DICTIONARY
OF
SCIENTIFIC BIOGRAPHY

PUBLISHED UNDER THE AUSPICES OF
THE AMERICAN COUNCIL OF LEARNED SOCIETIES

The American Council of Learned Societies, organized in 1919 for the purpose of advancing the study of the humanities and of the humanistic aspects of the social sciences, is a nonprofit federation comprising forty-five national scholarly groups. The Council represents the humanities in the United States in the International Union of Academies, provides fellowships and grants-in-aid, supports research-and-planning conferences and symposia, and sponsors special projects and scholarly publications.

MEMBER ORGANIZATIONS

AMERICAN PHILOSOPHICAL SOCIETY, 1743
AMERICAN ACADEMY OF ARTS AND SCIENCES, 1780
AMERICAN ANTIQUARIAN SOCIETY, 1812
AMERICAN ORIENTAL SOCIETY, 1842
AMERICAN NUMISMATIC SOCIETY, 1858
AMERICAN PHILOLOGICAL ASSOCIATION, 1869
ARCHAEOLOGICAL INSTITUTE OF AMERICA, 1879
SOCIETY OF BIBLICAL LITERATURE, 1880
MODERN LANGUAGE ASSOCIATION OF AMERICA, 1883
AMERICAN HISTORICAL ASSOCIATION, 1884
AMERICAN ECONOMIC ASSOCIATION, 1885
AMERICAN FOLKLORE SOCIETY, 1888
AMERICAN DIALECT SOCIETY, 1889
AMERICAN PSYCHOLOGICAL ASSOCIATION, 1892
ASSOCIATION OF AMERICAN LAW SCHOOLS, 1900
AMERICAN PHILOSOPHICAL ASSOCIATION, 1901
AMERICAN ANTHROPOLOGICAL ASSOCIATION, 1902
AMERICAN POLITICAL SCIENCE ASSOCIATION, 1903
BIBLIOGRAPHICAL SOCIETY OF AMERICA, 1904
ASSOCIATION OF AMERICAN GEOGRAPHERS, 1904
HISPANIC SOCIETY OF AMERICA, 1904
AMERICAN SOCIOLOGICAL ASSOCIATION, 1905
AMERICAN SOCIETY OF INTERNATIONAL LAW, 1906
ORGANIZATION OF AMERICAN HISTORIANS, 1907
AMERICAN ACADEMY OF RELIGION, 1909
COLLEGE ART ASSOCIATION OF AMERICA, 1912
HISTORY OF SCIENCE SOCIETY, 1924
LINGUISTIC SOCIETY OF AMERICA, 1924
MEDIAEVAL ACADEMY OF AMERICA, 1925
AMERICAN MUSICOLOGICAL SOCIETY, 1934
SOCIETY OF ARCHITECTURAL HISTORIANS, 1940
ECONOMIC HISTORY ASSOCIATION, 1940
ASSOCIATION FOR ASIAN STUDIES, 1941
AMERICAN SOCIETY FOR AESTHETICS, 1942
AMERICAN ASSOCIATION FOR THE ADVANCEMENT OF SLAVIC STUDIES, 1948
METAPHYSICAL SOCIETY OF AMERICA, 1950
AMERICAN STUDIES ASSOCIATION, 1950
RENAISSANCE SOCIETY OF AMERICA, 1954
SOCIETY FOR ETHNOMUSICOLOGY, 1955
AMERICAN SOCIETY FOR LEGAL HISTORY, 1956
AMERICAN SOCIETY FOR THEATRE RESEARCH, 1956
SOCIETY FOR THE HISTORY OF TECHNOLOGY, 1958
AMERICAN COMPARATIVE LITERATURE ASSOCIATION, 1960
AMERICAN SOCIETY FOR EIGHTEENTH-CENTURY STUDIES, 1969
ASSOCIATION FOR JEWISH STUDIES, 1969

DICTIONARY

OF

SCIENTIFIC BIOGRAPHY

CHARLES COULSTON GILLISPIE

EDITOR IN CHIEF

Volume 1

PIERRE ABAILARD—L. S. BERG

CHARLES SCRIBNER'S SONS · NEW YORK

Copyright © 1970, 1971, 1972, 1973, 1974, 1975, 1976, 1978, 1980
American Council of Learned Societies.
First publication in an eight-volume edition 1981.

Library of Congress Cataloging in Publication Data

Main entry under title:

Dictionary of scientific biography.

"Published under the auspices of the American Council
of Learned Societies."
Includes bibliographies and index.
1. Scientists—Biography. I. Gillispie, Charles
Coulston. II. American Council of Learned Societies
Devoted to Humanistic Studies.
Q141.D5 1981 509'.2'2 [B] 80-27830
ISBN 0-684-16962-2 (set)

ISBN 0-684-16963-0 Vols. 1 & 2   ISBN 0-684-16967-3 Vols. 9 & 10
ISBN 0-684-16964-9 Vols. 3 & 4   ISBN 0-684-16968-1 Vols. 11 & 12
ISBN 0-684-16965-7 Vols. 5 & 6   ISBN 0-684-16969-X Vols. 13 & 14
ISBN 0-684-16966-5 Vols. 7 & 8   ISBN 0-684-16970-3 Vols. 15 & 16

Published simultaneously in Canada
by Collier Macmillan Canada, Inc.
Copyright under the Berne Convention.

7 9 11 13 15 17 19   B/C   20 18 16 14 12 10 8

Printed in the United States of America

Panel of Consultants

Preface

The *Dictionary of Scientific Biography* is designed to make available reliable information on the history of science through the medium of articles on the professional lives of scientists. All periods of science from classical antiquity to modern times are represented, with the exception that there are no articles on the careers of living persons. In many instances the articles are either the first or the most considerable study yet made of an individual body of work, for the purpose of the *Dictionary* is not only to draw upon existing scholarship but to constitute scholarship where none exists.

The work is published under the sponsorship of the American Council of Learned Societies with the endorsement of the History of Science Society. The undertaking has been made possible by a generous grant from the National Science Foundation.

In planning the work, the Editorial Board has intended it to perform for the historiography of science services like those that the *Dictionary of National Biography* and the *Dictionary of American Biography* have long rendered in British and American historiography. During the compilation of those indispensable instruments of study and research, their editors enlisted the efforts of virtually all members of the historical profession concerned with national history in the respective countries. So it has been in the creation of this *Dictionary,* which has depended throughout upon large participation by professional historians of science all over the world. Theirs is a discipline which has now become a well-defined field of historical scholarship in recognition of the importance that science has held in the evolution of modern civilization. At the same time the Board has been equally fortunate to be able to rely (as, indeed, does the historiography of science generally) upon the contributions of many professional scientists who, in consequence of their commitment to science itself, take a lively and scholarly interest in its past.

Authors of articles were asked to place emphasis upon the scientific accomplishments and careers of their subjects. In some instances it has seemed wise to divide among several authors the account of figures whose work was manifold. Personal biography has intentionally been kept to the minimum consistent with explaining the subject's place in the development of science. Each article concludes with a bibliography that will guide the reader to the original scientific work and also to the main biographical items concerning personal and public life. Authors of articles on prolific scientists have included in the bibliographies the most important papers and indicated in what publications others may be found. In many cases, the citation of a systematic and authoritative bibliography makes it unnecessary to duplicate readily available information.

The aim has been to include articles on those figures whose contributions to science were sufficiently distinctive to make an identifiable difference to the profession or community of knowledge. The Board has felt reasonable confidence in the information available to it in selecting the more important figures and doubts that notable bodies of scientific work have been excluded. For the subjects of marginal significance, however, an important and often decisive limiting factor in reaching decisions about their inclusion or omission has been the availability of scholarship. The Board has not wished to commission any articles unless satisfied that they could be based on first-hand study of the sources. Given the way in which the selection of the lesser subjects had to be made, it does not appear that either the configuration of the list as a whole, or that of various distributions that might be discerned within it, could properly serve for statistical analysis of the population of science in various times, places, or disciplines. One exception to that caveat may be admitted. The high proportion of persons who worked in the biological sciences, amounting to almost a third of the total, does not appear to be an accident of the selection procedure. Their number reflects the historical situation. Until the nineteenth century only the sciences now pertaining to biology formed any part of the responsibility of a well-defined and populous profession, that of medicine. Perhaps for that reason, among others, the manner of growth of the biological and descriptive sciences requires taking account of a larger number of discrete investigations than does that of the physical and mathematical sciences.

Areas of science covered by the *Dictionary* are those that in modern times fall within the purview of mathematics, astronomy, physics, chemistry, biology, and the sciences of the earth. There are a few articles on historians of science. Technology, medicine, the behavioral and social sciences, and philosophy are included only in the instances of persons whose work was intrinsically related to the sciences of nature or to mathematics. Thus physiologists who were also physicians appear in virtue of their biological contributions. Authors of articles in these ancillary fields were asked to treat mainly the matters relevant to science, and if they have not always been as restrictive as the policy contemplated, there is a limit to the rigidity with which such distinctions can usefully be applied.

Several other disproportions in the selection of subjects must be noted. In the twentieth century, the choice has been held to relatively important figures. The justification might well be that historical perspective on the recent past is notoriously deceptive in science as elsewhere, but a more practical reason is the shortage of technically qualified scholars in the contemporary history of science. The same consideration has governed in certain decisions about whether to publish articles on the careers of notable scientists who have died while the *Dictionary* was in preparation. In a number of cases it has not been possible to obtain a worthy article.

The limited availability of scholarship is also the factor that has determined policy in respect to the treatment afforded the scientific knowledge and natural philosophy of early India, China, and Japan. In modern times the scientific culture of Asia has merged with that of the West, the continuous history of which goes back from Europe in large part through Islam to its origins in classical antiquity and the ancient Near East. Although this latter tradition was the one that developed the type of science that has become world-wide since the seventeenth century, we should not want to assert that it alone was scientific. Ancient Indian, Chinese, and Japanese sciences are vast and fascinating subjects. The difficulty is that they have been very little studied in the West. In this situation, we have decided that the most practical course is to publish a few articles that could be written to illustrate the kinds of work done in the ancient East and to leave any attempt at adequate representation of these subjects to the development of scholarship in the future.

In the case of Indian science, a biographical format is not well suited to the matter since many of the names associated with its schools of astronomy and natural philosophy are mythical or legendary. Professor David Pingree has undertaken to write brief articles identifying the Indian names at the appropriate places in the alphabetical sequence and further to write a series of essays giving accounts of the several schools and traditions to which they belonged. Those essays appear in the Supplement. Similar treatment is accorded the main topics of ancient Babylonian and Egyptian science, the identity of the scribes who founded it being largely unknown to history. Also published in the Supplement are certain articles that for one reason or another could not be printed in their proper places in the alphabetical sequence, where references to them will be found. The Supplement forms part of the final volume, containing also the Index.

The Index itself is organized to permit tracing the evolution of problems, concepts, and subjects through the articles about the persons who contributed to their development. During the initial planning discussions for the *Dictionary* some took the view that a basic work of reference in the history of science would better be organized by subject matter and topic, i.e., *Gravity, Natural Selection, Atomism,* etc., rather than by individual careers. The Board decided against it, however, on the grounds that history of science like other aspects of history is made by men and not by themes or abstractions, that in any case the specification of topics would prove difficult, arbitrary, and liable to obsolescence and that the purpose might be served by a suitably detailed analytical Index.

In producing this work, the Board of Editors has incurred immense obligations, which are a pleasure to acknowledge with the greatest of gratitude. The first and foremost is to the authors of the articles, who have taken time and energy from their own writing, research, and teaching in order to serve the interest of the subject as a whole. In establishing the list of subjects to be treated, we have enjoyed invaluable assistance from the Panel of Consultants, who advised us from the outset and who kindly put us in touch with many of the authors represented. The members both of that Panel and of the Advisory Committee established by the American Council of Learned Societies have answered innumerable inquiries and responded to all requests for advice. From the beginning, Mr. Charles Scribner's relation to the work has been that of colleague as much as publisher.

Finally, certain colleagues as well as certain members of the staff rendered services in the organization of the work at the outset that required a high degree of imagination and attention. The Board would like to record its particular gratitude to Professor H. G. Georgiadis and Professor Derek J. de Solla Price, and to Miss Jeanne Armel, Mr. Joseph G. E. Hopkins, Mr. Steven Pappayliou, and Miss Arlene Witt.

THE EDITORIAL BOARD

Contributors to Volume 1

HANS AARSLEFF

GIORGIO ABETTI

DEAN C. ALLARD

TORSTEN ALTHIN

PETER AMACHER

D. M. BALME

HANS BAUMGÄRTEL

WHITFIELD J. BELL, JR.

ENRIQUE BELTRÁN

JOHN A. BENJAMIN

ARTHUR BIREMBAUT

MARIA LUISA RIGHINI-BONELLI

FRANCK BOURDIÉR

MARJORIE NICE BOYER

J. MORTON BRIGGS

W. H. BROCK

THEODORE M. BROWN

K. E. BULLEN

VERN L. BULLOUGH

G. V. BYKOV

ALBERT V. CAROZZI

ETTORE CARRUCCIO

CARLO CASTELLANI

SEYMOUR L. CHAPIN

F. B. CHURCHILL

MARSHALL CLAGETT

THOMAS H. CLARK

EDWIN CLARKE

ERIC M. COLE

WILLIAM COLEMAN

ALBERT B. COSTA

PIERRE COSTABEL

RUTH SCHWARTZ COWAN

A. C. CROMBIE

M. P. CROSLAND

CHARLES A. CULOTTA

J. AL-DABBAGH

F. DAGOGNET

EUGENIO DALL'OSSO

GLYN DANIEL

ALLEN G. DEBUS

R. G. C. DESMOND

BERN DIBNER

SALLY H. DIEKE

J. G. DORFMAN

STILLMAN DRAKE

LOUIS DULIEU

A. HUNTER DUPREE

J. M. EDMONDS

OLIN J. EGGEN

H. ENGEL

GUNNAR ERIKSSON

V. A. ESAKOV

CHARLES L. EVANS

JOSEPH EWAN

V. A. EYLES

EDUARD FARBER

IVAN A. FEDOSEYEV

LUCIENNE FÉLIX

KONRADIN FERRARI D'OCCHIEPPO

BERNARD S. FINN

C. S. FISHER

MARCEL FLORKIN

MENSO FOLKERTS

GEORGE A. FOOTE

PAUL FORMAN

PIETRO FRANCESCHINI

EUGENE FRANKEL

HANS FREUDENTHAL

B. V. FREYBERG

KURT VON FRITZ

JUSTO GARATE

GERALD L. GEISON

RUTH ANNE GIENAPP

BERTRAND GILLE

JEAN GILLIS

CHARLES C. GILLISPIE

OWEN GINGERICH

HARRY GODWIN

STANLEY GOLDBERG

J. B. GOUGH

JOSEPH T. GREGORY

NORMAN T. GRIDGEMAN

N. GRIGORIAN

M. D. GRMEK

MORTON GROSSER

FRANCISCO GUERRA

NORBERT GÜNTHER

DOUGLAS GUTHRIE

IAN HACKING

ROGER HAHN

A. RUPERT HALL

OWEN HANNAWAY

BERT HANSEN

ROBERT H. HARDIE

WILLY HARTNER

JAGDISH N. HATTIANGADI

JOHN L. HEILBRON

MARY HESSE

FREDERIC L. HOLMES

WILLIAM T. HOLSER

R. HOOYKAAS

MICHAEL A. HOSKIN

PIERRE HUARD

G. L. HUXLEY

AARON J. IHDE

JEAN ITARD

CHARLES W. JONES

PHILLIP S. JONES

MIROSLAV KATĚTOV

ALAN S. KAY

MARTHA B. KENDALL

G. B. KERFERD

DANIEL J. KEVLES

PEARL KIBRE

JOHN S. KIEFFER

GEORGE KISH

MARC KLEIN

FRIEDRICH KLEMM

DAVID M. KNIGHT

HULDRYCH M. KOELBING

ZDENĚK KOPAL

EDNA E. KRAMER

CLAUDIA KREN

FRIDOLF KUDLIEN

P. G. KULIKOVSKY

V. I. KUZNETZOV

YVES LAISSUS

L. L. LAUDAN

MARTIN LEVEY

JAMES LONGRIGG

EDWARD LURIE

CONTRIBUTORS TO VOLUME 1

Eric McDonald

A. G. MacGregor

Robert M. McKeon

H. Lewis McKinney

Ernan McMullin

Michael McVaugh

M. S. Mahoney

C. L. Maier

Daniel Massignon

Seymour H. Mauskopf

Kenneth O. May

Robert M. Mengel

Philip Merlan

Wyndham Davies Miles

Lorenzo Minio-Paluello

Ernest A. Moody

J. E. Morère

Shigeru Nakayama

Henry Nathan

J. P. Nicolas

J. D. North

Herbert Oettel

Robert Olby

C. D. O'Malley

Jane Oppenheimer

Oystein Ore

G. E. L. Owen

Shin'ichi Oya

A. Pabst

Jacques Payen

Giorgio Pedrocco

Georges Petit

Shlomo Pines

David Pingree

Martin Plessner

Loris Premuda

P. M. Rattansi

Nathan Reingold

Gloria Robinson

Andrew Denny Rodgers III

Francesco Rodolico

Alfred Romer

Vasco Ronchi

Conrad E. Ronneberg

P. G. Roofe

Edward Rosen

George Rosen

Charles E. Rosenberg

Franz Rosenthal

Josef Sajner

William Schaaf

H. Schadewaldt

Rud. Schmitz

Bruno Schoeneberg

Dorothy V. Schrader

Dorothy M. Schullian

E. L. Scott

Paul Sentein

Aleksei Nikolaevich Shamin

Harold I. Sharlin

Kazuo Shimodaira

O. B. Sheynin

C. S. Smith

H. A. M. Snelders

E. Snorrason

William H. Stahl

Jerry Stannard

Johannes Steudel

Lloyd G. Stevenson

Bernhard Sticker

Dirk J. Struik

Charles Süsskind

L. Tarán

Juliette Taton

René Taton

George Taylor

Arnold Thackray

Jean Théodoridès

Phillip Drennon Thomas

V. V. Tikhomirov

G. J. Toomer

G. L'E. Turner

Juan Vernet

Kurt Vogel

William A. Wallace, O.P.

Charles Webster

C. E. Wegmann

D. T. Whiteside

Gweneth Whitteridge

L. Pearce Williams

Helmut M. Wilsdorf

Leonard G. Wilson

R. P. Winnington-Ingram

F. R. Winton

J. Witkowski

M. Wong

Helen Wright

A. P. Youschkevitch

DICTIONARY
OF
SCIENTIFIC BIOGRAPHY

DICTIONARY OF SCIENTIFIC BIOGRAPHY

ABAILARD—BERG

ABAILARD, PIERRE, also known as **Peter Abelard** (*b.* Le Pallet, or Palais, Brittany, France, 1079; *d.* near Chalon-sur-Saône, France, 21 April 1142), *logic, theology, philosophy.*

Abailard was the son of Berengar, lord of Le Pallet, but he abandoned the militaristic and governmental traditions of the nobility. He did preserve, however, a determination to impose his personality on the studies and intellectual polemics of his time, and often he dominated the entire field. Intolerant of what was not the best, he moved from school to school, fighting against his masters and colleagues and founding his own schools and a religious community. When he was forced as a punishment to reside in a monastery and when he accepted the leadership of another, he applied his exacting moral principles, his scholarship, and his energy to correcting and reforming mistakes and practices; if defeated, he prepared for further battle.

Of the subjects forming the basic curriculum for scholars, Abailard was interested only in those concerning language, especially grammar and dialectic. He confesses not to have mastered mathematics, although he shows himself competent to deal with the question of continuity. In astrology he follows the accepted views. At this time in France doctrinal conflicts centered largely on dialectic, both within its proper field and in its applications to the problems of human life, then usually presented in theological terms. As a discipline in its own right, it was expanding into the province of metaphysics. Combined with deeper inquiries into grammatical concepts, it was developing new distinctions, refining its procedures, and purifying itself from the sources of easy sophistry. In its applications, it would claim to be the method of clarifying ideas, organizing statements, even extending the province of knowledge, and producing statements normally accepted as having a supernatural origin as valid conclusions derived from nonrevealed truths.

Such was the background of Abailard's career. He was an uneasy pupil at the school of Roscelin in Loches (*ca.* 1094–1096). Roscelin's doctrines on significant words being merely words had appeared to endanger traditional views on knowledge and the dogma of the Trinity. Abailard soon passed on to William of Champeaux's school in Paris; but impatient of this master's opinions concerning the existence in our world (and possibly also in a Platonic world of ideas) of things referred to by general words, he began teaching in Melun and Corbeil. Perhaps, too, he was impatient of being just a pupil. About 1106 illness forced him to return to Brittany. Again in Paris (*ca.* 1110) he fought it out, victoriously, with William. The latter abandoned his chair, which soon after was given to Abailard. But intrigue had the better of learning; Abailard was dismissed. Undaunted, he opened a new school on the outskirts of the city, on the Montagne Ste. Geneviève.

So far, language, logic, and their metaphysical implications had dominated Abailard's mind; after a business sojourn in Brittany he was attracted to Anselme and Ralph of Laon's theological school (*ca.* 1114). Instead of clear words he found verbosity, instead of a scientific approach the smoke of traditional apologetics. The cathedral school of Paris now opened its doors to him as to an honored master of dialectic and theology. The disturbing love affair with Héloïse, physically concluded with Abailard's emasculation, turned into a friendship of a religious and intellectual character. He withdrew to the abbey of St. Denis outside Paris, and became a monk (*ca.* 1118): a bad choice for the abbey and for Abailard. He attacked the laxity of the monks; they attacked the dangers of his dialectical theology. The monks promoted the Council of Soissons (1121), where his doctrine of the Trinity was condemned. An attempt at demolishing, with the tools of historical criticism, the legend concerning the foundation of St. Denis by a pupil of St. Paul involved the *enfant terrible* in further trouble. He escaped, and finally obtained permission to settle at a place of his choice: a new convent was thus born under the symbolic name of Paraclete (the consoling Holy Spirit).

1

Peace was short-lived: too many people were attracted to the rebel. He accepted the position of abbot of St. Gildas in Brittany, leaving the Paraclete to Héloïse and her nuns, only to fight once more in vain against irreligiosity and immorality. By 1136 he was again on his Montagne Ste. Geneviève, again provoking hostility by his methods and doctrines. The unflinching St. Bernard was among the attackers. The Council of Sens (1140) dramatically—or theatrically—condemned, with the pope's support, the man for whom reason was a good companion of faith and intention rather than action the touchstone of sin. Abailard set out toward Rome for an appeal, but was persuaded by Peter the Venerable of Cluny to accept the verdict. From Cluny he moved to the priory of St. Marcel, where he died soon after.

Abailard's more strictly logico-philosophical works are partly documents of the elaborate development and preparation for his activity as a teacher and partly the systematic organization of his knowledge and critical evaluation of others' views concerning the whole of logico-philosophical studies. The *Introductiones parvulorum* is an elementary commentary on the three basic texts studied by every boy aiming at a career that required learning: Porphyry's *Isagoge* and Aristotle's *Categories* and *De interpretatione*. The more extensive commentaries on these same works and on Boethius' *De differentiis topicis* embody, both in the form of a penetrating analysis and in the form of constructive and destructive argument, much of Abailard's most original philosophical production. The *Dialectica* is the first full-scale attempt, in the Latin West, at producing a system of logic covering all the recognized sections of that discipline, until then dispersed in disconnected works composed by authors of different periods and treated without a uniform pattern and often without a clear plan. Abailard's own plan, however, depended too much on a traditional set of texts and on an old division of the parts of logic: his contribution is to be found more in details than in the general scheme.

Most of Abailard's philosophico-theological works, including sections of two biblical commentaries (on the beginning of Genesis and on St. Paul's Letter to the Romans) contain elaborations of the main themes of Christian doctrine from the point of view of the man of faith. But the elaborations are aimed at showing how much of this doctrine is accessible to the man without faith who uses his reason (itself, after all, of divine origin) both for directly establishing truth and for critically accepting non-Christian reasonable authorities, such as Plato and Aristotle. This is most evident in the successive editions of his *Theologia*. His *Scito te ipsum* (*Know Yourself*), the study

of the psychology of intention, volition, and action, as related to the concept of guilt, appeared to revolutionize the dogma of original sin. The *Sic et non* (*Yes and No*) is an analysis of texts chosen from works of the Fathers of the Church; in it, critical rules of interpretation of the written word are applied to show to what extent apparently contradictory statements can be seen to agree in their basic meaning.

Abailard's contributions that are of interest to science are more of a methodological character than discoveries of facts and laws of nature. From the introduction to the *Yes and No* the following principles or rules can be elicited: (1) methodical doubt (doubting is necessary [Aristotle]); search and you will find (the Gospels); (2) distinguish statements that compel assent from those on which free judgment must be exercised; (3) distinguish between the levels of language used (technical [proper] or common [vulgar, improper], explicit or metaphorical or rhetorical, stating the writer's views or quoting the opinions of others); (4) meanings of words change with time; (5) fallibility of human writers, however authoritative (mistakes even in Scripture); (6) fallibility of written tradition (textual criticism); (7) context affects meanings.

Abailard's discussion of "universals" in his longer commentary on Porphyry's *Isagoge* exemplifies his procedure. It can be schematized in this way: (*a*) be clear about the meaning that is ascribed to "universal," starting, as one normally does, from Aristotle's statement "universals are those that are predicated of many"; (*b*) properly used, the key term "predicated" applies only to words; (*c*) consequently "this kind of universals" can only be words, i.e., universal (or common) words; (*d*) these words have, in a proposition, the special function of "being predicated," not of "signifying"; (*e*) a more serious problem is this: What makes us invent and use universal words, i.e., what is the cause (common cause) of common words; is it a community in things or a community in our concept; (*f*) there is a common state of affairs ("status") for A and B such that each can be said to be man; this "status" is not a thing (*res*); (*g*) our mind "melts together" (*confundit*) into one image that which it elicits, abstracts, from things according to their "common status"; (*h*) the "common cause" of universal words is primarily to be found in the common "status" of things, secondarily in the *imago confusa*, i.e., in our concept; (*i*) extrapolating from the common status to the knowledge of it possessed by the maker of things (not by us, men), it is possible to conceive a knowledge of the common cause as *forma* (a Platonic idea in God's mind). In this way, Abailard surveys the linguistic, logical, naturalistic, gnoseological, metaphysical, and theological aspects of the problems of universals.

In the *Dialectica* as well as in the several commentaries there are many statements of importance to philosophy, theory of language, logic, methods of expression, and possibly of research in science, which were either first put forward clearly or strongly endorsed by Abailard. Some are to the effect that (*a*) "is," "are," etc., in sentences like *John is a man* and *John is,* have no existential import (the second sentence being elliptical = *John is an existing being*) but are connectives (*copulae*); (*b*) propositions in the future or past must be resolved into propositions in the present; (*c*) a self-referring word, e.g., "Man" (= "the word *Man*"), does not alter the nature of "is" qua *copula*; (*d*) "not every . . ." and "some . . . not" are not equivalent; (*e*) "all" implies both collectivity and exclusivity; (*f*) modal words ("possible," "necessary" . . . , "true," "false" . . . , etc.) have two different functions according to whether they affect the relationship between subject and predicate or the status of a proposition. In the study of conditional propositions, often called by Abailard *consequentiae* (possibly a new, systematically technical use of this term), a number of rules are made explicit in forms easily translatable into modern symbolism, e.g., the rules of transitivity of entailment, of incompatibility between a true affirmation and a true negation, and of entailments between modalities.

With his inquiries into the logic of language Abailard contributed possibly more than anyone else to the developments of the new logico-linguistic theories, especially those concerning *suppositio, copulatio,* and *appellatio.*

BIBLIOGRAPHY

I. ORIGINAL WORKS. All the known works of Abailard have been published at least once, but no single edition contains more than about half the extant texts. The largest collection is in J. P. Migne, ed., *Patrologia Latina,* 178 (Paris, 1855). This volume includes all the works ed. before, with the exception of the logical texts published by Victor Cousin as part of *Ouvrages inédits d'Abélard* (Paris, 1836). Migne gives, for each work, the necessary information on the eds. reproduced in his volume. We mention here only the two complementary collections older than Migne's. The first was ed. by F. d'Amboise and A. Duchesne, *Petri Abaelardi filosofi* [sic] *et theologi . . . et Heloissae coniugis . . . opera* (Paris, 1616); the second, ed. by Cousin, Ch. Jourdain, and E. Despois, *Abaelardi opera hactenus seorsim edita* (Paris, 1849–1859), includes little of what Cousin had published in 1836.

The following eds. contain, with minor exceptions, all the works that had not been published by 1855, or had

been published in an incomplete form: (*a*) the longer commentaries on Porphyry's *Isagoge* and Aristotle's *Categories* and *De interpretatione* (the latter incomplete and with an apocryphal last section), which came to be known as *Logica "Ingredientibus,"* together with an incomplete commentary on the *Isagoge,* now known as *Logica "Nostrorum petitioni,"* were ed. by B. Geyer, *Peter Abaelards philosophische Schriften* (Münster, 1919–1935), Vol. XXI of *Beiträge zur Geschichte der Philosophie und Theologie des Mittelalters*; (*b*) the shorter commentaries on Porphyry and Aristotle (*Introductiones parvulorum*) and the commentaries on Boethius' *De divisione* and *De differentiis topicis,* all published—incompletely—by Cousin in 1836, were published in full by M. Dal Pra, *Abelardo, Scritti filosofici* (Rome-Milan, 1954); (*c*) the *Dialectica,* also incompletely published by Cousin, was published in full by L. M. De Rijk (Assen, 1956); (*d*) the last section of the longer commentary on Aristotle's *De interpretatione,* missing from the Geyer ed., was published, together with two shorter texts, by L. Minio-Paluello, *Abaelardiana inedita* (Rome, 1958), Vol. II of *Twelfth Century Logic*; (*e*) *De unitate et trinitate,* R. Stölze, ed. (Freiburg, 1891); (*f*) "Ein neuaufgefundenes Bruchstück der Apologia Abaelards," P. Ruf and M. Grabmann, eds., in *Sitzungsberichte der Bayerischen Akademie der Wissenschaften,* Philos.-hist. Abt., **5** (1930); (*g*) *Theologia "Summi boni,"* H. Ostlender, ed. (Münster, 1939), Vol. XXXV. 2–3 of *Beiträge zur Geschichte der Philosophie und Theologie des Mittelalters.*

Mention should also be made of the new critical ed. of the first letter of Abailard, the autobiographical *Historia calamitatum,* prepared by J. Monfrin (Paris, 1959), and of the new ed. by J. T. Muckle of "The Personal Letters Between Abelard and Heloise," in *Mediaeval Studies,* **15** (1953), 47–94.

There are English trans. of the following: (*a*) the section on universals from the *Logica "Ingredientibus,"* in R. McKeon, ed., *Selections From Medieval Philosophers,* I (New York, 1929, 1957), 208–258; (*b*) C. K. Scott Moncrieff, *Letters to Heloise* (London, 1925–1926); (*c*) J. R. McCallum, *Scito te ipsum* (*Ethics*) (Oxford, 1935); (*d*) J. T. Muckle, *Historia calamitatum* (*The Story of Abelard's Adversities*) (Toronto, 1954).

II. SECONDARY LITERATURE. A quite extensive bibliography on Abailard down to 1928 is in B. Geyer, pp. 213–214 and 702–703 of the vol. mentioned below. More detailed bibliographical material for special fields will be found in several of the works listed here.

The following general histories contain important sections on Abailard: *Histoire littéraire de France,* XII (Paris, 1763), 86–152 (repr. in Migne, *Patrologia Latina,* CLXXVIII, 10–54); G. Robert, *Les écoles et l'enseignement de la théologie . . .* (Paris, 1909), pp. 149–211, reelaborated by G. Paré, A. Brunet, and P. Tremblay, *La Renaissance du XIIe siècle* (Paris-Ottawa, 1933), pp. 275–312; M. Grabmann, *Geschichte der scholastischen Methode,* II (Freiburg, 1911), 177–221; M. De Wulf, *History of Mediaeval Philosophy,* English trans. by E. C. Messenger of the 6th French ed. (London, 1935, 1952), pp. 194–205; L. Thorndike, *A History of Magic and Experimental Science,* II (New York,

1923), 3–8; B. Geyer, *Die patristische und scholastische Philosophie,* Vol. II of F. Ueberweg, *Grundriss der Geschichte der Philosophie,* 11th ed. (Berlin, 1928), 213–226; I. M. Bocheński, *Formale Logik* (Freiburg–Munich, 1956), English trans. by I. Thomas (South Bend, Ind., 1961); W. and M. Kneale, *The Development of Logic* (Oxford, 1962), pp. 202–224.

Comprehensive works on Abailard's life and doctrines include P. de Rémusat, *Abélard, sa vie, sa philosophie et sa théologie* (Paris, 1845, 1855); C. Ottaviano, *Pietro Abelardo, la vita, le opere, il pensiero* (Rome, 1931); J. G. Sikes, *Peter Abailard* (Cambridge, 1932); and H. Waddell, *Peter Abelard* (London, 1933).

Among the many books and articles on particular fields or problems, the following call for special attention: E. Kaiser, *Pierre Abélard critique* (Fribourg, 1901); J. Cottiaux, "La conception de la théologie chez Abélard," in *Revue d'histoire ecclésiastique,* **28** (1932), 247–295, 533–551, 788–828; H. Ostlender, "Die Theologia 'Scholarium' des Peter Abailard," in *Beiträge zur Geschichte der Philosophie und Theologie des Mittelalters,* Supp. III (Münster, 1935), 263–281; T. Reiners, *Der Nominalismus in der Frühscholastik, ibid.,* VIII.3 (1910), 41–59; B. Geyer, "Die Stellung Abaelards in der Universalienfrage," *ibid.,* Supp. I (1913), 101–127; E. A. Moody, *Truth and Consequence in Mediaeval Logic* (Amsterdam, 1953); J. R. McCallum, *Abelard's Christian Theology* (Oxford, 1949); G. Engelhardt, *Die Entwicklung der dogmatischen Psychologie in der mittelalterlichen Scholastik* in *Beiträge zur Geschichte der Philosophie und Theologie des Mittelalters,* Vol. XXX.4–6 (Münster, 1933); O. Lottin, "Le problème de la morale intrinsèque d'Abélard à S. Thomas," in *Revue Thomiste,* **39** (1934), 477–515; E. Gilson, *Héloïse et Abélard* (Paris, 1938, 1955, 1958), English trans. by L. K. Shook (Chicago, 1951).

The intros. to several of the eds. listed in Section I of this bibliography are of particular importance. The following deserve special mention: V. Cousin, *Ouvrages inédits*; L. M. De Rijk, *Dialectica*; L. Minio-Paluello, *Abaelardiana inedita*; P. Ruf and M. Grabmann, "Ein neuaufgefundenes Bruchstück"; H. Ostlender, *Theologia "Summi boni"*; J. Monfrin, *Historia calamitatum*; and the section "Untersuchungen" at the end of B. Geyer's ed. of *Peter Abaelards philosophische Schriften.*

                                          L. MINIO-PALUELLO

ABANO, PIETRO D' (*b.* Abano, Italy, 1257; *d.* Padua, Italy, *ca.* 1315), *medicine, natural history, alchemy, philosophy.*

D'Abano completed his early studies in Padua and later took many voyages which focused his attention upon nature studies and ethics. He lived in Constantinople and then, about 1300, went to Paris, where he attended the university and perhaps taught and composed his *Conciliator differentiarum philosophorum et praecipue medicorum.* In 1307 d'Abano returned to Padua, where for several years he taught philosophy and medicine, arousing the apprehension and the perplexity of the academic and ecclesiastical authorities. Although he was acquitted during his lifetime of the charge of heresy—of which he had been accused because of his attempt to interpret the birth and ministry of Christ as other than miraculous—his reputation as a sorcerer persisted. Some forty years after his death his writings were again put on trial; they were found to be heretical, and his bodily remains were disinterred and burned.

In his *Conciliator,* d'Abano undertook a superb synthetic program: the reconciliation of medicine with philosophy. In this he states 120 questions that give rise to as many controversies between physicians and philosophers. For their solution he adopts the method of didactic demonstration that is characteristic of the period, yet on the whole there are signs of a new intention and a new uncertainty.

The practice of medicine implies the necessity of resolving every problem in a natural manner. D'Abano maintained more or less that "the art of medicine must not consider only things that can be seen and felt." Hence he possessed a good knowledge of anatomy; he affirmed, in opposition to the authority of Aristotle (who thought the nerves originated in the heart) that the center of all sensation and motion resides in the brain. His notions of the central nervous system are probably derived from direct visualization. According to d'Abano, the doctor is the symbol of the zealous servant and the collaborator of nature. Considerable importance is attached to the relationship of trust that exists between the doctor and the patient. A good reputation is more useful to the doctor than rare drugs.

These concepts, as d'Abano developed them in his work, have considerable importance. The doctor must be free in his reasoning and must have no ties with scholastic authorities. Such ideas imply a revolt against established and wearisome tradition: they prepare for a rupture with the past and indicate a new path for scientific progress. D'Abano's voice was one of those that, at the dawn of humanism, announced the beginning of a scientific revival.

The Paduan master acknowledged the dependence of every living being and of earthly events on planetary influences. The *Conciliator* gives an outline of astrology as a two-part science comprising one that deals with the laws of celestial movements (astronomy) and another, more important, that draws from these laws the judgments and predictions concerning the effects of those motions on our world—on all human events, on human conception, and even on religion.

D'Abano has been considered by such scholars as Ferrari and Troilo as the initiator of Latin Averroism in Italy. Others—Thorndike, Nardi, and Giacon—have maintained that d'Abano's thought bears no trace of Averroistic theses—above all, that dealing with the unity of the intellect, either as an agent or as a possibility.

BIBLIOGRAPHY

I. ORIGINAL WORKS. The most important works of d'Abano are *Additio in librum Joh. Mesue* (Venice, 1471); *Conciliator differentiarum philosophorum et praecipue medicorum* (Mantua, 1472); *De venenis* (Mantua, 1473); *Liber compilationis physiognomiae* (Padua, 1474); *Expositio problematum Aristotelis* (Mantua, 1475); *Expositiones in Dioscoridem* (Colle [Tuscany], 1478); *Quaestiones de febribus* (Padua, 1482); *Hippocratis libellus de medicorum astrologia* (Venice, 1485); and *Geomantia* (Venice, 1549).

II. SECONDARY LITERATURE. The most important works on d'Abano are M. T. d'Alverny, "Pietro d'Abano et les 'naturalistes' à l'époque de Dante," in Leo S. Olschki, Vittore Branca, Giorgio Pedoan, eds., *Dante e la cultura veneta* (Florence, 1966), pp. 207–219; G. Della Vedova, *Biografia degli scrittori padovani,* I (Padua, 1832), 25–33; P. Duhem, *Le système du monde. Histoires des doctrines cosmologiques de Platon à Copernic,* IV (Paris, 1916), 229–263; S. Ferrari, *I tempi, la vita, le dottrine di Pietro d'Abano* (Genoa, 1900), which contains considerable information on d'Abano, and "Per la biografia e per gli scritti di Pietro d'Abano," in *Atti Regale Accademia Lincei, Memorie Classe Scienzi Morali, Storiche e Filologiche,* 5th ser., **15** (1915), 629–725; C. Giacon, "Pietro d'Abano e l'averroismo padovano," in *Atti XXVI riunione S.I.P.S.* (Rome, 1938), pp. 334–339; B. Nardi, "La teoria dell'anima e la generazione delle forme secondo Pietro d'Abano," in *Rivista filosofica neoscolastica,* **4** (1912), 723–737; "Intorno alle dottrine filosofiche di Pietro d'Abano," in *Nuova rivista storica,* **4** (1920), 81–97, and **5** (1921), 300–313; and *Dante e Pietro d'Abano, saggi di filosofia dantesca* (Milan, 1930), pp. 43–65; L. Norpoth, "Zur Bio-Bibliographie und Wissenschaftslehre des Pietro d'Abano, Mediziners, Philosophen und Astronomen in Padua," in *Kyklos,* **3** (1930), 292–353, which contains considerable information on d'Abano; J. H. Randall, Jr., *The School of Padua and the Emergence of Modern Science* (Padua, 1961); G. Saitta, *Il pensiero italiano nell'umanesimo* (Bologna, 1949), pp. 32–39; L. Thorndike, *A History of Magic and Experimental Science,* II (New York, 1947), 874–947; and E. Troilo, "Averroismo o aristotelismo 'alessandrista' padovano," in *Rendiconti classe scienze morali, storiche e filologiche, Accademia Nazionale Lincei,* 8th ser., **9,** nos. 5-6 (1954), 188–244.

LORIS PREMUDA

ʿABBĀS IBN FIRNĀS (*b.* Ronda, Spain; *d.* 274/A.D. 887), *humanities, technology.*

ʿAbbās ibn Firnās, who was of Berber origin, is sometimes confused with the poet ʿAbbās ibn Nāṣiḥ (*d.* 240/A.D. 844). He was the court poet and astrologer of the emirs ʿAbd al-Raḥmān II and Muḥammad I, but he attracted the attention of his compatriots because of his inventions and his dissemination of oriental science in the West. Ibn Firnās was the first Andalusian to understand the prosodic rules first laid down by al-Khalīl ibn Aḥmad in the eighth century, and he made known the tables of the *Sind Hind,* which later had great influence on the development of astronomy in Europe. He also attempted to fly—and actually managed to glide for a distance—but the landing was rough because, according to his critics, he did not devote enough study to the way birds use their tails when they land. This flight was often mentioned in Spanish and Arabic writings.

Ibn Firnās constructed a planetarium, a clock, and an armillary sphere; and he is often credited with the discovery of rock crystal. The texts now available are not explicit, however, and one cannot judge on the basis of the statements of E. Lévi-Provençal, for he had access to the *Muqtabis* manuscript, which has been lost. In any case, the historians Ibn Saʿid and Maqqarī state that "he was the first in al-Andalus [Andalusia] to invent (discover) stone crystal." The statement can be interpreted in various ways, but it seems clear that rather than inventing or discovering "stone crystal," Ibn Firnās introduced the cutting of rock crystal, an industry already known in other regions, into the Islamic West. This would have brought about a reduction in the export of quartz to the east, especially to Egypt, for it could now be worked where it was mined. The technique of making glass, known at least since the third millennium before Christ, does not seem to have undergone any change at this time.

BIBLIOGRAPHY

None of Ibn Firnās' original works is extant. His biography can be reconstructed only from a few verses and from the information given by the chroniclers, which can be found in the monograph by Elias Terés, in *Al-Andalus,* **25** (1960), 239–249. Also of value is that of E. Lévi-Provençal, in *Encyclopaedia of Islam,* I, 11. For a discussion of crystal, see P. Kahle, "Bergkristall, Glas und Glasflüsse nach dem Steinbuch von al-Beruni," in *Zeitschrift der Deutschen morgenländischen Gesellschaft,* **90** (1936), 322–356; and *Libri Eraclii de coloribus et artibus Romanorum,* in *Quellenschriften für Kunstgeschichte,* IV (1873).

JUAN VERNET

AL-ʿABBĀS IBN SAʿĪD AL-JAWHARĪ. See al-Jawharī.

ABBE, CLEVELAND (*b.* New York, N.Y., 3 December 1838; *d.* Washington, D.C., 28 October 1916), *meteorology.*

As the first regular official weather forecaster of the U.S. government and a promoter of research in atmospheric physics, Abbe served as a symbol of what a meteorologist should be. Unlike many of his colleagues, he was well trained. After studying under Oliver Wolcott Gibbs at the City College of New York, Abbe worked with the German astronomer F. F. E. Brünnow, then at the University of Michigan (1859–1860), and later (1860–1864) with B. A. Gould, who was on detached duty with the Coast Survey, at Cambridge, Massachusetts. In Cambridge he came in contact with the group of astronomers and mathematicians in the *Nautical Almanac* office, notably William Ferrel. Desiring better preparation in astronomy, Abbe spent two years (1864–1866) at Pulkovo, Russia, working under Otto Struve.

Abbe's Russian stay had two consequences. First, Pulkovo provided a model of the symbiotic relationship between theory and practice, ironically like the one that had obtained in Cambridge with the Coast Survey. Second, through his translations and personal connections Abbe provided a point of contact between the American and Russian scientific communities.

Today we would characterize Abbe as a geophysicist, for he sought to apply the methods of astronomy to the development of a physics of the earth. Outlets for this ambition were scarce in nineteenth-century America, and after failing to establish in New York an observatory modeled on Pulkovo, Abbe served from 1868 to 1870 as director of the Cincinnati Observatory before joining the Weather Service of the Signal Corps, the predecessor of the present Weather Bureau, in 1871. Under his aegis the Corps established a laboratory and a "study room," a center for basic research. Although not a notable discoverer, Abbe insisted on mathematical rigor and a close following of new developments in the physical sciences.

BIBLIOGRAPHY

I. PRIMARY SOURCES. Abbe's personal papers are in the Library of Congress. They are described in Nathan Reingold, "A Good Place to Study Astronomy," in Library of Congress, *Quarterly Journal of Current Acquisitions,* **20** (Sept. 1963), 211–217. Other documents bearing on Abbe's career are in the U.S. Weather Bureau records in the National Archives, Washington, D.C., and in the papers of his son, Cleveland Abbe, Jr., in the library of the City University of New York. His reprint collection is in the library of the Johns Hopkins University, Baltimore, Md.

A good bibliography of Abbe's writings is in W. J. Humphreys, "Biographical Memoir of Cleveland Abbe, 1838–1916," in U.S. National Academy of Sciences, *Biographical Memoirs,* **8** (1919), 469–508.

II. SECONDARY WORKS. The Humphreys memoir is still the best single account of Abbe's career. The only full biography was written by Abbe's son Truman: *Prof. Abbe and the Isobars* (New York, 1965). Although a work of filiopietism, it is quite charming and still useful because of the son's liberal use of his father's papers, then in his possession. Nathan Reingold interprets one aspect of Abbe's career in "Cleveland Abbe at Pulkowa: Theory and Practice in the Nineteenth Century Physical Sciences," in *Archives internationales d'histoire des sciences,* **17** (April–June 1964), 133–147. Useful background information is in D. R. Whitnah, *A History of the United States Weather Bureau* (Urbana, Ill., 1961).

NATHAN REINGOLD

ABBE, ERNST (*b.* Eisenach, Germany, 23 January 1840; *d.* Jena, Germany, 14 January 1905), *physics.*

Abbe's importance for the development of scientific and practical optics can be comprehended only in connection with the founding and rise of the Zeiss Works.

In 1846 Carl Zeiss, a thirty-year-old mechanic, established his shop in Jena; in 1866, he began a technical and scientific collaboration with Abbe, who was then a lecturer at the university there. Abbe's fortunes grew with those of the Zeiss company; he had become a partner in 1876 and held a professorship at the university. Within ten years, the once small Zeiss workshop developed into an internationally famous industrial enterprise. The company's apochromatic lens was the greatest advance in technical optics made to that date. At the same period, Abbe began to manifest that interest in social welfare that soon led to the creation of the Carl Zeiss Foundation.

Abbe, according to Jena University curator M. Seebeck, "was born of lowly station, but with predestined claim to scientific fame." His father, Adam Abbe, a spinning-mill worker, would never have been able to send his son through high school and university if his employers had not provided a scholarship for the intelligent and industrious youth.

Upon graduating from the Eisenach Gymnasium in 1857, Abbe studied physics in Jena and subsequently in Göttingen, where he received the doctorate on 23 March 1861. Among the Göttingen professors who exerted a lasting influence on him were the mathematician Riemann, the famous exponent of the theory of functions, and the physicist Wilhelm Weber, former assistant to Gauss and one of the "Göttingen Seven," who had been temporarily suspended because of their protest against the king of Hannover's violation of the constitution.

Abbe's decision to apply for the position of lecturer at Jena University must not have been an easy one to make, since there would be a two-year interval, with its inevitable economic hardships, between his doctorate and his inauguration. He managed to make ends meet by accepting a poorly paid teaching position with the Physikalischer Verein in Frankfurt am Main, a group founded by local citizens for the propagation of the natural sciences. He also did some private tutoring. On 8 August 1863, at the age of twenty-three, Abbe finally achieved his ambition and was admitted to the faculty of Jena University as lecturer in mathematics, physics, and astronomy.

Abbe's straitened circumstances did not improve until he was made associate professor in 1870. On 24 September 1871 he married Elise Snell, the daughter of Karl Snell, head of the physics department at the University of Jena. The marriage was an extremely happy one from the start. The couple had two daughters. In 1876 Abbe's economic difficulties were resolved when Zeiss offered him a partnership. During the preceding ten years Abbe had contributed eminently to the phenomenal rise of Zeiss's company; he now shared in the quite considerable profits.

Zeiss had early begun experiments to convert the production of his microscope, consisting of an objective and an ocular lens, into a scientific process; whereas formerly he had relied on trial and error to find the best lenses, he now wished to use scientific methods. In this effort, Zeiss had met with as little success as his teacher Friedrich Körner; he had also attempted to use the knowledge of the mathematician Friedrich Barfuss. After the latter's death, Zeiss remained unable to solve this problem because of his limited scientific training. He therefore turned to Abbe in 1866 and succeeded in interesting the young physicist in the systematic production of microscopes. During the following decade they constructed the machinery required for industrial production and turned out many commercially marketed instruments (illuminating apparatus for the microscope, known in England as "the Abbe," the Abbe refractometer, and others). Abbe also solved their main problem so completely and ingeniously that his theoretical findings became the basis for the further development of practical optics for decades to come. For example, in 1934 Frits Zernike derived from these findings the

phase-contrast process, for which he was awarded the Nobel Prize in physics (1953). Somewhat earlier, Hans Busch, on the basis of Abbe's theory, had seen the possibility of developing electron microscopes.

Abbe's two most important scientific achievements were in radiation optics (the "sine condition") and undulatory optics ("Beiträge zur Theorie des Mikroskops und der mikroskopischen Wahrnehmung," 1873). The latter led Helmholtz to offer Abbe a professorship at the University of Berlin, but Abbe declined, mainly because of his ties to Zeiss.

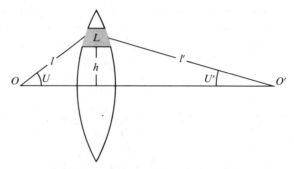

Figure 1. For the Determination of the Sine Condition.

The sine condition is easily derived with the aid of Figure 1. By imaging object point O in the image point O' through the decentered lens L, we obtain the image scale

$$(1) \qquad M = l'/l,$$

which, in accordance with the known image formula, equals the quotient of the image distance l' divided by the real distance l. The relation

$$(2) \qquad h = l \cdot \sin u = l' \cdot \sin u'$$

can also be derived from Figure 1. If the decentered lens L is regarded as the zone of a microscope objective at distance h from the optical axis, then it follows from the above equations that in the case of

$$(3) \qquad \sin u/\sin u' = M = \text{constant},$$

the image scale M is constant for any zone of the lens, that is, over the entire aperture of the objective. Abbe applied the term "aplanatic" to optical systems where the spherical aberration has been corrected, i.e., the axis point O is accurately refracted in the axis point O' and the sine condition (3) has been fulfilled, so that the surface element around O is imaged by all lens zones on the surface element at the image point O' at the same scale. These two corrective conditions can be simultaneously fulfilled only for a single object and image distance; therefore, in the case of the microscope we are limited to a tube of a certain length.

In examining a number of hit-or-miss microscope objectives, Abbe found that in operational position they all fulfilled the sine condition, thus solving the mystery of the success of the hit-or-miss method.

Now Abbe was in a position—through application of the sine condition—to undertake accurate corrections of the aberrations of image systems that did not have too large a divergence. His pupil, Siegfried Czapski, remarked that despite the vastly superior ray union, the images of fine microscopic objects produced by these objectives were duller, showed fewer details, and had less resolving power than the old, poorly corrected systems with larger divergence. After many strenuous and often vain attempts—at tremendous expense to Zeiss—Abbe finally came upon the solution to the problem, as follows.

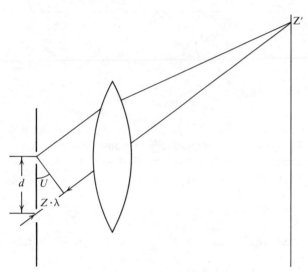

Figure 2. Diffraction on the Grid.

If—as shown in Figure 2—a graticule (heavy vertical lines in figure) is illuminated with, say, red filtered sunlight, then the light rays passing the edges of the graticule gap are deflected. Thus the lower ray in the figure, when it meets the upper ray in point Z' of the rear focal plane of the objective, has covered a distance longer by a fraction or multiple z of the wavelength λ than the upper ray has. Consequently, the two rays are out of phase by

$$(4) \qquad z \cdot \lambda = d \cdot \sin u.$$

The rays are optimally intensified when z is an integer, i.e., the wave crests coincide. The resulting images of the light source are designated as diffraction images of the order of $z = 0, 1, 2, \cdots$. According to Abbe, the accuracy of the image reproduction is a function of the number of diffraction patterns received by the microscope. For the resolution of a graticule the diffraction image of the first order suffices.

If the space between object and objective is filled with a substance (immersion fluid) having the refractive index n, the wavelength λ is reduced to λ/n. Accordingly, the diffraction equation is

$$(5) \qquad d = \lambda/(n \cdot \sin u)$$

limited to the diffraction maximum of the first order ($z = 1$). This renders possible calculation of the grating constants d, that is, the smallest still separable structures of the sample, provided the first-order diffraction maxima of the light source are recognizable in the microscope without the eyepiece. Abbe called the denominator of equation (5) "numerical aperture":

$$(6) \qquad A = n \cdot \sin u.$$

Thus it is possible to separate microscopically such structures as

$$(7) \qquad d = \lambda/A$$

which become finer as the wavelength of the rays used for illumination diminishes and the numerical aperture of the objective used increases. But even if the aperture of the microscope is too small to accommodate diffraction images of the first order, the grating constant can be calculated if, in addition to the diffraction image of 0 order of magnitude of the light source, at least one of the two diffraction images of the first order is accommodated in the microscope. In the extreme case, given oblique illumination, grating structures of the order

$$(8) \qquad d = \lambda/2A$$

can still be calculated and thus resolved.

With the setting up of equations (3) and (7) or (8) the last difficulty was cleared and the reason found for the previously puzzling observation that a poorly corrected objective with large aperture revealed more details in the sample than did a well-corrected objective with small divergence. This peculiarity, derived from Abbe's theory of image resolution, had a powerful effect upon electron microscopy, which developed half a century later. According to (7), the extraordinarily short equivalent wavelength of the electrons should have made the resolution in the electron microscope 100,000 times greater than that in the light microscope. The numerical apertures (6) achieved to date, however, amount to only a fraction of a percent of those of the light microscope, resulting in a much smaller superiority—hardly 100 times—of the electron microscope in the resolution of the smallest objects.

Today it is difficult to realize the magnitude of the effect that the above theoretical considerations exerted on the optical production of the Jena workshop.

The effort to discover a better chromatic correction of the microobjective is also noteworthy. In his report on his visit to the South Kensington Exposition in London (1876), which had an excellent optical section, Abbe points out the causes for this shortcoming: the refusal of the glassworks to consider not only economic but also scientific interests in the application of glass smelting. Nevertheless, his report led one glass chemist, Otto Schott, to undertake this task. Joining forces with Zeiss and Abbe, Schott perfected production methods in the Jena glassworks of Schott & Associates by refining a great number of new optical glasses to high perfection. Ten years after the London Exposition, the Zeiss Works celebrated its greatest triumph to that date with the development of an apochromatic system in which not only the primary but also the secondary color spectrum had been eliminated.

Of no lesser importance is the change in Ernst Abbe's personal outlook that occurred at this time and turned the physicist into a social reformer of equal stature. Having become sole owner of the optical plant and its share in the glassworks, after the death of Carl Zeiss in 1888 and the departure from the firm of the latter's son Roderich, in 1891 he created the Carl Zeiss Foundation, to which he bequeathed his personal fortune, with his wife's approval. In the foundation's charter—which in some respects later served the Prussian state as the model for its progressive social legislation (then generally admired) and turned over the larger part of the profits to the University of Jena—Abbe originated an economic system that unites socialism and capitalism. The economist Alfred Weber, in Volume I of *Schriften der Heidelberger Aktionsgruppe zur Demokratie und zum freien Sozialismus* (1947), proposed a voluntary socialization of German industry modeled after the Carl Zeiss Foundation.

BIBLIOGRAPHY

Abbe's work is collected in the *Gesammelte Abhandlungen von Ernst Abbe*, 5 vols. (Jena, 1904–1940). Further bibliographical information may be obtained from Firma Carl Zeiss, Oberkochen, or Volkseigener Betrieb, Jena.

N. GÜNTHER

'ABD AL-, 'ABDALLAH. See last element of name.

ABEL, JOHN JACOB (*b.* near Cleveland, Ohio, 19 May 1857; *d.* Baltimore, Maryland, 26 May 1938), *pharmacology, biochemistry.*

Abel, the son of George M. and Mary Becker Abel, was born on a farm. His parents were moderately prosperous, and their Rhenish origin may have helped in shaping his receptivity to German academic values. Abel attended high school in Cleveland and entered the University of Michigan in 1876, graduating with a Ph.B. in 1883. He overcame the obstacle of a financially dictated three-year interruption, during which he served first as principal in and then as superintendent of the La Porte, Indiana, public schools. During that period he met Mary Hinman, who was also teaching in La Porte and whom he married in 1883. After graduation Abel went to the Johns Hopkins University, and spent part of a year in Newell Martin's laboratory.

In 1884 Abel sailed for Germany, where he remained until late 1890—the longest German apprenticeship served by any prominent American scientist of his generation. The first two years were spent in Leipzig, studying the basic medical sciences—physiology, histology, pharmacology, and chemistry; like many other Americans he worked with Carl Ludwig. Not surprisingly, Abel undertook an electrophysiological problem. Finally titled "Wie verhalt sich die negative Schwankung des Nervenstroms bei Reizung der sensiblen und motorischen Spinal-surzeln des Frosches?," it was presented as a doctoral thesis at Strasbourg in 1888.

The next two years were spent in Strasbourg, with short periods of clinical study at Würzburg and Heidelberg. Abel received his M.D. from Strasbourg in 1888, then spent the winter semester of 1888–1889 in Vienna, taking postgraduate clinical courses. His interests had already turned to biochemistry and experimental pharmacology; in 1889–1890 he worked in M. von Nencki's laboratory in Berne and in the fall of 1890 returned to Leipzig to work with Drechsel, the physiological chemist at Ludwig's institute. He completed his first essentially biochemical studies in Berne, one on the composition of melanin and another on the determination of the molecular weights of cholic acid, cholesterol, and hydrobilirubin.

While studying in Europe, Abel was aware that when he returned to the United States, he would in all probability have to depend upon clinical work for a livelihood, although he hoped to find a position in which he would have the opportunity to conduct research. He was fortunate to be offered a full-time teaching position at the University of Michigan school of medicine just as his funds were running low. This offer came through the good offices of Victor Vaughan, who had taught Abel as an undergraduate and shared his conviction that chemistry was to play an increasingly central role in the future of medical research.

Abel was appointed lecturer in materia medica and therapeutics in January 1891. Classroom teaching, of course, and not research or laboratory instruction, at first made up the bulk of Abel's duties. By his third year, however, he was able to offer a graduate course on "the influence of certain drugs in the metabolism of tissue" and another on "the methods of modern pharmacology." In 1893 the Johns Hopkins University offered Abel its first professorship of pharmacology, a position that he accepted and occupied continuously until his retirement in 1932. Until 1908 he was also nominally in charge of instruction in physiological chemistry. Abel's research was his life: in his mature years he evinced little concern for the formal routine of teaching. From his retirement until his death, Abel remained steadfastly and constructively at work in his laboratory.

Through this ascetic dedication to research, Abel made one of his most significant contributions to the development of biochemistry and pharmacology in the United States. It was his way of life, his students and colleagues agreed, that influenced them—not his teaching of particular techniques. Almost all his memorialists, for example, mention the intellectual stimulation they received at the austere lunch the laboratory staff shared each day at a plain table, the legs of which were immersed in cans of kerosene to discourage Baltimore's predatory cockroaches. Photographs show Abel's never-changing laboratory attire— white operating-room cap, gray laboratory coat, and long white apron. In values and style of life, Abel seemed to embody the idealized figure of the German professor.

The character of Abel's work played a significant and distinct influence in the reshaping of his discipline. The key to this influence lay in his complete and farsighted commitment to the importance of chemistry in medicine and physiology. Few other biological scientists of his generation had had the prescience to undertake the high-level chemical training necessary in a world of medical and biological research that was increasingly dependent upon chemical and physical sophistication. In the analysis of vital phenomena, Abel warned in 1915, "the investigator must associate himself with those who have labored in fields where *molecules and atoms rather than multi-cellular tissues or even unicellular organisms are the units of study.*" In his own work, Abel was consistently motivated by the chemist's concern

with determining the composition and structure of the substances with which he worked; such concerns were, of course, highly atypical in the medical world of the 1890's. At Johns Hopkins, Abel sought, for example, to introduce microanalytic techniques and even the formal teaching of biophysics. His correspondence and programmatic statements indicate his assumption of what might, if formally articulated, be called a biochemical and biophysical reductionism.

Despite an occasional disdain for mere matters of administration, Abel also played a significant role in the institutional development of American science. He was instrumental in the founding of the *Journal of Experimental Medicine* in 1896 and, in conjunction with Christian Herter, the *Journal of Biological Chemistry* in 1905. Abel was one of the principal organizers of the American Society of Biological Chemists in 1906 and the American Society for Pharmacology and Experimental Therapeutics in 1908. A year later he led in establishing the *Journal of Pharmacology and Experimental Therapeutics,* a publication that he edited until his retirement. Abel's statements and actions indicate a rare mixture of insight and practicality in his understanding of the conditions favoring institutional growth. He of course also figured prominently in shaping the evolution of the traditionally didactic and empirical field of materia medica into modern, laboratory-oriented pharmacology.

After his physiological interlude with Ludwig, Abel's work was, for the half century of his active scientific life, essentially biochemical. His first significant work related to the metabolism of sulfur, a problem he had begun in Leipzig with Drechsel and had continued in Ann Arbor. Abel succeeded in demonstrating the presence of ethyl sulfide in dog urine and, in a related project, in explaining the presence of ammonia in the urine of children who had been given large quantities of limewater. He suggested that it was a product of the breakdown of carbamic acid, a substance Abel had previously studied in alkaline horse urine.

Soon after coming to Johns Hopkins, Abel turned to work with the physiologically active substance found in extracts from the adrenal medulla; this became his all-consuming interest from 1895 to 1905. He published his first article on the substance he had isolated in 1897 (with A. C. Crawford), and in 1899 he christened this blood-pressure-raising hormone "epinephrine." The substance he described, however, was not the free hormone, but a monobenzoyl derivative. This work was the first to give Abel international prominence, although in 1900 Jokichi Takamine was able to isolate the hormone without the benzoyl radical. Both assumed, of course, that they were dealing with a unitary substance. In the years after 1905, Abel completed less elaborate studies of the physiological effects of alcoholic beverages, isolated epinephrine from the parotid secretions of the South American toad *Bufo agua,* studied the poisons of the mushroom *Amanita phalloides*—and even published on the pharmacology of several new chemotherapeutic agents. Work on the pharmaceutical action of phthalein derivatives led to the elaboration of a test for kidney function.

In 1912 and the years immediately following, Abel became deeply interested in work with the protein constituents of the blood. He suggested in 1912 that an "artificial kidney" might be utilized in the removal and study of diffusible substances of the blood. An apparatus of coiled collodion tubes surrounded by a saline solution was soon devised and used for this purpose; arterial blood was shunted through these tubes and then returned to the experimental animal's vein. Using this technique, Abel succeeded in demonstrating the existence of free amino acids in the blood. Even at this time (1913), Abel seems to have been aware of the clinical potential of what he called his "vividiffusion" apparatus; it might, he suggested, prove useful in managing renal failure. A second and related aspect was Abel's demonstration that large quantities of blood could be removed from the circulation if the washed and centrifuged corpuscles were returned. Abel also showed remarkable foresight in his suggestion that "plasmaphaeresis"—his term for this procedure—might ultimately be used to create "blood banks" for use in traumatic and surgical emergencies.

A natural extension of this work led Abel's laboratory to a concern with amino acids and protein degradation products in the blood. A related study of histamines, however, soon revived his earlier interest in hormones. (A resemblance between histamine and the active principle of the posterior pituitary seemed at first to exist.) From the publication of his first paper on the pituitary in 1917 until 1924, when he turned abruptly to work with insulin, Abel labored singlemindedly, although fruitlessly, to isolate a unitary hormone with the protean physiological characteristics associated with pituitary extracts.

Abel's interest in insulin resulted from the explicit invitation of his personal friend A. A. Noyes of the California Institute of Technology. Noyes had acquired funds to subsidize an attempt to isolate pure insulin from the expensive, although readily available, commercial preparation. Abel accepted Noyes' invitation, arrived in Pasadena in October 1924, and was soon able to report encouraging findings. A key step forward lay in his insight that amounts of labile sulfur in his fractions of commercial insulin were directly correlated with physiological activity. Not only did this have ultimate structural implications but—more immediately—it allowed Abel to save a great deal of time; he could now separate out the more active fractions by the use of this criterion without resorting to as yet unstandardized bioassay procedures, in which his laboratory had never excelled and which were far more time-consuming.

Late in 1925, Abel succeeded in forming crystals that, according to the chemical criteria he instinctively employed—crystallization, optical rotation, melting point, and elementary analysis—seemed to be the pure hormone. Despite early scientific enthusiasm at the announcement of this finding in 1926, Abel spent much of the next four years in defending his discovery. The reasons for skepticism were several. One was an initial difficulty in reproducing his crystals. Perhaps more important were certain theoretical implications. It seemed apparent that the substance isolated was a protein, and it was difficult for Abel's contemporaries to believe that the immense protein molecule, of which the regularity of structure was still very much in doubt, could be capable of performing the precise physiological functions of a hormone. The protein, many biochemists believed, must necessarily be an inert carrier, the active principle an adsorbent on its surface. (Similar objections greeted the parallel findings of J. B. Sumner and of J. H. Northrop and M. Kunitz when they announced the protein nature of enzymes they had crystallized.)

By the mid-1930's, however, it was becoming generally assumed that proteins could act as enzymes. The isolation of insulin and its attendant publicity had, of course, helped to sharpen the debate. Abel's insulin work and that of his students played an important part in the line of research culminating in Frederick Sanger's identification in 1955 of the complete primary structure of insulin, the first protein structure to be thus elucidated. After his crystallization of insulin, Abel turned back to his earlier work on the pituitary. When he retired in 1932, he was placed at the head of a laboratory of endocrinology created especially for him at Johns Hopkins. With a touch of characteristic individuality, he then abandoned his hormone work and devoted the remaining years of his life to a study of tetanus toxin and the pathological mechanism through which it acts.

BIBLIOGRAPHY

I. ORIGINAL WORKS. A complete bibliography of Abel's work is available in William deB. MacNider, "Biographical

Memoir of John Jacob Abel 1857–1938," in National Academy of Sciences, U.S.A., *Biographical Memoirs,* **24** (1946), 231–257. Several of Abel's own papers provide valuable synthetic and expository accounts of his work. See *Experimental and Chemical Studies of the Blood With an Appeal for More Extended Chemical Training for the Biological and Medical Investigator,* the first Mellon lecture, given at the University of Pittsburgh under the auspices of the Society for Biological Research (Pittsburgh, 1915), also in *Science,* **42** (1915), 135–147, 165–178; "Some Recent Advances in Our Knowledge of the Ductless Glands," in *Bulletin of the Johns Hopkins Hospital,* **38** (January 1926), 1–32; and "Chemistry in Relation to Biology and Medicine With Especial Reference to Insulin and Other Hormones," in *Science,* **66** (1927), 307–319, 337–346.

The most important source for Abel's life and scientific career is his papers, deposited at the Welch Medical Library, the Johns Hopkins University. This extensive collection, including correspondence, notebooks, and other memoranda, constitutes an important source of information on the development of American biochemistry and pharmacology as well as for the history of the specific research areas that concerned Abel.

II. SECONDARY LITERATURE. There is no full-length biography of Abel available, but among the most useful of numerous biographical sketches are Carl Voegtlin, "John Jacob Abel. 1857–1938," in *Journal of Pharmacology and Experimental Therapeutics,* **67** (1939), 373–406, a detailed account of Abel's scientific work; H. H. Swain, E. M. K. Geiling, and A. Heingartner, "John Jacob Abel at Michigan. The Introduction of Pharmacology Into the Medical Curriculum," in *University of Michigan Medical Bulletin,* **29** (1963), 1–14; E. K. Marshall, Jr., "Abel the Prophet," in *The Johns Hopkins Magazine,* **1** (1950), 11–14; Paul D. Lamson, "John Jacob Abel—A Portrait," in *Bulletin of the Johns Hopkins Hospital,* **68** (1941), 119–157, an engagingly detailed personal portrait; McNider's "Biographical Memoir," cited above; H. H. Dale, "John Jacob Abel. 1857–1938," in *Obituary Notices of Fellows of the Royal Society,* **2** (1939), 577–581; and E. M. K. Geiling, "John Jacob Abel," in *Dictionary of American Biography,* XXII, Supp. 2 (New York, 1958), 4–5. *John Jacob Abel, M.D., Investigator, Teacher, Prophet. 1857–1938* (Baltimore, 1957) is a useful commemorative volume that includes the Lamson and Marshall sketches cited above, as well as a number of Abel's most important papers. In 1957 the Johns Hopkins University celebrated the centenary of Abel's birth with a symposium to which contributions were made by Torald Sollman, Samuel Amberg, Carl Voegtlin, L. G. Rowntree, E. K. Marshall, Jr., E. M. K. Geiling, and Warfield M. Firor; the proceedings appeared in the *Bulletin of the Johns Hopkins Hospital,* **101** (1957), 297–328. An excellent recent study of Abel's insulin work by Jane Murnaghan and Paul Talalay that succeeds in placing this research in the broader context of twentieth-century biochemistry is "John Jacob Abel and the Crystallization of Insulin," in *Perspectives in Biology and Medicine,* **10** (1967), 334–380.

CHARLES E. ROSENBERG

ABEL, NIELS HENRIK (*b.* Finnöy, an island near Stavanger, Norway, 5 August 1802; *d.* Froland, Norway, 6 April 1829), *mathematics.*

Abel's father, Sören Georg Abel, was a Lutheran minister and himself the son of a minister. He was a gifted and highly ambitious theologian, educated at the University of Copenhagen, which was at that time the only such institution in the united kingdom of Denmark-Norway. He had married Ane Marie Simonson, the daughter of a wealthy merchant and shipowner in the town of Risör, on the southern coast. Finnöy was the first parish for pastor Abel; it was small and toilsome, comprising several islands. The couple had seven children, six sons and a daughter; Niels Henrik was their second child.

In 1804 Sören Georg Abel was appointed successor to his father in the parish of Gjerstad, near Risör. The political situation in Norway was tense. Because of its alliance with Denmark the country had been thrown into the Napoleonic Wars on the side of France, and a British blockade of the coast created widespread famine. Pastor Abel was prominent in the nationalistic movement, working for the creation of separate Norwegian institutions—particularly a university and a national bank—if not for outright independence. At the conclusion of the peace treaty of Kiel, Denmark ceded Norway to Sweden. The Norwegians revolted and wrote their own constitution, but after a brief and futile war against the Swedes under Bernadotte, they were compelled to seek an armistice. A union with Sweden was accepted, and Abel's father became one of the members of the extraordinary Storting called in the fall of 1814 to write the necessary revision of the new constitution.

Niels Henrik Abel and his brothers received their first instruction from their father, but in 1815 Abel and his older brother were sent to the Cathedral School in Christiania (Oslo). This was an old school to which many public officials in the province sent their children; some fellowships were available. The Cathedral School had been excellent, but was then at a low ebb, because most of its good teachers had accepted positions at the new university, which began instruction in 1813.

Abel was only thirteen years old when he left home, and it seems probable that deteriorating family life expedited his departure. During the first couple of years his marks were only satisfactory; then the quality of his work declined. His brother fared even worse; he began to show signs of mental illness and finally had to be sent home.

In 1817 an event took place at the school that was destined to change Abel's life. The mathematics teacher mistreated one of the pupils, who died shortly afterward, possibly as a consequence of the punishment. The teacher was summarily dismissed and his place was taken by Bernt Michael Holmboe, who was only seven years older than Abel. Holmboe also served as an assistant to Christoffer Hansteen, professor of astronomy and the leading scientist at the university.

It did not take Holmboe long to discover young Abel's extraordinary ability in mathematics. He began by giving him special problems and recommending books outside the school curriculum. The two then started to study together the calculus texts of Euler, and later the works of the French mathematicians, particularly Lagrange and Laplace. So rapid was Abel's progress that he soon became the real teacher. From notebooks preserved in the library of the University of Oslo one sees that even in these early days he was already particularly interested in algebraic equation theory. By the time he finished school, he was familiar with most of the important mathematical literature. Holmboe was so delighted by the mathematical genius he had discovered that the rector of the school made him moderate his statements about Abel in the record book. But the professors at the university were well informed by Holmboe about the promising young man and made his personal acquaintance. Besides Hansteen, who also taught applied mathematics, there was only one professor of mathematics, Sören Rasmussen, a former teacher at the Cathedral School. Rasmussen, a kindly man, was not a productive scholar; his time was largely taken up by tasks assigned to him by government, particularly in his post as an administrator of the new Bank of Norway.

During his last year at school Abel, with the vigor and immodesty of youth, attacked the problem of the solution of the quintic equation. This problem had been outstanding since the days of del Ferro, Tartaglia, Cardano, and Ferrari in the first half of the sixteenth century. Abel believed that he had succeeded in finding the form of the solution, but in Norway there was no one capable of understanding his arguments, nor was there any scientific journal in which they could be published. Hansteen forwarded the paper to the Danish mathematician Ferdinand Degen, requesting its publication by the Danish Academy.

Degen could not discover any fault in the arguments, but requested that Abel illustrate his method by an example. Degen also found the topic somewhat sterile and suggested that Abel turn his attention to a topic "whose development would have the greatest consequences for analysis and mechanics. I refer to the elliptic transcendentals [elliptic integrals]. A serious investigator with suitable qualifications for re-

search of this kind would by no means be restricted to the many beautiful properties of these most remarkable functions, but could discover a Strait of Magellan leading into wide expanses of a tremendous analytic ocean" (letter to Hansteen).

Abel began constructing his examples for the solution of the fifth-degree equation, but discovered to his dismay that his method was not correct. He also followed Degen's suggestion about the elliptic transcendentals, and it is probable that within a couple of years he had in the main completed his theory of the elliptic functions.

In 1818 pastor Abel was reelected to the Storting, after an unsuccessful bid in 1816. But his political career ended in tragedy. He made violent unfounded charges against other representatives and was threatened with impeachment. This, together with his drunkenness, made him the butt of the press. He returned home in disgrace, a disillusioned man. Both he and his wife suffered from alcoholism, and the conditions at the vicarage and in the parish became scandalous. It was generally considered a relief when he died in 1820. His widow was left in very straitened circumstances, with a small pension barely sufficient to support her and her many children.

The penniless Abel entered the university in the fall of 1821. He was granted a free room at the university dormitory and received permission to share it with his younger brother Peder. But the new institution had no fellowship funds, and some of the professors took the unusual measure of supporting the young mathematician out of their own salaries. He was a guest in their houses and became particularly attracted to the Hansteen home, and to Mrs. Hansteen and her sisters.

Abel's first task at the university was to satisfy the requirements for the preliminary degree, *Candidatus Philosophiae*. Once this was achieved, after a year, Abel was entirely on his own in his studies. There were no advanced courses in mathematics and the physical sciences, but this does not seem to have been a handicap; in a letter from Paris a little later he stated that he had read practically everything in mathematics, important or unimportant.

He devoted his time to advanced research and his efforts received a strong impetus when Hansteen started a scientific periodical, *Magazin for Naturvidenskaben*. In 1823 this journal published Abel's first article, in Norwegian, a study of functional equations. Mathematically it was not important, nor was his second little paper. The subscribers to the magazine had been promised a popular review, however, and Hansteen, probably after criticism, felt obliged to apologize for the character of these papers: "Thus

I believe that the *Magazin* in addition to scientific materials should also further the tools serving for their analysis. It will be reckoned to our credit that we have given the learned public an opportunity to become acquainted with a work from the pen of this talented and skillful author" (*Magazin*, **1**). Abel's next paper, "Opläsning afet Par Opgaver ved bjoelp af bestemte Integraler" ("Solution of Some Problems by Means of Definite Integrals"), is of importance in the history of mathematics, since it contains the first solution of an integral equation. The paper, which went unnoticed at the time, in part because it was in Norwegian, deals with the mechanical problem of the motion of a mass point on a curve under the influence of gravitation. During the winter of 1822–1823 Abel also composed a longer work on the integration of functional expressions. The paper was submitted to the university Collegium in the hope that that body would assist in its publication. The manuscript has disappeared, but it seems likely that some of the results obtained in it are included in some of Abel's later papers.

Early in the summer of 1823 Abel received a gift of 100 daler from Professor Rasmussen to finance a trip to Copenhagen to meet Degen and the other Danish mathematicians. His letters to Holmboe reveal the mathematical inspiration that he received. He stayed in the house of his uncle and here made the acquaintance of his future fiancée, Christine Kemp.

Upon his return to Oslo, Abel again took up the question of the solution of the quintic equation. This time he took the reverse view and succeeded in solving the centuries-old problem by proving the impossibility of a radical expression that represents a solution of the general fifth- or higher-degree equation. Abel fully realized the importance of his result, so he had it published, at his own expense, by a local printer. To reach a larger audience, he wrote it in French: "Mémoire sur les équations algébriques où on démontre l'impossibilité de la résolution de l'équation générale du cinquième degré." To save expense the whole pamphlet was compressed to six pages. The resulting brevity probably made it difficult to understand; at any rate, there was no reaction from any of the foreign mathematicians—including the great C. F. Gauss, to whom a copy was sent.

It had become clear that Abel could no longer live on the support of the professors. His financial problems had been increased by his engagement to Christine Kemp, who had come to Norway as a governess for the children of a family living near Oslo.

Abel applied for a travel grant, and after some delays the government decided that Abel should receive a small stipend to study languages at the university to prepare him for travel abroad. He was then to receive a grant of 600 daler for two years of foreign study.

Abel was disappointed at the delay but dutifully studied languages, particularly French, and used his time to prepare a considerable number of papers to be presented to foreign mathematicians. During the summer of 1825 he departed, together with four friends, all of whom also intended to prepare themselves for future scientific careers; one of them later became professor of medicine, and the three others became geologists. Abel's friends all planned to go to Berlin, while Abel, upon Hansteen's advice, was to spend his time in Paris, then the world's principal center of mathematics. Abel feared being lonely, however, and also decided to go to Berlin, although he well knew that he would incur the displeasure of his protector.

Abel's change of mind turned out to be a most fortunate decision. On passing through Copenhagen, Abel learned that Degen had died, but he secured a letter of recommendation from one of the other Danish mathematicians to Privy Councilor August Leopold Crelle. Crelle was a very influential engineer, intensely interested in mathematics although not himself a strong mathematician.

When Abel first called upon Crelle, he had some difficulty in making himself understood, but after a while Crelle recognized the unusual qualities of his young visitor. The two became lifelong friends. Abel presented him with a copy of his pamphlet on the quintic equation, but Crelle confessed that it was unintelligible to him and recommended that Abel write an expanded version of it. They talked about the poor state of mathematics in Germany. In a letter to Hansteen, dated from Berlin, 5 December 1825, Abel wrote:

> When I expressed surprise over the fact that there existed no mathematical journal, as in France, he said that he had long intended to edit one, and would presently bring his plan to execution. This project is now organized, and that to my great joy, for I shall have a place where I can get some of my articles printed. I have already prepared four of them, which will appear in the first number.

Journal für die reine und angewandte Mathematik, or *Crelle's Journal,* as it is commonly known, was the leading German mathematical periodical during the nineteenth century. The first volume alone contains seven papers by Abel and the following volumes contain many more, most of them of preeminent importance in the history of mathematics. Among the first is the expanded version of the proof of the impossibility of the solution of the general quintic equation by radicals. Here Abel develops the necessary algebraic background, including a discussion of algebraic field extensions. Abel was at this time not aware that he had a precursor, the Italian mathematician Paolo Ruffini. But in a posthumous paper on the equations which are solvable by radicals Abel states: "The only one before me, if I am not mistaken, who has tried to prove the impossibility of the algebraic [radical] solution of the general equations is the mathematician Ruffini, but his paper is so complicated that it is very difficult to judge on the correctness of his arguments. It seems to me that it is not always satisfactory." The result is usually referred to as the Abel-Ruffini theorem.

After Abel's departure from Oslo an event took place that caused him much concern. Rasmussen had found his professorship in mathematics too burdensome when combined with his public duties. He resigned, and shortly afterward the faculty voted to recommend that Holmboe be appointed to fill the vacancy. Abel's Norwegian friends found the action highly unjust, and Abel himself probably felt the same way. Nevertheless, he wrote a warm letter of congratulation to his former teacher, and they remained good friends. But it is evident that from this moment Abel worried about his future and his impending marriage; there was no scientific position in sight for him in his home country.

During the winter in Berlin, Abel contributed to *Crelle's Journal*; among the notable papers are one on the generalization of the binomial formula and another on the integration of square root expressions. But one of his main mathematical concerns was the lack of stringency in contemporary mathematics. He mentioned it repeatedly in letters to Holmboe. In one of these, dated 16 January 1826, he wrote:

My eyes have been opened in the most surprising manner. If you disregard the very simplest cases, there is in all of mathematics not a single infinite series whose sum has been stringently determined. In other words, the most important parts of mathematics stand without foundation. It is true that most of it is valid, but that is very surprising. I struggle to find the reason for it, an exceedingly interesting problem.

A result of this struggle was his classic paper on power series which contains many general theorems and also, as an application, the stringent determination of the sum of the binomial series for arbitrary real or complex exponents.

During the early spring of 1826, Abel felt obliged to proceed to his original destination, Paris. Crelle had promised to accompany him, and on the way they intended to stop in Göttingen to visit Gauss. Unfortunately, pressure of business prevented Crelle from leaving Berlin. At the same time, Abel's Norwegian friends were planning a geological excursion through central Europe, and, again reluctant to be separated from them, he joined the group. They traveled by coach through Bohemia, Austria, northern Italy, and the Alps. Abel did not reach Paris until July, low on funds after the expensive trip.

The visit to Paris was to prove disappointing. The university vacations had just begun when Abel arrived, and the mathematicians had left town. When they returned, he found that they were aloof and difficult to approach; it was only in passing that he met Legendre, whose main interest in his old age was elliptic integrals, Abel's own specialty. For presentation to the French Academy of Sciences Abel had reserved a paper that he considered his masterpiece. It dealt with the sum of integrals of a given algebraic function. Abel's theorem states that any such sum can be expressed as a fixed number p of these integrals, with integration arguments that are algebraic functions of the original arguments. The minimal number p is the genus of the algebraic function, and this is the first occurrence of this fundamental quantity. Abel's theorem is a vast generalization of Euler's relation for elliptic integrals.

Abel spent his first months in Paris completing his great memoir; it is one of his longest papers and includes a broad theory with applications. It was presented to the Academy of Sciences on 30 October 1826, under the title "Mémoire sur une propriété générale d'une classe très-étendue de fonctions transcendantes." Cauchy and Legendre were appointed referees, Cauchy being chairman. A number of young men had gained quick distinction upon having their works accepted by the Academy, and Abel awaited the referees' report. No report was forthcoming, however; indeed, it was not issued until Abel's death forced its appearance. Cauchy seems to have been to blame; he claimed later that the manuscript was illegible.

Abel's next two months in Paris were gloomy; he had little money and few acquaintances. He met P. G. L. Dirichlet, his junior by three years and already a well-known mathematician, through a paper in the Academy sponsored by Legendre. Another acquaintance was Frédéric Saigey, editor of the scientific revue *Ferrusac's Bulletin,* for whom Abel wrote a few articles, particularly about his own papers in *Crelle's Journal.* After Christmas he spent his last resources to pay his fare to Berlin.

Shortly after his return to Berlin, Abel fell ill; he seems to have then suffered the first attack of the tuberculosis that was later to claim his life. He borrowed some money from Holmboe, and Crelle probably helped him. Abel longed to return to Norway but felt compelled to remain abroad until his fellowship term had expired. Crelle tried to keep him in Berlin until he could find a position for him at a German university; in the meantime he offered him the editorship of his *Journal*.

Abel worked assiduously on a new paper: "Recherches sur les fonctions elliptiques," his most extensive publication (125 pages in the *Oeuvres complètes*). In this work he radically transformed the theory of elliptic integrals to the theory of elliptic functions by using their inverse functions corresponding in the most elementary case to the duality

$$y = \arcsin x = \int \frac{dx}{\sqrt{1 - x^2}} \quad x = \sin y.$$

The elliptic functions thereby become a vast and natural generalization of the trigonometric functions; in the wake of Abel's work they were to constitute one of the favorite research topics in mathematics during the nineteenth century. Abel had already developed most of the theory as a student in Oslo, so he was able to present the theory of elliptic functions with a great richness of detail, including double periodicity, expansions in infinite series and products, and addition theorems. The theory led to the expressions for functions of a multiple of the argument with the concomitant determination of the equations for fractional arguments and their solution by radicals, much in the way that Gauss had treated the cyclotomic equations; Abel's letters to Holmboe (from Paris in December 1826 and from Berlin on 4 March 1827) indicate that he was particularly fascinated by a determination of the condition for a lemniscate to be divisible into equal parts by means of compass and ruler, analogous to Gauss's construction of regular polygons. The last part deals with the so-called theory of complex multiplication, later so important in algebraic number theory.

Abel returned to Oslo on 20 May 1827, to find that the situation at home was as gloomy as he had feared. He had no position in prospect, no fellowship, and an abundance of debts. His application to have his fellowship prolonged was turned down by the Department of Finance, but the university courageously awarded him a small stipend out of its meager funds. This action was criticized by the department, which reserved the right to have the amount deducted from any future salary he might receive.

Abel's fiancée found a new position with friends of Abel's family, the family of the owner of an iron-works at Froland, near Arendal. During the fall Abel eked out a living in Oslo by tutoring schoolboys and probably with the help of friends. At the new year the situation became brighter. Hansteen, a pioneer in geomagnetic studies, received a large grant for two years to examine the earth's magnetic field in unexplored Siberia. In the meantime Abel became his substitute both at the university and at the Norwegian Military Academy.

The first part of the "Recherches" was published in *Crelle's Journal* in September 1827, and Abel completed the second part during the winter. He lived in isolation at Oslo; there was no package mail during the winter, and he had no inkling of the interest his memoir had created among European mathematicians. Nor did he know that a competitor had appeared in the field of elliptic functions until early in 1828, when Hansteen showed him the September issue of the *Astronomische Nachrichten*. In this journal a young German mathematician, K. G. J. Jacobi, announced without proofs some results concerning the transformation theory of elliptic integrals. Abel hurriedly added a note to the manuscript of the second part of the "Recherches," showing how Jacobi's results were the consequence of his own.

Abel was keenly aware that a race was at hand. He interrupted a large paper on the theory of equations that was to contain the determination of all equations that can be solved by radicals; the part that was published contained the theory of those equations that are now known as Abelian. He then wrote, in rapid succession, a series of papers on elliptic functions. The first was "Solution d'un problème général concernant la transformation des fonctions elliptiques." This, his direct response to Jacobi, was published in *Astronomische Nachrichten*; the others appeared in *Crelle's Journal*. In addition, Abel prepared a book-length memoir, "Précis d'une théorie des fonctions elliptiques," which was published after his death. Jacobi, on the other hand, wrote only brief notices which did not reveal his methods; these were reserved for his book, *Fundamenta nova theoriae functionum ellipticarum* (1829).

Much has been written about the early theory of elliptic functions. There seems to be little doubt that Abel was in possession of the ideas several years before Jacobi. On the other hand, it is also an established fact that Gauss, although publishing nothing, had discovered the principles of elliptic functions long before either Abel or Jacobi.

The European mathematicians watched with fascination the competition between the two young mathematicians. Legendre noticed Jacobi's announcements and also received a letter from him. In a meeting of the French Academy in November 1827, he praised the new mathematical star; the speech was reproduced in the newspapers and Legendre sent the clipping to Jacobi. In his reply Jacobi, after expressing his thanks, pointed out Abel's "Recherches" and its general results. Legendre responded: "Through these works you two will be placed in the class of the foremost analysts of our times." He also expressed his disappointment over Jacobi's method of publication and was irritated when Jacobi confessed that in order to derive some of his results he had had to rely on Abel's paper. About this time also, Abel began a correspondence with Legendre and poured out his ideas to him.

All that the European mathematicians knew about Abel's condition in Norway was that he had only a temporary position and had recently been compelled to tutor schoolboys to make a living. The main source of their information was Crelle, who constantly used his influence to try to obtain an appointment for Abel at a new scientific institute to be created in Berlin. Progress was very slow, however. In September 1828 four prominent members of the French Academy of Sciences took the extraordinary step of addressing a petition directly to Bernadotte, now Charles XIV of Norway-Sweden, calling attention to Abel and urging that a suitable scientific position be created for him. In a meeting of the Academy, on 25 February 1829, Legendre also paid tribute to Abel and his discoveries, particularly to his results in the theory of equations.

In the meantime Abel, in spite of his deteriorating health, wrote new papers frantically. He spent the summer vacation of 1828 on the Froland estate with his fiancée. At Christmas he insisted on visiting her again, notwithstanding that it required several days' travel in intense cold. He was feverish when he arrived, but enjoyed the family Christmas celebration. He may have had a premonition that his days were numbered, however, and he now feared that the great paper submitted to the French Academy had been lost forever. He therefore wrote a brief note, "Demonstration d'une propriété générale d'une certaine classe de fonctions transcendantes," in which he gave a proof of the main theorem. He mailed it to Crelle on 6 January 1829.

While waiting for the sled that was to return him to Oslo, Abel suffered a violent hemorrhage; the doctor diagnosed his illness as tuberculosis and ordered prolonged bed rest. He died in April, at the age of twenty-six, and was buried at the neighboring Froland church during a blizzard. The grave is marked by a monument erected by his friends. One of them, Baltazar Keilhau, wrote to Christine Kemp,

without ever having seen her, and made her an offer of marriage which she accepted. Two days after Abel's death Crelle wrote jubilantly to inform him that his appointment in Berlin had been secured.

On 28 June 1830, the French Academy of Sciences awarded its Grand Prix to Abel and Jacobi for their outstanding mathematical discoveries. After an intensive search in Paris the manuscript of Abel's great memoir was rediscovered. It was published in 1841, fifteen years after it had been submitted. During the printing it again disappeared, not to reappear until 1952 in Florence.

Crelle wrote an extensive eulogy of Abel in his *Journal* (**4** [1829], 402):

> All of Abel's works carry the imprint of an ingenuity and force of thought which is unusual and sometimes amazing, even if the youth of the author is not taken into consideration. One may say that he was able to penetrate all obstacles down to the very foundations of the problems, with a force which appeared irresistible; he attacked the problems with an extraordinary energy; he regarded them from above and was able to soar so high over their present state that all the difficulties seemed to vanish under the victorious onslaught of his genius. . . . But it was not only his great talent which created the respect for Abel and made his loss infinitely regrettable. He distinguished himself equally by the purity and nobility of his character and by a rare modesty which made his person cherished to the same unusual degree as was his genius.

BIBLIOGRAPHY

I. ORIGINAL WORKS. Abel's complete works are published in two editions, *Oeuvres complètes de N. H. Abel, mathématicien*, ed. and annotated by B. Holmboe (Oslo, 1839), and *Nouvelle édition*, M. M. L. Sylow and S. Lie, eds., 2 vols. (Oslo, 1881).

II. SECONDARY LITERATURE. Materials on Abel's life include *Niels Henrik Abel: Mémorial publié à l'occasion du centenaire de sa naissance* (Oslo, 1902) which comprises all letters cited in the text; and O. Ore, *Niels Henrik Abel; Mathematician Extraordinary* (Minneapolis, Minn., 1957).

OYSTEIN ORE

ABEL, OTHENIO (*b.* Vienna, Austria, 20 June 1875; *d.* Pichl am Mondsee, Austria, 4 July 1946), *paleontology, paleobiology.*

Abel's greatest scientific achievement, the founding of paleobiology, undoubtedly grew out of his background. For several generations his ancestors on his father's side had been gardeners. His grandfather had taken his examination in botany under Nicolaus Jaquin and had received *summa cum eminentia;* his father was a teacher at the horticulture school and *Dozent* at the Institute of Agriculture in Vienna.

As a sixteen-year-old Gymnasium student Abel eagerly collected fossils. To please his parents, however, he continued his education at the Faculty of Law of the University of Vienna. Soon, though, he devoted more time to the natural sciences, especially botany: his first publications were several papers on orchids.

In 1898, while still a student, Abel became an assistant at the Geological Institute of the University of Vienna under Eduard Suess. He took his major examination in geology and paleontology in 1899 and received the Ph.D. Abel then attended the school of mining in Leoben for a short while, and in 1900 he accepted a position as *Mitarbeiter* at the Imperial–Royal Geological State Institute in Vienna, where he was active until 1907. At the very beginning of this period he published the result of his studies on Cetacea, "Untersuchungen über die fossilen Platanistiden des wiener Beckens," which he had begun while still a student. This paper brought him an invitation to investigate the fossil whales of Belgium. From 1900 on, Abel was *collaborateur étranger* of the Royal Museum of Natural History of Belgium, and several times he traveled to that country. There he met Louis Dollo, who became his teacher of paleontology and his friend. Abel was stimulated by Dollo's teachings, as well as by the writings of Vladimir Kovalevsky and Henry Fairfield Osborn, which were influential in the founding of paleobiology.

In 1901 Abel was appointed *Privatdozent* in paleontology at the University of Vienna; in 1907, associate professor; and in 1912, full professor. He became professor of paleobiology and director of the paleobiology department, later the Paleobiological Institute of the university, in 1917.

Abel's important work *Grundzüge der Paläobiologie der Wirbeltiere* appeared in 1912, *Die vorzeitlichen Säugetiere* in 1914, and *Paläobiologie der Cephalopoden* in 1916. By then he had published over 100 papers. The stream of writings continued. *Die Stämme der Wirbeltiere* (1919) was followed by *Lehrbuch der Paläozoologie* (1920), *Lebensbilder aus der Tierwelt der Vorzeit* (1922), and *Geschichte und Methode der Rekonstruktion vorzeitlicher Wirbeltiere* (1925). Many shorter works were published as well. During this period Abel was in charge of extensive paleozoological excavations in the Drachenhöhle near Mixnitz, Styria; he had previously (1912) undertaken similar excavations at Pikermi, Greece.

In 1911 Abel was awarded the Bigsby Medal of the Geological Society of London; in 1922, the Elliot Medal of the National Academy of Sciences in Washington and the Rainer Medal of the Zoological-Botanical Society of Vienna. He became president of the Paleontological Society in 1921, and was subsequently a member or an honorary member of numerous scientific academies and learned societies, as well as honorary doctor of the universities of Cape Town and Athens. He was also dean (1927/1928) and rector (1932/1933) of the University of Vienna.

Abel had been at the University of Vienna for twenty-eight years and had published more than 250 papers when, in 1935, he accepted a post at the University of Göttingen. He spent five years at Göttingen, amassing new collections and turning his attention to newer subdivisions of paleontology that had not yet been extensively investigated. In 1940 he retired and returned to Austria, where, in Salzburg, he founded a short-lived institute for biological natural history. Despite this failure, he continued his scientific activities until very shortly before his death.

BIBLIOGRAPHY

I. ORIGINAL WORKS. Among the most important of Abel's 275 publications are *Grundzüge der Paläobiologie der Wirbeltiere* (Stuttgart, 1912); *Die vorzeitlichen Säugetiere* (Jena, 1914); *Paläobiologie der Cephalopoden* (Jena, 1916); *Die Stämme der Wirbeltiere* (Berlin–Leipzig, 1919); *Lehrbuch der Paläozoologie* (Jena, 1920); *Lebensbilder aus der Tierwelt der Vorzeit* (Jena, 1922); *Eroberungszüge der Wirbeltiere in die Meere der Vorzeit* (Jena, 1924); *Geschichte und Methode der Rekonstruktion vorzeitlicher Wirbeltiere* (Jena, 1925); *Amerikafahrt* (Jena, 1926); *Paläobiologie und Stammesgeschichte* (Jena, 1929); *Die Drachenhöhle bei Mixnitz in Steiermark,* 3 vols. (Vienna, 1931); *Die Stellung des Menschen im Rahmen der Wirbeltiere* (Jena, 1931); and *Vorzeitliche Lebensspuren* (Jena, 1935).

In 1928 Abel founded the journal *Palaeobiologica,* and most of his shorter papers appeared in it.

II. SECONDARY LITERATURE. Of several obituary notices giving information on Abel's life and works, the most extensive one is written by his son-in-law, Kurt Ehrenberg: "Othenio Abel, sein Werden und Wirken," in *Neues Jahrbuch für Mineralogie, Geologie und Paläontologie,* Mitteilungshefte no. 11/12 (1949), 325–328. See also E. Fischer, "Othenio Abel 1875–1946," in *Die Naturwissenschaften,* 6, no. 33 (1946); and K. Leuchs, "Othenio Abel," in *Almanach der Österreichischen Akademie der Wissenschaften, 1947* (Vienna, 1948).

An oil portrait of Abel by P. F. Gsur, painted in 1933, is in the possession of the University of Vienna.

HANS BAUMGÄRTEL

ABELARD, PETER. See **Abailard, Pierre.**

ABENARE. See **Ibn Ezra.**

ABENGUEFITH. See **Ibn Wāfid.**

ABETTI, ANTONIO (*b.* S. Pietro di Gorizia, Italy, 19 June 1846; *d.* Arcetri [Florence], Italy, 20 February 1928), *astronomy.*

Although Abetti received a degree in civil engineering from the University of Padua in 1867, he abandoned engineering the following year in order to devote himself to astronomy. He was appointed astronomer of the observatory of the University of Padua in 1868, and he remained there until 1893. In 1894, following a public competitive examination, he was appointed director of the astronomical observatory of Arcetri and professor of astronomy at the University of Florence; he held this post until 1921, when he was obliged to retire because he had reached the compulsory retirement age. Nonetheless, from 1921 to 1928, Abetti continued his researches in astronomy at the observatory. He was a member of the Accademia Nazionale dei Lincei (Rome), associate member of the Royal Astronomical Society (London), and a member of several other Italian academies.

Abetti's scientific activity was devoted essentially to positional astronomy. At Padua, with a modest equatorial telescope, he made many observations on the positions of small planets, comets, occultation of stars, and eclipses. In 1874, as a member of the Italian expedition directed by P. Tacchini, he went to Muddapur in Bengal, to observe the transit of Venus over the disc of the sun; it was the first time that this transit was observed through the spectroscope. Abetti also determined the geographic coordinates of the station.

In addition, Abetti determined the differences in longitude between various Italian localities, a project sponsored by the Italian Geodetic Commission. These studies resulted in the perfection and simplification of determinations of time. When he took over the directorship of the Arcetri observatory, which had been founded by G. B. Donati in 1872 and had been practically abandoned after Donati's death, Abetti set about reconstructing it. He provided the observatory with an equatorial telescope, which he had built in the shops of the observatory of Padua. To it he adapted the well-known objective previously constructed by G. B. Amici, with a diameter of twenty-eight centimeters, and a "Bamberg's small meridian circle." With these instruments he was able to carry out, and to encourage others to carry out, many observations on the positions of small planets, comets, and fixed stars.

All these observations and his scientific studies on the precision of observations, and the solution of equations that are met with in the method of least squares, are published in various issues of *Memoirs and Observations of the Observatory of Arcetri.*

BIBLIOGRAPHY

Obituary notices are G. Armellini, *Rendiconti Accademia dei Lincei,* **9** (1929), 13; L. Carnera, *Vierteljahrschrift der Astronomischen Gesellschaft,* **64** (1929), 2; and G. Silva, *Memorie della Società Astronomica Italiana,* **4** (1928), 193.

Giorgio Abetti

ABICH, OTTO HERMANN WILHELM (*b.* Berlin, Germany, 11 December 1806; *d.* Vienna, Austria, 1 July 1886), *geology.*

Abich's interest in natural science and travel was formed under the influence of his father, Wilhelm, an official in the Department of Mines; and two uncles, Martin Klaproth, a chemist, and Julius Klaproth, an ethnographer, orientalist, traveler, and expert on the Caucasian peoples.

He began his higher education at the Faculty of Law of Heidelberg University but later transferred to the department of physics and mathematics at the University of Berlin, from which he graduated in 1831. Having defended a dissertation on the minerals of the spinel group, he received the Ph.D. At Heidelberg, Abich's teachers included Hegel, Humboldt, and Buch. On Buch's advice, Abich traveled to Italy, where he took great interest in the problems of recent volcanicity and published a series of articles that brought him widespread fame.

Shortly thereafter Abich was invited to become extraordinary professor of mineralogy at the University of Dorpat (now Tartu, Estonian S.S.R.), and in 1843 he moved to Russia. As an outstanding expert on volcanic phenomena, he was immediately sent to Transcaucasia in order to determine the causes of the catastrophic earthquake that had destroyed part of the mountain peak Great Ararat in 1840. Abich was captivated by the Caucasus, and for almost thirty-five years he studied this complex, mountainous country. His fundamental papers on the geology of the region, which previously had been totally ignored, made him world famous.

In 1853 the Academy of Sciences in St. Petersburg elected Abich to membership, and in 1866 he was elevated to honorary membership. Since Abich was continuously involved in work far from Dorpat, it became necessary for him to resign his post there. He was then assigned to the headquarters of the Corps of Mining Engineers and was detailed to the command of the viceregent of the Caucasus. Spending the greater part of his time in Armenia, Azerbaijan, and Georgia, Abich went to St. Petersburg or western Europe only to work on the specimens he collected or to publish important monographs. During one such trip in the late 1850's he met and married Adelaide Hess, daughter of the renowned chemist Hermann Heinrich Hess.

Abich was a proponent of the volcanistic theory that assigned the decisive role in geologic processes to hypogene forces and to magma. In the first years of his career Abich's interests were concentrated on mineralogy and petrography. In his paper on spinels he was the first to establish that a characteristic of this group is the constancy of a crystallographic form that is preserved regardless of the wide range of isomorphous replacements. While conducting his petrographic investigations, Abich uncovered the important role of feldspar in the composition of igneous rocks.

In the years that followed, Abich continued his intensive study of various Caucasian magmatic formations, conducting chemical investigations simultaneously with his mineralogic-petrographic analysis. Abich's work in the Caucasus demanded that he answer many stratigraphic questions. The frequent change of facies and the complexity of the tectonics seriously hampered investigation, and Abich therefore paid serious attention to the collecting of fossils, which he determined himself. He studied the fauna of the whole geological sequence exposed in the Caucasus and in Transcaucasia—from the Paleozoic to the Quaternary. Despite the specific features of the Caucasian fauna, Abich successfully established the geological age of the enclosing strata. For some years his stratigraphic conclusions served as the basis of all geological research in the Caucasus, and only in the middle of the twentieth century have his data been subjected to more or less important revision.

Studying the tectonic structure of the areas in which he carried out his investigations, Abich interpreted this structure from the standpoint of ideas then current. He thought that all parallel mountain ranges arose concurrently; thus, in the Caucasus he distinguished four major directions of tectonic lines and, consequently, four stages of tectogenesis. Adhering to Buch's ideas, he considered all mountains to be elevation craters that originated from the intrusion of magmatic masses, and classified canyon-like mountain passes as tectonic fissures.

On the whole, Abich correctly plotted the most important tectonic lines in the Caucasus and established that the highest seismicity is associated with these lines. In the southeastern and northwestern ends of the main Caucasian range Abich discovered a regularity in the location of mud volcanoes according to a definite geometrical grid. He established that the discharge of combustible gases and mineral springs are associated with zones of large faults.

Abich was especially interested in the practical aspects of his geological investigations, and he allotted much time to the study of mineral deposits. He located and described a large number of ore deposits and sources of both ferrous and nonferrous metals, gypsum, rock salt, sulfur, alum, combustible minerals, and mineral waters. He made the important discovery of the Chiatura manganese deposit, one of the largest on earth. Studying the oil shows in the Baku region, Abich established that large accumulations of petroleum are associated with the more elevated parts of tectonic structures, and, from 1847 on, he developed and successfully applied the anticlinal theory in the search for oil fields. He apparently made this discovery independently of American geologists who, in the same period (1840–1860), began prospecting for oil in the arched parts of the uplifts. On the question of the origin of petroleum, Abich adhered to the distillation (inorganic) theory and thought petroleum was the product of sublimation, emanating at a substantial depth from coal beds affected by volcanic heat.

Because he possessed an exceptional capacity for work and the ability to understand the peculiarities of a complex geological structure, Abich had great success in conducting his regional investigations and in mapping the geology of large areas. In addition, he completed a series of papers generalizing data collected by the many Russian geologists then working in the Caucasus; these led to fundamental summaries on the geology of various parts of this mountainous country. The maps published by Abich over a period of many decades have been widely used by geologists.

The range of Abich's scientific interests was wide and varied. Among his printed works, the total number of which exceeds 200, in addition to geological subjects there are papers on geomorphology, glaciology, meteorology, geography, geobotany, meteoritics, and archaeology.

By the time Abich reached the age of seventy, he found it difficult to continue his expeditionary research. Retiring with a pension in 1876, he went to Vienna, where, working intensively, he prepared important summaries on Caucasian geology for publication. He died in his eightieth year from what was tardily diagnosed as acute appendicitis. In accordance with his wishes, Abich's body was cremated and an urn containing his ashes was buried in his mother's grave in Koblenz.

BIBLIOGRAPHY

I. ORIGINAL WORKS. Abich's most important works are *De spinello. Dissertatio inauguralis chemicaquam ad consequendos ab amplissimo Universitatis Berolinensis philosophorum rodine summos in philosophia honores* (Berlin, 1831); "Vergleichende geologische Grundzüge der kaukasischen, armenischen und nordpersischen Gebirge als Prodromus einer Geologie der kaukasischen Länder," in *Mémoires de l'Académie des Sciences de St. Pétersbourg,* 6th ser., **7** (1859), 359–534; "Über eine im Caspischen Meer erschienene Insel nebst Beiträgen zur Kenntniss der Schlammvulkane der caspischen Region," *ibid.,* 7th ser., **6** (1863), i–viii, 1–151, with 4 maps; *Geologische Beobachtungen auf Reisen in der Gebirgsländern zwischen Kur und Araxes* (Tiflis, 1867); "Geologische Beobachtungen auf Reisen im Kaukasus im Jahre 1873," in *Bulletin de la Société des Naturalistes de Moscou,* **48**, pt. 1, no. 2 (1874), 278–342; pt. 2, no. 3 (1874–1875), 63–107, and no. 4, 243–272; *Über krystallinische Hagel im unteren Kaukasus und seine Beziehung zu der Physik des Bodens* (Vienna, 1879); *Geologische Forschungen in den kaukasischen Ländern,* pt. 2, *Geologie des armenischen Hochlands. 1. Westhälfte* (Vienna, 1882); and *Geologische Forschungen in kaukasischen Ländern,* pt. 3, *Geologie des armenischen Hochlands. 2. Östhälfte* (Vienna, 1887).

II. SECONDARY LITERATURE. Works on Abich are E. Suess and A. Abich, "Herman Abich. Biographie und Verzeichnis seiner Werke," in H. Abich's *Geologische Forschungen in den kaukasischen Ländern,* pt. 3, pp. i–xii; and V. V. Tikhomirov and S. P. Volkova, "Zhizn' i trudy Germana Vil'gelmovicha Abikha" ("The Life and Works of Hermann Wilhelm Abich"), in *Ocherki po istorii geologicheskikh znanij* ("Essays on the History of Geological Knowledge"), **8** (Moscow, 1959), pp. 177–238, which includes a portrait of Abich and a bibliography of works by him and about him.

V. V. TIKHOMIROV

ABNEY, WILLIAM DE WIVELESLIE (*b.* Derby, England, 24 July 1843; *d.* Folkestone, England, 2 December 1920), *photography, astronomy.*

Abney was one of the founders of modern photography, combining a scientific approach, ingenuity, and manipulative skill with a talent for popularization. In later life his interest in color photography led him to investigate theories of color vision.

The eldest son of a clergyman, Abney graduated from the Royal Military Academy and served several years with the Royal Engineers in India before being invalided home. In 1869 he was made chemical assistant to the instructor in telegraphy at the Chatham School of Military Engineering, where he was able to pursue a boyhood interest in photography. His first book was *Chemistry for Engineers* (1870); his second, *Instruction in Photography* (1871), rapidly became a standard text.

Abney pioneered in the quantitative sensitometry of photographic images: his studies of how negatives blacken in response to varying amounts of incident light (1874, 1882) preceded the "D-log E" curves of F. Hurter and V. C. Driffield in use today. His early work tended to confirm the photochemical law of Robert Bunsen and Sir Henry Roscoe, which states that intensity of light and the time of exposure to it are reciprocally responsible for the effect produced; but when Julius Scheiner showed, beginning in 1888, that the density of stellar images did not follow this law, Abney was quick to confirm the "failure of reciprocity" in the laboratory, and himself discovered the related intermittency effect (1893).

Abney's first astronomical publications were reports on an expedition he led to Egypt in December 1874, to photograph a transit of Venus across the face of the sun. In preparation for this expedition Abney—by then a captain—invented a dry photographic emulsion (1874): this, his "albumen beer" process, remained in use for general as well as solar photography until superseded by commercial gelatin products. He went on to study the chemistry of latent image developing (1877) and to introduce hydroquinone (1880), still one of the best developing agents known.

Extending his interests to spectroscopy, Abney was the first to suggest (1877) that stars with rapid axial rotation could be detected by broadened lines in their spectra—an idea later to have wide application. He then devised a red-sensitive emulsion and with it made the first spectroscopic analyses of the structure of organic molecules (1882) and the first photographs of the solar spectrum in the infrared (1887). This was followed by comparative studies of how sunlight is altered in passing through our atmosphere, made at sea level and in the Swiss Alps (1888, 1894).

In 1877 Abney began a long supervisory career with the Board of Education for England and Wales. He was already a member of the Royal Photographic Society (which he served as president in 1892–1894, 1896, and 1903–1905), of the Royal Astronomical Society (president 1893–1895), and of the Physical Society of London (president 1895–1897). He had been made a fellow of the Royal Society in 1876 and was awarded its Rumford Medal in 1882, for his spectroscopic work. He was knighted in 1900.

BIBLIOGRAPHY

I. ORIGINAL WORKS. The books and papers by Abney referred to above are *Chemistry for Engineers* (Chatham, 1870); *Instruction in Photography* (Chatham, 1871; 11th ed., London, 1906); "On the Opacity of the Developed Photographic Image," in *Philosophical Magazine,* 4th ser.,

48 (1874), 161–165; "Dry Plate Process for Solar Photography," in *Monthly Notices of the Royal Astronomical Society,* 34 (1874), 275–278; "Photography in the Transit of Venus," *ibid.,* 35 (1875), 309–310, with Abney's report from the scene on 208; "Effect of a Star's Rotation on Its Spectrum," *ibid.,* 37 (1877), 278–279; "On the Alkaline Development of the Photographic Image," in *Philosophical Magazine,* 5th ser., 3 (1877), 46–51; "A New Developer" [hydroquinone], in *Photographic News,* 24 (1880), 345–346; "On the 'Sensitometric' Sensitiveness of Gelatine and Other Plates," in *British Journal of Photography,* 29 (1882), 243–244; "On the Influence of the Molecular Grouping in Organic Bodies on Their Absorption in the Infra-Red Region of the Spectrum," in *Philosophical Transactions of the Royal Society of London,* 172 (1882), 887–918, written with E. R. Festing; "The Solar Spectrum, From λ7150 to λ10,000," *ibid.,* 177 (1887), 457–469; "Transmission of Sunlight Through the Earth's Atmosphere," *ibid.,* A178 (1888), 251–283 and A184 (1894), 1–42; and "On a Failure of the Law in Photography That When the Products of the Intensity of the Light Acting and the Time of Exposure Are Equal, Equal Amounts of Chemical Action Will Be Produced," in *Proceedings of the Royal Society* (London), 54 (1893), 143–147.

Other books by Abney include *Thebes, and its Five Greater Temples* (London, 1875); *A Treatise on Photography* (London, 1878; 10th ed., 1916); *The Pioneers of the Alps* (London, 1888), written with C. D. Cunningham; *A Facsimile of Christian Almer's Führerbuch 1856–1894* (London, 1896), edited with C. D. Cunningham; and *Researches in Colour Vision and the Trichromatic Theory* (London, 1913), which summarizes his work in this field.

No complete list of Abney's publications is available: the 117 items in the Royal Society of London's *Catalogue of Scientific Papers,* VII (London, 1877), 5; IX (London, 1891), 7–8; and XIII (Cambridge, 1914), 15–16, do not include anything published after 1900, or any of his numerous short papers in *Photographic News,* and only part of his publications in the *Photographic Journal* (London), of which Abney was editor from 1876 until his death. Many of these additional works are mentioned in Chapman Jones's memorial lecture (see below).

II. Secondary Literature. Henry Chapman Jones wrote the article on Abney in *Dictionary of National Biography,* 3rd supp. (London, 1927), pp. 1–2, and also delivered a memorial lecture, printed in *Photographic Journal,* 61 [n.s. 45] (1921), 296–310. An obituary notice, by Col. Edmund Herbert Grove-Hills, with portrait, appeared in *Proceedings of the Royal Society* (London), A99 (1921), i–v; and another, by W. B. Ferguson, appeared in *Photographic Journal,* 61 [n.s. 45] (1921), 44–46.

Sally H. Dieke

ABOALY. See Ibn Sīnā.

ABRAHAM. See also Ibrāhīm.

ABRAHAM BAR ḤIYYA HA-NASI, also known as **Savasorda** (*fl.* in Barcelona before 1136), *mathematics, astronomy.*

In Arabic he was known as Ṣāḥib al-Shurṭa, "Elder of the Royal Suite," denoting some type of official position; this title later gave rise to the commonly used Latin name of Savasorda. He was also known as Abraham Judaeus. Savasorda's most influential work by far is his Hebrew treatise on practical geometry, the *Ḥibbūr ha-meshīḥah we-ha-tishboret.* Translated into Latin as *Liber embadorum* by Plato of Tivoli, the work holds an unusual position in the history of mathematics. It is the earliest exposition of Arab algebra written in Europe, and it contains the first complete solution in Europe of the quadratic equation, $x^2 - ax + b = 0$.

The year the *Ḥibbūr* was translated (1145) also saw the Robert of Chester translation of al-Khwārizmī's algebra and so may well be regarded as the birth year of European algebra. Thus the *Ḥibbūr* was among the earliest works to introduce Arab trigonometry into Europe, and it was also the earliest to treat of Euclid's *Book of Divisions.* Leonardo Fibonacci was influenced by Savasorda and devoted an entire section of his *Practica geometriae* to division of figures. Savasorda made a novel contribution when he included the division of geometric figures in a practical treatise, thus effecting a synthesis of Greek theory with the pragmatic aspects of mathematics.

Savasorda himself recommended Euclid, Theodosius of Bithynia, Menelaus, Autolycus, Apollonius of Perga, Eudemus of Rhodes, and Hero of Alexandria for study in geometry. He knew well al-Khwārizmī and al-Karajī. Following Hero and not Euclid he did not accept the Pythagorean figurate numbers in his explanation of plane and square numbers. In general, Savasorda preferred those definitions and explanations that may be aligned more easily and closely with reality.

To understand this approach, it is necessary to go back to the earliest known Hebrew geometry, the *Mishnat ha-Middot* (ca. A.D. 150). This work may be considered as a link in the chain of transmission of mathematics between Palestine and the early medieval Arab civilization. The Arab mathematicians al-Khwārizmī and al-Karajī, and later Savasorda, followed the methodological lines of this old *Mishna.* Savasorda himself provided a new cross-cultural bridge a thousand years after the *Mishna.* In his *Encyclopedia* there is the same teaching of both theory and practice, including not only the art of practical reckoning and business arithmetic but also the theory of numbers and geometric definition. This book is probably the earliest algorismic work written in Western Europe, but knowledge of the work is not apparent in the arithmetical works of either Abraham ibn Ezra or Levi ben Gerson, although they may have had a common origin.

In the history of decimal theory and practice, the two mainstreams of development in the Middle Ages came from the Jewish and Christian cultures. Savasorda, however, did not belong definitely to any one mathematical group. He spent most of his life in Barcelona, an area of both Arab and Christian learning, and was active in translating the masterpieces of Arab science. In an apologetic epistle on astrology to Jehuda ben Barsillae al-Barceloni, he deplored the lack of knowledge of Arab science and language among the people of Provence. He wrote his own works in Hebrew, but he helped translate the following works into Latin: al-ʿImrānī's *De horarum electionibus* (1133–1134), al-Khayyāt's *De nativitatibus* (1136), and Almansori's *Judicia seu propositiones* . . . (1136). Savasorda may have worked on translations of the *Quadripartitum* of Ptolemy, the *Spherics* of Theodosius, the *De motu stellarum* of al-Battānī, and others, with Plato of Tivoli. It is also possible that he worked with Rudolf of Bruges on the *De astrolabia.*

BIBLIOGRAPHY

I. ORIGINAL WORKS. Savasorda's *Ḥibbūr,* Michael Guttman, ed., was published in Berlin (1912–1913); a Catalan translation was done by J. Millás-Vallicrosa (Barcelona, 1931). Three treatises of Savasorda constitute a complete astronomical work: *Zurat ha-ereẓ* ("Form of the Earth," Bodleian MS 2033) is concerned with astronomical geography and general astronomy; *Ḥeshbon mahlakot ha-kokhabim* ("Calculation of the Movement of the Stars," Leiden MS 37 Heb.) covers astronomical calculations; and *Luhot ha-nasi* ("Tables of the Prince," Berlin MS 649; Bodleian MS 443, 437) follows al-Battānī's work. His work on the calendar, *Sefer ha-'ibbur* ("Book on Intercalation," H. Filipowski, ed., London, 1851), written 1122–1123, exerted great influence on Maimonides and Isaac Israeli the Younger. Savasorda's philosophical works include *Megillat ha-megalleh* ("Scroll of the Revealer," A. Poznanski and J. Guttmann, eds., Berlin, 1924) and *Hegyon ha-nefesh* ("Meditation of the Soul," E. Freimann, ed., Leipzig, 1860). See also J. Millás-Vallicrosa, ed., *Llibre de Geometria* (Barcelona, 1931), *Llibre revelador* (Barcelona, 1929), and *La obra enciclopédica yĕsodé ha-tĕbuná u-migdal ha-ĕmuná* (Madrid-Barcelona, 1952); and A. Poznanski and J. Guttmann, eds., *Megillat ha-nefesh* (Berlin, 1924).

II. SECONDARY LITERATURE. For other studies of mathematics and Savasorda's contributions, see F. Baer, *Die Juden im Christlichen Spanien,* I (Berlin, 1929), 81, n. 1; M. Curtze, "Der *Liber Embadorum* des Abraham bar Chijja Savasorda in der Übersetzung des Plato von Tivoli," in *Abhandlungen zur Geschichte der mathematischen Wissenschaften,* 12 (1902), 1–183; I. Efros, "Studies in Pre-Tibbonian Philosophical Terminology," in *Jewish Quarterly Review,* 14 (1926–1927), 129–164, 323–368; S. Gandz, "The Invention of Decimal Fractions . . . ," in *Isis,* 25 (1936), 17, and *"Mishnat ha-Middot,"* in *Quellen und Studien zur Geschichte der Mathematik, Astronomie und Physik,* Abt. A, 2 (Berlin, 1932). See also B. Goldberg and L. Rosenkranz, eds., *Yesod Olam* (Berlin, 1846–1848); J. M. Guttmann, *Chibbur ha-Meschicha weha-Tischboreth* (Berlin, 1913); M. Levey, "Abraham Savasorda and His Algorism: A Study in Early European Logistic," in *Osiris,* 11 (1954), 50–63, and "The Encyclopedia of Abraham Savasorda: A Departure in Mathematical Methodology," in *Isis,* 43 (1952), 257–264. Additional studies are by P. A. Sorokin and R. K. Merton, "The Course of Arabian Intellectual Development, 700–1300 A.D., A Study in Method," in *Isis,* 22 (1935), 516–524; M. Steinschneider, *Gesammelte Schriften* (Berlin, 1925), p. 345; H. Suter, "Die Mathematiker und Astronomen der Araber und Ihre Werke," in *Abhandlungen zur Geschichte der mathematischen Wissenschaften,* 10 (1900); J. Tropfke, "Zur Geschichte der quadratischen Gleichungen ueber dreieinhalb Jahrtausend," in *Jahresbericht der Deutschen Mathematiker-vereinigung,* 43 (1933), 98–107; 44 (1934), 95–119.

MARTIN LEVEY

ABRAHAM BEN JACOB. See **Ibrāhīm ibn Yaʿqūb.**

ABRAHAM JUDAEUS. See **Ibn Ezra.**

ABRAHAM BEN MEIR IBN EZRA. See **Ibn Ezra.**

ABRAHAM, MAX (*b.* Danzig, Germany, 26 March 1875; *d.* Munich, Germany, 16 November 1922), *physics.*

Abraham was born to a wealthy Jewish merchant family. He studied under Max Planck and completed his doctoral dissertation in 1897. He then assisted Planck at Berlin and in 1900 assumed the position of *Privatdozent* at Göttingen.

Abraham's lifework amounted to the explication of Maxwell's theory. He exhibited a virtuosity in the handling of Maxwell's equations like few others before him. In spite of his many original contributions, however, he was repeatedly passed over for academic appointments. This was due to the fact that he had no patience with what he considered to be silly or illogical argumentation. Abraham had a penchant for being critical and had no hesitation in publicly chastising his colleagues, regardless of their rank or position. His sharp wit was matched by an equally sharp tongue, and as a result he remained a *Privatdozent* at Göttingen for nine years. In 1909, he accepted a professorship at the University of Illinois, but he did not like the atmosphere at a small American university and returned to Göttingen after one semester. He then took the post of professor of rational mechanics at the University of Milan, where he remained until 1914. When World War I broke out, he was forced to return to Germany. He spent the war years investigating theoretical problems in radio transmission for the Telefunkengesellschaft. After the war, unable to return to Milan, he substituted as professor of physics at the Technische Hochschule at Stuttgart. Finally, in 1921, he received the call to a chair of theoretical physics at Aachen. On the trip to Aachen in April 1921, he was stricken with a fatal brain tumor. Abraham died after six painful months in a hospital in Munich. "Just as his life was suffering, his end was full of agony" (Born and von Laue).

Abraham is best remembered for his two-volume textbook, *Theorie der Elektrizität,* which went through five editions during his lifetime. Volume I, first published in 1904, was an adaptation of Föppl's *Einfuhrung in die Maxwellsche Theorie der Elektrizität.* Volume II, subtitled "Der Elektromagnetische Theorie der Strahlung" ("The Electromagnetic Theory of Radiation") contained Abraham's theory of electrons. It appeared in 1905. Subsequent to Abraham's death the book was revised under the authorship of Abraham and Becker. Today the modern counterpart of Abraham's text, R. Becker and F. Sauter's *Electromagnetic Fields and Interactions,* is in use.

The Abraham textbook was the standard work in electrodynamics in Germany for several generations of physicists. His consistent use of vectors was a significant factor in the rapid acceptance of vector notation in Germany. But one of the most noteworthy features of the text was that in each new edition Abraham saw fit to include not only the latest experimental work but also the latest in theoretical contributions, even if these contributions were in dispute. Furthermore, he had no hesitation, after explicating both sides of a question, in using the book to argue his own point of view. This was especially true with regard to theories of the electron as well as with regard to rival views of space and time.

Abraham's theory of the electron was developed in 1902 shortly after a close friend, Wilhelm Kaufmann, had published his first tentative experimental results on the variation of the transverse mass of the electron as a function of its velocity. The basic underlying assumptions of Abraham's theory were, first, that the conception of an ether in which electromagnetic phenomena took place was valid and,

second, that the differential equations of the electromagnetic field (Maxwell's equations) are applicable to the dynamics of electrons. In Abraham's view, the central question that had to be answered before any other was to what extent the mass of the electron was electromagnetic. If all the mass of the electron could be ascribed to the interaction of the electron's charge with electromagnetic fields, then one could hope to build a consistent and universal physics based on electrodynamics. Abraham's approach was to calculate the inertia due to the self-induction of the electron as it moved through its own field and the induction due to any external field in which the electron found itself. One could compare the results thus obtained with Kaufmann's results, and if agreement was substantial, then it could be said with some assurance that the mass of the electron was purely electromagnetic. The analysis was difficult, sophisticated, and lengthy. However, making the assumption that the electron was a perfectly rigid sphere and that the charge is distributed uniformly on the surface of the sphere, Abraham calculated the transverse electromagnetic mass of the moving electron to be

$$m = m_0 \cdot \frac{1}{\beta^2}\left(\frac{1+\beta^2}{2\beta}\ln\left[\frac{1+\beta}{1-\beta}\right] - 1\right)$$

where m_0 is the electron's rest mass and $\beta = v/c$, the ratio of the velocity of the electron to the velocity of light. Expressed in terms of powers of β this equation becomes

$$m = m_0(1 + \tfrac{2}{5}\beta^2 + \tfrac{3}{70}\beta^4 + \cdots +).$$

Noting that the data were very difficult to obtain, and that there was a high degree of uncertainty in Kaufmann's results, Abraham was pleased to find agreement between his own predictions and Kaufmann's data.

In 1904 H. A. Lorentz published his second-order theory of the electrodynamics of moving bodies. His expression for the mass of the moving electron, based on the conception of a deformable sphere which contracted in the direction of motion was

$$m = m_0(1 - \beta^2)^{-1/2}$$

which when expressed in terms of powers of β becomes

$$m = m_0(1 + \tfrac{1}{2}\beta^2 + \tfrac{3}{8}\beta^4 + \cdots +).$$

Of course, Einstein obtained the same result as a kinematic consequence of his special theory of relativity.

Finally, in 1906, Kaufmann undertook a new set of measurements in the hopes of distinguishing between Abraham's theory and those of Lorentz and Einstein. He reported that his experiments supported the Abraham theory. Although Kaufmann's work was later criticized on methodological grounds and later experiments vindicated the Lorentz and Einstein predictions, opponents of the theory of relativity often cited Abraham's theory and Kaufmann's data as evidence against Einstein's special theory.

Abraham himself remained unalterably opposed to Einstein's theory throughout his life. Early (ca. 1906–1910) he not only was convinced that the data did not support the theory, but he was unwilling to accept the postulates of the theory. By 1912 Abraham admitted that he had no objection to the logic of Einstein's theory; however, he expressed the hope that astronomical observations would contradict it, paving the way for the resurrection of the old absolute ether. "He loved his absolute ether, his field equations, his rigid electron just as a youth loves his first flame, whose memory no later experience can extinguish" (Born and von Laue). But throughout, Abraham's objections were not based on misunderstanding of the theory of relativity. He understood it better than most of his contemporaries. He was simply unwilling to accept postulates he considered contrary to his classical common sense.

BIBLIOGRAPHY

I. ORIGINAL WORKS. *Theorie der Elektrizität,* 1st ed. (Leipzig, 1904–1905), 2nd ed. (1908), 3rd ed. (1912), 4th ed. (1914), 5th ed. (1918); "Energie electrische Drahtwellen," in *Annalen der Physik,* **6** (1901), 217–241; "Prinzipien der Dynamik des Elektrons," *ibid.,* **10** (1903), 105–179; "Die Grundhypothesen der Elektronentheorie," in *Physikalische Zeitschrift,* **5** (1904), 576–579; "Zur Elektrodynamik bewegter Körper," in *Rendiconti del Circolo matimatico di Palermo,* **28** (1909), 1–28; "Zur elektromagnetische Mechanik," in *Physikalische Zeitschrift,* **10** (1909), 731–741; "Relativität und Gravitation; Erwiderung an Herrn A. Einstein," in *Annalen der Physik,* **38** (1912), 1056–1058; "Erhaltung der Energie und die Materie im Schwerkraftfelde," *Physikalische Zeitschrift,* **13** (1912), 311–314; "Theorie der Gravitation," *ibid.,* **13** (1912), 1–4; "Zur Theorie der Drahtwellen in ein leitenden Medium," *ibid.,* **20** (1919), 147–149.

II. SECONDARY LITERATURE. R. Becker and F. Sauter, *Electromagnetic Fields and Their Interactions,* 3 vols. (New York, 1964); Max Born and Max von Laue, "Max Abraham," in *Physikalische Zeitschrift,* **24** (1923), 49–53; Albert Einstein, "Über Relativitätsprinzip und die aus demselben gezogene Folgerungen," in *Jahrbuch der Radioactivität und Elektronik,* **4** (1907), 411–462, esp. 436–439; Stanley Goldberg, "Early Response to Einstein's Special Theory of Relativity," unpublished doctoral thesis (Harvard University, 1969); Gerald Holton, "Influence on and Reception of Einstein's Early Work in Relativity Theory" (mimeo, 1965); Wilhelm Kaufmann, "Die Elektromagnetische Masse des Elektrons," in *Physikalische Zeitschrift,* **3** (1902), 54–57; and "Die Konstitution des Elektrons," in *Annalen der Physik,* **19** (1906), 487–553; Max Planck, "Die Kaufmannschen Messungen der Ablenkbarkeit der Strahlen und ihren Bedeutung für die Dynamik der Elektronen," in *Physikalische Zeitschrift,* **7** (1906), 753–761; Arnold Sommerfeld, "Abraham, Max," *Neue Deutsche Biographie,* I (Berlin, 1953), 23–24; E. T. Whittaker, *A History of the Theories of Aether and Electricity,* 2 vols. (New York, 1960).

STANLEY GOLDBERG

ABREU, ALEIXO DE (*b.* Alcáçovas, Alentejo, Portugal, 1568; *d.* Lisbon, Portugal, 1630), *tropical medicine.*

Abreu was named for his grandfather, captain of an India galleon, who was killed in Malacca in 1500. He entered Évora University in 1577 and graduated as bachelor of arts about 1583. Afterward, against his parents' wishes, he studied medicine at Coimbra University on a royal scholarship, and seven years later graduated as a licentiate of medicine. Abreu practiced medicine in Lisbon with little success until, thanks to his father's friendship with Count Duarte de Castelo Branco, in 1594 he was appointed physician to the governor of Angola, João Furtado de Mendonça, with an annual salary of 24,000 reis. In São Paulo de Loanda he was both physician and colonizer, helping with slaves and horses in the conquest of that territory.

From 1604 to 1606 Abreu was in Brazil with the governor, Diogo Botelho, and served as surgeon during the Dutch attack on Bahía de Todos os Santos. During his tropical sojourn he contracted amoebiasis and yellow fever. In 1606 Abreu returned, ill, to Lisbon, and in 1612 was appointed physician to the treasury officials, a position he had to relinquish in July 1629 because of his poor health. Abreu was of sickly constitution and was very seriously ill in 1605, 1614, and 1621. On this last occasion he began to write a book describing his illness—which lasted five months—and including clinical reports of a case of malaria and of a kidney ailment, a discussion of phlebotomy, and an account of certain tropical diseases he had suffered from or had observed in Angola and Brazil.

Abreu's *Tratado de las siete enfermedades* (1623), the first text on tropical medicine, was written partly in Spanish and partly in Latin, but because of its corrupt language, archaic terminology, and particularly its extraordinary rarity, no full appraisal of its text has ever been made. In it Abreu described his own case of liver involvement in his recurrent amoebiasis (ff. 11–71). His early description of scurvy, which he called *mal de loanda* (ff. 150–192), emphasized gingivitis gum ulcers. He had first observed the disease in slaves, sailors, and travelers arriving in Lisbon after long sea voyages from Brazil and the Orient; his postmortem examinations and his therapeutic preparations made of green vegetables are noteworthy.

The name Abreu gave to yellow fever, *enfermedad del gusano* (ff. 193v.–199v.), or *bicho* in Brazil, has led to confusion because he incorporated in that description the *gusano Trichuris trichiura* (f. 195), a worm found in the rectum in some cases. His account of headache, pain in the lumbar region and the thighs, fever, vomiting, ulcers, and sudden death in otherwise strong young people is clearly recognizable. Abreu also described the *Tunga penetrans* or Brazilian *tungiasis,* the flea that penetrates the skin of the foot, usually around the toenail (f. 199v.), and the Guinea worm, the macrofilaria *Dracunculus medinensis,* which grows to the thickness of a violin string, together with the techniques used by the natives to extract the parasites (ff. 199v.–200).

BIBLIOGRAPHY

Aleixo de Abreu published only one work, the *Tratado de las siete enfermedades, de la inflammación universal del higado, zirbo, pyloron, y riñones, y de la obstrución, de la satiriasi, de la terciana y febre maligna, y passión hipocondriaca. Lleva otros tres tratados, del mal de loanda, del guzano, y de las fuentes y sedales* (Lisbon, 1623). In the preliminary leaves is an autobiography of Abreu that remains the best biographical source.

There are two standard references on Abreu, both published in the nineteenth century: A. Chinchilla, *Anales de la medicina . . . española,* II (Valencia, 1845), 325–328; and A. Hernández Morejón, *Historia bibliográfica de la medicina española,* V (Madrid, 1846), 51–54. More recent is M. Ferreira de Mira, *Historia da medicina portuguesa* (Lisbon, 1947), pp. 132, 178–180. All three are based on Abreu's autobiography. F. M. Sousa Viterbo, "O licenciado Aleixo de Abreu," in *Archivos de historia da medicina portugueza,* n. s. **2** (1911), 121–124, reproduces some documents about Abreu's life; a good discussion of his role in tropical medicine appears in G. Osorio de Andrade and E. Duarte, *Morão, Rosa & Pimenta* (Pernambuco, 1956). See also F. Guerra, "Aleixo de Abreu (1568–1630), Author of the Earliest Book on Tropical Medicine," in *Journal of Tropical Medicine,* **71** (1968), 55–69.

FRANCISCO GUERRA

ABŪ BAKR MUḤAMMAD IBN AL-ḤASAN AL-KARAJĪ, AL-ḤĀSIB. See **al-Karajī.**

ABU'L-BARAKĀT AL-BAGHDĀDĪ, HIBAT ALLAH (*b.* Iraq, *ca.* 1080; *d.* Baghdad, Iraq, after 1164/ 1165, aged eighty or ninety), *physics, psychology, philosophy.*

Abu'l-Barakāt was physician to the caliph of Baghdad. Of Jewish origin, he was converted to Islam late in life—according to one report, in reaction to a social slight inflicted upon him because of his Judaism; according to another, in order to counter a threat to his life. His writings include *Kitāb al-Muʿtabar,* his main work (the title may be translated as "The Book of What Has Been Established by Personal Reflection"); a philosophical commentary on the Ecclesiastes, written in Arabic in Hebrew characters; and the treatise "On the Reason Why the Stars Are Visible at Night and Hidden in Daytime."

According to Abu'l-Barakāt's own account, which is on the whole quite plausible, *Kitāb al-Muʿtabar* consists in the main of critical remarks jotted down by him over the years while reading philosophical texts and published, at the insistence of his friends, in the form of a philosophical work. From the formal point of view, its composition closely follows that of the *Logic, Naturalia,* and *Metaphysics* of Avicenna's voluminous *Kitab al-Shifāʾ* (the *Sufficientia* of the Latins), which seems to have been the principal philosophical text studied in Abu'l-Barakāt's time in the Islamic East. The genesis of *Kitāb al-Muʿtabar* as an accumulation of notes may account for various doctrinal inconsistencies in the work; Abu'l-Barakāt's many bold deviations from Avicenna's physics and metaphysics appear to be at variance with his complete acceptance of considerable portions of his predecessor's views.

In his psychology as well as in his physics, Abu'l-Barakāt bases his views on what he regards as immediate certainties rather than on an assessment, made by discursive reasoning, of empirical data. The use of this method clearly renders both the Aristotelian approach and many Aristotelian theories unacceptable to him, and Abu'l-Barakāt is not chary of proclaiming his disagreement with the then dominant philosophical tradition, which he declares to be a corruption of the true doctrines of ancient philosophers. Nonetheless, the starting point of his psychology is identical with that of Avicenna and is obviously taken over from the latter. Like Avicenna, Abu'l-Barakāt considers that immediate self-awareness, the awareness of one's own existence and of one's own actions, constitutes an unchallengeable proof of the existence and activity of the soul (identified with the ego). But unlike Avicenna, he does not try to fit this insight into the categories of Peripatetic psychology.

According to Abu'l-Barakāt, man is aware that his intellectual, imaginative, sensory, and motor activities, and any other psychic activities he may have, are due to one and the same agent: namely the soul. This awareness, accompanied as it is by a sense of certainty, may be relied upon to provide the truth. Abu'l-Barakāt uses this intuition in order to deny the existence of a variety of psychic faculties. He goes even further; he rejects the distinction (fundamental to Aristotelianism) between the intellect and the soul. There is no place in his doctrine for the speculations of the Peripatetics concerning the active, the passive, and the other intellects.

Abu'l-Barakāt regards the soul as an incorporeal entity, linked with the body but not located in it or anywhere else. Not being restricted by position in space, it is able to perceive anything that exists or occurs anywhere in the universe—but only one thing at a time. Thus, according to a conception that is reminiscent of Bergson, it has to choose among the multitudinous external impressions liable to impinge upon it; this choice, the sifting of these potential impressions, is done by the body—more precisely, by the sense organs, which circumscribe the perceptive activities of the soul.

The primordial role played by consciousness and self-awareness in Abu'l-Barakāt's conception of the soul impels him to try to explain the existence of unconscious psychic activities, *inter alia,* the organic ones (for instance, digestion) and latent memories (which, contrary to the Aristotelians, he does not regard as preserved in a part of the brain in the form of corporeal impressions, but as being incorporeal). One of these explanations is centered upon the notion of attention; some of the unconscious activities of the soul are considered as those to which the soul pays no attention.

Abu'l-Barakāt's conception of God seems to be modeled to a considerable extent upon his view of the soul. His God, unlike the Aristotelian one, is not a pure intellect but an entity that, like the human ego but with much greater powers, is engaged in many different activities and has knowledge of particulars (but not of an infinity of particulars at the same time, because the notion is self-contradictory). When His attention is engaged, He may intervene in the course of events. In other cases, this course may be regarded as causally determined, if one envisages only one chain of causes and effects. In fact, however, a large proportion of events is determined by chance, the latter notion being defined by Abu'l-Barakāt as

an encounter of two mutually independent chains of causes and effects. He gives as an example the encounter of a scorpion and a man crossing a street; the direction and the speed of both are strictly determined; yet their meeting, which may lead to the man killing the scorpion or to the scorpion stinging the man, is due to chance. A similar theory of chance is set forth by Boethius (see, e.g., *De consolatione philosophiae* V, 1), who could not have influenced Abu'l-Barakāt, and is hinted at by Plotinus (see *Enneads* VI, 8, 10, where coming-about-by-chance appears to be explained as coming-about-through-encounter). These seem to be the only known precursors of Abu'l-Barakāt on this point.

In his physics Abu'l-Barakāt employs a method similar to the one he utilized in his psychology. Like his tenth-century predecessor Abū Bakr al-Rāzī (by whom he may have been influenced, although there is no evidence either way) and in contrast with the Aristotelians, he relies in his physical theories, just as he does in his doctrine of the soul, on what he regards as self-evident, i.e., immediately perceived truths that are not dependent upon empirical data. Applying this method, he rejects the Aristotelian contention that time is the measure of movement. According to him, the notion of time is ontologically prior to the notion of movement. Nor does he regard time as being merely a subjective phenomenon. It is in fact the measure of Being, and as such it should not be regarded as external to Being. Comparisons between the lengths of two or more durations are, however, due to a mental comparison between the two.

The fundamental connection that Abu'l-Barakāt establishes between time and Being leads to his denying the existence of the two other higher modes of temporality postulated in Avicenna's philosophy: according to him there is no eternity (*sarmad*) and no aevum (*dahr*). Time is real even with regard to God.

Abu'l-Barakāt's theory of space, or of place—in the medieval philosophical vocabulary the two notions are designated by one and the same term—resembles his doctrine of time in its rejection of the Aristotelian conception, which was based on empirical data. For the Aristotelian view, according to which place (or space) is the inner surface of the surrounding body (and consequently bidimensional), Abu'l-Barakāt substitutes the conception that there exists a tridimensional space that in itself is empty; in physical reality it is generally (although, according to some passages, which assert in certain cases the existence of vacuums, not always) occupied by bodies. In the mind, however, the conception of a tridimensional empty space is prior to the conception of a plenum. Abu'l-Barakāt also refutes the arguments used by the Aristotelians to prove that infinite space is impossible. According to him, space is infinite because it is impossible for man to conceive a space that has a limit.

In his explanation of the movement of projectiles, Abu'l-Barakāt, like Avicenna, subscribes to the doctrine positing a "violent inclination" (*mayl qasrī;* the notion is similar to, or identical with, the *impetus* of the Schoolmen). "Violent inclination"—opposed to the "natural inclination," in virtue of which bodies removed from their natural place tend to return—is regarded as having been imparted by the mover to a body in a state of violent motion (for instance, to a stone thrown upward or to an arrow shot from a bow). The notion of violent inclination is used to account for the continuation of violent motion after the separation of the projectile from the mover. Contrary to Avicenna, Abu'l-Barakāt regards "violent inclination" as self-expending; it is used up in the very process of violent motion.

The acceleration of the motion of falling bodies is attributed by Abu'l-Barakāt to two causes:

(1) He holds that a violent and a natural inclination can simultaneously coexist in a projectile. Thus, when a body begins to fall, a residue of violent inclination still subsists in it and opposes the natural inclination that causes the body to descend, slowing down its fall. The acceleration of the fall is due to the gradual weakening of the violent inclination.

(2) The second cause of the acceleration of the motion of falling bodies is that the force (i.e., gravity) generating natural inclination resides in the falling body and produces a succession of natural inclinations in such a way that the strength of the inclination increases throughout the fall.

Abu'l-Barakāt's conception of the second cause seems to anticipate in a vague way the fundamental law of classical mechanics, according to which a continually applied force produces acceleration. According to Aristotelian mechanics such a force produces a uniform motion.

While there is no concrete evidence to show that Abu'l-Barakāt exercised a significant influence on Jewish philosophers, Fakhr al-Dīn al-Rāzī (*d.* 1210), a celebrated Muslim author, was his professed disciple. His influence appears to have extended also to other Muslim philosophers.

BIBLIOGRAPHY

The Arabic text of *Kitāb al-Muʿtabar* was published in 3 vols., Serefettin Altkaya, ed. (Hyderabad, 1937–1940).

E. Wiedemann, "Über den Grund warum die Sterne bei Nacht sichtbar und bei Tag verborgen sind," in *Jahrbuch für Photographie* (Halle, 1909), pp. 49–54, is a translation of a treatise by Abu'l-Barakāt.

Works on Abu'l-Barakāt or his writings are Shlomo Pines, "Études sur Awḥad al-Zamān Abu'l-Barakāt al-Baghdādī," in *Revue des études juives,* **103** (1938), 4–64, and **104** (1938), 1–34; "La conception de la conscience de soi chez Avicenne et Abu'l-Barakāt al-Baghdādī," in *Archives d'histoire doctrinale et littéraire du moyen âge,* **29** (1954), 21–98; *Nouvelles études sur Awḥad al-Zamān Abu'l-Barakāt al-Baghdādī,* Vol. I of *Mémoires de la Société des études juives* (Paris, 1955); "Studies in Abu'l-Barakāt al-Baghdādī's Poetics and Metaphysics," in *Studies in Philosophy,* Vol. VI of Scripta Hierosalymitana (Jerusalem, 1960), pp. 120–198; and "A Study of Abu'l-Barakāt's Commentary on the Ecclesiastes" (in Hebrew), in *Tarbiz,* **33** (1964), 198–213.

SHLOMO PINES

ABU'L-FIDĀ' ISMĀ'ĪL IBN 'ALĪ IBN MAḤMŪD IBN . . . AYYŪB, 'IMĀD AL-DĪN (*b.* Damascus, Syria, 1273; *d.* Ḥamā, Syria, 1331), *history, geography.*

A prince of the Ayyūbid family, Abu'l-Fidā' participated from the age of twelve in the Muslim campaigns against the Crusaders and the Mongols. During his youth his family lost its estates, which he recovered in 1312 through his fidelity to the Mamelukes, whom he served, as a vassal prince, until his death.

Biographical dictionaries have preserved information on a number of Abu'l-Fidā''s historical-literary and scientific works. Among the most distinguished of the former is the *Mukhtaṣar taʾrīkh al-bashar,* a continuation of the *Kāmil fi 'l-taʾrīkh* of Ibn al-Athīr. It is a historical treatise that begins with pre-Islamic Arabia and becomes most interesting when it deals with happenings during the author's lifetime. Written in 1315, it was continued by Abu'l-Fidā' himself until 1329 and was the object of attention of a number of fourteenth-century Arabic historians, who kept it up-to-date until 1403 (among them Ibn al-Wardī until 1348, and Ibn al-Shiḥna al-Ḥalabī until the beginning of the fifteenth century). This work was translated into Western languages and became the basis for several historical syntheses by eighteenth-century Orientalists, which explains the strong influence it exerted on nineteenth-century Western historiography.

Abu'l-Fidā''s outstanding scientific work is the *Taqwīm al-buldān* ("A Sketch of the Countries"), written between 1316 and 1321. This is a general geography of twenty-eight chapters of varying lengths, with a prologue containing interesting observations: the gain or loss of a day according to the direction in which one goes around the earth, and the assertion that three-fourths of the earth's surface is covered with water. The descriptions of rivers, lakes, oceans, and mountains are interesting and instructive. The text contains some tables—suggested to Abu'l-Fidā᾽ by his reading of the *Taqwīm al-abdān* ("The Cure of Bodies") of Ibn Jazla—that recapitulate the written variants for each place name, its geographical coordinates, the sources utilized, the climate or zone to which it belongs, and the natural region in which it is located. The order followed in the presentation has often been argued.

The sources of the work are the Arabic translation of Ptolemy and the works of Idrīsī, Ibn Ḥawqal, Isṭakhri, al-Bīrūnī, and above all the *Geography* or *Kitāb Basṭ al-arḍ fi'l-ṭūl wa'l-ʿard* of Ibn Saʿīd al-Magribī. The latter book is frequently quoted, and Abu'l-Fidā᾽ took from it the information about the trip of one Ibn Fāṭima (very possibly a Berber from the Sahara), who explored in detail the Atlantic and western Mediterranean coasts of Africa. The longitudes recorded in the *Taqwīm al-buldān* often contain obvious errors that are a result of their having been taken from sources that did not adopt the same prime meridian (some used the western coast of Africa, others the Canary Islands); poor conversion of the distance between extreme points on an itinerary into degrees and minutes of latitude and longitude; and faulty reading of the canvas maps in use in the Near East.

The *Taqwīm al-buldān* underwent a number of critical abridgments, among which that in Turkish by Muḥammad ibn ʿAlī Sipāhīzādé (*d.* 1589) should be noted.

BIBLIOGRAPHY

I. ORIGINAL WORKS. For a list of MSS, see C. Brockelmann, *Geschichte der arabischen Litteratur*, II, 44–46, and supp. II, 44. There is a complete edition of the *Mukhtaṣar*, 2 vols. (Istanbul, 1869–1870) and a number of partial translations. The *Taqwīm al-buldān* was edited by Joseph Toussaint Reinaud and Baron de Slane (Paris, 1840). The French translation, Reinaud and Stanislas Guyard, eds., *Géographie d'Aboulféda*, 2 vols. (Paris, 1848–1883), contains a lengthy prologue on Arab geography by Reinaud.

II. SECONDARY LITERATURE. For the Arabic sources and works connected with the book, consult Carra de Vaux, *Les penseurs de l'Islam* (Paris, 1921), I, 139–146, and II, 13–14; H. A. R. Gibb, in *Encyclopaedia of Islam*, 2nd ed., I, 122; Joseph Needham, *Science and Civilization in China*, III (Cambridge, 1959), 561–565; George Sarton, *Introduction to the History of Science*, III (Baltimore, 1947), 793–799; and J. Vernet, "Marruecos en la Geografía de Ibn Saʿīd al-Magribī," in *Tamuda*, **1** (1955), 123–157, and *Kitāb basṭ al-arḍ fi'l-ṭūl wa'l-ʿarḍ* (Tetuán, 1957), a critical edition of the Arabic text.

J. VERNET

ABŪ ḤĀMID AL-GHARNĀṬĪ, also known as **Abū ʿAbdallāh Muḥammad ibn ʿAbd al-Raḥīm . . . al-Māzinī al-Andalusī** (*b.* Granada, Spain, 1080; *d.* Damascus, Syria, 1169), *geography.*

Born of Hispano-Arabic parents, Abū Ḥāmid emigrated to the Orient by sea. In 1117 he was in Alexandria, where he studied with Abū Bakr al-Ṭurṭūshī. He visited Cairo and Baghdad, traveled through Persia, crossed the Caucasus, and went up the Volga as far as Bulgār (55° north). He later visited Bashgird (an area then occupied by Hungarian tribes), and then returned to Baghdad by way of Persia. He later undertook the pilgrimage to Mecca. The latter part of his life was dedicated to the preparation of two works: the *Muʿrib* and the *Tuḥfa*. Both influenced later Arabic cosmographers, such as al-Qazwīnī, and they must have undergone many re-elaborations. Other books that appear in general bibliographies have also been attributed to him. Among these works of secondary importance are *Nukhbat al-adhhān fī ʿajāʾib al-ʿajab* ("Selected Memories Concerning the Greatest Marvels") and *ʿAjāʾib al-makhlūqāt* ("Marvels of Creatures"), both of which appear to be revisions or adaptations of the *Muʿrib* and the *Tuḥfa.*

The work *Muʿrib ʿan baʿḍ ʿajāʾib al-Maghrib* ("Anthology of the Marvels of the Maghrib") has been partially translated into Spanish (folios 96a–114b of the MS of the Royal Academy of History, Madrid). The manuscript contains a description of some of the marvels of Andalucía and, above all, some long dissertations on astronomical, astrological, and chronological matters. The part that has been published contains references to his travels through Eurasia and a series of quite interesting observations about physical geography and ethnography, such as one of the oldest descriptions of the skis used by the so-called Yura peoples of the Arctic, complete with diagram. The first information about this text was provided by Emilio García Gómez in *A.B.C.* (6 January 1947). The description of the flora and fauna of northern Russia is also of great interest.

The *Tuḥfat al-albāb wa-nukhbat al-ʿajāʾib* ("Gift From the Heart and Selection of Marvelous Things") is in some ways similar to the *Muʿrib*, as in the wording of passages dealing with the same material.

But while the *Muᶜrib* may be considered authentic (at least the part that has been published), the same cannot be said for the *Tuḥfa*, the last part of which, as it appears in the Ferrand edition, may contain interpolations by other authors. Also, in the *Tuḥfa* the material referring to Abū Ḥāmid's homeland is relatively sparse, and it contains common fables of the period. The work consists of four parts:

(1) A description of the world and the men and spirits who inhabit it. It includes interesting details about the gold and salt trade in the Sudan, as well as information about how the merchants of the time crossed the Sahara, guiding themselves by the stars. (Texts later than Abū Ḥāmid's mention use of the compass.) On the other hand, the description he gives of some quasi-human beings, a description repeated in the Latin cosmographic manuscripts of the early Middle Ages, is completely fanciful.

(2) A description of strange countries and interesting monuments, such as the pyramids of Egypt and the lighthouse at Alexandria.

(3) A description of the seas and the animals that inhabit them, scientifically the most interesting part. He tells us with some degree of realism about flying fish, squid, octopuses, torpedoes, pumice, oil wells, and India paper.

(4) A discussion of caves and tombs. This part contains, in an incidental fashion, a description of fossils, data about the utilization of ivory from the Siberian mammoths, and a mention of fireproof asbestos cloth.

BIBLIOGRAPHY

I. ORIGINAL WORKS. A general bibliography is in George Sarton, *Introduction to the History of Science,* II (Baltimore, 1931), 412. There are lists of MSS in C. Brockelmann, *Geschichte der arabischen Litteratur* (Weimar, 1898; Leiden, 1937), I, 477 and Supp. I, 877; and in F. Pons Boigues, *Ensayo bio-bibliográfico sobre los historiadores y geógrafos arabigoespañoles* (Madrid, 1898), pp. 229–231. A bilingual version of his writings on Eurasia is César E. Dubler, *Abū Ḥāmid el Granadino y su relación de viaje por tierras euroasiáticas. Texto árabe, traducción e interpretación* (Madrid, 1953).

II. SECONDARY LITERATURE. Works dealing with Abū Ḥāmid or his writings are Gabriel Ferrand, "Le Tuḥfat al-albāb de Abū Ḥāmid al-Andalusī al Garnāṭī," in *Journal asiatique,* 2 (1925), 1–148, 193–304; I. Hrbek, in *Archiv Orientální,* 23 (1955), 109–135; and F. Tauer, *ibid.,* 18 (1950), 298–316.

J. VERNET

ABŪ JAᶜFAR AL-KHĀZIN. See **al-Khāzin.**

ABŪ KĀMIL

ABŪ KĀMIL SHUJĀᶜ IBN ASLAM IBN MUḤAMMAD IBN SHUJĀᶜ (*b. ca.* 850, *d. ca.* 930), *mathematics.*

Often called al-Ḥāsib al-Miṣrī ("the reckoner of Egypt"), Abū Kāmil was one of Islam's greatest algebraists in the period following the earliest Muslim algebraist, al-Khwārizmī (*fl. ca.* 825). In the Arab world this was a period of intellectual ferment, particularly in mathematics and the sciences.

There is virtually no biographical material available on Abū Kāmil. He is first mentioned by al-Nadīm in a bibliographical work, *The Fihrist* (987), where he is listed with other mathematicians under "The New Reckoners and Arithmeticians," which refers to those mathematicians who concerned themselves with the practical algorisms, citizens' arithmetic, and practical geometry (see Bibliography). Ibn Khaldūn (1322–1406) stated that Abū Kāmil wrote his algebra after the first such work by al-Khwārizmī, and Ḥajjī Khalīfa (1608–1658) attributed to him a work supposedly concerned with algebraic solutions of inheritance problems.

Among the works of Abū Kāmil extant in manuscripts is the *Kitāb al-ṭarāʾif fiʾl-ḥisāb* ("Book of Rare Things in the Art of Calculation"). According to H. Suter[1] this text is concerned with integral solutions of some indeterminate equations; much earlier, Diophantus (*ca.* first century A.D.) had concerned himself with rational, not exclusively integral, solutions. Abū Kāmil's solutions are found by an ordered and very systematic procedure. Although indeterminate equations with integral solutions had been well known in ancient Mesopotamia, it was not until about 1150 that they appeared well developed in India. Aryabhata (*b.* A.D. 476) had used continued fractions in solutions, but there is uncertain evidence that this knowledge had been passed on in any ordered form to the Arabs by the time of Abū Kāmil.

A work of both geometric and algebraic interest is the *Kitāb . . . al-mukhammas waʾal-muᶜashshar . . .* ("On the Pentagon and Decagon"). The text is algebraic in treatment and contains solutions for a fourth-degree equation and for mixed quadratics with irrational coefficients. Much of the text was utilized by Leonardo Fibonacci (1175–*ca.* 1250) in his *Practica geometriae.*[2] Some of the equations solved by Abū Kāmil in this work read as follows:

$$s_{15} = \sqrt{\frac{s}{32}d^2 \sqrt{\frac{s}{1024}d^4} + \sqrt{\frac{3}{64}d^2} - \sqrt{\frac{15}{64}d^2}}$$

$$= \frac{r}{4}\left(\sqrt{10 + 2\sqrt{5}} + \sqrt{3} - \sqrt{15}\right),$$

with s the side of a regular polygon inscribed in a circle. Also

$$S_5 = \sqrt{5d^2 - \sqrt{20d^4}} = 2r\sqrt{5 - 2\sqrt{5}},$$

with S the side of a regular polygon (pentagon here) which has an inscribed circle.[3]

The outstanding advance of Abū Kāmil over al-Khwārizmī, as seen from these equations, is in the use of irrational coefficients.[4] Another manuscript, which is independent of the *Ṭarā'if*, mentioned above, is the most advanced work on indeterminate equations by Abū Kāmil. The solutions are not restricted to integers; in fact, most are in rational form. Four of the more mathematically interesting problems are given below in modern notation. It must be remembered that Abū Kāmil gave all his problems rhetorically; in this text, his only mathematical notation was of integers.

(1) $x^2 - 8x - 30 = y^2$

(2) $x + x^2 = y^2$
             $x - x^2 = z^2$

(3) $20 + x = y^2$
             $50 - (10 - x) = z^2$

(4) $10 + x^2 = y^2$
             $10 - x^2 = z^2$

Many of the problems in *Kitāb fi'l-jabr wa'l-muqābala* had been previously solved by al-Khwārizmī. In Abū Kāmil's work, a solution[5] for x^2 was worked out directly instead of first solving for x. Euclid had taken account of the condition x less than $p/2$ in $x^2 + q = px$, whereas Abū Kāmil also solved the case of x greater than $p/2$ in this equation.

Abū Kāmil was the first Muslim to use powers greater than x^2 with ease. He used x^8 (called "square square square square"), x^6 (called "cube cube"), x^5 (called "square square root"), and x^3 (called "cube"), as well as x^2 (called "square"). From this, it appears that Abū Kāmil's nomenclature indicates that he added "exponents." In the Indian nomenclature a "square cube" is x^6, in contradistinction. Diophantus (*ca.* A.D. 86) also added "powers," but his work was probably unknown to the Arabs until Abu'l Wafā' (940–998) translated his work into Arabic (*ca.* 998).

Abū Kāmil, following al-Khwārizmī, when using *jadhr* ("root") as the side of a square, multiplied it by the square unit to get the area ($x \cdot 1^2$). This method is older than al-Khwārizmī's method and is to be found in the *Mishnat ha-Middot*, the oldest

Hebrew geometry, which dates back to A.D. 150.[6] This idea of root is related to the Egyptian *khet* ("cubit strip").[7]

The Babylonians stressed the algebraic form of geometry as did al-Khwārizmī. However, Abū Kāmil not only drew heavily on the latter but he also derived much from Heron of Alexandria and Euclid. Thus he was in a position to put together a sophisticated algebra with an elaborated geometry. In actuality, the resulting work was more abstract than al-Khwārizmī's and more practical than Euclid's. Thus Abū Kāmil effected the integration of ancient Mesopotamian practice and Greek theory to yield a wider approach to algebra.

Some of the more interesting problems to be found in the *Algebra*, in modern notation, are:[8]

(No. 57) $$\frac{x \cdot \sqrt{10}}{2 + \sqrt{3}} = x - 10$$

(No. 60) $x + \sqrt{x} + \sqrt{2x} + \sqrt{5x^2} = 10$

(No. 61) $x + y + z = 10; x < y < z$
                  $x^2 + y^2 = z^2$
                  $xz = y^2$

(No. 63) $$\frac{10}{x} + \frac{10}{10 - x} = 6\frac{1}{4}$$

It is possible that Greek algebra was known to Abū Kāmil through Heron of Alexandria, although a direct connection is difficult to prove. The influence of Heron is, however, definite in Abraham bar Ḥiyya's work.[9] That Abū Kāmil influenced both al-Karajī and Leonardo Fibonacci may be demonstrated from the examples they copied from his work. Thus through Abū Kāmil, mathematical abstraction, elaborated together with a more practical mathematical methodology, impelled the formal development of algebra.

NOTES

1. "Das Buch der Seltenheiten."
2. See also Suter, "Die Abhandlung des Abū Kāmil."
3. *Ibid.,* p. 37. Levey will soon publish the Arabic text of "On the Pentagon and Decagon," discovered by him.
4. At least twenty of these problems from this text may be found in Leonardo Fibonacci, *Scritti*, Vol. I, sect. 15; Vol. II.
5. Tropfke, *Geschichte der Elementar-Mathematik*, pp. 74–76; 80–82; Weinberg, "Die Algebra des abu Kamil."
6. S. Gandz, "On the Origin of the Term 'Root.'"
7. M. Levey, *The Algebra of Abū Kāmil*, pp. 19–20. P. Schub and M. Levey will soon publish Abū Kāmil's advanced work on indeterminate equations, newly discovered in Istanbul.
8. *Ibid.,* pp. 178, 184, 186, 202.
9. M. Levey, "The Encyclopedia of Abraham Savasorda" and "Abraham Savasorda and His Algorism."

BIBLIOGRAPHY

I. ORIGINAL WORKS. The following manuscripts of Abū Kāmil are available: *Kitāb fi'l-jabr wa'l-muqābala* ("Book on Algebra," Paris BN MS Lat. 7377A; Munich Cod. MS Heb. 225; Istanbul-MS Kara Mustafa 379), trans. into Hebrew by Mordecai Finzi *ca.* 1460. See also *Kitāb al-ṭarāʾif fi'l ḥisāb* ("Book of Rare Things in the Art of Calculation," Leiden, MS Arabic 1003, ff. 50r–58r; translations are found in Munich Cod. MS Heb. 225 and in Paris BN MS Lat. 7377A); *Kitāb . . . al-mukhammas wa'l-muʿashshar . . .* ("On the Pentagon and Decagon," Paris BN MS Lat. 7377A; Munich Cod. MS Heb. 225; Istanbul-MS Kara Mustafa 379, ff. 67r–75r); *Al-wāṣāyā bi'l-judhūr* (MS Mosul 294) discusses the ordering of roots. Works of Abū Kāmil listed in *The Fihrist* of al-Nadīm (p. 281) include *Kitāb al-falāḥ* ("Book of Fortune"), *Kitāb miftāḥ al-falāḥ* ("Book of the Key to Fortune"), *Kitāb fi'l-jabr wa'l-muqābala* ("Book on Algebra"), *Kitāb al-misāḥa wa'l-handasa* ("Book on Surveying and Geometry"), *Kitāb al-kifāya* ("Book of the Adequate"), *Kitāb al-ṭayr* ("Book on Omens"), *Kitāb al-ʿasīr* ("Book of the Kernel"), *Kitāb al-khaṭaʾayn* ("Book of the Two Errors"), *Kitāb al-jamʿ wa'l-tafrīq* ("Book on Augmentation and Diminution").

II. SECONDARY LITERATURE. For works on both Arab mathematics and Abū Kāmil see the following: H. T. Colebrooke, *Algebra with Arithmetic and Mensuration from the Sanskrit* (London, 1817); G. Fluegel, ed. and trans., *Lexicon bibliographicum et encyclopedicum a Haji Khalfa compositum* (Leipzig, 1835–1858); W. Hartner, "Abū Kāmil Shudjāʿ," in the *Encyclopedia of Islam,* 2nd ed., I (Leiden, 1960), 132–133; II (Leiden, 1962), 360–362; *Ibn Khaldūn The Muqaddimah,* Franz Rosenthal, trans., 3 vols. (New York, 1958); Leonardo Fibonacci, *Scritti di Leonardo Pisano,* 2 vols: Vol. I, *Liber abaci;* Vol. II, *Practica geometriae;* S. Gandz, "On the Origin of the Term 'Root,' " in *American Math. Monthly,* **35** (1928), 67–75; M. Levey, *The Algebra of Abū Kāmil (Kitāb fi'l-jabr wa'l-muqābala) in a Commentary by Mordecai Finzi* (Madison, Wisc., 1966), "The Encyclopedia of Abraham Savasorda: A Departure in Mathematical Methodology," in *Isis* **35** (1952), 257–264; and "Abraham Savasorda and His Algorism: A Study in Early European Logistic," in *Osiris,* **11** (1954), 50–63.

For additional material see G. Libri, *Histoire des sciences mathématiques en Italie* (Paris, 1938), pp. 253–297; 2nd ed. (Paris, 1865), pp. 304–369; al-Nadīm, *Fihrist al-ʿulūm,* G. Fluegel, ed. (Leipzig, 1871–1872); M. Steinschneider, *Die Hebraeischen Uebersetzungen des Mittelalters und die Juden als Dolmetscher,* a reprint (Graz, 1956), pp. 584–588; and H. Suter, "Die Abhandlung des Abū Kāmil Šojaʿ b. Aslam über das Fünfeck und Zehneck," in *Bib. Math.,* **10** (1909–1910), 15–42; "Das Buch der Setenheiten der Rechenkunst von Abū Kāmil el-Misrī," in *Bib. Math.,* Ser. 3, **11** (1910–1911), 100–120; "Die Mathematiker und Astronomen der Araber und ihre Werke," in *Abhandlungen z. Gesch. d. Math. Wissenschaften,* **10** (1900). See also J. Tropfke, *Geschichte der Elementar-Mathematik,* Vol. III (Berlin, 1937); J. Weinberg, "Die Algebra des abū Kāmil Šoğaʿ ben Aslam" (doctoral diss., Munich, 1935); A. P. Youschkevitch, *Geschichte der Mathematik im Mittelalter* (Basel, 1964).

MARTIN LEVEY

ABŪ MAʿSHAR AL-BALKHĪ, JAʿFAR IBN MUḤAMMAD, also known as **Albumasar** (*b.* in or near Balkh in Khurasan, 10 August 787; *d.* al-Wāsiṭ, Iraq, 9 March 886), *astrology.*

The ancient city of Balkh, where Abū Maʿshar grew up, had once been an outpost of Hellenism in central Asia, and then had become a center for the mingling of Indians, Chinese, Scythians, and Greco-Syrians with Iranians during the Sassanian period; when it was conquered by Aḥnaf ibn Qays during the caliphate of ʿUthmān (644–656), its religious communities included Jews, Nestorians, Manichaeans, Buddhists, and Hindus, as well as Zoroastrians. In the revolution of the middle of the eighth century, the people of Khurasan provided the Abbasids with their army, their general, and many of their intellectuals.

These intellectuals, like those from other frontier areas of the former Sassanian empire, were politically inclined toward pro-Iranism and against their Arab masters, and religiously inclined toward heresy, especially the Shīʿa sect. They were called upon, despite these tendencies, to play a large role in the activities of the libraries and translation institutes established at Baghdad by the early Abbasids; and they succeeded in making a generous portion of their Sassanian heritage of syncretic science and philosophy an integral part of the Muslim tradition.

Abū Maʿshar was a member of the third generation of this Pahlavi-oriented intellectual elite. He retained a strong commitment to the concept of Iranian intellectual superiority (expressed most vehemently in his *Kitāb ikhtilāf al-zījāt* and *Kitāb al-ulūf*), but he himself relied entirely on translations for his knowledge of Sassanian science. He mingled his already complex cultural inheritance with various intellectual trends current in Baghdad in his time, and became a leading exponent of the theory that all different national systems of thought are ultimately derived from a single revelation (thus, in a sense, paralleling in intellectual history the Neoplatonic doctrine of emanation, which he accepted philosophically in its Harrānian guise). This theory could be used to justify the most astonishing and inconsistent eclecticism; it also permitted an advocate to adopt wildly heretical views while maintaining strict adherence to the tenets of Islam. Abū Maʿshar's great reputation and usefulness as the leading astrologer of the Muslim world also helped to preserve him from persecution; there are reports of only one unfortunate incident, a whipping administered because of his practice of astrology, during the caliphate of al-Mustaʿīn (862–866).

Abū Maʿshar began his career in Baghdad, probably at the beginning of the caliphate of al-Maʾmūn (813–833), as an expert in *ḥadīth*, the sayings traditionally ascribed to Muḥammad and his companions. It was undoubtedly in studying this subject that he developed his proficiency in such subjects as the pre-Islamic Arabic calendar and the chronology of the early caliphs. But, in his forty-seventh year (832–833), according to the biographical tradition, but actually in about 825, an event occurred that completely changed his scholarly career. He became involved in a bitter quarrel with the Arabs' first "philosopher," Abū Yūsuf Yaʿqūb ibn Isḥāq al-Kindī (*ca.* 796–873), who was interested at once in Plato, in Aristotle and his commentators, in various Neoplatonists, in the works that the "Sabaeans" of Ḥarrān attributed to Hermes and Agathodemon, and, in general, in "mathematics" (arithmetic, geometry, music, astronomy, and astrology). It was his urging that made Abū Maʿshar realize the necessity of studying "mathematics" in order to understand philosophical arguments. He henceforth devoted his energies to expounding the philosophical and historical justifications of astrology, and to discoursing on and exemplifying the practical efficacy of this science. In this effort he drew upon elements of all the diverse intellectual traditions to which he was almost uniquely heir: upon the Pahlavi Greco-Indo-Iranian tradition in astrology, astronomy, and theurgy as preserved in Buzurjmihr, Andarzghar, Zaradusht, the *Zīj al-Shāh,* Dorotheus, and Valens; upon a Sanskrit Greco-Indian tradition in astrology and astronomy from Varāhamihira, Kanaka, the *Sindhind,* the *Zīj al-Arkand,* and Āryabhaṭa; upon the Greek tradition in philosophy, astrology, and astronomy through Aristotle, Ptolemy, and Theon; upon the Syriac Neoplatonizing philosophy of astral influences and theurgy from al-Kindī and the books of the Ḥarrānians; and upon the earlier, less complete attempts at such vast syntheses among Persian scholars writing in Arabic as represented by those of Māshāʾallāh, Abū Sahl al-Faḍl ibn Nawbakht, ʿUmar ibn al-Farrukhān al-Ṭabarī, and Abū Yūsuf Yaʿqūb al-Qaṣrānī.

Abū Maʿshar's renown as an astrologer was immense, both among his contemporaries and in later times. He cast the horoscope of an Indian (Rāṣṭrakūṭa?) prince who was born 11 January 826; he advised several rebels against the authority of the caliph; and he accompanied al-Muwaffaq on his expedition against the Zanj in Basra in 880–883. To Ibn al-Qifṭī, as to most students of Islamic astrology,

he was "the teacher of the people of Islam concerning the influences of the stars."

Abū Maʿshar's philosophical proof of the validity of astrology was probably most elaborately presented in his lost *Kitāb ithbāt ʿilm al-nujūm* ("Book of the Establishment of Astrology"), but it is also discussed at length in the first *maqāla* of his *Kitāb al-madkhal al-kabīr* ("Great Introduction"), which was written in 849/850. The argument, as has been pointed out by Lemay, is largely Aristotelian, with some Neoplatonic elements; but Lemay, working only with the Latin translations, failed to realize that the immediate sources of Abū Maʿshar's Aristotelianism were not the Arabic translations of the *De caelo,* the *Physica,* and the *De generatione et corruptione,* but the purported writings of the Ḥarrānian prophets, Hermes and Agathodemon. That the "Sabaeans" of Ḥarrān depended on Aristotle's *Physica, De caelo, De generatione et corruptione,* and *Meteorologica* for their theories regarding the material universe is clearly stated by Aḥmad ibn al-Ṭayyib al-Sarakhsī (*ca.* 835–899), another student of al-Kindī (fr. I B1 in F. Rosenthal). Since the Ḥarrānians were interested in the laws of perceptible nature precisely because they saw the same relationships between the ethereal spheres and the sublunar world of change that Abū Maʿshar seeks to prove (as well as a further relationship between the ethereal spheres and the One which Abū Maʿshar only hints at), it is an easy step to the conclusion that this justification of astrology is, in its main outline, taken by Abū Maʿshar from the books of the Ḥarrānians, and is thus only a part of a much more elaborate universal philosophy of emanation.

That philosophy, closely similar to doctrines common to a number of religious movements of the first half-millennium of the Christian era (they are found, for example, in the *Corpus Hermeticum,* in the *Chaldaean Oracles,* and in the writings of various Neoplatonists), and not unlike the philosophical background of Jābirian alchemy, posits three levels of being analogous to three concentric spheres: the divine (the sphere of light), the ethereal (the eight celestial spheres), and the hylic (the sublunar core, in which matter is involved in a constant process of change due to the motions of the four Empedoclean elements).

This view of the universe gains religious content when there is added to it the idea that man's soul has descended from the sphere of light to the hylic sphere, and now must strive to return to union with the divine. But, according to "Sabaean" doctrine, it cannot leap over the ethereal sphere and attain this union without the assistance of intermediaries, which are the celestial spheres; therefore, man's religion—

his liturgy and his ritual—must be addressed to the deities of the planets and of the constellations rather than to the One. The form of this worship is determined by the attributes, qualities, and conditions of the intermediaries; these are known by the study of astrology and astronomy.

The religious view of the Ḥarrānians, then, assumes an Aristotelian physical universe in which the four Empedoclean elements are confined to the sublunar world, and the celestial spheres consist of a fifth element. The normal astrological view is concerned to some extent with schematic correlations between celestial figures and (*a*) the four Empedoclean elements and (*b*) the various Pythagorean contrasting principles. Primarily, however, it works with somewhat arbitrary associations of planets, zodiacal signs, decans, and so on; with the psychological factors governing man's behavior; with the attributes and characteristics apparent in material objects; and with various selected species of plants, animals, stones, fish, and so on. The Ḥarrānians, followed by Abū Maʿshar, attempted to validate the scientific basis of these arbitrary associations between the celestial and sublunar worlds in astrology by casting over the whole system a peculiar interpretation of Aristotelian physics. According to this interpretation, the nature of the influence of the superior spheres on the inferior is not restricted to the transmission of motion alone; terrestrial bodies each possess the potentiality of being moved by particular celestial bodies, and the celestial bodies similarly each possess the possibility of influencing particular terrestrial bodies. The precise details of the mode of this influence need not detain us here; suffice it to say that the practical effect of this elaborate development of theory in the *Kitāb al-madkhal al-kabīr* was the reassertion of the truth of the astrological doctrines already long current.

For the Ḥarrānians the *Kitāb al-madkhal al-kabīr* also provided the justification for the elaboration of a theory of talismans and planetary theurgy which made them the recognized masters of these esoteric practices (although they had, of course, been popular for centuries in the Roman and Sassanian empires). Abū Maʿshar is among those who helped, in his *Kitāb al-ulūf* and *Kitāb fī buyūt al-ʿibādāt,* to establish their reputation. From time to time he refers to talismans (see especially his *Kitāb ṣuwar al-daraj*), but in general he is interested more in predicting the future than in manipulating it.

For Abū Maʿshar, however, the validity of astrology is determined not only by the Neoplatonizing Aristotelianism of the Ḥarrānians; it also rests on an elaborate world history of the transmission of science which permits one to trace back the fragments of truth about nature scattered among the peoples of the earth to a pristine divine source: it is a sort of prophetology of science.

Man's knowledge of the relationship between the three spheres comes not from his own powers of reasoning, but from revelation. For the Ḥarrānians, the prophet of revelation was Hermes Trismegistus. Abū Maʿshar, however, desired to universalize the personality of the prophet and to demonstrate the essential unity of human thought, and identifies a first Hermes with the Iranian Hūshank and the Semitic Enoch-Idrīs; following this composite figure are a succession of pupils of various nations (including two more named Hermes) who spread the revealed truth among the nations of the *oecumene.* Abū Maʿshar's cultural background helps to explain this universalism, although it must be noted that in certain details his elaborate history of science had been anticipated by Persian scholars of the preceding generation. It was his theory of an original "Sabaeanism" followed by all of mankind, however, which became the basis of much of Muslim historiography of philosophy and religion.

In conformity with this theory as expounded in his *Kitāb al-ulūf,* and on the alleged basis of a manuscript said to have been buried at Isfahan before the Flood, Abū Maʿshar produced his *Zīj al-hazārāt,* which was to restore to mankind the true astronomy of the prophetic age. The mean motions of the planets are computed in this *zīj* by the Indian method of the *yuga* and by using Indian parameters; in this section Abū Maʿshar depended largely on the *Zīj al-Sindhind* of al-Fazārī and the *Zīj al-Arkand* (both of Indian origin), although his *yuga* of 360,000 years, while Indian, was also used by the Ismāʿīlīs. His prime meridian and the parameters for his planetary equations were taken from the Persian *Zīj al-Shāh,* which is greatly indebted to Indian sources. His planetary model, however, was evidently Ptolemaic. Thus this "antediluvian" *zīj* proves by its mixture of Indian, Persian, and Greek elements that the theory of the original unity of the intellectual traditions of mankind is a true one; each has preserved a bit of the revelation.

Accompanying this astronomical work and history of science was an elaborate astrological interpretation of history expounded in the *Kitāb al-qirānāt,* which, being originally of Sassanian (Zoroastrian) origin, reached Abū Maʿshar through the works of Māshāʾallāh, ʿUmar ibn al-Farrukhān al-Ṭabarī, and al-Kindī. This theory, based on periods of varying length under the influence of the several planets and zodiacal signs, on the recurring conjunctions at regular intervals of Saturn and Jupiter and of Saturn and

Mars, on the horoscopes of year transfers, and on transits, postulated the inherent impermanence of all human institutions—including the religion of Islam and the rule of the Arab caliphate. It was particularly popular among the Iranian intellectuals of the eighth, ninth, and tenth centuries, who delighted in predicting the imminent downfall of the Abbasids and restoration of the royal house of Iran to the throne of world empire. And it is one other element in Abū Maʿshar's system that links him with the Ismāʿīlīs.

Parallel to these methods of universal astrological history, Sassanian scientists had developed similar techniques of progressive individual genethlialogy based on periods, the horoscopes of birthdays, and transits. Their sources had been Hellenistic, the primary one being the fourth book of Dorotheus. Abū Maʿshar, like many other Muslim astrologers, has elaborately dealt with this type of astrology (in his *Kitāb taḥāwīl sinī al-mawālīd*). He also composed a number of other works on nativities, some mere compilations of the sayings of the wise men of India, Persia, Greece, Egypt, and Islam, intended to demonstrate again their fundamental unity (the *Kitāb al-jamhara* and the *Kitāb aṣl al-uṣūl*), and some more orthodox compositions modeled on the Hellenistic textbooks that had been translated into Arabic (the two versions of the *Kitāb aḥkām al-mawālīd*).

In these writings, as in his other works listed in the critical bibliography (see below), Abū Maʿshar did not display any startling powers of innovation. They are practical manuals intended for the instruction and training of astrologers. As such, they exercised a profound influence on Muslim intellectual and social history and, through translations, on the intellectual and social history of western Europe and of Byzantium. Abū Maʿshar's folly as a scientist has been justly pointed out by al-Bīrūnī (*Chronology*, ed. C. E. Sachau, repr. Leipzig, 1923, pp. 25–26; trans. *idem*, London, 1879, pp. 29–31). One gains the strong impression from his pupil Shādhān's *Mudhākarāt* that even as an astrologer he was not intellectually rigorous or honest (no matter what the situation may be now, it certainly was possible to be an intellectually honest astrologer in the ninth century). He is an interesting and instructive phenomenon, but is not to be ranked among the great scientists of Islam.

BIBLIOGRAPHY

I. ORIGINAL WORKS. There are two old lists of Abū Maʿshar's works. The first and most complete is that in the *Fihrist* of Ibn al-Nadīm, who wrote *ca.* 987, G. Flügel, ed. (Leipzig, 1871–1872), p. 277; this I call N. The second is a shorter catalog preserved in the *Taʾrīkh al-ḥukamāʾ* of Ibn al-Qifṭī, who wrote before 1248, J. Lippert, ed. (Leipzig, 1903), p. 153; this I call Q I. Ibn al-Qifṭī (p. 154) adds a list of those works found in Ibn al-Nadīm's list which he could not identify in Q I; this I call Q II. There are some repetitions of titles in Q I and Q II where Ibn al-Qifṭī has been led by differences in wording to believe in the existence of separate works. Note that Ibn al-Nadīm (p. 275; copied by Ibn al-Qifṭī, p. 154) claims, on the authority of Ibn al-Jahm (is this Muḥammad ibn al-Jahm al-Barmakī? See *Fihrist*, p. 277) that Abū Maʿshar plagiarized nos. 1, 4, 8, and 16 from Sanad ibn ʿAlī, who flourished under al-Maʾmūn; but this seems to be a mistaken allegation (cf. Ibn Yūnis, *Zīj al-ḥākimī*, Caussin de Perceval, ed., in *Notices et extraits des manuscrits*, VII [Paris, 1803], 58).

In this catalog, I refer in general only to manuscripts which I have personally examined, and not to all of these. The reader should realize that this list is as exhaustive as it can be made at present, and he should be particularly aware that all of the *genuine* works listed by Brockelmann (*Geschichte der arabischen Literatur*, I²–II² [Leiden, 1943–1949]; and *Supplementum* I–III [Leiden, 1937–1942], henceforth referred to as *GAL*) are included.

(1) *Kitāb al-madkhal al-kabīr ʿalā ʿilm aḥkām al-nujūm* ("Great Introduction to the Science of Astrology"). N 1; Q I, 3; Ḥājjī Khalīfa (*Kashf al-ẓunūn*, ed. G. Flügel [London, 1835–1858], henceforth referred to as Ḥājjī Khalīfa), V, 475. I have examined Leiden Or. 47 and NO 2806. This is a work in eight *maqālāt* covering the following: (1) the philosophical and historical justifications of astrology; (2) the numbers and characteristics of the fixed stars and the zodiacal signs; (3) the influence of the seven planets, and particularly of the two luminaries, on the sublunar world; (4) the astrological natures of the planets; (5) the lordships of the planets over the zodiacal signs and their parts; (6) the zodiacal signs in relation to each other and to man; (7) the strengths of the planets, their relations to each other, and their chronocratories; and (8) astrological lots. It was written in 849/850 or shortly thereafter. Only one chapter (6,1—on the decans) has been published of the original Arabic version: see K. Dyroff, in F. Boll, *Sphaera* (Leipzig, 1903), pp. 490–539; cf. D. Pingree, "The Indian Iconography of the Decans and Horās," in *Journal of the Warburg and Courtauld Institutes*, **26** (1963), 223–254.

Lengthy selections from the "Great Introduction" were translated into Greek *ca.* 1000; they form the bulk of the third book of the *Mysteries of Abū Maʿshar*, of which the present writer is preparing an edition. The whole of the "Great Introduction" was translated into Latin by John of Seville in 1133 and (with some abridgments) by Hermann of Carinthia in 1140; the latter translation was printed by Erhard Ratdolt at Augsburg in 1489 and 1495, and by Jacobus Pentius Leucensis (de Leucho) at Venice in 1506. From the Latin versions were derived a Hebrew version by Yakob ben Elia in the late thirteenth century, which is referred to in M. Steinschneider, *Die hebräischen Übersetzungen des Mittelalters* (Berlin, 1893), pp. 567–571 (henceforth referred to as Steinschneider), and the *Liber Albumazarus,* written by Zothorus Zaparus Fendulus in the fourteenth century, as well as German and English translations. The most recent discussion of the Latin translations, in which a strong case has been made for their influence on western European philosophy in the twelfth century, is R. Lemay, *Abū Maʿshar and Latin Aristotelianism in the Twelfth Century* (Beirut, 1962); cf. also J. C. Vadet, "Une défense de l'astrologie dans le madhal d'Abū Maʿsar al Balḫī," in *Annales islamologiques,* **5** (1963), 131–180. The Peripatetic and Neoplatonic background of Abū Maʿshar's theory of tides is discussed by P. Duhem, *Le système du monde,* II (Paris, 1914), 369–386 (henceforth referred to as Duhem). For the date of the "Great Introduction," see H. Hermelink, "Datierung des Liber Introductorius von Albumasar (Kitāb al-mudḫal al-kabīr von Abū Maʿsar)," in *Sudhoffs Archiv,* **46** (1962), 264–265.

(2) *Kitāb al-madkhal al-ṣaghīr,* also called *Kitāb mukhtaṣar al-madkhal* ("Little Introduction"). N 2; Q II, 1. I have consulted British Museum Additional Manuscript 7490, pt. 4 (Yeni Cami 1193, pt. 6, listed by Brockelmann, is the *Kitāb al-madkhal fī ʿilm al-aḥkām al-falakiyya,* in seventy-three chapters, of Abu'l-Qāsim ʿAlī ibn Aḥmad al-Balkhī, also known as Abū Maʿshar, but this has nothing to do with the "Little Introduction"). This work was written after the "Great Introduction," which it epitomizes at the expense of all philosophical and historical passages. It consists of seven *fuṣūl:* (1) on the natures, conditions, and indications of the zodiacal signs; (2) on the conditions of the planets alone and with respect to the sun; (3) on the twenty-five conditions of the planets; (4) on the strength and goodness of the planets and their dodecatemoria; (5) on the natures of the planets and their indications; (6) on lots; and (7) on the planetary chronocratories. It was translated into Latin by Adelard of Bath in the early twelfth century.

(3) *Zīj al-Hazārāt* ("Tables of the Thousands"). N 3; Q I, 12; Q II, 2; Ḥājjī Khalīfa, III, 558–559. This work, composed between 840 and 860 in "sixty and some" chapters, is now lost. However, an attempt at recovering its planetary parameters and some of its astronomical theories has been made by the present writer in his *The Thousands of Abū Maʿshar* (London, 1968). Note that the phrase *et ego Albumasar in tabulis nostris maioribus in fine richene elchebir* [*zījinā al-kabīr*] *celestium discursus persecutus sum,* which is found in Hermann of Carinthia's translation of the "Great Introduction" (1,1), does not appear in the Arabic manuscripts I have examined.

(4) *Kitāb al-mawālīd al-kabīr* ("Great Book of Nativities"). N 4; Q II, 3. According to Ibn al-Nadīm, Abū Maʿshar never finished this book. Perhaps it is identical with the "Book of the Multitude"; cf. also the "Book of Judgments About Nativities."

(5) *Kitāb hayʾat al-falak wa-ikhtilāf ṭulūʿihi* ("Form of the Sphere and Differences in Rising-times"), N 5; Q I, 7 (?); Q II, 4. Ibn al-Nadīm informs us that this book, of which no copies have survived, was in five *fuṣūl.* Its subject is clear.

(6, 7) *Kitāb al-kadkhudāh* ("Book of the Kadkhudāh")

and *Kitāb al-haylāj* ("Book of the Haylāj"). N 6 and N 7; Q I, 8. Ibn al-Nadīm treats these as two separate books; Ibn al-Qifṭī, more naturally, as one. The *Haylāj* ("Prorogator") and *Kadkhudāh* ("Lord of Life") are frequently discussed in Abū Maʿshar's other works (e.g., in the "Book of Judgments About Nativities," chs. 4 and 5), and were often treated by his predecessors—most notably by Dorotheus in his third book. The Persian terminology, of course, indicates a Sassanian background, and we know that the Arabic version of Dorotheus was translated from Pahlavi *ca.* 800. It is at least possible that Abū Maʿshar's work was the source of the *Kitāb al-zāʾirjāt fī istikhrāj al-haylāj wa 'l-kadkhudāh,* which forms the fourth part of the *Al-Jāmiʿ al-Shāhī* of Aḥmad ibn Muḥammad ibn ʿAbd al-Jalīl al-Sijzī, which was written in the second half of the tenth century (I have used British Museum Or. 1346, Esad Ef. 1998, and Hamidiye 837). The original of al-Sijzī's work relied on Hermes, Ptolemy, Dorotheus, and "the moderns."

(8) *Kitāb al-qirānāt* ("Book of Conjunctions"), also known as the *Kitāb al-milal wa 'l-duwal*). N 8; Q I, 4; Q I, 5; Ḥājjī Khalīfa, V, 136. I have used British Museum Or. 7716 and Escorial 937. This work, in eight *maqālāt,* as is the "Great Introduction," was written after 869 or perhaps even after 883 (cf. 1,3, which mentions events in Basra predicted for the fifteen years after the sixtieth year following the conjunction of 809; in 2,7 he refers to the murder of al-Mutawakkil, which occurred in December 861). The subjects of the eight *maqālāt* are as follows: (1) the appearance of prophets and their laws; (2) the rise and fall of dynasties and kings; (3) the effects of planetary combinations; (4) the effects of each zodiacal sign's being in the ascendant; (5) the lordships of the planets; (6) transits; (7) each zodiacal sign as *muntahā* and as ascendant of the revolution of the year; and (8) the revolutions of the years and the *intihāʾāt.*

Ibn al-Nadīm claims that this work was dedicated to Ibn al-Bāzyār, a pupil of Ḥabash al-Ḥāsib (*fl.* 829–864); this statement is perhaps supported by the fact that one manuscript of the "Book of Conjunctions" ascribes it to Ibn al-Bāzyār (*GAL Suppl.* I, 394; cf. also al-Bīrūnī, *Chronology,* ed. C. E. Sachau, repr. Leipzig, 1923, p. 21; trans. *idem,* London, 1879, p. 25; and *Fihrist,* p. 276). It was translated into Latin by John of Seville; this translation was printed by Erhard Ratdolt at Augsburg in 1489, and reprinted by Jacobus Pentius de Leucho at Venice in 1515 (these are the same two printer-scholars to whom we owe the editions of Hermann's translation of the "Great Introduction"). One chapter of this Latin translation (2, 8), which Abū Maʿshar had plagiarized from al-Kindī, was reprinted by O. Loth in his article "Al-Kindī als Astrolog," in *Morgenländische Forschungen* (Leipzig, 1875), pp. 261–310. For Abū Maʿshar's reference to trepidation in this work, see Duhem, II, 503–504.

(9) *Kitāb taḥāwīl sinī al-ʿālam,* or *Kitāb al-nukat* ("Book of Revolutions of the World-years," or "Book of Subtleties"). N 9; Q I, 10; Ḥājjī Khalīfa, I, 171. I have used Bodleian Marsh 618, Escorial 938, and Fatih 3426. This is a relatively short work on the nature of a year (or

month or day) as determined by the horoscope of its beginning. It was translated into Latin by John of Seville under the title *Flores;* cf. J. Vernet, "Cuestiones catalográficas referentes a autores orientales: Problemas bibliográficas en torno a Albumasar," in *Biblioteconomia* (Barcelona, 1952), 12–17. This is undoubtedly identical with the *De revolutionibus annorum mundi seu liber experimentorum,* also translated by John of Seville; for this, see F. J. Carmody, *Arabic Astronomical and Astrological Sciences in Latin Translation* (Berkeley–Los Angeles, 1956), p. 94 (henceforth referred to as Carmody). *Flores* was published by Erhard Ratdolt at Augsburg in 1488, 1489, and 1495, and by the house of Sessa in Venice in 1488 and 1506.

(10) *Kitāb al-ikhtiyārāt* ("Book of Elections"). N 10; Q II, 5; Ḥājjī Khalīfa, I, 198. This may be the fifth text in British Museum Additional Manuscript 7490, which is entitled *Kitāb al-ikhtiyārāt* and which follows Abū Maʿshar's "Little Introduction"; it contains fifty-five chapters quoting from many sources that were favorites of Abū Maʿshar (e.g., Dorotheus). There is also a *Kitāb al-ikhtiyārāt* which is the eighth component of al-Sijzī's *Al-Jāmiʿ al-Shāhī,* but its relation to Abū Maʿshar remains obscure. There are also many chapters on elections in the first book of the Byzantine *Mysteries of Abū Maʿshar* and, in Latin, an *Electiones planetarum* and a *De mode eligendi* (Carmody, p. 96). Cf. also Steinschneider, p. 571.

(11) *Kitāb al-ikhtiyārāt ʿalā manāzil al-qamar* ("Book of Elections According to the Lunar Mansions"). N 11; Q II, 6. This is perhaps different from the preceding work. There is in Latin a *Flores de electionibus* which is based on the moon (Carmody, p. 97) and a *De electionibus lunae* (Carmody, p. 101; cf. Steinschneider, p. 571), both ascribed to Abū Maʿshar. Compare also the *Kitāb masāʾil al-qamar* in Berlin oct. 1617 (not seen by me).

(12) *Kitāb al-ulūf* ("Book of the Thousands"). N 12; Q I, 2; Ḥājjī Khalīfa, V, 50; cf. I, 22. This, one of Abū Maʿshar's most important works, is lost; but we do have summaries of it by al-Sijzī (part 9 of *Al-Jāmiʿ al-Shāhī;* cf. the *Dastūr al-munajjimīn* in Paris Bibliothèque Nationale 5968), al-Tanūkhī (in British Museum Or. 3577), and an anonymous author (in Berlin 5900). Unfortunately, the epitome by Abū Maʿshar's pupil Ibn al-Māzyār (Ibn al-Bāzyār?) is lost. All the available material has been assembled and discussed by the present writer in his *The Thousands of Abū Maʿshar.* The *Kitāb al-ulūf* is not to be confused with the *Kitāb fī buyūt al-ʿibādāt* ("Book of Temples") mentioned by al-Bīrūnī in the *Chronology,* despite what Ḥājjī Khalīfa, who had no copy, says of its nature.

(13) *Kitāb al-ṭabāʾiʿ al-kabīr* ("Great Book of Natures"). N 13; Q I, 1; Q II, 7. According to Ibn al-Nadīm, this apparently lost work was divided into five *ajzāʾ.* According to at least one manuscript, the "Book of the Foundation of Foundations" was called the *Kitāb al-ṭabāʾiʿ,* but this identification is probably not to be taken seriously. If it refers to any part of that work, it could only be to the first section, which precedes the quotations from ancient authorities.

(14) *Kitāb al-sahmayn wa-aʿmār al-mulūk wa 'l-duwal* ("Book of the Two Lots and the Lives of Kings and Dynasties"). N 14; Q II, 8. The two lots must be the Lot of Fortune and the Lot of the Demon; their relevance to astrological history is not yet clear. The text of this book has not been found.

(15) *Kitāb zāʾirjāt [wa] al-intihāʾāt wa 'l-mamarrāt* ("Book of Tables of the Intihāʾāt and of the Transits"). N 15. This work, too, must have been on astrological history; cf. books 6 and 8 of the "Book of Conjunctions." No manuscripts of it are known.

(16) *Kitāb iqtirān al-naḥsayn fī burj al-Saraṭān* ("Book of the Conjunction of the Two Malefics in Cancer"). N 16; Q II, 9. The particularly maleficent effects of conjunctions of Saturn and Mars in Cancer are also treated extensively in the "Book of Conjunctions" (2, 8, which is largely copied from al-Kindī). There seem to be no copies of this book extant.

(17, 18) *Kitāb al-ṣuwar wa 'l-ḥukm ʿalayhā* ("Book of the Images and Their Influences"). N 17; Q II, 10. *Kitāb [al-] ṣuwar [wa] al-daraj wa 'l-ḥukm ʿalayhā* ("Book of the Images of the Degrees, and Their Influences"). N 18. This is most likely one work, as Ibn al-Qifṭī assumes, and seems to be on talismans ("Zaradusht's" work on talismans, which forms the thirteenth part of *Al-Jāmiʿ al-Shāhī,* is entitled *Kitāb ṣuwar darajāt al-falak*). Again, the Arabic is lost; but there is in Latin a *De ascensionibus imaginum* ascribed to Abū Maʿshar (Carmody, p. 100). Cf. the first part of the "Small Book of Nativities."

(19) *Kitāb taḥāwīl sinī al-mawālīd* ("Book of the Revolutions of the Years of Nativities"). N 19; Q I, 11; Ḥājjī Khalīfa, VI, 242. I have consulted Escorial 917. This work contains nine *maqālāt* rather than just eight, as Ibn al-Nadīm claims: (1) introductory; (2) on the various astrological lords as signifiers; (3) on the direction and the division; (4) on the planetary periods; (5) on the transits of the planets; (6) on various planetary and zodiacal signifiers; (7) on the effects of the planetary motions; (8) on the effects of the planets being in each other's houses and terms; and (9) on casting monthly and daily horoscopes. Al-Sijzī summarized this work in his *Al-Jāmiʿ al-Shāhī* (part five; cf. Ḥājjī Khalīfa, II, 46); he also translated this summary into Persian, cf. C. Storey, *Persian Literature,* II, 1 (London, 1958), 39 (henceforth referred to as Storey). The original Arabic was also translated into Greek; the first five books survive (see the present writer's edition of them [Leipzig, 1968]). These five books were translated from Greek into Latin and were published by H. Wolf at Basel in 1559.

(20) *Kitāb al-mizājāt* ("Book of Mixtures"). N 20; Q II, 11. Ibn al-Nadīm states that this work is rare, and so it is. But perhaps it is identical with the *Kitāb mizājāt al-kawākib* summarized by al-Sijzī (*Al-Jāmiʿ al-Shāhī,* part 6); this deals with combinations of two, three, four, five, six, and seven planets.

(21) *Kitāb al-anwāʾ* ("Star-calendar"). N 21; Q II, 12. This work of Abū Maʿshar I find mentioned nowhere else. The contemporary *Kitāb al-anwāʾ* of Ibn Qutayba (d. 879) has been edited by Hamidullah and Pellat (Hyderabad-

Deccan, 1956); a list of twenty-four authors of *Kutub al-anwāʾ* in the ninth and tenth centuries will be found on p. 14 of their introduction.

(22) *Kitāb al-masāʾil* ("Book of Interrogations"). N 22; Q II, 13. Ibn al-Nadīm calls this a compendium; it is probably, then, identical with the "Perfect Book." There is a work entitled *Abwāb al-masāʾil wa-mā baʿdahā min al-ikhtiyārāt* in Mingana 922.

(23) *Kitāb ithbāt ʿilm al-nujūm* ("Book of the Proof of Astrology"). N 23; Q II, 14. This work, which presumably expounded in detail the Ḥarrānian theories found in the "Great Introduction," was perhaps written against ʿAlī ibn ʿĪsā al-Ḥarrānī's *Risāla fī ibṭāl ṣināʿat aḥkām al-nujūm,* which is mentioned by al-Qabīṣī (*d.* 967) in the preface to his *Al-madkhal ilā ṣināʿat aḥkām al-nujūm;* ʿAlī ibn ʿĪsā participated in the measurement of a terrestrial degree carried out at Sinjar under al-Maʾmūn, and made observations in Baghdad in 843–844.

(24) *Kitāb al-kāmil* or *Kitāb al-masāʾil* ("Perfect Book" or "Book of Interrogations"). N 24; Q II, 15. This unfinished compendium is perhaps identical with the "Book of Interrogations."

(25) *Kitāb al-jamhara* ("Book of the Multitude"). N 25; Q I, 9; Q II, 16. Ibn al-Nadīm informs us that this was a collection of sayings of earlier astrologers concerning nativities. It was, then, perhaps the original form of the second part of the "Book of the Foundation of Foundations."

(26) *Kitāb aṣl al-uṣūl* ("Book of the Foundation of Foundations"). N 26; Q II, 17; Ḥājjī Khalīfa, I, 282–283 (?). As Ibn al-Nadīm states, this compendium of sayings about genethlialogy is also attributed to Abu'l-ʿAnbas al-Ṣaymarī (828–888/889). Most manuscripts (e.g., Hamidiye 829 and British Museum Or. 3540) ascribe the work to al-Ṣaymarī (apparently correctly), but its title in Hamidiye 824 is *Al-aṣl fī ʿilm al-nujūm* and *Sirr al-asrār* by Abū Maʿshar, which is the *Kitāb al-ṭabāʾiʿ* (cf. also Mingana 921); this confusion of three titles does not inspire confidence in the manuscript's accuracy. The "Book of the Foundation of Foundations" is an extremely valuable work, especially its second part, which contains extensive excerpts from such authorities as Antiochus, Teucer, Dorotheus, Valens, Democritus, Zeno, Jina (?) the Indian, Ṛṣi (?) the Indian, Buzurjmihr, and Zaradusht.

(27) *Kitāb tafsīr al-manāmāt min al-nujūm* ("Book of the Explanation of Dreams From the Stars"). N 27; Q II, 18. This work, whose purpose was probably to predict dreams from astrological indications rather than to expound oneiromancy, is not mentioned in the inventory of oneirocritical treatises drawn up by T. Fahd, *La divination arabe* (Leiden, 1966), pp. 329–363. The attribution to Abū Maʿshar of Muḥammad ibn Sīrīn's *Tafsīr al-manāmāt* by J. Leunclavius in his Latin translation of the Greek version (Frankfurt, 1577) is, of course, false.

(28) *Kitāb al-qawāṭiʿ ʿala 'l-haylājāt* ("Book of Severances [of Life] According to the Haylājāt"). N 28; Q I, 6 (?); Q II, 19. This book on what the Greeks call the ἀφαιρέτης is lost.

(29) *Kitāb al-mawālīd al-ṣaghīr* ("Small Book of Na-

tivities"). N 29; Q II, 20. According to Ibn al-Nadīm this work consists of two *maqālāt* and thirteen *fuṣūl.* It is not, then, identical with the "Book of Judgments About Nativities," but it does coincide with the state of the so-called "Book of the Meticulous Investigator, the Greek Philosopher Known as Abū Maʿshar al-Falakī" (*Kitāb al-muḥaqqiq al-mudaqqiq al-Yūnānī al-faylasūf al-shahīr bi-Abī Maʿshar al-Falakī*). This curious work has several times been published in Cairo, and J.-M. Faddegon has given a brief description of it in "Notice sur un petit traité d'astrologie attribué à Albumasar (Abu-Maʿšar)," in *Journal asiatique,* **213** (1928), 150–158. The first four *fuṣūl* are on magic and astrology related to various times; the next five *fuṣūl* are on the science of prediction from the numerical equivalents of proper names which classical antiquity commonly ascribed to Pythagoras or Petosiris; and the last four *fuṣūl* are concerned with nativities. *Faṣl* 12 is a zodiologion for men, and *faṣl* 13 a zodiologion for women; this last is introduced with a *basmala,* and therefore represents the second *maqāla* mentioned by Ibn al-Nadīm.

(30) *Kitāb zīj al-qirānāt wa 'l-ikhtirāqāt* ("Tables of Conjunctions and Transits"). N 30; Q I, 13; Q II, 21. Ibn al-Qifṭī, who also calls this work *Kitāb zīj al-ṣaghīr* ("Small Tables"), asserts that it gives the mean longitudes of the planets at the times of the (mean) conjunctions of Saturn and Jupiter since the epoch of the Flood (17 February 3102 B.C.). The work, then, is closely related to the "Tables of the Thousands," the "Book of the Conjunctions," and the "Book of the Thousands." The editions' *iḥtirāfāt* I take to be a scribal error for *ikhtirāqāt.* This was not a *zīj* in the normal sense of the term.

(31, 32) *Kitāb al-awqāt* ("Book of Times"). N 31. *Kitāb al-awqāt ʿalā ithnā ʿashariyyat al-kawākib* ("Book of Times According to the Dodecatemoria of the Planets"). N 32; Q II, 22. This is clearly one work, as Ibn al-Qifṭī perceived, and was presumably concerned with the proper times for commencing various activities as determined from the ascendant dodecatemorion (a well-known Greek technique of καταρχαί). This may be the *Kitāb al-masʾala* [*ʿalā*] *al-ithnā ʿashariyya* in Aya Sofya 2672 (not seen by me).

(33) *Kitāb al-sihām* ("Book of Lots"). N 33; Q II, 23. This work covers the special lots governing the material objects utilized by man; it must, then, to a large extent duplicate the contents of the eighth *maqāla* of the "Great Introduction." It is possible that at least the first tract in the Latin *De partibus et eorum causis* (Carmody, p. 98) is translated from this work. The work entitled *Liber Albumazar de duodecim domibus astrorum* (Carmody, pp. 98–99) also seems to be a translation of the "Book of Lots," and much of this sort of material is found in the first book of the Byzantine *Mysteries.* See also al-Bīrūnī, *Book of Instruction in the Elements of the Art of Astrology,* ed. R. R. Wright (London, 1934), pp. 282–289.

(34) *Kitāb al-amṭār wa 'l-riyāḥ wa-taghayyur al-ahwiya* ("Book of Rains and Winds and of Changes in the Weather"). N 34; Q II, 25; Ḥājjī Khalīfa I, 147, and V, 94. This is probably the *Kitāb al-sirr* ("Book of the Secret") which is found in Escorial 938 (ff. 1v–28) and in Bodleian

Marsh 618 (ff. 162v–173v and 198v et seq.); its first part deals with meteorological astrology, its second with the astrology of prices. This work includes a horoscope cast by Abū Maʿshar in Nīshāpūr on 5 March 832.

(35) *Kitāb ṭabāʾiʿ al-buldān wa-tawallud al-riyāḥ* ("Book of the Natures of Places and the Generation of Winds"). N 35; Q II, 24. This title brings to mind those of two older books: Hippocrates' *Airs, Waters and Places,* and the *Book of the Laws of the Regions* of Bardesanes' pupil Philip. But Abū Maʿshar's work was probably a technical astrological discussion of why the same celestial influences simultaneously cause different meteorological phenomena in the various regions of the world.

(36) *Kitāb al-mayl fī taḥwīl sinī al-mawālīd* ("Book of the Obliquity [of the Ecliptic] in the Revolution of the Years of Nativities"). N 36. In this lost work Abū Maʿshar must have tried to explain the differences between the lives of several individuals born at the same time as due in part to the effect of different terrestrial latitudes on the interpretation of the revolutions of their birth anniversaries.

(37) *Kitāb fī buyūt al-ʿibādāt* ("Book of Temples"). This work, which is mentioned by al-Bīrūnī (*Chronology,* C. E. Sachau, ed. [Leipzig, 1923], p. 205; trans. *idem* [London, 1879], p. 187), described the curious planetary temples of the "Sabaeans"; cf. the "Book of Conjunctions," 1,4. One wonders about the extent to which al-Dimashqī has relied on Abū Maʿshar's work in his description of the temples of Ḥarrān. The "Book of Temples," it should be noted, is different from the "Book of Thousands."

(38) *Kitāb ikhtilāf al-zījāt* ("Book of the Differences Between Tables"). The fragments of this work have been discussed in the present writer's *The Thousands of Abū Maʿshar.*

(39, 40) *Kitāb aḥkām al-mawālīd* ("Book of Judgments About Nativities"). Two versions of this work were written by Abū Maʿshar. The first is found on the last twenty-four ff. of Hamidiye 856, and consists of thirty-one *abwāb;* the second is preserved on ff. 1–64v of Bodleian Huntington 546, and originally contained eighteen *maqālāt* (only 1–15 and the beginning of 16 survive). Both works are traditionally Hellenistic; the second is based on the opinions of Dorotheus, Ptolemy, and Valens, and gives as examples a nativity of 128 of the era of Diocletian (A.D. 412) and the nativity of Paulus of Alexandria in 145 of the same era (A.D. 429). This second work is summarized by al-Sijzī as the third part of *Al-Jāmiʿ al-Shāhī.* The first work has apparently been translated into Persian (Storey, p. 39).

(41) *Kitāb qirānāt al-kawākib fī 'l-burūj al-ithnā ʿashara* ("Book of Conjunctions of the Planets in the Twelve Signs"). Ḥājjī Khalīfa, V, 136. I have used Bodleian Hyde 32. This work—differing from the "Book of Conjunctions"—discusses the effects of combinations of the planets in each of the zodiacal signs. This work has evidently been translated into Persian (Storey, p. 40). It is also apparently the work included in the first book of the Byzantine translation of the *Introduction to Astrology* of Ahmad the Persian, and published in *Catalogus codicum astrologorum Graecorum,* II (Brussels, 1900), 123–130.

(42) *Mudhākarāt Abī Maʿshar fī asrār ʿilm al-nujūm* ("Sayings of Abū Maʿshar on the Secrets of Astrology"). I have used Bodleian Huntington 546 and Cambridge University Gg. 3, 19. (I have not yet seen the manuscript in Ankara.) The *Mudhākarāt* was not written by Abū Maʿshar himself, but by his pupil Abū Saʿīd Shādhān. It contains much valuable information on the practice of astrology in ninth-century Baghdad, and therefore is frequently cited by Muslim historians of the period. It was translated into Greek (it constitutes most of the second book of the *Mysteries*) and into Latin (see L. Thorndike, "Albumasar in Sadan," in *Isis,* **45** [1954], 22–32, which is very inadequate). The present author is preparing an edition of the Arabic to accompany his edition of the Byzantine *Mysteries.*

II. SECONDARY LITERATURE. Besides the evidence of his own writings and the rich anecdotes of his pupil Shādhān, neither of which has yet been adequately explored, biographical information about Abū Maʿshar comes from two Muslim sources. The most important of these is the *Fihrist* of Ibn al-Nadīm (G. Flügel, ed. [Leipzig, 1871–1872], p. 277). Much of this was copied by Ibn al-Qifṭī (*Taʾrīkh al-ḥukamāʾ.* J. Lippert, ed. [Leipzig, 1903], pp. 152–154), but with some important additions taken from Shādhān and other sources, including the allegations that he was a drunkard and an epileptic. Ibn al-Qifṭī's biography was partially copied by Abu'l-Faraj in *Taʾrīkh mukhtaṣar al-duwal* (Beirut, 1958), p. 149.

Modern discussions of Abū Maʿshar are generally unreliable compilations based on secondary sources. The most authoritative of these is that by H. Suter, "Die Mathematiker und Astronomen der Araber und ihre Werke," in *Abhandlungen zur Geschichte der mathematischen Wissenschaften* (Leipzig, **10** (1900), 28–30. His list of Abū Maʿshar's works, however, is extremely unreliable. Most recent is a brief article by J. M. Millás in the *Encyclopaedia of Islam,* I (Leiden, 1960), 139–140.

DAVID PINGREE

ABŪ NAṢR AL-FĀRĀBĪ. See **al-Fārābī.**

ABŪ'L-RAYḤĀN MUHAMMAD IBN AHMAD AL-BĪRŪNĪ. See **al-Bīrūnī.**

IBN ABŪ'L-SHUKR. See **Muḥyī al-Dīn al-Maghribī.**

ABŪ ʿUBAYD AL-BAKRĪ. See **al-Bakrī.**

ABŪ'L-WAFĀʾ AL-BŪZJĀNĪ, MUHAMMAD IBN MUHAMMAD IBN YAHYĀ IBN ISMĀʿĪL IBN AL-ʿABBĀS (*b.* Būzjān [now in Iran], 10 June 940; *d.* Baghdad [now in Iraq], 997 or July 998), *mathematics, astronomy.*

Abū'l-Wafāʾ was apparently of Persian descent. In 959 he moved to Baghdad, which was then the capital of the Eastern Caliphate. There he became the last great representative of the mathematics-astronomy school that arose around the beginning of the ninth century, shortly after the founding of Baghdad. With his colleagues, Abū'l-Wafā' conducted astronomical observations at the Baghdad observatory. He continued the tradition of his predecessors, combining original scientific work with commentary on the classics—the works of Euclid and Diophantus. He also wrote a commentary to the algebra of al-Khwārizmī. None of these commentaries has yet been found.

Abū'l-Wafā''s textbook on practical arithmetic, *Kitāb fī mā yaḥtaj ilayh al-kuttāb wa'l-'ummāl min 'ilm al-ḥisāb* ("Book on What Is Necessary From the Science of Arithmetic for Scribes and Businessmen"), written between 961 and 976, enjoyed widespread fame. It consists of seven sections (*manāzil*), each of which has seven chapters (*abwāb*). The first three sections are purely mathematical (ratio, multiplication and division, estimation of areas); the last four contain the solutions of practical problems concerning payment for work, construction estimates, the exchange and sale of various grains, etc.

Abū'l-Wafā' systematically sets forth the methods of calculation used in the Arabic East by merchants, by clerks in the departments of finance, and by land surveyors in their daily work; he also introduces refinements of commonly used methods, criticizing some for being incorrect. For example, after indicating that surveyors found the area of all sorts of quadrangles by multiplying half the sums of the opposite sides, he remarks, "This is also an obvious mistake and clearly incorrect and rarely corresponds to the truth." Abū'l-Wafā' does not introduce the proofs here "in order not to lengthen the book or to hamper comprehension," but in a series of examples he defines basic concepts and terms, and also defines the operations of multiplication and division of both whole numbers and fractions.

Abū'l Wafā''s book indicates that the Indian decimal positional system of numeration with the use of numerals—which Baghdad scholars, acquainted with it by the eighth century, were quick to appreciate—did not find application in business circles and among the population of the Eastern Caliphate for a long time. Considering the habits of the readers for whom the textbook was written, Abū'l-Wafā' completely avoided the use of numerals. All numbers and computations, often quite complex, he described only with words.

The calculation of fractions is quite distinctive. Operation with common fractions of the type m/n, where m, n are whole numbers and $m > 1$, was uncommon outside the circle of specialists. Merchants and other businessmen had long used as their basic fractions—called *ra's* ("principal fractions") by Abū'l-Wafā'—those parts of a unit from 1/2 to 1/10,

and a small number of *murakkab* ("compound fractions") of the type m/n, with numerators, m, from 2 to 9 and denominators, n, from 3 to 10, with the fraction 2/3 occupying a privileged position. The distinction of principal fractions was connected with peculiarities in the formation of numerical adjectives in the Arabic language of that time. All other fractions m/n were represented as sums and products of basic fractions; businessmen preferred to express the "compound" fractions, other than 2/3, with the help of principal fractions, in the following manner:

$$\frac{2}{5} = \frac{1}{3} + \frac{2}{3} \cdot \frac{1}{10}, \frac{9}{10} = \frac{1}{2} + \frac{1}{3} + \frac{2}{3} \cdot \frac{1}{10}.$$

Any fraction m/n, the denominator of which is a product of the sort $2^P\ 3^Q\ 5^R\ 7^S$, can be expanded into basic fractions in the above form. In the first section of his book, Abū'l-Wafā' explains in detail how to produce such expansions with the aid of special rules and auxiliary tables. Important roles in this operation are played by the expansion of fractions of the type $a/60$ and the preliminary representations of the given fraction m/n in the form $m \cdot 60/n \div 60$ (see below). Since usually for one and the same fraction one can obtain several different expansions into sums and products of basic fractions, Abū'l-Wafā' explains which expansions are more generally used or, as he wrote, more "beautiful."

If the denominator of a fraction (after cancellation of the fraction) contains prime factors that are more than seven, it is impossible to obtain a finite expansion into basic fractions. In this case approximate expansions of the type $3/17 \approx (3 + 1) \div (17 + 1) = 2/9$ or $3/17 \approx 3\frac{1}{2} \div 17\frac{1}{2} = 1/5$—or still better, $3/17 \approx 3\frac{1}{7} \div 17\frac{1}{7} = \frac{1}{6} + \frac{1}{6} \cdot \frac{1}{10}$—were used.

Instead of such a method, which required the skillful selection of a number to be added to the numerator and denominator of a given fraction, Abū'l-Wafā' recommended the regular method, which enables one to obtain a good approximation with reasonable speed. This method is clear from the expansion

$$\frac{3}{17} = \frac{180}{17} \div 60 = \frac{10 + \frac{10}{17}}{60} \approx \frac{11}{60} = \frac{1}{6} + \frac{1}{6} \cdot \frac{1}{10}.$$

Analogously, one can obtain

$$\frac{3}{17} \approx \frac{1}{10} + \frac{1}{2} \cdot \frac{1}{9} + \frac{1}{6} \cdot \frac{1}{8}$$

or

$$\frac{3}{17} \approx \frac{1}{10} + \frac{1}{2} \cdot \frac{1}{9} + \frac{1}{6} \cdot \frac{1}{8} + \frac{1}{2} \cdot \frac{1}{6} \cdot \frac{1}{10} \cdot \frac{1}{10} \cdot \frac{1}{10}.$$

The error of this last result, as Abū'l-Wafā' demonstrates, equals

$$\frac{1}{4} \cdot \frac{1}{9} \cdot \frac{1}{10} \cdot \frac{1}{10} \cdot \frac{1}{10} \cdot \frac{1}{17}.$$

The calculation described somewhat resembles the Egyptian method, but, in contrast with that, it (1) is limited to those parts of a unity $1/q$, for which $2 \le q \le 10$; (2) uses products of the fractions $1/q_1 \cdot 1/q_2$ and $2/3 \cdot 1/q$; and (3) does not renounce the use of compound fractions m/n, $1 < m < n \le 10$. Opinions differ regarding the origin of such a calculation; many think that its core derives from ancient Egypt; M. I. Medovoy suggests that it arose independently among the peoples living within the territory of the Eastern Caliphate.

In the second section is a description of operations with whole numbers and fractions, the mechanics of the operations with fractions being closely connected with their expansions into basic fractions. In this section there is the only instance of the use of negative numbers in Arabic literature. Abū'l-Wafā' verbally explains the rule of multiplication of numbers with the same ten's digit:

$$(10a + b)(10a + c) = [10a + b - \{10(a + 1) -$$
$$(10a + c)\}]10(a + 1) + [10(a + 1) -$$
$$(10a + b)] \cdot [10(a + 1) - (10a + c)].$$

He then applies it where the ten's digit is zero and $b = 3$ and $c = 5$. In this case the rule gives

$$3 \cdot 5 = [3 - (10 - 5)] \cdot 10 + [10 - 3] \cdot [10 - 5]$$
$$= (-2) \cdot 10 + 35 = 35 - 20.$$

Abū'l-Wafā' termed the result of the subtraction of the number $10 - 5$ from 3 a "debt [*dayn*] of 2." This probably reflects the influence of Indian mathematics, in which negative numbers were also interpreted as a debt (*kśaya*).

Some historians, such as M. Cantor and H. Zeuthen, explain the lack of positional numeration and "Indian" numerals in Abū'l-Wafā''s textbook, as well as in many other Arabic arithmetic courses, by stating that two opposing schools existed among Arabic mathematicians: one followed Greek models; the other, Indian models. M. I. Medovoy, however, shows that such a hypothesis is not supported by fact. It is more probable that the use of the positional "Indian" arithmetic simply spread very slowly among businessmen and the general population of the Arabic East, who for a long time preferred the customary methods of verbal expression of whole numbers and fractions, and of operations dealing with them. Many authors considered the needs of these people; and, after Abū'l-Wafā', the above computation of fractions,

for example, is found in a book by al-Karajī at the beginning of the eleventh century and in works by other authors.

In the third section Abū'l-Wafā' gives rules for the measurement of more common planar and three-dimensional figures—from triangles, various types of quadrangles, regular polygons, and a circle and its parts, to a sphere and sectors of a sphere, inclusive. There is a table of chords corresponding to the arcs of a semicircle of radius 7, which consists of $m/22$ of the semicircumference ($m = 1, 2, \cdots, 22$), and the expression for the diameter, d, of a circle superscribed around a regular n-sided polygon with side a:

$$d = \sqrt{a^2 \left[(n - 1) \frac{n}{2} + 3 \right] \frac{2}{g}}$$

Abū'l-Wafā' thought this rule was obtained from India; it is correct for $n = 3, 4, 6$, and for other values of n gives a good approximation, especially for small n. At the end of the third section, problems involving the determination of the distance to inaccessible objects and their heights are solved on the basis of similar triangles.

Another practical textbook by Abū'l-Wafā' is *Kitāb fī mā yaḥtaj ilayh al-ṣāniʿ min al-aʿmāl al-handasiyya* ("Book on What is Necessary From Geometric Construction for the Artisan"), written after 990. Many of the two-dimensional and three-dimensional constructions set forth by Abū'l-Wafā' were borrowed mostly from the writings of Euclid, Archimedes, Hero of Alexandria, Theodosius, and Pappus. Some of the examples, however, are original. The range of problems is very wide, from the simplest planar constructions (the division of a segment into equal parts, the construction of a tangent to a circle from a point on or outside the circle, etc.) to the construction of regular and semiregular polyhedrons inscribed in a given sphere. Most of the constructions can be drawn with a compass and straightedge. In several instances, when these means are insufficient, intercalation is used (for the trisection of an angle or the duplication of a cube) or only an approximate construction is given (for the side of a regular heptagon inscribed in a given circle, using half of one side of an equilateral triangle inscribed in the same circle, the error is very small).

A group of problems that are solved using a straightedge and a compass with an invariable opening deserves mention. Such constructions are found in the writings of the ancient Indians and Greeks, but Abū'l-Wafā' was the first to solve a large number of problems using a compass with an invariable opening. Interest in these constructions was probably aroused by the fact that in practice they give more exact results than can be obtained by changing the compass opening. These constructions were widely circulated in Renaissance Europe; and Lorenzo Mascheroni, Jean Victor Poncelet, and Jakob Steiner developed the general theory of these and analogous constructions.

Also in this work by Abū'l-Wafā' are problems concerning the division of a figure into parts that satisfy certain conditions, and problems on the transformation of squares (for example, the construction of a square whose area is equal to the sum of the areas of three given squares). In proposing his original and elegant constructions, Abū'l-Wafā' simultaneously proved the inaccuracy of some methods used by "artisans."

Abū'l-Wafā''s large astronomical work, al-majisṭī, or Kitāb al-kāmil ("Complete Book"), closely follows Ptolemy's Almagest. It is possible that this work, available only in part, is the same as, or is included in, his Zīj al-Wāḍiḥ, based on observations that he and his colleagues conducted. The Zīj seems not to be extant. Abū'l-Wafā' apparently did not introduce anything essentially new into theoretical astronomy. In particular, there is no basis for crediting him with the discovery of the so-called variation of the moon (this was proved by Carra de Vaux, in opposition to the opinion expressed by L. A. Sédillot). E. S. Kennedy established that the data from Abū'l-Wafā''s observations were used by many later astronomers.

Abū'l-Wafā''s achievements in the development of trigonometry, specifically in the improvement of tables and in the means of solving problems of spherical trigonometry, are undoubted. For the tabulation of new sine tables he computed sin 30′ more precisely, applying his own method of interpretation. This method, based on one theorem of Theon of Alexandria, gives an approximation that can be stated in modern terms by the inequalities

$$\sin \frac{15°}{32} + \frac{1}{3}\left(\sin \frac{18°}{32} - \sin \frac{15°}{32}\right) < \sin 30'$$

$$< \sin \frac{15°}{32} + \frac{1}{3}\left(\sin \frac{15°}{32} - \sin \frac{12°}{32}\right).$$

The values sin 15°/32 and sin 18°/32 are found by using the known values of sin 60° and sin 72°, respectively, with the aid of rational operations and the extraction of a square root, which is needed for the calculation of the sine of half a given angle; the value sin 12°/32 is found as the sine of the difference 72°/32 − 60°/32. Setting sin 30′ equal to half the sum of the quantities bounding it above and below,

with the radius of the circle equal to 60, Abū'l-Wafā' found, in sexagesimal fractions, sin 30′ = 31^I 24^{II} 55^{III} 54^{IV} 55^V. This value is correct to the fourth place, the value correct to five places being sin 30′ = 31^I 24^{II} 55^{III} 54^{IV} 0^V.

In comparison, Ptolemy's method of interpolation, which was used before Abū'l-Wafā', showed error in the third place. If one expresses Abū'l-Wafā''s approximation in decimal fractions and lets $r = 1$ (which he did not do), then sin 30′ = 0.0087265373 is obtained instead of 0.0087265355—that is, the result is correct to 10^{-8}. Abū'l-Wafā' also compiled tables for tangent and cotangent.

In spherical trigonometry before Abū'l-Wafā', the basic means of solving triangles was Menelaus' theorem on complete quadrilaterals, which in Arabic literature is called the "rule of six quantities." The application of this theorem in various cases is quite cumbersome. Abū'l-Wafā' enriched the apparatus of spherical trigonometry, simplifying the solution of its problems. He applied the theorem of tangents to the solution of spherical right triangles, priority in the proof of which was later ascribed to him by al-Bīrūnī. One of the first proofs of the general theorem of sines applied to the solution of oblique triangles also was originated by Abū'l-Wafā'. In Arabic literature this theorem was called "theorem which makes superfluous" the study of complete quadrilaterals and Menelaus' theorem. To honor Abū'l-Wafā', a crater on the moon was named after him.

BIBLIOGRAPHY

I. ORIGINAL WORKS. The text on practical arithmetic has never been published in any modern language; however, discussions of it may be found in Woepcke, Luckey, and Medovoy (see below). Manuscripts of this work are preserved at the library of the University of Leiden (993) and the National Library, Cairo (^{I}V, 185). Besides these, there exist the manuscript of a work containing the fundamental definitions of theoretical arithmetic: "Risāla fī'l-aritmāṭīqī," at the Institute of Oriental Studies of the Academy of Sciences of the Uzbek S.S.R. in Tashkent (4750/8), which is described by G. P. Matvievskaja (see below); an unstudied arithmetical manuscript at the Escorial (Casiri, 933); and an unstudied arithmetical manuscript at the Library Raza, Rampur (I, 414). The text on geometric constructions has been studied in a Persian variant (Paris, Bibliothèque Nationale, pers. anc., 169) by Woepcke; Suter has studied the Milan manuscript (Biblioteca Ambrosiana, arab. 68); and a Russian translation by Krasnova of the Istanbul manuscript (Aya Sofya, 2753) has appeared. Eleven of the thirteen chapters of the latter are extant. The manuscript of the al-Majisṭī, only part of which has survived, is in Paris (Bibliothèque Nationale, ar. 2497) and has been studied by Carra de Vaux (see below). MS Istanbul, Carullah, 1479, is unstudied.

II. SECONDARY LITERATURE. General works concerning Abū'l-Wafā' are A. von Braunmühl, *Vorlesungen über Geschichte der Trigonometrie,* I (1900), 45–61; C. Brockelmann, *Geschichte der arabischen Litteratur,* I, 2nd ed. (Leiden, 1943), 255; Supp. I (Leiden, 1937), p. 400; M. Cantor, *Vorlesungen über Geschichte der Mathematik,* 2nd ed., I (Leipzig, 1894), 698–704, Index; Ibn al-Nadīm (Abū'l-Farāj Muḥammad Ibn Isḥāq), *Kitāb al-Fihrist,* G. Flügel, Y. Rödiger, and A. Müller, eds., I (Leipzig, 1871), 266, 283; H. Suter's translation of the *Fihrist,* "Das Mathematikerverzeichnis im Fihrist des Ibn Abī Ya'kūb al-Nadīm," in *Abhandlungen zur Geschichte der mathematischen Wissenschaften,* 6 (1892), 39; A. Youschkevitch, *Geschichte der Mathematik im Mittelalter* (Leipzig, 1964), Index; G. Sarton, *Introduction to the History of Science,* I, 666–667; H. Suter, *The Encyclopaedia of Islam,* new ed., I (Leiden–London, 1954), 159, and *Die Mathematiker und Astronomen der Araber* (Leipzig, 1900–1902), 71–72; Supp., 166–167; and Joh. Tropfke, *Geschichte der Elementar-Mathematik,* 2nd ed., VII (Leipzig, 1921–1924), Index.

The first attention to Abū'l-Wafā"s work was F. Woepcke's "Analyse et extraits d'un recueil de constructions géométriques par Aboûl Wefâ," in *Journal asiatique,* 5th ser., 5 (1855), 218–256, 309–359, which deals with the Paris (Persian) manuscript. For an analysis of Abū'l-Wafā"s practical arithmetic, see P. Luckey, *Die Rechenkunst bei Ğamšid b. Mas'ūd al-Kāši mit Rückblicken auf die ältere Geschichte des Rechnens* (Wiesbaden, 1951). There are two detailed investigations of the arithmetic text by M. I. Medovoy: "Ob odnom sluchae primenenija otritsatel'nykh chisel u Abu-l-Vafy" ("On One Case of the Use of Negative Numbers by Abū'l-Wafā'"), in *Istoriko-matematicheskie issledovanija* ("Studies in the History of Mathematics"), 11 (1958), 593–598, and "Ob arifmeticheskom traktate Abu-l-Vafy" ("On the Arithmetic Treatise of Abū'l-Wafā'"), *ibid.,* 13 (1960), 253–324; both articles constitute the first detailed investigation of this work. On the Tashkent manuscript, see G. P. Matvievskaja, *O matematicheskikh mkopissiakh iz sobranija instituta vostokovedenija AN Uz. S.S.R.* ("On the Mathematical Manuscripts in the Collection of the Institute of Oriental Studies of the Academy of Sciences of the Uzbek S.S.R."), Publishing House of the Academy of Sciences of the Uzbek S.S.R. Physical and Mathematical Sciences Series, pt. 9 (1965), no. 3, and *Uchenije o chisle na siednevekovom Vostoke* ("Number Theory in the Orient During the Middle Ages"); (Tashkent, 1967). An exposition of the Milan geometric manuscript is H. Suter, "Das Buch der geometrischen Konstruktionen des Abûl Wefâ," in *Abhandlungen zur Geschichte der Naturwissenschaften und Medizin* (1922), pp. 94–109. A Russian translation of the Istanbul geometric manuscript, with commentary and notes, has been done by S. A. Krasnova: "Abu-l-Vafa al-Buzdzhani, Kniga o tom, chto neobkhodimo remeslenniku iz geometricheskikh postroenij" ("Abū'l-Wafā' al-Būzjānī, 'Geometrical Constructions for the Artisan' "), in *Fiziko-matematicheskie*

nauki v stranakh vostoka ("Physics and Mathematics in the Orient"), I (IV) (Moscow, 1966), 42–140.

The astronomical and trigonometrical works of Abū'l-Wafā' are discussed in Carra de Vaux, "L'*Almageste* d'Abū-l-Wéfā al-Būzjānī," in *Journal asiatique,* 8th ser., 19 (1892), 408–471; E. S. Kennedy, "A Survey of Islamic Astronomical Tables," in *Transactions of the American Philosophical Society,* n.s. 46 (1956), 2; and F. Woepcke, "Sur une mesure de la circonférence du cercle due aux astronomes arabes et fondée sur un calcul d'Aboûl Wefâ," in *Journal asiatique,* 5th ser., 15 (1860), 281–320.

A. P. YOUSCHKEVITCH

ACCUM, FRIEDRICH CHRISTIAN (*b.* Bückeburg, Germany, 29 March 1769; *d.* Berlin, Germany, 28 June 1838), *applied chemistry.*

The son of a converted Jew, Herz Marcus, and a Huguenot, Judith La Motte, Accum was the sixth of seven children and the youngest of the three who lived to maturity. A small family soap-boiling business begun with the aid of his mother's dowry continued to provide income after his father's death in 1772; Friedrich was thus able to follow the usual classical curriculum at the more than usually respected local Gymnasium. It was, however, an interest in chemistry that led to his move to England in 1793, as assistant in the Hannover and London firm of Brande, apothecaries to George III. He married Mary Simpson of London in 1798 and had eight children, of whom one son and one daughter lived to adulthood.

Accum was early indebted to the patronage of Anthony Carlisle and the friendship of William Nicholson. About 1800 he established his own laboratory; he was also "assistant chemical operator" to Humphry Davy, resigning in September 1803. In 1802 he began public lecturing, and the steady stream of laboratory pupils included the Americans Benjamin Silliman (in 1805) and William Peck. Orders for chemical apparatus for Harvard, Yale, even Pondicherry, India, were a natural consequence.

In December 1820 the scandal that followed Accum's arrest for mutilating books in the Royal Institution library led to his flight to Germany. Some of the proceeds from his technological and publishing successes went with him, and he was soon reestablished in Berlin, with two technical professorships. In England he was a subscriber to the Royal Institution, a member of the Royal Irish Academy and the Linnean Society, and a corresponding member of the Royal Academy of Sciences in Berlin. The last connection continued after his move to Germany.

The value of Accum's work lies in the way he saw and exploited the technological possibilities of the rapidly advancing science of chemistry. His activities as lecturer, author, laboratory instructor, merchant, consultant, and technical adviser epitomize the opportunities that the industrial revolution opened to the emerging class of professional chemists. His pioneer work on gas-lighting and food adulteration was of fundamental importance.

Accum was intimately concerned with the application of F. A. Winsor's 1804 patent of a gas-lighting process. He undertook the experimental work necessary to overcome the complaints of Winsor's rival William Murdoch and the scruples of Parliamentary committees. As a result his name appeared as "practical chymist" on the 1812 list of the first Corporation of London's highly successful Gas-Light and Coke Company. Profiting from his experience, Accum advised other fledgling gas companies and wrote the 1815 treatise that became the classic text of gas technology. Equal fame, although not success, surrounded his work on food adulteration. He was long aware of this problem and his deliberately sensational 1820 work (motto "There is death in the pot") did much to awaken that public concern that eventually resulted in the Adulteration Act of 1860. Not surprisingly, Accum's outspoken attacks, and his naming of offending individuals, antagonized powerful interests. This antagonism may well explain the harsh line the Royal Institution took toward his own shortcomings, despite the pleadings of his former patron Carlisle.

BIBLIOGRAPHY

I. Original Works. In addition to many articles Accum wrote at least fifteen books, from *System of Theoretical and Practical Chemistry*, 2 vols. (London, 1803), to *Physische und chemische Beschaffenheit der Baumaterialien*, 2 vols. (Berlin, 1826). The two most important are his *Practical Treatise on Gas Light; Exhibiting a Summary Description of the Apparatus and Machinery* (London, 1815; 4th ed., 1817), which was trans. into French, German, and Italian, and the *Treatise on the Adulterations of Food, and Culinary Poisons* (London, 1820; 2nd ed., 1820).

II. Secondary Literature. C. A. Browne, "The Life and Chemical Services of Fredrick Accum," in *Journal of Chemical Education*, **2** (1925), 829–851, 1008–1034, 1140–1149, and "Recently Acquired Information Concerning Fredrick Accum," in *Chymia*, **1** (1948), 1–9, contain detailed data on Accum's life and work, surviving MSS, and chemical samples; R. J. Cole's largely derivative "Friedrich Accum. A Biographical Study," in *Annals of Science*, **7** (1951), 128–143, contains some information on eds. and trans. of Accum's books.

Arnold Thackray

ACHARD, FRANZ KARL (*b.* Berlin, Germany, 20 April 1753; *d.* Kunern, Germany, 20 April 1821), *chemistry, experimental physics.*

Achard, baptized François-Charles, was born of French Protestant émigré parents. His father, Guillaume Achard, a minister, died when Franz Karl was only two years old. In 1759 his mother, whose maiden name was Margarette Henriette Roupert, married a Charles Vigne. Of Achard's upbringing and early education, virtually nothing is known. At the age of twenty he began his scientific career in association with the botanist J. G. Gleditsch and with the renowned chemist A. S. Marggraf. It was Marggraf, especially, who trained young Franz Karl, gave him entrée into Berlin scientific circles, and finally obtained, in June 1776, his admission to the Berlin Academy. When Marggraf died in 1782, Achard succeeded him as director of the "Class of Physics" of the Berlin Academy.

In his youth, Achard was a prolific writer. He published articles on the thermal expansion of gases and liquids, the effects of soluble and insoluble fluids on the freezing point of water, cooling by evaporation, and various electrical phenomena, as well as numerous papers on miscellaneous chemical and physical subjects. Most of these articles were published in French in the *Nouveaux mémoires* of the Berlin Academy, in Rozier's journal (*Observations sur la physique. . .*), and in the *Journal littéraire* (Berlin). In addition, Achard published German articles in Crell's *Chemische Annalen* and in two volumes of collected essays (1780, 1784).

None of Achard's early papers was of great importance. In general, he eschewed theoretical discussion and concentrated his efforts on the detailed investigation of facts. Typical of this procedure was his paper on cooling by evaporation, which amounted to little more than a short history of previous investigations of the topic, followed by a description of his experimental method and long lists of various liquids and their relative abilities to cool. The few general remarks that he appended to this discourse had been established years, even decades, before. Achard is known for several minor achievements in applied chemistry, most notably his description of a workable alloy of platinum and arsenic and his process for fabricating lye from common salt and litharge. In addition, he was one of the first to conduct detailed investigations of galvanism.

Achard is best known for his development of a method of extracting sugar from beets in large quantity. In 1747 Marggraf had published a paper in which he showed that crystallizable sugar could be extracted from various plants native to Europe. The most promising of these plants appeared to be the common beetroot. At his estate near Berlin, Achard began experiments in 1786, in an effort to develop a process for extracting sugar from beets in quantities large enough and at cost low enough to be commercially valuable. In 1799, after having tried various methods of cultivation and extraction, he had a loaf of beet sugar, along with a description of the process by which it had been made, presented to Frederick William III. The king appointed a special committee to investigate this new process, and when it issued a favorable report, Achard was given financial aid to build a beet-sugar refinery in Kunern, a Silesian village near Breslau. The factory was completed in 1801, and it began operations the following year.

Achard's new process of obtaining sugar was simple but costly. It consisted of boiling specially cultivated, white Silesian beets and then pressing them to extract a sugary liquid. This liquid, added to that obtained from a second pressing, was boiled to remove excess water and then placed in an oven at moderate temperature to allow crystallization. After a crust had formed, the liquid was cooled and the sugar was separated by filtration. The muscovado, or raw sugar, could be refined to any desired degree of purity by recrystallization. The by-products were beet pulp, which could be used for cattle fodder, and molasses, which could be made into spirits.

News of Achard's accomplishment spread quickly throughout Europe. In France the first details were communicated by Van Mons in an article written for the *Annales de chimie*. The Institut formed a committee of chemists and agriculturalists to examine Achard's process and suggest possible improvements. In June 1800 it issued an encouraging report and made several valuable recommendations, the most important of which was that the beets be pressed without cooking them. Achard afterward adopted this technique in order to reduce the considerable expenditures for fuel.

Under the artificial stimulus of the Continental System, which reduced the supply of West Indian sugar, and with the aid of liberal governmental financing, the beet-sugar industry prospered briefly. In 1813 an investigating commission of the French government reported that there were 334 beet-sugar factories in France, with a combined annual production of 7,700,000 pounds. Although no such specific figures exist for Germany, it seems evident that the beet-sugar industry was also widely established there. However, when the Napoleonic Wars came to an end and normal markets were reestablished, the industry all but disappeared. It revived about the middle of the nineteenth century, and made its greatest strides

with the development of the diffusion process of extraction and with the cultivation of new strains of beets containing over twice as much sugar as those Achard had used.

After a five-week illness, Achard died only a week before his sixty-eighth birthday. He was survived by his second wife and several children. His death passed virtually unnoticed in academic circles, from which he had long since retired.

BIBLIOGRAPHY

I. ORIGINAL WORKS. Achard's writings include *Chymisch-physische Schriften* (Berlin, 1780); *Sammlung physikalischer und chymischer Abhandlungen,* Vol. I only (Berlin, 1784); and *Vorlesungen über die Experimentalphysik,* 4 vols. (Berlin, 1790–1792). Among Achard's many works on the beet-sugar industry, three are especially worthy of mention: *Kurze Geschichte der Beweise welche ich von der Ausführbarkeit im Grossen und den vielen Vortheilen der von mir angegebenen Zuckerfabrication aus Runkelrüben geführt habe* (Berlin, 1800); *Anleitung zum Anbau der zur Zuckerfabrication anwendbaren Runkelrüben und zur vortheilhaften Gewinnung des Zuckers aus denselben* (Breslau, 1809), repr. as Ostwalds Klassiker der exacten Wissenschaften, no. 159 (Leipzig, 1907); and *Die europäische Zuckerfabrication aus Runkelrüben* (Leipzig, 1809).

II. SECONDARY LITERATURE. A detailed and reliable account of Achard's life is found in Wilhelm Stieda, "Franz Karl Achard und die Früzeit der deutschen Zuckerindustrie," in *Abhandlungen der philologisch-historischen Klasse der Sächsischen Akademie der Wissenschaften,* **39**, no. 3 (Leipzig, 1928). See also Adolf von Harnack, *Geschichte der Königlich Preussischen Akademie der Wissenschaften zu Berlin,* IV (Berlin, 1900), 3–7, 389 f.; and Noël Deerr, *The History of Sugar,* II (London, 1950), 471–500.

J. B. GOUGH

ACHARIUS, ERIK (*b.* Gavle, Sweden, 10 October 1757; *d.* Vadstena, Sweden, 14 August 1819), *botany.*

Acharius was the last to defend his thesis (*De planta Aphyteia,* 1776) under Linnaeus, and all his life he pursued the Linnaean tradition of research in his botanical work. Like most of the university-trained botanists of his time, Acharius studied medicine, first at the University of Uppsala and then at Lund, where he received the M.D. in 1782. After that, he practiced medicine and spent the major part of his life in the province of Ostergotland, in the small town of Vadstena. He worked on botany only in his leisure time, but his scientific achievement was nevertheless considerable. He devoted himself almost exclusively to the study of lichens, and his description and classification of them laid the foundation for later scholarship.

Linnaeus had concentrated on the classification of the higher plants, and at the time of his death was the recognized authority on world flora. The generation of botanists who followed him sought to enlarge on his work by defining new areas of the plant kingdom. Some turned to new areas of the world as sources of plants for study, and others devoted themselves to the previously neglected cryptogams, which despite their large numbers and enormous variety were all placed in the twenty-fourth class of the Linnaean sexual system. As for the study of lichens, German botanists, among them J. Hedwig and H. A. Schrader, had considerably extended Linnaeus' findings, but only Acharius laid the rational foundations of their classification.

In his first important publication, *Lichenographiae suecicae prodromus* (1798), Acharius still classified lichens according to the appearance of the thallus, which had been observed earlier, but he soon developed a new system based on structure. His terminology for the morphological description of lichens is still, to a large extent, valid. Using this method, he described a considerable number of new families and species, both Scandinavian and tropical. Some of the tropical specimens were collected from the bark and tissues of tropical plants that came to him in the form of botanical drugs. His advanced views on the taxonomy of lichens were presented in *Methodus* (1803), *Lichenographia universalis* (1810), and *Synopsis methodica lichenum* (1814).

For a long time Acharius thought that lichens were not really plants, but animals, most closely related to polyps. While this shows the vagueness of the conception still held at the beginning of the nineteenth century with regard to the "lower" organisms and their reproduction, it may also serve to illustrate the problems the lichen group posed before it was established that they are composed of an alga and a fungus living symbiotically. Since only inferior microscopes were available, the structure of lichens remained obscure to Acharius. During the decades following his death, Acharius' scientific work was severely criticized, especially by the German botanists H. G. Floerke, G. F. W. Meyer, and K. F. W. Wallroth. Both his terminology and his classification of species were considered defective, and he was thought to have distinguished too many species. In contemporary lichenology, however, Acharius is highly respected, and many of his species are still recognized.

BIBLIOGRAPHY

I. ORIGINAL WORKS. Acharius' complete bibliography is in T. O. B. N. Krok, *Bibliotheca botanica suecana* (Uppsala–Stockholm, 1925), pp. 2–4; and R. Sernander, "Acharius," in *Svenskt biografiskt lexikon,* I (Stockholm, 1917–1918), 39–40. Among his works are *Lichenographiae suecicae prodromus* (Linköping, 1798); *Methodus qua omnes detectos lichenes secundum organa carpomorpha ad genera, species et varietates redigere atque observationibus illustrare tentavit* (Stockholm, 1803); *Lichenographia universalis* (Göttingen, 1810); and *Synopsis methodica lichenum* (Lund, 1814). His lichen herbarium is at the botanical museum of Helsingfors University; his correspondence is in the library of the University of Uppsala.

II. SECONDARY LITERATURE. There is no complete modern biography of Acharius. Information on his life is in *Svenskt biografiskt lexikon* (see above). The best résumé of his work is A. von Krempelhuber, *Geschichte und Literatur der Lichenologie,* I–II (Munich, 1867–1869): I, 96–98, 112–114, 194–196, and *passim;* II, 61–68, 79–88, and *passim.*

GUNNAR ERIKSSON

ACHILLINI, ALESSANDRO (*b.* Bologna, Italy, 29 October 1463; *d.* Bologna, 2 August 1512), *anatomy.*

Alessandro Achillini was the son of Claudio Achillini, of an old family of Bologna. He was graduated doctor of philosophy and of medicine from the University of Bologna in 1484, whereupon he was appointed substitute lecturer of philosophy. After 1495 he taught medicine also. In 1506 he was obliged to leave Bologna, owing to the expulsion of the powerful Bentivoglio family, of whom he was a partisan. He went to Padua, where he was appointed teacher of philosophy. (It is possible that he had taught at Padua before 1506.) In 1508 the regents of the University of Bologna requested Achillini to return. He did so and taught there for three years. He is buried in the Church of Saint Martin, in Bologna. Achillini never married.

During his lifetime, Achillini was known mainly as a philosopher. Today, however, he is remembered for his considerable activity in research on human anatomy. He gave a good description of the veins of the arm, and he described the seven bones of tarsus, the fornix of the brain, the cerebral ventricles, the infundibulum, and the trochlear nerve. He also described, exactly, the ducts of the submaxillary salivary glands—a discovery generally attributed to the Englishman Thomas Wharton (1614–1673)—and the ileocecal valve, described later by Costanzo Varolio and Gaspard Bauhin. Finally, to Achillini is attributed the first description of the two ossicles of the ear, the malleus and incus.

A famous teacher (Magnus Achillinus) and a man of independent critical judgment, Achillini first demonstrated some mistakes of Galen. Among the Italian pre-Vesalian anatomists (Alessandro Benedetti, Gabriele Zerbi, Berengario da Carpi, and Niccolo Massa) the influence of Achillini's work was slight, chiefly because his works were not illustrated and he used an obscure medieval terminology derived largely from Mondino's texts and heavily sprinkled with Arabic terms.

BIBLIOGRAPHY

I. ORIGINAL WORKS. Achillini's major philosophical writings are to be found in *Clarissimi Achillini opera.* I. *Aristotelis, philosophorum maximi, de secretis secretorum ad Alexandrum opusculum;* II. *Eiusdem de regum regimine;* III. *De universalibus et de elementis libri tres;* IV. *De orbibus libri quatuor;* V. *De signis tempestatum;* VI. *De mineralibus;* VII. *Alexandri Aphrodisei de intellectu;* VIII. *Averrois de animae beatitudine;* IX. *Alexandri Macedonis de mirabilibus Indiae ad Aristotilem* (Venice, 1508, 1516, 1545, 1551, 1568). His major anatomical writings include *Alexandri Achillini de humani corporis anatomia* (Venice, 1516; Bologna, 1520); "In Mundini anatomiam adnotationes," in *Fasciculus medicinae* by John of Ketham (Venice, 1522); *Adnotationes anatomicae* (Bologna, 1520).

II. SECONDARY LITERATURE. See also P. Capparoni, *Profili di medici e naturalisti celebri italiani dal secolo XV° al XVIII°,* I (Rome, 1925), 11–14; S. De Renzi, *Storia della medicina in Italia,* II (Naples, 1845), 358–360. The best essay on Achillini's work is L. Münster, "Alessandro Achillini anatomico e filosofo nello Studio di Bologna," in *Rivista di storia delle scienze mediche e naturali,* **24** (1933), 7–22, 54–77. Further material can be found in B. Nardi, *Studi su Pietro Pomponazzi* (Florence, 1965); C. D. O'Malley, "The Discovery of the Auditory Ossicles," in *Bulletin of the History of Medicine,* **35** (1961), 419–441; A. Pazzini, *Storia della medicina,* I (Milan, 1945), 614; V. Putti, *Berengario da Carpi* (Bologna, 1937); G. Rath, "Pre-Vesalian Anatomy in the Light of Modern Research," in *Bulletin of the History of Medicine,* **35** (1961), 142–148.

PIETRO FRANCESCHINI

ACOSTA, CRISTÓBAL (*b.* Bõa Ventura, Santo Antão, Cape Verde Islands, *ca.* 1525; *d.* La Peña de Tharsis [?], Huelva, Spain, *ca.* 1594), *natural history, medicine.*

Born Christovão da Costa—according to Portuguese usage—Acosta moved to Lisbon and lived in Setúbal and Peniche, but his excellent Spanish and broad education indicate that he studied arts and medicine, most probably in Salamanca. He went to the East Indies before 1550 as a soldier and visited Persia, India, Malaya, and perhaps China. There is record of his actions against the Arabs in Hormuz; against the Abexins near Daman, where he lost two horses in battle; and on the Malabar Coast, where he was taken prisoner on the way to Bengal; and of his meetings in Goa with García d'Orta.

Acosta returned to Portugal but soon rejoined his former captain, Luiz de Ataíde, who had been appointed viceroy of India, and landed at Goa in October 1568, a few months after the death of García d'Orta. In 1569 he was appointed physician to the Royal Hospital in Cochin, but by 1571 he was collecting botanical specimens in Tanor, Cranganor, and other parts of India. Luiz de Ataíde ended his term of office in 1572, and Acosta sailed from Cochin back to Lisbon via the Cape of Good Hope. He practiced medicine in Burgos and was that city's physician and surgeon from 1576 to 1587. After his wife died, he retired to the hermitage of La Peña de Tharsis, which was probably where he died.

Like his other books, Acosta's *Tractado de las drogas, y medicinas de las Indias orientales* was written in a fluid and concise style. It offers systematic, first-hand observations of the Oriental drugs and is illustrated by woodcuts made from his own accurate drawings. This book clearly surpasses that of d'Orta, whose contributions Acosta readily acknowledges. His *Tratado in loor de las mujeres* followed Boccaccio's work but was not influenced by Espinosa's *Dialogo en laude de las mugeres* (1580). His other printed work, *Tratado en contra y pro de la vida solitaria,* incorporated two other independent treatises, *Tratado de la religión y religioso* and *Collación á los mohateros, usureros, aparceros, tratantes y seducadores,* both moral works. In all, Acosta wrote thirteen works, but the manuscripts of his *Discurso del viaje á las Indias orientales y lo que se navega en aquellos mares, Tres diálogos teriacales,* and above all his great *Tratado de las yerbas, plantas, frutas y animales, así terrenos como aquatiles que en aquellas partes y en la Persia y en la China hay, no dibujadas al natural hasta agora,* are not extant.

BIBLIOGRAPHY

Acosta's first printed book was the *Tractado de las drogas, y medicinas de las Indias orientales* (Burgos, 1578), trans. into Italian by F. Ziletti (Venice, 1585); into French by A. Colin (Lyons, 1602, 1619); and into Latin by C. L'Ecluse (Antwerp, 1582, 1593, 1605). His two other published books were the *Tratado en loor de las mujeres* (Venice, 1592) and the *Tratado en contra y pro de la vida solitaria* (Venice, 1592).

J. Olmedilla y Puig, *Estudio histórico de . . . Cristóbal*

Acosta (Madrid, 1899), is comprehensive and offers two documents on Acosta's practice in Burgos. Unfortunately, the information that Cristóbal Acosta was born of Jewish parents in Tangier, Ceuta, or Mozambique *ca.* 1515, was a brother of José Acosta, studied at Coimbra or Burgos, returned from India via Jerusalem, was shipwrecked near Italy, and died in Burgos *ca.* 1580 is in conflict with the facts.

FRANCISCO GUERRA

ACOSTA, JOSÉ DE (*b.* Medina del Campo, Spain, 1539; *d.* Salamanca, Spain, 15 February 1600), *geography.*

Acosta was one of the first Europeans to provide a detailed image of the physical and human geography of Latin America; his studies of the Indian civilizations of the New World were a major source of information for several centuries. At the age of fifteen he entered the Jesuit order in his native city and underwent rigorous theological and literary training. He early displayed a strong interest in the New World, and although he was offered a chair of theology in Rome, he asked to be sent to the Americas. He left Spain in 1570 and sailed via Panama to Peru, where he remained for fourteen years. For many years Acosta lived at the Jesuit college on the shore of Lake Titicaca and learned enough of both Aymara and Quechua to produce a trilingual catechism (1583). After acting as historian of the third council of the church at Lima (1582–1583), he embarked for Mexico, where he spent three years. Next he went to Rome, and then to Spain, where he filled several important posts for the Jesuits. He died while serving as rector of the Jesuit college of Salamanca.

Acosta was a prolific writer on both sacred and profane subjects, but his most important scientific work was *Historia natural y moral de las Indias.* It provides firsthand observations on such diverse phenomena as altitude sickness, the nature and uses of coca, and the crops, farm techniques, and domesticated animals of America. Equally important are his descriptions of Inca and Aztec history, religious observances, folk customs, and statecraft. He was the first to describe in detail Mexican ideograms and Peruvian quipu, and the Inca postal system. He may, indeed, be called the first of the true Americanists.

BIBLIOGRAPHY

I. ORIGINAL WORKS. Acosta's most important work on the Americas was *Historia natural y moral de las Indias, en que se tratan las cosas más notables del cielo, y elementos, metales, plantes y animales dellas* (Seville, 1590). The sole English trans., by Edward Grimston, is *The Natural and Moral History of the Indies* . . . (London, 1604), repr. with notes and intro. by C. R. Markham as Vols. LX and LXI of the Hakluyt Society Series (London, 1880).

II. SECONDARY LITERATURE. References to Acosta's writings may be found in Marcelino Menéndez y Pelayo, *La ciencia española,* III (Madrid, 1887); and Carlos Sommervogel, *Bibliothèque de la Compagnie de Jésus,* I (Paris, 1890), cols. 31–38.

GEORGE KISH

ACTEDIUS, PETRUS. See **Artedi, Peter.**

ACYUTA PIṢĀRAṬI (*b.* Tṛkkaṇṭiyūr [Sanskrit, Kuṇḍapura], Kerala, India, *ca.* 1550; *d.* Kerala, 7 July 1621), *astronomy.*

Acyuta was a member of the Piṣāraṭi community, which is a section of the Ampalavāsi community and is traditionally employed in looking after the external affairs of temples. He studied astronomy under Jyeṣṭhadeva (*ca.* 1500–1600), a Nambūtiri Brāhmaṇa of the Paraṅṅoṭṭu family in the village called Ālattūr, who wrote the *Yuktibhāṣā* based on the *Tantrasaṅgraha* of Nīlakaṇṭha Somasutvan. Jyeṣṭhadeva and Nīlakaṇṭha were pupils of Dāmodara, the son and student of Parameśvara, who founded the *Dṛggaṇita* school of astronomy (for an account of this school, see my essay, V A 2). Acyuta's patron was Ravivarman, King of Veṭṭattunād (Sanskrit, Prakāśa; 1595–1607). He was a scholar in grammar (his famous pupil Nārāyaṇa Bhaṭṭatiri [1560–*ca.* 1646] refers to him in his *Prakriyāsarvasva*), poetics, and medicine, as well as in astronomy and astrology. In the field of astrology, there existed in his school a long line of scholars writing in Malayalam.

Acyuta wrote the following works dealing largely with astronomy.

1. *Praveśaka.* This is an introduction to Sanskrit grammar in about six hundred *anuṣṭubh* stanzas. It was edited, with a commentary, by P. S. Anantanarayana Sastri, as Vol. II in Cochin Sanskrit Series (Trippunithura, 1938).

2. *Karaṇottama.* This is a work on astronomy in five chapters and about one hundred verses; it deals with the computation of the mean and true longitudes of the planets, with eclipses, and with the *vyatīpātas* of the sun and moon. Acyuta himself wrote a commentary on it. It was published as Vol. 213 of Trivandrum Sanskrit Series (Trivandrum, 1964).

3. *Uparāgakriyākrama.* This is a treatise in four chapters on lunar and solar eclipses which was completed in 1593. There is a commentary on it in Malayalam.

4. *Sphuṭanirṇaya.* This is a work on astronomy in six chapters, written before the *Rāśigolasphuṭānīti.*

5. *Chāyāṣṭaka.* This is a short astronomical text in eight verses.

6. *Uparāgaviṃśati.* This is a manual in twenty verses on the computation of eclipses. It was published with a Malayalam commentary in Vol. II of the Ravivarma Sanskrit Series (Trippunithura, 1954).

7. *Rāśigolasphuṭānīti.* This work in fifty verses is concerned with the reduction of the moon's true longitude in its own orbit to the ecliptic. Since in this work Acyuta quotes not only the *Sphuṭanirṇaya* but also the *Uparāgakriyākrama,* it is clear that the *Rāśigolasphuṭānīti* was written after 1593. It was edited and translated into English by K. V. Sarma as Adyar Library Series, Paper 29 (Madras, 1955); reprinted from *Brahmavidyā,* **18** (1954), 306–335.

8. *Veṇvārohavyākhyā.* This Malayalam commentary on the *Veṇvāroha* of Mādhava of Saṅgamagrāma (*ca.* 1340–1425) was written at the request of Netranārāyaṇa, a spiritual head of the Nambūtiri Brāhmaṇas. The *Veṇvāroha* deals with the calculation of the *tithis* and *nakṣatras.* The text and its commentary have been edited by K. V. Sarma in Vol. III of the Ravivarma Sanskrit Series (Trippunithura, 1956).

9. *Horāsāroccaya.* This is an adaptation in seven chapters of the *Jātakapaddhati* of Śrīpati. The relationship to it of a Malayalam commentary on the *Jātakapaddhati* entitled *Horātantraṃ Paribhāṣa* remains uncertain.

BIBLIOGRAPHY

Further discussion of Acyuta Piṣārati may be found in S. Venkitasubramonia Iyer, "Acyuta Piṣāroṭi: His Date and Works," in *JOR Madras,* **22** (1952–1953), 40–46; and K. Kunjunni Raja, *The Contribution of Kerala to Sanskrit Literature* (Madras, 1958), pp. 122–125, and "Astronomy and Mathematics in Kerala," in *Brahmavidyā,* **27** (1963), 158–162.

DAVID PINGREE

ADAM OF BODENSTEIN (*b.* 1528; *d.* Basel, Switzerland, 1577), *medicine, alchemy.*

Adam was doctor of arts and medicine at Basel, where he studied and practiced medicine. He was a follower of the doctrines of Paracelsus, who had taught medicine at Basel and had been known especially for his emphasis upon the relationship between medicine and minerals, being noted particularly for his advocacy of the use of metallic compounds in medicine. Adam participated with other scholars of his time, among whom were Michael Toxites, Adam Schröter, George Forberger, Balthasar Flüter, and Gerard Dorn, in the translation, editing and publication of works by Paracelsus still in manuscript. He preceded these editions with prefatory remarks of his own. His chief contemporary rival in the interpretation of Paracelsus was Leo Suavus (Jacques Gohory).

In close association with Paracelsus' special predilection for the use of metallic compounds, Adam developed an interest in minerals, particularly in the traditional alchemical process of transmuting baser metals into gold. In the *Epistola* addressed to the Fuggers, he related the circumstances of the change in his opinion of alchemy from one of scorn and contempt for it as a suspect art and for those who wrote on it as evil men, to a belief in the verity of alchemy and of the philosophers' stone. This change he attributed to his discovery of the famous alchemical tract, the *Rosarium,* of Arnald of Villanova. On reading that work and taking cognizance of the author's orderly procedure and the presentation of the variety of theories, persons, and scientific paraphernalia involved in the art of alchemy, Adam was convinced that the contentions of the alchemists were valid and that the transmutation of baser metals into gold was possible. He strongly affirmed this conviction in the *Isagoge,* or introduction to Arnald of Villanova's *Rosarium,* which he paraphrased or edited. Adam of Bodenstein went on to expound the traditional views set forth by Arnald that mercury (quicksilver) is the primary matter of metallic bodies and that sulfur and mercury, the constituents of gold, are found in the viscera of the earth. Hence, since art follows nature, one may learn to discern the causes of the transmutation of sulfur and mercury into gold by a close observation of the process in nature.

Besides the editions of Paracelsus' works and the introduction to the paraphrase or edition of Arnald of Villanova's *Rosarium,* Adam is credited with the composition of other tracts. His chief biographer, Melchior Adam, ascribed to him further tracts entitled *De podagra* ("On Gout") and *De herbis duodecim zodiaci signis dicatis* ("On the Relation of Herbs to the Twelve Signs of the Zodiac"). Furthermore, the noted Swiss naturalist Conrad Gesner, who was Adam of Bodenstein's contemporary, reported that he had learned about salmon from Adam.

On the whole, Adam of Bodenstein's works contain little that is novel. Rather, as Lynn Thorndike pointed out, they demonstrate the strength of tradition in both medicine and alchemy in the sixteenth century. They do, however, exemplify the proclivity of the scholars of the time to carry on an active correspondence by means of which they exchanged views and the results of their scientific activities and discoveries.

BIBLIOGRAPHY

I. ORIGINAL WORKS. The following printed works are all at the British Museum in London and the Bibliothèque Nationale in Paris. Editions of Paracelsus are: *Libri V de vita longa, cum dedicatoria epistola* ([1562] Basel, 1566); *Drei Bücher von Wunden und Schäden sampt allen iren Zufellen und derselben vollkommener Cur* (Frankfurt, 1563); *Spittal Büch* (Mühlhausen, 1562; Frankfurt, 1566); *Dess erfarnesten Fürsten aller Artzeten Aureoli Theophrasti Paracelsi von ersten dreyen principijs...* (Basel, 1563); *Weyssagung Sibylle Tyburtine von ... Lucas Gauricus ... ausgelegt für das 1557 Jar,* trans. by Adam of Bodenstein (Nuremberg [?], 1556 [?]); *Baderbüchlin... Mit fleyss Adams von Bodenstein publicirt* (Mülhausen, 1562; Frankfurt, 1566, 1576); *Libri quinque de causis, signis et curationibus morborum ex tartaro utilissimi Opera...* (Basel, 1563); *De gradibus, de compositionibus et dosibus receptorum ac naturalium... libri septem* (Mylau, 1562; Basel, 1568); *Das Buch Paragranum ... Item, von Aderlassens, Schrepffens und Purgirens rechtem gebrauch* (Frankfurt, 1562); *Das Buch Paramirum* (Mühlhausen, 1562); *Praeparationum libri duo* (Strasbourg, 1569); *Metamorphosis ... der zerstörten ... Artzney restauratoris ...* (Basel, 1572); *Opus chyrurgicum* (Strasbourg, 1566; Basel, 1581); *Operum latine redditorum tomus I (II)*, with preface by Adam of Bodenstein (Basel, 1575); *Schreiben von tartarischen Kranckheiten,* trans. into German and ed. by Adam of Bodenstein (n.p., 1563); *Labyrintus und Irrgang der vermeinten Artzet* (Basel, 1574); *Libri duo; I. Defensiones septem adversus aemulos suos; II. De tartaro sive morbis tartareis* (Strasbourg, 1566, 1573); *Drey Schreiben ... Von tribus principiis aller Generaten, Libro vexationum und thesauro alchimistarum,* in B. G. Penot, *Theophrastisch Vade mecum* (Magdeburg, 1608); *Pyrophilia vexationumque liber* (Basel, 1568); *Schreyben von den Kranckheyten so die Vernunfft berauben als da sein S. Veyts Thantz hinfallender Siechtage; Melancholia und Unsinnigkeit* (n.p., 1567); *De tartaro libri septem ... nunc vero auctiores et castigatiores denuo excusi* (Basel, 1570); *Schreiben von warmen oder Wildbäden* (Basel, 1576); *Zwey Bücher ... von der Pestilentz und ihren Zufallen* (Strasbourg, 1559); *Herrlicher philosophischer Rhatschlag zu curirn Pestilentz Brustgeschwer, Carfunckl: Dardurch auch andere Gyfft ... aussgetriben mögen werden* (Basel, 1577).

Works on Arnald of Villanova and alchemy are: *Isagoge in excellentissimi philosophi Arnaldi de Villanova Rosarium chymicum, paraphrastice et magna diligentia tradita, Epistola operi praefixa ad amplissimos et generosos dominos, dominos Fuggeros in qua argumenta alchymiam infirmantia et confirmantia adducuntur, quibus et eam artem esse certissimam demonstratur* (Basel, 1559); also available in MS: British Museum Sloane 3737, 17th cent., ff. 96r–106r, "Ex Isagoge Adam à Bodenstein in Rosarium Arnaldi."

The two tracts *De podagra* and *De herbis duodecim*

zodiaci signis dicatis, noted by Melchior Adam (see below), have not been otherwise identified or located.

II. SECONDARY LITERATURE. The principal modern account is Lynn Thorndike, *History of Magic and Experimental Science,* V (New York, 1959), 619, 636; VI, 267. Dr. Thorndike utilized for the biographical information Melchior Adam, *Vitae medicorum Germanorum,* 3rd ed. (Frankfurt, 1706), pt. 4, p. 104, the sole extended source of information on Adam of Bodenstein; also Conrad Gesner, *Historia animalium* (Frankfurt, 1604), p. 829, "lib. IV, qui est de piscium et aquatilium animalium natura."

PEARL KIBRE

ADAMS, FRANK DAWSON (*b.* Montreal, Canada, 17 September 1859; *d.* Montreal, 26 December 1942), *geology.*

Adams' father, Noah Adams, belonged to the distinguished Adams family of New England; his mother, Frances Tait Dawson, was a United Empire Loyalist from Northern Ireland. At nineteen he graduated with first rank honors in natural science from McGill University, where he came under the influence of the scholarly principal, J. W. Dawson (no relation), and the versatile and magnetic B. J. Harrington of the department of chemistry and mineralogy. Comfortable family financial circumstances allowed Adams to study chemistry and mineralogy at Yale University (1878–1879), and later to attend several sessions at Heidelberg University.

After spending 1880 to 1889 in government service, Adams joined the staff at McGill as lecturer (1890–1893); he followed Dawson as holder of the Logan chair of geology in 1893. Adams became dean of the Faculty of Applied Science in 1908 and dean of the Faculty of Graduate Studies in 1922. He retired in 1924, after thirty-five years of arduous service. He was deeply but quietly religious. His *History of Christ Church Cathedral* (1941) was a tribute to the church to which he was devoted and to his wife, Mary Stuart Finley, to whom the book was dedicated. Adams also found time to devote to many philanthropic and social benevolences.

Adams belonged to numerous scientific societies, including the Royal Society of London, the Geological Society of America (of which he was president in 1918), the Royal Society of Canada (president in 1913; Flavelle Gold Medal in 1937), and the Geological Society of London (Lyell Medal in 1906; Wollaston Medal in 1939). He was also president of the Twelfth International Geological Congress (Montreal, 1913). He received honorary degrees from McGill, Toronto, Queen's, and Mount Allison universities, and Bishop's College.

In 1880 Adams joined the Geological Survey of Canada as assistant chemist and lithologist. One of the first tasks given him by the director, A. R. C. Selwyn, was to determine the nature and origin of certain rocks from southern Quebec. In order to work out these derivations and associations, Adams requested, and was granted, leave of absence to master the new petrographic technique being developed by H. Rosenbusch at Heidelberg. He completed this work to Selwyn's satisfaction and was next assigned to study areas of partly foliated anorthosites in southwestern Quebec that William Logan had considered to be the upper and stratified portion of the Laurentian series. To this difficult task he applied his newly won skill in the use of the petrographic microscope, an instrument with which he had become familiar in Heidelberg and which he was probably the first in Canada to use. He was able to demonstrate conclusively not only the igneous origin of the anorthosites but also the sedimentary origin of some of the Grenville crystalline rocks upon which Logan had supposed the anorthosites rested. The presentation of the results of this study gained him the Ph.D. *summa cum laude* at Heidelberg. The publication of his thesis (1893) established Adams as one of the North American experts in the use of the petrographic microscope. Several publications resulted from his work with the Geological Survey, among the more important being descriptions of Precambrian rocks north of Montreal and St. Jerome (1896), based on field work carried out from 1885 to 1891.

The Laurentian system had been considered by Logan to consist of two divisions, the lower a complex of metamorphosed sedimentary rocks, which he named Grenville and Ottawa, and an upper, or Norian, division, made up largely of anorthosite, which was considered to be an altered and crystallized sediment. This anorthosite, well exposed around Morin, Quebec, was very carefully inspected by Adams, who showed that it was composed largely of plagioclase feldspar, with few accessories. He established its igneous nature both by his petrographic determinations in the laboratory and by the intrusive contacts with the Grenville rocks that he was able to demonstrate in the field. He attributed the marked differences in grain size to crystal fracturing, which in places was carried to granulation. He also recognized that the Laurentian granites, then supposed to be the oldest rocks of the Canadian shield and possibly part of the original crust, were intrusive into metamorphic rocks, which therefore must have preceded them in time and space. These conclusions were among the foundations upon which the modern classification of the Precambrian series rests.

After leaving the Geological Survey and joining the staff of McGill, Adams devoted the summers of 1902 to 1908 to the mapping and description of the forbiddingly difficult Haliburton and Bancroft areas of southern Ontario. In his report (1910), written with A. E. Barlow, who had collaborated with him during the later years of the project, he showed that the oldest rocks were highly metamorphosed sediments, now gneisses, schists, quartzites, and marbles, and assigned them to the Grenville series. He found widespread intrusions of granite, diorite, and gabbro penetrating the Grenville rocks, and correctly attributed most of the metamorphism to the thermal effects of the abundant granite bodies. Adams noted that the metamorphism of the stratified rocks became more intense and the sedimentary rocks were "fretted away and [ultimately] represented only by occasional shreds and patches of amphibolite," as the intrusive granites were approached. His discovery of nepheline syenite adjacent to granite and marble bodies was of great scientific importance, and paved the way for their later industrial exploitation. Because of the clarity of the writing, the painstaking carefulness of the descriptions, and the logical deductions, Adams' report has become one of the classics of Canadian geology.

At McGill, Adams could not fail to be impressed by the peculiar rocks of Mount Royal. He found that the same general rock types prevailed in the half-dozen prominent hills dotted across the Paleozoic plain between Montreal and the Appalachian front. He announced the occurrence of these remarkable rock types in his paper "The Monteregian Hills: A Canadian Petrographical Province" (1903).

Close study of the deformed foliated gneisses and schists of the Grenville area had stimulated Adams' curiosity concerning the causes of such structures and the possibility of their being duplicated in laboratory experiments. Aided by colleagues in the engineering laboratories at McGill, he started a sequence of experiments, spread over the first decade of the present century, utilizing a gigantic (for that time) press in which he could subject rocks to enormous pressures. High pressures had long been used to test the strength of cubes of rock to determine architectural suitability; but to duplicate the conditions within the earth's crust, Adams subjected cylinders of rock encased in metallic tubes to compression under high confining pressures—for the most part less than 20,000 pounds per square inch but on one occasion 296,725 pounds per square inch. Manipulation of the apparatus allowed Adams to develop differential stresses and presumably to imitate the conditions under which some of the foliated rocks may have originated; he was also able to correlate some of the experimentally developed structures with natural ones observed in the field. The influence of this work upon our understanding of metamorphic processes is profound, and it has contributed in no small measure to the development of modern ideas of mountain building.

Adams' first paper in this field, written with J. T. Nicholson, concerned a thumb-size cylinder of Carrara marble that he exposed to a confining pressure of 18,000 pounds per square inch; after 124 days the column had shortened by 11.4 percent of its original height. Examined microscopically, it showed many of the characteristics peculiar to the Grenville marbles. Never before had properties of metamorphic rocks been imitated in controlled experiments. Adams returned to this topic several times, and by using the highest available pressure, 296,725 pounds per square inch, he developed in his samples a schistose structure essentially similar to that of some highly metamorphic calcareous rocks. One interesting result was his discovery that quick-loading techniques caused calcite to yield to stress along intergranular slip planes, giving a cataclastic structure, whereas slow increase in loading produced intracrystalline polysynthetic twinning.

Other experiments were designed to record the plasticity of rocks under high pressures, up to 200,000 pounds per square inch. Most soft materials were easily deformed, but the harder rocks, such as granite, failed along fracture lines, yielding zones of granulation; this corroborated Adams' own early ideas of the granulation of anorthosite by crystal fracturing. Other studies were directed to the determination of the depth at which pressure would close cavities in rocks. In granite, one of the least plastic rocks, Adams determined that cavities could exist as deep as eleven miles below the surface. One of his last papers on experimentation (1917), written with J. A. Bancroft, showed that the strength of rocks increases with pressure, and hence with depth in the crust, the conclusion being that rocks at great depth have great strength. To Adams must go the credit for establishing this phase of geological investigation upon a sure engineering foundation. His reputation as a pioneer in the field is secure.

In the decade following his retirement from active participation in university affairs, he and Mrs. Adams traveled widely, and following their third visit to Ceylon, he published the first complete geological report and map of that island (1929). Adams had always been intrigued by the beginnings of geological thinking, and during his travels he visited most of the Old World universities whose libraries held a wealth of early geological treatises. Wherever he could, he acquired early writings and amassed what was certainly the greatest such collection in private hands (now kept, intact, at McGill). This formed the basis for his scholarly work *Birth and Development of the Geological Sciences* (1938), a text that will long remain a standard treatment of the subject.

BIBLIOGRAPHY

I. ORIGINAL WORKS. Complete listings of Adams' geological papers relating to North America are in United States Geological Survey Bulletins 746, 823, 937, 1049, and 1195. His most important works are "Notes on the Microscopic Structures of Some Rocks of the Quebec Group," in *Geological Survey of Canada, Report of Progress for 1880–1882,* Part A (1883), 8–23; "Ueber das Norian oder Oberlaurentian von Canada," in *Neues Jahrbuch,* **8** (1893), 419–498; trans. in *Canadian Record of Science* (1895–1896), 169–198, 277–305, 416–443; "Report on the Geology of a Portion of the Laurentian Area Lying to the North of the Island of Montreal," in *Geological Survey of Canada, Annual Report (New Series),* **8,** Part J (1896); "The Monteregian Hills: A Canadian Petrographical Province," in *Journal of Geology,* **11** (1903), 239–282; "An Experimental Investigation Into the Flow of Marble," in *Royal Society of London, Philosophical Transactions* (Section A), **195** (1901), 363–401, with J. J. Nicholson; *An Investigation Into The Elastic Constants of Rocks, More Especially With Reference to Cubic Compressibility,* Carnegie Institute Publication 46 (Washington, 1905), with E. G. Coker; "An Experimental Investigation Into the Action of Differential Pressure of Certain Minerals and Rocks," in *Journal of Geology,* **18** (1910), 489–525; "An Experimental Investigation Into the Flow of Rocks—the Flow of Marble," in *American Journal of Science,* ser. 4, **29** (1910), 465–487; "An Experimental Investigation Into the Flow of Rocks" [with discussion], in Eleventh International Geological Congress, *Comptes rendus,* **14** (1912), 911–945; "On the Amount of Internal Friction Developed in Rocks During Deformation and on the Relative Plasticity of Different Types of Rocks," in *Journal of Geology,* **25** (1917), 597–637, with J. A. Bancroft; "Earliest Use of the Term Geology," in *Geological Society of America Bulletin,* **43** (1932), 121–123; "Geology of Ceylon," in *Canadian Journal of Research,* **1** (1929), 425–511; "Origin and Nature of Ore Deposits, an Historical Study," in *Geological Society of America Bulletin,* **45** (1934), 375–424; *The Birth and Development of the Geological Sciences* (Baltimore, 1938); and *History of Christ Church Cathedral* (Montreal, 1941).

II. SECONDARY LITERATURE. Of the many biographies the following are selected: H. M. Tory, "Frank Dawson Adams (1859–1942)," in *Royal Society of Canada Proceedings,* ser. 3, **37** (1943), 69–71; J. A. Dresser, "Memorial to Frank Dawson Adams," in *Geological Society of America Proceedings 1944* (1945), 143–150; and J. W. Flett, "Frank

Dawson Adams 1859–1942," in *Royal Society of London, Obituary Notices of Fellows*, **4**, no. 12 (1943), 381–393.

THOMAS H. CLARK

ADAMS, JOHN COUCH (*b.* Laneast, Cornwall, England, 5 June 1819; *d.* Cambridge, England, 21 January 1892), *astronomy, mathematics.*

John Couch Adams was born at Lidcot farm, seven miles from Launceston. He was the eldest son of Thomas Adams, a tenant farmer and a devout Wesleyan, and Tabitha Knill Grylls. The family circumstances were modest but respectable: Tabitha Adams' cousin was the headmaster of a private school in Devonport, and in 1836 her adoptive mother left her some property and a small income which helped support John's education.

Adams had his first schooling in a Laneast farmhouse. In 1827 he was tutored in calligraphy, Greek, and mathematics, but quickly outpaced his teacher. He developed an early interest in astronomy, inscribing a sundial on his window sill and observing solar altitudes with an instrument he built himself. In 1831 he was sent to his cousin's academy, where he distinguished himself in classics, spending his spare time on astronomy and mathematics. Teaching himself, he finished the standard texts on conic sections, differential calculus, theory of numbers, theory of equations, and mechanics. Adams' precocity convinced his parents that he should be sent to a university, and in October 1839 he sat for examinations at St. John's College, Cambridge University, and won a sizarship. He went on to win the highest mathematical prizes in his college and took first prize in Greek testament every year that he was at Cambridge.

In July 1841, Adams, having read about the irregularities in the motion of the planet Uranus, decided to investigate them as soon as he had taken his degree. He graduated from Cambridge in 1843 as senior wrangler in the mathematical tripos and first Smith's prizeman; shortly afterward he became a fellow and tutor of his college. At the beginning of the next long vacation he returned to Lidcot and began the long-deferred investigation of Uranus.

By October 1843 Adams had arrived at a solution of the inverse perturbation problem: given the mass of a body and its deviations from the path predicted for it by Newtonian mechanics, find the orbit and position of another body perturbing it through gravitational attraction. This problem required, among other procedures, the solution of ten simultaneous equations of condition for as many unknowns. Although Adams' first result was approximate, it con-

vinced him that the disturbances of Uranus were due to an undiscovered planet.

In February 1844, Adams applied through James Challis to the astronomer royal, Sir George Biddell Airy, for more exact data on Uranus. Using figures supplied by Airy, Adams computed values for the elliptic elements, mass, and heliocentric longitude of the hypothetical planet. He gave his results to Challis in September 1845, and after two unsuccessful attempts to present his work to Airy in person, he left a copy of it at the Royal Observatory on 21 October 1845. Although Airy wrote to Adams a few weeks later criticizing his paper, he did not institute a search for the planet until July 1846.

In the meantime a French astronomer, Urbain Jean Joseph Leverrier, independently published several papers on the theory of Uranus and reached the same conclusions as Adams had regarding an exterior planet. Although Leverrier began his investigation later, he pressed his case more aggressively, and on 23 September 1846 the perturbing body—Neptune—was discovered as a result of his efforts. Johann Gottfried Galle, an astronomer at the Berlin Observatory, found the planet less than one degree distant from the point where Leverrier predicted it would lie.

Leverrier was immediately showered with honors and congratulations. Adams' earlier prediction, which agreed closely with Leverrier's, was thus far unpublished. It was first publicized in a letter from Sir John Herschel to the London *Athenaeum* on 3 October 1846 and provoked a long and bitter controversy over priority of discovery. The two principals took little part in the feud, but the issue became a public sensation. It still seems remarkable that Airy suppressed Adams' work for so long and that Adams was so reticent about pressing his claims. This behavior was, however, characteristic of Adams. The modesty that temporarily cost him some glory endeared him to colleagues and friends throughout his life.

The disparity between the credit accorded to Leverrier and that accorded to Adams was not made up for some years, but the two men met at Oxford in 1847 and became good friends. Adams was offered a knighthood by Queen Victoria in 1847 but declined it; the following year the Adams Prize, awarded biennially for the best essay in physics, mathematics, or astronomy, was instituted at Cambridge. The Royal Society gave Adams its highest award, the Copley Medal, in 1848.

In 1851 Adams was elected president of the Royal Astronomical Society and shortly afterward began to work on lunar theory. After much laborious calculation he finished new tables of the moon's parallax which corrected several errors in lunar theory and gave more accurate positions. In the meantime, since he had not taken holy orders, his fellowship at St. John's expired in 1852. He was elected a fellow of Pembroke College in 1853, and shortly afterward he presented to the Royal Society a remarkable paper on the secular acceleration of the moon's mean motion. This quantity was thought to have been definitively investigated by Pierre Simon de Laplace in 1788, but Adams showed that Laplace's solution was incorrect. In particular, Laplace had ignored a variation in solar eccentricity that introduces into the differential equations for the moon's motion a series of additional terms. Adams calculated the second term of the series, on which the secular acceleration depends, as $3771/64 \, m^4$; the value computed from Laplace's work was $2187/128 \, m^4$. The effect of the correction was to reduce the figure for the moon's secular acceleration by about half, from $10''.58$ to $5''.70$.

This paper caused a sharp scientific controversy, marked by angry chauvinism on the part of several French astronomers. Their attacks stimulated a number of independent investigations of the subject, all of which confirmed Adams' result. The matter was definitely settled in his favor by 1861, but not without hard feelings.

In 1858 Adams occupied the chair of mathematics at the University of St. Andrews, vacating it the following year to accept the appointment as Lowndean professor of astronomy and geometry at Cambridge. In 1861 he succeeded James Challis as director of the Cambridge Observatory, and in 1863, when he was forty-four, he married Eliza Bruce of Dublin. In 1866 the Royal Astronomical Society awarded Adams a gold medal for his work on lunar theory.

The brilliant Leonid meteor shower of November 1866 stimulated Adams to investigate the elements of the Leonid system. By dividing the orbit into small segments, he calculated an analysis of perturbations for the meteor group, resulting in improved values for its period and elements. This work provided another demonstration of Adams' extraordinary ability to manipulate equations of great length and complexity without error.

In 1870 the Cambridge Observatory acquired a Simms transit circle. In order to exploit it fully, Adams undertook—a rarity for him—the direction of a program of observational astronomy. The circle was used to map a zone lying between 25° and 30° of north declination for the *Astronomische Gesellschaft* program. This work was first published in 1897.

In 1874 Adams was elected to a second term as president of the Royal Astronomical Society. His scientific interest at this time turned to mathematics.

Like Euler and Gauss, Adams enjoyed the calculation of exact values for mathematical constants. In 1877 he published thirty-one Bernoullian numbers, thus doubling the known number. With sixty-two Bernoullian numbers available, he decided to compute a definitive value of Euler's constant; this required the calculation of certain logarithms to 273 decimal places. Using these terms, Adams extended Euler's constant to 263 decimal places. This result was published in the *Proceedings* of the Royal Society in 1878; in the same year Adams published expressions for the products of two Legendrian coefficients and for the integral of the product of three.

Adams was a fervent admirer of Isaac Newton. In 1872, when Lord Portsmouth presented Newton's scientific papers to Cambridge University, Adams willingly undertook to arrange and catalog those dealing with mathematics. He was also an omnivorous reader in other fields, especially botany, history, and fiction. He usually kept a novel at hand when working on long mathematical problems.

In retrospect Adams' many mathematical and astronomical achievements pale in comparison to his analysis of the orbit of Uranus and his prediction of the existence and position of Neptune at the age of twenty-four. Much of his later work has been superseded, but as the co-discoverer of Neptune he occupies a special, undiminished place in the history of science.

BIBLIOGRAPHY

I. ORIGINAL WORKS. Works by Adams include MSS on the perturbations of Uranus, 1841–1846, St. John's College Library, Cambridge, England; *Lectures on the Lunar Theory* (Cambridge, England, 1900); and William Grylls Adams, ed., *The Scientific Papers of John Couch Adams*, 2 vols. (Cambridge, England, 1896–1900).

II. SECONDARY LITERATURE. See Morton Grosser, *The Discovery of Neptune* (Cambridge, Mass., 1962); Urbain Jean Joseph Leverrier, MS of the memoir "Recherches sur le mouvement de la planète Herschel (dite Uranus)," in the library of the Paris Observatory; W. M. Smart, "John Couch Adams and the Discovery of Neptune," in *Occasional Notes of the Royal Astronomical Society* (London), **2** (1947), 33–88.

MORTON GROSSER

ADAMS, WALTER SYDNEY (*b.* Kessab, near Antioch, Syria, 20 December 1876; *d.* Pasadena, California, 11 May 1956), *astrophysics.*

The son of Lucien Harper Adams and Nancy Dorrance Francis, missionaries in Syria, Adams was brought up in a strict, though broad-minded, environment. He received his early schooling from his mother and from his father's classical and historical library. His childhood was spent near Antioch, crossroads of the Crusades, and at the age of six he knew more of the history of Athens and Rome and the campaigns of Alexander the Great and Hannibal than of the United States. Here, too, his interest in astronomy was aroused when his father pointed out the constellations in the clear Syrian skies.

In 1885 the family returned to Derry, New Hampshire. Adams graduated with the A.B. from Dartmouth College in 1898 and, on the advice of Edwin B. Frost, his teacher of astronomy, entered the University of Chicago. There he earned a reputation as a skillful mathematician. He gained his first practical observing experience under George Ellery Hale, founder and director of the Yerkes Observatory, which, with its forty-inch refracting telescope, was then the largest in the world. After receiving his M.A. in 1900, he went to the University of Munich; the following year he returned to Yerkes as computer and general assistant.

Adams was married twice, first to Lillian Wickham, who died in 1920, and in 1922 to Adeline L. Miller, by whom he had two sons, Edmund and John.

Inspired by Hale's vision of the future of astrophysics and by his belief in the importance of an observatory as a physical laboratory, Adams followed his path enthusiastically. In 1947, fifty years after the dedication of the Yerkes Observatory, he described the revolution in astronomy:

> It opened at a time when visual observations were still a major factor in observatory activities, photographic methods were in their infancy, the spectrum was studied empirically and cosmogony was almost a completely sealed book. The period ends with visual observations greatly reduced in amount, although still holding an important place, with photographic methods applied almost universally and enormously improved and extended, with the spectrum analyzed and used as an extraordinarily powerful tool to seek out physical processes in the sun and stars, and with a clear and logical picture of a physical universe beyond the imagination of the astronomer of fifty years ago.[1]

In this transformation Adams played a leading role, first at the Yerkes, then at the Mount Wilson Observatory. As acting director of Mount Wilson at various periods, then as director from 1923 to 1946, he contributed significantly to the design of instruments, especially of the 100-inch and of the 200-inch Hale telescopes on Palomar Mountain. Through his leadership he helped to make the Mount Wilson and Palomar observatories preeminent, so that astronomers the world over came to use these, the most powerful astronomical instruments on earth, to push back the frontiers of the universe.

Adams' observations ranged from planetary atmospheres to interstellar gases, from sunspot spectra to his greatest achievement—the discovery of a method for determining stellar distances. His research was characterized by notable skill in observation and precision in measurement.

His influence on astronomical development, both nationally and internationally, was reflected in his positions in the American Astronomical Society (president, 1931–1934), the Astronomical Society of the Pacific (president, 1923), and the International Astronomical Union (vice-president, 1935–1948, acting secretary, 1940–1945). His broader interests in science were reflected in his membership in the American Philosophical Society (elected 1915) and the National Academy of Sciences (1917) and by his election as foreign associate to many academies of science, including those of France, Sweden, and the Soviet Union, and the Royal Society of London (foreign member).

His earliest work, on the polar compression of Jupiter, was followed by research with Frost on radial velocities in B-type, or helium, stars. By 1903, despite the difficulty of observing the diffuse spectral lines, measurement of the velocities in twenty such stars showed the average motion to be exceptionally small. This result would prove important in discussions of stellar motions, especially in the recognition by W. W. Campbell of the so-called K-term. In April 1904, Adams joined Hale on the Yerkes expedition to Pasadena, California; this led to the establishment by the Carnegie Institution of Washington of the Mount Wilson Solar Observatory on 20 December 1904. In his vivid "Early Days at Mount Wilson," he described the wild and primitive conditions, the joys and difficulties of a pioneer time when transportation of equipment was wholly by pack train and the only means of reaching the peak was by mule or burro, or on foot. He had a wiry, athletic build and an indefatigable spirit, and often climbed the steep, twisting, eight-mile trail; prepared the telescope for observing; worked the night through; then walked down the mountain the following morning.

At Mount Wilson he joined Hale in an intensive study of that "typical star," our sun, first with the horizontal Snow reflector, then with the sixty-foot and 150-foot tower solar telescopes. Visually it had been observed that the spectrum of a sunspot differs from that of the solar disk, but little was known of the nature of the spot spectrum, and nothing of its cause. Now, for the first time, it became possible to study spot spectra photographically with adequate apparatus. In 1906 Adams and Hale took the first photograph of a spot spectrum at Mount Wilson and undertook a detailed comparison of spot spectra with those of the solar disk. Simultaneously experiments were begun in the primitive laboratory on the mountain to imitate the conditions observed in the sun. Working with Henry Gale on arc spectra, Adams and Hale were able to show that temperature must be the cause of the differences observed between spots and disk, and thus to prove that sunspots are cooler than the surrounding solar surfaces. From further laboratory studies of pressure and density, they also showed that "enhanced lines" (those lines identified and named by Norman Lockyer to denote lines that are much stronger in the electric spark spectrum than in that of the arc) are the result of a lower density of the gases, while in sunspots the density proved to be higher. These and other results found a rational explanation when, in 1920, M. N. Saha published his theory of ionization.

In the course of this investigation Adams became interested in the problem of solar rotation. From a study of the minute Doppler displacements at various solar levels, he found that higher levels in the sun showed a higher rate of rotation and a smaller equatorial acceleration than the lower levels. In 1909, with Hale, he succeeded in photographing the flash spectrum without eclipse.

Gradually Adams turned from studies of the sun to other, larger stars. Yet, as he was often to show, the early sunspot studies played an important role in the understanding of other stars, and especially in the determination of their distances. In "Sunspots and Stellar Distances" he described the fascinating chain of events that led from the classification of groups of spectral lines according to temperature, to show how unexpected the ramifications of scientific investigation can be. "The study attained its primary objectives, but in addition it provided in the field of physics the first clues to the analysis of complex spectra according to energy levels in the atom, in solar physics the discovery of magnetism in the sun, and in astrophysics a new and fundamental method for determining the distances of the stars."[2]

In 1906, using the Snow telescope, Adams had succeeded in taking a twenty-three-hour exposure, on five successive nights, of the line spectrum of the cool star Arcturus. When he compared this spectrum with that of a sunspot, he found them to be similar. In 1908, after the sixty-inch reflecting telescope was set up on Mount Wilson, he extended his comparative studies to other stars—their motions, spectral classifications, magnitudes, and the distances of those too

far away to be measured trigonometrically. These studies included the first thorough investigation of the differences in the spectra of the large and massive stars of high luminosity called giants and the comparatively dense bodies of very low luminosity known as dwarfs.

In 1914, working at first with Arnold Kohlschütter, Adams compared pairs of stars of nearly the same spectral type, and therefore of nearly the same temperature, but of very different luminosity.

> It soon appeared that a few lines were stronger in the spectrum of the highly luminous star, and others in that of the intrinsically faint star. With the use of all the available material for well-determined luminosities it then became possible to establish numerical correlations between luminosity and the intensities of these sensitive lines. The process could then be reversed and in the case of a star of unknown luminosity its value could be derived from the intensities of the lines; the distance of the star is then readily calculated from the simple relationship connecting apparent brightness, luminosity and distance.[3]

This ingenious method of obtaining "spectroscopic parallaxes," applied to thousands of stars, has become a fundamental astronomical tool of immense value in gaining knowledge of giant and dwarf stars and of galactic structure. Otto Struve commented, "It is not an exaggeration to say that almost all our knowledge of the structure of the Milky Way which has developed during the past quarter of a century has come from the Mount Wilson discovery of spectroscopic luminosity criteria."[4]

In 1917 the 100-inch telescope went into operation. Adams had made the Hartmann tests of the mirror which were vital to its successful figuring. For it he built the powerful Coudé spectrograph that would provide higher dispersion and make the penetration of hitherto unknown regions possible. The following year, in studies that stemmed from his investigation of giants and dwarfs, he became interested in Sirius B, the companion of Sirius that he had first identified as a tiny white-hot star, or white dwarf, in 1915. He found that while the companion is small, it has a mass not much less than that of the sun (actually four-fifths of that mass). It proved, almost incredibly, to be about 50,000 times as dense as water. A ton of such material could be squeezed into a matchbox. Sir Arthur Eddington predicted that, since the Einstein effect is proportional to the mass divided by the radius of the star and the radius of the companion of Sirius is very small, the relativity effect should be large. In 1925 Adams performed the difficult feat of taking a spectrogram of the faint companion, which is 10,000 times fainter than its neighbor, yet only twelve arc seconds away. He confirmed Eddington's prediction when he found a displacement to the red of 21 km./sec., a result he later modified to 19 km./sec. Eddington wrote: "Prof. Adams has thus killed two birds with one stone. He has carried out a new test of Einstein's general theory of relativity, and he has shown that matter at least 2,000 times denser than platinum is not only possible, but actually exists in the stellar universe."[5]

In the 1920's and 1930's Adams also applied the spectrograph to studies of the atmospheres of Venus and Mars—these observations were difficult because of the problem of identifying such substances as oxygen, carbon dioxide, and water vapor, which are also contained in the earth's atmosphere. In 1932, with Theodore Dunham, Jr., he identified carbon dioxide in the infrared spectrum of Venus. In 1934 similar observations of Mars indicated that the amount of free oxygen above a given area of the surface of Mars cannot exceed one-tenth of one percent.

Over the years other investigations included Cepheids, spectroscopic binaries, and, from 1901 to 1936, the spectra of novae that he felt might be explained by an expanding shell or succession of shells. In his last extensive research on the clouds of interstellar gas he had the arduous task of sorting out the lines in a star's own spectrum from those belonging to the tenuous interstellar gases. He found double or multiple interstellar lines in 80 percent of the stars examined, identified two classes of clouds, and observed four clouds moving with radial velocities up to 100 km./sec. The highly accurate velocities provided good values for the relative motions caused by the rotation of the galaxy.

These, then, were the far-ranging programs through which Adams contributed to our knowledge of the nature of the universe and profoundly influenced the development of cosmogony.

NOTES

1. "Some Reminiscences of the Yerkes Observatory," p. 196.
2. *Cooperation in Solar Research*, pp. 135–137.
3. "Biographical Notes—Walter S. Adams," written for the National Academy of Sciences (Jan. 1954), p. 9 (unpublished).
4. "Fifty Years of Progress in Astronomy," p. 6.
5. *Stars and Atoms*, p. 52.

BIBLIOGRAPHY

I. ORIGINAL WORKS. Adams' bibliography in Joy's biographical article (see below) includes 270 papers, in addi-

tion to his annual Mount Wilson reports. Among these papers are "The Polar Compression of Jupiter," in *Astronomical Journal*, **20** (1899), 133, written while he was still a graduate student; "Radial Velocities of Twenty Stars Having Spectra of the Orion Type," in *Publications of the Yerkes Observatory*, **2** (1904), 143–250, written with E. B. Frost; "Photographic Observations of the Spectra of Sunspots," in *Astrophysical Journal*, **23** (1906), 11–44, written with G. E. Hale; "Preliminary Paper on the Cause of the Characteristic Phenomena of Sunspot Spectra," *ibid.*, **24** (1906), 185–213, written with G. E. Hale and H. G. Gale; "Sunspot Lines in the Spectrum of Arcturus," *ibid.*, 69–77; "Spectroscopic Observations of the Rotation of the Sun," *ibid.*, **26** (1907), 203–224; "Photography of the Flash Spectrum Without an Eclipse," *ibid.*, **30** (1909), 222–230, written with G. E. Hale; "The Radial Velocities of 100 Stars With Measured Parallaxes," *ibid.*, **39** (1914), 341–349, written with A. Kohlschütter; "The Spectrum of the Companion of Sirius," in *Publications of the Astronomical Society of the Pacific*, **27** (1915), 236–237; "A Spectroscopic Method of Determining Parallaxes," in *Proceedings of the National Academy of Sciences*, **2** (1916), 147–152; "Address of the Retiring President . . .," in *Publications of the Astronomical Society of the Pacific*, **36** (1924), 2–9; "The Relativity Displacement of the Spectral Lines in the Companion of Sirius," in *Proceedings of the National Academy of Sciences*, **11** (1925), 382–387; "The Past Twenty Years of Physical Astronomy," in *Publications of the Astronomical Society of the Pacific*, **40** (1928), 213–228; "The Astronomer's Measuring Rods," *ibid.*, **41** (1929), 195–211; "Absorption Bands in the Spectrum of Venus," *ibid.*, **44** (1932), 243–245, written with Theodore Dunham, Jr.; "The B-Band of Oxygen in the Spectrum of Mars," in *Astrophysical Journal*, **79** (1934), 308–316, written with Theodore Dunham, Jr.; "The Planets and Their Atmospheres," in *Scientific Monthly*, **39** (1934), 5–19; "The Sun's Place Among the Stars," in *Annual Report of the Smithsonian Institution for 1935*, pp. 139–151; "Sunspots and Stellar Distances," in *Cooperation in Solar Research*, Carnegie Institution of Washington pub. no. 506 (1938), pp. 135–147; "George Ellery Hale," in *Biographical Memoirs of the National Academy of Sciences*, **21** (1940), 181–241; "Newton's Contributions to Observational Astronomy," in *The Royal Society Newton Tercentenary Celebrations* (Cambridge, 1946), pp. 73–81; "Early Days at Mount Wilson," in *Publications of the Astronomical Society of the Pacific*, **59** (1947), 213–231, 285–304; the Henry Norris lecture of the American Astronomical Society (29 Dec. 1947), *ibid.*, **60** (1948), 174–189; "Some Reminiscences of the Yerkes Observatory," in *Science*, **106**, no. 2749 (5 Sept. 1947), 196–200; "The History of the International Astronomical Union," in *Publications of the Astronomical Society of the Pacific*, **61** (1949), 5–12; "The Founding of the Mount Wilson Observatory," *ibid.*, **66** (1954), 267–303; and "Early Solar Research at Mount Wilson," in Arthur Beer, ed., *Vistas in Astronomy*, I (London, 1955), 619–623. See also the *Annual Report* of the Mount Wilson Observatory (1923–1945).

The bulk of Adams' correspondence, original manuscripts, and other source materials are (as of 1968) in the

Hale Solar Laboratory and in the director's files of the Mount Wilson and Palomar observatories.

II. SECONDARY LITERATURE. Biographical articles on Adams include Alfred H. Joy, "Walter S. Adams, a Biographical Memoir," in *Biographical Memoirs of the National Academy of Sciences,* **31** (1958); Paul W. Merrill, "Walter S. Adams, Observer of Sun and Stars," in *Science,* **124** (13 July 1956), 67; Harlow Shapley, "A Master of Stellar Spectra," in *Sky and Telescope,* **15** (1956), 401; and F. J. M. Stratton, "Walter Sydney Adams (1876–1956)," in *Biographical Memoirs of the Royal Society,* **2** (Nov. 1956), 1–18.

Additional works that contribute to a picture of the development of astronomy in this period, and of Adams' role in that development, are Charles G. Abbot, *Adventures in the World of Science* (Washington, D.C., 1958), esp. ch. 6; Giorgio Abetti, "Solar Physics," in *Handbuch der Astrophysik,* IV (Berlin, 1929), 161–168, and VII (Berlin, 1936), 184; *The History of Astronomy,* Betty B. Abetti, trans. (New York, 1952), pp. 255, 258, 259, 274, 291–292, 298, 306, 327; and *The Sun,* J. B. Sidgwick, trans. (New York, 1957), pp. 103, 142, 143, 144, 145, 147, 148, 164–165, 206, 231, 232; Herbert Dingle, "The Message of Starlight," in T. E. R. Phillips and W. H. Steavenson, eds., *Splendour of the Heavens,* II (London, 1924), 479–499; Sir Arthur Eddington, *Stars and Atoms* (Oxford, 1927), pp. 48–53; Edwin B. Frost, *An Astronomer's Life* (Boston, 1933); George Ellery Hale, *The Study of Stellar Evolution* (Chicago, 1908); Caryl Haskins, *The Search for Understanding* (Washington, D.C., 1967), which includes a large part of "Early Days at Mount Wilson" (pp. 301–326) and other material relating to the observatory (pp. 234–277); Gerard P. Kuiper, ed., *The Sun* (Chicago, 1953), especially the excellent introduction by Leo Goldberg, pp. 7–22; Knut Lundmark, "Luminosities, Colours, Diameters, Densities, Masses of the Stars," in *Handbuch der Astrophysik,* V (Berlin, 1932), ch. 4; A. Pannekoek, *A History of Astronomy* (New York, 1961); M. N. Saha, "Ionization in the Solar Chromosphere," in *Philosophical Magazine,* **40** (1920), 479; Harlow Shapley, "Brief Historical Analysis on the Spectra of Stars and Nebulae," in *Source Book in Astronomy, 1900–1950* (Cambridge, Mass., 1960), pp. 159–161, 162–164, in which Shapley discusses Adams' work on star distances and reprints two of his papers; Otto Struve, "The Story of an Observatory (The Fiftieth Anniversary of the Yerkes Observatory)," in *Popular Astronomy,* **55**, nos. 5–6 (May–June 1947); and "Fifty Years of Progress in Astrophysics," in *The Science Counselor* (Mar. 1948), 4–6, esp. 6, 26–27; Otto Struve and Velta Zebergs, *Astronomy of the 20th Century* (New York, 1962); and Helen Wright, *Palomar, the World's Largest Telescope* (New York, 1952), and *Explorer of the Universe, a Biography of George Ellery Hale* (New York, 1966).

HELEN WRIGHT

ADANSON, MICHEL (*b.* Aix-en-Provence, France, 7 April 1727; *d.* Paris, France, 3 August 1806), *natural history, philosophy.*

Adanson belonged to an Auvergne family that moved to Provence at the beginning of the eighteenth century and to Paris about 1730. He was educated at the Plessis Sorbon, the Collège Royal, and the Jardin du Roi. Among his *maîtres* were Pierre Le Monnier, Réaumur, G.-F. Rouelle, and Antoine and Bernard de Jussieu. He made his first four-year scientific expedition to Senegal on behalf of the Compagnie des Indes and brought back a large group of natural history specimens; a few of these later became part of the royal collection, then under the care of Buffon. While traveling in Africa, Adanson was elected (24 July 1750) a corresponding member of the Académie des Sciences. His travel journal (1757) was accompanied by a general survey of the living mollusks he had found in Senegal. His classification of mollusks was an original one; based on the anatomical structure of the living animals inside the shells, it appeared the same year as the work of Argenville, who claimed to have originated such a scheme.

In 1761 Adanson was elected a foreign member of the Royal Society of London, and in 1763–1764 he published *Familles des plantes.* In this book he proclaimed his contempt for "systems" and proposed a natural classification based upon all characters rather than upon a few arbitrarily selected ones, an attempt that brought him into conflict with Linnaeus. Recent historical studies have shown that Adanson's views were shared by many Parisian botanists and that he was responsible for the maintenance of Joseph Tournefort's system at the Jardin du Roi until 1774, when A. L. de Jussieu's system was adopted. Adanson owed much to Bernard de Jussieu's plant families as they were developed in his manuscript plan for the Trianon garden in which he arranged the plants in beds in an order corresponding to his system of classification. He soon recognized that his *Familles des plantes* was only an outline of his general conception, and in 1769 he prepared a new edition that was never published.

Adanson knew Diderot but did not collaborate on the *Encyclopédie,* although he played an important role in the publication of the supplement (1776) by Panckoucke, to whom he sent more than 400 articles. He had his own views about encyclopedias, and in 1775 he presented a plan for one to the Académie des Sciences. By that time he had amassed a collection of documents, observations of his own, and natural history specimens. Nothing came of this plan, however, and he spent the rest of his life in futile attempts to publish his own encyclopedia. On 23 July 1759 Adanson had been elected *adjoint botaniste;* on 25 February 1773, *associé botaniste;* and on 6 December 1782, *académicien pensionnaire.* Upon the creation of the Institut de France, he was immediately selected a member of the first college. Later Napoleon made him a member of the Legion of Honor.

In many respects Adanson played a hidden role in the development of science, for he was in touch with most of the learned people of Europe. He studied static electricity in the torpedo fish, the tourmaline, and various plants; agricultural problems concerning corn, wheat, barley, and fruits; microscopic animalcules; and the circulation of sap in lower plants. He also experimented on regeneration of the limbs and head of frogs and snails. Although he kept most of his materials for his own use, we know that he was an important contributor to Buffon's *Histoire naturelle générale,* where he is quoted more than a hundred times. He had sent several hundred new plant species from Senegal to Bernard de Jussieu, and before his controversy with Linnaeus, he had sent to Sweden a number of African plants that Linnaeus said he included with those of Hasselquist. His general herbarium, now in the Muséum National d'Histoire Naturelle, contains about 30,000 specimens, many of which have been studied; the plants he sent to Jussieu were used by A. L. de Jussieu for his *Genera plantarum* and by later botanists. Lamarck used Adanson's articles in the *Encyclopédie* supplement for his *Dictionnaire de botanique.*

Adanson survived the Revolution without political difficulties, but suffered much from the financial crash. His whole life, however, was one of periodic financial insecurity, alleviated by the patronage obtained for him by his friends and by the life annuity granted in the 1760's when his natural history collection became part of the Cabinet du Roi. Intellectually, he was perhaps equally insecure, admired by many of his contemporaries and disliked by others for both scientific and personal reasons. It is only recently that his historical influence and his role in introducing modern statistical methods into systematic botany have received proper recognition.

BIBLIOGRAPHY

I. ORIGINAL WORKS. Adanson's first book was *Histoire naturelle du Sénégal. Coquillages. Avec la relation abrégée d'un voyage fait en ce pays pendant les années 1749 . . . 1753* (Paris, 1757); the travel portion was translated by "an English gentleman" as *A Voyage to Senegal, the Isle of Goree and the River Gambia* (London-Dublin, 1759), and into German by Martini as *Reise nach Senegall* (Brandenburg, 1772) and by Schreber as *Nachricht von seiner Reise nach Senegall* (Leipzig, 1773); a review of the *Histoire naturelle* is in G. R. Boehmer, *Bibliotheca scriptorum historiae naturalis oeconomiae, aliarum que artium ac scientiarum,* Vol. I (Leipzig, 1785). His other book is *Familles*

des plantes, 2 vols. (Paris, 1763–1764), reviewed in G. R. Boehmer, *op. cit.*

See also "Marées de l'Ile de Gorée," in *Mémoires de mathématiques et de physique, présentés à l'Académie royale des sciences, par divers sçavans,* **2** (1755), 605–606; "Plan de botanique," in *Collection académique* (*Savants français*), **8** (n.d. [after 1759]), appendix p. 59; "Description . . . du baobab," in *Mémoires de l'Académie des sciences* (1761 [1763]), pp. 218–243; "Description d'une nouvelle espèce de vers . . . ," *ibid.* (1759 [1765]), pp. 249–279; "Remarques sur les bleds appelés de miracle," *ibid.* (1765 [1768]), pp. 613–619; "Mémoire sur un mouvement particulier . . . de la tremelle," *ibid.* (1767 [1770]), pp. 564–572; "Examen de la question: si les espèces changent parmi les plantes . . . ," *ibid.* (1769 [1772]), pp. 31–48; "Premier mémoire sur l'acacia des anciens," *ibid.* (1773 [1777]), pp. 1–17; "Deuxième mémoire sur le gommier blanc . . . ," *ibid.* (1778 [1781]), pp. 20–35; and "Observations météorologiques . . . ," *ibid.,* p. 425.

Published after Adanson's death were *Cours d'histoire naturelle fait en 1772,* 2 vols. (Paris, 1845), and *Histoire de la botanique et plan des familles naturelles des plantes* (Paris, 1864), A. Adanson and J. B. Payer, eds.

Manuscripts on botanical subjects and a large part of Adanson's library, with many annotated books, are in the Hunt Botanical Library of Carnegie Institute of Technology, Pittsburgh, Pa. Letters are in the Royal Society of London; Wellcome Library, London; Académie des Sciences, Paris; Institut de France, Paris; Bibliothèque Nationale, Département des Manuscrits, Paris; Bibliothèque Centrale du Muséum National d'Histoire Naturelle, Paris; Bibliothèque Publique et Universitaire de Genève; and Bibliothèque de la Bourgeoisie, Berne.

II. SECONDARY LITERATURE. Although rather superficial in its judgments, Cuvier's *éloge* of Adanson, in *Recueil des éloges historiques,* new ed. (Paris, 1861), I, 173–204, was the basic source used by nineteenth-century biographers. More recent is *Adanson,* 2 vols., Hunt Monograph Series, G. H. M. Lawrence, ed. (Pittsburgh, 1963–1964), containing several original papers, a biography, a bibliography, and notes. A general review of J. P. Nicolas's studies of Adanson is the pamphlet "Adanson et les Encyclopédistes," Lecture D.104, Palais de la Découverte, Paris (3 April 1965).

J. P. NICOLAS

ADDISON, THOMAS (*b.* Long Benton, England, *ca.* April 1793; *d.* Brighton, England, 29 June 1860), *medicine.*

Although the birth date generally assigned to Thomas Addison is April 1793, the tablet in Guy's Hospital Chapel in London and that in Lanercost Abbey in Cumberland, where he is buried, state that he died on 29 June 1860, at the age of sixty-eight. The Long Benton church baptismal register has the following entry: "1795, Oct. 11. Thomas s. of Joseph and Sarah Addison, Lg. Benton." The same register gives 13 April 1794 as the baptismal date of John, the second son of Joseph and Sarah Addison. Since it is unlikely that if Thomas had been born in 1793 his baptism would have been deferred until after that of his younger brother, it is reasonable to believe that in the course of transcription a five has become a three, as Hale-White suggested.

Addison married Elizabeth Catherine Hauxwell at Lanercost Church in September 1847. They were childless, although she had two children by her first marriage.

Addison was first sent to school near Long Benton, and then went to a grammar school at Newcastle-on-Tyne. He learned Latin so well that he made notes in that language and spoke it fluently. His father had wished him to become a lawyer, but in 1812 he entered the University of Edinburgh as a medical student. He graduated in 1815, at the age of twenty-two, as a Doctor of Medicine. The title of his thesis was "De syphilide et hydrargyro" ("Concerning Syphilis and Mercury"). Guy's Medical School book records his entrance: "Dec. 13, 1817, from Edinburgh, T. Addison, M.D., paid £22-1s. to be a perpetual Physician's Pupil."

Addison became house surgeon at Lock Hospital in London in 1815 and was appointed assistant physician to Guy's Hospital on 14 January 1824. He became lecturer on materia medica three years later. He was joint lecturer on medicine with Richard Bright in 1835, and in 1837 he became physician to Guy's Hospital. In 1840 Bright retired from the lectureship, and Addison became sole lecturer. He held this position until either 1854 or 1855. He obtained his licentiateship in the Royal College of Physicians on 22 December 1819 and was elected a fellow on 4 July 1838.

Addison's numerous clinical studies include works on the clinical signs of a fatty liver (1836), appendicitis (1839), pneumonia (1843), phthisis (1845), and xanthoma (1851). In 1849 he described Addison's anemia before a meeting of the South London Medical Society: "For a long period I had from time to time met with a remarkable form of general anemia. . . ." His clinical findings fit with both Vitamin B_{12} and folic acid deficiency states. One feature peculiar to Vitamin B_{12} deficiency is: ". . . the bulkiness of the general frame and the obesity often present, a most striking contrast."

In the absence of a separate formal report, it is not surprising that the world overlooked this excellent description of pernicious anemia. That description, good as it was, was quite overshadowed by Addison's spectacular discovery of the disturbance of the suprarenal capsules. In 1855, in a paper entitled "On the

Constitutional and Local Effects of Disease of the Suprarenal Capsules," he described what is now known as Addison's disease, a condition characterized by progressive anemia, bronze skin pigmentation, severe weakness, and low blood pressure. It is now known that in Addison's disease the blood sodium and chloride are lowered, potassium and nitrogen are increased, and there is a diminution in the blood volume. The intravenous administration of a physiologic solution of sodium chloride helps the patient to recover from these conditions. This work laid the foundation for modern endocrinology.

At Guy's Hospital both conditions became increasingly familiar and were recorded separately from time to time in Guy's Hospital Reports, but elsewhere Addison's description of anemia was forgotten until his pupils Samuel Wilks and Thomas Daldy published his collected work and made it clear that Addison had described the disease in 1849, although A. Biermer reported it as a new disease in 1872.

In 1839 Bright and Addison published Elements of Practical Medicine. Only Volume I (two volumes were planned) appeared, and the work is incomplete and very rare.

Probably the best evaluation of Addison comes from Wilks, who said: "The personal power which he possessed was the secret of his position, much superior to what Bright could ever claim, and equal, if not greater, than that of Sir Astley Cooper."

On 7 July 1860 the Medical Times and Gazette published a notice of Addison's death on 29 June 1860, but neither Lancet nor the British Medical Journal recorded it.

BIBLIOGRAPHY

I. ORIGINAL WORKS. Addison's writings include "Observations on Fatty Degeneration of the Liver," in Guy's Hospital Reports, 1st Series, 1 (1836), 476–485; Elements of the Practice of Medicine (London, 1839), written with Richard Bright; "Observations on the Anatomy of the Lungs" (1840), in his Collected Writings (London, 1868), pp. 1–6; "Observations on Pneumonia and Its Consequences," in Guy's Hospital Reports, 2nd Series, 1 (1843), 365–402; "On the Pathology of Phthisis," ibid., 3 (1845), 1–38; "Disease: Chronic Suprarenal Insufficiency, Usually due to Tuberculosis of Suprarenal Capsule. 1st Announcement," in London Medical Gazette, n.s. 43 (1849), 517–518, reprinted in his Collected Writings (London, 1868), pp. 209–239, and in Medical Classics, 2 (1937), 239–244; "On a Certain Affection of the Skin, Vitiligoidea—a. plana, b. tuberosa, With Remarks," in Guy's Hospital Reports, 2nd Series, 7 (1851), 265–276, written with William Gull; On the Constitutional and Local Effects of Disease of the Suprarenal Capsules (London, 1855), also in Medical Classics,

2 (1937), 244–280; and *A Collection of the Published Writings of the Late Thomas Addison,* Samuel Wilks and Thomas M. Daldy, eds. (London, 1868).

II. Secondary Literature. More on Addison and his work may be found in Thomas Bateman, in *The Roll of the Royal College of Physicians of London,* 2nd ed. (London, 1878), III, 19–22; A. Biermer, "Form von progressiver perniciöser Anämie," in *Korresp.-Bl. schweizer Ärtze,* **2** (1872), 15; Herbert French, "Pernicious Anemia," in Clifford Allbutt and Humphrey Davy Rolleston, eds., *System of Medicine* (London, 1909), V, 728–757; William Hale-White, "Biography by Sir William Hale-White," in *Guy's Hospital Reports,* **76** (July 1926), 253–279; Victor Herbert, "The Megaloblastic Anemias," in *Modern Medical Monographs* (New York and London, 1959), p. 63; and E. R. Long, "Addison and His Discovery of Idiopathic Anemia," in *Annals of Medical History,* **7** (1935), 130–132.

See also "Obituary," in *Medical Times and Gazette,* **2** (1860), 20; and "Biography," in *The Roll of the Royal College of Physicians of London,* 2nd ed. (London, 1878), III, 205.

John A. Benjamin

ADELARD OF BATH (*b.* Bath, England; *fl.* 1116–1142), *mathematics, astronomy.*

Among the foremost of medieval English translators and natural philosophers, Adelard of Bath was one of the translators who made the first wholesale conversion of Arabo-Greek learning from Arabic into Latin. He traveled widely, first journeying to France, where he studied at Tours and taught at Laon. After leaving Laon, he journeyed about for seven years, visiting Salerno, Sicily (before 1116, perhaps before 1109), Cilicia, Syria, and possibly Palestine. It seems probable that he spent time also in Spain, on the evidence of his manifold translations from the Arabic (particularly his translation of the astronomical tables of al-Khwārizmī, from the revised form of the Spanish astronomer Maslama al-Majrīṭī).

It may be, however, that he learned his Arabic in Sicily and received Spanish-Arabic texts from other Arabists who had lived in or visited Spain, for example, Petrus Alphonsus and Johannes Ocreatus. He is found in Bath once more in 1130 when his name is mentioned in the Pipe Roll for 31 Henry I as receiving 4*s.* 6*d.* from the sheriff of Wiltshire. There are several indications in his writings of some association with the royal court. The dedication of his *Astrolabe* to a young Henry (*regis nepos*) seems to indicate a date of composition for that work between 1142 and 1146, and no later date for his activity has been established. F. Bliemetzrieder[1] has attempted to show that Adelard made a later trip to Salerno and Sicily, where he undertook the translation from the Greek of the *Almagest* of Ptolemy (completed about 1160), but a lack of any positive evidence and an improbable chronology militate against acceptance of this theory.

Adelard's modest contributions to medieval philosophy are found in two of his works: *De eodem et diverso* (1), written prior to 1116 and dedicated to William, bishop of Syracuse, and *Quaestiones naturales* (6), certainly written before 1137 and probably much earlier. [The numbers assigned here to the works of Adelard are those used by Haskins.[2] The author of this article has divided no. (5) into three parts, (5*a*), (5*b*), and (5*c*), and also has added a no. (15), which may reflect a further possible work.]

In the first work no trace of Arabic influence is evident, and he speaks as a quasi Platonist. From the *Timaeus,* he drew the major theme of *Philosophia* as representing "the same" and *Philocosmia* "the diverse." To the problem of universals, Adelard proposed as a kind of harmonizing of Plato and Aristotle his theory of *respectus,* that is, that the names of individuals, species, and genus are imposed on the same essence but under different aspects. ("Nam si res consideres, eidem essentiae et generis et speciei et individui nomina imposita sunt, sed respectu diverso."[3])

Both in *De eodem et diverso* and *Quaestiones naturales,* Adelard exhibits eclectic tendencies rather than strictly Platonic views. The *Natural Questions,* a dialogue with his unnamed nephew, comprises seventy-six chapters covering such manifold subjects as the nature and growth of plants (with attention to the doctrine of the four elements and four qualities); the nature of animals (including the question of whether animals have souls, which is answered in the affirmative); the nature of man (including his psychology and physiology); and meteorology, physics, and astrology.

Although professedly written to reveal something of his recent Arabic studies, no Arabic author is mentioned by name or quoted directly. Still the work shows traces of Arabic influence. The nephew describes a pipette-like vessel with holes in both ends. Water is prevented from flowing out of the holes in the lower end by covering the holes in the upper end with the thumb; "but with the thumb removed from the upper perforations the water [is] wont to flow immediately through the lower holes."[4] This is not unlike the vessel described in Hero's *Pneumatica* or in Philo of Byzantium's *Pneumatica,* which was translated from the Arabic in the twelfth century. Adelard explains this phenomenon by using a theory of the continuity of elements; no element will leave its place unless another element succeeds it; but with the upper holes covered and a vacuum formed, no air can enter the tube to replace the water. Hence the water cannot fall from the open holes below until the upper holes are uncovered and air can enter and replace it.

While there is some tendency to exaggerate Adelard's use of observation and experiment, it is clear that the *Natural Questions* exhibits a naturalistic trend, a tendency to discuss immediate natural causation rather than explain natural phenomena in terms of the supernatural.[5] This was also to become the practice of later writers such as William of Auvergne and Nicole Oresme. Adelard expressly prefers reason to authority, calling authority a *capistrum* ("halter") like that used on brutes.[6] He claims in the final chapter of the *Natural Questions* that he will write (7) on pure elements, simple forms, and the like, which lie behind the composite things treated in the *Natural Questions;* but no such work has been found.

There is extant, however, the tract *On Falcons* (8), which harkens back to the *Natural Questions*. According to Haskins, it is the "earliest Latin treatise on falconry so far known."[7] Perhaps also indicative of his interest in natural phenomena is the enlarged edition of the work on chemical recipes, *Mappae clavicula* (12), which is attributed to him.[8] However, the pristine version of that work is far earlier than Adelard. It is possible that some miscellaneous notes (14) that appear in a manuscript at the British Museum are by Adelard.[9] These are philosophical, astronomical, cosmological, and medical notes that seem to conform to Adelard's wide naturalistic interests, and the lunar cycle therein is that of 1136–1154.

Adelard's chief role in the development of medieval science lay, as has been noted, not so much in his contributions to natural philosophy as in the various translations he made from the Arabic. His translations were of a crucial and seminal nature in several areas.

Adelard gave the Latin Schoolmen their first example of the work of one of the most important Arabic astrologers with his *Ysagoga minor Iapharis matematici in astronomicam per Adhelardum bathoniensem ex arabico sumpta* (10), a translation of Abū Maʿshar's *Shorter Introduction to Astronomy*.[10] Consisting of some astrological rules and axioms, it was abridged by Abū Maʿshar from his longer *Introductorium maius*. Adelard's translation may well have served to whet the appetite of the Schoolmen for the longer work, which was twice translated into Latin: by John of Seville in 1135 and five years later by Hermann of Carinthia. Adelard also translated an astrological work of Thābit ibn Qurra on images and horoscopes, *Liber prestigiorum Thebidis* (*Elbidis*) *secundum Ptolomeum et Hermetem per Adelardum bathoniensem translatus* (11).[11]

In astronomy Adelard's most significant achievement was his translation of the *Astronomical Tables* of al-Khwārizmī, *Ezich Elkauresmi per Athelardum bathoniensem ex arabico sumptus* (3). At the end of chapter 4, the Arabic date A.H. 520 Muḥarram 1 is said to be 26 January 1126,[12] and this has usually been taken as the approximate date of translation. However, a manuscript at Cambridge gives examples for 1133 and 1134 and mentions a solar eclipse in 1133, throwing some doubt on the date.[13] These additional examples may, of course, be accretions not present in the original translation. How dependent this translation was on a possible earlier translation of the *Tables* by Petrus Alphonsus cannot definitely be determined from the available evidence. Millás-Vallicrosa has proposed that Petrus composed an earlier translation or adaptation of al-Khwārizmī's work, which Adelard then retranslated in 1126 with the assistance or collaboration of Petrus himself.[14]

At any rate, the *Tables* (comprising some 37 introductory chapters and 116 tables in the edition published by Suter) provided the Latin West with its initial introduction (in a considerably confused form) to the complex of Hellenistic-Indian-Arabic tabular material, including, among others, calendric tables; tables for the determination of the mean and true motions of the sun, moon, and planets; and trigonometric tables. (Tables 58 and 58*a* were very probably the first sine tables to appear in Latin.) In addition to this basic translation, Adelard also composed a tract on the *Astrolabe* (9),[15] continuing a line of work that began with translations from the Arabic as early as the middle of the tenth century. It is in this work that he cites his *De eodem et diverso*, his translation of the *Tables* of al-Khwārizmī, and his rendering of the *Elements* of Euclid.

Adelard's earliest efforts in arithmetic appear in a work entitled *Regule abaci* (2), which was apparently a work composed prior to his study of Arabic mathematics, for it is quite traditional and has Boethius and Gerbert for its authorities. But another work, the *Liber ysagogarum Alchorismi in artem astronomicam a magistro A. compositus* (4), based in part on Arabic sources, might well have been composed by him. Manuscript dates and internal evidence point to a time of composition compatible with the period in which Adelard worked. Hence the "magister A." is usually thought to be Adelard. The first three books of this work are concerned with arithmetic; the remaining two consider geometry, music, and astronomy. The subject of Indian numerals and the fundamental operations performed with them is introduced as follows: ". . . since no knowledge (*scientia*) goes forth if the doctrine of all the numbers is neglected, our tract begins with them, following the reasoning of the Indians."[16] (The section on geometry is, however, based on the Roman-Latin tradition rather than the Arabic-Indian tradition. The astronomical section returns to Arabic and Hebrew sources.) It has been suggested that the first three books on Indian reckoning have been drawn from an early Latin translation of al-Khwārizmī's *De numero Indorum* (not extant in its pristine state) or from a version of that translation revised sometime before 1143, which is preserved in an incomplete state at Cambridge and which has the incipit "Dixit algorizmi laudes deo rectori. . . ."[17] This work has been published three times: in transcription by B. Boncompagni,[18] in transcription and facsimile by K. Vogel,[19] and in facsimile only by A. P. Youschkevitch.[20] It has been suggested by Vogel[21] and Youschkevitch,[22] without any decisive evidence, that the original Latin translation of the *De numero Indorum* was executed by Adelard.

Adelard of Bath in all likelihood was the first to present a full version, or versions, of the *Elements* of Euclid in Latin and thus to initiate the process that led to Euclid's domination of high and late medieval mathematics. Prior to Adelard's translation (5a–5c) from the Arabic, the evidence exists that there were only grossly incomplete translations from the Greek, such as that of Boethius. Adelard's name is associated in twelfth-century manuscripts with three quite distinct versions. Version 1 (5a) is a close translation of the whole work (including the non-Euclidean Books XIV and XV) from the Arabic text, probably that of al-Hajjāj. No single codex contains the whole version, but on the basis of translating techniques and characteristic Arabicisms the text has been pieced together.[23] Only Book IX, the first thirty-five propositions of Book X, and the last three propositions of Book XV are missing.

The second treatment of the *Elements* bearing Adelard's name, Version II (5b), is of an entirely different character. Not only are the enunciations differently expressed but the proofs are very often replaced by instructions for proofs or outlines of proofs. It is clear, however, that this version was not merely a paraphrase of Version I but derives at least in part from an Arabic original since it contains a number of Arabicisms not present in Version I. It may be that Version II was the joint work of Adelard and his student Johannes Ocreatus or that Ocreatus revised it in some fashion since some manuscripts of Version II include a statement specifically attributed to "Joh. Ocrea," i.e., Ocreatus.[24] (In another work, addressed "to his master Adelard of Bath," Ocreatus' name is given as "N. Ocreatus.") It was Version II that became the most popular of the various translations of the *Elements* produced in the twelfth century. Apparently this version was the one most commonly studied in the schools. Certainly its enunciations provided a skeleton for many different commentaries, the most celebrated of which was that of Campanus of Novara, composed in the third quarter of the thirteenth century. Version II also provided the enunciations for Adelard's Version III (5c).

Version III does not appear to be a distinct translation but a commentary. Whether or not it is by Adelard, it is attributed to him and distinguished from his translation in a manuscript at the Bibliothèque National in Paris;[25] and judging from a twelfth-century copy at Oxford,[26] it was written prior to 1200. This version enjoyed some popularity and was quoted by Roger Bacon, who spoke of it as Adelard's *editio specialis.* Still another quasi commentary, consisting of a hodgepodge of geometrical problems, is found in a Florence manuscript, *Bachon Alardus in 10 Euclidis* (15).[27] It may be based in some way on a work of Adelard. Incidentally, the set of proofs for the *Elementa de ponderibus,* which were almost certainly composed by Jordanus de Nemore, is assigned in one manuscript to "Alardus."[28] Finally, in the area of geometry, note should be made of a thirteenth-century reference to a commentary on the *Spherica* of Theodosius, *Dicti Theodosii liber de speris, ex commentario Adelardi* (13), in the *Biblionomia* of Richard de Fournival.[29] No such work has been found, and the fact that the *Spherica* was translated only later by Gerard of Cremona makes it quite unlikely that Adelard did a commentary. The foregoing is an impressive list of geometrical translations and compositions; and, if by any chance, Bliemetzrieder should be proven correct concerning Adelard's role as the translator of the *Almagest* of Ptolemy, then the recently discovered translation from the Greek of the *Elements*[30] would also have to be assigned to Adelard since both translations exhibit identical translating techniques and styles.

The conclusion that must be drawn from the widespread translating activity described above is that Adelard should be considered, along with Gerard of Cremona and William of Moerbeke, as one of the pivotal figures in the conversion of Greek and Arabic learning into Latin.

NOTES

1. Bliemetzrieder, *Adelhard von Bath,* pp. 149–274.
2. Haskins, *Studies in Medieval Science,* ch. 2.
3. *De eodem,* edit. of Willner, p. 11, ll. 20–21.
4. *Quaestiones naturales,* edit. of Müller, ch. 58, p. 53.
5. *Ibid.,* ch. 4, p. 8.
6. *Ibid.,* ch. 6, p. 11.
7. Haskins, p. 28.

8. Brit. Mus., Royal MS 15.C.iv., Table of Contents.
9. Brit. Mus., Old Royal and King's Collections, MS 7.D.xxv.
10. Oxford, Bodleian Lib. MS Digby 68, 116r. The opening paragraphs are published in Richard Lemay, *Abu Ma'shar*, p. 355.
11. MS Lyons 328, 70r–74r, is among the extant MSS.
12. Edit. of Suter in Björnbo et al., ch. 4, p. 5.
13. Oxford, Corpus Christi Coll. MS 283, f. 142r.
14. Millás-Vallicrosa, *Nuevos estudios*, p. 107.
15. Cf. Cambridge, Fitzwilliam Mus., McClean MS 165, ff. 81r–88r, and Brit. Mus. Arundel MS 377, ff. 69r–74r.
16. *Liber ysagogarum*, edit. of Curtze, p. 18.
17. Cambridge Univ. Lib. MS Ii.6.5.
18. *Trattati d'aritmetica*, pp. 1–23.
19. *Mohammed ibn Musa Alchwarizmi's Algorismus.*
20. "Über ein Werk," pp. 1–63; cf. his earlier paper, in Russian, cited on p. 22, n. 2.
21. *Op. cit.*, p. 43.
22. *Op. cit.*, p. 22.
23. Clagett, "The Medieval Latin Translations," p. 18.
24. *Ibid.*, p. 21.
25. Paris, BN MS Lat. 16648, f. 58r.
26. Oxford, Balliol Coll. MS 257.
27. Biblioteca Nazionale Centrale Conv. Soppr. J.IX.26, 46r–55r.
28. Oxford, Corpus Christi Coll., MS 251, 10r–12v.
29. Haskins, p. 31.
30. Cf. Paris, BN MS Lat. 7377 and Florence, Biblioteca Nazionale Centrale Conv. Soppr. C.I.448.

BIBLIOGRAPHY

Among the works of Adelard of Bath available in modern editions and in manuscript form are the following: *De eodem et diverso*, edit. of H. Willner, in *Beiträge zur Geschichte der Philosophie des Mittelalters*, **4**, Heft 1 (1903); (?) *Liber ysagogarum Alchorismi in artem astronomicam a magistro A. compositus*, MSS Paris, BN Lat. 16208, ff. 67r–71r; Milan, Ambrosian Lib., A. 3 sup., ff. 1r–20r; Munich, Staatsbibliothek, Cod. 13021, ff. 27r–68v, Cod. 18927, ff. 31r *seq.*; Vienna, Nationalbibliothek, Cod. 275, f. 27r; first three books, edit. of M. Curtze, in *Abhandlungen zur Geschichte der Mathematik*, Heft 8 (1898), 1–27; *Quaestiones naturales*, edit. of M. Müller, in *Beiträge zur Geschichte der Philosophie und Theologie des Mittelalters*, **31**, Heft 2 (1934); *Regule abaci*, edit. of B. Boncompagni, in *Bullettino di bibliografia e di storia delle scienze matematiche e fisiche*, **14** (1881), 1–134.

For the texts of Adelard's translations and studies on his activities, see A. Björnbo, R. Besthorn, and H. Suter, *Die astronomischen Tafeln des Muhammed ibn Mūsā al-Khwārizmī in der Bearbeitung des Maslama ibn Ahmed al-Madjrītī* (Copenhagen, 1914); F. Bliemetzrieder, *Adelhard von Bath* (Munich, 1935); B. Boncompagni, *Trattati d'aritmetica, I. Algoritmi de numero Indorum* (Rome, 1857), pp. 1–23; M. Clagett, "The Medieval Latin Translations from the Arabic of the *Elements* of Euclid, with Special Emphasis on the Versions of Adelard of Bath," in *Isis*, **44** (1953), 16–42; C. H. Haskins, *Studies in the History of Mediaeval Science*, 2nd ed. (Cambridge, Mass., 1927), pp. 20–42; R. Lemay, *Abu Ma'shar and Latin Aristotelianism in the Twelfth Century* (Beirut, 1962), p. 355; and J. M. Millás-Vallicrosa, "La aportación astronómica de Pedro Alfonso," in *Sefarad*, **3** (1943), 65–105, and *Nuevos estudios*

sobre historia de la ciencia española (Barcelona, 1960), pp. 105–108; O. Neugebauer, *The Astronomical Tables of al-Khwārizmī. Translation with Commentaries of the Latin Versions edited by H. Suter supplemented by Corpus Christi College MS 283* (Copenhagen, 1962); T. Phillipps, "The *Mappae Clavicula*; a Treatise on the Preparation of Pigments During the Middle Ages," in *Archaeologia*, **32** (1847), 183–244; G. Sarton, *Introduction to the History of Science*, II (Baltimore, 1931), 167–169; L. Thorndike, *A History of Magic and Experimental Science*, II (New York, 1923), 19–49; K. Vogel, *Mohammed ibn Musa Alchwarizmi's Algorismus* (Aalen, 1963); A. P. Youschkevitch, "Über ein Werk des 'Abdallah Muhammad ibn Mūsā al-Huwārizmī al-Maǧusī zur Arithmetik der Inder," in *Beiheft 1964 zur Schriftenreihe Geschichte der Naturwissenschaften, Technik und Medizin*, pp. 1–63.

Marshall Clagett

ADET, PIERRE-AUGUSTE (*b.* Nevers, France, 17 May 1763; *d.* Paris, France, 19 March 1834), *chemistry.*

Although Adet was *docteur-régent* of the Faculty of Medicine in Paris, his life was devoted to politics rather than to science. He was deeply interested in chemistry but it was, nevertheless, only a spare-time pursuit, and he made no important contributions to it. In 1789, however, he did participate in the founding of the *Annales de chimie*, which was designed to permit easy publication of papers on antiphlogistic chemistry since, at that time, the *Journal de physique* was opposed to the new doctrines. Adet was one of the editors for several years and in this capacity published a number of translations of English papers in the journal and a few original works. He was, then, a keen supporter of the "new chemistry" from his early years; further evidence of this may be found in the appendix that he added to his translation of Priestley's *Considerations on the Doctrine of Phlogiston and the Decomposition of Water*, in which he replied to a number of Priestley's arguments (1797).

Adet was also interested in other reforms in chemistry and, to supplement the new system of chemical nomenclature that was being developed, he and Hassenfratz, Lavoisier's assistant, proposed a new system of chemical symbols. In it, a symbol indicated not merely the identity of the substance but its physical state, the proportion of oxygen it contained (as in sulfurous and sulfuric acids) and, if it was a salt, the extent to which the acid had been neutralized by the base. The system was never generally adopted, however, perhaps because of its complexity.

Adet's last published work before the Revolution, when he became much more involved in politics, was on stannic chloride (1789). He then became a colonial administrator and, while in Santo Domingo in 1791, investigated pineapple juice, in which he believed he had found both citric and malic acids. He could not confirm the presence of citric acid, however, because someone threw away his liquids.

In 1798 Adet investigated "acetous" and acetic acids. When verdigris is heated strongly, one of the products is a very concentrated acetic acid, which was known as "radical vinegar." No acid as concentrated as this could be obtained from vinegar, so it was thought that two acids existed—acetous acid (vinegar) and acetic acid (radical vinegar), which contained a higher proportion of oxygen. Adet was unable to oxidize "acetous acid" to acetic acid, but obtained acetic acid when he distilled "acetites" with concentrated sulfuric acid. He therefore concluded that the acids differed only in the proportion of water they contained. Although this conclusion was not widely accepted, it was confirmed by Proust in 1802.

After 1803, when he became prefect of the Nièvre, Adet seems to have published nothing of consequence apart from his textbook, *Leçons élémentaires de chimie* (1804). This had the distinction of being translated into modern Greek, but nonetheless was not a work of outstanding merit.

BIBLIOGRAPHY

All the articles translated from English by Adet are omitted from the following selection of his works, except those written by Priestley and Kirwan, to which Adet appended notes which show his attitude to the new doctrines of chemistry. The new system of chemical symbols devised by Adet and Hassenfratz is in De Morveau, Lavoisier, Berthollet, and Fourcroy, *Méthode de nomenclature chimique* (Paris, 1787), pp. 253–287 (N.B., some of the pages within this range are incorrectly numbered), immediately followed by a report by the Academy on this work, pp. 288–312. The English translation, *A Translation of the Table of Chemical Nomenclature* (London, 1799), contains as an appendix "Explanation of the Table of Symbols of Messrs. Hassenfratz and Adet; With the Additions and Alterations of the Editor." His text is *Leçons élémentaires de chimie* (Paris, 1804).

Adet's articles are "Lettre à M. Ingenhouz sur la décomposition de l'eau," in *Observations sur la physique*, **28** (1786), 436–439; "Lettre à M. de La Métherie," *ibid.*, **30** (1787), 215–218, written with Hassenfratz, and a reply by La Métherie, pp. 218–226; "Sur le muriate d'étain fumant ou liqueur de Libavius," in *Annals de chimie*, **1** (1789), 5–18; "'An Essai on Phlogiston and the Constitution of Acids,' Kirwan. Extrait de l'anglois avec des notes par P. A. Adet," *ibid.*, **7** (1790), 194–237; "Essai sur l'analyse du suc acide de l'ananas," *ibid.*, **25** (1798), 32–36; "Mémoire sur l'acide acétique," *ibid.*, **27** (1798), 299–319; and "Réflexions sur la doctrine du phlogistique et la décomposition de l'eau par J. Priestley etc., traduit de l'anglais et suivi d'une réponse, par P. Adet," *ibid.*, **26** (1798), 302–309; according to Partington (III, 244), this French translation was also published in Philadelphia in 1797.

Nothing in any detail exists on Adet as a chemist, but J. Balteau's article in *Dictionnaire de biographie française,* I (1933), 574–575, gives the background of his life as a politician.

E. McDonald

ADRAIN, ROBERT (*b.* Carrickfergus, Ireland, 30 September 1775; *d.* New Brunswick, New Jersey, 10 August 1843), *mathematics.*

Adrain was a teacher in Ireland and took part in the rebellion of 1798. With his wife, Ann Pollock, he escaped to America, where he first served as a master at Princeton Academy, then moved to York, Pennsylvania, as principal of the York County Academy. In 1805 he became principal of the academy in Reading, Pennsylvania. From 1809 to 1813 Adrain was professor of mathematics at Queen's College (now Rutgers), New Brunswick, New Jersey, and from 1813 to 1826 at Columbia College, New York. He then returned to Queen's College for a short while. He taught from 1827 to 1834 at the University of Pennsylvania in Philadelphia, where in 1828 he became vice-provost. From 1836 to 1840 he taught at the grammar school of Columbia College, after which he returned to New Brunswick. It is reported that in the classroom he often showed impatience with ill-prepared students. He had seven children, one of whom, Garnett Bowditch Adrain (1815–1878), was a Democratic member of Congress from New Brunswick between 1857 and 1861.

Adrain's first mathematical contributions were in George Baron's *Mathematical Correspondent* (1804), in which he solved problems and wrote on the steering of a ship and on Diophantine algebra. He continued the latter subject in *The Analyst* (1808), a short-lived periodical that he published himself. Here we find Adrain's most interesting mathematical paper, a study of errors in observations with the first two published demonstrations of the normal (exponential) law of errors. Gauss's work was not published until 1809. This volume also contains Adrain's paper on what he calls isotomous curves, inspired by Rittenhouse's hygrometer. If a family of curves (e.g., circles or parabolas) are all tangent at a point A, then an isotomous curve cuts these curves at equal arcs measured from A. Another article deals with the *catenaria volvens,* the form taken by a homogeneous, flexible, nonelastic string uniformly revolving about two points, without gravity.

Adrain shares with his contemporary Nathaniel Bowditch the honor of being the first creative mathematician in America. Like Bowditch, he was an ardent student of Laplace, and his paper on errors is in the spirit of Laplace.

Adrain became a member of the American Philosophical Society in 1812, and six years later he published in its *Transactions* a paper on the figure of the earth, in which he found 1/319 as its ellipticity (Laplace had 1/336; the modern value is 1/297). In the same issue of the *Transactions* he also published a paper on the mean diameter of the earth. Both papers were inspired by Laplace.

BIBLIOGRAPHY

I. ORIGINAL WORKS. Adrain's papers include "A Disquisition Concerning the Motion of a Ship Which Is Steered in a Given Point of the Compass," in *Mathematical Correspondent,* **1** (1804), 103–114; "Research Concerning the Probabilities of the Errors Which Happen in Making Observations," in *The Analyst,* **1** (1808), 93–109; "Researches Concerning Isotomous Curves," *ibid.,* 58–68; "Investigation of the Figure of the Earth and of the Gravity in Different Latitudes," in *Transactions of the American Philosophical Society,* n.s. **1** (1818), 119–135; and "Research Concerning the Mean Diameter of the Earth," *ibid.,* 352–366. He also contributed to *Portico,* **3** (1817); *Scientific Journal and Philosophical Magazine* (1818–1819); *Ladies and Gentleman's Diary* (1819–1822); and *The Mathematical Diary* (1825–1833), of which he edited the first six issues. In addition, Adrain prepared American editions of T. Keith, *A New Treatise on the Use of Globes* (New York, 1811); and C. Hutton, *Course in Mathematics* (New York, 1812).

II. SECONDARY LITERATURE. The most easily available source of information on Adrain is J. L. Coolidge, "Robert Adrain and the Beginnings of American Mathematics," in *American Mathematical Monthly,* **33** (1926), 61–76, with an analysis of Adrain's mathematical work. On his theory of errors, see also O. R. Seinin, "R. Adrain's Works in the Theory of Errors and Its Applications," in *Istoriko-matematicheskie issledovaniya,* **16** (1965), 325–336 (in Russian). An early source is an article in *United States Magazine and Democratic Review,* **14** (1844), 646–652, supposedly written by Adrain's son Garnett. See also G. E. Pettengill, in *Historical Review of Berks County* (*Penna.*), **8** (1943), 111–114; and D. E. Smith, in *Dictionary of American Biography,* I (1928), 109–110.

Coolidge mentions the existence of manuscript material of Adrain's on which M. J. Babb of Princeton was working. These papers seem to have been lost after Babb's death in 1945. The library of the American Philosophical Society has some letters by and concerning Adrain to John Vaughan in Philadelphia, and a letter written to Adrain by M. Roche in 1831.

D. J. STRUIK

ADRIAANSON, ADRIAAN. See **Metius, Adriaan.**

AEGIDIUS. See **Giles of Rome.**

AEPINUS, FRANZ ULRICH THEODOSIUS (*b.* Rostock, Germany, 13 December 1724; *d.* Dorpat, Russia [now Tartu, Estonian Soviet Socialist Republic], 10 August 1802), *mathematics, electricity, magnetism.*

Aepinus came from a family long distinguished for its learning. His great-grandfather, who had translated the family name, Hoeck, into Greek, had been an important evangelical theologian. His father held the chair of theology and his elder brother that of oratory at the University of Rostock. Aepinus studied medicine and mathematics at Jena, particularly under the guidance of G. E. Hamberger, and at Rostock, where he took his M.A. in 1747 with a dissertation on the paths of falling bodies. Until 1755 he taught mathematics at Rostock, as a junior lecturer, and published only on mathematical subjects: the properties of algebraic equations, the integration of partial differential equations, the concept of negative numbers. In 1751–1752 one of his auditors was J. C. Wilcke, who had come to Rostock to study under Franz's brother. With Franz's encouragement and instruction, Wilcke concentrated on physics and mathematics, and soon decided against the clerical career for which his father had intended him. A few years later Wilcke played an equally important role in reorienting his mentor's professional career.

In the spring of 1755 Aepinus became director of the observatory in Berlin and a member of the Academy of Sciences there. These appointments were apparently merely a device for establishing Aepinus, who had begun to acquire a reputation, in Frederick's capital: he was neither especially interested nor experienced in astronomy, and his closest published approach to the subject during his Berlin sojourn was a mathematical analysis of a micrometer adapted to a quadrant circle. His main preoccupation at the time was the study of the tourmaline, to which he was introduced by Wilcke, who had followed him to Berlin. Aepinus' first researches on the thermoelectric properties of this stone, which was then of extreme rarity, were fundamental. He recognized the electrical nature of the attractive power of a warmed tourmaline and attempted, not altogether successfully, to reduce its apparent capriciousness to rule. He was particularly struck by the formal similarity between the tourmaline and the magnet in regard to polarity, which inspired him to reconsider the possibility, then occasionally discussed, that electricity and magnetism were basically analogous. This thought became the theme for his masterwork, *Tentamen theoriae electricitatis et magnetismi* (1759).

In experimenting on the tourmaline Aepinus was often assisted by Wilcke, who was then preparing a dissertation on electricity. Their closeness made it natural for Wilcke to bring to Aepinus' attention certain phenomena he had discovered that apparently conflicted with Franklin's principles. In seeking an explanation, Aepinus came to the anti-Franklinian idea of a Leyden jar without the glass. The success of this air condenser eventually helped to persuade many to abandon Franklin's special assumptions about electrical atmospheres and the electricity of glass, and to prepare the ground for more general views of the kind Aepinus urged in his *Tentamen*.

In October 1756 Aepinus asked to be relieved of his positions in Berlin in order to accept the directorship of the observatory and the professorship of physics, vacant since the death of Richmann, at the Imperial Academy of St. Petersburg. Euler, with whom he boarded in Berlin, warmly recommended him for the job and interceded with Frederick to procure his release, which occurred in the spring of 1757. The Petersburg academicians expected that Aepinus, as befitted Richmann's successor, would continue to work on electricity. They were not disappointed. Late in 1758 Aepinus completed the lengthy *Tentamen*, which the Academy rushed into print before its author could finish his polishing.

The *Tentamen* is one of the most original and important books in the history of electricity. It is the first reasoned, fruitful exposition of electrical phenomena based on action-at-a-distance. Aepinus emphatically rejects the current notion of electrical atmospheres. Not that he believes that bodies act where they are not: he merely takes literally Newton's precepts about natural philosophy, and deduces the phenomena from certain assumed forces, without inquiring into the manner in which the forces themselves might be effected. Three such forces, according to him, create all the appearances of electricity: a repulsion between the particles of the electric fluid, an attraction between them and the corpuscles of common matter, and a repulsion between the corpuscles. This last is necessary to prevent unelectrified bodies—bodies with their normal complement of electrical fluid—from attracting one another. Aepinus observes that although such a repulsion might appear to conflict with universal gravitation there is no reason not to suppose several types of forces between matter corpuscles, and in fact the phenomena require it. As for the law of force, it is proportional to the excess or deficiency of fluid, and the same for all pairs of particles and corpuscles. Aepinus does not pretend

to know its precise form. Analogy, he thinks, favors the inverse square, which he uses in one numerical application; but generally he leaves the matter open, the great unanswered question in electrical theory.

Aepinus does not need the precise law, however, to explain the phenomena qualitatively. He is particularly successful with induction effects, which had puzzled philosophers since Canton's experiments of 1752; his explanations, with appropriate terminological changes, are essentially those used in elementary electrostatics today. Although his exposition is not quantitative, it is mathematical, with symbols used to indicate the excess or deficiency of fluid and the associated forces. Assuming that the forces decrease with distance, he is able to anticipate the direction of electrical interactions. In this way he predicts apparently paradoxical phenomena, e.g., that if two bodies with like charges of greatly different strengths are pushed together, their repulsion will at some point change to attraction. The magnetic theory of the *Tentamen* operates on the same principles, except that the magnetic fluid can freely penetrate all substances but iron, in which it is so tightly held that it can neither increase nor decrease. A piece of iron is thus to the magnetic fluid what a perfect insulator would be to the electric. All magnetic phenomena depend on the displacement of the magnetic fluid within iron. Aepinus' analysis of magnetization is exactly analogous to his treatment of electrical induction; it is adequate to all problems he considers except the formation of two magnets by the halving of one. Most notably it leads him to improve on Canton's and Michell's method of preparing artificial magnets, and on the usual disposition of armatures.

In 1760 or 1761 Aepinus became instructor to the Corps of Imperial Cadets, a position that left him too little time to fulfill his duties at the academy. The observatory was seldom used, and the equipment in the physics laboratory deteriorated. These circumstances gave Lomonosov the opportunity for a furious attack on Aepinus, whose haughtiness toward Russian scientists and quick preferment at court had already irritated him. Despite such unfavorable conditions, Aepinus continued for a few years to produce papers on various mathematical and physical subjects. He published the most important and coherent of these, several dissertations on the tourmaline, along with some criticism and corrections of his earlier work, as *Recueil des différents mémoires sur la tourmaline* (1762). Among the more occasional pieces, perhaps the most interesting are a masterful discussion of the mercurial phosphorus and a critical examination of Mayer's theory of magnetism, both of which appeared in the *Novi commentarii* of the Petersburg Academy for 1766–1767. About that time Aepinus' scientific activity ceased almost entirely. He became preceptor to the crown prince, a member of the prestigious Order of St. Anne, an educational reformer, a diplomat, a courtier, and finally a privy councillor. In 1798, after forty years in Russia, he resigned his offices and retired to Dorpat.

Except for his work on the tourmaline, which established a new subject, it is difficult to assess Aepinus' immediate influence. He had no distinguished students besides Wilcke. His contributions to mathematics, astronomy, and optics were competent but not outstanding. The *Tentamen* was at first not widely read. It was not easy to find (Beccaria had not seen a copy as late as 1772), and it was not easy to read (it demanded greater mathematical facility than most physicists then possessed). Although it was known and praised by Volta, Cavendish, and Coulomb, those physicists appear largely to have developed their own views before they came across it. But, less directly, the *Tentamen* was of great importance. Most of its content became easily available in 1780 in the excellent nonmathematical epitome composed by R. J. Haüy, who managed to preserve the spirit and clarity of the original. A much less adequate notice appeared in Priestley's *History*. Through such means the message of the *Tentamen* became widely diffused. Those who returned to the original then discovered in it a model for the application of mathematics to electricity and magnetism, and a store of apposite experiments. As one can see from P. T. Riess's *Die Lehre von der Reibungselektricität* (1853), the *Tentamen* remained an important source until the middle of the nineteenth century.

BIBLIOGRAPHY

I. ORIGINAL WORKS. Aepinus' most important works are "Mémoire concernant quelques nouvelles expériences électriques remarquables," in *Histoire de l'Académie Royale des Sciences de Berlin* (1756), 105–121; *Tentamen theoriae electricitatis et magnetismi* (St. Petersburg, 1759); *Recueil des différents mémoires sur la tourmaline* (1762); and the discussions of phosphorus and Mayer's theory of magnetism in *Novi commentarii* of the Imperial Academy (1766–1767). The best bibliography is in Poggendorff, to which should be added *Commentatio de notatione quantitatis negativae* (Rostock, 1754); and "Two Letters on Electrical and Other Phenomena," in *Transactions of the Royal Society of Edinburgh,* **2** (1790), 234–244. In addition, there are a few essays, in Russian, listed in Ia. G. Dorfman, ed., *Teoriia elektrichestva i magnetizma* (Moscow, 1951), a modern translation of the *Tentamen* and of Aepinus' contributions to the *Recueil*. Notes on Aepinus' lectures in Rostock, taken by Wilcke, are preserved in the library of the Swedish Academy of Sciences; other manuscripts may exist in the Soviet Union.

II. SECONDARY LITERATURE. Biographical information about Aepinus is sparse and scattered. The older, standard biographical entries are summarized and slightly expanded in W. Lorey's notice in *Allgemeine deutsche Biographie* and in H. Pupke, "Franz Ulrich Theodosius Aepinus," in *Naturwissenschaften,* **37** (1950), 49–52. For other data, see Euler's correspondence, particularly A. P. Youschkevitch and E. Winter, eds., *Die Berliner und die Petersburger Akademie der Wissenschaften im Briefwechsel Leonhard Eulers. I. Der Briefwechsel L. Eulers mit G. F. Müller* (Berlin, 1959); A. A. Morosow, *Michail Wassilyewitsch Lomonossow 1711–1765* (Berlin, 1954); and E. Winter, ed., *Die Registres der Berliner Akademie der Wissenschaften 1746–1766* (Berlin, 1957).

For Aepinus' work, see Dorfman's essay in *Teoriia* (above); Haüy's abridgment, *Exposition raisonée de la théorie de l'électricité et du magnetisme d'après les principes de M. Aepinus* (Paris, 1787); C. W. Oseen, *Johan Carl Wilcke. Experimental-fysiker* (Uppsala, 1939); Joseph Priestley, *The History and Present State of Electricity,* 2 vols., 3rd ed. (London, 1775); and P. T. Riess, *Die Lehre von der Reibungselektricität,* 2 vols. (Berlin, 1853).

JOHN L. HEILBRON

AËTIUS OF AMIDA (*b.* Amida, Mesopotamia [now Diyarbakir, Turkey], *fl. ca.* A.D. 540), *medicine.*

Aëtius had the title *comes obsequii,* which indicates that he had a relatively high rank, possibly of a military nature, at court. Since this title seems not to have been introduced until the reign of Justinian I, Aëtius cannot have lived before the sixth century. It is sometimes supposed that he was physician in ordinary at the Byzantine court, and this is occasionally stated as a fact both in books dealing with antiquity and in books on medical history. In any case, Aëtius lived after Oribasius, for the latter's medical encyclopedia is one of his main sources. Several times in his work Aëtius speaks of a sojourn in Alexandria. It cannot be proved that he was a Christian, for he does no more than mention Christian institutions and customs several times. In any event, he ought not to be confused with the physician and Arian Christian Aëtius who lived in the fourth century and is mentioned in Philostorgios' church history, as well as in Gregory of Nyssa's *Contra Eunomium.*

Aëtius wrote a large medical encyclopedia that is called either *Sixteen Medical Books* or *Tetrabibloi* (i.e., four volumes, each containing four parts or books). This form of medical encyclopedia, typical of late antiquity and the Byzantine period, corresponds to that of the known encyclopedias of Oribasius and Paul of Aegina. They are all collections of more or less verbatim excerpts from the works of previous medical authors, primarily Galen.

Aëtius' originality has often been questioned, but since there exists only an incomplete critical edition of his work (with proof of sources), the question cannot be answered conclusively. The Byzantine Photius stated that Aëtius had "added nothing and left out much" from his original sources, but this must be viewed skeptically, for there are indications that Aëtius evaluated his sources, using his own experiences and his own thoughts.

BIBLIOGRAPHY

The first eight books of Aëtius' encyclopedia have been critically edited by Alexander Olivieri in *Corpus medicorum Graecorum*, VIII, Part 1 (Berlin-Leipzig, 1935), and VIII, Part 2 (Berlin, 1950). The few notices on Aëtius' biography are in Olivieri, VIII, 1, p. 8, 11.14–15; his sojourn in Alexandria, *ibid.*, p. 65, 1.4, and p. 67, 1.1. Photius' statement on Aëtius' originality, *ibid.*, p. 8, 11.12 ff. For Aëtius' criticism of his predecessors see, e.g., *ibid.*, VIII, 1, p. 153, 11.16 ff. The most usable appreciation of Aëtius is by Ivan Bloch, in Max Neuburger and Julius Pagel, *Handbuch der Geschichte der Medizin*, I (Jena, 1902), 529 ff.

FRIDOLF KUDLIEN

IBN AFLAḤ. See **Jābir ibn Aflaḥ.**

AGARDH, CARL ADOLPH (*b.* Bastad, Sweden, 23 January 1785; *d.* Karlstad, Sweden, 28 January 1859), *botany.*

Agardh's fame is based on his contributions to the taxonomy of algae, but his scientific interests covered a far wider area. In many ways he reflects the philosophic romanticism that flourished when he was a professor at the University of Lund (1812–1835).

In Sweden, where Linnaeus had been active until 1778, as well as abroad, knowledge of algae and their classification was still rudimentary at the beginning of the nineteenth century. Linnaeus had divided the algae known to him into three families (*Fucus, Ulva, Conferva*), and after his death botanists continued to incorporate new forms into the same groups. In 1812 the French botanist Lamouroux took an important step toward a more comprehensive and natural differentiation, especially among the red algae, but a new understanding of the relationships existing within the larger groups of algae was first presented in Agardh's *Synopsis algarum Scandinaviae* (1817). Although it dealt basically with only one limited regional flora, the introduction presented an entirely new systematic survey of everything then considered algae. Agardh's broad outline became the *Species algarum* (1821–1828), which was never finished, and the more concentrated *Systema algarum* (1824), which summarized the state of algology at that time with precise groupings and clearly defined descriptions. In these works, with a collection of illustrations, he presented theories that are still considered nodal points in the development of algology. He achieved eminence partly through fieldwork, but he acquired a thorough knowledge of the literature and an extensive knowledge of various collections (among others, the herbaria of algae in Paris, which he examined in 1820–1821). It was not until 1827 that he undertook an extensive field trip to the north shore of the Adriatic Sea, where he became familiar with the little-known algal flora.

Although Agardh is remembered mainly as an algologist, he represents several of the main trends in botany at that time. He took an active part in contemporary discussions of the natural system of plant classification. Agardh presented his outline of the plant kingdom in his *Aphorismi botanici,* in the form of sixteen academic dissertations (1817–1826), and *Classes plantarum* (1825), in which he characterized several new plant families, some of which are still considered valid. His opinions reflect the views of nature developed by German Romantic *Naturphilosophen:* Schelling, Oken, and Nees von Esenbeck. Agardh, however, opposed the deductive, speculative method of the Romantics. He insisted that all study of nature had to be approached inductively, that it is not possible to establish a few groups within which all species can be classified. He believed that attention must be focused on the individual species and genera, which step by step, and with great care, might be arranged in larger groupings whose mutual relationships could be established only by further research. But in his plea for caution he expounded the Romantic *Naturphilosophie*. Nature is freedom; therefore it does not obey human logic, but its own logic, which cannot be penetrated by reason. Thus, no deductive, logically functioning system of classification can conform to the laws of nature.

Agardh's romanticism was even more pronounced in his writings on plant anatomy and plant physiology. On his way home from the Adriatic he stopped at the mineral springs of Karlsbad, where he met the Romantic philosopher Schelling. Together they studied algal forms in the hot springs, and Agardh demonstrated their life cycle. He later called his visit to Karlsbad "the most interesting days of my life"; it is evident that his interests here shifted from taxonomy to the problems of plant life. He first published his views on this subject in several articles in French, and then in a more extensive form in *Lärobok i botanik* (1830–1832), which was translated into German and dedicated to Schelling. The general tenor of the manual, and in particular the importance that Agardh attributes to chemistry, led to a violent disagreement with his friend Berzelius. Their animated correspondence on this subject reveals a strong contrast between a romantic, speculative temperament and an empirical one.

Agardh had divided personal aims and was often quite disturbed. He found himself increasingly at odds with his academic surroundings. He was, however, politically active, made important contributions to economics, and participated in pedagogical and theological debates.

In 1835 Agardh was offered the bishopric of the Karlstad diocese in western Sweden, which he accepted, and thereupon gave up writing on botany.

BIBLIOGRAPHY

I. ORIGINAL WORKS. A complete bibliography of Agardh's writings can be found in both J. E. Areschoug and A. B. Carlsson (see below). His works include *Synopsis algarum Scandinaviae* (Lund, 1817); *Aphorismi botanici* (Lund, 1817–1826); *Species algarum,* I, pt. 1 (Greifswald, 1821); I, pt. 2 (Lund, 1822); II, pt. 1 (Greifswald, 1828); *Systema algarum* (Lund, 1824); *Classes plantarum* (Lund, 1825); and *Lärobok i botanik* (Malmö, 1830–1832). His correspondence with Berzelius is in H. G. Soderbaum's ed. of Berzelius' letters, *Jac. Berzelius' brev,* X (Uppsala, 1925). His unpublished correspondence is mainly in the library of the University of Lund.

II. SECONDARY LITERATURE. Two articles on Agardh, both in Swedish, are J. E. Areschoug, "Carl Adolph Agardh," in *Levnadsteckningar över Kungliga Svenska Vetenskapsakademiens ledamöter,* I (Stockholm, 1869–1873); and A. B. Carlsson, *Svenskt biografiskt lexikon,* I (Stockholm, 1917–1918).

GUNNAR ERIKSSON

AGARDH, JACOB GEORG (*b.* Lund, Sweden, 8 December 1813; *d.* Lund, 17 January 1901), *botany.*

As the son of the prominent botanist Carl Adolph Agardh, Jacob had exceptional opportunities to acquire scientific experience at an early age. He was only fourteen when he accompanied his father on an important algological expedition to the Adriatic, and he showed both keen powers of observation and a marked aptitude for collecting. He also began to do research on phanerogams and cryptogams. Agardh followed in his father's footsteps and soon gained international renown as an algologist, specializing in sea algae. Agardh was basically a taxonomist, but his scientific approach was different. When Agardh began his scientific career, the conditions for research were better (through the improvement of microscopy) than in his father's time.

Agardh's first research on algae dealt with the germination process in some species. He studied their development and explained the nature of the swarm spores, which had previously been unclear. At the same time he began to observe how such external conditions as depth of water and currents influenced the appearance of the various species. This was of primary importance for the understanding of the taxonomic characteristics.

Agardh soon acquired firsthand knowledge of numerous types of algae by undertaking an extensive field trip to the Mediterranean and through herbarium studies in the large collections at the Muséum d'Histoire Naturelle in Paris, among others, which had also been studied by his father. His important *Algae maris Mediterranei et Adriatici* (1842) dealt with his new findings and contained his first work on the taxonomy of the *Florideae,* which was to become his most important field of research. The point of departure for his research on the differentiation of the classes of the *Florideae* was the structure of the reproductive organ and the structure of the cystocarp. Although his criteria are no longer considered definitive, many of the groups he differentiated are still valid. His *magnum opus* as an algae taxonomist was *Species, genera et ordines algarum,* in six volumes, published during the course of more than half a century (1848–1901). It contains all the then known species of the *Florideae* and all the known species of *Fucaceae* (brown algae) as well as their description and a general morphological survey, all in accordance with the Swedish Linnaean tradition. When the plant physiologist Julius Sachs published a critical appraisal of Linnaeus' contribution in his history of botany, Agardh was one of Linnaeus' most ardent defenders. Agardh's interest in taxonomy encompassed the entire plant kingdom, and he developed his ideas in *Theoria systematis plantarum* (1858). Here he reveals himself as an idealist, as were many leading scientists during the pre-Darwinian era. He interprets the natural relationship between various genera, as well as between other taxa, not as a phylogenetic one but as one dependent on the premise that all genera within a family reflect the same prototype, that is, a pattern according to which the Creator worked when He created the various species.

Following the trend of his contemporaries, Agardh combined idealism with certain evolutionary beliefs. Hence, he considered that each species had evolved from a lower to a higher state and had developed through the ages into different and progressively more perfect forms. He definitely dismissed the thought that one species could develop into another, and thereby denied the theory of the origin of species that has constituted the nucleus of the philosophy of evolution that originated with Darwin. Still less did he concur with the materialistic approach to life that was often expressed by Darwin's followers, for nature, to him, was a harmonious whole, the development of which had been planned from the beginning and had been directed by an omnipotent and omniscient Creator.

As have those of most other taxonomists, many of Agardh's concepts have become obsolete, especially the more general ones. For his descriptions of species, he had only pressed and dried material in herbaria; thus, his ideas about species have often had to be revised. He introduced many new ideas to algology, however, and was active in developing an increasingly keener systematic and morphological approach. His algae collection, which had been started by his father, was given to the University of Lund, where Agardh had been active as a teacher since 1834 and professor of botany from 1854 to 1879. The collection is one of the most varied in the world and contains many type specimens—an indication of the importance of his work.

BIBLIOGRAPHY

I. ORIGINAL WORKS. Complete bibliographies of Agardh's writings are in J. Ericksson, "Jacob Georg Agardh," in *Levnadsteckningar över Kungliga Svenska Vetenskapsakademiens ledamöter,* V, pt. 2 (Stockholm, 1915–1920); and N. Svedelius, in *Svenskt biografiskt lexikon,* I (Stockholm, 1917–1918), 268–274. Among his works are *Algae maris Mediterranei et Adriatici* (Paris, 1842); *Species, genera et ordines algarum,* 6 vols. (Lund, 1848–1901); and *Theoria systematis plantarum* (Lund, 1858). Most of his letters and MSS are in the library of the University of Lund.

II. SECONDARY LITERATURE. Writings on Agardh are the articles of Ericksson and Svedelius cited above and G. B. de Toni, "G. G. Agardh e la sua opera scientifica," in *La nuova notarisia,* **17** (1902), 1–28.

GUNNAR ERIKSSON

AGASSIZ, ALEXANDER (*b.* Neuchâtel, Switzerland, 17 December 1835; *d.* mid-Atlantic, 27 March 1910), *zoology, oceanography, engineering.*

Alexander Agassiz was the son of Louis Agassiz and Cécile Braun Agassiz, the sister of the botanist Alexander Braun. From 1847, after his father departed for

America, until 1849, when he went to Cambridge, Massachusetts, following the death of his mother, he lived at Freiburg im Breisgau, where he came under the influence of his uncle. In America he soon formed a lasting bond with his stepmother, Elizabeth Cary Agassiz, and moved naturally into a scientific career. Agassiz graduated from Harvard College in 1855, from the Lawrence Scientific School with a degree in engineering in 1857, and again from the Lawrence Scientific School with a degree in zoology in 1862. After a short career in the U.S. Coast Survey in 1859, he became his father's assistant at the Museum of Comparative Zoology, which he continued to serve for the rest of his life, chiefly as its director.

In 1866 Agassiz undertook, on behalf of himself and a brother-in-law, the management of the Calumet and Hecla copper mines in the Upper Peninsula of Michigan. By 1869, although he had impaired his health, he had laid the basis for a fortune that he plowed into scientific research, both by gifts to Harvard and the museum and by freeing himself from a conventional career in either teaching or business. After 1873 (when his wife, Anna Russell, whom he had married in 1860, died within a few days of his father) his life consisted of a regular round of research in the tropics in the winter, summers at his laboratory near Newport, Rhode Island, and stays in Cambridge and Michigan each fall and spring. Although his fortune and his benefactions place him first among those late nineteenth-century captains of industry who supported science in the United States, he was distinguished as both a zoologist and an oceanographer. He died while crossing the Atlantic from England to America.

Although usually reticent about large theoretical schemes, in 1860 Agassiz spoke in private letters in terms that were closer to the theories of his father about the geographical distribution of animals than to the ideas of Charles Darwin which were sweeping through the American scientific community (including among their adherents most of Louis Agassiz's own students). By 1872, when Agassiz visited the British exploring ship *Challenger* at Halifax, Nova Scotia, he impressed its naturalists, including Sir John Murray, as holding views quite different from his father's. His work from 1860 to the late 1870's was largely concerned with the study of zoology, beginning with the animals of the New England shore, especially the echinoderms, and culminating in his *Revision of the Echini* (1872–1874). Using the embryological and paleontological approach of his father, he produced a masterly work that belonged to the era of Darwin, writing that it "is astonishing that so little use has been made of the positive data furnished by embryology in support of the evolution hypothesis." He also worked up the echinoderms from the *Challenger* expedition.

In 1877 Agassiz's interest began to shift to deep-sea dredging for abyssal fauna. Using his engineering background to good advantage and his wealth to support both operations and publications, he began with three cruises of the Coast Survey steamer *Blake* in the Caribbean. In 1891 he explored the deep water of the Pacific from the Galápagos Islands to the Gulf of California in the Fish Commission steamer *Albatross.* His aim in this period was to make a comparative study of marine fauna on both sides of the Isthmus of Panama. His interest from 1892 onward shifted strongly to the problem of the formation of coral atolls. Questioning the universality of Darwin's theory of atoll formation by subsidence, he used his knowledge of the Caribbean and Hawaiian islands as a basis of comparison. In 1893 and 1894 he explored the Bahama and Bermuda islands, in 1896 the Great Barrier Reef, in 1897 the Fijis, in 1898–1900 the central Pacific, and in 1900–1902 the Maldives. The publications of his later years were usually reports of the various voyages; a general work on coral reefs was never finished. Agassiz's later work is as close to modern oceanography and marine zoology as his earlier work was to that of his father.

BIBLIOGRAPHY

A list of Alexander Agassiz's published writings appears in George Lincoln Goodale, "Biographical Memoir of Alexander Agassiz 1835–1910," in National Academy of Sciences, *Biographical Memoirs,* VII (Washington, D. C., 1912), 291–305. Manuscripts, letter books, incoming letters, and many photographs are preserved at the Museum of Comparative Zoology, Harvard University.

The preeminent biographical source is George R. Agassiz, ed., *Letters and Recollections of Alexander Agassiz With a Sketch of His Life and Work* (Boston-New York, 1913).

A. HUNTER DUPREE

AGASSIZ, JEAN LOUIS RODOLPHE (*b.* Motier-en-Vuly, Switzerland, 28 May 1807; *d.* Cambridge, Massachusetts, 14 December 1873), *ichthyology, geology, paleontology.*

Louis Agassiz, the son of Rodolphe and Rose Mayor Agassiz, grew to manhood enjoying the prosperity and status of his family and the natural beauty of the Swiss cantons of Fribourg, Vaud, and Neuchâtel. He never identified with a sectarian religious persuasion. He did embrace the Protestant pietism of his minister father, but was more fundamentally devoted to an idealistic romanticism that saw the power of the Creator exemplified in all flora and fauna. The Agassiz and Mayor families were anxious to see Louis succeed in the world of commerce or medicine, but he triumphed over their opposition and entered the larger world of European scholarship and cosmopolitanism by attending the universities of Zurich, Heidelberg, and Munich. In 1829 he earned his doctorate in philosophy at the universities of Munich and Erlangen and published a monograph on the fishes of Brazil that brought him to the attention of Baron Georges Cuvier. In 1830 he earned the doctor of medicine degree at Munich. After studying under Cuvier's tutelage in Paris, Agassiz accepted a professorship at the newly established College of Neuchâtel in 1832. In the same year he married Cécile Braun, the sister of his Heidelberg classmate Alexander Braun. In 1846 he accepted an invitation to lecture at the Lowell Institute in Boston. On the death of his wife in 1847, he accepted a professorship at the Lawrence Scientific School of Harvard University, where he continued to teach until his death. Agassiz's decision to make the United States his permanent home—despite attractive offers to return to Europe—was influenced by his love for and marriage to Elizabeth Cabot Cary. From 1850 until 1873 she raised Agassiz's three children by his first wife and acted as a constant companion in the writing, exploration, and interpretation of natural history.

Agassiz's career had two distinct geographic and intellectual aspects. As a European, he published monographs on ichthyology, paleontology, and geology whose promise earned him the admiration of such established savants as Cuvier, Alexander von Humboldt, and Sir Charles Lyell. As an American, Agassiz made nature study popular and appealing, explored the American environment with great enthusiasm, and established lasting institutions of research and education. His robust attitude toward life and nature study was a perpetual passion that tolerated no opposition to plans he deemed vital. Agassiz demanded unquestioning loyalty, and repaid such dedication by deep love and devotion. His dedication to science and culture won him the admiration of statesman and commoner alike, although his reputation among fellow scientists diminished with the passing of time. His exceptionally strong constitution sustained him on journeys of exploration through central Europe, the Swiss Alps, the eastern United States and the trans-Mississippi West, and South America. In 1873, shortly after an expedition through the Strait of Magellan, Agassiz died of a cerebral hemorrhage. Among his numerous awards and honors were the Wollaston Medal of the Geological Society of London and the Copley Medal of the Royal Society of London.

Agassiz thought of himself primarily as a naturalist, generalizing about the entire range of organic creation. Nevertheless, it is the modern sciences of ichthyology, geology, and paleontology that bear the stamp of his contributions. In the middle decades of the nineteenth century, when the natural sciences were in transition from classical to evolutionary biology, Agassiz's work and career were typical. He had an insatiable desire to record data; he described and analyzed material significant for the study of marine biology, freshwater fishes, embryology, and fossil fishes. In this last realm, his *Poissons fossiles,* written directly in the tradition of his mentor Cuvier, contained precise descriptions of more than 1,700 ancient species, together with illustrated reconstructions based on principles of comparative anatomy. This pioneer effort was a model of exactitude, providing future students with primary data relating zoology to geology and paleontology.

Agassiz never viewed his work in paleoichthyology as providing a framework for conceptions of natural history related to the development of lower forms into higher ones. He insisted that ancient and modern species were permanent representations of a divine idea, and bore no genetic relationship to each other. While employing techniques of close empirical study learned from such teachers as Cuvier and Ignaz Döllinger, Agassiz affirmed a view of the world above and beyond experience. In this sense, he reflected the teaching of Lorenz Oken and Friedrich Schelling. These diverse influences in Agassiz's intellectual history make it impossible to separate his contributions to exact science from his philosophy of nature. He worked in two divergent traditions, and his efforts reflected the virtues and deficiencies of each. This is why evolutionists found Agassiz so mystifying an opponent and why the Swiss naturalist found their views to be mere restatements of ideas absorbed and partly rejected in his youth.

These divergent qualities were reflected in Agassiz's geological investigations. From 1835 to 1845, while still serving as a professor at Neuchâtel, Agassiz studied the glacial formations of Switzerland and compared them with the geology of England and central Europe. The resulting concept of the "Ice Age" was remarkable for its breadth of generalization and for the exacting field study represented. Agassiz held that in the immediately recent past there had been an era during which large land masses over much of northern Europe were covered with ice. With the onset of warming periods, the recession of the ice was responsible for upheaval and subsidence. The marks of glaciers could be discerned in the scratched and polished rocks as well as in the configurations of the earth in glaciated regions. Glacial movement was responsible for modern geological configurations, and could be traced in such areas as Switzerland. Agassiz was not the first to observe the phenomena of glaciation, but he was innovative in the wide-ranging character of his research, his measurement of ice formations, and his elaboration of local geology into a theory explaining Continental natural history. Such events, now known to have been of greater cyclical duration than Agassiz asserted, were still sufficient to convince such naturalists as Darwin and Lyell that Pleistocene glaciation was a primary mechanism in causing the geographical distribution and consequent genetic relationship of flora and fauna otherwise inexplicably separated by land and water masses. But Agassiz could never accept such a conclusion. He interpreted glaciation in metaphysical terms. To him, the Deity had been responsible for the Ice Age, a catastrophe that provided a permanent physical barrier separating the species of the past from those of the present era. There were as many as twenty separate creations in the history of the earth, each distinguished by animal and plant forms bearing no relationship to present types. At best, paleontology could only provide a glimpse of those "prophetic types" that suggested the course of future development, while those forms that remained unchanged over time were evidence of the wisdom of the Creator in inspiring perfect creatures from the beginning. Agassiz extended his conception of natural history to include mankind, asserting that men, like other animals, were of distinct types or species and were marked by different physical and intellectual traits. In the United States of the pre-Civil War years, such ideas provided convenient rationalizations for defenders of the slave system.

Agassiz's visit to the United States in 1846 was a notable success, for the brilliant young naturalist described his adventures and communicated his love of nature to lecture hall audiences in Boston and other eastern cities. He had also come to compare the natural history of the Old World with that of America, but this temporary purpose soon vanished in the adulation he received from all classes of Americans. Agassiz found the natural environment fascinating, and after accepting the Harvard professorship, he determined to explore it and interpret it to his new countrymen. In 1855 he announced a grand plan for the publication of a monumental ten-volume study, *Contributions to the Natural History of the United States,* that would depict the full scope of the

American natural environment. Only four volumes appeared; and these, although magnificently illustrated, were valuable only for their descriptions of North American turtles. The work was at once too complicated for the general public and too descriptive for those naturalists increasingly interested in new theoretical conceptions identifiable with the work of Charles Darwin.

Agassiz was philosophically and scientifically unprepared to meet the challenge of the theory of evolution as it was propounded in 1859. During his early years in the United States he extended his glacial theory to North America, he explored large portions of the country, and conducted some potentially valuable research in marine biology. More than all these efforts, it was the collection of the raw data of nature that drove Agassiz ever onward, so that Harvard University became a center for natural history instruction and research. The capstone of such efforts was the establishment at Harvard College in 1859 of the Museum of Comparative Zoology, an institution made possible by private gifts and funds supplied by the state of Massachusetts. The museum always bore the impress of Agassiz's conception of the relationship between graduate instruction, research, fieldwork, and publication, centered in an institution of higher learning and supported by private philanthropy and public funds.

It was inevitable that Agassiz became the leading American opponent of Darwin, but regrettable that his public activity left little time for reflection on the data he had collected or on alternate interpretations of its significance. Agassiz had become a public man in the fullest sense, but even had he devoted more time to intellectual labor, it is doubtful that he could have accepted an interpretation of nature that seemed to deny permanence and immaterialism. Some of his critiques of evolution were trenchant ones, but in the main his attacks were inconclusive efforts that failed to convince his scientific colleagues. Many of these appeared in popular journals, reflecting Agassiz's conviction that this "error" had to be opposed with the full power of his public position. While Agassiz's opposition to evolution was inconsequential, the years from 1859 to his death were nevertheless a period of notable public accomplishment. He was able to obtain more than $600,000 in public and private support for the Harvard museum, and to convince fellow scientists to establish the National Academy of Sciences in 1863. This achievement, coupled with his earlier efforts to advise the federal government on the operations of the U.S. Coast Survey and the Smithsonian Institution, revealed Agassiz in the prime of his American influence and international prestige.

By 1873, despite Darwin, Agassiz's name was synonymous with the study of natural history. It was fitting that in that last year of his life he established the Anderson School of Natural History on Penikese Island, off the Massachusetts coast, as a combined summer school and marine biological station. In testimony to Agassiz's American influence, the faculty of the school was entirely composed of his former students. The *Poissons fossiles* and *Études sur les glaciers* were high points of Agassiz's career in Europe; in America, the life and work of such students as William James, David Starr Jordan, Alexander Agassiz, Frederick Ward Putnam, and Nathaniel Southgate Shaler exemplify his role and cultural significance.

BIBLIOGRAPHY

I. ORIGINAL WORKS. Bibliographies of Agassiz's writings are in his *Bibliographia zoologiae et geologiae,* 4 vols. (London, 1848–1854), I, 98–103; Jules Marcou, *Life, Letters, and Works of Louis Agassiz* (see below), II, 258–303; and Max Meisel, *A Bibliography of American Natural History,* 3 vols. (New York, 1924–1929), *passim.* Among his significant works are *Selecta genera et species piscium quas in itinere per Brasiliam 1817–1820 . . .* (Munich, 1829); *Recherches sur les poissons fossiles,* 5 vols. (Neuchâtel, 1833–1844); *Monographies d'échinodermes vivans et fossiles . . .,* 4 vols. (Neuchâtel, 1838–1842); *Études sur les glaciers* (Neuchâtel, 1840); *Twelve Lectures on Comparative Embryology* (Boston, 1849); *Lake Superior* (Boston, 1850); *Contributions to the Natural History of the United States,* 4 vols. (Boston, 1857–1862); *Essay on Classification* (London, 1859), also ed., with intro., by Edward Lurie (Cambridge, Mass., 1962); *Geological Sketches* (Boston, 1866); "Evolution and Permanence of Type," in *Atlantic Monthly,* **33** (Jan. 1874), 94–101; and *Geological Sketches, Second Series* (Boston, 1876).

II. SECONDARY LITERATURE. Works on Agassiz are Elizabeth Cary Agassiz, ed., *Louis Agassiz, His Life and Correspondence,* 2 vols. (Boston, 1885); Lane Cooper, *Louis Agassiz as a Teacher,* rev. ed. (Ithaca, N.Y., 1945); Edward Lurie, *Louis Agassiz: A Life in Science* (Chicago, 1960); Jules Marcou, *Life, Letters, and Works of Louis Agassiz,* 2 vols. (New York, 1896); and Ernst Mayr, "Agassiz, Darwin and Evolution," in *Harvard Library Bulletin,* **13** (Spring 1959), 165–194.

EDWARD LURIE

AGATHINUS, CLAUDIUS (*fl. ca.* A.D. 50), *medicine.*

Agathinus was a Spartan physician who lived in Rome, where he was connected with the family of the Stoic philosopher L. Annaeus Cornutus. His association with known Stoics was certainly not fortuitous, since he is known to have been a practitioner of the Pneumatic school of medicine founded by Athenaeus of Attalia under the influence of Poseidonius, another Stoic philosopher. Although the identity of Agathinus' medical teacher is not known, it was certainly not Athenaeus of Attalia (*fl.* 50 B.C.) himself. Agathinus founded his own school, which he called episynthetic (i.e., eclectic). In direct opposition to the schismatic spirit of Imperial Roman medicine, the episynthetic school championed the intellectual unity of medicine as interpreted by Galen.

Agathinus had many medical disciples, of whom the best known was the celebrated Archigenes, who described his teacher's scientific attitude: "Therefore Agathinus—who was particular about everything and never relied on mere eclecticism, but for safety's sake always required empirical verification—administered hellebore (elleborus) to a dog, which therewith vomited" (see Oribasius in *Corpus medicorum Graecorum*, VI, part 1, 1 [Leipzig–Berlin, 1928], 252).

None of Agathinus' original writings survive intact, but antique sources mention a work on the pulse, dedicated to his pupil Herodot; a work on fever, especially the kind he called "semitertian fever"; and, finally, a work on hellebore. All of these show, therefore, that Agathinus was neither a narrow specialist nor a dubious charlatan, as were so many of his colleagues in Imperial Rome. On the contrary, although only scant information about Agathinus' life and a few fragments of his writings remain, it is apparent that he was among the really important and influential physicians of the intellectually rich first century A.D.

BIBLIOGRAPHY

Information on Agathinus' personality and the episynthetic school can be found in G. Kaibel, ed., *Epigrammata Graeca* (Berlin, 1878), no. 558; and in Galen's *Works*, ed. C. G. Kühn, XIX (Leipzig, 1830), 353. His associations in Rome are discussed in Suetonius; see *Suetonii Reliquiae*, ed. A. Reifferscheid (Leipzig, 1860), p. 74, with textual correction by Osann. Agathinus' disciples are discussed in Galen; see *Corpus medicorum Graecorum*, V, part 10, 2, 2 (Berlin, 1956), 86. His writing on the pulse is mentioned in Kühn's edition of Galen's *Works*, VIII (Leipzig, 1824), 749 ff.; on fever, in *Corpus medicorum Graecorum*, V, part 10, 1 (Leipzig, 1934), 62; and on the hellebore, in Caelius Aurelianus, *Celeres vel acutae passiones*, Bk. 3, sec. 135. Agathinus' influence is assessed in Galen, *Über die medizinischen Namen*, M. Meyerhof and J. Schacht, eds. (Berlin, 1931), p. 10. To be used, with some reservations, is

M. Wellmann, *Die pneumatische Schule bis auf Archigenes* (Berlin, 1895), pp. 9 ff.

Fridolf Kudlien

AGNESI, MARIA GAETANA (*b.* Milan, Italy, 16 May 1718; *d.* Milan, 9 January 1799), *mathematics.*

Maria Gaetana Agnesi, the first woman in the Western world who can accurately be called a mathematician, was the eldest child of Pietro Agnesi and Anna Fortunato Brivio. Her father, a wealthy Milanese who was professor of mathematics at the University of Bologna, encouraged his daughter's interest in scientific matters by securing a series of distinguished professors as her tutors and by establishing in his home a cultural salon where she could present theses on a variety of subjects and then defend them in academic disputations with leading scholars. Agnesi invited both local celebrities and foreign noblemen to his soirées. During the intermissions between Maria Gaetana's defenses, her sister, Maria Teresa, a composer and noted harpsichordist, entertained the guests by playing her own compositions.

In all her discourses at these gatherings, Maria Gaetana demonstrated her genius as a linguist. At age five she spoke French fluently. At age nine, she translated into Latin, recited from memory, and released for publication a lengthy speech advocating higher education for women. By age eleven, she was thoroughly familiar with Greek, German, Spanish, and Hebrew. The disputations were conducted in Latin, but during the subsequent discussions a foreigner would usually address Maria in his native tongue and would be answered in that language. The topics on which she presented theses covered a wide range—logic, ontology, mechanics, hydromechanics, elasticity, celestial mechanics and universal gravitation, chemistry, botany, zoology, and mineralogy, among others. Some 190 of the theses she defended appear in the *Propositiones philosophicae* (1738), her second published work.

Although the 1738 compilation does not contain any of Agnesi's purely mathematical ideas, various other documents indicate her early interest in mathematics and her original approach to that subject. At fourteen she was solving difficult problems in analytic geometry and ballistics. Her correspondence with some of her former tutors indicates that, as early as age seventeen, she was beginning to shape her critical commentary on the *Traité analytique des sections coniques* of Guillaume de L'Hospital, a leading mathematician of the Newtonian era. The manuscript material that she prepared, although judged excellent by all the professors who examined it, was never published.

In 1738, after the publication of the *Propositiones philosophicae,* Agnesi indicated that the constant public display of her talents at her father's gatherings was becoming distasteful to her, and she expressed a strong desire to enter a convent. Persuaded by her father not to take that step, she nevertheless withdrew from all social life and devoted herself completely to the study of mathematics. In the advanced phases of the subject she was guided by Father Ramiro Rampinelli, a member of the Olivetan order of the Benedictines, who later became professor of mathematics at the University of Pavia. A decade of concentrated thought bore fruit in 1748 with the publication of her *Istituzioni analitiche ad uso della gioventù italiana,* which she dedicated to Empress Maria Theresa of Austria. This book won immediate acclaim in academic circles all over Europe and brought recognition as a mathematician to Agnesi.

The *Istituzioni analitiche* consisted of two huge quarto volumes containing more than a thousand pages. Its author's objective was to give a complete, integrated, comprehensible treatment of algebra and analysis, with emphasis on concepts that were new (or relatively so) in the mid-eighteenth century. In this connection one must realize that Newton was still alive when Agnesi was born, so that the development of the differential and integral calculus was in progress during her lifetime. With the *gioventù* (youth) in mind, she wrote in Italian rather than in Latin and covered the range from elementary algebra to the classical theory of equations, to coordinate geometry, and then on to differential calculus, integral calculus, infinite series (to the extent that these were known in her day), and finally to the solution of elementary differential equations. She treated finite processes in the first volume and infinitesimal analysis in the second.

In the introduction to the *Istituzioni analitiche,* Agnesi—modest as she was, with too great a tendency to give credit to others—had to admit that some of the methods, material, and generalizations were entirely original with her. Since there were many genuinely new things in her masterpiece, it is strange that her name is most frequently associated with one small discovery which she shared with others: the formulation of the *versiera,* the cubic curve whose equation is $x^2 y = a^2 (a - y)$ and which, by a process of literal translation from colloquial Italian, has come to be known as the "witch of Agnesi." She was apparently unaware (and so were historians until recently) that

Fermat had given the equation of the curve in 1665 and that Guido Grandi had used the name *versiera* for it in 1703.

Agnesi's definition of the curve may be stated as follows: If C is a circle of diameter a with center at $(O, 1/2\ a)$, and if the variable line OA through the origin O intersects the line $y = a$ at point A and the circle at point B, then the *versiera* is the locus of point P, which is the intersection of lines through A and B parallel to the Y axis and X axis, respectively. The curve, generated as the line OA turns (Latin *vertere,* hence the name *versiera*), is bell-shaped with the X axis as asymptote. There are interesting special properties and some applications in modern physics, but these do not completely explain why mathematicians are so intrigued by the curve. They have formulated a *pseudo-versiera* by means of a change in the scale of ordinates (a similarity transformation). Even Giuseppe Peano, one of the most formidable figures in modern axiomatics and mathematical logic, could not resist the temptation to create the "*visiera* of Agnesi*,*" as he called it, a curve generated in a fashion resembling that for the *versiera.*

The tributes to the excellence of Agnesi's treatise were so numerous that it is impossible to list them all, but those related to translations of the work will be noted. The French translation (of the second volume only) was authorized by the French Academy of Sciences. In 1749 an academy committee recorded its opinion: "This work is characterized by its careful organization, its clarity, and its precision. There is no other book, in any language, which would enable a reader to penetrate as deeply, or as rapidly, into the fundamental concepts of analysis. We consider this treatise the most complete and best written work of its kind."

An English translation of the *Istituzioni analitiche* was made by John Colson, Lucasian professor of mathematics at Cambridge, and was published in 1801 at the expense of the baron de Masères. In introducing the translation, John Hellins, its editor, wrote: "He [Colson] found her [Agnesi's] work to be so excellent that he was at the pains of learning the Italian language at an advanced age for the sole purpose of translating her book into English, that the British Youth might have the benefit of it as well as the Youth of Italy."

The recognition of greatest significance to Agnesi was provided in two letters from Pope Benedict XIV. The first, dated June 1749, a congratulatory note on the occasion of the publication of her book, was accompanied by a gold medal and a gold wreath adorned with precious stones. In his second letter, dated September 1750, the pope appointed her to the chair of mathematics and natural philosophy at Bologna.

But Agnesi, always retiring, never actually taught at the University of Bologna. She accepted her position as an honorary one from 1750 to 1752, when her father was ill. After his death in 1752 she gradually withdrew from all scientific activity. By 1762 she was so far removed from the world of mathematics that she declined a request of the University of Turin to act as referee for the young Lagrange's papers on the calculus of variations.

The years after 1752 were devoted to religious studies and social work. Agnesi made great material sacrifices to help the poor of her parish. She had always mothered her numerous younger brothers (there were twenty-one children from Pietro Agnesi's three marriages), and after her father's death she took his place in directing their education. In 1771 Agnesi became directress of the Pio Albergo Trivulzio, a Milanese home for the aged ill and indigent, a position she held until her death.

BIBLIOGRAPHY

I. ORIGINAL WORKS. Agnesi's main works are *Propositiones philosophicae* (Milan, 1738) and *Analytical Institutions,* an English translation of the *Istituzioni analitiche* by the Rev. J. Colson (London, 1801).

II. SECONDARY LITERATURE. Further information about Agnesi and her work may be found in L. Anzoletti, *Maria Gaetana Agnesi* (Milan, 1900); A. F. Frisi, *Elogio storico di Domina Maria Gaetana Agnesi milanese* (Milan, 1799); and A. Masotti, "Maria Gaetana Agnesi," in *Rendiconti del seminario matematico e fisico di Milano,* **14** (1940), 1–39.

EDNA E. KRAMER

AGRICOLA, GEORGIUS, also known as **Georg Bauer** (*b.* Glauchau, Germany, 24 March 1494; *d.* Chemnitz, Germany [now Karl-Marx-Stadt, German Democratic Republic], 21 November 1555), *mining, metallurgy.*

Agricola's father was probably Gregor Bauer, a dyer and woolen draper. His youngest son, Hans, Georg's favorite brother, who joined Georg at Chemnitz in 1540, followed the same profession. The eldest son, Franciscus, became a priest at Zwickau and later at Glauchau. Georg attended various schools in Glauchau, Zwickau, and Magdeburg (1511), and in 1514—rather late, since the average age at matriculation was between twelve and fifteen—he entered Leipzig University. In 1515 he received the B.A. and remained at the university as lecturer in elementary Greek until he was chosen *ludi moderator* at Zwickau in 1517. In 1519, as *rector extraordinarius,* he organized the new Schola Graeca and wrote his first work, *De prima ac simplici institutione grammatica* (1520). This short booklet is an excellent specimen of the new humanistic pedagogy, with interesting examples taken from a schoolboy's experiences.

Zwickau was a center of the Reformation, and although Agricola believed a reformation was necessary, he did not approve of its revolutionary aspects. He therefore returned to Leipzig in 1523 to study medicine under Heinrich Stromer von Auerbach; to support himself, he had been endowed with the prebend of the St. Erasmus altar for three years by the council of Zwickau. This enabled him to visit Italy, and on his way he stopped in Basel to pay his respects to Erasmus. Agricola spent three years at Bologna and Venice as a member of the editorial staff for the Aldina editions of Galen and Hippocrates. He also joined the English group headed by Edward Wotton and John Clement, son-in-law of Sir Thomas More. This group may have aroused Agricola's interest in politics and economics.

Following the route through the mining districts in Carinthia, Styria, and the Tyrol, Agricola returned to Germany in the fall of 1526 with the M.D. and a wife, the widow of Thomas Meiner, director of the Schneeberg mining district. The following spring he was elected town physician and apothecary of St. Joachimsthal (now Jáchymov), Czechoslovakia. Here he continued his studies on the pharmaceutical use of minerals and smelting products, with a view to compiling comments on Galen and Hippocrates.

In those days St. Joachimsthal was the most important mining center in Europe besides Schwaz in the Tyrol. Miners and smelters, some of whom suffered from occupational diseases, were crowded together. Agricola studied not only their ailments but also their life, labor, and equipment. Day and night he visited the mines and the smoky smelting houses, and soon he had an excellent knowledge of mining and metallurgy. He recorded his impressions in *Bermannus sive de re metallica dialogus* (1530).

The success of this pioneer delineation of mining and metallurgy was assured by Erasmus, who contributed a letter of recommendation. Agricola was now a well-known author, and he indefatigably sustained his reputation with a flow of important books. The next ones were political and economic: *Oratio de bello adversus Turcam suscipiendo* (1531) and *De mensuris et ponderibus* (1533).

Since there were too many demands on his time in St. Joachimsthal, Agricola decided to return to Chemnitz, to be town physician in this quieter, smaller town on the northern slope of the Erzgebirge. Chemnitz had a copper smelter which was used to extract silver from the ore. Agricola's knowledge of mining enabled him to profit from mining shares. He always seemed to enter into the right partnership and to avoid profitless ventures. By 1542 he was one of the twelve richest inhabitants of Chemnitz. After fifteen years of hard work he succeeded in finishing a complete series of inquiries concerning the principles of geology and mineralogy.

This series must be considered his greatest scientific achievement. It had not yet been published when Agricola became involved in the war of Emperor Charles V against the Protestant Schmalkaldic League: he was elected mayor of Chemnitz, appointed a councillor to the court of Saxony, and sent as an ambassador to the emperor and his younger brother Ferdinand, king of Bohemia. For more than three years Agricola was with the councillors of Moritz, duke of Saxony, as one of the few Roman Catholic representatives at the Protestant court. He never wrote about the diplomatic missions he was charged with, but we may assume that his parleys with the Catholic emperor's commanders and diplomats were effective. He was not able to return to his scientific work until 1548, but new books appeared soon after: *De animantibus subterraneis* (1549) and an enlarged edition of *De mensuris et ponderibus* (1550).

In 1550 Agricola returned to St. Joachimsthal for some weeks. He saw a very changed situation: the prosperity was gone, nearly all of the ruling family had been deposed or expelled, and some of the new royal officials had not the slightest idea of the needs of the town and its inhabitants. Agricola gave a 5,000-thaler credit—worth 2,000 cows in those days—to the counts Schlick to promote prospecting for new deposits, a search that was successful. He went home to Chemnitz conscious of having done a good deed, and with him he took the finished text of his chief work, *De re metallica libri XII,* begun twenty years before in St. Joachimsthal. During his visit to St. Joachimsthal he had met the expert designer Blasius Weffring, who spent the next three years illustrating the text.

When the black plague spread through Saxony in 1552–1553, Agricola worked day and night, going from the pesthouses to his family, always fearing that he would bring the contagion with him; one daughter did die of the plague. His first wife had died in 1541, and the following year he had married Anna Schütz, daughter of the guild master and smelter owner Ulrich

Schütz, who had entrusted his wife and children to Agricola's guardianship when he died in 1534. His studies during the plague led Agricola to publish *De peste libri III* (1554).

Agricola could not retire until another work was finished. In 1534 Georg the Whiskered, duke of Saxony and a patron of the Catholic church, had nominated Agricola as historiographer of the court of Saxony, probably with the hope of discovering genealogical claims on territories by heirs-at-law. For twenty years Agricola studied yellowed parchments and old chronicles. His honesty forbade him to conceal the rulers' mistakes uncovered during his research: he was a scholar, not a courtier. He recorded his findings very frankly—much to the disappointment of Augustus, third duke after Georg the Whiskered. It is no wonder, then, that the *Sippschaft des Hausses zu Sachssen,* an evaluation of all the rulers of Saxony, remained unpublished until 1963.

Augustus ignored the dedication dated 9 August 1555, but more important is that of 18 March 1555, for the second, enlarged edition of the mineralogical works (1558). It contains Agricola's most quoted words on peace and war, written before the Peace of Augsburg (September 1555), when war between the Catholic and Protestant confessions seemed imminent. For Agricola, that decisive agreement was the end of all his hopes for a reunion in faith. He fell ill soon after and died after suffering a relapse in November.

After Agricola's death the religious struggle renewed over his corpse. The Protestant clergy refused to allow his being buried in the parish church at Chemnitz, an honor traditionally accorded to mayors; and it was only through the intervention of his old friend Julius von Pflug, bishop of Zeitz-Naumburg, that he was interred in the cathedral at Zeitz.

Four months after his death, *De re metallica libri XII,* illustrated with 292 woodcuts, appeared. A year later an Old German translation by Philippus Bech was published using the same woodcuts, which were used for 101 years in seven editions.

BIBLIOGRAPHY

I. ORIGINAL WORKS. Agricola's writings include *De prima ac simplici institutione grammatica* (Leipzig, 1520); *Bermannus sive de re metallica dialogus* (Basel, 1530; Paris, 1541); *Oratio de bello adversus Turcam suscipiendo*, original Latin ed. (Basel, 1538), in Old German as *Oration, Anrede und Vermanung . . . widder den Türcken*, Lorenz Bermann, trans. (Dresden–Nuremberg, 1531); *De mensuris et ponderibus* (Basel-Paris, 1533; Venice, 1535), reissued as *De mensuris et ponderibus Romanorum atque Graecorum libri*

V (Basel, 1550), with the following additions: *De externis mensuris et ponderibus, Brevis defensio, De mensuris quibus intervalla metimur, De restituendis mensuris atque ponderibus,* and *De precio metallorum et monetis;* the big foliant, containing *De ortu et causis subterraneorum, De natura eorum quae effluunt e terra, De natura fossilium, De veteribus et novis metallis, Bermannus sive de re metallica dialogus* (revised), and *Interpretatio Germanica vocum rei metallicae addito indice foecundissimo* (Basel, 1546; 2nd ed., rev. and enl., Basel, 1558)—*De natura fossilium* was translated into English by Mark C. Bandy and Jean A. Bandy (New York, 1955); *De animantibus subterraneis* (Basel, 1549); *De peste libri III* (Basel, 1554); *Sippschaft des Hausses zu Sachssen* (1555), in *Ausgewählte Werke,* VII, 77–416; and *De re metallica libri XII* (Basel, 1556, 1561, 1621, 1657), translated into English, with biographical introduction, annotations, and appendixes by Henry Clark Hoover and Lou Henry Hoover (London, 1912; new ed. [unchanged], New York, 1950).

Nearly all of Agricola's works are brought together in *Ausgewählte Werke,* 12 vols., incl. supps., Hans Prescher, ed. (Berlin, 1955–   ).

II. SECONDARY LITERATURE. Works on Agricola are Bern Dibner, *Agricola on Metals,* Burndy Library Publication no. 15 (Norwalk, Conn., 1958); Erwin Herlitzius, *G. A. Seine Weltanschauung und seine Leistung als Wegbereiter einer materialistischen Naturauffassung,* Freiberger Forschungsheft no. D32 (Berlin, 1960); William B. Parsons, *Engineers and Engineering in the Renaissance* (Baltimore, 1939); Georg Spackeler, ed., *Georgius Agricola 1555–1955* (Berlin, 1955); and Helmut M. Wilsdorf, *Präludien zu Agricola,* Freiberger Forschungsheft no. D5 (Berlin, 1954); *Georg Agricola und seine Zeit* (Berlin, 1956); and "Dr. Georgius Agricola und die Begründung des Bergbaumedizin," in *Jahrbuch des Museums für Mineralogie und Geologie Dresden,* **5** (1959), 112–154.

HELMUT M. WILSDORF

AGRIPPA, HEINRICH CORNELIUS, also known as **Agrippa von Nettesheim** (*b.* near Cologne, Germany, 14 September 1486; *d.* Grenoble, France, *ca.* 18 February 1535), *magic, alchemy, philosophy, medicine.*

Agrippa's father, Heinrich von Nettesheim, was a citizen of Cologne; nothing is known of his mother. Agrippa's surname and epithet indicate both his birthplace (Cologne was formerly Colonia Agrippina) and the origin of his family (Nettesheim, a village near Cologne); his given names suggest a Dutch or Flemish influence. Agrippa married three times. His first wife, who came from Pavia and was married to him in 1514, died in 1518 in Metz. They had a son, Theodoricus, who was born in 1515 and died in 1522. Six children were born to his second wife, Jeanne Loyse Tissie, whom he married in Geneva in 1521; she died in 1528. A third union, apparently unhappy, took place the following year.

Agrippa enrolled at the University of Cologne on 22 July 1499. While there he studied law, medicine, magic sciences, and theology—particularly under Peter Ravenna. He also served in the army of Emperor Maximilian I for several years. At the age of twenty he made his first trip to Paris to study; he then went, again in military service, to Catalonia, and finally to Dôle, where he gave lectures on Johann Reuchlin's *De verbo mirifico.* In 1510 he spent a short time in London where he stayed with John Colet, the friend of Erasmus, and then he returned to Cologne, where he held theological disputations. That same year, in Würzburg, he met Johannes Trithemius, the abbot of St. Jacob's monastery. This was probably the most important meeting of Agrippa's life, for Trithemius encouraged him to finish the *De occulta philosophia.* Following this, Agrippa led a restless, roving life throughout Europe, especially in Italy. Among the places he visited were Milan, Pisa, Pavia (where in 1515 he expounded Hermetic writings), and Turin (where he taught theology), sometimes as an independent rhetorician, sometimes in military service.

In 1518 Agrippa was a public advocate in Metz and the defense lawyer in a sorcery trial; the latter service aroused such opposition that he had to leave town. He then went to Geneva via Cologne and became a physician. During 1523 and 1524 he was a salaried town physician in Fribourg, Switzerland. After 1524 he was at the court of Francis I in Lyons, where he was personal physician to the queen mother and court astrologer. He was always in monetary difficulties and constantly being dunned by his creditors.

In 1528 Margaret of Austria, the regent of the Netherlands, summoned Agrippa to become historian and librarian in Antwerp. Two years later he published his polemic *De incertitudine et vanitate scientiarum atque artium declamatio et de excellentia verbi Dei,* which he had begun to draft while still in Lyons. In 1531 he published the first of the three books of the *De occulta philosophia* (the fourth book is apocryphal), which had probably been written around 1510–1515. After the death of Margaret he returned, via Brussels and Cologne, to Lyons, where he was often persecuted because of his writings. He died in great poverty.

Agrippa's personality and *curriculum vitae* are still open to dispute, as is the authorship of his works. He has been described as an "honest, fearless, and generous man, . . . but somewhat vainglorious . . . , whereby he himself several times spoiled his chances at success" and also as a scientific swindler. Today Agrippa's importance is considered to lie in the social criticism that is embodied in his works on magic as well as in his polemic against the vanity and uncertainty of science. He has his *De occulta* and *De incertitudine* to thank not only for his fame but also for the doubt cast upon his having been a scientist. For a long time historians lumped him together with Reuchlin and even with Ramón Lull, for he attempted to combine Neoplatonic mysticism and magic—subject to nature—with Renaissance skepticism. Recent historical investigation does not support this view, however, and assigns him a central place in the history of ideas of the Middle Ages; he is seen as characterizing the main line of intellectual development from Nicholas of Cusa to Sebastian Franck. Modern opinion evaluates him on the basis of his Platonic, Neoplatonic, and Hermetic influences—primarily in the *De occulta philosophia*—without insisting on his skepticism.

The basic idea of Agrippa's *De occulta philosophia* is that from the void God had created several worlds, three of which constitute the All: the domain of the elements, the heavenly world of the stars, and the intelligible cosmos of the angels. These and the things existing in them are endowed with the *spiritus mundi* (the soul, the fifth element, the *quinta essentia* in the sense of the Aristotelian "ether"), which is set above the four classical elements. This spirit of the world represents the all-germinating force (comparable to the "germ-form" of the Stoics). At the center of these three worlds is man, who, because he is a microcosm and thus represents a mirror image of the macrocosm, can obtain knowledge of everything. The effectiveness of magic, according to Agrippa, is based on the connection of the three worlds. Only the human spirit can uncover the hidden forces present in matter, and by the latter's aid man can also call on greater forces to serve him. What Agrippa meant by this becomes evident in his small work *De triplici ratione cognoscendi Deum* (1516), in which the role of the cabala as intermediary step in his system signifies that true knowledge is to be found only in the love of God.

Although Agrippa was an admirer of Luther, he understood the *verbum Dei* as a Catholic; in one letter to Melancthon he called Luther the invincible heretic. Although this aspect of his thought is often neglected, it occupies the key position in his polemic on the arts and sciences, *De incertitudine*. This work gives emphasis to the tension between the *verbum Dei* and human knowledge, without providing any basis for the skepticism of which Agrippa has often been accused. Rather, at the beginning of the era of natural science, it is one of the first testimonials to knowledge of the limits of human understanding. *Incertitudo* here means a real uncertainty of existence, based on the concept of the human being as a created entity.

The question of why the otherwise critical Agrippa published nearly simultaneously two such opposing works as *De occulta philosophia* and *De incertitudine* remains open. In the former he appears to follow the metaphysical and speculative tradition of natural philosophy, while in the latter he attempts to overcome the magic of the *verbum mirificum*. There is no satisfactory explanation for this, a fact of which even Agrippa himself was aware. With a Faustian restlessness (he is considered the historical prototype of Goethe's Faust) he always returns to this theme in his letters; posterity has often considered this a fault in his character. Such a conflict is representative of Agrippa's age, however, and demonstrates a point of view widely held in Germany during the Renaissance.

BIBLIOGRAPHY

I. ORIGINAL WORKS. Agrippa's writings are collected in his *Opera omnia,* 2 vols. (Lyons, n.d.; 2nd ed., Lyons, 1600). During his lifetime thereof was edited: *De occulta philosophia,* 3 vols. (Vol. I, Antwerp, 1531; complete ed., Cologne, 1533); *De incertitudine et vanitate scientiarum atque artium declamatio* (Antwerp, 1530; Cologne, 1531); *Liber de triplici ratione cognoscendi Deum* (1516); *In artem brevem Raymundi Lulli commentaria* (Cologne, 1533). Not in *Opera omnia:* "Contra pestem antidoton," in P. Poitier, *Insignes curationes . . . et observationes centum,* Vol. I (Cologne, 1625).

II. SECONDARY LITERATURE. Works on Agrippa are M. H. Morley, *The Life of Henry Cornelius Agrippa von Nettesheim,* 2 vols. (London, 1856); Auguste Prost, *Les sciences et les arts occultes aux XVIe siècle: Corneille Agrippa, sa vie et ses oeuvres,* 2 vols. (Paris, 1881–1882); J. Orsier, *Henri Cornelis Agrippa, sa vie et son oeuvre d'après sa correspondance* (Paris, 1911); A. Reichl, "Goethes Faust und Agrippa von Nettesheim," in *Euphorion,* **4** (1897) 287–301; G. Ritter, "Ein historisches Urbild zu Goethes Faust (Agrippa von Nettesheim)," in *Preussische Jahrbuecher,* **141**, no. 2 (1910), 300–324; J. Meurer, "Zur Logik und Metaphysik des Heinrich Cornelius Agrippa von Nettesheim," in *Renaissance und Philosophie, Beiträge zur Geschichte der Philosophie,* Adolf Dryoff, ed., Vol. XI (Bonn, 1920); E. Hahn, "Die Stellung des H. C. Agrippa von Nettesheim in der Geschichte der Philosophie," diss. (Munich-Leipzig, 1923); R. Stadelmann, "Zweifel und Verzweiflung bei Agrippa von Nettesheim," pp. 80–86 of *Vom Geist des ausgehenden Mittelalters,* Vol. XV of the series *Deutsche Vierteljahrsschrift für Literaturwissenschaft und Geistesgeschichte* (Halle, 1929); E. Cazalas, "Les sceaux planétaires de C. Agrippa," in *Revue de l'histoire des religions* **110** (1934), 66–82; E. Metzke, "Die 'Skepsis' des Agrippa von Nettesheim," in *Deutsche Vierteljahrsschrift für Literaturwissenschaft und Geistesgeschichte,* **13** (1935), 407–420; L. Thorndike, *A History of Magic and Experimental*

Science, V (New York, 1941), 127–138; H. Grimm, *Neue deutsche Biographie,* I (Berlin, 1953), 105–106; Charles G. Nauert, Jr., *Agrippa von Nettesheim, His Life and Thought,* diss. (University of Illinois, 1955), and "Agrippa in Renaissance Italy; the Esoteric Tradition," in *Studies in the Renaissance,* **6** (1959), 195–222; P. Zambelli, "Umanesimo magico-astrologico et raggruppamenti segreti nei platonici della preriforma," in *Umanesimo e esoterismo,* Enrico Castelli, ed. (Padua, 1960), 141–174; R. Schmitz, and K. U. Kuhlmay, "Zum Handschriftenproblem bei Agrippa von Nettesheim," in *Sudhoffs Archiv für Geschichte der Medizin und der Naturwissenschaften,* **46** (1962), 350–354; R. Schmitz, "Agrippa von Nettesheim und seine Bemerkungen ueber die Wirkungen der Magie in Medizin und Pharmazie," in *Pharmazeutische Zeitung,* **110** (1965), 1131–1138; Charles G. Nauert, Jr., "Agrippa and the Crisis of Renaissance Thought," in *Illinois Studies in the Social Sciences,* **55** (Urbana, Ill., 1965); and G. Rudolph, "'De incertitudine et vanitate scientiarum,' Tradition und Wandlung der wissenschaftlichen Skepsis von Agrippa von Nettesheim bis zum Ausgang des 18. Jahrhunderts," in *Gesnerus,* **23**, no. 3/4 (1966), 247–265.

R. SCHMITZ

AGUILON, FRANÇOIS D' (*b.* Brussels, Belgium, 1546; *d.* Antwerp, Belgium, 1617), *physics, mathematics.*

The son of the secretary to Philip II, Aguilon became a Jesuit in 1586. After having taught syntax and logic, then theology, he was charged with organizing in Belgium the teaching of the exact sciences, which were useful in commerce, geography, navigation, and architecture, as well as military activities. This project led to the composition of a master treatise on optics that synthesized the works of Euclid, Ibn al-Haytham (Alhazen), Vitellion, Roger Bacon, Pena, Ramus (Pierre de la Ramée), Risner, and Kepler. Its organization into three sections was determined by the manner in which the eye perceives objects (directly, by reflection on polished surfaces, and by refraction through transparent bodies). Aguilon's death prevented the publication of the second and third sections, on catoptrics, dioptrics, and telescopes. Only the first part exists, with six frontispieces drawn by Rubens: *Francisci Aguilonii e Societate Jesu Opticorum libri sex juxta ac mathematicis utiles* (1613).

Aguilon treated, successively, the eye, the object, and the nature of vision; the optic ray and horopter; the general ideas that make possible the knowledge of objects; errors in perception; luminous and opaque bodies; and projections.

The sixth book, on orthographic, stereographic, and scenographic projections, remains important in the history of science. It accounts for a third of the treatise and was meant for the use of astronomers, cosmographers, architects, military leaders, navigators, painters, and engravers. It places particular emphasis on stereographic projection—a type of projection, used by Ptolemy, in which the portion of the sphere to be represented is projected from the pole onto the plane of the equatorial circle.

The balance of the treatise is of interest for the history of optics: description of the eye; controversies on the nature of light and its action; the application of mathematics to optics; the analysis of the concepts of distance, quantity, shape, place, position, continuity, discontinuity, movement, rest, transparency, opacity, shadow, light, resemblance, beauty, and deformity; and explanation of the various errors of perception linked to distance, size, position, shape, place, number, movement, rest, transparency, and opacity.

Book 5, in spite of an Aristotelian concept of light, studies the propagation of light, the limit of its action, the phenomena produced by the combinations of light sources, and the production of shadows. Aguilon proposes an experimental apparatus, drawn by Rubens, that made it possible to study the variations of intensity according to variations in distance and to compare lights of different intensities. This attempt to apply mathematics to the intensity of light was continued by Mersenne, then by Claude Milliet de Chales, and resulted in Bouguer's photometer.

BIBLIOGRAPHY

Aguilon's only work is *Francisci Aguilonii e Societate Jesu Opticorum libri sex juxta ac mathematicis utiles* (Antwerp, 1613; Würzburg, 1685; Nuremberg, 1702).

Writings on Aguilon or his work are P. Alegambe, *Bibliotheca scriptorum Societatis Jesu* (Antwerp, 1643), p. 112; A. de Backer, *Bibliothèque des écrivains de la Compagnie de Jésus* (Liège, 1853); Michel Chasles, *Aperçu historique sur l'origine et le développement des méthodes en géométrie* (Brussels, 1837), pp. 222, 517; F. V. Goethals, *Histoire des lettres, des sciences et des arts en Belgique et dans les pays limitrophes,* I (Brussels, 1840), 149, 153; J. E. Morère, "La photométrie: Les sources de l'Essai d'optique sur la gradation de la lumière de Pierre Bouguer, 1729," in *Revue d'histoire des sciences,* **18**, no. 4 (1965), 337–384; L. Moréri, *Dictionnaire historique* (Paris, 1749), I, 231; V. G. Poudra, *Histoire de la perspective ancienne et moderne* (Paris, 1864), pp. 68–70; Adolphe Quetelet, *Histoire des sciences mathématiques chez les Belges* (Brussels, 1864), pp. 192–198; E. Quetelet, "Aïguillon" [*sic*], in *Biographie nationale,* I (Brussels, 1866), 140–142; and C. Sommervogel, *Bibliothèque de la Compagnie de Jésus,* I (Louvain, 1890), 90.

J. E. MORÈRE

AḤMAD IBN IBRĀHĪM AL-UQLIDĪSĪ. See **al-Uqlidīsī.**

AḤMAD IBN MUḤAMMAD IBN ʿABD AL-JALĪL AL-SIZJĪ. See **al-Sizjī.**

AḤMAD IBN MUḤAMMAD IBN AL-BANNĀʾ. See **Ibn al-Bannāʾ.**

AḤMAD IBN MŪSĀ IBN SHĀKIR. See **Banū Mūsā.**

AḤMAD IBN YŪSUF (*b.* Baghdad, Iraq [?]; *fl. ca.* 900–905; *d.* Cairo, Egypt, 912/913 [?]), *mathematics.*

Aḥmad ibn Yūsuf ibn Ibrāhīm ibn al-Dāya al-Miṣrī was the son of an Arab scholar, Yūsuf ibn Ibrāhīm. Yūsuf's home was in Baghdad, but in 839/840 he moved to Damascus, and later to Cairo; hence his son was known as an Egyptian. Aḥmad's birth date is not known, although it seems probable that he was born before the move to Damascus. His death date is likewise in doubt, although the most probable date is 912/913.

Aḥmad's father, sometimes referred to as *al-ḥāsib* ("the reckoner"), was one of a group of learned and influential men. A work on the history of medicine, another on the history of astronomy, and a collection of astronomical tables are attributed to him, although no written work of his survives today.

In Egypt, Aḥmad ibn Yūsuf was a private secretary to the Ṭūlūn family, which ruled Egypt from 868 to 905. In his writing, Aḥmad made several references to one Hudā ibn Aḥmad ibn Ṭūlūn. This was probably Abu'l-Baqāʾ Hudā, the thirteenth son of Aḥmad ibn Ṭūlūn, and probably Aḥmad ibn Yūsuf's employer.

Aḥmad ibn Yūsuf wrote a treatise on ratio and proportion, a work on similar arcs, a commentary on Ptolemy's *Centiloquium,* and a work on the astrolabe. All the works survive in Arabic manuscript, and all but the work on the astrolabe exist in Latin translation. While it is impossible to distinguish absolutely the work of the father from that of the son, there seems to be little doubt of Aḥmad's authorship of the above four works. A number of other works are attributed to him, but these cannot be authenticated.

Aḥmad's most significant work is the treatise on ratio and proportion. This was translated from the Arabic into Latin by Gerard of Cremona and then extensively copied. Manuscript copies of the Latin version exist today in at least eleven libraries in England, Spain, Austria, France, and Italy, thus testifying to the wide interest in the treatise in medieval times. Arabic versions of the work are in manu-

script form in Cairo and Algiers libraries. The work is largely an expansion of and commentary on Book V of Euclid's *Elements.* Aḥmad developed and expanded Euclid's definitions of ratio and proportion in a long dialectic argument. Having clarified the meaning of these terms, he went on to show in great detail various methods for finding unknown quantities from given known quantities when the knowns and unknowns existed in certain proportional relationships.

By applying the Euclidean definitions of composition, separation, alternation, equality, and repetition to the given proportional relationships, Aḥmad found eighteen different cases: six when there are three different quantities in the proportion, eight when there are four quantities, and four when there are six. The discussion and geometrical interpretation of these eighteen cases form the nucleus of the treatise. Since many of his proofs referred to variations on a single triangular figure, later authors have referred to his work as the eighteen cases of the divided figure.

Besides his obvious dependence on Euclid, Aḥmad acknowledged his indebtedness to Ptolemy. The latter part of the treatise on ratio and proportion is actually an extension of two lemmas from Book I, chapter 13, of Ptolemy's *Almagest.* Aḥmad also made reference to, and quoted from, Archimedes, Hero, Plato, Empedocles, and Apollonius, indicating that he was acquainted with at least some of their works.

Writing as he did at the beginning of the tenth century, not only was Aḥmad ibn Yūsuf profoundly influenced by his Greek predecessors, but also in his turn he exerted an influence on the works of several medieval mathematicians. Leonardo Fibonacci, in his *Liber abacci,* mentioned the work of Aḥmad (Ametus in the Latin form) in the eighteen cases of proportion, and he used Aḥmad's methods in the solution of tax problems. Some traces of Aḥmad's influence have been seen in the work of Jordanus de Nemore, *Arithmetica in decem libris demonstrata.* Aḥmad was cited as an authority by Thomas Bradwardine in his differentiating between continuous and discontinuous proportions. Pacioli listed Aḥmad (Ametus), along with such well-known scholars as Euclid, Boethius, Jordanus, and Bradwardine, as one of those whose work on proportions was of major significance.

On the somewhat negative side, Aḥmad was guilty of a grave logical error. Campanus of Novara, in his commentary on the definitions of Book V of Euclid's *Elements,* devoted considerable attention to Aḥmad's method of proof and pointed out a subtle but real bit of circular reasoning. In his eagerness to establish definitions and postulates, Aḥmad did, at one point in his treatise, accept as a postulate a principle that he later was to prove as a theorem. This logical error does not detract from the value of his careful classification and solution of the various cases of proportional quantities. In fact, it is for this that he is remembered: his eighteen cases of the divided figure.

BIBLIOGRAPHY

Latin MSS of the *Epistola de proportione et proportionalitate* are in Paris, Bibliothèque Nationale, MS Lat. 9335, ff. 64r–75v; Florence, Biblioteca Medicea-Laurenziana, MS San Marco 184, ff. 90r–112v; and Vienna, Oesterreichische Nationalbibliothek, MS 5292, ff. 158r–179v. Arabic MSS, with the title *Risāla fi 'l-nisba wa 'l-tanāsub,* are in Algiers, MS 176 R. 898e + 684, ff. 54r–73r; and Cairo, National Library, MS 39 Riyāḍa mīm, ff. 1–25r.

Works containing information on Aḥmad ibn Yūsuf are Abū Muḥammad ʿAbd Allāh ibn Muḥammad al-Madīnī al-Balawī, *Sīrat Aḥmad ibn Ṭūlūn,* Muḥammad Kurd ʿAlī, ed. (Damascus, 1939); C. Brockelmann, *Geschichte der arabischen Litteratur,* supp. I (1937), 229, and I, 2nd ed. (1943), 155; George Sarton, *Introduction to the History of Science,* I, 598; M. Steinschneider, "Iusuf ben Ibrahim und Ahmed ben Iusuf," in *Bibliotheca mathematica* (1888), 49–117, esp. 52, 111; H. Suter, "Die Mathematiker und Astronomen der Araber und ihre Werke," in *Abhandlungen zur Geschichte der mathematischen Wissenschaften,* **10** (1900), 42–43; and Yāqūt, *Irshād al-arīb ilā maʿrifat al-adīb,* D. S. Margoliouth, ed., II (Leiden, 1909), 157–160.

Dorothy V. Schrader

AḤMAD IBN YŪSUF AL-TĪFĀSHĪ. See **al-Tīfāshī.**

AIDA YASUAKI, also known as **Aida Ammei** (*b.* Yamagata, Japan, 10 February 1747; *d.* Edo [now Tokyo], Japan, 26 October 1817), *mathematics.*

Aida studied mathematics under Yasuyuki Okazaki in Yamagata when he was fifteen. When he was twenty-two, he went to Edo, determined to become the best mathematician in Japan, and worked as a field supervisor of engineering, river improvement, and irrigation under the Edo shogunate. His co-workers in the civil service included Teirei Kamiya, who was one of the ablest disciples of the famous mathematician Sadasuke Fujita. Aida wanted to become Fujita's pupil, and asked Kamiya for an introduction. Fujita did not receive Aida as a pupil, however, perhaps because of a falling-out occasioned by Fujita's pointing out mistakes in the problems inscribed on a tablet donated to a temple by Aida (these tablets, called *sangaku,* were hung on the walls of shrines and temples by recognized mathematicians as votive offerings—they further served as an exhibition of scholarship and as a supplement to textbooks).

Aida then devoted his efforts to composing and publishing his *Kaisei sampo* (1781), in which he criticized and revised Fujita's highly regarded *Seiyo sampo* of 1781. Kamiya accordingly lost face, because he had introduced Aida to Fujita who then was insulted by him; he retaliated by publicly pointing out the faults in Aida's book. Kamiya's criticism of Aida initiated a series of polemics that, conducted in private correspondence and in more than ten published mathematical works, lasted for the next twenty years.

In this dispute Naonobu Ajima, who was a friend of Fujita, sided with Kamiya. Ajima and Fujita had both been pupils of Nushizumi Yamaji, a master of the Seki school, and the private feud was thus transformed into a rivalry between the Seki school of mathematicians and the Saijyo school established by Aida. The Seki school was the most popular of the many schools of mathematics in Japan. Yoriyuki Arima (1714–1783), Lord of Kurume, was one of its leaders and was the first to publish its secret theories of algebra. Arima personifies the anomaly of a member of a hereditary warrior class drawn, in a time of enforced peace, to mathematics of the mostly highly abstract and purely aesthetic sort; he, too, had been a pupil of Yamaji, and he took Fujita under his protection and assisted him in the publication of *Seiyo sampo.* (Arima's own *Shuki sampo* was as popular in its time as Fujita's work, and Aida drew heavily upon both books.)

In 1788 Aida published *Sampo tensei shinan,* a collection of conventional geometry problems which were, however, presented in a new and simplified symbolic notation. The same year saw the coronation of a new shogun, and Aida was released from his post to face the social and cultural dislocation of the masterless samurai. He then decided that it was heaven's will that he concentrate on mathematics; he would live on his savings and devote himself to the perfection of his studies. He also took pupils, including many from the northeastern provinces; these returned to teach in their native regions, where Aida is still revered as a master of mathematics.

In *Sampo tensei shinan,* Aida compiled the geometry problems presented in Arima's *Shuki sampo* and Fujita's *Seiyo sampo* and *Shinpeki sampo.* These were largely the problems of *yo jutsu,* the inscribing in circles or triangles of other circles, a mainstay of traditional Japanese mathematics. In his book, Aida also showed how to develop formulas for ellipses, spheres, circles, regular polygons, and so on, and explained the use of algebraical expressions and the construction of equations.

Aida was well acquainted with the mathematical literature of his time, and edited several other books.

In the course of his research he developed a table of logarithms, transmitted from China, that differed substantially from that of Ajima, being calculated to the base of two.

Aida also worked in number theory and gave an explanation of approximate fractions by developing a continued fraction (a simplification of the methods of Seki and Takebe). And, by expanding $x_1^2 + x_2^2 + x_3^2 + \cdots x_n^2 = y^2$, he obtained the integral solutions of $x_1^2 + k_2 x_2^2 + \cdots + k_n x_n^2 = y^2$.

Aida was hard-working and strong-willed and produced as many as fifty to sixty works a year. Nearly 2,000 works survived him, including many on nonmathematical subjects. He was a distinguished teacher of traditional mathematics and a successful popularizer of that discipline.

KAZUO SHIMODAIRA

AILLY, PIERRE D', also known as **Petrus de Alliaco** (*b.* Compiègne, France, 1350; *d.* Avignon, France, 1420), *theology, cosmography.*

D'Ailly studied at the College of Navarre of the University of Paris, where he received his doctorate of theology in 1381. He was grand master of Navarre from 1384 to 1389, and from 1389 to 1395 he was chancellor of the University of Paris. In 1395 d'Ailly became bishop of LePuy, and in 1397 bishop of Cambrai. He was made a cardinal in 1411. D'Ailly wrote commentaries on Aristotle (the *De anima* and the *Meteorologica*), as well as a number of astronomical and astrological works, including a commentary on the *De sphaera* of Sacrobosco. In his treatises on astrology, he reflects a more lenient attitude than that of either Nicole Oresme or Henry of Hesse. He was also concerned with the problem of calendar reform, and wrote a work on this subject for the Council of Constance (1414).

His most significant scientific work is a collection of cosmographical and astronomical treatises with the collective title *Imago mundi*. The *Imago* includes sixteen treatises on geography and astronomy, and the concordance of astrology, astronomy, and theology with historical events; only the first of these is the *Imago mundi* properly speaking. In the geographical portion of the *Imago* (the first treatise), d'Ailly makes use of the newly translated *Geography* of Ptolemy. It was thought that d'Ailly's work caused Columbus to underestimate his distance from the supposed coast of Asia, but it is now known that Columbus did not read the *Imago* until after his first voyage.

In his philosophical and scientific outlook, d'Ailly is considered a nominalist; however, his scientific

writing shows little originality and much unacknowledged borrowing. He has a more significant claim to historical prominence as a leader of the conciliar movement.

BIBLIOGRAPHY

I. ORIGINAL WORKS. There is a modern edition of d'Ailly's *Imago mundi* in Latin-French, edited by Edmond Buron (Paris, 1930). This is only the first of the sixteen treatises in the medieval *Imago mundi*.

II. SECONDARY LITERATURE. Louis Selembier, *Pierre d'Ailly* (Tourcoing, 1931), a French version of a Latin dissertation, *Petrus de Alliaco* (1886). There is a list of secondary material on d'Ailly following the entry on Pierre d'Ailly in the *Lexikon für Theologie und Kirche,* VIII (Freiburg, 1963), 330; also in Francis Oakley, *The Political Thought of Pierre d'Ailly* (New Haven-London, 1964), bibliographical note, pp. 350–356. Other works that discuss d'Ailly are Pierre Duhem, *Le système du monde,* IV (Paris, 1953), 168–183; George H. T. Kimble, *Geography in the Middle Ages* (London, 1938), pp. 92, 208–212, 218, n. 4; Lynn Thorndike, *A History of Magic and Experimental Science,* IV (New York, 1934), 101–113, 322, and *The Sphere of Sacrobosco and Its Commentators* (Chicago, 1949), pp. 38–40, 49–51.

CLAUDIA KREN

AIRY, GEORGE BIDDELL (*b.* Alnwick, Northumberland, England, 27 July 1801; *d.* Greenwich, England, 2 January 1892), *astronomy.*

George Airy was the eldest of four children of William Airy, a farmer who through self-education acquired posts in the Excise, and of Ann Biddell, daughter of a well-to-do farmer. At the age of ten he took first place at Byatt Walker's school at Colchester but, as he himself records, because he had very little animal vitality, he was not a favorite with his schoolmates. In the school he thoroughly learned arithmetic, double-entry bookkeeping, and the use of the slide rule. An introverted but not shy child, Airy was, even for the time and especially for his circumstances, a young snob. Nevertheless, he overcame some of the dislike of his schoolmates by his great skill and inventiveness in the construction of peashooters and other such devices.

At the age of twelve Airy came to know his uncle Arthur Biddell, a well-educated and highly respected farmer near Ipswich. He recognized in his uncle an opportunity to escape what he considered unpromising surroundings, and secretly requested that he be removed from his family. Arthur Biddell almost literally kidnapped him, without any word to his parents, but because of financial difficulties caused by William Airy's loss of his Excise post, the escape was not blocked. From 1814 to 1819 Airy spent nearly half of his time with his uncle. In later life he put great value on this connection, especially because of the resulting acquaintances, including Thomas Clarkson, the abolitionist, who could help his career. It was through Clarkson and Charles Musgrave, fellow of Trinity College, Cambridge, that he was entered as sizar of Trinity College in October 1819.

Airy entered Cambridge with the determination to get on, and he was certainly equipped to do so. Although his own assessment of his abilities was immodestly high, it was nevertheless matched, albeit sometimes reluctantly, by his tutors and college friends. He graduated as a Senior Wrangler in 1823 after far outdistancing all the men of his year, although beginning in his second term he had the burden of supporting himself by taking pupils. He was elected a fellow of Trinity College in 1824.

Three incidents from this period illustrate the care and foresight with which Airy planned his life. The first concerns the habit he adopted, as an undergraduate, of always keeping by him a quire of large-sized scribbling paper, sewn together, upon which everything was entered: translations into Latin and out of Greek, several lines of which he attempted every day, no matter how pressing other business might be; mathematical problems; and nearly every thought he had, complete with date. The sheets, even after the more important items were transferred to exercise books or diaries, were kept, together with nearly every communication received and a copy of those sent throughout his life, and are still extant. He seems not to have destroyed a document of any kind whatever: stubs of old checkbooks, notes for tradesmen, circulars, bills, and correspondence of all sorts were carefully preserved in chronological order from the time that he went to Cambridge. This material provides possibly the best existing documentation of a truly Victorian scientist.

The second illustrative incident involves Airy's courtship of his future wife, Ricarda Smith, the eldest daughter of the Rev. Richard Smith, private chaplain to the duke of Devonshire. He met Miss Smith while he was on a walking tour in Derbyshire, and within two days of first seeing her he made an offer of marriage. Neither his means nor his prospects at the time permitted an immediate marriage, and the Rev. Smith would not permit an engagement. Undaunted, Airy renewed his suit from time to time, and six years after his first proposal they were married.

A similar singleness of purpose is shown in Airy's approach to a prospective position at the Royal Greenwich Observatory. In 1824 an attempt was made to improve the educational level of assistants at the Royal Observatory by hiring one or two Cambridge graduates. Airy was proposed as one of these assistants and traveled to Greenwich to investigate the possibility. However, in his own words, "when I found that succession to the post of Astronomer Royal was not considered as distinctly a consequence of it, I took it cooly [sic] and returned to Cambridge the next night."

Airy applied for and won the Lucasian professorship in 1826. In doing this, he exchanged an assistant tutorship worth £150 per annum, and the prospect of succeeding to a tutorship, for the £99 per annum of the professorship, supplemented by a somewhat uncertain £100 per annum as *ex officio* member of the Board of Longitude. Other considerations were that "my prospects in the law or other profession might have been good if I could have waited but marriage would have been out of the question and I much preferred a moderate income in no long time. I had now in some measure taken science as my line (though not irrevocably) and I thought it best to work it well for a time at least and wait for accidents."

The Plumian professorship, which involved the care of the Cambridge Observatory, became vacant in 1828, and Airy "made known that I was a candidate and nobody thought it worthwhile to oppose me. . . . I told everybody that the salary (about £300) was not sufficient and drafted a manifesto to the University for an increase. . . . the University had never before been taken by storm in such a manner and there was some commotion about it. I believe very few people would have taken the same step . . . I had no doubt of success." He was appointed Plumian professor and director of the observatory on 6 February 1828, with a salary of £500 per annum. Although he accepted the post of astronomer royal in 1835, when he moved from Cambridge to Greenwich, Airy's considerable influence on British astronomy stretches without break from his appointment at Cambridge in 1828 to his retirement as astronomer royal in 1881. He was knighted in 1872, after thrice refusing on the basis that he could not afford the fees.

The ruling feature of Airy's character was undoubtedly order, and from the time he went up to Cambridge until the end of his life his system of order was strictly maintained. He wrote his autobiography up to date as soon as he had taken his degree, and made his first will as soon as he had any money to leave. His accounts were personally kept by double entry, and he regarded their keeping as one of his greatest joys. The effect of this sense of order on British observational astronomy is the only reason that Airy is included in this volume, for he was an organizer rather than a scientist. To realize his importance, it is necessary to understand the astronomy of the nineteenth century and the role played by such institutions as the Royal Greenwich Observatory.

The rise of astronomy in the seventeenth and eighteenth centuries took the form of careful observations of stellar positions made to provide a framework within which planetary motions could be measured. The first astronomer royal, John Flamsteed, provided the earliest observations of this kind that are still useful today. Although the emphasis in modern astronomy has shifted beyond the planets to the stars and external galaxies, these early observations provide us with a three-hundred-year base line for measuring the motions of the stars themselves, and knowledge of these motions is vital to the understanding of the origin and evolution of the stars. Observations of this kind are not only necessary in large numbers but they must be extremely exacting if the results are to be of general use. They are therefore best made in a routine way by those more interested in the technological problems of their procurement than in their scientific use. The Royal Greenwich Observatory, following Flamsteed's early lead, became the primary producer of such observations, mainly because the Admiralty was interested in the more immediate need of them for navigational purposes. Partly because the utilitarian purpose was stressed, scientific supervision of the observations eventually decreased and was refocused only in the nineteenth century, when it became obvious that their lack of accuracy was adversely affecting their use in navigation. The situation was ripe for Airy with his scientific training and his sense of order. The reforms he introduced were copied by other countries that, because they were expanding their navies to protect their expanding merchant fleets, needed the navigational aids.

The secret of Airy's long and successful official career was that he was a good servant who thoroughly understood his position. He never set himself in opposition to his masters, the Admiralty. He recognized the task for which he was appointed and transformed the Royal Greenwich Observatory into a highly efficient institution. The cost, however, was high. No independent thought could be tolerated, and as a result no scientists were trained there. The often slipshod methods that lead to scientific discovery were carried on outside, by John Herschel, John Adams, and many others. Airy himself would not understand this criticism. He wrote, in 1832, ". . . in those parts of astronomy which depend principally on the assistance of Governments, requiring only method and judgement, with very little science in the persons employed, we have done much; while in those which depend exclusively on individual effort we have done little. . . . our principal progress has been made in the lowest branches of astronomy while to the higher branches of science we have not added anything." He needed only to add that he had done *his* job.

In any article on Airy mention must be made of the controversy accompanying the discovery of the planet Neptune. It is ironical that the kind of order Airy restored to the observational work at Greenwich should coincide with the greatest need for the results since Newton had put Flamsteed's observations to such good use in the *Principia*—and then be unfairly blamed in nearly every subsequent article on the discovery of Neptune for withholding these observations. In fact, Airy supplied all the major participants in this discovery with the observational data they requested, and the only basis for the subsequent attacks upon him was that he was not at home when John Adams, then a young Cambridge mathematician, called unannounced to present one of his early predictions that such a planet as Neptune had to exist in order to account for the motions of the other planets. Airy's great efficiency in the observatory was noted by other government services and he rapidly became the prototype of the modern government scientist. This kept him from the observatory a large amount of time.

Always of medium stature and not powerfully built, Airy seemed to shrink as he aged, mainly because of an increasing stoop. His constitution, even at eighty-five, was remarkably sound. He took not the least interest in athletic sports or competition, but he was always a very active walker and could endure a great deal of fatigue. His eyesight was peculiar, and he studied it thoroughly all his life, correcting the astigmatism with a cylindrical lens, a method that he invented and is still used. As his powers failed with age, he was tyrannized by his ruling passion for order, and his efforts went into correctly filing his correspondence rather than understanding its contents. He was by nature eminently practical, and his dislike of mere theoretical problems and investigations put him continually in dissent with some of the resident Cambridge mathematicians. This practical bent led him to undertake, in 1872, the preparation of a numerical lunar theory. This work consisted, essentially, of obtaining from observations numerical values of the 320 periodic terms in Delaunay's equations for the moon's motion. His difficulties are summed up in a note of 29 September 1890:

> I had made considerable advance (under official difficulties) in calculations on my favourite Numerical Lunar Theory, when I discovered that, under the heavy

pressure of unusual matters (two Transits of Venus and some eclipses) I had committed a grievous error in the first stage of giving numerical value to my Theory. My spirit in the work was broken, and I have never heartily proceeded with it since.

Airy was not a great scientist, but he made great science possible. It is true that he was indirectly responsible for guiding British observational astronomy into a cul-de-sac from which it took many years to retreat, but it was not his fault that the methods he devised to provide a particular service at a particular time were so efficiently contrived and completely implemented that weaker successors continued to apply them, unchanged, to changing conditions.

BIBLIOGRAPHY

Airy's bibliography contains over 500 printed papers and the following books: *Mathematical Tracts on Physical Astronomy, the Figure of the Earth, Precession and Nutation, and the Calculus of Variations* (Cambridge, 1826); 2nd ed. (London, 1831), with the *Undulatory Theory of Optics* added; 4th ed. (London, 1858); *Undulatory Theory of Optics* also published separately (London, 1877); *Gravitation: An Elementary Explanation of the Principal Perturbations in the Solar System* (London, 1834, 1884); *Six Lectures on Astronomy* (London, 1849); *A Treatise in Trigonometry* (London, 1855); *On the Algebraical and Numerical Theory of Errors of Observations and the Combination of Observations* (London, 1861; 3rd ed., 1879); Essays on the invasion of Britain by Julius Caesar, the invasion of Britain by Plautius, and by Claudius Caesar; the early military policy of the Romans in Britain; the Battle of Hastings; and correspondence were published in *Essays* (London, 1865); *An Elementary Treatise on Partial Differential Equations* (London, 1866); *On Sound and Atmospheric Vibrations, With the Mathematical Elements of Music* (London, 1868, 1871); *A Treatise on Magnetism* (London, 1870); *Notes on the Earlier Hebrew Scriptures* (London, 1876); and *Numerical Lunar Theory* (London, 1886).

The complete list of printed papers is given in Wilfred Airy's edition of the *Autobiography of Sir George Airy* (London, 1896). They can be divided into four main categories: optics, both practical and theoretical; practical astronomy, including reports of progress and final publication of results obtained by the observers at Cambridge and Greenwich; government science, concerning the many tasks other than astronomy for which the government claimed his time; and contributions to the many polemics that marked nineteenth-century British science.

The extensive biographical data are housed mainly in the new Royal Greenwich Observatory at Herstmonceux Castle, Sussex. Some of those covering Airy's pre- and post-Greenwich careers are in the hands of the writer, and the remainder are scattered between the archives of the Royal Astronomical Society, the Royal Society, and the Royal Greenwich Observatory. As already noted, Airy apparently never discarded a piece of paper. His son Wilfred (as I have been informed by Airy's granddaughter) had no such inhibitions, and, after including a few extracts in his edition of the *Autobiography,* destroyed the voluminous correspondence between Sir George and Lady Airy.

Olin J. Eggen

AITKEN, ROBERT GRANT (*b.* Jackson, California, 31 December 1864; *d.* Berkeley, California, 29 October 1951), *astronomy.*

During forty years at the Lick Observatory, Aitken was an outstanding observer of double stars, and his *New General Catalogue of Double Stars Within 120° of the North Pole* (1932) is still a standard work.

Aitken at first intended to become a minister, but his studies in biology and astronomy at Williams College (1883–1887) diverted his interests to science. In 1891 he became professor of mathematics at the University of the Pacific, where there was a small observatory with a six-inch Clark refractor. In 1894 he visited Mt. Hamilton, and the following year returned there as assistant astronomer at the Lick Observatory. He was successively promoted to astronomer (1907), associate director (1923), and director (1930). He retired in 1935.

Aitken's early work at Lick was routine and varied, but double stars soon came to take up more and more of his time. In 1899 he embarked on a systematic survey of double stars that would provide the basis for statistical investigations. He was soon joined by W. J. Hussey, and together they examined stars given in the *Bonner Durchmusterung* as not fainter than 9.0 (Aitken) or 9.1 (Hussey) down to 22° southern declination. When Hussey left Lick in 1905, Aitken took over his share of the work and completed the survey in 1915. It resulted in the discovery of over 4,400 new pairs with separations mostly below 5″, and over two-thirds of these were found by Aitken; he published statistical investigations of this material in 1918 in *The Binary Stars.* Yet Aitken's real vocation was the observation of doubles; and in 1920, on the death of Eric Doolittle, he took over the compilation of material for a revision of S. W. Burnham's 1906 catalog of double stars, a revision that he published in 1932 as complete to 1927. Meanwhile, he devoted much of his time to the popularization of astronomy, and this became his main interest in retirement.

BIBLIOGRAPHY

I. Original Works. Aitken's works include *The Binary Stars* (New York, 1918; 2nd ed., rev., 1935) and *A New*

General Catalogue of Double Stars Within 120° of the North Pole, 2 vols., Carnegie Institute of Washington Publication No. 417 (Washington, D.C., 1932). A full bibliography of his articles is in the biographical memoir by Van den Bos (see below).

II. SECONDARY LITERATURE. Works on Aitken are mainly the numerous obituary notices in astronomical and other journals, of which by far the most comprehensive is that by Willem H. Van den Bos, in *National Academy of Sciences, Biographical Memoirs,* **32** (1958), 1–30.

MICHAEL A. HOSKIN

AITON, WILLIAM (*b.* Avondale, Lanarkshire, Scotland, 1731; *d.* Kew, Surrey, England, 2 February 1793), *horticulture.*

According to the parish baptismal roll, Aiton was the eldest of eleven children of William Aiton of Wailsely, whose occupation is not revealed. The son was a trained gardener when he left in 1754 for London, where in the following year he became an assistant to Philip Miller, curator of Chelsea Physic Garden, who was the most eminent gardener of his time and author of the celebrated and immensely important *Gardener's Dictionary.* Miller's influence on young Aiton greatly enlarged the latter's botanical knowledge and to a very large extent determined his future career.

Aiton's aptitude and proficiency led to his engagement in 1759 by Princess Augusta to plant a botanical garden at Kew House, under the supervision of John Haverfield; this was the inauguration of the present-day Royal Botanic Gardens. The princess also engaged Sir William Chambers as landscape architect to lay out the grounds in the fashionable mode; he is responsible for the orangery, the pagoda, and several temples.

At this time the princess depended for scientific direction on John Stuart, third earl of Bute, a most accomplished and knowledgeable botanist, who introduced many new species to Kew. Chambers has recorded Bute's assiduity in the assembling of plants from many parts of the globe to make the collection at Kew the largest in Europe, so Aiton's responsibility for its care was indeed a heavy one. A generous patron of botanical science, Bute encouraged Aiton in every possible way.

Upon the death of Princess Augusta in 1772, the gardens came into the possession of George III, and Bute was replaced by Sir Joseph Banks as the royal adviser on Kew. Banks was undoubtedly the greatest scientific impresario of the day and spared no effort in building up the collection at Kew. He used his connections with naval officers, merchants, doctors, and travelers to obtain plants; these specimens were entrusted to Aiton, who assumed control of the

garden in 1783. Like Bute, Banks befriended Aiton; and the two men, together with Daniel Carl Solander and Jonas Dryander, Banks's librarian, enabled Aiton to publish his *Hortus Kewensis* in 1789. This three-volume work is of fundamental importance as a catalog of some 5,500 plants under cultivation at the time, and records their provenance and the date of their introduction. It also contains descriptions of new species. It is unlikely that Aiton, who was a gardener rather than a botanist, had the scholarship required to produce, entirely by himself, a work of this nature. Indeed, in the Preface he acknowledges in general terms the assistance from those more learned than himself, without mentioning them by name. For the strictly botanical content, especially the Latin descriptions, Aiton depended heavily on Solander, but even more on Dryander, who was largely responsible for editing the work and seeing it through the press. The *Hortus* was well received and was sold out in two years.

Aiton is commemorated by the interesting monotypic South African genus *Aitonia,* which was described by Carl Peter Thunberg in 1780. The plant was introduced into cultivation by Aiton's fellow Scot, Francis Masson, whom he had trained as a gardener before Masson went in 1772, on behalf of Kew, to the Cape of Good Hope. The species was featured in *The Botanical Magazine* in 1791 by William Curtis, who remarked:

> The great length of time Mr. Aiton has been engaged in the cultivation of plants, the immense numbers which have been the constant objects of his care through every period of their growth, joined to his superior discernment, give him a decided superiority in the *prima facie* knowledge of living plants over most Botanists of his day; his abilities in the other line of his profession, are displayed in the eulogies of all who have seen the royal collections at Kew, which he has the honour to superintend.

There is little evidence available of his personal affairs. His wife was named Elizabeth, and she lies in the family tomb in Kew churchyard with her husband, four daughters, and two sons. A measure of the high esteem in which Aiton was held is shown by the presence as pallbearers at his funeral of Sir Joseph Banks, Bishop Goodenough, Jonas Dryander, and the famous artist John Zoffany, who lived in nearby Strand-on-the-Green.

BIBLIOGRAPHY

Aiton's only work is *Hortus Kewensis; or, a Catalogue of the Plants Cultivated in the Royal Botanic Garden at Kew,* 3 vols. (London, 1789), with color illustrations.

There is a good deal of minor reference to Aiton in various journals. The most significant sources are under the pseudonym Kewensis in *The Gentleman's Magazine,* **63** (1793), 389–391; James Britten and Edmund G. Baker, *Journal of Botany,* **35** (1897), 481–485; J. Britten, *ibid.,* **50,** supp. 3 (1912), 1–16; and W. Botting Hemsley, *Journal of the Kew Guild,* **2,** no. 10 (1902), 87–90.

GEORGE TAYLOR

AITON, WILLIAM TOWNSEND (*b.* Kew, Surrey, England, 2 February 1766; *d.* Kensington, London, England, 9 October 1849), *horticulture.*

Aiton was the eldest son of William Aiton, who was in charge of the living collections in the Royal Botanic Gardens, Kew, and who described himself as "Gardener to His Majesty." He entered school at Chiswick, and when he was thirteen transferred to one at Camberwell. He remained there until, at the age of sixteen, he entered Kew as assistant to his father. Apart from gaining practical knowledge of horticulture and experience with a wide range of living plants, Aiton became greatly interested in landscaping and acquired a considerable reputation as a landscape gardener. Indeed, he received commissions from many eminent people, including the duke of Kent and a number of noblemen.

In 1793, upon the death of his father, Aiton succeeded to the control of the gardens at Kew and Richmond, and on the accession of George IV he also had charge of other royal gardens, including Kensington, Buckingham Palace, those around the bizarre Royal Pavilion at Brighton, and certain areas at Windsor. He was styled director-general of the royal gardens, but there is evidence that he was more concerned with improvements elsewhere. He left the cultivation of the rich assemblage of plants at Kew to his subordinates (particularly John Smith, the first Kew curator), although he assiduously continued his father's close association with Sir Joseph Banks and sought his advice and support on how best to enrich the Kew collections by sending suitably trained men overseas.

Aiton was one of the seven gentlemen who, on the initiative of John Wedgwood, son of the potter Josiah Wedgwood, assembled "at Mr. Hatchard's House for the purpose of instituting a Society for the improvement of Horticulture" on 7 March 1804. The Royal Horticultural Society was founded at this meeting.

Largely through the encouragement of Banks, a second and much enlarged edition of the senior Aiton's *Hortus Kewensis* was published. Nominally the work of the son, who certainly had considerable knowledge of botany and its literature, it was in fact largely revised by Jonas Dryander, who was mainly responsible for the botanical matter in the original

edition, and by Robert Brown, who became Banks's librarian upon the death of Daniel Carl Solander. An epitome of the *Hortus* was published in 1814, but neither this nor the parent work enjoyed the success of the first edition. Aiton prepared a second edition of the *Epitome* but the manuscript was probably destroyed with his voluminous collection of letters, although James Britten states in the *Journal of Botany* (1912) that two versions appeared in 1814.

Aiton's responsibilities were drastically curtailed when William IV became king and his authority was restricted to the gardens at Kew. After Banks's death in 1820 cut off invaluable patronage, royal funds for Kew's maintenance were severely restricted. In spite of the efforts of Aiton and Smith, the collections and the gardens suffered from so much neglect that in 1838 the Treasury was obliged to appoint a committee of inquiry to investigate the condition of the Royal Gardens at Kew. The report, signed by Professor John Lindley on 28 February 1838, noted "That it is little better than a waste of money to maintain it in its present state, if it fills no intelligible purpose except that of sheltering a large quantity of rare and valuable plants." Lindley suggested that Kew might become a national botanic garden, so the garden was transferred to the nation in 1840 and William Jackson Hooker was appointed the first official director in 1841. For a time Aiton remained in charge of the pleasure grounds at Kew, but he resigned in 1841, taking with him his library, records, and drawings. When he died, apparently unmarried, his enormous and rich correspondence, containing a vast amount of information on Kew affairs over a period of nearly fifty years, was burned by his brother John. His collection of drawings and plant record books were retrieved for Kew after John's death.

BIBLIOGRAPHY

I. ORIGINAL WORKS. Aiton's writings are *Delineations of Exotick Plants Cultivated in the Royal Gardens at Kew* (London, 1796); *Hortus Kewensis; or a Catalogue of the Plants Cultivated in the Royal Botanic Garden at Kew,* 2nd ed., enl., 5 vols. (London, 1810–1813); and *An Epitome of the Second Edition of Hortus Kewensis, for the Use of Practical Gardeners; to Which Is Added a Selection of Esculent Vegetables and Fruits Cultivated in the Royal Gardens at Kew* (London, 1814).

II. SECONDARY LITERATURE. Writings on Aiton are *Proceedings of the Linnean Society,* **2** (1850), 82–83 (anon.); W. T. Thiselton-Dyer, in *Bulletin of Miscellaneous Information, Royal Botanic Gardens, Kew* (1891), pp. 304–305; *ibid.,* (1910), pp. 306–308 (anon.); J. Britten, in *Journal of Botany,* **50,** supp. 3 (1912), 1–16; and E. Nelmes, in *Curtis's Botanical Magazine Dedications* (1931), pp. 7–8, with a portrait. R. G. C. Desmond, "John Smith, Kew's First Curator," in *Kew Guild Journal* (1965), pp. 576–587, has many references to Aiton. See also John L. Gilbert, "The Life and Times of William Townsend Aiton," *ibid.* (1966), pp. 688–693.

GEORGE TAYLOR

AJIMA NAONOBU, also known as **Ajima Chokuyen** (common name, **Manzo;** pen name, **Nanzan**) (*b.* Shiba, Edo [now Tokyo], Japan, *ca.* 1732; *d.* Shiba, 1798), *mathematics.*

Ajima was born at the official residence of the Shinjo family, and was later stationed in Edo as a retainer of that clan. He remained there until his death, and is buried in the Jorin-ji Temple, Mita, Tokyo.

Ajima first studied mathematics under Masatada Irie of the Nakanishi school, and later he studied both mathematics and astronomy under Nushizumi Yamaji, who initiated him into the secret mathematical principles of the Seki school. He was apparently over thirty when he began his studies with Yamaji; his career before then (save for his studies with Masatada) is largely a matter of conjecture. Ajima wrote several works on astronomy soon after becoming Yamaji's pupil; it is presumed that during this time he was also engaged in helping his master to compile an almanac. After Yamaji's death, Ajima began to write on mathematics.

In the traditional succession of the Seki school, Ajima is in the fourth generation of masters. None of his books were published in his lifetime; they existed solely as copies handwritten by his students, perhaps because of the esoteric nature of the discipline. Most of the essential points of his work are summarized in his *Fukyu sampo,* a book that Ajima intended as an emendation of Sadasuke Fujita's *Seiyo sampo,* which was then a popular textbook. His pupil Makoto Kusawa wrote a preface to this book in 1799, a year after Ajima's death: although Kusawa planned to publish the work, he did not do so. Kusawa succeeded Ajima as a master of the Seki school. Masatoda Baba and Hiroyasu Sakabe were also students of Ajima; as did Kusawa, they had their own pupils, many of whom became first-rate mathematicians and continued the tradition of the Seki school until the Meiji restoration (an arithmetic book in the European style was published in Japan in 1856, and marked the end of the native forms of mathematics).

The mathematics originated by Takakazu Seki was refined by his successive pupils and tentatively completed and systematized by Yoshisuke Matsunaga and Yoriyuki Arima, who was the first to publish it. Upon this base Ajima began to develop a new mathematics;

his works reflect an innovative trend toward geometry within a tradition that was basically algebraic and numerical in approach.

This trend is exemplified in the development of *yenri,* a method for determining the area of a circle, of a sphere, or of plane figures composed of curved lines.

Seki's technique for calculating the length of an arc of a circle depended upon giving a fixed number to the diameter of the circle, and was not much more sophisticated a method than that of Archimedes. His pupil Takebe used letters instead of numbers to represent a diameter and found infinite series, expressed exponentially. Matsunaga improved Takebe's method and increased the number of types of infinite series capable of representing the different elements of circles.

The *yenri* process began with the inscription in the circle of a regular polygon to divide the circle or arc into equal parts; this method was, however, by definition limited, and could not be expanded to include curves other than circles and their arcs. Ajima expanded the process by dividing the diameter or chord into equal, small segments, initiating a technique somewhat similar to the definite integration of European mathematics. The earlier Japanese mathematicians had been concerned with subdividing the circle or arc directly; it was Ajima's contribution to proceed from the subdivision of the chord. He introduced his method in his *Kohai jutsu kai,* and used it for the basis of further calculations.

Japanese mathematicians were accustomed to using exponential notation for convenience in dealing with large numbers. Integration was also easier if an exponent were used, and double integrals were thus easily obtained. In the same year that Ajima developed his *yenri* method, he discovered a way to obtain the volume common to two intersecting cylinders by using double integration, which he presented in *Enchu kokuen jutsu.* Ajima's new method was a logical outgrowth of the method he described in *Kohai jutsu kai,* and required the application of the earlier technique.

For his work with logarithms, Ajima drew upon *Suri seiran,* a book published in China in 1723 that almost certainly incorporated some of the Western principles brought to China by the Jesuits. *Suri seiran* introduced the seven-place logarithmic table into Japan, and also showed how to draw up such a table. It is apparent that Ajima knew this book, since he used the same terminology in setting up his own table of logarithms (actually antilogarithms). Ajima's table and its uses are described in *Fukyu sampo,* and there is also a copybook of the table only. The Chinese logarithmic table was useful for multiplication and division; Ajima's was not. Ajima used his table to find the tenth root, and it was also useful in finding the power of a number. Before this table could be used, however, it was necessary to find the logarithm by division. The setting up of Ajima's table, as explained in *Fukyu sampo,* is based upon $\log 10 = 1$, $\log^{10} \sqrt{10} = 0.1$; therefore, the logarithm of $^{10}\sqrt{10} = 1$, while the value of 0.1 is 258925411. Ajima's tables permitted the calculation of a logarithm to twelve places.

Ajima drew upon Japanese mathematical tradition for *yo jutsu,* problems involving transcribing a number of circles in triangles and squares. Ajima wrote a major work on this subject, which included the problem described by Malfatti in 1803: in a given triangle, inscribe three circles, each tangent to the other and to two sides of the triangle. Although this problem became known as "Malfatti's question," it is obvious that Ajima's work preceded Malfatti's, although it is not known when Ajima published his problem. Malfatti approached the problem analytically, while Ajima was concerned with finding the diameters of the circles, but it is apparent that the problems are essentially identical.

BIBLIOGRAPHY

Ajima's works include *Fukyu sampo*; *Kohai jutsu kai*; and *Enchu kokuen jutsu.*

SHIN'ICHI OYA

AL-. See next element of name.

ALAIN DE LILLE, also known as **Alanus de Insulis** (*b.* Lille, France, first half of the twelfth century; *d.* Cîteaux, France, 1203), *theology, philosophy.*

Although he was popular and enjoyed a reputation for wide learning during subsequent centuries, the birthdate of Alain de Lille is a matter of conjecture, the most reasonable surmise being 1128. He taught theology at Paris and Montpellier, and subsequently became a member of the Cistercian order. He is the author of numerous theological works and was among the first to write against the Albigensians (in his *Contra haereticos*). Influenced by the *Quomodo substantiae* (better known as the *De hebdomadibus*) of Boethius, he attempted to construct a deductive theology derived from axioms in the manner of mathematics (in his *Regulae caelestis iuris*). His theological treatises reveal Neoplatonic influences; in addition to Boethius, these works employ such Neoplatonic materials as the *Liber de causis* and the pseudo-Hermetic *Liber XXIV philosophorum.*

Alain's reputation rests largely on two literary works, the *De planctu naturae,* possibly composed between 1160 and 1170, and the very famous *Anticlaudianus,* written around 1182–1184. The *De planctu* is extant in few early manuscripts, and was apparently not the subject of commentary. The work is an exposition of Neoplatonic Christian naturalism. As an allegorical portrayal of Nature and attendant Virtues, it exerted extensive influence on the part of the *Roman de la Rose* written by Jean de Meun.

The popular *Anticlaudianus* survives in countless manuscripts and was a frequent subject of commentary. Consisting of a prose prologue followed by more than 4,000 lines of classic hexameter, the *Anticlaudianus* was intended as a refutation of the *In Rufinum* of Claudian. The Roman poet had portrayed Rufinus as a creation of evil Nature. Alain's theme concerns Nature's wish to atone for previous errors and to create a perfect man. She and the Virtues send Prudence on a celestial journey to the throne of God to seek the soul of the perfect man. Prudence travels in a chariot constructed by the seven liberal arts and drawn by the five senses. Guided by Reason, the chariot ascends through the heavens. As it approaches the throne of God, Reason falters, and Prudence is then guided through these exalted regions by Theology and Faith. Her petition is successful; God grants her request.

The *Anticlaudianus* is a mélange of Neoplatonisms. One can detect the influences of such Chartrain masters and disciples as Bernard and Thierry of Chartres, Bernard Silvester, and Gilbert de la Porrée. There are also traces of the pseudo-Dionysian corpus and its interpreter, John Scotus Erigena. Above all, Alain is indebted to Boethius' *Consolation of Philosophy* and to the *De nuptiis philologiae et Mercurii* of Martianus Capella. In his description of the personification of Astronomy, Alain alludes to the eccentric, and possibly to the equant, and mentions Ptolemy and Abū Ma'shar by name. His treatment of Arithmetic is limited to a few propositions taken possibly from either Boethius or Nicomachus; he mentions the latter by name. Geometry fares even more poorly.

Alain de Lille, although known in the middle ages as "doctor universalis," was not an original thinker. However, he wove together successfully many of the Neoplatonic traditions available to twelfth-century humanism.

BIBLIOGRAPHY

I. ORIGINAL WORKS. Thomas Wright, *The Anglo-Latin Poets and Epigrammatists of the Twelfth Century,* in *Rerum britannicarum medii aevi scriptores,* **59**[2] (London, 1872), 268–428, contains the texts of the *De planctu naturae* and the *Anticlaudianus;* R. Bossuat, *Alain de Lille: Anticlaudianus* (Paris, 1955); J. Huizinga, *Über die Verknüpfung des Poetischen mit dem Theologischen bei Alanus de Insulis* (Amsterdam, 1932), with an edition of the *De virtutibus et vitiis* in an appendix; English versions are D. M. Moffat, *The Complaint of Nature by Alain de Lille* (New York, 1908); W. H. Cornog, *The Anticlaudian of Alain de Lille* (Philadelphia, 1935); M.-T. d'Alverny, *Alain de Lille, Textes inédits* (Paris, 1965) contains two small works by Alain (pp. 185ff).

II. SECONDARY LITERATURE. Étienne Gilson, *History of Christian Philosophy in the Middle Ages* (New York, 1955), lists many of Alain's theological works and gives their *Patrologia latina* entries, p. 635; P. Duhem, *Le système du monde,* 10 vols. (Paris, 1954–1959), III, 223–230, discusses Alain's scientific importance; R. de Lage, *Alain de Lille, poète du XIIᵉ siècle* (Paris, 1951), has an extensive bibliography on pp. 169–173 and an appendix on pp. 175–186 listing the manuscripts of Alain's works; see also bibliographical material in R. Bossuat, *Anticlaudianus* (see above), and on pp. 322–348 in M.-T. d'Alverny, *Textes inédits* (see above).

CLAUDIA KREN

ALBATEGNI. See **al-Battānī.**

ALBERT I OF MONACO (HONORÉ CHARLES GRIMALDI) (*b.* Paris, France, 13 November 1848; *d.* Paris, 26 June 1922), *oceanography.*

Albert was the son of Charles III of Monaco (Honoré Grimaldi) and Antoinette Ghislaine, countess of Mérode. He succeeded his father on 10 September 1889. In 1870 he fought against Germany as a lieutenant-commander in the French navy.

His career as a navigator actually began in 1873, when he bought a 200-ton schooner, the *Pleiad,* and renamed it the *Hirondelle.* By 1885 he had decided to devote himself to the study of the sea, and each following year, for nearly forty years, he made voyages in the North Atlantic, taking soundings wherever he went. He made four cruises in the *Hirondelle;* six, between 1892 and 1897, in the *Princesse Alice I;* and twelve, between 1898 and 1910, in the *Princesse Alice II.* He used the *Hirondelle II* until his death, making five cruises between 1911 and 1915. He may truly be considered one of the founders of oceanography.

In physical oceanography, Albert studied currents, especially the Gulf Stream (1885). He set out floating mines to study drift in both the North Atlantic and the Arctic. Using the Richard bottle, he took samples of water at various depths in order to determine the differences in temperature. Albert also established three observation centers in the Azores in order to study the meteorology of the ocean regions. One of his major achievements was a general atlas to the millionth, which had twenty-four plates and illustrated the bathymetry of all the oceans; it represented a synthesis of all previous findings.

Albert conducted valuable physiological research as well. Interested in the venom of *Physalia,* a pelagic coelenterate, he crushed its tentacles, filtered the product, and injected it into experimental animals. The result was a deep state of anesthesia, and the toxin was therefore called hypnotoxin. This was a first step toward the discovery of anaphylaxis.

Plankton was another concern of Albert's—from the surface to depths as great as 5,000 meters. His huge, baited polyhedral nets brought forth abundant evidence of a rich and varied bathypelagic fauna. In the waters off the Cape Verde Islands he broke the previous record of 5,800 meters by dredging at a depth of 6,035 meters. Some of his ideas, accepted today, were far ahead of his time: his protests that the depths were being overfished, the use of airplanes to spot schools of fish, and the creation of underwater preserves.

In order to display his collections, Albert founded the Musée Océanographique de Monaco in 1910 and the Institut Océanographique, Paris, in 1911. He also established publications for these institutions: *Bulletin du Musée Océanographique* in 1904, which became *Bulletin de l'Institut Océanographique* in 1906, and *Annales de l'Institut Océanographique* in 1910. Since he was also interested in man's origin and evolution, Albert founded the Musée Anthropologique de Monaco and the Institut de Paléontologie Humaine in Paris.

Albert was a corresponding member of the Académie des Sciences, Paris, and later a foreign associate member (1909), succeeding Lord Kelvin; a foreign associate member of the Académie de Médecine; and a member of the Académie d'Agriculture (for the model farming practices on his property at Marchais, Aisne).

BIBLIOGRAPHY

I. ORIGINAL WORKS. Between 1885 and 1915 Albert published numerous reports in the *Comptes rendus de l'Académie des sciences* (Paris), *Comptes rendus des séances de la Société de Biologie* (Paris), *Bulletin de la Société de Géographie* (Paris), *Bulletin du Musée Océanographique,* and *Revue scientifique* (*Revue rose*), among others. His works include *Sur le Gulf-Stream. . . .* (Paris, 1886); "La pêche de la Sardine sur les côtes d'Espagne," in *Revue scientifique,* 3rd ser., **13,** no. 17 (1887), 513–519—a very similar work with the title *L'industrie de la Sardine sur les côtes de la Galice* was published separately with the notation "taken from the *Revue scientifique"* (Paris, 1887); "Sur les filets fins de profondeur employés à bord de l'*Hirondelle,*" in *Comptes rendus des séances de la Société de Biologie,* 8th ser., **4,** no. 37 (1887), 661–664; "Sur l'alimentation des naufragés en pleine mer," in *Comptes rendus de l'Académie des sciences,* **107** (Dec. 1888), 980–982; "Recherche des animaux marins. . . ," in *Congrès International de Zoologie,* **1** (1889), 133–159; "Sur le développement des tortues (*T. caretta*)," in *Comptes rendus des séances de la Société de Biologie,* 10th ser., **5,** no. 1 (1898), 10–11; *La carrière d'un navigateur* (Monaco, 1901, 1951, 1966), the 1966 ed. with intro. by J. Y. Cousteau and preface by J. Rouch; "Les progrès de l'océanographie," in *Bulletin du Musée Océanographique,* no. 6 (1904), 1–13; "Progrès de la biologie marine," *ibid.,* no. 14 (1904), 1–7; "Sur le lancement de ballons pilotes au-dessus des océans," in *Comptes rendus de l'Académie des sciences,* **141** (1905), 492–493; "L'outillage moderne de l'océanographie," in *Bulletin du Musée Océanographique,* no. 25 (1905), 1–12; "Vingt-cinquième campagne scientifique (*Hirondelle II*)," in *Bulletin de l'Institut Océanographique de Monaco,* **10,** no. 268 (1913), 1–4; "Marche des mines flottantes dans l'Atlantique nord et l'océan Glacial pendant et après la guerre," *ibid.,* **16,** no. 357 (1919), 1–8; *Sur les résultats partiels des deux premières expériences pour déterminer la direction des courants de l'Atlantique nord* (Paris, n.d.); and "Discours sur l'océan" (delivered 25 Apr. 1921 to National Academy of Sciences, Washington), in *Bulletin du Musée Océanographique,* no. 392 (1921), 1–16.

II. SECONDARY LITERATURE. Works on Albert are C. Carpine, "Les navirés océanographiques dont les noms ont été choisis par S.A.S. le Prince Albert Ier pour figurer sur la façade du Musée Océanographique de Monaco . . .," in *Bulletin de l'Institut Océanographique,* spec. no. 2 (1966), 627–638; R. Damien, *Albert Ier Prince Souverain de Monaco, précédé de l'historique des origines de Monaco et de la dynastie des Grimaldi* (Villemomble, 1964); M. Fontaine, "La découverte de l'anaphylaxie," in *Bulletin de l'Institut Océanographique,* no. 997 (1951), 3–9; "Liste des campagnes scientifiques de S.A.S. Prince Albert Ier de Monaco," in *Bulletin des amis du Musée Océanographique,* no. 5 (1948), 9–14; L. Mayer, "S.A.S. Albert Ier, Prince de Monaco. L'homme et l'oeuvre," in *Bulletin de l'Institut Océanographique,* no. 421 (1922), 1–8; P. Portier, "La carrière scientifique du Prince de Monaco," in *Revue générale des sciences pures et appliquées,* **33,** no. 19 (1922), 542–544, and in *Bulletin des amis du Musée Océanographique,* no. 6 (1948), 2–7; J. Richard, *Les campagnes scientifiques de S.A.S. le Prince Albert Ier de Monaco* (Monaco, 1900), and many articles on Albert's cruises and the apparatus used on board his vessels in *Bulletin du Musée Océanographique* and *Bulletin de l'Institut Océanographique* between 1904 and 1941; J. Rouch, "Le Prince Albert Ier et Jean Charcot . . .," in *Bulletin de l'Institut Océanographique,* spec. no. 2 (1967); and J. Thoulet, "S.A.S. le Prince Albert Ier de Monaco," in *Bulletin des amis du Musée Océanographique,* no. 5 (1948), 1–6.

GEORGES PETIT

ALBERT OF BOLLSTÄDT. See **Albertus Magnus.**

ALBERT OF SAXONY (*b.* Helmstedt, Lower Saxony, *ca.* 1316; *d.* Halberstadt, Saxony, 8 July 1390), *physics, logic, mathematics.*

The family name of Albert of Saxony was de Ricmestorp; his father, Bernard de Ricmestorp, was a well-to-do burgher of Helmstedt. A brother, John, was a master of arts at the University of Paris in 1362, while Albert himself was still there. Of Albert's youth and early schooling nothing is known, although there is some evidence to indicate that he studied at Prague before going to Paris, where he obtained the degree of master of arts in 1351.

He quickly achieved renown as a teacher on the faculty of arts at Paris and was made rector of the university in 1353. During most of the period of Albert's study and teaching at Paris, the most influential figure on the faculty of arts was Jean Buridan, and Albert's own lectures on natural philosophy, represented by his books of questions on Aristotle's *Physics* and *De caelo et mundo,* were modeled closely on those of Buridan. Nicole Oresme, another pupil of Buridan, also taught at Paris at this time, and there is evidence that he influenced Albert in the direction of mathematical studies. Albert apparently studied theology also but never received a theological degree.

It is believed that he left Paris by the end of 1362, going to Avignon and spending the next two years carrying out various commissions for Pope Urban V. The pope obtained for him a benefice at Mainz, later made him parochial priest at Laa, and shortly afterward canon of Hildesheim. Albert played a major role in obtaining the authorization of the pope for the establishment of a university at Vienna and in drawing up its statutes. When the university was established in June 1365, Albert was its first rector. But he held this position for only a year; at the end of 1366 he was appointed bishop of Halberstadt and his academic career came to an end. His twenty-four years as bishop were marred by political and financial difficulties, and at one point he was even accused of heresy by some inimical clergy of his own region who intimated that he was "more learned in human science than in divine wisdom," and that he had openly taught an astrological determinism with denial of human freedom of choice. Surviving these vicissitudes, he held the bishopric until he died at the age of seventy-six. He was buried in the cathedral of Halberstadt.

Albert's writings, which were probably composed during the years when he was teaching at Paris, consist mostly of books of questions on Aristotle's treatises and of some treatises of his own on logic and mathematical subjects. Extant in early printed editions are questions on Aristotle's *Physics, De caelo et mundo, De generatione et corruptione, Posterior Analytics,* and on the "old logic" (Porphyry's *Predicables* and Aristotle's *Categories* and *De interpretatione*); a complete textbook of logic published in 1522, under the title *Logica Albertutii;* an extensive collection of logical puzzles, entitled *Sophismata;* and a treatise on the mathematical analysis of motion, entitled *Tractatus proportionum.* In unpublished manuscripts there are sets of questions on Aristotle's *Meteora, Ethics, De sensu et sensato,* and *Oeconomica;* a book of questions on John of Sacrobosco's *De sphaera;* and two short treatises on the mathematical problems of "squaring the circle" and of determining the ratio of the diameter of a square to its side. Suter's ascription of the second of these mathematical treatises to Albert has been questioned by Zoubov (see Bibliography), who attributes it to Oresme. It does in fact echo passages found in one of Oresme's known works, but since Albert often paraphrased the content of works whose ideas he borrowed, this does not prove that the work was not written by Albert. There is much uncertainty concerning the attribution of a number of these manuscript works to Albert. It has been shown that his *Questions on the Ethics,* although written by Albert as his own work, is an almost literal plagiarism of the corresponding work of Walter Burley.

Albert's significance in the history of science is primarily that of a transmitter and an intelligent compiler of scientific ideas directly drawn from the works of Buridan, Thomas Bradwardine, William of Ockham, Burley, Oresme, and other writers in the medieval scientific tradition. His works in physics are heavily dependent on the corresponding works of Buridan, to the extent that all but a few of the questions devoted to the *Physics* and the *De caelo et mundo* correspond directly to those of Buridan's works of similar title, both in form and in content. Most of the questions that Albert adds, and which are not found in Buridan's works, draw their materials from the Oxford tradition of Bradwardine and his Mertonian pupils, or, in a few cases, from the early thirteenth-century works on statics and hydrostatics associated with Jordanus de Nemore. Albert's *Tractatus proportionum* is modeled directly on Bradwardine's treatise *De proportionibus velocitatum in motibus,* although it adds some refinements in terminology and in the analysis of curvilinear motions that reflected the later Mertonian developments and probably also the influence of Oresme.

Despite his lack of originality Albert contributed many intelligent discussions of aspects of the problems dealt with, and he had the particular merit of seeing the importance of bringing together the mathematical treatments of motion in its kinematic aspect, stemming from the Oxford tradition of Bradwardine, with the dynamical theories that Buridan had developed without sufficient concern for their mathematical formulation. As a transmitter of Buridan's work, Albert played an important part in making known the explanations of projectile motion and of gravitational acceleration provided by Buridan's theory of impetus, although he tended to blur the distinction between Buridan's quasi-inertial concept of impetus and the older doctrine of the self-expending "impressed virtue." Unlike Buridan, he introduced an error into the analysis of projectile motion, by supposing that there is a short period of rest between the ascent of a projectile hurled directly upward and its descent. Yet this led him to initiate a fruitful discussion by raising the question of the trajectory that would be followed by a projectile shot horizontally from a cannon. He supposed that it would follow a straight horizontal path until its *impetus* ceased to exceed the force of its gravity, but that it would then follow a curved path for a short period in which its lateral impetus would be compounded with a downward impetus caused by its gravity, after which it would fall straight down. Leonardo da Vinci took up the problem, but it remained for Nicolò Tartaglia to show that the entire trajectory would be a curve determined by a composition of the two forces.

Albert's textbook of logic is one of the best organized of the late medieval works in the field. In its first three sections it presents the analysis of the signification and supposition of terms, and the internal analysis and classification of propositional forms, provided by the work of Ockham and Buridan. The fourth section, on "consequence," shows influence by Burley and Buridan, developing the theory of inference on the foundation of the logic of unanalyzed propositions, exhibiting the syllogism as a special type of consequence, and ending with a very full treatment of modal syllogisms and a shorter formulation of the rules of topical argumentation. The last two sections deal with logical fallacies, with the "insoluble" (or paradox of self-reference), and with the rules of disputation known as *Obligationes.* There is little that is not directly traceable to the sources Albert used, but these materials are skillfully integrated, reduced to a uniform terminology, and presented with systematic elegance.

Despite its excellence as a textbook, this work did not achieve the popularity or influence attained by Albert's *Tractatus proportionum* and by his questions on the physical treatises of Aristotle. These, printed in many editions at Venice, Padua, and Pavia, became the principal means by which the contributions of the northern Scholastics of the fourteenth century to the science of mechanics were made known to the physicists and mathematicians of Italy, from Leonardo da Vinci to Galileo himself.

BIBLIOGRAPHY

I. ORIGINAL WORKS. *Expositio aurea et admodum utilis super artem veterem . . . cum quaestionibus Alberti parvi de Saxonia* (Bologna, 1496); *Quaestiones subtilissimae Alberti de Saxonia super libros Posteriorum* (Venice, 1497); *Logica Albertutii* (Venice, 1522); *Sophismata Alberti de Saxonia* (Paris, 1490, 1495); *Tractatus obligationum* (Lyons, 1498; with Albert's *Insolubilia,* Paris, 1490, 1495); *Subtilissimae quaestiones super octo libros Physicorum* (Venice, 1504, 1516); *Quaestiones in libros de caelo et mundo* (Pavia, 1481; Venice, 1492, 1497, 1520); *Quaestiones in libros de generatione et corruptione* (Venice, 1504, 1505, 1518); *Quaestiones et decisiones physicales insignium virorum . . .,* Georgius Lockert, ed. (Paris, 1516, 1518), contains Albert's questions on the *Physics* and the *De caelo et mundo; Tractatus proportionum* (Bologna, 1502, 1506; Padua, 1482, 1484, 1487; Venice, 1477, 1494, 1496; Paris, *s.a.*).

II. SECONDARY LITERATURE. Philotheus Boehner, *Medieval Logic* (Chicago, 1952); B. Boncompagni, "Intorno al Tractatus proportionum di Alberto de Sassonia," in *Bolletino di bibliografia e di storia delle scienze matematiche e fisiche,* **4** (1871), 498 ff.; Maximilian Cantor, *Vorlesungen über die Geschichte der Mathematik,* II, 2nd. ed. (1900), 137–154; Marshall Clagett, *The Science of Mechanics in the Middle Ages* (Madison, Wis., 1959); Pierre Duhem, *Études sur Léonard de Vinci,* Vols. I–III (Paris, 1906–1913); A. Dyroff, "Ueber Albertus von Sachsen," in *Baeumker-Festgabe* (Münster, 1913), pp. 330–342; G. Heidingsfelder, "Albert von Sachsen: Sein Lebensgang und sein Kommentar zur Nikomachischen Ethik des Aristoteles," in *Beiträge zur Geschichte der Philosophie und Theologie des Mittelalters,* **22,** 2nd ed. (Münster, 1926); M. Jullien, "Un scolastique de la décadence: Albert de Saxe," in *Revue Augustinienne,* **16** (1910), 26–40; Anneliese Maier, *Zwei Grundprobleme der scholastischen Naturphilosophie* (Rome, 1951), pp. 259–274; C. Prantl, *Geschichte der Logik im Abendlande,* **4** (Leipzig, 1870), 60–88; H. Suter, "Der Tractatus 'De quadratura circuli' des Albertus de Saxonia," in *Zeitschrift für Mathematik und Physik,* **29** (1884), 81–102 (reedited and translated in M. Clagett, *Archimedes in the Middle Ages* [Madison, Wis., 1964], pp. 398–432); H. Suter, "Die Quaestio 'De proportione dyametri quadrati ad costam eiusdem' des Albertus de Saxonia," in *Zeitschrift für Mathematik und Physik,* **32** (1887), 41–56; V. P. Zoubov, "Quelques Observations sur l'Auteur du Traité Anonyme 'Utrum dyameter alicuius quadrati sit commensurabilis costae ejusdem,'" in *Isis,* **50** (1959), 130–134.

ERNEST A. MOODY

ALBERTI, FRIEDRICH AUGUST VON (*b.* Stuttgart, Germany, 4 September 1795; *d.* Heilbronn, Germany, 12 September 1878), *geology, mining.*

Alberti's father, Karl Franz, was a colonel in Württemberg and a teacher at the well-known Karlsschule in Stuttgart; a member of the middle class, he was ennobled in 1807. His mother, Christiane Friederike, also came from the middle class; she had family connections with the princely court and was the aunt of the novelist and short-story writer Wilhelm Hauff.

In 1809 Alberti entered the Bergkadettenkorps in Stuttgart, in which he received instruction in general scientific subjects as well as special training in mineralogy, geology, and mining. In 1815 he went to the saltworks at Sulz; in 1818 he supervised drilling experiments near Jagstfeld; and in 1820 he was appointed inspector of the saltworks at Friedrichshall. The first proof of his abilities came in 1823, when he drilled a rock salt deposit near Schwenningen and established the saltworks at Wilhelmshall; he became manager in 1828. In his book *Über die Gebürge des Königreiches Württemberg, in besonderer Beziehung auf Halurgie* (1825) he also demonstrated his scientific abilities.

Alberti was appointed mining counselor in 1836. From 1852 to 1870 he was again manager of the saltworks at Friedrichshall. There, under his direction, between 1854 and 1859 the Friedrichshall shaft was bored, and the center of Württemberg's salt production was shifted from Wilhelmshall to Friedrichshall. His most important technical improvement was the introduction of steam heating into salt processing. Alberti was considered one of the foremost salt-mining engineers, but like other German mining officials of his era, he was not only a good manager with technical capabilities but also a scientist of significant achievement.

After Quenstedt, Alberti must be reckoned one of the founders of the geology of southwest Germany. His investigations of the Triassic period and its fossils were of fundamental significance. He coined the name Triassic for the oldest formations of the Mesozoic era and thoroughly investigated the three divisions—variegated sandstone, shell limestone, and Keuper sandstone—dividing them into groups characterized by petrographic and paleontologic features. In 1834 he published his most important results in *Beiträge zu einer Monographie des Bunten Sandsteins, Muschelkalks und Keupers und der Verbindung dieser Gebilde zu einer Formation.* He also investigated and described crystalline slate and the eruptive rocks, as well as their superimposed formations in the Black Forest.

BIBLIOGRAPHY

I. Original Works. Alberti's main writings are *Über die Gebürge des Königreiches Württemberg, in besonderer Beziehung auf Halurgie* (Stuttgart–Tübingen, 1825); *Beiträge zu einer Monographie des Bunten Sandsteins, Muschelkalks und Keupers und der Verbindung dieser Gebilde zu einer Formation* (Stuttgart–Tübingen, 1834); *Halurgische Geologie,* 2 vols. (Stuttgart–Tübingen, 1852); and *Überblick über die Trias* (Stuttgart, 1864).

II. Secondary Literature. Biographical notices are in *Schwäbische Kronik* (1878), 2165; and in W. Serlo, *Männer des Bergbaues* (1937). The best recent summary is Erich Krenkel, in *Neue deutsche Biographie,* I (1953), 140–141.

Hans Baumgärtel

ALBERTI, LEONE BATTISTA (*b.* Genoa, Italy, 18 February 1404; *d.* Rome, Italy, April 1472), *mathematics, physics, natural history, technology.*

In the twelfth century Alberti's ancestors were feudal lords of Valdarno who settled in Florence, where they became judges and notaries and were members of the wealthy bourgeoisie. In the fourteenth century they engaged in commercial and banking enterprises, organizing a firm with branches scattered all over Europe; their wealth enriched Florence. At the same time, the Albertis became involved in politics. Toward the end of the fourteenth and the beginning of the fifteenth centuries, this led to the family's exile; they sought refuge in the foreign branches of their firm. Thus Leone Battista Alberti, the son of Lorenzo Alberti, came to be born in Genoa. It is possible that he was illegitimate.

From his early childhood Alberti is said to have been precocious; little else is known about his youth. Fleeing the plague, his father went to Venice, the site of perhaps the most important branch of the house of Alberti. The father died suddenly, leaving his children in the care of their uncle, who disappeared soon thereafter. It is possible that unscrupulous relatives liquidated the Venice branch in order to make themselves rich at the orphans' expense.

Alberti seems to have started his advanced education at Padua. At any rate, after 1421 he continued it at Bologna, where he began the study of law. Overwork caused him to fall ill, and he had to interrupt his studies; nevertheless, he received a doctorate in canon law. For relaxation he took up the study of mathematics, natural sciences, and physics, subjects that he pursued to a rather advanced level. Subsequently, the decrees of exile against his family having been revoked, Alberti undoubtedly returned to Florence, or at least to Tuscany. In Florence he met Brunelleschi, who became a good friend. Between

1430 and 1432 he was in the service of a cardinal, who took him with his entourage to France, Belgium, and Germany.

In 1432 Alberti arrived in Rome, where he became a functionary at the papal court. In Rome he discovered antiquity and became the artist we know today—painter, sculptor, and then architect. His paintings and sculptures, however, have never been found or identified. As part of the papal court, he necessarily shared all its tribulations. In 1437 he was in Bologna and Ferrara with Pope Eugene IV, who was roaming all over northern Italy. He was often in Rome, yet he also served those humanistic families who ruled small, more or less independent principalities. Thus he certainly spent some time at the court of Rimini, with the Malatesta family. Here Alberti conceived and partially executed his most important architectural work, the Malatesta Temple, a chapel designed to shelter their tombs.

Alberti was, we are told, amiable, very handsome, and witty. He was adept at directing discussions and took pleasure in organizing small conversational groups. Alberti represented, perhaps even better than Brunelleschi, the first scholar-artists of the Renaissance, more inquisitive than given to realization, more collectors of facts and ideas than imaginative and creative. Still close to the expiring Middle Ages, Alberti had trouble freeing himself of its shackles on the scientific level. He was possessed of a perpetual need to know—and a perpetual need to expound his ideas—as well as a desire to mingle with intellectual equals. It is certain that from these encounters at the courts of rulers like the Malatestas, a new scientific spirit arose. In this sense Alberti occupies a place of particular importance in the history of thought. At the end of his life, aside from architectural works or such engineering projects as the attempt to refloat the Roman galleys in Lake Nemi in 1447 (on which he wrote a short treatise, now lost), he was occupied with these meetings and with the editing of his written works, which were numerous.

Unfortunately, a large part of Alberti's scientific work has been lost. It is not impossible, however, that some of his works may be submerged in the scientific literature of the age without being known. Like all of his contemporaries, Alberti inherited a fragmentary science. He seems to have been interested in isolated problems which furnished subjects for discussion but which individually could not result in anything important. It was difficult to give them a personal emphasis, for these questions had already been debated, discussed, and restated many times.

Alberti's mathematics is exactly that of his times.

He wrote, at least on an advanced level, only a small treatise, the *Ludi matematici,* dedicated to his friend Meliadus d'Este, himself an accomplished mathematician. Only twenty problems were involved, some of which had to do with mathematics only remotely. Only one of them touched on an abstract question— lunules in "De lunularum quadratura," in which he furnished an elegant solution to the problem but lost his way in the squaring of the circle. On all other points he shared the preoccupation of a great number of fifteenth-century scholars, considering mathematics as a tool rather than an independent science. Often he merely applied formulas. Thus, geometry was used to calculate the height of a tower, the depth of a well, the area of a field. In this work we find notions of the hygrometer which is simply the hygrometer of Nicholas of Cusa. Alberti wrote a book of mathematical commentaries that may have contained more precise ideas, but unfortunately the manuscript has never been found.

Not much is known about Alberti's physics. He wrote *De motibus ponderis,* which has been lost also. In some of his works we can find some references to physics, but they are rather elusive ones. Some years ago the *Trattati dei pondi, lieve e tirari,* long attributed to Leonardo da Vinci, was reattributed to Alberti. It concerns gravity, density (harking back to the works of Archimedes), hydrostatics, and heat. There are only vague, undoubtedly traditional ideas on the preservation of labor. His optics is more pragmatic than theoretical, although he sets forth a theory of vision. In his opinion bodies, even dark ones, emit in all directions rays that move in a straight line. They converge toward the eye and together form a visual pyramid. This theory is also completely traditional. The camera obscura, which may be his greatest discovery, deeply impressed his contemporaries, although he perhaps borrowed this device from Brunelleschi, to whom he was greatly indebted for his studies on perspective. In his *Elementa picturae,* however, he contributed nothing more than applied geometry. He worked from the idea that the construction of similar figures was the basis for all figure representation.

Alberti displayed the same attitude in his writings on the natural sciences, in which he speculated on nature rather than on scientific data. Like many others, he admitted the roundness of the earth, and also wrote briefly on the development of its crust. He seemingly spoke knowledgeably of earthquakes, atmospheric erosion, water circulation, the action of plants on soil, plant decomposition and formation of humus, sedimentary layers, and the formation of deltas. He considered fossils merely a freak of nature.

Alberti's best-known work, containing many of his scientific ideas, is his *De re aedificatoria,* which was presented to Pope Nicolas V about 1452. The work was printed in 1485 and exerted a certain influence. It was to be a treatise on the art of engineering, but this aim was not completely achieved. Alberti dealt with lifting devices, grain bins and "other conveniences that albeit of little esteem nevertheless bring profit," water supply, ways of quarrying rock and cutting through mountains, the damming of the sea or of rivers, the drying up of swamps, machines of war, and fortresses. In this work he was concerned less with architecture per se and architectural techniques than with an actual attempt at town planning. His ideas of a city were still largely inspired by the Middle Ages, but they also contained elements clearly belonging to the Renaissance, such as the respect for urban aesthetics, perspective, and orderly arrangement. Something that certainly seems new—but we hardly know his predecessors—is the application of the entire range of scientific knowledge to town planning and architectural practice. Alberti applied his knowledge of the natural sciences to building materials; his knowledge of physics was applied to equilibrium of buildings, the flexibility of beams, and the construction of engines; and that of mathematics (still very simple mathematics) was shown in the very Pythagorean layout of cities and the arrangement of fortresses.

As was typical of his time, Alberti was preoccupied with various machines and apparatuses, some in current use and some the subject of scattered and almost confused observation which made it impossible to draw the parallels and comparisons necessary to develop a technology. He spoke of balances, clocks, sundials, pulleys, water mills and windmills, and canal locks. He developed topographical instruments and envisaged the odometer and the "sulcometer," which measured distances traveled by ships. He studied the methods of sounding in deep waters. In all of this work he manifested more interest in manual crafts than in true science.

Alberti is difficult to place in both the history of science and the history of technology. Contemporary works in these fields almost invariably cite him in their lists of scholars, but he is not credited with anything really new. He contributed no new principles, but he seems to have had a very profound knowledge. In short, he seems to have regarded science as a means for action rather than as a system of organized knowledge. On many occasions he admitted his interest in knowledge, but more for reasons of efficiency than as an abstract science, as power rather than as intellectuality. He knew only the perspective and natural science that serve the artist or the architect, and only the mathematics and physics of use to the engineer and the technician. Nevertheless, he perceived certain directions for research. He was well aware of the difference between sensation (common observation) and scientific ideas: "Points and lines are not the same for the painter as for the mathematician." Observation was a point of departure for scientific hypothesis, which must be verified by systematic observation. In the last analysis, although Alberti contributed nothing but a supplementary collection of special cases to scientific progress, he nevertheless outlined some promising avenues for future work.

BIBLIOGRAPHY

I. ORIGINAL WORKS. In most cases only very old editions of Alberti's works are extant. *De re aedificatoria* was first published in Florence in 1485; there were many subsequent editions in Italian, and a French version appeared in Paris in 1553. *Opere volgari dei L. B. Alberti,* IV (Florence, 1847), contains *Ludi matematici. Opera inedita et pauca separatim impressa* (Florence, 1890) includes *Elementa picturae;* it also contains a treatise on perspective incorrectly attributed to Alberti. *Trattati dei pondi, lieve e tirari* was published as an appendix to Vasari (Florence, 1917).

II. SECONDARY LITERATURE. There are few works on Alberti. The essential work is P. H. Michel, *La pensée de L. B. Alberti* (Paris, 1930), with an exhaustive bibliography of works published until then. There is a good chapter on Alberti in L. Olschki, *Geschichte der neuspralichen Literatur* (Leipzig, 1919). The technological aspects of Alberti's work are discussed by B. Gille in *Les ingénieurs de la Renaissance* (Paris, 1967), pp. 80–84.

BERTRAND GILLE

ALBERTI, SALOMON (*b.* Naumburg, Germany, October 1540; *d.* Dresden, Germany, 29 March 1600), *medicine.*

Although he is usually associated with Nuremberg, where his family moved in 1541 and where he received his elementary education, Alberti studied medicine at Wittenberg (M.D., 1574) and taught in the medical faculty there for many years. He was chiefly interested in anatomy. As early as 1579, he began public demonstrations of the venous valves; his study of these valves was his most noteworthy achievement. A knowledge of the venous valves was essential to the formation of Harvey's concept of a systemic circulation of the blood, fifty years later. First referred to in 1546, they were apparently forgotten after about 1560; they were rediscovered in 1574 by Girolamo Fabrizio (Fabrizio d'Acquapendente) at Padua. Although Alberti acknowledged his indebtedness to Fabrizio for rediscovery of these valves, he deserves recognition as being the first to provide illustrations of venous valves in his *Tres orationes* (Nuremberg, 1585), which also included the first extensive printed account devoted solely to their structure.

Alberti also studied and described the lacrimal apparatus (*De lacrimis,* Wittenberg, 1581), as well as such then curious but rational problems as why boys ought not to be forbidden to cry, why sobbing usually accompanies weeping, and whether asthma might be ameliorated by breathing the fumes of various minerals burned on coals (*Orationes quatuor,* Wittenberg, 1590). In addition, he provided an extended account of the ileocecal valve, or Bauhin's valve (mentioned by Mondino in 1316 and described briefly by Laguna in 1535), the cochlea (described in detail by Fallopio in 1561), and, as an original contribution, the renal papillae. (See *Orationes duae,* Wittenberg, 1575–1576; and *Historia plerarunque partium humani corporis,* a textbook for medical students, Wittenberg, 1583, and later editions.) Alberti discussed the problem of deafness and muteness in *Oratio de surditate et mutitate* (Nuremberg, 1591). He emphasized the difference between hardness of hearing and deafness, which latter condition he considered as possibly being caused by a defect in the development of the fetus.

In 1592 Alberti became physician to Duke Friedrich Wilhelm of Saxony. A year earlier, his interest in the problem of scurvy had led to the treatise *De schorbuto* (Wittenberg, 1591). Alberti made a survey of the incidence of the deficiency disease in the ducal territory, and the result was his *Schorbuti historia* (Wittenberg, 1594), which for the most part is of no great significance except for its demonstration of the prevalence of the complaint and the recommendation of citrus fruit as part of a preventive diet. The book was known by James Lind and referred to by him in his celebrated treatise of 1753.

BIBLIOGRAPHY

For bibliographies of Alberti's writings see Georg Andreas Will, *Nürnbergisches Gelehrten-Lexicon,* **1** (Nuremberg, 1755); and Claudius F. Mayer, "Bio-bibliography of XVI. Century Medical Authors. Fasciculus 1, Abarbanel-Alberti, S.," in *Index Catalogue of the Library of the Surgeon General's Library,* ser. 4, 3rd supp. (Washington, D. C., 1941), which contains an exhaustive list but without indication of imprint. See also Lynn Thorndike, *A History of Magic and Experimental Science, The Sixteenth Century,* VI (New York, 1941), 229–230.

C. D. O'MALLEY

ALBERTUS MAGNUS, SAINT, also known as **Albert the Great** and **Universal Doctor** (*b.* Lauingen, Bavaria, *ca.* 1200; *d.* Cologne, Prussia, 15 November 1280). *Proficient in all branches of science, he was one of the most famous precursors of modern science in the High Middle Ages.*

Albert was born in the family castle and probably spent his childhood at the family manor in nearby Bollstädt—whence he is variously referred to as Albert of Lauingen and Albert of Bollstädt. His birth date could have been as early as 1193 or as late as 1206 or 1207. His family was wealthy and powerful, of the military nobility, and he received a good education.

He studied liberal arts at Padua, where, over strong opposition from his family, he was recruited into the Dominican Order by its master general, Jordan of Saxony—identified by some (but probably falsely) as Jordanus de Nemore, the mechanician. He likely studied theology and was ordained a priest in Germany, where he also taught in various priories before being sent to the University of Paris *ca.* 1241. In Paris he was the first German Dominican to become a master of theology and to lecture in the chair "for foreigners" (1245–1248). In the summer of 1248 he went to Cologne to establish a *studium generale*: among his students were Thomas Aquinas, Ulrich of Strassburg, and Giles (Aegidius) of Lessines.

He began the administrative phase of his career as provincial of the German Dominicans (1253–1256). Subsequently he became bishop of Regensburg (1260), a post he resigned in 1262. The latter part of his life was spent in preaching and teaching, mainly at Cologne. He took part in the Council of Lyons (1274) and journeyed to Paris in an unsuccessful attempt to block the famous condemnation of 1277, where some of Aquinas' teachings were called into question. His health was good and he had great powers of physical endurance, even to old age, although his eyesight failed during the last decade of his life. Albert was canonized by Pope Pius XI on 16 December 1931 and was declared the patron of all who cultivate the natural sciences by Pope Pius XII on 16 December 1941.

Albert's principal importance for the history of modern science derives from the role he played in rediscovering Aristotle and introducing Greek and Arab science into the universities of the Middle Ages. Before his time, what was to become the subject matter of modern science was usually treated in encyclopedias, which assembled a curious mélange of fact and fable about nature, or in theological treatises, which described the cosmos in terms of the six days of creation, as recounted in Genesis and variously analyzed by the church fathers. Aristotle, of course, had already made his entry into the Latin West through the translations of Gerard of Cremona and James of Venice, among others; but Christendom was generally hostile to the teachings of this pagan philosopher, particularly as contained in his *libri naturales* ("books on natural science"). In 1210, the ecclesiastical authorities at Paris had condemned Aristotle's works on natural philosophy and had prohibited their being taught publicly or privately under pain of excommunication. Although this condemnation was revoked by 1234, it had a general inhibiting effect on the diffusion of Greek science in the schools of the Middle Ages.

Albert seems to have become acquainted with the Aristotelian corpus while at the Paris priory of St. Jacques in the 1240's. Here too he probably began his monumental paraphrase of all the known works of Aristotle and Pseudo-Aristotle, to which are allotted seventeen of the forty volumes in the Cologne critical edition of Albert's works (see Bibliography). The project was undertaken by Albert, then studying and teaching theology, at the insistence of his Dominican brethren, who wished him to explain, in Latin, the principal physical doctrines of the Stagirite so that they could read his works intelligently. Albert went far beyond their demands, explaining not only the natural sciences but also logic, mathematics, ethics, politics, and metaphysics, and adding to Aristotle's exposition the discoveries of the Arabs and of whole sciences that were not available to him. The gigantic literary production that this entailed was recognized as one of the marvels of his age and contributed in no small measure to Albert's outstanding reputation. Roger Bacon, a contemporary who was not particularly enamored of the German Dominican, complained of Master Albert's being accepted as an authority in the schools on an equal footing with Aristotle, Avicenna, and Averroës—an honor, he protested, "never accorded to any man in his own lifetime."

Like all medieval Aristotelians, Albert incorporated considerable Platonic thought into his synthesis, and even commented on a number of Neoplatonic treatises. In several places he represents himself as merely reporting the teachings of the Peripatetics and not as proposing anything new; some historians charge him, on this basis, with being a compiler who was not too judicious in his selection of source materials. Those who have studied his works, however, detect there a consistent fidelity to Aristotle's basic theses, a clear indication of his own views when he thought Aristotle in error, a repudiation of erroneous interpretations of Aristotle's teaching, and an explicit rejection of Platonic and Pythagorean physical doctrines—all of which would seem to confirm his Aristotelianism. J. A. Weisheipl, in particular, has stressed the differences between thirteenth-century Oxford masters such as Robert Grosseteste, Robert Kilwardby, and Roger Bacon (all of whom were more pronouncedly Platonist in their scientific views) and Paris masters such as Albert and Aquinas (who were more purely Aristotelian). Whereas the former held that there is a successive subalternation between physics, mathematics, and metaphysics (so that the principles of natural science are essentially mathematical, and the principle of mathematics is the unity that is identical with Being), the latter held for the autonomy of these sciences, maintaining that each has its own proper principles, underived from any other discipline.

Albert's early identification as a precursor of modern science undoubtedly stemmed from his empiricist methodology, which he learned from Aristotle but which he practiced with a skill unsurpassed by any other Schoolman. From boyhood he was an assiduous observer of nature, and his works abound in descriptions of the phenomena he noted, usually in great detail. Considering that his observations were made without instruments, they were remarkably accurate. Some of the "facts" he reported were obviously based on hearsay evidence, although he was usually at pains to distinguish what he had himself seen from what he had read or been told by others. *Fui et vidi experiri* ("I was there and saw it happen") was his frequent certification for observations. Sometimes, as Lynn Thorndike has well illustrated in his *A History of Magic and Experimental Science,* even these certifications test the reader's credulity; what is significant in them, however, is Albert's commitment to an empiricist program. He stated that evidence based on sense perception is the most secure and is superior to reasoning without experimentation. Similarly, he noted that a conclusion that is inconsistent with the evidence cannot be believed and that a principle that does not agree with sense experience is really no principle at all. He was aware, however, that the observation of nature could be difficult: much time, he remarked, is required to conduct an experiment that will yield foolproof results, and he suggested that it be repeated under a variety of circumstances so as to assure its general validity.

On the subject of authority, he pointed out that science consists not in simply believing what one is told but in inquiring into the causes of natural things. He had great respect for Aristotle, but disagreed with the Averroists of his day on the Stagirite's infallibility. "Whoever believes that Aristotle was a god, must also believe that he never erred. But if one believes that he was a man, then doubtless he was liable to error just as we are." His *Summa theologica,* for example, contains a section listing the errors of Aristotle, and

in his *Meteorology* he observes that "Aristotle must have spoken from the opinions of his predecessors and not from the truth of demonstration or experiment."

Albert recognized the importance of mathematics for the physical sciences and composed treatises (unfortunately lost) on its pure and applied branches. Yet he would not insist that the book of nature is written in the language of mathematics, as Galileo was later to do, and as Roger Bacon intimated in his own lifetime. Rather, for Albert, mathematics had only a subsidiary role to play in scientific activity, insofar as it assisted in the discovery of physical causes. Mathematics is itself an abstract science, prescinding from motion and sensible matter, and thus its applications must be evaluated by the science that studies nature as it really exists, *in motu et inabstracta* ("in motion and in concrete detail").

The mechanics of Albert was basically that of Aristotle, with little innovation in either its kinematical or its dynamical aspects. One part of Albert's teaching on motion, however, did assume prominence in the late medieval period and influenced the emerging new science of mechanics. This was his use of the expressions *fluxus formae* and *forma fluens* to characterize the scholastic dispute over the entitative status of local motion. Arab thinkers such as Avicenna and Averroës had pursued the question whether this motion, or any other, could be located in the Aristotelian categories; the question quickly led to an argument whether motion is something really distinct from the terminus it attains. Local motion, in this perspective, could be seen in one of two ways: either it was a *fluxus formae* (the "flowing" of successive forms, or locations) or a *forma fluens* (a form, or absolute entity, that is itself a process). Although Albert made no clear dichotomy between these two views and allowed that each described a different aspect of motion, later writers came to be sharply divided over them. Nominalists, such as William of Ockham, defended the first view: this equivalently denied the reality of local motion, equating it simply with the distance traversed and rejecting any special causality in its production or continuance—a view that stimulated purely kinematical analyses of motion. Realists, such as Walter Burley and Paul of Venice, on the other hand, defended the second view: for them, local motion was an entity really distinct from the object moved and from its position, and thus had its own proper causes and effects—a view that stimulated studies of its more dynamical aspects.

Albert mentioned the term *impetus* when discussing projectile motion, but spoke of it as being in the medium rather than in the projectile, thus defending the original Aristotelian teaching; certainly he had no treatment of the concept to match that found in the work of fourteenth-century thinkers. His analysis of gravitational motion was also Aristotelian: he regarded the basic mover as the generator of the heavy object, giving it not only its substantial form but also its gravity and the motion consequent on this. He knew that bodies accelerate as they fall, and attributed this to their increasing propinquity to their natural place.

The cause of sound, for Albert, is the impact of two hard bodies, and the resulting vibration is propagated in the form of a sphere whose center is the point of percussion. He speculated also on the cause of heat, studying in detail how light from the sun produces thermal effects; here his use of simple experiments revealed a knowledge of the method of agreement and difference later to be formulated by J. S. Mill. He knew of the refraction of solar rays and also of the laws of refraction of light, although he employed the term *reflexio* for both refraction and reflection, as, for example, when discussing the burning lens and the burning mirror. His analysis of the rainbow was diffuse in its historical introduction, but it made an advance over the theory of Robert Grosseteste in assigning individual raindrops a role in the bow's formation, and undoubtedly prepared for the first correct theory of the rainbow proposed by another German Dominican, Dietrich von Freiberg, who was possibly Albert's student. In passing, he corrected Aristotle's assertion that the lunar rainbow occurs only twice in fifty years: "I myself have observed two in a single year."

Although he had no telescope, he speculated that the Milky Way is composed of stars and attributed the dark spots on the moon to configurations on its surface, not to the earth's shadow. His treatise on comets is notable for its use of simple observation to verify or falsify theories that had been proposed to explain them. He followed Grosseteste in correlating the occurrence of tides with the motion of the moon around its deferent. He favored the mathematical aspects of the Ptolemaic theory of the structure of the solar system, contrasting it with that of al-Biṭrūjī, although he acknowledged the superiority of the latter's theory in its physical aspects. Albert accepted the order of the celestial spheres commonly taught by Arabian astronomers; he knew of the precession of the equinoxes, attributing knowledge of this (falsely) to Aristotle also. Like most medieval thinkers, Albert held that heavenly bodies are moved by separated substances, but he denied that such substances are to be identified with the angels of Christian revelation, disagreeing on this point with his celebrated disciple Thomas Aquinas.

On the structure of matter, when discussing the presence of elements in compounds, Albert attempted to steer a middle course between the opposed positions of Avicenna and Averroës, thereby preparing for Aquinas' more acceptable theory of "virtual" presence. In a similar vein, he benignly viewed Democritus' atoms as equivalent to the *minima naturalia* of the Aristotelians. He seems to have experimented with alchemy and is said to have been the first to isolate the element arsenic. He compiled a list of some hundred minerals, giving the properties of each. During his many travels, he made frequent sidetrips to mines and excavations in search of specimens. He was acquainted with fossils, and made accurate observations of "animal impressions" and improved on Avicenna's account of their formation. Albert suggested the possibility of the transmutation of metals, but he did not feel that alchemists had yet found the method to bring this about.

Extensive as was Albert's work in the physical sciences, it did not compare with his contributions to the biological sciences, where his powers of observation and his skill at classification earned for him an unparalleled reputation. Some aspects of his work have been singled out by A. C. Crombie as "unsurpassed from Aristotle and Theophrastus to Cesalpino and Jung." His *De vegetabilibus et plantis,* in particular, is a masterpiece for its independence of treatment, its accuracy and range of detailed description, its freedom from myth, and its innovation in systematic classification. His comparative study of plants extended to all their parts, and his digressions show a remarkable sense of morphology and ecology. He drew a distinction between thorns and prickles on the basis of their formation and structure, classified flowers into the celebrated three types (bird-form, bell-form, and star-form), and made an extensive comparative study of fruits. His general classification of the vegetable kingdom followed that proposed by Theophrastus: he ranged plants on a scale reaching from the fungi to the flowering types, although, among the latter, he did not explicitly distinguish the monocotyledons from the dicotyledons. He seems to have been the first to mention spinach in Western literature, the first to note the influence of light and heat on the growth of trees, and the first to establish that sap (which he knew was carried in veins—like blood vessels, he said, but without a pulse) is tasteless in the root and becomes flavored as it ascends.

On plant evolution, Albert proposed that existing types were sometimes mutable and described five ways of transforming one plant into another; he believed, for example, that new species could be produced by grafting. Here he registered an advance over most medieval thinkers, who accounted for the succession of new species not by modification but by generation from a common source such as earth.

Albert's *De animalibus* includes descriptions of some fabulous creatures, but it also rejects many popular medieval myths (e.g., the pelican opening its breast to feed its young) and is especially noteworthy for its sections on reproduction and embryology. Following Aristotle, Albert distinguished four types of reproduction; in sexual reproduction among the higher animals he taught that the material produced by the female was like a seed (a *humor seminalis*), differentiating it from the catamenia (*menstruum*) in mammals and the yolk of the egg in birds, but incorrectly identifying it with the white of the egg. The cause of the differentiation of the sexes, in his view, was that the male "vital heat" could "concoct" semen out of surplus blood, whereas the female was too cold to effect the change.

He studied embryology by such simple methods as opening eggs at various intervals of time and tracing the development of the embryo from the appearance of the pulsating red speck of the heart to hatching. He was acquainted, too, with the development of fish and mammals, and understood some aspects of fetal nutrition. His studies on insects were especially good for their descriptions of insect mating, and he correctly identified the insect egg. He showed that ants lose their sense of direction when their antennae are removed, but concluded (wrongly) that the antennae carry eyes.

Among the larger animals, he described many northern types unknown to Aristotle, noting changes of coloration in the colder climates, and speculating that if any animals inhabited the poles they would have thick skins and be of a white color. His knowledge of internal anatomy was meager, but he did dissect crickets and observed the ovarian follicles and tracheae. His system of classification for the animal kingdom was basically Aristotelian; occasionally he repeated or aggravated the Stagirite's mistakes, but usually he modified and advanced Aristotle's taxonomy, as in his treatment of the ten genera of water animals. His anthropology was more philosophical than empirical in intent, but some have detected in it the adumbration of methods used in experimental psychology.

Apart from these more speculative concerns, Albert made significant contributions also to veterinary and medical science, dentistry included. In anatomy, for example, he took the vertebral column as the basis for structure, whereas in his day and for long afterward most anatomists began with the skull. He was reported to have cures for all manner of disease, and despite his own repudiation of magic and astrology came to be regarded as something of a magician. Many spurious works, some utterly fantastic, were attributed to him or published under his name to assure a wide diffusion—among these are to be included the very popular *De secretis mulierum* ("On the Secrets of Women") and other occult treatises.

Albert's productivity in science was matched by a similar output in philosophy and theology. In these areas his teachings have been overshadowed by those of his most illustrious disciple, Thomas Aquinas. The latter's debt to Albert is, of course, considerable, for Aquinas could well attribute the extent of his own vision to the fact that he stood on the shoulders of a giant.

BIBLIOGRAPHY

I. Major Works and Writings. Standard editions include *Omnia opera*, B. Geyer, ed. (Cologne, 1951–), a critical edition, in progress, 40 vols.; Vol. XII (1955) is the only work of direct scientific interest to appear thus far; it contains the *Quaestiones super de animalibus* and other treatises related to Albert's work in zoology; *Omnia opera*, A. Borgnet, ed. (Paris, 1890–1899), 38 quarto vols.; *Omnia opera*, P. Jammy, ed. (Lyons, 1651), 21 folio vols., available on microfilm positives from the Vatican Library; his *Book of Minerals* is translated from the Latin by Dorothy Wychoff (Oxford, 1967). Special texts include H. Stadler, ed., "Albertus Magnus De animalibus libri XXVI," in *Beiträge zur Geschichte der Philosophie des Mittelalters*, **15-16** (Münster, 1916; 1921); L. Thorndike, *Latin Treatises on Comets Between 1238 and 1368 A.D.* (Chicago, 1950), pp. 62–76; J. A. Weisheipl, "The Problema Determinata XLIII ascribed to Albertus Magnus (1271)," in *Mediaeval Studies*, **22** (1960), 303–354.

II. Secondary Literature. For a compact summary of Albert's life and works, with bibliography, see J. A. Weisheipl, "Albert the Great (Albertus Magnus), St.," in the *New Catholic Encyclopedia* (New York, 1967). Biographies include S. M. Albert, *Albert the Great* (Oxford, 1948) and T. M. Schwertner, *St. Albert the Great* (Milwaukee, 1932), a fuller biography with indication of sources. Works concerned with scientific teachings include H. Balss, *Albertus Magnus als Biologe* (Stuttgart, 1947); M. Barbado, *Introduction à la psychologie expérimentale*, P. Mazoyer, trans. (Paris, 1931), pp. 114–189; C. B. Boyer, *The Rainbow: From Myth to Mathematics* (New York, 1959), esp. pp. 94–99; A. C. Crombie, *Medieval and Early Modern Science*, I (New York, 1959), esp. 147–157; A. C. Crombie, *Robert Grosseteste and the Origins of Experimental Science* (Oxford, 1953), esp. pp. 189–200; E. J. Dijksterhuis, *The Mechanization of the World Picture*, C. Dikshoorn, trans. (Oxford, 1961); P. Duhem, *Le système du monde*, III (Paris, 1914; reprinted, 1958), 327–345; A. Maier, *Die Vorläufer Galileis im 14. Jahrhundert*, Edizioni di Storia e Letteratura, **22** (Rome, 1949), 11–16, 183–184; L. Thorndike, *A History of Magic* *and Experimental Science*, II (New York, 1923), esp. pp. 517–592; J. A. Weisheipl, *The Development of Physical Theory in the Middle Ages* (London, 1959); J. A. Weisheipl, "Celestial Movers in Medieval Physics," in *The Thomist*, **24** (1961), 286–326. See also *Serta Albertina*, a special issue of the Roman periodical *Angelicum*, **21** (1944), 1–336, devoted to all branches of Albert's science; includes a bibliography classified by fields.

WILLIAM A. WALLACE, O. P.

ALBRECHT, CARL THEODOR (*b.* Dresden, Germany, 30 August 1843; *d.* Potsdam, Germany, 31 August 1915), *surveying, astronomy.*

Albrecht's father, Friedrich Wilhelm Albrecht, and both grandfathers were soap boilers. Indeed, his maternal grandfather, Christian Friedrich Pohle, was a senior official of the soap boilers' guild of Dresden. Carl, however, did not continue the family tradition. His parents recognized the boy's intelligence, and set him on quite another path in that era when technology and the exact sciences flowered. As a student his major fields were mathematics and the natural sciences, but he occupied himself independently with astronomy and meteorology. About 1865, after passing his examinations at the Polytechnicum in Dresden, which at that time was an engineering school, Albrecht studied astronomy at the University of Leipzig in order to follow his special inclinations and to enlarge his theoretical knowledge.

From 1866 on, Albrecht was an assistant in the central European degree measurement project while continuing his studies. In 1869 he graduated from Leipzig and was immediately accepted at the newly founded Geodetic Institute in Potsdam, an indication that he already had a good scientific reputation. In 1873 he was appointed director of the astronomy department of the Geodetic Institute, a post he held until his death. In 1875 he became professor; in the same year he married Marie Stiemer.

The Geodetic Institute quickly became one of the leading research institutes in astronomy and geodesy. From 1895 on, Albrecht also directed the International Latitude Service, a cooperative group of various research institutes in many countries that sought the precise determination of the geographic degree of latitude.

BIBLIOGRAPHY

I. ORIGINAL WORKS. Albrecht published a large number of scientific writings, most of which appeared in various astronomical and geodetic journals. The most notable are "Über die Bestimmung von Längendifferenzen mit Hilfe des elektrischen Telegraphen" (Leipzig, 1869), his dissertation; "Genauigkeit der telegraphischen Ortsbestimmung," in *Astronomische Nachrichten,* **89** (1877); "Ausgleichungen des deutschen Längenbestimmungsnetzes," *ibid.,* **95** (1879); "Provisorische Resultate der Beobachtungsreihen Berlin, Potsdam und Prag betr. der Veränderlichkeit der Polhöhe," in *Internationale Erdmessung, Publikationen* (Berlin, 1890); "Stand der Erforschung der Breitenvariation," in *Internationale Erdmessung, Verhandlungen* (Berlin, 1894–1896); "Bestimmung der Längendifferenz Potsdam–Pulkovo im Jahre 1901," *ibid.* (Berlin, 1901); "Bestimmung der Längendifferenz Potsdam–Greenwich im Jahre 1903," *ibid.* (Berlin, 1904); "Bestimmung der Polhöhe und des Azimutes in Memel im Jahre 1907. Telegraphische Längenbestimmung Potsdam–Jena, Jena–Gotha und Gotha–Göttingen im Jahre 1909," *ibid.* (Berlin, 1910); and "Ergebnisse der Breitenbeobachtungen auf dem Observatorium in Johannesburg von März 1910 bis März 1913. Bearbeitet von Theodor Albrecht," *ibid.* (Berlin, 1915). For the *Astronomisch-geodätische Arbeiten für die Gradmessung im Königreich Sachsen* he wrote the third section, "Die astronomischen Arbeiten" (Berlin, 1883–1885). He also wrote *Formeln und Hilfstafeln für geographische Ortsbestimmungen* (Leipzig, 1869; 1873; 5th ed., Berlin, 1967).

II. SECONDARY LITERATURE. Obituary notices with *curriculum vitae* are A. Galle, in *Vierteljahresschrift der Astronomischen Gesellschaft,* **50** (1915), 170–175; and F. R. Helmert, in *Astronomische Nachrichten,* **201** (1915), 269. A short biography is Hans-Ulrich Sandig, in *Neue deutsche Biographie,* I (1953), 183.

HANS BAUMGÄRTEL

ALBUMASAR. See **Abū Maʿshar.**

ALCABITIUS. See **al-Qabīṣī.**

ALCMAEON OF CROTONA (*b.* Crotona, Magna Graecia, *ca.* 535 B.C.), *medicine, natural philosophy.*

Alcmaeon, the son of Peirithoos and a pupil of Pythagoras, is often reported to have been a physician. There is no support for this in ancient sources, however, although Diogenes Laertius stated that Alcmaeon "wrote mostly about medical affairs." As far as we can judge, he also wrote about meteorological and astrological problems and about such philosophical questions as the immortality of the soul. It may therefore be best to call him a natural philosopher, deeply versed in medicine, who was in close contact with both the Pythagoreans and the physicians in Crotona (in this connection we may also think of his contemporary, the physician Democedes of Crotona). One must also keep in mind that at that time the "physiological" side of medicine was treated predominantly by philosophers, Hippocrates being the first to "separate medicine from philosophy," as Celsus states in the preface to *De re medicina.* Aristotle's lost writing *Against Alcmaeon* apparently concerned Alcmaeon as a philosopher.

In the history of science Alcmaeon is especially important for two reasons: he may have written the very first Greek prose book, a *physikos logos;* and he furnished medicine with the first material for a fundamental intellectual mastery of the nontraumatic internal diseases. He defined health as "the isonomy [balance] of forces" (that is, a balance of the opposite bodily qualities of cold and warm, bitter and sweet, and so forth) and internal disease as the "monarchy" of one of these "forces." He further divided the causes of disease into disorders of environment (climatic factors and the like), of nutrition, and of physical mode of living (exertion and such). From these definitions he formulated the bases of a general pathophysiology of internal diseases; similar hypotheses were made by the Hippocrateans. Apparently Alcmaeon clearly recognized the conjectural character of his formulae; they constituted, for him, an "opinion about the invisible."

Alcmaeon also seems to have engaged in dissection, especially ocular dissection for the investigation of the visual process. Obviously, the word *exsectio* in Chalcidius' report is to be taken in this sense; it could hardly refer to a surgical operation on a man since human dissection in a systematic form was, for religious reasons, neither then nor until much later possible in Greece. Among the pre-Socratic philosophers of around 500 B.C., Alcmaeon is the one most closely connected with medicine and therefore had the greatest significance for medicine *per se,* although he himself did not practice as a physician.

BIBLIOGRAPHY

Information on Alcmaeon and fragments of his writings are most accessible in H. Diels and W. Kranz, eds., *Die Fragmente der Vorsokratiker,* I (Berlin, 1951), 210 ff., which covers his statement on nontraumatic internal diseases; his description of his mode of thought as "an opinion about the invisible"; and his references to dissection and Chalcidius' report. Another work of value is Johannes Wachtler, *De Alcmaeone Crotoniata* (Leipzig, 1896). Also of value is Diogenes Laërtius, *Lives of Eminent Philosophers,* V, §25, and VIII, §83, which deal, respectively, with Aristotle's *Contra Alkmaion* and with Alcmaeon's early life.

Fridolf Kudlien

ALCUIN OF YORK (*b.* York, England, *ca.* 735; *d.* Tours, France, 19 May 804), *education.*

Alcuin is not famous for contributing to a specific scientific discipline; rather, his reputation and renown are based upon more general accomplishments. As Charlemagne's educational advisor, he brought Anglo-Saxon learning and teaching methods to the Franks.

Alcuin was born of a noble Northumbrian family. His English name was Ealhwine (Alchvine), but he preferred the Latin form, Albinus; at the court of Charlemagne he acquired the surname Flaccus. Educated at the cathedral school of York under the supervision of the archbishops Egbert and Aelbert, he was exposed to the best traditions of the early English schools. The school of York was heir to the rich pedagogical legacy of the Venerable Bede, and by the beginning of the eighth century its library was the finest in England. The methods and curriculum developed at York brought vitality to early medieval learning.

Alcuin's abilities attracted the attention of his teachers, and he became the protégé of Aelbert. At the death of Egbert in 766, Aelbert became archbishop and Alcuin assumed a major role in the leadership of the school; in 778 he became head of the school and library. When Eanbald became archbishop in 780, Alcuin was sent to Rome to receive the *pallium.* On his return journey the following year, he met Charlemagne at Parma. By this time Alcuin's fame as an educator and scholar had spread to the Continent. The Frankish king needed a competent educational advisor, for education in his kingdom was in a state of decline; he therefore invited Alcuin to become his minister of education. Upon accepting the offer in 782, Alcuin initiated a reform of the Frankish schools. He now became the guiding force behind Charlemagne's educational policies and the leading spirit of the palace school. Charlemagne rewarded Alcuin well for his services: he was granted the abbeys of Ferrières, Troyes, and St. Martin at Tours.

Alcuin popularized the study of the seven liberal arts in France and wrote elementary textbooks on these subjects. While these works do not demonstrate brillant philosophical insight, they do reflect the mind of a creative teacher. His dialogue method of instruction brought needed vitality to teaching; there was now more give and take between teacher and pupil. The emphasis on the elementary subjects of the *trivium* and *quadrivium* encouraged both secular and sacred learning—indeed, the schools themselves were opened to both clerics and laymen, for both church and state needed educated servants.

The knowledge of science imparted by the schools was restricted, and Alcuin's works show only a limited awareness of the physical world. In his *Disputation of the Royal and Most Noble Youth Pepin with Albinus, the Scholastic,* there is a very general discussion of man, the universe, and the natural world. This work is presented in the form of 101 questions, problems, and riddles, with symbolic answers. There are almost no natural or scientific answers; the explanations are in terms of effects rather than causes:

> Pepin: What is the sun?
> Albinus: The splendor of the universe, the beauty of the sky, the glory of the day, the divider of the hours.

Alcuin expressed some interest in astronomy, but it was an interest based on the need for an understanding of calendrical calculations. He helped to develop the Continental interest in the *computus,* and to aid the development of the skills needed to establish the date of Easter, he encouraged the study of mathematics.

In a work ascribed to him, *Propositions for Sharpening the Minds of Youth,* Alcuin presents fifty-three mathematical puzzles. While some can be solved through elaborate and ingenious calculations, many of them require geometrical and algebraic solutions. His encouragement of education was a valuable stimulant to the culture of Charlemagne's realm, and thus he left a lasting legacy to both the culture and the science of Europe.

BIBLIOGRAPHY

I. ORIGINAL WORKS. Alcuin's writings, both in verse and in prose, cover a wide range of subjects. They can be topically classified as educational texts, philosophical and theological treatises, historical works, and letters. Collected editions of his writings have been made by A. Quercetanus (Paris, 1617) and Frobenius Forster, *Alcuini opera* (Regensburg, 1777). The latter edition is reprinted in J. P. Migne, *Patrologia Latina,* C and CI; the letters are in W. Wattenbach and E. L. Duemmler, eds., *Monumenta Alcuiniana,* Vol. VI of Bibliotheca Rerum Germanicarum, P. Jaffe, ed. (Berlin, 1873).

II. SECONDARY LITERATURE. General surveys of Alcuin's life and accomplishments are in E. S. Duckett, *Alcuin, Friend of Charlemagne* (New York, 1951); C. J. B. Gaskoin, *Alcuin: His Life and Work* (London, 1904); A. Kleinclausz, *Alcuin* (Lyons, 1948); J. B. Laforet, *Alcuin restaurateur des sciences en Occident* (Louvain, 1851); Luitpold Wallach, *Alcuin and Charlemagne* (Ithaca, N.Y., 1959); and K. Werner, *Alcuin und sein Jahrhundert* (Paderborn, 1876).

PHILLIP DRENNON THOMAS

ALDER, KURT (*b.* Königshütte, Germany [now Chorzów, Poland], 10 July 1902; *d.* Cologne, Germany, 20 June 1958), *organic chemistry.*

Alder, the son of a schoolteacher in the heavily industrialized area around Kattowitz (now Katowice) in Upper Silesia, received his early education in the German schools of Königshütte. When the region became a part of the new Polish nation after the end of World War I, his family left in order to remain in Germany. After completing the Oberrealschule in Berlin, Alder studied chemistry at the University of Berlin and later at the University of Kiel, where he received the doctorate in 1926. His dissertation, "On the Causes of the Azoester Reaction," was carried out under the direction of Otto Diels. Alder continued his work at Kiel, being made a reader in organic chemistry in 1930 and extraordinary professor of chemistry in 1934. He became a research director at the Bayer Werke in Leverkusen, a branch of I. G. Farbenindustrie, in 1936. In 1940 he returned to academic life as ordinary professor of chemistry and director of the chemical institute at the University of Cologne, where he served until his death. In 1949–1950 he was dean of the Faculty of Philosophy. With Diels, he received the Nobel Prize for chemistry in 1950.

Alder's principal contributions to organic chemistry are associated with the diene synthesis, which grew out of his studies in Diels's laboratory and was first reported in 1928. The synthetic method, frequently referred to as the Diels-Alder reaction, involves the addition of dienes (compounds with conjugated unsaturation, i.e., double bonds on adjacent carbon atoms) to dienophiles (compounds having a double bond activated by nearby carbonyl or carboxyl groups). A simple example is the addition of butadiene to maleic anhydride:

Diene   Dienophile   Diels–Alder adduct

Although a few reactions of this type had been reported over a period of more than 30 years, Diels and Alder recognized the widespread and general nature of the reaction and subsequently spent much of their lives in developing the consequences. They called particular attention to the ease with which such reactions take place and the high yield of adduct.

Their earliest work involved the addition of cyclopentadiene (I) to *p*-quinone (II). The nature of the product (III) of this reaction was the subject of controversy from the time of its preparation by Walter Albrecht in 1893. Diels and Alder, utilizing the corresponding addition of cyclopentadiene to azoester

(IV), were able to identify the structure of Albrecht's compound correctly.

(It will be recalled that azoester had been the subject of Alder's doctoral dissertation.) The two investigators were able to show that, besides azoester and *p*-quinone, they could obtain a reaction of cyclopentadiene with the double bonds in maleic, citraconic, and itaconic acids. They also demonstrated that the adduct is always a six-membered ring, with the addition taking place between the double bond of the dienophile and the carbon atoms at the 1 and 4 positions in the diene.

At first in association with Diels, and then independently with his own students, Alder studied the general experimental conditions of the diene synthesis and the overall scope of the method for synthetic purposes. He was a particularly able stereochemist and showed that diene addition took place at double bonds with a *cis* configuration. In his Nobel Prize address he listed more than a dozen diene types of widely differing structures that had been shown to participate in the reaction. Similarly, he showed that the reaction was equally general with respect to dienophiles, provided the double bond was properly activated by nearby carbonyl, carboxyl, cyano, or nitro groups. Unsaturated compounds without such properly placed activating groups failed to participate in an addition reaction. Many of the compounds studied were prepared in Alder's laboratory for the first time. The Diels-Alder reaction also became useful in structural studies because it provided an analytical means for the detection of conjugated double bonds.

The bridged-ring compounds formed by the use of cyclic dienes were closely related to such naturally occurring terpenes as camphor and norcamphor. The diene synthesis stimulated the understanding of ter-

pene chemistry by providing a synthetic method for preparing such compounds. The ease with which such reactions took place suggested that the diene synthesis might occur in biosynthetic reactions in nature. This role in biosynthesis was also found relevant in connection with anthraquinone-type dyes and a compound that could substitute for vitamin K in stimulating blood coagulation.

The diene synthesis proved to have broad applicability, not only in laboratory syntheses but in commercial operations as well. Commercial products prepared by Diels-Alder reactions include dyes, drugs, insecticides (e.g., dieldrin, aldrin, chlordane), lubricating oils, drying oils, synthetic rubber, and plastics.

During his period of industrial research Alder was involved in the study of polymerization processes connected with the production of Buna-type synthetic rubbers by polymerization of butadiene with such suitable compounds as styrene.

In 1955 he joined seventeen other Nobel laureates in issuing a declaration requesting the nations of the world to renounce war.

BIBLIOGRAPHY

I. ORIGINAL WORKS. Most of Alder's papers were published in *Berichte der Deutschen chemischen Gesellschaft, Liebig's Annalen der Chemie,* and *Angewandte Chemie.* For a full bibliography, see Poggendorff. The original paper on the diene synthesis is in *Liebig's Annalen der Chemie,* **460** (1928), 98–122. Alder's Nobel Prize address, "Diensynthese und verwandte Reaktionstypen," appears in *Les Prix Nobel in 1950* (Stockholm, 1951), pp. 157–194. An English translation is available in the Nobel Foundation's *Nobel Lectures Including Presentation Speeches and Laureates' Biographies, Chemistry, 1942–1962* (Amsterdam, 1964), pp. 266–303. His two main works are "Die Methoden der Diensynthese," in *Handbuch der biologischen Arbeitsmethoden,* sec. 1, II, pt. 2 (1933); and *Neuere Methoden der präparativen organischen Chemie* (Berlin, 1944).

II. SECONDARY LITERATURE. There is no lengthy biography of Alder. Short sketches are Eduard Farber, *Nobel Prize Winners in Chemistry* (New York, 1953), pp. 205–207; M. Günzl-Schumacher, in *Chemikerzeitung,* **82** (1958), 489–490; H. Hauptman, in *Boletim da Associação química do Brasil,* **9** (1951), 1–6; M. Lora-Tamayo, in *Revista de ciencia aplicada* (Madrid), **14** (1960), 193–205; *McGraw-Hill Encyclopedia of Science and Technology* (New York, 1966), I, 6–7; *Les Prix Nobel in 1950* (Stockholm, 1951), pp. 117–118; and *Nobel Lectures Including Presentation Speeches and Laureates' Biographies, Chemistry, 1942–1962* (Amsterdam, 1964), pp. 304–305.

AARON J. IHDE

ALDEROTTI, TADDEO, also known as **Thaddaeus Florentinus** (*b.* Florence, Italy, 1223; *d.* Bologna, Italy, *ca.* 1295), *medicine.*

Biographical information about Alderotti is for the most part based on references to himself in his writings. From these it is known that he was born and brought up in extreme poverty and was an adult before he began his education. Once started on his studies at Bologna, however, he made rapid progress, and within a few years (*ca.* 1260) he was teaching at the university. Indeed, he was one of the founders of medical study at Bologna and was held in such esteem in the city that he was accorded citizenship in 1289.

Alderotti's commentaries on various classical and Islamic writers established the dialectical method of teaching in the medical school, a method that was used until the sixteenth century. He also developed a new form of medical literature, the *Consilia,* a collection of clinical cases with advice on how to treat them. Besides being a teacher of medicine, Alderotti was a well-known and successful practitioner; Pope Honorius IV was one of his patients. The extent of his reputation is attested to by the fact that he is mentioned by Dante in *Paradiso,* XII, 83. He also had a reputation for charging very high fees. His pupils included such persons as Bartolomeo da Varignana, Henri de Mondeville, and Mondino dei Luzzi.

In his commentaries on the works of Hippocrates, Galen, Hunayn ibn Isḥāq, Avicenna, and others, Alderotti utilized the translations of Burgundio of Pisa in preference to those of Constantine the African. He is unique in that he urged his readers to read the original as well as his commentary. At the same time he encouraged more and better translations of classical and Arabic works.

BIBLIOGRAPHY

I. Original Works. In spite of Alderotti's influence, his *Consilia* was not published until the twentieth century (Turin, 1937, G. M. Nardil, ed.). Other works include *In Claudii Galeni artem parvam commentarii* (Naples, 1522); *Expositiones in arduum aphorismorum Hippocratis volumen, in divinum prognosticorum Hippocratis volumen, in praeclarum regiminis acutorum Hippocratis opus, in subtilissimum Joanniti Isagogarum libellum* (Venice, 1527). His treatise *Sulla conservazione della salute* is one of the oldest medical texts in Italian, although it was not published until the nineteenth century, G. Manuzzi and L. Razzolini, eds. (Florence, 1863). A Latin version, *De conservatione sanitatis,* was published much earlier (Bologna, 1477).

II. Secondary Literature. Much of the information about Alderotti was brought together by George Sarton, *Introduction to the History of Science,* II (Baltimore, 1927–1948), 1086–1087. There have been additions and corrections, however, and these have been incorporated in this article. See also Lynn Thorndike, *A History of Magic and Experimental Science,* III (New York, 1923–1958), 14; and "Further Incipits," in *Speculum,* **26** (1951), 675. For a bibliography that gives some of the most recent Italian scholarship, see L. Belloni and L. Vergnano, "Alderotti," in *Enciclopedia Italiana,* II (Rome, 1960), 85. Further information can be found in H. Adelmann, *Marcello Malpighi and the Evolution of Embryology,* I (Ithaca, N. Y., 1966), 76–78.

Vern L. Bullough

ALDINI, GIOVANNI (*b.* Bologna, Italy, 10 April 1762; *d.* Milan, Italy, 17 January 1834), *physics.*

The most significant single event in the history of the development of electricity was the discovery by Alessandro Volta in 1797 of the continuous-flow electric current from a voltaic pile. Next in importance to Volta stood Luigi Galvani, the uncle of Giovanni Aldini.

In the controversy over Galvani's "animal electricity" and Volta's "galvanic current," it was not the modest Galvani but his lusty nephew who wrote, lectured, and published in Italian, French, and English on the theories and experiments of both his uncle and himself. Aldini added notes and a commentary to the second edition of Galvani's important *De viribus electricitatis in motu musculari* (1792). An ardent partisan of his uncle's cause, he followed this supplement with *De animale electricitate, dissertatione duae* (1794) and his best-known work, *Essai théorique et expérimentale sur le galvanisme* (1804). This appeared in two volumes and also, in the same year, as a single quarto volume dedicated to Napoleon. The *Dissertatione duae* resulted from Aldini's galvanic experiments, including those on warm-blooded animals, and generally followed suggestions made by Galvani. A paper on the results of these experiments was read before the Accademia delle Scienze di Bologna; an English translation appeared in 1803 and a French one in 1804.

While Galvani (with one exception) remained silent during the growing controversy over the true nature of his animal electricity, the effervescent Aldini became his uncle's champion—so much so that Volta addressed his arguments to Aldini instead of Galvani. Aldini also probably joined Galvani in the preparation of the anonymous *Dell'uso e dell'attività dell'arco conduttore* (1794). This contained an important experiment, intended to demonstrate the contraction of a dissected frog's leg without the use of any metal, that established the existence of electrical forces within living tissue. Early in 1803 he attempted to determine the velocity of an electric current across the harbor of Calais.

Aldini became professor of physics at the University of Bologna in 1794 and earnestly investigated galvanism. He helped organize a society at Bologna to foster the practices of galvanism in opposition to a Volta society established at the University of Pavia. In 1802 Aldini lectured before the Société Galvanique of Paris and in the following year demonstrated galvanic action in England. Some of his more dramatic experiments involved motion in the anatomical members of a just-executed murderer and induced muscular contraction in dissected parts of sheep, oxen, and chickens. His final writings concerned lighthouses, fire fighting, and quarrying. For his work he was knighted by the emperor of Austria and made councillor of state in Milan.

BIBLIOGRAPHY

Aldini's first known writing is his contribution of notes and a commentary to the second edition of Galvani's *De viribus electricitatis in motu musculari* (Modena, 1792). He and Galvani probably prepared the anonymous *Dell'uso e dell'attività dell'arco conduttore* (Bologna, 1794). Other works by Aldini are *De animale electricitate, dissertatione duae* (Bologna, 1794) and *Essai théorique et expérimentale sur le galvanisme* (Paris, 1804).

BERN DIBNER

ALDROVANDI, ULISSE (*b*. Bologna, Italy, 11 September 1522; *d*. Bologna, May 1605), *natural sciences.*

Aldrovandi is a typical representative of those "universal" and multifaceted minds which seem to have been characteristic of the Renaissance. He was the son of a nobleman, Teseo Aldrovandi, a notary who served as secretary of the Senate of Bologna, and of Veronica Marescalchi, also of a noble family. His mother was a first cousin of Pope Gregory XIII, a circumstance that was helpful to Aldrovandi later in his life, for Bologna was then a papal state.

As a young man, Aldrovandi first studied mathematics under Annibale della Nave, a famous mathematician of the period. Restless by nature and eager to see new things, new countries, and new people, he ran away from home on several occasions. During one of these escapades he went as far as Spain.

After the voyage to Spain, which had been replete with adventures and perils, Aldrovandi returned to Bologna, where he enthusiastically studied Latin under Giovanni Gandolfo, one of the most distinguished humanists of the period.

Aldrovandi's mother, now a widow, wanted him to become a jurist, and he readily applied himself to studying law. Within seven years he was on the verge of receiving his degree, which would have qualified him to practice law, but instead of completing the work he dedicated himself to philosophy. After having studied under the best philosophers of Bologna, he decided, about 1545, to go to Padua to complete his preparation there. This decision had a major influence on his life, for at Padua he began to study medicine and, with the aid of Pietro Catena, again took up mathematics.

On his return to Bologna, Aldrovandi and some of his friends were charged with heresy, probably because at that time the University of Padua was reputed to be one of the main centers for the teaching of Averroës' doctrines. He was obliged to go to Rome to exonerate himself, and there, after proving his innocence, he became interested in the archaeological discoveries in which the city abounded. Later he collected his observations in a book, but perhaps more important, at Rome he met Guillaume Rondelet, who was there as the personal physician of Cardinal Tournon.

Rondelet was then gathering material for his work on fishes. Aldrovandi, who accompanied the French physician to fish markets in order to study the various species, finally decided to study natural history, and began collecting specimens for his own museum.

Upon his return to Bologna, Aldrovandi met Luca Ghini, who then held the professorship of pharmaceutical botany at the university. When Ghini moved to Pisa, Aldrovandi followed him in order to attend his lectures.

The need to earn his living obliged Aldrovandi to take his medical degree, which he received on 23 November 1553. On 14 December, at a solemn ceremony, he was admitted to the Collegio dei Dottori of Bologna, a membership that entitled him not only to practice medicine but also to teach in the university. Thanks to the support given him by an uncle who was a senator, he was also appointed a teacher of "logic" in the University of Bologna. Teaching, however, was merely an easy way of earning an income that would enable him to devote himself entirely to the study of the natural sciences. During vacation periods, Aldrovandi went on long trips, to study nature firsthand and to enrich his knowledge and collections. In 1551 he went as far as Monte Baldo, which he climbed with Luigi dell'Anguillara and Luigi Alpago, who were well-known botanists of the period. In later years he was frequently accompanied on these expeditions by his pupils, who went with him to study botany and to collect samples of fossils and minerals to enrich his "museum" with specimens from every part of Italy.

As a direct result of his intense scientific activity, the Senate appointed Aldrovandi professor of the history of "simples" (which study Aldrovandi had extended to embrace what would now be called natural sciences, including animals and minerals, as well as plants, whether they were of medicinal value or not). His appointment to this professorship was important for the development of natural history, for until then, lectures had been confined to the concise illustration of some single specimen of medicinal value. He was so successful in arousing a lively interest in the more systematic study of natural science, however, that his lectures were attended by an increasing number of students. At the request of the students themselves, the chair was finally declared a full professorship on 11 February 1561.

In the wake of his first success, Aldrovandi, after long and bitter battles, also established at Bologna a botanical garden, of which he was named curator. This new appointment aroused further opposition and envy, and shortly afterward new quarrels arose when he was assigned the task of preparing an *Antidotario,* an official pharmacopoeia. It was to be authoritative in the state of Bologna and would fix the exact characteristics of the drugs and medicinal substances that druggists would be required to use in filling prescriptions.

The variety of tasks, the public and semipublic positions he held, and the conflicts and disputes (which his somewhat obstinate character served only to embitter) were responsible for Aldrovandi's recurrent disagreements with his colleagues on the medical faculty of Bologna. They did not, however, seriously interfere with his truly prodigious studies in natural history. Aldrovandi also had the support of Pope Gregory XIII, who granted him, as a token of his benevolence and esteem, a large sum of money to aid him in the publication of his works.

At his death, Aldrovandi bequeathed to the city of Bologna his museum, his library, and the manuscripts of his unpublished works. During his life he had been able to publish only four folio volumes, illustrated with beautiful copperplates; other volumes were published after his death. His manuscripts are preserved in the libraries of Bologna.

Aldrovandi carried out studies in several fields of natural history: botany, teratology, embryology, icthyology, and ornithology. He has been criticized for having included in his works information and legends devoid of any scientific basis—material that he derived largely from the works of Pliny and that would have been better confined to a medieval bestiary than included in scholarly works.

The period in which Aldrovandi lived and studied was one of transition, however. Science was then being born through the labors of men who, like Aldrovandi, wrote of distant lands but were still obliged to base their accounts almost entirely on secondhand information, gleaned from texts and accounts of travelers. Very often the authors of these accounts were not men of science, but merchants and adventurers whose chief interests had nothing in common with science.

On the other hand, science assumes the existence of a critical, experimental mind, which the men of the Renaissance (Aldrovandi among them) were striving to achieve; it also assumes the inheritance of knowledge, already critically evaluated and classified, with which to compare and test new knowledge as it is acquired. It would therefore be mistaken to ridicule the minute descriptions that Aldrovandi gives us of the sirens, or of other fabulous animals and things.

In embryology, Aldrovandi was able to carry out, within certain limitations, studies in which he excelled and which influenced the work of Volcher Coiter, the Flemish scientist considered one of the founders of embryology. He and Coiter were the first to examine, as Aristotle had suggested, the development of the chick in the egg day by day, opening the eggs successively on each day of the incubation period, in order to describe minutely the changes that take place in the embryo. By this method it became possible for him to show that the heart of the embryo is formed in the "sacco vitellino" and not in the albumen, as other writers had maintained. He also showed that, just as Aristotle had correctly stated, the formation of the heart in the embryo precedes that of the liver, which Galen had incorrectly stated as taking place at the start of the embryonic development.

Aldrovandi also deserves credit for having carried out, in this area of studies, keen observations of a teratological nature, tracing the cause of the morphological changes of the chick to corresponding chemicophysical changes in the substance of the egg yolk.

Even if, from a practical viewpoint, his work and his observations did not contribute greatly to the progress of embryology, they unquestionably had the merit of recalling to the attention of scholars the method of direct observation of natural phenomena. Aldrovandi's studies in this field paved the way for work along the same lines by Fabrizio (Fabricius ab Aquapendente), Malpighi, and Harvey.

Although he did not practice medicine, Aldrovandi's efforts to place botany and pharmacology on a scientific plane and the lucidness and modernity of the legislation he suggested for public health and the civic sanitation of Bologna (found in his unpublished works) suggest that he was a pioneer in hygiene and pharmacology.

Although Aldrovandi is not identified with any revolutionary discoveries, his work as a teacher and as the author of volumes that constitute an irreplaceable cultural patrimony earns him a place among the fathers of modern science. Perhaps most importantly, he was among the first to attempt to free the natural sciences from the stifling influence of the authority of textbooks, for which he substituted, as far as possible, direct study and observation of the animal, vegetable, and mineral worlds.

BIBLIOGRAPHY

I. ORIGINAL WORKS. Works by Aldrovandi, all published at Bologna and all in folio, are *Ornithologiae, hoc est, de avibus historiae libri XII. Agunt de avibus rapacibus* (1600); *Ornithologiae tomus alter de avibus terrestribus, mensae inservientibus et canoris* (1600); *De animalibus insectis libri VII* (1602); *Ornithologiae tomus tertius et ultimus de avibus aquaticis et circa quas degentibus* (1603); *De reliquis animalibus exanguibus, utpote de mollibus, crustaceis, testaceis et zoophytis, libri IV* (1606); *Quadrupedum omnium bisulcorum historia* (1613); *De piscibus libri V et de cetis liber unus* (1613); *De quadrupedibus digitatis viviparis libri III, et de quadrupedibus oviparis libri II* (1637); *Historiae serpentum et draconum libri duo* (1640); *Monstruorum historia* (1642); *Museum metallicum* (1648); and *Dendrologiae naturalis, scilicet arborum historiae libri duo* (1668).

II. SECONDARY LITERATURE. Works on Aldrovandi are H. B. Adelmann, *Marcello Malpighi and the Evolution of Embryology* (Ithaca, N.Y., 1966), for Aldrovandi's contributions to the advancement of embryology; G. Fantuzzi, *Memorie sulla vita e sulle opere di U. Aldrovandi* (Bologna, 1774); L. Frati, *Catalogo dei manoscritti di Ulisse Aldrovandi* (Bologna, 1907); "La vita di U. Aldrovandi," in *Intorno alla vita e alle opere di U. Aldrovandi* (Bologna, 1907); and *La vita di U. Aldrovandi scritta da lui medesimo* (Imola, 1907); L. Samoggia, *Ulisse Aldrovandi medico e igienista* (Bologna, 1962), containing an extensive and up-to-date bibliography of Aldrovandi's manuscripts; and A. Sorbelli, "Contributi alla bibliografia delle opere di Ulisse Aldrovandi," in *Intorno alla vita e alle opere di Ulisse Aldrovandi* (Bologna, 1907), which lists the published works of Aldrovandi and gives information on the various editions.

CARLO CASTELLANI

ALEMBERT, JEAN LE ROND D' (*b.* Paris, France, 17 November 1717; *d.* Paris, 29 October 1783), *physics, mathematics.*

Jean Le Rond d'Alembert was the illegitimate child of Madame de Tencin, a famous salon hostess of the eighteenth century, and the Chevalier Destouches-Canon, a cavalry officer. His mother, who had renounced her nun's vows, abandoned him, for she feared being returned to a convent. His father, however, located the baby and found him a home with a humble artisan named Rousseau and his wife. D'Alembert lived with them until he was forty-seven years old. Destouches-Canon also saw to the education of the child. D'Alembert attended the Collège de Quatre-Nations (sometimes called after Mazarin, its founder), a Jansenist school offering a curriculum in the classics and rhetoric—and also offering more than the average amount of mathematics. In spite of the efforts of his teachers, he turned against a religious career and began studies of law and medicine before he finally embarked on a career as a mathematician. In the 1740's he became part of the *philosophes,* thus joining in the rising tide of criticism of the social and intellectual standards of the day. D'Alembert published many works on mathematics and mathematical physics, and was the scientific editor of the *Encyclopédie.*

D'Alembert never married, although he lived for a number of years with Julie de Lespinasse, the one love of his life. A slight man with an expressive face, a high-pitched voice, and a talent for mimicry, he was known for his wit, gaiety, and gift for conversation, although later in life he became bitter and morose. D'Alembert spent his time much as the other *philosophes* did: working during the morning and afternoon and spending the evening in the salons, particularly those of Mme. du Deffand and Mlle. de Lespinasse. He seldom traveled, leaving the country only once, for a visit to the court of Frederick the Great. D'Alembert was a member of the Académie des Sciences, the Académie Française, and most of the other scientific academies of Europe. He is best known for his work in mathematics and rational mechanics, and for his association with the *Encyclopédie.*

D'Alembert appeared on the scientific scene in July 1739, when he sent his first communication to the Académie des Sciences. It was a critique of a mathematical text by Father Charles Reyneau. During the next two years he sent the academy five more *mémoires* dealing with methods of integrating differential equations and with the motion of bodies in resisting media. Although d'Alembert had received almost no formal scientific training (at school he had studied Varignon's work), it is clear that on his own he had become familiar not only with Newton's work, but also with that of L'Hospital, the Bernoullis, and the other mathematicians of his day. His communications to the academy were answered by Clairaut, who although only four years older than d'Alembert was already a member.

After several attempts to join the academy, d'Alembert was finally successful. He was made *adjoint* in astronomy in May 1741, and received the title of *associé géometre* in 1746. From 1741 through 1743 he worked on various problems in rational mechanics and in the latter year published his famous *Traité de dynamique.* He published rather hastily (a pattern he was to follow all of his life) in order to forestall the loss of priority; Clairaut was working along similar lines. His rivalry with Clairaut, which continued until Clairaut's death, was only one of several in which he was involved over the years.

The *Traité de dynamique,* which has become the most famous of his scientific works, is significant in many ways. First, it is clear that d'Alembert recognized that a scientific revolution had occurred, and he thought that he was doing the job of formalizing the new science of mechanics. That accomplishment is often attributed to Newton, but in fact it was done over a long period of time by a number of men. If d'Alembert was overly proud of his share, he was at least clearly aware of what was happening in science. The *Traité* also contained the first statement of what is now known as d'Alembert's principle. D'Alembert was, furthermore, in the tradition that attempted to develop mechanics without using the notion of force. Finally, it was long afterward said (rather simplistically) that in this work he resolved the famous *vis viva* controversy, a statement with just enough truth in it to be plausible. In terms of his own development, it can be said that he set the style he was to follow for the rest of his life.

As was customary at the time, d'Alembert opened his book with a lengthy philosophical preface. It is true that he was not always faithful to the principles he set down in the preface, but it is astonishing that he could carry his arguments as far as he did and remain faithful to them. D'Alembert fully accepted the prevailing epistemology of sensationalism. Taken from John Locke and expanded by such men as Condillac, sensationalism was to be d'Alembert's metaphysical basis of science. The main tenet of this epistemology was that all knowledge was derived, not from innate ideas, but from sense perception. In many ways, however, d'Alembert remained Cartesian. The criterion of the truth, for example, was still the clear and simple idea, although that idea now had a different origin. In science, therefore, the basic concepts had to conform to this ideal.

In developing his philosophy of mechanics, d'Alembert analyzed the ideas available to him until he came to those that could be analyzed no further; these were to be his starting points. Space and time were such. So simple and clear that they could not even be defined, they were the only fundamental ideas he could locate. Motion was a combination of the ideas of space and time, and so a definition of it was necessary. The word "force" was so unclear and confusing that it was rejected as a conceptual building block of mechanics and was used merely as a convenient shorthand when it was properly and arbitrarily defined. D'Alembert defined matter as impenetrable extension, which took account of the fact that two objects could not pass through one another. The concept of mass, which he defined, as Newton had done, as quantity of matter, had to be smuggled into the treatise in a mathematical sense later on.

In the first part of the *Traité,* d'Alembert developed his own three laws of motion. It should be remembered that Newton had stated his laws verbally in the *Principia,* and that expressing them in algebraic form was a task taken up by the mathematicians of the eighteenth century. D'Alembert's first law was, as Newton's had been, the law of inertia. D'Alembert, however, tried to give an a priori proof for the law, indicating that however sensationalistic his thought might be he still clung to the notion that the mind could arrive at truth by its own processes. His proof was based on the simple ideas of space and time; and the reasoning was geometric, not physical, in nature. His second law, also proved as a problem in geometry, was that of the parallelogram of motion. It was not until he arrived at the third law that physical assumptions were involved.

The third law dealt with equilibrium, and amounted to the principle of the conservation of momentum in impact situations. In fact, d'Alembert was inclined to reduce every mechanical situation to one of impact rather than resort to the effects of continual forces; this again showed an inheritance from Descartes. D'Alembert's proof rested on the clear and simple case of two equal masses approaching each other with equal but opposite speeds. They will clearly balance one another, he declared, for there is no reason why one should overcome the other. Other impact situations were reduced to this one; in cases where the masses or velocities were unequal, the object with the greater quantity of motion (defined as mv) would prevail. In fact, d'Alembert's mathematical definition of mass was introduced implicitly here; he actually assumed the conservation of momentum and defined mass accordingly. This fact was what made his work a mathematical physics rather than simply mathematics.

The principle that bears d'Alembert's name was introduced in the next part of the *Traité*. It was not so much a principle as it was a rule for using the previously stated laws of motion. It can be summarized as follows: In any situation where an object is constrained from following its normal inertial motion, the resulting motion can be analyzed into two components. One of these is the motion the object actually takes, and the other is the motion "destroyed" by the constraints. The lost motion is balanced against either a fictional force or a motion lost by the constraining object. The latter case is the case of impact, and the result is the conservation of momentum (in some cases, the conservation of *vis viva* as well). In the former case, an infinite force must be assumed. Such, for example, would be the case of an object on an inclined plane. The normal motion would be vertically downward; this motion can be resolved into two others. One would be a component down the plane (the motion actually taken) and the other would be normal to the surface of the plane (the motion destroyed by the infinite resisting force of the plane). Then one can easily describe the situation (in this case, a trivial problem).

It is clear that the use of d'Alembert's principle requires some knowledge beyond that of his laws. One must have the conditions of constraint, or the law of falling bodies, or some information derived either empirically or hypothetically about the particular situation. It was for this reason that Ernst Mach could refer to d'Alembert's principle as a routine form for the solution of problems, and not a principle at all. D'Alembert's principle actually rests on his assumptions of what constitutes equilibrium, and it is in his third law of motion that those assumptions appear. Indeed, in discussing his third law (in the second edition of his book, published in 1758) d'Alembert arrived at the equation $\phi = dv/dt$, which is similar to the standard expression for Newton's second law, but which lacks the crucial parameter of mass. The function ϕ was to contain the parameters for specific problems. For example (and this is d'Alembert's example), should the assumption be made that a given deceleration is proportional to the square of the velocity of an object, then the equation becomes $-gv^2 = dv/dt$. The minus sign indicates deceleration, and the constant g packs in the other factors involved, such as mass. In this fashion d'Alembert was able to avoid dealing with forces.

It has often been said that d'Alembert settled the *vis viva* controversy in this treatise, but such a view must be qualified. In the preface d'Alembert did discuss the issue, pointing out that in a given deceleration the change in velocity was proportional to the time. One could therefore define force in terms of the velocity of an object. On the other hand, if one were concerned with the number of "obstacles" that had to be overcome to stop a moving body (here he probably had in mind 'sGravesande's experiments with objects stopped by springs), then it was clear that such a definition of force depended on the square of the velocity and that the related metric was distance, not time. D'Alembert pointed out that these were two different ways of looking at the same problem, that both methods worked and were used with success by different scientists. To use the word "force" to describe either mv or mv^2 was therefore a quarrel of words; the metaphysical notion of force as a universal causal agent was not clarified by such an argument. In this way d'Alembert solved the controversy by declaring it a false one. It involved convention, not reality, for universal causes (the metaphysical meaning of the idea of force) were not known, and possibly not even knowable. It was for this reason that d'Alembert refused to entertain the possibility of talking of forces in mechanics. He did not throw the word away, but used it only when he could give it what today would be called an operational definition. He simply refused to give the notion of force any metaphysical validity and, thus, any ontological reality.

In this way d'Alembert was clearly a precursor of positivistic science. He employed mathematical abstractions and hypothetical or idealized models of physical phenomena and was careful to indicate the shortcomings of his results when they did not closely match the actual events of the world. The metaphysician, he warned in a later treatise, too often built systems that might or might not reflect reality, while the mathematician too often trusted his calculations, thinking they represented the whole truth. But just as metaphysics was suspect because of its unjustified claim to knowledge, so mathematics was suspect in its similar claim. Not everything could be reduced to calculation.

> Geometry owes its certainty to the simplicity of the things it deals with; as the phenomena become more complicated, the results become less certain. It is necessary to know when to stop, when one is ignorant of the thing being studied, and one must not believe that the words *theorem* and *corollary* have some secret virtue so that by writing QED at the end of a proposition one proves something that is not true [*Essai d'une nouvelle théorie de la résistance des fluides,* pp. xlii–xliii].

D'Alembert's instincts were good. Unfortunately, in this case they diverted him from the path that was eventually to produce the principle of the conservation of energy.

A major question that beset all philosophers of the Enlightenment was that of the nature of matter. While d'Alembert's primary concern was mathematical physics, his epistemology of sensationalism led him to speculate on matter theory. Here again, he was frustrated, repeating time after time that we simply do not know what matter is like in its essence. He tended to accept the corpuscular theory of matter, and in Newton's style; that is, he conceived of the ideal atom as perfectly hard. Since this kind of atom could not show the characteristic of elasticity, much less of other chemical or physical phenomena, he was sorely perplexed. In his *Traité de dynamique,* however, he evolved a model of the atom as a hard particle connected to its neighbors by springs. In this way, he could explain elasticity, but he never confused the model with reality. Possibly he sensed that his model actually begged the question, for the springs became more important that the atom itself, and resembled nothing more than a clumsy ether, the carrier of an active principle. Instead of belaboring the point, however, d'Alembert soon returned to mathematical abstraction, where one dealt with functional relations and did not have to agonize over ontology.

In 1744 d'Alembert published a companion volume to his first work, the *Traité de l'équilibre et du mouvement des fluides.* In this work d'Alembert used his principle to describe fluid motion, treating the major problems of fluid mechanics that were current. The sources of his interest in fluids were many. First, Newton had attempted a treatment of fluid motion in his *Principia,* primarily to refute Descartes's *tourbillon* theory of planetary motion. Second, there was a lively interest in fluids by the experimental physicists in the eighteenth century, for fluids were most frequently invoked to give physical explanations for a variety of phenomena, such as electricity, magnetism, and heat. There was also the problem of the shape of the earth: What shape would it be expected to take if it were thought of as a rotating fluid body? Clairaut published a work in 1744 which treated the earth as such, a treatise that was a landmark in fluid mechanics. Furthermore, the *vis viva* controversy was often centered on fluid flow, since the quantity of *vis viva* was used almost exclusively by the Bernoullis in their work on such problems. Finally, of course, there was the inherent interest in fluids themselves. D'Alembert's first treatise had been devoted to the study of rigid bodies; now he was giving attention to the other class of matter, the fluids. He was actually giving an alternative treatment to one already published by Daniel Bernoulli, and he commented that both he and Bernoulli usually arrived at the same conclusions. He felt that his own method was superior. Bernoulli did not agree.

In 1747 d'Alembert published two more important works, one of which, the *Réflexions sur la cause générale des vents,* won a prize from the Prussian Academy. In it appeared the first general use of partial differential equations in mathematical physics. Euler later perfected the techniques of using these equations. The pattern was to become a familiar one: d'Alembert, Daniel Bernoulli, or Clairaut would pioneer a technique, and Euler would take it far beyond their capacity to develop it. D'Alembert's treatise on winds was the only one of his works honored by a prize and, ironically, was later shown to be based on insufficient assumptions. D'Alembert assumed that wind patterns were the result of tidal effects on the atmosphere, and he relegated the influence of heat to a minor role, one that caused only local variations from the general circulation. Still, as a work on atmospheric tides it was successful, and Lagrange continued to praise d'Alembert's efforts many years later.

D'Alembert's other important publication of 1747 was an article in the *Mémoirs* of the Prussian Academy dealing with the motion of vibrating strings, another problem that taxed the minds of the major mathematicians of the day. Here the wave equation made its first appearance in physics. D'Alembert's mathematical instincts led him to simplify the boundary conditions, however, to the point where his solution, while correct, did not match well the observed phenomenon. Euler subsequently treated the same problem more generally; and although he was no more correct than d'Alembert, his work was more useful.

During the late 1740's, d'Alembert, Clairaut, and Euler were all working on the famous three-body problem, with varying success. D'Alembert's interest in celestial mechanics thus led him, in 1749, to publish a masterly work, the *Recherches sur la précession des équinoxes et sur la nutation de la terre.* The precession of the equinoxes, a problem previously attacked by Clairaut, was very difficult. D'Alembert's method was similar to Clairaut's, but he employed more terms in his integration of the equation of motion and arrived at a solution more in accord with the observed motion of the earth. He was rightly proud of his book.

D'Alembert then applied himself to further studies in fluid mechanics, entering a competition announced by the Prussian Academy. He was not awarded the prize; indeed, it was not given to anybody. The

Prussian Academy took this action on the ground that nobody had submitted experimental proof of the theoretical work. There has been considerable dispute over this action. The claim has been made that d'Alembert's work, although the best entered, was marred by many errors. D'Alembert himself viewed his denial as the result of Euler's influence, and the relations between the two men deteriorated further. Whatever the case, the disgruntled d'Alembert published his work in 1752 as the *Essai d'une nouvelle théorie de la résistance des fluides.* It was in this essay that the differential hydrodynamic equations were first expressed in terms of a field and the hydrodynamic paradox was put forth.

In studying the flow lines of a fluid around an object (in this case, an elliptical object), d'Alembert could find no reason for assuming that the flow pattern was any different behind the object than in front of it. This implied that whatever the forces exerted on the front of the object might be, they would be counteracted by similar forces on the back, and the result would be no resistance to the flow whatever. The paradox was left for his readers to solve. D'Alembert had other difficulties as well. He found himself forced to assume, in order to avoid the necessity of allowing an instantaneous change in the velocity of parts of the fluid moving around the object, that a small portion of the fluid remained stagnant in front of the object, an assumption required to prevent breaking the law of continuity.

In spite of these problems, the essay was an important contribution. Hunter Rouse and Simon Ince have said that d'Alembert was the first "to introduce such concepts as the components of fluid velocity and acceleration, the differential requirements of continuity, and even the complex numbers essential to modern analysis of the same problem." Clifford Truesdell, on the other hand, thinks that most of the credit for the development of fluid mechanics must be granted to Euler; thus historians have continued the disputes that originated among the scientists themselves. But it is often difficult to tell where the original idea came from and who should receive primary recognition. It is certain, however, that d'Alembert, Clairaut, Bernoulli, and Euler were all active in pursuing these problems, all influenced one another, and all deserve to be remembered, although Euler was no doubt the most able of the group. But they all sought claims to priority, and they guarded their claims with passion.

D'Alembert wrote one other scientific work in the 1750's, the *Recherches sur différens points importants du systême du monde.* It appeared in three volumes, two of them published in 1754 and the third in 1756.

Devoted primarily to the motion of the moon (Volume III included a new set of lunar tables), it was written at least partially to guard d'Alembert's claims to originality against those of Clairaut. As was so often the case, d'Alembert's method was mathematically more sound, but Clairaut's method was more easily used by astronomers.

The 1750's were more noteworthy in d'Alembert's life for the development of interests outside the realm of mathematics and physics. Those interests came as a result of his involvement with the *Encyclopédie.* Denis Diderot was the principal editor of the enterprise, and d'Alembert was chosen as the science editor. His efforts did not remain limited to purely scientific concerns, however. His first literary task was that of writing the *Discours préliminaire* of the *Encyclopédie,* a task that he accomplished with such success that its publication was largely the reason for his acceptance into the Académie Française in 1754.

The *Discours préliminaire,* written in two parts, has rightly been recognized as a cardinal document of the Enlightenment. The first part is devoted to the work as an *encyclopédie,* that is, as a collection of the knowledge of mankind. The second part is devoted to the work as a *dictionnaire raisonnée,* or critical dictionary. Actually, the first part is an exposition of the epistemology of sensationalism, and owes a great deal to both John Locke and Condillac. All kinds of human knowledge are discussed, from scientific to moral. The sciences are to be based on physical perception, and morality is to be based on the perception of those emotions, feelings, and inclinations that men can sense within themselves. Although d'Alembert gives lip service to the truths of religion, they are clearly irrelevant and are acknowledged only for the sake of the censors. For this reason, the *Discours préliminaire* came under frequent attack; nevertheless, it was generally well received and applauded. It formed, so to say, the manifesto of the now coalescing party of *philosophes;* the body of the *Encyclopédie* was to be the expression of their program.

The second part of the *Discours préliminaire* is in fact a history of science and philosophy, and clearly shows the penchant of the *philosophes* for the notion of progress through the increased use of reason. As a history, it has often quite properly been attacked for its extreme bias against the medieval period and any form of thought developed within the framework of theology, but this bias was, of course, intentional. At the end of this history, the *philosophes'* debt to Francis Bacon is clearly acknowledged in the outline of the organization of knowledge. A modified version of Bacon's tree of knowledge is included and briefly explained. All knowledge is related to three functions of the mind: memory, reason, and imagination. Reason is clearly the most important of the three. Bacon's emphasis on utility was also reflected in the *Encyclopédie,* although more by Diderot than by d'Alembert. D'Alembert's concept of utility was far wider than that of most people. To him, the things used by philosophers—even mathematical equations—were very useful, even though the bulk of the public might find them mysterious and esoteric.

In the midst of this activity, d'Alembert found time to write a book on what must be called a psychophysical subject, that of music. In 1752 he published his *Élémens de musique théorique et pratique suivant les principes de M. Rameau.* This work has often been neglected by historians, save those of music, for it was not particularly mathematical and acted as a popularization of Rameau's new scheme of musical structure. Yet it was more than simply a popularization. Music was still emerging from the mixture of Pythagorean numerical mysticism and theological principles that had marked its rationale during the late medieval period. D'Alembert understood Rameau's innovations as a liberation; music could finally be given a secular rationale, and his work was important in spreading Rameau's ideas throughout Europe.

As time went on, d'Alembert's pen was increasingly devoted to nonscientific subjects. His articles in the *Encyclopédie* reached far beyond mathematics. He wrote and read many essays before the Académie Française; these began to appear in print as early as 1753. In that year he published two volumes of his *Mélanges de littérature et de philosophie.* The first two were reprinted along with two more in 1759; a fifth and last volume was published in 1767. The word *mélanges* was apt, for in these volumes were essays on music, law, and religion, his treatise on the *Élémens de philosophie,* translations of portions of Tacitus, and other assorted literary efforts. They make an odd mixture, for some are important in their exposition of Enlightenment ideals, while others are mere polemics or even trivial essays.

In 1757 d'Alembert visited Voltaire at Ferney, and an important result of the visit was the article on Geneva, which appeared in the seventh volume of the *Encyclopédie.* It was clearly an article meant to be propaganda, for the space devoted to the city was quite out of keeping with the general editorial policy. In essence, d'Alembert damned the city by praising it. The furor that resulted was the immediate cause of the suspension of the license for the *Encyclopédie.* D'Alembert resigned as an editor, convinced that the enterprise must founder, and left Diderot to finish

the task by himself. Diderot thought that d'Alembert had deserted him, and the relations between the men became strained. Rousseau also attacked d'Alembert for his view that Geneva should allow a theater, thus touching off another of the famous controversies that showed that the *philosophes* were by no means a totally unified group of thinkers.

D'Alembert's chief scientific output after 1760 was his *Opuscules mathématiques,* eight volumes of which appeared from 1761 to 1780. These collections of mathematical essays were a mixed bag, ranging from theories of achromatic lenses to purely mathematical manipulations and theorems. Included were many new solutions to problems he had previously attacked—including a new proof of the law of inertia. Although the mathematical articles in the *Encyclopédie* had aired many of his notions, these volumes provide the closest thing to a collection of them that exists.

As Carl Boyer has pointed out, d'Alembert was almost alone in his day in regarding the differential as the limit of a function, the key concept around which the calculus was eventually rationalized. Unfortunately, d'Alembert could never escape the tradition that had made geometry preeminent among the sciences, and he was therefore unable to put the idea of the limit into purely algorithmic form. His concept of the limit did not seem to be any more clear to his contemporaries than other schemes invented to explain the nature of the differential.

It has often been said that d'Alembert was always primarily a mathematician and secondarily a physicist. This evaluation must be qualified. No doubt he sensed the power of mathematics. But, as he once said, "Mathematics owes its certainty to the simplicity of the things with which it deals." In other words, d'Alembert was never able to remove himself to a world of pure mathematics. He was rather in the tradition of Descartes. Space was the realization of geometry (although, unlike Descartes, d'Alembert drew his evidence from sense perception). It was for this reason that he could never reduce mathematics to pure algorithms, and it is also the reason for his concern about the law of continuity. In mathematics as well as physics, discontinuities seemed improper to d'Alembert; equations that had discontinuities in them gave solutions that he called "impossible," and he wasted no time on them. It was for this reason that the notion of perfectly hard matter was so difficult for him to comprehend, for two such particles colliding would necessarily undergo sudden changes in velocity, something he could not allow as possible.

It was probably the requirement of continuity that led d'Alembert to his idea of the limit, and it also led him to consider the techniques of handling series. In Volume V of the *Opuscules* he published a test for convergence that is still called d'Alembert's theorem. The mathematical statement is:

If $\lim_{n \to \infty} |S_{n+1}/S_n| = r$, and $r < 1$, the series $\sum_{n=1}^{\infty} S_n$ converges. If $r > 1$, the series diverges; if $r = 1$, the test fails.

But in spite of such original contributions to mathematical manipulation, d'Alembert's chief concern was in making this language not merely descriptive of the world, but congruent to it. The application of mathematics was a matter of considering physical situations, developing differential equations to express them, and then integrating those equations. Mathematical physicists had to invent much of their procedure as they went along. Thus, in the course of his work, d'Alembert was able to give the first formulation of the wave equation, to express the first partial differential equation, and to be the first to solve a partial differential equation by the technique of the separation of variables. But probably the assignment of "firsts" in this way is not the best manner of evaluating the development of mathematics or of mathematical physics. For every such first, one can find other men who had alternative suggestions or different ways of expressing themselves, and who often wrote down similar but less satisfactory expressions.

More important, possibly, is the way in which these ideas reflect the mathematicians' view of nature, a view that was changing and was then very different from that of a mathematical physicist today. D'Alembert's very language gives a clue. He used, for example, the word *fausse* to describe a divergent series. The word to him was not a bare descriptive term. There was no match, or no useful match, for divergence in the physical world. Convergence leads to the notion of the limit; divergence leads nowhere—or everywhere.

D'Alembert has often been cited as being oddly ineffective when he considered probability theory. Here again his view of nature, not his mathematical capabilities, blocked him. He considered, for example, a game of chance in which Pierre and Jacques take part. Pierre is to flip a coin. If heads turns up on the first toss, he is to pay Jacques one *écu.* If it does not turn up until the second toss, he is to pay two *écus.* If it does not turn up until the third toss, he is to pay four *écus,* and so on, the payments mounting in geometric progression. The problem is to determine how many *écus* Jacques should give to Pierre before the game begins in order that the two men have equal chances at breaking even. The solution seemed to be that since the probability on each toss was one-half,

and since the number of tosses was unlimited, then Jacques would have to give an infinite number of *écus* to Pierre before the game began, clearly a paradoxical situation.

D'Alembert rebelled against this solution, but had no satisfactory alternative. He considered the possibility of tossing tails one hundred times in a row. Metaphysically, he declared, one could imagine that such a thing could happen; but one could not realistically imagine it happening. He went further: heads, he declared, must *necessarily* arise after a finite number of tosses. In other words, any given toss is influenced by previous tosses, an assumption firmly denied by modern probability theory. D'Alembert also said that if the probability of an event were very small, it could be treated as nothing, and therefore would have no relevance to physical events. Jacques and Pierre could forget the mathematics; it was not applicable to their game.

It is no wonder that such theorizing caused d'Alembert to have quarrels and arguments with others. Moreover, there were reasons for interest in probability outside games of chance. It had been known for some time that if a person were inoculated with a fluid taken from a person having smallpox, the result would usually be a mild case of the disease, followed by immunity afterward. Unfortunately, a person so inoculated occasionally would develop a more serious case and die. The question was posed: Is one more likely to live longer with or without inoculation? There were many variables, of course. For example, should a forty-year-old, who was already past the average life expectancy, be inoculated? What, in fact, was a life expectancy? How many years could one hope to live, from any given age, both with and without inoculation? D'Alembert and Daniel Bernoulli carried on extensive arguments about this problem. What is significant about d'Alembert's way of thinking is that he expressed the feeling that the laws of probability were faint comfort to the man who had his child inoculated and lost the gamble. To d'Alembert, that factor was as important as any mathematical ratio. It was not, as far as he was concerned, irrelevant to the problem.

Most of these humanitarian concerns crept into d'Alembert's work in his later years. Aside from the *Opuscules,* there was only one other scientific publication after 1760 that carried his name: the *Nouvelles expériences sur la résistance des fluides* (published in 1777). Listed as coauthors were the Abbé Bossut and Condorcet. The last two actually did all of the work; d'Alembert merely lent his name.

In 1764 d'Alembert spent three months at the court of Frederick the Great. Although frequently asked by Frederick, d'Alembert refused to move to Potsdam as president of the Prussian Academy. Indeed, he urged Frederick to appoint Euler, and the rift that had grown between d'Alembert and Euler was at last repaired. Unfortunately, Euler was never trusted by Frederick, and he left soon afterward for St. Petersburg, where he spent the rest of his life.

In 1765 d'Alembert published his *Histoire de la destruction des Jésuites.* The work was seen through the press by Voltaire in Geneva, and although it was published anonymously, everyone knew who wrote it. A part of Voltaire's plan *écraser l'infâme,* this work is not one of d'Alembert's best.

In the same year, d'Alembert fell gravely ill, and moved to the house of Mlle. de Lespinasse, who nursed him back to health. He continued to live with her until her death in 1776. In 1772 he was elected perpetual secretary of the Académie Française, and undertook the task of writing the eulogies for the deceased members of the academy. He became the academy's most influential member, but, in spite of his efforts, that body failed to produce anything noteworthy in the way of literature during his pre-eminence. D'Alembert sensed his failure. His later life was filled with frustration and despair, particularly after the death of Mlle. de Lespinasse.

Possibly d'Alembert lived too long. Many of the *philosophes* passed away before he did, and those who remained alive in the 1780's were old and clearly not the vibrant young revolutionaries they had once been. What political success they had tasted they had not been able to develop. But, to a large degree, they had, in Diderot's phrase, "changed the general way of thinking."

BIBLIOGRAPHY

I. Original Works. There have been no collections made of d'Alembert's scientific works, although reprints of the original editions of his scientific books (except the *Opuscules mathématiques*) have recently been issued by Éditions Culture et Civilisation, Brussels. There are two collections of d'Alembert's *Oeuvres* which contain his literary pieces: the Bélin ed., 18 vols. (Paris, 1805); and the Bastien ed., 5 vols. (Paris, 1821). The most recent and complete bibliographies are in Grimsley and Hankins (see below).

II. Secondary Literature. The following works are devoted primarily to d'Alembert or accord him a prominent role: Joseph Bertrand, *D'Alembert* (Paris, 1889); Carl Boyer, *The History of the Calculus and Its Conceptual Development* (New York, 1949), ch. 4; René Dugas, *A History of Mechanics* (Neuchâtel, 1955), pp. 244–251, 290–299; Ronald Grimsley, *Jean d'Alembert* (Oxford,

1963); Maurice Müller, *Essai sur la philosophie de Jean d'Alembert* (Paris, 1926); Hunter Rouse and Simon Ince, *A History of Hydraulics* (New York, 1963), pp. 100–107; Clifford Truesdell, *Continuum Mechanics,* 4 vols. (New York, 1963–1964); and Arthur Wilson, *Diderot: The Testing Years* (New York, 1957). Of the above, Boyer, Dugas, Rouse and Ince, and particularly Truesdell, deal specifically and in detail with d'Alembert's science.

Three recent doctoral dissertations on d'Alembert are J. Morton Briggs, *D'Alembert: Mechanics, Matter, and Morals* (New York, 1962): Thomas Hankins, *Jean d'Alembert, Scientist and Philosopher* (Cornell University, 1964); and Harold Jarrett, *D'Alembert and the Encyclopédie* (Durham, N. C., 1962).

J. Morton Briggs

ALEXANDER OF APHRODISIAS (*fl.* second–third century A.D.), *philosophy.*

Alexander was a Peripatetic philosopher of the second–third century among whose masters were Herminus, Sosigenes, and Aristocles. His fame rests mainly on his interpretation of Aristotle's doctrines, the scholarly qualities of which earned him the sobriquet of "the interpreter" (ὁ ἐξηγητής). Of his works other than commentaries, four have survived in Greek manuscripts: *On the Soul; On Fate;* a writing going under the title *On Mixture;* and another, in four books, going under the title *Natural Questions,* of which the fourth book, however, deals mainly with ethical problems. Additional material is likely to be found in Arabic and Armenian.

Of these, the second part of *On the Soul* and the *Natural Questions* are collections of short pieces (some twenty-five in *On the Soul,* sixty-nine in the *Questions*) dealing with a great variety of topics and representing different literary forms. It is rather certain that these collections were not arranged or edited by Alexander, and that some pieces are inauthentic. But few of the problems thus posed have yet been sufficiently explored; the following presentation of Alexander's doctrines will be based indiscriminately on texts handed down to us under his name.

Of his commentaries on Aristotle, those on *Analytica priora I, Topics, Meteorologica,* and *On Sense and Sensibilia* survive in their entirety. Of the commentary on *Metaphysics* under his name, only the part dealing with *Metaphysics A–Λ* is genuine; the rest, usually referred to as a work by Pseudo-Alexander (his identity is not known), is not his, although it does contain some genuine passages. Of other commentaries, only fragments in the form of quotations in other commentators on Aristotle survive. Of his interpretations of Aristotle (either in formal commentaries or in other writings), two are particularly famous. Whereas Aristotle made Plato's ideas immanent in the sensible individuals but insisted that only this "ideal" (i.e., universal) aspect of sensibles can be known, so that with regard to us the individual (object of sensation) is prior to the universal—although the universal is actually prior to the individual—Alexander went one step further and declared that only individuals actually exist, the universals existing only as products of our mental (noetic) activity (νοεῖν), which abstracts them from the individuals (or the individual existing only in one exemplar, e.g., the phoenix). Therefore, the universals exist only as long as they are perceived. Alexander calls them νοητά (usually translated "intelligibles," in which case νοεῖν would best be translated "to intelligize"; we could then translate the noun νοῦς, the agent of intelligizing, as "intelligence"; one of the Latin translations of νοῦς is *mens,* to which, unfortunately, only the adjective "mental" corresponds in English).

But in addition to these intelligibles (corresponding to Plato's transcendent ideas made immanent by Aristotle) existing only as the results of our mental acts, Alexander admits the existence of intelligibles existing outside the realm of the sensible. Roughly, they correspond to Aristotle's "pure" forms, of which the best-known example is his supreme deity, the Unmoved Mover. These "higher" intelligibles (κυρίως νοητά) have one thing in common with the lower ones: they exist only as objects of mental (noetic) acts, but the νοῦς (intelligence) that intelligizes ("perceives") them is not our human intelligence. Rather, it is a (or the) divine intellect, one of whose marks is that its activity is eternal and incessant, so that these "higher" intelligibles also exist eternally and incessantly. They are "caused" by the highest intelligible, in the description of whose causality Alexander anticipates some Neoplatonic categories. The mental act perceiving them does not "abstract" them from matter, for they are not embodied.

Connected with this piece of noetics is another, with the help of which Alexander interprets a most difficult aspect of Aristotle's psychology in his *On the Soul,* Book III, chapters 4 and 5. According to Alexander, Aristotle teaches the existence of human intelligence (Alexander calls it passive, or potential, or material intelligence), which is different for different individuals and is part of everybody's soul, and of another intelligence, which is identical with the Supreme Deity, called active intelligence (it is this intelligence that incessantly and eternally perceives itself by perceiving the "higher" intelligibles). This intelligence–Deity is unique; it "enters" man from without (i.e., it is not connected in any way with his body); it is active also in the sense of activating the human

intelligence, thus enabling this intelligence to perceive intelligibles of both the lower and the higher order. Human intelligence thus activated (we could also say "transformed," and Alexander almost says "divinized") in different aspects of its activity is called by Alexander intelligence "in action," or "acquired as habit," or "acquired as disponible skill" (ἐνεργείαι, ἐπίκτητος, καθ᾽ ἕξιν). The most conspicuous result of this theory is the denial of any kind of personal immortality. Man's soul perishes with his body; his intelligence, qua transformed by the active intelligence, survives by being reabsorbed into that unique, impersonal, divine intelligence. It is remarkable that Pseudo-Alexander describes the experience of "transformation," after which human intelligence becomes capable of perceiving the "higher" intelligibles, as a mystical (ineffable) experience. The assertion or the denial of the correctness of Alexander's interpretation of Aristotle and, even more, the correctness of the doctrine (denial of personal immortality) became one of the great controversies of the Middle Ages and early modern times.

Only a few other doctrines of Alexander can be mentioned here.

(1) In Aristotle's writings all change is ultimately reduced to locomotion, and prime locomotion is attributed to the celestial bodies (fixed stars and planets, and their spheres). Three explanations are given of the cause of this locomotion. One is that all celestial bodies are moved by being attracted to their Unmoved Mover as lovers are attracted by the objects of their love; the second is that they consist of an element, the ether, which by nature moves eternally, incessantly, and circularly; the third is that they are animated and moved by their souls. Alexander tried to reconcile these three explanations. Ether is animated and the soul is its nature. This soul desires to imitate the Unmoved Mover, which it does by eternally circling him.

(2) Alexander undertakes to prove that man's will is free (or, as the Greek has it, that there are things in our power, ἐφ᾽ ἡμῖν). One of his main arguments is that nature distinguished man from other animals by endowing him with the faculty of deliberation, which mediates between stimuli (φαντασίαι) and actions, whereas animals simply react to stimuli. And since nature does nothing in vain, the exercise of this deliberative faculty results in reasonable assent to (or dissent from) stimuli, which proves that we are free to choose.

(3) Another theory explaining the freedom of will is based on the assertion that whereas the realm of the eternal and immutable is, if we may say so, full of being, the realm of the changeable (of becoming

and perishing) is permeated by nonbeing. In fact, this nonbeing is responsible for such things as chance and freedom of the will; there is no cause of these phenomena.

(4) Connected with the free-will theory is Alexander's treatment of the problem of fate or destiny ($\varepsilon \dot{\iota} \mu \alpha \rho \mu \acute{\varepsilon} \nu \eta$), i.e., the doctrine of an unbroken causal chain. The fact of human freedom proves this doctrine wrong. The meaning of the word "fate" should be taken to indicate that everything acts according to its own, individual, nongeneric nature ($\phi \acute{\upsilon} \sigma \iota s$); in fact, "fate" and "nature" coincide without abridgment of man's freedom.

(5) Alexander discusses the problem of providence. He denies that the divine provides in a direct way ($\pi \rho o \eta \gamma o \upsilon \mu \acute{\varepsilon} \nu \omega s$), i.e., the way a shepherd provides for his flock; such providence, says Alexander, would amount to saying that the divine (superior) exists for the sake of or profits from the inferior. He also denies that the effects of divine providence are merely accidental ($\kappa \alpha \tau \grave{\alpha} \sigma \upsilon \mu \beta \varepsilon \beta \eta \kappa \acute{o} s$), but he does insist that there are other manners of divine providence, and promises to prove that contrary to what others have asserted, Aristotle recognizes providence. Alexander himself at least tentatively identifies the sum total of effects emanating from the everlasting circular movement of the celestial bodies with providence, its main effect being the *generic* immortality of perishable individuals of which the world of becoming consists.

(6) Alexander asserts the existence of natural justice. His main proof is that nature created man to live in community; that there can be no community without justice; that therefore justice is natural ($\phi \acute{\upsilon} \sigma \varepsilon \iota$).

(7) The object of man's fundamental desire ($\tau \grave{o} \pi \rho \tilde{\omega} \tau o \nu \ o \dot{\iota} \kappa \varepsilon \tilde{\iota} o \nu$) is pleasure (the apparent good) rather than, e.g., self-preservation.

(8) Moral perfection ($\dot{\alpha} \rho \varepsilon \tau \acute{\eta}$) does not guarantee a happy life, as can be seen from the fact that a morally perfect man is justified in committing suicide for good reasons ($\varepsilon \ddot{\upsilon} \lambda o \gamma o s \ \dot{\varepsilon} \xi \alpha \gamma \omega \gamma \acute{\eta}$), which he would never do if his life were a happy one.

(9) Nobody can possess one moral perfection ($\dot{\alpha} \rho \varepsilon \tau \acute{\eta}$), such as courage, without possessing all others.

(10) Alexander devotes a comparatively large amount of space to the problem of vision and related problems.

(11) Alexander refutes in great detail the Stoic doctrine of total interpenetration of bodies ($\kappa \rho \tilde{\alpha} \sigma \iota s \ \delta \iota$' $\ddot{o} \lambda o \upsilon$), which he feels is the foundation of the main tenets of the whole Stoic system.

(12) The magnet attracts because iron desires it, just as other things, although inanimate, desire that which nature has destined for them.

In any history of the problem of squaring the circle, Alexander is likely to be mentioned as he commented on all passages in which Aristotle criticized the methods used for this purpose by Hippocrates of Chios, Bryson, and Antiphon, always briefly to the point of obscurity. It seems that Alexander, probably misled by Aristotle, falsely assumed that Hippocrates did not distinguish lunules formed on quadrants (sides of a square inscribed in a circle) from sextants (sides of a hexagon inscribed in a circle), and also assumed that Antiphon violated the principle that a curve and a straight line can have only a point in common, which probably implies that Antiphon asserted the existence of atomic lengths of which both curves and straight lines would consist. On this basis Alexander rejected Hippocrates' and Antiphon's methods of squaring the circle.

BIBLIOGRAPHY

I. ORIGINAL WORKS. All commentaries mentioned in the text are available in the collection *Commentaria in Aristotelem Graeca*, 23 vols. (Berlin, 1882–1909), Vols. II and III in 2 parts each. All his other writings preserved in Greek are in *Supplementum Aristotelicum*, II, pts. 1 and 2 (Berlin, 1887–1892). The content of a writing on providence, translated from Greek into Arabic, has been translated into French in P. Thillet, "Un traité inconnu d'Alexandre d'Aphrodise sur la Providence dans une version arabe inédite," in *Actes du Premier Congrès International de Philosophie Médiévale* (Louvain–Paris, 1960), pp. 313–324. See also A. Dietrich, "Die arabische Version einer unbekannten Schrift des Alexander von Aphrodisias über die differentia specifica," in *Nachrichten der Akademie der Wissenschaften in Göttingen*, **1** (1964), 90–148; E. G. Schmidt, "Alexander von Aphrodisias in einem altarmenischen Kategorien-Kommentar," in *Philologus*, **110** (1966), 277–286; J. van Ess, "Über einige neue Fragmente des Alexander von Aphrodisias und des Proklos in arabischer Übersetzung," in *Der Islam*, **42** (1966), 148–168. Translations of Alexander's works include *On Destiny*, A. FitzGerald, ed. and trans. (London, 1931); and *Commentary on Book IV of Aristotle's Meteorologica*, V. C. B. Coutant, trans. (New York, 1936).

II. SECONDARY LITERATURE. A brief but comprehensive presentation is E. Zeller, *Die Philosophie der Griechen*, 5th ed., III, pt. 1 (Leipzig, 1923; repr. 1963), 817–830. Still briefer are A. Tognolo, "Alessandro di Aphrodisia" and "Alessandrismo," in *Enciclopedia filosofica*, I (Venice–Rome, 1957), 136–139; and F. Ueberweg and K. Praechter, *Die Philosophie des Altertums*, 12th ed. (Berlin, 1926; repr. Basel, 1953). Special problems are discussed in I. Bruns, "Studien zu Alexander von Aphrodisias," in *Rheinisches Museum*, **44** (1889), 613–630; **45** (1890), 138–145, 223–235; and Preface to his ed. of the *Natural Questions* in *Supplementum Aristotelicum* (see above), pp. v–xiv; E. Freudenthal, "Die durch Averroes erhaltenen Fragmente Alexanders zur Metaphysik des Aristoteles," in *Abhandlungen der Berliner Akademie vom Jahre 1884* (Berlin, 1885); P. Merlan, "Ein Simplikios-Zitat bei Ps. Alexandros und ein Plotinos-Zitat bei Simplikios," in *Rheinisches Museum,* **84** (1935), 154–160; *Philologische Wochenschrift,* **58** (1938), 65–69; and *Monopsychism, Mysticism, Metaconsciousness* (The Hague, 1963), Index, under "Alexander" and "Pseudo-Alexander"; P. Moraux, *Alexandre d'Aphrodise, exégète de la noétique d'Aristote* (Paris, 1942); and J. Zahlfleisch, "Die Polemik Alexanders von Aphrodisias gegen die verschiedenen Theorien des Sehens," in *Archiv für Geschichte der Philosophie,* **8** (1895), 373–386, 498–509; **9** (1896), 149–162. Additional literature is listed in Ueberweg and Praechter (see above).

Virtually all presentations of Aristotle's noetics deal with Alexander; a recent example is L. Barbotin, *La théorie aristotelicienne de l'intellect d'après Théophraste* (Louvain, 1954). Other recent literature includes O. Becker, "Formallogisches und Mathematisches in griechischen philosophischen Texten," in *Philologus,* **100** (1956), 108–112; R. Hackforth, "Notes on Some Passages of Alexander Aphrodisiensis *De fato,*" in *Classical Quarterly,* **40** (1946), 37–44; F. P. Hager, "Die Aristotelesinterpretation des Alexander von Aphrodisias und die Aristoteleskritik Plotins bezüglich der Lehre vom Geist," in *Archiv für Geschichte der Philosophie,* **46** (1964), 174–187; P. Henry, "Une comparaison chez Aristote, Alexandre et Plotin," in *Les sources de Plotin* (Geneva, 1960), pp. 429–444; H. Langerbeck, "Zu Alexander von Aphrodisias' *De fato,*" in *Hermes,* **64** (1936), 473–474; S. Luria, "Die Infinitesimaltheorie der antiken Atomisten," in *Quellen und Studien zur Geschichte der Mathematik, Astronomie und Physik, Abteilung B: Studien,* **2** (1932), 106–185; P. Moraux, "Alexander von Aphrodisias *Quaestiones* 2, 3," in *Hermes,* **95** (1967), 159–169; R. A. Pack, "A Passage in Alexander of Aphrodisias Relating to the Theory of Tragedy," in *American Journal of Philology,* **58** (1937), 418–436; S. Pines, "Omne quod movetur necesse est ab aliquo moveri: A Refutation of Galen by Alexander of Aphrodisias and the Theory of Motion," in *Isis,* **52** (1961), 21–54; J. M. Rist, "On Tracking Alexander of Aphrodisias," in *Archiv für Geschichte der Philosophie,* **48** (1966), 82–90.

PHILIP MERLAN

ALEXANDER OF MYNDOS (*b.* Myndos, Caria; *fl. ca.* A.D. 25–50), *biology.*

Nothing certain is known of Alexander's life and his dates are conjectural, but internal evidence indicates that he flourished in the first half of the first century. None of his writings has survived intact, although he seems to have been widely read in antiquity and is cited by Aelian, Athenaeus, Diogenes Laertius, and Photius. He was essentially a compiler, but with wide interests ranging from animal lore and medicine to dream analysis and mythology.

Alexander's principal work in natural history was entitled Περὶ ζῴων (*On Animals*). A second work,

Περὶ τῆς τῶν πτηνῶν ἱστορίας (*Inquiry on Birds*) may have been an alternate title of Book II of his *On Animals.* The extant fragments from his zoological writings are a mixture of fact and fancy, in which the strange, unusual, or fabulous behavior of land animals and birds is emphasized. Unable to explain certain observed and authenticated data, such as the annual migration of birds, Alexander resorted to analogies with human behavior or to religious and mythological symbolism. Despite his apparent lack of originality, he wisely followed Aristotle in reporting on zoological matters and provided, in turn, one of the principal sources for the account of birds in Book IX of Athenaeus' *Deipnosophistae.* There the size, behavior, and feeding habits of about a dozen identifiable species of birds are recounted, although most of the passages derive ultimately from one of Aristotle's lost writings. Additional remarks deal with the color of the plumage and the external differences between males and females of the same species. The description of the internal organs of a female quail (*Coturnix* sp.) probably derives from Aristotle as well. The only evidence of independent research concerns Alexander's inability to hear the legendary song of a dying swan. Another zoological fragment describes an unusual animal that has been tentatively identified as a gnu. His interest in animal behavior tends to merge with the fabulous in his moralizing tales about the transformations of storks after death and the semihuman intelligence of chameleons and goats.

Only one identifiable fragment exists from Alexander's Περὶ θηριακῶν (*On Theriac*), which may have been an account of the miraculous curative properties of theriac as a drug and as a universal protection against poisons. He may have written a separate book on plants, but the title is not known.

The combination of natural history and an interest in miracles is further evidenced in Alexander's "Dream Book," whose exact title is not recorded. In the few surviving fragments, predictions are based upon the behavior and properties of plants and birds.

No certain opinion can be formed of Alexander's study of early myths. His Τὰ μυθικά (*Mythical Stories*), originally in nine books, is represented by only two fragments. He may also be the author of Περίπλους τῆς ᾿ερυθρᾶς θαλάττης (*Voyage Around the Red Sea*), of which a fragment that deals with monstrous snakes and their symbolic associations with Poseidon is preserved by Aelian.

BIBLIOGRAPHY

For further information on Alexander, see Eugen Oder, "Das Traumbuch des Alexander von Myndos," in *Rhein-*

isches Museum für Philologie, **45** (1890), 637–639; Max Wellmann, "Alexander von Myndos," in *Hermes,* **26** (1891), 481–566, a fundamental study containing an edition of the thirty-four identifiable fragments, and "Alexandros von Myndos" [Alexandros 100], in Pauly-Wissowa, *Real-Encyclopädie der classischen Altertumswissenschaft.*

JERRY STANNARD

ALEXANDER OF TRALLES (*b.* Tralles, in Lydia, first half of the sixth century A.D.; *fl.* in the time of Justinian), *medicine.*

Alexander of Tralles was the son of Stephanus, a physician. He had four brothers: Anthemius, a famous mechanician who was involved in rebuilding Hagia Sophia; Metrodorus, a grammarian; Olympius, a jurist; and Dioscorus, another physician. As we know by his own dedicatory preface, Alexander was the protégé of the father of a certain Cosmas; to this Cosmas he dedicated his work, which he says that he wrote, at the behest of Cosmas, at an advanced age, when he was no longer able to practice medicine. It is unlikely, however, that this Cosmas was the famous geographer Cosmas Indicopleustes, as his modern biographer Theodor Puschmann hypothesizes. The historian Agathias, a contemporary of Alexander, indicates that Alexander's life was beset with hardships. Agathias also records that Alexander practiced in Rome for some time, while Alexander's own writings mention his travels in Gaul and Spain.

The writings of Alexander that have survived have been subjected to a thorough and critical examination by Puschmann, whose findings may well be accepted. According to him, the dedicatory preface to the works as a whole and the book *Concerning Fever* were written during Alexander's last years; the other eleven books are either hastily sketched or more elaborate notes for a handbook on internal medicine, in accordance with a plan set forth at the beginning of the work. A letter about intestinal worms, directed to an unknown Theodorus, is extant, and Puschmann has collected a few additional fragments; all of Alexander's other writings are apparently lost. The entirety of Alexander's work is known to have been available in Greek in numerous codices and in equally numerous Latin translations; his work was also much read and translated by the Arabs.

Puschmann is perhaps biased in favor of his subject. One must remember that Alexander's importance lies within the framework of Byzantine medicine—a rather sterile, literary tradition. Alexander is praised for his self-reliance; this independence should, however, be more precisely formulated as deriving from the consideration that Alexander did not simply edit a medical anthology composed of other people's texts, as did Oribasius or Aetios of Amida, but wrote a work of his own. He was not the only Byzantine author to do so, however—compare especially the work of Johannes Actuarios. Alexander indeed had an extensive practice, made many original observations, and knew the value of empiricism; but this may also be said of other Byzantine physicians (again, especially Johannes Actuarios). While Alexander sometimes dared to criticize even Galen, Johannes Actuarios, too, had a self-confident sense of the value of his own work as compared to the work of his predecessors, as is especially apparent in his book *About the Urine.* Alexander's style is justly praised for its comprehensibility and clarity; not all other Byzantine physicians wrote pompously, however. Moreover, Alexander—like all other Byzantine physicians, and like all those of late antiquity—was uncritical of a great deal of the older medical literature that he cited, and was by no means entirely free from superstition. Finally, very few remnants of Byzantine medical texts and practice are available to us to provide a measure by which Alexander may be objectively evaluated. In summary, one may state that Alexander was, as a representative of Byzantine medicine, rather refreshing, not uninteresting, and not, perhaps, altogether unimportant.

BIBLIOGRAPHY

Text, German translation, and an extensive introduction may be found in Theodor Puschmann, *Alexander von Tralleis,* 2 vols. (Vienna, 1878–1879). It includes Alexander's mention of his father (II, 139), Alexander's preface to his work on medicine (I, 289), mention of Cosmas Indicopleustes (I, 83), Alexander's accounts of his travels (I, 565), and descriptions of his works (I, 101 ff.). Also by Puschmann is "Nachträge zu *Alexander von Tralleis,*" in *Berliner Studien für classische Philologie und Archaeologie,* **5,** part 2 (1887).

For a French translation with an extensive introduction, see F. Brunet, *Oeuvres medicales d'Alexandre de Tralles,* 4 vols. (Paris, 1933–1937); this does not go far beyond Puschmann, however.

A good source of material on Alexander is Agathias; see "Agathias Histor.," in *Historici Graeci Minores,* W. Dindorf, ed., II (Leipzig, 1871), 357. Reference to Greek codices and Latin translations may be found in H. Diels, *Die Handschriften der griechischen Ärzte,* II (Berlin, 1906), 11–13; the Arabic versions are mentioned by I. Bloch in Theodor Puschmann, *Handbuch der Geschichte der Medizin,* I (Jena, 1902), 537–538. This work also covers Alexander's criticism of Galen (p. 539).

FRIDOLF KUDLIEN

ALEXIS OF PIEDMONT. For a study of his life and work, see **Ruscelli, Girolamo,** in Supplement.

ALFONSO EL SABIO (*b.* Toledo, Spain, 23 November 1221; *d.* Seville, Spain, 24 April 1284), *astronomy, dissemination of science and learning.*

Alfonso el Sabio, "the learned," was the son of Ferdinand III of Castile and León, and Beatrice of Swabia, granddaughter of Frederick Barbarossa. Upon the death of his father in 1252, he became Alfonso X. Descended from the Hohenstaufens through his mother, he sought between 1256 and 1272 to become Holy Roman Emperor by pressing the Swabian claims to that position. This continued preoccupation alienated the Castilian nobility and depleted his treasury. Upon the death of his eldest son, Ferdinand, in 1275, his second son, Sancho, sought to dethrone him and gain power. Seville remained loyal to Alfonso, but the majority of his subjects opposed him and the Cortes declared him deposed in 1282.

Although Alfonso's domestic policies threatened the stability of the state, his patronage of science and learning sowed the seeds of later Castilian greatness. In the tradition of his predecessors, he supported the translation of Arabic works into Latin and Castilian. Alfonso gave Spain a great legal code, *Las siete partidas,* a compilation of the legal knowledge of his time, and sponsored important scientific translations. These translations of Arabic astronomical, astrological, and magical treatises reveal Alfonso's active interest in science. He gained his most lasting scientific fame by supporting a new edition of the Toledan Tables of the Cordoban astronomer al-Zarqālī (Arzachel, *ca.* 1029–*ca.* 1087). This new edition, the *Tablas alfonsinas,* was not an original work. Although new observations were made from 1262 to 1272, it still followed the general format of al-Zarqālī's earlier compilation and, with only minor qualifications, retained the Ptolemaic system for explaining celestial motion. It utilized mean solar, lunar, and planetary orbits and equations; declination of stars; ascension, opposition, and conjunction of the sun and moon; visibility of the moon and of eclipses; and a trigonometrical theory of sines and chords to predict the motion of celestial bodies. The original Spanish edition of the *Tablas* has been lost. Its popularity in the medieval period was based on the Latin versions.

Alfonso also supported the translation of a series of Arabic astronomical studies known collectively as the *Libros del saber de astronomia.* The fifteen treatises in this collection are either based on or translated from Arabic astronomical works written in the ninth through the twelfth centuries. The collection, which includes a catalog of stars and a study of the celestial globe, spherical astrolabe, quadrants, clocks, and other assorted astronomical instruments, was prepared while observations were being made for the Alfonsine Tables. Never as popular as the latter, it was not published until 1863–1867. Alfonso may have also encouraged Rabi Zag of Toledo to prepare a treatise on the quadrant. Entitled *Tratado del cuadrante "sennero,"* it exists today in an incomplete manuscript; only eight of its thirteen chapters are extant. Alfonso is said also to have sponsored a vernacular translation of an Arabic work on magic, the Latin *Liber picatrix.* His reputation is based not on his occult endeavors but on the royal patronage he gave so willingly to astronomy.

BIBLIOGRAPHY

The Alfonsine Tables may be found in numerous Latin editions, none of which is, in any sense, a critical edition. The 1st ed. of the Latin version is *Alfonti . . . celestium motuum tabule: nec non stellarum fixarum longitudines ac latitudines Alfontii tempore ad motus veritatem mira diligentia reducte* (Venice, 1483; 2nd ed., with *canones* of Joannes Lucilius Santritter, 1492). The *Libros del saber de astronomia* was ed. by Manuel Rico y Sinobas, 5 vols. (Madrid, 1863–1867). A beautiful color facsimile of Alfonso's famous study on lapidaries has been reproduced by José Fernandez Montaña in *Lapidario, reproducción fotolitográfica* (Madrid, 1881). The fragmented *Tratado del cuadrante "sennero"* was ed. by José M.ª Millás Vallicrosa in his *Nuevos estudios sobre historia de la ciencia española* (Barcelona, 1960). The work on magic was trans. by H. Ritter and H. Plessner in their *Picatrix: Das Ziel des Weisen* (London, 1962).

The definitive portrait of Alfonso's political activities is Antonio Ballesteros y Baretta, *Alfonso el Sabio* (Barcelona, 1963). Excellent brief biographies are John Esten Keller, *Alfonso X, El Sabio* (New York, 1967), and Evelyn S. Proctor, *Alfonso X of Castile, Patron of Literature and Learning* (Oxford, 1951). For his scientific works and translations see Evelyn S. Proctor, "The Scientific Works of the Court of Alfonso X of Castile," in *Modern Language Review,* **40** (1945), 12–29; and Moritz Steinschneider, "Die europäischen Uebersetzungen aus dem Arabischen," in *Sitzungsberichte der Akademie der Wissenschaften in Wien,* **4, 9, 40, 55, 60, 61, 69, 87, 93, 97,** and **108** (1904–1905). For his astronomy and its influence see J. L. E. Dreyer, "The Original Form of the Alfonsine Tables," in *Monthly Notices of the Royal Astronomical Society,* **80** (1920), 243–267; Jose Soriano Viguera, *La astronomia de Alfonso X* (Madrid, 1926); Pierre Duhem, *Le système du monde,* II, (Paris, 1914), 259–266; and Alfred Wegener, *Die Alfonsinischen Tafeln für den Gebrauch eines modernen Rechners* (Berlin, 1905), and "Die astronomischen Werke Alfons X," in *Bibliotheca mathematica,* **6** (1905), 129–185.

PHILLIP DRENNON THOMAS

ALFRAGANUS. See **al-Farghānī.**

ALHAZEN. See **Ibn al-Haytham.**

ʿALI. See last element of name.

ALKINDUS. See **al-Kindī.**

ALLEN, EDGAR (*b.* Canyon City, Colorado, 2 May 1892; *d.* New Haven, Connecticut, 3 February 1943), *endocrinology.*

Edgar Allen discovered estrogen and investigated the hormonal mechanisms that control the female reproductive cycle. In so doing he helped to create the science of endocrinology, one of the most significant branches of modern biology.

Allen, the son of a physician, received his early education in the public schools of Pawtucket and Cranston, Rhode Island. He attended Brown University and earned all of his higher degrees at that institution: Ph.B. (1915), M.A. (biology, 1916), Ph.D. (biology, 1921). During World War I his studies were interrupted for a short time by military service. In 1918 he married Marion Pfieffer of Providence; the couple had two daughters.

Allen's distinguished academic career began in 1919 when he was appointed instructor in anatomy at Washington University, St. Louis, Missouri. In 1923 he moved to the University of Missouri, where he served initially as professor of anatomy and subsequently as dean of the medical school and director of university hospitals. In 1933 he returned to the east coast as professor of anatomy and chairman of the department at the Yale School of Medicine, a position he held until his death.

Allen's Ph.D. thesis on the estrous cycle of the mouse (published in 1922) is a thorough description of the histological changes that occur in primary and secondary sex organs during the reproductive cycle. During the early 1920's several investigators had suggested that the ovary might be the control center for this cycle; it was thought that the fluid content of the *corpus luteum* might be the active agent of control. Allen's Ph.D. thesis cast doubt on this latter hypothesis. He noticed that at any given time during the cycle *corpora lutea* can be found in many different stages of degeneration, making it highly unlikely that they could be controlling a continuous series of progressive histological changes.

In 1923 Allen undertook a study of ovogenesis during sexual maturity and discovered that females are not born with a complete complement of ova; ova are continually formed in the germinal epithelium and the follicles that develop around them have a cycle of growth and decay not unlike that of the *corpus luteum.* This fact led Allen to suspect that the ovarian follicle, not the *corpus luteum,* might be the focus of control. In collaboration with a biochemist, E. A. Doisy, Allen proceeded to test this hypothesis by extracting a fluid from the follicle and determining its effects. He found that repeated injections of the follicular fluid produced histological changes that were identical to the early stages of normal estrus; when the injections were halted, the later phases ensued.

Allen and Doisy had discovered the existence and the effects of estrogen. Within fifteen years the other hormones that influence estrus were also discovered (see the work of Zondek, Aschheim, H. M. Evans and J. A. Long on pituitrin and Hartmann, Corner, Hisaw and Zuckerman on progesterone) and the relations between them were becoming clear. All of Allen's subsequent investigations were concerned, in some way, with these sex hormones. He proved, for example, that estrogen causes the onset of puberty in immature female animals and demonstrated that the hormonal mechanisms of primates (including man) are very similar to those of rodents, on which the original studies had been done. Allen also studied the relation between estrogen and malignancy, in order to determine whether there is any similarity between rapid cell growth caused by estrogenic stimulation and rapid cell growth that is characteristic of cancerous tissue. In addition, Allen's publications contain a wealth of methodological information that was of great value to subsequent researches in endocrinology.

Allen was a member, and president, of two scientific societies, the American Association of Anatomists and the Association for the Study of Internal Secretions. Brown, Yale, and Washington universities awarded him honorary degrees. The French government made him a member of the Legion of Honor (1937), and the Royal College of Physicians awarded him its Baly Medal (1941). He was also an advisory trustee (1939–1943) and member of the Scientific Advisory Committee of the International Cancer Research Foundation.

A devoted sailor, Allen joined the Coast Guard Auxiliary at the onset of World War II and died of a heart attack while commanding a patrol boat on Long Island Sound.

BIBLIOGRAPHY

Allen's total bibliography contains more than 140 items; it can be found in the *Yale Journal of Biology and Medicine,* **17**, part 1 (1944–1945), 2–12, along with his *curriculum vitae.* The *Yale Journal,* **15** (1943), 641–644, contains an excellent biographical sketch of Allen.

For an understanding of the development of follicular control, see, in the following order, "The Estrous Cycle in the Mouse," in *American Journal of Anatomy,* **30** (1922), 297 (Allen's doctoral dissertation); "Ovogenesis During Sexual Maturity," in *American Journal of Anatomy,* **31** (1923), 439; and "The Hormone of the Ovarian Follicle; Its Localization and Action in Test Animals, and Additional Points Bearing Upon the Internal Secretion of the Ovary," in *American Journal of Anatomy,* **34** (1924), 133. The last study was written in collaboration with E. A. Doisy.

Some of the extensions of his original idea can be found in "The Induction of a Sexually Mature Condition in Immature Females by Injection of the Ovarian Follicular Hormone," in *American Journal of Physiology,* **69** (1924), 577; "The Menstrual Cycle of the Monkey, *Macacus Rhesus* . . .," in *Contributions to Embryology,* Carnegie Institute of Washington, No. 380 (1927), p. 98; and "The Estrous Cycle of Mice During Growth of Spontaneous Mammary Tumors and the Effects of Ovarian Follicular and Anterior Pituitary Hormones," in *American Journal of Cancer,* **25** (1935), 291.

RUTH SCHWARTZ COWAN

ALLIACO, PETRUS DE. See **Ailly, Pierre D'.**

ALLONVILLE, J. E. D'. See **Louville, J. E. d'A., Chevalier de.**

ALPETRAGIUS. See **al-Biṭrūjī.**

ALPHARABIUS. See **al-Fārābī.**

ALPINI, PROSPERO (*b.* Marostica, Italy, 23 November 1553; *d.* Padua, Italy, 23 November 1616), *botany.*

Alpini was among the first of the Italian physician-botanists of the sixteenth century to examine plants outside the context of their therapeutic uses. Although he shared his contemporaries' reverence for the past, he helped to advance the frontiers of botanical science by taking advantage of knowledge gained through his travels.

The oldest of the four children of Francesco Alpini, a physician, and Bartolomea Tarsia, Alpini studied medicine at the University of Padua, from which he received his degree on 28 August 1578. His master was Melchiore Guilandino (originally Melchior Wieland of Königsberg), the second director of the botanical garden at Padua, who acted as respondent in Alpini's dialogue *De plantis Aegypti.* For a short time Alpini practiced medicine in Camposampiero, near Padua. In 1580 he became physician to Giorgio Emo, the Venetian consul to Cairo, and in September of that year he accompanied Emo to Egypt. En route, he botanized on the island of Crete. After three years in Egypt, he returned to Venice. In 1594 the Venetian

Senate elected him *lettore dei semplici* ("reader in simples") at the University of Padua. He succeeded to the directorship of the botanical garden at Padua in 1603, assuming both the title and the duties of *prefetto ed ostensore dei semplici* ("prefect and demonstrator of simples"). Alpino, a son by his first wife, Guadagnina Guadagnini, later became the seventh prefect of the botanical garden. Alpini is said to have died of a kidney infection contracted during his stay in Egypt. He is buried in the Church of St. Anthony, Padua.

Alpini's major contributions are the outgrowth of his travels. From a scientific point of view, the *De plantis Aegypti* (1592) is his most important work. The pioneer study of Egyptian flora, it introduced exotic plants to the still-parochial European botanical circles. Obviously incomplete, this small book later was used by such systematists as F. Hasselquist and P. Forskål as a basis for their more complete studies. Moreover, some of Alpini's original descriptions were included in the writings of Linnaeus, who regarded him with sufficient esteem to name the genus *Alpinia* (*Zingiberaceae*) in his honor.

Fifty-seven plants and trees are described in the *De plantis Aegypti,* and forty-nine are illustrated. Alpini's medical training led him to approach the new flora in the traditional manner of attempting to correlate these plants with the names and descriptions found in classical sources. When this proved impossible, he described the plant under its local name. The descriptions are based upon specimens that Alpini personally examined, either cultivated in gardens or growing wild. This in itself provided a much-needed corrective to the fables and vague reports associated with Eastern plants. Among the plants previously undescribed in a European botanical text were the coffee bush (*Coffea arabica* L.), banana (*Musa* sp.), and baobab (*Adansonia digitata* L.). Perhaps because of the dialogue form of the book, there is no discernible system. There is, however, a wide range of miscellaneous information based upon observation. Alpini observed that the fertilization of the date palm was a sexual process, described the phototropic movements of the leaves of the tamarind (*Tamarindus indica* L.), speculated that the tree cotton (*Gossypium arboreum* L.) was the *byssos* of the ancients, and noted the edibility of plants unknown in Europe, such as bammia or okra (*Hibiscus esculentus* L.). Evidently puzzled by the treelike banana plant, he accepted the story that it was the result of a sugar cane grafted onto the root of colocasia (*Colocasia esculenta* Schott.) and supplied a good description of the latter.

Another product of Alpini's study of Egyptian plants is the *De balsamo dialogus* (1591). In the form of a dialogue involving the author, an Egyptian physician, and a Jew, the source of balsam (*Commiphora* spp.) is discussed and questions are raised concerning its identity, ancient names, and medical uses, and the possibility that the true balsam has become extinct. Closely related in form and method is the *De rhapontico* (1612). The source and therapeutic properties of rhubarb (*Rheum* sp.) are discussed with a show of classical scholarship controlled by a personal examination of specimens grown in the botanical garden under Alpini's supervision.

The material for the *De plantis exoticis,* which was published posthumously, also derived from Alpini's travels. With Onorio Belli he carefully studied the flora of Crete. Information on plants from other areas was later incorporated into the manuscript, which was edited by his son Alpino and completed in 1614. Data concerning some of these plants were obtained by examining specimens grown from seeds sent to Alpini. A total of 145 plants, each illustrated by a woodcut, formed a notable contribution to Mediterranean floristics. This is especially true of the flora of Crete, many of whose plants were described for the first time. The accuracy of Alpini's descriptions was demonstrated by A. Baldacci and P. A. Saccardo, who identified seventy-one of the eighty-five Cretan plants on which he reported.

Alpini's interest in medicine was expressed in several books, of which the most important were *De medicina Aegyptiorum* (1591) and *De praesagienda vita* (1601). The former, like his study of Egyptian plants, was based upon personal experience. Primarily an examination of contemporary Egyptian (i.e., Turkish) medicine, it ranks as one of the earliest studies of non-European medicine. Although he took a dim view of local customs, Alpini was sufficiently impressed by novel therapeutic practices to introduce the technique of *moxa* into European medicine. The *De praesagienda vita* is a detailed study of prognostics in which attention is devoted to the patient's mental state and its bearing on health, as well as to the usual physical and diagnostic signs.

Mention should be made of the *Rerum Aegyptiarum* (Volume I of *Historiae Aegypti naturalis*), a pioneer and undeservedly neglected contribution to Egyptology. Edited after Alpini's death by Bartolomeo Cellari, it contains a wealth of information on the natural history (including zoology and mineralogy), customs, and ancient monuments of Egypt.

BIBLIOGRAPHY

I. ORIGINAL WORKS. Principal editions of Alpini's works include *De balsamo dialogus* (Venice, 1591; Padua, 1639), also in *Medicina Aegyptiorum* (Leiden, 1719) and in Blasius Ugolino, *Thesauro antiquitatum sacrarum* (Venice, 1750), also translated by Antoine Colin as "Histoire du baulme

. . ., version françoise . . .," in Garcia d'Orta, *Histoire des drogues, espisceries, et de certains medicamens simples* (Lyons, 1619); *De medicina Aegyptiorum libri quatuor* (Venice, 1591; Paris, 1646), also published as *Medicina Aegyptiorum. Accedunt huic editioni ejusdem auctoris libri de balsamo et rhapontico* (Leiden, 1719; 1745); *De plantis Aegypti liber . . . Accessit etiam liber de balsamo, alias editus* (Venice, 1592), later edited by Johannes Vesling (Padua, 1638; 1640); *De praesagienda vita et morte aegrotantium libri septem* (Venice-Frankfurt, 1601; Leiden, 1733; Hamburg, 1734), later edited by H. Boerhaave (Leiden, 1710; Venice, 1735; 1751; Bassano, 1774) and by J. B. Friedreich, 2 vols. (Nördlingen, 1828), and translated by R. James as *The Presages of Life and Death in Disease,* 2 vols. (London, 1746); *De medicina methodica libri tredecim* (Padua, 1611; Leiden, 1719); *De rhapontico—disputatio in gymnasio Patavino habita* (Padua, 1612), also in *De plantis Aegypti liber* (Padua, 1640) and *Medicina Aegyptiorum* (Leiden, 1719); "Trattato della teriaca egittia," in Ippolito Ceccarelli, *Antidotario romano latino e volgare tradotto . . .,* A. Manni, ed. (Rome, 1619); *De plantis exoticis libri duo* (Venice, 1627; 1629); *Historiae Aegypti naturalis,* 2 vols. (Leiden, 1735), Vol. I, *Rerum Aegyptiarum libri quattuor . . .*; Vol. II, *De plantis Aegypti liber auctus et emendatus, cum observationibus et notis Johannis Veslingii*; and *De longitudine et brevitate morborum, libri duo,* introduction, translation, and notes by Giuseppe Ongaro (Marostica, 1966).

II. SECONDARY LITERATURE. Works dealing with Alpini and his contributions to botany are A. Baldacci and P. A. Saccardo, "Onorio Belli e Prospero Alpino e la flora dell'isola di Creta," in *Malpighia,* 14 (1900), 140–163; Augusto Béguinot, "Prospero Alpini," in Aldo Mieli, ed., *Gli scienziati italiani·dall'inizio del medio evo ai nostri giorni,* I, pt. 1 (Rome, 1921), 84–90; Pietro Capparoni, "Prosper Alpini (1553–1616)," in *Bulletin de la Société Française d'Histoire de la Médecine,* 23 (1929), 108–115, and *Profili bio-bibliografici di medici e naturalisti celebri italiani dal sec. XV al sec. XVIII* (Rome, 1932), pp. 20–23; G. Fasoli and C. Cappelletti, "Prospero Alpino (1553–1616)," in *Rassegna trimestrale di odontoiatria,* 41 (1960), 597–613; Ludwig Keimer, "Quelques détails oubliés ou inconnus sur la vie et les publications de certains voyageurs européens venus en Égypte pendant les derniers siècles," in *Bulletin de l'Institut d'Égypte,* 31 (1949), 121–175; Giuseppe Ongaro, "Contributi alla biografia di Prospero Alpini," in *Acta medicae historiae Patavina,* 8–9 (1961–1963), 79–168—the most complete study to date, based on Alpini's unpublished manuscripts at Padua; John Ray, *A Collection of Curious Travels and Voyages,* 2 vols. (London, 1693), II, 92–98; P. A. Saccardo, "Contribuzioni alla storia della botanica italiana," in *Malpighia,* 8 (1894), 476–539; and Kurt Sprengel, *Geschichte der Botanik,* 2 vols. (Altenburg-Leipzig, 1817–1818), I, 356–359.

JERRY STANNARD

ALSTED, JOHANN HEINRICH (*b.* Ballersbach, Germany, 1588; *d.* Weissenburg, Transylvania [after 1715 Karlsburg; now Alba Iulia, Rumania], 9 November 1638), *natural philosophy.*

Alsted was the second son of Jacob Alsted (*d.* 1622), a Reformed Church minister, and Rebecca Pincier, the daughter of a Reformed Church minister and sister of Johannes Pincier, humanist scholar and professor of medicine and philosophy at the Herborn Academy. After an elementary education at Ballersbach, Alsted entered the lower school of the Herborn Academy in 1602. This academy, founded in 1584, had achieved considerable prominence as a center of Calvinist and Ramist influence.

After completing his studies at Herborn, Alsted undertook the customary *peregrinatio academicae,* visiting Frankfurt am Main, Heidelberg, Strasbourg, and Basel, where he met Amandus Polanus von Polandsdorf. In 1608 he returned to Herborn and was appointed teacher and examiner at the high school of the academy. Two years later he became professor of philosophy, and in 1619 rector and professor of theology. He made brief visits to other parts of Europe, attending the Synod of Dordrecht in 1618.

Alsted married Anna Katherine Rab (1593–1648), daughter of the Herborn printer Christoph Rab (Corvinus), who was to print the majority of Alsted's works. They had four children.

Alsted attracted students from numerous German and Slavic states; the most famous was Jan Amos Komenský (Comenius), who taught for a short time at Herborn before embarking upon his pansophic missions. The Thirty Years' War upset the continuity of Alsted's work, and he reluctantly decided to leave Herborn in 1629, to become the first rector of the new high school at Stuhlweissenburg, which had been established by the Protestant prince Gabriel Bethlen von Siebenbürgen.

The majority of Alsted's writings were on theology, and in them he displayed the same logical and encyclopedic approach found in the philosophical writings. Throughout the areas of Calvinist influence, from Transylvania to New England, Alsted's systematic treatises on educational theory, theology, and philosophy exerted great influence in the universities during most of the seventeenth century. His writings covered the whole spectrum of natural philosophy: commentaries on the cabala, the *Ars magna* of Lull, mnemonics, traditional and Ramist logic, physics, mathematics, and astronomy.

Alsted's major monographs—compendia or harmonia of logic, physics, the Scriptures, and education—display a strikingly uniform organization, at the cost of oversimplification. This was an inherent danger of the Ramist approach. The *Systema physicae harmonicae* (1612) is typical. It analyzes the principles of "physics" derived according to four conflicting systems: *physicum Mosaicum; rabbinica et cabbalis-tica; peripateticam;* and *chemicam.* These systems are based, respectively, on the Old Testament, Jewish mystical writings, Aristotle, and Paracelsus. The principles of each system are discussed in a clear logical sequence that draws upon a wide range of sources, from the humanist editors of the cabala to the late sixteenth-century neo-scholastic commentators Magirus and Scaliger, and the Paracelsian or mystical authors John Dee and Oswald Croll. Throughout, Alsted gives his own judgments on the physical principles, favoring a "Christianized" Peripatetic philosophy, the description of which occupies more than half the book.

The *Methodus admirandorum* contains information about improved techniques of surveying and physical astronomy, and discusses the merits of the Copernican hypothesis. Copernicus is admired, but his system is deemed unacceptable, for it is refuted by the Scriptures and common sense.

Alsted's ultimate fame rests upon his conception of the encyclopedia as a universal system of knowledge. He believed in the fundamental unity of divine and secular knowledge, the nature of which unity could be displayed by the use of *logica-mnemonica,* the art of directing the mind and perfecting the memory. Also prominent was his logical analysis of the nature and divisions of the parts of knowledge, or *technologia,* which provided the basis for the organization of his encyclopedia.

These systematic writings had an immediate but ephemeral appeal in institutions of higher education, his *Encyclopaedia* being to such students as Cotton Mather the "North-West Passage to all the sciences." More important, they influenced the educational theories of Comenius, as well as his pansophia, and the encyclopedic philosophies of Leibniz and Morhof.

BIBLIOGRAPHY

I. ORIGINAL WORKS. Alsted's writings most directly related to natural philosophy are *Clavis artis Lullianae et verae logices* (Strasbourg, 1609); *Methodus formandorum studiorum, continens commonfactiones concilia, regulas . . . de ratione bene discendi et ordine studiorum recte instituendo* (Strasbourg, 1610); *Panacea philosophica, id est . . . Methodus docendi et discendi universam encyclopaediam* (Herborn, 1610), to which was appended *Harmonico philosophiae Aristotelicae, Lullianae et Rameae; Systema mnemonicum duplex* (Frankfurt, 1610); *Theatrum scholasticum . . . I Systema et gymnasium mnemonicum . . . II Gymnasium logicum . . . III Systema et gymnasium oratorium* (Herborn, 1610); *Compendium I Systematis logici . . . II Gymnasii logici* (Herborn, 1611); *Elementale mathematicum in quo mathesis methodice traditur* (Frankfurt, 1611); *Philosophia digne restituta: libros quatuor* (Herborn, 1612); *Systema physicae harmonicae* (Herborn, 1612), entitled *Physica harmonica* in later editions; *Compendium logicae harmonicae . . . accedit nucleus logicae* (Herborn, 1613), which appeared in a slightly different form in 1614; *Methodus admirandorum mathematicorum* (Herborn, 1613), an expanded version of the *Elementale mathematicum* that was often reprinted; *Theologia naturalis exhibens augustissimam naturae scholam* (Frankfurt, 1615); *Cursus philosophici encyclopaedia libris XXVII* (Herborn, 1620), also in an expanded edition with a new section entitled *Compendium lexici philosophici* (Herborn, 1626); and *Encyclopaedia septem tomis distincta* (Herborn, 1630; Leiden, 1649; Stuttgart, 1663, abridged).

II. SECONDARY LITERATURE. Works on Alsted are *Allgemeine deutsche Biographie*, I, 354–355; C. G. Jöcher, *Allgemeines Gelehrten-Lexicon* (Leipzig, 1750), I, 302–303; Johann Kvacsala, "Johann Heinrich Alsted," in Ungarische *Revue*, **9** (1889), 628–642, and *J. A. Comenius* (Leipzig, 1892), pp. 98–104; Max Lippert, *J. H. Alsted pädagogisch-didactische Reform Bestrebungen* (Meissen, 1899); L. E. Loemker, "Leibniz and the Herborn Encyclopaedists," in *Journal of the History of Ideas*, **22** (1961), 323–338; *Neue deutsche Biographie*, I, 206; Walter J. Ong, *Ramus, Method, and the Decay of Dialogue* (Cambridge, Mass., 1958); Wilhelm Risse, *Die Logik der Neuzeit*, I (Stuttgart–Bad Canstatt, 1964), 477–485, and *Bibliographia logica*, I (Hildesheim, 1965); E. W. E. Roth, "Johann Heinrich Alsted (1588–1638), sein Leben und sein Schriften," in *Monatshefte der Comenius-Gesellschaft*, **4** (1895), 29–44, the most complete bibliography of Alsted's works; and Max Wundt, *Die deutsche Schulmetaphysik des 17 Jahrhunderts* (Tübingen, 1939), pp. 80–83, 236–237.

CHARLES WEBSTER

ALZATE Y RAMÍREZ, JOSÉ ANTONIO (*b.* Ozumba, Mexico, 1738; *d.* Mexico City, Mexico, 1799), *natural history, mathematics, geography, astronomy.*

Born into a wealthy country family, Alzate attended San Ildefonso College and graduated in 1753 with a bachelor of arts degree. In 1756, he received a bachelor of divinity degree from the University of Mexico, and was subsequently ordained as a Roman Catholic priest.

An enthusiastic naturalist and man of letters, Alzate was a member of the Sociedad Económica Vascongada, the Real Jardín Botánico de Madrid, and the Académie Royale des Sciences de Paris. He embraced the ideas of the Enlightenment and devoted his life to the study of all branches of natural science. On various occasions, he was commissioned by the colonial government to solve problems affecting the public interest. His principal aim was to transcend the Aristotelian philosophy of his day and to promote the development of technology in New Spain. The value of his scientific production was not consistent, however, for his work covered a great many fields and was often conducted in an unfavorable atmosphere.

Aggressive by nature, Alzate was continually involved in scientific polemics, and his sarcasm aroused the animosity of his colleagues. He struggled to contradict the European opinions regarding the inferiority of American scientific knowledge. When Charles III of Spain sent a botanical expedition to New Spain, Alzate touched off a lengthy controversy by defending the advanced botanical knowledge of the ancient Mexicans and criticizing the Spaniards' application of Linnaean methods and principles.

Using his own limited economic resources, Alzate founded several scientific periodicals: *Diario literario de México* (1768); *Asuntos varios sobre ciencias y artes* (1772); *Observaciones sobre la física, historia natural y artes utiles* (1787); and *Gazeta de literatura* (1788–1795). On the basis of these journals, all of which were designed to improve the country's welfare through technology, Alzate is considered to be one of the pioneers of scientific journalism in the western hemisphere.

As a result of his continuous efforts to promote the scientific advancement of his countrymen and his successful fight to abolish the scholastic systems used in the colonial institutions, Alzate is regarded as one of the forerunners of Mexican independence. In 1884, the Sociedad Científica Antonio Alzate (now known as the Academia Nacional de Ciencias) was founded in Mexico City. Many Mexican intellectuals consider Alzate to be the father of modern natural science in Mexico.

BIBLIOGRAPHY

For information on Alzate's life and work, see F. Fernández del Castillo, "Apuntes para la biografía del Presbítero Bachiller J. A. F. de Alzate y Ramírez," in *Memorias de la Sociedad científica "Antonio Alzate,"* **48** (1927), 347–375; J. Galindo y Villa, "El Pbro. J. A. Alzate y Ramírez. Apuntes biográficos y bibliográficos," *ibid.,* **3** (1889–1890), 125–183, and "El enciclopedista Antonio Alzate," in *Memorias de la Academia nacional de ciencias "Antonio Alzate,"* **54** (1934), 9–14; A. Gómez Orozco, "Don Antonio Alzate y Ramírez," in *Humanidades,* **1** (1943), 169–177. See also R. Moreno Montes de Oca, "Alzate y la conciencia nacional," in *Memorias de la Academia nacional de ciencias "Antonio Alzate,"* **57** (1955), 561–572, and "Alzate y su concepción de la ciencia," in *Memorias del primer coloquio mexicano de historia de la ciencia,* **2** (1965), 185–200; B. Navarro, "Alzate, símbolo de la cultura ilustrada mexicana," in *Memorias de la Academia nacional de ciencias "Antonio Alzate,"* **57** (1952), 176–183.

ENRIQUE BELTRÁN

AMAGAT, ÉMILE (*b.* Saint-Satur, Cher, France, 2 January 1841; *d.* Saint-Satur, 15 February 1915), *physics.*

Amagat became *docteur-es-sciences* at Paris in February 1872 and was then, successively, *agrégé*, professor of physics at the Faculté Libre des Sciences of Lyons, and examiner at the École Polytechnique. He was elected a corresponding member of the Académie des Sciences on 5 May 1890, received from the Institut the Prix Lacaze pour la Physique in 1893, and became a full member of the Académie on 9 June 1902.

Amagat's work dealt with fluid statics. At the time he began his work, Andrews had just announced the structure of a few isotherms of carbon dioxide at between 10° and 50°C. and up to 110 atmospheres, the region in which the product of pressure times volume is at a minimum for this compound. Andrews' report was published in France in 1870, and his research was the only one, at the time, to use varying temperatures. From 1869 to 1872 Amagat studied the effect of temperatures up to 320° on the compressibility and expansion of gases. These studies led to his doctoral thesis.

Knowledge of liquids was even more limited. It was known that, except for water, the coefficient of compressibility increases with the temperature. Amagat published an extensive report on this subject in 1877, showing that this coefficient clearly decreases when pressure increases, which was contrary to the results reached by other experimenters.

There remained the search for the laws of the coefficients of compressibility, the coefficients of expansion under constant pressure and constant volume, the coefficients of pressure when both pressure and temperature are varied, and the limits toward which these laws tend when matter is more and more condensed by pressure. It was to this research that Amagat devoted the active phase of his career.

His first works were published between 1879 and 1882. They display the isotherms of a number of gases for temperatures between 0° and 100° and for pressures up to slightly more than 400 atmospheres (the limit that could be borne by the sturdiest glass tubes available).

The experimental data used as a base for this research were furnished by measurements taken in 1879 in a mine shaft at Verpilleux, near Saint-Étienne. Amagat determined by means of a column of mercury 327 meters high the compressibility of nitrogen up to 430 atmospheres. His results were universally adopted for the calculation of gas manometers. Until then only Regnault's results had been available, and they went only to thirty atmospheres. As for Andrews, he had

been content to apply Mariotte's (Boyle's) law up to 110 atmospheres, using a compressed-air manometer.

At the end of this first stage of his research Amagat realized that certain questions vital to the theory of fluids could be elucidated only by greatly increasing the limit of pressures. But having already gone beyond the possible resistance of glass tubes, he had to invent new experimental methods. At this point Amagat created his most ingenious apparatus, the manometer with free-moving pistons in viscous liquids; thus he was able to measure with certainty and regularity pressures above 3,000 atmospheres. Using an idea that originated with Gally-Cazolat, Amagat constructed an apparatus that was a back-acting hydraulic press made gas-tight by means of a viscous liquid such as molasses or castor oil.

Amagat's manometer served his own research and was also adopted in many military and firearms laboratories. The same apparatus, put in reverse, made it possible to create considerable pressures while measuring them. A remarkable application of this device is credited to Vieille in the adjustment of crushing cylinders for the creation of pressure in firearms.

Amagat's own work, which had to do with gases and a fairly large number of liquids, was published between 1886 and 1893. After having constructed his network of isotherms (1887–1891), he spent the next two years extracting the experimental laws of fluid statics as they emerged from these results. Without the experimental data contributed by Amagat, many important theoretical works would have been impossible: research on liquids, Tait's kinetic theory of gases, the works of Van Der Waals, the computation of coefficients in Sarrau's formula (pressure and temperature in the detonation of explosives), and Sarrau's application of his formula to oxygen, which made it possible to determine the critical temperature of this gas before it was liquefied by Wroblewski.

Having been forced to study the elasticity of glass in order to gauge the variations in volume of the containers holding the fluids studied, in 1889–1890 Amagat broadened his study until it embraced the elasticity of solids in general, starting with a verification of the general formulas of elasticity. As a correlative study he had to determine the compressibility of mercury. As early as 1876 and 1882 he had done research on the elasticity of rarefied gases. Amagat found that air, hydrogen, and carbon dioxide follow Mariotte's law at pressures as low as 1/10,000 atmosphere.

Amagat then turned to various special studies related to his general research: interior pressure of fluids; negative interior pressure; laws regulating the specific heats of fluids at different temperatures and pressures, and their relationship; verification of Van Der Waals' law of corresponding states, according to which the equation of the state of all gases can be represented by a single function in which critical temperature, pressure, and volume appear as parameters; solidification of liquids by pressure; determination of the density of liquefied gases and of their saturated vapor; the atomic volume of oxygen and of hydrogen; the interaction of oxygen and mercury; the differential equation of the speed of sound in gases; and the relations between the coefficients of the formulas of Coulomb (magnetism), Laplace, and Ampère: "The Laplace formula connects those of Coulomb and of Ampère but only by employing an almost constant factor that experience alone can determine."

BIBLIOGRAPHY

Amagat presented all accounts of his work to the Académie des Sciences; his papers are listed in the following indexes to the *Comptes rendus* of the Académie: for Vols. **62–91** (12 papers); **92–121** (48 papers); **122–151** (18 papers); and **152–181** (4 papers). Individual papers are "Compressibilité de l'air et de l'hydrogène à des températures élevées," in *Annales de chimie et de physique*, 4th series, **28** (1873), 274–279; "Dilatation et compressibilité des gaz à diverses températures," *ibid.*, **29** (1873), 246–285; "Recherches sur l'élasticité de l'air sous de faibles pressions," *ibid.*, 5th series, **8** (1876), 270–278; "Recherches sur la compressibilité des liquides," *ibid.*, **11** (1877), 520–549; "Sur l'équation différentielle de la vitesse du son," in *Journal de physique*, 1st series, **9** (1880), 56–59; "Mémoire sur la compressibilité des gaz aux fortes pressions. Influence de la température," in *Annales de chimie et de physique*, 5th series, **22** (1881), 353–398; "Mémoire sur la compressibilité de l'air et de l'acide carbonique de 1 atm. à 8 atm. et de 10° à 300°," *ibid.*, **28** (1883), 456–464; "Mémoire sur la compressibilité de l'air, de l'hydrogène et de l'acide carbonique raréfiés," *ibid.*, 464–480; "Sur une forme nouvelle de la fonction φ (pvt) = 0," *ibid.*, 480–507; "Sur la détermination du rapport $\frac{C}{c}$," in *Journal de physique*, 2nd series, **4** (1885), 174–177; "Mémoires sur l'élasticité des solides et la compressibilité du mercure," in *Annales de chimie et de physique*, 6th series, **22** (1891); "Sur la détermination de la densité des gaz et de leur vapeur saturée, éléments du point critique de l'acide carbonique," in *Journal de physique*, 3rd series, **1** (1892), 288–298; "Sur le déplacement de la température du maximum de densité de l'eau par la pression et le retour aux lois ordinaires sous l'influence de la pression et de la température," *ibid.*, 3rd series, **2** (1893), 449–459; "Mémoires sur l'élasticité et la dilatation des fluides jusqu'aux très hautes pressions," in *Annales de chimie et de physique*, 6th series, **29** (1893), 68–137 and 505–574, 2 articles;

"Pressions intérieures dans les fluides et forme de la fonction φ (pvt) = 0," in *Journal de physique*, 3rd series, **3** (1894), 307–316; and "Sur les chaleurs specifiques des gaz et les propriétés des isothermes," *ibid.*, 3rd series, **5** (1896), 114–123.

No general study has been made of Amagat's life or his work. See, however, *Notice sur les travaux scientifiques de M. E. H. Amagat . . .*(Paris, 1896), which, although published anonymously, was written by Amagat himself to be presented to the Académie des Sciences at the time of his candidature, in accordance with traditional practice.

JACQUES PAYEN

AMATUS LUSITANUS. See **Lusitanus, Amatus.**

AMEGHINO, FLORENTINO (*b.* Moneglia, Liguria, Italy, 19 September 1853; *d.* La Plata, Argentina, 6 August 1911), *paleontology, prehistory, anthropology, geology.*

The son of Antonio Ameghino, a mason and a warehouse keeper, and Maria Dina Armanino, both of whom had come to Argentina from Italy, Florentino received only a scanty formal education. One of his teachers, Carlos D'Aste, awed by Ameghino's intelligence, took him to his home in Buenos Aires in 1868 so that he could attend the Escuela Normal de Preceptores. He was thus able to teach school in Mercedes in the early 1870's.

At fourteen Ameghino read Lyell in French and was imbued with the spirit of evolutionism. The geology and geography of the Luján area, particularly the exposed strata and the fauna fossils in the ravines of the Luján River, had aroused his interest in natural science. In 1869 he explored the area around Luján, collecting fossil bones of extinct fauna and Indian relics. His studies at this time included anthropology, Argentine geology, and paleontology. He also frequented the German naturalist Karl Burmeister's natural history museum and library. Ironically, in later life Ameghino and Burmeister became bitter enemies through the combination of Burmeister's professional jealousy and Ameghino's bold advocacy of his doctrines.

In 1873 Ameghino contracted an unidentified illness and was directed to take long walks as part of the cure. His interest in and love of nature were thus increased, as was his endurance; in 1882, short of funds, he walked from the capital to Luján. His health was again threatened when he was poisoned by a mushroom of the type identified by R. Singer as *Amanita ameghinoi.* From 1890 on, Ameghino suffered from diabetes and other disabilities, especially in the four years after 1898. He died of gangrene of the foot.

Soon after he turned twenty, Ameghino published his first important geological work, "El Tajamar y sus futuras consecuencias y el origen de la Tosca." On 31 October 1875 he announced to Paul Gervais, director of the *Journal de zoologie,* that he had made a discovery in the Frías brook near Mercedes, which had been verified by Ramorino, and another in the Luján River. Gervais also helped him to publish a summary of his years in Mercedes. In it Ameghino claimed that the Argentine Amerindian was contemporary with the extinct fauna of the Pampas, a position that encountered stiff opposition. Ameghino sent a work on fossil man, espousing Darwin's transformist view, to the Scientific Society of Argentina. It was not published, but it put him in touch with Francisco Moreno and Estanislao Zeballos. Articles on the Pampean formation, written in 1875, were incorporated into "Los terrenos de transporte cuaternario de la provincia de Buenos Aires," which was presented to the Scientific Society of Argentina in 1876 but was never published. In this work he defended his Pliocene chronology (now accepted as Pleistocene) and Lyell's uniformitarianism against catastrophism.

Nogaro and Salomones, two friends from Mercedes, provided Ameghino with the means to go to Paris. "El hombre de la formación pampeana," presented at the French Exposition early in 1878, was published in *The American Naturalist.* During his three years in Europe, Ameghino did much to increase his knowledge: he traveled to Copenhagen, where he examined the Brazilian fossils collected by Lund; he attended the Zoological Congress in Bologna; he visited museums in Belgium and England; and he took courses at the Museum of Natural Sciences and the School of Anthropology in Paris. Ameghino supported himself at this time mainly through selling fossils he had brought with him. He became a friend of Paul Gervais, who helped him classify many specimens, and of Henri Gervais, with whom he collaborated on the publication of a work detailing seventy new species of fossil mammals. With the help of Adolph and Oskar Doering he published his principal work, *Contribución al conocimiento de los mamíferos fósiles de la República Argentina* (1880). The two-volume work was so expensive to produce that he suffered financially, but it brought him gold medals at the Paris Exposition of 1889 and the Chicago Exposition of 1892. In 1880–1881 he published *La antigüedad del hombre en el río de La Plata,* the first work on Argentine prehistory in both Spanish and French. He acquired direct knowledge of the first European fossil discoveries and, with Mortillet, he visited the Chelles site near Paris.

While he was in Paris, Ameghino lost his teaching position in Mercedes. When he returned to Buenos Aires he started a bleaching business, which soon failed. He then opened a secondhand bookshop in 1882 and brought his brother Juan into the business. Later Ameghino gave the shop to his mother, who ran it until 1908, and then to his brother, who kept it going until his death in 1932.

In 1884 Ameghino published *Filogenia,* in which he proposed to find irrefutable proof of transformism. This gave rise to a great quarrel with the Linnaean-Cuvierist wing of the clergy. In the same year, with the help of Juárez Celman, he obtained the chair of zoology at the University of Córdoba. Through the intercession of Adolph Doering, he was also offered the directorship of the Anthropological Museum, but he remained there for only a year. Two years later Francisco Moreno nominated him for secretary of the La Plata Museum, and he was appointed to the post in July 1886. Ameghino sold to the museum at least part of his collection of fossils and archaeological finds for $16,500. The collection was thought to number about 15,000 items, although Ameghino later said that there were only 8,000, of which half would go to the museum. He was named professor of geology and mineralogy at the University of La Plata in 1887. Unfortunately, his friend Moreno had an exaggerated concept of his authority, and they quarreled. Ameghino resigned as secretary of the museum in 1890.

The saddest period of Ameghino's life began in 1890, when he lost his professorship and was barred from entering the museum housing his collection. He attempted to support himself by opening a bookstore in La Plata and by founding the *Revista argentina de historia natural,* which lasted for only six issues. What little money he did accumulate was lost in the depression of 1893. Ameghino therefore sent a collection to Zittel in Munich, and was too poor to refuse the pittance offered for it. The American paleontologist William B. Scott, who benefited from Ameghino's hospitality and scholarly materials during this period, wrote: "I do not know of a finer example of courage and abnegation under the most distressing circumstances in the history of science. . . . He has made a vow of humility and poverty in the name of science and he is one of the greatest civil heroes in Argentina."

Despite the hardships, the period from 1895 to 1902 was Ameghino's most fruitful; the *Proceedings of the Zoological Society of London* published his account of the plexodontal molars of mammals in 1899. His fortunes improved considerably in 1902, when, thanks to the minister of education, Joaquín González, he was made director of the Museum of Natural Sciences in Buenos Aires. During his nine years as director he added 71,000 objects to its collections and published fifteen issues of its *Anales*. In 1906 he assumed the added duties of professor of geology at the University of La Plata.

Ameghino's brothers were of great help to him. Carlos assisted him on sixteen classification explorations, and his accurate observations resulted in Florentino's modification of his identification of a number of orders and genera. Juan, although an accomplished botanist, was too timid to publish any of his work; he preferred to help in the family bookstore, thereby freeing Florentino for science.

Ameghino was one of the most eminent geologists and paleontologists of his day and discovered many fossil fauna. On the other hand, his anthropological works are of doubtful value today. For instance, it is now known that his Pampas finds were more recent than he and his disciples thought. His errors were unavoidable, however, because of his early isolation and because the sciences themselves were so new. As so often happens, his fame was greater outside his own country.

In geology, Ameghino placed the Guaraní and Chubut formations in the Cretaceous period of the Secondary; six Argentine formations in the Tertiary period; the post-Pampean in the Pleistocene period of the Quaternary; and upper alluvial deposits in the Holocene. He then divided each of these formations into subaerial and freshwater stages with marine parallels, detailing and dividing them further in 1908. This paleontological geochronology has been disputed by geologists relying upon tectonic and mineralogical characteristics. Ingenieros, however, believes that Ameghino's stratigraphy itself is correct, very valuable, and based upon a great deal of material.

In paleontology, his first specialty, Ameghino achieved the same results that Haeckel had achieved in embryology by utilizing mathematical zoology with the seriation procedure. In 1889 he determined 450 species of fossil mammals, an astounding year's work. By 1906 he had classified thirty-five suborders; in all, he discovered over 6,000 species. His most controversial theory was that these fossils were older than those of other countries and that Argentina was the center from which those creatures had spread.

It is very possible that Ameghino had no knowledge of the work in prehistory that occurred during his early life, and therefore, according to Schobinger, he re-created and rediscovered human prehistory in Argentina. This brought him into conflict with Euro-

peans over the characteristics he attributed to Pampean man, although his contacts with Gervais lead one to suppose that he soon came to know the work of his French contemporaries. In Luján and Mercedes he found worked stones (chips), arrowheads, boleadora stones, and other more modern items, as well as ceramics and objects made of bone. Near the Córdoba observatory he found an authentic Pleistocene site with a fire site, some stone chips, and animal bones. He was not infallible, however: he attributed two rough artifacts from the Buenos Aires seashore to the Tertiary period, when they actually belonged to the beginning of the Holocene.

In his later years Ameghino had four theories on the evolution of the higher primates: (1) He excluded *Homo heidelbergensis* (Mauer's mandible) and *Pithecanthropus erectus* from the direct line of human phylogeny, considering them extinct lateral branches. (2) The primitive hominids of the early Miocene were derived from the anthropoid apes. (3) *Homo simius*, or *Homo australopithecus,* of Africa, was derived from the *Tetraprothomo*. (4) The Neanderthal man (*Homo primigenus*) was derived laterally from *Homo sapiens.*

All was not success, however, for Ameghino's construction of Eocene ancestors and his stubborn identification of remains he found with various stages of protohumanity (his *prothomo*) have since been repudiated. His anthropological knowledge of the *Tetraprothomo* was based only on an atlas and a femur he found in Monte Hermoso in 1907, and the latter was accepted as human by only a few scholars. Ameghino provided no physical evidence for the *Triprothomo*, and the *Diprothomo* of 1909 was also very weak, for it was based only on a piece of cranial cap. Ever since Father Blanco there has been criticism of the circumstances of the latter discovery, which took place in terrain of doubtful antiquity in the port of Buenos Aires, and was based upon an erroneous bone orientation.

This was not the only criticism of Ameghino's work. Márquez considered his four phylogenetic links irritatingly simple, but he accepted the contemporaneity of the extinct large mammals of the loess and Pampean man. He also felt Ameghino's philosophical elucidations were colored by a markedly candid materialism. Frenguelli criticized his geological stratigraphy because the Pampas were more modern than Ameghino claimed, but nevertheless recognized Ameghino's great perceptivity, his profound powers of observation, and his great talent. His theory that man originated during the Tertiary period on the Pampas has been rejected, and the bones and archaeological items he uncovered have proved, upon investigation, to be the weakest part of his work in prehistory.

BIBLIOGRAPHY

I. ORIGINAL WORKS. Ameghino's writings were collected as *Obras completas y correspondencia científica,* Alfredo Torcelli, ed., 24 vols. (La Plata, 1913–1936). Individual works are "El Tajamar y sus futuras consecuencias y el origen de la Tosca," in *El pueblo* of Mercedes (2 June 1875) and in *Obras completas,* I, 11; "Nouveaux débris de l'homme et de son industrie melés à des ossements quaternaires recueillis auprès (près) de Mercedes," in *Journal de zoologie,* **4** (1875), 527; **5** (1875), 27; "The Man of the Pampean Formation," in *The American Naturalist,* **12** (1880), 828; *Contribución al conocimiento de los mamíferos fósiles de la República Argentina,* 2 vols. (1880); *Los maníferos fósiles de la América Meridional* (Paris–Buenos Aires, 1880), text in both Spanish and French, written with H. Gervais; "Un recuerdo a la memoria de Darwin. El transformismo considerado como ciéncia exacta," in *Boletín del Instituto geográfico argentino,* **3,** no. 12 (1882), 205–213; *Filogenia* (Buenos Aires, 1884, 1915); "New Discoveries of Fossil Mammalians of Southern Patagonia," in *The American Naturalist,* **27** (1893), 445 ff.; the account of the primitive types of plexodont molars of mammals, in *Proceedings of the Zoological Society of London* (1899), 555–571 (or 575), also trans. into French in *Anales del Museo nacional de historia natural de Buenos Aires* (16 Dec. 1902); and *La antigüedad del hombre en el río de La Plata,* Vol. III of Ameghino's *Obras completas* (La Plata, 1915), also published separately in 2 vols. (Buenos Aires, 1918).

II. SECONDARY LITERATURE. Among the works on Ameghino or his contributions are Juan B. Ambrosetti, "Florentino Ameghino," in *Anales del Museo nacional de historia natural de Buenos Aires* (1912); José María Blanco, *La evolución antropológica y Ameghino* (Buenos Aires, 1916); Ángel Cabrera, *El pensamiento vivo de Ameghino* (Buenos Aires, 1944); Arturo Capdevila, "Ameghino el vidente," in *La prensa* of Buenos Aires (25 May 1932); Alfredo Castellanos, *Homenaje a Florentino Ameghino* (Rosario, 1937); Pedro Daniels, "Presencia y actualidad de Ameghino," in *La hora médica argentina* (Sept. 1954), pp. 167–169; Joaquín Frenguelli, *La personalidad y la obra de Florentino Ameghino* (La Plata, 1934); Max Friedmann, "Vorlage eines Gipsabgusses des Schädeldaches von *Diprothomo Platensis* Ameghino," in *Zeitschrift für Ethnologie* (1910); Bernardo González Arrili, *Vida de Ameghino* (Santa Fe, 1954); Mario Graci Larravide, *Florentino Ameghino* (Mendoza, 1944); Alex Hrdlicka, *Early Man in South America* (Washington, D.C., 1912), see index; José Ingenieros, *Las doctrinas de Ameghino,* Vol. XVIII of his *Obras completas* (Buenos Aires, 1939); Arthur Keith, *The Antiquity of Man* (London, 1916), ch. 17; Robert Lehmann-Nitsche, "Ameghino como antropólogo," in *Renacimiento,* **3** (31 Aug. 1911); P. G. Mahoudeau, "Les primates et les prosimiens fossiles de la Patagonie d'après les travaux de M. Florentino Ameghino," in *Revue de l'École d'anthropologie de Paris,* **11** (1907), 354–361; Osvaldo Menghin, *Origen y desarrollo racial de la especie humana* (Buenos Aires, 1957), p. 70; Fernando Márquez Miranda, *Ameghino. Una vida heróica* (Buenos Aires, 1951); Victor Mercante, "Dr. Florentino Ameghino. Su vida y sus obras," in Ameghino's *Obras completas,* I, 148–170; Aldobrandino Mocchi, "Nota preventiva sul *Diprothomo platensis,*" in *Revista del Museo de La Plata,* **17** (1910–1911), 70; Ricardo Rojas, *Historia de la literatura argentina,* VII (Buenos Aires, 1960), 53–60; Alberto Rovero and Victor Delfino, "La obra antropológica de Florentino Ameghino," in *La semana médica,* no. 18 (1914); V. G. Ruggeri, "Die Entdeckungen Florentino Ameghino's und der Ursprung des Menschen," in *Globus,* **94,** no. 11 (1908), 21–26; Carlos Rusconi, *Florentino Ameghino. Rasgos de su vida y su obra* (Mendoza, 1965); and *Animales extinguidos de Mendoza y de la Argentina* (Mendoza, 1967), see index; Antonio Santiana, *La personalidad creadora de Ameghino* (Quito, 1954); Domingo Sarmiento, "El señor Ameghino," in *El nacional* (10 July 1883) and in his *Obras completas,* XLII (Buenos Aires, 1900), 140; Juan Schobinger, *Cincuentenario de la muerte de Ameghino* (Mendoza, 1961); G. Schwalbe, "Studien zur Morphologie der südamerikanischen Primatenformen," in *Zeitschrift für Morphologie und Anthropologie,* **13** (1910); William B. Scott, *A History of Land Mammals in the Western Hemisphere* (New York, 1937), pp. 114, 500, 504, 544; Rodolfo Senet, *Ameghino. Su vida y su obra* (Buenos Aires, 1934); George Simpson, "The Beginning of Mammals in South America," in *Bulletin of the American Museum of Natural History,* **91** (1948), which lists 19 works by Ameghino; Miguel Soria, "Intoxicación accidental por *Chlorophyllum molybdites,*" in *Prensa universitaria* of Buenos Aires (28 Nov. 1966), pp. 2427–2428; Kasimiers Stolihwo, "Contribución al estudio del hombre fósil sudamericano y su pretendido precursor el *Diprothomo platensis,*" trans. from the Polish by Victor Delfino, in *Semana médica* (15 Aug. 1912); Herbert Wenk, *Tras las huellas de Adán* (Barcelona, 1958), pp. 545, 547, 548; and Karl von Zittel, *Grundzüge der Paläontologie. Vertebrata* (Munich–Berlin, 1911), see index.

JUSTO GARATE

AMES, JOSEPH SWEETMAN (*b.* Manchester, Vermont, 3 July 1864; *d.* Baltimore, Maryland, 24 June 1943), *physics.*

Ames's father, George L. Ames, a physician, died in 1869; in 1874 his mother, Elizabeth Bacon Ames, married Dr. James Dobbin, rector of the Shattuck School in Fairbault, Minnesota, where Ames was a student. At home, Ames, who was raised an Episcopalian, acquired a lasting taste for classical education, books, and good society. At Johns Hopkins University, where he went after the Shattuck School, he developed an enthusiasm for physics. After graduating in 1886, he spent two years in Helmholtz' laboratory in Berlin. He returned to Johns Hopkins to take his Ph.D. in 1890 under Henry A. Rowland, the inventor of the curved spectral grating, and then joined the Johns Hopkins faculty. He was director of the physical laboratory from 1901 to 1926, provost of the university from 1926 to 1929, and president from 1929 to 1935. In 1899 he married Mary B. Harrison, a widow from Maryland.

Ames's research was limited in quantity and largely confined to the field of spectroscopy. Working closely with Rowland in the 1890's, he struggled with the problem of finding relationships among the lines of particular spectra. Johann Balmer had already advanced his formula for the hydrogen lines; Ames, measuring with great exactitude the spectra of more complex atoms, tried empirically to find relationships among the wave numbers (the reciprocals of the wave lengths in vacuo). He concluded that the answer could come only from theoretical considerations. But by 1913, when Bohr's theory was published, Ames had given up research and had turned to administration.

Ames, a man of courtly manner and executive talent, was an able administrator. He kept the physics department at Johns Hopkins alive despite persistent budgetary problems by cooperating with the National Bureau of Standards. He encouraged the faculty to offer courses there and graduate students to do their research at the Bureau's well-equipped laboratories.

In World War I Ames, a member of the National Academy of Sciences, was drawn into the affairs of government research. He served on the Academy's National Research Council and on the National Advisory Committee for Aeronautics. The NACA, created by Congress in 1915 to promote the scientific study of flight, was deeply enmeshed in policy matters. Ames got into trouble for writing publicly that the government's ambitious aircraft program was far behind schedule; Secretary of Commerce William C. Redfield considered the statement a treasonous act. But Ames, as it turned out, was right. His judgment won increasing respect, and he became chairman of the NACA's Executive Committee in 1919, the year that he was elected president of the American Physical Society.

Ames held the chairmanship of the Executive Committee until 1936 and served as chairman of the entire NACA from 1927 to 1939. During his administration the agency's technical publications won the respect of aeronautical experts all over the world. In 1935 the Smithsonian Institution awarded Ames the Langley Gold Medal for his leadership of the NACA. In 1939 the Committee decided to name its new Moffet Field, California, research facility the Ames Aeronautical Laboratory.

BIBLIOGRAPHY

Aside from a scattering of Ames's papers in the Archives of the Johns Hopkins University, Baltimore, Maryland, the principal body of his extant correspondence is in the records of the National Advisory Committee for Aeronautics in the National Archives, Washington, D. C. In the *Annual Reports* of the NACA during his administration, Ames discussed the work of the Committee and provided a running commentary on the development of aeronautics.

Henry Crew, "Joseph Sweetman Ames," in *National Academy of Sciences Biographical Memoirs,* **23** (1945), 181–201, is a shallow, sentimental essay but contains a complete bibliography of Ames's scientific work.

DANIEL J. KEVLES

AMES, WILLIAM (*b.* Ipswich, Suffolk, England, 1576; *d.* Rotterdam, Netherlands, 1 November 1633), *theology, natural philosophy.*

Ames was the son of a prosperous Ipswich merchant, William Ames, and his wife, Joan Snelling. His parents died during his childhood, and he was brought up by an uncle, Robert Ames, at Oxford. He proceeded to Cambridge, where he matriculated as a pensioner at Christ's College in 1593/1594, obtaining his B.A. in 1597/1598 and his M.A. in 1601. As a fellow of Christ's College from 1601 until 1610, Ames took up the controversial theological position of his tutor, the celebrated Puritan William Perkins (1558–1602). Ames became a central figure in Puritan agitation at Cambridge, with the result that, in 1609, he was suspended from his degrees and expelled from his college and the university.

For a short time Ames preached at Colchester, but the opposition of George Abbot, bishop of London, caused him to emigrate to the Netherlands in 1610. There he occupied minor clerical positions at Leiden and The Hague, and became chaplain to Sir Horace Vere, English governor at Brill, succeeding another well-known Puritan, John Burgess, whose daughter became Ames's first wife. She died childless shortly after the marriage.

Ames achieved prominence at the Synod of Dordrecht (1618), where he advised the Calvinist faction. As a result of his success in the theological debates emanating from the synod, he was appointed to the chair of divinity at the University of Franeker in Friesland, in 1622. There, his erudition and abilities as a teacher attracted students from many parts of Europe, one of them the future first head of Harvard College, Nathaniel Eaton. Ames was rector of the university from 1626 to 1632. During this Franeker period he wrote most of his philosophical and theological works. He was not entirely settled in Holland, however, for the climate did not favor his health.

Plans were made for emigration to New England (1629), but instead he became minister and lecturer to the English congregation at Rotterdam (1633). Among his colleagues there were Hugh Peter and Thomas Hooke, both of whom eventually settled in America. Soon after his arrival at Rotterdam, Ames died of a fever contracted after his house was flooded. His second wife, Joan Fletcher Ames, a relation of Governor John Winthrop, and their three children went to New England in 1637. Two sons were educated at Harvard College.

As the numerous editions of Ames's works indicate, he occupied a prominent role in the Protestant theology of the first half of the seventeenth century, systematizing and developing certain aspects of the Calvinist theology of Perkins. Two particular points—practical divinity and Ramist philosophy—make Ames significant, not only for theology but also for the general intellectual history of the seventeenth century.

First, he stressed the role of "practical divinity," in reaction against the tendency of contemporary Dutch philosophers to divorce ethics from theology. By analyzing the nature of conscience, it could be shown that the tenets of theology and ethics had the same origin. The unity of theology and ethics was also proved by their mutual reliance on the Scriptures. Ames's resultant system of divinity paid the greatest attention to the rules of personal behavior and organization of the community. Like Bacon, Ames directed all Christians to the practical reform of society. This practical divinity certainly influenced the scientific outlooks of such figures as John Winthrop, Jr., Samuel Hartlib, and Robert Boyle, directing them to problems of social usefulness.

One of the reasons for the enormous popularity of Ames's writings was their strict logical organization. Through Alexander Richardson, George Downham, and Perkins, Ames had become an enthusiastic exponent of Ramist philosophy: his *Demonstratio logicae verae* and *Theses logicae* were commentaries on Ramus' *Dialecticae libri duo*. His theological works were organized according to the Ramist dichotomies, and the Ramist logic was applied to the interpretation of the dictates of conscience and the Scriptures. In this system, the scholastic boundaries of knowledge were broken down. Theology impinged on natural philosophy, and mathematics and physics were seen as having inherent moral and spiritual value. These three subjects, the *artes speciales,* took their place alongside the three *artes generales*—dialectic, grammar, and rhetoric. All were subject to logical analysis, and all amalgamated into an encyclopedic system, the *Technometria.*

The apprehension of the principles of the arts was seen as an important moral duty. Ames adopted an empirical approach to this problem, believing that these principles were derived from a knowledge of the objects of nature, by a process of observation and experiment that he recognized as akin to the philosophy of induction announced in Bacon's *Novum organum* (*Technometria,* §§ 69, 70). This encyclopedic and empirical view of nature appealed greatly to the Puritan educationalists of New England, and Ames's works became the dominant forces in the curriculum of the newly founded college of Harvard, as well as in Cambridge, the Low Countries, and Transylvania. Because of the influence of such authors as Ames, there was constant interaction between theology, ethics, and natural philosophy in the areas of Calvinist influence during the seventeenth century.

BIBLIOGRAPHY

I. ORIGINAL WORKS. *Philosophemata* (Leiden, 1643; Cambridge, 1646; Amsterdam, 1651) is the collected edition of Ames's philosophical writings. Certain theological works are included in *The Works of the Reverend and Faithfull Minister of Christ William Ames* (London, 1643). Ames's collected works are *Opera, quae Latine scripsit omnia,* Matthias Nethemus, ed., 5 vols. (Amsterdam, 1658), with a biographical account in the Preface; the philosophical writings are in Vol. V.

Some of Ames's individual works are *Medulla theologia* (Franeker, 1623; Amsterdam, 1627, 1659), translated into English as *The Marrow of Sacred Divinity, Drawne Out of the Holy Scriptures, and the Interpreters Thereof* (London, ca. 1638, 1642, 1643); *De Conscientia et eius iure vel casibus libri quinque* (Amsterdam, 1631, 1670), translated into English as *Conscience With the Power and Cases Thereof* (London [?], 1639 [?], 1643); *Demonstratio logicae verae* (Leiden, 1632; Cambridge, 1646); *Disputatio theologica adversus metaphysicam* (Leiden, 1632; Hanau, 1640; Cambridge, 1646); *Technometria, omnium & singularum artium fines adaequate circumscribens* (Amsterdam–Leiden, 1632, 1633); and *Theses logicae* (Cambridge, 1646).

II. SECONDARY LITERATURE. See "William Ames," in *Biographia Britannica,* A. Kippis, ed., I (London, 1747), 135–137; J. Bass Mullinger, "William Ames," in *Dictionary of National Biography,* I (London, 1885), 355–357; Paul Dibon, *La philosophie néerlandaise au siècle d'or,* I (Amsterdam, 1954), 151–154; G. L. Kittredge, "A Note on Dr. William Ames," in *Transactions of the Colonial Society of Massachusetts,* **13** (1910/1911), 60–69; Perry Miller, *The New England Mind* (New York, 1939); S. E. Morison, *The Founding of Harvard College* (Cambridge, Mass., 1935), and *Harvard College in the Seventeenth Century,* 2 vols. (Cambridge, Mass., 1936), I, 164–165; George L. Mosse, *The Holy Pretence* (Oxford, 1957), ch. 5; *Nieuw Nederlandsch biographisch Woordenboek* (Amsterdam, 1911–

1937), VI, 36; J. Piele, *Biographical Register of Christ's College,* 2 vols. (Cambridge, 1910), I, 211–212; Karl Reuter, *Wilhelm Amesius der führende Theologe des erwachenden reformierten Pietismus* (Neukirchen, 1940); Keith L. Sprunger, "Technometria: A Prologue to Puritan Theology," in *Journal of the History of Ideas,* **29** (1968), 115–122; and Hugo Visscher, *Gulielmus Amesius. Zijn Leven en Werken* (Haarlem, 1894), theology thesis, Leiden University.

CHARLES WEBSTER

AMICI, GIOVAN BATTISTA (*b.* Modena, Italy, 23 March 1786; *d.* Florence, Italy, 10 April 1868), *optics, microscopy, natural sciences.*

Amici was the son of Giuseppe Amici, a ministerial official, and Maria Dalloca, a member of a well-to-do family. In 1806 he married Teresa Tamanini, the daughter of a wealthy Tyrolean bookseller. The following year he graduated as an engineer-architect from the University of Bologna and immediately became a mathematics teacher at the Modena *liceo.* In 1831 Amici was invited to Florence by the grand duke of Tuscany to head the astronomical observatory and the Royal Museum of Physics and Natural History. He held this office until 1859, when, because of his advanced age of seventy-three, he accepted the less demanding office of director of microscopic research at the museum.

From early youth, Amici was interested in optical instruments, particularly microscopes, and it was in this field that he achieved his fame. In spite of the efforts of many pioneers during the previous two centuries to refine the microscope, in the early nineteenth century scientific microscopy was carried out exclusively with the simple microscope because it performed better than the compound microscope, particularly in resolving power. A comparative study of the microscopes of the past centuries has been made by P. Harting and the Van Citterts, who have thoroughly examined the microscopes in the collection of the Museum of the History of Science at Utrecht. Their study clearly indicates Amici's decisive contribution to the use of the compound microscope.

In the early nineteenth century, compound microscopes were much less accurate than the simple microscopes, not only because their objectives caused strong aberrations, especially chromatic ones, but above all because the numerical aperture was not yet known to be the determining factor of the resolving power; therefore construction was governed by the erroneous conviction that enlargement was the most important factor. In 1791 the first achromatic lens for compound microscopes was built as the result of the work of an Amsterdam amateur, F. Beeldsnijder; the images were quite good, but the resolution was only 0.01

mm., while the resolution of simple microscopes reached 0.0015 mm. Only around 1806 did microscopes with achromatic objectives appear on the market, through the efforts of Harmanus Van Deyl, also of Amsterdam; these were instruments that could magnify 150 times and could resolve to 0.005 mm.

This lead was immediately followed by scientists all over the world, especially Chevalier and Nachet; Oberhaüser and Plössl; Tulley, Pritchard, and Ross; and Amici, who was just embarking on his career. Within thirty years, the efforts of these and many other scientists succeeded in making the resolving power of the compound microscope equal to that of the simple microscope. For example, as early as 1818 Amici, after having built a type of catadioptric microscope that was free of chromatic aberrations, succeeded in appreciably improving the knowledge of the circulation of protoplasm in *Chara* cells, thereby becoming immediately famous not only as an optician but also as a microscopic biologist.

His most sensational innovation was achieved in 1837, when Amici arrived at a resolving power of 0.001 mm. with a new type of microscope that had a numerical aperture of 0.4 and was capable of magnifying up to 6,000 times. This device consisted of a hemispheric frontal lens applied to the objective; only through this contrivance was it possible to increase the numerical aperture to an appreciable extent. It was the hemispheric lens that permitted maximum use of the compound microscope. All opticians, with the exception of the French, immediately adopted this new design.

The improvement introduced by Amici also led to a significant theoretical clarification. The moderate influence of the magnification in the operation of the microscope was soon recognized and values ranging around 1,000X were generally achieved. It was emphasized, however, that resolution was due mainly to the numerical aperture and to the optical correction by the objective. Amici clearly stated his discovery in a letter of 25 October 1855 to his friend Ottaviano Mossotti, professor at the University of Pisa and a famous optical mathematician; the following passage is particularly interesting:

> The objectives of my microscopes consisting of six lenses, three crown lenses and three flint lenses, happen to be achromatic. I found out, however, that the sets consisting of three pairs of lenses, as I mentioned, were not the best suitable for achieving greater magnification, particularly due to the fact that the lower pair closer to the object are too large and make it impossible to obtain a very short focal distance of the system and its very large aperture. I then had the idea of replacing the lower pair with a simple lens, i.e., a half-sphere of

any transparent substance, either crown, flint, low-grade ruby, diamond, molten rock crystal, etc., and thus eliminating its aberrations by means of the two suitably processed upper pairs. In order to accomplish this, I needed a flint with very high dispersion, which I was able to obtain from Faraday thanks to Airy. The English opticians laughed at this request but, when I showed them the superiority of the new construction in London in 1844, they soon tried to imitate it, and the Americans followed suit. The French, who were not interested and who were unable to understand the improvement, were outdistanced by the others.

Amici was so convinced that the numerical aperture was the theoretical factor determining a microscope's power that he continued to do everything possible to increase it, not only by manufacturing objectives with ever greater numerical aperture (Harting states that in 1856 he purchased from Amici a microscope objective with a numerical aperture of 0.985) but also by inventing the technique of immersion microscopy; he first used water, then olive oil, and finally sassafras oil. He even became aware of the influence that the thickness of the cover glass has on the quality of the image in the microscope.

Amici moved to Florence, where he devoted his attention to other optical processes, although his interest in microscopy never flagged. He invented widely used prisms that still bear his name, and reconsidered the direct-vision prism, which had been forgotten. He also built concave mirrors and astronomical lenses. His masterpiece was a lens with a diameter of 285 mm.: this was so far in advance of his time that it was presented as a rarity to the Third Congress of Italian Scientists, held in Florence in 1841. Only one lens larger than this existed at that time. Amici's lens was so successful that it is still in use at the Arcetri Astronomical Observatory. Amici also invented new micrometers to improve the accuracy of astronomical measurements, and new types of distance-measuring telescopes.

While Amici kept increasing the power of his instruments, through his theoretical know-how and his skill in optical processing, he also used them in astronomical and microscopical observations. His findings in microscopy attracted the attention of the entire biological world of his time. Besides studying the circulation of protoplasm, Amici made remarkable observations in anatomy, physiology, histology, and plant pathology, as well as in leaf morphology (discovering palisade parenchyma) and animal biology. The discovery that made him famous, however, was that of the fertilization of phanerogams, particularly the travel of the pollen tube through the pistil of the flower (1821). His early observations were

followed by a heated controversy with the best-known botanists of the world, who for thirty years disputed Amici's ideas. However, by making ever finer microscopic observations, he finally won over his rivals.

The splendid wax models by which Amici illustrated the results of his microscopic observations can still be seen in many museums of natural history.

BIBLIOGRAPHY

I. ORIGINAL WORKS. Amici's main works in optics and astronomy are "Descrizione di un nuovo micrometro," in *Memorie di matematica e fisica della Società Italiana delle Scienze,* **17** (1816), 344–349; "Dei microscopii catadiottrici," *ibid.,* **18,** no. 1 (1818), 107–118; "Memoria sopra un cannocchiale iconantidiptico," in *Memorie della Società Italiana delle Scienze,* **1** (1821), 113–125; *Description d'une nouvelle lunette micrométrique* (Genoa, 1823), IX, 517–521; *Osservazioni sopra i satelliti di Giove* (Genoa, 1825), XII, 539–560; "Descrizione di un nuovo strumento per livellare," in *Atti della R. e I. Accademia dei Georgofili,* **15** (1837), 129–137; "Lettura relativa a due macchine ottiche da lui inventate," in *Atti della Prima Riunione degli Scienziati Italiani* (Pisa, 1839), pp. 49–50; and "Di alcuni perfezionamenti recentemente ottenuti in fotografia," in *Atti della Sesta Riunione degli Scienziati Italiani* (Milan, 1844), pp. 87–97.

Amici's principal publications in biology are "Osservazioni sulla circolazione del succhio nella Chara," in *Memorie della Società Italiana delle Scienze,* **18** (1818), 183–198; "Osservazioni microscopiche sopra varie piante," *ibid.,* **19** (1823), 234–255; "Descrizione di alcune specie nuove di Chara," in *Memorie della R. Accademia di Scienze, Lettere ed Arti di Modena,* **1** (1833), 199–207; "Opinione relativa all'ascensione della linfa nelle piante," in *Atti della Prima Riunione degli Scienziati Italiani* (Pisa, 1839), pp. 165–167; "Sulla presenza dei pori nei casi delle conifere e sulla loro struttura," *ibid.,* pp. 162–164; "Sul processo col quale gli ovuli vegetabili ricivono l'azione fecondante del polline," *ibid.,* pp. 134–139; "Sull'uredo rosae," *ibid.,* p. 157; "Sulla fecondazione delle piante (cucurbita pepo)," in *Atti della Quarta Riunione degli Scienziati Italiani* (Padua, 1842), pp. 279–283; "Nuove osservazioni sugli stomi del *Cereus Peruvianus," ibid.,* pp. 327–333; "Sulla struttura degli stomi," in *Atti della Sesta Riunione degli Scienziati Italiani* (Milan, 1844), pp. 513–518; "Sulla fecondazione delle orchidee," in *Atti della Ottava Riunione degli Scienziati Italiani* (Genoa, 1846), pp. 542–550; "Nota in risposta al primo articolo dello Schleiden," *ibid.,* pp. 89–90; "Sulla malattia dell'uva," in *Atti della I. e R. Accademia dei Georgofili,* **30** (1852), 454–460; "Sulla malattia della foglia del gelso, detta fersa o seccume," *ibid.,* n.s. **1** (1853), 72–79; "Sulla malattia del frumento detta rachitide," *ibid.,* 570–577; and "Sulla fibra muscolare," in *Il tempo,* **1** (1858), 2–4.

Amici's correspondence, which consists of more than 6,000 letters and has not yet been published, is in the Biblioteca Estense of Modena.

II. Secondary Literature. The principal publications on Amici are *Celebrazione del centenario della morte di G. B. Amici* (Modena, 1963); G. B. Donati, "Elogio del Prof. G. B. Amici," in *Reale Accademia dei Georgofili,* **1** (1864), 1–23; H. von Mohl, "Giovambattista Amici," in *Botanische Zeitung,* **15** (1863), 21–32; P. Pagini, "L'ottica geometrica in Italia nella prima metà del secolo XIX e l'opera di G. B. Amici," in *Rassegna nazionale* (1917), 1–35; F. Palermo, "Sulla vita e le opere di G. B. Amici," in *Bollettino di bibliografia e storia delle scienze matematiche e fisiche,* **28** (1870), 3–18; V. Ronchi, "Sopra gli obbiettivi astronomici dell'Amici," in *Rivista d'ottica e meccanica di precisione,* **2** (1922), 3–21; and "Giovan Battista Amici, Optician," in *Atti della Fondazione Giorgio Ronchi,* **18** (1963), 481–504; and P. H. Van Cittert and J. G. Van Cittert-Eymers, "The Amici Microscopes About 1850 in Possession of the University of Utrecht," in *Proceedings of the Koninklijke Nederlandsche Akademie van Wetenschappen,* **50** (1947), 5–10.

Also of value for an understanding of the microscope and its development are P. Harting, *Das Mikroskop,* 2nd German ed. (Brunswick, 1866); Maria Roseboom, *Microscopium* (Leiden, 1956); and P. H. Van Cittert and J. G. Van Cittert-Eymers, "Some Remarks on the Development of the Compound Microscope in the 19th Century," in *Proceedings of the Koninklijke Nederlandsche Akademie van Wetenschappen,* **54** (1951), 1.

Vasco Ronchi

AMMONIUS, SON OF HERMIAS (*d.* Alexandria, Egypt, *ca.* A.D. 517–526), *philosophy.*

Ammonius, surnamed "son of Hermias" to distinguish him from namesakes, was head of the Platonic school in Alexandria from 485. He is one of the characters in Zacharias Scholasticus' dialogue *Ammonius* (probably historical in essence), in which Zacharias refutes Ammonius' assertion that the cosmos is coeternal with God and explains to him the doctrine of Trinity. A sober and scholarly interpreter of Plato and Aristotle, Ammonius "harmonized" them. His works and lecture courses were frequently edited and fully utilized by three generations of his students, including Johannes Philoponus, Simplicius, Olympiodorus, David, Elias, and perhaps Boethius. He and his school managed to come to terms with Christian authorities; he himself perhaps became a Christian, albeit in name only: the faculty and students of the school were partly pagan, partly Christian. The school survived until the Arab conquest of Alexandria (*ca.* 641/642), whereas Plato's Academy was closed in 529. Its Platonism was in many respects pre-Plotinian, as has been shown especially by Praechter (*pace* Lloyd's doubts).

Damascius praised Ammonius as an accomplished mathematician and astronomer. Having witnessed his astronomical observations, Simplicius (and probably Ammonius himself) deduced the existence of a starless sphere (the *primum mobile*) enveloping the sphere of fixed stars and imparting its own motion to that sphere. Ammonius agreed with Aristotle that mathematical objects do not subsist, although they can be abstracted from physical objects. He divided theoretical philosophy into theology, mathematics, and physics (and mathematics into arithmetic, geometry, music, and astronomy), thereby again combining Aristotelian and Platonic points of view, as did Boethius later.

BIBLIOGRAPHY

I. Original Works. Ammonius' works include *In Porphyrii Isagogen,* in *Commentaria in Aristotelem Graeca,* 23 vols., Vol. IV, pt. 3 (Berlin, 1891); *In Aristotelis De interpretatione, ibid.,* Vol. IV, pt. 5 (1897) (the part defending both free will and providence in J. C. von Orelli, ed., *Alexandri Aphrodisiensis, Ammonii Hermiae filii, Plotini, Bardesanis Syri, et Georgii Gemisti Plethonis de fato quae supersunt Graece* [Zurich, 1824]); *In Aristotelis Anal. pr. 1, I,* in *Commentaria in Aristotelem Graeca,* Vol. IV, pt. 6 (1890); and lectures on Aristotle's *Metaphysics,* edited with additions by Asclepius, *ibid.,* Vol. VI, pt. 2 (1888).

See also Damascius, *Vita Isidori,* C. Zintzen, ed. (Hildesheim, 1967); Simplicius, *In De caelo,* in *Commentaria in Aristotelem Graeca,* Vol. VII (1894); and Zacharias Scholasticus, *Ammonius,* in J. P. Migne, ed., *Patrologiae cursus completus, Series graeca,* 162 vols. (Paris, 1857–1866), Vol. LXXXV.

II. Secondary Literature. Writings on Ammonius or his work are J. Freudenthal, "Ammonios 15," in Pauly-Wissowa, *Real-Encyclopädie;* P. Duhem, *Le système du monde,* II (Paris, 1914); E. Zeller and R. Mondolfo, *La filosofia dei Greci,* Vol. VI, pt. 3, G. Martano, ed. (Florence, 1961); P. Courcelle, *Les lettres grecques en Occident* (Paris, 1948); P. Merlan, *From Platonism to Neoplatonism,* 2nd ed. (Berlin, 1960), pp. 75 f.; K. Kremer, *Der Metaphysik-Begriff in den Kommentaren der Ammonius-Schule* (Berlin, 1961); L. G. Westerink, *Anonymous Prolegomena to Platonic Philosophy* (Amsterdam, 1962); A. H. Armstrong, ed., *The Cambridge History of Late Greek and Early Medieval Philosophy* (1967), the contributions by A. C. Lloyd, J. P. Sheldon-Williams, and H. Liebeschütz; and P. Merlan, "Ammonius Hermiae, Zacharias Scholasticus and Boethius," in *Greek-Roman-Byzantine Studies,* **9** (1968).

For the character of the school of Alexandria in general, see K. Praechter, "Richtungen und Schulen im Neuplatonismus," in *Genethliakon für C. Robert* (Berlin, 1910), pp. 147–156; "Christlich-neuplatonische Beziehungen," in *Byzantinische Zeitschrift,* **21** (1912), 1–27; and "Simplicius 10," in Pauly-Wissowa, *Real-Encyclopädie,* III (1927), 1.

Philip Merlan

AMONTONS, GUILLAUME (*b.* Paris, France, 31 August 1663; *d.* Paris, 11 October 1705), *physics.*

Amontons's father was a lawyer from Normandy who settled in Paris. The boy became almost deaf during adolescence, and his interest then turned toward mechanics. After vain efforts to develop a perpetual motion machine, he decided, despite his family's opposition, to study physical sciences and mathematics. After studying drawing, surveying, and architecture, he was employed on various public works projects that gave him practical knowledge of applied mechanics. Later he studied celestial mechanics and applied himself to the improvement of hygrometers, barometers, and thermometers.

His first scientific production was a hygrometer in 1687. The apparatus consisted of a ball of beechwood, horn, or leather filled with mercury; it varied in size according to the humidity of the atmosphere. In 1688 he developed his shortened barometer, composed of several parallel tubes connected alternately at the top and bottom, with only alternate tubes containing mercury.

Sometime between 1688 and 1695, Amontons tried out his optical telegraph in the presence of the royal family. He published no data on this experiment, but the device is known to have consisted of a series of stations, each equipped with a spyglass, for the rapid transmission of signals. The nature of the signals to be transmitted is not known, however.

In 1695 Amontons sought to renew the use of the clepsydra as a timing apparatus on ships in order to solve the problem of determining longitude at sea. In his paper on this, he described two apparatuses that became well known by his name in the eighteenth century, although their use was never common. One was a cisternless barometer consisting of a tube narrow enough for the column of mercury to remain suspended. In his experiments with this, Amontons gradually broadened the tube into the shape of an inverted funnel. The mercury column then became shortened as atmospheric pressure decreased and lengthened as it increased.

The other was an air thermometer independent of the atmospheric pressure. Air occupied the top of one of the branches of a U-shaped tube, and by its dilation it pushed down one of the mercury columns so that the other end of the branch formed a barometric chamber.

As early as 1699 Amontons proposed a thermic motor: a machine using hot air and external combustion with direct rotation. The experiments carried on in connection with this machine led him to note that ordinary air going from the temperature of ice to that of boiling water increases its volume by about one third.

In the same year Amontons produced the first known study on the question of losses caused by friction in machines. He then established the laws of proportionality between the friction and the mutual pressure of the bodies in contact.

In 1702 Amontons returned to thermometry. Having noted that water ceases to increase its temperature from the boiling point, he proposed that the latter be the fixed thermometric point. He also observed that for an equal elevation of temperature, the increase of pressure of a gas always exists in the same proportion, no matter what the initial pressure.

The following year Amontons indicated practical ways of graduating ordinary alcohol thermometers. Also, returning to his observations of 1702, he proposed an explanation for certain natural catastrophes, such as earthquakes: If there is air very deep within the earth, it is extremely compressed and could reach an irresistible pressure as the result of a relatively small increase in temperature.

Among Amonton's last works was a barometer with a U-tube, without an open surface of mercury, to be used on shipboard. Using the same receptacle and liquids whose coefficients of expansion differed, Amontons was able to establish as false the theory that liquids "condense and cool first, before expanding with approaching heat." The observed results were due only to the expansion of the containers. Also, using a barometer as an altimeter, he tried to verify the exactitude of Mariotte's (Boyle's) law at low pressures.

One really cannot understand what has led certain authors to attribute to Amontons the creation of an air thermometer of unvarying volume. As for the idea of absolute zero, he barely implies it in his memoir of 1703 ("Le thermomètre réduit à une mesure fixe," pp. 52–54); this brief notice nevertheless presented Johann Heinrich Lambert with a point of departure for his explication of this idea (1779).

BIBLIOGRAPHY

I. ORIGINAL WORKS. Excerpts from letters are in *Journal des savants* (8 March 1688), 245–247, and (10 May 1688), 394–396. His only book is *Remarques et expériences physiques sur la construction d'une nouvelle clepsydre . . .* (Paris, 1695). Amontons's papers, all in *Mémoires de l'Académie des Sciences*, are "De la résistance causée dans les machines . . ." (1699), 206–227, and *Histoire . . .*, 104, 109; "Moyen de substituer commodément l'action du feu . . .," *ibid.*, 112–126, and *Histoire . . .*, 101; "Discours sur quelques propriétes de l'air . . ." (1702), 155–174, and *Histoire* . . ., 1; "Que les nouvelles expériences que nous avons du poids et du ressort de l'air . . ." (1703), 101–108, and *Histoire* . . ., 6; "Remarques sur la table des degrés de chaleur . . .," *ibid.,* 200–212, and *Histoire* . . ., 9; "Le thermomètre réduit à une mesure fixe et certaine . . .," *ibid.,* 50–56 and *Histoire* . . ., 9; "Discours sur les baromètres" (1704), 271–278, and *Histoire* . . ., 1; "Que tous les baromètres tant doubles que simples . . .," *ibid.,* 164–172, and *Histoire* . . ., 1; "Baromètres sans mercure à l'usage de la mer" (1705), 49–54, and *Histoire* . . ., 1; "De la hauteur du mercure dans les baromètres" (four articles), *ibid.,* 229–231, 232–234, 234–236, 267–272, and *Histoire* . . ., 10; "Expériences sur la raréfaction de l'air," *ibid.,* 119–124 and *Histoire* . . ., 10; "Expériences sur les solutions et sur les fermentations froides . . .," *ibid.,* 83–84, and *Histoire* . . ., 68; and "Que les expériences sur lesquelles on se fonde pour prouver que les liquides se condensent et se refroidissent . . .," *ibid.,* 75–80, and *Histoire* . . ., 4.

II. SECONDARY LITERATURE. Works that discuss Amontons and his instruments are Maurice Daumas, *Les instruments scientifiques aux XVIIe et XVIIIe siècles* (Paris, 1953); [Bernard le Bovier de Fontenelle] "Éloge de M. Amontons," in *Histoire de l'Académie Royale des Sciences* (1705), 150–154; René Taton, *Histoire générale des sciences,* II, *La science moderne (de 1450 à 1800)* (Paris, 1958), pp. 258, 472, 516; and W. E. Knowles Middleton, *The History of the Barometer* (Baltimore, 1964).

JACQUES PAYEN

AMPÈRE, ANDRÉ-MARIE (*b.* Lyons, France, 22 January 1775; *d.* Marseilles, France, 10 June 1836), *mathematics, chemistry, physics.*

Ampère's father, Jean-Jacques, was a merchant of independent means who, soon after his son's birth, moved the family to the nearby village of Poleymieux, where André-Marie grew up. The house is today a national museum. Jean-Jacques Ampère had been greatly influenced by the educational theories of Rousseau and was determined to educate his son along the lines laid down in *Émile.* The method he seems to have followed was to expose his son to a considerable library and let him educate himself as his own tastes dictated. One of the first works Ampère read was Buffon's *Histoire naturelle,* which stimulated his lifelong interest in taxonomy. Probably the most important influence on him was the great *Encyclopédie*—even thirty years later he could recite many of the articles from memory. In his father's library he also discovered Antoine Laurent Thomas's eulogy of Descartes, which convinced him of the nobility of a life in science. It also introduced him to metaphysics, the one passion he sustained throughout his life.

Almost incidentally Ampère discovered and perfected his mathematical talents. As an infant, he was fascinated by numbers and taught himself the elements of number theory. Like the young Pascal, having been forbidden the rigors of geometry because of his tender years, he defied parental authority and worked out the early books of Euclid by himself.

When the librarian in Lyons informed him that the works by Euler and Bernoulli that he wished to read were in Latin, Ampère rushed home to learn this language. He soon became adept enough to read the books that interested him, but continued his studies to the point where he could write quite acceptable Latin verse.

Ampère's early education was also conducted in a deeply religious atmosphere. His mother, the former Jeanne Desutières-Sarcey, was a devout woman who saw to it that her son was thoroughly instructed in the Catholic faith. Throughout his life, Ampère reflected the double heritage of the *Encyclopédie* and Catholicism. He was almost constantly assailed by the doubts sown by the Encyclopedists and, just as constantly, renewed his faith. From this conflict came his concern for metaphysics, which shaped his approach to science.

Ampère's childhood ended in 1789 with the outbreak of the French Revolution. Although Poleymieux was a rural backwater, the events in Lyons soon involved the Ampère family. Jean-Jacques was called upon by his fellow citizens to assume the post of *juge de paix,* a post with important police powers. He met the threat of a Jacobin purge head-on by ordering the arrest of Joseph Chalier, the leading Jacobin of Lyons. Chalier was executed. When Lyons fell to the troops of the Republic, Jean-Jacques Ampère was tried and guillotined on 23 November 1793. The event struck André-Marie like a bolt of lightning. The world had always been remote; now it had moved to the very center of his life, and this sudden confrontation was more than he could immediately bear. For a year he retreated within himself, not speaking to anyone and trying desperately to understand what had happened. His contact with the outside world was minimal; only an interest in botany, stimulated by a reading of Rousseau's letters on the subject, seemed to survive.

It was in this extremely vulnerable emotional state that Ampère met the young lady who was to become his wife. Julie Carron was somewhat older than Ampère and as a member of a good bourgeois family must have seen Ampère's suit in a somewhat unfavorable light. Although the Ampères and the Carrons lived in neighboring villages and shared a common economic and social background, marriage seemed impossible. At twenty-two, Ampère had only a small patrimony and no trade or other special skill. He was also homely and rustic, characteristics that were hardly likely to attract someone accustomed to the society and usages of Lyons. Ampère's courtship, carefully documented in his journal, reveals an essential aspect of his character: he was an incurable romantic whose emotional life was both intense and simple. Having lost his heart to Julie, he had no choice but to pursue her until she finally consented to marry him. His joy, like his despair at the death of his father, was immoderate. So, in his science, Ampère was possessed by his own enthusiasm. He never laid out a course of experiments or line of thought; there would be a brilliant flash of insight that he would pursue feverishly to its conclusion.

On 7 August 1799 Ampère and Julie were wed. The next four years were the happiest of Ampère's life. At first he was able to make a modest living as a mathematics teacher in Lyons, where on 12 August 1800 his son, Jean-Jacques, was born. In February 1802 Ampère left Lyons to become professor of physics and chemistry at the *école centrale* of Bourg-en-Bresse, a position that provided him with more money and, more important, with the opportunity to prepare himself for a post in the new *lycée* that Napoleon intended to establish at Lyons. In April of that year he began work on an original paper on probability theory that, he was convinced, would make his reputation. Thus, everything concurred to make him feel the happiest of men. Then tragedy struck. Julie had been ill since the birth of their son, and on 13 July 1803 she died. Ampère was inconsolable, and began to cast about desperately for some way to leave Lyons and all its memories.

On the strength of his paper on probability, he was named *répétiteur* in mathematics at the École Polytechnique in Paris. Again his emotional state was extreme, and again he fell victim to it. Bored by his work at the École Polytechnique, lonely in a strange and sophisticated city, Ampère sought human companionship and was drawn into a family that appeared to offer him the emotional warmth he so desperately craved. On 1 August 1806 he married Jeanne Potot. The marriage began under inauspicious circumstances: his father-in-law had swindled him out of his patrimony and his wife had indicated that she was uninterested in bearing children. The marriage was a catastrophe from the very beginning. After the birth of a daughter, Albine, his wife and mother-in-law made life so unbearable for Ampère that he realized that his only recourse was a divorce. Albine joined Jean-Jacques in Ampère's household, now presided over by his mother and his aunt, who had come to Paris from Poleymieux.

In 1808 Ampère was named inspector general of the newly formed university system, a post he held, except for a few years in the 1820's, until his death. On 28 November 1814 he was named a member of the class of mathematics in the Institut Impérial. In September 1819 he was authorized to offer a course in philosophy at the University of Paris, and in 1820 he was named assistant professor (*professeur suppléant*) of astronomy. In August 1824 Ampère was elected to the chair of experimental physics at the Collège de France.

During these years, Ampère's domestic life continued in turmoil. His son, for whom he had great hopes, fell under the spell of Mme. Recamier, one of the great beauties of the Empire, and for twenty years was content to be in her entourage. His daughter, Albine, married an army officer who turned out to be a drunkard and a near maniac. There was, too, a constant anxiety about money. In 1836 Ampère's health failed and he died, alone, while on an inspection tour in Marseilles.

Ampère's personal misery had an important effect on his intellectual development. His deep religious faith was undoubtedly strengthened by the almost constant series of catastrophes with which he was afflicted. Each successive tragedy also reinforced his desire for absolute certainty in some area of his life. His son later remarked on this characteristic of his father's approach; he was never content with probabilities but always sought Truth. It is no coincidence that his first mathematical paper, "Des considérations sur la théorie mathématique du jeu" (1802), proved that a single player inevitably would lose in a game of chance if he were opposed by a group whose financial resources were infinitely larger than his own. The outcome was certain.

In science Ampère's search for certainty and the exigencies of his faith led him to devise a philosophy that determined the form of his scientific research. The dominant philosophy in France in the early years of the nineteenth century was that of the Abbé de Condillac and his disciples, dubbed *Idéologues* by Napoleon. It maintained that only sensations were real, thus leaving both God and the existence of an objective world open to doubt. Such a position was abhorrent to Ampère, and he cast about for an alternative view. He was one of the earliest Frenchmen to discover the works of Immanuel Kant. Although Kant's philosophy made it possible to retain one's religious faith, Ampère felt that his treatment of space, time, and causality implied the doubtful existence of an objective reality at a fundamental level.

Space and time, as Ampère interpreted Kant, became subjective modes of the human understanding, and Ampère, as a mathematician, could not accept this.

He therefore constructed his own philosophy. Its foundation was provided by his friend Maine de Biran, who felt he had successfully refuted David Hume's conclusion that *cause* simply meant succession of phenomena in time. The act of moving one's arm provided a firm proof that a cause explained an act and was not simply a description of succession. One wills the arm to move and one is conscious of the act of willing; the arm then moves. Therefore the arm moves *because* one wills it to. Ampère used this argument to prove the existence of an external world. If one's arm cannot move because it is, say, under a heavy table, then one becomes conscious of causes external to oneself. The arm does not move because the table prevents it from doing so. Thus Ampère carried causation from the psychological world to the physical world. Moreover, the resistance of the table proved, to Ampère's satisfaction, that matter does exist, for this external cause must be independent of our sensation of it. With similar arguments, Ampère was able to prove that the soul and God also must exist.

Ampère's philosophy permitted him to retain both a belief in God and a belief in the real existence of an objective nature. The next step was to determine what could be known about the physical world. Here again, Ampère's analysis contained highly idiosyncratic views on the nature of scientific explanation which were to be clearly illustrated in his own work. There are (and here the influence of Kant is obvious) two levels of knowledge of the external world. There are phenomena, presented to us directly through the senses, and there are noumena, the objective causes of phenomena. Noumena, according to Ampère, are known through the activity of the mind, which hypothesizes certain real, material entities whose properties can be used to account for phenomena. These two aspects of reality, however, are not all that we can know. We also can know relations (*rapports*) between phenomena and relations between noumena, and these relations are just as objectively real as the noumena. One example may suffice to illustrate this. It had been known since the end of the eighteenth century that two volumes of hydrogen combined with one volume of oxygen to form two volumes of water vapor. This is knowledge of a specific phenomenon.

In 1808 Gay-Lussac discovered that all gases combine in simple ratios, and thus was able to announce his law of combining volumes. The law states a relationship between phenomena and thereby extends our knowledge of the phenomenal world. In 1814

Ampère published his "Lettre de M. Ampère à M. le comte Berthollet sur la détermination des proportions dans lesquelles les corps se combinent d'après le nombre et la disposition respective des molécules dont leurs particules intégrantes sont composées."[1] It was an attempt to provide the noumenal, and therefore deeper, explanation of the phenomenal relations. From the theory of universal attraction used to account for the cohesion of bodies and the fact that light easily passes through transparent bodies, Ampère concluded that the attractive and repulsive forces associated with each molecule hold the ultimate molecules of bodies at distances from one another that are, as Ampère put it, "infinitely great in comparison to the dimensions of these molecules." This is knowledge of the noumena. It explains certain basic qualities of the observable world in terms of theoretical entities whose properties can be hypothesized from phenomena.

From the science of crystallography Ampère borrowed the idea of the integral particle, that is, the smallest particle of a crystal that has the form of the crystal. Ampère's molecules now were assumed to group themselves in various ways to form particles that had specific geometric forms. Thus there would be particles composed of four molecules that formed **tetrahedrons (oxygen, nitrogen, and hydrogen), of six molecules that formed an octahedron (chlorine),** etc. These geometrical forms were of the greatest importance in Ampère's theory, for they allowed him to deal with the problem of elective affinity and also to deduce Avogadro's law. Ampère's particles were compound and could, therefore, be broken down into smaller parts. Thus, oxygen was composed of four molecules that could, and did, separate under certain conditions, with two molecules going one way and two the other. The rule was that only compounds whose molecules were regular polyhedrons could be formed. If a tetrahedron met an octahedron, there could not be a simple combination, for the result would be a bizarre (in Ampère's terms) geometrical figure. Two tetrahedrons could combine with one octahedron, however, since the result would be a dodecahedron.

Ampère's philosophy and its influence on his science are obvious here. The relations of noumena, in this case the association of molecules to form a geometrically regular form, are simply assumed. If Ampère had been asked what evidence he had for the existence of such forms, he would have replied that no evidence could be offered. One hypothesizes noumena and relations between them in order to give causal explanations of phenomena. There can be no "evidence" for the noumena; there can be only the greater or lesser success of the noumenal hypothesis in explaining what can be observed. The point is of central importance, for it permitted Ampère to assume whatever he wished about the noumena. His assumption of an electrodynamic molecule followed this pattern exactly.

Ampère's philosophical analysis also provided him with the key for his classification of the sciences, which he considered the capstone of his career. Like Kant, he was concerned with relating precisely what man could know with the sciences that dealt with each part of man's ability to know. The chart appended to the first volume of his *Essai sur la philosophie des sciences* (1834) seems, at first glance, to be a fantastic and uncorrelated list of possible objects of investigation. If Ampère's philosophical views are attended to, however, they all fall into a rather simple pattern. We may use general physics as an example. In Ampère's classification this is divided into two second-order sciences—elementary general physics and mathematical physics. Each of these, in turn, has two divisions. Elementary general physics consists of experimental physics and chemistry; mathematical physics is divided into stereonomy and atomology (Ampère's neologisms). Experimental physics deals with phenomena, i.e., with the accurate description of physical facts. Chemistry deals with the noumenal causes of the facts discovered by experimental physics. Stereonomy concerns the relations between phenomena, e.g., laws of the conduction of heat through a solid. Atomology explains these laws by demonstrating how they may be deduced from relations between the ultimate particles of matter. All other sciences are treated in this fashion and have exactly the same kind of fourfold division.

This classification reveals Ampère's far-ranging mind and permits us to understand his occasional excursions into botany, taxonomy, and even animal anatomy and physiology. He was, in large part, seeking confirmation for his philosophical analysis, rather than setting out on new scientific paths. By the time of his death, Ampère had found, to his great satisfaction, that his scheme did fit all the sciences and, in his *Essai sur la philosophie des sciences,* he maintained that the fit was too good to be coincidence; the classification must reflect truth. Once again he had found certainty where his predecessors had not.

Although the one continuing intellectual passion of Ampère's life from 1800 to his death was his philosophical system, these years were also devoted to scientific research of considerable originality. From 1800 to about 1814, he devoted himself primarily to mathematics. As his mathematical interests declined, he became fascinated with chemistry and, from 1808

to 1815, spent his spare time in chemical investigations. From 1820 to 1827, he founded and developed the science of electrodynamics, the scientific work for which he is best known and which earned him his place in the first rank of physicists.

Ampère was not a truly outstanding mathematician. His first paper showed considerable originality and, more revealingly, great ability as an algorist. Like Leonhard Euler, Ampère had the uncanny ability, found only in the born mathematician, to discover mathematical relations. His largest mathematical memoir, "Mémoire sur l'intégration des équations aux différences partielles" (1814), was on various means of integrating partial differential equations. Although one should not underestimate the utility of such works, they should not be put in the same class as, say, the invention of quaternions by Sir William Rowan Hamilton or the laying of rigorous foundations of the calculus by Augustin Cauchy.

Ampère's failure to achieve the early promise he had shown in mathematics was undoubtedly the result of his passion for metaphysics and the necessity of earning a living. But there was also the fact, worth noting here, that the French scientific system forced him to do mathematics when his interests were focused elsewhere. Having been classified as a mathematician, Ampère found himself unable to gain recognition as anything else until after his epoch-making papers on electrodynamics. The security that came with election to the Academy of Sciences was achieved by Ampère only at the cost of putting aside his chemical interests and writing a mathematical memoir for the express purpose of gaining entry to the Academy. Original mathematical work is rarely done under such conditions.

Ampère's interest in chemistry had been aroused in the days when he gave private lessons at Lyons. This interest continued to grow at Bourg-en-Bresse, where he mastered the subject, and it became most intense about 1808, when Humphry Davy was shaking the foundations of the orthodox chemistry of the French school. Ampère once described himself as credulous in matters of science, and this again reflected his philosophy. A new scientific idea could be immediately accepted as a hypothesis even if there was no evidence for it, just as a fundamental assumption could be made without evidence: the main criterion was whether it worked or not. Ampère was not committed to Lavoisier's system of chemistry.

When Davy announced the discovery of sodium and potassium, the orthodox were startled, if not dismayed. How could oxygen be the principle of acidity, as Lavoisier had insisted, if the oxides of potassium and sodium formed the strongest alkalies?

For Ampère there was no problem; he simply accepted the fact. If this fact were true, however, then the oxygen theory of acids was probably wrong. And if this were so, then the great riddle of muriatic acid could easily be solved. The green gas that was given off when muriatic acid was decomposed need not be a compound of some unknown base and oxygen; it could be an element. Thus, at the same time that Davy was questioning the compound nature of chlorine, Ampère had also concluded that it was an element. Unfortunately, and much to his later regret, he had neither the time nor the resources to prove this point, and the credit for the discovery of chlorine as an element went to Davy. Ampère was forestalled once again by Davy in 1813, when he brought Davy a sample of a new substance that Bernard Courtois had isolated from seaweed. Ampère had already seen its similarities to chlorine, but it was Davy who first publicly insisted upon its elemental character and named it iodine.

The noumenal aspect of chemistry fascinated Ampère. Although his derivation of Avogadro's law came three years after Avogadro had enunciated it, the law is known today in France as the Avogadro-Ampère law. This was Ampère's first excursion into molecular physics, and was followed almost immediately by a second. In 1815 he published a paper demonstrating the relation between Mariotte's (Boyle's) law and the volumes and pressures of gases at the same temperature. The paper is of some interest as a pioneer effort, along with Laplace's great papers on capillarity, in the application of mathematical analysis to the molecular realm.

In 1816 Ampère turned to the phenomenal relations of chemistry in a long paper on the natural classification of elementary bodies ("Essai d'une classification naturelle pour les corps simples"). Here he drew attention to the similarities between Lavoisier's and his followers' classification of elements in terms of their reactions with oxygen and Linnaeus' classification of plants in terms of their sexual organs. Bernard de Jussieu had successfully challenged Linnaeus with his natural system that took the whole plant into account and sought affinities between all parts of the plant, not just the flowers, as the basis of classification. Ampère now wished to do the same thing for chemistry. By discovering a natural classification, i.e., one that tied the elements together by real rather than artificial relations, Ampère hoped to provide a new insight into chemical reactions. His classificatory scheme, therefore, was not merely an ordering of the elements but, like the later periodic table of Dmitri Mendeleev, a true instrument of chemical research. Unfortunately, Ampère's system was as artificial as

Lavoisier's. Although he looked for more analogies among elements than Lavoisier had, the ones he selected offered little insight into the relations between the groups founded on them. The paper may be noted, however, as an early attempt to find relationships between the elements that would bring some order into the constantly growing number of elementary bodies.

By 1820 Ampère had achieved a certain reputation as both a mathematician and a somewhat heterodox chemist. Had he died before September of that year, he would be a minor figure in the history of science. It was the discovery of electromagnetism by Hans Christian Oersted in the spring of 1820 which opened up a whole new world to Ampère and gave him the opportunity to show the full power of his method of discovery. On 4 September 1820 François Arago reported Oersted's discovery to an astonished and skeptical meeting of the Académie des Sciences. Most of the members literally could not believe their ears; had not the great Coulomb proved to everyone's satisfaction in the 1780's that there could not be any interaction between electricity and magnetism? Ampère's credulity served him well here; he immediately accepted Oersted's discovery and turned his mind to it. On 18 September he read his first paper on the subject to the Académie; on 25 September and 9 October he continued the account of his discoveries. In these feverish weeks the science of electrodynamics was born.

There is some confusion over the precise nature of Ampère's first discovery. In the published memoir, "Mémoire sur l'action naturelle de deux courants électriques . . ." (1820), he stated that his mind leaped immediately from the existence of electromagnetism to the idea that currents traveling in circles through helices would act like magnets. This may have been suggested to him by consideration of terrestrial magnetism, in which circular currents seemed obvious. Ampère immediately applied his theory to the magnetism of the earth, and the genesis of electrodynamics may, indeed, have been as Ampère stated it. On the other hand, there is an account of the meetings of the Académie des Sciences at which Ampère spoke of his discoveries and presented a somewhat different order of discovery. It would appear that Oersted's discovery suggested to Ampère that two current-carrying wires might affect one another. It was this discovery that he announced to the Académie on 25 September.[2] Since the pattern of magnetic force around a current-carrying wire was circular, it was no great step for Ampère the geometer to visualize the resultant force if the wire were coiled into a helix. The mutual attraction and repulsion of two helices was also announced to the Académie on 25 September. What Ampère had done was to present a new theory of magnetism as electricity in motion.

From this point on, Ampère's researches followed three different but constantly intertwining paths. They conform exactly to his ideas on the nature of science and scientific explanation. The phenomenon of electromagnetism had been announced by Oersted; the relations of two current-carrying wires had been discovered by Ampère. It remained to explore these relations in complete and elaborate detail. Then, following his own philosophy, it was necessary for Ampère to seek the noumenal causes of the phenomena, which were found in his famous electrodynamic model and theory of the nature of electricity. Finally, Ampère had to discover the relations between the noumena from which all the phenomena could be deduced. Between 1820 and 1825 he successfully completed each of these tasks.

Ampère's first great memoir on electrodynamics was almost completely phenomenological, in his sense of the term. In a series of classical and simple experiments, he provided the factual evidence for his contention that magnetism was electricity in motion. He concluded his memoir with nine points that bear repetition here, since they sum up his early work.

1. Two electric currents attract one another when they move parallel to one another in the same direction; they repel one another when they move parallel but in opposite directions.

2. It follows that when the metallic wires through which they pass can turn only in parallel planes, each of the two currents tends to swing the other into a position parallel to it and pointing in the same direction.

3. These attractions and repulsions are absolutely different from the attractions and repulsions of ordinary [static] electricity.

4. All the phenomena presented by the mutual action of an electric current and a magnet discovered by M. Oersted . . . are covered by the law of attraction and of repulsion of two electric currents that has just been enunciated, if one admits that a magnet is only a collection of electric currents produced by the action of the particles of steel upon one another analogous to that of the elements of a voltaic pile, and which exist in planes perpendicular to the line which joins the two poles of the magnet.

5. When a magnet is in the position that it tends to take by the action of the terrestrial globe, these currents move in a sense opposite to the apparent motion of the sun; when one places the magnet in the opposite position so that the poles directed toward the poles of the earth are the same [S to S and N to N, not south-seeking to S, etc.] the same currents are found in the same direction as the apparent motion of the sun.

6. The known observed effects of the action of two magnets on one another obey the same law.

7. The same is true of the force that the terrestrial globe exerts on a magnet, if one admits electric currents in planes perpendicular to the direction of the declination needle, moving from east to west, above this direction.

8. There is nothing more in one pole of a magnet than in the other; the sole difference between them is that one is to the left and the other is to the right of the electric currents which give the magnetic properties to the steel.

9. Although Volta has proven that the two electricities, positive and negative, of the two ends of the pile attract and repel one another according to the same laws as the two electricities produced by means known before him, he has not by that demonstrated completely the identity of the fluids made manifest by the pile and by friction; this identity was proven, as much as a physical truth can be proven, when he showed that two bodies, one electrified by the contact of [two] metals, and the other by friction, acted upon each other in all circumstances as though both had been electrified by the pile or by the common electric machine [electrostatic generator]. The same kind of proof is applicable here to the identity of attractions and repulsions of electric currents and magnets.[3]

Here Ampère only hinted at the noumenal background. Like most Continental physicists, he felt that electrical phenomena could be explained only by two fluids and, as he pointed out in the paper, a current therefore had to consist of the positive fluid going in one direction and the negative fluid going in the other through the wire. His experiments had proved to him that this contrary motion of the two electrical fluids led to unique forces of attraction and repulsion in current-carrying wires, and his first paper was intended to describe these forces in qualitative terms. There was one problem: how could this explanation be extended to permanent magnets? The answer appeared deceptively simple: if magnetism were only electricity in motion, then there must be currents of electricity in ordinary bar magnets.

Once again Ampère's extraordinary willingness to frame *ad hoc* hypotheses is evident. Volta had suggested that the contact of two dissimilar metals would give rise to a current if the metals were connected by a fluid conductor. Ampère simply assumed that the contact of the molecules of iron in a bar magnet would give rise to a similar current. A magnet could, therefore, be viewed as a series of voltaic piles in which electrical currents moved concentrically around the axis of the magnet. Almost immediately, Ampère's friend Augustin Fresnel, the creator of the wave theory of light, pointed out that this hypothesis simply would not do. Iron was not a very good conductor of the electrical fluids and there should, therefore, be some heat generated if Ampère's views were correct. Magnets are not noticeably hotter than their surroundings and Ampère, when faced with this fact, had to abandon his noumenal explanation.

It was Fresnel who provided Ampère with a way out. Fresnel wrote in a note to Ampère that since nothing was known about the physics of molecules, why not assume currents of electricity around each molecule. Then, if these molecules could be aligned, the resultant of the molecular currents would be precisely the concentric currents required. Ampère immediately adopted his friend's suggestion, and the electrodynamic molecule was born. It is, however, a peculiar molecule. In some mysterious fashion, a molecule of iron decomposed the luminiferous ether that pervaded both space and matter into the two electrical fluids, its constituent elements. This decomposition took place *within* the molecule; the two electrical fluids poured out the top, flowed around the molecule, and reentered at the bottom. The net effect was that of a single fluid circling the molecule. These molecules, when aligned by the action of another magnet, formed a permanent magnet. Ampère did not say why molecules should act this way; for him it was enough that his electrodynamic model provided a noumenal foundation for electrodynamic phenomena.

There was no doubt that Ampère took his electrodynamic molecule seriously and expected others to do so too. In an answer to a letter from the Dutch physicist Van Beck, published in the *Journal de physique* in 1821, Ampère argued eloquently for his model, insisting that it could be used to explain not only magnetism but also chemical combination and elective affinity. In short, it was to be considered the foundation of a new theory of matter. This was one of the reasons why Ampère's theory of electrodynamics was not immediately and universally accepted. To accept it meant to accept as well a theory of the ultimate structure of matter itself.

Having established a noumenal foundation for electrodynamic phenomena, Ampère's next steps were to discover the relationships between the phenomena and to devise a theory from which these relationships could be mathematically deduced. This double task was undertaken in the years 1821–1825, and his success was reported in his greatest work, the *Mémoire sur la théorie mathématique des phénomènes électrodynamique, uniquement déduite de l'expérience* (1827). In this work, the *Principia* of electrodynamics, Ampère first described the laws of action of electric currents, which he had discovered from four extremely ingenious experiments. The measurement of

electrodynamic forces was very difficult, although it could be done, as J.-B. Biot and Félix Savart had shown in their formulation of the Biot-Savart law. Ampère realized, however, that much greater accuracy could be achieved if the experiments could be null experiments, in which the forces involved were in equilibrium.

The first experiment, to quote Ampère, "demonstrated the equality of the absolute value of the attraction and repulsion which is produced when a current flows first in one direction, then in the opposite direction in a fixed conductor which is left unchanged as to its orientation and at the same distance from the body on which it acts." The second "consists in the equality of the actions exerted on a mobile rectilinear conductor by two fixed conductors situated at equal distances from the first of which one is rectilinear and the other bent or contorted in any way whatsoever. . . ." The third case demonstrated "that a closed circuit of any form whatsoever cannot move any portion of a conducting wire forming an arc of a circle whose center lies on a fixed axis about which it may turn freely and which is perpendicular to the plane of the circle of which the arc is a part." Ampère rather casually mentioned at the end of the *Mémoire* that he had not actually performed the fourth experiment, which was intended to determine certain constants necessary for the solution of his mathematical equation. These constants, it would appear, had been found by measuring the action of a magnet and a current-carrying wire upon one another and were sufficiently accurate to permit Ampère to continue his researches.

From these cases of equilibrium, Ampère was able to deduce certain necessary consequences that permitted him to apply mathematics to the phenomena. It was time to turn to the noumena once again and to complete the edifice by deducing from the noumenal elements those mathematical relationships that had been indicated by experiment. The flow of an electrical current, it will be remembered, was a complicated process in Ampère's theory. Positive electricity was flowing in one direction in the wire while negative electricity flowed in the opposite direction. The luminiferous ether was a compound of these two fluids, so that it was constantly being formed from their union, only to be decomposed as each fluid went its way. Thus, at any moment in the wire there were elements of positive electricity, negative electricity, and the ether. Ampère's current element (*ids*), therefore, was not a mathematical fiction assumed out of mathematical necessity, but a real physical entity. What his experiments had done was to tell him the basic properties of this element. The force associated with the element is a central force, acting at a distance at right angles to the element's direction of flow. From this fact it was easy to deduce that the mutual action of two lengths of current-carrying wire is proportional to their length and to the intensities of the currents. Ampère was now prepared to give precise mathematical form to this action. As early as 1820, he had deduced a law of force between two current elements, ids and $i'ds'$. He gave the formula

$$F = \frac{i \cdot i' \cdot ds \cdot ds'}{r^2}$$
$$[\sin \theta \cdot \sin \theta' \cdot \cos \omega + k \cos \theta \cdot \cos \theta']$$

for the force between two current elements, making angles θ and θ' with the line joining them and the two planes containing this line and the two elements respectively making an angle ω with each other. At that time he had been unable to evaluate the constant k. By 1827 he was able to show that $k = -1/2$. The formula above could now be written

$$F = \frac{i \cdot i' \cdot ds \cdot ds'}{r^2}$$
$$[\sin \theta \cdot \sin \theta' \cdot \cos \omega - 1/2 \cos \theta \cdot \cos \theta'].$$

When integrated around a complete circuit (as in practice it must be), this formula is identical with that of Biot and Savart.

It was now possible for Ampère to attack the theory of magnetism quantitatively. He could show that his law of action of current elements led to the conclusion that the forces of a magnet composed of electrodynamic molecules should be directed toward the poles. He was also able to deduce Coulomb's law of magnetic action. In short, he was able to unify the fields of electricity and magnetism on a basic noumenal level. The theory was complete.

Not everyone accepted Ampère's theory. His primary opponent was Michael Faraday, who could not follow the mathematics and felt that the whole structure was based on *ad hoc* assumptions for which there was no evidence whatsoever. The phenomenal part was accepted; even in France the electrodynamic molecule was regarded with considerable suspicion. The idea, however, did not die with Ampère. It was accepted later in the century by Wilhelm Weber and became the basis of his theory of electromagnetism.

After 1827 Ampère's scientific activity declined sharply. These were the years of anxiety and fear for his daughter's well-being, as well as years of declining health. He produced an occasional paper but, by and large, after the great 1827 memoir Ampère's days as a creative scientist were ended. He turned instead to the completion of his essay on the philosophy of science and his classification of the sciences. He must have derived some satisfaction from the fact that he had, almost single-handedly, created a new science to be placed in his taxonomic scheme.

NOTES

1. *Annales de chimie,* **90** (1814), 43 ff.
2. See *Bibliothèque universelle des sciences, belles-lettres, et arts,* **17** (1821), 83.
3. *Mémoires sur l'électrodynamique,* I (Paris, 1885), 48.

BIBLIOGRAPHY

I. ORIGINAL WORKS. The most important source for the life and work of Ampère is the forty cartons of documents in the archives of the Académie des Sciences in Paris. This material has been catalogued but never used. For Ampère's correspondence, see Louis de Launay, ed., *Correspondance du Grand Ampère,* 3 vols. (Paris, 1936–1943). It should be used with care, for there are many errors of transcription and it is not complete. It does, however, have a complete bibliography of Ampère's works at the end of the second volume. One should also consult *André-Marie Ampère et Jean-Jacques Ampère: Correspondance et souvenirs (de 1805 à 1864) recuellis par Madame H. C.[heuvreux],* 2 vols. (Paris, 1875), and Mme. Cheuvreux's *Journal et correspondance d'André-Marie Ampère* (Paris, 1872). These volumes should be used with great caution, for Mme. Cheuvreux was not a scholar and changed the order of whole passages, sometimes inserting part of one letter in another for artistic reasons. Ampère's papers on electrodynamics were reprinted by the Société Française de Physique as *Mémoires sur l'électrodynamique,* 2 vols. (Paris, 1885–1887). The *Mémoire sur la théorie mathématique des phénomènes électrodynamiques, uniquement déduite de l'expérience* (Paris, 1827) was republished with a foreword by Edmond Bauer (Paris, 1958). Portions of this work and others of Ampère's papers on electrodynamics have been translated and appear in R. A. R. Tricker, ed., *Early Electrodynamics: The First Law of Circulation* (Oxford, 1965). This volume contains a long commentary by the editor that is of great value in explaining Ampère's theory.

For Ampère's philosophical development, see *Philosophie des deux Ampère publiée par J. Barthélemy Saint-Hilaire* (Paris, 1866), which contains a long essay by Jean-Jacques Ampère on his father's philosophy.

II. SECONDARY LITERATURE. There is no adequate biography of Ampère. C. A. Valson, *André-Marie Ampère* (Lyons, 1885), and Louis de Launay, *Le grand Ampère* (Paris, 1925), are the standard biographies, but neither discusses Ampère's work in any detail. The eulogy by François Arago provides a survey of Ampère's scientific achievement from the perspective of a century ago. See his *Oeuvres,* 17 vols. (Paris, 1854–1862), II, 1 ff. For some modern appreciations of Ampère's work, see the *Revue général de l'électricité,* **12** (1922), supplement. The entire issue is devoted to Ampère's work. For an interesting account of Ampère's early career, see Louis Mallez, *A.-M. Ampère, professeur à Bourg, membre de la Société d'Émulation de l'Ain, d'après des documents inédits* (Lyons, 1936).

For various aspects of Ampère's career, see Borislav Lorenz, *Die Philosophie André-Marie Ampères* (Berlin, 1908); and Maurice Lewandowski, *André-Marie Ampère. La science et la foi* (Paris, 1936). The *Bulletin de la Société des Amis d'André-Marie Ampère,* which appears irregularly, contains much Ampère lore. Two interesting sketches of Ampère are Henry James, "The Two Ampères," in *French Poets and Novelists* (London, 1878); and C. A. Sainte-Beuve, "M. Ampère," in *Portraits littéraires,* 3 vols. (Paris, 1862), I.

A discussion of Ampère's electrodynamic molecule is to be found in L. Pearce Williams, "Ampère's Electrodynamic Molecular Model," in *Contemporary Physics,* **4** (1962), 113 ff. For Ampère's relations with England, see K. R. and D. L. Gardiner, "André-Marie Ampère and His English Acquaintances," in *The British Journal for the History of Science,* **2** (1965), 235 ff.

L. PEARCE WILLIAMS

AMSLER (later **AMSLER-LAFFON**), **JAKOB** (*b.* Stalden bei Brugg, Switzerland, 16 November 1823; *d.* Schaffhausen, Switzerland, 3 January 1912), *mathematics, precision instruments.*

The son of a farmer, Amsler was educated at local schools before going on to study theology at the universities of Jena and Königsberg. At Königsberg he came under the influence of Franz Neumann, whose lectures and laboratory sessions he attended for seven semesters. After earning his doctorate in 1848, Amsler spent a year with Plantamour at the Geneva observatory; he went from there to Zurich, where he completed his *Habilitation* and began his teaching career. For four semesters he lectured on various topics in mathematics and mathematical physics, then in 1851 accepted a post at the Gymnasium in Schaffhausen. From this he hoped to gain some financial independence as well as an opportunity for more research. In 1854 Amsler married Elise Laffon, the daughter of a Schaffhausen druggist who was well known in Swiss scientific circles. Henceforth he used the double form Amsler-Laffon. The change applied to Jakob alone and was not adopted by his children.

Until 1854 Amsler's interests lay in the area of mathematical physics; he published articles on magnetic distribution, the theory of heat conduction, and the theory of attraction. One result of his work was a generalization of Ivory's theorem on the attraction of ellipsoids and of Poisson's extension of that theorem.

In 1854 Amsler turned his attention to precision mathematical instruments, and his research resulted in his major contribution to mathematics: the polar planimeter, a device for measuring areas enclosed by plane curves. Previous such instruments, most notably that of Oppikofer (1827), had been based on the Cartesian coordinate system and had combined bulkiness with high cost. Amsler eliminated these drawbacks by basing his planimeter on a polar coordinate system referred to a null circle as curvilinear axis. The instrument, described in "Ueber das Polarplanimeter" (1856), adapted easily to the determination of static and inertial moments and of the coefficients of Fourier series; it proved especially useful to shipbuilders and railroad engineers.

To capitalize on his inspiration, Amsler established his own precision tools workshop in 1854. From 1857 on, he devoted full time to the venture. At his death, the shop had produced 50,000 polar planimeters and 700 momentum planimeters. The polar planimeter marked the height of Amsler's career. His later research, mostly in the area of precision and engineering instruments, produced no comparable achievement, although it did bring Amsler recognition and prizes from world exhibitions at Vienna (1873) and Paris (1881, 1889), as well as a corresponding membership in the Paris Academy (1892). From 1848 until his death, Amsler was an active member of the Naturforschende Gesellschaft in Zurich.

BIBLIOGRAPHY

I. ORIGINAL WORKS. Amsler's writings include: "Zur Theorie der Verteilung des Magnetismus im weichen Eisen," in *Abhandlungen der naturforschenden Gesellschaft in Zürich* (1847), reprinted in *Neue Denkschriften der allgemeinen schweizerischen Gesellschaft für die gesammten Naturwissenschaften,* **10** (1849); "Methode, den Einfluss zu kompensieren, welchen die Eisenmassen eines Schiffes infolge der Verteilung der magnetischen Flüssigkeiten durch den Erdmagnetismus auf die Kompassnadel ausüben," in *Verhandlungen der schweizerischen naturforschenden Gesellschaft* (1848); "Ueber die klimatologischen Verhältnisse der Polargegenden" and "Ueber die Anwendung von Schwingungsbeobachtungen zur Bestimmung der spezifischen Wärme fester Körper bei konstantem Volumen," in *Mitteilungen der naturforschenden Gesellschaft in Zürich,* **2** (1850–1852), 314–315; "Neue geometrische und mechanische Eigenschaft der Niveauflächen," "Zur Theorie der Anziehung und der Wärme," and "Ueber die Gesetze der Wärmeleitung im Innern fester Körper, unter Berücksichtigung der durch ungleichförmige Erwärmung erzeugten Spannung," in *Crelle's Journal,* **42** (1851), the last reprinted in *Neue Denkschriften,* **12** (1852); "Ueber das Polarplanimeter," in *Dingler's Journal,* **140** (1856); "Ueber die mechanische Bestimmung des Flächeninhaltes, der statischen Momente und der Trägheitsmomente ebener Figuren, insbesondere über einen neuen Planimeter," in *Vierteljahrsschrift der naturforschenden Gesellschaft in Zürich,* **1** (1856), also printed separately (Schaffhausen, 1856); "Anwendung des Integrators (Momentumplanimeters) zur Berechnung des Auf- und Abtrages bei Anlage von Eisenbahnen, Strassen und Kanälen," a pamphlet (Zurich, 1875); "Der hydrometrische Flügel mit Zählwerk und elektrischer Zeichengebung," a pamphlet (Schaffhausen, 1877), reprinted in *Carls Repertorium,* **14** (1878); "Neuere Planimeterkonstruktionen," in *Zeitschrift für Instrumentkunde,* **4** (1884); and "Die neue Wasserwerksanlage in Schaffhausen und einige darauf bezügliche technische Fragen," in *Schweizerische Bauzeitung,* **16** (1890).

II. Secondary Literature. See Poggendorf, Vols. III, IV, and V. The present article is based on the necrology by Ferdinand Rudio and Alfred Amsler in *Vierteljahrsschrift der naturforschenden Gesellschaft in Zürich,* **57** (1912), 1–17, and on the extensive study by Fr. Dubois, "Die Schöpfungen Jakob und Alfred Amsler's auf dem Gebiete der mathematischen Instrumente anhand der Ausstellung im Museum Allerheiligen systematisch dargestellt," in *Mitteilungen der naturforschenden Gesellschaft Schaffhausen,* **19** (1944), 209–273.

M. S. Mahoney

ANARITIUS. See **al-Nayrīzī.**

ANATOLIUS OF ALEXANDRIA (*b.* Alexandria; *d.* Laodicea; *fl. ca.* A.D. 269), *mathematics, philosophy.*

The historian Eusebius, whose *Ecclesiastical History* provides what we know of Anatolius' life, says, "For his learning, secular education and philosophy [he] had attained the first place among our most illustrious contemporaries." Learned in arithmetic, geometry, astronomy, and other sciences both intellectual and natural, Anatolius was also outstanding in rhetoric. The Alexandrians deemed him worthy of heading the Aristotelian school in that city.

Bishop Theotecnus of Caesarea consecrated Anatolius as his successor, and he held office for a while in Caesarea. About A.D. 280, however, as he passed through Laodicea on his way to Antioch, he was retained by the inhabitants as their bishop, the previous bishop, also called Eusebius, having died. He remained bishop of Laodicea until his death some years later.

Anatolius' Christian and humanitarian character was much admired. During a siege of the Greek quarter of Alexandria by the Roman army, he attempted to make peace between the factions. He failed, but he succeeded in winning safe conduct from the besieged quarter for all noncombatants.

Anatolius put his knowledge of astronomy at the service of his religion in a treatise on the date of Easter. Eusebius gives the title of the work as *The Canons of Anatolius on the Pascha* and quotes several paragraphs that display Anatolius' grasp of astronomy in the discussion of the position of the sun and moon in the zodiac at the time of Easter. According to Eusebius, Anatolius did not write many books; but those that he did write were distinguished for eloquence and erudition, which is evident through his quotation of Philo, Josephus, and two of the seventy who translated the Old Testament into Greek during the third and second centuries B.C.

The only other work of Anatolius known to us by name is his *Introduction to Arithmetic.* In ten books, it seems to have been excerpted by the author of the curious writing entitled *Theologoumena arithmetica.* A Neoplatonic treatise, uncertainly attributed to Iamblichus, it is a discussion of each of the first ten natural numbers. It mixes accounts of truly arithmetical properties with mystical fancies. Many parts of the discussion are headed "of Anatolius." The character of its arithmetical lore may be illustrated by the following quotation from a part attributed to Anatolius: "[Four] is called 'justice' since its square is equal to the perimeter [i.e., $4 \times 4 = 16 = 4 + 4 + 4 + 4$]; of the numbers less than four the perimeter of the square is greater than the area, while of the greater the perimeter is less than the area."

In contrast with the flights of fancy preserved in *Theologoumena arithmetica,* some paragraphs of a writing of Anatolius are found in manuscripts of Hero of Alexander in which Anatolius deals soberly and sensibly, and in Aristotelian terms, with questions about mathematics, its name, its philosophical importance, and some of its methods. The structure of *Theologoumena arithmetica* and its selection of material from Anatolius suggest that Anatolius' *Introduction to Arithmetic* may have dealt with each of the first ten natural numbers. The Pythagoreanism or Neoplatonism manifested here was in the spirit of the times. Despite the number mysticism, however, Anatolius' competence in mathematics is clear and justifies the esteem in which Eusebius says he was held in Alexandria.

BIBLIOGRAPHY

No individual works of Anatolius' are known to exist today. Some paragraphs of a work by him are found in *Heronis Alexandrini geometricorum et stereometricorum reliquiae,* F. Hultsch, ed. (Berlin, 1864), pp. 276–280. A seeming use of excerpts from Anatolius' *Introduction to Arithmetic* is *Theologoumena arithmetica,* V. De Falco, ed.

148

(Leipzig, 1922). Two sources of information on the life of Anatolius are Eusebius, *The Ecclesiastical History*, H. J. Lawlor, trans., II (Cambridge-London, 1942), 228–238; and Pauly-Wissowa, eds., *Real-Enzyklopädie der Klassischen Altertumswissenschaft*, XII (Stuttgart, 1894–　　), col. 2073 f.

JOHN S. KIEFFER

ANAXAGORAS (*b.* Clazomenae, Lydia, 500 B.C.[?]; *d.* Lampsacus, Mysia, 428 B.C.[?]), *natural philosophy.*

Although he was born of wealthy parents, Anaxagoras neglected his inheritance to devote himself to natural philosophy. At the age of twenty, he traveled to Athens, where he spent the next thirty years. There he became a friend of Pericles and brought Ionian physical speculation to Athens at the height of its intellectual development. Subsequently he was prosecuted for impiety and banished[1] because, it was alleged, he held the sun to be a mass of red-hot stone. This charge doubtless was instigated by the political opponents of Pericles, who sought to attack him through his friendship with an atheistic scientist. Anaxagoras wrote only one treatise, completed after 467 B.C.

Like Empedocles, Anaxagoras sought to reconcile Parmenides' logic with the phenomena of multiplicity and change. Each maintained that there was never a unity in either the qualitative or the quantitative sense and postulated instead a plurality of eternal, qualitatively different substances that filled the whole of space. They accepted Parmenides' negation of coming-into-being and passing-away but replaced the former with the aggregation of their indestructible elements and the latter with their segregation. Motive forces were introduced to account for motion—a phenomenon whose validity had, prior to Parmenides, been taken for granted.

Anaxagoras evidently did not consider that Empedocles had fully satisfied the demands of Eleatic logic.[2] Empedocles had seen no objection to making secondary substances come into being as various combinations of his elements. A piece of flesh, according to him, consisted of the four elements juxtaposed in almost equal quantities. Theoretically, if it were divided, one would arrive at a minimum piece of flesh and thereafter at particles of the constituent elements. Thus, flesh originally came into being from the elements and, strictly speaking, from what is not flesh. Anaxagoras' own formulation of the problem is preserved: "How," he asks, "could hair come to be from what is not hair and flesh from what is not flesh?" (Diels and Kranz, B10). His answer was to claim that everything preexisted in our food. Thus, he denied the existence of elements simpler than and

prior to common natural substances and maintained that every natural substance must itself be elementary, since it cannot arise from what is not itself. Furthermore, to avoid being confuted by Zeno's paradoxes against plurality, he held that matter was infinitely divisible; that however far any piece of matter might be divided, there always resulted smaller parts of the same substance, each of which always contained portions of every other substance[3] and was itself capable of further division. Its predominant ingredients were responsible for its most distinctive features.

Initially, Anaxagoras held, all things were together in an apparently uniform, motionless mixture. Then Mind (Noûs) instituted a vortex, causing the dense, wet, cold, and dark matter to settle at the center and the rare, hot, and dry matter to take up peripheral positions as the sky. From the former, the disklike earth was compacted (Diels and Kranz, B15–16). The sun, moon, and stars, however, were torn from the earth and carried around, ignited by friction.[4]

Although strikingly rational, Anaxagoras' astronomy was not fruitful because it provided no stimulus to discover the laws of planetary motion. A more important contribution was his concept of a separate, immaterial moving cause, which paved the way for a fully teleological view of nature.[5] His theory of matter, however, was not influential, doubtless as much because of its subtlety and sophistication as because of its lack of economy.

NOTES

1. For the chronology of Anaxagoras' life see Taylor, Davison, and Guthrie.
2. For the relative dating of the works of these two see Longrigg, p. 173, n. 49.
3. The interpretation of Anaxagoras' theory is highly controversial. Certain scholars, most cogently Vlastos, reject this so-called naïve interpretation on the grounds that it involves a redundancy and an infinite regress. Their solution, although plausible, is, however, less in accordance with the fragments. On the question of the regress see especially Strang, pp. 101 ff.
4. The fall of the meteorite at Aegospotami in 467 B.C. probably suggested this theory. (It might be observed here that although Anaxagoras is commonly stated to have been the first to discover the true explanation of eclipses, there is evidence against his priority.)
5. For the reaction of Plato and Aristotle see *Phaedo* 97B and *Metaphysics* 985a18 ff. (Diels and Kranz, A47).

BIBLIOGRAPHY

The collected fragments and later testimony are in H. Diels and W. Kranz, *Die Fragmente der Vorsokratiker*, 6th ed. (Berlin, 1951–1952), II, 5–44.

Secondary literature includes C. Bailey, *The Greek*

Atomists and Epicurus (Oxford, 1928), pp. 537–556; D. Bargrave-Weaver, "The Cosmogony of Anaxagoras," in *Phronesis,* **4,** no. 2 (1959), 77–91; J. Burnet, *Early Greek Philosophy,* 4th ed. (London, 1930), pp. 251–275; W. Capelle, "Anaxagoras," in *Neue Jahrbücher für das klassische Altertum* (1919), 81–102, 169–198; F. M. Cleve, *The Philosophy of Anaxagoras* (New York, 1949); F. M. Cornford, "Anaxagoras' Theory of Matter," in *Classical Quarterly,* **24** (1930), 14–30, 83–95; J. A. Davison, "Protagoras, Democritus and Anaxagoras," *ibid.,* n.s. **3** (1953), 33–45; O. Gigon, "Zu Anaxagoras," in *Philologus,* **91** (1936–1937), 1–41; W. K. C. Guthrie, *A History of Greek Philosophy,* II (Cambridge, England, 1965), 266–338; G. S. Kirk and J. E. Raven, *The Presocratic Philosophers* (Cambridge, England, 1957), pp. 362–394; J. Longrigg, "Philosophy and Medicine: Some Early Interactions," in *Harvard Studies in Classical Philology,* **67** (1963), 147–175; R. Mathewson, "Aristotle and Anaxagoras: An Examination of F. M. Cornford's Interpretation," in *Classical Quarterly,* n.s. **8** (1958), 67–81; C. Mugler, "Le problème d'Anaxagore," in *Revue des études grecques,* **69** (1956), 314–376; A. L. Peck, "Anaxagoras: Predication as a Problem in Physics," in *Classical Quarterly,* **25** (1931), 27–37, 112–120; J. E. Raven, "The Basis of Anaxagoras' Cosmogony," *ibid.,* n.s. **4** (1954), 123–137; C. Strang, "The Physical Theory of Anaxagoras," in *Archiv für Geschichte der Philosophie,* **45,** 2 (1963), 101–118; P. Tannery, *Pour l'histoire de la science hellène,* 2nd ed. (Paris, 1930), pp. 275–303; A. E. Taylor, "On the Date of the Trial of Anaxagoras," in *Classical Quarterly,* **11** (1917), 81–87; G. Vlastos, "The Physical Theory of Anaxagoras," in *Philosophical Review,* **59** (1950), 31–57; and M. L. West, "Anaxagoras and the Meteorite of 467 B.C.," in *Journal of the British Astronomical Association,* **70** (1960), 368–369.

JAMES LONGRIGG

ANAXILAUS OF LARISSA (described by Eusebius as a Pythagorean magician who was banished from Italy by Augustus in Olympiad 188, 1 [28 B.C.]).

He wrote about the "magical" or peculiar effects of some minerals, herbs, and animal substances and of the drugs made with them; in this connection he is cited several times by Pliny. He seems to have been famous for his magical tricks, since he is mentioned as the author of *ludicra* ("sports" or "tricks") by Irenaeus and by Epiphanius. Anaxilaus may have been a source for Pliny in other instances where Pliny does not mention him by name, but it is impossible to prove it.

Because, in a magical papyrus now at Stockholm, Anaxilaus is reported as quoting a recipe of Pseudo-Democritus, Diels assumed that through Anaxilaus the author of the papyrus knew at least part of the treatise of Pseudo-Democritus on alchemy. Wellmann, through a comparison of Pliny's *Natural History* (XXV, 154), where Anaxilaus is mentioned, and

Dioscorides' *De materia medica* (IV, 79), infers that Anaxilaus was one of the authors used by Sextius Niger. Other passages, in Sextus Empiricus, Psellus, and others, that are parallel to passages where Anaxilaus is mentioned suggest that he was one of the important sources for magical and alchemical authors. The Anaxilaus mentioned by Diogenes Laertius (I, 107) may or may not be Anaxilaus of Larissa, but there is no reason to identify either with the Anaxilaïdes cited by Diogenes Laertius (III, 2) as the author of a work entitled *On Philosophers,* as Schwartz has suggested.

BIBLIOGRAPHY

H. Diels provides a review of Lagercrantz's book (see below) in *Deutsche Literaturzeitung,* **34** (1913), cols. 901–906. See also O. Lagercrantz, *Papyrus Graecus Holmiensis. Recepte für Silber, Steine und Purpur* (Uppsala-Leipzig, 1913); E. Schwartz, "Anaxilaides," in Pauly-Wissowa, *Real-Encyclopädie der klassischen Altertumswissenschaft,* I, 2 (Stuttgart, 1894), col. 2083; H. Stadler, *Die Quellen des Plinius im 19. Buche der naturalis historia,* inaugural dissertation (Munich, 1891), pp. 29–30; and M. Wellmann, "Sextius Niger. Eine Quellenuntersuchung zu Dioscorides," in *Hermes,* **24** (1889), 530–569; "Anaxilaos," in Pauly-Wissowa, *Real-Encyclopädie der klassischen Altertumswissenschaft,* I, 2 (Stuttgart, 1894), col. 2084; "Die ΦΥΣΙΚΑ des Bolos Demokritos und der Magier Anaxilaos aus Larissa, Teil I," in *Abhandlungen der Preussischen Akademie der Wissenschaften* (1928), philosophisch-historische Klasse, no. 7; see pp. 77–80 for a collection of the fragments of Anaxilaus.

Anaxilaus is mentioned in the following classical works: Epiphanius, *Contra haereses,* 34, 1; Eusebius, *Chronikon,* Schoene ed., II, 141; Irenaeus, *Contra haereses,* I, 13, 1; and Pliny, *Natural History,* XIX, 20; XXV, 154; XXVIII, 181; XXX, 74; XXXII, 141; XXXV, 175.

L. TARÁN

ANAXIMANDER, *astronomy, natural philosophy.*

There were at least two Anaximanders, both citizens of Miletus: the elder, who was born about 610 B.C. and who is said to have died shortly after 547/546 B.C. (i.e., in 546/545 B.C., the year of the fall of Sardis), and the younger, who is said to have been a historian, the author of *Interpretation of Pythagorean Symbols,* who lived toward the end of the fifth century B.C. We are concerned here with the older Anaximander. Of him we possess only one verbatim quotation, which is difficult to interpret with certainty, and several reports by later authors. Most of this indirect evidence, however, cannot be taken at its face value because it ultimately goes back to Theophrastus, who, like his teacher Aristotle, had the tendency to see and interpret the doctrines of pre-Aristotelian philosophers in the light of the problems, the terminology, and the positive doctrines of Aristotle's philosophy. Needless to say, the later doxographical reports also contain mistakes of their own making; and, in the case of the astronomical and mathematical data, later authors transferred the knowledge of what to them were obvious notions to the heroes of early Greek thought. Thus, for example, Anaximander is credited with the discovery of the equinoxes and of the obliquity of the ecliptic, attributions that are anachronistic and that contradict other notions that he is said to have held.

Fortunately, for Anaximander we possess information from a different tradition, the geographical. Thus, we know that he drew a map of the inhabited world and that he wrote a book in which he tried to explain the present state of the earth and of its inhabitants, especially the human race. For this purpose he advanced a cosmogony. According to Anaximander, at any given time there are an infinite number of worlds that have been separated off from the infinite, τὸ ᾿ἄπειρον, which is the source and reservoir of all things. These worlds come into being, and when they perish, they are reabsorbed into the infinite, which surrounds them and is eternal and ageless. Our world came into being when a mass of material was separated off from the infinite; a rotatory motion in a vortex caused the heavy materials to concentrate at the center, while masses of fire surrounded by air went to the periphery and later constituted the heavenly bodies. The sun and the moon are annular bodies constituted of fire surrounded by a mass of air. This mass of air has pipelike passages through which the light produced by the fire inside escapes, and this is the light the earth receives. In this way Anaximander perhaps accounted for the different shapes of the moon's face and also for eclipses. The earth, at the center of this world, has the shape of a rather flat cylinder. Animals originated from inanimate matter, by the action of the sun on water, and men originated from fish.

What is significant in all this is that Anaximander tried to explain all these different phenomena as the result of one law that rules everything; and it is this law that is preserved in the only verbatim quotation from Anaximander that we possess (here paraphrased): All things pass away into that from which they took their origin, the infinite, as it is necessary; for they make reparation to one another for their injustice in the fixed order of time. The extent and exact meaning of this quotation are controversial, but there can be no question that here we have an imper-

sonal law according to which all occurrences in the universe are explained. This all-inclusive, immanent law of nature is Anaximander's lasting contribution to human thought.

BIBLIOGRAPHY

The ancient sources are collected in H. Diels and W. Kranz, eds., *Die Fragmente der Vorsokratiker,* 6th ed., I (Berlin, 1951), 81–90.

Modern works dealing with Anaximander are J. Burnet, *Early Greek Philosophy,* 4th ed. (London, 1930), pp. 50–71; H. Cherniss, *Aristotle's Criticism of Presocratic Philosophy* (Baltimore, 1935), *passim*; and "The Characteristics and Effects of Presocratic Philosophy," in *Journal of the History of Ideas,* **12** (1951), 319–345; D. R. Dicks, "Solstices, Equinoxes, and the Presocratics," in *Journal of Hellenic Studies,* **86** (1966), 26–40; F. Dirlmeier, "Der Satz des Anaximandros von Milet," in *Rheinisches Museum,* **87** (1938), 376–382; W. A. Heidel, "The ΔINH in Anaximenes and Anaximander," in *Classical Philology,* **1** (1906), 279–282; "On Anaximander," *ibid.,* **7** (1912), 212–234; "On Certain Fragments of the Pre-Socratics," in *Proceedings of the American Academy of Arts and Sciences,* **48** (1913), 681–734, esp. 682–691; and "Anaximander's Book, the Earliest Known Geographical Treatise," *ibid.,* **56** (1921), 237–288; C. H. Kahn, *Anaximander and the Origins of Greek Cosmology* (New York, 1960); G. S. Kirk and J. E. Raven, *The Presocratic Philosophers* (Cambridge, 1957), pp. 99–142; A. Maddalena, *Ionici, testimonianze e frammenti* (Florence, 1963), pp. 76–157; J. B. McDiarmid, "Theophrastus on the Presocratic Causes," in *Harvard Studies in Classical Philology,* **61** (1953), 85–156; R. Mondolfo, *L'infinito nel pensiero dell'antichità classica* (Florence, 1956), pp. 188 ff.; and in his edition of E. Zeller's *Philosophie der Griechen, La filosofia dei greci nel suo sviluppo storico,* II (Florence, 1938), 135–205; and E. Zeller, *Die Philosophie der Griechen,* W. Nestle, ed., I (Leipzig, 1923), 270–315.

Additional bibliographies may be found in the works by Kahn and Maddalena cited above.

L. Tarán

ANAXIMENES OF MILETUS (*fl.* 546/545 B.C.), *philosophy.*

The year in which Anaximenes is said to have flourished is that of the fall of Sardis, and therefore his chronology, which comes from Apollodorus, may be arbitrary. Anaximenes may or may not have been a student of Anaximander, but there can be no question that he was acquainted with Anaximander's book, since his cosmological and astronomical views are very close to those of Anaximander. His interests, however, seem to have been more restricted than those of Anaximander; but like the latter, Anaximenes thought that the source from which all things come into being is infinite. He further qualified this original substance by saying that it is air, for he had discovered a mechanism that could account for the transformation of one thing into another: the mechanism of condensation and rarefaction. When air is evenly distributed, it is invisible; when it is condensed, it becomes water; and when it is condensed further, it becomes earth and then stone. When, on the other hand, air is hot, it becomes rarefied and eventually becomes fire. Anaximenes seems to have been satisfied that the following "experiment," which proved to him that cold air is condensed and hot air is rarefied, corroborated his theory: When we expel air, it becomes cold if we press our lips, whereas it becomes hot if we open our mouths. This experiment, which is recounted—perhaps in Anaximenes' own words—by Plutarch, shows that for Anaximenes air is a substance composed of small, discrete particles: when the particles are compressed, we have water; when they are expanded by heat, we have hot air, fire, etc. This substance thus composed of discrete particles is not more air than water, earth, or fire, but Anaximenes seems to have named it "air" because air is the most widely distributed body in the universe and because breath is identified with the soul, which he believed holds living beings together.

On the basis of this analogy, Anaximenes seems to have believed that the cosmos breathes by inhaling the surrounding air. Consequently, what Anaximenes calls air should not be identified with a single substance that by qualitative alteration becomes all the things we see around us—as Aristotle, Theophrastus, and the doxographers believed he meant. Anaximenes seems also to have believed in an infinite number of worlds that come into being and pass away, to be reabsorbed into the infinite air that surrounds them and is in perpetual motion. We also find in Anaximenes the use of rotatory motion to explain the formation of our world: the big masses of air and water and the heavenly bodies are formed through the process of condensation and rarefaction; the earth, a flat disk at the center, is supported by air; and the same annular form is attributed to the heavenly bodies, which are carried around and supported by air. Since for Anaximenes the sun and the moon are formed out of fire, he must have been ignorant of the fact that the moon reflects the light of the sun. The heavenly bodies turn around the earth and become invisible because they are so far away and because the northern parts of the earth are elevated.

BIBLIOGRAPHY

The ancient sources are collected in Diels and Kranz, *Die Fragmente der Vorsokratiker,* 6th ed., I (Berlin, 1951), 90–96.

Modern works dealing with Anaximenes are J. Burnet, *Early Greek Philosophy,* 4th ed. (London, 1930), pp. 72–79; H. Cherniss, *Aristotle's Criticism of Presocratic Philosophy* (Baltimore, 1935); W. A. Heidel, "The ΔINH in Anaximenes and Anaximander," in *Classical Philology,* **1** (1906), 279–282; J. B. McDiarmid, "Theophrastus on the Presocratic Causes," in *Harvard Studies in Classical Philology,* **61** (1953), 85–156; A. Maddalena, *Ionici, testimonianze e frammenti* (Florence, 1963), with bibliography; and R. Mondolfo, in E. Zeller and R. Mondolfo, *La filosofia dei greci nel suo sviluppo storico,* pt. 1 (Florence, 1938), 206–238, with bibliography.

L. Tarán

ANCEL, PAUL ALBERT (*b*. Nancy, France, 21 September 1873; *d*. Paris, France, 27 January 1961), *biology.*

Ancel became an intern in the Nancy hospitals in 1898 and received the M.D. in 1899 and the *docteur ès sciences* from the University of Nancy in 1903. The following year he was made *professeur agrégé* of anatomy at the Faculté de Médecine in Lyons. He then became *professeur titulaire* of anatomy at Nancy in 1908 and of embryology at Strasbourg in 1919. The chair of embryology in Strasbourg was the first of its kind in France. It demanded, both in teaching and in research, a complete reorientation, which Ancel accomplished superbly.

Ancel's numerous honors include corresponding membership in the Académie des Sciences of Paris, national corresponding membership in the Académie de Médecine, honorary foreign membership in the Académie Royale de Médecine Belge of Brussels, the Prix du Prince de Monaco of the Académie de Médecine (1937), and the Prix de la Fondation Singer Polignac (1950), both shared with Pol Bouin.

Ancel's work can be divided into three sections:

(1) His first publications sum up the observations of an anatomist trained in the operating room, but at the same time he carried on his research in cytology, which furnished the material for an important thesis on the hermaphroditic genital gland of the snail.

(2) Beginning in 1903, he collaborated for twenty-five years with Pol Bouin, first at Nancy and then at Strasbourg, in investigations on the physiology of reproduction in mammals. Their essential discoveries can be summed up as follows: In the male it is the

interstitial gland of the testis that produces a hormone responsible for the secondary sexual characteristics. In the female the internal secretion of the *corpus luteum* determines the preparation of the uterine mucosa for the nidation of the fertilized ovum, as well as the morphogenetic development of the mammary gland. These basic facts are still accepted, and Ancel and Bouin must figure in any history of sexual endocrinology. This science began, not around 1930 with the isolation of sex hormones, but in the first years of the twentieth century, through experimental investigations, among which those of Ancel and Bouin are of primary importance.

(3) Studies in experimental embryology and teratogenesis, started in Strasbourg as early as 1919, and continued by Ancel until his death, sometimes with various collaborators, constitute an important part of modern embryology. Ancel elucidated the determinism of bilateral symmetry in the amphibian embryo and gave a new scope to experimental teratogenesis, which had remained stagnant for scores of years, by trying out new and fruitful methods, such as the experimental production of monstrosities by precise and localized lesions and the use of chemistry in applying to the whole embryo, at a precise stage of development, substances having a specific action determined by previous tests. Ancel thus became one of the creators of present experimental embryology, and particularly of teratogenesis, by using physical and chemical means.

Of a determined and at times vehement character, Ancel did not fear discussion in order to defend university policy or conclusions reached through his investigations. He was most meticulous in preparing his courses, using blackboard diagrams as well as practical demonstrations in embryology. His scientific writings exceeded 300 memoranda, reports, and books, and his research, begun as early as his medical studies, covered nearly sixty years. After his retirement from teaching, Ancel continued his research at the Institut de Physico-Chimie Biologique in Paris until the eve of his death.

BIBLIOGRAPHY

For a complete list of Ancel's publications and references to his obituary notices, see Étienne Wolff, "Le Professeur Paul Ancel," in *Archives d'anatomie, d'histologie, d'embryologie normales et expérimentales,* supp. **44** (1961), 5–27.

MARC KLEIN

ANDERNACH, GUNTHER. See Guenther, Johann.

ANDERSON, ERNEST MASSON (*b.* Falkirk, Scotland, 9 August 1877; *d.* Edinburgh, Scotland, 8 August 1960), *geology.*

Anderson, a structural geologist, tectonist, and mathematical geophysicist, was a son of Rev. John Anderson and Annie Masson, daughter of a minister. He was educated at Dundee High School and Edinburgh University, from which he received the B.Sc. in 1897, the M.A. with first class honours in mathematics and natural philosophy in 1898, and the D.Sc. in 1933. In 1915 he married Alice Catherine Esson, by whom he had two daughters.

Anderson joined the Geological Survey of Great Britain in 1903. Except during 1916–1917, when he served in the army and was wounded in France, he worked in Scotland. Temporary ill health forced him to retire as a senior geologist in 1928.

He was a Christian free-thinker; self-effacing in spite of outstanding mathematical ability, he was characterized by his innate courtesy. Anderson was awarded three medals: by the Royal Society of Edinburgh, the Geological Society of London, and the Geological Society of Edinburgh.

Anderson contributed to many Geological Survey memoirs. His main scientific work stemmed from official problems but was done in leisure time and published almost entirely in other than governmental journals. This work concerned the dynamics of faulting and igneous intrusion, the lineation of schists, crustal heat and structure, and volcanism. After his retirement he expanded all these studies by mathematical analysis.

In 1905 Anderson gave the first explanation of the dynamic basis of the three main classes of faults—reversed, normal, and wrench (a term he preferred to "transcurrent" or "strike-slip"). He later extended his structural studies in discussions of the dynamics of intrusion of igneous sheets and dikes and of the formation of cauldron subsidences (1924, 1936, 1937, 1938). *The Dynamics of Faulting and Dyke Formation* (1942, 1951) has been acclaimed by W. B. Harland as a landmark in structural geology. E. S. Hills, however, has criticized some applications of Anderson's theories; H. Jeffreys has questioned certain of his views on fracture; and G. R. Robson and K. G. Barr have rejected a postulate in his theory of igneous intrusion.

Anderson's theory regarding lineation in schists of the Scottish Highlands was inspired by the work of his Geological Survey supervisor C. T. Clough. From kinematic analysis Anderson inferred that lineations, which may be due to subcrustal convection currents, are parallel to the direction of transport or shear, and

Anuchin's studies of man's racial types were also significant. He thought that all human physical types (races) were essentially transitions from some types toward others: "Mankind represents properly one form. . . . In other words, all mankind proceeds from the same ancestors, whose descendants only gradually formed different races" ("O zadachakh i metodakh antropologii," p. 69). Anuchin decisively rejected the view that the human races originated from distinct apelike ancestors. He wrote several fundamental works on the physical types of various nationalities, among them a monograph on the Ainu people (1876).

Anuchin's works on ethnography were distinguished by originality and strict adherence to scientific method. Such works as the comparative study of bows and arrows (1881), "Sani, lad'ja i koni, kak prinadlezhnosti pokhoronnogo obrjada" ("Sleighs, Large Boats, and Horses as Appurtenances of Funeral Rites," 1890), and "K istorii oznakomlenija s Sibir'ju do Ermaka" ("Toward the History of the Acquaintance with Siberia Before Ermak," 1890) are regarded as models of scientific creativity. Anuchin was also interested in the origin of domestic animals as the basis of one of the most important branches of agricultural economics.

Anuchin did work in the history of science, writing original studies of Lomonosov, Darwin, Humboldt, Miklukho-Maklaja, and others. He published approximately a thousand papers.

BIBLIOGRAPHY

I. ORIGINAL WORKS. Anuchin's writings include "Antropomorfnye obez'jany i nizshie tipy chelovechestva" ("Anthropoid Apes and Lower Types of Man"), in *Priroda*, no. 1 (1874), 185–280; no. 3 (1874), 220–276; no. 4 (1874), 81–141; "Materialy dlja antropologii Vostochnoj Azii. Plemja Ajnov" ("Materials for the Anthropology of Eastern Asia. The Ainu Tribe"), in *Izvêstīya Imperatorskago obshchestva lyubiteleĭ estestvoznanīya, antropologīi i étnografīi. Trudȳ antropologicheskago otdela,* **20**, no. 2 (1876), 79–204; *Antropologija, ee zadachi i metody* ("Anthropology, Its Problems and Methods"; Moscow, 1879); "O nekotorykh anomalijakh chelovecheskogo cherepa i preimushchestvenno ob ikh rasprostranenii po rasam" ("On Several Anomalies of the Human Cranium, Primarily on Their Prevalence According to Race") in *Izvêstīya Imperatorskago obshchestva lyubiteleĭ estestvoznanīya, antropologīi i étnografīi. Trudȳ antropologicheskago otdela,* **6** (1880); *Kurs lektsij po istorii zemlevedinija* ("Lecture Course on the History of Geography"; Moscow, 1885); *Kurs lektsij po obshchej geografii* ("Lecture Course in General Geography"; Moscow, 1887); *O geograficheskom raspredelenii rosta muzhskogo naselenija Rossii* ("On the Geographic Distribution of the Height of Russia's Male Population"; St. Petersburg, 1889); "Rel'ef poverkhnosti Evropejskoj Rossii v posledovatel'nom razvitii o nem predstavlenii" ("The Surface Relief of European Russia in the Chronological Development of Its Mapping"), in *Zemlevedenie,* **2,** no. 1 (1895), 77–126, and no. 4 (1895), 65–124, also in *Rel'ef Evropejskoj chasti SSSR* (see below), pp. 35–147; *Verkhnevolzhskie ozera i verkhov'ja Zapadnoj Dviny. Rekognostsirovki i issledovanija 1894–1895 gg.* ("The Upper Volga Lakes and the Upper Reaches of the Western Dvina. Reconnaissance and Study in 1894–1895"; Moscow, 1897); "O zadachakh i metodakh antropologii" ("On the Problems and Methods of Anthropology"), in *Russkij antropologicheskij zhurnal,* **9,** no. 1 (1902), 62–88; *Japonija i japontsy* ("Japan and the Japanese"; Moscow, 1907); "Proiskhozdhenie cheloveka i ego istoricheskie predke" ("The Origin of Man and His Historical Predecessors"), in *Itogi nauki v teorii i pratike,* VI (Moscow, 1912), 691–784; *Lektsii po fizicheskoj geografii* ("Lectures on Physical Geography"; Moscow, 1916); *Proiskhozhdenie cheloveka* ("The Origin of Man"; Moscow, 1922; 3rd ed., Moscow-Leningrad, 1927); *Rel'ef Evropejskoj chasti SSSR* ("Relief of the European Sector of the U.S.S.R."; Moscow, 1948), written with A. A. Borzov; *Izbrannye geograficheskie raboty* ("Selected Geographical Works"), L. S. Berg, gen. ed. (Moscow, 1949); *O ljudjakh russkoj nauki i kul'tury* ("On Men of Russian Sciences and Culture"; Moscow, 1950, 1952); *Geograficheskie raboty* ("Geographical Works"), A. A. Grigor'ev, ed. (Moscow, 1954), with bibliography; and *Ljudi zarubezhnoj nauki i kul'tury* ("Men of Foreign Science and Culture"; Moscow, 1960).

II. SECONDARY LITERATURE. Works on Anuchin are V. V. Bogdanov, *D. N. Anuchin. Antropolog i geograf* ("D. N. Anuchin, Anthropologist and Geographer"; Moscow, 1941), with bibliography; *Sbornik v chest semidesjatiletija prof. D. N. Anuchina* ("Collection in Honor of the Seventieth Birthday of Professor D. N. Anuchin"; Moscow, 1913); V. A. Esakov, *D. N. Anuchin i sozdanie russkoj universitetskoj geograficheskoj shkoly* ("D. N. Anuchin and the Creation of the Russian University School of Geography"; Moscow, 1955); A. A. Grigor'ev, "D. N. Anuchin," in *Ljudi russkoj nauki* ("Men of Russian Science"), I (Moscow-Leningrad, 1948), 599–605; 2nd ed., II (1962), 508–515; G. V. Karpov, *Put' uchenogo* ("The Path of a Scientist"; Moscow, 1958); *Pamjati D. N. Anuchina (1843–1923)* ("In Memory of D. N. Anuchin [1843–1923]"), in *Trudȳ Instituta étnografii imeni N. N. Miklukho-Maklaja. Akademiya nauk SSR,* n.s. **1** (1947); and A. I. Solov'ev, "D. N. Anuchin, ego osnovnye geograficheskie idei i ego rol' v razvitii russkoj geografii" ("D. N. Anuchin, His Basic Geographical Ideas and his Role in the Development of Russian Geography"), in *Voprosy geografii,* **9** (1948), 9–28.

V. A. ESAKOV

ANVILLE, JEAN-BAPTISTE BOURGUIGNON D' (*b.* Paris, France, 11 July 1697; *d.* Paris, 28 January 1782), *cartography.*

D'Anville was the son of Hubert Bourguignon and

Charlotte Vaugon. About 1730 he married Charlotte Testard, who bore him two daughters. He was secretary to the duke of Orléans (regent during the minority of Louis XV) and was named royal geographer as early as 1717. Having been elected to the Académie des Inscriptions et Belles-Lettres in 1754, he succeeded Buache as chief royal geographer in 1773 and as assistant geographer to the Académie Royale des Sciences.

D'Anville contributed greatly to the renaissance of geography and cartography in France in the eighteenth century. He was the author of 211 maps or plans, and from his first publications he demonstrated great intellectual honesty by leaving unknown territories (the interior of Africa, America, and Asia) blank, contrary to the practice of many cartographers, who used ornaments—the less their knowledge of a territory, the greater the ornamentation.

Furthermore, his maps were solidly based on triangulation nets. As a result, the maps of China, drawn at the request of the Jesuits for Father du Halde's history, after reports of their missionaries, were the first to give an accurate indication of the Pacific coastline (the Yellow Sea and its gulfs and headlands and the Molucca Islands). D'Anville also drew maps of France for Longuerue's *Description de la France* (1719), and maps of Africa and Santo Domingo.

He made studies of ancient measurements in several memoirs and in his *Traité des mesures itinéraires anciennes et modernes* (1769). Comparing them with modern figures, he established remarkably accurate maps for Rollin's *Histoire ancienne,* Rollin and Crevier's *Histoire romaine,* and the *Histoire des empereurs romains.* His maps were greatly appreciated by geographers and navigators, and their exactness was confirmed for Italy by the geodesic operations carried out under the pontificate of Benedict XIV, and for Egypt during Napoleon's campaign in 1799. D'Anville was less successful in his studies of the figure of the earth, and his two memoirs on that subject (1735, 1736) contain some erroneous conclusions. In 1780 d'Anville gave the king his remarkable collection of 10,000 maps, which were both engraved and in manuscript. This collection is now in the Bibliothèque Nationale. In spite of delicate health d'Anville lived until his eighty-fifth year, having devoted his life almost wholly to his work and having published, besides his maps, numerous articles on geography and cartography.

BIBLIOGRAPHY

I. Original Works. The complete bibliography is in L.-C.-J. de Manne, *Notice des ouvrages de M. d'Anville* (see below). The works most important to understanding his achievements in cartography are *Mémoire instructif pour dresser sur lieux des cartes particulières et topographiques d'un canton de pays . . .* (Paris, 1743); *Géographie ancienne abrégée* (Paris, 1769); *Traité des mesures itinéraires anciennes et modernes* (Paris, 1769); and *Considérations générales sur l'étude et les connaissances que demande la composition des ouvrages de géographie* (Paris, 1777). Manne prepared an ed. of the *Oeuvres de d'Anville* in 6 vols., only 2 of which appeared, with an atlas (Paris, 1834).

II. Secondary Literature. Works on d'Anville are Condorcet, "Éloge de d'Anville," in *Oeuvres de Condorcet,* II (Paris, 1847), 528 ff.; M. Dancier, "Éloge de d'Anville," in *Oeuvres de d'Anville,* I, i–xvii; C. Du Bus, "La collection d'Anville à la Bibliothèque Nationale," in *Bulletin de géographie historique,* **41** (1926–1929), 93 ff.; F. Hoefer, in *Nouvelle biographie générale,* II (1852), cols. 368–370; L.-C.-J. de Manne, *Notice des ouvrages de M. d'Anville* (Paris, 1802); Michaud, in *Biographie universelle,* II, 97–98; N. Nielsen, *Géomètres français du dix-huitième siècle* (Paris, 1935), pp. 24–25; Poggendorff, I, 51; and M. Prévost, in *Dictionnaire de biographie française,* III, cols. 84–86.

Juliette Taton

APÁTHY, STEPHAN (*b.* Budapest, Austria-Hungary, 4 January 1863; *d.* Szeged, Hungary, 27 September 1922), *medical science, histology, zoology.*

Apáthy's father, István, was a professor at the University of Budapest and a famous expert in international law. Apáthy attended high school in Budapest and then studied medicine at the university. As a student he worked from 1883 to 1884 at the Institute of Pathology, where his interest in histology was awakened; in 1884 he published a paper dealing with the microscopic anatomy of naiades. He was known as the leader of a progressive student movement, and wrote poetry and essays on social themes. After obtaining his doctor's degree in 1885, he became assistant to Theodor Margó at the Institute of Zoology and Comparative Anatomy in Budapest, where he improved his knowledge of histology.

From 1886 to 1889 Apáthy had a scholarship to the marine biological laboratory in Naples, where he worked under Anton Dohrn. During this period he published seventeen papers and traveled in several European countries. In 1890 he was appointed professor of zoology, and a few years later became professor of histology and embryology at Kolozsvár in Transylvania (now Cluj, Rumania), where he established a modern institute that became a famous international histological research center. In 1895 Apáthy was elected a corresponding member of the Hungarian Academy of Sciences, and in 1905 he was appointed an honorary foreign member of the Royal Medicinal Academy of Belgium.

From his youth Apáthy had opposed the official policies of Vienna and Budapest. Unfortunately, after

World War I he devoted more time to politics than to science. When Transylvania became part of Rumania, the government of the newly created Hungarian Republic appointed Apáthy as government commissioner of Transylvanian affairs to the Rumanian National Committee in Sibiu because of his progressive ideas and scientific reputation. He was, as a commissioner and representative spokesman of a defeated country, imprisoned by the Rumanians in Sibiu, but they, respecting his scientific reputation and patriotism, set him free again. He was even allowed to take all of his research material, collections, microscopes, and such. Apáthy returned to Hungary as professor of zoology at Szeged University, where he founded another modern institute. His health was impaired, however, and he died soon after.

Apáthy's scientific achievements were in three fields: zoology, neurohistology, and microscopic technique. In zoology his most important contributions concerned the systematic and comparative anatomy of Hirudinea, especially of their nervous system.

In neurohistology Apáthy defended the concept of a continuous network of neurofibrils, passing from one neuron to another. The first research worker in this field, he found, with the aid of his original impregnation containing gold chloride, a network of finest neurofibrils in the intestine wall of marine Hirudines. According to his interpretation of his microscopic examination of the whole thickness of the intestinal wall—not, unfortunately, in microscopic sections—these neurofibrils are closely connected with all other cells of the intestinal wall tissues and pass from one neuron to another.

Apáthy considered the neurofibrils to be a connected system, forming an intimate unity with the various tissues penetrated by them. Consequently, he advanced the assumption of the continuity of neurofibrils in the whole animal body. To this idea Apáthy applied all his interpretations, and his further scientific work was given over, for the most part, to the proof and the polemic defense of his view. He published a series of papers opposing the views of Santiago Ramón y Cajal, Joseph Gerlach, Wilhelm Waldeyer, and others, in which he perseverantly tried to prove that the neurons are not merely apposed to each other and that the nervous irritation from one individual neuron to the other is not realized through its transfer *per contiguitatem,* but that the whole nervous system is connected continuously by means of reticularly arranged neurofibrils. Unfortunately he extended his observations, made on invertebrate nervous systems with a specific microscopic feature and function, to the whole class of vertebrates, thus leaving his hypothesis open to attack.

The prolonged controversy contributed, by the response in the world and the tough bilateral polemics lasting many years, to many new and important findings in the field of neurology. This controversy has recently been settled in favor of the neuron theory through electron microscopy. Apáthy always interpreted his morphological findings functionally, and thanks to him many scientists in other branches of biology (e.g., physiology, pathophysiology, neurology, pharmacology) were induced to extend his research. Apáthy is for these reasons to be considered the founder of the modern trends in neurohistology.

Apáthy's work in microscopic technique constituted a great contribution to the development of modern histological techniques. His improvements in the fixation of tissues—in embedding them in paraffin, gelatine, or celloidin; in sectioning and staining them, or impregnating them with gold salts, or both—meant a great advance in histology and furnished many good results. Apáthy's *Die Mikrotechnik der thierischen Morphologie* was an indispensable handbook for two generations of histologists. His greatest achievement is that he made microscopic technique a scientific method, based not only on empirical knowledge but also on systematic comparisons and investigations.

BIBLIOGRAPHY

I. ORIGINAL WORKS. Lists of Apáthy's works may be found in Ambrus Ábráham, "Stephan von Apáthy" (see below), and in Ferenc Kiss, "Stephan von Apáthy als Neurolog" (see below).

II. SECONDARY LITERATURE. There is no biography of Apáthy, but information on his life and work may be found in the following papers: Ambrus Ábráham, "Apáthy István," in *Communicationes ex Bibliotheca Historiae Medicae Hungarica,* 25 (1962), 13–24, and "Stephan von Apáthy," in H. Freud and A. Berg, *Geschichte der Mikroskopie* (Frankfurt am Main, 1963), pp. 65–75; Ferenc Kiss, "Stephan von Apáthy als Neurolog," in *Communicationes ex Bibliotheca Historiae Medicae Hungarica,* 3 (1956), 1–64, and "Apáthy István nehézségei" ("The Difficulties of István Apáthy"), *ibid.,* 36–41; Gábor Kolosváry, "Apáthy mint rendszerezö elme a zoológiában források és személyes kapcsolat alapján" ("Apáthy's Contribution to Zoology, Based on Works of Reference and Personal Recollections"), *ibid.,* 25 (1962), 29–35; Maria Koszoru, "Apáthy István korának társadalma és a tudós szociálpolitikai munkássága" ("István Apáthy, the Society of His Time and the Social-Political Activities of the Scientist"), *ibid.,* 53–57; Endre Réti, "Apáthy István emberi jelentosege" ("The Human Importance of István Apáthy"), *ibid.,* 42–49, and "Darwinista humanizmus Apáthy és Lenhossék szemléletében"—"Darvinističeskij gumanism vo vzljadach Apati i Lenchoseka" ("Apáthy's and Lenhossék's Opinion About Darwinian Humanism"), *ibid.,* 27 (1963), 111–116, 117–122; and Gylyás Pál, *Magyar irók élete és munkái* (Budapest, 1939).

JOSEF SAJNER

APELT, ERNST FRIEDRICH (*b*. Reichenau bei Zittau, Germany, 3 March 1812; *d*. Oppelsdorf, Germany, 27 October 1859), *philosophy, history of science.*

Following a conventional education at a Dorfschule in Reichenau from 1818 to 1822 and at a Bürgerschule in Zittau from 1822 to 1824, Apelt was admitted to the Gymnasium in Zittau, where he studied natural science. An early interest in mathematics led him to speculate on cosmological and philosophical questions, and his teachers directed him to Jakob Fries's *Neue Kritik der Vernunft*. With characteristic aplomb, he wrote to Fries, commenting on several passages in the *Neue Kritik;* as a result, he was invited to study at the University of Jena.

Under Fries's tutelage there from 1831 to 1833, Apelt read mathematics and philosophy. He continued his studies at Leipzig until 1835, specializing in astronomy and physics under the direction of A. F. Moebius. After a brief period in his father's business, in 1839 he became a *Privatdozent* at Jena, where he spent the rest of his life. Apelt lectured initially on mathematics and astronomy, later turning to philosophy. A prolific writer, his inaugural dissertation (*De viribus naturae primitivis,* 1839) was followed in quick succession by a number of important works on metaphysics, epistemology, and history of science: *Die Epochen der Geschichte der Menschheit* (1845–1846), *Johann Keplers Weltansicht* (1849), *Die Reformation der Sternkunde* (1852), *Theorie der Induction* (1854), *Metaphysik* (1857), and the posthumously published *Religionsphilosophie* (1860).

Although his direct contributions to science were slight, Apelt's influence on German history and philosophy of science was far-reaching. His general approach was much like that of Fries and other first-generation Neo-Kantians: philosophy was conceived of as an analysis of knowledge and reason by the critical method. Apelt's special contribution to the Kantian tradition was his insistence, both by precept and by example, that scientific knowledge should be the epistemologist's touchstone and that a careful historical scrutiny of the sciences should be absolutely essential for epistemology. His most important work in the philosophy of science was his *Theorie der Induction,* which was prompted in part by the controversy then raging among Comte, Mill, Herschel, and Whewell about scientific method in general and induction in particular. Apelt's views are probably closest to those of Whewell, which is hardly surprising in view of their common Kantian orientation. Like Whewell, he drew extensively on case studies from the history of science to support his methodological analyses. Moreover, his *Theorie der Induction* in-

cludes a lengthy analysis of mathematical induction and the foundations of mathematical inference. Apelt must be regarded as one of the two leading German methodologists of the period, Justus von Liebig being the other. His historical writings were concerned primarily with astronomy and cosmology; his books and monographs dealt with the Copernican revolution, Greek astronomy, Kepler's natural philosophy, and the scientific work of F. H. Jacobi.

BIBLIOGRAPHY

I. ORIGINAL WORKS. Apelt's works are *De viribus naturae primitivis* (Jena, 1839); "Die Conjunktion des Jupiter und Saturn und ihr Zusammenhang mit den Prophezeihungen des Peter d'Ailln und Nostradamus," in *Zeitschrift Minerva* (July 1840), pp. 1–37; "Ernst Reinhold und die Kantische Philosophie," in *Kritik der Erkenntnistheorie,* **1** (1840), 74–103; *Die Epochen der Geschichte der Menschheit,* 2 vols. (Jena, 1845–1846); *Wie muss das Glaubensbekenntnis beschaffen sein, das zur Vereinigung aller Confessionen führen soll?* (Jena, 1846); "Die Sternhimmel," in *Zeitschrift Minerva* (Jan. 1847), 145–174; "Untersuchungen über die Philosophie und Physik der Alten," in *Abhandlungen der Fries'schen Schule,* I (Leipzig, 1847), 31–144; "Erwiederung auf H. Ritters Recension des ersten Hefts," *ibid.,* II (Leipzig, 1849), 1–26; "Die Sphärentheorie des Eudoxus und Aristoteles," *ibid.,* 27–49; "Die Entdeckung von Amerika: eine historische Skizze," *ibid.,* 51–78; "Bemerkungen über F. H. Jacobi und seine Lehre," *ibid.,* 79–88; "Metaphysische Betrachtungen über die Welt," *ibid.,* 89–105; *Parmenides et Empedoclis doctrina de mundi structura* (Jena, 1847); *Johann Keplers astronomische Weltansicht* (Leipzig, 1849); *Die Reformation der Sternkunde* (Jena, 1852); *Die Theorie der Induction* (Jena, 1854); *Metaphysik* (Leipzig, 1857); and *Religionsphilosophie* (Leipzig, 1860).

II. SECONDARY LITERATURE. The only detailed recent studies of Apelt's work are by W. Gresky: *Die Ausgangspunkte der Philosophie Ernst Friedrich Apelts* (Würzburg, 1936); and "21 Briefe von Hermann Lotze an Ernst Friedrich Apelt (1835–1841)," in *Blätter für deutsche Philosophie,* **10** (1936), 319–331; **11** (1937), 184–203.

L. L. LAUDAN

APIAN, PETER, also known as **Petrus Apianus, Peter Bienewitz** (or **Bennewitz**) (*b*. Leisnig, Germany, 16 April 1495; *d*. Ingolstadt, Germany, 21 April 1552), *astronomy, geography.*

Apian was a pioneer in astronomical and geographical instrumentation, and one of the most successful popularizers of these subjects during the sixteenth century. He studied mathematics and astronomy at Leipzig and Vienna, and quickly established a reputation as an outstanding mathematician. His first work was *Typus orbis universalis,* a world map, based on the work of Martin Waldseemüller, which illustrated the 1520 Vienna edition of Solinus' *Polyhistor seu de mirabilibus mundi.* The following year he published the *Isagoge,* a commentary on the *Typus* and on geography.

Apian's first major work, *Cosmographia seu descriptio totius orbis* (1524), was based on Ptolemy. Starting with the distinction between cosmography, geography, and chorography, and using an ingenious and simple diagram, the book defines terrestrial grids; describes the use of maps and simple surveying; defines weather and climate; and provides thumbnail sketches of the continents. In its later form, as modified by Gemma Frisius, the *Cosmographia* was one of the most popular texts of the time and was translated into all major European languages. The success of this and his previous works led to Apian's appointment as professor of mathematics at the University of Ingolstadt, where he remained until his death. He was knighted by Charles V.

In his *Cosmographia,* Apian suggests the use of lunar distances to measure longitude; in his second major work, *Astronomicon Caesareum* (1540), he supports the use of solar eclipses for that purpose. The *Astronomicon* is notable for Apian's pioneer observations of comets (he describes the appearances and characteristics of five comets, including Halley's) and his statement that comets point their tails away from the sun. Also important is his imaginative use of simple mechanical devices, particularly valvelles, to provide information on the position and movement of celestial bodies. Of greater scientific significance, however, is Apian's *Instrumentum sinuum sive primi mobilis* (1534), where he calculates sines for every minute, with the radius divided decimally. These are the first such tables ever printed.

Apian's contribution to cartography was as a compiler and publisher, rather than as a mapmaker. His cordiform world map and maps of Hungary and France survive; his large-scale map of Europe (1534), the first of its kind, is lost. He also designed a quadrant and an armillary sphere that were popular in his day.

BIBLIOGRAPHY

The best list of Apian's writings is F. Van Ortroy, "Bibliographie de l'oeuvre de Pierre Apian," in *Bibliographie moderne* (Mar.–Oct. 1901).

The standard biographical source is S. Günther, "Peter und Phillipp Apian: Zwei deutsche Mathematiker und Kartographen," in *Abhandlungen der Königlich böhmischen Gesellschaft der Wissenschaften,* 6th ser., **11** (1882).

GEORGE KISH

APOLLONIUS OF PERGA (*b.* second half of third century B.C.; *d.* early second century B.C.), *mathematical sciences.*

Very little is known of the life of Apollonius. The surviving references from antiquity are meager and in part untrustworthy. He is said to have been born at Perga (Greek Πέργη), a small Greek city in southern Asia Minor, when Ptolemy Euergetes was king of Egypt (i.e., between 246 and 221 B.C.)[1] and to have become famous for his astronomical studies in the time of Ptolemy Philopator, who reigned from 221 to 205 B.C.[2] Little credence can be attached to the statement in Pappus that he studied for a long time with the pupils of Euclid in Alexandria.[3] The best evidence for his life is contained in his own prefaces to the various books of his *Conics.* From these it is clear that he was for some time domiciled at Alexandria and that he visited Pergamum and Ephesus.

The prefaces of the first three books are addressed to one Eudemus of Pergamum. Since the Preface to Book II states that he is sending the book by the hands of his son Apollonius, he must have been of mature age at the time of its composition.[4] We are told in the Preface to Book IV that Eudemus is now dead;[5] this and the remaining books are addressed to one Attalus. The latter is commonly identified with King Attalus I of Pergamum (reigned 241–197 B.C.); but it is highly unlikely that Apollonius would have neglected current etiquette so grossly as to omit the title of "King" (βασιλεύς) when addressing the monarch, and Attalus was a common name among those of Macedonian descent. However, a chronological inference can be made from a passage in the Preface to Book II, where Apollonius says, ". . . and if Philonides the geometer, whom I introduced to you in Ephesus, should happen to visit the neighborhood of Pergamum, give him a copy [of this book]."[6] Philonides, as we learn from a fragmentary biography preserved on a papyrus and from two inscriptions, was an Epicurean mathematician and philosopher who was personally known to the Seleucid kings Antiochus IV Epiphanes (reigned 175–163 B.C.) and Demetrius I Soter (162–150 B.C.). Eudemus was the first teacher of Philonides. Thus the introduction of the young Philonides to Eudemus probably took place early in the second century B.C. The *Conics* were composed about the same time. Since Apollonius was then old enough to have a grown son, it is reasonable to accept the birth date given by Eutocius and to place the period of Apollonius' activity in the late third and early second centuries B.C. This fits well with the internal evidence which his works provide on his relationship to Archimedes (who died an old man in 212–211 B.C.); Apollonius appears at times to be developing and improving on ideas that were originally conceived by Archimedes (for examples see p. 189). It is true that Apollonius does not mention Archimedes in his extant works; he does, however, refer to Conon, an older (?) contemporary and correspondent of Archimedes, as a predecessor in the theory of conic sections.[7]

Of Apollonius' numerous works in a number of different mathematical fields, only two survive, although we have a good idea of the content of several others from the account of them in the encyclopedic work of Pappus (fourth century A.D.). But it is impossible to establish any kind of relative chronology for his works or to trace the development of his ideas. The sole chronological datum is that already established, that the *Conics* in the form that we have them are the work of his mature years. Thus the order in which his works are treated here is an arbitrary one.

The work on which Apollonius' modern fame rests, the *Conics* (κωνικά), was originally in eight books. Books I–IV survive in the original Greek, Books V–VII only in Arabic translation. Book VIII is lost, but some idea of its contents can be gained from the lemmas to it given by Pappus.[8] Apollonius recounts the genesis of his *Conics* in the Preface to Book I[9]: he had originally composed a treatise on conic sections in eight books at the instance of one Naucrates, a geometer, who was visiting him in Alexandria; this had been composed rather hurriedly because Naucrates was about to sail. Apollonius now takes the opportunity to write a revised version. It is this revised version that constitutes the *Conics* as we know it.

In order to estimate properly Apollonius' achievement in the *Conics,* it is necessary to know what stage the study of the subject had reached before him. Unfortunately, since his work became the classic textbook on the subject, its predecessors failed to survive the Byzantine era. We know of them only from the scattered reports of later writers. It is certain, however, that investigation into the mathematical properties of conic sections had begun in the Greek world at least as early as the middle of the fourth century B.C., and that by 300 B.C. or soon after, textbooks on the subject had been written (we hear of such by Aristaeus and by Euclid). Our best evidence for the content of these textbooks comes from the works of Archimedes. Many of these are concerned with problems involving conic sections, mostly of a very specialized nature; but Archimedes makes use of a number of more elementary propositions in the theory of conics, which he states without proof. We may assume that these propositions were already well known. On occasion Archimedes actually states that such and such a proposition is proved "in the

Elements of Conics" ('εν τοῖς κωνικοῖς στοιχείοις).[10] Let us leave aside the question of what work(s) he is referring to by this title; it is clear that in his time there was already in existence a corpus of elementary theorems on conic sections. Drawing mainly on the works of Archimedes, we can characterize the approach to the theory of conics before Apollonius as follows.

The three curves now known as parabola, hyperbola, and ellipse were obtained by cutting a right circular cone by a plane at right angles to a generator of the cone. According to whether the cone has a right angle, an obtuse angle, or an acute angle at its vertex, the resultant section is respectively a parabola, a hyperbola, or an ellipse. These sections were therefore named by the earlier Greek investigators "section of a right-angled cone," "section of an obtuse-angled cone," and "section of an acute-angled cone," respectively; those appellations are still given to them by Archimedes (although we know that he was well aware that they can be generated by methods other than the above). With the above method of generation, it is possible to characterize each of the curves by what is known in Greek as a σύμπτωμα, i.e., a constant relationship between certain magnitudes which vary according to the position of an arbitrary point taken on the curve (this corresponds to the equation of the curve in modern terms). For the parabola (see Figure 1), for an arbitrary point K, $KL^2 = 2 \, AZ \cdot ZL$ (for suggested proofs of this and the σύμπτωμα of hyperbola and ellipse, see Dijksterhuis, *Archimedes,* pp. 57–59, whom I follow closely here). In algebraic notation, if $KL = y$, $ZL = x$, $2 \, AZ = p$, we get the characteristic equation of the parabola $y^2 = px$. Archimedes frequently uses this relationship in the parabola and calls the parameter p "the double of the distance to the axis" ('α διπλασία τᾶς μέχρι τοῦ 'άξονος)[11] exactly describing $2 \, ZA$ in Figure 1 ("axis" refers to the axis of the *cone*). For the hyperbola and ellipse the following σύμπτωμα can be derived (see Figures 2 and 3):

$$\frac{KL^2}{ZL \cdot PL} = \frac{2 \, ZF}{PZ};$$

in algebraic notation, if $KL = y$, $ZL = x_1$, $PL = x_2$, $2 \, ZF = p$, $PZ = a$,

$$\frac{y^2}{x_1 x_2} = \frac{p}{a} = \text{constant}.$$

This is found in Archimedes in the form equivalent to

$$\frac{y^2}{x_1 x_2} = \frac{y'^2}{x_1' x_2'}.^{12}$$

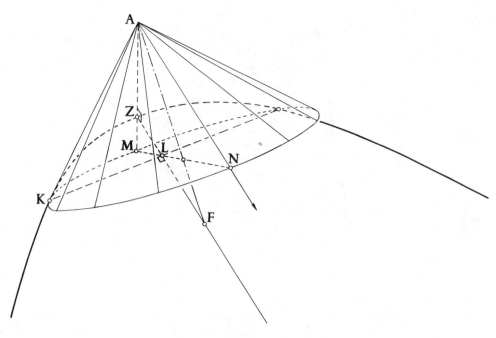

FIGURE 1

It is to be noted that in this system *ZL* always lies on the axis of the section and that *KL* is always at right angles to it. In other words, it is a system of "orthogonal conjugation."

Apollonius' approach is radically different. He generates all three curves from the double oblique circular cone, as follows: in Figures 4, 5, and 6 *ZDE* is the cutting plane. We now cut the cone with another plane orthogonal to the first and passing through the axis of the cone; this is known as the axial triangle (*ABG*); the latter must intersect the base of the cone in a diameter (*BG*) orthogonal to the line in which

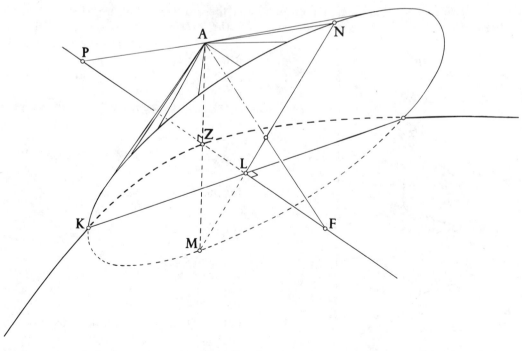

FIGURE 2

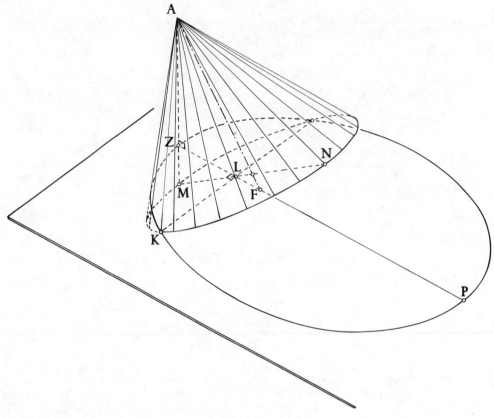

FIGURE 3

the cutting plane intersects it (or its extension); it intersects the cutting plane in a straight line *ZH*. Then, if we neglect the trivial cases where the cutting plane generates a circle, a straight line, a pair of straight lines, or a point, there are three possibilities:

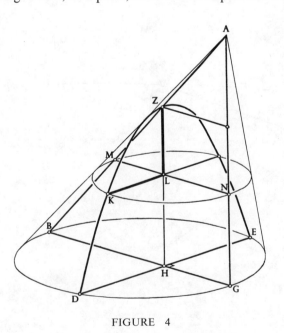

FIGURE 4

(*a*) The line *ZH* in which the cutting plane intersects the axial triangle intersects only one of the two sides of the axial triangle, *AB, AG*; i.e., it is parallel to the other side (Figure 4).

(*b*) *ZH* intersects one side of the axial triangle below the vertex *A* and the other (extended) above it (Figure 5).

(*c*) *ZH* intersects both sides of the axial triangle below *A* (Figure 6).

In all three cases, for an arbitrary point *K* on the curve,

(1) $$KL^2 = ML \cdot LN.$$

Furthermore, in case (*a*),

(2) $$\frac{ML}{LZ} = \frac{BG}{AG} \quad \text{and} \quad \frac{LN}{BG} = \frac{AZ}{AB}.$$

If we now construct a line length Θ such that $\Theta = BG^2 \cdot AZ/AB \cdot AG$, it follows from (1) and (2) that $KL^2 = LZ \cdot \Theta$. Since none of its constituent parts is dependent on the position of *K*, Θ is a constant. In algebraic terms, if $KL = y$, $LZ = x$, and $\Theta = p$, then $y^2 = px$. In cases (*b*) and (*c*),

182

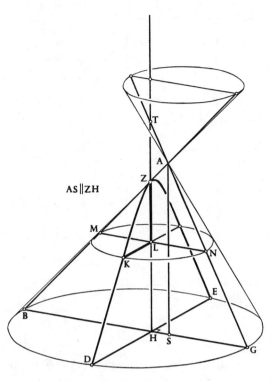

FIGURE 5

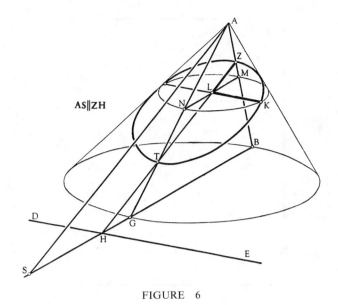

FIGURE 6

(3) $$\frac{ML}{LZ} = \frac{BS}{AS} \text{ and } \frac{LN}{LT} = \frac{SG}{AS}.$$

If we now construct a line length Ξ such that $\Xi = BS \cdot SG \cdot ZT/AS^2$, it follows from (1) and (3) that

$$KL^2 = \Xi \cdot \frac{LZ \cdot LT}{ZT}.$$

Thus

$$KL^2 = \Xi \cdot \frac{LZ(LZ + ZT)}{ZT} \text{ for case } (b)$$

and

$$KL^2 = \Xi \cdot \frac{LZ(LZ - ZT)}{ZT} \text{ for case } (c).$$

In algebraic terms, if $KL = y$, $LZ = x$, $\Xi = p$, and $ZT = a$,

$$y^2 = x\left(p + \frac{p}{a}x\right) \text{ for case } (b),$$

$$y^2 = x\left(p - \frac{p}{a}x\right) \text{ for case } (c).$$

The advantage of such formulation of the συμπτώματα of the curves from the point of view of classical Greek

geometry is that now all three curves can be determined by the method of "application of areas," which is the Euclidean way of geometrically formulating problems that we usually express algebraically by equations of second degree. For instance, Euclid (VI, 28) propounds the problem "To a given straight line to apply a parallelogram equal to a given area and falling short of it by a parallelogrammic figure similar to a given one." (See Figure 7, where for simplicity rectangles have been substituted for parallelograms.) Then the problem is to apply to a line of given length b a rectangle of given area A and side x such that the rectangle falls short of the rectangle bx by a rectangle similar to another with sides c, d. This is equivalent to solving the equation

$$bx - \frac{c}{d}x^2 = A.$$

Compare the similar problem Euclid VI, 29: "To a given straight line to apply a parallelogram equal to a given area and exceeding it by a parallelogrammic figure similar to a given one."

This method is used by Apollonius to express the συμπτώματα of the three curves, as follows (see Figures 8–10).

For case (a) (Figure 8), a rectangle of side x (equal to the abscissa) is applied (παραβάλλεται) to the line-length p (defined as above): this rectangle is equal to the square on the ordinate y. The section is accordingly called parabola (παραβολή, meaning "exact application").

For case (b) (Figure 9), there is applied to p a rectangle, of side x, equal to y^2 and exceeding

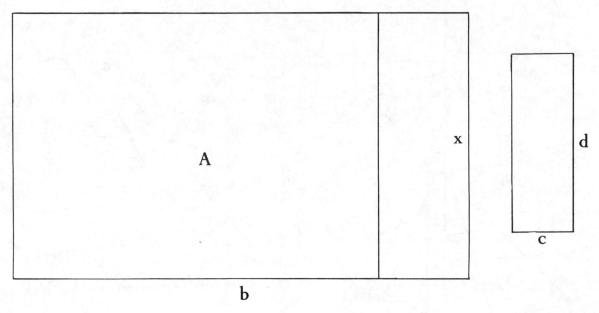

FIGURE 7

($\dot{v}\pi\epsilon\rho\beta\acute{a}\lambda\lambda o\nu$) p by a rectangle similar to p/a. The section is accordingly named hyperbola ($\dot{v}\pi\epsilon\rho\beta o\lambda\acute{\eta}$, meaning "excess").

For case (c) (Figure 10) there is applied to p a rectangle of side x, equal to y^2 and falling short ('$\epsilon\lambda\lambda\epsilon\hat{i}\pi o\nu$) of p by a rectangle similar to p/a. The section is accordingly named ellipse ('$\acute{\epsilon}\lambda\lambda\epsilon\iota\psi\iota s$, meaning "falling short").

This approach has several advantages over the older one. First, all three curves can be represented by the method of "application of areas" favored by classical Greek geometry (it has been appropriately termed "geometrical algebra" in recent times); the older approach allowed this to be done only for the parabola. In modern terms, Apollonius refers the equation of all three curves to a coordinate system of which one axis is a given diameter of the curve and the other the tangent at one end of that diameter. This brings us to a second advantage: Apollonius' method of generating the curves immediately pro-

duces oblique conjugation, whereas the older method produces orthogonal conjugation. As we shall see, oblique conjugation was not entirely unknown to earlier geometers; but it is typical of Apollonius' approach that he immediately develops the most general formulation. It is therefore a logical step, given this approach, for Apollonius to prove (I,50 and the preceding propositions) that a $\sigma\acute{v}\mu\pi\tau\omega\mu\alpha$ equivalent to those derived above can be established for any diameter of a conic and its ordinates: in modern terms, the coordinates of the curves can be transposed to any diameter and its tangent.

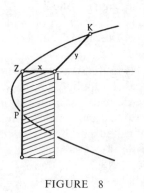

FIGURE 8

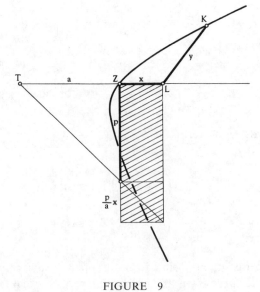

FIGURE 9

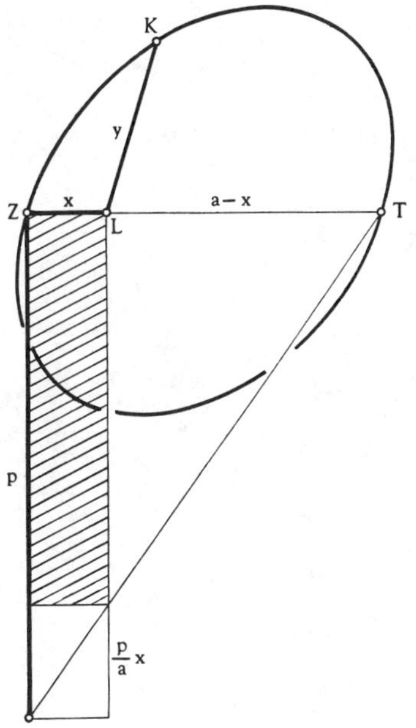

FIGURE 10

We cannot doubt that Apollonius' approach to the generation and basic definition of the conic sections, as outlined above, was radically new. It is not easy to determine how much of the *content* of the *Conics* is new. It is likely that a good deal of the nomenclature that his work made standard was introduced by him; in particular, the terms "parabola," "hyperbola," and "ellipse" make sense only in terms of Apollonius' method. To the parameter which we have called *p* he gave the name ʾορθία ("orthogonal side" [of a rectangle]), referring to its use in the "application": this term survives in the modern *latus rectum*. He defines "diameter" as *any* line bisecting a system of parallel chords in a conic, in accordance with the new generality of his coordinate system: this differs from the old meaning of "diameter" of a conic section (exemplified in Archimedes), which is (in Apollonian and modern terminology) the axis. But though this new terminology reflects the new approach, it does not in itself exclude the possibility that many of Apollonius' results in the *Conics* were already known to his predecessors. That this is true at least for the first four books is suggested by his own Preface to Book I. He says there:[13]

> The first four books constitute an elementary introduction. The first contains the methods of generating the three sections and their basic properties (συμπτώματα), developed more fully and more generally (καθόλου μᾶλλον) than in the writings of others; the second contains the properties of the diameter and axes of the sections, the asymptotes, and other things . . . ;

the third contains many surprising theorems useful for the syntheses of solid loci and for determinations of the possibilities of solutions (διορισμούς); of the latter the greater part and the most beautiful are new. It was the discovery of these that made me aware that Euclid has not worked out the whole of the locus for three and four lines,[14] but only a fortuitous part of it, and that not very happily; for it was not possible to complete the synthesis without my additional discoveries. The fourth book deals with how many ways the conic sections can meet one another and the circumference of the circle, and other additional matters, neither of which has been treated by my predecessors, namely in how many points a conic section or circumference of a circle can meet another. The remaining books are particular extensions (περιουσιαστικώτερα); one of them [V] deals somewhat fully with minima and maxima, another [VI] with equal and similar conic sections, another [VII] with theorems concerning determinations (διοριστικῶν), another [VIII] with determinate conic problems.

From this one gets the impression that Books I–IV, apart from the subjects specifically singled out as original, are merely reworkings of the results of Apollonius' predecessors. This is confirmed by the statement of Pappus, who says that Apollonius supplemented the four books of Euclid's *Conics* (which Pappus *may* have known) and added four more books.[15]

Apollonius also claims to have worked out the methods of generating the sections and setting out their συμπτώματα "more fully and more generally" than his predecessors. The description "more generally" is eminently justified by our comparison of the two methods. However, it is not clear to what "more fully" (ʾεπὶ πλέον) refers. Neugebauer suggests that Apollonius meant his introduction of conjugate hyperbolas (conjugate diameters in ellipse and hyperbola are dealt with in I, 15–16).[16] At least as probable is a more radical alternative, rejected by Neugebauer, that Apollonius is referring to his treatment of the two branches of the hyperbola as a unit (exemplified in I, 16 and frequently later). It is true that Apollonius applies the name "hyperbola" only to a single branch of the curve (he refers to the two branches as the "opposite *sections*" [τομαὶ ʾαντικείμεναι]); it is also true that in his own Preface to Book IV[17] he reveals that at least Nicoteles among his predecessors had considered the two branches together; but Apollonius' very definition of a conic surface[18] as the surface *on both sides* of the vertex is significant in this context; and it is unlikely that any of his predecessors had *systematically* developed the theory of both branches of the hyperbola. Here again, then, we may reasonably regard Apollonius as an innovator in his *method*. But we are not justified in assuming that any of the

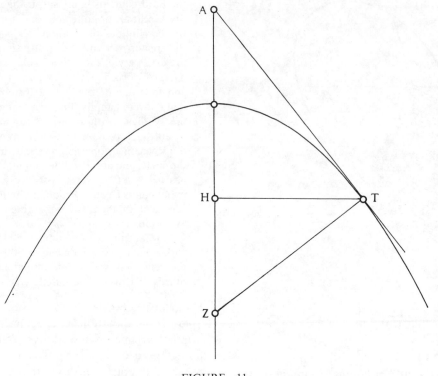

FIGURE   11

results stated in the first four books were unknown before Apollonius, except where he specifically states this. In this part of the work we must see him rather as organizing the results of his predecessors, consisting in part of haphazard and disconnected sets of theorems, into an exposition ordered rationally according to his own very general method. His mastery is such that it seems impossible to separate different sources (as one can, for instance, in the comparable work of Euclid on elementary geometry).

Nevertheless, we may suspect that Archimedes, could he have read Books I–IV of the *Conics,* would have found few results in them that were not already familiar to him (although he might well have been surprised by the order and mutual connection of the theorems). The predecessors of Archimedes were already aware that conic sections could be generated by methods other than that described on p. 180. Euclid states that an ellipse can be produced by cutting a cylinder by a plane not parallel to the base.[19] Archimedes himself certainly knew that there were many different ways of generating the sections from a cone. The best proof of this is *De conoidibus et sphaeroidibus* VII–IX, in which it is shown that for any ellipse it is possible to find an *oblique* circular cone from which that ellipse can be generated. Furthermore, it is certain that the essential properties of the oblique conjugation of at least the parabola were known to

Apollonius' predecessors; for that is the essence of propositions I–III of Archimedes' *Quadrature of the Parabola,* which he states are proved "in the Elements of Conics."[20] In *De conoidibus et sphaeroidibus* III, Archimedes states "If two tangents be drawn from the same point to *any* conic, and two chords be drawn inside the section parallel to the two tangents and intersecting one another, the product of the two parts of each chord [formed by the intersection] will have the same ratio to one another as the squares on the tangents. . . . this is proved in the Elements of Conics."[21] It is plausible to interpret this, with Dijksterhuis, as treatment of all three sections in oblique conjugation.[22]

It is probable then that much of the contents of Books I–IV was already known before Apollonius. Conversely, Apollonius did not include in the "elementary introduction" of Books I–IV all theorems on conics known to his predecessors. For example, in a parabola the subnormal to any tangent formed on the diameter (*HZ* in Figure 11) is constant and equal to half the parameter *p*. This is assumed without proof by Archimedes[23] and by Diocles (perhaps a contemporary of Apollonius) in his proof of the focal property of the parabola in his work *On Burning Mirrors.*[24] We can therefore be sure that it was a well-known theorem in the *Elements of Conics.* Yet in Apollonius it can be found only by combining the results of propositions 13 and 27 of Book V, one of the "particular extensions."

If Apollonius omitted some of his predecessors' results from the elementary section, we must not be surprised if he omitted altogether some results with which he was perfectly familiar: his aim was not to compile an encyclopedia of all possible theorems on conic sections, but to write a systematic textbook on the "elements" and to add some more advanced theory which he happened to have elaborated. The question has often been raised in modern times why there is no mention in the *Conics* of the focus of the parabola. The focal properties of hyperbola and ellipse are treated in III, 45–52: Apollonius proves, *inter alia,* that the focal distances at any point make equal angles with the tangent at that point and that their sum (for the ellipse) or difference (for the hyperbola) is constant. There is no mention of directrix, and from the *Conics* we might conclude that Apollonius was totally ignorant of the focus-directrix property of conic sections. However, it happens that Pappus proves at length that if a point moves in such a way that the ratio of its distance from a fixed point and its orthogonal distance from a fixed straight line is constant, then the locus of that point is a conic section; and that according as the ratio is equal to, greater than, or less than unity, the section will be respectively a parabola, a hyperbola, or an ellipse.[25] This amounts to the generation of the sections from focus and directrix. Pappus gives this proof as a lemma to Euclid's (lost) book *On Surface Loci;* hence, it has been plausibly concluded that the proposition was there stated without proof by Euclid.[26] If that is so, here is a whole topic in the theory of conics that must have been completely familiar to Apollonius, yet which he omits altogether. Thus the lack of any mention of the focus of the parabola in the *Conics* is not an argument for Apollonius' ignorance of it. I agree with those who argue on a priori grounds that he must have known of it.[27] Since he very probably dealt with it in his work(s) on burning mirrors (see p. 189) there was all the more justification for omitting it from the *Conics*. In any case, we now have a proof of it, by Diocles, very close to the time of Apollonius. Since Diocles further informs us that a parabolic burning mirror was constructed by Dositheus, who corresponded with Archimedes, it is highly probable that the focal property of the parabola was well known *before* Apollonius.

For a detailed summary of the contents of the *Conics,* the reader is referred to the works of Zeuthen and Heath listed in the Bibliography. Here we will only supplement Apollonius' own description quoted above by noting that Book III deals with theorems on the rectangles contained by the segments of intersecting chords of a conic (an extension to conics of that proved by Euclid for chords in a circle), with the harmonic properties of pole and polar (to use the modern terms: there are no equivalent ancient ones), with focal properties (discussed above), and finally with propositions relevant to the locus for three and four lines (see n. 14). Of Books V–VII, which are, to judge from Apollonius' own account, largely original, Book V is that which has particularly evoked the admiration of modern mathematicians: it deals with normals to conics, when drawn as maximum and minimum straight lines from particular points or sets of points to the curve. Apollonius finally proves, in effect, that there exists on either side of the axis of a conic a series of points from which one can draw only one normal to the opposite side of the curve, and shows how to construct such points: these points form the curve known, in modern terms, as the *evolute* of the conic in question. Book VII is concerned mainly with propositions about inequalities between various functions of conjugate diameters. Book VIII is lost, but an attempt at restoration from Pappus' lemmas to it was made by Halley in his edition of the *Conics.* If he is right, it contained problems concerning conjugate diameters whose functions (as "determined" in Book VII) have given values.

For a modern reader, the *Conics* is among the most difficult mathematical works of antiquity. Both form and content are far from tractable. The author's rigorous rhetorical exposition is wearing for those used to modern symbolism. Unlike the works of Archimedes, the treatise does not immediately impress the reader with its mathematical brilliance. Apollonius has, in a way, suffered from his own success: his treatise became canonical and eliminated its predecessors, so that we cannot judge by direct comparison its superiority to them in mathematical rigor, consistency, and generality. But the work amply repays closer study; and the attention paid to it by some of the most eminent mathematicians of the seventeenth century (one need mention only Fermat, Newton, and Halley) reinforces the verdict of Apollonius' contemporaries, who, according to Geminus, in admiration for his *Conics* gave him the title of The Great Geometer.[28]

In Book VII of his mathematical thesaurus, Pappus includes summaries of and lemmas to six other works of Apollonius besides the *Conics.* Pappus' account is sufficiently detailed to permit tentative reconstructions of these works, all but one of which are entirely lost. All belong to "higher geometry," and all consisted of exhaustive discussion of the particular cases of one or a few general problems. The contrast with

Apollonius' approach in the *Conics,* where he strives for generality of treatment, is notable. A brief indication of the problem(s) discussed in these works follows.

(1) *Cutting off of a Ratio* (λόγου ᾿αποτομή), in two books, is the only surviving work of Apollonius apart from the *Conics.* However, it is preserved only in an Arabic version which, by comparison with Pappus' summary, appears to be an adaptation rather than a literal translation. Pappus describes the general problem as follows: "To draw through a given point a straight line to cut off from two given straight lines two sections measured from given points on the two given lines so that the two sections cut off have a given ratio."[29] Apollonius discusses particular cases before proceeding to the more general (e.g., in every case discussed in Book I the two given lines are supposed to be parallel) and solves every case by the classical method of "analysis" (in the Greek sense). That is, the problem is presumed solved, and from the solution is deduced some other condition that is easily constructible. Then, by "synthesis" from this latter construction, the original condition is constructed. We may presume that Apollonius followed the same method in all six of these works, especially since Book VII of Pappus was named ᾿αναλυόμενος ("Field of Analysis"). In the *Cutting off of a Ratio* the problem was reduced to one of "application of an area." Zeuthen[30] points out the relevance of this work to *Conics* III, 41: If one regards the theorem proved there as a method of drawing a tangent to a given point in a parabola by determining the intercepts it makes on two other tangents to the curve, that is exactly the problem discussed by Apollonius in this work. Although there is no mention in it of conic sections, the connection is surely not a fortuitous one. In fact, many of the problems discussed by Apollonius in the six works summarized in Book VII of Pappus can be reduced to problems connected with conics. (This helps to explain the great interest shown in this part of Pappus' work by mathematicians of the sixteenth and seventeenth centuries.)

(2) *Cutting off of an Area* (χωρίου ᾿αποτομή), in two books, has a general problem similar to that of the preceding work. But in this case the intercepts cut off from the two given lines must have a given product (in Greek terms, contain a given rectangle) instead of a given proportion.[31] Here again Zeuthen has shown that *Conics* III, 42 and 43, which concern tangents drawn to ellipse and hyperbola, are equivalent to particular cases of the problem discussed by Apollonius in this work.[32]

(3) *Determinate Section* (διωρισμένη τομή) deals with the following general problem: Given four points—A, B, C, D—on a straight line *l,* to determine a point P on that line such that the ratio $AP \cdot CP / BP \cdot DP$ has a given value.[33] Since this comparatively simple problem was discussed at some length by Apollonius, Zeuthen conjectured—plausibly—that he was concerned to find the limits of possibility of a solution for the various possible arrangements of the points (e.g., when two coincide).[34] We know from Pappus' account that it dealt, among other things, with maxima and minima. Whether, as Zeuthen claims, the work amounted to "a complete theory of involution" cannot be decided on existing evidence. But it is a fact that the general problem is the same as determining the intersection of the line *l* and the conic that is the "locus for four lines," the four lines passing through A, B, C, and D; and Apollonius must have known this. Here again, then, is a connection with the theory of conics.

(4) *Tangencies* (᾿επαφαί), in two books, deals with the general problem characterized by Pappus[35] as follows: "Given three elements, either points, lines or circles (or a mixture), to draw a circle tangent to each of the three elements (or through them if they are points)." There are ten possible different combinations of elements, and Apollonius dealt with all eight that had not already been treated by Euclid. The particular case of drawing a circle to touch three given circles attracted the interest of Vieta and Newton, among others. Although one of Newton's solutions[36] was obtained by the intersection of two hyperbolas, and solutions to other cases can also be represented as problems in conics, Apollonius seems to have used only straight-edge and compass constructions throughout. Zeuthen provides a plausible solution to the three-circle problem reconstructed from Pappus' lemmas to this work.[37]

(5) *Inclinations* (νεύσεις), in two books, is described by Pappus on pages 670–672 of the Hultsch edition. In Greek geometry, a νεῦσις problem is one that consists in placing a straight line of given length between two given lines (not necessarily straight) so that it is inclined (νεύει) toward a given point. Pappus tells us that in this work Apollonius restricted himself to certain "plane" problems, i.e., ones that can be solved with straight-edge and compass alone. The particular problems treated by Apollonius can be reconstructed with some probability from Pappus' account.

(6) *Plane Loci* (τόποι ᾿επίπεδοι), in two books, is described on pages 660–670 of Pappus. "Plane loci" in Greek terminology are loci that are either straight lines or circles. In this work, Apollonius investigated

certain conditions that give rise to such plane loci. From them one can easily derive the equation for straight line and circle in Cartesian coordinates.[38]

A number of other works by Apollonius in the field of pure mathematics are known to us from remarks by later writers, but detailed information about the contents is available for only one of these: a work described by Pappus in Book II of his *Collectio*.[39] Since the beginning of Pappus' description is lost, the title of the work is unknown. It expounds a method of expressing very large numbers by what is in effect a place-value system with base 10,000. This way of overcoming the limitations of the Greek alphabetic numeral system, although ingenious, is not surprising, since Archimedes had already done the same thing in his ψαμμίτης (or "Sand Reckoner").[40] Archimedes' base is 10,000[2]. It is clear that Apollonius' work was a refinement on the same idea, with detailed rules of the application of the system to practical calculation. Besides this we hear of works on the cylindrical helix (κοχλίας);[41] on the ratio between dodecahedron and eicosahedron inscribed in the same sphere;[42] and a general treatise (καθόλου πραγματεία).[43] It seems probable that the latter dealt with the foundations of geometry, and that to it are to be assigned the several remarks of Apollonius on that subject quoted by Proclus in his commentary on the first book of Euclid (see Friedlein's Index).

Thus Apollonius' activity covered all branches of geometry known in his time. He also extended the theory of irrationals developed in Book X of Euclid, for several sources mention a work of his on unordered irrationals (περὶ τῶν ἀτάκτων ἀλόγων).[44] The only information as to the nature of this work comes from Pappus' commentary on Euclid X, preserved in Arabic translation;[45] but the exact connotation of "unordered irrationals" remains obscure. Finally, Eutocius, in his commentary on Archimedes' *Measurement of a Circle*,[46] informs us that in a work called ὠκυτόκιον, meaning "rapid hatching" or "quick delivery," Apollonius calculated limits for π that were closer than Archimedes' limits of 3-1/7 and 3-10/71. He does not tell us what Apollonius' limits were; it is possible to derive closer limits merely by extending Archimedes' method of inscribing and circumscribing regular 96-gons to polygons with an ever greater number of sides (as was frequently done in the sixteenth and seventeenth centuries).[47] Very probably this was Apollonius' procedure, but that cannot be proved.

In applied mathematics, Apollonius wrote at least one work on optics. The evidence comes from a late Greek mathematical work preserved only fragmentarily in a palimpsest (the "Bobbio Mathematical Fragment"). Unfortunately, the text is only partly legible at the crucial point,[48] but it is clear that Apollonius wrote a work entitled *On the Burning Mirror* (περὶ τοῦ πυρ⟨ε⟩ίου), in which he showed to what points parallel rays striking a spherical mirror would be reflected. The same passage also appears to say that in another work, entitled *To the Writers on Catoptrics* (πρὸς τοὺς κατοπτρικούς), Apollonius proved that the supposition of older writers that such rays would be reflected to the center of sphericity is wrong. The relevance of his work on conics to the subject of burning mirrors is obvious. We may conjecture with confidence that Apollonius treated of parabolic as well as of spherical burning mirrors. But the whole history of this subject in antiquity is still wrapped in obscurity.

Several sources indicate that Apollonius was noted for his astronomical studies and publications. Ptolemaeus Chennus (see n. 2) made the statement that Apollonius was called Epsilon, because the shape of the Greek letter ϵ is similar to that of the moon, to which Apollonius devoted his most careful study. This fatuous remark incidentally discloses some valuable information. "Hippolytus," in a list of distances to various celestial bodies according to different authorities, says that Apollonius stated that the distance to the moon from the earth is 5,000,000 stades (roughly 600,000 miles).[49] But the only specific information about Apollonius' astronomical studies is given by Ptolemy (fl. A.D. 140) in the *Almagest*.[50] While discussing the determination of the "station" of a planet (the point where it begins or ends its apparent retrogradation), he states that Apollonius proved the following theorem. In Figure 12, O is the observer (earth), the center of a circle on the circumference of which moves an epicycle, center C, with (angular) velocity v_1; the planet moves on the circumference of the epicycle about C with velocity v_2, and in the same sense as C moves about O. Then Apollonius' theorem states that if a line $OBAD$ is drawn from O to cut the circle at B and D, such that

$$\frac{\frac{1}{2}BD}{BO} = \frac{v_1}{v_2},$$

B will be that point on the epicycle at which the planet is stationary. Ptolemy also indicates that Apollonius proved it both for the epicycle model and for an equivalent eccenter model (depicted in Figure 13; here the planet P moves on a circle, center M, eccentric to the earth O, such that $OM/MP = CD/OC$ in Figure 12; M moves about O with speed $[v_1 + v_2]$,

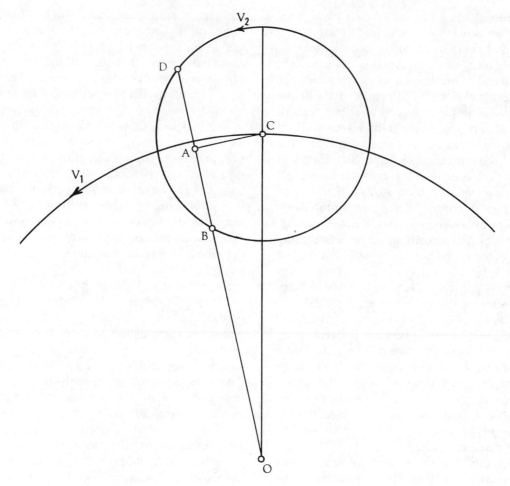

FIGURE   12

P about *M* with speed v_1). Even this much information is valuable, for it shows that Apollonius had already gone far in the application of geometrical models to explain planetary phenomena, and that he must have been acquainted with the equivalence of epicyclic and eccentric models (demonstrated by Ptolemy in *Almagest* III, 3); yet he was still operating with a simple epicycle/eccentric for the planets, although this would, for instance, entail that the length of the retrograde arc of a planet is constant, which is notoriously not the case. Neugebauer (see Bibliography) supposes, however, that the whole of the passage in which Ptolemy himself proves the above theorem is taken from Apollonius. That proof combines the two models of epicycle and eccenter in one by the ingenious device of using the same circle as both epicycle and eccenter; in other words, the epicycle model is transformed into the eccentric model by inversion on a circle. The procedure is worthy of Apollonius, and is indeed a particular case of the pole-polar relationship treated in *Conics* III, 37. But

Ptolemy (who of all ancient authors is most inclined to give credit where it is due) seems to introduce this device as his own,[51] and to return to Apollonius only later.[52] Fortunately, this uncertainty does not affect the main point: that Apollonius represents an important stage in the history of the adaptation of geometrical models to planetary theories. His real importance may have been much greater than we can ever know, since not only his astronomical works, but also those of his successor in the field, Hipparchus (*fl.* 130 B.C.), are lost.

It is not clear how far Apollonius applied his theoretical astronomical models to practical prediction (i.e., assigned sizes to the geometrical quantities and velocities). For the fact that he "calculated" the absolute distance of the moon need imply no more than imitation of the crude methods of Aristarchus of Samos (early third century B.C.); for "Hippolytus" also lists figures for distances in stades between the spheres of the heavenly bodies as given by Archimedes which cannot be reconciled with any rational astronomical

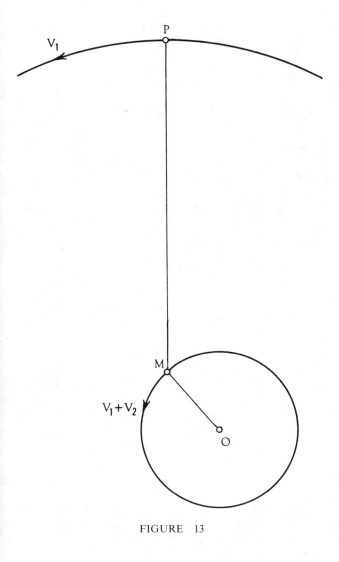

FIGURE 13

Conics became the standard treatise on the subject, and were duly provided with elementary commentaries and annotations by succeeding generations. We hear of such commentaries by Serenus (fourth century A.D.?) and Hypatia (*d.* A.D. 415). The commentary of Eutocius (early sixth century A.D.) survives, but it is entirely superficial. Of surviving writers, the only one with the mathematical ability to comprehend Apollonius' results well enough to extend them significantly is Pappus (*fl.* A.D. 320), to whom we owe what knowledge we have of the range of Apollonius' activity in this branch of mathematics. The general decline of interest in the subject in Byzantium is reflected in the fact that of all Apollonius' works only *Conics* I–IV continued to be copied (because they were used as a textbook). A good deal more of his work passed into Islamic mathematics in Arabic translation, and resulted in several competent treatises on conics written in Arabic; but so far as is known, no major advances were made. (Ibn al-Haytham discusses the focus of the parabola in his work on parabolic burning mirrors;[56] but this, too, may be ultimately dependent on Greek sources.) The first real impulse toward advances in mathematics given by study of the works of Apollonius occurred in Europe in the sixteenth and early seventeenth centuries. The *Conics* were important, but at least as fruitful were Pappus' reports on the lost works, available in the excellent Latin translation by Commandino, published in 1588. (We must remember in this context that Books V–VII of the *Conics* were not generally available in Europe until 1661,[57] too late to make a real impact on the subject.) The number of "restorations" of the lost works of Apollonius made in the late sixteenth and early seventeenth centuries, some by outstanding mathematicians (e.g., Vieta, whose *Apollonius Gallus* [1600] is a reconstruction of the *Tangencies,* and Fermat, who reconstructed the *Plane Loci*) attests to the lively interest that Pappus' account excited. It is hard to overestimate the effect of Apollonius on the brilliant French mathematicians of the seventeenth century, Descartes, Mersenne, Fermat, and even Desargues and Pascal, despite their very different approach. Newton's notorious predilection for the study of conics, using Apollonian methods, was not a chance personal taste. But after him the analytic methods invented by Descartes brought about a lack of interest in Apollonius which was general among creative mathematicians for most of the eighteenth century. It was not until Poncelet's work in the early nineteenth century, picking up that of Desargues, Pascal, and la Hire, revived the study of projective geometry that the relevance of much of Apollonius' work to some basic modern theory was

system.[53] We should not assume without evidence that Apollonius had any better basis for his lunar distance. There is, however, a passage in the astrologer Vettius Valens (*fl.* A.D. 160) that has been taken to show that Apollonius actually constructed solar and lunar tables.[54] The author says that he has used the tables of Hipparchus for the sun; of Sudines, Kidenas, and Apollonius for the moon; and also of Apollonius for both. But there is no certainty that "Apollonius" here refers to Apollonius of Perga. At least as likely is the suggestion of Kroll that it may be Apollonius of Mynda, who is known to us only from a passage of Seneca, from which it appears that he claimed to have studied with the "Chaldaeans" and that he was "very experienced in the examination of horoscopes."[55] The Apollonius of the Vettius Valens passage is also associated with Babylonian names and practices.

Although the mathematical stature of Apollonius was recognized in antiquity, he had no worthy successor in pure mathematics. The first four books of realized. It is no accident that the most illuminating accounts of Apollonius' geometrical work have been written by mathematicians who were themselves leading exponents of the revived "synthetic" geometry, Chasles and Zeuthen.

The contribution of Apollonius to the development of astronomy, although far less obvious to us now, may have been equally important but, unlike his geometrical work, it had an immediate effect on the progress of the subject. Hipparchus and Ptolemy absorbed his work and improved on it. The result, the Ptolemaic system, is one of the most impressive monuments of ancient science (and certainly the longest-lived), and Apollonius' work contributed some of its essential parts.

NOTES

1. Eutocius, *Commentary*, Heiberg, II, 168, quoting one Heraclius.
2. Photius, *Bibliotheca*, p. 151b18 Bekker, quoting the dubious authority Ptolemaeus Chennus of the second century A.D.
3. Pappus, *Collectio* VII, Hultsch, p. 678.
4. Heiberg, I, 192.
5. *Ibid.*, II, 2.
6. *Ibid.*, I, 192.
7. *Ibid.*, Preface to Bk. IV, II, 2, 4.
8. Pappus, *Collectio* VII, Hultsch, p. 990 ff.
9. Heiberg, I, 2.
10. *De quadratura parabolae* III, Heiberg, II², 268; *cf. De conoidibus et sphaeroidibus* III, Heiberg, I², 270.
11. E.g., *De conoidibus et sphaeroidibus* III, Heiberg, I², 272.
12. For the ellipse, see, e.g. *De conoidibus et sphaeroidibus* VIII, Heiberg, I², 294, 22–26; for the hyperbola, *ibid.* XXV, Heiberg, I², 376, 19–23.
13. Heiberg, I, 2, 4.
14. In modern terms, the locus for four lines is the locus of a point whose distances x, y, z, u from four given straight lines, measured along a given axis, satisfy the equation $xz/yu = $ constant. This locus is a conic. (The locus for three lines is just a particular case of the above: for the distances x, y, z from three lines, $xz/y^2 = $ constant. This is, in modern terms, an *anharmonic* ratio: it can be shown that the theorem that this locus is a conic is equivalent to some basic theorems of projective geometry. (See Michel Chasles, *Aperçu historique*, pp. 58, 354 ff.)
15. *Collectio* VII, Hultsch, p. 672.
16. "Apollonius-Studien," p. 219.
17. Heiberg, II, 2.
18. *Ibid.*, I, 6.
19. *Phaenomena*, ed. H. Menge (*Euclidis Opera Omnia* VIII) (Leipzig, 1916), p. 6.
20. Heiberg, II², 266–268.
21. *Ibid.*, I², 270.
22. Dijksterhuis, *Archimedes*, pp. 66, n.1, 106.
23. *De corporibus fluitantibus* II, 4, Heiberg, II², 357.
24. Chester Beatty MS. Ar. 5255, f.4v.
25. *Collectio* VII, Hultsch, pp. 1006–1014.
26. See, e.g., Zeuthen, *Kegelschnitte*, p. 367 ff.
27. For a method of proving the focal property of the parabola exactly parallel to Apollonius' procedure for those of hyperbola and ellipse, see Neugebauer, "Apollonius-Studien," pp. 241–242.
28. Eutocius, *Commentary*, Heiberg, II, 170.
29. Hultsch, p. 640.
30. *Kegelschnitte*, p. 345.

31. See Pappus, ed. Hultsch, pp. 640–642.
32. *Kegelschnitte*, p. 345 ff.
33. Pappus, ed. Hultsch, pp. 642–644.
34. *Kegelschnitte*, p. 196 ff.
35. Hultsch, p. 644.
36. *Principia*, Bk., I, Lemma XVI (Motte-Cajori trans., pp. 72–73).
37. *Kegelschnitte*, p. 381 ff.
38. See T. L. Heath, *A History of Greek Mathematics*, II, 187–189.
39. Hultsch, p. 2 ff.
40. Heiberg, II², 216 ff.
41. Proclus, *Commentary on Euclid*, ed. Friedlein, p. 105.
42. "Euclid," Bk. XIV, ed. Heiberg, V, 2: the problem is solved by the author of this part of the *Elements*, a man named Hypsicles (*fl. ca.* 150 B.C.), but we cannot tell exactly how much he owes to Apollonius.
43. See the commentary on Euclid's *Data* by Marinus (fifth century A.D.), in *Euclidis opera*, ed. Heiberg-Menge, VI, 234.
44. Proclus, *op. cit.*, p. 74.
45. Ed. Junge-Thomson, p. 219.
46. *Archimedis opera*, ed. Heiberg, III², 258.
47. See E. W. Hobson, *Squaring the Circle* (Cambridge, 1913), pp. 26–28.
48. *Mathematici Graeci Minores*, ed. Heiberg, p. 88.
49. *Refutation of all Heresies* IV, 8, ed. Wendland, III, 41.
50. XII, 1, ed. Heiberg, II, 450 ff.
51. *Ibid.*, pp. 451, 22.
52. *Ibid.*, pp. 456, 9.
53. *Refutation*, ed. Wendland, pp. 41–42.
54. *Anthologiae* XI 11, ed. Kroll, 354.
55. *Quaestiones naturales* VII, 4, 1, ed. Oltramare, II, 304.
56. Ed. Heiberg-Wiedemann, in *Bibliotheca mathematica*, **10** (1910), 201–237.
57. 1661 is the date of the publication at Florence of Abraham Ecchellensis' unsatisfactory version. Some knowledge of it had trickled out before, for Mersenne mentions some of the propositions in a book published in 1644 (see Introduction, xlvi, of ver Eecke's translation of the *Conics*).

BIBLIOGRAPHY

Ancient sources for Apollonius' life include the Prefaces to Books I, II, IV, V, VI, and VII of the *Conics* (in editions of Heiberg and Halley); Eutocius, *Commentary on Apollonius* I (in Heiberg, II, 168, 170); Pappus, *Collectio* VII (Hultsch, p. 678); Photius, *Bibliotheca*, ed. Bekker (Berlin, 1824–1825), p. 151b18. The fragmentary papyrus containing the life of Philonides is edited by Wilhelm Crönert in "Der Epikureer Philonides," in *Sitzungsberichte der Königlich Preussischen Akademie der Wissenschaften zu Berlin*, Jahrgang 1900.2, pp. 942–959. Crönert there points out the importance of this text for dating Apollonius. See further R. Philippson, article "Philonides 5," in *Real-Encyclopädie*, XX.1 (Stuttgart, 1941), cols. 63 ff. A convenient summary of the evidence is given by George Huxley in "Friends and Contemporaries of Apollonius of Perge," in *Greek, Roman and Byzantine Studies*, **4** (1963), 100–103.

A critical text of books I–IV of the *Conics* (with Latin translation) and Eutocius' commentary was published by J. L. Heiberg, *Apollonii Pergaei quae Graece exstant cum commentariis antiquis*, 2 vols. (Leipzig, 1891–1893). Of the Arabic version, only part of Book V has been published, with German translation, by L. Nix, *Das Fünfte Buch der Conica des Apollonius von Perga in der Arabischen Uebersetzung des Thabit ibn Corrah* (Leipzig, 1889). For the rest of Books V–VII the basis is still Edmund Halley's Latin translation from the Arabic in the first edition of the Greek text (Oxford, 1710). The most influential translation was Commandino's Latin version of the first four books (Bologna, 1566). For other editions and early versions and a history of the text, see Heiberg, II, lvii ff. The best modern translation is the French version of all seven books (from the Greek for I–IV and from Halley's Latin for V–VII) by Paul ver Eecke, *Les coniques d'Apollonius de Perge* (Bruges, 1923; reprinted Paris, 1963); the introduction gives a good survey of the work of Apollonius. T. L. Heath's *Apollonius of Perga* (Cambridge, 1896; reprinted 1961) is a free adaptation of the *Conics* rather than a translation. The fundamental modern work on Apollonius (and the ancient theory of conics in general) is H. G. Zeuthen, *Die Lehre von den Kegelschnitten im Altertum* (Copenhagen, 1886; reprinted Hildesheim, 1966), originally published in Danish. It is indispensable for anyone who wishes to make a serious effort to understand the methods underlying the *Conics.* The Introduction of Heath's *Apollonius* is valuable for those who cannot read Zeuthen. A useful summary of the contents of the *Conics* is provided by T. L. Heath, *A History of Greek Mathematics* (Oxford, 1921), II, 126–175. O. Neugebauer's "Apollonius-Studien," in *Quellen und Studien zur Geschichte der Mathematik,* Abteilung B: Studien Band 2 (1933), pp. 215–253, a subtle analysis of some parts of the *Conics,* attempts to trace certain "algebraic" procedures of Apollonius.

On the theory of conic sections before Apollonius, Zeuthen is again the best guide. On Archimedes in particular, J. L. Heiberg, "Die Kenntnisse des Archimedes über die Kegelschnitte," in *Zeitschrift für Mathematik und Physik,* 25 (1880), Hist.-lit. Abt., 41–67, is a careful collection of the relevant passages. In English, an account of pre-Apollonian conic theory is provided by Heath, *A History of Greek Mathematics,* II, 110–126; and E. J. Dijksterhuis, *Archimedes* (Copenhagen, 1956), ch. 3, gives an illuminating comparison between the Apollonian and Archimedean approaches. Another relevant work is Diocles' "On Burning Mirrors," which is extant only in Arabic translation. The sole known manuscript is Chester Beatty Arabic no. 5255, ff. 1–26, in the Chester Beatty Library, Dublin. An edition is being prepared by G. J. Toomer.

The Arabic text of *Cutting off of a Ratio* has never been printed. Halley printed a Latin version, together with a restoration of *Cutting off of an Area,* in *Apollonii Pergaei De sectione rationis libri duo* (Oxford, 1706); see also W. A. Diesterweg, *Die Bücher des Apollonius von Perga De Sectione Rationis* (Berlin, 1824), adapted from Halley's Latin.

Ancient texts giving information on lost mathematical works of Apollonius are the commentary of Proclus (fifth century A.D.) on Euclid Book I, edited by G. Friedlein, *Procli Diadochi in primum Euclidis Elementorum librum* (Leipzig, 1873); and *The Commentary of Pappus on Book X of Euclid's Elements,* ed. G. Junge and W. Thomson (Cambridge, Mass., 1930); but the most important is in Book VII of Pappus' *Collectio,* ed. Fr. Hultsch, *Pappi Alexandrini Collectionis quae supersunt,* 3 vols. (Berlin,

1876–1878). There is a good French translation of this work by P. ver Eecke, 2 vols. (Paris-Bruges, 1933). In modern times many attempts have been made at restoration of lost works of Apollonius on the basis of Pappus' account. Here we mention only the following: for the *Determinate Section,* Willebrordus Snellius, *Apollonius Batavus* (Leyden, 1608), and Robert Simson, in *Opera quaedam reliqua* (Glasgow, 1776); for the *Tangencies*—apart from Vieta's *Apollonius Gallus* (Paris, 1600)—J. Lawson, *The Two Books of Apollonius Pergaeus Concerning Tangencies* (Cambridge, 1764); for the *Inclinations,* Samuel Horsley, *Apollonii Pergaei inclinationum libri duo* (Oxford, 1770); for the *Plane Loci,* Pierre de Fermat, *Oeuvres,* P. Tannery and C. H. Henry, eds., I (Paris, 1891), 3–51, and Robert Simson, *Apollonii Pergaei locorum planorum libri II restituti* (Glasgow, 1749).

For other restorations of all the above see the Introduction to ver Eecke's translation of the *Conics,* pp. xxii–xxxiv. A good account of the probable contents of all six works is given by Heath, *A History of Greek Mathematics,* II, 175 ff. This is heavily dependent on Zeuthen's *Kegelschnitte;* the Index to the 1966 reprint of the latter is the most convenient guide to Zeuthen's scattered treatment of these lost works. F. Woepcke, "Essai d'une restitution de travaux perdus d'Apollonius sur les quantités irrationelles," in *Mémoires présentées à l'Académie des Sciences,* 14 (Paris, 1856), 658–720, is devoted to the work on unordered irrationals; see also T. L. Heath, *The Thirteen Books of Euclid's Elements Translated,* III (2nd ed., Cambridge, 1925), 255–259.

Ancient texts relevant to Apollonius' astronomical works are "Hippolytus," *Refutatio omnium haeresium,* ed. P. Wendland, Hippolytus Werke III (Leipzig, 1916), IV 8–10; Vettius Valens, *Anthologiarum libri,* ed. W. Kroll (Berlin, 1908), IX, 11; Seneca, *Quaestiones naturales,* ed. P. Oltramare, 2 vols. (Paris, 1961), VII, 4, 1; and especially Ptolemy, *Almagest* XII, 1, ed. J. L. Heiberg, in *Claudii Ptolemaei syntaxis mathematica,* 2 vols. (Leipzig, 1898–1903).

For Apollonius' astronomical work, see O. Neugebauer, "Apollonius' Planetary Theory," in *Communications on Pure and Applied Mathematics,* 8 (1955), 641–648, and "The Equivalence of Eccentric and Epicyclic Motion According to Apollonius," in *Scripta mathematica,* 24 (1959), 5–21.

No detailed account of the influence of Apollonius on later mathematics exists. Much interesting information can be found in ver Eecke's introduction to his translation. The best guide is Michel Chasles, *Aperçu historique sur l'origine et le développement des méthodes en géométrie* (Paris, 1837; reprinted 1875), a work which is also remarkable for its treatment of Apollonius in the light of nineteenth-century synthetic geometry.

G. J. TOOMER

APPELL, PAUL (-ÉMILE) (*b.* Strasbourg, France, 27 September 1855; *d.* Paris, France, 24 October 1930), *mathematics, mathematical physics.*

Appell's parents, Jean-Pierre Appell and Elizabeth

Müller, were Catholic Alsatians ardently loyal to revolutionary France. The family lived in a corner of the great Ritterhus, formerly a knightly lodge, where the master-dyer father and two sons by a previous marriage managed production while the mother, her sister, and a stepdaughter tended the store. Paul accepted the family ambition and patriotism but rejected Catholic piety. His character was forged by a forced move from the Ritterhus in 1866, his father's death in 1867, transfer from a religious school to the *lycée* at his own insistence in 1869, bitter experiences in the siege of Strasbourg in 1870, and a close relationship with the younger of his half brothers, Charles, who served in the Foreign Legion, fought as an irregular in 1870–1871, and in 1889 was sentenced to ten years' confinement for anti-German activities. When Appell went to Nancy in 1871 to prepare for the university and to assume French citizenship in 1872, he was carrying the hopes of his family, who remained behind in Strasbourg as German subjects.

Blessed with unbounded energy, this attractive outsider with an accent moved rapidly toward the inner circles of French mathematics. At Nancy, he and Henri Poincaré formed a friendship that lasted until the latter's death. In 1873 he entered the École Normale, from which he graduated first in the class of 1876, three months after earning his doctorate. From this time on, Appell maintained an amazing level of activity in teaching, research, editing, and public service. He typically held several teaching posts at the same time, including the chair of mechanics at the Sorbonne from 1885. He was elected to the Académie des Sciences in 1892. He served as dean of the Faculty of Science of the University of Paris from 1903 to 1920 and as rector from 1920 to 1925. In various government posts, including membership in the Conseil Supérieure d'Instruction Publique, he was an exponent of educational reform and initiator of numerous large-scale projects, including the Cité Universitaire.

In 1881 he married Amelie, daughter of the archaeologist Alexandre Bertrand, niece of the mathematicians Joseph Bertrand and Charles Hermité, and a cousin of Appell's classmate and friend Émile Picard. Their son became a deputy and undersecretary of state. Two of their three daughters married the academicians Émile Borel and J. E. Duclaux. The household included Paul's mother, who had joined him in 1877 and remained until her death in 1902. In his *Souvenirs* (p. 180) he described his life as "flowing tranquilly between teaching, mathematical work and vacations in Alsace" at the maternal home

in Klingenthal, but he found energy to support vigorously the movement for women's rights, to carry from Alsace his brother's reports destined for the French War Office, and to defend his fellow Alsatian Dreyfus and serve on an expert commission whose ruling played a key role in his final rehabilitation. During World War I he founded and led the Secours National, a semiofficial organization uniting all religious and political groups to aid civilian victims. He described the return of the tricolor to Alsace as the fulfillment of his "lifelong goal" and felt that Germany had been treated too easily. He served as secretary-general of the French Association for the League of Nations.

Appell's first paper (1876) was his thesis on projective geometry in the tradition of Chasles, but at the suggestion of his teachers he turned to algebraic functions, differential equations, and complex analysis. He generalized many classical results (e.g., the theories of elliptic and of hypergeometric functions) to the case of two or more variables. From the first his work was close to physical ideas. For example, in 1878 he noted the physical significance of the imaginary period of elliptic functions in the solution of the pendulum problem, and thus showed that double periodicity follows from physical considerations. In 1880 he wrote on a sequence of functions (now called the Appell polynomials) satisfying the condition that the derivative of the nth function is n times the previous one.

In 1885 Appell was awarded half the Bordin Prize for solving the problem of "cutting and filling" (*deblais et remblais*) originally posed by Monge: To move a given region into another of equal volume so as to minimize the integral of the element of volume times the distance between its old and new positions. In 1889 he won second place (after Poincaré) in a competition sponsored by King Oscar II of Sweden: To find an effective method of calculating the Fourier coefficients in the expansion of quadruply periodic functions of two complex variables.

The flow of papers continued, augmented by treatises, textbooks, and popularizations and seemingly unaffected by other responsibilities. Although Appell never lost his interest in "pure" analysis and geometry, his activity continued to shift toward mechanics, and in 1893 Volume I of the monumental *Traité de mécanique rationnelle* appeared. Volume V (1921) included the mathematics required for relativity, but the treatise is essentially an exposition of classical mechanics of the late nineteenth century. It contains many of Appell's contributions, including his equations of motion valid for both holonomic and nonholonomic systems, which have not displaced the classical Lagrangian system in spite of undoubted advantages.

It is difficult to do justice to Appell's work because it lacks central themes, seminal ideas, and dramatic results. In 1925 he wrote: "I always had little taste for developing general theories and preferred to study limited and precise questions that might open new paths" ("Notice," p. 162). Indeed, his scientific work consists of a series of brilliant solutions of particular problems, some of the greatest difficulty. He was a technician who used the classical methods of his time to answer open questions, work out details, and make natural extensions in the mainstream of the late nineteenth century; but his work did not open new doors, as he hoped. On the contrary, he does not seem to have looked down any of the new paths that were leading to a period of unbridled abstraction and generalization. During the last half of his career he was a pillar of a backward-looking establishment that was to give way to Nicolas Bourbaki, a namesake of a general who was one of his boyhood heroes.

BIBLIOGRAPHY

I. ORIGINAL WORKS. Appell's "Notice sur les travaux scientifiques," in *Acta mathematica,* **45** (1925), 161–285, describes 140 publications in analysis, 30 in geometry, and 87 in mechanics. The most notable are *Notice sur les travaux* (Paris, 1884, 1889, 1892), written to support his candidacy for the Académie; "Sur les intégrales des fonctions à multiplicateurs," in *Acta mathematica,* **13** (1890); *Traité de mécanique rationnelle,* 5 vols. (Paris, 1893–1921 and later eds.); *Théorie des fonctions algébriques et de leurs intégrales* (Paris, 1895, 1922), written with E. Goursat; and *Principes de la théorie des fonctions elliptiques et applications* (Paris, 1897), written with E. Lacour.

Not listed in the "Notice" are numerous elementary textbooks, popularizations, addresses and papers on history and education, and several later publications, including *Sur une forme générale des équations de la dynamique,* Mémorial des Sciences Mathématiques (Paris, 1925); *Sur les fonctions hypergéométriques de plusieurs variables, les polynomes d'Hermité et autres fonctions sphériques dans l'hyperspace, ibid.* (Paris, 1925); *Henri Poincaré* (Paris, 1925); *Fonctions hypergéométriques et hypersphériques. Polynomes d'Hermité* (Paris, 1926), written with M. J. Kampé de Feriet; *Le problème géométrique des deblais et remblais,* Mémorial des Sciences Mathématiques (Paris, 1928); *Sur la décomposition d'une fonction en éléments simples, ibid.* (Paris, 1929); and "Sur la constante d'Euler," in *Enseignement mathématique,* **29** (1930), 5–6, apparently his last paper, a follow-up to one on the same subject, *ibid.,* **26** (1927), 11–14, which had been welcomed by the editor with a note observing that "a great source of light is still burning."

II. SECONDARY LITERATURE. Appell's life and work are unusually well documented by his four *notices* mentioned above; his charming and revealing autobiography, *Souvenirs d'un alsacien 1858–1922* (Paris, 1923); and E. Lebon, *Biographie et bibliographie analytique des écrits de Paul Appell* (Paris, 1910), which gives many biographical details and seems to have been written with Appell's collaboration. Other biographical articles rely on these sources, but some of them contain personal recollections or other interesting information, notably *Cinquantenaire scientifique de Paul Appell* (Paris, 1927); "Centenaire de la naissance de Paul Appell," in *Annales de l'Université de Paris,* **26,** no. 1 (1956), 13–31; A. Buhl, in *Enseignement mathématique,* **26** (1927), 5–11; **30** (1931), 5–21; **33** (1934), 229–231; T. Levi-Civita, in *Rendiconti Accademia dei Lincei,* 6th ser., **13** (1931), 241–242; and Raymond Poincaré, in *Annales de l'Université de Paris,* **5** (1930), 463–477.

KENNETH O. MAY

APPLETON, EDWARD VICTOR (*b.* Bradford, Yorkshire, England, 6 September 1892; *d.* Edinburgh, Scotland, 21 April 1965), *radio physics.*

Appleton showed exceptional promise as a boy, matriculating at the University of London at sixteen and winning a scholarship to St. John's College, Cambridge, at eighteen. He graduated with first class honours in physics in 1913 and started postgraduate work in crystallography under William Henry Bragg. Shortly after the outbreak of World War I he became a signals officer in the Royal Engineers, an assignment that aroused his interest in radio.

Upon his return to Cambridge after the war, Appleton first investigated vacuum tubes at the Cavendish Laboratory with Balthazar van der Pol, Jr., and later wrote a monograph on the subject, *Thermionic Vacuum Tubes* (1932). He next turned to the study of the fading of radio signals. In 1924, when he was only thirty-two, he became Wheatstone professor of physics at King's College of the University of London, where he remained for twelve years. During the first year, he and Miles Barnett, a graduate student from New Zealand, performed a crucial experiment that led to a measurement of the height of the reflecting atmospheric layer of ionized gases, which had been postulated by Oliver Heaviside and A. E. Kennelly in explanation of the first transatlantic radio transmission by Guglielmo Marconi in 1901. In the experiment, the frequency of the new British Broadcasting Company transmitter at Bournemouth was periodically varied after broadcasting hours at a constant rate, so that the interference between the direct (ground) and reflected (sky) waves resulted in a regular fading in and out at a site about 100 kilometers away, in a manner analogous to the behavior of optical interference fringes.

Appleton's experimental proof that the Kennelly-Heaviside, or E, layer really existed, a scientific accomplishment of the highest order, was honored by his election as fellow of the Royal Society (1927), a knighthood (1941), and the Nobel Prize in physics (1947), "for his investigation of the physics of the upper atmosphere, especially for the discovery of the so-called Appleton layers." The last refers to the later discoveries of a second (F) layer at more than twice the height of the E layer and a third (D) layer below the E layer. Besides discovering these layers, Appleton and his co-workers showed that the sky wave generally was elliptically polarized, and calculated the reflection coefficients and electron densities of the layers and their diurnal and seasonal variations. His work may also be considered to be of prime technological significance, not only in regard to radio transmission but also as a milestone in the development of radar; the determination of the height of the E layer was the first distance measurement made by radio, a technique that was closely followed by Robert Alexander Watson-Watt, the British radar pioneer, who had collaborated with Appleton in atmospheric research and had many subsequent professional contacts with him. The rest of Appleton's life was spent in research flowing from his own discoveries, an endeavor in which he continued to maintain a degree of involvement that was astonishing in view of the many other duties thrust upon him.

Following a three-year tenure as Jacksonian professor of natural philosophy at the University of Cambridge, Appleton was appointed secretary of the government's Department of Scientific and Industrial Research. In 1949 he was made principal and vice-chancellor of the University of Edinburgh, where he remained until his death. Appleton was a great international figure: his first paper on the ionosphere was published in a Dutch journal (in Dutch); he served as vice-president of the (U.S.) Institute of Radio Engineers in 1932; he was president of the International Scientific Radio Union (URSI) from 1934 to 1952; and he was instrumental in organizing the first International Geophysical Year in 1957, a year of maximum sunspot activity. After moving to Edinburgh, he founded the *Journal of Atmospheric Research* (affectionately known as "Appleton's Journal") and served as its editor-in-chief for the rest of his life.

Appleton married Jessie Longson in 1915; they had two daughters, Marjery and Rosalind. A month before his death, Appleton, a widower since 1964, married Mrs. Helen F. Allison, who had been his private secretary for thirteen years.

BIBLIOGRAPHY

A list of Appleton's honors, decorations, medals, and the most important papers he wrote and collaborated in (a total of 140) appears in the obituary by his long-time associate, J. A. Ratcliffe, in *Biographical Memoirs of Fellows of the Royal Society*, **12** (1966), 1–21. See also obituaries in *The Times* (London), 23 April 1965; and in *Science and Culture*, **31** (1965), 348–350.

CHARLES SÜSSKIND

AQUINAS, SAINT THOMAS (*b.* Roccasecca, near Monte Cassino, Italy, *ca.* 1225; *d.* Fossanuova, near Maenza, Italy, 7 March 1274), *not a scientist in the modern sense, but a philosopher and theologian whose synthesis of Christian revelation with Aristotelian science has influenced all areas of knowledge—including modern science, especially in its early development.*

Thomas, the youngest of nine children, was born in the castle of the Aquino family. His father, Landolfo, and his older brothers served the Holy Roman Emperor, Frederick II, then warring against the papacy; his mother, Teodora of Chieti, was a Lombard. The family's political situation was precarious, and in 1231 Thomas was placed in the abbey of Monte Cassino for his elementary education. When the abbey was occupied by Frederick's troops in 1239, Thomas was sent to finish his studies at the recently founded University of Naples; his teachers there were Master Martin in grammar and logic and Peter of Ireland in natural science.

Thomas entered the Dominican order at Naples in 1244, against his family's wishes, and was sent to Paris, and then to Cologne, for further studies (1245–1252). The Dominicans at the time were in the forefront of intellectual life; in natural science, groups of friars were synthesizing the heritage of Greece and Rome, which soon appeared in the encyclopedias of Thomas of Cantimpré and Vincent of Beauvais. Albertus Magnus, earlier recruited into the order by Jordan of Saxony, was himself paraphrasing in Latin all of the works of Aristotle that had just been brought to the West, thus rendering them intelligible to the younger friars. Studying under Albert, possibly at Paris and certainly at Cologne (1248–1252), Thomas was soon abreast of the most advanced scholarship of his time, including the major Greek, Arab, and Latin sources that were to revivify the intellectual life of the Middle Ages.

Sent to Paris "to read the *Sentences*" at the priory of Saint-Jacques in 1252, Thomas quickly demonstrated his proficiency as a theologian. There was, however, growing jealousy and antipathy toward the friars (both Dominican and Franciscan) among the secular masters at the University of Paris, and in 1256 the intervention of Pope Alexander IV was required before Thomas and the Franciscan Bonaventure were accepted as masters at the university. During or before this, his first Paris professorship (1256–1259), Thomas composed his commentary on the *Sentences,* some smaller treatises—including the highly original *De ente et essentia* ("On Being and Essence")—and the disputed question *On Truth;* he also began work on the *Summa contra gentiles,* of special importance for its evaluation of Arab thought.

From 1259 to 1268 he was back in Italy, first at Anagni and at Orvieto, where he was associated with the papal courts of Alexander IV and Urban IV, respectively; then at Rome (1265–1267), where he taught at the Dominican priory of Santa Sabina and began his famous *Summa theologiae*; and finally at Viterbo, where he served at the court of Clement IV.

Then, in 1268 or 1269, possibly because of disputes at the University of Paris over the Aristotelianism which he and Albert had introduced, Thomas returned to Paris for a somewhat unusual second professorship (1269–1272). Here he combated both the traditional Augustinian orthodoxy being fostered by such Franciscans as Bonaventure and John Peckham and the heterodox Aristotelianism of Siger of Brabant and his associates, who are usually referred to as Latin Averroists. One of the key issues in the dispute was Thomas' teaching that the world's creation in time cannot be demonstrated by reason alone, since there is no philosophical repugnance in a created universe's having existed from eternity—a thesis with important ramifications for later medieval concepts of infinity.

The condemnation, in 1270, of certain Averroist theses by Étienne Tempier, bishop of Paris, is regarded by some scholars as directed, at least implicitly, against Aquinas' teaching. Of the later condemnation, in 1277, there can be no doubt that two propositions concern matters taught by Thomas, including his thesis on the unicity of the substantial form in man, which bears on the problem of the presence of elements in compounds. Such controversies drew a series of polemical treatises from Aquinas' pen, including *De aeternitate mundi contra murmurantes* ("On the Eternity of the Universe, Against the 'Murmurers' " [i.e., the traditionalist Augustinians]) and *De unitate intellectus contra Averroistas* ("On the Unity of the Intellect, Against the Averroists"). The intellectual ferment also stimulated him to further efforts at philosophical and theological synthesis. During these years he elaborated most of his detailed commentaries on Aristotle and worked steadily on the *Summa theologiae.*

After his second Paris professorship was concluded in 1272, Thomas returned to Italy, this time to Naples,

to erect a Dominican *studium* near the university there. He lectured, directed disputations, and continued writing; but the pace of his work slowed noticeably, partly because of failing health. He suspended all writing activity late in 1273 and died a few months later, while en route to the second Council of Lyons. He was canonized on 18 July 1323 and subsequently was approved by the Roman Catholic Church as its most representative teacher.

Today the name of Thomas is so associated with Catholic orthodoxy that one tends to forget that he was an innovator. In an atmosphere dominated by faith, especially at the University of Paris, he took the leadership in championing the cause of reason. Almost single-handedly he turned the theologians of that university to a study of the pagan Aristotle, to the use of what was then a rigorous scientific method, learned from investigating the world of nature, for probing the mysteries of revelation. Opposing the popular teaching that all knowledge comes by divine illumination, he allowed that man, by sense observation and through the use of unaided reason, could arrive at truth and certitude.

It would be a mistake, of course, to urge that Aquinas' main concern was with the physical universe. Rather, he was preoccupied with questions about God, the angels, and man; first and foremost he was a metaphysician and a theologian. Yet there can be no doubt that, like Aristotle, his approach to metaphysical problems was through the physical sciences. Like St. Paul, he firmly believed that the invisible things of God are seen through his visible creation, provided it is rightly understood (Romans 1:20). So convinced of this was he that in his later life he turned from his unfinished *Summa theologiae* to comment on all the physical works of Aristotle. He probably completed his exposition of *De caelo et mundo,* one of his best works as a commentator, at Naples (1272–1273) and ceased commenting on *De generatione et corruptione* and the *Meteorologica* only shortly before his death.

Furthermore, for a man not usually recognized as a scientist, he made noteworthy contributions to medieval science. These can best be indicated by summarizing his more significant teachings relating to the medieval counterparts of physics, astronomy, chemistry, and the life sciences.

In the high scholastic period, foundations were laid for later medieval discussions that adumbrated the distinction in modern mechanics between kinematics and dynamics. The kinematical content of Thomas' teaching is meager, although he did hold that velocity is a mode of continuous quantity and thus is capable of intensification in the same manner as qualities, thereby allowing for the type of comparison between qualitative change and local motion later made by Nicole Oresme.

In dynamics, he inaugurated some new directions in the study of causality affecting gravitational and projectile motions. Aquinas would probably look askance at the tendency of present-day historians of science to identify Aristotle's motive powers and resistances with forces and to represent Aristotle's teaching with precise dynamical equations. His own exegesis of the relevant Aristotelian texts, as opposed to that of Averroës and Avempace, discounts any demonstrative intent on Aristotle's part and interprets his statements as dialectical efforts to confute his atomist opponents.

Thus, on the disputed question whether motion through a vacuum would take place instantaneously, Thomas did not follow Aristotle literally. He insisted that if, by an impossibility, a vacuum were to exist, motion through it would still take time—that the temporal character of the motion does not arise uniquely from external resistance but, rather, from the proportion of the mover to the moved (which prevents the movements of the heavens from being instantaneous, although they are not impeded by resistance) and also from the continuity of the distance being traversed. The latter reason, particularly, provoked speculation among fourteenth-century thinkers; some, such as Oresme, saw no necessary connection between spatial continuity and velocity limitation and were led on this account to seek some type of resistance internal to the moving body—thereby foreshadowing the modern concept of inertia.

Thomas' analysis of gravitation is basically Aristotelian, yet it differs in significant respects from that of other commentators. Like all medieval thinkers, he regarded gravitation as the natural motion of a heavy body to its proper place. For Thomas, however, nature was a relational concept, and thus he disagreed with those who defined it as a *vis insita* or as something absolute; it is a principle of motion, either actively or passively, depending on the particular motion that results.

Aquinas held that the body's gravity is the proximate cause of its falling, but only in the manner of a "passive principle." He rejected Averroës' teaching that the medium through which the body falls plays an essential role in its motion and that there is an active source of such motion within the body, whether this be its gravity or its substantial form. In this respect, Aquinas was also at variance with the later teaching of Walter Burley and the Paris terminists, all of whom saw the cause of falling as some type of active force within the body itself, thereby fore-

shadowing animist theories of gravitation such as subsequently proposed by William Gilbert. Again, Thomas disagreed with Bonaventure and Roger Bacon, who regarded place as exerting some type of repulsive or attractive influence on the falling body; for Thomas, there was no repulsion involved, and the attractive aspect of place was sufficiently accounted for by its being the end, or final cause, of the body's movement. Here Thomas implicitly rejected the absolute space and attractive forces later proposed by the Newtonians; his own analysis, it has been remarked, shows more affinity with the ideas behind Einstein's theory of general relativity, although the two are so remote in thought context as to defy any attempt at detailed comparison.

On the subject of *impetus,* authors are divided as to Thomas' teaching. Certainly he has no treatment of the concept to match that found later in Franciscus de Marcia, Jean Buridan, and Oresme, nor does he use it to explain any details of projectile or gravitational motion. In his later writings, particularly the commentaries on the *Physics* and the *De caelo,* Thomas clearly defends the original Aristotelian teaching on the proximate cause of projectile motion. In some earlier writings, on the other hand, he speaks of a *virtus* in the projectile, and, in one text of the *Physics* commentary, discussing the case of a ball that bounces back from a wall, he mentions that the *impetus* is given not by the wall but by the thrower. Later Thomists, such as Joannes Capreolus and Domingo de Soto, had no difficulty in assimilating a fully developed *impetus* theory to Thomas' teaching, evidently regarding the Aristotelian element in his expositions as reflecting his role as a commentator more than his personal views.

Aquinas took up the problems of the magnet, of tidal variations, and of other "occult" phenomena in a letter entitled *De occultis operationibus naturae* ("On the Occult Workings of Nature"), whose very title shows his preoccupation with reducing all of these phenomena to natural, as opposed to supramundane, causes. Significantly, Thomas' analysis of magnetism was known to Gilbert and was praised by him.

Commenting as he did on the *De caelo* and also, in his theological writings, on the cosmogony detailed in Genesis, Thomas could not help but evaluate the astronomical theories of his contemporaries. He contributed nothing new by way of observational data, nor did he evolve any new theories of the heavens, but his work has an importance nonetheless, if only to show the care with which he assessed the current state of astronomical science. His view of the structure of the universe was basically Aristotelian; he knew of two theories to account for the phenomena of the heavens, both geocentric in the broad sense: that of Eudoxus, Callippus, and Aristotle, and that of Ptolemy. Aquinas generally employed the Eudoxian terminology; he mentions the Ptolemaic system at least eleven times, and five of these are in his late commentary on the *De caelo*. In most of his references to the Eudoxian or Ptolemaic systems, he refrains from expressing any preference; clearly, he was aware of the hypothetical character of both. At least once, commenting on Ptolemy's cumbersome theory of eccentrics and epicycles, he voices the expectation that this theory will one day be superseded by a simpler explanation.

The astronomical data reported by Aquinas, according to an extensive analysis by Thomas Litt, were those of a well-informed thirteenth-century writer; he errs in one or two particulars, but on matters of little theoretical consequence. His treatise on comets, included in a work by Lynn Thorndike, is one of the most balanced in the high Middle Ages, rejecting fanciful explanations and pointing out how little is actually known about these occurrences.

In his more philosophical views, however, Thomas was not so fortunate. He believed in the existence of spheres that transport the heavenly bodies and, with his contemporaries, regarded such bodies as incorruptible. He was convinced also of the existence of an empyrean heaven, the dwelling place of the blessed and known only through revelation, but nonetheless included in the corporeal universe. He accorded an extensive causality to the heavenly bodies, while excepting from this all actions that are properly human (i.e., that arise from man's intelligence and deliberate will) and completely fortuitous events, so as to discourage any naïve credence in the astrologers of his day.

Aquinas has no treatment of alchemy to match that of his teacher Albertus Magnus, but he does discuss one topic that had important bearing on later views of the structure of matter, that of the presence of elements in compounds. Earlier thinkers, attempting to puzzle out Aristotle's cryptic texts, favored one of two explanations of elemental presence offered by Avicenna and Averroës, respectively. Dissatisfied with both, Aquinas formulated a third position, which soon became the most popular among the Schoolmen. He taught that the elements do not remain actually in the compound, but that their qualities give rise to "intermediate qualities" that participate somewhat in each extreme; these intermediate qualities are in turn proper dispositions for a new substantial form, that of the compound, which is generated through the alteration that takes place. Since the elemental qualities remain "in some way" in the compound, one can

say also that the substantial forms of the elements are present there, too—not actually, but virtually.

The subsequent influence of Thomas' teaching has been traced in considerable detail by Anneliese Maier, who characterizes it as inaugurating a modern direction that dominated treatments of the problem in later Scholasticism (*Studien* III, pp. 89–140). Duns Scotus and his school, particularly, became enamored of the theory and attempted a consistent development of its ramifications. Nominalists such as William of Ockham and Gregory of Rimini took it up, too, as did such Paris terminists as Buridan, Oresme, Albert of Saxony, and Marsilius of Inghen. The basic explanation continued to be taught through the sixteenth century and, coupled with Aristotelian teaching on *minima naturalia,* became the major alternative to a simplistic atomist view of the structure of matter before the advent of modern chemistry.

In biology and psychology, Aquinas followed Aristotle, Galen, and the medieval Arab tradition; his work is noteworthy more for its philosophical consistency than for its scientific detail. He wrote commentaries on Aristotle's *De anima, De sensu et sensato,* and *De memoria et reminiscentia,* all based on the texts of William of Moerbeke, his fellow Dominican. Also, *ca.* 1270, he composed a letter to a Master Philippus, who seems to have been a physician and professor at Bologna and Naples, on the motion of the heart (*De motu cordis*), explaining how the principle "Whatever is moved, is moved by another" is saved in this phenomenon. Like his contemporaries, he believed in spontaneous generation and countenanced a qualified type of evolution in the initial formation of creatures. Catholic thinkers, on the basis of his philosophy, have been more open to evolutionary theories than have fundamentalists, who follow a strict, literal interpretation of the text of Genesis.

Thomas was a mild man, objective and impersonal in his writing, more cautious than most in giving credence to reported facts. He showed neither the irascible temperament of Roger Bacon, nor the subtle questioning of Duns Scotus, nor the pious mysticism of Bonaventure. Calm and methodical in his approach, proceeding logically, step by step, he offered proof where it could be adduced, appealing to experience, observation, analysis, and (last of all) authority. He appreciated the importance of textual criticism, and possibly was one of the instigators of Moerbeke's many Latin translations of scientific treatises from the original Greek. He had a penetrating intellect and a strong religious faith, both of which led him to seek a complete integration of all knowledge, divine as well as human. Working with the science of his time, he succeeded admirably in this attempt, thus providing a striking example for all who were to be similarly motivated in the ages to come.

BIBLIOGRAPHY

I. ORIGINAL WORKS. Standard editions of Aquinas are the Leonine edition, *S. Thomae Aquinatis opera omnia, iussu Leonis XIII edita* (Rome, 1882–   ), a critical edition still in process (American section of the Leonine Commission at Yale University, New Haven, Conn.); Parma edition, *S. Thomae opera omnia,* 25 vols. (Parma, 1852–1873; photographically reproduced, New York, 1948–1949); Vivès edition, *D. Thomae Aquinatis opera omnia,* S. E. Fretté and P. Maré, eds., 32 vols. (Paris, 1871–1880).

English translations of major works include *Summa theologiae,* trans. and commentary, T. Gilby et al., eds., 60 vols. planned (New York-London, 1964–   ); *Summa theologiae,* English Dominicans trans., 22 vols., 2nd ed. (New York-London, 1912–1936); *Summa contra gentiles* (*On the Truth of the Catholic Faith*), A. C. Pegis et al., trans., 5 vols. (New York, 1955–1956); *De veritate* (*Truth*), R. W. Mulligan et al., trans., 3 vols. (Chicago, 1952–1954).

English translations of scientific writings include *Commentary on Aristotle's Physics,* R. J. Blackwell et al., trans. (New Haven, Conn., 1963); *Exposition of Aristotle's Treatise On the Heavens,* R. F. Larcher and P. H. Conway, trans. (Columbus, Ohio, 1963–1964), mimeographed, available from College of St. Mary of the Springs, Columbus; *Exposition of Aristotle's Treatise On Generation and Corruption,* Bk. I, chs. 1–5, R. F. Larcher and P. H. Conway, trans. (Columbus, Ohio, 1964)—as above; *Exposition of Aristotle's Treatise On Meteorology,* Bks. I–II, chs. 1–5, R. F. Larcher and P. H. Conway, trans. (Columbus, Ohio, 1964)—as above. Excerpt on comets in Lynn Thorndike, *Latin Treatises on Comets Between 1238 and 1368 A.D.* (Chicago, 1950), pp. 77–86. *Aristotle's De Anima With the Commentary of St. Thomas Aquinas,* K. Foster and S. Humphries, trans. (New Haven, Conn., 1951); *Exposition of the Posterior Analytics of Aristotle,* P. Conway, trans. (Quebec, 1956)—mimeographed, available from La Librairie Philosophique M. Doyon, Quebec, Canada; *The Letter of St. Thomas Aquinas De Occultis Operibus Naturae,* with a commentary, J. B. McAllister, trans., The Catholic University of America Philosophical Studies, **42** (Washington, D. C., 1939).

II. SECONDARY LITERATURE. General works on Aquinas include V. J. Bourke, *Aquinas's Search for Wisdom* (Milwaukee, 1965), an excellent biography; M. D. Chenu, *Toward Understanding St. Thomas,* A. M. Landry and D. Hughes, trans. (Chicago, 1964), a good introduction to Thomas' intellectual milieu; K. Foster, ed. and trans., *The Life of Saint Thomas Aquinas: Biographical Documents* (Baltimore, 1959); W. A. Wallace and J. A. Weisheipl, "Thomas Aquinas, St.," in *The New Catholic Encyclopedia* (New York, 1967), a compendious survey of his life and works.

Aquinas' scientific work is discussed in Thomas Litt, *Les corps célestes dans l'univers de saint Thomas d'Aquin,* Phi-losophes Médiévaux, VII (Louvain-Paris, 1963), the best on Thomas' astronomy; Anneliese Maier, *Studien zur Naturphilosophie der Spätscholastik,* I, *Die Vorläufer Galileis im 14. Jahrhundert,* Edizioni di storia e letteratura, **22** (Rome, 1949); II, *Zwei Grundprobleme der scholastischen Naturphilosophie,* Edizioni . . ., **37** (Rome, 1951); III, *An der Grenze von Scholastik und Naturwissenschaft,* Edizioni . . ., **41** (Rome, 1952); IV, *Metaphysische Hintergründe der spätscholastischen Naturphilosophie,* Edizioni . . ., **52** (Rome, 1955); V. *Zwischen Philosophie und Mechanik,* Edizioni . . ., **69** (Rome, 1958). These are the most complete sources; consult the index of each volume under "Thomas von Aquin." Some of this material is summarized in English in E. J. Dijksterhuis, *The Mechanization of the World Picture,* C. Dikshoorn, trans. (Oxford, 1961); W. A. Wallace, ed. and trans., *Cosmogony,* Vol. X of Aquinas' *Summa theologiae* (New York-London, 1967), with notes and appendices on Thomas' science and its background.

WILLIAM A. WALLACE, O. P.

ARAGO, DOMINIQUE FRANÇOIS JEAN (*b.* Estagel, France, 26 February 1786; *d.* Paris, France, 2 October 1853), *physics, astronomy.*

Arago was the eldest son of Marie Roig and François Bonaventure Arago, a modest landowner of Catalonian origin who became mayor of Estagel in 1789. The family moved to Perpignan in 1795, when Arago's father was named cashier at the mint. There Arago completed the usual classical education and set his sights on a military career in the artillery. He prepared for admission to the École Polytechnique by mastering the works of Euler, Lagrange, and Laplace, and passed the entrance examination with great distinction in 1803. After two years at the head of his class, he was named secretary of the Bureau des Longitudes and sent to Spain with Biot on a geodetic expedition. After being held prisoner in Spain and Algeria, he returned in June 1809 to France, where he was welcomed into the Société d'Arcueil. He was elected to the Institut de France as an astronomer on 18 September 1809 and in that year also succeeded Monge as professor of descriptive geometry at the École Polytechnique, where he taught a variety of subjects until his resignation in 1830. At the request of the Bureau des Longitudes, which placed the Paris Observatory under his direction, Arago also taught astronomy to the general public at the observatory from 1813 to 1846. He was the main contributor to the *Annuaire du Bureau des Longitudes* for more than forty years and coeditor, with Gay-Lussac, of the *Annales de chimie et physique* from 1816 to 1840. He was a member of most of the important scientific societies, receiving the Royal Society's Copley Medal in 1825 and being elected perpetual secretary of the

Académie des Sciences, replacing Fourier, on 7 June 1830.

Arago was at once volatile and warm-hearted in his personal relations. He either forged strong bonds with fellow scientists or engaged in sharp polemics that often were provoked by priority controversies. Among his closest friends were Alexander von Humboldt, with whom he shared a room in Paris from 1809 to 1811, Gay-Lussac, and Malus; and among his relatives, the physicist Alexis Petit and the astronomer Claude L. Mathieu. He had a stormy relationship with Biot, Thomas Young, and Brewster, but it did not blind him to their scientific merits. In both his writings and his public appearances, Arago conveyed a contagious sense of excitement that won him a large following. His personal style, which spilled over into his work habits, was that of a romantic—restless, inquisitive, volatile, and constantly bubbling with enthusiasm and optimism. Married in 1811, Arago had three sons and lived in an apartment at the observatory. In his later years he gradually lost his eyesight, went blind, and was reduced to dictating to his students.

Arago's most important original work in science was carried out before 1830, for his younger brothers, particularly Étienne, drew him into politics following the July Revolution of 1830. He was repeatedly elected deputy for his native department (Pyrénées-Orientales) and for Paris between 1830 and 1852, and sat on the left in the Chamber of Deputies, delivering influential speeches on educational reform, freedom of the press, and the application of scientific knowledge to technological progress, particularly concerning canals, steam engines, railroads, the electric telegraph, and photography. He also was twice named president of the Paris Municipal Council. The peak of Arago's political career came after the February Revolution of 1848, when he was made a member of the provisional government and was named, successively, minister of the navy and the army and president of the Executive Committee. As minister he signed decrees outlawing corporal punishment and improving the rations of sailors on the high seas, and abolishing slavery in the French colonies. His politics were those of a constitutional liberal, passionately concerned with social reform (he helped found La réforme in 1843), freedom of association, and education of the lower classes. He was, however, violently opposed to mob rule and to the socialistic programs espoused by Louis Blanqui, Alexandre Ledru-Rollin, and Louis Blanc. Arago's effective political career ended following his loss of control over the revolutionaries during the June days of 1848.

Arago's scientific life was dominated by a persistent interest in physical phenomena related to electricity, magnetism, and, above all, to light. His earliest investigations with Biot in 1805 and 1806 continued the work of Borda on the factors affecting the refraction of light passing through the atmosphere of the earth. They helped to verify the formulas given in Laplace's Mécanique céleste, which were based on the assumption that the atmosphere is composed of concentric rings of a mixture of oxygen and nitrogen, with density as a function of altitude. Biot and Arago showed experimentally that temperature and pressure were significant variables, whereas humidity and the traces of carbon dioxide in the atmosphere could be disregarded. But when Arago extended his investigations to refraction in liquids and solids—with Petit in 1813 and Fresnel in 1815—he recognized the failure of the current theory of emission and particulate attraction to account for the empirical formulas he derived. After his return from the geodetic expedition to extend meridian triangulations from Barcelona to the Balearic Islands, Arago became a vocal critic of the Newtonian emission theory and, by 1816, an ardent supporter of the undulatory theory.

The original source of Arago's interest was Thomas Young's classic paper of 1801 on the color of thin glass plates and the discovery of polarization by Malus in 1808. Arago continued their independent investigations by passing beams of polarized light through a variety of gaseous and crystalline substances at various degrees of incidence to study the light's properties. His results, which suggested the usefulness of the undulatory theory, included the discovery of chromatic polarization by the use of thin mica plates (1811), rediscovered independently by Brewster; the elaboration of the conditions necessary to produce Newton's rings (1811); and the observation of special cases of rotary polarization (1812), which were shortly thereafter made a general law of optics by Biot.

It was this series of disparate experiments that caused Fresnel to write to Arago in 1815 to announce his theory of stellar aberration and the explanation of diffraction phenomena by undulatory principles. Although Fresnel's "discoveries" had retraced the work of Bradley and Thomas Young, Arago urged him to pursue his investigations and agreed to collaborate with the young engineer. Together they published a series of papers advocating the undulatory theory of light, answering one by one the criticisms of the partisans of emission theory, especially Arago's colleagues and former friends, Laplace and Biot. In this collaborative enterprise Fresnel supplied the crucial mathematical analyses and the seminal concept of transverse waves, while Arago contributed his encyclopedic command of the current literature in optics, his critical powers, and a significant number of experimental insights and actual experiments.

Above all, Arago functioned as a catalytic agent and public defender of the new theory, and eventually as its major historian. In 1824 he wrote an important article on polarization, translated by Young for the *Encyclopaedia Britannica,* and later wrote detailed and moving biographical notices of Fresnel (1830), Young (1832), and Malus (1853), sprinkled with personal anecdotes of great significance. It was Arago who, in 1838, borrowing and amplifying the idea and apparatus from Wheatstone's experiments for measuring the speed of electricity, suggested the "crucial experiment" to decide between the corpuscular and undulatory theories of light by comparing the speed of light in water and in air. The experiment, which vindicated the undulatory position, was carried out by Foucault in 1850 and announced to the Academy in Arago's presence.

Arago was also concerned with optical instruments that proved useful for a variety of purposes, in physics and meteorology as well as in astronomy, for which they were mainly devised. In 1811 he invented the polariscope to determine the degree of polarization of light rays by passing them successively through a mica or rock-crystal polarizer and an Iceland spar analyzer. With the addition of a series of properly graduated plates that could be inclined at will with reference to the incident ray, Arago transformed his polariscope into a polarimeter, which he used to verify one of the few mathematically expressed laws he discovered: the cosine-squared law for calculating the intensity of the ordinary ray in double refraction. In 1833 he derived from it the ratio of the amount of polarized light to neutral light: $\cos 2i$ for the ordinary ray and $2 \sin^2 i$ for the extraordinary ray, where i is the angle between the rock-crystal polarizer and the plane of polarization of the incident ray. With the polarimeter, he was able to differentiate between light emanating from solid and liquid surfaces, polarized by reflection and from incandescent gases, and to determine that the edge of the sun is gaseous. The polarimeter also suggested to him ways to determine polarization of the corona during total eclipses, to determine that rays from the sun's halo are refracted but not reflected, to observe the nature of a comet's tail, and to calculate the height of isolated clouds.

In 1815 Arago built a primitive cyanometer to measure the degree of blueness of the atmosphere, which was later adapted for use in hydrographical determinations of the depth of the sea. In 1833 he proposed a photometer to measure comparative intensities of stellar light; his student Paul Ernest Laugier later employed it. He also perfected an ocular micrometer for measuring small angles, which was erroneously attributed to William Pearson. The workings of all these instruments, based upon polarization phenomena, were expounded with great clarity and enthusiasm in Arago's public lectures at the observatory, published posthumously as *Astronomie populaire.*

As a young astronomer and member of the Bureau des Longitudes, Arago made numerous observations and important theoretical proposals. Among them were the explanation of the scintillation of stars by the use of interference phenomena and the realization of the asymmetry of the layers of atmosphere with reference to the observer. In his later years he made some important remarks on solar appendages noticed during the 1842 eclipse, which he observed with Laugier and Mauvais. But it was even more by the stimulus he gave younger astronomers—including Paul Laugier, Félix Mauvais, Jean Goujon, Jules Jamin, Hervé Faye, and Charles Mathieu—that Arago made his reputation as an astronomer. It also was Arago who urged Leverrier, his successor as director of the observatory, to take up Bouvard's work on the tables of Uranus. These investigations eventually led to the prediction of the existence and position of Neptune. Arago was also attentive to instrument makers, being responsible for promoting the precision work of Henri Gambey and Louis Bréguet. He was proud that during his tenure at the observatory most of its late eighteenth-century, English-made instruments were gradually replaced by better, French apparatus.

In 1820 Arago interrupted his optical work to play a significant role in the elaboration of electrodynamic and electromagnetic theories. Invited to the La Rive laboratory in Geneva to witness the verification of Oersted's experiments linking electricity to magnetism, he immediately acquired a passionate interest in the subject, displaying what Humboldt characterized as "the intolerance of a new convert." Arago repeated the Geneva experiments at the Paris Academy on 11 September 1820, thereby inspiring Ampère to elaborate his electrodynamic theory of electricity and magnetism. Although the two scientists did not write joint papers, they were in constant and friendly communication, often working in each other's laboratories. Just as Arago had been the champion of Fresnel's theories in 1815, so now did he propagandize Ampère's new theory and vehemently support his novel views. Because of his loyalty to Ampère, Arago was never fully able to appreciate or accept the rival theory of Faraday.

Arago also made several important contributions to electromagnetism on his own. On 20 September 1820 he announced the discovery of the temporary magnetization of soft iron by an electric current, which suggested to Ampère a theory about the nature of magnetic "currents" and provided the technological key to the electric telegraph. Ampère calculated that the magnetic power could be multiplied by twisting the current-carrying wire into a helix, and with Arago he carried out the first experiments on primitive solenoids. In his historical articles Arago was always careful to credit Ampère with the major share of this discovery, which ultimately depended upon Ampère's mathematical theory. In 1822, while he and Humboldt were measuring the magnetic intensity of a hill at Greenwich, Arago casually noticed the dampening effect that metallic substances had on the oscillations of the compass needle. After a delay of several years, during which he worked on the speed of sound and the crystalline nature of ice, and wrote up his observations on the chemical and thermal effects of light, Arago recognized the importance of his original observation at Greenwich. He announced that the rotation of nonmagnetic metallic substances (especially copper) created a magnetic effect on a magnetized needle. Known as Arago's "disc" or "wheel," it was the discovery of this effect that won him the Copley Medal in 1825. John Herschel and Babbage attempted to explain the phenomenon on the basis of Ampère's theory, but it was Faraday who in 1831 explained it by his theory of induction. By this time Arago had abandoned electrical research and had turned to other, more eclectic concerns.

In 1824 Arago was a member of an academic commission to study steam pressure, with the aim of reducing the dangers of explosion in steam engines. He and Dulong prepared elaborate apparatus for measuring pressure under high temperatures, verifying Boyle's law for values up to 24 atmospheres. Through his long-standing friendship with Humboldt, Arago was led to write popular articles on meteorology and physical geography, which ranged from discussions of the temperature of the earth, the seas, and the atmosphere to earthquakes and magnetic variations on the earth. He was particularly influential in propagating Humboldt's concept of isothermal lines and in setting down the purposes of and data required from scientific expeditions. In 1839 Arago took a personal interest in announcing and popularizing the inventions of Niepce and Daguerre, who were awarded government pensions as a result of Arago's recognition of their inventions' potential significance.

In his last years, while his sight was failing him, Arago continued to discharge his duties as perpetual secretary of the Academy by summarizing the achievements of other scientists and by suggesting new experiments that he himself could not carry out. Surrounded by a group of devoted younger scientists who wrote, observed, and experimented for him, Arago never lost his mental energies and his ability to stimulate his colleagues and excite the public about the progress of science.

BIBLIOGRAPHY

I. ORIGINAL WORKS. Arago published no single scientific treatise of major significance during his lifetime. After the June days of 1848 he began arranging his papers in preparation for a complete edition. It was published posthumously as *Oeuvres de François Arago,* J. A. Barral, ed., 17 vols. (Paris, 1854–1862; 2nd ed., 1865), with an introduction by Humboldt. It contains most of his published articles (somewhat edited) and reports, the revised portions of his lectures (*Astronomie populaire*), and many previously unpublished notes. Missing is the *Recueil d'observations géodésiques, astronomiques et physiques en Espagne, en France, en Angleterre et en Écosse, pour déterminer la variation de la pesanteur et des degrés terrestres sur le prolongement du méridien de Paris* (Paris, 1821), written with Biot, and a number of reports prepared for the Academy, which are printed in the *Procès-verbaux des séances de l'Académie,* 10 vols. (Hendaye, 1910–1922), for the period until 1835 and unpublished in the archives of the Academy thereafter. Arago also collaborated with Bouvard, Mathieu, and Nicollet in preparing *Observations astronomiques, faites à l'Observatoire royal de Paris,* 2 vols. (Paris, 1825–1838). He annotated and edited Alexandre Bertrand, *Lettres sur les révolutions du globe* (5th ed., Paris, 1839); Jacques Étienne Victor Arago, *Souvenirs d'un aveugle,* 2 vols. (Paris, 1842–1843); and Condorcet, *Oeuvres,* 12 vols. (Paris, 1847–1849).

Of Arago's voluminous correspondence, only a small portion has been published: *Correspondance d'Alexandre de Humboldt avec François Arago (1809–1853),* E. T. Hamy, ed. (Paris, 1908). The library of the Paris Observatory has over 50 unpublished letters (B4 9–12) and most of his MS notes and observations related to astronomy (C6 8–11, E1 19, E3 4–13). An important correspondence with Thomas Young is at the British Museum (Add. MSS 34613) and at the Royal Society Library.

II. SECONDARY LITERATURE. The best biography is still Maurice Daumas, *Arago* (Paris, 1943), despite its meager scholarly apparatus. In addition to the works cited in Daumas's bibliographical essay, pp. 273–275, and in Horace Chauvet, *François Arago et son temps* (Perpignan, 1954), consult Maurice Crosland, *The Society of Arcueil* (London, 1967), *passim.*

ROGER HAHN

ARANZIO

ARANZIO, GIULIO CESARE (*b*. Bologna, Italy, *ca.* 1529/1530; *d*. Bologna, 7 April 1589), *surgery, anatomy.*

Since Aranzio's parents, Ottaviano di Jacopo and Maria Maggi, were poor, he was aided in his medical education by his maternal uncle, Bartolomeo Maggi (1477–1552), lecturer in surgery at the University of Bologna and principal court physician of Julius III. He was a favorite pupil of this uncle, whom he loved and esteemed so highly that he assumed his surname, calling himself Giulio Cesare Aranzio Maggio. He studied at the University of Padua, where in 1548, at nineteen, he made his first anatomical discovery: the elevator muscle of the upper eyelid. He received his degree at Bologna on 20 May 1556, and shortly thereafter, at the age of twenty-seven, he became lecturer in medicine and surgery at the same university.

The excellent scientific and practical preparation Aranzio had received from his uncle immediately brought him fame. He discovered the *pedes hippocamp*; the cerebellum cistern; and the fourth ventricle, the arterial duct between the aorta and the pulmonary duct, which discovery was erroneously attributed to Leonardo Botallo.

In 1564 Aranzio published *De humano foetu opusculum*, and fifteen years later his *Observationes anatomicae* appeared. In these he presented the new direction of anatomy, based not merely on simple description of the organs of the body but also on experimental investigations of their functions.

Aranzio was the first lecturer at the University of Bologna to hold a separate professorship of anatomy; prior to him, instruction was given by lecturers in surgery. He himself began as a lecturer in surgery, but in 1570 he was able to have the two subjects separated so that each would have its own professorship. He held both professorships all his life, beloved and esteemed by his students.

Aranzio's *De tumoribus secundum locus affectum* (1571) is devoted to surgical subjects and gives a very good idea of the quality of his surgical lectures. He performed rhinoplastic surgery several years before Gaspare Tagliacozzi, but he wrote nothing on these operations. One of his pupils, Oczok Wojciech, who graduated from Bologna in 1569, did publish *Przymiot* (Cracow, 1581), a treatise on syphilis, however. In this treatise, in discussing the loss of the nose as the result of an attack of syphilis, he mentions rhinoplastic surgery and then states that in Bologna he frequently saw Aranzio perform such surgery successfully by using the "skin of the arm." It was Tagliacozzi, though, who gave the first scientific description of facial plastic surgery, illustrating the account with splendid charts.

BIBLIOGRAPHY

I. ORIGINAL WORKS. Aranzio's writings are *De humano foetu opusculum* (Rome, 1564; Venice, 1571; Basel, 1579); *De tumoribus secundum locus affectum* (Bologna, 1571); and *Observationes anatomicae* (Basel, 1579; Venice, 1587, 1595).

II. SECONDARY LITERATURE. Works on Aranzio are A. Malati Benedicenti, *Medici e farmacisti,* 2 vols. (Milan, 1947); T. G. Benedict, *Collectanea ad historiam rinoplastices Italorum* (Bratislava, 1843); *Biographie medicale: Dizionario delle scienze mediche* (Paris, 1820–1825); A. Castiglioni, "La scuola bolognese e la rinascita dell'anatomia," in *Annali Merck* (1931); U. Cesarano, "Giulio Cesare Aranzi," in *Comune di Bologna,* **1** (1929); E. Dall'Osso, "Giulio Cesare Aranzio e la rinoplastica," in *Annali di medicina navale e tropicale,* **61**, no. 5 (Sept.–Oct. 1956), and "Un contributo al pensiero scientifico di Giulio Cesare Aranzio: la sua opera chirurgica," *ibid.,* no. 6 (Nov.–Dec. 1956); O. Dezeimeris and Rainge-Delorme, *Dizionario storico della medicina antica e moderna* (Paris, 1828–1829); *Dizionario classico di medicina interna ed esterna* (Milan, 1838–1847); G. Fantuzzi, "Aranzio," in *Notizie degli scrittori bolognesi,* I (Bologna, 1790); A. Gallassi, "Chirurgia plastica. Ars medica per saecula," in *Collana di studi e ricerche* (Bologna, 1950–1951); G. Marini, *Degli archiatri pontifici* (Rome, 1784); G. Martinotti, *L'insegnamento dell'anatomia a Bologna prima del secolo XIX* (Bologna, 1910); G. M. Mazzucchelli, *Gli scrittori d'Italia* (Brescia, 1753); M. Medici, *Compendio storico della scuola anatomica di Bologna* (Bologna, 1857); L. Münster, "Un precursore bolognese della rinoplastica del '400," in *Atti del 1º Convegno Società Medica Italo-Svizzera* (Bologna, 1953); A. Pazzini, *Storia della medicina* (Milan, 1947), *Bio-bibliografia di storia della chirurgia* (Rome, 1948), and "Breve storia della rinoplastica," in *La chirurgia plastica,* **1**, no. 1; A. Sorbelli and L. Simeoni, *Storia dell'Università di Bologna* (Bologna, 1949); and J. P. Webster and M. Tesch Gnudi, *The Life and Times of M. Gaspare Tagliacozzi* (New York-Bologna, 1953).

EUGENIO DALL'OSSO

ARATUS OF SOLI

ARATUS OF SOLI (*b*. Soli, Cilicia, *ca.* 310 B.C.; *d*. *ca.* 240/239 B.C.), *astronomy.*

Although we possess four anonymous "lives" and a biography by Suidas, we are poorly informed about Aratus' life. The letters supposedly by him that are in the first life, edited by Westermann, are most probably spurious. Aratus went to Athens as a young man and there became acquainted with Stoicism. He then spent some time in Macedon at the court of Antigonus Gonatas (276–239 B.C.) and in Syria with Antiochus I. He is said to have prepared an edition of Homer's *Odyssey* and of the *Iliad*. Aratus was the author of several poems that are now lost: *Hymn to Pan,* celebrating the marriage of Antigonus Gonatas to Phile, half sister of Antiochus; *epideceia* addressed to his friends (and one to his brother); *Ostologia,* which seem to have been poems on medical subjects; and *Catalepton,* a collection of poems from which Strabo quotes two hexameters. Two of his epigrams are in *Anthologia Palatina* (XI, 437; XII, 129).

Aratus' only extant work is *Phaenomena,* a poem in 1,154 hexameters. After a prelude (lines 1–18) consisting of a hymn to Zeus, he describes the northern (19–320) and the southern (320–453) constellations. He refrains from giving an explanation of the planetary movements (454–461), apparently because of their complicated nature and the difficulty of calculating their conjunction (an allusion to the great year). Next (462–558) Aratus describes the circles of the celestial sphere and then (559–757) deals with the calendar: the hours of the risings and settings of stars (559–732), the days of the lunar month (733–739), the seasons (740–751), and the Metonic cycle (752–757). The second part of the poem (758–1154) deals with weather signs and is an integral part of it even though some ancient commentators give it a separate title (*Prognosis*). After a transitional part (758–772), in which he again emphasizes the power of all-pervading Zeus, Aratus deals with the signs derived from the observation of the different celestial phenomena (the stars, the sun, etc.); he ends with a description of the signs that depend on terrestrial phenomena. He concludes his poem (1142–1154) with an invitation to observe all these signs during the whole year, certain that we will not, by doing so, reach unwarranted conclusions.

The *Phaenomena* became famous as soon as it was published, as may be seen from the epigrams that Callimachus (*Anthologia Palatina* IX, 507) and Leonidas of Tarentum (*Anthologia Palatina* IX, 25) dedicate to Aratus. The poem was translated into Latin by Cicero and by Germanicus; Avienus translated it in the fourth century, and there is extant a seventh-century translation into barbarous Latin. The *Phaenomena* is cited by many authors, both Greek and Latin, and remained fashionable until the sixteenth century, as may be seen by the numerous manuscripts that have come down to us. It possesses some literary value and is indebted mainly to Hesiod and Homer for vocabulary and syntax. Aratus' adherence to Stoicism is patent throughout the poem, especially in the opening hymn to Zeus, who stands for the Stoic pantheistic divinity. From an astronomical standpoint the poem contains many errors, more than were in its source, the *Phaenomena* of Eudoxus of Cnidos. That this work of Eudoxus' was the source for Aratus, at least for the first part of the poem, we know from the commentary that Hipparchus devoted to both works. Aratus' source for the second part may

have been the same or another work by Eudoxus, but a work by Theophrastus dealing with meteorology (now lost) has also been suggested. Two manuscripts now at the Vatican give the names of twenty-seven commentators on Aratus.

BIBLIOGRAPHY

The most important editions of Aratus are those of E. Maass, *Arati Phaenomena* (Berlin, 1893; repr. 1955); and J. Martin, *Arati Phaenomena* (Florence, 1956); the latter contains an extensive commentary and a translation, both in French. For a translation into English, see that of G. R. Mair in the Loeb Classical Library's *Callimachus, Lycophron, Aratus* (London, 1921). On Aratus' life, text tradition, and influence, see J. Martin, *Histoire du texte des Phénomènes d'Aratos* (Paris, 1956). The texts of the four lives and of Suidas' biography are in A. Westermann, *BIOGRAPHOI. Vitarum scriptores Graeci minores* (Brunswick, 1845; repr. Amsterdam, 1964), pp. 52–61; see also Suidas, *s.n.* Aratus (A. Adler, ed.). For the commentaries and scholia, see E. Maass, *Aratea* (Berlin, 1892), and *Commentariorum in Aratum reliquiae* (Berlin, 1898; repr. 1958); and C. Manitius, *Hipparchi in Arati et Eudoxi Phaenomena commentariorum libri tres* (Leipzig, 1894). See also W. von Christ and W. Schmid, *Geschichte der griechischen Literatur,* pt. 2, 1st half (Munich, 1920), 163–167; G. Knaack, "Aratos," no. 6, in Pauly-Wissowa, *Real-Encyclopädie,* II, pt. 1 (1894), cols. 391–399; F. Susemihl, *Geschichte der griechischen Literatur in der Alexandrinerzeit,* I (Leipzig, 1891), 284–299; and M. Erren, *Die Phainomena des Aratos von Soloi* (Wiesbaden, 1967).

L. TARÁN

ARBER, AGNES ROBERTSON (*b.* London, England, 23 February 1879; *d.* Cambridge, England, 22 March 1960), *botany.*

Agnes Robertson's interest in botany developed strongly at school in northern London and later at University College, London, and the Botany School, Cambridge. She was particularly influenced by her friend and teacher, Ethel Sargant, who excited her interest in comparative plant anatomy. In 1909 she married E. A. N. Arber, demonstrator in paleobotany at Cambridge University.

At the suggestion of A. C. Seward, Mrs. Arber engaged in the study of early printed herbals, and her book on them (1912) became a standard work. It was revised and largely rewritten for the 1938 edition.

She published many papers on comparative anatomy, especially of the monocotyledons, and coordinated her results in three books, the first of which was *Water Plants: A Study of Aquatic Angiosperms.*

The second, *Monocotyledons: A Morphological Study*, was noted for its illumination of the so-called phyllode theory of the origin of the monocotyledonous leaf. The third, *The Gramineae*, was like *Water Plants* in that it embraced a very wide botanical approach.

In the later years of her life Mrs. Arber published three books that reflect the way she had then turned to consideration of scientific thought in relation to philosophy and metaphysics.

Agnes Arber was in the vanguard of the movement of women into scientific research: she was the first woman botanist to be made a fellow of the Royal Society and she received the Gold Medal of the Linnean Society in 1948. Throughout her life she had few formal contacts with college or university, choosing to work largely by herself. She was nonetheless kindly, helpful, and gracious, and certainly deserved A. G. Tansley's tribute (1952): "Dr. Agnes Arber is the most distinguished as well as the most erudite of British plant morphologists."

BIBLIOGRAPHY

I. Original Works. Between 1902 and 1957 Agnes Arber published in scientific periodicals some eighty-four original papers, the bulk of them in the field of comparative plant anatomy. They are listed in full in H. Hamshaw Thomas' obituary notice (see below). In addition she published *Herbals: Their Origin and Evolution* (Cambridge, 1912, 2nd rev. ed. 1938); *Water Plants: A Study of Aquatic Angiosperms* (Cambridge, 1920); *Monocotyledons: A Morphological Study* (Cambridge, 1925); *The Gramineae: A Study of Cereal, Bamboo and Grass* (Cambridge, 1934); *The Natural Philosophy of Plant Form* (Cambridge, 1950); *The Mind and the Eye: A Study of the Biologist's Standpoint* (Cambridge, 1954); and *The Manifold and the One* (London, 1957).

II. Secondary Literature. The most comprehensive biography is that by H. Hamshaw Thomas, in *Biographical Memoirs of Fellows of the Royal Society*, **6** (1960), 1–11. Shorter obituaries appeared in *Taxon*, **9** (1960), 261–263; *Proceedings of The Linnean Society of London*, **172** (1961), 128; and *Phytomorphology*, **11** (1961), 197–198.

Harry Godwin

ARBOGAST, LOUIS FRANÇOIS ANTOINE (*b*. Mutzig, Alsace, 4 October 1759; *d*. Strasbourg, France, 18 April 1803), *mathematics*.

There is no exact information on Arbogast's early years nor on his studies. He is registered as a nonpleading lawyer to the Sovereign Council of Alsace about 1780, and it is known that he taught mathematics at the Collège de Colmar about 1787. In 1789 he moved to Strasbourg, where he taught the same subject at the École d'Artillerie. He also was professor of physics at the Collège Royal, and after it was nationalized he served as director from April to October 1791. He then became rector of the University of Strasbourg. In 1790 he joined the society known as the Amis de la Constitution. He was a noted person in the Commune of Strasbourg, and in 1791 was elected a deputy to the Assemblée Législative and, in the following year, deputy from Haguenau to the Convention Nationale.

At the first of these assemblies, he and Gilbert Romme, Condorcet's closest collaborator, were on the committee of public instruction. Arbogast was the author of the general plan for public schools at all levels, which was brought before the convention but not adopted. He was responsible for the law introducing the decimal metric system in the whole of the French Republic.

Arbogast and his Alsatian colleagues were responsible for making the two assemblies aware of the efforts in Alsace toward building up a teaching force, as well as introducing the methods of pedagogy used in Germany. This information was useful in the establishment of the École Normale in the year III (1795).

Although he had been made *instituteur d'analyse* (probably professor of calculus) at the École Centrale de Paris (now École Polytechnique) in 1794, Arbogast taught only at the École Préparatoire. In this temporary institution an accelerated course of three months was given to 392 students before they were divided into three groups, which then proceeded to finish their studies in one, two, or three years.

In July 1795 Arbogast was entrusted with the planning of the École Centrale du Bas-Rhin, which replaced the abolished university. There he held the chair of mathematics from 1796 until 1802.

Arbogast was elected corresponding member of the Académie des Sciences in 1792 and an associate nonresident member of the Institut National (mathematics section, first class) four years later.

Arbogast's interest in the history of mathematics led to his classification of papers left by Marin Mersenne. He also amassed an important collection of manuscripts that are for the most part copies, in his writing, of the originals of memoirs or letters of Pierre Fermat, René Descartes, Jean Bernoulli, Pierre Varignon, Guillaume de L'Hospital, and others. At Arbogast's death these manuscripts were collected by his friend Français. They were bought in 1839 by Guglielmo Libri, the inspector of libraries, from a bookseller in Metz. After Libri's committal for trial on charges of malfeasance, his escape, and the seizure of his property, some of Arbogast's copies were deposited at the Bibliothèque Nationale in Paris. Other documents sold by the unscrupulous historian of science to Lord Ashburnham have also come to rest there. Other copies are now in the Laurenziana Library, Florence. The collection gathered by Arbogast became extremely valuable when definitive editions of the complete works of Fermat and Descartes, and of Mersenne's correspondence, were published.

In 1787 Arbogast took part in a competition organized by the Academy of St. Petersburg on "the arbitrary functions introduced by the integration of differential equations which have more than two variables," the question being "Do they belong to any curves or surfaces either algebraic, transcendental, or mechanical, either discontinuous or produced by a simple movement of the hand? Or shouldn't they legitimately be applied only to continuous curves susceptible of being expressed by algebraic or transcendental equations?"

The Academy was thus requesting a drastic settlement of the dispute between Jean d'Alembert, who adopted the second point of view, and Leonhard Euler, partisan of the first.

Arbogast won the prize and was even bolder than Euler in his conclusions. He showed that arbitrary functions may tolerate not only discontinuities in the Eulerian sense of the term, but also "combinations of several portions of different curves or those drawn by the free movement of the hand," that is, discontinuities in the sense afterward used by Augustin Cauchy.

Two years later, Arbogast sent a report to the Académie des Sciences de Paris on the new principles of differential calculus. This was never published, but Joseph Lagrange mentions it in 1797 as setting forth the same idea that he had developed in 1772, an idea that is the fundamental principle of his theory of analytic functions, "with its own developments and applications."

In speaking of his report in the Preface to *Calcul des dérivations,* Arbogast recalled the general ideas that anticipate Cauchy's and Niels Abel's ideas on the convergence of series. He added, "It caused me to reflect on fundamental principles . . . I then foresaw the birth of the first inkling of the ideas and methods which, when developed and extended, formed the substance of calculus of derivatives."

The principal aim of the calculus of derivatives, as Arbogast understood it, was to give simple and precise rules for finding series expansions. In order not to stay in the domain of pure theory, he used his rapid methods to find important formulas that were reached more laboriously by some of the great geometers.

Arbogast's work is dominated by a general idea that has become increasingly important in science and that until then had barely been anticipated: operational calculus. His only followers in this field were the brothers Français, then François Servois. But he was part of a vast mathematical movement that later included such names as Cauchy, George Boole, Sir William Rowan Hamilton, and Hermann Grassmann.

Arbogast clearly saw the difference that should be made between function and operation. When he defined his method of the "separation of the scale of operations," he said (*Traité des dérivations,* Preface):

> This method is generally thought of as separating from the functions of variables when possible, the operational signs which affect this function. Then of treating the expressions formed by these signs applied to any quantity whatsoever, an expression which I have called a scale of operation, to treat it, I say, nevertheless as if the operational signs which compose it were quantities, then to multiply the result by the function.

Arbogast appears in his mathematical work as a philosophical thinker whose ideas prefigured many mathematical notions of modern times, such as the introduction into analysis of discontinuous functions, the limitation of certain methods of algebra to what are today known as holomorphic functions, the necessity for care in the use of infinite series, and the conception of calculus as operational symbols, disregarding the quantities or functions on which they are based.

BIBLIOGRAPHY

I. ORIGINAL WORKS. Arbogast's works include "Essai sur de nouveaux principes du calcul différentiel et intégral indépendant de la théorie des infiniment petits et de celle des limites; mémoire envoyé à l'Académie des Sciences de Paris au printemps 1789" (unpublished); *Mémoire sur la nature des fonctions arbitraires qui entrent dans les intégrales des équations aux dérivées partielles. Présenté a l'Académie Impériale de Pétersbourg pour concourir au Prix proposé en 1787 et couronné dans l'Assemblée du 29 novembre 1790. Par M. ARBOGAST, professeur de mathématiques à Colmar* (St. Petersburg, 1791); and *Du calcul des dérivations et de ses usages dans la théorie des suites et dans le calcul différentiel* (Strasbourg, 1800).

II. SECONDARY LITERATURE. Works concerning Arbogast or his work are Paul Dupuy, *L'École Normale de l'an III* (Paris, 1895), p. 28; Maurice Fréchet, "Biographie du mathematicien alsacien Arbogast," in *Thales,* **4** (1937–1939), 43–55; Joseph Lagrange, *Théories des fonctions analytiques* (Paris, 1797), Introduction; Niels Nielsen, *Géomètres français sous la Révolution* (Copenhagen, 1929), pp. 1–5; Paul Tannery, *Mémoires scientifiques,* VI (Paris,

1926), 157; and K. Zimmermann, dissertation (Heidelberg, 1934).

JEAN ITARD

ARBOS, PHILIPPE (*b.* Mosset, France, 30 July 1882; *d.* Andancette, Drôme, France, 28 October 1956), *geography.*

The son of schoolteachers, Arbos was admitted in 1904, after secondary schooling at Perpignan, to the École Normale Supérieure. Among his teachers there was Vidal de la Blache, who was trying to free geography from its subordination to history (it was restricted to mere enumeration) and transform it into an autonomous science, and to link the study of the physical environment to that of human activities. Before Arbos, he had taught Raoul Blanchard, Jean Brunhes, Emmanuel de Martonne, and others who were to be among the masters of the French school of geography.

In 1907, after passing his *agrégation* in both history and geography, Arbos became a professor at the Grenoble *lycée.* Under the direction of Raoul Blanchard, in 1912 he began a doctoral thesis on pastoral life in the French Alps, a work that took ten years to complete (he visited all the valleys and nearly all the townships of the Alps) and became a model for similar studies. Indeed, he clarified the general principles of pastoral life, little known and poorly understood until then. The newness of the subject, the soundness of the information, and the depth of treatment accorded this thesis inspired a whole school of research.

Arbos made the University of Clermont-Ferrand, where he became lecturer in 1919 and served as professor from 1922 to 1952, a center for geographical studies. His study of the urban geography of Clermont-Ferrand (1930) has not become outdated, and his book on the Auvergne (1946) is a classic of regional geography. His great interest was the geography of the Massif Central, and toward the end of his life he contributed a lengthy article on it to Larousse's *La France* (1951). Arbos's scientific writing also includes a great many articles and notes in *Revue de géographie alpine* and *Annales de géographie.* Greatly admired abroad, he lectured in Belgium, the United States, and Brazil.

An excellent teacher, Arbos made a point of being available to his students every day. He developed methods of study in the field by increasing the number of field trips, and taught his students the need for solid groundwork and concern for precision, as well as breadth of view and the need to synthesize. Having had excellent literary training, he accustomed his students to the skill and the rigor of his presentation. Many of his students were teachers, and besides influencing them, Arbos had the pleasure of seeing his methods of instruction spread to the secondary and even to the primary level of instruction.

BIBLIOGRAPHY

I. ORIGINAL WORKS. Arbos's writings include *La vie pastorale dans les Alpes françaises* (Paris, 1922); *Étude de géographie urbaine: Clermont-Ferrand* (Clermont-Ferrand, 1930); *L'Auvergne* (Paris, 1946); and "Le Massif Central," in *La France* (Paris, 1951), pp. 25–111. *Mélanges géographiques offerts à Ph. Arbos* (Clermont-Ferrand, 1953) contains three articles by Arbos.

II. SECONDARY LITERATURE. Obituaries of Arbos are R. Blanchard, in *Revue de géographie,* **32** (1957), 57–58; D. Faucher, in *Revue de géographie des Pyrénées et du Sud-ouest,* **27** (1956), 418; and M. Sorre, in *Annales de géographie,* **32** (1957), 182–183.

JULIETTE TATON

ARBUTHNOT, JOHN (*b.* Arbuthnot, Kinkardineshire, Scotland, 29 April 1667; *d.* London, 27 February 1735), *mathematical statistics.*

The son of a Scottish Episcopal clergyman, Arbuthnot studied at Aberdeen, took his doctor's degree in medicine at St. Andrews in 1696, and settled in London in 1697. He was elected a fellow of the Royal Society in 1704 and was appointed a physician extraordinary to Queen Anne in 1705 (he became ordinary physician in 1709). The Royal College of Physicians elected him a fellow in 1710.

Arbuthnot wrote a few scientific and medical essays, but he became especially famous for his political satires. He was a close friend of the wits and literary men of his day: with Swift, Pope, John Gay, and Thomas Parnell he was a member of the Scriblerus Club. Of the characters in his political novels, the one that has survived is John Bull.

Arbuthnot was well acquainted with the theory of probability. It is certain that he published an English translation of Christian Huygens' *De ratiociniis in ludo aleae* (probably to be identified with a work said to have appeared in 1692, and with the first edition of part of an anonymous work that appeared in a fourth edition in 1738 in London under the title *Of the Laws of Chance . . .*). His scientific importance, however, resides in a short paper in the *Philosophical Transactions of the Royal Society,* which has been taken as the very origin of mathematical statistics. Entitled "An Argument for Divine Providence, Taken From the Constant Regularity Observ'd in the Birth of Both Sexes," it begins:

Among innumerable footsteps of divine providence to be found in the works of nature, there is a very remarkable one to be observed in the exact balance that is maintained, between the numbers of men and women; for by this means it is provided, that the species never may fail, nor perish, since every male may have its female, and of proportionable age. This equality of males and females is not the effect of chance but divine providence, working for a good end, which I thus demonstrate.

He first shows by numerical examples that if sex is determined by a die with two sides, M and F, it is quite improbable that in a large number of tosses there will be as many M as F. However, it is also quite improbable that the number of M will greatly exceed that of F. Nevertheless, there are more male infants born than female infants—clearly through divine providence—to make good the greater losses of males in external accidents. In every year from 1629 to 1710, there were more males christened in London than females—as if 82 tosses of the die would all show M. Such an event has a very poor probability: 2^{-82}. Therefore it cannot have been produced by chance; it must have been produced by providence.

Arbuthnot's argument is the first known example of a mathematical statistical inference and, in fact, is the ancestor of modern statistical reasoning. It immediately drew the attention of Continental scientists, particularly the Dutch physicist 's Gravesande, as is shown by contemporary correspondence. Daniel Bernoulli used it in 1732 to show that it could not be by chance that the planetary orbits are only slightly inclined to the ecliptic. In 1757 John Michell proved the existence of double stars by showing that stars are found close to each other more often than mere chance would allow.

Condorcet applied the argument to test the veracity of the tradition of Roman history that seven kings had reigned for a total of 257 years. Laplace, in his classic work, reconsidered such applications and added many new ones. This crude argument, although now greatly refined, is still the basis of statistical inference.

BIBLIOGRAPHY

Arbuthnot's major paper, "An Argument for Divine Providence, Taken from the Constant Regularity Observ'd in the Birth of Both Sexes," is found in *Philosophical Transactions of the Royal Society,* **27** (1710–1712), 186–190.

See also G. A. Aitken, *The Life and Works of John Arbuthnot* (Oxford, 1892); L. M. Beattie, *John Arbuthnot* (Cambridge, Mass., 1935); Daniel Bernoulli, in *Recueil des pièces qui ont remporté le prix double de l'Académie Royale des Sciences,* **3** (1734), 95–144; M. J. A. N. C. le Marquis de Condorcet, in *Histoire de l'Académie* (Paris, 1784), pp. 454–468; Hans Freudenthal, "Introductory Address," in *Quantitative Methods in Pharmacology* (Amsterdam, 1961), and "De eerste ontmoeting tussen de wiskunde en de sociale wetenschappen," in *Verhandelingen van de Koninklijke Vlaamse Akademie, Klasse Wetenschappen,* **28** (1966), 3–51; W. J. 's Gravesande, *Oeuvres philosophiques et mathématiques,* II (Amsterdam, 1774), 221–248; P. S. Laplace, *Théorie analytique de la probabilité,* 2nd ed. (Paris, 1820); John Michell, in *Philosophical Transactions of the Royal Society,* **57**, no. 1 (1767), 234–264; and I. Todhunter, *A History of the Mathematical Theory of Probability* (Cambridge, 1865; repr. New York, 1949).

HANS FREUDENTHAL

ARCHIAC, ÉTIENNE-JULES-ADOLPHE DESMIER (or **DEXMIER**) **DE SAINT-SIMON, VICOMTE D'** (*b.* Rheims, France, 24 September 1802; *d.* Paris, France, 24 or 25 December 1868), *geology, paleontology.*

Registered as the natural son of Marie-Elisabeth-Françoise Commelin, Adolphe was acknowledged by his father, Étienne-Louis-Marie Dexmier d'Archiac de Saint-Simon, a former cavalry captain who was an impoverished descendant of an ancient noble family from Angoumois.

After a childhood spent at Mesbrecourt, Aisne, d'Archiac entered the school for pages at Versailles and prepared for the military academy at Saint-Cyr, to which he was admitted in November 1819. He was appointed second lieutenant in the cavalry in 1821 and promoted to first lieutenant in 1827. In 1830, shortly before the July Revolution, d'Archiac published a pamphlet that showed his attachment to the *ancien régime* and expressed his aversion to the nobility of the Empire and to the bourgeoisie. He was placed on leave pay (about half the pay of a soldier on active duty) on 1 October 1830 because of this attachment to the Bourbon cause. The amount he received enabled him to subsist without working for a living.

How he came to study the geology of sedimentary formations is unknown. On 4 September 1832, at the annual extraordinary meeting of the Geological Society at Caen, he arranged to be introduced by Graves, secretary-general of the prefecture of Oise, and by Mutel-Delisle, a Paris attorney—a choice of sponsorship that reveals his lack of connections in the teaching profession. D'Archiac then declared that he had renounced his military career in order to devote himself to geology.

Leaving the study of eruptive and volcanic formations to the mining engineers, d'Archiac concentrated on sedimentary formations. His curiosity first led him to study two regions through which he had traveled in his youth: Aisne and Charente. From the beginning of his surveys in Aisne he undertook to make detailed sections and to collect fauna, which he carefully catalogued layer by layer. In 1835 he published a summary of his research as well as "Note sur la position du calcaire de Château-Landon." Following these publications, the Ministry of Public Works commissioned him to make a geological map of Aisne. His search for correlations led him to England for two months in 1837 and, the following year, to Brussels and Düsseldorf. He followed the lignite deposits from the Soissons region to Berkshire and established that they were of the same age and belonged at the level of the plastic clay of the Paris basin, as had already been indicated by Alexandre Brongniart and Élie de Beaumont.

His "Essais sur la coordination des terrains tertiaires du nord de la France, de la Belgique et de l'Angleterre" appeared in 1839. In 1843 he published a geological map of Aisne on the scale of 1:160,000 in a serial monograph, in which he described the physical features of the department as well as its hydrography, industries, and meteorology. Above all, he classified its formations into five groups: Transitional, Oolitic, Cretaceous, Tertiary, and Diluvial. In these he distinguished separate strata and determined their fossil contents. The "Essais" was followed by another work dealing with the Cretaceous formation of the southwestern, northern, and northwestern slopes of the French central plateau. D'Archiac subdivided the Cretaceous formation into four groups (Neocomian, green sandstone, tufa chalk, white chalk), which were further subdivided into stages.

Approaching problems of narrower scope with the same meticulous care, d'Archiac published notes on the petrographic features of Silurian and Carboniferous limestones, on the fossilization of echinoderms, the serpentine rocks of Limousin, pelagic formations, and the comparative geographic distribution of Paleozoic and contemporary mollusks. His last surveying trips during the summers of 1853–1856 and 1858 were devoted to the Corbières region, little known until then. In 1854 he published *Coupe géologique des environs des Bains de Rennes* (*Aude*), *suivie de la description de quelques fossiles de cette localité.* On 16 March 1857 he presented to the Geological Society his newly completed geological map of the regions adjacent to Aude and Pyrénées-Orientales.

D'Archiac did not restrict himself to fauna that he himself had collected. In 1842, in collaboration with his friend Edouard de Verneuil, he published a descrip-

tion of Devonian fossils of the Rhineland. Three years later the two naturalists described the fauna brought back by Adrien Paillette, who had prospected the Primary formations of Asturias. When the Geological Society received an important collection of "tourtia" (a glauconite conglomerate of quartz pebbles, deposited by the Cenomanian transgression in Hainaut and Flanders) from Belgium, d'Archiac undertook a study of it that was published in 1846–1847.

He was particularly interested in *Nummulites.* In 1846 and 1850 he described and illustrated finds made by geologists at Bayonne and Dax. In 1853 he published, in collaboration with Jules Haime, *Description des animaux fossiles du groupe nummulitique de l'Inde, précédée d'un résumé géologique et d'une monographie des Nummulites.* This work, containing illustrations of 352 species or varieties received by the Geological Society of London, has remained the basic work for every paleontological laboratory interested in the determination of Foraminifera. In 1866 d'Archiac, in collaboration with Verneuil and Paul Fischer, described the fossils brought back from Asiatic Turkey by the Russian geographer Pëtr Tschichatschew. The following year he made public the findings of Auguste Viquesnel in European Turkey.

D'Archiac published most of his articles and memoirs in the *Bulletin* or the *Mémoires* of the Geological Society. Named the society's deputy secretary in 1836 and its secretary in 1838, he served as president in 1844, 1849, and 1854. In March 1842 the society commissioned him to present an analysis of papers published since 1834, the review for the previous publications having been Ami Boué's work. In 1847 d'Archiac began publication of *Histoire des progrès de la géologie* with Volume I (*Cosmogénie et géogénie, physique du globe, géographie physique, terrain moderne*) and announced that three additional volumes would complete the work. Geology developed so rapidly, however, that it soon outdistanced his plans. D'Archiac's abstracts make up eight other volumes published between 1848 and 1860 and terminate with the Triassic, leaving out older formations. These nine volumes are a remarkable collection of learning that is still highly appreciated by geologists.

D'Archiac was elected a member of the Académie des Sciences, Section for Mineralogy, on 27 April 1857, to replace Constant Prevost. When d'Orbigny's death left vacant the chair of paleontology at the Muséum d'Histoire Naturelle, d'Archiac, supported by the assembly of the professors of the museum and by the Academy of Sciences, offered himself as a candidate on 28 September 1857 and wrote:

> In summing up, the Museum's course in paleontology, in order to be truly useful and to respond to present

needs and clearly evident scientific tendencies, should have as its aim not so much making known the zoological or anatomical characteristics of the organic fossil forms . . . as demonstrating the relationships of these forms to the earth strata and the manner in which they are distributed therein, and searching for the overall laws that have governed the succession of beings in the course of time, as well as the causes that have modified these laws for short periods of time [Archives Nationales, F17.20036].

Despite his exceptional qualifications, the authorities were slow to decide. A first appeal by d'Archiac served merely to add another document to his file. A second appeal brought his appointment on 14 June 1861. "Précis de l'histoire de la paléontologie stratigraphique" was the subject of his first course.

Among the subjects d'Archiac treated was a lively critique of Darwin's *On the Origin of Species,* the first French translation of which had just appeared and in which the term *élection* had been preferred to *sélection.* Accustomed to accuracy, d'Archiac stressed the vagueness and gratuitousness of numerous claims made by the English author. "The sad impressions of fatalism prevail from beginning to end in Darwin's book," he wrote (*Cours de paléontologie stratigraphique,* II, 113), but by 1866 he had become an evolutionist:

> The present state of the earth is only the consequence of its past—and this holds true for the organic as well as the inorganic realm. The animals and plants surrounding us and among which we live are only the descendants or the representatives of those which have preceded us. The living forms, as well as those which are extinct, are all part of a continuous chain [*Géologie et paléontologie,* p. 345].

D'Archiac, who remained single, lived alone and had no close ties. The frantic work that enabled him to rank among the leading scientists was his consolation. His physical stamina, which had sustained him as long as he engaged in surveying trips, finally began to fail. On 24 December 1868, apparently in the grip of a severe depression, he submitted his resignation as academician and professor to Victor Duruy, minister of public education, and left his home. It is believed that within the next few hours he threw himself into the Seine, from which his body was recovered at Meulan on 30 May 1869.

BIBLIOGRAPHY

I. Original Works. D'Archiac's major writing is *Histoire des progrès de la géologie de 1834 à 1845,* in 9 vols., all pub. in Paris: I, *Cosmogénie et géogénie, physique du globe, géographie physique, terrain moderne* (1847); II, pt. 1, *Terrain quaternaire ou diluvien* (1848); II, pt. 2, *Terrain tertiaire* (1849); III, *Formation nummulitique. Roches ignées ou pyrogènes des époques quaternaire et tertiaire* (1850); IV, *Formation crétacée, 1re partie* (1851); V, *Formation crétacée, 2e partie* (1853); VI, *Formation jurassique, 1re partie* (1856); VII, *Formation jurassique, 2e partie* (1857); and VIII, *Formation triasique* (1860).

His other works include "Résumé d'un mémoire sur une partie des terrains tertiaires inférieurs du département de l'Aisne," in *Bulletin de la Société géologique de France,* **6** (1835), 240–247; "Note sur la position du calcaire de Château-Landon," *ibid.,* **7** (1835), 30–35; "Mémoire sur la formation crétacée du sud-ouest de la France," in *Mémoires de la Société géologique de France,* **2,** pt. 2 (1837), 157–192; "Note sur les sables et grès moyens tertiaires," in *Bulletin de la Société géologique de France,* **9** (1837), 54–73; "Observations sur les lignites tertiaires du nord de la France et de l'Angleterre," *ibid.,* **9** (1838), 103–106; "Observations sur le groupe moyen de la formation crétacée," in *Mémoires de la Société géologique de France,* **3,** pt. 1 (1838), 261–311; "Essais sur la coordination des terrains tertiaires du nord de la France, de la Belgique et de l'Angleterre," in *Bulletin de la Société géologique de France,* **10** (1839), 168–225, a German translation is in Leonhard and Bronn's *Neues Jahrbuch für Mineralogie* (Stuttgart, 1839); *Discours sur l'ensemble des phénomènes qui se sont manifestés à la surface du globe, depuis son origine jusqu'à l'époque actuelle* (Paris, 1840); "Description géologique du département de l'Aisne," in *Mémoires de la Société géologique de France,* **5,** pt. 2 (1843), 129–421; "Études sur la formation crétacée des versants sud-ouest, nord et nord-ouest du plateau central de la France," pt. 1 in *Annales des sciences géologiques,* **2** (1843), 121–143, 169–191; pt. 2 in *Mémoires de la Société géologique de France,* 2nd ser., **2,** pt. 1 (1846), 1–148; "Description des fossiles recueillis par M. Thorent, dans les couches à *Nummulines* des environs de Bayonne," *ibid.,* 189–217; "Rapport sur les fossiles du Tourtia, légués par M. Léveillé à la Société géologique de France," *ibid.,* pt. 2 (1847), 291–351; *Notice sur les travaux géologiques de M. d'Archiac* (Paris, 1847); "Description des fossiles du groupe nummulitique recueillis par M. S.-P. Pratt et M. J. Delbos aux environs de Bayonne et de Dax," in *Mémoires de la Société géologique de France,* 2nd ser., **3,** pt. 2 (1850), 397–456; *Liste bibliographique par ordre de dates des travaux géologiques de M. d'Archiac* (Paris, 1856); "Les Corbières. Études géologiques d'une partie des départements de l'Aude et des Pyrénées-Orientales," in *Mémoires de la Société géologique de France,* 2nd ser., **6,** pt. 2 (1859); *Cours de paléontologie stratigraphique professé au Muséum d'Histoire Naturelle,* 2 vols.: I, *Précis de l'histoire de la paléontologie stratigraphique* (Paris, 1862); II, *Connaissances générales qui doivent précéder l'étude de la paléontologie stratigraphique et des phénomènes organiques qui s'y rattachent* (Paris, 1864); *Leçons sur la faune quaternaire* (Paris, 1865); *Géologie et paléontologie* (Paris, 1866); "Paléontologie," in Auguste Viquesnel, *Voyage dans la Turquie d'Europe,* II (Paris, 1868), 449–481; and *Paléontologie de la France* (Paris, 1868).

Works written with others are "On the Fossils of the Older Deposits in the Rhenish Provinces; Preceded by a General Survey of the Fauna of the Paleozoic Rocks, and Followed by a Tabular List of the Organic Remains of the Devonian System in Europe," in *Transactions of the Geological Society of London,* **6** (1842), 303–410, written with Edouard de Verneuil, trans. into German by G. von Leonhard (Stuttgart, 1844); *Description des animaux fossiles du groupe nummulitique de l'Inde, précédée d'un résumé géologique et d'une monographie des Nummulites* (Paris, 1853), written with Jules Haime; and "Paléontologie de l'Asie Mineure," written with Verneuil and Paul Fischer, in Pëtr Tschichatschew, *Asie Mineure. Description physique de cette contrée,* IV (Paris, 1869), 83–234, 393–420.

II. Secondary Literature. An article on d'Archiac is Albert Gaudry, "Notice sur les travaux scientifiques de d'Archiac," in *Bulletin de la Société géologique de France,* 3rd ser., **2** (1874), 230–244. Source material may be found in Archives de la Guerre, Vincennes, d'Archiac's file; Archives Nationales, Paris, F17.13566 and F17.20036; and the archives of the Académie des Sciences, which has d'Archiac's papers that were turned over to it in 1960 by the Société Géologique de France, including 987 letters written to d'Archiac.

Arthur Birembaut

ARCHIGENES (*b.* Apameia, Syria, *ca.* A.D. 54; *fl.* Rome, 98–117), *medicine.*

Archigenes was the son of Philippos and pupil of Agathinos; he practiced medicine in Rome during the reign of Trajan and achieved sufficient popularity to be mentioned by the poet Juvenal (VI, 236; XIII, 98; XIV, 252).

Nearly all of his many and varied writings have been lost, and their contents can be reconstructed only approximately from the fragments preserved in Galen and later medical writers. According to Galen, Archigenes belonged to the eclectic school; the surviving fragments indicate, however, that he was influenced considerably by the doctrines of the pneumatic school. His main contributions were in pathology, surgery, and therapeutics.

Archigenes' main work in general medicine was the semidiagnostic Περὶ τόπων πεπονθότων ("On Places Affected"), in which he sought to explain the causes of diseases by concentrating upon their localized manifestations. Although he seems to have had a vague idea of the difference between a generalized, systemic disease and the locally painful injury or infection, it is difficult, as Galen noted, to understand him because of his tendency to designate the types of pain associated with local inflammations by separate names. Certain inconsistencies in Archigenes' medical theories were probably the result of his reliance on Stoic doctrines. Thus, his belief that the

hegemonikon (ruling principle) was located in the heart did not prevent him from treating loss of memory by local applications to the head. Because he discussed the role of the nerves and arteries in the pneuma's transmission of pain throughout the body, it is not surprising that he devoted a special treatise to the pulse. His Περὶ τῶν σφυγμῶν ("On the Pulses") was frequently cited by Galen, who also wrote a commentary on it and probably took some of his ideas on the pulse from it.

Among the titles preserved of Archigenes' other writings, several pertain to more specific diseases. Acute and chronic diseases were distinguished, and a separate treatise was written on the signs of fevers. At least some of the conditions underlying the feverish symptoms were treated by surgery. Archigenes described an amputation of a gangrenous limb that used both ligatures and cauterization. He noted the importance and location of tendons, and perhaps of nerves, in surgical repair. He described cancer of the breast and employed a speculum in the examination of uterine tumors.

In antiquity, Archigenes was highly regarded for his writings on therapeutics and materia medica. Numerous fragments are preserved from his Περὶ τῶν κατὰ γένος φαρμάκον ("On Drugs According to Their Nature"). Two other writings are known by title: Περὶ καστορίου χρήσεως ("On the Use of Castoreum") and Περὶ τῆς δόσεως τοῦ 'ελλβόρου ("On the Giving of Hellebore"), but these may have been portions of a larger work. Some of his prescriptions have been preserved in Galen and Alexander of Tralles. For epilepsy he resorted to amulets, and he relied heavily on animal substances in his compound drugs. In accordance with the prevailing humoral-pneumatic doctrines, the purpose of therapy was to control the δυσκρασίαι ("bad temperaments") by neutralizing them.

BIBLIOGRAPHY

Cesare Brescia, ed., *Frammenti medicinali di Archigene* (Naples, 1955), consists of three short tracts edited from MS Vat. Pal. 199.

Additional writings on Archigenes are E. Gurlt, *Geschichte der Chirurgie und ihrer Ausübung,* I (Berlin, 1898), 411–414; Alessandro Olivieri, "Frammenti di Archigene," in *Memorie dell'Accademia di Archeologia, Lettere e Belle Arti di Napoli,* **6** (1942), 120–122; Erwin Rohde, "Aelius Promotus," in *Rheinisches Museum für Philologie,* **28** (1873), 264–290, an analysis of a tract on poisonous drugs, part of which has been attributed to Archigenes; and Max Wellmann, "Die pneumatische Schule bis auf Archigenes," in *Philologische Untersuchungen,* **14** (1895), a fundamental study that lists sources for and titles of thirteen of Archigenes' lost writings, and "Archigenes," in Pauly-Wissowa, *Real-Encyclopädie der classischen Altertumswissenschaft,* II (1896), cols. 484–486.

JERRY STANNARD

ARCHIMEDES (*b.* Syracuse, *ca.* 287 B.C.; *d.* Syracuse, 212 B.C.), *mathematics, mechanics.*

Few details remain of the life of antiquity's most celebrated mathematician. A biography by his friend Heracleides has not survived. That his father was the astronomer Phidias we know from Archimedes himself in his *The Sandreckoner* (Sect. I. 9). Archimedes was perhaps a kinsman of the ruler of Syracuse, King Hieron II (as Plutarch and Polybius suggest). At least he was on intimate terms with Hieron, to whose son Gelon he dedicated *The Sandreckoner.* Archimedes almost certainly visited Alexandria, where no doubt he studied with the successors of Euclid and played an important role in the further development of Euclidian mathematics. This visit is rendered almost certain by his custom of addressing his mathematical discoveries to mathematicians who are known to have lived in Alexandria, such as Conon, Dositheus, and Eratosthenes. At any rate Archimedes returned to Syracuse, composed most of his works there, and died there during its capture by the Romans in 212 B.C. Archimedes' approximate birth date of 287 B.C. is conjectured on the basis of a remark by the Byzantine poet and historian of the twelfth century, John Tzetzes, who declared (*Chiliad* 2, hist. 35) that Archimedes "worked at geometry until old age, surviving seventy-five years." There are picturesque accounts of Archimedes' death by Livy, Plutarch, Valerius Maximus, and Tzetzes, which vary in detail but agree that he was killed by a Roman soldier. In most accounts he is pictured as being engaged in mathematics at the time of his death. Plutarch tells us (*Marcellus,* Ch. XVII) that Archimedes "is said to have asked his friends and kinsmen to place on his grave after his death a cylinder circumscribing a sphere, with an inscription giving the ratio by which the including solid exceeds the included." And indeed Cicero (see *Tusculan Disputations,* V, xxiii, 64–66), when he was Quaestor in Sicily in 75 B.C.,

> . . . tracked out his grave. . . . and found it enclosed all around and covered with brambles and thickets; for I remembered certain doggerel lines inscribed, as I had heard, upon his tomb, which stated that a sphere along with a cylinder had been put upon the top of his grave. Accordingly, after taking a good look all around (for there are a great quantity of graves at the Agrigentine Gate), I noticed a small column arising a little above the bushes, on which there was the figure of a sphere

and a cylinder. . . . Slaves were sent in with sickles. . . and when a passage to the place was opened we approached the pedestal in front of us; the epigram was traceable with about half of the lines legible, as the latter portion was worn away.

No surviving bust can be certainly identified as being of Archimedes, although a portrait on a Sicilian coin (whatever its date) is definitely his. A well-known mosaic showing Archimedes before a calculating board with a Roman soldier standing over him was once thought to be a genuine survival from Herculaneum but is now considered to be of Renaissance origin.

Mechanical Inventions. While Archimedes' place in the history of science rests on a remarkable collection of mathematical works, his reputation in antiquity was also founded upon a series of mechanical contrivances which he is supposed to have invented and which the researches of A. G. Drachmann tend in part to confirm as Archimedean inventions. One of these is the water snail, a screwlike device to raise water for the purpose of irrigation, which, Diodorus Siculus tells us (*Bibl. hist.,* V, Ch. 37), Archimedes invented in Egypt. We are further told by Atheneus that an endless screw invented by Archimedes was used to launch a ship. He is also credited with the invention of the compound pulley. Some such device is the object of the story told by Plutarch in his life of *Marcellus* (Ch. XIV). When asked by Hieron to show him how a great weight could be moved by a small force, Archimedes "fixed upon a three-masted merchantman of the royal fleet, which had been dragged ashore by the great labors of many men, and after putting on board many passengers and the customary freight, he seated himself at a distance from her, and without any great effort, but quietly setting in motion a system of compound pulleys, drew her towards him smoothly and evenly, as though she were gliding through the water." It is in connection with this story that Plutarch tells us of the supposed remark of Archimedes to the effect that "if there were another world, and he could go to it, he could move this one," a remark known in more familiar form from Pappus of Alexandria (*Collectio,* Bk. VIII, Prop. 11): "Give me a place to stand on, and I will move the earth." Of doubtful authenticity is the oft-quoted story told by Vitruvius (*De architectura,* Bk. IX, Ch. 3) that Hieron wished Archimedes to check whether a certain crown or wreath was of pure gold, or whether the goldsmith had fraudulently alloyed it with some silver.

> While Archimedes was turning the problem over, he chanced to come to the place of bathing, and there,

as he was sitting down in the tub, he noticed that the amount of water which flowed over by the tub was equal to the amount by which his body was immersed. This indicated to him a method of solving the problem, and he did not delay, but in his joy leapt out of the tub, and, rushing naked towards his home, he cried out in a loud voice that he had found what he sought, for as he ran he repeatedly shouted in Greek, *heurēka, heurēka.*

Much more generally credited is the assertion of Pappus that Archimedes wrote a book *On Sphere-making,* a work which presumably told how to construct a model planetarium representing the apparent motions of the sun, moon, and planets, and perhaps also a closed star globe representing the constellations. At least, we are told by Cicero (*De re publica,* I, XIV, 21–22) that Marcellus took as booty from the sack of Syracuse both types of instruments constructed by Archimedes:

> For Gallus told us that the other kind of celestial globe [that Marcellus brought back and placed in the Temple of Virtue], which was solid and contained no hollow space, was a very early invention, the first one of that kind having been constructed by Thales of Miletus, and later marked by Eudoxus of Cnidus . . . with the constellations and stars which are fixed in the sky. . . . But this newer kind of globe, he said, on which were delineated the motions of the sun and moon and of those five stars which are called the wanderers . . . contained more than could be shown on a solid globe, and the invention of Archimedes deserved special admiration because he had thought out a way to represent accurately by a single device for turning the globe those various and divergent courses with their different rates of speed.

Finally, there are references by Polybius, Livy, Plutarch, and others to fabulous ballistic instruments constructed by Archimedes to help repel Marcellus. One other defensive device often mentioned but of exceedingly doubtful existence was a burning mirror or combination of mirrors.

We have no way to know for sure of Archimedes' attitude toward his inventions. One supposes that Plutarch's famous eulogy of Archimedes' disdain for the practical was an invention of Plutarch and simply reflected the awe in which Archimedes' theoretical discoveries were held. Plutarch (*Marcellus,* Ch. XVII) exclaims:

> And yet Archimedes possessed such a lofty spirit, so profound a soul, and such a wealth of scientific theory, that although his inventions had won for him a name and fame for superhuman sagacity, he would not consent to leave behind him any treatise on this subject, but regarding the work of an engineer and every art

that ministers to the needs of life as ignoble and vulgar, he devoted his earnest efforts only to those studies the subtlety and charm of which are not affected by the claims of necessity. These studies, he thought, are not to be compared with any others; in them, the subject matter vies with the demonstration, the former supplying grandeur and beauty, the latter precision and surpassing power. For it is not possible to find in geometry more profound and difficult questions treated in simpler and purer terms. Some attribute this success to his natural endowments; others think it due to excessive labor that everything he did seemed to have been performed without labor and with ease. For no one could by his own efforts discover the proof, and yet as soon as he learns it from him, he thinks he might have discovered it himself, so smooth and rapid is the path by which he leads one to the desired conclusion.

Mathematical Works. The mathematical works of Archimedes that have come down to us can be loosely classified in three groups (Arabic numbers have been added to indicate, where possible, their chronological order). The first group consists of those that have as their major objective the proof of theorems relative to the areas and volumes of figures bounded by curved lines and surfaces. In this group we can place *On the Sphere and the Cylinder* (5); *On the Measurement of the Circle* (9); *On Conoids and Spheroids* (7); *On Spirals* (6); and *On the Quadrature of the Parabola* (2), which, in respect to its Propositions 1–17, belongs also to the second category of works. The second group comprises works that lead to a geometrical analysis of statical and hydrostatical problems and the use of statics in geometry: *On the Equilibrium of Planes,* Book I (1), Book II (3); *On Floating Bodies* (8); *On the Method of Mechanical Theorems* (4); and the aforementioned propositions from *On the Quadrature of the Parabola* (2). Miscellaneous mathematical works constitute the third group: *The Sandreckoner* (10); *The Cattle-Problem;* and the fragmentary *Stomachion.* Several other works not now extant are alluded to by Greek authors (see Heiberg, ed., *Archimedis opera,* II, 536–554). For example, there appear to have been various works on mechanics that have some unknown relationship to *On the Equilibrium of Planes.* Among these are a possible work on *Elements of Mechanics* (perhaps containing an earlier section on centers of gravity, which, however, may have been merely a separate work written before *Equilibrium of Planes,* Book I), a tract *On Balances,* and possibly one *On Uprights.* Archimedes also seems to have written a tract *On Polyhedra,* perhaps one *On Blocks and Cylinders,* certainly one on *Archai* or *The Naming of Numbers* (a work preliminary to *The Sandreckoner*), and a work on *Optics* or *Catoptrics.* Other works are attributed to Archimedes by Arabic authors, and, for the most part, are extant in Arabic manuscripts (the titles for which manuscripts are known are indicated by an asterisk; see Bibliography): *The Lemmata**, or *Liber assumptorum* (in its present form certainly not by Archimedes since his name is cited in the proofs), *On Water Clocks**, *On Touching Circles**, *On Parallel Lines*, *On Triangles**, *On the Properties of the Right Triangle**, *On Data*, and *On the Division of the Circle into Seven Equal Parts**.

But even the genuine extant works are by no means in their original form. For example, *On the Equilibrium of Planes*, Book I, is possibly an excerpt from the presumably longer *Elements of Mechanics* mentioned above and is clearly distinct from Book II, which was apparently written later. A solution promised by Archimedes in *On the Sphere and the Cylinder* (Bk. II, Prop. 4) was already missing by the second century A.D. *On the Measurement of the Circle* was certainly in a much different form originally, with Proposition II probably not a part of it (and even if it were, it would have to follow the present Proposition III, since it depends on it). The word *parabolēs* in the extant title of *On the Quadrature of the Parabola* could hardly have been in the original title, since that word was not yet used in Archimedes' work in the sense of a conic section. Finally, the tracts *On the Sphere and the Cylinder* and *On the Measurement of the Circle* have been almost completely purged of their original Sicilian-Doric dialect, while the rest of his works have suffered in varying degrees this same kind of linguistic transformation.

In proving theorems relative to the area or volume of figures bounded by curved lines or surfaces, Archimedes employs the so-called Lemma of Archimedes or some similar lemma, together with a technique of proof that is generally called the "method of exhaustion," and other special Greek devices such as *neuseis*, and principles taken over from statics. These various mathematical techniques are coupled with an extensive knowledge of the mathematical works of his predecessors, including those of Eudoxus, Euclid, Aristeus, and others. The Lemma of Archimedes (*On the Sphere and Cylinder*, Assumption 5; cf. the Preface to *On the Quadrature of the Parabola* and the Preface to *On Spirals*) assumes "that of two unequal lines, unequal surfaces, and unequal solids the greater exceeds the lesser by an amount such that, when added to itself, it may exceed any assigned magnitude of the type of magnitudes compared with one another." This has on occasion been loosely identified with Definition 4 of Book V of the *Elements* of Euclid (often called the axiom of Eudoxus): "Magnitudes are said to have a ratio to one another which are capable, when multiplied, of exceeding one another."

But the intent of Archimedes' assumption appears to be that if there are two unequal magnitudes capable of having a ratio in the Euclidian sense, then their difference will have a ratio (in the Euclidian sense) with any magnitude of the same kind as the two initial magnitudes. This lemma has been interpreted as excluding actual infinitesimals, so that the difference of two lines will always be a line and never a point, the difference between surfaces always a surface and never a line, and the difference between solids always a solid and never a surface. The exhaustion procedure often uses a somewhat different lemma represented by Proposition X.1 of the *Elements* of Euclid: "Two unequal magnitudes being set out, if from the greater there be subtracted a magnitude greater than its half, and from that which is left a magnitude greater than its half, and if this process be repeated continually, there will be left some magnitude which will be less than the lesser magnitude set out." This obviously reflects the further idea of the continuous divisibility of a continuum. One could say that the Lemma of Archimedes justifies this further lemma in the sense that no matter how far the procedure of subtracting more than half of the larger of the magnitudes set out is taken (or also no matter how far the procedure of subtracting one-half the larger magnitude, described in the corollary to Proposition X.1, is taken), the magnitude resulting from the successive division (which magnitude being conceived as the difference of two magnitudes) will always be capable of having a ratio in the Euclidian sense with the smaller of the magnitudes set out. Hence one such remainder will some time be in a relationship of "less than" to the lesser of the magnitudes set out.

The method of exhaustion, widely used by Archimedes, was perhaps invented by Eudoxus. It was used on occasion by Euclid in his *Elements* (for example, in Proposition XII.2). Proof by exhaustion (the name is often criticized since the purpose of the technique is to avoid assuming the complete exhaustion of an area or a volume; Dijksterhuis prefers the somewhat anachronistic expression "indirect passage to the limit") is an indirect proof by reduction to absurdity. That is to say, if the theorem is of the form $A = B$, it is held to be true by showing that to assume its opposite, namely that A is not equal to B, is impossible since it leads to contradictions. The method has several forms. Following Dijksterhuis, we can label the two main types: the compression method and the approximation method. The former is the most widely used and exists in two forms, one that depends upon taking decreasing differences and one that depends on taking decreasing ratios. The fundamental procedure of both the "difference" and the "ratio" forms starts with the successive inscription and circumscription of regular figures within or without the figure for which the area or volume is sought. Then in the "difference" method the area or volume of the inscribed or circumscribed figure is regularly increased or decreased until the difference between the desired area or volume and the inscribed or circumscribed figure is less than any preassigned magnitude. Or to put it more specifically, if the theorem is of the form $A = B$, A being the curvilinear figure sought and B a regular rectilinear figure the formula for the magnitude of which is known, and we assume that A is greater than B, then by the exhaustion procedure and its basic lemma we can construct some regular rectilinear inscribed figure P such that P is greater than B; but it is obvious that P, an included figure, is in fact always less than B. Since P cannot be both greater and less than B, the assumption from which the contradiction evolved (namely, that A is greater than B) must be false. Similarly, if A is assumed to be less than B, we can by the exhaustion technique and the basic lemma find a circumscribed figure P that is less than B, which P (as an including figure) must always be greater than B. Thus the assumption of A less than B must also be false. Hence, it is now evident that, since A is neither greater nor less than B, it must be equal to B. An example of the exhaustion procedure in its "difference" form is to be found in *On the Measurement of the Circle:*[1]

Proposition 1

The area of any circle is equal to a right-angled triangle in which one of the sides about the right angle is equal to the radius, and the other to the circumference, of the circle.

Let $ABCD$ be the given circle, K the triangle described.

Then, if the circle is not equal to K, it must be either greater or less.

I. If possible, let the circle be greater than K.

Inscribe a square $ABCD$, bisect the arcs AB, BC, CD, DA, then bisect (if necessary) the halves, and so on, until the sides of the inscribed polygon whose angular points are the points of division subtend segments whose sum is less than the excess of the area of the circle over K.

Thus the area of the polygon is greater than K.

Let AE be any side of it, and ON the perpendicular on AE from the centre O.

Then ON is less than the radius of the circle and therefore less than one of the sides about the right angle in K. Also the perimeter of the polygon is less than the circumference of the circle, i.e. less than the other side about the right angle in K.

Therefore the area of the polygon is less than K; which is inconsistent with the hypothesis.

Thus the area of the circle is not greater than K.

II. If possible, let the circle be less than K.

Circumscribe a square, and let two adjacent sides, touching the circle in E, H, meet in T. Bisect the arcs between adjacent points of contact and draw the tangents at the points of bisection. Let A be the middle point of the arc EH, and FAG the tangent at A.

Then the angle TAG is a right angle.

Therefore
$$TG > GA$$
$$> GH.$$

It follows that the triangle FTG is greater than half the area $TEAH$.

Similarly, if the arc AH be bisected and the tangent at the point of bisection be drawn, it will cut off from the area GAH more than one-half.

Thus, by continuing the process, we shall ultimately arrive at a circumscribed polygon such that the spaces intercepted between it and the circle are together less than the excess of K over the area of the circle.

Thus the area of the polygon will be less than K.

Now, since the perpendicular from O on any side of the polygon is equal to the radius of the circle, while the perimeter of the polygon is greater than the circum-

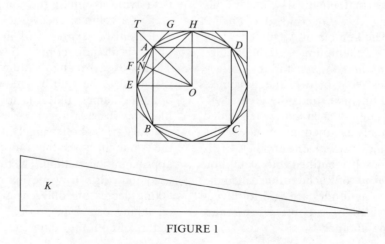

FIGURE 1

ference of the circle, it follows that the area of the polygon is greater than the triangle K; which is impossible.

Therefore the area of the circle is not less than K.

Since then the area of the circle is neither greater nor less than K, it is equal to it.

Other examples of the "difference" form of the exhaustion method are found in *On Conoids and Spheroids* (Props. 22, 26, 28, 30), *On Spiral Lines* (Props. 24, 25), and *On the Quadrature of the Parabola* (Prop. 16).

The "ratio" form of the exhaustion method is quite similar to the "difference" form except that in the first part of the proof, where the known figure is said to be less than the figure sought, the ratio of circumscribed polygon to inscribed polygon is decreased until it is less than the ratio of the figure sought to the known figure, and in the second part the ratio of circumscribed polygon to inscribed polygon is decreased until it is less than the ratio of the known figure to the figure sought. In each part a contradiction is shown to follow the assumption. And thus the assumption of each part must be false, namely, that the known figure is either greater or less than the figure sought. Consequently, the known figure must be equal to the figure sought. An example of the "ratio" form appears in *On the Sphere and the Cylinder* (Bk. I):[2]

Proposition 14

The surface of any isosceles cone excluding the base is equal to a circle whose radius is a mean proportional between the side of the cone [a generator] and the radius of the circle which is the base of the cone.

Let the circle A be the base of the cone; draw C equal to the radius of the circle, and D equal to the side of the cone, and let E be a mean proportional between C, D.

Draw a circle B with radius equal to E.

Then shall B be equal to the surface of the cone (excluding the base), which we will call S.

If not, B must be either greater or less than S.

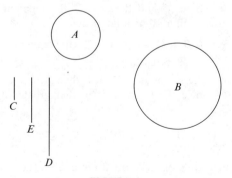

FIGURE 2

I. Suppose $B < S$.

Let a regular polygon be described about B and a similar one inscribed in it such that the former has to the latter a ratio less than the ratio $S:B$.

Describe about A another similar polygon, and on it set up a pyramid with apex the same as that of the cone.

Then (polygon about A):(polygon about B)
$= C^2:E^2$
$= C:D$
$=$ (polygon about A):(surface of pyramid excluding base). Therefore

(surface of pyramid) $=$ (polygon about B).

Now (polygon about B):(polygon in B) $< S:B$.

Therefore

(surface of pyramid):(polygon in B) $< S:B$,

which is impossible (because the surface of the pyramid is greater than S, while the polygon in B is less than B).

Hence $B \not< S$.

II. Suppose $B > S$.

Take regular polygons circumscribed and inscribed to B such that the ratio of the former to the latter is less than the ratio $B:S$.

Inscribe in A a similar polygon to that inscribed in B, and erect a pyramid on the polygon inscribed in A with apex the same as that of the cone.

In this case

(polygon in A):(polygon in B) $= C^2:E^2$
$= C:D$

$>$ (polygon in A):(surface of pyramid excluding base).

This is clear because the ratio of C to D is greater than the ratio of the perpendicular from the center of A on a side of the polygon to the perpendicular from the apex of the cone on the same side.

Therefore

(surface of pyramid) $>$ (polygon in B).

But (polygon about B):(polygon in B) $< B:S$.

Therefore, *a fortiori*,

(polygon about B):(surface of pyramid) $< B:S$;
which is impossible.

Since therefore B is neither greater nor less than S,

$B = S$.

Other examples of the "ratio" form of the exhaustion method are found in *On the Sphere and the Cylinder*, (Bk. I, Props. 13, 33, 34, 42, 44.)

As indicated earlier, in addition to the two forms of the compression method of exhaustion, Archimedes used a further technique which we may call the approximation method. This is used on only one occasion, namely, in *On the Quadrature of the Parabola* (Props. 18–24). It consists in approximating from below the area of a parabolic segment. That is to say, Archimedes continually "exhausts" the parabola by drawing first a triangle in the segment with the same base and vertex as the segment. On each side of the triangle we again construct triangles. This process is continued as far as we like. Thus if A_1 is the area of the original triangle, we have a series of inscribed triangles whose sum converges toward the area of parabolic segment: $A_1, 1/4 A_1, (1/4)^2 A_1, \cdots$ (in the accompanying figure A_1 is $\triangle PQq$ and $1/4 A_1$ or A_2 is the sum of triangles Prq and PRQ and A_3 is the sum of the next set of inscribed triangles—not shown on the diagram but equal to $[1/4]^2 A_1$). In order to prove that K, the area of the parabolic segment, is equal to $4/3 A_1$, Archimedes first proves in Proposition 22 that the sum of any finite number of terms of this series is less than the area of the parabolic segment. He then proves in Proposition 23 that if we have a series of terms $A_1, A_2, A_3, \cdots$ such as those given above, that is, with $A_1 = 4A_2, A_2 = 4A_3, \cdots$, then

$$A_1 + A_2 + A_3 + \cdots + A_n + \frac{1}{3} \cdot A_n = \frac{4}{3} \cdot A_1,$$

or

$$A_1 \left[1 + \frac{1}{4} + \left(\frac{1}{4}\right)^2 + \cdots + \left(\frac{1}{4}\right)^{n-1} + \frac{1}{3} \cdot \left(\frac{1}{4}\right)^{n-1} \right]$$
$$= \frac{4}{3} \cdot A_1$$

With modern techniques of series summation we would simply say that as n increases indefinitely $(1/4)^{n-1}$ becomes infinitely small and the series in brackets tends toward 4/3 as a limit and thus the parabolic segment equals $4/3 \cdot A_1$. But Archimedes followed the Greek *reductio* procedure. Hence he showed that if we assume $K > 4/3 \cdot A_1$ on the basis of a corollary to Proposition 20, namely, that by the successive inscription of triangles "it is possible to inscribe in the parabolic segment a polygon such that the segments left over are together less than any assigned area" (which is itself based on Euclid, *Elements* X.1), a contradiction will ensue. Similarly, a contradiction results from the assumption of $K < 4/3 \cdot A_1$. Here in brief is the final step of the proof (the reader is reminded that the terms $A_1, A_2, A_3, \cdots, A_n$, which were used above, are actually rendered by $A, B, C, \cdots, X$):[3]

Proposition 24

Every segment bounded by a parabola and a chord Qq is equal to four-thirds of the triangle which has the same base as the segment and equal height.

Suppose $\qquad K = \frac{4}{3} \triangle PQq,$

where P is the vertex of the segment; and we have then to prove that the area of the segment is equal to K.

For, if the segment be not equal to K, it must either be greater or less.

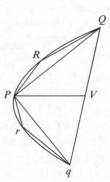

FIGURE 3

I. Suppose the area of the segment greater than K.

If then we inscribe in the segments cut off by PQ, Pq triangles which have the same base and equal height, i.e. triangles with the same vertices R, r as those of the segments, and if in the remaining segments we inscribe triangles in the same manner, and so on, we shall finally have segments remaining whose sum is less than the area by which the segment PQq exceeds K [Prop. 20 Cor.].

Therefore the polygon so formed must be greater than the area K; which is impossible, since [Prop. 23]

$$A + B + C + \cdots + Z < \frac{4}{3} A,$$

where $\qquad A = \triangle PQq.$

Thus the area of the segment cannot be greater than K.

II. Suppose, if possible, that the area of the segment is less than K.

If then $\triangle PQq = A$, $B = 1/4 A$, $C = 1/4 B$, and so on, until we arrive at an area X such that X is less than the difference between K and the segment, we have

$$A + B + C + \cdots + X + \frac{1}{3} X = \frac{4}{3} A \qquad \text{[Prop. 23]}$$
$$= K.$$

Now, since K exceeds $A + B + C + \cdots + X$ by an area less than X, and the area of the segment by an area greater than X, it follows that

$$A + B + C + \cdots + X > \text{(the segment)};$$

which is impossible, by Prop. 22. . . .

Hence the segment is not less than K.

Thus, since the segment is neither greater nor less than K,

$$(\text{area of segment } PQq) = K = \frac{4}{3} \triangle PQq.$$

In the initial remarks on the basic methods of Archimedes, it was noted that Archimedes sometimes used the technique of a *neusis* ("verging") construction. Pappus defined a *neusis* construction as "Two lines being given in position, to place between them a straight line given in length and verging towards a given point." He also noted that "a line is said to verge towards a point, if being produced, it reaches the point." No doubt "insertion" describes the mathematical meaning better than "verging" or "inclination," but "insertion" fails to render the additional condition of inclining or verging toward a point just as the name *neusis* in expressing the "verging" condition fails to render the crucial condition of insertion. At any rate, the *neusis* construction can be thought of as being accomplished mechanically by marking the termini of the linear insertion on a ruler and shifting that ruler until the termini of the insertion lie on the given curve or curves while the ruler passes through the verging point. In terms of mathematical theory most of the Greek *neuseis* require a solution by means of conics or other higher curves. *Neusis* constructions are indicated by Archimedes in *On Spirals* (Props. 5–9). They are assumed as possible without any explanation. The simplest case may be illustrated as follows:[4]

Proposition 5

Given a circle with center O, and the tangent to it at a point A, it is possible to draw from O a straight line OPF, meeting the circle in P and the tangent in F, such that, if c be the circumference of any given circle whatever,

$$FP:OP < (\text{arc } AP): c.$$

Take a straight line, as D, greater than the circumference c. [Prop. 3]

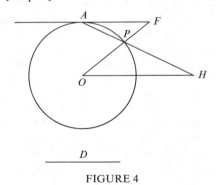

FIGURE 4

Through O draw OH parallel to the given tangent, and draw through A a line APH, meeting the circle in P and OH in H, such that the portion PH intercepted between the circle and the line OH may be equal to D [literally: "let PH be placed equal to D, verging toward A"]. Join OP and produce it to meet the tangent in F.

Then $FP:OP = AP:PH$, by parallels,
$\qquad\qquad = AP:D$
$\qquad\qquad < (\text{arc } AP):c.$

With the various methods that have been described and others, Archimedes was able to demonstrate a whole host of theorems that became a basic part of geometry. Examples beyond those already quoted follow: "The surface of any sphere is equal to four times the greatest circle in it" (*On the Sphere and the Cylinder,* Bk. I, Prop. 23); this is equivalent to the modern formulation $S = 4\pi r^2$. "Any sphere is equal to four times the cone which has its base equal to the greatest circle in the sphere and its height equal to the radius of the sphere" (*ibid.,* Prop. 34); its corollary that "every cylinder whose base is the greatest circle in a sphere and whose height is equal to the diameter of the sphere is 3/2 of the sphere and its surface together with its base is 3/2 of the surface of the sphere" is the proposition illustrated on the tombstone of Archimedes, as was noted above. The modern equivalent of Proposition 34 is $V = 4/3\ \pi r^3$. "Any right or oblique segment of a paraboloid of revolution is half again as large as the cone or segment of a cone which has the same base and the same axis" (*On Conoids and Spheroids,* Props. 21–22). He was also able by his investigation of what are now known as Archimedean spirals not only to accomplish their quadrature (*On Spirals,* Props. 24–28), but, in preparation therefore, to perform the crucial rectification of the circumference of a circle. This, then, would allow for the construction of the right triangle equal to a circle that is the object of *On the Measurement of a Circle* (Prop. I), above. This rectification is accomplished in *On Spirals* (Prop. 18): "If a straight line is tangent to the extremity of a spiral described in the first revolution, and if from the point of origin of the spiral one erects a perpendicular on the initial line of revolution, the perpendicular will meet the tangent so that the line intercepted between the tangent and the origin of the spiral will be equal to the circumference of the first circle" (see Fig. 5).

It has also been remarked earlier that Archimedes employed statical procedures in the solution of geometrical problems and the demonstration of theorems. These procedures are evident in *On the Quadrature of the Parabola* (Props. 6–16) and also in *On the Method.* We have already seen that in the latter part of *On the Quadrature of the Parabola* Archimedes demonstrated the quadrature of the parabola by purely geometric methods. In the first part of the tract he demonstrated the same thing by means of a balancing method. By the use of the law of the lever and a knowledge of the centers of gravity of triangles and trapezia, coupled with a *reductio* procedure, the quadrature is demonstrated. In *On the Method* the same statical procedures are used; but, in addition, an entirely new assumption is joined with them, namely, that a plane figure can be considered as the summation of its line elements (presumably infinite in number) and that a volumetric figure can be

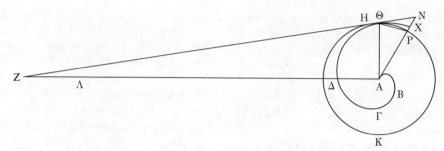

FIGURE 5

considered as the summation of its plane elements. The important point regarding this work is that it gives us a rare insight into Archimedes' procedures for discovering the theorems to be proved. The formal, indirect procedures that appear in demonstrations in the great body of Archimedes' works tell us little as to how the theorems to be proved were discovered. To be sure, sometimes he no doubt proved theorems that he had inherited with inadequate proof from his predecessors (such was perhaps the case of the theorem on the area of the circle, which he proved simply and elegantly in *On the Measurement of the Circle* [Prop. 1], as has been seen). But often we are told by him what his own discoveries were, and their relation to the discoveries of his predecessors, as, for example, those of Eudoxus. In the Preface of Book I of *On the Sphere and the Cylinder,* he characterizes his discoveries by comparing them with some established theorems of Eudoxus:[5]

> Now these properties were all along naturally inherent in the figures referred to . . ., but remained unknown to those who were before my time engaged in the study of geometry. Having, however, now discovered that the properties are true of these figures, I cannot feel any hesitation in setting them side by side both with my former investigations and with those of the theorems of Eudoxus on solids which are held to be most irrefragably established, namely, that any pyramid is one third part of the prism which has the same base with the pyramid and equal height, and that any cone is one third part of the cylinder which has the same base with the cone and equal height. For, though these properties also were naturally inherent in the figures all along, yet they were in fact unknown to all the many able geometers who lived before Eudoxus, and had not been observed by anyone. Now, however, it will be open to those who possess the requisite ability to examine these discoveries of mine.

Some of the mystery surrounding Archimedes' methods of discovery was, then, dissipated by the discovery and publication of *On the Method of Mechanical Theorems.* For example, we can see in Proposition 2 how it was that Archimedes discovered by the "method" the theorems relative to the area and volume of a sphere that he was later to prove by strict geometrical methods in *On the Sphere and the Cylinder:*[6]

Proposition 2

We can investigate by the same method the propositions that

(1) Any sphere is (in respect of solid content) four times the cone with base equal to a great circle of the sphere and height equal to its radius; and

(2) the cylinder with base equal to a great circle of the sphere and height equal to the diameter is 1-1/2 times the sphere.

(1) Let *ABCD* be a great circle of a sphere, and *AC*, *BD* diameters at right angles to one another.

Let a circle be drawn about *BD* as diameter and in a plane perpendicular to *AC*, and on this circle as base let a cone be described with *A* as vertex. Let the surface of this cone be produced and then cut by a plane through *C* parallel to its base; the section will be a circle on *EF* as diameter. On this circle as base let a cylinder be erected with height and axis *AC*, and produce *CA* to *H*, making *AH* equal to *CA*.

Let *CH* be regarded as the bar of a balance, *A* being its middle point.

Draw any straight line *MN* in the plane of the circle *ABCD* and parallel to *BD*. Let *MN* meet the circle in *O*, *P*, the diameter *AC* in *S*, and the straight lines *AE*, *AF* in *Q*, *R* respectively. Join *AO*.

Through *MN* draw a plane at right angles to *AC*; this plane will cut the cylinder in a circle with diameter *MN*, the sphere in a circle with diameter *OP*, and the cone in a circle with diameter *QR*.

Now, since $MS = AC$, and $QS = AS$,

$$MS \cdot SQ = CA \cdot AS$$
$$= AO^2$$
$$= OS^2 + SQ^2.$$

And, since $HA = AC$,

$$HA:AS = CA:AS$$
$$= MS:SQ$$
$$= MS^2:MS \cdot SQ$$
$$= MS^2:(OS^2 + SQ^2), \text{ from above,}$$
$$= MN^2:(OP^2 + QR^2)$$
$$= (\text{circle, diam. } MN):(\text{circle, diam. } OP$$
$$+ \text{ circle, diam. } QR).$$

220

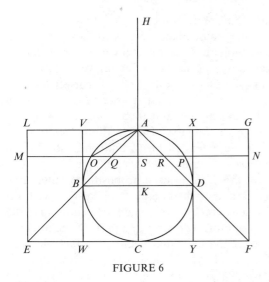

FIGURE 6

That is,

$HA : AS =$ (circle in cylinder) : (circle in sphere + circle in cone).

Therefore the circle in the cylinder, placed where it is, is in equilibrium, about A, with the circle in the sphere together with the circle in the cone, if both the latter circles are placed with their centers of gravity at H.

Similarly for the three corresponding sections made by a plane perpendicular to AC and passing through any other straight line in the parallelogram LF parallel to EF.

If we deal in the same way with all the sets of three circles in which planes perpendicular to AC cut the cylinder, the sphere and the cone, and which make up those solids respectively, it follows that the cylinder, in the place where it is, will be in equilibrium about A with the sphere and the cone together, when both are placed with their centers of gravity at H.

Therefore, since K is the center of gravity of the cylinder,

$HA : AK =$ (cylinder) : (sphere + cone AEF).

But $HA = 2AK$;
therefore cylinder $= 2$ (sphere + cone AEF).

Now   cylinder $= 3$ (cone AEF);   [Eucl. XII. 10]

therefore   cone $AEF = 2$ (sphere).

But, since $EF = 2BD$,

        cone $AEF = 8$ (cone ABD);
therefore   sphere $= 4$ (cone ABD).

(2) Through B, D draw VBW, XDY parallel to AC; and imagine a cylinder which has AC for axis and the circles on VX, WY as diameters for bases.

Then

    cylinder $VY = 2$ (cylinder VD)
            $= 6$ (cone ABD)   [Eucl. XII. 10]
            $= \frac{3}{2}$ (sphere), from above.

                                        Q.E.D.

From this theorem, to the effect that a sphere is four times as great as the cone with a great circle of the sphere as base and with height equal to the radius of the sphere, I conceived the notion that the surface of any sphere is four times as great as a great circle in it; for, judging from the fact that any circle is equal to a triangle with base equal to the circumference and height equal to the radius of the circle, I apprehended that, in like manner, any sphere is equal to a cone with base equal to the surface of the sphere and height equal to the radius.

It should be observed in regard to this quotation that the basic volumetric theorem was discovered prior to the surface theorem, although in their later formal presentation in *On the Sphere and the Cylinder,* the theorem for the surface of a sphere is proved first. By using the "method" Archimedes also gave another "proof" of the quadrature of the parabola—already twice proved in *On the Quadrature of the Parabola*—and he remarks in his preface (see the quotation below) that he originally discovered this theorem by the method. Finally, in connection with *On the Method,* it is necessary to remark that Archimedes considered the method inadequate for formal demonstration, even if it did provide him with the theorems to be proved more rigorously. One supposes that it was the additional assumption considering the figures as the summation of their infinitesimal elements that provoked Archimedes' cautionary attitude, which he presents so lucidly in his introductory remarks to Eratosthenes:[7]

Seeing moreover in you, as I say, an earnest student, a man of considerable eminence in philosophy, and an admirer [of mathematical inquiry], I thought fit to write out for you and explain in detail in the same book the peculiarity of a certain method, by which it will be possible for you to get a start to enable you to investigate some of the problems in mathematics by means of mechanics. This procedure is, I am persuaded, no less useful even for the proof of the theorems themselves; for certain things first became clear to me by a mechanical method, although they had to be demonstrated by geometry afterwards because their investigation by the said method did not furnish an actual demonstration. But it is of course easier, when we have previously acquired, by the method, some knowledge of the questions, to supply the proof than it is to find it without any previous knowledge. This is a reason why, in the case of the theorems the proof of which Eudoxus was the first to discover, namely that the cone is a third part of the cylinder, and the pyramid of the prism, having the same base and equal height, we should give no small share of the credit to Democritus who was the first to make the assertion with regard to the said figure though he did not prove it. I am myself in the position of having first made the discovery of the theorem now to be published [by the method indicated], and I deem it necessary to expound the method partly because I have already spoken of it and I do not want to be thought to have uttered vain words, but equally because I am persuaded that it will be of no little service to mathematics; for I apprehend that some, either of my contemporaries or of my successors, will, by means of the method when once established, be able to discover other theorems in addition, which have not yet occurred to me.

While Archimedes' investigations were primarily in geometry and mechanics reduced to geometry, he made some important excursions into numerical calculation, although the methods he used are by no means clear. In *On the Measurement of the Circle* (Prop. 3), he calculated the ratio of circumference to diameter (not called π until early modern times) as being less than 3-1/7 and greater than 3-10/71. In the course of this proof Archimedes showed that he had an accurate method of approximating the roots of large numbers. It is also of interest that he there gave an approximation for $\sqrt{3}$, namely, $1351/780 > \sqrt{3} > 265/153$. How he computed this has been much disputed. In the tract known as *The Sandreckoner,* Archimedes presented a system to represent large numbers, a system that allows him to express a number P^{10^8}, where P itself is $(10^8)^{10^8}$. He invented this system to express numbers of the sort that, in his words, "exceed not only the number of the mass of sand equal in magnitude to the earth . . ., but also that of a mass equal in magnitude to the universe." Actually, the number he finds that would approximate the number of grains of sand to fill the universe is a mere 10^{63}, and thus does not require the higher orders described in his system. Incidentally, it is in this work that we have one of the few antique references to Aristarchus' heliocentric system.

In the development of physical science, Archimedes is celebrated as the first to apply geometry successfully to statics and hydrostatics. In his *On the Equilibrium of Planes* (Bk. I, Props. 6–7), he proved the law of the lever in a purely geometrical manner. His weights had become geometrical magnitudes possessing weight and acting perpendicularly to the balance beam, itself conceived of as a weightless geometrical line. His crucial assumption was the special case of the equilibrium of the balance of equal arm length supporting equal weights. This postulate, although it may ultimately rest on experience, in the context of a mathematical proof appears to be a basic appeal to geometrical symmetry. In demonstrating Proposition 6, "Commensurable magnitudes are in equilib-

rium at distances reciprocally proportional to their weights," his major objective was to reduce the general case of unequal weights at inversely proportional distances to the special case of equal weights at equal distances. This was done by (1) converting the weightless beam of unequal arm lengths into a beam of equal arm lengths, and then (2) distributing the unequal weights, analyzed into rational component parts over the extended beam uniformly so that we have a case of equal weights at equal distances. Finally (3) the proof utilized propositions concerning centers of gravity (which in part appear to have been proved elsewhere by Archimedes) to show that the case of the uniformly distributed parts of the unequal weights over the extended beam is in fact identical with the case of the composite weights concentrated on the arms at unequal lengths. Further, it is shown in Proposition 7 that if the theorem is true for rational magnitudes, it is true for irrational magnitudes as well (although the incompleteness of this latter proof has been much discussed). The severest criticism of the proof of Proposition 6 is, of course, the classic discussion by Ernst Mach in his *Science of Mechanics,* which stresses two general points: (1) experience must have played a predominant role in the proof and its postulates in spite of its mathematical-deductive form; and (2) any attempt to go from the special case of the lever to the general case by replacing expanded weights on a lever arm with a weight concentrated at their center of gravity must assume that which has to be proved, namely, the principle of static moment. This criticism has given rise to an extensive literature and stimulated some successful defenses of Archimedes, and this body of literature has been keenly analyzed by E. J. Dijksterhuis (*Archimedes,* pp. 289–304). It has been pointed out further, and with some justification, that Proposition 6 with its proof, even if sound, only establishes that the inverse proportionality of weights and arm lengths is a sufficient condition for the equilibrium of a lever supported in its center of gravity under the influence of two weights on either side of the fulcrum. It is evident that he should also have shown that the condition is a necessary one, since he repeatedly applies the inverse proportionality as a necessary condition of equilibrium. But this is easily done and so may have appeared trivial to Archimedes. The succeeding propositions in Book I of *On the Equilibrium of Planes* show that Archimedes conceived of this part of the work as preparatory to his use of statics in his investigation of geometry of the sort that we have described in *On the Quadrature of the Parabola* and *On the Method.* In his *On Floating Bodies,* the emphasis is once more largely on geometrical analysis. In Book I, a somewhat obscure concept of hydrostatic pressure is presented as his basic postulate:[8]

> Let it be granted that the fluid is of such a nature that of the parts of it which are at the same level and adjacent to one another that which is pressed the less is pushed away by that which is pressed the more, and that each of its parts is pressed by the fluid which is vertically above it, if the fluid is not shut up in anything and is not compressed by anything else.

As his propositions are analyzed, we see that Archimedes essentially maintained an Aristotelian concept of weight directed downward toward the center of the earth conceived of as the center of the world. In fact, he goes further by imagining the earth removed and so fluids are presented as part of a fluid sphere all of whose parts weigh downward convergently toward the center of the sphere. The surface of the sphere is then imagined as being divided into an equal number of parts which are the bases of conical sectors having the center of the sphere as their vertex. Thus the water in each sector weighs downward toward the center. Then if a solid is added to a sector, increasing the pressure on it, the pressure is transmitted down through the center of the sphere and back upward on an adjacent sector and the fluid in that adjacent sector is forced upward to equalize the level of adjacent sectors. The influence on other than adjacent sectors is ignored. It is probable that Archimedes did not have the concept of hydrostatic paradox formulated by Stevin, which held that at any given point of the fluid the pressure is a constant magnitude that acts perpendicularly on any plane through that point. But, by his procedures, Archimedes was able to formulate propositions concerning the relative immersion in a fluid of solids less dense than, as dense as, and more dense than the fluid in which they are placed. Proposition 7 relating to solids denser than the fluid expresses the so-called "principle of Archimedes" in this fashion: "Solids heavier than the fluid, when thrown into the fluid, will be driven downward as far as they can sink, and they will be lighter [when weighed] in the fluid [than their weight in air] by the weight of the portion of fluid having the same volume as the solid." This is usually more succinctly expressed by saying that such solids will be lighter in the fluid by the weight of the fluid displaced. Book II, which investigates the different positions in which a right segment of a paraboloid can float in a fluid, is a brilliant geometrical tour de force. In it Archimedes returns to the basic assumption found in *On the Equilibrium of Planes, On the Quadrature of the Parab-*

ola, and *On the Method,* namely, that weight verticals are to be conceived of as parallel rather than as convergent at the center of a fluid sphere.

Influence. Unlike the *Elements* of Euclid, the works of Archimedes were not widely known in antiquity. Our present knowledge of his works depends largely on the interest taken in them at Constantinople from the sixth through the tenth centuries. It is true that before that time individual works of Archimedes were obviously studied at Alexandria, since Archimedes was often quoted by three eminent mathematicians of Alexandria: Hero, Pappus, and Theon. But it is with the activity of Eutocius of Ascalon, who was born toward the end of the fifth century and studied at Alexandria, that the textual history of a collected edition of Archimedes properly begins. Eutocius composed commentaries on three of Archimedes' works: *On the Sphere and the Cylinder, On the Measurement of the Circle,* and *On the Equilibrium of Planes.* These were no doubt the most popular of Archimedes' works at that time. The *Commentary on the Sphere and the Cylinder* is a rich work for historical references to Greek geometry. For example, in an extended comment to Book II, Proposition 1, Eutocius presents manifold solutions of earlier geometers to the problem of finding two mean proportionals between two given lines. *The Commentary on the Measurement of the Circle* is of interest in its detailed expansion of Archimedes' calculation of π. The works of Archimedes and the commentaries of Eutocius were studied and taught by Isidore of Miletus and Anthemius of Tralles, Justinian's architects of Hagia Sophia in Constantinople. It was apparently Isidore who was responsible for the first collected edition of at least the three works commented on by Eutocius as well as the commentaries. Later Byzantine authors seem gradually to have added other works to this first collected edition until the ninth century when the educational reformer Leon of Thessalonica produced the compilation represented by Greek manuscript A (adopting the designation used by the editor, J. L. Heiberg). Manuscript A contained all of the Greek works now known excepting *On Floating Bodies, On the Method, Stomachion,* and *The Cattle Problem.* This was one of the two manuscripts available to William of Moerbeke when he made his Latin translations in 1269. It was the source, directly or indirectly, of all of the Renaissance copies of Archimedes. A second Byzantine manuscript, designated as B, included only the mechanical works: *On the Equilibrium of Planes, On the Quadrature of the Parabola,* and *On Floating Bodies* (and possibly *On Spirals*). It too was available to Moerbeke. But it disappears after an early fourteenth-century reference. Finally, we can mention a third Byzantine manuscript, C, a palimpsest whose Archimedean parts are in a hand of the tenth century. It was not available to the Latin West in the Middle Ages, or indeed in modern times until its identification by Heiberg in 1906 at Constantinople (where it had been brought from Jerusalem). It contains large parts of *On the Sphere and the Cylinder,* almost all of *On Spirals,* some parts of *On the Measurement of the Circle* and *On the Equilibrium of Planes,* and a part of the *Stomachion.* More important, it contains most of the Greek text of *On Floating Bodies* (a text unavailable in Greek since the disappearance of manuscript B) and a great part of *On the Method of Mechanical Theorems,* hitherto known only by hearsay. (Hero mentions it in his *Metrica,* and the Byzantine lexicographer Suidas declares that Theodosius wrote a commentary on it.)

At about the same time that Archimedes was being studied in ninth-century Byzantium, he was also finding a place among the Arabs. The Arabic Archimedes has been studied in only a preliminary fashion, but it seems unlikely that the Arabs possessed any manuscript of his works as complete as manuscript A. Still, they often brilliantly exploited the methods of Archimedes and brought to bear their fine knowledge of conic sections on Archimedean problems. The Arabic Archimedes consisted of the following works: (1) *On the Sphere and the Cylinder* and at least a part of Eutocius' commentary on it. This work seems to have existed in a poor, early ninth-century translation, revised in the late ninth century, first by Isḥāq ibn Ḥunayn and then by Thābit ibn Qurra. It was reedited by Nasīr ad-Dīn al-Ṭūsī in the thirteenth century and was on occasion paraphrased and commented on by other Arabic authors (see Archimedes in Index of Suter's "Die Mathematiker und Astronomen"). (2) *On the Measurement of the Circle,* translated by Thābit ibn Qurra and reedited by al-Ṭūsī. Perhaps the commentary on it by Eutocius was also translated, for the extended calculation of π found in the geometrical tract of the ninth-century Arabic mathematicians the Banū Mūsā bears some resemblance to that present in the commentary of Eutocius. (3) A fragment of *On Floating Bodies,* consisting of a definition of specific gravity not present in the Greek text, a better version of the basic postulate (described above) than exists in the Greek text, and the enunciations without proofs of seven of the nine propositions of Book I and the first proposition of Book II. (4) Perhaps *On the Quadrature of the Parabola*—at least this problem received the attention of Thābit ibn Qurra. (5) Some indirect material from *On the Equilibrium of Planes* found in other mechanical works translated into Arabic (such as

Hero's *Mechanics,* the so-called Euclid tract *On the Balance,* the *Liber karastonis,* etc.). (6) In addition, various other works attributed to Archimedes by the Arabs and for which there is no extant Greek text (see list above in "Mathematical Works"). Of the additional works, we can single out the *Lemmata (Liber assumptorum),* for, although it cannot have come directly from Archimedes in its present form, in the opinion of experts several of its propositions are Archimedean in character. One such proposition was Proposition 8, which employed a *neusis* construction like those used by Archimedes:[9]

Proposition 8

If we let line *AB* be led everywhere in the circle and extended rectilinearly [see Fig. 7], and if *BC* is posited as equal to the radius of the circle, and *C* is connected to the center of the circle *D,* and the line (*CD*) is produced to *E,* arc *AE* will be triple arc *BF.* Therefore, let us draw *EG* parallel to *AB* and join *DB* and *DG.* And because the two angles *DEG, DGE* are equal, $\angle GDC = 2\angle DEG$. And because $\angle BDC = \angle BCD$ and $\angle CEG = \angle ACE$, $\angle GDC = 2\angle CDB$ and $\angle BDG = 3\angle BDC$, and arc $BG = $ arc AE, and arc $AE = 3$ arc BF; and this is what we wished.

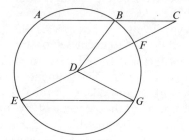

FIGURE 7

This proposition shows, then, that if one finds the position and condition of line *ABC* such that it is drawn through *A,* meets the circle again in *B,* and its extension *BC* equals the radius, this will give the trisection of the given angle *BDG.* It thus demonstrates the equivalence of a *neusis* and the trisection problem—but without solving the *neusis* (which could be solved by the construction of a conchoid to a circular base).

Special mention should also be made of the *Book on the Division of the Circle into Seven Equal Parts,* attributed to Archimedes by the Arabs, for its remarkable construction of a regular heptagon. This work stimulated a whole series of Arabic studies of this problem, including one by the famous Ibn al-Haytham (Alhazen). Propositions 16 and 17, leading to that construction, are given here in toto:[10]

Proposition 16

Let us construct square *ABCD* [Fig. 8] and extend side *AB* directly toward *H.* Then we draw the diagonal

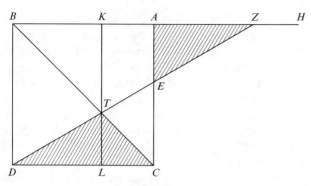

FIGURE 8

BC. We lay one end of a rule on point *D.* Its other end we make meet extension *AH* at a point *Z* such that △*AZE* = △*CTD.* Further, we draw the straight line *KTL* through *T* and parallel to *AC.* And now I say that *AB* · *KB* = *AZ*² and *ZK* · *AK* = *KB*² and, in addition, each of the two lines *AZ* and *KB* > *AK.*

Proof:

(1) $CD \cdot TL = AZ \cdot AE$ [given]   Hence

(2) $\dfrac{CD(=AB)}{AZ} = \dfrac{AE}{TL}$

Since △*ZAE*~△*ZKT*~△*TLD*, hence

(3) $\dfrac{AE}{TL} = \dfrac{AZ}{LD(=KB)}, \dfrac{AB}{AZ} = \dfrac{AZ}{KB},$ and

$\dfrac{TL(=AK)}{KT(=KB)} = \dfrac{LD(=KB)}{ZK}.$ Therefore

(4) $AB \cdot KB = AZ^2$ and
$ZK \cdot AK = KB^2$

and each of the lines *AZ* and *KB* > *AK.*   Q.E.D.

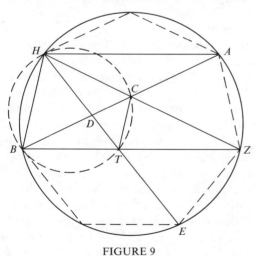

FIGURE 9

Proposition 17

We now wish to divide the circle into seven equal parts (Fig. 9). We draw the line segment *AB,* which we set out as known. We mark on it two points *C* and *D,* such that *AD* · *CD* = *DB*² and *CB* · *BD* = *AC*² and in addition each of the two segments *AC* and *DB* > *CD,*

following the preceding proposition [i.e., Prop. 16]. Out of lines *AC, CD* and *BD* we construct △*CHD.* Accordingly *CH* = *AC, DH* = *DB* and *CD* = *CD.* Then we circumscribe about △*AHB* the circle *AHBEZ* and we extend lines *HC* and *HD* directly up to the circumference of the circle. On their intersection with the circumference lie the points *Z* and *E.* We join *B* with *Z.* Lines *BZ* and *HE* intersect in *T.* We also draw *CT.* Since *AC* = *CH,* hence ∠*HAC* = ∠*AHC,* and arc *AZ* = arc *HB.* And, indeed, *AD* · *CD* = *DB*² = *DH*² and [by Euclid, VI.8] △*AHD*~△*CHD;* consequently ∠*DAH* = ∠*CHD,* or arc *ZE* = arc *BH.* Hence *BH, AZ* and *ZE* are three equal arcs. Further, *ZB* is parallel to *AH,* ∠*CAH* = ∠*CHD* = ∠*TBD; HD* = *DB, CD* = *DT, CH* = *BT.* Hence, [since the products of the parts of these diagonals are equal], the 4 points *B, H, C* and *T* lie in the circumference of one and the same circle. From the similarity of triangles *HBC* and *HBT,* it follows that *CB* · *DB* = *HC*² = *AC*² [or *HT*/*HC* = *HC*/*HD*] and from the similarity of △*THC* and △*CHD,* it follows that *TH* · *HD* = *HC*². And further *CB* = *TH* [these being equal diagonals in the quadrilateral] and ∠*DCH* = ∠*HTC* = 2∠*CAH.* [The equality of the first two angles arises from the similarity of triangles *THC* and *CHD.* Their equality with 2∠*CAH* arises as follows: (1) *AHD* = 2∠*CAH,* for ∠*CAH* = ∠*CHD* = ∠*CHA* and ∠*AHD* = ∠*CHA* + ∠*CHD;* (2) ∠*AHD* = ∠*BTH,* for parallel lines cut by a third line produce equal alternate angles; (3) ∠*BTH* = ∠*DCH,* from similar triangles; (4) hence ∠*DCH* = 2∠*CAH.*] [And since ∠*HBA* = ∠*DCH,* hence ∠*HBA* = 2∠*CAH.*] Consequently, arc *AH* = 2 arc *BH.* Since ∠*DHB* = ∠*DBH,* consequently arc *EB* = 2 arc *HB.* Hence, each of arcs *AH* and *EB* equals 2 arc *HB,* and accordingly the circle *AHBEZ* is divided into seven equal parts. Q.E.D. And praise be to the one God, etc.

The key to the whole procedure is, of course, the *neusis* presented in Proposition 16 (see Fig. 8) that would allow us in a similar fashion to find the points *C* and *D* in Proposition 17 (see Fig. 9). In Proposition 16 the *neusis* consisted in drawing a line from *D* to intersect the extension of *AB* in point *Z* such that △*AZE* = △*CTD.* The way in which the *neusis* was solved by Archimedes (or whoever was the author of this tract) is not known. Ibn al-Haytham, in his later treatment of the heptagon, mentions the Archimedean *neusis* but then goes on to show that one does not need the Archimedean square of Proposition 16. Rather he shows that points *C* and *D* in Proposition 17 can be found by the intersection of a parabola and a hyperbola.[11] It should be observed that all but two of Propositions 1–13 in this tract concern right triangles, and those two are necessary for propositions concerning right triangles. It seems probable, therefore, that Propositions 1–13 comprise the so-called *On* the *Properties of the Right Triangle* attributed in the *Fihrist* to Archimedes (although at least some of these propositions are Arabic interpolations). Incidentally, Propositions 7–10 have as their objective the formulation $K = (s - a) \cdot (s - c)$, where K is the area and a and c are the sides including the right angle and s is the semiperimeter, and Proposition 13 has as its objective $K = s(s - b)$, where b is the hypotenuse. Hence, if we multiply the two formulations, we have

$$K^2 = s(s - a) \cdot (s - b) \cdot (s - c)$$

or $\quad K = \sqrt{s(s - a) \cdot (s - b) \cdot (s - c)},$

Hero's formula for the area of a triangle in terms of its sides—at least in the case of a right triangle. Interestingly, the Arab scholar al-Bīrūnī attributed the general Heronian formula to Archimedes. Propositions 14 and 15 of the tract make no reference to Propositions 1–13 and concern chords. Each leads to a formulation in terms of chords equivalent to $\sin A/2 = \sqrt{(1 - \cos A)/2}$. Thus Propositions 14–15 seem to be from some other work (and at least Proposition 15 is an Arabic interpolation). If Proposition 14 was in the Greek text translated by Thābit ibn Qurra and does go back to Archimedes, then we would have to conclude that this formula was his discovery rather than Ptolemy's, as it is usually assumed to be.

The Latin West received its knowledge of Archimedes from both the sources just described: Byzantium and Islam. There is no trace of the earlier translations imputed by Cassiodorus to Boethius. Such knowledge that was had in the West before the twelfth century consisted of some rather general hydrostatic information that may have indirectly had its source in Archimedes. It was in the twelfth century that the translation of Archimedean texts from the Arabic first began. The small tract *On the Measurement of the Circle* was twice translated from the Arabic. The first translation was a rather defective one and was possibly executed by Plato of Tivoli. There are many numerical errors in the extant copies of it and the second half of Proposition 3 is missing. The second translation was almost certainly done by the twelfth century's foremost translator, Gerard of Cremona. The Arabic text from which he worked (without doubt the text of Thābit ibn Qurra) included a corollary on the area of a sector of a circle attributed by Hero to Archimedes but missing from our extant Greek text.

Not only was Gerard's translation widely quoted by medieval geometers such as Gerard of Brussels, Roger Bacon, and Thomas Bradwardine, it also served as the point of departure for a whole series of emended versions and paraphrases of the tract in the course of the thirteenth and fourteenth centuries. Among these are the so-called Naples, Cambridge, Florence, and Gordanus versions of the thirteenth century; and the Corpus Christi, Munich, and Albert of Saxony versions of the fourteenth. These versions were expanded by including pertinent references to Euclid and the spelling-out of the geometrical steps only implied in the Archimedean text. In addition, we see attempts to specify the postulates that underlie the proof of Proposition I. For example, in the Cambridge version three postulates (*petitiones*) introduce the text:[12] "[1] There is some curved line equal to any straight line and some straight line to any curved line. [2] Any chord is less than its arc. [3] The perimeter of any including figure is greater than the perimeter of the included figure." Furthermore, self-conscious attention was given in some versions to the logical nature of the proof of Proposition I. Thus, the Naples version immediately announced that the proof was to be *per impossibile*, i.e., by reduction to absurdity. In the Gordanus, Corpus Christi, and Munich versions we see a tendency to elaborate the proofs in the manner of scholastic tracts. The culmination of this kind of elaboration appeared in the *Questio de quadratura circuli* of Albert of Saxony, composed some time in the third quarter of the fourteenth century. The Hellenistic mathematical form of the original text was submerged in an intricate scholastic structure that included multiple terminological distinctions and the argument and counterargument technique represented by initial arguments ("principal reasons") and their final refutations.

Another trend in the later versions was the introduction of rather foolish physical justifications for postulates. In the Corpus Christi version, the second postulate to the effect that a straight line may be equal to a curved line is supported by the statement that "if a hair or silk thread is bent around circumference-wise in a plane surface and then afterwards is extended in a straight line, who will doubt—unless he is hare-brained—that the hair or thread is the same whether it is bent circumference-wise or extended in a straight line and is just as long the one time as the other." Similarly, Albert of Saxony, in his *Questio*, declared that a sphere can be "cubed" since the contents of a spherical vase can be poured into a cubical vase. Incidentally, Albert based his proof of the quadrature of the circle not directly on Proposition X.1 of the *Elements*, as was the case in the other medieval versions of *On the Measurement of the Circle*, but rather on a "betweenness" postulate: "I suppose that with two continuous [and comparable] quantities proposed, a magnitude greater than the 'lesser' can be cut from the 'greater.'" A similar postulate was employed in still another fourteenth-century version of the *De mensura circuli* called the Pseudo-Bradwardine version. Finally, in regard to the manifold medieval versions of *On the Measurement of the Circle*, it can be noted that the Florence version of Proposition 3 contained a detailed elaboration of the calculation of π. One might have supposed that the author had consulted Eutocius' commentary, except that his arithmetical procedures differed widely from those used by Eutocius. Furthermore, no translation of Eutocius' commentary appears to have been made before 1450, and the Florence version certainly must be dated before 1400.

In addition to his translation of *On the Measurement of the Circle*, Gerard of Cremona also translated the geometrical *Discourse of the Sons of Moses* (*Verba filiorum*) composed by the Banū Mūsā. This Latin translation was of particular importance for the introduction of Archimedes into the West. We can single out these contributions of the treatise: (1) A proof of Proposition I of *On the Measurement of the Circle* somewhat different from that of Archimedes but still fundamentally based on the exhaustion method. (2) A determination of the value of π drawn from Proposition 3 of the same treatise but with further calculations similar to those found in the commentary of Eutocius. (3) Hero's theorem for the area of a triangle in terms of its sides (noted above), with the first demonstration of that theorem in Latin (the enunciation of this theorem had already appeared in the writings of the *agrimensores* and in Plato of Tivoli's translation of the *Liber embadorum* of Savasorda). (4) Theorems for the volume and surface area of a cone, again with demonstrations. (5) Theorems for the volume and surface area of a sphere with demonstrations of an Archimedean character. (6) A use of the formula for the area of a circle equivalent to $A = \pi r^2$ in addition to the more common Archimedean form, $A = 1/2\,cr$. Instead of the modern symbol π the authors used the expression "the quantity which when multiplied by the diameter produces the circumference." (7) The introduction into the West of the problem of finding two mean proportionals between two given lines. In this treatise we find two solutions: (*a*) one attributed by the Banū Mūsā to Menelaus and by Eutocius to Archytas, (*b*) the other presented by the Banū Mūsā as their own but similar to the solution attributed by Eutocius to Plato. (8) The first solution in Latin of the problem of the trisection of an angle. (9) A method of approximating cube roots to any desired limit.

The *Verba filiorum* was, then, rich fare for the geometers of the twelfth century. The tract was quite widely cited in the thirteenth and fourteenth cen-

turies. In the thirteenth, the eminent mathematicians Jordanus de Nemore and Leonardo Fibonacci made use of it. For example, the latter, in his *Practica geometrie*, excerpted both of the solutions of the mean proportionals problem given by the Banū Mūsā, while the former (or perhaps a continuator) in his *De triangulis* presented one of them together with an entirely different solution, namely, that one assigned by Eutocius to Philo of Byzantium. Similarly, Jordanus (or possibly the same continuator) extracted the solution of the trisection of an angle from the *Verba filiorum*, but in addition made the remarkably perspicacious suggestion that the *neusis* can be solved by the use of a proposition from Ibn al-Haytham's *Optics*, which solves a similar *neusis* by conic sections.

Some of the results and techniques of *On the Sphere and the Cylinder* also became known through a treatise entitled *De curvis superficiebus Archimenidis* and said to be by Johannes de Tinemue. This seems to have been translated from the Greek in the early thirteenth century or at least composed on the basis of a Greek tract. The *De curvis superficiebus* contained ten propositions with several corollaries and was concerned for the most part with the surfaces and volumes of cones, cylinders, and spheres. This was a very popular work and was often cited by later authors. Like Gerard of Cremona's translation of *On the Measurement of the Circle*, the *De curvis superficiebus* was emended by Latin authors, two original propositions being added to one version (represented by manuscript D of the *De curvis superficiebus*)[13] and three quite different propositions being added to another (represented by manuscript M of the *De curvis*).[14] In the first of the additions to the latter version, the Latin author applied the exhaustion method to a problem involving the surface of a segment of a sphere, showing that at least this author had made the method his own. And indeed the geometer Gerard of Brussels in his *De motu* of about the same time also used the Archimedean *reductio* procedure in a highly original manner.

In 1269, some decades after the appearance of the *De curvis superficiebus*, the next important step was taken in the passage of Archimedes to the West when much of the Byzantine corpus was translated from the Greek by the Flemish Dominican, William of Moerbeke. In this translation Moerbeke employed Greek manuscripts A and B which had passed to the pope's library in 1266 from the collection of the Norman kings of the Two Sicilies. Except for *The Sandreckoner* and Eutocius' *Commentary on the Measurement of the Circle*, all the works included in manuscripts A and B were rendered into Latin by William. Needless to say, *On the Method, The Cattle*

Problem, and the *Stomachion,* all absent from manuscripts A and B, were not among William's translations. Although William's translations are not without error (and indeed some of the errors are serious), the translations, on the whole, present the Archimedean works in an understandable way. We possess the original holograph of Moerbeke's translations (MS Vat. Ottob. lat. 1850). This manuscript was not widely copied. The translation of *On Spirals* was copied from it in the fourteenth century (MS Vat. Reg. lat. 1253, 14r–33r), and several works were copied from it in the fifteenth century in an Italian manuscript now at Madrid (Bibl. Nac. 9119), and one work (*On Floating Bodies*) was copied from it in the sixteenth century (MS Vat. Barb. lat. 304, 124r–141v, 160v–161v). But, in fact, the Moerbeke translations were utilized more than one would expect from the paucity of manuscripts. They were used by several Schoolmen at the University of Paris toward the middle of the fourteenth century. Chief among them was the astronomer and mathematician John of Meurs, who appears to have been the compositor of a hybrid tract in 1340 entitled *Circuli quadratura.* This tract consisted of fourteen propositions. The first thirteen were drawn from Moerbeke's translation of *On Spirals* and were just those propositions necessary to the proof of Proposition 18 of *On Spirals,* whose enunciation we have quoted above. The fourteenth proposition of the hybrid tract was Proposition 1 from Moerbeke's translation of *On the Measurement of the Circle.* Thus this author realized that by the use of Proposition 18 from *On Spirals,* he had achieved the necessary rectification of the circumference of a circle preparatory to the final quadrature of the circle accomplished in *On the Measurement of the Circle,* Proposition 1. Incidentally, the hybrid tract did not merely use the Moerbeke translations verbatim but also included considerable commentary. In fact, this medieval Latin tract was the first known commentary on Archimedes' *On Spirals.* That the commentary was at times quite perceptive is indicated by the fact that the author suggested that the *neusis* introduced by Archimedes in Proposition 7 of *On Spirals* could be solved by means of an *instrumentum conchoydeale.* The only place in which a medieval Latin commentator could have learned of such an instrument would have been in that section of the *Commentary on the Sphere and the Cylinder* where Eutocius describes Nicomedes' solution of the problem of finding two mean proportionals (Bk. II, Prop. 1). We have further evidence that John of Meurs knew of Eutocius' *Commentary* in the Moerbeke translation when he used sections from this commentary in his *De arte mensurandi* (Ch. VIII, Prop. 16), where three

of the solutions of the mean proportionals problem given by Eutocius are presented. Not only did John incorporate the whole hybrid tract *Circuli quadratura* into Chapter VIII of his *De arte mensurandi* (composed, it seems, shortly after 1343) but in Chapter X of the *De arte* he quoted verbatim many propositions from Moerbeke's translations of *On the Sphere and the Cylinder* and *On Conoids and Spheroids* (which latter he misapplied to problems concerning solids generated by the rotation of circular segments). Within the next decade or so after John of Meurs, Nicole Oresme, his colleague at the University of Paris, in his *De configurationibus qualitatum et motuum* (Part I, Ch. 21) revealed knowledge of *On Spirals,* at least in the form of the hybrid *Circuli quadratura.* Further, Oresme in his *Questiones super de celo et mundo,* quoted at length from Moerbeke's translation of *On Floating Bodies,* while Henry of Hesse, Oresme's junior contemporary at Paris, quoted briefly therefrom. (Before this time, the only knowledge of *On Floating Bodies* had come in a thirteenth-century treatise entitled *De ponderibus Archimenidis sive de incidentibus in humidum,* a Pseudo-Archimedean treatise prepared largely from Arabic sources, whose first proposition expressed the basic conclusion of the "principle of Archimedes": "The weight of any body in air exceeds its weight in water by the weight of a volume of water equal to its volume.") Incontrovertible evidence, then, shows that at the University of Paris in the mid-fourteenth century six of the nine Archimedean translations of William of Moerbeke were known and used: *On Spirals, On the Measurement of the Circle, On the Sphere and the Cylinder, On Conoids and Spheroids, On Floating Bodies,* and Eutocius' *Commentary on the Sphere and the Cylinder.* While no direct evidence exists of the use of the remaining three translations, there has been recently discovered in a manuscript written at Paris in the fourteenth century (BN lat. 7377B, 93v–94r) an Archimedean-type proof of the law of the lever that might have been inspired by Archimedes' *On the Equilibrium of Planes.* But other than this, the influence of Archimedes on medieval statics was entirely indirect. The anonymous *De canonio,* translated from the Greek in the early thirteenth century, and Thābit ibn Qurra's *Liber karastonis,* translated from the Arabic by Gerard of Cremona, passed on this indirect influence of Archimedes in three respects: (1) Both tracts illustrated the Archimedean type of geometrical demonstrations of statical theorems and the geometrical form implied in weightless beams and weights that were really only geometrical magnitudes. (2) They gave specific reference in geometrical language to the law of the lever (and in the *De canonio* the law of the lever is connected directly to Archimedes). (3) They indirectly reflected the centers-of-gravity doctrine so important to Archimedes, in that both treatises employed the practice of substituting for a material beam segment a weight equal in weight to the material segment but hung from the middle point of the weightless segment used to replace the material segment. Needless to say, these two tracts played an important role in stimulating the rather impressive statics associated with the name of Jordanus de Nemore.

In the fifteenth century, knowledge of Archimedes in Europe began to expand. A new Latin translation was made by James of Cremona in about 1450 by order of Pope Nicholas V. Since this translation was made exclusively from manuscript A, the translation failed to include *On Floating Bodies,* but it did include the two treatises in A omitted by Moerbeke, namely, *The Sandreckoner* and Eutocius' *Commentary on the Measurement of the Circle.* It appears that this new translation was made with an eye on Moerbeke's translations. Not long after its completion, a copy of the new translation was sent by the pope to Nicholas of Cusa, who made some use of it in his *De mathematicis complementis,* composed in 1453–1454. There are at least nine extant manuscripts of this translation, one of which was corrected by Regiomontanus and brought to Germany about 1468 (the Latin translation published with the *editio princeps* of the Greek text in 1544 was taken from this copy). Greek manuscript A itself was copied a number of times. Cardinal Bessarion had one copy prepared between 1449 and 1468 (MS E). Another (MS D) was made from A when it was in the possession of the well-known humanist George Valla. The fate of A and its various copies has been traced skillfully by J. L. Heiberg in his edition of Archimedes' *Opera.* The last known use of manuscript A occurred in 1544, after which time it seems to have disappeared. The first printed Archimedean materials were in fact merely Latin excerpts that appeared in George Valla's *De expetendis et fugiendis rebus opus* (Venice, 1501) and were based on his reading of manuscript A. But the earliest actual printed texts of Archimedes were the Moerbeke translations of *On the Measurement of the Circle* and *On the Quadrature of the Parabola* (*Tetragonismus, id est circuli quadratura etc.*), published from the Madrid manuscript by L. Gaurico (Venice, 1503). In 1543, also at Venice, N. Tartaglia republished the same two translations directly from Gaurico's work, and, in addition, from the same Madrid manuscript, the Moerbeke translations of *On the Equilibrium of Planes* and Book I of *On Floating Bodies* (leaving the erroneous impression that he had made these translations

from a Greek manuscript, which he had not since he merely repeated the texts of the Madrid manuscript with virtually all their errors). Incidentally, Curtius Trioianus published from the legacy of Tartaglia both books of *On Floating Bodies* in Moerbeke's translation (Venice, 1565). The key event, however, in the further spread of Archimedes was the aforementioned *editio princeps* of the Greek text with the accompanying Latin translation of James of Cremona at Basel in 1544. Since the Greek text rested ultimately on manuscript A, *On Floating Bodies* was not included. A further Latin translation of the Archimedean texts was published by the perceptive mathematician Federigo Commandino in Bologna in 1558, which the translator supplemented with a skillful mathematical emendation of Moerbeke's translation of *On Floating Bodies* (Bologna, 1565) but without any knowledge of the long lost Greek text. Already in the period 1534–1549, a paraphrase of Archimedean texts had been made by Francesco Maurolico. This was published in Palermo in 1685. One other Latin translation of the sixteenth century by Antonius de Albertis remains in manuscript only and appears to have exerted no influence on mathematics and science. After 1544 the publications on Archimedes and the use of his works began to multiply markedly. His works presented quadrature problems and propositions that mathematicians sought to solve and demonstrate not only with his methods, but also with a developing geometry of infinitesimals that was to anticipate in some respect the infinitesimal calculus of Newton and Leibniz. His hydrostatic conceptions were used to modify Aristotelian mechanics. Archimedes' influence on mechanics and mathematics can be seen in the works of such authors as Commandino, Guido Ubaldi del Monte, Benedetti, Simon Stevin, Luca Valerio, Kepler, Galileo, Cavalieri, Torricelli, and numerous others. For example, Galileo mentions Archimedes more than a hundred times, and the limited inertial doctrine used in his analysis of the parabolic path of a projectile is presented as an Archimedean-type abstraction. Archimedes began to appear in the vernacular languages. Tartaglia had already rendered into Italian Book I of *On Floating Bodies,* Book I of *On the Sphere and the Cylinder,* and the section on proportional means from Eutocius' *Commentary on the Sphere and the Cylinder.* Book I of *On the Equilibrium of Planes* was translated into French in 1565 by Pierre Forcadel. It was, however, not until 1670 that a more or less complete translation was made into German by J. C. Sturm on the basis of the influential Greek and Latin edition of David Rivault (Paris, 1615). Also notable for its influence was the new Latin edition of Isaac Barrow (London,

1675). Of the many editions prior to the modern edition of Heiberg, the most important was that of Joseph Torelli (Oxford, 1792). By this time, of course, Archimedes' works had been almost completely absorbed into European mathematics and had exerted their substantial and enduring influence on early modern science.

NOTES

1. Heath, *The Works of Archimedes,* pp. 91–93. Heath's close paraphrase has been used here and below because of its economy of expression. While he uses modern symbols and has reduced the general enunciations to statements concerning specific figures in some of the propositions quoted below, he nevertheless achieves a faithful representation of the spirit of the original text.
2. *Ibid.,* pp. 19–20.
3. *Ibid.,* 251–252.
4. *Ibid.,* pp. 156–57.
5. *Ibid.,* pp. 1–2.
6. *Ibid.,* Suppl., pp. 18–22.
7. *Ibid.,* pp. 13–14.
8. Dijksterhuis, *Archimedes,* p. 373.
9. Clagett, *Archimedes in the Middle Ages,* pp. 667–668.
10. Schoy, *Die trigonometrischen Lehren,* pp. 82–83.
11. *Ibid.,* pp. 85–91.
12. Clagett, *op. cit.,* p. 27. The succeeding quotations from the various versions of *On the Measurement of the Circle* are also from this volume.
13. *Ibid.,* p. 520.
14. *Ibid.,* p. 530.

BIBLIOGRAPHY

I. Original Works.

 1. *The Greek Text and Modern Translations.* J. L. Heiberg, ed., *Archimedis opera omnia cum commentariis Eutocii,* 2nd ed., 3 vols. (Leipzig, 1910–1915). For the full titles of the various editions cited in the body of the article as well as others, see E. J. Dijksterhuis, *Archimedes* (Copenhagen, 1956), pp. 40–45, 417. Of recent translations and paraphrases, the following, in addition to Dijksterhuis' brilliant analytic summary, ought to be noted: T. L. Heath, *The Works of Archimedes,* edited in modern notation, with introductory chapters (Cambridge, 1897), which together with his *Supplement, The Method of Archimedes* (Cambridge, 1912) was reprinted by Dover Publications (New York, 1953); P. Ver Eecke, *Les Oeuvres complètes d'Archimède, suivies des commentaires d'Eutocius d'Ascalon,* 2nd ed., 2 vols. (Paris, 1960); I. N. Veselovsky, *Archimedes. Selections, Translations, Introduction, and Commentary* (in Russian), translation of the Arabic texts by B. A. Rosenfeld (Moscow, 1962). We can also mention briefly the German translations of A. Czwalina and the modern Greek translations of E. S. Stamates.

 2. *The Arabic Archimedes* (the manuscripts cited are largely from Suter, "Die Mathematiker und Astronomen" [see Secondary Literature], and C. Brockelmann, *Geschichte der arabischen Literatur,* 5 vols., Vols. I–II [adapted to Suppl. vols., Leiden, 1943–1949], Suppl. Vols. I–III [Leiden, 1937–1942]). *On the Sphere and the Cylinder* and *On the Measurement of the Circle;* both appear in Nāṣir al-Dīn al-Ṭūsī, *Majmūʿ al-Rasāʾil,* Vol. II (Hyderabad, 1940). Cf. MSS Berlin 5934; Florence Palat. 271 and 286; Paris 2467; Oxford, Bodl. Arabic 875, 879; India Office 743; and M. Clagett, *Archimedes in the Middle Ages,* I, 17, n. 8. The al-Ṭūsī edition also contains some commentary on Bk. II of *On the Sphere and the Cylinder. Book of the Elements of Geometry* (probably the same as *On Triangles,* mentioned in the *Fihrist*) and *On Touching Circles;* both appear in *Rasāʾil Ibn Qurra* (Hyderabad, 1947, given as 1948 on transliterated title page). *On the Division of the Circle into Seven Equal Parts* (only Props. 16–17 concern heptagon construction; Props. 1–13 appear to be the tract called *On the Properties of the Right Triangle;* Props. 14–15 are unrelated to either of other parts). MS Cairo A.-N.8 H.-N. 7805, item no. 15. German translation by C. Schoy, *Die trigonometrischen Lehren des persischen Astronomen Abu ʾl-Raihân Muh. ibn Ahmad al-Bîrûnî* (Hannover, 1927), pp. 74–84. The text has been analyzed in modern fashion by J. Tropfke, "Die Siebeneckabhandlung des Archimedes," in *Osiris,* **1** (1936), 636–651. *On Heaviness and Lightness* (a fragment of *On Floating Bodies*); Arabic text by H. Zotenberg in *Journal asiatique;* Ser. 7, **13** (1879), 509–515, from MS Paris, BN Fonds suppl. arabe 952 bis. A German translation was made by E. Wiedemann in the *Sitzungsberichte der Physikalisch-medizinischen Sozietät in Erlangen,* **38** (1906), 152–162. For an English translation and critique, see M. Clagett, *The Science of Mechanics in the Middle Ages* (Madison, Wis., 1959, 2nd pr., 1961), pp. 52–55. *Lemmata* (*Liber assumptorum*), see the edition in al-Ṭūsī, *Majmūʿ al-Rasāʾil,* Vol. II (Hyderabad, 1940). MSS Oxford, Bodl. Arabic 879, 895, 939, 960; Leiden 982; Florence, Palat. 271 and 286; Cairo A.-N. 8 H.-N 7805. This work was first edited by S. Foster, *Miscellanea* (London, 1659), from a Latin translation of I. Gravius; Abraham Ecchellensis then retranslated it, the new translation being published in I. A. Borelli's edition of *Apollonii Pergaei Concicorum libri V, VI, VII* (Florence, 1661). Ecchellensis' translation was republished by Heiberg, *Opera,* II, 510–525. See also E. S. Stamates' effort to reconstruct the original Greek text in *Bulletin de la Societé Mathématique de Grèce,* new series, **6 II,** Fasc. 2 (1965), 265–297. *Stomachion,* a fragmentary part in Arabic with German translation in H. Suter, "Der Loculus Archimedius oder das Syntemachion des Archimedes," in *Abhandlungen zur Geschichte der Mathematik,* **9** (1899), 491–499. This is one of two fragments. The other is in Greek and is given by Heiberg, *Opera,* II, 416. Eutocius, *Commentary on the Sphere and the Cylinder,* a section of Bk. II. MSS Paris, BN arabe 2457, 44°; Bibl. Escor. 960; Istanbul, Fatīh Mosque Library Ar. 3414, 60v–66v; Oxford, Bodl. Arabic 875 and 895. Various tracts and commentaries *On the Sphere and the Cylinder,* Bk. II, in part paraphrased and translated by F. Woepcke, *L'Algebra d'Omar Alkhayâmmî* (Paris, 1851), pp. 91–116.

 3. *The Medieval Latin Archimedes.* A complete edition and translation of the various Archimedean tracts arising

from the Arabic tradition have been given by M. Clagett, *Archimedes in the Middle Ages,* Vol. I (Madison, Wis., 1964). Vol. II will contain the complete text of Moerbeke's translations and other Archimedean materials from the late Middle Ages. Moerbeke's translation of *On Spirals* and brief parts of other of his translations have been published by Heiberg, "Neue Studien" (see below). See also M. Clagett, "A Medieval Archimedean-Type Proof of the Law of the Lever," in *Miscellanea André Combes,* II (Rome, 1967), 409–421. For the Pseudo-Archimedes, *De ponderibus (De incidentibus in humidum),* see E. A. Moody and M. Clagett, *The Medieval Science of Weights* (Madison, 1952; 2nd printing, 1960), pp. 35–53, 352–359.

II. SECONDARY LITERATURE. The best over-all analysis is in E. J. Dijksterhuis, *Archimedes* (Copenhagen, 1956), which also refers to the principal literature. The translations of Heath and Ver Eecke given above contain valuable evaluative and biographical materials. In addition, consult C. Boyer, *The Concepts of the Calculus* (New York, 1939; 2nd printing, 1949; Dover ed. 1959), particularly ch. 4 for the reaction of the mathematicians of the sixteenth and seventeenth centuries to Archimedes. M. Clagett, "Archimedes and Scholastic Geometry," in *Mélanges Alexandre Koyré,* Vol. I: *L'Aventure de la science* (Paris, 1964), 40–60; "The Use of the Moerbeke Translations of Archimedes in the Works of Johannes de Muris," in *Isis,* **43** (1952), 236–242 (the conclusions of this article will be significantly updated in M. Clagett, *Archimedes in the Middle Ages,* Vol. II); and "Johannes de Muris and the Problem of the Mean Proportionals," in *Medicine, Science and Culture, Historical Essays in Honor of Owsei Temkin,* L. G. Stevenson and R. P. Multhauf, eds. (Baltimore, 1968), 35–49. A. G. Drachmann, "Fragments from Archimedes in Heron's Mechanics," in *Centaurus,* **8** (1963), 91–145; "The Screw of Archimedes," in *Actes du VIIIe Congrés international d'Histoire des Sciences Florence-Milan 1956,* **3** (Vinci-Paris, 1958), 940–943; and "How Archimedes Expected to Move the Earth," in *Centaurus,* **5** (1958), 278–282. J. L. Heiberg, "Neue Studien zu Archimedes," in *Abhandlungen zur Geschichte der Mathematik,* **5** (1890), 1–84; and *Quaestiones Archimedeae* (Copenhagen, 1879). Most of the biographical references are given here by Heiberg. S. Heller, "Ein Fehler in einer Archimedes-Ausgabe, seine Entstehung und seine Folgen," in *Abhandlungen der Bayerischen Akademie der Wissenschaften. Mathematisch-naturwissenschaftliche Klasse,* new series, **63** (1954), 1–38. E. Rufini, *Il "Metodo" di Archimede e le origini dell'analisi infinitesimale nell'antichita* (Rome, 1926; new ed., Bologna, 1961). H. Suter, "Die Mathematiker und Astronomen der Araber und ihre Werke," in *Abhandlungen zur Geschichte der mathematischen Wissenschaften,* **10** (1892), *in toto;* "Das Mathematiker- Verzeichniss im Fihrist des Ibn Abī Jaʿkûb an-Nadîm," *ibid.,* **6** (1892), 1–87. B. L. Van der Waerden, *Erwachende Wissenschaft,* 2nd German ed. (Basel, 1966), pp. 344–381. See also the English translation, *Science Awakening,* 2nd ed. (Groningen, 1961), pp. 204–206, 208–228. E. Wiedemann, "Beiträge zur Geschichte der Naturwissenschaften III," in *Sitzungsberichte der Physikalisch-medizinischen Sozietät in Er-langen,* **37** (1905), 247–250, 257. A. P. Youschkevitch, "Remarques sur la méthode antique d'exhaustion," in *Mélanges Alexandre Koyré,* I: *L'Aventure de la science* (1964), 635–653.

MARSHALL CLAGETT

ARCHYTAS OF TARENTUM (*fl.* Tarentum [now Taranto], Italy, *ca.* 375 B.C.), *philosophy, mathematics, physics.*

After the Pythagoreans had been driven out of most of the cities of southern Italy by the Syracusan tyrant Dionysius the Elder at the beginning of the fourth century B.C., Tarentum remained their only important political center. Here Archytas played a leading role in the attempt to unite the Greek city-states against the non-Greek tribes and powers. After the death of Dionysius the Elder, he concluded, through the agency of Plato, an alliance with his son and successor, Dionysius the Younger.

Archytas made very important contributions to the theory of numbers, geometry, and the theory of music. Although extant ancient tradition credits him mainly with individual discoveries, it is clear that all of them were connected and that Archytas was deeply concerned with the foundations of the sciences and with their interconnection. Thus he affirmed that the art of calculation (λογιστική) is the most fundamental science and makes its results even clearer than those of geometry. He also discussed mathematics as the foundation of astronomy.

A central point in Archytas' manifold endeavors was the theory of means (μεσότητες) and proportions. He distinguished three basic means: the arithmetic mean of the form $a - b = b - c$ or $a + c = 2b$; the geometric mean of the form $a:b = b:c$ or $ac = b^2$; and the harmonic mean of the form $(a - b):(b - c) = a:c$. Archytas and later mathematicians subsequently added seven other means.

A proposition and proof that are important both for Archytas' theory of means and for his theory of music have been preserved in Latin translation in Boethius' *De musica.* The proposition states that there is no geometric mean between two numbers that are in "superparticular" (ἐπιμόριος) ratio, i.e., in the ratio $(n + 1):n$. The proof given by Boethius is essentially identical with that given for the same proposition by Euclid in his *Sectio canonis* (Prop. 3). It presupposes several propositions of Euclid that appear in *Elements* VII as well as VIII, Prop. 8. Through a careful analysis of Books VII and VIII and their relation to the above proof, A. B. L. Van der Waerden has succeeded in making it appear very likely that many of the theorems in Euclid's *Elements* VII and their proofs existed before Archytas, but that a considerable part of the propositions and proofs of VIII were added by Archytas and his collaborators.

Archytas' most famous mathematical achievement was the solution of the "Delian" problem of the duplication of the cube. A generation before Archytas, Hippocrates of Chios had demonstrated that the problem can be reduced to the insertion of two mean proportionals between the side of the cube and its double length: If a is the side of the cube and $a:x = x:y = y:2a$, then x is the side of the doubled cube. The problem of the geometrical construction of this line segment was solved by Archytas through a most ingenious three-dimensional construction. In the figure below, everything, according to the custom of the ancients, is projected into a plane.

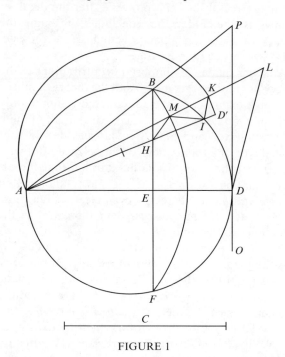

FIGURE 1

Let AD and C be the two line segments between which the two mean proportionals are to be constructed. Let a circle $ABDF$ be drawn with AD as diameter and $AB = C$ as a chord. Then let the extension of AB cut the line tangent to the circle at D at point P. Let BEF be drawn parallel to PDO. Next, imagine a semicylinder above the semicircle ABD (i.e., with ABD as its base) and above AD another semicircle located perpendicularly in the rectangle of the semicylinder (i.e., with AD as diameter in a plane perpendicular to plane $ABDF$). Let this semicircle be rotated around point A in the direction of B. Being rotated in this way, it will cut the surface of the semicylinder and will describe a curve.

If $\triangle APD$ is turned around the axis AD in the direction opposite to that of the semicircle, its side AP will describe the surface of a circular cone and in doing so will cut the aforementioned curve on the surface of the semicylinder at a point. At the same time, point B will describe a semicircle in the surface of the cone. Then, at the moment in which the aforementioned curves cut one another, let the position of the moving semicircle be determined by points AKD' and that of the moving triangle by points ALD, and let the point of intersection be called K. The semicircle described by B will then be BMF (namely, B at the moment of intersection of the curves being in M). Then drop a perpendicular from K to the plane of semicircle $ABDF$. It will fall on the circumference of the circle, since the cylinder is a right cylinder. Let it meet the circumference in I, and let the line drawn from I to A cut BF at H, and let line AL meet semicircle BMF in M (see above). Also let the connecting lines KD', MI, and MH be drawn.

Since each of the two semicircles AKD' and BMF is then perpendicular to the underlying plane ($ABDF$), their common intersection MH is also perpendicular to the plane of that circle, i.e., MH is perpendicular to BF. Hence the rectangle determined by BH, HF, and likewise that determined by AH, HI, is equal to the square on HM. Hence $\triangle AMI$ is similar to $\triangle HMI$ and AHM and $\angle AMI$ is a right angle. But $\angle D'KA$ is also a right angle. Hence $KD' \parallel MI$. Therefore $D'A:AK = KA:AI = AI:AM$ because of the similarity of the triangles. The four line segments $D'A$, AK, AI, and AM are therefore in continuous proportion. Thus, between the two given line segments AD and C (i.e., AB) two mean proportionals, AK and AI, have been found.

Tannery suspected and Van der Waerden, through an ingenious interpretation of *Epinomis* 990E, tried to show how the theory of means also served Archytas as a basis for his theory of music. Starting from the octave 1:2 or 6:12, one obtains the arithmetic mean 9 and the harmonic mean 8. The ratio 6:9 = 8:12 is 2:3 or, in musical terms, the fifth. The ratio 6:8 = 9:12 is 3:4 or, in musical terms, the fourth. Forming in like manner the arithmetic and harmonic means of the fifth, one obtains the ratios 4:5 and 5:6 or, in musical terms, the intervals of the major third and the minor third. Using the same procedure with the fourth, one obtains 6:7, or the diminished minor third, and 7:8, or an augmented whole tone. On these intervals Archytas built his three musical scales: the enharmonic, the chromatic, and the diatonic. The theory is also related to the theorem (mentioned above) that there is no geometric mean between two numbers in superparticular ratio. Since all the basic musical intervals are in that ratio, they can be sub-divided by means of the arithmetic and the harmonic means but not by the geometric mean.

Archytas also elaborated a physical theory of sound, which he expounded in the longest extant fragment of his works. He starts with the observation that arithmetic, geometry, astronomy, and the theory of music are all related, but then proceeds to draw conclusions from empirical observations that are not subjected to mathematical analysis within the fragment. The fundamental observation is that faster motion appears to produce higher sounds. Thus, the bull-roarers used at certain religious festivals produce a higher sound when swung around swiftly than when swung more slowly. There are other easy experiments that confirm this observation. For instance, when the holes of a flute that are nearest the mouth of the flutist are opened, a higher sound is produced than when the farther holes are opened. Archytas reasoned that the air pressure in the first case ought to be higher, and therefore the air motion should be faster than in the second case. This is true of the frequencies of the air impulses produced, but Archytas appears to have concluded that the higher sounds reach the ear of the listener more quickly than the lower ones. Thus, he can hardly have applied his arithmetical theory of music consistently to his theory of the production of sound, or he would almost certainly have discovered his error.

Archytas is also credited with the invention of a wooden dove that could fly.

BIBLIOGRAPHY

The surviving fragments of Archytas' work are in H. Diels and W. Kranz, *Fragmente der Vorsokratiker,* 6th ed. (Berlin, 1951). The following are most relevant to this article: 47A16, on the musical scales; 47A19, on the super-particular ratio (also in a Latin translation in Boethius' *De musica,* III, 11); 47B1, on the mathematical foundations of astronomy; 47B2, on the three basic means; Pappus, *Synagogê,* VIII, 13 (F. Hultsch, ed., I, 85 ff.), on other means; and 47B4, on the art of calculation. A three-dimensional drawing for the doubling of the cube is in Diels and Kranz, 6th ed., I, 426.

Writings related to Archytas or his work are T. L. Heath, *A History of Greek Mathematics,* I (Oxford, 1921), 215; P. Tannery, *Mémoires scientifiques,* III (Paris, 1915), 105; M. Timpanaro Cardini, *Pitagorici. Testimonianze e frammenti,* II (Florence, 1962), 226–384; and A. B. L. Van der Waerden, "Die Harmonielehre der Pythagoreer," in *Hermes,* **78** (1943), 184 ff.; *Mathematische Annalen,* **130** (1948), 127 ff.; and *Science Awakening* (Groningen, 1954), pp. 151, 153 f.

KURT VON FRITZ

ARDUINO (or **ARDUINI**), **GIOVANNI** (*b.* Caprino Veronese, Italy, 16 October 1714; *d.* Venice, Italy, 21 March 1795), *geology.*

Arduino was born of poor parents, and it was only through the interest of the Marchese Andrea Carlotti, who was impressed by the boy's intelligence and by his aptitude for the exact sciences, that he was able to complete a good literary and mathematical education at Verona. He did not, however, take a degree. Arduino spent much of his youth working in the mines of Chiusa di Bressanone (in the Upper Adige valley), and was soon able to establish himself as a mining expert.

Today Arduino is counted among the founders of geology: by applying Galilean methodology to investigations of the earth's structure and composition for the first time, he achieved results of lasting validity. He was, above all, a typical representative of the utilitarian rationalism characteristic of the eighteenth-century European mind. In Italy many governments turned to natural scientists for aid in solving the serious economic problems that resulted from a great increase in population. Scientists were employed to search for new or neglected resources, and to study and develop technological procedures that would be of economic advantage. In the Republic of Venice, of which Arduino was a citizen, the excellent reputation that he had earned as a scientist and technician, both as an independent mining expert and in the service of the city of Venice, led the Senate, in 1769, to entrust him with an important office concerned with the development of agriculture and industry throughout the republic.

The diversity of problems faced by Arduino in agriculture and industry is striking. He made each problem the subject of thorough study, most often accompanied by original experimental research, especially in chemistry and metallurgy. A few of his concerns were the reclamation of marshy lands, the fattening of cattle for slaughter, the construction of agricultural equipment for cultivating grain and rice, a chemical study of the ashes of marine plants to be used in glassmaking, and a metallurgical study of the best way of working iron to obtain the best possible castings. He sometimes had the collaboration of his brother Pietro (1728–1805), a botanist who was appointed to the first chair of agriculture at the University of Padua.

Arduino's lifelong passion, however, was mining. Following his early work in the mines of the Upper Adige valley, Arduino worked for an English copper and lead mining company in Montieri and Boccheggiano (Massa Marittima) in 1773 and for a mercury mining operation in Monte Amiata in 1775.

Even during his stay in Vicenza and in Venice, he did not miss an opportunity to visit and study mine sites; and in view of this interest, the Venetian Senate commissioned him to make a complete study of all the mines in the republic's territory.

Arduino's contribution to mining was not merely economic, for he was also receptive to the new discipline of geology. He used to say, "I have always loved to begin with facts, to observe them, to walk in the light of experiment and demonstration as much as possible, and to discuss the results." By observing the phenomena of nature without prejudice or consideration of the opinions of contemporary scientists, from John Woodward to William Whiston, he was able to identify four very distinct geological units of successively later periods in the Atesine Alps, the foothills of the Alps, the subalpine hills, and the plains of the Po. Discussing his observations, he established the bases for modern stratigraphic chronology: "From whatever I have been able to observe up to this time, the series of strata which form the visible crust of the earth appear to me classified in four general and successive orders. These four orders can be conceived to be four very large strata, as they really are, so that wherever they are exposed, they are disposed one above the other, always in the same way." He called these orders Primary, Secondary, Tertiary, and Quaternary, specifying that each was constituted of innumerable minor strata of different materials formed at different times under different conditions. He proposed a further subdivision only for the Primary order, distinguishing a lower part constituted of metamorphic rocks and an upper part of formerly sedimentary, calcareous, and arenaceous rocks. In consequence, he attributed to the Primary order the Paleozoic formations of the Atesine Alps; to the Secondary order, the Mesozoic prealpine formations from Lombardy to Venezia; and to the Tertiary order, the band of subalpine hills from Lombardy to Venezia and the Pliocene hills of Tuscany. In the Quaternary order he considered the alluvial deposits of the plain that stretches to the foot of the Alps.

Arduino's vision of paleontology was correspondingly clear. Not only did he understand that fossil species change according to variations in the age of the terrains in which they occur, but he also realized that in the Secondary order "unrefined and imperfect" species are found, while in the Tertiary order the species are "very perfect and wholly similar to those that are seen in the modern sea." He affirmed, as a result, that "as many ages have elapsed during the elevation of the Alps, as there are races of organic fossil bodies embedded within the strata." Arduino's investigations extended to magmatic rocks; he iden-

tified, among other things, the trachytic origin of the Euganean Hills and of the basaltic rocks in the area of Verona and Vicenza, rocks he linked to "ancient extinct volcanoes." He also realized the transformations that magmatic rocks can work upon preexistent sedimentary rocks. In these special researches, as in his general study of rocks and minerals, he utilized his skill as a chemist.

Arduino was one of the most brilliant precursors of actualism: "With the sole guidance of our practical knowledge of those physical agents which we see actually used in the continuous workings of nature, and of our knowledge of the respective effects induced by the same workings, we can with reasonable basis surmise what the forces were which acted even in the remotest times."

Arduino's fundamental ideas were diffused throughout Europe by means of his publications, his frequent letters to and conferences with Italian and foreign scholars, and, above all, the active interest of Ignaz de Born and Johann Jacob Ferber. De Born translated a collection of Arduino's writings into German, and Ferber fervently expounded Arduino's geological results in his well-known work on the natural history of Italy.

BIBLIOGRAPHY

I. Original Works. Arduino's geological writings, of which there is no collected critical edition, include "Due lettere del Signor Giovanni Arduino sopra varie sue osservazioni naturali," in *Nuova raccolta d'opuscoli scientifici e filologici* (*dell' abate Calogerà*), **6** (1760), 97, 133; *Raccolta di memorie chimico-mineralogiche, metallurgiche e orittografiche* (Venice, 1775), also translated into German by Ignaz de Born (Dresden, 1778); and "Effetti di antichissimi vulcani estinti . . . nei monti della villa di Chiampo, ed in altri luoghi del territorio di Vicenza e di quello di Verona," in *Nuovo giornale d'Italia,* **7** (1783), 163.

II. Secondary Literature. There is no definitive biography, but see T. A. Catullo, *Elogio di Giovanni Arduino* (Padua, 1835), which has an incomplete bibliography of Arduino's works; G. B. Ronconi, *Giovanni Arduino e le miniere di Toscana* (Padua, 1865); and G. Stegagno, *Il veronese Giovanni Arduino e il suo contributo alla scienza geologica* (Verona, 1929), in which sketches and sections are reproduced from unpublished manuscripts that show Arduino to be a precursor of geological cartography.

Francesco Rodolico

ARETAEUS OF CAPPADOCIA (*fl. ca.* A.D. 50), *medicine.*

Aretaeus is known for his text on the causes, symptoms, and treatment of acute and chronic diseases.

The dates of this physician, who was rarely cited in antiquity, have long been a matter of dispute, but it should now be clear that Aretaeus belongs to the middle of the first century and that he was a contemporary of the famous pharmacologist Pedanius Dioscorides, who cites him once. Further, since Aretaeus probably knew Andromachos, Nero's personal physician (to whom Dioscorides dedicated his work), we may conclude that at some time he had resided in Rome. Nothing more is known about his life.

Aretaeus belonged to the so-called Pneumatic school of physicians, which had been founded in the first century B.C. by Athenaeus of Attalia, who had studied under the Stoic Posidonius. Since Aretaeus is the only one of the Pneumatic school whose work has come down to us intact, he is a valuable source in several respects: (1) One can trace in his work several specific influences of the Stoic-Posidonian ideas and influences in medicine (for instance, the idea of the soul's ability to predict). (2) Aretaeus' position on the physician's compassion for the patient, for instance, reveals that the Pneumatics took a position close to that of the early Christians, and thus forms an important link between medicine and early Christianity (which actually was not very affable to medicine). (3) Aretaeus reveals that the "orthodox" Pneumatic school of the first century led to a strong revival of the doctrine of Hippocrates as well as of the Ionic dialect within medical literature. Since Aretaeus also used Homeric words and modes of expression, he represents an important example of the Greek style of prose in imperial times, especially that of the so-called second sophistic.

BIBLIOGRAPHY

Aretaeus' text on acute and chronic diseases, the original title of which is unknown, is edited by C. Hude in *Corpus medicorum Graecorum*, 2nd ed., II (Berlin, 1958): for the physician's compassion, see p. 7, ll. 17 ff.; for the soul's mantic power, see p. 22, ll. 26 ff.

Dioscorides' reference to Aretaeus is *Pedanii Dioscuridis Anazarbei De materia medica*, M. Wellmann, ed., 2nd ed., III (Berlin, 1958), 298, l. 19. For the dating of Aretaeus' life and for the Stoic-Posidonic and Christian features of the Pneumatics and of Aretaeus, see F. Kudlien, "Untersuchungen zu Aretaios von Kappadokien," in *Abhandlungen der Akademie der Wissenschaften und Literatur in Mainz*, no. 11 (1963), ch. 1. See also "Pneumatische Ärzte," in Pauly-Wissowa, *Real-Encyclopädie der classischen Altertumswissenschaft*, supp. XI (in press).

Fridolf Kudlien

ARGAND, ÉMILE (*b.* Geneva, Switzerland, 6 January 1879; *d.* Neuchâtel, Switzerland, 14 September 1940), *geology.*

Argand's father was a clerk in government service; his mother, a remarkable woman, was from Morzine, Savoy. Mountain climbing was the favorite sport of the young people of Geneva, so Argand was familiar with the Alps long before he studied geology. His father apprenticed him to an architect, but his mother had other plans. She took him with her for long stays in Italy, France, and Greece, and encouraged him to read medicine. Argand began his studies in Geneva but soon transferred to Lausanne, where he met the young professor Maurice Lugeon and became interested in the problems of Alpine structure. He soon came to believe that these problems had been put into the world especially for him to solve, and he turned to the study of geology.

Argand had the talent to be a geologist but, as his friends remarked, he also could have been an architect, an artist, a linguist, a writer, or a businessman. He would have been outstanding in any of these professions. His extraordinary ability to think in three dimensions allowed him to visualize and to represent not only very complicated solids but also their movements and deformations. A gifted artist, he also could sketch these solids as seen from different angles. Argand had his own recognizable style, not only in his illustrations but also in his writing—and, as a result, in the kind of geology he developed. Argand's visual recall was uncommonly good and enabled him to sketch landscapes, maps, and stereograms from memory (and even portraits that were more or less caricatures). His students and colleagues used to compare the maps he drew on the blackboard during lectures with the atlas, and they were always astonished by his accuracy.

These gifts enabled Argand to solve some of the most complex problems of terrestrial architecture. It was also his good fortune to enter science at a crucial stage in the evolution of Alpine geology and at one of the best possible places. Lugeon, his teacher, not only had shown the existence of the Helvetic nappes but also was the first to analyze these structures over an extended area rather than a restricted cross section. He had studied the entire zone from the Lake of Geneva to the Rhine, clarifying and extending the new methods of structural interpretation used earlier by Marcel Bertrand, Hans Schardt, and Pierre Termier. These involved not only the recognition of the nappe folding but also the extension from the detailed geology of the simple parts to the complex structure on the regional scale. The work had to be extended to the interior of the Alps, but there the metamorphism and the paucity of fossils meant that the details of the stratigraphy would have to be worked out from the structure. In the outer zones, a well-developed stratigraphy had made the unraveling of the structure possible. To unravel the structure of the inner zone Argand had to develop geometric methods, for example, in proceeding from the Alpine surface to the missing structure that was the origin of the projection.

Several years of intensive mapping in the highest central part of the Pennine Alps resulted in the now classic map of the Dent Blanche massif and the memoir explaining his results (1908). A new world of forms was revealed by this work. Argand spoke of the strata deformed into *faisceaux* of families of cylindrical surfaces, of the deformation of these cylinders, of their ruin in depth and space by the processes of erosion, and of the concomitant sedimentation. Of this work, Ulrich Grubenmann said that it was the first time that the inversion of his metamorphic zones had been proved and used for structural explanation. Argand had shown that the top of the Matterhorn was in the Grubenmann katazone, or deepest zone of metamorphism, while the roots of the mountain, as revealed in the rock of the valley below, were in the shallow, or metamorphic, epizone. These geometric methods were the key that enabled Argand to unveil the whole of the Pennic zone, establishing the axial culminations and depressions (i.e., the deformation of the generatrices of the cylindrical nappe structures). He integrated consecutive series of vertical sections with surface geology in complex block diagrams and traced the architecture of the nappes from one sector to the next. His concept of grouping tectonic elements in preferred directions, inclinations, and orientations became a basic notion of structural geology and the foundation of many of the more advanced techniques of structural analysis.

In this way the "Pennic" type, a new type of deformation characteristic of the central zones of many other orogenic belts, was brought to light. Argand's methods of representation and the accompanying full chronological and dynamic interpretation soon became classic, and were adopted by the students and professors who made up a new generation of Alpine geologists. Unfortunately, the methods were not always applied with the caution of their author, who understood their limits. Some of the rigid applications of these heuristic principles were later attacked as "cylindrism" by the school of Grenoble, which nevertheless made use of the principles stated by Argand.

Having not only the keys for the solution of regional problems but also the necessary methods and techniques, Argand extended his task and attempted a synthetic picture of the structure of the arc of the western Alps. The results were condensed in his famous four plates of 1911, with a brief commentary showing the relationships between the major lines and significant details. From regional problems Argand advanced to geology on the planetary scale. He knew by heart the fundamental five-volume work of Eduard Suess, *Das Antlitz der Erde,* and his linguistic skill allowed him to collect further information from many sources. He tried to synthesize this wealth of information as concretely as possible in a tectonic map of Eurasia conceived in a new style. In 1911 Argand had succeeded Hans Schardt as professor at Neuchâtel. Ordinarily systematic, almost to the point of compulsiveness, Argand, during the preparation of the map of Eurasia, worked in the midst of great disorder, driving himself and his assistants as much as thirty-two hours at a stretch, living on coffee and catching naps on an old sofa among the piles of books and papers. This map, although it exists only in the original manuscript, became famous, and in 1913 it received the Spendiaroff Prize. The map is preserved at the University of Neuchâtel, Switzerland.

Until then, Argand's research had been three-dimensional. In order to understand the genesis of the Alps, the dimension of time would have to be added. Lecturing to the Swiss Geological Society in 1915, Argand announced a new branch of research, kinematic analysis and synthesis, which he called embryotectonics. It was the sequential analysis of the evolving structure back through time to the original sedimentary terrain. His heuristic methods were intended to explain the succession of events leading to the present complex structures, and to connect the subsurface movements with the changing sedimentary landscapes of the successive former surfaces and with the emplacement of ophiolites. Argand's picture showed the importance of the sedimentology and stratigraphy of the western Alps and their place in a synthetic study. In this way he reconstructed what was then regarded as the typical evolution of a mountain chain. This "type biography" approach was later completed by a paper on the precursor phases and late phases of an orogenic segment (1920).

In 1915 Alfred Wegener published his fundamental paper containing his hypothesis of continental drift. Fascinated by the possibilities, Argand believed that continental drift offered a new motor for orogeny consistent with his evolutionary model. The Wegenerian hypothesis became the frame within which he created a new concept of Eurasian structural development. This work led to Argand's being chosen to deliver the inaugural address of the Brussels International Geological Congress in 1922. His revised tectonic map of Eurasia, *La tectonique de l'Asie,* was reproduced with text in 1924. This is not only a fundamental text of structural geology but also a work of art. One of the most important of the many new concepts introduced is the notion of basement folds. Contemporary models of the formation of mountain chains showed the upfolding of geosynclinal accumulations; against this simple pattern Argand set the concept of the warping and thrusting of old crystalline platforms without reactivation of the ancient structures and with the resulting sedimentary formations, drawing examples from Asia, Europe, and America. The paper had great influence in the French-speaking world, where Argand's vocabulary and manner of reasoning became a part of geological thinking.

In 1931 Argand was called upon to preside over the first congress for the study of the Precambrian and ancient mountain chains, convened by J. J. Sederholm.

Before World War II, Argand became interested in philosophy and linguistics, but this work was never published. He died suddenly, not long after the death of his mother.

BIBLIOGRAPHY

Argand's works are "Carte géologique du Massif de la Dent Blanche, 1:50.000," in *Matériaux pour la carte géologique de la Suisse,* n.s. **23** (1908), carte spéc. no. 52; "L'exploration géologique des Alpes pennines centrales," in *Bulletin de la Société vaudoise des sciences naturelles,* **45** (1909), 217–276; "Les nappes de recouvrement des Alpes pennines et leurs prolongements structuraux," in *Matériaux pour la carte géologique de la Suisse,* n.s. **31** (1911), 1–26; "Les nappes de recouvrement des Alpes Occidentales et les territoires environnents. Essai de carte structurale. 1:500.000," *ibid.,* n.s. **27** (1911), carte spéc. no. 64; "Sur l'arc des Alpes Occidentales," in *Eclogae geologicae Helvetiae,* **14** (1916), 145–191; "Plissements précurseurs et plissements tardifs des chaines de montagne," in *Actes de la Société helvétique des sciences naturelles,* sess. 101 (1920), 13–39; "La géologie des environs de Zermatt," *ibid.,* sess. 104 (1923), pt. 2, 96–110; "La tectonique de l'Asie," in *Comptes rendus de la XIIIᵉ Congrès international de géologie* (1924), 171–372; *Carte tectonique de l'Eurasie, 1922, 1:8.000.000. Réduction photographique à l'échelle 1:25.000.000* (Brussels, 1928); "La zone pennique," in *Guide géologique de la Suisse* (Basel, 1934), pp. 149–189; "Carte géologique du Grand Combin, 1:50.000," in *Matériaux pour la carte géologique de la Suisse* (1934), carte spéc. no. 93; "Feuille Saxon-Morcles, 1:25.000," in *Atlas géologique de la Suisse* (Bern, 1937), with Lugeon, Reinhard, Poldini, et al.; "Note explicative de la feuille 485, Saxon-Morcles avec annexes de la feuille 526 Martigny (feuille 1° de l'Atlas)," in *Atlas géologique de la Suisse,* pp. 25–53.

For a biography of Argand see Maurice Lugeon, "Emile Argand," in *Bulletin de la Société neuchâteloise des sciences naturelles,* **65** (1940), 25–53.

C. E. WEGMANN

ARGAND, JEAN ROBERT (*b.* Geneva, Switzerland, 18 July 1768; *d.* Paris, France, 13 August 1822), *mathematics.*

Biographical data on Argand are limited. It is known that he was the son of Jacques Argand and Èves Canac; that he was baptized on 22 July (a date given by some for his birth); that he had a son who lived in Paris and a daughter, Jeanne-Françoise-Dorothée-Marie-Élizabeth, who married Félix Bousquet and lived in Stuttgart.

Argand, a Parisian bookkeeper, apparently never belonged to any group of mathematical amateurs or dilettantes. His training and background are so little known that he has often been confused with a man to whom he probably was not even related, Aimé Argand, a physicist and chemist who invented the Argand lamp.

It is remarkable that Argand's single original contribution to mathematics, the invention and elaboration of a geometric representation of complex numbers and the operations upon them, was so timed and of such importance as to assure him of a place in the history of mathematics even among those who credit C. F. Gauss with what others call the Argand diagram.

Other circumstances make Argand's story unusual. His system was actually anticipated by Caspar Wessel, a Norwegian, in 1797, but Wessel's work was without significant influence because it remained essentially unknown until 1897. Argand's own work might have suffered the same fate, for it was privately printed in 1806 in a small edition that did not even have the author's name on the title page. He received proper credit for it through a peculiar chain of events and the honesty and generosity of J. F. Français, a professor at the École Impériale d'Artillerie et du Génie, who published a similar discussion in 1813.

Argand had shown his work to A. M. Legendre before its publication, and Legendre mentioned it in a letter to Français's brother. Français saw the letter among his dead brother's papers, and was so intrigued by the ideas in it that he developed them further and published them in J. D. Gergonne's journal *Annales de mathématiques.* At the end of his article Français mentioned the source of his inspiration and expressed the hope that the unknown "first author of these ideas" would make himself known and publish the work he had done on this project.

Argand responded to this invitation by submitting an article that was published in the same volume of the *Annales*. In it he recapitulated his original work (with a change in notation) and gave some additional applications. A key to his ideas may be presented by a description and analysis of Figures 1 and 2. Figure 1 accompanies his initial discussion of a geometric representation of $\sqrt{-1}$. His motivation for this can be traced back to John Wallis' *Treatise of Algebra* (1685). In it Wallis suggested that since $\sqrt{-1}$ is the mean proportional between $+1$ and -1, its geometric representation could be a line constructed as the mean proportional between two oppositely directed unit segments.

Argand began his book, *Essai sur une manière de représenter les quantités imaginaires dans les constructions géométriques,* with a brief discussion of models for generating negative numbers by repeated subtraction; one used weights removed from a pan of a beam balance, the other subtracted francs from a sum of money. From these examples he concluded that distance may be considered apart from direction, and that whether a negative quantity is considered real or "imaginary" depends upon the kind of quantity measured. This initial use of the word "imaginary" for a negative number is related to the mathematical-philosophical debates of the time as to whether negative numbers were numbers, or even existed. In general, Argand used "imaginary" for multiples of $\sqrt{-1}$, a practice introduced by Descartes and common today. He also used the term "absolute" for distance considered apart from direction.

Argand then suggested that "setting aside the ratio of absolute magnitude we consider the different possible relations of direction" and discussed the proportions $+1:+1::-1:-1$ and $+1:-1::-1:+1$. He noted that in them the means have the same or opposite signs, depending upon whether the signs of the extremes are alike or opposite. This led him to consider $1:x::x:-1$. In this proportion he said that x cannot be made equal to any quantity, positive or negative; but as an analogy with his original models he suggested that quantities which were imaginary when applied to "certain magnitudes" became real when the idea of direction was added to the idea of absolute number. Thus, in Figure 1, if KA taken as positive unity with its direction from K to A is written $\overline{KA}$ to distinguish it from the segment KA, which is an absolute distance, then negative unity will be $\overline{KI}$. The classical construction for the geometric mean would determine $\overline{KE}$ and $\overline{KN}$ on the unit circle with center at K. Argand did not mention the

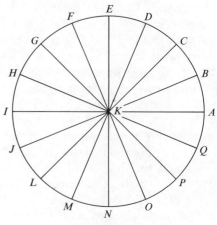

FIGURE 1

geometric construction, but merely stated that the condition of the proportion will be met by perpendiculars $\overline{KE}$ and $\overline{KN}$, which represent $\sqrt{-1}$ and $-\sqrt{-1}$, respectively. Analogously, Argand inserted $\overline{KC}$ and $\overline{KL}$ as the mean proportionals between $\overline{KA}$ and $\overline{KE}$ by bisecting angle AKE.

Argand's opening paragraphs included the first use of the word "absolute" in the sense of the absolute value of a positive, negative, or complex number; of the bar over a pair of letters to indicate what is today called a vector; and of the idea that $\overline{AK} = -\overline{KA}$. Later in the *Essai* Argand used the term "modulus" (*module*) for the absolute value or the length of a vector representing a complex number. In this Argand anticipated A. L. Cauchy, who is commonly given credit for originating the term.

Argand's notation in his original essay is of particular interest because it anticipated the more abstract and modern ideas, later expounded by W. R. Hamilton, of complex numbers as arbitrarily constructed new entities defined as ordered pairs of real numbers. This modern aspect of Argand's original work has not been generally recognized. One reason for this, no doubt, is that in later letters and journal articles he returned to the more standard $a + b\sqrt{-1}$ notation. In his book, however, Argand suggested omitting $\sqrt{-1}$, deeming it no more a factor of $a\sqrt{-1}$ than is $+1$ in $+a$. He wrote $\sim a$ and $\dashv a$ for $a\sqrt{-1}$ and $-a\sqrt{-1}$, respectively. He then observed that both $(\sim a)^2$ and $(\dashv a)^2$ were negative. This led him to the rule that if in a series of factors every curved line has a value of 1 and every straight line a value of 2—thus $\sim = 1$, $- = 2$, $\dashv = 3$, $+ = 4$—then the sign of the product of any series of factors can be determined by taking the residue modulo, 4, of the sum of the values of the symbols associated with the factors. Here he recognized the periodicity of the powers of the imaginary unit.

238

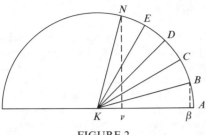

FIGURE 2

Argand generalized the insertion of geometric means between two given vectors to the insertion of any number of means, n, between the vectors $\overline{KA}$ and $\overline{KB}$ by dividing the angle between them by n. He noted that one could also find the means between $\overline{KA}$ and $\overline{KB}$ by beginning with the angles $AKB + 360°$, and $AKB + 720°$. This is a special case of de Moivre's theorem, as is more clearly and completely shown in Argand's explanation of Figure 2. In it AB, BC, $\cdots$ EN are n equal arcs. From the diagram Argand reasoned that $\overline{KN} = \overline{KB^n}$, $\overline{KN} = \overline{K\nu} + \overline{\nu N}$, and $\overline{KB} = \overline{K\beta} + \overline{\beta B}$; hence $\overline{K\nu} + \overline{\nu N} = (\overline{K\beta} + \overline{\beta B})^n$, which leads to $\cos na + \sqrt{-1}$ $\sin na = (\cos a + \sqrt{-1} \sin a)^n$.

This result was well known before Argand, as were the uses he made of it to derive infinite series for trigonometric and logarithmic functions. As noted earlier, we know nothing of Argand's education or contacts with other mathematicians prior to 1813. It seems highly probable, however, that he had direct or indirect contact with some of the results of Wallis, de Moivre, and Leonhard Euler. Nevertheless, the purely geometric-intuitive interpretation and reasoning leading to these results seem to have been original with Argand. This geometric viewpoint has continued to be fruitful up to the present day. Argand recognized the nonrigorous nature of his reasoning, but he defined his goals as clarifying thinking about imaginaries by setting up a new view of them and providing a new tool for research in geometry. He used complex numbers to derive several trigonometric identities, to prove Ptolemy's theorem, and to give a proof of the fundamental theorem of algebra.

Argand's work contrasts with Wessel's in that the latter's approach was more modern in its explicit use of definitions in setting up a correspondence between $a + b\sqrt{-1}$ and vectors referred to a rectangular coordinate system (which neither Wessel nor Argand ever explicitly mentioned or drew). Wessel stressed the consistency of his assumptions and derived results without regard for their intuitive validity. He did not present as many mathematical consequences as Argand did.

Just as it seems clear that Argand's work was entirely independent of Wessel's, so it also seems clear that it was independent of the algebraic approach published by Suremain de Missery in 1801. Argand refuted the suggestion that he knew of Buée's work published in the *Philosophical Transactions of the Royal Society* in 1806 by noting that since academic journals appear after the dates which they bear, and that his book was printed in the same year the journal was dated, he could not have known of Buée's work at the time he wrote the book. Buée's ideas were not as clear, extensive, or well developed as Argand's.

There are obvious connections between Argand's geometric ideas and the later work of Moebius, Bellavitis, Hermann Grassmann, and others, but in most cases it is as difficult to establish direct outgrowths of his work as it is to establish that he consciously drew on Wallis, de Moivre, or Euler.

Two of the most important mathematicians of the early nineteenth century, Cauchy and Hamilton, took care to note the relationship of Argand's work to some of their own major contributions, but claimed to have learned of his work only after doing their own. Gauss probably could have made a similar statement, but he never did. Cauchy mentioned Argand twice in his "Mémoire sur les quantités géométriques," which appeared in *Exercices d'analyse et de physique mathématique* (1847). He cited Argand as the originator of the geometric interpretation of imaginary quantities, which he suggested would give clarity, a new precision, and a greater generality to algebra than earlier theories of imaginary quantities had. He also cited Argand and A. M. Legendre as authors of proofs of what Gauss termed the "fundamental theorem of algebra." Argand's proof involved considering the modulus of

$$P(x) = a_0 x^n + a_1 x^{n-1} + \cdots + a_{n-1} x^1 + a_n$$

when $x = a + bi$. He noted that if $|P(x)| = 0$ the theorem was true, and argued geometrically that if $|P(x)| > 0$ one could find $x' = a' + b'i$ such that $|P(x')| < |P(x)|$. Servois objected that this only showed that $P(x)$ was asymptotic to 0 for some sequence of x's. Argand replied that such behavior was associated with hyperbolas having zeros at infinity, not with polynomials. Cauchy asserted that a proof proposed by Legendre reduced to Argand's but left much to be desired, while his own method for approximating roots of $P(x) = 0$ could be used to demonstrate their existence. Gauss had published a proof of this existence in his thesis (1799). Although the geometric representation of complex numbers was implicit in this thesis, Gauss did not actually publish a discussion of it until 1832 in his famous paper "Theoria residuorum biquadraticorum." Argand, however, was the first mathematician to assert that the fundamental theorem also held if the coefficients of $P(x)$ were complex.

Hamilton used lengthy footnotes in the first edition of his *Lectures on Quaternions* (1853) to assert the priority and quality of Argand's work, especially with respect to the "multiplication of lines." He traced the roots of his own development of the algebra of couples and of quaternions, however, to John Warren's *A Treatise on the Geometrical Representation of the Square Roots of Negative Quantities* (1828). This, like C. V. Mourrey's *La vraie théorie des quantités négatives et des quantités prétendues imaginaires* (1828), seems to have been free of any dependence on Argand's work.

Argand's later publications, all of which appeared in Gergonne's *Annales,* are elaborations of his book or comments on articles published by others. His first article determined equations for a curve that had previously been described in the *Annales* (**3**, 243). Argand went on to suggest an application of the curve to the construction of a thermometer shaped like a watch. His analysis of probable errors in such a mechanism showed familiarity with the mechanics of Laplace, as presented in *Exposition du système du monde.*

His fifth article in the *Annales,* defending his proof of the fundamental theorem of algebra, showed his familiarity with the works of Lagrange, Euler, and d'Alembert, especially their debates on whether all rational functions of $(a + bi)$ could be reduced to the form $A + Bi$ where a, b, A, and B are real. Argand, oddly enough, did not accept this theorem. He apparently was not familiar with Euler's earlier reduction of $\sqrt{-1}^{\sqrt{-1}}$, for he cited this as an example of an expression that could not be reduced to the form $A + Bi$.

His last article appeared in the volume of *Annales* dated 1815–1816 and dealt with a problem in combinations. In it Argand devised the notation (m,n) for the combinations of m things taken n at a time and the notation $Z(m,n)$ for the number of such combinations.

Argand was a man with an unknown background, a nonmathematical occupation, and an uncertain contact with the literature of his time who intuitively developed a critical idea for which the time was right. He exploited it himself. The quality and significance of his work were recognized by some of the geniuses of his time, but breakdowns in communication and the approximate simultaneity of similar developments by other workers force a historian to deny him full credit for the fruits of the concept on which he labored.

BIBLIOGRAPHY

I. ORIGINAL WORKS. There have been three editions of Argand's book (his first publication), *Essai sur une manière de représenter les quantités imaginaires dans les constructions géométriques.* The first edition (Paris, 1806) did not bear the name of the author; the second edition, subtitled *Précédé d'une préface par M. J. Hoüel et suivie d'une appendice contenant des extraits des Annales de Gergonne, relatifs à la question des imaginaires* (Paris, 1874), cites the author as "R. Argand" on the title page but identifies him as Jean-Robert Argand on page xv. The *Essai* was translated by Professor A. S. Hardy as *Imaginary Quantities: Their Geometrical Interpretation* (New York, 1881). Argand's eight later publications all appeared in Vols. **4, 5,** and **6** (1813–1816) of J. D. Gergonne's journal *Annales de mathématiques pures et appliquées.* Hoüel lists them at the end of his preface to the second edition of the *Essai.*

II. SECONDARY LITERATURE. Data on Argand's life were included by Hoüel with the second edition of the *Essai.* Verification of the dates of his birth and death is given by H. Fehr in *Intermédiaire des mathématiciens,* **9** (1902), 74. Niels Nielsen, *Géomètres français sous la Révolution* (Copenhagen, 1929), pp. 6–9, discusses Argand with reference to Wessel, Français, and others. William Rowan Hamilton gives a comparative analysis of contemporary work with complex numbers while praising Argand in *Lectures on Quaternions* (Dublin, 1853), pp. 31–34, 56, 57. Augustin Louis Cauchy's appraisal is found in "Mémoire sur les quantités géométriques," in *Exercices d'analyse et de physique mathématique,* IV (Paris, 1847), and in *Oeuvres,* 2nd series, XIV (Paris, 1938), 175–202. J. F. Français's development of Argand's ideas contained in a letter to his brother, "Nouveaux principes de géométrie de position, et interpretation géométrique des symboles imaginaires," is *Annales de mathématiques,* **4** (1813–1814), 61–71.

PHILLIP S. JONES

ARGELANDER, FRIEDRICH WILHELM AUGUST (*b.* Memel, Prussia, 22 March 1799; *d.* Bonn, Germany, 17 February 1875), *astronomy.*

Argelander's father, Johann Gottlieb Argelander, was a wealthy merchant from Finland; his mother, Dorothea Wilhelmine Grünlingen, was German. The boy grew up in the easternmost corner of Prussia, and the Napoleonic Wars brought him into close contact with the major political events of his day. After the defeat of Prussia in the campaign of 1806, Queen Louise and her sons fled from Berlin to Memel. The crown prince, who later became Friedrich Wilhelm IV, and Prince Friedrich lived in the Argelander home, and young Argelander formed a lasting friendship with both of them. After attending the secondary school in Elbing and the Collegium Fridericianum in Königsberg, Argelander entered the University of Königsberg in 1817 in order to study political economy and political science. At this time he attended the lectures of the astronomer Friedrich Wilhelm Bessel and took an active part in the calculating work of the observatory at Königsberg. When the results of this work were published, Bessel referred to Argelander as one of his "most outstanding students."

When he was twenty-one, Argelander was completely won over to astronomy by Bessel and was appointed an assistant in the observatory. He took part in a newly initiated project, designed to determine the exact positions of all the brighter stars of the northern sky between 15° southern declination and 45° northern declination; this was the first part of Bessel's projected survey of the entire northern sky. In 1822 Argelander earned his doctorate with a study of the older Greenwich observations (which were made with sextant and quadrant), *De observationibus astronomicis a Flamsteedio institutis,* and in the same year he earned his lectureship with a study entitled *Untersuchung über die Bahn des grossen Kometen von 1811* ("Investigation of the Orbit of the Great Comet of 1811"). In both these works he demonstrated his great proficiency in the critical evaluation of astronomical observations.

Bessel recommended him for the post of observer at the small, just rebuilt observatory in Åbo (now Turku), Finland. Thus Argelander, who had just turned twenty-four, went into the Russian civil service. Equipped with excellent instruments from the best German factories, so that the Åbo observatory could compete with Königsberg and Dorpat, Argelander made it famous as the northernmost station of the "three-point constellation" of east European observatories. Here he demonstrated his ability, later tested again and again, to achieve optimum results in spite of limited resources and an unfavorable climate. While Bessel continued to contribute to the knowledge of the positions of stars through observations of zones, Argelander dedicated his observatory to the study of the alteration of the positions in time—the "proper motions" of the fixed stars.

In 1718 Halley, by the comparison of older and more recent positions, had discovered that some brighter stars, such as Sirius, Arcturus, and Aldebaran, showed deviations that could be explained only through proper spatial motions and not through errors in observation. In 1760 Johann Tobias Mayer found that of eighty stars he investigated, fifteen or twenty showed proper motions of several arc-seconds in 100 years. Argelander devoted all his energies to this problem. In a few years he collected over 10,000 observations with the transit circle of several hundred stars that were suspected of proper motion. The catalog of 560 stars, which was based on these observations (1835), is incontestably the most exact of the contemporary catalogs.

Argelander was forced to abandon these observations in 1827 because of a fire that destroyed most of Åbo. Even if the observatory itself had been saved, a transfer of the entire university to Helsinki would still have been unavoidable. Argelander, who in 1828 was named full professor at the new university, was assigned the task of planning and building a new observatory. In 1832 he was finally able to move to Helsinki. Here his earlier work proved fruitful. The ingenious investigation *Über die eigene Bewegung des Sonnensystems, hergeleitet aus den eigenen Bewegungen der Sterne* ("Concerning the Peculiar Movement of the Solar System as Deduced From the Proper Motions of the Stars," 1837) is one of the few theoretical works in which Argelander found the basis of his observations conclusive enough to make certain deductions.

Through all his other works Argelander has become renowned as a master of practical investigations of fundamental importance. It would, however, be wrong to disregard his theoretical ability. His foresight and critical strength protected him from the danger of drawing premature and insufficiently proved conclusions from imperfect material. Lalande had already concluded from theoretical considerations that the sun, like the other fixed stars, has a progressive motion in space in addition to its rotation. All other bodies of the solar system, the planets and comets, participate in this movement; therefore the peculiar motion of the sun is only to be recognized as relative to the fixed stars outside the solar system. The motions of the fixed stars consist of the apparent perspectivic changes in position, which are caused by the motion of the sun, and of the real, progressive motions of the fixed stars through space. Only the most exact observations of a greater number of stars can allow the separation of these two aspects.

In 1783 Sir William Herschel had deduced, on the basis of the proper motions of only seven bright stars, that the apex of the movement of the sun was somewhat north of the star λ Herculis. A repetition of his calculations in 1805/1806 by the aid of a somewhat larger number of stars yielded data that did not essentially deviate. While Herschel himself had made no observations of his own for this purpose and had frequently encountered doubt and contradiction, Argelander saw in his observations an opportunity to derive anew the motion of the sun. He felt this would be worthwhile because of the progress both in the construction of instruments and in critical calculation methods that, as far as possible, freed observations from all accidental and systematic errors.

His result, based on no fewer than 390 proper motions, verified the accuracy of Herschel's pioneering work. The spatial movement of the sun has since then been proved beyond doubt.

It would have been obvious to a confirmed theorist that a further question should be asked: What is the law that governs the motion of the sun and the other stars through space? Are all bodies, Argelander asked, subject only to their mutual attractions, or do all of them obey the attractive force of a large central body? In other words, is the solar system only the smaller model of a larger, similarly constructed stellar system? Argelander did not address himself to this question; he did not indulge in idle speculation, as did other astronomers who claimed that now Sirius, or again the Pleiades, or even a central sun as yet unknown was the center of the realm of the stars. His merit lies much more in the recognition of the significance of the dynamics of the stellar system and in his provisions that enabled later astronomers to solve the problem. Everything he did in the next four decades served to increase the knowledge of the positions and proper motions of the stars, in the hope that someday a second Kepler might be successful in revealing the more complex laws of the orbits of the fixed stars. In November 1836 Argelander was appointed to the professorship of astronomy at the Prussian University of Bonn. He was also promised a new, richly fitted observatory, and his old friend the crown prince, who ascended the throne several years later, supported the project.

During his first years at Bonn, Argelander worked in an old bastion situated directly on the Rhine, where a provisional observatory was erected. Under these limitations Argelander again showed himself to be the ingenious improviser who, even with modest resources, could achieve lasting results. Although without measuring instruments, he created in these years one of his finest works, the *Uranometria nova* (1843). The main feature of this work was not the determination of exact positions, but the recording of all stars visible to the naked eye and a settlement of the nomenclature that had been used arbitrarily up to that time, as well as a demarcation of the constellations of the stars. At the same time, this atlas and the accompanying catalog fulfilled the task of a reliable representation of the magnitudes of the stars.

The exact observation of stellar magnitudes was not yet possible in Argelander's time because there was no suitable photometer. By means of the method of "estimation by steps," developed by Argelander, it was nevertheless possible, even without instruments, to obtain reliable magnitude data with an exactness

of about one-third of a magnitude class. For this purpose the magnitudes were, by means of mutual comparison with neighboring stars of slightly different magnitudes, arranged in an arbitrary scale which could be gauged with the aid of stars of known magnitudes. This simple method, which is based on the ability of man's eyesight to perceive very slight differences in brightness, proved itself especially useful in the investigation of the changing brightness of variable stars, of which only eighteen were known at that time. By qualitatively determining the changing brightness of these stars, Argelander opened a completely new field of research which soon earned an important place in the working program of many astronomers; and, at his suggestion, it became of interest to many amateurs as well. It has to a considerable degree enriched our knowledge of the physical nature of the fixed stars. Argelander devoted himself to the continuous observation of these stars and inspired a similar zeal in his students. At the time of his death, the number of variable stars with known period of changing light had increased to almost 200.

When a five-foot transit instrument, the first larger instrument ordered for the new observatory, arrived from Ertel, it was immediately set up at the provisional observatory. With it Argelander extended the work begun by Bessel at Königsberg (which reached only to 45°) up to 80° northern declination. In more than 200 nights he measured the exact positions of 22,000 stars. The work was later continued on the same instrument at the new observatory to the south from 15° southern declination up to 31° southern declination with 17,000 additional stars. With this work, in conjunction with the Besselian zones, a first, although not wholly complete, inventory of the entire sky visible in the northern latitudes was made. This inventory included many faint stars which were invisible to the naked eye, but it was not completed to a limited magnitude, which is necessary to permit the investigation of certain questions of stellar statistics. Argelander stands out above all other astronomers of his century not only because of his perception of this important problem but also because he followed with all his energy the only practicable way to its solution. In order not to stretch this undertaking to an impossible extreme, he set the limit of completeness at stars of the ninth magnitude. His goal was a uniform registration of all stars up to this magnitude and the cataloging of their positions and magnitudes with an exactness sufficient for further identification.

From these considerations there arose the *Bonner Durchmusterung*, which has since provided the working basis for every observatory. It consists of a three-volume catalog of stars and a forty-plate atlas in which are recorded the positions (exactness $\pm 0.1'$) and magnitudes ($\pm 0.3^M$) of 324,198 stars between the northern celestial pole and 2° southern declination. The observations were made with a very small instrument, a so-called comet seeker with only a 7.8-centimeter aperture and a 63-centimeter focal distance with a ninefold magnification. The observations, which lasted over a period of 625 nights, from 1852 until 1859, extended for terms of five, six, seven, eight, and occasionally twelve hours, were mostly made by his assistants Eduard Schönfeld and Adalbert Krüger. The plan of observation, thought out to the smallest detail, comprised not only the tremendous number of individual observations, which exceeded a million, but also the extensive revisions of doubtful cases, which were carried out by Argelander himself on the meridian circle. In addition, it included the detailed calculations for the final catalog and for the plates of the atlas. Thus the results, despite the small circle of collaborators, were published after a few years and became the foundation for all future astronomical work.

The next step was the improvement of the preliminary determination of the positions of stars, utilizing the high development of the art of measurement, but it could no longer be the task of a single observatory.

Therefore, in 1867 Argelander proposed to the Astronomische Gesellschaft that several observatories acting as a team undertake according to a uniform plan the observation of the most exact positions possible on meridian circles of all stars in the *Bonner Durchmusterung* up to the ninth magnitude. Argelander lived to see the project, in which seventeen observatories took part, set in motion in the 1870's.

Some idea of Argelander's renown as an astronomer may be gained from the fact that he was a member of the academies of St. Petersburg, London, Berlin, Stockholm, Paris, Vienna, Boston, and Brussels; of the Societas Fennica in Helsinki; of the Royal Astronomical Society of London; of the National Academy of Sciences of the United States; and a charter member of the Astronomische Gesellschaft, of which he was a member of the governing body from 1863 to 1871 and chairman from 1864 to 1867.

Argelander's achievement does not lie in substantial discoveries, but in the single-minded and systematic planning of his lifework and his unique resoluteness and skill in its execution. His main endeavor consisted in offering extensive and complete numerical data for the study of the construction of the heavens and the motions of the fixed stars in the stellar system. Without his work, which was con-tinued at the Bonn Observatory, by Schönfeld and Küstner, the knowledge attained in our century about the structure of the universe would not have been imaginable.

BIBLIOGRAPHY

I. ORIGINAL WORKS. Argelander's writings include *De observationibus astronomicis a Flamsteedio institutis,* his dissertation (Königsberg, 1822); *Untersuchung über die Bahn des grossen Kometen von 1811* (Königsberg, 1822); *DLX stellarum fixarum positiones mediae ineunte anno 1830* (Helsinki, 1835); *Über die eigene Bewegung des Sonnensystems, hergeleitet aus den eigenen Bewegungen der Sterne* (St. Petersburg, 1837); *Uranometria nova* (Berlin, 1843), with seventeen charts and a catalog of stars; *Bonner Sternverzeichnis,* 3 vols. (Bonn, 1859–1862), Vols. III–V in the series Astronomische Beobachtungen auf der Königlichen Sternwarte zu Bonn; and *Atlas des nördlichen gestirnten Himmels für den Anfang des Jahres 1855* (Bonn, 1863), with forty charts.

II. SECONDARY LITERATURE. Works providing information about Argelander include B. Sticker, *Fr. W. Argelander und die Astronomie vor hundert Jahren* (Bonn, 1944); E. Schönfeld, obituary, in *Vierteljahrsschrift der Astronomische Gesellschaft,* **10** (1875), 150–178; articles on him in *Allgemeine deutsche Biographie,* Vol. XLVI; *Neue deutsche Biographie,* Vol. I; and *Westermanns Monatshefte,* **51** (1906), which concerns Argelander's family.

BERNHARD STICKER

ARGENVILLE, ANTOINE-JOSEPH DEZALLIER D' (*b.* Paris, France, 1 July 1680; *d.* Paris, 29 November 1765), *natural history, engraving, art history.*

D'Argenville was the son of Antoine and Marie Mariette Dezallier. His father owned the d'Argenville estate, near Bezons and Versailles, from which he took his name. After studying at the Collège du Plessis, he devoted himself to the fine arts under the direction of the engraver Bernard Picart, the painter Roger de Piles, and the architect Alexandre Le Blond. He also became interested in natural science. In 1709 his first work, *Traité sur la théorie et la pratique du jardinage,* was published. D'Argenville went to Italy in 1713, and upon his return in 1716 he purchased the post of secretary to the king; he was later named *maître des comptes* (2 July 1733) and counsellor to the king (1748). After settling in Paris, he soon acquired a deserved reputation as an expert collector of objects of art and curiosities of nature. Trips to Germany, Holland, and England (1728) made it possible for him to enrich both his knowledge and his collections.

Today d'Argenville is known particularly through his *Abrégé de la vie des plus fameux peintres* (1745–

1752), a mediocre work. He also produced *L'histoire naturelle éclaircie dans deux de ses parties principales, la lithologie et la conchyliologie* (1742). This work, illustrated with beautiful plates, was a great success. D'Argenville profited from the eighteenth century's infatuation with the natural sciences and, indeed, contributed to this vogue by publishing descriptions of the most notable exhibits of natural history in Paris and the provinces. His *L'histoire naturelle* was reissued in two volumes (1755, 1757), and there was also a third edition of *La conchyliologie* (1780).

In 1718 d'Argenville married Françoise-Thérèse Hémart; they had one son, Antoine-Nicolas. He became a member of the Société Royale des Sciences de Montpellier in 1740, of the Royal Society of London in 1750, and of the Académie de La Rochelle in 1758.

BIBLIOGRAPHY

I. ORIGINAL WORKS. D'Argenville's writings include *Traité sur la théorie et la pratique du jardinage* (Paris, 1709); *L'histoire naturelle éclaircie dans deux de ses parties principales, la lithologie et la conchyliologie* (Paris, 1742), reissued in 2 vols. (Paris, 1755–1757), and, for *La conchyliologie*, a 3rd ed., 2 vols. (Paris, 1780); and *Abrégé de la vie des plus fameux peintres,* 3 vols. (Paris, 1745–1752; 1762).

The Bibliothèque Nationale, Paris, Département des Manuscrits, has a catalog of d'Argenville's paintings, prints, and curiosities (n.a. 1564); corrections and additions to the *Abrégé de la vie des plus fameux peintres* (19.094); and letters (n.a. 4814). The Bibliothèque de l'Arsenal, Paris, has an autograph manuscript of the *Histoire générale des coquilles* (2807); and the Bibliothèque d'Avignon, two letters addressed to Esprit-Claude-François Calvet (2358, 3050). At the Royal Society of London are a letter to the Royal Society (MM. 3.85), and a copy and translation of another letter (L&P.II.160).

II. SECONDARY LITERATURE. Other writings on d'Argenville appear under his name in *Biographie universelle Michaud,* new ed., X (Paris, 1855), 598–599; *Dictionnaire de biographie française,* III (Paris, 1939), cols. 581–583; *Nouvelle biographie générale,* XIV (Paris, 1855), 10–11; see also de Ratte, "Éloge de M. Desallier d'Argenville," in d'Argenville's *La conchyliologie,* 3rd ed. (Paris, 1780), I, ix–xxiv.

YVES LAISSUS

ARGOLI, ANDREA (*b.* Tagliacozzo, Italy, 15 March 1570 [1568?]; *d.* Padua, Italy, 27 September 1657), *astrology, astronomy.*

Andrea was the son of Octavio Argoli, a lawyer, and Caterina Mati; his own son, Giovanni (*b.* 1609), achieved considerable celebrity as a precocious poet.

From 1622 to 1627 Argoli held the chair of mathematics at the Sapienza in Rome; evidence suggests that he lost this post because of his enthusiasm for astrology. In 1632 he became professor of mathematics in Padua, where he spent the remainder of his life. If the reports that he studied with Magini and taught Albrecht Wallenstein astrology are correct, then he must also have been in Padua earlier, around 1600.

Argoli's extensive astronomical ephemerides, based first on the *Prutenic Tables* (1620–1640) and later on his own tables (1630–1700), which were based on the observations of Tycho Brahe, gave a permanence to his reputation that his other writings would scarcely have achieved.

In his *Astronomicorum* (1629), Argoli proposed his own geocentric system of the world: the orbits of Mercury and Venus are centered on the sun but those of Mars, Jupiter, and Saturn are centered on the earth (in contradistinction to the Tychonic hypothesis). This scheme is essentially the same as that of Martianus Capella, with the addition of the rotation of the earth on its own axis. A communication to Galileo from F. Micanzio gives evidence that Argoli later planned a defense of Galileo's *Dialogi;* but if the work was ever completed, it no longer survives.

In his *Pandosion,* Argoli devotes chapter 41 to an accurate and succinct exposition of Harvey's doctrine of the circulation of the blood—without, however, mentioning Harvey's name. His principal astrological text, *De diebus criticis . . .,* which concerned astrology in general and astrological medicine in particular, has been described by Thorndike.

BIBLIOGRAPHY

I. ORIGINAL WORKS. Argoli's writings are *Problemata astronomica* (Rome, 1604); *Tabulae primi mobilis* (Rome, 1610; Padua, 1644, 1667); *Ephemerides . . . ab anno 1621 ad 1640 ex Prutenicis tabulis supputatae* (Rome, 1621; Venice, 1623); *Novae caelestium motuum ephemerides . . . et anno 1620 ad 1640* (Rome, 1629), which includes *Astronomicorum libri tres; Secundorum mobilium tabulae juxta Tychonis Brahe et novas e coelo deductas observationes* (Padua, 1634, 1650); *Ephemerides . . . juxta Tychonis Brahe hypotheses* (for 1630–1680, Venice–Padua, 1638; for 1631–1680, Padua, 1638, 1642)—the following editions, for 1641–1700, begin *Exactissimae coelestium motuum* and include the *Astronomicorum libri tres* (Padua, 1648, 1652; Lyons, 1659, 1677); *De diebus criticis et de aegrorum decubitu libri duo* (Padua, 1639, 1652); *Pandosion sphaericum* (Padua, 1644; 2nd ed., enl., 1653); *Ptolemaus parvus in genethliacis junctus Arabibus* (Padua–Lyons, 1652; Lyons, 1654, 1659, 1680); and *Brevis dissertatio de cometa 1652,*

1653 et aliqua de meteorologicis impressionibus (Padua, 1653).

II. SECONDARY LITERATURE. The most complete modern treatment is M. Gliozzi, in *Dizionario biografico degli italiani,* IV (Rome, 1964), 122–124. See also B. J. B. Delambre, *Histoire de l'astronomie moderne,* II (Paris, 1821), 514–517; R. P. Niceron, *Mémoires pour servir à l'histoire des hommes illustres,* XXXIX (Paris, 1738), 325–331; and Lynn Thorndike, *A History of Magic and Experimental Science,* VII (New York, 1958), 122–124.

OWEN GINGERICH

ARISTAEUS (*fl. ca.* 350–330 B.C.), *mathematics.*

Aristaeus lived after Menaechmus and was an older contemporary of Euclid. Nothing is known of his life, but he was definitely not the son-in-law of Pythagoras mentioned by Iamblichus. Pappus, whose *Collectio* (Book VII) is our chief source, calls him Aristaeus the Elder, so presumably there was a later mathematician with the same name.

None of Aristaeus' writings have been preserved. On the other hand, Pappus (some 650 years later) had in his possession Aristaeus' treatment of conic sections as loci, the *Five Books Concerning Solid Loci.* He mentions the work in his *Treasury of Analysis.* In another place, he speaks of the solid loci of Aristaeus as "standing in relation to conic sections."[1] For that reason, and because in a scholium there is mention of *Five Books of the Elements of Conic Sections,* some have concluded that Aristaeus wrote another work of this nature. According to Heiberg's investigations, this is untrue. Pappus also reports that Aristaeus introduced the terms "section of the acute-angled, right-angled, and obtuse-angled cone."

As for its contents, it can be determined from the passages by Pappus and Apollonius that the "locus with respect to three or four lines" was treated by Aristaeus.[2] Well-founded suppositions have been expressed concerning other loci treated by him, in connection with vergings ($\nu\epsilon\upsilon\sigma\epsilon\iota\varsigma$) by means of conic sections, trisection of an angle with the aid of a hyperbola, and above all the focus-directrix property of conic sections; Pappus establishes them in a lemma to the lost *Surface Loci* of Euclid. Since Euclid evidently supposes that the principle necessary for an understanding of surface loci is well known, it could have originated with Aristaeus.

In any case, Aristaeus played a major part in the development of the conic section theory, which began with Menaechmus.[3] Zeuthen and Heath give a comprehensive presentation of his accomplishments. In 1645 Viviani undertook a revision of the *Solid Loci,* starting from the same general interpretation of the contents that was used later by Zeuthen.

Hypsicles (*ca.* 180 B.C.), the editor of Book XIV of Euclid's *Elements,* reports another work, *Concerning the Comparison of Five Regular Solids,* from which he quotes a proposition.[4] It is not certain whether Hypsicles had in mind here the author of the *Solid Loci.* One would suppose a younger Aristaeus; in that case, Euclid's dependence (in Book XIII) on Aristaeus, which has been maintained by Heath but has been denied with good reasons by Sachs, would not be a point of contention.

After Aristaeus, Euclid was the next to deal with conic sections; his work was rendered out-of-date by Apollonius and became superfluous. From a statement by Pappus one must assume that Euclid had no intention of developing further the treatment of conic sections as loci, but that he sought—as did Apollonius—to give a general, synthetic construction. The solution of the problem of the "locus with respect to three or four lines" must have been incomplete in Aristaeus' works, for the "proposition of powers" (prop. Euclid III, 36) extended to conics and the second branch of the hyperbola were still missing. Apollonius, who must have had the complete solution,[5] notes additionally that the problem could not be completely solved without the propositions he discovered. He does not give the solution, but in the third book Apollonius does prove the converse of the proposition.[6]

After Apollonius and Pappus became well known again during the Renaissance, the problem of loci again began to attract interest. It appears in letters from Golius to Mydorge (after 1629) and Descartes (1631), and in letters from Descartes to Mersenne (1632, 1634). Fermat gives (before 1637) a solution in the ancient manner; in a letter to him (4 August 1640), Roberval reports that he completely reconstructed the *Solid Loci.*[7] In Descartes's *Géométrie* (1637) the "locus with respect to three and four lines" forms the starting point for the new analytic treatment of conic sections. Descartes cites the pertinent passages in Pappus and, going beyond him, expands the problem to arbitrarily many straight lines. By means of this, he laid the foundation for a theory of the general properties of algebraic curves.

NOTES

1. $\sigma\upsilon\nu\epsilon\chi\hat{\eta}$ $\tau o\hat{\iota}\varsigma$ $\kappa\omega\nu\iota\kappa o\hat{\iota}\varsigma$ (Pappus VII, 672). Hultsch translates it freely as *supplementum conicorum doctrinae.*

2. The proposition is as follows: If from a given point arbitrarily directed lines *a, b, c* (or *a, b, c, d*) are drawn to meet at given angles three (or four) straight lines given in position, and then if $ac:b^2$ (or $ac:bd$) is a given value, then the point lies on a conic section. Cf. Pappus VII, 678.

3. Sarton (*Introduction*, I, 125) calls him the greatest mathematician of the second half of the fourth century.
4. "The same circle circumscribes both the pentagon of the dodecahedron and the triangle of the icosahedron when both are inscribed in the same sphere." (Euclid, Heiberg ed., V, 6 f.)
5. Descartes is of still another opinion. See also Zeuthen, *Lehre*, pp. 127 ff.
6. Heath, *Apollonius*, pp. cxxxviii ff. The proposition is this: For a point of a conic section the relationship given in note 2 holds true.
7. Tannery, "Note," pp. 46 ff. From Fermat only the solution for the locus of three straight lines survived.

BIBLIOGRAPHY

Aristaeus is also discussed in J. L. Coolidge, *A History of the Conic Sections and Quadric Surfaces* (Oxford, 1945, 1947), chs. 1 (para 2), 3, 5; P. ver Eecke, *Les coniques d'Apollonius de Perge* (Paris, 1932, 1959), pp. x, xvii, 2; and *Pappus d'Alexandrie,* I (Paris–Bruges, 1933), pp. lxxxix–ci; Euclid, *Elementa,* J. L. Heiberg, ed., V (Leipzig, 1888), 6 f.; T. L. Heath, *Apollonius of Perga* (Cambridge, 1896), pp. xxxviii, cxxxviii ff.; and *A History of Greek Mathematics,* I (Oxford, 1921), 143 f., 420; J. L. Heiberg, *Apollonii Pergaei quae Graece exstant,* I (Leipzig, 1891), 4; and *Geschichte der Mathematik und Naturwissenschaften im Altertum* (Leipzig, 1925, 1960), p. 13; F. Hultsch, trans., *Pappi Alexandrini Collectionis quae supersunt,* II (Berlin, 1877), 634, 636, 672 ff., 1004 ff.; Pauly-Wissowa, *Real-Encyclopädie,* supp. III (1918), 157 f.; E. Sachs, *Die fünf Platonischen Körper* (Berlin, 1917), pp. 107 ff.; G. Sarton, *Introduction to the History of Science,* I, 125; P. Tannery, "Note sur le problème de Pappus," in *Mémoires scientifiques,* 3 (1915), 42–50; V. Viviani, *De locis solidis secunda divinatio geometrica in quinque libros injuria temporum amissos Aristaei senioris geometrae* (Florence, 1673, 1701); and H. G. Zeuthen, *Die Lehre von den Kegelschnitten im Altertum,* R. von Fischer-Benzon, ed. (Copenhagen, 1886), repr. with foreword and index by J. E. Hofmann (Hildesheim, 1966), pp. 127 ff., 276, and *Geschichte der Mathematik im Altertum und Mittelalter* (Copenhagen, 1890), pp. 197 ff.

KURT VOGEL

ARISTARCHUS OF SAMOS (*ca.* 310–230 B.C.), *mathematics, astronomy.*

Aristarchus is celebrated as being the first man to have propounded a heliocentric theory, eighteen centuries before Copernicus. He was born on the island of Samos, close by Miletus, cradle of Ionian science and philosophy. Little is known of Aristarchus' subsequent habitation. He was a pupil of Strato of Lampsacos, third head of the Lyceum founded by Aristotle. It is more likely that he studied under Strato at Alexandria than at Athens after the latter's assumption of the headship of the Lyceum in 287 B.C. Aristarchus' approximate dates are determined by Ptolemy's record (*Syntaxis* 3.2) of his observation of the summer solstice in 280 B.C. and by Archimedes' account of his heliocentric theory in a treatise, *The Sand-Reckoner,* which Archimedes composed before 216 B.C. The sole surviving work of Aristarchus is the treatise *On the Sizes and Distances of the Sun and Moon.*

To his contemporaries Aristarchus was known as "the mathematician"; the epithet may merely have served to distinguish him from other men of the same name, although *On Sizes and Distances* is indeed the work of a highly competent mathematician. The Roman architect Vitruvius lists him with six other men of rare endowment who were expert in all branches of mathematics and who could apply their talents to practical purposes. Vitruvius also credits him with inventing the *skaphē,* a widely used sundial consisting of a hemispherical bowl with a needle erected vertically in the middle to cast shadows. Speculations as to why a reputable mathematician like Aristarchus should interest himself in the true physical orientation of the solar system thus appear to be idle. Some have pointed to the possible influence of Strato, who was known as "the physical philosopher." There is no evidence, however, to indicate that Aristarchus got his physical theories from Strato. A more likely assumption is that *On Sizes and Distances* gave him an appreciation of the relative sizes of the sun and earth and led him to propound a heliocentric system.

The beginnings of heliocentrism are traced to the early Pythagoreans, a religiophilosophical school that flourished in southern Italy in the fifth century B.C. Ancient tradition ascribed to Pythagoras (*ca.* 520 B.C.) the identification of the Morning Star and the Evening Star as the same body. Philolaus (*ca.* 440 B.C.) gave the earth, moon, sun, and planets an orbital motion about a central fire, which he called "the hearth of the universe." According to another tradition, it was Hicetas, a contemporary of Philolaus, who first gave a circular orbit to the earth. Hicetas was also credited with maintaining the earth's axial rotation and a stationary heavens. More reliable ancient authorities, however, associate the hypothesis of the earth's diurnal rotation with Heraclides of Pontus, a pupil of Plato, who is also explicitly credited with maintaining (*ca.* 340 B.C.) an epicyclic orbit of Venus—and presumably that of Mercury also—about the sun. Some Greek astronomer may have taken the next logical step toward developing a complete heliocentric hypothesis by proposing the theory advanced in modern times by Tycho Brahe, which placed the five visible planets in motion about the sun, and the sun, in turn, in motion about the earth. Several scholars have argued that such a step was indeed taken, the most notable being the Italian astronomer Schiaparelli, who ascribed the Tychonic system to Heraclides; but evidence of its existence in antiquity is lacking.

Ancient authorities are unanimous in attributing the heliocentric theory to Aristarchus. Archimedes, who lived shortly afterward, says that he published his views in a book or treatise in which the premises that he developed led to the conclusion that the universe is many times greater than the current conception of it. Archimedes, near the opening of *The Sand-Reckoner,* gives a summary statement of Aristarchus' argument:

> His hypotheses are that the fixed stars and the sun are stationary, that the earth is borne in a circular orbit about the sun, which lies in the middle of its orbit, and that the sphere of the fixed stars, having the same center as the sun, is so great in extent that the circle on which he supposes the earth to be borne has such a proportion to the distance of the fixed stars as the center of the sphere bears to its surface.

Plutarch (*ca.* A.D. 100) gives a similar brief account of Aristarchus' hypothesis, stating specifically that the earth revolves along the ecliptic and that it is at the same time rotating on its axis.

After reporting Aristarchus' views, Archimedes criticizes him for setting up a mathematically impossible proportion, pointing out that the center of the sphere has no magnitude and therefore cannot bear any ratio to the surface of the sphere. Archimedes intrudes the observation that the "universe," as it is commonly conceived of by astronomers, is a sphere whose radius extends from the center of the sun to the center of the earth. Accordingly, as a mathematician he imputes to the mathematician Aristarchus a proportion that he feels is implicit in his statement, namely, that the ratio that the earth bears to the universe, as it is commonly conceived, is equal to the ratio that the sphere in which the earth revolves, in Aristarchus' scheme, bears to the sphere of the fixed stars.

Modern scholars have generally supposed that Aristarchus did not intend to have his proportion interpreted as a mathematical statement, that instead he was using an expression conventional with Greek mathematical cosmographers—"having the relation of a point"—merely to indicate the minuteness of the earth's orbit and the vastness of the heavens. Sir Thomas Heath points to similar expressions in the works of Euclid, Geminus, Ptolemy, and Cleomedes, and in the second assumption of Aristarchus' extant treatise *On Sizes and Distances* (see below). Heath feels that Archimedes' interpretation was arbitrary and sophistical and that Aristarchus introduced the statement to account for the inability to observe

stellar parallax from an orbiting earth. Neugebauer defends the proportion that Archimedes ascribes to Aristarchus,

$$r : R_e = R_e : R_f,$$

as mathematically sound and providing finite dimensions for the sphere of the fixed stars: the earth's radius (r) is so small in comparison with the sun's distance (R_e) that no daily parallax of the sun is discernible for determining R_e; according to Aristarchus' hypothesis, the earth moves in an orbit whose radius is R_e and no annual parallax of the fixed stars is discernible.

Why did the Greeks, after evolving a heliocentric hypothesis in gradual steps over a period of two centuries, allow it to fall into neglect almost immediately? Only one man, Seleucus of Seleucia (*ca.* 150 B.C.), is known to have embraced Aristarchus' views. The common attitude of deploring the "abandonment" of the heliocentric theory as a "retrogressive step" appears to be unwarranted when it is realized that the theory, however bold and ingenious it is to be regarded, never attracted much attention in antiquity. Aristarchus' system was the culmination of speculations about the physical nature of the universe that began with the Ionian philosophers of the sixth century, and it belongs to an age that was passing away. The main course of development of Greek astronomy was mathematical, not physical, and the great achievements were still to come—the exacting demonstrations and calculations of Apollonius of Perga, Hipparchus, and Ptolemy. These were based upon a geocentric orientation.

To a mathematician the orientation is of no consequence; in fact it is more convenient to construct a system of epicycles and eccentrics to account for planetary motions from a geocentric orientation. A heliocentric hypothesis neatly explained some basic phenomena, such as the stations and retrogradations of superior planets; but a circular orbit for the earth, about a sun in the exact center, failed to account for precise anomalies, such as the inequality of the seasons. In explanation of this inequality, Hipparchus determined the eccentricity of the earth's position as 1/24 of the radius of the sun's circle and he fixed the line of absides in the direction of longitude 65° 30′. Ptolemy adopted Hipparchus' solar data without change, unaware that the sun's orbit describes a revolving eccentric, the shift being 32′ in a century. The Arab astronomer al-Battānī (A.D. 858–929) discovered this shift. Epicyclic constructions had two advantages over eccentric constructions: they were applicable to inferior as well as superior planets and they palpably demonstrated planetary stations and retrograde motions. By the time of Apollonius it was understood that an equivalent eccentric system could be constructed for every epicyclic system. Henceforth, combinations of epicycles and eccentrics were introduced, all from a geocentric orientation. Aristarchus, too, had used a geocentric orientation in calculating the sizes and distances of the sun and moon.

It is not hard to account for the lack of interest in the heliocentric theory. The *Zeitgeist* of the new Hellenistic age was set and characterized by the abstruse erudition of the learned scholars and the precise researches of the astronomers, mathematicians, and anatomists working at the library and museum of Alexandria. Accurate instruments in use at Alexandria were giving astronomers a better appreciation of the vast distance of the sun. Putting the earth in orbit about the sun would lead to the expectation that some variation in the position of the fixed stars would be discernible at opposite seasons. Absence of displacement would presuppose a universe of vast proportions. The more precise the observations, the less inclined were the astronomers at Alexandria to accept an orbital motion of the earth. It is the opinion of Heath that Hipparchus (*ca.* 190–120 B.C.), usually regarded as the greatest of Greek astronomers, in adopting the geocentric orientation "sealed the fate of the heliocentric hypothesis for so many centuries."

The intellectual world at large was also disinclined to accept Aristarchus' orientation. Aristotle's doctrine of "natural places," which assigned to earth a position at the bottom or center among the elements comprising the universe, and his plausible "proofs" of a geocentric orientation, carried great weight in later antiquity, even with the mathematician Ptolemy. Religious minds were reluctant to relinquish the central position of man's abode. According to Plutarch, Cleanthes, the second head of the Stoic school (263–232 B.C.), thought that Aristarchus ought to be indicted on a charge of impiety for putting the earth in motion. Astrology, a respectable science in the eyes of many leading intellectuals, was enjoying an extraordinary vogue after its recent introduction. Its doctrines and findings were also based upon a geocentric orientation.

It is interesting to note in passing that Copernicus' disappointment at being anticipated by Aristarchus has recently come to light. Copernicus deliberately suppressed a statement acknowledging his awareness of Aristarchus' theory; the statement, deleted from the autograph copy of the *De revolutionibus,* appears in a footnote in the Thorn edition (1873) of that work. Elsewhere Copernicus tells of his search for classical precedents for his novel ideas about the heavens and of his finding in Plutarch the views of Philolaus, Heraclides, and Ecphantus; but he omits mention of the clear statement about Aristarchus' theory that appears a few pages earlier. Lastly, Copernicus' almost certain acquaintance with Archimedes' *The Sand-Reckoner,* the work containing our best account of Aristarchus' theory, has recently been pointed out.

His accomplishments as an astronomer have tended to detract attention from Aristarchus' attainments as a mathematician. Flourishing a generation after Euclid and a generation before Archimedes, Aristarchus was capable of the same sort of rigorous and logical geometrical demonstrations that distinguished the work of those famous mathematicians. *On Sizes and Distances* marks the first attempt to determine astronomical distances and dimensions by mathematical deductions based upon a set of assumptions. His last assumption assigns a grossly excessive estimate to the apparent angular diameter of the moon (2°). We are told by Archimedes in *The Sand-Reckoner* that Aristarchus discovered the sun's apparent angular diameter to be 1/720 part of the zodiac circle (1/2°), a close and respectable estimate. Aristarchus uses a geocentric orientation in *On Sizes and Distances* and concludes that the sun's volume is over 300 times greater than the earth's volume. For these reasons it is generally assumed that the treatise was an early work, antedating his heliocentric hypothesis.

Aristarchus argues that at the precise moment of the moon's quadrature, when it is half-illuminated, angle *SME* is a right angle; angle *SEM* can be measured by observation; therefore it is possible to deduce angle *MSE* and to determine the ratio of the distance of the moon to the distance of the sun (Figure 1). Two obvious difficulties are involved in his procedures: the determination with any exactitude (1) of the time of the moon's dichotomy and (2) of the measurement of angle *SEM*. A slight inaccuracy in either case would lead to a grossly inaccurate result. Aristarchus assumes angle *SEM* to be 87°, when in actuality it is more than 89° 50′, and he derives a distance for the sun of 18 to 20 times greater than the moon's distance (actually nearly 400 times greater). His mathematical procedures are sound, but his observational data are so crude as to make it apparent that Aristarchus was interested here in mathematical demonstrations and not in physical realities.

Aristarchus' treatise begins with six assumptions:

(1) That the moon receives its light from the sun.

(2) That the earth has the relation of a point and center to the sphere of the moon.

(3) That when the moon appears to us to be exactly

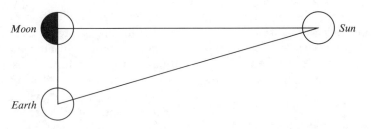

Figure 1

at the half the great circle dividing the light and dark portions of the moon is in line with the observer's eye.

(4) That when the moon appears to us to be at the half its distance from the sun is less than a quadrant by 1/30 part of a quadrant (87°).

(5) That the breadth of the earth's shadow (during eclipses) is that of two moons.

(6) That the moon subtends 1/15 part of a sign of the zodiac (2°).

He then states that he is in a position to prove three propositions:

(1) The distance of the sun from the earth is more than eighteen times but less than twenty times the moon's distance (from the earth); this is based on the assumption about the halved moon.

(2) The diameter of the sun has the same ratio to the diameter of the moon (i.e., assuming that the sun and moon have the same apparent angular diameter).

(3) The diameter of the sun has to the diameter of the earth a ratio greater than 19:3, but less than 43:6; this deduction follows from the ratio between the distances thus discovered, from the assumption about the shadow, and from the assumption that the moon subtends 1/15 part of a sign of the zodiac.

Then follow eighteen propositions containing the demonstrations. Heath has edited and translated the complete Greek text, together with Pappus' comments on the treatise, in his *Aristarchus of Samos* (pp. 352–414), and presents a summary account of the treatise in *A History of Greek Mathematics* (Vol. II).

Anticipating trigonometric methods that were to come, Aristarchus was the first to develop geometric procedures for approximating the sines of small angles. He deals with angles expressed as fractions of right angles and ratios of the sides of triangles, determining limits between which actual values lie. In Proposition 7, demonstrating that the distance of the sun is more than eighteen times but less than twenty times the distance of the moon, which would be expressed trigonometrically $1/18 > \sin 3° > 1/20$, he uses in his proof certain inequalities that he assumes to be known and accepted. These may be expressed trigonometrically. If α and β are acute angles and $\alpha > \beta$, then

$$\tan \alpha / \tan \beta > \alpha / \beta > \sin \alpha / \sin \beta.$$

If Aristarchus had had a correct measurement of the angle *SEM*—89 5/6° instead of 87°—his result would have been nearly correct. A century later Hipparchus was able to obtain a very close approximation of the moon's distance, expressed in terms of earth radii, by measuring the earth's shadow during lunar eclipses; but an appreciation of the vast distance of the sun had to wait upon the development of modern precision instruments.

Other dimensions deduced by Aristarchus in his treatise, all of them grossly underestimated because of his poor observational data, are:

(Prop. 10) The sun has to the moon a ratio greater than 5,832:1 but less than 8,000:1.

(Prop. 11) The diameter of the moon is less than 2/45 but greater than 1/30 of the distance of the center of the moon from the observer.

(Prop. 16) The sun has to the earth a ratio greater than 6,859:27 but less than 79,507:216.

(Prop. 17) The diameter of the earth is to the diameter of the moon in a ratio greater than 108:43 but less than 60:19.

(Prop. 18) The earth is to the moon in a ratio greater than 1,259,712:79,507 but less than 216,000:6,859.

BIBLIOGRAPHY

Thomas W. Africa, "Copernicus' Relation to Aristarchus and Pythagoras," in *Isis,* **52** (1961), 403–409; Angus Armitage, *Copernicus, the Founder of Modern Astronomy* (London, 1938); John L. E. Dreyer, *A History of the Planetary Systems from Thales to Kepler* (Cambridge, England, 1906; repr., New York, 1953); Pierre Duhem, *Le système du monde,* Vols. I–II (Paris, 1954); Sir Thomas Heath, *Aristarchus of Samos* (Oxford, 1913) and *A History of Greek Mathematics,* 2 vols. (Oxford, 1921); Otto Neugebauer, "Archimedes and Aristarchus," in *Isis,* **39** (1942), 4–6; Giovanni V. Schiaparelli, "Origine del sistema planetario eliocentrico presso i Greci," in *Memorie del'Istituto lom-*

bardo di scienze e lettere, **18** (1898), fasc. 5; and William H. Stahl, "The Greek Heliocentric Theory and Its Abandonment," in *Transactions of the American Philological Association,* **77** (1945), 321–332.

WILLIAM H. STAHL

ARISTOTLE (*b.* Stagira in Chalcidice, 384 B.C.; *d.* Chalcis, 322 B.C.), *the most influential ancient exponent of the methodology and division of sciences; contributed to physics, physical astronomy, meteorology, psychology, biology.* The following article is in four parts: Method, Physics, and Cosmology; Natural History and Zoology; Anatomy and Physiology; Tradition and Influence.

Method, Physics, and Cosmology.

Aristotle's father served as personal physician to Amyntas II of Macedon, grandfather of Alexander the Great. Aristotle's interest in biology and in the use of dissection is sometimes traced to his father's profession, but any suggestion of a rigorous family training in medicine can be discounted. Both parents died while Aristotle was a boy, and his knowledge of human anatomy and physiology remained a notably weak spot in his biology. In 367, about the time of his seventeenth birthday, he came to Athens and became a member of Plato's Academy. Henceforth his career falls naturally into three periods. He remained with the Academy for twenty years. Then, when Plato died in 347, he left the city and stayed away for twelve years: his reason for going may have been professional, a dislike of philosophical tendencies represented in the Academy by Plato's nephew and successor, Speusippus, but more probably it was political, the new anti-Macedonian mood of the city. He returned in 335 when Athens had come under Macedonian rule, and had twelve more years of teaching and research there. This third period ended with the death of his pupil, Alexander the Great (323), and the revival of Macedon's enemies. Aristotle was faced with a charge of impiety and went again into voluntary exile. A few months later he died on his maternal estate in Chalcis.

His middle years away from Athens took him first to a court on the far side of the Aegean whose ruler, Hermeias, became his father-in-law; then (344) to the neighboring island Lesbos, probably at the suggestion of Theophrastus, a native of the island and henceforth a lifelong colleague; finally (342) back to Macedon as tutor of the young prince Alexander. After his return to Athens he lectured chiefly in the grounds of the Lyceum, a Gymnasium already popular with sophists and teachers. The Peripatetic school, as an institution comparable to the Academy, was probably not founded until after his death. But with some distinguished students and associates he collected a natural history museum and a library of maps and manuscripts (including his own essays and lecture notes), and organized a program of research which *inter alia* laid the foundation for all histories of Greek natural philosophy (see Theophrastus), mathematics and astronomy (see Eudemus), and medicine.

Recent discussion of his intellectual development has dwelt on the problem of distributing his works between and within the three periods of his career. But part of the stimulus to this inquiry was the supposed success with which Plato's dialogues had been put in chronological order, and the analogy with Plato is misleading. Everything that Aristotle polished for public reading in Plato's fashion has been lost, save for fragments and later reports. The writings that survive are a collection edited in the first century B.C. (see below, Aristotle: Tradition), allegedly from manuscripts long mislaid: a few items are spurious (among the scientific works *Mechanica, Problemata, De mundo, De plantis*), most are working documents produced in the course of Aristotle's teaching and research; and the notes and essays composing them have been arranged and amended not only by their author but also by his ancient editors and interpreters. Sometimes an editorial title covers a batch of writings on connected topics of which some seem to supersede others (thus *Physics* VII seems an unfinished attempt at the argument for a prime mover which is carried out independently in *Physics* VIII); sometimes the title represents an open file, a text annotated with unabsorbed objections (e.g., the *Topics*) or with later and even post-Aristotelian observations (e.g., the *Historia animalium*). On the other hand it cannot be assumed that inconsistencies are always chronological pointers. In *De caelo* I–II he argues for a fifth element in addition to the traditional four (fire, air, water, earth): unlike them, its natural motion is circular and it forms the divine and unchanging substance of the heavenly bodies. Yet in *De caelo* III–IV, as in the *Physics,* he discusses the elements without seeming to provide for any such fifth body, and these writings are accordingly sometimes thought to be earlier. But on another view of his methods (see below, on dialectic) it becomes more intelligible that he should try different and even discrepant approaches to a topic at the same time.

Such considerations do not make it impossible to reconstruct something of the course of his scientific thinking from the extant writings, together with what is known of his life. For instance it is sometimes said that his distinction between "essence" and "accident," or between defining and nondefining characteristics, must be rooted in the biological studies in which it plays an integral part. But the distinction is explored at greatest length in the *Topics,* a handbook of dialectical debate which dates substantially from his earlier years in the Academy, whereas the inquiries embodied in his biological works seem to come chiefly from his years abroad, since they refer relatively often to the Asiatic coast and Lesbos and seldom to southern Greece. So this piece of conceptual apparatus was not produced by the work in biology. On the contrary, it was modified by that work: when Aristotle tries to reduce the definition of a species to one distinguishing mark (e.g., *Metaphysics* VII 12, VIII 6) he is a dialectician, facing a problem whose ancestry includes Plato's theory of Forms, but when he rejects such definitions in favor of a cluster of differentiae (*De partibus animalium* I 2–3) he writes as a working biologist, armed with a set of questions about breathing and sleeping, movement and nourishment, birth and death.

The starting point in tracing his scientific progress must therefore be his years in the Academy. Indeed without this starting point it is not possible to understand either his pronouncements on scientific theory or, what is more important, the gap between his theory and his practice.

The Mathematical Model. The Academy that Aristotle joined in 367 was distinguished from other Athenian schools by two interests: mathematics (including astronomy and harmonic theory, to the extent that these could be made mathematically respectable), and dialectic, the Socratic examination of the assumptions made in reasoning—including the assumptions of mathematicians and cosmologists. Briefly, Plato regarded the first kind of studies as merely preparatory and ancillary to the second; Aristotle, in the account of scientific and philosophical method that probably dates from his Academic years, reversed the priorities (*Posterior Analytics* I; *Topics* I 1–2). It was the mathematics he encountered that impressed him as providing the model for any well-organized science. The work on axiomatization which was to culminate in Euclid's *Elements* was already far advanced, and for Aristotle the pattern of a science is an axiomatic system in which theorems are validly derived from basic principles, some proprietary to the science ("hypotheses" and "definitions," the second corresponding to Euclid's "definitions"), others having an application in more than one system ("axioms," corresponding to Euclid's "common notions"). The proof-theory which was characteristic of Greek mathematics (as against that of Babylon or Egypt) had developed in the attempt to show why various mathematical formulae worked in practice. Aristotle pitches

on this as the chief aim of any science: it must not merely record but explain, and in explaining it must, so far as the special field of inquiry allows, generalize. Thus mathematical proof becomes Aristotle's first paradigm of scientific explanation; by contrast, the dialectic that Plato ranked higher—the logical but free-ranging analysis of the beliefs and usage of "the many and the wise"—is allowed only to help in settling those basic principles of a science that cannot, without regress or circularity, be proved within the science itself. At any rate, this was the theory.

Aristotle duly adapts and enlarges the mathematical model to provide for the physical sciences. Mathematics, he holds, is itself a science (or rather a family of sciences) about the physical world, and not about a Platonic world of transcendent objects; but it abstracts from those characteristics of the world that are the special concern of physics—movement and change, and therewith time and location. So the nature and behavior of physical things will call for more sorts of explanation than mathematics recognizes. Faced with a man, or a tree, or a flame, one can ask what it is made of, its "matter"; what is its essential character or "form"; what external or internal agency produced it; and what the "end" or purpose of it is. The questions make good sense when applied to an artifact such as a statue, and Aristotle often introduces them by this analogy; but he holds that they can be extended to every kind of thing involved in regular natural change. The explanations they produce can be embodied in the formal proofs or even the basic definitions of a science (thus a lunar eclipse can be not merely accounted for, but defined, as the loss of light due to the interposition of the earth, and a biological species can be partly defined in terms of the purpose of some of its organs). Again, the regularities studied by physics may be unlike those of mathematics in an important respect: initially the *Posterior Analytics* depicts a science as deriving necessary conclusions from necessary premises, true in all cases (I ii and iv), but later (I xxx) the science is allowed to deal in generalizations that are true in most cases but not necessarily in all. Aristotle is adapting his model to make room for "a horse has four legs" as well as for "2 × 2 = 4." How he regards the exceptions to such generalizations is not altogether clear. In his discussions of "luck" and "chance" in *Physics* II, and of "accident" elsewhere, he seems to hold that a lucky or chance or accidental event can always, under some description, be subsumed under a generalization expressing some regularity. His introduction to the *Meteorologica* is sometimes cited to show that in his view sublunary happenings are inherently irregular; but he probably means that, while the laws of sublunary physics are commonly (though not always) framed to allow of exceptions, these exceptions are not themselves inexplicable. The matter is complicated by his failure to maintain a sharp distinction between laws that provide a necessary (and even uniquely necessary), and those that provide a sufficient, condition of the situation to be explained.

But in two respects the influence of mathematics on Aristotle's theory of science is radical and unmodified. First, the drive to axiomatize mathematics and its branches was in fact a drive for autonomy: the premises of the science were to determine what questions fell within the mathematician's competence and, no less important, what did not. This consequence Aristotle accepts for every field of knowledge: a section of *Posterior Analytics* I xii is given up to the problem, what questions can be properly put to the practitioner of such-and-such a science; and in I vii, trading on the rule "one science to one genus," he denounces arguments that poach outside their own field—which try, for instance, to deduce geometrical conclusions from arithmetical premises. He recognizes arithmetical proofs in harmonics and geometrical proofs in mechanics, but treats them as exceptions. The same impulse leads him to map all systematic knowledge into its departments—theoretical, practical, and productive—and to divide the first into metaphysics (or, as he once calls it, "theology"), mathematics, and physics, these in turn being marked out in subdivisions.

This picture of the autonomous deductive system has had a large influence on the interpreters of Aristotle's scientific work; yet it plays a small part in his inquiries, just because it is not a model for inquiry at all but for subsequent exposition. This is the second major respect in which it reflects mathematical procedure. In nearly all the surviving productions of Greek mathematics, traces of the workshop have been deliberately removed: proofs are found for theorems that were certainly first reached by other routes. So Aristotle's theoretical picture of a science shows it in its shop window (or what he often calls its "didactic") form; but for the most part his inquiries are not at this stage of the business. This is a piece of good fortune for students of the subject, who have always lamented that no comparable record survives of presystematic research in mathematics proper (Archimedes' public letter to Eratosthenes—the *Ephodos,* or "Method"—is hardly such a record). As it is, Aristotle's model comes nearest to realization in the systematic astronomy of *De caelo* I–II (cf., e.g., I iii, "from what has been said, partly as premises and partly as things proved from these, it

follows . . ."), and in the proof of a prime mover in *Physics* VIII. But these constructions are built on the presystematic analyses of *Physics* I–VI, analyses that are expressly undertaken to provide physics with its basic assumptions (cf. I i) and to define its basic concepts, change and time and location, infinity and continuity (III i). *Ex hypothesi* the latter discussions, which from Aristotle's pupils Eudemus and Strato onward have given the chief stimulus to physicists and philosophers of science, cannot be internal to the science whose premises they seek to establish. Their methods and data need not and do not fit the theoretical straitjacket, and in fact they rely heavily on the dialectic that theoretically has no place in the finished science.

Dialectic and "Phenomena." Conventionally Aristotle has been contrasted with Plato as the committed empiricist, anxious to "save the phenomena" by basing his theories on observation of the physical world. First the phenomena, then the theory to explain them: this Baconian formula he recommends not only for physics (and specifically for astronomy and biology) but for ethics and generally for all arts and sciences. But "phenomena," like many of his key terms, is a word with different uses in different contexts. In biology and meteorology the phenomena are commonly observations made by himself or taken from other sources (fishermen, travelers, etc.), and similar observations are evidently presupposed by that part of his astronomy that relies on the schemes of concentric celestial spheres proposed by Eudoxus and Callippus. But in the *Physics* when he expounds the principles of the subject, and in many of the arguments in the *De caelo* and *De generatione et corruptione* by which he settles the nature and interaction of the elements, and turns Eudoxus' elegant abstractions into a cumbrous physical (and theological) construction, the data on which he draws are mostly of another kind. The phenomena he now wants to save—or to give logical reasons (rather than empirical evidence) for scrapping—are the common convictions and common linguistic usage of his contemporaries, supplemented by the views of other thinkers. They are what he always represents as the materials of dialectic.

Thus when Aristotle tries to harden the idea of location for use in science (*Physics* IV 1–5) he sets out from our settled practice of locating a thing by giving its physical surroundings, and in particular from established ways of talking about one thing taking another's place. It is to save these that he treats any location as a container, and defines the place of X as the innermost static boundary of the body surrounding X. His definition turns out to be circular:

moreover it carries the consequence that, since a point cannot lie within a boundary, it cannot strictly have (or be used to mark) a location. Yet we shall see later that his theories commit him to denying this.

Again, when he defines time as that aspect of change that enables it to be counted (*Physics* IV 10–14), what he wants to save and explain are the common ways of *telling* the time. This point, that he is neither inventing a new vocabulary nor assigning new theory-based uses to current words, must be borne in mind when one encounters such expressions as "force" and "average velocity" in versions of his dynamics. The word sometimes translated "force" (*dunamis*) is the common word for the "power" or "ability" of one thing to affect or be affected by another—to move or be moved, but also to heat or to soften or to be heated, and so forth. Aristotle makes it clear that this notion is what he is discussing in three celebrated passages (*Physics* VII 5, VIII 10, *De caelo* I 7) where later critics have discerned laws of proportionality connecting the force applied, the weight moved, and the time required for the force to move the weight a given distance. (Two of the texts do not mention weight at all.) A second term, *ischus,* sometimes rendered "force" in these contexts, is the common word for "strength," and it is this familiar notion that Aristotle is exploiting in the so-called laws of forced motion set out in *Physics* VII 5 and pre-supposed in VIII 10: he is relying on what a nontech-nical audience would at once grant him concerning the comparative strengths of packhorses or (his example) gangs of shiphaulers. He says: let A be the strength required to move a weight B over a distance D in time T; then (1) A will move $1/2$ B over $2D$ in T; (2) A will move $1/2$ B over D in $1/2$ T; (3) $1/2$ A will move $1/2$ B over D in T; and (4) A will move B over $1/2$ D in $1/2$ T; but (5) it does not follow that A will move some multiple of B over a proportion-ate fraction of D in T or indeed in any time, since it does not follow that A will be sufficient to move that multiple of B at all. The conjunction of (4) with the initial assumption shows that Aristotle takes the speed of motion in this case to be uniform; so com-mentators have naturally thought of A as a force whose continued application to B is just sufficient to overcome the opposing forces of gravity, friction, and the medium. In such circumstances propositions (3) and (4) will yield results equivalent to those of New-tonian dynamics. But then the circumstances described in (1) and (2) should yield not just the doubling of a uniform velocity which Aristotle supposes, but ac-celeration up to some appropriate terminal velocity. Others have proposed to treat A as prefiguring the later idea not of *force* but of *work,* or else *power,* if

these are defined in terms of the displacement of weight and not of force; and this has the advantage of leaving Aristotle discussing the case that is central to his dynamics—the carrying out of some finite task in a finite time—without importing the notion of action at an instant which, for reasons we shall see, he rejects. But Aristotle also assumes that, for a given type of agent, A is multiplied in direct ratio to the size or quantity of the agent; and to apply this to the work done would be, once more, to overlook the difference between conditions of uniform motion and of acceleration. The fact is that Aristotle is appealing to conventional ways of comparing the strength of haulers and beasts of burden, and for his purposes the acceleration periods involved with these are negligible. What matters is that we measure strength by the ability to perform certain finite tasks before fatigue sets in; hence, when Aristotle adduces these proportionalities in the *Physics,* he does so with a view to showing that the strength required for keeping the sky turning for all time would be immeasurable. Since such celestial revolutions do not in his view have to overcome any such resistance as that of gravity or a medium we are not entitled to read these notions into the formulae quoted. What then is the basis for these proportionalities? He does not quote empirical evidence in their support, and in their generalized form he could not do so; in the *Physics* and again in the *De caelo* he insists that they can be extended to cover "heating and any effect of one body on another," but the Greeks had no thermometer nor indeed any device (apart from the measurement of strings in harmonics) for translating qualitative differ-ences into quantitative measurements. Nor on the other hand does he present them as technical defini-tions of the concepts they introduce. He simply com-ments in the *Physics* that the rules of proportion require them to be true (and it may be noticed that he does not frame any of them as a function of more than two variables: the proportion is always a simple relation between two of the terms, the others remain-ing constant). He depends on this appeal, together with conventional ways of comparing strengths, to give him the steps he needs toward his conclusion about the strength of a prime mover: it is no part of the dialectic of his argument to coin hypotheses that require elaborate discussion in their own right.

It is part of the history of dynamics that, from Aristotle's immediate successors onward, these formulae were taken out of context, debated and refined, and finally jettisoned for an incomparably more exact and powerful set of concepts which owed little to dialectic in Aristotle's sense. That he did not intend his proportionalities for such close scrutiny becomes even clearer when we turn to his so-called laws of natural motion. Aristotle's universe is finite, spherical, and geocentric: outside it there can be no body nor even, therefore, any location or vacuum or time (*De caelo* I 9); within it there can be no vacuum (*Physics* IV 6–9). Natural motion is the unimpeded movement of its elements: centripetal or "downward" in the case of earth (whose place is at the center) and of water (whose place is next to earth), centrifugal or "upward" in the case of fire and (next below fire) air. These are the sublunary elements, capable of changing into each other (*De generatione et corruptione* II) and possessed of "heaviness" or "lightness" according as their natural motion is down or up. Above them all is the element whose existence Aristotle can prove only by a priori argument: ether, the substance of the spheres that carry the heavenly bodies. The natural motions of the first four elements are rectilinear and terminate, unless they are blocked, in the part of the universe that is the element's natural place; the motion of the fifth is circular and cannot be blocked, and it never leaves its natural place. These motions of free fall, free ascent, and free revolution are Aristotle's paradigms of regular movement, against which other motions can be seen as departures due to special agency or to the presence of more than one element in the moving body. On several occasions he sketches some proportional connection between the variables that occur in his analysis of such natural motions; generally he confines himself to rectilinear (i.e., sublunary) movement, as, for example, in *Physics* IV 8, the text that provoked a celebrated exchange between Simplicio and Salviati in Galileo's *Dialoghi*. There he writes: "We see a given weight or body moving faster than another for two reasons: either because of a difference in the medium traversed (e.g., water as against earth, water as against air), or, other things being equal, because of the greater weight or lightness of the moving body." Later he specifies that the proviso "other things being equal" is meant to cover identity of shape. Under the first heading, that of differences in the medium, he remarks that the motion of the medium must be taken into account as well as its density relative to others; but he is content to assume a static medium and propound, as always, a simple proportion in which the moving object's velocity varies inversely with the density of the medium. Two comments are relevant. First, in this as in almost all comparable contexts, the "laws of natural motion" are dispensable from the argument. Here Aristotle uses his proportionality to rebut the possibility of motion in a vacuum: such motion would encounter a medium of nil density and hence would have infinite velocity, which is impossible. But this

is only one of several independent arguments for the same conclusion in the context. Next, the argument discounts acceleration (Aristotle does not consider the possibility of a body's speed in a vacuum remaining finite but increasing without limit, let alone that of its increasing to some finite terminal speed); yet he often insists that for the sublunary elements natural motion is always acceleration. (For this reason among others it is irrelevant to read his proportionalities of natural motion as an unwitting anticipation of Stokes's law.) But it was left to his successors during the next thousand years to quarrel over the way in which the ratios he formulated could be used to account for the steady acceleration he required in such natural motion; and where in the passage quoted he writes "we see," it was left to some nameless ancient scientist to make the experiment recorded by Philoponus and later by Galileo, of dropping different weights from the same height and noting that what we see does not answer to Aristotle's claim about their speed of descent. It was, to repeat, no part of the dialectic of his argument to give these proportionalities the rigor of scientific laws or present them as the record of exact observation.

On the other hand the existence of the natural motions themselves is basic to his cosmology. Plato had held that left to themselves, i.e., without divine governance, the four elements (he did not recognize a fifth) would move randomly in any direction: Aristotle denies this on behalf of the inherent regularity of the physical world. He makes the natural motions his "first hypotheses" in the *De caelo* and applies them over and again to the discussion of other problems. (The contrast between his carelessness over the proportionalities and the importance he attaches to the movements is sometimes read as showing that he wants to "eliminate mathematics from physics": but more on this later.)

This leads to a more general point which must be borne in mind in understanding his way of establishing physical theory. When he appeals to common views and usage in such contexts he is applying a favorite maxim, that in the search for explanations we must start from what is familiar or intelligible to us. (Once the science is set up, the deductions will proceed from principles "intelligible in themselves.") The same maxim governs his standard way of introducing concepts by extrapolating from some familiar, unpuzzling situation. Consider his distinction of "matter" and "form" in *Physics* I. He argues that any change implies a passage between two contrary attributes—from one to the other, or somewhere on a spectrum between the two—and that there must be a third thing to make this passage, a substrate which changes but survives the change. The situations to which he appeals are those from which this triadic analysis can be, so to speak, directly read off: a light object turning dark, an unmusical man becoming musical. But then the analysis is extended to cases progressively less amenable: he moves, via the detuning of an instrument and the shaping of a statue, to the birth of plants and animals and generally to the sort of situation that had exercised earlier thinkers—the emergence of a new individual, the apparent coming of something from nothing. (Not the emergence of a new *type:* Aristotle does not believe that new types emerge in nature, although he accepts the appearance of sports within or between existing types. In *Physics* II 8 he rejects a theory of evolution for putting the random occurrence of new types on the same footing with the reproduction of existing species, arguing that a theory that is not based on such regularities is not scientific physics.) *Ex nihilo nihil fit;* and even the emergence of a new individual must involve a substrate, "matter," which passes between two contrary conditions, the "privation" and the "form." But one effect of Aristotle's extrapolation is to force a major conflict between his theories and most contemporary and subsequent physics. In his view, the question "What are the essential attributes of matter?" must go unanswered. There is no general answer, for the distinction between form and matter reappears on many levels: what serves as matter to a higher form may itself be analyzed into form and matter, as a brick which is material for a house can itself be analyzed into a shape and the clay on which the shape is imposed. More important, there is no answer even when the analysis reaches the basic elements—earth, air, fire, and water. For these can be transformed into each other, and since no change can be intelligibly pictured as a mere succession of discrete objects these too must be transformations of some residual subject, but one that now *ex hypothesi* has no permanent qualitative or quantitative determinations in its own right. Thus Aristotle rejects all theories that explain physical change by the rearrangement of some basic stuff or stuffs endowed with fixed characteristics. Atomism in particular he rebuts at length, arguing that movement in a vacuum is impossible (we have seen one argument for this) and that the concept of an extended indivisible body is mathematically indefensible. But although matter is not required to identify itself by any permanent first-order characteristics, it does have important second-order properties. Physics studies the regularities in change, and for a given sort of thing at a given level it is the matter that determines what kinds of change are open to it. In some respects the idea has

more in common with the field theory that appears embryonically in the Stoics than with the crude atomism maintained by the Epicureans, but its chief influence was on metaphysics (especially Neoplatonism) rather than on scientific theory. By contrast, the correlative concept of *form,* the universal element in things that allows them to be known and classified and defined, remained powerful in science. Aristotle took it from Plato, but by way of a radical and very early critique of Plato's Ideas; for Aristotle the formal element is inseparable from the things classified, whereas Plato had promoted it to independent existence in a transcendent world contemplated by disembodied souls. For Aristotle the physical world is all; its members with their qualities and quantities and interrelations are the paradigms of reality and there are no disembodied souls.

The device of extrapolating from the familiar is evident again in his account of another of his four types of "cause," or explanation, viz. the "final," or teleological. In *Physics* II 8 he mentions some central examples of purposive activity—housebuilding, doctoring, writing—and then by stages moves on to discerning comparable purposiveness in the behavior of spiders and ants, the growth of roots and leaves, the arrangement of the teeth. Again the process is one of weakening or discarding some of the conditions inherent in the original situations: the idea of purposiveness sheds its connection with those of having a skill and thinking out steps to an end (although Aristotle hopes to have it both ways, by representing natural sports and monsters as *mistakes*). The resultant "immanent teleology" moved his follower Theophrastus to protest at its thinness and facility, but its effectiveness as a heuristic device, particularly in biology, is beyond dispute.

It is worth noting that this tendency of Aristotle's to set out from some familiar situation, or rather from the most familiar and unpuzzling ways of describing such a situation, is something more than the general inclination of scientists to depend on "explanatory paradigms." Such paradigms in later science (e.g., classical mechanics) have commonly been limiting cases not encountered in common observation or discourse; Aristotle's choice of the familiar is a matter of dialectical method, presystematic by contrast with the finished science, but subject to rules of discussion which he was the first to codify. This, and not (as we shall see) any attempt to extrude mathematics from physics, is what separates his extant work in the field from the most characteristic achievements of the last four centuries. It had large consequences for dynamics. In replying to Zeno's paradox of the flying arrow he concedes Zeno's claim that nothing can be said to be moving at an instant, and insists only that it cannot be said to be stationary either. What preoccupies him is the requirement, embedded in common discourse, that any movement must take a certain time to cover a certain distance (and, as a corollary, that any stability must take a certain time but cover no distance); so he discounts even those hints that common discourse might have afforded of the derivative idea of motion, and therefore of velocity, at an instant. He has of course no such notion of a mathematical limit as the analysis of such cases requires, but in any event this notion came later than the recognition of the cases. It is illuminating to contrast the treatment of motion in the *Mechanica,* a work which used to carry Aristotle's name but which must be at least a generation later. There (*Mechanica* 1) circular motion is resolved into two components, one tangential and one centripetal (contrast Aristotle's refusal to assimilate circular and rectilinear movements, notably in *Physics* VII 4). And the remarkable suggestion is made that the proportion between these components need not be maintained for any time at all, since otherwise the motion would be in a straight line. Earlier the idea had been introduced of a point having motion and velocity, an idea that we shall find Aristotle using although his dialectical analysis of movement and location disallows it; here that idea is supplemented by the concept of a point having a given motion or complex of motions at an instant and not for any period, however small. The *Mechanica* is generally agreed to be a constructive development of hints and suggestions in Aristotle's writings; but the methods and purposes evident in his own discussions of motion inhibit him from such novel constructions in dynamics.

It is quite another thing to say, as is often said, that Aristotle wants to debar physics from any substantial use of the abstract proofs and constructions available to him in contemporary mathematics. It is a common fallacy that, whereas Plato had tried to make physics mathematical and quantitative, Aristotle aimed at keeping it qualitative.

Mathematics and Physics. Plato had tried to construct the physical world of two-dimensional and apparently weightless triangles. When Aristotle argues against this in the *De caelo* (III 7) he observes: "The principles of perceptible things must be perceptible, of eternal things eternal, of perishable things perishable: in sum, the principles must be homogeneous with the subject-matter." These words, taken together with his prescriptions for the autonomy of sciences in the *Analytics,* are often quoted to show that any use of mathematical constructions in his physics must be adventitious or presystematic, dis-

pensable from the science proper. The province of physics is the class of natural bodies regarded as having weight (or "lightness," in the case of air and fire), heat, and color and an innate tendency to move in a certain way. But these are properties that mathematics expressly excludes from its purview (*Metaphysics* K 3).

In fact, however, the division of sciences is not so absolute. When Aristotle contrasts mathematics and physics in *Physics* II he remarks that astronomy, which is one of the "more physical of the mathematical sciences," must be part of physics, since it would be absurd to debar the physicist from discussing the geometrical properties of the heavenly bodies. The distinction is that the physicist must, as the mathematician does not, treat these properties as the attributes of physical bodies that they are; i.e., he must be prepared to explain the application of his model. Given this tie-line a good deal of mathematical abstraction is evidently permissible. Aristotle holds that only extended bodies can strictly be said to have a location (i.e., to lie within a static perimeter) or to move, but he is often prepared to discount the extension of bodies. Thus in *Physics* IV 11, where he shows an isomorphic correspondence between continua representing time, motion, and the path traversed by the moving body, he correlates the moving object with points in time and space and for this purpose calls it "a point—or stone, or any such thing." In *Physics* V 4, he similarly argues from the motion of an unextended object, although it is to be noticed that he does not here or anywhere ease the transition from moving bodies to moving points by importing the idea of a center of gravity, which was to play so large a part in Archimedes' *Equilibrium of Planes.* In his meteorology, explaining the shape of halos and rainbows, he treats the luminary as a point source of light. In the biological works he often recurs to the question of the number of points at which a given type of animal moves; these "points" are in fact the major joints, but in *De motu animalium* 1 he makes it clear that he has a geometrical model in mind and is careful to explain what supplementary assumptions are necessary to adapting this model to the actual situation it illustrates. In the cosmology of the *De caelo* he similarly makes use of unextended loci, in contrast to his formal account of any location as a perimeter enclosing a volume. Like Archimedes a century later, he represents the center of the universe as a point when he proves that the surface of water is spherical, and again when he argues that earth moves so as to make its own (geometrical) center ultimately coincide with that of the universe. His attempt in *De caelo* IV 3 to interpret this in terms of perimeter locations is correct by his own principles, but confused.

This readiness to import abstract mathematical arguments and constructions into his account of the physical world is one side of the coin whose other face is his insistence that any mathematics must be directly applicable to the world. Thus, after arguing (partly on dialectical grounds, partly from his hypothesis of natural movements and natural places) that the universe must be finite in size, he adds that this does not put the mathematicians out of business, since they do not need or use the notion of a line infinite in extension: what they require is only the possibility of producing a line n in any required ratio with a given line m, and however large the ratio n/m it can always be physically exemplified for a suitable interpretation of m. The explanation holds good for such lemmata as that applied in Eudoxus' method of exhaustion, but not of some proportionalities he himself adduces earlier in the same context or in *De caelo* I. (These proportionalities are indeed used in, but they are not the subject of, *reductio ad absurdum* arguments. In the *De caelo* Aristotle even assumes that an infinite rotating body would contain a point at an infinite distance from its center and consequently moving at infinite speed.) The same concern to make mathematics applicable to the physical world without postulating an actual infinite is evident in his treatment of the sequence of natural numbers. The infinity characteristic of the sequence, and generally of any countable series whose members can be correlated with the series of numbers, consists just in the possibility of specifying a successor to any member of the sequence: "the infinite is that of which, as it is counted or measured off, it is always possible to take some part outside that already taken." This is true not only of the number series but of the parts produced by dividing any magnitude in a constant ratio; and since all physical bodies are in principle so divisible, the number series is assured of a physical application without requiring the existence at any time of an actually infinite set of objects: all that is required is the possibility of following any division with a subdivision.

This positivistic approach is often evident in Aristotle's work (e.g., in his analysis of the location of A as the inner static boundary of the body surrounding A), and it is closely connected with his method of building explanations on the familiar case. But here too Aristotle moves beyond the familiar case when he argues that infinite divisibility is characteristic of bodies below the level of observation. His defense and exploration of such divisibility, as a defining characteristic of bodies and times and mo-

tions, is found in *Physics* VI, a book often saluted as his most original contribution to the analysis of the continuum. Yet it is worth noticing that in this book as in its two predecessors Aristotle's problems and the ideas he applies to their solution are over and again taken, with improvements, from the second part of Plato's *Parmenides*. The discussion is in that tradition of logical debate which Aristotle, like Plato, called "dialectic," and its problems are not those of accommodating theories to experimentally established facts (or vice versa) but logical puzzles generated by common discourse and conviction. (But then Aristotle thinks of common discourse and conviction as a repository of human experience.) So the argument illustrates Aristotle's anti-Platonic thesis that mathematics—represented again in this case by simple proportion theory—has standing as a science only to the extent that it can be directly applied to the description of physical phenomena. But the argument is no more framed as an advance in the mathematical theory itself than as a contribution to the observational data of physics.

Probably the best-known instance of an essentially mathematical construction incorporated into Aristotle's physics is the astronomical theory due to Eudoxus and improved by Callippus. In this theory the apparent motion of the "fixed stars" is represented by the rotation of one sphere about its diameter, while those of the sun, moon, and the five known planets are represented each by a different nest of concentric spheres. In such a nest the first sphere carries round a second whose poles are located on the first but with an axis inclined to that of the first; this second, rotating in turn about its poles, carries a third similarly connected to it, and so on in some cases to a fourth or (in Callippus' version) a fifth, the apparent motion of the heavenly body being the resultant motion of a point on the equator of the last sphere. To this set of abstract models, itself one of the five or six major advances in science, Aristotle makes additions of which the most important is the attempt to unify the separate nests of spheres into one connected physical system. To this end he intercalates reagent spheres designed to insulate the movement of each celestial body from the complex of motions propelling the body next above it. The only motion left uncanceled in this downward transmission is the rotation of the star sphere. It is generally agreed that Aristotle in *Metaphysics* XII 8 miscalculates the resulting number of agent and reagent spheres: he concludes that we need either fifty-five or forty-seven, the difference apparently representing one disagreement between the theories of Eudoxus and Callippus, but on the latest computation (that of Hanson) the figures should be sixty-six and forty-nine. The mistake had no effect on the progress of astronomy: within a century astronomers had turned to a theory involving epicycles, and Aristotle's physical structure of concentric nonoverlapping spheres was superseded. On the other hand his basic picture of the geocentric universe and its elements, once freed from the special constructions he borrowed and adapted from Eudoxus, retained its authority and can be seen again in the introductory chapters of Ptolemy's *Syntaxis*.

Conclusion. These arguments and theories in what came to be called the exact sciences are drawn principally from the *Posterior Analytics, Topics, Physics, De caelo* and *De generatione,* works that are generally accepted as early and of which the first four at least probably date substantially from Aristotle's years in the Academy or soon after. The influence of the Academy is strong on them. They are marked by a large respect for mathematics and particularly for the techniques and effects of axiomatizing that subject, but they do not pretend to any mathematical discoveries, and in this they are close in spirit to Plato's writings. Even the preoccupation with physical change, its varieties and regularities and causes, and the use of dialectic in analyzing these, is a position to which Plato had been approaching in his later years. Aristotle the meticulous empiricist, amassing biological data or compiling the constitutions of 158 Greek states, is not yet in evidence. In these works the analyses neither start from nor are closely controlled by fresh inspections of the physical world. Nor is he liable to think his analyses endangered by such inspections: if his account of motion shows that any "forced" or "unnatural" movement requires an agent of motion in constant touch with the moving body, the movement of a projectile can be explained by inventing a set of unseen agents to fill the gap—successive stages of the medium itself, supposed to be capable of transmitting movement even after the initial agency has ceased acting. In all the illustrative examples cited in these works there is nothing comparable to even the half-controlled experiments in atomistic physics and harmonics of the following centuries. His main concerns were the methodology of the sciences, which he was the first to separate adequately on grounds of field and method; and the meticulous derivation of the technical equipment of these sciences from the common language and assumptions of men about the world they live in. His influence on science stemmed from an incomparable cleverness and sensitiveness to counterarguments, rather than from any breakthrough comparable to those of Eudoxus or Archimedes.

BIBLIOGRAPHY

Aristotle is still quoted by reference to the page, column, and line of Vols. I, II of I. Bekker's Berlin Academy edition (1831–1870, recently repr.). The later texts published in the Oxford Classical Texts and in the Budé and Loeb series are generally reliable for the works quoted. The standard Oxford translation of the complete works with selected fragments occupies 12 volumes (1909–1952). The ancient commentaries are still among the best (*Commentaria in Aristotelem Graeca,* Berlin, 1882–1909). Of recent editions with commentaries pride of place goes to those by Sir David Ross of the *Metaphysics* (2nd ed., 1953), *Physics* (1936), *Analytics* (1949), *Parva naturalia* (1955), and *De anima* (1961). Others are T. Waitz, *Organon* (1844–1846); H. H. Joachim, *De generatione et corruptione* (1922).

Important modern works are W. Jaeger, *Aristotle* (2nd English ed., Oxford, 1948); W. D. Ross, *Aristotle* (4th ed., London, 1945). On the mathematics and physics, T. L. Heath, *Mathematics in Aristotle* (Oxford, 1949); P. Duhem, *Le système du monde,* I (Paris, 1913); H. Carteron, *La notion de force dans le système d'Aristote* (Paris, 1924); A. Mansion, *Introduction à la physique aristotélicienne* (2nd ed., Louvain–Paris, 1945); F. Solmsen, *Aristotle's System of the Physical World* (Ithaca, N.Y., 1960); W. Wieland, *Die aristotelische Physik* (Göttingen, 1962); I. Düring, *Aristoteles* (Heidelberg, 1966).

On the so-called laws of motion in Aristotle, I. Drabkin, *American Journal of Philology,* **59** (1938), pp. 60–84.

G. E. L. OWEN

ARISTOTLE: Natural History and Zoology.

It is not clear when Aristotle wrote his zoology, or how much of his natural history was his own work. This is unfortunate, for it might help us to interpret his philosophy if we knew whether he began theorizing in biology before or after his main philosophical formulations, and how many zoological specimens he himself collected and identified. Some believe that he began in youth, and that his theory of potentiality was directed originally at the problem of growth. Others (especially Jaeger) hold that his interest in factual research came late in life and that he turned to biology after founding the Lyceum. Most probably, however, it was in middle life, in the years 344–342 B.C., when he was living on Lesbos with Theophrastus; many of his data are reported from places in that area. This would imply that he wrote the zoology with his philosophical framework already established, and on the whole the internal evidence of the treatises bears this out. It follows that in order to understand his zoological theory, we must keep his philosophy in mind. Yet it may also be true that in thinking out his philosophy, he was conscious of biological problems in a general way.

The zoological treatises must represent many years' work, for they make up a fourth of the whole corpus, and both data and discussion are concisely presented. They owe little to Herodotus, Ctesias, Xenophon, or other extant literature; their possible debt to Democritus cannot be assessed, however, because his three zoological books are lost. Comparing the quality of Aristotle's data with previous writings, we must conclude that he sifted and rejected a great deal; even by modern standards of natural history his reports are cautious. The chief collection of data is the *Historia animalium.* Out of 560 species mentioned in all his zoology, 400 appear only in this work and only five are not included. The treatises, as we now have them, form a course of instruction in which the *Historia* is referred to as the descriptive textbook, intended to be studied first and then kept at hand. Internal evidence suggests, however, that it was in fact written after the others, and that most of it was not written by Aristotle himself. This implies that he wrote the theoretical treatises before the main collection of data. Not that the treatises lack supporting data, but most of the information was common knowledge, whereas the reports that read like new, firsthand observation are nearly all confined to the later parts of the *Historia.*

Biological data were normally quoted in cosmological arguments, not least in the Academy. The Academicians' interest was not so much in the animals for their own sake, but rather in using them as evidence for—and giving them a place within—a rational cosmology. There were two issues: to identify the formal groups of animals, and thus to classify them, and to explain their functioning as part of nature. Plato and Speusippus opposed the materialism of those like Democritus, whose lost books, entitled *Causes Concerning Animals,* were probably intended to explain biology in terms of atomism. Aristotle would have been familiar with these discussions since his youth, and his writings follow this essentially etiological approach. His earliest zoology is probably in the *De partibus animalium,* the *De incessu animalium,* and the *Parva naturalia* (all of which in their present form show signs of revision and editing), in which he sets out the "causes" of tissues and structures, and of such significant functions as locomotion, respiration, aging, and death. Here the a priori element in his theory appears strongly: for example, right is superior to left, and hence the right-hand side is the natural side to lead off with; organs properly exist in pairs, and hence the spleen (for which he found no function) exists as the partner of the liver. On the other hand, the teleological explanation, which is the main theme of

De partibus animalium, is argued in a mature fashion with evidential support. This scientific maturity is even clearer in the next great treatise, *De generatione animalium,* in which he applies his concepts of form and matter, actuality and potentiality, to the problems of reproduction, inheritance, and growth of such inessential characters as color. On the question of classification he remains tentative and critical, as we would expect of one who rejected Plato's theory of Forms. He often returns to the problem in both early and late writings, but states no clear position.

His teleology differs from others. He argues it in *De partibus* I on the same grounds as in *Physics* B, where he states more of his opponents' case. He makes it clear that the "natural philosophers" (Empedocles, Anaxagoras, Democritus) were combating a popular teleology which presented the gods as purposive powers intervening in nature, so that "rain falls in order to make the crops grow." Against it they had argued that the "necessity" of natural causes was sufficient to explain events and that the crops happened to grow because the rain happened to fall, the real cause being the automatic interactions of the hot, the cold, and the other elements. In reply, Plato had posited a world soul and a creative "Demiurge." Aristotle, however, does not invoke a supernatural agency (for the relation between the cosmos and the Unmoved Mover is different), nor does he present nature as a quasi-conscious entity capable of purpose: his personification of nature "who does nothing in vain" is no more than a rhetorical abbreviation for "each natural substance." Neither does he posit an extra factor in nature, as modern teleologists posit a *conatus* that is not reducible to physics.

The directiveness that Aristotle sees in nature is part of the natural interactions, so that the teleological explanation coexists with the causal explanation. But he bases the teleology not primarily on directiveness but on the existence of forms. To explain an organ, he says, we must first grasp the complete animal's form and functions, what it means to be that animal, its *ousia.* Our explanation will include both the "necessary" causes and the "end" toward which development tends. This is not the temporal end or a state of equilibrium between phases of activity; indeed, it may never be reached. It is the perfect condition of the whole animal, "for the sake of which" each part develops. Thus, Empedocles was wrong to suppose that the spine is vertebrated because it gets bent: on the contrary, vertebration is necessary to the animal's functioning, and was contained potentially in the parent's seed before the embryo's vertebrae were formed. He was also wrong to think that random necessity could be a primary cause, for it could not produce the general regularity of nature, let alone the absolute regularity of the stars. Necessity in nature is secondary, or, as Aristotle calls it, "hypothetical": on the hypothesis that an animal is coming into existence, certain materials must interact, but these materials do not of themselves produce the animal any more than bricks produce a house. As the house needs a builder and a plan, the animal needs a soul and a form—factors ignored by the materialists. But whereas builder and plan are separate, soul and form are identical. The final cause of the animal is the actualization of its form, and its primary efficient cause is its soul, which "uses" the necessary movements of the materials. Aristotle's teleology therefore rests upon his theory of substantial form. The definition of a substance is logically prior to the definition of its parts, and so the final cause is prior to the necessary cause. It is prior temporally as well as logically, for Aristotle believed that the world never began—so that hen has forever preceded egg.

Although he used Plato's language ("existence is prior to coming-into-existence" and necessity is "the concomitant cause"), Aristotle did not follow Plato in positing an overall teleology or in the dualism that the *Timaeus* set up between creator and material. The few passages where Aristotle seems to imply that some species exist for the sake of man, or act for the general good as opposed to their own, cannot be meant literally. What he probably meant was a balance of nature, in which species are interdependent. The final cause of each animal is its own complete state, and nothing more. And instead of Plato's dualism, Aristotle places finality within natural interactions, not as something imposed upon them.

Within sublunary nature there are continual fresh beginnings of movement for which there are no sufficient external causes. They may be stimulated from outside, but the source of the movements in plants and animals is their souls. Only in a general way is the Unmoved Mover the prime cause. As a final and formal cause it presents the perfection that lesser beings desire to imitate. It can therefore be argued, although it is never clearly stated by Aristotle, that nature's tendency toward actualization and the *orexis* within souls are ultimately oriented toward the Unmoved Mover's perfection. As an efficient cause, the Unmoved Mover promotes general growth and decay on earth because it elicits the sun's movements in the ecliptic, and these movements cause the alternation of summer and winter. These general causes, however, do not bring about the particular starts of motion in nature. Nor, again, are souls regarded as separable entities that inhabit bodies and direct them, as Plato thought and as Aristotle may once have

thought but later rejected. In his mature view, found in his biology as well as in the *Metaphysics* and *De anima,* the soul (except, possibly, for the intellect) is not an independent substance but is the form of the body. On the other hand, it is not merely a resultant form, as in the "harmonia" theory, which Aristotle refuted; rather, it is both form and source of action. In plants it causes growth and reproduction; in animals it also causes sensation (here he differs from Plato, who thought that plants had sensation); in man the soul has a third faculty, intellect, and this is its only faculty that is not the form of body and could therefore be separable.

The concept of soul as both form and efficient cause may reflect a trace of ancient hylozoism. In Aristotle's view, finality pervades nature. If there is a cosmos, this implies that the elements not only have simple motions but also combine with modified motions. Both the simple motions and their modifications are hypothetically necessary and are natural. An animal contains many motions, all natural, that by a natural coordination tend toward a specific pattern. Its soul is both the tendency and the pattern. In nonliving substances, which have no soul, the tendency to form complexes is in their nature. Aristotle accepts as his data both the observable materials and the observable forms and species; therefore the movement of nature is simultaneously necessitated and endlike.

According to the *Metaphysics,* the form toward which animals grow is their species: individual differences arise from matter and consequently are unknowable to science. In Aristotle's earlier zoology we cannot tell whether he maintains this strict view, but in *De generatione animalium* his theory of reproduction implies that individuals differ in form to some extent. He does not say so, but repeats the doctrine of *De generatione et corruptione* that sublunary beings, which cannot achieve eternity as individuals, instead achieve it as species by reproduction. Nevertheless, Aristotle's discussion is in fact about an individual's reproduction of another animal "like itself." He starts from the long-standing controversy about the origin of seed. Do both male and female contribute seed? From what part or parts of the body does it come, and what does it contain? He analyzes the problem in terms of form and matter. The male alone makes seed from his blood; it contains potentially the sensitive soul and the adult form, but actually it contains no bodily parts (here he ridicules preformism and pangenesis). The female contributes only material (the *catamenia*), whose form is nutritive soul. When the male's form has been imposed upon the female material, the somatic part of the seed is sloughed away: all that is transmitted is soul, the source of form and motion. If the fetus develops regularly, the father's form will be actualized; failing that, the mother's; failing that again, more distant ancestors successively, until eventually the form may be merely that of the species, or even just the genus *Animal* (that is, a monstrous birth).

This long and careful argument, which is supported by observed evidence, gives a brilliant impression of maturity and originality, and in several points goes beyond the biological arguments that we occasionally find in the philosophical works. Aristotle's view that the father's form is reproduced, as distinct from the species, can only mean that some individual differences are formal and apodictic. He also brings to scientific account other differences due to "necessity"—not only monstrous births but differences of coloration, voice, or sharpness of senses. Since he calls them "concomitants" arising from irregularities in the material, he may have regarded them as unpredictable, but they seem to be accountable after the event. He now argues not from the fixity of species but from the reproduction of forms. True, he does not contemplate the obsolescence or alteration of existing species (for he had no paleontology); but he does accept, within limits, the evidence for miscegenation's resulting in new forms. In fact, the emphasis on species becomes less, while the concept of necessity as hypothetical becomes more important and sophisticated than in the philosophical works, where necessity is either "simple" (axiomatic) or brute (material). The one exception among the biological works is the *Historia animalium,* from which the teleological explanation is absent. Although a discussion of causes is not to be expected here, nevertheless the account of characters and life histories involves some causal explanation; and it is noteworthy that this explanation is given only in material terms. No doubt this is because the *Historia* was mainly the work of Aristotle's successors, among whom Theophrastus ignores the final cause even in his *Causes of Plants.*

In explaining the "necessary" causes—the interaction of materials—Aristotle does not innovate so much as rationalize theories that were already current. He accepts from Plato's *Timaeus* the four elements—fire, air, water, and earth—that were common to the medical writers and can be traced back through Empedocles into popular tradition. But the tradition had confused two notions: the cosmic regions of fire, air, water, and earth, and the seasonal powers of hot, cold, wet, and dry. The two sets do not exactly match, as is obvious in the ambiguous reports of Empedocles. Aristotle systematizes them by means of a formula that survived through the Middle Ages, treating fire, air, water, and earth as combinations of hot, cold, wet, and dry: fire is hot plus dry, air is hot plus wet, and so on. In his system hot, cold, wet, and dry are the primitive qualities of matter, but cannot exist in isolation. Fire, air, water, and earth are the simplest separable bodies, and are transformable into each other.

Like his predecessors, Aristotle regards the hot as the chief active power; its characteristic action is *pepsis* ("concoction"), which transforms food into blood and blood into flesh. By its opposite, the cold, he sometimes means merely the absence of hot, but more often a power in itself. The hot means more than temperature, which he calls "the hot according to touch." Another sort of hot is that possessed by pine wood, which is not hotter to the touch than other timber but contains more heat and therefore burns better. Animals have an innate heat upon which life depends. Their droppings still contain some of it, which generates flies. While the hot is the soul's chief agent in bringing about growth, cold is also needed to solidify things. Like the medical writers, Aristotle attaches importance to the due mingling (*krasis*) of hot and cold, which does not mean a point on a temperature scale but a mixture of two powers. He follows them in extending this notion to a general "right proportion" (*symmetria*) necessary for growth and health.

The other elements—the wet or watery, and the dry or earthy—are needed to provide the fluid and the solid parts of plants and animals. Whether Aristotle really intended a fifth element, *pneuma,* is debatable. The notion was current, and soon after him it became the chief element for the Pneumatic school of medicine and the Stoics. Aristotle had his own fifth substance in the outer heaven, the *aither,* and in *De generatione animalium* he compares it with the bodily *pneuma: pneuma* is the material of the animal seed, and conveys soul and the generative warmth, which he says is different from other heat. Yet he defines *pneuma* merely as warmed air, and since warmth has various powers for him, it is probable that he means no more. So he explains spontaneous generation by the presence of a warm soul-source in the materials.

The four elements combine to form the tissues, which Aristotle calls "made of like parts" (as flesh is divisible into flesh); and the tissues form the organs, which are "made of unlike parts" (hand is not divisible into hands). Taking this distinction from Plato, he uses it in finding homologies, but he makes only general statements about the processes. The hot concocts blood into flesh here, fat there, marrow or seed somewhere else; skin, hair, bone, nails, and horn all come from the earthy. He does not explain how. Medical literature of the time contains some practical investigations, such as the action of heat upon blood, and Aristotle occasionally refers to such evidence. In *Meteorologica* IV he goes further and analyzes the actions of hot and cold into evaporating, emulsifying, dissolving, condensing, and coagulating, and differentiates many types of earthy material. But this is a late work, and may not even be his. It seems, therefore, that in his biology Aristotle is content to take these theories in a general form from current tradition, although he is careful to rationalize them. For example, he will not allow Empedocles to say that spontaneous generation results from rottenness: new life comes not from disintegration but from concoction. The heart—not the brain, as many held—is the center of sensation and of the soul's motor impulses; as the first part to develop (observed in daily openings of a clutch of eggs), it is the source of the vital heat and innate *pneuma*. In it the blood is pneumatized and then flows out to nourish the tissues. (The distinction between arteries and veins is post-Aristotelian.) The lungs admit air to replenish the *pneuma* and to moderate the heat, an excess of which brings on senescence and death. Animals without lungs are cooled by the surrounding air or water: this suffices because they are "less perfect" and therefore cooler; also, their innate store of *pneuma* is sufficient.

Classification of animals remained a difficulty, and Aristotle suggested a solution by taking an animal's vital heat as an index of its superiority. Plato had proposed *diaeresis* (division), in which a major group is progressively divided by differentiae into genera and species. This method, used by Aristotle in his early logic and later by his successors, became the basis of Linnaean systematics. In his zoology, however, Aristotle criticizes it for splitting natural groups. He shows how groupings based on habitat and locomotion, and such characters as horns and rumination, cut across each other, while many animals belong to both sides of a formal division. He also criticizes the emphasis on morphology, which he holds subordinate to function. He prefers to start from the natural genus, as defined by multiple characters, then to arrange it with other types, not in a genus-species hierarchy but in a *scala naturae* ranging from man through less perfect animals down through plants to lifeless compounds. In this he emphasizes the continuity of nature and the many borderline or overlapping types, such as the seal, the bat, and the testaceans. The degree of vital heat is indicated by method of reproduction, state at birth, respiration, posture, and other signs. But he does not produce an actual scheme, nor does he finally reject genus-species classification. For practical purposes Aristotle discusses the animals by major groups: the "blooded" (i.e., red-blooded)—man, viviparous quadrupeds, oviparous quadrupeds, cetaceans, fishes, birds; and the "bloodless"—mollusks, crustaceans, testaceans, and insects. But he points out that even these groups exclude many types, such as snakes and sponges. In fact, before any classification could succeed, far more information was needed. He may have felt this, for the *Historia animalium* was begun as a comparative study of characters, arranged under the headings *parts, activities, lives, dispositions* (i.e., psychology). Major groups were to be compared by "analogy" (as wing to fin), while within a group each structure would vary by "the more and the less" (as wings are longer or shorter).

This project, however, was not carried through; instead, the treatise became a running collection of data. As new information came in and new significant characteristics were distinguished, they were inserted at convenient places, as if into a filing cabinet. Book I gives a program of the characters to be discussed, and by comparing this with the later books, we can see that many of those proposed are never mentioned again while many more new characters come to be recognized, so much so that the whole plan of the treatise is altered. The latest additions, which can be identified in all books from the second onward, consist of dossiers or even complete descriptions of single animals, no doubt awaiting breakdown under appropriate character headings. Thus the work eventually begins to approximate a descriptive zoology, and this is how it has been taken ever since. But in judging Aristotle as a natural historian, we should remember that we are judging him as something that he never set out to be. Although the classificatory intention of the *Historia animalium* came to nothing, it remained essentially an analysis of differentiae, the ways in which animals "are like to and different from each other," in the words of the introduction. The data about animals are put there to illustrate characteristic differences, and except in the late and unassimilated additions there is no description of an animal for its own sake. The statements about a given animal are spread through the nine books of the treatise, which is arranged not by animals but by characters. It has repeated signposts helping the reader to find his way among characters, but there are none to help him find animals, and there is no index. Some animals are cited frequently to illustrate but one point—for example, the mole's blindness: Aristotle obviously examined the mole, for he describes a dissection of its concealed eyes, which is of great interest; but this is all he tells us of the mole. In fact, like all his treatises, the *Historia animalium* is a theoretical study. It is not so much about animals as about *Animal*—and the various ways it is differentiated in nature.

Aristotle names about 500 "kinds" of animals. Some of these comprise several varieties, which his reports sometimes distinguish but sometimes confuse. Altogether, between 550 and 600 species can be distinguished, and of these as many as 200 are mentioned in connection with only one character. He includes some thirty from such distant places as Libya, Ethiopia, the Red Sea, and even India. A very few are taken from travelers' tales, especially from Herodotus and Ctesias, and of these some are fabulous—for example, the flying snake and the martichoras, or manticore (a monster, perhaps derived from a garbled account of the Indian tiger, which became a favorite of the Middle Ages), of which he plainly indicates his suspicion. But most were to be seen in Greece in menageries and shows—certainly the bear, monkeys and apes, elephant, camel, and lion. Aristotle gives much information about all of these, for the very reason that they exhibited interesting differences. Some information is evidently hearsay: for example, he reports that the lion has no cervical vertebrae, which shows that he never examined a dead lion. But his remarks about the lion's appearance and gait show equally that he observed it in life. He describes the elephant's leg joints in order to contradict a popular belief that it sleeps standing against a tree.

However, the great majority of Aristotle's reports concern animals native to Greece, its islands, and the Greek colonies in Asia Minor. It is incorrect to accuse him of showing more interest in exotics than in what was at his own doorstep. If we compare the variety of information given on each animal, we find not only that the nearest animals are the most fully reported but also that he covers most of what was available to him. Among mammals, of which he mentions some eighty, by far the most information is given about the horse, dog, sheep, ox, and pig; next comes a group including the goat, donkey, mule, hare, deer, elephant, bear, camel, seal, and dolphin. Of 180 birds mentioned, the best-reported are the domestic fowl, the pigeons, and the partridge, and there is a good deal on the sparrow, swallow, blackbird, crows, larks, eagles, hawks, quail, and stork. On the other hand, over 100 birds are mentioned only once or twice, as examples of differences in feeding or nesting, and so on. The information on marine animals is especially good, although out of 130 fishes only twenty are cited in connection with more than a very few characters. Among over eighty insects, he gives considerable information about the flies, ants, wasps, and cicadas, and three long, separate discussions of the honeybee;

there is a fair amount about the grasshoppers, gadflies, spiders, beetles, and chafers. It is true that he has relatively little on the gnats and mosquitoes, common though they were; but he reports their external structures, reproduction from larvae, feeding, and habitat—and there is, after all, little more that he could know, having no optical apparatus. Aristotle often complains that the smallness of some insects makes it impossible to discern their structures, especially the internal ones. Many features, in all groups of animals, are reported in a generalized form—"all two-winged insects have a proboscis and no rearward sting," "all fishes except selachians have gill covers"—so that if one is to assess what he knew about a given animal, these general statements have to be broken down and included. In some of them he generalizes further than the facts warrant, through faulty or deficient information.

The tests that Aristotle applies to reports are primarily observational checks, made either on the same type of animal or on "analogous" types. He shows himself well aware of the need for repeated observations, but he has not developed the refined technique of provoked and controlled observations that later (very much later) scientists learned to demand. Where observational checks are not available, he tests by inherent probability—that is, by reference to theory. The accusation that he relies on a priori argument, and not on observation, is not well founded; on the contrary, like most Greek philosophers, with the exception of Plato, he is overready to accept uncontrolled observation and to jump to large conclusions.

His chief sources of information are fishermen, farmers, stockbreeders, and hunters; to a lesser extent travelers, menageries, augurs, and drug manufacturers; and he owes a very little to such previous writers as Herodotus, Ctesias, Xenophon, Empedocles, and Democritus. There are many faulty reports that he corrects from observation. His favorite method is the counterinstance. He refutes a report that the viper does not slough its skin simply by describing an observation of the sloughing. The legend that the hyena has the genitalia of both sexes (which in fact it can appear to have externally) is refuted by inspection and dissection, and here he indicates that many specimens were examined. Fishermen said that all mullets are generated spontaneously, but he has examples of mullets with eggs and with sperm (although he allows that one kind of mullet is spontaneous).

Where such direct checks are not possible, he refers to analogous examples or to theory. He denies that the cuckoo is a metamorphosed hawk on the grounds that the hawk preys on the cuckoo, a thing never seen done by one bird to another of its own kind. Fishermen believed what Herodotus also said, that fishes are impregnated by swallowing the sperm; Aristotle denies this because there is no connection between stomach and uterus, and because fishes have been observed in coition—which, he remarks, is difficult to observe, and fishermen have missed it because they are not interested in acquiring knowledge. Here he has been misled by faulty observation that, unluckily, agreed with theory—a coincidence that accounts for many of the mistakes in his reports. He held that where there are separate male and female, there must be coition. He knew that the male fish sprinkles the eggs with sperm after spawning, but thought this an additional process of fertilization. Another famous example is the fishermen's report of hectocotylization—the extraordinary method by which a sperm-carrying tentacle is inserted into the female's mantle cavity and then completely detached from the male (eventually proved true): Aristotle denies that the tentacle assists reproduction, because it is not connected with the body and the spermatic channel—he was wrong because his theory could not accommodate what is, after all, a surprising fact. But in another context he makes it clear that theory must always yield to reliable observation: after his long discussion of the reproduction of bees he makes a statement that fairly represents his own practice (*De generatione animalium* 760b27):

> This, then, appears to be the method of reproduction of bees, according to theory together with the apparent facts. But the facts have not been satisfactorily ascertained, and if ever they are, then credence must be given to observation rather than to theory, and to theory only in so far as it agrees with what is observed.

Many of the reports, however, are from firsthand observation. He refers sometimes to "the dissections," evidently a collection of drawings and diagrams of internal organs; unfortunately nothing survives of them. Some of his data clearly come from deliberate dissection, while others come as clearly from casual observations in the kitchen or at augury. One of the best is a full-scale vivisection of a chameleon; and the internal organs of crabs, lobsters, cephalopods, and several fishes and birds are described from direct observation. Many of the exterior observations also presuppose a prolonged study. He speaks of lengthy investigations into the pairing of insects. He satisfies himself that birds produce wind eggs entirely in the absence of the cock. There are graphic accounts of courtship behavior, nest-building, and brood care. He records tests for sense perception in scallops, razor fish, and sponges. He watches the cuttlefish anchor itself to a rock by its two long arms when it is stormy. The detailing of structures in some crustaceans and shellfishes vividly suggests that the author is looking at the animal as he dictates. The sea urchin's mouth parts are still known as "Aristotle's lantern" from his description, and his statement that its eggs are larger at the full moon has only recently been confirmed for the Red Sea urchin. He is able to assert that two kinds of *Serranidae* are "always female" (they are in fact hermaphrodite). All such data require deliberate and patient observation. How much Aristotle himself did is not known, but it is clear enough that he caused reports to be collected and screened with great care.

The first main heading in the *Historia animalium* is "Parts of the Body." Aristotle methodically lists the external and internal structures, noting the significant differences between animal types. Through drawing an analogy between legs and fins, he holds that fishes are moved primarily by their fins; this error creates difficulties for his theory of locomotion, whereby the blooded animals are moved by two or four points and the bloodless by more than four. He classifies the forms of uterus by position: rearward and ventral in the viviparous quadrupeds, forward and dorsal in the birds and oviparous quadrupeds, rearward and dorsal in the oviparous fishes, and "in both ways" in the ovoviviparous fishes—that is, extending from a forward dorsal to a rearward ventral position, because they first produce eggs and then hatch them within the uterus. There are various mistakes, mostly concerning man (where dissection was impossible) or the rarer animals. He is prone to accept them when they fall in with theory, thus accepting that men have more sutures in the skull than women (possibly based on an unlucky observation of a female skull with sutures effaced in pregnancy), for it fits his theory that men need more heat regulation in the brain. He reports that if one blows down the windpipe, the air reaches the heart: again a faulty observation that agreed with theory (that the *pneuma* in the heart is replenished from the lungs). His account of the heart's three intercommunicating chambers, disastrous for later anatomy, was due to wrong observation in a difficult field, but it fell conveniently into his theory of the blood system.

Nevertheless, Aristotle is aware how easily observations can mislead. For example, he remarks that those who believed the lungs to be devoid of blood were misled by observing dissected animals from which the blood had escaped. Much of what he says of the lion is mistaken, as is his statement that the crocodile moves the upper jaw: in these cases external appearances have not been tested by inspection of the dead body. Some could have been better tested—for example, his reports of the incidence of the gall bladder are unreliable, probably because he trusted the augurs. But the great majority of data in this section are accurate and shrewdly observed, especially the details of alimentary canal and reproductive organs, in which he took special theoretical interest.

Under "Lives and Activities" Aristotle compares differences in reproduction, feeding, migration, hibernation, and sloughing, and variations due to season, breeding, disease, age, and habitat. His theory of reproduction, applied to all groups of animals, is argued in *De generatione animalium;* the *Historia animalium* summarizes this and adds much more information about sexual behavior, breeding methods and seasons, gestation, incubation, and brood care. He distinguishes the viviparous quadrupeds theoretically by the degree of perfection in the young at birth, and he has many details of seal and dolphin as well as land animals. The next step down is to the ovoviviparous, such as the vipers, sharks, and dogfishes. In them he describes the egg's development and its movement rearward to the position where the young are released within the uterus; in one dogfish (*Mustelus laevis*) he notes the placentoid structure, like that of mammals, which was not rediscovered until comparatively modern times. He mistakenly generalizes that all cartilaginous fishes are ovoviviparous. He divides the ovipara into those that lay perfected eggs (birds and quadrupeds) and those whose eggs develop after laying, requiring what he took to be a second fertilizing by the male. He describes minutely the development of the eggs of birds, fishes, cephalopods, and others by opening eggs at intervals during the whole incubation period. He records many special cases: for example, the way that *Syngnathus acus* carries its eggs in a pouch, which then splits to release them (although he does not observe that it is the male which carries them). The lowest mode of reproduction in his scale of "perfectedness" is spontaneous generation, which he attributes to all testaceans, many insects, the eel, and a few fishes. He describes the spawn of whelks, but judges it to be a budding-off comparable with that of plants, not a mass of eggs; otherwise, testaceans originate from various mixtures of mud and rotting substances, the type of animal being determined by the mixture. He considers that insects (except for one butterfly) produce grubs, not eggs; although one speaks of spiders' or bees' eggs, and so on, he says that what at first looks like an egg is really a motionless larva, on the (mistaken) grounds that the subsequent animal is formed out of the whole of it. The grubs of spiders, bees, cicadas, and others develop into the parental type, but those of flies and beetles do not develop

further, and originate spontaneously from a variety of materials, which he lists. Gnats and mosquitoes do not even produce grubs, but themselves arise from grubs that are spontaneously generated. He describes many types of larval development through pupa to imago, including the change of the bloodworm into the gnat. His conclusion about the honeybee (which he says is a puzzle) is tentatively that the queen produces queens and workers, the workers produce the drones, and the drones produce nothing. His view here is not exactly parthenogenesis: he holds that bees contain both male and female principles, and therefore generate without coition.

The final section on "Characters," that is, animal psychology and intelligence, contains little imputation of motives: he records strictly the observed behavior. He compares animals in compatibility, rivalry, nesting and homemaking, and miscellaneous habits of defense and self-support. Among many, for example, he reports the nests made by the octopus and the wrasse, and the brood care by the male river catfish—recently rediscovered and named after him (*Parasilurus aristotelis*). He notes that the partridge makes two nests, on one of which the male sits; and his report that some partridges cackle and others whistle led to the discovery in 1962 that two populations (rock partridge and chukar) live side by side in Thrace. Among the honeybee's habits he seems to refer to the "dance language." The section is unfinished, and the treatise in its present form ends abruptly with a distinction between birds that take dust baths and those that take water baths.

The more complete descriptions, which have been inserted throughout the treatise and seem to be the latest additions, include those of the ape, chameleon, and wryneck, and extracts from Herodotus and Ctesias on the crocodile, hippopotamus, and martichoras. But most of the fabulous or unauthenticated reports are in a separate work called *Mirabilia*, where they were perhaps held awaiting corroboration: some of them—for example, the bison—are in both treatises. For entirely new animals, Aristotle no doubt required reliable eyewitnesses. But when it comes to details reported of known animals, which is the subject matter of most of his reports, his first point of reference is the adult living animal in its natural environment. His standard of judgment is function rather than morphology, as he makes clear in *De partibus animalium*. The "analogies" that he seeks, and from which he constantly argues, are not structural but functional; and, wherever possible, his identification of differentiae is based on function. Because this is his aim in the *Historia,* he picks out the significant details better, for instance, than does

Xenophon (whose excellent accounts of the hare and of horses provide the best contemporary comparison with Aristotle's reports). Its change of plan and lack of revision make the treatise seem incoherent and bewildering, but its comprehensiveness and acumen made it the outstanding descriptive zoology of ancient times, even though it was not intended to be primarily descriptive. It outlasted the work of such later encyclopedic compilers as Pliny, and combined with Aristotle's other zoological works it became—through the Arabic version translated into Latin by Michael Scot—the major ingredient in Albertus Magnus' *De animalibus,* which dominated the field until the sixteenth century.

BIBLIOGRAPHY

The standard text is Bekker's *Corpus Aristotelicum* with Latin trans. (Berlin, 1831–1870). There is also the text with English trans., intro., and brief notes in the Loeb Classical Library; see especially A. L. Peck's eds. of *De partibus animalium* (rev. 1955), *De generatione animalium* (rev. 1953), and *Historia animalium,* I (1965; remaining 2 vols. in press). *Parva naturalia* was ed. with full English commentary by W. D. Ross (Oxford, 1955). The Loeb and Ross eds. contain bibliographies of previous eds. and full accounts of the MSS.

There are also lesser works with zoological content included in the Bekker ed., but not all are by Aristotle—*De incessu animalium, De motu animalium, De spiritu, Mirabilia,* and *Problemata.* See also *The Works of Aristotle Translated Into English,* W. D. Ross, ed.: III, *De spiritu* (1931); V, *De incessu animalium* and *De motu animalium* (1912); VI, *Mirabilia* (1913); and VII, *Problemata* (1927).

D. M. BALME

ARISTOTLE: Anatomy and Physiology.

In his discussion of animals Aristotle gives great importance to the heart, the blood vessels, and the blood, making the possession of blood the basis for distinguishing one great class of animals, those with blood, from those without blood (roughly the vertebrates and invertebrates). In giving this fundamental position to the heart and blood Aristotle departs from the physiological ideas of the Hippocratic writers; in doing so he seems to have been influenced by the ideas of the Italo-Sicilian-Greek medical thinkers. The stopping of the heartbeat was a certain sign of death and thereafter the body rapidly cooled and became stiff and lifeless. In the developing chick Aristotle saw the beating heart as the first manifestation of life. From this beating heart he saw blood vessels grow out over the yolk, and within the skein

of blood vessels thus formed, the body of the young chick gradually emerged. Aristotle emphasized that the heart is the center and the origin of all the blood vessels. He considered that the blood was formed in the heart and passed out from it, because from the moment that the heart became visible it was seen to contain blood and as the network of blood vessels spread out from it, in the embryo chick, the blood accompanied them.

Since the heart, blood, and blood vessels were so fundamental to the bodies of animals Aristotle undertook to discuss them first in his *Historia animalium.* Possibly because of his belief in their fundamental importance he gave one of the earliest accurate descriptions of the blood vessels as a system extending throughout the body, but with its center in the heart. References to the blood vessels by Greek writers before Aristotle emphasized superficial veins, most easily visible in emaciated men, which might be used in bloodletting. Their accounts of the internal arrangement of the blood vessels were extremely vague and fragmentary. By his full and accurate account of the cardiovascular system Aristotle may be considered a founder of detailed anatomical study.

The basis for Aristotle's success in the dissection of the blood vessels was that instead of stunning the animal and bleeding it, in the manner of butchers, he first allowed it to starve to emaciation and then strangled it, thereby retaining in the dead animal all of the blood within the blood vessels. This treatment of the animal had, however, certain physiological consequences which were to influence the character of his observations. The animal killed by strangulation dies in a state of shock which produces a constriction of the small arteries and arterioles in the lungs, thereby cutting off the supply of blood to the left side of the heart. The left ventricle of the heart contracts to empty itself of blood and cannot be refilled. Moreover, the elastic muscle walls of the arterial system contract to squeeze the blood they contain through the capillaries into the veins. Almost all the blood in the body, therefore, accumulates in the venous system, leaving the left side of the heart and the arteries nearly empty. The right side of the heart, on the other hand, is enormously swollen and engorged with blood. When the heart relaxes in death the pressure of blood in the veins will keep open the right auriculoventricular aperture. The flaps of the tricuspid valve will be pressed back against the wall of the ventricle and will be relatively inconspicuous. As a result of these circumstances the right auricle and ventricle will appear as one large chamber continuous with the superior and inferior venae cavae. Instead of four cavities, the heart will appear to have only three, the largest of which will be the united right auricle and ventricle, while the two others will be the left ventricle and the left auricle. Thus to Aristotle the vena cava or "great blood vessel" appeared as a single continuous vessel that broadened in the heart "as a river that widens out in a lake" (*Historia animalium* 513b5, Thompson, trans.). The aorta he saw arising from the middle chamber of the heart and noted that it was more sinewy than the "great blood vessel."

Aristotle did not distinguish between arteries and veins and applied the same term, *phleps* (φλεψ), to both. Neither did he describe the heart valves. He saw the pulmonary artery extending from the "largest chamber on the right" (the right ventricle) upward toward the lung, and he described how in the lung the branches of the pulmonary artery are distributed throughout its flesh and everywhere lie alongside the branches of the tubes (bronchioles) that extend from the windpipe. He traced the main branches of both the venous and arterial systems and described the blood vessels, at least in outline, as a system coextensive with the body, having a shape "like a sketch of a manikin" (*ibid.,* 515a34–515b2).

Aristotle interpreted the pulsation of the heart as the result of a kind of boiling movement in the blood which caused it to press against the walls of the heart and to pour out into the blood vessels. The heart walls were thick in order to contain the innate heat generated in it and the heat of the heart produced respiration by causing the lungs to expand and cool air to rush in. The entering air cooled the lungs so that they again subsided and the air, warmed now by the heat taken up from the blood, was expired. Thus for Aristotle respiration served the purpose of cooling and moderating the heat of the blood and the heart.

Aristotle considered the brain to be cold and to exert a cooling influence on the body in opposition to the heating influence of the heart. Since he did not know of the existence of the nervous system as a system extending throughout the body in a manner similar to the blood vessels, he could not conceive of the brain as having the same kind of central role as the heart.

BIBLIOGRAPHY

See T. H. Huxley, "On Certain Errors Respecting the Heart Attributed to Aristotle," in *Nature,* **21** (1880), 1–5; William Ogle's note to 667b in his translation of Aristotle, *De partibus animalium* (Oxford, 1910); and Arthur Platt, "Aristotle on the Heart," in C. Singer, *Studies in the History and Method of Science,* 2 vols. (Oxford, 1921), II, 520–532.

LEONARD G. WILSON

ARISTOTLE: Tradition and Influence.

An account of the Aristotelian tradition would cover, without any interruption, the whole of the intellectual history of the Western world and, in recent times, of other areas as well. On the other hand, the influence of Aristotle's works and doctrines on the cultural developments of civilization is, in most fields, elusive and undefinable. Especially in the province of science—if we use "science" in the stricter, modern sense—it may be found that Aristotle's influence is very limited, or effective only in the sense that mistakes, eliciting opposition, criticism, and new solutions to old and new problems, are the starting point of scientific progress. Positive influence and starting points for positive developments are found, for the different sciences, much more frequently in the works of Euclid and Ptolemy; of Hippocrates and Galen; of Archimedes; of al-Fārābī, Ibn Sīnā (Avicenna), and Ibn Rushd (Averroës); possibly of Boethius; and, back through Boethius, of Nicomachus of Gerasa.

Still, there are two aspects in this progress that bear the Aristotelian imprint and justify an extensive account of the spread of Aristotle's works and of their study: the methodical aspect and the conceptual-linguistic aspect. These two cannot always be separated, but they must not be confused if Aristotle's influence is to be clearly seen and properly assessed. This section will, therefore, be devoted first and foremost to such an account. We shall then consider a set of concepts and words that became essential for the elaboration of scientific problems and, indeed, for making scientific discoveries clearly expressible and understandable in the technical and, at the same time, the common language. Some exemplification will be given of the methodical aspect, insofar as it can be traced back to Aristotle's influence, and of the actual contributions derived from his works, mainly by discussion, rejection, and positive substitution of anti-Aristotelian views. In this connection it must be recorded that a very limited amount of the literature that developed around the works of Aristotle in later antiquity, in the Middle Ages, and even into the eighteenth century has been properly edited, much less has been critically read, and only a minimal proportion of it has been examined from the point of view that interests us here.

The transmission and spread of Aristotle's works can best be followed by considering the different languages or groups of languages in which it took place: basic, of course, was the Greek tradition, from which all others sprang, directly or indirectly (fourth century B.C. to our times); most important and permanent in value was the Latin (fourth century A.D. to sixteenth and seventeenth centuries); very influential, especially through elaborations and translations into Latin, was the Semitic (first Syriac, then [and mainly] Arabic, finally Hebrew [fifth century A.D. to sixteenth century]); only occasionally effective in its own right and more valuable as a help in the rebirth of the study of Greek civilization was the tradition in German, Neo-Latin, English, and, more recently, many other modern languages (tenth century to our times); limited to very narrow cultural units was the Armenian and possibly the Georgian (*ca.* fifth century A.D. to tenth century and later).

The Transmission of Aristotle's Works in Greek. Compared with the impact of what constitutes the traditional Aristotelian corpus, typically represented by the Berlin Academy edition of 1835, the influence of the other works of Aristotle—preserved, if at all, in a number of more or less extensive fragments—can be considered negligible; we cannot pursue their tradition here. The corpus, based mainly, it seems, on lectures, preparations for lectures, accounts of lectures, and elaboration of collected material (*De animalibus*), must have begun to be organized in Aristotle's own time, by Aristotle himself and his pupils (Theophrastus, Eudemus, and others). The process continued in his school, with vicissitudes, for 250 years after his death. The quasi-final organization of Aristotle's available material seems to have been accomplished by Andronicus of Rhodes (*ca.* 70 B.C.). It may be assumed that from Andronicus' edition there derived, with minor changes and developments, the transmitted texts as we know them in Greek. From Andronicus to the middle of the sixth century, the spread of the corpus or parts of it is continuously testified by the activities in the several philosophical schools, whether mainly Peripatetic in character, or eclectic, or more purely Neoplatonic. Andronicus' pupil Boëthus of Sidon commented on Aristotle's works, making the *Physics* the basis of Aristotelian philosophy; a century after, Nicholas of Damascus expounded Aristotle's philosophy and wrote (in the mood of Aristotle's *De animalibus*) a *De plantis,* which came to be ascribed to Aristotle; and *ca.* A.D. 100, Ptolemy Chennos of Alexandria wrote a work on the life and works of Aristotle. In the second half of the second century A.D., Galen, famous for his medical work, was a critical popularizer of Aristotle's logic, physics, and metaphysics, and many other authors commented on this or that work.

The texts of Aristotle were, obviously, already popular over a wide area. When, *ca.* A.D. 200, Alexander of Aphrodisias became professor of philosophy in Athens, as a "second Aristotle," he commented upon a large proportion of the corpus and left in his works abundant evidence of the variety of readings

that had been infiltrating the nearly 300-year-old transmission of the basic edition. Although only minor fragments of papyri containing Aristotle's texts from the corpus and no manuscript older than the ninth century exist, the expanding study of the works in Athens, Constantinople, Alexandria, and Pergamum justifies the statement that many manuscripts were available in many centers. The sixth century adds new evidence, since, at least in the case of some logical works, we possess not only the quotations of many Greek commentators but also the literal translations into Latin, Syriac, and Armenian: these testify to the variety of the Greek tradition, a variety that continued and became more complicated in later centuries.

The ban on pagan schools in 529 led to a reduction, if not to a halt, in the production of Greek copies of the works of Aristotle until the revival of the late eighth and ninth centuries. Then really "critical" editions of some works, and transcriptions of many, if not all, started again. The University of Constantinople became a center of studies of some of these works; the old libraries still possessed among them at least one copy of each of the writings of Aristotle. And it is possible to surmise that in form (some of them were rich in scholia extracted from the old commentaries) they were like the manuscripts of the sixth or earlier centuries. The number of extant manuscripts of the ninth and tenth centuries is very small, and does not cover the whole corpus; but the stronger revival of the eleventh century was the beginning of the uninterrupted transcription and transmission of the more popular works. This gathered momentum, not only in Constantinople but also in the numerous centers where lay and theological schools were flourishing.

By the thirteenth and fourteenth centuries publication had expanded to such an extent that about 150 manuscripts from that period still survive. There are only a few exceptions to show that not all of Aristotle was dominating the higher philosophical studies, side by side with Plato: the *Politics,* unearthed perhaps in the eleventh century and turned into a fruitful career by the Latin translator William of Moerbeke, does not appear in our collections in any manuscript older than the thirteenth century. The *Poetics* appears in late manuscripts, except for one of the eleventh century and one of the thirteenth. But the bigger collections, especially of the logical works, are relatively numerous. A new impetus to the dissemination was given in the fifteenth century by the migration of scholars from the Greek world to Italy and by the interest in Greek studies in Florence, Venice, and other cities. In the fifteenth century the number of copies of the several parts of the corpus, including the rarest works, multiplied, and the way was prepared for the printed editions, from the Aldine of 1495–1498 to those of the seventeenth century. There was then about a century of interruption: Aristotle was "out" from most points of view. By the end of the eighteenth century the new interests of learning brought about the new wave of Greek editions of Aristotle—a process that is still in full swing.

The Transmission of Aristotle's Works in Latin. No evidence has come to light to show that any work by Aristotle or any extensive paraphrase was available in Latin before *ca.* A.D. 350. Cicero's claim that his *Topica* was based directly on Aristotle's work of the same title is false. His model was the work of a rhetorician, not of a logician, and bears only vague, occasional, accidental resemblances to what Aristotle wrote. The latinization of Aristotle took place through different channels: by far the most important was the direct translation from the Greek originals; second in importance was the translation of Greek paraphrases and commentaries; third, the translation of some of Aristotle's works from direct or indirect Arabic versions, whether alone or accompanied by Arabic commentaries; fourth, the versions of Arabic works based, in various measure, on Aristotelian texts; finally, some translations from the Hebrew renderings of Arabic versions, commentaries, and paraphrases. All this happened in the course of four identifiable stages, very different in length, between the middle of the fourth century and the end of the sixteenth: (*a*) the first stage probably lasted only a few years and involved a few individuals belonging to two groups working in Rome; (*b*) the second corresponds to a few years in the first quarter or first half of the sixth century, with Boethius as the only person concerned with this activity in Italy, and possibly some minor contributors in Constantinople; (*c*) the third stage covers about 150 years, from *ca.* 1130 to *ca.* 1280, when the work was carried out probably in Constantinople and certainly in Sicily, Italy, Spain, Greece, England, and France by at least a score of people of many nationalities and callings—by the end of this period the whole of the Aristotelian corpus as it has reached us in Greek, with very minor exceptions, could be read and studied in Latin; (*d*) the fourth stage extended from shortly after 1400 to *ca.* 1590. Only in the third stage did the Arabic tradition contribute directly to the Latin one; and only in the fourth did it do so through the Hebrew.

(*a*) The intellectual intercourse between Greek and Latin in the third and fourth centuries, of which the most striking example outside religion was the spread of the knowledge of Plotinus' doctrines, led to the need for Latin texts of some of the works considered basic by the Greeks. It was in this Neoplatonic atmosphere (tempered by Porphyry with more Aristotelianism than Plotinus had accepted, rather than discussed and criticized) that the African Marius Victorinus, a pagan converted to Christianity, popularized the contents of Porphyry's introduction to logic, the *Isagoge*; if we accept Cassiodorus' testimony, he also translated Aristotle's *Categories* and *De interpretatione.* He certainly included Aristotelian views in his *De definitionibus,* the only work by Victorinus that contains some Aristotle and that has reached us in full (only sections of his version of the *Isagoge* survive in one of Boethius' commentaries). The attraction exercised by Themistius' school in Constantinople led to another, possibly purer, wave of Aristotelianism among the pagan revivalists, so vividly depicted in Macrobius' *Saturnalia.* Vettius Agorius Praetextatus, one of their leaders, rendered into Latin Themistius' teaching on the *Analytics.* Agorius' work was probably lost very soon, and there was no Latin text of Themistius' work on the *Analytics* until the second half of the twelfth century. This was based on an Arabic translation of part of that work (which was not translated from the Greek before the end of the fifteenth century). But Themistius' teaching of the *Categories*—a detailed exposition with additions and modernizations—found its Latin popularizer in a member of the same circle (perhaps Albinus). It is from this work, later ascribed to St. Augustine, under the title of *Categoriae decem,* that the Latin Aristotelianism of the Middle Ages started its career, never since interrupted.

(*b*) The middle and late fourth-century Aristotelianism, and much else of the cultural life of that time, was a faded, but not a lost, memory when, in the first decade of the sixth century, Boethius married a descendant of one of the prominent intellectual families, Symmachus' daughter Rusticiana. He took up what remained of that tradition, and was encouraged by his father-in-law to renew it. Cultural relations with the Greeks were not as active around 505 as around 370, but Boethius managed to obtain some Greek books, among them a copy of the collection of Aristotle's logical texts with an ample selection of notes from the greater masters of the past (Alexander, Themistius, and, mainly, Porphyry). So he probably managed to achieve what he had planned, to translate as much of Aristotle as he could get hold of: at least, we still preserve, in more or less original form, his translations of the *Categories, De interpretatione, Prior Analytics, Topics,* and *Sophistici elenchi;* he also claims to have produced a now lost translation of the

Posterior Analytics. Since, by the fifth century, Aristotle's logical works were prefaced by Porphyry's *Isagoge,* Boethius also translated this text. He wrote that he intended to comment upon the works of Aristotle accessible to him; as it turned out, he commented on only the two shortest texts, the *Categories* and *De interpretatione*—or, better, he translated, adapted, and coordinated passages from Greek commentaries that he must have found on the margins of his Greek volume. The existence of a double recension for many sections of the *Categories, Prior Analytics,* and one short section of the *Topics;* the existence of a Latin version of a considerable collection of scholia to the *Prior Analytics* translated from the Greek and connected with one of the two recensions of this work; and a variety of evidence pointing to some editorial activity in Constantinople centering on Boethius' work in the first half of the sixth century suggest that Boethius' work as a translator in Italy had some continuation in the circle of Latin culture in Constantinople.

(c) The third stage is by far the most impressive, representing as it does a variety of interests, of cultural backgrounds, of centers of progressive attitude toward the renewal, on the basis of older traditions, of the intellectual life in Europe and, to a certain extent, also representing one further step in a continuity of Aristotelian studies, hardly interrupted from the first century B.C. to the thirteenth century A.D. It is here necessary to consider separately the translators from the Greek and those from the Arabic, as well as some of the centers and people connected with this transmission of Aristotle. First of all, it cannot be emphasized too strongly that Aristotle was latinized from the Greek much more than from the Arabic and, with very few exceptions, earlier from the Greek than from the Arabic. Although competent scholars have tried to make this fact known, the commonly held view of historians of ideas and of people in general is the wrong view: that the Latin Middle Ages owed their knowledge of Aristotle first and foremost to the translations from the Arabic.

(c-1) The Aristotelian revival of the ninth and eleventh centuries in the higher schools of Constantinople—particularly the second revival, due to such people as Michael Psellus, Ioannes Italus, Eustratius of Nicaea, and Michael of Ephesus—brought its fruits to the Latin revival (or, better, discovery) in the twelfth and thirteenth centuries. In the second quarter of the twelfth century James (Iacobus), a cleric with philosophical, theological, and juridical interests who seems to describe himself as Venetian-Greek, was in Constantinople and in touch with the Aristotelian corpus. He translated, either in Constantinople itself, or possibly in Italy, at least the *Posterior Analytics,* the *Sophistici elenchi,* the *Physics,* the *De anima,* parts of the *Parva naturalia,* and the *Metaphysics.* Of the translation of the last work only Books I–III and the beginning of Book IV remain; of the translation of the *Elenchi* only fragments have been recovered, mainly in contaminated texts of Boethius' version. He also translated some Greek notes to the *Metaphysics,* a short introduction to the *Physics* (known, in much of the Latin tradition, as *De intelligentia Aristotelis*), and probably *Commentaries to the Posterior Analytics* and *Elenchi* ascribed to Alexander of Aphrodisias. Finally, he himself commented at least on the *Elenchi.* James's translations, in spite of their extreme literalness, reveal a considerable knowledge of the learned Greek language of his time and interests in a variety of fields. Conscious of his limitations, which seem to be more marked when the technical language of mathematics and some philosophical terminology in Latin are concerned, he transcribes some key words in Greek letters, occasionally attempting an approximate translation. Some of his versions remained the basis, directly and through revisions, of the knowledge and study of much of Aristotle until the fifteenth and sixteenth centuries.

In 1158 Henry, nicknamed Aristippus, a Norman dignitary of the church and court in Sicily, was on an embassy at Constantinople, from which he brought back several books. With its combination of a recent Arabic past, enlightened Norman rule, and refined cultural life, Sicily was, in its own right, one of the best training grounds for a man like Henry, interested in problems of human life and death (he translated Plato's *Phaedo* and *Meno*) and curious about the workings of nature (like Empedocles, he climbed Mt. Etna to observe the volcano firsthand). He, and others around him, were conscious of the scientific tradition of Sicily; books of mechanics, astronomy, optics, and geometry were available, and attracted people from as far as England. Henry contributed to this tradition with a translation of at least Book IV of the *Meteorologics.* With less pedantry than James, he varied his vocabulary more than a work of science could admit; still, his translation remained indispensable for about a century, and what may be called Aristotle's physical chemistry was known primarily through his text.

(c-2) At approximately the same time, and presumably drawing on the same Greek sources of Aristotelian studies, a number of scholars with quite a good knowledge of Greek produced either new versions of texts already translated—whether the older translations were known to them cannot always be established—and versions of works previously unknown in Latin. These scholars remain anonymous, with the possible exception of a certain John, who produced, after the Venetian James, another translation of the *Posterior Analytics;* a second scholar translated anew the *Topics* and the *Prior Analytics;* a third, the *De sensu;* a fourth, the short treatises *De somno* and *De insomniis;* a fifth, the *De generatione et corruptione* and the *Nicomachean Ethics* (of which only Book I ["Ethica nova"], Books II and III ["Ethica vetus"], and fragments of Books VII and VIII ["Ethica Borghesiana"] remain); a sixth, again after James, the *Physics* (only Book I ["Physica Vaticana"] remains) and the *Metaphysics* without Book XI (the first chapter is lost); and a seventh, probably the *Rhetoric.* Some of these translations had little or no success (*Prior* and *Posterior Analytics, Topics, Rhetoric, Physics*); the others, within the limits of their survival (*De generatione et corruptione, De sensu, De somno, De insomniis, Nicomachean Ethics, Metaphysics*), remained in use, in the original form or in revisions, for three or four centuries. They all testify to the vast interest in the recovery of Aristotle in the twelfth century.

(*c*-3) While Constantinople, possibly together with minor Greek centers, was giving the Aristotelian material to the Latin scholars, the intense cultural activity of the Arab world had spread to northwestern Africa and Spain, providing Latin scholarship, especially in the part of Spain freed from Arab domination, with a vast amount of scientific and philosophical material and the linguistic competence for this to be rendered into Latin. Leaving aside for the moment the spreading of Aristotelian ideas through works of Arabic writers, mention must be made of the one translator of Aristotelian work from the Arabic, the Italian Gerard of Cremona, active in Toledo from *ca.* 1150 to his death in 1187. Being a scientist, he translated from the Arabic what was accessible to him of the more scientific works of Aristotle: the *Posterior Analytics* (theory of science by induction and deduction), *Physics, De generatione et corruptione, De caelo,* and *Meteorologics* (most of Book IV of this was either not translated or was soon lost). He also translated Themistius' paraphrase of the *Posterior Analytics.* The two of these works that did not exist in translation from the Greek (*Meteorologics* I–III and *De caelo*) were often transcribed and not infrequently studied for about sixty years in these versions from the Arabic. The others were occasionally used as terms of comparison or as additional evidence where the texts from the Greek were considered basic. It should also be mentioned that Gerard translated, under the name of Aristotle, thirty-one propositions from Proclus' *Elements of Theology* accompanied by an Arabic commentary, which formed the text

(occasionally ascribed to Aristotle, more frequently left anonymous by the Latins) known under the title *Liber de causis.* Toward the end of the twelfth century, Alfred of Sareshel translated, again under the name of Aristotle (which attribution remained unchallenged for several centuries), Nicholas of Damascus' *De plantis.*

By the end of the twelfth century most of Aristotle had, therefore, found its way into Latin, but that does not mean that his works were soon widely accessible. To make them so, activity was still necessary in both transcription and translation. Some works had not yet been translated, and versions of others had been partly or completely lost; it was also realized that new versions made directly from the Greek would be necessary where only translations from the Arabic or inadequate versions from the Greek were available, and that revisions were necessary for almost every text; finally, it was felt that in order to achieve a more complete understanding of the words of Aristotle, translated by people whose knowledge of Greek was based mainly on the modernized, Byzantine usage, it was useful or necessary to give the reader of Latin access to many of the commentaries, Greek or Arabic, that linked the present with the past.

(*c*-4) The work done with these aims in view, on the basis of Greek texts, was carried out almost completely in the thirteenth century by two outstanding northerners: Robert Grosseteste, bishop of Lincoln and chancellor of Oxford University, and the Flemish Dominican William of Moerbeke, later archbishop of Corinth. A minor contribution came from a Sicilian, Bartholomew of Messina. Grosseteste, philosopher and theologian, linguist and scientist, politician and ecclesiastic, grew up at a time when it was already known how much Aristotle could help in the promotion of that Western European culture of which the foundations had been laid in the twelfth century. He was well aware of the contributions that the fading Greek renaissance could now offer, at least in books and teachers of the language. Grosseteste encouraged other Englishmen to go to Greece, southern Italy, and Sicily to collect books and men of learning. With their help, in the second quarter of the thirteenth century, he learned the language and, what concerns us here, thoroughly revised what remained of the older version of the *Nicomachean Ethics;* translated anew the major part of it, of which the older translation had been lost; and translated a large collection of commentaries on the several books of this work, some of them dating as far back as the third century, some as recent as the eleventh and twelfth. He also replaced with a translation from the Greek the *De caelo,* available until then only in a version from the

Arabic, and added the translation of at least part of the vast commentary by Simplicius on the same work. Finally, he translated as Aristotelian the short treatise *De lineis insecabilibus* ("On Lines Not Made of Points").

William of Moerbeke, also a philosopher, theologian, scientist, and ecclesiastic, but in these fields a lesser man than Grosseteste, traveled from the Low Countries to Italy, Greece, and Asia Minor, widening the scope of his discoveries and of his translations to include Neoplatonic philosophy, geometry, mechanics, and medicine. His activity as an Aristotelian translator was enormous and covered approximately the third quarter of the century. He was the first to translate from Greek into Latin the Aristotelian zoological encyclopedia, the *De animalibus,* and Books I–III of the *Meteorologics;* he can almost be considered the discoverer, for our civilization, of the *Politics;* he was the first to translate into Latin the *Poetics* and Book XI of the *Metaphysics;* he translated anew the *De caelo,* the *Rhetoric* (he probably did not know of the existence of the Greco-Latin translations of these two works), and Book IV of the *Meteorologics;* he accompanied his versions of Greek commentaries with new translations of the *Categories* and *De interpretatione;* and he revised, with different degrees of thoroughness but always having recourse to Greek texts, James's versions of *Posterior Analytics, Physics, De anima, De memoria* and other minor texts of the *Parva naturalia,* Boethius' version of the *Sophistici elenchi,* and the anonymous versions of the *De generatione et corruptione,* of Books I–X and XII–XIV of the *Metaphysics,* and of the *De sensu, De somno,* and *De insomniis.* He also translated the extensive commentaries by Simplicius on the *Categories* and (again, after Grosseteste) the *De caelo,* by Alexander of Aphrodisias on the *De sensu* and *Meteorologics,* by Themistius on the *De anima,* by Ammonius on the *De interpretatione,* and by Philoponus on one part of Book III of the *De anima.* With the possible exception of the *De coloribus* (one fragment seems to be translated by him), he avoided all the works wrongly ascribed to Aristotle.

In contrast, Bartholomew of Messina, working for King Manfred around 1260, specialized in the pseudepigrapha: *De mundo, Problemata, Magna moralia, Physionomia, De mirabilibus auscultationibus, De coloribus,* and *De principiis* (Theophrastus' *Metaphysics*). The only translation of a possibly genuine Aristotelian text made by Bartholomew is that of the *De Nilo.* To complete the picture of the translations from the Greek of "Aristotelian" works before the end of the thirteenth century (or possibly a little after), we should add a second translation of the *De mundo,*

by one of Grosseteste's collaborators, Nicholas of Sicily, two anonymous translations of the *Rhetorica ad Alexandrum,* and two partial translations of the *Economics.* Finally, an anonymous revision of Books I–II and part of Book III of James's translation of the *Metaphysics* was made around 1230, and an equally anonymous revision of the whole of Grosseteste's version of the *Nicomachean Ethics* was carried out probably between 1260 and 1270.

(*c*-5) The work of translating Aristotle or Aristotelian commentaries from the Arabic in the thirteenth century centered, again, mainly in Toledo and to a smaller extent in southern Italy. Most of this work was carried out by Michael Scot; other contributors were William of Luna and Hermann the German. Michael Scot was the first to make known to the Latins the *Books on Animals,* and it was his translation of most of the *Metaphysics* (parts of Books I and XII and the whole of Books XI, XIII, and XIV were not included), together with Averroës' *Great Commentary,* that provided many students of Aristotle with the bulk of this complex of Aristotelian texts: most of James's translation had probably been lost before anybody took any real interest in this work, and the anonymous Greco-Latin version (*Media*) made in the twelfth century emerged from some isolated repository *ca.* 1250. Under the title *Metaphysica nova,* Michael's version, isolated from Averroës' commentary, held its ground for about twenty years and was quite widely used for another twenty. The following translations must be ascribed to Michael Scot, some with certainty, some with great probability: the *De anima, Physics,* and *De caelo* with Averroës' *Great Commentary,* the *Middle Commentary* of the *De generatione et corruptione* and of Book IV of the *Meteorologics,* and Averroës' *Summaries* of the *Parva naturalia.*

William of Luna translated, in or near Naples, the *Middle Commentaries* to Porphyry's *Isagoge* and Aristotle's *Categories, De interpretatione,* and *Prior and Posterior Analytics.* Hermann the German translated Averroës' *Middle Commentaries* on the *Nicomachean Ethics, Rhetoric,* and *Poetics.* The last-mentioned was, in fact, the only source from which Latin readers acquired what knowledge they had—and that was mainly distorted—of Aristotle's *Poetics:* under the title *Poetria* (*Averrois* or *Aristotelis*) it was read quite widely; William of Moerbeke's translation from the Greek remained unknown until 1930, and the next translation from the Greek was not made until shortly before 1500.

By the end of the thirteenth century, the whole of the Aristotelian corpus as we know it, and as it has been known—if we except the relatively few fragments of early works—since the first century B.C., was available in Latin to practically everybody who cared to have access to it. The only exception consisted of the four books of the *Ethics* that are not common to the *Nicomachean Ethics* (which appears with the full complement of ten books) and to the *Eudemian Ethics* (which normally contains only the four that differ from those of the *Nicomachean*); only a small portion of this seems to have been translated, and is connected with passages of the *Magna moralia* in the so-called *De bona fortuna*. The general picture of the diffusion of Aristotle in these translations until the beginning of the sixteenth century is provided by the survival to our times of no fewer than 2,000 manuscripts containing from one to about twenty works, and by the fact that the most complete catalog of early printings (down to 1500) lists over 200 editions, without counting a large number of volumes that contain some of these translations with commentaries.

The detailed picture, when properly drawn, will show the difference in the popularity of the several works; but the difficulty in drawing such a picture derives from the fact that many works, especially minor ones, were transcribed as parts of general, mainly Aristotelian, collections without being actually taken into detailed account. Still, it may be significant that one of these collections, *Corpus Vetustius*—containing the *Physics, Meteorologics, De generatione et corruptione, De anima, Parva naturalia, De caelo,* and *Metaphysics* in the translations made before 1235—remains in slightly fewer than 100 manuscripts, all of the thirteenth (or very early fourteenth) century; a similar collection, including the same works in the new or revised translations in a more complete form (*Corpus recentius*) is preserved in about 200 manuscripts of the thirteenth, fourteenth, and fifteenth centuries. This shows that the more scientific of the works of Aristotle became indispensable in all centers of study and in private libraries. A statistical study of their provenance has not been made: it is, however, clear that France and England are most prominent in this respect for the *Corpus Vetustius;* and France, Italy, Germany, England, and Spain for the *Corpus recentius.*

If we consider the translations that most influenced Western culture and ascribe the authorship to those who produced them in the basic form, a quite accurate assessment of the individual abilities in transmitting Aristotle's works, and thus in shaping some of the philosophical, scientific, and common language of modern civilization, can be made. Their success in presenting formulations that, although not always carefully and strictly Aristotelian, have contributed

a basis for discussion and polemics, and have thus led, in the dialectic of history, to much progress, can be suggested by the following list:

(1) Boethius: *Categories, De interpretatione, Prior Analytics, Topics, Sophistici elenchi;*

(2) James the Venetian-Greek: *Posterior Analytics, De anima, Physics, De memoria* (perhaps *Metaphysics* I–III);

(3) Twelfth-century anonymous translators from the Greek: *Metaphysics* IV–X, XII–XIV (perhaps I–III), *De generatione et corruptione, Nicomachean Ethics* I–III, *De sensu, De somno, De insomniis;*

(4) Michael Scot: *Metaphysics* I–X, XII, *De animalibus;*

(5) Robert Grosseteste: *Nicomachean Ethics* IV–X;

(6) William of Moerbeke: *Meteorologics, Politics, Rhetoric, De animalibus, Metaphysics* XI, *De caelo.*

An important, if sometimes misleading, role in the Latin transmission of Aristotle must be ascribed to the translators of commentaries. All of them contributed to the transmission and improvement of the technique of interpretation, as developed in the Greek schools of the second through sixth centuries. From this point of view, the greatest influence was probably exercised by the commentaries adapted from the Greek by Boethius and those of Averroës, which are linked, through an almost continuous line of scholastic discipline, with the tradition of the Greek schools. From the point of view of the contributions to the actual critical understanding of Aristotle, probably the most important of Averroës' commentaries were those on the *Metaphysics, Physics,* and *De anima.*

(*d*) The last stage in the Latin transmission of Aristotle—if we disregard the occasional translations of the seventeenth to twentieth centuries—covers what is normally called the humanistic and Renaissance period. This is the period beginning with and following the reestablishment of a more intimate collaboration between Greeks and western European scholars, which extended and deepened the understanding of the "old" Greek through a wider knowledge of the history, literature, science, etc., of the ancient world and a much more accurate understanding of the language as it was understood in ancient times. Another aspect that was soon presented as typical of the new movement in translations was the purity and perspicuity of the Latin language (purity ought to have carried with it the elimination of technical words that were not yet technical in classical Latin); but a closer study of many translations shows that the standards of knowledge of the ancient Greek background and of the Greek language were not consistently higher than in the Middle Ages, and that the need for very literal translations and technical usages of a medieval or of a new kind could not be avoided. In fact, very many new versions of Aristotle are hardly distinguishable, in their essential features, from those of the twelfth and thirteenth centuries. And what there was of a new philosophy of language applied to translations—the philosophy of meanings of contexts as against the meanings of individual words—was not always conducive to a better understanding of the original.

A complete survey of new translations down to the last quarter of the sixteenth century is impossible here. Although some of the later versions may still have exercised some influence in their own right, it seems that greater influence was exercised by some of those of the fifteenth century. And it is questionable how much even the latter ousted the medieval translations, or substituted something of great importance for them. We shall confine ourselves to a quick survey of the new versions of the fifteenth century, which were due in almost equal measure to Greek scholars attracted to Italy and to the Italians whose Greek scholarship resulted from contact with them.

The first Italian translator was a pupil of Manuel Chrysoloras, Roberto de' Rossi, who in 1406 translated the *Posterior Analytics*. Probably the greatest and most influential translator at the beginning of this movement was Leonardo Bruni of Arezzo, translator of the *Nicomachean Ethics, Politics,* and *Economics* (1416–1438). Gianozzo Manetti added to new translations of the *Nicomachean Ethics* and *Magna moralia* the first version of the *Eudemian Ethics* (1455–1460), an effort soon followed by Gregorio of Città di Castello (or Tifernate). Giovanni Tortelli again translated (*ca.* 1450) the *Posterior Analytics;* and in the 1480's Ermolao Barbaro translated, if his statements are to be taken literally, the whole of the logical works, the *Physics,* and the *Rhetoric* (only some of his versions remain). Before 1498 Giorgio Valla produced new translations of the *De caelo, Magna moralia,* and *Poetics,* and Lorenzo Laurenziano one of the *De interpretatione.*

In the meantime, from the early 1450's, the Greeks who had entered into the heritage of Latin culture were competing, or leading the way, in translation. The greatest of all, as a man of culture, collector of books, theologian, ecclesiastic, and philosopher, was Iohannes Bessarion, who translated the *Metaphysics.* His vast collection of manuscripts, among them many Greek volumes of Aristotle, was the basis of the Library of St. Mark in Venice. The most productive were John Argyropulos, translator of the *Categories, De interpretatione, Posterior* (and part of the *Prior*) *Analytics, Physics, De anima, De caelo, Metaphysics,*

and *Nicomachean Ethics* (and the pseudo-Aristotelian *De mundo,* also translated shortly before by Rinucio Aretino), and George of Trebizond, translator of the *De animalibus, Physics, De caelo, De generatione et corruptione, De anima, Problemata,* and *Rhetoric.* Theodore of Gaza translated the *De animalibus* and *Problemata,* and Andronicus Callistus the *De generatione et corruptione.*

What had been done to a very limited extent in the fifteenth century was done on a large scale in the first half of the sixteenth, mainly by Italian scholars: the translation of Greek commentaries from the second to the fourteenth centuries. In this field the Renaissance obscured almost completely what had been done in the Middle Ages, something that, with a few exceptions, it failed utterly to do with the entrenched translations of Aristotle.

The Oriental Transmission of Aristotle's Works. The Greek philosophical schools of the fifth and sixth centuries were attended by people of the various nations surrounding the Mediterranean. Greek was the language of learning, but new languages were emerging to a high cultural level, especially as a consequence of the development of theology from the basic tenets and texts of the Christian faith. What had become necessary for the Greek-speaking theologian, a lay cultural basis, was necessary for the Syrian and for the Armenian. Apart from this, most probably, pure philosophical interest was spreading to other nations that were becoming proud of their nationhood. Thus, probably from the fifth century, and certainly from the sixth, Aristotelian texts started to be translated, and commentaries to be translated into, or originally written in, these languages.

The Armenian tradition, to some extent paralleled by or productive of a more limited Georgian tradition, has not been sufficiently investigated. Armenian culture continued in several parts of the world through the centuries—Armenia itself, India, Europe, and recently America—obviously depending on the culture of the surrounding nations but probably with some independence. A vast amount of unexplored manuscript material, stretching from the eighth century or earlier to the nineteenth century, is now concentrated in the National Library of Manuscripts in Yerevan, Armenian Soviet Socialist Republic. What is known in print is confined to translations of Porphyry's *Isagoge,* the *Categories* and *De interpretatione,* the apocryphal *De mundo,* and Helias' commentary to the *Categories.* A semimythical David the Unconquered (David Invictus) of the fourth or fifth century is mentioned as the author of some of these translations.

The Syriac tradition, more limited in time and space, apparently was richer both in translations of works of Aristotle and in original elaboration; apart from this, it formed the basis of a considerable proportion of the Arabic texts of Aristotle and, through them, of some of the Latin versions. The Nestorian Probus (Probha), of the fifth century, is considered the author of the surviving translations of *De interpretatione* and of *Prior Analytics* I.1–7, which may well belong to an eighth-century author. But there is no reason to doubt the ascription of translations and commentaries to Sergius of Theodosiopolis (Reshʿayna). He was a student in Alexandria and later active in Monophysite ecclesiastical and political circles in Antioch and in Constantinople, where he died *ca.* 535. He translated into Syriac the *Categories* with the *Isagoge,* and the *De mundo* (all still preserved), and possibly an otherwise unknown work by Aristotle, *On the Soul.* Toward the end of the seventh century, the Jacobite Jacob of Edessa translated the *Categories;* shortly after, George, bishop of the Arabs (*d.* 724), produced a new version of this book, of the *De interpretatione,* and of the entire *Prior Analytics.* Probably the most influential Syriac translators were two Nestorians, Ḥunayn ibn Isḥāq (*d.* 876) and his son Isḥāq ibn Ḥunayn (*d.* 910 or 911). Ḥunayn translated into Syriac the *De interpretatione, De generatione et corruptione, Physics* II (with Alexander of Aphrodisias' commentary), *Metaphysics* XI, and parts of the *Prior* and *Posterior Analytics;* his son possibly finished the version of these last two works, and translated the *Topics* into Syriac. ʿAbd al-Masih ibn Naʿima and Abū Bishr Matta translated the *Sophistici elenchi.* Isḥāq and Abū Bishr Matta also are among the translators from Greek into Arabic. Other translations into Syriac, which cannot be assigned to a definite author, include the *Poetics* (probably by Isḥāq ibn Ḥunayn), the *De animalibus,* possibly the *Meteorologics,* and a number of Greek commentaries to Aristotelian works. Not the least important feature of these translations into Syriac is the fact that numerous Arabic versions were made from the Syriac, rather than from the Greek.

Arabic translations from Aristotle were made in the ninth and tenth centuries, some by Syriac scholars, among whom the most prominent was Isḥāq ibn Ḥunayn. They were done in the latter part of the ninth century and at the beginning of the tenth, when Baghdad had become the great center of Arabic culture under al-Mamun. Of the many translations listed in the old Arabic bibliographies we shall mention only those that still exist. Those made by Isḥāq ibn Ḥunayn, presumably directly from the Greek, are

Categories, De interpretatione, Physics, De anima, and *Metaphysics* II; by Yaḥyā ibn Abī-Manṣūr, Isa ben Zura, and ibn Naim, the *Sophistici elenchi* (Yaḥyā also translated part of *Metaphysics* XII); Abū ʿUthman ad-Dimashki and Ibrahim ibn ʿAbdallāh, the *Topics;* Abū Bishr Matta, the *Posterior Analytics* and the *Poetics* (perhaps both through the lost Syriac version by Isḥāq ibn Ḥunayn); Yaḥyā ibn al Bitriq, the *De caelo, Meteorologics,* and *De animalibus;* Astat (Eustathius), *Metaphysics* III–X; Theodorus (Abū Qurra [?]), the *Prior Analytics;* unknown translators, the *Rhetoric* and *Nicomachean Ethics* VII–X. Of the apocrypha, we have two translations of the *De mundo,* one of which was made by ʿUsa ibn Ibrahim al-Nafisi from the Syriac of Sergius of Theodosiopolis (Reshʿayna). Finally, it must be mentioned that it was in the Arab world that sections of Plotinus' work (or notes from his conversations) were edited under the title *Theology of Aristotle,* and thirty-one propositions from Proclus' *Elements of Theology* were commented upon and edited as Aristotle's *Book of Pure Goodness* (generally known under the title *De causis,* which it acquired in the Latin tradition).

Elaborations of Aristotle's Works. The transcriptions of the Greek texts, the translations into the several languages, and the multiplication of the copies of these translations were obviously only the first steps in the spread of Aristotle's pure or adulterated doctrines. The more permanent influence of those doctrines was established in the schools, through oral teaching, or on the margin of and outside the schools, through writings of different kinds at different levels. There would be, at the most elementary level, the division into chapters, possibly with short titles and very brief summaries; then occasional explanations of words and phrases in the margins or between the lines in the manuscripts of the actual Aristotelian texts (glosses or scholia), or more extensive summaries and explanations of points of particular interest at some moment or other in the history of thought.

At a higher level there would be systematic expositions or paraphrases, adhering closely to the original text but adapting the diction, the language, and the articulation of the arguments to the common scholastic pattern of this or that time, place, or school; then, expository commentaries, section by section, with or without introductory surveys and occasional recapitulations. The commentaries could aim at clarifying Aristotle's doctrine or adding doctrinal developments, criticisms, or digressions. The discussions would then take on an independent status: "questions about the *Physics,*" "questions about the *De anima,*" and so on. These would normally represent the most marked transition from the exposition of Aristotle's views—however critically they might be treated—to the original presentation of problems arising from this or that passage. Very often such *quaestiones* would not have more than an occasional, accidental connection with Aristotle: the titles of Aristotle's works would become like the headings of one or another of the main branches of philosophy, of the encyclopedia of knowledge, or of sciences. This soon led to the abandonment of the pretense of a connection with the "Philosopher's" works and doctrines or, in many cases, to the pretense of abandoning him and being original while remaining, in fact, under the strongest influence of what he had said.

Systematic works covering a wide province of philosophy, or even aiming at an exhaustive treatment of all its provinces, could take the form of a series of expositions or commentaries on the works of Aristotle, or organize the accumulated intellectual experience of the past and the original views of the author with great independence at many stages, but with explicit or implicit reference to Aristotle's corpus as it had been shaped into a whole—to a small extent by him and to a larger extent by his later followers.

Much of the philosophical literature from the first to the sixteenth centuries could be classified under headings corresponding to the ways in which Aristotle was explained, discussed, taken as a starting point for discussions, used as a model for great systematizations containing all kinds of details, or abandoned—either with or without criticism. In the Greek-speaking world, the vast commentaries by Alexander of Aphrodisias (third century) on the *Metaphysics,* the *Analytics, Topics,* and *Meteorologics;* those by Simplicius (sixth century) on the *Categories,* the *De caelo,* and the *Physics;* and those by John Philoponus (the Grammarian) on the *De anima* were among the most prominent examples of the developed, systematic, and critical commentaries of Aristotle's texts. They were matched in the Latin world of the sixth century by Boethius' commentaries on the *Categories* and *De interpretatione,* in the Arab world of the twelfth century by the "great" commentaries of Averroës, and in the Latin world of the twelfth and thirteenth centuries by those of Abailard, Robert Grosseteste, Aquinas, Giles of Rome, and many others. Themistius' paraphrases (fourth century) of the logical works and of the *De anima,* partly imitated or translated into Latin in his own time, had their counterparts in works by Syriac-, Armenian-, and Arabic-writing philosophers: al-Kindī in the ninth century, the Turk al-Fārābī in the tenth, the Persian Ibn Sīnā (Avicenna) in the eleventh, and Averroës in the

twelfth contributed in this way much-needed information on Aristotle to those who would not read his works, but would like to learn something of his doctrines through simplified Arabic texts. *Summae* or *summulae* of the *Elenchi,* of the *Physics,* and of other works appeared in Latin in the twelfth and thirteenth centuries, under such names as that of Grosseteste, or have remained anonymous. The collections of scholia of Greek manuscripts were continued by such genres as *glossae* and *notulae:* such collections on the *Categories,* written in the ninth century, and on the *Posterior Analytics,* the *De anima,* and the *Meteorologics,* written between the end of the twelfth and the middle of the thirteenth centuries, became in many cases almost standard texts accompanying the "authoritative" but difficult texts of the great master. At the level of philosophical systems we find the great philosophical encyclopedia of Avicenna (eleventh century), organized on the basis of the Aristotelian corpus but enriched by the philosophical experience of Aristotelians, Platonists, and other thinkers of many centuries, and above all by the grand philosophical imagination and penetration of its author. On the other hand, in the Latin world Albertus Magnus (thirteenth century), a man of inexhaustible curiosity, and with a frantic passion for communicating as much as he knew or thought he knew as quickly as possible, followed up his discoveries in the books of others with his own cogitations and developments, and presented his encyclopedia of knowledge almost exclusively as an exposition-cum-commentary of the works by Aristotle or those ascribed to him. What he had learned from others—he was one of the most learned men of his times, and much of his reading derived from the Arabic—finds its place in this general plan.

Quaestiones (ζητήσεις) are found in the Greek philosophical literature, and one might be tempted to include in this class much of Plotinus' *Enneads.* But it is when impatience with systematic explanatory commentary (mildly or only occasionally critical) leads to independent treatment of problems that the *quaestio* comes into its own—first, perhaps, as in Abailard, in the course of the commentary itself; then, in the second half of the thirteenth and much more in the fourteenth and fifteenth centuries, independently of the commentaries. It is in many of these collections of *quaestiones* that we find the minds of philosophers, impregnated with Aristotelian concepts and methods, searching more deeply the validity of accepted statements, presenting new points of view, and inserting in the flow of speculation new discoveries, new deductions from known principles, and corrected inferences from ambiguous formulations.

Aristotle's Influence on the Development of Civilization. The influence exercised by Aristotle's writings varied from work to work and often varied for the several sections of one and the same work. It would be relatively easy to select those short writings which, in spite of their inferior and confused nature or their incompleteness—the *Categories* and the *De interpretatione* from the first century B.C. to the sixteenth, and the *Poetics* from the early sixteenth to the nineteenth—penetrated more deeply and widely into the minds of intelligent people than did the more extensive, organized, and imaginative works, such as *De animalibus, De anima,* and the *Physics.* Moreover, one could possibly select a limited number of passages that left their permanent mark because they were repeatedly quoted, learned by heart, and applied, rightly or wrongly, as proverbs, slogans, and acquired "truths" are applied. Most of all, it is possible, and essential for our purpose, to select those concepts that became common property of the civilized mind, however much they may have been elaborated and, in the course of time, transformed. And if these concepts are not all originally Aristotelian, if they have found their way into the several fields of culture in more than one (the Aristotelian) way, it is our contention that pressure of continuous study and repetition and use of those concepts in Aristotelian contexts, in the ways sketched above, are responsible more than anything else for their becoming so indispensable and fruitful.

It is enough to try to deprive our language of a certain number of words in order to see how much we might have to change the whole structure of our ways of thinking, of expressing, even of inquiring. A conceptual and historico-linguistic analysis of a definition like "mass is the quantity of matter" would show us that whatever was and is understood by these words owes much to the fact that the concepts of "quantity" and of "matter" were for two millennia inculcated into the minds of men and into their languages, more than in any other way, through the agency of Aristotle's *Categories, Physics,* and *Metaphysics.* If "energy" means something when we read it in the formula $e = mc^2$, we may forget that this "linguistic" tool is the creation of Aristotle and that it traveled through the ages with all its appendages of truths, half-truths, and hypotheses, which affected its meaning in different ways through the centuries, stimulating thoughts, experiments, and interpretations of facts, because some bits of the *Metaphysics* and of the *Physics* were the *sine qua non* condition of men's "knowledge" of the world. And if "potential" has assumed so many uses—from social and military

contexts to electricity, dynamics, and what not—is it not because we have been trained to handle this term as an indispensable instrument to describe an infinite variety of situations that have something in common, as Aristotle repeated *ad nauseam,* when making "potency" (δύναμις) one of the basic concepts for the understanding of the structure of the world? We have used, misused, abused, eliminated, and reinstated the concepts of "substance" and "essence." "Relation" and "analogy," "form," "cause," "alteration of qualities," and "development from potentiality to actuality" are all terms that have not yet stopped serving their purpose. A writer of a detailed history of science would be hard put if he tried to avoid having recourse to Aristotle for his understanding of how things progressed in connection with them. At the very root of much of our most treasured scientific development lies the quantification of qualities; this started in the form of a general problem set by the distinction between two out of the ten "Aristotelian categories" in conjunction with Aristotle's theory of the coming into being of new "substances." It may be contended that, by his very distinction, Aristotle created difficulties and slowed progress. Perhaps there is something in that complaint; nevertheless, in this way he stimulated the search for truth and for formulations of more satisfactory hypotheses to fit, as he would say, τὰ φαινόμενα—to fit what we see.

His exemplification of continuous and discontinuous quantities in the *Categories* may elicit an indulgent smile from those who lack any historical sense; and it would be impertinent to skip over twenty-two and a half centuries and say that here we are, faced by the same problems that worried Aristotle, but with more sophistication: continuous waves or discontinuous quanta? But how did it happen that the problems came to be seen in this way, with this kind of alternative? No doubt Aristotle was not the only ancient sage who taught the concept of continuity to the millennia to come, but no text in which the distinction—and the problems it brought with it—appeared was learned by heart, discussed and commented upon, or became the text for examinations and testing as often and as unavoidably as the *Categories.* Do things happen by chance, or through a chain of causality? Can we determine how and why this happens—is it "essential" that it should happen or is it "accidental"? Much scientific progress was achieved by testing and countertesting, under *these,* Aristotelian, headings, what the world presents to our perception and to our mind.

Again: classification, coordination, and subordination have been and are instruments of clear thinking, of productive procedures, of severe testing of results. The terms "species" and "genus" may be outmoded in some fields, but the fashion is recent; the words have changed, yet the concepts have remained. And with them we find, not even outmoded, "property" and "difference." We have been conditioned by these distinctions, by these terms, because we come from Aristotelian stock.

It is, in conclusion, significant of Aristotle's impact on the development of culture, and particularly of science, that among the more essential elements in our vocabulary there should be the following terms, coming directly from his Greek (transliterated in the Latin or later translations) or from the Latin versions, or from texts where some of these terms had to be changed in order to preserve some equivalence of meaning when they proved ambiguous: (*a*) *category* (*class, group,* etc.) and the names of the four categories actually discussed in the *Categoriae*—*substance* (*essence*), *quantity, quality, relation;* (*b*) *universal* and *individual,* and the *quinque voces* (another title for Porphyry's *Isagoge,* which developed a passage of Aristotle's *Topics* and was studied as the introduction to his logic)—*genus, species, difference, property, accident* (in the sense of accidental feature); (*c*) *cause* and the names or equivalents applied to the four causes until quite recent times—*efficient, final, material,* and *formal;* (*d*) couples of correlative terms, like *matter-form* (structure), *potency-act* (energy), *substance-accident.*

Terms like "induction" and "deduction," "definition" and "demonstration" have certainly become entrenched in our language from many sources apart from Aristotle's *Analytics.* But again, the extent of their use, the general understanding of their meaning and implications, and the application in all fields of science of the methods of research and exposition that those terms summarize depend possibly more on the persistent study of Aristotle than on any other single source. All the wild anti-Aristotelianism of the seventeenth century would have been more moderate if people had realized then, as it had been realized, for instance, in the thirteenth century, how aware Aristotle was that experience, direct perception and knowledge of individual facts, is the very basis of scientific knowledge. The anti-Aristotelians were much more Aristotelian than they thought in some aspects of their methods; and that was because they had, unconsciously, absorbed Aristotle's teaching, which had seeped through from the higher level of philosophical discussion to the common attitude of people looking for truth.

It has become a truism that observation of facts was recognized as the necessary beginning of science through a revolutionary attitude which had as its pioneers such people as Roger Bacon and Robert Grosseteste. One wonders whether many realize that —because he thought Aristotle to be very often right on important matters—Aquinas insisted that a problem which, for him and his contemporaries, was of the utmost importance—the problem of the existence of God—could be solved only by starting from the observation of facts around us. If, as it happened, Aquinas was going to carry the day with his very awkward "five ways," he was also going to boost very widely the value of the basic principle on which so much depended in the development of science: observe first, collect facts, and draw your conclusions after. And it is in the course of the discussion of the *Posterior Analytics* that probably one of the main steps forward in the methodology of science was made by Grosseteste around 1230: probably not so much—as has been maintained—in passing from "experience" to "experiment" as in the discrimination of the contributory factors of a certain effect, in the search for the really effective causes, as against the circumstantial, accidental state of affairs.

One further example of the permanence of Aristotle's teaching is provided by his insistence on the old saying that nature does nothing in vain. The development from this principle of the wrongly called "Ockham's razor" is the result of a series of refinements; it may be possible (or has it already been done?) to see through which steps this principle of finality and economy of nature has established itself in all but the most independent or anarchic scientific minds.

Above all, probably, Aristotle's explicitly stated methodical doubt as a condition for the discovery of truth and his exhaustive accumulation of "difficulties" (ἀπορίαι) have trained generation after generation in the art of testing statements, of analyzing formulations, of trying to avoid sophistry. The picture of an Aristotelianism confined to teaching how to pile up syllogisms that either beg the question or, at best, make explicit what is already implicit in the premises is very far from the Aristotelianism of Aristotle, and hides most of what Aristotle has meant for the history of culture and science. It is through observation, ἀπορίαι, reasoned and cautious argument, that he thought our statements should fit the phenomena (φαινόμενα): no wonder that Aquinas himself was not troubled by the possibility that geocentrism might prove to be less "valid" than heliocentrism.

It is much more difficult to discover, isolate, and follow up the influence of Aristotle's writings on the advancement of science considered in the several fields and, what counts more, in the solution of particular problems. It is also difficult to locate exactly in time and space the several steps by which methods of inquiry, learned directly or indirectly at the Aristotelian school, have been successfully applied as Aristotelian. Out of the vast amount of evidence existing, only a small fraction has been studied. Influences have hardly ever been the result of isolated texts or of individual authors; the accumulation of interpretations, refinements, new contributions, and variations in the presentation of problems has continued for centuries, and the more striking turning points are those at which the influence has been *a contrario*. Whether it is Simplicius (sixth century) commenting on the *De caelo,* and thus contributing to the methodical transformation of the study of the heavens, or William Harvey (eleven centuries later) taking as one of his basic texts for the study of the mechanics of the living body the *De motu animalium,* there is no doubt that we can rightly speak of Aristotle's influence on the advancement of astronomy and of physiology. But determining the exact point at which that influence can be located, in what precise sense it can be interpreted, and in what measure it can be calculated would require much more than a series of textual references.

It might be suggested that one precise point in history at which Aristotle's deductive theory in the *Posterior Analytics* contributed to the mathematization of nonmathematical sciences can be found in Robert Grosseteste's commentary on that work (*ca.* 1230). Aristotle had considered optics as a science dependent on mathematics (geometry), and in his discussion of two types of demonstration, the *demonstratio quia* and the more penetrating and valuable *demonstratio propter quid,* he had used optical phenomena to exemplify the general rule that it is the higher-level science (in those particular cases, obviously, mathematics) that holds the key to the *demonstratio propter quid.* For Grosseteste the whole of nature was fundamentally light, manifesting itself in different states. It could be argued, therefore, that Grosseteste would have inferred that Aristotle's examples revealed, more than he imagined, the mathematical structure of all natural (and supernatural) sciences. One can go further and, magnifying Grosseteste's influence, state that quantification in natural sciences has its roots in the *Posterior Analytics* as interpreted by Grosseteste in the frame of his metaphysics of light. This is the kind of fallacy that results from not realizing how difficult it is to discover and assess Aristotelian influences. Nothing has so far been

shown—although much has been said—to prove that statement.

Among the few fields in which many necessary inquiries have been made (through commentaries to Aristotle, *quaestiones* arising from the *Physics,* and independent treatises with an Aristotelian background) to show how (by appropriate or forced interpretation, by intelligent criticism or the process of development) modern science has to some extent come out of the study of Aristotle are those of the theories of rectilinear movement (constant velocity and acceleration), of "essential" transformations consequent to quantitatively different degrees of qualities, and of the nature and basic qualities of matter in connection with gravity. The temptation must, of course, be resisted to see Aristotle's influence wherever some connection can be established, whether *prima facie* or after detailed consideration of chains of quotations, repetitions, and slight transformations. But the pioneering studies of Pierre Duhem, the detailed analyses and historical reconstructions by Anneliese Maier, Nardi, Weisheipl; the attempts at wider historical systematizations by Thorndike, Sarton, and Crombie; and the contributions by many scholars of the last thirty years confirm more and more the view that the debt of scientists to the Aristotelian tradition is far greater than is generally accepted.

Setbacks in the Aristotelian Tradition. The progress in the spread of Aristotelian studies had its obstacles and setbacks, at different times in different spheres and for a variety of reasons. These ranged from purely philosophical opposition to purely theological convictions and prejudices, and to the interference of political and political-ecclesiastical powers with the free flow of speculation and debate. The story of the setbacks could be considered as diverse and rich as that of the actual progress; we shall mention only some of the most famous, or notorious, examples.

In 529 Justinian ordered the closing down of all philosophical schools in Athens; such people as Simplicius and Damascius became political-philosophical refugees in the "unfaithful" Persian kingdom. Greek Aristotelian studies then had over two centuries of almost total eclipse.

A similar attack on philosophy, at a very "Aristotelian" stage, was carried out in 1195 by Caliph Ya'ūb al-Manṣūr in southern Spain; one of the exiled victims was the great Averroës, who had, among other things, strongly defended philosophy against the religious mystical onslaught by al-Ghazali, the author of the *Destruction of Philosophers.* Whatever the reasons for the centuries-long eclipse of

Arabic philosophy, the blow of 1195 was certainly one of the most effective contributions to it.

Much has been made by the historians of philosophy, and particularly of science, of the Roman Church's hostility to Aristotelianism, as made manifest by the decrees of 1210, 1215, and 1231—also confirmed later—"prohibiting" the study of Aristotle's works on natural philosophy and then of those on metaphysics. The prohibitions, confined first to Paris and then to a few other places, and soon limited in scope (the works in question were to be examined by a committee of specialists and, where necessary, revised), turned out to be probably one of the most important factors in the most powerful and permanent expansion of Aristotelian studies in the whole of history. Interest was intensified, obstacles were avoided or disregarded, and witch-hunting did not succeed in doing much more than alerting philosophers and scholars to the danger of expressing Aristotle's views as their own views, and of describing developments based on Aristotle's works as *the* truth rather than as logically compelling inferences from authoritative statements.

The real setbacks to the spread of Aristotelian studies—not necessarily of the kind of Aristotelian influence sketched above—came in the seventeenth and eighteenth centuries, when progress in scientific and historical knowledge; the interplay of the new interests with a sterilized, scholastic "Aristotelianism"; a passion for grand philosophical systems; refined, systematic criticism of current beliefs; and the impact of new theological disputes filled the minds of thoughtful people with problems that either were not present in Aristotle's works or had now to be expressed in a differently articulated language.

BIBLIOGRAPHY

I. ORIGINAL WORKS. This section will be limited to the more essential references. The others will be found in works cited below under "Secondary Literature."

The tradition of the Greek texts of Aristotle is documented mainly in their critical editions; for these see the article on his "Life and Works." For the medieval Latin tradition see, above all, the *Corpus philosophorum medii aevi, Aristoteles Latinus* (Bruges-Paris, 1952–   ), of which the following vols. have appeared: I.1–5, *Categoriae,* L. Minio-Paluello, ed. (1961); I.6–7, *Supplementa Categoriarum* (Porphyry's *Isagoge* and Pseudo-Gilbertus' *Liber sex principiorum*), L. Minio-Paluello, ed. (1966); II.1–2, *De interpretatione,* L. Minio-Paluello, ed. (1965); III.1–4, *Analytica priora,* L. Minio-Paluello, ed. (1962); IV.1–4, *Ana-*

lytica posteriora, L. Minio-Paluello and B. G. Dod, eds. (1968); VII.2, *Physica* I ("Physica Vaticana"), A. Mansion, ed. (1957); XI.1–2, *De mundo,* 2nd ed., W. L. Lorimer *et al.,* eds. (1965); XVII.2.v, *De generatione animalium,* trans. Guillelmi, H. J. Drossaart Lulofs, ed. (1966); XXIX.1, *Politica* I–II.11, 1st vers. by William of Moerbeke, P. Michaud-Quantin, ed. (1961); and XXXIII, *Poetica,* 2nd ed., trans. Guillelmi, with Hermann the German's version of Averroës' *Poetria,* L. Minio-Paluello, ed. (1968). V.1–3, *Topica,* L. Minio-Paluello, ed., is to appear in 1969. Older eds. of most of the translations or revisions of the thirteenth century appeared from 1475 on. Among other more recent eds., the following should be recorded: *Politics,* in F. Susemihl's ed. of Greek text (Leipzig, 1872); *Rhetoric,* in L. Spengel's ed. of Greek text (Leipzig, 1867); *Metaphysica media,* in *Alberti Magni Opera omnia,* XVI, B. Geyer, ed. (Münster, 1960–   ); *Metaphysica,* trans. Iacobi ("Metaphysica Vetustissima"), in *Opera . . . Rogeri Baconi,* XI, R. Steele, ed. (Oxford, 1932).

The best ed. of the Armenian texts of the *Categoriae, De interpretatione,* and *De mundo* was produced by F. C. Conybeare in *Anecdota Oxoniensia,* Classical Series I.vi (Oxford, 1892). George's Syriac version of *Categoriae, De interpretatione,* and *Prior Analytics* was edited by G. Furlani in *Memorie dell'Accademia . . . dei Lincei,* Classe scienze morali, VI.5,i and iii, and VI.6.iii (Rome, 1933–1937). Most of the surviving Arabic translations of the Middle Ages were first edited or reedited by Abdurrahman Badawi in the collection Studii Islamici (then Islamica) (Cairo 1948–   ): these include all the works of logic, the *Rhetoric, Poetics, De anima, De caelo,* and *Meteorologics.* Of other eds. the following should be mentioned: *Metaphysics* (missing parts of Bks. I and XII, and the whole of Bks. XI and XIII–XIV), M. Bouyges, ed. (Beirut, 1938–1952); and *Poetics,* J. Tkatsch, ed. (Vienna, 1928–1932).

The extant Greek commentaries were edited by H. Diels and his collaborators in *Commentaria in Aristotelem Graeca* (Berlin, 1882–   ); the medieval Latin trans. are being published in the *Corpus Latinum commentariorum in Aristotelem Graecorum* (Louvain, 1957–   ), thus far consisting of I. *Themistius on De anima,* II. *Ammonius on De interpretatione,* III. *Philoponus on De anima,* and IV. *Alexander on De sensu*—all ed. by G. Verbeke.

The one major commentary by Averroës that is preserved in Arabic, on the *Metaphysics,* was published with the Aristotelian text by Bouyges (see above). Many of the Latin medieval trans. of the longer and shorter commentaries by Averroës were printed several times in the fifteenth and sixteenth centuries (1st ed., Venice, 1483); new trans. from the Hebrew of some of the same commentaries and of others (most importantly, the long commentary on *Posterior Analytics*) were published in the sixteenth century (first comprehensive ed., Venice, 1551–1561). Critical eds. of the medieval Latin and Hebrew trans. of Averroës' commentaries are being published in the *Corpus philosophorum medii aevi, Corpus commentariorum Averrois in Aristotelem,* the most important of which is Michael

Scot's trans. of the long commentary on *De anima,* in Vol. VI.1, F. Stuart Crawford, ed. (Cambridge, Mass., 1953).

II. SECONDARY LITERATURE. A list of Greek MSS of Aristotle's works and of those of his commentators, based mainly on printed catalogs, was ed. by A. Wartelle, *Inventaire des manuscrits grecs d'Aristote et de ses commentateurs* (Paris, 1963), and supplemented by D. Harlfinger and J. Wiesner in *Scriptorium,* **18,** no. 2 (1964), 238–257. A descriptive catalog of all the known MSS of Aristotle's works is being prepared by P. Moraux and his collaborators of the Aristotelian Archive at the University of Berlin. The best sources for knowledge of the printed tradition are still the general catalogs of the British Museum and of the Prussian libraries; for recent times, see also the catalog of the U.S. Library of Congress.

Nearly all the available basic information for the Latin tradition in the Middle Ages is collected in the three vols. of G. Lacombe, E. Franceschini, L. Minio-Paluello, *et al., Aristoteles Latinus, Codices:* I., Rome, 1939; II., Cambridge, 1955; *Supplem. Alt.,* Bruges, 1961. The bibliography that is in these vols. includes all the works of importance on the subject. Additional information on individual works will be found in the intros. to the eds. of texts in the *Aristoteles Latinus* series. Special mention should be made of E. Franceschini, "Roberto Grossatesta, vescovo di Lincoln, e le sue traduzioni latine," in *Atti della Reale Istituto Veneto,* **93,** no. 2 (1933–1934), 1–138; G. Grabmann, *Guglielmo di Moerbeke, il traduttore delle opere di Aristotele* (Rome, 1946); J. M. Millás Vallicrosa, *Las traducciones orientales en los manuscritos de la Biblioteca Catedral de Toledo* (Madrid, 1942); L. Minio-Paluello, "Iacobus Veneticus Grecus, Canonist and Translator of Aristotle" in *Traditio,* **8** (1952), 265–304; "Note sull'Aristotele Latino medievale," in *Rivista di filosofia neo-scolastica,* **42** ff. (1950 ff.). For the printed eds. of medieval Latin trans., see the *Gesamtkatalog der Wiegendrucke* and the library catalogs cited above.

For the humanistic and Renaissance trans. into Latin, see E. Garin, *Le traduzioni umanistiche di Aristotele nel secolo XV,* Vol. VIII in Accademia Fiorentina La Colombaria (Florence, 1951), and the *Gesamtkatalog* and the library catalogs.

For the study of Aristotle in the Middle Ages, M. Grabmann's *Mittelalterliches Geistesleben,* 3 vols. (Munich, 1926–1956), and his earlier *Geschichte der scholastischen Methode* (Freiburg im Breisgau, 1909–1911) are of fundamental importance. Among the many works of a more limited scope, see F. Van Steenberghen, *Siger de Brabant d'après ses oeuvres inédites, II: Siger dans l'histoire de l'Aristotélisme,* Vol. XII of Les Philosophes Belges (Louvain, 1942).

For the Armenian tradition, see Conybeare's ed. mentioned above; the catalogs of the more important collections of Armenian MSS (Vatican Library, British Museum, Bodleian Library, Bibliothèque Nationale); and G. W. Abgarian, *The Matenadaran* (Yerevan, 1962).

For the Syriac tradition, see A. Baumstark, *Geschichte der syrischen Literatur* (Bonn, 1922); and many articles by G. Furlani, listed in the bibliog. of his writings in *Rivista degli studi orientali,* **32** (1957).

For the Arabic tradition, see C. Brockelmann, *Geschichte der arabischen Literatur,* 2nd ed., 2 vols. (Leiden, 1943–1949) and 3 vols. of supps. (Leiden, 1937–1942); R. Walzer, "Arisṭūṭālīs," in *Encyclopaedia of Islam,* 2nd ed., I, 630–635; Abdurrahman Badawi, *Aristu ʿinda 1-ʿArab* (Cairo, 1947); M. Steinschneider, "Die arabischen Uebersetzungen aus dem Griechischen," in *Zentralblatt für Bibliothekswesen,* **8** (1889) and **12** (1893), and "Die europäischen Uebersetzungen aus dem Arabischen bis Mitte des 17 Jahrhunderts," in *Sitzungsberichte der Kaiserliche Akademie der Wissenschaften,* philos.-hist. Klasse, **149,** no. 4, and **151,** no. 1.

For the Hebrew tradition, see M. Steinschneider, *Die hebraïschen Uebersetzungen des Mittelalters und die Juden als Dolmetscher* (Berlin, 1893); and H. A. Wolfson, "Plan for the Publication of a *Corpus commentariorum Averrois in Aristotelem,*" in *Speculum* (1931), 412–427.

No comprehensive study of Aristotle's influence through the ages has ever been published. The standard histories of philosophy and science, general or specialized, contain much useful information, including bibliographies, e.g.: F. Ueberweg, *Geschichte der Philosophie,* 5 vols., 11th–13th eds. (Berlin, 1924–1928); E. Zeller, *Die Philosophie der Griechen,* 4th–7th eds. (1882–1920); I. Husik, *A History of Medieval Jewish Philosophy* (Philadelphia, 1916; 6th ed., 1946); G. Sarton, *Introduction to the History of Science,* 3 vols. (Baltimore, 1927–1948); Lynn Thorndike, *A History of Magic and Experimental Science,* 8 vols. (New York, 1923–1958); C. Singer, *Studies in the History and Method of Science* (Oxford, 1921); and A. C. Crombie, *Augustine to Galileo* (London, 1952).

Special problems, periods, or fields have been surveyed and analyzed in, e.g., P. Duhem, *Le système du monde,* 8 vols. (Paris, 1913–1916, 1954–1958), and *Études sur Léonard de Vinci* (Paris, 1906–1913); A. Maier, *Metaphysische Hintergründe der spätscholastischen Naturphilosophie* (Rome, 1951), *Zwei Grundprobleme der scholastischen Naturphilosophie,* 2nd ed. (Rome, 1951), *An der Grenze von Scholastik und Naturwissenschaft,* 2nd ed. (Rome, 1952), and *Zwischen Philosophie und Mechanik* (Rome, 1958); A. C. Crombie, *Robert Grosseteste and the Origins of Experimental Science* (Oxford, 1953); M. Clagett, *The Science of Mechanics in the Middle Ages* (Madison, Wis., 1959); and R. Lemay, *Abu Maʿshar and Latin Aristotelianism in the Twelfth Century* (Beirut, 1962).

L. MINIO-PALUELLO

ARISTOXENUS (*b.* Tarentum, *ca.* 375–360 B.C.; *d.* Athens [?]), *harmonic theory.*

Aristoxenus was a native of Tarentum, a Greek city in southern Italy. He flourished in the time of Alexander the Great (reigned 336–323), and can hardly have been born later than 360. His father's name was either Mnaseas or Spintharus; the latter was certainly his teacher, a musician whose wide acquaintance included Socrates, Epaminondas, and Archytas. Aristoxenus studied at Mantinea, an Arcadian city that had a strong conservative musical tradition, and later became a pupil of Aristotle in Athens. His position in the Lyceum was such that he hoped to become head of the school upon the death of Aristotle (322 B.C.); he is said to have vented his disappointment in malicious stories about his master. The date of his death is unknown, but the attribution to him of 453 published works (even if many were spurious) suggests a long life.

Of this vast production little has survived except for three books that have come down under the title *Harmonic Elements*. Modern scholars are agreed, however, that these represent two or more separate treatises. There is a substantial fragment of the second book of *Rhythmical Elements*. Aristoxenus' numerous other writings on music are all lost, except for quotations, but much of our scattered information on early Greek musical history must derive from him. He also wrote biographies (he was one of those who established this kind of writing as a tradition of the Peripatetic school); treatises on educational and political theory and on Pythagorean doctrine; miscellanies; and memoranda of various kinds.

It is proper, if paradoxical, that Aristoxenus should be included in a dictionary of scientific biography. It is proper because music under the form of "harmonics" or the theory of scales was an important branch of ancient science from the time of Plato onward, and because Aristoxenus was the most famous and influential musical theorist of antiquity. It is paradoxical because he turned his back upon the mathematical knowledge of his time to adopt and propagate a radically "unscientific" approach to the measurement of musical intervals.

When the Pythagorean oligarchs were expelled from Tarentum, Archytas, the celebrated mathematician and friend of Plato, remained in control of the new democracy and may still have been alive when Aristoxenus was born. From Archytas' pupils in Tarentum—or from the exiled Xenophilus in Athens—Aristoxenus must have become familiar with Pythagorean doctrine. The Pythagoreans recognized that musical intervals could be properly measured and expressed only as ratios (of string lengths or pipe lengths). Pythagoras himself is said to have discovered the ratios of the octave, fifth, and fourth; and the determination of the tone as difference between fifth and fourth must soon have followed. In each case the ratio is superparticular and incapable, without the aid of logarithms, of exact division in mathematical terms. Thus, "semitone" (Greek, *hemitonion*) is a

misnomer; when two tones were subtracted from the fourth, the Pythagoreans preferred to call the resulting interval "remainder" (*leimma*). The Pythagorean diatonic scale, consisting of tones (9:8) and *leimmata* (256:243), was known to Plato. Archytas worked out mathematical formulations for the diatonic, chromatic, and enharmonic scales upon a different basis.

Aristoxenus, however, turned his back upon the mathematical approach and stated that the ear was the sole criterion of musical phenomena. To the ear, he held, the tone was divisible into halves (and other fractions); the octave consisted of six tones, the fifth of three tones and a half, the fourth of two tones and a half, and so on. His conception of pitch was essentially linear; the gamut was a continuous line that could be divided into any required fractions, and these could be combined by simple arithmetic. It was for the cultivated ear to decide which intervals were "melodic," i.e., capable of taking their places in the system of scales.

The division of the octave into six equal tones and of the tone into two equal semitones recalls the modern system of "equal temperament," and it is held by some authorities that Aristoxenus envisaged such a system or sought to impose it upon the practice of music. If that were so, he might well have rejected a mathematics that was still incapable of expressing it. Equal temperament, however, was devised in modern times to solve a specific problem: how to tune keyboard instruments in such a way as to facilitate modulation between keys. No comparable problem presented itself to Greek musicians: although modulation was exploited to some extent by virtuosi of the late fifth century B.C. and after, there is no reason to suppose that it created a need for a radical reorganization of the system of intervals or that such could have been imposed upon the lyre players and pipe players of the time. Furthermore, such a "temperament" would distort all the intervals of the scale (except the octave) and, significantly, the fifths and fourths; but Aristoxenus always speaks as though his fifths and fourths were the true intervals naturally grasped by the ear and his tone the true difference between them. It seems more likely, then, that he took up a dogmatic position and turned a blind eye to facts that were inconsistent with it; this would be in keeping with the rather truculent tone he sometimes adopted.

Writers on harmonics, from Aristoxenus on, fall into two schools: his followers, who reproduced and simplified his doctrines in a number of extant handbooks, and the "Pythagoreans" such as Eratosthenes, Didymus the musician, and—notably—Ptolemy, who elaborated ratios for the intervals of the scale. It is perhaps doubtful whether any writer of the mathematical school prior to Ptolemy produced a comprehensive theory of scales in relation to practical music, and it may well have been the inadequacy and limited interests of the Pythagoreans that set Aristoxenus against this approach. Nor would it be fair to deny Aristoxenus' scientific merits because of his disregard of mathematics. He was not in vain a pupil of Aristotle, from whom he had learned inductive logic and the importance of clear definition; and what he attempted was, in the words of M. I. Henderson, "a descriptive anatomy of music." His arguments are closely reasoned, but, lacking his master's breadth and receptivity, he can be suspected of sacrificing musical realities to logical clarity.

The details of his system, which can be found in standard textbooks and musical encyclopedias, are not in themselves of primary interest to the historian of science. The quality of his thinking at its best can, however, be illustrated from his work on rhythm. Earlier writers had tended to discuss rhythm in terms of poetic meters, but, since rhythm also manifests itself in melody and in the dance, there was some confusion of thought and terminology. Aristoxenus drew a clear distinction between rhythm, which was an organized system of time units expressible in ratios, and the words, melodies, and bodily movements in which it was incorporated (*ta rhythmizomena*) and from which it could be abstracted. This was a much-needed piece of clarification worthy of Aristotle.

BIBLIOGRAPHY

I. ORIGINAL WORKS. Modern editions of Aristoxenus' works are *The Harmonics of Aristoxenus,* edited, with translation, notes, introduction, and index of words, by H. S. Macran (Oxford, 1902); and *Aristoxeni elementa harmonica,* Rosetta da Rios, ed. (Rome, 1954).

II. SECONDARY LITERATURE. Works dealing with Aristoxenus include Ingemar Düring, *Ptolemaios und Porphyrios über die Musik* (Göteborg, 1934), which contains, in German translation, Ptolemy's criticisms of Aristoxenus; C. von Jan, "Aristoxenos," in Pauly-Wissowa, *Real-Encyclopädie,* II (Stuttgart, 1895), 1057 ff.; L. Laloy, *Aristoxène de Tarente* (Paris, 1904), an outstanding work; F. Wehrli, *Die Schule des Aristoteles, Texte und Kommentar,* Vol. II, *Aristoxenos* (Basel, 1945), for the shorter fragments; and R. Westphal, *Aristoxenos von Tarent* (Leipzig, 1883–1893), which includes the fragment on rhythm.

R. P. WINNINGTON-INGRAM

ARISTYLLUS (*fl. ca.* 270 B.C.), *astronomy.*

We may infer when Aristyllus flourished from the information provided by Ptolemy and from the date of Hipparchus' observations.

He is mentioned by Plutarch in *De Pythiae oraculis* (402 F)—with Aristarchus, Timocharis, and Hipparchus—as an astronomer who wrote in prose. His name also occurs in two catalogs of commentators on Aratus; although two persons named Aristyllus are mentioned, the reference may be to one and the same person, as Maass maintains. Aristyllus' name also occurs in a catalog of astronomers who wrote about "the pole," i.e., about the polar stars. More important is the information we find in Ptolemy's *Syntaxis mathematica.* He mentions Aristyllus and Timocharis as two astronomers whose observations of the fixed stars (declinations and differences of longitude) were used by Hipparchus. The latter, partly because of the differences between his own observations and those made a hundred years earlier by Aristyllus and Timocharis, discovered the precession of the equinoxes and calculated the amount of retrogression of the equinoctial points. According to Ptolemy, the observations of Aristyllus were not very accurate.

BIBLIOGRAPHY

Aristyllus' observations are recorded in Ptolemy, *Syntaxis mathematica,* Heiberg ed., VII, 1 and 3. See also "Aristyllos," in Pauly-Wissowa, *Real-Encyclopädie der Altertumswissenschaft,* II, 1 (Stuttgart, 1895), cols. 1065–1066; and E. Maass, *Aratea* (Berlin, 1892), pp. 121, 123, 151.

L. TARÁN

ARKADIEV, VLADIMIR KONSTANTINOVICH (*b.* Moscow, Russia, 21 April 1884; *d.* Moscow, U.S.S.R., 1 December 1953), *physics.*

Arkadiev's father died when the boy was young, leaving him to be raised by his mother, who worked in a library. These circumstances disposed him to study from his earliest years. While still at the Gymnasium he became interested in physics; he met N. A. Umov—who at that time was a professor at Moscow University—and even tested his own apparatus in Umov's laboratory. Upon graduating from the Gymnasium in 1904, Arkadiev entered the Physics and Mathematics Faculty of Moscow University, where Umov and P. N. Lebedev lectured. In 1907, under Lebedev's direction, Arkadiev began an experimental study of the magnetic properties of ferromagnetic substances in high-frequency fields of which the wavelength was on the order of one centimeter. In 1908 he obtained new results—ferromagnetic properties of iron and nickel disappeared when the wavelength was on the order of three centimeters—a discovery for which Arkadiev was awarded the Society of Lovers of Natural Science Prize in 1908.

Arkadiev's work at Moscow University was interrupted in 1911 when many progressive professors and lecturers—including Lebedev, Umov, and Arkadiev himself—left the university to protest the arbitrariness of the administration of L. A. Kasso, the czarist minister of national education. Lebedev and his colleagues transferred to Shanyavsky Municipal University, which was privately run. After the October Revolution, Arkadiev returned to Moscow University, where he organized a large laboratory for electromagnetic research that he headed until his death. In 1927 he was chosen an associate member of the Academy of Sciences of the U.S.S.R.

The basic direction of Arkadiev's research was the development of his previous work on ferromagnetism. He was the first to determine experimentally the exact relationship between wavelength and the complex magnetic permeability of iron and nickel (1912). On the basis of his results, Arkadiev arrived at two important conclusions. First, he suggested that the magnetic parameters of a substance depend on the frequency of the field. Second, he proposed the introduction of an additional parameter—"permeance"—to calculate the influence of the lag between changes of magnetic induction and changes in the field, in connection with the field loss of energy during remagnetization. Thus he gave symmetric form to Maxwell's equations:

$$\nabla x\ \boldsymbol{H} = \frac{\varepsilon}{c}\frac{\delta \boldsymbol{E}}{\delta t} + 4\pi\delta \boldsymbol{E}$$

$$-\nabla x\ \boldsymbol{E} = \frac{\mu}{c}\frac{\delta \boldsymbol{H}}{\delta t} + 4\pi\rho \boldsymbol{H}.$$

Into this Arkadiev introduced the concept of complex magnetic permeability $\mu = \mu' - i\mu''$, where $\mu = 2\rho T$ (T is the period of the wave), by analogy to the concept of the complex dielectric constant, $\varepsilon = \varepsilon' - i\varepsilon''$, $\varepsilon = 2\delta T$.

In 1913, starting from the classic conceptions, Arkadiev was the first to indicate the possibility that natural oscillations of elementary magnets exist in ferromagnetic substances, resulting in the appearance of resonance; he confirmed this experimentally. Thus the honor of the discovery of ferromagnetic resonance (1913) belongs to him.

In 1922 Arkadiev proposed a generator of original construction that enabled A. A. Glagoleva-Arkadieva (his wife) to obtain the first electromagnetic spectra in the wavelength range of a few centimeters to 0.080 millimeters.

In 1934 Arkadiev developed a photographic plate sensitive to centimeter-length waves. Between 1934 and 1936 he thoroughly developed the theory of the behavior of ferromagnetic conductors in rapidly changing fields. In this work he introduced the concept of "magnetic viscosity," which proved to be exceedingly fruitful. He named the realm of these phenomena "magnetodynamics." In addition, he developed the theory of the magnetization and demagnetization of bodies with various shapes.

Among Arkadiev's other work was his 1913 study in which it was first proved that the diffraction of light can be observed on large objects as well as on small ones. He also was the first to construct a high-tension impulse generator, subsequently called a lightning generator (1925). Finally, in 1947 Arkadiev was the first to perfect a refined demonstration experiment—the levitation of a small permanent magnet above a superconducting dish.

BIBLIOGRAPHY

Arkadiev's writings are in his *Izbrannye trudy* ("Selected Works"; Moscow, 1961).

Articles on Arkadiev are N. N. Malov, "Vladimir Konstantinovitch Arkad'ev," in *Uspekhi fizicheskikh nauk* ("Successes of the Physical Sciences"), **52**, no. 3 (1954), 459–469; E. I. Miklashevskaja, "Kratkij ocherk nauchnoj i pedagogicheskoj dejatel'nosti V. K. Arkad'ev" ("Short Essay on the Scientific and Pedagogical Career of V. K. Arkadiev"), in *Izbrannye trudy,* pp. 11–16; and B. A. Vvedensky and N. N. Malov, "O nauchnom znachenii rabot V. K. Arkad'eva" ("On the Scientific Significance of V. K. Arkadiev's Work"), *ibid.,* pp. 5–10.

J. G. Dorfman

ARKELL, WILLIAM JOSCELYN (*b.* Highworth, Wiltshire, England, 9 June 1904; *d.* Cambridge, England, 18 April 1958), *geology, paleontology.*

Arkell was the youngest of seven children of James Arkell, a partner in the family brewery at Kingsdown, Wiltshire, and of Laura Jane Rixon, daughter of a London solicitor. He married Ruby Lilian Percival of Boscombe, Hampshire, in 1929; they had three sons.

After a boarding-school education, he entered New College, Oxford, in 1922. He graduated in geology three years later and was awarded the Burdett-Coutts research scholarship. After appointment as a lecturer at New College in 1929, he was elected a senior research fellow of the college from 1933 to 1940.

Arkell became a principal in the Ministry of War Transport in 1941. In 1943 he contracted a serious chest illness, and for several years was unable to undertake strenuous activity. In 1947 he was elected a fellow of the Royal Society of London and became a senior research fellow at Trinity College, Cambridge. With improved health he was once more able to travel abroad, and his studies on the Jurassic ammonites and stratigraphy were intensified. In August 1956, Arkell suffered a stroke that left him severely paralyzed. Thereafter he was confined to his house, although he eventually managed, with the cooperation of Cambridge friends, to resume his writing.

Arkell received the Mary Clark Thompson Gold Medal of the National Academy of Sciences, Washington, in 1944; the Lyell Medal of the Geological Society of London in 1949; and the von Buch Medal of the German Geological Society in 1953. He was an honorary member or correspondent of the Linnean Society of Normandy; the Geological Societies of France, Germany, and Egypt; and the Paleontological Society of America.

Arkell's first research culminated in the publication of two works: "The Corallian Rocks of Oxford, Berkshire, and North Wiltshire" (1927) and "A Monograph of British Corallian Lamellibranchia." He had spent his youth among these rocks and so had come to know them well, but he also felt that these strata merited attention because in England they had been neglected, in comparison with the Middle and Lower Jurassic, where other workers had long been active. During the following years he extended his studies to other parts of the Jurassic, in "The Stratigraphical Distribution of the Cornbrash" (1928, 1932), and to the Continent, in "A Comparison Between the Jurassic Rocks of the Calvados Coast and Those of Southern England" (1930). Field study and museum visits in Europe greatly increased his knowledge of Jurassic rocks and their fossil fauna.

Early in his career Arkell conceived the plan of revising Albert Oppel's *Die Juraformation Englands, Frankreichs und des sudwestlichen Deutschlands* (1856–1858) and of extending its coverage throughout Europe and the rest of the world. Seen in this perspective, Arkell's book *The Jurassic System in Great Britain* (1933) was the first stage of a larger program. This work, published before he was thirty, is arranged in chapters under the traditional formation names. Arkell stated his difficulties in translating these "haphazard terms" into the systematized stage names that had evolved in Europe, and concluded that the problems involved in extended correlation were beyond solution by any one man.

Nevertheless, Arkell included a searching analysis of the means whereby the Jurassic rocks might be more precisely dated and correlated over wider areas. In particular he discussed the contribution of Alcide

d'Orbigny, whose concept of "stages"—a systematic classification based on a combination of paleontology and stratigraphy—Arkell later revived in Great Britain as the basis of Jurassic classification. He then dealt with Oppel's ideas aimed at providing a more detailed biostratigraphical time scale, with units that could be recognized independently of local lithological considerations. These "zones" were documented by guide fossils, which were characteristic of the beds they defined, and they enabled correlation planes to be established over considerable distances.

In the years following, Arkell made numerous contributions to the corpus of knowledge of Jurassic stratigraphy, and by detailed classification of the ammonites of the Middle and Upper Jurassic he gradually stabilized many stratigraphically significant zonal assemblages that had been in doubt or little known. Over a period of twenty years the Palaeontographical Society published his monographs on the Corallian (Upper Oxfordian) and Bathonian ammonites. In 1946 Arkell published a paper, "Standard of the European Jurassic," advocating a commission to formulate a code of rules for stratigraphical nomenclature analogous to that which had brought order into zoological terminology.

He himself put forward a draft code and showed how it could be applied to the European, including central Russian, Jurassic. Stages and zones were precisely defined, the latter being characterized by particular ammonite assemblages. Arkell further asserted that since the limits of usefulness of species are circumscribed by the ascertained facts of their distribution, there must be separate zonal tables for each faunal province. His proposals were widely accepted as a suitable framework and were applied by investigators into Jurassic stratigraphy in many countries.

In 1956 Arkell published *Jurassic Geology of the World*, designed as a guide to the Jurassic of particular areas and to each individual stage over the whole world. In it he brought together and reviewed critically the information dispersed throughout the enormous literature on the world's Jurassic stratigraphy; for the first time, a comprehensive picture began to emerge, forming the framework for further elaboration. By way of introduction, he set down his final judgment on the principles of stratigraphical classification, especially on the basic notion of the zone. Arkell used the concept of a zone as being any bed, stratum, or formation deposited in any part of the world that could be recognized to be of a particular age on the strength of the fossils it contained. For the Jurassic, the ammonites are most frequently used as the significant fauna. He also denied any necessity to construct a parallel terminology to express the time units to which the zones would correspond.

Whatever modifications may subsequently have emerged from theoretical or philosophical refinement of this definition of zone as a unit measure of strata defined by the special fauna as the time factor, and of stage as the grouping of zones capable of recognition over wide areas, Arkell's synthesis in *Jurassic Geology of the World* must stand as a unique contribution. By this work he brought about a vigorous revival of interest in the principles of stratigraphical classification, which has led to numerous proposals for a basic set of rules applicable to all geological systems and periods. In 1957 Arkell contributed the section on the Jurassic Ammonoidea to the *Treatise on Invertebrate Paleontology;* this section is to a large extent complementary to his *Jurassic Geology* and is in itself a major contribution to paleontology.

Other interests included the elucidation of structural and tectonic problems within the Jurassic. In 1936 Arkell published "Analysis of the Mesozoic and Cainozoic Folding in England," which was an important assessment of knowledge in this field. From 1926 to 1929, he and K. S. Sandford investigated the Pliocene and Pleistocene deposits of the Nile Valley and Red Sea coast of Egypt. The work was organized by the Oriental Institute of the University of Chicago, and four monographs were published under Sandford and Arkell's joint authorship.

BIBLIOGRAPHY

I. ORIGINAL WORKS. A complete list of Arkell's publications, amounting to more than 200 items, together with a memoir of his life and a portrait, is given in *Biographical Memoirs of Fellows of the Royal Society,* **4** (1958), 1–14. Details of those mentioned in the text are "The Corallian Rocks of Oxford, Berkshire, and North Wiltshire," in *Philosophical Transactions of the Royal Society,* **B216** (1927), 67–181; "A Monograph of British Corallian Lamellibranchia," in *Palaeontographical Society Monographs* (1929–1937); "A Comparison Between the Jurassic Rocks of the Calvados Coast and Those of Southern England," in *Proceedings of the Geologists' Association* (London), **41** (1930), 396–411; *The Jurassic System in Great Britain* (Oxford, 1933); "A Monograph on the Ammonites of the English Corallian Beds," in *Palaeontographical Society Monographs* (London, 1935–1948); "Analysis of the Mesozoic and Cainozoic Folding in England," in *Report of the 16th International Geological Congress,* **2** (Washington, D.C., 1936), 937–952; "Standard of the European Jurassic," in *Bulletin of the Geological Society of America,* **57** (1946), 1–34; "A Monograph of the English Bathonian Ammonites," in *Palaeontographical Society Monographs* (London,

1951–1958); and *Jurassic Geology of the World* (Edinburgh–London, 1956).

Works written with collaborators include the following: with J. A. Douglas, "The Stratigraphical Distribution of the Cornbrash. I. The South-western Area," in *Quarterly Journal of the Geological Society of London,* **84** (1928), 117–178; with K. S. Sandford: "Palaeolithic Man and the Nile-Faiyum Divide," in *Oriental Institute Publications* (Chicago), **10** (1929); "Palaeolithic Man and the Nile Valley in Nubia and Upper Egypt," *ibid.,* **17** (1933); "Paleolithic Man and the Nile Valley in Upper and Middle Egypt," *ibid.,* **18** (1934); "Paleolithic Man and the Nile Valley in Lower Egypt, With Some Notes Upon a Part of the Red Sea Littoral," *ibid.,* **46** (1939).

Arkell's library and manuscripts are deposited in the University Museum, Oxford, and his fossil collections are divided between Oxford and the Sedgwick Museum, Cambridge.

II. SECONDARY LITERATURE. See Alcide d'Orbigny, *Paléontologie française. Terrains jurassiques* (1842–1849); and Albert Oppel, *Die Juraformation Englands, Frankreichs und des sudwestlichen Deutschlands* (1856–1858).

J. M. EDMONDS

ARMSTRONG, EDWARD FRANKLAND (*b.* London, England, 5 September 1878; *d.* London, 14 December 1945), *chemistry.*

The eldest son of Henry Edward Armstrong, a fellow of the Royal Society, Edward went to Germany in 1898 for training in organic chemistry with Claisen and in physical chemistry with van't Hoff at the University of Berlin, where he received the Ph.D. in 1901. He conducted research with Emil Fischer, work that aroused his lifelong interest in carbohydrates. He returned to London to continue work on disaccharides, glucosides, and enzymes at Central Technical College, for which he received the first D.Sc. by research of the University of London.

In 1905 Armstrong entered the chemical industry but continued with his research; his monograph on carbohydrates—one of the first—appeared in 1910. He was intimately involved in the negotiations that resulted in the formation of Imperial Chemical Industries, Ltd., one of the largest industrial concerns in Great Britain. During World War I, he and associates solved a problem of great wartime importance—the large-scale catalytic production of acetic acid and acetone from ethyl alcohol. This was an important application of his extensive research on heterogeneous catalysis. During World War II, Armstrong served as scientific adviser to several important governmental agencies. At his untimely death he was a principal adviser to the British delegates and a delegate to the founding conference of UNESCO. He was instrumental in ensuring the inclusion of science in the programs of this organization.

From 1928 Armstrong was a consultant to the chemical industry. He was elected a fellow of the Royal Society of London and was active in its affairs.

BIBLIOGRAPHY

Among Armstrong's writings are *The Simple Carbohydrates and Glucosides* (London, 1910; 1912; 1919; 1924; 1935); a series of 13 papers, "Catalytic Actions at Solid Surfaces," in *Proceedings of the Royal Society* (1920–1925); a paper on heterogeneous catalysis, in *Proceedings of the Royal Society*, **97A** (1920), 259–264; *Chemistry in the 20th Century* (London, 1924); *The Glycosides* (London, 1931), written with his son Kenneth; "The Chemistry of the Carbohydrates and the Glycosides," in *Annual Review of Biochemistry*, **7** (1938), 51; and *Raw Materials From the Sea* (Leicester, 1944), written with W. MacKenzie Miall.

A discussion of Armstrong's career and of the significance of his research and of his lifework in industry and public service is in C. S. Gibson and T. P. Hildreth, "Edward Frankland Armstrong," in *Obituary Notices of Fellows of the Royal Society*, **5** (1948), 619. This includes references to nearly 100 publications.

CONRAD E. RONNEBERG

ARMSTRONG, EDWIN HOWARD (*b.* New York, N.Y., 18 December 1890; *d.* New York, 1 February 1954), *radio engineering*.

Armstrong's father, John Armstrong, was a publisher who became vice-president in charge of the American branch of Oxford University Press; his mother, Emily Smith, graduated from Hunter College and taught for ten years in New York public schools before her marriage in 1888. When Armstrong was twelve, the family moved to Yonkers, New York, where he attended high school and became interested in radiotelegraphy. He entered Columbia University at nineteen and studied electrical engineering under Michael Idvorsky Pupin, the inventor of the Pupin loading coil used in long-distance telegraphy and telephony, graduating in 1913.

While still an undergraduate, Armstrong made the first of his many inventions, one of four that proved to be particularly significant: the triode feedback (regenerative) circuit. That invention, and the negative-bias grid circuit invented by Frederick Löwenstein, ultimately led to wide utilization of the as yet little-exploited triode (invented in 1906 by Lee De Forest), but Armstrong became embroiled in patent litigation and received only modest royalties.

In 1917, after serving as an assistant at Columbia for some years, Armstrong became a U.S. Army Signal Corps officer when the United States entered World War I. He was sent to France and while there developed his second important invention, the superheterodyne circuit, an improvement on the heterodyne circuit that was invented in 1905 by Reginald Aubrey Fessenden. In the heterodyne circuit, the received signal is mixed with a locally generated signal to produce an audible "beat" note at a frequency equal to the difference between those of the two signals; Armstrong's method, which greatly improved the sensitivity and stability of radio receivers, extended the technique to much higher frequencies and shifted the beat note above the audible range.

Upon returning to America, Armstrong was once again beset by patent interference proceedings, although his personal fortunes took a turn for the better: he sold his feedback and superheterodyne patents to Westinghouse Electric & Manufacturing Company (retaining royalty-earning licensing rights for the use of amateurs); he resumed his position at Columbia University; and he married Marion MacInnis, secretary to David Sarnoff, then general manager of the Radio Corporation of America.

In 1921 Armstrong made his third important discovery, superregeneration—a method of overcoming the regenerative receiver's principal limitation, the tendency to burst into oscillations just as the point of maximum amplification was reached. RCA purchased the patent, but it did not yield the company much in royalties, since it was unsuited for broadcast receivers; it did not come into its own until special applications were developed many years later. However, RCA profited greatly from the "superhet," to which it had acquired the rights through a cross license with Westinghouse. Armstrong found himself a millionaire.

The next decade of his life was marred by the long battle with De Forest over the feedback patents. The case was taken to the U.S. Supreme Court but Armstrong lost on a legal technicality. Before that decision had been handed down, however, Armstrong had completed and patented his greatest invention, frequency modulation (FM). Once again he was beset by difficulties: the U.S. radio industry resisted the introduction of FM broadcasting, FM production was interrupted when the United States entered World War II, and the Federal Communications Commission dealt FM a stunning blow in 1945 when it relegated it to a new frequency band and put restrictions on transmitter power, thus making over fifty existing transmitters and half a million receivers obsolete. At the same time, FM came to be widely used in military and other mobile communications, radar, telemetering, and the audio portion of television; but widespread adoption of FM broadcasting came only after Armstrong's death. Exhausted by a five-year suit for patent infringement against RCA and almost destitute as his FM patents began to expire, Armstrong committed suicide in 1954.

He had received many honors, including the highest awards of the two U.S. electrical engineering societies, the American Institute of Electrical Engineers (Edison Medal, 1942) and the Institute of Radio Engineers (Medal of Honor, 1918, reaffirmed in 1934 when he tried to return it after losing the legal fight against De Forest); the Franklin Medal (1941); and, for his war work, the U.S. Medal for Merit (1945). No inventor contributed more profoundly to the art of electronic communication. Armstrong is one of the two dozen honored in the Pantheon of the International Telecommunications Union in Geneva.

BIBLIOGRAPHY

Armstrong received forty-two patents and wrote twenty-six papers; the papers are listed in his biography by Lawrence Lessing, *Man of High Fidelity* (Philadelphia, 1956). See also obituaries in *New York Times* (2 Feb. 1954), p. 27; and in *Proceedings of the Institute of Radio Engineers,* **42** (1954), 635.

CHARLES SÜSSKIND

ARMSTRONG, HENRY EDWARD (*b.* Lewisham, London, England, 6 May 1848; *d.* Lewisham, 13 July 1937), *chemistry, science education.*

Armstrong entered the Royal College of Chemistry, London, in 1865. Here he studied with A. W. Hofmann, then in his last year in Britain, and John Tyndal, whose lectures greatly impressed him. In his third year he started research under Sir Edward Frankland, at whose suggestion he went to Leipzig to study under Hermann Kolbe. He was awarded the Ph.D. in 1870 and returned to London in that year to start his lifelong career in research and teaching. Armstrong's first positions were largely "bread and butter" ones: teaching combined with research, often under primitive conditions. His first paper to the Chemical Society of London, "On the Formation of Sulpho-Acids," was submitted in 1870. In 1884 he became professor of chemistry at the then new Central Institution, South Kensington, which merged with the Imperial College of Science and Technology in 1907. He retired from teaching in 1911.

Armstrong had a very vigorous mind and independent spirit, which, coupled with great breadth of interests and personal drive, enabled him to assume leadership in the development of technological chemistry in Great Britain. He became a fellow of the Royal Society of London when only twenty-eight, and he served the Chemical Society of London as secretary, president, and vice-president. He had an extraordinary impact on his contemporaries, his accomplishments and influence being such that between 1890 and 1935 he was regarded as the doyen of British chemists. Armstrong's mission was to advance chemistry in order to improve society, and he carried it out with the fervor of a prophet. His published papers, many written with students, totaled more than 250. He did pioneer work on the structure of benzene, devising the centric formula independently of Baeyer, and his work with W. P. Wynne on the structure and reactions of naphthalene helped establish the dye industry. Many of his papers were concerned with the quinonoid theory for the color of organic compounds, the reactions of camphor and terpenes, and the mechanism of enzyme reactions.

Armstrong's views were often prophetic. He was among the first to base instruction and writing in chemistry upon Mendeleev's periodic table, and he early emphasized that molecules must have spatial configurations that determine crystal structures. His researches in crystallography were significant and antedated X-ray diffraction methods.

The vigor of Armstrong's advocacy of unorthodox views at times caused intense controversy and often made enemies. On the basis of the unique properties of water and the hypothetical structure of water molecules, he castigated and ridiculed the theories of Arrhenius, van't Hoff, and Ostwald because they ignored the solvent and complex character of water. He believed that the process of solution of electrolytes had to consider and include the unique properties of water. Thus, his ideas and researches were an important step toward the present concepts of ion complexes and ion atmospheres in aqueous solutions of electrolytes.

He was the first to devise curricula to relate chemistry and engineering, and he came to be regarded as the father of chemical engineering. Armstrong helped organize the Education Section of the British Association for the Advancement of Science and served as its president in 1902. He vigorously opposed the didactic method of teaching and championed the view that the best method of teaching science was experimental. Appointed to the Committee on Management for Rothamsted Experiment Station at Harpenden, Armstrong served the rest of his life as member, vice-chairman, or chairman of this committee. Here his guidance and vision in the application of chemistry and scientific procedures to the problems of agriculture did much to advance this basic economic activity.

In 1877 Armstrong married Frances Louisa Lavers of Plumstead, Kent, who bore him four sons and three daughters. His eldest son, Edward Frankland, achieved prominence as an industrial chemist.

During his long career Armstrong's zeal and pioneering spirit brought him acknowledged leadership in many areas of organic research, in science and technical education, and in agriculture. But in the minds of his colleagues his chief contribution to chemistry was the teaching and inspiration of his many distinguished pupils.

BIBLIOGRAPHY

I. ORIGINAL WORKS. Armstrong's most important books are *The Teaching of the Scientific Method and Other Papers on Education* (London, 1903, 1925) and *The Art and Principles of Chemistry* (London, 1927). His first paper to the Chemical Society was "On the Formation of Sulpho-Acids," in *Journal of the Chemical Society of London* (1871), p. 173.

II. SECONDARY LITERATURE. J. Vargas Eyre, *Henry Edward Armstrong 1848–1937* (London, 1958), a biography by a former colleague and intimate, gives titles and journal sources for many of Armstrong's papers on benzene, naphthalene, crystallography, enzymes, and electrolytes. Many of these papers were influential. For example, his sixty papers on laws of substitution in naphthalene materially helped to establish naphthalene chemistry and the dye industry. A long and very detailed obituary notice is "Henry Edward Armstrong," in *Journal of the Chemical Society of London* (1940), p. 1418.

CONRAD E. RONNEBERG

ARNALD OF VILLANOVA (*b.* Aragon, Spain, *ca.* 1240; *d.* at sea off Genoa, Italy, 6 September 1311), *medical sciences.*

The family of Arnald of Villanova may originally have been Provençal, but he himself was Catalan by birth, probably from Valencia. We have no information about his parents (except that they may have been converted Jews) and little about his professional education. It is certain only that he was a student at Montpellier *ca.* 1260; the ascription to him of training under John Casamicciola at the University of Naples between 1267 and 1276 rests on the doubtful authenticity of the *Breviarium practice* traditionally included among his works. By 1281 Arnald had become physician to Peter III of Aragon, and he subsequently served Peter's son and heir, Alfonso III, in the same capacity. The death of the latter in 1291

and the succession of his brother as James II brought Arnald further into prominence.

By 1291 Arnald had taken up residence in Montpellier as a medical master at its newly chartered *studium generale,* but he was repeatedly called back to Spain during the next several years for professional consultations. Moreover, he seems to have been able to win considerable support from the royal family for his developing religious views. While Arnald's teaching and writings at Montpellier were of the first importance in establishing the content of scholastic medicine there, his own commitment to medicine was gradually being replaced by a concern for theological matters. He studied for six months with the Montpellier Dominicans, but he seems to have been most strongly influenced by the Joachimite Peter John Olivi and the spiritual-Franciscans, a group of rigorists within that order. By 1299 Arnald had completed a number of mystical, prophetic works—notably the *De adventu antichristi* (begun in 1288), which announced that the world would end and the Antichrist appear in 1378 and insisted upon the need for a drastic reform of the Church. When in 1299 Arnald was sent by James II on a diplomatic mission to Philip IV of France, he took the opportunity to explain his beliefs to the Parisian theologians. The *De adventu* was condemned, however, and Arnald was spared imprisonment only through the intervention of Philip's minister, William of Nogaret.

Arnald spent the next five years traveling between Provence and Rome. He continued to write while defending himself against his critics in the Church. Two popes, Boniface VIII and Benedict XI, were willing to tolerate him as a physician but were not receptive to his theology. With the accession in 1305 of Clement V, Arnald's friend for some time, he finally had both papal and royal patronage and was repeatedly given the opportunity to express his views, in which his concern for Church reform soon became predominant over his interest in eschatology. Then, in 1309, before a large assembly at Avignon, Arnald made statements that called James II's orthodoxy into question, thereby losing the king's favor. Thereafter Arnald attached himself to James's younger brother, Frederick I of Trinacria, a much more willing instrument for Arnald's plans for the reform of Christian society. He died while on a mission for Frederick.

In his medical practice Arnald seems to have been remarkable only for his success. There are many striking testimonies to his abilities, two of which are James II's demand that Arnald come from Montpellier to attend the queen's second pregnancy (1297) and Boniface VIII's delight at his relief from the stone (1301). From the various regimens and practical guides composed for such patients—for example, the *Regimen sanitatis* written for James II in 1307, and the *Parabole medicationis* dedicated to Philip IV (1300)—it is clear that his methods continued to be those in which he had been trained at Montpellier. His diagnostic and therapeutic principles appear quite conservative, with none of the dependence on uroscopy or extreme polypharmacy that marks the work of some of his colleagues. As a practicing physician, Arnald was evidently committed to experience rather than to theory or to authority.

Yet Arnald was also the principal figure in Montpellier's fusion of the Western empirical tradition with the systematic medical philosophy of the Greeks and Arabs. He was well placed for this role. Living in Valencia shortly after its reconquest by James I, it was natural for him to learn Arabic, and at the Aragonese court in the 1280's he translated, from Arabic into Latin, Avicenna's *De viribus cordis,* which subsequently proved immensely popular, and Galen's *De rigore*; he also translated a work on drugs by Albuzale and one on regimen by Avenzoar. There is very little in the choice of these texts to suggest any interest in the natural-philosophical aspects of medicine, but Arnald's stay at Montpellier seems to have deepened his medical as well as his theological interests. One indication of this is the unusual number of different Hippocratic and Galenic works on which he is known to have lectured. The classical authors had been commented upon by earlier masters at Montpellier, notably Cardinalis, but apparently not so extensively; they had preferred to pursue the more purely empirical methods of such moderns as Gilbert the Englishman and Walter Agilon. By the fourteenth century, however, Arnald's approach had become more widespread. The program of the university was unsettled in the 1290's, and Arnald certainly encouraged its subsequent Scholasticism by his example; moreover, it was his advice that shaped the papal bull of 8 September 1309, which regularized medical education at Montpellier, defining a set of fifteen Greek and Arab texts as the basis for future study at the school.

Arnald himself went beyond commentary to try to develop a coherent, systematic science of medicine on the Galenic foundations. It is possible to trace the development of his ideas in a continuous series of works datable to Montpellier and the 1290's, bound together by cross-references and a remarkably consistent interest in the philosophical aspects of medicine. In the earlier works (e.g., the *De intentione medicorum*) he is primarily concerned to defend pragmatically the presence of a rational element in medicine. Against his colleagues, Arnald holds that the

physician can and should draw on theory insofar as it is meaningful to his practice; he requires simply that it save the medical phenomena, not that it be absolutely, philosophically true. Arnald repeatedly makes an analogy with the astronomers' epicycles.

Gradually, however, Arnald became more preoccupied with problems of philosophy apart from any practical applications, and the later Montpellier treatises are highly technical discussions of sophisticated medical theory. They most often treat the *complexio* (the set of sensible qualities, principally hot and cold, that characterizes a man's state of health), which provides Arnald with a basis for both medical practice and theory. In the *Aphorismi de gradibus* he develops this qualitative medicine most fully: drawing on material from al-Kindī and Averroës, he establishes a mathematical pharmacy upon the empirical law that qualitative intensity increases arithmetically with a geometric increase in the ratio of the opposing forces that produce it. For medicinal qualities, the law would have the form (in modern terms) intensity $= \log_2$ hot/cold. The *Aphorismi* continued in use at Montpellier for over fifty years, and may have provided the stimulus for Bradwardine's Law, the Merton College dynamic rule that, in local motion, velocity $\sim\log_n$ force/resistance. But this success was exceptional. For the most part, Arnald's writings—even his final synthesis of medical theory, the *Speculum medicine*—were too elaborate and too abstract to have any vogue among professional physicians.

Arnald's unusual attention to philosophical medicine coincided with the development of his theological position, and one concern may well have inspired the other. Certainly he should not be understood as a simple rationalist in medical matters. Although Boniface VIII once told him, "Occupy yourself with medicine and not with theology and we will honor you," Arnald did not find it easy to make such a distinction, for to him the two disciplines were continually overlapping in subject matter. The outstanding instance of this is his *De esu carnium* (1304), a defense of the Carthusians' abstension from meat, which gave equal weight to citations from Hippocrates and from St. Paul. More profoundly, his epistemology shows the same blurred dualism, for it is the mystical element in Arnald's thought that restricts his rationalism. He believes that while all physical events may be natural and ordered, their causes are not all easily and directly accessible to the understanding and men wise in these more occult matters will prove to have been guided by chance—or, like himself, by divine illumination. This is the spirit behind his scientific writings on astrology and oneiromancy, as well as his prophetic works; it is also the rationale behind his acceptance of medical intuition as a tool complementary to reason, which was quite in agreement with the view of his more orthodox colleagues that medicine is an art as much as (or more than) a science. The scholastic writings by themselves do not make plain this complexity of his scientific thought, the coexistence in it of rationalism and mysticism.

Arnald's heterodoxy made a great impression upon the fourteenth and fifteenth centuries, and during that period his name became associated with a number of alchemical texts. Some of these, notably the *Flos florum* and the *Rosarius philosophorum,* were until recently accepted as genuine. At present the authenticity of all appears doubtful, the more so because Arnald himself considered alchemists "ignorant" and "foolish." More recently the empirical element in Arnald's medicine has been emphasized by critics, who cite the *Breviarium practice* in support. But the authenticity of this work is dubious as well. From internal evidence, it seems that it must have been written in the first years of the fourteenth century, at just the time when Arnald was deeply concerned with medical theory and theology, and there is no hint of these subjects in the *Breviarium.* The most famous of the medical writings once attributed to him, the *Commentum super regimen sanitatis Salernitanum,* is now also thought to be apocryphal.

BIBLIOGRAPHY

I. ORIGINAL WORKS. Virtually all of Arnald's Latin medical works were collected in various sixteenth-century editions of his *Opera* (Lyons, 1504, 1509, 1520, 1532; Venice, 1505, 1527; Basel, 1585). In general, the later editions are the more complete, although the collection entitled *Praxis medicinalis* (Lyons, 1586) omits his discussions of medical theory. None of these works has been published in a modern edition, but one not printed earlier, *De conservatione visus,* was published by P. Pansier in *Collectio ophtalmologica veterum auctorum,* I (Paris, 1903), 1–25.

Arnald's translations from the Arabic are not included in his *Opera.* His versions of Aristotle and Galen were included with other writings of the same authors in editions of the fifteenth and sixteenth centuries; his translation of the *De medicinis simplicibus* attributed to Albuzale or Albumasar (MSS Paris, Univ. 128, 165–168; Erfurt, Amplon. F. 237, 63–66; Q. 395, 138–160) and the translation of the *De regimine sanitatis* of Avenzoar less certainly ascribed to him (MS Oxford, Corpus Christi 177, 261–265v; cf. MS Paris, Univ. 131, 54v–59v [among others], where it is attributed to Profatius) have not been published. He has also been credited with two translations that are demonstrably not by him: Costa ben Luca's *De physicis ligaturis* (which exists in a twelfth-century manuscript, MS Brit. Mus. Add. 22719, fol. 200v) and al-Kindī's treatise *De gradibus* (for which Arnald's own *Aphorismi de gradibus* were mistaken).

A good bibliographical guide to the printed and manuscript sources is Juan Antonio Paniagua Arellano, "La obra médica de Arnau de Vilanova. Estudio 1°: Introducción y fuentes," in *Estudios y notas sobre Arnau de Vilanova* (Madrid, 1962), pp. 1–51, reprinted from *Archivo iberoamericano de historia de la medicina y antropología médica,* **11** (1959), 351–402.

A satisfactory bibliography of Arnald's theological writings, indicating those that have been published in whole or in part, will be found in Joaquín Carreras y Artau, "Les obres teològiques d'Arnau de Vilanova," in *Analecta sacra Tarraconensia,* **12** (1936), 217–231, although this is by now somewhat incomplete. The article by Paniagua cited above lists a number of more recent editions. It might also be mentioned that MSS Vat. Borg. 205 and Vat. Lat. 3824 are collections of Arnald's theological writings made at his own direction, in 1302 and 1305, respectively; the former manuscript may be annotated in Arnald's hand (see Anneliese Maier, "Handschriftliches zu Arnaldus de Villanova und Petrus Johannis Olivi," in *Analecta sacra Tarraconensia,* **21** [1948], 53–74).

Arnald's Catalan writings have been edited by Miguel Batllori as *Obres catalanes* (Barcelona, 1947) in two volumes, the first of which contains religious texts and the second, medical texts.

II. SECONDARY LITERATURE. The Arnaldian literature up to 1947 is thoroughly covered in the bibliographies of the volumes edited by Batllori, cited above; of this early material, Paul Diepgen's articles go most deeply into Arnald's scientific work. Since then a few other important studies of Arnald's science and medicine have appeared: Michael McVaugh, "Arnald of Villanova and Bradwardine's Law," in *Isis,* **58** (1967), 56–64. Juan Antonio Paniagua Arellano, "La patología general en la obra de Arnaldo de Vilanova," in *Archivos iberoamericanos de historia de la medicina,* **1** (1948), 49–119; Jacques Payen, "*Flos florum* et *Semita semite,* deux traités d'alchimie attribués à Arnaud de Villeneuve," in *Revue d'histoire des sciences,* **12** (1959), 288–300; René Verrier, *Études sur Arnaud de Villeneuve,* 2 vols. (Leiden, 1947–1949). The many biographical and theological studies that have appeared since 1947 can best be found by following up the references given in Miguel Batllori, "Dos nous escrits espirituals d'Arnau de Vilanova," in *Analecta sacra Tarraconensia,* **28** (1955), 45–70; or in Joaquín Carreras y Artau, *Relaciones de Arnau de Vilanova con los reyes de la casa de Aragón* (Barcelona, 1955).

MICHAEL MCVAUGH

ARNAULD, ANTOINE (*b.* Paris, France, 1612; *d.* Brussels, Belgium, 6 August 1694), *mathematics, linguistics.*

Arnauld was the youngest of the twenty children

of Antoine Arnauld, a lawyer who defended the University of Paris against the Jesuits in 1594. He was ordained a priest and received the doctorate in theology in 1641, and entered the Sorbonne in 1643, after the death of Richelieu. In 1656 he was expelled from the Sorbonne for his Jansenist views, and spent a good part of the rest of his life in more or less violent theological dispute. He died in self-imposed exile.

Although in many of his nontheological writings Arnauld is identified with the Port-Royal school, his voluminous correspondence—with Descartes and Leibniz, among others—bears witness to his own influence and acumen. His philosophical contributions are to be found in his objections to Descartes's *Méditations,* in his dispute with Malebranche, and in the *Port-Royal Logic,* which he wrote with Pierre Nicole. The latter, a text developed from Descartes's *Regulae,* elaborates the theory of "clear and distinct" ideas and gives the first account of Pascal's *Méthode.* It had an enormous influence as a textbook until comparatively recent times.

The profound influence of the *Regulae* is shown in both the *Logic* and the *Port-Royal Grammar,* where it is assumed that linguistic and mental processes are virtually identical, that language is thus to be studied in its "inner" and "outer" aspects. This point of view underlies the project for a universal grammar and the notion of the "transparency" of language: mental processes are common to all human beings, although there are many languages. The *Grammar* and the *Logic* are based on a common analysis of signs that has brought the Port-Royal school to the attention of modern linguistic theorists, who see in it an anticipation of their own point of view.

The *Élémens* (1667) undertakes a reworking and reordering of the Euclidean theorems in the light of the contemporary literature (in which he was widely read) and Pascal's influence. It bases its claim to originality and influence on the new order in which the theorems, many of them adapted from contemporary sources, are arranged. As mathematics, it is characterized by the mastery of the contemporary literature and by its clear and fresh exposition; its virtues are pedagogical. It is interesting to compare Arnauld's order of theorems with such recent ones as that of Hilbert and Forder, whose aims are quite different. If Arnauld's pedagogical concerns are insufficiently appreciated, it may be because the role of what are properly pedagogical concerns in the habits and "methods" of modern science is insufficiently understood: its preoccupation with clarity and procedure, with formal exercises and notation, and the use of these as instruments of research.

BIBLIOGRAPHY

I. ORIGINAL WORKS. Arnauld's writings include *Grammaire générale et raisonnée* (*Port-Royal Grammar;* Paris, 1660); *La logique ou l'art de penser* (*Port-Royal Logic;* Paris, 1662), crit. ed. by P. Clair and F. Girbal (Paris, 1965); and *Nouveaux élémens de géométrie* (Paris, 1667). Collections of his work are *Oeuvres,* 45 vols. (Lausanne, 1775–1783); and *Oeuvres philosophiques,* J. Simon and C. Jourdain, eds. (Paris, 1893).

II. SECONDARY LITERATURE. Works on Arnauld are K. Bopp, "Arnauld als Mathematiker," in *Abhandlundgen zur Geschichte der mathematischen Wissenschaften,* **14** (1902), and "Drei Untersuchungen zur Geschichte der Mathematik," in *Schriften der Strassburger wissenschaftlichen Gesellschaft in Heidelberg,* no. 10 (1929), pt. 2, 5–18; H. L. Brekle, "Semiotik und linguistische Semantik in Port-Royal," in *Indogermanische Forschungen,* **69** (1964), 103–121; Noam Chomsky, *Cartesian Linguistics* (Cambridge, Mass., 1966); J. Coolidge, *The Mathematics of Great Amateurs* (Oxford, 1949); Leibniz, "Remarques sur les nouveaux Élémens de Géométrie Antoine Arnaulds," in *Die Leibniz-Handschriften zu Hannover,* E. Bodemann, ed., I (Hannover, 1895), no. 21, 287; and H. Scholz, "Pascals Forderungen an die mathematische Methode," in *Festschrift Andreas Speiser* (1945).

HENRY NATHAN

ARNOLD, HAROLD DE FOREST (*b.* Woodstock, Connecticut, 3 September 1883; *d.* Summit, New Jersey, 10 July 1933), *electronics.*

Arnold received bachelor's and master's degrees from Wesleyan University in Connecticut and the doctorate from the University of Chicago (1911), where he studied physics under Robert Andrews Millikan. When the Bell System needed someone to develop repeaters for its projected transcontinental line, Millikan recommended Arnold, who thus became one of the scientists who later laid the foundation of the Bell Telephone Laboratories. Arnold first developed a mercury-arc repeater, but the device saw only limited use before his attention turned to the triode, which had been invented six years earlier by Lee de Forest (no kin). Its operation was still not entirely understood; the inventor himself did not appreciate the need for the highest attainable vacuum. Arnold was among the first to recognize the importance of high vacuum, and quickly developed designs that utilized reliable triodes and thus made long-distance telephony possible for the first time.

After World War I, part of which he spent as a captain in the U.S. Army Signal Corps, Arnold returned to research work for Bell and made a number of important contributions to the development of new magnetic alloys used in sound reproduction and to electroacoustics generally. He was named the first director of research when the Bell Telephone Laboratories were formed in 1925, a post he occupied until his death. He helped to lead that organization to its preeminent position among the industrial laboratories of the world.

BIBLIOGRAPHY

A biography containing extensive quotations from Arnold's writings on the organization of research appears in *Bell Laboratories Record,* **11** (1933), 351–359. See also his obituary in *New York Times* (11 July 1933), p. 17.

CHARLES SÜSSKIND

AROMATARI, GIUSEPPE DEGLI (*b.* Assisi, Italy, 25 March 1587; *d.* Venice, Italy, 16 July 1660), *embryology.*

Aromatari was the son of Favorino Aromatari and Filogenia Paolucci. He was brought up by his paternal uncle, Renier Aromatari, a learned and wealthy physician. He studied philosophy and medicine in Perugia, Montpellier, and Padua, where he attended Fabricius' lectures. After graduating M.D. in 1605, Aromatari remained at Padua until 1610 when he settled at Venice to practice medicine. His fame soon caused him to be requested as personal physician by King James I of England and by Pope Urban VIII; but Aromatari declined. He died of a stone in the urinary bladder and was buried in the Church of Saint Luke in Venice.

Aromatari was famous as a man of letters as well as a physician. From 1609 until 1613, he was involved in a literary debate with the poet Alessandro Tassoni over the work of Petrarch. At an advanced age, he wrote and published an anthology of passages from the classics. Aromatari is remembered today, however, for his hypothesis of the preformation of the germ. He also investigated the so-called permeability of the interventricular septum of the heart; but on this subject no writing exists.

In 1625 Aromatari published at Venice his famous *Epistola de generatione plantarum ex seminibus* ("Letter on the Generation of Plants from Seeds"). Addressed to a friend, Bartholomeo Nanti, the work was only four pages long, but it immediately made Aromatari famous. It was reprinted in 1626 at Frankfurt; and it was included by Richter in his *Epistolae selectae* in 1662, and by Junge in *Opuscula botanico-physica* in 1747.

Aromatari affirmed that the seeds of plants are composed of two parts: a smaller part, the germ, which

contains, in miniature, all parts of the future plant; and a larger part, which is destined as nourishment for the germ and therefore comparable to the yolk of an egg. He was also explicit on two other fundamental points: he denied, absolutely, the spontaneous generation of all living species (animal and vegetable) and postulated that each living kind is born from the seed (plants) or the egg (animals) of the same kind.

In the beginning of the seventeenth century, animal generation was studied in Padua; it is possible that Aromatari was induced to study the problems of generation by reading the works of Volcher Coiter or Ulisse Aldrovandi on generation.

Aromatari was unaware of the sexuality of plants, demonstrated in 1694 by Rudolph Camerarius. Aromatari thought that a single plant, as a hermaphrodite, produced both the *ova* (seeds) and the *semen prolificum* (pollen). However, on the precise problem of embryogenesis, he first advanced the hypothesis of the preformation of the germ: "In the aforesaid seeds, the plant exists already made . . . the plant arises from the seed, but it is not generated in the seed; we think that likely the chick is sketched in the egg, before it is brooded by the hen." Aromatari's priority on the doctrine of the preformation of the germ was acknowledged by William Harvey in his *De generatione animalium.*

Aromatari's hypothesis of germinal preformation became the new idea of the seventeenth century, and was developed later in the famous works of Marcello Malpighi, Jan Swammerdam, and Charles Bonnet, the greatest theorists of preformation in animal generation. Therefore, Aromatari's *Epistola* marks the origin of an idea of great importance in the history of embryology.

BIBLIOGRAPHY

I. ORIGINAL WORKS. Works by Aromatari include *Disputatio de rabie contagiosa, cui praeposita est epistola de generatione plantarum ex seminibus, qua detegitur in vocatis seminibus plantas contineri vere confirmatas, ut dicunt, actu* (Venice, 1625) and *Autori del bel parlare* (Venice, 1643).

II. SECONDARY LITERATURE. Works relating to Aromatari include *Memorie di Giuseppe Aromatari letterato medico e naturalista, pubblicate per cura dell'Accademia Properziana del Subasio* (Assisi, 1887), which contains a portrait engraved on copper, a bibliography, and the complete text of the *Epistola de generatione plantarum*; these bibliographical notes, collected by Leonello Leonelli, were extracted from the paper *Vita dell'Eccellentissimo Gioseffe degli Aromatari* (Venice, 1661), written by Father Giovanni Battista De Fabris; W. Harvey, Exercitatio XI, "Ovum esse primordium commune omnibus animalibus, ut et semina plantarum omnium . . . Ita olim mihi, Venetiis cum essem, Aromatarius, Medicus clarissimus, ostendit," of *De generatione animalium* (London, 1651); R. Herrlinger, *Volcher Coiter* (Nuremberg, 1952), p. 72; J. Rostand, *La formation de l'être: histoire des idées sur la génération* (Paris, 1930), p. 50; and *Esquisse d'une histoire de la biologie* (Paris, 1945), p. 24.

PIETRO FRANCESCHINI

ARONHOLD, SIEGFRIED HEINRICH (*b.* Angerburg, Germany [now Węgorzewo, Poland], 16 July 1819; *d.* Berlin, Germany, 13 March 1884), *mathematics.*

Aronhold attended the Angerburg elementary school and the Gymnasium in Rastenburg (now Kętrzyn, Poland). Following the death of his father, his mother moved to Königsberg, where the boy attended a Gymnasium. He graduated in 1841 and then studied mathematics and natural sciences at the University of Königsberg from 1841 to 1845. Among his teachers were Bessel, Jacobi, Richelot, Hesse, and Franz Neumann. When Jacobi went to Berlin, Aronhold followed him and continued his studies under Dirichlet, Steiner, Gustav Magnus, and Dove. He did not take the state examinations, but in 1851 the University of Königsberg awarded him the *Doctor honoris causa* for his treatise "Über ein neues algebraisches Prinzip" and other studies.

From 1852 to 1854 Aronhold taught at the Artillery and Engineers' School in Berlin and, from 1851, at the Royal Academy of Architecture in Berlin, where he was appointed professor in 1863. In 1860 he joined the Royal Academy for Arts and Crafts, where, when Weierstrass became ill in 1862, he took over the entire teaching schedule. He was appointed professor in 1864. In 1869 Aronhold became a corresponding member of the Academy of Sciences in Göttingen. He was considered an enthusiastic and inspiring teacher, and was held in high esteem everywhere.

Aronhold was particularly attracted by the theory of invariants, which was then the center of mathematical interest, and was the first German to do research in this area. The theory of invariants is not, however, connected with Aronhold alone—others who worked on it were Sylvester, Cayley, and Hesse—but he developed a special method that proved to be extremely successful. In 1863 he collected his ideas in a treatise entitled "Über eine fundamentale Begründung der Invariantentheorie."

In this treatise, Aronhold offers solid proof of his theory, which he had welded into an organic entity. His method refers to functions that remain unchanged under linear substitutions. He stresses the importance of the logical development of a few basic principles so that the reader may find his way through other papers. Aronhold establishes his theory in general and does not derive any specific equations. He derives the concept of invariants from the concept of equivalency for the general linear theory of invariants. Special difficulties arise, of course, if not only general but also special cases are to be considered. His efforts to obtain equations independent of substitution coefficients led to linear partial differential equations of the first order, which also have linear coefficients. These equations, which are characteristic for the theory of invariants, are known as Aronhold's differential equations.

With these equations "Aronhold's process" can be carried out. This process permits the derivation of additional concomitants from one given concomitant. Aronhold investigates the characteristics of these partial differential equations and expands the theory to include the transformation of a system of homogeneous functions, furnishes laws for simultaneous invariants, and investigates contravariants (relevant forms), covariants, functional invariants, and divariants (intermediate forms).

Aronhold stresses that he arrived at his principles as early as 1851, citing his doctoral dissertation and the treatise "Theorie der homogenen Funktionen dritten Grades . . ." (1858). Since the subsequent theory and terminology did not yet exist, he claimed priority.

Before Aronhold developed his theory, he had worked on plane curves. The problem of the nine points of inflection of the third-order plane curve, which had been discovered by Plücker, was brought to completion by Hesse and Aronhold. Aronhold explicitly established the required fourth-degree equation and formulated a theorem on plane curves of the fourth order. Seven straight lines in a plane always determine one, and only one, algebraic curve of the fourth order, in that they are part of their double tangents and that among them there are no three lines whose six tangential points lie on a conic section.

BIBLIOGRAPHY

I. ORIGINAL WORKS. Aronhold's writings are "Zur Theorie der homogenen Funktionen dritten Grades von drei Variabeln," in *Journal für die reine und angewandte Mathematik* (Crelle), **39** (1849); "Bemerkungen über die Auflösung der biquadratischen Gleichung," *ibid.,* **52** (1856), trans. into French as "Remarque sur la résolution des équations biquadratiques," in *Nouvelles annales de mathématiques,* **17** (1858); "Theorie der homogenen Funktionen dritten Grades von drei Veränderlichen," in *Journal für die reine und angewandte Mathematik* (Crelle), **55** (1858); "Algebraische Reduktion des Integrals ∫F(x,y) dx, wo F(x,y) eine beliebige rationale Funktion von x,y bedeutet und zwischen diesen Grössen eine Gleichung dritten Grades von der allgemeinsten Form besteht, auf die Grundform der elliptischen Transzendenten," in *Berliner Monatsberichte* (1861); "Form der Kurve, wonach die Rippe eines T-Konsols zu formen ist," in *Verhandlung der Polytechnischen Gesellschaft* (Berlin), **22** (1861); "Über eine neue algebraische Behandlungsweise der Integrale irrationaler Differentiale von der Form Π (x,y) dx, in welcher Π (x,y) eine beliebige rationale Funktion ist, und zwischen x und y eine allgemeine Gleichung zweiter Ordnung besteht," in *Journal für die reine und angewandte Mathematik* (Crelle), **61** (1862); "Über eine fundamentale Begründung der Invariantentheorie," *ibid.,* **62** (1863); "Über den gegenseitigen Zusammenhang der 28 Doppeltangenten einer allgemeinen Kurve vierten Grades," in *Berliner Monatsberichte* (1864); "Neuer und direkter Beweis eines Fundamentaltheorems der Invariantentheorie," in *Journal für die reine und angewandte Mathematik* (Crelle), **69** (1868); and "Grundzüge der kinetischen Geometrie," in *Verhandlungen des Vereins für Gewerbefleiss,* **52** (1872).

II. SECONDARY LITERATURE. More detailed information on mathematics in Berlin can be found in E. Lampe, *Die reine Mathematik in den Jahren 1884–1899 nebst Aktenstücken zum Leben von Siegfried Aronhold* (Berlin, 1899), pp. 5 ff. For the theory of invariants, see Weitzenböck, *Invariantentheorie* (Groningen, 1923); *Enzyklopädie der mathematische Wissenschaften,* I, pt. 1 (Leipzig, 1898), 323 ff.; Felix Klein, *Vorlesungen über die Entwicklung der Mathematik im 19. Jahrhunderts* (Berlin, 1926–1927), I, 157, 166, 305; II, 161, 195; and Enrico Pascal, *Repertorium der höheren Analysis,* 2nd ed. (Leipzig–Berlin, 1910), ch. 5, pp. 358–420.

HERBERT OETTEL

AROUET, FRANÇOIS-MARIE. See **Voltaire.**

ARREST, HEINRICH LOUIS D' (*b.* Berlin, Germany, 13 August 1822; *d.* Copenhagen, Denmark, 14 June 1875), *astronomy.*

A diligent investigator of comets, asteroids, and nebulae, d'Arrest is known today chiefly for his role in the discovery of the planet Neptune, and for the periodic comet that bears his name—this comet, which he discovered in 1851, was last seen in October 1963 and is significant because its orbit is gradually getting larger through the action of some nongravitational force.

D'Arrest, whose father was an accountant of Huguenot descent, attended the Collège Français in Berlin before entering the University of Berlin in 1839. He was a promising graduate student, with half a dozen publications and a medal from the King of Denmark (for discovering the comet 1845 I), when Johann Gottfried Galle got permission from Johann Franz Encke, director of the Berlin Observatory, to look for the trans-Uranian planet predicted by Urbain Leverrier. D'Arrest volunteered to help, and suggested the star chart to use: Hora XXI of the *Berliner Akademische Sternkarten,* completed by Carl Bremiker but not yet published. The search was successful that same night (23 September 1846), owing in large part to the excellence of the chart, but in making the initial announcement Encke mentioned only his staff member Galle and himself; it was not until 1877 that Galle set the record straight.

In 1848 d'Arrest was elected a foreign associate of the Royal Astronomical Society of London and chosen to fill a new post at the Leipzig observatory, where he worked under August Ferdinand Möbius—whose daughter he subsequently married. He received his Ph.D. degree from the University of Leipzig in 1850, and in 1851 published his first book, *Ueber das System der kleinerer Planeten zwischen Mars und Jupiter,* a study of the thirteen asteroids then known. His interest in comets and asteroids continued, as shown by his discovery of two more comets (1851 II, mentioned above, and 1857 I) and of the asteroid (76) Freia in 1862, but now d'Arrest began the studies of nebulae for which he received the Gold Medal of the Royal Astronomical Society in 1875.

Although many nebulae had already been observed, notably by William Herschel and his son John, their nature and particularly their distances were still unknown. To improve the situation d'Arrest made, and published in 1857, accurate measurements of the positions and appearances of two hundred and sixty-nine selected nebulous objects; after he became, in 1858, professor in the University of Copenhagen and director of its new observatory, he extended these observations to 1,942 nebulae, published as *Siderum nebulosorum observationes Hafniensis 1861–1867,* but gave up this approach when he realized that even those nebulae bright enough to be detected by his eleven-inch telescope were too numerous for any one man to observe in a lifetime.

Just before his untimely death d'Arrest began spectroscopic observations, following the lead of Sir William Huggins, and was the first to point out, in 1873, that the gaseous nebulae (those with bright line spectra) were preferentially located near the plane of the Milky Way and therefore probably relatively nearby objects in our own galaxy.

BIBLIOGRAPHY

I. ORIGINAL WORKS. D'Arrest's discovery of comet 1845 I was announced in the "Cometen-Circular" of the *Astronomische Nachrichten,* **22** (1845), cols. 343–344, and the orbit he calculated for it was published in a letter to the editor written by Encke, *ibid.,* **23** (1846), cols. 81–82. His early publications included three letters to the editor, on Colla's comet of June 1845: *ibid.,* cols. 231–234, 275–278, and 349–352; "Bestimmung der Elemente der Astraea, mit Rücksicht auf die Störungen aus der ganzen Reihe der Beobachtungen," *ibid.,* **24** (1846), cols. 277–288; "Elemente und Ephemeride der Astraea. 1846–47," *ibid.,* cols. 349–358; and "Ueber die Bahn des vom Dr. Peters entdeckten Cometen," *ibid.,* cols. 387–390.

During his stay in Leipzig d'Arrest's publications included: "Neue Verbesserung der Elemente der Hygiea-Bahn," in *Berichte über die Verhandlungen der königlich sächsischen Gesellschaft der Wissenschaften,* **2** (1850), 1–9; "Nachricht von der Entdeckung und den ersten Beobachtungen des Planeten Victoria, des Cometen von Bond und des dreizehnten Hauptplaneten," *ibid.,* 105–108; "Über die Gruppirung der periodischen Cometen," *ibid.,* **3** (1851), 31–38; the discovery of comet 1851 II [comet d'Arrest] in the "Cometen-Circular" of *Astronomische Nachrichten,* **32** (1851), cols. 327–328, with further observations of it in cols. 341–342; *Ueber das System der kleineren Planeten zwischen Mars und Jupiter* (Leipzig, 1851); "Resultate aus Beobachtungen der Nebelflecken und Sternhaufen. Erste Reihe," in *Abhandlungen der mathematisch-physischen Classe der königlich sächsischen Gesellschaft der Wissenschaften,* **3** (1857), 293–377, with errata on 378; and his discovery of comet 1857 I, announced in the "Cometen-Circular" of *Astronomische Nachrichten,* **45** (1857), cols. 223–224, with more observations in cols. 253–254 and 365–368.

While in Copenhagen d'Arrest's works included the discovery of asteroid (76) Freia, announced in *Astronomische Nachrichten,* **59** (1863), cols. 16–17, with further observations and orbital elements in cols. 77–78 and 91–92; *Siderum nebulosorum observationes Havnienses institutae in secula universitatis per tubum sedecimpedalem Merzianum, ab anno 1861 ad annum 1867* (Copenhagen, 1867); "Om Beskaffenheden og Ubbyttet af de spektral analytiske Undersogelser indenfor Solsystemet, sete i forbindelse med Kometernes Udviklingshistorie," in *Forhandlingerne ved de Skandinaviske Naturforskeres* (1873), 145–161; and a letter to the editor concerning the location of gaseous nebulae, in *Astronomische Nachrichten,* **80** (1873), cols. 189–190.

There is a list of 127 papers by d'Arrest in the Royal Society of London's *Catalogue of Scientific Papers*: I (London, 1867), 101–103; VII (London, 1877), 49–50; IX (London, 1891), 72; and XII (London, 1902), 24. A number of additional items can be found among the more than 400 entries under d'Arrest's name in the cumulative indices to *Astronomische Nachrichten:* **21–40** (Hamburg, 1856), 123–128; **41–60** (Hamburg, 1866), 122–124; and **61–80** (Leipzig, 1875), 80–81.

II. SECONDARY WORKS. An address delivered by the president of the Royal Astronomical Society, John Couch Adams, when d'Arrest received—*in absentia*—that society's Gold Medal, was printed in *Monthly Notices of the Royal Astronomical Society,* **35** (1875), 265–276, and gives a contemporary evaluation of his accomplishments. For further details on his life, see the biographical memoir by John Louis Emil Dreyer in *Vierteljahrsschrift der Astronomischen Gesellschaft,* **11** (1876), 1–14 and an unsigned obituary in *Monthly Notices of the Royal Astronomical Society,* **36** (1876), 155–158.

The discovery of the planet Neptune was announced by Encke, in a letter to the editor dated 26 Sept. 1846, that appeared in *Astronomische Nachrichten,* **25** (1847), cols. 49–52. The first printed mention of d'Arrest's contribution was in Dreyer's memoir (see above); Galle's first public acknowledgment of the part played by d'Arrest in this discovery was "Ein Nachtrag zu den in Band 25 und dem Ergänzungshefte von 1849 der Astr. Nachrichten enthaltenen Berichten über die erste Auffindungen des Planeten Neptun," in *Astronomische Nachrichten,* **89** (1877), cols. 349–352. In his "Historical Note Concerning the Discovery of Neptune," published in *Copernicus,* **2** (1882), 63–64, Dreyer called attention to this paper by Galle, remarking upon the fact that neither Encke nor Galle had previously seen fit to give proper credit to d'Arrest. Galle agreed to the justice of this rebuke in "Ueber die erste Auffindung der Planeten Neptun," *ibid.,* 96–97.

SALLY H. DIEKE

ARRHENIUS, SVANTE AUGUST (*b.* Vik, Sweden, 19 February 1859; *d.* Stockholm, Sweden, 2 October 1927), *chemistry, physics.*

Svante August Arrhenius, one of the founders of modern physical chemistry, came from a Swedish farming family. His father, Svante Gustav Arrhenius, was a surveyor and later a supervisor of the University of Uppsala. He also was employed as overseer on the ancient estate of Vik (Wijk), on Lake Mälar near Uppsala. In 1855 he married Carolina Christina Thunberg; Svante August was their second son. By the beginning of 1860, the father's position had improved enough so that the family moved to Uppsala, where he could devote full time to his university position.

After attending the Cathedral School in Uppsala, Arrhenius entered the University of Uppsala at the age of seventeen. He studied mathematics, chemistry, and physics, and passed the candidate's examination in 1878. Arrhenius chose physics as the principal subject for his doctoral study, but he was not satisfied with his chief instructor, Tobias Robert Thalén. Although Thalén was an eminent and competent experimental physicist and lecturer, he was interested only in spectral analysis. Arrhenius went to Stockholm in 1881 with the intention of working under Erik Edlund, physicist of the Swedish Academy of Sciences. The results of his first independent research, entitled "The Disappearance of Galvanic Polarization in a Polarization Vessel, the Plates of Which Are Connected by Means of a Metallic Conductor," was published in 1883. During the winter of 1882–1883 Arrhenius determined the conductivity of electrolytes; this resulted in his doctoral dissertation (1884), in which he discussed the electrolytic theory of dissociation. He presented it to the University of Uppsala and defended it in May 1884, but his dissertation was awarded only a fourth class (*non sine laude approbatur,* "approved not without praise") and his defense a third (*cum laude approbatur,* "approved with praise"). According to the then prevailing custom, this was not sufficient to qualify him for a docentship, which was a bitter disappointment to Arrhenius.

The chemist Sven Otto Pettersson, professor of chemistry at the Technical High School of Stockholm, reviewed Arrhenius' dissertation in the journal *Nordisk Revy* and praised it very highly, however: "The faculty have awarded the mark *non sine laude* to this thesis. This is a very cautious but very unfortunate choice. It is possible to make serious mistakes from pure cautiousness. There are chapters in Arrhenius' thesis which alone are worth more or less all the faculty can offer in the way of marks." [1] Pettersson referred here to the discovery of the connection between conductivity and speed of reaction. Per Theodor Cleve, speaking to Ostwald during the latter's visit to Uppsala, remarked, "But it is nonsense to accept with Arrhenius that in a solution of potassium-chloride chlorine and potassium are separated from each other," and in his speech honoring Arrhenius at the Nobel banquet in 1903 he said: "These new theories also suffered from the misfortune that nobody really knew where to place them. Chemists would not recognize them as chemistry; nor physicists as physics. They have in fact built a bridge between the two."

Arrhenius sent copies of his thesis to a number of prominent scientists: Rudolf Clausius in Bonn, Lothar Meyer in Tübingen, Wilhelm Ostwald in Riga, and Jacobus Henricus van't Hoff in Amsterdam. Ostwald, a physical chemist and professor at the Polytechnikum in Riga, was deeply impressed by the paper. He visited Arrhenius in Uppsala in August 1884, and offered him a docentship in Riga. Thanks to this, Arrhenius was appointed lecturer in physical chemistry at the University of Uppsala in November of that year. The English physicist Oliver Lodge was also impressed by Arrhenius' paper, and wrote an abstract and critical analysis of it for the *Reports of the British Association for the Advancement of Science* in 1886.

Through the influence of Edlund, Arrhenius re-

ceived a travel grant from the Swedish Academy of Sciences which made it possible for him to work in the laboratories of Ostwald in Riga (later in Leipzig), Kohlrausch in Würzburg, Ludwig Boltzmann in Graz, and van't Hoff in Amsterdam. During these *Wanderjahre* (1886–1890), he further developed the theory of electrolytic dissociation. Arrhenius' theory was, however, slowly accepted at first, but because of neglect rather than active opposition. It was the enthusiasm and influence of Ostwald and van't Hoff that helped to make it widely known. In 1887 Arrhenius met Walther Nernst in Kohlrausch's laboratory. There, too, he carried out an important investigation on the action of light on the electrolytic conductivity of the silver salts of the halogens. In 1891 Arrhenius received an invitation from the University of Giessen, but he preferred the post of lecturer at the Technical High School in Stockholm, where he was appointed professor of physics in 1895 and was rector from 1896 to 1905. After refusing an offer from the University of Berlin, he became director of the physical chemistry department of the newly founded Nobel Institute in Stockholm, a post which he held until his death.

One of Arrhenius' first honors was election as honorary member of the Deutsche Elektrochemische Gesellschaft in 1895. In 1901 he was appointed to the Swedish Academy of Sciences, over strong opposition. The following year he received the Davy Medal from the Royal Society of London, and in 1903 he won the Nobel Prize for chemistry "in recognition of the extraordinary services he has rendered to the advancement of chemistry by his theory of electrolytic dissociation." Arrhenius was elected an honorary member of the Deutsche Chemische Gesellschaft in 1905; he became a foreign member of the Royal Society of London six years later. During his visit to the United States in 1911, Arrhenius was awarded the first Willard Gibbs Medal. In 1914 he received the Faraday Medal.

Arrhenius was married twice: in 1894 to his best pupil and assistant, Sofia Rudbeck, and in 1905 to Maria Johansson. By the first marriage he had one son, Olof, and by the second, a son, Sven, and two daughters, Ester and Anna-Lisa.

Arrhenius' aim, during his study of the conductivity of electrolytic solutions at Edlund's laboratory, was to find a method for determining the molecular weight of dissolved nonvolatile compounds by measuring electric conductivity. Soon he recognized that the state of the electrolyte was the matter of primary importance. Arrhenius completed his experimental work in the spring of 1883 and submitted a long memoir (in French) to the Swedish Academy of Sciences on 6

June 1883, with the results of his experiments and the conclusions he deduced from them. The memoir was published in 1884 under the title "Recherches sur la conductibilité galvanique des électrolytes." The first part ("La conductibilité des solutions aqueuses extrêmement diluées" and "Recherches sur la conductibilité galvanique des électrolytes") contains his findings on the conductivity of many extremely dilute solutions. Instead of measuring the conductivities with the exact alternating-current method, which Kohlrausch had introduced in 1876, Arrhenius used a "depolarizer," devised by Edlund in 1875, which corresponded roughly to a hand-driven rotating commutator.

In the first part of his memoir, Arrhenius gave an account of his experimental work: He measured the resistance of many salts, acids, and bases at various dilutions to 0.0005 normal (and sometimes to even lower concentrations), and gave his results so as to show in what ratio the resistance of an electrolyte solution is increased when the dilution is doubled. It is true that Heinrich Lenz and Kohlrausch had made similar measurements, but they did not use such great dilutions. Like Kohlrausch, Arrhenius found that for very dilute solutions the specific conductivity of a salt solution is in many cases nearly proportional to the concentration (thesis 1) when the conditions are identical. The conductivity of a dilute solution of two or more salts is always equal to the sum of the conductivities that solutions of each of the salts would have at the same concentration (thesis 2). Furthermore, the conductivity of a solution equals the sum of the conductivities of salt and solvent (thesis 3).

If these three laws are not observed, it must be because of chemical action between the substances in the solution (theses 4 and 5). The electrical resistance of an electrolytic solution rises with increasing viscosity (thesis 7), complexity of the ions (thesis 8), and the molecular weight of the solvent (thesis 9). Thesis 9 is an example of a proposition that is not correct. In addition to the viscosity of the solvent, its dielectric constant, not the molecular weight, is significant. Arrhenius worked, however, with a limited number of solvents (water, several alcohols, ether) for which the dielectric constant decreases approximately as the molecular weight rises. Arrhenius summarized Part I of his memoir as follows:

> In the first six sections of the present work we have described a new method of measuring the resistance of electrolytic solutions. In this method we made use of rapidly alternating currents, produced by a depolarizer constructed for the purpose by M. Edlund. We have tried to show the use of this method, and to make clear the practical advantages which it possesses.

The main importance of Arrhenius' memoir, however, does not lie in the experimental measurements or in the thirteen detailed deductions of Part I, but in his development of general ideas. These contain the germ of the theory of electrolytic dissociation (which received its definitive statement only three years later).

In Part II ("Théorie chimique des électrolytes"), Arrhenius gave a theoretical treatment of his experimental work, which he based on the hypothesis of the British chemist Alexander William Williamson and the German physicist Rudolf Clausius. In his famous article "Theory of Aetherification," Williamson suggested that in a chemical system a molecule continually exchanges radicals or atoms with other molecules, so that there is a state of dynamic equilibrium between atoms and molecules. Thus, in hydrochloric acid "each atom of hydrogen does not remain quietly in juxtaposition with the atom of chlorine with which it first united, but, on the contrary, is constantly changing places with other atoms of hydrogen, or, what is the same thing, changing chlorine." Williamson, however, did not assume that the radicals or atoms were electrically charged. Clausius advanced the hypothesis that a small fraction of a dissolved salt is dissociated into ions even when no current is passing through the solution. He did not state or calculate how much of the salt is thus affected.

Arrhenius stated that the dissolved molecules of an electrolyte are partly "active," partly "inactive": "The aqueous solution of any hydrate [by *hydrates* Arrhenius always meant hydrogen compounds like acids and bases] is composed, in addition to the water, of two parts, one active, electrolytic, the other inactive, non-electrolytic. These three substances, viz. water, active hydrate, and inactive hydrate, are in chemical equilibrium, so that on dilution the active part increases and the inactive part diminishes" (thesis 15). Arrhenius gave no precise account of the nature of the active and inactive parts, however; he only indicated what they might be. He extended his hypothesis to other dissolved electrolytes (salts) and defined the "coefficient of activity of an electrolyte" (corresponding to our notion of degree of electrolytic ionization) as "the number expressing the ratio of the number of ions actually contained in the electrolyte to the number of ions it would contain if the electrolyte were completely transformed into simple electrolytic molecules." In 1890 Arrhenius said that he chose the name "activity coefficient" instead of "degree of electrolytic dissociation" on grounds of prudence![2]

After thesis 16 we find a number of chemical applications. Arrhenius asserted that "the strength of an acid is the higher, the greater its activity coefficient. The same holds for bases." The dissociation becomes complete at infinite dilution of the solution (thesis 31); and in solutions of salts of weak acids, strong acids displace the weak acids (thesis 34). From a chemical point of view, thesis 23 is important: "When the relative amounts of ions A, B, C, and D are given, the final result is independent of their original form of combination, whether AB and CD, or AD and BC." The principle of the calculation of the degree of hydrolyzation by means of the law of mass action is given in thesis 29: "Every salt, dissolved in water, is partly dissociated in acid and base. The amount of the decomposition products is greater the weaker the acid and the base and the greater the amount of water."

In the last thesis (56), Arrhenius clearly stated the constancy of the heat of neutralization of a strong acid with a strong base: "The heat of neutralisation, set free by the transformation of a perfectly active base, and perfectly active acid, into water and simple salt, is only the heat of activity of the water," where "heat of activity" is the heat used in transforming a body from the inactive to the active state. Arrhenius ended his memoir with a long summary, which begins as follows:

> In the present part of this work we have first shown the probability that electrolytes can assume two different forms, one active, the other inactive, such that the active part is always, under the same exterior circumstances (temperature and dilution), a certain fraction of the total quantity of the electrolyte. The active part conducts electricity, and is in reality the electrolyte, not so the inactive part.[3]

Although Arrhenius discussed electrolytic dissociation in his memoir of 1884, he nowhere used the word "dissociation," nor is there any explicit identification of the "active part" of the electrolyte with free ions in the solution. It is not so surprising that the acceptance of his theory was slow at first, above all because it had to overcome preconceived ideas that oppositely charged ions could not exist separately in solution. The influence and enthusiasm of Ostwald and van't Hoff were consequently needed to make it widely known and accepted.

The next step toward a definite and clear electrolytic dissociation theory came from a famous memoir of van't Hoff, "The Role of Osmotic Pressure in the Analogy Between Solutions and Gases" (1887). Van't Hoff recognized in this memoir an analogy between dilute solutions and gases: "The pressure which a gas exerts at a given temperature, if a definite number of molecules is contained in a definite volume, is equal to the osmotic pressure which is produced by most substances under the same conditions, if they are dissolved in any given liquid." He showed that it was possible to write for solutions an equation $PV = iRT$, analogous to the gas equation, where P is the osmotic pressure instead of the gaseous pressure, R the gas constant, V the volume, T the absolute temperature, and i a coefficient that is sometimes equal to unity but for the salts is greater than unity. Thus, van't Hoff concluded that the law was valid only for the "great majority of substances," but he could not explain the fact that solutions of salts, acids, and bases possess greater osmotic pressure, higher vapor tension, and greater depression of the freezing point than the results calculated from Raoult's experiments. Van't Hoff made no attempt to explain this exception, but Arrhenius identified the number of ions in solution with the value of i. In a letter to van't Hoff, dated 30 March 1887, Arrhenius wrote: "Your paper has cleared up for me to a remarkable degree the constitution of solutions. . . . Since . . . electrolytes decompose into their ions, the coefficient i must lie between unity and the number of ions." He continued with a statement of the theory of electrolytic dissociation in a clear and definite form: "In all probability all electrolytes are completely dissociated at the most extreme dilution."

In 1887 Arrhenius published a much revised, extended, and consolidated version of his theory of electrolytic dissociation in its quantitative formulation under the title "Ueber die Dissociation der im Wasser gelösten Stoffe." He wrote:

> In a previous communication . . . I have designated those molecules whose ions have independent motion, active molecules, and those whose ions are bound together, inactive molecules. I have also maintained it probable that at the most extreme dilution all the inactive molecules of an electrolyte are converted into active molecules. On this assumption I will base the calculations now to be carried out. The ratio of the number of active molecules to the total number of molecules, active and inactive, I have called the activity coefficient. The activity coefficient of an electrolyte at infinite dilution is therefore taken as unity. At smaller dilutions it is less than unity. . . .[4]

Arrhenius calculated the degree of electrolytic dissociation quantitatively as the ratio of the actual molecular conductivity of the solution and the limiting value to which the molecular conductivity of the same solution approaches with increasing dilution. He then gave the relationship between van't Hoff's constant i and the degree of ionization or activity coefficient α in the form $i = 1 + (k - 1)\alpha$, where k is the number of ions into which the molecule of the electrolyte dissociates. He compared the values of i calculated from Raoult's freezing-point data of solutions in water with the values obtained from the molecular conductivity for twelve nonconductors, fifteen bases, twenty-three acids, and forty salts, and found a very satisfactory agreement. He concluded that van't Hoff's law holds good, not merely for the majority but for all substances, including electrolytes in aqueous solution. "Every electrolyte in aqueous solution consists in part of molecules electrolytically and chemically active and in part of inactive molecules, which, however, on dilution change into active molecules, so that at infinite solution only active molecules are present." [5] With this publication, the full statement of the theory of electrolytic dissociation was given, and soon received substantial confirmation.

Among Arrhenius' most important contributions to this theory are his publications on isohydric solutions, solutions of two acids that can be mixed without any change in the degree of dissociation (1888); the relation between osmotic pressure and lowering of vapor tension (1889); the heat of dissociation of electrolytes and the influence of temperature on the degree of dissociation (1889); the condition of equilibrium between electrolytes (1889); the determination of electrolytic dissociation of salts through solubility experiments (1892); the hydrolysis of salts and weak acids and weak bases (1894); and the alteration of the strength of weak bases by the addition of salts (1899).

A problem that had always held Arrhenius' attention was the abnormality of strong electrolytes that do not follow Ostwald's law of dilution, which can be obtained by applying the law of mass action to the equilibrium between the dissociated and undissociated parts of an electrolyte. Arrhenius stated clearly that the law of mass action is not applicable to strong electrolytes, even when they are very diluted. A theory for the modern treatment of strong electrolytes was given by the Danish chemist Niels Bjerrum, by the Dutch-American scholar Peter Joseph Debye, and by the German Erich Hückel, who based their treatment on electrical interactions between the ions in solution.

Among the other physical-chemical works of Arrhenius, his important theoretical contribution, "Ueber die Reaktionsgeschwindigkeit bei der Inversion von Rohrzucker durch Säuren" (1889) must be mentioned. In this publication, Arrhenius studied the influence of an increase in temperature on the reaction velocity. Using the equilibrium equation deduced by van't Hoff in 1884, which gives mathematically the relation between the velocity coefficient and the temperature, Arrhenius realized that the study of the temperature coefficients of reaction velocity is important from the point of view of the general mechanism of chemical change. From the observation that the reaction velocity shows an abnormal increase of 10 to 15 per cent for one degree in temperature, Arrhenius supposed that active cane sugar molecules are formed;

these activated molecules (with much greater than average energy) are more susceptible to reaction. In a reaction system there are only a certain number of "active" molecules that can undergo reaction. This idea that molecules require a certain critical energy in order to react, as well as the concept of activation energy, is of great significance in modern chemistry.

During the last twenty-five years of his life, Arrhenius' interests were diverted to other fields of science, especially to the physics and chemistry of cosmic and meteorological phenomena. His contribution to these subjects consists mostly in the application of the laws of theoretical chemistry to existing astronomical, geophysical, and geological observations. Besides a short treatise on ball lightning (1883) and a publication on the influence of the rays of the sun on the electric phenomena of the earth's atmosphere (1888), Arrhenius and the meteorologist Nils Ekholm investigated the influence of the moon on the electric state of the atmosphere, on the aurora, and on thunderstorms (1887). Arrhenius supposed in the 1888 article that electric charges originate from ionization of the air by ultraviolet rays.

In 1896 he published a long memoir "On the Influence of Carbonic Acid in the Air Upon the Temperature of the Ground," in which he developed a theory for the explanation of the glacial periods and other great climatic changes, based on the ability of carbon dioxide to absorb the infrared radiation emitted from the earth's surface. Although the theory was based on thorough calculations, it won no recognition from geologists. In 1898 Arrhenius wrote a remarkable paper on the action of cosmic influence on physiological processes.

"Zur Physik des Vulkanismus," published in 1901, was also based on physical-chemical facts. Although at normal temperature, water is an acid about a hundred times weaker than silicic acid, increasing ionic dissociation with increased temperature would at a few hundred degrees make water a stronger acid than silicic acid. Arrhenius calculated by extrapolation that water at 1000°C. is eighty times, and at 2000°C. 300 times, stronger than silicic acid. In the magma, water penetrates at a temperature of between 1000°C. and 2000°C., and decomposes silicates. The magma expands, its volume increases, and it penetrates into the fissures of volcanoes. When the rising magma is cooled, the reverse process takes place, water is liberated, and under low pressure violent explosions occur, leading to volcanic eruptions. However, the hypothetical reaction between the molten silicate and water was not tenable, and Arrhenius' theory was soon forgotten.

In 1903 Arrhenius published his *Lehrbuch der kosmischen Physik,* the first textbook on cosmic physics. His work on the cosmic effects of the pressure of light rays attracted deserved attention in professional circles. With the aid of very light mirrors in a vacuum, the Russian physicist Pëtr Nikolajevich Lebedev and the American physicists Ernest Fox Nichols and Gordon Ferrie Hull proved in 1901 that a ray of light that meets material particles exerts a pressure on them, as James Clerk Maxwell had predicted in his electromagnetic theory of light. Arrhenius applied the radiation pressure to various phenomena even before its experimental confirmation. He calculated that we might expect streams of minute particles to be shot out from the sun in all directions. Arrhenius explained phenomena of the solar corona, comets, the aurora, and the zodiacal light by these charged particles, many of which, he said, would be electrically charged by ionization in the gaseous atmosphere of the sun. In 1905 he applied this concept to the problem of the origin of life by assuming that living seeds, spores, and so forth could be transported from interstellar space by the pressure of light (panspermic theory). Since Arrhenius' basic idea of the universe was its infinity in time, he did not have any need for a hypothesis involving a singular event like the creation of life. His concept that there was no beginning and no end of the universe follows from his inability to resolve by any other means the paradox in the application of the first and second laws of thermodynamics to the universe. According to Clausius, the energy of the world is constant and the entropy approaches a maximum, so that the universe is tending to what he called the *Wärmetod* ("heat death") through exhaustion of all sources of heat and motion. Now, if the universe were assumed to have a finite lifetime, the creation of energy at some time would be required—and this is contrary to the first law of thermodynamics. On the other hand, if the universe were assumed to have existed for an infinite time, according to the second law of thermodynamics, the maximum entropy would have been achieved. To solve the paradox, Arrhenius assumed that it is possible that there are galaxies in the universe where processes take place with decreasing entropy. His last paper (1927) was on thermophilic bacteria and the radiation pressure of the sun. In it he stated that on earth there are thermophilic bacteria that exist in volcanic areas at temperatures between 40°C. and 80°C. The temperature of the surface of the planet Venus is 50°C., and Arrhenius thought that it was possible that these bacteria are transported from Venus to earth by radiation pressure. Of course, he did not know of the existence of cosmic radiation, which makes it physically impossible for unprotected living things to survive transportation through interplanetary space.

In addition to his cosmic researches, Arrhenius was concerned with the theory of immunity, an interest that resulted in two textbooks: *Immunochemistry* (1907) and *Quantitative Laws in Biological Chemistry* (1915). After working during the summer of 1902 in the Frankfurt laboratory of the German bacteriologist Paul Ehrlich, Arrhenius and the Danish bacteriologist Thorvald Madsen (later founder and director of the Danish State Serum Institute at Copenhagen) published a paper on physical chemistry applied to toxins and antitoxins (1902). Against Ehrlich, who in his "side-chain theory" regarded the mutual relationship of toxins and antitoxins as a phenomenon of chemical neutralization, Arrhenius postulated a chemical equilibrium between toxin and antitoxin which follows the ordinary mass action law. The immunological phenomenon of antitoxin action was linked to the interaction of a weak acid and a weak base.

Besides the textbooks on immunochemistry and cosmic physics mentioned above, Arrhenius wrote a number of scientific books: *Lärobok i teoretisk elektrokemi* (1900), *Theorien der Chemie* (1906), and *Theories of Solution* (the Silliman Lectures of 1911, published in 1912). Arrhenius devoted most of his later years to popularizing science. His books and articles had a simple but always scientific approach, and were immediate worldwide successes. They were translated into several languages and appeared in numerous editions. Among these are *Världnarnas utveckling* (1906), *Människan inför världsgåtan* (1907), *Das Schicksal der Planeten* (1911), and *Stjärnornas Öden* (1915). His *Kemien och det moderna livet* (1919) contains a popular scientific treatment of the significance and the problems of technical chemistry. These books give a good idea of Arrhenius' aptitude for scientific speculation, a penchant also exhibited in his original ideas in cosmic physics, meteorology, immunology, and in his greatest contribution to chemistry, the theory of electrolytic dissociation.

NOTES

1. *Nordisk revy* (15 December 1884); cf. *Svensk kemisk tidskrift* (1903), 208.
2. *Svensk kemisk tidskrift* (1890), 9.
3. *Bihang till K. Svenska Vet.-Akad. Handlingar,* **8,** no. 14 (1884), 87.
4. *Zeitschrift für phys. Chemie,* **1** (1887), 632.
5. *Ibid.,* p. 637.

BIBLIOGRAPHY

For a bibliography of Arrhenius' works and writings, see E. H. Riesenfeld, *Svante Arrhenius* (Leipzig, 1931), pp.

93–110. In the bibliography given below, the following abbreviations are used: *Bihang* (*Bihang till kungliga vetenskapsakademiens handlingar*); *Öfversigt* (*Öfversigt af kungliga vetenskapsakademiens för handlingar*); *Meddelanden* (*Meddelanden från kungliga vetenskapsakademiens Nobelinstitut*); *Z. phys. Chem.* (*Zeitschrift für physikalische Chemie*).

I. ORIGINAL WORKS. Articles that are autobiographical or deal with the history of the theory of electrolytic dissociation are "The Development of the Theory of Electrolytic Dissociation," in *Les prix Nobel en 1903* (Stockholm, 1905); *Proceedings of the Royal Institute*, **17**, pt. 3 (1906); and *Nobel Lectures Chemistry 1901–1921* (Amsterdam-New York-London, 1966), pp. 45–58; "Electrolytic Dissociation," in *Journal of the American Chemical Society*, **34** (1912), 353–364; "Aus der Sturm- und Drangzeit der Lösungstheorien," in *Chemisch weekblad*, **10** (1913), 584–599; "The Theory of Electrolytic Dissociation," in *Lectures Delivered Before the Chemical Society* (London, 1928), pp. 237–249.

Articles by Arrhenius concerning the theory of electrolytic dissociation are "Recherches sur la conductibilité galvanique des électrolytes," in *Bihang*, **8**, no. 13 (1884) and no. 14 (1884), translated as *Untersuchungen über die galvanische Leitfähigkeit der Elektrolyte,* in Ostwald's *Klassiker der exakten Wissenschaften*, no. 160 (Leipzig, 1907); "Ueber die Dissociation der im Wasser gelösten Stoffe," in *Z. phys. Chem.*, **1** (1887), 631–648, expanded from two papers published in *Öfversigt* (1887), pp. 405–414, 561–575, and translated in the Alembic Club Reprints, no. 19 (Edinburgh, 1929); "Theorie der isohydrischen Lösungen," in *Öfversigt* (1888), pp. 233–247, and *Z. phys. Chem.*, **2** (1888), 284–295; "Einfache Ableitung der Beziehung zwischen osmotischem Druck und Erniedrigung der Dampfspannung," in *Z. phys. Chem.*, **3** (1889), 115–119; "Ueber die Dissociationswärme und den Einfluss der Temperatur auf den Dissociationsgrad der Elektrolyte," *ibid.*, **4** (1889), 96–116; "Ueber die Gleichgewichtsverhältnisse zwischen Elektrolyten," in *Öfversigt* (1889), pp. 619–645, and *Z. phys. Chem.*, **5** (1890), 1–22; "Ueber die Bestimmung der elektrolytischen Dissociation von Salzen mittelst Löslichkeitsversuchen," in *Öfversigt* (1892), pp. 481–494, and *Z. phys. Chem.*, **11** (1893), 391–402; "Ueber die Hydrolyse von Salzen schwacher Säuren und schwacher Basen," in *Z. phys. Chem.*, **13** (1894), 407–411; "Ueber die Aenderung der Stärke schwacher Säuren durch Salzzusatz," *ibid.*, **31** (1899), 197–229; "Zur Berechungsweise des Dissociationsgrades starker Elektrolyte," *ibid.*, **36** (1901), 28–40.

Articles on other physical-chemical subjects are "Ueber die Einwirkung des Lichtes auf das elektrische Leitungsvermögen der Haloïdsalze des Silbers," in *Zeitschrift der Wiener Akademie der Wissenschaft*, **96** (1887), 831–837; "Ueber die Reaktionsgeschwindigkeit bei der Inversion von Rohrzucker durch Säuren," in *Z. phys. Chem.*, **4** (1889), 226–248; and "Zur Theorie der chemischen Reaktionsgeschwindigkeit," in *Bihang*, **24**, no. 2 (1898), and *Z. phys. Chem.*, **28** (1899), 317–335.

Articles about meteorology and cosmic physics are "Ueber den Einfluss der Sonnenstrahlen auf die elektrischen Erscheinungen in der Erdatmosphäre," in *Meteorologische Zeitschrift*, **5** (1888), 297–304, 348–360; "Ueber den Einfluss des atmosphärischen Kohlensäuregehalts auf die Temperatur der Erdoberfläche," in *Bihang*, **22**, no. 1 (1896), 102 ff., excerpted in *Philosophical Magazine*, **41** (1896), 237–276; "Die Einwirkung kosmischer Einflüsse auf die physiologischen Verhältnisse," in *Skandinavisches Archiv für Physiologie*, **8** (1898), 367–426; "Zur Physik des Vulkanismus," in *Geologiska föreningens i Stockholm förhandlingar*, **22**, no. 5 (1901), 26 ff.; "Ueber die Wärmeabsorption durch Kohlensäure," in *Öfversigt* (1901), pp. 25–58, and *Drudes Annalen*, **4** (1901), 689–705; "Lifvets utbredning genon världsrymden," in *Nordisk tidskrift* (1905), pp. 189–200, and *The Monist* (1905), pp. 161 ff.; "Die vermutliche Ursache der Klimaschwankungen," in *Meddelanden*, **1**, no. 2 (1906); "Physikalisch-chemische Gesetzmässigkeiten bei den kosmisch-chemischen Vorgängen," in *Zeitschrift für Elektrochemie*, **28** (1922), 405–411; and "Die thermophilen Bakterien und der Strahlungsdruck der Sonne," in *Z. phys. Chem.*, **130** (1927), 516–519.

An article on serum therapy is "Anwendung der physikalischen Chemie auf das Studium der Toxine und Antitoxine," in *Festskrift v. inv. af Stat. Serum-Inst.* (Copenhagen, 1902), and *Z. phys. Chem.*, **44** (1903), 7–62, written with Thorvald Madsen.

Books by Arrhenius are *Lärobok i teoretisk elektrokemi* (Stockholm, 1900), translated as *Text-Book on Theoretical Electrochemistry* (London-New York, 1902); *Lehrbuch der kosmischen Physik,* 2 vols. (Leipzig, 1903); *Theorien der Chemie* (Leipzig, 1906), translated as *Theories of Chemistry* (London-New York, 1907); *Immunochemistry* (New York, 1907); *Theories of Solution* (London-New Haven, Conn., 1912); and *Quantitative Laws in Biological Chemistry* (New York-London, 1915).

II. SECONDARY LITERATURE. Works on Arrhenius include W. Ostwald, "Svante August Arrhenius," in *Z. phys. Chem.*, **69** (1909), v–xx; J. Walker, "Arrhenius Memorial Lecture," in *Journal of the Chemical Society* (1928), pp. 1380–1401; W. Palmaer, "Arrhenius," in G. Bugge, *Buch der grossen Chemiker* (Weinheim, 1929), II, 443–462, translated and abridged by R. E. Oesper in E. Farber, ed., *Great Chemists* (New York, 1961), pp. 1093–1109; E. H. Riesenfeld, "Svante Arrhenius," in *Berichte der Deutschen Chemischen Gesellschaft,* **63** (1930), 1–40, and *Svante Arrhenius* (Leipzig, 1931); and A. Olander, O. Arrhenius, A. L. Arrhenius-Wold, and G. O. S. Arrhenius, in *Svante Arrhenius till 100-arsminnet av hans födelse* (Stockholm, 1959).

H. A. M. SNELDERS

ARSONVAL, ARSÈNE D' (*b.* Chateau de la Borie, St. Germain-les-Belles, La Porcherie, France, 8 June 1851; *d.* Chateau de la Borie, 31 December 1940), *biophysics.*

The d'Arsonval family was part of France's ancient nobility, having held land and wealth in Limoges for centuries. D'Arsonval studied classics at the Lycée Impérial de Limoges and later at the Collège Ste.-Barbe. By the time he received a baccalaureate degree from the Université de Poitiers (1869), d'Arsonval had decided upon a career in medicine. He was the fourth generation to make this decision. His studies began at Limoges, but after the war of 1870 he continued them in Paris. A chance social encounter with Claude Bernard at the Salon de Lachard altered the course of the young physician's career. Drawn to Bernard's lectures, d'Arsonval on one occasion was able to correct the faulty wiring in Bernard's equipment, permitting the completion of a classroom demonstration. Thereafter d'Arsonval became Bernard's *préparateur* from 1873 to 1878. After Bernard's death he assisted C. Brown-Séquard, eventually replacing him at the Collège de France. With Paul Bert's assistance as Minister of Public Education, the Collège de France was able to establish a laboratory for biophysics at rue St.-Jacques in 1882. D'Arsonval directed the laboratory until 1910, when he moved to the new laboratory at Nogent-sur-Marne, erected with funds raised by public subscription. He directed this laboratory until his retirement in 1931.

Bernard's influence led d'Arsonval formally to give up a medical career for a life of physiological research. His thesis (1876) was on pulmonary elasticity and circulation. The young scientist adopted Bernard's organismic philosophy, adding little to it but a belief that electrical potential was one of the physicochemical characteristics of cells.[1] D'Arsonval believed life was vital but completely deterministic. The primary manifestation of life was the conversion of various forms of energy for work. As Bernard's assistant, d'Arsonval's first projects were on animal heat and body temperature.

In 1882 d'Arsonval was awarded the Prix Montyon of the Académie des Sciences for his ingenious apparatus for studying these problems. His double-chambered calorimeter was remarkably accurate and based upon a new approach. He maintained a constant temperature within its inner chamber by circulating ice water through tubes surrounding the inner chamber. The temperature and quantity of water exchanged was a measure of the heat produced. The constant interior temperature increased the accuracy of gas volume measurements and insured more constant rates of breathing. He also devised thermoelectric needles which allowed Bernard to measure simultaneously the temperature of tissue and blood in adjacent vessels. In 1894 d'Arsonval invented a simplified but less accurate calorimeter for hospital tests.

While assisting Brown-Séquard, d'Arsonval became involved in the former's famous experiments on endocrine extracts. D'Arsonval took personal charge of preparing the extracts and sterilizing them by immersion in high-pressure carbon dioxide atmospheres.[2] Their investigations of the therapeutic properties of animal extracts revealed clues to the later controversial hormone theory of wound healing. They found that testicular extracts from guinea pigs had definite antiseptic properties.[3]

D'Arsonval's most outstanding scientific contributions involved the biological and technological applications of electricity. His early studies dealt with the electrical properties of muscle contraction. He recognized that Bell's new invention, the telephone, provided a perfect device for detecting the current in muscle tissue. Telephones operate on extremely feeble currents similar to animal electricity. Galvanometers then in use drew too much current for sensitive tests. D'Arsonval used a frog muscle to join the mouthpiece of a phone and an induction coil with the receiving portion, which completed the circuit of a functional telephone. This interest in muscle current led to a series of practical inventions in the early 1880's. They included nonpolarizable silver chloride electrodes for biological research, refinement of carbon-rod microphones, and the invention with Marey of myographic equipment. D'Arsonval, in cooperation with Deprez, invented the mobile circuit galvanometer in 1882.[4]

Muscle contraction continued to interest him, especially Ranvier's histological studies of striated muscle and L. Hermann's discovery that a negative potential reading characterized the point of direct excitation of a muscle and was followed by a positive variation throughout the body of the muscle. D'Arsonval's own research on contraction led to the same conclusions about the negative activation potential. Using his highly sensitive galvanometer, d'Arsonval found a feeble positive current during normal rest or tonus, which vanished during the act of direct excitation. Experiments showed that the electrical changes were surface phenomena of approximately the same strength as is needed to induce contraction in a muscle adjacent to an excited nerve or muscle. E. DuBois-Reymond explained the negative action potential by assuming a basic bipolarity in muscles. D'Arsonval believed the contractile elements were Ranvier's disks and that the surface production of electrical charge had a physical explanation. Lippmann had shown that a globule of mercury in acidified water produced a measurable current flow when mechanically deformed. D'Arsonval reversed the argument, stating that every current causes a physical deformation. He theorized that a positive stimulus should cause elongation and a negative variation should cause contraction; he built a model to test his concept. A thin rubber tube was filled with porous plugs impregnated with mercury surrounded by acidified water. When subjected to electrical charges the model acted exactly according to d'Arsonval's predictions.[5] Later studies of the electrical organs of the torpedo fish substantiated his suggestions, which were highly plausible in the absence of an alternate chemical theory.

D'Arsonval also found that high voltage shocks did not always lead to sudden or inevitable death. Artificial respiration could frequently revive victims of accidental electric shock.[6] Gradually d'Arsonval's interests shifted from pure biological research to technological problems. For example, as an aside to his calorimetric work, d'Arsonval designed the first electrically controlled constant temperature incubator for embryological and bacterial research.[7] The d'Arsonval incubator was used well into the twentieth century. He was consulted frequently by E. Marey, and he aided Ferrie in constructing the first triode.[8] D'Arsonval became known as the foremost authority on laboratory apparatus, especially in regard to electrical equipment.

In later years d'Arsonval became increasingly involved in the application of electricity to industry, a role which he clearly enjoyed and fostered. He was instrumental in founding national (1881) and international (1897) societies for electrical science, a government supported laboratory for electrical research (1888), the École Supérieure d'Électricité (1894), an international society for cryogenic studies (1908), and La Compagnie Générale d'Électro-Céramique (1923), to name only a few. He worked with Georges Claude on industrial methods for the liquefaction of gases (1902), was consulted on high energy electrical transmission equipment, served as government science consultant during World War I, and was a constant promoter of the automobile and airplane.

His contribution to medicine, now overshadowed by the antibiotic era, created a minor revolution in clinical therapeutics. D'Arsonval literally founded the paramedical field of physiotherapy. In 1918 he was elected president of the Institute for Actinology. H. Herz, a physicist, built the first high frequency oscillator, and shortly thereafter d'Arsonval used it to experiment upon the effects of high frequency (500,000–1,500,000 c.p.s.), low voltage alternating current on animals. This led him in 1891 to report that no sensory or motor responses were evoked by high frequency currents. As Herz had noted earlier, the only effect was the production of heat.[9] The heating effect could be applied to muscle aches, spasms, tetanus, tumors, arthritis, and circulatory and gynecological problems. D'Arsonval correlated the frequency with expected temperatures for a given period of time.

The first high frequency heat therapy unit was established under d'Arsonval's direction at the Hôtel-Dieu Hospital in 1895. Indeed, electrotherapy was called *d'Arsonvalization* until the broader term *diathermy* came into use after 1920. The applications of high frequency treatment were highly successful. D'Arsonval helped to develop apparatus for electrocoagulation which was widely used for surgical excisions and tumor treatments.[10] High energy procedures were favored because these wavelengths were antibacterial. By 1910 methods of physiotherapy utilizing high frequency waves, X rays, and radium had become a professional discipline.

D'Arsonval's international reputation was closely associated with physiotherapy and industrial applications of electricity. He was an active member of societies for electrotherapy, physics, electronics, civil engineering, electroceramics, and soldering, in addition to being a member of the Society of Biologists, the Academy of Medicine (1888), and the Academy of Sciences (1894). In 1933 the Ministry of Education held an official jubilee for d'Arsonval at the Sorbonne. He was created knight of the Legion of Honor in 1884 and received the Grand Cross in 1931.

NOTES

1. *Lumière électrique,* **7** (1882), 302.
2. *Bulletin de l'Académie de médecine.* Paris, 3rd ser., **27** (1892), 250–261.
3. *Comptes rendus des séances de la Société de biologie.* Paris, 9th ser., **3** (1891), 235, 248–250; O. Glasser, *Medical Physics* (1947), pp. 1582–1583.
4. *Comptes rendus de l'Académie des sciences,* **94** (1882), 1347–1350.
5. *Archives de Physiologie,* 5th ser., **1** (1889), 246–252, 460–472; Chauvois (1937), 160–161.
6. *Comptes rendus de l'Académie des sciences,* **104** (1887), 978–981.
7. *Archives de physiologie,* 5th ser., **2** (1890), 83–86.
8. Chauvois (1937), p. 377.
9. Glasser, p. 414.
10. Chauvois (1937), pp. 270–271, 359.

BIBLIOGRAPHY

I. ORIGINAL WORKS. A bibliography of d'Arsonval's works is *L'oeuvre scientifique du Prof. A. d'Arsonval,* compiled by the Institut d'Actinologie (Paris, 1933). D'Arsonval's technological work may be best approached through his *Traité de physique biologique,* 2 vols. (Paris, 1903); "Nouveaux appareils destinés aux recherches d'électrophysiologie," in *Archives de physiologie,* 5th ser., **1** (1889), 423–437, and "Appareils a température fixe pour embryologie et cultures microbiennes," *ibid.,* **2** (1890), 83–88.

The following articles are only a few of the hundreds of articles written by d'Arsonval: "Les nouvelles applications et les perfectionnements du téléphone," in *Revue scientifique,* **1** (1879), 200–212; "Les sciences physiques en biologie," in *Lumière électrique,* **6** (1882), 174–177, 329–331, 394–395, 415–416, 512–513, 546–547; **7** (1882), 43–45, 64–65, 222–224, 302–303, 352–353, 421–422, 495–497, 519–522, 543–544, 567–570, 595–598; "Recherches sur le téléphone," in *Comptes rendus de l'Académie des sciences,* **95** (1882), 290–292; "Nouvelle méthode calorimétrique applicable a l'homme," in *Comptes rendus de la Société de biologie,* 8th ser., **1** (1884), 651–654; "La mort par l'électricité dans l'industrie . . . Moyens préservateurs," in *Comptes rendus de l'Académie des sciences,* **104** (1887), 978–981; "Relations entre la forme de l'excitation électrique et la réaction néuro-musculaire," in *Archives de physiologie,* **1** (1889), 246–252; "Recherches d'électrophysiologie," *ibid.,* 460–472; **2** (1890), 156–167; "De l'injection des extraits liquides provenant des différent tissus de l'organisme . . .," in *Comptes rendus de la Société de biologie,* 9th ser., **4** (1891), 248–250, or *Bulletin de l'Académie de médecine,* 3rd ser. (1892), pp. 250–261. "Galvanomètre apériodique," written with Deprez, is in *Comptes rendus de l'Académie des sciences,* **94** (1882), 1347–1350. Some of d'Arsonval's letters may be found in L. Delhome, *De Claude Bernard et une correspondance Brown-Séquard-d'Arsonval* (Paris, 1939).

II. SECONDARY LITERATURE. Useful but not always reliable are two works by Louis Chauvois, *D'Arsonval, soixante-cinq ans à travers la science* (Paris, 1937) and *D'Arsonval; une vie, une époque 1851–1940* (Paris, 1945). Other useful articles are J. Belot, "Jubilé du professeur d'Arsonval," in *La presse médicale,* **41,** no. 44 (1933), 899–901, and "D'Arsonval (1851–1940)," in *Journal de radiologie et d'électrologie,* **24** (1941), 49–60; G. Blech, "D'Arsonval's Service to Surgery," in *Archives of Physical Therapy,* **13** (1932), 775–779; G. Bourguignon, "Professor d'Arsonval," *ibid.,* 717–726; H. Bordier, "L'Oeuvre scientifique de d'Arsonval," in *Paris médical,* **88** (1933), v–viii; and Otto Glasser, ed., *Medical Physics* (Chicago, 1947).

CHARLES A. CULOTTA

ARTEDI, PETER (*b.* Anundsjö, Angermanland, Sweden, 27 February/10 March 1705; *d.* Amsterdam, the Netherlands, 27 September 1735), *biology.*

Artedi was the son of a curate, Olaus Arctaedius, and his second wife, Helena Sidenia, the daughter of a court chaplain. In September 1716 the father was appointed to the living of Nordmaling, on the Gulf of Bothnia. Artedi had early shown a strong interest in animals, especially fishes; and that same autumn he was sent to school at Hernösand, where he did well, using out-of-school hours for dissecting fishes

and collecting plants. As soon as he could read Latin, he greedily devoured the writings of the medieval alchemists. With the highest certificate he matriculated at Uppsala University on 30 October 1724 as Petrus Arctelius Angerm., but he used the signature Petrus Arctaedius Angermannus; some years later he assumed the variant Artedi.

Although he was expected to succeed his father (and grandfather) as a clergyman, Artedi devoted his time to chemistry and natural history in the medical faculty, where only Lars Roberg and Olof Rudbeck were interested in natural sciences. Both were old: Roberg, the anatomist, had practically stopped teaching, and lectured privately on the *Problemata* of Aristotle in the light of the principles of Descartes; Rudbeck was engaged in philological investigations and not until 1727 did he start a two-year course on the birds of Sweden. As Linnaeus, who arrived in Uppsala in 1728, wrote, "No one ever heard or saw any anatomy, nor chemistry or botany." Absent, because of the death of his father, when Linnaeus arrived, Artedi returned about March 1729 and their lifelong friendship began. They compared notes and at last divided their interests, Artedi studying fishes, amphibians, mammals, minerals, alchemy, and, in botany, the *Umbelliferae.* Linnaeus describes Artedi as lofty of stature and spare of figure, with long black hair and a face that reminded him of John Ray's: humble-minded, cautious, firm, mature, a man of Old World honor and faith.

In September 1734 Artedi, with the financial aid of his brothers-in-law, set sail for England. He stayed there for nearly a year, studying collections and writing on the literature and taxonomy of fishes, amphibians, and mammals (*Trichozoologia*), and on mineralogy. Returning through Leiden, where he hoped to qualify for the doctor's degree in medicine, Artedi unexpectedly met Linnaeus. He told Linnaeus of the excellent opportunities he had had in England for studying ichthyology and of the kind help of the president of the Royal Society, Sir Hans Sloane. Since Artedi was short of money, Linnaeus introduced him to the chemist Albert Seba of Amsterdam, a well-known collector. Seba entrusted Artedi with the study of the fishes for the third part of his *Thesaurus.* Artedi hired a room in Amsterdam and Linnaeus lived at "Hartekamp," the estate of his Maecenas, George Clifford. The two friends met only once more to discuss Artedi's scientific manuscripts. One evening, returning from a convivial evening with Seba and some friends, Artedi lost his way in the darkness and drowned in one of Amsterdam's many canals.

Thanks to Linnaeus, Artedi's manuscripts were

saved and, according to a vow they had made concerning their scientific papers, Linnaeus published, without altering anything, *Petri Artedi sueci, medici, ichthyologia sive opera omnia de piscibus* (1738). This taxonomically most important work assured Artedi the honor of being the father of the science of ichthyology. It contains an analytical review of the literature and a philosophical dissertation on a natural classification, proving that he and Linnaeus agreed on the principles that were to govern the new systematics. Artedi applied these to the system of fishes and résuméd their synonyms from older literature. In the last section he describes the seventy-two species of fishes he dissected and examined alive, an important comparative anatomy of fishes.

Artedi was the first to settle definitely the notion of genus in zoology; the distinction between species and variety; and the classification into classes, orders, and maniples (families). Most of the descriptions of the fishes in Part III of Seba's *Thesaurus* were prepared by Artedi. Artedi's classification of the *Umbelliferae* was incorporated in the first edition of Linnaeus' *Systema naturae* (1735). In 1905 a manuscript of 1729 on the plants near Nordmaling was published in Uppsala to commemorate the two-hundredth anniversary of Artedi's birth.

BIBLIOGRAPHY

I. Original Works. Artedi's writings include *Petri Artedi sueci, medici, ichthyologia sive opera omnia de piscibus* (Leiden, 1738), which contains a biography by Linnaeus and was reprinted with an introduction by A. C. Wheeler as No. 15 of Hist. Nat. Class. (Weinheim, 1962); and "Kort förteckning på de träen, buskar åg örter såmm wäxa sponté wid Nordmalings prästebord" ("List of Trees, Bushes and Plants That Are Indigenous in the Glebe-Lands in Nordmaling and the Villages Lying in Its Immediate Vicinity"), in *Yearbook of the Swedish Academy of Science* (Uppsala, 1905).

II. Secondary Literature. Works on Artedi are H. Engel, "Some Artedi Documents in the Amsterdam Archives," in *Svenska linnésällskapets årsskrift,* **33–34** (1950–1951), 51–66; E. Lönnberg, "Petrus Artedi, a Bicentenary Memoir," in *Yearbook of the Swedish Academy of Science* (Uppsala, 1905), also translated into English by W. E. Harlock; "Linne og Artedi," in *Svenska linnésällskapets årsskrift,* **2** (1919), 30–43; and "Artedi," in *Svensk Biografisk Lexikon,* II (1920); D. Merriman, "A Rare Manuscript Adding to Our Knowledge of the Work of Peter Artedi," in *Copeia,* **2** (1941), 65–69; and O. Nybelin, "Tvenne opublicerade Artedi manuscript," in *Svenska linnésällskapets årsskrift,* **17** (1934), 35–90.

H. Engel

ARTIN, EMIL (*b.* Vienna, Austria, 3 March 1898; *d.* Hamburg, Germany, 20 December 1962), *mathematics.*

Artin was the son of the art dealer Emil Artin and the opera singer Emma Laura-Artin. He grew up in Reichenberg, Bohemia (now Liberec, Czechoslovakia), where he passed his school certificate examination in 1916. After one semester at the University of Vienna he was called to military service. In January 1919 he resumed his studies at the University of Leipzig, where he worked primarily with Gustav Herglotz, and in June 1921 he was awarded the Ph.D.

Following this, he spent one year at the University of Göttingen, and then went to the University of Hamburg, where he was appointed lecturer in 1923, extraordinary professor in 1925, and ordinary professor in 1926. He lectured on mathematics, mechanics, and the theory of relativity. In 1929 he married Natalie Jasny. Eight years later they and their two children emigrated to the United States, where their third child was born. Artin taught for a year at the University of Notre Dame, then from 1938 to 1946 at Indiana University in Bloomington, and from 1946 to 1958 at Princeton. He returned to the University of Hamburg in 1958, and taught there until his death. He was divorced in 1959. His avocations were astronomy and biology; he was also a connoisseur of old music and played the flute, the harpsichord, and the clavichord.

In 1962, on the three-hundredth anniversary of the death of Blaise Pascal, the University of Clermont-Ferrand, France, conferred an honorary doctorate upon Artin.

In 1921, in his thesis, Artin applied the arithmetical and analytical theory of quadratic number fields over the field of rational numbers to study the quadratic extensions of the field of rational functions of one variable over finite constant fields. For the zeta function of these fields he formulated the analogue of the Riemann hypothesis about the zeros of the classical zeta function. In 1934 Helmut Hasse proved this hypothesis of Artin's for function fields of genus 1, and in 1948 André Weil proved the analogue of the Riemann hypothesis for the general case.

In 1923 Artin began the investigations that occupied him for the rest of his life. He assigned to each algebraic number field k a new type of L-series. The functions

$$L(s,\chi) = \Sigma\chi(n)(Nn)^{-s}$$

—generalizations of the Dirichlet L-series—in which χ is the character of a certain ideal class group and n traverses certain ideals of k were already known.

These functions play an important role in Teiji Takagi's investigations (1920) of Abelian fields K over k. Artin started his L-series from a random Galois field K over k with the Galois group G; he utilized representations of the Frobenius character χ by matrices. Further, he made use of the fact that, according to Frobenius, to each unbranched prime ideal, p, in K, a class of conjugated substitutions σ from G, having the character value $\chi(\sigma)$, can be assigned in a certain manner. Artin made $\chi(p^h) = \chi(\sigma^h)$ and formulated $\chi(p^h)$ for prime ideals p branched in K; he also defined his L-series by the formula

$$\log L(s,\chi,K/k) = \sum_{p,h} h^{-1}\chi(p^h)(Np^h)^{-s}.$$

Artin assumed, and in 1923 proved for special cases, the identity of his L-series formed of simple character and the functions $L(s,\chi)$ for Abelian groups, if at the same time χ were regarded as a certain ideal class character. The proof of this assumption led him to the general law of reciprocity, a phrase he coined. Artin proved this in 1927, using a method developed by Nikolai Chebotaryov (1924). This law includes all previously known laws of reciprocity, going back to Gauss's. It has become the main theorem of class field theory.

With the aid of the theorem, Artin traced Hilbert's assumption, according to which each ideal of a field becomes a principal ideal of its absolute class field, to a theory on groups that had been proved in 1930 by Philip Furtwaengler.

Artin had often pointed to a supposition of Furtwaengler's according to which a series k_i ($i = 1,2, \ldots$) is necessarily infinite if k_{i+1} is an absolute class field over k_i. This was disproved in 1964 by I. R. Safarevic and E. S. Gold.

In 1923 Artin derived a functional equation for his L-series that was completed in 1947 by Richard Brauer. Since then it has been found that the Artin L-series define functions that are meromorphic in the whole plane. Artin's conjecture—that these are integral if χ is not the main character—still remains unproved.

Artin had a major role in the further development of the class field theory, and he stated his results in *Class Field Theory*, written with John T. Tate (1961).

In 1926 Artin achieved a major advance in abstract algebra (as it was then called) in collaboration with Otto Schreier. They succeeded in treating real algebra in an abstract manner by defining a field as real—today we say formal-real—if in it -1 is not representable as a sum of square numbers. They defined a field as real-closed if the field itself was real but none of the algebraic extensions were. They then demonstrated that a real-closed field could be ordered in one exact manner and that in it typical laws of algebra, as it had been known until then, were valid.

With the help of the theory of formal-real fields, Artin in 1927 solved the Hilbert problem of definite functions. This problem, expressed by Hilbert in 1900 in his Paris lecture "Mathematical Problems," is related to the solution of geometrical constructions with ruler and measuring standard, an instrument that permits the marking off of a single defined distance.

In his work on hypercomplex numbers in 1927, Artin expanded the theory of algebras of associative rings, established in 1908 by J. H. Maclagan Wedderburn, in which the double-chain law for right ideals is assumed; in 1944 he postulated rings with minimum conditions for right ideals (Artin rings). In 1927 he further presented a new foundation for, and extension of, the arithmetic of semisimple algebras over the field of rational numbers. The analytical theory of these systems was treated by his student Käte Hey, in her thesis in 1927.

Artin contributed to the study of nodes in three-dimensional space with his theory of braids in 1925. His definition of a braid as a tissue made up of fibers comes from topology, but the method of treatment belongs to group theory.

Artin's scientific achievements are only partially set forth in his papers and textbooks and in the drafts of his lectures, which often contained new insights. They are also to be seen in his influence on many mathematicians of his period, especially his Ph.D. candidates (eleven in Hamburg, two in Bloomington, eighteen in Princeton). His assistance is acknowledged in several works of other mathematicians. His influence on the work of Nicholas Bourbaki is obvious.

BIBLIOGRAPHY

I. ORIGINAL WORKS. Artin's works are in *The Collected Papers of Emil Artin*, Serge Lang and John T. Tate, eds. (Reading, Mass., 1965), and in the books and lecture notes that are listed there, including "Einführung in die Theorie der Gammafunktion" (1931); "Galois Theory" (1942); "Rings With Minimum Condition," written with C. J. Nesbitt and R. M. Thrall (1944); *Geometric Algebra* (1957); and *Class Field Theory*, written with J. T. Tate (1961). Missing from the list is "Vorlesungen über algebraische Topologie," a mathematical seminar given with Hel Braun at the University of Hamburg (1964).

II. SECONDARY LITERATURE. Works on Artin are R. Brauer, "Emil Artin," in *Bulletin of the American Mathematical Society*, **73** (1967), 27–43; H. Cartan, "Emil Artin," in *Abhandlungen aus dem Mathematischen Seminar der Hamburgischen Universität,* **28** (1965), 1–6; C. Chevalley, "Emil Artin," in *Bulletin de la Société mathématique de France,* **92** (1964), 1–10; B. Schoeneberg, "Emil Artin zum Gedächtnis," in *Mathematisch-physikalische Semesterberichte,* **10** (1963), 1–10; and H. Zassenhaus, "Emil Artin and His Work," in *Notre Dame Journal of Formal Logic,* **5** (1964), 1–9, which contains a list of Artin's doctoral candidates.

BRUNO SCHOENEBERG

ĀRYABHAṬA I (*b.* A.D. 476).

Āryabhaṭa I clearly states his connection with Kusumapura (Pāṭaliputra, modern Patna in Bihar), which had been the imperial capital of the Guptas for much of the fourth and fifth centuries. The assertion of Nīlakaṇṭha Somasutvan (*b.* 1443) that Āryabhaṭa was born in the Aśmakajanapada (this presumably refers to the Nizamabad district of Andhra Pradesh) is probably the result of a confusion with his predecessor, Bhāskara I, as commentator on the *Āryabhaṭīya.* Āryabhaṭa I wrote two works: the *Āryabhaṭīya* in 499 (see Essay V), and another, lost treatise in which he expounded the *ārddharātrika* system (see Essay VI).

The *Āryabhaṭīya* consists of three parts and a brief introduction: *Daśagītikā,* introduction with parameters (ten verses); *Gaṇitapāda,* mathematics (thirty-three verses); *Kālakriyāpāda,* the reckoning of time and the planetary models (twenty-five verses); *Golapāda,* on the sphere, including eclipses (fifty verses). It was translated into Arabic in about 800 under the title *Zīj al-Arjabhar,* and it is to this translation that all the quotations in al-Bīrūnī refer, including those that led Kaye to conclude—mistakenly—that the *Gaṇitapāda* was not written by Āryabhaṭa I.

The *Āryabhaṭīya* has been commented on many times, especially by scholars of south India, where it was particularly studied. The names of those commentators who are known are as follows:

1. Prabhākara (*ca.* 525). His commentary is lost.
2. Bhāskara I (629). His *Bhāṣya* is being edited by K. S. Śukla of Lucknow.
3. Someśvara (*fl.* 1040). His *Vāsanābhāṣya* is preserved in two manuscripts in the Bombay University Library.
4. Sūryadeva Yajvan of Kerala (*b.* 1191). There are many manuscripts of his *Bhaṭaprakāśa,* in south India.
5. Parameśvara (*fl.* 1400–1450). His *Bhaṭadīpikā,* based on Sūryadeva's *Bhaṭaprakāśa,* was published by H. Kern (see below).
6. Nīlakaṇṭha Somasutvan (*b.* 1443). His *Bhāṣya* is published in *Trivandrum Sanskrit Series* (see below).

7. Yallaya (*fl.* 1482). His *Vyākhyāna* is based on Sūryadeva's *Bhaṭaprakāśa;* there is one manuscript of it in Madras and another among the Mackenzie manuscripts in the India Office Library.
8. Raghunātha (*fl.* 1590). His *Vyākhyā* is dealt with by K. Madhava Krishna Sarma, "The *Āryabhaṭīyavyākhyā* of Raghunātharāja—A Rare and Hitherto Unknown Work," in *Brahmavidyā,* **6** (1942), 217–227.
9. Kodaṇḍarāma of the Koṭikalapūḍikula, a resident of Bobbili in the Godāvarī district of Andhra Pradesh (*fl.* 1854). Besides an *Āryabhaṭatantragaṇita,* he wrote a Telugu commentary on the *Āryabhaṭīya* entitled *Sudhātaraṅga;* it was edited by V. Lakshmi Narayana Sastri, in *Madras Government Oriental Series,* **139** (Madras, 1956).
10. Bhūtiviṣṇu. There is apparently only one manuscript (in Berlin) and its apograph (in Washington, D.C.) of his commentary (*Bhāṣya*) on the *Daśagītikā.*
11. Ghaṭīgopa. There are two manuscripts of his *Vyākhyā* in Trivandrum.
12. Virūpākṣa Sūri. There is a manuscript of his Telugu commentary in Mysore.

There also exists a Marāṭhī translation of the *Āryabhaṭīya* in a manuscript at Bombay.

There are several editions of the *Āryabhaṭīya.* That by H. Kern (Leiden, 1874) is accompanied by the commentary of Parameśvara. Kern's text and commentary were reprinted and translated into Hindi by Udaya Nārāyana Singh (Madhurapur, Etawah, 1906). A new edition of the text, with the commentary of Nīlakaṇṭha Somasutvan (who does not include the *Daśagītikā*), was published in three volumes: Vols. I and II by K. Sāmbaśiva Śāstrī and Vol. III by Suranad Kunjan Pillai, in *Trivandrum Sanskrit Series,* **101, 110,** and **185** (Trivandrum, 1930, 1931, 1957). The text has also been published accompanied by two new commentaries, one in Sanskrit and one in Hindi, by Baladeva Mishra (Patna, 1966). The *Gaṇitapāda* was translated into French by Léon Rodet, in *Journal Asiatique,* **7,** no. 13 (1879), 393–434; and into English by G. R. Kaye, in *Journal of the Asiatic Society of Bengal,* **4** (1908), 111–141. Complete English translations have been made by Baidyanath Rath Sastri (Chicago, 1925; unpub.); P. C. Sengupta, *Journal of the Department of Letters of Calcutta University,* **16** (1927), 1–56; and W. E. Clark (Chicago, 1930).

BIBLIOGRAPHY

It is intended here to include references only to those books and articles that are primarily concerned with Āryabhaṭa I and his works; the many other papers and volumes that mention and/or discuss him can be found listed in David Pingree, *Census of the Exact Sciences in India*. Listed chronologically, the references are F.-E. Hall, "On the Ārya-siddhānta," in *Journal of the American Oriental Society*, **6** (1860), 556–559, with an "Additional Note on Āryabhaṭṭa and his Writings" by the Committee of Publication (essentially W. D. Whitney), *ibid.*, 560–564; H. Kern, "On Some Fragments of Āryabhaṭa," in *Journal of the Royal Asiatic Society*, **20** (1863), 371–387; repr. in Kern's *Vespreide Geschriften*, I (The Hague, 1913), 31–46; Bhāu Dājī, "Brief Notes on the Age and Authenticity of the Works of Āryabhaṭa, Varāhamihira, Brahmagupta, Bhaṭṭotpala, and Bhāskarāchārya," in *Journal of the Royal Asiatic Society* (1865), pp. 392–418 (Āryabhaṭa only pp. 392–406, 413–414); L. Rodet, "Sur la véritable signification de la notation numérique inventée par Āryabhaṭa," in *Journal Asiatique*, ser. 7, **16** (1880), 440–485; Sudhākara Dvivedin, *Gaṇakataraṅgiṇī* (Benares, 1933; repr. from *The Pandit*, **14** [1892]), 2–7; Ś. B. Dīkṣita, *Bhāratīya Jyotiḥśāstra* (Poona, 1931; repr. of Poona ed., 1896), pp. 190–210; G. Thibaut, *Astronomie, Astrologie und Mathematik, Grundriss der indo-arischen Philologie und Altertumskunde*, III, pt. 9, (Strasbourg, 1899), 54–55; T. R. Pillai, *Ārybhaṭa or the Newton of Indian Astronomy* (Madras, 1905—not seen—reviewed in *Indian Thought* [1907], pp. 213–216); G. R. Kaye, "Two Āryabhaṭas," in *Bibliotheca mathematica*, **10** (1910), 289–292; J. F. Fleet, "Āryabhaṭa's System of Expressing Numbers," in *Journal of the Royal Asiatic Society* (1911), pp. 109–126; N. K. Mazumdar, "Āryyabhatta's Rule in Relation to Indeterminate Equations of the First Degree," in *Bulletin of the Calcutta Mathematical Society*, **3** (1911/1912), 11–19; J. F. Fleet, "Tables for Finding the Mean Place of Saturn," in *Journal of the Royal Asiatic Society* (1915), pp. 741–756; P. C. Sengupta, "Āryabhaṭa's Method of Determining the Mean Motions of Planets," in *Bulletin of the Calcutta Mathematical Society*, **12** (1920/1921), 183–188.

See also R. Sewell, "The First Arya Siddhanta," in *Epigraphia Indica*, **16** (1921/1922), 100–144, and **17** (1923–1924), 17–104; A. A. Krishnaswami Ayyangar, "The Mathematics of Āryabhaṭa," in *Quarterly Journal of the Mythic Society*, **16** (1926), 158–179; B. Datta, "Two Āryabhaṭas of al-Biruni," in *Bulletin of the Calcutta Mathematical Society*, **17** (1926), 59–74; S. K. Ganguly, "Was Āryabhaṭa Indebted to the Greeks for His Alphabetical System of Expressing Numbers?," *ibid.*, 195–202, and "Notes on Āryabhaṭa," in *Journal of the Bihar and Orissa Research Society*, **12** (1926), 78–91; B. Datta, "Āryabhaṭa, the Author of the *Gaṇita*," in *Bulletin of the Calcutta Mathematical Society*, **18** (1927), 5–18; S. K. Ganguly, "The Elder Āryabhaṭa and the Modern Arithmetical Notation," in *American Mathematical Monthly*, **34** (1927), 409–415; P. C. Sengupta, "Āryabhaṭa, the Father of Indian Epicyclic Astronomy," in *Journal of the Department of Letters of Calcutta University*, **18** (1929), 1–56; S. K. Ganguly, "The Elder Āryabhaṭa's Value of π," in *American Mathematical Monthly*, **37** (1930), 16–29; P. C. Sengupta, "Āryabhaṭa's Lost Work," in *Bulletin of the Calcutta Mathematical Society*, **22** (1930), 115–120; B. Datta, "Elder Āryabhaṭa's Rule for the Solution of Indeterminate Equations of the

First Degree," *ibid.*, **24** (1932), 19–36; P. K. Gode, "Appayadīkṣita's Criticism of Āryabhaṭa's Theory of the Diurnal Motion of the Earth (*Bhūbhramavāda*)," in *Annals of the Bhandarkar Oriental Research Institute*, **19** (1938), 93–95, repr. in Gode's *Studies in Indian Literary History*, II, *Singhi Jain Series*, **38** (Bombay, 1954), 49–52; S. N. Sen, "Āryabhaṭa's Mathematics," in *Bulletin of the National Institute of Sciences of India*, **21** (1963), 297–319; Satya Prakash, *Founders of Sciences in Ancient India* (New Delhi, 1965), pp. 419–449.

DAVID PINGREE

ĀRYABHAṬA II (*fl.* between *ca.* A.D. 950 and 1100).

Of the personality of Āryabhaṭa II, the author of the *Mahāsiddhānta* (or *Āryasiddhānta*), virtually nothing is known. His date can be established only by his alleged dependence on Śrīdhara, who wrote after Mahāvīra (*fl.* 850) and before Abhayadeva Sūri (*fl.* 1050); and by his being referred to by Bhāskara II (*b.* 1114). He must be dated, then, between *ca.* 950 and 1100. Kaye's strange theories about the two Āryabhaṭas, which would have placed Āryabhaṭa II before al-Bīrūnī (963–after 1048), have been refuted by Datta. Nothing further can be said of Āryabhaṭa II; manuscripts of his work are found in Mahārāṣṭra, Gujarat, and Bengal.

The *Mahāsiddhānta* (see Essay VII) consists of eighteen chapters:

1. On the mean longitudes of the planets.

2. On the mean longitudes of the planets according to the (otherwise unknown) *Parāśarasiddhānta*.

3. On the true longitudes of the planets.

4. On the three problems relating to diurnal motion.

5. On lunar eclipses.

6. On solar eclipses.

7. On the projection of eclipses and on the lunar crescent.

8–9. On the heliacal risings and settings of the planets.

10. On the conjunctions of the planets.

11. On the conjunctions of the planets with the stars.

12. On the *pātas* of the sun and moon.

Chapters 13–18 form a separate section entitled *Golādhyāya* ("On the Sphere").

13. Questions on arithmetic, geography, and the mean longitudes of the planets.

14–15. On arithmetic and geometry.

16. On geography.

17. Shortcuts to finding the mean longitudes of the planets.

18. On algebra.

The *Mahāsiddhānta* was edited, with his own Sanskrit commentary, by MM. Sudhākara Dvivedin, in *Benares Sanskrit Series* **148–150** (Benares, 1910).

BIBLIOGRAPHY

Works dealing with Āryabhaṭa II, listed chronologically, are F. Hall, "On the Ārya-Siddhānta," in *Journal of the American Oriental Society,* **6** (1860), 556–559; G. R. Kaye, "Two Āryabhaṭas," in *Bibliotheca mathematica,* **10** (1910), 289–292; J. F. Fleet, "The Katapayadi Notation of the Second Arya-Siddhanta," in *Journal of the Royal Asiatic Society* (1912), 459–462; B. Datta, "Two Āryabhaṭas of al-Biruni," in *Bulletin of the Calcutta Mathematical Society,* **17** (1926), 59–74, and "Āryabhaṭa, the Author of the *Gaṇita," ibid.,* **18** (1927), 5–18; Ś. B. Dīkṣita, *Bhāratīya Jyotiḥśāstra* (Poona, 1931; repr. from Poona, 1896), pp. 230–234.

DAVID PINGREE

ARZACHEL. See **al-Zarqālī.**

ASADA GŌRYŪ (*b.* Kizuki, Bungo Province, Japan, 10 March 1734; *d.* Osaka, Japan, 25 June 1799), *astronomy.*

Asada was instrumental in turning Japanese astronomy and calendrical science away from the traditional Chinese style and toward Western models. His given name was Yasuaki, but he is better known by his pen name, Gōryū. He was the fourth son of Ayabe Yasumasa, a Confucian scholar-administrator of the Kizuki fief government.

Asada taught himself medicine and astronomy. Because Japanese books written on the basis of the Chinese *Shou-shih* calendrical system (promulgated in 1281) were abundant and popular in his youth, his first steps in the study of astronomy must have been to read some of these works. He may also have had direct access to some Chinese writings of the seventeenth- and early eighteenth-century Jesuit missionaries.

Asada placed great weight upon empirical verification, and every time he came across a new theory, he determined its value by observation. The earliest record of an observation by him is that of a lunar eclipse in 1757.

A year before the current official ephemeris caused a crisis of confidence in the official techniques by miscalculating a solar eclipse in 1763, Asada had already pointed out the systematic error and had shown the results of his calculations to his friends. When the day arrived, the eclipse coincided exactly with his calculations. It is apparent that his capacity as a student of astronomy was far superior to that

of the official astronomers of the shogunate. Because the position of official astronomer was hereditary, it often happened that the incumbent lacked the ability required to produce a sound revision of the calendar. Such untalented officials did no more than concentrate on preserving their sinecures. Within this hereditary bureaucracy conservatism naturally prevailed, and the spirit of free inquiry was stifled; innovations were dangerous. But in the eighteenth century astronomical knowledge was diffused by various means to private scholars, who then openly criticized the failure of the official ephemeris. Asada was most prominent among those amateur astronomers.

Asada was financially dependent upon his father until 1767; the freedom from money worries left him free to devote himself to the study of astronomy and medicine. In that year, however, he was appointed a physician of the fief government, and since he was ever on the move accompanying his feudal lord to Edo (Tokyo) or Osaka, he found it impossible to pursue his favorite study. He repeatedly implored his lord to excuse him from service, but in vain. He made up his mind at last to desert his fief and in 1772 went to Osaka, where he resumed the study of astronomy, making his living by practicing medicine. It was during his residence in Osaka that he changed his surname from Ayabe to Asada, because normal relations with his fief could no longer be openly maintained.

Osaka was the right choice for his residence, for the city was the focus of nationwide commercial activities; and the wealthy Osaka merchants, whose financial power often surpassed that of the fief governments, could afford expensive imported books and could support instrument-making and astronomical observation. Such was the case of the wealthy merchant Hazama Shigetomi (1756–1816), one of Asada's most able pupils.

Asada and his school introduced modern instruments and observational methods into Japan. Traditional fieldwork was limited largely to solstitial observations of gnomon shadows, eclipses, and occultations. It was customary to make regular observations during the few years preceding an anticipated calendar reform; beyond this, only occasional checks were made. Earlier astronomers had done little more than make minor amendments to the *Shou-shih* calendar.

Long before, the shogun Tokugawa Yoshimune (1684–1751) had intended to carry out observations with new instruments, but even at the time of the *Hōryaku* calendar reform in 1755, astronomers employed instruments of the traditional Chinese type, such as the gnomon. Japanese observations up to this period had been much inferior to those of Kou Shouching in thirteenth-century China; the data could be used only to check a calendar, not to make significant improvements.

Now the Asada school began to gather more reliable data. Asada himself initiated techniques for precise observation. He ground lenses and made a telescope, which he used to observe the movements of Jupiter's satellites. Hazama showed the greatest talent of his time for inventing and improving instruments; expending his wealth freely, he also sponsored the training of talented instrument-makers and conducted systematic observations with the assistance of his employees.

Only a few of Asada's treatises exist. One of them, *Jikkenroku* ("Records Based on Observations," 1786), gives the essentials of his calendar; taking the winter solstice of 1781 as the temporal origin and Osaka as the standard station, he gives the fundamental constants and his method for determining the positions of the sun and the moon, and calculating solar and lunar eclipses. He adopted the method described in the first volume of the Chinese *Li-hsiang k'ao-ch'eng* ("Compendium of Calendrical Science," 1713)—essentially Tychonian in content—which was his chief means of studying Western astronomy. The constants employed for calculation were mostly new ones that Asada had worked out from his own data.

In the spring of 1793 he made a considerable correction in the constant for the distance between the sun and the earth, and corrected other constants affected by it. That the corrected constant is almost identical with that given in the partly Keplerian sequel to the *Li-hsiang k'ao-ch'eng* (*hou p'ien,* compiled in 1737) shows that Asada had access to the sequel about this time, when he came across the theory of the elliptic orbit and became aware of the gross error in the constant for the distance of the sun given in the first volume. It seems, remarkably enough, that Asada retained the old theory concerning the mode of the motions of the sun and the moon while he made radical corrections in the important constants, evidently for the purpose of testing their adaptability. From the time that Asada obtained the sequel to the *Li-hsiang k'ao-ch'eng,* through the efforts of Hazama, he and his best pupils occupied themselves with studying the theory of the elliptic orbit.

The sequel employs Kepler's first and second laws without reference to the heliocentric system. Dynamics, as an approach, is absent, and the name of Newton is associated only with observational data, most of which are J. D. Cassini's. The arrangement of the treatise is to a great extent that dictated by traditional Chinese calendar-making practice. Within this framework it was unnecessary to relate Kepler's laws

to heliocentric coordinates. Lack of interest in planetary motion in traditional calendar-making seems to have made adoption of the third law unnecessary.

Asada's pupils attributed two major innovations to him—cyclic variation of astronomical parameters and independent discovery of Kepler's third law.

He mistakenly claimed the discovery of the Antarctic continent by means of lunar eclipse observations. After repeated observations of the shadow of the earth projected on the surface of the moon, Asada came to believe that the parts of the shadow corresponding to the South Pole and the Asian continent are somewhat more upthrust than the other parts of the shadow, and identified part of the shadow outline with the Antarctic continent, which appeared on a world map newly imported from the West.

Asada, after settling down in Osaka, was busily engaged for twenty years in making observations and in educating his pupils. Consequently, his influence increased and led to the formation of an important school of calendrical scientists. For this reason, when the shogunate proposed to revise the current Japanese calendar by use of the new theories of Western astronomy and found that its own official astronomers were not equal to the task, it turned to him. Instead of accepting the appointment himself, Asada recommended his best pupils, Takahashi Yoshitoki (1764–1804) and Hazama. Takahashi, since he belonged to the Samurai class (although he was only a minor official), was appointed an official astronomer; Hazama became a consultant or assistant.

When Asada deserted his fief and hid in Osaka, his feudal lord, to whom his whereabouts was known, was generous enough not to charge him with the crime of desertion, and even permitted him to communicate with his relatives at home, because he was eager for him to succeed in his pursuit of learning. This made it possible for Asada to receive a gift of some money each year from his eldest brother and thus to study astronomy without being destitute. Asada was so grateful to his former master for this that when he became a well-known scholar and was offered a high position by other feudal lords—and even by the shogunate—he always refused, saying that he could not turn his back on his former lord.

In Asada's later years a limited number of Japanese pioneers, the *rangakusha* (scholars of Dutch learning), undertook the prodigious and almost completely unaided labor of translating Dutch scientific works into Japanese. In the 1770's a notable expansion of the study of the Dutch language and of science led to a movement for the translation of Dutch scientific works—or retranslation of Dutch translations of European works. This task was begun by two groups, the official interpreters at Nagasaki, who alone were authorized access to foreign books, and the physicians of Edo.

The interpreters concentrated on introducing the core of genuine Western science—particularly of elementary astronomy, navigation, and geography—but this material did not have much interest for practical astronomers, except when directly applicable to traditional calendar-making. Men like Asada wanted observational data and astronomical constants. For this reason the writings of the Jesuit missionaries in China, although cosmologically obsolete, were much more useful to him. He himself had found no time to learn the Dutch language.

In studying medicine, as in studying astronomy, Asada collected and read the very best literature. He also dissected dogs and cats, and thus became acquainted with internal organs. From Asada's expounding of Western-style astronomy, some people conclude that his medical art was copied from some Dutch school of medicine. This is entirely wrong. At the time that Asada taught himself there were a few Japanese physicians who taught a type of very elementary Western-style surgery, but there was no proper literature for students. As for the European style of internal treatment, no one in Japan knew about it, and naturally there was nothing written on it. Asada could have had no access to what knowledge was available. Thus, his style of medicine was not Western, but the positivistic and clinical koihō (ancient medical learning) school that flourished during the mid-eighteenth century in Japan.

In Osaka, Asada practiced medicine, but it was only a means to make his living. All his energy had been devoted to the study of astronomy, but now his research was temporarily completed. When, in addition, Takahashi and Hazama, on whom his mantle had fallen, accomplished the task of revising the official calendar in 1797 and Asada himself was rewarded by the shogunate, he could say, "I have found men who will develop my astronomy. I have henceforth to devote myself to the study of medicine."

His health failed about the beginning of 1798 and he suffered a stroke, of which he died in the summer of the following year.

In praise of Asada, his distinguished pupil Takahashi stated in his Zōshū shōchō hō ("Variations of Astronomical Parameters," rev. and enl., 1798):

> Laboring over Chinese and Western works, Asada Gōryū at Osaka discovered the shōchō law. Although Western astronomy is most advanced, we have not heard of its mentioning this law, known only in our country. Therefore I have said that although we are unable to boast about our achievements in comparison with those

of the Westerners, my country should be proud of this man and his discovery.

This is perhaps the only notable originality to be found in the entire history of Japanese astronomy; it therefore merits critical examination.

In adopting the idea of shōchō (hsiao-ch'ang in Chinese, the secular diminution of tropical-year length), astronomers at the time of the Shou-shih and Jōkyō (promulgated in 1685) calendars were required only to account for the ancient records and modern data of Chinese solstitial observations by a single formula. While it is true that neither the Jesuits nor the Chinese had incorporated the concept of hsiao-ch'ang into their calendars during the Ming and Ch'ing periods, the Jesuit compilation Ch'ung-chen li-shu ("Ch'ung-chen Reign Period Treatise on Calendrical Science," 1635) pointed out three possible causes of variation in tropical-year length: (1) rotation of the center of the solar orbit in reference to the earth (perhaps referring to the progressive motion of the solar perigee); (2) variation of the eccentricity of the solar orbit; (3) variable precession (trepidation). Numerical values were not given, however, because such a minute parameter was not determinable within a single lifetime.

Classical Western data, such as those listed in the Almagest of Ptolemy, became available to Asada through the Jesuit treatises. He dared endeavor to synthesize Western and Chinese astronomy and to give a numerical explanation, by means of a single principle, of all the observational data available to him—old and new, Eastern and Western.

It seems that Asada did not fully comprehend the epicyclic system, based on that of Tycho Brahe, which appeared in the Jesuit works. In Western astronomy, only observed data and numerical parameters interested him. These he could utilize for his purely traditional approach, that of obtaining an algebraic representation that corresponded as closely as possible to the observed phenomena.

Copernicus appears in the Ch'ung-chen li'shu, not as an advocate of heliocentrism but as an observational astronomer and the inventor of the eighth sphere of trepidation. He is said in that work to have believed that the ancient tropical year was longer than that of the Middle Ages, which in turn was shorter than the contemporary constant. Asada, perhaps struck by this passage, formulated a modified conception in which the length of the ancient tropical year tended to decrease until it reached a minimum in the Middle Ages and to grow longer afterward, varying in a precession cycle of 25,400 years. The minimum was not associated with the solar perigee, but was arbitrarily chosen in order to fit the recorded data. He also presumed that the only perpetual constant was the length of the anomalistic (sidereal) year. Other basic parameters, such as the length of the synodic, nodical, and anomalistic months, were assumed to be subject to variation in a precession cycle. This idea seems to have originated in the Chinese *T'ung-t'ien* calendar (1199) of the Sung period. In the West, the first systematic study of the variation of basic astronomical parameters was carried out by Laplace on the basis of the perturbation theory. Although superficially similar, Asada's approach was by no means comparable with Laplace's well-founded theoretical considerations.

Although the mathematical derivation is quite complicated, the length of the tropical year, T, in Asada's formula is essentially expressed in terms of the equations

$$T = 365.250469717756 - 1.038645 \times 10^{-5}t,$$

where t is years elapsed since the epoch of 720 B.C. (this equation is valid up to A.D. 133), and

$$T = 365.2416204385 + 0.0435370 \times 10^{-5}t,$$

where t is years elapsed since the epoch of A.D. 133 (this equation is valid up to A.D. 11981).

These two equations together cover only half of the precession cycle since 720 B.C. In the other half of the cycle, t is expressed in the dotted line of Figure 1. Applying this formula to historical observations, we see in Figure 2 the extent to which it reconciles the data. After A.D. 133, the year of the epoch, the formulas of Simon Newcomb, Asada, and the *Shou-shih* calendar roughly coincide. Before the epoch, Asada's formula appears as a parabola of deep curvature, which comprehends the Greek observations as well as the ancient Chinese records. It is apparent

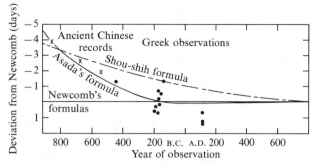

The X-axis represents the calculated value following Newcomb's formula. Each observation is plotted with the indication of the amount of deviation from calculation in the Y-axis direction (negative values represent times earlier than the calculated date, and positive later).

FIGURE 2

that what Asada really intended to do was account for the newly acquired Western data. His basic goal, that of "saving the ancient records" by numerical manipulation, differs not at all from that of the traditional approach. His consideration of the precession cycle was theoretical decoration.

In spite of resistance from the conservative hereditary official astronomers, Asada's pupils finally succeeded in applying Asada's variation term to the *Kansei* calendar promulgated in 1798. In the same year, Takahashi wrote the *Zōshū shōchō hō* in order to provide a theoretical foundation for his teacher's method. Takahashi had attained a mastery of the theory of spherical geometry and epicycles. Furthermore, rejecting the authority of Tycho, he revived the old idea of trepidation as contained in the *Ch'ung-chen li-shu,* in which trepidation was somewhat vaguely mentioned in order to contrast it with Tycho's more accurate view. Unlike Alphonsine trepidation, which had a 7,000-year cycle, however, Takahashi's cycle of trepidation had the same period as the cycle of precession.

The falsity of Asada's variation concept soon became apparent. In the 1830's it was realized that observations did not agree with the *Kansei* calendar; removal of Asada's variation factors gave better agreement. Asada's idea was doomed. In the next calendar reform, that of *Tenpo* (1843), it was entirely neglected.

During the Tokugawa period (seventeenth to early nineteenth centuries) Japanese astronomers were continually preoccupied by the contrast between Chinese and Western astronomy. While they generally followed Chinese astronomy in the first half of the period, Western astronomy became dominant during the latter half. During the period of transition there appeared mental attitudes like those of Asada,

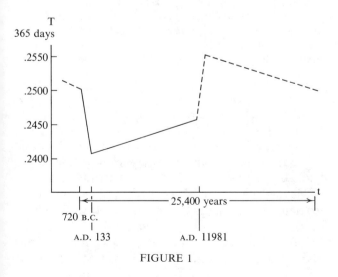

FIGURE 1

who tended to syncretize and synthesize Chinese and Western astronomy. His originality proved to be rather anachronistic, however, in view of the rapid contemporary development of Western astronomy.

His pupils claimed for Asada the honor of having independently discovered the relationship between the distances of planets from the sun and the periods of their revolution (in other words, Kepler's third law), although he did not publish it. Ōtani Ryōkichi maintained, however, that the law was first known in Japan in 1800, after Asada's death, when his pupils obtained the Dutch translation of J. J. L. de Lalande's *Astronomie*. Kepler's third law had, as a matter of fact, been described in the *Tenmon kanki* ("Astronomical Collection," 1782), one of Shizuki Tadao's draft translations of the Dutch version of John Keill's *Introductiones ad veram astronomiam*. One suspects that Asada somehow had the chance to become acquainted with Shizuki's works.

There is another possible interpretation. Prior to Asada's time, Chinese and Japanese observations of planetary motions were infrequent and imprecise. Even if neither the Keplerian planetary theory nor adequate observational data were available to him, he was fully acquainted with the first two volumes of the *Li-hsiang k'ao-ch'eng,* in which values for the relative sizes of planetary orbits are mentioned. These are close to the modern values, as they are calculated by trigonometry in the post-Copernican fashion. Thus, Asada might have tried a blind search for some numerical relationship between these values and the planets' revolution periods, and finally reached Kepler's third law independently.

Asada was not geometry-oriented, as were Western astronomers, but he was accustomed to the traditional algebraic approach and was fascinated by numerical manipulation. It would have been impossible for him to recognize the potential importance of Kepler's third law, which led to the discovery of Newton's inverse-square law and the establishment of modern mechanics.

Although Asada missed the meaning of his alleged discovery, it is interesting that his pupils believed that Asada had suggested a crude analogy between a balance and the solar system in a pseudomechanistic interpretation of Kepler's third law. Hazama elaborated it in his own fashion, as follows:

If we express a weight in a balance as the area of a square, the side of which is x, the relationship between arm length a and weight x^2 is $ax^2 = $ constant. A similar relationship holds between length l and frequency v of a pendulum: $lv^2 = $ constant.

The radius of planetary orbit r seems to correspond to balance arm length a, which in turn corresponds to pendulum length l. The velocity of the planet v is taken to correspond to the frequency of the pendulum v. An analogous relationship between r and v thus should hold in the planetary domain: $rv^2 = $ constant. Substituting $2r/T$ ($T = $ period of revolution) for v in the above equation, $r^3/T^2 = $ constant is obtained. In this way, we arrive at Kepler's third law.

The argument is, of course, quite misleading, but at a time when modern mechanics was not well understood, this crude analogy between planetary motion and simple mechanical laws gave an explanation satisfactory to Hazama's contemporaries.

BIBLIOGRAPHY

I. ORIGINAL WORKS. Asada has left unusually few works. Almost all of his ideas appear in his pupils' writings and compilations. Notable among his own works are *Jichūhō* ("The Method of the Jichū Calendar" [1786]), MS preserved in Mukyūkai Library; *Shōchō hō* ("Variation of Astronomical Parameters" [1788]), MS preserved in Tohoku University Library; *Gekkei o motte nisshoku o osu hō* ("A Method to Predict Solar Eclipses by Means of Observing the Moon's Shadow" [n.d.]), MS preserved in the Sonkeikaku Library; *Gosei kyochi no kihō* ("Remarkable Law of Planetary Distance" [n.d.]), MS preserved in the Sonkeikaku Library; and Takahashi Yoshitoki, *Zōshū shōchō hō* ("Variation of Astronomical Parameters," enl. and rev. [1798]), MS preserved in the Tokyo Astronomical Observatory.

II. SECONDARY LITERATURE. Works on Asada in English are Shigeru Nakayama, "Cyclic Variation of Astronomical Parameters and the Revival of Trepidation in Japan," in *Japanese Studies in the History of Science*, no. 3 (1964), 68–80, and *Outline History of Japanese Astronomy* (Cambridge, Mass., in press); and Ryōkichi Ōtani, *Tadataka Inō* (Tokyo, 1932).

A work on Asada in Japanese is Nishimura Tachū (one of Asada's four major pupils), "Asada sensei gyōjōki" ("Achievements of Master Asada") in Nomura Jun *et al.,* eds., *Nishimura Tachū jiseki* ("Achievements of Nishimura Tachū" [Toyama, 1934]). Watanabe Toshio is now preparing a comprehensive biography of Asada.

SHIGERU NAKAYAMA

ASCLEPIADES (*b.* Prusa, Bithynia, *ca.* 130 B.C.; *d.* Rome, *ca.* 40 B.C.), *medicine.*

Trained originally as a philosopher and orator, Asclepiades achieved fame as a physician in Rome. He had a large practice and wealthy patients, and was befriended by Cicero, Crassus, and other influential Romans. At the height of his fame he was invited to become the personal physician of Mithridates Eupator, king of Pontus, but he declined and remained in Rome until his death. His forceful personality, his clinical successes, and the simplicity of his therapeutic counsels aided in the acceptance of his doctrines and did much to overcome the Roman prejudice against Greek medicine. None of his many writings has survived intact.

Asclepiades rejected the teachings of Hippocrates and other advocates of humoralism in favor of his own original system of solidism. Thoroughly and consistently materialistic, his doctrines derived partially from Epicurean atomism and partially from the now lost, quasi-atomistic teachings of Heracleides Ponticus. The most important ideas derived from his predecessors were a theory of knowledge based upon sensory appearances alone and the rejection of teleology. These, plus his atomism, provided the philosophical basis of his medical theories. All diseases, Asclepiades held, resulted from an abnormal arrangement of the atoms relative to the "pores" that constituted the physical basis of the human body. Since he also denied the healing power of nature, it followed that diseases could be cured by human intervention. His therapy was simple but effective, and capable of innumerable modifications, depending upon the patient's condition and his purse. He relied principally upon diet, exercise, massage, and baths, avoiding, whenever possible, powerful drugs and surgery. The purpose of therapy was to restore the atomic constituents of the human body to their normal state of unimpeded movement. By means of a controlled regimen, he claimed that he was capable of achieving the goal of a good physician—*curare tuto, celeriter et jucunde* ("to cure safely, swiftly, and pleasantly").

Despite the rigid outlines of his mechanistic system, Asclepiades wisely was not always consistent in his adherence to it or in its practical application. His writings on the pulse show the influence of Stoic pneumatic theory. He recognized a limited role for enemas, surgery, bloodletting, and cupping; but he was violently opposed to the study of anatomy. For this reason, and perhaps because of his extravagant claims of success, he was much maligned and criticized for charlatanry. His influence in antiquity, however, was great, and he is often cited by Caelius Aurelianus, Celsus, and Galen. The philosophical and theoretical foundations of the methodist school can, in part, be traced back to Asclepiades' pupil, Themison of Laodicea.

Among the several hundred fragments that survive (many are in Wellmann), the following titles of his writings are recorded: "On Acute Diseases," "On the Preservation of Health," "Common Aids," "Practices," "To Geminius on Hygiene," "On Enemas," "On Periodic Fevers," "On Baldness," "On Pestilence," "On Dropsy," "On the Use of Wine," "On the Elements," "Definitions," "On Respiration and the Pulse," "On Ulcers," and commentaries on the *Aphorisms* and *In the Surgery,* both attributed to Hippocrates.

BIBLIOGRAPHY

Works dealing with Asclepiades are E. Gurlt, *Geschichte der Chirurgie und ihrer Ausübung,* I (Berlin, 1898), 329–330; W. A. Heidel, "The ἄναρμοι ὄγκοι of Heraclides and Asclepiades," in *Transactions of the American Philological Association,* **40** (1909), 5–21; T. Meyer-Steineg, *Das medizinische System der Methodiker* (Jena, 1916), pp. 5–18; and Max Wellmann, "Asklepiades aus Prusa" ["Asklepiades 39"], in Pauly-Wissowa, *Real-Encyclopädie der classischen Altertumswissenschaft,* and "Asklepiades aus Bithynien von einem herrschenden Vorurteil befreit," in *Neue Jahrbücher für das klassische Altertum,* **21** (1908), 684–703, the best modern study.

JERRY STANNARD

ASELLI, GASPARE (*b.* Cremona, Italy, 1581; *d.* Milan, Italy, 9 September 1625), *anatomy.*

A descendant of an ancient patrician family, Aselli revealed a marked propensity for the natural sciences early in his schooling. He studied medicine at the University of Pavia, where he soon distinguished himself among his fellow students. His teacher of anatomy was Giambattista Carcano-Leone, a pupil of Fallopio and author of *De cordis vasorum in foetu unione* (1574).

Later Aselli moved to Milan, where he gained recognition in his profession. Since his scientific preparation was essentially in anatomy, he distinguished himself in his practice of surgery. He was appointed first surgeon of the Spanish army in Italy from 1612 to 1620. In Milan he had the opportunity to continue his anatomical researches, which won him honorary citizenship of that city and an outstanding position in the history of anatomy. He died at the age of forty-four from an acute and malignant fever.

Aselli's scientific activity occurred during the first decades of the seventeenth century, in an atmosphere that was particularly sympathetic to anatomical studies, especially in northern Italy. At the end of the sixteenth century the study of descriptive human anatomy had made considerable progress. The early decades of the seventeenth century felt the effects of the baroque attitude toward science and the influences of the new mechanical concepts of Galileo. Anatomy, which had been essentially static in the sixteenth century, now assumed a dynamic character. It was enlivened by a new consideration of physiology.

Aselli discovered the chylous vessels, although it would perhaps be more correct to refer to his work as a rediscovery rather than a discovery. According to information that has come down to us from Galen, Herophilus, and Erasistratus, both Hippocrates and Aristotle had with considerable clarity already pointed to the existence of the so-called absorbent vessels. Nevertheless, not even Eustachi and Fallopio, who in the sixteenth century had noted and described the thoracic duct in the horse and the deep lymphatics in the liver, respectively, had succeeded in clarifying the functional significance of these vessels.

On 23 July 1622, during a vivisection performed on a dog that had recently been fed, Aselli, in the course of removing the intestinal tangle to reveal the abdominal fasciae of the diaphragm, noticed numerous white filaments ramified throughout the entire mesentery and along the peritoneal surface of the intestine. The most obvious interpretation was that these filaments were nerves. The incision of one of the larger of these "nerves" released a whitish humor similar to milk. Therefore, Aselli interpreted these formations as a multitude of small vessels, proposing to call them "aut lacteas, sive albas venas." The vivisection had been performed at the request of four of his friends, among whom were Senator Settala and Quirino Cnogler, to demonstrate the recurrent nerves and the movements of the diaphragm.

Immediately following his discovery, Aselli began a systematic study of the significance of these vascular structures. He recognized the chronological relationship that existed between their turgidity and the animal's last meal. His experimental findings enabled Aselli to observe the chylous vessels in different species of animals. The results of these investigations were collected in *De lactibus sive Lacteis venis quarto vasorum mesaraicorum genere novo invento Gasparis Asellii Cremonensis anatomici Ticinensis dissertatio* (1627), which is divided into thirty-five chapters that are followed by an index and preceded by four charts with accompanying commentaries and a portrait of the author. Besides their intrinsic scientific value, the importance of the charts lies in the new technique used in composing them: they were the first anatomical illustrations to appear in color. Aselli used color because he felt that several tints were needed in order to distinguish the various types of vessels more clearly.

Aselli traced the course of the chylous vessels to the mesenteric glands and probably confused them with the lymphatics of the liver; therefore he did not follow their course to the thoracic duct. (His discovery occurred six years before the publication of Harvey's *De motu cordis,* and so the Galenic concept of the liver as the center of the venous system still appeared valid.) Harvey himself believed that the absorption of the chyle took place through the mesenteric veins and that the liver generated the blood. With the discovery of the thoracic duct in the dog by Jean Pecquet in 1651, the old Galenic error, according to which the vessels of the intestine carried the chyle to the liver, was corrected. In addition to noting and describing the valvular apparatus of the chylous vessels, Aselli attempted to interpret their functional significance in health and in disease.

BIBLIOGRAPHY

I. ORIGINAL WORKS. Aselli's only published work is *De lactibus sive Lacteis venis quarto vasorum mesaraicorum genere novo invento Gasparis Asellii Cremonensis anatomici Ticinensis dissertatio* (Milan, 1627; Basel, 1628; Leiden, 1640; Amsterdam, 1645), which was issued at the insistence of his friends Senator Settala and Alessandro Tadino.

Two earlier works by Aselli, *De venenis* and *Observationes chirurgicae,* are presumed lost. Two series of manuscripts are extant: the more important is preserved in the Archives of the Civic Museum of Pavia. In addition to other lectures and charts, it contains the full text of the lectures on the chylous vessels that Aselli delivered in 1625. The other series concerns the *Observationes chirurgicae;* formerly in the possession of the noble Belgioioso family, it was turned over in 1920 to the Trivulziana Library of Milan.

II. SECONDARY LITERATURE. Writings on Aselli are F. Argelati, *Biblioteca scriptorum Mediolan* (Milan, 1755), II, 2058; H. Boruttau, "Geschichte der Physiologie . . . ," in Theodor Puschmann's *Handbuch der Geschichte der Medizin,* II (Jena, 1903), 335–336; C. A. Calderini, *Storia della letteratura e delle arti in Italia* (Milan, 1836), II, 379; P. Capparoni, *Profili bio-bibliografici di medici e naturalisti celebri italiani dal sec. XV al sec. XVIII* (Rome, 1928), II, 70–72; B. Corte, *Notizie istoriche intorno a' Medici Scrittori Milanesi* (Milan, 1718), p. 176; V. Ducceschi, "I manoscritti di Gaspare Aselli," in *Archivio di storia della scienza,* 3 (1922), 125–134; J. F. Fulton, "The Early History of the Lymphatics," in *Bulletin of the Hennepin County Medical Society,* 9 (1938), 5; J. I. Mangeti, *Bibliotheca scriptorum medicorum* (Geneva, 1731), I, 185; G. Mazzucchelli, *Gli scrittori d'Italia* (Brescia, 1753), I, part 2, 1159–1160; L. Premuda, *Storia dell'iconografia anatomica* (Milan, 1957), pp. 163–164; R. von Töply, "Geschichte der Anatomie," in Puschmann's *Handbuch (supra),* pp. 215–216; and G. Zoia, *Cenno sulla vita di Gaspare Aselli anatomico del secolo XVII* (Pavia, 1875).

LORIS PREMUDA

ASHMOLE, ELIAS (*b.* Lichfield, England, 23 May 1617; *d.* London, 18 [or 19] May 1692), *natural philosophy.*

Elias Ashmole was the only child of Simon Ashmole, a Lichfield saddler, and Anne Ashmole (née Bowyer). In keeping with the family's humble social position, Elias was educated at Lichfield Grammar School, in the expectation that he would enter a craft. Through the intervention of James Pagit, a relative of the Bowyer family, he settled in London, however, and there obtained legal training. He established a law practice in 1638 but practiced only sporadically, for he was involved in the Royalist faction during the first part (from 1641 to 1646) of the Civil War. A fortunate marriage then provided him with independent means that reduced his reliance on the income from his law practice.

In 1660 Ashmole's loyalty to the crown was rewarded by Charles II, who granted him the offices of comptroller of the excise and Windsor herald. He was successful in both offices; the former provided a large income, and he devoted his intellectual energies to the latter. He contributed greatly to the revival of English heraldry and in the course of his work gained an encyclopedic knowledge of its history and complex rules.

His chambers at the Middle Temple and South Lambeth house were used to display substantial and famous antiquarian collections. Ashmole's office at the College of Arms brought him considerable social prestige, and he became well known in court circles. In religious and political matters he was excessively conservative and orthodox. His closest friends, however, were drawn from the eccentric and varied class who shared his obsession with alchemy and astrology. These interests showed no diminution in his later life, despite his association with the Royal Society, whose view of nature undoubtedly did not accord with his own.

Ashmole married three times but had no children. His first wife was Eleanor Manwaring; the second was a rich widow, Lady Mary Manwaring (née Forster); the third was Elizabeth Dugdale, daughter of Sir William Dugdale, Ashmole's friend and a prominent antiquarian.

Ashmole's scientific outlook was deeply influenced by the mathematicians and astrologers with whom he associated during the Civil War. Both factions in the war made use of astrology, which became a fashionable and respectable scientific pursuit; its influence affected various disciplines from mathematics to medicine. While serving in the Oxford garrison in 1645, Ashmole became acquainted with the Royalist astrologers George Wharton and Sir John Heydon. After retiring from the conflict in 1646, he formed his most significant and lasting astrological friendship, with the Parliamentarian astrologer William Lilly, who had greatly contributed to the revival of astrology in England. Ashmole's first published writings were two short translations of astrological works included in Lilly's *Worlds Catastrophe* (1647).

From astrology, Ashmole gradually extended his interests to botany, medicine, and stenography. Inevitably, alchemy attracted him, and he became the eager pupil of William Backhouse of Swallowfield, Berkshire. Alchemy appealed to Ashmole's mystical and antiquarian instincts. As a practical and contemplative study, it appeared to offer the key to the secrets of nature; its antiquarian aspect lay in the collection and publication of the rare and often corrupt texts that formed the basis for its theory. Ashmole aspired to publish the "choicest flowers" of alchemical literature, and his first book, the *Fasciculus chemicus* (1650), was a modest translation of works by Arthur Dee and Jean d'Espagnet. By this time he had embarked upon a more ambitious enterprise— the restoration of English astrology. He hoped to produce a comprehensive collection of English verse and prose alchemical works, drawn from manuscript sources. This project had an auspicious beginning with the publication of the *Theatrum chemicum Brittanicum* (1652), a collection of verse alchemical works that displayed Ashmole's industry, erudition, and editorial skill. The *Prolegomena* indicate familiarity and agreement with the leading themes of Hermetic philosophy.

After the publication of this work, Ashmole's alchemical activities diminished. The only other English alchemical work he published was *The Way to Bliss* (1658). Ashmole saw no incompatibility between the occult sciences that he favored and the experimental natural philosophy that was becoming a dominant influence among his contemporaries. He regarded them as complementary means of discovering the fundamental principles of natural philosophy. Consequently, he was mildly interested in experimental science, became a founding fellow of the Royal Society in 1660, and made significant bequests to the society's museum.

Ashmole's other scientific activities diminished considerably after 1660, as he concentrated on the duties connected with his crown appointments. His major intellectual energies were increasingly absorbed by the compilation of the history of the Order of the Garter, begun in 1655 and completed in 1672. This book has become a major reference work on many aspects of heraldry. At the same time he continued collecting books, manuscripts, and archaeological and scientific "rarities." This was not an uncommon avocation among the gentry of Restoration England, but Ashmole exceeded others in his zeal for collection and in his desire to replace ephemeral personal "cabinets" with permanent public museums.

In 1675 he offered his collections to Oxford University, on the condition that suitable housing was provided for it. The university gladly complied, and the elegant museum, designed by Thomas Wood, was opened in 1683. This was the first English public museum, and Ashmole actively supervised its affairs. He added to the collections and persuaded other collectors, notably the zoologist Martin Lister, to make donations. His protégé, Robert Plot, who shared his patron's alchemical interests, was appointed keeper of the museum and first professor of chemistry. The museum was equipped with a laboratory that became the focus for scientific activities at Oxford. Thus, paradoxically, Ashmole created the institutional basis for the growing tradition of experimental science while himself representing the declining magical and astrological outlook.

BIBLIOGRAPHY

I. ORIGINAL WORKS. Ashmole's works include *Fasciculus chemicus: Or, Chymical Collections. Expressing the Ingress, Progress, and Egress, of the Secret Hermetick Science, out of the Choicest and Most Famous Authors* (London, 1650), written under the name of James Hasolle and presenting translations of Arthur Dee's *Fasciculus chemicus* (1629) and Jean d'Espagnet's *Arcanum hermeticae philosophiae* (1623); *Theatrum chemicum Britannicum. Containing Severall Poeticall Pieces of Our Famous English Philosophers, Who Have Written the Hermetique Mysteries in their owne Ancient Language* (London, 1652), reprinted with an introduction by A. G. Debus as No. 39 in the series *Sources of Science* (New York, 1967); *The Way to Bliss. In Three Books* (London, 1658); *Sol in Ascendente: Or, The Glorious Appearance Of Charles the Second, Upon The Horizon of London* (London, 1660), an anonymously published poem celebrating the Restoration; *A Catalogue of the Peers of the Kingdome of England* (London, 1661); *A Brief Narrative of His Majestie's Solemn Coronation* (London, 1662); "The Worlds Catastrophe, or the Miraculous Changes, and Alterations, . . . in several Kingdoms, and Commonwealths of Europe . . . ," translated from Franciscus Spina's *De mundi catastrophe* (1625), and "A Prophecie of Ambrose Merlin . . . ," from Geoffrey of Monmouth's *Historia regum Britanniae,* in William Lilly's *The Worlds Catastrophe, Or, Europes many Mutations untill 1666* (London, 1647); and *The Institution, Laws & Ceremonies of the Most Noble Order of the Garter* (London, 1672).

II. SECONDARY LITERATURE. More information on Ashmole and his work can be found in Mea Allen, *The Tradescants* (London, 1964), pp. 200–207; Thomas Birch, *History of the Royal Society,* 4 vols. (London, 1756/7), I, 4,8; J. Campbell, in *Biographia Britannica,* 2nd ed. (London, 1778), I, 293–307; J. Ferguson, *Bibliotheca chemica,* 2 vols. (Glasgow, 1906), II, 52–53; C. H. Josten, *Elias Ashmole (1617–1692). His Autobiographical and Historical Notes, His Correspondence, and Other Contemporary Sources Relating to His Life and Work,* 5 vols. (Oxford, 1966), the most comprehensive work on Ashmole, superseding the partial editions of the *Diary* by C. Burnan (1774) and R. T. Gunther (1927), and "William Backhouse of Swallowfield," in *Ambix,* 4 (1949), 1–33; R. Rawlinson, "Some Memoires of the Life of Elias Ashmole," prefixed to *The Antiquities of Berkshire* (London, 1719), which is incorrectly attributed to Ashmole; Taylor E. Sherwood, "Alchemical Papers of Dr. Robert Plot," in *Ambix,* 4 (1949), 61–79; and A. Wood, *Athenae Oxonienses,* P. Bliss, ed., 4 vols. (London, 1817), III, 354–364.

CHARLES WEBSTER

ASSALTI, PIETRO (*b.* Acquaviva Picena, Italy, 23 June 1680; *d.* Rome, Italy, 29 April 1728), *medicine.*

A descendant of a family who had held public office in the small town of Fermo in the fifteenth and sixteenth centuries, Assalti received his early education, which included Latin, in Acquaviva Picena. At fifteen he went to Fermo, where he studied Greek, Hebrew, Syriac, and Arabic. Four years later, following his father's wishes, he went to Rome to study law; he continued, however, to show increasing interest in languages and natural history. Assalti soon earned a reputation as a scholar and was chosen by Pope Clement XI to be one of the "writers" of the Vatican Library, where he had the means to broaden the scope of his learning.

In 1709 Assalti was appointed professor of botany at the University of Rome, succeeding Giovanni Battista Trionfetti; his teaching of the natural sciences led to his appointment as professor of anatomy in 1719 and, two years later, of theoretical medicine, with a stipend of 234 *scudi.*

Assalti collected the works of Lancisi and published them in two volumes (1718), and he was almost certainly the author of the scholarly and elegant annotations to Michele Mercati's *Metallotheca* (1717). His letter to Morgagni regarding Lancisi was published as a preface to Lancisi's *De motu cordis et aneurysmatibus* (1728). Also worthy of mention is his unpublished "Oratio de incrementis anatomicae in hoc saeculo XVIII," which, among other topics, deals with the function of the spleen.

In *De incrementis anatomie* Lorenz Heister writes most enthusiastically of Assalti, and the anatomist G. B. Bianchi (1681–1761) describes him as "eximium Medicum virum longe doctissimum, et in linguis eruditis peritissimum."

BIBLIOGRAPHY

I. ORIGINAL WORKS. Assalti collected and edited Lancisi's work as *Opera omnia in duos tomos distributa*, 2 vols. (Geneva, 1718). The year before, his annotations to Michele Mercati's *Metallotheca* (Rome, 1717) had appeared. He was also responsible for the preface to Lancisi's *De motu cordis et aneurysmatibus* (Rome, 1728).

II. SECONDARY LITERATURE. Works on Assalti are A. Bacchini, *Vita ed opere di G. M. Lancisi* (Rome, 1920), p. 32; Biblioteca Comunale di Fermo, MSS 8/3, in *Notizie raccolte da Rodolfo Emiliani sulla famiglia Assalti e su Pietro Assalti;* V. Curia, *L'Università degli Studi di Fermo* (Ancona, 1880), p. 71; G. Natalucci, *Medici insigni italiani antichi e contemporanei nati nelle Marche* (Falerona, 1934), pp. 49–50; G. Panelli, *Memorie degli uomini illustri e chiari in medicina del Piceno*, II (Ascoli, 1767), 364–383; A. Pazzini, *La storia della Facoltà Medica di Roma*, II (Rome, 1961), 458–459; S. de Renzi, *Storie della medicina in Italia*, IV (Naples, 1846), 53; and F. Vecchietti, *Biblioteca Picena o sia Notizie storiche delle opere e degli scrittori Piceni*, I (Osimo, 1790), 228–232.

LORIS PREMUDA

ASTBURY, WILLIAM THOMAS (*b.* Longton [now part of Stoke-on-Trent], England, 25 February 1898; *d.* Leeds, England, 4 June 1961), *X-ray crystallography, molecular biology.*

Astbury was brought up in the pottery town of Longton. He read chemistry at Cambridge and began research under Sir William Bragg at University College, London, in 1921, moving with him to the Royal Institution in 1923. In 1928 Astbury was appointed lecturer in textile physics at the University of Leeds, then reader, and finally professor of biomolecular structure.

In London, Astbury worked on the structure of tartaric acids, measured diffraction intensities photometrically, and, with Kathleen Lonsdale, produced the first tables of space groups. His assignment at Leeds was the structure of natural and synthetic fibers, especially wool. In 1930 he discovered that two diffraction patterns can be produced from the same wool fiber by exposing it to the X-ray beam when relaxed and when under tension. On the meridian the relaxed fiber showed a prominent spot at 5.1 A., and the stretched fiber a spot at 3.4 A. Astbury concluded that the long polypeptide chains which make up the keratin fiber are folded into a series of hexagons resembling diketopiperazines in structure, and are spaced 5.1 A. apart. When the wool fiber is stretched, these hexagon folds are pulled out into long chains in which the residue repeat is at 3.4 A. He noted the similarity of this stretched, or beta, form

of keratin to silk, and proposed for both a two-dimensional grid structure, the cross-links between the polypeptide chains being through salt linkages and sulfur bridges.

Henceforth, Astbury's work was dominated by this theory of the reversible transformation of keratin. In 1940 Hans Neurath showed that the amino acid side chains could not be accommodated in his structures, so in 1941 Astbury, with Florence Bell, modified them; but when the alpha helix and pleated sheets were proposed by Linus Pauling in 1951, Astbury's models were discarded.

In 1937 Astbury and Florence Bell took the first good X-ray pictures of sodium thymonucleate and discovered the strong meridional spot at 3.4 A. In 1946 Astbury assigned this spot to the eighth layer line, and Mansel M. Davies built a single-chain "pile of plates" model with a chemical repeat at 27 A. From model-building, Davies recognized that the planes of sugar and purine or pyrimidine must be at right angles and that the most likely phosphate ester linkages are between C3 and C5.

Although Astbury's structures for proteins were all wrong in detail, they represent the first attempt at molecular models in which specific cross-linkages hold the polypeptide chains in a characteristic conformation. His suggestion of the folding and unfolding of these chains as the basis of extensibility of fibrous proteins and of the denaturation process of globular proteins was correct in essence. Nowhere did he utilize helical models, but his picture of DNA as a dense molecule with the bases stacked one above another 3.4 A. apart was the first step toward the elucidation of its structure.

As a pioneer, Astbury was bound to see his work superseded, but by his enthusiasm and mastery of the art of lecturing he drew others into the then young field of molecular biology. He provided the stimulus to the more detailed and reliable work of his successors.

BIBLIOGRAPHY

I. ORIGINAL WORKS. Astbury wrote over 100 papers and two books. J. D. Bernal has published an almost complete bibliography in his Royal Society memoir on Astbury (see below). Astbury's most important papers are the three he published under the title "The X-ray Studies of the Structure of Hair, Wool, and Related Fibres," in *Philosophical Transactions of the Royal Society*, **230A** (1931), 75–101; **232A** (1933), 333–394; and *Proceedings of the Royal Society*, **150A** (1935), 533–551. His best papers on nucleic acids are "X-ray Studies of Nucleic Acids," in *Symposium of the*

Society for Experimental Biology, **1** (1947), 66–76; and "Protein and Virus Studies in Relation to the Problem of the Gene," in *Proceedings of the 7th. International Congress of Genetics* [held at Edinburgh in 1939] (Cambridge, 1941), pp. 49–51. Biographical information will be found in "The Fundamentals of Fibre Research: A Physicist's Story," in *Journal of the Textile Institute,* **51** (1960), 515–526. His two books are *Fundamentals of Fibre Structure* (London, 1933); and *Textile Fibres Under the X-Rays* (London, 1940).

II. SECONDARY LITERATURE. The most detailed biography is J. D. Bernal's, in *Biographical Memoirs of Fellows of the Royal Society,* **9** (1963), 1–35, with bibliography and portrait. For obituary notices see K. Lonsdale, in *Chemistry and Industry* (1961), 1174–1175; and I. MacArthur, in *Nature,* **191** (1961), 331–332.

ROBERT OLBY

ASTON, FRANCIS WILLIAM (*b.* Harbonne, Birmingham, England, 1 September 1877; *d.* Cambridge, England, 20 November 1945), *experimental chemistry, physics.*

Aston was the second son of William Aston, a metal merchant and farmer, and Fanny Charlotte Hollis, the daughter of a Birmingham gunmaker. After primary education at Harbonne vicarage school, Aston spent four years at Malvern College. In 1893 he entered Mason's College, Birmingham, where he studied for the London intermediate science examination with the chemists W. A. Tilden and P. F. Frankland and the physicist J. H. Poynting. In 1898 he obtained a Forster Scholarship to work with Frankland on the stereochemistry of dipyromucyltartaric acid esters. Simultaneously he took a course in fermentation chemistry, and from 1900 to 1903 he earned a living as a brewery chemist at Wolverhampton. He returned to Birmingham University (formed from Mason's College in 1900) from 1903 to 1908 as a physics research student with Poynting, and after a world tour in 1909 he spent a term at Birmingham as an assistant lecturer. From 1910 to 1919 Aston worked with J. J. Thomson at the Cavendish Laboratory, Cambridge, and the Royal Institution, London; first as a personal assistant, then from 1913 as a Clerk Maxwell Scholar. This period was interrupted by the war, during which Aston returned to chemistry as a technical assistant at the Royal Aircraft Establishment, Farnborough. In 1919 he was elected a fellow of Trinity College, Cambridge, where he spent the remainder of his life.

Conservative in politics and of no decided religious views, Aston was an animal lover, a keen traveler, a varied and skilled sportsman, a technically brilliant photographer, and an accomplished amateur musician. Like most of J. J. Thomson's associates, he acquired an interest in finance, and in consequence of skilled investment he was able to leave a large estate to Trinity College and several scientific beneficiaries. Aston was a bachelor, a poor teacher and lecturer, and a lone worker who detested the thought of experimental collaboration. (Only six out of 143 papers were collaborative.) He recognized his own fallibility as a theorist, and frequently sought the aid of such mathematical physicists as F. A. Lindemann (Lord Cherwell), R. H. Fowler, and W. W. Sawyer. Aston received the Nobel Prize for chemistry in 1922. He held several honorary degrees, and was elected to, and received medals from, many scientific institutions. From 1936 to 1945 he was chairman of the Committee on Atoms of the International Union of Chemistry.

Although Aston liked to recall that his first two publications were on organic chemistry, these two papers broke no new ground although they did exhibit his talent for devising ingenious apparatus. The appearance of Thomson's *Conduction of Electricity Through Gases* in 1903 opened up, for Aston the chemist, the physicist's world of cathode rays, positive rays, and X rays. Already an expert glassblower, and trained by Frankland in "extreme care and meticulous accuracy," he began to work under Poynting on the variable structure of the phenomena observed during gaseous conduction at low pressures. He was particularly interested in the variation, with pressure and current, of the length of the dark space between the cathode and the negative glow named after W. C. Crookes. By making special Geissler discharge tubes with movable aluminum cathodes, Aston was able to obtain a sufficiently well-bounded Crookes space to demonstrate that its length was proportional to $1/P + 1/\sqrt{C}$, where P is the pressure and C is the current. In 1908, while using hydrogen and helium, he detected a new "primary cathode dark space," about a millimeter thick and directly adjoining the cathode. This phenomenon now bears Aston's name. Research on the relationship between the Crookes dark space and current, voltage, pressure, and electrode nature and design continued intermittently until 1923. Aston then abandoned it in order to devote all his attention to isotopes.

When Aston became Thomson's assistant in 1910, he was given the task of improving the apparatus in which a beam of positively charged particles (positive rays), which emerged through a perforated cathode in a discharge tube, were deflected by perpendicularly arranged electric and magnetic fields into sharp, visible parabolas of constant e/m (charge over mass). Aston produced an improved spherical discharge tube, finely engineered cathode slits, an improved pump, a coil for detecting vacuum leaks, and an ingenious camera for photographing the parabolas. In 1912 he thought this apparatus for positive ray analysis gave a rigorous proof that all the individual molecules of any given substance had the same mass. This Daltonian belief was rudely shattered in the same year when Thomson obtained two parabolas, of mass 20 and 22, for neon. There were two obvious possibilities: if neon had a true atomic weight of 20 (instead of 20.2), then either mass 22 was an unknown hydride, NeH_2, or a new element, meta-neon. Thomson investigated the first possibility and left Aston to check the unlikely alternative.

Aston, who was sympathetic toward F. Soddy's contemporary ideas on radioactive isotopes, tried to separate the meta-neon by fractional distillation, and later by diffusion. He invented a quartz microbalance, which was sensitive to 10^{-9} gram, to measure the density of the minute heavier fraction. The partial separation of a new element, with the same properties as neon, was announced in 1913; Thomson, however, remained doubtful. During the war Aston had time to think over the problem and to debate the possibility of the existence of natural isotopes with the skeptical F. A. Lindemann.

In 1919, to test the neon isotope hypothesis, Aston built a positive ray spectrograph, or mass spectrograph, with a resolving power of 1 in 100 and an accuracy of 1 part in 10^3. The design was based upon an optical analogy. Just as white light can be analyzed into an optical spectrum by a prism, so an electric field will disperse a beam of heterogeneous positive rays. By arranging a magnetic field to deflect the dispersed rays in the opposite direction, but in the same plane, rays of uniform mass can be focused into a mass spectrum on a photographic plate, irrespective of their velocities. This was a great advance on Thomson's apparatus, where the arrangement of the fields produced parabolas that were dependent on the velocities of the positive rays. Aston adopted several methods to calculate the masses of the particles, including comparison with a calibration curve of reference lines of known masses. In the case of neon, the intensities of the 20 and 22 mass lines implied a relative abundance of about 10 : 1, enough to produce an average mass of 20.2, the known atomic weight of neon. Neon was isotopic.

Two larger mass spectrographs were built. The second (1927) had five times more resolving power and an accuracy of 1 in 10^4; the third (1935) had a resolving power of 1 in 2000 and a claimed accuracy of 1 in 10^5. The latter instrument proved difficult to adjust, and World War II intervened before any significant work could be done with it. By then, however, Aston's instruments had been surpassed by the mass spectrometers developed by A. J. Dempster (1918), K. T. Bainbridge (1932), and A. O. Nier (1937).

Aston's personal motto, "Make more, more, and yet more measurements," led him to analyze successfully all but three of the nonradioactive elements in the periodic table. But since the mass spectrograph was unsuitable for detecting minute amounts of isotopes, he missed finding those of oxygen and hydrogen. In 1930 Aston showed how his instrument could be used photometrically to determine and correct chemical atomic weights. Here much depended on his brilliant development of photographic plates that were highly sensitive to positive ions.

In December 1919 Aston announced the "whole-number rule" that atomic masses were integral on the scale O^{16} (a notation introduced by Aston in 1920). Fractional atomic weights were merely "fortuitous statistical effects due to the relative quantities of the isotopic constituents," and the elements were to be defined physically by their atomic numbers, rather than in terms of isotopic mixtures. Prout's hypothesis (1816), that all elements were built up from atoms of a common substance, appeared to be vindicated at last.

Aston's work, therefore, provided important insights into the structure of the atom and the evolution of the elements. At first, only hydrogen appeared to violate the whole-number rule. Aston explained this seeming violation as due to the "loss" of mass within this atom by binding energy; mass was additive only when nuclear charges were relatively distant from one another. This concept of "packing" had been proposed on theoretical grounds by W. D. Harkins (1915), and derived ultimately from J. C. G. Marignac (1860). However, it soon became clear that all elements deviated slightly from whole numbers. In 1927, with his second machine, Aston measured and codified the deviations in terms of the "packing fraction" (the positive or negative deviation of an atomic mass from an integer divided by its mass number). By plotting these fractions against mass numbers, Aston obtained a simple curve which gave valuable information on nuclear abundance and stability.

Aston's achievements were kept continually before the scientific public by revised editions of his excellent book *Isotopes* (1922). This included observations on the abundance and distribution of isotopes and a clear forecast of the power and dangers of harnessed atomic energy.

BIBLIOGRAPHY

I. ORIGINAL WORKS. A full list of Aston's papers can be found in Hevesy (see below); to it should be added "The Mass-spectra of the Elements (Part II)," in *The London, Edinburgh and Dublin Philosophical Magazine,* **40** (1920),

628–636; reports of the Committee on Atoms of the International Union of Chemistry (Paris, 1936–1941); and the obituary of J. J. Thomson, in *The Times* (London, 4 Sept. 1940).

The most important papers are "Experiments on the Length of the Cathode Dark Space," in *Proceedings of the Royal Society of London,* **79A** (1907), 80–95; "Experiments on a New Cathode Dark Space," *ibid.,* **80A** (1908), 45–49; "Sir J. J. Thomson's New Method of Chemical Analysis," in *Science Progress,* **7** (1912), 48–65; "A New Elementary Constituent of the Atmosphere," in *British Association for the Advancement of Science Reports,* **82** (1913), 403; "A Micro-balance for Determining Densities," in *Proceedings of the Royal Society of London,* **89A** (1914), 439–446; "A Positive Ray Spectrograph," in *The London, Edinburgh and Dublin Philosophical Magazine,* **38** (1919), 707–714; "The Possibility of Separating Isotopes," *ibid.,* **37** (1919), 523–534, written with F. A. Lindemann; "Problems of the Mass-spectrograph," *ibid.,* **43** (1922), 514–528, written with R. H. Fowler; "Photographic Plates for the Detection of Mass Rays," in *Proceedings of the Cambridge Philosophical Society,* **22** (1923–1925), 548–554; "A New Mass-spectrograph," in *Proceedings of the Royal Society of London,* **115A** (1927), 487–514, a Bakerian lecture; "The Photometry of Mass-spectra," *ibid.,* **126A** (1930), 511–525; and "A Second-order Focusing Mass-spectrograph," *ibid.,* **163A** (1937), 391–404.

Also of interest are his Nobel lecture, "Mass-spectra and Isotopes," in *Nobel Lectures in Chemistry (1922–1941)* (Amsterdam-London-New York, 1966); and "Forty Years of the Atomic Theory," in J. Needham and W. Pagel, eds., *Background to Modern Science* (Cambridge, 1938).

Aston's books are *Isotopes* (London, 1922; 2nd ed., 1924); and *Mass-spectra and Isotopes* (London, 1933; 2nd ed., 1942), on which he was assisted by C. P. Snow.

A few of Aston's original instruments are displayed in various museums: a quartz microbalance, 1913 (Science Museum, London); fragments from the neon diffusion apparatus, 1913 (Cavendish Laboratory Museum, Cambridge); the first mass spectrograph, 1919 (Science Museum, London); the third mass spectrograph, 1935 (Cavendish Laboratory Museum). The mass spectrograph of 1927 appears to have been broken up.

II. Secondary Literature. Works concerning Aston include N. Feather, "F. W. Aston," in *Dictionary of National Biography (1941–1950)* (Oxford, 1959), pp. 24–26; G. C. de Hevesy, "F. W. Aston," in *Obituary Notices of Fellows of the Royal Society,* **5** (1945–1948), 635–651, which includes a photograph and a bibliography; F. M. Green, "The Chudleigh Mess," in *R. A. E. News* (Jan. 1958), pp. 4–7, available only at Farnborough; G. P. Thomson, "F. W. Aston," in *Nature,* **157** (1946), 290–292, and *J. J. Thomson and the Cavendish Laboratory* (London, 1964), pp. 93, 136; J. J. Thomson, *Rays of Positive Electricity* (London, 1913), passim; and *Who Was Who (1941–1950)* (London, 1952), p. 38, which includes a complete list of honors.

W. H. Brock

ASTRUC, JEAN (*b.* Sauve, Gard, France, 19 March 1684; *d.* Paris, France, 5 May 1766), *medicine.*

Astruc was the son of a Protestant clergyman who probably had Jewish ancestors who chose to renounce their religion rather than leave France. After receiving a doctorate in medicine at Montpellier in 1703, he temporarily occupied Pierre Chirac's chair of medicine in 1706. He passed the competitive examination of the Faculty of Medicine of Toulouse in 1711, and then returned to Montpellier to occupy the chair of medicine of Jacques Chastelain from 1716 to 1728, at which time he became general physician to the duke of Orléans. In 1720 he received a pension from the king and in the following year was named inspector general of the mineral waters of Languedoc. In 1729 Astruc became the chief physician of Augustus II of Poland, and he was named municipal magistrate of Toulouse in 1730. In that same year he became the king's counsellor and physician, and in 1751 he occupied E. F. Geoffroy's chair of pharmacy at the Collège Royal.

In 1743 Astruc was elected regent doctor of the Faculty of Medicine of Paris, the first time that this exceptional honor was awarded to a doctor from a provincial medical school—despite the fact that the statutes of the university forbade it. This honor was followed by a second one, that of having his bust placed in the amphitheater of the Faculty. These honors were a recompense for his constant fight against his two enemies, the surgeons and the *variolisateurs* (users of a primitive form of vaccination against variola), as well as for his assiduous attendance at meetings. He died after twenty-three years in Paris, years made painful by a tumor of the bladder that Georges de la Faye, a member of the Académie de Chirurgie, perforated with a metal probe in order to overcome chronic retention of urine.

Astruc was a born teacher, extremely methodical and clear in his instruction. In a series of courses lasting six years he covered all phases of medicine: anatomy, physiology, psychology, gerontology, pathology, therapy, venereology, gynecology, neurology, and pediatrics. Even in American libraries there are manuscript copies of these carefully prepared and highly appreciated courses; during Astruc's lifetime some of them were used for editions printed without his knowledge in England, Switzerland, and Holland. His works were translated into English and German and were widely known in Europe. An iatrochemist and iatrophysicist, Astruc had no personal doctrine. His philosophy, influenced by Descartes and Malebranche, was only mildly opposed to the cold reason of Locke. His place in the history of medicine was somewhat behind Haller, Morgagni, and Boerhaave, rather than ahead of them.

Although a mediocre practitioner, Astruc was a scientist of note, a solitary and erudite scholar who often worked through the winter nights until three in the morning, without a fire in the library. Among his several thousand volumes were many works on theology, history, geography, and literature. He called this his "militant life." Around 1730 Astruc began to frequent the mansion of Mme. de Tencin, who was a patient. The famous hostess was helpful in arranging the marriage of Astruc's daughter to Daubin de Silhouette and also remembered him in her will. For some time Astruc, Bernard de Fontenelle, Pierre Marivaux, Jean de Mairan, the Marquis Victor de Mirabeau, Claude de Boze, and Charles Duclos were considered the seven sages of Mme. de Tencin's salon.

In his *Traité sur les maladies des femmes* (1761) Astruc described septicemia caused by uterine infections and puerperal fever, ovarian cysts, tubal pregnancies, abdominal pregnancies, and lithopedions, of which he reported four cases. He advised operating on extrauterine pregnancies, and the use of Caesarean sections only in emergencies.

In 1743 Astruc compared the transformation of an impression or sensation into a motor discharge to a ray of light reflected on a surface; he called it *reflex*. He thus had an intuition of reflex action, which was described in 1833 by Marshall Hall.

Astruc's best-known work is his treatise on venereology, *De morbis venereis*, the fourth French edition of which (1773–1774) contains "Dissertation sur l'origine, la dénomination, la nature et la curation des maladies vénériennes à la Chine," in which for the first time Chinese medical terminology was reproduced in an Occidental work in correctly printed Chinese characters.

His family background led Astruc to consider the exegesis of the Old Testament as one of the elements of his personal inner life. This work appeared as *Conjectures sur la Genèse* (1753), in which the different names of God (Elohim or Jehovah) gave him the key to dating various parts of the Bible. Both Catholic and Protestant theologians frowned on these discoveries, for they were incapable of appreciating Astruc's quick mind, his constructive criticism, and his remarkable philological knowledge. Now, after more than two centuries of discussion of these ideas, often minute and often passionate, historical criticism always comes back to them. It still cannot be proved that the thesis is true, but Astruc's conception is in line with our present knowledge and, according to

Lods, gives the best account of the formation of Jewish historiography. Sir William Osler judged the *Conjectures* worthy of inclusion in his *Bibliotheca prima* as a remarkable example of scientific criticism.

BIBLIOGRAPHY

I. ORIGINAL WORKS. The complete bibliography of Astruc's works has been drawn up by Janet Doe (see below). The principal items are *De motus fermentativi causa* (Montpellier, 1702); *Responsio critica animadversionibus F. R. Vieussens in tractatum de causa motus fermentativi* (Montpellier, 1702); *Dissertatio physico medica de motu musculari* (Montpellier, 1710); *Mémoire sur la cause de la digestion des aliments* (Montpellier, 1711); *Traité de la cause de la digestion . . .* (Toulouse, 1714); *Epistolae Joannès Astruc quibus respondetur epistolari dissertationi Thomae Boerü, de concoctione* (Toulouse, 1715); *Dissertatio de ani fistula* (Montpellier, 1718); *De sensatione* (Montpellier, 1719); *Dissertatio medica de hydrophobia* (Montpellier, 1720); *Quaestio medica de naturali et praeternaturali judicii exercitio* (Montpellier, 1720); *De phantasia et imaginatione* (Montpellier, 1723); *Cinq lettres contre les chirurgiens* (Paris, ca. 1731); *De morbis venereis libri sex* (Paris, 1736, 1738, 1740), translated into French by Augustin Jault with notes by Antoine Louis (Paris, 1743; 4th ed., 1773–1774); *Mémoires pour servir à l'histoire naturelle de la province du Languedoc* (Paris, 1737); *An sympathia partium a certa nervorum positura in interno sensorio* (Paris, 1743); *Conjectures sur les mémoires originaux . . . de la Genèse* (Paris, 1753); *Traité sur l'immatérialité et l'immortalité de l'âme* (Paris, 1755); *Doutes sur l'inoculation de la petite vérole, proposés à la Faculté de Médecine de Paris* (Paris, 1756); *Traité des tumeurs et des ulcères* (Paris, 1759); *Traité sur les maladies des femmes* (Paris, 1761; 1765); *L'art d'accoucher* (Paris, 1766, 1771); and *Mémoire pour servir ici l'histoire de la Faculté de Médecine de Montpellier* (Paris, 1767).

II. SECONDARY LITERATURE. Writings on Astruc are P. Astruc, "Une bibliothèque médicale au XVIIIème siècle," in *Progrès médicale,* 2 (1934), 94–95; Louis Barbillion, "Un ancien syphiligraphe," in *Paris médical,* 52 (1924), 196–198; C. S. Butler, "Hero Worship and the Propagation of Fallacies," in *Annals of Internal Medicine,* 5 (1932), 1033–1038; Georges Canguilhem, *La formation du concept de réflexe au XVIIème et XVIIIème siècles* (Paris, 1955), which includes Hall's description; F. Chaussier and N. P. Adelon, "Astruc," in *Biographie universelle ancienne et moderne* (Paris, 1811), pp. 486–487; A. Chéreau, "Astruc," in *Dictionnaire encyclopédique des sciences médicales,* VII (Paris, 1867), 31–34; Paul Delauney, *Le monde médical parisien au XVIIIème siècle* (Paris, 1935), *passim;* J. E. Dézeimeris, Claude Ollivier, and J. Raige-Delorme, *Dictionnaire historique de la médecine ancienne et moderne,* I (Paris, 1828), 200–203; Janet Doe, "Jean Astruc (1684–1766), a Biographical and Bibliographical Study," in *Journal of the*

History of Medicine, **20** (Apr. 1960), 184–197; Charles Fiessinger, "La thérapeutique de Jean Astruc," in *Thérapeutique des vieux maîtres* (Paris, 1897), pp. 226–232; Fischer, *De senio eiusque gradibus* (Erfurt, 1754); Pierre Huard and Ming Wong, "Montpellier et la médecine chinoise," in *Montpelliensis Hippocrates,* no. 2 (Dec. 1958), pp. 13–20; "Antonio Nunes Ribeiro Sanchès," in *Société Française de l'Histoire de la Médecine,* spec. no. 4 (13 Jan. 1962) pp. 96–103; and "J. Astruc Scholar and Biblical Critic," in *Journal of the American Medical Association,* **92** (1965), 249 ff.; A. M. Lautour, "Astruc," in *Dictionnaire de biographie française* (Paris, 1939); A. C. Lorry, "Éloge de J. Astruc," in *Mémoires pour servir à l'histoire de la Faculté de Médecine de Montpellier* (Paris, 1767), pp. xxxiii–li; A. Lods, *Jean Astruc et la critique biblique au XVIIIème siècle* (Strasbourg–Paris, 1924); P. M. Masson, *Madame de Tencin* (Paris, 1909); R. O. Moreau, "L'oeuvre d'Astruc dans son traité de maladies des femmes," thesis (Univ. of Paris, 1930); Sir William Osler, "Jean Astruc and the Higher Criticism," in *Canadian Medical Association Journal,* **2** (1912), 151–152; E. Ritter, "Jean Astruc, auteur des *Conjectures sur la Genèse,*" in *Bulletin de l'histoire du protestantisme français,* **15** (1916), 274–287; John Ruhrah, "J. Astruc the Pediatrician," in *American Journal of Diseases of Children,* **39** (1930), 403–408; F. M. Scapin, "L'inchiesta di Giovanni Astruc sulla sifilide cinese," in *Acta medicae historiae Patavina,* **3** (1956), 57–61; Sir A. Simpson, "Jean Astruc and his *Conjectures,*" in *Edinburgh Medical Journal,* **14** (1915), 461–475, and *Proceedings of the Royal Society of Medicine,* no. 8 (1914/15), 59–71; and F. D. Zeman, "Jean Astruc on Old Age," in *Journal of the History of Medicine,* **20** (1965), 52–57.

PIERRE HUARD

ATHENAEUS OF ATTALIA (*b.* Attalia in Pamphylia [?] [now Antalya, Turkey]), *medicine.*

Athenaeus was a physician who practiced in Rome, apparently during the reign of Claudius I. Biographical details are lacking, but it is known that he founded the Pneumatic school of medicine. His successors in the school included his pupil Agathinos and Herodotus, Magnus, and Archigenes, who flourished during the reign of Trajan.

The name of Athenaeus' school came from a new term, *pneuma* ("breath" or "spirit"), that he introduced into medical theory from Stoic philosophy. Chrysippus, the third head of the Stoic school, had defined soul as "a spirit [pneuma] innate in us, continuously penetrating the whole body." Athenaeus taught that the body was composed, ultimately, of the traditional four qualities—hot, cold, wet, dry—but that these were held together and governed by pneuma, which permeated the entire body. Although it owed something to Aristotle's view of pneuma as associated with semen in generation, the Stoic doctrine was more

general—and, indeed, cosmic in extent. Athenaeus, accepting the materialism of the Stoics, could identify with pneuma the ruling part of the soul (*hegemonikon*), which for Aristotle was immaterial. Athenaeus is often linked with Aristotle in Galen's criticism of previous teachings. Following Aristotle and Chrysippus, Athenaeus located the *hegemonikon* in the heart, and from this belief he drew certain conclusions about the medical treatment of mental illness. He explained disease in general as a pathological affection of the pneuma caused by the putrefaction or rotting of pneuma.

The doctrine of the proper mixture of the qualities or the humors composed of them was a staple of ancient medical tradition. The Pneumatic school felt the need of a governing spirit to maintain a proper temperament of the components of the physical body. Athenaeus, according to Galen, considered hot and cold as efficient causes and wet and dry as material ones, but apparently he was not able to conceive of them as operating on their own. Galen also asserts that Athenaeus was not clear as to whether the four qualities were potencies or bodies.

Athenaeus was an original thinker who enriched medical theory with a consistent philosophical position. He was also a teacher, as his founding of a school proves. He is known to have written a comprehensive treatise on medicine that ran to at least thirty books. He treated in detail the principal branches of medicine: physiology, pathology, embryology, therapeutics, and dietetics, as well as the medical aspects of meteorology and geography.

Practically all of our information about Athenaeus comes from Galen, who cites him frequently and seems to have accepted some of his teachings, and from Oribasius, who was the physician of Julian the Apostate and who was known to his contemporaries as the second Galen. Galen cites Athenaeus in many of his writings. The diverse subjects of the citations testify to the breadth of Athenaeus' knowledge. Galen, in general, states the opinion of Athenaeus relevant to the point he is discussing and tries either to refute it or to interpret it in the light of his own beliefs. Whether Galen agrees with Athenaeus or not, he treats him with obvious respect. Oribasius' extensive quotations from Athenaeus show us the latter's care and thoroughness in handling a subject as well as giving us some of the content of his teaching.

BIBLIOGRAPHY

Some idea of Athenaeus' works and views may be found in Galen, *Claudii Galeni opera omnia,* C. Kühn, ed., 20 vols. (Leipzig, 1821–1833), Index; and Oribasius, *Oribasii collectionum reliquiae,* J. Raeder, ed. (Leipzig–Berlin, 1928–　　), and *Medici Graeci varia opuscula,* CH. F. Matthai, ed. (Moscow, 1908).

Modern accounts are Pauly-Wissowa, eds., *Realencyclopädie der classischen Altertumswissenschaft;* and M. Wellman, *Die pneumatische Schule* (Berlin, 1895).

JOHN S. KIEFFER

ATWATER, WILBUR OLIN (*b.* Johnsburg, New York, 3 May 1844; *d.* Middletown, Connecticut, 22 September 1907), *agricultural chemistry, physiology, scientific administration.*

The son of William Warren Atwater, a Methodist clergyman, and Eliza Barnes Atwater, Wilbur Atwater studied for two years at the University of Vermont and received his bachelor's degree at Wesleyan College in 1865. Interested in both agriculture and chemistry, he then went for postgraduate work to Yale University's Sheffield Scientific School, where he studied under the chemist Samuel W. Johnson, a Leipzig graduate and America's leading authority on agricultural chemistry. Atwater earned his doctorate in 1869, with a thesis on the analysis of the composition of several varieties of American maize. He then spent two years at Leipzig and Berlin. After brief teaching periods at the University of Tennessee and Maine State College, in 1873 he was appointed professor of chemistry at Wesleyan, a position he held until his death.

In 1875 the Connecticut legislature—with the encouragement and financial aid of agricultural editor Orange Judd—established America's first agricultural experiment station, patterned largely after German stations, institutions admired by both Johnson and Atwater. The establishment of the station was a goal to which Johnson had been dedicated since the mid-1850's. From 1875 until 1877, the station was at Middletown and under Atwater's direction. In 1877 it moved to New Haven and the guidance of Johnson. Like most of his contemporary agricultural chemists, Atwater became increasingly involved in fertilizer investigation and testing, using this work partly as a means of gaining agricultural support for scientific research generally. (In the course of it he was able to demonstrate independently the role of leguminous plants in the fixation of atmospheric nitrogen.) Even after the directorship of the Connecticut station had passed into Johnson's hands, Atwater continued to organize fertilizer experiments and to write regularly for farm readers on the application of science to agriculture.

In 1887, with the passage of the Hatch Act, a measure providing federal funds for the establishment of an agricultural experiment station in each state, Atwater was appointed chief of the Office of Experiment Stations, established within the U.S. Department of Agriculture to oversee and coordinate the work of the state experiment stations. Although he occupied this post for only two years—during which he continued his academic duties at Wesleyan— Atwater exerted a decisive influence on administrative policy in regard to the stations. His basic policies were elaborated and implemented through the next quarter century by his successor, A. C. True, a Wesleyan classicist who depended heavily on Atwater's advice. The influence of True and Atwater upon the development of agricultural research in the United States was both positive and surprisingly pervasive, extending to many aspects of basic biological investigation.

In 1887 Atwater also visited Europe, where at Munich he became deeply interested in the calorimetric work of Carl Voit and Max Rubner. On his return to Wesleyan, Atwater sought the aid of E. B. Rosa, his physicist colleague, in the design and construction of what came to be called the Atwater-Rosa calorimeter. Begun in 1892, the calorimeter was in operation by 1897 (preliminary studies had appeared in 1896). Atwater was concerned not only with metabolism as a problem in physiology, but also with the use of his new techniques for the determination of improved dietary standards for the working class, standards that might prescribe a diet providing optimum food value at lowest cost. An adroit manipulator of political and business support, Atwater was able to demonstrate the value of such nutritional investigations to the Committee on Agriculture of the House of Representatives, which in 1894 began to support nutrition research (a program directed by Atwater until his death). Graham Lusk, Francis Benedict, and H. P. Armsby were among the other American students of metabolism who used Atwater's calorimetric techniques.

By the first decade of the twentieth century, calorimetric work had become an extremely popular, almost fashionable field, with broad implications for public policy and popular health education. It is ironic that the total impact of Atwater's nutrition work was somewhat clouded: his emphasis on caloric values—in the absence of knowledge of vitamin and amino acid requirements—led to recommendations that the working class purchase carbohydrates and avoid such "luxuries" as green vegetables. Vigorous though Atwater's scientific work was, his greatest contribution to the development of science in the United States was organizational and administrative— especially his efforts to establish scientific standards for experiment station research. Indeed, his forceful-

325

ness in such matters provided an occasional source of disquietude to certain of his colleagues and his career was marked at times by friction with contemporaries.

BIBLIOGRAPHY

The basic source for Atwater's life is his extensive collection of papers and correspondence, including documents recording his tenure as first director of the Office of Experiment Stations. The papers are in the possession of the University Archives, Wesleyan University, Middletown, Conn.; microfilm copies are available at Cornell University and the University of Pennsylvania. There is no full-length biography of Atwater, but for sketches of his life see Benjamin Harrow, "Wilbur Olin Atwater," in *Dictionary of American Biography;* Leonard A. Maynard, "Wilbur O. Atwater—A Biographical Sketch (May 3, 1844–September 22, 1907)," in *Journal of Nutrition,* **78** (1962), 3–9. For useful studies placing Atwater's work, both administrative and scientific, in perspective see Graham Lusk, *The Elements of the Science of Nutrition,* 2nd ed. (Philadelphia-London, 1909); E. V. McCollum, *A History of Nutrition* (Boston, 1957); and A. C. True, *A History of Agricultural Experimentation and Research in the United States, 1607–1925,* U.S. Department of Agriculture, Misc. Pub. No. 251 (Washington, D.C., 1937).

CHARLES E. ROSENBERG

ATWOOD, GEORGE (*b.* England, 1745; *d.* London, England, July 1807), *mathematics, physics.*

Atwood attended Westminster School and was awarded a scholarship to Trinity College, Cambridge, at the age of nineteen. He graduated with a B.A. in 1769, received his M.A. in 1772, then became a fellow and tutor at his college. His lectures were well attended and well received because of their delivery and their experimental demonstrations. These experiments, published in 1776, the year he was elected a fellow of the Royal Society, consisted of simple demonstrations to illustrate electricity, optics, and mechanics.

Among his admirers was William Pitt, who in 1784 gave Atwood an office in the treasury, at £500 a year, so that, according to an obituary in the *Gentleman's Magazine,* he could "devote a large portion of his time to financial calculation" in which he was apparently employed "to the great advantage of revenue." His only published work in this connection was *A Review of the Statutes . . .* (1801), in which he analyzed the cost of bread. The price that the baker could charge for a loaf of bread was governed by statute and was determined by the cost of grain plus an allowance for profit. Central to the problem was how much grain was required to make a loaf of bread. Atwood's work,

an attempt to rationalize the standards, was based on computation as well as on the results of experiments carried out by Sir George Young in 1773.

The work for which Atwood is best known and which bears his name—Atwood's machine—is described in *A Treatise on the Rectilinear Motion . . .* (1784), which is essentially a textbook on Newtonian mechanics. Atwood's machine was designed to demonstrate the laws of uniformly accelerated motion due to gravity and was constructed with pulleys, so that a weight suspended from one of the pulleys descends more slowly than a body falling freely in air but still accelerates uniformly.

Most of Atwood's other published work consisted of the mathematical analysis of practical problems. In "A General Theory for the Mensuration . . ." (1781), he derived equations for use in connection with Hadley's quadrant; and in "The Construction and Analysis . . ." (1796) and "A Disquisition on the Stability of Ships" (1798), he extended the theories of Euler, Bougier, and others to account for the stability of floating bodies with large angles of roll. For "The Construction and Analysis . . ." he was awarded the Copley Medal of the Royal Society. His work on arches, *A Dissertation on the Construction and Properties of Arches* (1801), based on the assumption that the material of an arch is perfectly hard and rigid and that the only critical forces are those relating to the wedging action of the individual arch units, is now totally superseded. It was published with a supplement containing Atwood's questions about the proposed new London Bridge over the Thames, which was to be of iron.

BIBLIOGRAPHY

Atwood's works are *A Description of the Experiments Intended to Illustrate a Course of Lectures on the Principles of Natural Philosophy* (London, 1776); "A General Theory for the Mensuration of the Angle Subtended by Two Objects, of Which One Is Observed by Rays After Two Reflections From Plane Surfaces, and the Other by Rays Coming Directly to the Spectator's Eye," in *Philosophical Transactions of the Royal Society,* **71,** part 2 (1781), 395–434; *An Analysis of a Course of Lectures on the Principles of Natural Philosophy* (Cambridge, 1784), a revised version of *A Description of the Experiments . . .* ; *A Treatise on the Rectilinear Motion and Rotation of Bodies With a Description of Original Experiments Relative to the Subject* (Cambridge, 1784); "Investigations Founded on the Theory of Motion for Determining the Times of Vibration of Watch Balances," in *Philosophical Transactions,* **84** (1794), 119–168; "The Construction and Analysis of Geometrical Propositions Determining the Positions Assumed by Homogeneal

Bodies Which Float Freely, and at Rest, on the Fluid's Surface; Also Determining the Stability of Ships and of Other Floating Bodies," *ibid.*, **86** (1796), 46–130; "A Disquisition on the Stability of Ships," *ibid.*, **88** (1798), 201–310; *A Dissertation on the Construction and Properties of Arches* (London, 1801); and *Review of the Statutes and Ordinances of Assize Which Have Been Established in England From the Fourth Year of King John, 1202 to the Thirty-seventh of His Present Majesty* (London, 1801).

ERIC M. COLE

AUBERT DUPETIT-THOUARS, L. M. See **Dupetit-Thouars, L. M. Aubert.**

AUBUISSON DE VOISINS, JEAN-FRANÇOIS D' (*b.* Toulouse, France, 19 April 1769; *d.* Toulouse, 20 August 1841), *mining, geology, hydraulics.*

Son of a squire, Jean-François d'Aubuisson de Voisins, and of Jeanne-Françoise Dassié, d'Aubuisson entered the Benedictine College of Sorèze in 1779. In 1786, having won prizes in infinitesimal calculus, physics, and natural history, he was admitted to the Artillery School in Metz. D'Aubuisson was commissioned a second lieutenant on 1 January 1791 and spent the next six years in military service in Spain.

After the Treaty of Campo Formio, d'Aubuisson retired to Freiberg, where he taught mathematics. His civilian vocation was determined by Werner's courses in mineralogy, geology, and mining, which he attended at the Bergakademie in 1800–1801, and field trips to Saxon mines. In 1802 he published the translation of a work by Werner, which he enriched with personal annotation. In the same year d'Aubuisson also published *Des mines de Freiberg en Saxe et de leur exploitation.*

Returning to France in 1802, d'Aubuisson attempted to enter the mining administration, although since 1795 it had been reserved for highly qualified graduates of the École Polytechnique who had spent two years at the École des Mines. He had an old Sorèze colleague, Antoine-François Andréossy, intervene for him. On Andréossy's very first recommendation to the minister of the interior, d'Aubuisson was appointed to the mining administration on 20 January 1803 as assistant to Tonnelier, the curator of the mineralogical collections and library.

Like Werner, d'Aubuisson attributed the formation of the Saxon basalts to precipitates from the primeval sea. Since volcanists, as a result of the observations made in Auvergne by Guettard and Desmarest, regarded the basalts as the product of volcanic eruptions, d'Aubuisson decided to study the question by spending the summer of 1803 in Auvergne. He returned a convert to volcanism.

Andréossy again aided d'Aubuisson by proposing to the minister of the interior that d'Aubuisson would be of greater service if he were appointed a mining engineer. In March 1807 d'Aubuisson became mining engineer for the departments of Doire and Sesia, in the Piedmont, now part of Italy. He was promoted to chief engineer, second class, in March 1811 and was appointed chief of the mineralogical district of Toulouse less than two months later. There he served for more than thirty years in an administrative post covering the entire Pyrenees region.

In 1819 d'Aubuisson published *Traité de géognosie, ou Exposé des connaissances actuelles sur la constitution physique et minérale du globe terrestre,* the first competent treatment of general geology published in France. As a result of this book, on 5 February 1821 the Académie des Sciences named him correspondent for the Mineralogical Section, replacing the Abbé Palassou, whose death had been erroneously announced. When the mistake was discovered, the Academy decided that d'Aubuisson should retain his title.

Publications by English, German, and French authors soon considerably enlarged the scope of geology, and d'Aubuisson undertook a second edition of his *Traité.* The first volume appeared in 1828. Pressed for time, he entrusted the task of publishing the subsequent volumes to Amédée Burat, the first professor of geology at the École Centrale; Volume II appeared in 1834 and Volume III in 1835.

During the latter part of his career d'Aubuisson was compelled gradually to abandon his geological studies in order to attend to the major tasks that resulted from his living in Toulouse. He had to devote a great deal of time to the "miners' mine" of Rancié (Ariège), which he administered for thirty years. Intelligent, tenacious, self-assertive, and industrious, he was the prototype of the technical civil servant in a young state. Against the changes of the political regime, he symbolized the continuity of viewpoint of this state, working untiringly for the adaptation of the miners' mine to the new economic situation. His work led to a pamphlet entitled *Observations sur les mines et les mineurs de Rancié, et sur l'administration de ces mines* (1818).

Aided by the engineer Marrot, in July 1825 d'Aubuisson found that in an airshaft the resistance to passage of the air is directly proportional to the length of the pipe and the square of the speed of the air flow, and inversely proportional to the diameter. He published these findings in two articles in the *Annales des mines* in 1828. These results served as the starting point for Théophile Guibal's establishment of the temperature constant of a mine. D'Aubuisson also found that the volumes of two gases streaming from equal openings and under the same pressure are inversely proportional to the square roots of their densities. A further accomplishment was the establishment of a simple formula that made it possible to use cast-iron water pipes with walls half as thick as those previously used.

The nature of the technical problems he dealt with as municipal councillor induced d'Aubuisson to publish *Traité du mouvement de l'eau dans les tuyaux de conduite* (1827) and "Histoire de l'établissement des fontaines à Toulouse" (1830). His professional activities had caused him to reflect on many problems of hydraulics, and he summarized his experiments and thoughts on these problems in *Traité d'hydraulique à l'usage des ingénieurs* (1834).

On 2 August 1828, d'Aubuisson was appointed chief engineer, first class. Since he preferred to remain at Toulouse, he refused a transfer to Paris and thus lost all chance of rising to the rank of inspector general.

BIBLIOGRAPHY

I. ORIGINAL WORKS. D'Aubuisson's writings include *Des mines de Freiberg en Saxe et de leur exploitation,* 3 vols. (Leipzig, 1802); *Nouvelle théorie de la formation des filons. Application de cette théorie à l'exploitation des mines, particulièrement de celles de Freiberg* (Freiberg, 1802), his trans. of Werner's *Neue Theorie von der Entstenung der Gänge; Observations sur les mines et les mineurs de Rancié, et sur l'administration de ces mines* (Toulouse, 1818); *Traité de géognosie, ou Exposé des connaissances actuelles sur la constitution physique et minérale du globe terrestre,* 2 vols. (Paris, 1819), 2nd ed., completed by Burat, 3 vols. (Paris, 1828–1835); *Considérations sur l'autorité royale en France depuis la Restauration et sur les administrations locales* (Paris, 1825); *Traité du mouvement de l'eau dans les tuyaux de conduite* (Paris, 1827; 2nd ed., 1836); "Expériences sur la résistance que l'air éprouve dans des tuyaux de conduite, faites aux mines de Rancié, en 1825," in *Annales des mines,* 2nd ser., **3** (1828), 367–486; "Expériences sur la trompe du ventilateur des mines de Rancié, suivies de quelques observations sur les trompes en général," *ibid.,* **4** (1828), 211–244; "Histoire de l'établissement des fontaines à Toulouse," in *Mémoires de l'Académie royale des sciences, inscriptions et belles-lettres de Toulouse* (1823–1827), **2,** pt. 1 (1830), 159–400; *Traité d'hydraulique à l'usage des ingénieurs* (Paris, 1834; 2nd ed., 1840, 1846; 3rd ed., 1858); and *Tables à l'usage des ingénieurs et des physiciens. Tables de logarithmes* (Paris, 1842).

D'Aubuisson's many articles in the *Journal des mines* are listed in P.-X. Leschevin's *Table analytique des matières* (Paris, 1813), which covers the first 28 vols., and in A.-C.-L. Peltier's work of the same title (Paris, 1821), which covers the last 10 vols. (the name is spelled Daubuisson in the two works). The articles in the *Annales des mines* (the title of the *Journal* after 1815) are listed in the indexes for the 1st and 2nd ser. (Paris, 1831) under "D'Aubuisson" and in the index for the 3rd ser. (Paris, 1847) under "Aubuisson (d')."

II. SECONDARY LITERATURE. Works on d'Aubuisson are E. Brassinne, "Éloge de M. d'Aubuisson de Voisins," in *Mémoires de l'Académie royale des sciences, inscriptions et belles-lettres de Toulouse,* 3rd ser., **1** (1845), 265–284; and René Garmy, *La "mine aux mineurs" de Rancié 1789–1848* (Paris, 1943), pp. 117–133. Original materials may be found in Archives Nationales, Paris, F14.2712[2], and the Archives de la Guerre at Vincennes, in the dossier "D'Aubuisson de Voisins."

ARTHUR BIREMBAUT

AUDOUIN, JEAN VICTOR (*b.* Paris, France, 27 April 1797; *d.* Paris, 9 November 1841), *zoology.*

Audouin was the second child of Victor Joseph Audouin, a notary, and Jeanne Marie Pierrette Enée. He began his studies at Rheims in 1807, and continued them at Paris in 1809 and at Lucca from 1812 to 1814; in Lucca he stayed with a relative who was an official in the household of Princess Elisa Baciocchi. He then returned to Paris and attended the Lycée Louis-le-Grand. Later Audouin began to study law, but he soon abandoned it for medicine, pharmacy, and the natural sciences. In 1816 he met the mineralogist Alexandre Brongniart, who hired him as his secretary and had a notable influence on Audouin's scientific career. In the same year Audouin published his first entomological work, and in 1820 he read his important paper, "Recherches anatomiques sur le thorax des animaux articulés et celui des insectes en particulier," before the Académie des Sciences. This report won him the praise of Cuvier and was printed in 1824. He and other naturalists founded the Société d'Histoire Naturelle de Paris in 1822, and in 1824 he began publication of the *Annales des sciences naturelles,* an important scientific periodical still in existence. His collaborators on the journal were Adolphe Brongniart and Jean-Baptiste Dumas, both of whom later became his brothers-in-law. In 1825 he became assistant to Lamarck and Latreille at the Museum of Natural History in Paris; and the following year he was asked to finish the work of Savigny (then incapacitated by serious eye trouble) on the invertebrates collected during the Egyptian expedition of 1798–1799.

The same year, 1826, Audouin and Henri Milne Edwards began a course of anatomical, physiological, and biological research on the marine invertebrates of the Breton and Norman coast. These were pioneer studies of this type in France. It was also in 1826 that Audouin presented his doctoral thesis on the chemical, pharmaceutical, and medical natural history of the cantharides. On 6 December 1827, he married Mathilde Brongniart, who became his collaborator because of her talent as a draftsman; in 1827 and 1828 he and Milne Edwards published the results of their anatomical and physiological research on crustacea. In 1830 he replaced Latreille as assistant naturalist at the Natural History Museum. He and Milne Edwards then published, in 1832, the first volume of their *Recherches pour servir à l'histoire naturelle du littoral de la France.* This work initiated the bionomic classification of coastal marine invertebrates. Also in 1832, Audouin and other entomologists founded the Société Entomologique de France, and in 1833 he succeeded Latreille as professor of zoology (in the chair dealing with crustacea, arachnids, and insects) at the Natural History Museum. From 1834 on, he specialized in agricultural entomology, and in 1836–1837 he confirmed the observations of Agostino Bassi concerning the muscardine of the silkworm (*Beauveria bassiana*). On 5 February 1838, Audouin was elected a member of the Académie des Sciences (Section d'Économie Rurale). He died after a short illness, and his *Histoire naturelle des insectes nuisibles à la vigne* was finished by his collaborator, Émile Blanchard, and his friend Henri Milne Edwards, in 1842.

Audouin's work is both that of a scrupulously careful morphologist and anatomist of Cuvier's school, and that of a biologist who has left behind important observations on the physiology of crustacea as well as on the ethology of various insects harmful to cultivated plants. This last phase of his research marks Audouin's work as the precursor of modern applied entomology.

BIBLIOGRAPHY

I. ORIGINAL WORKS. Audouin's published works include "Recherches anatomiques sur le thorax des animaux articulés et celui des insectes en particulier," in *Annales des sciences naturelles,* **1** (1824), 97–135, 416–432; "Recherches pour servir à l'histoire naturelle des Cantharides," *ibid.,* **9** (1826), 31–61; "Recherches anatomiques et physiologiques sur la circulation dans les crustacés," written with Henri Milne Edwards, *ibid.,* **11** (1827), 283–314, 352–393; "Recherches anatomiques sur le système nerveux des crustacés," written with Henri Milne Edwards, *ibid.,* **14** (1828), 77–102; *Recherches pour servir à l'histoire naturelle du littoral de la France ou Recueil de mémoires sur l'anatomie, la physiologie, la classification et les moeurs des animaux de nos côtes,* written with Henri Milne Edwards, 2 vols. (Paris, 1832–1834); "Recherches anatomiques et physiologiques sur la maladie contagieuse qui attaque les vers à soie, et qu'on désigne sous le nom de muscardine," in

Annales des sciences naturelles (zoologie), 8th series, **2** (1837), 229–245; "Nouvelles expériences sur la nature de la maladie contagieuse qui attaque les vers à soie et qu'on désigne sous le nom de muscardine," *ibid.,* 257–270; and *Histoire des insectes nuisibles à la vigne et particulièrement de la pyrale* (Paris, 1842), completed by Émile Blanchard and Henri Milne Edwards.

Numerous manuscripts, drawings, and letters of Audouin's are in the Archives de la Famille Audouin, Archives de l'Académie des Sciences, Bibliothèque de l'Institut, Bibliothèque du Muséum National d'Histoire Naturelle, and Archives de la Société Entomologique de France—all in Paris; and the Wellcome Historical Medical Library and the Library of the British Museum of Natural History—both in London.

II. SECONDARY LITERATURE. More on Audouin and his work can be found in M. Duponchel, "Notice sur la vie et les travaux de Jean-Victor Audouin," in *Annales de la Société Entomologique de France,* **11** (1842), 95–164; H. Milne Edwards, *Notice sur la vie et les travaux de Victor Audouin* (Paris, 1850); and Jean Théodoridès, "Jean-Victor Audouin, Journal d'un étudiant en médecine et en sciences à Paris sous la Restauration (1817–1818)," texte inédit, in *Histoire de la médecine,* **9** (Nov. 1958), 4–63; *ibid.* (Dec. 1958), 5–56; **10** (Jan. 1959), 5–48; "La Rhénanie en 1835 vue par un naturaliste français," unpublished letters, in *Sudhoffs Archiv,* **43** (1959), 233–253; and "Les débuts de la biologie marine en France: Jean-Victor Audouin et Henri Milne Edwards, 1826–1829," in *Actes du 1ᵉʳ Congrès International d'Histoire de l'Océanographie (Monaco, 1966),* 1968.

JEAN THÉODORIDÈS

AUDUBON, JOHN JAMES (*b.* Les Cayes, Santo Domingo [now Haiti], 26 April 1785; *d.* New York, N.Y., 27 January 1851), *ornithology.*

Audubon's father was Jean Audubon, a French sea captain and planter of moderate substance in Santo Domingo; his mother was a Mlle. Jeanne(?) Rabin(e?), who died soon after his birth. In 1791, he and a half sister were sent to Nantes, where their father had already arrived, to join him and Mme. Audubon (Anne Moynet), who graciously accepted the children of her husband's island sojourn. They were formally adopted in 1794, the boy as Jean Jacques Fougère Audubon.

Audubon's youth at Nantes and Coüeron, where he received a minimal elementary education, was comfortable and unexceptional. In 1803 he was sent to a farm, owned by his father, in eastern Pennsylvania and entrusted to the care of good friends. There his boyhood interest in birds—especially in drawing them—was intensified. In 1808 he married Lucy Bakewell, daughter of a prosperous neighbor, and moved to the new settlement of Louisville, Kentucky, where Audubon was to share in running a store.

Audubon had no formal training in natural history, having had only a brief acquaintance (upon revisiting France in 1805) with the obscure naturalist Charles d'Orbigny and a period in New York as a taxidermist under the many-faceted Samuel L. Mitchell (later founder of the Lyceum of Natural History). As an artist he was equally untutored (a persistent legend that he had briefly studied under Jacques Louis David seems to lack foundation). Marginally literate, Audubon had only hunting skill, undisciplined curiosity, great latent artistic power, and unfailing energy. He worked hard on his bird drawings, however, and developed a useful method of mounting dead birds on wires as an aid to delineation—a technique invaluable in a day without binoculars or cameras.

Between 1808 and 1819 Audubon failed as merchant and miller in both Louisville and Henderson, Kentucky, but in these formative years he ranged widely, from Pittsburgh as far west as Ste. Genevieve (now in Missouri). The country, if not untouched, was mostly unspoiled wilderness teeming with birds as little known to science as to him. He hunted and drew, sporadically at first, innocent of such patchy and uncertain knowledge as the few extant, relevant books would have given him. In common with not a few better-educated naturalists of the time, Audubon lacked formal method. He merely sought birds new to him, and shot and painted them, sometimes repeatedly, often substituting improved efforts for old.

Audubon briefly met the distinguished ornithologist Alexander Wilson at Louisville in 1810, and saw the first two (of nine) volumes of the artist–author's pioneer *American Ornithology* (he later implied, perhaps correctly, that his own drawings were, even at that time, better than Wilson's). Perhaps the idea of publication first entered his mind on this occasion, yet not until 1820, after going bankrupt, did Audubon set out by flatboat for Louisiana, with the single goal of enriching his portfolio of bird pictures. He would support himself precariously as itinerant artist and tutor, leaving much of the burden of supporting herself and their two sons to Lucy.

For the first time Audubon began a regular journal, some of which is extant. The journal of 1820 (the original of which survives) is a disorderly, semiliterate document. Like all the rest—the later ones are more articulate—it combines daily events, impressions of people and countryside, and random notes on birds encountered. These journals are valuable to ornithologists as checks on the formal texts that followed. Often more informative than the latter, they are nevertheless marked by lack of detail, imprecision,

and not infrequent discrepancies. Audubon never kept a full, orderly record of his observations on birds, and in formal writing he obviously relied as often on memory as on the sketchy notes he kept.

In 1821–1824, chiefly in Louisiana and Mississippi, Audubon came into his full powers as a gifted painter of birds and master of design. There would be many more pictures, but he would never improve upon the best of those years. Neither, although he would acquire a modicum of worldliness and a veneer of zoological sophistication, would his working methods and descriptive skills be basically changed. Whatever Audubon in essence was to be, he was by 1824.

In that year Audubon sought publication of his work in Philadelphia and New York. This failing, he traveled to England in 1826. There, finding support, subscribers, and skilled engravers, he brought out the 435 huge, aquatint copperplates of *The Birds of America,* in many parts, over the next twelve years.

The dramatic impact of his ambitious, complex pictures and a romantic image as "the American woodsman" secured Audubon entry into a scientific community much preoccupied with little-known lands. He met the leaders of society and science and was elected to the leading organizations, including the Royal Society of London. Among his friends were the gifted ornithologist William Swainson, from whom he learned some niceties of technical ornithology, and the orderly, brilliant Scottish naturalist-anatomist William MacGillivray. The text for Audubon's pictures, separately produced at Edinburgh, emerged as the five-volume *Ornithological Biography*. MacGillivray edited this for grammatical form, and he also contributed extensive anatomical descriptions to the later volumes.

Audubon's remaining efforts were devoted to the hopeless task of including all the birds of North America in his work. To this end he made increasing efforts to obtain notes and specimens from others and to cull the growing literature. Thus, much more than the early ones, the last volumes of his work have an element of compilation. He returned several times from his publishing labors in Scotland and England for more fieldwork, visiting the Middle Atlantic states in 1829, the Southeast as far as the Florida keys in 1831–1832, part of Labrador in 1833, and as far southwest as Galveston, Texas, in 1837. After his final return to the United States in 1839, Audubon journeyed up the Missouri River to Fort Union (the site of which is now in North Dakota) in 1843, obtaining birds treated in a supplement to the small American edition of his *Birds,* as well as some of the mammals discussed in his *Viviparous Quadrupeds,* which he wrote with John Bachman. In this, his last major effort, he was considerably assisted by his sons Victor and John.

Much, if not most, of Audubon's singularly enduring fame, which tends to cloud scientific and popular thought alike, rests on his much-debated but obviously significant efforts as an artist. (The relevancy of his established artistic stature to his scientific contribution is critical and difficult to assess, but can scarcely be ignored.) The illustration of new and little-known animals, as part of their zoological descriptions, was a characteristic and important part of eighteenth- and nineteenth-century natural history. Certainly Audubon kindled wide and enduring interest in this aspect of zoology—more, indeed, than would have been necessary for the strictly scientific appreciation of the subjects; his birds were portrayed with a flair, a concern for the living, acting animal in a suggested environment that was undreamed of before, and with a vigorous sense of drama, color, and design rarely equaled since. He had few significant predecessors and no debts in this area (only Thomas Bewick had earlier drawn—in simple woodcuts—birds as authentic). That Audubon's pictures contained innumerable technical errors seems to be comprehended only by specialists. The facial expressions and bodily attitudes of his birds are often strikingly human, rather than avian, but this is natural enough, considering his emotional nature and lack of optical equipment; paradoxically, this kind of error may have much to do with his enduring popularity with the general public.

Other than his art—aside from the inevitable accumulation of general knowledge of the kinds, habits, and distribution of birds—Audubon produced little that was new. Even the grand scale of his work had been anticipated by Mark Catesby a century earlier and by François Levaillant a generation earlier. Essentially, he built on Wilson's descriptive–anecdotal model (name the bird; say something general of its ways, habits, and haunts; and flesh out the account with a story or two of encounters with it in nature), as he states in the introduction to Volume V of the Biography. He went beyond Wilson in scope because he lived longer and had greater vigor; in point-for-point comparison, he tends to come off second best—where Wilson is dry and factual, even acerbic (but not artless), Audubon is grandiose, often irrelevant, romantic at best and florid at worst. His work, nevertheless, was the most informative available to American ornithologists between that of Alexander Wilson (as supplemented 1825–1833 by C. L. Bonaparte) and the beginning of Spencer Fullerton Baird's vast in-

fluence around 1860. He influenced such American successors as Baird, Elliott Coues, and Robert Ridgway, however, more by kindling interest than by procedural example.

Although he possessed a good eye for specific differences and inevitably discovered a number of new forms, Audubon was not basically a systematist; the classification of his *Synopsis* (1839), which ordered the randomly discussed birds of the *Biography*, is routine. As a theoretician he fared little better, being distinctly inferior to Gilbert White, who wrote half a century earlier and without pretension (see Audubon's curiously labored and undistinguished discussion of why birds do not need to migrate, in *Biography*, V, 442–445).

That Audubon possessed an original mind is shown, however, by a penchant (unfortunately little exploited) for experiment. As a young man in Pennsylvania he marked some phoebes with colored thread and recovered individuals after a year, thus anticipating bird banding by more than half a century. With Bachman in 1832, he conducted experiments designed to test the ability of the turkey vulture to locate its food by smell. The ingenious experiments lacked adequate controls and produced erroneous (though long credited) results.

In assessing Audubon, whose firm grip on the popular imagination has scarcely lessened since 1826, we must as historians of science seriously ask who would remember him if he had not been an artist of great imagination and flair. Not only does Audubon's artistic stature seem to dwarf his scientific stature, but the latter would probably be still less had he not been a painter expected to provide text for his paintings. The chances seem to be very good that had he not been an artist, he would be an unlikely candidate for a dictionary of scientific biography, if remembered to science at all.

BIBLIOGRAPHY

I. ORIGINAL WORKS. Nearly all the paintings for *The Birds of America* are at the New-York Historical Society, and have been reproduced by modern methods in *The Original Water-color Paintings by John James Audubon for The Birds of America* (New York, 1966). Miscellaneous additional paintings are cited by biographers listed below.

Audubon's books are *The Birds of America,* 435 aquatint copperplate engravings, 4 vols. without text (Edinburgh–London, 1827–1838); *Ornithological Biography,* 5 vols. (Edinburgh, 1831–1839); *Synopsis of the Birds of North America* (Edinburgh, 1839); *The Birds of America,* 7 vols. (New York–Philadelphia, 1840–1844), which combines the text of *Ornithological Biography* with inferior, much reduced, and sometimes altered copies of the plates of the 1st ed. of *The Birds of America;* and *Viviparous Quadrupeds of North America,* 3 vols. plates (New York, 1845–1848) and 3 vols. text (New York, 1846–1854), subsequent eds. (to at least 1865) combine text and reduced plates (plates by J. J. and J. W. Audubon; text by J. J. Audubon and John Bachman).

Audubon's comparatively few short articles in periodical literature are cited by biographers listed below.

The Life and Adventures of John James Audubon, written by Charles Coffin Adams from materials provided by Mrs. Audubon, Robert Buchanan, ed. (London, 1868), and its variant text, *The Life of John James Audubon,* Lucy Audubon, ed. (New York, 1869), contain the sole (but doubtless considerably modified) surviving record of Audubon's trip to Ste. Genevieve in 1810–1811 (pp. 25–33 in the 1868 version; 1869 version not seen) and other matter; Maria R. Audubon's *Audubon and His Journals,* E. Coues, ed. (New York, 1897), presents the only surviving version of the Labrador, Missouri River, and European journals. There is also a painstaking transcript of the extant *Journal of John James Audubon Made During His Trip to New Orleans in 1820–21* (Boston, 1929).

II. SECONDARY LITERATURE. The biographies cited below all contain extensive bibliographies that collectively provide detailed collations of Audubon's major works, elucidate the complexities of later editions and imprints, and give access to all but the most recent literature on the subject.

Lesser biographies and much miscellany are cited in F. H. Herrick's *Audubon the Naturalist,* 2nd ed., rev. (New York, 1938), still the basic and most extensive source but now outdated in some particulars by Ford; in S. C. Arthur's *Audubon, an Intimate Life of the American Woodsman* (New Orleans, 1937), which includes some sources not cited elsewhere; and in A. Ford's *John James Audubon* (Norman, Okla., 1964), which contains extensive new information on his parentage and early life in France. The popular *John James Audubon* by A. B. Adams (New York, 1966) contains some new sources and insights.

Searching appraisal of Audubon as an ornithologist may be found in the historical introduction to A. Newton's *A Dictionary of Birds* (London, 1896), p. 24; in E. Stresemann's preeminent *Die Entwicklung der Ornithologie* (Berlin, 1951), pp. 407–409; and in W. E. C. Todd's exhaustive *Birds of the Labrador Peninsula* (Toronto–Pittsburgh, 1963), pp. 731–732, 742, a detailed evaluation of his work in Labrador. An extensive critique of Audubon as a bird painter is given in R. M. Mengel's "How Good Are Audubon's Bird Pictures in the Light of Modern Ornithology?," in *Scientific American,* **216,** no. 5 (1967), 155–159.

ROBERT M. MENGEL

AUENBRUGGER, JOSEPH LEOPOLD (*b.* Graz, Austria, 19 November 1722; *d.* Vienna, Austria, 18 May 1809), *medicine.*

The son of a wealthy innkeeper, Auenbrugger received his medical education at the University of Vienna, where Gerard van Swieten had, through a series of reforms, made the Faculty of Medicine the leading one in Europe. Soon Auenbrugger came under van Swieten's influence; the extent of this influence is shown by his dedication to van Swieten of a work suggesting camphor as a treatment for a special form of mania (*Experimentum nascens de remedio specifico sub signo specifico in mania virorum,* 1776). He graduated on 18 November 1752.

From 1751 to 1758 Auenbrugger worked as assistant physician at the Spanish Hospital, but did not receive a salary until 1755. Because of his work in the hospital, Empress Maria Theresa in 1757 ordered the Faculty of Medicine to admit him as a member without charging him any fees. From 1758 to 1762 he was chief physician at the Spanish Hospital, obtaining experience in the diagnosis of chest diseases. After leaving the Spanish Hospital, Auenbrugger was a prominent practitioner in Vienna. For his medical achievements he was ennobled in 1784 by Emperor Joseph II.

Auenbrugger is considered the founder of chest percussion. He was undoubtedly aided in developing this diagnostic technique by his musical knowledge (he wrote the libretto for a comic opera by Antonio Salieri), which enabled him to perceive differences in tone when the chest was tapped. For seven years he had observed the changes in tone caused by diseases of the lungs or the heart in patients at the Spanish Hospital, checking and controlling his findings by dissections of corpses and by experiments. In the *Inventum novum* (1761) he presented his findings. If one taps with the fingertips on a healthy chest wall, one will perceive a sound like that of a drum. Diseases in the chest cavity change the normal tone of the tapping to a *sonus altior* (high or tympanitic sound), a *sonus obscurior* (indistinct sound), or a *sonus carnis percussae* (dull sound). Auenbrugger's method permitted the determination of disease-caused changes in the lungs and heart of a live patient and thus gave a new, dependable foundation to the diagnosis of chest diseases. Even with the development of X rays, this method still has diagnostic value.

In the first few years after its publication the *Inventum novum* was reviewed in several journals, the first mention probably being that of Oliver Goldsmith in the London *Public Ledger* (27 August 1761). In 1762 Albrecht von Haller drew attention to "this important work" in his lengthy review in the *Göttingische Anzeigen von gelehrten Sachen.* More influential than these positive references, however, was the opinion of Rudolf August Vogel, who could not find anything new in the *Inventum novum;* rather, he claimed to recognize in it only the *successio Hippocratis.* Van Swieten and Anton de Haen, the chief of the Vienna Clinic, never mentioned Auenbrugger's percussion, not even when discussing diseases of the chest, but Maximilian Stoll, de Haen's successor, described it in his publications and systematically taught it at the bedside. The spread of Auenbrugger's technique was interrupted by Stoll's premature death; his successors Jacob Reinlein and Johann Peter Frank did nothing to carry on his work.

Nevertheless, chest percussion was used as a diagnostic tool before 1800. Heinrich Callisen, a surgeon in Copenhagen, reported several observations obtained by percussion in his *System der Wundarzneikunst* (1788); and the Parisian surgeon Raphael Bienvenu Sabatier used it to advantage for the diagnosis of empyema. Percussion was practiced and taught at several German universities, including Halle, Wittenberg, Würzburg, and Rostock. About 1797 Jean-Nicolas Corvisart learned of chest percussion by reading Stoll. He investigated the method for several years and soon taught it to his students. In his classic book on heart diseases, *Essai sur les maladies et les lésions organiques du coeur* (1806), he based numerous diagnoses on percussion. Since there was only Rozière de la Chassagne's inadequate translation of the *Inventum novum* (1770), Corvisart published a new one in 1808, enriching it with a large number of his own observations and thus ending any question of the applicability of the new method.

BIBLIOGRAPHY

I. ORIGINAL WORKS. Auenbrugger's major work was *Inventum novum ex percussione thoracis humani ut signo abstrusos interni pectoris morbos detegendi* (Vienna, 1761, 1763, 1775). French translations were made by Rozière de la Chassagne, in *Manuel des pulmoniques* (Paris, 1770), and by Jean-Nicolas Corvisart (Paris, 1808). An English translation by John Forbes, in *Original Cases With Dissections and Observations . . . Selected from Auenbrugger, Corvisart, Laennec and Others* (London, 1824), was reprinted in F. A. Willius and T. E. Keys, *Cardiac Classics* (St. Louis, Mo., 1941), pp. 193–213, and in C. N. B. Camac, *Classics of Medicine and Surgery* (New York, 1959), pp. 120–147; it also appears, with an introduction by Henry E. Sigerist, in *Bulletin of the Institute of the History of Medicine,* **4** (1936), 373–403. A facsimile of the Latin text, together with the French, English, and German translations and a biographical sketch, has been published by Max Neuburger (Vienna–Leipzig, 1922). Auenbrugger also wrote *Von der stillen Wuth oder dem Triebe zum Selbstmorde als einer wirklichen Krankheit, mit Original-Beobachtungen und Anmerkungen* (Dessau, 1783) and the libretto for Salieri's *Der Rauchfangkehrer* (Vienna, 1781).

II. SECONDARY LITERATURE. Writings on Auenbrugger or his work are P. James Bishop, "A List of Papers, etc., on Leopold Auenbrugger (1722–1809) and the History of Percussion," in *Medical History,* **5** (1961), 192–196, a bibliography listing all important works pertinent to the subject with short critical remarks that facilitate orientation; H. L. Blumgart, "Leopold Auenbrugger. His 'Inventum novum'—1761," in *Circulation,* **24** (1961), 1–4; C. Costa, "Sobre la vicisitud creadora—Auenbrugger y Morgagni frente a frente," in *Anales chilenos de historia de la medicina,* **5** (1963), 63–223; Charles Coury, "Auenbrugger, Corvisart et les origines de la percussion," in *J. N. Corvisart, Nouvelle méthode pour reconnaître les maladies internes de la poitrine par la percussion de cette cavité, par Auenbrugger* (Paris, 1968), pp. 109–160; M. Jantsch, "200 Jahre 'Inventum novum,'" in *Wiener medizinische Wochenschrift,* **111** (1961), 199–202; R. G. Rate, "Leopold Auenbrugger and the 'Inventum novum,'" in *Journal of the Kansas Medical Society,* **67** (1966), 30–33; and J. J. Smith, "The 'Inventum novum' of Joseph Leopold Auenbrugger," in *Bulletin of the New York Academy of Medicine,* **38** (1962), 691–701.

JOHANNES STEUDEL

AUGUSTINE OF HIPPO, SAINT, also known as **Aurelius Augustinus** (*b.* Tagaste, North Africa, 13 November 354; *d.* Hippo, North Africa, 28 August 430), *theology, philosophy.*

Augustine was the son of Patricius, a minor official in the Roman province of Numidia, and his Christian wife, Monica. A thorough education in the classics of Roman rhetoric and philosophy led to his becoming at the age of twenty-one a teacher of rhetoric in nearby Carthage. For a time he was greatly attracted by the Manichaean religious doctrines, then at the height of their popularity in Africa, but their promise of a true "science of all things" proved illusory, and he turned to Stoic, Pythagorean, and Aristotelian sources for a surer light. His enormous success as a teacher led him to think of a wider field for his talents, and in 383 he embarked for Rome. After a year there, he was appointed to a professorship of rhetoric in Milan. The influence of Ambrose, bishop of Milan, led him to realize that the Manichaean objections to the Christian Scriptures were based on a simplistic, literalist mode of interpretation of the Bible. This realization and the reading of Plotinus and Porphyry led him to reject Manichaeism and begin the formulation of a highly personal Neoplatonism. His final step to Christianity was brought about by a dramatic conversion experience, centered around the letters of St. Paul. At this point, he gave up his teaching career, and with a group of friends retired to a monastic life of seclusion and study.

After his baptism in 387, he decided to return to Africa. There he began a prodigious writing career which led to almost one hundred books and countless treatises in sermon or letter form. In 395, he was consecrated bishop of Hippo, one of the major centers of Christian influence in North Africa, and was forced to forsake his monastic retirement for a life of constant travel and demanding pastoral duties. Despite this, he somehow found time for a stream of powerful polemics against those who seemed to him to threaten the doctrinal structure of Christianity; he opposed Donatists, Pelagians, and Arians in turn. In addition, he wrote such great creative works as *The City of God, On the Trinity,* and his incomparable *Confessions.* But the Empire was crumbling around him, and as he lay dying in 430, the Vandal armies had crossed the Strait of Gibraltar, and Roman Africa (once the most prosperous and fertile province of the Empire) lay in ruins. The Church of Africa, which Augustine had done so much to shape, was wiped out within a few years of his death.

But Augustine's influence had not been confined to the administration of the affairs of the church in North Africa. His was unquestionably the most powerful mind the Christian church had known until then, perhaps the most creative it has ever known. The success of his efforts to isolate and define the heretical elements in the work of Pelagius, Arius, and a legion of others gave to Roman Christianity a self-understanding, a methodology, a philosophical power and scope it had never before possessed. The categories in which the medieval church thought of man, of the world, of God, were largely those developed by Augustine. His metaphysics resembled those of Neoplatonism, but with *creation* replacing *emanation* as the focal concept. But what shaped his thought was not primarily the categorial systems of the Greek and Roman philosophers; rather, it was the overwhelming experience of sin and of conversion he had known in his own life. To an extent not to be seen again until the rise of existentialism in our own day, he built his philosophy around the certainties and realities of a profound inner experience. The experience was first and foremost one of weakness: naturally weak both in understanding and in will, man needs the help of God if he is to accomplish anything of value. Augustine's theory of knowledge is built around the notion of a Divine illumination which is integral to any genuine human act of understanding, and his theory of will centers on the idea of grace, that is, the aid God freely gives man to strengthen his will in pursuit of the good. The universe (and man

with it) is thus dependent on God in two significantly different ways: first in its being, because God, the Creator, is responsible for all that it is; second, in its activity, because it does not have within it a sufficiency of power to bring it to the goals that God designated for it and so He has to intervene to help it to completion.

Few men have influenced human thought as Augustine did Western religion and philosophy. But does he have any claim to notice in the history of natural science? At first sight it might not seem so. In the sixteen enormous volumes of his collected works, there is not a single treatise on what would be called today a "scientific" topic. Yet, in point of fact, Augustine's work can be seen as marking the second crucial stage in the development of the peculiar matrix of thought and value within which natural science, as we know it, emerged in the West. Greek philosophers had already made the bold and hitherto unthinkable claim that the universe can be grasped in a "true knowledge," a "science," of purely human making, and Greek geometers had produced an impressive ideal of what such a science might aspire to. But after an unparalleled burst of creative activity, Greek science had just as rapidly declined, until by the period of Roman dominance in the first century A.D. it had become little more than a memory. At that point, Christianity made its first appearance and rapidly swept through the Empire. Its message was primarily a religious one, but gradually this message was seen to have far-reaching philosophical consequences.

The major one—and Augustine was the first to draw it clearly—was that if all that is depends totally on God for its being, then it must be good through and through: "Everything which You made was good, and there is nothing at all which You did not make. . . . Each thing by itself is good, and the sum of them all is very good."[1] There is nothing independent of God, therefore, no positive principle of evil on which all the defects of the world can be blamed. At one stroke, the dualism of Manichaeism and of Oriental religions was rejected. The universe is henceforth to be regarded in the Christian West as the work of an intelligent Creator, itself therefore both intelligible and good. Christian faith thus began to point the way, dimly at first, to the possibility of a science of nature of the sort that the Greeks had dreamed of, but a science whose pursuit could now be construed in religious terms, something the Greeks had never succeeded in doing. A second consequence of Christian belief, one that equally permeated Augustine's thinking, would take even longer for Christian thinkers to explore, namely that human history has a meaning, a direction, a beginning, and an end. The way was thus opened to developmental thinking, to an understanding of things in terms of their origins and of the steps that have led to their present state. A shift away from the cyclic concept of time that dominated Greek and Oriental ways of thinking about the world had begun.

Augustine stands, therefore, at a fateful parting of the ways between West and East, one whose importance was often overlooked by early historians of science because it lay at a deeper level of thought than they were wont to regard as relevant. Yet it was at this deeper level that really crucial changes were occurring over the next millennium, so that by the seventeenth century the West was prepared in both attitude and motivation for a giant new effort of understanding.

Augustine's influence on the growth of the approach to nature and the knowledge of nature that science would one day demand cannot be represented as in all respects a positive one, however. There was a tension in his attitude toward the project of a "science of Nature," a tension deriving from many sources. He thought of the universe as a "sign" of God, and elaborated one of the most detailed treatments of the notion of "sign" before Peirce's work on signs in our own day. As a sign, the world should somehow be transparent; we should see God through it. It ought not become an object of interest entirely in its own right; if our gaze terminates at the sign instead of what the sign points to, it has failed to function as a sign for us. We have not really understood it, and our reputed knowledge is vain curiosity instead of true science. The knowledge of nature can never be a proper end in itself, then, unless "I proceed by occasion thereof to praise Thee."[2] Even though the universe is eminently worthy of study as the handiwork of God and the domain in which we can best come to grasp His power and wisdom, such study is always to be subordinate, a means to a knowledge of something other than the universe itself.

Some of Augustine's suspicion of science derived, however, from a much more specific cause. One of the features of Manichaeism that had most attracted him as a young man was its claim to a special knowledge of past and future, the former through an elaborate mythology of the origins of the cosmos and the latter through astrology. He rapidly came to see through both of these claims; in an incisive critique of astrology,[3] he points to instances of people born at the same hour of the same day whose subsequent fate is altogether different. What particularly irked him about the "rash promises of a scientific knowl-

edge" made by the Manichaean astrologers was the denial of human freedom implicit in such claims, and their use of "creatures of God" (the stars) to turn men away from God. His comments on astronomy, by far the most developed science of his day, are always tinged by the current misuse of astronomy on the part of astrologers: "I do not care to know the courses of the stars, nor does my soul ever demand an answer from a departed spirit, for I detest all such sacrilegious superstitions."[4] The sciences with which he was most familiar, instead of fulfilling their function of revealing the Creator in His creation, tended instead, it seemed, to satisfy men's lower instincts. He felt constrained to warn his readers of the dangers inherent in their pursuit: "Some men dive toward the discovery of secrets in Nature, whereof the knowledge, though not beyond our ken, doth profit nothing, yet men desire to know it for the sake of knowing. From this perverse desire of knowledge also it groweth that men enquire into things by magical arts."[5] Faith brought men to God and thus to the fullest satisfaction of their natures as free intelligent creatures. If a question of priorities arose, faith had therefore to rank before knowledge, although in principle the two ought to work together. "A faithful man . . . although he knows not the circles of the Great Bear is much better than another who can weigh out the elements and number the stars and measure the skies, if withal he neglects Thee, O Lord, who disposest of all things in number, weight and measure."[6]

Augustine's theory of knowledge dominated all discussions of scientific method for almost a millennium, until challenged by Aristotelian doctrines in the schools of Paris and Oxford in the mid-thirteenth century. Since all human knowledge (he asserted) needed the help of an intellectual illumination from God, the knowledge in which this illumination showed itself most clearly would obviously have to be the supreme type of knowledge. This pointed immediately to the revealed word of God in the Bible, and to its systematic explication in theology. When Augustine asserted that theology is the "queen of the sciences," this was no arbitrary claim on his part; it was a logical and quite inevitable consequence of his theory of knowledge. The Greek philosophers had seen something suprahuman in man's power of insight, the power by which men recognize eternal truths. These truths so far transcend the realm of sense and of the mutable in which man himself lives and moves that they can be grasped only with the aid of a power that equally transcends the normal material and human order. The notion of the Bible as the "revealed word of God" gave the Christian a warrant and model for such a theory of knowledge, a warrant that carried even more conviction than did the geometrical analogies of the Platonists.

Because the primary instance of Divine illumination is the Bible, and because all human sciences have to depend ultimately on a Divine illumination (rather than upon automatic modes of deduction or of empirical generalization), it follows that the word of the Bible carries immensely more weight than any claim of human science ever could. "That which is supported by Divine authority ought to be preferred over that which is conjectured by human infirmity."[7] Augustine favored the literal interpretation of Scripture, except where there was internal evidence that it was intended to be taken metaphorically or where the passage, literally taken, conflicted with a strictly demonstrated conclusion of a human science. In this literal emphasis, he differed from other Scriptural scholars of his day, especially those of the Eastern church who favored a more allegorical approach. His influence swung the Latin church heavily toward a "literalist" theory of inspiration, the view most in keeping with Augustine's overall theory of knowledge.

Augustine himself did not hesitate to depart from the literal interpretation of Scripture when he felt that scientific orthodoxy excluded the possibility of such an interpretation. The problem that preoccupied him by far the most in this regard was that of cosmic origins, about which the Neoplatonists had had much to say. The idea of a series of discrete creations spread over six days was absolutely unacceptable from the Neoplatonic point of view. It is interesting to note that Augustine assumed without further question that this conflict legitimized a partly metaphorical interpretation of the disputed *Hexaemeron,* the account of creation given in the first three chapters of Genesis. His commentary on this narrative, the *De Genesi ad litteram,* clearly gave him much trouble in the writing. It appeared in different versions, and the final one was written and rewritten over fourteen years. His purpose in writing it was to resolve what he clearly took to be the major dispute between Scripture and the science of his day.

He argued that the days mentioned in the Creation story have to be understood as lengthy periods of time, not as days in the strict sense. This was plausible enough, but he went much further than this. Instead of innumerable successive acts of creation, God implanted in the primal nebulous (*nebulosa*) matter the seminal potentialities (or *rationes seminales,* as the Neoplatonists called them) from which all the species of the world we know would successively develop in the order described in Genesis. There was very little internal evidence in support of this ingenious mode

of interpretation and the detailed theory of origins based on it; despite appearances, therefore, the warrant of the theory was really nothing more than its original philosophicoscientific one. But then methodological distinctions of this sort were not part of the Augustinian theory of knowledge to begin with.

Despite the liberties Augustine took in this particular instance, he much more often showed himself unexpectedly literalist, even when a metaphorical reading of Scripture would have seemed the obvious one to adopt. In one passage he discusses the Biblical phrase, "the waters above the firmament," the implications of which appeared to run counter to the contemporary belief that the "natural" place of water is below that of air. After an ingenious speculation about "waters" here perhaps meaning water vapor, he concludes: "In whatever manner the waters may be (above the firmament of air), and of whatsoever kind they may be, that they *are* there we cannot doubt. The authority of Scripture is far greater than the strength of man's reasoning."[8]

The influence of Augustine was thus a dangerously ambiguous one in this regard. Although in his view the knowledge of nature is worth attaining, it can apparently be discovered in the phrases of Scripture, as well as in the elaborations of science. And if a conflict should arise between a literal reading of the Bible and some finding of science, the former must take precedence unless and until the latter has been demonstrated beyond all possibility of error, in which case the theologian will have grounds for assuming that a metaphorical interpretation is the proper one. The potential for conflict in such a theory of interpretation is sufficiently obvious today, but it was more than a millennium before scientific claims would depart sufficiently from the commonsense Hebrew world view for the problem to become a disruptive one. It was no accident that both Galileo and his opponents called upon Augustine when the question arose whether the Copernican doctrine was invalidated by the frequent Scriptural mentions of the sun's motion.

Three Augustinian doctrines are worth noting because of the part they played in the early history of the natural sciences. The first of these is the doctrine of *rationes seminales,* which has already been mentioned and which has often been said to prefigure the theory of an evolutionary origin of species. The *De Genesi ad litteram* became the definitive work in the Middle Ages on origins and greatly encouraged developmental ways of understanding cosmic history, despite the counter influence of Aristotle in the later medieval period. But the Augustinian theory of *rationes seminales* was a developmental and not in any genuine sense an evolutionary view. Augustine believed in the fixity of species as strongly as did any of his Neoplatonist precursors. He did not suppose that one species could give rise to another; his theory was in no sense intended to suggest that every present species developed from a chain of earlier different species, back to the primal chaos of first matter. Rather, he held that God had implanted within this matter the germ of each separate species that would later develop; at some later appropriate moment, each germ would be activated, and the adult species would appear. The germs were not, therefore, earlier species, themselves destined to be replaced by the new species descending from them (as evolutionary theories hold); they were invisibly small seeds, each one carrying within it the potentiality for only one species. Furthermore, the activation of the seeds in some cases (as in the case of man's origin) required a further Divine intervention; the potentialities were not sufficient of themselves to bring about the new species unaided. Thus it was not a true theory of evolution. Yet it cannot be denied that it was a lot closer to evolutionary modes of thought than were the major cosmologies of Greek origin.

A second characteristically Augustinian emphasis had a more direct effect on the direction taken by earlier medieval science. Augustine made much of a passing Biblical phrase about God's having formed the universe "in number, weight and measure" in order to reinforce the mathematicism of Plato with an additional Biblical sanction. "These three things, measure, form, and order . . . are as it were generic good things to be found in all that God has created. . . . Where the three are present in a high degree, there are great goods. Where they are absent, there is no goodness."[9] Being, goodness, and mathematical intelligibility are one in the world God created. The Forms of Plato become for Augustine Ideas in the mind of God, and he is more sanguine than is Plato about the success of the Demiurge-Creator in incarnating them in matter. Since mathematical truths come closest to the immutability characteristic of God, mathematical ideas provide the most appropriate patterns for the Creator. Augustine emphasized, even more than Plato did, the preeminence of mathematics in the constructing of a science of nature. This assurance about the type of concept appropriate to physics came to be taken for granted during the long period of dominance of Augustinian theology, so that even when Aristotelian natural science was rediscovered in the mid-thirteenth century, the old reluctance about the use of mathematics in physics had almost vanished, and the new mechanics of Merton and Paris was a nascent mathematical physics.

A final feature of Augustine's thought that laid its impress upon medieval science was his basic metaphor of illumination, the principal causal mode relating God and man.[10] It was clearly all-important to understand the nature of illumination, since this would lead to a grasp of causal action generally. Such was quite explicitly the motivation that led to the extraordinary developments in mathematical and experimental optics in the thirteenth century, for instance to the work of Grosseteste and Theodoric of Fribourg on reflection and refraction.

When the third, and decisive, stage in the development of natural science occurred in the seventeenth century, Augustine was not one of those (like Archimedes and Aristotle) with whose views the pioneers of the new science would have to reckon directly. He is not quoted by them, nor would they have been likely to be aware of what they owed him. His contributions lay far beneath the surface; they were of the sort that once made are later taken for granted, so obvious do they seem. But when one tries to grasp what it was that came about in western Europe in the centuries between Grosseteste and Newton, the specific new theories of the workings of nature are less relevant than are the underlying slow changes in attitude toward nature, man, and God that made the search for such theories seem worth making and suggested the general lines along which the search would take place. In this more fundamental development, Augustine played a not inconsiderable role.

NOTES

1. *Confessions,* Bk. VII, ch. 12.
2. *Ibid.,* Bk. X, ch. 35. See also Bk. V, ch. 3.
3. *Ibid.,* Bk. VII.
4. *Ibid.,* Bk. X, ch. 35.
5. *Ibid.,* Bk. X, ch. 35.
6. *Ibid.,* Bk. V, ch. 4.
7. *De Genesi ad litteram,* Bk. II, ch. 9.
8. *Ibid.,* Bk. II, ch. 5.
9. *De natura boni,* Bk. I, ch. 3.
10. A. C. Crombie, *Augustine to Galileo* (London, 1952), pp. 71–80.

BIBLIOGRAPHY

Augustine's works fill sixteen vols. of Migne's standard *Patrologia Latina* collection (Vols. 32–47). There is an enormous amount of secondary literature, but Augustine's role in the history of science is scarcely mentioned in it. See, however, H. Pope, "St. Augustine and the World of Nature," ch. 6 of his *Saint Augustine of Hippo* (London, 1937); and F. van der Meer, *Augustine, the Bishop* (London, 1961), ch. 4. For a good general treatment of Augustine's work and a useful bibliography, see E. Portalié, *A Guide to the Thought of St. Augustine* (New York, 1960).

ERNAN MCMULLIN

AUSTEN, RALPH A. C. See **Godwin-Austen, Ralph A. C.**

AUSTIN, LOUIS WINSLOW (*b.* Orwell, Vermont, 30 October 1867; *d.* Washington, D.C., 27 June 1932), *radio physics.*

Austin graduated from Middlebury College in Vermont in 1889 and received the doctorate from the University of Strasbourg in 1893. He served as a member of the physics faculty of the University of Wisconsin from 1893 to 1901, and then returned to Germany for research work at Charlottenburg. In 1904 he entered U.S. government service in the Bureau of Standards and began a series of researches on radio transmission that made him world famous. The scope of his research was considerably enhanced by the establishment of a naval radiotelegraphic laboratory at the Bureau of Standards in 1908. This arrangement continued until 1923, when the laboratory and two others were merged into the radio division of a new unit (at the Navy Department), the Naval Research Laboratory.

Austin was in charge throughout this period, and thus had many opportunities for long-range transmission experiments. The most important was a test made in 1910, when U.S. cruisers en route to and from Liberia maintained radio contact with America and helped Austin and his collaborator Louis Cohen (1876–1948) to establish the Austin-Cohen formula, a semiempirical method for predicting the strength of radio signals at remote locations. The formula remained in use for many years and played an important role in the design and manufacture of improved apparatus. (Prior to World War I, the U.S. Navy had so much difficulty in making American firms meet its specifications that it began to design and manufacture its own radio receivers.)

In 1923 Austin resumed his work for the Bureau of Standards proper and became head of its laboratory for special radio transmission research. He contributed significantly to the understanding of the sources of radio atmospheric disturbances ("static"), a field in which he remained fruitfully active until his death.

As the doyen of U.S. government radio scientists, Austin exerted considerable influence and leadership in the development of radio engineering. He was one of the first members of the Institute of Radio Engineers (he joined on 22 January 1913), and one of the

few with a doctorate; he served as IRE's third president in 1914 and in 1927 received its Medal of Honor. Just before his death, Austin was unanimously nominated for the presidency of the International Radio Scientific Union (URSI), of whose U.S. national committee he had served as president.

BIBLIOGRAPHY

A list of Austin's major publications appears in Poggendorff, *Biographisch-literarisches Handwörterbuch,* Vols. V and VI. An obituary by Lyman J. Briggs is in *Science,* **76** (1932), 137. For Austin's contributions to marine radio communications, see L. S. Howeth, *History of Communications-Electronics in the U.S. Navy* (Washington, D.C., 1963).

CHARLES SÜSSKIND

AUTOLYCUS OF PITANE (*fl. ca.* 300 B.C.), *astronomy, geometry.*

Autolycus came from Pitane in the Aeolis, Asia Minor, and was an instructor of Arcesilaus, also of Pitane, who founded the so-called Middle Academy. It is reported by Diogenes Laertius (4.29) that Arcesilaus accompanied his master on a journey to Sardis.

Autolycus was a successor to Eudoxus in the study of spherical astronomy, but was active somewhat later than the successors of the Cnidian, Callippus, and Polemarchus. It appears that he attempted to defend the Eudoxian system of concentric rotating spheres against critics, notably Aristotherus, the teacher of the astronomer-poet Aratus. The critics had pointed out that Venus and Mars seem brighter in the course of their retrograde arcs and that eclipses of the sun are sometimes annular and sometimes total, so that not all heavenly bodies remain at fixed distances from the earth. Autolycus acknowledged the difficulty in his discussion with Aristotherus and, inevitably, was unable to account for the variations by means of the Eudoxian system.

The two treatises of Autolycus, *On the Moving Sphere* and *On Risings and Settings,* are among the earliest works in Greek astronomy to survive in their entirety. *On the Moving Sphere* is almost certainly earlier than Euclid's *Phaenomena,* which seems to make use of it (*Sphere,* Ch. 11, may be compared with *Phaenomena* 7). Pappus tells us it was one of the works forming the "Little Astronomy"—the "Small Collection," in contrast with the "Great Collection" of Ptolemy.

Both Euclid and Autolycus make use of an elementary textbook, now lost, on the sphere, from which they take several propositions without proof, because the proofs were already known. In *On the Moving Sphere,* a sphere is considered to move about an axis extending from pole to pole. Four classes of circular sections through the sphere are assumed: (1) great circles passing through the poles; (2) the equator and other, smaller, circles that are sections of the sphere formed by planes at right angles to the axis—these are the "parallel circles"; (3) great circles oblique to the axis of the sphere. The motion of points on the circles is then considered with respect to (4) the section formed by a fixed plane through the center of the sphere. A circle of class (3) is the ecliptic or zodiac circle, and (4) is equivalent to the horizon circle, which defines the visible and invisible parts of the sphere. In Euclid the great circle (4) has already become a technical term, "horizon," separating the hemisphere above the earth; Autolycus' treatment is more abstract and less overtly astronomical.

On Risings and Settings is strictly astronomical and consists of two complementary treatises or "books." True and apparent morning and evening risings and settings of stars are distinguished. Autolycus assumes that the celestial sphere completes one revolution during a day and a night; that the sun moves in a direction opposite to the diurnal rotation and traverses the ecliptic in one year; that by day the stars are not visible above the horizon owing to the light of the sun; and that a star above the horizon is visible only if the sun is 15° or more below the horizon measured along the zodiac (i.e., half a zodiacal sign or more).

The theorems are closely interrelated. Autolycus explains, for example, that the rising of a star is visible only between the visible morning rising and the visible evening rising, a period of less than half a year; similarly, he shows that the setting of a star is visible only in the interval from the visible morning setting to the visible evening setting, again a period of less than half a year. Another theorem states that the time from visible morning rising to visible morning setting is more than, equal to, or less than half a year if the star is north of, on, or south of the ecliptic, respectively.

Autolycus' works were popular handbooks and were translated into Arabic, Latin, and Hebrew.

BIBLIOGRAPHY

In addition to works cited in the text, see J. Mogenet, *Autolycus de Pitane* (Louvain, 1950); and O. Schmidt, "Some Critical Remarks About Autolycus' 'On Risings and Settings,'" in *Transactions of Den 11te Skandinaviske Matematikerkongress i Trondheim* (Oslo, 1952), pp. 202–209. See also the Mogenet edition of the Latin translation by Gerard of Cremona of *De sphaera mota,* in *Archives internationales d'histoire des sciences,* **5** (1948), pp. 139–164; and T. L. Heath, *Aristarchus of Samos* (Oxford, 1913), pp. 221–223.

G. L. HUXLEY

AUVERGNE, WILLIAM OF. See **William of Auvergne.**

AUWERS, ARTHUR JULIUS GEORG FRIEDRICH VON (*b.* Göttingen, Germany, 12 September 1838; *d.* Lichterfelde bei Berlin, Germany, 24 January 1915), *astronomy.*

The son of Gottfried Daniel Auwers, the riding master of Göttingen University, Auwers attended the gymnasia in Göttingen and Schulpforta from 1847 to 1857. He made a great number of observations and calculations of planets, comets, and variable stars as early as 1857–1859, his first years at Göttingen University. During his term as assistant at Königsberg (1859–1862) he made heliometric observations of double stars, which led him to his dissertation *Untersuchungen über veränderliche Eigenbewegungen* (1862). Bessel's assumption that certain changes in the proper motions of the stars Sirius and Procyon are based on the presence of invisible companion stars had opened a new field in celestial mechanics. Auwers was able to derive the orbits for both Sirius and Procyon as weakly eccentric ellipses, corresponding to the motion of the visible principal stars around the center of gravity of the double star system in forty or fifty years, respectively. A few years later it became possible also to observe their very faint companions by optical means.

Auwers then spent four years with Hansen at the Gotha observatory, for the most part determining parallaxes of the fixed stars. On Hansen's recommendation, in 1866 Auwers was appointed astronomer of the Berlin Academy, where he displayed great scientific and organizational ability during the next five decades. In 1878 the Academy appointed him as its permanent secretary, a position he held until his death. He was elected secretary of the Astronomical Society (founded in 1865) at its first meeting, and from 1881 to 1889 he was its president. Later he was awarded the Ordre pour le Mérite, grade of chancellor, and in 1912 was elevated to the hereditary nobility.

Auwers' lifework was the meticulous observation and calculation necessary to draw up star catalogs with highly accurate positions of stars. Since this required exact knowledge of the proper motions of the fixed stars, he also made new reductions of previous observations. Therefore, from 1865 to 1883 he undertook the laborious task of making a completely new reduction of Bradley's Greenwich observations, the oldest measurements of tolerable precision. The result was the publication of three volumes of Bradley's observations (1882–1903), the basis of all modern star positions and proper motions.

Auwers participated in the Zonenunternehmen der Astronomischen Gesellschaft, a project which had as its aim the observation and cataloging of all stars in the *Bonner Durchmusterung* up to the ninth magnitude. He not only observed the half of the zone for which Berlin was responsible but also took over the secretaryship of the zone commission appointed by the Astronomical Society to coordinate this task and to perform the final reduction. He also demonstrated his great organizational talents when he was in charge of the German expeditions that observed the transits of Venus in 1874 and 1882 in order to determine the sun's parallaxes. On the first expedition Auwers observed in Luxor, and on the second one in Punta Arenas, Chile. He published the results of both expeditions in six volumes (1887–1898).

In 1889 the opposition of the minor planet Victoria gave Auwers the opportunity to undertake the determination of the sun's parallax once again. He made the complete reduction of the meridian observations done in this connection by twenty-two other observatories, and obtained a surprisingly good determination of the sun's distance, considering the relatively great distance of the planet from the earth.

His incomparable experience and his remarkable endurance enabled Auwers to establish extremely accurate fundamental catalogs of positions of selected bright stars that had been observed for hundreds of years. These efforts led to the *Neue Fundamentalcatalog der Astronomischen Gesellschaft,* the foundation of all present precise measurements. Here his talent for the detection and elimination of systematic errors in observations was brought to the fore. Auwers also initiated the *Geschichte des Fixsternhimmels,* an extensive reduction and listing of all meridian observations of fixed stars from 1743 to 1900. Thus he must be considered the one who completed Bessel's epoch of classical astronomy in the second half of the nineteenth century. His participation in the founding of the astrophysical observatory in Potsdam bears witness also to his receptivity to the new science of astrophysics.

BIBLIOGRAPHY

I. ORIGINAL WORKS. Auwers' writings include *Untersuchungen über veränderliche Eigenbewegungen,* 2 vols. (Königsberg, 1862; Leipzig, 1868); *Bericht über die Beobachtung des Venusdurchgangs 1874 in Luxor* (Berlin, 1878); *Fundamentalcatalog für die Zonenbeobachtungen am nördlichen Himmel* (Leipzig, 1879); *Neue Reduktion der Bradleyschen Beobachtungen 1750 bis 1762,* 3 vols. (St. Petersburg, 1882–1903); *Die Venusdurchgänge 1874 und 1882,* 6 vols. (Berlin, 1887–1898); *Fundamentalcatalog für die Zonenbeobachtungen am südlichen Himmel* (Kiel, 1897); *Bearbeitung der Bradleyschen Beobachtungen an der Greenwicher Sternwarte* (Leipzig, 1912–1914); and many other works.

II. SECONDARY LITERATURE. Articles on Auwers are in *Neue deutsche Biographie,* I, 462; *The Observatory,* **38** (1915), 177–181; Poggendorff, Vols. III, IV, V; and *Vierteljahrschrift der Astronomischen Gesellschaft,* **53** (1918), 15–23.

BERNHARD STICKER

AUWERS, KARL FRIEDRICH VON (*b.* Gotha, Germany, 16 September 1863; *d.* Marburg, Germany, 3 May 1939), *chemistry.*

Auwers was a master organic chemist who investigated problems in structural theory for more than fifty years. He was a student of A. W. von Hofmann at Berlin, an assistant to Victor Meyer at Göttingen and Heidelberg, and then director of the chemical institute at the University of Marburg. On joining Victor Meyer, he became involved in stereochemical studies. In 1888 Auwers and Meyer substituted the name "stereochemistry" for van't Hoff's "chemistry in space." At this time the concept of geometrical isomerism had not been extended to structures other than those involving carbon atoms. Auwers and Meyer found that there were three isomeric forms of benzildioxime

$$C_6H_5-C\quad C-C_6H_5$$
$$N\quad N$$
$$OH\quad OH$$

and proposed, contrary to van't Hoff's principle of the free rotation of single bonds, that the isomerism resulted from the restricted rotation about the singly bound carbon atoms. Arthur Hantzsch and Alfred Werner elaborated the presently accepted theory in 1890 by extending geometric isomerism from double-bonded carbon atoms to double-bonded carbon and nitrogen atoms. Auwers and Meyer themselves aided in the confirmation of the Hantzsch and Werner theory.

Auwers' studies on isomerism led him into a lifelong investigation of stereochemical problems. He mastered the difficult art of assigning configuration to stereoisomers, establishing the configuration of the crotonic acids as well as many other geometric isomers. Auwers also investigated the spectrochemistry of organic compounds, the relation of physical properties to molecular structure, and molecular rearrangements, the most outstanding result of these studies being the determination of the keto-enol proportions in many tautomeric mixtures.

BIBLIOGRAPHY

Auwers was a prolific organic chemist, publishing over 500 papers in German journals from 1884 to 1938. A bibliography of his periodical publications, with a biographical notice by Hans J. Meerwein, is in *Berichte der Deutschen chemischen Gesellschaft*, **72** (1939), 111–121. His only book is *Die Entwicklung der Stereochemie* (Heidelberg, 1890). Among his important early papers with Victor Meyer on the stereochemistry of oxime compounds are "Untersuchungen über die zweite van't Hoffsche Hypothese," in *Berichte . . .*, **21** (1888), 784–817; "Weitere Untersuchungen über die Isomerie der Benzildioxime," *ibid.*, 3510–3529; and "Über die isomeren Oxime unsymmetrischer Ketone und die Konfiguration des Hydroxylamins," *ibid.*, **23** (1890), 2403–2409. His unraveling of the spatial configuration of the crotonic acids is found in "Über die Konfiguration der Crotonsäuren," in *Berichte . . .*, **56** (1923), 715–791, written with H. Wissebach. Among the many important papers on the relation between physical properties and chemical constitution are "Zur Spektrochemie und Konstitutionsbestimmung tautomerer Verbindungen," in *Annalen der Chemie*, no. 415 (1918), 169–232; "Über Beziehungen zwischen Konstitution und physikalischen Eigenschaften hydroaromatischer Stoffe," *ibid.*, no. 420 (1920), 84–111; and "Zur Bestimmung der Konfiguration raumisomerer Äthylenderivate," in *Zeitschrift für physikalische Chemie*, **143** (1929), 1–20, written with L. Harres.

ALBERT B. COSTA

AUZOUT, ADRIEN (*b.* Rouen, France, 28 January 1622; *d.* Rome, Italy, 23 May 1691), *astronomy, physics, mathematics.*

Auzout approached science with instruments rather than with mathematics. In the fall of 1647 he designed an ingenious experiment—creating one vacuum inside another—in order to prove that the weight of a column of air pressing on a barometer causes the mercury to rise inside. Auzout did not neglect mathematics, however; he criticized the treatise of François-Xavier Anyscom on the quadrature of the circle and prepared a treatise of reasons and proportions. By 1660 his career centered on astronomical instruments. He made a significant contribution to the final development of the micrometer and to the replacement of open sights by telescopic sights.

Not until after Christiaan Huygens discovered that there exists a special point (the focus) inside the Keplerian telescope, which has only convex lenses, could there be a breakthrough leading to the micrometer and to radically superior instruments. (Well before Huygens, William Gascoigne, after discovering the focal point, invented the micrometer and telescopic sights, but his inventions remained unknown to Continental astronomers until 1667.) Since an object can be superimposed on the image without distortion at the focus, precise measurements of the size of the image can be made. To do this, Huygens fashioned a crude micrometer. Cornelio Malvasia's lattice of fine wires was for Auzout, and probably Jean Picard, who worked with him, the jumping-off point for the perfection of the micrometer. They were dissatisfied with the accuracy of the lattice because images never covered exactly an integral number of squares, so they modified it. Two parallel hairs were separated by a distance variable according to the size of the image; one hair was fixed to a mobile chassis that was displaced at first by hand and later by a precision screw. By the summer of 1666 Auzout and Picard were making systematic observations with fully developed micrometers.

Soon after Huygens' discovery, Eustachio Divini and Robert Hooke replaced open sights with telescopic sights, and during the period 1667–1671 Auzout, Picard, and Gilles Personne de Roberval developed the systematic use of telescopic sights. An incomplete concept of focus caused the delay between discovery and systematic use, for Auzout, reasoning by analogy with open sights, suggested at the end of 1667 that each hair in the crosshairs be placed in a separate plane so that a line of sight might be assured. Unfortunately, after Auzout withdrew from the Académie des Sciences in 1668, he drifted into obscurity.

BIBLIOGRAPHY

I. ORIGINAL WORKS. A complete bibliography of Auzout's work is in R. M. McKeon (see below), pp. 314–324; the claims to priority in development of the micrometer of Auzout, Picard, and Pierre Petit are discussed on pp. 63–68. Most of Auzout's published works are reprinted in the *Mémoires de l'Académie Royale des Sciences, depuis 1666 jusqu'à 1699* (Paris, 1729), **6**, 537–540; **7**, Part 1, 1–130; **10**, 451–462. The major part of his correspondence is published in E. Caillemer, *Lettres de divers savants à l'Abbé Claude Nicaise* (Lyons, 1885), pp. 201–226; Christiaan Huygens, *Oeuvres*, 22 vols. (La Haye, 1888–1950), IV, 481–482; V, *passim;* VII, 372–373 (two letters are incorrectly attributed to Auzout in VI, 142–143, 580); H. Oldenberg, *Correspondence*, A. Rupert Hall and Marie Boas, eds. and trans. (Madison, Wis., 1965-   ); S. J. Rigaud and S. P. Rigaud, *Correspondence of Scientific Men* of the Seventeenth Century (Oxford, 1841), I, 206–210. The manuscript copies of Auzout's letters to Abbé Charles that were in the possession of the Ginori-Venturi family of Florence—called Épreuves in McKeon, p. 228, n.1—were listed in the spring 1966 catalog of Alain Brieux, Paris, and were sold. (The editors of Huygens' Oeuvres identify Abbé Charles as Charles de Bryas [IV, 72, n.4], but Abbé Charles's horoscope, which relates that he was born at Avignon in March 1604 and formerly was employed by Cardinal Mazarin [Bibliothèque Nationale, MSS fonds français, 13028, fol. 323], rules out their identification.)

II. Secondary Literature. Writings on Auzout or his work are Harcourt Brown, Scientific Organizations in Seventeenth Century France (1620–1680) (Baltimore, 1934); C. Irson, Nouvelle méthode pour apprendre facilement les principes et la pureté de la langue française (Paris, 1660), pp. 317–318; and Robert M. McKeon, "Établissement de l'astronomie de précision et oeuvre d'Adrien Auzout," unpublished dissertation (University of Paris, 1965); "Le récit d'Auzout au sujet des expériences sur le vide," in Acts of XI International Congress of the History of Science (Warsaw, 1965), Sec. III; and "Auzout," in Encyclopaedia universalis, in preparation.

Robert M. McKeon

AVEBURY, BARON. See **Lubbock, John.**

AVEMPACE. See **Ibn Bājja.**

AVENARE. See **Ibn Ezra.**

AVERROËS. See **Ibn Rushd.**

AVERY, OSWALD T. (b. Halifax, Nova Scotia, Canada, 21 October 1877; d. Nashville, Tennessee, 20 February 1955), biology.

Avery began his career as a physician. His father, a Canadian clergyman, had moved his family to New York in 1887, and Avery spent the next sixty-one years of his life there. He attended Colgate University, graduating A.B. in 1900, and received his medical degree from the Columbia University College of Physicians and Surgeons in 1904. He worked for a short time in the field of clinical medicine and then joined the Hoagland Laboratory in Brooklyn, New York, as a researcher and lecturer in bacteriology and immunology (his lectures at Hoagland won him the appellation "The Professor," by which he was known throughout his career). In 1913 he became a member of the staff of the Rockefeller Institute Hospital, where he remained until 1948.

At the time that Avery came to the hospital, an investigation of lobar pneumonia was in progress, and he joined with A. Dochez in work on the immunological classification of the pneumococcus bacterium.

This research led to the announcement, in 1917, that the pneumococcus produced an immunologically specific soluble chemical substance during growth in a culture medium.[1] Dochez and Avery further established that the substance was not a disintegration product of cell mortality but a true product of its metabolic processes, and that the substance was also present in the serum and urine of animals and men suffering from lobar pneumonia.

Beginning in 1922, Avery and his colleagues studied the chemical nature of these "soluble specific substances," which Avery believed were closely related to the immunological specificity of bacteria. They were soon shown to be polysaccharides derived from the capsular envelopes of the bacteria and to be specific to each pneumococcal type. By relating differences in bacterial specificity to these chemical differences in the capsular substances produced, Avery was able to explain in chemical terms many of the anomalies of immunological specificities. He showed that immunology could be analyzed biochemically, in terms of specific cellular components rather than the entire cellular complex, and thus contributed to the development of the study of immunochemistry.

Because of his concern with the extracellular substances of the pneumococcus, Avery began work in 1932 on a phenomenon first reported by F. Griffith in 1928. Griffith had observed that heat-killed virulent pneumococci could convert a nonvirulent strain to a disease-producer in vivo. Later investigations showed that this change in immunological specificity could be brought about in vitro, and that the alteration was permanent. Avery set out to isolate and analyze the active factor in the transformation, and his conclusions were reported in a well-known 1944 paper.[2]

A culture of an unencapsulated, and therefore nonvirulent, variant (designated R) of Type II pneumococcus, when exposed to an extract derived from encapsulated, virulent (S) Type III pneumococcus, was found to be converted to a Type III S culture. The isolated transforming material, upon examination, tested strongly positively for desoxyribonucleic acid (DNA) by the diphenylamine reaction. Elementary analysis revealed close resemblances in element ratios between the transforming substance and sodium desoxyribonucleate. Transforming ability was not inhibited by protein-destroying and ribonucleate-destroying enzymes, but only by those enzyme-containing preparations also capable of depolymerizing DNA. Treatment with desoxyribonucleodepolymerase also produced complete inactivation, at all temperatures up to the point of inactivation of the enzyme. Serologically it was found that transforming activity increased with increasing purity while the purified substance itself exhibited little or no immunological reactivity. On the basis of these and other tests, Avery and his collaborators concluded that the active fraction consisted principally, if not solely, of a highly polymerized form of desoxyribonucleic acid.

Avery thus showed that, in one instance at least, DNA was the active causative factor in an inherited variation in bacterial cells. The experiments showed that the preparations most active in bringing about transformation were those purest and most protein-free, thereby effectively casting doubt on the widespread and commonly accepted belief that proteins were the mediators of biological specificity and cellular inheritance. It was to a great extent through this work that the stage was set for the rapidly ensuing elaboration of the structure, function, and importance of DNA. Avery himself speculated about the mechanism of specificity determination and pointed out that "There is as yet relatively little known of the possible effect that subtle differences in molecular configuration may exert on the biological specificity of these substances,"[3] a situation that was well on the way to being remedied within ten years with the development of the Watson-Crick model for the DNA molecule.

NOTES

1. *J. Exp. Med.,* **26** (1917), 477–493; *Proc. Soc. Exp. Biol. Med.,* **14** (1917), 126–127.
2. *J. Exp. Med.,* **79** (1944), 137–158.
3. *Ibid.,* p. 153.

BIBLIOGRAPHY

I. ORIGINAL WORKS. Avery's major early work, on the immunological specificity of the pneumococcus involved in lobar pneumonia, is to be found in his two 1917 papers with A. Dochez, "The Elaboration of Specific Soluble Substance by Pneumococcus During Growth," in *Journal of Experimental Medicine,* **26** (1917), 477–493, also published in the *Transactions of the Association of American Physicians,* **32** (1917), 281–298, and "Soluble Substance of Pneumococcus Origin in the Blood and Urine During Lobar Pneumonia," in *Proceedings of the Society for Experimental Biology and Medicine,* **14** (1917), 126–127. The majority of his work on the chemical nature of the soluble specific substance appeared in a series of papers with M. Heidelberger: "Soluble Specific Substance of Pneumococcus," in *Journal of Experimental Medicine,* **38** (1923), 73–79; "The Specific Soluble Substance of Pneumococcus," in *Proceedings of the Society for Experimental Biology and Medicine,* **20** (1923), 434–435; and "The Soluble Specific Substance of Pneumococcus," in *Journal of Experimental Medicine,* **40** (1924), 301–316. The one paper for which Avery is best

known reported his work with MacLeod and McCarty, "Studies on the Chemical Nature of the Substance Inducing Transformation of Pneumococcal Types. Induction of Transformation by a Desoxyribonucleic Acid Fraction Isolated from Pneumococcus Type III," in *Journal of Experimental Medicine,* **79** (1944), 137–158.

II. SECONDARY LITERATURE. Complete bibliographies of Avery's work can be found appended to two major biographical sketches written by former colleagues: R. J. Dubos, in *Biographical Memoirs of Fellows of the Royal Society,* **2** (1956), 35–48; and A. R. Dochez, in *National Academy of Sciences, Biographical Memoirs,* **32** (1958), 32–49.

ALAN S. KAY

AVICENNA. See **Ibn Sīnā.**

AVOGADRO, AMEDEO (*b.* Turin, Italy, 9 August 1776; *d.* Turin, 9 July 1856), *physics, chemistry.*

He was the son of Count Filippo Avogadro and Anna Maria Vercellone. His father was a distinguished lawyer and higher civil servant who had become a senator of Piedmont in 1768 and had been appointed advocate general to the senate of Vittorio Amedeo III in 1777. His subsequent important administrative work led to his being chosen under the French rule of 1799 to reorganize the senate, of which he was made president. Amedeo Avogadro received his first education at home but went to the grammar school in Turin for his secondary education. Coming from a family well established as ecclesiastical lawyers (the name *Avogadro* itself probably being a corruption of *Advocarii*), Avogadro was guided toward a legal career and in 1792 he became a bachelor of jurisprudence. In 1796 he gained his doctorate in ecclesiastical law and began to practice law. In 1801 he was appointed secretary to the prefecture of the department of Eridano. Avogadro also showed interest in natural philosophy, and in 1800 he began to study privately mathematics and physics. Probably he was particularly impressed by the recent discoveries of his compatriot Alessandro Volta, since the first scientific research undertaken by Avogadro (jointly with his brother Felice) was on electricity in 1803.

In 1806 he was appointed demonstrator at the college attached to the Academy of Turin, and on 7 October 1809 he became professor of natural philosophy at the College of Vercelli. In 1820 when the first chair of mathematical physics (*fisica sublime*) in Italy was established at Turin with a salary of 600 lire, Avogadro was appointed. The political changes of 1821 led to the suppression of this chair, which Avogadro lost in July 1822. In 1823 he was given the purely honorary title of professor emeritus by way of compensation. When the chair was reestablished in 1832 it was first given to Cauchy. At the end of 1833 Cauchy went to Prague, and on 28 November 1834 Avogadro was reappointed. He held this position until his retirement in 1850.

In 1787 Avogadro succeeded to his father's title. He married Felicita Mazzé and they had six children. Avogadro was described in the *Gazzetta Piemontese* after his death as "religious but not a bigot."

Avogadro led an industrious life. His modesty was one of the factors contributing to his comparative obscurity, particularly outside Italy. Unlike his great contemporaries Gay-Lussac and Davy, he worked in isolation. Only toward the end of his life do we find letters exchanged with leading men of science in other countries. Avogadro's isolation cannot be attributed to language difficulties. He wrote good French, understood English and German, and kept abreast of all developments in physics and chemistry.

On 5 July 1804 Avogadro was elected a corresponding member of the Academy of Sciences of Turin, and on 21 November 1819 he became a full member of the Academy. His name is conspicuously absent from the foreign membership of the Paris Academy of Sciences and the Royal Society of London. Avogadro was a member of a government commission on statistics and served as president of a commission on weights and measures. In this latter capacity he was largely responsible for the introduction of the metric system in Piedmont. After 1848 Avogadro served on a commission on public instruction.

Avogadro is known principally for Avogadro's hypothesis, which provided a much-needed key to the problems of nineteenth-century chemistry by distinguishing between atoms and molecules. Dalton had considered the possibility that equal volumes of all gases might contain the same number of atoms but had rejected it. The source of Avogadro's inspiration was not however Dalton but Gay-Lussac, whose law of combining volumes of gases was published in 1809. In Avogadro's classic memoir of 1811 he wrote:

> M. Gay-Lussac has shown in an interesting memoir . . . that gases always unite in a very simple proportion by volume, and that when the result of the union is a gas, its volume also is very simply related to those of its components. But the quantitative proportions of substances in compounds seem only to depend on the relative number of molecules which combine, and on the number of composite molecules which result. It must then be admitted that very simple relations also exist between the volumes of gaseous substances and the numbers of simple or compound molecules which form them. The first hypothesis to present itself in this connection, and apparently even the only admissible one,

is the supposition that *the number of integral molecules in any gas is always the same for equal volumes, or always proportional to the volumes*. . . . The hypothesis we have just proposed is based on that simplicity of relation between the volumes of gases on combination, which would appear to be otherwise inexplicable.[1]

Avogadro, therefore, modestly presented his hypothesis as no more than an extension of Gay-Lussac's law.

From Avogadro's hypothesis there immediately follows the inference that the relative weights of the molecules of any two gases are the same as the ratios of the densities of these gases under the same conditions of temperature and pressure. Molecular weights could thus be determined directly. The hypothesis also enabled the chemist to deduce atomic weights without recourse to Dalton's arbitrary rule of simplicity. The molecular weight of water would be calculated in the following way:

$$\frac{\text{Weight of molecule of oxygen}}{\text{Weight of molecule of hydrogen}}$$
$$= \frac{\text{Density of oxygen}}{\text{Density of hydrogen}} = \frac{1.10359}{0.07321} = \frac{15.074}{1}.$$

As water is produced by the combination of two volumes of hydrogen with one of oxygen, this would give for the weight of two molecules of water vapor: $15 + 2 = 17$. The weight of one molecule would therefore be 8.5 (or, more accurately, 8.537). This agreed well with the known density of water vapor referred to the hydrogen standard. It should be noted that Avogadro's molecular weights are values based on the comparison with the weight of a molecule of hydrogen rather than an atom of hydrogen. The molecular weights given in Avogadro's paper of 1811 are therefore half the modern values. However expressed, they were a vast improvement on Dalton's values.

The superiority of Avogadro's method of deriving the molecular weights of compounds over that of Dalton is seen not only with water but with many other compounds. Where the values given by Avogadro were of the same order as those given by Dalton, the Italian pointed out that this resulted from the canceling out of errors. Avogadro was, however, completely fair in his criticism of Dalton, and at the end of his memoir he modestly concluded that his hypothesis was "at bottom merely Dalton's system furnished with a new means of precision from the connection we have found between it and the general fact established by Gay-Lussac."

It is necessary to comment on Avogadro's use of

the term *molecule*. Although it has been suggested that he used the term inconsistently, a close examination of his memoir enables us to distinguish four uses of the term: *molécule* ("molecule"), a general term denoting either what today would be called an atom or a molecule; *molécule intégrante* ("integral molecule"), corresponding to the present-day usage of molecule, particularly in relation to compounds; *molécule constituante* ("constituent molecule"), denoting a molecule of an element; and *molécule élémentaire* ("elementary molecule"), denoting an atom of an element. Although Avogadro deserves credit for his application of these expressions, he did not invent them. The terms *partie intégrante, molécule primitive intégrante,* and *partie constituante* are to be found in Macquer's *Dictionnaire de chimie* of 1766 (article on "Agrégation"); and the terms *integrant, constituent,* and *elementary* molecules are to be found in Fourcroy's textbook of 1800.

Avogadro had a solution to the problem that arose when the hypothesis of equal volumes was applied to compound substances. Gay-Lussac had shown that (above 100°C.) the volume of water vapor was twice the volume of oxygen used to form it. This was possible only if the molecule of oxygen was divided between the molecules of hydrogen. Dalton had seen this difficulty and, as it was to him inconceivable that the particles of oxygen could be subdivided, he had rejected the basic hypothesis that Avogadro was now defending. Avogadro overcame the difficulty by postulating compound molecules. This is the second and most important part of Avogadro's hypothesis. It may be regarded as his second hypothesis; and, unlike the first, it seems completely original. Avogadro wrote: "We suppose . . . that the constituent molecules of any simple gas whatever . . . are not formed of a solitary elementary molecule, but are made up of a certain number of these molecules united by attraction to form a single one." Compound molecules of gases must therefore be composed of two or more atoms. Avogadro implies that there are always an even number of atoms in the molecule of a gas. For nitrogen, oxygen, and hydrogen it is two, "but it is possible that in other cases, the division might be into four, eight, etc." Avogadro is not at his clearest in this part of the memoir, but clarity is achieved when he gives an example: "Thus, for example, the integral molecule of water will be composed of a half-molecule of oxygen with one molecule, or, what is the same thing, two half-molecules, of hydrogen."

Avogadro's reasoning about the divisibility of the integrant molecule raises the question of atomicity. Gases for Avogadro were usually diatomic, but certain substances could be tetratomic in the vapor state, as indeed phosphorus is. In later memoirs he allowed for the possibility of monatomic molecules, as, for example, in gold. In his work there is implicit the idea of equivalence, e.g., that one atom of oxygen is equivalent to two atoms of hydrogen, which in turn is equivalent to two atoms of chlorine. The concept of valency, was, however, not developed until the time of Frankland (1852).

In a second memoir, which he sent in January 1814 for publication in Lamétherie's *Journal de physique,* Avogadro developed his earlier ideas. He began by pointing out that no alternative explanation had been offered to the one published by him three years previously to account for Gay-Lussac's law of combining volumes of gases. He suggested that his hypothesis could be used to correct the theory of definite proportions, which he now saw as "the basis of all modern chemistry and the source of its future progress." It was therefore important "to establish by facts, or in default, by probable conjectures, the densities which the gases of different substances would have at a common pressure and temperature." The recent experimental work of Gay-Lussac, Davy, Berzelius, and others now gave Avogadro scope to apply his principle to a greater number of substances. In 1811 Avogadro had been able to give the modern formulas (in words) for water vapor, nitric oxide, nitrous oxide, ammonia, carbon monoxide, and hydrogen chloride. In 1814 he was able to give the correct formulas for several compounds of carbon and sulfur, including carbon dioxide, carbon disulfide, sulfur dioxide, and hydrogen sulfide. Falling back on analogy when experimental evidence was lacking, he reasoned correctly that silica was SiO_2 by comparison with CO_2. He also extended his earlier treatment of metals in the hypothetical gaseous state. He gave molecular weights for (using modern symbols) Hg, Fe, Mn, Ag, Au, Pb, Cu, Sn, Sb, As, K, Na, Ca, Mg, Ba, Al, and Si, based on analyses of the compounds of these elements. His mention of "gaz métalliques" may have done more harm than good to the reception of his hypothesis.

In 1821 Avogadro was able to state the correct formulas of several other compounds including those of phosphorus and the oxides of nitrogen. After deducing the formulas of such inorganic compounds from combining volumes, densities, or merely by analogy, he turned to organic chemistry. He gave the correct empirical formula for turpentine, $C:H = 1.6:1$, (in the symbols of Berzelius $C_{10}H_{16}$) and the correct molecular formulas for alcohol (C_2H_6O) and ether ($C_4H_{10}O$). Berzelius did not arrive at the correct formulas for alcohol and ether until 1828. Berzelius, however, who had developed a theory deriving from both Dalton and Gay-Lussac, took little account of Avogadro.

Avogadro's claim to the hypothesis that is named after him rests on more than his mere statement of it. We have seen that Dalton earlier considered a similar hypothesis but rejected it. Ampère in 1814 independently arrived at a similar conclusion, and later Dumas (1827), Prout (1834), and others formulated the same hypothesis. To Avogadro alone, however, belongs the distinction of applying his hypothesis to the whole field of chemistry.

Ampère's ideas on the constitution of molecules were published in 1814 in the form of a letter to Berthollet. He stated that it was only after writing his memoir that he had "heard that M. Avogadro had used the same idea." Ampère was not primarily concerned with providing an explanation of Gay-Lussac's law. He was interested in the structure of crystals and introduced geometrical considerations which led him to suppose that molecules of, for example, hydrogen, oxygen, and nitrogen each contained four atoms, whereas Avogadro had already shown correctly that these gases contained two atoms per molecule. Despite the inferiority in many respects of Ampère's memoir to that of Avogadro and its clear lack of priority, Avogadro's hypothesis was usually attributed to Ampère until the Italian Cannizzaro called the attention of chemists to the publication of his fellow countryman. Certainly by being published in the *Annales de chimie* Ampère's paper had the greatest possible publicity, and Ampère himself became famous throughout the scientific world after his work on electromagnetism in the 1820's.

In view of the lack of interest shown by chemists in Avogadro's hypothesis, it is all the more noteworthy that he himself continually drew attention to the significance of his ideas. If Avogadro was not heard, therefore, it was not because he had made an isolated pronouncement. He repeated it in a memoir published in 1816 and 1817; and again in two different memoirs published in 1818 and 1819 in Italian publications, he drew attention to his earlier work. The 1819 memoir was later translated into French and the relevant passage reads:

> In considering the matter theoretically and supposing in conformity with what I have established elsewhere (*Journal de Phys. de La Metherie,* July 1811 and February 1814) that in gases reduced to the same temperature and pressure the distances between the centers of the integrant molecules is constant for all gases, so that the density of a gas is proportional to the mass of its molecules. . . .[2]

Even more important was Avogadro's long memoir of 1821: "Nouvelles considerations sur . . . la détermination des masses des molécules des corps." This was published in the memoirs of the Turin Academy of Science, but a summary was published in France in 1826 in the *Bulletin . . . de Ferussac,* and this extract contains a reassertion of Avogadro's hypothesis with the claim that its introduction would do much to simplify and generalize chemistry. Avogadro insisted in this memoir on the necessity of correlating the data obtained by Gay-Lussac on the combining volumes of gases with the theory of fixed proportions—a problem that had been considered only superficially by Berzelius.

From the above it is clear that Avogadro stated his new hypothesis repeatedly over the period 1811–1821. It is true that the later memoirs were particularly long, and only the most relevant parts are included here. More important for the reception of Avogadro's work, however, is that these later memoirs were published in Italy, then at the periphery of the scientific world. When these memoirs were translated into French they appeared not in the influential *Annales de chimie et de physique* but in the comparatively obscure and second-rate *Bulletin . . . de Ferussac.* As regards translation into other languages, not more than three of Avogadro's papers were translated into English and not more than two into German, and the memoirs chosen are not those that we should today regard as significant. Reports on Avogadro's later work were, however, included in Berzelius' *Jahresberichte* from 1832 onward. Some part of the blame for the lack of attention given to Avogadro's hypothesis by his contemporaries must attach to the editors of the influential British, French, and German scientific periodicals.

One of the most remarkable features of Avogadro's hypothesis was the way in which it was neglected by the vast majority of chemists for half a century after its initial publication. The following are some of the reasons for this delay:

(1) It was not clearly enough expressed, particularly in the 1811 memoir. Avogadro did not coin a new term for the "solitary elementary molecules" nor did he use the term *atom.* In any case there was general looseness in the use of the terms *atom* and *molecule,* and they were often used synonymously. It would not have been clear to everyone in the early nineteenth century that Avogadro's hypothesis was quite different from that of Berzelius. Berzelius had substituted atom for volume in cases of combining gases and had ended with the logical contradiction of half an atom. Avogadro overcame this difficulty by introducing a polyatomic molecule, but this concept was quite novel.

(2) Avogadro did not support his hypothesis with any impressive accumulation of experimental results.

He never acquired, nor did he deserve, a reputation for accurate experimental work. Thus, for Regnault, for example, Avogadro was not a brilliant theoretician but merely a careless experimenter.

(3) From the beginning Avogadro applied his hypothesis to solid elements. When experimental evidence was not available, he relied only on analogy. He was correct in considering oxygen and hydrogen as diatomic, but he had little justification in coming to a similar conclusion in the cases of carbon and sulfur. His speculative treatment of metals in the vapor state in his second paper of 1814 cannot have helped his cause. He was, therefore, overly ambitious in extending his hypothesis. (Yet Berzelius too used a "volume theory" not restricted to the gaseous state, and his work did not suffer any eclipse because of this.) There were only a comparatively small number of well-defined substances that existed in the gaseous state and to which Avogadro's hypothesis could be applied. Only with Gerhardt was its full relevance to organic substances appreciated.

(4) The half century after the publication of Avogadro's hypothesis was a time when most attention was paid to organic chemistry, where the primary need was for analysis and classification. Organic analysis was based on weights, not volumes.

(5) Avogadro's idea of a diatomic molecule conflicted with the dominant dualistic outlook of Berzelius. According to the principles of electrochemistry, two atoms of the same element would have similar charges and therefore repel rather than attract each other.

(6) Avogadro, a modest and obscure physicist on the wrong side of the Alps, remained intellectually isolated from the mainstream of chemistry.

Despite the failure of chemists to appreciate the full significance of Avogadro's hypothesis, he could claim in 1845 that his statement that the mean distances between the molecules of all gases were the same under the same conditions of temperature and pressure and the consequence that the molecular weight was proportional to the density was generally accepted by physicists and chemists either explicitly or implicitly.

What was ignored, however, was the use of the hypothesis to determine atomic weights. It was Cannizzaro who, in a paper published in 1858 but not made widely known until 1860 at the Karlsruhe Congress, showed how the application of Avogadro's hypothesis would solve many of the major problems of chemistry. In particular he clarified the relation between atom and molecule which Avogadro had not made explicit. Gerhardt and Laurent had applied the hypothesis with some success to organic chemistry.

Cannizzaro performed a roughly complementary task in systematizing inorganic chemistry on the basis of Avogadro's hypothesis. He determined the molecular weights of many inorganic substances and hence the atomic weights. For the first time there began to exist among chemists substantial agreement on atomic weights. Cannizzaro had been able to make use of techniques unknown in Avogadro's time for the determination of the molecular weights of solid elements, including sulfur, phosphorus, and mercury.

Avogadro's first two memoirs to be published, in 1806 and 1807, were on electricity. He considered the state of a nonconductor placed between two oppositely charged elementary layers. If there was air between two charged bodies, it would become charged. In the later terminology of Faraday the claim could be made that Avogadro had some conception of the polarization of dielectrics. In 1842 Avogadro himself claimed that some of Faraday's treatment of condensers was to be found in his own earlier work. At the end of the same memoir Avogadro suggested that the capacity of a condenser was independent of the gas between the plates and that there would be the same process of induction even in a vacuum—a phenomenon later verified by James Clerk Maxwell.

Avogadro's 1822 memoir "Sur la construction d'un voltimètre multiplicateur" was given publicity by Oersted. Avogadro's "multiplier" was one of the most sensitive instruments of the time, and by using it he found that when certain pairs of metals are plunged into concentrated nitric acid the direction of the electric current is momentarily reversed. This happens for the pairs of metals: Pb/Bi, Pb/Sn, Fe/Bi, Co/Sb. This phenomenon had actually been observed in the case of Pb/Sn by Pfaff in 1808, but it was sometimes referred to as "Avogadro's reversal." Avogadro used a succession of pairs of metals with an electrolyte to establish the order Pt, Au, Ag, Hg, As, Sb, Co, Ni, Cu, Bi, Fe, Sn, Pb, Zn, a list that showed several differences from the order found by Volta using a condenser.

Avogadro published his first article dealing only with chemistry in 1809. This memoir, on acids and alkalies, is interesting for several reasons. In the first place it illustrates his abiding concern with chemical affinity and incidentally the great influence exerted on him by Berthollet. Second, in the opening paragraph, which criticizes the oxygen theory of acidity, it illustrates his radical approach to post-Lavoisier chemistry. He postulated a relative scale of acidity in which oxygen and sulfur were placed toward the acid end of the scale, neutral substances in the middle, and hydrogen at the alkali end. A significant feature of this scale was that it was continuous. Avogadro would not allow any absolute distinctions. He was not, for example, prepared to agree with Berzelius that oxygen was absolutely electronegative. Davy, in 1807, had suggested a connection between acidity and alkalinity and electricity; Avogadro developed this idea. Another feature of Avogadro's interests found in this memoir is the subject of nomenclature. He was to give more detailed attention to this in the 1840's.

Avogadro might claim to share with Berthollet the honor of having been one of the founders of physical chemistry in the early nineteenth century. Certainly he saw no boundary between physics and chemistry and made constant use of a mathematical approach. This is exemplified by his various studies on heat; his thermal studies of gases provided a useful approach to physical chemistry. In 1813 Delaroche and Bérard suggested that there was a simple relationship between the specific heat of a compound gas and its chemical composition. Particles of a gas were then envisaged as surrounded by an "atmosphere" of caloric. But the amount of caloric between particles of a gas governed the mutual attraction between its particles. Avogadro thus saw a means of relating chemical affinity to specific heat. Assuming that (specific heat)^m is the attractive power of each molecule and that these attractive powers are additive, Avogadro obtained a number of equations from which m was found to be rather less or rather greater than 2. Taking the deviation from $m = 2$ to be due to experimental error (Avogadro never claimed any high degree of accuracy in his experimental work), he derived the general formula:

$$c^2 = p_1c_1{}^2 + p_2c_2{}^2 + \text{etc.},$$

where c, c_1, c_2, etc. are the specific heats at constant volume of the compound gas and its constituents, respectively, and p_1, p_2, etc. are numbers of molecules of the components taking part in the reaction.

In 1822 Avogadro considered himself justified in making the generalization that the specific heats at constant volume of gases were proportional to the square root of the attractive power of their molecules for caloric. In 1824 he made further progress toward an evaluation of a "true affinity for heat," to which he could assign a numerical value. This value was obtained by taking the square of the specific heat determined by experiment and dividing by the density of the gas, which by Avogadro's hypothesis was proportional to its molecular weight. He thus obtained a series of values ranging from oxygen = 0.8595 to hydrogen = 10.2672. This confirmed his conjecture that oxygen and substances similar to it had the least affinity for heat. He concluded triumphantly that the

order so obtained, representing the affinity for heat, coincided with the order in the electrochemical series. He obtained further confirmation by comparing his results with those obtained by Biot and Arago for the relationship between affinity for heat and the refractive indices of gases. Avogadro concluded with a table of twenty-nine substances, headed by acids and terminating with bases. By dividing each affinity for heat by that of oxygen, he obtained a series of what he called "affinity numbers" (*nombres affinitaires*), and in his next memoir he attempted to determine further affinity numbers.

By 1828 he doubted the validity of much of his earlier work on assigning numerical values to affinities, and he therefore reverted to a purely chemical method, developing Berthollet's idea of combining proportions as a measure of affinity. At the end of his life Avogadro claimed that he had succeeded in deriving affinity numbers from atomic volumes and "by a method independent of all chemical considerations"[3]— reminding us of his predominantly physical attitude. By trying to derive chemical information from nonchemical sources, however, he contributed to a situation in which his work lay outside the sphere of interest of contemporary chemists.

In 1819 Dulong and Petit announced that there was a simple relationship between specific heats and atomic weights. Although they suggested that their law might be extended to compounds, it was F. E. Neumann who, in 1831, first applied the law practically to solid compounds. Avogadro, who began his research in this field in 1833, investigated both liquids and solids.

He decided that the formula of a compound in the liquid or solid state could not be the same as that in the gaseous state. He therefore introduced the arbitrary division of molecules and considered, for example, that a molecule of water or ice contained only a quarter as many atoms as one of steam. Thus since H_2O represented steam, water would have been $HO_{1/2}$. (Actually Avogadro says a compound of "1/4 *atom* of hydrogen and 1/4 *atom* of oxygen," but if we interpret "atom" as "molecule" and consider these to be diatomic this would have been the resultant formula.) Mercuric oxide was considered to be 1/2 atom of metal + 1/4 atom of oxygen; aluminum oxide was considered to be 1/2 atom of metal + 3/8 atom of oxygen; ferric oxide was considered to be 1/4 atom of metal + 3/16 atom of oxygen. By such arbitrary division he was able to fit a fairly wide selection of solid compounds into a "law" that he devised to relate specific heat to molecular weight. His 1838 memoir on this subject probably shows Avogadro at his very worst. Fractional atoms were introduced in the most irresponsible way although they were related to the supposed specific heats. The latter soon proved to be inaccurate. Because the memoir was published so long after the first announcement of Avogadro's hypothesis, it was not mentioned earlier as one of the reasons for the rejection of the hypothesis by his contemporaries. Yet the publication of the memoir can only have weakened Avogadro's scientific reputation.

The inaccuracies of the values obtained by Avogadro for specific heats were later criticized by Regnault. We are reminded by such criticism of the opposite poles represented by such men of science as Avogadro and Regnault—the one intuitive, speculative, and theoretical and the other empirical, precise, and practical. Avogadro's research on the vapor pressure of mercury cannot stand comparison for accuracy with Regnault's later work, yet Avogadro deserves credit for being the first to make this determination over the temperature range 100°C.-360°C.

Toward the end of his life Avogadro devoted a total of four memoirs to the subject of atomic volumes. In the first (1843) he pointed out the connection with his classic memoir of 1811—the mean distance between the molecules of all gases is the same under the same conditions of temperature and pressure. In 1824 he had read to the Turin Academy a memoir in which he had pointed out that the atomic volumes (i.e., the volume occupied by the molecule together with its surrounding caloric) of all substances in the liquid or solid state would be the same if it were not for certain factors and in particular the different affinities of bodies for caloric. But the latter factor was directly related to the electronegativity of the element. Comparing the densities of the elements with their atomic weights, he now concluded that the distances between the molecules of solids and liquids, and consequently their volumes, were greater, and hence their densities compared with their atomic weights were less as the body became more electropositive. Alternatively expressed, the atomic volume (atomic weight/density) is greater for the more electropositive elements, and this is now accepted.

Avogadro's work on atomic volumes differs from that of his contemporaries in his insistence on its connection with the position of elements in the electrochemical series. Also his "atomic" volumes were really molecular volumes. In this later work he was as isolated as he had been in his earlier speculations. By his habit of consistently giving references to his own earlier work Avogadro established the lineage of his research with any corresponding priority claim, but the practice had the disadvantage of revealing him as a solitary worker perhaps born a generation too soon.

NOTES

1. Alembic Club Reprint, No. 4, pp. 28-30 (the italics are the author's).
2. *Bulletin . . . de Ferussac*, **5** (1826), 39.
3. *Annales de chimie et de physique*, **29** (1850), 248.

BIBLIOGRAPHY

I. ORIGINAL WORKS. Articles by Avogadro include "Considérations sur l'état dans lequel doit se trouver une couche d'un corps non-conducteur de l'électricité, lorsqu'elle est interposée entre deux surfaces douées d'électricités de différente espèce," in *Journal de physique*, **63** (1806), 450–462; "Second mémoire sur l'électricité," in *Journal de physique*, **65** (1807), 130–145; "Idées sur l'acidité et l'alcalinité," in *Journal de physique*, **69** (1809), 142–148; "Essai d'une manière de déterminer les masses relatives des molécules élémentaires des corps, et les proportions selon lesquelles elles entrent dans ces combinaisons," in *Journal de physique*, **73** (1811), 58–76; "Réflexions sur la théorie électro-chimique de M. Berzelius," in *Annales de chimie*, **87** (1813), 286–292; "Mémoire sur les masses relatives des corps simples, ou densités présumées de leurs gaz, &c.," in *Journal de physique*, **78** (1814), 131–156; "Memoria sul calore specifico de' gaz composti paragonato a quello de' loro gaz componenti," in *Biblioteca italiana*, **4** (1816), 478–491; **5** (1817), 73–87; "Osservazioni sulla legge di dilatazione dell'acqua pel calore," in *Giornale di fisica*, **1** (1818), 351–377; "Sopra la relazione che esiste tra i calori specifici e i poteri refringenti delle sostanze gazose" [1817], in *Memorie di matematica e di fisica*, **18** (1818), 154–173; "Sulla determinazione delle quantità di calorico che si sviluppano nelle combinazioni per mezzo de' poteri refringenti" [1817], *ibid.*, pp. 174–182; "Osservazioni sulla forza elastica del vapor acqueo a diverse temperature," in *Giornale di fisica*, **2** (1819), 187–199; "Memoria sulle leggi della dilatazione de' diversi liquidi pel calore," *ibid.*, pp. 416–427; "Memoria sopra lo stabilimento d'una relazione tra le densità e dilatabilità de' liquidi e la densità dei vapori che essi formano," *ibid.*, pp. 443–456; "Memoria sulla legge della dilatazione del mercurio dal calore," *ibid.*, **3** (1820), 24–38; "Nouvelles considérations sur la théorie des proportions déterminées dans les combinaisons, et sur la détermination des masses des molécules des corps," in *Memorie della Reale Accademia delle scienze*, **26** (1821), 1–162; "Sur la manière de ramener les composés organiques aux lois ordinaires des proportions déterminées, *ibid.*, pp. 440–506; "Nuove considerazioni sull' affinità dei corpi pel calorico, calcolate per mezzo de' loro calori specifici" [1822], in *Memorie di matematica e di fisica*, **19** (1823), 83–137; "Sur la construction d'un voltimètre multiplicateur" [1822], in *Memorie della Reale Accademia delle scienze*, **27** (1823), 43–82; "Sur l'affinité des corps pour le calorique, et sur les rapports d'affinité qui en résultent entre eux:-1er Mém." [1823], *ibid.*, **28** (1824), 1–122; **29** (1825), 79–162; "Osservazioni sopra un articolo del Bollettino delle Scienze del sig. B. di Ferussac relativo alle Memorie sull' affinità dei corpi pel calorico, e sui rapporti d'affinità che ne risultano tra oro," in *Giornale di fisica,* **8** (1825), 432–438; "Mémoire sur la densité des corps solides et liquides, comparée avec la grosseur de leurs molécules et avec leurs nombres affinitaires" [1824], in *Memorie della Reale Accademia della scienze,* **30** (1826), 80–154; **31** (1827), 1–94; "Comparaison des observations de M. Dulong sur les pouvoirs réfringens des corps gazeux, avec les formules de relation entre ces pouvoirs et les affinités pour le calorique" [1826], *ibid.,* **33** (1829), 49–112; "Sur la loi de la force élastique de l'air par rapport à sa densité dans le cas de compression sans perte de calorique" [1828], *ibid.,* pp. 237–274; "Sur les pouvoirs neutralisants des différents corps simples, déduits de leurs proportions en poids dans les composés neutres qui en sont formés," *ibid.,* **34** (1830), 146–216; "Mémoire sur les chaleurs spécifiques des corps solides et liquides," in *Annales de chimie et de physique,* **55** (1833), 80–111; "Sur la force élastique de la vapeur du mercure à différentes températures" [1831], **49** (1832), 369–392; "Nouvelles recherches sur la chaleur spécifique des corps solides et liquides," *ibid.,* **57** (1834), 113–148; "Fisica de' corpi ponderabili, ossia trattato della costituzione generale de' corpi," in *Giornale Arcadio di Scienze,* **79** (1839), 104–107; "Diversi gradi della facoltà elettronegativa ed elettropositiva dei corpi semplice," in *Reunione degli scienziati italiana atti* (1840), 64–65; "Note sur la chaleur spécifique des différents corps, principalement à l'état gazeux," in *Bibliothèque universelle,* **29** (1840), 142–152; "Note sur la nature de la charge électrique," in *Archives de l'électricité,* **2** (1842), 102–110; "Saggio di teoria matematica della distribuzione della elettricità sulla superficie dei corpi conduttori nell'ipotesi della azione induttiva esercitata dalla medesima sui corpi circostanti, per mezzo delle particelle dell'aria frapposta" [1842], in *Memorie di matematica e di fisica,* **23** (1844), 156–184; "Proposizione di un nuovo sistema di nomenclatura chimica" [1843], *ibid.,* pp. 260–304; "Mémoire sur les volumes atomiques, et sur leur relation avec le rang que les corps occupent dans la série électro-chimique" [1843], in *Memorie della Reale Accademia delle scienze,* **8** (1846), 129–194; "Mémoire sur les volumes atomiques des corps composés" [1845], *ibid.,* pp. 293–532; "Note sur la nécessité de distinguer les molécules intégrantes des corps de leurs équivalents chimiques dans la détermination de leurs volumes atomiques," in *Archives des sciences physiques et naturelles,* **11** (1849), 285–298; "Sopra un sistema di nomenclatura chimica" [1847], in *Memorie di matematica e di fisica,* **24** (1850), 166–211; "Troisième mémoire sur les volumes atomiques. Détermination des nombres affinitaires des différents corps élémentaires par la seule considération de leur volume atomique et de celui de leurs composés" [1849], in *Memorie della Reale Accademia delle scienze,* **11** (1851), 231–318; "Quatrième mémoire sur les volumes atomiques. Détermination des volumes atomiques des corps liquides à leur température d'ébullition" [1850], *ibid.,* **12** (1852), 39–120; "Mémoire sur les conséquences qu'on peut déduire des expériences de M. Regnault sur la loi de compressibilité des gaz" [1851], *ibid.,* **13** (1853), 171–242.

In addition Avogadro wrote the major treatise: *Fisica de' corpi ponderabili ossia trattato della costituzione generale de' corpi,* 4 vols. (Turin, 1837–1841).

II. SECONDARY LITERATURE. V. Cappelletti and A. Alippi, "Avogadro," in *Dizionario biografico degli Italiani,* IV (Rome, 1962), 689–707; N. C. Coley, "The Physico-Chemical Studies of Amedeo Avogadro," in *Annals of Science,* **20** (1964), 195–210; I. Guareschi, "Amedeo Avogadro e la sua opera scientifica. Discorso storico-critico," in *Opere scelte di Amedeo Avogadro pubblicata dalla R. Accademia delle scienze di Torino* (Turin, 1911), 1–140; A. N. Meldrum, *Avogadro and Dalton. The Standing in Chemistry of Their Hypotheses* (Edinburgh, 1904); L. K. Nash, "The Atomic Molecular Theory," Case 4 of *Harvard Case Histories in Experimental Science,* J. B. Conant, ed., Vol. I (Cambridge, Mass., 1957); J. R. Partington, *A History of Chemistry,* Vol. IV (London, 1964).

M. P. CROSLAND

IBN AL-ʿAWWĀM ABŪ ZAKARIYYĀ YAḤYĀ IBN MUḤAMMAD (*fl.* Spain, second half of the twelfth century; nothing more is known of his life), *agronomy.*

Ibn Khaldūn mentions Ibn al-ʿAwwām in his *Muqaddima* as the author of a treatise on agriculture, the *Kitāb al-filāḥa,* which was, according to him, a summary of the *Nabatean Agriculture* of Ibn Waḥshiyya. The work of Ibn al-ʿAwwām, published in Spanish at the beginning of the nineteenth century, consists of thirty-five chapters, of which thirty are devoted to agronomy and the rest to related matters. It deals with 585 plants and more than fifty fruit trees, and is generally limited to a repetition of the doctrines of his predecessors, although there are a few observations, made by Ibn al-ʿAwwām in the Aljarafe of Seville, that are introduced by the term *lī.*

Among the classical writers he mentions Democritus, the Pseudo Aristotle, Theophrastus, Vergil, Varro, and especially Columella (the format of the *Kitāb al-filāḥa* is similar to that of the *De re rustica*). The Oriental Arabs are represented by Abū Ḥanīfa al-Dīnawarī (the tenth-century botanist who wrote the *Kitāb al-nabāt*) and the *Nabatean Agriculture.* The most extensive quotations, however, are from the Hispano-Arab agriculturists, among them Albucasis, possibly the author of a *Mukhtaṣar kitāb al-filāḥa;* the Sevillian Abū ʿUmar ibn Ḥajjāj (d. *ca.* 1073), author of the *Muqniʿ;* Ibn Baṣṣāl, a Toledan who was the director of the botanical garden of al-Maʾmūn and later that of al-Muʿtamid, as well as author of the *al-Qaṣd waʾl-bayān;* Abūʾl-Khayr al-Shajjār of Seville; and Abū ʿAbd Allāh Muḥammad al-Tijnarī, author of the *Zahr al-bustān wạ-nuzhat al-adhhān.* Most of the works of these authors were known, until very recently, only through the quotations of Ibn al-ʿAwwām.

The *Kitāb al-filāḥa* is an excellent manual that was designed to increase the value of land through the education of the farmer. Therefore the Spanish governments of the Enlightenment, advised by Pedro Campomanes (1723–1802), incited the Arabists of the period to publish its translation.

BIBLIOGRAPHY

The first Spanish edition of the *Kitāb al-filāḥa* was prepared by José Antonio Banqueri, a canon from Tortosa who was a disciple of Casiri: *Libro de agricultura, su autor el doctor excelente Abu Zacaria . . .,* 2 vols. (Madrid, 1802). There is also a résumé of this translation by D. C. Boutelou, 2 vols. (Madrid, 1878), and a translation into French by J. J. Clément Mullet (Paris, 1864–1867) that is not definitive. Interesting observations about the text have been made by J. J. Clément Mullet, in *Journal asiatique,* **1** (1860), 449–454; E. M. Chehabi, in *Revue de l'Académie arabe,* **11** (1931), 193 ff.; and C. C. Moncada, "Sul taglio della vita di'Ibn al-ʿAwwām," in *Actes du VIIIᵉ Congrès des orientalistes* (Stockholm, 1889), Sec. I, 215 ff.

There are no Arabic sources for Ibn al-ʿAwwām's life, except for the quotation by Ibn Khaldūn in his *Muqaddima,* trans. de Slane, III (Paris, 1863–1868), 165. For the manuscripts, consult Ziriklī, *Aʿlām,* IX (1954–1959), 208; Brockelmann, *Geschichte der arabischen Litteratur,* I (Weimar, 1898), 494, and Supp. I (Leiden, 1937), 903; and George Sarton, *Introduction to the History of Science,* II, 424.

JUAN VERNET

AZARA, FÉLIX DE (*b.* Barbuñales, Huesca, Spain, 18 May 1742; *d.* Huesca, 20 October 1821), *mathematics, geography, natural history.*

Azara was the third son of Alejandro de Azara y Loscertales and María de Perera. At the University of Huesca he studied philosophy, arts, and law from 1757 to 1761 and in 1764 became an infantry cadet. The following year he continued his mathematical training in Barcelona, and by 1769, as a second lieutenant, he was assisting in the hydrographic surveys being carried out near Madrid; afterward he taught mathematics in the army until 1774. During the assault on Argel in 1775 he received a serious chest wound.

In 1781 Azara received a commission to establish the frontier between Brazil and the neighboring Spanish colonies. Upon his arrival in Montevideo, Uruguay, he was appointed captain of a frigate by the Spanish viceroy, who then sent him to Rio Grande and later to Asunción, Paraguay; this was the area

Azara was to explore as both a geographer and a naturalist for thirteen years. Félix de Azara never married but, according to Walckenaer, while on his travels he was fond of female company, particularly that of mulattoes.

Between 1784 and 1796 Azara prepared at least fifteen maps of the Brazilian frontier; the Paraná, Pequeri, and Paraguay rivers; and the territory of Mato Grosso, Uruguay, Paraguay, and Buenos Aires. During those years he filled several diaries with accounts of travels in Paraguay and the Buenos Aires viceroyalty, the geography of Paraguay and the Río de la Plata, and the natural history of the birds and quadrupeds in those areas, relying on direct observation because he was practically without books or reference collections.

Azara returned to Spain in 1801, but soon afterward he moved to Paris, where his brother José Nicolás—a man greatly admired by Napoleon—was the Spanish ambassador. He was welcomed by the French naturalists because his *Essais sur l'histoire naturelle des quadrupèdes . . . du Paraguay* had just appeared, but on the death of José Nicolás in 1804 he returned to Madrid. Azara, with his liberal ideas, declined an appointment as viceroy of Mexico, and after 1808 during the Napoleonic War in Spain he was torn between his political and patriotic beliefs. Azara retired to Barbuñales and, as mayor of Huesca, ended his days there.

Azara enlarged natural history by discovering a large number of new species. He also visualized great biological concepts expanded by Cuvier and Darwin, both of whom quoted and accepted his views; for instance, on the variation undergone by horses under domestication.

BIBLIOGRAPHY

I. ORIGINAL WORKS. Azara's first published work was *Essais sur l'histoire naturelle des quadrupèdes de la province du Paraguay,* M. L. E. Moreau de Saint Méry, trans. (Paris, 1801), which shortly afterward was much improved and corrected (Madrid, 1802). Other works by Azara are *Apuntamientos para la historia natural de los pájaros del Paraguay y Río de la Plata* (Madrid, 1802); *Voyages dans l'Amérique méridionale,* C. A. Walckenaer, trans. (Paris, 1809), with notes by Cuvier; *Descripción e historia del Paraguay y del Río de la Plata* (Madrid, 1847); and *Memorias sobre el estado rural del Río de la Plata en 1801 . . .* (Madrid, 1847), published by his nephew Agustín de Azara. Several other works have subsequently appeared in Madrid and Buenos Aires.

II. SECONDARY LITERATURE. Both Moreau de Saint Méry's translation of the *Essais* (1801) and C. E. Walckenaer's translation of the *Voyages* give details of Azara's life. An excellent bibliographical survey, including the manuscript material in Madrid, Rome, Rio de Janeiro, and Buenos Aires, is Luis M. de Torres, "Noticias biográficas de D. Félix de Azara y exámen general de su obra," in *Anales de la Sociedad científica argentina*, **108** (1929), 177–190. The biography of Azara and a discussion of his role as a precursor of Darwin's ideas on the origin of species and the hypothesis of successive creations is E. Álvarez López, *Félix de Azara* (Madrid, 1935).

FRANCISCO GUERRA

BAADE, WILHELM HEINRICH WALTER (*b.* Schröttinghausen, Westphalia, Germany, 24 March 1893; *d.* Bad Salzuflen, Westphalia, Germany, 25 June 1960), *astronomy.*

Baade's father, Konrad, was a schoolteacher; he and his wife, Charlotte Wulfhorst, were Protestants and planned a career in theology for their son. But at the Friedrichs-Gymnasium in Herford, which he attended from 1903 to 1912, Baade decided that astronomy appealed to him more than the church: his choice resulted in a lifetime devoted to telescopic observations that have seldom been equaled, either in technical skill or in theoretical significance.

From the Gymnasium, Baade went briefly to the University of Münster, transferring to Göttingen in the Easter term of 1913. There he remained as a student throughout World War I, gaining competence in the observatory under Leopold Ambronn and serving for three years as assistant to the famous mathematician Felix Klein. Because of his congenitally dislocated hip, Baade was exempt from active military service, but beginning in 1917 had to spend eight hours a day on war work, in an installation for testing airplane models.

Shortly after passing his doctoral examination in July 1919, Baade became scientific assistant to Richard Schorr, director of the University of Hamburg's observatory at Bergedorf, located about ten miles southeast of the city. His main interest was in astrophysical problems—as shown by his dissertation, written under Johannes Hartmann, on the spectroscopic binary star β Lyrae (1921)—but his job required him to confirm the positions of many comets and asteroids. This he did conscientiously, soon dis-

covering a comet of his own (1923) and providing an explanation for the shape of comets' tails, published in 1927 with the first of a series of distinguished collaborators—the theoretical physicist Wolfgang Pauli, who was then a *Privatdozent* at Hamburg. Baade also discovered two remarkable asteroids (1920, 1949): (944) Hidalgo, which recedes farther from the sun than any other asteroid, and (1566) Icarus, noted for its close approaches both to the sun and to the earth, having passed within 4,000,000 miles of us as recently as 14 June 1968.

A meeting with Harlow Shapley in 1920 aroused Baade's interest in globular star clusters and the pulsating variables they contain. Despite the relatively small size of his telescope, Baade began observing them and in 1926 suggested a way to prove that the radiating surface of such variable stars actually rises and falls. That same year a Rockefeller fellowship enabled him to realize his dream of visiting the big telescopes in California.

Returning to Bergedorf in 1927, Baade was promoted to observer, but got no action on his plea that their 39-inch telescope be moved to a more favorable site. He turned down a job offered him at Jena, because it provided even less in the way of good observing facilities, and became a *Privatdozent* at Hamburg in 1928. The following year he married Johanna Bohlmann and went to the Philippine Islands to observe a total solar eclipse on 9 May. Clouds covered the sun during the crucial time, but the long sea voyage served to cement his friendship with the Estonian Bernard Voldemar Schmidt, who at Baade's urging designed the optical system used in the Schmidt wide-angle telescopes.

The year 1931 brought Baade an invitation to join the staff at Mt. Wilson Observatory near Pasadena, California; he accepted immediately, resigning his position at Hamburg and moving to the clearer skies and larger instruments in California as to the Promised Land. When his friend Rudolf Minkowski was forced to resign his professorship at Hamburg in 1933, Baade helped him to emigrate, thus enabling the two men to continue in California an already productive collaboration.

Now that Baade had realized his dream of working with the best telescopes in the world, he made the most of it. With Minkowski he continued spectroscopic work begun in Germany (1937), and with Edwin Powell Hubble he studied distant galaxies (1938). His work with Fritz Zwicky on supernovae (1938) led to papers on the Crab Nebula (1942) and on Nova Ophiuchi 1604 (1943).

When the United States entered World War II, Baade was classified as an enemy alien, but this favored rather than hampered his astronomical career: with many astronomers engaged in war work, and the lights of Los Angeles dimmed to protect coastal shipping, Baade had increased access to the telescopes and the added advantage of darker skies. Under these favorable circumstances he was able to photograph stars in the hitherto unresolved inner portions of the Andromeda galaxy, M31, and in both of its satellite galaxies, M32 and NGC205 (1944). This achievement was both technically difficult— thought, indeed, to be unattainable with the 100-inch telescope he used—and theoretically significant, because the brightest stars in the nucleus of M31 turned out to be red—not blue, as in the surrounding spiral arms. Baade realized he was dealing with two different stellar populations, which he called Type I (found in dusty regions, brightest stars blue) and Type II (found in dust-free regions, brightest stars red).

At a ceremony dedicating the 200-inch Palomar telescope on 1 July 1948, Baade outlined the ways he thought this great new instrument should be used to explore the universe. He suggested that since distance measurements depend critically on a reliable standard sequence of stellar magnitudes, the first thing to do was to replace the North Polar Sequence with a better one based on photoelectric techniques. His concern with distance criteria soon had spectacular results in another way, when he looked for cluster-type variables in the Andromeda galaxy. At the accepted distance of 750,000 light-years, these stars should have appeared in photographs taken with the 200-inch telescope; when they did not, Baade reasoned correctly that the galaxy was more distant than had previously been thought. Announced in 1952 at the Rome meeting of the International Astronomical Union, this conclusion cleared up several inconsistencies, but also essentially doubled the size of the universe. At a symposium on stellar evolution held during that same meeting, Baade presented color-magnitude diagrams for star clusters that his young collaborators Halton C. Arp, William A. Baum, and Allan R. Sandage had just completed, in support of his theory that Type I stars are younger—on a cosmological time scale—than Type II stars. This was an oversimplification, as suggested by Boris V. Kukarkin during the ensuing discussion, but it represented a major step toward today's understanding of the life cycles of stars.

Baade left his mark on yet another aspect of astronomy when he identified, in photographs he had taken with the 200-inch telescope, several objects first detected by radiotelescopes (1954). One of them, the radio source Cygnus A, had a twinned appearance: Baade correctly placed it far outside our galaxy— which seemed unbelievable—but his conclusion, based on theoretical work done in 1950 with Lyman Spitzer, Jr., that it represented two galaxies in collision, has not stood the test of time. In a final contribution to the understanding of objects later called quasars, Baade showed that a jet issuing from the galaxy M87 emitted strongly polarized light (1956).

Upon his retirement in 1958 Baade gave a series of lectures at Harvard University (published as *Evolution of Stars and Galaxies*), and then spent six months in Australia, where he used the 74-inch telescope at Mt. Stromlo near Canberra, before returning to Göttingen as Gauss professor. An operation on his hip and six months' convalescence in bed preceded his death from respiratory failure. Much of his work remained unpublished—in some cases because he was dissatisfied with it, in others because he preferred active research to the tedious job of writing about it—but of what did appear, the British astronomer Fred Hoyle has commented: "Almost every one of Baade's papers turned out to have far-reaching consequences."

Considering his accomplishments, Baade received few honors during his lifetime. He was awarded the Gold Medal of the Royal Astronomical Society in 1954, for his observational work on galactic and extragalactic objects, and in 1955 received the Bruce Medal of the Astronomical Society of the Pacific. He was a member of the American Philosophical Society, and of scientific academies in Amsterdam, Göttingen, Lund, Munich, and Mainz.

BIBLIOGRAPHY

I. Original Works. Baade's dissertation, "Bahnbestimmung des spektroskopischen Doppelsterns β Lyrae nach Spectrogrammen von Prof. Hartmann," was presented at Göttingen 3 Aug. 1921. Works cited above are "Über die Möglichkeit, die Pulsationstheorie der δ Cephei-Veränderlichen zu prüfen," in *Astronomische Nachrichten,* **228** (1926), 359–362; "Über den auf die Teilchen in den Kometenschweifen ausgeübten Strahlungsdruck," in *Die Naturwissenschaften,* **15** (1927), 49–51, written with W. Pauli, Jr.; "The Trapezium Cluster of the Orion Nebula" and "Spectrophotometric Investigations of Some O- and B-type Stars Connected With the Orion Nebula," in *Astrophysical Journal,* **86** (1937), 119–122 and 123–135, both written with R. Minkowski; "The New Stellar Systems in Sculptor and Fornax," in *Publications of the Astronomical Society of the Pacific,* **51** (1938), 40–44, written with E. Hubble; "Photographic Light Curves of the Two Supernovae in IC4182 and NGC1003," in *Astrophysical Journal,* **88** (1938), 411–421, written with F. Zwicky; "The Crab Nebula," *ibid.,* **96** (1942), 188–198; "Nova Ophiuchi of 1604 as a Supernova," *ibid.,* **97** (1943), 119–127; "The Resolution of Messier 32, NGC205, and the Central Region of the Andromeda Nebula," *ibid.,* **100** (1944), 137–146; "A Program of Extragalactic Research for the 200-inch Hale Telescope," in *Publications of the Astronomical Society of the Pacific,* **60** (1948), 230–234; "Stellar Populations and Collisions of Galaxies," in *Astrophysical Journal,* **113** (1950), 413–418, written with L. Spitzer, Jr.; "Basic Facts on Stellar Evolution," in *Transactions of the International Astronomical Union VIII* (Cambridge, 1954), pp. 682–688 (discussion on pp. 688–689); "Identification of the Radio Sources in Cassiopeia, Cygnus A, and Puppis A," and "On the Identification of Radio Sources," in *Astrophysical Journal,* **119** (1954), 206–214, and 215–231, both written with R. Minkowski; "Polarization in the Jet of Messier 87," *ibid.,* **123** (1956), 550–551; and *Evolution of Stars and Galaxies* (Cambridge, Mass., 1963), ed. by Cecilia Payne-Gaposhkin from tape recordings of Baade's lectures at Harvard in 1958.

A list that includes seventy-three papers by Baade and ninety short communications is appended to Heckmann's obituary notice (see below); references to Baade's book (mentioned above) and several other contributions to symposia can be found in Poggendorff, VIIb (1967), 166. Baade's notebooks and other unpublished material are divided between the Mt. Wilson–Palomar Observatories and the Leiden Observatory.

II. SECONDARY LITERATURE. The citation delivered by John Jackson when Baade received (*in absentia*) the Gold Medal of the Royal Astronomical Society was printed in *Monthly Notices of the Royal Astronomical Society* (London), **114** (1954), 370–383; the one by Olin Chaddock Wilson that accompanied the Bruce Medal appeared in *Publications of the Astronomical Society of the Pacific,* **67** (1955), 57–61, and includes a portrait. Obituary notices on Baade include those by Fred Hoyle, in *Nature,* **187** (1960), 1075; Erich Schoenberg, in *Bayerische Akademie der Wissenschaften. Jahrbuch 1960,* pp. 177–181, plus a portrait facing p. 184; Otto Hermann Leopold Heckmann, in *Mitteilungen der Astronomischen Gesellschaft* [Hamburg] *1960* (1961), 5–11, with portrait and list of publications; Allan R. Sandage, in *Quarterly Journal of the Royal Astronomical Society,* **2** (1961), 118–121; and Halton C. Arp, in *Journal of the Royal Astronomical Society of Canada,* **55** (1961), 113–116.

The role Baade played in the early days of quasar research is described in Ivor Robinson, Alfred Schild, and E. L. Schücking, eds., *Quasi-Stellar Sources and Gravitational Collapse* (Chicago, 1965), pp. xi–xiv; and an essay on Baade's life and works by a longtime associate, Robert S. Richardson, constitutes ch. 16 (pp. 260–294) of Richardson's *The Star Lovers* (New York, 1967).

Further accounts of Baade's work are Fred Hoyle, "Report of the Meeting of Commission 28," in *Transactions of the International Astronomical Union VIII* (Cambridge, 1954), pp. 397–399; Hermann Kobold, "Komet 1922c (Baade)," in *Astronomische Nachrichten,* **217** (1923), cols. 175–176, and an unsigned report on Planet 1920 HZ [later named Hidalgo], in *Nature,* **106** (1920), 482; and R. S. Richardson, "A New Asteroid With Smallest

Known Distance" [Icarus], in *Publications of the Astronomical Society of the Pacific,* **61** (1949), 162–165.

SALLY H. DIEKE

BABBAGE, CHARLES (*b.* Teignmouth, England, 26 December 1792; *d.* London, England, 18 October 1871), *mathematics, computer logic, computer technology.*

Babbage's parents were affluent. As a child, privately educated, he exhibited unusually sharp curiosity as to the how and why of everything around him. Entering Cambridge University in 1810, he soon found that he knew more than his teachers, and came to the conclusion that English mathematics was lagging behind European standards. In a famous alliance with George Peacock and John Herschel, he began campaigning for a revitalization of mathematics teaching. To this end the trio translated S. F. Lacroix's *Differential and Integral Calculus* and touted the superiority of Leibniz's differential notation over Newton's (then widely regarded in England as sacrosanct).

After graduation, Babbage plunged into a variety of activities and wrote notable papers on the theory of functions and on various topics in applied mathematics. He inquired into the organization and usefulness of learned societies, criticizing the unprogressive ones (among which he included the Royal Society) and helping found new ones—in particular the Astronomical Society (1820), the British Association (1831), and the Statistical Society of London (1834). He became a fellow of the Royal Society in 1816, and in 1827 was elected Lucasian professor of mathematics at Cambridge. He had not sought this prestigious chair (he described his election as "an instance of forgiveness unparalleled in history") and, although he held it for twelve years, never functioned as professor. This is a little surprising, in that the position could have been used to further the pedagogic reforms he advocated. But Babbage was becoming absorbed, if not obsessed, by problems of the mechanization of computation. He was to wrestle with these for decades, and they were partly responsible for transforming the lively, sociable young man into an embittered and crotchety old one, fighting all and sundry, even the London street musicians, whose activities, he figured, had ruined a quarter of his working potential.

Babbage had a forward-looking view of science as an essential part of both culture and industrial civilization, and he was among the first to argue that national government has an obligation to support scientific activities, to help promising inventors, and even to give men of science a hand in public affairs.

Few eminent scientists have had such diversified interests as Babbage. A listing of them would include cryptanalysis, probability, geophysics, astronomy, altimetry, ophthalmoscopy, statistical linguistics, meteorology, actuarial science, lighthouse technology, and the use of tree rings as historic climatic records. Two deserve special mention: the devising of a notation that not only simplified the making and reading of engineering drawings but also helped a good designer simplify his "circuits"; and his insightful writings on mass production and the principles of what we now know as operational research (he applied them to pin manufacture, the post office, and the printing trade).

Computational aids began to haunt Babbage's mind the day he realized that existing mathematical tables were peppered with errors whose complete eradication was all but infeasible. As a creature of his era—the machine-power revolution—he asked himself, at first only half in earnest, why a table of, say, sines could not be produced by steam. Then he went on to reflect that maybe it could. He was at the time enthusiastic about the application of the method of differences to tablemaking, and was indeed using it to compile logarithms. (His finished table of eight-figure logarithms for the first 108,000 natural numbers is among the best ever made.) While still engaged in this work, Babbage turned to the planning of a machine that would not only calculate functions but also print out the results.

To understand his line of thought, we must take a close look at the method of differences—a topic in what later became known as the calculus of finite differences. The basic consideration is of a polynomial $f(x)$ of degree n evaluated for a sequence of equidistant values of x. Let h be this constant increment. We next take the corresponding increments in $f(x)$ itself, calling these the *first* differences; then we consider the differences between consecutive first differences, calling these the *second* differences. And so forth. An obvious recursive definition of the rth difference for a particular value of x, say x_i, is

$$\Delta^r f(x_i) = \Delta^{r-1} f(x_i + h) - \Delta^{r-1} f(x_i),$$

and it is not difficult to show that, specifically,

$$\Delta^r f(x_i) = \sum_{m=0}^{r} (-1)^m \binom{r}{m} f[x_i + (r - m)h].$$

As r increases, the differences become smaller and more nearly uniform, and at $r = n$ the differences are constant (so that at $r = n + 1$, all differences are

zero). A simple example—one that Babbage himself was fond of using—is provided by letting the function be the squares of the natural numbers. Here $n = 2$, and we have

x	1	2	3	4	5	6	$\cdots$
$f(x)$	1	4	9	16	25	36	$\cdots$
Δ^1		3	5	7	9	11	$\cdots$
Δ^2			2	2	2	2	$\cdots$

Two propositions follow. The first, perhaps not obvious but easily demonstrated, is that the schema can be extended to most nonrational functions (such as logarithms), provided that we take the differences far enough. (This is linked to the fact that the calculus of finite differences becomes, in the limit, the familiar infinitesimal calculus.) The second, originated by Babbage, is that the inverse of the schema is readily adaptable to mechanization. In other words, a machine can be designed (and it will be only slightly more sophisticated than an automobile odometer or an office numbering machine) that, given appropriate initial values and nth constant differences, will accumulate values of any polynomial, or indeed of almost any function. (For nonrational functions the procedure will be an approximation conditioned by the choice of r and h and the accuracy required, and will need monitoring at regular checkpoints across the table.)

This is what Babbage set out—and failed—to do. As the work progressed, he was constantly thinking up new ideas for streamlining the mechanism, and these in turn encouraged him to enlarge its capacity. In the end his precepts ruined his practice. The target he set was a machine that would handle twenty-decimal numbers and sixth-order differences, plus a printout device. When he died, his unfinished "Difference Engine Number One" had been a museum piece for years (in the museum of King's College, at Somerset House, London, from 1842 to 1862, and subsequently in the Science Museum, London—where it still is). What is more revealing and ironic is that, during his own lifetime, a Swedish engineer named Georg Scheutz, working from a magazine account of Babbage's project, built a machine of modest capacity (eight-decimal numbers, fourth-order differences, and a printout) that really worked. It was used for many years in the Dudley Observatory, Albany, New York.

Aside from technicalities, two factors militated against the production of the difference engine. One was cost (even a generous government subsidy would not cover the bills), and the other was the inventor's espousal of an even more grandiose project—the

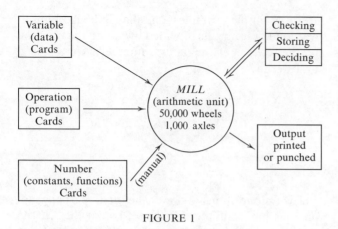

FIGURE 1

construction of what he called an analytical engine.

Babbage's move onto this new path was inspired by his study of Jacquard's punched cards for weaving machinery, for he quickly saw the possibility of using such cards to code quantities and operations in an automatic computing system. His notion was to have sprung feeler wires that would actuate levers when card holes allowed them access. On this basis he drew up plans for a machine of almost unbelievable versatility and mathematical power. A simplified flow diagram of the engine is shown in the accompanying figure. The heart of the machine, the mill, was to consist of 1,000 columns of geared wheels, allowing up to that many fifty-decimal-digit numbers to be subjected to one or another of the four primary arithmetic operations. Especially remarkable was the incorporation of decision-making units of the logical type used in today's machines.

Although the analytical engine uncannily foreshadowed modern equipment, an important difference obtains: it was decimal, not binary. Babbage, not having to manipulate electronics, could not have been expected to think binarily. However, his having to use wheels meant that his system was not "purely" digital, in the modern sense.

All who understood the plans expressed unbounded admiration for the analytical engine and its conceiver. But material support was not forthcoming, and it remained a paper project. After Babbage's death his son, H. P. Babbage, sorted the mass of blueprints and workshop instructions, and, in collaboration with others, built a small analytical "mill" and printer. It may be seen today in the Science Museum, London.

BIBLIOGRAPHY

I. ORIGINAL WORKS. Babbage appended a list of eighty of his publications to his autobiographical *Passages From the Life of a Philosopher* (London, 1864), and it is repro-

duced in P. and E. Morrison's *Charles Babbage and His Calculating Engines* (New York, 1961). It is a poor list, with reprinted papers and excerpts separately itemized. Apart from translations and the autobiography and a few small and minor works, the only books of substance that Babbage published were *Reflections on the Decline of Science in England* (London, 1830); *Economy of Manufactures and Machinery* (London, 1832); and *The Exposition of 1851* (London, 1851). His logarithms deserve special mention: they were originally published in stereotype as *Table of the Logarithms of the Natural Numbers From 1 to 108,000* (London, 1827), with a valuable introduction dealing with the layout and typography of mathematical tables. A few years later he published *Specimen of Logarithmic Tables* (London, 1831), a 21-volume, single-copy edition of just two of the original pages printed in a great variety of colored inks on an even greater variety of colored papers, in order "to ascertain by experiment the tints of the paper and colors of the inks least fatiguing to the eye." In the same "experiment" about thirty-five copies of the complete table were printed on "thick drawing paper of various tints." In 1834 regular colored-paper editions were published in German at Vienna and in Hungarian at Budapest, by C. Nagy. Babbage's formal scientific articles number about forty. The first publication dealing with his main subject is "A Note Respecting the Application of Machinery to the Calculation of Mathematical Tables," in *Memoirs of the Astronomical Society,* **1** (1822), 309; the last is chs. 5–8 of his entertaining autobiography (see above).

II. SECONDARY LITERATURE. Practically all the significant material is either reproduced or indexed in the Morrisons' book, the only one entirely devoted to Babbage (see above). The symposium *Faster Than Thought* (London, 1953) has a first chapter (by the editor, B. V. Bowden) that is largely concerned with Babbage. Both of these books carry reprints of a translation and annotation of an article on the analytical engine written by the Italian military engineer L. F. Menabrea (Geneva, 1842). The translator was Lady Lovelace, Lord Byron's mathematically gifted daughter, and her detailed annotations (especially a sketch of how Bernoulli numbers could be computed by the engine) are excellent. It is in the course of this commentary that she finely remarks that "the Analytical Engine *weaves algebraic patterns,* just as the Jacquard-loom weaves flowers and leaves." The sectional catalog *Mathematics, I. Calculating Machines and Instruments,* The Science Museum (London, 1926), contains much useful illustrated information about Babbage's engines, as well as about allied machines, such as the Scheutz difference engine.

NORMAN T. GRIDGEMAN

BABCOCK, STEPHEN MOULTON (*b.* Bridgewater, New York, 22 October 1843; *d.* Madison, Wisconsin, 1 July 1931), *agricultural chemistry.*

Babcock, the son of Pelig and Mary Scott Babcock, received the B.A. from Tufts College in 1866. His engineering studies at Rensselaer Polytechnic Institute were cut short when he had to manage the family farm after his father's death. However, he was soon taking chemistry courses at Cornell University, and in 1875 was made an instructor in the subject. In 1877 Babcock began graduate studies at the University of Göttingen under Hans Hübner, receiving the Ph.D. in 1879. He resumed his instructorship at Cornell in 1881 but left in 1882 to become chief chemist at the New York Agricultural Experiment Station in Geneva. He moved to an equivalent position at the Wisconsin Agricultural Experiment Station in 1888 and was also appointed professor of agricultural chemistry at the University of Wisconsin. Both positions were held until his retirement as emeritus professor in 1913. He married May Crandall in 1896.

Babcock is best known for his test for butterfat in milk, introduced in 1890. By using sulfuric acid to release the fat from its normal suspension and centrifuging and diluting, it was possible to measure directly the percentage of fat by observing it in the neck of a specially designed test bottle. The simplicity of the test permitted its use by persons without scientific training. Its use altered the economics of dairying and stimulated growth of the dairy industry.

With the bacteriologist Harry L. Russell, Babcock developed the process for cold curing of cheese in 1900. The great improvement in the quality of cheese led to wide acceptance of the process in the dairy industry.

Babcock's most important contribution arose from his skepticism regarding the biological equivalency of chemically similar feeds from different crops. In 1907 four of his younger associates—E. B. Hart, E. V. McCollum, H. Steenbock, and G. Humphrey— began a cattle-feeding experiment using chemically equivalent rations, each derived from a different plant. The experiment not only confirmed Babcock's skepticism but led to studies that helped develop the vitamin concept.

Babcock also studied metabolic water in insects and, in his later years, sought to investigate the structure of matter and its relation to energy.

BIBLIOGRAPHY

I. ORIGINAL WORKS. Babcock had an aversion to writing and published very little. Most of his work is described in bulletins and annual reports of the New York and Wisconsin Agricultural Experiment Stations. The fat test was originally published as "A New Method for the Estimation of Fat in Milk, Especially Adapted to Creameries and Cheese Factories," in *Bulletin of the University of Wisconsin Experiment Station,* no. 24 (July 1890). The studies with H. L. Russell on cold curing of cheese ap-

peared in *Bulletin,* no. 94 (1902). Research Bulletin no. 22 (1912) is entitled *Metabolic Water: Its Production and Role in Vital Phenomena.* The single-grain feeding experiments, under the authorship of E. B. Hart, E. V. McCollum, H. Steenbock, and G. Humphrey, are reported in Research Bulletin no. 17 (1911).

Babcock's unpublished papers are held by the State Historical Society of Wisconsin.

II. SECONDARY LITERATURE. Paul de Kruif has a perceptive biographical sketch of Babcock in *Hunger Fighters* (New York, 1928), ch. 9; and Aaron J. Ihde has a short sketch in Eduard Farber, ed., *Famous Chemists* (New York, 1961), pp. 808–813, with a bibliography of other biographical sketches on pp. 828–829. The history of the fat test is described by J. L. Sammis, *The Story of the Babcock Test,* Circular 172, Extension Services of the College of Agriculture, University of Wisconsin (Madison, 1924).

AARON J. IHDE

BABINET, JACQUES (*b.* Lusignan, France, 5 March 1794; *d.* Paris, France, 21 October 1872), *physics, meteorology.*

Jacques Babinet was a French physicist who did important work in the theory of diffraction, meteorological optics, and optical instrumentation. His parents—Jean Babinet, the mayor of Lusignan, and Marie-Anne Félicité Bonneau du Chesne, daughter of a lieutenant-general—hoped that he would become a magistrate and gave him a strong literary education. Jacques preferred science, however, and entered the École Polytechnique in 1812. After graduation he held professorships at Fontenay-le-Comte and Poitiers. In 1820 he became professor of physics at the Collège Louis-le-Grand in Paris. Babinet was elected a member of the physics section of the Académie des Sciences in 1840 and a year later he was appointed librarian at the Bureau of Longitudes. He married Adelaide Laugier, by whom he had two sons.

An early proponent of the wave theory of light which had recently been perfected by Fresnel and Young, Babinet devoted much of his research to extending its applications. His first published paper on optics, "Sur les couleurs des réseaux," dealt with Fraunhofer's discovery that white light viewed through a wire grating produces a series of continuous spectra. Babinet derived the formula that relates the deviation of rays of a given color to the ratio of their wavelength divided by the sum of the diameter of the wires plus the distance between any two wires. He also showed that diffraction experiments could yield an improved measure of wavelengths and presented a table of new values.

Babinet realized that the grating was only one of a number of means of producing diffraction effects, so he extended his theoretical work to include other systems. The result was a concept known today as Babinet's principle: "If parallel rays fall normally on a diffraction system formed from a large number of openings . . . the diffraction phenomena will remain identically the same if the transparent parts become opaque, and reciprocally."

This theoretical work was accompanied by experimental investigations of several unexplained optical phenomena in mineralogy and meteorology. Here Babinet showed his talent as an inventor. Finding Malus' goniometer (a device for measuring angles of refraction) too cumbersome for his experiment on minerals, he constructed a portable one, which was considerably easier to use and was capable of making measurements that previously required separate instruments. To aid in his study of polarization he invented the Babinet compensator, which could produce or analyze arbitrarily polarized light.

Babinet's interests in physics transcended laboratory work and included all phenomena in nature. Thus, the study of meteorology, particularly meteorological optics, occupied much of his career. He began his work in this field with an investigation of interference phenomena produced in the atmosphere: rainbows and "coronas," or colored rings surrounding the sun or moon under certain weather conditions. Later work included modifications of the theory of atmospheric refraction and a study of polarization of skylight, especially the mysterious existence of neutral or unpolarized points in the sky.

While engaged in this original research, Babinet also achieved considerable fame as a popularizer of science, explaining natural phenomena to the layman in public courses and in articles in popular journals. In these exercises, whose subjects ranged over geology, mineralogy, astronomy, and meteorology, Babinet exhibited his rare ability to reduce complex phenomena to an easily comprehensible level.

BIBLIOGRAPHY

I. ORIGINAL WORKS. Among Babinet's works are "Sur les couleurs des réseaux," in *Annales de chimie,* **60** (1829), 166–176; "Mémoires d'optique météorologique," in Académie des Sciences, *Comptes rendus,* **4** (1837), 638–648; "Sur les caractères optiques des minéraux," *ibid.,* 758–766; "Sur la réfraction terrestre," *ibid.,* **52** (1861), 394–395, 417–425, 529–535. A study of polarization of skylight, especially the mysterious existence of neutral or unpolarized points in the sky, may be found in Académie des Sciences, *Comptes rendus,* **11** (1840), 618–620; **20** (1845), 801–804; **23** (1846), 233–235. Many of Babinet's treatises on science for the layman were collected in a volume entitled *Études et expériences sur les sciences d'observation* (Paris, 1855).

II. SECONDARY LITERATURE. Biographical material on Babinet is found in the following sources: *Revue des cours scientifiques de la France et de l'étranger,* **3** (1872), 409–410; L. Figuier, *L'année scientifique,* **16** (1872), 533; the speech by Becquerel at Babinet's funeral; E. Beauchat-Filleau, *Dictionnaire historique et généalogique de Poitou,* 2nd ed., I (Poitiers, 1891), 120, especially for family background. A complete bibliography of Babinet's original papers is given in the *Royal Society Catalogue of Scientific Papers,* I, 134–136 and VII, 72. Other publications are listed in J. C. Poggendorff's *Biographisch-literarisches Handwörterbuch . . .,* I (Leipzig, 1863; repr. Amsterdam, 1965), 82.

EUGENE FRANKEL

BABINGTON, CHARLES CARDALE (*b.* Ludlow, England, 23 November 1808; *d.* Cambridge, England, 22 July 1895), *botany.*

The only child of Joseph Babington and Catherine Whitter, Charles married Anna Maria Walker in 1866. His father, a physician, was a keen amateur botanist and doubtless influenced his son's inclination to natural history.

After a succession of private schools and a brief interlude at Charterhouse, Babington entered St. John's College, Cambridge, in 1826, graduating B.A. in 1830 and receiving the M.A. in 1833. In his first year at Cambridge he established an enduring friendship with J. S. Henslow, professor of botany, whose enthusiasm confirmed Babington's lifelong devotion to botany. Completely involved in the natural history activities of Cambridge for more than forty years, Babington was a leading member of the Ray Club, which developed into the Ray Society (founded 1844); and a number of its publications, such as *Memorials of John Ray* and *Correspondence of John Ray,* owed much to his help. A man of wide intellectual interests, he was a founding member of the Entomological Society in Cambridge and the Cambridge Antiquarian Society.

Babington's first work, *Flora Bathoniensis* (1834), with critical notes and references to Continental floras, adumbrated the direction of his future taxonomic work. Two visits to the Channel Islands, in 1837 and 1838, resulted in his *Primitiae florae Sarnicae* (1839). The Napoleonic Wars had isolated the British Isles from botanical research in the rest of Europe, where the natural system of plant classification was generally accepted, and therefore Linnaeus' artificial arrangement was perpetuated in such standard English works as J. E. Smith's *English Botany* and the earlier editions of W. J. Hooker's *British Flora.* Consequently, it was difficult for English botanists to identify the new plants published in Continental floras, a defect remedied by Babington in successive editions of his *Manual of British Botany.* Considered to be his *magnum opus,* it made its first appearance in 1843 and, with the exception of the fifth edition of Hooker's *British Flora,* was the first complete guide to British plants arranged according to a natural system. Accurate and clear in its descriptions, meticulous in its assignment of genera and species, the *Manual* soon established itself as an indispensable field companion.

Babington differed from many of his contemporaries in insisting upon a more critical delimitation of species; this was well demonstrated in his *British Rubi* (1869), which described in impressive detail some forty-five species.

On the death of his friend Henslow in 1861, Babington was elected to the chair of botany at Cambridge, which he held until his death. He was an indifferent and infrequent lecturer; his interests were mainly in research, and during his professorship many additions were made to the Cambridge Herbarium, the most notable being John Lindley's collection. His own collection of nearly 55,000 sheets was bequeathed to Cambridge, together with his library. At the time of his death Babington was the senior fellow of the Linnean Society, having been elected in 1830; in 1851 he was elected a fellow of the Royal Society.

BIBLIOGRAPHY

I. ORIGINAL WORKS. Babington's writings include *Flora Bathoniensis* (Bath, 1834); *Primitiae florae Sarnicae* (London, 1839); *Manual of British Botany* (London, 1843; 8th ed., 1881); *Synopsis of British Rubi* (London, 1846); *Flora of Cambridgeshire* (London, 1860); and *British Rubi* (London, 1869).

II. SECONDARY LITERATURE. A. M. Babington, ed., *Memorials, Journal and Botanical Correspondence of Charles Cardale Babington* (Cambridge, 1897), contains reprinted obituaries and reminiscences of Babington and a bibliography of his periodical articles. Obituaries include James Britten, in *Journal of Botany,* **33** (1895), 257–266; and J. E. B. Mayor, in *Cambridge Chronicle* (30 Aug. 1895), p. 4.

R. G. C. DESMOND

BABINGTON, WILLIAM (*b.* Port Glenone, Antrim, Northern Ireland, 21 May 1756; *d.* London, England, 29 April 1833), *mineralogy, geology.*

After apprenticeship to a medical practitioner in Londonderry, Northern Ireland, and completion of his medical training at Guy's Hospital, London, Babington was appointed assistant surgeon at Haslar Naval Hospital, Portsmouth, in 1777. He became apothecary (1781), assistant physician (1795), and full physician (1802) at Guy's Hospital, where he also lectured on chemistry. In 1795 he was awarded an M.D. by Aberdeen University. The following year he qualified as licentiate of the College of Physicians, London, and was created fellow of the college by "special grace" in 1827. Also in 1796 he commenced private practice, and was so successful that in 1811 he resigned from Guy's Hospital. Although prominent as a practicing physician for the rest of his life, Babington made no conspicuous contribution to medical science. He was elected a fellow of the Royal Society in 1805.

Babington's interest in mineralogy arose through the purchase of a cabinet of minerals—the collection of minerals was at that time a fashionable pursuit. His study of these minerals, to which he applied his knowledge of chemistry, led to the publication of his *Systematic Arrangement of Minerals* (1795). In 1799 he published a greatly enlarged *New System of Mineralogy.* He classified minerals by the Linnaean system into orders, genera, and species, the main subdivisions being based on chemical composition; crystal form was used only to establish subdivisions. This constituted an advance on the then widely used Wernerian system, which was based principally on external characters.

Babington was a member of the British Mineralogical Society, which was formed in 1799 to carry out analyses of British minerals. Some of the members of the society, including Babington and a few others with scientific inclinations, founded the Geological Society of London on 13 November 1807. As an influential member, Babington played an important part in establishing the new society on a firm basis and in promoting mineralogical and geological science in general. He served as president from 1822 to 1824.

BIBLIOGRAPHY

I. ORIGINAL WORKS. Works by Babington are *A Syllabus of a Course of Lectures Read at Guy's Hospital on Chemistry* (London, 1789); *A Systematic Arrangement of Minerals, Founded on the Joint Consideration of Their Chemical, Physical, and External Characters, &c.* (London, 1795, 2nd ed. 1796); and *A New System of Mineralogy in the Form of a Catalogue, &c.* (London, 1799).

II. SECONDARY LITERATURE. See G. B. Greenough, "President's Anniversary Address, for 1834," in *Proceedings of the Geological Society of London,* **2** (1838), 42–44; W. Munk, *The Roll of the Royal College of Physicians,* II (London, 1878), 451; S. Wilks and G. T. Bettany, *A Biographical History of Guy's Hospital* (London, 1892), pp.

199–204; and H. B. Woodward, *The History of the Geological Society of London* (London, 1907), pp. 6–11.

V. A. EYLES

BACCELLI, GUIDO (*b.* Rome, Italy, 25 November 1830; *d.* Rome, 10 January 1916), *clinical medicine.*

Baccelli, the son of Antonio Baccelli and Adelaide Leonori, studied in Rome. He graduated surgeon (1852) and physician (1853) from the Roman School of Medicine; he was appointed director of the Medical Clinic of Rome in 1863, and remained in that chair until his death. Of wealthy family, Baccelli had a youthful interest in politics, and from 1870 was active in public life. In 1875 he was elected a deputy of the Italian parliament. Loyal to the house of Savoy, Baccelli was always staunchly liberal in his political beliefs. He was Minister of Public Education three times (1881, 1893, 1898) and Minister of Agriculture, Commerce, and Industry (1901). In Rome, he advocated the repair of the Capitol, the Pantheon, and the Palatine, and conceived the beautiful Archeological Walk; he also founded the Gallery of Modern Art and promoted the draining of the Pontine Marshes. On the national level, Baccelli restored the autonomy of the Italian universities and reformed the system of primary education. He promoted agriculture, resolutely supported the idea of an armed nation, and used his political power to advance the cause of public health through legislation setting up programs to eliminate malaria and pellagra. In 1893, in collaboration with the noted surgeon Francesco Durante, he founded the widely read medical weekly *Il policlinico*. Baccelli's greatest achievement, however, was the founding of the Policlinico Umberto I, a medical center which was opened in Rome on 9 April 1906. A fascinating orator, Baccelli presided at the first congress of the Italian Society of Internal Medicine (Rome, 1888), and the Eleventh International Congress of Medicine (Rome, 1894).

Baccelli's medical writings covered a variety of subjects. From 1863 to 1877 he studied heart diseases and pleural diseases, and in 1876 illustrated Gaucher's disease. From 1866 to 1894 he studied malarial infection, demonstrating that such infection acts upon the erythrocytes.

Baccelli appreciated traditional therapeutics, but he was also one of the first supporters of oxygen therapy (1870). However, his major merit was "to have opened the way of the veins to the heroic drugs," as he wrote. The first introduction of drugs into the veins of a living animal (1656, in a dog) was performed in Oxford by Sir Christopher Wren; previously, in the sixteenth century, Alessandro Massaria, teacher of practical medicine at Padua University, had proposed the same operation in man. But these facts were practically unknown when, in 1890, Baccelli first injected quinine chlorhydrate into a vein of a woman, thereby saving her from death by pernicious malaria. In 1906 Baccelli also announced positive results from the intravenous injection of strophantine in the treatment of grave heart failure.

Baccelli gave great impetus to the modern study of medicine, but he also acknowledged the importance of the ancient masters; in 1907, he was elected chairman of the committee that drafted the constitution of the Italian Society of the History of Medicine.

BIBLIOGRAPHY

I. ORIGINAL WORKS. Baccelli's works include *Ascoltazione e percussione nella Scuola Romana* (Rome, 1857); *Patologia del cuore e dell'aorta,* 3 vols. (Rome, 1863–1866); *La pettiroloquia afonetica e la diplofonia* (Rome, 1864); *Il plessimetro lineare della Scuola Clinica di Roma* (Rome, 1866); *De primitivo splenis carcinomate* (Rome, 1876); *Sulla trasmissione dei suoni attraverso i liquidi endopleurici di differente natura* (Rome, 1877); "Le iniezioni endovenose di sali di chinino nella infezione malarica," in *La riforma medica,* **6** (1890), 14–16; *L'infezione malarica. Studi di Guido Baccelli 1866–1894 raccolti dal Dott. Zeri* (Rome, 1894); "I discorsi inaugurali pronunciati dal Baccelli ai Congressi Nazionali di Medicina Interna," in *La clinica medica italiana,* **45** (1906), 317–338; *La via delle vene aperta aimedicamenti eroici* (Rome, 1907).

II. SECONDARY LITERATURE. Works relating to Baccelli include "Numero speciale in occasione delle onoranze a Guido Baccelli," in *Il policlinico,* **14** (8 April 1906); A. Baccelli, *Guido Baccelli. Ricordi* (Naples, 1931); L. Belloni, *Da Asclepio ad Esculapio* (Milan, 1959); A. Castiglioni, *Storia della Medicina,* II (Milan, 1948), 699–700; M. Di Segni, "Il contributo italiano alle origini della trasfusione del sangue e della iniezione di medicamenti nelle vene," in *Bollettino dell'Istituto Storico Italiano dell'Arte Sanitaria,* **10** (1930), 66–90 and 179–199; G. Gorrini, *Guido Baccelli. La vita, l'opera, il pensiero* (Turin, 1916), which contains a portrait of Baccelli and a complete inventory of his medical works; E. Maragliano, "Guido Baccelli clinico," in *Il policlinico,* **23** (1916), 217–262; A. Pazzini, *La storia della Facoltà Medica di Roma,* 2 vols. (Rome, 1961); and G. Sanarelli, "Guido Baccelli uomo politico e medico sociale," in *Il policlinico,* **23** (1916), 263–280.

PIETRO FRANCESCHINI

BACH, ALEKSEI NIKOLAEVICH (*b.* Zolotonosha [now in Ukrainian S.S.R.], Russia, 17 March 1857; *d.* Moscow, U.S.S.R., 13 May 1946), *biochemistry, physical chemistry.*

Bach's father, a technician in a distillery, stimulated his son's interest in science. During his years in a

Gymnasium in Kiev (1867–1875), Bach, under the influence of the ideas of the writer and revolutionary Dmitrii Ivanovich Pisarev, decided to dedicate himself to the natural sciences and to their dissemination for the good of society. In 1875 he enrolled in the natural science section of the Department of Physics and Mathematics of Kiev University, where he specialized in chemistry. In 1878 he was expelled from the university and exiled to Belozersk for three years for participating in student political activities. Allowed to resume his studies at the university in 1882, Bach devoted even more energy than before to the political struggle, and soon became one of the leaders of the Kiev branch of the Narodnaja Volja ("People's Will") party. Inspired by revolutionary propaganda, Bach wrote the pamphlet "Tsar'-golod" ("Tsar-hunger"; 1883), one of the first statements of Marxist economic theory to be published in Russian. (Before 1917 this pamphlet had sold 100,000 copies, including translations into the languages of several Russian peoples.) In 1883, fearing arrest, Bach went into hiding; the utter defeat of the Narodnaja Volja by the tsarist government forced him to flee to France in 1885.

From 1885 to 1894 Bach lived in Paris, where he collaborated on the journal *Moniteur scientifique,* a collaboration that enabled him to establish close contact with French scientific circles. As a result, Paul Schützenberger invited him in 1890 to work in his laboratory at the Collège de France. There Bach completed his first investigations of the chemical mechanism of the assimilation of carbon by plants, the results of which Schützenberger regularly reported to the Paris Academy of Sciences. In the same year Bach married Aleksandra Aleksandrovna Cherven-Vodali, a teacher who had later become a pediatrician. In 1891 Bach traveled to the United States at the behest of the Brussels Society of Maltose, in order to introduce an improved method of fermentation into Chicago's distilleries.

Bach moved to Switzerland in 1894, residing there until he returned to Russia in 1917. In the laboratory that he built in his home in Geneva, Bach conducted his investigations into slow oxidation and biological oxidation, investigations culminating in the peroxide theory of oxidation and the theory of biological oxidation.

In 1916 Bach was elected chairman of the Société de Physique et d'Histoire Naturelle de Genève, and in the following year he was awarded the doctorate *honoris causa* by Lausanne University.

After the February 1917 revolution Bach returned to his homeland. In Moscow he met Maxim Gorky, at whose request he wrote a series of articles that were later published as *Istorija narodnogo khozjajstva* ("The History of the National Economy"). Bach renewed his scientific research in the spring of 1918 in Moscow, first at Blumenthal's chemical and bacteriological institute and later at the chemical laboratory of the chemical industry section of the Higher Soviet National Economy. He was made director of this laboratory upon the recommendation of Lev Yakovlevich Karpov, a renowned chemist and prominent Communist and statesman. Owing to a significant degree to Bach's efforts, the laboratory became a major scientific center and, in 1922, was renamed the L. Y. Karpov Institute of Physics and Chemistry. Many prominent chemists, including Aleksandr Ivanovich Oparin, began their scientific careers there under Bach's guidance. At this institute—as well as at the National Commissariat of Health's Biochemistry Institute, which was founded in 1920 on Bach's recommendation—wide-ranging investigations of oxidation and the problems of biological catalysis, as well as many applied studies into the biochemical methods of processing raw material, were carried out.

In 1927 Bach was chosen a member of the Central Executive Committee of the U.S.S.R. and was awarded the Lenin Prize. The following year the All-Union Association of Scientists and Technicians was organized at his urging. He was chosen a member of the Academy of Sciences of the U.S.S.R. in 1929, and in 1935 he and Oparin organized the Institute of Biochemistry, with Bach as director, within the Academy of Sciences. This institute became the coordinating center of Soviet biochemistry. Bach participated in the preparation of the scientific publications *Zhurnal fizicheskoj khimii* ("Journal of Physical Chemistry"), *Acta physicochimica URSS,* and *Biokhimija* ("Biochemistry") and of the *Bol'shaja sovetskaja èntsiklopedija* ("Great Soviet Encyclopedia") and the *Bol'shaja meditsinskaja èntsiklopedija* ("Great Medical Encyclopedia"). In 1939 Bach was chosen the first academician-secretary of the newly created chemical sciences section of the Academy of Sciences. From 1932 until his death, he was the permanent president of the Dmitrii Ivanovich Mendeleev All-Union Chemical Society.

During World War II Bach organized the research of the evacuated institutes of the Academy of Sciences. In 1942–1943, in Frunze, Kirghiz S.S.R., he directed the study of the exploitational possibilities of local biological raw materials. Bach was awarded the State Prize for his work in biochemistry in 1941 and, in 1945, the title Hero of Socialist Labor—the highest distinction for civilians in the U.S.S.R.—was conferred upon him. In 1943 Bach was chosen an honorary member of the Imperial Society of the Chemical Industry (London) and of the American Chemical Society. Annual memorial "Bach readings" on the fundamental questions of biochemistry are held in the Soviet Union on his birthday.

Bach's most important works are studies of carbon assimilation by plants and of slow oxidation and biological oxidation. These investigations are the basis of his works on enzymology (oxidizing and hydrolytic enzymes) and on technical biochemistry (the regulating principles of industrial biochemical processes). Research into the assimilation of carbonic acid preceded the development of the theory of slow oxidation: Bach proposed that in plants carbonic acid decomposes into H_2CO_3, and not $CO_2 + H_2O$; that this process is completed in several stages; and that during this process hydrogen peroxide and peroxides of organic compounds must be formed as intermediary products. On this basis he formed the conclusion that the oxygen formed during photosynthesis is the oxygen of water, and not of carbonic acid, as would follow from the hypotheses concerning the assimilation of carbonic acid proposed by Adolph von Baeyer and E. Erlenmeyer.

The supposition of the existence of conjugate oxidation-reduction reactions and of the formation of intermediary peroxides during the assimilation of carbonic acid led Bach to deal with biological oxidation. Inasmuch as there was no satisfactory general theory of oxidation, he first developed one. His work on slow oxidation, and later on biological oxidation, continued the development of the idea concerning the necessary activation of oxygen in the process of oxidation that had been stated in 1845 by Christian Friedrich Schönbein. "O roli perekisej v protsessakh medlennogo okislenija" ("On the Role of Peroxides in the Process of Slow Oxidation") stated the foundations of a new theory of slowly proceeding oxidation processes and was published in the *Zhurnal Russkago fiziko-khimicheskago obshchestva* in 1897. This paper was presented to the Paris Academy of Sciences by Schützenberger in 1897 and published in the *Comptes rendus* the same year. Almost simultaneously there appeared, in the *Berichte der Deutschen chemischen Gesellschaft,* an article by Karl Engler and Wilhelm Wild, "Die sogennante Activirung des Oxygens und über Superoxydbildung," which developed a theory analogous to Bach's and independent of it. The theory of slow oxidation was later named the Bach-Engler peroxide theory of oxidation.

At the heart of the peroxide theory lies the notion that the energy necessary for the activation of oxygen is introduced by the oxidized substance itself. Before Bach, the activation process was considered either the result of the break-up of an oxygen molecule into separate atoms, as was envisaged by Rudolph Clausius, Felix Hoppe-Seyler, and Jacobus van't Hoff, or the result of the severance of only one bond in the oxygen molecule and the formation of hydrogen peroxide, as Moritz Traube proposed. Bach significantly widened the limits of Traube's hypothesis, thus formulating a new theory of slow oxidation.

(1) In slow combustion processes the oxygen molecule $O=O$ partially dissociates under the influence of the free energy of the substance being oxidized and enters into the reaction in the form $-O-O-$.

(2) All substances capable of being oxidized, independent of their chemical nature, annex such $-O-O-$ groups, initially forming a peroxide, either

$$R'-O-O-R'' \text{ or } R\begin{matrix} O \\ | \\ O \end{matrix}.$$

(3) Peroxides formed in such a manner contain half of the annexed oxygen in a weakly bonded, "active" state, and therefore easily lose this to other substances.

Thus, slow oxidation can be represented schematically as a two-stage process:

(1) $A + O_2 = AO_2$

(2) $AO_2 + A = 2AO$.

Having advanced a correct presentation of the influence of the oxidized substance on oxygen, Bach could not explain the nature of this influence. An explanation became possible only after the development of the theory of free organic radicals. It was subsequently demonstrated that during oxidation the free valence of such a radical influences the oxygen molecule. With this correction, Bach's views are at the base of the modern chain theory of oxidation processes.

Bach subsequently extended the peroxide theory to processes of biological oxidation, focusing attention on the following facts: oxidation processes in organisms proceed with great intensity; the organic substances that serve as food or enter into the composition of the body are practically indifferent to free oxygen; and the oxygen that is separated from the blood's oxyhemoglobin is found in a passive state. According to Bach, these facts attested to the existence in organisms of the sources—biological catalysts—of the activation of oxygen: enzymes (whose existence was considered far from indisputable at the time) and unsaturated organic compounds. In order successfully to apply the peroxide theory in the explanation of biological oxidation, it was necessary to prove not only the presence of peroxides in living cells but also the existence of mechanisms that would defend the cell from their harmful influence. Bach's most important work on biological oxidation during the Geneva period of his career was specifically directed to the solution of these problems. In 1902, with R. Chodat, he investigated the action of hydrogen peroxide on a cell, a new reaction for the assay of hydrogen peroxide in green plants (the system potassium bichromate-aniline-oxalic acid) that he had proposed. During the next two years he investigated the formation of peroxides in plants. According to the peroxide theory of biological oxidation proposed by Bach in 1904, the oxidation of biological substances proceeds according to the following schema:

$$A + O_2 \xrightarrow{\text{oxygenase}} AO_2$$

$$AO_2 + B \xrightarrow{\text{peroxidase}} AO + BO \quad \text{or} \quad A + BO_2.$$

The proposition that at the base of these processes are two consecutively coordinated enzymatic reactions is the most important statement in such an interpretation of biological oxidation. Such an approach was altogether new in the description of metabolic chemical processes. The representation introduced by Bach made note of new means for elucidating the general order of chemical processes in organisms. Although the peroxide mechanism has been confirmed only for the action of lipoxydase, and although it has been found that the more important oxidation-reduction processes are achieved with the aid of mechanisms other than peroxide mechanisms, Bach's theory played a positive role in the development of the concepts involved in biological oxidation by stimulating Otto Warburg's successful search for the enzymatic mechanisms of oxidation. Experimental investigations of the conditions under which many oxidation processes are achieved enabled Bach to draw conclusions that were subsequently used in various industrial processes.

Bach carried out fundamental investigations in industrial biochemistry after his return to the Soviet Union. These were inspired by work on the reconstruction and development of the Soviet food industry in the 1920's and 1930's. Many recommendations for the improvement of the baking of bread and for the drying and preservation of grain and flour were worked out with Bach's help. Original investigations aimed at the creation of new methods for the processing of tea, wine, tobacco, and raw-vitamin materials (tea leaves, pine needles, etc.) were carried out. The methods of industrial biochemistry developed by Bach are widely used in many countries.

BIBLIOGRAPHY

I. ORIGINAL WORKS. Articles by Bach are "O roli perekisej v protsessakh medlennogo okislenija" ("On the Role of Peroxides in the Process of Slow Oxidation"), in *Zhurnal Russkago fiziko-khimicheskago obshchestva,* **29,** no. 2 (1897), 375; "Untersuchungen über die Rolle der Peroxyde in der Chemie der lebenden Zelle. I. Ueber das Verhalten der lebenden Zelle gegen Hydroperoxyde," in *Berichte der Deutschen chemischen Gesellschaft,* **35** (1902), 1275, with R. Chodat; and "Khimizm dykhatel'nykh protsessov" ("The Chemical Mechanism of Respiratory Processes"), in *Zhurnal Russkago fiziko-khimicheskago obshchestva,* **44,** no. 2 (1912), 1–74. His *Sobranie trudov po khimii i biokhimii* ("A Collection of Works on Chemistry and Biochemistry"; Moscow, 1950), A. I. Oparin and A. N. Frumkin, eds., includes a detailed bibliography of his works.

II. SECONDARY LITERATURE. Works on Bach are *Aleksei Nikolaevich Bach,* in the series Materialy k bibliografii uchenykh SSSR ("Materials Toward the Bibliography of Scientists of the U.S.S.R."), A. N. Nesmejanov, ed. (Moscow–Leningrad, 1946); L. A. Bach and A. I. Oparin, *Aleksei Nikolaevich Bach. Biograficheskij ocherk* ("Aleksei Nikolaevich Bach. Biographical Essay"; Moscow, 1957); B. E. Bykhovsky, *Razvitie biologii v SSSR* ("The Development of Biology in the U.S.S.R."; Moscow, 1957); J. I. Gerassimov, ed., *Razvitie fizicheskoj khimii v SSSR* ("The Development of Physical Chemistry in the U.S.S.R."; Moscow, 1967); L. A. Orbeli, ed., *Trudy soveshchanija, posvjashchennoogo 50-letiju perekisnoj teorii medlennogo okislenija i roli Bacha v razvitii otechestvennoj biokhimii* ("Works of the Conference Dedicated to the Fiftieth Anniversary of the Peroxide Theory of Slow Oxidation and to Bach's Role in the Development of National Biochemistry"; Moscow–Leningrad, 1946); and "Trudy konferentsii, posvjashchennoj 40-letiju perekisnoj teorii Bacha-Ènglera" ("Works of the Conference Dedicated to the Fortieth Anniversary of the Bach-Engler Peroxide Theory"), in M. B. Neyman, ed., *Problemy kinetiki i kataliza* ("Problems of Kinetics and Catalysis"), IV (1940).

ALEKSEI NIKOLAEVICH SHAMIN

BACHE, ALEXANDER DALLAS (*b.* Philadelphia, Pennsylvania, 19 July 1806; *d.* Newport, Rhode Island, 17 February 1867), *physics.*

Bache was a great-grandson of Benjamin Franklin and related to leading families of Philadelphia, facts not insignificant in his future career. After graduating from West Point at the head of his class in 1825, he served for two years in the Corps of Engineers before accepting a professorship of natural philosophy and chemistry at the University of Pennsylvania, a position he held until his resignation in 1836 to organize Girard College. Upon returning from a two-year sojourn in Europe (1836–1838), where he studied primary and secondary education, Bache wrote a report on his findings for Girard College that exerted considerable influence on the development of education in the United States, by proposing adoption, in American high schools, of features from the German Gymnasium and the French *lycée* (1839). He put his views into practice by organizing Central High School of Philadelphia. In 1842 Bache returned to the University of Pennsylvania, but left for Washington at the end of 1843 to succeed F. R. Hassler as head of the Coast Survey, a position he held for the rest of his life.

In Philadelphia, Bache's scientific career followed many of the conventional paths for that period. He dabbled in chemical analysis and experimented on the effects of color on the radiation and absorption of heat. Like many of his contemporaries, he tried his hand at electromagnetism and astronomy, but with no particular success, his dispute with Denison Olmsted on meteoric showers being a notably poor showing. He was, however, outstanding in assuming leading roles in the affairs of both the American Philosophical Society and the Franklin Institute.

At the latter he directed a significant investigation of the explosion of steam boilers for the federal government (*Journal of the Franklin Institute,* **17,** 1836). Not only was this notable for experimental virtuosity; it was also one of the first deliberate uses of science by the government for the solution of a practical problem. It also established a pattern for Bache's subsequent career and a precedent for the later development of federal policy toward science and technology.

Increasingly during the Philadelphia period, Bache became involved in studies of the physics of the earth, particularly terrestrial magnetism and meteorology. After returning from Europe, where he had made observations of declination and inclination for comparison with American readings, he attempted to establish an American system that would fit into Sir Edward Sabine's world network of magnetic observatories. All that resulted, however, was an observatory at Girard College, the first of its kind in the United States. Perhaps the most interesting work in this vein was Bache's unsuccessful attempt with Humphrey Lloyd to determine longitude by simultaneous magnetic observations (*Proceedings, Royal Irish Academy,* **1,** 1839).

When Bache assumed direction of the Coast Survey, it was a small, insecurely established body with high scientific standards. In less than two decades it became entrenched, the largest employer of physical scientists in the United States, and active in many scientific fields. First-order triangulation was expanded along the Atlantic, Gulf, and Pacific coasts. Under Bache's direction, Sears Walker and W. C. Bond developed the use of telegraphy in the determination of longitude. Bache and the Survey supported astronomical research, including study of the solar eclipses of 26 May 1854 and 18 July 1860. Survey vessels amassed the most extensive series of observations of

the Gulf Stream up to that time. Continuing and broadening Hassler's tidal observations, Bache became embroiled in a dispute with Whewell in 1851, when the Survey's findings deviated from the latter's theory. Using deflections on tide staffs on the Pacific Coast, Bache studied waves from an earthquake in Japan (*American Journal of Science,* **21,** 1855), work that foreshadowed the Survey's later work in seismology. Bache also succeeded Hassler as head of the Office of Weights and Measures, the predecessor of the National Bureau of Standards. During all this time, while he was successfully administering a large research program, Bache was able to spend several months in the field with a survey party and to continue doing his own investigations.

Bache is clearly one of the founders of the scientific community in the United States. His administration of the Coast Survey established a model for large-scale scientific organization that was followed either implicitly or explicitly by later groups. Bache and his close friend Joseph Henry established many of the patterns of interaction of science and the federal government. Perhaps most significant of all was the way pure science, in Bache's scheme of things, became the necessary antecedent and companion of applied science, rather than purely a philosophical endeavor. Around him gathered a small, changing group of followers, the *Lazzaroni,* or scientific beggars. Bache clearly had their admiration, but what they specifically wanted is hazy in many respects. The group included nonscientists; some of the scientists, such as Dana and Henry, later split with Bache. The *Lazzaroni* wanted to form a true professional scientific community to reform higher education so that more young people would be interested in science, and to find administrative means of increasing governmental support of science. It was Bache's misfortune that his warm admirers in Cambridge—including Louis Agassiz, B. A. Gould, and Benjamin Peirce—lacked his diplomatic talents, thus embroiling the *Lazzaroni* "program" in irrelevant personal squabbles.

The culmination of Bache's influence and of the outlook he represented came during the Civil War. Because of his knowledge of the coasts, in 1861 he served with the informal Commission on Conference planning the naval campaign against the Confederacy. As vice-president of the U.S. Sanitary Commission, Bache became involved in a notable medical and welfare program. A member of the Permanent Commission in 1863–1864, Bache advised the navy on technical matters. Linked to the Permanent Commission was the formation of the National Academy of Sciences in 1863, with Bache as its first president. The Academy was the concrete culmination of the attitudes he enunciated in his 1851 presidential address to the American Association for the Advancement of Science.

Bache was incapacited by a stroke early in the summer of 1864. After Bache's death, Henry kept the Academy alive largely because his friend had left his estate to it. The Bache Fund was a small but important source of support for research in the United States before 1900. The Michelson-Morley experiment, for example, was conducted with its aid.

BIBLIOGRAPHY

I. ORIGINAL WORKS. Joseph Henry's memoir of Bache in *Biographical Memoirs of the National Academy of Sciences,* I (1869), has a usable bibliography of Bache's writings. The best single source of information on his career is the *Annual Reports of the U.S. Coast Survey* (1844–1866). For his pre-Washington experiences, see *Journal of the Franklin Institute* and *Transactions of the American Philosophical Society.*

An extensive body of manuscripts from Bache's tenure as head of the Coast Survey is preserved in the U.S. National Archives. They are described in Nathan Reingold, *Records of the Coast and Geodetic Survey* (Washington, D.C., 1958), Preliminary Inventory No. 105 of the National Archives. A discussion of their significance is in Reingold's "Research Possibilities in the U.S. Coast and Geodetic Survey Records," in *Archives internationales d'histoire des sciences,* **11** (Oct.-Dec. 1958), 337–346. These records include a considerable amount of private correspondence. Substantial portions of Bache's private papers are also found in the Rhees Collection of the Huntington Library in San Marino, Calif., the Smithsonian Archives, and the Library of Congress. Other Bache letters are in the American Philosophical Society, the Franklin Institute, and the Benjamin Peirce Papers at Harvard. For hostile comments on Bache's work by a scientific amateur, see the John Warner Papers at the American Philosophical Society. For hostile comments of a professional scientist, see the letters of C. H. F. Peters in the Harvard Observatory records.

II. SECONDARY LITERATURE. The most recent biography of Bache, M. M. Odgers, *Alexander Dallas Bache, Scientist and Educator, 1806–1867* (Philadelphia, 1947) is a slight advance over the necrologies published by friends shortly after Bache's death. It suffers, however, from the absence of work in many of the major manuscript sources and a nearly total lack of insight into or knowledge of Bache's scientific environment. *Proceedings of the American Philosophical Society,* **84,** no. 2 (1941), "Commemoration of the Life and Work of Alexander Dallas Bache and Symposium on Geomagnetism," is still quite useful for an introduction to Bache's career. Two recent publications discuss the steam boiler investigations: John G. Burke, "Bursting Boilers and the Federal Power," in *Technology and Culture,* **7** (Winter 1966), 1–23; and Bruce Sinclair, *Early Research at the Franklin Institute, the Investigation Into the Causes of*

Steam Boiler Explosions, 1830–1837 (Philadelphia, 1966). For Bache's career in more general frames of reference, see Nathan Reingold, *Science in Nineteenth Century America, a Documentary History* (New York, 1964), pp. 127–161, 200–225; and A. H. Dupree, *Science in the Federal Government . . .* (Cambridge, Mass., 1957), pp. 216–232, 316–336. The activities of the *Lazzaroni* are treated in E. Lurie, *Louis Agassiz, a Life in Science* (Chicago, 1960), pp. 166–211, 303–350; and A. H. Dupree, *Asa Gray* (Cambridge, Mass., 1959).

NATHAN REINGOLD

BACHELARD, GASTON (*b.* Bar-sur-Aube, France, 27 June 1884; *d.* Paris, France, 16 October 1962), *philosophy of science, epistemology.*

Bachelard became a philosopher late in life. He had previously taught physics and chemistry in the *collège* of his native city. His knowledge of physics later enabled him to determine the epistemological change brought about by modern science and, particularly, to gauge the growing distance between it and classical physics, which had suddenly become only relative.

As early as 1928, in the *Essai sur la connaissance approchée,* Bachelard penetrated to the heart of the new mathematical physics and began to simplify its methods of measuring (calculus of errors), experimenting, and generalizing. A complementary study, *Étude sur l'évolution d'un problème de physique* (1928), was designed to show how thermodynamics was both established by and liberated from its early, very poor intuitions (such as that a metal bar heated at one end will become longer). His study of conductibility in anisotropic media, and the mathematical theory of Poisson and Fourier, facilitated the integration of thermodynamics with mechanics, especially since heat was not linked to an isolated molecule but was determined within a rather large volume (the quantitative view).

As early as 1930 Bachelard's work branched out in several directions. One was pedagogical, concerned with how to arouse the mind and modernize it so that it could participate in the formulation of scientific concepts. This led to *La formation de l'esprit scientifique* (1938), in which Bachelard examines mental resistances and prejudices anthropologically and methodologically, in order to expose them. His original objective was to apply Freud's psychoanalytic method to epistemology and to science itself, and to draw up a highly systematic list of complexes that paralyze intelligence.

Bachelard's work was also historical: he became the historian of scientific changes. He did not give a chronological description of the evolutionary stages of thought as it grappled with reality, but clarified its tensions and expressed its breakdowns. These tensions and breakdowns occurred because modern physics attacked what had previously been considered basic principles.

A vast prescientific or, as it were, prehistoric universe underlies the period of concentrated scientific discovery—the revolution that substituted the surrationalism of the twentieth century for the rationalism of the nineteenth. If this rationalism conditions our view of the science of the eighteenth and, *a fortiori,* the preceding centuries, it by no means invalidates it. Bachelard used earlier, naive physics as a subject for a new study of elemental psychology. Alchemy represented for him the projection of the soul's desires, the magic lantern that enlarged dreams. From this belief came his celebrated analyses of the classical four elements and of dreams, space, time, and imagination.

Bachelard was never satisfied merely to point out obstructive archaisms. He went further, introducing the reader to the most abstract categories of contemporary science. From generalities, he moved on to make specialized analyses on the point, the corpuscle, movement, space, and the simultaneous. Bachelard also simplified the dialectic contained in apparatus that embodied a theory: the Wilson cloud chamber, the spectroscope, and the particle accelerator. "Phenomenotechnology" replaced instruments based on calculating systems; probabilism replaced realism; the discursive replaced the intuitive.

Bachelard attempted to renew the traditional questions of philosophy and physical reality. He sought to base metaphysics upon the new physics and to resume the work of Descartes, Newton, and Leibniz.

All these avenues of endeavor cut across each other. With Bachelard the past became the poetry of the world; classical science, a basis for exploration and enlargement; and modern science, a phenomenon that must constantly be reviewed.

BIBLIOGRAPHY

I. ORIGINAL WORKS. Bachelard's writings are *Essai sur la connaissance approchée* (Paris, 1928); *Étude sur l'évolution d'un problème de physique: La propagation thermique dans les solides* (Paris, 1928); *La valeur inductive de la relativité* (Paris, 1929); *Le pluralisme cohérent de la chimie moderne* (Paris, 1932); *Le nouvel esprit scientifique* (Paris, 1934); *Les intuitions atomistiques* (Paris, 1935); *L'intuition de l'instant* (Paris, 1935); *La dialectique de la durée* (Paris, 1936); *L'expérience de l'espace dans la physique contemporaine* (Paris, 1937); *La formation de l'esprit scientifique. Contribution à une psychanalyse de la connaissance objective* (Paris, 1938); *La psychanalyse du feu* (Paris, 1938); *La phi-losophie du non* (Paris, 1940); *Lautréamont* (Paris, 1940); *L'eau et les rêves. Essai sur l'imagination de la matière* (Paris, 1942); *L'air et les songes. Essai sur l'imagination du mouvement* (Paris, 1943); *La terre et les rêveries de la volonté. Essai sur l'imagination des forces* (Paris, 1948); *La terre et les rêveries du repos. Essai sur l'imagination de l'intimité* (Paris, 1948); *Le rationalisme appliqué* (Paris, 1949); *L'activité rationaliste de la physique contemporaine* (Paris, 1951); *Le materialisme rationnel* (Paris, 1953); *La poétique de l'espace* (Paris, 1957); *La flamme d'une chandelle* (Paris, 1961); and *La poétique de la rêverie* (Paris, 1961).

II. SECONDARY LITERATURE. Three essential studies on Bachelard's epistemology are in Vrin, ed., *Problèmes d'histoire et de philosophie des sciences* (Paris, 1968); Georges Canguilhem, "L'histoire des sciences dans l'oeuvre epistémologique de Gaston Bachelard," originally in *Annales de l'Université de Paris* (1963); "Gaston Bachelard et les philosophes," originally in *Sciences* (March 1963); and "Dialectique et philosophie du non chez Gaston Bachelard," originally in *Revue internationale de philosophie* (1963).

See also François Dagognet, *Gaston Bachelard* (Paris, 1965); *Hommage à Gaston Bachelard* (Paris, 1957), consisting mainly of articles by Georges Canguilhem and Jean Hyppolite; Jean Hyppolite, "L'épistémologie de Gaston Bachelard," in *Revue d'histoire des sciences* (January 1964); R. Martin, "Dialectique et esprit scientifique chez Gaston Bachelard," in *Les études philosophiques* (October 1963); and Pierre Quillet, *Bachelard* (Paris, 1964).

F. DAGOGNET

BACHELIER, LOUIS (*b.* Le Havre, France, 11 March 1870; *d.* Saint-Servan-sur-Mer, Ille-et-Villaine, France, 28 April 1946), *mathematics.*

Bachelier obtained the bachelor of sciences degree in 1898, and two years later he defended his doctoral dissertation on the theory of speculation. Encouraged by Henri Poincaré, he published three memoirs on probability: "Théorie mathématique des jeux" (1901), "Probabilités à plusieurs variables" (1910), and "Probabilités cinématiques et dynamiques" (1913). After being named lecturer at Besançon in 1919, he taught at Dijon and Rennes before returning to Besançon as professor from 1927 to 1937, when he retired.

Under the name "théorie de la spéculation" Bachelier introduced continuity into problems of probability. Making time the variable, he considered the movements of probabilities and chains of probabilities, then applied these ideas to the fluctuations of market rates and to the propagation of light and radiance. The relation between haphazard movement and diffusion was the first attempt at a mathematical theory of Brownian movement, a classic theory known today as the Einstein–Wiener theory; Wiener's rigorous study was made in 1928. Bachelier was the first to examine the stochastic methods of the Markovian type, on which A. N. Kolmogorov's theory is based.

Yet lack of clarity and precision, certain considerations of doubtful interest, and some errors in definition explain why, in spite of their originality, his studies exerted no real scientific influence.

Bachelier participated in the diffusion of probabilistic thought through his articles "La périodicité du hasard" and "Quelques curiosités paradoxales du calcul des probabilités" and the book *Le jeu, la chance et le hasard.*

BIBLIOGRAPHY

Bachelier's writings include "Théorie de la spéculation," in *Annales de l'École normale supérieure,* 3rd ser., **17** (1900), 21–86; "Théorie mathématique des jeux," *ibid.,* **18** (1901), 143–210; "Théorie des probabilités continues," in *Journal de mathématiques pures et appliquées,* 6th ser., **2,** pt. 3 (1906), 259–327; "Probabilités à plusieurs variables," in *Annales de l'École normale supérieure,* 3rd ser., **27** (1910), 340–360; "Mouvement d'un point ou d'un système soumis à l'action de forces dépendant du hasard," in *Comptes rendus de l'Académie des sciences,* **151** (1910), 852–855; *Calcul des probabilités* (Paris, 1912); "Probabilités cinématiques et dynamiques," in *Annales de l'École normale supérieure,* 3rd ser., **30** (1913), 77–119; *Le jeu, la chance et le hasard* (Paris, 1924); *Les lois du grand nombre du calcul des probabilités* (Paris, 1937); *La spéculation et le calcul des probabilités* (Paris, 1938); and *Les nouvelles méthodes du calcul des probabilités* (Paris, 1939).

LUCIENNE FÉLIX

BACHET DE MÉZIRIAC, CLAUDE-GASPAR (*b.* Bourg-en-Bresse, France, 9 October 1581; *d.* Bourg-en-Bresse, 26 February 1638), *mathematics.*

One of the ablest men of the seventeenth century, Bachet came from an ancient and noble family. His grandfather, Pierre Bachet, seigneur de Meyzériat, was counselor to King Henry II. His father was the honorable Jean Bachet, appeals judge of Bresse and counselor to the duke of Savoy; his mother was the noblewoman Marie de Chavanes. Orphaned at the age of six, the precocious Claude-Gaspar received his early education in a house of the Jesuit order that belonged to the duchy of Savoy. Presumably he studied in Padua as a young man, and he may have taught in a Jesuit school in Milan or Como. Bachet also spent a few years in Paris and in Rome, where, with his friend Claude Vaugelas, he composed a great deal of Italian verse. He was a prolific reader of poetry, history, commentary, scholarship, and mathematics.

In his fortieth year Bachet married Philiberte de Chabeu, by whom he had seven children. He suffered greatly from rheumatism and gout; when the Académie Française was founded in 1634, he was too ill to attend the inaugural ceremony. He was made a member of the Académie in the following year. His early literary works, humanist in outlook, consisted of poems in Latin, French, and Italian. Between 1614 and 1628 he published a Latin epistle from the Virgin Mary to her Son, canticles, brief sacred and profane Latin poems, translations of psalms, and a metrical translation of seven of Ovid's *Epistulae heroïdum.* He also published an anthology of French poetry, *Délices,* and the *Epistles of Ovid* (1626); the latter assured his reputation as a mythologist.

Bachet claims our attention today, however, chiefly for his contributions to the theory of numbers and to the field of mathematical recreations, in which he was one of the earliest pioneers. The two mathematical works for which he is remembered are his first edition of the Greek text of Diophantus of Alexandria's *Arithmetica,* accompanied by prolix commentary in Latin (1621), and his *Problemes plaisans et delectables qui se font par les nombres* (1612).

Diophantus had anticipated the advent of algebra and the theory of numbers. His work, which was known to the Arabs, was not appreciated until it was rediscovered in Europe during the latter part of the sixteenth century. Prior to Bachet's translation, only a few scholars had written on the work of Diophantus: Maximus Planudes, who gave an incomplete commentary on the first two books of the *Arithmetica* (ca. 1300); Raphael Bombelli, who embodied all of the problems of the first four books in his *Algebra* (1572); Wilhelm Holzmann, better known as Xylander, who gave a complete Latin translation (1575); and Simon Stevin, who gave a French translation of the first four books (1585). Bachet's translation, *Diophanti Alexandrini Arithmeticorum libri sex,* was based largely on the writings of Bombelli and Xylander, particularly the latter, although he admitted this with reluctance. Indeed, it is the opinion of T. L. Heath that although Bachet generally has been regarded as the only writer to interpret the contributions of Diophantus effectively, perhaps as much—if not more—of the credit is due to Xylander.

It is noteworthy that Bachet struggled with this work while suffering from a severe fever. He asserted that he corrected many errors in Xylander's version; filled in numerous omissions, such as proofs of porisms and abstruse theorems that Diophantus merely mentioned; and clarified much of the exposition. He apparently added few original contributions to number theory or Diophantine analysis, however, except for a generalization of the solution of the system $ax + v = u^2$, $cx + d = w^2$. Yet despite its imperfections, Bachet's work is commendable for being the first and only edition of the Greek text of Diophantus. It was subsequently reprinted, with the addition of Fermat's notes, in 1670; and while Fermat's notes are significant, the Greek text is inferior to that of the first edition.

Bachet's penchant for arithmetical rather than geometric problems is also obvious in the contents of his *Problemes plaisans et delectables*. These problems fall into several readily recognizable categories. The most elementary are those mildly amusing but mathematically unimportant parlor tricks of finding a number selected by someone, provided the results of certain operations performed on the number are revealed. Variations include problems involving two or three numbers, or two persons; problems depending upon the scale of notation; and tricks with a series of numbered objects, such as watch-dial puzzles and card tricks—many of these have appeared ever since in most collections of mathematical recreations. Somewhat more sophisticated is the famous problem of the Christians and the Turks, which had previously been solved by Tartaglia. (In a storm, a ship carrying fifteen Christians and fifteen Turks as passengers could be saved only by throwing half the passengers into the sea. The passengers were to be placed in a circle, and every ninth man, beginning at a certain point, was to be cast overboard. How should they be arranged so that all the Christians would be saved? Answer: CCCCTTTTTCCTCCCTCTTCCTTTCTTCCT.)

Of greater mathematical significance was Bachet's problem of the weights: to determine the least number of weights that would make possible the weighing of any integral number of pounds from one pound to forty pounds, inclusive. Bachet gave two solutions: the series of weights 1, 2, 4, 8, 16, 32; and the series 1, 3, 9, 27—depending upon whether the weights may be placed in only one scale pan or in each of the two scale pans. Last, there is the celebrated prototype of ferrying problems or difficult crossings, the problem of the three jealous husbands and their wives who wish to cross a river in a boat that can hold no more than two persons, in such a manner as never to leave a woman in the company of a man unless her husband is also present. Eleven crossings are required, but Bachet gave a solution that asserts "Il faut qu'ils passent en six fois en cette sorte." The analogous problem with four married couples cannot be solved; Bachet stated this fact without proof. It should also be noted that Bachet gave a method for constructing magic squares which is essentially that of Moschopulous (*ca.* 1300), although Bachet appears to have discovered it independently.

BIBLIOGRAPHY

I. ORIGINAL WORKS. Bachet's translation of Diophantus' *Arithmetica* is *Diophanti Alexandrini Arithmeticorum libri sex, et de numeris multangulis liber unus. Nunc primum Graecè et Latinè editi, atque absolutissimis commentariis illustrati* (Paris, 1621); it was reprinted with the subtitle *Cum commentariis C. G. Bacheti V. C. & observationibus D. P. de Fermat senatoris Tolosani accessit doctrinae analyticae inventum novum, collectum ex variis D. de Fermat epistolis* (Toulouse, 1670). His other major work is *Problemes plaisans et delectables, qui se font par les nombres; Partie recueillis de diuers autheurs, & inuentez de nouueau auec leur demonstration. Tres-utiles pour toutes sorte de personnes curieuses, qui se seruent d'arithmetique* (Lyons, 1612). It was reprinted in several editions: *2ᵉ édition, revue, corrigée, et augmentée de plusieurs propositions, et de plusieurs problems, par le meme autheur* (Lyons, 1624; reprinted 1876); *3ᵉ édition, revue, simplifiée, et augmentée par A. Labosne* (Paris, 1874; abridged version, 1905); and *Cinquième édition, revue, simplifiée et augmentée par A. Labosne. Nouveau tirage augmenté d'un avant-propos par J. Itard* (Paris, 1959), a paperback edition.

II. SECONDARY LITERATURE. Works dealing with Bachet are W. W. R. Ball and H. S. M. Coxeter, *Mathematical Recreations and Essays* (London, 1942), pp. 2–18, 28, 30, 33, 50, 116, 313, 316; Pierre Bayle, *Dictionnaire historique et critique* (Amsterdam, 1734), pp. 553–556; Moritz Cantor, *Vorlesungen über die Geschichte der Mathematik* (Leipzig, 1913), II, 767–780; C.-G. Collet and Jean Itard, "Un mathématicien humaniste: Claude-Gaspar Bachet de Méziriac (1581–1638)," in *Revue d'histoire des sciences*, **1** (1947), 26–50; and T. L. Heath, *Diophantos of Alexandria; A Study in the History of Greek Algebra* (Cambridge, 1885), pp. 49–54.

WILLIAM SCHAAF

BACHMANN, AUGUSTUS QUIRINUS, also known as **Augustus Quirinus Rivinus** (*b.* Leipzig, Germany, 9 December 1652; *d.* Leipzig, 30 December 1723), *botany.*

Bachmann was a son of the physician Andreas Bachmann (1600–1656), professor of poetry and then of physiology in the University of Leipzig. The father had already adopted the latinized form of his family name, and the son consistently used it. Young Rivinus studied humanities and medicine in Leipzig but took his degree at the then flourishing University of Helmstedt in 1676. He settled in his native town as a medical practitioner and became a lecturer in medicine in the University of Leipzig in 1677. He was appointed to the chair of physiology and botany in 1691; became professor of pathology in 1701; and was made professor of therapeutics and dean perpetual in 1719. His last years were clouded by failing vision and urolithiasis, and he died of what was

Rivinus' System of Plant Classification (1690)

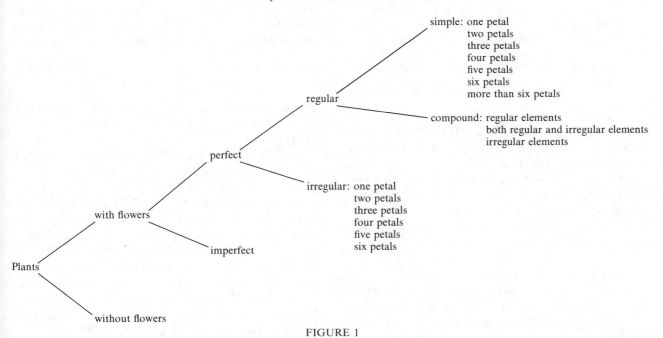

FIGURE 1

diagnosed as a pleurisy. Rivinus was married four times and had one son, Johannes Augustus Rivinus (1692–1725), who also became a physician.

Rivinus wrote many medical dissertations, among them *De dyspepsia* (1678), in which he describes the excretory ducts of the sublingual salivary glands. He also tried to remove from the *materia medica* of his time such squalid substances as feces and urine, as well as all superstitious, counterfeit, or otherwise useless drugs.

Rivinus' main scientific interest, however, was botany, particularly botanical taxonomy. In 1690 he published *Introductio generalis in rem herbariam,* with 125 tables of "plants with irregular flowers of one petal" (*Labiatae* and others). Atlases of "irregular" flowers of four petals (mostly *Leguminosae,* 121 tables) and of five petals (mostly *Umbelliferae,* 139 tables) followed in 1691 and in 1699, respectively. Rivinus published these tables at his own expense; and it is therefore no wonder that he could not afford to bring out the last volume he had prepared, which dealt with "irregular" flowers of six petals (orchids). He anticipated Tournefort and Linnaeus in devising an artificial system of plant classification. His system comprised eighteen classes (*ordines*) based on the number of petals of a flower and on its regularity or irregularity.

Rivinus' standards of regularity were extremely high. For instance, he included all the *Umbelliferae* in the class of *irregulares pentapetalae* because they have not only one style exactly in the center of the flower but also two eccentric ones. His attempt to base classification on a single part of a plant led Rivinus into a controversy with the famous English naturalist John Ray, who held the sound view that the whole plant must be considered.

Since he emphasized the need for short plant names of no more than two words, Rivinus was a pioneer of modern binomial nomenclature.

BIBLIOGRAPHY

I. ORIGINAL WORKS. Among Rivinus' works are *De dyspepsia* (Leipzig, 1678); *Introductio generalis in rem herbariam* and *Ordo plantarum, quae sunt flore irregulari monopetalo* (Leipzig, 1690), *Introductio* also published separately (Leipzig, 1696, 1720); *Ordo plantarum, quae sunt flore irregulari tetrapetalo* (Leipzig, 1691); *Epistola botanica ad Johannem Raium* (Leipzig, 1694; repr. London, 1696, with Ray's answer); *Ordo plantarum quae sunt flore irregulari pentapetalo* (Leipzig, 1699); and *De medicamentorum officinalium censura* (Leipzig, 1701, 1707), repr. with forty-six other papers in *Dissertationes medicae, diversis temporibus habitae, nunc vero in unum fasciculum collectae* (Leipzig, 1710).

Universal-Lexicon (see below) gives a complete list of Rivinus' writings, and Jourdan (see below) lists his more important medical publications.

II. SECONDARY LITERATURE. Articles on Rivinus are L.-M. Dupetit-Thouars, in *Biographie universelle ancienne et moderne,* new ed., XXXVI (*ca.* 1860), 90–94, which gives a very careful account of Rivinus' botanical work; A. von Haller, in *Bibliotheca botanica,* I (Zurich, 1771), 551, para.

651; C. G. Jöcher, in *Allgemeines Gelehrten-Lexicon,* III (Leipzig, 1751; repr. Hildesheim, 1961); A.-J.-L. Jourdan, in *Dictionnaire des sciences médicales—Biographie médicale,* VIII (1825); and Zedler's *Universal-Lexicon,* XXXI (Leipzig–Halle, 1742). See also C. Rabl, *Geschichte der Anatomie an der Universität Leipzig* (Leipzig, 1909).

HULDRYCH M. KOELBING

BACHMANN, PAUL GUSTAV HEINRICH (*b.* Berlin, Germany, 22 June 1837; *d.* Weimar, Germany, 31 March 1920), *mathematics.*

Bachmann's father was pastor at the Jacobi Kirche, and from his home the young Paul inherited a pious Lutheran view of life, coupled with modesty and a great interest in music. During his early years in the Gymnasium he had some trouble with his mathematical studies, but his talent was discovered by the excellent teacher Karl Schellbach. After a stay in Switzerland for his health, presumably to recover from tuberculosis, Bachmann studied mathematics at the University of Berlin until he transferred to Göttingen in 1856 to attend Dirichlet's lectures. Here he became a close friend of Dedekind, who was a fellow student.

From 1856 on, Bachmann's interests were centered almost exclusively upon number theory. He completed his studies in Berlin, where in 1862 he received his doctorate under the guidance of Ernst Kummer for a thesis on group theory. Two years later he completed his habilitation in Breslau with a paper on complex units, a subject inspired by Dirichlet. After some years as extraordinary professor in Breslau, Bachmann was appointed to a professorship in Münster.

Around 1890 Bachmann divorced his wife and resigned his professorship. With his second wife he settled in Weimar, where he combined his mathematical writing with composing, playing the piano, and serving as music critic for various newspapers. His main project, however, was a complete survey of the state of number theory, *Zahlentheorie. Versuch einer Gesamtdarstellung dieser Wissenschaft in ihren Hauptteilen* (1892–1923). It includes not only a review of known results but also an evaluation of the various methods of proof and approach, labors for which his close association with Dirichlet, Kummer, Dedekind, and Hensel made him ideally suited.

BIBLIOGRAPHY

Bachmann's writings are *Vorlesungen über die Natur der Irrationalzahlen* (Leipzig, 1892); *Zahlentheorie. Versuch einer Gesamtdarstellung dieser Wissenschaft in ihren Haupt-teilen,* 5 vols. (Leipzig, 1892–1923); *Niedere Zahlentheorie,* 2 vols. (Leipzig, 1902–1910); and *Das Fermat-Problem in seiner bisherigan Entwicklung* (Leipzig, 1919), which has stimulated much research in this field.

OYSTEIN ORE

BACK, ERNST E. A. (*b.* Freiburg im Breisgau, Germany, 21 October 1881; *d.* Munich, Germany, 20 July 1959), *physics.*

Back, a scion of an old Alsatian family, entered the University of Strasbourg in 1902. There and at Berlin and Munich he studied law for four years. After passing his first state examination he served as a law clerk from 1906 to 1908. In 1908 Back abandoned the law and began to study experimental physics in Friedrich Paschen's institute at the University of Tübingen, which specialized in atomic spectroscopy. After completing his doctoral dissertation, *Zur Prestonschen Regel* (accepted February 1913, published 1921), Back stayed on at Tübingen as Paschen's assistant until the outbreak of World War I.

Preston's rule—that the magnetic splittings (Zeeman effects) of the lines in a given series (e.g., triplet diffuse series, doublet sharp series, and so forth) are identical—suffered many apparent exceptions. Clarification required better—above all, sharper—spectrographs, and Back's entire scientific career was devoted to meeting this need. His most important technical innovations were in the design and fabrication of the light source, because this was the component of the apparatus chiefly responsible for the failure of Zeeman-effect spectrographs to match the quality of those made in the absence of a magnetic field. By the late 1920's Back's unique vacuum arc, packed into the four millimeters between the pole pieces of Tübingen's electromagnet, included extraordinarily elaborate features: the electrode containing the element to be investigated was advanced (through a vacuum seal) by clockwork, while the other electrode, oscillated (also through a vacuum seal) by a motor, intermittently struck and extinguished the arc precisely on the axis of the field.

Although Back was able to rule out a number of apparent exceptions to Preston's rule, others were confirmed. The latter were usually very narrow or unresolved doublets or triplets showing "normal" Lorentz triplets in a magnetic field rather than the anticipated superposition of the anomalous splittings of the individual lines. It was Paschen who saw the solution to the difficulty: in sufficiently strong magnetic fields the several splitting patterns characteristic of the different types of series are transformed into the normal triplet (the Paschen-Back effect, 1912).

370

During the following years this effect was regarded by leading theorists as both one of the most important problems in atomic physics and one of the most promising sources of information on the subject of atomic structure.

A lieutenant in the reserve, Back was employed in the service corps (*Train*) on the Western Front from the outbreak of World War I until 1918. In 1918 he became head (*Leiter*) of the physical laboratory of Veifa-Werke, a manufacturer of electrical and X-ray equipment in Frankfurt am Main. Yet Back never abandoned his interest in the Zeeman effect, and in the spring of 1920 he returned to Tübingen as an assistant in Paschen's institute. There Back accumulated spectrographs of unparalleled quality, including the Zeeman effects of elements in whose spectra Miguel Catalán (working at the University of London) and Hilda Gieseler (working at Tübingen) were then discovering the existence of higher multiplicities (quartets, quintets, and so forth). When Alfred Landé arrived in Tübingen in October 1922, Back made this data available to him. Six months later, in consecutive articles in the *Zeitschrift für Physik,* Landé published the well-known formula expressing the "*g* factors" for all multiplicities as a function of the quantum numbers of the stationary state of the atom, and Back published the data upon which it was based.

In the spring of 1923 Back became *Privatdozent* at Tübingen. An appointment as extraordinary professor at the Landwirtschaftliche Hochschule in Hohenheim in 1926 (ordinary professor in 1929) provided financial security, but his scientific work was still carried on in the Tübingen institute. There, in the fall of 1926, he showed Samuel Goudsmit photographs of the bismuth spectrum that he had made over the years. Together they effected an analysis of the hyperfine structure of several lines, interpreting it as the result of the interaction of the total electronic angular momentum with an angular momentum that they attributed to the bismuth nucleus. Back's spectrographs (1927) fixed this nuclear angular momentum at $9/2 \cdot h/2\pi$. They thus provided the first firm evidence of a counterexample to an almost inescapable inference from the contemporary assumption that nuclei consisted of protons and electrons—namely, that a nucleus of even atomic number must have integral angular momentum (in units of $h/2\pi$).

In 1936 Back became professor at Tübingen; he retired in 1948. After the acceptance of electron spin (1926) the Zeeman effect lost most of its intrinsic importance, and in the 1930's other techniques vied with spectroscopy in determinations of angular momenta and magnetic moments. Moreover, after 1930 Back lacked the perceptive collaborators who had previously kept his very special talents directed toward significant problems in atomic physics.

BIBLIOGRAPHY

I. ORIGINAL WORKS. A bibliography of Back's works is in J. C. Poggendorff, *Biographisch-literarisches Handwörterbuch zur Geschichte der exakten Naturwissenschaften* (Leipzig, 1926–1962), V, 47–48; VI, 100–101; VIIA, pt. 1, 74. Some twenty-five letters to Sommerfeld, Landé, and Goudsmit are listed in T. S. Kuhn, *et al., Sources for History of Quantum Physics. An Inventory and Report* (Philadelphia, 1967), p. 13.

II. SECONDARY LITERATURE. For the Paschen-Back effect, see J. B. Spencer, "An Historical Investigation of the Zeeman Effect (1896–1913)," Ph.D. dissertation (University of Wisconsin, 1964). Back's work in the early 1920's is discussed in P. Forman, "Environment and Practice of Atomic Physics in Weimar Germany," Ph.D. dissertation (University of California, Berkeley, 1967), ch. 3; the later 1920's, in S. A. Goudsmit, "Pauli and Nuclear Spin," in *Physics Today* (June 1961), pp. 18–21. The author of the present article is preparing a fuller account of Back's work and career.

PAUL FORMAN

BACKLUND, JÖNS OSKAR (*b.* Långhem, Sweden, 28 April 1846; *d.* Pulkovo, Russia, 29 August 1916), *astronomy.*

The main object of Backlund's research was comet 1786 I, known as Encke's comet or (in the U.S.S.R.) comet Encke-Backlund. Despite its forty-eight observed returns to perihelion between 1786 and 1967, this comet still puzzles astronomers: Backlund was the third man, following Johann Franz Encke and Friedrich Emil von Asten, to devote a major portion of his life to studying it, and the first to show that the long-term acceleration in its motion is subject to irregular changes, attributable to nongravitational forces.

Born in poverty, Backlund left school at an early age but nevertheless managed to prepare himself for entrance to the University of Uppsala, where he received a Ph.D. in 1875. After three years at the observatory of Dorpat, he went to Pulkovo Observatory, located about ten miles south of Petrograd (now Leningrad), to work under Otto von Struve. In 1886 he left Pulkovo for the Imperial Academy of Sciences of Petrograd, to which he had been elected in 1883, but returned in 1895 to serve for the rest of his life as director at Pulkovo. He also traveled throughout Europe and the United States to attend scientific meetings. He received worldwide recognition for his work on Encke's comet, including the award in 1909 of the Gold Medal of the Royal Astronomical Society.

Although Backlund favored Encke's idea that the comet's accelerated motion was due to a thin interplanetary medium, he nevertheless wanted to make sure no gravitational accelerations had been overlooked. He therefore decided to recalculate all planetary perturbations since 1819, the year Encke first obtained an orbit for the comet. By 1886 Backlund had got the financial backing necessary to carry out this tremendous task and had hired a number of people as computers.

The results, published between 1892 and 1898, were impressive, but the comet continued to depart from predictions based upon them in a way that Backlund was unable to explain, although he considered two different possibilities: first, a patchy stream of meteoroids lying along the inner part of the comet's orbit, then electrical forces related to the sunspot cycle. Until the advent of electronic computers the efforts of later workers served mainly to confirm Backlund's suspicion that no single explanation would suffice. Current thinking tends toward the idea, first suggested in 1836 by Friedrich Wilhelm Bessel, that loss of mass by the comet itself is responsible.

BIBLIOGRAPHY

I. ORIGINAL WORKS. Backlund's first paper on Encke's comet was written with A. Bonnsdorf and appeared as "Allmänna störingar, som af jorden förorsakas i Enckeska Kometens rörelse i en viss del af dess bana," in *Bihang till Kongliga Svenska Vetenskaps-Akademiens Handlingar,* 4th series, **3,** no. 16 (1876), 39 pp.; another early work on the subject was "Ueber die Berechnung der allgemeinen Jupiterstörung des Encke'schen Cometen," in *Vierteljahrsschrift der Astronomischen Gesellschaft* (Leipzig), **12** (1877), 313–323. His recalculation of the perturbations affecting Encke's comet between 1819 and 1891 was published in six parts in *Mémoires de l'Académie Impériale des Sciences de Saint-Pétersbourg* under the general title "Calculs et Recherches sur la Comète d'Encke": 7th series, **38,** no. 8 (1892); 7th series, **41,** nos. 3, 7 (1893); 7th series, **42,** nos. 7, 8 (1894); 8th series, **6,** no. 13 (1898).

Three papers by Backlund summarize his attempts to reconcile the observed motion of Encke's comet with theory: "Sur la Masse de la Planète Mercure et sur l'Accélération du Mouvement Moyen de la Comète d'Encke," in *Bulletin astronomique* (Paris), **11** (1894), 473–485; "Vergleichung der Theorie des Encke'schen Cometen mit den Beobachtungen 1894–95," in *Mémoires de l'Académie Impériale des Sciences de Saint-Pétersbourg,* 8th series, **16,** no. 3 (1904); and "Encke's Comet, 1895–1908," in *Monthly Notices of the Royal Astronomical Society* (London), **70** (1910), 429–442.

Approximately fifty other papers that Backlund wrote between 1874 and 1900 are listed in the Royal Society of London's *Catalogue of Scientific Papers,* IX (London, 1891), 93, and XIII (Cambridge, 1914), 228–229. After 1900 he wrote about thirty more, mainly for various publications of the Pulkovo Observatory and of the St. Petersburg Academy. His last paper, "On Chandler's Period in the Latitude Variation," appeared posthumously in *Monthly Notices of the Royal Astronomical Society* (London), **77** (1917), 2–6.

II. SECONDARY LITERATURE. Further details about Backlund's life can be found in an obituary notice, signed H. H. T. [Herbert Hall Taylor], in *Proceedings of the Royal Society* (London), **94A** (1918), xx–xxiv. The citation delivered by Hugh Frank Newall when Backlund received the Gold Medal of the Royal Astronomical Society provides an assessment of Backlund's work by one of his contemporaries; it appeared in *Monthly Notices of the Royal Astronomical Society* (London), **69** (1909), 324–331. A retrospective evaluation, pointing out the shortcomings of Backlund's methods, appeared in a paper by S. G. Makover and N. A. Bokhan, of the Leningrad Institute of Theoretical Astronomy, entitled "The Motion of Comet Encke-Backlund during 1898–1911 and a New Determination of the Mass of Mercury," in *Soviet Physics-Doklady,* **5** (1961), 923–925.

For other recent material on Encke's comet, see: K. Wurm, "Die Kometen," in *Handbuch der Physik,* Vol. 52, ed. S. Flügge (Berlin, 1959), p. 476; F. L. Whipple, "On the Structure of the Cometary Nucleus," chap. 19 of *The Moon, Meteorites and Comets,* ed. B. M. Middlehurst and G. P. Kuiper (Chicago, 1963), pp. 648 and 659; and B. G. Marsden, "Comets and Nongravitational Forces," in *Astronomical Journal,* **73** (1968), 367–379.

SALLY H. DIEKE

BACON, FRANCIS (*b.* London, England, 22 January 1561; *d.* London, 9 April 1626), *philosophy of science.*

Bacon was the son of Sir Nicholas Bacon, lord keeper of the great seal, and Ann, daughter of Sir Anthony Cooke. He was educated at Trinity College, Cambridge, from 1573 to 1575, when he entered Gray's Inn; he became a barrister in 1582. Bacon's life was spent in court circles, in politics, and in the law; in religion he adhered to the middle road of the Church of England, neither authoritarian nor sectarian. In 1606 he married Alice Barnham. He was knighted on the accession of James I in 1603, became lord chancellor in 1618, and was made viscount St. Albans in 1621. Bacon was dismissed from the chancellorship in 1621 after being convicted of bribery, a strain under which his health broke down. He then lived in retirement near St. Albans, devoting his remaining years to natural philosophy.

Bacon's writings in history, law, politics, and morals are extensive; but his place in the history of science rests chiefly upon his natural philosophy, his philosophy of scientific method, his projects for the practical organization of science, and the influence of all these upon the science of the later seventeenth century. During and immediately following his lifetime his principal publications in these areas were *The Advancement of Learning* (1605), expanded and latinized as *De augmentis scientiarum* (1623); *De sapientia veterum* (1609); *Novum organum* (1620); and *Sylva sylvarum* and *New Atlantis* (1627). Many of his shorter works, some of them fragmentary and published posthumously, are of equal scientific and philosophical interest.

Although Bacon was a contemporary of William Gilbert, Galileo, Johannes Kepler, and William Harvey, he was curiously isolated from the scientific developments with which they were associated. His knowledge of and contribution to the natural sciences were almost entirely literary; and, indeed, it has been shown that much of the empirical material collected in his "histories" is not the result of his own firsthand observation, but is taken directly from literary sources. Furthermore, most of Bacon's comments on both his scientific contemporaries and his philosophical predecessors are critical. For example, he never accepted the Copernican "hypothesis," attacking both Ptolemy and Copernicus for producing mere "calculations and predictions" instead of "philosophy . . . what is found in nature herself, and is actually and really true."[1] On similar grounds he attacked the theory of "perspective" as not providing a proper theory of the nature of light because it never went further than geometry. Mathematics was, he thought, to be used as a tool in natural philosophy, not as an end, and he had no pretensions to mathematical learning. He was not unsympathetic to Gilbert's magnetic philosophy, but he criticized him for leaping too quickly to a single unifying principle without due regard for experiment.

Bacon's closest associations with contemporary science were with atomism and with the Renaissance tradition of natural magic. His views on atomism underwent considerable change during the period of his philosophical writings, from a sympathetic discussion of Democritus in *De sapientia veterum* and *De principiis atque originibus*,[2] to outright rejection of "the doctrine of atoms, which implies the hypothesis of a vacuum and that of the unchangeableness of matter (both false assumptions)" in *Novum organum*.[3] There were both philosophical and scientific reasons for this change of mind. Even in his earlier works, Bacon posed the fundamental dilemma of atomism: either

the atom is endowed with some of the qualities that are familiar to sense, such as "matter, form, dimension, place, resistance, appetite . . .,"[4] in which case it is difficult to justify taking these qualities rather than any other sensible qualities as primary; or the atom is wholly different from bodies apprehended by the senses, in which case it is difficult to see how we come to know anything about them. On the other hand, empirical phenomena of cohesion and continuity are impossible to understand in terms of inert atoms alone; and the existence of spirituous substances, even in space void of air, seems to cast doubt upon the existence of the absolute void demanded by atomism.

In any case, Bacon was never an orthodox atomist, for as early as *De sapientia veterum* he insisted that the atom has active powers other than mere impenetrability—it has "desire," "appetite," and "force that constitutes and fashions all things out of matter."[5] In *Novum organum* these qualities are ascribed to bodies in general. All bodies have powers to produce change in themselves and in other bodies; they have "perceptions" that, although distinct from the "sensations" of animals, nevertheless enable them to respond to other bodies, as iron does in the neighborhood of a magnet. That virtues seem thus to emanate from bodies through space is an argument for suspecting that there may be incorporeal substances: "Everything tangible that we are acquainted with contains an invisible and intangible spirit, which it wraps and clothes as with a garment." It "gives them [bodies] shape, produces limbs, assimilates, digests, ejects, organises and the like." It "feeds upon" tangible parts and "turns them into spirit."[6]

Commentators have seen in this dualism of tangible, inert matter and active, intangible spirits a legacy of Renaissance animism, and have tended to apologize for it as being out of harmony with Bacon's other, more progressive views. Indeed, in seventeenth-century writings and later, Bacon was most often listed with the revivers of the Democritan philosophy, in company with those to whom his "active spirits" might be an embarrassment. It would be a mistake, however, to suppose that Bacon necessarily thought of his view on spirits as opposed to a mechanical theory of nature. There are many passages in which he objects to his predecessors' purely verbal ontologies of spirits and describes his own view as essentially unitary: "this spirit, whereof I am speaking, is not a virtue, nor an energy, nor an actuality, nor any such idle matter, but a body thin and indivisible, and yet having place and dimension, and real . . . a rarefied body, akin to air, though greatly differing from it."[7] Both in the description of heat as a species of motion in *Novum organum* and in the discussion of transmission of light and sound in *Sylva sylvarum,*[8] Bacon showed his sympathy with explanations in terms of mechanical analogues.

Bacon's natural philosophy is indecisive, and also a good deal more subtle than that of his corpuscularian successors. It is not surprising, therefore, that it did not lead to any detailed theoretical developments (as did, for example, Descartes's) and had, in fact, little direct influence. It can be argued, however, that, unlike Descartes, Bacon was not attempting to reach theoretical conclusions but, rather, to lay the necessary foundations for his inductive method. To that method we now turn.

Bacon's method is foreshadowed in the early *Valerius terminus* (1603), but was not developed until the last few years of his life. *De augmentis scientiarum* and *Novum organum* are the first and second parts of his projected *Great Instauration,* and the applications of the method that were to have constituted four further parts reached only a fragmentary stage in the *Histories,* most of which were published posthumously.

In *De augmentis scientiarum,* which is concerned primarily with the classification of philosophy and the sciences, Bacon develops his influential view of the relation between science and theology. He distinguishes in traditional fashion between knowledge by divine revelation and knowledge by the senses, and divides the latter into natural theology, natural philosophy, and the sciences of man. Natural philosophy is independent of theology; but in a sense its end is also knowledge of God, for it seeks the "footprints of the Creator imprinted on his creatures."[9] Indeed, Bacon sees both speculative and practical science as religious duties, the first for the understanding of creation and the second for the practice of charity to men. We should not read back into Bacon's separation of science and theology any implication that theology is depreciated or superseded by science. Such a view was hardly influential until the Enlightenment, whereas seventeenth-century natural philosophers generally followed Bacon in claiming a religious function for their investigations; this was undoubtedly one important factor in the public success of the scientific movement.

Having placed his project within the complete framework of knowledge in true Aristotelian fashion, Bacon proceeds to demolish all previous pretensions to natural philosophy. His aim is to lay the foundations of science entirely anew, neither leaping to unproved general principles in the manner of the ancient philosophers nor heaping up unrelated facts in the manner of the "empirics" (among whom he counts contemporary alchemists and natural magicians). "Histories," or collections of data, are to be drawn up systematically and used to raise an ordered system of axioms that will eventually embrace all the phenomena of nature. Many commentators, in the seventeenth century and later, have been misled, by the apparently unorganized collections of facts that fill Bacon's works, into supposing that his method was a merely empirical one, with no concern for theoretical interpretation. Such an impression is easily dispelled, however, by a closer reading of the text of *Novum organum.* We shall follow his account of the method in that work.

The first step in making true inductions is, as in a religious initiation, a purging of the intellect of the "idols" that, in man's natural fallen state, obstruct his unprejudiced understanding of the world. Bacon holds that we must consciously divest our minds of prejudices caused by excessive anthropomorphism (the "idols of the tribe"), by the particular interests of each individual (the "idols of the cave"), by the deceptions of words (the "idols of the market place"), and by received philosophical systems (the "idols of the theater").[10] Only in this way can the mind become a *tabula abrasa* on which true notions can be inscribed by nature itself. The consequences of the Fall for the intellect will then be erased, and man will be able to return to his God-given state of dominion over creation.

The aim of scientific investigation is to discover the "forms of simple natures." What Bacon means by a "form" is best gathered from his example concerning the form of heat (which is the only application of his method that he works out in any detail): "The Form of a thing is the very thing itself, and the thing differs from the form no otherwise than as the apparent differs from the real, or the external from the internal, or the thing in reference to man from the thing in reference to the universe." Hence, when the "form or true definition of heat" is defined as *"Heat is a motion, expansive, restrained, and acting in its strife upon the smaller parts of bodies,"* Bacon means "Heat itself, or the *quid ipsum* of Heat, is Motion and nothing else."[11] Thus, the form is not to be understood in a Platonic or Aristotelian sense but, rather, as what was later called an "explanation" or "reduction" of a secondary quality (heat) to a function of primary bodies and qualities (matter in motion). In order to discover what primary qualities are relevant to the form, Bacon prescribes his Tables of Presence, Absence, and Comparison: "[the form] is always present when the nature is present. . . . absent when the nature is absent" and "always decrease[s] when the nature in question decreases, and . . . always increase[s] when the nature in question increases." [12]

Therefore, we are to draw up a Table of Instances that all agree in the simple nature, heat—such as rays of the sun, flame, and boiling liquids—and then to look for other natures that are copresent with heat and therefore are candidates for its form. To ensure that as many irrelevant natures as possible are eliminated at this stage, these instances should be as unlike each other as possible except in the nature of heat. Second, a Table of Absence should be drawn up, in which as far as possible each instance in the Table of Presence should be matched by an instance similar to it in all respects *except* heat, such as rays of the moon and stars, phosphorescence, and cool liquids. This is the method of *exclusion* by negative instances, which will at once test a putative form drawn from the Table of Presence; if it is not the true form, it will not be absent in otherwise similar instances when heat is absent. The tables are the precursors of Mill's "Joint Method of Agreement and Difference," and clearly are more adequate than the method of induction by simple enumeration of positive instances, with which Bacon has so often been wrongly identified. Construction of the tables demands not a passive observation of nature, but an active search for appropriate instances; and it therefore encourages artificial experiment. Nature, Bacon says, must be "put to the question." [13]

Inference of the form from the tables is, however, only the beginning of the method. Bacon speaks often of raising a "ladder of axioms" by means of the forms, until we have constructed the complete system of natural philosophy that unifies all forms and natures. The conception seems to be something like an Aristotelian classification into genus, species, and differentia, in which every nature has its place. It also has some affinity with the later conception of a theoretical structure that yields observation statements by successive deductions from theoretical premises. But it would be misleading to press these parallels too closely, for the essence of Bacon's ascent to the axioms is that it is the result of a number of inductive inferences whose conclusions are infallible if they have been properly drawn from properly contrived Tables of Instances. The axioms are emphatically not the result of a leap to postulated premises from which observations may be deduced, for this is not an infallible method and gives no guarantee that the axioms arrived at are unique, let alone true. This deductive method is, in fact, what Bacon calls the method of "anticipation of nature," which, he thinks, may be useful in designing appropriate Tables of Instances, but is to be avoided in inductive inference proper.

Bacon is not unaware that the infallibility of his method depends crucially on there being only a finite number of simple natures and on our ability to enumerate all those present in any given instance. His faith that nature is indeed finite in the required respects seems to rest upon his natural philosophy. Although he rejected atomism, he retained the belief that the primary qualities are few in number and regarded the inductive method as the means to discover which qualities they are. Forms are the "alphabet of nature" [14] that suffice to produce the great variety of nature from a small stock of primary qualities, just as the letters of the alphabet can generate a vast literature. The whole investigation is further complicated, as Bacon also sees, by the fact that some natures are "hidden" and cannot be taken account of in the tables unless we employ "aids to the senses" to bring them within reach of sensation. Much of the later part of *Novum organum* is taken up with this problem, which leads Bacon to commend not only instruments such as the telescope, but also "fit and apposite experiments" that bring hidden and subtle processes to light. [15]

Complementary to Bacon's ascent to axioms is his insistence on subsequent descent to works. The aim is not merely passive understanding of nature, but also practical application of that understanding to the improvement of man's condition; Bacon holds that each of these aspects of his method is sterile without the other. Furthermore, he claims to have given in his method a means whereby anyone who follows the rules can do science—he has "levelled men's wits." [16] Thus, with proper organization and financial support, it should be possible to complete the edifice of science in a few years and to gather all the practical fruit that it promised for the good of men. Such a vision inspired Bacon as early as 1592, when he described in a letter to Cecil his "vast contemplative ends . . . I hope I should bring in industrious observations, grounded conclusions, and profitable inventions and discoveries." [17] Throughout his life he used his status and influence in a succession of frustrated attempts to obtain the Crown's support for this enterprise. In 1605 it was advertised in *The Advancement of Learning*—the only work Bacon ever published in English. His unfinished account of the ideal scientific society was published posthumously in *New Atlantis,* which ranks among the best-known and most delightful Utopian writings in the world and has been perhaps the most influential.

New Atlantis contains a description of the island of Bensalem, on which there is a cooperative college of science called Salomon's House. Bacon's account of it begins with a concise expression of his whole vision of science: "The End of our Foundation is the knowledge of Causes, and secret motions of things; and the enlarging of the bounds of Human Empire, to the effecting of all things possible."[18] The house is essentially a religious community, having "certain hymns and services, which we say daily, of laud and thanks to God."[19] It contains all kinds of laboratories and instruments for the pursuit of science, and is organized on the principle of a division of labor among those who perform experiments and collect information from various sources; those who determine the significance of the information and experiments, and direct and perform new and more penetrating experiments; and the "Interpreters," who "raise the former discoveries by experiments into greater observations, axioms, and aphorisms."[20] It is noticeable that while there are said to be thirty-three men assigned to the experimental parts of this task, only three are assigned to interpretation—a proportion that seems to reflect neither the status Bacon gives to the raising of axioms in his explicit accounts of method, nor the ease with which he thought this part of the task would be completed. Unfortunately, however, it does reflect the way Bacon's ideas were subsequently understood.

Bacon's immense prestige and influence in later seventeenth-century science does not rest upon positive achievements in either experiment or theory but, rather, upon his vision of science expressed in *Novum organum* and *New Atlantis,* and in particular upon his fundamental optimism about the possibilities for its rapid development. Now that the true method had been described, he thought all that was required was the purgation of the intellect to make a fit instrument for the method, and the human and financial resources to carry it out. When patronage and manpower for the organization of science were eventually forthcoming in the form of the Royal Society, its *Philosophical Transactions* was soon full of just the sort of "histories" Bacon had prescribed. His program for the raising of axioms, however, was taken less seriously than his strictures against "anticipations" and hypotheses, so that the weight of his influence was toward empiricism rather than toward theoretical system-building. At the time this did provide a useful corrective to Cartesianism, as can even be seen in Newton's insistence on the inductive "ascent" to the law of gravitation, in contrast with the merely imagined hypotheses of Descartes. But although most leading members of the Royal Society took every opportunity to proclaim themselves Bacon's loyal disciples, they tacitly adopted a more tolerant attitude toward hypotheses than his; and subsequent theoretical developments took place in spite of, rather than

as examples of, his elaboration of method. His successors in this area should be sought among the inductive logicians, beginning with Hume and Mill, and not among the scientists.

NOTES

1. *Descriptio globi intellectus* (1612), in *Works,* III, 734; V, 511.
2. Probably written before 1620, published 1653.
3. *Novum organum,* in *Works,* I, 234; IV, 126.
4. *De principiis atque originibus,* in *Works,* III, 111; V, 492.
5. *De sapientia veterum,* in *Works,* VI, 655, 729.
6. *Novum organum,* I, 310; IV, 195.
7. *Historia vitae et mortis* (1623), in *Works,* II, 213; V, 321.
8. *Sylva sylvarum,* in *Works,* II, 429 ff.
9. *De augmentis scientiarum,* in *Works,* I, 544; IV, 341. Also *Novum organum,* I, 145; IV, 33.
10. *Novum organum,* I, 169; IV, 58 ff.
11. *Ibid.,* I, 248, 262, 266; IV, 137, 150, 154.
12. *Ibid.,* I, 230, 248; IV, 121, 137.
13. *Ibid.,* I, 403; IV, 263.
14. *Abecedarium naturae,* in *Works,* II, 85; V, 208.
15. *Novum organum,* I, 168; IV, 58.
16. *Ibid.,* I, 217; IV, 109.
17. *Letters and Life,* I, 109.
18. *New Atlantis,* in *Works,* III, 156.
19. *Ibid.,* III, 166.
20. *Ibid.,* III, 165.

BIBLIOGRAPHY

I. ORIGINAL WORKS. The standard edition of Bacon's works is *The Works of Francis Bacon,* J. Spedding, R. L. Ellis, and D. D. Heath, eds., 7 vols. (London, 1857–1859), which contains valuable prefaces and notes. The philosophical and scientific works are in Vols. I–III and VI, with English translations in Vols. IV and V. Also of value is *The Letters and Life of Francis Bacon, Including All His Occasional Works,* J. Spedding, ed., 7 vols. (London, 1861–1874). All page references in the Notes are to these editions. I have modified some of the translations. For further original works and secondary literature, see R. Gibson, *Francis Bacon: A Bibliography of His Works and of Baconiana to the Year 1750* (Oxford, 1950). A useful edition of an individual work with notes, introduction, and bibliography is *Novum organum,* Thomas Fowler, ed. (2nd ed., Oxford, 1889).

II. SECONDARY LITERATURE. There are many biographies, and their quality varies greatly. The most valuable recent examples are F. H. Anderson, *Francis Bacon, His Career and His Thought* (Los Angeles, 1962); J. G. Crowther, *Francis Bacon* (London, 1960); and B. Farrington, *Francis Bacon; Philosopher of Industrial Science* (London, 1961). The last two interpret Bacon mainly as a "philosopher of works."

The literature on Bacon's philosophy and science is enormous, and there is no attempt at completeness here. The recent books and articles listed give references to further material whose absence from this list does not imply any value judgment: B. Farrington, *The Philosophy of*

Francis Bacon, an Essay on Its Development From 1603 to 1609 With New Translations of Fundamental Texts (Liverpool, 1964)—neither the commentary nor the newly translated texts throw much additional light on Bacon's scientific ideas during this period; Kuno Fischer, *Francis Bacon of Verulam, Realistic Philosophy and Its Age,* John Oxenford, trans. (London, 1857), which is still a useful analysis of Bacon's method but, like most nineteenth-century works (except those of Ellis and Spedding), underestimates the significance of Bacon's doctrine of spirits; Thomas Fowler, *Bacon* (London–New York, 1881), which deals with Bacon's philosophy and scientific method and their influence; W. Frost, *Bacon und die Naturphilosophie* (Munich, 1927); Mary B. Hesse, "Francis Bacon," in D. J. O'Connor, ed., *A Critical History of Western Philosophy* (New York, 1964), pp. 141–152; C. W. Lemmi, *The Classic Deities in Bacon* (Baltimore, 1933), not always accurate on Bacon's science; A. Levi, *Il pensiero di F. Bacone considerato in relazione con le filosofie della natura de Rinascimento e col razionalismo cartesiano* (Turin, 1925), one of the first detailed accounts of Bacon's relation to his immediate predecessors; G. H. Nadel, "History as Psychology in Francis Bacon's Theory of History," in *History and Theory,* 5 (1966), 275–287; M. Primack, "Outline of a Reinterpretation of Francis Bacon's Philosophy," in *Journal of the History of Philosophy,* 5 (1967), 123–132; P. Rossi, *Francesco Bacone, dalla magia alla scienza* (Bari, 1957; English trans. by S. Rabinovitch, London, 1968), which interprets Bacon's science as being heavily indebted to the natural magic tradition and his logic as indebted to Ramist rhetoric; P. M. Schuhl, *La pensée de lord Bacon* (Paris, 1949); and K. R. Wallace, *Francis Bacon on Communication and Rhetoric* (Chapel Hill, N.C., 1943).

Recent general works with extensive references to Bacon are R. M. Blake, C. J. Ducasse, and E. H. Madden, *Theories of Scientific Method: The Renaissance Through the Nineteenth Century* (Seattle, Wash., 1960), ch. 3; C. Hill, *Intellectual Origins of the English Revolution* (Oxford, 1965), esp. ch. 3; R. H. Kargon, *Atomism in England From Hariot to Newton* (Oxford, 1966), which contains a good bibliography of recent scholarly articles on Bacon, pp. 150–153; R. McRae, *The Problem of the Unity of the Sciences: Bacon to Kant* (Toronto, 1961); and Margery Purver, *The Royal Society: Concept and Creation* (Cambridge, Mass., 1967), esp. ch. 2.

MARY HESSE

BACON, ROGER (*b.* England, *ca.* 1219; *d. ca.* 1292), *natural philosophy, optics, calendar reform.*

Apart from some brief references in various chronicles, the only materials for Roger Bacon's biography are his own writings. The date 1214 for his birth was calculated by Charles, followed by Little, from his statements in the *Opus tertium* (1267) that it was forty years since he had learned the alphabet and that for all but two of these he had been "in studio."[1] Taking this to refer to the years since he entered the

university—the usual age was then about thirteen—they concluded that in 1267 Bacon was fifty-three and thus was born in 1214. But Crowley has argued that his statements more probably refer to his earliest education, beginning about the age of seven or eight, which would place his birth about 1219 or 1220. Of his family the only good evidence comes again from Bacon himself. He wrote in the *Opus tertium* that they had been impoverished as a result of their support of Henry III against the baronial party, and therefore could not respond to his appeal for funds for his work in 1266.[2]

After early instruction in Latin classics, among which the works of Seneca and Cicero left a deep impression, Bacon seems to have acquired an interest in natural philosophy and mathematics at Oxford, where lectures were given from the first decade of the thirteenth century on the "new" logic (especially *Sophistici Elenchi* and *Posterior Analytics*) and *libri naturales* of Aristotle as well as on the mathematical *quadrivium.* He took his M.A. either at Oxford or at Paris, probably about 1240. Probably between 1241 and 1246 he lectured in the Faculty of Arts at Paris on various parts of the Aristotelian corpus, including the *Physics* and *Metaphysics,* and the pseudo-Aristotelian *De vegetabilibus* (or *De plantis*) and the *De causis,* coincident with the Aristotelian revival there. In arguing later, in his *Compendium studii philosophie,* for the necessity of knowledge of languages,[3] he was to use an incident in which his Spanish students laughed at him for mistaking a Spanish word for an Arabic word while he was lecturing on *De vegetabilibus.* He was in Paris at the same time as Albertus Magnus, Alexander of Hales (*d.* 1245),[4] and William of Auvergne (*d.* 1249).[5]

The radical intellectual change following Bacon's introduction to Robert Grosseteste (*ca.* 1168–1253) and his friend Adam Marsh on his return to Oxford about 1247 is indicated by a famous passage in the *Opus tertium:*

> For, during the twenty years in which I have laboured specially in the study of wisdom, after disregarding the common way of thinking [*neglecto sensu vulgi*], I have put down more than two thousand pounds for secret books and various experiments [*experientie*], and languages and instruments and tables and other things; as well as for searching out the friendships of the wise, and for instructing assistants in languages, in figures, in numbers, and tables and instruments and many other things.[6]

Grosseteste's influence is evident in Bacon's particular borrowings, especially in his optical writings, but above all in the devotion of the rest of his life to the promotion of languages and of mathematics, optics (*perspectiva*), and *scientia experimentalis* as the essential sciences.

He was in Paris again in 1251, where he says in the *Opus maius*[7] that he saw the leader of the Pastoreaux rebels. This story and some later works place him there for long periods as a Franciscan. He entered the Franciscan order about 1257 and, soon afterward, he also entered a period of distrust and suspicion—probably arising from the decree of the chapter of Narbonne, presided over by Bonaventure as master general in 1260, which prohibited the publication of works outside the order without prior approval. Bonaventure had no time for studies not directly related to theology, and on two important questions, astrology and alchemy, he was diametrically opposed to Bacon. He held that only things dependent solely on the motions of the heavenly bodies, such as eclipses of the sun and moon and sometimes the weather, could be foretold with certainty. Bacon agreed with the accepted view that predictions of human affairs could establish neither certainty nor necessity over the free actions of individuals, but he held that nevertheless astrology could throw light on the future by discovering general tendencies in the influence of the stars, acting through the body, on human dispositions, as well as on nature at large. In alchemy Bonaventure was also skeptical about converting base metals into gold and silver, which Bacon thought possible.

Whatever the particular reasons for Bacon's troubles within the order, he felt it necessary to make certain proposals to a clerk attached to Cardinal Guy de Foulques; as a result, the cardinal, soon to be elected Pope Clement IV (February 1265), asked him for a copy of his philosophical writings. The request was repeated in the form of a papal mandate of 22 June 1266.[8] Bacon eventually replied with his three famous works, *Opus maius, Opus minus,* and *Opus tertium,* the last two prefaced with explanatory *epistole* in which he set out his proposals for the reform of learning and the welfare of the Church. It is reasonable to suppose that after twenty years of preparation he composed these *scripture preambule* to an unwritten *Scriptum principale* between the receipt of the papal mandate and the end of 1267. In that year he sent to the pope, by his pupil John, the *Opus maius* with some supplements, including *De speciebus et virtutibus agentium* in two versions[9] and *De scientia perspectiva,*[10] followed (before the pope died in November 1268) by the *Opus minus* and *Opus tertium* as résumés, corrections, and additions to it. The pope left no recorded opinion of Bacon's proposals.

Perhaps at this time Bacon wrote his *Communia*

naturalium and *Communia mathematica,* mature expressions of many of his theories. These were followed in 1271 or 1272 by the *Compendium studii philosophie,* of which only the first part on languages remains and in which he abused all classes of society, and particularly the Franciscan and Dominican orders for their educational practices. Sometime between 1277 and 1279 he was condemned and imprisoned in Paris by his order for an undetermined period and for obscure reasons possibly related to the censure, which included heretical Averroist propositions, by the bishop of Paris, Stephen Tempier, in 1277. The last known date in his troubled life is 1292, when he wrote the *Compendium studii theologii.*[11]

Scientific Thought. The *Opus maius* and accompanying works sent to the pope by Bacon as a *persuasio* contain the essence of his conception of natural philosophy and consequential proposals for educational reform. He identified four chief obstacles to the grasping of truth: frail and unsuitable authority, long custom, uninstructed popular opinion, and the concealment of one's own ignorance in a display of apparent wisdom. There was only one wisdom, given to us by the authority of the Holy Scriptures; but this, as he explained in an interesting history of philosophy, had to be developed by reason, and reason on its part was insecure if not confirmed by experience. There were two kinds of experience, one obtained through interior mystical inspiration and the other through the exterior senses, aided by instruments and made precise by mathematics.[12] Natural science would lead through knowledge of the nature and properties of things to knowledge of their Creator, the whole of knowledge forming a unity in the service and under the guidance of theology. The necessary sciences for this program were languages, mathematics, optics, *scientia experimentalis,* and alchemy, followed by metaphysics and moral philosophy.

Bacon leaves no doubt that he regarded himself as having struck a highly personal attitude to most of the intellectual matters with which he dealt, but his writings are not as unusual as the legends growing about him might suggest. They have, on the whole, the virtues rather than the vices of Scholasticism, which at its best involved the sifting of evidence and the balancing of authority against authority. Bacon was conscious of the dangers of reliance on authority: Rashdall draws attention to the irony of his argument against authority consisting chiefly of a series of citations. Most of the content of his writings was derived from Latin translations of Greek and Arabic authors. He insisted on the need for accurate translations. When it was that he learned Greek himself is not certain, but his Greek grammar may be placed after 1267, since in it he corrected a philological mistake in the *Opus tertium*. He also wrote a Hebrew grammar to help in the understanding of Scripture.

One of the most interesting and attractive aspects of Bacon is his awareness of the small place of Christendom in a world largely occupied by un-believers, "and there is no one to show them the truth." [13] He recommended that Christians study and distinguish different beliefs and try to discover common ground in monotheism with Judaism and Islam, and he insisted that the truth must be shown not by force but by argument and example. The resistance of conquered peoples to forcible conver-sion, such as practiced by the Teutonic knights, was "against violation, not to the arguments of a better sect." [14] Hence the need to understand philosophy not only in itself but "considering how it is useful to the Church of God and is useful and necessary for direct-ing the republic of the faithful, and how far it is effective for the conversion of infidels; and how those who cannot be converted may be kept in check no less by the works of wisdom than the labour of war." [15] Science would strengthen the defenses of Christendom both against the external threat of Islam and the Tartars and against the methods of "fascina-tion" that he believed had been used in the Children's Crusade and the revolt of the Pastoreaux, and would be used by the Antichrist.

Bacon's mathematics included, on the one hand, astronomy and astrology (discussed later) and, on the other, a geometrical theory of physical causation related to his optics. His assertions that "in the things of the world, as regards their efficient and generating causes, nothing can be known without the power of geometry" and that "it is necessary to verify the matter of the world by demonstrations set forth in geometrical lines" [16] came straight from Grosseteste's theory of *multiplicatio specierum*, or propagation of power (of which light and heat were examples), and his account of the "common corporeity" that gave form and dimensions to all material substances. "Every multiplication is either according to lines, or angles, or figures." [17] This theory provided the effi-cient cause of every occurrence in the universe, in the celestial and terrestrial regions, in matter and the senses, and in animate and inanimate things. In thus trying to reduce different phenomena to the same terms, Grosseteste and Bacon showed a sound physi-cal insight even though their technical performance remained for the most part weak. These conceptions made optics the fundamental physical science, and it is in his treatment of this subject that Bacon appears most effective. Besides Grosseteste his main optical sources were Euclid, Ptolemy, al-Kindī, and Ibn

al-Haytham (Alhazen). He followed Grosseteste in emphasizing the use of lenses not only for burning but for magnification, to aid natural vision. He seems to have made an original advance by giving constructions, based on those of Ptolemy for plane surfaces and of Ibn al-Haytham for convex refracting surfaces, providing eight rules (*canones*) classifying the properties of convex and concave spherical sur-faces with the eye in various relationships to the refracting media. He wrote:

> If a man looks at letters and other minute objects through the medium of a crystal or of glass or of some other transparent body placed upon the letters, and this is the smaller part of a sphere whose convexity is towards the eye, and the eye is in the air, he will see the letters much better and they will appear larger to him. For in accordance with the truth of the fifth rule [Fig. 1] about a spherical medium beneath which is the object or on this side of its centre, and whose convexity is towards the eye, everything agrees towards magnifica-tion [*ad magnitudinem*], because the angle is larger under which it is seen, and the image is larger, and the position of the image is nearer, because the object is between the eye and the centre. And therefore this instrument is useful for the aged and for those with weak eyes. For they can see a letter, no matter how small, at sufficient magnitude. [18]

According to the fifth rule, [19] if the rays leaving the object, *AB,* and refracted at the convex surface of the lens meet at the eye, *E,* placed at their focus, a magnified image, *MN,* will be seen at the intersections of the diameters passing from the center of curvature, *C,* through *AB* to this surface and the projections of the rays entering the eye. As he did not seem to envisage the use of combinations of lenses, Bacon got no further than Grosseteste in speculating about magnifications such that "from an incredible distance we may read the minutest letters and may number the particles of dust and sand, because of the magni-tude of the angle under which we may see them." [20]

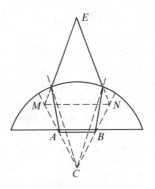

FIGURE 1

But he did make an important contribution to the history of physiological optics in the West by his exposition of Ibn al-Haytham's account of the eye as an image-forming device, basing his ocular anatomy on Ḥunayn ibn Isḥāq and Ibn Sīnā. In doing so, he seems to have introduced a new concept of laws of nature (a term found in Lucretius and numerous other authors more widely read, such as St. Basil) by his reference to the "laws of reflection and refraction" as *leges communes nature.*[21] His meaning is clarified by his discussion elsewhere of a *lex nature universalis*[22] requiring the continuity of bodies and thus giving a positive explanation, in place of the negative *horror vacui,* which he rejected, of such phenomena as water remaining in a clepsydra so long as its upper opening remained closed—an explanation comparable to one found in Adelard of Bath's *Natural Questions.* Universal nature constituted from these common laws, including those *de multiplicatione specierum,* was superimposed on the system of particular natures making up the Aristotelian universe—not yet the seventeenth-century concept but perhaps a step toward it.

"Having laid down the roots of the wisdom of the Latins as regards languages and mathematics and perspective," Bacon began Part VI of the *Opus maius,* "I wish now to unfold the roots on the part of *scientia experimentalis,* because without experience [*experientia*] nothing can be known sufficiently."[23] This science, "wholly unknown to the general run of students," had "three great prerogatives with respect to the other sciences."[24] The first was to certify the conclusions of deductive reasoning in existing speculative sciences, including mathematics. As an example he gave an investigation of the shape and colors of the rainbow involving both theoretical reasoning and the collection of instances of related phenomena in order to discover their common cause. The second prerogative was to add to existing sciences new knowledge that they could not discover by deduction. Examples were the discovery of the properties of the magnet, the prolonging of human life by observing what plants produced this effect naturally in animals, and the purification of gold beyond the present achievements of alchemy. The third prerogative was to investigate the secrets of nature outside the bounds of existing sciences, opening up knowledge of the past and future and the possibility of marvelous inventions, such as ever-burning lamps and explosive powders.

It is clear that Bacon's *scientia experimentalis* was not exactly what this term might now suggest, but belonged equally to "natural magic" aimed at producing astonishing as well as practically useful effects by harnessing the hidden powers of nature. His approach had been profoundly influenced by the pseudo-Aristotelian *Secretum secretorum,* of which he had produced an annotated edition variously dated between 1243 and sometime before 1257, but he also insisted that his new science would expose the frauds of magicians by revealing the natural causes of effects. The "dominus experimentorum" of the *Opus tertium,*[25] who may have been Pierre de Maricourt, the pioneer investigator of magnetism, is praised for understanding all these essential characteristics. In the *Opus minus,*[26] Bacon described possibly original experiments of his own with a lodestone held above and below a floating magnet, and argued that it was not the Nautical (Pole) Star that caused its orientation, or simply the north part of the heavens, but all four parts equally. It was in this work, and in the *Opus tertium,*[27] that he inserted his main discussion of alchemy, including the conversion of base metals into gold and silver. There is a further discussion in the *Communia naturalium,*[28] together with sketches of the sciences of medicine and agriculture. In the *Communia mathematica*[29] and the *Epistola de secretis operibus artis et naturae et de nullitate magiae,*[30] he described more wonderful machines for flying, lifting weights, and driving carriages, ships, and submarines, and so on, which he believed had been made in antiquity and could be made again.

Despite his occasional references to them, Bacon in his accredited writings deals with neither instruments nor mathematical tables in any but a superficial way. For this reason it is hard to measure his stature by comparison with that of his contemporaries whom we should call astronomers and mathematicians. We are not encouraged to set great store by the stories that while in Paris he constructed astronomical tables and supplied the new masters with geometrical problems that none of their audiences could solve.[31] His mathematics and astronomy were in fact almost wholly derivative, and he was not always a good judge of competence, preferring, for instance, al-Biṭrūjī to Ptolemy.

Bacon is often held to have achieved a deep and novel insight in regard to the role of mathematics in science, an insight that to the modern mind is almost platitudinous. In this connection it is easy to forget the large numbers of astronomers of antiquity and the Middle Ages for whom mathematics was an essential part of the science, and the smaller numbers of natural philosophers who had made use of simpler mathematical techniques than those of astronomy. It is more to the point to notice that Bacon argued for the usefulness of mathematics in almost every realm of academic activity. Part IV of the *Opus maius* is devoted to the usefulness of mathematics (1) in human affairs (this section was published separately as the *Specula mathematica*); (2) in divine affairs, such as chronology, the fixing of feasts, natural phenomena, arithmetic, and music; (3) in ecclesiastical affairs, such as the certification of faith and the emendation of the calendar; and (4) in affairs of state, under which heading are included geography and astrology. When Bacon sang the praises of mathematics, "the first of the sciences," "the door and key of the sciences," "the alphabet of philosophy," it has to be remembered that he used the word in an unusually wide sense. Bacon seemed to fear that mathematics would be dismissed as one of the blacker arts, as when arithmetic was applied to geomancy. He sought "per vias mathematice verificare omnia que in naturalibus scientias sunt necessaria"; and yet in the last resort, experience was still necessary, and in a sense supreme.[32]

So loud and long were Bacon's praises of mathematics that it is hard to avoid the conclusion that his love of the subject was unrequited. He could compose his *De communibus mathematice* and mention, in geometry, nothing beyond definitions, axioms, and methods. Apart from mathematically trivial results in such practical contexts as engineering, optics, astronomy, and the like, his works apparently contain not a single proof, not a single theorem; and we must take on trust the story of the difficult problems he devised for the young Paris masters. As for his analytical skills and his views on the citation of authority, rather than try to resolve the geometrical paradox of the doctrine of atomism—that it can make the hypotenuse and side of a square commensurable—he preferred simply to dismiss it as being contrary to Euclid.

The standard discussion of ratios in Euclid, Book V, did not include a numerical treatment of the subject, for which the standard medieval authority was the *Arithmetica* of Boethius. There the different species of ratio are tediously listed and subdivided, and the absence of a similar logical division of ratio in Euclid was complained of by Bacon in *Communia mathematica*.[33] He was not to carry out the program at which he might seem to have hinted, and not until Bradwardine's *Geometria speculativa* did the Schoolmen make any progress toward a numerical description of irrational ratios, except perhaps in some halting attempts to elucidate Proposition III of Archimedes' *De mensura circuli*.

As for the relation of logic to mathematics, Bacon inverted, in a sense, the logistic thesis of our own century: without mathematics, for instance, the categories were unintelligible.[34] Mathematics alone gave

absolute certainty. Bacon was unusual in that he generally named his sources, citing such authors as Theodosius, Euclid, Ptolemy, Boethius, al-Fārābī, and—among modern writers—Jordanus de Nemore (*De triangulis* and *Arithmetica*) and Adelard. Despite his criticism of Jordanus, by any reckoning a better mathematician than Bacon, he had praise for "the only two perfect mathematicians" (of his time), John of London and Pierre de Maricourt. He also condescended to praise Campanus of Novara and a "Master Nicholas," teacher of Amauri, son of Simon de Montfort. In the last analysis, almost everything Bacon wrote under the title of mathematics is best regarded as being at a metaphysical level. His view that in mathematics we have perfect demonstration reinforced his theory of natural action. His philosophy of science, however, was inherently empiricist: rational argument may cause us to dismiss a question, but it neither gives us proof nor removes doubt.

It was held in the *Opus maius* that a more accurate knowledge of the latitudes and longitudes of places was needed for (1) knowledge of mankind and the natural world; (2) facilitation of the spiritual government of the world—missionaries, for example, would be saved from danger and from much wasted labor; (3) knowledge of the whereabouts of the ten tribes and even of the Antichrist. His geography was nevertheless a compilation of works on descriptive geography (in which he gave, as it were, an extended verbal map of the world) by such writers as Ptolemy and al-Farghānī, supplemented by the reports of Franciscan travelers, especially to the East.

In the *Opus maius*[35] he stated the possibility of voyaging from Spain to India. The passage was inserted, without reference to its source, in the *Imago mundi*[36] of Cardinal Pierre d'Ailly (d. 1420). Humboldt argued that this passage, quoted by Columbus in a letter of 1498 to Ferdinand and Isabella, was more important in the discovery of America than the Toscanelli letters. Thorndike suggests that Columbus probably did not read the vital work until his return from the first voyage of 1492.[37] It is immaterial, as Thorndike points out, whether Bacon was merely optimistically citing Aristotle, Seneca, Nero, and Pliny on the distance of Spain from India. In fact Bacon argued as cogently from such longitudes and latitudes as were available in the Toledan tables as he did from classical authors.

For the radius of the earth Bacon took a figure of 3,245 miles (al-Farghānī). He stated that the earth's surface was less than three-quarters water. In both cases he selected good figures from a great many authoritative but bad ones. It is clear, nevertheless, from his repetition of the method of determining the size of the earth—a method he took from al-Farghānī—that he had no appreciation whatsoever of the practical difficulties it involved.

Bacon appears to have sent a map to the pope with his *Opus maius*. Although it is now lost, from the description he gave it appears to have included the better-known towns of the world plotted by their latitudes and longitudes as found in many contemporaneous lists.[38] We have no knowledge of the projection adopted, but the description is compatible with the use of a rectangular coordinate system.

Bacon used the words "astronomia" and "astrologia" in a typically ambiguous manner, but there is no doubt that he believed in the reasonableness of what we would call astrology. In the *Opus tertium* he spoke of astrology as the most important part of mathematics, dividing it into a speculative, or theoretical, part, presumably of the sort included in Sacrobosco's *Sphere,* and a practical part, "que dicitur astronomia,"[39] concerned with the design of instruments and tables.[40] A remark in the *Opus maius,*[41] written in 1267, confirms a similar remark made four years later by Robertus Anglicus,[42] to the effect that conscious efforts were being made to drive what amounts to a clock (in Bacon's example the spherical astrolabe was to be driven) at a constant rate. This seems to confirm approximately the *terminus ante quem non* previously determined for the mechanical clock.

On many occasions Bacon emphasized at length that the two sorts of "astrology" were essential if man was to learn of the celestial influences on which terrestrial happenings depended. By reference to Ptolemy, Haly, Ibn Sīnā, Abū Maʿshar, Messahala, and others, he showed that the best astrologers had not held that the influence of the stars subjugated the human will, and that the Fathers who objected to astrology on these grounds had never denied that astrology could throw light on future events. It was possible to predict human behavior statistically but not with certainty in individual cases. Astrology might strengthen faith in the stability of the Church and foretell the fall of Islam and the coming of the Antichrist; and all these things "ut auctores docent et experiencia certificat."[43] On occasion he likened astrological influence to the influence of a magnet over iron.

In his main works Bacon did not discuss the technicalities of astronomy or astrology, but in both of the works ascribed to him with the title *De diebus creticis*[44] the standard medical astrology of the time is rehearsed. These works are not merely compilations of older authorities. Although technically they are in no sense new, they have a rational cast and even include the testimony of medical men of the time.

The first of these two works is interesting because it incorporates the whole of the *De impressione aeris* attributed to Grosseteste and printed among his works by Baur. Little[45] suggests that Grosseteste (*d.* 1253) collaborated with Bacon. Internal evidence suggests a date of composition of about 1249. Some planetary positions quoted for that year are sufficiently inaccurate to suggest that the work was written before 1249 rather than after, and that the author was by no means as skilled as the best astronomers of the time.

The *Speculum astronomie,* of doubtful authorship (see below), is inconsistent with certain of Bacon's accredited writings. It is essentially a criticism of Stephen Tempier's decree of 1277 attacking 219 errors, several involving a belief in astrology. As already seen, Bacon's prison sentence was probably related to the bishop's decrees.

Bacon's astronomical influence was slight in all respects, although through Paul of Middelburg he is said to have influenced Copernicus.[46] His writings on the calendar were frequently cited.[47] Theologians treated the calendar with a respect it did not deserve, regarding it as a product of astronomy, while astronomers would have treated it with more disdain had they been detached enough to perceive it in a historical context. Here Bacon's skepticism was useful, and whatever the depth of his astronomical knowledge, he wrote on calendar reform with as much insight as anyone before Regiomontanus—Nicholas of Cusa notwithstanding. In discussing the errors of the Julian calendar, he asserted that the length of the Julian year (365 $\frac{1}{4}$ days) was in excess of the truth by about one day in 130 years, later changing this to one day in 125 years. The length of the (tropical) year implied was better than Ptolemy's, and indeed better than that accepted in the Alphonsine tables compiled a few years after the *Opus maius.* (The correct figure for Bacon's time was one day in a little over 129 years.) The Alphonsine tables imply that the Julian error is one day in about 134 years. There is no reason whatsoever to suppose, as many have done following Augustus De Morgan, that Bacon's data were his own. Thābit ibn Qurra made the length of the year shorter than the Julian year by almost exactly one day in 130 years, and according to a curious passage in the *Communia naturalium,* Thābit was "maximus Christianorum astronomus." In the *Computus,* however, Thābit is grouped with al-Battānī and others who are said to have argued for one day in 106 years, while Asophus (ʿAbd al-Raḥmān ibn ʿUmar al-Sūfī) appears to have been the most probable source of influence, with his one day in 131 years.[48]

As a means of reforming the calendar, Bacon seems finally to have recommended the removal of one day in 125 years (cf. the Gregorian method of ignoring three leap years in four centuries), and in connection with Easter, since the nineteen-year cycle is in error, the astronomical calculation of the feast; otherwise a lunisolar year like that of the eastern nations should be adopted. (Grosseteste had previously made this proposal.) He tempered this rash suggestion with the pious qualification that if an astronomical calculation of Easter was to be adopted, Hebrew astronomical tables should be used. His proposals may be compared with the much less radical ones of Nicholas of Cusa, who in his *Reparatio calendarii* (pre 1437?) merely suggested a temporary patching up of the calendar, eliminating a number of days to alter the equinox suitably (Gregorian reform, supervised by Clavius, took the same superfluous step) and changing the "golden number" so as to make the ecclesiastical moon correspond for a time with reality. These solutions were inferior to Bacon's, including fewer safeguards against a future state of affairs in which Church usage and the ordinances of the Fathers might differ appreciably. It is worth noting that Stöffler proposed to omit one day in 134 years (an obviously Alphonsine parameter), while Pierre d'Ailly followed Bacon explicitly in advocating a lunisolar cycle. Again, in connection with a proposal for calendar reform in England, we find that in 1582 John Dee commended Bacon to Queen Elizabeth as one who had "instructed and admonished" the "Romane Bishopp," who was now "contented to follow so neare the footsteps of veritye."[49] Judging by the speed of English legislation in the matter of calendar reform, it seems that Bacon was a little less than five centuries ahead of most of his countrymen.

Little wrote in 1914, "The extant manuscripts of Bacon's works show that the 'Doctor mirabilis never wanted admirers,'"[50] and cited as evidence the existence of twenty-seven manuscripts of the *Perspectiva*[51] alone, dating from the thirteenth to the seventeenth centuries. Apart from his proposals for the calendar it was on Bacon's optics that most scientific value was placed, by his contemporary Witelo as well as by Francesco Maurolico, John Dee, Leonard Digges, Hobbes, and the first editors of his works. At the same time his accounts of alchemy and natural magic gave him more dubious fame, varying from the sixteenth to the nineteenth centuries with current popular prejudices.

NOTES

1. *Opus tertium,* Brewer ed., p. 65.
2. *Ibid.,* p. 16.
3. *Compendium studii philosophie,* Brewer ed., pp. 467–468.
4. *Opus minus,* Brewer ed., p. 325; *Opus tertium,* Brewer ed., p. 30; *Compendium studii philosophie,* p. 425.
5. *Opus tertium,* Brewer ed., pp. 74–75.
6. *Ibid.,* p. 59.
7. *Opus maius* (1266–1267), Bridges ed., I, 401.
8. Brewer, p. 1.
9. Cf. *Opus maius,* Bridges ed., pt. IV, dist. ii–iv; and *De multiplicatione specierum,* Bridges ed.
10. Cf. *Opus maius,* pt. V.
11. Rashdall, pp. 3, 34.
12. *Opus maius,* VI, 1.
13. *Ibid.,* Bridges ed., III, 122.
14. *Ibid.,* II, 377.
15. *Opus tertium,* Brewer ed., pp. 3–4.
16. *Opus maius,* Bridges ed., I, 143–144.
17. *Ibid.,* p. 112.
18. *Ibid.,* V.iii.ii.4 (Bridges ed., II, 157).
19. Figure 1 is redrawn and relettered from *Opus maius,* V.iii.ii.3, British Museum MS Royal 7.f.viii, 13th cent., f. 93r.
20. *Ibid.,* Bridges ed., II, 165.
21. *Opus tertium,* Duhem ed., pp. 78, 90; *Opus maius,* Bridges ed., II, 49.
22. *Ibid.,* I, 151; *De multiplicatione specierum, ibid.,* II, 453; *Communia naturalium,* Steele ed., fasc. 3, pp. 220, 224.
23. *Opus maius,* Bridges ed., II, 167.
24. *Ibid.,* p. 172.
25. Brewer ed., pp. 46–47.
26. *Ibid.,* pp. 383–384.
27. Little ed., pp. 80–89.
28. Steele ed., fasc. 2, pp. 6–8.
29. Steele ed., fasc. 16, pp. 42–44.
30. Brewer ed., p. 533.
31. *Opus tertium,* Brewer ed., pp. 7, 36, 38.
32. See, e.g., *Opus maius,* Bridges ed., II, 172–173.
33. Steele ed., fasc. 16, p. 80.
34. *Opus maius,* Bridges ed., I, 102; cf. *Communia mathematica,* Steele ed., fasc. 16, p. 16.
35. Bridges ed., I, 290 ff.
36. *Imago mundi* was first published at Louvain in 1480 or 1487.
37. *A History of Magic and Experimental Science,* II, 645.
38. Bridges ed., I, 300.
39. Cf. *Communia mathematica,* Steele ed., fasc. 16, p. 49.
40. Brewer ed., p. 106. Since in ch. XII of the same work he seems to have used the word "tables" to refer primarily to almanacs, i.e., ephemerides, and to have spoken of instruments only as a means of verifying tables, it is probable that here he meant to refer only to the astrolabe and the equatorium.
41. Bridges ed., II, 202–203.
42. See L. Thorndike, *The Sphere of Sacrobosco and Its Commentators* (Chicago, 1949), p. 72.
43. *Opus maius,* I, 385.
44. Steele ed., fasc. 9, appendices ii and iii, ed. Little.
45. Little, *ibid.,* p. xxx.
46. Bridges ed., I, xxxiii, 292.
47. See bibliography. Note that the same passage occurs, word for word, in *Opus tertium,* Brewer ed., pp. 271–292; and in *Opus maius,* Bridges ed., I, 281. Notice, however, that the *Computus,* written 1263–1265, does not contain any passage from either of these works, and that it acknowledges Arabic, rather than paying lip service to Hebrew, sources.
48. Steele ed., fasc. 6, pp. 12–18.
49. Corpus Christi College, Oxford, MS C. 254, f. 161r.
50. Pp. 30–31.
51. *Opus maius,* pt. V.

BIBLIOGRAPHY

I. ORIGINAL WORKS. A number of Baconian problems must remain unsolved until there is a complete critical edition of his works: see the bibliography by Little in *Roger Bacon: Essays* (Oxford, 1914), pp. 375–426; compare

G. Sarton, *Introduction to the History of Science,* II (Baltimore, 1931), 963–967; and L. Thorndike and P. Kibre, *A Catalogue of Incipits of Mediaeval Scientific Writings in Latin* (2nd ed., Cambridge, Mass., 1963).

The earliest of Bacon's authentic works to be printed was the *Epistola de secretis operibus artis et naturae* (*De mirabili potestate artis et naturae*) (Paris, 1542; Basel, 1593); in the *Opera,* J. Dee, ed. (Hamburg, 1618); in French (Lyons, 1557; Paris, 1612, 1629); in English (London, 1597, 1659); in German (Eisleben, 1608); and other eds. After this appeared the *De retardandis senectutis accidentibus et de sensibus conservandis* (Oxford, 1590; in English, London, 1683); and *Specula mathematica* (part of *Opus maius* IV); *in qua De specierum multiplicatione earumdemque in inferioribus virtute agitur* and *Perspectiva* (*Opus maius* V), both ed. J. Combach (Frankfurt, 1614). There were other early eds. of the doubtful *Speculum alchemiae* (Nuremburg, 1541; in French, 1557; English, 1597; German, 1608; with later reissues) and the collection *De arte chymiae scripta* (Frankfurt, 1603, 1620).

The 1st ed. of the *Opus maius* was by S. Jebb (London, 1733), followed by an improved ed. (Venice, 1750), both including only pts. I–VI. Pt. VII was included in the new ed. by J. H. Bridges, 2 vols. (Oxford, 1897), with a supp. vol. (III) of revisions and additional notes (London, 1900). This ed. was trans. into English by R. B. Burke (Philadelphia, 1928). Pt. VII of the actual MS sent to the pope has been ed. by E. Massa, *Rogeri Baconi Moralis philosophia* (Zurich, 1953). The eds. of Jebb and Bridges (Vols. II and III, pp. 183–185) both include *De multiplicatione specierum,* a separate treatise forming part of a larger work; a further section of this has been ed. with a discussion of its date and associations by F. M. Delorme, "Le prologue de Roger Bacon à son traité De influentiis agentium," in *Antionianum,* **18** (1943), 81–90.

The 1st eds. of the *Opus minus* and the *Opus tertium,* together with the *Compendium studii philosophie* and a new ed. of the *Epistola de secretis operibus,* were by J. S. Brewer in *Fr. Rogeri Bacon Opera quaedam hactenus inedita* (London, 1859). Further sections of the first two works have been ed. by F. A. Gasquet, "An Unpublished Fragment of Roger Bacon," in *The English Historical Review,* **12** (1897), 494–517, a prefatory letter and other parts of *Opus minus;* P. Duhem, *Un fragment inédit de l'Opus tertium de Roger Bacon* (Quaracchi, 1909), on optics, astronomy, and alchemy; and A. G. Little, *Part of the Opus tertium of Roger Bacon,* British Society of Franciscan Studies, IV (Aberdeen, 1912). The last two items include Bacon's *De enigmatibus alkimie.* For further parts of the *Opus minus,* including discussions of alchemy, still unpublished, see A. Pelzer, "Une source inconnue de Roger Bacon, Alfred de Sareshel, commentateur des Météorologiques d'Aristote," in *Archivium Franciscanum historicum,* **12** (1919), 44–67.

Other works have been ed. by E. Nolan and S. A. Hirsch, *The Greek Grammar of Roger Bacon, and a Fragment of His Hebrew Grammar* (Cambridge, 1902); H. Rashdall, *Fratris Rogeri Baconi Compendium studii theologii,* British Society of Franciscan Studies, III (Aberdeen, 1911); S. H. Thomson, "An Unnoticed Treatise of Roger Bacon on

Time and Motion," in *Isis,* **27** (1937), 219–224; and in *Opera hactenus inedita Rogeri Baconi,* R. Steele, ed. (unless otherwise stated), 16 fasc. (Oxford, 1905–1940): (1) *Metaphysica: De viciis contractis in studio theologie* (1905); (2–4) *Communia naturalium* (1905–1913); (5) *Secretum secretorum cum glossis et notulis* (1920); (6) *Computus* (1926); (7) *Questiones supra undecimum prime philosophie Aristotelis* (*Metaphysica,* XII) (1926); (8) *Questiones supra libros quatuor physicorum Aristotelis,* F. M. Delorme, ed. (1928); (9) *De retardatione accidentium senectutis cum aliis opusculis de rebus medicinalibus,* A. G. Little and E. Withington, eds. (1928); (10) *Questiones supra libros prime philosophie Aristotelis* (*Metaphysica,* I, II, V–X) (1930); (11) *Questiones altere supra libros prime philosophie Aristotelis* (*Metaphysica,* I–IV), *Questiones supra de plantis* (1932); (12) *Questiones supra librum de causis* (1935); (13) *Questiones supra libros octo physicorum Aristotelis,* F. M. Delorme, ed. (1935); (14) *Liber de sensu et sensato, Summa de sophismatibus et distinctionibus* (1937); (15) *Summa grammatica, Sumule dialectices* (1940); and (16) *Communia mathematica* (1940). The *Chronica XXIV generalium ordinis minorum* (*ca.* 1370) was pub. in *Analecta Franciscana,* **3** (1897).

II. SECONDARY LITERATURE. The best critical study of Bacon's life is T. Crowley, *Roger Bacon: The Problem of the Soul in His Philosophical Commentaries* (Louvain–Dublin, 1950). The pioneering study by E. Charles, *Roger Bacon: Sa vie, ses ouvrages, ses doctrines d'après des textes inédits* (Paris, 1861), is now mostly of historical interest. Essential general studies are A. G. Little, ed., *Roger Bacon: Essays Contributed by Various Writers* (Oxford, 1914), especially contributions by Little (life and works); L. Baur (Grosseteste's influence); Hirsch (philology); E. Wiedemann, S. Vogl, and E. Würschmidt (optics); Duhem (vacuum); M. M. P. Muir (alchemy); E. Withington (medicine); and J. E. Sandys (English literature); Little, *Franciscan Letters, Papers and Documents* (Manchester, 1943); L. Thorndike, *A History of Magic and Experimental Science,* II (New York, 1929), 616–691; S. C. Easton, *Roger Bacon and His Search for a Universal Science* (Oxford, 1952), with bibliography; and F. Alessio, *Mito e scienza in Ruggero Bacone* (Milan, 1967).

Studies of particular aspects are E. Schlund, "Petrus Peregrinus von Maricourt: Sein Leben und seine Schriften," in *Archivum Fransiscanum historicum,* **4** (1911), 445–449, 636–643; L. Baur, "Die philosophischen Werke des Robert Grosseteste," in *Beiträge zur Geschichte der Philosophie des Mittelalters,* **9** (1912), 52–63; and "Die Philosophie des Robert Grosseteste," *ibid.,* **18** (1917), 92–120; P. Duhem, *Le système du monde* (Paris, 1916–1958), III, 260–277, 411–442; V, 375–411; VIII, 121–168; A. Birkenmajer, "Études sur Witelo, i–iv," in *Bulletin international de l'Académie polonaise des sciences et des lettres,* Classe d'histoire et de philosophie (1920), 354–360; and "Robert Grosseteste and Richard Fournival," in *Mediaevalia et humanistica,* **5** (1948), 36–41; R. Carton, *L'expérience physique chez Roger Bacon, L'expérience mystique de l'illumination intérieure chez Roger Bacon, La synthèse doctrinale de Roger Bacon,* nos. 2, 3, 5 in the series

Études de philosophie médiévale (Paris, 1924); C. B. Vandewalle, *Roger Bacon dans l'histoire de la philologie* (Paris, 1929); G. Meyer, "En quel sens peut-on parler de 'méthode scientifique' de Roger Bacon," in *Bulletin de littérature ecclésiastique* (Toulouse), **53** (1952), 3–25, 77–98; A. C. Crombie, *Robert Grosseteste and the Origins of Experimental Science 1100–1700*, 3rd imp. (Oxford, 1969), pp. 41, 139–162, 204–207, 213–218, 278–281, with bibliography; and "The Mechanistic Hypothesis and the Scientific Study of Vision," in *Proceedings of the Royal Microscopical Society*, **2** (1967), 20–30, 43–45; M. Schramm, "Aristotelianism: Basis and Obstacle to Scientific Progress in the Middle Ages," in *History of Science*, **2** (1963), 104–108; and A. Pacchi, "Ruggero Bacone e Roberto Grossatesta in un inedito hobbesiano del 1634," in *Rivista critica di storia della filosofia*, **20** (1965), 499–502.

A. C. CROMBIE
J. D. NORTH

BADĪᶜ AL-ZAMĀN AL-JAZARĪ. See **al-Jazarī.**

BAEKELAND, LEO HENDRIK (*b.* Ghent, Belgium, 14 November 1863; *d.* Beacon, New York, 23 February 1944), *chemistry.*

Baekeland graduated with honors from the Municipal Technical School of Ghent in 1880, received the degree of Doctor of Natural Science from the University of Ghent in 1884, and stayed on to serve as professor of chemistry and physics at the Government Normal School of Science. In 1887 he received first prize in the chemistry division of a competition among the graduates of the four Belgian universities. Baekeland's travels took him to the United States in 1889, and he settled there, first as an employee of a photographic firm and then as the head of his own company, which manufactured a photographic paper he named Velox. This was a "gaslight paper" like that invented by Josef M. Eder for making, developing, and handling prints from negatives by artificial light. Baekeland perfected the process and sold it in 1899. He then became a consultant in electrochemistry and obtained patents on dissolving salt in spent electrolytes and on making more durable diaphragms from asbestos cloth treated with gummy iron hydroxides.

In 1905 Baekeland began his third enterprise, which he developed with great success and pursued until his death: the manufacture of condensation products from phenol and formaldehyde. This condensation had first been described in 1872 by Adolf von Baeyer. In a lecture before the American Chemical Society on 8 February 1909, Baekeland surveyed the previous attempts at industrial utilization of the reaction, which resulted in slow processes and brittle products; then he continued: ". . . by the use of small amounts of bases, I have succeeded in preparing a solid initial condensation product, the properties of which sim-

plify enormously all molding operations. . . ." He distinguished three stages of reaction, with a soluble intermediate product. Manufacture of Bakelite resins started in 1907; by 1930, the Bakelite Corporation occupied a 128-acre plant at Bound Brook, New Jersey.

In 1914, Baekeland received the first Chandler Medal. During World War I, he was active in the National Research Council. He was elected president of the American Chemical Society in 1924 and received many other honors.

BIBLIOGRAPHY

Baekeland's writings include "The Synthesis, Constitution, and Uses of Bakelite," in *Industrial and Engineering Chemistry*, **1** (1909), 149–161; and "Some Aspects of Industrial Chemistry," in *Science*, **40** (1914), 179–198.

Further information on Baekeland and his work can be found in Helmut and Alison Gernsheim, *The History of Photography From the Earliest Use of the Camera Obscura in the 11th Century up to 1914* (London, 1955), pp. 284 ff.; see also J. Gillis, *Leo Hendrik Baekeland* (Brussels, 1965).

EDUARD FARBER

BAER, KARL ERNST VON (*b.* Piep, near Jerwen, Estonia, 28 February 1792; *d.* Dorpat, Estonia [now Tartu, Estonian S.S.R.], 28 November 1876), *biology, anthropology, geography.*

During the earlier years of his professional life, Baer concentrated his principal efforts on what is now known as embryology. He is known to Western biologists for his discovery of the mammalian egg in 1826 and for his treatise *Ueber die Entwickelungsgeschichte der Thiere, Beobachtung und Reflexion* (1828, 1837), the publication of which provided a basis for the systematic study of animal development. Baer was a professor in Germany when he carried out his embryological investigations. When he was about forty-two years old, he left Germany for Russia; there, during the remaining years of his long, active life, he devoted his attention primarily to anthropology, both physical and ethnographic, to geography, and, to a lesser degree, to archaeology. But this gives only a bare indication of how widely and how philosophically his mind ranged through nature.

Baer, whose complete style was Karl Ernst Ritter von Baer, Edler von Huthorn, was descended from an originally Prussian family. One of his ancestors, Andreas Baer, emigrated from Westphalia to Reval, Livonia, in the mid-sixteenth century; a collateral descendant of Andreas bought an estate in Estonia during the mid-seventeenth century and was made a member of the nobility. Karl's father, Magnus Johann von Baer, was an Estonian landholder whose estate, Piep, was modest in size. He had been trained in law and, after Karl's birth, served a term as a district official (*Landrat*) and as an official of the Estonian knighthood, in which the family had gained membership during the late eighteenth century.

Magnus Johann von Baer married his first cousin Juliane Louise von Baer. Karl was one of ten children, of whom three were sons. His parents, because of the large size of the family, entrusted Karl during his early years to his father's brother Karl and his wife, Baroness Ernestine von Canne, from Coburg, who lived on a neighboring estate and were childless. Here Karl acquired the love of plants that later drew his interest to botany and natural history. He returned to his own family when he was seven. His first formal instruction was from tutors at home; then, from 1807 to 1810, he attended a cathedral school for members of the nobility in Reval.

Baer's uncle Karl had enjoyed a military life and hoped that his nephew would follow a similar career, as did the boy's father. When Karl decided to enter a university instead, his father encouraged him to go to Germany, but he insisted on entering the University of Dorpat, opened six years earlier. He matriculated in August 1810 as a medical student, perhaps to prepare himself for a career in natural science. He later said that he did not know why he decided to take a medical degree.

Dorpat was a small university, and a provincial one. Baer was interested, however, in his work in botany, physics, and physiology. The professor of physiology was Karl Friedrich Burdach, who later exerted further influence on Baer's career. Baer received the M.D. at Dorpat in September 1814. He was dissatisfied with his medical training, however, and continued his studies in Berlin and Vienna in 1814–1815 and in Berlin during the winter semester 1816–1817.

Baer went to Würzburg in 1815 to further his medical studies, but as an indirect result of his interest in botany he met Ignaz Döllinger, one of the great teachers of the nineteenth century. During the academic year 1815–1816 he studied comparative anatomy with Döllinger, an experience that was critical for his later career. At that time Döllinger tried to persuade him to study the development of the chick by improved methods, studying the blastoderm removed from the yolk, but Baer was not willing to spend the time or money that would have been necessary. He was instrumental, however, in bringing to Würzburg Christian Heinrich Pander, whom he

had known in Dorpat and Berlin, and Pander began the study.

In August 1817, Baer went to Königsberg as prosector in anatomy at the invitation of Burdach, who was professor there. In 1819 he became extraordinary professor of anatomy, and in 1826 ordinary professor of zoology. At various times during his years at Königsberg he taught zoology, anatomy, and anthropology. Baer founded a zoological museum, acted several times as director of the botanical gardens, and served terms as dean of the medical faculty and as rector of the university. He married Auguste von Medem of Königsberg on 1 January 1820. They had five sons and a daughter; the first son died in childhood; the second, Karl, who was interested in natural history, died of typhus at the age of twenty-one while a student at the University of Dorpat.

Most of Baer's contributions to embryology were made between 1819 and 1834, when he was in Königsberg. He made a number of specific discoveries in vertebrate morphogenesis relating to the development of particular organs or organ systems. These alone would have sufficed to warrant his inclusion among major contributors to embryology. He was the first to discover and describe the notochord. He was among the first to recognize that the neural folds represent the rudiment of the central nervous system and that they form a tube, although he did not understand the precise mechanism by which the folds form the substance of brain and spinal cord. He was the first to describe and name the five primary brain vesicles. He made considerable advances in the understanding of the development and function of the extraembryonic membranes (chorion, amnion, allantois) in the chick and the mammal. Incidentally, he was responsible for the introduction of the term "spermatozoa" for what were then known as animalcules in the seminal fluid (but he thought them parasites).

Baer's greatest contributions to embryology were of far wider general significance. In 1826 he discovered the egg of the mammal in the ovary, bringing to completion a search begun at least as early as the seventeenth century. William Harvey had unsuccessfully attempted to find eggs of the deer in the uterus; others, after Harvey's time, had mistaken ovarian follicles for mammalian eggs. Baer first found the true egg in Burdach's house dog, a bitch sacrificed for the investigation; subsequently he found eggs in a number of other mammals. Thus he concluded that "every animal which springs from the coition of male and female is developed from an ovum, and none from a simple formative liquid" (*De ovi mammalium et hominis genesi,* O'Malley trans., p. 149). This was a unifying doctrine whose importance cannot be over-emphasized.

Equally important were Baer's careful descriptions and thoughtful interpretations of the whole course of vertebrate development. He had been in Würzburg in 1815–1816 when his friend Pander had first analyzed the development of the chick in terms of germ layers, and he had participated in that work. In Königsberg, Baer continued the work begun by Pander, extending the observations and generalizing their meaning. Some of his ideas were first published in contributions to Volumes I and II (1826, 1827) of Burdach's *Die Physiologie als Erfahrungswissenschaft.* The first volume of Baer's own treatise was published in 1828, the second volume (unfinished) in 1837; part of what was lacking in Volume II was published posthumously in 1888. *Entwickelungsgeschichte* was the key word in his title and his thought; his great contribution rested on his ability to envisage the organism as a historical entity, as a being that undergoes observable change during its life. He described the development of vertebrates from conception to hatching or birth. Baer observed the formation of the germ layers and described the way in which they formed various organs by tubulation, and he knew this to be more or less similar in all vertebrates. Even more important, he emphasized that development is epigenetic, that it proceeds from the apparently homogeneous to the strikingly heterogeneous, from the general to the special. The old idea, long disputed, that embryonic parts might be preformed in the egg was no longer tenable after Baer's work.

In discussing his view that development proceeds from the general to the special, Baer emphasized that embryos resemble each other more than adults do, and he strongly opposed the opinion previously expressed by Johann Friedrich Meckel that embryos resemble adults of other species. As part of the heritage of German *Naturphilosophie* in which he had been trained, Baer had a great interest in symmetry. His embryological observations led him to believe that there are four fundamental animal types that differ from each other according to their symmetry: the peripheral or radial, the segmental, the massive, and the double symmetrical (vertebrate). These types were very similar to the four *embranchements* described at approximately the same time by Cuvier. Baer held some belief in limited transformationism, the idea that one kind of animal species might during the course of history be transformed into another, but when Darwin's *Origin of Species* was published (1859) Baer could not agree that all organisms could have evolved from a single or a few progenitors. Unfortunately, Baer's valid objections to Meckel's interpretations were ignored by Darwin's immediate followers, who made the recapitulation doctrine a key to evolution theory. But when the new analytical and experimental approach to the study of development began to be followed in the late nineteenth century, the new pathways led out not from the ideas of the recapitulationists, but from those of Baer.

Although Baer reached the peak of his career as an embryologist when at Königsberg, he was restless, and for reasons not yet fully understood, he was unwilling to remain there. He was elected a corresponding member of the Academy of Sciences in St. Petersburg in 1826, and in 1828 he refused an invitation to work at the academy, even though his friend Pander was already there as academician. Baer spent several months in St. Petersburg in 1829–1830, but found conditions for work less favorable than in Königsberg. Nonetheless, when his elder brother Louis, who had been managing the estate at Piep, died in 1834, Baer moved with his wife and children to St. Petersburg. His wish to retain the Piep estate for his family and broken health resulting from overwork may have been factors contributing to a move that had not seemed desirable earlier. Baer entered the academy as a full member in zoology in December 1834, and remained there for the rest of his working life. He performed many duties, his first appointment being as librarian for the foreign division. In 1846 Baer became academician for comparative anatomy and physiology, and from 1846 to 1852 he served as ordinary professor in those fields at the Medico-Chirurgical Academy in St. Petersburg. He retired from active membership in the Academy of Sciences in 1862 but continued work as an honorary member until 1867. He then returned to Dorpat, where he resided until his death.

When he became academician for comparative anatomy and physiology, Baer took charge of the academy's anatomical museum, a decision related to his long-standing interest in anthropology. His doctoral dissertation (1814), on diseases endemic among Estonians, was ethnographically inclined. He first lectured on anthropology as early as the winter of his first year in Königsberg (1817–1818), to students of all faculties, not only of medicine. One volume of these lectures was published in 1824; a second, although promised, did not appear.

In the 1820's, while he was still in Königsberg, Baer had contemplated a trip to Lapland and Novaya Zemlya. After becoming an academician in St. Petersburg, he was able to satisfy his desire for travel. In 1837 he headed an expedition to Novaya Zemlya under the auspices of the academy and was the first naturalist to collect specimens there. During more than twenty-five years he made a number of scientific expeditions, traveling widely: to Lapland, to the North Cape, to the Caspian Sea and the Caucasus, to the Sea of Azov, to Kazan, and to other far parts of Russia, as well as on the Continent and in England. Baer made a number of important discoveries in natural history and geology, and contributed, beginning in 1845, to the founding of the Russian Geographical Society. From 1839 he was coeditor of and contributor to the important *Beiträge zur Kenntniss des Russischen Reiches und der angränzender Länder Asiens.* His travels also increased his already considerable interest in ethnography.

Baer classified man into six categories, ranked according to the degree of primitiveness. His interpretations of some peoples as more primitive than others were similar to those of his contemporaries and immediate predecessors; he did not bring to this area of investigation the same vision that he had carried into embryology. Nonetheless, at least one of his contributions to modern anthropology was truly effective. After he became academician for comparative anatomy and physiology, one of his primary accomplishments, perhaps growing out of his earlier museum experience in Königsberg, was the establishment at the academy of a craniological collection. Attempts to classify skulls were based on measurements, and Baer thought it desirable that methods of cranial measurement be standardized. To this end, he called together a group of craniologists in Göttingen in 1861. The measurements were not standardized, but the meeting led to the founding of the German Anthropological Society and of the German *Archiv für Anthropologie.*

Baer's scientific and intellectual interests reached beyond the areas already enumerated. He did some work in entomology and was instrumental in the establishment of the Russian Entomological Society, of which he was the first president in 1860. He was deeply interested in pisciculture and in the Russian fisheries. He wrote on the origin of the tin found in ancient bronze, on the routes of Odysseus' voyages, and on the whereabouts of biblical Ophir.

Among his other abilities Baer had particular talents that distinguished him socially. He had great wit, which endeared him to those who knew him, and he was very loyal to his friends. One friend in particular may be singled out. In the winter of 1839–1840 Baer made the acquaintance of the Grand Duchess Helen Pavlovna, the former Princess Frederika Charlotte Marie of Württemberg. The wife of Grand Duke

Michael Pavlovitch, youngest brother of Czar Alexander I, she was an enlightened and intelligent patron of the arts and the sciences; Baer instructed her two daughters in natural history and enjoyed her friendship for many years.

Baer was a patriotic Russian, as is clear from the zeal with which he carried out his duties for the academy and from his evident interest in Russian geography and ethnography. But he was also an expressed enthusiast of Prussia. His true political views remain obscure, for some were expressed cryptically and others, we are told by his biographer Stieda, were probably eliminated from his publications by the censors.

As for his writing, Baer began more than he completed. The second volume of his great *Entwickelungsgeschichte* was never finished; he neglected even to read the proofs when the publisher decided to bring it out unfinished. He began, but failed to complete, other writings; he never completely described his collections from Novaya Zemlya. Nonetheless, he was a prolific lecturer and author. The second edition of Stieda's biography enumerates approximately 300 of his publications, and the list is incomplete.

Baer was not only an accomplisher but also a thinker, and the range and profundity of his thought are reflected in many of his writings. He was particularly interested in problems related to teleology; he lectured and wrote on living creatures and life as related to the wider cosmos. Baer saw nature as a whole, not merely as did the *Naturphilosophen,* who constructed elaborate ideas about natural schemes, but as an observer and discoverer who with his eyes as well as his mind searched deep into many of nature's realms.

Baer received many honors during his lifetime. An island in the Russian North was named for him; and in 1864 the Estonian Knights held a celebration for him on the golden jubilee of his doctorate. They also published his autobiography, which was especially prepared for that event. In 1872, Volume 5 of the *Archiv für Anthropologie* was dedicated to him. Baer was elected a member of the Royal Society of London and of the Paris Academy. He was awarded the Copley Medal by the Royal Society and also received a medal from the Paris Academy. Alexander von Humboldt personally brought him the medal from Paris, much to Baer's delight. Baer once wrote of Humboldt that he was "versatile, yet always accurate as an observer, deep and far-seeing as a thinker, exalted as a seer" (*Reden,* I, 296). He might well have been speaking of himself.

BIBLIOGRAPHY

The most important sources of information about Baer's life and writings are his autobiography and an authoritative biography by Ludwig Stieda. Both contain extensive classified and annotated lists of publications by Baer.

I. ORIGINAL WORKS. Baer's writings include *De ovi mammalium et hominis genesi epistola* (Leipzig, 1827); German trans., B. Ottow, ed., *Ueber die Bilding des Eies der Säugetiere und des Menschen* (Leipzig, 1927); facsimile of Latin ed. in George Sarton, "The Discovery of the Mammalian Egg and the Foundation of Modern Embryology," in *Isis,* **16** (1931), 315–[378]; English trans. by Charles Donald O'Malley, *ibid.,* **47** (1956), 117–153; *Ueber die Entwickelungsgeschichte der Thiere. Beobachtung und Reflexion,* 2 vols. (Königsberg, 1828, 1837), Vol. III, L. Stieda, ed. (Königsberg, 1888); *Reden gehalten in wissenschaftlichen Versammlungen und kleinere Aufsätze vermischten Inhalts,* 3 vols. (St. Petersburg, 1864–1876); and *Nachrichten über Leben und Schriften des Geheimrathes Dr. Karl Ernst von Baer, mitgetheilt von ihm selbst. Veröffentlicht bei Gelegenheit seines fünfzigjährigen Doctor-Jubiläums, am 29. August 1864, von der Ritterschaft Ehstlands* (St. Petersburg, 1865), privately distributed ed. of 400 copies; trade eds. appeared under same title (St. Petersburg, 1866; 2nd ed., Brunswick, 1886).

II. SECONDARY LITERATURE. Works on Baer include B. Ottow, "K. E. von Baer als Kraniologe und die Anthropologen-Versammlung 1861 in Göttingen," in *Sudhoffs Archiv für Geschichte der Medizin und der Naturwissenschaften,* **50** (1966), 43–68; and L. Stieda, *Karl Ernst von Baer. Eine biographische Skizze* (Brunswick, 1878, 1886).

JANE OPPENHEIMER

BAEYER, ADOLF JOHANN FRIEDRICH WILHELM VON (*b.* Berlin, Germany, 31 October 1835; *d.* Starnberg, Oberbayern, Germany, 20 August 1917), *chemistry.*

Adolf von Baeyer was the eldest of the five children of Johann Jacob Baeyer and Eugenie Hitzig. His father, a lieutenant general in the Prussian army, took part in the government project to measure degrees of latitude and longitude directed by the astronomer W. F. Bessel. He subsequently participated in a general European measurement program and published these results, as well as other investigations on the shape of the earth. Baeyer's mother was the daughter of Julius Eduard Hitzig, a criminal judge. Her uncle, the art historian Franz Kugler, and her grandfather made Baeyer's first home on Friedrichstrasse a center for Berlin literary life by attracting E. T. A. Hoffman and others for weekly evenings of conversation. Baeyer's mother was Jewish but had been converted to the Evangelical faith and had been confirmed by Schleiermacher. Baeyer was of the same faith as his parents.

Baeyer's early education at the Friedrich-Wilhelms Gymnasium and at the University of Berlin under P. G. Dirichlet and G. Magnus emphasized mathematics and physics. In 1856, after a year's military service, he decided to study experimental chemistry with R. Bunsen in Heidelberg, where the emphasis was on applied physical chemistry. Dissatisfied with this approach, Baeyer in 1858 entered Kekulé's private laboratory in Heidelberg. Since he was pleased with Kekulé's tuition in organic chemistry, Baeyer followed him to Ghent and remained there until 1860, taking time out only in 1858 to receive his doctoral degree from Berlin for work on arsenic methyl chloride. After leaving Ghent, Baeyer returned to Berlin and held various teaching positions at the technical institute and the military academy. He married Lida, daughter of Emil Bendemann, in 1868; the couple had three children: Eugenie, Hans, and Otto. In 1872 Baeyer was appointed ordinary professor of chemistry at the new imperial university at Strasbourg. Three years later he moved to Munich as successor to Justus von Liebig.

Baeyer's contributions to science were widely recognized during his lifetime. He received the Liebig Medal of the Berlin Chemists Congress, the Royal Society's Davy Medal, and in 1905 the Nobel Prize for his work on dyes and hydroaromatic compounds. Among the many scientific societies to which he belonged were the Berlin Academy of Sciences and the German Chemical Association.

Most of Baeyer's scientific interests grew out of work he began at the technical institute in Berlin. His early work, in which he followed the lead of Liebig and Wöhler, centered on the derivatives of uric acid. He prepared various derivatives, including barbiturates. From his uric acid studies Baeyer turned to another physiological problem, that of assimilation in green plants. He thought that the ease with which a water molecule could be eliminated from a carbon atom with two hydroxyl groups would explain the formation of sugars in plants. Baeyer also saw that a study of condensation reactions where water was eliminated might be used to probe the structure of complicated organic molecules. He and his students conducted a series of investigations on condensation reactions of aldehydes and phenols which led to the discovery of phenolphthalein and other members of the phthalein group. Baeyer developed the dyestuffs gallein and coerulein from this class of compounds.

Connected with his work on phthaleins was his work with phthalic acid, which led Baeyer to investigate the structure of benzene. He started with mellitic acid, which he showed was a benzene derivative rather than an alkyne, as he had first believed. He then investigated the carboxylic acids of benzene and found one new tricarboxylic and two new tetracarboxylic acids. From the reduced forms of these acids, prepared by heating them with zinc dust—a method he had previously introduced—he tried to elucidate the structure of benzene. Failure to find cis-trans isomers led him to reject Kekulé's double-bond model and Ladenberg's prism model and to adopt his own model of benzene, in which one valence bond from each carbon was directed inward. But in 1892 he produced cis-trans derivatives of benzene, which made him reject any one theory of benzene structure.

Baeyer's best-known research, on indigo, was begun in 1865 and followed up earlier work on phthalein dyes. In 1841 Laurent had investigated indigo and had obtained isatin by oxidation. Baeyer reversed Laurent's process, and in 1870 he and Emmerling produced indigo by treatment of isatin with phosphorous trichloride, followed by reduction. In 1878, by synthesizing isatin from phenylacetic acid, Baeyer demonstrated the possibility of a complete synthesis of indigo. However, neither the exact process nor the structure of indigo was yet known. Baeyer had worked on isatin in 1865–1866 and had produced a series of related compounds, dioxindole and oxindole, with successively less oxygen. It occurred to him that this series bore a similarity to the alloxan series he had developed with uric acid. Consequently, he predicted and found a final member of the series, which he named indole, containing no oxygen. He found the structure of indole, the kernel of the indigo molecule, through work with pyrrole. To determine the structural formula of indigo, Baeyer examined the linkage of the two indole kernels. He prepared indigo from several other reagents, and finally in 1883 he showed, through a synthesis from o-dinitrodiphenyldiacetylene, the exact location of the link, and thus the exact structural formula.

Naturally the German dye industry was interested in the synthesis of indigo, and spent large sums on research to discover a practical process. Baeyer declined to take an active part in eliminating commercial difficulties, and the resulting ill feeling on the part of the dye industry caused him to discontinue further work on dyestuffs and to turn his attention elsewhere. He began an investigation of the polyacetylenes and prepared a number of highly explosive compounds. The properties of these compounds focused his attention on the resistance of double and triple carbon bonds. From these considerations he developed his strain theory, in which the stability of a ring structure is related to the amount of bending necessary to form a closed ring.

A final area of extensive research for Baeyer was that of oxonium compounds. In the course of his work on terpenes Baeyer had used peroxide compounds and had suggested that hydrogen peroxide had the formula HO · HO. Certain phthalein salts, derivatives of pyrone, also had unusual oxygen linkages. At first Baeyer opposed the idea of J. N. Collie and T. Tickle that the oxygen in these compounds becomes quadrivalent and basic, but work with the peroxides and perchloric acid convinced him of the weak basic properties of oxygen, and he called such compounds of oxygen "oxonium compounds."

Baeyer's chemical research was in many ways an extension of Kekulé's work on the tetravalency of carbon. He used this framework to elucidate the structure of compound after compound. His work on indigo indicated that he was not interested merely in synthesis, but also believed that understanding a compound required knowing its exact structure. Baeyer's work on the stability of ring structures showed his concern with the direction of valence bonds and the extent to which the direction can be changed. With the idea that valence bonds can be strained and with his criticism of existing benzene formulas, Baeyer went beyond Kekulé's earlier picture. His final unwillingness to assign a definite structure to benzene, his work on stable and unstable isomers of isatin and indoxyl, which he called lactim and lactam forms, as well as that on phloroglucinol, pointed up difficulties in determining exact structure, which were later resolved with the concept of tautomerism.

Baeyer's approach to chemistry, however, was different from that of Kekulé. Baeyer had little interest in theoretical statements and attacked most problems empirically. His work on benzene, he explained, was only an experimental investigation and not an attempt to prove a particular hypothesis. He was a master at test-tube analysis, and eschewed complicated apparatus.

Most of Baeyer's investigations of compounds grew out of his early work on uric acid derivatives, which Liebig had also investigated. He was instrumental in setting up a chemical laboratory at Munich in 1877 and thus was a successor of Liebig in several senses. As a teacher, Baeyer's strength lay in the laboratory rather than in the lecture hall. He gathered about him a large group of students and assistants, among whom were C. Lieberman, C. Graebe, V. and R. Meyer, and E. and O. Fischer. The enormous number of articles by Baeyer himself, those written with his students, and those written by his students indicates both the scope of the activities in his laboratory and the co-operative nature of the research. Baeyer filled out and extended existing theory by the vast amount of his empirical research, stimulated the chemical dye industry through his work on indigo, and laid the foundation for subsequent work in biochemistry through his investigation of complex ring structures.

BIBLIOGRAPHY

I. Original Works. Baeyer's works are collected in *Gesammelte Werke* (Brunswick, 1905). Also of interest are unpublished letters to H. Caro, which are in the library of the Deutsches Museum, Munich.

II. Secondary Literature. Works on Baeyer include Günther Bugge, "Adolf v. Baeyer," in *Prometheus,* **29** (1917), 1–5; W. Dieckmann, O. Dimroth, F. Friedländer, C. Harries, P. Karrer, R. Meyer, W. Schlenk, H. Wieland, and R. Willstätter, *Die Naturwissenschaften,* Sonderheft zum 80 Geburtstag von Adolf v. Baeyer, **44** (29 Oct. 1915); J. Gillis, "Lettres d'Adolf Baeyer à son ami Jean Servais Stas," in *Mémoires de l'Académie royale de Belgique. Classe des sciences,* **32** (1960), 1–45; J. Partington, "Adolf von Baeyer," in *Nature,* **136** (1935), 669; H. Rupe, *Adolf von Baeyer als Lehrer und Forscher* (Stuttgart, 1932); Karl Schmorl, *Adolf von Baeyer* (Stuttgart, 1952); Richard Willstätter, "Adolf von Baeyer," in *Das Buch der grossen Chemiker,* ed. Günther Bugge (Weinheim, 1955).

RUTH ANNE GIENAPP

AL-BAGHDĀDĪ. See **Ibn Ṭāhir.**

BAGLIVI, GEORGIUS (*b.* Republic of Dubrovnik [now a part of Yugoslavia], 8 September 1668; *d.* Rome, Italy, 15 June 1707), *biology.*

Baglivi's family name was Armeno, which probably indicates that his father was of Armenian origin. His parents were Blasius Armeno and Anna de Lupis, respectable but poor merchants who both died in 1670; Georgius and his younger brother Jacobus were educated first by their uncle and then in the Jesuit College of Dubrovnik. At the age of fifteen, Georgius left his native town and went to southern Italy, where he and his brother were adopted by Pietro Angelo Baglivi, a physician in Lecce. At college Georgius received very good classical training and was imbued with Peripatetic philosophy. He began medical practice with his foster father. After completing his medical studies in Naples and receiving the M.D. degree— probably at Salerno in 1688—Baglivi attended Bellini's lectures at Pisa; worked in hospitals in Padua, Venice, Florence, Bologna, and other Italian cities; traveled to Dalmatia; and finally decided to settle in

Bologna as a pupil and assistant of Marcello Malpighi in 1691.

In his student days Baglivi had been attracted by physiological experiments and by postmortem examinations. Therefore, in 1685 he began to experiment with the infusion of various substances into the jugular veins of dogs and to observe the life habits of tarantulas. From 1689 to 1691, he dissected such various animals as lions, tortoises, snakes, and deer, and made morphological and physiological discoveries, studied the function of *dura mater* by experiments on dogs and observations on wounded men, performed many autopsies, and experimented with toxic drugs. At the same time he served as a physician in many hospitals. In this way he became aware of a curious discrepancy between clinical practice and newly developing biological research. In his opinion, physicians were slaves to systems and hypotheses. Under Malpighi's direction Baglivi experimented on dogs (performing resection of the pneumogastric nerve, infusion of drugs in the veins and spinal canal, and so forth) and on frogs (experiments concerning circulation of blood) and continued his anatomical research, perfecting it by microscopic observations. He studied the fine structure of muscles and of the brain's membranous envelopes.

Accepting a position as the pope's archiater, Malpighi moved from Bologna to Rome, and in 1692 invited Baglivi to join him and to live in his house as a kind of scientific secretary. They collaborated closely until Malpighi's death in 1694. Baglivi performed an autopsy on his master's corpse and gave a very good description of his last illness, cerebral apoplexy. Introduced to the papal court, the friend of such influential scientists as Lancisi, Bellini, Redi, Tozzi, and Trionfetti, and praised as a highly competent practitioner, the young man was destined for a brilliant career. He became the pope's second physician in 1695 and in the following year was elected professor of anatomy at the Sapienza in Rome. In connection with this election, Baglivi published and dedicated to his protector, Innocent XII, a book entitled *De praxi medica* (1696). It was a lucid program of what medicine should be in the future, an attack against the medico-philosophical systems, and a claim for the Hippocratic principles of sound clinical observation. With the exception of some fine general statements in an aphoristic form and a small number of fairly good clinical descriptions (e.g., of typhoid fever and of cardiac decompensation), this book offers little to a modern reader; its style is somewhat baroque, and its factual medical content is often doubtful. In any case, Baglivi's treatise is representative of a stream of thought opposed to philosophical generalizations in medical practice. Baglivi became a member of the Royal Society in 1697, the Academia Naturae Curiosorum and the Arcadia in 1699, and the Accademia dei Fisiocritici in 1700.

The new pope, Clement XI, confirmed Baglivi in his position at court, and even named him professor of theoretical medicine at the Sapienza in 1701. Baglivi's research was now concentrated upon the microscopic structure of muscle fibers and the physical and physiological properties of saliva, bile, and blood. His lectures and anatomical demonstrations, as well as his medical consultations, acquired a very high reputation all over Europe.

His philosophical conception of life phenomena makes Baglivi a member of the iatrophysical school, a defender of biomechanicism. Strongly influenced by Santorio, he wrote a commentary to the latter's *De medicina statica.* Accepting Harvey's theory, Baglivi hoped to complete it with his own theory of fluid circulation in nerves propelled by contractions of the *dura mater.* He was able to distinguish between the smooth and striated muscles; and he discovered the histological distinction between two categories of fibers, which he called *fibrae motrices seu musculares* (with parallel fiber bundles) and *fibrae membranaceae* (with bundles running in various directions). He believed that all physiological and pathological processes should be explained by the living properties of *fibrae.* Baglivi gave new life to the ancient doctrine of the methodists that life and health are determined by the physical balance of the active solids (fibers) and the more passive fluids of the body. His fundamental research concerning the fibers made him one of the most important students of muscle physiology before Albrecht von Haller. Baglivi denied the possibility of spontaneous generation of intestinal parasites, described paralytic phenomena after infusion of certain substances into the spinal canal, and predicted the rise of specific chemotherapy. A strange conflict in his writings is his acceptance of biomechanistic doctrine as a guide for research work and his rejection of all speculative theoretical background in actual medical practice. He said that the iatromathematic physician must forget his theories when he appears at the bedside.

BIBLIOGRAPHY

I. ORIGINAL WORKS. Among Baglivi's writings are *De praxi medica ad priscam observandi rationem revocanda* (Rome, 1696), trans. as *The Practice of Physick* (London, 1704); *De fibra motrice, et morbosa, nec non de experimentis, ac morbis salivae, bilis et sanguinis* (Perugia, 1700); *Speci-*

men quatuor librorum de fibra motrice et morbosa (Rome, 1702); *Canones de medicina solidorum ad rectum statices usum* (Rome, 1704); *Opera omnia medico-practica et anatomica* (Lyons, 1704; new enlarged ed., 1710); and *Opera omnia medico-practica et anatomica,* C. G. Kuhn, ed., 2 vols. (Leipzig, 1827–1828). For a complete list of Baglivi's published works, see M. D. Grmek, *Hrvatska medicinska bibliografija,* I, pt. 1 (Zagreb, 1955), 32–34.

The main collections of Baglivi's papers and letters are in the National Library, Florence; the Waller Collection, Uppsala; and the Osler Collection, McGill University, Montreal.

II. SECONDARY LITERATURE. More information on Baglivi and his work can be found in E. Bastholm, *History of Muscle Physiology* (Copenhagen, 1950), pp. 178–189; A. Castiglioni, "Di un illustre medico raguseo del secolo decimosettimo," in *Rivista di storia critica delle scienze mediche e naturali,* **12** (1921), 1–11; P. Fabre, *Un médecin italien de la fin du XVII siècle: Georges Baglivi* (Paris, 1896); M. D. Grmek, "Osservazioni sulla vita, opera ed importanza storica di Giorgio Baglivi," in *Atti del XIV Congresso Internazionale di Storia della Medicina,* I (Rome, 1954), p. 423, and "Životni put dubrovačkog liječnika Gjure Baglivija," in *Liječnički vjesnik,* **79** (1957), 599–624; J. Jiménez Girona, *La Medicina de Baglivi* (Madrid, 1955); L. Münster, "Nuovi contributi alla biografia di Giorgio Baglivi," in *Archivio storico Pugliese,* **3**, nos. 1–2 (1950); M. Salomon, *Giorgio Baglivi und seine Zeit* (Berlin, 1889); F. Scalzi, *Giorgio Baglivi, altre notizie biografiche* (Rome, 1889); and F. Stenn, "Giorgio Baglivi," in *Annals of Medical History,* 3rd ser., **3** (1941), 183–194.

M. D. GRMEK

BAIER, JOHANN JACOB (*b.* Jena, Germany, 14 June 1677; *d.* Altdorf, Bavaria, 11 July 1735), *medicine, geology, paleontology.*

Baier was the son of Johann Wilhelm Baier, professor of Protestant theology at the University of Jena, and Anna Katharina Musaeus. After private tutoring he matriculated in 1693 at the University of Jena, where he dutifully studied philosophy, classical languages, mathematics, medicine, and natural science. During 1699 and 1700 he traveled in northern Germany and in the Baltic Sea provinces to Riga and Dorpat, enriching his knowledge by conversations with other scholars and by examining collections and visiting libraries. In 1700 he finished his studies and was awarded the degrees of M.A., Ph.D., and M.D. After another educational trip, to visit mining facilities in the Hartz Mountains and university towns in northern Germany, he settled in 1701 as a practicing physician in Nuremberg.

In 1703, during the War of the Spanish Succession, Baier was director of a field hospital for the soldiers of the Nuremberg contingent. The following year he was awarded a professorship at the medical faculty of the University of Altdorf in Nuremberg, which he held until his death. He was twice elected rector of the University. In 1708 he became a member of the Leopoldina (Academy of Natural Scientists) and in 1730 was chosen its president.

Baier's scientific fame today does not rest on his medical investigations, but on his studies of minerals and fossils. At that time the natural sciences could be pursued only within the framework of medicine: oryctography, comprising geology and paleontology, was at the very beginning of its development. Baier's *Oryctographia norica* (1708) was a new, systematic presentation based on his own studies. The work contributed much to disproving the idea that fossils were a mere sport of nature. By means of exact descriptions and good illustrations he laid the foundations for the investigation of Jurassic fauna and of scientific paleontology in general. Instead of theory, he clearly presented what could be observed. He believed that the earth had been created in one act and that the Deluge was the only great change since the Creation. His exact foundation work, however, helped to prepare the ground for the next generation to determine historically the geological structure of mountains and to transform oryctography into geology.

BIBLIOGRAPHY

I. ORIGINAL WORKS. Baier's most important publications (a complete list comprises about forty titles) are *Dissertatio de ambra* (Jena, 1698); *Dissertatio de necessaria salinae inspectione ad conservandum et restaurandum sanitatem* (Halle-Magdeburg, 1698); *Dissertatio de capillis* (Jena, 1700); *Oryctographia norica sive rerum fossilium et ad minerale regnum pertinentium in territorio norimbergensi eiusque vicinia observatarum succincta descriptio* (Nuremberg, 1708); *Wahrhaffte und gründliche Beschreibung der Nürnbergischen Universitätsstadt Altdorf samt dero fürnehmsten Denkwürdigkeiten kürtzlich entworfen und mit accuraten Kupferstichen gezieret* ("Real and True Description of the Nuremberg University Town Altdorf With Its Most Noble Places of Interest Recently Designed and Adorned With Accurate Copper Engravings," Altdorf, 1714); *Horti medici Academia Altorfina historia, curiose conquisita* (Altdorf, 1727); *Orationum varii argumenti, variis occasionibus in Academia Altorfina publice habitarum fasciculus* (Altdorf, 1727); *Biographiae professorum medicinae, qui in Academia Altorfina unquam vixerunt* (Nuremberg-Altdorf, 1728); "Sciagraphia Musei sui. Accendunt supplementa *Oryctographiae noricae*," in *Acta physico-medica Academiae Caesareae Leopoldina-Carolinae Naturae Curiosorum,* 2 (1730), Appendix, also published separately (Nuremberg, 1730); *Oryctographia norica sive rerum fossilium et ad minerale regnum pertinentium in territorio Norimbergensi eiusque vicinia observatorum succincta descriptio. Cum supplementis*

a. 1730 editis (Nuremberg, 1758); *Epistolae ad viros eruditos eorumque responsiones historiam literariam et physicam specialem explanantes, curante filio Ferdinando Jacobo Baiero* (Frankfurt-Leipzig, 1760); and "Oryctographia Norica and Supplements," translated from the Latin by Hermann Hornung, with paleontological explanations by Florian Heller, edited and explicated by Bruno von Freyberg, in *Erlanger geologische Abhandlungen,* 29 (1958).

II. SECONDARY LITERATURE. Works on Baier are Bruno von Freyberg, "250 Jahre geologische Forschung in Franken," in *Geologische Blätter für Nordost-Bayern und angrenzende Gebiete,* 8 (1958), 34–43; "Einführung (in Baiers wissenschaftliches Lebenswerk)," in *Erlanger geologische Abhandlungen,* 29 (1958), 7–12; and "Memoria viri perillustris, magnifici excellentissimique domini Joanni Jacobi Baieri," in *Acta physico-medica Academiae Caesareae Leopoldina-Carolinae Naturae Curiosorum,* 4, Appendix, Biographies (1737), 35–48; and Ernst Stromer von Reichenbach, "Johann Jakob Baier, einer der ersten deutschen Paläontologen, ein Beispiel der Willkür des Nachruhmes," in *Natur und Volk,* 75/76 (1946), 25–31.

B. V. FREYBERG

BAILAK AL-QABAJAQĪ. See **Baylak al-Qibjāqī.**

BAILEY, EDWARD BATTERSBY (*b.* Marden, England, 1 July 1881; *d.* London, England, 19 March 1965), *geology.*

Bailey was a son of John Battersby Bailey, medical practitioner, and Louise Florence Carr, daughter of a farmer. He was educated at Kendal Grammar School and at Clare College, Cambridge, where he graduated with First-Class Honours in geology and physics in 1902. In 1914 Bailey married Alice Meason, by whom he had a son and a daughter. Alice Meason died in 1956, and six years later he married Mary Young.

While at Cambridge, the reading of a translation of Suess's great synthesis, *Das Antlitz der Erde,* inspired Bailey with a passionate enthusiasm for geological research—an enthusiasm favored by his robust health, remarkable physique, and unusually fertile mind. Temperamentally gay, exuberant, and self-confident, he wrote in a clear, distinctive style characterized by picturesque phrases. In religious matters he was an atheist.

Upon graduation from Cambridge, Bailey became geologist, and later district geologist, with the Geological Survey of Great Britain (in Scotland). He held the latter post until 1929. He then became professor of geology at Glasgow University, and in 1937 returned to government service as director of the Geological Survey and Museum. He retired in 1945. Bailey's many honors included fellowship of the Royal Society, six honorary doctorates, and seven scientific medals. He was knighted in 1945.

His work on the Geological Survey was interrupted from 1915 to 1919, when he served as an artillery subaltern on the Western Front and was twice wounded, losing an eye. By his gallantry he gained the Military Cross, the Croix de Guerre, and the Legion of Honor.

Bailey made notable contributions to tectonics and metamorphism, to igneous and general geology and to the history of the development of geological ideas. His revolutionary tectonic reinterpretation of Dalradian schists in the Scottish Highlands was inspired by the work of his colleagues C. T. Clough and H. B. Maufe, and by excursions in Switzerland and Scandinavia. In Scotland from 1910 onward he introduced new or modified stratigraphical groupings and reinterpreted structure in terms of great recumbent folds and contemporaneous slides (fold faults) that had locally cut out parts of the succession and had then been refolded (1916, 1922, 1925, 1937). Bailey's syntheses met with criticism, details of which may be found in "Discussions" following his *Quarterly Journal* papers and in Read and MacGregor (1948). From time to time he modified his ideas (1930, 1934, 1938), notably because of evidence provided by the current-bedding criterion for stratigraphical order introduced to Scotland in 1930 by T. Vogt and T. L. Tanton. Bailey's mature views (e.g. 1916, revision of 1960) are, in the main, now accepted.

In northern Scotland, Bailey dealt with tectonics and metamorphism in the Moine Thrust areas of Skye and the northwest mainland (1951, 1955). He supported the correlation of Moinian and Torridonian, and the Caledonian age of Moinian and Dalradian metamorphism (1955).

In 1929 Bailey published an account of the Paleozoic mountain systems of Europe and North America, in which he suggested that the crossing of Caledonian and Hercynian chains, begun in southern Wales and Ireland and completed in New England, may be evidence of continental drift. His *Tectonic Essays* (1935) explain Alpine terminology and structure. In later years he produced tectonic reassessments of parts of Iran, Turkey (1953), Provence, Gibraltar, and Liguria (1963).

Bailey's main contributions to Scottish igneous geology stemmed from work with C. T. Clough and H. B. Maufe. In the Glen Coe area they demonstrated a Devonian ring fracture accompanied by a ring intrusion and a central 1,000-foot caldron subsidence of an oval block of schists and Devonian lavas several miles in diameter. A northeasterly dyke swarm was also mapped. These remarkable discoveries were described by Bailey (1909, 1916). Both editions of *The Geology of Ben Nevis* (1916, 1960) include detailed petrographic accounts of Devonian lavas, dykes, and granitic plutons. In a wider study, Bailey inferred that the hornblendic parental Devonian magma was relatively richer in water than the pyroxenic parental British Tertiary magma (1958).

Bailey's compilation of the memoir on the Tertiary volcanic complex of Mull (1924), and its accompanying intricate map, was recognized as a major scientific achievement. He and his colleagues found evidence of a caldera with pillow lavas, innumerable cone sheets, a northwesterly dyke swarm, and massive ring dykes locally modified by gravitative differentiation. Bailey and H. H. Thomas studied the petrology of the complex and introduced the concept of "magma type." There has been controversy over the ring-dyke differentiation, involving Holmes (1936) and Koomans and Kuenen (1938), as well as over the parental magma, the magma-type concept, and the use of terms "tholeiite" and "tholeiitic," in which Kennedy (1933), Wells and Wells (1948), Holmes (1949), Tomkeieff (1949), Tilley (1950), Chayes (1966), Dunham (1966), and Tilley and Muir (1967) participated.

Bailey also reassessed the intrusion tectonics of the Arran granite (1926) and of the volcanic complex of Rhum (1945), both of the Tertiary period. With W. J. McCallien, he later studied pillow lavas and serpentines associated with radiolarian cherts in Scotland, Turkey, and the Apennines, and advocated their submarine origin (1953, 1957, 1960, 1963). In Ireland he and McCallien inferred that dolerite can crack before its crystallization has gone very far and can then be locally chilled by contact with invading colder acid magma (1956); such ideas, originally developed with L. R. Wager (1953), have been criticized by Reynolds (1953) and supported by Skelhorn and Elwell (1966). Bailey's general Scottish work included glaciation studies (1908, 1916, 1924) and an inference that Chalk seas had desert shores (1924).

With L. W. Collet and R. M. Field, Bailey published a reinterpretation of the Quebec and Lévis conglomerates; these were regarded as the products of submarine landslips, detached during successive subsidences along the hinge of the Logan Slope (1928). He was thus led to infer that Kimmeridgian boulder beds in Sutherland originated from movements along a submarine fault scarp (1932). He later suggested that submarine landslips produce graded bedding and may locally merge into submarine mud rivers (1930, 1936, 1940). Kuenen and Migliorini (1950) were led thereby to develop the idea of submarine turbidity currents of high density.

Bailey's historical writings include an invaluable chronological summary of the development of world tectonic research (1935); a history of the Geological Survey (1952); a biography of Sir Charles Lyell (1962); and an account of James Hutton's life and work which includes a commentary on each chapter of Hutton's *Theory of the Earth* (1967).

Obituary notices of Bailey have been written by Stubblefield (1965) and by MacGregor (1966); they contain bibliographies much more comprehensive than that given below.

BIBLIOGRAPHY

I. ORIGINAL WORKS. Bailey's writings include "The Glaciation of East Lothian South of the Garleton Hills," in *Transactions of the Royal Society of Edinburgh,* **46** (1908), 1–31, written with P. F. Kendall; "The Cauldron-subsidence of Glen Coe and the Associated Igneous Phenomena," in *Quarterly Journal of the Geological Society of London,* **65** (1909), 611–678, written with C. T. Clough and H. B. Maufe; *The Geology of Ben Nevis and Glen Coe and the Surrounding Country,* a memoir of the Geological Survey (Edinburgh, 1916, 1960), written with H. B. Maufe *et al.;* "The Structure of the South-west Highlands of Scotland," in *Quarterly Journal of the Geological Society of London,* **78** (1922), 82–131; "The Desert Shores of the Chalk Seas," in *Geological Magazine,* **61** (1924), 102–116; *The Tertiary and Post-Tertiary Geology of Mull, Loch Aline and Oban,* a memoir of the Geological Survey (Edinburgh, 1924), written with C. T. Clough *et al.;* "Perthshire Tectonics; Loch Tummel, Blair Atholl and Glen Shee," in *Transactions of the Royal Society of Edinburgh,* **53** (1925), 671–698; "Domes in Scotland and South Africa: Arran and Vredefort," in *Geological Magazine,* **63** (1926), 481–495; "Palaeozoic Submarine Landslips Near Quebec City," in *Journal of Geology,* **36** (1928), 577–614, written with L. W. Collet and R. M. Field; "The Palaeozoic Mountain Systems of Europe and America," in *Report of the British Association for 1928* (London, 1929), pp. 57–76; "New Light on Sedimentation and Tectonics," in *Geological Magazine,* **78** (1930), 77–92; "Submarine Faulting in Kimmeridgian Times: East Sutherland," in *Transactions of the Royal Society of Edinburgh,* **57** (1932), 429–467, written with J. Weir; "West Highland Tectonics: Loch Leven to Glen Roy," in *Quarterly Journal of the Geological Society of London,* **90** (1934), 462–525; *Tectonic Essays, Mainly Alpine* (Oxford, 1935); "Sedimentation in Relation to Tectonics," in *Bulletin of the Geological Society of America,* **47** (1936), 1713–1726; "Perthshire Tectonics: Schichallion to Glen Lyon," in *Transactions of the Royal Society of Edinburgh,* **59** (1937), 79–117, written with W. J. McCallien; "Northwestern Europe: Caledonides," in *Regionale Geologie der Erde,* II, pt. 2 (Leipzig, 1938), 1–76, written with O. Holtedahl; "American Gleanings: 1936," in *Transactions of the Geological Society of Glasgow,* **20** (1940), 1–16, issued as an offprint in 1938; "Tertiary Igneous Tectonics of Rhum (Inner Hebrides)," in *Quarterly*

Journal of the Geological Society of London, **100** (1945), 165–191; "Scourie Dykes and Laxfordian Metamorphism," in *Geological Magazine,* **88** (1951), 153–165; *Geological Survey of Great Britain* (London, 1952); "Basic Magma Chilled Against Acid Magma," in *Nature,* **172** (1953), 68–69, written with R. L. Wager; "Serpentine Lavas, the Ankara Mélange and the Anatolian Thrust," in *Transactions of the Royal Society of Edinburgh,* **62** (1953), 403–442, written with W. J. McCallien; "Moine Tectonics and Metamorphism in Skye," in *Transactions of the Edinburgh Geological Society,* **16** (1955), 93–166; "Composite Minor Intrusions and the Slieve Gullion Complex, Ireland," in *Liverpool and Manchester Geological Journal,* **1** (1956), 466–501, written with W. J. McCallien; "The Ballantrae Serpentine," in *Transactions of the Edinburgh Geological Society,* **17** (1957), 33–53, written with W. J. McCallien; "Some Chemical Aspects of Southwest Highland Devonian Igneous Rocks," in *Bulletin of the Geological Survey,* **15** (London, 1958), 1–20; "Some Aspects of the Steinmann Trinity, Mainly Chemical," in *Quarterly Journal of the Geological Society of London,* **116** (1960), 365–395, written with W. J. McCallien; *Charles Lyell* (London, 1962); "Liguria Nappe: Northern Apennines," in *Transactions of the Royal Society of Edinburgh,* **65** (1963), 315–333, written with W. J. McCallien; and *James Hutton —The Founder of Modern Geology* (Amsterdam-London-New York, 1967).

II. SECONDARY LITERATURE. Writings dealing with Bailey or his work are F. Chayes, in *American Journal of Science,* **264** (1966), 128–145; K. C. Dunham, in *The Geology of Northern Skye,* a memoir of the Geological Survey (Edinburgh, 1966), p. 157; A. Holmes, in *Geological Magazine,* **73** (1936), 228–238, and *ibid.,* **86** (1949), 71–72; W. Q. Kennedy, in *American Journal of Science,* **25** (1933), 239–256; C. Koomans and P. H. Kuenen, in *Geological Magazine,* **75** (1938), 145–160; P. H. Kuenen and C. I. Migliorini, in *Journal of Geology,* **58** (1950), 91–127; A. G. MacGregor, in *Bulletin of the Geological Society of America,* **77** (1966), "Proceedings," 31–39; H. H. Read and A. G. MacGregor, *British Regional Geology: The Grampian Highlands,* 2nd ed. (Edinburgh, 1948), pp. 16–38; D. L. Reynolds, in *Nature,* **172** (1953), 69; R. R. Skelhorn and R. W. D. Elwell, in *Transactions of the Royal Society of Edinburgh,* **66** (1966), 294; C. J. Stubblefield, in *Biographical Memoirs of the Royal Society of London,* **11** (1965), 1–21; T. L. Tanton, in *Geological Magazine,* **67** (1930), 73–76; C. E. Tilley, in *Quarterly Journal of the Geological Society of London,* **106** (1950), 37–61; C. E. Tilley and I. D. Muir, in *Geological Magazine,* **104** (1967), 337–343; S. I. Tomkeieff, *ibid.,* **86** (1949), 130; T. Vogt, *ibid.,* **67** (1930), 68–73; and M. K. Wells and A. K. Wells, *ibid.,* **85** (1948), 349–357.

A. G. MACGREGOR

BAILEY, LIBERTY HYDE, JR. (*b.* South Haven Township, Michigan, 15 March 1858; *d.* Ithaca, New York, 25 December 1954), *botany, horticulture, agriculture.*

Bailey was the youngest son of Liberty Hyde Bailey, who had migrated from Townshend, Vermont, and Sarah Harrison Bailey, who came of a distinguished Virginia family. At an early age he manifested his precocity in the study of plants, birds, insects, and unusual rocks of the region. Since he grew up in an area abounding in orchards, he was also interested in the grafting of apple tree varieties.

At nineteen, Bailey entered Michigan State (Agricultural) College, where his genius for plant study was soon recognized by William James Beal, a former student of Asa Gray and a pioneer in the laboratory method of teaching botany. Bailey also studied Darwinian evolution, and in 1880 the *Botanical Gazette* (**5** [1880], 76–77) published his article "Michigan Lake Shore Plants." When Bailey received the B.S. degree in 1882, he had been trained in the use of compound microscopes and had begun experiments with *Rubus* and other plants. A brief stint as a reporter on the Springfield, Illinois, *Monitor* followed; but a visit to the herbarium of Michael Schuck Bebb testifies to his continued interest in plants. Late in 1882 Asa Gray of Harvard employed Bailey "at Cambridge . . . for a year or two" as assistant curator of the university herbarium. He was also assistant in physiological experiments and had charge of nomenclature for gardens, greenhouses, and the students' and garden herbaria. In June 1883 he married Annette Smith, of a farm near Lansing, Michigan.

From 1884 through 1900 Bailey published many papers on *Carex,* his first being a catalog that was presented in fuller form in 1887. In 1885 Michigan State (Agricultural) College called him to serve as professor of horticulture and landscape gardening. In that same year the American Pomological Society awarded Bailey its Wilder medal for an exhibit of native nuts and fruits. No later than 1886 he began making crosses and varietal studies in *Cucurbita,* another group on which he became an authority. His instruction in horticulture, in both classroom and field, embraced every facet of the subject as it was then conceived, and introduced such innovations as classification and nomenclature of fruits and vegetables, hybridization, and cross-fertilization of plant varieties. Also in 1886 Bailey was elected president of the Ingham County Horticultural Society, received the M.S. degree from Michigan State, and, with Joseph Charles Arthur, participated in a botanical survey in Minnesota. The first three of his more than sixty books appeared during this period.

In 1888 Bailey was summoned to Cornell University to occupy its chair of practical and experimental horticulture, the first such chair in an American university. He held it with distinction as horticulturist, botanist, rural sociologist, nature-study proponent,

editor, poet, philosopher, and world traveler until 1903, when he became the second director of the College of Agriculture. In May 1904 he was made first dean of the New York State College of Agriculture and director of its experiment station.

Bailey wrote hundreds of papers, but the magnitude of his work may best be indicated by enumerating some of his many honors. He was a founder and first president of the American Society for Horticultural Science (1903), a founder and president of the Botanical Society of America (1926), and twice president of the American Nature-Study Society (1914–1915). He also served as president of the American Association of Agricultural Colleges and Experiment Stations (1906), the American Pomological Society (1917), the American Association for the Advancement of Science (1926), the Fourth International Botanical Congress (1926), the American Country Life Association (1931), and the American Society of Plant Taxonomists (1939). Honorary degrees were conferred on him by Alfred University and the universities of Wisconsin, Vermont, and Puerto Rico, and he was a member of, among many organizations, the National Academy of Sciences, the American Philosophical Society, the American Academy of Arts and Sciences, and the Academy of Natural Sciences of Philadelphia.

In 1908 President Theodore Roosevelt appointed Bailey chairman of his Country Life Commission, which investigated the possibilities of improving rural conditions and, until 1911, submitted recommendations that were nationwide in scope. Bailey was also the founder, developer, and donor of the Bailey Hortorium of the New York State College of Agriculture, which publishes the quarterly journal *Baileya,* devoted to the botany of cultivated plants, their identification, nomenclature, classification, and history. Another periodical, *Gentes herbarum,* founded by Bailey, has attained worldwide influence.

Bailey was a recognized authority on *Carex, Rubus, Cucurbita,* the palms, *Vitis,* and certain of the cultivated groups, notably *Brassica.* He will be most remembered, however, for his great encyclopedias and important manuals of horticulture and agriculture; his summaries of progress; his texts; his books on principles of cultivation, harvesting, plant breeding, and evolution; his (and his daughter Ethel's) *Hortus;* his beautiful and informative "garden" books; and his so-called background books.

BIBLIOGRAPHY

Additional works on Bailey are addresses published in *Baileya,* **6,** and delivered at Cornell University, mid-March

1958, on the celebration of the Liberty Hyde Bailey centennial; E. Eugene Barker, "Liberty Hyde Bailey, Philosopher and Poet," in *Cornell Countryman,* **14,** no. 1 (1958), 13 f.; Lewis Knudson, "Liberty Hyde Bailey," in *Science,* **121** (1955), 322–323; George H. M. Lawrence, "Liberty Hyde Bailey, 1858–1954," in *Baileya,* **3** (1955), 27–40, with a list of degrees, honors, and societies and a bibliography; "Liberty Hyde Bailey, the Botanist," in *Bulletin of the Torrey Botanical Club,* **82** (1955), 300–305; "Professor L. H. Bailey," in *Nature,* **175** (1955), 451–452; and "The Bailey Hortorium, Its Past and Present," in *Baileya,* **4,** no. 1 (1956), 1–9; Andrew Denny Rodgers III, *Liberty Hyde Bailey, A Story of American Plant Sciences* (Princeton, 1949), see "Acknowledgments" for source materials—also a facsimile of 1949 ed. (New York, 1965), and *"Portrait Liberty Hyde Bailey,"* in *American Scholar* (Summer 1951), 336–340.

ANDREW DENNY RODGERS III

BAILEY, LORING WOART (*b.* West Point, New York, 28 September 1839; *d.* Fredericton, Canada, 10 January 1925), *geology.*

Loring Bailey was born at the United States Military Academy, where his father, Jacob Whitman Bailey, was professor of chemistry, mineralogy, and geology. His mother was Maria Slaughter Bailey. He received both his bachelor of arts degree (1859) and his master's degree (1861) from Harvard University. He thereupon moved to Canada, where from 1861 until his retirement in 1907 he held the chair of chemistry and natural science at the University of New Brunswick in Fredericton. In August 1863, Bailey married Laurestine Marie d'Avray.

The position at the university required that Bailey teach physics, chemistry, zoology, physiology, botany, and geology. After several years, however, the first two subjects were assigned to a colleague. At the beginning he had some knowledge of chemistry but little of the other fields—least of all geology. Yet geology captured his attention, primarily because of the complex stratigraphical relations of New Brunswick, which the previous studies by Abraham Gesner, J. William Dawson, and James Robb had done little to elucidate. Together with George F. Matthew and Charles F. Hartt, Bailey was the first to find fossils of Cambrian age in New Brunswick (near St. John), thus establishing the base for deciphering the chronostratigraphic interrelationships of a large area. Bailey's numerous studies of the region not only provided an extensive foundation for further work in New Brunswick but also contributed to the understanding of the geology of neighboring New England, especially Maine. His *Report on the Mines and Minerals of New Brunswick* and his *Observations on the Geology of Southern New Brunswick* are classics of regional

geology. Most of Bailey's numerous publications were in the form of official reports, lectures, and newspaper and magazine articles.

After his retirement from professorial duties and from summer field work, Bailey continued his biological researches, in particular the microscopic investigation of diatoms. (His first publication, in 1861, had been the completed form of his father's "Microscopical Organisms from South America," which the elder Bailey had left unfinished at his death in 1857.) During the course of his research, which was carried out in conjunction with the St. Andrews Marine Biological Station, the number of known New Brunswick diatoms increased from fifty to about four hundred. His work resulted in a catalog of Canadian diatoms, which was issued late in 1924. In 1888 and 1918 Bailey served as president of the geology section of the Royal Society of Canada, of which he was a charter member.

BIBLIOGRAPHY

The only published list of Loring Bailey's works is contained in *Loring Woart Bailey: The Story of a Man of Science* (St. John, 1925), his reminiscences, collected and edited by his son, Joseph Whitman Bailey. His most important regional studies are *Report on the Mines and Minerals of New Brunswick* (Fredericton, 1864); "Notes on the Geology and Botany of New Brunswick," in *Canadian Naturalist,* **1** (1864), 81–97; *Observations on the Geology of Southern New Brunswick . . . with a Geological Map* (Fredericton, 1865); and *The Mineral Resources of the Province of New Brunswick,* Geological Survey of Canada, Annual Report, n.s. **10,** part M (Ottawa, 1898). His microscopic biology begins with "Notes on New Species of Microscopical Organisms from the Para River, South America," in *Boston Journal of Natural History,* **7** (July 1861), 329–351, and concludes with "An Annotated Catalogue of the Diatoms of Canada showing their Geographical Distribution," in *Contributions to Canadian Biology,* n.s. **2** (1924), 31–68.

The biography by his son provides little insight into or information on Bailey's scientific achievements; for such see W. F. Ganong's memoir in *Proceedings and Transactions of the Royal Society of Canada* (May 1925), pp. xiv–xvii.

BERT HANSEN

BAILEY, SOLON IRVING (*b.* Lisbon, New Hampshire, 29 December 1854; *d.* Norwell, Massachusetts, 5 June 1931), *astronomy, meteorology.*

Bailey pioneered in studies in the southern hemisphere for the Harvard Observatory, at a time when there were no large telescopes south of the equator. His two main contributions to astronomy were studies of variable stars in globular clusters and his long-exposure photographs, which not only helped to elucidate the structure of our galaxy but also showed the value of photography in detecting extragalactic objects too faint to be observed visually.

Graduating from Boston University with an A.M. degree in 1884, Bailey began a career with the Harvard Observatory that was to last forty-four years. He earned a second A.M. degree from Harvard in 1888 and was then sent to South America by E. C. Pickering (director of the Observatory), to find a suitable location for Harvard's proposed southern observing post.

Bailey traveled throughout Peru and Chile, trying out various sites. No information on weather patterns was available to make his task easier, so he established a chain of meteorological stations—from sea level up to 19,000 feet. He published the data that he accumulated over the following thirty-seven years in his "Peruvian Meteorology" (1889 to 1930). As a result of his survey, Harvard's Boyden station, with the 24-inch Bruce doublet as its main telescope, was established near the city of Arequipa in Peru and remained in operation from 1890 until 1927.

Bailey took with him to Peru the visual photometer used for the original "Harvard Photometry," a compilation of standard magnitudes for stars located north of declination −30° (1884, 1885), and with it completed Harvard's coverage of the sky (1895). He then began photographing ω Centauri and other globular clusters, to get light curves for over a hundred of the stars now called cluster variables (1902, 1916); this work proved fundamental to estimates of astronomical distances subsequently made by Harlow Shapley and others.

Bailey's photographs of regions in and away from the plane of the Milky Way were made with exposures ranging up to nineteen hours and twenty-seven minutes (1908, 1913, 1917). They were used for relative star counts and also revealed more than 3,000 new extragalactic objects.

Bailey served as acting director of the Harvard Observatory from 1919 to 1921, between the tenures of Pickering and Shapley. During his fifth and final stay in Arequipa, in 1923, the University of San Agustín awarded him an honorary Sc.D. degree. His other honors included election to the National Academy of Sciences in Washington. After retiring in 1925 he completed his *History and Work of the Harvard Observatory, 1839–1927* (1931).

BIBLIOGRAPHY

I. Original Works. Except for the *History and Work of the Harvard College Observatory, 1839–1927,* which was published as Harvard Observatory Monograph No. 4 (Cambridge, Massachusetts, 1931), all works cited in the text appeared in *Annals of the Astronomical Observatory of Harvard College.* The original "Harvard Photometry" had appeared as "Observations with the Meridian Photometer during the years 1879–1882," by E. C. Pickering, Arthur Searle, and Oliver C. Wendell, in *Annals,* **14,** part I (1884), 1–324, and **14,** part II (1885), 325–512; Bailey's extension of it was "A Catalogue of 7922 Southern Stars Observed with the Meridian Photometer during the years 1889–1891" (reduced under the direction of E. C. Pickering), *ibid.,* **34** (1895), 1–259. The other papers by Bailey were "A Discussion of Variable Stars in the Cluster ω Centauri," *ibid.,* **38** (1902), 1–252; "Globular Clusters," *ibid.,* **76** (1916), 43–82; "A Catalogue of Bright Clusters and Nebulae," *ibid.,* **60** (1908), 199–229; "The Southern Milky Way," *ibid.,* **72** (1913), 71–78, with 9 half-tone plates; and "The Northern Milky Way," *ibid.,* **80** (1917), 83–89, with 9 half-tone plates.

Bailey's "Peruvian Meteorology," covering the period 1888–1925, appeared in the *Annals* in six parts: **39,** part I (1889), 1–153; **39,** part II (1906), 157–292; **49,** part I (1907), 1–103; **49,** part II (1908), 107–232; **86,** part III (1923), 123–194; and **87,** part 2A (1930), 179–217.

II. Secondary Literature. Annie Jump Cannon wrote two obituaries of Bailey: one in *Monthly Notices of the Royal Astronomical Society,* **92** (1932), 263–266; the other, with a portrait, appeared in *Publications of the Astronomical Society of the Pacific,* **43** (October, 1931), 317–323 and was reprinted, with the addition of a list of 95 publications by Bailey, in *Biographical Memoirs of the National Academy of Sciences* (Washington), **15** (1934), 193–203.

Sally H. Dieke

BAILLIE, MATTHEW (*b.* Shots Manse, Lanarkshire, Scotland, 27 October 1761; *d.* Duntisbourne, Gloucestershire, England, 23 September 1823), *medicine.*

Baillie's father was a Presbyterian minister in Shots, and later professor of divinity at Glasgow University. His mother, Dorothy Hunter Baillie, was the sister of the famous surgeons John and William Hunter. His sister, Joanna Baillie, was a well-known playwright and poet, who was an intimate friend of Sir Walter Scott. Baillie attended Hamilton Grammar School, Glasgow University, and, for a year, Balliol College, Oxford. He had won a scholarship to Oxford shortly after his father died, and before moving to Oxford he stayed in London with his uncle William Hunter, who operated the famous Windmill Street School of Anatomy. He soon was affiliated with the school, and when his uncle died in 1783, Baillie, then twenty-two, became master of the school. His uncle also left him the use of his medical museum for twenty years (after which it was to revert to Glasgow University) and a considerable fortune. In 1787 Baillie was elected physician to St. George's Hospital, a position he held for thirteen years. He received the M.D. from Oxford in 1789 and soon after became a fellow of the Royal College of Physicians as well as of the Royal Society. In 1791 he married Sophia Denman, sister of the Lord Chief Justice. The couple had two children.

Baillie's great importance is in the field of pathology. His *Morbid Anatomy of Some of the Most Important Parts of the Human Body* (1795) was the first English text on pathology, and the first systematic study in any language. Although Joseph Lieutaud and Giovanni-Battista Morgagni had earlier published studies in the field, Baillie was the first to take the various organs of the body serially and set forth the diverse morbid conditions of each. He believed that by studying the altered structure of morbid anatomy, it might be possible for the physician to learn what had caused the changes. Baillie was, however, careful not to go beyond his observations, and his modest goal prevented him from analyzing, correlating, or theorizing about his data. Most of his descriptions are based upon his own firsthand observations, many of them from specimens preserved by his uncles in their medical museum. Although explanations of morbid appearances have changed radically since his time, Baillie's descriptions are still of value. His work is limited primarily to thoracic and abdominal organs and the brain; he ignored changes observed in the skeleton, muscles, nerves, and spinal cord. In general, Baillie was quite logical in his procedures, and he carefully distinguished between inflamed states, thickenings, hardenings, softenings, ulcers, tubercles, aneurysms, and the like. The second and subsequent editions of his work included sections on symptoms to be observed. Most subsequent pathology textbooks have been modeled on his.

Once his study of morbid anatomy had been published, and then revised and illustrated, Baillie concentrated on the practice of medicine. By 1800 he had the largest medical practice in London, and he has been called the most popular physician of his time. His bust was later placed in Westminster Abbey; and when the London Pathological Society was founded in 1846, its members put Baillie's portrait on their seal. He left a fortune of over £80,000.

BIBLIOGRAPHY

I. ORIGINAL WORKS. Baillie's major work is *The Morbid Anatomy of Some of the Most Important Parts of the Human Body* (London, 1795; rev. 1797, 1799), the 1799 ed. illustrated with engravings. Earlier he had edited a MS by his uncle William Hunter that is often credited to Baillie:

Anatomy of the Gravid Uterus (London, 1794). Many of his essays, accompanied by a brief sketch of his life, were published in *The Works of Matthew Baillie,* James Wardrop, ed., 2 vols. (London, 1825).

II. SECONDARY LITERATURE. Information about Baillie's career is usually included in various works that discuss William and John Hunter. There is, for example, a brief biographical sketch of him in Richard Hingston Fox, *William Hunter, Anatomist, Physician, Obstetrician* (London, 1901), pp. 26–29. See also William Macmichael, *The Gold Headed Cane,* 7th ed. (Springfield, Ill., 1953), pp. 165–185; and Benjamin Ward Richardson, *Disciples of Aesculapius,* 2 vols. (London, 1900), II, 554–572. For an assessment of his importance, see Lester S. King, *The Medical World of the Eighteenth Century* (Chicago, 1958), pp. 277–281.

VERN L. BULLOUGH

BAILLOU, GUILLAUME DE, also known as **Baillon** and **Ballonius** (*b.* Paris, France, *ca.* 1538; *d.* Paris, 1616), *medicine.*

Baillou was the son of a mathematician and noted architect. He became a physician in 1570 and dean of the Faculty of Medicine in 1580. He had a reputation for dialectics and was known as "the bane of the bachelors," but he also was the most erudite and well-read doctor of his time—famous for his eloquence, the clarity of his courses, and his writings, almost all of which were published posthumously. Baillou fought against the tradition of Arab medicine and revived the Hippocratic tradition of clinical observation, human understanding, and macrocosmic concepts of illnesses and their treatment. Consequently, he thought that solar eclipses and the position of the stars were important in medical practice, and restored a little of the astrology completely eliminated by Jean Fernel—whose pupil he never was, despite claims to the contrary.

During the many epidemics in Paris between 1570 and 1579, Baillou developed the idea of the ephemerides, which influenced the work of Thomas Sydenham. He was thus the first Occidental epidemiologist since Hippocrates. He left excellent descriptions of the plague (and possibly of typhoid fever); of measles, which he distinguished from variola; and of diphtheria, whose choking false membranes he identified. Baillou is also credited with the first mention of adhesive pericarditis complicated by edema, and of whooping cough (*tussis quinta*). The latter is uncertain, however, for whooping cough seems to have been diagnosed in the *Khulasat-ul Tajjarib* (1501) of Baha'-ul-Douleh, an Iranian doctor.

Baillou brought rheumatism into nosology, and in this connection he gave a learned presentation of Fernel's, Lepois's, and Felix Platter's rather vague conceptions of the overflow of phlegm and aqueous humors. Fernel thought the *distillatio* was a superfluous humor running down from the head into the declivitous parts of the human organism and was the sole cause of chronic arthritis and crural neuralgia. He had located these ailments, two hundred years before Domenico Cotugno, in the sciatic notch, which was the seat of the diffusion of pain. Baillou attributed to rheumatism all pain in the external *habitus corporis,* that is, all areas situated between the skin and the internal parts of the body. The *habitus corporis* therefore included cellular tissues, muscles, tendons, aponeuroses, bones, and joints. He went on to describe a rheumatic diathesis (not necessarily articular) with painful localizations in the limbs. This was in the form of gout, sciatica, and arthritis or *morbus articularis* (rheumatic fever or subacute or chronic rheumatism). To this infinitely variable disease Baillou related all pains of the external body, from scurvy, smallpox, and lead poisoning to those thought to be caused by wind, worms, and other obscure agents.

BIBLIOGRAPHY

I. ORIGINAL WORKS. Baillou's writings are *Comparatio medici cum chirurgo ad castigandum quorumdam chirurgicorum audaciam* (Paris, 1577); *Consiliorum medicinalium libri duo* (Paris, 1635–1639), also published separately (Vol. II only) by Jacques Thévart (Paris, 1640); *Definitionum medicinalium liber* (Paris, 1639); *De convulsionibus libellus* (Paris, 1640); *Epidemiorum et ephemeridum libri duo* (Paris, 1640), also in Ralph Mayor, *Classic Descriptions of Disease* (Lawrence, Kans., 1932), pp. 94–96, 159–160, and translated into French by Prosper Yvaren as *Épidémies et éphémérides* (Paris, 1858); *Liber de rhumatismo et pleuritide dorsali* (Paris, 1642), trans. by Barnard in *British Journal of Rheumatism,* **2** (1940), 141–162; *Commentarius ad libellum Theophrasti de Vertigine. De virginum et mulierum morbis liber* (Paris, 1643); *Opuscula medica de arthritide, de calculo et urinarum hypostasi* (Paris, 1643); *Adversaria medicinalia* (Paris, 1644); and *Opera omnia medica,* 4 vols. (Paris, 1685), also Tronchin, ed., 4 vols. (Geneva, 1762).

II. SECONDARY LITERATURE. Works on Baillou are A. Chéreau, "Baillou," in *Dictionnaire encyclopédique des sciences médicales,* VIII (Paris, 1878); A. Delpeuch, *Histoire des maladies—La goutte et le rhumatisme* (Paris, 1900); Charles Fiessinger, *La thérapeutique des vieux maîtres* (Paris, 1897), pp. 110–115; E. W. Goodhall, "A French Epidemiologist of the Sixteenth Century," in *Annals of Medical History,* **7** (1935), 409–427, and "De Baillou Clinician and Epidemiologist," in *Journal of the American Medical Association,* **195** (1966), 957; and Philippe Very, "Il y a 350 ans mourait Guillaume de Baillou," in *Scalpel,* **119** (1966), 677–678.

PIERRE HUARD

BAILLY, JEAN-SYLVAIN (*b.* Paris, France, 15 September 1736; *d.* Paris, 12 November 1793), *astronomy.*

Bailly was born in the Louvre, where his father, Jacques, as keeper of the king's paintings, had an apartment. His father also owned a house in Chaillot, a fashionable suburb of Paris. Bailly later took up residence there, where he met Madame Gaye, a widow whom he married in 1787. Little is known of his youth, but it appears that he received schooling that would prepare him to assume his father's office, a position held by a member of the family since Bailly's great-grandfather. He succeeded to the title in 1768. Bailly was relieved of his functions in 1783 but continued to receive the pension accorded to the position and to be known as honorary keeper. When he was about twenty, he received mathematics lessons from Montcarville of the Collège Royal. Shortly thereafter he met Nicolas de Lacaille, France's greatest observational astronomer, and Alexis Clairaut, France's greatest theoretical astronomer. The further studies that he pursued with them directed him into his lifework.

Bailly undertook his first astronomical research in 1759—concerned, appropriately, with Halley's comet. His paper on the comet's theory, read to and published by the Académie des Sciences, pointed out that one could not conclude the duration of its revolution from that of its visibility. This paper was based upon the observations of Lacaille.

In 1760 Bailly established his own observatory in the Louvre. Although not ideal, the site served him well, and he felt it important for an astronomer to base his theories on his own observations. He now began to do this with Jupiter's satellites, reading two papers on this subject in 1762 and adding a third near the end of 1763.

By the time he read the third report, Bailly had become a member of the Académie. In January 1763 Bailly was elected to the vacancy created by the death of Lacaille, probably because he had been Lacaille's protégé and was engaged in editing an unfinished work of his. He had also shown greater promise than his competitors, Jeaurat and Messier, of significant theoretical researches.

Bailly approached the problem of inequalities in the motions of the four known satellites of Jupiter with Clairaut's lunar theory in mind. Improvements had been made in tables of the motions since Cassini's 1668 ephemerides, but these improvements had been made empirically. Bailly was the first to attempt to achieve better tables theoretically, by treating each satellite in turn as the third body in a three-body problem. His success was not complete—there were considerable discrepancies between his theoretical formulation of orbitary elements and their observed values—but he did demonstrate that the problem was amenable to solution by Newtonian principles.

Bailly's memoirs sparked interest in this subject, and the Académie made it the topic of its essay contest for 1766. As an Academician, Bailly could not win the prize; nevertheless, he considered himself in competition with the astronomers who entered, and therefore synthesized and extended his earlier researches to produce his major *Essai.* Although excellent, this work was greatly overshadowed by the prize-winning entry of Lagrange. As Bailly himself later admitted, Lagrange's use of the new method of the variation of parameters allowed him to resolve almost completely the problem of five simultaneously perturbed bodies, as opposed to his own continuing series of three-body treatments.

In 1766 there arose the possibility of Bailly's succeeding the ailing Grandjean de Fouchy as perpetual secretary of the Académie. To prepare for this position, Bailly wrote several *éloges.* He did not achieve his goal, however; in 1770 Fouchy named Condorcet as his assistant.

Bailly followed this setback with one of his best scientific papers, his 1771 memoir on the inequalities of the light of Jupiter's satellites. Fouchy had earlier noted that a satellite disappeared before its total immersion and that, at its emersion, it was observed only when a small segment had already emerged from the planet's shadow. He concluded that the size of the segment involved depended upon the amount of the particular satellite's light. Utilizing a new observational technique, Bailly confirmed Fouchy's theory while coupling his study of light intensities with determinations of the diameters of the satellites. The resultant paper added greatly to contemporary knowledge of Jupiter's satellites and suggested a standard observing method to reduce instrument and observer errors. Because the fourth satellite had not been eclipsing during the period of his observations, the memoir dealt only with the first three satellites. Bailly never completed the study, for this paper was his last theoretical effort.

Having moved to his father's house in Chaillot, Bailly turned his attention to literary pursuits. These were given direction by his scientific activity, as shown in the four-volume history of astronomy that he published between 1775 and 1782. These tomes represent his most lasting achievement and were responsible for additional honors. In 1783 he was elected to the Académie Française, and two years later he was named to the Académie des Inscriptions et Belles-Lettres. This made him a triple academician, the only Frenchman besides Bernard de Fontenelle to achieve this distinction.

Meanwhile, Bailly continued to work within the Académie des Sciences, although not at astronomical pursuits. In 1784 he was named to the commission appointed to investigate the extravagant claims then being made for "animal magnetism" by Mesmer and others, and he drafted the commission's damning report on that alleged phenomenon. It was probably for this service that, near the end of 1784, Bailly was made a supernumerary pensioner within the Académie. A year later he was appointed to a commission to investigate the Hôtel Dieu, the hospital of the poor of Paris; again he prepared the findings. Three reports submitted between 1786 and 1788 deplored the miserable conditions existing at the hospital and suggested means for their correction.

It was chiefly acclaim through these reports that catapulted Bailly into public affairs at the beginning of the French Revolution; the movement culminated on 15 July 1789 in his unanimous proclamation as the first mayor of Paris. He was reelected for a second term in August 1790, but in this second year he lost popularity, particularly after the unfortunate massacre of the Champ-de-Mars. Although he retired from office in November 1791 and left Paris in July 1792, he was not forgotten. Arrested in September 1793, Bailly was soon tried, found guilty, and condemned to the guillotine. Like those of his victims, his head fell on the Champ-de-Mars.

BIBLIOGRAPHY

I. ORIGINAL WORKS. Bailly's first paper, the "Mémoire sur la théorie de la comète de 1759," was published in the *Mémoires de mathématiques et de physique présentés à l'Académie Royale des sciences par divers savans et lûs dans ses assemblées,* **5** (1768), 12–18. The Académie employed this journal, generally known as its collection of memoirs *par savants étrangers,* as an outlet for papers by scientists who, like Bailly at that time, were not members of the Académie and, thus, were "foreign" to it. This would have been the case with his first two papers "sur la théorie des satellites de Jupiter," except that he reread them to the Académie after attaining membership; thus, these and the third were published in the regular *Mémoires de l'Académie Royale des sciences* for 1763. He also placed many subsequent papers there, including, most importantly, his "Mémoire sur les inégalités de la lumière des satellites de Jupiter, sur la mesure de leur diamètres, et sur un moyen aussi simple que commode de rendre les observations comparables, en remédiant à la différence des vues et des lunettes" (1771), pp. 580–667. His *Essai sur la théorie des satellites de Jupiter,* on the other hand, was issued as a separate volume with *privilège* granted by the Académie (Paris, 1766).

Two of Bailly's *éloges,* those of Corneille and Leibniz, won prizes. They were published separately by the appropriate institutions: the Académie des Sciences, Belles-Lettres & Arts of Rouen (Rouen, 1768) and the Académie Royale des Sciences et des Belles-Lettres of Berlin (Berlin, 1769), respectively. These were incorporated with others, including that of Lacaille, in a collection (Berlin, 1770). But all of these and some later ones may be most conveniently consulted in Vol. I of the two-volume *Discours et mémoires* (Paris, 1790).

The first of Bailly's historical works was the *Histoire de l'astronomie ancienne, depuis son origine jusqu'à l'établissement de l'école d'Alexandre* (Paris, 1775). Its argument for an antediluvean astronomy that prepared the way for the astronomers of recorded history brought him into conflict with Voltaire and resulted in the publication of his *Lettres sur l'origine des sciences* (Paris-London, 1777) and the similar *Lettres sur l'Atlantide de Platon* (Paris-London, 1778). These were sidelights, however, and he continued to work on the larger theme. His *Histoire de l'astronomie moderne* appeared in two stages. Two volumes, devoted to the period up to 1730, were published in 1779, and were followed by a third that brought the story up to 1782, the year of its publication. To these he added a special *Traité de l'astronomie indienne et orientale* (Paris, 1787).

The *Rapport des commissaires chargés par le roi de l'examen du magnétisme animal* was published separately by order of the king (Paris, 1784), as were the three hospital reports (Paris, 1786, 1787, 1788). The latter were subsequently reprinted in the *Histoire de l'Académie Royale des sciences,* which prefaces each volume of the regular *Mémoires,* but in association with incorrect years, as follows: (1785), 1–110; (1786), 1–9, 13–40. As with his earlier *éloges,* all of these and other reports may be consulted in his *Discours et mémoires,* Vol. II.

Bailly's own treatment of his political career is contained in the three-volume *Mémoires de Bailly, avec une notice sur la vie, des notes et des éclaircissemens historiques* (Paris, 1821), which constitute Vols. VIII–X of Berville and Barrière's *Collection des mémoires relatifs à la révolution française. . . .*

II. SECONDARY LITERATURE. A definitive biography has not yet been written. The most important general account by a contemporary is that by Bailly's friend Mérard de Saint-Just: *Éloge historique de Jean-Sylvain Bailly au nom de la république de lettres, par une société de gens de lettres; suivi de notes et de quelques pièces en prose et en vers* (London, 1794). Providing more insight into his astronomical work are J. J. Lalande's "Éloge de Bailly," originally published in the *Décade philosophique* of 30 pluviôse an III, **4** (1795), 321–330, but more easily consulted in his *Bibliographie astronomique . . .* (Paris, 1803), pp. 730–736; J. B. J. Delambre's treatment in his *Histoire de l'astronomie au dix-huitième siècle* (Paris, 1827), pp. 735–748; and D. F. Arago's "Bailly, biographie lue en séance publique de l'Académie des Sciences, le 26 février 1844," in his *Oeuvres complètes,* II (Paris, 1854), 247–426. On Bailly's observa-

tory, see G. Bigourdan, *Histoire de l'astronomie d'observation et des observatoires en France,* part 2 (Paris, 1930), 125–132.

Ignoring works that concentrate on his political career, the best modern study is E. B. Smith, "Jean-Sylvain Bailly; Astronomer, Mystic, Revolutionary, 1736–1793," in *Transactions of the American Philosophical Society,* n.s., **44** (1954), 427–538. Based entirely upon printed sources—and, in fact, providing a comprehensive checklist of Bailly's printed works and an extensive bibliography—Smith's work should be supplemented by an article devoted to demonstrating the existence of various as yet unused manuscript sources on Bailly: R. Hahn, "Quelques nouveaux documents sur Jean-Sylvain Bailly," in *Revue d'histoire des sciences,* **8** (1955), 338–353.

SEYMOUR L. CHAPIN

BAILY, FRANCIS (*b.* Newbery, Berkshire, England, 28 April 1774; *d.* London, England, 30 August 1844), *astronomy, metrology.*

Baily was one of the founders of the Astronomical Society of London (later the Royal Astronomical Society). His enthusiasm and organizing ability served to arouse interest in astronomy and to put its practical aspects on a firm footing. Today he is remembered (although frequently with his name misspelled) for "Baily's beads," an effect seen during solar eclipses by many men before Baily but never so vividly described.

Baily, the third son of a banker, received only an elementary education. At the age of fourteen he was apprenticed to a London mercantile firm. As soon as his seven years of apprenticeship were up, Baily set sail for the United States and two years of rugged adventures at sea and in the backwoods, which he described in his *Journal.* He returned to England in 1798, hoping to spend his life as an explorer. When all efforts to obtain backing for such a career proved fruitless, he became a stockbroker instead. Before the end of the Napoleonic wars he had published several actuarial tables, *An Epitome of Universal History,* and an astronomical paper (1811).

Having prospered on the stock exchange, Baily retired at the age of fifty to devote his full time to astronomy. He had become a fellow of the Royal Society in 1821 and was to serve the Astronomical Society as president during four two-year terms, the first beginning in 1825 and the last interrupted by his death.

Baily's first substantial astronomical work dealt with methods of determining latitude and time by the stars. Since no up-to-date star catalog was available for this purpose, Baily calculated the mean positions of 2,881 stars for the epoch 1 January 1830; this work, published in 1826, earned him his first Gold Medal from the Astronomical Society. (The second came in 1843, for his redetermination of the density of the earth.)

Work on the standard pendulum, the standard yard, and the ellipticity of the earth followed, interspersed among revisions of many star catalogs. In the course of preparing a new edition of the *Historica coelestis* of 1712, Baily found and published (1835) evidence that Edmund Halley, the second astronomer royal, had unduly maligned his predecessor, John Flamsteed.

It was during the annular eclipse of 15 May 1836, which he observed from Inch Bonney in Scotland, that Baily first saw the "beads." They are a transient phenomenon often seen at the beginning and end of totality in a solar eclipse, when the edge of the moon is close to inner tangency and a thin crescent of sunlight shines between mountains on the moon's limb. In Baily's own words, they appear as "a row of lucid points, like a string of bright beads . . . running along the lunar disc with beautiful coruscations of light." His report (1838) included a list of all previous observers, beginning with Halley in 1715, and aroused keen interest. For the solar eclipse of 8 July 1842, many astronomers accordingly journeyed to Italy, where the eclipse was to be total. The British astronomer royal, George B. Airy, who was in Turin, looked for, but did not see, "Mr. Baily's beads"; Baily himself, in Pavia, did see them but only at the beginning of totality (1846).

BIBLIOGRAPHY

I. ORIGINAL WORKS. Baily's works mentioned in the text are *Tables for the Purchasing and Renewing of Leases for Terms of Years Certain and for Lives* (London, 1802; 3rd ed., 1812); *The Doctrine of Life-Annuities and Assurances Analytically Investigated* (London, 1810; appendix 1813); "On the Solar Eclipse Said to Have Been Predicted by Thales," in *Philosophical Transactions of the Royal Society*, **101** (1811), 220–241; *An Epitome of Universal History*, 2 vols. (London, 1813); "On the Construction and Use of Some New Tables for Determining the Apparent Places of Nearly 3000 Principal Fixed Stars," in *Memoirs of the [Royal] Astronomical Society of London*, **2** (1826), whole appendix: Baily's Preface iii–xli, auxiliary tables xlii–liv, the general catalog lv–ccxxi, supplementary tables ccxx–ccxxiii, errata ccxxiv; *An Account of the Revd. John Flamsteed, the First Astronomer Royal* (London, 1835; facsimile reprint, omitting the star catalog, London, 1966); "On a Remarkable Phenomenon that Occurs in Total and Annular Eclipses of the Sun," in *Memoirs of the Royal Astronomical Society*, **10** (1838), 1–42; "Some Remarks on the Total Eclipse of the Sun on July 8th, 1842," *ibid.*, **15** (1846), 1–8.

Ninety-one publications are listed in the preface to Baily's travel diary, *Journal of a Tour in Unsettled Parts of North America in 1796 and 1797*, Augustus de Morgan, ed. (London, 1856), pp. 61–69; fifty appear in the Royal Society of London's *Catalogue of Scientific Papers*, I (London, 1867), 158–160. Baily's long articles constitute the bulk of the first fifteen volumes of the *Memoirs of the [Royal] Astronomical Society of London* (1822–1846)—his edition of Tobias Mayer's star catalog in **4** (1831), 391–445; of the Abbé de La Caille's catalog in **5** (1833), 93–124; of the catalogs of Ptolemy, Ulugh Beigh, Tycho Brahe, Halley, and Hevelius in **13** (1843), prefaces 1–48, tables (1)–(248), with errata facing p. 1; his work on the earth's ellipticity in **7** (1834), 1–378; on the standard yard in **9** (1836), 35–184; and on the earth's density in **14** (1843), 1–120 with tables on i–ccxlvii. His paper on correcting a pendulum to vacuum is in *Philosophical Transactions of the Royal Society of London*, **122** (1832), 399–492. Three more star catalogs were completed and published posthumously: a revision of Jérôme Lalande's *Histoire céleste française* and of La Caille's southern hemisphere stars (both London, 1847) and finally the ultimate evidence of Baily's industry, the British Association for the Advancement of Sciences' catalog of almost 10,000 stars (London, 1845).

II. SECONDARY LITERATURE. Baily's entry in *Dictionary of National Biography*, II (London, 1885), 427–432, written by Agnes M. Clerke, includes an appraisal of his achievements. An obituary by Sir John Herschel appeared in *Philosophical Magazine* (London), ser. 3, **26** (1845), 38–75, and was reprinted, with additions, as part of the preface to Baily's *Journal*.

SALLY H. DIEKE

BAIN, ALEXANDER (*b.* Aberdeen, Scotland, 11 June 1818; *d.* Aberdeen, 18 September 1903), *philosophy, psychology.*

Alexander Bain was the son of George Bain, a handloom weaver. Poverty forced the father to take young Alexander out of school at the age of eleven, and for the next seven years the boy had to work for a living. However, he spent his spare time studying Latin, Greek, mathematics (chiefly algebra), and some mechanics. He became very proficient in these subjects, and in 1836 he won a bursary to study at Marischal College, in Aberdeen. He excelled at his studies during his four years at Marischal, and when he received his degree of master of arts in 1840, he was adjudged the best candidate of the year. He had acquired a training both in the humanities and the natural sciences.

After receiving his degree, Bain taught mental and moral philosophy for five years at Marischal. During these years he did much of his thinking on psychology. For the next fifteen years he held various lectureships for short durations. It was also during this time that he became a close friend of John Stuart Mill. In 1860 he was elected to the chair of logic at the newly formed Aberdeen University (a union of Marischal and King's colleges). Bain remained in this post until he retired in 1880. In 1876 he founded, and for sixteen years edited, the philosophical journal *Mind,* which has remained an important journal in the field to the present day. He served as rector of the university from 1882 to 1886. Bain was married twice. His first wife, Frances, whom he married in 1855, died in 1892. In April 1893 he married Barbara Forbes.

After an early interest in the natural sciences, Bain turned his attention to philosophical psychology. Here his training in the philosophy of Reid and Beattie, the influence of his friends Mill and George Grote, and the writings of Comte and Whewell are apparent. He had a great respect for facts and a mistrust of speculative metaphysics. He was also impressed by the physiological theories of Johannes Müller and was convinced that they were essential to the study of psychology.

Two powerful and complementary ideas of Bain's philosophy concerned the unity of the mind and the active power of the mind: "The argument for the two substances have, we believe, now entirely lost their validity; they are no longer compatible with ascertained science and clear thinking. The one substance with two sets of qualities, the physical and mental—a double-faced unity—would comply with all the exigencies of the case" (*Mind and Body,* p. 196). Bain's study of the nervous system gave him a way of correlating every mental process with some physiological process. For example, the will is identified with surplus energy in the nervous system.

The active nature of the mind is emphasized not only with respect to the feelings and volitions of an individual but also to the sensations. The mind can discriminate between sensations and can retain some of them; and he found that greater retention of some sensations is connected closely with greater discriminations of these sensations. Bain also believed that instinct, which includes reflex actions and primitive combined movements, is another active principle of the mind. Thus Bain presented the mind as an active unity, which superseded the then reigning theory of the faculties of the mind.

In ethics Bain followed the utilitarian position set out by Mill. He also followed Mill in logic, even in criticizing the Aristotelian syllogism as fallacious; but later, in a note in *Mind* (reprinted in *Dissertations*), he renounced this position.

BIBLIOGRAPHY

For a complete bibliography of Bain's work, see his *Autobiography* (London, 1904). Articles of primary importance include "Electrotype and Daguerrotype," in *Westminster Review,* **34** (Sept. 1840), 434; "Constitution of Matter," *ibid.,* **36** (July 1841), 69; "An Attempt to Generalise and Trace to One Sole Cause—viz. the Liberation of Latent Heat—All Cases of Terrestrial Heat," read to the Aberdeen Philosophical Society (6 Jan. 1843), manuscript in the Aberdeen University Library; "On a New Classification of the Sciences," read to the Aberdeen Philosophical Society (1 Dec. 1843), manuscript in the Aberdeen University Library; and "On the Definition and Classification of the Human Senses," read to the Aberdeen Philosophical Society (9 Dec. 1844), manuscript in the Aberdeen University Library. See also "On the Impediments to the Progress of Truth from the Abuse of Language," an 1845 MS in the Aberdeen University Library. For major books by Bain see *The Senses and the Intellect* (London, 1855); *The Emotions and the Will* (London, 1859); *An English Grammar* (London, 1863); *Mental and Moral Science—a Compendium of Psychology and Ethics* (London, 1868); *Logic,* part I, "Deduction" (London, 1870), part II, "Induction" (London, 1870); *Mind and Body: The Theories of Their Relation,* International Scientific Series, Vol. IV (London, 1872); *Mental Science: Psychology and History of Philosophy* (London, 1872); *Moral Science: Ethical Philosophy and Ethical Systems* (London, 1872); *Education as a Science,* International Scientific Series, Vol. XXV (London, 1879); *James Mill: A Biography* (London, 1882); *John Stuart Mill: A Criticism With Personal Recollections* (London, 1882). Some of Alexander Bain's most important articles, chiefly from the journal *Mind,* can be found in *Dissertations on Leading Philosophical Topics* (London, 1903). Works edited by Bain include *Paley's Moral Philosophy,* with dissertations and notes, Chamber's Instructive and Entertaining Library (Edinburgh, 1852); *James Mill's Analysis of the Human Mind,* with notes, 2 vols. (London, 1869); *Grote's Aristotle,* with G. C. Robertson (London, 1872); and *Grote's Minor Works,* with critical remarks (London, 1873).

JAGDISH N. HATTIANGADI

BAIRD, SPENCER FULLERTON (*b.* Reading, Pennsylvania, 3 February 1823; *d.* Woods Hole, Massachusetts, 19 August 1887), *zoology, scientific administration.*

Baird's father was Samuel Baird, a lawyer of local prominence; his mother was the former Lydia MacFunn Biddle of Philadelphia. Upon Samuel's death in 1833, Lydia and her seven children moved to Carlisle, Pennsylvania, where Spencer Baird entered Dickinson College in 1836, receiving a B.A. in 1840 and an M.A. in 1843. In November 1841, Baird went to New York City to study medicine, but two months later he abandoned his studies and returned to Carlisle, determined to pursue a career in zoology despite the limited opportunities then available to American biologists. Since the United States lacked institutions offering professional training in science, Baird's education consisted of self-study and informal instruction from established naturalists, including James Dwight Dana, John James Audubon, and George N. Lawrence. In 1846 he married Mary Helen Churchill, daughter of a well-known army officer, and in the same year became professor of natural history at Dickinson College. Lucy Hunter Baird, the couple's only child, was born in 1848.

In 1850 Baird's writings in systematic zoology; his translation and revision of the four-volume *Iconographic Encyclopaedia of Science, Literature, and Arts,* compiled in cooperation with many leading American scientists; and recommendations from politicians and scientists secured his appointment as assistant to Joseph Henry at the Smithsonian Institution. For the next thirty-seven years Baird used numerous governmental expeditions, plus a network of private collectors, to bring distinguished zoological and anthropological collections to the Smithsonian's National Museum. He became a member of the National Academy of Sciences in 1864, over the opposition of Louis Agassiz, who had personal differences with Baird and also contended that, as a descriptive biologist, Baird contributed no new knowledge to science. In 1871 Congress established the U.S. Fish Commission under Baird's direction. This agency conducted basic research in marine biology, propagated food fishes, and aided the fishing industry. In 1878 he succeeded Joseph Henry as secretary of the Smithsonian, a position he held until his death. Baird was noted for his serenity and modesty, but in his private dealings he was forceful and persistent in pursuing his ambitions. He gradually drifted from his Protestant upbringing, and after 1875 did not attend religious services.

Baird's bibliography includes more than a thousand titles, of which about ninety were formal scientific contributions. The remaining writings were largely official reports or brief review articles in the *Annual Record of Science and Industry,* a semipopular series of eight volumes that he edited from 1871 to 1878. His most notable scientific papers were taxonomic studies of birds and mammals, but he also wrote on reptiles, amphibians, and fishes, usually in collaboration with Charles Girard.

His scientific writings earned Baird his reputation as the leading vertebrate zoologist of mid-nineteenth-century America. Four of his works were especially significant. The first two, *Mammals* (1857) and *Birds* (1858), were comprehensive monographs based on American collections taken north of Mexico by fifteen governmental surveys and by numerous individual naturalists. From this extensive material, Baird was able to describe seventy species of mammals and 216 bird species not previously known. Moreover, he completely recast the nomenclature and classification of the two classes. Because of the accuracy and originality of his descriptions, and his use of several specimens to establish the characteristics of a particular species, Baird's methods were in themselves an advance. According to David Starr Jordan, Baird's "minute exactness" thereby "departed widely from the loose and general type of description" of his predecessors, especially by "using a particular individual and then indicating with precision any deviations due to age, sex, geographical separation or other influences which might appear in other specimens" ("Spencer Fullerton Baird and the United States Fish Commission," p. 101). The two works replaced studies prepared in the preceding two decades by Audubon and John Bachman, and remained standard sources at the time of Baird's death. Many of the species identified by Baird are now considered to be synonyms or subspecies.

A third major work, "The Distribution and Migrations of North American Birds," was Baird's only important attempt at scientific generalization. This 1865 paper was notable for the definition of biological zones and the discussion of "general laws" showing "certain influences exerted upon species by their distribution . . . and by their association with each other" ("Distribution," p. 189). Baird noted that an understanding of environmental influences allowed taxonomists to reduce the number of nominal species. The paper's larger import was to reveal Baird's support of organic evolution. It was recognized by his contemporaries as one of the major commentaries on Darwinian evolution produced by American naturalists during this era.

In 1874 Baird, assisted by Thomas M. Brewer and Robert Ridgway, published his final major work, a three-volume study of land birds entitled *A History of North American Birds.* Baird and Ridgway prepared the technical descriptions while Brewer, working primarily from field notes forwarded to the Smithsonian by Baird's collectors, contributed accounts of the habits of individual species. The *History* was notable for presenting the first comprehensive information on the behavior of birds in Arctic breeding grounds. It remained a standard treatise on ornithological life history throughout the nineteenth century.

Baird's significance as a teacher and as a molder of scientific institutions was probably greater than his personal scientific work. He was the patron of numerous naturalists who collected for him or studied informally at the Smithsonian. Among his most important protégés were George Brown Goode, C. Hart Merriam, Robert Ridgway, and William Healey Dall. Baird's simultaneous direction of the U.S. Fish Commission, the U.S. National Museum, and the Smithsonian during the last ten years of his life indicates his influence on American scientific institutions. From 1871 through 1887, the Fish Commission's research ships and volunteer scientists undertook the first sustained biological study of American waters. The U.S. National Museum, along with Harvard's Museum of Comparative Zoology, were the leaders in American zoology during this period. The Smithsonian, under Baird's administration, was no longer restrained by its modest private income. In contrast with Joseph Henry, who feared that government funds would lead to political interference, Baird did not hesitate to seek congressional appropriations to expand the work of the Smithsonian.

BIBLIOGRAPHY

I. ORIGINAL WORKS. For a comprehensive bibliography of Baird's writings, see George Brown Goode, *The Published Writings of Spencer Fullerton Baird, 1843–1882* (Washington, D.C., 1883). Baird's two monographs on mammals and birds were initially published as Vols. VIII and IX of *Reports of Explorations and Surveys to Ascertain the Most Practicable and Economical Route for a Railroad from the Mississippi River to the Pacific Ocean* (Washington, D.C., 1857, 1858). An abstract of "The Distribution and Migrations of North American Birds," read to the National Academy of Sciences in 1865, appears in *The American Journal of Science and Arts*, 2nd ser., **41** (Jan.–May 1866), 78–90, 184–192, 337–347.

Two valuable sources presenting portions of Baird's scientific correspondence are Ruthven Deane, "Unpublished Letters of John James Audubon and Spencer F. Baird," in *The Auk*, **23** (Apr. and July 1906), 194–209, 318–334, and **24** (Jan. 1907), 53–70; and Elmer Charles Herber, ed., *Correspondence Between Spencer Fullerton Baird and Louis Agassiz, Two Pioneer American Naturalists* (Washington, D.C., 1963).

Baird's personal papers and letters documenting his work as a Smithsonian official are held by the Smithsonian Institution, Washington, D.C. The comprehensive records of the U.S. Fish Commission are in the U.S. National Archives, Washington, D.C.

II. SECONDARY LITERATURE. The only biography is William Healey Dall, *Spencer Fullerton Baird: A Biography* (Philadelphia, 1915). Dall presents much useful data, but the work is now dated. He slights the last seventeen years of Baird's career.

Among the most useful brief sketches of Baird are John Shaw Billings, "Memoir of Spencer Fullerton Baird,

1823–1887," in *Biographical Memoirs of the National Academy of Sciences,* III (Washington, D.C., 1895), 141–160; George Brown Goode, "The Three Secretaries," in *The Smithsonian Institution, 1846–1896,* George Brown Goode, ed. (Washington, D.C., 1897), pp. 115–234; and David Starr Jordan, "Spencer Fullerton Baird and the United States Fish Commission," in *The Scientific Monthly,* **17** (Aug. 1923), 97–107. A detailed assessment of Baird's personal scientific work is presented in T. D. A. Cockerell, "Spencer Fullerton Baird," in *Popular Science Monthly,* **68** (Jan. 1906), 63–83.

A recent and extended discussion of Baird's career, concentrating on his work with the Fish Commission, is Dean C. Allard, "Spencer Fullerton Baird and the U.S. Fish Commission: A Study in the History of American Science," unpublished Ph.D. diss. (Washington, D.C., 1967). This work contains a lengthy bibliography of published literature relating to Baird.

DEAN C. ALLARD

BAIRE, RENÉ LOUIS (*b.* Paris, France, 21 January 1874; *d.* Chambéry, France, 5 July 1932), *mathematics.*

Baire was one of three children in a modest artisan's family. His parents had to sacrifice in order to send him through high school. Having won a scholarship competition for the city of Paris in 1886, he entered the Lycée Lakanal as a boarding student; there he completed his advanced classes in 1890 after having won two honorable mentions in the Concours Général de Mathématiques.

In 1891 Baire entered the section for special mathematics at the Lycée Henri IV, and in 1892 was accepted at both the École Polytechnique and the École Normale Supérieure. He chose the latter, and during his three years there attracted attention by his intellectual maturity. He was a quiet young man who kept to himself and was profoundly introspective. During this period he was found to be in delicate health.

Although he placed first in the written part of the 1895 *agrégation* in mathematics, Baire was ranked third because of a mistake in his oral presentation on exponential functions, which the board of examiners judged severely. In the course of his presentation Baire realized that his demonstration of continuity, which he had learned at the Lycée Henri IV, was purely an artifice, since it did not refer sufficiently to the definition of the function. This disappointment should be kept in mind, because it caused the young lecturer to revise completely the basis of his course in analysis and to direct his research to continuity and the general idea of functions. While studying on a scholarship in Italy, Baire was strengthened in his decisive reorientation by Vito Volterra, with whom he soon found himself in agreement and who recognized the originality and force of his mind.

On 24 March 1899 Baire defended his doctoral thesis, on the theory of the functions of real variables, before a board of examiners composed of Appell, Darboux, and Picard. The few objections, which Baire fully appreciated, proved that he had embarked on a new road and would not find it easy to convince his listeners.

Baire began his teaching career in the *lycées* of Troyes, Bar-le-Duc, and Nancy, but he could not long endure the rigors of teaching the young. In 1902, as lecturer at the Faculty of Sciences of Montpellier, he wrote a paper on irrational numbers and limits. In 1904 he was awarded a Peccot Foundation fellowship to teach for a semester at the Collège de France. At that time this award went to young teachers to enable them to spend several months, free of routine duties, in developing their own specialties. Baire chose to work on a course in discontinuous functions, later edited by his pupil A. Denjoy and published in the Collection Borel, a series of monographs on the theory of functions.

Upon his return to Montpellier, Baire experienced the first violent attack of a serious illness that became progressively worse, manifesting itself in constrictions of the esophagus. After the crisis passed, he began drafting his paper "Sur la représentation des fonctions discontinues." Appointed professor of analysis at the Faculty of Science in Dijon in 1905, to replace Méray, he devoted himself to writing an important treatise on analysis (1907–1908). This work revivified the teaching of mathematical analysis. His health continued to deteriorate, and Baire was scarcely able to continue his teaching from 1909 to 1914. In the spring of 1914 he decided to ask for a leave of absence. He went first to Alésia and then to Lausanne. War broke out while he was there, and he had to remain—in difficult financial circumstances—until the war ended.

Baire was never able to resume his work, for his illness had undermined his physical and mental health. He now devoted himself exclusively to calendar reform, on which he wrote an article that appeared in the *Revue rose* (1921). While still at his retreat on the shores of Lake Geneva, Baire received the ribbon of the Legion of Honor, and on 3 April 1922 was elected corresponding member of the Academy of Sciences. A pension granted him in 1925 enabled him to live in comparative ease, but the devaluation of the franc soon brought money worries. His last years were a struggle against pain and worry.

Thus Baire was able to devote only a few periods, distributed over a dozen years, to mathematical research. In addition to the already mentioned works, of particular importance are "Sur les séries à termes continus et tous de même signe" and "Sur la non-applicabilité de deux continus à n et $n + p$ dimensions." His writings, although few, are of great value.

Baire's doctoral thesis solved the general problem of the characteristic property of limit functions of continuous functions, i.e., the pointwise discontinuity on any perfect aggregate. In order to imagine this characteristic, one needed very rare gifts of observation and analysis concerning the way in which the question of limits and continuity had been treated until then. In developing the concept of semicontinuity—to the right or to the left—Baire took a decisive step toward eliminating the suggestion of intuitive results from the definition of a function over a compact aggregate. But in order to obtain the best possible results, one needed a clear understanding of the importance of the concepts of the theory stemming from aggregates.

> Generally speaking, in the framework of ideas that here concerns us, every problem in the theory of functions leads to certain questions in the theory of sets, and it is to the degree that these latter questions are resolved, or capable of being resolved, that it is possible to solve the given problem more or less completely [*Sur les fonctions*].

In this respect, Baire knew how to use the transfinite in profoundly changing a method of reasoning that had been applied only once before—and this in a different field.

> From the point of view of derivative sets that interests us here, it may be said that if α is a number of the first type, then P^α is the set derived from $P^{\alpha-1}$, and if α is of the second type, P^α is by definition the set of points that belongs to all $P^{\alpha'}$ where α' is any number smaller than α. Independently of any abstract considerations arising from Cantor's symbolism, P^α represents a fully determined object. Nothing more than a convenient language is contained in the use we shall be making of the term "transfinite number" [*Ibid.*, p. 36].

Until the arrival of Bourbaki, his success greatly influenced the orientation of the French school of mathematics.

In line with his first results, Baire was led to approach the problem of integrating equations with partial derivatives at a time when their solution was not subjected to any particular condition of continuity. But it is in another, less specialized field that Baire's name is associated with lasting results.

Assigning to limit functions of continuous functions the name of Class 1 functions, Baire first endeavored to integrate functions of several variables into this class. Thus he considered those functions that are separately continuous in relation to each of the variables of which they are a function. Then he defined as Class 2 functions the limits of Class 1 functions, and as Class 3 functions the limits of Class 2 functions. Having established basic solutions for these three classes, he obtained a characteristic common to the functions of all classes.

Since the beginning of the nineteenth century, mathematicians had been interested in the development of the theory of functions of real variables only incidentally and in relation to complex variables. Baire's work completely changed this situation by furnishing the framework for independent research and by defining the subject to be studied.

If Baire did not succeed, even in France and despite intensive efforts, in overcoming the mistrust of the transfinite, he nevertheless put an end to the privileged status of continuity and gave the field to aggregate-oriented considerations for the definition and study of functions. Baire's work, held in high esteem by Émile Borel and Henri Lebesgue, exerted considerable influence in France and abroad while its author found himself incapable of continuing or finishing the task he had set himself.

There can be no doubt that the progress of modern mathematics soon made obsolete Baire's work on the concept of limit and the consequences of its analysis. But this work, written in beautiful French, has an incomparable flavor and merits inclusion in an anthology of mathematical thought. It marks a turning point in the criticism of commonplace ideas. Moreover, the class of Baire's functions, according to the definition adopted by Charles de la Vallée Poussin, remains unattainable as far as the evolution of modes of expression is concerned. This model of a brief and compact work is part of the history of the most profound mathematics.

BIBLIOGRAPHY

I. ORIGINAL WORKS. Baire's writings are *Sur la théorie analytique de la chaleur* (Paris, 1895), his ed. of Henri Poincaré's 1893–1894 lectures; *Sur les fonctions de variables réelles* (Milan, 1899), his doctoral thesis (Faculté des Sciences, Paris, no. 977); "Sur les séries à termes continus et tous de même signe," in *Bulletin de la Société mathématique de France,* no. 32 (1904), 125–128; *Leçons sur la théorie des fonctions discontinues,* A. Denjoy, ed. (Paris, 1905, 1930); *Théorie des nombres irrationnels, des limites et de la continuité* (Paris, 1905, 1912, 1920); "Sur la représentation des fonctions discontinues," in *Acta mathematica,* no. 30 (1906), 1–48, and no. 32 (1909), 97–176; "Sur la non-applicabilité de deux continus à n et $n + p$

dimensions," in *Bulletin des sciences mathématiques* (Darboux), no. 31 (1907), 94–99; *Leçons sur les théories générales de l'analyse,* 2 vols. (Paris, 1907–1908); and "Origine de la notion de semi-continuité," in *Bulletin de la Société mathématique de France,* no. 55 (1927), 141–142.

Baire's works are now being collected.

II. SECONDARY LITERATURE. Works dealing with Baire are Émile Borel, *L'espace et le temps* (Paris, 1922), p. 123; A. Buhl and G. Bouligand, "En mémoire de René Baire," in *L'enseignement mathématique,* **31,** nos. 1–3 (1932); Constantin Caratheodory, *Vorlesungen über reellen Funktionen* (Vienna, 1918), pp. 178, 393–401; Hans Hahn, *Reellen Funktionen* (Vienna, 1932), pp. 276–324; H. Lebesgue, notice in *Monographies de l'enseignement mathématique,* no. 4 (1958); Marijon, notice in *Annuaire de l'Association amicale des anciens élèves de l'École normale supérieure* (1933), pp. 82–87; Paul Montel, "Quelques tendances dans les mathématiques contemporaines," in *Science (L'encyclopédie annuelle;* 1 Mar. 1937); P. Sergescu, *Les sciences mathématiques (tableau du XXe siècle)* (Paris, 1933), pp. 77–79, 170; and C. de la Vallée Poussin, *Intégrales de Lebesgue, fonctions d'ensemble, classes de Baire* (Paris, 1916, 1934), pp. 113–160.

PIERRE COSTABEL

AL-BAIRŪNĪ. See **al-Bīrūnī.**

IBN AL-BAIṬĀR. See **Ibn al-Bayṭār.**

IBN BĀJJA, ABŪ BAKR MUḤAMMAD IBN YAHYĀ IBN AL-ṢĀ'IGH, also known as **Avempace** or **Avenpace** (*b.* Saragossa, Spain, end of the 11th century; *d.* Fez, Morocco, 1138/1139), *philosophy.*

Ibn Bājja was a Muslim philosopher who wrote in Arabic. Besides Saragossa and Fez, he lived and worked in Seville and Granada. He is said to have been a vizier serving an Almoravid prince and to have been poisoned by physicians who were jealous of his medical skill. Ibn Bājja is often described as the earliest Arabic Aristotelian in Spain, which is correct but rather less significant than a characterization made by a friend and editor of his work, Abu'l-Hasan 'Alī of Granada, who indicates that Ibn Bājja had a preeminent part in establishing in Spain a systematic method for the study of the philosophical sciences. Such a method was already in existence in the Muslim East, but not in the peripheral Muslim West.

It can be taken as certain that, in working out his curriculum, Ibn Bājja, like the philosophers of the East, attached the greatest importance to the study of the *corpus Aristotelicum.* In spite of far-reaching doctrinal divergences he appears to have modeled his philosophical method on that of al-Fārābī rather than on that of Ibn Sīnā (Avicenna), whose influence was at that time predominant in the eastern centers of learning. In this and in various other respects he seems to have been responsible for the distinctive character of Spanish Aristotelianism, which counts among its representatives Ibn Rushd (Averroës) and Moses Maimonides. Among his extant works are the following:

Tadbīr al-mutawaḥḥid ("The Regimen of the Solitary"). The work deals with the various categories of men—those interested only in the bodily functions, those swayed by such "spiritual" faculties as imagination, and those governed by reason—in relation to man's final end, which is intellectual perfection; with various regimens; and with the position that the philosopher should adopt in relation to the imperfect communities in which he has to live (the reference is to the Islamic states of the time). In the absence of any hope for the creation of a perfect philosophical body politic (the conception of which goes back to Plato), the philosopher should regard himself as a stranger in his own community and as a citizen of the ideal State constituted by the happy few, i.e., the men, in whatever country they live, who in the past or the present have attained intellectual perfection.

A treatise on the union of man with the Active Intellect.

Risālat al-widāʿ ("The Epistle of Farewell"). The work treats some of the themes discussed in the *Tadbīr.*

A book on the soul.

Commentaries and notes on some works or parts of works belonging to the *corpus Aristotelicum,* including treatises of the *Organon* (there is also a commentary on Porphyry's *Isagoge*), the *Physics, De generatione et corruptione,* the *Meteorologica,* and *The Book of Animals.* There is also a "Book of Plants."

In a letter to Abū Jaʿfar Yūsuf Ibn Ḥasday, Ibn Bājja gives some details concerning his intellectual biography. After having learned the art of music, he studied astronomy; he then went on to study Aristotle's *Physics.* In astronomy Ibn Bājja rejected the theory of epicycles as being incompatible with the (Aristotelian) physical doctrine (see Maimonides, *The Guide of the Perplexed,* II, 24). He thus seems to have been one of the initiators of the tendency—which, after him, came to the fore in Muslim Spain—to reject and attempt to replace the Ptolemaic system.

Ibn Bājja's dynamics, as set forth in his notes on the seventh book of Aristotle's *Physics,* may, *inter alia,* be regarded as an attempt to unify the Aristotelian theory of movement by replacing the multiform concept of cause with the notion of force. The Arabic term used by Ibn Bājja is a translation of the Greek *dynamis,* but in the context it seems to have an exclusively active sense (which the term also has in Book VII of Aristotle's *Physics,* in certain texts of pagan Neoplatonic philosophers, and in John Philoponus' writings); it is in no way a potentiality. The unifying function of Ibn Bājja's concept of force may to some extent be classified by a consideration of the notion of "fatigue" (a term used in somewhat similar but not identical contexts by Alexander of Aphrodisias and by John Philoponus), which in Ibn Bājja's doctrine is associated with it. According to Ibn Bājja, the force of a mover may become "fatigued" (1) by the mere fact that it is exerted in moving a body and (2) because of the reaction of the body that is moved, when this body is other than the mover.

In relation to the first factor Ibn Bājja's statements give rise to some perplexity. One of them seems to purport that the "natural" motions of simple bodies do not cause fatigue, the reason being that in these motions there is no opposition between the mover and the moved body. It would thus appear that factor (1) causes fatigue only when it is accompanied by factor (2). The latter factor produces fatigue because a body moved by another body (and also a living being moved by its soul) causes in its turn (apparently in proportion to its fatigue) motion in its mover and thus fatigues the latter. In other words, there are an action and a reaction, the two being essentially comparable. While this view can be regarded as a development of some conceptions implicitly contained in the seventh book of Aristotle's *Physics,* it does not agree with other Aristotelian texts, which tend to assimilate the relation between a mover and a moved body to the relation between an agent and the thing acted upon or between a cause and an effect; that is to say, to relations that from the Peripatetic point of view do not lend themselves easily, and perhaps do not lend themselves at all, to being quantified. As against this, there is no theoretical difficulty about the quantification of the relation of action and reaction postulated by Ibn Bājja.

This view could, as it seems, be expressed in the following formula (whose validity would, however, be restricted to the instant in which a given body is moved): $M = F_1 - F_2$, M being the motion, F_1 the force of the mover, and F_2 the force of the moved body. This formula does not take into consideration the progressive weakening, or in Ibn Bājja's terminology "fatigue," of the force of the mover, which in all probability (although this is not explicitly stated by Ibn Bājja) is directly proportional to the duration of the motion.

It is indicated in a rather vague way that in accordance with the formulas of Book VII of the *Phys-*

ics, the distance covered by a body in motion is directly proportional to the relation between the force of the mover and the force of the moved body. The formula breaks down, however, in the case of a moved body that weighs too little "to fatigue," i.e., to move, its mover.

In the case of descent on an inclined plane, the fatigue (perhaps the slowing down) of the falling body is proportional to the angle formed by the inclined plane and a perpendicular drawn to the surface of the earth from the point on the plane at which the falling body happens to be. Ibn Bājja is possibly the earliest author known to us who has outlined, admittedly in a very summary way, a dynamic theory of descent on an inclined plane.

In his explanation of the motion of projectiles, Ibn Bājja follows Aristotelian lines. According to him, the continued motion of an arrow shot from a bow and of a stone thrown by the hand is due to their being pushed by particles of air, the motion of which is, in the last analysis, due to air's being pushed by the action of the hand or the bow. He seems to ignore the theory of violent inclination, similar to the theory of impetus, which was propounded by such philosophers of the Muslim East as Ibn Sīnā. As an Aristotelian philosopher, Ibn Bājja believed that a projectile moves more quickly at the middle of its course than at its beginning.

In Ibn Bājja's view a magnet does not directly cause a piece of iron that is not immediately contiguous to it to move; the latter's motion is occasioned by the air, or some other body, such as a piece of copper or silver, that is placed between the magnet and the piece of iron. In another context Ibn Bājja refers to, but rejects, the belief that an attraction of the earth, similar to the attraction exerted upon a piece of iron by a magnet (rather than the tendency to reach their natural place), is the cause of the downward motion of heavy bodies.

In Latin Europe, the most influential of Ibn Bājja's physical theories was the one sometimes described as the doctrine concerning the original time of motion. This theory attacks the Peripatetic conceptions concerning the role of the resistance of the medium (air or water), regarded as one of the factors determining the velocity of a moving body. According to the Aristotelians, the function of this factor is such that, in the absence of any resistance on the part of a medium, i.e., in a vacuum, the velocity of a moving body must become infinite. The impossibility of this proves that there is no vacuum. Although Ibn Bājja did not hold a brief for the existence of a vacuum, he refuted the Aristotelian argumentation. His view on this question is similar to a doctrine set forth by

his contemporary Abu'l-Barakāt al-Baghdādī as well as by John Philoponus, by whom Ibn Bājja may have been influenced, but it also derives in a logical way from Ibn Bājja's theory concerning the relation between a mover and a moved body. As stated above, it is this relation that determines the velocity of a moved body. Ibn Bājja contends that, in the absence of a medium, the body would move with this original velocity, which is clearly finite. This velocity would decrease in proportion to the resistance of a medium. As has been shown by E. A. Moody, this theory, which was known in Latin Europe through the exposition of Ibn Rushd, who refuted it, influenced Thomas Aquinas, Duns Scotus, and other Schoolmen.

BIBLIOGRAPHY

I. ORIGINAL WORKS. Miguel Asin Palacios has provided the Arabic text and Spanish translation of the following works by Ibn Bājja: "Avempace Botánico" ("Book of Plants"), in *Al-Andalus,* **5** (1940), 259–299; "Tratado de Avempace sobre la unión del intelecto con el hombre" ("The Union of Man With the Active Intellect"), *ibid.,* **7** (1942), 1-47; "La carta de adiós de Avempace" ("The Epistle of Farewell"), *ibid.,* **8** (1943), 1–85; and *El régimen del solitario* ("The Regimen of the Solitary"; Madrid-Granada, 1946), English trans. by D. M. Dunlop in *Journal of the Royal Asiatic Society* (1945), 61–81.

II. SECONDARY LITERATURE. Further information on Ibn Bājja can be found in E. A. Moody, "Galileo and Avempace," in *Journal of the History of Ideas,* **12** (1951), 163–193, 375–422; and S. Pines, "La dynamique d'Ibn Bājja," in *Mélanges Alexandre Koyré,* I, *L'aventure de la science* (Paris, 1964), 442–468.

SHLOMO PINES

BAKER, HENRY (*b.* London, England, 8 May 1698; *d.* London, 25 November 1774), *microscopy.*

Henry Baker did valuable work on the teaching of the deaf and dumb, but he is especially noted for his popularization of the use of the microscope and for his contribution to the study of crystals. His father, William, was a Clerk in Chancery, and his mother, the former Mary Pengry, was "a midwife of great practice." At the age of fifteen Baker was apprenticed to a bookseller whose business later passed into the hands of Robert Dodsley, the printer of Baker's microscopical works. In 1720, at the close of his indentures, Baker went to stay with John Forster, a relative and an attorney, whose daughter had been born deaf. Baker felt inspired to teach the child to read and speak, and was so successful that he became in great demand as a teacher both of the deaf and dumb, and of those with speech defects. He amassed a considerable fortune, and possibly it was for financial reasons that he kept his teaching methods secret. Four manuscript volumes of exercises written by his pupils have, however, survived, and are in the library of the University of Manchester.

Baker's work with the deaf attracted the interest of Daniel Defoe, one of whose early novels, *The Life and Adventures of Duncan Campbell* (1720), was about a deaf conjurer. This book shows that Defoe was familiar with the methods for teaching the deaf used by John Wallis. Baker married Defoe's youngest daughter, Sophia, in 1729. The year before, Defoe and Baker had established the *Universal Spectator and Weekly Journal,* Baker using the pseudonym Henry Stonecastle. The magazine existed until 1746, and Baker was in charge of its production and a frequent contributor until 1733. His annotated volume containing copies of the journal, from the first copy until April 1735, is now in the Bodleian Library, Oxford. Baker's early literary efforts also included several volumes of verse, both original and translated. In 1727 he published *The Universe: a Philosophical Poem intended to restrain the Pride of Man,* which was much admired and reached a third edition, incorporating a short eulogy of the author, in 1805. This poem reveals Baker's keen interest in natural philosophy, as well as his pious approach to the wonders of nature, and includes a reference to the microscope.

In January 1740, Baker became a fellow of the Society of Antiquaries, and in March 1741 a fellow of the Royal Society. For the next seventeen years he was a frequent contributor to the *Philosophical Transactions,* on subjects as diverse as the phenomenon of a girl "able to speak without a tongue" and the electrification of a myrtle tree. Microscopical examinations of water creatures and fossils were, however, the subject of the majority of his papers and were included in his books on microscopy. In 1742 there was considerable interest among fellows of the Royal Society in the freshwater polyp (*Hydra viridis*) as a result of the recent discovery and description of this animal by Abraham Trembley, and with Martin Folkes Baker carried out experiments on this animalcule which he published in 1743 under the title *An Attempt towards a Natural History of the Polype.* The first edition of *The Microscope Made Easy* appeared in 1742; it ran to five editions in Baker's lifetime and was translated into several foreign languages. The Copley Medal of the Royal Society was awarded to Baker in 1744 "For his curious Experiments relating to the Crystallization or Configuration of the minute particles of Saline Bodies dissolved in a menstruum."

Baker was an indefatigable correspondent with scientists and members of philosophical societies all over Europe. He introduced the alpine strawberry into England with seeds sent to him from a correspondent in Turin, and the rhubarb plant (*Rheum palmatum*) sent from a correspondent in Russia. Eleven years after the appearance of *The Microscope Made Easy,* Baker published a second microscopical work, *Employment for the Microscope,* which was as successful as its predecessor. In 1754 the Society for the Encouragement of Arts Manufactures and Commerce was established, and Baker was the first honorary secretary. He died in his apartments in the Strand at the age of seventy-six, and left the bulk of his property to his grandson, William Baker, a clergyman. Neither of his two sons had a successful career, and both predeceased their father. He bequeathed the sum of £100 to the Royal Society for the establishment of an oration, which was called the Bakerian Lecture. Among notable Bakerian lecturers in the fifty years following Baker's death were Tiberius Cavallo, Humphry Davy, and Michael Faraday. Baker's considerable collection of antiquities and objects of natural history was sold at auction in the nine days beginning 13 March 1775.

Henry Baker was in many respects a typical natural philosopher of the eighteenth century. His interests ranged widely, and his skills were equally various; he was by no means dedicated to one branch of study, nor did he do research in the modern sense. Yet he deserved the title "a philosopher in little things"; and he had the rare gift of communicating his knowledge of, and above all his enthusiasm for, the microscope to others. This was what made his two books so widely popular. He regarded the microscope with reverence, as a means to the deeper appreciation of the wonders of God's world. "Microscopes," he wrote in the introduction to *The Microscope Made Easy,* "furnish us as it were with a new sense, unfold the amazing operations of Nature," and give mankind a deeper sense of "the infinite Power, Wisdom and Goodness of Nature's Almighty Parent." *The Microscope Made Easy* is divided into two parts, the first dealing with the various kinds of microscopes, how each may be best employed, the adjustment of the instrument, and the preparation of specimens. Part II has chapters devoted to the examination of various natural objects, in the manner of Robert Hooke's *Micrographia*—e.g., the flea, the poison of the viper, hairs, and pollen. Part I of *Employment for the Microscope* is devoted to the study of crystals, and Part II is a miscellany of Baker's microscopical discoveries. Both books were deliberately written for the layman.

Apart from his work as an instructor in the techniques of microscopy, Baker's most important scientific achievements were the observation under the microscope of crystal morphology, for which he received the Copley Medal, and his account of an examination of twenty-six bead microscopes bequeathed to the Royal Society by Antony van Leeuwenhoek. His measurements of these unique microscopes (lost during the nineteenth century) are most valuable historical material.

BIBLIOGRAPHY

I. Original Works. Baker's writings are *The Universe: a Philosophical Poem* (London, 1727); *Universal Spectator and Weekly Journal,* Henry Stonecastle [Henry Baker], ed. (London, 1728–1735), annotated volume Bodl. Hope F103; *The Microscope Made Easy* (London, 1742); *An Attempt towards a Natural History of the Polype in a Letter to Martin Folkes* (London, 1743); and *Employment for the Microscope* (London, 1753). Some twenty papers were published between 1740 and 1758 in *Philosophical Transactions of the Royal Society.* The following are manuscript materials relating to Baker: letters from Baker—Bodl.MS Montague D11, ff.70–75, 79; BM Add.MS4426, f.242; BM Add.MS4435, ff.255–259; indentures relating to Baker—BM MS Egerton 738, ff.2–5; four volumes of exercises written by Baker's pupils, formerly in Arnold Library—Special Collection, Manchester University, 371 924 B1, Vols. I–IV; legal agreements concerning teaching—Special Collection, Manchester University, 371 92092 f B4.

II. Secondary Literature. Works on Baker are A. Farrar, *Arnold on the Education of the Deaf: A Manual for Teachers,* Book 1, "Historical Sketch" (London, 1932), 33; William Lee, *Daniel Defoe: His Life and Recently Discovered Writings,* 3 vols. (London, 1869), I, 439, 441, 455–459; and John Nichols, *Biographical and Literary Anecdotes of William Bowyer* (London, 1782), pp. 413–416, 596, 645; and *Literary Anecdotes of the Eighteenth Century,* 9 vols. (London, 1812–1816), V, 272–278.

G. L'E. Turner

BAKER, JOHN GILBERT (*b.* Guisborough, England, 13 January 1834; *d.* Kew, England, 16 August 1920), *botany.*

Baker's parents, John Baker and Mary Gilbert, moved from Guisborough to Thirsk, Yorkshire, in 1834. A Quaker, he attended the Friends' School at Ackworth; when he was twelve, he was transferred to the Friends' School at Bootham, York, which then enjoyed a reputation for natural history study. His formal education ended in 1847, and he spent the next eighteen years in a drapery business in Thirsk. This uncongenial occupation did not impede Baker's enthusiasm for natural history; when only fifteen, he communicated a new record of a rare *Carex* to *The*

Phytologist. In 1854 he collaborated with J. Nowell in a supplement to Baines's *Flora of Yorkshire.* Baker's zeal helped to create the Botanical Exchange Club of the Thirsk Natural History Society; when the society was dissolved in 1865, the club moved to London, with Baker as one of the two curators. In May 1864 Baker's home and business premises were completely destroyed by fire and his entire herbarium and library were lost, including the stock of his book *North Yorkshire* (1863). This catastrophe caused him seriously to consider his future career, and when an opportunity was offered to engage in botanical research, he readily abandoned the drapery business.

The opportunity arose from an invitation by J. D. Hooker, director of the Royal Botanic Gardens, Kew, to join the staff of its herbarium. In 1866 Baker was appointed first assistant in the Kew Herbarium, with the initial task of finishing W. J. Hooker's *Synopsis Filicum,* left incomplete on his death in 1865. To supplement his slender salary, Baker lectured on botany at the London Hospital Medical School from 1869 to 1881; the following year he was appointed lecturer in botany at the Chelsea Physic Garden. In 1890 he became keeper of the herbarium and library at Kew, serving until his retirement in 1899.

His very able work on Hooker's *Synopsis Filicum* earned Baker wide recognition as an expert on vascular cryptogams, and he was invited by Martius to undertake the volume on ferns in his monumental *Flora Brasiliensis;* Baker later contributed the *Compositae* to the same work. An early interest in *Rosa,* manifested in a review of the genus in *The Naturalist* for 1864, was followed by a monograph on British roses in 1869; he also wrote the botanical descriptions for E. A. Willmott's *Genus Rosa* (1910–1914). He published monographic accounts of other plant families and genera, and made substantial contributions to *Flora of Tropical Africa, Flora Capensis,* and *Flora of British India.* Baker was one of the great English taxonomists and a pioneer investigator in plant ecology. Botany was his *raison d'être;* his enormous capacity for work and his output were impressive by any standard.

Baker was a notably effective lecturer, clear and concise, and had an instinctive sympathy for the problems of his students. His long, fruitful career was attended by numerous distinctions: fellowship of the Royal Society in 1878, the Victoria Medal of Honour of the Royal Horticultural Society in 1897, in acknowledgment of his valuable services to horticulture, and the Linnean Medal in 1899. His first child (he married Hannah Unthank in 1860), Edmund Gilbert Baker, emulated his father by choosing botany as his vocation.

BIBLIOGRAPHY

I. ORIGINAL WORKS. Baker's writings include *Supplement to Baines' Flora of Yorkshire* (London, 1854), written with J. Nowell; *North Yorkshire* (London, 1863); the completion of Sir W. J. Hooker's *Synopsis Filicum* (London, 1868, 1874); "A Monograph of the British Roses," in *Journal of the Linnean Society,* **11,** no. 52 (1869), 197–243; *Handbook of the Fern-allies* (London, 1887); *Handbook of the Amaryllideae* (London, 1888); *Handbook of the Bromeliaceae* (London, 1889); and *Handbook of the Irideae* (London, 1892). He wrote over 400 papers, most of which are listed in the *Royal Society Catalogue of Scientific Papers, 1867–1914:* I, 164–165; VII, 74–75; IX, 102–104; XIII, 253–254.

II. SECONDARY LITERATURE. Obituary notices are *Botanical Society and Exchange Club of the British Isles,* **6** (1920), 93–100; James Britten, in *Journal of Botany,* **58** (1920), 233–238; and Sir David Prain, in *Proceedings of the Royal Society of London,* **92B** (1921), xxiv–xxx.

R. G. C. DESMOND

BAKEWELL, ROBERT (*b.* England, 1768; *d.* Hampstead, England, 15 August 1843), *geology.*

Bakewell's parentage and birthplace are not known, although he may have come from Nottingham. He was not related to Robert Bakewell (1725–1795), the celebrated Leicestershire husbandman with whom he has often been confused. That he became interested in science, particularly geology, about the middle of his life perhaps explains the emphasis in his works on presenting science to his readers in clear, simple language, scorning merely technical distinctions and abstruse vocabulary. He wrote for the general reader but without undue popularization. During much of his life he lived in London, working as a mineralogical surveyor and teaching mineralogy and geology.

Bakewell's *Introduction to Geology* appeared in 1813. The work was widely read and appreciated, largely because it used examples and illustrations taken from the English countryside. A second, enlarged edition appeared in 1815, a third edition in 1828, a fourth edition in 1833, and a final edition in 1838. The second and third editions were translated into German. In 1829, Benjamin Silliman published an American edition, in which he included his own lecture notes.

Bakewell's brightly ironic style contributed greatly to the popularity of the *Introduction,* and perhaps accounts for the book's continued success despite its being outdated by advances in the subject. William Smith's ideas of using fossils for the correlation of strata, for example, were never included in the *Introduction.* Bakewell appreciated James Hutton's "plutonic" ideas while failing to grasp the principle of uniformity—Hutton's greatest contribution. He was highly critical of the geognosy of the Wernerian school, missing no opportunity to disparage it, and he rejected "neptunism," depending almost entirely on volcanic processes to account for rock formations. Like Baron Cuvier, a French contemporary, Bakewell found evidence in the rocks for geological revolutions of great magnitude with quiet intervals lasting tens of thousands of years.

In addition to the *Introduction,* Bakewell wrote many articles on geological and biological subjects. Most of these appeared in the *Philosophical Magazine,* although one was published by the Geological Society of London, to which Bakewell was never admitted as a member. One of his sons, also named Robert Bakewell, wrote on geology, and the three articles on the Falls of Niagara listed for the elder Bakewell in the *Royal Society Catalogue of Scientific Papers* were written by his son.

BIBLIOGRAPHY

In addition to *An Introduction to Geology, . . . and Outline of the Geology and Mineral Geography of England* (London, 1813; 5th ed. 1838), Bakewell wrote *An Introduction to Mineralogy* (London, 1819) and *Travels . . . in the Tarentaise,* 2 vols. (London, 1823). Further details of Bakewell's life may be found in the *Dictionary of National Biography.*

BERT HANSEN

BAKH, ALEKSEI NIKOLAEVICH. See **Bach, Aleksei Nikolaevich.**

AL-BAKRĪ, ABŪ ʿUBAYD ʿABDALLĀH IBN ʿABD AL-ʿAZĪZ IBN MUḤAMMAD (*b. ca.* 1010; *d.* 1094), *geography.*

Al-Bakrī was a Hispano-Arabic geographer who belonged to a family of landowners that took advantage of the fall of the caliphate of Córdoba to declare itself independent in Huelva and Saltes. When his father was deposed by al-Muʿtaḍid, al-Bakrī moved to Córdoba, where he studied with the historian Ibn Ḥayyān (*d.* 1075) and the geographer al-ʿUdhrī (*d.* 1085). For much of his life al-Bakrī was a member of al-Muʿtaṣim's court at Almería, and late in life he became acquainted with the writer Ibn Khāgān. He spent much time in Seville, and was there when El Cid arrived to collect the tributes due Alfonso VI.

A man of wide-ranging knowledge, al-Bakrī was a good poet and philologist who devoted much of his time to geography, even though it appears that he never traveled outside the Iberian Peninsula. His main scientific works are the following:

(1) *Muʿjam mā istaʿjam,* a collection of place names that are mainly Arabian and that are frequently misspelled in common use. The analytical part is preceded by an introduction detailing the geographic framework of the region under study.

(2) *Kitāb al-masālik wa'l-mamālik* (completed in 1068), only part of which has been preserved. As the title ("Book of the Roads and Kingdoms") implies, it is a description of land and sea routes written to facilitate travel. Aside from the purely geographical facts, al-Bakrī also includes historical and social data. The description of the coasts is in some instances so precise that it makes one wonder if the author's source was a navigation or prenavigation text of the western Mediterranean. The book is independent of the works of the same title written by Oriental geographers. Still extant is a general introduction that contains a description of the Slavic and Nordic peoples; it has been published only in fragments. There is also a description of North Africa and Spain (*al-Mughrib fī dhikr Ifrīqiyya wa'l-Maghrib*). The first part presents interesting data about the Sahara routes and the origins of the Almoravid movement. We have only fragments of the second part, which was used by Alfonso X (el Sabio). The main source of the work (aside from al-Bakrī's teacher al-ʿUdhrī), of which many fragments about Spain have been preserved, is the book about Ifrīqiyya written by Muḥammad ibn Yūsuf al-Warrāq (*d.* 973). Al-Bakrī also had access to Latin sources, such as the *Etimologiae* of St. Isidore, the *History* of Orosius, and many others known in Córdoba either in Arabic versions or in oral tradition. In addition, he utilized the description of the trip of Ibrāhīm ibn Yaʿqūb, a Jew from Tortosa, to the north of Europe in the tenth century. The *Kitāb al-masālik wa'l-mamālik* was influential in Arabic literature for centuries.

(3) Al-Ghafiqī and Ibn al-Bayṭār cite al-Bakrī as an authority several times in their pharmacological compilations. From this we can deduce that he wrote a treatise, since lost, entitled *Aʿyān al-nabāt* or *Kitāb al-nabāt,* about simple medicines, quoted by Ibn Khayr (*Fahrasa,* 377, Codera and Ribera, eds. [Zaragoza, 1894–1895]).

BIBLIOGRAPHY

I. ORIGINAL WORKS. There is a list of al-Bakrī's manuscripts in Brockelmann, *Geschichte der arabischen Literatur,* I (Weimar, 1898), 476, and Supp. I (Leiden, 1937), 875–876. The *Muʿjam mā istaʿjam* was edited by F. Wüs-tenfeld and published as *Das geographische Wörterbuch* (Göttingen–Paris, 1876; Cairo, 1945–1949). The manuscript of the *Kitāb al-masālik wa'l-mamālik* is in Paris (BN 5905); the part dealing with North Africa was published in French trans. by de Slane as "Description de l'Afrique septentrionale," in *Journal asiatique* (1858–1859), and as a book (2 vols., Algiers, 1911–1913; 1 vol., Paris, 1965). It has often been reprinted with original and translation combined. The part dealing with the Slavs was translated into Russian by A. Kunik and V. Rosen (St. Petersburg, 1878); it also appears in A. Seippel's *Rerum normanicorum fontes arabici* (Oslo, 1896–1928). The best edition with Polish and Latin versions is that of T. Kowalski (Krakow, 1946).

II. SECONDARY LITERATURE. The Arabic sources for the life of al-Bakrī are listed in Khayr al-Dīn al-Ziriklī, *al-Aʿlām,* 2nd ed., IV, 233; secondary listings may be found in the above sources and in George Sarton, *Introduction to the History of Science,* I, 768; and E. Lévi-Provençal, in *Encyclopédie de l'Islam,* 2nd ed., I, 159–161. See also the important text by his teacher al-ʿUdhrī, published by al-Ahwānī, in *Fragmentos geográficos-históricos de al-Masālik ilā Gamīʿ al-Mamālik* (Madrid, 1965), and the note by J. Vernet in *Revista del Instituto de estudios islámicos,* 13 (1965–1966), 17–24.

J. VERNET

BALANDIN, ALEKSEY ALEKSANDROVICH (*b.* Yeniseysk, Siberia, Russia, 20 December 1898; *d.* Moscow, U.S.S.R., 22 May 1967), *chemistry.*

Balandin was the founder of an important school of catalytic chemistry. He graduated from Moscow University in 1923. Professor of chemistry from 1934, he was elected a corresponding member of the Soviet Academy of Sciences in 1943 and a full member in 1946.

As a result of his quantitative studies of the kinetics of catalytic reactions of hydrogenation and dehydrogenation of cyclic hydrocarbons, Balandin established that two factors play a decisive role in all catalytic processes: (1) a similarity between the structure of the reacting molecules and the surface of the catalyst and (2) comparable energy values for the chemical bonds of the molecular reagents and the energy of chemical interaction of these molecules with the surface of the catalysts. In 1929 Balandin made this discovery the basis of his universal catalysis theory, which became known as the universal "multiplet" theory of catalysis.

For the first time in the history of chemistry, Balandin proposed and demonstrated physically the principle of an active transitional state characterized by the formation of an unstable complex containing not fully valent chemical bonds (i.e., bonds with less than two electrons). With some modifications, this principle has been incorporated into one of the most universal theories of chemical processes, the theory of the absolute reaction rates.

Balandin showed that the active, or multiplet, complex (formed as a transition state in hydrogenation and dehydrogenation processes of cyclic hydrocarbons) can take on an orientation that conjoins either the plane or a facet of a molecule with the surface of the catalyst. The practical outcome of this work was that Balandin was able to predict catalytic activity for a number of metals in group VIII of the periodic table, as well as for chromium and vanadium oxides. Chromium sesquioxide later became a commercial-type catalyst.

Having worked out quantitative methods to determine a catalyst's maximum activity, Balandin showed that the peak activity is located in the vicinity of an adsorption potential, q, equal to one half of the total energy of the reacting bonds, s, i.e., at the $q = s/2$ point. This relationship gives a solution to the problem of selecting catalysts by experimental determination of their absorption potential. In addition, Balandin determined the conditions for enhancing the activity of mixed catalysts and found explanations for poisoning—promoting and utilizing these conditions in the control of the activity. He was thus able to forecast the sequence of reactions in stepwise processes.

On the basis of the multiplet theory, Balandin proposed a classification of the organic catalytic reactions that reflected the positions of the atoms in the reacting molecule on the catalyst surface in relation to the active center. The importance of this classification for catalytic organic synthesis is equivalent to that of the system of forms used in crystallography. New types of processes can be forecast by means of this system of classification. For example, in 1935 Balandin predicted the then unknown types of paraffin and olefin dehydrogenation processes that now form the basis for the production of monomers used in the synthesis of rubber.

Balandin introduced a number of basic concepts and equations into chemical kinetics. In 1929–1930 he was one of the first (together with F. Constable) to indicate the importance of activation energy determinations in the study of heterogenous processes; he established a logarithmic dependence between ε and k_0 in the Arrhenius equation $k = k_0 \cdot e^{-\varepsilon/rt}$ (ε being the empirical value of activation energy). In 1930–1935 he gave a precise equation of reaction kinetics in a flow system and derived a general equation for monomolecular reactions complicated by adsorption equilibrium. Having studied the kinetics of the dehydrogenation of butane and butylene to butadiene and of ethylbenzene to styrene, Balandin was the first to find conditions under which the yield of these most valuable monomers approached the thermodynamically feasible maximum.

Balandin's work in catalysis is closely connected with petrochemical synthesis. Together with N. D. Zelinski and his aides, Balandin participated in the discovery and study of the alkane and cyclane (cycloparaffin) reactions and, in particular, in the dehydrocyclization of paraffins, aromatization, and hydrogenolysis. In 1950 he applied the multiplet theory to fermentation processes, demonstrating the special role of the matrix effect by imposing substituents on the surface of ferment (entropy factor) and by lowering the potential barrier of the reaction and of the heat of adsorption (energy factor).

BIBLIOGRAPHY

I. ORIGINAL WORKS. Among Balandin's writings are "Khimiya i struktura" ("Chemistry and Structure"), a series of articles in *Izvestiia Akademii nauk SSSR* (1940), 295–310; (1940), 571–584; (1942), 168–178, 286–296; (1943), 35–42; "Teoria slozhnykh reaktsy" ("Theory of Complex Reactions"), in *Zhurnal fizicheskoi khimii,* **15** (1941), 615–628; "Tochny metod opredelenia adsorbtsionnykh koeffitsientov" ("A Precise Method of Determining Adsorption Coefficients"), in *Izvestiia Akademii nauk SSSR* (1957), 882–884; "Printsipy strukturnogo i energeticheskogo sootvetstvia v fermentativnom katalize" ("Principles of Structural Energetic Conformity in Fermentative Catalysis"), in *Biokhimiya,* **23** (1958), 475–485; and *Multipletnaya teoria kataliza* ("Multiple Catalysis Theory"), 2 vols. (Moscow, 1963–1965).

II. SECONDARY LITERATURE. Works on Balandin are *Materialy k biobibliografy uchenykh SSSR. Aleksey Aleksandrovich Balandin* ("Bibliographical Material Concerning the Lives of USSR Scientists. Aleksey Aleksandrovich Balandin"; Moscow, 1958); A. M. Rubinstein, "Akademik Aleksey Aleksandrovich Balandin" ("Academician Aleksey Aleksandrovich Balandin"), in *Uspekhi khimii,* **18** (1949), 38 (issued on the occasion of his fiftieth birthday); H. S. Taylor, "Geometry in Heterogeneous Catalysis," in *Chemical Architecture* (New York, 1948), pp. 8–18; and B. M. W. Trapnell, "Balandin's Contribution to Heterogeneous Catalysis," in *Advances in Catalysis and Related Subjects,* III (New York, 1951), 1–25.

V. I. KUZNETZOV

BALARD, ANTOINE JÉROME (*b.* Montpellier, France, 30 September 1802; *d.* Paris, France, 30 March 1876), *chemistry.*

Balard was born into a family of modest circumstances; his parents were wine-growers. His godmother noticed his intelligence, however, and enabled him to attend the *lycée* in Montpellier. In 1819, upon his graduation, he began to train for a career in pharmacy. While training in Montpellier, he served as *préparateur* in chemistry to Joseph Anglada of the Faculté des Sciences and *préparateur* at the École de Pharmacie. At the latter he studied chemistry and physics under Jacques Étienne Bérard, who permitted Balard to do research at the chemical factory of which Bérard was the director. Balard received his degree in pharmacy in 1826, having written a thesis on cyanogen and its compounds.

It was in this period, about 1825 (the date is uncertain but was before 28 November), that Balard made his discovery of the element bromine. The discovery of a new chemical element by a young and obscure provincial pharmacist caused a sensation in Parisian, and subsequently in foreign, scientific circles. Balard's achievement was recognized by the Académie des Sciences and he was awarded a medal by the Royal Society of London. Professionally he advanced steadily, first in Montpellier, where he succeeded Anglada in 1834, and later in Paris, where he assumed Louis Jacques Thénard's chair at the Sorbonne in 1842 and that of Théophile Jules Pelouze at the Collège de France in 1851. For this last position, Balard's chief competitor was Auguste Laurent. In 1844 Balard was elected to the Académie des Sciences.

Balard's mode of life was modest, even in his successful years in Paris. Moreover, his abstemiousness influenced the style of his scientific research; both Jean Baptiste Dumas and Charles Adolphe Wurtz testified to Balard's preference for simple apparatus and homemade reagents over elaborate techniques and materials. He was amiable and generous both to his colleagues and to his students.

Balard was principally an experimental chemist, although his experimentation was guided by a keen awareness of the analogies between the substances he was investigating and others of which the chemistry was better known. He did not publish a great deal, but what did appear was of high quality and great interest.

The discovery of bromine, Balard's first and greatest achievement, actually was a by-product of his more general chemical investigations of the sea and its life forms. In the course of his studies, Balard devised a reliable test for the presence of iodine, the content of which he was determining in plants taken from the Atlantic and the Mediterranean. Chlorine water was added to the test solution, to which starch and sulfuric acid had already been added. The iodine was manifested by its characteristic blue color at the interface of the test solution and the chlorine water. Then Balard noticed that, in some samples, above the blue layer there appeared a yellow-orange layer, which had its own characteristic odor. He isolated the substance causing the yellow color, which proved to be a red liquid. He first collected it by simple distillation, but soon he found a more effective method: shaking the chlorinated sample with ether and treating the resultant orange layer with caustic potash. The crystallized precipitate was then distilled with manganese dioxide and sulfuric acid to produce the red vapor of elemental bromine, which Balard dried with calcium chloride and condensed to a liquid.

At first he thought the substance might be a combination of chlorine and iodine, but he was unable to detect the presence of iodine and the liquid did not decompose under electrolysis. Balard concluded that he had discovered a new element, which, at the suggestion of Anglada, he named muride, subsequently changed to bromine. He proceeded to study its properties, which he found to be analogous to those of chlorine and iodine.

A number of German chemists, including Justus Liebig, had in fact isolated bromine before Balard without realizing its elemental nature. Carl Jacob Löwig, while still a student, isolated it almost simultaneously with Balard. The discovery of this element had a significance greater than that of the isolation of most other elements, for it made manifest the most striking "family" of elements—the halogens—in which bromine possessed an atomic weight that was approximately the arithmetic mean between those of chlorine and iodine.

In 1834 Balard published the results of his study of the bleaching agent Javelle water. In the course of working out the chemistry of this chlorine bleach, he succeeded in preparing hypochlorous acid and chlorine monoxide.

The project to which Balard devoted the most time was the inexpensive extraction of salts from the sea. Beginning in 1824, he spent many years developing techniques for precipitating sodium sulfate and potassium salts, publishing his method in 1844. Unfortunately for Balard, cheaper techniques for producing sodium sulfate were developed and huge deposits of potassium sulfate were discovered at Stassfurt.

Two other chemical studies by Balard deserve mention: the discovery of oxamic acid from the decomposition by heat of ammonium hydrogen oxalate (ammonium bioxalate) and the study and naming of amyl alcohol.

Of perhaps greater importance than his chemical researches was the interest Balard took in his students' careers, particularly those of Louis Pasteur and Marcelin Berthelot. Balard petitioned to have Pasteur assigned to him as an assistant in 1846, and in 1851 he secured a similar position for Berthelot at the Collège de France. He maintained close friendships with both these pupils, coming to Pasteur's defense in the spontaneous-generation controversy and securing the creation of the chair in organic chemistry at the Collège de France for Berthelot.

BIBLIOGRAPHY

I. ORIGINAL WORKS. For a listing of Balard's scientific papers, see the Royal Society of London's *Catalogue of Scientific Papers,* I, 166–167, and VII, 76. His most important papers are "Note pour servir à l'histoire naturelle de l'iode," in *Annales de chimie,* 2nd ser., **28** (1825), 178–181; "Mémoire sur une substance particulière contenue dans l'eau de la mer (le brôme)," *ibid.,* **32** (1826), 337–384; "Recherches sur la nature des combinaisons décolorantes de chlore," *ibid.,* **57** (1834), 225–304; "Note sur la décomposition du bioxalate d'ammoniaque par la chaleur et les produits qui en résultent," *ibid.,* 3rd ser., **4** (1842), 93–103; "Mémoire sur l'alcool amylique," *ibid.,* **12** (1844), 294–330; and "Sur l'extraction des sulfates de soude et de potasse des eaux de la mer," in *Comptes-rendus de l'Académie des Sciences,* **19** (1844), 699–715.

II. SECONDARY LITERATURE. Works on Balard are J. B. Dumas, "Éloge de M. Antoine-Jérome Balard," in *Mémoires de l'Académie des Sciences,* 2nd ser., **41** (1879), lv–lxxx; M. Massol, "Centenaire de la découverte du brôme par Balard," in *Bulletin de la Société Chimique de France,* 4th ser., **41** (1927), 1–9; J. R. Partington, *A History of Chemistry,* IV (London, 1964), 96–97; M. E. Weeks, *Discovery of the Elements,* 4th ed. (Easton, Pa., 1939), 360–362; and C. A. Wurtz, "Discours qui M. Wurtz, membre de l'Académie des Sciences, se proposait de prononcer aux funérailles de M. Balard, le 3 Avril, 1876," in *Journal de pharmacie et chimie,* 4th ser., **23** (1876), 375–379.

SEYMOUR H. MAUSKOPF

BALBIANI, EDOUARD-GÉRARD (*b.* Port-au-Prince, Haiti, 31 July 1823; *d.* Meudon, France, 25 July 1899), *biology.*

Balbiani's father, German by birth but of Italian descent, married a French Creole and went to Haiti to set up a banking firm. While still young, Balbiani was sent to Frankfurt am Main, and about 1840 he went to Paris to finish his studies, his mother having

settled there. For a time Balbiani attended law school, but he was soon attracted by the natural sciences as they were being taught at the Muséum National d'Histoire Naturelle by de Blainville.

Balbiani became *licencié ès sciences naturelles* in 1845 and *docteur en médecine* on 30 August 1854. His thesis, "Essai sur les fonctions de la peau considérée comme organe d'exhalation, suivi d'expériences physiologiques sur la suppression de cette fonction," was on a far higher level than the usual medical thesis.

His financial situation made it possible for Balbiani not to practice medicine, but to devote himself entirely to microscopic studies. As early as 1858 he communicated the first results of his research to the Academy. Balbiani founded the Société de Micrographie and was a faithful member of the Société de Biologie and the secretary of the *Journal de physiologie.* He was already well known in 1867, when Claude Bernard, who often praised him, asked him to direct the histological research at the laboratory of general physiology at the Muséum. On 13 February 1874 he became professor of embryogeny at the Collège de France, a post he held for the rest of his life.

Balbiani's work was extensive. His early research concerned protozoa, which at the time were subject to various interpretations. Some naturalists, following Ehrenberg, considered the infusoria as complete organisms with differentiated groups of organs. Others, such as Dujardin, saw in the protozoa only a mass of "sarcoda" without any organization whatever. These extreme views eventually were reconciled, but it is not possible to consider them here without outlining the history of protozoology. Balbiani discovered sexual reproduction in the ciliata, a finding that aroused much controversy until Bütschli confirmed Balbiani's research and gave it its true interpretation. While studying the binary fission of the infusoria, Balbiani set forth its laws in 1861. This work led him to perform genuine microsurgical experiments that enabled him to specify the role of the nucleus. He also introduced the technique and the term "merotomy."

Balbiani's research on the formation of the sexual organs of the *Chironomus* demonstrated that the sexual cells derive directly from the egg and are differentiated before the blastoderm appears—and that consequently they precede the individual itself. This essential fact was later observed in other species and eventually was responsible for the general theory of the autonomy of the germ cell. Notice should also be made of Balbiani's investigations on the reproduction of aphids and of his work on pebrine, the disease of the silkworm that later attracted Pasteur's attention.

Balbiani is known eponymously through his research on cytoplasmic inclusions. He described in the yolks of young ovules a special formation that Milne-Edwards called the Balbiani vesicle (1867). This had already been pointed out by Julius Carus (1850) under the name "vitelline nucleus," but Henneguy proposed calling it "corps vitellin de Balbiani" (1893). This body, interpreted by turns as a derivative of the centrosome or as an equivalent of the chondriome, has received new attention with the development of the electron microscope.

A solitary worker, Balbiani did not seek recognition through his lectures at the Collège de France. His publications were numerous, and with Ranvier he founded *Archives d'anatomie microscopique,* which is still published.

BIBLIOGRAPHY

L. F. Henneguy has written two works that provide additional information on Balbiani: *Leçons sur la cellule* (Paris, 1896), and "Balbiani, E. G.—Notice biographique," in *Archives d'anatomie microscopique,* **3** (1900), i–xxxi, with a complete bibliography of Balbiani's publications and a portrait.

Marc Klein

BALBUS (BALBUS MENSOR [?]) (*fl. ca.* A.D. 100), *surveying, mathematics.*

For information on the life of Balbus, we must depend upon the scanty data appearing in his works. He served as an officer in the campaign that opened Dacia to the Romans (see Lachmann, p. 93). Since Balbus does not name the reigning emperor, this may have been either the war of Domitian, which started in A.D. 85, or one of the campaigns of Trajan, which terminated in A.D. 106 with the conversion of Dacia into a Roman province. The second possibility seems more likely. Immediately following his return from war, Balbus completed a treatise dedicated to a prominent engineer named Celsus, about whom nothing has come down to us.

Authenticated Works. Balbus' work in its oldest manuscript, the so-called Arcerianus B, is entitled "Balbi ad Celsum expositio et ratio omnium formarum."[1] Other manuscripts name Frontinus—or even Fronto—as the author. The work is a geometric manual for surveyors. Starting with a summary of standard measurements, it contains definitions of geometric concepts (point, line, area, types of angles, figures). The work, which refers back to the works

of Hero of Alexandria (Hultsch, pp. 103 f.; Cantor, pp. 101 f.), is distorted by gaps and interpolations (see Lachmann's ed.; and Bubnov, pp. 419 f.) and undoubtedly is not preserved in full, for its contents do not match its title. Probably the end of the preserved text (Lachmann, p. 107)[2] was followed by examples of the calculation of triangles, rectangles, and polygons, such as can be found in Hero's works.

Works of Uncertain Origin. The oldest surveying codices have handed down a list of cities (*liber coloniarum I,* Lachmann, I, 209–251). Repeatedly mentioned as the source is a Balbus *mensor* (surveyor), who is said to have lived at the time of Augustus. Mommsen has demonstrated that this Balbus could not have been a contemporary of Augustus, so it is possible that this surveyor Balbus is identical with the author of *Expositio.* Nevertheless, the *liber coloniarum* does not form part of *Expositio,* since the latter almost certainly contained only geometry and not surveying material.

In 1525 Fabius Calvus of Ravenna published a short treatise on fractions (*De asse minutisque eius portiunculis*) as a fragment derived from a larger work by Balbus. However, this Balbus has no connection with the author of *Expositio:* Christ (*Sitzungsberichte der Bayerischen Akademie der Wissenschaften zu München* [1863], p. 105) and Hultsch (*Metrologicorum scriptorum reliquiae,* II [Leipzig, 1866], 14) found that the treatise on fractions originated between the time of Alexander Severus (222–235) and Constantine I (306–337).

Whether other works in the *Corpus agrimensorum* must be attributed to Balbus is uncertain.

Some chapters in the *Collection of the Surveyors* go back to Balbus (for example, Lachmann, I, 295.17–296.3). Particularly clear references to Balbus appear in the two works on geometry ascribed to Boethius: similar formulations in the five-book work on geometry have been grouped by Bubnov, p. 426. In the two-book work on geometry edited by G. Friedlein almost the entire section pp. 393.12–395.2; pp. 401.15–403.24 relates back to Balbus. At one place (p. 402.28–29 of Friedlein's edition) its anonymous author credits Frontinus with a definition actually derived from Balbus.

NOTES

1. The translation of *forma* as "geometric figure" is corroborated by Balbus' own definition (p. 104), despite the contrary views of Lachmann (*forma* as *mensura,* or "measure"; Lachmann, II, 134) and Mommsen (*forma* as *Grundriss* or "plan"; Lachmann, II, 148).
2. The text (I, 107.10–108.8 of Balbus' text, edited by Lachmann),

which Lachmann himself considered spurious, is believed to be the work of Frontinus; compare C. Thulin, "Die Handschriften des *Corpus agrimensorum Romanorum*," in *Abhandlungen der Preussischen Akademie der Wissenschaften,* Phil.-hist. Kl., no. 2 (1911), p. 23. In the London MS, BM Add. 47679, the only prehumanistic codex containing the ending of Balbus' work, this text is preceded by excerpts from Balbus and followed by abstracts from Frontinus. There are no headings. On this codex, so far neglected but a must for the reconstruction of the Balbus text, see M. Folkerts, *Zur Überlieferung der Agrimensoren: Schrijvers bisher verschollener "Codex Nansianus,"* which is to be printed in the *Rheinisches Museum für Philologie.*

BIBLIOGRAPHY

There is an edition of the *Expositio* by K. Lachmann in K. Lachmann et al., *Die Schriften der römischen Feldmesser,* I (Berlin, 1848), 91–107.

Works on Balbus are N. Bubnov, *Gerberti opera mathematica* (Berlin, 1899), pp. 400, 419–421; M. Cantor, *Die römischen Agrimensoren* (Leipzig, 1875), pp. 99–103; Gensel, in Pauly-Wissowa, II, 2820–2822; F. Hultsch, in *Ersch und Grubers allgemeine Encyclopaedie,* **92** (1872), 102–104; K. Lachmann, "Über Frontinus, Balbus, Hyginus und Aggenus Urbicus," in K. Lachmann, Blume, and Rudorff, *Die Schriften der römischen Feldmesser,* II (Berlin, 1852), 131–136; T. Mommsen, "Die *libri coloniarum*," *ibid.,* 145–214; and Schanz and Hosius, *Geschichte der römischen Literatur,* 4th ed., II (1935), 802 f.

MENSO FOLKERTS

BALDI, BERNARDINO (*b*. Urbino, Italy, 5 June 1553; *d.* Urbino, 10 October 1617), *mechanics.*

After a classical education by private tutors at Urbino, Baldi studied mathematics with Guido Ubaldo del Monte, under Federico Commandino, beginning about 1570. At Commandino's suggestion he translated the *Automata* of Hero of Alexandria into Italian, but left it unpublished until 1589. He also translated the *Phenomena* of Aratus of Soli and wrote didactic poems on the invention of artillery and the nautical compass, but did not publish them. In 1573 he enrolled at the University of Padua, and when it was closed shortly afterward because of plague, Baldi returned to Urbino. He was with Commandino during the latter's final illness in 1575, and obtained from him an account of his life. Baldi's studies at Padua centered on philology and literature, but he obtained no degree.

In 1580 Baldi went to Mantua in the service of Ferrante II Gonzaga, who in 1585 secured him the post of abbot of Guastalla. He then put in order his biographies of some 200 mathematicians, a work conceived during the writing of Commandino's biography and completed in 1588–1589. About that time he also wrote his principal contribution to physics, a

commentary on the pseudo-Aristotelian *Questions of Mechanics,* posthumously published in 1621. In 1589 he published his translation of Hero's *Automata,* prefaced by a history of mechanics.

Baldi visited Urbino in 1601 to compile materials for a life of Federico di Montefeltro, and in 1609 he resigned his abbacy to enter the service of the duke of Urbino as historian and biographer, remaining there until his death. A vast work on geography on which he worked in his later years remains unpublished, although many of his poetic and literary compositions were printed in his lifetime. His final scientific contribution was a translation of Hero's *Belopoeica* into Latin, accompanied by the Greek text and by Baldi's Latin *Life of Hero* (1616). Brief extracts from his lives of the mathematicians were published in 1707, and about forty of those biographies have since been published in full. Most of the work, however, remains in manuscript.

Except for that of Henri de Monantheuil, with which Baldi was certainly unacquainted, Baldi's commentary on the *Questions of Mechanics* was the most important work of its kind to appear up to that time. He was probably influenced in his ideas on the continuance of motion by the earlier commentary of Alessandro Piccolomini. His account of dynamic equilibrium in spinning tops was superior to that of G. B. Benedetti, with whose principal work he appears to have been unfamiliar. The most significant aspect of Baldi's approach to mechanics lay in the development and application of the concept of centers of gravity, particularly with regard to stable and unstable equilibrium. It was the opinion of Pierre Duhem that Baldi drew his chief ideas from manuscripts of Leonardo da Vinci. Duhem accepted the year 1582 for the original composition of Baldi's commentary, as given by Baldi's first biographer. That date is, however, inconsistent with passages in Baldi's own preface and in the text of the work, which imply the year 1589. The latter year is also supported by textual indications that Baldi's principal inspiration was drawn from Commandino's translation of Pappus and from Guido Ubaldo's commentary on the *Plane Equilibrium* of Archimedes, both of which were published in 1588. The influence of Baldi's work was doubtless diminished by its delay in publication until after the Archimedean ideal had largely supplanted the Aristotelian among students of mechanics.

BIBLIOGRAPHY

I. ORIGINAL WORKS. An essentially complete bibliography of Baldi's published works is given in Pierre

Duhem, *Études sur Léonard de Vinci,* I (Paris, 1906; repr. 1955), 93–99; those of scientific interest are *Di Herone Alessandrino de gli automati* . . . (Venice, 1589; repr. 1601); *Scamilli impares Vitruviani* . . . (Augsburg, 1612); *De Vitruvianorum verborum significatione* . . . (Augsburg, 1612); *Heronis Ctesibii Belopoeeca* . . . *et* . . . *Heronis vita* . . . (Augsburg, 1616); *In mechanica Aristotelis problemata exercitationes* . . . (Mainz, 1621); *Cronica de' matematici* . . . (Urbino, 1707); "Vita di Federigo Commandino," in *Giornale de' letterati de'Italia,* **19** (1714), 140 ff.; "Vite inedite di matematici italiani," in *Bollettino di bibliografia e di storia delle scienze,* **19** (1886), 335–640; **20** (1887), 197–308; and *L'invenzione del bossolo da navigare,* G. Canerazzi, ed. (Leghorn, 1901).

II. SECONDARY LITERATURE. Principal biographical sources are Fabritio Scharloncini, *De vita et scriptis Bernardini Baldi Urbinatis,* prefaced to Baldi's *In mechanica Aristotelis* . . .; Ireneo Affò, *Vita di Bernardino Baldi* (Parma, 1783); and R. Amaturo, "Bernardino Baldi," in *Dizionario biografico degli Italiani,* V (Rome, 1963), 461–464. For Baldi's scientific work and its influence, see Duhem, *op. cit.,* pp. 89–156; for his relation to the central Italian mathematicians, see S. Drake and I. E. Drabkin, *Mechanics in Sixteenth-Century Italy* (Madison, Wis., 1968).

STILLMAN DRAKE

BALFOUR, FRANCIS MAITLAND (*b.* Edinburgh, Scotland, 10 November 1851; *d.* near Courmayeur, Switzerland, 19/20 July 1882), *embryology.*

The third son of James Maitland Balfour and Lady Blanche, daughter of the marquis of Salisbury, Francis Balfour came from an illustrious family, the outstanding member of which was his oldest brother, the prominent philosopher and statesman Arthur James Balfour. Francis spent his childhood at Whittingham. He attended the preparatory school at Hoddesdon, Hertfordshire, and in 1865 enrolled at Harrow. A keen naturalist as a boy, at Harrow Balfour wrote a prize-winning essay for the school's scientific society, "The Geology and Natural History of East Lothian," which was judged and highly praised by Thomas H. Huxley.

Balfour matriculated at Trinity College, Cambridge, in 1870; he took a first in the Natural Science Scholarships the following year, and by 1872 was working under the direction of the physiologist Michael Foster. During the winter of 1873/1874 he began his investigation of the embryology of elasmobranchs at the Stazione Zoologica in Naples, work he pursued intermittently, both in Italy and in Cambridge, until 1878, when he completed his outstanding monograph on elasmobranch development. He began lecturing on morphology and embryology at Trinity College in the fall term of 1873 and became director of the university's morphological laboratory, where

he attracted an extraordinarily enthusiastic group of students. Balfour was elected to the Royal Society in 1878, became vice-president of the biological section of the British Association for the Advancement of Science in 1880, and won the Royal Society's Royal Medal the following year upon completion of his two-volume text on comparative embryology.

Balfour's star had risen so quickly and his reputation as a remarkable teacher was so well known that it was not surprising that both Oxford and the University of Edinburgh tried to lure him away from Cambridge. In 1882, at the urging of Foster, Cambridge created a professorship of animal morphology that Balfour was to hold for his lifetime; but before the year was out, Balfour was dead. For several years he had spent his vacations mountain climbing in Switzerland and had become an experienced alpinist. On 18 July 1882 he set out with his Swiss guide to scale the unconquered peak Aiguille Blanche de Peteret in the Chamonix district. They failed to return; their bodies were recovered at the foot of an icefall several days later.

Balfour's embryological contributions exemplify the descriptive studies that characterized the two decades of embryology following the publication of Darwin's *Origin of Species.* He began his research in Naples at a time when Ernst Haeckel, Aleksandr Kovalevski, E. Ray Lankester, and Anton Dohrn, among others, were emphasizing comparative embryology and placing a premium on the search for homologies and phylogenetic links. Balfour's work was unquestionably directed by similar concerns, but his painstaking microscopic examinations and his crisp ordering of embryological details permitted him to draw certain sound generalizations about development that are far more impressive to the historian than the wilder speculations of his peers.

His monograph on elasmobranchs was a closely knit document of nearly 300 pages. At the outset, Balfour avoided the customary phylogenetic discussion; instead, he began by describing the development of the ovum and the segmentation of the fertilized egg. He carried the description forward by embryonic stages until, at the end of the treatise, he could give a detailed analysis of the major organ systems and interpret their development in relation to that in other animals. Balfour's analysis of germ-layer formation in elasmobranchs is a good example of his detailed microscopic examination and also demonstrates his ability to draw far-reaching generalizations. Carefully tracing the migration of lower-level cells during the formation of the segmentation cavity, he argued that both the segmentation cavity and the alimentary canal were formed by a delamination in the hypoblast. At first glance this conclusion had grave implications for Haeckel's *gastraea* theory, to which Balfour himself subscribed, because Haeckel had maintained that the alimentary canal of all vertebrates was formed by an involution of the epiblast—that is, in the same manner that the archenteron of his hypothetical *gastraea* had been formed in eons past.

At this juncture Balfour introduced quite a different question: If the segmentation cavity and alimentary canal in *amphioxus,* elasmobranchs, amphibians, and birds were to be considered homologous, how should one interpret their different germ-layer origins? He concluded that one must take into consideration the quantity of yolk in the egg and its degree of segmentation at the time of germ-layer formation. Thus, he argued that since there was little yolk in *amphioxus* and amphibian eggs, there existed perforce an anus of Rosconi and an involution of the epiblast at the dorsal lip of the blastopore; however, his argument continued, since there was a large quantity of yolk in elasmobranch eggs, there existed only a temporary anal opening and a delamination of the hypoblast. As for bird eggs, which held an extremely large quantity of yolk, Balfour claimed that no anus of Rosconi could possibly exist, nor could the epiblast invaginate. In this explanation Balfour indicated an appreciation for the mechanical influences of cellular movement that forced different embryos to develop along different lines, irrespective of their phylogenetic connections.

Balfour's recognition of the mechanical influence of the mass of yolk may have prompted his inquiry into the meaning of the primitive streak of the chick embryo. Comparing the positions of amphibian and selachian embryos, which are asymmetrically situated on the blastoderm, with the chick embryo, which lies on the center, Balfour argued that the primitive streak was the homologue of the blastopore in the lower vertebrates. In collaboration with one of his students, Balfour further explained that the mesoblast of the chick was derived through a simultaneous differentiation from both the hypoblast and the epiblast, which lay along the axial line of the primitive streak. Such evidence again implicitly challenged a too rigid belief in the specificity of the germ layers. Balfour, however, did not draw the ultimate conclusions but faithfully adhered to the germ-layer doctrine.

In his study of elasmobranchs, Balfour included a detailed and superb description of the development of the excretory system. Although this work was anticipated in many respects by Karl Semper's similar study, Balfour definitively pointed out the initial segmental character of the Wolffian duct ("segmental duct") and traced the manner in which the Müllerian duct arose from the Wolffian duct's ventral side. He then described in detail how the Müllerian duct in the female became converted to the oviduct and how the Wolffian duct in the male was utilized by the testes as the *vas deferens.* Considering the various modes of development in the abdominal opening of the Müllerian duct in elasmobranchs, amphibians, and birds, Balfour argued that there were sufficient differences to indicate three lines of urogenital evolution that diverged from a more general and primitive state. Although it is clear that the reconstruction of an evolutionary tree was a concern which constantly framed his detailed research, Balfour again minimized a discussion of the phylogenetic implications of his work.

When one surveys his many shorter articles, it is surprising to discover how strategically Balfour covered portions of the animal kingdom. A clear pattern emerges from his studies of elasmobranchs, *Lepidosteus* (a ganoid), *amphioxus, Peripatus* (an aberrant arthropod), and *Araneina* (a true spider), in which Balfour appears intent upon locating possible links between major taxa. Although he did not commit himself to any current theory concerning the link between invertebrates and vertebrates, he believed that any protochordate must have been segmental in character and must have had a suctorial mouth on the ventral surface. He believed, moreover, that any claimant of the title "missing link" must show undeniable signs of a primitive notochord.

Even taking these broader interests in phylogeny into consideration, it is still remarkable that within three years of the completion of his monograph on elasmobranchs, Balfour finished his two-volume *Treatise on Comparative Embryology* (1880–1881). According to Waldeyer, this was the first successful attempt at a complete comparative embryology text; it was translated into German the same year, and one finds important references to it for decades thereafter. Not only did Balfour survey developments peculiar to each phylum and give a comparative survey of the embryology of each organ system, but he also included introductory chapters on gamete formation, fertilization, and early cleavage, additions that demonstrated his keen interest in the most recent advances in cytology. By 1882 microscopy had made such rapid gains that many of Balfour's facts, particularly about the nucleus, were already dated. There are certain signs, however, that he would have become more involved in this direction of research; in fact, as indicated above, there are even signs that he would have appreciated the type of question which the experimental embryologists in Germany began asking of the embryo from about 1882. Reflecting upon these interests, one can only muse how the embryological sciences in England would have progressed had Balfour not died at the age of thirty.

BIBLIOGRAPHY

The Works of Francis Maitland Balfour, Michael Foster and Adam Sedgwick, eds., 4 vols. (London, 1885), contains all of Balfour's published works, including *A Monograph on the Development of the Elasmobranch Fishes* (London, 1878) and *A Treatise on Comparative Embryology,* 2 vols. (London, 1880–1881).

There are only a few scanty biographical notices on Balfour, and even fewer discussions of his work. Besides Michael Foster's introduction to the collected works, W. Waldeyer, "Francis Maitland Balfour, Ein Nachruf," in *Archiv für mikroskopische Anatomie,* **21** (1882), 828–835, and Henry Fairfield Osborn, "Francis Maitland Balfour," in *Science,* **2** (1883), 299–301, are useful. Balfour's work is placed in its historical context in E. S. Russell, *Form and Function, a Contribution to the History of Animal Morphology* (London, 1916), pp. 268–301.

F. B. CHURCHILL

BALFOUR, ISAAC BAYLEY (*b.* Edinburgh, Scotland, 31 March 1853; *d.* Haslemere, England, 30 November 1922), *botany.*

Balfour possessed a distinguished academic pedigree: his father, John Hutton Balfour, was professor of botany at Edinburgh University and his mother, Marion Spottiswood Bayley, was the daughter of Isaac Bayley, writer to the signet; his ancestry also included George H. Baird, principal of Edinburgh University, and geologist James Hutton.

At Edinburgh University he obtained his B.S. in 1873 and M.B. in 1877, then continued his botanical studies, particularly morphology and physiology, at the universities of Würzburg and Strasbourg. He accompanied the transit of Venus expedition to Rodriguez Island in 1874 as botanist and geologist. The botanical results of this expedition, published in the *Philosophical Transactions* in 1879, clearly indicated that Balfour was a taxonomist of considerable promise. In 1879–1880 he collected plants on the island of Socotra, and in 1888 published his description of the island's flora, which included many new species. He observed that the flora had affinities with that of the African mainland and argued that Socotra had once formed part of that continent.

After his appointment to the chair of botany at Glasgow University in 1879, Balfour revealed his capacity for organization. In his five years there he rebuilt the principal range of greenhouses, saved the herbarium from imminent destruction, and improved the laboratory facilities for students. In 1884 he was elected Sherardian professor of botany at Oxford, where his energy and administrative skill revitalized the ancient but neglected botanic garden. The herbaceous beds were remodeled, and the valuable herbarium and library thoroughly reorganized. Having established good relations with the Clarendon Press in Oxford, he persuaded them to undertake, under his editorship, the publication of translations of standard German botanical texts. The press also launched the *Annals of Botany* in 1887, with Balfour as joint editor with S. H. Vines and W. G. Farlow until 1912.

He moved to Edinburgh in 1888 as professor of botany, queen's botanist in Scotland, and Regius keeper of the Royal Botanic Garden, holding these offices, as did his father before him, for thirty-four years. During this time Balfour gradually accomplished many of his desired reforms: a massive wall that obstructed the redesigning of the botanic garden was removed; the plant collections were enriched, particularly with alpines, for which a splendid new rock garden was created; and the greenhouses were rebuilt. Under his direction Edinburgh became an exemplar of horticultural practice.

A capacity for unrelenting work brought Balfour distinction in several fields of activity: taxonomy, teaching, horticulture, and administration. The pressures of a full life never hindered his botanical research, in which he concentrated on *Rhododendron* and *Primula.* His taxonomic papers on these two genera appeared in *Notes From the Royal Botanic Garden, Edinburgh,* which he founded in 1900. He was an authority on the vegetation of the Himalayas and western China, and successfully grew at Edinburgh many of the new plants introduced by George Forrest from that region. A gifted lecturer, Balfour modernized the teaching methods of three universities, and for a generation Edinburgh was the main center for the teaching of taxonomy. In short, he was that *rara avis,* an all-round botanist with an aptitude for organization.

Balfour was elected a fellow of the Royal Society in 1884; awarded the Victoria Medal of Honour of the Royal Horticultural Society in 1897, and the Linnean Medal in 1919; and created K.B.E. in 1920 for services rendered during World War I. He married Agnes Boyd in 1884, and had one son and one daughter.

BIBLIOGRAPHY

I. ORIGINAL WORKS. Among Balfour's writings are "Botany [of Rodriguez]," in *Philosophical Transactions of the*

Royal Society of London, **168** (1879), 302–387; and "Botany of Socotra," which constitutes *Transactions of the Royal Society of Edinburgh,* **31** (1888).

II. SECONDARY LITERATURE. Obituaries include F. O. Bower, in *Proceedings of the Royal Society of Edinburgh,* **43**, no. 3 (1923), 230–236; J. B. Farmer, in *Annals of Botany,* **37**, no. 146 (1923), 335–339; *Kew Bulletin* (1923), 30–35; Sir David Prain, in *Proceedings of the Royal Society of London,* **96B**, no. 678 (1924), i–xvii; and W. Wright Smith, in *Transactions and Proceedings of the Botanical Society of Edinburgh,* **28**, no. 4 (1923), 192–196.

R. G. C. DESMOND

BALFOUR, JOHN HUTTON (*b.* Edinburgh, Scotland, 15 September 1808; *d.* Edinburgh, 11 February 1884), *botany.*

The eldest son of Andrew Balfour, an army surgeon who became a printer and publisher in Edinburgh, John Balfour received his basic education at the High School, Edinburgh, and the universities of Edinburgh and St. Andrews. He resisted his parents' wishes that he enter the church, and chose medicine for his future career. He graduated M.D. at Edinburgh in 1832, and after a period of further study in a medical school in Paris, returned to his native city to practice medicine in 1834.

Like so many other physicians before him, Balfour found himself irresistibly drawn to botany. He first became seriously interested in the subject when he was about eighteen, and he was allowed to attend Robert Graham's botanical lectures for four sessions. He was a founding member in 1836 of the Botanical Society of Edinburgh, and two years later he established the Edinburgh Botanical Club. In 1840 he enjoyed some success as a lecturer on botany in the extramural School of Medicine.

In 1841 Balfour abandoned medicine on succeeding Sir William J. Hooker as professor of botany at Glasgow University. His few years at Glasgow were not marked by any radical changes, although he arranged the removal of the botanic garden, which was becoming hemmed in by adjacent streets, to a new site at Kelvinside. Upon the death of Graham in 1845, he was elected to the chair of botany at Edinburgh and the associated posts of queen's botanist in Scotland and Regius keeper of the Royal Botanic Garden. At Edinburgh, Balfour was fortunate in having as successive curators William and James McNab, father and son, and through their horticultural skill marked improvements were effected in the botanic garden: the grounds were enlarged from fourteen to forty-two acres, an arboretum was established, and a fine palm house and botanical museum were erected.

It is, however, as a teacher that Balfour is remembered. He made extensive use of the microscope in his demonstrations—not a common practice at that time—and placed great emphasis on the value of botanical excursions. These excursions had always been a characteristic feature of the teaching of botany in Scottish universities, and Balfour, like his predecessor Graham, used them for the dual purposes of instruction and supplying the botanical garden with specimens suitable for cultivation. Through these excursions he explored the greater part of Scotland, inculcating in his students not only a knowledge of the flora but also an appreciation of ecological diversity. His manner of teaching, lucid and displaying an impressive wealth of detail, was distinguished by an enthusiasm that he communicated to his audience. A number of useful students' textbooks, which went through numerous editions, came from his pen, including *Manual of Botany* (1849) and *Class Book of Botany* (1852).

His father had been a strict Presbyterian, and Balfour himself was a deeply religious man who sought in nature confirmation of God's existence. Among his botanico-religious books were *Phyto-Theology* (1851), the title of which was changed to *Botany and Religion* in the third edition, *Plants of the Bible* (1857), and *Lessons From Bible Plants* (1870).

Balfour was an editor of the *Edinburgh New Philosophical Journal.* That his original work should have been slight was the inevitable consequence of the time he devoted to teaching and administration. Elected a fellow of the Royal Society of Edinburgh in 1835, he was its secretary for many years; he also acquired the secretaryship of the Royal Caledonian Horticultural Society and was dean of the Medical Faculty at Edinburgh for thirty years. He was elected a fellow of the Linnean Society in 1844, and of the Royal Society in 1856. On his retirement in 1879, the universities of Edinburgh, St. Andrews, and Glasgow each conferred on him the LL.D.

BIBLIOGRAPHY

Among Balfour's works are *Manual of Botany* (London, 1849); *Phyto-Theology* (Edinburgh, 1851); *Class Book of Botany* (Edinburgh, 1852); *Outlines of Botany* (Edinburgh, 1854); *Plants of the Bible* (London, 1857); *Flora of Edinburgh* (Edinburgh, 1863), written with J. Sadler; and *Lessons From Bible Plants* (Glasgow, 1870).

There is an article on Balfour by his son, Isaac, in F. W. Oliver, ed., *Makers of British Botany* (Cambridge, 1913), pp. 293–300.

R. G. C. DESMOND

BALIANI, GIOVANNI BATTISTA (*b.* Genoa, Italy, 1582; *d.* Genoa, 1666), *physics.*

Baliani, the son of a senator, was trained in the law and spent most of his life in public service. His scientific interests appear to have begun about 1611, when he was prefect of the fortress at Savona. There he noted the equal speed of fall of cannon balls differing greatly in weight. About the same time he devised an apparatus for cooking by frictional heat—an iron pot rotating on a concave iron base.

In 1613 Filippo Salviati met Baliani and wrote of him to Galileo, who began corresponding with Baliani concerning the experimental determination of the weight of air. In 1615 Baliani visited Galileo at Florence and met Benedetto Castelli. The intermittent correspondence that lasted for many years shows Baliani to have been a talented experimentalist and an ingenious speculator. In 1630 he wrote to Galileo of the failure of a siphon that had been expected to carry water over a rise of about sixty feet (eighty *piedi*). Baliani attributed the action of a lift pump to atmospheric pressure, but doubted that the total weight of a column of air many miles high could be less than that of a thirty-foot column of water, at which height Galileo had already noted the failure of lift pumps.

In astronomy, although Baliani preferred Tycho Brahe's system to that of Copernicus, he speculated on a terrestrial motion as the possible cause of tides.

In 1638 Baliani published a short treatise on the motions of heavy bodies, which he reprinted in 1646 with many additions. In the first edition he gave correct laws for free fall, motion on inclined planes, and pendulums. In the second edition he speculated on the possibility that in unmeasurably small successive finite times, the spaces traversed by a falling body might increase in proportion to the natural numbers, assuming that the body received very rapid successive impulses and retained them unimpaired. Baliani's argument embodied an important step toward the concept of mass and the analysis of acceleration, but it was widely misunderstood as intended to contradict the law that both he and Galileo had explicitly stated: that for successive equal measurable times, the spaces traversed by a falling body are as the odd numbers 1, 3, 5,···. The misunderstanding caused Baliani's name to become associated with the false hypothesis, specifically rejected by Galileo, that velocity in free fall increases in proportion to space traversed. Although Baliani did not uphold the false law of acceleration, he rejected the parabolic trajectory in such a way as to show that his idea of inertial motion was inexact.

In 1647 Baliani published a treatise on the plague, suggesting a chemical explanation of its nature and its contagious character. In this work he stated the principle that the rate of human population increase as related to arable land and food production would necessarily result in famine were it not for the occurrence of war and pestilence. The quantitative nature of his argument entitles him to be regarded as a predecessor of the Malthusian law.

Baliani returned to Savona in 1647 as governor of its fortress, a post he held until 1649. He was then elevated to membership in the principal governing body of Genoa, where he remained until his death.

Baliani's previously unpublished works were collected and printed in 1666. They include several philosophical dialogues and discussions of light, action at a distance, the existence of a vacuum and of motion therein, and some prismatic experiments. The works were republished in 1792 with an anonymous life of Baliani and a number of letters praising his achievements.

The direct influence of Baliani on other scientists probably was not great; he worked as an amateur with notable ability and success, but his principal fields of interest were those that were simultaneously receiving attention from Galileo and his disciples. He did not arrive at the laws of falling bodies independently of Galileo, and although their explanation in terms of incremental impulse was probably his own, a similar analysis had been published by G. B. Benedetti in 1585. His correct conception of atmospheric pressure remained unpublished, although it may have become known to Torricelli through conversations with Galileo, who rejected it. Baliani's most important contribution, a discussion of elastic shock, seems to have gone unnoticed until recently, and hence probably did not influence the development of the laws of impact. No final evaluation of Baliani's place in the history of physics is possible, however, without an exhaustive study of his extant correspondence, for many of his germinal ideas were not published and are known only through his letters.

BIBLIOGRAPHY

I. ORIGINAL WORKS. Known copies of Baliani's works in North America not listed in the Union Catalogue are designated as follows: California Institute of Technology, CIT; University of Toronto Library, ULT. His works are *De motu naturali gravium solidorum Ioannis Baptistae Baliani patritii genuensis* (Genoa, 1638); *De motu gravium solidorum et liquidorum Io. Baptistae Baliani patritii genuensis* (Genoa, 1646; ULT); *Trattato della pestilenza di Gio. Battista Baliano* (Savona, 1647; ULT); *Di. Gio. Batista Baliani. Opere diverse* (Genoa, 1666); and *Opere diverse di Gio. Battista Baliani, patrizio genovese; aggiuntovi nell'avviso a chi legge, una compendiosa notizia di lui vita* (Genoa, 1792; CIT).

The Baliani-Galileo correspondence is included in *Le opere di Galileo Galilei*, Ed. Naz. (Florence, 1934–1937), Vols. XII–XVIII, *passim*. Baliani's correspondence with Mersenne has been published to 1640 in *Correspondance du P. Marin Mersenne*, Cornelis De Waard *et al.*, eds. (Paris, 1945–1965). Unpublished Baliani letters at Milan are described in Moscovici's 1965 article.

II. Secondary Literature. A general discussion of Baliani and his published works is given in Alpinolo Natucci, "Giovan Battista Baliani letterato e scienzato del secolo XVII," in *Archives internationales d'histoire des sciences,* **12** (1959), 267–283. A critical examination of the misinterpretation by earlier writers of Baliani's law of acceleration occupies Ottaviano Cametti's *Lettera critico-meccanica* (Rome, 1758); its subsequent fate and its implications for the concepts of mass and inertia are discussed in S. Moscovici, *L'expérience de mouvement. Jean-Baptiste Baliani—disciple et critique de Galilée* (Paris, 1967), which discusses extensively the barometric correspondence with Mersenne and others and which has an appendix with many previously unpublished letters of Baliani. Other aspects of Baliani's physics are discussed in S. Moscovici, "Les développements historiques de la théorie galiléenne des marées," in *Revue d'histoire des sciences et de leurs applications,* **18**, no. 2 (1965), 193–220; and Cornelis De Waard, *L'expérience barométrique. Ses antécédents et ses explications* (Thouars, 1936), p. 95.

Stillman Drake

AL-BALKHĪ. See **Abū Ma'shar.**

BALLONIUS. See **Baillou, Guillaume.**

BALLOT, CHRISTOPH BUYS. See **Buys Ballot, Christoph.**

BALMER, JOHANN JAKOB (*b.* Lausen, Basel-Land, Switzerland, 1 May 1825; *d.* Basel, Switzerland, 12 March 1898), *mathematics, physics.*

Balmer was the oldest son of Chief Justice Johann Jakob Balmer and Elisabeth Rolle Balmer. He attended the district school in Liestal and the secondary school in Basel, studied mathematics at Karlsruhe and Berlin, and was granted a doctorate at Basel in 1849 with a dissertation on the cycloid. In 1868 he married Christine Pauline Rinck, who bore him six children. He taught at the girls' secondary school in Basel from 1859 until his death, and from 1865 to 1890 he also held a part-time lectureship at the University of Basel. His major field of professional interest was geometry; spectral series, the topic of his most noted contribution, was an area in which he became involved only late in life.

The earliest attempts to establish relationships

between the observed lines of an elementary spectrum were organized primarily within the theoretical context of a mechanical acoustical analogy. Many investigators attempted to establish simple harmonic ratios, but in 1881 Arthur Schuster demonstrated the inadequacy of this approach. The essentially successful mathematical organization of the data began in 1885 with Balmer's presentation of the formula $\lambda = hm^2/(m^2 - n^2)$ for the hydrogen series. This formula could be used to generate, with considerable accuracy, the wavelengths of the known characteristic spectral lines for hydrogen when $n = 2$, $h = 3645.6 \times 10^{-8}$ cm., and $m = 3, 4, 5, 6, \cdots$ successively.

Initially, Balmer knew only Ångström's measurements for the first four visible hydrogen lines, but he calculated the next, or fifth, result of the formula using $m = 7$, obtaining the wavelength of a line that, if it existed, would be barely on the edge of the visible spectrum. Jakob Edward Hagenbach-Bischoff, a friend and colleague at the University of Basel who had stimulated his interest in this topic, informed him that this fifth line had been observed, and that a number of other hydrogen lines had been measured. Comparisons between the calculated results obtained using Balmer's formula for these lines and the observed values showed close agreement, differences being at most approximately one part in one thousand. Later investigators demonstrated that the formula, with a slightly altered constant, represented the whole series, including additional lines, with unusual accuracy. Balmer speculated that values for other hydrogen series in the ultraviolet and infrared regions would be generated if n were assigned integer values other than two. Such predicted series were experimentally found later and are known as the Lyman, Paschen, Brackett, and Pfund series.

Balmer's relationship was so different from the simple harmonic ratios expected on the acoustical analogy, that investigators were bewildered as to what sort of mechanism could produce such lines. In spite of this disturbing feature, Balmer's work served as a model for other series formulas, especially the more generalized formulas of Johannes Robert Rydberg (1854–1919), Heinrich Kayser (1853–1940), and Carl Runge (1856–1927). Balmer published only one further paper on spectra, in which he extended his considerations to the spectra of several other elements (1897).

BIBLIOGRAPHY

I. Original Works. Balmer's major article on the hydrogen spectrum is "Notiz über die Spektrallinien des

Wasserstoffs," in *Verhandlungen der Naturforschenden Gesellschaft in Basel,* **7** (1885), 548–560, 750–752; also in *Annalen der Physik,* 3rd ser., **25** (1885), 80–87. His second and only other spectral article is "Eine neue Formel für Spektralwellen," in *Verhandlungen der Naturforschenden Gesellschaft in Basel,* **11** (1897), 448–463; and in *Annalen der Physik,* 3rd ser., **60** (1897), 380–391. It is also available in *Astrophysical Journal,* **5** (1897), 199–209. A short note of historical interest is Jakob Edward Hagenbach-Bischoff's "Balmer'sche Formel für Wasserstofflinien," in *Verhandlungen der Naturforschenden Gesellschaft in Basel,* **8** (1890), 242.

II. Secondary Literature. For discussions on aspects of Balmer's life and works, see August Hagenbach, "J. J. Balmer und W. Ritz," in *Die Naturwissenschaften,* **9** (1921), 451–455, and "Johann Jakob Balmer," in Edward Fueter, ed., *Grosse Schweizer Forscher* (Zurich, 1939), pp. 248–249; L. Hartmann, "Johann Jakob Balmer," in *Physikalische Blätter,* **5** (1949), 11–14; and Eduard His, "Johann Jakob Balmer," in *Basler Gelehrte des 19. Jahrhunderts* (Basel, 1941), pp. 213–217.

C. L. Maier

BAMBERGER, EUGEN (*b.* Berlin, Prussia, 19 July 1857; *d.* Ponte Tresa, Switzerland, 10 December 1932), *chemistry.*

In 1875 Bamberger entered the University of Breslau as a medical student. The next summer he studied at Heidelberg with Robert Bunsen. He later returned to Berlin, worked with Carl Liebermann, and completed his degree under A. W. Hofmann. He then became an assistant to Karl Rammelsberg at the Technische Hochschule of Berlin-Charlottenburg. In 1883 Bamberger went to Munich as Baeyer's assistant, first in the analytic, then in the organic, laboratory. He gave his inaugural lecture in 1885 and in 1891 became extraordinary professor of organic chemistry. In 1893 he accepted a professorship at the Eidgenössische Technische Hochschule in Zurich. He relinquished this position in 1905, however, because of a nervous condition that left his right arm paralyzed. He nevertheless continued private research with the aid of an assistant.

In his dissertation Bamberger worked on derivatives of guanine, then turned to similar substances. He investigated aromatic hydrocarbons and elucidated the structures of retene, chrysene, pyrene, and the glycoside picein. As he pursued his work with naphthalene derivatives, he noticed that some compounds in which hydrogen had been added to the substituted ring no longer retained their aromatic properties. He called these alicyclic because they behaved like aliphatic compounds but had a closed-ring structure. In connection with this work he also extended the centric formula of benzene to naph-

thalene and supported this view with his work on benzimidazole and isoquinoline. Turning to by-products of reactions previously studied, Bamberger examined such mixed nitrogen compounds as the hydrazones and formaryl derivatives. He then directed his attention to organic nitrogen compounds, and discovered the isodiazo compounds. He found that, unlike normal diazo compounds, in the isodiazo compounds the NO could be oxidized to NO_2. This led Bamberger to believe that a formula similar to that of nitrosamines could be used to represent those isodiazo compounds which on hydrolysis yielded normal diazo compounds. In 1894 this view was sharply criticized by Arthur Hantzsch. He suggested that the differences between the two types of diazo compounds could be explained by isomerism. As a classical organic chemist, Bamberger countered most of Hantzsch's arguments by studying the chemical activity of the supposed isomer, but he admitted the cogency of the physicochemical arguments by replacing his proposed formula with one containing the phenylazo radical. The controversy continued for years, and only with the rejection of pentavalent nitrogen could a new formula for diazo compounds be proposed.

Bamberger then turned to the oxidation and reduction of nitrogen compounds. He reduced nitrobenzene to nitrosobenzene and phenylhydrozylamine with zinc dust in a neutral solution, but refused to patent the process. His preparation of dimethylaniline oxide supported the idea of the pentavalence of nitrogen. Following the diazo controversy Bamberger paid more attention to physical properties, investigating the optical properties of anthranil derivatives and the photochemical properties of benzaldehyde derivatives. In spite of this, however, he remained a classical organic chemist devoted to theory. He investigated natural compounds, using minimum material and the simplest equipment, and studied all by-products thoroughly.

BIBLIOGRAPHY

I. Original Works. An extensive list of Bamberger's works is in Poggendorff. Among his writings are "Ueber die Constitution des Acenaphtens und der Naphthalsäure," in *Bericht der Deutschen chemischen Gesellschaft,* **20** (1887), 237–244, written with Max Philip; "Ueber α-tetrahydronaphtylene," *ibid.,* **21** (1888), 1786–1795, 1892–1904, written with Max Althausse; "Ueber Aethyl-α-naphtylamin," *ibid.,* **27** (1894), 2469–2472, written with Carl Goldschmidt; "Zur Kenntnis des Diazotirungsprocess," *ibid.,* 1948–1953; "Ueber das Phenylhydrozylamin," *ibid.,* 1548–1557; "Ueber die Reduction der Nitroverbindungen," *ibid.,*

1347–1350; "Ueber die 'Stereoisomeren' Diazoamidverbindungen von A. Hantzsch," *ibid.*, 2596–2601; "Weiteres über Diazo- und Isodiazoverbindungen," *ibid.*, 914–917; and "Zur Geschichte der Diazoniumsalze," *ibid.*, **32** (1899), 2043–2046, 3633–3635.

II. SECONDARY LITERATURE. Works dealing with Bamberger are "E. Bamberger zum 75. Geburtstag," in *Zeitschrift für angewandte Chemie*, **45** (1932), 514; Louis Blangey, "Eugene Bamberger," in *Helvetica chimica acta*, **16** (1933), 644–676; and Eduard Hjelt, *Geschichte der organischen Chemie von altester Zeit bis zur Gegenwart* (Brunswick, 1916). J. R. Partington, *A History of Chemistry*, IV (London, 1964), discusses Bamberger's contributions to organic chemistry on pp. 840–842 and gives an account of the controversy with Hantzsch under "Hantzsch," pp. 842–847.

RUTH ANNE GIENAPP

BANACH, STEFAN (*b.* Krakow, Poland, 30 March 1892; *d.* Lvov, Ukrainian S.S.R., 31 August 1945), *mathematics.*

Banach's father, a railway official, and mother turned their son over to a laundress, who became his foster mother and gave him her surname. From the age of fifteen he supported himself by giving private lessons. After graduating from secondary school in Krakow in 1910, Banach studied at the Institute of Technology in Lvov but did not graduate. He returned to Krakow in 1914, and from 1916, when he met H. Steinhaus, he devoted himself to mathematics. His knowledge of the field was already fairly extensive, although it probably was not very systematic. Banach's first paper, on the convergence of Fourier series, was written with Steinhaus in 1917 and was published two years later. Also in 1919 he was appointed lecturer in mathematics at the Institute of Technology in Lvov, where, in addition, he lectured on mechanics. In the same year he received his doctorate with an unusual exemption from complete university education. Banach's thesis, "Sur les opérations dans les ensembles abstraits et leur application aux équations intégrales," appeared in *Fundamenta mathematicae* in 1922. The publication of this thesis is sometimes said to have marked the birth of functional analysis.

In 1922 Banach became a *Dozent* on the basis of a paper on measure theory (published in 1923). Soon afterward he was made associate professor, and in 1927 he became full professor at the University of Lvov. In 1924 he was elected corresponding member of the Polish Academy of Sciences and Arts. Banach's research activity was intense, and he had a number of students who later became outstanding mathematicians: S. Mazur, W. Orlicz, J. Schauder, and S. Ulam, among others. Banach and Steinhaus

founded the journal *Studia mathematica,* but often Banach had little time left for scientific work because the writing of both college texts (of which the book on mechanics is of special importance) and secondary-school texts took most of his time and effort.

From 1939 to 1941 Banach was dean of the faculty at Lvov and was elected a member of the Ukrainian Academy of Sciences. In the summer of 1941, Lvov was occupied by the German army, and for three years Banach was compelled to feed lice in a German institute that dealt with infectious diseases. After the liberation of Lvov in the autumn of 1944, he resumed his work at the university. His health was shattered, however, and he died less than a year later.

Banach's scientific work comprises about fifty papers and the monograph *Théorie des opérations linéaires* (1932). Although he laid the foundations of contemporary functional analysis, most of his papers are closely connected with the field but are not precisely in it.

Banach made a significant contribution to the theory of orthogonal series, and his theorem on locally meager sets is of lasting importance in general topology. In the descriptive theory of sets and mappings, he extended to mappings some theorems previously known only for numerical functions. A number of results, many of which can now be found in textbooks, concern derivation and absolute continuity, as well as related properties. Banach made a substantial contribution to the theory of measure and integration, results that stimulated a great number of papers and, apparently, the discovery of the Radon-Nikodým theorem. The questions of the existence of measures investigated by Banach have proved to be closely connected with the axiomatic theory of sets.

Despite the great importance of these results and the unusual lucidity and force of mathematical thinking manifested in them, functional analysis is Banach's most important contribution. His work started, of course, from what was achieved during the decades following Vito Volterra's papers of the 1890's on integral equations. Before Banach there were either rather specific individual results that only much later were obtained as applications of general theorems, or relatively vague general concepts. Papers on the so-called general analysis (mainly by E. H. Moore) formed a significant trend, but these were none too comprehensible and, for that period at least, much too general. Ivar Fredholm's and David Hilbert's papers on integral equations marked the most substantial progress. The concepts and theorems they had discovered later became an integral part of functional analysis, but most of them concern only a single linear space (later called Hilbert space).

Later, more or less simultaneously with Banach, several mathematicians—O. Hahn, L. Fréchet, E. Helly, and Norbert Wiener, among others— attained many of the concepts and theorems forming the basis of Banach's theory. None of them, however, succeeded in creating as comprehensive and integrated a system of concepts and theorems and their applications as that of Banach, his co-workers, and his students.

From 1922 on, Banach introduced through his papers the concept of normed linear spaces and investigated them (and metric linear spaces), particularly regarding the assumption of completeness (complete normed linear spaces are now generally known as Banach spaces). The concept was introduced at almost the same time by Norbert Wiener (who about a quarter of a century later founded cybernetics), but he did not develop the theory, perhaps because he did not see its possible application. Banach proved three fundamental theorems of the theory of normed linear spaces: the theorem on the extension of continuous linear functionals, now called the Hahn–Banach theorem (they proved it independently, and Hahn actually did so first); the theorem on bounded families of mappings, now called the Banach–Steinhaus theorem; and the theorem on continuous linear mappings of Banach spaces. He also introduced and examined the concept of weak convergence and weak closure, and gave a series of applications of the general theorems on normed linear spaces.

Further development of functional analysis proved that metric linear spaces are not sufficient for the needs of analysis and that it is essential to use more general, and also richer and more special, objects and structures. Nevertheless, the theory of Banach spaces has—often in combination with other methods—numerous applications in analysis. The theory of these spaces is both an indispensable tool and the basis of contemporary theory of more general linear spaces; it also provided the stimulus and the starting point for other branches of functional analysis.

The fact that functional analysis originated as late as Banach and his school, although favorable conditions seemingly existed at the beginning of the century, is due largely to the way mathematics had developed until then. In fact, sufficiently detailed knowledge about the different concrete instances of linear spaces was not achieved until the 1920's. Also, by that time the applications of some methods of the theory of sets, such as transfinite construction, were clarified, and some theorems on general topology (e.g., Baire's theorem on complete metric spaces and some propositions from the descriptive theory of sets) became widely known and applied.

BIBLIOGRAPHY

I. ORIGINAL WORKS. Banach's major work is *Théorie des opérations linéaires* (Warsaw, 1932). Numerous papers are in *Fundamenta mathematicae* and *Studia mathematica*. See also "Sur le problème de la mesure," in *Fundamenta mathematicae*, **4** (1923), 7–33; and *Mechanika w zakresie szkol akademickich*, Vols. 8 and 9 in the series Monografie Matematyczne (Warsaw–Lvov–Vilna, 1938), translated into English as *Mechanics* (Warsaw–Breslau, 1951). The first volume of his collected works, *Oeuvres* (Warsaw, 1967), contains, besides papers by Banach, up-to-date comments on almost every article.

II. SECONDARY LITERATURE. A short biography, a practically complete list of scientific papers, and an analysis of Banach's work are in *Colloquium mathematicum*, **1**, no. 2 (1948), 65–102. Also see H. Steinhaus, "Stefan Banach," in *Studia mathematica*, special series, **1** (1963), 7–15; and S. Ulam, "Stefan Banach 1892–1945," in *Bulletin of the American Mathematical Society*, **52** (1946), 600–603.

MIROSLAV KATĚTOV

BANACHIEWICZ, THADDEUS (*b.* Warsaw, Poland, 13 February 1882; *d.* Krakow, Poland, 17 November 1954), *astronomy.*

The younger son of Arthur Banachiewicz, a landowner in the Warsaw district, and Sophia Rzeszotarska, Banachiewicz received his bachelor's degree in astronomy in 1904 from Warsaw University, where one of his astronomical papers had earlier won a gold medal. He continued his studies in Göttingen under Schwarzchild and in Pulkovo. After his return to Warsaw he was junior assistant at the Warsaw Observatory in 1908–1909. In 1910 he received the master's degree in astronomy from Moscow University and soon afterward was appointed assistant at the Engelhardt Observatory, near Kazan, where he stayed until 1915. For the next three years he taught at Dorpat. He returned to Warsaw in 1918 and for a short time was *Dozent* of geodesy at the Warsaw Polytechnic High School. Toward the end of that year he accepted the professorship of astronomy at the University of Krakow and directorship of the Krakow Observatory. He spent the rest of his life in Krakow, the only interruption occurring in the winter of 1939/ 1940, when the Krakow faculty was taken to the Gestapo concentration camp at Sachsenhausen, near Berlin.

Banachiewicz's work concerned many important problems of astronomy, geodesy, geophysics, mathematics, mechanics, and numerical calculus. His principle of repeated verification made his 240 published papers safe from errors. His most important astronomical and geodetical work was theoretical. As early as 1906 a paper of his that dealt with Lagrange's three-body problem was presented to the Paris Academy by Poincaré. The paper, "Über die Anwendbarkeit der Gyldén-Brendelschen Störungstheorie auf die Jupiternahen Planetoiden," gave a brilliant analysis of Gyldén-Brendel's theory, pointing out its illusiveness when applied to small planets in the vicinity of Jupiter.

He also published several papers on Gauss's equation and gave useful tables to facilitate its numerical solution. These tables have been reprinted in J. Bauschinger and G. Stracke's *Tafeln zur theoretischen Astronomie.*

Banachiewicz paid considerable attention to multiple solutions in the determination of parabolic orbits. Legendre, Charlier, Vogel, and others claimed to have solved the question in the sense that the two equations obtained in the process of determining a parabolic orbit from three observations lead to a single solution. Banachiewicz showed that they were basing their reasoning on Lambert's equation, which fails in certain circumstances. In these exceptional cases three solutions are possible, as he demonstrated with a fictitious numerical example.

One of Banachiewicz's great achievements in theoretical astronomy was the simplification of Olbers' method of determining parabolic orbits. These new methods used a much improved technique of computing, for which Banachiewicz had invented the cracovian calculus. The cracovians are related to Cayley's matrices but differ in the definition of the product, the cracovians being multiplied column by column. This change in the product rule leads to considerable differences between the theories of Cayley and Banachiewicz and makes cracovians more suitable for machine computation. The invention of rotary cracovians enabled Banachiewicz to obtain the solution of the general problem of spherical polygonometry, which had been sought for over a century. Having these new formulas at his disposal, he simplified Bessel's classic method of reducing heliometric observations of the moon's libration. He successfully applied the cracovians to the correction of orbits and gave a simple, practical, and elegant solution of the problem. Convenient cracovian formulas were introduced by Banachiewicz for computing the precessional effect of star coordinates, and his orthogonal transformation formulas facilitated the reduction of the vectorial elements of planets and comets from one epoch to another.

Banachiewicz's investigations into the theory of linear equations produced interesting applications of the cracovian method to such problems as the reduction of astrographic plates; here cracovian formulas led to a general solution that comprised the formulas of Turner and those of the dependency method. He also simplified the classical method of least squares. The cracovian method is well suited to numerical computation. Its importance lies in the fact that the unknowns and their weights are found simultaneously during the process of computation, which is not the case in Gauss's method.

Banachiewicz was not only a prominent theorist but also a gifted and assiduous observer. While a student he promoted observations of occultations of stars by the moon, insisting on their importance for the study of the moon's motion, and developed a purely mathematical method for predicting occultations of stars that had great advantages over the graphical methods. He was also interested in occultations of stars by planets. His ephemerides drew attention to the occultation of 6G Librae by Ganymede, Jupiter's III satellite, on 13 August 1911, which was unique in the history of astronomy because it was the only occultation of a bright star by a planet's satellite to be predicted and observed. From probability considerations one may conclude that an occultation of a star of magnitude $\leq 7^n$ by Ganymede occurs once in a thousand years. While in Kazan he carried out very precise observations of the moon's libration with a four-inch heliometer. Banachiewicz attached much importance to observations of eclipsing variables and introduced them into the working program of the Krakow Observatory. He considered eclipsing binaries the clue to many important questions of the sidereal universe and insisted on gathering observational material for "the future Kepler of eclipsing binaries."

Since he was interested in higher geodesy, Banachiewicz took part in gravimetric observations when he was at Kazan. He later organized gravimetric observations and first-order leveling in Poland. He was Poland's representative to the Baltic Geodetic Commission. At its 1928 meeting Banachiewicz reported the results obtained by the Polish expedition to Lapland (12 June 1927) in timing a solar eclipse by his "chrono-cinematographic" method. This method makes use of Baily's beads and thus greatly increases the number of observed contacts of the two heavenly bodies. Thus the difference in right ascension of moon–sun could be established for the Lapland eclipse with a mean error of $\pm 0.''04$. Banachiewicz proposed to use total eclipses for "lunar triangulations" capable of connecting distant points of the earth's surface.

In astrophysics Banachiewicz was especially interested in photometric problems. Besides the photometry of variable stars he was interested in the illumination of planetary disks and of our sky. He was the first in Poland to appreciate the value of radio signals for the time service and of phototubes in photometry. At his urging the first Polish radio telescope was installed at Fort Skala, a branch station of the Krakow Observatory. Polish astronomy is also indebted to him for his organizational activity. Because of the poor observing conditions at the old Krakow Observatory, he set up a branch station on Mount Lubomir, about nineteen miles south of Krakow. Many observations of variable stars were made there, and several new comets were discovered. The station was burned down by a Gestapo detachment in 1944, but shortly before his death Banachiewicz began the building of a new branch station at Fort Skala.

Banachiewicz was founder and editor of *Acta astronomica, Ephemerides of Eclipsing Binaries,* and *Circulaire de l'Observatoire de Cracovie.*

BIBLIOGRAPHY

I. ORIGINAL WORKS. Among Banachiewicz's numerous publications are "Sur un cas particulier du problème des *n* corps," in *Comptes rendus de l'Académie des sciences* (26 Feb. 1906); "Sur les occultations de l'étoile BD—12⁰ 4042 le 13 août 1911," in *A. N.* 4508; "Sur le mouvement d'un corps céleste à masse variable," in *Comptes-rendus de la Société des sciences de Varsovie,* **6,** fasc. 8 (1913), 657–666; "Sur la méthode d'Olbers et ses solutions multiples," in *Comptes rendus de l'Académie des sciences* (9 Aug. 1915); "Sur la théorie de l'erreur de fermeture dans le cas de la détermination de la latitude géographique" (in Russian), in Kazan, Université Impériale (1915), pp. 1–11; "Tri eskiza po teorii refrakcii," *ibid.;* "Sur la résolution de l'équation de Gauss dans la détermination d'une orbite planétaire," in *Bulletin de l'Académie impériale des sciences,* Petrograd (1916), 739–750; "Tables auxiliaires pour la résolution de l'équation de Gauss sin $(z - q) = m \sin^4 z$" (Paris, 1916), pp. 3–19; "Bemerkungen zu Teil V der Photometrie von Lambert," in *A. N.* **207** (1918), 113–118; "Sur les points d'inflexion des courbes généralisées de Cassini," in *Bulletin de l'Académie polonaise,* ser. A (1921); "Calcul de la précession en coordonnées orthogonales," *ibid.* (1923); "Sur un théorème de Legendre relatif à la détermination des orbites cométaires," *ibid.* (1924), 1–7; "Ueber die Anwendbarkeit der Gyldén-Brendelschen Störungstheorie auf die jupiternahen Planetoiden," in *Circulaire de l'Observatoire de Cracovie,* **22** (1926); "Les rélations fondamentales de la polygonométrie sphérique," *ibid.,* **25** (1927), and *Comptes rendus de l'Académie des sciences* (21 Nov. 1927); "Sur un théorème de Poincaré relatif aux marées océaniques" (in Polish), in *Comptes-rendus de l'Académie polonaise,* Krakow (7 Mar. 1927); "Voies nouvelles dans l'astronomie mathématique," in *Bulletin de l'Académie polonaise* (1927); "Determination of the Constants of the Position of an Orbit From Its Ecliptic Elements," in *Monthly Notices of the Royal Astronomical Society* (Dec. 1928), 215–217;

"Méthodes arithmométriques de la correction des orbites," *Acta astronomica,* **1** (1929), 71–86; "Coordonnées sélénographiques relatives aux occultations," *ibid.* (1930), 127–130; "Sur la détermination de l'orbite de Pluton," in *Comptes rendus de l'Académie des sciences,* **191** (1930), 1–3; "Sur l'amélioration d'une orbite elliptique," in *Bulletin astronomique,* **7,** fasc. 1 (1932), 1–11; "Ueber die Anwendung der Krakoviane in der Methode der kleinsten Quadrate," in *Verhandlungen der 6 Tagung d. Balt. Geod. Kommission* (1933); "Quelques points fondamentaux de la théorie des orbites," in *Acta astronomica,* **3** (1933), 53–56; "On the Prediction of Occultations by the Moon," *ibid.* (1934), 67–76; "Divers points de la théorie des étoiles à éclipses," in *CRM Cracovie* (1936); "Calcul des déterminants à l'aide de cracoviens," in *CRM de l'Académie polonaise des sciences,* **3,** no. 1 (1937); "Contrôle des opérations avec les cracoviens," in *Acta astronomica,* **3** (1938), 133–143; "Einfluss der Gewichte auf die Resultate einer Ausgleichung nach der Methode der kleinsten Quadrate," *ibid.,* 109–118; "Méthode de la résolution numérique des équations linéaires," in *Bulletin de l'Académie polonaise,* ser. A (1938), 393–404; "Principes d'une nouvelle technique de la méthode des moindres carrés," *ibid.,* 134–135; "Sur les rotations dans l'espace à 4 dimensions," *ibid.,* 127–133; "On the Computation of Inverse Arrays," in *Acta astronomica,* **4** (1948), 26–30; "An Outline of the Cracovian Algorithm of the Method of Least Squares," in *Astronomical Journal,* **50** (1942), 38–41; "Improvement of a 9 Cosine Table by Least Squares and Cracovians," in *Acta astronomica,* **4** (1948), 78–80; "La précision d'une orbite provisoire," *ibid.,* **5** (1950), 37–50; "On the General Least Squares Interpolation Formula," *ibid.,* **4** (1950), 123–128; "Sur la détermination du profil de la lune," *ibid.,* **5** (1952), 19–32; and *The Cracovian Calculus* (in Polish) (Warsaw, 1959).

The International Supplement to *Rocrnik astronomiczny Obserwatorium Krakowskiego* ("Astronomical Annual of the Krakow Observatory"), founded by Banachiewicz, is published yearly. Supp. no. 39 for 1968 (1967) contains ephemerides of 754 eclipsing binaries, a list of eclipsing binaries and bases of the ephemeris 1968, RR-Lyrae-type variables, and auxiliary tables, including geocentric ephemeris of the oppositions of the libration points L_4 and L_5 in the earth–moon system. It is published in English.

II. SECONDARY LITERATURE. Banachiewicz's tables are reprinted in Julius Bauschinger, *Tafeln zur theoretischen Astronomie,* 2nd ed. by G. Stracke (Leipzig, 1934); table 22, "Auflösung der Gauss'schen Gleichung," pp. 126–130, is identical with Banachiewicz's. See also Stracke's *Bahnbestimmung der Planeten und Kometen,* ch. 13, which concerns Banachiewicz's method of determining parabolic orbits; fourteen of Banachiewicz's papers are also cited.

J. WITKOWSKI

BANCROFT, WILDER DWIGHT (*b.* Middletown, Rhode Island, 1 October 1867; *d.* Ithaca, New York, 7 February 1953), *chemistry.*

Bancroft was descended from an early New England family, his grandfather being George Bancroft the historian, secretary of the navy under James K. Polk, and founder of the United States Naval Academy. He became interested in chemistry at Harvard, where he received the B.A. in 1888 and assisted in the chemistry department in 1888–1889. He then journeyed to Europe and studied at Strasbourg; Berlin; Leipzig, where he received the Ph.D. under Wilhelm Ostwald in 1892; and Amsterdam, where he studied under van't Hoff. On returning to the United States, Bancroft taught at Harvard from 1893 to 1895 and at Cornell from 1895 to 1937. During World War I he joined the Chemical Warfare Service with the rank of lieutenant colonel. His ability in science, coupled with his good judgment in administrative matters, led his professional colleagues to elect him president of the American Chemical Society in 1910, and of the American Electrochemical Society in 1905 and 1919. In 1937 on the Cornell campus an automobile struck him, injuring him so severely that he was a semi-invalid for the remainder of his life.

Bancroft was one of the early physical chemists in the United States. He taught physical chemistry at Cornell, trained graduate students, wrote a text entitled *The Phase Rule* (1897), and founded the *Journal of Physical Chemistry* in 1896. He supported the *Journal* financially, coedited it until 1909, and edited it until 1932, when he gave it to the American Chemical Society. The value of the *Journal* in its early years lay not only in the information it transmitted, but also in the stimulus it gave to the study of physical and colloid chemistry in the United States.

In research Bancroft did not follow one line exhaustively, but preferred to roam into many fields, letting his curiosity lead him. Generally he had a number of investigations going on at one time. His early investigations were in electrochemistry, a subject he pursued into the practical field of electrodeposition of metals. At the same time he and his students took up the study of heterogeneous equilibria, which was then in an early stage of development, and applied the phase rule to a great variety of systems. They investigated freezing-point equilibria in three-component systems; showed that the minimum in boiling-point curves of binary liquid mixtures was caused by association of one or both components; and made intelligible the heterogeneous solid–liquid equilibria encountered in dynamic isomerides.

Bancroft's second specialty was colloid chemistry. His early work on emulsions and the chemistry of photography led him into theories of dyeing, of the color of colloids, and, toward the end of his career, of the colloidal phenomena associated with anesthesia, asthma, insanity, and drug addiction.

Through his writings on contact catalysis, Bancroft clarified what was, around 1920, largely an empirical art and stimulated chemists to carry out investigations that placed the art on a firm theoretical foundation.

BIBLIOGRAPHY

I. ORIGINAL WORKS. Bancroft's writings include *The Phase Rule* (Ithaca, N.Y., 1897); *Applied Colloid Chemistry: General Theory* (New York, 1921; 3rd ed., 1932); and *Research Problems in Colloid Chemistry* (Washington, D.C., 1921). Bancroft's articles, of which there are scores, may be found by reference to *Chemical Abstracts.*

II. SECONDARY LITERATURE. Articles on Bancroft are A. Findlay, "Wilder Dwight Bancroft, 1867–1953," in *Journal of the Chemical Society* (London) (1953), 2506–2514, with portrait, also in Eduard Farber, ed., *Great Chemists* (New York, 1961), pp. 1245–1261; H. W. Gillett, "Wilder D. Bancroft," in *Industrial and Engineering Chemistry,* **24** (1932), 1200–1201; and C. W. Mason, "Wilder Dwight Bancroft, 1867–1953," in *Journal of the American Chemical Society,* **76** (1954), 2601–2602, with portrait.

WYNDHAM DAVIES MILES

BANISTER, JOHN (*b.* Twigworth, Gloucestershire, England, 1650; *d.* on Roanoke River, Virginia, May 1692), *botany, entomology, malacology, anthropology.*

The first resident, university-educated naturalist of what is now the eastern United States, Banister contributed to English horticulture, to Linnaeus' understanding of the American flora, to Martin Lister's iconography of mollusks, and to James Petiver's catalog of insects. He would have influenced the course of colonial natural history even more but for his early death. Approximately 340 plant descriptions, specimens, and at least eighty plant drawings by Banister are behind the citations in John Ray's *Historia,* Robert Morison's *Historia,* and Leonard Plukenet's *Phytographia* and texts cited in Linnaeus' *Species plantarum* (1753). Jan Gronovius, working with Linnaeus on *Flora Virginica* (1739–1743), compared John Clayton's specimens with Plukenet's—and thus Banister's—illustrations and the descriptions in Ray and Morison.

Lister reproduced, without acknowledgment, at least fourteen of Banister's drawings of Virginia shells in his *Historia conchyliorum,* and published material from four of Banister's letters in *Philosophical Transactions.* Banister's "Collectio insectorum [et arachnidorum]," dealing with about a hundred specimens, and sent probably to Lister in 1680, was published by Petiver in *Philosophical Transactions* in 1701. Banister was one of the first to describe the internal anatomy of a snail, and the first to explain the function of balancers of *Diptera.* He spent the last several years of his life composing his significant "Natural History of Virginia." His "Of the Natives," and undoubtedly some geographical and economic materials, were used by Robert Beverley in *History and Present State of Virginia* (1708): indeed Beverley lifted paragraphs and sentences intact from Banister. Banister's citations are a comprehensive bibliography of the natural history of the New World.

With James Blair as the other minister, Banister was appointed to the committee to establish the College of William and Mary and was named a trustee. He had assembled at least twenty-eight natural history and travel books, which were later acquired by William Byrd I, thus adding immeasurably to the luster of the library at Westover in the time of William Byrd II and III.

Banister's matriculation record at Magdalen College, Oxford, dated 21 June 1667, states that he was the son of John Bannister, "pleb." He was a chorister, receiving the B.A. in 1671 and the M.A. in 1674; was "clerk" (and/or librarian, *fide* his grandson) for two years; and chaplain from 1676 to 1678. That he was well acquainted with American plants growing at Oxford is attested by his carefully labeled *hortus siccus* with author citations. He probably took this with him to Virginia, leaving its catalog at Oxford.

Although Banister arrived as an Anglican minister in Virginia in 1678, after a pause probably in Barbados and Grenada, his ambition from the beginning was his "Natural History," modeled on Robert Plot's *Natural History of Oxfordshire* (1677). Banister's name is absent from the Bristol Parish register of ministers for 1680. Virginians were reluctant to assume the financial burden imposed by induction: thus Banister was financially insecure. He received at least encouragement and hospitality from William Byrd I, trader at the falls of the James River, and from Theodorick Bland, then of Westover. He probably received some assistance from the Temple Coffee House Botany Club in London: Bishop Compton, Lister, Plukenet, and Samuel Doody, to all of whom, and to John Ray, who called him "eruditissimus," he sent specimens.

Banister finally realized that in order to maintain his status in the colony, to have sufficient income, and to have time to complete his "Natural History," he must become a planter. In 1690 he was able to patent 1,735 acres near the Appomattox River in return for importing thirty-five persons, including two Negroes.

He married a young widow named Martha sometime before 16 April 1688, and had one son, John.

A fall from a horse and at least one "fit of sickness" impeded Banister's field excursions. One longer trip of several days, probably up the James River, took his party well toward the Appalachians before Indians discouraged their further progress. In May 1692 he joined a party going to the Roanoke River, where he was accidently shot while botanizing among the river rocks.

The originals of his catalogs (after being copied?), some drawings, dried plants, seeds, and shells were sent to Bishop Compton, who passed them on: the botanical material was given to Plukenet, who loaned much of it to Petiver. Upon Petiver's death his vast accumulation of natural history materials, still containing some of Banister's and Plukenet's specimens and letters, was purchased by Sir Hans Sloane.

Banister's devotion to natural history is well expressed in his letter of 1688: "Had I an Estate would bear out my Expense, There is no part of this, or any other Country that would afford new matter, though under ye Scorching Line, or frozen Poles my genius would not incline me to visit."

BIBLIOGRAPHY

Banister's known letters and manuscripts are in the British Museum, Sloane MSS 3321 and 4002; the Bodleian Library, Oxford, Sherard MSS B26 and B37; and Lambeth Palace, letter from "the Falls" (Apr. 1679). These are reproduced in Joseph and Nesta Ewan, *John Banister and His Natural History of Virginia* (Urbana, Ill., in press).

Printed excerpts and information come mainly from Martin Lister, "Extracts of Four Letters From Mr. John Banister to Dr. Lister . . .," in *Philosophical Transactions of the Royal Society,* **17** (1693), 667–672; Robert Morison, *Plantarum historiae universalis Oxoniensis,* Jacob Bobart the younger, ed., III (Oxford, 1699)—although not acknowledged in the Preface, Banister is mentioned in the discussion of many species; James Petiver, "Herbarium Virginianum Banisteri," in *Monthly Miscellany or Memoirs for the Curious,* Decad no. 7 (Dec. 1707), and "Some Observations Concerning Insects Made by Mr. John Banister in Virginia, A.D. 1680 . . .," in *Philosophical Transactions of the Royal Society,* **22** (1701), 807–814; Leonard Plukenet, *Opera,* especially the *Phytographia* (London, 1691–1705); John Ray, *Historia plantarum,* 3 vols. (London, 1686–1704), esp. II, Preface and 1926–1928, and III, Preface and *passim;* and Arthur Foley Winnington-Ingram, *The Early English Colonies* (London–Milwaukee, 1908), pp. 192–201.

JOSEPH EWAN

BANKS, JOSEPH (*b.* London, England, 13 February 1743; *d.* Isleworth, England, 19 June 1820), *botany.*

Joseph Banks was the only son of William Banks of Revesby Abbey, Lincolnshire, and his wife Sarah, the eldest daughter of William Bate of Derbyshire. The Banks family first became well known in the seventeenth century, and by the eighteenth century was firmly established among the landed gentry. Joseph's great-grandfather, a prosperous attorney of the same name, was a man who acquired property and rendered public service as a member of Parliament, first for Grimsby and later for Totnes. In the next generation another Joseph was sheriff of the county and a member of the Royal Society of London. His second son, Joseph's father, continued the tradition of public service and took particular interest in the drainage of the Fenland. Thus the inheritance of the family was a happy if not unusual one. In the early eighteenth century the Banks family elevated its social position by marrying into the Grenville family and the family of the earl of Exeter.

The early years of Banks's life were spent at Revesby Abbey. In 1752 he entered Harrow, transferring in 1756 to Eton. Henry Brougham, in a biographical sketch published after Banks's death, commented that the young Joseph was not particularly "bookish," and his school record bears this out. At the age of fifteen, while still at Eton, Banks became aware of his interests and found a goal for his life, the study of botany. One summer evening, after swimming with schoolmates, he remained behind his companions and returned to school alone. While he was wandering along a country lane, the beauty of the flowers and the solitude overwhelmed him. He told a friend, Sir Everard Home, of the experience, and Home repeated the story in his Hunterian Oration of 1822:

> He stopped and looking round, involuntarily exclaimed, How beautiful! After some reflection, he said to himself, it is surely more natural that I should be taught to know all these productions of Nature, in preference to Greek and Latin; but the latter is my father's command and it is my duty to obey him; I will however make myself acquainted with all these different plants for my own pleasure and gratification. He began immediately to teach himself Botany; and, for want of more able tutors, submitted to be instructed by the women, employed in culling simples, as it is termed, to supply The Druggists and Apothecaries shops, paying sixpence for every material piece of information.

When he next went home on holiday, Banks found a battered copy of John Gerarde's *The Herball or Historie of Plants* (1598); the woodcuts illustrating the text were of the very same plants he had been collecting at school. He proceeded to broaden his interest in natural history, turning to the study of insects, shells, and fossils as well as plants.

In 1760 Banks entered as a gentleman commoner at Christ Church, Oxford. The classical curriculum was far from his taste, but he was able to begin his formal training in botany. This was due to his own initiative, however, rather than to the challenge of the curriculum. The professor of botany at Oxford, Humphrey Sibthorp, never gave lectures. At Sibthorp's suggestion, Banks turned to the professor of botany at Cambridge, who found him an instructor, Israel Lyons. Lyons came to Oxford and gave instruction to Banks and other interested students. Lyons was supported by Banks with income from his estate, the estate that was to be the source of support for many scientific projects during Banks's long life. As Linnaeus somewhat chauvinistically remarked later: "I cannot sufficiently admire Mr. Banks who has exposed himself to so many dangers and has bestowed more money in the service of Natural Science than any other man. Surely none but an Englishman would have the spirit to do what he has done."

Banks's father had died in 1761, and in 1764, at his majority, Joseph came into his well-managed inheritance. During his lifetime he was spoken of as a man of large fortune, and estimates of his yearly income varied from £6,000 to £30,000. It was not his wealth that made him a prominent man; rather, it was his vision and interests that could be furthered by his income. A modest man when it came to his scientific prowess, he spoke of himself as a botanizer rather than as a botanist. His chief reputation was not based on his scientific ability as much as it was on his ability to organize and administer scientific affairs. He became a patron of science. Owing to his background, connections, interests, and pleasant personality, he became a person of importance and influence early in life.

Finding the atmosphere of Oxford not conducive to his interests, Banks left the university without a degree and settled in London. In 1766 he made the first of several voyages. These trips appealed to him because of the opportunity they presented to collect new botanical specimens. His first voyage, lasting from April to November 1766, took him to Labrador and Newfoundland. He indulged his interests to the full, bringing back specimens that marked the beginning of the famous Banks Herbarium. In the same year he was elected a fellow of the Royal Society of London.

In 1768 the governments of Europe, in cooperation with scientific academies, were planning a series of observations of the transit of Venus in 1769. The English government was to send an expedition to the South Seas; the Admiralty and the Royal Society planned the expedition. Banks asked for and gained permission to join the voyage as a naturalist. He became a participant in the famous first voyage of Captain James Cook on the *Endeavour*. Besides observing the transit of Venus, the expedition was to conduct explorations in the South Seas for the southern continent that was thought to exist. Although the two aims of the voyage, geography and astronomy, were far from Banks's interests, he easily accommodated himself to the ship and its company. His preparations were extensive. He was accompanied by his own party of eight men with their equipment, all paid for from his own pocket. One of his companions was a Swede, Daniel Carl Solander.

One of Linnaeus' outstanding pupils, Solander had come to England in 1760 at the request of Peter Collinson and John Ellis, English naturalists who corresponded with Linnaeus. He became well known in England and spread his master's teachings in the British Isles. As early as 1762 he was invited to attend meetings of the Royal Society. In the same year the St. Petersburg Academy of Sciences offered him a professorship in botany. Solander did not wish to leave England, for he found it much to his liking, both professionally and personally. By 1764 he was made an assistant at the British Museum and in the same year was elected a member of the Royal Society. Banks's other companions, aside from personal servants, were Armon Sporing, a naturalist; Sydney Parkinson, a skilled artist; and A. Buchan, another artist, who died early in the voyage.

The preparations for the expedition by Banks and Solander were extensive. With these preparations there developed a close personal and professional relationship between the two men that ended only with the death of Solander in 1782. A letter from James Ellis to Linnaeus, written on the eve of the voyage, reveals the careful planning:

> No people ever went to sea better fitted out for the purpose of Natural History. They have got a fine library of Natural History; they have all sorts of machines for catching and preserving insects; all kinds of nets, trawls, drags and hooks for coral fishing, they have even a curious contrivance of a telescope, by which, put into the water, you can see the bottom at a great depth, where it is clear. They have many cases of bottles with ground stoppers of several sizes, to preserve animals in spirits. They have the several sorts of salts to surround the seeds; and wax, both beeswax and that of Myrica; besides there are many people whose sole business it is to attend them for this very purpose. They have two

painters and draughtsmen, several volunteers who have a tolerable notion of Natural History; in short Solander assured me this expedition would cost Mr. Banks £10,000. All this is owing to you and your writings.

Banks was also well informed about the problem of scurvy on long voyages. Cook provided quantities of sauerkraut for the crew. Banks preferred to use lemon juice, which he brought along on the voyage. Both of these remedies worked well in keeping the crew free of the disease.

Fortunately, Banks was a man of easy disposition, and from the beginning he and Cook were friendly. Cook was the navigator, Banks the botanizer. After the astronomical aspects of the voyage were accomplished, the expedition turned to exploring for the southern continent. Although the hope of discovery was not realized, Cook did conduct explorations in New Zealand and Australia. The gathering of specimens was of marked importance for descriptive botany and zoology. During the voyage over 800 previously unknown specimens were collected. Banks took great interest in the native languages and customs, and was the only person on the trip who gained any mastery of Tahitian, although he had little facility in foreign languages. The voyage acted as an intellectual stimulus to Banks, then in the full vigor of youth. Most of all, this expedition, one of the first to carry a professional naturalist, made a reputation for Banks that would greatly aid him in his later career. On their return in 1771, both Banks and Cook were acclaimed as heroes.

Banks kept an extensive journal of this three-year period. The publication of the journal was delayed for over a century, however, and not until 1962 was an accurate and well-edited version published. Banks was not himself a man of letters, and seems to have felt hesitation in putting his journal into a publishable form. He had hoped to publish the journals with handsome illustrations, but the death of his associate and librarian Solander in 1782 delayed publication for the moment. This delay continued until his death. His herbarium and library were willed to his last librarian and custodian, Robert Brown, a noted botanist in his own right. In his will Banks provided that the bequest would go to the British Museum at Brown's death, unless Brown chose to deposit the collection at the museum during his lifetime. Brown elected the latter course and became keeper of the botanical department of the British Museum in 1827, his chief job being the supervision of the collections left by Banks. Because his journal and private papers were ignored by his heirs for some time after his death, Banks was not very well known in the fields of his greatest competence and interest until several generations after his death. Instead, his name and reputation were associated with the Royal Society of London. Banks dominated the scientific community of England and presided over its affairs for many years in the same way that Samuel Johnson presided over the literary community. In 1772 Banks made a brief visit to Iceland, the last of his voyages.

Banks has been best remembered for his long tenure of office as president of the Royal Society. Elected at the age of thirty-five in 1778, he continued in that office until his death in 1820. At the time of his election nonscientists made up the majority of the society membership, a situation that had existed since its founding. Although boasting of the sovereign as its patron, the society lacked any regular income from royal or parliamentary grants. Often membership was given to men of influence who might bring benefaction, if not scientific knowledge, to the society. During Banks's long tenure the proportion of scientific to nonscientific members did not increase significantly. Early in his presidency he was criticized for packing the council of the society with his favorites, for being dictatorial in selection of new members, and for not possessing an appropriate mathematical bent. Only the last was true. Although somewhat high-handed in introducing nominations for membership and appointments to the council, this was not necessarily favoritism on his part, but may have represented an attempt to strengthen the membership rolls of the society. Yet Banks was not an innovator and made no attempt to redress the lack of balance between professional scientists and amateurs in the society. In this respect he maintained the status quo.

Banks considered the presidency of the Royal Society the greatest honor bestowed upon him in his lifetime and was faithful in attending to the duties of his office. During his forty-one years of service there were 450 meetings of the council, and he presided over 417 of them, although later in life he was crippled with gout and could move only with difficulty. He made the society much better known both at home and abroad. After 1777 his house in Soho Square became an unofficial headquarters for scientists in London. His weekly receptions, his famous breakfast parties for noted guests, and his large library, available to students, all contributed to the enhancement of the reputation of the society. And Banks did not hesitate to use the prestige of his office to further causes connected with science. He was one of the founders of the Linnean Society, assisted in the founding of the Royal Institution, and took an active interest in the affairs of the Board of Longitude.

Banks was made a baronet in 1781, a knight of the Order of the Bath in 1795, and a member of the Privy Council in 1797. Oxford University had already awarded him an honorary degree on his return from his voyage to the South Seas. One of the great pleasures of his life came in 1771, when he was introduced to King George III. The two men immediately struck up a friendship that grew with their common interest in horticulture and agriculture. It was Banks who persuaded the king to turn Kew Gardens, already noted for their beauty, into a botanical research center. Banks became the king's chief adviser in connection with Kew, and plants from all over the world were brought there for study and cultivation. Sir Joseph had the foresight to recognize that various plants could be adapted for cultivation in the broad reaches of the British Empire. He suggested the growing of Chinese tea in India, and his desire to cultivate the breadfruit tree in the West Indies led to the ill-fated voyage of the *Bounty*. A later expedition succeeded in bringing plants to the West Indies, but unfortunately the tree did not adapt itself very successfully to its new environment.

Kew Gardens became a bond between Banks and George III. The king was interested, as were others, in the possibility of bringing merino sheep from Spain in order to improve the quality of English wool. However, the sheep, long valued for the fine quality of their wool, were carefully protected and Spanish law forbade their sale to foreigners. Earlier in the century a few of the sheep had been obtained in France and Silesia. Banks managed the delicate transaction of obtaining sheep for England; in later years, during the Peninsular War, merino sheep became more available.

The friendship of the king for Banks undoubtedly assisted his election to the presidency of the Royal Society. The king had been severely displeased with Sir John Pringle, Banks's predecessor. In 1775 a committee of the society, among whom was Benjamin Franklin, had conducted, at the request of the government, a series of experiments to determine the most effective type of lightning rod. In its report the committee had advocated the use of lightning rods with pointed conductors, as opposed to blunt or knobbed conductors. The knobbed conductors had their advocates, however, and as a result, a spirited argument arose. With the outbreak of the American Revolution in 1776, Franklin's name became associated with rebellion as well as with pointed conductors, and a scientific question thus became a political question as well. Pringle backed Franklin, and on being urged by the king to change his opinion, is supposed to have said: "Sire, I cannot reverse the laws and operations of nature." Pringle's decision to retire in 1778 made it possible for the society to elect as president a person who, in addition to other qualifications, was a friend of the king.

Banks's interest in the South Seas did not lag after his return from the *Endeavour* voyage. Throughout the remainder of his life he continued to be concerned about the prospects of obtaining botanical specimens from these areas. And both Banks and Cook had been impressed with the east coast of Australia, which in both climate and soil was appealing as a prospective site for European colonization. As early as 1779 Banks appeared before a committee of the House of Commons to urge the use of Botany Bay as a place to send English convicts who had long terms to serve. The defection of the American colonies, which in the past had received some of these convicts, made a new destination urgent. Not until 1783 did the government begin to make plans for a convict settlement in New South Wales. In 1786, when the formal plans were made, Banks was consulted, and most probably was the author of a document entitled "Heads of the Plan," which described the proposed voyage and the aims of the settlement. The plan pointed out that New Zealand flax would undoubtedly thrive in Australia and would give an additional and superior supply of material for making canvas for both the navy and the merchant marine.

From the beginning of the colony in 1788 there was a constant problem for the military governors in dealing with the convicts, problems of supply, and the general lack of interest of the government at London—due in large part to the French Revolution. Throughout the extended period of war following the Revolution, the various governors sent not only official dispatches and pleas to London but also private letters to Banks outlining their problems and asking for help and advice. He did much to aid the struggling colony. At first Banks did not encourage the raising of sheep in the new land because the coastal area that Banks had explored had not seemed good for pasturage. An early governor had also reported that a flock of seventy sheep had died shortly after their arrival from the Cape Colony. As sheep raising became successful, Banks slowly changed his mind.

One of the most outstanding contributions that Banks made to the scientific world was his interest in maintaining close contacts in the international scientific community. He had been impressed with Franklin's protection of Captain Cook during the American Revolution; Franklin had made it possible for Cook to sail unmolested by ships of the new republic. The outbreak of war with France in 1793 provided a great challenge to the free exchange of ideas between French and English scientists. Throughout the long war Banks was a constant advocate of maintaining contact with French scientists. He also spent much time in attempting to maintain a free exchange of scientific publications at a time when communications with the enemy were strictly forbidden.

Scientists frequently became prisoners of war, and when appeals were made to Banks, his efforts to gain their release were often successful. Enemy ships, captured on their return from scientific expeditions, often carried valuable botanical specimens, and Banks would arrange for the release and return of this valuable scientific cargo to French ports. The French geologist Dolomieu, returning from the Egyptian expedition of Napoleon, was imprisoned by the king of Naples. When Banks found that it was impossible to obtain his release, he used all of the influence at his command to make Dolomieu's confinement easier. His correspondence with Frenchmen in this period is impressive. Such names as Lalande, Du Pont de Nemours, Delambre, and Cuvier appear; they wrote to Banks concerning such diverse matters as the release of a hostage, a copy of the *Nautical Almanac,* and letters of safe passage for scientific workers. In 1802 Banks was elected to membership in the Institut, an honor he accepted with pride.

Until his death Banks maintained his interest in science and his encouragement of scientific pursuits. His own scientific writings are few and of no importance; and he would have been the first to admit that such was the case. His great collection of specimens and his library are now in the British Museum. Long remembered by many as an autocrat in his rule of the Royal Society and often associated with the patronizing characterization of him by Sir Humphry Davy, Banks is assessed much more fairly in Cuvier's *éloge* of 1821:

> The works which this man leaves behind him occupy a few pages only; their importance is not greatly superior to their extent; and yet his name will shine out with lustre in the history of the sciences. In his youth, resigning the pleasures which an independent fortune had placed at his command, he devoted himself to Science and in her cause braved the dangers of the sea and the rigors of the most opposite climates. During a long series of years, he has done good service to the cause of Science by exciting in its favour all the influences arising from his fortunate position and friendship with men in power; but his special claim to our homage rests on the fact that he always considered the labourers in the field of science as having a rightful claim on his interest and protection.

BIBLIOGRAPHY

Little recognition was given to Banks by writers until many years after his death. A convenient listing of early memorials and reminiscences of Banks, along with the titles of Banks's scientific treatises, can be found in B. D. Jackson, "Sir Joseph Banks," in *Dictionary of National Biography.* The first biography, a typical life-and-letters biography of the Edwardian period, is Edward Smith, *The Life of Sir Joseph Banks* (London, 1911). George Mackaness, *Sir Joseph Banks, His Relations With Australia* (Sydney, 1936), is of some help, although the title is somewhat misleading. The best biography is Hector Charles Cameron, *Sir Joseph Banks, K.B., P.R.S., The Autocrat of the Philosophers* (London, 1952). Halldor Hermannsson, "Sir Joseph Banks and Iceland," in *Islandica,* **18** (1928), gives a good account of the voyage to Iceland and Banks's continued interest in the island.

Warren R. Dawson, *The Banks Letters; A Calendar of the Manuscript Correspondence of Sir Joseph Banks* (London, 1958), is an invaluable guide to Banks's extensive correspondence. Sir Gavin de Beer, *The Sciences Were Never at War* (London, 1960), supplies interesting correspondence of Banks in connection with his attempts to maintain communication with France during the French Revolution and the Napoleonic Wars. A good, brief account of Banks's presidency of the Royal Society can be found in Henry Lyons, *The Royal Society, 1660–1940* (Cambridge, 1944). The work of J. C. Beaglehole in the study of Banks's career is most important. His editing of *The Endeavour Journal of Joseph Banks, 1768–1771,* 2 vols. (Sydney, 1962), is not only a model of editing but also supplies, in the authoritative introduction to the journals, by far the best account of Banks's life in this formative period.

GEORGE A. FOOTE

IBN AL-BANNĀ' AL MARRĀKUSHĪ, also known as **ABŪ'L-ʿABBĀS AHMAD IBN MUHAMMAD IBN ʿUTHMĀN AL-AZDĪ** (*b.* Marrakesh, Morocco, 29 December 1256; *d.* Marrakesh [?], 1321), *mathematics.*

Some authors, following Casiri, say Ibn al-Bannā' was a native of Granada. In any case, he studied all the literary and scientific subjects that had cultural value in Fez and Marrakesh. Muhammad ibn Yahyā al-Sharīf taught him general geometry and Euclid's *Elements;* Abū Bakr al-Qallūsī, nicknamed al-Fār ("the Mouse"), introduced him to fractional numbers; and Ibn Hajala and Abū ʿAbd Allāh ibn Makhlūf al-Sijilmāsī rounded out his training in mathematics. He also studied medicine with al-Mirrīkh, but he did not delve deeply into the subject. The mystic al-Hazmirī was responsible for directing a great part of Ibn al-Bannā''s work to the study of the magic properties of numbers and letters.

He taught arithmetic, algebra, geometry, and astronomy in the *madrasa* al-ʿAttārīn in Fez. Among his disciples were Abū Zayd ʿAbd al-Rahmān . . . al-Lajā'ī (*d. ca.* 771/1369), teacher of Ibn Qunfudh, who left us an excellent biographical sketch of Ibn al-Bannā'; Muhammad ibn Ibrāhīm al-Abūlī (*d.* 770/1368); Abu'l-Barakāt al-Balāfiqī (*d.* 771/1370), who had Ibn al-Khatīb and Ibn Khaldūn as disciples; and Ibn al-Najjār al-Tilimsānī.

H. P. J. Renaud lists eighty-two works by Ibn al-Bannā'. The most important scientific ones are an introduction to Euclid; a treatise on areas; an algebra text dedicated to Abū ʿAlī al-Hasan al-Milyānī; a book about acronical risings and settings (*Kitāb al-anwā'*), which is not as good as his other works on astronomy, such as the *Minhāj;* and an almanac that is possibly the earliest known, in which the word *manākh* appears for the first time in its Arabic form. The works of greatest merit, however, are the *Talkhīs* and the *Minhāj.*

The *Talkhīs,* as its title indicates, is a summary of the lost works of the twelfth- or thirteenth-century mathematician al-Hassār. It was later summarized in verse by Ibn al-Qādi (*d.* 1025/1616) and was often commented on and glossed. Outstanding commentaries are the *Rafʿ al-hijāb* by Ibn al-Bannā' himself, with notes by Ibn Haydūr, and that of al-Qalasādī of Granada. These works contain a type of fraction that corresponds to what are today called continuing ascending fractions and an approximate method for extracting square roots that corresponds, more or less, to the third or fourth reduction in the development of the continuous fraction, and is similar to al-Qalasādī's

$$a + \frac{r}{2a} - \frac{\left(\frac{r}{2a}\right)^2}{2a + \frac{r}{2a}}.$$

The possible connection between this formula and that of Juan de Ortega seems evident, but the transmission has not been sufficiently proved. The works also contain sums of cubes and squares according to the formulas

$$1^3 + 3^3 + 5^3 + \cdots + (2n - 1)^3 = n^2(2n^2 - 1)$$

$$1^2 + 3^2 + \cdots + (2n - 1)^2 = \left(\frac{2n + 1}{6}\right)2n(2n - 1).$$

One cannot be sure that Ibn al-Bannā' was responsible for introducing a system of mathematical notation.

The *Kitāb minhāj al-tālib li taʿdīl al-kawākib* is a very practical book for calculating astronomical

437

ephemerals, thanks to the attached tables that are based upon those that Ibn Isḥāq al-Tūnisī calculated for the year 1222. The theoretical part does not contribute anything new and sometimes gives incorrect relationships between contradictory theories.

Ibn al-Bannā' is credited with a *Risāla* ("epistle") on the astrolabe called *ṣafīḥa shakāziyya,* a variation of the *ṣafīḥa zarqāliyya,* or "al-Zarqālī's plate," which is the topic of many manuscripts in the libraries of north Africa. An examination of some of these manuscripts does not show the differences that should, in theory, exist between the two instruments.

BIBLIOGRAPHY

I. ORIGINAL WORKS. Manuscripts of works by Ibn al-Bannā' are listed in Brockelmann, *Geschichte der arabischen Litteratur,* II (Berlin, 1902), 255, 710; Supp. II (Leiden, 1938), 363–364; J. Vernet, "Los manuscritos astronómicos de Ibn al-Bannā'," in *Actes du VIIIe Congrès International d'Histoire des Sciences* (1956), 297–298; and Griffini in *RSO,* 7 (1916), 88–106. A. Marre published a French translation of the *Talkhīṣ* in *Atti dell'Accademia pontificia de Nuovi Lincei,* 17 (5 July 1864); the commentary of al-Qalaṣādī was translated by M. F. Woepcke, *ibid.,* 12 (3 April 1859). The *Minhāj* has been edited, translated into Spanish, and studied by J. Vernet (Tetuán, 1951). The *Kitāb al-anwā'* was edited, translated into French, and commented on by H. J. P. Renaud (Paris, 1948).

II. SECONDARY LITERATURE. The Arabic sources for Ibn al-Bannā''s life are listed in al-Ziriklī, *al-Aʿlām,* 2nd ed., I, 213–214; especially in H. P. J. Renaud, "Ibn al-Bannā' de Marrakech, ṣūfī et mathématicien," in *Hesperis,* 25 (1938), 13–42. The *Muqaddima* of Ibn Khaldūn is fundamental; see the English translation by F. Rosenthal, 3 vols. (New York, 1958), indexes and esp. III, 121, 123, 126, 137. Also consult H. P. J. Renaud, "Sur les dates de la vie du mathématicien arabe marocain Ibn al-Bannā'," in *Isis,* 27 (1937), 216–218; and "Sur un passage d'Ibn Khaldoun relatif à l'histoire des mathématiques," in *Hesperis,* 31 (1944), 35–47.

Additional information can be found in George Sarton, "Tacuinum, taqwīm. With a digression on the word 'Almanac,' " in *Isis,* 10 (1928), 490–493, and *Introduction to the History of Science,* II, 998–1000; H. Suter, *Die Mathematiker und Astronomen der Araber und ihre Werke* (Leipzig, 1900), 162–164, 220, 227; J. A. Sánchez Pérez, *Biografías de matemáticos árabes que florecieron en España,* no. 44, pp. 51–54; M. Cantor, *Vorlesungen ueber Geschichte der Mathematik,* I (Leipzig, 1907), 805–810; *Encyclopaedia of Islam,* II, 367; M. Steinschneider, "Rectification de quelques erreurs relatives au mathématicien Arabe Ibn al-Banna," in *Bulletino di bibliografia e di storia delle scienze matematiche e fisiche,* 10 (1877), 313; and F. Woepcke, "Passages relatifs à des sommations de séries de cubes," in *Journal des mathématiques pures et appliquées,*

2nd ser., 10 (1865), reviewed by M. Chasles in *Comptes rendus des séances de l'Académie des Sciences* (27 March 1865).

J. VERNET

BANTI, GUIDO (*b.* Montebicchieri, Italy, 8 June 1852; *d.* Florence, Italy, 8 January 1925), *pathology.*

Banti, the most eminent Italian pathologist of the early twentieth century, was born in a typical village of Tuscany, in the lower valley of the Arno River. He was the son of Dr. Scipione Banti, a physician, and Virginia Bruni. He studied medicine at the University of Pisa, but was graduated in 1877 from the Medical School of Florence. He was then appointed assistant at the Archihospital of Santa Maria Nuova in Florence and, at the same time, assistant at the Laboratory of Pathological Anatomy. Banti was a tireless worker. Chief of the hospital medical service from 1882, in 1890 he became temporary professor and, in 1895, ordinary professor of pathological anatomy at the Medical School of Florence. His medical service at the hospital ended in 1924, after forty-seven years; he died the following year, his thirty-fifth year of teaching.

As a result of then existing arrangements, Banti could observe patients in bed and later study their corpses through autopsy as well as through laboratory tests: he wrote that clinical observation, anatomical report, and laboratory examination are three links in the same chain. Banti's numerous writings are original, and few men of science have spoken or written with such conciseness and clarity.

Banti was a perspicacious clinician, as evidenced by his study on heart enlargement (1886) and his notes for the surgical treatment of hyperplastic gastritis (1898) and acute appendicitis (1905). He was also a precise histologist who studied cancer cells (1890–1893) and a capable bacteriologist who published the first Italian textbook of bacteriological technique, *Manuale di Tecnica Batteriologica* (Florence, 1885). Thus, he contributed decisively to the advancement of the study of human pathology.

As a bacteriologist, Banti integrated bacteriology with the pathogenesis of infectious diseases. His works on typhoid fever (1887, 1891) and his paper *Le setticemie tifiche* (1894) contained the first observations of typhoid without intestinal localizations. Of fundamental importance were his studies (1886–1890) on *Diplococcus pneumoniae Fraenkelii.* In particular, Banti analyzed the characteristics of the types *hemolytic* and *viridans.* In 1890 he affirmed the hematogenic pathogenesis of acute pneumonia. In his remarkable experimental work on the destruction of bacteria in organisms (1888), Banti contributed to the develop-

(1839), Barrande recognized the similarity between the rocks of central Bohemia and those Murchison described in Britain.

From 1840 to his death Barrande collected, described, and drew the fossils of the central Bohemian basin—this area has been called the Barrandian ever since. The strata of this basin are Proterozoic–early Paleozoic in age. At that time "Silurian" might be applied to strata that today are in the Cambrian, Ordovician, Silurian, or Devonian system. The results formed the outstanding monograph *Système silurien du centre de la Bohême* (Prague, 1852–1902), which appeared in eight parts forming thirty quarto volumes. Because this work is so comprehensive, its drawings so accurate, and its descriptions so fine, it is still used as a reference book by paleontologists. Barrande's enormous collection, with all his scientific manuscripts and his library, was given to the Prague Museum at his death. His will also provided 10,000 florins to defray the costs of the remaining volumes of the *Système silurien,* which the museum was to publish from the extensive notes he left.

The *Système silurien* would have been a significant contribution to the geology of the mid-nineteenth century if it had offered no more than the naming and analysis of over 4,000 new fossil species. Yet these investigations had even greater impact on geological thought. In 1851 Charles Lyell pointed out in the *Quarterly Journal of the Geological Society of London* (**7,** xxiii), on the basis of some of Barrande's preliminary studies:

> . . . it is not the least interesting circumstance attending these discoveries, to learn that all these fossils were obtained from a superficial area, not more extensive than one-sixtieth part of the Adriatic; and they certainly show that the Silurian Fauna was not only as rich, but as much influenced by geographical conditions, or as far from being uniform throughout the globe, as that of any subsequent era.

Barrande's meticulous efforts distinguished metamorphosis in several species of trilobites, and he even identified embryonic states. He described *Sao hirsuta,* of which twenty naturally occurring forms had previously been identified as eighteen species in ten genera.

Barrande never abandoned the Cuvierian conception of the constancy of species, which he acquired during his early training in Paris. Similarly, he persevered in his conception of "colonies," assemblages of more recent fauna found intercalated among older strata. They supposedly resulted from migrations, although his opponents argued that these anomalies were due to tectonic disturbances of the strata.

In an international controversy over the status of the Taconic system, proposed by the American geologist Ebenezer Emmons, Barrande actively joined Jules Marcou in supporting Emmons' claim to be the true discoverer of the primordial fauna.

BIBLIOGRAPHY

I. ORIGINAL WORKS. The *Système silurien du centre de la Bohême,* VII, pt. 1 (Prague, 1887), ix-xvi, contains a bibliography of Barrande's many articles. His *Défense des colonies* (Prague, 1861–1881) is a series of his articles, letters, and essays relevant to that controversy.

II. SECONDARY LITERATURE. A bibliography of secondary materials on Barrande is in K. Lambrecht, W. Quenstedt, and A. Quenstedt, eds., *Fossilium catalogus I: Animalia, pars 72: Palaeontologi catalogus bio-bibliographicus* (The Hague, 1938), p. 22. An interesting modern biography is Josef Svoboda and Ferdinand Prantl, *Barrandium: Geologie des mittelböhmisches Silur und Devon* (Prague, 1958), pp. 47–67.

BERT HANSEN

BARRÉ DE SAINT VENANT. See **Saint Venant, A. J. C. Barré de.**

BARRELL, JOSEPH (*b.* New Providence, New Jersey, 15 December 1869; *d.* New Haven, Connecticut, 4 May 1919), *geology.*

Barrell was well prepared for a fruitful and scholarly career in the earth sciences: he studied engineering at Lehigh University, graduating with high honors in 1892, and continued there for E.M. and M.S. degrees in 1893 and 1897. He received the Ph.D. in geology from Yale in 1900 and the D.Sc. from Lehigh in 1916. To defray the expense of such extended education, Barrell took part-time positions teaching mathematics, mining and metallurgy, geology, zoology, and astronomy; he considered this experience invaluable when he became professor of structural geology at Yale (1908–1919).

Although he was concerned with a variety of problems, from mining technology to the evolution of protoman, Barrell's chief contributions were in isostasy, sedimentology, and metamorphism. He preferred building theories to collecting data, and his most impressive papers were deductive interpretations of the work of others. When writing on a particular topic, Barrell followed what he called the method of multiple working hypotheses: beginning with a set of carefully considered assumptions, he attempted to derive a number of hypothetical explanations that could later be checked against existing facts and theories. His own conclusions were typically prefaced with lengthy and critical presentations.

Barrell's ideas on metamorphism grew out of fieldwork during the summers of 1897–1901. In 1901 he joined the U.S. Geological Survey in Montana, to study the Marysville mining district and large Marysville and Boulder granite batholiths. Dissatisfied with contemporary theories of their origin, he directed his efforts to an accurate description of the characteristics of igneous intrusions occurring in nature, and developed a theory of magmatic stoping. The method of invasion was by subsidence of roof blocks and the rise of magma. Superheated magma confined at great depths shows a maximum of marginal assimilation. In the zone of flowage, magmatic intrusions crowd aside their containing wall rock, with the development of peripheral schistosity, and in the zone of fracture they force strata apart, forming sheets, laccoliths, and dikes. Greatly confined magmas expand and give rise to volcanism. Charles Schuchert called the *Geology of the Marysville Mining District, Montana* (1907) a geological classic.

Many of Barrell's 150 published papers deal with topics in paleoclimatology and sedimentology. He was a pioneer dry-land geologist, greatly influencing the manner in which stratigraphic problems were subsequently approached and conceptualized. Before his work, it was generally held that most sedimentary strata were of marine origin. Barrell was convinced by examination of the floodplains of western deserts and Triassic deposits of New Jersey that at least a fifth of all land surfaces are mantled by continental, fluvial, or eolian sediments.

He early recognized causal relationships between climatic variation and sedimentation, emphasizing that the ratio of terrestrial to littoral and marine deposits fluctuates markedly through time. Barrell proposed that sedimentation is a complex repetition of many compound rhythms and that such cyclic events as diastrophism, erosion, temperature change, and rainfall variation influence and are influenced by topography and physical geography, by the depth and streaming force of water bodies. He outlined numerous covarying factors of deposition in "Criteria for the Recognition of Ancient Delta Deposits" (1912) and pointed out that the heterogeneity of stratified deposits is the result.

Barrell read the history of the earth in its strata and interpreted their irregularities as meaning that geological processes are halting and discontinuous. Anti-uniformitarian arguments are forcefully presented in "Rhythms and the Measurements of Geologic Time" (1917), one of Barrell's more philosophical works. Here he tried to estimate the age of the earth by calculating the rates of denudation, sedimentation, uplifts and subsidences, deposition of salt in the sea, and emission of radioactivity. The figure he obtained, 1,400 million years to the Precambrian, was more than ten times the usual uniformitarian estimates.

Geodesic theory occupied Barrell's interest for a portion of his career. He published eight papers in the *Journal of Geology* (1914–1915), under the series title "Strength of the Earth's Crust," that present his views on isostasy and terrestrial dynamics. Positing two crustal layers, an outermost and stronger *lithosphere* (50–70 miles thick and of varying density) and a zone of flowage that he named the *aesthenosphere* (70–300 miles thick), he sought to explain geological phenomena by their dynamic interaction. Barrell said that isostatic equilibrium obtains in general, despite the effects of erosion and sedimentation. He wrote that the lithosphere is capable of supporting limited loads, uncompensated, however, if the vertical anomaly is inversely proportional to the area. Anomalies in excess of this proportion are compensated for by vertical displacement of the lithosphere against the foundation aesthenosphere (isostatic adjustment).

Although Barrell's concerns were seemingly diverse, they were actually variations on a common theme: the effects of physical agents on the evolution of the earth and its inhabitants. "The Origin of the Earth" (1916), a lecture delivered to Yale's Sigma Xi Society, discussed the conditions required for the genesis of the solar system and the development of the earth. His papers on sedimentology always relate sedimentological processes to the larger problems of historical geology, as do his treatments of structural geology. Barrell even maintained that biological evolution was the result of physical and chemical agents, in that these are the factors determining the environment of organisms.

BIBLIOGRAPHY

I. ORIGINAL WORKS. Barrell's writings include *Geology of the Marysville Mining District, Montana . . .*, U.S. Geological Survey, Professional Paper no. 57 (Washington, D.C., 1907); "Criteria for the Recognition of Ancient Delta Deposits," in *Bulletin of the Geological Society of America,* **23** (1912), 377–446; the series on the strength of the earth's crust: "Geologic Tests of the Limits of Strength," in *Journal of Geology,* **22** (1914), 28–48; "Regional Distribution of Isostatic Compensation," *ibid.,* 145–165; "Influence of Variable Rate of Isostatic Compensation," *ibid.,* 209–236; "Heterogeneity and Rigidity of the Crust as Measured by Departures From Isostasy," *ibid.,* 289–314; "The Depth of Masses Producing Gravity Anomalies and Deflection Residuals," *ibid.,* 441–468, 537–555; "Relation of Isostatic Movements to Sphere of Weakness," *ibid.,* 655–683; "Variation of Strength With Depth," *ibid.,* 729–741, and **23** (1915), 27–44; and "Physical Conditions Controlling the Nature of the Lithosphere and Aesthenosphere," *ibid.,* **23** (1915), 425–443, 449–515; "Rhythms and the Measurements of Geologic Time," in *Bulletin of the Geological Society of America,* **28** (1917), 745–904; and "The Origin of the Earth," in Sigma Xi Society, ed., *The Evolution of the Earth and Its Inhabitants* (New Haven, 1918), pp. 1–44.

II. Secondary Literature. Works on Barrell are Herbert E. Gregory, "Memorial to Joseph Barrell," in *Bulletin of the Geological Society of America,* **34** (1923), 18–28; G. P. Merrill, "Joseph Barrell," in *Dictionary of American Biography,* I, 642–644; and Charles Schuchert, "Biographical Memoir of Joseph Barrell," in *Biographical Memoirs of the National Academy of Science,* **12** (1927), 1–40.

Martha B. Kendall

BARRESWIL, CHARLES-LOUIS (*b.* Versailles, France, 13 December 1817; *d.* Boulogne-sur-Mer, France, 22 November 1870), *chemistry.*

Barreswil studied chemistry in Paris under the guidance of Robiquet and Bussy, and later under Jules Pelouze, whose laboratory on the Rue Dauphine he directed. It was there, from 1844 to 1849, that Barreswil worked side by side with Claude Bernard and gave the young physiologist, then far from his future eminence, extremely valuable assistance. They conducted experiments on themselves, such as subsisting only on gelatin and attempting to discover the influence of food on the chemical reaction in the urine.

At first Barreswil was especially interested in the improvement of chemical analysis; starting in 1843 he published several articles in this field, notably on a new process for separating cobalt from manganese (1846) and on the cupropotassic solution as a reagent facilitating the detection of sugar (1844). Rayer introduced the "blue liquid of Barreswil" into the systematic clinical detection of diabetes; only slightly altered, it is the Fehling solution used today.

Barreswil participated in Bernard's researches on digestion (1844); on the chemical composition of gastric juices, bile, and pancreatic juice (1844–1846); on the means of eliminating urea after the removal of the kidneys (1847); and especially on the role of sugar in the animal organism. Both Bernard and Barreswil signed the two basic communications on the presence of sugar in the liver (1848) and in egg whites (1849).

While carrying on his own research, Barreswil discovered a new chrome compound, blue chromic acid (1847), and prepared quinine tannate.

Having become professor of chemistry at the École Municipale Turgot, and later at the École Supérieure de Commerce in Paris, Barreswil abandoned physiological chemistry and became more interested in the problems of industrial chemistry, particularly those of coloring, of photographic chemistry, and of the manufacture of sulfuric acid. He also carried out useful and humanitarian work as inspector of child labor in the factories of the Department of the Seine and as secretary of a philanthropic organization for the protection of young workers.

BIBLIOGRAPHY

I. Original Works. Barreswil's books include *Appendice à tous les traités d'analyse chimique* (Paris, 1848), written with A. Sobrero; *Chimie photographique* (Paris, 1854; 4th ed., 1864), written with Davanne; *Répertoire de chimie pure et appliquée,* 6 vols. (Paris, 1858–1866); and *Dictionnaire de chimie industrielle,* 3 vols. (Paris, 1861–1864), written with A. Girard *et al.* For the bibliography of his articles see *Royal Society of London, Catalogue of Scientific Papers,* I, 191–192.

II. Secondary Literature. There is no proper biographical study of Barreswil. Short notices are in G. Vapereau, *Dictionnaire universel des contemporains,* 4th ed. (Paris, 1870), and *Dictionnaire de biographie française,* V (Paris, 1951).

M. D. Grmek

BARROIS, CHARLES (*b.* Lille, France, 21 April 1851; *d.* Ste.-Geneviève-en-Caux, Seine Maritime, France, 5 November 1939), *geology, paleontology.*

Barrois belonged to a family of rich industrialists who had been liberals under the Restoration but became conservatives and ardent Catholics under the Second Empire. After studying with the Jesuits in Lille, Charles and his brother Jules wanted to be zoologists, an ambition encouraged by their parents. Jules became a specialist in invertebrate embryology, and Charles wrote his second doctoral thesis on the development of living *Spongiae* (1876). As early as 1871, however, he was caught up in Jules Gosselet's contagious enthusiasm for geology. He remained Gosselet's devoted disciple for half a century.

In 1871 Barrois was named assistant at the Faculté des Sciences of Lille, where he spent the rest of his career. He thereupon began a doctoral thesis on the Cretaceous formations of England, which he explored during four successive summers. His keen powers of observation were focused particularly on fossil fauna, which his training as a zoologist helped him to analyze. His thesis, presented in Paris in 1876 under Hébert's supervision, became the first volume of the *Mémoires de la Société géologique du Nord,* which was founded by Gosselet and Barrois himself. In this work Barrois extended to England the same paleontologic and stratigraphic zones found by Hébert in the French Cretaceous formation. Following Godwin-Austen, he observed that the axes of the deformations that affected the English Cretaceous were those active during the Primary era. This discovery was expanded by Marcel Bertrand in 1892 and by Eduard Suess as the theory of posthumous flections.

The English geologists were enthusiastic over Barrois's thesis. In 1908 Arthur Rowe declared his work to be that of a genius. According to E. Bailey (1940), Barrois was the most illustrious stratigrapher since Murchison. The compliment was slightly excessive, since Barrois owed a great deal to Gosselet and to Hébert. But his reputation benefited from the friendships that he made abroad, facilitated by his thorough command of languages and his personal fortune. His rather cold and haughty manner, the result of his education as a member of the high bourgeois, was mitigated by a gentle spirit and perfect manners. It was impossible to do other than like such a man, who was always as good as his word.

Once his thesis was finished, Barrois began a study of the Cretaceous formation of northern and eastern France. The most ancient formations of the globe had attracted him for many years, however, and after a period of study in the laboratory of Fouqué and Michel-Lévy at the Collège de France (where he became acquainted with the new methods of optical petrography), he undertook research on the Primary formations of northern Spain in 1877. He went on to make other field studies in Asturias and in Galicia that were the subject of an imposing memoir (1882). He also undertook a study of the Sierra Nevada with A. Offret in 1889; their stratigraphic results will long remain authoritative.

In 1878 Barrois had met the American geologist James Hall, who developed a paternal affection for him and took him back to the United States. Hall conducted him on a tour of the most important stratigraphic sequences in New York State, and Barrois wrote to Hébert, "One learns more in one hour in Albany than in a week anywhere else."

Back in France, Barrois was named lecturer at the Faculté des Sciences at Lille (1878) and embarked on his great work: mapping the geological formations of Brittany. This was a difficult task, for the Precambrian and Primary formations, overthrown by Hercynian flections and made unrecognizable by meta-

morphism, are practically always without fossils; moreover, they are often concealed by the present vegetation. Barrois hunted for outcrops in road cuts, small quarries, or barren land where the bedrock was visible. From 1880 to 1909 he published twenty geological maps on the scale of 1:80,000, representing more than 25,000 square kilometers. After his retirement from teaching in 1926, he returned to his beloved Brittany, where until his death he tried to improve his previous observations.

Although Brittany did not suggest any original hypotheses to Barrois (who rather dreaded them), it gave him the opportunity to verify new theories put forward by others. He showed the great diversity in the age of granites, and in the field he followed the transformation, through metamorphism, of bands of sandstone into veins of quartzite or quartz, and that of carbonaceous sediments into graphitic layers. He dated the numerous volcanic eruptions that affected Brittany during the Primary epoch and established the existence of a marine transgression during the Lower Carboniferous. He also collected specimens of Lamballe's *phtanites,* which L. Cayeux found to be one of the oldest microfauna of Europe, possibly going back to the Precambrian era.

As early as 1888 Barrois was made *professeur-adjoint* at the University of Lille, and in 1902 he became director of the Institute of Geology of Lille, following Gosselet. He also was an ingenious museologist, as is shown by the Musée Houiller, which he founded in 1907 to help train students, engineers, and technicians for the coal-mining industry. Barrois chose valuable collaborators, such as Pierre Pruvost, his successor; Paul Bertrand, a specialist in fossil flora; and André Duparc, who pioneered the microscopic study of coal. In 1905 Léon Bertrand's views on the structure of the Franco-Belgian coal basin were prevalent. It was conceived as a mildly folded depression where enormous reserves of coal had been deposited, the thick beds at the bottom and the thinner coals on top. Using a detailed stratigraphic and paleontological study, Barrois and his disciples designed a completely different tectonics for the basin, seeing it as crosshatched by faults. This led them to anticipate far smaller coal reserves. It was the triumph of minute observation over simplistic theory.

Barrois, who had a remarkable ability to discover and describe paleontological species, became interested in the fossil corals, the Spongiae, the Bryozoa, the brachiopods, and especially the graptolites. He also wrote an extensive monograph on the Devonian fauna of Erbray (1888). Pierre Pruvost said, however, that Barrois "defined his species in terms of their value as characteristic [guide and index] fossils, and he avoids . . . philosophical meditation on the origin and relationships of living beings. Paleontology is for him a chronology of organic life, as exact as possible . . . for the use of the geologist." When Barrois undertook the unrewarding but useful task of translating into French the five large volumes of Zittel's *Handbuch der Palaeontologie* (1883–1894), he made it difficult for others to publish treatises on paleontology that openly praised theories of evolution. Although he wrote "All scientific effort broadens our freedom of action," he was inhibited from speaking freely by his religious convictions: he and his friend Albert de Lapparent had been among the militants of what he called the Catholic party. He did not scorn honors and became a corresponding member of numerous academies; although not a resident of Paris, he was successful in being elected a titular member of the Académie des Sciences, after several defeats. During the last year of his life the pope made him a member of the Pontifical Scientific Academy.

Barrois was one of the last geologists to be *complet,* as Pruvost put it—that is, to be capable of carrying his research almost to perfection in both the field and the laboratory, whether the research was paleontological or petrographic. Barrois would have been outstanding among geologists had he been a bit more daring and had he had more of a feeling for synthesis.

BIBLIOGRAPHY

I. ORIGINAL WORKS. A list of Barrois's scientific works includes 268 titles, of which twenty-five are geological maps; it may be found on pp. 251–262 of Pruvost's article in *Bulletin de la Société géologique de France* (see below). His principal works are *Recherches sur le terrain crétacé supérieur de l'Angleterre et de l'Irlande,* his thesis (Paris, 1876), *Mémoires de la Société géologique du Nord de la France,* **1;** "Mémoire sur le terrain crétacé des Ardennes . . .," in *Annales de la Société géologique du Nord de la France,* **5** (1878), 227–287; *Recherches sur les terrains anciens des Asturies et de la Galice, Mémoire de la Société géologique du Nord de la France,* **2** (1882); *Mémoire sur la faune du Calcaire d'Erbray (Loire-inférieure), Mémoire de la Société géologique du Nord de la France,* **3** (1889); "Mémoire sur la constitution géologique du Sud de l'Andalousie . . .," in *Académie des sciences, Mémoires des savants étrangers,* 2nd ser., **30** (1889), 79–169, written with Albert Offret; "Étude des strates marines du terrain houiller du Nord," in *Gîtes minéraux de la France* (1912), pt. 1; "Description de la faune siluro-dévonienne de Liévin," in *Mémoires de la Société géologique du Nord de la France,* **6,** fasc. 2 (1922), 67–223, written with P. Pruvost and Georges Dubois; and "Les grandes lignes de la Bretagne,"

in *Livre jubilaire de la Société géologique de France* (1930), pp. 83 ff.

II. SECONDARY LITERATURE. Barrois's pupil and friend Pierre Pruvost analyzed the man and his work in *Bulletin de la Société géologique de France,* 5th ser., **10** (1950), 231–262, with a portrait. Charles Jacob attempted a sketch of the social milieu in which Barrois lived in *Obituary Notices of Fellows of the Royal Society,* **5** (1947), 287–293, with a portrait.

FRANCK BOURDIER

BARROW, ISAAC (*b.* London, England, October 1630; *d.* London, 4 May 1677), *geometry, optics.*

Barrow's father, Thomas, was a prosperous linen-draper with court connections; his mother, Anne, died when Isaac was an infant. A rebel as a dayboy at Charterhouse, Barrow came later, at Felsted, to accept the scholastic disciplining in Greek, Latin, logic, and rhetoric imposed by his headmaster, Martin Holbeach. In 1643, already as firm a supporter of the king as his father was, he entered Trinity College, Cambridge, as pensioner. There he survived increasingly antiroyalist pressure for twelve years, graduating B.A. in 1648, being elected a college fellow (1649), and receiving his M.A. (1652), the academic passport to his final position as college lecturer and university examiner. In 1655, ousted by Cromwellian mandate from certain selection as Regius professor of Greek (in succession to his former tutor, James Duport), he sold his books and set out on an adventurous four-year tour of the Continent. On his return, coincident with the restoration of Charles II to the throne in 1660, he took holy orders and was promptly rewarded with the chair previously denied him. In 1662 he trebled his slender income by concurrently accepting the Gresham professorship of geometry in London and acting as locum for a fellow astronomy professor; he was relieved of this excessive teaching load when, in 1663, he was made first Lucasian professor of mathematics at Cambridge.

During the next six years, forbidden by professorial statute to hold any other university position, Barrow devoted himself to preparing the three series of *Lectiones* on which his scientific fame rests. In 1669, however, increasingly dissatisfied with this bar to advancing himself within his college, he resigned his chair (to Newton) to become royal chaplain in London. Four years later he returned as king's choice for the vacant mastership of Trinity, becoming university vice-chancellor in 1675. Barrow never married and, indeed, erased from his master's patent the clause permitting him to do so. Small and wiry in build, by conventional account he enjoyed excellent health, his early death apparently being the result of an overdose of drugs. He was remembered by his contemporaries for the bluntness and clarity of his theological sermons (published posthumously by Tillotson in 1683–1689), although these were too literary and long-winded to make him a popular preacher. His deep classical knowledge resulted in no specialized philological or textual studies. Although he was one of the first fellows of the Royal Society after its incorporation in 1662, he never took an active part in its meetings.

As an undergraduate, Barrow, like Newton a decade later, endured a traditional scholastic course, centered on Aristotle and his Renaissance commentators, which was inculcated by lecture and examined by disputation; but from the first he showed great interest in the current Gassendist revival of atomism and Descartes's systematization of natural philosophy. (His 1652 M.A. thesis, *Cartesiana hypothesis de materia et motu haud satisfacit praecipuis naturae phaenomenis,* is based on a careful study of Descartes and Regius.) That, also like Newton, he mastered Descartes's *Géométrie* unaided is unlikely. The elementary portion of Euclid's *Elements* was part of Barrow's college syllabus, but some time before 1652 he went on to read not only Euclidean commentaries by Tacquet, Hérigone, and Oughtred, but also more advanced Greek works by Archimedes and possibly Apollonius and Ptolemy. His first published work, his epitomized *Euclidis Elementorum libri XV* (probably written by early 1654), is designed as a quadrivium undergraduate text, with emphasis on its deductive structure rather than on its geometrical content, its sole concessions to contemporary mathematical idiom being its systematic use of Oughtred's symbolism and a list "ex P. Herigono" of numerical constants relating to inscribed polyhedra. To its reedition in 1657, Barrow added a similar epitome of Euclid's *Data,* and in his 1666 Lucasian lectures expounded a likewise recast version of Archimedes' method in the *Sphere and Cylinder;* a full edition, in the same style, of the known corpus of Archimedes' works, the first four books of Apollonius' *Conics,* and the three books of Theodosius' *Spherics* appeared in 1675. Overloaded with marginal references, virtually bare of editorial amplification, and fussy in their symbolism, these texts can hardly have been easier to read than their Latin originals, and only the conveniently pocket-sized *Euclid* reached a wide public. Barrow himself commented that his Apollonius had in it "nothing considerable but its brevity." His early attempt at a modern approach to Greek mathematics was a short, posthumously edited *Lectio* in which he analyzed the Archimedean quadrature method in terms of indivisibles on the style of Wallis' *Arithmetica infinitorum.*

Barrow's Gresham inaugural, still preserved, tells little of the content of his lost London lectures: perhaps they were similar to works of his on "Perspective, Projections, Elem^ts of Plaine Geometry" mentioned by Collins. The first of his Lucasian series, the *Lectiones mathematicae* (given in sequence from 1664 to 1666), discourse on the foundations of mathematics from an essentially Greek standpoint, with interpolations from such contemporaries as Tacquet, Wallis, and Hobbes (usually cited only to be refuted). Such topics as the ontological status of mathematical entities, the nature of axiomatic deduction, the continuous and the discrete, spatial magnitude and numerical quantity, infinity and the infinitesimal, and proportionality and incommensurability are examined at length. Barrow's conservatism reveals itself in his artificial preservation of the dichotomy between arithmetic and geometry by classifying algebra as merely a useful logical (analytical) tool which is not a field of mathematical study in itself. The *Lectiones geometricae* were, no doubt, initially intended as the technical study of higher geometry for which the preceding course had paved the way, and the earlier lectures may indeed have been delivered as such.

About 1664, having heard (as he told Collins) that "Mersennus & Torricellius doe mencōn a generall method of finding ye tangents of curve lines by composition of motions; but doe not tell it us," he found out "such a one" for himself, elaborating an approach to plane geometry in which the elements were suitably compounded rotating and translating lines. In his first five geometrical lectures he took some trouble to define the uniformly "fluent" variable of time which is the measure of all motion, and then went on to consider the properties of curves generated by combinations of moving points and lines, evolving a simple Robervallian construction for tangents. Later lectures (6–12), evidently thrown together in some haste, are in large part a systematic generalization of tangent, quadrature, and rectification procedures gathered by Barrow from his reading of Torricelli, Descartes, Schooten, Hudde, Wallis, Wren, Fermat, Huygens, Pascal, and, above all, James Gregory; while the final *Lectio,* 13, is an unconnected account of the geometrical construction of equations. We should (despite Child) be careful not to overemphasize the originality of these lectures: the "fundamental theorem of the calculus," for example, and the *compendium pro tangentibus determinandis* in *Lectio 10* are, respectively, restylings, by way of propositions 6 and 7 of Gregory's *Geometriae pars universalis* (1668), of William Neil's rectification method (in Wallis' *De cycloide,* 1659) and of the tangent algorithm thrashed out by Descartes and Fermat in their 1638 correspondence (published by Clerselier in 1667). In theory, as Jakob Bernoulli argued in 1691, Barrow's geometrical formulations could well have been the basis on which systematized algorithmic calculus structures were subsequently erected; but in historical fact the *Lectiones geometricae* were little read even by the few (Sluse, Gregory, Newton, Leibniz) qualified to appreciate them, and their impact was small. Perhaps only John Craige (1685) based a calculus method on a Barrovian precedent, and then only in a single instance (*Lectio 11,*1).

Barrow's optical lectures, highly praised on their first publication by Sluse and James Gregory, had an equally short-lived heyday, being at once rendered obsolete by the Newtonian *Lectiones opticae,* which, both in methodology and in subject matter, they inspired. In his introduction he lays down the scarcely novel mechanical hypothesis of a lucid body (a "congeries corpusculorum ultra pene quam cogitari potest minutorum" or "collection of particles minute almost beyond conceivability") as the propagating source of rectilinear light rays. His hypothesis of color (in *Lectio 12*) as a dilution in "thickness" and swiftness, of white light through red, green, and blue to black, is no less shadowy than the Cartesian explanation to which it is preferred. Structurally, the technical portion of the *Lectiones* is developed purely mathematically from six axiomatic "Hypotheses opticae primariae et fundamentales [seu] leges . . . ab experientiâ confirmatae," notably the Euclidean law of reflection and the sine law of refraction, and presents a reasonably complete discussion of the elementary catoptrics and dioptrics of white light. Not unexpectedly, the organization and mathematical detail are Barrow's, but his topics are mostly taken from Alhazen, Kepler, Scheiner, Descartes, and others: thus, his improvement of the Cartesian theory of the rainbow (*Lectio 12,* 14) derives from Huygens by way of Sluse. The most original contributions of the work are his method for finding the point of refraction at a plane interface (*Lectio 5,* 12) and his point construction of the diacaustic of a spherical interface (*Lectio 13,* 24): both were at once subsumed by Newton into his own geometrical optics, and the latter (in ignorance) was triumphantly rediscovered by Jakob Bernoulli in 1693.

Barrow's relationship with Newton, although of considerable historical importance, has never been clarified. That Newton was Barrow's pupil at Trinity is a myth, and Barrow's name does not appear in the mass of Newton's extant early papers; nor is there good evidence for supposing that any of Newton's early mathematical or optical discoveries were in any way due to Barrow's personal tutelage. In his old age, the furthest that Newton would go in admitting a

mathematical debt to Barrow was that attendance at his lectures "might put me upon considering the generation of figures by motion, tho I not now remember it." It may well be that Barrow came to know Newton intimately only after his election to senior college status in 1667. Certainly by late 1669 there was a brief working rapport between the two which, if it did not last long, at least resulted in Newton's consciously choosing to continue the theme of his predecessor's lectures in his own first Lucasian series.

BIBLIOGRAPHY

I. Original Works. The contents of Barrow's library at the time of his death are recorded in "A Catalogue of the Bookes of Dr Isaac Barrow Sent to S.S. by Mr Isaac Newton . . . July 14. 1677" (Bodleian, Oxford, Rawlinson D878, 33^r–59^r). His *Euclidis Elementorum libri XV breviter demonstrati* appeared at Cambridge in 1655; to its 1657 reedition (reissued in 1659) was appended his edition of Euclid's *Data.* Both reappeared in 1678, together with Barrow's *Lectio . . . in qua theoremata Archimedis De sphaera & cylindro per methodum indivisibilium investigata exhibentur* (Royal Society, London, MS XIX). An English edition of the *Elements; the Whole Fifteen Books* (London, 1660) was reissued half a dozen times in the early eighteenth century, and an independent English version by Thomas Haselden of the *Elements, Data,* and *Lectio* together appeared there in 1732. The manuscript of Barrow's *Archimedis opera: Apollonii Pergaei Conicorum libri IIII: Theodosii Sphaerica: Methodo novo illustrata, & succincte demonstrata* (London, 1675) is now in the Royal Society, London (MSS XVIII–XX): a proposed appendix epitomizing Apollonius' *Conics,* 5–7 (from Borelli's 1661 edition) never appeared. His *Lectiones mathematicae XXIII; In quibus Principia Matheseôs generalia exponuntur: Habitae Cantabrigiae A.D. 1664, 1665, 1666* was published posthumously at London in 1683 (reissued 1684 and 1685); an English version by John Kirkby came out there in 1734. The rare 1669 edition of Barrow's *Lectiones XVIII Cantabrigiae in scholis publicis habitae; In quibus opticorum phaenomenωn genuinae rationes investigantur, ac exponuntur* was speedily followed (1670) by his *Lectiones geometricae: In quibus (praesertim) generalia curvarum linearum symptomata declarantur*: these were issued (both together and separately) at London in 1670, 1672, and 1674. Unpublished variant drafts of geometrical lectures 10, 11, and 13 are extant in private possession.

The optical lectures were reprinted, none too accurately, in C. Babbage and F. Maseres' *Scriptores optici* (London, 1823), and all three Lucasian series were collected, together with Barrow's inaugural, in W. Whewell's *The Mathematical Works of Isaac Barrow D.D.* (Cambridge, 1860). A mediocre English translation of the geometrical lectures by Edmund Stone (London, 1735) is more accurate than J. M. Child's distorted abridgment, *The Geometrical Lectures of Isaac Barrow* (Chicago-London, 1916). Alexander

Napier's standard edition of Barrow's *Theological Works* (9 vols., Cambridge, 1859), based on original manuscripts in Trinity College, Cambridge, and otherwise restoring the text from Tillotson's "improvements," is scientifically valuable for the *Opuscula* contained in its final volume: here will be found the texts of Barrow's early academic exercises and college orations, as well as of his professorial inaugurals. The extant portion of Barrow's correspondence with Collins has been published several times from the originals in possession of the Royal Society, London, and the Earl of Macclesfield, notably in Newton's *Commercium epistolicum D. Johannis Collins, et aliorum de analysi promota* (London, 1712) and in S. P. Rigaud's *Correspondence of Scientific Men of the Seventeenth Century*, II (Oxford, 1841), 32–76.

II. SECONDARY LITERATURE. Existing sketches of Barrow's life (by Abraham Hill, John Aubrey, John Ward, and, more recently, J. H. Overton) are mostly collections of unsupported anecdote, both dreary and derivative. Percy H. Osmond's *Isaac Barrow, His Life and Times* (London, 1944) has few scientific insights but is otherwise a lively, semipopular account of Barrow's intellectual achievement. In "Newton, Barrow and the Hypothetical Physics," in *Centaurus,* **11** (1965), 46–56, and *Atomism in England From Hariot to Newton* (Oxford, 1966), p. 120, Robert H. Kargon argues that Barrow's scientific methodology, as expounded in the *Lectiones mathematicae,* should be interpreted as a rejection of hypothetical physics rather than as Archimedean classicism, but he is uncritical in his acceptance of Barrow's early influence on Newton.

D. T. WHITESIDE

BARRY, MARTIN (*b.* Fratton, Hampshire, England, 28 March 1802; *d.* Beccles, Suffolk, England, 27 April 1855), *embryology, histology.*

Barry was trained for a career in his father's Nova Scotia-based mercantile concern, which sent ships to various parts of North America and the West Indies. After a short time in business, Barry began medical studies to prepare himself for work in science. Before receiving the M.D. at Edinburgh in 1833 he studied medicine at London, Paris, Erlangen, and Berlin. After graduation he specialized in anatomy and physiology for about a year under Friedrich Tiedemann at Heidelberg. On his return to Edinburgh, Barry attended the Royal Infirmary to further his medical knowledge.

Since he had a private income, Barry never had to practice medicine and did not hold any permanent appointment. As a result, he was able to divide his residency between Scotland, England, and Germany. In 1843 he presented a course of physiological lectures at St. Thomas' Hospital, London. The following year Barry became house surgeon at the new Royal Maternity Hospital, Edinburgh, and made observations on the position of the fetus both before and during delivery. These observations were noted by Sir James Young Simpson in various papers. Barry had a partiality for his obstetric work, of which Simpson spoke highly, but he soon had to resign this position because of recurring ill health.

Barry received the Royal Society's Royal Medal in 1839 for his 1838 and 1839 papers on embryology, and the following year was elected a fellow of the Royal Society. Other societies in which he was active were the Royal Society of Edinburgh and the Wernerian Society.

In 1835 Barry began to study the embryological literature, having been led into microscopical researches by an embryological work given to him by Robert Jameson. This was soon after K. E. von Baer and his fellow workers in Germany had stimulated the study of embryology. His interest in embryology led him into general histological studies at about the time that the cell theory was being formulated. The bulk of microscopic research was then being published in German, and Barry was one of few British scientists interested in and conversant with the German microscopic literature of the 1830's and 1840's.

Barry's first embryological paper was "On the Unity of Structure in the Animal Kingdom," in which he started from the assumption that all of nature is part of the same grand design. Barry recognized that the germ cells of several species of animals were essentially the same and that there is a general law directing the development of animal structure from a homogeneous or general state to a heterogeneous or special one. He adapted his description of the general development of animals from Baer, on whom he relied quite heavily.

In 1837 Barry was in Germany again, this time using the facilities of Johannes Müller, C. G. Ehrenberg, Rudolph Wagner, and Theodor Schwann. After his return to England he presented his results to the Royal Society as "Researches in Embryology" (1838–1840). In this three-part series Barry tried to follow the history of the mammalian ovum from its first appearance within the ovary through its early stages of development. His numerous observations (mostly on rabbits) resulted in a series of descriptions and illustrations (drawn by himself) that give a good account of that development.

Barry made two notable embryological observations: the segmentation of the yolk in the fertilized mammalian ovum and the penetration of the spermatozoon into the mammalian ovum. In 1839 he pictured the two-, four-, eight-, and sixteen-cell stages in mammals and described as similar to a mulberry that stage which Ernst Haeckel later named the morula. Barry concluded that the process he described in the mammal was similar to that already recognized in fishes and batrachians, thus strengthening his belief in the unity of the animal world.

In 1839 Barry avoided discussing the problem of whether contact of the seminal fluid with the ovum was necessary for impregnation. He did mention, however, that in some instances he had found spermatozoa on the surface of the ovary. The following year Barry reported that he had seen a spermatozoon within the *zona pellucida* with its head directed toward the interior of the ovum. Barry read a note to the Royal Society on 8 December 1842 in which he announced that he had recently seen spermatozoa within the ova of a rabbit. As far as he knew, this was an original observation. He showed one or more of these ova to Richard Owen, William Sharpey, and Richard Grainger, all of whom, he said, agreed that the spermatozoa were within the ova. Theodor Bischoff refused to believe that Barry had seen spermatozoa within ova, and Barry answered him in 1844, pointing out Bischoff's errors in obtaining and preserving ova. Only in 1851 did Alfred Nelson confirm Barry's observations; further proof came from George Newport and Georg Meissner, and finally from Bischoff himself.

Barry presented a second series of papers, "On the Corpuscles of the Blood," to the Royal Society in 1840–1841, and concluded it in a fourth part, "On Fibre," in 1842. A later, expanded version of the 1842 paper was translated into German by J. E. Purkinje, with whom Barry had been living, and was published in Müller's *Archiv* (1850). The histological work in these papers was an outgrowth of Barry's embryological studies, being prompted particularly by the appearance of the rabbit's generative organs when they were in a highly vascular condition. Barry was interested in the origin of the red blood corpuscles and the changes they undergo. The basic idea in his hypothesis of red corpuscle formation was that the nucleus of a parent cell somehow broke into several "discs," all but two of which eventually disappeared. A cell then formed around each of these two "discs," and when the new cells were complete with their own membranes, the membrane of the parent cell disappeared, leaving two new cells. Barry based these ideas on the observations he had made on the fertilized ovum and its development to the morula stage, during which time there is no appreciable increase in the total organic mass.

Barry thought that all of the animal structures which he had studied arose from red blood corpuscles, or corpuscles very similar to them. The bulk of his four-part series of papers was devoted to his observations on many types of tissues and his arguments that

they had all arisen in the same manner from the same basic elements, thus developing further his ideas about the unity of all animal life. He also thought that a knowledge of the formation of various tissues might lead to an understanding of the role of the blood corpuscles in nutrition. In a later paper he tried to show that John Goodsir's work in nutrition supported his ideas.

The greatest portion of Barry's later work was aimed at promoting his conclusions on the origin of blood corpuscles and the formation of animal tissues from them. Often he was overly concerned with his own priority and tried to show that other workers had merely confirmed his theories, even when there were scant grounds for such a claim.

Barry's contemporaries generally thought highly of, and accepted, his observations and the illustrations accompanying his papers. His conclusions, however, were often considered highly speculative and were sometimes assailed. In addition to a number of valuable observations, Barry should also be credited with stimulating microscopic studies in Great Britain and increasing his countrymen's acquaintance with the German literature in his field.

BIBLIOGRAPHY

I. ORIGINAL WORKS. Barry's first embryological paper was "On the Unity of Structure in the Animal Kingdom," in *Edinburgh New Philosophical Journal,* **22** (1837), 116–141, 345–364. His main embryological results are in "Researches in Embryology. First [Second and Third] Series," in *Philosophical Transactions of the Royal Society,* **128** (1838), 301–341; **129** (1839), 307–380; **130** (1840), 529–593; and in a short note, "Spermatozoa Observed Within the Mammiferous Ovum," *ibid.,* **133** (1843), 33. The important histological papers are the series "On the Corpuscles of the Blood," *ibid.,* **130** (1840), 595–612; **131** (1841), 201–216, 217–268; and the continuation "On Fibre," *ibid.,* **132** (1842), 89–135. Many of these observations and conclusions, with some new material, were summarized in a paper read before the Wernerian Society, "On the Nucleus of the Animal and Vegetable 'Cell,'" in *Edinburgh New Philosophical Journal,* **43** (Apr.–Oct. 1847), 201–229. The 1842 paper, "On Fibre," was expanded, translated into German by J. E. Purkinje, and published as "Neue Untersuchungen über die schraubenförmige Beschaffenheit der Elementarfasern der Muskeln, nebst Beobachtungen über die muskulöse Natur der Flimmerhärchen," in Müller's *Archiv* (1850), 529–596. See the Royal Society's *Catalogue of Scientific Papers,* 1st ser., I, 194–195, for Barry's other papers, most of which treat the same material.

II. SECONDARY LITERATURE. The most complete account of Barry's life and work is in the *Edinburgh Medical*

Journal, **1** (1856), 81–91. This consists of a memoir of Barry by J. B., pp. 81–87, and a letter to the editor by Allen Thomson, pp. 87–91. Thomson knew Barry and his scientific endeavors, having at times worked with him, and he gives a short comment on Barry's achievements.

WESLEY C. WILLIAMS

BARTELS, JULIUS (*b.* Magdeburg, Germany, 17 August 1899; *d.* Göttingen, Germany, 6 March 1964), *geophysics.*

Bartels was educated at the University of Göttingen, graduating Ph.D. in 1923 and then working in close association with the distinguished geomagnetician Adolph Schmidt for four years. He was head of the Meteorological Institute at the Fortliche Hochschule in Eberswalde from 1927 to 1941, professor of geophysics at the University of Berlin from 1941 to 1945, and professor of geophysics and director of the Geophysical Institute of the University of Göttingen from 1945 on. From 1956 he was also director of the Max Planck Institute of Aeronomy at Lindau and, from 1954 to 1957, president of the International Association of Geomagnetism and Aeronomy.

At the time Bartels began his research, the mathematical theory of statistics was emerging as a major scientific tool. Bartels saw how it could be used to improve the quality of inferences in important sections of geomagnetism. Accordingly, he developed rigorous statistical procedures that both served as a pattern in subsequent geomagnetic analyses and led to new results of much importance. He himself applied the procedures skillfully and fruitfully.

Bartels' statistical analyses led him to make the first clear discrimination between the geomagnetic variations caused by wave and particle radiation from the sun, and from this to develop reliable measures, based on geomagnetic observations, of the two types of radiation. Some of the indexes he introduced in his treatment of geomagnetic variations came to be used internationally, e.g., a sensitive geomagnetic index of the influx of solar particles into the auroral region and his planetary indexes K_p. His procedures further enabled Bartels to elucidate features of tides in the earth's atmosphere that are caused by the moon's gravitational attraction. In investigating twenty-seven-day variations in geomagnetic activity, which are connected with the sun's rotation, he was led to postulate the existence in the sun of certain magnetically active regions (*M* regions), which astronomers later connected with the development of sunspots. He also showed that the sun's surface is never wholly active or wholly quiet.

BIBLIOGRAPHY

Among Bartels' works are "Statistical Methods for Research on Diurnal Variations," in *Terrestrial Magnetism and Atmospheric Electricity,* **37** (1932), 291–302; "Twenty-Seven-Day Recurrence in Terrestrial Magnetic and Solar Activity, 1932–33," *ibid.,* **39** (1934), 201–202; "Random Fluctuations, Persistence and Quasi-persistence in Geophysical and Cosmical Periodicities," *ibid.,* **40** (1935), 1–60; "The Eccentric Dipole Approximating the Earth's Magnetic Field," *ibid.,* **41** (1936), 225–250; "Geophysical Lunar Almanac," *ibid.,* **43** (1938), 155–158, written with G. Fanselau; "Harmonic Analysis of Diurnal Variations for Single Days," *ibid.,* **44** (1939), 137–156; "Some Problems of Terrestrial Magnetism and Electricity," in J. A. Fleming, ed., *Physics of the Earth VIII: Terrestrial Magnetism and Electricity* (New York–London, 1939), pp. 385–430; "The Three-Hour-Range Index Measuring Geomagnetic Activity," in *Terrestrial Magnetism and Atmospheric Electricity,* **44** (1939), 411–454; *Geomagnetism,* 2 vols. (Oxford, 1940, 1951, 1962), written with Sydney Chapman; "Geomagnetic Data on Variations of Solar Radiation. Part 1—Wave-radiation," in *Terrestrial Magnetism and Atmospheric Electricity,* **51** (1946), 181–242; "Geomagnetic and Solar Data," in *Journal of Geophysical Research,* **54** (1949), 295–299; "Geomagnetically Detectable Local Inhomogeneities in Electrical Conductivity Below the Surface," in *Nachrichten. Akademie der Wissenschaften in Göttingen,* **5** (1954), 95–100; "Solar Influences on Geomagnetism," in *Proceedings of the National Academy of Sciences,* **43** (1957), 75–81; and "Discussion of Time-variations of Geomagnetic Activity Indices K_p and A_p, 1932–1961," in *Annales de géophysique,* **19** (1963), 1–20.

Bartels was also editor of Vols. XLVII and XLVIII of *Handbuch der Physik* (Berlin, 1957).

K. E. BULLEN

BARTHEZ, PAUL-JOSEPH (*b.* Montpellier, France, 11 December 1734; *d.* Paris, France, 15 October 1806), *physiology.*

Barthez was an extraordinarily influential French physician who helped to popularize vitalistic doctrines at a time when the most unsophisticated forms of mechanism still held sway in biology and medicine. Succeeding generations have regarded Barthez's theories as hopelessly naïve, but the general point of view that he advocated was adopted by many important biologists of the early nineteenth century.

The son of Guillaume Barthez de Marmorières, chief engineer of Languedoc, Barthez received his early schooling in Narbonne and Toulouse and entered the medical school at Montepellier at the age of sixteen. Completing his degree in three years, he went to Paris, where he became a protégé of Falconnet, physician to Louis XV. He frequented the intellectual circles of the capital and became particularly intimate with d'Alembert. From 1755 to 1757 he served as a physician with the French army. Upon returning to Paris, he was employed first as a royal censor and then as an editor of the *Journal des savants*. During this time he also contributed to the *Encyclopédie* edited by his friend d'Alembert.

Barthez returned to Montpellier about 1760 as professor of medicine and remained there for twenty years, becoming chancellor of the medical school in 1773. During this period he developed his vitalistic doctrines, expounding them in three books: *De principio vitali hominis* (1773), *Nova doctrina de fonctionibus naturae humanae* (1774), and his most important work, *Nouveaux éléments de la science de l'homme* (1778).

Barthez's vitalism is based on the distinction between three different types of phenomena—matter (*la matière*), life (*la vie*), and soul (*l'âme*). He argued that even if like effects follow from like causes, we cannot assume that the laws which govern one type of phenomenon will be meaningful when applied to another; life cannot be subsumed under the laws that govern matter, and it cannot be described in the same manner as the behavior of the soul. Barthez denied both the mechanistic doctrines of Borelli and Boerhaave and the vitalism of Stahl and van Helmont. He thought that their approaches to physiology were illogical and—even worse—totally useless, since they yielded results that were either obviously wrong or incapable of being tested. In his view, life was a valid subject for investigation, but it needed its own distinctive science with unique principles and techniques.

One important aspect of this new science was the development of clinical teaching and research. Barthez thought that physicians should return to the inductive method of Hippocrates, learning the principles of physiology as they manifest themselves in the whole, living body. Although such clinical research later became the basis for the superiority of French medical science in the nineteenth century, the idea was opposed by Barthez's colleagues, and the controversy that it aroused caused him to resign his position at Montpellier in 1781.

Once again Barthez returned to the literary and intellectual life of Paris. He was awarded many honors, among them membership in the Académie des Sciences and the Société Royale de Médecine. The Revolution stripped him of these honors and sent him into retirement in southern France, where he spent the last two decades of his life studying and writing. In 1798 he published *Nouvelle mécanique des mouvements de l'homme et des animaux,* in which he demonstrated, through very intricate anatomical analysis,

that the simple hydraulic explanations offered by the iatromechanists (particularly Borelli) would never explain the delicate balance and control of muscles that are needed for such motions as walking and swimming. During these years Barthez also published several practical medical handbooks and revised his *Nouveaux éléments,* which had been very popular in its earlier version.

With the coming of the Directorate, Barthez regained some of his former prominence. At the time of his death in 1806 he was personal physician to Napoleon and honorary professor of medicine at Montpellier. He had been appointed to the Légion d'Honneur and the Institut National, and had served with Corvisart as a medical member of Napoleon's consular government.

Unfortunately, Barthez did not really practice what he had preached on the subject of clinical research. His works are almost wholly theoretical, and he was particularly adept at producing the teleological explanations which were anathema to later generations of physiologists. Despite these failings his influence was widely felt; and his attitude toward physiology is mirrored in the work of Xavier Bichat and Johannes Müller, and even in that of the antivitalists François Magendie and Claude Bernard.

BIBLIOGRAPHY

I. ORIGINAL WORKS. The two most important works for an understanding of Barthez's physiological doctrines are *Nouveau éléments de la science de l'homme* (Montpellier, 1778; 2nd ed., Paris, 1806) and *Nouvelle méchanique des mouvements de l'homme et des animaux* (Carcassonne, 1798). Other works include *De principio vitali hominis* (Montpellier, 1773); *Nova doctrina de fonctionibus naturae humanae* (Montpellier, 1774).

II. SECONDARY LITERATURE. For reliable discussions of Barthez's life and theories see Jacques Lordat, *Exposition de la doctrine medical de Barthez et mémoire sur la vie de Barthez* (Paris, 1818); A. C. E. Barthez, *Sur la vitalisme de Barthez* (Paris, 1864).

RUTH SCHWARTZ COWAN

BARTHOLIN, CASPAR (*b.* Malmö, Denmark [now Sweden], 12 February 1585; *d.* Sorø, Denmark, 13 July 1629), *theology, anatomy.*

Bartholin's father, Bertel Jespersen, was court chaplain; his mother, Ane Rasmusdotter Tinckel, the daughter of a clergyman from Skåne. Because of his aptitude for languages Caspar was sent to grammar school when he was only three; at eleven he delivered lectures in Greek and Latin. In 1603 he matriculated at the University of Copenhagen, but transferred the following year to Wittenberg, where he studied philosophy and theology for the next three years. He was *respondens* at theses, lectured, and was elected master of philosophy and humane arts in 1608 with the thesis *Exercitatio physica de natura.* In 1606 Bartholin traveled through Germany to Holland, France, and England, visiting the universities and meeting learned physicians and philosophers. During a stay at Leiden he began to study medicine, but without giving up theology. After a short visit home he returned to Wittenberg, where in 1606 he published *Exercitatio de stellis,* which was reissued seven times, with queries and corrections, as *Astrologia sive de stellarum naturae, emendiator et auctior.*

In 1607 Bartholin went to Basel, where he lectured and worked with Felix Platter, Gaspard Bauhin, and Jacob Zwinger. While there he was offered the M.D. degree, but declined it. From 1608 to 1610 Bartholin was in Italy, where he studied anatomy and performed dissections at Padua with Fabricius ab Aquapendente and Casserio, and at Naples, where he was offered a professorship in anatomy. He helped to prepare the engravings for Casserio's work on the sense organs, *Pentaesthesicon* (1609), and his anatomical studies here formed the basis for the manual *Anatomicae institutiones corporis humani* (1611), which made him famous. During this time he also published several manuals of logic, physics, and ethics: *Enchiridion logicum ex Aristotele* (1608), *Janitores logici bini* (1609), *Disp. physica Basileensis* (1610), and *Enchiridion metaphysicum ex philosophorum coryphaei* (1610).

During a visit to Basel in 1610 Bartholin was made doctor of medicine after defending his *Paradoxa CCXL pathologica, simiotica, diaetetica.* When he returned to Denmark in 1611 he was appointed *professor eloquentia* at the University of Copenhagen. His marriage in 1612 to Anna Fincke, daughter of the professor of medicine Thomas Fincke, strengthened his ties to the university. He became professor of medicine in 1613, inaugurating his lectures with a speech on the use of philosophy in medicine. During the next ten years he wrote prolifically and lectured on medicine, physics, and religion, but he performed no dissections at Copenhagen. By 1622 his health had failed and, tortured by renal stones and rheumatism, he sought recovery at Carlsbad.

Vowing to continue his theological studies, upon his cure in 1624 Bartholin accepted a professorship of theology at the University of Copenhagen. Two years later he was made a doctor of divinity, and in the following years he published such works as *De natura theologiae* (1627), *De auctoritate Sacrae Scripturae* (1627), *Benedictio Aharonis* (1628), and

Systema physicum (1628). A mild religious tone is found in his writings, especially in *De studio theologico ratione inchoando et continuando* (1628), where he urges the young to study Holy Writ in both the vernacular and the original language. In 1629, when he was dean of the University of Copenhagen for the second time, Bartholin went to visit his children, who had been removed to Sorø for fear of the plague at Copenhagen. He died there, of renal failure, in the home of his friend J. Burser (1583–1639), a botanist.

Bartholin's fame is due not to his originality, but to his learning and reputation as a teacher; as a strict Aristotelian he clarified the essential points in the doctrines of his time, eliminating obsolete and superfluous theories. As a theologian his personal life was marked by piety and Lutheran orthodoxy. His anatomical manual *Institutiones,* well arranged and handy but without illustrations, was reprinted five times. It became still more famous when his son Thomas brought out an enlarged and illustrated edition, *Novis recentiorum opinionibus* (1641). In addition to four main sections on the abdomen, thorax, head, and extremities there were four *libelli* on the blood vessels, nerves, and skeleton. It was the first manual to describe the olfactory nerves, found by Casserio, as the first pair of cerebral nerves. Bartholin called the suprarenal glands, recently discovered by Bartolomeus Eustachius, the *capsulae atrabiliares,* in the belief that they were hollow and the source of black bile. In 1628 Bartholin published an excellent textbook, *De studio medico,* for his sons.

BIBLIOGRAPHY

I. ORIGINAL WORKS. A full catalog of Bartholin's works is in H. Ehrencron-Müller, *Forfatterlexikon,* I (Copenhagen, 1924), 258–264. Individual works are *Exercitatio de stellis* (Wittenberg, 1606), reissued as *Astrologia sive de stellarum naturae, emendiator et auctior* (Wittenberg, 1606; 6th ed., 1627); *Enchiridion logicum ex Aristotele* (Augsberg, 1608); *Exercitatio physica de natura* (Wittenberg, 1608); *Janitores logici bini* (Wittenberg, 1609); *Disp. physica Basileensis* (Basel, 1610); *Enchiridion metaphysicum ex philosophorum coryphaei* (Augsburg, 1610); *Paradoxa CCXL pathologica, simiotica, diaetetica* (Basel, 1610); *Anatomicae institutiones corporis humani* (Wittenberg, 1611); *De auctoritate Sacrae Scripturae* (Copenhagen, 1627); *De natura theologiae* (Copenhagen, 1627); *Benedictio Aharonis* (Copenhagen, 1628); *De studio medico* (Copenhagen, 1628); *De studio theologico ratione inchoando et continuando* (Copenhagen, 1628); and *Systema physicum* (Copenhagen, 1628). The *Institutiones* was translated by Simon Paulli as *Künstliche Zerlegung menschlichen Leibes* (Copenhagen, 1648) and also into Italian (Florence, 1651).

II. SECONDARY LITERATURE. V. Ingerslev, *Danmarks*

Laeger og Laegevaesen, I (Copenhagen, 1873), 270–274, and G. A. Sommelius, *Lexicon eruditor. Scanensium,* I (Lund, 1776), 133–146, give biographical information, while E. Gotfredsen, *Medicinens historie* (Copenhagen, 1964), p. 230, treats Bartholin's time.

E. SNORRASON

BARTHOLIN, ERASMUS (*b*. Roskilde, Denmark, 13 August 1625; *d*. Copenhagen, Denmark, 4 November 1698), *mathematics, physics.*

Erasmus Bartholin was the son of Caspar Bartholin (1585–1629) and the brother of Thomas (1616–1680). He matriculated at the University of Leiden in 1646 and remained in Holland for several years, studying mathematics. Later he traveled in France, Italy (he received his M.D. at Padua in 1654), and England. Upon his return to Copenhagen, Bartholin was appointed professor of mathematics in 1656 but transferred to an extraordinary chair of medicine in 1657 and to an ordinary one in 1671. He served the University of Copenhagen as dean of the faculty of medicine, librarian, and rector, and was appointed royal physician and privy councilor.

Bartholin wrote little on medicine, although he and his brother Thomas played some part in introducing cinchona bark to Denmark; he also contributed to the journal founded and edited by Thomas, *Acta medica et philosophica Hafniensia.*

His publications in pure mathematics were fairly numerous, but not of great importance. As an exponent of the Cartesian tradition, Bartholin's main interest was in the theory of equations; in this he was directly influenced by Frans Van Schooten. Besides his own works, he issued in almost every year from 1664 to at least 1674 a *Dissertatio de problematibus geometricis* consisting of theses propounded by himself and defended by his students.

Bartholin also worked in astronomy. Like many others, he observed the comets of 23 December 1664–9 April 1665. In this effort he was assisted by Ole Rømer. He did not reach a conclusion about the true orbits, for he was skeptical of all statements about either the place of comets in the heavens (including Tycho Brahe's) or their physical nature (including Descartes's).

Also in 1664 Bartholin began, at the direction of Frederick III of Denmark, to prepare for publication the collected manuscript observations of Tycho Brahe, which the king had bought from Ludwig Kepler. In this task he was again assisted by Rømer. The king's death prevented the project's completion, and its only result was Bartholin's critique of the imperfect *Historia coelestis* of Albert Curtz.

Bartholin's major contribution to science was undoubtedly his study of Icelandic spar (specially collected by an expedition sent to Helgusta ir in Reyðarfyorðr, Iceland, in 1668). In physics, as in mathematics, Bartholin was a fervent admirer of Descartes (of whom he wrote: "Miraculum reliquum solus in orbe fuit"), as is evident in his attempt to deal with the newly discovered phenomenon of double refraction. Having shown that both rays (*solita* and *insolita*) are produced by refraction, and given a construction for determining the position of the extraordinary image, he argued that double refraction could be explained in the Cartesian theory of light by assuming that there was a double set of "pores" in the spar. This puzzling phenomenon proved to be of great theoretical interest to both Huygens and Newton. Bartholin was in fairly close touch with both French and German scientists and, through the latter (initially), with the Royal Society. The copy of his *Experimenta crystalli Islandici* that he sent to Henry Oldenburg is now in the British Museum; from it Oldenburg prepared an excellent English précis.

BIBLIOGRAPHY

I. ORIGINAL WORKS. Bartholin's works are in *Francisci à Schooten Principia matheseos universalis* (Leiden, 1651); *Dissertatio mathematica qua proponitur analytica ratio inveniendi omnia problemata proportionalium* (Copenhagen, 1657), a monograph on harmonic proportionals ($2ac = ab + bc$) leading into a short discussion of the resolution of equations; a translation of a minor Greek optical text by Damianos, or Heliodorus of Larissa (Copenhagen, 1657); a completion of two papers of Florimond de Beaune as *De aequationum natura, constitutione & limitibus,* in Descartes's *Geometria* (Amsterdam, 1659); *Auctarium trigonometriae ad triangulorum sphaericorum et rectilineorum solutiones* (Copenhagen, *ca.* 1663/1664); and *Dioristice, seu Aequationum determinationes duabus methodis propositae* (Copenhagen, 1663). The *Dissertatio, Dioristice,* and *Auctarium* (and perhaps others) were issued under the title *Selecta geometrica* (Copenhagen, 1664). Also see *De cometis anni 1664 et 1665 opusculum* (Copenhagen, 1665); his critique of Albert Curtz's *Historia coelestis* (Augsberg, 1668) in *Specimen recognitionis nuper editarum observationum astronomicarum N. V. Tychonis Brahe* (Copenhagen, 1668); and *Experimenta crystalli Islandici disdiaclastici quibus mira & insolita refractio detegitur* (Copenhagen, 1669). *De naturae mirabilibus quaestiones academicae* (Copenhagen, 1674) is a collection of reprinted essays and addresses (original dates in brackets): "The Study of the Danish Language" [1657], "The Shape of Snow" [1660], "The Pores of Bodies" [1663], "On Cartesian Physics" [1664], "On Attraction" [1665], "On Custom" [1666], "On Nature" [1666], "On Judgment and Memory" [1667], "On Experiment" [1668], "On Physical Hypotheses" [1669], "On the Shapes of Bodies" [1671], and "Secrets of the Sciences" [1673]; the article on the Danish language has attracted some interest from modern Danish scholars. *De aere Hafniensi dissertatio* (Frankfurt, 1679) is a pamphlet on climatology that alludes to medieval Iceland. There may well be other tracts by Bartholin.

II. SECONDARY LITERATURE. Works dealing with Bartholin include Axel Garboe, "Nicolaus Steno and Erasmus Bartholinus," in *Danmarks geologiske undersøgelse,* 4th ser., **3,** no. 9 (1954), 38–48; V. Maar, *Den første anvendelse af kinabark i Danmark* (Leiden, 1925); Kirstine Meyer, *Erasmus Bartholin. Et Tidsbillede* (Copenhagen, 1933); and Henry Oldenburg's précis of the *Experimenta crystalli Islandici,* in *Philosophical Transactions of the Royal Society,* **6,** no. 67 (16 Jan. 1671), 2039–2048.

A. RUPERT HALL

BARTHOLIN, THOMAS (*b.* Copenhagen, Denmark, 20 October 1616; *d.* Copenhagen, 4 December 1680), *physiology, anatomy.*

Thomas Bartholin was the second of the six sons in the famous family produced by Caspar and Anna (Fincke) Bartholin. He entered the University of Copenhagen in 1634 and in 1637 went to Leiden, where he decided on medicine as his vocation. Like his renowned father, who successively was professor of anatomy and religion at the University of Copenhagen, he retained a strong interest in the humanities. In Leiden, with the help of Sylvius (Franciscus de le Boë) and Johannes de Wale, he produced in 1641 the first of the many revised editions of his father's *Institutiones anatomicae* (1611). Notably, the new edition recognized the work of Aselli and Harvey. In 1640, threatened by pulmonary tuberculosis, Bartholin went to Paris, then to Orléans and Montpellier, and finally to Padua, where he regained his health, only to develop chronic renal stones. In Padua, he studied with a fellow countryman, Johan Rhode, and the anatomist Johann Vesling. The latter assisted him with a second revision of the *Institutiones,* published in 1645.

In the winter of 1643–1644, Bartholin visited Rome and Naples, where he gained the enduring friendship of Marco Aurelio Severino; in the following spring he visited Sicily and Malta. At Messina he was offered, but declined, a professorship in philosophy. During this time he wrote a thesis (never published) on fossil sharks' teeth (*glossopetrae*), which were thought to have value as medicine. When he returned to Padua, he produced a related treatise, *De unicornu* (1645). The peripatetic Bartholin then moved to Basel, where he obtained a mèdical degree, and in October 1646 he returned to Copenhagen and joined the faculty of the university. Three years later he married Else Christoffersdatter. Of their children, the most notable was Caspar Bartholin, known eponymously

for Bartholin's gland (*Glandula vestibularis major*) and Bartholin's duct (*Ductus sublingualis major*).

After teaching mathematics for a time, in 1649 Bartholin was chosen to succeed Simon Paulli in the chair of anatomy. Thus Paduan anatomy was introduced to Copenhagen. Hitherto, human dissection had been performed infrequently, was somewhat restricted, and was performed only at the discretion of the king, who sometimes observed from a concealed position. Bartholin's most famous student was Niels Stensen. In his new post, Bartholin was able to study anatomy extensively and regularly, and this is reflected in the third revised edition of the *Institutiones* (1651), an edition much superior in text to the second, which had been severely criticized, both justly and unjustly, by Caspar Hoffmann and Jean Riolan. The third edition was also noteworthy for the illustrations from Casserius and Vesling that replaced the earlier Vesalian figures.

After being informed by his brother Erasmus Bartholin of Pecquet's discovery in dogs of the thoracic duct (*ductus thoracicus*) and the cisterna chyli (*receptaculum chyli*), Bartholin undertook a search for them in the cadavers of two criminals, donated for the purpose by the king. He found the duct, which he reported in *De lacteis thoracis in homine brutisque nuperrime observatis* (1652), but apparently he overlooked the cisterna chyli and declared that it is not always present in man. Bartholin's greatest contribution to physiology was his discovery that the lymphatic system is an entirely separate system. At first he sought to explain the lymphatics, already recognized as anatomical structures, as providing the liver with chyle for the manufacture of blood. On 28 February 1652, working with his assistant, Michael Lyser, Bartholin concluded that the lymphatics formed a hitherto unrecognized physiological system. This was reported in *Vasa lymphatica nuper hafniae in animalibus inventa et hepatis exsequiae* (1653). Failure, in this edition, to indicate the date of discovery by more than the term "28 February" and the inclusion of the further date "1652" in the second edition led to the belief by many that the true year of discovery was 1653. Such was the opinion of Olof Rudbeck, who claimed priority of discovery by reason of his demonstration of the lymphatics in April 1652. Although there was extended controversy, there is now little doubt of Bartholin's priority. In *Vasa lymphatica in homine nuper inventa* (1654), he confirmed the existence of the human lymphatic system.

Continuing attacks of renal stones forced Bartholin to give up his anatomical duties in 1656, after which he turned his attention to a wider range of medical problems. His *Dispensatorium hafniense*

(1658) was the first Danish pharmacopeia. The *Historarium anatomicarum rariorum centuria I–VI* (1654–1661) dealt with numerous limited problems of human and comparative anatomy, and *Cista medica hafniensis* (1662) was a medical miscellany. Bartholin immediately recognized the significance of Malpighi's work on the lungs, *De pulmonibus* (Bologna, 1661)—not least because it provided the first account and illustration of the capillaries, the link between arteries and veins hypothesized by Harvey as a requirement for a systemic circulation of the blood and now proved to exist. Consequently, he included these two celebrated letters in *De pulmonum substantia et motu* (1663), their second publication in Europe. In 1661, Bartholin was elected *professor honorarius,* which freed him from all academic duties, and in 1663 he bought the estate of Hagestedgaard, forty-five miles from Copenhagen. There he devoted himself largely to literary, historical, antiquarian, and medicophilosophical studies, such as *De insolitis partus humani viis* (1664), *De medicina danorum domestica* (1666), *De flammula cordis epistola* (1667), *Orationes et dissertationes omnino varii argumenti* (1668), and *Carmina varii argumenti* (1669). Many unpublished works were lost in 1670 in a disastrous fire at Hagestedgaard, described in *De bibliothecae incendio* (1670).

As the most distinguished physician in Denmark and held in high esteem by the king, Bartholin was responsible for the royal decree of 1672 that decided the organization of Danish medicine for the next hundred years. In 1673 he established the first examination in midwifery at Copenhagen, and in the same year he began publication of the first Danish scientific journal, *Acta medica et philosophica hafniensa.* His health continued to decline, and in 1680 Bartholin sold Hagestedgaard and returned to Copenhagen, where he died. He is buried in the cathedral, the Vor Frue Kurke (Church of Our Lady), but unfortunately the location of his grave is not known.

BIBLIOGRAPHY

I. ORIGINAL WORKS. Bartholin's works, which are discussed in the text, are *Institutiones anatomicae* (Leiden, 1641, 1645, 1651), revised editions of his father's work; *De unicornu* (Padua, 1645); *De lacteis thoracis in homine brutisque nuperrime observatis* (Copenhagen, 1652); *Vasa lymphatica nuper hafniae in animalibus inventa et hepatis exsequiae* (Copenhagen, 1653); *Vasa lymphatica in homine nuper inventa* (Copenhagen, 1654); *Historarium anatomicarum rariorum centuria I–VI* (Copenhagen, 1654–1661); *Dispensatorium hafniense* (Copenhagen, 1658); *Cista medica hafniensis* (Copenhagen, 1662); *De pulmonum substantia* *et motu* (Copenhagen, 1663); *De insolitis partus humani viis* (Copenhagen, 1664); *De medicina danorum domestica* (Copenhagen, 1666); *De flammula cordis epistola* (Copenhagen, 1667); *Orationes et dissertationes omnino varii argumenti* (Copenhagen, 1668); *Carmina varii argumenti* (Copenhagen, 1669); *De bibliothecae incendio* (Copenhagen, 1670).

II. SECONDARY LITERATURE. Despite the destruction of a number of his manuscripts in the fire of 1670, Bartholin's bibliography remains extensive. The fullest list is H. Ehrencron-Müller, *Forfatterlexikon omfattende Danmark, Norge og Island indtil 1814,* I (Copenhagen, 1924), 276–290. Bartholin's work on the lymphatics has been translated from Latin into Danish by F. Lützhøft and G. Tryde as *Lymfekarrene af Thomas Bartholin* (Copenhagen, 1936); and by G. Tryde as *Thomas Bartholins skrifter om opdagelsen af lymfekarsystemet hos dyrene og mennesket* (Copenhagen, 1940). There is an English version of Bartholin's youthful travels and of the burning of Hagestedgaard, translated by C. D. O'Malley: *Thomas Bartholin on the Burning of His Library and on Medical Travel* (Lawrence, Kans., 1961).

The fullest biographical account is that of Axel Garboe, *Thomas Bartholin et bidrag til dansk natur-og laegevidenskabs historie i det 17. aarhundrede,* 2 vols. (Copenhagen, 1949–1950).

C. D. O'MALLEY

BARTOLI, DANIELLO (*b.* Ferrara, Italy, 12 February 1608; *d.* Rome, Italy, 12 January 1685), *physics.*

Bartoli entered the Society of Jesus at the age of fifteen and studied with the Jesuits at Piacenza and Parma. After teaching rhetoric at Parma for a time, he proceeded to Milan and Bologna for theological studies. At Bologna he studied under G. B. Riccioli. He took his vows in 1643. In 1650, after extensive travel, he was made historian of the Jesuits, and thereafter resided principally in Rome. From 1671 to 1673 he was rector of the Collegio Romano (now the Gregoriana), the principal Jesuit university.

Bartoli's major published works comprise histories of the first century of Jesuit activity in England, Italy, China, and Japan. He also wrote extensively on Italian literary matters and on morals. In the scientific field he did much to expound and popularize the work of contemporary physicists, particularly barometric experiments and the concept of atmospheric pressure (1677). He also wrote works on the physical analysis of sound, sound waves, and the sense of hearing (1679), and on the phenomena of freezing (1681). The last-mentioned work was severely criticized by Giuseppe Del Papa, professor of philosophy at the University of Pisa, in a published letter to Francesco Redi.

Although not distinguished by valuable original contributions, Bartoli's scientific expositions were generally objective, clear, and attractively written; they evidence wide reading and a spirit of true inquiry. Bartoli sought to link the speculative and experimental approaches in science. He did not hesitate to mention Galileo with praise and to quote from his works (which were still on the *Index*) and from those of such foreign writers as William Harvey, Marin Mersenne, and Pierre Gassendi. Occasionally he disputed Galileo's opinions, especially on harmonic theory. To judge by their numerous editions and translations, his books were widely read and did much to render the scientific debates of his day interesting and accessible to the general reader, as well as to inculcate a taste for impartial consideration of scientific evidence.

Thomas Salusbury, the English translator of Galileo's works, published in 1660 a translation of Bartoli's *Dell'huomo di lettere difeso, et emendato*. The book was designed to encourage the pursuit of scholarship under such difficulties as poverty, hostile criticism, and neglect of one's published work; to stimulate public appreciation of original ideas; and to discourage plagiarism or the worship of authority by students. Bartoli's work was undoubtedly effective, both in Italy and abroad, in increasing interest among lay readers in the burgeoning physics of the seventeenth century.

BIBLIOGRAPHY

Bartoli's collected works (*Opere*) were published in fifty volumes (Florence, 1829–1837). His letters were published with a biography by G. Boero in *Lettere edite ed inedite del p. Daniello Bartoli* (Bologna, 1865). The works of scientific interest include *Dell'huomo di lettere difeso, et emendato* (Rome, 1645), translated by Thomas Salusbury as *The Learned Man Defended and Reformed* (London, 1660); *La ricreazione del savio* (Rome, 1659); *La tensione, e la pressione* (Rome, 1677); *Del suono, de' tremori armonici, e dell'udito* (Rome, 1679); and *Del ghiaccio e della coagulazione* (Rome, 1681). An exhaustive bibliography appears in C. Sommervogel, S. J., *Bibliothèque de la compagnie de Jésus* (Louvain, 1960).

The biographical article by A. Asor-Rosa in *Dizionario biografico degli italiani,* VI (Rome, 1964), includes a modern critical appraisal of Bartoli's literary and historical activities.

STILLMAN DRAKE

BARTOLOTTI, GIAN GIACOMO (*b*. Parma, Italy, *ca.* 1470; *d*. after 1530), *medicine, history of medicine.*

Bartolotti came from a family of doctors; his father, Pellegrino, was competent in both pharmacy and surgery. Gian Giacomo studied philosophy and medicine at the universities of Bologna and Ferrara; at the latter he was a pupil of Antonio Cittadini and of Sebastiano Dell'Aquila, both of secondary importance in the medical ranks of Ferrara. He was there in 1494, among those who witnessed the conferring of degrees, and in 1497 he attended an anatomical dissection. A year later, although not included on the list of professors holding regular appointments at Ferrara, he was assigned to teach a course on the fourth "fen" of the first book of the *Canon* of Ibn Sīnā (Avicenna). He prefaced the course with a series of historical lessons that he also used as an introduction to his *Opusculum de antiquitate medicinae*. Toward the close of the century he was practicing medicine, and in the early sixteenth century he was doing so at Venice.

In 1758 Mazzuchelli revealed an Italian translation that Bartolotti had made of the *Table* (*Pinax*), a dialogue attributed to Cebes, the Theban philosopher who was a disciple of Socrates and Philolaus.

In his *Opusculum de antiquitate medicinae*, a brief treatise on the history of ancient medicine, Bartolotti exhibits praiseworthy zeal, even though the work is based on obvious critical naïveté regarding historical-scientific orientation of the ancient period. Nevertheless, he reveals a good knowledge of classical medical literature, and in the early chapters he analyzes the origins of medical thought.

BIBLIOGRAPHY

I. ORIGINAL WORKS. MS of Cebe's *Table,* translated into the vernacular at the request of Niccolò Maria d'Este, bishop of Adria, is preserved in the Library of the Somaschi Fathers of the Salute in Venice, Codex Vaticanus 288, and bears the notation "Ferrariae 1498, die 28. Aprilis." *Opusculum de antiquitate medicinae*, Codex Vaticanus 5376, was translated, along with introduction, into English by Dorothy M. Schullian and, in the same volume, into Italian by Luigi Belloni (Milan, 1954). Mention must also be made of *Tractatus de natura daemonum* (1498) and *Tractatus complessionum* (1520), which are in the MS Vaticanus latinus 5376.

II. SECONDARY LITERATURE. Mention of Bartolotti may be found in I. Affo, *Memorie degli scrittori e letterati parmigiani* (Parma, 1791), pp. 178–179; and G. M. Mazzuchelli, *Gli scrittori d'Italia,* II, pt. 1 (Brescia, 1758), 479.

LORIS PREMUDA

BARTON, BENJAMIN SMITH (*b.* Lancaster, Pennsylvania, 10 February 1766; *d.* Philadelphia, Pennsylvania, 19 December 1815), *botany, zoology, ethnography, medicine.*

Barton was the author of the first botanical textbook published in the United States, *Elements of*

Botany (1803), which ran to six editions (three during his lifetime); an influential teacher at the University of Pennsylvania (his students included William Darlington, William Baldwin, Charles Wilkins Short, Thomas Horsfield, and Meriwether Lewis); the patron of Frederick Pursh and Thomas Nuttall, with whose specimens he hoped to produce a flora of North America; and the owner of the largest private natural history library of his time. Between 1797 and 1807 he assembled what was then the largest herbarium of native plants (1,674 specimens).

His father, Thomas Barton, attended Trinity College, Dublin; came to Pennsylvania in 1750; and married David Rittenhouse's older sister, Esther, in 1753. Ordained in the Anglican Church, he worked with the Indians around Carlisle, and settled on Conestoga Creek. Benjamin's mother died in 1774, and his father in 1780, leaving Benjamin an orphan at fourteen. Educated by his older brother, William, and at an academy in York, Pennsylvania, he early showed a liking for history, natural history, and drawing, but recurrent gout soon plagued his health. He commenced medical training at eighteen under Dr. William Shippen of Philadelphia. During 1785 he joined his uncle David Rittenhouse, commissioned to survey the western boundary of Pennsylvania. During the first of his two years (1786–1787) of medical studies at the University of Edinburgh, although twice ill, he won for his dissertation on black henbane the Harveian Prize, which he did not receive until about 1813. He had been made one of four annual presidents of the Royal Medical Society of Edinburgh and had been entrusted with a sum of the Society's money, which he was unable to return before his departure.

Barton did not receive a medical degree from Edinburgh, nor is there confirmation that he did from Göttingen, but he received a diploma from Lisbon Academy, Portugal. He took no pride in the Lisbon diploma, however, and in 1796 he wrote (but may not have sent) a letter to D. Christopher Ebeling of Hamburg University, seeking a degree from that or another German university. In 1801 *A Memoir concerning the disease of Goitre, as it prevails in different parts of North America* (Philadelphia, 1800) was published in Portuguese in Lisbon (his doctoral thesis?) and in 1802 in German in Göttingen.

Barton's "irritable and even cholerick" disposition with his colleagues, influenced by his gout, was no doubt worsened by the consciousness of the sum unreturned to the Medical Society. As his nephew, W. P. C. Barton, wrote (p. 283): " . . . the struggles he made in early life through the most discouraging, nay appalling influence of want, added to the direful ravages of disease—his subsequent elevation appears astonishing. . . . He whose mental exertions survive such a fate, and who perseveres through it, is not, believe me, a common man!"

In 1789, at the age of twenty-three, Barton returned to Philadelphia to become professor of natural history and botany; in 1795 he succeeded to the professorship of materia medica, and in 1813 he added the professorship of the practice of physic to his already too busy life. For ten years (1790–1800) he served as one of the American Philosophical Society's curators, and from 1802 to 1815 as one of its vice-presidents. In 1797 he married Mary Pennington, by whom he had two children, Thomas Pennant and Hetty. He visited Thomas Jefferson at Monticello in 1802, and revisited Virginia in 1805.

Barton was continually publishing short papers on his observations, or those related to him by his associates, not always with permission or acknowledgment. His prose was diffuse and sometimes redundant; he seldom revised but expostulated in an intimate style. His interpretations of complex phenomena generally followed the views of the antiquarians. His bibliography is a farrago, for he continually announced projected works, varying their titles. He began a revision of Gronovius' *Flora Virginica,* and a "Flora of Pennsylvania." In 1808 he began editing successful European works, relating the content to American readers. Meanwhile, to the dismay of Jefferson and others, his natural history of the Lewis and Clark expedition lay unfinished.

In 1815 Barton revisited France and England, then died within two weeks of his return to Philadelphia. Barton had, said Elliott Coues, "every qualification of a great naturalist except success, his actual achievements being far from commensurate with his eminent ability and erudition."

BIBLIOGRAPHY

I. ORIGINAL WORKS. Barton's papers are preserved in the Academy of Natural Sciences of Philadelphia, the American Philosophical Society, the Pennsylvania Historical Society, the Boston Public Library, the Library of Congress, and the possession of John R. Delafield. Barton launched the *Philadelphia Medical and Physical Journal* (1804–1809), to which he contributed numerous articles on medicine, natural history, physical geography, and lives of naturalists. Lists of his writings are found in Max Meisel, *Bibliography of American Natural History,* 3 vols. (New York, 1924–1929); *Elements of Botany,* 6th ed., William P. C. Barton, ed. (Philadelphia, 1836), pp. 27–30, illustrated, as were all editions, with drawings by William Bartram; Whitfield J. Bell, Jr., *Early American Science, Needs and*

Opportunities for Study (Williamsburg, Va., 1955), p. 45; and Francis W. Pennell, "Benjamin Smith Barton as Naturalist," in *Proceedings of the American Philosophical Society,* **86** (1942), 108–122. W. L. McAtee edited "Journal of Benjamin Smith Barton on a Visit to Virginia, 1802," in *Castanea,* **3** (1938), 85–117.

II. SECONDARY LITERATURE. Additional sources, any of which must be read with caution as to the known facts and/or the bias of the writer in mind, are Jeannette E. Graustein, "The Eminent Benjamin Smith Barton," in *Pennsylvania Magazine of History and Biography,* **85** (1961), 423–438; John C. Greene, "Early Scientific Interest in the American Indian: Comparative Linguistics," in *Proceedings of the American Philosophical Society,* **104** (1960), 511–517; [John E. Hall] "Life of Benjamin S. Barton M.D.," extracted from the paper read by W. P. C. Barton before the Philadelphia Medical Society shortly after his uncle's death, in *The Port Folio,* 4th ser., **1** (1816), 273–287; Francis Harper, "Proposals for Publishing Bartram's *Travels,"* in *Library Bulletin of the American Philosophical Society* (1945), 27–38; Francis W. Pennell, "The Elder Barton—His Plant-Collection and the Mystery of His Floras," in *Bartonia,* no. 9 (1926), 17–34; and Edgar Fahs Smith, "Benjamin Smith Barton," in *Papers of the Lancaster County Historical Society,* **28** (1924), 59–66. Barton presents his future biographer a challenge matched by few men of his period. The acidulous estimates of his students Charles Caldwell and James Rush may be balanced by the regard of DeWitt Clinton.

JOSEPH EWAN

BARTRAM, JOHN (*b.* Marple, Pennsylvania, 23 May 1699; *d.* Kingsessing, Pennsylvania, 22 September 1777), *botany.*

The eldest child of William Bartram and his first wife, Elizabeth Hunt, both members of the Society of Friends, John Bartram was born on his father's farm near Darby, Pennsylvania. He had only a common country schooling; but, from the age of twelve, as he later said, he had "a great inclination to Botany and Natural History," although for a time medicine and surgery were his "chief study." On reaching manhood, Bartram inherited from an uncle a farm on which he established himself and his young family; he sold it in 1728 and bought another, of 102 acres, on the banks of the Schuylkill River at Kingsessing, four miles from Philadelphia. Here he converted the marshy lands into productive meadows by draining them; and, through intelligent use of fertilizer and crop rotation, he was soon reaping more abundant crops than most of his neighbors. By 1730 he had laid out a small garden where he cultivated plants, shrubs, and trees from different parts of America. As Bartram's interest in scientific botany grew, James Logan, chief justice of the province and a learned amateur of science, encouraged him with loans and gifts of books. About 1734 Bartram was introduced to the London mercer and enthusiast of science Peter Collinson as "a very proper person" to provide specimens of the products of American fields and forests; and thus his career was launched.

Collinson ordered seeds, plants, and shrubs; got Bartram other customers; advised him on what would sell in England; and instructed him how to pack and ship the specimens and even how to behave toward his patrons. The dukes of Richmond, Norfolk, Argyll, and Bedford; Lord Petre; Philip Miller, author of *The Gardener's Dictionary*; Sir Hans Sloane; and Thomas Penn all enriched their gardens and greenhouses with plants obtained from Bartram. In this way Bartram introduced more than a hundred American species into Europe. Collinson and his friends annually raised a fund to pay for their purchases and thus to underwrite Bartram's collecting. Sometimes they sent him botanical works as gifts so that he could identify the plants he found, and Collinson persuaded the Library Company of Philadelphia to give Bartram a membership so that he might use its collections. In addition, Collinson introduced Bartram by letter to Linnaeus, Gronovius, G. L. Buffon, and other European naturalists, and also to Americans who shared his interests.

With a market for his plants thus assured, Bartram began to make a series of botanical journeys to distant parts of the country. The first, in 1736, was to the sources of the Schuylkill River. In 1738 he traveled to Virginia and the Blue Ridge, covering 1,100 miles in five weeks and spending but a single night in any town. He made shorter expeditions to the New Jersey coast and pine barrens and to the cedar swamps of southern Delaware. The yield to science from these explorations was so great that in 1742 Benjamin Franklin and other Philadelphians opened a subscription to enable Bartram "wholly to spend his Time and exert himself" in discovering and collecting plants, trees, flowers, and other natural products. The subscription was abandoned when Logan opposed it, and Bartram never had the kind of financial independence he repeatedly sought. Nonetheless, in the summer of 1742 he tramped over the Catskill Mountains, and in 1743, with Conrad Weiser, the province interpreter, and the cartographer Lewis Evans, he traveled through Pennsylvania into the Indian country of New York as far as Oswego and Lake George. "Our way . . . lay over rich level ground," Bartram wrote of one day's journey, in a good example of both his life in the woods and his prose style,

> . . . but when we left it, we enter'd a miserable thicket of spruce, opulus, and dwarf yew, then over a branch

of Susquehannah, big enough to turn a mill, came to ground as good as that on the other side of the thicket; well cloathed with tall timber of sugar birch, sugar maple, and elm. In the afternoon it thunder'd hard pretty near us, but rained little: We observed the tops of the trees to be so close to one another for many miles together, that there is no seeing which way the clouds drive, nor which way the wind sets: and it seems almost as if the sun had never shone on the ground since the creation.

By 1750 Bartram was famous. Copies of his journals circulated in manuscript in London, and that of the trip to Onondago was published there in 1751. Such American naturalists as Dr. John Mitchell of Virginia and such philosophers as Cadwallader Colden of New York sought him out. Peter Kalm spent so much time at Kingsessing that Logan complained that during an eight-month visit to Philadelphia the Swedish botanist had seen no one but Franklin and Bartram. Dr. Alexander Garden of Charleston wrote of a visit to Bartram in 1754:

His garden is a perfect portraiture of himself, here you meet with a row of rare plants almost covered over with weeds, here with a Beautifull Shrub, even Luxuriant Amongst Briars, and in another corner an Elegant & Lofty tree lost in common thicket. On our way from town to his house he carried me to severall rocks & Dens where he shewed me some of his rare plants, which he had brought from the Mountains &c. In a word he disclaims to have a garden less than Pensylvania & Every den is an Arbour, Every run of water, a Canal, & every small level Spot a Parterre, where he nurses up some of his Idol Flowers & cultivates his darling productions. He had many plants whose names he did not know, most or all of which I had seen & knew them. On the other hand he had several I had not seen & some I never heard of [Earnest, p. 21].

Bartram's observations were not limited to things botanical. He collected shells, insects, hummingbirds, terrapin, and wild pigeons. He described the mussels of the Delaware River, rattlesnakes, wasps, and the seventeen-year locust; and from his letters about them Collinson fashioned communications that were printed in the Royal Society's *Philosophical Transactions.* Except for the introduction and notes for a Philadelphia edition of Thomas Short's *Medicina Britannica* and two short pieces on snakeroot and red cedar for *Poor Richard's Almanack,* Bartram wrote almost nothing for publication; even the Onondago journal, in Kalm's estimate, contained not "a thousandth part of the great knowledge which he has acquired in natural philosophy and history."

Bartram was also interested in every scheme to promote scientific inquiry in America, and he offered several of his own. In 1739 he suggested that "ingenious & curious men" be organized into a society or college for "the study of natural secrets arts & syances." The suggestion was premature, but in 1743 Franklin succeeded in establishing for a few years a less ambitious group, the American Philosophical Society, to which Bartram was especially devoted. In a letter to Garden in 1756 he proposed a kind of geological survey of the mineral resources of the North American continent. Bartram was always ready to become a traveling naturalist, supported by the government or private patrons and reporting his discoveries to his sponsors. As increasing numbers of explorers and collectors uncovered and carried away mammoth fossils from Big Bone Lick, Kentucky, in the 1750's and 1760's, Bartram expressed the hope that wealthy "curiosos" would "send some person that will take pains to measure every bone exactly, before they are broken and carried away, which they soon will be, by ignorant, careless people, for gain."

Single-minded and untiring in pursuit of natural history, Bartram grumbled and complained when anything impeded his work. On the other hand, he was friendly and open to those who shared his devotion. Kalm gratefully acknowledged that he owed Bartram much, "for he possessed that great quality of communicating everything he knew." Everyone who knew him was impressed with Bartram's industry, his capacity for accurate observation and recall, and his independence. When Kalm repeated Mark Catesby's theory that trees and plants decrease in size and strength when they are taken north, Bartram answered that if the question were "more limited," Catesby's answer "would prove more worthwhile"—some trees grow better in the south, others in the north.

This independent turn of mind also showed itself in Bartram's religious views, which were deeply influenced by his studies of nature. He was contemptuous of ecclesiastical formalities and theological points; he was scornful of Quaker pacifism, which he believed made men hypocrites, "for they can't banish freedom of thought"; and, unlike his coreligionists, he judged that the only way to establish lasting peace with the Indians was to "bang them stoutly." He was in fact a deist, and when he persisted in expressions of disbelief in the divinity of Jesus, the Friends disowned him in 1757. He continued to attend Quaker meetings, however—and to express his unorthodox views. "My head runs all upon the works of God, in nature," he wrote in 1762. "It is through that telescope I see God in the sky." Over the window of his house in 1770 he carved the words, "'Tis God alone, Almighty Lord,/The Holy One, by me ador'd."

487

The sentiment, his son William remembered, gave offense to pious neighbors.

The close of the French and Indian War brought Great Britain a vast increase of territory in North America, and Bartram set out to explore it. He traveled in 1761 to the forks of the Ohio River and to the springs of western Virginia, crawling "over many deep wrinkles in the face of our antient mother earth"; and in 1764 he appealed to Collinson to raise funds to send him on an exploration of Florida. In consequence of his friend's representations, in 1765 Bartram was named king's botanist with an annual stipend of fifty pounds. Although he complained that it was not enough, Bartram set out all the same, accompanied by his son William. Entertained by governors and other officials, he traveled from Charleston through Georgia into Florida, visiting plantations, noting the quality of the soils, and recording trees, plants, and fossils. During these journeys he discovered the lovely *Franklinia altamaha,* which has never since been found in its native soils and survives today only in descent from a specimen Bartram brought back to his garden. In Florida, Bartram went up the St. John's River to Fort Picolata, where he and William witnessed an impressive Indian treaty ceremonial.

This was Bartram's last trip. He was aging, and his sight was failing. He spent his remaining years at Kingsessing, surrounded by family and friends, tending his garden, visited by the great and the curious. One visitor, J. Hector St. John de Crèvecoeur, later published a pleasing, although romanticized, account of Bartram's life and manner of living. Inevitably honors came to Bartram. The Royal Academy of Sciences of Sweden made him a member in 1769, and in 1772 "A Society of Gentlemen in Edinburgh" who were interested in propagating arts and sciences, awarded him a gold medal for his services. Before he died, his fellow citizens and European friends of America had ranked Bartram with Franklin and David Rittenhouse as one of the country's authentic natural geniuses.

Bartram was married twice: in 1723 to Mary Maris of Chester Monthly Meeting, by whom he had two sons; and in 1729 to Ann Mendenhall, of Concord Monthly Meeting, who bore him five sons and four daughters. His sons Isaac and Moses became apothecaries, John inherited the farm and famous garden, and William achieved lasting fame as a botanical traveler, artist, and author. John Bartram, in William's words, was "rather above the middle size, and upright," with a long face and an expression that was at once dignified, animated, and sensitive. A painting by John Wollaston in the National Portrait Gallery in Washington is thought to be a portrait of Bartram.

BIBLIOGRAPHY

I. ORIGINAL WORKS. Bartram's published writings are *Observations on the Inhabitants, Climate, Soil, Rivers, Productions, Animals, and Other Matters . . . From Pensilvania to Onondago, Oswego and the Lake Ontario, in Canada* (London, 1751); and "Diary of a Journey Through the Carolinas, Georgia, and Florida . . . ," Francis Harper, ed., in American Philosophical Society, *Transactions,* **33,** part 1 (1942–1944), 1–120.

II. SECONDARY LITERATURE. Works on Bartram are William Bartram, "Some Account of the Late Mr. John Bartram, of Pennsylvania," in *Philadelphia Medical and Physical Journal,* **1,** part 1 (1804), 115–124; William Darlington, *Memorials of John Bartram and Humphry Marshall* (Philadelphia, 1849); Ernest Earnest, *John and William Bartram: Botanists and Explorers* (Philadelphia, 1940); Peter Kalm, *Travels in North America,* A. B. Benson, ed., 2 vols. (New York, 1937); Leonard W. Larabee et al., eds., *The Papers of Benjamin Franklin* (New Haven, Conn., 1959–), II, 298–299, 355–357, 378–380, *et passim*; Francis D. West, "John Bartram's Journey to Pittsburgh in the Fall of 1761," in *Western Pennsylvania Historical Magazine,* **38** (1955), 111–115; "John Bartram and the American Philosophical Society," in *Pennsylvania History,* **23** (1956), 463–466; and "The Mystery of the Death of William Bartram, Father of John Bartram the Botanist," in *Pennsylvania Genealogical Magazine,* **20** (1956–1957), 253–255.

WHITFIELD J. BELL, JR.

BARTRAM, WILLIAM (*b.* Kingsessing, Pennsylvania, 9 April 1739; *d.* Kingsessing, 22 July 1823), *botany, ornithology.*

William Bartram, third son of the botanist John Bartram and his second wife, Ann Mendenhall, was born on his father's farm at Kingsessing, four miles from Philadelphia. In 1752 he was sent to the Academy of Philadelphia; his studies there included Latin and French, but botany and drawing, his father wrote, were "his darling delight." The father encouraged his son in these directions, taking him on botanizing trips to the Catskill Mountains in 1753 and to Connecticut in 1755, and letting him sketch on Saturday afternoons and Sundays, instead of working and going to meeting. Samples of William's work were sent to Peter Collinson, his father's London friend and patron, who was much pleased with them. Before he left the Academy in 1756, William was making natural history drawings for Collinson and for George Edwards, who used them and William's descriptions of Pennsylvania birds in his *Gleanings of Natural History.* Two of William's drawings of turtles were printed in the *Gentleman's Magazine.*

William had no idea what profession or trade he should enter. His father, sure only that he did not "want him to be what is called a gentleman," consulted Collinson and Benjamin Franklin. He considered medicine, surveying, printing, and engraving for the lad; but finally, in 1757, apprenticed him to a merchant. In 1761 William moved to Cape Fear, North Carolina, where he opened a trading store that soon failed. Nothing, in fact, went well for him, and by 1764 he seemed to Collinson, who liked him, to be "lost in indolence and obscurity."

In 1765 John Bartram, setting out on his trip to Florida, asked William to accompany him. When the journey was over, William chose to remain in the south. Turning down an invitation from Major John G. W. DeBrahm to join him as a draftsman on his survey of Florida, he settled as a planter of rice and indigo, living in a mean hovel on the St. John's River. His father did not approve and could not understand—Billy, he lamented, was "so whimsical and so unhappy." The young man succeeded no better as a planter than as a merchant. He returned to Pennsylvania, eked out a poor living by various means, and made a few drawings for Dr. John Fothergill, the London Quaker physician who had succeeded Collinson as the Bartrams' patron. But the lure of the Carolinas and Florida was strong, and in 1771 William was in the south again. Fothergill encouraged him to visit Florida, offered him fifty pounds a year, and agreed to purchase any drawings he made. With this assurance William set out in March 1773 on the travels that were to make his reputation.

For four years he traveled through South Carolina, Georgia, and Florida. Much of the way he went alone; and so patent were his innocence and devotion to nature that colonial governors—even in wartime—let him come and go freely, and the Indians everywhere welcomed him, naming him *Puc-puggy,* the "Flower Hunter." Bartram made notes on birds, animals, fishes, and plants, which were everywhere in profusion, and he recorded the life of the Indians. He wandered along meandering streams, refreshed himself from crystal springs, took refuge in dark caves from crashing thunderstorms, marveled at the terrifying roar and thrash of angry alligators, and stood in awe before forests of azalea so bright that the very mountains seemed on fire. In this world, which was as much of his own making as nature's, Bartram found peace. He returned to Kingsessing in January 1778, however, and never left again. His brother John, who had inherited their father's farm and garden, made him welcome; and so, after John's death, did their sister and her husband. William never married.

He worked in the garden, planting, grafting, and performing the many other chores the nursery required. In 1782 the University of Pennsylvania offered him the professorship of botany, but he declined it.

He wrote up his notes and sometimes shyly showed them to visitors; but he was in no hurry to publish, and the *Travels* did not appear until 1791. The book was reprinted in London in 1792 and 1794; and before the decade closed, other editions appeared in Paris, Dublin, Vienna, Berlin, and Holland. Most reviewers praised Bartram's descriptions of natural products and of the Indians; but almost all criticized his style as florid, luxuriant, imaginative, imprecise, and personal. Yet it is for its style that the *Travels* is best known, and for its influence on romantic literature that it has its most lasting fame. Coleridge, Wordsworth, and Chateaubriand filled pages of their notebooks with excerpts from the *Travels;* and from these notebooks Bartram's figures and sense of nature passed through the poets' richly imaginative minds and reappeared, still recognizable, in *Ruth,* "Kubla Khan," "The Rime of the Ancient Mariner," *Atala,* and other works. Bartram, in short, was one of the earliest voices, and one of the authentic sources, of the nineteenth-century romantic movement.

The fame of the *Travels,* in addition to John Bartram's reputation, brought a procession of visitors to Kingsessing; otherwise, William's life was unchanged. He wrote down his observations of the Creek and Cherokee Indians for Benjamin Smith Barton, made most of the drawings for Barton's *Elements of Botany* (1803), and wrote a few papers for Barton's *Philadelphia Medical and Physical Journal,* among them a biographical sketch of his father. To the Philadelphia Society for Promoting Agriculture, which elected him a member in 1785, he sent "Observations on the Pea-Fly or Beetle." At its founding in 1812, the Academy of Natural Sciences made him a member. He was also a member of the American Philosophical Society, but took no interest in its work and never attended a meeting.

Gentle, loving, and well loved, "the happiest union of moral integrity with original genius and unaspiring science," William Bartram lived out his life in harmony with the numberless lives in nature all about him. Without formal education in science, he was nonetheless recognized, as his father had been, for his unequaled knowledge of the natural world. Many sought his help, and he cheerfully shared all he knew. Thomas Nuttall, F. A. Michaux, Thomas Say, and Palisot de Beauvois felt they could hardly begin their work without his advice and blessing. In particular, the ornithologist Alexander Wilson owed much of his training to Bartram, lived for a time at the garden, and urged his benefactor to accompany him on his ornithological journeys. On 22 July 1823, having completed in his study a note on the natural history of a plant, Bartram took a few steps, collapsed, and died in the house where he was born. A portrait by Charles Willson Peale, done in 1808, is in the Independence National Historical Park collection, Philadelphia.

BIBLIOGRAPHY

I. ORIGINAL WORKS. William Bartram's works are "Observations on the Creek and Cherokee Indians, 1789," in American Ethnological Society, *Transactions,* **3,** part 1 (1853), 1-81; "Travels in Georgia and Florida, 1773-74: A Report to Dr. John Fothergill," Francis Harper, ed., in American Philosophical Society, *Transactions,* **33,** part 2 (1943), 121-242; and *Travels Through North & South Carolina, Georgia, East & West Florida, . . .* (Philadelphia, 1791). Of many subsequent printings and editions of the *Travels,* that edited by Francis Harper (New Haven, Conn., 1958) is most useful for scholars.

II. SECONDARY LITERATURE. Works concerning Bartram are William Darlington, *Memorials of John Bartram and Humphry Marshall* (Philadelphia, 1849); Ernest Earnest, *John and William Bartram: Botanists and Explorers* (Philadelphia, 1940); N. Bryllion Fagin, *William Bartram, Interpreter of the American Landscape* (Baltimore, 1933); John Livingston Lowes, *The Road to Xanadu* (Boston, 1927).

The drawings Bartram made for Fothergill, now in the British Museum (Natural History), have been published, in color and with an introduction and notes by Joseph Ewan, by the American Philosophical Society (Philadelphia, 1968).

WHITFIELD J. BELL, JR.

BARUS, CARL (*b.* Cincinnati, Ohio, 19 February 1856; *d.* Providence, Rhode Island, 20 September 1935), *physics.*

Barus was the son of German immigrants. His father, Carl Barus, Sr., was a musician and choirmaster; his mother, Sophia Moellmann, was a clergyman's daughter. Despite persistent financial difficulties, he attended Columbia University's School of Mines from 1874 to 1876 and from 1876 to 1879 studied at Würzburg, where he took his Ph.D. under Friedrich Kohlrausch. Barus worked for the United States Geological Survey (1880-1892), the United States Weather Bureau (1892-1893), and the Smithsonian Institution (1893-1894), where he assisted Samuel P. Langley with the development of the flying machine. Appointed to the Hazard professorship of physics at Brown University in 1895, he became dean of the graduate school in 1903 and taught until his retirement in 1926. Barus was married in 1887 to Annie G. Howes of Massachusetts, a graduate of Vassar College who was active in civic and social reform.

While at the Geological Survey, Barus worked in collaboration with the geologist Clarence King, an advocate of aiding the study of geological evolution with the laboratory analysis of rocks and minerals under high temperatures and pressures. Perhaps Barus' most significant achievement was the development, independently of Châtelier, of methods of measuring, with thermocouples, temperatures over a range of some $1,000°$ C. Internationally known as an authority on pyrometry, he was elected to the National Academy of Sciences in 1892 and awarded the Rumford Medal of the American Academy of Arts and Sciences in 1900. A founder of the American Physical Society in 1899, he served as its president from 1905 to 1906.

At the Weather Bureau, Barus' research shifted to the condensation of water vapor on nuclei, but his departure interrupted his experiments. Resuming this work about 1900, he studied the effects of X rays and radioactivity on condensation in a fog chamber. The chamber, a cylinder made of wood impregnated with resin, was filled with moist air at atmospheric pressure and connected through a stopcock to another chamber at lower pressure. When the stopcock was opened, the air became supersaturated.

Barus concluded that ionization was relatively unimportant in condensation, but C. T. R. Wilson pointed out that his apparatus was unreliable. Wilson had published two authoritative papers in 1899 showing that ions produced by X rays and radioactivity were significant in condensation. He had used a chamber made almost wholly of glass; supersaturation was achieved by expansion at the sudden drop of a piston. Wilson sharply criticized Barus for ignoring the contaminating effect of the resin in his chamber and for failing to recognize that the stopcock method of expansion allowed condensation to occur before maximum supersaturation. In addition, Barus had failed to shield the chamber from the strong electric field of his X-ray apparatus.

Rutherford, who disliked the fragility of Wilson's glass equipment, encouraged Barus to improve the fog chamber so as to explore the nature of X rays. But Barus soon dropped this work entirely, and by the 1920's, despite continued publication, he had fallen into professional obscurity.

BIBLIOGRAPHY

Barus' papers are in the Brown University Archives, the John Hay Library, Brown University, Providence, R.I. The collection consists of his 269-page unpublished autobiography, a frank account of his personal and professional life; a file of about 750 letters, mostly to Barus from personal and scientific correspondents, covering 1879–1935; ten letterpress books for the years 1882–1897, particularly useful for his work at the Geological Survey; and a small group of miscellaneous scientific papers and notes.

R. Bruce Lindsay, "Carl Barus," in *National Academy of Sciences Biographical Memoirs*, **22** (1943), 171–213, is far better than the usual official memoir by a former student. Based on the Barus papers, it is a sympathetic and perceptive account and contains an apparently complete bibliography of Barus' many publications.

DANIEL J. KEVLES

BARY, HEINRICH ANTON DE. See **DeBary, Heinrich Anton.**

BASIL VALENTINE. See **Valentine, Basil.**

BASSANI, FRANCESCO (*b*. Thiene, Vicenza, Italy, 29 October 1853; *d*. Capri, Italy, 26 April 1916), *paleontology.*

The son of Antonio Bassani and Anna Brolis, Bassani studied geology at Padua with Giuseppe Meneghini and, especially, Giovanni Omboni, who introduced him to fossil fishes. After assisting Omboni for two years, he continued his studies at Paris in 1877 under Hébert, Gaudry, and Sauvage, and in Vienna in 1878 with E. Suess and M. Neumayr. He also studied in Munich with Zittel. From 1879 to 1887 Bassani taught natural history, geology, and mineralogy in secondary schools in Padua, Modena, and Milan, and actively investigated and published on fossil fishes of northern Italy and adjacent regions. In 1887 he was called to the chair of geology at Naples, which he occupied until his death.

Bassani was an outstanding teacher, combining the zest and talent for research with the gift of communicating his discoveries and aspirations clearly and understandably to others. As director of the Geological Institute in Naples he brought about extensive improvements in laboratory facilities and collections. He married Everdina Dowkes Dekker, a native of the Netherlands, whose drawings of fossils illustrate his monographs; they had two sons. In his last years he was afflicted with diabetes, in spite of which he continued his research with the aid of assistants.

Bassani was primarily a student of fossil fishes. He also investigated other geological problems, particularly those of the stratigraphic relations and geologic age of various formations in southern Italy, volcanic phenomena at Vesuvius and Solfatara, the contemporaneity of man and extinct animals on Capri, and marine mammals. Many of his short papers pointed out solutions to problems that led to rapid advances

in regional geology. While still a student, he collaborated in translating Charles Darwin's *Expression of Emotions in Man and Animals* into Italian. In 1885 he published a textbook of zoology for secondary schools.

Bassani's earliest studies were of the Eocene fishes of Bolca (Verona), and he wrote a total of eight papers describing and interpreting these famous fossils; a monographic review commenced shortly before his death was completed by his student D'Erasmo in 1922. The slightly later deposit at Chiavòn (Vicenza) was written up in 1889. His studies of the Cretaceous fishes of Comen (Istria) in 1880 and the island of Lesina (Hvar) in 1882 are more important for the extensive comparisons between numerous Cretaceous faunas of the Mediterranean basin than for their systematic or descriptive portions. In 1886 he described a collection of fishes and reptiles from the Triassic bituminous shales of Besano, near the southern tip of Lake Lugano.

Soon after his appointment at Naples, Bassani investigated the fishes from limestones of the Sorrento Peninsula; these proved to belong to the alpine Triassic fauna, and he was able to show that the main dolomitic limestones of southern Italy, which up to that time had been regarded as Jurassic or Cretaceous, were actually Triassic. About the same time he wrote a monograph on the Miocene fishes of Sardinia; in 1899, on the Eocene fishes of Gassino, in Piedmont; in 1905, those of the Pleistocene marls of Taranto and Nardo; and in 1915, on the fishes of the Miocene of Lecce. At least nine of Bassani's papers are devoted exclusively to fossil sharks, and studies of this group of fishes form prominent parts of other reports. He became expert at discriminating shark teeth and eventually concluded that all Pliocene sharks belonged to existing species.

Bassani's most significant work consisted of faunal revisions; he reviewed nearly all important fossil fishes of the Mesozoic and Cenozoic in Italy. Characteristically, he studied entire fossil assemblages rather than reviewing genera or other systematic groups. His reports include full surveys of earlier studies, carefully organized and detailed descriptions of the fossils, and logically developed conclusions about the environmental implications of the fauna and its geological age. He made extensive use of comparative range charts and based his correlations upon the indications of the bulk of the fauna rather than a few "index" species. As his familiarity with faunas of many ages grew, he developed a concept of fish species evolving slowly through geologic time; in accordance with this, he believed that many of the differently named fossils from faunas of different ages actually belonged to the same species, but in the absence of adequate proof of identity he refrained from relegating them to synonymy. He was widely regarded as one of the leading paleoichthyologists of his day.

BIBLIOGRAPHY

I. ORIGINAL WORKS. Lists of Bassani's publications are in the memorial by D'Erasmo, pp. lxviii–lxxvi, and those by Lorenzo and Parona. See also A. S. Romer, N. E. Wright, T. Edinger, and R. van Frank, "Bibliography of Fossil Vertebrates Exclusive of North America, 1509–1927," in *Memoirs of the Geological Society of North America,* no. 87, I (1962), 100–103.

Individual works by Bassani are *L'espressione dei sentimenti nell'uomo e negli animali* (Turin, 1878), a translation of Darwin's *Expression of Emotions in Man and Animals,* with G. Canestrini; *Elementi di zoologia descrittiva al uso delle scuole secondarie* (Milan, 1885, 1889); "Sui fossili e sull'età degli scisti bituminosi triasici di Besano in Lombardia," in *Atti della Società italiana di scienze naturale,* **29** (1886), 15–72; "Ricerche sui pesci fossili di Chiavòn (strati di Sotzka, Miocene inferiore)," in *Atti della Reale Accademia delle scienze, scienze fisiche e matematiche di Napoli,* 2nd ser., **3,** no. 6 (1889), 1–103; "Sui fossili e sull'età degli scisti bituminosi di Monte Pettine presso Giffoni Valle Piana in provincia di Salerno," in *Memorie della Società italiana delle scienze detta dei XL,* 3rd ser., **9,** no. 3 (1892), 1–27; "Contributo alla paleontologia della Sardegna. Ittioliti miocenici," in *Atti della Reale Accademia delle scienze, scienze fisiche e matematiche di Napoli,* 2nd ser., **4,** no. 3 (1892); "La ittiofauna del calcare eocenico di Gassino in Piemonte," *ibid.,* **9,** no. 13 (1899), 1–41; "La ittiofauna delle argille marnose pleistoceniche di Taranto e di Nardò," *ibid.,* **12,** no. 3 (1905), 1–59; "La ittiofauna del calcare cretacico di Capo d'Orlando presso Castellammare," in *Memorie della Società italiana delle scienze,* 3rd ser., **17** (1912), 185–243, written with G. D'Erasmo; and "La ittiofauna della pietra leccese (Terra d'Otranto)," in *Atti della Reale Accademia delle scienze, scienze fisiche e matematiche di Napoli,* 2nd ser., **16,** no. 4 (1915), 1–52.

II. SECONDARY LITERATURE. Articles on Bassani are G. D'Erasmo, "Commemorazione di F. Bassani," in *Bollettino della Società geologica italiana,* **35** (1916), xlix–lxxvi, including a bibliography of 105 titles; G. de Lorenzo, "Commemorazione di F. Bassani," in *Rendiconti della Reale Accademia delle scienze, scienze fisiche e matematiche di Napoli,* 3rd ser., **22** (1916), 69–88; C. F. Parona, "Cenno necrologico," in *Atti dell'Accademia delle scienze di Torino* (Classe di scienze fisiche, matematiche e naturali), **51** (1915–1916), 945–950, and "A ricordo di F. Bassani," in *Bollettino. Comitato geologico d'Italia,* 5th ser., **6** (1919), 89–102; and C. Stefani, "Commemorazione del Prof. F. Bassani," in *Atti dell'Accademia nazionale dei Lincei. Rendiconti,* 5th ser., **26** (1917), 335–336.

JOSEPH T. GREGORY

BASSI, AGOSTINO MARIA (*b.* Mairago, Italy, 25 September 1773; *d.* Lodi, Italy, 6 February 1856), *law, agriculture, natural science.*

Bassi experimented over many years to determine the cause of the *mal del segno,* or muscardine, the silkworm disease then prevalent. His discovery of the microscopic fungus parasite that caused this malady, commonly believed to be of spontaneous origin, was published in 1835 and contributed importan ..y to the understanding of contagious disease.

He was one of twins born to Rosa Sommariva and Onorato Bassi. After his early schooling in Lodi, he entered the University of Pavia. He studied law, but at his parents' desire, he later wrote. His study of physics, chemistry, mathematics, natural science, and some medicine is evidence of his interest in science. Among his teachers were Antonio Scarpa, the anatomist, Alessandro Volta, the physicist, and Giovanni Rasori, professor of pathology and a strong proponent of the *contagium vivum* theory at a time when this was in dispute. Bassi was able to attend the lectures of Lazzaro Spallanzani, who had opposed the doctrine of spontaneous generation and carried out experiments to disprove it. He received his doctorate in jurisprudence in 1798 and was named provincial administrator and police assessor in Lodi, newly under French administration. Subsequently he held various positions in the civil service. Failing eyesight forced his return to Mairago, to his father's farm, where, except for a brief period in public service in 1815, he remained for the rest of his life.

As early as 1807 Bassi's attention was drawn to the silkworm disease, but while he conducted his experiments over the years, he maintained a number of agricultural interests, ranging from the importation and breeding of merino sheep and the publication of *Il pastore bene istruito* (1812) to potato cultivation and winemaking, on which he also published.

The silkworm disease, known in Italy as *mal del segno,* or as *calcino* or *calcinaccio,* because of the white efflorescence and calcined appearance that developed after the worms had died of its ravages, had caused heavy losses to the silkworm industry in Italy and in France, where it was called *la muscardine.* Bassi at first accepted the opinion of the silkworm breeders that the disease arose spontaneously. He experimented, therefore, upon various premises: considering the possibility that environmental factors of food, atmospheric conditions, or methods of breeding were responsible, he maltreated the worms in many ways; they died, but not of the muscardine. Then, suspecting excessive acidity to be the cause, he used phosphoric acid, but was unsuccessful in producing the disease. When Bassi obtained worms that seemed to be calcified after death, he found that they lacked the contagious property that characterized *mal del segno.*

After some years, Bassi concluded that the disease was due to external agents and was transmitted by food, by contact with the dead worms that showed white efflorescence, and by the hands and clothing of the silkworm breeders. The germs were also carried by animals and flies. The contagion contaminated the air in rooms where the worms had become infected with the disease. Bassi reproduced the muscardine by inoculating healthy worms with the white dust, or with matter from diseased worms. He infected caterpillars of other species and then, in turn, produced the same disease in silkworms again. Bassi's microscopic investigations showed that the disease was caused by a cryptogam, a fungus parasitic upon the silkworm. The fine efflorescence, which usually appeared following the death of the animal, was composed of a multitude of minute plants bearing the "seeds," and only when they developed did the disease become infectious. The seeds, transmissible in numerous ways, penetrated the bodies of healthy worms and there nourished themselves. The worms eventually died, and, upon maturation, the small plants produced new seeds. The latent life of the seeds was somewhat under two years, but under certain circumstances it was three.

Bassi presented his results at the University of Pavia in 1833, and in the following year repeated his experiments to the satisfaction of a nine-member committee of the faculties of philosophy and medicine. He reported his experiments and conclusions in *Del mal del segno, calcinaccio o moscardino, malattia che afflige i bachi da seta e sul modo di liberarne le bigattaje anche le più infestate* (1835–1836). He was reviving the *contagium vivum* theory of Rasori, he remarked; it had already been suggested that the disease was caused by a fungus, but only on the basis of odor. It was Bassi who first demonstrated through experiments that a living fungus parasite was the cause of a disease in animals. We know the white efflorescence actually is a mass of spores. During the life of the silkworm the mycelium grows at the expense of the animal and the worm, in time, dies. The hyphae, which penetrate the skin, bear the fine, white, dustlike spores which only then appear.

The first part of the *Del mal del segno* contained a theory proposing that some contagions of plants and animals had their source in the "germs" of plant or animal parasites, and that possibly certain diseases of man were caused by vegetable organisms. The second part was devoted to practical methods of preventing and eliminating the silkworm disease through the avoidance of contamination, the disinfection of rooms where the disease had occurred, and the boiling of implements. He included fresh air and sunlight among his disease preventives and disinfectants.

The minute fungus parasite discovered by Bassi was examined by the botanist Giuseppe Balsamo-Crivelli, professor at the University of Milan. He described it as the vegetable parasite *Botrytis paradoxa,* of the family *Mucedinaceae,* and later named it for Bassi, *Botrytis bassiana* (today *Beauvaria bassiana*).

Confirmation of Bassi's discovery by Jean-Victor Audouin and others came soon after the publication of *Del mal del segno.* The significance of this work was recognized as far greater than its immediate value to practical agriculturists. Johann Lukas Schönlein of Zurich remarked upon the implications for pathogenesis in Bassi's discovery of the fungus origin of muscardine. He recounted his own repetition of Bassi's experiments, for with these in mind he had microscopically examined material from the pustules of favus, and had found the disease to be caused by a vegetable parasite, a fungus. Bassi's influence has been traced in the subsequent investigations of David Gruby, Miles J. Berkeley, and others who considered fungus parasites as a cause of disease in both animals and plants. Bassi's conviction that parasites were actively involved in most contagions was also reflected in Jakob Henle's classic paper of 1840, "Von den Miasmen und Contagien und von den miasmatisch-contagiösen Krankheiten." Over the ensuing years, the clearer understanding of the nature of parasitism affected concepts of infectious disease and contributed to the undermining of the doctrine of spontaneous generation. Bassi's use of experimental inoculation showed his understanding of the importance of this means of tracing the life history of the invading organism. It was to be utilized by the mycologist Anton de Bary in later years.

With time Bassi's increasing loss of eyesight precluded further microscopic observations. His later writings showed his continuing interest in contagion, for he suggested that parasites, animal or vegetable, were the cause of various diseases, including plague, smallpox, syphilis (1844), and cholera (1849). Since the minute germs spread, he advocated quarantine and various modes of prevention and disinfection, employing both asepsis and antisepsis.

Bassi's outstanding and enduring contribution to the understanding of the origin of infectious disease was his demonstration of the fungus cause of the silkworm disease.

BIBLIOGRAPHY

I. ORIGINAL WORKS. Bassi's outstanding publication was *Del mal del segno, calcinaccio o moscardino, malattia che afflige i bachi da seta e sul modo di liberarne le bigattaje anche le più infestate,* 2 pts. (Lodi, 1835–1836; 2nd ed., Milan, 1837; facsimile of 1st ed., Pavia, 1956). The American Phytopathological Society has published a translation of pt. 1 by P. J. Yarrow in Phytopathological Classics, no. 10, C. G. Ainsworth and P. J. Yarrow, eds. (Baltimore, 1958). Three medical works are *Sui contagi in generale e specialmente su quelli che affliggono l'umana specie* (Lodi, 1844); *Discorsi sulla natura e cura della pellagra* (Milan, 1846); and *Istruzioni per prevenire e curare il colera asiatico* (Lodi, 1849). *Opere di Agostino Bassi* (Pavia, 1925) contains a collection of his works. *Documenti Bassiani,* Luigi Belloni, ed. (Milan, 1956), including an autobiographical writing (1842), portraits, facsimiles, and manuscripts, was issued to commemorate the centenary of Bassi's death, and is a valuable source of material on his life and contributions. In the same year there appeared *Studi su A. Bassi,* Luigi Cremascoli, Luigi Belloni, Letizia Vergnano, and Attilio Zambianchi, eds. (Lodi, 1956), with the autobiographical fragment and lists of works, manuscripts, documents, portraits, and illustrations. It also contains a catalog of the 1956 commemorative exhibition relating Bassi's work to other writings, both earlier and later, on *contagium vivum* and the etiology of infectious disease.

II. SECONDARY LITERATURE. Works on Bassi are C. G. Ainsworth, "Agostino Bassi, 1773–1856," in *Nature,* **177** (1956), 255–257; Giovanni P. Arcieri, *Agostino Bassi in the History of Medical Thought, A. Bassi and L. Pasteur* (New York, 1938; Florence, 1956), which stresses his role as a founder of microbiology; Bajla, "Agostino Bassi da Lodi, 1773–1856," in *La clinica veterinaria,* **47** (1924), 186–189; William Bulloch, in *The History of Bacteriology* (London, 1938, 1960), which presents Bassi's doctrine of pathogenic microorganisms in its historical relation to concepts of disease and to the investigations of parasitic cryptogams that followed his work on the silkworm disease; Ralph H. Major, "Agostino Bassi and the Parasitic Theory of Disease," in *Bulletin of the History of Medicine,* **16,** no. 2 (1944), 97–107; Adalberto Pazzini, "The Influence of Agostino Bassi on the Evolution of Microbiology," in *Scientia medica italica,* 2nd ser., English ed., **3** (1954), 187–195; and G. C. Riquier, "Agostino Bassi, 1773–1856," in *Rivista di storia delle scienze mediche e naturali,* **6** (1924), 48–51. Gian Battista Grassi's "Commentario all'opera parasitologica (sui contagi) di Agostino Bassi" is included in the *Opere di Agostino Bassi,* pp. 11–48, and in the booklet accompanying the facsimile edition of *Del mal del segno,* pp. 5–33. Luigi Belloni, "La scoperta di Agostino Bassi nella storia del contagio vivo," is in *Studi su A. Bassi,* pp. 55–86, and in *Actes du VIIIe Congrès international d'histoire des sciences* (Florence–Milan, 1956), II, 897–909.

GLORIA ROBINSON

BASSLER, RAYMOND SMITH (*b.* Philadelphia, Pennsylvania, 22 July 1878; *d.* Washington, D.C., 31 October 1961), *paleontology.*

In his presidential address to the Paleontological Society in 1933, Bassler lamented the passing of an era when training was acquired more in the field than in the classroom: The modern student, attaining maturity before deciding on paleontology as his lifework, loses precious years of interest and experience. As a youth, Bassler had himself spent many hours collecting marine fossils in the dusty Ordovician outcrops near his Cincinnati home, and had worked after high school every day as technical assistant to E. O. Ulrich.

Bassler's close association with Ulrich lasted well beyond his high-school days. When the elder scientist moved to Washington, the younger withdrew from the University of Cincinnati and followed him. Ulrich's influence colored his entire career. He adopted Ulrich's methods and approaches to problems, his interest in paleontology, and his passion for collections and catalogs.

Bassler did graduate work at George Washington University, taking the M.S. and Ph.D., and during this time was employed by the United States National Museum. He was affiliated with these two institutions for nearly four decades.

A world authority on Silurian and Ordovician Bryozoa, Bassler assembled an outstanding collection of these fossils, now at the Smithsonian Institution. He also published valuable bibliographic indexes relating to them: "Bibliographical Index of American Ordovician and Silurian Fossils" (1915), "Early Bryozoa of the Baltic Provinces" (1911), and "Synopsis of American Fossil Bryozoa" (1900). His study of other phyla resulted in the compilation of the *Bibliographic Index of Paleozoic Ostracoda* (1934) and *Bibliographic and Faunal Index of Paleozoic Echinoderms* (1943), as well as shorter papers on tetra corals, Cystoidea, and conodonts.

In the 1933 presidential address, Bassler advocated the classical geological practice of retouching photographs of specimens, a position that has led modern paleontologists to characterize his work as "enthusiastic."

BIBLIOGRAPHY

I. ORIGINAL WORKS. Bassler's writings include "A Synopsis of American Fossil Bryozoa," in *Bulletin of the U.S. Geological Survey,* no. 173 (1900), 9–663, written with J. M. Nickles; "A Revision of Paleozoic Bryozoa," in *Smithsonian Miscellaneous Collections,* **45** (1904), 256–296; **47** (1904), 15–45, written with Ulrich; "The Early Bryozoa of the Baltic Provinces," in *Bulletin of the U.S. National Museum,* **77** (1911), 1–382; "Bibliographic Index of American Ordovician and Silurian Fossils," *ibid.,* **92** (1915), 1–1521 (in 2 vols.); "Development of Invertebrate Paleontology in America," in *Bulletin of the Geological Society of America*, **44** (1933), 265–286; *Bibliographic Index of Paleozoic Ostracoda*, Geological Society of America, special paper no. 1 (Washington, D.C., 1934); and *Bibliographic and Faunal Index of Paleozoic Echinoderms*, Geological Society of America, special paper no. 45 (Washington, D.C., 1943).

II. SECONDARY LITERATURE. Articles on Bassler are Kenneth Caster, "Memorial to Raymond S. Bassler," in *Bulletin of the Geological Society of America*, **76**, no. 11 (1965), 167–173; and *National Cyclopedia of American Biography*, XLIX (1966), 482.

MARTHA B. KENDALL

BASSO, SEBASTIAN (*fl.* second half of the sixteenth century), *natural philosophy.*

Although Basso was famous as a reviver of the atomic philosophy, and was mentioned by Descartes, Gassendi, and Jungius, there is little biographical information about him. He is known only as the author of *Philosophiae naturalis adversus Aristotelem*, which tells us that he was a physician and had studied at the new Academia Mussipontana (Pont à Mousson).

In the choice and arrangement of topics, Basso follows the Aristotelian tradition of the sixteenth century. The authors most discussed are Scaliger, Toletus, Piccolomini, Zabarella, and the Conimbricenses. But Basso's aim is to oppose Peripatetic opinions, and to prove that many puzzling problems in natural philosophy can be solved satisfactorily only by reviving the doctrines of such pre-Aristotelian thinkers as Empedocles, Democritus, Anaxagoras, and Plato. Basso's central doctrines are that all matter consists of very small atoms of different natures and that a very fine, corporeal ether extends throughout the universe and fills the pores between the atoms. The atomic doctrine is invoked, in a sustained polemic with Scaliger, to prove that the forms of the constituents really persist in the compound. Particles of the four elements combine into groupings of the second order, third order, and so on, and produce the characteristic properties of a specific body. The predominant element determines whether an inorganic, compound body is *solida, liquida, fusilia,* or *meteoria.* Transmutation of the elements is held to be impossible, since the ultimate particles are immutable. When water evaporates, it turns into steam, not air: the fiery particles penetrate the water, and drive the watery particles farther apart and into the air.

The ether, or *spiritus,* which Basso likens to the universal ether of the Stoics, governs the motion and arrangement of the atoms, and is responsible for all material change. The immediate instrument of God

in moving and directing all things, it is different from fire, which consists of *spiritus* plus very fine and sharp corpuscles. Basso is convinced that the vacuum of Democritus signifies nothing but the ether of the Stoics. He attempts to explain a wide range of phenomena by assigning to the ether the function of bringing together likes and driving apart unlikes. He opposes Zabarella's opinions when discussing the acceleration of falling bodies and the motion of projectiles: Acceleration of falling bodies is due to increasing pressure from the air above and diminishing resistance of the air beneath.

Basso considers such traditional questions as the influence of the planets on men, especially of the moon on critical days and tides, and tries to explain them (like Pomponazzi and Fracastoro, among the Aristotelians) in terms of contact-action: very fine and fiery spirits emanate from the planets. Their action begins at birth; before that the body of the mother shields the fetus from their influence.

In attacking Aristotelian philosophy, reviving the pre-Socratics and atomists, and offering corpuscular explanations of physical change, Basso was seen as a precursor by some of the founders of the mechanical philosophy in the seventeenth century.

BIBLIOGRAPHY

Basso's only printed work is *Philosophiae naturalis adversus Aristotelem libri XII. In quibus abstrusa veterum physiologia restauratur, & Aristotelis errores solidis rationes refelluntur* (Geneva, 1621; Elzevir ed., Amsterdam, 1649). It has not been possible to trace a 1574 edition that Jöcher claims was published at Rome.

Works on Basso are I. Guareschi, "La teoria atomistica e Sebastiano Basso con notizie e considerazioni su William Higgins," in *Memoria della Regale Accademia dei Lincei, Classe di Scienze Fisiche, Matematiche e Naturali*, **11** (1916), 289–388; Kurt Lasswitz, *Geschichte der Atomistik vom Mittelalter bis Newton*, I (Hamburg–Leipzig, 1890), 467–481, and *Vierteljahrschrift für wissenschaftlichen Philosophie*, **8** (1884), 18–55; Lynn Thorndike, *A History of Magic and Experimental Science*, VI (New York, 1941), 386–388; and J. R. Partington, *A History of Chemistry*, II (London, 1961), 387–388.

P. M. RATTANSI

BASTIAN, HENRY CHARLTON (*b.* Truro, England, 26 April 1837; *d.* Chesham Bois, England, 17 November 1915), *neurology, bacteriology.*

Little is known of Bastian's early life except that his father was named James and that the boy showed great interest in natural history. In 1856 he entered University College London, and seven years later was graduated M.B. At first he worked at St. Mary's Hospital, London, as assistant physician and lecturer on pathology, but returned to his *alma mater* in 1867 as professor of pathological anatomy, having received the M.D. a year before. He continued to practice clinical medicine, and in 1878 was promoted to physician to University College Hospital. From 1887 to 1898 Bastian held the chair of the principles and practice of medicine and also had an appointment at the National Hospital in Queen Square, London, from 1868 to 1902. Early in his career he worked on the problem of abiogenesis, or spontaneous generation, and the return of this interest determined his premature retirement from clinical neurology.

Bastian married Julia Orme in 1866, and had three sons and a daughter. He was elected a fellow of the Royal Society of London in 1868, at the age of thirty-one; served as a censor of the Royal College of Physicians of London from 1897 to 1898; and received honorary fellowship of the Royal College of Physicians of Ireland and an honorary M.D. from the Royal University of Ireland. From 1884 to 1898 Bastian was crown referee in cases of supposed insanity, and a few months before he died, he was awarded a Civil List pension of £150 a year in recognition of his services to science. He is said to have been a close friend of Herbert Spencer, but the philosopher makes only very slight reference to Bastian in his lengthy autobiography. As one of Spencer's executors, Bastian helped to publish the latter work.

Bastian's earliest scientific work was with guinea worms and other nematodes, but his investigations ended suddenly when he developed a strange allergy to these creatures. His clear analytical mind, sound reasoning, and acute powers of observation drew him to clinical neurology, and he spent the rest of his hospital and academic life in this discipline. Beginning in 1869, he published a series of papers on speech disorders. At a time when it was thought that speech was controlled by independent brain centers, Bastian was the most important of those who represented this view by means of diagrams almost akin to electrical circuits. He described a visual and an auditory word center, and in 1869 he gave the first account of word blindness (alexia) and of word deafness, which is known today as "Wernicke's aphasia." His views on aphasia, which were founded on an assumption that there is a direct relationship between psychological functions and localized areas of brain, were an oversimplification and are no longer accepted. They are presented, with supporting clinical data, in *A Treatise on Aphasia and Other Speech Defects* (1898).

The many publications on clinical and clinico-pathological neurology prepared by Bastian reveal his outstanding practical, philosophical, and literary skills, which he exercised to the full in a specialty of medicine that was then emerging as a more precise science. Nevertheless, his work was overshadowed by that of some of his contemporaries who excelled him. His book *The Brain as an Organ of Mind* (1880) and those on paralyses (1875, 1886, 1893), although important, were less popular than those of other writers; and the claim that he and his contemporaries John Hughlings Jackson and William Gowers founded neurological science in Britain is difficult to justify. Nevertheless, he was an important pioneer of neurology as a science; in addition, his anatomical skill is revealed by the discovery in 1867[1] of the anterior spinocerebellar tract of the spinal cord—now known, however, as "Gowers' tract" because of the more detailed investigation of it made by Gowers in 1880. In 1887 Bastian published his important paper on "muscular sense," which he thought was represented in the cerebral cortex, whereas, in contrast with Ferrier and Beevor, in particular, he considered the cortical motor centers to be unnecessary. In 1890 he showed for the first time that complete section of the upper spinal cord abolishes reflexes and muscular tone below the level of the lesion; this has been known occasionally as Bastian's law.

Bastian himself claimed that his studies on abiogenesis were more significant; and there were two periods of activity in this field, approximately 1868–1878 and 1900–1915. Contrary to accepted biological and bacteriological opinion, he believed that there was no strict boundary between organic and inorganic life. He denied the doctrine of *omne vivum ex ovo* and argued that since living matter must have arisen from nonliving matter at an early stage in evolution, such a process could still be taking place. His battle was fought mostly alone; and eventually he was the last scientific opponent of Pasteur, Tyndall, Koch, and the other pioneers of bacteriology. This was an important role, for he pointed out many of their mistakes; and thus, in a negative way and quite against his purpose, he helped to advance the germ theory of fermentation and of disease. As Pasteur's main opponent, he was responsible for the development of some of the techniques that advanced bacteriology.

Thus, Bastian denied that boiling destroyed all bacteria, as Pasteur claimed, and thereby opened the way for the discovery of heat-resistant spores. On the whole, his criticisms of Pasteur's logic were more effective than the multitude of experiments he cunningly conceived and carried out, for the techniques he used are now known to have been frequently defective. His views and supporting experimental data were set forth in a large book of over 1,100 pages, *The Beginnings of Life* (1872). Concerning the role of bacteria in fermentation, he concluded that "These lowest organisms are, in fact, to be regarded as occasional concomitant products rather than invariable or necessary causes of all fermentative changes."[2] A controversy with Tyndall concerning the existence of airborne germs closed his first period of interest in bacteriology. His opponent's comments summarize the situation in 1878: "Neither honour to the individual nor usefulness to the public is likely to accrue from its [the controversy's] continuance, and life is too serious to be spent in hunting down in detail the Protean errors of Dr. Bastian."[3] He began again to devote all his time to his neurological practice and writings.

Bastian returned to his biological studies in 1900 and devoted the last fifteen years of his life to the fundamental problem of the origin of life. He was the last scientific believer in spontaneous generation, for he succeeded in converting no one to his cause. He had been self-instructed in the field of biological research, and although he learned such recording procedures as photomicrography, he was eventually left behind as the technology of the new science of microbiology developed apace, mainly in the hands of the French and German pioneers. Bastian thought that abiogenesis included "archebiosis," living things arising from inorganic matter, or from dead animal or plant tissues, through new molecular combinations, and "heterogenesis," the interchangeability of the lowest forms of animal and vegetable life, both among themselves and with each other; thus ciliates and flagellae could arise from amoebae.

In this regard Bastian was criticized by biologists who accused him of insufficient knowledge of lower forms of life. He attempted to substantiate these phenomena in *Studies in Heterogenesis* (1904), illustrated by 815 photomicrographs, and in *The Nature and Origin of Living Matter* (1905). Each contains similar arguments, in which he compares the unit of living matter with crystals and suggests that the persistence of lower organisms can be adequately ascribed to their successive evolution. Bastian made solutions containing colloidal silica and iron, and put them in sealed glass tubes that were heat-sterilized and stored under regulated conditions of light and temperature. His claim was that after a time, such organisms as bacteria, torulae, and molds grew, but no one could or would repeat these observations. *The Evolution of Life* (1907) dealt exclusively with archebiosis. Many of his papers were rejected by the Royal Society of London and by other learned bodies; this

was the case with the material of his last book, *The Origin of Life* (1911). Two years later *The Lancet* summarized the scientists' judgment of this losing battle: "To our mind the position is quite unchanged, and we ourselves still remain unconvinced, save, of course, of the courage and good faith of Dr. Bastian."[4]

After Bastian's death his son challenged bacteriologists and others to disprove his father's work on abiogenesis. The correspondence thus stimulated[5] reveals that several investigators thought they had done so, but certain findings are still unexplained. Careful repetition of crucial experiments and a dialectical analysis of his results are still needed. Of Bastian's industry, tenacity, logic, and experimental versatility—although perhaps misplaced—there can be no question.

NOTES

1. "On a Case of Concussion-Lesion . . .," in *Medical-Chirurgical Transactions*, 2nd ser., **32** (1867), 499–537.
2. *Modes of Origin*, p. 108.
3. "Spontaneous Generation. A Last Word," in *Nineteenth Century*, **3** (1878), 497–508, see 508.
4. *The Lancet* (1913), **i**, 970.
5. *Ibid.* (1919), **i**, 951–952, 1000–1001, 1044–1045, 1133–1134; **ii**, 216–217, 458.

BIBLIOGRAPHY

I. ORIGINAL WORKS. Bastian's writings include *The Modes of Origin of Lowest Organisms* (London, 1871); *The Beginnings of Life*, 2 vols. (London, 1872); *On Paralysis From Brain Disease in Its Common Forms* (London, 1875); *The Brain as an Organ of Mind* (London, 1880), also trans. into French and German; *Paralyses, Cerebral, Bulbar and Spinal* (London, 1886); "The 'Muscular Sense'; Its Nature and Cortical Localisation," in *Brain*, **10** (1888), 1–137; *Various Forms of Hysterical or Functional Paralysis* (London, 1893); *A Treatise on Aphasia and Other Speech Defects* (London, 1898), also trans. into German and Italian; *Studies in Heterogenesis* (pub. in 4 pts., London, 1901–1903; pub. together, 1904); *The Nature and Origin of Living Matter* (London, 1905; rev. and abbr. ed., 1909); *The Evolution of Life* (London, 1907), also trans. into French; and *The Origin of Life* (London, 1911, 1913). There is a complete list of Bastian's writings in an unpublished biography by Mercer Rang, "The Life and Work of Henry Charlton Bastian 1837–1915" (1954), MS in Library of University College Hospital Medical School, London.

II. SECONDARY LITERATURE. There is no published work comparable with the Rang MS. Obituaries are *British Medical Journal* (1915), **2**, 795–796; *The Lancet* (1915), **2**, 1220–1224 (including portrait and brief bibliography); and F. W. M[ott], in *Proceedings of the Royal Society of London,*

89 (1917), xxi–xxiv. See also J. K. Crellin, "The Problem of Heat Resistance of Micro-organisms in the British Spontaneous Generation Controversies of 1860–1880," in *Medical History,* **10** (1966), 50–59; Lothar B. Kalinowsky, "Henry Charlton Bastian," in Webb Haymaker, ed., *The Founders of Neurology* (Springfield, Ill., 1953), pp. 241–244; and Mercer Rang, "Henry Charlton Bastian (1837–1915)," in *University College Hospital Magazine,* **39** (1954), 68–73.

EDWIN CLARKE

BATAILLON, JEAN EUGÈNE (*b.* Annoire, Jura, France, 22 October 1864; *d.* Montpellier, France, 1 November 1953), *biology, zoology.*

Bataillon was the son of a stone mason. Although he had a scholarship to the *petit séminaire* of Vaux-sur-Poligny, he refused to prepare for an ecclesiastic career. After he had passed the first part of his *baccalauréat ès lettres* in 1882, he became a *surveillant* at the Collège d'Arbois; after passing the second, he was made assistant master at the Belfort *lycée* and started work on his *licence* in philosophy. Bataillon next went to Lyons as master at the *lycée* and, when he had passed his *baccalauréat ès sciences,* he became a student under the physiologist Fernand Arloing. As a *licencié ès sciences naturelles,* he became an assistant in zoology at the University of Lyons in 1887 and learned the elementary techniques of experimental embryology under Laurent Chabry.

On 17 April 1891, Bataillon presented a thesis for the doctorate in science, *La métamorphose des amphibiens anoures.* Seconded by Pasteur, he was made acting lecturer at the Faculté des Sciences at Lyons and deputy lecturer in zoology and physiology at Dijon in 1892. There he became professor of general biology (the first chair so named in France) in 1903, and dean in 1907. It was in Dijon that he discovered the traumatic parthenogenesis of the batrachians. Appointed professor at the University of Strasbourg, which was being reorganized in 1919, he first set up a new laboratory and then temporarily served as rector. He became rector of the University of Clermont-Ferrand in 1921 but left this administrative position in 1924 to accept a professorship of zoology and comparative anatomy at Montpellier. There he met his principal collaborator, Chou Su, who thereafter helped him in his research. In 1932 Chou Su had to return to China to teach, and Bataillon decided to retire ahead of time. He went to live with one of his sons at Castelnau-le-Lez and later on the outskirts of Montpellier, where he died. He had become a corresponding member of the Académie des Sciences in 1916 and a full member in 1946. He was awarded the Osiris Prize of the Academy in 1951.

The experimental attempts to produce parthenogenesis with sodium chloride and sugar hypertonic solutions or butyric acid had failed, resulting only in an abortive segmentation of the egg (simple activation). During an experimental attempt to fertilize *Bufo calamita* with sperm of *Triturus alpestris,* Bataillon, struck by the resemblance of the spermatozoa to fine needles, thought of substituting glass or platinum stylets for them. On a Sunday in March 1910 the experiment performed on *Rana temporaria* was successful and resulted in 90 percent of the eggs being segmented, of which about 10 percent evolved into normal larvae able to survive.

The analysis of these results demonstrated that besides the simple pricking (first activating factor) there is a second regulating factor, which can be identified as a cell or an inoculated cellular fragment. Operating upon eggs freed of their gelatin coating by potassium cyanide, Bataillon demonstrated, in an elegant experiment, using horse blood that had been defibrinated and sedimented, that the element inoculated was a leukocyte. Later, J. Shaver revealed that the active element is a cytoplasmic fragment rich in ribonucleoproteins.

By this interpretation Bataillon corrected Guyer's error. Guyer had attributed to the nucleus of the leukocyte, released inside the egg by the pricking, the power to divide itself and to direct the embryogenesis. The activator thus provoked the "proper reaction" of the egg membrane, making any further fertilization impossible; there followed a special monocentric rhythm, a succession of monopolar mitoses that were, of course, abortive. The second regulating factor was necessary to determine the sequence of the bipolar mitoses (dicentric or amphiastral rhythm), the only ones capable of bringing about segmentation and, consequently, normal development. An intermediate type of mitosis, bipolar but anastral, appeared when the regulation was insufficient.

Electric shocks, pricking by galvanocautery, and the action of fat solvents could all be the activating factor, but only a biological factor could be the regulator. The latter, however, was not a specific: blood of other anurans, urodeles, fish, or mammals could replace the blood of *Rana.* The action of potassium cyanide as a chemical unsheathing agent was a fortuitous discovery which confirmed that unfertilized eggs pricked, but not smeared with blood, never undergo morphogenesis; it also made possible the experiment with defibrinated and sedimented horse's blood, which could be performed only on naked eggs. Eggs sheathed in serum before pricking showed no regular cleavage; those sheathed in red corpuscles showed only 1 percent; and those sheathed in leukocytes showed 75 percent. The extract of hepatopancreas from the *Astacus* was useful in distinguishing the nonactivated eggs, which were cytolyzed, from the activated ones; the latter, through strengthening of the activating membrane, were able to resist cytolysis. This membrane had to be "modified in thickness," a concept confirmed by modern electronic microscopic research (the bursting of the cortical granules liberated their contents, which then reinforced the vitelline membrane).

Bataillon emphasized this paradox: the spermatozoa of *Rana*, when penetrating the eggs of *Bufo*, caused fewer normal developments than did the *Rana* blood inoculated in *Bufo* eggs. He gave the explanation that the inoculated leukocyte acts only as regulator and does not cause any intervention due to chromosome incompatibilities. G. Hertwig, causing irradiated spermatozoa of *Rana* to act upon eggs of *Bufo*, obtained the same results and came to absolutely the same conclusion. Traumatic parthenogenesis generally leads to haploid larvae, but occasionally diploids may result either from an ovule whose polar mitosis aborted or from a secondary regulation through the fusion of divided nuclei. In such cases a few diploid larvae undergo metamorphosis. The regulation factor therefore can substitute dicentric mitosis for monocentric mitosis but cannot double the number of chromosomes that remain haploid. The role of this factor was confirmed by research done with Chou Su on immature and overmature eggs, and subsequently on "false hybrids," where incompatible nuclei act only as regulation factors.

Bataillon's work, although extremely original, is little known, and his role as a precursor is underestimated. His own modesty and deliberate reticence have lessened his reputation. His unchallenged discovery of traumatic parthenogenesis is important, but even more important is his remarkable analysis of the process, which clarified the complex phenomena of fertilization.

BIBLIOGRAPHY

I. ORIGINAL WORKS. Among Bataillon's writings are "L'embryogenèse complète provoquée chez les amphibiens par piqûre de l'oeuf vierge, larves parthénogénésiques de *Rana fusca,*" in *Comptes rendus de l'Académie des sciences,* **150** (1910), 996–998; "Les deux facteurs de la parthénogenèse traumatique chez les amphibiens," *ibid.,* **152** (1911), 920–922; and *Une enquête de trente cinq ans sur la génération,* J. Rostand, S.E.D.E.S., ed. (Paris, 1955), which sums up his work.

II. SECONDARY LITERATURE. Writings on Bataillon include R. Courrier, "Notice sur la vie et les travaux d'Eugène Bataillon," in *Notices et discours de l'Académie des sciences,* **3** (1954), 1–26, which sums up his work; M. F. Guyer, "The Development of Unfertilized Frog Eggs Injected With Blood," in *Science,* **25** (1907), 910–911; G. Hertwig, "Parthenogenesis bei Wirbeltieren hervorgerufen durch artfremden radiumbestrahlten Samen," in *Archiv für mikroskopische Anatomie,* **81** (1913), 87; and J. Shaver, "The Role of Cytoplasmic Granules in Artificial Parthenogenesis," in *Journal de cyto-embryologie belgonéerlandaise* (1949), 61–66.

PAUL SENTEIN

BATE, HENRY. See **Henry Bate of Malines.**

BATEMAN, HARRY (*b.* Manchester, England, 29 May 1882; *d.* Pasadena, California, 21 January 1946), *mathematics, mathematical physics.*

The son of Marnie Elizabeth Bond and Samuel Bateman, a druggist and commercial traveler, Bateman became interested in mathematics while at Manchester Grammar School. He attended Trinity College, Cambridge, where he received a B.A. in 1903 and an M.A. in 1906. After a tour of Europe in 1905–1906 he taught for several years, first at Liverpool University and then at Manchester. In 1910 he emigrated to the United States, where he taught for two years at Bryn Mawr College, then held a three-year research fellowship and lectured at Johns Hopkins from 1912 to 1917, taking his Ph.D. there in 1913. Although by this time he had an international reputation as a mathematician, he worked part-time on meteorology at the Bureau of Standards. In 1917 Bateman was appointed professor of mathematics, theoretical physics, and aeronautics at Throop College (which later became California Institute of Technology) in Pasadena, California, where he taught until his death. He was made a fellow of the Royal Society in 1928 and a member of the U.S. National Academy of Sciences in 1930, was elected vice-president of the American Mathematical Society in 1935, and delivered the Society's Gibbs lecture in 1943.

General theories had little attraction for Bateman; he was a master of the special instance. Much of his work consisted of finding special functions to solve partial differential equations. After some geometrical studies, he used definite integrals to extend E. T. Whittaker's solutions of the potential and wave equations to more general partial differential equations (1904). These and later results he applied to the theory of electricity and, with Ehrenfest, to electromagnetic fields (1924).

While in Göttingen in 1906, Bateman became

familiar with the work of D. Hilbert and his students on integral equations. He applied integral equations to the problem of the propagation of earthquake waves, determining, from the time of contact at various surface points, the velocity of the motion at interior points. In 1910 he published a comprehensive report on research concerning integral equations.

Bateman's most significant single contribution to mathematical physics was a paper (1909) in which, following the work of Lorentz and Einstein on the invariance of the equations of electromagnetism under change of coordinates of constant velocity and constant acceleration, he showed that the most general group of transformations which preserve the electromagnetic equations and total charge of the system and are independent of the electromagnetic field is the group of conformal maps of four-dimensional space.

Bateman was one of the first to apply Laplace transform methods to integral equations (1906), but he felt that he never received recognition for this. In 1910 he solved the system of ordinary differential equations arising from Rutherford's description of radioactive decay. From 1915 to 1926 Bateman worked on problems of electromagnetism and classical atomic models that were solved by the quantum theory (1926). His interests shifted to hydrodynamics and aerodynamics; in 1934 he completed a monumental 634-page report on hydrodynamics for the National Academy of Sciences.

In 1930 Bateman set out to find complete systems of fundamental solutions to the most important equations of mathematical physics, and he wrote a text describing many of the methods of solving these equations (1932). Much of the remainder of his life was dedicated to completing the task of collecting special functions and integrals that solve partial differential equations. He developed many of these, such as Bateman's expansion and Bateman's function. Bateman kept references to the various functions and integrals on index cards stored in shoe boxes—later in his life these began to crowd him out of his office. His memory for special facts was phenomenal. Mathematicians both telephoned and wrote to him, asking about particular integrals; and after consulting his files, Bateman supplied the questioner with formulas and extensive references. After his death the Office of Naval Research assembled a team of mathematicians headed by Arthur Erdelyi to organize and publish Bateman's manuscripts. Only parts of the resulting volumes on transcendental functions and integral transforms made extensive use of Bateman's files.

BIBLIOGRAPHY

For bibliographies of Bateman's books and papers and biographical material, see Arthur Erdelyi, "Harry Bateman 1882–1946," in *Obituary Notices of Fellows of the Royal Society,* **5** (London, 1948), 591–618; and F. D. Murnaghan, "Harry Bateman," in *Bulletin of the American Mathematical Society,* **54** (1948), 88–103. The volumes resulting from the Bateman Manuscript Project are Arthur Erdelyi, ed., *Higher Transcendental Functions,* 3 vols. (New York, 1953–1955), and *Tables of Integral Transforms,* 2 vols. (New York, 1954).

C. S. FISHER

BATES, HENRY WALTER (*b.* Leicester, England, 8 February 1825; *d.* London, England, 16 February 1892), *natural history.*

The eldest son of Henry Bates, a hosiery manufacturer at Leicester, Henry received his elementary education in the schools there before attending Mr. H. Screaton's boarding school at Billesden, a village about nine miles from Leicester. Although Bates was an excellent student, his formal education was terminated in midsummer 1838, and he was apprenticed to a local hosiery manufacturer, for whom he labored thirteen hours a day. His capacity for work was prodigious, however, and he also took night classes at the local Mechanic's Institute, where he excelled in Greek, Latin, French, drawing, and composition. (Later, while in South America, he taught himself German and Portuguese.) Throughout his life Bates read very widely and was particularly fond of reading Homer in the original and Gibbon's monumental *Decline and Fall of the Roman Empire.* Music also engaged some of his time: he sang in the local glee club, played the guitar, and maintained a strong interest in classical music throughout his life.

In addition to these numerous activities, Bates was an avid entomologist and, with his brother Frederick, scoured the woods of nearby Charnwood Forest for specimens during his holidays. His earliest scientific work, a short paper on beetles (1843), was published in the first issue of *The Zoologist* when he was eighteen years old.

His supervisor died several years before Bates's apprenticeship was completed, and Bates assumed management of the firm for the son before becoming a clerk in nearby Burton-on-Trent. He strongly disliked his work, however; entomology was a much more congenial occupation, and he habitually spent much of his time writing detailed accounts of his expeditions and captured treasures. In 1844 (or 1845) Bates befriended Alfred Russel Wallace, a mutually beneficial act that profoundly influenced both their lives. Wallace was then a master at the Collegiate School at Leicester and in his spare time enthusiastically pursued his own amateurish interests in botany. Bates introduced him to entomology, and the two friends continued to correspond and exchange specimens after Wallace moved from Leicester early in 1845.

While exchanging specimens in 1847, Wallace audaciously suggested to Bates that they should travel to the tropical jungles to collect specimens, ship them home for sale, and gather facts "towards solving the problem of the origin of species"—a frequent topic of their conversations and correspondence. Wallace's attention had been directed to South America by the vivid prose of William H. Edwards' *Voyage up the River Amazon, Including a Residence at Pará* (1847); conversations with the author increased their interest, as did Edward Doubleday of the British Museum, who showed them some exquisite new species of butterflies collected near Pará (Belém), Brazil, and offered other encouragement. Arrangements were soon made, and after a swift voyage of thirty-one days, the two amateur naturalists disembarked on 28 May 1848 at Pará, near the mouth of the Amazon River. Although Wallace returned to England in July 1852, Bates remained for a total of eleven years, exploring and collecting within four degrees of the equator. Frequently discouraged by his chronically destitute condition and his isolation, he was nevertheless held there by an intense passion for collecting: the "exquisite pleasure of finding another new species of these creatures supports one against everything." Bates conservatively estimated that he had collected 14,712 animal species (primarily insects) while in South America; more than 8,000 of these were new to science. Despite the richness of his collections, he received a profit of only about £800 for his efforts—or about £73 a year.

From Pará Bates traveled almost 2,000 miles deep into the wilderness. He resided in Pará for a total of nearly eighteen months, returning there periodically for a few months after each of his shorter excursions to the interior. On 26 August 1848 Bates and Wallace embarked on a journey up the Tocantins River; they arrived in October 1849 at what was to be Bates's headquarters for three years—Santarém, a small town of 2,500 inhabitants some 475 miles from the sea at the mouth of the Tapajós River. By mutual agreement Bates and Wallace parted company on 26 March 1850 at Manaus, the latter departing for the Rio Negro and the Uaupés. Manaus, at the confluence of the Rio Negro and the upper Amazon (Solimões), was a classic hunting ground for naturalists, having

been a favorite spot for the celebrated travelers J. B. von Spix and K. F. P. von Martius, who had stayed there in 1820. On 6 November 1851 Bates set out to explore the Tapajós and Solimões river basins, spending a total of seven and one-half years there. His headquarters during his four and one-half years in the Solimões area was at Ega (Tefé), at the foot of the Andes. In September 1857 he plunged deep into the wilderness to São Paulo, some 1,800 miles from Pará. Finally, on 11 February 1859, Bates left Ega for Pará and England, his chronically poor health at its nadir.

After arriving in England in the summer of 1859, Bates began work on his enormous collections under the influence of a specific biological concept. In the previous year the now famous papers of Charles Darwin had been presented to the Linnean Society of London, and in November 1859 Darwin published *On the Origin of Species by Means of Natural Selection.* Bates was an immediate convert, and had some substantial and impressive evidence of his own to contribute to Darwin's arguments. (Bates's Unitarian religious views did not hinder his acceptance of natural selection.) On 21 November 1861 Bates first expounded his ideas on mimicry in his famous paper "Contributions to an Insect Fauna of the Amazon Valley. *Lepidoptera: Heliconidae,*" which he read before the Linnean Society. With typical enthusiasm for works by his followers, Charles Darwin commented that Bates's article was "one of the most remarkable and admirable papers I have ever read in my life."

Responding to Darwin's exhortations, Bates published early in 1863 a two-volume narrative of his travels in South America, *The Naturalist on the River Amazon.* This splendid work was one of the finest scientific travel books of the nineteenth century. The *Naturalist* went through many editions and was translated into several languages; nevertheless, popular and remunerative as it was, Bates remarked that he would rather spend another eleven years on the Amazon than write another book.

In 1862 Bates failed to secure a position in the zoology department at the British Museum—a post that went instead to A. W. E. O'Shaughnessy, a poet who, Bates said, was probably sponsored by Richard Owen. In 1864, however, Bates was appointed assistant secretary of the Royal Geographical Society of London, serving with distinction for twenty-eight years. Besides editing the *Journal* and *Proceedings* of the Society, as well as carrying on an immense correspondence with travelers and others throughout the world, Bates actually managed the Society and made arrangements for various meetings, including those held by the Geographical Section of the British Association.

During his tenure as assistant secretary, Bates's published works were devoted almost exclusively to entomology, primarily systematics, and numerous editions of travel works, such as Peter E. Warburton's *Journey Across the Western Interior of Australia* (1875), Thomas Belt's *The Naturalist in Nicaragua* (1873), and six volumes of Cassell's *Illustrated Travels: A Record of Discovery, Geography and Adventure* (1869–1875). His greatest contribution to systematic entomology appeared in various volumes of the *Biologia Centrali-Americana,* edited by F. D. Godman and O. Salvin (see his works on Coleoptera: Adephaga, Pectinicornia [Passalidae and Lucanidae], Lamellicornia [Scarabaeidae], and Longicornia [Cerambycidae]), and his fame as a preeminent authority on Coleoptera was worldwide. He contributed more than one hundred scientific papers to scholarly journals, corresponded with many entomologists, and served as consultant or assistant to various scientific journals, including the *The Entomologist.* (He also published anonymously purely literary works, and was at one time on the staff of the *Atheneum.*)

Many honors were bestowed on Bates, although he was strongly disinclined to discuss them. In 1861 he was elected to the Entomological Society of London and served as president in 1868, 1869, and 1878. He was elected fellow of the Linnean Society in 1871, fellow of the Zoological Society in 1863, and fellow of the Royal Society in 1881. Although he was secretary of the Geographical Section of the British Association, he declined the office of president. For his work in Brazil, the emperor of that country bestowed on Bates the Order of the Rose, a distinction rarely conferred upon foreigners. In 1863 he married Sara Ann Mason of Leicester, who bore him one daughter and three sons, two of whom emigrated to New Zealand.

That Bates was an immediate, early convert to Darwinian natural selection is understandable, for he had long been an evolutionist, as had his friend and traveling companion Alfred Russel Wallace. However, Bates appears not to have been converted as quickly as Wallace was by the evolutionary explanations adduced by Robert Chambers in his heretical, but quite popular, *Vestiges of the Natural History of Creation* (1844 *et seq.*). In a letter to Wallace in 1845, Bates apparently had described Chambers' ideas as hasty generalizations, a charge that Wallace felt compelled to answer at some length. By March 1850,

when the two parted company to explore by themselves, they had often discussed the species question and had reached some tentative conclusions. Nevertheless, Bates expressed surprise at Wallace's paper "On the Law Which Has Regulated the Introduction of New Species."[1] In this important work Wallace argued that every species has arisen "coincident in both space and time with a pre-existing closely allied species," thus supporting species creation through natural laws—as opposed, by implication, to numerous special creations by God.

Bates wrote Wallace from the Amazon that he was at first startled to see that he was "already ripe for the enunciation of the theory," commenting, however, "the theory I quite assent to, and, you know, was conceived by me also, but I profess that I could not have propounded it with so much force and completeness." Wallace replied from the Malay Archipelago that his paper naturally would appear more clear to Bates than "to persons who have not thought much on the subject," adding that the "paper is, of course, only the announcement of the theory, not its development. . . ."[2] Precisely what Bates meant by "the theory . . . was conceived by me also" is a matter for conjecture, for Wallace apparently was the earlier convert to the evolutionary hypothesis and, in fact, pleaded the evolutionist's case quite forcefully to Bates. Furthermore, Wallace certainly had a far more pregnant imagination than Bates, and it was he who perceived in 1858 the solution to the problem. Bates's original contribution to evolutionary biology appeared shortly after the *Origin* was published and fully corroborated Darwin's theory of natural selection.

Before 1861, naturalists had observed that certain butterflies belonging to distinct groups inhabiting the same geographical area bear remarkable superficial resemblances in appearance, shape, and color. In 1836 J. A. Boisduval had described the African swallowtailed butterfly, which was copied by quite distinct butterflies. The distinguished English entomologists William Kirby and William Spence had observed in 1817 that the flies of the genus *Volucella* enter bees' nests to deposit their eggs so that their larvae may feed on the bee larvae and that the flies curiously resemble the bees on which they are parasitic. They concluded that the resemblance exists to protect the flies from possible attacks by the bees. In an article published in 1861, the German entomologist A. Rössler enumerated many examples of mimicry, which he explained as a device to protect insects from their enemies. In general, however, belief in the handiwork of God sufficed to explain the phenomenon, although some thought that there might be an innate tendency in insects to vary in a particular direction.

Bates was the first naturalist to venture a comprehensive scientific explanation for the phenomenon that he labeled "mimicry" (Batesian mimicry). Batesian mimicry should here be differentiated from Müllerian mimicry. According to Cott:

> In Batesian mimicry a relatively scarce, palatable, and unprotected species resembles an abundant, relatively unpalatable or well-protected species and so becomes disguised. In Müllerian mimicry, on the other hand, a number of different species all possessing aposematic attributes and appearance, resemble one another, and so become easily recognized [*Adaptive Coloration in Animals,* p. 398].

One therefore leads to deception of enemies and the other to the education of enemies by warning colors. Bates, and others, at first confused the two kinds of mimicry.

Bates's discussion of mimicry was unobtrusively buried in his classic article on the heliconid butterflies of the Amazon, which were frequently mimicked by counterfeits so perfect that even Bates was unable to distinguish them in flight. The Leptalides. (*Dismorphia*) butterflies, although quite different structurally from the Heliconidae, are especially proficient mimics. Other examples abound. In various parts of the world beetles, spiders, flies, and grasshoppers mimic ants. While spiders may have a body configuration resembling that of ants, other mimics may use optical illusions to produce the appearance of a narrow antlike waist. Certain bees on the banks of the Amazon, as Bates observed, are also mimicked; indeed, many moths and longicorn beetles in the tropics mimic bees, wasps, and other hymenopterous insects. (Bates also adduced numerous examples of insects imitating inanimate objects.) In every case the mimic benefited to some extent from protection afforded by the original. The Heliconidae, which fly slowly and in plain sight of many potential enemies, are rendered foul-smelling and unpalatable by a glandular secretion. Birds and reptiles, to a lesser extent than insects, also mimic protected species. Bates observed an extraordinary example of a very large caterpillar that imitated a small, venomous snake. The first three segments behind the head could be dilated at will by the insect, and on each side of its head was a large pupillated spot resembling the eye of the poisonous snake. The general consternation produced among the natives when Bates displayed his specimen attests to the excellence of the imitation.

Granted that mimics are adaptations to their environment, the important question was why such remarkably close analogies exist. Bates ruled out direct action of physical conditions because in limited districts where these conditions were the same, the most widely contrasting varieties may be found coexisting. Likewise, sports (mutations) did not explain mimicry. To Bates it was quite clear that natural selection had produced these phenomena, "the selecting agents being insectivorous animals which gradually destroy those sports or varieties which are not sufficiently like [the protected species] to deceive them." The closer the resemblance of the mimic to the original, the greater will be its protection. Imperfect copies will be eliminated slowly, unless they have some supplementary protection of their own. (In fact, as V. Grant observed, "most animals in general have several alternative means of defense against their enemies" [*The Origin of Adaptations,* p. 112].) By observing the different forms of the mimic as it approximated (or diverged from) the original, species change itself could be carefully observed: "Thus, although we are unable to watch the process of formation of a new race as it occurs in time, we can see it, as it were, at one glance, by tracing the changes a species is simultaneously undergoing in different parts of the areas of distribution."[3] As Bates so aptly observed, it was "a most beautiful proof of natural selection."

Darwin was thoroughly delighted with Bates's paper, for it fully corroborated his theory and presented him with an excellent opportunity to rebut his critics in a short, unsigned review in the *Natural History Review* for 1863. He confronted the creationists with the embarrassing case of the mimicking forms of Leptalides (*Dismorphia*) butterflies, which could be shown through a graduated series to be varieties of one species; other mimickers clearly were distinct varieties, species, or genera. To be logically consistent—and they were not always—the creationists had to admit that some mimickers had been formed by commonly observed variations, while others were specially created. They would further be required to admit that some of these forms were created in imitation of forms known to arise through the ordinary processes of variation. The difficulties of such a position were insurmountable, and the arguments for Batesian mimicry were widely accepted. Moreover, A. R. Wallace soon extended these arguments in two excellent articles: "On the Phenomena of Variation and Geographical Distribution as Illustrated by the *Papilionidae* of the Malayan Region" (1865) and "Mimicry and Other Protective

Resemblances Among Animals" (1867). Thereafter, the literature was rich with references to mimicry, reaching a high point toward the end of the century with extensive discussions by Wallace, Poulton, and Beddard.

On the other hand, Bates, the author of the theory of mimicry, never published an extensive review of the subject, nor did he ever again write a philosophical or interpretative paper comparable to his article on mimicry. Unable to spin hypotheses as easily as did Darwin and Wallace, he devoted himself instead to numerous works on systematic entomology, such as his catalog of the Erycinidae (Riodinidae) butterflies, the foundation upon which subsequent authors worked. After selling his collections of Lepidoptera, Bates concentrated on Coleoptera, especially Adephaga, Lamellicornia and Longicornia.

That Bates failed to produce further works comparable to those on mimicry and his *Travels* has generally been attributed to the press of his many duties at the Royal Geographical Society. However, the character of his later work on entomological systematics may have been strongly influenced by other factors as well. In his presidential address to the Entomological Society in 1878, Bates discussed the "prevailing exclusively descriptive character of the entomological literature of the day," which he thought resulted primarily from the difficulty of describing the prodigious influx of newly discovered species. Nature was proving to be far more prolific, and her products more varied, than had previously been thought. "Thus our best working Entomologists are led to abandon general views from lack of time to work them out, and the consciousness that general views on the relations of forms and faunas are liable to become soon obsolete by rapid growth of knowledge." This, he felt, did not totally excuse the systematists' excessive narrowness, for it led them to neglect natural affinities that greatly illuminated evolution, which in his opinion was the greatest problem in biology. While Bates himself searched for natural affinities and focused attention on the important problems of the geographical distribution of animals, his general observation about systematists may also have applied to him, as his friend D. Sharp observed. Nevertheless, his erudite papers formed the substance for those who were able to develop theories, while his paper on mimicry and his *Travels* stand as classics in their own right.

NOTES

1. *Annals and Magazine of Natural History,* 2nd ser., **16** (1855), 184–196.

2. Bates to Wallace, 19 November 1856, and Wallace to Bates, 4 January 1858, in Marchant, *Alfred Russel Wallace: Letters and Reminiscences,* pp. 52–55.
3. "Contributions to an Insect Fauna . . .," in *Transactions of the Linnean Society of London,* **23** (1862), 512–513.

BIBLIOGRAPHY

I. ORIGINAL WORKS. Bates's best-known works are "Contributions to an Insect Fauna of the Amazon Valley. *Lepidoptera: Heliconidae,*" in *Transactions of the Linnean Society of London,* **23** (1862), 495–566, in which he announced his theory of mimicry, and *The Naturalist on the River Amazon,* 2 vols. (London, 1863; repr. London, 1892). Many other contributions to the insect fauna of the Amazon appeared in entomological journals. His presidential addresses to the Entomological Society of London, in its *Proceedings* for 1868, 1869, and 1878, are valuable for understanding his general opinions on entomology and biology.

II. SECONDARY LITERATURE. There are a number of informative obituaries on Bates; his neighbor and friend Edward Clodd had direct access to his papers (which have disappeared), which makes his account particularly valuable. See his memoir of Bates prefaced to the 1892 reprint of the first edition of Bates's *Travels.* Other obituaries are R. McLachlan, "Obituary of Henry Walter Bates, F. R. S.," in *The Entomologist's Monthly Magazine,* 2nd ser., **3** (1892), 83–85; *Proceedings of the Royal Geographical Society,* **14** (1892), 245–257; D. Sharp, "Henry Walter Bates," in *The Entomologist,* **25** (1892), 77–80; and Alfred Russel Wallace, "H. W. Bates, the Naturalist of the Amazons," in *Nature,* **45** (1892), 398–399.

Except for letters to Darwin at Cambridge, scarcely any of Bates's correspondence remains. Fortunately, however, some of his important letters have been published in the following works: Francis Darwin, ed., *The Life and Letters of Charles Darwin,* II (London, 1887), 378–381, 391–393; Francis Darwin and A. C. Seward, eds., *More Letters of Charles Darwin,* Vol. I (New York, 1903), *passim;* James Marchant, ed., *Alfred Russel Wallace: Letters and Reminiscences* (New York, 1916), pp. 52–59; and A. R. Wallace, *My Life: A Record of Events and Opinions,* I (London, 1905), 350–379.

On mimicry see Frank E. Beddard, *Animal Coloration* (London, 1892); Hugh B. Cott, *Adaptive Coloration in Animals* (London, 1940, 1957); Verne Grant, *The Origin of Adaptations* (New York–London, 1963); E. B. Poulton, *The Colours of Animals, Their Meaning and Use* (London, 1890); Charles L. Remington, "Mimicry, a Test of Evolutionary Theory," in *Yale Scientific Magazine,* **22**, no. 1 (October 1957), 10–11, 13–14, 16–17, 19, 21, written with Jeanne E. Remington; and "Historical Backgrounds of Mimicry," in *Proceedings of the XVI International Congress of Zoology,* IV (Washington, D. C., 1963), 145–149; and A. R. Wallace, *Darwinism* (London, 1889), pp. 232–267.

H. LEWIS MCKINNEY

BATESON, WILLIAM (*b.* Whitby, England, 8 August 1861; *d.* Merton, London, England, 8 February 1926), *morphology, genetics.*

Bateson was the son of William Henry and Anna Aiken Bateson. His father, a classical scholar, had become master of St. John's College, Cambridge (1857), and Bateson lived in that university town until 1910. In that year he moved to London and remained there until his death.

Marked as "a vague and aimless boy" while at Rugby, Bateson confounded the doubters by taking first-class honors in the natural science tripos at Cambridge, from which he received the B.A. in 1883. A student and later (1885–1910) a fellow of St. John's College, his interest early came to focus on zoology and morphology. Contributing largely to his scientific development were A. Sedgwick and W. F. R. Weldon at Cambridge and, while he was in Maryland and Virginia during 1883 and 1884, W. K. Brooks at Johns Hopkins. Bateson was ill-trained in physics and chemistry and knew virtually no mathematics. His abilities as a classicist were outstanding.

Bateson's career may be divided into three periods. During the first period (1883–1900) he turned from orthodox embryological Darwinizing to the rigorous study of heredity and variation. Dissatisfied with old truths, he was seeking a new approach to evolutionary studies. His views were distinctly out of favor with conventional zoologists, and he won no regular university teaching appointments. Bateson's Mendelian period opened in 1900 and continued until about 1915. His contributions toward the establishment of the Mendelian conception of heredity and variation were enormous, and he finally began to harvest popular and professional esteem. The third period (*ca.* 1915–1926) reintroduced the discordant element. Bateson continually attacked the new glory, chromosome theory, and concentrated upon problems of somatic segregation. Once a radical, furiously disturbing his zoological elders, he now seemed conservative. His science, genetics (so named by Bateson in 1905–1906), turned largely toward chromosomes and genes. As a result, Bateson, who had been a central figure, became isolated from current developments.

Bateson received the Balfour Studentship at Cambridge in 1887. His other Cambridge positions included deputy in zoology to Alfred Newton in 1899, reader in zoology in 1907, and finally professor of genetics in 1908. His was the first such chair in Britain. In 1910 he became the first director of the John Innes Horticultural Institution, then at Merton. He was elected a fellow of the Royal Society in 1894 and belonged to numerous other societies. Among his many honors were the Darwin Medal (1904), presi-

dency of the British Association for the Advancement of Science (1914), the Royal Medal (1920), and election as a trustee of the British Museum (1922).

Bateson's Cambridge training had emphasized genealogical reconstruction. Such work promised the recovery of a generalized view of the history of life, but it could tell very little about how such great changes had actually come to pass. True progress in evolutionary studies must, Bateson believed, reject customary phylogenetic ambitions and turn to the process wherein evolutionary novelty emerges, that is, the phenomena of heredity and, particularly, variation. From about 1887 such detailed investigation dominated Bateson's activity. He searched the literature and began experimental hybridization, demanding exact knowledge of the transmission of heritable features from parent to immediate offspring. Then, in May 1900, he read Mendel's report of 1866 on breeding experiments with peas.

The rediscovery of Mendel's work transformed Bateson's career. He reinterpreted experimental data already available in Mendelian terms. Simultaneously with L. Cuénot he proved that Mendelian behavior holds for animals as well as plants. His research group, which included R. C. Punnett, E. R. Saunders, and L. Doncaster, demonstrated the existence of epistatic modifications of simple Mendelian ratios and the phenomena of reversion and (as it was subsequently designated) linkage. As spokesman for the new discipline Bateson knew no peer, yet his work was financed with great difficulty and only by personal funds, private gifts, and the Evolution Committee of the Royal Society. Reports to the latter body were the main vehicle of Mendelian publication in Britain until Bateson and Punnett began the *Journal of Genetics* in 1910.

Behind all technical achievements persisted Bateson's concern for the "salt of biology," evolution theory, and its indispensable adjunct, the determinative mechanism(s) of inheritance. The chromosome theory of inheritance, rising from bare enunciation in 1902 to evident triumph around 1920, offered a formed, material element in the cell nucleus as the agent of heredity. But Bateson demanded a direct—indeed, a visible—correlation between any chromosome and consequent bodily feature, and he found none. Moreover, chromosome theory was patently materialistic, and this, for numerous and complex reasons, he could not tolerate. In place of the chromosome theory Bateson devised an ostensibly nonmaterialistic vibratory theory of inheritance, founded on force and motion. This theory and the phenomena that appeared consistent with it (especially plant variegation) were utterly unacceptable to most geneticists, and in its defense Bateson became increasingly an advocate without adversaries. Significant discussion of his views largely disappeared from leading genetics journals.

An ardent evolutionist, Bateson vigorously opposed the doctrine of natural selection. Like Fleeming Jenkin and St. G. Mivart, he failed to see how few and slight variations, the raw material for Darwinian selection, could be truly useful. He rejected the entire "utility" aspect of orthodox Darwinism. What, then, could be the source of undeniable evolutionary change? The answer was found in the organismic quality of any significant biological variation. Genuine evolutionary novelty was, therefore, not the product of gradual accumulation of seemingly trivial variations but the consequence of saltatory or major modifications. Variation was the driving force of evolutionary change, and Bateson always hoped to find its explanation in his vibratory theory of heredity. Perhaps the disputatious Bateson's most famous public battle was against the biometrical school led by W. F. R. Weldon and Karl Pearson. At issue was the nature of evolutionary change and how it might best be studied. Bateson emphasized the imprecision in individual instances of statistical analyses and decried the biometricians' ardent defense of natural selection. In straightforward Mendelism he found both support for evolutionary discontinuities and a dependable experimental approach to matters of descent.

Bateson's broad interests and remarkable personality greatly influenced his scientific thought. A man of deep and well-cultivated aesthetic sensibilities (he was a major collector of Oriental works of art and the creations of William Blake), he despaired before the harsh spiritual and economic realities of late Victorian England. He was at once a radical empiricist, wanting his facts absolutely secure, and a rather unconscious intuitionist. From his intuition, fed by insights taken from transcendental morphology, he developed his conception of the functionally integral organism. The same view informs his conception of human society, in which an intellectual elite is struggling desperately against a rising and insensitive middle-class horde bent on leveling all men to their own mediocrity. To Bateson this was not life, but mere existence. This complex of ideas and intuitions set Bateson firmly against utilitarian rationalism and its near neighbor, materialism. In natural selection and the chromosome theory he saw both at work, and would have none of them.

BIBLIOGRAPHY

I. ORIGINAL WORKS. With few and unimportant exceptions Bateson's scientific contributions were collected and reissued by R. C. Punnett as *Scientific Papers of William Bateson,* 2 vols. (Cambridge, 1928). His often remarkable general biological and social essays are available in B. Bateson, ed., *William Bateson, F.R.S. Naturalist. His Essays and Addresses* (Cambridge, 1928). Bateson's major publications are *Materials for the Study of Variation Treated with Especial Regard to Discontinuity in the Origin of Species* (London, 1894); *Reports to the Evolution Committee of the Royal Society, I–V* (London, 1902–1909); *Mendel's Principles of Heredity. A Defence* (Cambridge, 1902); *Mendel's Principles of Heredity* (Cambridge, 1909, 1913); and *Problems of Genetics* (New Haven, 1913). Punnett has provided a bibliography of Bateson's writings in *Scientific Papers,* II, 487–494.

The Bateson papers, in the possession of his family, constitute an extensive and exceptionally valuable archive. A comprehensive selection of scientific items from this collection has been microfilmed (a copy is deposited in the Library of the American Philosophical Society, Philadelphia) and roughly catalogued. All aspects of Bateson's career are represented, including his large and interesting correspondence, unpublished speculative essays, and abundant documentation to published works. These papers are described in *The Mendel Newsletter,* no. 2 (Philadelphia, 1968). Lesser collections of Bateson MSS are held by the John Innes Institute (to join the University of East Anglia, Norwich, *ca.* 1970), the Royal Society of London, and the American Philosophical Society.

II. SECONDARY LITERATURE. The principal biographical source is B. Bateson, "Memoir," in *William Bateson, F.R.S. Naturalist,* pp. 1–160. Other notices, particularly concerned with his scientific endeavor, are R. C. Punnett, "William Bateson," in *Edinburgh Review,* **244** (1926), 71–86; and T. H. Morgan, "William Bateson," in *Science,* **63** (1926), 531–535. See also O. Renner, "William Bateson and Carl Correns," in *Sitzungsberichte der Heidelberger Akademie der Wissenschaften,* Mathematische-naturwissenschaftliche Klasse, **6** (1960–1961), 159–184; L. C. Dunn, *A Short History of Genetics* (New York, 1965), pp. 55–87; and C. D. Darlington, *The Facts of Life* (London, 1953), pp. 95–116. An excellent discussion of Bateson's peculiar later conception of the heredity mechanism is R. C. Swinburne, "The Presence-and-Absence Theory," in *Annals of Science,* **18** (1962), 131–145. Bateson's refusal to pursue phylogenetic reconstruction is recorded in W. Coleman, "On Bateson's Motives for Studying Variation," in *Actes du XIᵉ Congrès International d'Histoire des Sciences,* V (Warsaw-Cracow, 1968), 335–339.

WILLIAM COLEMAN

BATHER, FRANCIS ARTHUR (*b.* Richmond, Surrey, England, 17 February 1863; *d.* Wimbledon, England, 20 March 1934), *paleontology.*

Francis Arthur Bather, the son of Arthur H. and Lucy Bloomfield Bather, attended Winchester School and New College, Oxford, where he earned the bachelor of arts degree in 1886. Bather received the master of arts degree in 1890 and the doctor of science in 1900, both from Oxford. In 1887 he took a position as assistant geologist in the British Museum, and in 1924 he was promoted to keeper of the geology department. Bather enjoyed his curatorial duties and took great interest in museum administration, finding ways to improve exhibits, corresponding with other curators around the world, and visiting natural history museums on the Continent.

In addition to his curatorial work, Bather carried out research in paleontology and contributed many fundamental studies on echinoderm morphology. Among the echinoderms the Pelmatozoa were his specialty. In 1893 his *Crinoidea of Gotland* (part 1) was published by the Royal Swedish Academy. His *The Echinoderma,* one volume in E. R. Lankester's *Treatise of Zoology* (1900), remained the leading exposition of echinoderm morphology for decades, with new discoveries fitting easily into the scheme that Bather established. Intensive study of fossil morphology did not hinder Bather's appreciation of the environment and an animal's place in it. The study of paleoecology interested him, and he particularly recommended this study to others. In such ways he brought biological concepts to bear on fossil studies. Although Bather began his museum work nearly thirty years after the publication of Darwin's *Origin of Species,* the classification of echinoderms had been untouched by evolutionary ideas.

Bather retired from the British Museum in 1928. Among the many honors he received for his scientific contributions were the Rolleston Prize of Oxford and Cambridge (1892) and the Wollaston Fund (1897) and the Lyell Medal (1911) of the Geological Society of London. He also served as president of the Geological Society, the Museums Association, and the British Association for the Advancement of Science.

BIBLIOGRAPHY

In addition to the works cited above, see Bather's presidential address to the Geological Society, "Biological Classification: Past and Future," in *Quarterly Journal of the Geological Society of London,* **83** (1927), lxii–civ. Besides his general views, the address contains a lucid historical review of classificatory schemes. Further biographical material is to be found in memoirs by W. D. Lang, in *Nature,* **133** (1934), 485–486; and Percy E. Raymond, in *Proceedings of the Geological Society of America* (June 1935), pp. 173–186. The latter contains a complete bibliography of Bather's writings, compiled by Thomas H. Withers.

BERT HANSEN

AL-BATTĀNĪ,[1] **ABŪ ʿABD ALLĀH MUḤAMMAD IBN JĀBIR IBN SINĀN AL-RAQQĪ AL-ḤARRĀNĪ AL-ṢĀBIʾ,** also **Albatenius, Albategni** or **Albategnius** in the Latin Middle Ages; *astronomy, mathematics.*

One of the greatest Islamic astronomers, al-Battānī was born before 244/858,[2] in all probability at or near the city of Ḥarrān (ancient Carrhae) in northwestern Mesopotamia, whence the epithet al-Ḥarrānī. Of the other two epithets, al-Raqqī, found only in Ibn al-Nadīm's *Fihrist,*[3] refers to the city of al-Raqqa, situated on the left bank of the Euphrates, where al-Battānī spent the greater part of his life and carried out his famous observations; al-Ṣābiʾ indicates that his ancestors (al-Battānī himself was a Muslim; witness his personal name Muḥammad and his *kunya* Abū ʿAbd Allāh) had professed the religion of the Ḥarranian Ṣabians,[4] in which a considerable amount of the ancient Mesopotamian astral theology and star lore appears to have been preserved and which, tolerated by the Muslim rulers, survived until the middle of the eleventh century. The fact that al-Battānī's elder contemporary, the great mathematician and astronomer Thābit ibn Qurra (221/835–288/901) hailed from the same region and still adhered to the Ṣabian religion, seems indicative of the keen interest in astronomy that characterized even this last phase of Mesopotamian star idolatry. As for the cognomen (*nisba*) *al-Battānī,* no reasonable explanation of its origin can be given. Chwolsohn's conjecture[5] that it derives from the name of the city of Bathnae (or Batnae; Gr., Βάτναι; Syr., Baṭnān) near the ancient Edessa, was refuted by Nallino[6] with the perfectly convincing argument that the possibility of a transition of Syriac *t* into Arabic *t* (Baṭnān to Battān) has to be strictly excluded; since there is no evidence of the existence of a city or town named Battān, Nallino suggests that this name, rather, refers to a street or a district of the city of Ḥarrān.

Nothing is known about al-Battānī's exact date of birth and his childhood. Since he made his first astronomical observations in 264/877, Nallino is on safe ground assuming the year 244/858 as a *terminus ante quem* for his birth. His father, in all probability, was the famous instrument maker Jābir ibn Sinān al-Ḥarrānī mentioned by Ibn al-Nadīm,[7] which would explain not only the son's keen astronomical interest but also his proficiency at devising new astronomical instruments, such as a new type of armillary sphere.

On al-Battānī's later life too the information is scanty. According to the *Fihrist*[8] and to Ibn al-Qifṭī's *Taʾrīkh al-Ḥukamāʾ,*[9] al-Battānī was

> . . . one of the illustrious observers and foremost in geometry, theoretical and practical [lit., computing]

astronomy, and astrology. He composed an important *zīj* [i.e., work on astronomy with tables] containing his own observations of the two luminaries [sun and moon] and an emendation of their motions as given in Ptolemy's *Almagest*. In it, moreover, he gives the motions of the five planets in accordance with the emendations which he succeeded in making, as well as other necessary astronomical computations. Some of the observations mentioned in his *Zīj* were made in the year 267 H.[10] [A.D. 880] and later on in the year 287 H. [A.D. 900]. Nobody is known in Islam who reached similar perfection in observing the stars and in scrutinizing their motions. Apart from this, he took great interest in astrology, which led him to write on this subject too; of his compositions in this field [I mention] his commentary on Ptolemy's *Tetrabiblos*.

He was of Ṣabian origin and hailed from Ḥarrān. According to his own answer to Jaᶜfar ibn al-Muktafī's question, he set out on his observational activity in the year 264 H. [A.D. 877] and continued until the year 306 H. [A.D. 918]. As an epoch for his [catalog of] fixed stars in his *Zīj* he chose the year 299 H. [A.D. 911].[11]

He went to Baghdad with the Banu'l-Zayyāt, of the people of al-Raqqa, on account of some injustice done them.[12] On his way home, he died at Qaṣr al-Jiṣṣ in the year 317 H. [A.D. 929].[13]

He wrote the following books: *Kitāb al-Zīj* [*Opus astronomicum*], in two recensions;[14] *Kitāb Maṭāliᶜ al-Burūj* ["On the Ascensions of the Signs of the Zodiac"];[15] *Kitāb Aqdār al-Ittiṣālāt* ["On the Quantities of the Astrological Applications"], composed for Abu'l-Ḥasan ibn al-Furāt; *Sharḥ Kitāb al-Arbaᶜa li-Baṭlamiyūs* ["Commentary on Ptolemy's *Tetrabiblos*"].[16]

It seems to have been a widespread belief among Western historians that al-Battānī was a noble, a prince, or even a king of Syria. Not the slightest allusion to it can be found in Arabic writers, so the source of this misunderstanding must be sought in Europe. The earliest reference quoted by Nallino[17] is Riccioli's *Almagestum novum*,[18] where al-Battānī is called "dynasta Syriae." J. F. Montucla[19] makes him a "commandant pour les califes en Syrie" and J. LaLande[20] a "prince arabe," as does J.-B. Delambre,[21] probably on LaLande's authority, since he expressly says that he used a copy, formerly in LaLande's possession, of the 1645 Bologna edition of al-Battānī's *Zīj*, although its title contains no reference to the author's alleged nobility.[22]

From al-Battānī's work, only one additional fact on his life can be derived: he mentions in his *Zīj*[23] that he observed two eclipses, one solar and one lunar, while in Antioch, on 23 January and 2 August A.D. 901, respectively.

The book on which al-Battānī's fame in the East and in the West rests is the *Zīj*, his great work on

astronomy. Its original title, in all probability, was that indicated by Ibn al-Nadīm and Ibn al-Qifṭī: *Kitāb al-Zīj*, or just *al-Zīj*. Later authors also often call it *al-Zīj al-Ṣābiʾ* ("*The Ṣabian Zīj*").[24] The word *zīj*, derived from the Middle Persian (Pahlavi) *zīk* (modern Persian, *zīg*), originally meant the warp of a rug or of an embroidery. As Nallino points out,[25] by the seventh century this had become a technical term for astronomical tables. In Arabic, it soon assumed the more general meaning "astronomical treatise," while for the tables themselves the word *jadwal* ("little river") came into use.[26]

Of the two recensions mentioned by Ibn al-Qifṭī, the first must have been finished before 288/900 because Thābit ibn Qurra, who died in February 901, mentions one of its last chapters.[27] Since the manuscript preserved in the Escorial[28] and the Latin version by Plato of Tivoli (Plato Tiburtinus) contain the two observations of eclipses mentioned above, the first of which occurred immediately before, and the second six months after, Thābit's death, Nallino concludes[29] that they must both have been copied (or translated) from the second recension.

In the preface to the *Zīj*,[30] al-Battānī tells us that errors and discrepancies found in the works of his predecessors had forced him to compose this work in accordance with Ptolemy's admonition to later generations to improve his theories and inferences on the basis of new observations, as he himself had done to those made by Hipparchus and others.[31] The Arabic version of the *Almagest*, on which he relied, seems to have been a translation from the Syriac, which Nallino shows on several occasions was not free from errors. All quotations from the *Almagest* are carefully made and can be verified.

A comparison of the *Zīj* with the *Almagest* at once reveals that it was far from al-Battānī's mind to write a new *Almagest*. To demonstrate this, it suffices to point out a few striking differences:

The arrangement of the fifty-seven chapters is dictated by practical rather than by theoretical considerations. Thus, contrary to al-Farghānī, who, writing half a century before al-Battānī, devotes his nine first chapters[32] to the same questions that are treated in *Almagest* I, 2–8 (spherical shape of the heavens and of the earth; reasons for the earth's immobility; the earth's dimensions and habitability; the two primary motions; etc.), al-Battānī starts his *Zīj* with purely practical definitions and problems: the division of the celestial sphere into signs and degrees, and prescriptions for multiplication and division of sexagesimal fractions. In chapter 3, corresponding to *Almagest* I, 11, he develops his theory of trigonometrical functions (see below); in chapter 4 he presents his own observations that resulted in a value for the obliquity of the ecliptic (23°35') that is more than 16' lower than Ptolemy's (23°51'20"; *Almagest* I, 12);[33] the next chapters (5–26), corresponding roughly to *Almagest* I, 13–16 and the whole of Book II, contain a very elaborate discussion of a great number of problems of spherical astronomy, many of them devised expressly for the purpose of finding solutions for astrological problems.

The Ptolemaic theory of solar, lunar, and planetary motion in longitude is contained in chapters 27–31. Then follows a discussion of the different eras in use and their conversion into one another (chapter 32),[34] serving as an introduction to the next sixteen chapters (33–48), in which detailed prescriptions for the use of the tables are given (chapters 39 and 40 deal with the theory of lunar parallax and the moon's distance from the earth, necessary for the computation of eclipses). Chapters 49–55 treat of the chief problems in astrology: chapter 55 has the Arabic title "Fī maʿrifat maṭāliʿ al-burūj fī-mā bayna 'l-awtād fī arbāʿ al-falak" ("On the Knowledge of the Ascensions of the Signs of the Zodiac in the Spaces Between the Four Cardinal Points of the Sphere"),[35] which is identical with Ibn al-Nadīm's title of one of al-Battānī's minor works. It is possible that this chapter actually existed as a separate treatise, but it is also possible that it was only due to an error that we find it listed separately in the *Fihrist* and in later biographies.

Of the two last chapters, 56 deals with the construction of a sundial indicating unequal hours (*rukhāma*, "marble disk"), and 57 with that of a novel type of armillary sphere, called *al-bayḍa* ("the egg"), and of two more instruments, a mural quadrant and a *triquetrum* (Ptolemy's τεταρτημόριον, *Almagest* I, 12, and ὄργανον παραλλακτικόν, *Almagest* V, 12).

Contrary to Ptolemy's procedure in the *Almagest,* the practical aspect of the *Zīj* is so predominant that it sometimes impairs the clarity of exposition and even evokes a totally wrong impression. This is felt more than anywhere else in chapter 31, which deals with the theory of planetary motion. The Arabic text consists of little more than five pages, only three of which deal with the theoretical (i.e., kinematic) aspect of the problem. Here the reader trained in Ptolemy's careful and sometimes slightly circumstantial way of exposing his arguments, and familiar with al-Farghānī's excellent *epitome,* will needs be struck by the brevity and—this is worse—by the insufficiency and inaccuracy of al-Battānī's outline. To point out some particularly bewildering features:[36] No distinction is made between, on the one hand, the theory of the three superior planets and Venus, and, on the other, the ingenious and intricate mechanism devised by

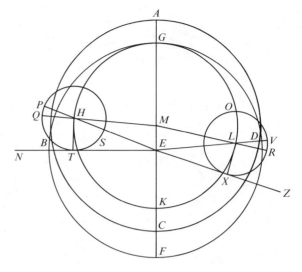

FIGURE 1

Ptolemy to represent Mercury's perplexing motion. With the aid of only one figure, which is wholly defective and misleading, al-Battānī tries—and of course fails—to demonstrate the motion of all of the five planets alike. In this figure, reproduced here (Fig. 1), the equant (*punctum aequans*), the essential characteristic of the Ptolemaic theory, is not indicated, nor is it referred to in the accompanying text, according to which the center, *M*, of the deferent itself is to be regarded as the center of mean motion (!). Moreover, the nodes of the planetary orbit are placed at right angles to the line of apsides (which of course is not true of any of the five planets), and for the planet's position in the epicycle, in the two cases indicated in the figure, the very special points are chosen in which the line earth-planet is tangent to the epicycle.

It is easy to point out all these errors and, as G. Schiaparelli has done at Nallino's request,[37] to show how the figure ought to look, were it drawn in accordance with Ptolemy's theory (Fig. 2). But Schiaparelli's surmise that al-Battānī's correct figure was distorted by some unintelligent reader or copyist does not exhaust the question. For, if so, who would dare at the same time to mutilate the text in such a way that the equant disappears from it altogether? And which uninitiated reader might have had the courage to suppress in this context the theory of Mercury, without which Ptolemy's system of planetary motions remains a torso? Since the Escorial manuscript and Plato of Tivoli's translation[38] have both the same erroneous figure and text, the alleged mutilation must have occurred, at the latest, in the eleventh century, during the lifetime of the great Spanish-Muslim astronomer al-Zarqālī (Azarquiel) or of one of his renowned

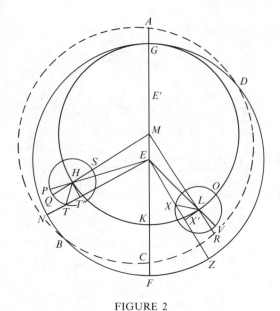

FIGURE 2

predecessors. To me it seems utterly improbable that an arbitrary disfiguring of one of the crucial chapters of al-Battānī's great *Zīj* could have escaped their attention and that no measures should have been taken to delete such faulty copies and to restore the original text. I am inclined, thus, to ascribe the matter to al-Battānī's carelessness rather than to anything else, in view of the circumstance that no other reasonable conjecture seems to square with the facts. Similar examples can be adduced from the writings even of the greatest astronomers; in this context I refer to a grave slip (although of lesser consequence) committed by al-Bīrūnī, which I have pointed out in an earlier paper.[39] It need not be emphasized that al-Battānī actually knew better; to prove this, it suffices to mention that his tables of planetary motion, far more elaborate than Ptolemy's, could not have been drawn up without a thorough familiarity with the Ptolemaic system, including all its finesses and intricacies.

While al-Battānī takes no critical attitude toward the Ptolemaic kinematics in general, he evidences, as said before, a very sound skepticism in regard to Ptolemy's practical results. Thus, relying on his own observations, he corrects—be it tacitly, be it in open words—Ptolemy's errors. This concerns the main parameters of planetary motion no less than erroneous conclusions drawn from insufficient or faulty observations, such as the invariability of the obliquity of the ecliptic or of the solar apogee.

The Islamic astronomers' interest in the question of the variability of the obliquity of the ecliptic started astonishingly early. This is the more remarkable because the effect, being on the order of magnitude of 0.5″ a year, is definitely of no practical use. According

to Ibn Yūnus (*d.* 399/1009),[40] the first measurement since Ptolemy was made shortly after 160/776, yielding 23°31′, which was 4′–5′ too low;[41] after this he reports quite a number of different values, all in the vicinity of 23°33′, made during and after the time of al-Ma'mūn (*d.* 215/830). Hence, al-Battānī's determination is nothing exceptional, but it is important for us because he gives a careful description of the procedure by which his value (23°35′), which squares perfectly with the modern formula, was obtained.

In chapter 28 of the *Zīj*, al-Battānī discusses his observations of the lengths of the four seasons, from which, employing Hipparchus' method as described in *Almagest* III, 4, he infers that the sun's apogee and its eccentricity have both changed since antiquity: the apogee, found at 65°30′ by Hipparchus and erroneously claimed to be invariable by Ptolemy, had moved to 82°17′, and the eccentricity had diminished from $2^p29′30″$ to $2^p4′45″$ ($1^p = 1/60$ of the radius).

Despite contrary assertions, however, al-Battānī was not the first since Ptolemy to check on these values. According to al-Bīrūnī,[42] who relies on Abū Ja'far al-Khāzin's (*d. ca.* 350/961) commentary on the *Almagest,* the first observations serving the purpose of a new determination of the apogee were carried out, on the basis of the specially devised new "method of the four *fuṣūl*,"[43] in the Shammāsiyya quarter of Baghdad in A.D. 830. In spite of this methodological improvement, the result was extremely poor: it yielded a value no less than 20° too small. One year after, Thābit ibn Qurra or the Banū Mūsā, using the old Ptolemaic method, obtained the excellent value 82°45′.[44] Comparing this with Hipparchus' value (65°30′) and rejecting Ptolemy's obviously wrong confirmation of the latter, he (or they) found that the motion of the apogee amounts to 1° in sixty-six years. Then, from the fact that he had also found the same value for the motion of precession, Thābit concluded that they must of necessity be identical—in other words, that the apogee of the sun remains fixed once and for all in regard to the fixed stars (for this type of reasoning and concluding, European scholasticism has invented the term "Ockham's razor").

Al-Battānī's value for the sun's apogee (82°17′) is not quite as good as Thābit's (or the Banū Mūsā's), although the perfect accordance of the latter with the one resulting from the modern formula must be considered to some degree accidental: for A.D. 831, Leverrier's formula yields 82°50′22″ (Thābit, 82°45′); for A.D. 884, 83°45′10″[45] (al-Battānī, 82°17′). It is of interest in this context that Hipparchus' value, 65°30′ (*ca.* 140 B.C.), also squares exceedingly well with the modern, 65°21′.

Thus, it is evident that al-Battānī has no special

claim to the discovery of the motion of the solar apogee. Apart from this, he was no more able than Thābit to decide whether this motion was identical with that of precession. It was only 150 years later that al-Bīrūnī furnished the theoretical foundation for such a distinction,[46] but even he had to admit that the data at his disposal did not allow him to make a conclusive statement. The first who actually made a clear (and very correct) numerical statement concerning the proper motion—1° in 299 Julian years, or 12.04″ in one year (modern: 1.46″), was al-Zarqālī (second half of the eleventh century) of Toledo. But his result is impaired by his belief in the reality of trepidation, which he shared with Thābit.

Al-Battānī's value for the eccentricity of the solar orbit ($2^p4'45''$), corresponding to an eccentricity in the modern sense of 0.017326 (instead of 0.016771, according to our modern formula, for A.D. 880) must be called excellent, while Ptolemy's value (0.0208 instead of 0.0175) is much too high.[47]

Among al-Battānī's many other important achievements is his improvement of the moon's mean motion in longitude;[48] his measurements of the apparent diameters of the sun and of the moon and their variation in the course of a year, or of an anomalistic month, respectively, from which he concludes that annular solar eclipses (impossible, according to Ptolemy) must be possible;[49] and his new and elegant method of computing the magnitude of lunar eclipses.[50]

For the precession of equinoxes, he accepts and confirms Thābit's value (1° in sixty-six years), far better than Ptolemy's (1° in 100 years), but about 10 percent too fast (correct, 1° in seventy-two years). Accordingly, his tropical year ($365^d5^h46^m24^s$) is too short by 2^m22^s (correct, $365^d5^h48^m46^s$), while Ptolemy's ($365^d5^h55^m12^s$) is too long by 6^m26^s.

Al-Battānī's catalog of fixed stars[51] is far less comprehensive than Ptolemy's (489 as against 1,022 stars). The latitudes and magnitudes are taken over (perhaps with a few corrections) from the *Almagest,* while the longitudes are increased by the constant amount of 11° 10′, which corresponds, for the interval of 743 years between the epochs of the two catalogs (A.D. 137 and 880), to the motion of precession indicated, viz., 1° in sixty-six years.

While Ptolemy's *Almagest* is often cited, his *Tetrabiblos* is mentioned on only one occasion (end of chapter 55). It is uncertain whether al-Battānī knew and used Ptolemy's *Geography.*[52] Ptolemy's *Hypotheses* (called by later authors *Kitāb al-Iqtiṣāṣ* or *Kitāb al-Manshūrāt*)[53] are made use of in chapter 50, which deals with the distances of the planets, but al-Battānī ascribes the underlying theory of contiguous spheres,

according to which the distances are computed, to "more recent scientists [who lived] after Ptolemy." Since al-Farghānī mentions no name at all in this connection,[54] it seems probable that al-Battānī's reference to "scientists after Ptolemy" reflects a passage from Proclus' *Hypotyposis,*[55] in which Ptolemy's name also does not occur, and that Ptolemy's authorship became known only when, some time before al-Bīrūnī, the *Hypotheses* were translated into Arabic.

Of other astronomical works from antiquity, only Theon of Alexandria's *Manual Tables* are mentioned. In the section of chapter 6 dealing with geographical questions,[56] al-Battānī refers to "the ancients" without further specification. Nallino[57] has shown that this expression, there, means Greco-Syriac sources.

In spite of the circumstance that al-Battānī, as demonstrated before, has a good deal in common with the Banū Mūsā, Thābit, and al-Farghānī, no reference whatever appears in his *Zīj* to his Islamic predecessors. In his terminology he abstains from using foreign (Persian or Indian) words, as found in earlier writings of his countrymen, such as *awj* for the apogee of the eccentric (circumstantially called by al-Battānī *al-buʿd al-abʿad min al-falak al-khārij al-markaz,* "the [point having] maximum distance in the eccentric"), *jayb* for the sine (al-Battānī: *watar munaṣṣaf,* "half-chord," or just *watar,* "chord"), *buht* for the (unequal) motion of a planet in the course of one day (a concept not used by al-Battānī), *jawzahar* for the ascending node (al-Battānī: *al-ra's,* "the head [of the dragon]"), *haylāj* for the astrological "significator" or "aphet" (Gr.: ἀφέτης; al-Battānī: *dalīl*), and so on.[58] His aversion to foreign terms, however, certainly springs not from any "purism of language" but, rather, from the circumstance that the words in question did not occur in the *Almagest* translations at his disposal; this explains why in some cases he does not hesitate to employ clumsy transliterations of Greek terms, such as *afījiyūn* for ἀπόγειον ("apogee") and *farījiyūn* for περίγειον ("perigee").

Al-Battānī uses the sine instead of the chord (of twice the angle), following the example of his Arab predecessors who had fused into one whole the new Indian notion (*Siddhānta*) and the old Greek notion. Besides the sine he also employs the cosine (*watar mā yabqā li-tamām . . . ilā tisʿīn,* "the sine of the complement of . . . to 90°") and the versine (R − cosine, called *watar rājiʿ,* "returning sine"), for which later authors also employed the term *jayb maʿkūs* ("inverted sine"), as opposed to *jayb mustawī* ("plain sine") or *sahm* ("arrow"), whence the medieval Latin *sagitta.* Tangents and cotangents do not occur in al-Battānī's formulas, which therefore often become as clumsy as Ptolemy's. He uses them only in his gnomonics, where they refer, as in the *Siddhāntas,* to a twelve-partite gnomon. For the cotangent, he employs the term *ẓill mabsūṭ* ("umbra extensa"; called by others also *ẓill mustawī,* "umbra recta"); for the tangent, *ẓill muntaṣib* ("umbra erecta"; called by others *ẓill maʿkūs,* "umbra versa").[59] By applying considerations based on the principle of orthographic projection, al-Battānī introduced new and elegant solutions into spherical trigonometry. In Europe, this principle was adopted and developed by Regiomontanus (1436–1476).

The epoch of al-Battānī's chief era (*Taʾrīkh Dhiʾl-Qarnayn,* "Epoch of the Two-Horned [Alexander]") is Saturday (mean noon, Raqqa), 1 September 312 B.C., which precedes by thirty days the epoch *Dhuʾl-Qarnayn* used by all other Arabic authors: Monday, 1 October 312 B.C. It is combined with the Julian year; for the months he uses the "Syrian" names: *aylūl* (September), *tishrīn* I and II, *kanūn* I and II, *subāṭ, ādhār, nīsān, ayyār, ḥazīrān, tammūz,* and *āb.* The epoch of the Coptic era (*Taʾrīkh al-Qibṭ*) is Friday, 29 August 25 B.C., while all other Arabs used this term to denote one of the following three: the era of Nabunassar (1 Thoth = 26 February 747 B.C.), the era of Philippus Arrhidaeus (12 November 324 B.C., mentioned in the *Almagest* as the "era of Alexander's death"), or the era of Diocletian (29 August 284, also called *Taʾrīkh al-Shuhadāʾ,* "era of the martyrs").[60]

A Latin translation of the *Zīj* made by the English Robertus Retinensis (also cited as R. Ketenensis, Castrensis, or Cestrensis; Nallino believes the correct form to be Cataneus),[61] who flourished about the middle of the twelfth century, has not survived. The only Latin version extant is the one by Plato of Tivoli, who flourished in Barcelona in the first half of the twelfth century. This translation was printed twice. The *editio princeps* (Nuremberg, 1537) carries the title *Rudimenta astronomica Alfragrani* [sic].[62] *Item Albategnius peritissimus de motu stellarum ex observationibus tum propriis tum Ptolemaei, omnia cum demonstrationibus geometricis et additionibus Ioannis de Regiomonte;* the title of the second edition (Bologna, 1645), printed without al-Farghānī's *Elements,* is *Mahometis Albatenii de scientia stellarum liber cum aliquot additionibus Ioannis Regiomontani. Ex Bibliotheca Vaticana transcriptus.*

A Spanish translation, made at the order of King Alfonso el Sabio (1252–1284), is preserved in the Bibliothèque de l'Arsenal in Paris.[64]

Although no Hebrew translation of the *Zīj* seems to have existed, its impact on Jewish scholarship was great. It was mentioned and praised by Abrāhām bar Ḥiyyāʾ (d. *ca.* 1136) and Abrāhām ibn ʿEzrā (*ca.* 1090–1167). Maimonides (1135–1204) follows al-Battānī closely, but without mentioning his name, in the eighth treatise of Book III of his *Mishne Tōrā,* which bears the title "Hilkōt qiddūsh ha-ḥōdesh."[65] In chapters 12–14 the parameters used (mean motion in longitude of the sun and of the moon, mean anomalistic motion of the moon, equation of the sun) are exactly the same as in al-Battānī's tables, except that the values for the solar equation are rounded off to minutes.[66] In his prescriptions for determining the limits of visibility of the new moon, too, Maimonides closely follows the elegant method devised by al-Battānī.

Among Islamic astronomers and historians, al-Battānī holds a place of honor. The great al-Bīrūnī composed a book entitled *Jalaʾ al-Adhhān fī Zīj al-Battānī* ("Elucidation of Genius in al-Battānī's *Zīj*"),[67] and Ibn Khaldūn (1332–1406)[68] counts his works among the most excellent in Islamic astronomy.

In Byzantine writings, al-Battānī's name is mentioned as δΠατανής, but Greek translations apparently have not existed. A great many medieval Latin authors who knew the *Zīj* or at least mention the name of its author can be enumerated. Among them are Henry Bate (1246–*ca.* 1310), who, in composing his *Magistralis compositio Astrolabii anno 1274 scripta,* makes ample use of the contents of the *Zīj*—not without giving its author due credit for it; Gerard of Sabbionetta; Albertus Magnus; Lēvi ben Gersōn (in the Latin translation of his *Astronomy*);[69] and, not least, Regiomontanus, whose keen interest in the work is evident from the great number of annotations in his handwriting found in his copy of Plato of Tivoli's version and printed as an appendix to the Nuremberg and the Bologna editions. His teacher Georg Peurbach's (1423–1461) *Theoricae planetarum,* printed and edited by Regiomontanus,[70] mentions al-Battānī's name on only one occasion (fol. 18r),[71] where he relates that "Albategni," contrary to those who defended the theory of trepidation, claimed that the stars move 1° in sixty years and four months,[72] and always toward the east. Since all Arab astronomers, in accordance with the text and the translations extant, state that al-Battānī accepted a motion of 1° in sixty-six years,[73] it is a mystery how this erroneous value, which cannot be due to a copyist's slip, could have found its way into Peurbach's book. As for his masterful presentation of planetary kinematics according to Ptolemy, it is a matter of course—in view of what I have said about al-Battānī's chapter 31[74]—that it could not be modeled after al-Battānī's fallacious chapter, while the influence of al-Farghānī seems perceptible in many places.

The indebtedness of Copernicus to al-Battānī is well known. He quotes him fairly often, especially—as does Peurbach—in the chapters dealing with the problems of solar motion and of precession.[75] Much more frequent references to him are found in Tycho Brahe's writings and in G. B. Riccioli's *New Almagest;*[76] in addition, Kepler and—only in his earliest writings—Galileo evidence their interest in al-Battānī's observations.

From the point of view of the history of astronomy, the names of two men are to be mentioned, although with a totally different weight. In 1819 Delambre published his *Histoire de l'astronomie du moyen âge.*[77] In chapter 2 he devotes fifty-three pages to a very thorough analysis of the *Zīj,* on the basis of the Bologna edition of Plato of Tivoli's translation. Even to the modern reader the chapter is of interest, in spite of the fact that a certain superciliousness, characteristic of all Delambre's historical works, is sometimes embarrassing. For it is, of course, not as interesting to learn how one problem or the other could have been solved in a less circumstantial way as it would be to get an insight into the historical situation in which al-Battānī's work came into being. This, however, was beyond Delambre, for even if he had possessed a sufficient knowledge of Arabic (he had none at all), the only extant manuscript would not have been within his reach. Thus he had to rely on Plato of Tivoli's version, whose errors and misunderstandings naturally led him astray in more than one case.

Eighty years after Delambre, in 1899, the young Italian orientalist C. A. Nallino published his model edition of the complete Arabic text of al-Battānī's *Zīj.*[78] The two other volumes, containing a Latin translation and exceedingly detailed and learned commentaries, followed during the next eight years. In a time like ours, characterized by the abuse of superlatives, it is hard to describe Nallino's work in appropriate terms. Al-Battānī's Arabic style, which at first sight looks simple and straightforward, but which reveals itself difficult and even obscure on many occasions, is rendered here in a Latin whose purity and clarity deserve the highest praise. In reading this book, which is Nallino's *magnum opus,* one understands that it was not due purely to a whim that he decided to compose it in Latin. As for the technical aspect, Nallino's work bears witness to a great familiarity with the mathematical and astronomical problems occurring in al-Battānī's *Zīj,* and no less with the historical facts that form its background. This third Latin translation, written eight centuries after the first two, will always stand as one of the masterpieces of the history of science.

Until recently it was believed that none of the three minor works (all of astrological content) listed in the *Fihrist* and in Ibn al-Qifṭī's *Ta'rīkh al-ḥukamā'*[79] had survived, since, on the one hand, the authenticity of

a manuscript preserved in Berlin seemed dubious[80] and, on the other, the only extant manuscript expressly titled "Commentary on Ptolemy's *Tetrabiblos,*" which figures still in Casiri's catalog,[81] could no longer be found in the Escorial library, as stated by H. Derenbourg[82] in 1884 and confirmed by Nallino in 1894 and by Father Pedro Blanco Soto in 1901.[83] Fortunately, however, the lost manuscript seems to have been recovered: in H.-P.-J. Renaud's new catalog,[84] it is listed as no. 969, 2 (not 966, as in Casiri), under the title *Kitāb al-arbaʿ maqālāt fī aḥkām ʿilm al-nujūm,* the copy dating from 939/1533 and comprising sixty folios. The word *sharḥ* ("commentary") does not appear in the title, but Renaud's and Casiri's descriptions read "Commentary on Ptolemy's *Quadripartitum*" (i.e., *Tetrabiblos*). It will require a special study to establish whether the Berlin and the Escorial manuscripts are identical in text. The fact that the latter also contains tables (which the Greek original does not have) suggests the existence in the text of rules and prescriptions for their use that might justify calling it a "commentary."

In this context,[85] Nallino mentions that the Egyptian ʿAlī ibn Riḍwān (latinized as Haly Heben Rodan, *d.* 453/1061) states that he has never come across any paraphrase (*glossa*) of the *Tetrabiblos* at all, whereas Abu'l-Ḥasan ʿAlī ibn Abi'l-Rijāl (Albohazen Haly filius Abenragel, *fl. ca.* 1050) counts al-Battānī among those who, like Ptolemy, attributed special importance to astrological prognostications made on the basis of eclipses that occur during the years of planetary conjunctions. Nallino evidently believes this refers to *Tetrabiblos* II, 6,[86] which deals with the same subject matter. The case, however, is different. In the Saib (Ismāʿīl Sāʾib) Library at Ankara there is preserved a volume (no. 1/199) containing three different works, the second of which (fols. 27r–42v) bears the title *Kitāb [Muḥammad ibn] Jābir b. Sinān al-Ḥarrānī al-Battānī fī dalāʾil al-qirānāt wa'l-kusūfāt* ("Jābir . . . al-Battānī's Book on the Significations of Conjunctions and Eclipses").[87] It is undoubtedly this book, not listed in any of the great oriental bibliographies, on which bears Albohazen's remark. Judging from a cursory inspection of a photostatic copy in the possession of the Institute for the History of Science of the University of Frankfurt, I see no reason to doubt its authenticity.

Another manuscript, entitled *Tajrīd uṣūl tarkīb al-juyūb*[88] ("Construction of the Principles of Establishing [Tables of] Sines"), also carries al-Battānī's name. From the fact that al-Battānī, at least in his *Zīj,* avoids using the term *jayb* (plural, *juyūb*) for "sine,"[89] it might be inferred that this manuscript is spurious.

For a number of other definitely spurious works existing only in Latin translations, see the list (with comprehensive discussions) found in Nallino.[90]

NOTES

1. The transliteration system used in this article is that of the *Encyclopaedia of Islam,* 2nd ed., with the following simplifications: *j* instead of *dj*; *q* instead of *k* (NB: *qu* is pronounced *ku*, not as English *qu*); no underlinings to indicate compound consonants: *kh* instead of <u>*kh*</u> for the Scottish *ch*-sound, <u>*th*</u> and <u>*dh*</u> for the English voiceless and voiced *th*, respectively.

2. Of two numbers separated by a slash, the first indicates the year according to the Muslim calendar and the second its beginning according to the Christian calendar. In quotations from Arabic texts, Muslim years are denoted by H. (Hegira), and the corresponding Christian years are added in brackets. Note that the Muslim year is 3 percent shorter than the Julian.

3. *Kitāb al-Fihrist* (composed *ca.* A.D. 987 by Ibn al-Nadīm), G. Flügel, ed. (Leipzig, 1871–1872), I, 279. See also C. A. Nallino, *Al-Battānī sive Albatenii Opus astronomicum, ad fidem codicis escurialensis arabice editum, latine versum, adnotationibus instructum,* I (Milan, 1903), viii ff. This *magnum opus* (Vol. II, 1907; Vol. III, 1899), cited hereafter as *O.A.,* will always remain the chief source of information in Arabic astronomy and on al-Battānī in particular.

4. See B. Carra de Vaux, "al-Ṣābiʾa," in *Encyclopaedia of Islam,* 1st ed., IV; and, for comprehensive information (although obsolete in certain parts), D. Chwolsohn, *Die Ssabier und der Ssabismus,* Vols. I/II (St. Petersburg, 1856).

5. *Die Ssabier,* I, 611.

6. *O.A.,* I, xiii.

7. *Fihrist,* p. 285.

8. p. 280.

9. J. Lippert, ed. (Leipzig, 1903), p. 280. Ibn al-Qifṭī, the author of this "History of Learned Men," died in 646/1248. His chapter on al-Battānī (which I follow in my translation), according to his own words, relies on Ṣāʿid al-Andalusī. It contains information not found in the *Fihrist.*

10. Owing to a scribal error, the *Fihrist* and Qifṭī have 269.

11. Instead of 299/911, read 267/880 (scribal error). The epoch of the catalog is actually 267/880.

12. The meaning, evidently, is "because unjust taxes had been requested of them." The text leaves open whether "them" includes al-Battānī. Cf. Nallino, *O.A.,* I, viii. As for the Banu'l-Zayyāt, Nallino (*ibid.,* pp. xvii f.) considers it almost certain that they are the descendants of the famous poet and vizier ʿAbd al-Malik ibn Abān al-Zayyād (executed by Caliph Mutawakkil in 233/847). It was to his great-grandson's son, Abū Tālib Aḥmad al-Zayyād, that Ibn Waḥshiyya dictated, in 318/930, his alleged "translation from the Syriac" of his book on the Nabataean agriculture.

13. In Ibn Khallikān's (*d.* 681/1282) biographical dictionary (Eng. trans. by Mac Guckin de Slane, Paris–London, 1843–1871, IV, 317–320; Arabic original: *Ibn Challikan, Vitae illustrium virorum,* F. Wüstenfeld, Göttingen, 1835–1842, no. 719 [cited after *O.A.,* I, ix, n. 6]), the place of al-Battānī's death is called Qaṣr al-Ḥaḍr. Nallino has shown (*O.A.,* p. xviii) that Jiṣṣ (of which Ḥaḍr is nothing but a graphical corruption) is the correct form.

14. Ibn al-Nadīm and Ibn Khallikān add the words "a first and a second; the second is better."

15. Ibn al-Nadīm adds the words *fī mā bayna arbāʿ al-falak* ("in the spaces between the four cardinal points of the sphere"). The book gives mathematical solutions of the astrological problem of finding the direction of the *aphet* (*tasyīr al-dalīl*).

16. Not listed by Ibn al-Nadīm.

17. *O.A.,* I, xvii, n. 1.

18. Bologna, 1651, II, xxix.

19. *Histoire des mathématiques,* new ed. (Paris, 1797–1800), I, 363.

20. *Astronomie,* 3rd ed. (Paris, 1792), I, 123.

21. *Histoire de l'astronomie du moyen âge* (Paris, 1819; repr. New York–London, 1965), pp. 4, 10.

22. In this title, the author is called "Mahometus, filius Geber, filius Crueni, qui vocatur Albategni." The strange name Cruenus is obviously due to a misreading of Sinanus, which may have been found spelled Cinenus.

23. Ch. 30, *O.A.,* I, 56.

24. Thus Ibn Khallikān and Ḥājjī Khalīfa (1017/1609–1067/1657); see *Haji Khalfae Lexicon bibliographicum et encyclopaedicum,* ed. and trans., with commentary, by G. Flügel (Leipzig–London, 1835–1858), III, 564, no. 6946.

25. *O.A.,* I, xxxi, n. 3.

26. In Byzantine Greek, the word is found as ζῆσι and identified with σύγταξις; see *O.A.,* I, xxxi, n. 5.

27. Ch. 57, which treats the theory of trepidation, refuted by al-Battānī but accepted by Thābit. The reference to al-Battānī is found in Thābit's letter to Isḥāq ibn Ḥunayn, preserved by Ibn Yūnus. See *O.A.,* I, 298.

28. Originally no. 903 (M. Casiri, *Bibliotheca arabico-hispanica Escurialensis,* Madrid, 1760, I, 342–343), now no. 908. Unbelievable as it sounds, only this copy of one of the most important books written in the Middle Ages has survived in the Arabic original.

29. *O.A.,* I, xxxii.

30. Ch. 1, *O.A.,* I, 5.

31. *Almagest,* III, 1 (German trans. by Karl Manitius, *Des Claudius Ptolemäus Handbuch der Astronomie,* Leipzig, 1912, p. 141). The wording there is a little different and contains no such express "admonition" or "order" (*amr*).

32. Except for ch. 1, which deals with the various eras and their mutual conversions, practically identical with al-Battānī's ch. 32. Cf. J. Golius' ed. and Latin trans.: *Muhammedis fil. Ketiri Ferganensis, qui vulgo Alfraganus dicitur, Elementa astronomica* (Amsterdam, 1669).

33. See W. Hartner, "The Obliquity of the Ecliptic According to the Hou-Han Shu and Ptolemy," in *Silver Jubilee Volume of the Zinbun-Kagaku-Kenkyusyo* (Kyoto, 1954), pp. 177–183; repr. in Hartner's *Oriens-Occidens* (Hildesheim, 1968), pp. 208–214.

34. See n. 32.

35. See p. 508 and n. 15.

36. See the figure and the accompanying text, *O.A.,* III, 96 f. (Arabic) and *O.A.,* I, 64 f. (Latin), from which Figure 1 in the text is reproduced.

37. *O.A.,* I, 237 f. Figure 2 in the text is reproduced from Schiaparelli's.

38. Nallino, *O.A.,* I, lxii, states that the Arabic original, from which Plato of Tivoli translated, and the Escorial MS (written, according to Nallino, about 1100) must both have belonged to the same archetype.

39. "Mediaeval Views on Cosmic Dimensions and Ptolemy's *Kitāb al-Manshūrāt,*" in *Mélanges Alexandre Koyré,* I (Paris, 1964), 254–282; repr. in W. Hartner, *Oriens-Occidens* (Hildesheim, 1968), pp. 319–348.

40. Bibliothèque Nationale, MS Ar. 2495, p. 222 (cited after *O.A.,* I, 157).

41. For an exact verification and comparison with modern formulas (Newcomb, de Sitter), the effect of refraction and of solar parallax has to be taken into account; by this the values derived from observation are reduced by about 40″, which of course is of no interest here (see n. 33).

42. Masʿūdic Canon, VI, chs. 7 and 8 (*al-Qānūnal-Masʿūdī,* pub. by The Dāʾirat al-Maʿārif Oṣmānia, II (Hyderabad-Dn., 1374/1955), pp. 650–685. Cf. W. Hartner and M. Schramm, "Al-Bīrūnī and the Theory of the Solar Apogee: An Example of Originality in Arabic Science," in A. C. Crombie, ed., *Scientific Change* (London, 1963), pp. 206–218.

43. Observation of the sun's passage through the points 15° Taurus, 15° Leo, 15° Scorpio, and 15° Aquarius.
44. According to the *Kitāb fī sanat al-shams bi'l-arṣād,* MS London India Office no. 734, fol. 6r, ll. 13 ff.
45. Schiaparelli (*O.A.,* I, 215) gives the erroneous value 83°50'51" for A.D. 884.
46. Cf. Hartner and Schramm (n. 42), pp. 216–218.
47. Cf. *O.A.,* I, 213 f. For comparing the ancient with the modern values (elliptic eccentricity), the former must of course be halved: $2^p4'45''/60 = 0.034653 = 2 \cdot 0.017326$.
48. Cf. *O.A.,* I, 225 f.
49. Cf. *ibid.,* 58, 236. He either is unaware or avoids mentioning that his own observations of the moon's apparent diameter at apogee ($d_1 = 29.5'$) and at perigee ($d_2 = 35.3'$) are in the ratio 5:6. According to Ptolemy's and his own theory, they ought to be in the ratio 17:33, or nearly 1:2.
50. Cf. *ibid.,* 99 f.
51. Arabic text, *O.A.,* III, 245–274; Latin, *O.A.,* II, 144–177.
52. Cf. *O.A.,* I, xli and 20 (end of ch. 4).
53. See W. Hartner, "Mediaeval Views on Cosmic Dimensions" (n. 39), and B. R. Goldstein, "The Arabic Version of Ptolemy's *Planetary Hypotheses,*" in *Transactions of the American Philosophical Society,* n.s. **57,** pt. 4 (1967).
54. Ch. 21, pp. 80–82; see n. 32.
55. *Procli Diadochi Hypotyposis astronomicarum positionum,* ed. and trans. into German by K. Manitius (Leipzig, 1909), ch. 7, 19 (p. 220). As I have shown in "Mediaeval Views . . ." (see n. 39), it can no longer be doubted that the idea of contiguous spheres was conceived by Ptolemy. The final proof for the correctness of my assertion has been furnished by Goldstein (see n. 53), who found that the extant Arabic and Hebrew versions of the *Hypotheses* contain the part missing from J. L. Heiberg's edition (*Claudii Ptolemaei opera,* II, *Opera astronomica minora,* Leipzig, 1907, 69–145) at the end of Book I. It has exactly the same parameters and ratios as indicated in my paper.
56. *O.A.,* I, 17–19.
57. *Ibid.,* 165–177.
58. Cf. *ibid.,* xlii f.
59. On the back of astrolabes, preference is given to the terms *ẓill mabsūṭ* for the shadow cast by a vertical gnomon on a horizontal plane, and *ẓill maʿkūs* (sometimes *mankūs*) for the shadow of a horizontal gnomon on a vertical plane.
60. For further information, see *O.A.,* I, 242–246.
61. See *O.A.,* I, xlix f. There can be no doubt that the name means Robert of Chester, a friend of Hermannus Dalmata. He was the first to translate the Koran into Latin (1143) and also one of the first translators of Muḥammad ibn Mūsā al-Khwārizmī's *Algebra.* See L. C. Karpinski, "Robert of Chester's translation of al-Khowarizmi," in *Bibliotheca mathematica,* **11** (1911), 125–131. Robertus Retinensis is not identical with Robertus Anglicus, who lived in the thirteenth century.
62. Cf. the other edition of al-Farghānī's *Elements,* quoted in n. 32. The division into chapters is different in the two editions.
63. For C. A. Nallino's Latin translation and edition, see p. 513.
64. Described by Rico y Sinobas in *Libros del saber de astronomía del Rey D. Alfonso X de Castilla,* V, pt. 1 (Madrid, 1867), 19 f.
65. Cf. *O.A.,* I, xxxiv; and S. Gandz, trans., "The Code of Maimonides, Book III, Treatise 8, Sanctification of the New Moon: With Supplementation and an Introduction by J. Obermann and an Astronomical Commentary by Otto Neugebauer," in J. Obermann, ed., *Yale Judaica Series,* XI (New Haven, 1956), 47–56.
66. *O.A.,* II, 20, 22, 75, and 78 ff. Note that for the sun's mean motion, the values given in the tables on pp. 22 and 75 differ by 1" and 2" for the arguments 9^d and 10^d. The figures on p. 75 are the correct ones: 9^d, 8°52'15"; 10^d, 9°51'23".
67. *Jalāʾ al-adhhān fī Zīj al-Battānī,* according to al-Bīrūnī's own bibliography, published in E. Sachau, ed., *Chronologie orientalischer Völker von Albêrûnî* (Leipzig, 1878), p. xxxxvi.

68. Cf. de Slane, trans., *Les prolégomènes d'Ibn Khaldūn* (Paris, 1863–1868), III, 148; cf. F. Rosenthal, trans., *Ibn Khaldûn, The Muqaddimah,* III (New York, 1958), 136.
69. Cod. Vat. Lat. 3098 (cited after *O.A.,* I, xxxvi).
70. Nuremberg, *ca.* 1473 (the exact year of this incunabulum cannot be established).
71. In the chapter "De motu octavae sphaerae." The folios carry no numbers.
72. "Albategni vero dicebat eas moveri uno gradu in sexaginta annis et quatuor mensibus semper versus orientem."
73. Cf. p. 510.
74. See p. 509.
75. See e.g., Copernicus, *De revolutionibus,* III, 13.
76. See n. 18.
77. See n. 21.
78. See n. 3.
79. See p. 508.
80. Staatsbibliothek, MS no. 5875; see W. Ahlwardt, *Verzeichniss der arabischen Handschriften der Kgl. Bibliothek zu Berlin,* V, 273 f. The MS, written *ca.* 800/1397 by Aḥmad ibn Tamīm and comprising 62 folios, lacks the title page and the first pages of the text. At the end is the grammatically incorrect phrase *tamma kitāb al-arbaʿa* [*sic!*] *sharḥ al-Battānī* ("Here ends the Book of the Four [Maqālas] of al-Battānī's commentary"), which obviously alludes to Ptolemy's *Tetrabiblos.* According to Ahlwardt, no division into four *maqālas* and no sections marked as commentaries (*sharḥ*) are recognizable. As Nallino (I, xxi ff.) has shown, the subjects treated and their sequence are, with few exceptions, those of the *Tetrabiblos.*
81. See n. 28. I, 399: "Cod. CMLXVI, nr. 2°: Commentarius in Quadripartitum Ptolemaei de astrorum iudiciis: subiectis tabulis. Auctor est vir clarissimus Mohammad Ben Geber Albategnius."
82. *Les manuscrits arabes de l'Escurial* (Paris, 1884), I, xxiv.
83. *O.A.,* I, xx.
84. "Les manuscrits arabes de l'Escurial décrits d'après les notes de Hartwig Derenbourg," in *Publications de l'École Nationale des Langues Orientales Vivantes,* 6th ser., **5,** 2, fasc. 3 (1941), 116.
85. *O.A.,* I, xxiii.
86. Cited after the anonymous Latin translation *C. Ptolemaei de praedictionibus astronomicis, cui titulum fecerunt Quadripartitum, libri IV. Ed. posterior* (Frankfurt, 1622).
87. This important MS was discovered by Fuat Sezgin, Institut für Geschichte der Naturwissenschaften, University of Frankfurt. It dates from the sixth/thirteenth century.
88. Istanbul, Carullah 1499, fol. 81v, written in 677/1278, consisting of only one page. This MS was discovered by Fuat Sezgin.
89. See p. 511.
90. *O.A.,* I, xxiii–xxxi.

BIBLIOGRAPHY

Consult the following works by C. A. Nallino: *Al-Battānī . . . Opus astronomicum* (see note 3); "Al-Battānī," in *Encyclopaedia of Islam,* Vol. I, repr. with augmented bibliography, *ibid.,* 2nd ed., Vol. I; "Astronomy," *ibid.;* "Astrologia e astronomia presso i Musulmani," in his *Raccolta di scritti editi e inediti,* Maria Nallino, ed., V (Rome, 1944), 1–87, esp. 52; "Storia dell'astronomia presso gli Arabi nel Medio Evo," *ibid.,* pp. 88–329, trans. by Maria Nallino from the Arabic original, *ʿIlm al-Falak . . .* (Rome, 1911–1912); and "Albatenio," in *Enciclopedia italiana,* repr. in his *Raccolta,* V, 334–336.

See also H. Suter, "Die astronomischen Tafeln des Muḥammad ibn Mūsā al-Khwārizmī," in *Det Kgl. Danske Videnskabernes Selskabs Skrifter,* 7. Række, Historisk og Filologisk Afdeling, **3,** no. 1 (1914); J. M. Millás-Vallicrosa,

Estudios sobre Azarquiel (Madrid–Granada, 1943–1950); E. Honigmann, "Bemerkungen zu den geographischen Tabellen al-Battānī's," in *Rivista degli studi orientali,* **11** (1927), 169–175; and E. S. Kennedy and Muhammad Agha, "Planetary Visibility Tables in Islamic Astronomy," in *Centaurus,* **7** (1960), 134–140.

WILLY HARTNER

IBN BAṬṬŪṬA (*b.* Tangier, Morocco, 24 February 1304; *d.* Morocco, *ca.* 1368–1369), *Muslim traveler.*

Muḥammad, the son of ʿAbdallāh, whose family name was Ibn Baṭṭūṭa and who was descended from members of the Berber tribe of the Lawāta, left his native Tangier on 14 June 1325 in order to make the pilgrimage to Mecca. He returned to Morocco, to Fez, almost a quarter of a century later, in November 1349, after having visited much of Asia, going as far east as China. He went through northern Africa to Egypt, which he reached in the spring of 1326. He traveled through Egypt, Palestine and Syria, the Ḥijāz, South Arabia, and down the east coast of Africa from Mogadishu to Kilwa. Ibn Baṭṭūṭa visited southwestern Persia and Iraq, Anatolia, and southern Russia in the region of the Crimea. From there, he paid an imaginary visit to the city of Bulghār on the Volga and undertook an overland trip to Constantinople that has occasionally been doubted but seems to have taken place. He then went through central Asia to India, where he took up residence in Delhi. He spent about eight years in India, traveling along the coasts and through various sections of the subcontinent. He stayed in the Maldives for more than a year and a half, then visited Ceylon, Sumatra, and China.

Soon after his return to Morocco, Ibn Baṭṭūṭa left on a trip to Spain and then turned south to visit the mighty Mālī-Mandinka state, in particular, the cities of Timbuktu and Gawgaw (Gao). He returned to Morocco early in 1354. He dictated the story of his travels to Ibn Juzayy—a Spanish scholar and official at the court of Sultan ʿInān of Fez—who edited it in a very short time. No more is heard of Ibn Baṭṭūṭa after this.

All the details of Ibn Baṭṭūṭa's life and accomplishments come from his *Travels* (in Arabic, *Riḥla*). His family in Tangier belonged to the legal establishment, and he received the usual legal-religious education. Going on the pilgrimage meant visiting and, on occasion, studying with the famous scholars to be met en route. These customary study trips often extended over many years; but in the case of Ibn Baṭṭūṭa, although he always kept in contact with scholars and mystics, the original purpose was soon forgotten and travel became an end in itself. He thoroughly enjoyed what for most other medieval Muslims was a necessary evil, and seized every opportunity to see new cities and regions. Only a very small part of his time abroad was spent outside the realm of Islam; and when he stayed in places not under Muslim political control, he preferred to move in the circles of his fellow expatriate coreligionists, where his upbringing and his linguistic background stood him in good stead. The judicial functions he exercised for the first time on his way from Morocco to Egypt and then as a Mālikite judge in India suggest a career not unlike the one he would have pursued had he never left Morocco. He made his living from gifts and stipends—some, he says, very generous—from rulers or high officials wherever he went. He apparently undertook occasional commercial ventures on the side. Some of his marriages may have procured helpful family connections for him. He speaks often of the women he married in the various places he visited and the many slave girls who always accompanied him. He also refers to children born to his women, children whom he did not see again or who did not live to maturity.

If the summary of his travels has been left rather vague, it is not for a lack of precise dates, which are amply present in his work. It is because these dates have been recognized as frequently incorrect. Attempts to harmonize them and to establish a coherent chronology of his travels necessarily involve repeated choices between indications furnished by the author for reasons that are never absolutely cogent. Even the date he gives for his arrival in India, 12 September 1333, has been questioned. This date cannot be correct if one chooses to credit his statement that he was in Mecca during the pilgrimage a year earlier. His itinerary often cannot be followed in exact detail. He lists places, in seemingly chronological succession, that he can be assumed to have visited on different trips. He reports abruptly on side excursions that do not fit into the main itinerary. On occasion he uses literary sources, reports on events of contemporary history that he did not witness, and can be shown to present as having seen himself what he can only have been told by others. All this does not detract from the credibility of his information. It is explained not so much by slips of memory (the question of whether he had been able to keep written notes to help him in his recollections is much debated) as by the spirit and purpose of medieval Muslim literature.

Although exceptionally rich in personal data, the *Travels* was not meant to constitute a personal record in terms of the circumstances of the author's life or the proper sequence of his itinerary. Its purpose was to enlighten the reader about remarkable and often marvelous things and events that could be observed in other countries and to deepen his understanding of human society and his respect for the divine handiwork in all its richness and variety. This purpose was uniquely achieved and has given the *Travels* its lasting greatness.

Ibn Baṭṭūṭa was not a scientist. He was not even a productive scholar in the traditional Muslim sense. While his general qualifications in the legal field must have been acceptable, the fact that he did not give the final form to his work clearly shows that he did not belong to the scholarly sector of Muslim society. His travels did not involve any scholarly or scientific research pursuits. In the traditional manner he reported the things that struck him as unusual and remarkable. He deeply believed in and reported on supernatural events, the miracles of saintly men, and their dreams and predictions. He was interested in political conditions and the glories of foreign rulers; in economic factors; in all sorts of strange customs, such as those of marriage and burial; in the construction of Indian beds and the kind of fuel used in China; in strange inventions, such as wagons in the Crimea or a supposed way of getting rid of vermin; in remarkable animals, minerals, and, to a greater degree, trees and plants, especially those useful to man (for example, the coconut, on which he lived in the Maldive Islands). The only systematic treatment of observed facts concerns the trees, fruits, and grains of South Arabia, India, and the Maldives. His curiosity about buildings and ancient ruins was not particularly great. He believed in alchemy; but while he was credulous of religious stories, he showed himself quite skeptical of marvelous natural phenomena, such as fables about the origin of the coconut, the strange shape of the vagina of certain Turkish women, or the purported age of some old man.

Ibn Baṭṭūṭa's work is a source of unmatched importance for fourteenth-century India (to which about one-fifth of it is devoted), and even more so for the Maldives, southern Russia, and especially Negro Africa. Ibn Baṭṭūṭa often is the only medieval author to give us information on these areas; and where there is additional material, the value of his observations remains undisputed. He also contributes considerably to our knowledge of comparatively well-known areas, such as contemporary Anatolia. Whether observed fact, hearsay, or legend, whether indubitably true or suspect, his data are often the only ones we have to fill voids in our knowledge. His personal contribution lies in the single-mindedness with which he traveled in order to learn more and more about man and nature, in the skill he showed for ferreting out meaningful facts, and in his realization of the importance of these facts for the growth of human knowledge. In

contrast with Christian explorers, whose journeys took them out of their own limited world and who thus brought back to it an entirely new experience, he remained spiritually and for the most part physically within the boundaries of the world of Islam. Yet he clearly belongs in the select circle of the men who, often misunderstood or not understood by their contemporaries, paved the way for the modern age of discovery.

BIBLIOGRAPHY

I. ORIGINAL WORKS. The original edition of the *Travels* was published in four volumes, with a French translation, by C. Defrémery and B. R. Sanguinetti (Paris, 1853–1858). There is an English translation of the first two volumes by H. A. R. Gibb (Cambridge, 1958–1962), Hakluyt Society, 2nd ser., nos. 110 and 117. For a translation of the section on India, the Maldives, and Ceylon, see Mahdi Husain, *The Reḥla of Ibn Baṭṭūṭa* (Baroda, 1953). For the section dealing with Negro Africa, see G. S. P. Freeman-Grenville, *The East African Coast* (Oxford, 1962), pp. 27–32; and R. Mauny et al., *Textes et documents relatifs à l'histoire de l'Afrique* (Dakar, 1966), University of Dakar, Section d'Histoire, Publication 9.

II. SECONDARY LITERATURE. Works on Ibn Baṭṭūṭa include G.-H. Bousquet, "Ibn Baṭṭūṭa et les institutions musulmanes," in *Studia Islamica,* 24 (1966), 81–106; H. A. R. Gibb, "A Provisional Chronology of Ibn Baṭṭūṭa's Travels in Asia Minor and Russia," II, 528–537, of his translation; I. Hrbek, "The Chronology of Ibn Baṭṭūṭa's Travels," in *Archiv Orientální,* 30 (1962), 409–486; H. F. Janssens, *Ibn Batouta, le voyageur de l'Islam* (Brussels, 1948); É. Lévi-Provençal, "Le voyage d'Ibn Baṭṭūṭa dans le royaume de Grenade," in *Mélanges offerts à William Marçais* (Paris, 1950), pp. 205-224; and George Sarton, *Introduction to the History of Science,* III, pt. 2 (Baltimore, 1948), 1614–1623.

FRANZ ROSENTHAL

BAUDRIMONT, ALEXANDRE ÉDOUARD (*b.* Compiègne, France, 25 February 1806 [Partington; 7 May in Micé and Poggendorff]; *d.* Bordeaux, France, 24 January 1880), *chemistry, physiology.*

Baudrimont had an impressive range of scientific interests, but he accomplished nothing of major importance in any one of them. He was the son of Marie Victor Baudrimont, an inspector of bridges and roads, and Adélaïde Sauvage. At the age of twelve, Baudrimont was apprenticed to a pharmacist, to be trained in that profession, although he himself hoped to become a doctor. In 1823 he went to Paris to continue his training in pharmacy, taking up the study of medicine as well in 1825. He received his medical degree in 1831 and his first-class degree in pharmacy in 1834.

In the meantime, Baudrimont had developed his abilities as a chemist, first as a research assistant to E. R. A. Serres, with whom he published his first paper, in 1828, and then as an industrial chemist in Valenciennes, where he went in 1830 to earn money for his education.

After returning to Paris to receive his medical degree in 1831, Baudrimont went back to Valenciennes to practice medicine. But Paris was where he wished to live and work. During the 1830's and 1840's, he attempted to obtain a post in one of the scientific or medical centers in the capital, but he never advanced beyond minor positions at the Collège de France and the Faculté de Médecine. It was nevertheless a creative period for Baudrimont: he published the *Introduction à l'étude de la chimie par la théorie atomique* (1833), a mineralogy and geology textbook (1840), and the two-volume *Traité de chimie générale et expérimentale* (1844–1846). In addition, he wrote various theses either to obtain scientific degrees or to compete for academic posts in Paris; the most important of these was that on organic chemistry which he submitted in 1838 in competition for the chair of organic chemistry at the École de Médecine.

His "Recherches anatomiques et physiologiques sur le développement du foetus et en particulier l'évolution embryonnaire des oiseaux et des batraciens," written with Gaspard Martin Saint-Ange, won the Académie des Sciences's grand prize in physical sciences for 1846. His interest in applied chemistry was shown in contributions he made to the *Dictionnaire de l'industrie commerciale et agricole* in agricultural and industrial chemistry. In addition, Baudrimont obtained his degrees in pure science—his licentiate in 1839 and his doctorate in 1847. He was probably impelled to secure them by his failure to obtain a post teaching medical chemistry.

In 1847 he accepted a post as assistant to Auguste Laurent at the University of Bordeaux. Two years later he was awarded the chair of chemistry, and he remained in Bordeaux for the rest of his life. He had married in 1842, and had one son, Édouard. His nephew, Marie Victor Ernest Baudrimont, became a distinguished professor of pharmacy.

Baudrimont worked primarily in chemistry and physiology. In chemistry, he was active both as a theorist and as an experimentalist. In chemical theory, Baudrimont devoted his attention to molecular structure. He was greatly influenced by Ampère's ideas on this subject. Like Ampère, he thought that molecules, at least in the solid state, were polyhedral, and

like both Ampère and Gaudin, he thought that these molecular shapes could be determined from chemical and crystallographical data. Moreover, he came to believe that in many cases atoms were grouped together in submolecular units within the molecule. He worked out a hierarchy of such units: merons, which roughly corresponded to chemical radicals; merules, which made up merons; and mericules, which composed merules. In his late publications, Baudrimont attempted to explain all physical phenomena in terms of the oscillations or rotations of one or another of these units.

Baudrimont's interest in molecular structure led, him to adumbrate a theory of chemical types. He himself claimed to have used chemical types for classificatory purposes as early as 1835, but he did not develop his ideas on them until his 1838 thesis on organic chemistry. He engaged in a priority dispute with both Dumas and Laurent over the type theory, although Laurent did give him credit for first employing the term *type chimique* (*Méthode de chimie,* p. 358). Baudrimont claimed that his ideas on chemical types were developed independently of the theory of substitution propounded by Dumas and Laurent.

As a theorist, Baudrimont was also one of the few proponents of Avogadro's gas hypothesis in the 1830's and 1840's.

In experimental chemistry, Baudrimont made an important study of aqua regia in 1843, which he published in 1846. Distilling this substance, he produced and condensed a red gas (most likely nitrosyl chloride mixed with some free chlorine) that he thought was the active ingredient in aqua regia. Gay-Lussac disproved Baudrimont's hypothesis but used his method of collecting the gaseous constituent. Baudrimont and F. J. M. Malaguti discovered the presence of sulfur in cystine in 1837.

Throughout his career Baudrimont was interested in physiological research and wrote extensively on physiology. The work that won him the greatest recognition was that done with Martin Saint-Ange. They studied the chemical changes that took place in embryonic development in chickens and amphibia and the toxological effects of various gaseous substances on the embryo.

Baudrimont's other interests can only be indicated here, since they were so varied. His concern with material structure extended into physics; in the 1830's he performed experiments testing the tenacity, ductibility, and malleability of metals. His interest in agricultural chemistry was lifelong and figured importantly in his work at Bordeaux. He also published works on geometry, philosophy of science, music theory, and linguistics.

BIBLIOGRAPHY

I. ORIGINAL WORKS. Baudrimont was a prolific writer. Numerous articles in many fields were published in the *Annales de chimie et de physique,* the *Comptes-rendus hebdomadaires de l'Académie des sciences,* and, especially after 1847, in the *Mémoires de la Société des sciences physiques et naturelles de Bordeaux* and the *Actes de l'Académie des sciences, belles-lettres et arts de Bordeaux.* The works listed here are the ones most relevant to the discussion of his scientific career: *Introduction à l'étude de la chimie par la théorie atomique* (Paris, 1833); "Quel est l'état actuel de la chimie organique, et quel secours a-t-elle reçu des recherches microscopes?," thesis submitted in competition for the chair of organic chemistry and pharmacy of the École de Médecine (Paris, 1838); "Théorie de substitution, discussion de M. Dumas et de M. Pelouze; nouvelle classification des corps moléculaires définis; réclamations de M. Baudrimont," in *Revue scientifique et industrielle,* **1** (1840), 5-60; *Traité de chimie générale et expérimentale,* 2 vols. (Paris, 1844–1846); "Recherches sur l'eau régale et sur un produit particulier auquel elle doit ses principales propriétés," in *Annales de chimie,* 3rd ser., **17** (1846), 24–42; and "Recherches anatomiques et physiologiques sur le développement du foetus et en particulier sur l'évolution embryonnaire des oiseaux et des batraciens," in *Mémoires présentés par divers savants à l'Académie des sciences,* 2nd ser., **11** (1851), 469–692, written with Martin Saint-Ange.

II. SECONDARY LITERATURE. Works on Baudrimont are L. Micé, "Discours d'ouverture de la séance publique du 19 mai 1881 (éloge de M. Baudrimont)," in *Actes de l'Académie de Bordeaux,* 3rd ser., **42** (1880), 729–766; and "Éloge de M. Baudrimont, notes complémentaires," *ibid.,* **44** (1882), 557–624, a typically laudatory *éloge,* very detailed in its account of Baudrimont's life and publications, that is his only extensive biography; and J. R. Partington, *A History of Chemistry,* IV (London, 1964), 391–393.

SEYMOUR H. MAUSKOPF

BAUER, EDMOND (*b.* Paris, France, 26 October 1880; *d.* Paris, 18 October 1963), *physics, physical chemistry.*

The son of a businessman, Bauer was educated at the *lycées* Condorcet and Janson de Sailly, and graduated from the University of Paris. At Paris he was first assistant to Jean Perrin; he worked for some time under Rubens and Nernst; and he engaged in theoretical and experimental research on luminescence and blackbody radiation under Langevin, writing his thesis on the latter work (1912). Bauer was married in 1911 and had four children. Severely wounded at Charleroi in 1914, he was a prisoner of war in Germany for three years. Once Alsace had been returned to France in 1919, he was appointed to the University of Strasbourg under Pierre Weiss; in 1928 Langevin asked him to be *sous-directeur* of his laboratory at

the Collège de France. During World War II, Bauer did not accept an invitation to go to the United States, and stayed in Vichy France; after the Gestapo arrested his daughter, who was fighting in the Resistance with her brothers, he had to seek refuge in Switzerland. His daughter was sent to Ravensbrück, but survived the war. His eldest son also was captured, and died in Neuengamme. Bauer's last position was as professor of physical chemistry at the University of Paris.

Bauer's main experimental and theoretical research dealt with radiation emitted by flames and metallic vapors, which he showed to be thermal; precise determination of the Stefan constant; diffusion of light by high-altitude atmosphere at the Mont Blanc observatory, involving Avogadro's number, the coming of night, and second twilight; the ferromagnetic equation of state; group theory and quantum mechanics; hydrogen bonding and the structure of water and ice, determined from vibration spectra; differential infrared detection of impurities in gas; dielectric dispersion and phase transformation, involving relaxation time and lifetime of hydrogen bonds; and chemical kinetics.

Throughout his life Bauer was keenly interested in the origin and development of the fundamental notions of physics. He contributed to treatises on the history of science and conducted much original research in that field.

Bauer's activity as a professor and an initiator of research stimulated a generation of French physicists. A man of great culture and inexhaustible generosity, he made himself always available to any scientist asking for help or direction in research. Thus, the sum of his work was more than what has been published under his name.

BIBLIOGRAPHY

I. ORIGINAL WORKS. Bauer's books on physics include *La théorie de Bohr* (Paris, 1922); *Les bases expérimentales immédiates de la théorie de quanta* (Paris, 1933), written with Pierre V. Auger, Louis de Broglie, and M. Courtines; *Introduction à la théorie des groupes et à ses applications à la physique quantique* (Paris, 1933), trans. by Paul E. H. Meijer as *Group Theory, the Application to Quantum Mechanics* (Amsterdam, 1962, 1965), and by N. D. Ershovoï as *Vvedenie v teoriu grup i ee prilozhenie k kvantovoï fizike* (Kiev, 1937); *La théorie de l'observation en mécanique quantique* (Paris, 1939), written with Fritz London (analysis, more concise than von Neumann's, of the modification of information on a system by the act of observation; the difference in statistical meaning between representation of information by a wave function and by a density matrix is clearly outlined); and *Champs de vecteurs et de tenseurs* (Paris, 1955).

He also wrote more than fifty original papers on his experimental and theoretical research in physics and physical chemistry. They include "Les coefficients d'aimantation des gaz paramagnétiques et la theorie du magnétisme," in *Journal de physique,* 6th ser., **1** (1920), 97–128, written with Auguste Piccard; "Les propriétés thermoélastiques des métaux ferromagnétiques et le champ moléculaire," *ibid.,* **10** (1929), 345–359; "Sur la déformation des molécules en phase condensée et la liaison hydrogène," *ibid.,* 7th ser., **9** (1938), 319–330, written with M. Magat; "Sur la théorie des diélectriques polaires," in *Cahiers de physique,* no. 20 (1944), 1–20; no. 21 (1944), 21–37; no. 27 (1945), 33–38; and "A Theory of Ultrasonic Adsorption in Unassociated Liquids," in *Proceedings of the Physical Society,* **52** (1949), 141–154.

Bauer's contributions to the history of science include *L'électromagnétisme, hier et aujourd'hui* (Paris, 1929); "L'oeuvre scientifique de Coulomb," in *Bulletin de la Société française des radioélectriciens,* 5th ser., **7** (1937), 1–18; and the chapters on electricity and magnetism in R. Taton, ed., *Histoire générale des sciences: II, La science moderne* (1958), 520–539; III, pt. 1, *Le XIXe siècle* (1961), 201–260; and III, pt. 2, *Le XXe siècle* (1964), 239–253. He also wrote all the chapters on the history of science in the seven vols. of Malet and Isaac's *Cours d'histoire,* the official textbook in all French secondary schools since 1928.

II. Secondary Literature. Information on Bauer's life and work is in the obituary by K. K. Darrow in *Physics To-day,* **17** (June 1964), 86–87; and in M. Letort, G. Champetier, J. Wyart et al., "Hommage à Edmond Bauer," in *Journal de chimie physique,* **61** (1964), 955–984.

Daniel Massignon

BAUER, FERDINAND LUCAS (*b.* Feldsberg, Lower Austria, 20 January 1760; *d.* Hietzing, near Vienna, Austria, 17 March 1826), *scientific illustration.*

Bauer's fame rests chiefly on his illustrations for John Sibthorp's *Flora Graeca,* the most beautiful flora of all time. With his brothers Joseph and Franz, he was educated by Father Boccius, prior of the monastery of the Merciful Brothers in Feldsberg, then worked under Nikolaus von Jacquin, professor of botany in Vienna. When Sibthorp, Oxford's professor of botany, came to Vienna in search of the *Codex vindobonensis* of Dioscorides, he learned of Bauer's work and traveled to Feldsberg to see Boccius' fourteen-volume manuscript, *Hortus botanicus,* the 2,750 plates of which were chiefly the work of Ferdinand and his brothers. So impressed was Sibthorp that he immediately engaged Bauer to serve as artist on his expedition to Greece and the Levant. In the ensuing twenty-two months (March 1786 to December 1787) Bauer painted nearly 1,000 watercolors of plants, 363 of animals, and 131 sepia landscapes. Only the plant watercolors were eventually published in the *Flora Graeca,* a herculean work that took from 1806 to 1840 and cost over £30,000.

Bauer worked on the *Flora* in Oxford until 1801, when he joined Matthew Flinders on his voyage to Australia. There Bauer and the botanist Robert Brown made a rich collection of material that included over 2,000 sketches by Bauer, 1,541 being of plants. Few of these were ever published, although Endlicher used some for his *Prodromus florae Norfolkicae.* Bauer attempted to use them in his *Illustrationes florae Novae Hollandiae,* a project that failed commercially, and in 1814 he left England for Hietzing, which was close to the gardens of Schönbrunn.

Bauer sketched rapidly and colored little on the spot, but noted gradations with a number code so that he could finish his illustrations later. Untroubled by the rigors of travel, he worked hard and achieved an incredible output. Goethe admired the three-dimensional quality of his work, and all botanists respected his accuracy. The overthrow of Linnaean taxonomy was partly due to the Australian work of Bauer and Brown.

BIBLIOGRAPHY

I. Original Works. Bauer's only writing was *Illustrationes florae Novae Hollandiae* (London, 1813). Plates are in Boccius, *Hortus botanicus,* Vols. I–IX, MS at Vaduz, Liechtenstein; Stephan Endlicher, *Prodromus florae Norfolkicae* (Vienna, 1833); Matthew Flinders, *A Voyage to Terra Australis* (London, 1814); A. B. Lambert, *A Description of the Genus Pinus* (London, 1803–1824); John Lindley, *Digitalium monographia* (London, 1821); and John Sibthorp, *Flora Graeca* (London, 1806–1840).

II. Secondary Literature. There is no detailed biography of Bauer, few of his letters have survived, and no portrait is extant. The best accounts are Wilfred Blunt, *The Art of Botanical Illustration* (London, 1950), and William Stearn, "Franz and Ferdinand Bauer, Masters of Botanical Illustration," in *Endeavour,* **19**, no. 73 (Jan. 1960), 27–35. For details of Bauer manuscripts see C. Nissen, *Die botanische Buchillustration,* 2nd ed. (Stuttgart, 1966).

R. Olby

BAUER, FRANZ ANDREAS (*b.* Feldsberg, Lower Austria, 4 October 1758; *d.* Kew, England, 11 December 1840), *botanical illustration, microscopy.*

Bauer was educated by Father Boccius in Feldsberg and worked under Jacquin in Vienna. In 1788 he accompanied Jacquin's son, Joseph, on his travels through Europe. In London they met Sir Joseph Banks and worked in his magnificent library and herbarium. Banks then engaged Bauer as artist at the

Royal Garden, Kew, and at the attractive salary of £300 a year Bauer remained there for the rest of his life. His output was not so great as that of his brother Ferdinand, for his services were not adequately utilized and much of the time he followed his own fancy. This led him into the intricacies of flower structure in strelitzias and orchids, the nature of red snow, and the structure of pollen grains. He also illustrated the works of friends with microscopical and anatomical drawings: Banks's works on cereal diseases and apple blight, Robert Brown's description of *Rafflesia* and *Woodsia,* Home's *Lectures on Comparative Anatomy,* and John Smith's account of his discovery of the apomictic plant dovewood.

Bauer achieved recognition as a microscopist but made no lasting contribution to the field. For instance, while others correctly attributed the color of red snow to an alga, he believed the causal organism was a fungus. In his study of the rye ergot he inclined to the view that nutritional factors were responsible until John Smith showed him hyphae of the fungal agent in the infected ears. On the difficult subject of the fertilization mechanism in orchids Bauer, like Robert Brown, came to the wrong conclusions because he did not know that they are cross-pollinated by insects.

Bauer's plant paintings are of outstanding beauty and scientific accuracy. From the beginning he tended to use a pure wash, whereas his brother Ferdinand relied more on body color; yet much of Ferdinand's later work is indistinguishable from that of Franz. Since the cell theory and staining and fixing techniques had not yet been developed, it is understandable that Bauer's microscopical paintings have little scientific worth.

BIBLIOGRAPHY

I. ORIGINAL WORKS. Bauer's writings include *Delineations of Exotic Plants Cultivated in the Royal Garden at Kew* (London, 1796); *Strelitzia depicta* (London, 1818); "Microscopical Observations on the Red Snow," in *Quarterly Journal of Literature, Science and the Arts,* **7** (1819), 222–229; "Some Experiments on the Fungi Which Constitute the Colouring Matter of the Red Snow Discovered in Baffin's Bay," in *Philosophical Transactions of the Royal Society,* **110** (1820), 165–172; "Croonian Lecture," *ibid.,* **113** (1823), 1–16; *Illustrations of Orchidaceous Plants* (London, 1830–1838); and "The Ergot of Rye," in *Transactions of the Linnean Society of London,* **18** (1841), 509–512. Plates by Bauer are in E. Home, *Lectures on Comparative Anatomy,* 4 vols. and 2 supps. (London, 1814–1828); and W. J. Hooker, *Genera Filicum* (London, 1842).

II. SECONDARY LITERATURE. Obituaries are in *Proceedings of the Linnean Society of London,* **1** (1849), 101–104; and *Proceedings of the Royal Society,* **4** (1843), 342–344. On Bauer's art see W. Blunt, *The Art of Botanical Illustration* (London, 1950), pp. 195–202.

R. OLBY

BAUER, GEORG. See **Agricola.**

BAUER, LOUIS AGRICOLA (*b.* Cincinnati, Ohio, 26 January 1865; *d.* Washington, D.C., 12 April 1932), *geophysics.*

Bauer, the founder of the department of terrestrial magnetism of the Carnegie Institution of Washington, was of that line of students of magnetism starting with Halley, and including Gauss and Sabine, who have considered this topic one of the great scientific problems. Originally trained as a civil engineer, he worked from 1887 to 1892 as a computer, primarily on geomagnetism, for the U.S. Coast and Geodetic Survey. From 1892 to 1895 he studied at the University of Berlin and earned the Ph.D. for a thesis on the secular variation of the earth's magnetism.

On his return from Europe, Bauer taught at the University of Chicago (1895–1896) and the University of Cincinnati (1897–1899). While at the latter he also directed the magnetic work of the Maryland Geological Survey, and from 1899 to 1906 he supervised the expanded geomagnetic work of the Coast and Geodetic Survey. During this period he became convinced that determination of the physical nature of terrestrial magnetism and the explanation of its secular variation required a great increase in observational data. When the Carnegie Institution of Washington was founded in 1902, Bauer proposed a worldwide program of research on the magnetic and electrical condition of the earth and its atmosphere and was appointed head of the new department in 1904.

Undoubtedly the most heroic aspect of Bauer's work at the Carnegie Institution was the extensive mapping of the earth's magnetic field on a vast area of the seas, quite like the efforts of Halley and Sabine. Like Gauss and Sabine, he also promoted the establishment of magnetic stations and was active in the mathematical analysis of observations and in the development of a mathematical theory. Bauer paid considerable attention to instrumentation. Although much of his work was devoted to the main magnetic field, he was also interested in atmospheric electricity and extraterrestrial influences on the earth's magnetism. An example of the latter is his work on the effects of solar eclipses.

Bauer was very active in various international geophysical organizations, such as the International Union of Geodesy and Geophysics. In 1896 he founded the periodical *Terrestrial Magnetism* (later

Terrestrial Magnetism and Atmospheric Electricity), which he edited until 1927. Now known as the *Journal of Geophysical Research,* the periodical was influential, highly regarded, and played an important role in stimulating geophysical investigations at a time when interest in classical physics was being overshadowed by newer intellectual pursuits.

Bauer commited suicide at the age of sixty-seven.

BIBLIOGRAPHY

I. ORIGINAL WORKS. Bauer has over 300 titles to his credit. While no comprehensive bibliography exists, H. D. Harradon, "Principal Published Papers of Louis A. Bauer," in *Terrestrial Magnetism and Atmospheric Electricity,* **37** (Sept. 1932), 220–224, is a good introduction to Bauer's writings.

Bauer documents are in the Archives of the Carnegie Institution of Washington and the records of the U.S. Coast and Geodetic Survey in the U.S. National Archives, Washington, D.C. The W. F. G. Swann Papers at the American Philosophical Society Library, Philadelphia, and the George Ellery Hale Papers at the Millikan Library, California Institute of Technology, Pasadena, California, contain interesting Bauer correspondence.

II. SECONDARY LITERATURE. *Terrestrial Magnetism and Atmospheric Electricity,* **37** (Sept. 1932), 203–420, was the Bauer Memorial Number; it contains a scattering of comments on Bauer, as well as many articles showing the state of development of fields he worked in. Both the *Annual Report* of the U.S. Coast and Geodetic Survey and the *Yearbook* of the Carnegie Institution of Washington for the years when Bauer was with them are replete with information on his career. The best general secondary source on Bauer's principal intellectual interest is Sydney Chapman and Julius Bartels, *Geomagnetism,* 2 vols. (Oxford, 1940), which has many specific references to Bauer. An additional virtue of the work is that the authors have, implicitly and explicitly, attempted to place recent works in a historical perspective.

NATHAN REINGOLD

BAUHIN, GASPARD (*b.* Basel, Switzerland, 17 January 1560; *d.* Basel, 5 December 1624), *anatomy, botany.*

Bauhin was the younger son of Jean Bauhin, a French Protestant physician and surgeon from Amiens, who sought refuge from religious persecution by settling in Basel in 1541 and became attached to the university. From childhood, Bauhin was taught anatomy by his father and botany by his brother Jean (almost twenty years his senior), who became a botanist of some repute. In 1572 Gaspard entered the University of Basel, where Felix Plater and Theodore Zwinger were among his teachers. He received the degree of Bachelor of Philosophy in 1575, and conducted his first medical disputation in 1577. In October of that year he went to Padua, where for eighteen months he studied anatomy with Girolamo Fabrizio (Fabricius ab Aquapendente), saw seven bodies dissected, "and even assisted myself in the private dissections." He also attended the teaching of Marco de Oddi and Emilio Campolongo at the Hospital of St. Francis, and probably that of Melchior Wieland (Guillandinus) in the botanical garden. At some time during this period he visited Bologna and received instruction in anatomy from Giulio Cesare Aranzio. In the spring of 1579 he signed the register of the University of Montpellier, but by his own account he spent more time in Paris attending the anatomies conducted by Sévérin Pineau, professor of anatomy and surgery, "whom I assisted in dissecting at his request." In May 1580 he was in Tübingen. Early the following year he returned to Basel, where "at the urgent request of the College of Physicians, I began to dissect bodies." He held his first public anatomy on 27 February. The disputation for his doctor's degree took place on 19 April and had as its subject *De dolore colico.* He received the doctorate on 2 May, and on 13 May he was made a member of the Faculty of Medicine.

In April 1582 Bauhin was made professor of Greek; two years later he became *consiliarius* in the Faculty of Medicine, an office he held until his death. He was dean of the faculty nine times (beginning in 1586), and four times rector of the university (1592, 1598, 1611, 1619). Despite his professorship of Greek, he continued to teach both anatomy and botany, holding public anatomies in the winter and taking the students on botanical expeditions in the summer. As a result of his efforts, work was begun on a permanent theater for anatomical demonstrations, and a botanical garden was laid out. In September 1589 he was rewarded by the creation of a special chair in anatomy and botany. During these years Bauhin's private medical practice grew, and in 1597 he was associated with his brother Jean as physician to Duke Frederick of Württemberg. From 1588 on, he was occupied with the writing and publishing of a series of books on anatomy. His first major botanical work was *Phytopinax* (1596). When his *Pinax* appeared in 1623, it was said to be the result of forty years' work. On the death of Felix Plater in 1614, Bauhin succeeded him as archiater to the city of Basel. The following year he was appointed professor of the practice of medicine. He was married three times: in 1581 to Barbara Vogelmann of Montbéliard, by whom he had one daughter; in 1596 to Maria Bruggler of Bern;

and sometime after 1597 to Magdalena Burckhardt, by whom he had two daughters and one son, Jean Gaspard, who succeeded his father as professor of anatomy and botany in 1629 and became professor of the practice of medicine in 1660. In 1658 Jean Gaspard published the first volume, all that was ever published of the intended twelve, of his father's *Theatrum botanicum.*

No great anatomical discoveries can be ascribed to Bauhin. He himself believed that he was the first to describe the ileocecal valve, which was long known as the *valvula Bauhini;* and in a number of his anatomical writings he gives an account of how he first found it during a private dissection that he performed as a student in Paris in 1579. His greatest contribution to anatomy was the reform he introduced into the nomenclature, particularly into that of muscles. Because it is very easy to make mistakes in the enumeration of muscles if they are merely called first, second, third, etc., and because different anatomists had named different muscles in this way, not agreeing on the order of the enumeration, Bauhin decided that it was better to use another kind of terminology. He therefore named some muscles according to their substance (semimembranosus, etc.), others according to their shape (deltoid, scalene, etc.), some according to their origin (arytenoideus, etc.), and others according to their origin and insertion (styloglossus, crycothyroideus, etc.). Some he named according to the number of their heads (biceps, triceps), some according to their amount (vastus, gracilis), some according to their position (pectoralis, etc.), and others according to their use (supinator, pronator, etc.). He also decided that veins and arteries should be named according to their use or course, and nerves according to their function. The system had so many obvious advantages over the old method that it was adopted by all subsequent anatomists.

There is no doubt that Bauhin's contribution as a teacher of anatomy was considerable, particularly through his many books on the subject. His first textbook was *De corporis humani partibus externis* (1588), written at the request of his students after the public anatomy he had given two years before. It was intended to be a succinct, methodical account of the external parts of the body suitable for beginners and to be used in conjunction with the *Tabulae* of Vesalius. The second volume, dealing with "similar and spermatic" parts, was published in 1592. Meanwhile, in 1590, Bauhin published his first complete textbook, *De corporis humani fabrica: Libri IIII.* It was a systematic account written from the point of view of dissection, and intended for students rather than professors of anatomy. It respected ancient findings

and theories, but did not hesitate to correct them when the results of actual dissections made it necessary. Its appearance provoked a storm of abuse from the Galenists. Corrected and enlarged by a description of female anatomy, it was republished in 1597 as *Anatomica corporis virilis et muliebris historia.* In 1605 all these anatomical writings were brought together, revised and enlarged, and published in Bauhin's most celebrated anatomical textbook, *Theatrum anatomicum,* which was accompanied by copper engravings based on the drawings of Vesalius and entitled *Vivae imagines partium corporis.* The *Theatrum anatomicum* soon acquired the reputation of being the best anatomical textbook available. It was systematic, gave adequate consideration to the ancient authorities, did not go into too much detail over the controversies, had a series of eminently useful footnotes, and mentioned anatomical anomalies and pathological findings. Its illustrations, although poor in comparison with those of Vesalius, were adequate for anyone using the book to accompany an actual dissection. It was this work that William Harvey chose as the basis for his Lumleian Lectures to the College of Physicians in London in 1616. It was translated into English in 1615 by Helkiah Crooke and, conflated with the textbook of Laurentius, was published under the title of *Microcosmographia, A Study of the Body of Man.*

Although Bauhin said that his anatomical works contained few novelties, he did include new anatomical findings that were obviously true, for he believed firmly that truth demonstrable to the beholder outweighs the opinions of the authorities. In this way, in his *Libri IIII* of 1590 he included a description of the valves in the veins, as demonstrated sixteen years previously by Girolamo Fabrizio in Padua. (Fabrizio's own account was not published until 1603.) In the same work he pointed out that there was no need to suppose the existence of pores in the interventricular septum of the heart, for the venous blood could more easily go to the lungs from the right ventricle through the pulmonary artery, there be refined and mixed with air, and return to the left ventricle through the pulmonary vein. He gives no authorities for this view, and it is open to conjecture whether he formulated it independently. In the end, it is a view he seems to have rejected, for in his later works he repeats Galen's traditional teaching, supported by his own findings of conspicuous pores in the septum of an ox's heart, which had been prepared by boiling.

Bauhin's medical works include treatises on the bezoar stone, on Caesarean section, on hermaphrodites and other monstrous births, and on the pulse.

His two pharmacological writings are designed as handbooks for young physicians.

As in anatomy, Bauhin's great contribution to botany was to nomenclature. He was primarily a taxonomist, concerned with the collecting and classifying of a great variety of plants. He is also remembered for separating botany from materia medica. His botanical fame rests chiefly on two works, *Prodromos* (1620) and *Pinax* (1623). In the latter he discarded the old alphabetical manner of enumeration and stated that any sound method of classification must be based on affinities. Consequently, he distinguished between genus and species and introduced a system of binomial nomenclature. He made little or no progress in classifying the genera into orders and classes. Although the system was not his invention, he vastly improved it, and thereafter it was generally adopted. His botanical work was commemorated by L. P. C. Plumier, who gave the name *Bauhinia* to a family of tropical trees; and Linnaeus, in memory of both Gaspard and his brother Jean, called one species of this family *Bauhinia bijuga*.

Although in a final assessment of his work in botany and anatomy it can be said that little was truly original, Bauhin's influence in both fields lasted for well over a century. His great merit was his ability to treat his subjects in an orderly and methodical manner, for he had a capacity to think clearly and an ability to work without tiring. Quiet and reserved, he can be remembered in William Harvey's words concerning him: "a rare industrious man."

BIBLIOGRAPHY

I. ORIGINAL WORKS. Bauhin's anatomical and medical writings, presented in the order that best indicates their relationships, are ΑΠΟΘΕΡΑΠΕΙΑ ΙΑΤΡΙΚΗ *Quam medicae laureae causa Casparus Bauhinus . . . subibit* (Basel, 1581); ΥΣΤΕΡΟΤΟΜΟΤΟΚΙΑ *Francisci Rousseti, Gallicè primum edita, nunc vero Caspari Bauhini . . . Latine reddita* (Basel, 1588), which contains Bauhin's account of Caesarean section in the "Appendix varias et novas historias continens . . . a Casparo Bauhino addita," repr. many times in various eds. of Rousset's work; *De corporis humani partibus externis Tractatus* (Basel, 1588), repr. as *Anatomes . . . liber primus* (Basel, 1591); *Anatomes liber secundus partium similarium spermaticarum tractationem . . . continens* (Basel, 1592), repr. with *Anatomes . . . liber primus* (Basel, 1597); *De corporis humani fabrica: Libri IIII. Methodo anatomico in praelectionibus publicis proposita . . .* (Basel, 1590), enl. and repr. as *Anatomica corporis virilis et muliebris historia* (Lyons, 1597) and repr. as *Institutiones anatomicae corporis virilis et muliebris historiam exhibentes* (Lyons, 1604; Basel, 1609; Frankfurt, 1616);

Praeludia anatomica (Basel, 1601), also known as *Disputatio prima; Disputatio secunda de partibus humani corporis* (Basel, 1602); *Disputatio tertia. De ossium natura* (Basel, 1604); "Introductio in doctrinam pulsuum ad tyrones," in *Ars sphygmica . . . à Josepho Struthio . . . conscripta . . .* (Basel, 1602), pp. 1–23; *Theatrum anatomicum novis figuris aenis illustratum . . .* (Frankfurt, 1605; new and enl. ed., 1621); *Appendix ad Theatrum anatomicum . . . sive Explicatio characterum omnium . . .* (Frankfurt, 1600), frequently bound with the *Theatrum anatomicum* of 1605; *Vivae imagines partium corporis humani aeneis formis expressae & ex Theatro anatomico . . . desumptae* (Frankfurt, 1620, 1640), frequently bound with the 1621 ed. of *Theatrum anatomicum; De compositione medicamentorum sive medicamentorum componendorum ratio et methodus* (Offenbach, 1610); *De lapidis Bezaar . . . ortu, natura, differentiis, veròque usu* (Basel, 1613, 1624, 1625); *De homine oratio* (Basel, n.d.), an oration delivered to the Faculty of Medicine 16 Nov. 1614; *De hermaphroditorum monstrosorumque partuum natura* (Oppenheim, 1614); and *De remediorum formulis Graecis, Arabibus & Latinis usitatis . . . libri duo, Iuniorum medicorum usum editi* (Frankfurt, 1619).

Bauhin's botanical works are *De plantis a divis sanctisve nomen habentibus . . . Ioanni Bauhini* (Basel, 1591, 1595), which contains letters from Gaspard Bauhin to Conrad Gesner; ΦΥΤΟΠΙΝΑΞ *seu enumeratio plantarum ab herbariis nostro seculo descriptarum. . . . Additis aliquot hactenus non sculptarum plantarum vivis iconibus* (Basel, 1596); *Animadversiones in historiam generalem plantarum Lugduni editam. Item catalogus plantarum* (Frankfurt, 1601); ΠΡΟΔΡΟΜΟΣ *Theatri botanici in quo plantae supra secentae ab ipso primum descriptae cum plurimis figuris proponuntur* (Frankfurt, 1620; Basel, 1671); *Catalogus plantarum circa Basileam spontè nascentium . . . in usum scholae medicae* (Basel, 1622; 3rd ed., 1671; 2nd ed. unknown); ΠΙΝΑΞ *Theatri botanici sive Index in Theophrasti, Dioscoridis, Plinii et botanicorum qui à seculo scripserunt opera* (Basel, 1623, 1671, 1740); and *Theatri botanici sive Historiae plantarum . . . liber primus*, Jean Gaspard Bauhin, ed. (Basel, 1658).

In addition, Bauhin edited a number of anatomical and botanical works by other writers, of which the most notable is his ed. of P. A. Mathioli's *Opera quae extant omnia* (Frankfurt, 1598). Letters written by Bauhin on medical and anatomical subjects are in J. Hornung, *Cista medica* (Nuremberg, 1625); and Wolfgang Wedel, *Ephemeriden Academiae Naturae Curiosorum* (1673). Unpublished letters in holograph, both Latin and vernacular, are in the library of the University of Basel.

II. SECONDARY LITERATURE. Albrecht Burckhardt, *Geschichte der Medizinischen Fakultät zu Basel, 1460–1900* (Basel, 1917), pp. 95–123, provides a good account of Bauhin's life and a bibliography of his writings. Older studies are J. von Hess, *Bauhins Leben* (Basel, 1860); and Ludovic Legré, *Les deux Bauhins* (Marseilles, 1904). References to Bauhin's botanical works are in R. J. Harvey-Gibson, *Outlines of the History of Botany* (London, 1919); G. A. Pritzel, *Thesaurus literaturae botanicae* (Leipzig, 1851; new ed.,

Milan, 1871); and *Sachs' History of Botany 1530–1860,* trans. Garnsey and Balfour (Oxford, 1890).

GWENETH WHITTERIDGE

BAUHIN, JEAN (*b.* Basel, Switzerland, 12 February 1541; *d.* Montbéliard, principality of Württemberg–Montbéliard, 26 October 1613), *botany.*

Bauhin's father, Jean, a Protestant from Amiens, was physician to Margaret of Navarre; while in Paris he married Jean Fontaine. The revival of religious persecution under Francis I resulted in the exile of the Bauhin family to Basel in 1541; there the younger Jean, eldest of seven children, was born shortly afterward. His younger brother was Gaspard Bauhin.

Bauhin's basic education was in Basel; his early teachers were the religious reformer Coelius Secundus Curione, an Italian exile who was professor of belles lettres at Basel University, and Conrad Gesner of Zurich. Bauhin completed his studies by paying short visits to foreign universities between 1560 and 1563. His most prolonged stay was in Montpellier, from 1561 to 1562. It has generally been assumed that he obtained an M.D. at Montpellier, but this is not supported by the university records. Nevertheless, he established a medical practice at Lyons in 1563 and married Denyse Bornand (Bernhard) in 1565. Of the six children born to them, three daughters lived to maturity, all of whom married doctors.

Renewed religious persecution forced Bauhin into exile in Geneva, where he again practiced medicine, in 1568; two years later he was appointed professor of rhetoric at Basel. Finally, in 1571 he became physician to Duke Frederick of Württemberg, the ruler of the dual principality of Württemberg–Montbéliard and an enlightened and sympathetic patron. Bauhin established botanical gardens, in which he grew both native and exotic plants, at Montbéliard and Stuttgart. He excavated the ruins of Mandurium (Mandeure) and displayed his archaeological collections in a museum at Duke Frederick's chateau. Bauhin was also appointed archiater by Frederick. In 1675 he was instrumental in establishing the College of Medical Practitioners in Montbéliard, which regulated the duties of all practitioners and provided free medical services to the poor. His second marriage, in 1598, was to Anne Grégoire, a Protestant refugee from Besançon.

Botany was Bauhin's primary interest from his youth. His teacher and friend, Conrad Gesner, spent his last years compiling a *Historia plantarum,* a work on descriptive botany paralleling his published *Historia animalium;* the first documentary evidence relating to Bauhin's concern with botany is his correspondence with Gesner from 1560 to 1565. He collected plants and sent the herbarium specimens to Gesner, whose artist and engraver prepared appropriate illustrations. Gesner's death left this great collaborative enterprise incomplete, and the unfinished *Historia plantarum* was not published until the eighteenth century. Bauhin's *Historia plantarum* may be regarded as a reconstruction of Gesner's work, carrying on the latter's humanistic learning and enthusiasm for exploration and description. Yet Bauhin realized that a comprehensive description of flora could not be undertaken single-handed, and he relied on informants from many countries.

Bauhin's longest journeys were made during his medical studies. In 1561 he accompanied Gesner on a tour of the Rhaetian Alps, during which they pioneered the study of the alpine flora, particularly in the Mount Albula region. Upon Gesner's advice, in the same year he traveled to Tübingen to meet Leonhard Fuchs. He next spent more than a year at Montpellier under Rondelet, one of the most gifted Renaissance naturalists and medical teachers. He conducted further systematic explorations in the Montpellier region, and in 1562 he reported to Gesner that he was compiling a *Catalogum stirpium Monspeliensium.* This was never published separately, but information from it was incorporated into the *Historia plantarum.* References in this work show that Bauhin and his colleagues Jacques Raynaudet and Leonard Rauwolff explored Provence and Languedoc. Most of 1563 was spent studying the flora of north Italy and the Apennines. He also visited botanical gardens in Padua and Bologna, which later provided him with materials for his own gardens. His first garden was established about 1564 at Lyons, where he collaborated with Jacques Daleschamps in studying the local flora. It is not certain whether Bauhin played any significant role in the compilation of the anonymous *Historia plantarum generalis* (1586), which is generally attributed to Daleschamps. Bauhin's duties at Montbéliard did not interrupt his explorations, which were often undertaken in connection with the missions of Duke Frederick. In his later years Bauhin's closest collaborator was his son-in-law Jean-Henri Cherler, who had married his youngest daughter, Geneviève. Cherler contributed a knowledge of the floras of Belgium and England, and was named as joint author of the *Historia plantarum*—an honor he scarcely deserved.

The two botanical works published by Bauhin in 1591 and 1593 give no intimation of the *Historia plantarum,* for each is a very specific exercise. The first, *De plantis a divis sanctisve nomen* (1591), is an alphabetical list of plants named after saints, with full citations to the botanical literature. Bauhin showed that commonly more than one species—and often quite unrelated ones—were named after a particular saint. The other work, *De plantis Absynthii nomen* (1593), underlines the great confusion that existed over a single type. He quoted a list, covering twelve pages, of authors who had used the name *Absynthium* and then attempted to identify each type, indicating the synonyms used by botanists. Although these works may seem obscure, they illustrate Bauhin's critical talents, his great knowledge of the botanical literature, and his ability to solve problems of nomenclature.

Bauhin's reputation as a botanist rests upon the encyclopedic *Historia plantarum universalis* (1650–1651), which was not published until thirty-seven years after his death. It completely overshadows the works published during his lifetime, which give only a limited indication of his originality. He was reluctant to publish, and most of his works were concentrated in the years 1593–1598. Most of them dealt with subjects peripheral to his main botanical interests, although they displayed the erudition and caution of his later botanical works. A specifically medical work described the remedies for diseases contracted as a result of animal bites (*Histoire notable de la rage*, 1591). More significant is the description of swarms of locusts and grasshoppers that Felix Platter and François Valleriol had recorded at Arles in 1553, and the swarms of insects at Porrentruy, Montbéliard, in 1590 (*Traicté des animauls, aians aisles*, 1593). He suggested that these phenomena were a manifestation of the power of Providence and asserted that insects were a cause of disease. His longest and most popular medical work was a description of European mineral waters and baths, based on a study of the springs at Bad Boll, near Groppingen, Württemberg (*Historia novi et admirabilis fontis*, 1598). This was the most detailed work on the subject written in the sixteenth century, and contains a lengthy appendix that gives an intimation of Bauhin's abilities as a naturalist. It consists of a series of illustrations, most of his fossil collections, that was probably inspired by Gesner's *De rerum fossilium* (1565). The most original were the illustrations of sixty varieties of apples and thirty-nine of pears, all collected in the alpine region. These large and distinctive woodcuts show the value of illustration for depicting fine morphological distinctions.

From 1600 until his death, Bauhin was engaged in compiling the *Historia plantarum*. The virtually complete manuscript passed to Dominic Chabrey, an Yverdon physician, who published a summary of the work in 1619. Eventually, François Louis de Graffen-ried of Yverdon financed the publication of the work, which appeared in three volumes. Despite the delay in publication, the work was not obsolete. It contained the description and synonyms of 5,226 plants, primarily from Europe, but with some Eastern and American floras. This represented the fruits of the explorations of Bauhin and his informants, and compilation from ancient and contemporary literature. It also indicates the great progress of botany in the sixteenth century: Brunfels had described 240 plants in 1532; the less accurate *Historia plantarum generalis* of 1587 reached 3,000; the only works to describe more than this number in the succeeding century were by Jean Bauhin and his brother Gaspard. Jean's work was paralleled by his brother's *Pinax theatri botanici* (1623). This contained 6,000 plants but gave only names and synonyms. Hence the particular value of Jean's work was its concise and accurate description of each type.

BIBLIOGRAPHY

I. ORIGINAL WORKS. Bauhin's works are *Histoire notable de la rage des loups advenue l'an MDXC, avec les remèdes pour empescher la rage* . . . (Montbéliard, 1591), trans. into Latin and German the same year; *De plantis a divis sanctisve nomen habentibus . . . Additae sunt Conradi Gesneri epistolae hactenus editae a Casparo Bauhino* (Basel, 1591); *De plantis Absynthii nomen habentibus, caput desumptum ex clarissimi ornatissimique viri D. Doct. Ioannis Bauhin . . . Tractatus item de absynthiis Claudii Rocardi* (Montbéliard, 1593; repr. 1599); *Traicté des animauls, aians aisles, qui nuisent par leurs piqueures ou morsures, auec les remèdes* . . . (Montbéliard, 1593); *Historia novi et admirabilis fontis balneique Bollensis in Ducatu Wirtembergico* . . . (Montbéliard, 1598), trans. into German (Stuttgart, 1602) and also reissued under two other titles: *De thermis aquisque medicatis Europae praecipuis opus succinctum* (Montbéliard, 1600) and *De aquis medicatis nova methodus libri quatuor* (Montbéliard, 1607/1608, 1612); *Histoire ou plustot un simple et véritable récit des merveilleux effects qu'une salubre fontaine située au comté de Montbéliard* . . . (Montbéliard, 1601), also trans. into German (Montbéliard, 1602); *De auxiliis adversus pestem* (Montbéliard, 1607), reported to have been trans. into German by Thiebaud Noblot in the same year; *Joh. Bauhini et Joh. Henr. Cherleri, Historiae plantarum generalis novae et absolutissimae, quinquaginta annis elaboratae jam prelo commissae, Prodromus* (Yverdon, 1619); and *Historia plantarum universalis, nova et absolutissima, cum consensu et dissensu circa eas. Auctoribus Joh. Bauhino et Joh. Henr. Cherlero philos. et med. doctoribus Basiliensibus; Quam recensuit et auxit Dominicus Chabraeus, med. doct. Genevensis; Juris vero publici fecit Franciscus Lud. a Graffenried*, 3 vols. (Yverdon, 1650–1651).

II. Secondary Literature. Works dealing with Bauhin and his accomplishments are A. Arber, *Herbals,* 2nd ed. (Cambridge, 1953), pp. 70, 85, 93, 113–119; P. Delaunay, *La zoologie au seizième siècle* (Paris, 1962), pp. 71, 225, 229, 230, 238, 302; L.-M. Dupetit-Thouars, in *Biographie universelle,* III, 556–559; C. Duvernay, *Notices sur quelques médecins, naturalistes et agronomes nés ou établis à Montbéliard dès le seizième siècle* (Besançon, 1835), pp. 1–24; É. and É. Haag, *La France protestante,* 2nd ed., I (Paris, 1887), 1016–1023; C. Jenssen, in *Allgemeine deutsche Biographie,* II, 149–151; L. Legré, *La botanique en Provence au XVIe siècle. Les deux Bauhin, Jean Henri Cherler et Valerand Dourez* (Marseille, 1904); J. E. Planchet, *Rondelet et ses disciples ou la botanique à Montpellier au XVIme siècle* (Montpellier, 1866); C. Roth, "Stammtafeln einiger ausgestorbener Gelehrtenfamilien," in *Basler Zeitschrift für Geschichte und Altertumkunde,* **15** (1916), 47–55; and C. P. J. Sprengel, *Geschichte der Botanik* (Leipzig, 1817–1818), pp. 364–369.

Charles Webster

BAUMÉ, ANTOINE (*b.* Senlis, France, 26 February 1728; *d.* Paris, France, 15 October 1804), *chemistry, pharmacy.*

The son of Guillaume Baumé, who ran two inns in Senlis, Antoine Baumé seems to have had the advantages of a fairly well-to-do home. About 1743 he was apprenticed to a pharmacist in Compiègne, but by 1745 he was working in C. J. Geoffroy's dispensary in Paris, where he was able to indulge and develop his interest in theoretical chemistry as well as in pharmacy. In 1752 he became *maître apothicaire,* and opened his own dispensary in the Rue St.-Denis, Paris, the following year; it was moved to the Rue Coquillère in 1762.

In addition to its role as a local pharmacy, Baumé's dispensary supplied drugs in bulk to pharmacies and hospitals over a very wide area and manufactured drugs and other chemicals in large quantities. In 1767 he began the first large-scale production of sal ammoniac in France; he doubtless owed something in this connection to Geoffroy, who had been the first in France to make sal ammoniac. Baumé also supplied industrial and laboratory apparatus, some of which he designed himself. In particular, his areometer (1768) was an important step forward, in that it possessed a scale having two fixed points (the density of distilled water and that of a salt solution of known concentration), thus enabling properly calibrated instruments to be produced. In 1777, Baumé won first prize for an essay on the best furnaces, alembics, and other apparatus to be used in the distillation of wine.

In 1757, Baumé and Macquer began a series of courses in chemistry and pharmacy that continued

for sixteen years. Baumé equipped the laboratory, supplied the funds, and prepared all the experiments to be carried out. An outline of this course, *Plan d'un cours de chymie expérimentale et raisonnée . . .,* was published in 1757. Apart from this, Baumé published a number of works on chemistry and pharmacy that ran into several editions. He also contributed to the *Dictionnaire des arts et métiers* (1766).

From 1755 until the end of his life, Baumé produced many memoirs, the first of these being a dissertation on ether. (One of the many bitter wrangles he entered into with his fellow scientists was a dispute with L. C. Cadet de Gassicourt concerning the best and cheapest method of preparing ether [1775].) Baumé's many disputes nearly cost him membership in the Académie des Sciences, but he was finally elected as *adjoint-chimiste* on 25 December 1772; he became *associé* in 1778 and *pensionnaire* in 1785.

Despite his perceptiveness in some matters—he pointed out that two affinity tables were necessary (*Chymie expérimentale et raisonnée,* **1** [1773], 22), one for reactions done in the wet way and one for reactions done in the dry way—Baumé resisted the new theories in chemistry until the end of his life. He remained a determined phlogistonist—as late as 1797 he stated quite categorically that water could not be decomposed. With the coming of the Revolution, Baumé lost his fortune, including his pension as a member of the Academy, but in 1795, as a result of an appeal to the Committee of Public Instruction, he received a sum of money that enabled him to open another dispensary in 1796. At the newly formed Institut de France, he obtained only a place as *associé non résidant,* which carried a minimum of privileges.

BIBLIOGRAPHY

An excellent and detailed bibliography of Baumé's works is given in René Davy, *L'apothicaire Antoine Baumé (1728–1804)* (Cahors, 1955), pp. 141–147.

Other works on Baumé are L. C. Cadet de Gassicourt, *Éloge de Baumé, apothicaire* (Brussels, an XIII [1805]); Delunel, "Éloge de Baumé," in *Société de pharmacie de Paris, séance publique* (an XIV [1806]), p. 27; and N. Deyeux, "Éloge de Baumé," in *Annales de chimie et de physique,* **4** (30 messidor an XIII [1805]), 105.

E. McDonald

BAUMHAUER, EDOUARD HENRI VON (*b.* Brussels, Belgium, 18 September 1820; *d.* Haarlem, Netherlands, 18 January 1885), *chemistry.*

The son of a solicitor-general at the High Court of Justice, Brussels, Baumhauer attended the Latin

School of Brussels, then studied classical literature and natural sciences at Utrecht, graduating with a degree in classical literature in 1843 and one in natural sciences in 1844. From 1845 to 1847 he was professor of physics and chemistry at the Royal Athenaeum in Maastricht, and from 1848 to 1865, professor of chemistry at the Athenaeum Illustre in Amsterdam. From 1865 until his death Baumhauer was perpetual secretary of the Dutch Society, Haarlem, and edited the *Archives néerlandaises des sciences exactes et naturelles.*

Baumhauer was primarily a teacher. During his stay in Amsterdam, however, he did a great deal of work in practical chemistry. In 1853 and the years following, he developed a method for the quantitative determination of oxygen in organic substances. His investigations on drinking water led in 1854 to the establishment of a waterworks in Amsterdam which drew its water from the coastal dunes. Baumhauer analyzed milk in 1857, and the following year he effected the passage of a municipal regulation calling for the inspection of food. In 1859 he began work on the accurate determination of the strength of alcohol, which was important for the levying of taxes.

Baumhauer published papers on a great variety of subjects, including diamonds, marine pileworms, and meteorology. After oil had been discovered in great quantities, he analyzed a number of samples from the Dutch East Indies and gave advice to those drilling for oil in that territory.

As a theoretical chemist Baumhauer was of minor significance. His analysis of meteoric stone, the subject of his thesis (1844), led him to theoretical speculations. He supposed, as Nördenskjöld did later, that meteorites were condensed nebulae, which, as small planets surrounded by noncondensed matter, described their own paths before they were attracted by the earth. He also held that when noncondensed matter, consisting of loose atoms, came within the earth's atmosphere, it gave rise to shooting stars, fireballs, and other forms of light phenomena associated with thermochemical effects. He believed that the aurora borealis originated from the oxidation of elementary iron and nickel particles attracted into the atmosphere by the earth's magnetic field. This theory was not accepted.

BIBLIOGRAPHY

I. Original Works. Almost all of Baumhauer's work was published in Dutch journals, and some has been translated in German and French periodicals. A complete bibliography is in Gunning, pp. 49–57, and is supplemented in Van der Beek, p. xv. Individual works are *Specimen inaugurale continens sententias veterum philosophorum Graecorum de visu, lumine et coloribus* (Utrecht, 1843); *Specimen meteorologico-chemicum de ortu lapidum meteoricorum, annexis duorum lapidum analysibus chemicis* (Utrecht, 1844); "Ueber den muthmasslichen Ursprung der Meteorsteine," in *Annalen der Physik und Chemie,* **66** (1845), 465–503; and *Beknopt leerboek der onbewerktuigde scheikunde,* 3rd ed. (Amsterdam, 1864).

II. Secondary Literature. Works on Baumhauer are J. W. Gunning, "Levensbericht van E. H. von Baumhauer," in *Jaarboek van de Koninklijke Akademie van Wetenschappen, 1887* (Amsterdam, 1887), pp. 1–57; and J. H. Van der Beek, *E. H. von Baumhauer. Zijn betekenis voor de wetenschap en de Nederlandse economie* (Leiden, 1963).

H. A. M. Snelders

BAUMHAUER, HEINRICH ADOLF (*b.* Bonn, Germany, 26 October 1848; *d.* Fribourg, Switzerland, 1 August 1926), *chemistry, mineralogy.*

Baumhauer, the son of a lithographer and merchant, studied chemistry under Kekulé at Bonn and in 1869 earned his degree with the dissertation "Die Reduction des Nitrobenzols durch Chlor- und Bromwasserstoff." After a further year of study at Göttingen he briefly held three teaching posts before serving as teacher of chemistry at the agricultural school of Lüdinghausen, Westphalia, from 1873 to 1896. In that year he was called to the young University of Fribourg as its first professor of mineralogy; during his first ten years there he also taught inorganic chemistry.

Although Baumhauer was trained as a chemist, crystallography and mineralogy were his principal interests and he is generally regarded as a mineralogist. In spite of a heavy work load—he was a conscientious teacher—Baumhauer always had time for research, which at first was chemical. In 1870 he wrote a tract on the relation between atomic weights and the properties of elements. In it he commented on the work of Mendeleev and Lothar Meyer, and proposed a spiral arrangement for the periodic table. He also wrote a textbook of inorganic chemistry (1884) and one of organic chemistry (1885).

At Lüdinghausen, Baumhauer's interest turned to crystallography and mineralogy. He wrote a short text on mineralogy (1884) and, soon thereafter, a popular work on crystallography. Baumhauer is best known for his study of the etch figures produced on crystal faces by various solvents, a method of study that for a long time provided one of the principal means for establishing the symmetry of crystals. His *Die Resultate der Aetzmethode* (1894) was the standard and only work on the subject until it was supplemented in 1927 by A. P. Honess' book.

Soon after the appearance of *Resultate,* Baumhauer added other fields to the scope of his research. He became the foremost authority on the sulfosalts of the Binnental occurrence, and made extensive morphological studies of minerals. During his last decades his chief interest was comparative crystal morphology. He gathered material for a book on this subject, which was edited by his successor, Leonhard Weber, and published as an extended article.

BIBLIOGRAPHY

Complete bibliographies of Baumhauer's works are in Poggendorff, III, 84; IV, 78-79; and V, 75; in *Schweizerische mineralogische und petrographische Mitteilungen,* **6** (1926), 391-397, following a short autobiography dated July 1896; and in *Verhandlungen der Schweizerischen Naturforschenden Gesellschaft,* **107** (1926), Anhang, 3-15, following a fine biographical sketch by Leonhard Weber.

Among his books are *Die Beziehungen zwischen dem Atomgewichte und der Natur der chemischen Elemente* (Brunswick, 1870); *Kurzes Lehrbuch der Mineralogie (einschliesslich Petrographie)* (Freiburg im Breisgau, 1884); *Leitfaden der Chemie. I. Anorganische Chemie* (Freiburg im Breisgau, 1884); *Leitfaden der Chemie. II. Organische Chemie* (Freiburg im Breisgau, 1885); *Das Reich der Kristalle* (Leipzig, 1889); and *Die Resultate der Aetzmethode in der krystallographischen Forschung* (Leipzig, 1894). His articles include "Über die Einwirkung von Brom- und Chlorwasserstoff auf Nitrobenzol," in *Zeitschrift für Chemie,* n.s. **5** (1869), 198; "Über die Einwirkung von Bromwasserstoff auf Nitrobenzol," in *Berichte der Deutschen chemische Gesellschaft,* **2** (1869), 122; and "Beitrag zur vergleichenden Kristallographie," Leonhard Weber, ed., in *Schweizerische mineralogische und petrographische Mitteilungen,* **5** (1925), 348-426.

Arthur P. Honess, *The Nature, Origin and Interpretation of Etch Figures on Crystals* (New York, 1927), is the principal work continuing the studies summarized in the *Resultate* and contains many references to Baumhauer.

A. PABST

BAYEN, PIERRE (*b.* Châlons-sur-Marne, France, 7 February 1725; *d.* Paris, France, 15 February 1798), *chemistry, pharmacy.*

Bayen received his early education at the Collège de Troyes. After deciding on a career in pharmacy, he was placed with the pharmacist Faciot in Rheims. In 1749 he went to Paris and studied under P. de Chamousset and G. F. Rouelle.

With G. F. Venel, Bayen was commissioned by the government in 1753 to analyze all the mineral waters of France, but this work had to be suspended when he was sent as chief pharmacist on an army expedition in 1755. Later, after continuing the analyses alone for many years, he finally had to abandon the work for lack of funds. From 1763 to 1793 Bayen was *apothicaire-major des camps et armées du roi,* and he became *pharmacien inspecteur* to the Conseil de Santé in the latter year. He was never elected to the Academy but became a member of the Institut de France when it was founded in 1795.

Bayen's first publication was an analysis of the mineral waters of Bagnères-de-Luchon (1765), a long and detailed work published only incompletely in his *Opuscules.* However, his most important contribution to chemistry was a series of four memoirs on the precipitates of mercury (1774–1775). His observations led him to doubt the phlogiston theory, for he noticed that when red precipitate (mercuric oxide) was heated alone (in which case, there was no reducing agent to supply phlogiston to it), it was reduced to mercury and a large quantity of an elastic fluid was produced. He made careful quantitative observations of all that took place and concluded that when mercury was calcined, it did not lose phlogiston but combined intimately with the elastic fluid, and that this addition of elastic fluid to the mercury was responsible for its increase in weight. He failed to identify the elastic fluid (oxygen), however, and did not even apply the simple test of plunging a lighted taper into it.

Bayen helped to make known the work of Jean Rey, the seventeenth-century Périgord doctor whose theory of calcination was very similar to his own.

In 1781, Bayen produced a report, commissioned by the Paris College of Pharmacy, on the alleged presence of arsenic in tin. Henckel and Marggraf had reported that commercial tin contained arsenic, which would have made its use in kitchen equipment dangerous. Bayen concluded that, at the most, the tin contained only a trace of arsenic.

In addition to these researches, Bayen published analyses of minerals (1776, 1778, 1785, 1798). Some of his work, however, cannot be traced, because he burned all his manuscripts during the Terror.

Bayen was a thorough and careful worker, and often spent years examining a material, seeking the least destructive method of analysis in an attempt to imitate nature and preserve the constituents intact. That he was forty when he published his first memoir was doubtless a consequence of this reluctance to commit himself to paper before he had conducted the most exhaustive research.

BIBLIOGRAPHY

I. ORIGINAL WORKS. A detailed bibliography of Bayen's writings is in A. Balland, ed., *Travaux scientifiques des pharmaciens militaires français* (Paris, 1882), pp. 6–8. His works were collected in *Opuscules chimiques de Pierre Bayen,* P. Malatret, ed., 2 vols. (Paris, 1798), in which the memoir "Analyse des eaux de Bagnères-de-Luchon, par ordre du ministre de la guerre" is incomplete. A complete version of this memoir is in Richard, *Recueil d'observations de médecine des hôpitaux militaires,* II (Paris, 1772), 633–778; it was originally published, complete, in an octavo volume (Paris, 1765).

II. SECONDARY LITERATURE. Works on Bayen are A. Balland, "En mémoire de Bayen," in *Journal de pharmacie et chimie,* 8th ser., **1** (1925), 83–94; P. A. Cap, "Étude biographique sur Pierre Bayen," *ibid.,* 4th ser., **1** (Jan. 1865); P. Lassus, "Notice historique sur Bayen," in *Annales de chimie,* **26** (an VI [1798]), 278–288, also published as "Notice sur la vie et les ouvrages du C^en Bayen," in *Mémoires de l'Institut national des sciences mathématiques et physiques,* **2** (fructidor, an VII [1799]), hist., 144–152; A. A. Parmentier, "Eloge de Pierre Bayen," in Bayen's *Opuscules,* I, xxxiii–lxxiv; see also "Discours préliminaire," *ibid.,* pp. iii–xxxii, which contains sections of another *éloge,* not published elsewhere, in which some of Bayen's ideas are discussed; M. B. Reber, "Une lettre inédite de Pierre Bayen, suivie de quelques observations," in *Bulletin de la Société française d'histoire de la médecine,* **9** (1910), 50–63; and A. F. Silvestre, "Notice sur le Citoyen Bayen," in *Rapports généraux des travaux de la Société philomathique de Paris,* I (1793), 233–240.

E. MCDONALD

BAYER, JOHANN (*b.* Rain, Germany, 1572; *d.* Augsburg, Germany, 7 March 1625), *astronomy.*

Bayer, a lawyer who was an amateur astronomer, established the modern nomenclature of stars visible to the naked eye. He enrolled at Ingolstadt University on 19 October 1592 as a student of philosophy, but later moved to Augsburg. There, on 1 September 1603, he dedicated his *Uranometria,* a popular guide to the starry heavens, to two leading citizens and the city council, which promptly rewarded him with an honorarium of 150 gulden. On 13 December 1612, Bayer was appointed legal adviser to the city council of Augsburg, at an annual salary of 500 gulden. He died a bachelor.

The oldest surviving star catalogue, contained in Ptolemy's *Syntaxis,* lists forty-eight constellations, with each star located in reference to a constellation by means of a verbal description. This was often cumbersome, and did not always direct every observer to the same star, particularly after the original Greek expression had been translated into one or more languages. This lack of compactness and of precision had long prevailed when Bayer undertook his reform by unambiguously and succinctly identifying every star visible to the naked eye.

In essence, the novelty of Bayer's method consisted of assigning to each star in a constellation one of the twenty-four letters of the Greek alphabet. For constellations with more than two dozen stars, he resorted to the Latin alphabet after exhausting the Greek letters. He placed these Greek and Latin letters on his star charts, beautiful copper-plate engravings by Alexander Mair. In addition, Bayer reproduced the traditional numeration of the stars in the constellations, as well as the many and very different names used by Ptolemy and his successors, whose works Bayer studied with painstaking care. In this way he sought to facilitate the identification of any star in his *Uranometria* with the same star as it had been recorded by his various predecessors. Thus, just a few years before the invention of the telescope enormously increased the number of stars visible with the aid of an optical instrument, Bayer produced a stellar nomenclature that astronomers still use for most stars visible to the naked eye.

In Bayer's own time the popularity of his work was further enhanced by his forty-ninth plate, which displayed twelve new southern constellations. These had recently been defined by the Dutch navigator Pieter Dirckszoon Keyzer (Petrus Theodori) of Emden, who corrected the older observations of Amerigo Vespucci and Andrea Corsali, as well as the report of Pedro de Medina.

On the other side of the ledger, two aspects of Bayer's *Uranometria* created difficulties. In the first place, what his predecessors had called the right side of a constellation's figure, he labeled the left side (Table 4, verso). Second, in bracketing stars of the same magnitude in each constellation (Table 1, recto), Bayer failed to indicate on what basis he assigned the letters within each bracket. The widespread assumption that he used the order of decreasing magnitude led to considerable confusion in the study of variable stars. The alternative assumption that he used some spatial arrangement is likewise open to serious objection.

BIBLIOGRAPHY

I. ORIGINAL WORKS. Bayer's main work is *Uranometria, omnium asterismorum continens schemata, nova methodo delineata, aereis laminis expressa* (Augsburg, 1603). *Explicatio characterum aeneis Uranometrias imaginum tabulis insculptorum* reprints the text of Bayer's *Uranometria* but omits its plates. Bayer also collaborated on Julius Schiller's *Coelum stellatum Christianum* (Augsburg, 1627).

II. SECONDARY LITERATURE. There is no biography of Bayer. Some information about his life is provided in Franz Babinger, "Johannes Bayer, der Begründer der neuzeitlichen Sternbenennung," in *Archiv für die Geschichte der*

Naturwissenschaften und der Technik. **5** (1915). 108–113. Bayer's reliability is analyzed in F. W. A. Argelander, *De fide Uranometriae Bayeri* (Bonn, 1842). Bayer's collaboration with Schiller is discussed by Ernst Zinner in *Vierteljahrsschrift der astronomischen Gesellschaft,* **72** (1937), 64–68.

EDWARD ROSEN

BAYES, THOMAS (*b.* London, England, 1702; *d.* Tunbridge Wells, England, 17 April 1761), *probability.*

Bayes was a member of the first secure generation of English religious Nonconformists. His father, Joshua Bayes, F.R.S., was a respected theologian of dissent; he was also one of the group of six ministers who were the first to be publicly ordained as Nonconformists. Privately educated, Bayes became his father's assistant at the presbytery in Holborn, London; his mature life was spent as minister at the chapel in Tunbridge Wells. Despite his provincial circumstances, he was a wealthy bachelor with many friends. The Royal Society of London elected him a fellow in 1742. He wrote little: *Divine Benevolence* (1731) and *Introduction to the Doctrine of Fluxions* (1736) are the only works known to have been published during his lifetime. The latter is a response to Bishop Berkeley's *Analyst,* a stinging attack on the logical foundations of Newton's calculus; Bayes's reply was perhaps the soundest retort to Berkeley then available.

Bayes is remembered for his brief "Essay Towards Solving a Problem in the Doctrine of Chances" (1763), the first attempt to establish foundations for statistical inference. Jacques Bernoulli's *Ars conjectandi* (1713) and Abraham de Moivre's *The Doctrine of Chances* (1718) already provided great textbooks of what we now call probability theory. Given the probability of one event, the logical principles for inferring the probabilities of related events were quite well understood. In his "Essay," Bayes set himself the "converse problem": "*Given* the number of times in which an unknown event has happened and failed: *Required* the chance that the probability of its happening in a single trial lies somewhere between any two degrees of probability that can be named." "By chance," he said, "I mean the same as probability."

In the light of Bernoulli's *Ars conjectandi,* and of a paper by John Arbuthnot (1710), there was some understanding of how to reject statistical hypotheses in the light of data; but no one had shown how to measure the probability of statistical hypotheses in the light of data. Bayes began his solution of the problem by noting that sometimes the probability of a statistical hypothesis is given before any particular

events are observed; he then showed how to compute the probability of the hypothesis after some observations are made. In his own example:

> Postulate: 1. I suppose the square table or plane ABCD to be so made and levelled that if either of the balls O or W be thrown upon it, there shall be the same probability that it rests upon any one equal part of the plane as another, and that it must necessarily rest somewhere upon it.

> 2. I suppose that the ball W shall be first thrown, and through the point where it rests a line *os* shall be drawn parallel to AD, and meeting CD and AB in *s* and *o,* and that afterwards the ball O shall be thrown $p + q$ or *n* times, and that its resting between AD and *os* after a single throw be called the happening of the event M in a single trial.

For any fractions *f* and *b* (between zero and one), Bayes was concerned with the probability of assertions of the form

$$f \leq \text{probability of M} \leq b.$$

From his physical assumptions about the table ABCD, he inferred that the prior probability (i.e., the probability before any trials have been made) is $b - f$. He proved, as a theorem in direct probabilities, that the posterior probability (i.e., the probability, on the evidence, that M occurred *p* times and failed *q* times) is

$$\int_f^b x^p(1 - x)^q \, dx \Big/ \int_0^1 x^p(1 - x)^q \, dx.$$

A generalization on this deduction is often, anachronistically, called Bayes's theorem or Bayes's formula. In the case of only finitely many statistical hypotheses, $H_1, \cdots, H_n$, let there be prior data *D* and some new observed evidence *E*. Then the prior probability of H_i is Prob (H_i/D); the posterior probability is Prob $(H_i/D$ and $E)$; the theorem asserts:

Prob $(H_i/D$ and $E)$
$$= \frac{\text{Prob } (H_i/D) \text{ Prob } (E/H_i \text{ and } D)}{\sum_{j-1}^{j=n} \text{Prob } (H_j/D) \text{ Prob } (E/H_j \text{ and } D)}.$$

A corresponding theorem holds in the continuum; Bayes's deduction essentially involves a special case of it.

The work so far described falls entirely within probability theory, and would now be regarded as a straightforward deduction from standard probability axioms. The striking feature of Bayes's work is an argument found in a *scholium* to the paper, which does not follow from any standard axioms. Suppose we have no information about the prior probability of a statistical hypothesis. Bayes argues by analogy that, in this case, our ignorance is neither more nor less than in his example where prior probabilities are known to be entirely uniform. He concludes, "I shall take for granted that the rule given concerning the event M . . . is also the rule to be used in relation to any event concerning the probability of which nothing at all is known antecedently to any trials made or observed concerning it."

If Bayes's conclusion is correct, we have a basis for the whole of statistical inference. Richard Price, who sent Bayes's paper to the Royal Society, seems to imply in a covering letter that Bayes was not satisfied with his argument by analogy and, hence, had declined to publish it. Whatever the case with Bayes, Laplace had no qualms about Bayes's argument; and from 1774 he regularly assumed uniform prior probability distributions. His enormous influence made Bayes's ideas almost unchallengeable until George Boole protested in his *Laws of Thought* (1854). Since then, Bayes's technique has been a constant subject of controversy.

Today there are two kinds of Bayesians. Sir Harold Jeffreys, in his *Theory of Probability,* maintains that, relative to any body of information, even virtual ignorance, there is an objective distribution of degrees of confidence appropriate to various hypotheses; he often rejects Bayes's actual postulate, but accepts the need for similar postulates. Leonard J. Savage, in his *Foundations of Statistics,* rejects objective probabilities, but interprets probability in a personal way, as reflecting a person's personal degree of belief; hence, a prior probability is a person's belief before he has made some observations, and his posterior probability is his belief after the observations are made. Many working statisticians who are Bayesians, in the sense of trying to argue from prior probabilities, try to be neutral between Jeffreys and Savage. In this respect they are perhaps close to Bayes himself. He defined the probability of an event as "the ratio between the value at which an expectation depending upon the happening of the event ought to be computed, and the value of the thing expected upon its happening." This definition can be interpreted in either a subjective or an objective way, but there is no evidence that Bayes had even reflected on which interpretation he might prefer.

BIBLIOGRAPHY

I. ORIGINAL WORKS. Bayes's works published during his lifetime are *Divine Benevolence, or an Attempt to Prove That the Principal End of the Divine Providence and Government Is the Happiness of His Creatures* (London, 1731); and *An Introduction to the Doctrine of Fluxions, and a Defence of the Mathematicians Against the Objections of the Author of* The Analyst (London, 1736). "An Essay Towards Solving a Problem in the Doctrine of Chances" was published in *Philosophical Transactions of the Royal Society of London,* **53** (1763), 370–418, with a covering letter written by Richard Price; repr. in *Biometrika,* **45** (1958), 296–315, with a biographical note by G. A. Barnard. Also of interest is "A Letter on Asymptotic Series from Bayes to John Canton," in *Biometrika,* **45** (1958), 269–271; repr. with the paper from the *Philosophical Transactions* in *Facsimiles of Two Papers by Bayes* (Washington, D.C., n.d.), with a commentary on the first by Edward C. Molina and on the second by W. Edwards Deming.

II. SECONDARY LITERATURE. Supplementary information may be found in John Arbuthnot, "An Argument for Divine Providence Taken From the Constant Regularity of the Births of Both Sexes," in *Philosophical Transactions of the Royal Society of London,* **23** (1710), 186–190; George Berkeley, *The Analyst* (London–Dublin, 1734); Jacques Bernoulli, *Ars conjectandi* (Basel, 1713); George Boole, *An Investigation of the Laws of Thought* (London, 1854); Harold Jeffreys, *Theory of Probability* (Oxford, 1939; 3rd ed., 1961); Pierre Simon Laplace, "Mémoire sur la probabilité des causes par les événements," in *Mémoires par divers savants,* **6** (1774), 621–656, and *Théorie analytique des probabilités* (Paris, 1812); Abraham de Moivre, *The Doctrine of Chances* (London, 1718; 3rd ed., 1756); and Leonard J. Savage, *The Foundations of Statistics* (New York–London, 1954).

IAN HACKING

BAYLAK AL-QIBJĀQĪ (fl. Cairo, Egypt, *ca.* 1250), *mineralogy,* probably also *mathematics, technology.*

Baylak's period of activity is determined from his signature, dated 1282, in the *Kanz al-tujjār.* An autobiographical note there states that in 1242/1243 he undertook a trip by sea from Tripoli, Syria, to Alexandria. He must have had connections at the court of Ḥamā, Syria, for the *Kanz* is dedicated to either the Ayyūbid ruler al-Malik al-Muẓaffar II (1229–1244) or his son al-Malik al-Manṣūr II (1244–1284). If Baylak is identical with the mathematician—and this can be assumed—who in 1260 prepared a copy of a work on the knowledge and use of the clocks of Riḍwān al-Khurāsānī, and if he is also the one who made a handwritten mark of ownership on another mathematical manuscript (1269/1270), both preserved in Istanbul, then he was concerned with at least three areas of mathematics and natural science. However, in the *Kanz* his father apparently is named Muḥammad, and in the manuscript on clocks he is definitely ʿAbdallāh. The authenticity of these details is open to investigation, but in any case, all dates mentioned fall within the lifetime of one particular mature man.

Baylak's only known book is his mineralogical work *Kanz al-tujjār* [or *al-tijār*] *fī maʿrifat al-ahjār* ("Treasure of the Merchants on the Knowledge of Minerals"). If this was dedicated to al-Malik al-Muzaffar II, then the autograph of 1282 must be a copy of the original prepared by the author himself, for by 1282 al-Malik al-Manṣūr II was the ruler of Hamā. Apparently, however, 1282 was the year in which the original was written, for the note on the sea voyage is composed in such a manner that the *Kanz* can hardly have been written before 1244, the year of al-Malik al-Muzaffar's death. If one is to judge by excerpts of the book published by Clément-Mullet in 1868, the book contains little that is original; and this little is further reduced if one compares it with other texts additional to those preserved in the Paris manuscripts used by Clément-Mullet, and especially if it is compared with the older mineralogical books that have been discovered and published within the past hundred years.

Baylak was the first author writing in Arabic to treat the use of the magnetic needle as a ship's compass. On the trip to Alexandria, he watched one starless night as the captain stuck a needle through a straw so that the two objects formed a cross. This apparatus was then floated in a vessel of water. The needle responded to the circular movement of a nearby magnet. At the sudden removal of the magnet, the needle came to rest with a north-south orientation. Baylak had also heard that seamen in the Indian Ocean used hollow iron fish as compasses. His report on the magnetic needle was translated into French by Klaproth, then by Clément-Mullet (with the original Arabic text), and finally by De Saussure; Ferrand emended the latter translation at the time of its reprinting. Wiedemann translated the report into German in 1904.

BIBLIOGRAPHY

C. Brockelmann, *Geschichte der arabischen Literatur,* I (1898), 495; Supp. I (1937), 904; 2nd ed., I (1943), marginal 495—with contradictory details in each place; J. Clément-Mullet, "Essai sur la minéralogie arabe," in *Journal asiatique,* 6th ser., **11** (1868), 5–81, 105–253, 502–522, esp. 13 f., 146 ff., also published separately (1869); L. De Saussure, "L'origine de la rose des vents et l'invention de la boussole," repr. in Gabriel Ferrand, *Introduction à l'astronomie nautique arabe* (1928), pp. 80 ff.; F. Kern, in *Mitteilungen des Seminars für orientalische Sprachen in Berlin, Westasiatische Studien,* **11** (1908), 268 f.; Jules Klaproth, *Lettre à A. de Humboldt sur l'invention de la boussole* (1834), p. 59; M. Krause, "Stambuler Handschriften islamischer Mathematiker," in *Quellen und Studien zur Geschichte der Mathematik, Astronomie und Physik,* Abt. B, **3** (1936), 490 f.; G. Sarton, *Introduction to the History of Science,* II (1931), 1072, with several mistakes; MG. de Slane, *Bibliothèque Nationale, Catalogue des manuscrits arabes* (Paris, 1883–1895), MS No. 2779; G. Vajda, *Index général des manuscrits arabes musulmans de la Bibliothèque Nationale de Paris* (1953), p. 420, which differs from de Slane; and E. Wiedemann, "Beiträge zur Geschichte der Naturwissenschaften (II)," in *Sitzungsberichte der physikalisch-medizinischen Sozietät in Erlangen,* **39** (1904), 330 f. (the quotation from an Arabic author of 854 does not refer to the naval compass), and "Maghnaṭīs," in *Encyclopaedia of Islam,* III (1936).

MARTIN PLESSNER

BAYLISS, LEONARD ERNEST (*b.* London, England, 15 November 1900; *d.* London, 20 August 1964), *physiology.*

The youngest son of Sir William Maddock Bayliss and Gertrude Starling Bayliss, sister of the physiologist Ernest H. Starling, Bayliss grew up near Hampstead Heath, on the northern fringe of London. His parents entertained generously and attracted many overseas physiologists to meet their British colleagues in an atmosphere of tennis, good fellowship, and unending, searching arguments about the fundamentals of physical, biological, and medical sciences. Here his father had his private laboratory and workshop although, like Starling, he was also a professor in the physiological laboratory of University College, London. In such a home Bayliss developed his "do-it-yourself" attitude toward apparatus and his capacity for exacting and informed analysis of scientific problems.

After graduating from University College School, a day school for boys fairly near his home, Bayliss entered Trinity College, Cambridge, and graduated in physiology in 1922. He stayed on in the physiology department as Michael Foster Student and in 1925 received the Ph.D. with the thesis "Tone in Plain Muscle." From 1926 to 1929 he was Sharpey Scholar at University College, and in 1928–1929 he worked under A. N. Richards at the University of Pennsylvania on a Rockefeller Fellowship; he returned in 1929 to University College, where he was awarded the Schäfer Prize (given for original work by a young physiologist associated with University College). In 1933 Bayliss was appointed physiologist at the Marine Biological Station, Plymouth, and the following year he became lecturer in physiology at the University of Edinburgh, from which he "retired" in 1939, hoping to be free to prepare a revised version of his father's famous book *Principles of General Physiology,* the last edition of which had appeared in 1924, soon after his father's death. Bayliss returned to University College as honorary part-time research assistant in the same year.

Bombing and fire interrupted Bayliss' work in physiology in 1940, and he joined a group of pioneering scientists in the Army Operational Research Group, where he was later put in charge of research on anti-aircraft gunnery and accuracy of fire against aircraft at various heights and against guided missiles.

Bayliss returned to physiology and to battered University College as full-time reader from 1945 to 1950. He resigned in 1950 in order to concentrate on rewriting *Principles of General Physiology* but retained his room near the Medical Sciences Library at the college. His accessibility to students and colleagues continued, and he pursued his research and writing with the title of honorary research associate. He also made an important contribution to advanced teaching of physiology. By 1960 he had completed his project, entirely rewriting the *Principles* and expanding it to two volumes.

His bibliography reveals Bayliss' wide interests in physical approaches to physiology and in relevant instrumentation, but he was a reluctant and unambitious publisher; much work never reached the journals at all or did so only in the papers of friends. A great deal of his time was always given to helping others, and he was sought after as perhaps the most learned, ingenious, and cooperative of English physiologists. His early papers on the pH of blood (1923, 1926) were written when the newly developed glass electrode, still of very high resistance, was connected to a delicate electrometer—quite a challenge even to skilled experimenters in the sulfurous atmospheres of towns. While at University College he also worked with Starling on the metabolism of dog heart-lung preparations; devised pump-oxygenator perfusion techniques for kidneys; and studied serotonins in defibrinated blood, water diuresis, and glomerular permeability in both cold-blooded and warm-blooded animals. His interest in invertebrates, as shown in a 1930 paper, was strengthened at the Plymouth Marine Biological Station.

The rheology of blood was a major but intermittent interest during Bayliss' later years. It began about 1930 with Fahraeus' papers showing that the viscosity of blood, relative to water in the same tube, diminished as the diameter diminished, down to 0.03 millimeters, at which point cells blocked the tube. Bayliss had been making and using micropipettes, and had encountered the problem of cell clumping in his blood perfusion experiments with pumps and tubes. He extended the range of tube diameters down as far as that of arterioles and found the apparent viscosity nearly halved. This finding was published as a "personal communication" by colleagues (1933) and again in the second edition of *Human Physiology* (1936). Bayliss' first definitive paper on blood viscosity did not appear until 1952, and many experimental findings have remained unpublished.

In 1939 Bayliss married M. Grace Palmer Eggleton, a physiologist at University College, whom he had known since about 1930. They took an apartment at the top of a tall old building near the college and, except for the early war years, lived there happily and frugally until his death. Also in the year of their marriage they bought a weekend cottage, isolated in the East Sussex countryside, and Bayliss spent much energy and ingenuity in improving and enlarging both house and garden, in anticipation of their retirement in 1966. He was physically active and enjoyed good health. Among his avocations were mountain climbing and playing the piano. He was quietly courteous to students and friends but unyielding in essentials.

Bayliss was a fellow of the Royal Society of Edinburgh and a member of the Physiological Society, the Society for Experimental Biology, and the Marine Biological Association.

BIBLIOGRAPHY

Bayliss' earliest book was *Human Physiology* (London, 1930; 3rd ed., 1948; 5th ed., 1962), written with F. R. Winton. His major work was the complete revision and rewriting of his father's *Principles of General Physiology*, 2 vols. (London, 1959–1960). His *Living Control Systems* appeared in the New Science Series (London, 1966).

His papers are "A Comparison Between the Colorimetric and the Electrometric Methods of Determining the Hydrogen Ion Concentration of Blood," in *Journal of Physiology*, **58** (1923), 101–107, written with Ruth Conway-Verney; "Reversible Haemolysis," *ibid.*, **59** (1924), 48–60; "The Determination of the Hydrogen-ion Concentration of the Blood," *ibid.*, **61** (1926), 448–454, written with Phyllis Tookey Kerridge and Ruth Conway-Verney; "A Conductivity Method for the Determination of Carbon Dioxide," in *Biochemical Journal*, **21** (1927), 662–664; "The Action of Insulin and Sugar on the Respiratory Quotient and Metabolism of the Heart-lung Preparation," in *Journal of Physiology*, **65** (1928), 33–47, written with E. A. Müller and E. H. Starling; "The Energetics of Plain Muscle," *ibid.*, i–ii; "A Method of Oxygenating Blood," *ibid.*, **66** (1928), 443–448, written with A. R. Fee and E. Ogden; "A Simple High-speed Rotary Pump," in *Journal of Scientific Instruments*, **5** (1928), 278–279, written with E. A. Müller; "The Adductor Mechanism of *Pecten*," in *Proceedings of the Royal Society*, **106B** (1930), 363–376, written with E. Boyland and A. D. Ritchie; "A Combined Flow Recorder and Pump Control," in *Journal of Physiology*, **69** (1930), v, written with A. R. Fee; "The Electrical Conductivity of Glomerular Urine From the Frog and From Necturus," in *Journal of Biological Chemistry*, **87** (1930), 523–540; "Studies in Water Diuresis. Part III. A Comparison of the Excretion of Urine by Innervated and Denervated Kidneys Perfused With the Heart-lung Preparation," in *Journal of Physiology*, **69** (1930), 135–143, written with A. R. Fee; "Studies in Water Diuresis. Part IV. The Changes in the Concentration of Electrolytes and Colloids in the Plasma of Decerebrate Dogs Produced by the Ingestion of Water," *ibid.*, **70** (1930), 60–66, written with A. R. Fee; "The Action of Cyanide on the Isolated Mammalian Kidney," *ibid.*, **74** (1932), 279–293, written with E. Lundsgaard; "Fourteenth International Physiological Congress," in *Nature*, **130** (1932), 705–707, written with P. Eggleton; "The Excretion of Protein by the Mammalian Kidney," in *Journal of Physiology*, **77** (1933), 386–398, written with Phyllis Tookey Kerridge and Dorothy S. Russell; "'Vaso-tonins' and the Pump-oxygenator-kidney Preparation," *ibid.*, 34P–35P, written with E. Ogden; "The Action Potentials in *Maia* Nerve Before and After Poisoning With Veratrine and Yohimbine Hydrochlorides," *ibid.*, **83** (1935), 439–454, written with S. L. Cowan and D. Scott, Jr.; "Digestion in the Plaice (*Pleuronectes platessa*)," in *Journal of the Marine Biological Association of the United Kingdom*, **20** (1935), 73–91; "A Drop Recorder for A.C. Mains," in *Journal of Physiology*, **84** (1935), 57P; "A Stimulation Unit for A.C. Mains," *ibid.*, 58P–59P, written with P. Eggleton; "Recent Developments in Physical Instruments for Biological Purposes," in *Journal of Scientific Instruments*, **12** (1935), 1–5, written with Phyllis Tookey Kerridge; "Some New Forms of Visual Purple Found in Sea Fishes With a Note on the Visual Cells of Origin," in *Proceedings of the Royal Society*, **120B** (1936), 95–113, written with R. J. Lythgoe and Katharine Tansley; "The Photographic Method for Recording Average Illuminations," in *Journal of the Marine Biological Association of the United Kingdom*, **23** (1938), 99–118; "The Visco-elastic Properties of the Lungs," in *Quarterly Journal of Experimental Physiology*, **29** (1939), 27–47, written with G. W. Robertson; "A Circulation Model," in *Journal of Physiology*, **97** (1940), 429–432; "The Part Played by the Renal Nerves in the Production of Water Diuresis in the Hypophysectomized and Decerebrate Dog," *ibid.*, **98** (1940), 190–206, written with A. Brown; "Schack, August Steenberg Krogh," in *Yearbook of the Royal Society of Edinburgh* (1948–1949); "Rheology of Blood and Lymph," in A. Frey-Wyssling, ed., *Deformation and Flow in Biological Systems* (Amsterdam, 1952), pp. 355–418; "A Mechanical Model of the Heart," in *Journal of Physiology*, **127** (1955), 358–379; "The Use of a Roller Pump for Long-continued Infusion," *ibid.*, **128** (1955), 29P–30P; "The Process of Secretion," in F. R. Winton, ed., *Modern Views on the Secretion of Urine: The Cushny Memorial Lectures* (London, 1956), pp. 96–127; "The 'Brown Dog' Affair," in *Potential: Journal of the Physiological Society of University College London*, no. 2 (1957), 11–22; "The Axial Drift of the Red Cells When Blood Flows in a Narrow Tube," in *Journal of Physiology*, **149** (1959), 593–613; "The Anomalous Viscosity of Blood," in A. L. Copley and G. Stainsby, eds., *Flow Properties of Blood and Other Bio-*

logical Systems (Oxford, 1960), pp. 29–62; "William Maddock Bayliss, 1860–1924: Life and Scientific Work," in *Perspectives in Biology and Medicine*, **4** (1961), 460–479; "The Rheology of Blood," in *Handbook of Physiology*, sec. 2, W. F. Hamilton, ed., I (Washington, D.C., 1962), 137–150; "Some Factors Concerned in Temperature Regulation in Man," in *Journal of Physiology*, **172** (1964), 8P–9P, written with S. E. Dicker and M. Grace Eggleton; and "The Flow of Suspensions of Red Blood Cells in Capillary Tubes. Changes in the 'Cell-free' Marginal Sheath With Changes in the Shearing Stress," *ibid.*, **179** (1965), 1–25.

F. R. WINTON

BAYLISS, WILLIAM MADDOCK (*b.* Wednesbury, Staffordshire, England, 2 May 1860; *d.* London, England, 27 August 1924), *general physiology.*

W. M. Bayliss was the only son of Moses Bayliss and Jan Maddock to survive infancy. His father was a manufacturer of galvanized goods and founder of the Wolverhampton firm of Bayliss, Jones and Bayliss.

William received his early education at a private school in Wolverhampton, Staffordshire, and was then apprenticed to a local general practitioner.

When, in 1880, the elder Bayliss moved to London, William continued his studies at University College, London, where in 1881 he gained a medical entrance exhibition and a science exhibition at the preliminary scientific examination. In 1882 he took the B.Sc. degree with a scholarship in zoology and physics, and began to study physiology and anatomy, but at the second M.B. examination he failed in anatomy. He then relinquished medical studies to concentrate on physiology.

Two men who influenced him at this stage were John Burdon-Sanderson, Jodrell professor of physiology at University College, and Ray Lankester, professor of zoology. Physiology was at that time emerging as a subject separate from anatomy, and the Jodrell chair, founded in 1874, was the first separate chair of physiology in Great Britain. When the Waynflete chair of physiology was established at Oxford in 1882, Burdon-Sanderson was elected to it, and in 1885 Bayliss followed him to Oxford as an undergraduate of Wadham College. He took a first-class degree in physiology in 1888 and then returned to University College, where, in the meantime, E. A. Schäfer (afterward Sir Edward Sharpey-Schafer) had succeeded to the Jodrell chair. Bayliss worked for the remainder of his life at University College, first as assistant, then in 1903 as assistant professor, and from 1912 onward as professor of general physiology.

He attended an Anglo-Catholic church, but was largely tolerant in religious matters. Politically, he was somewhat to the left of liberal, with a sympathy for reasonable minority movements, but had little interest in politics as such.

In 1893 he married Gertrude Starling, the sister of E. H. Starling; and, after his father's death in 1895, he returned to live in the family home in Hampstead, London. He had three sons and one daughter; the youngest son, Leonard (1900–1964), was also a physiologist. Bayliss was well-off financially, his married life was happy, and until shortly before his death from a blood dyscrasia his health was excellent. In their large house, set in a four-acre garden with two tennis courts and a private laboratory, the Baylisses entertained often and liberally, mainly colleagues and visiting physiologists.

Bayliss was elected to membership in the Physiological Society in 1890 and served as secretary from 1900 to 1922, and treasurer from 1922 to 1924. He was editor of *Physiological Abstracts* from 1923 to 1924, and joint editor of the *Biochemical Journal* from 1913 to 1924.

In 1903 he was elected a fellow of the Royal Society, serving on its council from 1913 to 1915. He was Croonian lecturer (jointly with Starling) in 1904 and was awarded a Royal Medal in 1911 and the Copley Medal in 1919. He also received honorary degrees from the universities of Oxford, St. Andrews, and Aberdeen and was elected to the Royal Danish Academy of Science, the Royal Academy of Belgium, and the Société de Biologie of Paris. In 1917 he was awarded the Baly Medal of the Royal College of Physicians and in 1918 gave the Oliver-Sharpey lectures, and the Sylvanus Thompson lectures in 1919. He was created a knight bachelor in 1922.

During World War I he served on the Food (War) Committee of the Royal Society, and on the Wound Shock Committee of the Medical Research Committee; he visited the front in France in 1917. From 1917 to 1924 he served on the Medical Research Council's committee to study the biological action of light.

In 1922 he visited the United States to deliver the Herter lectures at Baltimore, a Harvey lecture at New York, and to talk to the Research Club at Harvard.

The placid tenor of his life was interrupted in 1903 by the urgent necessity to take out an action for libel against Stephen Coleridge, secretary of the National Antivivisection Society. The trial, before the lord chief justice, occupied four days and Bayliss won the day, with £2,000 damages. He presented the money to University College for furtherance of research in physiology. The interest on the capital sum is still used for that purpose.

Bayliss devoted his life almost solely to his work in physiology. He was fond of music and played the violin. He was of a merry and gentle disposition, good with his hands, and would accept little help from others, although he gladly gave his own. Photography was more than a hobby to him; the illustrations for his books and papers were his own work.

Bayliss' generation entered the scene at the beginning of a period of exponential growth of science, and his outlook can be seen as a projection into the twentieth century of that of Claude Bernard (1814–1878), Ludwig (1816–1895), and Helmholtz (1821–1894). To him all science was one, although fundamentally he was a biologist who was deeply interested in the emergence of physical chemistry and biochemistry and in their relations to the problems of general physiology. The foundations of Hoppe-Seyler's *Zeitschrift für physiologische Chemie* (1877) and Ostwald's *Zeitschrift für physikalische Chemie* (1887) were harbingers of what was to follow, and Bayliss availed himself of the tools offered by these new subjects.

The frequent coupling together of the names of Bayliss and Starling in important publications might lead to the inference that their partnership was an essential condition for either or both of them, but nothing could be further from the truth. They were both great physiologists and in their collaboration were largely complementary. Bayliss was the more fundamental and erudite, but of a retiring disposition; Starling was a pragmatic and forceful extrovert, with an essentially medical outlook. Bayliss preferred to work alone, without even technical help, but collaboration was imposed upon them for a time; until a new Institute of Physiology was built there was inadequate space and equipment in the department. After 1909 they never collaborated, although they frequently conferred at the daily tea meetings.

Bayliss' researches can be divided into roughly the following six phases:

Electrophysiology. When Bayliss started at University College, electrophysiology was the central feature of the research work in the department. Bayliss worked with Dr. (afterward Sir John) Rose Bradford in studying, with the aid of a Thomson galvanometer, the changes in electric potential involved in the act of secretion, first in the salivary glands of a frog and then in the skin. In his second paper, "The Electromotive Phenomena of the Mammalian Heart," he collaborated with Starling, partly at Oxford; they were able to use the relatively rapidly responding capillary electrometer, which Lippmann had introduced in 1873. The hearts of mammals, frog, tortoise, and man (the two authors) were studied, the electrical changes shown to be triphasic, and the time relations of the cardiac cycle observed.

Vascular System. This work, which was begun in 1892 with Starling, resulted in two papers, one on a new form of optically registered blood pressure apparatus and one "On Some Points in the Innervation of the Mammalian Heart." He also worked with Bradford on the innervation of the vessels of the limbs. At about this same time, he began his own investigations, starting with the physiology of the depressor nerve, which led on to an examination of vasomotor reactions in general and of their central coordination. The nature of the antidromic reaction, which he showed to be an axon reflex, was another important outcome of this work. By 1894 Bayliss and Starling were working on circulatory problems, including venous and capillary pressures and intraventricular and aortic pressure curves.

Intestinal Movements. The partners now turned their attention to the study of the movements of the small and large intestines, and of their innervation. The most important result of this work was the elucidation of the peristaltic wave.

Pancreatic Secretion. Bayliss' interest in secretion led him, with Starling, to study the anomalous secretion of pancreatic juice which occurred when acid was introduced into the duodenum. At that time the only known means by which glands were excited to secrete was by the mediation of secretory nerves, as exemplified by the pioneer work of Ludwig (1851) on the submaxillary gland. Pavlov had shown that secretion of pancreatic juice was evoked on stimulation of the vagus nerves, but it could also be produced by the introduction of acid into the upper intestine, even after section of both vagi; this acid-provoked secretion, he concluded, could only be brought about by a local reflex from the intestinal mucosa. Bayliss and Starling showed, however, that introduction of acid into the duodenum still induced secretion of pancreatic juice after division of both vagi and of all the nerves to the upper intestine, so that the only means of communication of the pancreas with the intestine was via the bloodstream. The excitant must, therefore, be not a nervous but a chemical stimulus, a substance derived from the intestine and carried in the bloodstream to the pancreas, which it excited to secrete. This they proved by showing that the intravenous injection of an acid extract of intestinal mucosa resulted in the secretion of pancreatic juice. They called the excitant substance *secretin,* and thus recognized the existence of a new class of chemical messengers, the hormones (from the Greek *horman,* "to set in motion"), a name first promulgated by Starling in his Croonian lecture of 1905, in which he foreshadowed other hormone actions.

Enzyme Action. Bayliss and Starling next studied the activation of trypsin in the pancreatic juice. As secreted, the juice has no proteolytic action, but it becomes activated by contact with enterokinase in the intestinal juice. Pavlov regarded this activation to be the result of an enzyme action by which the precursor, trypsinogen, was converted into the active form, trypsin, by the separation from it of part of its molecule. Opponents had thought of the action as a *combination* between inactive enzyme and activator. Bayliss and Starling gave support to Pavlov's view by showing that the reaction was catalytic and not stoichiometric.

Bayliss now turned to study enzyme action, in the light of physical chemistry. He used trypsin as the enzyme, caseinogen or gelatin as substrate, and followed the course of the digestion by measurement of the electrical conductivity, correlated with other measurements such as viscosity, refractive index, optical rotation, osmotic pressure, etc. These studies led on to further work on the velocity of enzyme action, final equilibria, reversibility, and the effect of temperature, in the course of which he showed that enzyme action is largely influenced by adsorption effects in which there is combination between enzyme and substrate, a viewpoint developed by Michaelis and Menten in 1913.

Colloids. Bayliss' study of the osmotic pressure of Congo red, showing that this colloid exerted the same osmotic pressure that it would if it were in true solution, led on to the study of the electrical equilibrium set up when Congo red and sodium chloride solutions are separated by a semipermeable membrane. This work gave support to the studies of Donnan on the theory of membrane equilibria, and its relation to Nernst's equation for a concentration cell.

War Problems (1914–1918). Bayliss' most valuable contribution to the war effort was his work on wound shock, which led to the practice of replacing lost blood with a gum saline solution. He had shown the importance of maintaining an adequate colloidal osmotic pressure in the circulating blood.

Books. Bayliss' great book, on which he spent many years, was his *Principles of General Physiology,* which first appeared in 1914. It can be regarded as an extension into the twentieth century of Claude Bernard's *Phénomènes de la vie commune aux animaux et aux végétaux* (1878–1879). The *Principles* stands as a landmark in the history of biological literature. When Bayliss visited the United States in 1922, he was pleased, but embarrassed, to find that at some universities there were Bayliss clubs, formed to discuss the contents of the book.

BIBLIOGRAPHY

I. ORIGINAL WORKS. *Articles.* "On the Physiology of the Depressor Nerve," in *Journal of Physiology,* **14** (1893),

303-325; "On the Origin From the Spinal Cord of the Vaso-Dilator Fibres of the Hind Limb and on the Nature of These Fibres," *ibid.,* **26** (1901), 173-209; "Further Researches on Antidromic Nerve Impulses," *ibid.,* **28** (1902), 276-299; "The Kinetics of Tryptic Action," in *Archives des sciences biologiques,* **11** (1904), 261-296, supp., *Pavlov Jubilee Volume;* "On Some Aspects of Adsorption Phenomena, with Especial Reference to the Action of Electrolytes, and to the Ash Constituents of Proteins," in *Biochemical Journal,* **1** (1906), 175-232; "On Reciprocal Innervation in Vaso-Motor Reflexes, and on the Action of Strychnine and of Chloroform Thereon," in *Proceedings of the Royal Society,* **80B** (1908), 339-375; "The Excitation of Vaso-Dilator Fibres in Depressor Reflexes," in *Journal of Physiology,* **37** (1908), 264-277; "The Osmotic Pressure of Congo Red, and of Some Other Dyes," in *Proceedings of the Royal Society,* **81B** (1909), 269-286; "On Adsorption as Preliminary to Chemical Reaction," *ibid.,* **84B** (1911), 81-98; "The Osmotic Pressure of Electrolytically Dissociated Colloids," *ibid.,* pp. 229-254; "Researches on the Nature of Enzyme Action. II. The Synthetic Properties of Anti-Emulsin," in *Journal of Physiology,* **43** (1912), 455-466; "Researches on the Nature of Enzyme Action. III. The Synthetic Action of Enzymes," *ibid.,* **46** (1913), 236-266; "The Action of Insoluble Enzymes," *ibid.,* **50** (1915), 85-94; "Methods of Raising a Low Arterial Pressure," in *Proceedings of the Royal Society,* **89B** (1916), 380-393, and in *British Medical Journal* (1917); "The Action of Gum Acacia on the Circulation," in *Journal of Pharmacology and Experimental Therapeutics,* **15** (1920), 29-74; "Reversible Gelation in Living Protoplasm," in *Proceedings of the Royal Society,* **91B** (1920), 196-201.

Collaborations. J. R. Bradford: "The Electrical Phenomena Accompanying the Process of Secretion in the Salivary Glands of the Dog and Cat," in *Internationale Monatsschrift für Anatomie und Physiologie,* **4** (1885), 109-136; "Proc. Physiol. Soc.," in *Journal of Physiology,* **6** (1889), xiii-xvi; and "The Innervation of the Vessels of the Limbs," *ibid.,* **16** (1894), 10-22.

W. B. Cannon: "Note on Muscle Injury in Relation to Shock," in *Special Report of the Medical Research Committee,* no. 26 (1919), 19-23.

Leonard Hill: "On the Intra-Cranial Pressure and the Cerebral Circulations," in *Journal of Physiology,* **18** (1895), 334-362.

E. H. Starling: "On the Electromotive Phenomena of the Mammalian Heart," in *Internationale Monatsschrift für Anatomie und Physiologie,* **9** (1892), 256-281; "On Some Points in the Innervation of the Mammalian Heart," in *Journal of Physiology,* **13** (1892), 407-418; "On the Form of the Intraventricular and Aortic Pressure Curves Obtained by a New Method," in *Internationale Monatsschrift für Anatomie und Physiologie,* **11** (1894), Heft 9; "The Movements and Innervation of the Small Intestine," in *Journal of Physiology,* **24** (1899), 99-143; "The Mechanism of Pancreatic Secretion," *ibid.,* **28** (1902), 325-353; "The Proteolytic Activities of the Pancreatic Juice," *ibid.,* **30** (1903), 61-83; and "The Chemical Regulation of the Secretory Process," Croonian Lecture pub. in *Proceedings of the Royal Society,* **73B** (1904), 310-322.

Books. The Nature of Enzyme Action, 5 eds. (London, 1908-1925); *An Introduction to General Physiology* (London, 1919); *Principles of General Physiology,* 4 eds. (London, 1919-1924); *The Physiology of Food and Economy in Diet* (London, 1917); *Intravenous Injections in Wound Shock,* Oliver-Sharpey Lectures (London, 1918); *The Vasomotor System* (London, 1923); *The Colloidal State in Its Medical and Physiological Aspects* (Oxford, 1923); *Interfacial Forces and Phenomena in Physiology,* Herter Lectures (London, 1923).

II. SECONDARY LITERATURE. L. E. Bayliss, "William Maddock Bayliss (1860-1924)," in *Perspectives in Biology and Medicine,* **4** (1961), 460-479; C. Lovatt Evans, First Bayliss-Starling Memorial Lecture, *Reminiscences of Bayliss and Starling* (Cambridge, 1964), p. 17; E. H. Starling, obit. of W. M. Bayliss in *The Times* (28 Aug. 1924).

CHARLES L. EVANS

IBN AL-BAYṬĀR AL-MĀLAQĪ, ḌIYĀʾ AL-DĪN ABŪ MUḤAMMAD ʿABDALLĀH IBN AḤMAD (*b.* Málaga, Spain, *ca.* 1190; *d.* Damascus, Syria, 1248), *pharmacology, botany.*

Ibn al-Bayṭār may have belonged to the Bayṭār family of Málaga, on which considerable information is available in the biographical dictionaries of the period. His Hispano-Roman ancestry, suggested by Francisco Javier Simonet, has not been proved. He studied in Seville, and while there he gathered herbs with his teachers Abu'l-ʿAbbās al-Nabātī, ʿAbdallāh ibn Ṣāliḥ, and Abu'l-Hajjāj. We know he preferred to study the works of al-Ghāfiqī, al-Zahrāwī, al-Idrīsī, Dioscorides, and Galen.

Around 1220 Ibn al-Bayṭār migrated to the Orient, crossing North Africa and possibly sailing from there to Asia Minor and Syria in 1224. He finally settled in Cairo, where the Ayyūbid Sultan al-Kāmil named him chief herbalist, a post he continued to occupy under the sultan's successor, al-Ṣāliḥ. He traveled sporadically through Arabia, Palestine, Syria, and part of Iraq, accompanied most of the time by his disciples. The most outstanding of his followers was Ibn Abī Usaybiʿa, who in his *ʿUyūn* has left us a eulogistic passage about his teacher.

Ibn al-Bayṭār's *Al-Mughnī fi'l-adwiya al-mufrada* is dedicated to Sultan al-Ṣāliḥ and deals with the simple medicines appropriate for various illnesses. *Al-Jāmiʿ li-mufradāt al-adwiya wa'l-aghdhiya* enumerates alphabetically some 1,400 animal, vegetable, and mineral medicines, relying on his own observations as well as some 150 authorities, including, besides those already mentioned, al-Rāzī and Ibn Sīnā (Avicenna). The main contribution of Ibn al-Bayṭār was the systematization of the discoveries made by Arabs during the Middle Ages, which added between 300 and 400 medicines to the thousand known since antiquity. One must also point out his preoccupation with synonymy (technical equivalents between Arabic and Persian, Berber, Greek, Latin, Arab dialects, and Romance, taken in part from the *Sharḥ asmāʾ al-ʿuqqār* by Maimonides, which he knew well because he had translated it).

Meyerhof and Sobhy have cast doubt on the originality of this work, suspecting that it is a plagiarism of the pharmacopoeia of al-Ghāfiqī, which Ibn al-Bayṭār quotes more than 200 times. This hypothesis is difficult to prove, because the concept of intellectual property was very different among the medieval Arabs from what it is now and because al-Ghāfiqī's work has been preserved only in a résumé by Barhebraeus. The *Jāmiʿ* had great influence upon later pharmacopoeias in the Near East, both in and out of the Islamic world—as, for example, on the Armenian Amir Dowlat. On the other hand, his influence in the West was less, for the era of great translations from Arabic to Latin ended in the middle of the thirteenth century. Nevertheless, Andrea Alpago used the *Jāmiʿ* in his works on Ibn Sīnā, and later it was the object of attention by such Arabists as William Portel and Antoine Galland, who published both a résumé and a manuscript in French.

The other works of Ibn al-Bayṭār have received much less attention than the aforementioned two. They are *Mīzān al-ṭabīb; Risāla fi'l-aghdhiya wa'l-adwiya; Maqāla fi'l-laymūn,* also attributed to Ibn Jumaʿ, which exists in a Latin version by Alpago; and *Tafsīr kitāb Diyusqūrīdis,* a commentary on Dioscorides, a manuscript of which has recently been found. In it he inventories 550 medicines that are to be found in the first four books of Dioscorides and frequently gives synonyms for them.

BIBLIOGRAPHY

I. ORIGINAL WORKS. An inventory of the MSS is in C. Brockelmann, *Geschichte der arabischen Literatur,* I, 492, and Supp. I, 896; and Albert Dietrich, *Medicinalia Arabica* (Göttingen, 1966), p. 147. Ibn al-Bayṭār's second work, the *al-Jāmiʿ,* is not available in a good edition, although there is a defective German translation by J. Sontheimer, 2 vols. (Stuttgart, 1840–1842), and a very useful French edition by Lucien Leclerc in the series Notices et Extraits, XXIII, XXV, and XXVI (Paris, 1877–1883).

II. SECONDARY LITERATURE. Arabic sources for the biography of Ibn al-Bayṭār may be found in César E. Dubler, "Ibn al-Bayṭār en armenio," in *Al-Andalus,* **21** (1956), 125–130; Max Meyerhof, "Esquisse d'histoire de la pharmacologie et botanique chez les musulmans d'Espagne," *ibid.,* **3** (1935), 31–33; George Sarton, *Introduction to the History of Science,* II, pt. 2 (Baltimore, 1931), 663–664; and Ziriklī, *Aʿlām,* IV (Cairo, n.d.), 192.

A résumé of the work by al-Ghāfiqī may be found in *The Abridged Version of the "Book of Simple Drugs"*

of . . . al-Ghāfiqī by Gregorius Abu-l-Farag (*Barhebraeus*), Max Meyerhof and G. P. Sobhy, eds., fasc. 1 (Cairo, 1932), 32–33. An article on the *Tafsīr kitāb Diyusqūrīdis* is M. al-Shihābī, in *Majallat Maʿhad al-Makhṭūṭāt al-ʿArabiyya,* **3,** no. 1 (1957), 105–112.

J. VERNET

BEALE, LIONEL SMITH (*b.* London, England, 5 February 1828; *d.* London, 28 March 1906), *microscopy.*

Beale's father, Lionel John Beale (1796–1871), was a prominent London surgeon and medical author; his mother was Frances Smith. After attending Highgate, a private school, Beale entered King's College School, London, in 1840. From then on, his life and career were intimately connected with institutions under the direction of the council of King's College. He studied at King's College School until 1844, and during the last three years was also apprenticed to an apothecary-surgeon, Joseph Ross, of Islington. In 1845 Beale entered the medical department of King's College, University of London, and two years later matriculated with honors in chemistry and zoology. He then spent two years at Oxford as assistant to Henry Acland (1815–1900) in the Anatomical Museum.

Beale was licensed by the Society of Apothecaries in 1849, and in 1851 he graduated M.B. from the University of London, having already served (1850–1851) as resident physician at King's College Hospital. In 1852 he established a private laboratory in Carey Street, near the hospital, where he pioneered in teaching the use of the microscope in pathological anatomy. In 1853, at the age of twenty-five, he succeeded Robert Bentley Todd as professor of physiology and general and morbid anatomy at King's College, sharing the duties for two years with William Bowman, who had been Todd's assistant. Beale later served King's College as professor of pathological anatomy (1869–1876) and as professor of the principles and practice of medicine (1876–1896). Upon his retirement in 1896, he was nominated professor emeritus and honorary consulting physician to King's College Hospital.

Conspicuous among the many honors conferred upon Beale were those of the Royal College of Physicians: membership (1856), fellowship (1859), the Baly Medal (1871), and the post of Lumleian lecturer (1875). Elected fellow of the Royal Society in 1857, he delivered the Croonian lecture for 1865. He also was president of the Microscopical Society from 1879 to 1880.

In 1859 Beale married Frances Blakiston; their son, Peyton Todd Bowman Beale, served as assistant surgeon to King's College Hospital. Beale's death, in his seventy-eighth year, was attributed to pontine hemorrhage.

Beale's contemporary reputation derived primarily from his practical books on the microscope and from his vocal opposition to the mechanistic interpretation of life. The books are basic laboratory manuals, a natural and useful outgrowth of the Carey Street course on the use of the microscope. Beale's opposition to T. H. Huxley and the other "physicalists" commanded attention because his scientific credentials were impressive. He had, in fact, been elected to his first teaching position at King's College despite Huxley's candidacy.

The high quality of Beale's histological studies is demonstrated by his series of lectures, "On the Structure of the Simple Tissue of the Human Body," delivered before the Royal College of Physicians in April and May 1861, and published in book form later that year under the same title. These lectures reveal a careful microscopist at work and demonstrate Beale's mastery of the recently introduced techniques of vital staining. Like most of his works, they are illustrated by excellent plates executed by Beale himself. To a great extent, his plates and conclusions confirmed the work of Ferdinand Cohn, Max Schultze, and others who were working toward a redefinition of the cell and a deemphasis of the cell wall. But Beale, a strong-willed and independent figure, proposed a unique terminology, insisting upon an absolute distinction between what he called "germinal matter" (essentially equivalent to protoplasm) and "formed matter" (all other tissue constituents). Germinal matter was the living substance; formed matter had ceased to live.

Beale was also committed to a special interpretation of the living substance, and after Huxley had in 1868 delivered a mechanistic manifesto on the nature of protoplasm,[1] Beale countered with *Protoplasm,* a polemical work in which he asserted that a vital force was necessary for life. Generally regarded as the defender of religion and morality against the inroads of the mechanistic heresy, he wrote often, especially near the end of his life, on religious and moral considerations.

Beale seems to have been polemical by nature. Besides his spectacular attack against Huxley, he was also suspicious of Darwinian evolution, and was involved in an extended controversy with the German histologists Wilhelm Kühne and Rudolf von Kölliker over the termination of the nerve endings in voluntary muscles. Beale denied their belief that the nerves terminate in ends, insisting instead that the cells and fibers of each nervous apparatus form a continuous loop, "an uninterrupted circuit." In his book *On Disease Germs* (1870), Beale attacked both those who believed that the agents of contagion were bacteria (i.e., plants) and those who believed that disease germs did not exist; he proposed instead the unique theory that disease germs are minute particles of degraded protoplasm derived by direct descent from the normal protoplasm of the diseased organism. It is typical of Beale that he attacked Darwin's now discredited hypothesis of pangenesis "with much acerbity and some justice."[2]

Beale was an amazingly prolific writer, but many of his works cover essentially the same ground. Between 1857 and 1870 he edited a periodical called *Archives of Medicine.* Two eponyms were awarded to Beale: "Beale's solution," carmine in ammonia, is an effective histological stain, and "Beale's cells" are the pyriform nerve ganglion cells.

NOTES

1. T. H. Huxley, "On the Physical Basis of Life," in *Fortnightly Review,* **5** (1869), 129–145. Later published in his *Lay Sermons, Addresses, and Reviews* (London, 1870), pp. 132–161.
2. Charles Darwin, *The Variation of Animals and Plants Under Domestication,* auth. ed. (New York, 1899), p. 339, n. 1.

BIBLIOGRAPHY

I. Original Works. Among Beale's writings are *The Microscope, and Its Application to Clinical Medicine* (London, 1854; 4th ed., 1878), later editions varying slightly in title; *How to Work With the Microscope. A Course of Lectures on Microscopical Manipulation, and the Practical Application of the Microscope to Different Branches of Investigation* (London, 1857; 5th ed., 1880); "Some Points in Support of Our Belief in the Permanence of Species," in *Edinburgh New Philosophical Journal,* **11** (1860), 233–242; "Remarks on the Recent Observations of Kühne and Kölliker Upon the Terminations of the Nerves in Voluntary Muscle," in *Archives of Medicine,* **3** (1862), 257–265; *Kidney Diseases, Urinary Deposits, and Calculous Disorders; Their Nature and Treatment,* 3rd ed., enl. (London, 1869), earlier eds. (1861, 1864) under somewhat different titles; *Protoplasm; or Life, Matter, and Mind,* 2nd ed., rev. (London, 1870; 4th ed., 1892); and *On Slight Ailments: Their Nature and Treatment* (London, 1880; 4th ed., 1896). Nearly all of Beale's scientific papers are listed in the *Royal Society Catalogue of Scientific Papers,* I, 221–222; VII, 111–112; IX, 152; and XIII, 370.

II. Secondary Literature. Sketches of Beale's life and work can be found in *Bulletin de l'Académie royale de médecine de Belgique,* 4th ser., **20** (1906), 348–351; *Dictionary of National Biography,* Supp. 2, I, 118–120; *Journal of the American Medical Association,* **46** (1906), 1392; *Lancet* (7 Apr. 1906), 1004–1007; *Medical History,* **2** (1958),

269–273; *Nature*, **73** (1905–1906), 540; and *Proceedings of the Royal Society*, **79B** (1907), lvii–lxiii.

GERALD L. GEISON

BEAUGRAND, JEAN (*b*. Paris [?], France, *ca.* 1595 [?]; *d*. Paris [?], *ca.* 22 December 1640), *mathematics*.

In spite of the important role he played in the mathematics of the 1630's, what little is known or surmised about Beaugrand has had to be pieced together from sources dealing with his friends and enemies, and only rarely with him directly. There are few manuscripts or letters, and no records. He may have been the son of Jean Beaugrand, author of *La paecilographie* (1602) and *Escritures* (1604), who was chosen to teach calligraphy to Louis XIII. He studied under Viète and became mathematician to Gaston of Orléans in 1630; in that year J. L. Vaulezard dedicated his *Cinq livres des zététiques de Fr. Viette* to Beaugrand, who had already achieved a certain notoriety from having published Viète's *In artem analyticam isagoge*, with scholia and a mathematical compendium, in 1631. Some of the scholia were incorporated in Schooten's edition of 1646.

Beaugrand was an early friend of Fermat[1] and Étienne Despagnet (the son of Jean Despagnet); later of Mersenne and his circle; and for a time, before their bitter break, of Desargues. He seems to have been an official Paris correspondent to Fermat and was replaced in that function by Carcavi.[2] In 1634 he was one of the scientists who officially examined Morin's method for determining longitudes.[3] The following year he assumed the functions of *sécretaire du roi*, possibly under Pierre Séguier, who was appointed chancellor in the same year.

Sometime before 1630 Beaugrand visited England;[4] he met Hobbes in Paris, at the home of Mersenne, in 1634 and 1637.[5] He spent a year in Italy, from February 1635, as part of Bellièvre's entourage.[6] While there, he visited Castelli in Rome,[7] Cavalieri in Bologna,[8] and Galileo in Arcetri,[9] and communicated to them some of Fermat's results in a conversation alluded to in his *Géostatique*. All of them, especially Cavalieri, appear to have been impressed with Beaugrand as a mathematician, and he continued to correspond with them after his return to Paris in February 1636.[10] He conveyed results of the French mathematicians without always bothering about provenance, a habit that resulted in misunderstandings.

Although Beaugrand's *Géostatique* (1636) was well received by Castelli and Cavalieri, it was a disappointment in France; and his violent polemical exchanges with Desargues, his anonymous pamphlets against Descartes, and the disdain that characterizes

Descartes's references to him, as well as the cooling of his relations with Fermat, seem to stem from the period of its publication. Its main thesis is that the weight of a body varies as its distance from the center of gravity. Fermat[11] had adopted this law, and sought to demonstrate it in a satisfactory manner by arguing from a thought-experiment in which Archimedean arguments were applied to a lever with its fulcrum at the earth's center. Thus he defended a law of gravity later taken up independently by Saccheri in his *Neo-Statica* (1703).

Fermat's proposition gave rise to a long debate involving Étienne Pascal, Roberval, and Descartes.[12] Desargues appended a text inspired by this controversy to his *Brouillon projet*. Beaugrand in turn claimed that the proposition which occupies most of the *Brouillon projet* is nothing but a corollary to Apollonius, *Conics* III, prop. 17.[13] This attack was preserved by Desargues's enemies and occasioned Poncelet's rediscovery of Desargues's work 150 years later. Beaugrand's attacks on Descartes[14] took a similar form, including a charge of plagiarism from Harriot,[15] and are to be found in three anonymous pamphlets and a letter to Mersenne claiming that Viète's methods were superior and that Descartes had derived his *Géométrie* from them.[16]

NOTES

D, F, and *M* below refer to the respective standard editions (see bibliography) of the correspondence of Descartes, Fermat, and Mersenne.

1. *M*, V, 466 f.
2. Cf. C. de Waard, in *Bulletin des sciences mathématiques*, 2nd ser., **17** (June 1918).
3. Bigourdan, "La conférence des longitudes de 1634," in *Comptes rendus de l'Académie des sciences*, **163** (1916), 229–233.
4. *M*, II, 514.
5. *D*, III, 342.
6. *M*, V, 271.
7. Paris, B.N., f. fr. 15913, 15914.
8. *M*, V, 429. Cf. C. de Waard, in *Bollettino di bibliografia di storia delle scienze matematiche* (1919), 1–12.
9. *M*, V, 454; Galileo, Edizione nazionale, XVI (1905), 335–337, 340–344.
10. *Lettres de Chapelain*, Tamizey de Larroque, ed., I (1880), 109.
11. *F*, V, 100–103.
12. Cf. Descartes to Mersenne, 13 July 1638.
13. R. Taton, *L'oeuvre mathématique de Desargues* (Paris, 1937), *passim*.
14. P. Tannéry, *Oeuvres scientifiques*, V, 503–512; VI, 206 ff.
15. *M*, VII, 201 f.
16. *Ibid.*, 87–104.

BIBLIOGRAPHY

References to Beaugrand are scattered throughout the standard eds. of the correspondence of Fermat (P. Tannéry and C. Henry, eds. [Paris, 1891–1912], supp. vol., N. de Waard, ed. [Paris, 1922]), Descartes (Adam and P. Tannéry, eds. [Paris, 1897–1913]), and Mersenne (Paris, 1933–   ).

The extant writings by Beaugrand, besides his ed. of Viète's *In artem analyticam isagoge* (Paris, 1631) and *Géostatique* (Paris, 1636), pub. in Latin as *Geostatica* (Paris, 1637), are a letter on tangents in Fermat's *Oeuvres,* supp. vol. (1922), pp. 102–113; a letter to Desargues, in R. Taton, *L'oeuvre mathématique de Desargues* (Paris, 1951), pp. 186–190; four writings against Descartes, "Lettre de M. de Beaugrand (10 Aug. 1640)," in Descartes, *Oeuvres,* V, 503–512; letter to Mersenne (Apr. [?] 1638), in Descartes, *Correspondance* (1903); three anonymous pamphlets, in P. Tannéry, *Mémoires scientifiques,* VI (Paris, 1896), 202–229 ("La correspondance de Descartes dans les inédits du fonds Libri"); and Mersenne, *Harmonie universelle,* II, *Livre I des instrumens,* props. 14, 15, 31, and his *Correspondance,* IV, 429–431.

See also Guy de Brosse, *Éclaircissement d'une partie des paralogismes . . .* (1637); P. Costabel, "Centre de gravité et équivalence dynamique," in *Les conférences du Palais de la Découverte,* ser. D, no. 34 (Paris, 1954); and P. Duhem, *Les origines de la statique* (Paris, 1906), pp. 178 ff. For the confused priority controversy over the cycloid, see Pascal, *Oeuvres,* L. Brunschvicq and E. Boutroux, eds., VIII (Paris, 1914), 181–223.

HENRY NATHAN

BEAUMONT, ÉLIE DE. See **Élie de Beaumont, Jean B.**

BEAUMONT, WILLIAM (*b.* Lebanon, Connecticut, 21 November 1785; *d.* St. Louis, Missouri, 25 April 1853), *physiology.*

William Beaumont was the son of a farmer. Deciding not to follow his father's occupation, he left home in the winter of 1806. He set out with a horse and cutter, a barrel of cider, and $100 and drove northward, arriving at Champlain, New York, in the spring of 1807. There he became the village schoolmaster and taught for three years. During this period, Beaumont employed his spare time to read medical books borrowed from Dr. Seth Pomeroy, a Burlington, Vermont, physician. Having decided to study medicine, he became an apprentice to Dr. Benjamin Chandler, a general practitioner in St. Albans, Vermont. In June 1812, after two years of study, he received a license to practice from the Medical Society of Vermont. That month war broke out between Great Britain and the United States, and in September, Beaumont joined the army as acting surgeon's mate. After resigning from the army in 1815, he practiced at Plattsburg for five years, but then took a commission as a post surgeon and was ordered to Fort Mackinac, now in Michigan.

Here, on 6 June 1822, an accident occurred. A shotgun had been carelessly loaded and had gone off. A nineteen-year-old French Canadian, Alexis St. Martin, a trapper employed by the American Fur Company, had been standing about two or three feet from the muzzle of the gun and received the entire charge in his left side. Beaumont was called to care for the trapper. Despite a very poor prognosis, St. Martin recovered. After ten months his wound was to a large extent healed. However, extensive parts of the injured lung and of the stomach sloughed off, and as a result of the union between the lacerated edges of the stomach and the intercostal muscles, a gastric fistula developed. Touched by the young trapper's misfortune, Beaumont took him into his own home, where he remained almost two years, until he was completely recovered. But the large fistula into his stomach remained.

The idea of carrying out experiments on digestion in a human subject occurred to Beaumont in the spring of 1825, and in May he began his investigations. From 1825 to 1833 Beaumont carried out four groups of experiments while stationed at several army posts (Mackinac, Niagara, Crawford, St. Louis). These studies were interrupted for varying periods, ranging from several months to four years, because of St. Martin's trips to Canada. It was not by accident that Beaumont recognized the opportunity presented by St. Martin's gastric fistula. Although he was self-educated, he had a curious mind and was eager to learn. From his writings it is clear that he was abreast of the scientific literature concerned with gastric physiology and pathology, and he employed the knowledge available to him to organize his experiments and observations. Aware of his limited knowledge of chemistry, he requested the assistance of Robley Dunglison, professor of physiology at the University of Virginia, and of Benjamin Silliman, professor of chemistry at Yale University. Both of them analyzed specimens of St. Martin's gastric juice, and established the presence in it of free hydrochloric acid. A portion of the gastric juice was sent by Silliman to the Swedish chemist Berzelius, but his investigation was incomplete and arrived too late to be included in Beaumont's book. In 1834, Beaumont also sent some gastric juice to Charles F. Jackson, one of the co-discoverers of ether anesthesia.

The experiments were concluded by November 1833, and in December of that year his book was published at Plattsburg in an edition of 1,000 copies. The work is an octavo volume of 280 pages entitled *Experiments and Observations on the Gastric Juice and the Physiology of Digestion.* The book is divided into two parts, a general discussion of the physiology of digestion and a detailed description of the experiments. The presentation concludes with fifty-one inferences drawn by the author from his researches. The material is presented in an unpretentious, matter-of-fact style. Beaumont restricted himself to a relation of his experiments and observations, and was very cautious, avoiding all unsupported assertions.

In 1834, after the publication of his book, Beaumont was posted at St. Louis, where he remained in service until his resignation in 1839, when he went into general practice. As a general practitioner he exhibited the strong common sense that characterizes his experimental work, earning the esteem of his fellow citizens. In March 1853, he suffered a fall, resulting in a fractured hip. Several weeks later a carbuncle appeared on his neck, and on April 25, he succumbed to his illness.

Not only did Beaumont's experiments soon become known to his friends, but before long, reports of his work also spread across the Atlantic to Europe. As early as 1826 two articles dealing with the case of St. Martin appeared in Germany, and it was soon noticed in Scotland and France. The publication of Beaumont's book created considerable interest among European scientists and physicians, and in general was favorably received. A German translation was published at Leipzig in 1834, and at about the same time, the American edition was noted in the English, German, and French literature. The most significant response came from Germany, where Beaumont's experiments influenced various investigators, among them Johannes Müller, Theodor Schwann, and J. B. Purkinje. In France, Nicolas Blondlot, professor of chemistry at Nancy, impressed by the reports of St. Martin's accident, his recovery, and its consequences, hit upon the idea of imitating in animals the fistula that the Canadian had retained after his recovery. Until the 1860's, Beaumont's work was referred to by physiologists, pathologists, and clinicians. Because of a number of factors, however, interest in his investigations declined during the last third of the century. The introduction of the stomach pump by Kussmaul (1867) gave research workers a simpler instrumentality with which to carry out investigations. This circumstance, together with the rapid development of physiological chemistry during the later nineteenth century, shifted the emphasis in research to the chemical aspects of digestion. Moreover, it was found that the secretions of the salivary glands, of the liver, of the pancreas, and of the intestine played a very much greater role than had previously been assumed. Nonetheless, the value of William Beaumont's contributions to gastric physiology and pathology had been established, and it is clear that he was the first American physiologist to make a major contribution to the development of the science.

In the eighteenth century Réaumur and Spal-lanzani had shown digestion to be a chemical process. Nevertheless, a good deal of confusion prevailed concerning various aspects of the digestive process in the stomach; it is to Beaumont that we owe the clarification of this subject. The results of his investigations attracted the attention of the scientific world, and within a short time gained admission to the literature dealing with the stomach.

Beaumont contributed to the development of gastric physiology by substantiating doubtful observations of other scientists and by stimulating new investigations. He demonstrated once again, both in the stomach and outside the body, that digestion is the result of a chemical process. At the time Beaumont was carrying out his experiments, the nature of the effective agent of the gastric juice was still unknown. A number of experiments on artificial digestion with dilute acids convinced him that such an effort "to imitate the gastric juice . . . was not satisfactory. Probably the gastric juice contains some principles inappreciable to the senses, or to chemical tests, besides the alkaline substances already discovered in it." Schwann's discovery of pepsin in 1836 showed that Beaumont's surmise had been correct.

Several details of gastric physiology still to be found in modern textbooks were established by Beaumont. He observed that gastric juice was not found in the stomach in the absence of food and that water and other fluids passed very rapidly and directly from the stomach through the pylorus. He also observed a retrograde passage of the duodenal contents. Beaumont ascertained by direct observation that psychic influences are to a considerable degree able to affect gastric secretion and digestion. Some investigators saw the significance of these observations, which, however, were not pursued until Pavlov began his researches.

Beaumont studied the digestibility of various foods, as well as the action of stimulants such as coffee, tea, and alcohol on gastric digestion, thus creating a foundation for practical dietetics, which even today remains one of the most important contributions to this subject. He knew very little about the functions of the rest of the digestive system, and believed that the food was almost completely digested in the stomach. Consequently, he assumed that the time that elapsed between the entry of food into the stomach and its departure therefrom represented an index of its digestibility. Moreover, he also believed that several other conditions that might possibly influence digestion—for instance, the size of the meal—could be disregarded. Despite these limitations, however, his average determinations of the time required for digestion generally agree with those later obtained by means of the stomach tube.

On the basis of his observations *in vivo*, Beaumont was the first to attempt an exact description of the way the gastric contractions move and mix the chyme within the stomach. By fixing his glance on an easily recognizable object as soon as it passed through the cardia, he was able to follow its further movements. In the course of his experiments Beaumont frequently introduced a thermometer into the stomach and convinced himself that it was subject to the same movements. Except for slight modifications by William Brinton in 1848, Beaumont's observations on gastric mobility remained unchanged until they were superseded by Walter B. Cannon's roentgenological investigations between 1898 and 1911.

The technique of animal experimentation was also greatly influenced by Beaumont's experiments. During the seventeenth century, Regnier de Graaf had established a pancreatic fistula in a dog, but the method was not developed and soon lapsed into oblivion. St. Martin's history and Beaumont's experiments stimulated Blondlot to establish artificial gastric fistula in animals. This successful endeavor led other scientific investigators to develop and to extend the fistula method. Schwann in 1844 reported the experimental creation of a gallbladder fistula, and in 1864, Thiry introduced the intestinal fistula. Klemensiewicz, in 1875, used Thiry's approach to isolate the pyloric part of the stomach, and in 1879 R. Heidenhain employed it for the isolation of a portion of the corpus. In turn, Heidenhain, under whom Pavlov worked in 1877 and 1884, greatly influenced him in his study of gastric secretion and in the experimental development of an innervated gastric pouch. As these methods developed, it became possible to study with greater precision the activity of the digestive glands, their dependence on the nervous system, and the details of the digestive process.

Beaumont's work led to a more accurate conception of gastritis and contributed to the development of gastric pathology. He described the occurrence of acute gastritis as a consequence of different causes, most frequently the excessive ingestion of alcoholic beverages or overloading the stomach with irritating foods. He also observed a reddening of the mucous membrane and cessation of gastric secretion in febrile states. These observations in turn led to therapeutic regimens for such conditions.

BIBLIOGRAPHY

I. ORIGINAL WORKS. In 1825 the *Medical Recorder* (**8**, no. 1) published a report of the case of St. Martin; the following year a report of four experiments performed on the same subject appeared in this periodical (Jan. 1826, no. 33). Beaumont published his final results as *Experiments and Observations on the Gastric Juice and the Physiology of Digestion* (Plattsburg, 1833), which was reissued in Boston in 1834, and the same year saw the appearance of a German translation at Leipzig as *Neue Versuche und Beobachtungen über den Magensaft und die Physiologie der Verdauung*, trans. by Dr. Bernhard Luden. A Scottish edition was published at Edinburgh in 1838. A second edition with corrections by Samuel Beaumont, a cousin of Beaumont, appeared in 1847 at Burlington, Vt. Facsimiles of the original edition of 1833, together with a biographical essay, "William Beaumont. A Pioneer American Physiologist," by Sir William Osler, were issued in 1929, 1941, and 1959.

Two important collections of Beaumont papers, documents, and relics are located at the Washington University School of Medicine, St. Louis, Mo., and at the University of Chicago. A Beaumont memorial on Mackinac Island, Mich., recreates some of the circumstances under which Beaumont carried on his investigations.

II. SECONDARY LITERATURE. The chief source for Beaumont's life and work besides his own writings is Jesse S. Myer, *Life and Letters of Dr. William Beaumont Including Hitherto Unpublished Data Concerning the Case of Alexis St. Martin* (St. Louis, 1912; 2nd ed., 1939). Two of Beaumont's early notebooks that give a picture of his medical training and early medical practice have been edited by Genevieve Miller: *William Beaumont's Formative Years: Two Early Notebooks 1811–1821* (New York, 1946). William Osler's essay on Beaumont in *An Alabama Student, and Other Biographical Essays* (New York, 1909) is still useful and in addition is delightful to read. Another review of Beaumont's career that should be consulted is W. S. Miller, "William Beaumont and His Book. Elisha North and His Copy of Beaumont's Book," in *Annals of Medical History*, **1** (1929), 155–179. The extensive biographical literature dealing with Beaumont is for the most part filiopietistic, derivative, repetitive, dreary, and not worth listing.

Descriptions of the Beaumont papers and artifacts are available in Arno B. Luckhardt, "The Dr. William Beaumont Collection of the University of Chicago, Donated by Mr. and Mrs. Ethan Allen Beaumont," in *Bulletin of the History of Medicine,* **7** (1939), 535–563; Alfred M. Whittaker, "The Beaumont Memorial on Mackinac Island," *ibid.,* **28** (1954), 385–389; and Phoebe A. Cassidy and Robert S. Sokol, *Index to the William Beaumont M.D. (1785–1853) Manuscript Collection* (St. Louis, 1968).

For an evaluation of the reception of Beaumont's work by his contemporaries and successors, and its influence on physiology and medicine, see George Rosen, *Die Aufnahme der Entdeckung William Beaumont's durch die europäische Medizin. Ein Beitrag zur Geschichte der Physiologie im 19. Jahrhundert* (Berlin, 1935). An English translation of this work, with a foreword by John F. Fulton, appeared as *The Reception of William Beaumont's Discovery in Europe* (New York, 1942). For further discussion of some points in the reception of Beaumont's work, see George Rosen, "Notes on the Reception and Influence of William Beaumont's Discovery," in *Bulletin of the History of Medicine,* **13** (1943), 631–642; F. N. L. Poynter, "The Reception of William Beaumont's Discovery in England. Two Additional Early References," in *Journal of the History of Medicine and Allied Sciences,* **12** (1957), 511–512; George Rosen, "The Reception of William Beaumont's Discovery. Some Comments on Dr. Poynter's Note," *ibid.,* **13** (1958), 404–406; F. N. L. Poynter, "New Light on the Reception of William Beaumont's Discovery in England," *ibid.,* 406–409.

GEORGE ROSEN

BECCARI, NELLO (*b.* Bagno a Ripoli, near Florence, Italy, 11 January 1883; *d.* Florence, 20 March 1957), *anatomy.*

Beccari, son of Odoardo Beccari, a learned naturalist who explored the interior regions of Borneo, and of Nella Goretti-Flamini, studied in Florence. After receiving his M.D. in 1907, he dedicated himself to anatomical research at the Institute of Human Anatomy of Florence, at that time directed by Giulio Chiarugi. In 1911 he obtained a teaching diploma in human anatomy and was a substitute teacher of anatomy until 1915. He served as a major in the Italian army during World War I and was wounded at the front. In 1919 he returned to civilian life, and two years later was appointed director of the Institute of Human Anatomy at Catania University. In 1925 he agreed to transfer to the University of Florence as director of the Institute of Comparative Anatomy, a position he held until his seventieth year, the legal age limit for teaching.

Beccari was a very learned anatomist. Following the belief of his teacher Chiarugi that "comparative anatomy is the key to the comprehension of human anatomy," he developed his research essentially on the plan of comparative anatomy: his last book, *Anatomia comparata dei vertebrati* (only the first two volumes of the five projected were published, in 1951 and 1955), would have been his own definitive statement of his comparative approach. Nevertheless, he also did some work in more traditional descriptive anatomy; for instance, his precise observations on the rhinencephalon in man: on the olfactory gyri (1911), on the anterior perforated substance (1911), and on the hippocampal gyrus (1911).

Beccari was a fine cytologist, as is shown by his little-known observations on the cutaneous glands of sheep (1909), and particularly those on the suborbital glands of *Gazella dorcas* (1910): he revealed new, interesting details on sebaceous secretion and demonstrated a functional synergism between epithelial cells and melanophores. But Beccari best displayed his ability as a cytologist in his great work on the

genesis of the germ cells and on the early cytological differentiation of sex cells. From 1920 to 1925 he worked systematically on the *Bufo viridis,* aiming at the discovery of the origin of the germ cells and the morphological description of the initial differentiation of sex cells. Beccari confirmed that in vertebrates the germ cells spring from entodermal elements that appear very early in the embryo's development, but he could not affirm absolutely that in the fertilized egg of vertebrates a genital blastomere exists, already differentiated, in the earlier stages of segmentation. However, the primary germ cells of *Bufo* are, in the beginning, all alike, not differentiated cytologically at all. But these primary germ cells already have a sex and, contrary to the view of Ernst Haeckel, who called them hermaphroditic, are female. From the systematic study of Bidder's organ, Beccari could demonstrate that a constant ovogenesis appears before a doubtful abortive spermiogenesis. In *Bufo* the earlier morphological appearance of cellular sexuality manifests itself as an initial evolution of primary germ cells in a female direction. In addition, the fact that in *Bufo* all the males come from hermaphrodite larvae confirmed that, in *Bufo,* the female is the fundamental sex. Sexual determination results, in *Bufo,* from later embryonic development: the persistence of the primary sex cells to the periphery of the genital crest and the formation, within this, of a cavity—the so-called ovarian pocket—are signs of female development; the migration within the genital crest of primary sex cells and their proliferation, with precocious appearance of interstitial cells, are signs of masculinity.

Beccari's principal work was done, however, in the field of neurology. Undoubtedly one of the most eminent neurologists of his generation, he based his conclusions on a thorough knowledge of comparative neurology. Beccari collected and arranged his own wide neurological work in his *Nevrologia comparata anatomo-funzionale dei vertebrati e dell'uomo* (1943). In his book *Il problema del neurone* (1945) he gives us the measure of his wide critical knowledge. He defended the fundamental principle of the neuron, underlining the fact that the recent discoveries had not invalidated its individuality, even if they modified its character. Beccari also stated (1948) that in vertebrates metamery begins only when the nerves appear; that the arrangement in the spinal cord of neurons for segmental knots is consequent upon the formation of nerves, and not at all the reason for their metameric arrangement.

But Beccari's most important contributions related to the rhombencephalon. From his first note of 1906 on the Mauthner's fibers in the *Salamandrina perspicillata* to his further works (1911, 1913, 1915, 1919),

545

Beccari persistently analyzed and clarified their functional significance in the structure of the rhombencephalon. In 1921 and 1922 he demonstrated the importance of the tegmental centers of the medulla oblongata and of the pons. Beccari's research also revealed the role of the rhombencephalon in the determination of static extrapyramidal motility.

Beccari was a very able technician, a precise draftsman, and an excellent teacher. He was secretary of the Italian Society of Anatomy from 1929, associate founder of the Italian Society of Experimental Biology, and a member of the National Academy of the Lincei. Because his democratic beliefs were well known, in October 1944, following Florence's liberation from German occupation, he was chosen to preside over the Faculty of Natural and Physical Sciences of the University of Florence—a position he held until his death.

BIBLIOGRAPHY

Beccari's books include *Elementi di tecnica microscopica* (Milan, 1915, 1927, 1940); *La inversione sperimentale del sesso nei vertebrati* (Pavia, 1930); *Nevrologia comparata anatamo-funzionale dei vertebrati e dell'uomo* (Florence, 1943); *Genetica* (Florence, 1945); *Il problema del neurone,* 2 vols. (Florence, 1945–1947); *I problemi biologici del sesso* (Rome, 1950); and *Anatomia comparata dei vertebrati,* 2 vols. (Florence, 1951–1955).

There is no complete list of Beccari's many articles. For his scientific production between 1906 and 1922, see the bibliographical index of anatomical publications of the Institute of Human Anatomy at Florence for the years 1891–1921, in *Archivio italiano di anatomia e di embriologia,* **18** (1922), Supp.

Among Beccari's more important articles are "Sullo sviluppo delle ghiandole sudoripare e sebacee nella pecora," in *Archivio italiano di anatomia e di embriologia,* **8** (1909), 271–295; "Ricerche intorno alle tasche ed ai corpi ghiandolari suborbitali in varie specie di ruminanti: anatomia, struttura e sviluppo," *ibid.,* **9** (1910), 660–690; "La sostanza perforata anteriore e suoi rapporti col rinencefalo nel cervello dell'uomo," *ibid.,* **10** (1911), 261–275; "Le strie olfattorie nel cervello dell'uomo," in *Monitore zoologico italiano,* **22** (1911), 255–270; "La superficie degli emisferi cerebrali dell'uomo nelle regioni prossime al rinencefalo," in *Archivio italiano di anatomia e di embriologia,* **10** (1911), 482–500; "Studi sulla prima origine delle cellule genitali nei vertebrati. I. Storia delle indagini e stato attuale della questione," *ibid.,* **18** (1920), 157–190; "II. Ricerche nella *Salamandrina perspicillata,*" *ibid.,* **18** (1921), Supp. 29; "III. e IV. Ovogenesi larvale, organo del Bidder e differenziamento dei sessi nel *Bufo viridis,*" *ibid.,* **21** (1924) and **22** (1925); and "Il problema del differenziamento del sesso negli anfibi," in *Archivio di fisiologia,* **23** (1925), 385–410. See also "Dimostrazione delle fibre di Mauthner nella

Salamandrina perspicillata," in *Lo sperimentale,* **60** (1906), 456–460; "Ricerche sulle cellule e fibre del Mauthner e sulle loro connessioni in pesci ed anfibi," in *Archivio italiano di anatomia e di embriologia,* **6** (1907), 660–700; "Le cellule dorsali o posteriori dei ciclostomi. Ricerche nel *Petromyzon marinus,*" in *Monitore zoologico italiano,* **20** (1909), 308–325; "Sopra alcuni rapporti del fascicolo longitudinale posteriore con i nuclei di origine dei nervi oculomotore e trochleare nei teleostei," *ibid.,* 242–255; "La costituzione, i nuclei terminali e le vie di connessione del nervo acustico nella *Lacerta muralis* Merr.," in *Archivio italiano di anatomia e di embriologia,* **10** (1911), 646–660; "Sulla spettanza delle fibre del Lenhossek al sistema del nervo accessorio. Osservazioni in *Lacerta muralis,*" *ibid.,* **11** (1912), 299–315; "Il IX, X, XI e XII paio di nervi cranici ed i nervi cervicali negli embrioni di *Lacerta muralis,*" *ibid.,* **13** (1915), 1–25; "Duplicità delle cellule e delle fibre del Mauthner in un avanotto di trota," in *Monitore zoologico italiano,* **30** (1919), 88–100; "Peculiari modalità nelle connessioni di alcuni neuroni del sistema nervoso centrale dei pesci. Ulteriori ricerche sulle collaterali delle fibre del Mauthner," in *Archivio italiano di anatomia e di embriologia,* **17** (1919), 239–360; "Lo scheletro, i miotomi e le radici nervose nella regione occipitale degli avanotti di trota," *ibid.,* **19** (1921), 1–45; "Studi comparativi sulla struttura di rombencefalo. I. Nervi spino-occipitali e nervo ipoglosso. II. Centri tegmentali," *ibid.,* 122–210; "Il centro tegmentale o insterstiziale ed altre formazioni poco note nel mesencefalo e nel diencefalo di un rettile," *ibid.,* **20** (1923), 560–573; "La costituzione del nucleo del fascio ottico basale dei rettili, e la sua probabile importanza nella produzione del riflesso pupillare," in *Atti del VI° Congresso della Società Italiana di Neurologia* (Naples, 1923); "Primo differenziamento dei nuclei motori di nervi cranici," in *Monitore zoologico italiano,* **34** (1923), 161–170; "I centri tegmentali dell'asse cerebrale dei selaci," in *Archivio zoologico italiano,* **14** (1930), 411–455; "Studi comparativi sopra i nuclei terminali del nervo acustico nei pesci," *ibid.,* **16** (1931), 732–760; "Intorno all'esistenza di uno strato sinaptico nelle connessioni di alcuni neuroni dei pesci," in *Monitore zoologico italiano,* **45** (1934), 220–235; and "Morfogenesi filetica e morfogenesi embrionale del sistema nervoso dei vertebrati," report to the Ninth Congress of the Società Italiana di Anatomia (Bologna, 24–27 October 1947), *ibid.,* Supp. **56** (1948), 22–32.

PIETRO FRANCESCHINI

BECCARIA, GIAMBATISTA (*b.* Mondovì, Italy, 3 October 1716; *d.* Turin, Italy, 27 May 1781), *electricity.*

Beccaria, who is not to be confused with his younger contemporary, the Milanese publicist Cesare Beccaria, was christened Francesco, and became Giambatista when he joined the Piarists (the Clerks Regular of the Pious Schools), a teaching order established in the seventeenth century. He entered into his novitiate in 1732, studied at Rome and at Narni, and began his own pedagogical career in 1737. During the next decade he taught in Narni, Urbino, Palermo, and Rome; in 1748 he was appointed to the chair of physics at the University of Turin, which he occupied with great distinction for over twenty years.

Beccaria's predecessor at Turin, a Minim priest and a convinced Cartesian, had many followers in the university who did not appreciate the English science the new professor fancied. A battle ensued, for which Beccaria's ability, energy, and testiness admirably suited him. This struggle, which mixed local jealousies and the rivalry between Minims and Piarists with philosophical issues, was of critical importance in shaping Beccaria's career, for out of it grew his concern with electricity. One of those who had recommended Beccaria's appointment, the marchese G. Morozzo, learning of the lightning experiment of Marly, suggested to his protégé that Franklin's ideas on electricity might make an excellent weapon against the Cartesians. Electricity was a promising field for confrontation even without the glamorous new discoveries about lightning, as the Cartesian program of reducing electrical phenomena to the vortical motion of a special matter had never been very successful.

The results of Beccaria's brief, vigorous study of electricity appeared in his first book, *Dell'elettricismo artificiale e naturale* (1753). The volume, which Franklin praised, presents the elements of the new theory clearly and logically; illustrates them with variations of Franklin's experiments, to which Beccaria primarily added observations of the different appearances of discharges from positively and negatively electrified points; modifies secondary aspects of the theory and applies it to new territory; and seeks to explain meteorological and geophysical phenomena as manifestations of "natural" electricity. Apart from its new experiments and ingenious meteorological hypotheses, the book is notable for Beccaria's love of system, which caused him occasionally to oversimplify the phenomena; for his reluctance, characteristic of the Franklinian physicist, to discuss the detailed mechanical causes of electrical motions; for his courageous assertion, against the prevailing view, that air is a better conductor than pure water; and for his skill in designing apparatus, including such measuring devices as the electrical thermometer, whose invention is usually wrongly ascribed to Franklin's colleague, Ebenezer Kinnersley. The book also contains a long letter to the Abbé J. A. Nollet, who had raised objections against Franklin's system. The Parisian *franklinistes* thought the letter successful, translated it into French, and thus temporarily made Beccaria the leading champion of the new system.

In 1758 Beccaria published a second book on electricity, in the form of sixteen letters to G. B. Beccari. F. A. Eandi, Beccaria's biographer, and Priestley thought the *Lettere al Beccari* to be Beccaria's best work, although the modern reader is apt to find it annoyingly prolix. Two-thirds of it is devoted to natural electricity, particularly that of the atmosphere, which Beccaria and his students tirelessly probed with metal poles, kites, and even rockets. Here are many valuable observations of cloud formation, thunderstorms, and the accompanying electrical states of the lower air. The first seven letters, on man-made or artificial electricity, summarize, extend, and modify Franklin's system. The most successful extensions deal with insulators other than glass, from which Beccaria constructed parallel plate condensers whose relative powers he tried to estimate. Less happily, ignoring his earlier strictures, he sought a mechanical explanation of the apparent suspension of electrical attractions in the highly conducting "vacuums" produced by eighteenth-century air pumps. His investigation led him to revive Cabeo's theory, which attributes such attractions to the action of air displaced by electrical matter issuing from charged bodies. This view, although decidedly retrograde, was not necessarily anti-Franklinist, since it did not touch the characteristic Philadelphian doctrines relating to positive and negative electricity and to the perfect insulation of glass. Franklin himself had tacitly ascribed an important task to the air, that of opposing the natural tendency toward dissipation of the "electrical atmospheres" formed about charged bodies. Beccaria concurred in this view, which he tried (not altogether successfully) to combine with the hypothesis of Cabeo.

From the *Lettere,* Beccaria turned to other subjects, particularly astronomical and geophysical ones, in connection with measuring a degree of longitude in Piedmont. His interest was reawakened by the work of a former student, G. F. Cigna, who in 1765 published an account of experiments that developed Nollet's elaborations of Symmer's famous manipulations with electrified silk stockings. Beccaria pursued these experiments with all his skill, inventiveness, and energy, largely because they seemed to favor the anti-Franklinian double-fluid theory. But the more remarkable and delicate of the phenomena he investigated, which depend upon induction in the coatings from residual charges left on the dielectric after the discharge of a condenser, required something more than Franklin's system for their explanation. Beccaria, who would not admit action-at-a-distance, supplied this deficiency with a complicated scheme of electrical atmospheres and "vindicating," or regenerating, electricity. These ideas, which found their clearest expression in Beccaria's *Electricitas vindex* (1769), subsequently led Volta, while seeking alternatives to them, to the invention of the electrophorus.

Beccaria's last major contribution to the science of electricity was *Elettricismo artificiale* (1772), a difficult, verbose compendium of Beccaria's work on the subject, explained with the help of Franklin's principles, vindicating electricities and the doctrine of atmospheres. In 1775 he brought out a short volume on atmospheric electricity in fair weather, *Dell'elettricità a cielo sereno.* Thereafter, he produced a few small pieces on electricity, intermixed with short contributions on other physical and chemical subjects. In 1778 his health began to fail, and, although his painful affliction—hemorrhoids and tumors—sometimes remitted, he was thereafter unable to do much laboratory work.

Beccaria's success derived as much from his aggressive, vigorous character as from his native intelligence. He began the study in which he was to win fame as a new front in a larger battle. His thirst for fame, and his recurrent depressions when he felt unsure of it, were prominent features of his personality and his primary source of energy. He was jealous of his discoveries and quick to claim priority; Franklin was perhaps the only person he did not consider a competitor. The connection with Franklin, with whom he corresponded intermittently, was very precious to him: letters of praise from Franklin, besides the gratification they conferred, resulted in increases in financial support for himself and his work. His personality apparently did not impair his effectiveness as a teacher. Eandi thought him inspirational, an opinion strengthened by the success of his students, Cigna, Lagrange, and G. A. Saluzzo. His chief pedagogical achievement, however, may well have been his indirect, literary influence on the early studies of Volta.

BIBLIOGRAPHY

I. ORIGINAL WORKS. Beccaria's main works on electricity are *Dell'elettricismo artificiale e naturale libri due* (Turin, 1753); *Dell'elettricismo lettere . . . al chiarissimo Signor Giacomo Bartolommeo Beccari* (Bologna, 1758); *Experimenta atque observationes quibus electricitas vindex late constituitur atque explicatur* (Turin, 1769); *Elettricismo artificiale* (Turin, 1772), which also appeared in an English version sponsored by Franklin (London, 1776); and *Dell' elettricità terrestre atmosferica a cielo sereno* (Turin, 1775), which is included in the English version of *Elettricismo artificiale*. A bibliography of Beccaria's work, both published and in manuscript, drawn up by Prospero Balbo, appears in F. A. Eandi, *Memorie istoriche intorno gli studi del padre Giambatista Beccaria* (Turin, 1783), and in *Biblioteca degli eruditi e dei bibliofili*, no. 69 (Florence, 1961). The present location and nature of the extant manuscripts are described in A. Pace, "The Manuscripts of Giambatista Beccaria, Correspondent of Benjamin Franklin," in *Proceedings of the American Philosophical Society,* **97** (1952), 406–416.

II. SECONDARY LITERATURE. The standard biography of Beccaria is Eandi's *Memorie.* Further information and references are given in A. Pace's excellent notice of Beccaria in *Dizionario biografico degli italiani,* VII (Rome, 1965). For Beccaria's work, see Joseph Priestley, *The History and Present State of Electricity,* 3rd ed. (London, 1775), *passim;* M. Gliozzi, "Giambatista Beccaria nella storia dell'elettricità," in *Archeion,* **17** (1933), 15–47; and C. W. Oseen, *Johan Carl Wilcke experimental-fysiker* (Uppsala, 1939).

JOHN L. HEILBRON

BECHER, JOHANN JOACHIM (*b.* Speyer, Germany, 6 May 1635; *d.* London, England, October 1682), *chemistry, economics.*

Becher's father, Joachim, was a Protestant pastor who died when the boy was only eight years old. His mother, Anna Margaretha Gauss, came from a prominent Speyer family; her father also was a Protestant minister. Left with four sons at the death of her husband, she married a man who squandered the family's remaining resources. Young Becher was thus forced at an early age to help support his mother and his brothers, all of whom were younger than he. He was for the most part self-educated; in his later years he recalled only one teacher who had helped him, Debus (*Konrektor* of the Speyer Retscher-Gymnasium from 1644).

At the age of thirteen Becher began his *Wanderjahren.* Residing first in Sweden, he then traveled through Holland, Germany, and Italy. His earliest dated work (1654) deals with alchemy, but his farflung interests also included medicine, theology, politics, economics, and even the formulation of a universal language. By 1657 Becher had settled at Mainz, where he was converted to Roman Catholicism. His first published book, *Naturkündigung der Metallen,* appeared in 1661, and he was soon well established as an iatrochemist. He received the M.D. from the University of Mainz on 16 November 1661, and married Maria Veronika, the daughter of the influential jurist and imperial councillor Ludwig von Hörnigk, on 13 June 1662. In the same year Becher published his *Parnassus medicinalis illustratis,* and in 1663 he was appointed professor of medicine at Mainz and physician to the elector.

Nevertheless, the next year Becher left for Munich,

where he was named *Hofmedicus und Mathematicus* to Ferdinand Maria, elector of Bavaria; in 1666 he accepted the post of imperial commercial counsellor to Emperor Leopold I in Vienna. Always interested in problems of law, politics, and commerce, in 1668 he published his *Politischer Discurs,* which shows him to have been the leading German mercantilist of the seventeenth century. The years at Munich and Vienna were ones of great activity. In his official position it was Becher's duty to introduce profitable businesses and new industries. With this in mind, he built and organized an imperial arts and crafts center that included a glassworks and facilities for the manufacture of textiles, as well as a chemistry laboratory. Aware also of the importance of technical education for the advancement of the domestic economy of the state, he proposed important educational reforms, such as the institution of schools that gave practical instruction in civil and military engineering and statics. At the same time, eager to increase trade, Becher organized the Eastern Trading Company and proposed colonial settlements in South America. Neither did he neglect his role as alchemical advisor to the emperor. His important *Physica subterranea* appeared in 1669, and was followed by supplements in 1671 and 1675. At this time Becher also wrote his chief philological works and treatises on theology and moral philosophy.

Becher's economic policies included, in 1677, edicts against French imports that proved to be unsuccessful in the southern German cities. This failure resulted in his dismissal, and after a short imprisonment in 1678, he made his way to Holland, where he sold the plans for a machine that would spool silk cocoons to the city of Haarlem. More important, he submitted to the Dutch Assembly a plan for the extraction of gold from sea sand through smelting. This had been proposed as early as 1673, but it had been abandoned then because of the outbreak of the war with France. In 1679 a small-scale test of his process proved successful, but Becher suddenly left for England, without his family, when the process was scheduled to be repeated on a larger scale. In London, in March 1680, he completed the third and final supplement to the *Physica subterranea* in which he described the gold extraction process. He examined mines in Scotland and Cornwall; and his prefaces give evidence that he completed books at Falmouth and the Isle of Wight (the *Laboratorium portabile* and the *Centrum mundi concatenatum seu magnorum duorum productorum nitri & salis textura & anatomia,* both parts of the *Tripus Hermeticus fatidicus,* 1689, 1690). Shortly before his death he was back in London, where on 22 March 1682 he completed his *Chymischer Glücks-*

hafen, which gives 1,500 chemical processes, including detailed recipes for making the philosophers' stone. While in England, Becher unsuccessfully sought membership in the Royal Society. He had for years cited the works of Robert Boyle, and in 1680 he dedicated a short book on clock design to the Society. It was considered to be of little value, however, and he was not elected.

Becher's views on chemistry have much in common with standard seventeenth-century Paracelsian and Helmontian treatments of the subject. His major work, the *Physica subterranea,* begins in a fashion reminiscent of most theoretical iatrochemical texts. After stating the need for observations and experiments as a guide to a true understanding of the universe, Becher turned to the Mosaic account of the Creation. Here he argued that the universe gravitated from the initial Chaos into five regions (ranging from the sidereal to the mineral). Then motion was added through the rarefaction that followed the creation of light (heat).

As with all theoretical iatrochemists, the problem of the elements was a basic one for Becher. He had little respect for the four Aristotelian elements as they were commonly taught, and he felt that the efforts of Helmont and Boyle to show the elemental nature of water through the growth of vegetable substances were little better. Becher rejected their affirmation that water could be changed into earth in this fashion, and he explained that the willow tree experiment could best be understood in terms of earth being carried by the water into the substance of the tree. Similarly, he argued that observations show that the philosophical attributes of the Paracelsian triad have little in common with ordinary salt, sulfur, and mercury, so they could not really be "principles." He felt that on practical grounds—because of their familiarity—their use might be defended. Nevertheless, in the *Oedipus chimicus* (1664) Becher suggested that sulfur was analogous to earth, and salt to water, while earth and water, in more subtle form, were mercurial in nature.

Becher believed that air, water, and earth were the true elementary principles. The last two, however, form the real basis of all material things, since air is primarily an instrument for mixing. The essential substance of subterranean bodies (metals and stones) is earthy, and there is a need for three types of earth in metals and minerals. One type is needed for substance, another for color and combustibility, and a third, more subtle, for form, odor, and weight. These are, respectively, the *terra vitrescible,* the *terra pinguis,* and the *terra fluida,* earths that have been improperly identified with the Paracelsian principles: salt, sulfur, and mercury.

The concept of *terra pinguis* as a fatty, oily, and

combustible earth occurs in the older alchemical literature and Becher also calls it sulfur φλογιστòς. Here again, he was following an old tradition, since as an adjective meaning "inflammable" φλογιστòς may be found in the works of Sophocles and Aristotle. More recently φλογιστòν had been used in adjectival form by Hapelius (1609), Poppius (1618), and Helmont. In *The Sceptical Chymist* (1661), a work often referred to by Becher, Boyle cited Sennert's use of φλογιστòν in 1619.

Although Becher insisted that each combustible body must contain the cause of its combustibility within itself, he had no clearly defined position on the role to be played by this substance in the burning process. Generally he spoke of the rarefaction of the burning substance through the dissolving power of flame, and he gave relatively little attention to any part that air might play in combustion. He was aware that metals grew heavier when calcined, and, like Boyle, he credited this to the addition of ponderable fire particles.

Becher wrote in support of spontaneous generation, and he believed firmly in metallic transmutation. He wrote at length on fermentation, which, in typical alchemical fashion, he considered to be a basic natural process of great value for the chemist. For Becher, fermentation was a rarefaction leading to perfection, and it could not continue in closed vessels without a fresh supply of air. As such, fermentation may be clearly distinguished from putrefaction, in which a mixed body is completely broken down and destroyed.

Becher was a thoroughgoing vitalist who accepted the common belief that metals grow in the earth. Similarly, he compared the flow of subterranean waters in the living earth to the flow of blood in man. He believed in a perpetual circularity in nature, and he felt that earthly reactions may properly be compared to the laboratory manipulations of the chemists. He rejected the widely held hypothesis of an internal fire in the earth and sought another explanation of the origin of mountain streams and hot springs. Arguing that it is known that surface waters fall by gravity toward the center of the earth, he noted that laboratory experience gives us the further information that vapors arise when liquids are transferred from hot containers to cold ones. Becher reasoned, therefore, that hot waters from the surface are rapidly cooled at the center of the earth, where they vaporize and return to the surface. There mountains act as alembics, and condense the waters as mountain springs; in other places, separate vapors from the interior reach higher temperatures through mixing before they erupt as hot springs or medicinal spas.

Becher's importance for the history of science rests less in major innovations than in his influence on Georg Ernst Stahl. Stahl republished the *Physica subterranea* in 1703, along with his own lengthy *Specimen Beccherianum*. In this work and in others he lauded his predecessor while developing the concept of the *terra pinguis* into the phlogiston theory with the aid of experimental evidence, which was largely lacking in the writings of Becher.

BIBLIOGRAPHY

I. ORIGINAL WORKS. Becher's complex bibliography is discussed in considerable detail by Herbert Hassinger, in his *Johann Joachim Becher 1635–1682. Ein Beitrag zur Geschichte des Merkantilismus,* Veröffentlichungen der Kommission für Neuere Geschichte Österreichs, no. 38 (Vienna, 1951), pp. 254–272. J. R. Partington gives bibliographic details of Becher's more important chemical works in his *A History of Chemistry,* II (London, 1961), 640–643.

The *Physica subterranea* appeared first as *Actorum laboratorii chymici Monacensis, seu Physicae subterraneae libri duo* (Frankfurt, 1669)—Partington states that an earlier ed. appeared in 1667. There is only one volume. The three supplements were published separately in 1671, 1675, and 1680, and are included in the 2nd ed. (Frankfurt, 1681). Becher's own German translation exists as *Chymisches laboratorium Oder Unter-erdische Naturkündigung . . .* (Frankfurt, 1680), which includes the first two supplements and the *Oedipus chimicus.* Stahl's Latin edition was published with his *Specimen Beccherianum* (Leipzig, 1703), and a final composite edition (containing original text, supps., and Stahl's *Specimen*) was printed at Leipzig in 1738. It should be noted that Becher's *Theoria & experientia de nova temporis dimetiendi ratione, & accurata horologiorum constructione, ad Societatem Regiam Anglicanum in Collegio Greshamensi* (London, 1680) forms part of the third supplement to the *Physica subterranea.*

Institutiones chimicae prodromae i.e. . . . . Oedipus chimicus obscuriorum terminorum & principiorum mysteria aperiens et resolvens (Frankfurt, 1664) rapidly became a standard text on elements, principles, and chemical processes. Besides going through numerous separate Latin editions, it was included in J. J. Manget's *Bibliotheca chemica curiosa,* I (Geneva, 1702), 306–336. It appeared in German with Becher's translation of the *Physica subterranea,* and also as part of Friedrich Roth-Scholtz's *Deutsches theatrum chemicum,* II (Nuremberg, 1728), 620 ff.

Among Becher's many other works on chemistry are *Naturkündigung der Metallen,* or *Metallurgia Becheri* (Frankfurt, 1661), which discusses metallurgical problems on the basis of the four elements; *Tripus Hermeticus fatidicus, pandens oracula chymica* (Frankfurt, 1689), which was completed at Truro, Cornwall, in 1682 and includes an account of Becher's visit to the Cornish mines in 1680; *Psychosophia oder Seelen-Weisheit* (Güstrow, 1673), which gives details of Becher's life; *Parnassus medicinalis illustratis,* 4 pts. (Ulm, 1662–1663), which contains reworked cuts from the earlier herbal by Mathiolus; *Chymischer Glücks-hafen* (Frankfurt, 1682), which contains 1,500 chemical and alchemical processes; and *Magnalia naturäe, or the Philosophers-Stone Lately Expos'd to Publick Sight and Sale . . . Published at the Request . . . Especially of Mr. Boyl, &c.* (London, 1680), which describes a transmutation at Vienna by Wenzel Seiler.

Notable among Becher's other works are *Character pro notitia linguarum universali. Inventum steganographicum* (Frankfurt, 1661); *Politischer Discurs von den eigentlichen Ursachen, des Auf- und Abnehmens der Städt, Länder und Republicken* (Frankfurt, 1668); and *Moral Discurs von den eigentlichen Ursachen des Glücks und Unglücks* (Frankfurt, 1669).

The Rostock University Library has three volumes of Becher's papers. These are briefly described in Hassinger, p. 271.

II. SECONDARY LITERATURE. Recent research on Becher has emphasized his economic and administrative policies; see F. A. Steinhüser, *Johann Joachim Becher und die Einzelwirtschaft* (Nuremberg, 1931). Although Hassinger does not ignore Becher's scientific and medical contributions, there is no question that he considers these to be distinctly less important than his administrative accomplishments. He discusses the secondary literature on this topic in his introduction, pp. 1–9.

Becher's scientific work has received far less attention. Interest in the phlogiston theory has for the most part centered on the mid-eighteenth century rather than on its origins—see especially the four papers by J. R. Partington and Douglas McKie entitled "Historical Studies on the Phlogiston Theory," which appeared in *Annals of Science,* **2** (1937), 361–404; **3** (1938), 1–58, 337–371; **4** (1939), 113–149—and although Partington's chapter on Becher—*A History of Chemistry,* II (London, 1961), 637–652—is based on extensive reading in the sources, it is somewhat less than satisfactory because it does not clearly show the close connection of Becher's views with their alchemical–Paracelsian background. Other accounts, generally with far less detail, are in most histories of chemistry and alchemy. The best of these are those of Hermann Kopp, who points out Becher's often inconsistent views on the elements: *Beiträge zur Geschichte der Chemie,* 3 vols. (Brunswick, 1869–1875), III, 201–210; *Geschichte der Chemie,* 4 vols. (Brunswick, 1843–1847; repr. 1966), I, 178–180; II, 277–278; and *Die Alchemie in älterer und neurer Zeit. Ein Beitrag zur Culturgeschichte,* 2 vols. (Heidelberg, 1886), I, 65–68.

ALLEN G. DEBUS

BECKE, FRIEDRICH JOHANN KARL (*b.* Prague, Czechoslovakia, 31 December 1855; *d.* Vienna, Austria, 18 June 1931), *mineralogy, petrography.*

Becke's father, Friedrich, originally a bookseller in Prague, became a railway employee at Pilsen in 1866 and later at Vienna. After attending several schools, Becke enrolled in 1874 at the University of Vienna, where he studied under the geologist Gustav Tscher-

mak, who also published the journal *Mineralogische und petrographische Mitteilungen.* Becke soon devoted himself to mineralogy, and in 1878 he became Tschermak's assistant. He had already gained notice in 1877 with his studies on Greek rock formations and his inaugural dissertation, *Krystalline Schiefer des niederösterreichischen Waldviertels* (1882), the first modern petrographical study of the metamorphic rocks of Austria. In 1882, at the age of twenty-seven, Becke was appointed associate professor (and later full professor) of mineralogy at the university of the Polish town of Czernowitz (now Chernovtsy, Ukrainian S.S.R.). While there, he discovered the differential solubility of dextrose crystals (1889).

In 1890 Becke was called to the University of Prague. Three years later he developed the method for the relative determination of the refraction of light by means of what since 1896 has been known as the Becke line. The procedure for measuring the angle of optical axes, which had been developed by François E. Mallard, was extended by Becke in 1895 for application to microscopic preparations. The following year he presented the Becke volume rule, which states that—assuming isothermal conditions—with increased pressure, the formation of minerals with the smallest molecular volume (the greatest density) will be favored.

In 1898 Becke went to the University of Vienna, where he attempted to classify the solidified rocks (plutonic rocks and lava). In 1903 he put forth the differentiation—based on their chemical composition and still used today—between the Atlantic group (of alkalic origin, mainly sodium and potassium) and the Pacific group (of calc-alkalic origin, characterized by high calcium and aluminum content). The mineral composition and the chemistry of both groups were correlated by Becke with E. Suess's geotectonic concepts of an Atlantic Coast type (stratum) and a Pacific Coast type (mountain chain).

Becke's work on crystalline schists was especially extensive. For the elucidation of the crystallization stratification he used Riecke's rule, according to which mineral grains are relatively soluble in the direction of the applied pressure, whereas crystallization will proceed more rapidly perpendicular to the direction of applied pressure; thus the parallel planar texture of many rocks ("foliation") develops independent of their original layering, and is characteristic for many metamorphic rocks. Becke's explanation for the process in general considers only pressure as the cause, i.e., the process is static. According to more recent research, this is controversial.

For transformations in the solid state Becke introduced the terms *crystalloblastic, granoblastic, por-*

phyroblastic, blastophytic, blastogranitic, and *blastoporphyric.* With the aid of the graphic representation of rock components that he developed, it was possible, by utilizing the Si-U-L (Si = Si + Ti, U = Al + Fe + Mg, L = Ca + Na + K) triangle, to differentiate between orthorocks (solidified rocks transformed by metamorphosis) and pararocks (transformed sedimentary rocks). In 1909 Becke originated the term *diaphthoresis* for the adjustment of highly metamorphosed rocks to the conditions prevailing at lesser depths; the process is also known as regressive metamorphism. In later works he also dealt with the classification of the facies of metamorphic rocks, the mass movement during metamorphosis, and the graphic representation of rock analyses. His last publications dealt with the crystal systems and the nomenclature of the thirty-two symmetry point groups.

From 1899 on, Becke was the editor of Tschermak's *Mineralogische und petrographische Mitteilungen,* Volume 38 of which was dedicated to him in honor of his seventieth birthday. He was director of the Mineralogical Institute of the University of Vienna from 1906 until 1927, and in 1911 he was appointed secretary-general of the Viennese Academy of Sciences. Immediately after World War I, Becke was rector of the University of Vienna, and in 1929 he received the Wollaston Medal of the Geological Society of London.

Becke did fundamental work in the elucidation of metamorphism, combining exact observations with bold and sophisticated theoretical considerations. Many students came to his institute to learn the methods of the "Vienna School." His most important findings, especially those on metamorphic rocks, are still included in textbooks of mineralogy and geology.

BIBLIOGRAPHY

I. ORIGINAL WORKS. Becke published many articles. Among the most important are "Beziehungen zwischen Dynamometamorphose und Molekularvolumen," in *Neues Jahrbuch für Mineralogie,* II (1896), 182–183; "Fortschritte auf dem Gebiete der Metamorphose," in *Fortschritte der Mineralogie,* no. 2 (1911), 221-256, and no. 5 (1916), 210-264; *Das Wachstum und der Bau der Kristalle,* his inaugural lecture as rector of the University of Vienna (1918); "Struktur und Klüftung," in *Fortschritte der Mineralogie,* no. 9 (1924), 185-220; and "Vorschläge zur Systematik und Nomenklatur der 32 Symmetrie-klassen," *ibid.,* nos. 11 (1927), 289-292, and 12 (1927), 97–103. He also published many works in *Mineralogische und petrographische Mitteilungen* between 1883 and 1928.

II. SECONDARY LITERATURE. Of most value are A. Himmelbauer's obituary of Becke in *Mineralogische und petrographische Mitteilungen,* **42** (1932), i-viii; and the obituary in *Verhandlungen der Geologischen Bundesanstalt Wien* (1931), 239-241. A short but comprehensive article is by Walther Fischer, in *Neue deutsche Biographie,* I (1953), 708-709.

HANS BAUMGÄRTEL

BECKER, GEORGE FERDINAND (*b.* New York, N.Y., 5 January 1847; *d.* Washington, D.C., 20 April 1919), *geology.*

The son of Alexander Christian and Sarah Tuckerman Becker, George Becker grew up in a home where serious intellectual endeavor was rewarded. His mother encouraged his early inclination toward science, introducing him to her friends in the Boston and Cambridge academic communities: Asa Gray, Louis Agassiz, Benjamin Peirce, Jeffries Wyman, and Benjamin Gould. It was only natural that he studied at Harvard. He earned advanced degrees in chemistry and mathematics at the University of Heidelberg and the Royal Academy of Mines, Berlin.

His formal education completed, Becker worked in the German Royal Iron Works for a year before returning to the United States. From 1874 to 1879 he was employed at the University of California at Berkeley as an instructor of mining and metallurgy; there he met Clarence King, first director of the U.S. Geological Survey. Becker joined the Survey and contributed detailed studies of mining districts—the monograph *Geology of the Comstock Lode and Washoe District* (1882) and the article "Reconnaissance of San Francisco, Eureka and Bodie Districts" (1880)—as well as broader accounts of the Pacific Coast ranges and the Sierra Nevada.

Becker's great concern for mathematical, geophysical, and geochemical approaches to mining geology are evident in his field research. Several of his more general theoretical papers are devoted to geophysical problems, including the structure of the globe: "An Elementary Proof of the Earth's Rigidity" (1890), "The Finite Elastic Stress-Strain Function" (1893), and "Finite Homogeneous Strain, Flow and Rupture of Rocks" (1893). Becker maintained that subsidence phenomena followed from the theory that the earth is solid throughout.

Becker's interest in geophysics was manifested in deed as well as in published word. He was instrumental in establishing the Carnegie Geophysical Laboratory and served as director. He also left a sizable portion of his estate to the Smithsonian Institution to promote the advancement of geophysics.

BIBLIOGRAPHY

Becker's writings include "Reconnaissance of San Francisco, Eureka and Bodie Districts Nevada," in U.S. Geological Survey, *First Annual Report* (Washington, D.C., 1880), pp. 34–37; "A Summary of the Geology of the Comstock Lode and Washoe District," in U.S. Geological Survey, *Second Annual Report* (1880–1881); *Geology of the Comstock Lode and Washoe District. Monographs of the U.S. Geological Survey*, III (Washington, 1882), 1–422; "The Relation of Mineral Belts of the Pacific Slope to the Great Upheavals," in *American Journal of Science*, 3rd ser., **28** (1884), 209–212; "An Elementary Proof of the Earth's Rigidity," *ibid.*, **39** (1890), 336–352; "The Finite Elastic Stress-Strain Function," *ibid.*, **46** (1893), 337–356; "Finite Homogeneous Strain, Flow and Rupture of Rocks," in *Bulletin of the Geological Society of America*, **4** (1893), 13–90; and "The Witwatersrand and the Revolt of the Uitlanders," in *National Geographic Magazine*, **7** (1896), 349–367. A five-page bibliography, compiled by Isabel P. Evans, may be found in *Memoirs of the National Academy of Sciences*, **21** (1926), 15–19.

An article on Becker is George P. Merrill, "Biographical Memoir of George Ferdinand Becker," in *Memoirs of the National Academy of Sciences*, **21** (1926), 1–13.

MARTHA B. KENDALL

BECKMANN, ERNEST OTTO (*b*. Solingen, Germany, 4 July 1853; *d*. Berlin, Germany, 12 July 1923), *chemistry*.

Beckmann's ancestors were originally from Hannover and settled as farmers near Solingen. His father, Johannes Friedrich Wilhelm, established a small dye and emery works in Solingen and in 1814 independently discovered Paris green. His mother, Julie, was the daughter of a tanner. After five years of laboratory activity, including a year with Remigius Fresenius in Wiesbaden, Beckmann entered the University of Leipzig in 1875 and completed his degree in 1878 under Hermann Kolbe and Ernest S. C. Meyer. He then became an assistant to the toxicologist Robert Otto at the Technische Hochschule in Brunswick. In 1883 he returned to Leipzig, where he studied briefly with Kolbe and Johannes Wislicenus, and, after 1887, with Wilhelm Ostwald. In 1891 he became an extraordinary professor at Giessen, and from 1892 to 1897 was ordinary professor at Erlangen. In 1897, as ordinary professor, he organized the applied chemistry laboratory at Leipzig. After refusing offers from Munich and Berlin, in 1912 he became director of the newly founded Kaiser Wilhelm Institute of Applied and Pharmaceutical Chemistry. Here, except for the war years, when the institute was concerned with military work, he continued his research.

By gentle oxidation methods Beckmann was able to produce menthone and thymol and from early investigations on the oximes of menthone he was led to consider the nature of isonitroso compounds. In 1886 during this study he obtained a rapid reaction now known as the Beckmann transformation, in which ketoximes are converted into amides by the action of acids, acid chlorides, or phosphorous pentachloride. It was believed that this reaction could establish the *syn*-isomeric form of the original ketoxime until Meisenheimer showed in 1921 that it was not the *syn* but the *anti* form which underwent intermolecular arrangement. The exact mechanism of this reaction has had considerable theoretical interest for chemists and drew attention to isomerism in nitrogen bonding. Beckmann's work on isomerism of benzoldoxime contributed to an understanding of the valence properties of nitrogen. Interest in the physical properties of oximes led Beckmann to develop a thermometer in which it is possible to vary the amount of mercury in the bulb and thereby change the range of the thermometer. The Beckmann thermometer is a practical device for determining molecular weight based on the theoretical work of Raoult on the freezing point depression and boiling point elevation of solutions. Beckmann also identified SCl_4, and investigated the properties of furfurol.

BIBLIOGRAPHY

I. ORIGINAL WORKS. Beckmann's writings include "Zur Kenntniss der Isonitrosoverbindungen," in *Berichte der Deutschen chemischen Gesellschaft*, **19** (1886), 988–993; **20** (1887), 1507–1510, 2580–2585, 2766–2768; **21** (1888), 766–769; "Ueber die Methode der Molekulargewichtsbestimmung durch Gefrierpunktserniedrigung," in *Zeitschrift für physikalische Chemie*, **2** (1888), 638–645, 715–743; "Ueber die Bestimmung von Molekulargewichten aus Siedepunktserhöhungen," *ibid.*, **3** (1889), 603–604; "Zur Isomerie der Oximidoverbindungen. Isomere monosubstitutirte Hydroxylamine," in *Berichte der Deutschen chemischen Gesellschaft*, **22** (1889), 429–440, 514–517; "Untersuchungen in der Campherreihe. Erste Abhandlung," in *Annalen der Chemie*, no. 250 (1889), 322–375; *Das Laboratorium für angewandte Chemie der Universität Leipzig* (Leipzig, 1908); and "Ueber die Verbindung des Schwefels mit Chlor," in *Zeitschrift für physikalische Chemie*, **55** (1923), 81–82; **61** (1928), 87–130, written with T. Klopfer and F. Junker.

II. SECONDARY LITERATURE. Works on Beckmann are G. Lockemann, "Ernst Beckmann, zum siebzigsten Geburtstage am 4. Juli 1923," in *Zeitschrift für angewandte Chemie*, **26** (1923), 341–344, and *Ernst Beckmann, sein Leben und Werken* (Berlin, 1927); O. Ostwald, "Ernst Beckmann's Anfange als Physico-Chemiker," in *Zeitschrift* *für angewandte Chemie,* **26** (1923), 344; C. Paal, "Nachruf auf Ernest Otto Beckmann," in *Bericht über die Verhandlungen der Sächsischen Akademie der Wissenschaften zu Leipzig,* **76** (1924), 83–94; W. Schenk, "Gedachtnisrede," in *Sitzungsberichte der Deutschen Akademie der Wissenschaften zu Berlin,* Kl. Phil.-hist., **104** (1924); and Alfred Stock, "Forschungsinstitut," in *Zeitschrift für angewandte Chemie,* **26** (1923), 344–345.

RUTH ANNE GIENAPP

BECKMANN, JOHANN (*b.* Hoya, Germany, 4 June 1739; *d.* Göttingen, Germany, 3 February 1811), *economy, technology.*

Beckmann was the oldest son of the director of taxation and custodian of postal services, Nicolaus Beckmann, who died in 1745. His mother, the daughter of a Protestant parson, devoted herself to the education of the children. In 1754, at the age of fifteen, Beckmann enrolled in the Gymnasium at Stade, and in 1759 he entered the University of Göttingen, where he studied theology. He soon turned, however, to mathematics, the natural sciences, public finance and administration, and philology. He fancied languages just as much as scientific subjects.

In 1762, after completing his studies, Beckmann traveled to Brunswick and Holland, where he visited factories, mines, and natural history museums. In 1763 he accepted a teaching position at St. Peter's Gymnasium, a Lutheran school in St. Petersburg, which had been founded by Anton Friedrich Büsching. He taught there for two years. Aside from his teaching, Beckmann occupied himself with projects in the natural sciences, such as meteorological observations, and with the history of the natural sciences. In St. Petersburg he became friendly with August Ludwig Schloezer, who was later active in Göttingen as a historian and a political scientist; this friendship greatly influenced Beckmann's later historical researches. After his sojourn in St. Petersburg, he took an educational trip, which lasted most of the year 1765–1766, through Sweden and Denmark. Again he inspected factories, mines, and foundries, as well as collections of art and of natural objects. His love for natural history led him to Linnaeus, with whom he studied. In the fall of 1766 Beckmann was appointed extraordinary professor of philosophy in Göttingen. At this time he published his first larger work, *De historia naturali veterum libellus I,* which admirably combines aspects of natural science with philology.

In 1767 Beckmann married the daughter of a parson. Since his labors turned more and more to applied botany, agriculture, and public economy, an ordinary professorship of economic sciences was estab-lished in 1770 for him, and he occupied this post until his death. He lectured on mineralogy, agriculture, technology, materials science, commerce, and general public administration. Aside from his teaching duties, he devoted himself to writing.

Beckmann's scientific importance lies especially in the agricultural sciences, technology, and the history of technology and invention. He can be called the most important representative of German agricultural economy in the second half of the eighteenth century. He founded the independent science of agriculture with his textbook *Grundsätze der teutschen Landwirthschaft* (1769), and he stressed that practical agriculture needed a scientific foundation: natural history, mineralogy, chemistry, physics, and mathematics were recognized as necessary auxiliary sciences of agriculture. Beckmann's agricultural science followed the empirical treatment of agriculture practiced by the so-called *Hausväter* and the older economists. In his general economic treatment of agriculture, however, their special treatments of vegetable raising and of animal husbandry received short shrift. Yet his textbook remained prominent in the field for nearly half a century, until superseded by the work of Albrecht Thaer.

Since agriculture concerns the production of natural products and mining technology leads one into the production of metals, Beckmann became interested in the processing of raw materials by the individual trades. By 1769 he was calling this science of trades "technology," and in 1777 his *Anleitung zur Technologie* appeared, the first advanced textbook in this field. It is noteworthy for its systematic approach to the various vocations and for its descriptions of a number of trades. The book was addressed primarily to governmental economic officials, in order to make them cognizant of the problems of trade and manufacture within the framework of public affairs. Beckmann was not without precursors in his attempts to spread technological knowledge, but he was the only one to succeed in introducing technology as a separate subject into the high school curriculum. He enlivened his lectures by using models and by demonstrations, as well as by conducting inspections of manufacturing establishments. His attempt in 1806 (*Entwurf der algemeinen Technologie*) to compare the processes that are utilized in the various areas of technology and that are based on the same objectives also deserves special attention. Thus, for example, the various methods of crushing or grinding were examined with a view toward profiting from the transfer of an especially efficient procedure from one field to another.

Beckmann should also be credited with being the

first reliable historian of inventions, with his *Beyträge zur Geschichte der Erfindungen* (1782 *et seq.*), which is not a complete history of technology but, rather, an admirable collection of historical descriptions of individual inventions. For sources he used primarily literary material, and his excellent philological and good technological knowledge served him well.

BIBLIOGRAPHY

I. ORIGINAL WORKS. The Staatsbibliothek of the Stiftung Preussischer Kulturbesitz in Berlin has forty-two letters written by Beckmann, among them thirty-eight to F. Nicolai; the Library of the Deutsches Museum in Munich has nine letters. Six letters written by Beckmann to Linnaeus between 1765 and 1770 are in *Carl von Linné, Bref och skrifvelser,* II, pt. 1 (Uppsala, 1916), 254–268. Handwritten material used by W. Exner for his biography of Beckmann cannot be traced.

The most important published works by Beckmann are *De historia naturali veterum libellus I* (St. Petersburg–Göttingen, 1766); *Anfangsgründe der Naturhistorie* (Göttingen, 1767); *Grundsätze der teutschen Landwirthschaft* (Göttingen, 1769; 6th ed., 1806); *Physikalisch-ökonomische Bibliothek,* 23 vols. (Göttingen, 1770–1806); *Anleitung zur Technologie* (Göttingen, 1777; 7th ed., 1823); *Grundriss zu Vorlesungen über die Naturlehre* (Göttingen, 1779, 1785); *Beyträge zur Oekonomie, Technologie, Polizey und Cameralwissenschaft,* 12 pts. (Göttingen, 1779–1791), of which Beckmann is primarily the editor; *Beyträge zur Geschichte der Erfindungen,* 5 vols. (Leipzig, 1782–1805; 2nd ed. of Vol. I, 1786; English trans., London, 1797; 4th ed., 1846); *Sammlung auserlesener Landesgesetze, welche das Policey- und Cameralwesen zum Gegenstande haben,* 10 pts. (Frankfurt, 1783–1793); *Vorbereitung zur Waarenkunde,* 2 pts. (Göttingen, 1794–1800); *Vorrath kleiner Anmerkungen über mancherley gelehrte Gegenstände,* 3 pts. (Göttingen, 1795–1806), of which the third part is *Entwurf der algemeinen Technologie* (also pub. separately); *Lexicon botanicum* (Göttingen, 1801); *Litteratur der älteren Reisebeschreibungen,* 2 vols. (Göttingen, 1808–1810); and *Schwedische Reise in den Jahren 1765–1766* (Uppsala, 1911), a diary.

II. SECONDARY LITERATURE. Works on Beckmann are W. Exner, *Johann Beckmann, Begründer der technologischen Wissenschaft* (Vienna, 1878); G. Grundke, "Johann Beckmann als Begründer der Technologie in Deutschland," in *Wissenschaftliche Zeitschrift der Karl-Marx-Universität, Leipzig,* Mathematical-Natural Sciences Section, no. 3/4 (1954/1955), 343–352; H. J. Herpel, *Entwicklung des landwirtschaftlichen Studiums an der Universität Göttingen* (Göttingen, 1932); C. G. Heyne, "Memoria Jo. Beckmann," in *Commentationes Societatis Regiae Scientiarum Gottingensis recentiores,* 1 (1808–1811), appendix, 1–15; *Ideengeschichte der Agrarwirtschaft und Agrarpolitik im deutschen Sprachgebiet* (Munich, 1957);

Carl, Count of Klinckowstroem, "Johann Beckmann," in *Neue deutsche Biographie,* I (Berlin, 1953), 725–726; G. Schmid, "Linné im Urteil Johann Beckmanns," in *Svenska Linné-sällskapets årsskrift,* **20** (1937), 47–70; and U. Troitzsch, *Ansätze technologischen Denkens bei den Kameralisten des 17. and 18. Jahrhunderts* (Berlin, 1966), pp. 150–165.

FRIEDRICH KLEMM

BECQUEREL, ALEXANDRE-EDMOND (*b.* Paris, France, 24 March 1820; *d.* Paris, 11 May 1891), *experimental physics.*

Becquerel was the second son of Antoine-César Becquerel, experimental physicist and professor at the Muséum d'Histoire Naturelle. At the age of eighteen, Edmond was admitted to both the École Polytechnique and the École Normale Supérieure. He decided, however, to refuse both these opportunities in order to assist his father at the museum. After having served as assistant at the University of Paris and then as professor of physics at the short-lived Institut Agronomique de Versailles, Becquerel was appointed, in December 1852, to the chair of physics at the Conservatoire des Arts et Métiers. From 1860 to 1863, he taught chemistry at the Société Chimique de Paris; and after having served a term as *aide-naturaliste* at the museum, he succeeded his father as director of that institution in 1878. Becquerel received a Doctor of Science degree from the University of Paris in 1840 and was elected to the Académie des Sciences in 1863.

Becquerel's most important achievements in science were in electricity, magnetism, and optics. In electricity he measured the properties of currents and investigated the conditions under which they arose. In 1843 he showed that Joule's law governing the production of heat in the passage of an electrical current applied to liquids as well as to solids. In 1844 he rectified Faraday's law of electrochemical decomposition to include several phenomena that had not been taken into account, and in 1855 he discovered that the mere displacement of a metallic conductor in a liquid was sufficient to produce a current of electricity. Becquerel's measurement of the electromotive force of the voltaic pile was achieved through the use of his father's ingeniously devised electrostatic balance, an instrument that, in effect, allowed him to weigh the relative force of an electric current. He also studied the separate effects of the liquid, the metal, the temperature, and the polarization of the electrodes on the functioning of voltaic piles.

From 1845 to 1855, Becquerel devoted most of his attention to the investigation of diamagnetism. Anxious to preserve the simplicity of Ampère's electrical theory of magnetic action, he was unwilling to accept Faraday's contention that diamagnetic phenomena were fundamentally different from those of ordinary magnetism. To explain the repulsion of certain substances by the poles of a magnet, he conceived an "Archimedean law" of magnetic action, so called because of its resemblance to Archimedes' hydrostatic principle: "A body placed away from a magnetic center is attracted toward that center with a force equal to the difference which exists between the specific magnetism of the body and that of the milieu in which it is immersed" (*Traité d'électricité* . . . , III, 52). Just as specifically heavier bodies sink and lighter bodies rise in a liquid, so substances less magnetic than their surroundings are pushed away from a magnet, whereas the more magnetic ones are attracted to it.

Becquerel set out to measure the magnetic properties of oxygen, one of the substances composing the milieu in which magnetic action most often took place. He condensed a large volume of the gas in a glass tube filled with highly absorbent charcoal. When placed in a magnetic field, the tube containing oxygen was found to be much more magnetic than one containing only charcoal. This discovery might have provided conclusive proof of the validity of Becquerel's theory had it not been for the highly embarrassing fact that diamagnetic action took place in the vacuum as well as in the air. Becquerel attempted in vain to overcome this objection by assigning magnetic (and thus electrical) properties to the ether.

Becquerel's early investigations of light phenomena were closely related to his interest in electricity. In 1840 he demonstrated that electrical currents arose from certain light-induced chemical reactions. On the basis of this discovery, he constructed an instrument, called an "actinometer," that measured the intensity of light by measuring the intensity of the electrical currents produced in photochemical reactions. Later, he showed that by relating light intensity to heat intensity one could use this device to determine optically the temperatures of extremely hot bodies.

In spectroscopy Becquerel revealed in 1843 the presence of Fraunhofer lines in photographs of the ultraviolet portion of the spectrum. Earlier he had shown that rays at the red end of the spectrum reinforced or "continued" the chemical action initiated by rays at the violet end.

Becquerel did his most important work in optics on the phenomena of luminescence. In the middle years of the nineteenth century, he virtually monopolized the significant discoveries made in this field. His researches began in 1839, when he published a paper dealing primarily with the effects of temperature on the duration of phosphorescent light emission. In 1843 he demonstrated that phosphorescence was stimulated in different substances by specific frequencies of light and that at some frequencies the phosphorescent glow seemed to stop immediately after the cutting off of incident light rays. Between 1857 and 1859, Becquerel produced three pioneering studies on luminescent phenomena; these were collected and published in 1859 under the title *Recherches sur les divers effets lumineux qui resultent de l'action de la lumière sur les corps.* It was in these studies that Becquerel first described the phosphoroscope, an instrument of his own invention consisting of a box sealed with two disks mounted on the same axis and pierced with holes arranged in such a way that light could not at any one time pass through the entire apparatus. By rapidly revolving these perforated disks, an observer could continuously view substances in the dark only fractions of a second after they had been exposed to brilliant light; and by regulating the speed of revolution of the disks, one could measure the length of time that substances continued to glow after exposure to light. Using this device, Becquerel was able to identify many new phosphorescent substances and to show that the phenomenon G. G. Stokes had named fluorescence in 1852 was in reality only phosphorescence of an extremely short duration. By attaching a prism to his phosphoroscope, Becquerel was able to examine the spectra of light emitted from luminescent bodies. In this manner substances could be analyzed without physical or chemical alteration.

BIBLIOGRAPHY

I. ORIGINAL WORKS. Becquerel's *Traité d'électricité et de magnétisme, leurs applications aux sciences physiques, aux arts et à l'industrie,* 3 vols. (Paris, 1855–1856), was written in collaboration with his father. Besides the *Recherches sur les divers effets lumineux* . . . (Paris, 1859), Becquerel wrote *La lumière, ses causes et ses effets,* 2 vols. (Paris, 1867–1868). The first volume of this work is devoted to the study of luminescent phenomena.

II. SECONDARY LITERATURE. Works on Becquerel are Henri Becquerel, "La chaire de physique du Muséum," in *Revue scientifique,* **49,** No. 22 (28 May 1892), 673–678; and Jules Violle, "L'oeuvre scientifique de M. Edmond Becquerel," *ibid.,* No. 12 (19 March 1892), 353–360. See also E. N. Harvey, *A History of Luminescence From the Earliest Times Until 1900* (Philadelphia, 1957), 207–219, 351–360.

J. B. GOUGH

BECQUEREL, ANTOINE-CÉSAR (*b.* Châtillon-sur-Loing, Loiret, France, 7 March 1788; *d.* Paris, France, 18 January 1878), *electrochemistry.*

Becquerel's father was the royal lieutenant at Châtillon-sur-Loing. Becquerel himself entered the École Polytechnique in 1806, and on graduation embarked on a military career in the Corps of Engineers, being promoted to captain in 1812. The following year he was appointed *sous inspecteur* of the École Polytechnique; he returned to active service in 1814, but after the fall of Napoleon in 1815 devoted himself entirely to science. Becquerel was elected to the Académie des Sciences in 1829; was awarded the Copley Medal of the Royal Society of London in 1837; and, when the chair of physics at the Muséum d'Histoire Naturelle was founded in 1838, became its first occupant. In 1813 he married Aimée-Cécile Darlui, a relative of the duc de Feltre, and became the founder of a dynasty of distinguished scientists. His son Alexandre Edmond succeeded him at the Museum, and his grandson Antoine Henri was the discoverer of radioactivity. Becquerel lived to a ripe old age, and his vivacity amazed his younger colleagues in the Academy, in the proceedings of which he took a lively and regular part to the end of his life.

Becquerel's first studies were in mineralogy, and in association with Alexandre Brongniart, whose disciple he was, he discovered in 1819 a collection of important recent deposits at Auteuil that contained previously unknown crystalline forms of calcium phosphate. The study of minerals led easily to electrical experiments. Following up Haüy's observation that Iceland spar became electrified when compressed, Becquerel showed that this phenomenon was general, provided that the crystals under investigation were insulated; his studies on tourmaline became classics. This effect is called the piezoelectric effect, and obtains only in crystals which possess no center of symmetry. From studying the electrical effects of compression, Becquerel moved on to thermoelectricity, investigating the electrical effects of heating minerals. He discovered various definite transition temperatures, at which the electrical state of a substance changes discontinuously. These researches led to experiments on the electrical measurement of temperature.

Of perhaps greater interest was Becquerel's work on the voltaic cell. In the early decades of the nineteenth century, it was not clear whether the production of electricity in the cell was due to the mere contact of dissimilar bodies, or whether it depended upon a chemical reaction. The principle of conservation of energy had not yet been clearly enunciated; but Becquerel believed that there was a close relationship between electricity on the one hand, and heat, light, and chemical forces on the other. In a series of careful experiments, he put it beyond doubt that electricity could be generated only by the contact of dissimilar bodies when they reacted together chemically, or differed in temperature, or were rubbed together. Conversely, he established that all chemical reactions can generate electricity. This was related to earlier work, notably by Humphry Davy, and in 1829 Becquerel employed Davy's discovery that a battery could be made of two liquids separated by a solid barrier in the construction of the first battery that, not being polarized, could supply current at a reasonably constant EMF.

Using these cells, Becquerel performed elegant small-scale experiments on the synthesis of mineral substances. Since he was able to control his EMF's, through secondary electrolysis he succeeded in obtaining crystals of various substances—notably sulfides—that had previously been produced only in an amorphous condition. He suggested that the presence of crystalline substances in mineral veins might be accounted for by their having been deposited by electric currents operating over a very long period of time. Becquerel remarked that chemical synthesis had lagged behind analysis, and hoped that through his techniques many naturally occurring crystals would be synthesized and the balance thus redressed. It was for this work that he received the Copley Medal. These researches found industrial applications in the treatment of silver-bearing ores and the extraction of potassium chloride from the sea.

Becquerel wrote a great number of papers; of his joint papers, the majority were written in collaboration with his son, but his co-workers also included Ampère and Biot. He corresponded with Michael Faraday over diamagnetism; he had noticed examples of it before Faraday but had failed to generalize from them. He also invented an electromagnetic balance and a differential galvanometer.

BIBLIOGRAPHY

I. ORIGINAL WORKS. Becquerel's papers are listed in the *Royal Society Catalogue of Scientific Papers,* I (1867), 233–239; VII (1877), 118–121; IX (1891), 164–166. See also *Analyse succint des travaux de M. Becquerel* (Paris, 1829), published by the Académie des Sciences. His most important books are *Traité expérimental de l'électricité et du magnétisme, et de leur rapports avec les phénomènes naturels,* 7 vols. (Paris,

1834–1840); *Traité de physique considérée dans ses rapports avec la chimie et les sciences naturelles,* 2 vols. (Paris, 1842–1844); *Éléments d'électro-chimie appliquée aux sciences naturelles et aux arts* (Paris, 1843; 1864), also translated into German (Erfurt, 1845; 1848; 1857); and *Résumé de l'histoire de l'électricité et du magnétisme, et des applications de ces sciences à la chimie, aux sciences naturelles et aux arts* (Paris, 1858), written with Alexandre Edmond Becquerel. There is an extensive collection of MSS of A. C. Becquerel at the Académie des Sciences in Paris.

II. Secondary Literature. Obituaries of Becquerel are in *Comptes rendus de l'Académie des sciences,* **86** (1878), 125–131. See also *Abstracts of the Philosophical Transactions,* **4** (1843), 22–23; and T. Cooper, ed., *Men of the Time,* 9th ed. (London, 1875), p. 87.

David M. Knight

BECQUEREL, [ANTOINE-] HENRI (*b.* Paris, France, 15 December 1852; *d.* Le Croisic, Brittany, France, 25 August 1908), *physics.*

Becquerel is known for his discovery of radioactivity, for which he received the Nobel Prize for physics jointly with the Curies in 1903, and for other contributions to that field which he made during the half-dozen years when he was most active in it. He was a member of the Academy of Sciences, became its president, and was elected to the far more influential post of permanent secretary. He held three chairs of physics in Paris—at the Museum of Natural History, at the École Polytechnique, and at the Conservatoire National des Arts et Métiers—and attained high rank as an engineer in the National Administration of Bridges and Highways (Ponts et Chaussées).

Henri's father, Alexandre-Edmond Becquerel, and his grandfather, Antoine-César Becquerel, were renowned and prolific physicists, both members of the Academy of Sciences and each in his turn professor of physics at the Museum of Natural History. When Henri was born in the professor's house at the Museum, he was born almost literally into the inner circles of French science. He was educated at the Lycée Louis-le-Grand, from which he went to the École Polytechnique (1872–1874) and then to the École des Ponts et Chaussées (1874–1877), where he received his engineering training and from which he entered the Administration of Bridges and Highways with the rank of *ingénieur.* On leaving the Polytechnique, he married Lucie-Zoé-Marie Jamin, daughter of J.-C. Jamin, academician and professor of physics in the Faculty of Sciences of Paris. Before the end of his schooling he had begun both his private research (1875) and his teaching career (1876) as *répétiteur* at the Polytechnique. In January 1878 his grandfather died; his wife died the following March,

a few weeks after the birth of their son Jean. At this time Becquerel succeeded to the post of *aide-naturaliste,* which his father had hitherto held at the Museum, and from then on, his professional life was shared among the Museum, the Polytechnique, and the Ponts et Chaussées.

Becquerel's early research was almost exclusively optical. His first extensive investigations (1875–1882) dealt with the rotation of plane-polarized light by magnetic fields. He turned next to infrared spectra (1883), making visual observations by means of the light released from certain phosphorescent crystals under infrared illumination. He then studied the absorption of light in crystals (1886–1888), particularly its dependence on the plane of polarization of the incident light and the direction of its propagation through the crystal. With these researches Becquerel obtained his doctorate from the Faculty of Sciences of Paris (1888) and election to the Academy of Sciences (1889), after two preparatory nominations (1884, 1886), in the second of which he polled twenty of the fifty-one votes. He had in the meantime been promoted to *ingénieur de première classe* in the Ponts et Chaussées.

With his doctorate achieved, Becquerel became substantially inactive in research. In 1890 he married his second wife, the daughter of E. Lorieux, an inspector general of mines. Following the death of Edmond Becquerel in 1891, he succeeded in the following year to his father's two chairs of physics, at the Conservatoire National des Arts et Métiers and at the Museum. In the same year Alfred Potier withdrew from active teaching because of illness, and Becquerel took over his lectures in physics at the École Polytechnique. Two years later (1894) he became *ingénieur en chef* with the Ponts et Chaussées and the next year (1895) was named to succeed Potier at the Polytechnique.

Thus the beginning of 1896 found Becquerel, at the age of forty-three, established in rank and responsibility, his years of active research behind him and everything for which he is now remembered still undone. In the very opening days of the year, Roentgen had announced his discovery of X rays by a mailing of preprints and photographs, but Becquerel's personal knowledge of the discovery dates to 20 January, when two physicians, Paul Oudin and Toussaint Barthélemy, submitted an X-ray photograph of the bones of a living hand to the Academy. From Henri Poincaré, who had received a preprint, Becquerel learned that in Roentgen's tubes the X rays arose from the fluorescent spot where a beam of cathode rays played on the glass wall. Thus a natural, if perhaps not plausible, inference arose that the visible light and invisible X rays might be produced by the same mechanism, and that X rays might accompany all luminescence.

Becquerel's researches had been essentially descriptive, with a primary commitment to observation and a careful avoidance of theorizing. Nevertheless, this X-ray hypothesis caught his fancy. He had some personal acquaintance with luminescent crystals, he was familiar with his father's researches on them, and he began to hunt for a crystalline emitter of penetrating radiation. On 24 February 1896 he reported to the Academy that fluorescent crystals of potassium uranyl sulfate had exposed a photographic plate wrapped in black paper while they both lay for several hours in direct sunlight. On 2 March he reported comparable exposures when both crystals and plate lay in total darkness. By his working hypothesis, that would have been impossible because the luminescence of potassium uranyl sulfate ceases immediately when the ultraviolet radiation that excites it is withdrawn. One might speculate, nevertheless, that the penetrating rays persisted longer than the visible fluorescence when their common excitation was cut off. Becquerel did so, conscientiously condemned the speculation as unjustified, and then proceeded to act upon it.

He did not neglect his general studies. He showed that, like X rays, the penetrating rays from his crystals could discharge electrified bodies (in modern terms, could ionize the air they passed through). He found evidence to suggest that the rays were refracted and reflected like visible light, although later he attributed these effects to secondary electrons ejected from his glass plates and mirrors. Nevertheless, he devoted a substantial effort to searching out the radiation that had first excited his penetrating rays. He kept some of his crystals in darkness, hoping that their pent-up energy might dissipate itself and make them ready for reexcitation. He tried other luminescent crystals and found that only those containing uranium emitted the penetrating radiation. He tried ingeniously but unsuccessfully to release the energy of uranyl nitrate by warming its crystals in darkness until they dissolved in their own water of crystallization. He tested nonluminescent compounds of uranium and found that they emitted his penetrating rays. Finally, he tried a disk of pure uranium metal and found that it produced penetrating radiation three to four times as intense as that he had first seen with potassium uranyl sulfate.

With this last announcement, on 18 May, Becquerel's discovery of radioactivity was complete, although he continued with ionization studies of his penetrating radiation until the following spring. What he had accomplished at the most general level was to establish the occurrence and the properties of that radiation, so that it could be identified unambiguously. Of more importance, he had shown that the power of emitting penetrating rays was a particular property of uranium. However, the implications of this second conclusion were by no means clear at the time. Becquerel characterized his own achievement as the first observation of phosphorescence in a metal. His immediate successors, G. C. Schmidt and Marie Curie, started with quite conventional views about the rays and came only gradually to realize that such radiation might also be emitted by other elements. Both then searched among the known elements, finding that only thorium was also a ray-emitter. Marie Curie and her husband, Pierre, pushed on to search for unknown elements with the same property, however, and so discovered polonium and radium. With these discoveries, the field of radioactivity (a term that the Curies coined) was fully established.

Nothing that Becquerel subsequently accomplished was as important as this discovery, by which he opened the way to nuclear physics. Nevertheless, there were two other occasions on which he stood directly on the path of history: when he identified electrons in the radiations of radium (1899–1900) and when he published the first evidence of a radioactive transformation (1901).

Marie Curie's work, which attracted Becquerel's attention, brought the Curies within the circle of his acquaintance and turned him back to radioactive studies. He became the intermediary through whom their papers reached the Academy, and they lent him radium preparations from time to time. Toward the end of 1899 (his first report is dated 11 December), he began to investigate the effects on the radiation from radium of magnetic fields in various orientations to the direction of its propagation (in modern terms, the magnetic deflection of the beta rays from short-term decay products in equilibrium with the radium). In this work he united two descriptive traditions, the magneto optics of his own experience and a line of qualitative studies of the discharge of electricity through gases. He soon moved from these to J. J. Thomson's more radical program of quantitative observations on collimated beams, in which Thomson had shown (1897) that the cathode rays were corpuscular and consisted of streams of swiftly moving, negatively charged particles whose masses were probably subatomic. By 26 March 1900, Becquerel had duplicated those experiments for the radium radiation and had shown that it too consisted of negatively charged ions, moving at 1.6×10^{10} cm./sec. with a ratio of $m/e = 10^{-7}$ gm./abcoul. Thus Thomson's "corpuscles" (electrons) constituted a part of the radiations of radioactivity.

At this period an idea was current, although seldom formally expressed, that radioactivity should be a property only of rare substances like radium, and not of ordinary chemical elements. Perhaps under the impulse of such a notion, Becquerel undertook to remove from uranium a magnetically deviable (or beta) radiation he had recently identified. His method was borrowed from André Debierne, who had found it effective with actinium. To a solution of uranium chloride, he added barium chloride and precipitated the barium as the sulfate. The precipitate entrained something, for the deviable radiation of the uranium was diminished; by a long repetition of such operations he succeeded in July 1900 in reducing that radiation, in one specimen, to one-sixth of its original value. In confirmation of this result, he found that earlier that spring, Crookes had succeeded, by more effective chemical procedures, in separating from uranium the photographically active radiation, which he now attributed to a substance provisionally named uranium X. Something over a year later, Becquerel realized the logical incongruity of these two successes. It had been relatively easy to remove the apparent radioactivity from uranium by chemical purification, yet no one who had investigated uranium over the last five years had ever observed a nonradioactive specimen. It followed, then, that whatever radioactivity was lost in purification must always regenerate itself; and he verified this logical conclusion on his own earlier specimens. The uranium had regained its lost radioactivity, and the barium sulfate precipitates had lost all that they had carried down. The explanations he attempted were thoroughly confusing, but the facts remained.

These facts were brought squarely to the attention of Ernest Rutherford and Frederick Soddy, who had just succeeded in separating a thorium X analogous to Crookes's uranium X. Their subsequent tests showed a similar regeneration of the lost radioactivity of the thorium. From this they inferred, and immediately verified, a regeneration of the thorium X in those specimens and then came to realize that this chemically distinct thorium X could have been formed there only by a transformation of the thorium. On the basis of these and similar experiments, in a few months they formulated the transformation theory, which became the basic theory of radioactivity.

In 1903 the Nobel Prize for physics was divided between Henri Becquerel and Pierre and Marie Curie. It was an appropriate division. Becquerel's pioneer investigations had opened the way to the Curies'

discoveries, and their discoveries had validated and shown the importance of his. On 31 December 1906, Becquerel was elected vice-president of the Academy of Sciences, serving in that capacity during 1907 and succeeding to the presidency in 1908. On 29 June 1908 he was elected as one of the two permanent secretaries of the Academy, following the death of Lapparent. On confirmation by the president of the republic, he was installed in that office on 6 July, taking his seat beside Darboux, who had taught him mathematics nearly four decades before at the Lycée Louis-le-Grand. He died soon after at Le Croisic in Brittany, the ancestral home of his wife's family, the Lorieux.

In an assessment of Becquerel's scientific powers, it should be noted that he had little taste for physical theories, either his own or those of others, and much of his research effort was dissipated on observations of no great significance. Against this, he displayed an admirable versatility in experiment in unfamiliar as well as familiar fields. His greatest asset, however, was a strong, persistent power of critical afterthought. On those rare occasions when Becquerel did pursue a hypothesis, this critical power continually corrected his enthusiasms and redirected his line of investigation; so that, for example, while he persistently searched for X rays in phosphorescence, he managed to discover the inherent radioactivity of uranium.

BIBLIOGRAPHY

I. ORIGINAL WORKS. Becquerel wrote an account of his radioactive investigations in an extended *mémoire,* "Recherches sur une propriété nouvelle de la matière. Activité radiante spontanée ou radioactivité de la matière," in *Mémoires de l'Académie des sciences, Paris,* **46** (1903). Aside from this, his scientific work is to be found in some 150 papers and notes in various journals. No single list of them exists, but those published up to 1900 may be found in the Royal Society's *Catalogue of Scientific Papers,* IX, 166–167, and XIII, 395–396. His papers on radioactivity are listed in the bibliography of 214 items included with his *mémoire.* Among these papers on radioactivity, all of them in the *Comptes rendus de l'Académie des sciences, Paris,* are "Émission de radiations nouvelles par l'uranium métallique," **122** (1896), 1086–1088; "Sur quelques propriétés nouvelles des radiations invisibles émises par divers corps phosphorescents," *ibid.,* 559–564; "Sur les radiations émises par phosphorescence," *ibid.,* 420–421; "Sur les radiations invisibles émises par les corps phosphorescents," *ibid.,* 501–503; "Sur les radiations invisibles émises par les sels d'uranium," *ibid.,* 689–694; "Sur diverses propriétés des rayons uraniques," **123** (1896), 855–858; "Sur la loi de décharge dans l'air de l'uranium électrisé," **124** (1897), 800–803; "Recherches sur les rayons uraniques," *ibid.,* 444; "Influence d'un champ magnétique sur le rayonnement des corps radio-actifs," **129** (1899), 996–1001; "Sur le rayonnement des corps radio-actifs," *ibid.*, 1205–1207; "Contribution à l'étude du rayonnement du radium," **130** (1900), 206–211; "Déviation du rayonnement du radium dans un champ électrique," *ibid.*, 809–815; "Sur la dispersion du rayonnement du radium dans un champ magnétique," *ibid.*, 372–376; "Note sur le rayonnement de l'uranium," *ibid.*, 1583–1585, and **131** (1900), 137–138; and "Sur la radioactivité de l'uranium," **133** (1901), 977–980.

II. SECONDARY LITERATURE. There is a somewhat padded biography by Albert Ranc, *Henri Becquerel et la découverte de la radioactivité,* Sciences et Savants, no. 3 (Paris, 1946). Important material is in G. Darboux, E. Perrier, M. Vieille, and L. Passy, "Discours prononcés aux funerailles de M. Henri Becquerel," in *Comptes rendus de l'Académie des sciences, Paris,* **147** (1908), 443–451; and in the sketch included in *Les Prix Nobel en 1903* (Stockholm, 1906), pp. 62–63.

Critical studies include Sir O. Lodge, "Becquerel Memorial Lecture," in *Journal of the Chemical Society,* **101** (1912), 2005–2042, for a contemporary assessment; and L. Badash, "'Chance Favors the Prepared Mind': Henri Becquerel and the Discovery of Radioactivity," in *Archives internationales d'histoire des sciences,* **18** (1965), 55–66, for a modern one.

ALFRED ROMER

BECQUEREL, PAUL (*b.* Paris, France, 14 April 1879; *d.* Évian, France, 22 June 1955), *biology.*

Through his researches Paul Becquerel did much to explain the physiological nature of the plant seed and the reactions of protoplasm to freezing and dehydration. Becquerel, grandson of Edmond Becquerel and nephew of Henri Becquerel, received his licentiate in natural sciences at the Faculty of Sciences at Paris in 1903 and his doctorate in 1907. In that same year he was named an assistant at the Faculty of Sciences. For a period of time after World War I, Becquerel was a member of the Faculty of Sciences at Nancy, following which, in 1927, he was appointed professor of general botany at the University of Poitiers. He remained at Poitiers until his retirement.

Becquerel began his career with researches on the nature of the life of seeds. Were they, as some physiologists from Spallanzani through Claude Bernard had suggested, actually nonliving, in a state of suspended or latent life? Or were they merely in a state in which the life processes were greatly slowed?—an idea suggested by, among others, Leeuwenhoek and Gaston Bonnier, one of Becquerel's teachers in Paris. After studying the structure and function of the various parts of seeds, Becquerel concluded that a portion of the seed always lived, protected from hostile external environments by an impermeable protective layer. Shorn of this layer, seeds could easily be killed by toxic liquids and gases. However, Becquerel went

on to show that plant and animal tissue could in fact attain a state of suspended or latent life. He produced this condition in seeds; in fern, moss, bacteria, and mushroom spores; and in algae, rotifers, and infusoria by dehydrating them as much as possible in a vacuum. Living objects so treated could withstand exposure to liquid hydrogen (*ca.* −253°C.) and liquid helium (*ca.* −269°C.) and still be revived by warming and rehydration.

Becquerel then turned his attention to naturally occurring cases of "suspended life." Working with seeds of known ages—from 25 to 135 years old—he was able to produce a small percentage of germinations; and, in later experiments (1933), he was able to bring about the germination of two 158-year-old seeds of *Cassia multijuga.* He found in these results confirmation of his ideas, maintaining that these seeds had retained their germinative ability because they had undergone a strong dehydration and their seed coats were quite thick.

In 1937 Becquerel began a study of the effects of freezing on vegetable protoplasm, showing in 1939 that plant cells are not killed by plasmolysis upon freezing; rather, the lowered temperature induces synaeresis. This was elaborated upon in a 1949 paper.

BIBLIOGRAPHY

Becquerel's work appeared in a large number of short papers, most of them published in the *Comptes rendus de l'Académie des sciences, Paris.* Among his most important papers were the following: "Action de l'air liquide sur les graines décortiquées," in *Comptes rendus de l'Académie des sciences, Paris,* **140** (1905), 1652–1654; "Recherches expérimentales sur la vie latente des spores des *Mucorinées* et des *Ascomycétes* aux basses températures de l'hydrogène liquide," *ibid.*, **150** (1910), 1437–1439; "Action abiotique des rayons ultraviolets sur les spores sèches aux basses températures et l'origine cosmique de la vie," *ibid.*, **151** (1910), 86–88; "La suspension de la vie des graines dans le vide à −271°C.," *ibid.*, **181** (1925), 805–807; "La vie latente des graines de pollen dans le vide à 271°C. au-dessous de zéro," *ibid.*, **188** (1929), 1308–1310; "La vie latente des spores des Fougères dans le vide aux basses températures de l'hélium liquide," *ibid.*, **190** (1930), 1134–1136; "La vie latente des spores des bactéries et des moisissures," in *Travaux cryptogamiques, dédiés à Louis Mangin* (Paris, 1931), pp. 303–307; "La vie latente des spores des Mousses aux basses températures," in *Comptes rendus de l'Académie des sciences, Paris,* **194** (1932), 1378–1380; "L'anhydrobiose des tubercules des Renoncules dans l'azote liquide," *ibid.*, **194** (1932), 1974–1976; "La reviviscence des plantules desséchées soumises aux actions du vide et des très basses températures," *ibid.*, 2158–2159; "Sur la résistance de certains organismes végétaux aux actions des basses températures de l'azote et de l'hélium liquides, réalisées au laboratoire cryogène de Leiden," in *Actes du VIème Congrès International du Froid* (1932), IV, 23–27; "Role de la synérèse dans le mécanisme de la congélation cellulaire," in *Chronica botanica,* **5** (1939), 10–11, a brief summary article; "Reviviscence du *Xanthoria parietina* desséché avec sa faune, six ans dans le vide et deux semaines à −189°C. Ses conséquences biologiques," in *Comptes rendus de l'Académie des sciences, Paris,* **226** (1948), 1413–1415; and "L'action du froid sur la cellule végétale," in *Botaniste,* **34** (1949), 57–74.

A short biographical obituary of Becquerel, without bibliographical material, appeared in *Comptes rendus de l'Académie des sciences, Paris,* **241** (1955), 137–140.

ALAN S. KAY

BEDDOE, JOHN (*b.* Bewdley, Worcestershire, England, 21 September 1826; *d.* Bradford-on-Avon, England, 19 July 1911), *physical anthropology.*

Beddoe suffered from ill health in his youth. Although he was intended for a career in the law, he studied medicine at University College, London, and the University of Edinburgh; in 1853 he received the M.D. from the latter institution, with a thesis entitled "On the Geography of Phthisis." Beddoe next served as a house physician in the Royal Edinburgh Infirmary. During the Crimean War he was on the medical staff of a civilian hospital at Renkioi in the Dardanelles. After the war he went to Vienna to complete his medical training, then made an extensive tour of Austria, Hungary, Italy, and France, studying medicine and physical anthropology. Upon his return to England in 1857 he set up a medical practice in Clifton, a suburb of Bristol, where he remained until his retirement in 1891.

Beddoe was elected a fellow of the Royal Society in 1873. He was a founding member, in 1857, of the Ethnological Society; was president of the Anthropological Society in 1869–1870; and was president of the Royal Anthropological Institute in 1889–1891. In 1905 he gave the Huxley lecture, "Colour and Race," and received the Huxley Medal. Beddoe received an honorary LL.D at Edinburgh in 1891 and there delivered the Rhind lectures. The University of Bristol elected him honorary professor of anthropology in 1908.

From the age of twenty Beddoe was a keen observer of the physical variations in man, beginning with observations on hair and eye color in the west of England, then extending his work to the Orkneys, which resulted in his *Contributions to Scottish Ethnology* (1853). In 1864 A. Johnes, of Garthmyl, Wales, contributed a prize of 100 guineas, to be awarded at the Welsh National Eisteddfod, for the best essay on the origin of the English nation. For four years, although there were contenders, the prize was not awarded. In 1868 it was won by Beddoe, and the essay was published as *The Races of Britain: A Contribution to the Anthropology of Western Europe* (1885). In the preface Beddoe says: "The successful work, however, though composed expressly for the occasion, was really the outcome of a great part of the leisure of fifteen years devoted to the application of the numerical and inductive method to the ethnology of Britain and of Western Europe."

Beddoe's Rhind lectures extended the theme of his Eisteddfod essay to cover all Europe. The year after his death, they appeared in a revised form: *The Anthropological History of Europe, Being the Rhind Lectures for 1912* (1912).

Beddoe loved traveling, and set down some of the incidents of his widespread journeys in *Memories of Eighty Years* (1910). This book also describes his friendship and correspondence with all the major figures of the time who were interested in the physical and cultural history of early man in Europe: Broca, Davis, Topinard, Virchow, Darwin, Retzius, Rhys, Boyd-Dawkins, Cartailhac, Deniker, John Evans, Pitt-Rivers, and Ridgeway. His *Races of Britain* is dedicated "to Rudolf Virchow and Paul Topinard, and to the memory of Paul Broca and Joseph Barnard Davis."

Beddoe was first of all a descriptive physical anthropologist, noting and measuring the varieties of man's hair color, skin color, height, cephalic index, and so forth, throughout the British Isles and Europe. But he was not content with description; he wanted to explain the observed physical variations in terms of the spread of language and culture, and against the picture produced by prehistorians in the nineteenth century and the known movements of people in historical times. He continually sought to produce a coherent story in which the movements of, for example, speakers of the Celtic languages, or the people of the great migration period of post-Roman times, or the Jews and the Gypsies were integrated into the anthropological history of Europe.

The following passage, taken from the end of chapter 3 of *The Races of Britain,* shows the quality of Beddoe's thinking:

> The natives of South Britain, at the time of the Roman conquest, probably consisted mainly of several strata, unequally distributed, of Celtic-speaking people, who in race and physical type, however, partook more of the tall blond stock of Northern Europe than of the thick-set, broad-headed, dark stock which Broca has called Celtic, and which those who object to this attribution of that much-contested name may, if they like,

denominate Arvernian. Some of these layers were Gaelic in speech, some Cymric; they were both superposed on a foundation principally composed of the long-headed dark races of the Mediterranean stock, possibly mingled with the fragments of still more ancient races, Mongoliform or Allophylian. This foundation-layer was still very strong and coherent in Ireland and the north of Scotland, where the subsequent deposits were thinner, and in some parts wholly or partially absent. The most recent layers were Belgic, and may have contained some portion or colouring of Germanic blood: but no Germans, recognisable as such by speech as well as person, had as yet entered Britain.

Beddoe was a pioneer in this sort of synthetic writing about the physical, linguistic, and cultural history of Europe. His works were widely read, and greatly influenced the thinking of a wide range of people in the late nineteenth and early twentieth centuries, from pure physical anthropologists to historians like Rice Holmes, archaeologists like Ridgeway, and linguistic scholars like Sir John Rhys.

Even though today, over half a century after Beddoe's death, it is no longer considered possible to lump together in any meaningful way the facts and hypotheses about the physical, linguistic, and cultural history of Britain and Europe, we should not forget that Beddoe's work was instrumental in the development of anthropology. Prior thereto much speculation about the origins of the physical varieties of man in Europe had been largely guesswork. Beddoe put this matter on a firm basis of observed fact, of which his interpretations were in keeping with the evidence.

BIBLIOGRAPHY

I. ORIGINAL WORKS. Among Beddoe's writings are *Contributions to Scottish Ethnology* (Edinburgh, 1853); *The Races of Britain: A Contribution to the Anthropology of Western Europe* (Bristol–London, 1885); "Colour and Race," in *Journal of the Royal Anthropological Institute,* **35** (1905), 98–140; *Memories of Eighty Years* (Bristol–London, 1910); and *The Anthropological History of Europe, Being the Rhind Lectures for 1912* (Paisley, 1912).

GLYN DANIEL

BEDDOES, THOMAS (*b.* Shifnal, Shropshire, England, 13 April 1760; *d.* Clifton, England, 24 December, 1808), *medicine, chemistry.*

Beddoes' father was a tanner and wished his son to follow the same trade, but the boy proved bookish. His grandfather, a man of parts, recognized his abilities and insisted that he be trained for a profession. Beddoes therefore attended Bridgnorth Grammar School, Shropshire, and in 1776 went to Pembroke College, Oxford. There, while reading medicine, he became interested in chemistry and in modern languages. On going down from Oxford, he translated, or edited translations of, works by Bergman, Scheele, and Spallanzani. In 1784 he went to Edinburgh for three years and then paid a visit to France, where he met Lavoisier, to whose antiphlogistic chemistry he adhered. In 1788 Beddoes was appointed reader in chemistry at Oxford and attracted audiences larger than any lecturer since the thirteenth century. He supported the French Revolution, and his strong political views partly explain his resignation from Oxford in 1792. In 1794 he married Anna Edgeworth, sister of the novelist Maria Edgeworth; their eldest son, Thomas Lovell Beddoes, became a famous poet.

Beddoes' name is not perpetuated in any important discovery in chemistry or in medicine: while perhaps his most important contribution to science was his discovery of Humphry Davy, he was well known in his own day for his popular works on preventive medicine and his investigations of the use of gases in treating diseases. Among his publications are *Isaac Jenkins,* a moral tale (1792); *Essay on Consumption* (1799); and *Hygëia,* a series of essays describing the regimen necessary for avoiding disease (1802–1803). Beddoes considered it deplorable that although about one person in a million died annually of hydrophobia, and one in a hundred of tuberculosis, everybody dreaded hydrophobia and took tuberculosis for granted. He observed that some occupations carried great risks of consumption and declared that in civilized societies victims were thus sacrificed to alcohol, to fashion, and to commerce. Beddoes agreed that it was foolish to encourage people to doctor themselves; but to persuade them to adopt a healthy way of life, and to learn some biology, seemed eminently reasonable.

He called on doctors to produce more case studies and censured hospitals particularly for their failure to develop adequate statistics; when an adequate supply of genuine facts was available, medicine could be brought to the level of chemistry or astronomy, where charlatans stood no chance of success. Chemistry interested Beddoes primarily because he believed that true medical science must have a chemical basis. With this in mind, in 1798 he set up his Pneumatic Institution at Clifton for treating diseases by the administration of gases. He appointed Davy, then age nineteen, superintendent, and published his first essays on heat and light; like Davy, Beddoes adhered to the kinetic theory of heat. After Davy's departure in 1801, the Pneumatic Institution became a clinic where advice was given on preventive medicine.

Beddoes was associated with the Lunar Society of

Birmingham during its last years. His mind was restless, and he died unhappy at his lack of solid achievement.

BIBLIOGRAPHY

I. Original Works. *The History of Isaac Jenkins, and of the Sickness of Sarah his wife, and their three children* (Madeley, Shropshire, 1792); *An Essay on the Causes, Early Signs, and Prevention of Pulmonary Consumption* (Bristol, 1799); *Hygëia; or Essays Moral and Medical on the Causes Affecting the Personal State of Our Middling and Affluent Classes,* 3 vols. (Bristol, 1802–1803). The following works were written by Beddoes and James Watt: *Considerations on the Medicinal Use of Factitious Airs, and on the Manner of Obtaining Them in Large Quantities* (Bristol, 1794; 2nd ed., 1795; 3rd ed., 1796); *Medical Cases and Speculations; Including Parts IV and V of Considerations on the Medicinal Powers, and the Production of Factitious Airs* (Bristol, 1796). With Beddoes as editor: *Chemical Experiments and Opinions, Extracted from a Work Published in the Last Century* [Mayow's experiments] (Oxford, 1790); *Contributions to Physical and Medical Knowledge, Principally From the West of England* (Bristol, 1799).

II. Secondary Literature. Works on Beddoes are J. E. Stock, *Memoirs of the Life of Thomas Beddoes, M.D.* (London, 1811), which contains a bibliography of Beddoes' writings; F. F. Cartwright, *The English Pioneers of Anaesthesia* (Bristol, 1952), pp. 49–164; E. Robinson, "Thomas Beddoes, M.D., and the Reform of Science Teaching in Oxford," in *Annals of Science,* **11** (1955), 137–141; Editorial, in *Endeavour,* **19** (1960), 123–124; and R. E. Schofield, *The Lunar Society of Birmingham* (Oxford, 1963), pp. 373–377.

David M. Knight

BEDE, THE VENERABLE (*b.* Northumbria, England, A.D. 672/673; *d.* Jarrow-on-Tyne, England, A.D. 735), *philosophy.*

At the canonical age of seven Bede was entrusted to the care of Benedict Biscop, founding abbot of the monasteries of Wearmouth and Jarrow (near Newcastle). In Bede's words, written in 731, "From that time, spending all the days of my life in residence at that monastery, I devoted myself wholly to Scriptural meditation. And while observing the regular discipline and the daily round of singing in the church, I have always taken delight in learning, or teaching, or writing." [1] A disciple, Cuthbert, described his death in loving but not hagiographical terms. [2] There are no other biographical events of record, but from the twelve octavo volumes of his extant Latin writings emerges a consistent picture of a dedicated scholar and scientist.

Apparently Bede never traveled farther than fifty miles from his monastery, but he had unusual resources there. The English settlement of Britain in the fifth to the seventh centuries suggests the European settlement of North America in the seventeenth to the nineteenth centuries. Benedict Biscop, who founded his monastery two centuries after the first English settlement, had studied at Lérins (the most famous Western school of the period) and at Rome. He brought Archbishop Theodore and Abbot Hadrian from Rome to England. Hadrian came from culturally rich North Africa, via Naples. Theodore, whose home had been the "university city" of Tarsus in Cilicia, was schooled in Greek, law, and philosophy. Later, Benedict brought John, the archchanter of St. Peter's in Rome, to teach and to compose musical texts. In Northumbria, Benedict was surrounded by students from Ireland and Frankland, including, for example, the famous Abbess Hilda of Whitby, an English princess who had been educated in Ireland and Paris. In all, he made five trips to the Continent, importing examples of all the arts and crafts and "a very rich library." His successor Ceolfrid, Bede's master and abbot, was an author and the creator of the famous *Codex Amiatinus.* Bede himself was the primary voice of this flowering English culture.

Half of Bede's volumes are scriptural exegesis, an art in which he excelled. Five of the remaining six volumes contain homilies, hagiography, history, a guide to holy places (derived from the pilgrim Arculf), religious and occasional verses, and letters. They include his renowned work, *Historia ecclesiastica gentis Anglorum.* Cuthbert reports that he composed Anglo-Saxon verse and prose, but none has survived.

The remaining volume contains Bede's several *opera didascalica,* textbooks designed for such courses in the emerging vocational curriculum of monastic schools as *notae* (scribal work), *grammar* (literary science), and *computus* (the art and science of telling time). For the study of *notae* Bede composed *De orthographia* (dealing with spelling, word formation, and so forth), in the tradition of Caper, Agroecius, and Cassiodorus. For *grammar* he composed *De arte metrica* (on versification) and *De schematibus et tropis* (on figurative language). The first contains the first known treatment of isosyllabic rhythm (*De rhythmo*), which was then supplanting quantity as the formative principle of Western verse, while the second is unique in maintaining, with evidence, that every Greco-Roman figure of speech had been anticipated in the Hebrew Scriptures.

The study of *computus* had arisen pragmatically to meet the needs of Christian monastic communities, in which time was of the essence. The migrations of different peoples, each of whom had a different mode of reckoning time in both short and long units, and the establishment among them of convents as models for living in which the residents emphasized cycles of psalms and prayers through ordained days and years, made the study of *computus* second only to that of the grammar necessary for studying the Scriptures. Eventually *computus* attracted, as part of its discipline, most of the content (although not in those categories) of what might later be deceptively called the *quadrivium* (arithmetic, geometry, astronomy, and music). As time passed, the science necessary to physicians (largely dietary regimens and periodic phlebotomy), to agriculturists, to mariners, to historians (chronology), to geographers and cosmologists, and to musicians and versifiers was incorporated. The basic pattern for Bede's primary texts was a theoretical section (rules and formulas with explanations) and a practical section (a chronicle of world history emphasizing such "timely" events as eclipses, earthquakes, human and natural calamities), to which were appended a wide variety of calendars, tables, and formularies.

This pattern had begun to develop with Julius Africanus about A.D. 200. The Eusebius-Jerome Chronicle, which largely determined the medieval view of history, probably originated as part of such a text. Bede's earlier text of this kind, *De temporibus,* was published in 703. At about the same time, he wrote *De natura rerum,* an ancillary text on the model of Isidore's *Liber rotarum,* using sections of Pliny's *Natural History* together with patristic lore. Cosmology and natural history were traditionally (after Origen, Basil, Ambrose) linked to hexamera, that is, to commentaries on the six days of creation as described in Genesis. Bede wrote such a commentary, in which he refined some earlier statements of his own and of the Fathers. In 725 he rewrote his earlier text *On Times* (*De temporum ratione*), lengthening it tenfold because, he said, "When I tried to present and explain it to some brothers, they said it was far more condensed than they wanted, especially respecting the calculation of Easter, which seemed most useful; and so they persuaded me to write somewhat more at length about the nature, course, and end of time."[3] Bede reproduced or created tables of calculation, Easter tables, calendars, formularies, mnemonic verses, and the like. A letter to Plegwin defended his chronology; another, to Wicthede, explained contradictions in a document later proved a forgery.[4]

In this area three particular contributions are noteworthy. He is the first to have created, or at least to have recorded, on the basis of the Metonic nineteen-year lunar cycle, a perpetual (532-year) cycle of Easters and to have tabulated it. True, Victorius of Aquitaine (fifth century) had created a 532-year table, and even earlier Christians had created an eighty-four-year cycle, but Bede practically reconciled the two. Bede built upon the work of Dionysius Exiguus (*ca.* 525) and took his anchor date, the *annus Domini.* Because annals were added in the margins of these tables and because Bede incorporated such annals with their anchor dates in his historical writing, he became the first historian to use the Christian era. His popular *Historia ecclesiastica* helped to spread the practice. Finally, he first stated the tidal principle of "establishment of port," which has been described (e.g., by Duhem) as the only original formulation of nature to be made in the West for some eight centuries. Pseudo-Isidore's *De ordine creaturarum,* almost certainly written in Bede's time and region, uses Bede's technical diction for tides, but not his principle. It is not surprising that scholars living by the North Sea, in or near Lindisfarne, an island at high tide and a peninsula at low tide, should have been concerned with observed tidal action.

In the eighth century, Boniface (Winfrid), apostle to the Germans, and innumerable other insular missionaries and wandering scholars (e.g., Alcuin), in cooperation with Carolingian rulers, developed Continental schools based on English Scholasticism. Bede's writings were the staple texts. Among Carolingian epithets for Bede were "Doctor Modernus" and "Venerabilis." From the eleventh century on, these texts were supplanted in diocesan schools by Boethian texts. Judging by extant manuscripts, the ninth century was clearly the Age of Bede on the Continent, whereas his works had virtually disappeared from England, a fact which may be explained in part by the raids of the Danes and the decline of English ecclesiastical vigor. An unknown author of the eleventh century thought Bede first a geographer, "Living in the very corner of the world, after death he lived renowned in every other corner through his books. In them he discriminatingly described at length the locations, resources, qualities of the different lands and provinces."[5] Bede's manuscripts were more in demand in the twelfth century, but only in line with the increase for all standard authors. Another peak in the fifteenth century may reflect the reformers' search for pristine purity. Bede's greatest practical effect was on the Western calendar. His decisions (beginning the year, calculation of Easter, names of days and months, calculations of eras, and so forth) in most instances finally determined usage that was only refined, not changed, by Gregorian reform.

NOTES

1. *Historia ecclesiastica,* V, 24.
2. *De obitu Bedae,* trans. Plummer, I, lxxii–lxxviii.
3. *Praefatio,* ed. Jones, p. 175.
4. The *Liber Anatholi de ratione paschali,* ed. Bruno Krusch, in *Studien zur christlich-mittelalterlichen Chronologie* (Leipzig, 1880), pp. 311–327. See E. Dekkers and A. Gaar, *Clavis patrum Latinorum,* Sacris Eruditi III, editio altera (Bruges, 1961), no. 2303, p. 514.
5. *Patrologia Latina,* XC, 37C.

BIBLIOGRAPHY

I. ORIGINAL WORKS. Collected editions of Bede's work include J. Hervagius, ed., *Opera Bedae Venerabilis omnia . . .,* 8 vols. (Basel, 1563), the first collected edition; J. P. Migne, ed., *Patrologia cursus completus . . . series Latina,* XC–XCIV (Paris, 1850), useful reprints of earlier editions; C. Plummer, ed., *Bedae opera historica,* 2 vols. (Oxford, 1896), containing editions of Bede's historical works and Cuthbert's *De obitu Bedae* (I, clx–clxiv; translated I, lxxii–lxxviii), with valuable information about all works; and Fratres Abbatiae Sancti Petri, O.S.B., Steenbrugis, Belgiae, eds., *Corpus Christianorum series Latina: Bedae opera omnia* (Turnholt, Belgium, 1955–   ).
Individual works include J. Sichardus, ed., *Bedae . . . de natura rerum et temporum ratione* (Basel, 1529); K. F. Halm, ed., *Rhetores Latini minores* (Leipzig, 1863), pp. 607–618 (*De schematibus*); H. Kiel, ed., *Grammatici Latini,* VII (Leipzig, 1880), 227–294 (includes *De orthographia* and *De arte metrica*); P. Geyer, ed., *Itinera Hierosolymitana saeculi IV-VIII,* Vol. XXXIX in the series Corpus Scriptorum Ecclesiasticorum Latinorum (Vienna, 1898), which contains *De locis sanctis;* Theodore Mommsen, ed., *Bedae chronica maiora, chronica minora,* XIII, 3 in the series Monumenta Germaniae Historica, Auctores Antiquissimi (Berlin, 1898), which contains *Chronica;* C. W. Jones, ed., *Bedae opera de temporibus* (Cambridge, Mass., 1943), pp. 175–303, 307–315, 317–325 (includes *De temporibus, De temporum ratione, Ep. ad Plegwin,* and *Ep. ad Wichtede*); and B. Colgrave and R. A. B. Mynors, *The Ecclesiastical History of the English People,* in the series Oxford Medieval Texts (Oxford, 1968).
II. SECONDARY LITERATURE. Studies are the following: W. F. Bolton, *A History of Anglo-Latin Literature, 597–1066,* I (Princeton, 1967), 101–185 and an excellent bibliography, 264–287; C. W. Jones, "The 'Lost' Sirmond MS. of Bede's *Computus,*" in *English Historical Review,* **52** (1937), 209–219, and "Manuscripts of Bede's *De natura rerum,*" in *Isis,* **27** (1937), 430–440; M. L. W. Laistner, *A Hand-list of Bede MSS.* (Ithaca, N.Y., 1943); W. Levison, *England and the Continent in the Eighth Century* (Oxford, 1946); R. B. Palmer, "Bede as a Textbook Writer: A Study of his *De arte metrica,*" in *Speculum,* **34** (1959), 573–584; F. Strunz, "Beda in der Geschichte der Naturbetrachtung und Naturforschung," in *Zeitschrift für deutsche Geschichte,* **1** (1935), 311–321; A. H. Thompson, ed., *Bede: His Life, Times, and Writings* (Oxford, 1935); and Beda Thum,

"Beda Venerabilis in der Geschichte der Naturwissenschaften," in *Studia Anselmiana,* **6** (1936), 51–71.

CHARLES W. JONES

BEECKMAN, ISAAC (*b.* Middelburg, Zeeland, Netherlands, 10 December 1588; *d.* Dordrecht, Netherlands, 19 May 1637), *physics, mechanics.*

In preparation for a career in the Reformed Church, Beeckman studied letters, philosophy, and theology. As a student at Leiden from 1607 to 1610 he came into contact with the Ramist philosopher Rudolph Snell and his son Willebrord. He continued his ministerial studies at Saumur in 1612; in the interim he had privately studied mathematics and nautical science and had learned Hebrew, in Amsterdam, from the Brownist Ainsworth. After he had completed his studies, Beeckman entered his father's factory, which made candles and water conduits for various purposes, as an apprentice. When his apprenticeship was over, he pursued the same trade in Zierikzee, on the isle of Schouwen, Zeeland, which provided him easy access to the equipment required for experiments in combustion, pumping, and hydrodynamics. In addition to all this, Beeckman found time to study medicine—he received the M.D. from Caen in 1618—although he never practiced it.

In 1618–1619 Beeckman was conrector (assistant headmaster) of the Latin school in Veere, on the island of Walcheren, his brother Jacob being headmaster. While there, he made astronomical observations with Philip van Lansbergen, a well-known Copernican astronomer. In 1619–1620 Beeckman was conrector in Utrecht; he later filled the same position in Rotterdam, once again under his brother's rectorship. In Rotterdam he founded a Collegium Mechanicum, a society of craftsmen and scholars who occupied themselves with scientific problems, especially those that had technological applications.

In 1627 Beeckman was appointed rector of the Latin school at Dordrecht, which under his direction grew to 600 students and became the best school in the Netherlands. There, in 1628, with the help of the magistrate he established the first meteorological station in Europe, recording wind velocity and direction, rainfall, and temperature, and making astronomical observations with his former pupil Martinus Hortensius and the Reformed minister Andreas Colvius. Among his students were Frederick Stevin, the son of Simon Stevin; Jan de Witt, who became Grand Pensionary of Holland; and George Ent, who was one of the first adherents of Harvey and one of the first members of the Royal Society. Beeckman also planned a series of lectures on physics and mathematics, to be given in the vernacular, "for the benefit of carpenters, bricklayers, skippers, and other burghers."[1] He made experiments in physics and was the unsalaried adviser to the burgomasters on water regulation. Some of the greatest French philosophers went to Dordrecht to see him—Descartes made the trip in 1628 and 1629, Gassendi (who called Beeckman "the best philosopher I have met up to now") in 1629, and Mersenne in 1630. (Beeckman had first met Descartes at Breda in 1618; Descartes wrote his *Compendium musicae* for him and recognized that he had never before met one who so closely connected physics with mathematics.) He was an early proponent of the application of a mathematical method in physics; in his inaugural address at Dordrecht (1627) he said that he would introduce his pupils into "the true, that is the mathematical-physical, philosophy." Nevertheless, he emphasized the necessity of using experiments to check deductions.

Beeckman was a progressive thinker. He accepted the Copernican hypothesis, although with due reservations, and also the conceptions of the infinity of the universe and of atoms and the void. He became an early adherent of the doctrine of circulation of the blood, of which he learned from Ent. His work in astronomy, mechanics, and engineering was strongly influenced by that of Simon Stevin and Willebrord Snell. His early espousal of the corpuscular theory of matter was inspired mainly by the ancient engineers Hero of Alexandria and Vitruvius, by the physician Asclepiades, and, to a lesser extent, by Lucretius—in keeping with these diverse models, Beeckman's approach to this subject was undogmatic.

As early as 1613, Beeckman, rejecting any internal cause of motion, put forward a principle of inertia for both circular and rectilinear movement: When there is no impediment, there is no reason why velocity and curvature of movement should be altered. He stated that in collision the total quantity of movement remains the same, and deduced different laws for elastic and nonelastic bodies. In 1618 he worked out the law of uniformly accelerated movement of bodies falling *in vacuo* by combining his law of inertia with the hypothesis that the earth discontinuously attracts falling bodies by tiny impulses. On this basis he found the correct relation between time and space traversed: the distances are as the squares of the times. ("It was Beeckman who went beyond Oresme and Galileo by introducing sound infinitesimal considerations into Oresme's 'triangular' proof" [Clagett, 1961].) When the first impulse imparts the velocity v, the space traversed in the first moment t will be vt; in the second mo-

ment the effect of another impulse will be added to the continuing motion acquired in the first moment, so that double the space, i.e., $2vt$, will be traversed. Consequently, in n moments the total space traversed will be (in algebraic notation)

$$vt(1 + 2 + \cdots + n) = \frac{n(n + 1)}{2} vt.$$

In a time twice as long, the space will be $2n(2n + 1)/2vt$. That is, the distances traversed will be as $n(n + 1)/2$ is to $2n(2n + 1)/2$, that is, as $(n + 1)$ is to $(4n + 2)$. If the moments are supposed to be infinitely small, and n is supposed to be infinitely great, the proportion will be as $1:4$.

Beeckman was a professional scholar as well as a craftsman; consequently his experimental work was as concerned with purely scientific inquiry as with practical applications. Many of the procedures he describes are thought experiments; in other instances, however, he carried out the actual manipulations and made the measurements requisite to formulating the phenomenal law. In 1615, for example, Beeckman determined experimentally the law of the velocity of the outflow of water (usually attributed to Torricelli): The quantity of water passing through a hole in the bottom of a vessel varies as the square root of the height of the water column. He also gave experimental proof that water cannot be changed into air (as was then generally believed) and added a discussion of the errors possible in the procedure that he used. In 1626 he determined the relation between pressure and volume in a measured quantity of air, and discovered that pressure increases in a degree slightly greater than the diminution of the volume. Beeckman attributed the ascent of water in a pump tube not to the "horror of a vacuum," nor yet to the "weight" of the atmosphere, but rather to the pressure of the air.

Francis Bacon's works were known to Beeckman as early as 1623. He made both appreciative and devastating comments on the experiments contained in the *Historia experimentalis:* he agreed that heat is a kind of motion, but thought that Bacon "talks nonsense when saying that water can be transformed into air" and that he "wrongly" believed that refrigeration was always followed by contraction, even when water is turned to ice.[2] *Sylva sylvarum* gave, in Beeckman's opinion, "dubious arguments" of the experiments described.[3] He stated the weakness of Bacon's natural philosophy, saying that Bacon did not know how to relate mathematics to physics.[4]

In 1627 Beeckman learned of Gilbert's work; he deemed much of it conformable to his own findings but rejected "internal magnetical force" as the motive power of the earth, holding that the earth is subject to inertial motion in empty space. Beeckman himself offered a purely mechanistic explanation of magnetism; believing only in action by contact, he rejected the notion of any attractive force. He deemed arguments based on simplicity or beauty (as those of Copernicus and Kepler) to be of "no value," and condemned the idea that the earth should possess intelligence and a soul (as set forth by Gilbert and Kepler) as "unworthy of a philosopher."[5] Beeckman found only mechanistic explanations satisfactory, since only they "put things as it were sensible before the imagination." When Colvius informed him of Galileo's theory of tides, Beeckman considered it a strong argument for the rotation of the earth, but suggested first making a mechanical model to test it.[6]

Beeckman was no more uncritical of his own atomistic philosophy than he was of the theories of others. He confessed to having an irrefutable "argument against all atomists, and accordingly against myself": while the absolute solidity and incompressibility of true atoms exclude their resilience after collision, their deflection is indeed observable in a multitude of phenomena, so that "it seems that the doctrine of atoms is fundamentally overthrown by these phenomena." It was therefore necessary to accept the idea of flexible primary particles—even though their properties might be unintelligible—since the scholastic explanations of elasticity were even more incomprehensible.[7]

Beeckman's approach to scientific theorizing therefore clearly demonstrates the difference between his method and the more rationalistic one of Descartes. Beeckman's conceptions were formalized before his meeting with Descartes, and the two were to differ on many questions (for instance, Beeckman held that light is propagated with a finite velocity). Descartes and Mersenne certainly read Beeckman's diary, however, and Gassendi knew his ideas. Beeckman may thus be considered "a link of the highest importance in the history of the evolution of scientific ideas."[8]

NOTES

1. *Journal*, II, 455.
2. *Ibid.*, 476.
3. *Ibid.*, III, 57.
4. *Ibid.*, 51.
5. *Ibid.*, 17.
6. *Ibid.*, 171, 206.
7. *Ibid.*, II, 100–101, 157.
8. Koyré, p. 101.

BIBLIOGRAPHY

I. ORIGINAL WORKS. Beeckman's dissertation for his M.D. is *Theses de febre tertiana intermittente . . .* (Caen, 1618). *D. Isaaci Beeckmanni medici et rectoris apud Dordracenos mathematico-physicarum meditationum, quaestionum, solutionum centuria* (Utrecht, 1644) is a selection of Beeckman's notes, which was published by his younger brother Abraham. Of major importance is *Journal tenu par Isaac Beeckman de 1604 à 1634,* with notes and introduction by C. de Waard, 4 vols. (The Hague, 1939–1953). This work (in Latin and Dutch) is valuable because de Waard's notes are based largely on sources destroyed in World War II. Vol. IV contains some of Beeckman's discourses and documents relating to his life and work.

II. SECONDARY LITERATURE. Beeckman's work is dealt with in E. J. Dijksterhuis, *Val en Worp* (Groningen, 1924), pp. 304–321; R. Hooykaas, "Science and Religion in the 17th Century; Isaac Beeckman (1588–1637)," in *Free University Quarterly,* **1** (1951), 169–183; A. Koyré, *Études galiléennes,* II (Paris, 1939), 25–40; and C. de Waard, *L'expérience barométrique* (Thouars, 1936), pp. 75–91, 145–168.

R. HOOYKAAS

BEER, WILHELM (*b.* Berlin, Germany, 4 January 1797; *d.* Berlin, 27 March 1850), *astronomy.*

A scion of a cultivated Jewish family (his brother was the poet Michael Beer; and his half-brother was the composer Jacob Meyer Beer, better known under the name of Meyerbeer), Beer was a banker by profession; by avocation, however, he was an amateur astronomer and the owner of a private observatory made famous by the work of Johann Mädler.

Beer's place in the history of astronomy is inseparably connected with the contributions that he and Mädler made between 1830 and 1840 to selenography and solar-system studies in general. Their first joint work was *Physikalische Beobachtungen des Mars in der Erdnähe* (1830), and their chef d'oeuvre was *Mappa selenographica totam lunae hemisphaeram visibilem complectens* (1836). This map, based on its author's observations made at Beer's private observatory with a Fraunhofer refractor of only 9.4-centimeter free aperture, constitutes a milestone in the development of selenographical literature. It was the first map to be divided into four quadrants (corresponding to a diameter of 97.5 centimeters for the apparent lunar disk) and contained a remarkably faithful representation of the moon's face as it is visible through a 4-inch refractor. Its topographic structure was based on the positions of 105 fundamental points measured micrometrically by Beer and Mädler (and related to the previous measurements by Wilhelm Lohrmann).

Moreover, an accompanying volume, *Der Mond nach seinen kosmischen und individuellen Verhält-nissen, oder allgemeine vergleichende Selenographie* (1837), contains the results of micrometric measurements of the diameters of 148 lunar craters and of the relative altitudes of 830 lunar mountains, determined by the shadow method. This book also contains a reduced version (to 32 centimeters) of the original map of 1836.

One more book appeared under the joint authorship of Beer and Mädler—*Beiträge zur physischen Kenntniss der himmlichen Körper im Sonnensystem* (1839)—before Mädler left Berlin to accept the professorship of astronomy and directorship of the astronomical observatory in Dorpat (now Tartu), Estonia. His departure ended the ten-year partnership with Beer. For the remainder of his life Beer was unimportant in the history of astronomy, and one may suspect that, in his selenographical work with Mädler, he largely played the role of a Maecenas. His name remains inseparably connected, however, with the best map of the moon produced in the first half of the nineteenth century, a map that remained unsurpassed until the publication of the famous *Charte der Gebirge des Mondes* by J. F. J. Schmidt in 1878.

ZDENĚK KOPAL

BEEVOR, CHARLES EDWARD (*b.* London, England, 12 June 1854; *d.* London, 5 December 1908), *neurology, neurophysiology.*

Beevor was the eldest son of Charles Beevor, a fellow of the Royal College of Surgeons, and Elizabeth Burrell. He was educated at Blackheath Proprietary School, London, and University College, London. He qualified as a member of the Royal College of Surgeons in 1878, M.B. in 1879, L.S.A. in 1880, received the M.D. in 1881, was elected a member of the Royal College of Physicians in 1882, and became a fellow of the Royal College of Physicians of London in 1888. From 1882 to 1883 he studied in Austria, Germany, and France with Carl Weigert, Julius Cohnheim, Wilhelm Heinrich Erb, and others. Beevor was appointed assistant physician to the National Hospital, Queen Square, London, in 1883 and to the Great (later Royal) Northern Central Hospital in 1885. He became full physician at each of these hospitals, but was able to carry out neurophysiological investigations in addition to practicing clinical neurology. On 5 December 1882 he married Blanche Adine Leadam, who bore him a son and a daughter.

Beevor was a man of pleasant yet simple personality and graceful courtesy, with a keen sense of humor and considerable musical talent. His powers of observation, his industry, and his precision as a recorder were unsurpassed, and he was intensely self-critical. He possessed such marked scientific caution that occasionally he would not publish the results of his investigations if they seemed to refute established opinion. He was retiring, modest, and quite unselfish.

Essentially, Beevor was an excellent clinical neurologist whose main ambition was to make possible more accurate diagnosis of diseases of the nervous system. He was also, however, a general physician and practiced internal medicine as well as neurology for many years. His *Diseases of the Nervous System: A Handbook for Students and Practitioners* (1898) revealed his clinical ability as well as his literary skill, and his work on the diagnosis and localization of cerebral tumors was outstanding.

Beevor's researches in neurophysiology encompassed three areas.

(1) Cerebral cortical function. From 1883 to 1887 Beevor collaborated with Victor Horsley at the Brown Institution in London. They extended the work of Gustav Theodor Fritsch, Eduard Hitzig, and David Ferrier on the representation of function in the cerebral cortex. In particular they studied the motor region in monkeys by means of electrical stimulation and then carried out similar investigations on the internal capsule. Minute representation of movement could be mapped at each site. This work was an important landmark in the development of the concept of cerebral localization, and Beevor became widely known after it was published (1887–1891).

(2) Muscle movements. Beevor meticulously observed the function of muscles and muscle groups both in health and in disease. His Croonian lectures given before the Royal College of Physicians of London in 1903 and published in 1904 embodied his findings, which he linked to the earlier studies he had made on the motor cortex of the cerebral hemisphere.

(3) Cerebral circulation. Stimulated by the work of Otto Heubner (1872) and Henri Duret (1874), Beevor carried out experiments on the human brain in order to discover the areas of distribution of the five main arteries. He injected colored gelatin into all five vessels simultaneously, under a constant pressure. His findings, which appeared in 1908 and were of outstanding importance, were the first accurate ones to be published. Unfortunately, Beevor agreed with Duret that the brain arteries were end-arteries, each with its own territory. This view dominated anatomical and pathological considerations of cerebral circulation for two decades and was finally disproved by

R. A. Pfeifer in 1928. It is now known that Beevor's experiments and the deductions derived from them were incorrect and that all parts of the cerebral cortex are linked by an anastomosing vascular network.

BIBLIOGRAPHY

I. ORIGINAL WORKS. With Victor Horsley, Beevor wrote "A Minute Analysis (Experimental) of the Various Movements Produced by Stimulating in the Monkey Different Regions of the Cortical Centre for the Upper Limb, as Defined by Professor Ferrier," in *Philosophical Transactions of the Royal Society, Part B,* **178** (1887), 153–167; "A Further Minute Analysis by Electrico Stimulation of the So-called Motor Region of the Cortex Cerebri in the Monkey (*Macacus sinicus*)," *ibid.,* **179** (1888), 205–256; and "An Experimental Investigation Into the Arrangement of the Excitable Fibres of the Internal Capsule of the Bonnet Monkey (*Macacus sinicus*)," *ibid.,* **181** (1890), 49–88. Also see *Diseases of the Nervous System: A Handbook for Students and Practitioners* (London, 1898); *The Croonian Lectures on Muscular Movements and Their Representation in the Central Nervous System* (London, 1904); and "On the Distribution of the Different Arteries Supplying the Human Brain," in *Philosophical Transactions of the Royal Society, Part B,* **200** (1908), 1–55.

II. SECONDARY LITERATURE. Obituaries of Beevor are in *The Lancet,* **2** (1908), 1854–1855; and *British Medical Journal,* **2** (1908), 1785–1786. A short account of his life can be found in *Dictionary of National Biography,* 2nd supp., (London, 1912). For Beevor and his contribution to our knowledge of cerebral blood supply, see E. Clarke and C. D. O'Malley, *The Human Brain and Spinal Cord* (Berkeley-Los Angeles, 1968), pp. 779–783.

EDWIN CLARKE

BÉGHIN, HENRI (*b.* Lille, France, 16 September 1876; *d.* Paris, France, 22 February 1969), *mechanics.*

Béghin studied at the École Normale Supérieure from 1894 to 1897 and placed first in the mathematics *agrégation* in 1897. For over twenty years he lived in Brest, where his inclination toward mechanics was strengthened. At the *lycée* there he gave courses that would prepare students for the École Navale (1899–1908) and later taught mechanics at the École Navale itself (1908–1921). Béghin organized the training of radio electricians during World War I. He began his university career in Montpellier (1921–1924) and later occupied the chair of mechanics at Lille (1924–1929). He lectured in fluid mechanics at Paris from 1929 to 1932 and was made professor of physical and experimental mechanics and director of the mechanics laboratory. Béghin also taught at the École des Beaux-Arts (1924–1930), the École Supérieure d'Aéronautique (1930–1939), and the École Polytechnique (from 1936). He was elected to the Académie des Sciences on 25 February 1946.

A remarkable teacher and inspirer of applied research, Béghin had a long and fruitful career. His accomplishments stem from his ability to rephrase classical mechanics and to enlarge its scope. Béghin is associated with the most general treatment of systems of nonholonomic linkage, through which he demonstrated, in his first works (1902–1903), how to use Lagrange's equations in percussion problems. He also provided an elegant extension of Carnot's theorem that leads to an exhaustive formulation. He extended the solution of Painlevé's paradox by increasing the examples of dynamic interference due to friction. Béghin also showed the significance of indeterminates, which sometimes appear in very simple cases, for theoretical solutions in a branch of mechanics whose principles do not involve any consideration of microscopic deformations. In elastic impact, he showed that the main characteristic of the end of the impact is that the loss of kinetic energy equals the work absorbed by friction during the impact, and that the reversal of the normal component of the relative speed, until then believed to occur in all such impacts, is realized only under specific conditions of symmetry.

The theory of the gyrostatic compass, which Béghin published in 1921 and later perfected, is significant for his fusion of the extensions of rational mechanics with technical and experimental data. He thus became a specialist in servomechanisms. In this field he studied and experimented with the elimination of vibrations and developed numerous applications, notably for automatic piloting of ships and planes.

Béghin, whose training was classical, was ahead of his time because of his new method of scientific work. He always worked with others, which at times makes it difficult to assess his contribution.

BIBLIOGRAPHY

Among Béghin's writings are "Sur les percussions dans les systèmes non holonomes," in *Journal de mathématiques pures et appliquées,* **9** (1903), 21–26, written with T. Rousseau; "Extension du théorème de Carnot au cas où certaines liaisons dépendent du temps," *ibid.; Étude théorique des compas gyrostatiques Anschütz et Sperry* (Paris, 1921); *Statique et dynamique,* 2 vols. (Paris, 1921; 7th ed., 1957); "Sur un nouveau compas gyrostatique," in *Comptes rendus de l'Académie des sciences,* **176** (1923), written with P. Monfraix; "Étude d'une machine locomotive à l'aide du théorème des travaux virtuels," in *Bulletin de l'élève ingénieur* (1924); "Étude théorique d'un compas zénithal gyrostatique à amortisseur tournant," in *Annales hydro-*

graphiques (1924), written with P. Monfraix; "Sur certains problèmes de frottement," in *Nouvelles annales de mathématiques,* **2** (1924) and **3** (1925); "Sur les transmissions par adhérence," in *Bulletin de l'élève ingénieur* (1925); *Leçons sur la mécanique des fils* (Tourcoing, 1927–1929), a course given at the École Technique Supérieure de Textile; "Sur les conditions d'application des équations de Lagrange à un système non holonome," in *Bulletin de la Société mathématique de France,* no. 57 (1929); "Sur le choc de deux solides en tenant compte du frottement," *ibid.; Exercices de mécanique,* 2 vols. (Paris, 1930; 2nd ed., 1946, 1951), written with G. Julia; *Cours de mécanique de l'École Polytechnique,* privately mimeographed (1942–1943), Vol. I later pub. (Paris, 1947); *Cours de mécanique théorique et appliquée à l'usage des ingénieurs et des étudiants de facultés,* 2 vols. (Paris, 1952); and "Les preuves de la rotation de la terre," in *Conférences du Palais de la Découverte,* ser. A, no. 207 (23 Apr. 1955).

A more complete bibliography and a lengthy list of studies and theses done under Béghin's supervision at the Laboratory of Physical Mechanics of the Faculty of Sciences, Paris, may be found in Béghin's *Notice sur les travaux scientifiques de M. Henri Béghin* (1943, 1945), at the secretariat of the Academy of Sciences, Paris.

PIERRE COSTABEL

BEGUIN, JEAN (*b.* Lorraine, France, *ca.* 1550; *d.* Paris, France, *ca.* 1620), *chemistry.*

Little is known of Beguin's family and early life, but he seems to have received a good classical education. When he arrived in Paris, possibly from Sedan, the influence of the royal physician, Jean Ribit, and of Turquet de Mayerne enabled him to obtain permission to set up a laboratory and give public lectures on the preparation of the new chemical medicaments of Quercetanus and others. His clear and lucid exposition and demonstration of chemical techniques won him large audiences. He complained that his unveiling of the mysteries of iatrochemistry so angered other "spagyrists" that they twice broke into his laboratory, plundering drugs and valuable manuscripts.

Beguin's first publication was an edition of Michael Sendivogius' *Novum lumen chymicum* with a preface (1608). The signature to the dedication shows that he was then almoner to Henry IV. Jeremias Barth, a Silesian who had studied medicine at Sedan, became Beguin's pupil at some time and encouraged him to publish a "little book," so that he would not have to dictate his lectures to his pupils. As a result, Beguin published the *Tyrocinium chymicum* (1610), a slim volume of seventy pages. The book was immediately pirated at Cologne, and Beguin published a revised edition with a long defense of chemical remedies (1612).

According to information in various editions of the *Tyrocinium,* Beguin had visited the mines of Hungary and Slovenia in 1604, and visited the Hungarian mines again in 1642. In a letter to Barth in 1613, he said he had earned 700 crowns by his skill, and could hardly earn more by teaching. The preface to the 1615 French edition says he was about to depart for Germany in search of the repose he had once found there, but yielded to the wishes of his friends and remained in France. In the same edition he acknowledged the authority and censorship of the Paris Faculty of Medicine, which may be the reason for his omission of a Paracelsian quotation.

For Beguin, chemistry was the art of separating and recombining natural mixed bodies to produce agreeable and safe medicines. The book opened with a Paracelsian quotation, followed by an exposition of the *tria prima* in a short theoretical section. The three elements were spiritual rather than corporeal, since they were impregnated by seeds emanating from the celestial bodies. Most of the work is concerned with chemical operations rather than with theory, and Beguin emphasized that the most effective therapy combined Galenic and Paracelsian remedies. The "quintessences" brought into prominence by the *Archidoxes* of Paracelsus occupy only a short third book. Beguin is credited with the first mention of acetone, which he called "the burning spirit of Saturn." Long sections on techniques and processes in Beguin parallel Libavius' *Alchymia* (1597), and both may have depended on a common source. Beguin cited Hermes Trismegistus, Lull, and the *Turba philosophorum* as authorities, and was a firm believer in transmutation.

Tyrocinium chymicum was immensely popular through the seventeenth century, and had swollen to nearly 500 pages in some later editions. It was translated into the major European languages and issued in many editions; it set the pattern for the notable series of French chemical textbooks in the later part of the century and was not really superseded until the appearance of Nicolas Lémery's *Cours de chymie* (1675).

BIBLIOGRAPHY

I. ORIGINAL WORKS. Beguin's writings include the dedication and preface to an edition of Michael Sendivogius' *Novum lumen chymicum* (Paris, 1608); *Tyrocinium chymicum* (Paris, 1610; rev. ed., 1612, the *editio princeps*); and *Les élémens de chymie* (Paris, 1615), English trans. by Richard Russell (London, 1669).

II. SECONDARY LITERATURE. Beguin and *Tyrocinium chymicum* are comprehensively discussed in T. S. Patterson, "Jean Beguin and His *Tyrocinium chymicum,*" in *Annals of Science,* **2** (1937), 243–298; the editions are conveniently listed in J. R. Partington, *A History of Chemistry,* III (London, 1962), 2–3. Various aspects are discussed in A. Kent and O. Hannaway, "Some New Considerations on Beguin and Libavius," in *Annals of Science,* **16** (1960), 241–250; Hélène Metzger, *Les doctrines chimiques en France, du début du XVIIIe siècle à la fin du XVIIIe siècle* (Paris, 1923), pp. 36–44; R. P. Multhauf, "Libavius and Beguin," in E. Farber, ed., *Great Chemists* (New York–London, 1961), pp. 65–74; John Read, *Humour and Humanism in Chemistry* (London, 1947), pp. 81–88; and Lynn Thorndike, *A History of Magic and Experimental Science,* VIII (New York, 1958), 106–113.

P. M. RATTANSI

BÉGUYER DE CHANCOURTOIS, ALEXANDRE-ÉMILE

BÉGUYER DE CHANCOURTOIS, ALEXANDRE-ÉMILE (*b.* Paris, France, 20 January 1820; *d.* Paris, 14 November 1886), *geology.*

A grandson of René-Louis-Maurice Béguyer de Chancourtois, a noted artist and architect, Alexandre-Émile entered the École Polytechnique in 1838 and the École des Mines in 1840. At the age of twenty-three he left the latter to travel through Armenia, Turkestan, the Banat, and Hungary, and became involved in the exploration and study of their mountainous regions.

He returned to the École des Mines in 1848 as professor of descriptive geometry and subsurface topology, and retained his affiliation with this institution until his death. Here he met Élie de Beaumont, whose geological theories greatly influenced him. Béguyer became *professeur suppléant* to Élie in 1852, succeeding him as professor of geology in 1875. Élie, who headed the French Geological Survey, had his protégé named assistant director, and together they undertook an exploration of the Haute-Marne regions. This venture resulted in Béguyer's publication in 1860 of a geological map of that region (drawn by M. Duhamel) as well as his collaboration with Élie on *Études stratigraphiques sur le départ de la Haute-Marne* (1862). The latter work contained a detailed study of the geological distribution of certain mineral deposits (sulfur, sodium, chlorine, the hydrocarbons); according to Élie's theories, they should have been found in particular mineralogical, petrographic, and geological association.

Generalizing further from Élie's ideas, Béguyer formulated a method for classifying chemical elements based "in the last analysis upon the distribution of these elements in the crust of the globe." His scheme, a precursor of the periodic table, was put forth in "Vis tellurique, classement des corps simples ou radicaux au moyen d'une système de classification helicoïdal et numérique" (1862). The model for his theory was the "telluric screw," a helical graph wound about a cylinder. The base of the cylinder was divided by sixteen equally spaced points, and the screw thread was similarly divided on each of its turns; the seventeenth point was on the second turn directly above the first, the eighteenth above the second, and so forth. Each point was supposed to represent the "characteristic number" of some element that could be deduced from its physical properties or chemical characteristics. Actually, Béguyer used unit equivalent weights as characteristic numbers, following Prout, who made hydrogen the unit. These weights were derived by measuring the specific heat of each element in a manner suggested by Regnault.

When placed on the telluric screw, elements whose equivalent weights differed by sixteen units were aligned in vertical columns. Sodium, for example, with a weight of 23, appeared one thread above lithium, with a weight of 7; potassium was above sodium, manganese above potassium. To the right of these columns was the group containing magnesium (24), calcium (40), iron (56), strontium (88), uranium (120), and barium (152). Directly opposite this group on the screw, oxygen (16) was aligned with sulfur (32), tellurium (128), and bismuth (224). One could, Béguyer believed, draw helices of any pitch through any two characteristic numbers of elements in different columns and find relationships among the elements so connected. Contrasts and analogies in mineralogical associations of the elements would become apparent through the sequences of their characteristic numbers along these helices. He showed that a line passed through magnesium and potassium (associated in the micas) would unite them, just as a line through sodium and calcium would show an association confirmed in the feldspars. If such lines were extended through or near the characteristic numbers of other elements, one would also be able to relate them to these mineral groups.

The name "telluric screw" was suggested to Béguyer by several circumstances, "especially the position of tellurium in the middle of the table and at the end of a characteristic series" and by the geognostic origin of his ideas, "since *tellus* refers to globe in the most positive and familiar sense, in the sense of Mother Earth."

Despite an unfortunate lapse into numerology and several errors in determining equivalent weights, "Vis tellurique" may be said to have anticipated Mendeleev's periodic table. When describing the theoretical origins of his work, Mendeleev mentioned that he was aware of Béguyer's ideas among others; he further acknowledged that his own periodic classification was influenced to some extent by his knowledge of previous systems.

Béguyer made still other contributions to geology. He worked with Le Play in organizing the Universal Exposition of 1855. Prince Napoleon, who had been interested in the exposition and was pleased by its success, invited Béguyer to participate in the voyage of the *Reine Hortense* to the polar regions the following year. In 1875 Béguyer was appointed director general of mines in France and initiated programs for the safety of miners and engineers. He also advocated the use of stereographic and gnomonic projections and campaigned for the adoption of a uniform system of cartographic gradation based on the metric system.

Béguyer served as secretary of Prince Napoleon's Imperial Commission for the 1867 Universal Exposition. He also organized the French geological exhibits at the expositions in Venice (1881) and Madrid (1883), and was *chef de cabinet* during Prince Napoleon's administration of Algeria and the African colonies.

A man of diverse interests, Béguyer attempted to develop a universal alphabet. He also studied human geography, trying to see if there was any consistent relationship between the geology of a country and the life style of its people. He devoted a great deal of time and effort to the improvement of the geological collections of the École des Mines and, finally, he toyed with ideas for using imaginary numbers in physics.

BIBLIOGRAPHY

I. ORIGINAL WORKS. Béguyer contributed nearly 75 memoirs and notes, which may be found in *Comptes rendus de l'Académie des sciences* (1844–1864); *Annales des mines* (1846); and *Bulletin de la Société géologique* (1874–1884). Among his works are "Sur la distribution des minéraux de fer," in *Comptes rendus de l'Académie des sciences,* **51** (1860), 414–417, the text accompanying Duhamel's map; *Études stratigraphiques sur le départ de la Haute-Marne* (Paris, 1862); and, most important, "Vis tellurique," in *Comptes rendus de l'Académie des sciences,* **54** (1862), 757–761, 840–843, 967–971.

II. SECONDARY LITERATURE. Works on Béguyer are Roman D'Amat, "Béguyer de Chancourtois," in *Dictionnaire de biographie française,* V (1951), 1279; and Léon Sagnet, "Chancourtois," in *La grande encyclopédie,* X (1890), 495.

MARTHA B. KENDALL

BEHAIM, MARTIN, also known as **Martin of Bohemia** (*b.* Nuremberg, Germany, 1459; *d.* Lisbon, Portugal, 29 July 1507), *geography.*

The mercantile interests of Behaim's family extended from Venice to Flanders. It is traditionally acknowledged that he was a disciple of Regiomontanus, from whom he may have learned how to use astronomical instruments. He quite certainly was influenced by the *Ephemerides* published by the latter between 1475 and 1506. As a youth he began in the textile business in Flanders. He was in business in Lisbon in 1480.

There has been much speculation about Behaim's role in Portugal. We know that he became a member of the Council of Mathematicians created by John II and that he became friendly with José Vicinho. Nevertheless, we can no longer defend the thesis that celestial navigation was possible only because of Behaim's teaching the Portuguese how to use the cross staff (Jacob's staff, or *ballestilla*) and the astronomical tables of Regiomontanus. The cross staff, invented by Levi ben Gerson, was already well known on the Iberian Peninsula; and the declination of the sun given in the *Tabula directionum* of Regiomontanus is different from that found in the *Regimento do astrolabio . . . Tractado da spera do mundo* prepared by the Council of Mathematicians for use by navigators. Joaquín Bensaúde has demonstrated that the numerical values in the *Regimento* are copies of those in the four-year-cycle sun table in the *Almanach perpetuum* (1473) of Abraham Zacuto, a Jew from Salamanca (*d. ca.* 1515), who had been a teacher of Vicinho. The latter was also involved in the scientific movement at the University of Salamanca (with Diego Ortiz and Juan de Salaya), and thus one must believe that the scientific elements that made celestial navigation possible were already present on the peninsula before the arrival of Behaim.

Behaim took part in the expedition of Diogo Cão (1485–1486) that followed the coast of Africa to Cape Cross. It is difficult to determine whether he carried out an observation of the latitude of the North Star while on this trip, or whether this should be attributed to Diogo Gomes (who dictated his memoirs to Behaim) and thus be dated a few years earlier. The text in question, which constitutes the birth certificate of celestial navigation in the Atlantic Ocean (although not in the East Indian), says: "When I went to these places, I took a quadrant. I marked the latitude of the North Pole on the table of the quadrant, and I noticed that the quadrant is more useful than the chart. It is true that on the latter one can see the sea route, but if it is incorrect [i.e., if one has committed an error in determining the course] one never arrives at the point previously designated."

During a landfall at Fayal, Azores, on the return voyage in 1486, Behaim married the daughter of Job Huerter de Moerbeke (in Portuguese, Joz Dutra or Jorge de Utra), the leader of a large Flemish colony there. From 1491 to 1493 he was in Nuremberg, where with the assistance of the painter Glockenthon he built his famous globe. Upon his return to Portugal, Behaim was entrusted with a number of official missions and was taken prisoner by the English while on a journey to Flanders. After he was released, he took up residence in Fayal. He died in the Hospice of Saint Bartholomew while on one of his trips to Lisbon.

A nautical chart, since lost, that showed the strait discovered years later by Magellan was attributed to Behaim. His most important work, which places him among the greatest geographers of the Renaissance, is the globe that has been preserved in Nuremberg. It is a sphere fifty-one centimeters in diameter, covered with parchment luxuriously adorned with many types of figures (111 miniatures, forty-eight banners, and fifteen coats of arms). Many inscriptions clarify and explain, from the most diverse points of view, the geography of 1,100 localities in many parts of the world. Actually, this globe constitutes an adaptation of a nautical chart onto a sphere. It lacks coordinates, but it does represent the equator and the tropics. The space between the western coast of Europe and the eastern coast of Asia is full of fanciful islands from medieval tradition, such as Antilia, which has no relation to the Antilles.

BIBLIOGRAPHY

Behaim's globe has been reproduced in almost every book on the history of geography.

For information on Behaim and his work, see G. Beaujouan, "Les aspects internationaux de la découverte océanique aux XVe et XVIe siècles," in *Actes du Cinquième Colloque International d'Histoire Maritime* (Paris, 1960–1966), pp. 69–73; J. Bensaúde, *Histoire de la science nautique portugaise à l'époque des grandes découvertes. Collection de documents . . .*, I, *Regimento do astrolabio e do quadrante. Tractado da spera do mundo*, facs. (Munich, 1914); S. Gunther, *Martin Behaim*, Bayerische Bibliothek, XIII (Bamberg, 1890); R. Henning, *Terrae incognitae*, IV (Leiden, 1939), 342–376, and indexes; E. G. Ravenstein, *Martin Behaim, His Life and His Globe* (London, 1908); A. Reichenbach, *Martin Behaim, ein deutscher Seefahrer aus den 15 Jahrhundert* (Wurzen–Leipzig, 1889); and R. Uhden, "Die Behaimsche Erdkugel und die Nürnberger Globen technik am Ende des 15 Jahrhunderts," in *Minutes of the Amsterdam Geography Congress, Works of Section IV: Géographie historique et histoire de la géographie*, IV (1938), 196–198.

JUAN VERNET

BEHRING, EMIL VON (*b.* Hansdorf, Germany, 15 March 1854; *d.* Marburg, Germany, 31 March 1917), *medicine, serology.*

Behring, one of twelve children of August Georg Behring, a teacher, and his second wife, Augustine Zech, grew up in simple circumstances in Hansdorf, a small town that is now under Polish administration. His father intended him to be a teacher or a minister, both traditional family professions, and in 1866 enrolled him in the Gymnasium of Hohenstein, in East Prussia. During his school years Behring discovered his interest in medicine, but he saw no hope of pursuing it. Accordingly, he planned to enter the University of Königsberg as a theology student.

Fortunately, one of Behring's teachers arranged for his acceptance at the Friedrich Wilhelms Institute in Berlin, where future military surgeons received a free medical education in return for promising to serve in the Prussian Army for ten years after passing their university examinations. Thus, in 1874 Behring became a cadet at the institute. In 1878 he received the M.D. and in 1880 passed his state board examinations. In the same year he was appointed intern at the Charité, a Berlin hospital, and in 1881 was attached to a cavalry regiment in Posen (now Poznan, Poland) as assistant surgeon. In between, he served for a short time as physician to a battalion stationed in Wohlau.

Behring, who had shown remarkable dedication at the Friedrich Wilhelms Institute, began to ponder scientific questions during his service in Wohlau and Posen. He was particularly interested in the possibility of combating infectious diseases through the use of disinfectants.

In 1881 Behring wrote his first paper on sepsis and antisepsis in theory and practice. In it he raised the question whether, in addition to external disinfection, the entire living organism could not be disinfected internally. He started investigations on iodoform (discovered as early as 1822 but introduced into wound treatment only in 1880) and the disinfecting effect of its derivatives. In 1882 he published his first treatise, "Experimentelle Arbeiten über desinficierende Mittel," which had been written in Posen. He had to admit that in many cases the disinfectant's toxic effect upon the organism was obviously much stronger than its disinfecting effect upon the bacteria. He concluded that the favorable results observed after the application of iodoform to infected wounds were not due to its being a parasiticide, but to its antitoxic effects. On the basis of later research, however, he came to reject its general use. In 1898 he wrote:

The fact that living animal and human body cells show much more sensitivity to disinfecting agents than any hitherto known bacteria may almost be considered a law of nature. As a result, before bacteria are killed by a disinfectant or their growth in the organs can be stunted, the infected animal body itself is killed by this same agent ["Über Heilprinzipien, insbesondere über das ätiologische und das isopathische Heilprinzip," in *Deutsche medizinische Wochenschrift,* **24,** no. 5 (1898), 67].

According to Behring's own statements, these iodoform experiments were the beginning of his preoccupation with antitoxic blood-serum therapy.

In 1883, at his own request, Behring was transferred from remote West Prussia to Winzig, Silesia. There he published another paper on iodoform poisonings and their treatment. At this time he prepared for the civil service medical examinations, since he planned to enter the Prussian Public Health Service after completing his service as a military surgeon. In 1887 he was promoted to captain and sent to the Pharmacological Institute in Bonn for further training. The director of this institute, Carl Binz, was especially interested in all problems concerning disinfectants. In the same year Behring published a report on new investigations concerning iodoform and acetylene. At the institute he acquired the knowledge and working habits necessary for accurate animal experiments and research in toxicology. In 1888 Behring was sent to Berlin, and after a brief service at the Academy for Military Medicine, in 1889 he joined the Institute for Hygiene of the University of Berlin, then presided over by Robert Koch. Here, between 1889 and 1895, Behring developed his pioneering ideas on serum therapy and his theory of antitoxins. Also in 1889, Behring finished his army service and became Koch's full-time assistant.

As early as 1887, in Bonn, Behring had ascertained that the serum of tetanus-immune white rats contained a substance that neutralized anthrax bacilli. This he saw as the cause of "resistance." Beginning in 1889, he worked in Berlin with Shibasaburo Kitasato on the isolation and definition of this agent. One of their goals was still the discovery of suitable systemic disinfecting agents, especially against anthrax, for which iodine, gold, and zinc compounds were tested. But of greater promise were experiments aimed at inhibiting the causative agents by using certain sera similar in effect to disinfectants, since the organism showed far greater tolerance to the sera. On 4 December 1890 Behring and Kitasato jointly published their first paper on blood-serum therapy, followed on 11 December by another report, signed by Behring alone, which discussed the blood-serum therapy not only in the treatment of tetanus but also of diphtheria. In it he stressed four points:

(1) The blood of tetanus-immune rabbits possesses tetanus toxin-destroying properties.

(2) These properties are also present in extravascular blood and in the cell-free serum obtained from the latter.

(3) These properties are so lasting that they remain effective when injected into other animals, thus making it possible to achieve excellent therapeutic effects with blood or serum transfusions.

(4) Tetanus toxin-destroying properties are not present in the blood of animals not immune to tetanus.

Behring immediately recognized that evidently a new principle of defense by the organism against infection had been discovered, one that clearly clashed with the then-prevalent cellular pathology of Virchow. Subsequently, Behring clashed with Virchow over the importance of his discoveries. One day before the publication of Behring's discovery, Ludwig Brieger and Carl Fränkel published a paper in the *Berliner klinische Wochenschrift* on the isolation of a protein substance—in their opinion a toxic substance—from bacteria; they called it "toxalbumine" and ascribed to it the severity of various infectious diseases. In the following years, however, Behring was able to show that the therapeutic principle in the serum, which he called "antitoxin," was ineffective against "toxalbumine" but acted against a specific toxin secreted by the bacteria. Incidentally, he succeeded in obtaining his new antitoxin-containing blood serum from guinea pigs treated not only with live diphtheria bacilli but also with diphtheria toxin alone in increasing dosages. Thus, in contrast with the hitherto prevailing phagocytosis theory of Élie Metchnikoff, he demonstrated the humoral defense capacities of the organism. Accordingly, he terminated his first work with the famous passage from Goethe's *Faust*: "Blood is a very special liquid."

When Paul Ehrlich demonstrated in 1891 that even vegetable poisons led to the formation of antitoxins in the organism, Behring's theory was confirmed and a lifelong friendship was formed. Both Behring and Ehrlich were then serving as assistants at the Koch Institute in Berlin. Behring immediately recognized the unusual importance of his discovery, and wrote:

> For hundreds and thousands of years the wisest physicians and scientists have studied the properties of blood and its relation to health and illness, without ever suspecting the specific antibodies appearing in the blood as a result of an infectious disease, which are capable of rendering infectious toxins harmless [Kleinschmidt, p. 347].

In 1891, at the Seventh International Hygiene Congress in London, Behring appeared for the first time before the public and delivered a lecture entitled

"Desinfektion am lebenden Organismus." He stressed that his method resulted above all in a natural increase of natural healing powers, leading in turn to increased resistance to nerve and cell toxins produced by pathogens. In 1892 he published his investigations in *Die praktischen Ziele der Blutserumtherapie und der Immunisierungsmethoden zum Zwecke der Gewinnung von Heilserum* and *Das Tetanusheilserum und seine Anwendung auf tetanuskranke Menschen.* Behring had to defend himself incessantly against all kinds of attacks. Failures due to the low antitoxin content of his first sera made his enthusiastic statements less credible. Furthermore, Behring—who at times used sharp language in his polemics—was forced to take issue with priority claims by other authors.

The legendary account of the first use of diphtheria serum on a patient on Christmas Eve 1891 has not been fully verified. A critical case of diphtheria is said to have been successfully treated with the serum by Behring's colleagues Geissler and Wernicke, in the infectious-disease ward of the surgical clinic of Berlin University. It is doubtful that sufficient serum was available at that time, since it was obtained exclusively from guinea pigs and then from sheep. When, in 1894, Roux and André Martin introduced the immunization of horses, Behring immediately adopted and extended this procedure. From 1892 on, he was backed by Farbwerke Meister, Lucius und Brüning, a dye works in Höchst, a suburb of Frankfurt. Until then Behring had put his own money in his research.

From 1893 on, serum therapy experimentation was conducted on a more extensive scale. In that year Behring became professor. Soon afterward there appeared the first publications by Hermann Kossel and Otto Heubner on results obtained with the new serum therapy, which reduced the mortality rate from 52 percent to 25 percent. In 1894 the first serum therapy experiments were carried out in France, England, and the United States. In the meantime, in 1893 Behring had written two important works on problems of great interest to him, *Die ätiologische Behandlung der Infektionskrankheiten* and *Geschichte der Diphtherie,* to which he added two books on the treatment of infectious diseases (1894, 1898).

Behring was quick to see that in order to obtain results in man, methods for standardizing the serum must be found. These methods were developed in 1897 by Ehrlich. Since 1895, however, standardization of the serum had been under state control. In 1896 the control authority became the Institute for Serum Research and Testing; today it is known as the Paul Ehrlich Institute and performs the same tasks on a more extensive scale.

In the fall of 1894 Behring, whose relationship with Koch had perceptibly cooled, was appointed associate professor of hygiene in Halle. He taught there for only a short time and with moderate success. The following year he was appointed professor of hygiene in Marburg—against the wishes of the Medical Faculty and thanks to determined efforts on his behalf by Friedrich Althoff, a powerful figure in the Prussian Ministry for Education. In Marburg, Behring carried on intensive research and organized what is now known as the Behring Institute. In the meantime, he had acquired great renown, especially in France, where Roux and Metchnikoff had become his close friends. He was made an officer of the Legion of Honor in 1895 and with Roux shared the 50,000-franc prize of the Académie de Médecine as well as the 50,000-franc prize of the Académie des Sciences. Also in 1895 he received the Prussian title of *Geheimrat* (privy councillor), and in 1901 his lifework was crowned with the first Nobel Prize in physiology and medicine, followed by his elevation to the hereditary nobility.

Beginning in 1889, Behring dedicated himself to a new task, the fight against tuberculosis. In competition with Koch he also attempted to find a substance suitable for tuberculosis vaccination. Finally, he felt he had succeeded with "tulase," an extract from tuberculosis bacilli treated with chloral hydrate, but his vaccination attempts failed. Nevertheless, we are indebted to Behring for his important findings on the spread of tuberculosis, which he ascribed mainly to the consumption by infants of milk containing tuberculosis bacilli. In contrast with Koch, he was convinced bovine and human tuberculosis were identical, a belief based on an understandable error. Nevertheless, his suggestions for combating bovine tuberculosis were of extreme importance and brought about vital changes in public health policy. In 1900, however, he realized that he was not achieving his objective and concluded one of his papers as follows: "Here I should like to say simply that I have definitely abandoned my hope for obtaining antitoxin for humans from cured and immunized tubercular cattle. Consequently, I have stopped searching for an antitoxin against tuberculosis."

Nevertheless, Behring's preoccupation with tuberculosis continued, and in 1903 and 1904 he devoted two monographs to this subject. Finally, in 1905 he suggested disinfection of milk for infants by adding Formalin and hydrogen peroxide, a process that proved impractical.

In 1913, in dogged pursuit of his theory of the origin of antitoxins as a result of insufficient toxin in the organism, Behring introduced active preventive vaccination against diphtheria. Its basis was a balanced toxin-antitoxin mixture, rendered stable by formaldehyde.

World War I, which separated Behring from his friends outside Germany, helped to substantiate his theories. The preventive, although still passive, tetanus vaccination saved the lives of millions of German soldiers. For his contributions Behring was awarded the Iron Cross, an unusual decoration for a non-combatant.

In 1896 Behring had married Else Spinola, daughter of one of the directors of the Charité Hospital in Berlin, who bore him six sons. The Villa Behring in Marburg, still standing today, was the gathering place of society. Behring also owned a house on Capri, where he was fond of vacationing. He liked to seclude himself in Switzerland, especially when suffering from the serious depressions that occasionally required sanatorium treatment. A fractured thigh, which initially seemed harmless, led to a pseudarthrosis that resulted in increasingly limited mobility. When Behring contracted pneumonia, his already weakened constitution was unable to withstand the multiple strain, and he died in Marburg on 31 March 1917.

For the discovery of antitoxins and the development of passive and active preventive vaccinations against diphtheria and tetanus, Behring was honored with the epithet "Children's Savior." By the same token, he could be called the "Soldier's Savior." His modern concepts raised humoral pathology to renewed importance, and he was certainly the equal of the other two pioneers in bacteriology, Pasteur and Koch. In his antitoxin theory Behring discovered a new principle in the fight against infections. He was able to realize his plan for an important and worthwhile lifework only by single-mindedly pursuing his original ideas. He thereby became involved in disputes with certain experts. Also, since he embraced the principle of "authority, not majority," he was not particularly adept at making friends or founding a school. He remained one of the great solitary figures in the history of medicine.

BIBLIOGRAPHY

I. ORIGINAL WORKS. Most of Behring's scientific papers may be found in two editions of collected works, the first covering the period 1882–1893 and the second the later period up to 1915: *Gesammelte Abhandlungen zur ätiologischen Therapie von ansteckenden Krankheiten* (Leipzig, 1893); and *Gesammelte Abhandlungen. Neue Folge* (Bonn, 1915). The most important papers and monographs are "Über Iodoform und Iodoformwirkung" in *Deutsche medizinische Wochenschrift,* **8** (1882), 146–148; "Die Bedeutung des Iodoforms in der antiseptischen Wundbehandlung," *ibid.,* 323–329; "Über das Zustandekommen der Diphtherie-Immunität und der Tetanus-Immunität bei Thieren," *ibid.,* **16** (1890), 113–114, written with S. Kitasato; "Untersuchungen über das Zustandekommen der Diphtherie-Immunität bei Thieren," *ibid.,* 1145–1148; "Über Immunisierung und Heilung von Versuchsthieren bei der Diphtherie," in *Zeitschrift für Hygiene und Infektionskrankheiten,* **12** (1892), 10–44, written with E. Wernicke; *Die praktischen Ziele der Blutserumtherapie und die Immunisierungsmethoden zum Zwecke der Gewinnung von Heilserum* (Leipzig, 1892); "Die Behandlung der Diphtherie mit Diphtherieheilserum," in *Deutsche medizinische Wochenschrift,* **19** (1893), 543–547, and **20** (1894), 645–646; *Die Geschichte der Diphtherie* (Leipzig, 1893); *Die Bekämpfung der Infektionskrankheiten* (Leipzig, 1894); *Allgemeine Therapie der Infektionskrankheiten* (Berlin–Vienna, 1898); *Diphtherie, Begriffsbestimmung, Zustandekommen, Erkennung und Verhütung* (Berlin, 1901); *Tuberkulosebekämpfung* (Marburg, 1903); "Tuberkuloseentstehung, Tuberkulosebekämpfung und Säuglingsernährung," in *Beiträge zur experimentellen Therapie,* **8** (1904); *Einführung in die Lehre von der Bekämpfung der Infektionskrankheiten* (Berlin, 1912); and "Über ein neues Diphtherieschutzmittel," in *Deutsche medizinische Wochenschrift,* **39** (1913), 873–876, and **40** (1914), 1139.

II. SECONDARY LITERATURE. The best biography, with many illustrations and references, is H. Zeiss and R. Bieling, *Behring. Gestalt und Werk* (Berlin, 1940). An exhaustive bibliography may be found in H. Dold, *In memoriam Paul Ehrlich und Emil von Behring zur 70. Wiederkehr ihrer Geburtstage* (Berlin, 1924). A biographical novel is H. Unger, *Emil von Behring* (Hamburg, 1948). Additional biographical articles are E. Bauereisen, "Emil von Behring," in *Neue deutsche Biographie,* II (Berlin, 1955), 14–15; H. von Behring, "Emil v. Behring," in *Lebensbilder aus Kurhessen und Waldeck,* **1** (1935), 10–14; and "Emil v. Behring zum 100. Geburtstag," in *Deutsches medizinisches Journal,* **5** (1954), 172–173; A. Beyer, "Zum 100. Geburtstag von Paul Ehrlich und Emil v. Behring," in *Deutsches Gesundheitswesen,* **9** (1954), 293–296; C. H. Browning, "Emil von Behring and Paul Ehrlich; Their Contributions to Science," in *Nature,* **175** (1955), 616–619; K. W. Clauberg, "Das immunologische Vermächtnis Emil von Behrings und Paul Ehrlichs," in *Deutsches medizinisches Journal,* **5** (1954), 138–146; C. Hallauer, "Emil von Behring und sein Werk," in *Schweizerische Zeitschrift für allgemeine Pathologie und Bakteriologie,* **17** (1954), 392–399; M. Jantsch, "Gemeinsames im wissenschaftlichen Werk Ehrlichs und Behrings," in *Wiener klinische Wochenschrift,* **66** (1954), 181–182; H. Kleinschmidt, "Zum 100. Geburtstag von Emil v. Behring," in *Medizinische* (1954), 347–348; A. S. Macnalty, "Emil von Behring," in *British Medical Journal* (1954), **1,** 668–670; and P. Schaaf, *Emil von Behring zum Gedächtnis. Herausgegeben von der Universität Marburg* (Marburg, 1944); *Robert Koch und Emil von Behring. Ursprung und Geist einer Forschung* (Berlin, 1944); and obituary notices in *British Medical Journal* (1917), **1,** 498; and *Lancet* (1917), **1,** 890.

H. Schadewaldt

BEILSTEIN, KONRAD FRIEDRICH (*b.* St. Petersburg, Russia, 17 February 1838; *d.* St. Petersburg, 18 October 1906), *chemistry.*

Beilstein's parents, Friedrich Beilstein and Catharine Margaret Rutsch, were German. His father, a salesman and tailor from Lichtenberg near Darmstadt, was related to Justus Liebig. His mother came from a farming and weaving family, and her uncle, Johannes Conrad Rutsch, was court tailor to the czar. Rutsch took an interest in Beilstein and sent him, at the age of fifteen, to Germany for study. Beilstein studied with leading figures in Germany—Bunsen and Kekulé at Heidelberg, Liebig at Munich, and Wöhler at Göttingen. In 1858 he received the doctorate at Göttingen, then continued his studies with Wurtz and Charles Friedel in Paris, and with Löwig in Breslau. In 1860 he began to teach at Göttingen, and in 1865 was appointed extraordinary professor. As a result of family difficulties following his father's death, Beilstein moved to the Technical Institute of St. Petersburg in 1866 and remained there the rest of his life. From 1865 to 1871 he, Hübner, and Fittig edited the *Zeitschrift für Chemie.* He belonged to the Deutsche Chemische Gesellschaft, and in 1881 was elected to the St. Petersburg Academy of Sciences.

Beilstein's work with many of the leading chemists of the day impressed him with the differences and difficulties in organic chemistry. Some chemists were guided by dualistic electrochemical notions and the older radical theory, while others viewed compounds as substitution products of several basic types of molecules. In addition, ideas about valence and the geometrical arrangement of bonds in structural formulas were being put forth. Beilstein was aware of all these views. He was attracted by Kekulé's ideas, for they seemed to provide a possible solution to the difficulties and to permit the writing of specific formulas for compounds. His own work heightened his awareness of the need for specific formulas. In 1860, for example, he showed that ethylidene chloride, obtained by Wurtz in 1858, was identical with the *éther hydrochlorique monochloridée* of Regnault. With F. Reichenbach he showed in 1864 that Kolbe and Lautemann's salicylic acid was only impure benzoic, and with J. Wilbrand he showed in 1863 that dracylic acid from nitrodracylic acid (i.e., p-nitro-benzoic acid) was also identical with benzoic acid. He himself had prepared pure hydracrylic acid in 1862, only to see the formula he had assigned to this compound over-turned two years later by the work of Moldenhauer. In 1866, through his work with P. Geitner on the chlorination of benzene compounds, he became more convinced of the importance of Kekulé's ideas. They had investigated when chlorination would occur on a side chain and when in the nucleus. He also worked on aromatic compounds in various petroleum fractions.

At Göttingen, Beilstein had attempted to bring order into organic chemistry and had started his *Handbuch der organische Chemie,* which superseded Gmelin's textbook. In Beilstein's handbook, material that had previously been included in textbooks in an abbreviated form was now expanded into a new form of chemical literature. The first edition (1880–1882) described 15,000 compounds. The rapid growth of organic chemistry made new editions imperative. By the end of the century the magnitude of the task was too great for one man. The Deutsche Chemische Gesellschaft agreed to prepare a supplement to the third edition. After Beilstein's death they began to prepare a fourth edition, which would survey the literature to 1910. This edition appeared in 1937. The following year a supplement of equal length appeared, covering the decade 1910–1919. Work continued on a second supplement.

BIBLIOGRAPHY

I. Original Works. Among Beilstein's writings are "Ueber die Einwirkung verschiedener Aetherarten auf Aether-Natron und über die Aethylkohlensäure," in *Justus Liebigs Annalen der Chemie,* **112** (1859), 121–125; "Ueber die Umwandlung des Acetals zu Alderhyd," *ibid.,* 239–240; "Ueber die Identität des Aethylidenchlorürs und des Chlorürs des gechlorten Aethyls," *ibid.,* **113** (1860), 110–112; "Ueber die Umwandlung der Glycerinsäure in Acrylsäure," *ibid.,* **122** (1862), 366–374; "Ueber die Zersetzung der Aldehyde und Acetone durch Zinkäthyl," *ibid.,* **126** (1863), 241–247, written with K. F. Rieth and R. Rieth; "Ueber eine neue Reihe isomerer Verbindungen der Benzoegruppe—Nitrodracylsäure und deren Derivate," *ibid.,* **128** (1863), 257–273, written with J. Wilbrand; "Ueber die Natur der sogenannten Salylsäure," in *Annalen der Chemie und Pharmacie,* **132** (1864), 309–321, written with E. Reichenbach; "Untersuchung über Isomerie in der Benzoëreihe: Ueber das Verhalten der Homologen des Benzols gegen Chlor," *ibid.,* **139** (1866), 331–342, written with P. Geitner; "Ueber die Scheidung des Zinks von Nickel," in *Berichte der Deutschen chemischen Gesellschaft,* **11** (1878), 1715–1718; "Ueber die Natur des Kaukasischen Petroleums," *ibid.,* **13** (1880), 1818–1821, written with K. F. Kurbatov and A. Kurbatov; and *Handbuch der organischen Chemie,* 2 vols. (1880–1883; 2nd ed., 1886–1889; 3rd ed., 1892–1899; 4th ed., 1916–1937).

Letters to Zincke and others are in the Kekulé Archiv and in the Preussische Staatsbibliothek. Letters to R. Anschutz are in the possession of Prof. L. Anschutz.

II. SECONDARY LITERATURE. Works on Beilstein are F. Hjelt, "Verzeichnis in der deutschen and franzoischen Zeitschriften erscheinen Abhandlungen Beilstein," in *Berichte der Deutschen chemischen Gesellschaft,* **40** (1907), 5074–5078; F. Richter, "Konrad Friedrich Beilstein, sein Werk und seine Zeit," *ibid.,* **71** (1938), Abt. A, 35–55; and F. Richter, "Zum 100. Geburtstag von Konrad Friedrich Beilstein," in *Forschungen und Fortschritte,* **14** (1938), 59–60.

RUTH ANNE GIENAPP

BEKETOV, NIKOLAI NIKOLAEVICH (*b.* Alferevka village, Penzensky district, Russia, 13 January 1827; *d.* St. Petersburg [now Leningrad], Russia, 13 December 1911), *chemistry.*

Beketov graduated from Kazan University in 1849 and then worked in Zinin's laboratory at the Academy of Medicine and Surgery in St. Petersburg. In 1855 he became a junior scientific assistant at Kharkov University, and from 1859 to 1886 he was professor of chemistry. In 1864 Beketov organized the department of chemistry and physics at Kharkov, with laboratory work in physical chemistry, and taught a course in physical chemistry. He was elected a member of the Petersburg Academy of Sciences in 1886.

Under Zinin's influence Beketov began his scientific activity with work in organic chemistry, studying esterification reactions. In his master's thesis (1853) he further developed the concepts of basicity and affinity, which had been worked out by C. F. Gerhardt and his followers. In Beketov's work one can find the sources of the study of "chemical value," which was later developed by the Butlerov school.

Beketov's later interests were physical and inorganic chemistry. As a result of his studies of the liberation of certain metals by hydrogen and by other metals, Beketov established an activity series of metals, demonstrating that the process of reduction is associated with the formation of galvanic pairs. In order to promote concentration of the reagent, Beketov subjected the hydrogen to pressures of 100 atmospheres and higher. "This action of hydrogen," he wrote, "depends on the gas pressure and the concentration of the metal [in acids], or in other words, depends on the chemical mass of the reducing body." In this work he closely approached the law of mass action and also studied the reversibility of the reaction:

$$(CH_3COO)_2Ca + CO_2 + H_2O \rightleftarrows$$
$$2CH_3COOH + CaCO_3$$

The results of all these investigations were stated in his doctoral dissertation (1865).

Beketov later discovered and substantiated the theoretical possibility that metals could be reduced from their oxides by using aluminum, thus opening the way to the creation of the method of aluminothermal reduction. His constant interest in the theory of chemical affinity led to a large series of thermochemical researches, begun in Kharkov and continued in St. Petersburg. Specifically, Beketov determined the heats of formation and hydration of the oxides of the alkali metals. Beketov established that the heats of formation of chlorides, bromides, iodides, and oxides are in parallel with the compression taking place during the reactions.

BIBLIOGRAPHY

I. ORIGINAL WORKS. Beketov's writings include *O nekotorykh novykh sluchayakh khimicheskogo sochetania i obshchie zamechania ob etikh yavleniakh* ("On Several New Cases of Chemical Combination and General Remarks About These Occurrences"; St. Petersburg, 1853); *Issledovania nad yavleniami vytesnenia odnikh elementov drugimi* ("Researches Into the Occurrences of Liberation of Certain Elements by Other Elements"; Kharkov, 1865); "Recherches sur la formation et les propriétés de l'oxyde de sodium anhydre," in *Mémoires de l'Académie des sciences de St. Pétersbourg,* 7th ser., **30**, no. 2 (1881), 1–16; *Dinamicheskaya storona khimicheskikh yavleny* ("Dynamic Site of Chemical Phenomena"; Kharkov, 1886); "De quelques propriétés physico-chimiques de sels haloïdes du césium," in *Bulletin de l'Académie des sciences de St. Pétersbourg,* n.s., **4** (1894), 197–199; *Rechi khimika 1862–1903* ("Speeches of a Chemist 1862–1903"; St. Petersburg, 1908); and N. A. Izmailov, ed., *Izbrannie proizvedenia po fizicheskoy khimy* ("Selected Works on Physical Chemistry"; Kharkov, 1955), with a bibliography of his scientific works.

II. SECONDARY LITERATURE. Works on Beketov are A. I. Belyaev, *Nikolai Nikolaevich Beketov—vydayushchiysya russky fiziko-khimik i metallurg, 1827–1911* ("Nikolai Nikolaevich Beketov—Prominent Russian Physical Chemist and Metallurgist"; Moscow, 1953), with a bibliography of Beketov's works, and Y. I. Turchenko, *Nikolai Nikolaevich Beketov* (Moscow, 1954), which includes a list of Beketov's works and literature on him.

G. V. BYKOV

BEKHTEREV, VLADIMIR MIKHAILOVICH (*b.* Sorali, Vyatskaya oblast, Russia, 20 January 1857; *d.* Leningrad, U.S.S.R., 24 December 1927), *neurology, psychology.*

After graduating from the Vyatskaya Gymnasium, Bekhterev enrolled at the Medical and Surgical

Academy of St. Petersburg in 1873. He graduated in 1878, then prepared for a teaching career in the clinic of I. P. Merzheevsky. In 1881 Bekhterev defended his dissertation for the M.D., which dealt with the possible relation between body temperature and some forms of mental illness, then began work with Flechsig and Meinert, Westfall and Charcot, Du Bois-Reymond, and Wundt. In 1885 he was made a professor of psychiatry at Kazan University, where he organized the first laboratory for research on the anatomy and physiology of the nervous system. While in Kazan, Bekhterev completed investigations of the role of the cortex in the regulation of the functions of internal organs and prepared the first edition of the classic monograph *Provodyashchie puti spinnogo i golovnogo mozga* ("Passages of the Spinal Cord and Cerebrum") and the two-volume *Nervnye bolezni v otdelnykh nablyudeniakh* ("Nervous Diseases in Separate Observations").

From 1893 to 1913, Bekhterev headed the department of nervous and psychic diseases of the Military Medical Academy in St. Petersburg. During this period he published *Osnovy uchenia o funktsiakh mozga* ("Foundations of Knowledge About the Functions of the Brain"; 1903–1907). The last ten years of Bekhterev's life were devoted to the study of "reflexology" (objective psychology) and other areas of psychology.

In the anatomy and physiology of the central nervous system, Bekhterev defined more precisely the path and separation of the posterior rootlets of the spinal cord and described a group of cells on the surface of the shaft of the posterior horn (Bekhterev cells) and the internal bundle of the lateral column. He also described the large bundles of the brain stem and the pia mater nodes of the base of the brain; studied in detail and described the reticular formation (*formatio reticularis*) in 1885; and established the precise location of the taste center within the brain cortex in 1900.

In nervous diseases, Bekhterev isolated a number of reflexes and symptoms that have important diagnostic significance and described new illnesses: numbness of the spine (Bekhterev's disease), apoplectic hemitonia (*hemitonia postapopletica*), syphilitic dissipating sclerosis, a special form of facial tic, severe motor ataxia (*tabes dorsalistrans*), acroerythrosis, chorenic epilepsy (*epilepsia choreica*), and new phobias and obsessive states.

Bekhterev was a great organizer who created centers for the study of the neurological sciences and psychic illnesses—the Psychoneurological Institute, in 1908, and the State Institute for the Study of the Brain in St. Petersburg, which today bears his name. He also was the founder of the first Russian journals on nervous and psychic diseases: *Nevrologicheskii vestnik* and *Obozrenie psikhiatrii, nevrologii i eksperimentalnoy psikhologii.*

BIBLIOGRAPHY

I. ORIGINAL WORKS. Among Bekhterev's 800 or so publications are "O prodolnykh voloknakh setevidnoy formatsy na osnovany issledovania ikh razvitia i o soedineniakh setchatogo yadra pokryshki" ("About the Longitudinal Fibers of the Reticular Formation, on the Basis of an Investigation of Their Development, and About the Junctions of the Retinal Nucleus of the Cover"), in *Vrach,* no. 6 (1885); *Provodyashchie puti spinnogo i golovnogo mozga. Rukovodstvo k izucheniyu vnutrennikh svyazey mozga* ("Passages of the Spinal Cord and Cerebrum. Handbook Toward the Study of the Internal Connections of the Brain"), 2nd ed., rev. and enl., 2 vols. (St. Petersburg, 1896–1898); *Osnovy uchenia o funktsiakh mozga* ("Foundations of Knowledge About the Functions of the Brain"), 7 vols. (St. Petersburg, 1903–1907); *Zadachi i metod obektivnoy psikhology* ("The Tasks and Method of Objective Psychology"; St. Petersburg, 1909); *Obshchaya diagnostika bolezney nervnoy sistemy* ("General Diagnostics of Diseases of the Nervous System"), 2 vols. (Petrograd, 1911–1915); *Obshchie osnovy refleksology cheloveka (rukovodstvo k obektivnomu izucheniyu lichnosti)* ("General Foundations of the Reflexology of Man [Handbook Toward the Objective Study of Personality]"), 3rd ed. (Petrograd, 1926); *Avtobiografia (posmertnaya)* ("Autobiography [Posthumous]"; Moscow, 1928); *Mozg i ego deyatelnost* ("The Brain and Its Activity"), L. V. Gerver, ed. (Moscow–Leningrad, 1928); and *Izbrannye proizvedenia (stati i doklady)* ("Selected Works [Articles and Reports]"), ed. with introductory articles by V. N. Myasishchev (Moscow, 1951).

II. SECONDARY LITERATURE. Works on Bekhterev are V. D. Dmitriev, *Vydayushchysya russky ucheny V. M. Bekhterev* ("The Outstanding Russian Scientist V. M. Bekhterev"; Cheboskary, Chuvash Autonomous SSR, 1960); N. I. Grashchenkov, *Rol V. M. Bekhtereva v razvity otechestvennoy nevrology* ("The Role of V. M. Bekhterev in the Development of National Neurology"; Moscow, 1959); V. N. Myasishchev and T. Y. Khvilivitsky, eds., *"V. M. Bekhterev i sovremennye problemy stroenia i funktsy mozga v norme i patology. Trudy vsesoyuznoy konferentsy, posvyashchennoy stoletiyu so dnya rozhdenia V. M. Bekhtereva* ("V. M. Bekhterev and Modern Problems of the Structure and Functions of the Brain in Norm and in Pathology. Works of the All-Union Conference on the One Hundredth Anniversary of the Birthday of V. M. Bekhterev"; Leningrad, 1959); V. P. Osipov, *Bekhterev* (Moscow, 1947); and *Sbornik, posvyashchenny Vladimiru Mikhailovichu Bekhterevu. K 40-letiyu professorskoy deyatelnosti (1885–1925)* ("A Collection Dedicated to Vladimir

Mikhailovich Bekhterev. On the 40th Anniversary of His Professorial Career [1885–1925]"; Leningrad, 1926).

N. GRIGORYAN

BELAIEW, NICHOLAS TIMOTHY, in Russian **Nikolai Timofeevich Beliaev** (*b.* St. Petersburg, Russia, 26 June 1878; *d.* Paris, France, 5 November 1955), *metallurgy.*

Belaiew was the son of Gen. T. M. Beliaev. From 1902 to 1905 he studied at the Mikhailovskoi Artilleriiskoi Akademii, a graduate school of military engineering in St. Petersburg. He remained at this academy until 1914, first as a tutor and later (from 1909) as professor of metallurgical chemistry. Wounded early in World War I, in 1915 Belaiew was sent to England in connection with munitions supply; he remained there after the Revolution, working as an industrial consultant. In 1934 he moved to Paris.

Belaiew's papers have a strong historical bent. He claimed inspiration from his famed teacher, D. K. Chernoff, and from P. P. Anosov, who had established the manufacture of Damascus steel swords in Russia in 1841. Belaiew himself wrote a classic paper on the history and metallurgy of Damascus steel (1918). In 1944 he studied, in engineering steels, the coalescence of iron carbide that the Oriental swordmakers had unknowingly achieved through their methods of forging. His scientific contributions are mainly in his first book, *Kristallizatsia, struktura i svoystva stali pri medlennom okhlazhdenii* (1909), which provided the basis for several later papers in French, German, and English as well as for a small book in English (1922).

He showed that the geometric Widmanstätten structure, which had been discovered in 1804 in meteorites, could also be produced in steel under certain conditions of cooling. (He achieved the right structure by accident, because a foreman, anxious to get on with production, disregarded instructions and removed the steel ingot from the furnace before it was fully transformed.) Belaiew's emphasis on the crystallographic basis of the change and his detailed analysis of the geometry in this one case strongly influenced an important decade of metallurgical thinking.

BIBLIOGRAPHY

Belaiew's principal works are *Kristallizatsia, struktura i svoystva stali pri medlennom okhlazhdenii* (St. Petersburg, 1909); "Damascene Steel," in *Journal of the Iron and Steel Institute,* **97** (1918), 417–437, and **104** (1921), 181–184; *The Crystallization of Metals* (London, 1922); "Swords and Meteors," in *Mining and Metallurgy,* **20** (1939), 69–70; ". . .

la coalescence dans les aciers eutectoïdes et hypereutectoïdes," in *Revue de métallurgie,* **41** (1944), 65 ff. (in 8 parts).

Also see Robert F. Mehl, "On the Widmanstätten Structure," in *The Sorby Centennial Symposium on the History of Metallurgy* (New York, 1965), pp. 245–269.

C. S. SMITH

BÉLIDOR, BERNARD FOREST DE (*b.* Catalonia, Spain, 1697/1698; *d.* Paris, France, 1761), *mechanics, ballistics, military and civil architecture.*

Bélidor's career belongs to the early stages of engineering mechanics.

When Bélidor was born, his father, Jean-Baptiste Forest de Bélidor, a French officer of dragoons, was on duty in Spain. Both his father and his mother, Marie Hébert, died within five months of his birth, and he was brought up by the family of the widow of his godfather, an artillery officer named de Fossiébourg. Grateful for the family's protection during his childhood, Bélidor married Fossiébourg's daughter (or possibly his granddaughter) two years before his own death.

His life mingled the scientific with the military. A flair for practical mathematics secured him a post in the field under Jacques Cassini and Philippe de La Hire in the survey of the meridian from Paris to the English Channel, which they completed in 1718 and published in Cassini's *De la grandeur et de la figure de la terre* (1720). Bélidor's talents came to the attention of the regent, the Duc d'Orléans, who discouraged him from taking holy orders and arranged his appointment as professor of mathematics at the new artillery school at La Fère. In this post he made himself known as an author of textbooks and technical manuals in the 1720's and 1730's. After an interval of active duty in Bavaria, Italy, and Belgium during the War of the Austrian Succession, Bélidor settled in Paris with the rank of brigadier, was named chevalier of the Order of Saint Louis, and was elected an *associé libre* of the Academy of Sciences in 1756.

The book that made his reputation was *Nouveau cours de mathématique,* a text for artillery cadets and engineers. A second, *Le bombardier françois,* was for use in combat and contained systematic firing tables. It was with two fuller works, however—*La science des ingénieurs* (1729) and *Architecture hydraulique* (1737–1739)—that Bélidor entered into the science of mechanics proper with a summons to builders to base design and practice on its principles. The first of these treatises was concerned primarily with fortifications, their erection and reduction (the term *génie* then referred mainly to military and naval enterprises). The second, *Architecture hydraulique,* embraced civil constructions. The choice of title was a reflection of the actual prominence of problems involving transport, shipbuilding, waterways, water supply, and ornamental fountains. Both books opened with formulations of the principles of mechanics in mathematical terms, in which there was nothing original. The discussion was elementary, for the putative marriage of mathematics to mechanics was a rite more often celebrated than consummated in the early eighteenth century.

Nevertheless, the practical contents of both works proved to be invaluable to architects, builders, and engineers. They amount to rationalized engineering handbooks in which the man in charge of a construction might look up model specifications for a foundation or a cornice, a pediment or an arch; find diagrams he could follow or adapt; and consult job analyses and work plans for dividing and directing the labor. Both works were reprinted so often that the copper plates wore out and had to be reengraved for the final editions, in 1813 and 1819 respectively. Those editions were republished with notes by Navier, who in order to conserve the practical value, found it wiser to correct Bélidor's theoretical faults by up-to-date annotation than to revise or rewrite. He chose this course despite the immense development, amounting almost to creation of analytical mechanics as a science, that had occurred since Bélidor's first edition.

In that interval, Bélidor's writings had instructed innumerable practitioners as well as the first two generations of engineers who were also intrinsically scientists: for example, Lazare Carnot, Coulomb, and Meusnier, followed by Coriolis, Navier, and Poncelet, all of whom, under the designation "science of machines," inaugurated engineering mechanics. Bélidor's influence, therefore, was the reciprocal of what he intended: rather than introducing mathematics into practical construction, he brought the problems of engineering to mechanics.

BIBLIOGRAPHY

I. ORIGINAL WORKS. Bélidor's important writings were *Nouveau cours de mathématique à l'usage de l'artillerie et du génie* (Paris, 1725); *La science des ingénieurs dans la conduite des travaux de fortification et d'architecture civile* (Paris, 1729), also ed. with notes by Louis Navier (Paris, 1813; repub. 1830); *Le bombardier françois, ou nouvelle méthode de jetter les bombes avec précision* (Paris, 1731); and *Architecture hydraulique, ou l'art de conduire, d'élever et de ménager les eaux pour les différens besoins de la vie,* 2 vols. (Paris, 1737–1739), also ed. by Navier (Paris, 1819)—the most successful eighteenth-century edition was that published in 5 vols. (Paris, 1739–1790). Bélidor also published two memoirs in the *Mémoires de l'Académie Royale des Sciences:* "Théorie sur la science des mines propres à la guerre, fondée sur un grand nombre d'expériences" (1756), pp. 1–25; and "Seconde mémoire sur les mines, servant de suite au précédent" (1756), pp. 184–202.

II. SECONDARY LITERATURE. Bélidor's *éloge* by Grandjean de Fouchy is in *Histoire de l'Académie Royale des Sciences, année 1761* (Paris, 1763), pp. 167–181. Further biographical tradition is recalled in the introduction to the Navier edition of *Architecture hydraulique.*

CHARLES C. GILLISPIE

BELL, ALEXANDER GRAHAM (*b.* Edinburgh, Scotland, 3 March 1847; *d.* Baddeck, Nova Scotia, 2 August 1922), *technology.*

Both Bell's grandfather, Alexander, and his father, Alexander Melville, were teachers of elocution; his father was well known as the inventor of Visible Speech (a written code indicating the position and action of throat, tongue, and lips in forming sounds). Bell had a lifelong interest in teaching the deaf to speak, an interest intensified because his mother and his wife were deaf. In 1870, after the second of Bell's two brothers died of tuberculosis, the family moved to Canada. Bell did his early telephone work in Boston and subsequently moved to Washington. He became a citizen of the United States in 1882.

Bell achieved fame as inventor of the telephone and fortune under a broad interpretation given to the patent granted him 10 March 1876. His early experimental work was spurred on by a persistent belief in its ultimate commercial value, an enthusiasm unshared by his predecessor Philip Reis and his contemporary Elisha Gray. Although the telephone is not properly called a scientific invention (Bell's knowledge of electricity at the time was extremely limited), a fair proportion of the wealth he received from it was used by Bell to pursue scientific researches of his own and to support those of others.

His interest in the deaf led Bell to publish several articles on hereditary deafness. This in turn led to studies on longevity and a long-term series of experiments in which he attempted to develop a breed of sheep with more than the usual two nipples. In 1909, after twenty years of selection, he had a flock consisting solely of six-nippled sheep. He found, as he had suspected, that twin production increased with the number of nipples. Bell made a number of suggestions on the medical use of electricity but performed few experiments himself. His approach to these areas was as an amateur, although one with an active, inquiring mind.

Bell's financial support of science took several

forms. In 1880 he used the 50,000 francs of the Volta Prize to establish the Volta Laboratory Association (later the Volta Bureau), largely devoted to work for the deaf, in Washington. In 1882 he conceived the idea of the journal *Science,* which began publication in 1883. In the first eight years of its existence, Bell and his father-in-law, G. G. Hubbard, subsidized this journal to the amount of about $100,000. To allay S. P. Langley's concern that his post as secretary of the Smithsonian Institution would be merely administrative, Bell and J. H. Kidder each gave $5,000 for Langley's personal research; this money was used in the establishment of the Smithsonian's astrophysical observatory. In 1891 Bell gave $5,000 to support Langley's flight experiments. He himself experimented with kites, and in 1907 he organized the Aerial Experimental Association, which lasted for a year and a half and was financed by his wife. Bell also helped to organize and finance the National Geographic Society, serving as its president from 1898 to 1903.

Bell was elected to membership in the National Academy of Sciences in 1883 and was appointed a regent of the Smithsonian Institution in 1898.

BIBLIOGRAPHY

I. ORIGINAL WORKS. A complete list of Bell's publications is given in Osborne's article (see below). His notebooks, letters, and other documentary material are nicely housed by the Bell family at the National Geographic Society; some of these have been reproduced on microfilm and are available at the Library of Congress and the Bell Telephone Company of Canada, Montreal. Bell's court testimony dealing with the telephone appears in *The Bell Telephone* (Boston, 1908). Most of the surviving pieces of apparatus are preserved at the Smithsonian Institution.

II. SECONDARY LITERATURE. No satisfactory biography of Bell exists. Basic details can be found in W. C. Langdon's article on Bell, in *Dictionary of American Biography,* II, 148–152; C. D. Mackenzie, *Alexander Graham Bell, the Man Who Contracted Space* (New York, 1928); and H. S. Osborne, "Alexander Graham Bell," in *National Academy of Sciences, Biographical Memoirs,* **23** (1945), 1–30. Part of the experimental telephone work is analyzed in B. S. Finn, "Alexander Graham Bell's Experiments With the Variable-Resistance Transmitter," in *Smithsonian Journal of History,* **1,** no. 4 (1966), 1–16.

BERNARD S. FINN

BELL, CHARLES (*b.* Edinburgh, Scotland, November 1774; *d.* Hallow, Worcestershire, England, 28 April 1842), *anatomy.*

Bell introduced new methods of determining the functional anatomy of the nervous system. For the spinal and cranial nerves he correlated anatomical division with functional differentiation by cutting or stimulating the anatomical divisions and observing the changes produced in the experimental animals' behavior. Bell's techniques and observations led to Johannes Müller's generalizations on the sensory functions of the nervous system.

Bell was the son of a minister of the Church of England. His father died when he was five, and he received his basic education from his mother. He was also tutored in art and attended Edinburgh High School for three years. Bell's older brother John was a surgeon who gave private classes in anatomy. Charles assisted him in his classes, learning medicine from him and from lectures at Edinburgh University. He was admitted to the Royal College of Surgeons in 1799. The success of John Bell's anatomy classes aroused the jealousy of the medical faculty of the university, who succeeded in barring him and Charles from practice in the Royal Hospital of Edinburgh.

Since his career in Edinburgh was blocked, in 1804 Bell moved to London, where he opened his own school of anatomy and gradually built up a surgical practice. He combined his skill in painting with his scientific interests in *Essays on the Anatomy of Expression in Painting* (1806). Besides being an exposition of the anatomical and physiological basis of facial expression for artists, the book included much philosophy and critical history of art. The book gained Bell some reputation and remained popular, going through several editions up to 1893. He was co-owner of and principal lecturer at the Great Windmill Street School of Anatomy, founded by William Hunter, from 1812 to 1825 and was instrumental in the founding of the Middlesex Hospital Medical School in 1828. He returned to Edinburgh University as professor of surgery in 1836. Bell was knighted in 1831, in recognition of his scientific achievement. Further recognition came when he was selected to write the fourth Bridgewater Treatise, in which series he published *The Hand* in 1833.

Bell developed his experimental techniques involving the peripheral nerves in order to discover how the brain functions. In 1811 he published *Idea of a New Anatomy of the Brain,* a book giving his views on the brain. He circulated one hundred copies to his acquaintances, then published nothing more on the subject for ten years. Bell's first concern in *Idea* was to establish that the different parts of the brain serve different functions, rather than that the entire organ was involved in all functions. His statement that the peripheral nerves are composed of divisions "united for convenience of distribution" but "distinct in office" was a concomitant of this view of the brain.

Each division of a peripheral nerve received its functional specificity from the part of the brain with which it was connected. This was the crucial element in what were to be Müller's laws of sensation, and in *Idea* Bell incidentally stated the central law, that of specific nerve energies. It was Bell's techniques, however, not his generalizations, that influenced Müller.

Idea included a description of an experiment that demonstrated the differing functions of each root of a spinal nerve. Bell cut the posterior roots and observed no convulsions of the muscles of the back; touching the anterior roots convulsed them. Bell did not deduce the Bell-Magendie law—that the anterior roots are motor, the posterior sensory—from the experiment. Rather, it supported his opinion that the cerebellum, which he thought was the origin of the posterior root filaments, was the locus of the involuntary nervous functions. The cerebrum, the origin of the anterior root filaments, was the locus of the voluntary nervous functions. Bell reasoned that filaments of involuntary nerves did not elicit convulsions because there was no conscious sensation of pain.

Magendie did his own experimental work, formulating and publishing the Bell-Magendie law (1822) after hearing of Bell's work from John Shaw, Bell's assistant at the Great Windmill School. The law was a special case of the general principle of nervous function that Bell had worked out, but it was the special case that was noted and became the subject of a bitter priority dispute between Bell and Magendie.

Bell's later experimental studies, which he correlated with clinical observations, involved the functions of the cranial nerves.

BIBLIOGRAPHY

I. ORIGINAL WORKS. Bell's major works are *Essays on the Anatomy of Expression in Painting* (London, 1806); *Idea of a New Anatomy of the Brain* (London, 1811), a rare work that is reprinted in Gordon-Taylor and Walls; *An Exposition of the Natural System of Nerves of the Human Body* (London, 1824), which includes revisions of papers first published in *Philosophical Transactions of the Royal Society; The Nervous System of the Human Body* (London, 1830); and *The Hand, Its Mechanism and Vital Endowments, as Evincing Design* (London, 1833).

II. SECONDARY LITERATURE. Gordon Gordon-Taylor and E. W. Walls, *Sir Charles Bell, His Life and Times* (London, 1958), the important work on Bell, is an accurate but discursive biography and includes a bibliography of Bell's writings and the literature on him. J. M. D. Olmsted, *François Magendie* (New York, 1944), pp. 93–122, details the Bell-Magendie priority dispute.

PETER AMACHER

BELL, ERIC TEMPLE (*b*. Aberdeen, Scotland, 7 February 1883; *d*. Watsonville, California, 21 December 1960), *mathematics.*

The younger son of James Bell, of a London commercial family, and Helen Lyndsay Lyall, whose family were classical scholars, he was tutored before entering the Bedford Modern School, where a remarkable teacher, E. M. Langley, inspired his lifelong interest in elliptic functions and number theory. Bell migrated to the United States in 1902 "to escape being shoved into Woolwich or the India Civil Service" (as he later explained) and was able to "cover all the mathematics offered" at Stanford and graduate Phi Beta Kappa in two years. A single year at the University of Washington netted an M.A. in 1908; another at Columbia sufficed for the Ph.D. in 1912. The years between he spent as a ranch hand, mule skinner, surveyor, school teacher, and partner in an unsuccessful telephone company. In 1910 he married Jessie L. Brown, who died in 1940. They had one son, Taine Temple Bell, who became a physician in Watsonville. Bell produced about 250 mathematical research papers, four learned books, eleven popularizations, and, as "John Taine," seventeen science fiction novels, many short stories, and some poetry. He was active in organizations of research mathematicians, teachers, and authors. In religion and politics he was an individualist and uncompromising iconoclast. He remained active in retirement and was writing his last book in the hospital when overtaken by a fatal heart attack.

At the University of Washington from 1912, Bell published a number of significant contributions on numerical functions, analytic number theory, multiply periodic functions, and Diophantine analysis. His "Arithmetical Paraphrases" (1921) won a Bôcher Prize. Other honors (e.g., the presidency of the Mathematical Association of America [1931–1933]), editorial duties, and invitations multiplied, but they did not reduce his output. After lecturing at Chicago and Harvard, he went, in 1926, to the California Institute of Technology, where he remained (emeritus after 1953) until hospitalized a year before his death. Bell will be longest known for his *Men of Mathematics* and other widely read books "on the less inhuman aspects of mathematics," and for *The Development of Mathematics,* whose insights and provocative style continue to influence and intrigue professional mathematicians—in spite of their historical inaccuracies and sometimes fanciful interpretations.

BIBLIOGRAPHY

I. ORIGINAL WORKS. Typical are his first publication, "An Arithmetical Theory of Certain Numerical Functions,"

University of Washington Publication, no. 1 (1915); "Arithmetical Paraphrases, Part I," in *Transactions of the American Mathematical Society,* **22,** no. 1 (Jan. 1921), 1–30, and no. 3 (Oct. 1921), 273–275, which won a Bôcher Prize; *Algebraic Arithmetic,* American Mathematical Society Colloquium Publication, no. 7 (1927), which was based on his invited lectures at the Eleventh Colloquium of the American Mathematical Society in 1927; *Before the Dawn* (Baltimore, Md., 1934), which was his favorite science fiction novel, the only one published under his own name and inspired, he said, by boyhood views of models of dinosaurs in Croydon Park near London; *Men of Mathematics* (New York, 1937), awarded the gold medal of the Commonwealth Club of California; *The Development of Mathematics* (New York, 1940; 2nd ed., 1945); *Mathematics, Queen and Servant of Science* (New York, 1951), his most ambitious popularization based on two previous books, *Queen of the Sciences* and *The Handmaiden of the Sciences;* and *The Last Problem* (New York, 1961), a study of Fermat's conjecture, which was unfinished at the time of Bell's death.

II. SECONDARY LITERATURE. There is no detailed biography. Only the following give more information than appears in *American Men of Science, Who's Who in America,* and *Who Was Who:* an autobiography in *Twentieth Century Authors,* supp. 1 (New York, 1955), 70–71, from which we have taken the quotations in the article; T. A. A. Broadbent, obituary in *Nature,* no. 4763 (11 Feb. 1961), 443; and a news release from the California Institute of Technology News Bureau (21 Dec. 1960).

KENNETH O. MAY

BELL, ROBERT (*b.* Toronto, Ontario, 3 June 1841; *d.* Rathwell, Manitoba, 19 June 1917), *geology.*

The son of Rev. Andrew Bell, at fifteen Robert was appointed junior assistant on a Canadian Geological Survey party on the Gaspé Peninsula and continued to work for the Survey during the summers of his high school and college years. When he received a civil engineering degree from McGill University in 1861, he also won the Governor General's Gold Medal. Following a year at Edinburgh, Bell was appointed professor of chemistry and natural history at Queen's University in 1863. He resigned four years later, in order to devote full time to the Survey, and served as its chief geologist and acting director from 1901 until his retirement in 1906. By adroit use of his spare time he gained the M.D.,C.M. degree in 1878. His knowledge of medicine and surgery proved useful in his travels and resulted in his being appointed medical officer on government ships visiting Hudson Bay.

Among Bell's honors were honorary D.Sc. degrees from Queen's and Cambridge universities; fellowship of the Geological Society of London (1862), the Royal Society of Canada (1882), the Geological Society of America (1889), the Royal Society of London (1897)

and the Royal Astronomical Society (Canada); the Imperial Science Order (1903); the King's Gold Medal, from the Royal Geographic Society (1906); and the Cullum Gold Medal, from the American Geographic Society (1906).

After preliminary surveying projects on the Gaspé Peninsula and the north shores of the Great Lakes, in 1870 Bell commenced the work for which he is justly famous—thirty continuous years of exploration of the territory from Lake Superior northward to the Arctic and from Saskatchewan to the east shore of Hudson Bay—work that today would be classified as reconnaissance mapping. He traveled mainly by canoe, and his only instruments were the compass, sextant, boat log, and Rochon micrometer. His most important contribution was the mapping of both shores of Hudson Bay and parts of the Nottaway, Churchill, and Nelson rivers, which flow into it, as a result of which he became a strong advocate of the Hudson Bay route to the Atlantic. Bell also explored other parts of Canada: he was the first to survey Great Slave, Nipigon, and Amadjuak lakes, the latter in the interior of Baffin Island. He produced more than thirty reports on the geology of the areas he surveyed, the balance of his more than 200 titles dealing with the geography, zoology and botany, resources, and Indian lore along his routes.

No man has accomplished so much pioneer exploration in territory where previously only Indians had been. He used Indian names to a very great extent, and he has been called the "place-name father of Canada."

BIBLIOGRAPHY

A complete list of Bell's writings on North American geology is in *Bulletin of the United States Geological Survey,* no. 746 (1923), 89–91.

Articles on Bell are F. J. Alcock, "Bell and Exploration," in *A Century in the History of the Geological Survey of Canada,* National Museum of Canada, Special Contribution no. 47.1 (Ottawa, 1947), 48–54; H. M. Ami, "Memorial of Robert Bell," in *Bulletin of the Geological Society of America,* **38** (1927), 18–34; Charles Hallock, "One of Canada's Explorers," in *Forest and Stream,* **53,** no. 41 (1901), 9–15; and "Robert Bell," in *Proceedings and Transactions of the Royal Society of Canada* (1918), x–xiv.

T. H. CLARK

BELLANI, ANGELO (*b.* Monza, Italy, 31 October 1776; *d.* Milan, Italy, 28 August 1852), *physics, chemistry.*

Bellani received his education in religious institutions, as was common. He studied rhetoric at the school of the Merate dei Somaschi; philosophy at the school of Monza, run by former Jesuits; and theology at the seminary of Milan, where he was ordained a priest. He was appointed honorary canon of the basilica of Monza, and was a member of the Istituto Lombardo Accademia di Scienze e Lettere and of the Società Italiana delle Scienze di Verona.

Bellani began his scientific activity with the construction and study of thermometers. He designed the "thermometergraph." He made an instrument with a double scale that served as an inspiration to the English physicist Six and was to prove helpful in the automatic registering of maximum and minimum temperatures. In 1808 Bellani demonstrated how "zero" on the thermometer's scale is subject to variations caused by deformations of the glass over the years. In that year he also established a factory in Milan for the manufacture of thermometers, a business that flourished from the outset.

Bellani was also interested in chemistry, especially as it related to the construction of physical instruments. In fact, in his research on combustion, on the temperature of liquefaction and of ebullition of phosphorus, his idea was to construct a phosphorus eudiometer that would not present the limitations of the one invented by the Piedmontese chemist Giovanni Antonio Giobert and used by Lazzaro Spallanzani in his biological experiments; it would permit an exact quantitative determination of the gas produced. In 1824 Bellani also distinguished the catalytic function of a platinum sponge saturated with hydrogen. After 1815 he widened the scope of his scientific studies to include meteorology and agriculture, and during the later years of his life he turned his attention to natural history.

In meteorology Bellani continued his interest in instrumentation. He made a hygrometer from a fish bladder (1836) and perfected Landriani's *atmidomètre*. In 1834 he defined his theory on the formation of hail, resuming the criticism, which dated back to 1817, of the ideas that Alessandro Volta had expressed on this subject. Basing his arguments on laboratory data and on direct meteorological observations, he demonstrated that Volta's theory was founded on the false and contradictory principle of the extreme cold produced by the warming action of the sun.

BIBLIOGRAPHY

I. ORIGINAL WORKS. Most of Bellani's works were published in scientific journals and reviews. In the *Nuova scelta d'opuscoli interessanti sulle scienze e sulle arti* are "Descrizione di un termometro ad indice" (1804); "Osser-

vazioni sull'uso dell'aerometro di Farenheit e di Baumé" (1804); "Sulla divisione decimale del pesa-licori" (1804); and "Osservazioni d'esperienze fatte colla pila di Volta sulla produzione dell'acido muriatico ossigenato" (1806).

In the *Giornale di fisica, chimica e storia naturale* are "Sopra la spiegazione di un fenomeno idrostatico dato dal Sign. Robinet" (1808); "Esame dell'ebollizione dei liquidi" (1809-1810); "Dell'ascensione del mercurio ne' tubi capillari" (1810); "Sopra un nuovo termografo" (1811); "Sulla deliquescenza de' corpi" (1812); "Sull'antichità dei telegrafi" (1812); "Sul fosforo come mezzo eudiometrico" (1813-1814); "Se la fosforescenza di alcune piante sia analoga alla lenta combustione del fosforo" (1814); "Riflessione intorno all'evaporizzazione" (1816-1817); "Nuova ipotesi sulla coda di comete" (1820); "Descrizione di un nuovo atmidometro" (1820); "Storia di un aerolito" (1822); "Sull'origine e la natura delle stelle cadenti" (1822); "Dell'incertezza nel determinare il punto del ghiaccio" (1822); "Di alcune proprietà del mercurio e del vetro" (1823); "Sulla proprietà che posseggono alcune sostanze di facilitare la combinazione del gas idrogeno con l'ossigenio" (1824); "Sopra la cristallizzazione e congelazione" (1827); "Sul termobarometro" (1827); and "Sulla salsedine dell'acqua del mare" (1823).

In the *Memorie dell'Istituto lombardo* are "Sopra la teoria della combustione del fosforo" (1813); "Esposizione di alcune esperienze relative alla dilatazione dei corpi" (1814); "Sul freddo prodotto dall'evaporazione" (1815); and "Intorno alla corona ferrea esistente nella basilica di Monza" (1818).

In the *Giornale e biblioteca dell'Istituto lombardo* are "Sull'arte di filare il vetro" (1840); "Sopra diversi argomenti fisico-chimici" (1842); "Sopra una supposta causa principale dell'utilità degli avvicendamenti agrari" (1842); "Sulle funzioni delle radici nei vegetali" (1843); "Sopra un fenomeno di sospesa evaporazione" (1844); "Sulla filatura dei bozzoli a freddo" (1844); and "Osservazioni di fisiologia vegetale" (1846-1847).

In the *Giornale dell'Istituto lombardo* are "Conseguenze della vigente teoria dei vapori applicabile alla ispirazione di quelli sviluppatisi dall'etere solforico" (1847); and "Del gelso in agricoltura" (1850).

In the *Memorie di matematica e fisica della Società italiana delle scienze* are "Sullo spostamento del mercurio osservato al punto del ghiaccio nella scala dei termometri" (1841); and "Della malaria vicino ai fontanini d'irrigazione" (1844).

His books include *Tentativi per determinare l'aumento di volume che acquista l'acqua prima e dopo la congelazione* (Pavia, 1808); *Riflessioni intorno ad alcune recenti opinioni riguardanti il fosforo* (Pavia, 1814); *Delle rotazioni agrarie* (Pesaro, 1833); *Sulla grandine* (Milan, 1834); *Della indefinibile durabilità della vita delle bestie* (Milan, 1836); *Revisione di alcuni supposti assioma fisiologici intorno l'assorbimento o la evaporazione delle foglie nelle piante* (Milan, 1837); and *Osservazioni sulla bacologia*, 3 vols. (Milan, 1852).

II. SECONDARY LITERATURE. The only works that deal in depth with Bellani's work are the obituary by G. Veladini, in *Giornale dell'Istituto lombardo,* **9** (1856), 485–489; and G. Cantoni, "Sopra due strumenti metereologici ideati da Angelo Bellani," in *Rendiconti dell'Istituto lombardo,* **10** (1877), 17–23; **11** (1878), 873–880.

The transactions of the congresses of Italian scientists, in which Bellani participated actively, contain extensive extracts of his remarks.

In agriculture he collaborated on the *Giornale agrario lombardo-veneto,* contributing articles on the cultivation of the silkworm, wood, malaria, and the function of the roots of plants.

GIORGIO PEDROCCO

BELLARDI, LUIGI (*b.* Genoa, Italy, 18 May 1818; *d.* Turin, Italy, 17 September 1889), *paleontology, entomology.*

Although a native of Genoa, Bellardi spent most of his life in Turin, where, following the wishes of his family, he studied law. Since his early youth, however, the natural sciences had attracted him; and in his leisure time he collected the Cenozoic Mollusca abundant in the hills around Turin, Superga, Asti, and Tortona. At the age of twenty he published his first paper on the gastropod genus *Borsonia,* and from that time on, his major scientific activity concentrated on the Cenozoic Mollusca of the Piedmont and of Liguria. He also visited the Middle East, particularly Egypt, bringing back extensive collections for comparative study.

Between 1854 and 1874 a variety of circumstances prevented Bellardi from dedicating himself entirely to paleontology, and he therefore undertook entomological research, mostly on the diptera of the Piedmont. He also took some interest in botany and agriculture, being the first in Italy to discuss the phylloxera and its relationship to viticulture.

Upon the introduction of evolutionary ideas into paleontology, Bellardi immediately understood their fundamental importance, and his last works show the relationships between the different forms of Mollusca, and their probable filiation through geological time. During the last twenty years of his life he returned entirely to paleontology, but never completed his extensive and important *I molluschi,* which is characterized by perfectionism in presentation and content. With a similar attitude, he taught natural history for thirty years at the Liceo Gioberti and was curator of the paleontological collection of the Royal Geological Museum of Turin, to which he made many contributions. His desire to increase interest in natural history led him to write several elementary textbooks, also characterized by clarity of expression and precision of data.

Bellardi was elected an honorary member of many academies and scientific societies, and King Victor Emmanuel requested him to teach the natural sciences to his sons, an assignment that Bellardi particularly enjoyed.

BIBLIOGRAPHY

Among Bellardi's works are "Saggio orittografico sulla classe dei gasteropodi fossili dei terreni Terziarii del Piemonte," in *Memorie della Reggia Accademia delle scienze di Torino,* 2nd ser., **3** (1841), 93–174, written with G. Michelotti; "Description des cancellaires fossiles des terrains Tertiaires du Piémont," *ibid.,* 225–264; "Monografia delle pleurotome fossili del Piemonte," *ibid.,* **9** (1848), 531–650; "Monografia delle columbelle fossili del Piemonte," *ibid.,* **10** (1849), 225–247; "Monografia delle mitre fossili del Piemonte," *ibid.,* **11** (1851), 357–390; and *I molluschi dei terreni Terziarii del Piemonte e della Liguria,* 30 vols. (Turin, 1873–1904), I–V by Bellardi and VI–XXX by F. Sacco—I–VIII, XI, and XIII also in *Memorie della Reggia Accademia delle scienze di Torino,* 2nd ser., **27–44** (1873–1894).

ALBERT V. CAROZZI

BELLARMINE, ROBERT (*b.* Montepulciano, Italy, 4 October 1542; *d.* Rome, Italy, 17 September 1621), *theology, philosophy.*

Third of the twelve children of Vincenzo Bellarmino and Cynthia Cervini, half-sister of Pope Marcellus II, Robert joined the newly founded Jesuit order in 1560 and took a master's degree in philosophy at the Spanish-staffed Roman College in 1563. Natural philosophy formed an important part of his studies there, but it appears to have been wholly and routinely Aristotelian in character. Ordained priest in 1570, he completed his theological studies in Louvain.

The struggle between the Catholic and Protestant wings of Christendom had by then attained an extraordinary ferocity all over Europe. One of the major theoretical issues separating the two groups concerned the norms for the proper interpretation of Scripture. Because of his profound scriptural scholarship and his thorough grasp of the major Protestant writers (both of these achievements very rare in the Catholic church of the day), Bellarmine soon became the leading theologian on the Catholic side of the debate. His three-volume *Disputationes de controversiis* was by far the most effective piece of Catholic polemic scholarship of the century. After its appearance in 1588, he was recognized as the leading defender of the papacy; successive popes forced on a man whose natural temperament was at once gentle and gay the uncongenial role of controversialist and apologist.

Made rector of the Roman College in 1592, cardinal in 1599, and archbishop of Capua in 1602, Bellarmine was never far from Rome, and in his last years lived at the Vatican as the pope's major theological adviser.

Bellarmine's relevance to the history of science comes only from his role in the Galileo story. In 1611 he was among the Roman dignitaries whom Galileo invited to see the new-found wonders in the sky. The old man was disturbed at the implications of what he saw, and asked the astronomers of his old college (among them Clavius) to test the accuracy of Galileo's claims. This they soon did. Galileo sent him a copy of his important and effectively anti-Aristotelian work on hydrostatics (1612), to which Bellarmine replied that "the affection you have thus shown me is fully reciprocated on my part; you will see that this is so, if ever I get an opportunity of doing you a service." The opportunity was not long in coming.

The Aristotelian cosmology was crumbling in the face of the new astronomical evidence, notably that of the phases of Venus and the sunspots. The Aristotelians of the universities fell back on the authority of Scripture as a last desperate expedient to save their case. Galileo answered them in his brilliant *Letter to the Grand Duchess Cristina of Lorraine* (1615). Two rather different, and ultimately incompatible, positions were argued by him. On the one hand, he argued with great cogency that the language in which the scriptural writers described physical phenomena could not possibly have been intended to carry any probative weight in questions of natural science. On the other hand, he also appeared to concede the traditional Augustinian maxim: So great is the weight of authority behind the words of revelation that the literal sense ought to be taken as the correct one in every case, except where such an interpretation could be strictly shown, on commonsense or philosophical grounds, to lead to falsity.

In a letter written at this time to Foscarini, a Carmelite who had defended similar views on the nonrelevance of scriptural phrases to problems of physical science, Bellarmine accepted the Augustinian maxim, but went on to emphasize that since the heliocentric theory of Copernicus could in no way be "strictly demonstrated," the troublesome scriptural phrases about the motion of the sun could not be regarded as metaphorical. If, of course, a "strict proof" of heliocentrism were to be found (a contingency he regarded as in the highest degree unlikely but, significantly, did not wholly exclude), the scriptural texts would have to be reexamined. To argue that the celestial appearances are "saved" by supposing the earth to go around the sun does not constitute

a strict proof that this is what really happens. When a vessel recedes from the shore, the illusion that the shore is moving is corrected by *seeing* the ship to be in motion. Likewise, the experience of the wise man "tells him plainly that the earth is standing still."

This is the sort of unshakable trust in the ultimacy of observation that had made Aristotle (who had once seemed so dangerous an intellectual threat to Christian beliefs) a congenial cosmologist for those who regarded the Hebrew turn of phrase about sun or stars as somehow carrying a special authority. In Bellarmine's view, Solomon's phrase about the sun "returning to its place" carried far more weight than did the Copernican theory. The latter was no more than a "hypothesis," whereas Solomon had his wisdom from God.

A year later, a specially appointed tribunal of eleven theologians went much further than Bellarmine had, and advised the Congregation of the Holy Office that the heliostatic view was formally heretical, because it called into question the inspiration of Scripture. No account of the tribunal's deliberations survives, but presumably its arguments were the standard ones summarized in Bellarmine's letter.

The consequences, both for science and for the church, of the ensuing decree (1616) suspending the work of Copernicus "until it be corrected" can scarcely be overestimated. Once this decree was put into effect, the die was cast; and although later incidents (notably Galileo's trial) would come to have a greater symbolic and dramatic significance, it was with the decree of 1616 that the parting of the ways really came. The disastrous potentialities for conflict that were latent in Augustine's theory of scriptural interpretation were now for the first time realized. If the literal sense of Scripture is to be retained unless and until its untenability be *strictly* demonstrated, an impossible burden is laid on theologian and scientist alike. Each will be called on to evaluate the arguments of the other. And the argument of the scientist will not be allowed *any* weight until it is conclusive, when all of a sudden it will be conceded. The notions of evidence and probability underlying this approach (which originally derived from Augustine's theory of divine illumination as a basis for all human knowledge) are ultimately inconsistent.

In his criticisms of this approach, Galileo showed himself a better theologian than Bellarmine and the consultors. He had a far keener appreciation of what language is, and what the conditions for communication are. That his opponents did not accept his arguments, cogent as they seem to us today, was due mainly to the fact that the norms for the proper interpretation of Scripture were one of the two main issues then dividing Protestants and Catholics. Any liberalizing suggestion in this quarter was hardly likely to meet with favor on either side. It was a far cry from the calmer days of Oresme and Cusa, two centuries before, when similar suggestions about the interpretation of Scripture scarcely caused a ripple.

In his *Système du monde,* Duhem suggests that in one respect, at least, Bellarmine had shown himself a better scientist than Galileo by disallowing the possibility of a "strict proof" of the earth's motion, on the grounds that an astronomical theory merely "saves the appearances" without necessarily revealing what "really happens." This claim has often been repeated, most recently by Karl Popper, who makes Bellarmine seem a pioneer of the nineteenth-century positivist theory of science. In point of fact, nothing could be further from the case. Bellarmine by no means denied that strict demonstrations of what is "really" the case could be given in astronomical matters. In his view, however, such demonstrations had to rest on "physical" considerations of the type used by Aristotle, not on the mathematical models of positional astronomy.

This distinction between two epistemologically different types of astronomy was a time-honored one, taking its origin in the medieval debates over the relative merits of the Aristotelian and Ptolemaic astronomies. The former clearly gave a good causal account of *why* the planets moved, but was quite unable to provide any practical aids for the compiling of calendars and the like. On the other hand, it appeared impossible to account for the complex and eccentric epicyclic motions of Ptolemaic astronomy in causal terms, even though they did provide a good descriptive and predictive account of apparent planetary positions. The orthodox Aristotelian reading of this situation, such as one will find in writers like Aquinas, was that *real* motion could be asserted only on the basis of demonstrations of a properly "physical" sort; the models of the mathematical astronomer did not lend themselves to dynamic explanation because their purpose was merely that of computation.

Although the phrase "saving the appearances" was often used in reference to mathematical astronomy, it is important in this context to distinguish between the Aristotelian and Platonic views of what physics in general could accomplish. Aristotle argued that a strict science of physics can be achieved, one that tells us how the world really is. Plato, on the other hand, held that physical inquiry could at best only "save the appearances." Admittedly, such a "saving" provided some sort of insight into the physical world because of the relation of image between it and the

domain of Form, but the insight is a limited and defective one because the imaging is such a flickering and uncertain affair.

Bellarmine's comments in his letter to Foscarini cannot be construed as a protopositivist declaration. He was, indeed, just as much of an "essentialist" in his theory of science (to use Popper's term) as either Aristotle or Galileo. Even if he had been a Platonist, and extended the notion of "saving the appearances" to all of physics and not just to mathematical astronomy, it still would not be correct to take this in the positivist sense favored by Duhem.

To refute the Aristotelian separation between the two types of astronomy, it would be necessary to construct a dynamical substructure for Copernican kinematics, something Galileo could not do. It was not until Newton that the new mathematical astronomy was given an adequate causal interpretation in terms of central forces. In his *Dialogo,* Galileo attempted to meet Bellarmine's challenge to provide a dynamical proof of the earth's motion, but the tidal arguments he used carried little conviction. Galileo's opponents could, therefore, claim that the Copernican theory was still no more than a "hypothesis," in the traditional sense of a fictional account, because it lacked the "physics" (i.e., the dynamics) that, in their view, it would need to transform it into a claim about the nature of the real.

When the decree outlawing Copernicanism was promulgated in 1616, Galileo was in Rome. He was not mentioned in the decree, probably because of the respect in which he was held and the support of his many friends in Rome. But since he was the main protagonist both of Copernicanism and of the views on the interpretation of Scripture that were disapproved, he obviously was the person most affected by it. Wishing to make sure that Galileo appreciated the gravity of the matter, the pope asked Bellarmine to call him in and notify him officially of the contents of the decree before it was made public. If he showed himself unwilling to accept it, he was to be enjoined personally not to support or even discuss Copernicanism in any fashion.

What happened at this famous interview has been the subject of endless controversy in the past century, since the documents of the "Galileo case" have been made public. According to a document introduced in evidence at Galileo's trial nearly twenty years later, the personal injunction apparently *was* given to him, and the prosecution made much of the fact that its existence had not been made known to the censors in charge of licensing the *Dialogo.* Galileo protested that he could recall no such formal injunction, although he remembered the interview with Bellarmine well enough. In addition, he produced an attestation drawn up by Bellarmine before Galileo left Rome in 1616, in which the aged cardinal affirmed that Galileo had not been forced to abjure Copernicanism, as rumor had claimed. Bellarmine's note, whose existence obviously had not been suspected by the prosecution, forced an alteration in the strategy of the trial; in its later stages, the personal injunction was not mentioned.

Had it actually been delivered? The record in the Holy Office file is not signed in the usual form, and Bellarmine's attestation strongly suggests that it could not be valid. Some have claimed that it was forged by enemies of Galileo, either in 1616 or in 1633, with a view to trapping him. Others have suggested that it *was* delivered in 1616, but that there were no legal grounds for doing so. Still others argue that a genuine injunction was given, and that Bellarmine's attestation was the action of a friend protecting Galileo's reputation. We shall never know for sure. And in any event, it makes little difference, since the trial verdict would very likely have been the same whether or not a special injunction had been given to Galileo in 1616. Once the decree of 1616 implied that the heliocentric view is formally heretical, the writing of a book like the *Dialogo* automatically gave grounds for the suspicion of heresy, if the pope or the Holy Office cared to press the charge. This was where Bellarmine and the theologians of 1616 failed. Beset by the polemics of Reformation and Counter-Reformation, they did not grasp the limits of scriptural inspiration that were already becoming evident to the pioneers of the new sciences. One can account for their failure easily enough, but it was to have disastrous consequences for their church and for religion in general.

BIBLIOGRAPHY

The relationship between Bellarmine and Galileo is fully covered in James Brodrick, *Robert Bellarmine* (Westminster, Md., 1961). Brodrick's earlier *Life and Work of Robert Francis Bellarmine* (London, 1928) is quite defective in its treatment of the Galileo case. The most colorful recent account of the Bellarmine–Galileo dispute is Giorgio de Santillana, *The Crime of Galileo* (Chicago, 1955); a quite different reconstruction of the circumstances under which the disputed injunction came to be entered into the file is given by Stillman Drake in the appendix to his trans. of Ludovico Geymonat, *Galileo Galilei* (New York, 1965). Popper's view of Bellarmine as an "instrumentalist," in contrast with Galileo, the "essentialist," is worked out in his "Three Views Concerning Human Knowledge," in H. D. Lewis, ed., *Contemporary British Philosophy* (New York, 1956), pp. 355–388.

ERNAN MCMULLIN

BELLAVITIS, GIUSTO (*b.* Bassano, Vicenza, Italy, 22 November 1803; *d.* Tezze, near Bassano, 6 November 1880), *mathematics.*

Bellavitis was the son of Ernesto Bellavitis, an accountant with the municipal government of Bassano, and Giovanna Navarini; the family belonged to the nobility but was in modest circumstances. He did not pursue regular studies but was tutored under the guidance of his father, who directed his interest toward mathematics. Soon he surpassed his tutor and diligently pursued his studies on his own, occupying himself with the latest mathematical problems.

From 1822 to 1843 Bellavitis worked for the municipal government of Bassano—without pay for the first ten years—and conscientiously discharged his duties, occupying his free time with mathematical studies and research. During this period he published his first major works, including papers (1835, 1837) on the method of equipollencies, which were hailed as one of his major contributions. On 26 September 1840 Bellavitis became a fellow of the Istituto Veneto, and in 1843 he was appointed professor of mathematics and mechanics at the *liceo* of Vicenza. He then married Maria Tavelli, the woman who for fourteen years had comforted and encouraged him in his difficult career.

On 4 January 1845, through a competitive examination, Bellavitis was appointed full professor of descriptive geometry at the University of Padua. On 4 July 1846, the university awarded him an honorary doctorate in philosophy and mathematics. He transferred in 1867 to the professorship of complementary algebra and analytic geometry. On 15 March 1850, Bellavitis became a fellow of the Società Italiana dei Quaranta, and in 1879 a member of the Accademia dei Lincei. In 1866 he was named a senator of the Kingdom of Italy.

Bellavitis' method of equipollencies belongs to a special point of view in mathematical thought: geometric calculus. According to Peano, geometric calculus consists of a system of operations to be carried out on geometric entities; these operations are analogous to those executed on numbers in classical algebra. Such a calculus "enables us to express by means of formulae the results of geometric constructions, to represent geometric propositions by means of equations, and to replace a logical argument with the transformation of equations." This approach had been developed by Leibniz, who intended to go beyond the Cartesian analytic geometry, by performing calculations directly on the geometric elements, rather than on the coordinates (numbers). Moebius' barycentric calculus finds its expression within this context, but Bellavitis made special reference to

Carnot's suggestion of 1803, when he wrote in 1854:

> This method complies with one of Carnot's wishes, i.e., he wanted to find an algorithm that could simultaneously represent both the magnitude and the position of the various components of a figure; with the immediate result of obtaining elegant and simple graphic solutions to geometrical problems ["Sposizione del metodo delle equipollenze," p. 226].

In order to indicate that two segments, AB and DC, are equipollent—i.e., equal, parallel, and pointing in the same direction—Bellavitis used the formula

$$AB \backsimeq DC.$$

Thus we are given a kind of algebra analogous to that of complex numbers with two units; it found its application in various problems of plane geometry and mechanics, and paved the way for W. R. Hamilton's theory of quaternions (1853), through which geometric calculus can be applied to space; it also led to Grassmann's "Ausdehnungslehre" (1844), and finally to the vector theory. With his barycentric calculus, Bellavitis created a calculus more general than Moebius' "baricentrische Calcul."

In 1834, in his formula expressing the area of polygons and the volume of polyhedra as a function of the distances between their vertexes, Bellavitis anticipated results that were later newly discovered by Staudt and published in 1842.

In algebraic geometry, Bellavitis introduced new criteria for the classification of curves, and then completed Newton's findings on plane cubic curves, adding six curves to the seventy-two already known; these six had not been mentioned by Euler and Cramer. He also began the classification of curves of class three. He offered a graphical solution of spherical triangles, based on the transformation—through reciprocal vector radia—of a spherical surface into a plane. This method finds application in the solution of crystallographic problems.

Bellavitis furthered the progress of descriptive geometry with his textbook on the subject. Considering mathematics to be based essentially upon physical facts and proved by sensible experience, Bellavitis looked down on geometry of more than three dimensions and on non-Euclidean geometry. He did, however, like Beltrami's research on the interpretation of Lobachevski's geometry of the pseudosphere, for he felt that this research would help to diminish the prestige of the new geometry, reducing it to geometry of the pseudosphere. He continued to pursue his research on geodetic triangles on such surfaces as the pseudosphere.

In algebra, Bellavitis thoroughly investigated and continued Paolo Ruffini's research on the numerical solution of an algebraic equation of any degree; he also studied the theory of numbers and of congruences. He furnished a geometric base for the theory of complex numbers. Several of Bellavitis' contributions deal with infinitesimal analysis. In this connection we should mention his papers on the Eulerian integrals and on elliptic integrals.

Bellavitis solved various mechanical problems by original methods, among them Hamilton's quaternions. He developed very personal critical observations about the calculus of probabilities and the theory of errors. He also explored physics, especially optics and electrology, and chemistry. As a young man, Bellavitis weighed the problem of a universal scientific language and published a paper on this subject in 1863. He also devoted time to the history of mathematics and, among other things, he vindicated Cataldi by attributing the invention of continuous fractions to him.

BIBLIOGRAPHY

I. ORIGINAL WORKS. Bibliographies of Bellavitis' works are in A. Favaro, in *Zeitschrift für Mathematik und Physik,* Historisch-literarische Abteilung, **26** (1881); and E. Nestore Legnazzi, *Commemorazione del Conte Giusto Bellavitis* (Padua, 1881), pp. 74–88, which gives the titles of 181 published works, together with numerous "Riviste di giornali," in which Bellavitis examines the results of other authors and uses his problem-solving methods, published in the *Atti dell'Istituto veneto* between 1859 and 1880.

Among his works are "Sopra alcune formule e serie infinite relative ai fattoriali ed agli integrali euleriani," in *Annali delle scienze del Regno Lombardo Veneto,* **4** (1834), 10–19; "Teoremi generali per determinare le aree dei poligoni ed i volumi dei poliedri col mezzo delle distanze dei loro vertici," in *Annali Fusinieri* (1834); "Saggio di applicazioni di un nuovo metodo di geometria analitica-calcolo delle equipollenze," *ibid.,* **5** (1835); "Memoria sul metodo delle equipollenze," in *Annali delle scienze del Regno Lombardo Veneto,* **7** (1837), 243–261; **8** (1838), 17–37, 85–121; "Sul movimento di un liquido che discende in modo perfettamente simmetrico rispetto a un asse verticale," in *Atti dell'Istituto veneto,* **3** (1844), 206–210, and *Memorie dell'Istituto veneto,* **2** (1845), 339–360; "Sul più facile modo di trovare le radici reali delle equazioni algebriche e sopra un nuovo metodo per la determinazione delle radici immaginarie," *ibid.,* **3** (1847), 109–220; "Considerazioni sulle nomenclature chimiche, sugli equivalenti chimici e su alcune proprietà, che con questi si collegano," *ibid.,* 221–267; *Lezioni di geometria descrittiva* (Padua, 1851, 1858); "Sulla classificazioni delle curve del terzo ordine," in *Memorie della Società italiana delle scienze (dei Quaranta),* ser. 1, **25** (1851), 1–50; "Classificazione delle curve della terza classe," in *Atti dell'Istituto veneto,* ser. 3, **4** (1853),

234–240; "Teoria delle lenti," in *Annali Tortolini,* **4** (1853), 26–269, and *Atti dell'Istituto veneto,* ser. 4, **2** (1872), 392–406; "Sposizione del metodo delle equipollenze," in *Memorie della Società italiana delle scienze,* ser. 1, **25** (1854), 225–309, trans. into French by C. A. Laisant, in *Nouvelles annales de mathématiques* (1874); "Calcolo dei quaternioni di W. R. Hamilton e sue relazioni col metodo delle equipollenze," in *Memorie della Società italiana delle scienze,* ser. 2, **1** (1858), 126–184, and *Atti dell'Istituto veneto,* ser. 3, **3** (1856–1857), 334–342; "Sulla misura delle azioni elettriche," *ibid.,* **9** (1863), 773–786, 807–818; "Pensieri sopra una lingua universale e sopra alcuni argomenti analoghi," in *Memorie dell'Istituto veneto,* **11** (1863), 33–74; and "Sopra alcuni processi di geometria analitica" in *Rivista periodica della R. Accademia di scienze lettere e arti di Padova,* **30** (1880), p. 73.

II. Secondary Literature. Works on Bellavitis are G. Loria, *Storia delle matematiche,* III (Turin, 1933), 500–501; E. Nestore Legnazzi (see above); G. Peano, *Calcolo geometrico secondo l'Ausdehnungslehre di H. Grassmann* (Turin, 1888), p. v; D. Turazza, "Commemorazione di G. Bellavitis," in *Atti dell'Istituto veneto,* ser. 5., **8** (1881–1882), 395–422; and N. Virgopia, "Bellavitis, Giusto," in *Dizionario biografico degli italiani* (Rome, 1965), VII.

Ettore Carruccio

BELLEVAL, PIERRE RICHER DE (Châlonssur-Marne, France, *ca.* 1564; *d.* Montpellier, France, 17 November 1632), *medicine.*

Nothing is known about Belleval's origins. He came to Montpellier to study medicine on 22 October 1584, but it was in Avignon that he received his physician's degree on 4 June 1587. He practiced medicine in Avignon and then in Pézenas, where he met Henri de Montmorency, the governor of Languedoc, who became his lifelong protector.

When Henri IV created a chair of anatomy and botanical studies at the medical college of Montpellier, Belleval was named to it in December 1593. On 10 July 1595 he received his doctorate from Montpellier.

Also in December 1593 the first botanical garden in France was established at Montpellier, and Belleval devoted all of his time and money to it. The garden comprised the King's Garden (medicinal plants), the Queen's Garden (mountain plants from many foreign countries), and the King's Square (plants of purely botanical interest). During the Wars of Religion all these gardens were destroyed; Belleval set to work once again, but the King's Garden was never restored.

Belleval was planning to publish a general herbarium of the Languedoc when he died. We possess a great number of the plates that were to illustrate

it; the work was to utilize a binary nomenclature in Latin or Greek.

BIBLIOGRAPHY

Works on Belleval are P. J. Amoreux, *Notice sur la vie et les ouvrages de Pierre Richer de Belleval, pour servir à l'histoire de cette faculté et à celle de la botanique* (Avignon, 1786); J. A. Dorthès, *Éloge historique de Pierre Richer de Belleval, instituteur du Jardin royal de botanique de Montpellier sous Henri IV* (Montpellier, 1854); L. Guiraud, *Le premier jardin des plantes de Montpellier. Essai historique et descriptif* (Montpellier, 1854); and J. E. Planchon, *Pierre Richer de Belleval, fondateur du Jardin des plantes de Montpellier et Appendice contenant les pièces justificatives* (Montpellier, 1869).

Louis Dulieu

BELLINGSHAUSEN, FABIAN VON. See **Bellinsgauzen, Faddei.**

BELLINI, LORENZO (*b.* Florence, Italy, 3 September 1643; *d.* Florence, 8 January 1704), *physiology, medical theory.*

Bellini, born to a family of small businessmen, benefited from the patronage of Duke Ferdinand II of Tuscany from his early youth. Under the duke's protection, he attended the University of Pisa, where he studied philosophy and mathematics with some of Italy's leading scientists, most notably Giovanni Alfonso Borelli. At the precocious age of twenty, Bellini was appointed professor of theoretical medicine at Pisa, and with Duke Ferdinand's assistance he acquired a chair of anatomy five years later. He resigned the chair when he was about fifty, to become first physician to Duke Cosimo III of Tuscany.

Considered a founder of Italian iatromechanism, Bellini was a pioneer in applying mechanical philosophy to the explanation of the functions of the human body. His earliest interests were anatomy and physiology. The first essay he published, *Exercitatio anatomica de usu renum* (1662), was an important study of the structure and function of the kidneys. Until Bellini's time, the kidneys were widely understood in terms laid down by Galen. Said to be composed of a dense, undifferentiated "parenchymatous" material, they were thought to form urine through the exercise of a special "faculty" that controlled the selective secretion of urine from the mass of the incurrent blood. But under the influence of the mechanical philosophy he learned from Borelli, Bellini rejected this Galenic theory as absurd. He sought a less verbal and more mechanical explanation, and when he reexamined renal anatomy in this frame of mind, he discovered in the supposedly unorganized parenchyma a complicated structure composed of fibers, open spaces, and densely packed tubules opening into the pelvis of the kidney. This unsuspected structure provided the key to true renal function.

Bellini posited that arterial blood brought into the kidneys is separated into urinous and sanguineous portions by a sievelike arrangement in the renal vessels and fibers; the urinous portion collected in the urinary tubules and the blood returned to the circulatory stream through the venules. He meant this mechanical explanation of urinary secretion, achieved by the "configuration alone" of the renal vessels, as a general substitute for Galen's incomprehensible "faculties."

Bellini's mechanical manner of physiological explanation won the admiration of contemporary scientists in Italy and abroad, and his anatomical discoveries were considered worthy enough in themselves to merit careful scrutiny by, among others, Marcello Malpighi. Nevertheless, after his initial triumph Bellini's direct interest in anatomy and physiology slackened. In 1665 he published an essay on taste, *Gustus organum,* which included sections on the functional anatomy of the tongue. The intention of this work, however, was mainly to account for the general phenomena of gustatory sensation according to corpuscular principles. For years afterward, Bellini published no other serious anatomical work, although he did maintain active friendships with Francesco Redi and, after 1676, with Malpighi. He kept abreast of scientific developments through his contacts and correspondence, and in 1683 he returned to print with *De urinis et pulsibus et missione sanguinis,* the first important attempt by an Italian systematically to apply the mechanical philosophy to medical theory.

In *De urinis,* Bellini seems to have been inspired by two earlier authors: the English physician and corpuscular philosopher Thomas Willis, whose medical works were published on the Continent in the early 1680's, and Bellini's old mentor Borelli, whose *De motu animalium* appeared posthumously in 1680. Both Willis and Borelli had tried, to some extent, to apply the mechanical philosophy to medical theory. Willis in particular had attempted to formulate corpuscular hypotheses, both to account for disease phenomena and to justify such traditional therapeutic methods as bleeding, purging, and vomiting. Bellini sympathized with Willis' intentions but felt that he had introduced too many unnecessary and *ad hoc* assumptions. Willis supposed, for example, that a "fermentation" of the blood's constituent particles was required to explain such diseases as fevers. By contrast, Bellini, under

Borelli's influence, preferred to think of the blood as a physical fluid with such simple mechanical and mathematicizable attributes as density, viscosity, and momentum. He postulated that health consisted in a well-ordered circulation and that disease implied some sort of circulatory imbalance or inefficiency. Thus, to arrive at the proper theory for a particular disease, one needed merely to set out the pathological phenomena and then to deduce them as consequences of an increased or diminished velocity of the blood. In this manner, Bellini thought, one could get beyond Willis' tentative hypotheses to the real and "necessary" mechanical causes of fevers. He therefore substituted a hydraulic iatromechanism for a corpuscular one.

First reactions to Bellini's hydraulic iatromechanism were uncertain. A few scattered admirers, such as Bohn in Germany and Baglivi in Italy, communicated their enthusiasm, but otherwise the *De urinis* seems to have made little real impact. Bellini turned seriously to medical practice, and in his letters he consoled Malpighi with the news that he too had many enemies and critics. But in 1692 Bellini's affairs took a dramatic turn for the better. The Scots mathematician and physician Archibald Pitcairne, then lecturer on theoretical medicine at Leiden, wrote to express his admiration. Pitcairne was then trying to construct a mathematical and "Newtonian" medical theory, and he saw in Bellini's work an example of the "methods of the Geometers" applied to medical problems. Pitcairne praised Bellini's general theory of disease and urged him to develop further some of the ideas contained only in compressed form in the *De urinis.*

Bellini reacted favorably to Pitcairne's encouragement; and he devoted the next few years to his last set of essays, which he dedicated to Pitcairne and published in 1695 as *Opuscula aliquot.* The *Opuscula* developed Bellini's earlier iatromechanical themes most fully. Organized into postulates, theorems, and corollaries, it treats problems ranging from the hydraulics of intrauterine and extrauterine circulation to the mechanics of the "contractile villi." For example, he considered the blood to be a fluid flowing through conical vessels, according to Borelli's laws of hydraulics, and he imagined that the "contractile villi" (which explain the body's reaction to stimuli) expand, contract, and vibrate like ordinary musical strings. In these carefully stated propositions Bellini hoped to explain all important physiological phenomena according to the laws of mechanics.

By the time the *Opuscula* was published, Bellini had already begun to enjoy an international reputation, largely through the influence of Pitcairne. During the early decades of the eighteenth century that reputation spread. Hermann Boerhaave, the most influential medical teacher of the time, studied with Pitcairne at Leiden and absorbed his enthusiasm for Bellini's medical science. In England, too, Bellini's theories enjoyed a considerable vogue from 1710 to 1730, when such physicians as George Cheyne and Richard Mead tried to build a "Newtonian" theory of the "animal oeconomy" and turned appreciatively to Bellini's writings. His influence was somewhat tempered in Italy by the Hippocratic revival encouraged by Giorgio Baglivi, who had once been Bellini's ardent admirer. Nevertheless, Bellini remained a major scientific authority until the mid-eighteenth century, when a general reaction to iatromechanism, in all its forms, set in.

BIBLIOGRAPHY

I. ORIGINAL WORKS. Bellini's total scientific production was collected in a handsome posthumous edition, 2 vols. (Venice, 1732). Correspondence between Bellini and Malpighi has been edited by Gaetano Atti as *Notizie edite ed inedite della vita e della opere di Marcello Malpighi e di Lorenzo Bellini* (Bologna, 1847).

His individual works include *Exercitatio anatomica de usu renum* (Florence, 1662); *De urinis et pulsibus et missione sanguinis* (Bologna, 1683); and *Opuscula aliquot* (Leiden, 1695).

II. SECONDARY LITERATURE. There is no full-length study of Bellini. A useful short biography can be found in *Dizionario biografico degli italiani,* VII (Rome, 1965), 713–716. Augusto Botto-Micco's account in *Rivista di storia delle scienze mediche e naturale,* **12** (1930), 38–49, is also helpful. Charles Daremberg presents a summary of some of Bellini's theories, along with a general account of the medical environment of the late seventeenth and early eighteenth centuries, in *Histoire des sciences médicales,* II (Paris, 1870), esp. 765–783. Useful information on Bellini's career, particularly on his connections with Malpighi, can be found throughout Howard Adelmann's mammoth *Marcello Malpighi and the Evolution of Embryology,* 5 vols. (Ithaca, N.Y., 1966).

THEODORE M. BROWN

BELLINSGAUZEN, FADDEI F. (*b.* Arensburg, on the island of Oesel, Russia [now Kingissepp, Sarema, Estonian Soviet Socialist Republic], 30 August 1779; *d.* Kronstadt, Russia, 25 January 1852), *navigation, oceanography.*

Bellinsgauzen began his naval career in 1789 as a cadet in Kronstadt. From 1803 to 1806 he participated in the first round-the-world voyage by a Russian ship, the *Nadezhda,* under the command of Ivan F. Kruzenstern. Following this trip, and until July 1819, Bellinsgauzen commanded various ships on the Baltic and Black seas. From 1819 to 1821 he headed a Russian Antarctic expedition, then commanded various naval units in the Mediterranean, Black, and Baltic seas. From 1839 until the end of his life Bellinsgauzen was the military governor of Kronstadt. He married in 1826 and had four daughters. Bellinsgauzen was also one of the founders of the Russian Geographic Society in 1845.

Bellinsgauzen's fame is a result of the discoveries made on the expedition he led to the Antarctic. When the Russian government decided to send an expedition to study the Antarctic, he was designated commander because of his solid knowledge of physics, astronomy, hydrography, and cartography, and because of his wide naval experience. He also was responsible for practically all the maps of the Kruzenstern expedition. During his Black Sea service he had determined the coordinates of the main points of the eastern shore and had corrected several inaccuracies on the map of the shoreline.

Two three-masted vessels, the *Vostok* and the *Mirny,* were outfitted for the trip to the Antarctic. Bellinsgauzen commanded the *Vostok;* the commander of the *Mirny* was Mikhail P. Lazarev. The expedition was to survey South Georgia Island, sail to the east of the Sandwich Islands, go as close to the pole as possible, and not turn back unless it met impassable obstacles.

The *Vostok* and the *Mirny* left Kronstadt on 15 July 1819. In December the expedition sighted the south shore of South Georgia Island, and at the beginning of January 1820 it came upon the Traverse Islands. Reaching the Sandwich Islands seven days later, the expedition found not a single island, as Cook had proposed, but several islands. Naming them the South Sandwich Islands, the expedition headed south and, in spite of ice barriers, on 27 January 1820, near the coast of Princess Martha's Land at a latitude of 69°25′ (according to Bellinsgauzen's report of 20 April 1820 to the Russian minister) came within twenty miles of the Antarctic mainland, the shoreline of which was plotted. Bellinsgauzen wrote on this day that ice, which appeared "like white clouds in the snow that was then falling," was sighted. This was the first view of the earth's sixth continent, and on the navigational map he noted, "Sighted solid ice." Lazarev, who arrived at approximately the same place several hours later, wrote:

> . . . we reached latitude 69°23′, there to meet continuous ice of an immense height; watched from the crosstrees on that beautiful evening . . . ice ranged as far as the eye could see; but we could not enjoy that wonderful sight for very long because the sky soon went dull again and it started to snow as usual. From this place we

proceeded to the east, venturing at every opportunity to the south, but meeting each time an ice-covered continent before we could reach 70° [*Dokumenty*, I, 150].

On 1 February, Bellinsgauzen's expedition was about thirty miles from the icy continent, at latitude 69°25′ south and longitude 1°11′ west. The third leaf of the navigational map shows the ice, which the expedition again approached on 17–18 February. In a report from Port Jackson [Sydney], Australia, to the war minister on 20 April 1820, Bellinsgauzen wrote of these approaches:

> Here behind ice fields of smaller ice and islands, a continent of ice is seen, the borders of which are broken off perpendicularly and which ranges as far as we could see. The flat ice islands are situated close to that continent rising toward the south . . . and show clearly that they are wreckage of that continent, for their borders and surface are like the continent [*Dvukratnye izyskania* ("Twofold Investigations"), 3rd ed., p. 37].

The expedition later discovered a series of islands in the Pacific. In January 1821 the southernmost point of the trip was reached (latitude 69°53′ south, longitude 92°19′ west) and an island (which was named Peter the Great) and a coastline (which was named Alexander I Coast) were sighted.

Bellinsgauzen expressed his firm conviction that there was an ice continent at the South Pole: "I call the great ice rising to form sloping mountains as one proceeds to the South Pole the inveterate ice, provided it is 4° cold in the midst of the most perfect summer day. I suppose it is no less cold farther to the south, and thus I conclude that the ice ranges over the Pole and must be stationary . . ." (*ibid.*, p. 420).

In 751 sailing days, of which 122 were spent below the sixtieth parallel, the *Vostok* and the *Mirny* covered over 55,000 miles. Besides discovering Antarctica and sailing around it, one of the great achievements of the Bellinsgauzen-Lazarev expedition was the first description of large oceanic regions adjoining the ice continent. It discovered twenty-nine islands and one coral reef and conducted valuable oceanographic and other scientific observations. A test of the water at a depth of 402 meters, using a bathometer with a thermometer placed inside, showed higher specific weight and lower temperature proportional to depth. The climatic peculiarities of the Antarctic as recorded by Bellinsgauzen are generally in agreement with modern observations. The generalization concerning the temperatures of the equatorial and tropic zones of the Pacific is also important: "Here I ought to note that the greatest heat is not to be expected directly at the equator, where the passing southern trade wind refreshes the air. The greatest heat is in a zone of calm between the southern and the northern trade winds" (*ibid.*, p. 110).

Bellinsgauzen made the first attempt to describe and classify the Antarctic ice and determined that the Kanarsky flow is a branch of the Floridsky. He explained the origin of coral islands and on the basis of 203 observations of compass variations determined with great precision the position of the South Magnetic Pole at that time—latitude 76° south, longitude 142° 30′ west. He wrote of this in a letter to Kruzenstern, asking him to pass on to Gauss, at the latter's request, his table of compass variations. This table was published in Leipzig by Gauss (1840).

A sea in the Antarctic was named for Bellinsgauzen, as well as a cape in the southern part of Sakhalin Island and an island in the Tuamotus.

BIBLIOGRAPHY

Bellinsgauzen's account of the Antarctic expedition is *Dvukratnye izyskania v Yuzhnom Ledovitom okeane i plavanie vokrug sveta v prodolzhenie 1819, 20 i 21 godov, sovershennoe na shlyupakh "Vostoke" i "Mirnom"* ("Twofold Investigations in the Antarctic Ocean and a Voyage Around the World During 1819, 1820 and 1821, Completed on the Sloops *Vostok* and *Mirny*"; St. Petersburg, 1821; 3rd ed., Moscow, 1961).

Works on Bellinsgauzen are M. I. Belov, ed., *Pervaya russkaya antarkticheskaya ekspeditsia 1819–1821 gg. i ee otchetnaya navigatsionnaya karta* ("The First Russian Antarctic Expedition 1819–1821 and Its Navigational Map"; Leningrad, 1963), and "Slavu pervogo tochnogo vychislenia mestopolozhenia yuzhnogo magnitnogo polyusa Ross dolzhen razdelit c Bellingauzenom" ("Ross Must Share With Bellinsgauzen the Fame for the First Precise Calculation of the Location of the South Magnetic Pole"), in *Nauka i zhizn* ("Science and Life," 1966), no. 8, 21–23; "Geograficheskie nablyudenia v Antarktike ekspeditsy Kuka 1772–1775 gg. i Bellingauzena-Lazareva 1819–1820–1821 gg." ("Geographic Observations in the Antarctic of the Cook Expedition, 1772–1775, and the Bellinsgauzen-Lazarev Expedition, 1819–1821"), in *Doklady Mezhduvedomstvennoy Komissy po izucheniyu Antarktiki za 1960 g.* ("Reports of the Joint Commission for the Study of Antarctica, 1960"; Moscow, 1961), pp. 7–23; and M. P. Lazarev, *Dokumenty*, I (Moscow, 1952), 150.

IVAN A. FEDOSEYEV

BELON, PIERRE (*b.* Soultière, near Cerans, France, 1517; *d.* Paris, France, 1564), *zoology, botany.*

Belon's birthplace—or the house traditionally considered as such—still stands at Soultière. Practically nothing is known of his ancestry, and his biographer, Paul Delaunay, has been unable to pierce the mystery surrounding his origins. We know that he came from an obscure family and that as a boy he was apprenticed to an apothecary at Foulletourte. About 1535 he became apothecary to Guillaume Duprat, bishop of Clermont. In the course of subsequent wanderings (in Flanders and England) and zoological research he came back to the Auvergne. About that time he became the protégé of René du Bellay, bishop of Le Mans, a situation that enabled him to go to the University of Wittenberg and study under the botanist Valerius Cordus. In 1542 he went to Paris, where Duprat recommended him as an apothecary to François Cardinal de Tournon. Belon never acquired the doctorate, but in 1560, when Brigard was dean, he obtained the licentiate in medicine from the Paris Faculty of Medicine.

In his *Histoire naturelle des estranges poissons marins* (1551), Belon presented an orderly classification of fish that included the sturgeon, the tuna, the *malarmat* (peristedion), the dolphin, and the hippopotamus. The last, incidentally, was drawn from an Egyptian sculpture. Nevertheless, Belon can be considered the originator of comparative anatomy. By the same token, he depicted a porpoise embryo and set forth the first notions of embryology. Belon enriched the biological sciences by new observations and contributed greatly to the progress of the natural sciences in the sixteenth century. His learning was not derived solely from books. He was one of the first explorer-naturalists; and between 1546 and 1550 he undertook long voyages through Greece, Asia, Judaea, Egypt, Arabia, and other foreign countries. He passed through Constantinople, and in Rome he met the zoologists Rondelet and Salviani.

Belon discarded the bases of the comparative method and was not at all afraid of drawing parallels between human and bird skeletons. He was the first to bring order into the world of feathered animals, distinguishing between raptorial birds, diurnal birds of prey, web-footed birds, river birds, field birds, etc.

Belon was also a talented botanist and recorded the results of his observations in a beautiful work adorned with woodcuts showing, for the first time, several plants of the Near East, including *Platanus orientalis, Umbilicus pendulinus, Acacia vera,* and *Caucalis orientalis.* He was more interested in the practical uses of plants than in their scientific description. He also advocated the acclimatization of exotic plants in France.

Belon dwelt at great length on the applications of medicinal substances. He clarified the use of bitumen, which the ancient Egyptians had used for mummifying corpses. Its agglutinative and antiputrefactive properties had induced physicians to use it therapeutically.

Belon's observations were generally correct. He looked at the world as an analyst devoted to detail. He succeeded in winning the confidence of the great and was famous during his lifetime. His works were translated by Charles de l'Escluse and Ulisse Aldrovandi, both of whom recognized his authority. Charles IX installed him at the Château de Madrid and granted him a pension. Belon was murdered in the Bois de Boulogne under mysterious circumstances. He was only forty-seven.

Charles Plumier dedicated the species *Bellonia* (Rubiaceae) to Belon. Charles Alexandre Filleul, a sculptor from Le Mans, erected a statue to him in Le Mans in 1887. Pavlov called him the "prophet of comparative anatomy."

BIBLIOGRAPHY

I. ORIGINAL WORKS. Belon's writings are *L'histoire naturelle des estranges poissons marins, avec la vraie peincture et description du daulphin, et de plusieurs autres de son espèce, observée par Pierre Belon* . . . (Paris, 1551); *De admirabili operum antiquorum et rerum suspiciendarum praestantia liber primus. De medicato funere seu cadavere condito et lugubri defunctorum ejulatione liber secundus. De medicamentis non-nullis servandi cadaveris vim obtinentibus liber tertius* (Paris, 1553); *De aquatilibus libri duo* . . . (Paris, 1553); *De arboribus coniferis resiniferis, aliis quoque non-nullis sempiterna fronde virentibus* . . . (Paris, 1553); *Les observations de plusieurs singularitez et choses mémorables trouvées en Grèce, Asie, Judée, Égypte, Arabie et autres pays estranges, rédigées en trois livres* . . . (Paris, 1553); *L'histoire de la nature des oyseaux, avec leurs descriptions et naïfs portraicts retirez du naturel, escrite en sept livres* . . . (Paris, 1555); *La nature et diversité des poissons, avec leurs pourtraicts représentez au plus près du naturel, par Pierre Belon* . . . (Paris, 1555); *Portraits d'oyseaux, animaux, serpens, herbes, arbres, hommes et femmes d'Arabie et d'Égypte, observez par P. Belon, . . . le tout enrichy de quatrains, pour plus facile cognoissance des oyseaux et autres portraits* . . . (Paris, 1557); and *Les remonstrances sur le défaut du labour et culture des plantes et de la cognoissance d'icelles* . . . (Paris, 1558).

II. SECONDARY LITERATURE. Works on Belon are "La cronique de Pierre Belon, du Mans, médecin," Paris, Bibliothèque de l'Arsenal, MS 4561, fols. 88–141; Paul Delaunay, *Les voyages en Angleterre du médecin naturaliste Pierre Belon* (Anvers, 1923); and *L'aventureuse existence de Pierre Belon du Mans* (Paris, 1926), with a bibliography; and R. J. Forbes, *Pierre Belon and Petroleum* (Brussels, 1958).

M. WONG

BELOPOLSKY, ARISTARKH APOLLONOVICH
(*b.* Moscow, Russia, 13 July 1854; *d.* Pulkovo, U.S.S.R., 16 May 1934), *astrophysics.*

Belopolsky was born into an intellectual but poor family; his father, Apollon Grigorievich, a teacher in a Gymnasium, had not graduated from the Faculty of Medicine at the University of Moscow because of a lack of funds. His mother, the daughter of a doctor, taught music in Moscow. Upon graduating from the Gymnasium in 1873, Belopolsky entered Moscow University, graduating from its Faculty of Physics and Mathematics in 1877.

F. A. Bredikhin, director of the Moscow Observatory, enlisted Belopolsky to work there on photographing the sun. Belopolsky thus became seriously interested in and decided to devote himself to astronomy. In 1879 he was named a supernumerary assistant at the observatory.

An investigation of the photographs that he had taken of the sun served as the topic of Belopolsky's master's thesis, "Pyatna na solntse i ikh dvizhenie" ("Spots on the Sun and Their Movements," 1886). In 1888 Belopolsky was appointed a junior assistant in astronomy at the Pulkovo Observatory, and in 1891 a staff astrophysicist. He was elected an extraordinary academician in 1903 and ordinary academician in 1906. From 1908 to 1916 he was the vice-director of Pulkovo Observatory, from 1917 to 1919 its director, and from 1933 its honorary director.

During his eleven years at Moscow Observatory, in addition to his photographic work, Belopolsky measured on the meridional circle the precise positions of a selected group of stars, planets, and asteroids, and the positions of comets. Besides systematically photographing the sun and studying the law of its rotation, Belopolsky was the first in Russia to attempt to photograph stars; he achieved notable success with low-sensitivity bromide plates, photographing stars up to magnitude 8.5. He photographed the total solar eclipse of 1887, at which time he obtained on one negative a photograph of the internal solar corona, taken simultaneously by four separate objectives.

The main area of astrophysics in which Belopolsky worked was the determination of the radial velocity of celestial bodies. The application of photography to the study of their spectra had much significance for the development of astrophysics at the end of the nineteenth century. Belopolsky built spectrographs and was a tireless observer who used these devices in conjunction with all the powerful refractors at the Pulkovo Observatory.

The first two years of Belopolsky's stay at Pulkovo were devoted not to astrophysics, however, but to astrometrical observation, using a transit instrument, and to the working up of prior observations by August Wagner, from which he obtained reliable parallaxes for a number of bright stars (1889). Only in 1891, after Bredikhin became director at Pulkovo, did Belopolsky's active and fruitful career in astrophysics begin. Belopolsky reestablished the abandoned astrophysics laboratory and ordered from the observatory's workshops several spectrographs, which he himself helped to build. He was the first to use dry photographic plates, instead of the previously used colloidal plates. Belopolsky quickly became, with Scheiner and Fogel of Potsdam, a leading specialist in solar and laboratory spectroscopy. In 1902 he was invited to join the editorial board of the American *Astrophysical Journal.*

Belopolsky began his spectrographic work in 1895 with a study of the large planets, his goal being to explain the peculiarities of their axial rotation. It turned out that the angular rate of rotation of Jupiter's equatorial zones is 4–5 percent greater than that at other jovicentric latitudes. This confirmed a conclusion drawn by Belopolsky while he was still in Moscow; in analyzing observations of Jupiter taken over a period of 200 years, he had established that the period of rotation of the planet's equatorial region ($9^h 50^m$) differs from the period of rotation of the remainder of the surface ($9^h 55^m$), which is separated from the equatorial region by two dark bands.

In 1895, using a spectrograph attached to a 30-inch refractor to study the radial velocities of various points on Saturn's rings, Belopolsky brilliantly confirmed the theoretical conclusion of James Clerk Maxwell and Sofia V. Kovalevskaya that Saturn's rings are not solid, but consist of a multitude of small satellites. In addition, he found that the spectrum of the ring was rich in ultraviolet rays, which he explained as the influence of the planet's atmosphere on the spectrum of the planet itself.

At Pulkovo he continued his study of the sun's surface, but he shifted from a study of the sun's rotation by observing its faculae—the method he himself had proposed—to systematic observations of its protuberances and eruptions and, later, to a study of the movements of matter within the sun, a problem that is still important. Belopolsky also studied the fine structure of spectral lines and their changes in shape over time.

In 1904 Belopolsky joined the International Union for Cooperation in Solar Research, organized by G. E. Hale. In Russia there was a branch of the Union, the Commission for the Investigation of the Sun, to which many eminent physicists and astronomers belonged. The Union decided to undertake an eleven-and-a-half-year study of the sun, and for this work Belopolsky ordered a special three-prism spectrograph. In 1915 he published Russia's first study in spectrophotometry, "O temperature solnechnykh pyaten" ("On the Temperature of Sunspots"), in which he obtained a temperature of 3,500° C. for these spots. This has been confirmed by the latest investigations.

Belopolsky participated in two more expeditions to observe solar eclipses. In 1896 he succeeded in determining from the inclination of the spectral lines in the spectrum of the solar corona that the corona does not rotate like a solid body. The second expedition, in 1907, was not successful because of bad weather.

The study of star spectra on the basis of the Doppler principle was begun with the effective determination of the rate of expansion of the shell of Nova Aurigae, which appeared in 1892; Belopolsky conducted investigations of the spectra of Nova Persei (1901), which manifested semiperiodic fluctuations and changes in its spectrum, as well as of Nova Geminorum (1912), Nova Aquilae (1918), Nova Cygni (1920), and several others.

Belopolsky began studying the spectra of common (i.e., not new) stars in 1890, primarily to measure their radial velocities, and gave special attention to spectroscopic binary stars. He discovered the spectral duality of many stars, especially the star α Lyrae, which, until then had been considered a standard star in all catalogs of radial velocities. For the majority of spectroscopic binary stars he studied, Belopolsky determined the elements of their orbits and, in a number of cases, their changes. These changes of the elements attested to the presence of a third component in these systems (for example, Algol's third component and the polestar's third component).

Belopolsky discovered the variability of the spectrum of α_2 Canum Venaticorum, which bespoke strong and irregular changes in the atmosphere of this star, the prototype of a special class of stars with strong perturbations in their atmospheres.

A study of the spectra of stars of the δ Cephei type—Cepheid variables—led Belopolsky to the discovery of the noncoincidence of the phases of changes in brightness and changes of radial velocities, as well as the discovery of periodic changes in the intensity of absorption lines. Belopolsky's determination of the parallaxes and linear dimensions of several visual binary stars is also of interest. It is worth noting that in 1896, at Belopolsky's defense of his doctoral dissertation, *Issledovanie spektra peremennoy zvezdy δ Cephei* ("An Investigation of the Spectrum of the Variable Star δ Cephei"), the physicist N. A. Umov

first explained the periodic changes of radial velocity (observed by Belopolsky) by the hypothesis of an individual star's pulsation. The pulsation theory of Cepheids was not completely developed until the work of the Soviet astrophysicist S. A. Zhevakin.

Belopolsky's laboratory investigations occupy a special place in the history of science. In his master's thesis, as a supplement to the theoretical analysis of the laws of motion of solar matter, Belopolsky conducted an original experiment using a glass flask filled with water, in which were suspended minute particles of stearin. A coordinate grid plotted on the surface of the flask facilitated registration of the behavior of the stearin particles, which revealed the particularities of the liquid's rotation. The angular velocity of the rotation declined monotonically from the equator to the latitude of 55°, while the meridional velocity increased up to this same latitude, recalling the corresponding particularities of the rotation of elements of the sun's surface.

His verification of the validity of Doppler's principle in optics is considered a classic. For this experiment Belopolsky constructed an exceedingly original device in which, by means of the multiple reflections of a beam of light between two oppositely rotating "water wheels" whose blades were strips of plane mirrors, he succeeded in obtaining a speed of the image's motion on the order of 1 km./sec., which was capable of being measured with the sufficient certainty to verify the Doppler principle convincingly.

BIBLIOGRAPHY

I. ORIGINAL WORKS. Belopolsky's major writings are "Pyatna na solntse i ikh dvizhenie" ("Spots on the Sun and Their Movements"), master's thesis (University of Moscow, 1886); "Issledovanie smeshchenia liniy v spektre Saturna i ego koltsa" ("An Investigation of the Shift of Lines in the Spectrum of Saturn and of Its Ring"), in *Izvestiya Imperatorskoi akademii nauk,* **3** (1895); *Issledovanie spektra peremennoy zvezdy δ Cephei* ("An Investigation of the Spectrum of the Variable Star δ Cephei"; St. Petersburg, 1895); "O zvezde α' Bliznetsov kak spektralno-dvoynoy" ("On the Star α' Geminorum as a Spectral-Binary Star"), in *Izvestiya Imperatorskoi akademii nauk,* **7**, no. 1 (1897); "Opyt issledovania printsipa Doplera-Fizo, ne pribegaya k kosmicheskim skorostyam" ("An Attempt to Investigate the Doppler-Fizo Principle, Without Resorting to Cosmic Velocities"), *ibid.,* **13**, no. 5 (1900), 461–474; "O vrashceny Yupitera" ("On the Rotation of Jupiter"), *ibid.,* 6th ser., **3** (1909); "Avtobiografia" ("Autobiography"), in *Materialy dlya biograficheskogo slovarya deystvitelnykh chlenov Akademii nauk. 1889–1914, chast pervaya,* I (Petrograd, 1915), pp. 121–122; "O spektre Novoy 1918 g. (predvaritelnoe soobshchenie)" ("On the

Spectrum of the Nova of 1918 [Preliminary Report]"), in *Izvestiya Imperatorskoi akademii nauk; Astrospektroskopia* ("Astrospectroscopy"; Petrograd, 1921); "Comets and Ionization," in *Observatory,* **46** (1923), 124–125; "Über die Intensitätsveränderlichtkeit der Spektrallinien einiger Cepheiden," in *Izvestiya Pulkovo Obs.,* 2nd ser., **2,** no. 101 (1927), 79–88; "Ob izmeny intensivnosti liny v spektrakh nekotorykh tsefeid," ("On Changes in the Intensity of Lines in the Spectra of Certain Cepheids"), in *Izvestiya Akademii nauk,* 7th ser., no. 1 (1928); and "Bestimmung der Sonnenrotation auf spektroskopischen Wege in den Jahren 1931, 1932 und 1933 in Pulkovo," in *Zeitschrift für Astrophysik,* **7** (1933), 357–363.

II. SECONDARY LITERATURE. Works on Belopolsky are S. N. Blazhko, "A. A. Belopolsky," in *Bolshaya sovetskaya entsiklopedia,* IV (1950), 462–464; V. G. Fesenkov, "Aristarkh Apollonovich Belopolsky," in *Lyudi russkoy nauki* ("People of Russian Science"; Moscow, 1961), pp. 185–192; B. P. Gerasimovich, "A. A. Belopolsky. 1854–1934," in *Astronomicheskii zhurnal* (*SSSR*), **2,** pt. 3 (1934), 251–254; O. A. Melnikov, "Aristarkh Apollonovich Belopolsky (1854–1934). Nauchno-biografichesky ocherk" ("Aristarkh Apollonovich Belopolsky [1854–1934]. A Scientific-biographical Essay"), in *A. A. Belopolsky. Astronomicheskie trudy* ("A. A. Belopolsky. Astronomical Works"; Moscow, 1954), pp. 7–58; Y. G. Perel, *Vydayushchiesya russkie astronomy* ("Outstanding Russian Astronomers"; Moscow, 1951), pp. 85–107; K. D. Pokrovsky, "A. A. Belopolsky (k 50-letiyu ego nauchnoy deyatelnosti 1877–1927)" ("A. A. Belopolsky [on the 50th Anniversary of His Scientific Career 1877–1927]"), in *Astronomicheskii kalendar na 1928 god* ("Astronomical Calendar for 1928"; Nizhni Novgorod, 1927), pp. 123–125, with illustrations; and D. A. Zhukov, "Spisok nauchnykh rabot akademika A. A. Belopolskogo 1877–1934" ("A List of the Scientific Papers of Academician A. A. Belopolsky 1877–1934"), in *Byulleten' Komissii po issledovaniyu solntsa akademii nauk SSSR,* nos. 10–11 (1934), 7–20.

P. G. KULIKOVSKY

BELTRAMI, EUGENIO (*b.* Cremona, Italy, 16 November 1835; *d.* Rome, Italy, 18 February 1900), *mathematics.*

Beltrami was born into an artistic family: his grandfather, Giovanni, was an engraver of precious stones, especially cameos; his father, Eugenio, painted miniatures. Young Eugenio studied mathematics from 1853 to 1856 at the University of Pavia, where Francesco Brioschi was his teacher. Financial difficulties forced Beltrami to become secretary to a railroad engineer, first in Verona and then in Milan. In Milan he continued his mathematical studies and in 1862 published his first mathematical papers, which deal with the differential geometry of curves.

After the establishment of the kingdom of Italy in 1861, Beltrami was offered the chair of complemen-

tary algebra and analytic geometry at Bologna, which he held from 1862 to 1864; from 1864 to 1866 he held the chair of geodesy in Pisa, where Enrico Betti was his friend and colleague. From 1866 to 1873 he was back in Bologna, where he occupied the chair of rational mechanics. After Rome had become the capital of Italy in 1870, Beltrami became professor of rational mechanics at the new University of Rome, but served there only from 1873 to 1876, after which he held the chair of mathematical physics at Pavia, where he also taught higher mechanics. In 1891 he returned to Rome, where he taught until his death. He became the president of the Accademia dei Lincei in 1898 and, the following year, a senator of the kingdom. A lover of music, Beltrami was interested in the relationship between mathematics and music.

Beltrami's works can be divided into two main groups: those before *ca.* 1872, which deal with differential geometry of curves and surfaces and were influenced by Gauss, Lamé, and Riemann, and the later ones, which are concerned with topics in applied mathematics that range from elasticity to electromagnetics. His most lasting work belongs to this first period, and the paper "Saggio di interpretazione della geometria non-euclidea" (1868) stands out. In a paper of 1865 Beltrami had shown that on surfaces of constant curvature, and only on them, the line element $ds^2 = E du^2 + 2F du dv + G dv^2$ can be written in such a form that the geodesics, and only these, are represented by linear expressions in u and v. For positive curvature R^{-2} this form is

$$ds^2 = R^2[(v^2 + a^2)du^2 - 2uvdudv + (u^2 + a^2)dv^2] \times (u^2 + v^2 + a^2)^{-2}.$$

The geodesics in this case behave, locally speaking, like the great circles on a sphere. It now occurred to Beltrami that, by changing R to iR and a to ia ($i = \sqrt{-1}$), the line element thus obtained,

$$ds^2 = R^2[(a^2 - v^2)du^2 + 2uvdudv + (u^2 + a^2)dv^2] \times (a^2 - u^2 - v^2)^{-2},$$

which defines surfaces of constant curvature $-R^{-2}$, offers a new type of geometry for its geodesics inside the region $u^2 + v^2 < a^2$. This geometry is exactly that of the so-called non-Euclidean geometry of Lobachevski, if geodesics on such a surface are identified with the "straight lines" of non-Euclidean geometry.

This geometry, developed between 1826 and 1832, was known to Beltrami through some of Gauss's letters and some translations of the work of Lobachevski. Few mathematicians, however, had paid attention to it. Beltrami now offered a representation of this geometry in terms of the acceptable Euclidean geometry: "We have tried to find a real foundation [*sub-*

strato] to this doctrine, instead of having to admit for it the necessity of a new order of entities and concepts." He showed that all the concepts and formulas of Lobachevski's geometry are realized for geodesics on surfaces of constant negative curvature and, in particular, that there are rotation surfaces of this kind. The simplest of this kind of "pseudospherical" surface (Beltrami's term) is the surface of rotation of the tractrix about its asymptote, now usually called the pseudosphere, which Beltrami analyzed more closely in a paper of 1872.

Thus Beltrami showed how possible contradictions in non-Euclidean geometry would reveal themselves in the Euclidean geometry of surfaces; and this removed for most, or probably all, mathematicians the feeling that non-Euclidean geometry might be wrong. Beltrami, by "mapping" one geometry upon another, made non-Euclidean geometry "respectable." His method was soon followed by others, including Felix Klein, a development that opened entirely new fields of mathematical thinking.

Beltrami pointed out that his representation of non-Euclidean geometry was valid for two dimensions only. In his "Saggio" he was hesitant to claim the possibility of a similar treatment of non-Euclidean geometry of space. After he had studied Riemann's *Über die Hypothesen welche der Geometrie zu Grunde liegen,* just published by Dedekind, he had no scruples about extending his representation of non-Euclidean geometry to manifolds of $n > 2$ dimensions in "Teoria fondamentale degli spazi di curvatura costante."

In a contribution to the history of non-Euclidean geometry, Beltrami rescued from oblivion the Jesuit mathematician and logician Giovanni Saccheri (1667–1733), author of *Euclides ab omni naevo vindicatus,* which foreshadowed non-Euclidean geometry but did not achieve it.

In his "Ricerche di analisi applicata alla geometria," Beltrami, following an idea of Lamé's, showed the power of using so-called differential parameters in surface theory. This can be considered the beginning of the use of invariant methods in differential geometry.

Much of Beltrami's work in applied mathematics shows his fundamental geometrical approach, even in his analytical investigations. This trait characterizes the extensive "Richerche sulle cinematica dei fluidi" (1871–1874) and his papers on elasticity. In these he recognized how Lamé's fundamental formulas depend on the Euclidean character of space, and he sketched a non-Euclidean approach (1880–1882). He studied potential theory, particularly that of ellipsoids and cylindrical discs; wave theory in connection with Huygens' principle; and further problems in thermo-dynamics, optics, and conduction of heat that led to linear partial differential equations. Some papers deal with Maxwell's theory and its mechanistic interpretation, suggesting a start from d'Alembert's principle rather than from that of Hamilton (1889).

BIBLIOGRAPHY

I. Original Works. Beltrami's works are collected in *Opere matematiche,* 4 vols. (Milan, 1902–1920). Important individual works are "Saggio di interpretazione della geometria non-euclidea," in *Giornale di matematiche,* **6** (1868), 284–312, and *Opere,* I, 374–405, also translated into French in *Annales scientifiques de l'École Normale Supérieure,* **6** (1869), 251–288; "Richerche di analisi applicata alla geometria," in *Giornale di matematiche,* **2** (1864) and **3** (1865), also in *Opere,* I, 107–206; a paper on surfaces of constant curvature, in *Opere,* I, 262–280; "Teoria fondamentale degli spazi di curvatura costante," in *Annali di matematica,* ser. 2 (1868–1869), 232–255, and *Opere,* I 406–429; "Richerche sulle cinematica dei fluidi," in *Opere,* II, 202–379; a paper on the pseudosphere, in *Opere,* II, 394–409; a non-Euclidean approach to space, in *Opere,* III, 383–407; "Sulla teoria della scala diatonica," in *Opere,* III, 408–412; an article on Saccheri, in *Rendiconti della Reale Accademia dei Lincei,* ser. 4, **5** (1889), 441–448, and *Opere,* IV, 348–355; and papers on Maxwell's theory, in *Opere,* IV, 356–361.

II. Secondary Literature. There is a biographical sketch by L. Cremona in *Opere,* I, ix–xxii. See also L. Bianchi, "Eugenio Beltrami," in *Enciclopedia italiana,* VI (1930), 581; G. H. Bryan, "Eugenio Beltrami," in *Proceedings of the London Mathematical Society,* **32** (1900), 436–439; and G. Loria, "Eugenio Beltrami e le sue opere matematiche," in *Bibliotheca mathematica,* ser. 3, **2** (1901), 392–440. On Beltrami's contribution to non-Euclidean geometry, consult, among others, R. Bonola, *Non-Euclidean Geometry* (Chicago, 1912; New York, 1955), pp. 130–139, 234–236.

D. J. STRUIK

BENEDEN, EDOUARD VAN (*b.* Louvain, Belgium, 5 March 1846; *d.* Liège, Belgium, 28 April 1910), *zoology, embryology.*

Van Beneden's father, the zoologist P. J. Van Beneden, was professor at the Catholic University in Louvain. Edouard was appointed professor at the University of Liège in 1870 with the qualification of *chargé de cours* and was promoted to *ordinarius* in 1874.

His scientific work is characterized by a great unity resulting from the character of the problems to which he devoted his attention. In 1872 Van Beneden went to Brazil, and from there he brought back a number of specimens, particularly *Annelida,* which were studied by Armauer Hansen. P. J. Van Beneden had organized a modest laboratory on the Belgian sea-coast at Ostend, and there Edouard had the opportunity to collect and study many specimens of fauna, particularly from Thornton Bank, near Ostend. His main interests at that time were directed toward the animal groups his father had studied: protozoa, hydraria, cestodes, nematodes, and tunicates. He soon extended these studies to dicyemida, and later to vertebrates, particularly mammals.

In a paper on the origin of the sexual cells of hydroids (1874), Van Beneden showed, besides great care in observation, his tendency to sometimes bold generalization. At that time, the gastrula theory of metazoan development was presented by Huxley, Lankester, and Haeckel; Van Beneden suggested that ectoderm and endoderm have opposed sexual significance, the ectoderm being the male layer and the endoderm the female layer. He thereafter recognized the hermaphrodite nature of the egg, an idea he later developed in his studies on cestodes.

Van Beneden's first publication on dicyemida dates from 1876. He made a thorough and careful study of their structure and observed that they derive from an epibolic gastrula, the hypoblast of which is formed by a long, central single cell. In 1877 he proposed the creation of the phylum Mesozoa, considering it an ideal transition between monocellular and multicellular animal forms. This idea, after a temporary eclipse, has regained popularity.

In 1864, by the application of the embryogenetic method and the biogenetic law, Kovalevski had shown, to everyone's surprise, that tunicates were chordates. They therefore were excellent material for studies in comparative morphology, with a view to establishing their phylogenetic relations with other animals. In collaboration with his colleague Charles Julin, Van Beneden showed, in a contribution to ascidian embryology that represents one of the first and best works on the segmentation of the egg, that the germ in formation shows a bilateral symmetry and a clear polarity. In 1884 Van Beneden and Julin were the first to follow, in a strictly bilateral egg, what is now called cell lineage.

In 1887, Van Beneden and Julin published an important paper on tunicate morphology in which they stated that the mesoblast derives from the enterocele. They also accepted that the cardiopericardic vesicle and the epicardia derive from common forms, the procardia, which they considered to be pharyngeal diverticula.

Van Beneden pursued these studies until his death and left notes that have been published by his disciple Marc de Selys-Longchamps, who, from drawings left by his master, concluded that the endoblast is not derived from the enterocele and showed that the pericardium of tunicates is the ultimate remainder of the chordate celom.

Van Beneden's most famous contributions to science are his papers on the maturation and fertilization of the egg of *Ascaris megalocephala,* the first of which was published in 1883. In this paper he revealed the essential nature of fertilization: the union of two half-nuclei, one female and the other male. He showed that in *Ascaris,* it is not until the male and female pronuclei have formed what we now call chromosomes (which are accurately represented in his plates) that the respective nuclear membranes break down. He described how each set of chromosomes moves to the equatorial plate. Van Beneden saw that in the variety of *Ascaris* he called *univalens,* which had only two chromosomes, each parent contributes one chromosome to the pair found in the zygote. Through this discovery, the individuality of the single chromosome was first demonstrated.

In *bivalens* (the variety with four chromosomes in the fertilized egg) Van Beneden showed the presence of two chromosomes in each of the pronuclei, and of four in the zygote. He also discovered that in giving off the polar bodies, the number of chromosomes within the egg nucleus is reduced from four to two.

Van Beneden's opinion on the mechanism of karyogamic reduction, which he had discovered, was that of the four chromosomes of *bivalens,* two entered the first polar body and two remained to form the female pronucleus. It was shown independently, by Theodor Boveri in 1887–1888, and by Oscar Hertwig in 1890, that the real nature of the "maturation" division leading to the formation of ripe gametes in both sexes is that each chromosome in the mother cell divides once, while the cell itself divides twice. In 1892 Boveri showed, in *Ascaris megalocephala bivalens,* that two of the eight daughter chromosomes of the egg's mother cell go into each of four cells: into each of the three polar bodies (if the first polar body divides after being given off) and into the ovum.

In a paper published in collaboration with the photographer Neyt (1887), Van Beneden described the centrosome and showed that it was a permanent cell organ which remained in the cell during the resting period and divided into two parts before the beginning of the next mitosis.

From the beginning of his scientific work, Van Beneden had been preoccupied with the problem of the origin of vertebrates. His first interpretation of the didermic mammalian embryo was immediately accepted, but in 1888 he concluded a study of the gastrulation of mammals that became a classic only after his death.

BIBLIOGRAPHY

I. ORIGINAL WORKS. A complete list of Van Beneden's publications is in A. Brachet, "Notice sur Edouard Van Beneden," in *Annuaire de l'Académie royale de Belgique,* **89** (1923), 167–242. The most important of his works are "De la distinction originelle du testicule et de l'ovaire. Caractère sexuel des deux feuillets primordiaux de l'embryon. Hermaphrodisme morphologique de toute individualité animale. Essai d'une théorie de la fécondation," in *Bulletin de l'Académie royale de Belgique. Classe des sciences,* 2nd ser., **37** (1874), 530–595; "Recherches sur les dicyémides, survivants actuels d'un embranchement des mésozoaires," *ibid.,* **41** (1876), 1160–1205; **42** (1876), 35–97; "L'appareil sexuel femelle de l'ascaride mégalocéphale," in *Archives de biologie,* **4** (1883), 95–142; "La segmentation chez les ascidiens dans ses rapports avec l'organisation de la larve," in *Bulletin de l'Académie royale de Belgique. Classe des sciences,* 3rd ser., **7** (1884), 431–447, written with C. Julin; "La segmentation chez les ascidiens et ses rapports avec l'organisation de la larve," in *Archives de biologie,* **5** (1884), 111–126, written with C. Julin; "Nouvelles recherches sur la fécondation et la division mitosique chez l'ascaride mégalocéphale. Communication préliminaire," in *Bulletin de l'Académie royale de Belgique. Classe des sciences,* 3rd ser., **14** (1887), 215–295, written with A. Neyt; and "Recherches sur la morphologie des tuniciers," in *Archives de biologie,* **6** (1887), 237–476, written with C. Julin.

II. SECONDARY LITERATURE. The best biography of Van Beneden is the one by Brachet (see above). Van Beneden had asked in his will that his biography be written by Walter Flemming and Karl Rabl, published by them in a German periodical, and translated into French in *Archives de biologie.* Flemming was dead by 1910, so Rabl wrote the biography alone: "Edouard Van Beneden und der gegenwärtige Stand der Wichtigsten von ihm behandelten Probleme," in *Archiv für mikroskopische Anatomie,* **88** (1915), 1–470. This extensive biography, in spite of all its good points, was considered by Van Beneden's disciples as too polemic and too replete in allusions to Rabl's own work. Its French translation was replaced, in the *Archives de biologie,* by a reproduction of Brachet's notice.

MARCEL FLORKIN

BENEDEN, PIERRE-JOSEPH VAN (*b.* Mechelen [French, Malines], Belgium, 19 December 1809; *d.* Louvain, Belgium, 8 January 1894), *zoology.*

After studying the humanities, Van Beneden was apprenticed to the pharmacist Louis Stoffels, a great collector of natural history specimens, whose house was a veritable museum that inspired Van Beneden to become a zoologist. At that time in the kingdom of the Low Countries, it was not necessary to attend a university in order to become a pharmacist, but

Stoffels recognized in Van Beneden the talents of a scientist and persuaded the boy's parents to send him to the University of Louvain (then still a state university), where he studied medicine. After obtaining the M.D., Van Beneden went to Paris to study zoology, intending to become a professor of zoology at one of the state universities of the new kingdom of Belgium. Since other professors had already been appointed to these chairs, he accepted the post of professor of zoology at the newly created Catholic University of Louvain on 10 April 1835.

Van Beneden's contributions to zoology concern most animal phyla and are characterized by the importance given to embryology in the recognition of systematic affinities. He was a naturalist whose curiosity extended to the broadest spectrum of animal species. His main contribution, however, was the discovery of the life cycle of the cestodes (1849). When he started this work in 1845, the cycle was entirely unknown. It was known that the digestive tract of certain animals contains certain taenias, and the presence of cysticerci in other forms was also known, but no relation had been established between these forms, which were considered to be distinct and autonomous organisms, the products of a spontaneous generation in the tissues of their hosts. Taenias were considered to be hypertrophied intestinal villi.

Dujardin had already noticed a similarity between the heads of certain cysticerci of the liver of bony fishes and the head of Tetrarhynchus living in the gut of cartilaginous fishes that fed on bony fishes. From an extensive study of the contents of the digestive tracts of a large number of fishes, Van Beneden concluded, in January 1849, that a cysticercus is an incomplete taenioid. Siebold adopted this view in 1850, and the following year Küchenmeister demonstrated experimentally that *Cysticercus cellulosae* from a rabbit's peritoneum, fed to a dog, becomes *Taenia serrata.*

Since the Parisian zoologist Valenciennes refused to accept his theory, Van Beneden left for Paris on 22 April 1858 with four dogs, one of which had been fed thirty-two cysticercus; another, seventy; and the other two, none. In Paris he met Milne-Edwards and Quatrefages in Valenciennes's laboratory; he predicted which dogs would contain the taenias and confirmed his predictions by autopsy.

In 1853 the Institut de France gave an important prize to a paper by Van Beneden on the mode of development and transmission of intestinal worms. This report, published in 1858, covers a wide range of data on parasites, concerning not only cestodes and trematodes, but also nematodes, *Gordiacea,* and Acanthocephala. The paper ends with a masterly treatment of the systematics of worms that is based on embryology.

After 1859 Van Beneden devoted himself to the study of Cetacea, both living and fossil. In 1878 he determined that the first fossil skeleton discovered in the coal mine of Bernissart belonged to the genus *Iguanodon*. During the last years of his life Van Beneden devoted his efforts to new studies on parasites. In 1898 his native town raised a statue to honor his memory.

BIBLIOGRAPHY

I. ORIGINAL WORKS. A complete list of Van Beneden's publications is in *Notices biographiques et bibliographiques de l'Académie royale de Belgique* (1886, 1896). The most important are "Les Helminthes cestoïdes, considérés sous le rapport de leurs métamorphoses, de leur composition anatomique et de leur classification, et mention de quelques espèces nouvelles de nos poissons Plagiostomes," in *Bulletin de l'Académie royale des sciences de Belgique*, **16** (1849), 269; "Notice sur un nouveau genre d'Helminthe cestoïde," *ibid.*, 182; "Faune littorale de Belgique. Les vers cestoïdes, considérés sous le rapport physiologique, embryogénique et zooclassique," *ibid.*, **17** (1850), 102; "Notice sur l'éclosion du *Tenia dispar* et la manière dont les embryons de cestoïdes pénétrent à travers les tissus, se logent dans les organes creux, et peuvent même passer de la mère au foetus," *ibid.*, **20** (1853), 287; *Mémoire sur les vers intestinaux, Comptes rendus de l'Académie des sciences* (Paris), supp. 2 (1858); "Sur la découverte de reptiles fossiles gigantesques dans le charbonnage de Bernissart près de Péruwelz," in *Bulletin de l'Académie royale des sciences de Belgique*, 2nd ser., **45** (1878), 578–579.

II. SECONDARY LITERATURE. The most detailed biography of Van Beneden is A. Kemna, *P. J. Van Beneden, la vie et l'oeuvre d'un zoologiste* (Antwerp, 1897). Other publications concerning his life and work are J. B. Abeloos, *Discours prononcé à la Salle des Promotions . . . après le service funèbre . . . de M. Pierre-Joseph Van Beneden* (Louvain, 1894); Charles Van Bambeke, "P. J. Van Beneden, 1809–1894," in *Annales de la Société belge de microscopie*, **20** (1896); Paul Brien, "Pierre-Joseph Van Beneden," in *Florilège des sciences en Belgique* (Brussels, 1967), pp. 825–851; P. Debaisieux, "Un siècle de biologie à l'Université de Louvain," in *Revue des questions scientifiques* (1937); H. Filhol, "Inauguration de la statue de Van Beneden à Malines le 24 juillet 1898," in *Comptes rendus de l'Académie des sciences de Paris*, **127** (1898), 91; A. Gaudry, "Statue de Van Beneden et les fêtes de Malines," in *La nature*, **26** (1898); Auguste Lameere, "Notice sur Pierre-Joseph Van Beneden," in *Annuaire de l'Académie royale de Belgique*, **107** (1941), 1–13, with portrait; and "Pierre-Joseph Van Beneden," in *Biographie nationale*, XXVI, col. 184; Dr. F. Lefebvre, "Discours prononcé aux funérailles de P. J. Van Beneden," in *Bulletin de l'Académie de médecine de Belgique* (1894); *Manifestation en l'honneur de M. le professeur P. J. Van Beneden, Louvain* (Ghent, 1877); *Manifestation en l'honneur de M. le professeur P. J. Van Beneden, à l'occasion de son cinquantenaire de professorat, Louvain* (Louvain, 1886); "Manifestation en l'honneur de Pierre-Joseph Van Beneden à l'occasion du cinquantième anniversaire de sa nomination comme membre titulaire de la Classe des Sciences (1842–1892)," in *Bulletin de l'Académie royale de Belgique*, 3rd ser., **23** (1892), 702–709; M. Mourlon, "Discours prononcé aux funérailles de P. J. Van Beneden," *ibid.*, **27** (1894), 201–208; P. Pelseneer, "P. J. Van Beneden malacologiste," in *Mémoires de la Société royale malacologique de Belgique*, **29** (1894); J. Van Raemdonck, "Souvenirs du professeur Van Beneden," in *Annales du Cercle Archéologique du Pays de Waes*, **14** (1894); and *Souvenir de l'inauguration de la statue de P. J. Van Beneden . . .* (Malines, 1898).

MARCEL FLORKIN

BENEDETTI, ALESSANDRO (*b.* province of Verona, Italy, *ca.* 1450; *d.* Venice, Italy, 30 October 1512), *anatomy, medicine.*

Alessandro, the son of Lorenzo Benedetti, came from a medical family (a maternal uncle was also a physician), and his works attest to an excellent education in both letters and medicine. One of his teachers was the historian and humanist Giorgio Merula. He lived in a transitional period of Venetian and Italian history, and his career was very much influenced by the military and political events of the time. From his works, from the portion of his correspondence that has been published, and from documents we know that he transferred early to Padua, where he studied and taught. He practiced medicine in Venice and was a physician in the war against Charles VIII of France in 1495. For many years he also served Venice abroad—Greece, Crete, Melos, and the Dalmatian coast are mentioned. Among his friends were not only families high in the Venetian nobility and government but also the humanist and physician Giorgio Valla, the poet Quinzio Emiliano Cimbriaco, the historian Marino Sanudo, and the writer Jacopo Antiquari. In his will his wife, Lucia, is mentioned along with a daughter, Giulia.

Benedetti's services to his patients, his thirty-book codification of diseases, his painstaking labors to improve the text of the thirty-seven books of Pliny's *Natural History,* and his contributions to medical and general knowledge of the Caroline War in his diary of that event have assured him a position of importance in both medical history and historiography. His place in natural science rests upon his contributions to the study of anatomy. His importance consisted not in startling discoveries on the nature of the human body but in dissection before his students and in his treatise *Historia corporis humani sive anatomice.* He had been assembling material for this treatise at least since 1483, and parts of it were probably circulated in the closing years of the fifteenth century; it appeared complete in Venice (1502), with a letter of dedication dated 1 August 1497, and appeared in later editions at Paris (1514), Cologne (1527), Strasbourg (1528), and in the collected editions of his medical and anatomical works that were published at Basel (1539, 1549). The *Historia* was known to Leonardo da Vinci; Leonhart Fuchs called attention to a number of errors in it; and Giovanni Battista Morgagni accorded importance in anatomy to its author. It therefore had, we may believe, something of the authority of a textbook in a period when anatomists in Italy were becoming increasingly concerned with observation of the human rather than the animal body. The work comprises five books, which describe the body roughly *a calce ad caput*. It concludes with a paragraph praising dissection, in which the author urges all students, together with established physicians and surgeons, to seek truth in the anatomical theater and not rely solely upon the written word. This paragraph is a natural extension of the introductions to the several books in which Benedetti describes an anatomical theater inspired in form by the Colosseum at Rome and the Roman arena at Verona, and invites various political and literary figures to dissections. This encouragement of dissection was probably Benedetti's greatest contribution to natural science.

Benedetti, while profiting from earlier writers, followed the same firsthand methods in practicing medicine and in his encyclopedic *Omnium a vertice ad calcem morborum signa, causae, indicationes et remediorum compositiones utendique rationes,* in which a number of his cases are cited. He was the author also of aphorisms, of a tract on the pest (1493), and of a work on Paul of Aegina, which was not published.

BIBLIOGRAPHY

I. ORIGINAL WORKS. Benedetti's works include *Collectiones medicinae* (Venice, ca. 1493); *De observatione in pestilentia* (Venice, 1493); *Diaria de bello Carolino* (Venice, after 26 August 1496), the modern edition and translation of which by Dorothy M. Schullian (New York, 1967) provides an extensive bibliography of original works of Benedetti as well as secondary works on him, and makes unnecessary any lengthy listing here; *Historia corporis humani sive anatomice* (Venice, 1502); his edition of C. Plinius Secundus, *Historia naturalis libri XXXVII* (Venice, 1513); and *Omnium a vertice ad calcem morborum signa, causae, indicationes et remediorum compositiones utendique rationes . . .* (Basel, 1539). The correspondence of Giorgio Valla with Benedetti and others was edited by Johan Ludvig Heiberg in *Centralblatt für Bibliothekswesen,* Beiheft XVI (Leipzig, 1896).

II. SECONDARY LITERATURE. References to Benedetti are contained in Nicolaus Comnenus Papadopoli, *Historia gymnasii Patavini* (Venice, 1726); and in Marino Sanudo, *La spedizione di Carlo VIII in Italia* (Venice, 1873), and *I diarii,* XV (Venice, 1886), col. 283. Also see Roberto Massalongo, "Alessandro Benedetti e la medicina veneta nel quattrocento," in *Atti del Reale Istituto Veneto di Scienze Lettere ed Arti,* **76,** pt. 2 (1916–1917), 197–259; Curt F. Bühler, "Stoppress and Manuscript Corrections in the Aldine Edition of Benedetti's *Diaria de bello Carolino,*" in *Papers of the Bibliographical Society of America,* **43** (1949), 365–373; Glauco De Bertolis, "Alessandro Benedetti; Il primo teatro anatomico padovano," in *Acta medicae historiae Patavina,* **3** (1956–1957), 1–13; Hans Dieter Kickartz, *Die Anatomie des Zahn-, Mund- und Kieferbereiches in dem Werk "HISTORIA CORPORIS HUMANI SIVE ANATOMICE" von Alessandro Benedetti,* doctoral dissertation (Düsseldorf, 1964); Wolfgang Matschke, *Die Zahnheilkunde in Alessandro Benedettis posthumen Werk "Omnium a vertice ad calcem morborum signa, causae, indicationes et remediorum compositiones utendique rationes, generatim libris XXX conscripta,"* doctoral dissertation (Freiburg, 1965); and Mario Crespi, "Benedetti, Alessandro," in *Dizionario biografico degli italiani,* VIII (Rome, 1966), 244–247.

For additional references see the 1967 edition of Benedetti's *Diaria* cited above.

DOROTHY M. SCHULLIAN

BENEDETTI, GIOVANNI BATTISTA (*b.* Venice, Italy, 14 August 1530; *d.* Turin, Italy, 20 January 1590), *mathematics, physics.*

Benedetti is of special significance in the history of science as the most important immediate forerunner of Galileo. He was of patrician status, but has not been definitely connected with any of the older known families of that name resident at Venice. His father was described by Luca Guarico as a Spaniard, philosopher, and *physicus,* probably in the sense of "student of nature" but possibly meaning "doctor of medicine." It was to his father that Benedetti owed most of his education, which Guarico says made him a philosopher, musician, and mathematician by the age of eighteen. One of the few autobiographical records left by Benedetti asserts that he had no formal education beyond the age of seven, except that he studied the first four books of Euclid's *Elements* under Niccolò Tartaglia, probably about 1546–1548. Their relations appear to have been poor, for Tartaglia nowhere mentions Benedetti as a pupil; Benedetti named Tartaglia in 1553 only "to give him his due" and severely criticized his writings in later years.

Benedetti's originality of thought and mathematical skill are evident in his first book, the *Resolutio,* published at Venice when he was only twenty-two. The

Resolutio concerns the general solution of all the problems in Euclid's *Elements* (and of some others) using only a compass of fixed opening. Benedetti's treatment was more comprehensive and elegant than that of Tartaglia or Ludovico Ferrara in the published polemics of 1546–1547, and was more systematic than Tartaglia's later attack on the same problem in his final work, the *Trattato generale di numeri e misure* of 1560, in which he ignored the work of his former pupil.

Benedetti had one daughter, who was born at Venice in 1554 and died at Turin in 1580, but there is no record of his marriage. In 1558 he went to Parma as court mathematician to Duke Ottavio Farnese, in whose service he remained about eight years. In the winter of 1559/60 he lectured at Rome on the science of Aristotle; Girolamo Mei, who heard him there, praised his acumen, independence of mind, fluency, and memory. At Parma, Benedetti gave instruction at the court, served as astrologer, and advised on the engineering of public works. He also carried out some astronomical observations and constructed sundials mentioned in a later book on that subject. It appears that his private means were considerable, so that he was not inconvenienced by long delays in the payment of his salary.

In 1567 he was invited to Turin by the duke of Savoy and remained there until his death. The duke, Emanuele Filiberto, had great plans for the rehabilitation of Piedmont through public works, military engineering, and the general elevation of culture. Benedetti's duties included the teaching of mathematics and science at court. Tradition places him successively at the universities of Mondovì and Turin, although supporting official records are lacking and Benedetti never styled himself a professor. He appears, however, to have served as the duke's adviser on university affairs; for instance, he secured the appointment of Antonio Berga to the chair of philosophy at the University of Turin in 1569. Benedetti later engaged in a polemic with Berga, and on the title page of his *Consideratione* (1579) he referred to Berga as professor at Turin, but to himself only as philosopher to the duke of Savoy.

While at Turin, Benedetti designed and constructed various public and private works, such as sundials and fountains. His learning and mathematical talents were frequently praised by the duke and were mentioned by the Venetian ambassador in 1570, when Benedetti was granted a patent of nobility. In 1585 he appears to have been married a second time or rejoined by the mother of his daughter. In the same year he published his chief work, the *Speculationum,* a collection of treatises and of letters written to various correspondents on mathematical and scientific topics.

Benedetti died early in 1590. He had forecast his death for 1592 in the final lines of his last published book. On his deathbed he recomputed his horoscope and declared that an error of four minutes must have been made in the original data (published in 1552 by Luca Guarico), thus evincing his lifelong faith in the doctrines of judiciary astrology.

Benedetti's first important contribution to the birth of modern physics was set forth in the letter of dedication to his *Resolutio.* The letter was addressed to Gabriel de Guzman, a Spanish Dominican priest with whom he had conversed at Venice in 1552. It appears that Guzman had shown interest in Benedetti's theory of the free fall of bodies, and had asked him to publish a demonstration in which the speeds of fall would be treated mathematically. In order to forestall the possible theft of his ideas, Benedetti published his demonstration in this letter despite its irrelevance to the purely geometrical content of the book. Benedetti held that bodies of the same material, regardless of weight, would fall through a given medium at the same speed, and not at speeds proportional to their weights, as maintained by Aristotle. His demonstration was based on the principle of Archimedes, which probably came to his attention through Tartaglia's publication at Venice in 1551 of a vernacular translation of the first book of the Archimedean treatise on the behavior of bodies in water. Benedetti's "buoyancy theory of fall" is in many respects identical with that which Galileo set forth in his first treatise *De motu,* composed at Pisa about 1590 but not published during his lifetime.

Although no mention of Benedetti's theory has been found in books or correspondence of the period, lively discussions appear to have taken place concerning it, some persons denying the conclusion and others asserting that it did not contradict Aristotle. In answer to those contentions, Benedetti promptly published a second book, the *Demonstratio* (1554), restating the argument and citing the particular texts of Aristotle that it contradicted. In the new preface, also addressed to Guzman, Benedetti mentioned opponents as far away as Rome who had declared that since Aristotle could not err, his own theory must be false. Such discussions may explain the otherwise remarkable coincidence that another book published in 1553 also contains a statement related to free fall. This was *Il vero modo di scrivere in cifra,* by Giovanni Battista Bellaso of Brescia, in which it was asked why a ball of iron and one of wood will fall to the ground at the same time.

Two editions of Benedetti's *Demonstratio,* which

was by no means a mere republication of the *Resolutio,* appeared in rapid succession. The first edition maintained, as did the *Resolutio,* that unequal bodies of the same material would fall at equal speed through a given medium. The second edition stated that resistance of the medium is proportional to the surface rather than the volume of the falling body, implying that precise equality of speed for homogeneous bodies of the same material and different weight would be found only in a vacuum. This correction of the original statement was repeated in Benedetti's later treatment of the question in *Speculationum* (1585).

Benedetti's original publication of his thesis in 1553 was designed to prevent its theft; perhaps he had in mind the fate of Tartaglia's solution of the cubic equation a few years earlier. But even repeated publication failed to protect it, and indeed became the occasion of its theft. Jean Taisnier, who pirated the work of Petrus Peregrinus de Maricourt in his *Opusculum . . . de natura magnetis* (1562), included with it—as his own—Benedetti's *Demonstratio.* Taisnier's impudent plagiarism enjoyed wider circulation than Benedetti's original, and was translated into English by Richard Eden about 1578. Simon Stevin cited the proposition as Taisnier's when he published his own experimental verification of it in 1586. But since Taisnier had stolen the *Demonstratio* in its earlier form, he was criticized by Stevin for the very fault which Benedetti had long since corrected in the second *Demonstratio* of 1554. Taisnier's appropriation of his book ultimately became known to Benedetti, who complained of it in the preface to his *De gnomonum* (1573). The relatively small circulation of Benedetti's works is evidenced by the fact that it was not until 1741 that general attention was first called to the theft, by Pierre Bayle.

Benedetti's ultimate expansion of his discussion of falling bodies in the *Speculationum* is of particular interest because it includes an explanation of their acceleration in terms of increments of impetus successively impressed *ad infinitum.* That conception is found later in the writings of Beeckman and Gassendi, but it appears never to have occurred to Galileo. Despite this insight, however, Benedetti failed to arrive at (or to attempt) a mathematical formulation of the rate of acceleration. The difference between Galileo's treatment and Benedetti's is perhaps related to the fact that Benedetti neglected the medieval writers who had attempted a mathematical analysis of motion and did not adopt their terminology, which is conspicuous in Galileo's early writings. Benedetti was deeply imbued with the notion of impetus as a self-exhausting force, a concept that may have prevented his further progress toward the inertial idea implicit in the accretion of impetus.

Benedetti's next contribution to physics was made about 1563, the most probable date of two letters on music written to Cipriano da Rore and preserved in the *Speculationum.* Those letters, in the opinion of Claude Palisca, entitle Benedetti to be considered the true pioneer in the investigation of the mechanics of the production of musical consonances. Da Rore was choirmaster at the court of Parma in 1561–1562, when he returned to Venice. Benedetti's letters probably supplemented his discussions with da Rore at Parma. Departing from the prevailing numerical theories of harmony, Benedetti inquired into the relation of pitch, consonance, and rates of vibration. He attributed the generation of musical consonances to the concurrence or cotermination of waves of air. Such waves, resulting from the striking of air by vibrating strings, should either agree with or break in upon one another. Proceeding thus, and asserting that the frequencies of vibration of two strings under equal tension vary inversely with the string lengths, Benedetti proposed an index of agreement obtained by multiplication of the terms of the ratio of a given consonance; by this means he could express the degree of concordance in a mathematical scale. Benedetti's empirical approach to musical theory, as applied to the tuning of instruments, anticipated the later method of equal temperament and contrasted sharply with the rational numerical rules offered by Gioseffo Zarlino. It is of interest that Zarlino was the teacher of Galileo's father, who in 1578 attacked Zarlino's musical theories on somewhat similar empirical grounds. But since Benedetti's letters were not published until 1585, they were probably not known to Vincenzio Galilei when he launched his attack.

Benedetti's first publication after his move to Turin was a book on the theory and construction of sundials, *De gnomonum* (1573), the most comprehensive treatise on the subject to that time. It dealt with the construction of dials at various inclinations and also with dials on cylindrical and conical surfaces. This book was followed by *De temporum emendatione,* on the correction of the calendar (1578). In 1579 he published *Consideratione,* a polemic work in reply to Antonio Berga, concerning a dispute over the relative volumes of the elements earth and water. As with Galileo's polemic on floating bodies, the dispute had arisen at court as a result of the duke's custom of inviting learned men to debate topics of philosophical or scientific interest before him. Of all Benedetti's books, this appears to be the only one to have received notice in a contemporary publication, Agostino Michele's *Trattato della grandezza dell'acqua et della terra* (1583).

Benedetti's final work, containing the most important Italian contribution to physical thought prior to Galileo, was the *Diversarum speculationum* (1585). Its opening section includes a number of arithmetical propositions demonstrated geometrically. Other mathematical sections include a treatise on perspective, a commentary on the fifth book of Euclid's *Elements,* and many geometrical demonstrations—including a general solution of the problem of circumscribing a quadrilateral of given sides; the development of various properties of spherical triangles, circles, and conic sections; discussion of the angle of contact between circular arcs; and theorems on isoperimetric figures, regular polygons, and regular solids.

The section on mechanics is largely a critique of certain parts of the pseudo-Aristotelian *Questions of Mechanics* and of propositions in Tartaglia's *Quesiti, et inventioni diverse.* Benedetti disputed Tartaglia's assertion that no body may be simultaneously moved by natural and violent motions, although he did not enter into a discussion of projectile motion giving effect to composition. He did, however, assert clearly and for the first time that the impetus of a body freed from rapid circular motion is rectilinear and tangential in character, a conception of fundamental importance to his criticisms of Aristotle and to his attempted explanation of the slowing down of wheels and of spinning tops.

Following the section on mechanics is an attack on many of Aristotle's basic physical conceptions. This section restates the "buoyancy theory of fall" as it was set forth in the second edition of the *Demonstratio.* For equality of speed of different weights falling *in vacuo,* Benedetti proposed a thought experiment that is often said to be identical with Galileo's, although the difference is considerable. Benedetti supposes two bodies of the same weight connected by a line and falling *in vacuo* at the same speed as a single body having their combined weight; he appeals to intuition to show that whether connected or not, the two smaller bodies will continue to fall at the same speed. In Galileo's argument, two bodies of different weight—and therefore of different speeds, according to Aristotle—are tied together; by the Aristotelian assumption, the slower would impede the faster, resulting in an intermediate speed for the pair. But the pair being heavier than either of its parts, it should fall faster than either, under the Aristotelian rule. Thus Galileo's argument, unlike Benedetti's, imputes self-contradiction to Aristotle's view. Bene-

detti's discussion of the ratios of speeds of descent in different media is also essentially different from Galileo's in *De motu,* for it includes both buoyancy and the effect of resistance proportional to the surface; the latter effect was neglected by Galileo in his earlier writings.

Again, Benedetti correctly holds that natural rectilinear motion continually increases in speed because of the continual impression of downward impetus, whereas Galileo wrongly believed that acceleration was an accidental and temporary effect at the beginning of fall only, an error which vitiated much of the reasoning in *De motu* and was corrected only in his later works. These differences create historical perplexities described below.

Another of Benedetti's contributions to mechanics is the description of hydrostatic pressure and the idea of a hydraulic lift, prior to Stevin's discussion of the hydrostatic paradox (1586). Benedetti also attributed winds to changes in density of air, caused by alterations of heat. In opposition to the view that clouds are held in suspension by the sun, he applied the Archimedean principle and stated that clouds seek air of density equal to their own; he also observed that bodies are heated by the sun in relation to their degree of opacity.

Benedetti published no separate work on astronomy, but his letters in the *Speculationum* show that he was an admirer of Copernicus and that he was much concerned with accuracy of tables and precise observation. His astronomical interests appear to have been astrological rather than physical and systematic, as were those of Kepler, Galileo, and Stevin. Benedetti offered a correct explanation of the ruddy color of the moon under total eclipse, however, based on refraction of sunlight in the earth's atmosphere.

Benedetti's scientific originality and versatility leave little doubt that his work afforded a basis for the overthrow of Aristotelian physics. The extent of its actual influence on others, however, presents very difficult questions. Stevin was certainly unaware of Benedetti when he published his basic contributions to mechanics and hydrostatics. He had seen Benedetti's *Speculationum* before 1605, when he published on perspective, but in that work he built more on Guido Ubaldo del Monte than on any other writer. Kepler mentioned Benedetti but once, and only in the most general terms. Willebrord Snell's attention was called to Benedetti by Stevin. The case of Galileo is the most perplexing. It is widely held that he was directly indebted to Benedetti for the ideas underlying *De motu,* but the resemblances of those ideas are easily accounted for by the Archimedean principle and the medieval impetus theory, easily accessible to both men independently, while the differences, particularly with respect to acceleration and the accumulation of impressed motion, are hard to explain if the young Galileo had the work of Benedetti before him. The absence of Benedetti's name in Galileo's books and notes, where other kindred spirits such as Gilbert and Guido Ubaldo are praised, is suggestive; much more so is the fact that Benedetti is not mentioned to or by Galileo in the vast surviving correspondence of his time. Jacopo Mazzoni has been proposed as a positive link—he was a colleague and friend of Galileo's at Pisa about 1590, and certainly knew Benedetti's work by 1597; but since Galileo left Pisa for Padua in 1592, the connection is uncertain. Benedetti appears to have remained unknown to Galileo's teacher at Pisa, Francesco Buonamico, who in 1591 published a treatise, *De motu,* of over a thousand pages. On the whole, it appears that Benedetti's *Speculationum* was not widely read by his contemporaries, despite its outstanding achievements in extending the horizons of mathematics, physics, and astronomy beyond the Peripatetic boundaries.

BIBLIOGRAPHY

I. ORIGINAL WORKS. No collection of Benedetti's work has been published. Because the original editions are very scarce, locations of known copies not listed in the Union Catalogue are indicated by the following abbreviations: BM—British Museum; BN—Bibliothèque Nationale; BNT—Biblioteca Nazionale di Torino; ULT—University of Toronto Library.

Resolutio omnium Euclidis problematum aliorumque ad hoc necessario inventorum una tantummodo circini data apertura, per Ioannem Baptistam de Benedictis inventa (Venice, 1553), BM, BN, BNT, ULT.

Demonstratio proportionum motuum localium contra Aristotelem et omnes philosophes (Venice, 1554), first ed. in Vatican Library, 2nd ed. in Biblioteca Universitaria, Padua. First edition pirated by Jean Taisnier, *Opusculum perpetua memoria dignissimum de natura magnetis . . . Item de motu continuo, demonstratio proportionum motuum localium contra Aristotelem etc.* (Cologne, 1562; facs. repr., London, 1966). English translation by Richard Eden is in *A very necessarie and profitable Booke concerning Navigation, compiled in Latin by Joannes Taisnierus . . .* (London, 1578?), BM, Brown University Library. Both eds. repr. in C. Maccagni, *Le speculazioni giovanili "de motu" di G. B. Benedetti* (Pisa, 1967).

De gnomonum umbrarumque solarium usu liber . . . nunc primum publicae utilitati, studiosorumque commoditati in lucem aeditus (Turin, 1573), BM, BN, BNT, ULT.

De temporum emendatione opinio . . . (Turin, 1578), BM,

MSS copies in BNT and Archivio di Stato, Turin. Repr. in *Speculationum,* pp. 205–210.

Consideratione di Gio. Battista Benedetti . . . d'intorno al discorso della grandezza della terra, et dell'acqua, del Ecelent. Sig. Antonio Berga . . . (Turin, 1579), BN, BNT, ULT; a Latin trans. by F. M. Vialardi is reported to be contained in Antonio Berga, *Disputatio de magnitudine terrae et aquae* (Turin, 1580).

Lettera per modo di discorso all'Ill. Sig. Bernardo Trotto intorno ad alcune nuove riprensioni et emendationi contra alli calculatori delle effemeridi . . . (Turin, 1581), BNT, Latin trans. in *Speculationum,* pp. 228–248.

Diversarum speculationum mathematicarum, et physicarum, liber (Turin, 1585), BM, BN, BNT, ULT. Reissued as *Speculationum mathematicarum, et physicarum, fertilissimus, pariterque utilissimus tractatus . . .* (Venice, 1586) and as *Speculationum liber: in quo mira subtilitate haec tractata continentur . . .* (Venice, 1599), BM, Biblioteca Marciana, Venice.

De coelo et elementis . . . (Ferrara, 1591), sometimes reported as Benedetti's, was in fact written by Giovanni Benedetto of Tirna (Orsova).

Two volumes of manuscript letters and astronomical observations by Benedetti, formerly in the Biblioteca Nazionale, Turin, were lost in 1904 in a fire that also destroyed the only known portrait of Benedetti.

Manuscript volumes reported to be privately owned in Italy are "Descrittione, uso et ragioni del triconolometro. Al serenissimo Prencipe di Piemonte. Trattato di Giovan Battista Benedetti. MDLXXVII," 94 unnumbered pages including 21 blanks; "La generale et necessaria instruttione per l'intelligentia et compositione d'ogni sorte horologij solari," 46 folio pages, unnumbered, without the name of the author but in the hand of same scribe as the above MS; and "Dechiaratione delle parti et use dell'instromento chiamato isogonio," 24 folio pages in the same hand as the foregoing.

II. SECONDARY LITERATURE. The principal monograph is Giovanni Bordiga, "Giovanni Battista Benedetti, filosofo e matematico veneziano del secolo XVI," in *Atti del Reale Istituto veneto di scienze, lettere ed arti,* **85,** pt. 2 (1925–1926), 585–754.

Apart from histories of mathematics or physics, such as those of Guillaume Libri, Kurt Lasswitz, Moritz Cantor, Rafaello Caverni, and Ernst Mach, discussions of Benedetti's work (listed chronologically) are in Emil Wohlwill, "Die Entdeckung des Beharrungsgesetzes," in *Zeitschrift für Völkerpsychologie und Sprachwissenschaft,* **14** (1885), 391–401, and *Galileo Galilei und sein Kampf . . .* (Hamburg–Leipzig, 1909), I, 111 ff.; Giovanni Vailati, "Le speculazione di Giovanni Battista Benedetti sul moto dei gravi," in *Atti dell'Accademia delle scienze di Torino,* **33** (1897–1898), 559 ff., reprinted in *Scritti di G. Vailati* (Leipzig–Florence, 1911), pp. 161–178; Pierre Duhem, *Les origines de la statique* (Paris, 1905), I, 226–235; "De l'accélération produite par une force constante," Congrès International d'Histoire des Sciences, IIIᵉ Session (Geneva, 1906), pp. 885 ff; and *Études sur Léonard de Vinci* (Paris, 1913), III, 214 ff.

See also E. J. Dijksterhuis, *Val en Worp* (Groningen, 1924), pp. 179–190, and *The Mechanization of the World Picture* (Oxford, 1961), pp. 269–271; Alexandre Koyré, *Études galiléennes* (Paris, 1939), I, 41–54, and "Jean-Baptiste Benedetti critique d'Aristote," in *Mélanges offerts à Étienne Gilson* (Toronto–Paris, 1959), pp. 351–372, reprinted in *Études d'histoire de la pensée scientifique* (Paris, 1966), pp. 122–146, an English trans. in *Galileo, Man of Science,* ed. E. McMullin (New York, 1967); Claude Palisca, "Scientific Empiricism in Musical Thought," in *Seventeenth Century Science and the Arts* (Princeton, 1961), pp. 104 ff.; I. E. Drabkin, "Two Versions of G. B. Benedetti's *Demonstratio,*" in *Isis,* **54** (June 1963), 259–262, and "G. B. Benedetti and Galileo's *De motu,*" in *Proceedings of the Tenth International Congress of the History of Science* (Paris, 1964), I, 627–630; and C. Maccagni, "G. B. Benedetti: *De motu,*" in *Atti del Symposium Internazionale di Storia, Metodologia, Logica e Filosofia della Scienza* (Florence, 1966), pp. 53–54; "Contributi alla biobibliografia di G. B. Benedetti," in *Physis,* **9** (1967), 337–364; and *Le speculazioni "de motu" di G. B. Benedetti* (Pisa, 1967). English translations of Benedetti's most important scientific writings by I. E. Drabkin are included in *Mechanics in Sixteenth-Century Italy* (Madison, Wis., 1968).

STILLMAN DRAKE

BENEDICKS, CARL AXEL FREDRIK (*b.* Stockholm, Sweden, 27 May 1875; *d.* Stockholm, 16 July 1958), *metallography.*

Benedicks' father, Edvard Otto Benedicks, and his mother, Sofia Elisabeth Tholander, came from families that had been involved in the Swedish steel industry for generations. Interested from boyhood in the theoretical study of minerals and metals, Benedicks concentrated on the natural sciences at the University of Uppsala and received the Ph.D. in 1904 with the dissertation *Recherches physiques et physico-chimiques sur l'acier au carbone.*

In 1910 Benedicks was appointed professor of physics at the University of Stockholm. He is regarded as the father of Swedish metallography because he pleaded for the establishment of a special research laboratory for metals and metallography. Sponsored by the iron and steel industry and the government, the Swedish Institute of Metallography in Stockholm became a reality in 1920, and Benedicks served as its director for fifteen years. He continued his scientific work for twenty years more at the Laboratorium Benedicks.

Benedicks personally performed many experiments, usually employing apparatus designed and constructed by himself or in collaboration with his assistants. One of the fields in which he pioneered was metal microscopy.

In 1908 Benedicks was awarded the Iron & Steel Institute's gold medal for his investigations that were published as "Experimental Researches on the Cooling Power of Liquids on Queendring Velocities and on Constituents Troostite and Austenite."

In the light of modern physics, some of Benedicks' theories and interpretations of thermoelectrical phenomena and some of his solutions to problems concerning physical properties of steel seem to be outdated and are subject to criticism. Nevertheless, during his lifetime they aroused great interest in international circles.

Benedicks donated his private library of rare books on alchemy, natural science, metallurgy, and iron and steel manufacturing, totaling about 1,500 volumes, to the Institute of Metallography, now reorganized and named the Swedish Metal Research Institute.

BIBLIOGRAPHY

I. ORIGINAL WORKS. Benedicks' scholarly writings, often published simultaneously in various European journals, reflect a wide range of interests and considerable originality. Among them are "Thalénit, ein neues Mineral aus Österby in Dalekarlien," in *Bulletin of the Geological Institute of Uppsala,* **4** (1898–1899), 1–15; "On Fragments of Cast Iron, Designated as Crystals," in *The Iron and Steel Metallurgist and Metallographist,* **7** (1904), 252–257; "Sur les fers météoriques naturels et synthétiques et leur conductibilité électrique," in *Arkiv för matematik, astronomi och fysik,* **12,** no. 17 (1917), 1–11; "Le nouvel effet thermoélectrique et le principe d'étranglement," in *Revue de métallurgie,* **15** (1918), 296–332; *Non-metallic Inclusions in Iron and Steel* (London, 1930), written with H. Löfquist; "Über den Mechanismen der Supraleitung," in *Annalen der Physik,* **17** (1933), 184–196; "Effect of Gas Ions on the Benedicks Effects in Mercury," in *Nature,* **141** (1938), 1097, written with P. Sederholm; "Experiments Regarding the Influence of an Adsorbed Layer in Cohesion," in *Arkiv för matematik, astronomi och fysik,* **30A,** no. 6 (1944), 1–34, written with P. Sederholm; and "Theory of the Lightning-balls and Its Application to the Atmospheric Phenomenon Called 'Flying Saucers,'" in *Arkiv för geofysik,* **2,** no. 1 (1954), 1–11. A MS by Benedicks on his 483 articles and books is at the Swedish Metal Research Institute, Stockholm.

II. SECONDARY LITERATURE. Biographies are in *Journal of the Iron & Steel Institute, London,* **160,** no. 3 (1948), and The Svedberg, in *Svenskt biografiskt lexikon,* III (Stockholm, 1922), 163–170, which lists Benedicks' published works up to 1920. Erik Rudberg, in *Jernkontorets annaler,* **142** (1958), 733–738, is an elaborate obituary.

TORSTEN ALTHIN

BENEDICT, FRANCIS GANO (*b.* Milwaukee, Wisconsin, 3 October 1870; *d.* Machiasport, Maine, 14 May 1957), *chemistry, physiology.*

Washington Gano Benedict, a businessman, and Harriet Emily Barrett were average middle-class parents with secondary-school backgrounds. They moved in 1881 to Boston, where Francis attended public high school. He studied the piano and his parents planned a musical career for him. Meanwhile, he dabbled in his basement chemical laboratory.

Upon graduation (1888), Benedict devoted a year to the study of chemistry at the Massachusetts College of Pharmacy. He continued at Harvard University and received a B.A. (1893) and M.A. (1894) in chemistry. An additional year at Heidelberg under Professor Victor Meyer led to the Ph.D., *magna cum laude*. On his return Benedict assisted W. O. Atwater at Wesleyan University, where he became lecturer in chemistry in 1905 and professor in 1907. He supplemented his income through research as a chemist at the Storrs Agricultural Experimental Station (1896–1900) and as physiological chemist for the Department of Agriculture (1895–1907).

Atwater transformed Benedict from a chemist into the world's foremost expert on animal calorimetry and respiratory gas analysis. Benedict applied his mechanical skill to improvement of the apparatus used in these studies and the establishment of rigorous experimental controls.[1] William Welch and John Shaw Billings were impressed with Benedict's early publications on animal heat and metabolism, and they convinced the Carnegie Foundation trustees to establish a nutrition laboratory under Benedict's direction.[2] The result was the Boston Nutrition Laboratory, where Benedict remained until his retirement (1907–1937).

Benedict the man was straightforward and eminently Victorian. His marriage in 1897 to a third cousin, Cornelia Golay, provided Benedict with a research assistant and eventually a daughter. He was against alcohol, trade unions, and the "decline" of culture. His friendships and professional associations were largely European. Magic tricks and the piano were his hobbies. He equated science with progress and "the spirit of service" to humanity.[3]

Benedict's European affiliations were the result of his chosen field of research. The study of metabolism through calorimetry and the analysis of body intake and output was developed in the nineteenth century and was still current in Europe. In 1897, when Benedict joined him, Atwater prepared an exhaustive compilation of European contributions to the technique.[4] Atwater and Benedict confirmed the validity of energy conservation in animal metabolism.

The concept "metabolism" was radically changing in the early decades of the twentieth century.[5] The keys to energy transformation clearly lay within the emerging theories of hormonal control, catalysis, enzyme chemistry, and the vitamins. Yet Benedict continued in the classic methods, perhaps because of his chemical rather than biological training. His acknowledgment of the newer methods shows neither enthusiasm nor personal conviction.[6]

Benedict took care to distinguish between an organism's loss of heat and production of heat. He began with the basic assumption that all animal tissue produces heat to the same degree (calories per kilogram body weight per unit time). He tested the assumption through extensive comparisons of various cold- and warm-blooded animals and searched for extremes to prove his point. The newborn, the young, the elderly, the obese, the thin, the dieter, the vegetarian, patients with overactive thyroids or one lung, diabetics, the tiniest mouse, even the elephant, came under his scrutiny. By 1910 he realized that organisms do not produce heat in any simple, mechanical fashion. Large animals produced greater amounts of heat but small animals yielded more heat per unit of body weight. He remained steadfast, however, in the belief that homeostasis was maintained by heat loss, not heat production.[7] Accurate data were his driving passion, and comparative physiology provided the basis for his judgments.

He established criteria of comparison for the variable rates of metabolism that he found: sex, age, condition, water content, and the idea that "active protoplasmic mass" differed from total body weight. Two standards of judging metabolic rates were available to him: the relation to body surface area and to body weight. Benedict unequivocally favored the latter.

His most important contribution was the invention of an apparatus to measure simultaneously, directly, and accurately oxygen consumption, expired air, and heat (1924). His respirator provided the foundation for the basal metabolic-rate test, which has only recently given ground to less involved methods. His standards for the basal metabolic rates of humans are still valuable (1919). Insensible perspiration was found proportional to basal metabolism and body weight (1926), variations in the temperature of anatomical structures were mapped out (1911), lipogenesis investigated, and the caloric values of foods established. Diabetics were thought to have a low metabolic rate. Benedict revised clinical treatment of these patients by demonstrating that their metabolic rate was higher than normal (1910).

He was a member of the National Academy of Sciences (1914), American Philosophical Society (1910), American Academy of Arts and Sciences (1930), Society of American Magicians (1930), and honorary member of medical and scientific societies across Europe. He received the National Institute of Social Sciences medal in 1917 (for his work on the alcohol problem), the gold honor medal of the University of Hamburg in 1929 (physiology), and was given an honorary M.D. by the University of Würzburg in 1932.

NOTES

1. The best descriptions of his methods and apparatus are found in Benedict and Carpenter (1910), Benedict and Talbot (1914), and Benedict and Burger, Abderhalden's *Handbuch,* Abt. 4 (1924).
2. DuBois and Riddle, p. 69.
3. Commencement address given at the University of Maine, 9 June 1924; repr. in *Science,* n.s., **60** (1924), 209, 212–213.
4. W. O. Atwater and C. F. Langworthy, *A Digest of Metabolism Experiments in which the Balance of Income and Outgo was Determined,* bull. no. 45, U.S.D.A. Office of Experimental Stations (Washington, 1897).
5. See O. Folin, "A Theory of Protein Metabolism," in *American Journal of Physiology,* **13** (1905), 117–133; Russell Chittenden, *The Nutrition of Man* (New York, 1907). Benedict was aware of the changes: Atwater and Benedict (1899), p. 6, and the latter's work on creatinine (1907).
6. Benedict and Talbot (1921), p. 1; Benedict, *Lectures on Nutrition* (1925), pp. 17, 20; Benedict (1938), p. 200; DuBois and Riddle, pp. 73–74.
7. Benedict (1938), pp. 203, 212.

BIBLIOGRAPHY

I. ORIGINAL WORKS. Benedict published extensively. A list drawn up by E. DuBois and O. Riddle (see below) should be consulted. W. O. Atwater and Benedict joint publications are not included. They are as follows: *Experiments on the Metabolism of Matter and Energy in the Human Body,* bull. no. 69, U.S.D.A. Office of Experimental Stations (Washington, 1899); *Experiments on the Metabolism of Matter and Energy in the Human Body, 1898–1900,* bull. no. 109 (Washington, 1902); *Experiments on the Metabolism of Matter and Energy in the Human Body, 1900–1902,* bull. no. 136 (Washington, 1903).

II. SECONDARY WORKS. O. Riddle, "Francis Gano Benedict (1870–1957)," in *The American Philosophical Society Yearbook, 1957* (Philadelphia, 1958), pp. 109–113; E. F. DuBois and O. Riddle, "Francis Gano Benedict," in *Biographical Memoirs. National Academy of Sciences,* **32** (1958), 66–98.

CHARLES A. CULOTTA

BENIVIENI, ANTONIO (*b.* Florence, Italy, 3 November 1443; *d.* Florence, 11 November 1502), *pathological anatomy.*

A contemporary and friend of Angelo Poliziano, Marsilio Ficino, and Lorenzo the Magnificent, Benivieni was a famous physician, yet we have little information about him as a person. His father, Paolo, a nobleman, was a notary; he himself was the eldest of five brothers, of whom two others achieved distinction—Domenico as a theologian at the University of Pisa, and Girolamo as a philosopher and poet in Florence. Benivieni studied medicine at Pisa and practiced in Florence; his house stood where modern Ricasoli Street is, near the present-day Academy of Fine Arts. He had as patients the most important families of the city—the Bardi, Brunelleschi, Guicciardini, Pazzi, Strozzi, and the Medici themselves. He was on excellent terms with his medical colleagues Bernardo Torni (1452–1497) and Lorenzo Lorenzi (known as Laurentianus, who died in 1502), who were teachers in the University of Pisa, and he was well regarded by both Ficino and Savonarola. He was buried in the chapel of St. Michael in the Church of the SS. Annunziata, Florence. No reliable portrait of him exists.

Antonio Benivieni left a number of works in manuscript—*De pestilentia ad Laurentium Medicem; Consilium contra pestem Magistri Antonii Benivieni; De virtutibus,* an essay on Galenic physiology; notes on *De opinionibus antiquorum;* the title of a chapter of a *Liber de cometa ad Julianum Medicem;* separate notes on fossils and minerals; and a treatise, *De chirurgia.* A learned humanist, Benivieni occupies a place in the history of science for his *De abditis nonnullis ac mirandis morborum et sanationum causis,* an early essay on pathological anatomy published posthumously by Filippo Giunti at Florence in 1507. Originally conceived of by the author as a treatise of 300 selected observations or chapters, the book was put forth by his brother Girolamo with the aid of a physician (Giovanni Rosati) in an incomplete form containing only 111 chapters; a copy of this first edition is to be found in the Bibliotheca Marucelliana, Florence. Later issues of this incomplete version appeared at Paris (1528), Basel (1529), Cologne (1581), Leiden (1585), and Antwerp (1585). Alfonso Corradi in 1885 recorded an edition printed at Naples in 1519.

Although Benivieni's book was noted by Morgagni and Haller, it was virtually forgotten until Carlo Burci (1813–1875), the first ordinary professor (Florence, 1840) of pathological anatomy in Italy, translated it into Italian; it has been translated into English by Charles Singer and Esmond Long under the title *The Hidden Causes of Diseases* (1954). Both translations are based on the edition of 1507, and a complete knowledge of the work requires study of the author's manuscript, which was discovered by Francesco Puccinotti in 1855. Puccinotti noted in his researches that Benivieni had performed at least twenty autopsies and had observed gallstones, urinary calculi, scirrhous cancer of the stomach, fibrous tumor of the heart, peritonitis arising from intestinal perforation, and transmission of syphilis to the fetus. Benivieni was also a competent surgeon, as Burci pointed out.

The autograph manuscript of the *De abditis* was given by Puccinotti to Cesare Guasti, a paleographer, in order that a transcription of the unpublished parts might be made. This was most provident, for the original autograph again disappeared; fortunately, Guasti's transcription survives in the National Library of Florence and is believed to represent the text of the previously unpublished parts, unmodified and unedited.

Benivieni possessed among his books a copy of *De medicina libri octo* of Celsus, and Antonio Costa and his pupil Weber clearly documented the Celsian humanism of the *De abditis;* indeed, the title of Benivieni's work would appear to have been suggested by Celsus' "abditae morborum causae" (cf. Celsus, *De medicina,* Bk. I, *Proemium*). The critical work on the *De abditis* published by Costa and Weber is without doubt the best study of Benivieni's accomplishment and establishes his importance on a firm basis. As Ralph Major noted in 1935, "Benivieni's *De abditis* marks the beginning of a new method of thinking in medical science. . . . He was unquestionably a pathfinder in medicine who blazed a new path which physicians waited for more than two centuries and a half to follow."

BIBLIOGRAPHY

Benivieni's works are cited in the text.

Literature on Benivieni is Luigi Belloni, "Antonii Benivieni, De regimine sanitatis ad Laurentium Medicem," in *Atti del Congresso di Torino: 8–10 Giugno 1951, della Società italiana di patologia* (Milan, 1951); C. Burci, *Di alcune ammirabili ed occulte cause di morbi e loro guarigioni. Libro di Antonio Benivieni con un elogio storico intorno alla vita e alle opere dell'autore* (Florence, 1843); A. Corradi, "Un libro raro di sifilografia, e un'edizione ignota del Benivieni," in *Annali universali di medicina e chirurgia,* **271** (1885), 228–240; A. Costa and G. Weber, "L'inizio dell'anatomia patologica nel quattrocento fiorentino," in *Archivio De Vecchi,* **39** (1963), 429–878, 939–993, which includes a rich bibliography on the subject; B. De Vecchi, "La vita e l'opera di Maestro Antonio Benivieni," in *Atti della Società Colombaria di Firenze,* **30** (1930–1931), 103–122; and "Les livres d'un médecin humaniste," in *Janus,* **37** (1933), 97–108; L. Landucci, *Diario fiorentino dal 1450 al 1516* (Florence, 1883); Ralph Major, "Antonio di Pagolo Benivieni," in *Bulletin of the History of Medicine,* **3** (1935), 739–755; M. G. Nardi, "Antonio Benivieni e un suo scritto inedito sulla peste," in *Atti e memorie dell'Accademia di storia dell'arte sanitaria,* **4** (1938), 124–133; F. Puccinotti, *Storia della medicina* (Leghorn, 1855), II, pt. 1, 232–255; and Lynn Thorndike, "A Physician of Florence: Antonio Benivieni," in *A History of Magic and Experimental Science,* IV, 586–592.

PIETRO FRANCESCHINI

BENOIT, JUSTIN-MIRANDE RENÉ (*b.* Montpellier, France, 28 November 1844; *d.* Dijon, France, 5 May 1922), *physics.*

As director of the Bureau International des Poids et Mesures from 1889 to 1915, René Benoit played a large role in standardizing units of length, temperature, and electrical resistance for the scientific world.

Benoit's early training was in medicine, in compliance with the wishes of his father, Justin Benoit, a distinguished surgeon and dean of the faculty at the medical school of Montpellier. No sooner had René received his medical doctorate (1869), however, than he turned to his first interest, physics, and entered the laboratory of the École des Hautes Études. Here he conducted a series of experiments on the temperature dependence of electrical resistance in metals and received his *doctorat ès sciences* in 1873. Four years of work in the electrical industry followed, after which the Bureau International des Poids et Mesures offered him an assistant directorship.

The bureau was hardly three years old when Benoit arrived, and its main concern was the preparation of prototype standard units for its member nations. The new appointee was charged with determining the best method of measuring lengths. This involved a detailed study of thermal expansion, the various devices for measuring dilation, and the relative merits of different thermometers. Once the standard lengths were set up, Benoit played a large role in their verification and in establishing their relation to other standard units, such as the English yard.

The distribution of the first set of prototype lengths in 1889 coincided with Benoit's appointment as director of the bureau. In this capacity he became deeply involved in research to determine the relationship between the standard meter and various wavelengths of light, work that was carried out primarily by A. A. Michelson in 1891 and by C. Fabry and A. Perot in 1906. He was also associated with the standardization of twenty-four-meter Jaderin surveying wires and with the determination of the standard ohm.

Along with the directorship of the bureau, Benoit held the honorary position of president of the Société Française de Physique, correspondent of the Académie des Sciences, honorary fellow of the Physical Society of London, and officer of the Legion of

Honor. On his retirement in 1915, due to failing health and eyesight, his colleagues at the bureau made him honorary director as a token of their esteem.

BIBLIOGRAPHY

I. ORIGINAL WORKS. Benoit's published works are cataloged in Poggendorff, III, 107; IV, 98; V, 93; VI, 278; and in the *Royal Society Catalogue of Scientific Papers,* VII, 139; IX, 191; XIII, 449.

His thesis for his *doctorat ès sciences* is *Études expérimentales sur la résistance électrique des métaux et sa variation sous l'influence de la température* (Paris, 1873). Devices for measuring dilation are discussed in "Mesures de dilation et comparaison des règles métriques," in *Mémoires et travaux du Bureau international des poids et mesures,* **2** (1883), C.1–C.174, C.i–C.clxvii; and **3** (1884), C.1–C.44, C.1–C.xlvi. Benoit's work in determining the standard ohm is presented in *Détermination de l'ohm* (Paris, 1884), written with E. E. N. Mascart and F. G. Nerville; and *Construction d'étalons prototypes de résistance électrique* (Paris, 1885). An account of the relative merits of different thermometers is "Études préliminaires sur les thermomètres . . .," in *Mémoires et travaux du Bureau international des poids et mesures,* **7** (1890), 10–13. Research on the relationship between the standard meter and various wavelengths of light is presented in "Nouvelles déterminations du mètre en longueurs d'ondes lumineuses," in *Comptes rendus de l'Académie des sciences,* **144** (1907), 1082. For an idea of his work as director of the bureau, see his "Rapports règlementaires du directeur du Bureau international des poids et mesures," *Procès-verbaux des séances du Comité international des poids et mesures,* in each volume from 1890 to 1915.

II. SECONDARY LITERATURE. The most complete biography of Benoit is Charles Édouard Guillaume, in *Procès-verbaux des séances du Comité international des poids et mesures,* **20** (1923). For shorter works, see *Nature,* **109** (1922), 820; and *Dictionnaire de biographie française,* V, 1442.

EUGENE FRANKEL

BENSLEY, ROBERT RUSSELL (*b.* Hamilton, Ontario, 13 November 1867; *d.* Chicago, Illinois, 11 June 1956), *anatomy.*

Bensley was the third of six children born to Robert Daniel Bensley, a prosperous farmer. From his mother, Caroline Vandeleur, Bensley acquired his Irish wit, his love of music and the fine arts, and his talent for languages (he mastered French, German, Italian, and Spanish).

In 1883 he graduated from the Collegiate Institute in Hamilton, a notable achievement in that he had walked the two and a half miles (each way) up and down the "mountain" (as the Niagara escarpment is called there) each day and had also managed to do his share of the farm chores. The following year, Bensley entered University College of the University of Toronto. His academic career was almost ended when, on a hunting trip, he received a shotgun wound in the left leg that severed an artery. The limb was amputated twice, first below the knee and later above the knee, when gangrene set in. After a year of convalescence, Bensley managed to return to University College, and upon his graduation he received the Governor General's Medal in both arts and sciences. During his convalescence, working in his mother's kitchen, he learned to stain tissues obtained from farm animals with dyes made from local barks and herbs. He prepared histological slides so well that he was not only excused from the histology laboratory at medical school but was also congratulated for his excellent preparations.

In 1889 Bensley entered the medical department of the University of Toronto; three years later he received the M.B. and became an assistant demonstrator in biology there upon graduation. Four years later, in a paper that won high praise from Sir William Osler, he showed clearly the replenishment of the cells of the gastric mucosa.

Bensley joined the anatomy department of the University of Chicago in 1901, becoming its acting head in the same year and its director in 1907. He differentially stained the cells of the islands of Langerhans in 1906, an achievement that led to Banting's discovery of insulin. Among his other accomplishments were confirmation of the presence of the Golgi apparatus of cells in 1910 and, with B. C. H. Harvey in 1912, the demonstration of the mechanism of the gastric secretion of hydrochloric acid.

Bensley pioneered in the study of cell organelles. I. Altman, A. Fisher, W. B. Hardy, and L. Michaelis had previously described mitochondria but, in 1934, with his student I. Gersh, Bensley modified Altman's freezing-drying technique so that mitochondria could be isolated and subjected to microchemical analysis.

During his twenty-six years at Chicago, Bensley brought to the department of anatomy C. Judson Herrick, George W. Bartelmez, Alexander Maximov, Charles Swift, and William Bloom.

Arthur Benjamin Bensley, Robert's younger brother, gained fame for his book *Practical Anatomy of the Rabbit,* as well as for his directorship of the Ontario Museum of Zoology in Toronto.

Bensley's daughter, Caroline May, was for many years technical assistant in the department of anatomy at the University of Chicago. His son, Robert (husband of Sylvia Holton Bensley), has contributed much to color photography as well as aiding his father in some of the latter's endeavors in photomicrography.

BIBLIOGRAPHY

I. ORIGINAL WORKS. A complete list of Bensley's publications follows Norman L. Hoerr's memoir (see below), pp. 15–18. Among his works are "The Cardiac Glands of Mammals," in *American Journal of Anatomy,* **2** (1902), 105–156; "Upon the Formation of Hydrochloric Acid in the Foveolae and on the Surface of the Gastric Mucous Membrane and the Non-acid Character of the Contents of Gland Cells and Lumina," in *Biological Bulletin,* **23** (1912), 225–249, written with B. C. H. Harvey; "The Formation of Hydrochloric Acid on the Free Surface and Not in the Glands of the Gastric Mucous Membrane," in *Transactions of the Chicago Pathological Society,* **19** (1913), 14–16, written with B. C. H. Harvey; "The Thyroid Gland of the Opossum," in *Anatomical Record,* **8** (1914), 431–440; "Structure and Relationship of the Islets of Langerhans: Criteria of Histological Control in Experiments on the Pancreas," in *Harvey Lectures,* **10** (1915), 250–289; "Functions of Differentiated Segments of Uriniferous Tubules," in *American Journal of Anatomy,* **41** (1928), 75–96, written with W. B. Steen; "The Gastric Glands," in *Cowdry's Special Cytology* (New York, 1928); "The Functions of the Differentiated Parts of the Uriniferous Tubule in the Mammal," in *American Journal of Anatomy,* **47** (1931), 241–275, written with S. H. Bensley; "Studies on Cell Structure by the Freezing-Drying Method. I. Introduction. II. The Nature of Mitochondria in the Hepatic Cell of Amblystoma," in *Anatomical Record,* **57** (1933), 205–235, written with I. Gersh; "Studies on Cell Structure by the Freezing-Drying Method. III. The Distribution in Cells of the Basophil Substances, in Particular the Nissl Substance of the Nerve Cell," *ibid.,* 369–385, written with I. Gersh; "Studies on Cell Structure by the Freezing-Drying Method. IV. The Structure of the Interkinetic and Resting Nuclei," *ibid.,* **58** (1933), 1–15; "Studies on Cell Structure by the Freezing-Drying Method. V. The Chemical Basis of the Organization of the Cell," *ibid.,* **60** (1934), 251–266, written with N. L. Hoerr; and "Studies on Cell Structure by the Freezing-Drying Method. VI. The Preparation and Properties of Mitochondria," *ibid.,* 449–455, written with N. L. Hoerr.

II. SECONDARY LITERATURE. Hoerr wrote on his work with Bensley in "Introduction to Cytological Studies with the Freezing-Drying Method. II. Section of HCl in the Stomach," in *Anatomical Record,* **65** (1936), 417–435. His memoir on Bensley, prepared with the help of Sylvia H. Bensley, Gordon Scott, *et al.,* is in *Anatomical Record,* **128** (1957), 1–15.

P. G. ROOFE

BENTHAM, GEORGE (*b.* Stoke, near Plymouth, England, 22 September 1800; *d.* London, England, 10 September 1884), *botany.*

George Bentham was the third child and second son of Samuel Bentham and Maria Sophia Fordyce. His father, inspector-general of naval works, was ennobled in 1809; his mother was the eldest daughter of Dr. George Fordyce, F.R.S., a noted physician. Jeremy Bentham, the well-known authority on jurisprudence and ethics, was his uncle.

Bentham's early education was rather sporadic, largely because of the peripatetic family life, and was provided mainly by private tutors. During his father's tour of duty in St. Petersburg from 1805 to 1807, his precocity was evidenced in the ease with which he acquired conversational proficiency in Russian, French, and German and a knowledge of Latin. Later, in France, Bentham attended the faculty of theology at Montauban, where he studied French and Latin literature, natural philosophy, mathematics, and Hebrew, while indulging his tastes for music and drawing. Later in life he was able to read botanical works in fourteen languages, and it is a tribute to his industry, concentration, and high powers of reception that he did so largely by his own efforts.

He first became interested in botany at the age of seventeen, during travels in France, where he and his parents lived for eleven years. It was the encouragement of his mother, who was an accomplished gardener and a knowledgeable botanist, and access to her copy of Alphonse de Candolle's *Flore française,* whose analytical keys for the identification of plants appealed to his orderly mind, that fostered a penchant for systematic botany which became his consuming interest for over fifty years and to which he made outstanding contributions.

It was at Montauban that Bentham made his first dried specimens and thus began a herbarium which through his own collecting, purchase, and gift amounted to over 100,000 specimens when he gave it to Kew in 1854. Throughout the course of his studies he consistently worked at botany during his limited leisure time. He made his debut as a botanical author in November 1826, with publication in Paris of his *Catalogue des plantes indigènes des Pyrénées et du bas Languedoc.* As further testimony to his precocity and breadth of intellectual achievement, in 1827 he published *Outlines of a New System of Logic,* of which only about sixty copies were sold. The rest, owing to a financial crisis at the publisher, were disposed of for wastepaper.

Bentham studied law at Lincoln's Inn, and in November 1831, was called to the bar. However, in 1833, possessed of adequate wealth inherited from his father and his uncle, he determined to give up the legal profession for botany.

On 11 April 1833 Bentham married Sarah, daughter of Harford Brydges, onetime British envoy to Persia. They had no children. Bentham received many honors in recognition of his scientific achievements. In 1828 he was admitted as a fellow of the Linnean Society, of which he became president in 1861; in 1863 he was elected F.R.S., having been awarded the Society's Royal Medal in 1859. He received the LL.D. from Cambridge in 1874, and five years later was admitted to the Companionship of the Order of St. Michael and St. George. From 1829 to 1840 he was an extremely energetic and conscientious honorary secretary of the Royal Horticultural Society and, with John Lindley, steered its affairs through a very critical period.

Bentham corresponded with most of the leading botanists of the day and gave unstinting help to the many who requested it. His first major work, which appeared between 1832 and 1836, was *Labiatarum genera et species,* a masterly treatment of a very important family; and in 1848, at the request of Candolle, he contributed a revision of the group for the classic *Prodromus.* Subsequently he made additional contributions to this fundamental work.

Once his collections and library were amalgamated, by a generous gift in 1854, with those of the Hookers at Kew, Bentham was given special facilities to continue his research; in the years of intense effort which followed, some of his greatest taxonomic publications were produced. The *Genera plantarum* (1862–1883), in every respect the fulfillment of complete collaboration with Joseph D. Hooker, was the culmination of his scientific career. This monumental synthesis was based on critical analysis of the material in the Kew Herbarium and Gardens and is a masterly and meticulous extension of Candolle's system.

Bentham was one of the most unassuming of men, always regarding himself as an amateur and reluctant to accept any honor, yet a man who in his lucid, concise, and accurate writing did immense service to systematic botany.

BIBLIOGRAPHY

I. ORIGINAL WORKS. Bentham's works include *Catalogue des plantes indigènes des Pyrénées et du bas Languedoc* (Paris, 1826); *Outlines of a New System of Logic* (London, 1827); *Labiatarum genera et species* (London, 1832–1836); *Handbook of the British Flora* (London, 1858); *Flora Hongkongensis* (London, 1861); *Genera plantarum* (London, 1862–1883); *Flora Australiensis* (London, 1863–1878); "On the Recent Progress and Present State of Systematic Botany," in *Report of the British Association for the Advancement of Science* (1875), pp. 27–54.

II. SECONDARY LITERATURE. Works relevant to Bentham are Leonard Huxley, *Life and Letters of Sir Joseph Dalton Hooker* (London, 1918); B. Daydon Jackson, *George Bentham* (London, 1906).

GEORGE TAYLOR

BENZENBERG, JOHANN FRIEDRICH (*b.* Schöller, near Düsseldorf, Germany, 5 May 1777; *d.* Bilk, near Düsseldorf, 7 June 1846), *astronomy, geodesy, physics.*

Benzenberg was the son of Heinrich Benzenberg, a Protestant clergyman and theological writer. He studied theology in Herborn and Marburg, but later, in Göttingen, became interested in the natural sciences. He received his doctorate from the University of Duisburg in 1800, with a thesis "De determinatione longitudinis per stellas transvolantes." In 1805 Benzenberg became professor of mathematics at the Lyceum in Düsseldorf; later he directed the land survey of the duchy of Berg (1805–1810), organized its land registry, established the teaching of surveying, and wrote several textbooks for surveyors. He gave up this position and lived in Switzerland during the French occupation of the Rhineland. After his return to Germany, Benzenberg devoted himself to studying science as a private citizen; he simultaneously began to gain influence as a writer on politics, particularly constitutional law and economics. Benzenberg's political views were those of an enlightened liberal who advocated a strong central monarchy and a concomitant restriction of the power of the diet.

Benzenberg had begun original scientific investigation as early as 1798, while he was still a student at Göttingen; his independent work there, in which his collaborator was H. W. Brandes, was encouraged by the physicists A. G. Kästner and Georg Lichtenberg. With Brandes, he made the first simultaneous observations of meteors that made use of terminal points on a basis of either ten or fifteen kilometers. They then used these measurements to determine for the first time the height and velocity of the meteors. These data corroborated Chladni's theory of the cosmic origin of meteorites and their relation to fireballs, which had been published in 1794 and which was still hotly disputed. Benzenberg himself used his data to formulate a theory that "the shooting-stars are stones from moon-volcanoes." As late as 1839 (by which time it was widely assumed that meteorites were small heavenly bodies that rotated around the sun or the earth), he defended this idea against Olbers and Arago, among others.

Benzenberg's experiments on falling bodies were as original as his work on meteors. In these investigations he determined the displacement toward the east of falling lead spheres—in the tower of the Michaelis Church in Hamburg in 1802, and in a mine shaft in Schlebusch, in the Earldom of Mark, in 1804. He thereby demonstrated the revolution of the earth some fifty years before Foucault did. (That such a demonstration might be possible had been suspected by Newton, to be sure, but Robert Hooke's experiments of 1679 were inconclusive and those of Domenico Guglielmini in 1791 were highly inaccurate.)

In his later years Benzenberg occupied himself with ballistic experiments and published further works on geodetical, astronomical, and physical subjects. Two years before his death he built a small observatory on his own property in Bilk. In his will, he left his observatory to the city of Düsseldorf, providing a grant to pay the salary of a resident astronomer. Johann Schmidt, F. Brünnow, and R. Luther all worked there and made important contributions to astronomy, particularly in observations of the minor planets.

BIBLIOGRAPHY

I. Original Works. Among Benzenberg's works are *Versuche die Entfernung, die Geschwindigkeit und die Bahnen der Sternschnuppen zu Bestimmen* (Hamburg, 1800), written with H. W. Brandes; *Über die Bestimmung der geographischen Länge durch Sternschnuppen* (Hamburg, 1802); *Versuche über das Gesetz des Falls, über den Widerstand der Luft und über die Umdrehung der Erde* (Dortmund, 1804); *Vollständiges Handbuch der angewandten Geometrie* (1813); *Über den Cataster* (1818); *Über die Daltonsche Theorie* (Düsseldorf, 1830); and *Die Sternschnuppen* (Hamburg, 1839).

II. Secondary Literature. More on Benzenberg and his work can be found in articles by C. Reinhertz, in *Zeitschrift für Vermessungswesen*, **32** (1903), 17–25, 52–57, 65–92; K. Ketter, in *Allgemeine Vermessungsnachrichten*, **39** (1927), 385 f; W. Lindemann, in *Sternenwelt*, **3** (1951), 163–166; Poggendorff, I, 145; and in *Allgemeine deutsche Biographie*, II, 384; and *Neue deutsche Biographie*, II, 60, with a complete list of Benzenberg's works and secondary literature.

Bernhard Sticker

BÉRARD, JACQUES ÉTIENNE (*b.* Montpellier, France, 12 October 1789; *d.* Montpellier, 10 June 1869), *chemistry.*

Bérard was the son of Étienne Bérard, a chemical manufacturer and associate of Chaptal, by whom he was recommended to the chemist Berthollet as a laboratory assistant. About 1807 young Bérard went to live in Berthollet's house at Arcueil, near Paris, and he soon became a member of the Society of Arcueil. He took advantage of his proximity to Paris to study at the newly reestablished university and was successively *bachelier-ès-lettres* (1811) and *licencié-ès-sciences.* He returned to Montpellier in 1813 and received the M.D. on 9 July 1817. He became professor of chemistry in the faculty of pharmacy and later in the faculty of medicine. He was elected a correspondent in the chemistry section of the Paris Académie des Sciences on 20 December 1819.

Bérard's first published work (1809–1810) was on the analysis of salts and a study of solubilities, research that he undertook at the behest of Berthollet. His value for the density of nitric oxide was used by Gay-Lussac as a datum in the establishment of his law of combining volumes of gases. Bérard collaborated with Malus in a study of infrared and ultraviolet radiation, finding that both can be polarized.

Bérard's best-known research is that carried out in collaboration with François Delaroche on the specific heats of gases. This won the prize offered by the First Class of the Institute in 1811. To compare the specific heats of gases, they passed them through a spiral tube immersed in a copper calorimeter filled with water. The rise in temperature of the water when a current of hot gas under constant pressure was passed through the spiral was taken to be proportional to the specific heat. Their method was essentially that used earlier by Lavoisier. The accuracy of their results, although fair, was impaired by their failure to ensure that the gases were dry.

In 1821 the Académie des Sciences awarded Bérard a prize for his work on the ripening of fruit. This was the first scientific investigation of the effect of different atmospheres on the ripening of fruit. Bérard recognized that harvested fruits use oxygen and give off carbon dioxide. He stated that fruit does not ripen in the absence of oxygen, at least for a limited period, and therefore suggested that some fruits could be stored for a period of up to three months by placing them in a sealed jar with a paste of lime and ferrous sulfate, which was supposed to remove the oxygen.

BIBLIOGRAPHY

I. Original Works. Bérard's works include "Sur les élémens de quelques combinaisons, et principalement des carbonates et souscarbonates alcalins," in *Annales de chimie*, **71** (1809), 41–69; "Sur l'eau contenu dans la soude fondue," *ibid.*, **72** (1809), 96–101; "Mémoire sur la détermination de la chaleur spécifique des différens gaz," *ibid.*, **85** (1813), 72–110, 113–182, written with François Delaroche; "Mémoire sur les propriétés physiques et chimiques des divers rayons qui composent la lumière solaire," *ibid.*, 309–325; "Observations sur les oxalates et suroxalates alcalins," *ibid.*, **73** (1810), 263–289; "Mémoire sur les propriétés des différentes espèces de rayons qu'on peut séparer au moyen du prisme de la lumière solaire," in *Mémoires de la Société d'Arcueil*, **3** (1817), 1–47; "Mémoire sur la maturation des fruits," in *Annales de chimie et de physique*, **16** (1821), 152–183, 225–252; and "Lettre à M. Gay-Lussac sur les usines de gaz inflammable de la houille," *ibid.*, **28** (1825), 113–128.

II. SECONDARY LITERATURE. Further information on Bérard is in Maurice Crosland, *The Society of Arcueil. A View of French Science at the Time of Napoleon I* (Cambridge, Mass., 1967). Bérard's work on the ripening of fruit is discussed in Dana G. Dalrymple, *The Development of Controlled Atmosphere Storage of Fruit* (Washington, D.C., 1967), pp. 3–4.

M. P. CROSLAND

BÉRARD, JOSEPH FRÉDÉRIC (*b*. Montpellier, France, 8 November 1789; *d*. Montpellier, 16 April 1828), *medicine*.

Bérard's M.D. thesis, "Plan d'une médecine naturelle" (1811), aroused interest, and he soon became a very successful private teacher. He abandoned teaching to go to Paris, where he collaborated on the *Dictionnaire des sciences médicales,* contributing articles on cranioscopy, the elements, trance, and muscle strength. Returning to Montpellier in 1816, he tried unsuccessfully to found a journal and then returned to teaching. In 1823 Bérard was appointed professor of public health at the Faculté de Médecine of his native city, and he took the chair in 1826. He died two years later, at the age of thirty-nine.

Bérard was essentially an analyst of ideas and a philosopher as well as a historian, and his books are of great interest to those who wish to learn at first hand about the fundamental ideas of the leaders of the Montpellier school of science. He wished to prove that all these scholars were eager to have experience and observation triumph, but he had to admit that most of them were snared into the very errors that they condemned. For example, he wrote about Barthez, the great theoretician of vitalism:

The term of vital principle inserts into physiological language a genuine obscurity; it turns attention away from direct observation of phenomena and their comparative analysis . . . and directs it toward the search for causes . . . and thus destroys science. . . .

Barthez usually proceeds by . . . the synthetic method . . . the shortest and the least sure . . . which seems to me dangerous. On the contrary, the analytic method . . . which in physiology starts with individual facts and goes on to general phenomena, is sure and easy. It permits a free examination of dogmas and allows an

opportune place for all improvements possible [*Doctrine médicale de l'École de Montpellier . . .*, pp. 110–112].

Because of his acute critical sense, his lucidity, and his remarkable gift of exposition, Bérard deserves to be rescued from oblivion.

BIBLIOGRAPHY

I. ORIGINAL WORKS. Bérard's writings are "Plan d'une médecine naturelle," thesis (Montpellier, 1811); *Essai sur les anomalies de la variole et de la varicelle* (Montpellier, 1818), written with de Lavit; *Doctrine médicale de l'École de Montpellier et comparaison de ses principes avec ceux des autres écoles de l'Europe* (Montpellier, 1819); *Mémoire sur les avantages politiques et scientifiques du concours en général* (Paris, 1820); the articles "Cranioscopie," "Éléments," "Extase," and "Forces musculaires," in *Dictionnaire des sciences médicales* (1821–1824); various writings in *Revue médicale*, **6** (1821), 341–370; **7** (1822), 185–223, 456–493; **8** (1822), 152–180, 282–299; **9** (1822), 240–262; n.s. **1** (1824), 1–32, 462–486; n.s. **3** (1824), 277–294; *Doctrines des rapports du physique et du moral . . .* (Paris, 1823); *Lettre posthume et inédite de Cabanis à Fouquet sur les causes premières* (Paris, 1824), with notes; and *Notes additionnelles à l'édition de la doctrine générale des maladies chroniques de Dumas,* 2 vols. (Paris, 1824).

Détermination expérimentale des apports du système nerveux en général . . . was announced for publication in 1823 but never appeared.

II. SECONDARY LITERATURE. Works on Bérard are E. Beaugrand, "Bérard," in *Dictionnaire Dechambre,* IX (1868), 100; and R. de Saussure, "Fr. Bérard historien de la médecine," in *Bulletin of the History of Medicine,* supp. **3** (1944), 309–317.

PIERRE HUARD

BERENGARIO DA CARPI, GIACOMO (*b*. Carpi, Italy, *ca*. 1460; *d*. Ferrara, Italy, 24 November 1530 [?]), *medicine*.

Giacomo Berengario (or correctly, Barigazzi) was one of the five children of Faustino Barigazzi, surgeon of Carpi. He received rudimentary training in anatomy and surgery from his father and possibly some classical education from Aldus Manutius, who was in Carpi between 1469 and 1477 as tutor to the children of Lionello Pio, prince of Carpi. Berengario apparently studied more systematically, perhaps in Ferrara or Bologna, before entering the medical school of the University of Bologna. He studied there under Girolamo Manfredi, Lionello dei Vittori, Gabriele Zerbi, and Alessandro Achillini, and received his degree on 4 August 1489.

Nothing definite is known of Berengario's activities

from the time he left the university until his appointment as a member of its faculty of medicine in the early sixteenth century. It is likely, however, that he returned to Carpi, at least for a time, to assist his father in his surgical practice; and he probably practiced surgery independently during the wars that plagued Italy at this time and were further aggravated by the French invasion of 1494. That Berengario had acquired considerable distinction as a surgeon, very possibly as a military surgeon, is indicated by his appointment in the latter part of 1502 as lecturer in surgery at Bologna, where he was one of the four "foreign" members of the faculty chosen for their achievements and designated the *eminenti*. In 1506 he became a Bolognese citizen through the mediation of Pope Julius II, and in 1508 the government of Bologna placed him in charge of treating victims of the plague epidemic.

Berengario enjoyed great popularity as a teacher, despite a somewhat fiery and quarrelsome nature. In 1511, for reasons not now known, he developed a bitter enmity toward one of his colleagues; armed and accompanied by his servant and sixteen other companions, he searched for the luckless colleague in order to do him bodily harm. Thwarted in this, he incited his companions with the cry "To his home, let us kill his father and mother," which the group seemed not loath to do. Happily, they could not get into the house, and had to be content with smashing its exterior. The university authorities seem to have been willing to put up with several such violent incidents, partly because of Berengario's support by powerful individuals and families, notably the Medici, and partly because of his popularity with the students. Indeed, a record in the university archives reveals Berengario's success by a comparison of the numbers in his class with the much smaller numbers in the class of the second lecturer in surgery.

At some time during the pontificate of Julius II, Berengario was called to Rome for medical consultation. In 1513, Pope Leo X, a Medici, was directly responsible for Berengario's leave of absence from his academic duties, thus enabling him to go to Florence and take over the medical care of Alessandro Soderini, who was related to the pope by marriage. On several occasions thereafter Berengario was of service to the Medici, who repaid him with their patronage and protection.

In 1514 Berengario produced an edition of Mondino da Luzzi's early fourteenth-century Galenico-medieval dissection manual, *Anothomia Mundini noviter impressa ac per Carpum castigata*. Presumably the work was prepared as an aid for his students, but

its publication also suggests that Berengario had by this time begun to give greater prominence to anatomical investigation, which in his original appointment as surgical lecturer was merely an ancillary discipline.

In 1517 Berengario was one of the physicians called to attend Lorenzo de' Medici, who had received a gunshot wound and an occipital skull fracture in battle. Although he was not responsible for the immediate treatment and trepanation of Lorenzo's skull, Berengario was placed in charge of the postoperative care; this assignment led to his second book, *Tractatus de fractura calvae sive cranei* (1518). The *Tractatus* was written in little more than two months, soon after Berengario's return to Bologna, and dedicated to Lorenzo de' Medici. It opens with a short discussion of various sorts of skull fractures, followed by a grouping of the consequent lesions according to their symptoms. This is the most interesting and valuable portion of the work, for Berengario was able to cite from contemporary knowledge or from his own direct observation the relationship between the location of the lesions and the resulting neurological effects. Next, he discusses prognosis, diagnosis, treatment, the instruments to be employed, and the technique of craniotomy. Berengario's book was the most original neurosurgical treatise until then and was not surpassed until the appearance of Ambroise Paré's similar work in 1562, in which Paré expressed his appreciation of his predecessor's efforts and made use of them.

In 1520 Berengario was called to Cremona as one of the medical consultants for Marchese Galeazzo Pallavicini, who, according to Berengario, died in that same year of a renal complaint.

Although Berengario was officially lecturer in surgery, he was also responsible for instruction in anatomy, which, in the course of time, appears to have become his major interest. In his *Isagogae breves* (1522) he declares that by that time he had dissected several hundred bodies, and in the *Commentaria* (1521) he refers to a number of specific dissections he had performed on adult cadavers and on fetuses. It is not known when Berengario's interest in anatomy became predominant, although this may be marked by his edition of Mondino's *Anothomia* in 1514; and somewhere toward the close of the second decade of the sixteenth century his dissatisfaction with traditional accounts of the human body became strong enough to cause him to write his own treatise. It seems to have been produced very quickly, in approximately two years, and was published under the title *Commentaria cum amplissimis additionibus super anatomia Mundini cum textu eiusdem in pristinum &*

verum nitorem redacto, with a dedication to Cardinal Giulio de' Medici, later Pope Clement VII.

The *Commentaria,* consisting of slightly more than a thousand pages, is so arranged that each topic is introduced by the appropriate section of Mondino's *Anothomia* and followed by Berengario's critical commentary. The commentary presents the opinions of ancient and medieval writers, as well as those of Berengario's contemporaries, and finally his own judgment of the problem; sometimes these judgments are illustrated from his own cases or from other, contemporary ones known to him. As one who had an unusually great knowledge of human anatomy, as well as a considerable degree of intellectual independence and an argumentative and even pugnacious spirit, Berengario not only extended the knowledge of human anatomy but also opposed traditional, often Galenic, beliefs and sought to replace them with his own. The *Commentaria* was the first work since the time of Galen to display any considerable amount of anatomical information based upon personal investigation and observation, and although Berengario's antitraditional attitude was not the result of a consistent principle of independent investigation, his *Commentaria* nevertheless must be considered the most important forerunner of Vesalius' *Fabrica.*

After a brief discussion of the value of anatomical studies, Berengario, following the sequence of Mondino's text, systematically describes the general characteristics of the major structures and divisions of the body and the properties of fat, membrane, flesh, fiber, ligament, sinew and tendon, nerve, and muscle. This description, more extensive than any until then, is followed by a fairly extensive account of the abdominal muscles. In accordance with Mondino's arrangement, the abdominal organs, the most susceptible to putrefaction, are next considered. The account begins with mention of the peritoneum, but only the visceral portion is described; and Berengario erroneously criticizes Gentile da Foligno for referring to a muscular portion as well. The description of the intestines contains the first mention of the vermiform appendix; the yellow staining of the duodenum, caused by bile, is noted; and attention is called to the fact that the common biliary duct opens into the duodenum. The stomach is described as being formed of three intrinsic coats, to which a fourth, arising from the peritoneum, is added, although Berengario mentions only two intrinsic coats in the *Isagogae breves* of 1522. In his description of the kidneys Berengario denies that they contain a filter or sieve for straining off the urine, and is especially critical of his former teacher Gabriele Zerbi, who had claimed actual observation of such

a filter. During a public dissection in 1521, but too late for mention in the *Commentaria,* Berengario observed a fused kidney with horseshoe configuration, which he described in the *Isagogae breves.*

The description of the male and female reproductive organs, of reproduction itself, and of the fetus is much more extensive than any earlier account. In the course of this description Berengario for the first time called attention to the greater proportional capacity of the female pelvis than the male pelvis, and elsewhere he noted the proportionately larger size of the male chest. He also denied the medieval belief in the seven-celled uterus, and in direct opposition to Galen he declared that the umbilical cord carried a single vein rather than two.

Turning his attention to the neck and thorax, Berengario recognized five cartilages in the larynx, describing the two arytenoid cartilages for the first time, and provided the first good account of the thymus. He declared that the lungs were a single organ, although he admitted that some said they were plural. He did, however, note the existence of three right lobes, in contrast with two on the left. He remarked upon the oblique position of the heart, described the pericardium in some detail, and declared that it contained pericardial fluid at all times. In his description of the ventricles of the heart he sought a compromise between the Aristotelian description of a three-chambered heart and the Galenic description of a two-chambered one by declaring that the third ventricle posited by Aristotle was in the traditionally accepted Galenic pores of the cardiac septum. Although erroneous, this assertion was of some significance, for Niccolo Massa answered it in 1536 with the declaration that the heart's midwall was solid. Although Massa seems not to have realized the implication of his statement, he and Berengario unwittingly preluded the dispute that developed later in the century over the correct course of the blood from the right to the left side of the heart. In addition, although Berengario correctly described the aorta as arising from the left ventricle, he displayed some confusion regarding the position and structure of the pulmonary artery (which he called *vena arterialis*) and pulmonary vein (which he called *arteria venalis*).

In his description of the skull and its contents, Berengario was the first to describe the sphenoidal sinus, which he considered to be the source of catarrh, but he declared that the ethmoid plate was impervious to the passage of cerebral fluids. He also called attention to two of the ossicles of the ear, the malleus and incus, although he did not provide them with these names or claim the discovery. In his description of the brain, the account of the ventricles is relatively clear and extensive. Contrary to the then common practice of locating the mental faculties in the four ventricles, however, he placed them all in the first two, or lateral, ventricles and thereby reduced the significance of the remainder. He considered the third ventricle as being merely a passage through which excess cerebral fluids were transported to the infundibulum for ultimate excretion; it also served as a route for the transmission of a portion of the animal spirit into the fourth ventricle. The fourth ventricle served only as the receptacle for that portion of the animal spirit which served to operate the spinal nerves.

Of the other cerebral structures, Berengario noted the corpus striatum, declared that the choroid plexus was formed of veins and arteries, and described the pituitary and pineal glands. His most noteworthy anti-Galenic heresy was denial of the existence of the *rete mirabile* in the human brain, for the excellent reason that he had never been able to find it. However, since the accepted theory of the day would not permit him to deny the existence of animal spirit, the motive force of the nervous system until then asserted to be manufactured in the *rete mirabile,* Berengario declared that the spirit was manufactured in the very small branches of the arteries dispersed throughout the *pia mater.* Consequently, it was Berengario who began the movement to give greater importance to the entire brain substance.

Despite the significance of Berengario's anatomical findings and his correction of traditional errors, as well as his citations, which reflect wide reading in classical and contemporary literature, his literary style was more medieval than Renaissance. Whatever his merits or deficiencies as a stylist, he was greatly hindered by the absence of any standard anatomical vocabulary. Numerous medieval Latin translations from the Arabic by a number of different translators, often not really familiar with the subject matter of their translations, had produced a confusion of terminology; an anatomical structure might have several names or a single term might apply to several different structures. Berengario was well aware of this problem, and on one occasion described it in respect to the graphic illustration of a vein:

> In these figures can be seen the course of the salvatella of Mondino and Rhazes, the sceylen of Avicenna, the salubris of Haly, and that branch of the basilica which ends between the little and ring finger which Rhazes [also] calls salvatella. It can furthermore be seen how the vein called funis brachii by Avicenna ends around the middle finger in a branch called sceylen by Avicenna and salubris by Haly and how also the funis brachii

is a branch of the cephalic and terminates between the index finger and thumb [*Isagogae breves,* 1522, fol. 65 *r*].

Such conditions recommend visual assistance, and Berengario appears to have been the first anatomist to recognize the significance of anatomical illustrations properly related to the text. The *Commentaria* contains twenty-one pages of illustrations, some of them with the names of the significant structures written on the figures, with a brief explanation of the meaning of the particular illustration printed along its border. The figures are rather crude woodcuts, but the crudity does not greatly impede the explanation of anatomical terminology. Infrequently there is reference from the main text to the illustration. The illustrations are less successful when, as in those of the uterus, the purpose is to reveal details of structure rather than to explain terminology. It is particularly apparent in these figures that Berengario had not completely broken away from the older idea that illustrations must serve a decorative purpose, an idea that interfered with the presentation of anatomical detail. One figure, presumably displaying the abdominal muscles (actually plagiarized from Pietro d'Abano's *Conciliator* of 1496), is utterly worthless because the muscles are wholly imaginary. However, the remaining sequence of figures in which the successive muscle layers of the abdomen are displayed, dimly suggests, despite anatomical errors, the later Vesalian "muscle men." Berengario's results, however, were more important for the direction they took than for their achievement.

In 1521 there also appeared an edition of Ulrich von Hutten's small treatise of 1519 on the use of guaiac wood in the treatment of syphilis, *Ulrichi de Hutten Eq. de guaiaci medicina et morbo gallico liber unus.* It is uncertain whether the publication was primarily a publisher's speculation, since guaiac wood had gained sudden and very considerable popularity as a specific, or occurred at the urging of Berengario, whose name, however, appears only on the final page in the colophon. Berengario himself preferred to treat syphilis with mercury. In 1522 he was responsible for publishing an edition of Galen's work on medical prognostication, *Habes in hoc volumine candide lector magni Galeni Pergamensis medicorum principis libros tres de crisi. i. de judicationibus interprete Laurentiano medico Florentino,* and dedicated it to one of his students, Ochoa Gonzáles. Despite his intermittent criticisms of Galenic anatomy, Berengario remained a supporter of the "prince of physicians," who was not opposed for basing human anatomy on that of animals until 1543.

A much more important publication, in 1522, was the compendium of the *Commentaria,* entitled *Isagogae breves perlucide ac uberrime in anatomia humani corporis a communi medicorum academia usitatam a Carpo in almo Bononiensi Gymnasio ordinariam chirurgiae publice docente ad suorum scholasticorum preces in lucem datae* and dedicated to Alberto Pio, prince of Carpi. As a condensed version of the more extensive *Commentaria,* but with the same arrangement of contents, the *Isagogae breves* was intended by Berengario as a manual for his students and a replacement of his edition of Mondino's *Anothomia,* which he had made obsolete. If we may believe the title, it was prepared at his students' request. Its success is indicated by the fact that a second edition was published in 1523, with some alteration and increase in the number of illustrations, emphasizing the anatomy of the heart and brain.

In 1525, at the command of Pope Clement VII, Berengario went to Piacenza to assist in the care of Giovanni dalle Bande Nere, a member of the Medici family, who had suffered a wound of the right leg in battle. During 1526 Berengario spent several months in Rome as physician to the ailing Cardinal Pompeo Colonna, and, according to Vasari, in lieu of his fee accepted a "St. John in the Desert" painted by Raphael. It was also at this time that he met the sculptor and goldsmith Benvenuto Cellini, who described Berengario as a clever surgeon and a man of much learning whose services were sought on a permanent basis by Clement VII. Cellini also referred to Berengario's lucrative but ultimately disastrous practice of venerology, declaring that "he was a person of great sagacity and did wisely to get out of Rome," since through the excessive use of mercurial fumigations in the treatment of syphilis, "all the patients . . . grew so ill . . . that he would certainly have been murdered if he had remained."

Berengario's return to Bologna was brief, however, since he was dismissed from his position in the medical faculty. According to Gabriele Falloppia, he was charged with human vivisection, of which, however, there is no documentary evidence. Whatever the reason, Berengario retired to Ferrara, where he edited a collection of Latin translations of Galen: *Libri anatomici, horum indicem versa pagina indicabit. De motu musculorum liber primus, Nicolao Leoniceno vicentino interprete. Anatomicarum aggressionum interprete Demetrio Chalcondylo. De arteriarum et venarum dissectione, interprete Antonio Fortolo Ioseriensi. De hirudinibus, revulsione etc., interprete Ferdinando Balamio Siculo* (1529). The date of Berengario's death is uncertain and the place of his burial is unknown, although possibly it is the church of San Francesco in Ferrara.

BIBLIOGRAPHY

Berengario's writings include his edition of Mondino da Luzzi's *Anothomia Mundini noviter impressa ac per Carpum castigata* (Bologna, 1514; Venice, 1529, *ca.* 1530, 1538, *ca.* 1575); *Tractatus de fractura calvae sive cranei* (Bologna, 1518; Venice, 1535; Leiden, 1629, 1651, 1715, 1728); *Commentaria cum amplissimis additionibus super anatomia Mundini cum textu eiusdem in pristinum & verum nitorem redacto* (Bologna, 1521); his edition of Galen's *Habes in hoc volumine candide lector magni Galeni Pergamensis medicorum principis libros tres de crisi. i. de judicationibus interprete Laurentiano medico Florentino* (Bologna, 1522); *Isagogae breves perlucide ac uberrime in anatomia humani corporis a communi medicorum academia usitatam a Carpo in almo Bononiensi Gymnasio ordinariam chirurgiae publice docente ad suorum scholasticorum preces in lucem datae* (Bologna, 1522, 1523; Strasbourg, 1530; Venice, 1535); and his edition of a collection of Latin translations of Galen, *Libri anatomici, horum indicem versa pagina indicabit. De motu musculorum liber primus, Nicolao Leoniceno vicentino interprete. Anatomicarum aggressionum interprete Demetrio Chalcondylo. De arteriarum et venarum dissectione, interprete Antonio Fortolo Ioseriensi. De hirudinibus, revulsione etc., interprete Ferdinando Balamio Siculo* (Bologna, 1529).

All editions of the above works are very rare, and none has been published since the last date given for each. There is, however, an Italian translation of *De fractura calvae* in Vittorio Putti, *Berengario da Carpi saggio biografico e bibliografico* (Bologna, 1937). There is also an English translation of the *Isagogae breves* (London, 1660, 1664) and in *Jacopo Berengario da Carpi. A Short Introduction to Anatomy* (*Isagogae breves*), translated with introduction and historical notes by L. R. Lind and with anatomical notes by Paul G. Roofe (Chicago, 1959), pp. 3–29; it suffers, however, from some misunderstanding of Berengario's original text.

The best biographical account, including the pertinent documents and records, is in Putti's work cited above. The introduction to Lind's translation contains a brief biographical account drawn from Putti.

C. D. O'MALLEY

BERG, LEV SIMONOVICH (*b.* Bendery, Bessarabia, Russia, 14 February 1876; *d.* Leningrad, U.S.S.R., 24 December 1950), *geography, ichthyology.*

The son of Simon Grigor'evich Berg, a notary, and Klara L'vovna Bernstein-Kogan, Berg was awarded the gold medal when he graduated from the Second Kishinev Gymnasium in 1894. He then enrolled in the natural science section of the Faculty of Physics and Mathematics at Moscow University. In 1898 he received the gold medal and a first-degree diploma in zoology for his paper "Droblenie yaitsa i obrazovanie parablasta u shchuki" ("The Breakdown of the Egg and the Formation of the Parablast in Pike").

Striving to broaden his theoretical knowledge through practical work, Berg rejected an offer to remain at Moscow University and, having obtained the position of inspector of fisheries in the Aral Sea and on the Syr-Dar'ya River, he studied the lakes and rivers of central Asia and Kazakhstan for four years.

In 1902–1903 he was sent by the Russian Department of Agriculture to study oceanography in Bergen, Norway. Upon his return in 1904, Berg obtained the position of zoologist and director of the ichthyology section of the Zoological Museum of the Academy of Sciences in St. Petersburg, where he remained for ten years. In 1909 the degree of doctor of geography was conferred on Berg for his dissertation *Aral'skoe more* ("The Aral Sea"), for which he was also awarded the P. P. Semenov Gold Medal by the Russian Geographical Society. The following year Berg married Polina Abramovna Kotlovker, the daughter of a teacher, who bore him a son and a daughter. They were divorced in 1913, and in 1922 Berg married Maria Mikhailovna Ivanova, the daughter of a ship's commander.

In 1913 the Moscow Agricultural Institute made Berg professor of ichthyology, and the following year he moved to Moscow. In 1917 Petrograd University appointed him professor of geography, and he spent the rest of his life there. In addition, Berg was simultaneously head of the applied ichthyology section of the State Institute of Experimental Agronomy, which was later renamed the Institute of the Fish Industry (1922–1934), and ichthyologist and head of the fossil fish section of the Zoological Institute of the Academy of Sciences of the U.S.S.R. (1934–1950). He also taught for several years at the Pedagogical Institute and the Geographical Institute.

In 1934 Berg was awarded the degree of doctor of zoology for his many important ichthyological investigations. In recognition of his outstanding scientific services, the Academy of Sciences elected him an associate member in 1928 and an academician in 1946. He was also awarded the titles Honored Scientist of the R.S.F.S.R. (1934) and Laureate of the State, first degree (1951). From 1940 to 1950 he was president of the All-Union Geographical Society. Among his awards were the Konstantinovsky Medal of the Russian Geographical Society (1915) and the Medal of the Asiatic Society of India (1936).

Berg's works deal with a wide range of topics in geography and zoology. His early investigations were concentrated on the study of the lakes of northern Kazakhstan: the salinity of the water, the temperature regimen, the peculiarities of the shores, the fauna inhabiting them, and the origin of the lakes and all their components. Studying the changes in the mineralization of lakes, he put forth the notion (subsequently confirmed) that such changes occur because salts are carried away by the wind when the lakes temporarily decrease in size.

These detailed investigations, the beginning of limnology in Russia, gave Berg the basis for opposing the hypothesis popular at the end of the last century concerning the drying up of the lakes of central Asia. He convincingly demonstrated that these lakes do not dry up, but experience periodic fluctuations in their level and size that correspond to secular changes in climate and hydrological regimen.

Berg's limnological investigations are recognized as classics. He considered a lake to be a geographical complex, taking into account the history of its formation, its connections with the surrounding territory, the peculiarities of climatic and hydrobiological regimens, and an evaluation of its economic importance as a reservoir. These works were intimately connected with observations of climate, for study of the history of lakes demonstrated the existence of a clear link between the hydrobiological regimen of a lake and paleoclimatic changes. Berg's diversified works played an important role in the development of climatology as a science.

Berg's contributions to ichthyology are extremely significant. His works contain valuable information on the anatomy, embryology, and paleontology, as well as on the systematics and geographical distribution, of all the fishes in the U.S.S.R. Of great scientific value are his studies of summer and winter spawning runs of migratory fish, the periodicity of fish reproduction and distribution, and the influence of climatic changes on fish migration. He also established the symbiotic relationship between the lamprey and salmon families, which arose as a result of similar conditions of existence.

Most significant of all Berg's works are those on geography. Classifying the earth's surface according to climatic, soil, biological, and other natural factors, he distinguished ten different geographical zones, several of which he divided into subzones. This system of zones followed a latitudinal sequence in lowland areas and a longitudinal one in mountainous areas.

It contributed to the explanation of the surface characteristics of each area, ensuring detailed geographic study and the full exploitation of natural resources.

Berg opposed the opinion of some scientists that the earth's climate was becoming drier, and, having analyzed a mass of geological, geographical, and historical material, he demonstrated that in the modern epoch the earth's arid zones are contracting and that a displacement of zones is observable in the direction of the equator: the tundra is encroaching upon forest land, the forest is encroaching upon the steppe. On the other hand, equatorial forests are moving apart in the direction of the poles. This process was brought about, in Berg's opinion, by general climatic fluctuations, determined by cosmic causes.

Berg distinguished geographical zones according to the principle of the uniformity of landscape: the relief, climate, plants, and soil combine into a unified, harmonious whole that typically is duplicated over an entire geographical zone. He examined landscapes as they developed and continually changed. By reconstructing the paleogeographic conditions, he sought to trace their transformation into the modern landscape zones.

During his study of landscape, Berg devoted much attention to a thorough study of soils, especially of loess. He concluded that loess originated as a result of the weathering of fine-grained calcareous rocks in a dry climate. This set of conditions, which arose in the southern part of European Russia after the Ice Age, led, in his opinion, to the rise of loess there. Analyzing the climatic peculiarities of the more remote geological past, Berg concluded that even before the Ice Age a geographical zonality had existed, and that there were seasonal rhythms and the alternation of dry and moist epochs. In any case, loess-type rocks had appeared in the Upper Paleozoic.

Berg coordinated his paleoclimatologic and paleogeographic reconstructions with the origin of various sedimentary rocks, the history of relief development, the formation of various soils, and the displacement of landscape zones. These investigations contributed to the intimate coordination of modern geography with historical geology. His works included separate statements about, and penetrating investigations of, the physicogeographical conditions on Precambrian continents, the origin of land and sea organisms, the development of floras and faunas, the influence of the Ice Age on the present distribution of plants, the origin of iron and bauxite ores, and some general laws of sediment formation.

Berg believed that with a change of geographical landscape there also occurred a change in the character of the sedimentary rocks; thus, no periodicity or cyclical recurrence in the formation of sedimentary rocks exists, inasmuch as landscapes are unique. He stressed the significance of living organisms in the migration of chemical elements and in the formation of sedimentary rocks. Since the organic world has been continuously transformed throughout the history of the earth, the sedimentary rocks associated with it can neither remain unchanged nor be cyclically repeated.

Berg took exception to the concept of a fiery liquid stage in the earth's origin and supported the idea that the earth was initially solid. He also proposed that the position of the poles has changed little throughout geological history. On the basis of a biogeographic analysis, he determinedly spoke against Wegener's theory and thought there are better data in favor of comparatively minor vertical movements of the earth's surface than of its horizontal shifts for thousands of kilometers.

In his early papers Berg stated that great uplifts of the earth's crust caused continental glaciations, but he later concluded that a change in solar activity or more remote cosmic processes were decisive factors in the origin of glacial epochs. In this way, regions of low temperature arose not only in mountainous or high-latitude regions but also in low-lying areas and in subtropic zones. This explains the bipolar distribution of many marine organisms.

The zoogeographical analyses completed by Berg, particularly that on the distribution of fish in the carp family, permitted him to determine with reasonable precision the time of the appearance of glacial epochs and the periods in which the climate grew warmer. Discussing the problem of the development of the organic world, Berg spoke about distinct boundaries between separate species and about a saltatory transition from some species to others.

In some of his papers Berg dealt with the origin of life on earth. He believed that the first organisms were autotrophic and needed neither oxygen nor light. They developed near the earth's surface or in marshland. Subsequently, through their life processes, soils and atmospheric oxygen appeared and conditions were created for the transfer of life to the earth's surface and the ocean. He believed that there were epochs when the formation of new species proceeded very intensively in all groups of living organisms.

In the last years of his life Berg devoted serious attention to the study of the history of geography and geographic discoveries, especially in Asia and the Antarctic.

Berg published about 700 works, including a number of monographs and textbooks. His name is perpetuated by a peak and a glacier in the southwestern Pamir, a glacier in the Dzhungarskiy Alatau Mountains, a volcano in the Kuril Islands, and a promontory in the Arctic.

BIBLIOGRAPHY

I. ORIGINAL WORKS. Among Berg's writings are *Aral'skoe more. Opyt fiziko-geograficheskoj monografii* ("The Aral Sea. A Physicogeographical Monograph"; St. Petersburg, 1908); *Osnovy klimatologii* ("Fundamentals of Climatology"), 2nd ed., rev. and enl. (Leningrad, 1938); *Geograficheskie zony Sovetskogo Sojuza* ("Geographical Zones of the Soviet Union"), 3rd ed., 2 vols. (Moscow, 1947–1952); *Ryby presnykh vod SSSR i sopredel'nykh stran* ("Freshwater Fish of the U.S.S.R. and Neighboring Countries"), 4th ed., rev. and enl., 3 vols. (Moscow–Leningrad, 1948–1949); and *Izbrannye trudy* ("Collected Works"), 5 vols. (Moscow–Leningrad, 1956–1962).

II. SECONDARY LITERATURE. Works on Berg are *Lev Semenovich Berg (1876–1950)*, Academy of Sciences of the U.S.S.R., Materials and Bibliography of Scientists of the U.S.S.R., Geographical Series, no. 2 (Moscow–Leningrad, 1952); and E. Pavlovsky, ed., *Pamjati akademika L. S. Berga. Sbornik rabot po geografii i biologii* ("In Memory of L. S. Berg. A Collection of Papers on Geography and Biology"; Moscow–Leningrad, 1955).

V. V. TIKHOMIROV

DICTIONARY
OF
SCIENTIFIC BIOGRAPHY

DICTIONARY

OF

SCIENTIFIC BIOGRAPHY

CHARLES COULSTON GILLISPIE

EDITOR IN CHIEF

Volume 2

HANS BERGER—CHRISTOPH BUYS BALLOT

CHARLES SCRIBNER'S SONS · NEW YORK

Editorial Staff

MARSHALL DE BRUHL, *MANAGING EDITOR*

SARAH FERRELL, *Assistant Managing Editor*

LELAND S. LOWTHER, *Associate Editor*

JOYCE D. PORTNOY, *Associate Editor*

ROSE MOSELLE, *Editorial Assistant*

ELIZABETH I. WILSON, *Copy Editor*

DORIS ANNE SULLIVAN, *Proofreader*

JOEL HONIG, *Copy Editor*

LINDA FISHER, *Secretary-Typist*

Panel of Consultants

MARIA LUISA RIGHINI-BONELLI
Istituto e Museo di Storia della Scienza, Florence

GEORGES CANGUILHEM
University of Paris

PIERRE COSTABEL
École Pratique des Hautes Études, Paris

ALISTAIR C. CROMBIE
University of Oxford

MAURICE DAUMAS
Conservatoire National des Arts et Métiers, Paris

ALLEN G. DEBUS
University of Chicago

MARCEL FLORKIN
University of Liège

JOHN C. GREENE
University of Connecticut

MIRKO D. GRMEK
Archives Internationales d'Histoire des Sciences, Paris

A. RUPERT HALL
Imperial College of Technology, London

MARY B. HESSE
University of Cambridge

BROOKE HINDLE
New York University

JOSEPH E. HOFMANN
University of Tübingen

REIJER HOOYKAAS
Free University, Amsterdam

MICHAEL HOSKIN
University of Cambridge

E. S. KENNEDY
American University of Beirut

STEN H. LINDROTH
University of Uppsala

ROBERT K. MERTON
Columbia University

JOHN MURDOCH
Harvard University

SHIGERU NAKAYAMA
University of Tokyo

CHARLES D. O'MALLEY
University of California at Los Angeles

DEREK J. DE SOLLA PRICE
Yale University

J. R. RAVETZ
University of Leeds

DUANE H. D. ROLLER
University of Oklahoma

KARL EDUARD ROTHSCHUH
University of Münster/Westfalen

S. SAMBURSKY
Hebrew University, Jerusalem

GIORGIO DE SANTILLANA
Massachusetts Institute of Technology

AYDIN SAYILI
University of Ankara

ROBERT E. SCHOFIELD
Case Institute of Technology

CHRISTOPH J. SCRIBA
Technische Universität Berlin

NATHAN SIVIN
Massachusetts Institute of Technology

BOGDAN SUCHODOLSKI
Polish Academy of Sciences

RENÉ TATON
École Pratique des Hautes Études, Paris

J. B. THORNTON
University of New South Wales

RICHARD S. WESTFALL
Indiana University

W. P. D. WIGHTMAN
King's College, Aberdeen

L. PEARCE WILLIAMS
Cornell University

A. P. YOUSCHKEVITCH
Soviet Academy of Sciences

Contributors to Volume 2

Hans Aarsleff

L. R. C. Agnew

A. F. O'D. Alexander

G. C. Amstutz

R. Arnaldez

Richard P. Aulie

Cortland P. Auser

Lawrence Badash

D. L. Bailey

Walter Baron

Hans Baumgärtel

Silvio A. Bedini

Yvon Belaval

Luigi Belloni

James D. Berger

Francis Birch

Arthur Birembaut

R. P. Boas, Jr.

Uno Boklund

Carl B. Boyer

G. E. Briggs

L. Brillouin

T. A. A. Broadbent

Stephen G. Brush

Gerd Buchdahl

O. M. B. Bulman

John G. Burke

Harold L. Burstyn

H. L. L. Busard

G. V. Bykov

William F. Bynum

André Cailleux

Ronald S. Calinger

Walter F. Cannon

Albert V. Carozzi

Ettore Carruccio

J. G. van Cittert-Eymers

Thomas H. Clark

Edwin Clarke

William Coleman

George W. Corner

Ruth Schwartz Cowan

Paul F. Cranefield

M. P. Crosland

Michael J. Crowe

Charles A. Culotta

Glyn Daniel

Sally H. Dieke

Claude E. Dolman

Sigalia Dostrovsky

Stillman Drake

John Dubbey

Louis Dulieu

M. V. Edds, Jr.

Carolyn Eisele

H. Engel

Joseph Ewan

Eduard Farber

Lucienne Félix

E. A. Fellmann

Eugene S. Ferguson

J. O. Fleckenstein

Eric G. Forbes

Paul Forman

Vincenzo Francani

Pietro Franceschini

Eugene Frankel

H. C. Freiesleben

Walter Fricke

Henri Galliard

Gerald L. Geison

Wilma George

Ruth Anne Gienapp

Bertrand Gille

C. Stewart Gillmor

Owen Gingerich

Harry Godwin

Stanley Goldberg

G. J. Goodfield

D. C. Goodman

T. A. Goudge

J. B. Gough

Judith V. Grabiner

Edward Grant

Joseph T. Gregory

S. L. Greitzer

A. T. Grigorian

N. A. Grigorian

M. D. Grmek

J. Gruber

Henry Guerlac

Laura Guggenbuhl

Marie Boas Hall

Bert Hansen

Thomas Hawkins

John L. Heilbron

C. Doris Hellman

Mayo Dyer Hersey

Erwin N. Hiebert

J. E. Hofmann

Olaf Holtedahl

Gerald Holton

Dora Hood

David Hopkins

Michael A. Hoskin

Pierre Huard

Wlodzimierz Hubicki

G. L. Huxley

Jean Itard

Melvin E. Jahn

S. A. Jayawardene

Edouard Jeauneau

Børge Jessen

Sheldon Judson

Satish C. Kapoor

George B. Kauffman

Alan S. Kay

Suzanne Kelly

Edwin C. Kemble

Martha B. Kendall

E. S. Kennedy

Hubert C. Kennedy

Günther Kerstein

Geoffrey Keynes

Pearl Kibre

Lester S. King

Paul A. Kirchvogel

Marc Klein

B. Knaster

Zdeněk Kopal

Claudia Kren

P. G. Kulikovsky

CONTRIBUTORS TO VOLUME 2

Louis I. Kuslan
Gisela Kutzbach
Henry M. Leicester
Jean-François Leroy
Erna Lesky
Jacques R. Levy
G. A. Lindeboom
Sten Lindroth
Jacob Lorch
Eric McDonald
Susan M. P. McKenna
Robert M. McKeon
H. Lewis McKinney
Nikolaus Mani
Željko Marković
Seymour H. Mauskopf
Kenneth O. May
Josef Mayerhöfer
Everett Mendelsohn
Philip Merlan
W. E. Knowles Middleton
G. H. Miller
S. R. Mikulinsky
Lorenzo Minio-Paluello
Ernest A. Moody
Edgar W. Morse
Marston Morse
Jean Motte
John E. Murdoch
Henry Nathan
A. Natucci
Axel V. Nielsen
W. Nieuwenkamp
Lowell E. Noland

Luboš Nový
C. D. O'Malley
Charles O'Neill
Jane Oppenheimer
Adolf Pabst
Jacques Payen
Olaf Pedersen
Giorgio Pedrocco
Georges Petit
Mogens Pihl
P. E. Pilet
David Pingree
J. B. Pogrebyssky
P. M. Rattansi
Nathan Reingold
Bernard Rochot
Jacques Roger
Colin A. Ronan
B. van Rootselaar
Leon Rosenfeld
K. E. Rothschuh
Alain Rousseau
M. J. S. Rudwick
Susan Schacher
H. Schadewaldt
Charles B. Schmitt
Rud. Schmitz
Herbert Schriefers
B. P. M. Schulte
E. L. Scott
J. F. Scott
Christoph J. Scriba
Jan Sebestik
Thomas B. Settle
Harold I. Sharlin

O. B. Sheynin
Diana Simpkins
W. A. Smeaton
C. S. Smith
Y. I. Soloviev
Jerry Stannard
William T. Stearn
Johannes Steudel
Bernhard Sticker
Hans Straub
Dirk J. Struik
A. H. Sturtevant
Edith Sylla
Charles Süsskind
C. H. Talbot
Juliette Taton
René Taton
George Taylor
Andrzej A. Teske
Arnold Thackray
Joachim Thiele
K. Bryn Thomas
Heinz Tobien
J. J. Verdonk
Stig Veibel
Jean Vieuchange
Fred W. Voget
William A. Wallace, O.P.
Deborah Jean Warner
Charles Webster
Adrienne R. Weill-Brunschvicg
Thomas Widorn
Frances A. Yates
A. P. Youschkevitch

DICTIONARY
OF
SCIENTIFIC BIOGRAPHY

DICTIONARY OF
SCIENTIFIC BIOGRAPHY

BERGER—BUYS BALLOT

BERGER, HANS (*b.* Neuses bei Coburg, Germany, 21 May 1873; *d.* Jena, Germany, 1 June 1941), *psychiatry, electroencephalography.*

The son of Paul Friedrich Berger and of Anna Rückert, Berger graduated from the Gymnasium at Coburg and then entered the University of Jena in 1892. After one semester in astronomy he transferred to medicine. In 1897 he became assistant to Otto Binswanger at the university's psychiatric clinic. He was appointed chief doctor in 1912 and director and professor of psychiatry in 1919; he retired in 1938. His associates described Berger as punctual, strict, demanding, and reserved.

The central theme in Berger's work was the search for the correlation between objective activity of the brain and subjective psychic phenomena. In his work on blood circulation in the brain (1901) he described his efforts to gain insight into this correlation through plethysmographic registration of the brain pulsations. He investigated the influence of the heartbeat, respiration, vasomotor functions, and position of the head and body on brain pulsations, which were measured through an opening, made by trephination, in the skull. Berger also studied the effects of a number of medications—such as camphor, digitoxin, caffeine, cocaine, and morphine—on brain pulsations. The results of these investigations were disappointing, yet Berger continued his search for measurable expressions of psychic conditions through experiments on blood circulation (1904, 1907).

After 1907 Berger tried to discover a correlation between the temperature of the brain and psychic processes. He postulated that through dissimilation in the cortex, psychic energy (*P-Energie*) develops, along with heat, electrical energy, and neural energy. These experiments also came to a dead end, according to Berger's publication of 1910. Nevertheless, in his lectures on psychophysiology, given from 1905 on and published in 1921, the problem of *P-Energie* continued to hold his interest. His tenaciousness in this matter is apparent from a memo in his journal dated

14 December 1921, in which he says that the goal of his research continues to be the correlation between the expressions of the human mind and the processes of dissimilation in the brain.

After his disappointing experiments measuring the blood circulation and temperature of the brain, Berger (following his return from World War I) devoted himself mainly to the measurement of the brain's electrical activity. In 1902 he had taken measurements of electrical activity above skull defects with the Lippmann capillary electrometer, and later with the Edelmann galvanometer. In 1910, however, Berger mentioned in his journal that the results of these measurements were not satisfactory. Therefore, until 1925 he followed two methods of research: stimulation of the motor cortex through a defect in the skull, measuring the time between stimulus and contralateral motor reaction, and registration of the spontaneous potential differences of the brain surface. After 1925 Berger no longer used the stimulation method. He specialized, with ever increasing skill, in registering the spontaneous fluctuations in electrical potential that could be recorded through the skull from the cortex. In his first publication on electroencephalography (1929), he called 6 July 1924 the date of discovery of the human electroencephalogram. The EEG, the curves of the electrical potentials measured again and again between two points of the skull, did not give him a closer insight into the correlation between the electrical activity of the brain and psychic energy. However, electroencephalography has proved to be of ever increasing importance in diagnosing and treating neurological diseases (epilepsy, brain tumors, traumata).

Berger's work was strongly influenced by the exact psychology of the nineteenth century. In developing his psychophysiology, Berger used the ideas of J. F. Herbart, R. H. Lotze, G. T. Fechner, W. Wundt, and the Danish psychologist A. Lehmann as a base. In the experimental field, Berger was in all aspects a follower of A. Mosso. Berger's experiments on brain circulation and brain temperature were identical with Mosso's, and his publications on these subjects bore the same titles as Mosso's papers.

In developing electroencephalography, Berger was influenced by Caton and by Nemminski. Caton had measured electrical potentials on the exposed cortex of experimental animals in 1875, but he was not able to record these phenomena graphically. Nemminski recorded the first electrocerebrogram on dogs with the skull intact by means of the Einthoven string galvanometer in 1913.

Berger's historical significance lies in his discovery of the electroencephalogram of man. Although he began publishing his many papers on electroencephalography in 1929, he did not receive international recognition until Adrian and Matthews drew attention to his work in 1934.

BIBLIOGRAPHY

I. ORIGINAL WORKS. Berger's writings include *Zur Lehre von der Blutzirkulation in der Schädelhöhle des Menschen namentlich unter dem Einfluss von Medikamenten* (Jena, 1901); *Ueber die körperlichen Äusserungen psychischer Zustände,* 2 vols. (Jena, 1904–1907); *Untersuchungen über die Temperatur des Gehirns* (Jena, 1910); *Hirn und Seele* (Jena, 1919); *Psychophysiologie in 12 Vorlesungen* (Jena, 1921); *Ueber die Lokalisation im Grosshirn* (Jena, 1927); "Ueber das Elektrenkephalogramm des Menschen," in *Archiv für Psychiatrie,* **87** (1929), 527–570; **94** (1931), 16–60; **97** (1932), 6–26; **98** (1933), 231–254; **99** (1933), 555–574; **100** (1933), 301–320; **101** (1934), 452–469; **102** (1934), 538–557; **103** (1935), 444–454; **104** (1936), 678–689; **106** (1937), 165–187, 577–584; **108** (1938), 407–431; and *Psyche* (Jena, 1940).

II. SECONDARY LITERATURE. Works on Berger are E. Adrian and B. Matthews, "The Berger Rhythm," in *Brain,* **57** (1934), 355–385; Mary A. B. Brazier, "The Historical Development of Neurophysiology," in *Handbook of Physiology,* I (Washington, D. C., 1959), 1–58; R. Caton, "The Electric Currents of the Brain," in *British Medical Journal* (1875), **ii,** 278; H. Fischgold, "Hans Berger et son temps," in *Actualités neurophysiologiques,* 4th ser. (1962), 197–221; R. Jung, "Hans Berger und die Entdeckung des EEG nach seinen Tagebüchern und Protokollen," in *Jenenser EEG-Symposion* (Jena, 1963), pp. 20–53; K. Kolle, "Hans Berger," in *Grosse Nervenärzte,* I (Stuttgart, 1956), 1–16; A. Mosso, *Ueber den Kreislauf des Blutes im menschlichen Gehirn* (Leipzig, 1881), pp. 104–197; and *Die Temperatur des Gehirns* (Leipzig, 1894), pp. 120–135; and W. Nemminski, "Ein Versuch der Registrierung der elektrischen Gehirnerscheinungen," in *Zentralblatt für Physiologie,* **27** (1913), 951–960.

B. P. M. SCHULTE

BERGER, JOHANN GOTTFRIED (*b.* Halle, Germany, 11 November 1659; *d.* Wittenberg, Germany, 2[?] October 1736), *physiology, medicine.*

Berger was the son of Valentin Berger, an important educator of the mid-seventeenth century. He studied mathematics and medicine at Jena from 1677 to 1680, chiefly under Georg Wolfgang Wedel, a physician who was especially interested in iatrochemistry. At Jena, Friedrich Hoffmann and Georg Ernst Stahl were among his fellow students. After a brief period at Erfurt, Berger returned to Jena and graduated in 1682 with a thesis entitled *De circulatione lymphae et catarrhis.* In connection with this work he traveled through France, Italy, and perhaps Holland. From 1684 to 1688 he worked at the University of Leipzig with Johannes Bohn, who influenced Berger to develop a critical attitude toward iatrochemistry. Berger wrote two dissertations at Leipzig: *De mania* (1685) and *De chylo* (1686). From 1688 until his death Berger worked at the University of Wittenberg; first as assistant professor and from 1689 as third-ranking professor of anatomy and botany. He celebrated his appointment to the latter with an inaugural address in the spirit of natural theology. In 1693 he assumed the chair of pathology, and in 1697 Friedrich August I, king of Poland and Saxony, appointed him physician in ordinary. From 1730 he was the *consiliarus aulae,* that is, the senior of the entire university.

Berger's chief scientific work is his *Physiologia medica* (1701). Written in perfect Latin, this work deals in a "modern" way with the physiological functions of the organs and organ systems. While the *Physiologia medica* of his teacher Wedel was, in spirit and structure, still modeled closely on Jean Fernel's neo-Aristotelian physiology, Berger's work is based on the recent discoveries in anatomy, physics, and chemistry; in this respect, it is similar to Bohn's *Circulus anatomico-physiologicus* of 1680. Berger tends heavily toward iatromechanics and the corpuscular theory of Descartes, whom he often quotes. The body is a natural machine connected with an immortal soul (*mens* is *substantia cogitans; corpus* is *substantia extensa*) and united harmoniously by divine design.

The first, larger part of the book deals with the physiology of the adult human being; the second with reproduction and development. He discusses extensively the circulation of the blood and argues that the flow of blood and the nerve fluid determine the rhythmical activity of the heart. He considers the arteries to be elastic, as Giovanni Borelli had suggested. Berger describes several experiments involving the injection of mercury and colored liquids into the blood vessels. He attributed body heat to the rapid movement of those fine particles mentioned by Descartes. It originates in a heavenly material, and not in the *calor innatus,* as the ancients had suggested. Berger confirmed the circulation of blood by exsanguinating a dog within a few minutes, and he determined the total amount of blood.

Berger deals very thoroughly with the function of nerves: they are porous and conduct a fluid that is distributed by arterial blood pressure from the brain by way of the nerves to the periphery. There is neither a *spiritus animalis* nor are there the *facultates* or *archei* of ancient physiology. Berger rejects the explosion theory of muscle contraction proposed by William Thomas and Borelli, stating that the intellect controls the nerve fluid during voluntary motions. The soul resides in the *corpus callosum,* not in the pineal gland. The soul does not, as Stahl maintained, influence the activity of the internal organs. Berger also states that complete section of the vagus nerve is not immediately followed by death, but only after some days; breathing does not serve to cool the blood, but to restore and refine it by means of contact with the air.

The *Physiologia* is a general, critical presentation of contemporary physiology. No particular discoveries are associated with Berger's name. The last of the four editions of the *Physiologia* was edited after Berger's death in 1737 by Cregut, who added a long introduction, "De Anthropologia," in which the then important literature of anatomy and physiology is discussed.

In addition to a short treatise concerning the large arterial branches, accompanied by a good chart (1698), Berger wrote about fifty treatises. One dealing with the springs at Karlsbad has been reprinted several times. An attack against Stahl's animism was published in 1702.

Berger was very well read, very critical, and an opponent of all obscurity; he fought against the weak points of the Galenists and Paracelsians, as well as against the students of Stahl.

BIBLIOGRAPHY

I. ORIGINAL WORKS. Berger's thesis is *De circulatione lymphae et catarrhis* (Jena, 1682); his main work, *Physiologia medica sive De natura humana liber bipartitus* (Wittenberg, 1701; Frankfurt, 1737). Albertus Haller, *Bibliotheca anatomica,* I (1774), 720–721, lists several physiological treatises: *De chylo* (1686); *De corde* (1688); *De ovo et pullo* (1689); *De polypo* (1689); *De homine* (1691); *De succi intrinseci per nervos transitu* (1695); *De respiratione* (1697); *De odoratu* (1698); *De somno* (1706); *De nutritione* (1708); *De vita longa* (1708); and *De secretione* (1712). Additional clinical treatises are listed in Haller's *Bibliography of Practical Medicine,* I (1779), 641–643. Additional works are *Dissertatio de natura morborum medico* (Wittenberg, 1702), a polemic against Stahl; and *De thermis carolinis commentatio qua omnium origo fontium calidorum itemque acidorum ex pyritide ostenditur* (Leipzig–Wittenberg, 1709), also published in German.

II. SECONDARY LITERATURE. Of especial value is E. A. Underwood, "Johann Gottfried von Berger (1659–1736) of Wittenberg and His Text-book of Physiology (1701)," in his *Science, Medicine and History,* II (Oxford, 1953), 141–172, which includes the locations of the *Physiologia medica* but no general bibliography. Biographical information can also be found in *Allgemeine deutsche Biographie,* II (1875), 375; Bayle and Thillaye, *Biographie medicale,* II (Paris, 1855), 94; *Biographie universelle ancienne et moderne,* IV (Paris, 1843), 15; *Biographisches Lexikon der hervorragenden Ärzte,* 2nd ed., I (Berlin–Vienna, 1929), 475; and J. C. Poggendorff, *Biographisch-literarisches Handwörterbuch,* I (1863), 148.

K. E. ROTHSCHUH

BERGIUS, FRIEDRICH (*b.* Goldschmieden, near Breslau, Germany, 11 October 1884; *d.* Buenos Aires, Argentina, 30 March 1949), *chemistry.*

Bergius became involved with chemistry in his very early years. His father was head of a chemical plant and his mother was the daughter of a classics professor; thus, he grew up in a home where learning was highly valued. Following high school and a practical course in the laboratory of a foundry, Bergius began his higher education in Leipzig. In 1907 he passed his doctorate examination and became, successively, assistant to Nernst, Haber, and Ernst Bodenstein. All three allowed him to participate in their research.

In 1909 Bergius qualified as a university lecturer in Hannover, and thereupon set up a private laboratory where he could conduct his own research. Systematically he began work on the influence of high pressure and high temperature, seeking to clarify the conditions in nature under which wood was transformed into coal. Within a short time, Bergius succeeded in developing a coal very similar to that produced in nature through his process of carbonization of peat and cellulose.

At the same time Bergius studied the origin of petroleum and conducted experiments in which he sought to make carboniferous materials react with hydrogen to yield liquid products. As early as 1913 he was granted (with John Billwiller) his first patent for the manufacture of liquid hydrocarbons from coal. Needless to say, research of this kind could not be conducted in a private laboratory. Bergius, who had acquired a fine reputation in scientific circles, became head of a new research laboratory at the Goldschmidt Company in Essen in 1914. The owner, Karl Goldschmidt, was an enthusiastic friend and promoter of Bergius. Soon afterward a private experimental plant was constructed for laboratory experiments, continuous processing of petroleum reserves (the daily output was twenty tons), and commercial experiments.

Bergius recognized that the hydrogenation-dehydrogenation balance was based on temperature, partial hydrogen pressure, and the size of the molecules in the hydrocarbon. For petroleum reserves, slight hydrogenation was found to be sufficient; hydrogen-poor coal, however, required greater hydrogenation before it could be thermally cracked. Hydrogenation of carbon was conducted in two stages, the first being to process the carbon into a paste with oil.

Conditions after World War I made it impossible for Bergius to continue his work, and in 1925 he sold his patent rights to the Badische Anilin- und Sodafabrik in Ludwigshafen, which had begun experimenting with hydrogenation of carbon. He then left the field and dedicated himself to a new problem: hydrolysis of wood by means of acid. With Hägglund, Bergius developed a process by which he obtained complete hydrolysis of wood cellulose by using concentrated hydrochloric acid. The product was either dextrose or, after transformation, ethanol or a nutrient yeast with a 50 percent albumin content.

In 1931 Bergius and Karl Bosch were awarded the Nobel Prize for chemistry. Bergius was one of the most individualistic of research scholars. He persistently attempted to go his own way and to remain completely independent, but the tasks to which he committed himself exceeded the ability of a single individual.

After World War II, no longer able to find work in Germany, Bergius founded a company in Madrid and in 1947 became a scientific adviser to the Argentine government.

BIBLIOGRAPHY

I. ORIGINAL WORKS. Bergius' works include *Die Anwendung hoher Drucke bei chemischen Vorgängen und eine Nachbildung des Entstehungsprozesses der Steinkohle* (Halle, 1913); "Die Verflüssigung der Kohle," in *Zeitschrift des Vereins deutscher Ingenieure,* **69** (1926); "Die Herstellung von Zucker aus Holz und ähnlichen Naturstoffen," in *Ergebnisse der angewandten physikalischen Chemie,* Vol. I (Leipzig, 1931); "Gewinnung von Alkohol und Glucose aus Holz," in *Chemical Age* (London), **29** (1933), 481–483; and "Chemische Reaktionen unter hohem Druck," in *Les prix Nobel en 1931* (Stockholm, 1933), which also includes an autobiographical notice.

II. SECONDARY LITERATURE. For a discussion of Bergius' work, see E. Farber, *Nobel Prize-winners in Chemistry, 1901–1950* (New York, 1953), pp. 123–131.

GÜNTHER KERSTEIN

BERGMAN, TORBERN OLOF (*b.* Katrineberg, Sweden, 9 March 1735; *d.* Medevi, Sweden, 8 July 1784), *chemistry, mineralogy, entomology, astronomy, physics, geography.*

The son of Barthold Bergman, sheriff on the royal estate at Katrineberg, and Sara Hägg, Bergman received a conventional early education in classical subjects at Skara, and was also given private instruction in natural history by Sven Hof, a teacher in the Gymnasium there. He entered Uppsala University in 1752 and graduated in 1756, after studying mathematics, philosophy, physics, and astronomy. In 1758 he gained his doctorate with a thesis entitled *De interpolatione astronomica,* published *De attractione universali,* and became a lecturer in physics at the university. Bergman was appointed associate professor of mathematics in 1761, and in 1767 succeeded J. G. Wallerius as professor of chemistry, a subject that was new to him but in which he became famous. He corresponded with scientists all over Europe and Frederick II offered him an appointment in the Berlin Academy, but he preferred to stay in Sweden. Often in poor health, he regularly visited the medicinal springs at Medevi, where he died. In 1771 he had married Catherine Trast, who survived him.

While a student, Bergman made important contributions to natural history, the most interesting being his discovery, praised by Linnaeus, that the "insect" *coccus aquaticus* was in fact a leech's egg from which ten to twelve young hatched. He studied and classified insect larvae, and in 1762 the Stockholm Academy of Science awarded him a prize for his research on the winter moths that damaged fruit trees. He observed that the female was wingless, and found that metamorphosis occurred in the ground and that after mating, the female climbed the tree and laid eggs around the buds. Damage could therefore be prevented by tying a wax-covered band around the trunk, and this became standard practice in orchards.

Bergman's early contributions to physical science included studies of the rainbow, twilight, and the aurora borealis, of which he estimated the height to be about 460 miles. More important was his discovery of an atmosphere on Venus during the transit of 6

June 1761. Others who took part in this early example of international scientific cooperation included Lomonosov, Dunn, and Chappe d'Auteroche. Like Bergman, they saw a luminous aureole around the planet when it entered and left the sun's disk, and interpreted it as being due to refraction in an atmosphere, but Bergman gave the clearest description of the phenomenon, which was overlooked by some other observers.

Bergman's inaugural address to the Stockholm Academy in 1764 was "The Possibility of Preventing the Harmful Effects of Lightning," and he was one of the first to support Franklin's belief in lightning conductors. However, he disagreed with Franklin's one-fluid theory of electricity, and developed a two-fluid theory similar to that of Wilcke in order to explain why a luminous discharge apparently flowed from both negative and positive conductors.

Following Aepinus, Bergman investigated the pyroelectricity of tourmaline. In 1766 he showed that when the temperature was raised, one end of the crystal became positive and the other negative, and that reducing the temperature reversed the polarity. This was discovered independently by B. Wilson and J. Canton, but none of the three offered an explanation. In 1785 R. J. Haüy was the first to relate crystal structure to electrical properties.

A comprehensive work on cosmography was published in Uppsala in three volumes. F. Mallet gave an account of astronomy; S. Insulin described the customs of the various races inhabiting the world; and in 1766 Bergman contributed *Physical Description of the Earth.* This was an important treatise on physical geography, but its influence was probably diminished by the lack of English and French translations at a time when British and French navigators were rapidly adding to knowledge of the globe. Bergman included a long account of minerals, and he soon became actively interested in mineralogy and chemistry.

Wallerius had lectured on chemistry without demonstrations, and Bergman reformed the teaching arrangements as soon as he succeeded to the chair. Since he believed in applying chemistry to mining and industry, he provided two displays of minerals, one arranged according to chemical composition and one according to geographical distribution, and an exhibition of models of industrial equipment. He taught his students, who came from many countries, not only theoretical chemistry but also new experimental methods, especially in mineral analysis.

The blowpipe had been used by Swedish analysts at least as early as the 1740's. A. Swab and S. Rinman were the pioneers, and A. F. Cronstedt introduced soda, borax, and microcosmic salt (sodium am-monium phosphate) as fluxes. In *De tubo feruminatoria* (1779) Bergman gave a full account of the instrument. He recorded the reactions of minerals with the three fluxes, fused in hollow charcoal supports or silver or gold spoons, and he distinguished between the oxidizing and reducing flames, as they are now called. The blowpipe was an excellent instrument for qualitative analysis, but he recognized that it was unsuitable for quantitative analysis, a branch of the art that he greatly improved by wet methods.

Bergman published analyses of many individual minerals and mineral waters, and dissertations on most of the metals. These frequently contained new qualitative and quantitative results, but three general treatises were more important. In *De analysi aquarum* (1778), he gave the first comprehensive account of the analysis of mineral waters. Dissolved gases, usually carbon dioxide or hydrogen sulfide, were driven out by heating and were collected over mercury. The water was then evaporated to dryness and the residue extracted by four solvents in succession: alcohol, water, acetic acid, and hydrochloric acid. The resulting solutions were each treated with about twenty-five reagents, but not in any particular order. Many were well known, but Bergman introduced important new reagents, notably oxalic acid as a test for lime, and barium chloride for sulfate. His method represented an advance in the analysis of mineral waters, but it was soon improved, notably by A. F. de Fourcroy, who in 1781 pointed out the desirability of using the reagents in a systematic order and found that several were redundant. However, Bergman determined the compositions of some important mineral waters, and from 1773 he successfully prepared artificial seltzer and Pyrmont waters by dissolving the necessary compounds in water saturated with carbon dioxide, which he called "aerial acid" after discovering its acidity.

In *De minerarum docimasia humida* (1780) Bergman described his procedures for qualitative and quantitative analysis of minerals by wet methods. The mineral was finely ground and dissolved in purified acids. Reagents were used for qualitative analysis, as in the case of mineral waters, but for quantitative analysis he introduced an entirely new procedure that soon was generally adopted. Previously it had been customary to attempt to isolate the substance being estimated (metal, earth, and so forth) in the pure state, but Bergman precipitated it as an insoluble compound of known composition, which was filtered through a previously weighed paper and then weighed after drying at the temperature of boiling water. It was necessary to ensure that the precipitate was not contaminated. Thus, iron was precipitated by potassium ferrocyanide; this reagent also formed an insoluble salt with manganese that could be removed by nitric acid.

The method depended on the purity and insolubility of precipitates of known composition. Bergman discussed these factors in a third treatise, *De praecipitatis metallicis* (1780), which contained a table listing the weights of precipitates obtained from 100 parts by weight of different metals by various reagents. Other chemists, notably R. Kirwan and C. F. Wenzel, obtained different results, but Bergman's prestige caused his figures to be generally accepted for many years. In 1789 L. B. Guyton de Morveau proved that Bergman's results were inconsistent, and most of them were abandoned (as were Kirwan's and Wenzel's), but his new analytical methods were of permanent value.

In *De praecipitatis metallicis* Bergman also considered the phenomena observed when metals dissolved in acids, and, as in other writings by him, he accepted the view that phlogiston was lost by the metal. He adopted the phlogistic explanation of combustion and calcination proposed by C. W. Scheele: phlogiston from the combustible or metal combined with "fire air" (oxygen) to form heat, a subtle material that escaped through the vessel. Bergman did, however, make an important original contribution to the phlogiston theory in 1782, when he attempted to measure the relative quantities of phlogiston in different metals by determining the weight of one metal that would precipitate another from solution. Thus, from solution in nitric acid, 100 parts of silver were precipitated by 135 of mercury, 234 of lead, 31 of copper, and so on; and these weights were considered to contain the same amount of phlogiston, for the reaction involved only its transfer. There were inconsistencies due to such effects as incomplete precipitation and the occasional evolution of hydrogen, and it would be reading too much into these results to say that Bergman had grasped the idea of equivalents. This work is, however, important as one of the few attempts ever made to put the phlogiston theory on a quantitative basis.

With his early mathematical training, Bergman was well equipped to seek a geometrical as well as a chemical explanation of the composition of matter, and he made a notable contribution to the early development of crystallography. He may have been influenced by C. F. Westfeld, who in 1767 expressed the opinion that all calcite crystals were composed of rhombohedra and that other shapes were built up from these. A similar suggestion was made by Gahn, but Bergman was the first to demonstrate this in the case of a definite form in which calcite occurred, the scalenohedron. In 1773 he showed how from a rhombohedral nucleus with angles of 101.5° and 78.5° the scalenohedron could be constructed by superimposing rhombic lamellae in sizes that decreased according to some law as the layers developed. This was not merely a geometrical construction on paper, but agreed with the results of cleavage experiments. Dodecahedral garnet crystals were similarly explained, but Bergman could not derive the hexagonal calcite prism with plane ends. This line of research was later developed by Haüy, whose earliest investigations were on garnet and the forms of calcite discussed by Bergman. Haüy's denial that he knew of Bergman's theories has been questioned, but of course there is no doubt about the originality of his subsequent work.

Crystallization was commonly believed to be caused by a "saline principle," but Bergman's analyses showed that many gems and crystalline minerals contained no saline material, and he rejected the concept. Like Guyton de Morveau, with whom he regularly corresponded, he accepted the Newtonian theory that crystals were formed by the mutual attraction of the molecules of matter, but he did not go so far as Guyton, who suggested that these molecules themselves had definite geometrical shapes that could be inferred from the shapes of the crystals.

His belief in attraction between molecules and the vast knowledge of chemical reactions acquired in the course of his analytical work put Bergman in a good position to study chemical affinity. Much had been written about this since E. F. Geoffroy published his table of affinities in 1718, after Newton expressed the view that chemical change was caused by a force acting between particles at very small distances.

Geoffroy's table contained sixteen columns, each with the symbol for one substance at the head and below it the symbols for the substances with which it combined, arranged so that each displaced only those below it. As chemical knowledge increased, affinity tables grew larger: Spielmann's table (1763) had twenty-eight columns, and Fourcy's (1773) had thirty-six. These tables were convenient for summarizing chemical knowledge, but their compilers generally did not speculate about the cause of reactions. Newton's followers were convinced that a modified gravitational attraction was the sole cause of affinity. While using the words "affinity" and "attraction" indiscriminately, Bergman made it clear that he was a Newtonian, although, unlike Guyton and some others, he made no attempt to calculate the strength of intermolecular forces.

Eighteenth-century chemists sometimes thought that the affinity between two substances was a vari-

6

able quantity, for the course of many reactions could be altered, particularly by the action of heat. A. Baumé pointed this out in 1763, and suggested that two affinity tables should be drawn up, one for reactions "in the wet way" (in aqueous solution) and the other for reactions "in the dry way" (at high temperature, in the absence of water). Bergman was the first to do this, but he did not accept the view that affinities were variable. He stated emphatically that reactions in aqueous solution showed the true affinities of substances, and that changes in the order of affinities in the dry way were due to the action of heat. The first version of his *Disquisitio de attractionibus electivis* (1775) included a table with fifty columns representing reactions in the wet way and thirty-six in the dry way; in 1783 he enlarged these to fifty-nine and forty-three columns, respectively. These tables were widely praised and were reprinted as late as 1808 in William Nicholson's *Dictionary of Chemistry,* but it must be doubted whether they were of great utility, for, like all previous tables, they summarized experimental information without explaining it. Further, Lavoisier pointed out that while separate tables for wet and dry reactions were useful, the effect of heat was so great that there should really be an individual table for each degree of temperature.

In his theoretical discussion Bergman advanced a simplified version of P. J. Macquer's classification of the types of attraction. The union of two or more particles of the same substance was an example of "attraction of aggregation"; when two different substances united, it was "attraction of composition." "Single elective attraction" occurred when a simple substance combined with one of the constituents of a binary compound and set the other free; and "double elective attraction," when two binary compounds reacted and their constituents were exchanged to form two new binary compounds. Affinity tables showed single elective attractions, and Bergman considered that from these it should be possible to find the results of reactions involving several substances, for double elective attractions were to be calculated from the algebraic sums of single attractions. Bergman did not himself attempt to do this, for he did not give numerical values for elective attractions, although soon after the publication of his work this was done by Kirwan, Guyton, and others.

Although Bergman believed that single elective attractions were constant, he admitted that this was not immediately obvious. The apparent exceptions could, however, be explained. For example, it was sometimes difficult to ascertain the exact number of reactants, phlogiston in particular being often overlooked, so that what was apparently an example of single elective attraction might in fact be double. Bergman also noted that the proportions as well as the nature of reactants sometimes seemed to affect the course of a reaction—a line of investigation developed fruitfully by C. L. Berthollet between 1798 and 1803.

Descriptions of reactions in conventional prose were cumbersome, and Bergman introduced diagrams to represent reactions involving single and double elective attractions. These served the same purpose as the earlier diagrams of J. Black and W. Cullen, but were differently constructed. The reaction in the wet way between muriate (chloride) of soda and nitrate of silver is an example.

FIGURE 1

The vertical and horizontal brackets show the constituents of the reactants and the products, respectively.

In tables and diagrams Bergman represented substances by symbols, but his successors frequently preferred words. Their binary nomenclature for salts was partly due to Bergman, an early critic of the old unsystematic nomenclature in which the name of a substance was usually derived from its appearance, its discoverer, or some other chemically irrelevant factor. From 1775 he began to coin names related to chemical composition, and he attempted a general reform in *Sciagraphia regni mineralis* (1782). Guyton de Morveau's proposals for a systematic nomenclature were also published in 1782, and their influence can be seen in Bergman's final system, presented in *Meditationes de systemate fossilium naturali* (1784). Guyton wrote in French, but Bergman preferred Latin, which could be translated into all modern languages. Following Linnaeus, Bergman divided inorganic substances into classes, genera, and species; and, as Linnaeus had done for plants and animals, he defined each class and genus by one word and each species by two. There were four classes: salts (including acids and alkalies as well as neutral salts), earths, metals, and phlogistic materials. In the important and numerous class of salts, each acid or alkali constituted a genus, and a neutral salt was a species belonging to the genus of its acid. The acids were named *vitriolicum, nitrosum,* and so on; and the alkalies became *potassinum, natrum,* and *ammoniacum.* Neutral salts received such names as *vitriolicum potassinatum* and *nitrosum argentatum.* Provision was made for unusual salts containing excess acid or base, for example, or having more than two constituents. This was more comprehensive than Guyton's nomenclature of 1782; had he lived, Bergman would have collaborated with Guyton in perfecting the system.

Bergman's nomenclature was closely related to his classification of minerals by composition. Some mineralogists based their classification on external form, but he insisted that the best method demonstrated the inner composition of each mineral, for only thus would we know its utility. However, different external appearances were sometimes associated with the same composition; so while classes, genera, and species had to be defined by composition, varieties could be distinguished by appearance. This was the basis of *Meditationes de systemate fossilium naturali* (1784), his last major work. In the preface he announced his latest discoveries. He had found two sulfides of tin, one containing twice as much sulfur as the other; and resemblances between baryta (barium oxide) and calx (oxide) of lead made him suspect that baryta contained a metal. These final examples of experimental skill and chemical insight show that Bergman deserved the high esteem in which he was held by his contemporaries.

BIBLIOGRAPHY

I. ORIGINAL WORKS. More than 300 items are described in Birgitta Moström, *Torbern Bergman, A Bibliography of His Works* (Stockholm, 1957), which includes reprints and translations published up to 1956.

Bergman's manuscripts are preserved in the University Library, Uppsala. The letters that he received from foreigners are published in G. Carlid and J. Nordström, eds., *Torbern Bergman's Foreign Correspondence,* I (with an introductory biography by H. Olsson), *Letters From Foreigners to Torbern Bergman* (Stockholm, 1965). Vol. II will contain letters from Swedes living abroad, and Bergman's letters to his foreign correspondents.

II. SECONDARY LITERATURE. Several short biographies were written soon after Bergman's death, and later biographical accounts are based on these: P. F. Aurivillius, *Åminnelse-tal öfver . . . Bergman* (Uppsala, 1785); P. J. Hjelm, *Åminnelse-tal öfver . . . Bergman* (Stockholm, 1786); F. Vicq d'Azyr, in *Histoire de la Société royale de médecine de Paris* (1782/1783, pub. 1787), 141–187, repr. in his *Éloges historiques,* I (Paris, 1805), 209–248; A. N. de Condorcet, in *Histoire de l'Académie royale des sciences*

(1784, pub. 1787), 31–47, repr. in his *Oeuvres* (Paris, 1847), III, 139–161.

Recent studies of several aspects of Bergman's scientific work can be listed in the order in which the topics are discussed above: D. Müller-Hillebrand, "Torbern Bergman as a Lightning Scientist," in *Daedalus, Tekniska museets årsbok, Stockholm* (1963), 35–76; A. J. Meadows, "The Discovery of an Atmosphere on Venus," in *Annals of Science,* **22** (1966), 117–127; F. Szabadvary, *History of Analytical Chemistry* (Oxford, 1966), pp. 71–81, 86–89; U. Boklund, "Torbern Bergman as a Pioneer in the Domain of Mineral Waters," in T. Bergman, *On Acid of Air . . .* (Stockholm, 1956), pp. 105–128; R. Hooykas, "Les débuts de la théorie cristallographique de R. J. Haüy, d'après les documents originaux," in *Revue d'histoire des sciences,* **8** (1955), 319–337; J. G. Burke, *Origins of the Science of Crystals* (Berkeley–Los Angeles, 1966), pp. 26–27, 79–84; A. M. Duncan, "Some Theoretical Aspects of Eighteenth-Century Tables of Affinity," in *Annals of Science,* **18** (1962), 177–194, 217–232; and "Introduction" to a facsimile repr. of Thomas Beddoes' trans. (London, 1785) of Bergman's *A Dissertation on Elective Attractions* (London, 1969); W. A. Smeaton, "The Contributions of P.-J. Macquer, T. O. Bergman and L. B. Guyton de Morveau to the Reform of Chemical Nomenclature," in *Annals of Science,* **10** (1954), 87–106; and M. P. Crosland, *Historical Studies in the Language of Chemistry* (London, 1962), pp. 144–167.

W. A. SMEATON

BERGSON, HENRI LOUIS (*b.* Paris, France, 18 October 1859; *d.* Paris, 4 January 1941), *philosophy.*

Henri Bergson was the son of a Polish musician and composer, Michael Bergson, and of an English mother (née Kate Lewison). The family settled in Paris, where Henri attended the Lycée Condorcet and the École Normale Supérieure. His main studies were mathematics, literature, and philosophy; and his academic record was brilliant. After becoming *agrégé* in 1881, Bergson began teaching philosophy at the Angers Lycée. He advanced steadily through a series of other posts, until in 1900 he was appointed professor at the Collège de France, where he eventually succeeded Gabriel Tarde in the chair of modern philosophy. By this time his books and lectures had given him an international reputation as the author of an original and impressive philosophical doctrine.

Bergson was elected to the Académie Française in 1914, was awarded the Nobel Prize for Literature in 1928, and received the Grand Cross of the Legion of Honor in 1930. After World War I, he devoted himself to the cause of the League of Nations and served as chairman of its Committee for Intellectual Cooperation. Ill health forced him to retire from academic life in 1921; he then lived in comparative seclusion with his wife (née Louise Neuburger) and daughter Jeanne. Bergson's parents were Jewish, but he himself had no formal religious affiliation, although toward the end of his life he expressed a sympathy for Roman Catholicism.

Bergson's philosophy offers a new interpretation of four main ideas: time, freedom, memory, and evolution. These ideas are construed so as to produce a doctrine that is opposed to materialism, as well as to mechanistic and deterministic theories of living things. The doctrine strongly emphasizes the phenomenon of change or process, which continually creates unpredictable novelties in cosmic history. The doctrine also emphasizes the importance of direct, conscious experience as the source of man's most reliable knowledge.

In his first book, *Essai sur les données immédiates de la conscience (Time and Free Will)*, Bergson drew a distinction between two kinds of time. There is the time that occurs in the theories of the natural sciences, and there is the time that we experience directly. Scientific time is a mathematical conception, symbolized geometrically by a line or algebraically by the letter *t*, and measured by clocks and chronometers. Because these measuring instruments are spatial bodies, scientific time is represented as an extended, homogeneous medium composed of standard units (seconds, hours, years, and so forth). Most of man's practical life in society is guided by reference to such units. But time thus represented neither "flows" nor "acts"; it is wholly passive. When we turn to our direct experience, however, we find nothing like scientific time. What we find, Bergson contended, is a flowing, irreversible succession of phenomena which melt into one another to form an indivisible process. This process is not homogeneous, but heterogeneous. It is concrete, not abstract. In short, it is "lived time" (*temps vécu*) or "real duration" (*durée réelle*), something immediately experienced as active and ongoing. If we try to represent it by a spatial image such as a line, we will only generate abstract mathematical time, which is a fiction. The mistake of mechanistic modes of thought is to regard this fiction as a reality.

The recognition of real duration provides a basis, Bergson held, for vindicating human freedom and disposing of the bogey of determinism. A determinist contends that freedom of choice is illusory. A man may feel that he is free to choose, but theoretical considerations show that he never is. To support this contention, a determinist may depict the case in which a man confronts an ostensible choice as a situation similar to arriving at a point on a line where branching occurs, and then taking one of the branches. The determinist asserts that in fact the branch taken could not *not* have been taken and, indeed, that the choice

made was fully predictable beforehand, given complete knowledge of the antecedent states of mind of the agent.

This argument has a spurious plausibility, according to Bergson, because it represents the case of choice by a spatial image. The image may serve to symbolize a choice already made, but it is totally inadequate to symbolize a choice in the making: in acting, we do not move along a linear path through time. Deliberating about a course of action is not like being at a point in space where we oscillate between paths laid out before us. Deliberation and choice are temporal, not spatial, acts. Moreover, the determinist makes the mistake of supposing that the way in which the agent acts is determined by the totality of antecedent states of mind, each atomically independent of the rest. This conception of mental life as made up of basic units was the misconception promulgated by associationistic psychology. It invalidates all the traditional arguments for determinism.

Freedom of choice, Bergson held, is fully certified by direct experience. A man knows that he is free as he acts. However, one qualification must be added. Strictly speaking, a man is free only when his act springs spontaneously from his total personality as it has evolved up to the moment of action. If this spontaneity is absent, his action will be stereotyped and mechanical. Hence, free acts are far from being universal. Most people behave like automata most of the time. But the point is that they need not do so, for freedom is always attainable.

Bergson's second book, *Matière et mémoire,* advocates a dualism of body and mind, of the material and the spiritual. Each of us is alleged to know his own body in two ways: from without, as an object among other objects, and from within, as a center of action. Likewise, each of us knows his own conscious processes directly. How are the body and mind—i.e., brain activity and mental activity—related? The answer lies in a proper understanding of memory.

Living organisms, unlike nonliving things, retain their past in the present. This phenomenon is manifested, Bergson affirmed, in two kinds of memory. One kind consists of bodily habits fixed in the organism and designed to ensure its adaptation to the contemporary world. The other kind of memory, which man alone possesses, records in the form of images each event of his daily life as it takes place. This is "pure memory," which is wholly spiritual. It is quite independent of the brain, whose structure resembles that of a telephone exchange and which therefore has no facilities for storing memory traces.

How is pure memory related to the brain? Bergson's

answer depends on the assumption that each person's memory retains the whole of his past experience. If this is so, something must prevent all of one's memories from crowding into consciousness at every moment. The brain is the mechanism that performs this function. It is a device evolved to facilitate action by ensuring that only what is relevant to a particular occasion of action will be recalled. Hence, it acts as a filter which excludes the vast majority of memories at each instant. It is a device to promote forgetting, not remembering.

The relation between body and mind is, then, to be understood temporally. It is not to be envisaged as a spatial or quasi-spatial connection between two entities. On an occasion of action, body and mind (including memory) converge in time. A typical case is an act of perception. Traditional philosophers have thought of perception as being like a photographic process which provides a passive, cognitive reflection of the world. But this is a mistaken view. Perception is an adaptive response to the world in which the body contributes sense receptors to register the influences of environing objects, and the mind contributes relevant memory images to give a meaningful form to what is perceived. The aim is to put the organism in a condition in which it can act successfully. Body and mind are thus united in real duration, for perception is an event in the present, which is not a geometrical point or "knife edge" separating past from future, but a continuous flowing. Perceptual acts in which body and mind are fused are intrinsically temporal and practical.

In his next work, *Introduction à la métaphysique,* Bergson modified his position by affirming that sometimes "pure perception," detached from memory and action, can occur. This process is also called "intuition" and is contrasted with conceptual thought, the product of the intellect. Both processes arose in human evolution, intuition being derived from instinct and conceptual thought being derived from man's social existence, his tool-making capacity, and his power of speech. Because his inherently limited intellect is the human animal's distinctive instrument which he employs in his interactions with the world, conceptual thought has certain limitations. It has an inherent tendency to use spatial notions, to analyze things mechanistically into ultimate units, and to interpret motion and change in static terms. Like a motion-picture camera, it translates everything into a series of discrete "frames." Intuition, on the other hand, is a type of consciousness which achieves a direct participation in, or an identification with, what is intuited. By means of it "one is transported into the interior of an object in order to coincide with what

is unique and consequently inexpressible about it."

In the case of ourselves, intuition is immersion in the flow of consciousness, a grasping of pure becoming and real duration. Unlike the intellect—which remains outside what it knows, requires symbols, and produces knowledge that is relative—intuition enters into what it knows, dispenses with symbols, and produces knowledge that is absolute.

The natural sciences were for Bergson a typical achievement of the intellect. Hence, they inevitably falsify time, motion, and change by interpreting these items in terms of static concepts. The sciences are equipped to deal with matter, but not with real becoming. Hence, there is need of a discipline to supplement them, if a full understanding of the universe is to be attained. Such a discipline, Bergson proposed, is metaphysics, not in the classical sense of rational speculation or system building, but redefined as a "true empiricism" that explores real becoming by participating directly in it. Thus, by adopting the method of intuition, metaphysics can provide a supplement to the sciences by giving a true account of duration, of becoming, and even of evolution.

This last topic was dealt with by Bergson in his most famous book, *L'évolution créatrice (Creative Evolution)*. He accepted the historical fact of the evolution of living things on the earth, but rejected all mechanistic or materialistic explanations of the evolutionary process. In place of the theories of Darwin, Lamarck, and Spencer, he advanced a doctrine which owes much to the tradition of European vitalism and also draws inspiration from Plotinus. The result was a theory of cosmic evolution that goes far beyond the domain of biology.

Bergson did accept one aspect of Lamarck's doctrine, "the power of varying by use or disuse" certain bodily organs, and the transmission of such acquired variations to descendants. He may have derived this notion from his early study of Herbert Spencer, rather than from the study of Lamarck himself. However, he does not link the notion with a doctrine of "racial memory" (as did, for example, Samuel Butler). Bergson hints that the Lamarckian idea of "effort" has to be understood not in an individualistic sense, but in "a deeper sense" as a manifestation of the *élan vital*.

Bergson argued against the Darwinian doctrine that the cause of evolution was natural selection acting on random variations. This doctrine fails to give a satisfactory explanation of the evolution of complex organs and functions, such as the eye of vertebrates, for it obliges us to suppose that at each stage all the parts of an animal and of its organs vary contemporaneously, since integral functioning has to be preserved to ensure survival. But it is utterly implausible to suppose that such coadapted variations could have been random. Some agency must have been at work to maintain continuity of functioning through successive alterations of form.

Another fact that points in the same direction is the evolution of complex organisms from relatively simple ones. The earliest living things were minute, unicellular entities well adapted to their environment. Why did the evolutionary process not stop at this stage? Why did life continue to complicate itself "more and more dangerously"? Random variations and selection pressures cannot provide a satisfactory explanation. Something must have driven life on to higher and higher levels of organization, despite the risks involved. Neither Spencer, with his appeal to mechanistic notions, nor Lamarck, with his appeal to the "effort" exerted by individual organisms, can account for the great diversity and complexity which evolution has created.

The clue to this problem, Bergson affirmed, is to be found not in biology, but in metaphysics. Human beings, with their capacity for intuition, are typical constituents of the universe, and hence the forces that work in them may be supposed to work in all things. Intuition not only discloses pure duration and becoming in our experience, but also gives us a consciousness of a vital impulse (*élan vital*) within us. We are thus led to the idea of "*un élan original de la vie*" which pervades the whole evolutionary process and accounts for its dominant features. Accordingly, the history of life is to be understood as a process of creative evolution which has resulted from this primordial impulse.

Bergson spoke of the vital impulse as a "current of consciousness" that has penetrated matter, produced living organisms, and made possible an ever-increasing freedom of action. Yet the impulse is not striving to attain a final goal. Bergson was as opposed to radical finalism as he was to mechanism. Both doctrines disregard the creativity by which unpredictable novelties have periodically "leaped into existence." One of these novelties is man, for his appearance was in no sense designed or prefigured. Terrestrial evolution might have produced some other being "of the same essence." Such beings doubtless have arisen elsewhere, for the vital impulse must be supposed to animate countless planets in the universe. Creative evolution is thus cosmic in its scope.

Bergson's interest in science was shown not only by his discussion of biology but also by his discussion of physics. This subject is treated in *Durée et simultanéité*, which deals with the view of time set forth in Einstein's theory of relativity. Here, as in the case

of evolution, Bergson sought to demonstrate how philosophy can supplement science by providing a more adequate account of phenomena. Einstein's theory has, of course, a precisely defined physical sense and stands in no need of support from metaphysics. But some persons have sought to derive from the theory certain "paradoxes" that have their source in the notion of multiple times which flow more rapidly or less rapidly, depending on the motion of the reference systems with which they are associated. Bergson contended that these paradoxes arise because of a philosophical misconception. They rest on an assumption that all of the times are real when the observers in the various systems disagree in their measurements. But from a philosophical standpoint only experienced or "lived" time (*temps vécu*) is real. Hence, the paradoxes can be avoided by considering just one observer and his time as real at each stage of the calculation. For that observer the times of all other reference systems are mathematical fictions, not realities. A similar treatment can be given to simultaneity. It is even possible in the light of this to reinstate the idea of universal time (*l'idée d'une durée de l'univers*) as a philosophical basis for physics.

The religious element in Bergson's thought became more pronounced in his later works. *Creative Evolution* contains a reference to the vital impulse as a "supraconsciousness" to which the name "God" might be applied. This was a use of the term quite alien to traditional Christian theology. For if God is the vital impulse, He is pure activity, limited by the material world through which He is struggling to manifest Himself and, hence, is evolving rather than complete and perfect. In Bergson's final book, *Les deux sources de la morale et de la religion*, this conception is qualified so that it moves closer to the Christian position. God is now affirmed to be love and the object of love. A divine purpose in evolution is also affirmed, for evolution is nothing less than God's "undertaking to create creators, that He may have, besides Himself, beings worthy of His love."

The discovery of this purpose and of the reality of God cannot be made by the intellect. It can be made only by the special sort of intuition that is the mystical experience. The vital impulse, Bergson declared, is "communicated in its entirety" to certain exceptional persons. These are the mystics who attain partial union with the creative effort that "is of God, if it is not God Himself." But the mystical experience does not lead to passive withdrawal from the world; it leads to intense activity. The true mystics are impelled to help the divine purpose to advance by helping mankind to advance beyond its present state.

An important step in this advance is the replacing

of a "closed" society with an "open" one. Bergson's analysis here is influenced by that of the French sociological school of Émile Durkheim. A closed society is one dominated by what is routine and mechanical. It is resistant to change, conservative and authoritarian. Its morality is static and absolutistic, and its religion is ritualistic and dogmatic. An open society is progressive, diversified, experimental, and continually growing. Its morality is flexible and spontaneous. Religion in this society will dispense with stereotyped dogmas formulated by the intellect. These dogmas will be replaced by the intuition and illumination now achieved by the mystics. Men in the open society will be free, integral, creative, and able to reflect in their lives the divine *élan* that is the basic reality in the universe.

Bergson's philosophy represents an impressive statement of the antimechanist position. His use of biological and psychological material to support his contentions, his capacity to invent striking metaphors, and above all his fluent, persuasive style gave his philosophy wide appeal. Philosophers such as William James, dramatists such as G. B. Shaw, and *littérateurs* such as Proust were all influenced by Bergsonian doctrines. Yet the absence of precise definition and of rigorous argument in his books leaves many doctrines obscure. It is far from clear, for example, how his theory of knowledge, which is a form of idealism, can be made compatible with his parabiological theory, which is a form of evolutionary realism. Intuition is said to disclose pure becoming, real duration, and the vital impulse. Yet the differences, if any, between these items are not clearly specified. Sometimes matter and spirit are treated as quite distinct from one another, and sometimes as the "inverse" of one another, matter being spirit that has become "devitalized" and uniform. These obscurities are perhaps to be expected in the work of a philosopher for whom intuition is superior to conceptual thought. Bergson displayed his greatest originality when he undertook to describe direct experience and the temporal dimension of life.

BIBLIOGRAPHY

I. ORIGINAL WORKS. Bergson's works include *Écrits et paroles,* R. M. Mossé-Bastide, ed. (I, Paris, 1957; II and III, Paris, 1959); and *Oeuvres,* Édition du centenaire, annotated by André Robinet, intro. by Henri Gouhier (Paris, 1959). This edition contains the following works: *Essai sur les données immédiates de la conscience* (Paris, 1889); *Matière et mémoire* (Paris, 1896); *Le rire* (Paris, 1900); *Introduction à la métaphysique* (Paris, 1903); *L'évolution créatrice* (Paris, 1907); *L'énergie spirituelle* (Paris, 1919); *Les*

deux sources de la morale et de la religion (Paris, 1932); *La pensée et la mouvant* (Paris, 1934). The only major work omitted is *Durée et simultanéité* (Paris, 1922; 3rd ed., 1926), for the reasons given by E. Le Roy in "Lettre-préface" to *Écrits et paroles,* I, vii–viii.

English translations of Bergson's writings include *Laughter, An Essay on the Meaning of the Comic,* trans. C. Brereton and F. Rothwell (New York, 1910); *Time and Free Will: An Essay on the Immediate Data of Consciousness,* trans. F. L. Podgson (New York, 1910); *Creative Evolution,* trans. A. Mitchell (New York, 1911); *Matter and Memory,* trans. N. M. Paul and W. S. Palmer (New York, 1911); *Introduction to Metaphysics,* trans. T. E. Hulme (New York, 1914, 1949); *Mind-Energy,* trans. H. W. Carr (New York, 1920); *The Two Sources of Morality and Religion,* trans. R. A. Audra and C. Brereton (New York, 1935); and *The Creative Mind,* trans. M. L. Andison (New York, 1946).

II. SECONDARY LITERATURE. Works on Bergson and his philosophy include H. W. Carr, *Henri Bergson, the Philosophy of Change* (London, 1912; new ed., rev., 1919); J. Chevalier, *Henri Bergson* (New York, 1928); J. Delhomme, *Vie et conscience de la vie: Essai sur Bergson* (Paris, 1954); M. V. Jankélévitch, *Henri Bergson* (Paris, 1959), a centenary completion of a work published in 1931 which undertakes to refute the view that Bergson's philosophy is dualistic and to relate it to the contemporary movement of European phenomenology; A. D. Lindsay, *The Philosophy of Bergson* (London, 1911); A. O. Lovejoy, *The Reason, the Understanding and Time* (Baltimore, 1961), contains a valuable discussion of Bergson's ideas in relation to post-Kantian thought in Germany; J. Maritain, *La philosophie bergsonienne* (Paris, 1930), trans. M. L. and G. Andison as *Bergsonian Philosophy and Thomism* (New York, 1955), a highly polemical work by an ex-Bergsonian turned Thomist; R. M. Mossé-Bastide, *Bergson éducateur* (Paris, 1955), a full and accurate account of Bergson's career, with a massive bibliography of primary and secondary sources; B. A. Sharfstein, *Roots of Bergson's Philosophy* (New York, 1943); and *Les études bergsoniennes* (Paris, 1948–1966), studies published at intervals under the editorship of various Bergsonian scholars, of which Vol. II (1949) contains the text of Bergson's doctoral thesis (1889), translated into French from Latin as *L'idée de lieu chez Aristote.*

T. A. GOUDGE

BÉRIGARD (in modern French, **Beauregard**), **CLAUDE GUILLERMET DE** (*b.* Moulins, France, 1578 [according to Niceron; possibly as late as 1591]; *d.* Padua, Italy, 1663/1664), *medicine, physics, philosophy.*

Bérigard studied both medicine and philosophy at Aix-en-Provence and lived quietly in Avignon, Lyons, and Paris. He could, therefore, have witnessed the condemnation by the parliament and the Sorbonne of Villon and de Claves for their support of atomism in 1624. Summoned to Tuscany in 1625, possibly by

Christine de Lorraine, mother of Ferdinand II, grand duke of Tuscany, he taught philosophy at Pisa in 1628. Hobbes met him there around 1635. In 1632 Bérigard published the *Dubitationes,* concerning Galileo's *Dialogue Concerning the Two Chief World Systems* (condemned in 1633). Galileo himself is quoted as saying to Elie Diodati (25 July 1634) that it was more out of obligation than conviction. Bérigard, who must have known Galileo personally, always praised him, but remained firmly convinced of the earth's immobility.

In 1640 Bérigard went to Padua, where he became well known as a teacher. He later published a synopsis of his courses as *Circulus pisanus* (1643). Since the preface to the chapter "De generatione et corruptione" named Galileo, Torricelli, Viviani, Cabeo, Bourdin, Boulliau, Mersenne, Descartes, Digby, Kircher, Kaspar Bartholin and his sons, and Borel as some of the contemporaries he admired, he was surely abreast of the intellectual movement of his time and was well disposed toward change. Yet the Scholasticism that he had to teach at times dominated his thought.[1]

The above list also includes Gassendi, although he is usually cited as following Bérigard in reviving atomism. In fact, the opposite was probably true. At first there was no reciprocal influence. Bérigard left Aix before Gassendi came to teach there, and as far as is known they did not meet in 1624–1625, when Gassendi came to Paris. But in his *Exercitationes paradoxicae* (1624) Gassendi criticized Aristotle after having taught his doctrines—as Bérigard did in the *Circulus.* Although Gassendi's great Epicurean works date from 1647 and 1649, he was already working on them in 1630, as Mersenne and Peiresc reveal in their correspondence. Between 1640 and 1643, in published works, Gassendi cited atomistic physics as evidence and corresponded with Liceti, whom Bérigard succeeded in 1653. Finally, the second edition of the *Circulus* shows the influence of Gassendi's *Syntagma philosophicum* (1658). It should be noted that as early as 1633 Mersenne and Peiresc called Gassendi's attention to the *Dubitationes.*[2]

On the other hand, the spirits of the two works are somewhat different. Both admit *corpuscula tenuia, a Deo creata,* while attributing essential qualities to them—but with far more caution on Gassendi's part than on Bérigard's. Moreover, Bérigard's corpuscles are not true atoms. As a humanist inclined to archaism, Bérigard went back to the pre-Socratics because Aristotle was against them; Gassendi, a historian above all, thought that Aristotle's successors had surpassed him. Bérigard was unable to tie the new scientific ideas to his own endeavor, while

Gassendi found new arguments in these ideas and tried to prepare the future for rational physics.

Actually, Bérigard found his inspiration less in Democritus, who was quite modern, than in the Ionians, who were still influenced by mythology: Anaximander, Empedocles, and Anaxagoras. In order to explain variations in being, these philosophers divided it into simple elements likely to combine. Because of this, the form of the elements must be that most suitable to movement: round and smooth. These elements achieved cohesion by all moving in the same direction or toward one center.

Around the masses thus formed there circulated an ether that consolidated them and, through the pores of these contiguous but not adherent elements, penetrated the masses and also escaped from them, sweeping particles in and out without leaving a vacuum.[3] However, the subject in question was a *qualitative atomism,* not a mechanical one, for the elements were not homogeneous; each had distinct properties, and the appearances resulting from their composition are infinitely different.[4] Only the local predominance of certain more or less similar qualities gave rise to the most common perceptible appearances and accounted for there being lead here, for instance, and gold there. But there was some of all things in each thing—gold in lead, lead in gold, with an infinite possibility of transmutations.

However, this vague relationship between a quantitative determination in the external cause and a qualitative appearance in the conscious effect is insufficient for making reality rational. The explanatory power of atomism is related to the supposition of perfect homogeneity in the components of bodies. Only the number, distance, and density of atoms in a given volume—that is, only the quantity measurable and expressible by mathematical means—enters into the action that the composites exert on each other and on those composites which are our sense organs. Atoms, then, are units capable of being combined, and these combinations can be calculated. Bérigard's qualified corpuscles end only in Heraclitus' perpetual flux, in which the same entities are never reproduced; thus no law can be formulated or imagined.

From the very beginning Bérigard and Gassendi held opposing views on the ontology of space and time. For Gassendi space (the Epicurean void) and time (which passes outside events) were "neither substance nor accidents."[5] Thus time and space escape in parallel ways from the Aristotelian categories and need not have been created in order to exist, because they make creation possible. Against this opinion, Bérigard absolutely refused to recognize that space and time are uncreated things, nor did he admit that they are neither substance nor mere accident (*Circulus,* 1661 ed., p. 51). But Bérigard here drew back, in spite of the fact that he renounced the Scholastic definition of space as "limite du corps environnant" (*ibid.,* p. 47).

Gassendi's view leads directly to the law of inertia, to the conservation of straight and uniform movement as long as nothing in inert space stops it. Newton found in Gassendi the ontological categories that made universal attraction conceivable. Yet Gassendi had not completely renounced qualitative atomism, and lacked the mathematical ability to found a mechanistic physics. Bérigard, even after having read Gassendi's works, remained deeply involved with qualitative physics. He was a great scholar; if, however, he did see the scientific promised land from afar, he did not move toward it. He was not aware of the implications of his own corpuscular philosophy or the importance of universal mechanism.

NOTES

1. For instance, although recognizing that sunspots prove that the sun is not incorruptible, he did not dare go so far as to say the same of the "heavens." (The list was taken from the 2nd ed., 1661.)
2. See *Correspondance de Mersenne,* III, 355 (5 January 1633, Mersenne to Gassendi) and 380 (2 March 1633, Gassendi to Peiresc).
3. Up to this point, the parallel with Descartes's physics could be easily noted, for Descartes would not be called an atomist and denied the void, although he admitted vortexes of subtle matter.
4. Here it is legitimate to speak of either *points-qualités* or *éléments-qualités,* as is done by Jean Zafiropoulo in his *Anaxagore de Clazomène* (Paris, 1948), pp. 313–315.
5. See *Animadversiones in X. librum D. Laertii* (Lyons, 1649), p. 614, and *Syntagma philosophicum* (Lyons, 1658), I 182*a,* 183*b,* 220*b.* In his controversy over the void with P. Noël in 1648, Pascal was led to use the same anti-Scholastic formulas as Gassendi; however, he may have found them only in Noël's writings (see *Oeuvres de Pascal,* Pléïade ed., pp. 383, 1450).

BIBLIOGRAPHY

I. ORIGINAL WORKS. Bérigard's major works are *Dubitationes in dialogum Galilaei Lyncei . . .* (Florence, 1632); and *Circulus Claudii Berigardi, de veteri et peripatetica philosophia . . .* (Udine, 1643); 2nd ed., *Circulus pisanus Claudii Berigardi, de veteri et peripatetica philosophia . . .* (Padua, 1661), in 1 vol., with considerable modification.

II. SECONDARY LITERATURE. Works on Bérigard are E. Brehier, *Histoire de la philosophie,* II (Paris, 1960), 13; J. Brücker, *Historia critica philosophiae* (Leipzig, 1742); David Clement, *Bibliothèque curieuse* (Göttingen, 1750), III, 182–185; Kurt Lasswitz, *Geschichte der Atomistik* (Leipzig, 1890; 1926), I, 488–498; D. G. Morhof, *Polyhistor* (1732); J. P. Niceron, *Mémoires pour servir à l'histoire des*

hommes illustres dans la république des lettres (Paris, 1727–1745), XXXI, 123–127; G. Sortais, *Philosophie moderne depuis Bacon jusqu'à Leibniz* (Paris, 1922), II, 71–75; and W. G. Tennemann, *Histoire de la philosophie* (Leipzig, 1798–1819). There is also a note on Bérigard in the Edizione Nazionale of Galileo (new ed., Florence, 1964), XX, 90. Franck's *Dictionnaire des sciences philosophiques* (1885) includes an article on Bérigard in which there is a serious error: Galileus Lynceus is not a pseudonym, but refers to Galileo Galilei in the full title of *Dubitationes.*

BERNARD ROCHOT

BERING, VITUS (*b.* Horsens, Denmark, summer 1681; *d.* Bering Island, Russia, 19 December 1741), *geography.*

Bering's parents were Jonas Svendsen and Anna Bering. There were many children in the family, and the need to earn a living sent Bering to sea at an early age. Returning from a voyage to the East Indies in 1703, he was recruited into the Russian navy, where he rose quickly in rank.

On 5 February 1725, Bering left St. Petersburg on his first expedition, which, under Peter the Great's orders, was to determine whether Asia and America were joined by land. He journeyed overland through Siberia and across to the Kamchatka Peninsula, which he was the first to map. On 9 July 1728 he sailed in a northerly direction along the east coast of Asia; on 15 August he reached latitude 67° 18' N. There he turned back, having convinced himself that there was no more land to the north. Owing to poor visibility in the strait that was to be named after him, he could not sight the American mainland. He was much criticized for not having proved whether East Cape (now Cape Dezhnev) was the end of Asia.

In spite of that, Bering led a second and even more ambitious expedition to investigate the Kurile Islands and Japan, to make a landing in America, and to map the north coast of Siberia. The expedition achieved the first two objectives. Bering himself left St. Petersburg in 1733 and sailed from Kamchatka for the New World in June 1741. Setting his course in accordance with Delisle's map, he searched for the land mass of Gamaland on the 47th parallel. Failing to find it, he turned northeast and on 16 July 1741 sighted Alaska; four days later he landed on Kayak Island. Bering's return was delayed by the lateness of the season, by scurvy, and, again, by the inaccuracy of the maps. He returned along the hitherto unknown Aleutian Islands and was forced to winter on an uninhabited island only 300 miles short of home. There he died, and the crew named the island Bering Island.

BIBLIOGRAPHY

The translated report of the first expedition by Bering is reproduced in F. A. Golder, *Bering's Voyages, an Account of the Efforts of the Russians to Determine the Relations of Asia and America,* 2 vols. (New York, 1922–1925).

There is no complete biography of Bering. Of value are an account of the first expedition in W. H. Dale, "Notes on an Original Chart of Bering's Expedition of 1725–1730 and an Original Manuscript Chart of His Second Expedition, Together With a Summary of a Journal of the First Expedition, Kept by Peter Chaplin," Appendix 19 of *U.S. Coast and Geodetic Survey Report* for 1890 (Washington, D.C., 1891), pp. 759–774; accounts by members of the second expedition S. Khitrov, K. Yushin, and G. W. Steller, in F. A. Golder, *Bering's Voyages* (see above), I, 36–49; 50–230; II, 9–158, respectively; and Sven Waxell, *The American Expedition,* M. A. Michael, trans. (London, 1952), pp. 11–135.

WILMA GEORGE

BERINGER, JOHANN BARTHOLOMAEUS ADAM, also known as **Johann Barthel Adam Behringer** (*b.* Würzburg, Germany, *ca.* 1667; *d.* Würzburg, 11 April 1738),[1] *medicine, natural history.*

There is a signal lack of information regarding Beringer's life and career. His father was Johann Ludwig Beringer, professor at the University of Würzburg and senior physician and dean of the Faculty of Medicine there from 1669 to 1671. In 1693 Beringer passed the final examination (*periculum*) for the doctorate in medicine and on 14 December 1694 was named *professor quartus seu extraordinarius* at the University of Würzburg—where he remained, as far as is known, for his entire career. Appointed keeper in 1695, Beringer reordered and enlarged the botanical gardens of the university and the Julian Hospital.[2] In 1700 he was elevated to the rank of *professor ordinarius* and dean of the Faculty of Medicine; adviser and chief physician to the prince-bishop of Würzburg, Christoph Franz von Hutten; and chief physician to the Julian Hospital (1700/1701–1728). He frequently lectured at the university on reform in education, and succeeded in introducing a program for the education (at public expense) of poor but gifted students from various orphan asylums. In medicine Beringer was particularly concerned with the malpractices of wandering physicians.

Typical of the curious and learned men of the seventeenth and eighteenth centuries, Beringer was caught up in virtuoso endeavor. Occasionally his natural history lectures turned to the petrifactions found in the Würzburg Muschelkalk, from which he collected fossil shells for his cabinet. It was this interest that led to his involvement in the famous Würzburg *Lügensteine* hoax. Stones of shell lime carved in a great variety of forms were hidden about Mount Eibelstadt by two of Beringer's colleagues— J. Ignatz Roderick (an ex-Jesuit and professor of geography, algebra, and analysis at the university) and Johann Georg von Eckhart (privy councillor and librarian to the court and the university)—with the assistance of three young boys.[3] The stones were subsequently uncovered by Beringer and placed in his cabinet. The hoaxers soon realized the enormity of their actions, but in spite of their best efforts to dissuade him, Beringer published a preliminary report on the stones in 1726: *Lithographiae Wirceburgensis . . . Specimen primum. . . .* The fraud, however, was soon discovered. Roderick and Eckhart were taken to court by Beringer for the "saving of his honor," and were duly punished.[4]

Despite this partially successful attempt to discredit him, Beringer remained on the staff of the university, where he was occupied with teaching and research until his death in 1738. He was survived by two sons, Georg Philipp (who matriculated under the direction of his father in 1701, with the dissertation *Thesis de phthisi*) and Johannes Ludwig Anton (a student of metaphysics).

NOTES

1. Beringer's dates, especially that of his birth, are open to conjecture. His death is assigned both to 1738 and, without month, to 1740.
2. A catalog of the gardens, issued in 1722, contained a list of 423 species of plants arranged according to Tournefort's system.
3. The three youths, employed by Beringer for fieldwork, were Christian Zänger (age seventeen) and two brothers, Niklaus and Valentin Hehn (age eighteen and fourteen, respectively). Of the three only Zänger was involved in the hoax, being in the employ of Roderick and von Eckhart as well as of Beringer.
4. The judicial process was in three steps: first, the hearing at the Würzburg Cathedral Chapter on 13 April 1726; second, third, the municipal trials of 15 April and 11 June 1726. The transcripts are in the Staatsarchiv, Würzburg, and appear in a complete English translation in Jahn and Woolf, *Lying Stones,* pp. 125–141.

BIBLIOGRAPHY

I. ORIGINAL WORKS. Beringer's writings include the following: *Connubium Galenico-Hippocraticum, sive Idaea institutionum medicinae rationalium . . .* (Würzburg, 1708); *Tractatus de conservanda corporis humani sanitate . . .* (Würzburg, 1710); *Dissertatio prima de peste in genere et lue epidemica modo grassante in specie . . .* (Nuremberg, 1714); *Plantarum quarundam exoticarum perennium in horto medico Herbipolensi . . .* (Würzburg, 1722), written with Laurentius Anton Dercum—it is uncertain whether this catalog was actually published or was presented to the university in MS; *Lithographiae Wirceburgensis...Specimen primum* . . . (Würzburg, 1726; 2nd ed., Frankfurt–Leipzig, 1767), trans. by M. E. Jahn and Daniel J. Woolf in *The Lying Stones of Dr. Beringer* . . . (Berkeley–Los Angeles, 1963); and *Gründlich und richtigste Untersuchung deren kissinger Heyl- und Gesundheits-Brunnen* . . . (Würzburg, 1738).

The 2nd ed. of the *Lithographiae Wirceburgensis* consisted of original sheets with a new title page. The sheets undoubtedly came from copies of the original ed. recalled from booksellers and the press by Beringer but not destroyed. The reissue of the work as a literary curiosity is often attributed to one of Beringer's sons.

A partial list of dissertations directed by Beringer is in Sticker, p. 487.

II. Secondary Literature. It would be impossible to cite more than a handful of the articles written about Beringer and the Würzburg *Lügensteine.* The following contain the correct story of the hoax and, in most cases, a bibliography of secondary sources: M. E. Jahn, "Dr. Beringer and the Würzburg 'Lügensteine,'" in *Journal of the Society for the Bibliography of Natural History,* **4,** no. 2 (Jan. 1963), 138–146, and "A Further Note on Dr. Johann Bartholomew Adam Beringer," *ibid.,* no. 3 (Nov. 1963), 160–161; Heinrich Kirchner, "Die würzburger Lügensteine im Lichte neuer archivalischer Funde," in *Zeitschrift der Deutschen geologischen Gesellschaft,* **87,** no. 9 (Nov. 1935), 607–615; P. X. Leschevin, "Notice sur l'ouvrage singulier intitulé: *Lithographia Wirceburgensis,* et sur la mystification qui y a donné lieu," in *Magasin encyclopédique,* **6** (Nov. 1808), 116–128; August Padtberg, "Die Geschichte einer vielberufenen paläontologischen Fälschung," in *Stimmen der Zeit,* **104** (1923), 32–48; and Georg Sticker, "Entwicklungsgeschichte der Medizinischen Fakultät an der Alma Mater Julia," in Max Bruchner, ed., *Festschrift zum 350 Jährigen bestehen der Universität* . . . (Würzburg, 1932), pp. 483–487.

The trial transcripts excerpted in Kirchner's article and included in full in Jahn and Woolf (pp. 125–141) exist in MS in the Staatsarchiv, Würzburg.

Melvin E. Jahn

BERKELEY, GEORGE (*b.* County Kilkenny, Ireland, March 1685; *d.* Oxford, England, 14 January 1753), *philosophy of science.*

Berkeley was a critic of seventeenth- and eighteenth-century philosophical, scientific, mathematical, moral, political, and theological ideas and an important link in the development of general philosophy between the period of Descartes and Locke and that of Hume and Kant. From his earliest days at Trinity College, Dublin (1700–1713), he came under the influence of Bacon, Boyle, Newton, Locke, and Malebranche. In 1705 he helped to found a society with the aim of pursuing the inquiry into their "new philosophy"; the extent of this inquiry may be gauged from Berkeley's *Commonplace Book,* kept during the first few years

of that period. Subsequently, particularly in London, Berkeley formed intellectual associations with such prominent figures as Clarke, Swift, Addison, Steele, and Pope. After a brief interlude in America, connected with his abortive attempt to found a college in Bermuda (1729–1731), he retired to the bishopric of Cloyne in 1734. He moved to Oxford in 1752.

Berkeley's interests (excluding political economy, and his epistemological and theological inquiries except insofar as they bear on science) ranged from those with a primarily scientific focus to the scientifico-philosophical. In the former category belongs *A New Theory of Vision* (1709), reckoned by *Brett's History of Psychology* to have been "the most significant contribution to psychology produced in the eighteenth century," being "the first instance of clear isolation and purely relevant discussion of a psychological topic" (Peters ed., p. 409). The main problem examined in this work is the factors that determine our ability to see things at a distance, the assumption being that the sense of vision itself is incapable of doing so. Rather, seeing distant objects requires the *suggestions* supplied by other senses, especially that of touch, as well as such other experiences as visual distortion caused by failure of eye accommodation. We do not "judge" by means of quasi-optical calculation of the distance of objects (the traditional account of Berkeley's predecessors); rather, we let one group of sensations suggest another, in virtue of experience and custom. Moreover, from saying that all visual sensations "seem to be in the eye," Berkeley moves to his basic contention, later generalized in his *Principles of Human Knowledge* (1710), that visual ideas are in our minds. Given his general doctrine that the "being" of things amounts to their being perceived, i.e., being ideas in a mind (the ultimate reference is to the divine mind), he infers that external space is not basic, but is "only suggested" to us by visual ideas, via tactile and other ideas.

This close interweaving of science with epistemology, as well as of metaphysics with theology, is also very prominent in Berkeley's last major work, *Siris* (1744), which begins as an investigation of the medicinal virtues of tar water and ends with a disquisition on Platonic philosophy. The body of the book consists, on the one hand, of a discussion of contemporary chemical theory and, on the other, of a critique of Newtonian principles of explanation, of space and time, and of the true interpretation of the concept of causation. The sections on chemistry are of particular interest, for they display considerable acquaintance with most of the major chemical doctrines of Berkeley's period (e.g., Boerhaave, Homberg, Hales, the younger Lemery, etc.), including a discussion of acids, salts, alkalies, and air that leads to a discussion of fire and light, the latter providing a "bridge" to a spiritual interpretation of all phenomena. *Siris* thus involves an attempt to assimilate Newtonian concepts to the more complex phenomena of chemistry and animal physiology.

Apart from his more specifically scientific preoccupations, Berkeley's more general aim in these writings is to show that the goal of science can be no more than describing phenomena through the laws and theories ("hypotheses") of science that govern them, and thus to trace the "grammar" or "language of nature" without intervening concepts, at least insofar as these concepts might be construed existentially or as sources of "active power," which in Berkeley's terminology would amount to giving an "explanation." The opposition to a positive construction of such intervening concepts is paramount in Berkeley's writing on mathematics, as exemplified in his critique of the foundations of the differential calculus, whether our concern be with Newtonian "fluxions" or with Leibnizian "infinitesimals." Both, as Berkeley points out in *The Analyst* (1734), suffer from the fatal defect of demanding that certain "increments" vanish in a result whose demonstration requires these increments to have a finite value.

Berkeley's basic objection is to a sequence that is imagined to continue indefinitely, yet at the same time is conceived as suddenly ending. This difficulty formed the starting point of many discussions of the foundations of mathematics that continued in England until the nineteenth century, and he himself initially participated in them through replies to objections made to *The Analyst*. Berkeley does not impugn the employment of the differential calculus for "practical" purposes; his objection is to the quasi-existential positing of the "differential" entities involved. In the *Principles* this had been stated as an opposition to "abstract ideas." His fundamental thought (although he lacks the notion of the limit) is "operationalist," a concentration on the imaginative *process* of dividing a finite line into finite parts *indefinitely,* by always letting the new parts "grow" so that they remain finite lines; this conception is meant to replace "infinite divisibility" into the "infinitely small" (*Principles,* sec. 128). At a more technical level Berkeley developed an ingenious theory of compensating errors that was meant to explain the "correct" results of the calculus of fluxions, whose "faulty" foundations alone he deplored.

Berkeley's opposition to abstract ideas is closely connected with a theory of meaning the most relevant component of which is the contention that we should not suppose that to every noun there corresponds a particular idea. In *De motu* this is applied with special emphasis to the Newtonian concepts of gravitational attraction, action and reaction, and motion in general. Basically, Berkeley regards all such concepts as elements in "mathematical hypotheses" (i.e., what would now be called theoretical terms implicitly defined by certain theoretical axioms). Sometimes he holds that theoretical concepts are simply reducible to individual laws of phenomena (reductionism); at other times he emphasizes their place in the systematic constructions of these laws in overarching theories (a forerunner of the modern instrumentalist position).

The instrumentalist approach affected Berkeley's theory of explanation and causation, which also drew upon the basic doctrine that all phenomena must be construed as ideas. Since they stand in an accusative relation to a perceiver, the ideas are held to be inactive; this is the doctrine of *esse-percipi.* The logical counterpart of the doctrine that no idea can act on any other idea is that no necessary connections exist between any such ideas. As a result, causal explanation cannot be reducible to the "action" of any phenomenal agents, be they "attraction" or "insensible corpuscles." Causal action reduces to uniform "lawlike" association between ideas that function as signs for things signified; the logical center of gravity being again the theoretical system of scientific laws, laws whose ultimate inductive foundation Berkeley places in the uniform operation of the "Author of nature" (*Principles,* sec. 107).

It follows that the doctrine of the distinction between primary and secondary qualities, so central to the thinking of the "Newtonian century," in Berkeley loses its metaphysical relevance, reducing at most to no more than a difference of degree, since the opposition to abstract general ideas and the unavailability of the theoretical "corpuscles" for explanation rendered the conception unimportant. Berkeley does not so much deny unobservable entities; once again he is opposed only to treating them as genuine sources of transeunt causal action, since they are in reality no more than abstractions.

These approaches more or less naturally lead to Berkeley's critique of the Newtonian concepts of absolute space, time, and motion. For it follows at once that all motion must be relative and referred to a physical (phenomenal) system, a contention that Berkeley also urges against Newton's example of rotatory motion, thus anticipating part of what is now called Mach's principle. The impossibility of absolute motion is one of Berkeley's arguments against absolute space; another is its being an abstract idea. Moreover, it is otiose if taken to be an entity "existing without the mind" (*Principles,* sec. 116). This (somewhat weakly) seems to fit in with the conclusion drawn from the theory that distance and space cannot be determined visually. At best, empty space denotes a mere "possibility" for a body to be in motion, and certainly it is nothing "given in itself," separate from or prior to body.

Berkeley's general influence extended to such writers as Hume, Maclaurin, and Kant in the eighteenth century, and Mill, Helmholtz, and Mach in the nineteenth. He also anticipated many of the ideas of twentieth-century philosophers of science.

BIBLIOGRAPHY

I. ORIGINAL WORKS. The standard edition of Berkeley's writings is *The Works of George Berkeley, Bishop of Cloyne,* A. A. Luce and T. E. Jessop, eds., 9 vols. (Edinburgh, 1948–1957).

Berkeley's major writings on science and mathematics and their philosophy are *An Essay Towards a New Theory of Vision* (Dublin, 1709); *A Treatise Concerning the Principles of Human Knowledge*, pt. 1 (Dublin, 1710), the only part published; *Three Dialogues Between Hylas and Philonous* (London, 1713); *De motu* (London, 1721); *Alciphron, or the Minute Philosopher* (London, 1732); *Theory of Vision, or Visual Language, Vindicated and Explained* (London, 1733); *The Analyst* (London, 1734); *A Defence of Free-Thinking in Mathematics* (London, 1735); *Siris: A Chain of Philosophical Reflexions and Inquiries Concerning the Virtues of Tar-Water* (London, 1744); *Further Thoughts on Tar-Water* (London, 1752); and *Philosophical Commentaries [Commonplace Book]*, A. A. Luce, ed. (London, 1944).

Collections that include scientific writings are *Selections From Berkeley Annotated,* A. C. Fraser, ed. (Oxford, 1874); *Berkeley: Philosophical Writings*, T. E. Jessop, ed. (Edinburgh, 1952); *Berkeley: Works on Vision*, C. M. Turbayne, ed., in Library of Liberal Arts (Indianapolis, 1963); and *Berkeley's Philosophical Writings*, D. M. Armstrong, ed., in Collier Classics in the History of Thought (New York, 1965).

II. SECONDARY LITERATURE. The standard biography is A. A. Luce, *The Life of George Berkeley* (Edinburgh, 1949).

Discussions of aspects of Berkeley's philosophy of science and mathematics since 1842 may be found in the following works: T. K. Abbott, *Sight and Touch: An Attempt to Disprove the Received (or Berkeleian) Theory of Vision* (London, 1864); G. W. Ardley, *Berkeley's Philosophy of Nature* (Auckland, 1962); D. M. Armstrong, *Berkeley's Theory of Vision* (Melbourne, 1960); S. Bailey, *A Review of Berkeley's Theory of Vision* (London, 1842); C. B. Boyer, *The History of the Calculus* (New York, 1959), ch. 6, pp. 224–229; *Brett's History of Psychology,* R. S. Peters, ed. (London, 1953), pp. 408–414; *British Journal for the Philosophy of Science,* **4** (May 1953), which honors the bicentenary of Berkeley's death; G. Buchdahl, *Metaphysics*

and the Philosophy of Science. The Classical Origins: Descartes to Kant (Oxford, 1969), ch. 5; F. Cajori, *A History of the Conceptions of Limits and Fluxions in Great Britain From Newton to Woodhouse* (Chicago, 1919), pp. 57–95; D. W. Hamlyn, *Sensation and Perception. A History of the Philosophy of Perception* (London, 1961), pp. 104–116; T. H. Huxley, *Hume: With Helps to the Study of Berkeley* (London, 1894); G. A. Johnston, *The Development of Berkeley's Philosophy* (London, 1923); A. A. Luce, *Berkeley and Malebranche* (Oxford, 1934); J. S. Mill, *Dissertations and Discussions,* IV (London, 1875), 154–187; A. D. Ritchie, *George Berkeley. A Reappraisal* (Manchester, 1967); G. Stammler, *Berkeleys Philosophie der Mathematik* (Berlin, 1922); C. M. Turbayne, *The Myth of Metaphor* (New Haven, 1962); and G. J. Warnock, *Berkeley* (London, 1953).

GERD BUCHDAHL

BERKELEY, MILES JOSEPH (*b.* Biggin Hall, Oundle, Northamptonshire, England, 1 April 1803; *d.* Sibbertoft, Market Harborough, Northamptonshire, 30 July 1889), *mycology.*

Miles was the second son of Charles Berkeley and of the sister of Paul Sandby Munn, the well-known watercolor artist. Berkeley received his early education at the grammar school at Oundle and later at Rugby. He entered Christ's College, Cambridge, as a scholar in 1821 and obtained the B.A. in 1825. In the following year he entered the clergy, beginning his career as curate of St. John's, Margate. In 1833 he became perpetual curate of Apethorpe and Woodnewton, Northamptonshire; in 1868 he moved, on his appointment as vicar, to Sibbertoft. In 1830 he married Cecelia Emma Campbell; the couple had fifteen children. Berkeley was a man of splendid presence and great refinement, and had a sound classical background. He read the proofs of Bentham and Hooker's *Genera plantarum* to ensure the correctness of the Latin text. To support his family, he depended almost entirely on his meager clerical stipend, which for many years he supplemented by keeping a small boarding school for boys. His continued straitened circumstances were reflected in the grant of a Civil List pension of £100 per year in 1867. He was elected to the Linnean Society in 1836 and to the Royal Society in 1879, having already received its Royal Medal in 1863.

While at school at Rugby, Berkeley became intensely interested in natural history, particularly in animals, and built up an extensive shell collection. His first publications were on zoology and were illustrated with his own fine colored drawings. Later his bent toward biology was greatly fostered by his close acquaintance with J. S. Henslow, professor of botany at Cambridge and a friend of Charles Darwin. Probably encouraged by three well-known contemporary cryptogamists—William Henry Harvey of Trinity College, Dublin; Robert Kaye Greville of Edinburgh; and Captain Dugald Carmichael of Appin—Berkeley gave up his zoological studies and began investigations on the lower plants. In 1833 he produced his *Gleanings of British Algae,* in which he described in detail and with color plates the structure of a number of marine and freshwater species. Soon he became engrossed in his studies of fungi. The work that established his preeminence as a mycologist was his account of fungi, which was prepared at the invitation of William Jackson Hooker for a volume of Sir James Edward Smith's *The English Flora.* The meticulously accurate descriptions, mostly drawn from living material, remain unsurpassed in their construction.

Between 1837 and 1883 he published, in the later years in collaboration with Christopher Edmund Broome, a series of papers entitled "Notices of British Fungi" in the *Annals and Magazine of Zoology and Botany* (later called the *Annals and Magazine of Natural History*). Over the years a vast amount of exotic material from the Royal Botanic Gardens, Kew, was referred to Berkeley, and he became the accepted authority for information on mycological matters. His herbarium, comprising some 10,000 specimens, including about 5,000 types that he had described, was presented to Kew.

Berkeley's significant contributions to cryptogamic botany were not by any means confined to the taxonomy of the fungi. Indeed, he can with some justification be regarded as the founder of plant pathology, for he was the first to appreciate the economic importance of the incidence of plant disease caused by fungi. His pioneer researches established that potato blight was the result of the ravages of *Phytophthora infestans,* but this was merely one of a series of investigations on pathogenic fungi that he undertook between 1854 and 1880; his important results are to be found in the articles that he published in the *Gardeners Chronicle.* Berkeley's most distinguished morphological investigations concerned the structure of the hymenium, and it was he who originally established the constant presence of basidia with apically borne spores in a large group of fungi, thus laying the basis of the primary classification into Basidiomycetes (with spores produced externally) and Ascomycetes (with spores formed within a sac, or ascus). In all he published over 400 papers on fungi, either alone or in collaboration.

BIBLIOGRAPHY

I. ORIGINAL WORKS. Berkeley's writings include "Fungi," in J. E. Smith, *The English Flora,* V, pt. 2 (1836); "On the Fructification of the Pileate and Clavate Tribes of the Hymenomycetous Fungi," in *Annals and Magazine of Natural History,* **1** (1838), 81; *Introduction to Cryptogamic Botany* (London, 1857); and *Outlines of British Fungology* (London, 1860).

II. SECONDARY LITERATURE. For further biographical information on Berkeley, see *Gardeners Chronicle,* 3rd ser., **6** (1889), 135 (portrait), 141; J. D. Hooker, in *Proceedings of the Royal Society,* **47** (1890), ix; and G. Massee, *Makers of British Botany,* F. W. Oliver, ed. (Cambridge, 1913).

GEORGE TAYLOR

BERNARD OF CHARTRES, also known as **Bernardus Carnotensis** (*d.* Chartres, France, *ca.* 1130), *philosophy.*

Bernard, who should not be mistaken for Bernardus Silvestris, was of Breton origin and an older brother of Thierry of Chartres. He taught the masters responsible for the glory of the school of Chartres during the first half of the twelfth century. William of Conches, Richard the Bishop, and Gilbert of Poitiers were his most famous disciples.

Bernard is known to have been studying at Chartres as early as 1114. From 1119 to 1126 he was chancellor of the episcopal schools of that city.

The information left concerning Bernard's doctrine is very fragmentary. According to John of Salisbury, he was the most perfect Platonist of his time; but this Platonism, influenced by the *Timaeus* (17A–53C), which Calcidius had made accessible to the Latins, is colored by many overtones. Bernard also attempted to reconcile Plato and Aristotle, an endeavor considered a vain one by John of Salisbury.

Bernard's pedagogical method is better known. Its aim was to obtain effective and continuous work from his students; as John of Salisbury expressed it: "Each passing day became the disciple of the previous day." Bernard is particularly well known for being the first to use the comparison of dwarfs and giants: "We are," he said, "like dwarfs sitting on the shoulders of giants. Our glance can thus take in more things and reach farther than theirs. It is not because our sight is sharper nor our height greater than theirs; it is that we are carried and elevated by the high stature of the giants" (*Metalogicon* III, 4, p. 136). This comparison, in which the giants stand for the ancients and the dwarfs for the moderns, should not be taken as an act of faith in the indefinite progress of the sciences and culture. Rather, Bernard modestly remained at the level of *grammatica*: the secrets of good writing are learned by reading and rereading the great works of the past, not in order to copy them slavishly but in order to be inspired by them, so that future generations may take us as models, as we ourselves took the ancients as our models.

BIBLIOGRAPHY

I. Original Works. None of Bernard's works has survived in its entirety. The only fragments remaining are found in John of Salisbury's *Metalogicon,* C. Webb, ed. (Oxford, 1929). These are the famous comparison of dwarfs and giants, attributed to him by John of Salisbury and also transmitted by William of Conches—see E. Jeauneau, "Nani gigantum humeris insidentes. Essai d'interprétation de Bernard de Chartres," in *Vivarium,* **5** (1967), 79–99; a quotation from the *Expositio Porphyrii,* in the *Metalogicon* IV, 35, p. 206, 11. 19–25; and some fragments of philosophical poems: two elegiac distichs on form (*idea*) and matter (*ile*), cited in the *Metalogicon* IV, 35, p. 205, 11. 24–27, and recurring in some twelfth-century glosses on Plato's *Timaeus* (MS Vatican, Archivio di San Pietro, H 51, fol. 11v); six hexameters on the clear opposition of the eternally indestructible world of ideas to the realm of matter, destined to perish in time, also quoted in the *Metalogicon* IV, 35, p. 206, 11. 26–31; and three hexameters on the conditions favorable to the work of the mind, quoted in John of Salisbury's *Policraticus* VII, 13 (C. Webb, ed. [Oxford, 1909], II, p. 145, 11. 12–14), and commented on by Hugh of Saint Victor in his *Didascalicon* III, 13–20.

II. Secondary Literature. Works containing further information on Bernard are A. Clerval, *Les écoles de Chartres au moyen âge, du Ve au XVe siècle* (Paris, 1895); E. Garin, *Studi sul platonismo medievale* (Florence, 1958), pp. 50–53; L. Merlet and R. Merlet, "Dignitaires de l'Église Notre-Dame de Chartres. Listes chronologiques," in *Archives du diocèse de Chartres,* V (Chartres, 1900), 103; A. Nelson, "Ett citat från Bernard av Chartres," in *Nordisk tidskrift för Bok-och Bibliotheksväsen,* **17** (1930), 41; and R. L. Poole, *Illustrations of the History of Medieval Thought and Learning,* 2nd ed. (London, 1920).

Edouard Jeauneau

BERNARD OF LE TREILLE (TRILIA) (*b.* near what is now Nîmes, France, *ca.* 1240; *d.* Avignon, France, 4 August 1292), *astronomy, philosophy.*

As a youth Bernard entered the Dominican Order in the province of Provence, possibly at Montpellier. Early catalogs of the order list him as a Spaniard; Quétif and Échard explain this by the supposition that Montpellier and Nîmes fell at the time under the king of Aragon, or had done so earlier under James I. Sometime between 1260 and 1265 Bernard was sent to Paris to study, and subsequently he taught at Montpellier (1266, 1268), Avignon (1267, 1274), Bordeaux (1271), Marseilles (1272), and Toulouse (1273, 1276). He returned to the priory of St.-Jacques in Paris to lecture on the *Sentences* in 1279, then continued teaching there as a bachelor (1282–1284) and as a master of theology (1284–1287). From 1288 until his death he held various administrative posts in his order in Provence.

Bernard is described by Duhem as a disciple of Albertus Magnus, and is thought by some to have attended the lectures of Thomas Aquinas. While he was undoubtedly well acquainted with the teachings of these fellow Dominicans, the chronology of his education rules out the possibility of his having studied under either. Bernard is the earliest known French Dominican to be identified as a Thomist, however, and his philosophical and theological writings—which constitute the bulk of his literary output—bear out this identification. He consistently explained and defended Aquinas' teachings on the real distinction between essence and existence, on the pure potentiality of primary matter, and on the unicity of substantial form. The last two points are of some importance in the history of medieval science, since they committed Bernard to a rejection of the *forma corporeitatis,* or "form of corporeity" (a teaching of Avicenna that led some later Scholastics to a mathematicist view of nature), and to an acceptance and elucidation of Thomas' distinctive thesis on the virtual presence of elements in compounds.

Bernard's principal interest in the history of science, however, derives from his having composed a series of questions (*Quaestiones*) on the *Sphere* of John of Sacrobosco, of which only two fourteenth-century manuscripts are known: one is in the municipal library at Laon, the other in St. Mark's in Venice. Early catalogs of the Dominican Order mention a treatise *Super totam astrologiam* ("On All of Astronomy"), which might be another variant of these same *Quaestiones;* more probably it is an erroneous listing of a work with the same title by the Franciscan Bernard of Verdun. Thorndike dates the work between 1263 and 1266, holding that it was composed at Nîmes for the instruction of young Dominicans there; the dating seems somewhat early in light of the chronology of Bernard's studies and teaching given above. The treatise, however, does bear the mark of classroom origin, being divided into *lectiones,* or lectures, and providing a rather philosophical commentary on Sacrobosco's work. Duhem furnishes some twenty pages of French translation of its text, dealing mainly with the concepts of epicycle and eccentric and with explanations of the movement of the fixed stars. They show Bernard favoring the Ptolemaic system of the universe over the stricter geocentric theories of more conservative Aristotelians, and attempting to reconcile the Hipparchian theory of continuous precession of the equinoxes with Thābit's erroneous theory of trepidation along lines suggested by Albertus Magnus, whom Bernard appears to have studied closely.

BIBLIOGRAPHY

I. ORIGINAL WORKS. Bernard's major work is *Bernardi Triliae Quaestiones de cognitione animae separatae a corpore*, Stuart Martin, ed., Vol. XI in the Pontifical Institute of Medieval Studies' series Studies and Texts (Toronto, 1965), a critical edition of the Latin text, with an introduction and notes. Frederick J. Roensch, *Early Thomistic School* (Dubuque, Iowa, 1964), pp. 84–88, 289–296, contains a complete listing of Bernard's other works, with a guide to sources and literature, and a summary of his philosophical teachings.

II. SECONDARY LITERATURE. Works on Bernard are Pierre Duhem, *Le système du monde*, III (Paris, 1915; repr. 1958), 363–383, 391, 417; Jacques Quétif and Jacques Échard, *Scriptores Ordinis Praedicatorum*, I (Paris, 1719; repr. New York, 1959), 432–434; Lynn Thorndike, *The Sphere of Sacrobosco and Its Commentators* (Chicago, 1949), pp. 23–26, 29, 49–51, 54; and George Sarton, *Introduction to the History of Science*, II, part 2 (Baltimore, 1931), 749, 758, 989.

WILLIAM A. WALLACE, O.P.

BERNARD SILVESTRE (BERNARDUS SILVESTRIS), also known as **Bernard de Tours (Bernardus Turonensis)** (*fl.* mid-twelfth century), *philosophy*.

The only certain date of Bernard's life is that of his *Cosmographia*, sometimes improperly called *De mundi universitate*, written between 1145 and 1148. He lived and taught in Tours, where he owned a house near the church of Saint Martin—evidence of which has been found by André Vernet among property titles. It is possible that Bernard also lived and taught in other cities, such as Orléans.

Bernard's most famous work is the *Cosmographia*, a poetical cosmogony alternately in prose and verse, dedicated to Thierry of Chartres; it describes the creation of the universe (*megacosmos*) and of man (*microcosmos*). The *Cosmographia* is less a treatise on cosmogony than a dramatic interpretation of philosophical thoughts drawn from many sources: from the *Timaeus*, translated and commented upon by Calcidius; from Asclepius, Apuleius, Boethius, Macrobius, Martianus Capella, Ovid, and Virgil.

The literary qualities of the *Cosmographia* are undeniable; its interest for the historian of science is perhaps less evident but not without importance. This work shows us, in fact, that the heritage of Greek science had not been entirely lost in the first half of the twelfth century and, further, that a man of letters could have mastered much of it simply by reading the Latin texts available to him. But Bernard's major concern was not to construct a rigorous doctrinal synthesis from such material. It is for this reason that it is pointless to ask, as modern historians often do, whether the *Cosmographia* is pagan or Christian in character.

Bernard was first of all a poet. His attitude toward nature was less "scientific" than that of his contemporary William of Conches. Thus while Bernard seems ready to admit that a state of chaos preceded the order found in the world, William believes that the chaotic state of matter is merely hypothetical. In his commentary on Martianus Capella, Bernard also admits the existence of the waters "above the firmament," while William rejects that idea as being contrary to the laws of physics and considers the biblical verse (Gen. 1:7) upon which it is based as purely allegorical.

BIBLIOGRAPHY

I. ORIGINAL WORKS. With the exception of works that are certainly apocryphal (*De cura rei familiaris, De forma honestae vitae*) and works that are probably apocryphal (*De gemellis, De paupere ingrato*), the literary output of Bernard can be divided into two groups: surely authentic and probably authentic.

Certainly authentic works are *Experimentarius* (a manual of geomancy of which Bernard was not the author but merely the editor), M. Brini Savorelli, ed., in *Rivista critica di storia della filosofia*, **14** (1959), 283–342; *Mathematicus* (or *De patricida*, or *De parricidali*), in J. P. Migne, ed., *Patrologia latina*, CLXXI (Paris, 1893), cols. 1365–1380, where it is erroneously attributed to Hildebert of Lavardin, and in B. Hauréau, *Le Mathematicus de Bernard Silvestris et la Passio sanctae agnetis de Pierre Riga* (Paris, 1895); and *Cosmographia* (or *De mundi universitate*), C. S. Barach and J. Wrobel, eds. (Innsbruck, 1876), a very defective edition; there is also an unpublished critical edition prepared by André Vernet and described in "Bernardus Silvestris et sa Cosmographia," in *École nationale des chartes. Positions des thèses . . . de 1937*, pp. 167–174.

Among works considered "probably authentic" must be classed Bernard's *Ars dictaminis;* although he almost certainly composed a treatise on this subject, so many have been attributed to him that we have no way of determining which represent later modifications. See Charles-V. Langlois, "Maître Bernard," in *Bibliothèque de l'École des chartes*, LIV (1893), 225–250; Ch. H. Haskins, "An Italian Master Bernard," in *Essays in History Presented to R. L. Poole* (Oxford, 1927), pp. 211–226; "The Early Artes dictandi in Italy," in *Studies in Mediaeval Culture* (Oxford, 1929), pp. 170–192; E. Faral, "Le manuscrit 511 du Hunterian Museum de Glasgow," in *Studi medievali*, **9** (1936), 18–121, esp. 80–88; H. Koller, "Zwei pariser Briefsammlungen," in *Mitteilungen des Instituts für österreichische Geschichtsforschung*, **59** (1951), 229–327; B. Berulfsen, "Et blad av en Summa dictaminum," in *Avhandlinger utgitt av det Norske Videnskaps-Akademi i Oslo*, II. *Historisk-Filosofisk Klasse* (1953), no. 3; F.-J.

Schmale, "Der Briefsteller Bernhards von Meung," in *Mitteilungen des Instituts für österreichische Geschichtsforschung*, **66** (1958), 1–28; and W. Zöllner, "Eine neue Bearbeitung der 'Flores dictaminum' des Bernhard von Meung," in *Wissenschaftliche Zeitschrift der Martin-Luther-Universität Halle-Wittemberg, Gesellschafts- und Sprachwissenschaftliche Reihe*, **13** (1964), 335–342.

Another "probably authentic" work is *Commentum super sex libros Eneidos Virgilii*, G. Riedel, ed. (Greifswald, 1924). For a better text, consult the manuscripts mentioned in the following studies: M. De Marco, "Un nuovo codice del commento di Bernardo Silvestre all' Eneide," in *Aevum*, **28** (1954), 178–183; and G. Padoan, "Tradizione e fortuna del commento all' Eneide di Bernardo Silvestre," in *Italia medioevale e umanistica*, **3** (1960), 227–240.

The third is *Commentum super Martianum Capellam*. Extracts are in E. Jeauneau, "Note sur l'École de Chartres," in *Studi medievali*, 3rd. ser., **5** (1964), 821–865, esp. 855–864.

II. SECONDARY LITERATURE. Works on Bernard are E. Gilson, "La cosmogonie de Bernardus Silvestris," in *Archives d'histoire doctrinale et littéraire du moyen âge*, **3** (1928), 5–24; L. Thorndike, *A History of Magic and Experimental Science During the First Thirteen Centuries of Our Era*, 3d ed., II (New York, 1947); T. Silverstein, "The Fabulous Cosmogony of Bernardus Silvestris," in *Modern Philology*, **46** (1948), 92–116; R. B. Woolsey, "Bernard Silvester and the Hermetic Asclepius," in *Traditio*, **6** (1948), 340–344; M. F. McCrimmon, "The Classical Philosophical Sources of the 'De mundi universitate' of Bernard Silvestris," dissertation (Yale University, 1952); F. Munari, "Zu den Verseinlagen in Bernardus Silvestris' De mundi universitate," in *Philologus*, **104** (1960), 279–285; J. R. O'Donnell, "The Sources and Meaning of Bernard Silvester's Commentary on the Aeneid," in *Mediaeval Studies*, **24** (1962), 233–249; and W. von den Steinen, "Bernard Silvestre et le problème du destin," ("Les sujets d'inspiration chez les poètes latin du XIIe siècle," III), in *Cahiers de civilisation médiévale*, **9** (1966), 363–383, esp. 373–383.

EDOUARD JEAUNEAU

BERNARD OF TREVISAN, also known as **Bernard of Treviso, Bernard of Treves** (*fl. ca.* 1378, although also thought to have flourished in the fifteenth or sixteenth century, in France, Italy, or Germany), *alchemy*.

Although it is uncertain whether two or even three persons are responsible for the tracts bearing the name of Bernard of Trevisan, his name first appears in manuscript texts of the fourteenth century; and the contents of all of these works fit well into fourteenth-century alchemical thought and practice, both in the nature of the alchemical doctrines expounded and in the authorities or authors cited. For example, in a reply to Thomas of Bologna, physician to King Charles V of France (*d.* 1380), Bernard maintained against Thomas the dominant fourteenth-century theory that gold is made solely from quicksilver or mercury, although the process might be hastened by the addition of a small amount of gold. Bernard rejected the sulfur-mercury theory of the preceding century. He asserted that mercury contained within itself the four elements—that is, the air and fire of sulfur in addition to the earth and water usually associated with mercury. All these elements, he reported, remain when the mercury turns to gold. He also rejected Thomas of Bologna's association of the planets with the alchemical process.

The alchemical doctrine of the composition of the philosophers' stone by mercury alone was reiterated in the tracts that were printed under Bernard's name in the sixteenth and seventeenth centuries, particularly in *A Singular Treatise on the Philosophers' Stone* and in the *Traicté de la nature de l'oeuvf*. In the latter, Bernard asserted that the elixir is made of pure mercury and that this purified substance, which has lost all its terrestrial and consumable feces and which the philosophers call the water of volatility, contains within itself the entire *magisterium*.

Bernard, in common with other alchemists of the fourteenth century, likened the production of the philosophers' stone to human generation. In this process, he explained, the sun is the male and is hot and dry, the moon is the female and is cold and moist, and both are essential because nothing can be generated and brought to the light of existence without a male and a female. In the philosophers' stone, however, is to be found everything that is required for the production of the stone. This is demonstrated by the fact that it is composed of both body and spirit or of fixed and volatile elements, which, although they do not appear to be so, are indeed one in substance, i.e., quicksilver.

Furthermore, to demonstrate or explain the alchemical process, Bernard utilized another symbol commonly found in the alchemical literature of the time. He likened the mercury of the philosophers to the philosophers' egg, which contains in itself two natures in one substance, the white and the yellow, and from itself produces another—the chicken—which has life and the power of generation. Mercury, he held, similarly contains within itself two natures in the one body and from itself produces a whole that has body, soul, and spirit. Moreover, on the authority of Albertus Magnus, whom he had cited for the preceding exposition of the philosophers' egg as one and many, Bernard likened this oneness of spirit, soul, and body to the Holy Trinity, who are one in God without diversity of substance. In his view, mercury, the egg, contains in itself everything required for the perfection of its own *magisterium*, without the addition of anything else and without any diminution of its own perfection. It has everything for the production of the chicken.

The works bearing Bernard's name also reveal the author's acquaintance with a number of alchemical writers, several of them from earlier centuries and others belonging to the thirteenth and fourteenth centuries. Among the earlier group are Geber, Rasis, Avicenna, Morienus, and Hermes. The later group comprised the Latin authors Albertus Magnus, Thomas Aquinas, Arnald of Villanova, and his brother Pierre of Villanova, as well as Hortulanus and Raymond Lull, John Dastin, and Christopherus Parisiensis. Furthermore, Bernard paraphrased Hippocrates' *Aphorisms* and cited Aristotle and Galen.

There are other interesting and engaging features in Bernard's works. For example, in the *Chemica miracula* there is a long autobiographical account of his quest for the philosophers' stone. In another tract he cites as his reason for departing from the usual admonitions to keep the alchemical art secret the fear that so noble an art or science might perish or be lost if it were not imparted to others. Possibly because the works attributed to Bernard reproduced in this attractive form alchemical doctrines and practices that were familiar to his contemporaries and were to become traditional in the centuries that followed, they were printed and reprinted not only in the sixteenth and seventeenth centuries but as late as the eighteenth century.

BIBLIOGRAPHY

I. ORIGINAL WORKS. For manuscript texts see Lynn Thorndike and Pearl Kibre, *A Catalogue of Incipits of Mediaeval Scientific Writings in Latin* (Cambridge, Mass., 1963). In addition to the MSS there noted are the following, written after 1500: British Museum, Sloane 299, 16c, ff. 10v–19r, in English; Sloane 3117, 17c, ff. 2r–84r; and Sloane 3737, 17c, ff. 93r–95r, extracts.

Printed editions of Bernard's "Responsio ad Thomam de Bononia" are found, in Latin, with Morienus, *De re metallica* (Paris, 1564), in *Artis auriferae* (Basel, 1610), II, 38, and in J. J. Manget, ed., *Bibliotheca chemica*, II (Geneva, 1702), 399; in English, as "Epistle to Thomas of Bononia," in *Aurifontina chemica* (London, 1680), pp. 187, 269; in German, as "Ein Antwort an Thomam de Bononia," with Philip Morgenstern, *Turba philosophorum*, II (Vienna, 1750); in German and Latin, as "Bernardi von Tervis, Vom Stein der Weisen . . . ," in J. Tanckius, *Opuscula chemica* (Leipzig, 1605), pp. 215–230; and in French, as "La response de Messire Bernard Conte de la Marche, Trevisane, à Thomas de Boulongne [*sic*] Medicin du roi Charles huictiesme," Gabriel Joly, trans., in *Trois anciens traictez de la philosophie naturelle* (Paris, 1626), pp. 27–89.

Chymica miracula quod lapidem philosophiae appellant (Strasbourg, 1567; Basel, 1583, 1600), also appeared in L. Zetzner, ed., *Theatrum chemicum,* I (Strasbourg, 1613), 148–776, and 2nd ed., I (Strasbourg, 1659), 683; and as "De secretissimo philosophorum opere chemico," in J. J. Manget, ed., *Bibliotheca chemica,* II (Geneva, 1702), 388.

De chemia. Opus historicum et dogmaticum ex Gallico in Latinum, Gulielmus Gratarolus, trans. (Strasbourg-Basel, 1567), pp. 139–223, also appeared with J. Franciscus Pico della Mirandola, *Libri III de auro* (Ursel, 1598), p. 149; and with D. Zacaire, *Opuscule* (Anvers, 1567).

Vom der hermetischen Philosophia (Strasbourg, 1574), a translation from the Latin, is also in *Hermetischer Rosenkrantz* (Hamburg, 1659, repr. 1682), pp. 98–110; and in *Hermetische Philosophia* (Frankfurt-Leipzig, 1709).

Tractatus singularis Bernhardi Comitis Treverensis. De lapide philosophorum is in *Tractatus aliquot chemici singulares* (Geismar, 1647), pp. 16–30, and *Ginaeceum chimicum,* I (Lyons, 1679), 503–509; an English translation is *A Singular Treatise of Bernard, Earl of Trevisan, Concerning the Philosophers' Stone* (London, 1683), and in *Collectanea chymica* (London, 1684), pp. 83–94.

Traicté de la nature de l'oeuvf des philosophes, composé par Bernard, Comte de Treves, Allemand (Paris, 1659), pp. 1–64, also appeared as "Des Herrn Bernhards, Grafens von der Mark und Tervis," in *Abhandlungen von der Natur des philosophischen Eije* (Hildesheim, 1780).

"La parole delaissée traicté de Bernard, Comte de la Marche Trevisano," in *Trois traitez de la philosophie naturelle* (Paris, 1618), pp. 1–52, was also translated as "Verbum dimissum," in *Taeda trifida chimica* (Nuremberg, 1674), p. 97.

II. SECONDARY LITERATURE. Bernard of Trevisan is associated with the fourteenth century in Lynn Thorndike, *History of Magic and Experimental Science,* III (New York, 1934), 611–627, and V (New York, 1959), 601, 622–623, where he surveys both manuscript and printed texts; and in George Sarton, *Introduction to the History of Science,* III (Baltimore, 1948), 1480.

John Ferguson, *Bibliotheca chemica* (Glasgow, 1906), I, 103–104, and II, 466–467, differentiates between Bernardus Trevisanus of Padua (1406–1409), Bernardo Trevisano of Venice (1652–1720), and Bernardinus Trivisanus of Padua (1506–1583). He also has an extensive bibliography on the three Bernards.

PEARL KIBRE

BERNARD OF VERDUN, also known as **Bernardus de Virduno** (France, *fl.* latter part of the thirteenth century), *astronomy.*

Nothing is known with certainty about the life and career of Bernard of Verdun, except his place of origin (not necessarily the city on the Meuse), that he was a Franciscan, and that he was a professor in his order. It is possible that he is the Bernard who carried on a correspondence with the French scholar Nicolas of Lyra (*ca.* 1270–1349). Bernard's contribution to medieval astronomy is his *Tractatus super totam astrologiam,* most likely written in the late thirteenth century. It discusses a turquet similar to one described by Francon de Pologne in a manuscript dated 1284. If we assume that Bernard's sketchy account of this instrument is derived from Francon's more adequate description, we might consider 1284 as a possible *terminus post quem* for Bernard's *Tractatus.* However, it is quite possible that Bernard's work preceded that of Francon. The *Tractatus,* both a defense and a description of the Ptolemaic system, contains no astrological allusions. Bernard was familiar with the alternative system of al-Biṭrūjī, and he considered it unfavorably. He also rejected the theory of trepidation, which he associated with Thābit ibn Qurra.

In design, Bernard's treatise is similar to the *Almagest* of Ptolemy. The first two sections are devoted to preliminary matters both descriptive and mathematical, such as the characteristics of the four elements and the celestial region, the spherical nature of the heaven and its circular motion, the uniqueness of the world, the insensible size of the earth relative to the heaven, the construction of a table of arcs and chords, and the determination of declinations and ascensions. The remaining sections (excluding the tenth) treat the motion of the sun and the moon, eclipses, and lunar parallax, as well as the motion of the five planets visible to the naked eye. The solar, lunar, and planetary models are all derived from Ptolemy.

Bernard followed the popular rationalization that combined solid spheres with epicycles and eccentrics. Since in this adaptation of Ptolemy the greatest distance of any celestial body is equal to the least distance of the body immediately above, Bernard provides tables for the relative sizes and distances of the sun, moon, and planets. The values given correspond to those in the *Theorica planetarum* of Campanus of Novara, and were undoubtedly canonical in the Middle Ages. The tenth section describes the turquet (turketum, torquetum), a complex instrument designed for a variety of uses, e.g., to find the positions of the fixed stars and the planets, the altitude of the sun, the hour of the day or night. This section also contains a brief account of another astronomical instrument, a kind of noctilabium or "star-clock" that could be used to determine the hour of night by observation of the pole star and two other bright stars, the date being known.

Bernard's sources are few: Ptolemy, al-Battānī, and Aristotle are his most quoted authorities. His intent in his treatise is to present the Ptolemaic system in a clear and concise, although simplified, manner. It is possible that the work was originally intended to familiarize Bernard's students with the main outlines of Ptolemaic astronomy while avoiding the complexities of the *Almagest.* Insofar as it is a technical treatise on astronomy, the *Tractatus* falls in the same medieval astronomical tradition as the "theory of the planets" literature. The only distinctive difference between Bernard's treatise and others in this genre is his introductory defense of Ptolemy as having provided the only explanation that will account for astronomical phenomena.

BIBLIOGRAPHY

I. ORIGINAL WORKS. A modern edition of Bernard's *Tractatus super totam astrologiam* is Polykarp Hartmann, ed., Vol. XV in the series Franziskanische Forschungen (Werl, 1961). This edition is based on two manuscripts in the Bibliothèque Nationale. Lynn Thorndike, however, mentions the following additional manuscripts: Erfurt, Wissenschaftliche Bibliothek, Amplonian Collection, F 393, f22–f43, and F 386, f1–f25 (where the work is entitled *Speculum celeste*); Vatican Palatine, 1380; Vatican (Bibliotheca Apostolica Vaticana) 3097, f51r–f71r. See also Thorndike, "Vatican Latin Manuscripts in the History of Science and Medicine," in *Isis,* **13** (1929–1930), 53–102; Thorndike and P. Kibre, *Incipits of Mediaeval Scientific Writings in Latin,* rev. and enl. ed. (Cambridge, Mass., 1963).

II. SECONDARY LITERATURE. Works concerning Bernard are Pierre Duhem, *Le système du monde,* III (Paris, 1958), 442–460; E. Littré, *Histoire littéraire de la France,* XXI (Paris, 1847), 317–320; and Emmanuel Poulle, "Bernard de Verdun et le turquet," in *Isis,* **55** (1964), 200–208, esp. 202, n. 3.

Three works dealing with the turquet are R. T. Gunther, *Early Science in Oxford,* II (Oxford, 1923), 35–36, 370–375; L. Thorndike, "Franco de Polonia and the Turquet," in *Isis,* **36** (1945–1946), 6–7; and Ernst Zinner, *Astronomische Instrumente des 11.-18. Jahrhunderts* (Munich, 1956), pp. 177–183, plate 11, no. 2. See Zinner, p. 164, plate 57, for information on the noctilabium.

CLAUDIA KREN

BERNARD, CLAUDE (*b.* St.-Julien, near Villefranche, Beaujolais, France, 12 July 1813; *d.* Paris, France, 10 February 1878), *physiology.*

Bernard's parents, Pierre François Bernard and Jeanne Saulnier, who were vineyard workers, lived in very modest circumstances. His father seems to have exerted so little influence that several biographers have erroneously asserted that he died when Bernard was an infant. On the other hand, Bernard always remained close to his mother, a gentle and pious woman. All his life he remained attached to the place of his birth, the hamlet of Chatenay at the outskirts of the village of St.-Julien. Every fall he returned home to relax and to help with the grape harvest. His entire life revolved about two poles of attraction: the laboratories of Paris and the vineyards of Beaujolais. As a child Bernard lived close to nature and maintained his deep love of it throughout his life. His education, first from the parish priest and then in religious schools in Villefranche and Thoissey, was humanistic rather than scientific. At the age of nineteen, he was apprenticed to an apothecary named Millet in Vaise, a suburb of Lyons. Thus he had occasion to observe the rude empiricism of the pharmacotherapy of that period. The apprentice pharmacist turned, however, not toward the sciences but toward the theater and belles-lettres. One of his comedies brought him some local success, which induced him to write a heroic drama entitled *Arthur de Bretagne*. (A first, posthumous edition of 1887 was suppressed by court decision upon the request of Bernard's widow; the work was republished in 1943.)

In 1834 Bernard went to Paris, where he planned to seek a career in literature. The illustrious critic Saint-Marc Girardin discouraged him, however, and urged him first to acquire a profession in order to earn a living. In the same year, with great difficulty, Bernard passed the baccalaureate and entered the Faculty of Medicine in Paris. Thus, as Renan remarked in his *Éloge,* by turning his back on literature, Bernard took the road that nevertheless led him to the Académie Française.

Bernard was an average student, conscientious but not really brilliant. In 1839 he passed the examinations for internship in the Paris municipal hospitals. A protégé of Pierre Rayer, he worked at the Charité and, as intern on the staff of François Magendie, at the Hôtel Dieu. What he admired in Magendie, however, was less the clinician than the physiologist, the bold experimenter, and the aggressive skeptic. It was in Magendie's laboratory at the Collège de France that Bernard, even before the end of his clinical studies, discovered his real vocation: physiological experimentation.

From 1841 to December 1844, Bernard worked as *préparateur* to Magendie at the Collège de France, assisting him in experiments concerning the physiology of nerves (especially the problem of "recurrent sensitivity" of the spinal nerve roots), the cerebrospinal fluid, the question of the seat of oxidation in the body of horses (by important experiments with cardiac catheterization), and the physiology of digestion. In order to carry out his own research, Bernard installed a very modest private laboratory in the Cour du Commerce de Saint-André-des-Arts. He also made use of the adjoining laboratory of Jules Pelouze,

where he enjoyed the intelligent help of his friend Charles-Louis Barreswil. It was Magendie who taught Bernard to use animal vivisection as the principal means of medical research and to be suspicious of generally accepted theories and doctrines. But Bernard knew how to go beyond the empiricism and skepticism of his master and to create an especially productive method of research on living creatures.

Although he had graduated M.D. at Paris on 7 December 1843, Bernard never practiced medicine and always entertained ambivalent feelings about physicians. Nevertheless, his work was such that it laid new foundations for the profession. His doctoral thesis, *Du suc gastrique et de son rôle dans la nutrition* (1843), was a work both useful to medicine and dedicated to pure science, since it furnished new facts on gastric digestion and the transformations of carbohydrates in the animal organism.

In 1844 Bernard failed to pass the examinations for a teaching post with the Faculty of Medicine. Nevertheless, he resigned his position with Magendie. After having tried vainly to organize a free course in experimental physiology (in collaboration with his friend Charles Lasègue), Bernard resigned himself to giving up scientific research and to setting up as a country doctor in his native village. Rather than resolve his economic embarrassments in this way, however, Bernard decided to take the advice of Pelouze, and in July 1845 he married Fanny Martin, daughter of a Paris physician (they were to have three children: a boy who died in infancy and two daughters, Jeanne-Henriette and Marie-Claude). This match was to become a source of unhappiness, but for the moment his wife's dowry enabled Bernard to continue his physiological research. He now entered the most fruitful and certainly the most hectic period of his scientific career.

In December 1847 Bernard was made *suppléant* to Magendie at the Collège de France. At first he gave the course in the winter term, while Magendie continued to teach experimental medicine during the summer semester. In 1852 Magendie retired completely and turned over his chair and his laboratory to Bernard. In 1848 the Société de Biologie was founded, and Bernard became its first vice-president. Named a *chevalier* of the Légion d'Honneur in 1849, he applied (unsuccessfully) for membership in the Académie des Sciences in 1850 and started to work on his thesis for the doctorate in science. On 17 March 1853 he received the doctorate in zoology at the Sorbonne after a brilliant presentation of his thesis, *Recherches sur une nouvelle fonction du foie.*

Bernard made his principal discoveries early in his scientific career, in the period between his first publication, "Recherches anatomiques et physiologiques sur la corde du tympan" (1843), and his thesis for the doctorate in science (1853). The discoveries on the chemistry and nerve control of gastric digestion (1843–1845) were followed by the first experiments with curare, the discovery of the role of bile in the digestion of proteins, and research on the innervation of the vocal cords and the functions of the cranial nerves (1844–1845). In 1846 he made his first observations on the mechanism of carbon monoxide intoxication, discovered the difference between the urine of herbivores and that of carnivores, began studies on absorption of fats and the functions of the pancreas, and observed the inhibitory action of the vagus nerve on the heart. He solved the problem of "recurrent sensitivity" in 1847. In August 1848 Bernard discovered the presence of sugar in the blood under fasting conditions (nonfood-connected glycemia) and the physiological presence of sugar in the liver—which led rapidly to the revolutionary theory attributing a glycogenic function to the liver (October 1848). In February 1849, he published an important paper on the role of the pancreas in digestion and, in the same month, observed for the first time the presence of sugar in the urine after artificial traumatization of some particular cerebral structures. The following year Bernard made other discoveries concerning the metabolism of carbohydrates and resumed fruitful experiments with curare. In 1852 came the discovery of the vasoconstrictor nerves and the description of the syndrome now called the Horner-Bernard syndrome. This period concluded with a critical examination of Lavoisier's theory on the seat of the production of heat in the animal and with the systematic presentation of discoveries concerning animal glycogenesis.

Bernard's reputation was further enhanced by these works, and soon extended beyond the borders of France. Honor followed honor in quick succession. The government created a chair of general physiology for him at the Faculty of Sciences in Paris, and on 1 May 1854 he delivered his inaugural lecture at the Sorbonne. On 26 June of the same year he was elected to the Académie des Sciences, and in 1855, following the death of Magendie, he became professor of medicine at the Collège de France. He became a member of the Académie de Médecine in March 1861.

Bernard consolidated and completed his physiological discoveries between 1854 and 1860: in 1855 he made the experiment of the perfused liver and discovered glycogen; in 1857 he isolated glycogen; in 1858 he discovered the vasodilating nerves; and in 1859 he made experiments on the glycogenic functions of the placenta and of fetal tissues. This period

of Bernard's work is further marked by the creation of new concepts that were to facilitate the generalization of the results of his experimentation: the concepts of "experimental determinism," and "internal secretion" (1855; it must be stated that for Bernard this term did not have its precise present meaning), the "milieu intérieur" (1857), "local circulation," "reciprocal innervation," "paralyzing reflex actions," and so on.

The transition from laboratory work to dogmatic synthesis was mirrored in Bernard's teaching and in his *Cahier de notes, 1850–1860* (also called the *Cahier rouge*). The *Cahier* clearly demonstrates a change of emphasis from the tenacious pursuit of concrete facts to a concentration on research methods and principles of biological science, and may be said to mark the junction between Bernard's analytical and philosophical work; his teaching led him to the formulation of a comprehensive and didactic theoretical elaboration of his laboratory experience. As early as 1858, Bernard conceived a "plan for a dogmatic work on experimental medicine" in consideration of the new direction indicated by his teaching.

From 1860 on, Bernard spent all his vacations at St.-Julien, where he had bought the manor house of the landlord on whose farm he had been born. In March 1860 he came here to recover from the first of a series of illnesses that were to mark his last years, and here, during the leisure enforced by a period of convalescence in 1862–1863, he drafted his principal theoretical work, *Introduction à l'étude de la médecine expérimentale*. This work was conceived of as the preface of a great treatise, *Principes de médecine expérimentale,* for which Bernard wrote the rough drafts of several chapters. The *Introduction* itself was rewritten in the course of the following two years, and was given its definitive form in the version published in August 1865. A grave illness in October 1865, from which he recovered after eighteen months, led Bernard to abandon the *Principes*. (In 1947 L. Delhoume published, under the same title, a reconstruction of the work based on the rough drafts and augmented by the unpublished text of Bernard's lecture course of 1865.)

During his convalescence of 1865–1867, Bernard turned his attention to philosophy, and read and annotated the philosophical works of Tenneman and of Comte; these notes revealed a subtle and critical attitude toward positivism. In 1866, at the request of the minister of public education, he prepared his *Rapport sur les progrès et la marche de la physiologie générale en France.* The *Rapport,* which was published on the occasion of the World Exposition of 1867, was to have been an objective, historico-encyclopedic treatment of physiology in France. Bernard used the opportunity, however, to issue a passionate statement of his personal opinions and presented a unified synthetic physiology, founded on the notion of the "milieu intérieur" and on the regulatory functions that, under the control of the nervous system, maintain the stability of the fluids and the living tissues.

On 12 December 1868 the chair of general physiology was transferred from the Sorbonne to the Muséum d'Histoire Naturelle; as the titular holder of the chair, Bernard succeeded Flourens (who had held the chair as professor of comparative physiology) on the council of professors of the museum. Flourens's chair was transferred to the Sorbonne, and was awarded to Paul Bert, one of Bernard's most faithful pupils.

In January 1869, after a hiatus of three years, Bernard resumed his courses in experimental medicine at the Collège de France. Although he was only a mediocre lecturer, he was able to hold the attention of his audience by the novelty and vividness of his arguments and by the experiments that he improvised in the amphitheater to support his statements. (At the beginning of his career, Bernard's audiences had been composed almost exclusively of physicians and physiologists, especially foreigners; gradually, however, they became larger, more varied, and more fashionable.)

Bernard's teaching at the College was analytical and dedicated to his own research—demonstrating, as he was wont to say, science in the making rather than science already made. His methods attracted such listeners and collaborators as d'Arsonval, Bert, Dastre, Gréhant, Jousset de Bellesme, Moreau, Pasteur (whose notes made from Bernard's lectures remain unpublished), Ranvier, and Tripier; the Germans Kühne and Rosenthal; the Russians E. de Cyon, Setchenov, and Tarkhanov; the Italians Mosso and Vella; the Dane Panum; the Englishmen Ball and Pavy; such Americans as J. C. Dalton, Austin Flint, W. E. Horner, and S. W. Mitchell; and Emperor Pedro II of Brazil. Even those physiologists and physicians who did not actually attend Bernard's lectures knew his ideas from the ten volumes of *Leçons* delivered at the Collège de France, publications that ranged from the *Leçons de physiologie expérimentale appliquée à la médecine* (1855) to the *Leçons de physiologie opératoire* (published posthumously, 1879).

The courses that Bernard taught at the Sorbonne were, from their inception, of a more general character. His *Leçons sur les propriétés des tissus vivants*, delivered in 1864 and published in 1866, illustrate these tendencies. In this course, it was Bernard's

aim to "determine the elementary conditions of the phenomena of life," that is, "to return to the elementary condition of the vital phenomenon, a condition that is identical in all animals." In contrast to comparative physiology, general physiology "does not seek to grasp the differences that separate beings, but the common points that unite them and which constitute the essence of the vital phenomena." It is obvious why, when Bernard went to the Muséum d'Histoire Naturelle in 1868, the name of the chair that he was to occupy was changed.

In all the courses that he taught at the Muséum, Bernard sought to demonstrate the vital unity of all organisms. In contrast to the naturalists, Bernard was interested only in vital manifestations that did not differ from species to species. Encouraged by the general development of cellular theory and by his own research on the nonspecificity of the nutritive processes, he extended his work into plant physiology. In the first volume of his *Leçons sur les phénomènes de la vie communs aux animaux et aux végétaux* (1878; Bernard corrected the proofs on his deathbed), he went beyond the framework of traditional physiology to treat problems of general biology. His last experimental researches dealt with anesthesia of animals, influence of the ether application on plants, embryonic development, and fermentation.

Bernard was showered with honors in the final years of his life: he was commander of the Légion d'Honneur (1867), president of the Société de Biologie (1867), senator of the Empire (6 May 1869), member of the Académie Française (27 May 1869) and its president (1869). His legal separation from his wife and the Franco-Prussian War affected him profoundly, but he took pleasure in long stays at St.-Julien and in a tender friendship with Marie Raffalovich, to whom his letters reveal a glimpse of his poetic sensibility.

Bernard died of what was probably a kidney disease. He received a national funeral, an honor reserved until then for France's military and political leaders.

Scientific Works. As much through concrete discoveries as through the creation of new concepts, the work of Claude Bernard constitutes the founding of modern experimental physiology. His scientific career started with two series of precise and well delimited researches: on the one hand, the chemical and physiological study of gastric digestion, and on the other, experimental section of nerves. In both cases, the responsibility for the choice of method and subject rested less with Bernard than with his teacher Magendie. But once the initial impetus had been given, the disciple quickly gave his work a completely new orientation—one that had not been foreseen at the start.

Despite some errors (for example, Bernard believed that the acidity of the gastric juice was caused by the presence of lactic acid), his experiments on the digestive action of saliva, gastric juice, and bile resulted in discoveries of undeniable value: the presence of an organic enzymatic factor in the gastric juice (1843), the nervous control of gastric secretion, the decomposition of all carbohydrates into monosaccharides prior to their absorption, the special defense mechanism of the gastric wall against the digestive activity of the gastric juice, the proteolytic properties of bile (1844), the exact localization of gastric secretion, and so on.

Bernard's most impressive discoveries in the field of digestion proper concern the functions of the pancreas, especially the importance of pancreatic juice in the digestion and absorption of fats. Two observations showed him the road to follow. First, he had noted that the urine of herbivores is alkaline, while that of carnivores is acid. Bernard showed that fasting brought about acidity of the urine in herbivores (they lived off their body fat) and that man and carnivorous animals put on a vegetarian diet excreted alkaline urine (1846). Bernard then applied himself to the comparative study of the phenomena of digestion in both carnivores and herbivores. He initiated experiments by which to follow the changes in the chyle in the various parts of the intestinal tract of a dog and a rabbit. Thereby he noted that the absorption of fat by the chyliferous vessels occurred at a rather considerable distance from the pylorus in the rabbit and immediately at the beginning of the duodenum in the dog. Bernard discovered that this difference coincided with an anatomical difference at the point of discharge of the pancreatic juice into the intestine. Thus the role of the pancreas in the first phase of fat metabolism was demonstrated ("Du suc pancréatique et de son rôle dans les phénomènes de la digestion," 1849). In order to collect pancreatic juice in its pure state and to study the regulation of its secretion, Bernard conceived and made the temporary pancreatic fistula, later improved by Pavlov. Bernard found that pancreatic juice acted on fats by a saponification process.

In studying the digestive properties of the gastric and pancreatic juices, Bernard did not intend to restrict himself to a narrow view of the problem of local digestion alone, or of the decomposition of food in the gastrointestinal tract. Although he studied intensively the chemical changes in food exposed, both *in vivo* and *in vitro,* to saliva, gastric juice, or pancreatic juice, this was to him only one, fragmentary aspect of a vast research subject. What interested him above all was what happened to the food in the animal organism, from its entry until its total assimilation or excretion. Thus the horizon of Bernard's research kept widening and, by going beyond the limits of simple "digestion," it made its true object "nutrition" (or, in modern terminology, "metabolism").

Never wavering, Bernard was to advance beyond the then prevailing notions of "animal statics" and to set up the first milestones on the road to the understanding of intermediate metabolism. To begin with, Bernard accepted the theory of his teachers that animals are incapable of synthesizing sugar, fat, and albumin. These three substances would always originate in plants, and their percentage in the blood would vary and would depend essentially on the food consumed. Nutrition would consist of three stages: digestion, transport of digested substances, and chemical incorporation or combustion.

Then he discovered that the alleged transport of absorbed substances is an extremely complicated process, more chemical than physical, more a series of transformations than a series of displacements. He also understood that nutrition is a phenomenon of synthesis as much as it is an analytical process. If food intake is an intermittent process, "nutrition" (in the sense of metabolism) is continuous and is stopped only by death. "Nutrition" is also indirect: prior to being integrated into the tissues, the organic alimentary substances must be broken down to a certain degree and then recombined. In formulating and demonstrating these ideas, Bernard was able to talk with pride of his work on nutrition: "I am the first one to have studied the intermediary stage. The two extremes were known and the rest was accomplished by means of the physiology of probability."

In his thesis on gastric juice (1843), Bernard published, marginally to the principal subject, the first results of his experiments on the ingestion of food substances by other than natural means. His thesis relates two important discoveries: (1) if so-called "type 1" sugar (sucrose) is injected directly into the blood, it is eliminated by the kidneys, while the so-called "type 2" sugar (glucose) is retained in the organism; (2) gastric juice transforms sucrose into assimilable sugar, that is, sucrose exposed to the action of gastric juice and then injected into the blood no longer appears in the urine. "Type 2" sugars (in modern terminology, sugars of the monosaccharide group) represent the only "physiological" form of carbohydrates in the animal organism. Gastric juice changes all other forms of carbohydrates into assimilable physiological sugar.

Blinded by prevalent theories, Bernard searched in

vain for the site and the manner of breakdown of sugar in the animal organism. He wished especially to give experimental proof to Lavoisier's ideas, according to which sugar is burned in the lungs. After four years of experimentation an apparently contradictory observation, a new experimental fact, upset the entire theoretical structure. In August 1848, Bernard noted the presence of sugar in the blood of an animal from which all solid food had been withheld for several days. Greatly surprised, he turned his research in a new direction. Thus, he was soon able to discover (1) that glycemia is a normal and constant phenomenon, independent of food intake and (2) that the liver produces sugar and empties it into the blood. Published in October of the same year in "De l'origine du sucre dans l'économie animale," the discovery of the glycogenetic function of the liver compelled physiologists to revise certain fundamental notions and threw new light on the understanding of diabetes. Bernard was sometimes contradicted. If the criticisms by Louis Figuier, Pavy, and several other adversaries today appear justified with respect to certain details, Bernard was nevertheless on the right track.

In 1849 Bernard believed he had found a method of causing "artificial diabetes" by means of a local lesion of the nervous system. There followed the discoveries of the presence of sugar in the allantoic and the amniotic fluids (1850) and in the cerebrospinal fluid (1855), the proposition to utilize the quantitative determination of sugar in the liver of a fresh corpse in order to establish whether death had been sudden (January 1855), and the astonishing observation that the liver manufactures sugar even after the death of an animal (September 1855). By forcing a stream of water through the hepatic vessels into the still-warm liver, as soon as possible after the death of the animal, the hepatic tissue is completely freed of its sugar content. But if the liver is kept at moderate temperature, several hours afterward, or even the next day, the tissue will once more contain a quantity of sugar, produced since the irrigation. From this experiment Bernard derived proof of the existence of a special "glycogenetic substance." This was, strictly speaking, the first artificial perfusion of an organ separated from the body.

His handwritten notes reveal that Bernard had perfectly understood the general implication of this process and that he wanted to study the artificial survival of certain organs by means of continuous perfusion with blood. The discovery of glycogen, a kind of "animal starch" that could be converted into sugar and was barely soluble in water, was communicated to the Académie des Sciences on 24 September 1855, but this substance was not extracted in a rela-

tively pure state until February 1857. Almost simultaneously, V. Hensen, a young German physiologist, isolated glycogen by a process different from Bernard's.

In the glycogenetic function of the liver Bernard distinguished henceforth two types of phenomena: the creation (or synthesis) of glycogen in the liver and the transformation of this substance into sugar. According to Bernard, the first phenomenon was a "vital function whose true beginning is still unknown," while the second phenomenon is "purely chemical" and consequently can also be produced after the death of the individual. Thus, a fundamental distinction is established between the "plastic or organically created" phenomena and the "phenomena of attrition," or vital destruction.

The true culmination of Bernard's work in the field of carbohydrate metabolism was shown in the chapter on extrahepatic glycogenesis: the role of the placenta (1859), the ontogenetic and phylogenetic aspects of glycogenesis, production of sugar in animals without a liver, carbohydrate metabolism in muscle (for example, lactic fermentation of muscle glycogen), the breakdown of sugar in the tissues and its relation to the release of heat, the role of glycogenetic ferments, and, above all, the explanation of glycogenesis as a cellular process.

Since the discovery of the formation of sugar in the liver, Bernard had been convinced that the latter was subject to control by the nervous system. This hypothesis found strong support in the discovery of the so-called *piqûre sucrée*. In February 1849, Bernard experimented with severing the cerebellar peduncle in rabbits, in order to determine the accuracy of certain observations on the behavior of animals thus traumatized—observations that had been reported by Magendie and contradicted by François Longet. To his great surprise, Bernard found that this type of trauma caused glycosuria. He then showed that the lesion of a specific spot in the brain (the floor of the fourth ventricle) was regularly accompanied by increased glycemia. Experiments involving section of the vagus nerve also showed the influence of the nervous system on the intensity of glycogenesis.

In order to study the functions of nerves Bernard often resorted to severing them and to local galvanic stimulation. If the first research experiments on the chorda tympani (1843) today appear to have been a step in the wrong direction, his other work in this field represents a series of extraordinary successes: the "destruction" experiments on the spinal and vagus nerves and the innervation of the vocal cords (1844), the observation of the change in the sense of taste in paralysis of the facial nerves (1845), his research on the pneumonia that occurred in animals whose vagus nerves had been severed (1853), the fine experiments on the influence of the different nerves on saliva secretion (1857), and, above all, the discovery of the vasomotor nerves. Bernard had clarified the functions of the accessory nerve, particularly its connection with the vagus nerve in the innervation of the larynx.

Bernard put an end to a long dispute between Magendie and Longet on the significance of "recurrent sensitivity," that is, that stimulation of the anterior root of a spinal nerve (motor root) can, in certain cases, produce sensibility phenomena. In explaining the apparent contradictions between Magendie's experiments and those of Longet, Bernard drew a general conclusion: contradictions in experimental results always stem from a difference in the conditions under which such conflicting experiments are performed.

At the beginning of his important discovery of the vasomotor nerves, Bernard presented a brief communication ("Influence du grand sympathique sur la sensibilité et sur la calorification") in 1851 to the Société de Biologie relating the observations of phenomena that occurred after section of the cervical sympathetic nerve in rabbits. He had expected a cooling of the animal's face, since the experiment had been based on the hypothesis that the sympathetic system exerts a direct influence on the nutritive and calorific processes of the tissues. To his surprise, he found a very sharp increase in the temperature of the entire region innervated by the severed nerve. Although he noted and described the increased blood circulation in the parts affected, Bernard did not realize at the time—not even in his notes on this subject that were dated March and October 1852—the relation between these phenomena and vascular paralysis.

In November 1852 Bernard informed the Société de Biologie that the galvanization of the peripheral end of the sympathetic nerve produced effects that are the exact opposite of those obtained by severing this nerve. At that time Bernard did not know that he had been anticipated by Brown-Séquard. In fact, the latter had published in the United States—in August 1852—the results of experiments with the galvanization of the sympathetic nerve that preceded those performed by Bernard but were certainly inspired by Bernard's observations made in 1851, of which Brown-Séquard learned just before his departure for America.

Through the work of Brown-Séquard, Schiff, and Bernard, the knowledge of the vasoconstrictor nerves was incorporated into science. But we are indebted

to Bernard alone for the second stage in the explanation of vasomotor function: the discovery of the vasodilator nerves and the establishment of the concept of the physiological equilibrium of the two antagonistic innervations. In analyzing the causes of changes in the color of venous blood in the salivary glands, Bernard discovered the active vasodilator reflex ("De l'influence de deux ordres de nerfs qui déterminent les variations de couleur du sang veineux dans les organes glandulaires," 1858). Research on vasomotor nerves was very closely connected with (1) the description of the so-called Horner-Bernard ocular syndrome (paralysis of the sympathetic nerve provokes miosis, narrowing of the palpebral fissure, and enophthalmos on the side of the lesion; see "Expériences sur les fonctions de la portion céphalique du grand sympathique," 1852); (2) the elaboration of the concept of "local circulation" subject to variations occurring in the various organs, depending on whether they are functioning or in a state of rest ("Sur la circulation générale et sur les circulations locales," 1859); and (3) the idea of double and reciprocal innervation that enables the organ to function not only as a result of stimulation but also as a result of an inhibitive mechanism (for example, according to Bernard, the chorda tympani determines salivary secretion by a "paralyzing action" on the tonus of the sympathetic nerve; see "Du rôle des actions paralysantes dans le phénomène des sécrétions," 1864).

Bernard was deeply involved in the problems of animal heat production and its regulation. While he accepted Lavoisier's theory, which attributed the origin of animal heat to a combustion process (i.e., oxidation), Bernard insisted on two fundamental modifications: (1) this vital combustion could not be direct oxidation, an immediate union of oxygen with tissue carbon; it had to be a particular organic process, an indirect combustion taking place with the aid of special ferments; (2) organic combustion could not occur in the lungs exclusively, as Lavoisier had taught, but in all tissues. In order to demonstrate the latter statement, Bernard used (particularly in June 1853) cardiac catheterization: comparison of blood temperature in the left and right ventricles furnished results that disproved Lavoisier's original theory of pulmonary combustion. Research concerning the site of sugar decomposition furnished additional proof for "respiration of the tissues." Bernard also conducted experiments on the lowering of body temperature either by severing the spinal cord and certain nerves or by prolonged exposure to cold. He connected these phenomena with those observed during the hibernation of certain animals, involving artificial transformation of a homoiothermic animal into a poikilothermic one.

Several other subjects of Bernard's research in this field deserve mention: the mechanism of death caused by exposure to high temperature, the slowing down of the vital processes in a cold environment, and the pathogenesis of fever. Bernard's experiments on rigor mortis and on the acidity and alkalinity of muscles after death represent an anticipation of twentieth-century discoveries.

Bernard was a true innovator in the study of the effects of toxic and medicinal substances. No one before him had understood so well the role of drug metabolization. He regarded poisoning as a local phenomenon, and advocated the use of certain poisons in physiological research. Curare and carbon monoxide had served him, he said, as "chemical bistoury," making it possible to destroy specific structures selectively.

As early as his first experiments with curare in 1844, Bernard had noted that this substance somehow isolated the contractile property of the muscle from the motor property of the nerve (observation published in 1850). But only ten years later he thought—practically at the same time as Albert von Kölliker—of an experiment that would prove that curare acted only upon the peripheral ends of the motor nerves. Contrary to what is generally believed, Bernard never wanted to accept the correct explanation of curare's action, that is, that it paralyzes the motor end plates described by his pupil Kühne.

In studying the mechanism of carbon monoxide intoxication, Bernard found that animals died of asphyxiation because this gas replaces oxygen in the red blood cells (1855–1856). At last Bernard's theory on organic combustion in the tissues was confirmed. In toxicology, we must not forget, moreover, Bernard's work on opium, on strychnine, and on anesthetics. According to him, anesthesia was a biological phenomenon common to all living things and caused by a reversible coagulation of protoplasm (1875). Etherization can eliminate the sensitivity reactions, temporarily arrest germination in grain, and suspend fermentation (1876).

Bernard's last works concentrated on the nature of alcoholic fermentation. In them he distinguished two types of fermentation, one produced by intervention of a "figurative" ferment, the other produced by soluble ferments. Nevertheless, he hoped to reduce the activities of the former (Pasteur's ferments) to the soluble chemical principles of the latter (Berthelot's ferments). After Bernard's death, a series of his notes on alcoholic fermentation were made public by Berthelot (see "La fermentation alcoolique. Dernières expériences de Claude Bernard," 1878). Pasteur, surprised and embittered, published a rather angry reply. There was truth and error on both sides. Today we know that Berthelot and Bernard were wrong in accepting spontaneous generation of yeast in a fermentable medium and that they were right, in contrast with Pasteur, in claiming the existence of a soluble ferment, not living but nevertheless capable of causing alcoholic fermentation.

Philosophical Opinions. Although Bernard stated that he had "no philosophical pretensions," his works—particularly the *Introduction à l'étude de la médecine expérimentale* (1865)—are of such general scope that they enter the domain of philosophy. In the *lycées* of France, the *Introduction* is one of the official philosophy textbooks. Almost paradoxically, the "philosophical" aspects of Bernard's work resulted in a bibliography much larger than that of his strictly scientific work.

Bernard's views on philosophy and religion are imbued with the idea that the essence of things inevitably escapes us. Phenomena have two kinds of causes: immediate, or secondary, and primary causes. Only secondary causes are accessible to scientific investigation. The others remain beyond all possibility of proof and scientific control. It is the duty of the scholar to determine, by observation and experimentation, the immediate conditions of the phenomena. Investigation of primary causes lies beyond science, and the scholar—insofar as he is a scientist—must abandon it.

Such an attitude readily reflects the influence of positivism. Yet although he was indebted to this philosophical trend, Bernard deviated from it on several points and did not refrain from criticizing Comte with pronounced rudeness. Reaching beyond narrow positivism, Bernard rediscovered certain topics of Kantian thought and accorded extrascientific legitimacy to metaphysical deliberations.

What interested Bernard first of all, however, was not so much the general theory of knowledge as the psychology and logic of scientific research. Primarily a man of the laboratory, he was interested in philosophical questions principally as a theoretician of the experimental method.

Bernard's "experimental rationalism" is opposed both to Descartes's rationalism and to Magendie's empiricism while somehow embracing both and synthesizing them in a wider doctrine. For Bernard, the experimental method proceeds by three stages: observation, hypothesis, and experimentation. Observation and experiment, he wrote, are two extreme terms of "experimental reasoning." They furnish the knowledge of "facts," but between them there extends like a bridge the "experimental idea" (also called the "idea a priori" or simply "hypothesis"). The hypothesis is the *primum movens* of all scientific reasoning and the essential part of every discovery, but it is worthless if it is not followed and confirmed by experimental verification. Experiment is precisely an observation elicited under certain conditions for the verification of a hypothesis. Bernard's rationalism implies constant recourse to a test of the "experimental facts." If he can change the conditions of an event, man can become its master. This is the difference between the sciences of observation—essentially passive—and those of experimentation:

> With the aid of these active experimental sciences man becomes an inventor of phenomena, a real foreman of creation; and in this respect no limits could be set to the power man can acquire over nature through future progress of the experimental sciences [*Introduction*, I, ch. 1, § V].

The conscious aim of all of Bernard's work was to give medicine the decisive push along the road of its transformation into an "experimental" and "conquering" science.

Bernard confirmed the primordial role of "feeling" or "intuition" as the point of departure for "creative" experimental research. Convinced that "method by itself produces nothing," he did not insist on positive practical precepts. Nevertheless, he carefully set forth a series of precautions for the experimental biologist. His principal advice concerned the "experimental doubt" and the necessity to avoid fixed ideas and to keep one's mind free of doctrinal preconceptions. A good experimenter must—as he was himself—be simultaneously theoretician and practitioner: "A skilled hand without the head to direct it is a blind instrument; the head without the hand to carry out an idea remains impotent."

The success of the *Introduction* is due, at least in part, to the glimpse that it affords of the personal adventures of a great biologist and its claims to the revelation of the secrets of his scientific success. In fact, almost all the examples cited by Bernard in support of his general concepts stem from his own work. A careful analysis of his original laboratory notes shows, however, that at times there were some rearrangements in the chronology of Bernard's discoveries. The decisive turning point was almost always his extraordinary capacity for noting, in the course of an experiment, a fact that was somewhat marginal and did not accord with the prevailing theory.

Against a strong vitalist current—harking back to Bichat and the school of Montpellier—Bernard stressed the necessity of assuming that vital phenomena are subject to a determinism of the same kind as that which governs inert matter. This amounted to saying that "a vital phenomenon has—like any other phenomenon—a rigorous determinism, and [that] such determinism could only be a physicochemical determinism." Application of the experimental method in physiology would not be justified without acceptance of this principle. To proclaim this kind of determinism in biology signified rejection of vitalism in its classic form. Furthermore, Bernard was convinced that determinism renders the use of statistics in physiological research illusory. This criticism was nevertheless meant only for the method of the arithmetic mean.

Bernard's position between the vitalism and the materialism of his contemporaries was complex. He opposed the former by virtue of the principle of physicochemical determinism, but by the same token he did not rally to the cause of the latter, since he attributed a "directive and creative idea" to life. According to Bernard, life phenomena fall into two groups: the phenomena of organization, of creation, or of organic synthesis, on the one hand; and the phenomena of organic destruction, on the other. If the latter can be explained only by the laws of physics and chemistry, the phenomena of the first group (that is, embryonic development, the anabolic processes of nutrition, psychic life, and regeneration) defy physicochemical explanations, although they obey all the laws governing inert matter. "Life is creation"; it has a sense, a direction.

Bernard did not seek to reconcile opposing theories but, in going beyond them, to bring them into accord by means of an antisystematic attitude. In Bernard's mind, once research into determinism of the phenomenon is accepted as the only aim of the experimental method, there is no longer materialism, nor spiritualism, nor inanimate matter, nor living matter; there are only natural phenomena, the conditions of which should be determined.

The notion of "milieu intérieur" occupies a central place in Bernard's thought. It took form gradually: beginning in 1851 he formed a group of ideas on an intermediate animal milieu that nourished and protected. The term itself was coined in 1857 and, little by little, it was enriched with new meanings. The concept arose from the generalization of Schwann's theory of blastemas, and of the collision between Schwann's theory and the new forms of the cellular theory (Virchow, Brücke). It was strengthened by his research on how to overcome the conflict between Bichat's vitalism and the epistemological necessity of absolute determinism. Onto all this were grafted reflections on the "aquatic" character of life's elementary form and the idea of regulatory mechanisms that watch over the stability of the internal conditions of an organism. Life is a phenomenon of relationship—or, still better—a permanent conflict between the living particles and the outer world. The stability of the "milieu intérieur," Bernard declared, is the precondition of a free, independent life. The notion of homeostasis (Cannon) and even the beginnings of cybernetics relate to Bernard's ideas on the "milieu intérieur" and the way in which the equilibrium between this milieu, the tissues, and the outside world is maintained.

Although Bernard contributed to the spread of the cellular theory in France, he remained attached for a long time to the ideas of Schwann and never completely accepted Virchow's reform. For him, life was a protoplasmic, and not really a cellular, phenomenon. He thought life was tied more closely to chemical compounds than to histological structures. His astonishing criticism of Pasteur's experiments on spontaneous generation appears to us today as an extraordinary anticipation of molecular biology.

For Bernard, physiology had to be the basis of "experimental medicine." There is no qualitative difference between normal and pathological functions. Diseases have no ontological existence; in disease one always deals merely with exaggerated, weakened, or abolished physiological functions. If this is not the view of the majority of modern pathologists, it must be recognized that Bernard must be considered the pioneer of the "positive" concept of health (the state of health is not only the absence of illness) that characterizes modern hygiene.

BIBLIOGRAPHY

I. ORIGINAL WORKS. Bernard's books published during his lifetime are *Du suc gastrique et de son rôle dans la nutrition* (Paris, 1843), his thesis for the M.D.; *Des matières colorantes chez l'homme* (Paris, 1844), his thesis for the agrégation; *Recherches expérimentales sur les fonctions du nerf spinal ou accessoire de Willis* (Paris, 1851); *Recherches sur une nouvelle fonction du foie considéré comme organe producteur de matière sucrée chez l'homme et les animaux* (Paris, 1853), thesis presented to the Faculty of Sciences; *Notes of M. Bernard's Lectures on the Blood*, W. F. Atlee, ed. (Philadelphia, 1854); *Précis iconographique de médecine opératoire et d'anatomie chirurgicale* (Paris, 1854), written with C. Huette; *Recherches expérimentales sur le grand sympathique, et spécialement sur l'influence que la section de ce nerf exerce sur la chaleur animale* (Paris, 1854); *Illustrated Manual of Operative Surgery and Surgical Anatomy,* trans. with notes and addition by W. H. Van Buren and C. E. Isaacs (New York, 1855); *Leçons de physi-ologie expérimentale appliquée à la médecine*, 2 vols. (Paris, 1855–1856); *Mémoires sur le pancréas et sur le rôle du suc pancréatique dans les phénomènes digestifs, particulièrement dans la digestion des matières grasses neutres* (Paris, 1856); *Leçons sur les effets des substances toxiques et médicamenteuses* (Paris, 1857); *Leçons sur la physiologie et la pathologie du système nerveux* (Paris, 1858); *Leçons sur les propriétés physiologiques et les altérations pathologiques des liquides de l'organisme* (Paris, 1859); *Introduction à l'étude de la médecine expérimentale* (Paris, 1865, and many later eds., including that of F. Dagognet, Paris, 1966), trans. into English by H. C. Greene (New York, 1927, 1957); *Leçons sur les propriétés des tissus vivants* (Paris, 1866); *Lectures on the Physiology of the Heart and Its Connections With the Brain,* J. S. Morel, trans. (Savannah, Ga., 1867); *Rapports sur les progrès et la marche de la physiologie générale en France* (Paris, 1867), repub. as *De la physiologie générale* (Paris, 1872); *Éloge de Flourens* (Paris, 1869), delivered before the Académie Française; *Leçons de pathologie expérimentale* (Paris, 1872); *Leçons sur les anesthésiques et sur l'asphyxie* (Paris, 1875); *Leçons sur la chaleur animale, sur les effets de la chaleur et sur la fièvre* (Paris, 1876); and *Leçons sur le diabète et la glycogenèse animale* (Paris, 1877).

Books published after Bernard's death are *Leçons sur les phénomènes de la vie communs aux animaux et aux végétaux,* A. Dastre, ed., 2 vols. (Paris, 1878–1879), Vol. I reed. by G. Canguilhem (Paris, 1966); *La science expérimentale* (Paris, 1878); *Leçons de physiologie opératoire,* M. Duval, ed. (Paris, 1879); *Pensées. Notes détachées,* L. Delhoume, ed. (Paris, 1937); *Philosophie,* J. Chevalier, ed. (Paris, 1937); *Le cahier rouge* (partial ed.), L. Delhoume, ed. (Paris, 1942); *Principes de médecine expérimentale,* L. Delhoume, ed. (Paris, 1947); *Lettres beaujolaises,* J. Godard, ed. (Villefranche, 1950); *Esquisses et notes de travail inédites,* L. Binet, ed. (Paris, 1952); *Cahier de notes 1850–1860* (complete ed.), M. D. Grmek, ed. (Paris, 1965); *Notes, mémoires et leçons sur la glycogenèse animale et le diabète,* selected by M. D. Grmek (Paris, 1965); and *Notes inédites de Claude Bernard sur les propriétés physiologiques des poisons de flèches (curare, upas, strychnine et autres),* M. D. Grmek, ed. (Paris, 1966).

Bernard's articles include "Recherches anatomiques et physiologiques sur la corde du tympan," in *Annales medico-psychologiques,* **1** (1843), 408–439; "De l'origine du sucre dans l'économie animale," in *Archives générales de médecine,* 4th ser., **18** (1848), 303–319; "Du suc pancréatique et de son rôle dans les phénomènes de la digestion," in *Mémoires de la Société de biologie,* **1** (1849), 99–115; "Recherches sur le curare," in *Comptes rendus hebdomadaires de l'Académie des sciences,* **31** (1850), 533–537, written with J. Pelouze; "Influence du grand sympathique sur la sensibilité et sur la calorification," in *Comptes rendus de la Société de biologie,* **3** (1851), 163–164; "Expériences sur les fonctions de la portion céphalique du grand sympathique," *ibid.,* **4** (1852), 155; "De l'influence de deux ordres de nerfs qui déterminent les variations de couleur du sang veineux dans les organes glandulaires," in *Comptes rendus hebdomadaires de l'Académie des sci-*

ences, **47** (1858), 245–253; "Études physiologiques sur quelques poisons américains. I. Curare," in *Revue des deux mondes,* **53** (1864), 164–190; "Du rôle des actions paralysantes dans le phénomène des sécrétions," in *Journal d'anatomie et de physiologie,* **1** (1864), 507–513; and "La fermentation alcoolique. Dernières expériences de Claude Bernard," M. Berthelot, ed., in *Revue scientifique,* **16** (1878), 49–56.

For a complete bibliography, see G. Malloizel, "Bibliographie des travaux scientifiques," in *L'oeuvre de Claude Bernard* (Paris, 1881); and M. D. Grmek, *Catalogue des manuscrits de Claude Bernard, avec la bibliographie de ses travaux imprimés et des études sur son oeuvre* (Paris, 1967).

II. SECONDARY LITERATURE. Works on Bernard are P. Bert, "Les travaux de Claude Bernard," in *Revue scientifique de la France,* 2nd ser., **16** (1879), 741–755; G. Canguilhem, *L'idée de médecine expérimentale selon Claude Bernard* (Paris, 1965); P. E. Chauffard, *Claude Bernard, sa vie et ses oeuvres* (Paris, 1878); L. Delhoume, *De Claude Bernard à d'Arsonval* (Paris, 1939); J. L. Faure, *Claude Bernard* (Paris, 1925); M. Foster, *Claude Bernard,* in the series Masters of Medicine (London, 1899); M. D. Grmek, "La conception de la maladie et de la santé chez Claude Bernard," in *Mélanges Koyré, I. L'aventure de la science* (Paris, 1964), pp. 208–227; "Les expériences de Claude Bernard sur l'anesthésie des plantes," in *Comptes rendus du 89° Congrès des sociétés savantes* (Lyons, 1964), pp. 65–80; "Examen critique de la genèse d'une grande découverte: La piqûre diabétique de Claude Bernard," in *Clio medica,* **1** (1966), 341–350; "First Steps in Claude Bernard's Discovery of the Glycogenic Function of the Liver," in *Journal of the History of Biology,* **1** (1968), 141–154; and *La glycogenèse et le diabète dans l'oeuvre de Claude Bernard* (Paris, 1968); B. Halpern, "Concepts philosophiques de Claude Bernard d'après l'*Introduction,*" in *Revue d'histoire des sciences,* **19** (1966), 97–114; H. Hermann, "À propos d'un centenaire. Comment se fit la découverte des nerfs vaso-moteurs," in *Biologie médicale,* **41** (1954), 201–230; G. L. Jousset de Bellesme, "Notes et souvenirs sur Claude Bernard," in *Revue internationale des sciences biologiques,* **10** (1882), 433–461; L. N. Karlik, *Klod Bernar* (Moscow, 1964), in Russian; P. Lamy, *Claude Bernard et le matérialisme* (Paris, 1939); N. Mani, *Die historischen Grundlagen der Leberforschung,* II (Basel, 1967), 339–369; P. Mauriac, *Claude Bernard,* 2nd ed., rev. (Paris, 1954); R. Millet, *Claude Bernard ou l'aventure scientifique* (Paris, 1945); G. Monod et Thyss-Monod, "Claude Bernard, l'homme, sa vie," in *Revue du mois,* **17** (1917), 222–242; J. M. D. Olmsted, *Claude Bernard, Physiologist* (New York–London, 1938); J. M. D. Olmsted and E. Harris Olmsted, *Claude Bernard and the Experimental Method in Medicine* (New York, 1952); E. Renan, *Éloge de Claude Bernard* (Paris, 1879); W. Riese, "Claude Bernard in the Light of Modern Science," in *Bulletin of the History of Medicine,* **14** (1943), 281–294; J. Rostand, *Hommes de vérité* (Paris, 1943), pp. 53–123; P. Van Tieghem, *Notice sur la vie et les travaux de Claude Bernard* (Paris, 1910); P. Vendryès, *Les "conditions déterminées" de Claude Bernard* (Paris, 1940); R. Virtanen, *Claude Bernard and His Place in the History of Ideas* (Lincoln, Neb., 1960).

Collections are *L'oeuvre de Claude Bernard* (Paris, 1881); *Centenaire de Claude Bernard* (Paris, 1914); *Claude Bernard and Experimental Medicine,* F. Grande and M. B. Visscher, eds. (Cambridge, Mass., 1967); and *Philosophie et méthodologie scientifique de Claude Bernard,* B. Halpern, ed. (Paris, 1967).

M. D. GRMEK

BERNARD, NOËL (*b.* Paris, France, 13 March 1874; *d.* Mauroc [near Poitiers], France, 26 January 1911), *botany.*

During his short but productive career, Noël Bernard shed much light on the nature of the endophytic fungi found in orchids and their importance to the plant. His active research covered a period of only eleven years, cut short by his untimely death at the age of thirty-six. Moreover, the majority of Bernard's work was done before he received a university professorship. He began in 1899 as a demonstrator at the École Normale, and in 1902 moved to Caen as a lecturer. Six years later he was called to take charge of the course in botany at the Faculté des Sciences at Poitiers, and in 1909 he was named professor of botany there. He was to have been the director of the experimental botany research institute he was planning at Mauroc for the university, but he did not live to see it established.

When he began his work it had already been known for some time that orchids were mycorhizally infected plants. But it was Bernard, in a 1900 paper and his doctoral thesis of 1901, who determined that the relationship was obligatory; the presence of the fungus, he found, had become necessary for the germination of the seed. Since this infection was chronic and always present, the morphological features characteristic of many orchids, such as a tuberous root and atrophied vegetal organs, were actually fungus-induced symptoms. In analyzing the life cycles of several orchids, Bernard found differing degrees of fungal infection present. In some, such as the Ophrydeae, periods of noninfection and, therefore, morphological elaboration alternated with periods of infection and tuberization of the roots. In others, such as the Neottia, the plant is never free of the fungus, and its vegetal apparatus is reduced to no more than a rhizome.

Bernard's experimental work began with the isolation in pure culture for the first time of the endophyte. From more than twenty different orchid species three new species of fungus were isolated: *Rhizoctonia repens,* widespread among the Orchidaceae, and two more localized species, *Rhizoctonia mucorides* and *Rhizoctonia lanuginosa.* As a verification, he inoculated previously sterile orchid seeds with the fungus and, in 1904, brought about germination in this artificially produced symbiont, inducing tuber formation. On the basis of these results, he was able to advise horticulturists as to how to ensure the germination in hothouses of orchids, until that time a very uncertain, seemingly capricious event. By contaminating the soil with *Rhizoctonia repens,* he was able to improve greatly the growers' success.

Bernard announced this successful method for the germination of orchids at the international congress of horticulture held in Paris in 1905, only to find that his results were not unanimously confirmed by other workers. This disappointment led him to a reexamination of his fungal cultures, and therefore to the discovery of the phenomenon of attenuation of the fungi after having been cultured for lengths of time *in vitro.*

Further investigations revealed the physiological mechanism of the "disease" caused by fungal infection. From experiments, Bernard concluded that the fungus converted starch into sugar, and it was the increased osmotic pressure that stimulated growth and germination. In apparent verification of this, he found that tuberization could be produced in the orchid *Bletilla* without infection if the orchid were placed in a medium of high carbohydrate concentration. Similar results were obtained with the germination of orchids that normally required the presence of a virulent fungus. (It was later shown that the essential function of the fungus was to convert complex carbohydrates to simple sugars, and not necessarily to provide increased osmotic pressure.)

These results corresponded closely with observations made early in his career on potato tuberization, which is also dependent on the concentration of the medium. He was involved in further work on the potato tuber at the time of his death.

BIBLIOGRAPHY

I. Original Works. Works by Bernard include "Sur quelques germinations difficiles," in *Revue générale de botanique,* **12** (1900), 108–120; "Études sur la tubérisation," doctoral dissertation (Paris, 1901), also published *ibid.,* **14** (1902), 5–25, 58–71, 101–119, 170–183, 219–234 (mispaginated 139–154), 269–279; "La germination des Orchidées," in *Comptes rendus de l'Académie des sciences, Paris,* **137** (1903), 483–485; "Recherches expérimentales sur les Orchidées," in *Revue générale de botanique,* **16** (1904), 405–451, 458–476; "L'évolution dans la symbiose. Les Orchidées et leur champignons commensaux," *Annales des sciences naturelles* (*Botanique*), ser. 9, **9** (1909), 1–196 (a large work treating many aspects of the problem); *La Matière et la Vie* (Paris, 1909). Many of Bernard's later researches can be found in *Principes de biologie végétale,* edited after his death by Mme. M. L. Bernard (Paris, 1921).

II. Secondary Literature. The only substantial biographical reference to Bernard is a memoir written immediately after his death by a friend, C. Pérez, in *La Revue du Mois,* **11** (1911), 641–657. No bibliographical detail is provided, and none is available elsewhere in any complete form. A portrait of Bernard can be found in Boissonade *et al., Histoire de l'Université de Poitiers Passé et Présent (1432–1932)* (Poitiers, 1932), facing p. 424.

Alan S. Kay

BERNHEIM, HIPPOLYTE (*b.* Mulhouse, France, 27 April 1840; *d.* Paris, France, 1919), *psychology.*

Bernheim was an intern in Strasbourg hospitals, but left Alsace after the Franco-Prussian War and became a professor at the Faculté de Médecine in Nancy after his *agrégation.* He was particularly interested in pulmonary localizations of the Bouillaud syndrome (rheumatic fever); forms of prolonged typhus that affect the cerebrospinal nerves; the effect of arteriosclerosis in the circle of Willis on the Cheyne-Stokes respiratory phenomenon; and in a special form of the right asystole without retrograde intervention of the pulmonary stasis (sinking of the poorly developed right ventricle caused by the enlarged left ventricle), called the Bernheim syndrome by the South American school of cardiology.

After teaching in the medical clinics for thirteen years, Bernheim heard of a practitioner named Liebault (1823–1904) in one of Nancy's suburbs. Liebault was a philosopher and philanthropist who successfully treated his patients through induced sleep. Bernheim, although very skeptical, went to call; this was the beginning of his study of hypnotism, of suggestion, and of hysteria, which also interested Charcot and the Salpêtrière school, as well as Émile Coué (1857–1926), a pharmacist in Nancy.

As early as 1884 Bernheim stated his opposition to Charcot's concepts regarding hypnosis. He criticized the Parisian idea of hypnosis in three stages, and was the first to have the courage to say that it was a "cultural hypnosis," entirely explicable by suggestion.

Likewise, Bernheim demonstrated in 1904 that the great four-phase hysteria described by Charcot was not an illness, but an emotional, psychoneurotic reaction brought about through suggestion and curable by the same process. He thus anticipated Joseph Babinski, although there remained important differences between the latter's pithiatism, based on a very precise semeiology, and Bernheim's concepts. Bernheim's fame brought Paul Dubois, Economo, and Freud to visit him. The latter was much impressed, and wrote:

I went to Nancy in the summer of 1889 where I spent several weeks. I witnessed the astonishing experiments performed by Bernheim on his hospital patients, and it is there that I experienced the strongest impressions relating to the possible use of powerful psychical processes which remained hidden from human consciousness. I had many interesting discussions with him and I undertook to translate into German his two works on suggestion and its therapeutic effects.

It is not correct, however, to classify Bernheim as the father of psychoanalysis, for he remained a classical psychologist—he was a master of psychotherapy and a precursor of psychosomatic thought. At a time when the latter are both in full use, Bernheim's ideas have lost nothing of their interest and value.

BIBLIOGRAPHY

I. ORIGINAL WORKS. Bernheim's writings include "De la myocardite aigüe," thesis (Strasbourg, 1867); *Des fièvres typhiques en général* (Strasbourg, 1868); *De l'état cireux des muscles* (Strasbourg, 1870); *Leçons de clinique médicale* (Paris, 1877), Spanish trans. by E. Sánchez de Ocana (Madrid, 1879); *Contributions à l'étude des localisations cérébrales* (Paris, 1878); *Études sur les râles* (Paris, 1878); *De la suggestion dans l'état hypnotique et dans l'état de veille* (Paris, 1884), also in *Revue médicale de l'est*, **15** (1884), 513–520; 545–559; 577–592; 610–619; 641–658; 674–685; 712–721; **16** (1884), 7–20; *De la suggestion et de ses applications à la thérapeutique* (Paris, 1886), German trans. by Sigmund Freud (Leipzig–Vienna, 1888, 1889, 1896), English trans. by Christian A. Herter (New York–London, 1889); *Recueil des faits cliniques* (Paris, 1890), written with P. Simon; *Hypnotisme, suggestion, psychothérapie, études nouvelles* (Paris, 1891), Dutch trans. by A. W. Van Renterghem (Amsterdam, 1891), German trans. by Sigmund Freud (Leipzig–Vienna, 1892); *L'hypnotisme et la suggestion dans leurs rapports avec la médecine légale* (Nancy, 1897); *Hypnotisme, suggestion, psychothérapie avec considérations nouvelles sur l'hystérie* (Paris, 1903, 1910); *Doctrine de l'aphasie, conception nouvelle* (Paris, 1907); and *L'aphasie, conception psychologique et clinique* (Paris, 1914).

II. SECONDARY LITERATURE. Works on Bernheim are E. H. Ackerknecht, *A Short History of Psychiatry* (New York–London, 1959); G. Amselle, *Conception de l'hystérie* (Paris, 1907); P. Blum, *Des anesthésies psychiques* (Paris, 1906); K. Kolle, *Grosse Nervenärzte* (Munich, 1959), II, 220; III, 136–165; P. Kissel and P. Barrucand, "Le sommeil hypnotique d'après l'École de Nancy," in *L'encéphale*, **53** (1964), 5371–5388; P. E. Levy, *L'éducation rationnelle de la volonté. Son emploi thérapeutique* (Paris, 1898), Preface, 11, 21, 64–68; and H. H. Walsehr, "L'école hypnologique de Nancy," in *Médecine et hygiène*, no. 685 (1965), 443.

PIERRE HUARD

BERNOULLI,* DANIEL (*b.* Groningen, Netherlands, 8 February 1700; *d.* Basel, Switzerland, 17 March 1782), *medicine, mathematics, physics.*

Life. Daniel Bernoulli was the second son of Johann I Bernoulli and Dorothea Falkner, daughter of the patrician Daniel Falkner. At the time of Bernoulli's birth his father was professor in Groningen, but he returned to Basel in 1705 to occupy the chair of Greek. Instead, he took over the chair of mathematics, which had been made vacant by the death of his brother Jakob (Jacques) I. In 1713 Daniel began to study philosophy and logic, passed his baccalaureate in 1715, and obtained his master's degree in 1716. During this period he was taught mathematics by his father and, especially, by his older brother Nikolaus II. An attempt to place young Daniel as a commercial apprentice failed, and he was allowed to study medicine—first in Basel, then in Heidelberg (1718) and Strasbourg (1719). In 1720 he returned to Basel, where he obtained his doctorate in 1721 with a dissertation entitled *De respiratione* (1). That same year he applied for the then vacant professorship in anatomy and botany (2), but the drawing of the lot went against him. Bad luck also cost him the chair of logic (3). In 1723 he journeyed to Venice, whence his brother Nikolaus had just departed and continued his studies in practical medicine under Pietro Antonio Michelotti. A severe illness prevented him from realizing his plan to work with G. B. Morgagni in Padua.

In 1724 Bernoulli published his *Exercitationes mathematicae* (4) in Venice, which attracted so much attention that he was called to the St. Petersburg Academy. He returned to Basel in 1725 and declared his readiness to go to the Russian capital with Nikolaus. That same year, he won the prize awarded by the Paris Academy, the first of the ten he was to gain. Bernoulli's stay in St. Petersburg was marred by the sudden death of his beloved brother and by the rigorous climate, and he applied three times for a professorship in Basel, but in vain. Finally, in 1732, he was able to obtain the chair of anatomy and botany there.

His Petersburg years (1725–1733 [after 1727 he worked with Euler]) appear to have been Bernoulli's most creative period. During these years he outlined the *Hydrodynamica* and completed his first important work on oscillations (23) and an original treatise on the theory of probability (22). In 1733 he returned to Basel in the company of his younger brother Johann II, after a long detour via Danzig, Hamburg, and Holland, combined with a stay of several weeks in Paris. Everywhere he went, scholars received him most cordially.

*See p. 56 for genealogy chart.

Although largely occupied with his lectures in medicine, Bernoulli continued to publish in mathematics and mechanics, which interested him much more intensely. His principal work, the *Hydrodynamica* (31), had been completed as early as 1734 but was not published until 1738. About the same time his father published *Hydraulica,* predated to 1732.[1] This unjustifiable attempt to insure priority for himself was one among many instances that exhibited Johann I Bernoulli's antagonism toward his second son.

In 1743 Daniel Bernoulli was able to exchange his lectures in botany for those in physiology, which were more to his liking. Finally, in 1750, he obtained the chair of physics, which was his by rights. For almost thirty years (until 1776) he delivered his lectures in physics, which were enlivened by impressive experiments and attended by numerous listeners. He was buried in the Peterskirche, not far from his apartment in the Kleine Engelhof.

Works. Daniel Bernoulli's works include writings on medicine, mathematics, and the natural sciences, especially mechanics. His works in these different areas were usually conceived independently of each other, even when simultaneous. As a consequence it is legitimate to distinguish them by subject matter and to consider them in chronological order within each subject.

Medicine. Bernoulli saw himself, against his inclination, limited to the field of medicine. Thus the future physicist promptly turned his interest to the mechanical aspects of physiology.[2] In his inaugural dissertation of 1721 (1), as a typical iatrophysicist under the decisive influence of Borelli and Johann Bernoulli, he furnished a comprehensive review of the mechanics of breathing. During the same year he applied for the then vacant chair of anatomy and botany, presenting pertinent theses (2) in support of his candidacy. In St. Petersburg in 1728 he published a strictly mechanical theory of muscular contraction (10), which disregarded the hypothesis of fermentation in the blood corpuscles assumed by Borelli and Johann Bernoulli. That same year he furnished a beautifully clear contribution to the determination of the shape and the location of the entrance of the optic nerve into the bulbus, or blind spot (11). Also of great importance is a lecture on the computation of the mechanical work done by the heart (*vis cordis*). Bernoulli gave this address in 1737 at the graduation exercises of two candidates in medicine, and it was thus that he first developed a correct method for such calculations. Because of its lasting significance, this lecture was published with its German translation in 1941 (75). Contributions to the phys-

iology of work, more particularly to the determination of the maximum work that a man can perform, are found in *Hydrodynamica* (sec. 9) and the prize-winning treatise of 1753 (47). (In this context, Bernoulli meant by "maximum work" the quantity that a man could do over a sustained period of time, e.g., a working day.)

Mathematics. Medical research, however, did not divert Bernoulli from his primary interest, the mathematical sciences. This is evidenced by the publication in 1724 of his *Exercitationes mathematicae,* which he wrote during his medical studies in Italy. This treatise combined four separate works dealing, respectively, with the game of faro, the outflow of water from the openings of containers, Riccati's differential equation, and the lunulae (figures bounded by two circular arcs). Ultimately, Bernoulli's talent proved to lie primarily in physics, mechanics, and technology, but his mathematical treatises originated partly from external circumstances (Riccati's differential equation) and partly from applied mathematics (recurrent series, mathematics of probability).

The discussions on Jacopo Riccati's differential equation were initiated in 1724 by the problem presented by Riccati in the *Supplementa* to the *Acta eruditorum.* Immediately thereafter Daniel Bernoulli offered a solution in the form of an anagram (5). In the two following papers, published in the *Acta eruditorum* (6, 7), as well as in the *Exercitationes mathematicae,* Bernoulli demonstrated that Riccati's special differential equation $ax^n\, dx + u^2\, dx = b\, du$ could be integrated through separation of the variables for the values $n = -4c/(2c \pm 1)$, where c takes on all integral values—positive, negative, and zero.

In the first part of the *Exercitationes* (4), dealing with faro, Bernoulli furnished data on recurrent series that later proved to have no practical application. According to De Moivre, these series result from the generative fraction

$$\frac{a + bz + cz^2 + \cdots + rz^m}{1 - \alpha z - \beta z^2 - \cdots - \sigma z^n}.$$

Bernoulli made use of these series in (16) for the approximate calculation of the roots of algebraic equations. For this purpose, the fraction is broken up into partial fractions, which are then developed into power series yielding, in the case of simple roots $1/p,\ 1/q,$ and so on, the general term

$$P = (Ap^n + Bq^n + \cdots)z^n$$

and the following member

$$Q = (Ap^{n+1} + Bq^{n+1} \cdots)z^{n+1}.$$

If p is considerably larger than $q \cdots$, etc., then, for sufficiently large n, P is approximated by Ap^n, Q by Ap^{n+1}, and thus the smallest root, $1/p$, is approximated by P/Q. In treatise (20) this method is applied to infinite power series.

Divergent sine and cosine series are treated by Bernoulli in three papers (62, 64, 66). The starting point is the thesis formulated by Leibniz and Euler that the equation $1 - 1 + 1 - 1 \cdots = 1/2$ is valid, which they base on the equation $1/(1 + x) = 1 - x + x^2 \pm \cdots$ for $x = 1$ and by observing that the arithmetic mean of the two possible partial sums of the series equals $1/2$. In reality, however, this divergent series can be summed to many values, depending on the expression from which it is derived. On the other hand, it can be demonstrated that the mean-value method for the equations found by Euler,

$$\sum_{n=1}^{\infty} \cos nx = -\frac{1}{2}$$

and

$$\sum_{n=1}^{\infty} \sin nx = \frac{1}{2} \frac{\sin x}{1 - \cos x} = \frac{1}{2} \cot \frac{1}{2}x,$$

leads to a correct result. For if x is commensurable with π, but not a multiple of π, then the terms of these series for a definite p and for each n satisfy the conditions $a_{n+p} = a_n$ and $a_1 + a_2 + a_3 + \cdots + a_p = 0$. For this case, according to the Leibniz-Bernoulli rule, the sum of the series $\sum_1^{\infty} a_n$ becomes equal to the arithmetic mean of the values a_1, $a_1 + a_2$, $a_1 + a_2 + a_3$, $\cdots a_1 + a_2 + a_3 + \cdots + a_p$.

Interestingly, in (64) the integration of the above cosine series, with application of Leibniz' series for $\pi/2$, yields the convergent series:

$$\sin x + \frac{1}{2} \sin 2x + \frac{1}{3} \sin 3x + \cdots = \frac{\pi - x}{2}.$$

In (66) Bernoulli let the formulas derived by Bossut[3] for the sums of the finite sine and/or cosine series n extend to the infinite. He assigned the value zero to the corresponding $\cos \infty x$ and $\sin \infty x$ and thereby obtained the correct sums.

In his later years Bernoulli contributed two additional papers (70, 71) to the theory of the infinite continued fractions.

Rational mechanics. In order to appreciate Daniel Bernoulli's contributions to mechanics, one must consider the state of this branch of science in the first half of the eighteenth century. Newton's great work was already available but could be rendered fruitful only by means of Leibniz' calculus. Collaterally there appeared Jakob Hermann's *Phoronomia* (1716), a sort of textbook on the mechanics of solids and liquids that used only the formal geometrical method. Euler's excellent *Mechanica* (1736) dealt only with the mechanics of particles. The first theory on the movement of rigid bodies was published by Euler in 1765. The fields of oscillations of rigid bodies and the mechanics of flexible and elastic bodies were new areas that Daniel Bernoulli and Euler dominated for many years.

In his earliest publication in mechanics (9), Bernoulli attempted to prove the principle of the parallelogram of forces on the basis of certain cases, assumed to be self-evident, by means of a series of purely logical extensions; this was in contrast with Newton and Varignon, who attempted to derive this principle from the composition of velocities and accelerations. Like all attempts at logical derivation, Bernoulli's was circular, and today the principle of the parallelogram of forces is considered an axiom. This was one of the rare instances when Bernoulli discussed the basic principles of mechanics. Generally he took for granted the principles established by Newton; only in cosmology or astronomy (gravity) and magnetism was he unable to break away completely from a modified vortex theory of subtle matter propounded by Descartes and Huygens. The deduction of gravity from the rotation of the subtle matter can be found in (79) and (31, ch. 11) and the explanation of magnetism in (41).

Treatise (13), inspired by Johann I Bernoulli's reports, is a contribution to the theory of rotating bodies, which at that time, considering the state of the dynamics of rigid bodies, was no trivial subject. The starting point was the simple case of a system consisting of two rigidly connected bodies rotating around a fixed axis. By means of geometric-mechanical considerations based on Huygens, Bernoulli solved a number of pertinent problems. Let us mention here only a special case of König's theorem (1751), derived by formal geometrical means. Written analytically, it states that

$$m_1 v_1^2 + m_2 v_2^2 = (m_1 + m_2)V^2 + m_1 v_1'^2 + m_2 v_2'^2,$$

where v_1 and v_2 represent velocities in a fixed system, V the velocity of the center of gravity, and v_1' and v_2' the velocities around the center of gravity.

The determination of a movement imparted to a body by an eccentric thrust and the calculation of the center of instantaneous rotation were accomplished by Bernoulli in 1737 (27). At his invitation Euler took up the problem simultaneously, with similar results. In this problem, Bernoulli limited himself to the simplest case, that involving rigid, infinitely thin rods. The motion caused by an impact on elastic rods was dealt with only much later (61).

The principle of areas and an extended version of the principle concerning the conservation of live force, both of which furnished integrals of Newton's basic equations, were published by Bernoulli, probably with Euler's assistance, in the *Berlin Mémoires* in 1745 and 1748 (40, 43). The principle of areas (40) was used and clearly formulated almost simultaneously by Bernoulli and Euler in their treatments of the problem involving the movement of a tube rotating around a fixed point and containing freely moving bodies.

The principle of conservation of live force (43) was developed by Bernoulli not only—as had been done before him—for the movements within a field of uniform gravity or within a field of one or several fixed centers of force, but also for a system of mobile, mutually attracting mass points. For example, given three centers with the masses m_1, m_2, m_3, whose mutual distances change from initial *a, b, c,* to *x, y, z,* Bernoulli finds that if the gravity constant equals ρ^2/μ for the difference of live forces,

$$\sum_{i=1}^{3} (m_i v_i^2 - m_i v_{io}^2) = 2\rho^2/\mu[m_1 m_2(1/x - 1/a) + m_1 m_3(1/y - 1/b) + m_2 m_3(1/z - 1/c)].$$

Most probably this is the first time that the double sum of $m_h m_k/r_{hk}$ appears. However, the force function for conservative systems was first discovered by Lagrange.

Bernoulli also investigated problems of friction of solid bodies (36, 57, 60). In his first such paper (36) he studied the movement of a uniformly heavy sphere rolling down an inclined plane and calculated the inclination at which the pure rotation changes into a motion composed of a rotatory and a sliding part.

The main problem of (60) consists in determining the progressive and rotatory motion of a uniformly heavy rod pressing upon a rough surface while a force oblique to the axis of the rod acts upon the rod.

A group of papers (14, 18, 21) dealing with the movement of solid bodies in a resisting medium is based on the presentation given by Newton in the *Principia.* The first two papers (14) deal with a rectilinear motion, the three subsequent ones (18, 21) with movement along a curve (pendulum swing). Here Bernoulli started with the usual premise that the resistance is largely proportional to the square of the velocity. At the same time he denied Newton's affirmation of a partial linear relation between resistance and velocity, but considered as probable the

assumption that part of the resistance, at least for viscous fluids, is proportional to time (i.e., independent of speed). The value of these five papers rests primarily on their consistent analytical presentation and on the treatment of certain special problems.

Hydrodynamics.[4] Traditionally, Bernoulli's fame rests on his *Hydrodynamica* (31)—a term he himself introduced. The first attempt at solving the problem of outflow as presented in the *Exercitationes mathematicae* was conceived in accordance with the concepts of the time, and did little to advance them. Essentially it contained a controversy with Jacopo Riccati over Newton's two different views on the force of a liquid issuing from an opening. But as early as 1727 Bernoulli succeeded in breaking through to an accurate calculation of the problem (12). Further progress was represented by the published experiments on the pressure exerted on the walls of a tube by a fluid flowing through it (19). In 1733 Bernoulli left behind in St. Petersburg a draft of the *Hydrodynamica* that agrees extensively in substance although not in form with the final version. Only the thirteenth chapter of the definitive work is missing (82).

The treatise opens with an interesting history of hydraulics, followed by a brief presentation of hydrostatics. The following three chapters contain formulas for velocity, duration, and quantity of fluid flowing out of the opening of a container. The author treats both the case of a falling level of the residual fluid and that of a constant level in the reservoir, and takes into consideration the starting process (nonstationary flow) and radial contraction of the stream. Bernoulli based these deductions on the principle of the conservation of live force or, as he says, the equality of the *descensus actualis* (actual descent) and *ascensus potentialis* (potential ascent), whereby these physical magnitudes, which pertain to the center of gravity, are obtained from the former through division by the mass of water in the container. If we equate the changes in *ascensus potentialis* and *descensus actualis* resulting from the water outflow, we obtain, in the case of a dropping water level, a linear differential equation. The kinematic principle used was the hypothesis of the parallel cross sections, which states that all particles of the liquid in a plane vertical to the flow have the same velocity, and that this velocity is inversely proportional to the cross section (principle of continuity).

Chapter 7 deals with the oscillations of the water in a tube immersed in a water tank and considers mainly the energy loss. Many years later Borda resumed these investigations, but arrived at another formula for the loss.

Chapter 9 contains a theory of machinery, lifting devices, pumps, and such, and their performance, as well as an extensive theory of the screw of Archimedes. A spiral pump related to the latter was discussed by Bernoulli much later (65). A theory of windmill sails concludes the chapter.

Chapter 10 is devoted to the properties and motions of "elastic fluids" (i.e., gases), and its main importance lies in its sketch of a "kinetic gas theory," which enabled Bernoulli to explain the basic gas laws and to anticipate—in incomplete form—Van der Waals' equation of state, which was developed some hundred years later. Further on, Bernoulli examined the pressure conditions in the atmosphere, established a formula for relating pressure to altitude, provided a formula for the total refraction of light rays from various stellar heights, and was the first to derive a formula for the flow velocity of air streaming from a small opening.

Chapter 12 contains the somewhat questionable derivation of a rather unusual form of the so-called Bernoulli equation for stationary currents. For the wall pressure p in a horizontal tube, connected to an infinitely wide container filled with water to the level a and having the cross section n and an outlet with the cross section 1, he determined the expression $p = [(n^2 - 1)/n^2]a$. Since $a/n^2 \sim u^2$ represents the height from which a body must fall to obtain the velocity u at the point observed, that expression becomes the equation $p + u^2 = a = $ const. More generally, for a current in a tube of any shape and inclination, u^2 must equal A/n^2, A or a being the distances between water surface and discharge opening or any cross section n. We then obtain the equation $p + A/n^2 = a$, and—with $A - z = a$ ($z = $ distance between n and opening)—the term $p + z + u^2 = A = $ const. for the stationary current. Because of the system of measures used by Bernoulli, the constant factors have values other than those customarily used.

Chapter 13 is concerned with the calculation of the force of reaction of a laterally discharged fluid jet as well as with the determination of its pressure upon a facing plate. With the aid of the impulse theorem, Bernoulli proved that both pressures p are equal to the weight of the cylinder of water whose base equals the area n of the opening for the discharge and whose length is double the height a of the water. It is thus $p = 2$ $gan = nu^2$. In contrast, Johann I Bernoulli advocated throughout his life the erroneous assumption of a cylinder length equal to the height of the water. A complicated calculation of the pressure of a water jet on an inclined plate is contained in (26). Toward the end of chapter 13 Daniel Bernoulli discusses the question of whether the traditional propelling forces of sail and oar could be replaced by such a force of reaction. This principle was converted to practice only many years later.

The weaknesses in the deduction of the so-called Bernoulli equation and Daniel's incomplete concepts of internal pressure can only be mentioned here. In this respect, Johann Bernoulli's *Hydraulica* represents a certain progress, which in turn inspired Euler in his work on hydrodynamics.

Vibrating systems. From 1728, Bernoulli and Euler dominated the mechanics of flexible and elastic bodies,[5] in that year deriving the equilibrium curves for these bodies. In the first part of (15) Bernoulli determined the shape that a perfectly flexible thread assumes when acted upon by forces of which one component is vertical to the curve and the other is parallel to a given direction. Thus, in one stroke he derived the entire series of such curves as the velaria, lintearia, catenaria, etc.

More original was the determination of the curvature of a horizontal elastic band fixed at one end—a problem simultaneously undertaken by Euler. Bernoulli showed that the total moment of a uniform band around point s, by virtue of the weight P at its free end and of its own weight p acting on the center of gravity, relates to the curvature radius R by means of the equation

$$Px + \frac{p}{l} \int s \, dx = \frac{m}{R},$$

whereby the arc length s and the abscissa x are to be taken starting from the free end, with m being the modulus of bending and l the length of the string. A case involving a variable density and an optionally directed final load is quite possible.

When he departed from St. Petersburg in 1733, Bernoulli left behind one of his finest works (23), ready for the printer. Here, for the first time, he defined the "simple modes" and the frequencies of oscillation of a system with more than one degree of freedom, the points of which pass their positions of equilibrium at the same time. The inspiration for this work must have been the reports made by Johann I, toward the end of 1727, on treatment of a similar problem. In the first part of the treatise, Daniel Bernoulli discussed an arrangement consisting of a hanging rope loaded with several bodies, determined their amplitude rates and frequencies, and found that the number of simple oscillations equals the number of bodies (i.e., the degrees of freedom).

For a uniform, free-hanging rope of length l he found the displacement, y, of the oscillations at distance x from the lower end by means of the equation

$$y = AJ_o\left(2\sqrt{\frac{x}{\alpha}}\right),$$

where α has to be determined from the equation $J_o\left(2\sqrt{\frac{l}{\alpha}}\right) = 0$ and J_o is the first appearance of Bessel's function. It shows that α is the length of the simple pendulum of equal frequency. The above equation has an infinite number of real roots. Thus the rope can perform an infinite number of small oscillations with the frequencies $v = \frac{1}{2\pi}\sqrt{g/\alpha}$. These theorems were demonstrated in (25) on the basis of a principle that is equivalent to that subsequently named after d'Alembert.

Immediately following Bernoulli's departure from St. Petersburg, there began between him and Euler one of the most interesting scientific correspondences of that time. In its course, Bernoulli communicated much important information from which Euler, through his analytical gifts and tremendous capacity for work, was able to profit within a short time.

The above results were corroborated by Bernoulli and Euler through additional examples. Thus Bernoulli, in extending paper (30), investigated small vibrations of a plate immersed in water (32) and those of a rod suspended from a flexible thread (34). Both works stress the difference between simple and composite vibrations. He investigated only the former, however, for composite vibrations ultimately change into the slower ones.

The following two papers (37, 38), dating from 1741–1743, deal with the transversal vibrations of elastic strings, with (37) discussing the motion of a horizontal rod of length l, fastened at one end to a vertical wall. In order to derive the vibration equation, whose form he had known since 1735 (35), Bernoulli used the relation between curvature and moment, as detailed in (15): $m/R = M$. The resulting differential equation is $f^4\, d^4y = y\, dx^4$, where y becomes the amplitude at distance x from the band end, and $f^4 = m^4L/g$, if L is the length of the simple pendulum isochronal with the band vibrations and g is the load per unit of length. Bernoulli used the solution $y = y(x/f)$ through infinite series as well as in closed expression by means of exponential and trigonometric functions. The series of the roots l/f is an example of nonharmonic oscillations. In (38) Bernoulli discusses the differential equation in the case of free ends.

Treatise (45), on vibrating strings, represented a reaction to the publications of d'Alembert and Euler, who calculated the form of the vibrating string from the partial differential equation

$$\partial^2 y/\partial t^2 = c^2 \partial^2 y/\partial x^2.$$

They thus moved the inference from the finite to the infinite up into the hypothesis, whereas Bernoulli always made this transition without thinking about it in the final, completed formula.

His deliberations in (45) started from the assumption that the single vibrations of a string of length a were furnished by $y = \alpha_n \sin n\pi x/a$ ($n =$ any integral number). From this and from his previous deliberations he deduced that the most general motion could be represented by the superposition of these single vibrations, i.e., by a series of the form

$$y = \sum_{n=1}^{\infty} A_n \sin \frac{n\pi x}{a} \cos \frac{cn\pi t}{a}.$$

This equation appears nowhere explicitly, but it can be derived from a combination of various passages of this work and is valid only with the assumption of an initial velocity equaling zero.

In (46) Bernoulli determined the vibrations of a weightless cord loaded with n weights. He shows that in the case of $n = 2$, two simple vibrations, either commensurable or incommensurable, are possible, depending on the position and value of the two weights.

Treatise (53) is a beautiful treatment of the oscillations inside organ pipes, using only elementary mathematics. It is assumed that the movement of the particles parallel to the axis, the velocities, and the pressure are equal at all points of the same cross section and that the compression at the open end of the pipes equals zero. Among other things, this work contains the first theory of conical pipes and an arrangement consisting of two coaxial pipes of different cross sections as well as a series of new experiments.

In paper (54), on the vibrations of strings of uneven thickness, Bernoulli inquires about cases where oscillations assume the form $y = Aq \sin p \sin vt$, where p and q are functions of x only and v is a constant. Here, for the first time, are solutions for the inverse problem, the determination of vibration curves from the distribution of density. In (63) he treats a special case in which the string consists of two parts of different thickness and length.

In treatise (67) Bernoulli compared the two possible oscillations of a body suspended from a flexible thread with the movement of a body bound with a rigid wire, and showed that one of the two oscillations of the first arrangement closely approximated the oscillation of the second arrangement. The method followed by Bernoulli is applicable only to infinitely small vibrations, and thus represents only a special case of the problem treated simultaneously by Euler by means of the Newtonian fundamental equations of mechanics.

In (68, 69) Bernoulli once more furnished a comprehensive presentation of his views on the superposition principle, which he clarified by means of the example of the frequently studied double pendulum. These last papers show that he had nothing new to add to the problem of vibrations.

Probability and statistics. A number of valuable papers were published by Bernoulli on probability theory and on population statistics.[6] True, his youthful work on faro within the framework of the *Exercitationes mathematicae* contributed hardly anything new, but it was evidence of his early interest in the work on the theory of probability done by his predecessors Montmort and De Moivre, which had been nourished by discussions with his cousin Nikolaus I. The most important treatise, and undoubtedly the most influential, was the *De mensura sortis* (22), conceived while he was in St. Petersburg, which contains an unusual evaluation of capital gains, and thus also contains the mathematical formulation of a new kind of value theory in political economy.[7]

The basic idea is that the larger a person's fortune is, the smaller is the moral value of a given increment in that fortune. If we assume, with Bernoulli, the special case, that a small increase of assets dx implies a moral value, $dy,$ that is directly proportional to dx and inversely proportional to the fortune a—i.e., $dy = b \, dx/a$—then it follows that the moral value y of the gains $x - a$ complies with the formula $y = \log x/a$.

If a person has the chances $p_1, p_2, p_3 \cdots$, to make the gains $g_1, g_2, g_3 \cdots$, where $p_1 + p_2 + \cdots = 1$, which reflects one and only one gain, then the mean value of the moral values of the gains is equal to

$$bp_1 \log a \, (a + g_1) + bp_2 \log a \, (a + g_2)$$
$$+ \cdots - b \log a$$

and the moral expectation (hope)

$$H = (a + g_1)^{p_1}(a + g_2)^{p_2} \cdots - a.$$

If gains are very small in comparison with the assets, then the moral hope converts to the mathematical expectation $H = p_1 g_1 + p_2 g_2 + \cdots$. There follow some applications of the preceding to risk insurance and a discussion of the Petersburg paradox.

Only in 1760 did Bernoulli again treat a problem of this sort: medical statistics concerning the rate of mortality resulting from smallpox in the various age groups (51). If ξ is the number of survivors and s the number of those who at age x have not yet had

smallpox, there results—given certain conditions—a differential equation containing three variables that defines the ratio s/ξ as a function of x. A table calculated on that basis contains the values of ξ, s, $\xi - s$, and so on valid for the first twenty-four years; ξ was taken from Halley's mortality table.[8] In (52) Bernoulli ardently advocated inoculation as a means of prolonging the average lifetime by three years.

In paper (55) Bernoulli treats, by means of urn models, problems of probability theory as applied to his treatise on population statistics (56). Their main purpose was to determine for every age the expected average duration of a marriage. Here and in his subsequent papers (58, 59) Bernoulli preferred to make use of infinitesimal calculus in probability theory by assuming continuously changing states. The problem treated in (58) is as follows: Given several urns, each of which contains n slips of the same color, but of a different color for each urn, one slip is taken from each urn and deposited in the next one, with the slip taken from the last urn deposited in the first. The question is, How many slips of each color do the various urns contain after a number r of such "permutations"? The problem treated in (59) belongs in the field of the theory of errors, and concerns the determination of the probability with which (expressed in modern terms) a random variable subject to binomial distribution would assume values between two boundaries on either side of the mean value.

In paper (72) Bernoulli seeks to deal with the theory of errors in observation as a branch of probability theory. He challenges the assumption of Simpson and Lagrange that all observations are of equal importance. Rather, he maintains that small errors are more probable than large ones. Thus Bernoulli approximates the modern concept, except that he selects the semicircle instead of Gauss's probability curve.

Treatise (73) deals with errors to be considered in pendulum clocks, which are calculated partially by means of the method presented in (59).

Prizes of the Paris Academy. Bernoulli was highly esteemed for clarifying problems for a general public interested in the sciences. Of his essays entered in the competitions of the Paris Academy, ten were awarded prizes. Most of them concerned marine technology, navigation, and oceanology; but astronomy and magnetism were also represented.

His prize-winning paper of 1725 (8) dealt with the most appropriate shape for and the installation of hourglasses filled with sand or water. The subject of the 1728 contest was the cause and nature of gravity, on which Bernoulli prepared a manuscript, but the prize went to the Cartesian G. B. Bilfinger (79). In his entry for the 1729 competition Bernoulli indicated several methods for determining the height of the pole, particularly at sea, when only one unknown star is visible, or when one or more known stars are visible. The essay did not win a prize (80), but the manuscript is extant.

The prize of 1734 (24) was shared with his father, who begrudged Daniel his share of success. Here Daniel postulated an atmosphere resembling air and rotating around the solar axis, resulting in an increasing inclination of the planetary orbits toward the equator of the sun.

Bernoulli shared the 1737 prize for the best form of an anchor with Poleni (28). The 1740 prize on the tides was shared with Euler and several others. This important paper (33) on the relationship, recognized by Newton, between the tides and solar and lunar attraction, respectively, is still of interest, inasmuch as it furnishes a complete equilibrium theory of these phenomena.

The prize-winning papers of 1743 (39) and 1746 (41) deal with problems of magnetism. In the first paper Bernoulli considered all possibilities for reducing the sources of error in the inclination compass by improving construction. According to his instructions, the Basel mechanic Dietrich constructed such needles (49). The 1746 paper, written with his brother Johann II, contains an attempt to establish a theory of magnetism. Both authors believed that there is a subtle matter which moves in the direction of the magnetic meridian and forms a vortex around the magnet.

The next prize, for the best method of determining the time at sea with the horizon not visible, was offered in 1745 for the first time. It was offered for a second time in 1747, and Bernoulli won (42). Included in the wealth of information contained in this paper are the proposals for improving pendulum and spring clocks and the description of a mechanism for holding a rod equipped with diopter in a vertical position, even in a turbulent sea. A detailed account of the determination of the time, with the position of a given star known, concludes this paper (see 17).

The 1748 prize, for the irregular movements of Saturn and Jupiter (81), went to Euler. Bernoulli's manuscript has been preserved. The prize essay for 1749–1751 (44) discussed the question of the origin and nature of ocean currents, and added suggestions for measuring current velocities.

The problem treated by the prize essay of 1753 (47), the effect on ships of forces supplementary to that of the wind (e.g., rudder forces), was answered by Bernoulli, mainly by means of detailed data on the maximum work that could be performed by a man in a given unit of time. Among other things, he calculated the number of oarsmen required for attaining a given ship velocity.

The subject of the prize essay for 1757 (48), proposals for reducing the roll and pitch of ships, gave Bernoulli the opportunity to air his views on the pertinent works of Bouguer and Euler, published several years earlier. Whereas Euler had limited himself to the free vibrations of a ship, Bernoulli extended his views to the behavior of ships in turbulent seas, i.e., to forced vibrations. His findings prevailed for almost a century.

Evaluation and Appreciation. In order to appreciate both Bernoulli's importance in science, as indicated by the above summaries of his published works, and his private life, it is necessary to consider his extensive correspondence.[9] This includes his exchange of letters with Christian Goldbach (1723–1730), Euler (1726–1767, especially 1734–1750), and his nephew Johann III (1763–1774). Also important are his contemporaries' evaluations of Bernoulli and of his work. Unfortunately, his extremely popular lectures on experimental physics, in which he often introduced unproved hypotheses that have since been confirmed, apparently are not extant. Among them was his assertion of the validity of the relation later known as Coulomb's law in electrostatics. All of these achievements brought Bernoulli considerable fame in intellectual circles during his lifetime. He was a member of the leading learned societies and academies, including Bologna (1724), St. Petersburg (1730), Berlin (1747), Paris (1748), London (1750), Bern (1762), Turin (1764), Zurich (1764), and Mannheim (1767).

We can now assert that Bernoulli was the first to link Newton's ideas with Leibniz' calculus, which he had learned from his father and his brother Nikolaus. He did not, however, attempt to solve the problems that confronted him by means of the fundamental Newtonian equations; rather, he preferred to use the first integrals of these equations, especially Leibniz' principle of the conservation of living force, which his father had emphasized. Like Newton, whose battles he fought on the Continent, Bernoulli was first and foremost a physicist, using mathematics primarily as a means of exploring reality as it was revealed through experimentation. Thus he was interested in physical apparatus as well as the practical application of the results of physics and other sciences.

Bernoulli's active and imaginative mind dealt with the most varied scientific areas. Such wide interests, however, often prevented him from carrying some of his projects to completion. It is especially unfortunate that he could not follow the rapid growth of mathematics that began with the introduction of partial differential equations into mathematical physics.

Nevertheless, he assured himself a permanent place in the history of science through his work and discoveries in hydrodynamics, his anticipation of the kinetic theory of gases, a novel method for calculating the value of an increase in assets, and the demonstration that the most common movement of a string in a musical instrument is composed of the superposition of an infinite number of harmonic vibrations (proper oscillations).

Otto Spiess instituted the publication of editions of the works and correspondence of the Bernoullis, a project that has continued since Spiess's death.

NOTES

1. *Johannis Bernoulli Hydraulica nunc primum detecta ac demonstrata directe ex fundamentis pure mechanicis,* in his *Opera omnia,* IV (Lausanne–Geneva, 1742), 387–488.
2. Friedrich Huber, *Daniel Bernoulli (1700–1782) als Physiologe und Statistiker,* Basler Veröffentlichungen zur Geschichte der Medizin und der Biologie, fasc. 8 (Basel, 1958).
3. Charles Bossut, "Manière de sommer les suites . . .," in *Mémoires de mathématiques et de physique de l'Académie royale des sciences, Paris,* 1769 (1772), 453–466.
4. For Daniel Bernoulli's hydrodynamic studies, see Clifford Truesdell, *Rational Fluid Mechanics,* intro. to Euler's *Opera omnia,* 2nd ser., XII, XIII (Zurich, 1954–1955).
5. Two excellent works are Clifford Truesdell, "The Rational Mechanics of Flexible or Elastic Bodies (1638–1788)"; his intro. to Euler's *Opera omnia,* 2nd ser., X, XI (Zurich, 1960); and H. Burkhardt, "Entwicklungen nach oscillierenden Functionen und Integration der Differentialgleichungen der mathematischen Physik," in *Jahresbericht der Deutschen Mathematiker-Vereinigung,* **10,** no. 2 (1908), 1–24.
6. I. Todhunter, *A History of the Mathematical Theory of Probability From the Time of Pascal to That of Laplace* (Cambridge–London, 1865; repr. New York, 1949), pp. 213–238.
7. An English trans. by Louise Sommer appeared in *Econometrica,* **22** (Jan. 1954). There is also a German trans. with extensive commentary by Alfred Pringsheim (Leipzig, 1896).
8. Edmund Halley, "An Estimate of the Degrees of the Mortality of Mankind," in *Philosophical Transactions of the Royal Society of London,* no. 196 (1694), 596–610.
9. Correspondence with Euler and Christian Goldbach—as far as available—appeared in *Correspondance mathématique et physique de quelques célèbres géomètres du XVIIIème siècle,* II (St. Petersburg, 1843). Letters exchanged by Bernoulli and his nephew Johann III have not yet been published.

BIBLIOGRAPHY

The following abbreviations of journal titles are used in the listing of Bernoulli's published works: *AE, Acta eruditorum; AP, Acta Academiae Scientiarum Imperialis Petropolitanae; CP* (or *NCP*), *Commentarii* (or *Norvi commentarii*) *Academiae Scientiarum Imperialis Petropolitanae; Prix, Pièces qui ont remporté les prix de l'Académie royale des sciences* (Paris); *Hist. Berlin, Histoire de l'Académie royale des sciences et belles lettres, Berlin; Mem. Berlin, Mémoires de l'Académie royale des sciences et belles lettres, Berlin; Mem. Paris, Mémoires de mathématiques et de physique de l'Académie royale des sciences, Paris.*

The first year following a journal title is the serial year; the second is the year of publication.

1. *Dissertatio inauguralis physico-medica de respiratione* (Basel, 1721).
2. *Positiones miscellaneae medico-anatomico-botanicae* (Basel, 1721).
3. *Theses logicae sistentes methodum examinandi syllogismorum validitatem* (Basel, 1722).
4. *Exercitationes quaedam mathematicae* (Venice, 1724).
5. "Notata in praecedens schediasma Ill. Co. Jacobi Riccati," in *AE,* supp. **8** (1724).
6. "Danielis Bernoulli explanatio notationum suarum, quae exstant Supplem. Tomo VIII Sect II," *ibid.,* 1725 (1725), also published in (4).
7. "Solutio problematis Riccatiani propositi in Act. Lips. Suppl. Tom. VIII p. 73," *ibid.,* 1725 (1725), also published in (4).
8. "Discours sur la manière la plus parfaite de conserver sur mer l'égalité du mouvement des clepsidres ou sabliers," in *Prix,* 1725 (1725).
9. "Examen principiorum mechanicae, et demonstrationes geometricae de compositione et resolutione virium," in *CP,* **1,** 1726 (1728).
10. "Tentamen novae de motu musculorum theoriae," *ibid.*
11. "Experimentum circa nervum opticum," *ibid.*
12. "Theoria nova de motu aquarum per canales quoscunque fluentium," *ibid.,* **2,** 1727 (1729).
13. "De mutua relatione centri virium, centri oscillationis et centri gravitatis," *ibid.*
14. "Dissertatio de actione fluidorum in corpora solida et motu solidorum in fluidis," *ibid.;* "Continuatio," *ibid.,* **3,** 1728 (1732).
15. "Methodus universalis determinandae curvaturae fili," *ibid.*
16. "Observationes de seriebus quae formantur ex additione vel subtractione quacunque terminorum se mutuo consequentium," *ibid.*
17. "Problema astronomicum inveniendi altitudinem poli una cum declinatione stellae ejusdemque culminatione," *ibid.,* **4,** 1729 (1735).
18. "Theorema de motu curvilineo corporum, quae resistentiam patiuntur velocitatis suae quadrato proportionalem," *ibid.;* "Additamentum," *ibid.,* **5,** 1730/1731 (1738).
19. "Experimenta coram societate instituta in confirmationem theoriae pressionum quas latera canalis ab aqua transfluente sustinent," *ibid.,* **4,** 1729 (1735).
20. "Notationes de aequationibus, quae progrediuntur in infinitum, earumque resolutione per methodum serierum recurrentium," *ibid.,* **5,** 1730/1731 (1738).
21. "Dissertatio brevis de motibus corporum reciprocis seu oscillatoriis, quae ubique resistentiam patiuntur quadrato velocitatis suae proportionalem," *ibid.*
22. "Specimen theoriae novae de mensura sortis," *ibid.*
23. "Theoremata de oscillationibus corporum filo flexili connexorum et catenae verticaliter suspensae," *ibid.,* **6,** 1732/1733 (1738).

24. "Quelle est la cause physique de l'inclinaison des plans des orbites des planètes par rapport au plan de l'équateur de la révolution du soleil autour de son axe," in *Prix,* 1734 (1735).

25. "Demonstrationes theorematum suorum de oscillationibus corporum filo flexili connexorum et catenae verticaliter suspensae," in *CP,* **7,** 1734/1735 (1740).

26. "De legibus quibusdam mechanicis, quas natura constanter affectat, nondum descriptis, earumque usu hydrodynamico, pro determinanda vi venae aqueae contra planum incurrentis," *ibid.,* **8,** 1736 (1741).

27. "De variatione motuum a percussione excentrica," *ibid.,* **9,** 1737 (1744).

28. "Réflexions sur la meilleure figure à donner aux ancres," in *Prix,* 1737 (1737).

29. "Commentationes de immutatione et extensione principii conservationis virium vivarum, quae pro motu corporum coelestium requiritur," in *CP,* **10,** 1738 (1747).

30. "Commentationes de statu aequilibrii corporum humido insidentium," *ibid.*

31. *Hydrodynamica, sive de viribus et motibus fluidorum commentarii* (Strasbourg, 1738). The following trans. exist: German, with extensive commentary by Karl Flierl, in the series Veröffentlichungen des Forschungsinstituts des Deutschen Museums für die Geschichte der Naturwissenschaften und der Technik, Reihe C: Quellentexte und Uebersetzungen, nos. 1a, 1b (Munich, 1965); English, *Hydrodynamics by Daniel Bernoulli,* trans. Thomas Carmody and Helmut Kobus (New York, 1968), bound with Johann I Bernoulli's *Hydraulics* (pp. 343–451); and Russian, *Daniel Bernoulli. Gidrodinamika ili zapiski o silakh i dvizheniakh zhidkostei,* trans. A. I. Nekrasov, K. K. Baumgart, and V. I. Smirnov (Moscow, 1959).

32. "De motibus oscillatoriis corporum humido insidentium," in *CP,* **11,** 1739 (1750).

33. "Traité sur le flux et reflux de la mer," in *Prix,* 1740 (1741).

34. "De oscillationibus compositis praesertim iis quae fiunt in corporibus ex filo flexili suspensis," in *CP,* **12,** 1740 (1750).

35. "Excerpta ex litteris ad Leonhardum Euler," *ibid.,* **13,** 1741–1743 (1751).

36. "De motu mixto, quo corpora sphaeroidica super plano inclinato descendunt," *ibid.*

37. "De vibrationibus et sono laminarum elasticarum," *ibid.*

38. "De sonis multifariis quos laminae elasticae diversimode edunt disquisitiones mechanico-geometricae experimentis acusticis illustratae et confirmatae," *ibid.*

39. "Mémoire sur la manière de construire les boussoles d'inclinaison," in *Prix,* 1743 (1748).

40. "Nouveau problème de mécanique," in *Mem. Berlin,* 1745 (1746), trans. into German in Ostwald's Klassiker der Exacten Wissenschaften, no. 191 (Leipzig, 1914), pp. 29–43.

41. "Nouveaux principes de mécanique et de physique, tendans à expliquer la nature & les propriétés de l'aiman," written with Johann II, in *Prix,* 1746 (1748).

42. "La meilleure manière de trouver l'heure en mer," *ibid.,* 1745 and 1747 (1750).

43. "Remarques sur le principe de la conservation des forces vives pris dans un sens général," in *Mem. Berlin,* 1748 (1750), trans. into German in Ostwald's Klassiker der Exacten Wissenschaften, no. 191 (Leipzig, 1914), pp. 67–75.

44. "Sur la nature et la cause des courans," in *Prix,* 1749 and 1751 (1769).

45. "Réflexions et éclaircissemens sur les nouvelles vibrations des cordes," in *Mem. Berlin,* 1753 (1755).

46. "Sur le mélange de plusieurs espèces de vibrations simples isochrones, qui peuvent coexister dans un même système de corps," *ibid.*

47. "Recherches sur la manière la plus avantageuse de suppléer à l'action du vent sur les grands vaisseaux," in *Prix,* 1753 (1769).

48. "Quelle est la meilleure manière de diminuer le roulis & le tangage d'un navire," *ibid.,* 1757 (1771).

49. "Sur les nouvelles aiguilles d'inclinaison," in *Journal des sçavans,* 1757 (1757).

50. "Lettre de monsieur Daniel Bernoulli à M. Clairaut, au sujet des nouvelles découvertes faites sur les vibrations des cordes tendues," *ibid.,* 1758 (1758).

51. "Essai d'une nouvelle analyse de la mortalité causée par la petite vérole, & des avantages de l'inoculation pour la prevenir," in *Mem. Paris,* 1760 (1766).

52. "Réflexions sur les avantages de l'inoculation," in *Mercure de France* (June 1760).

53. "Sur le son & sur les tons des tuyaux d'orgues," in *Mem. Paris,* 1762 (1764).

54. "Mémoire sur les vibrations des cordes d'une épaisseur inégale," in *Mem. Berlin,* 1765 (1767).

55. "De usu algorithmi infinitesimalis in arte coniectandi specimen," in *NCP,* **12,** 1766/1767 (1768).

56. "De duratione media matrimoniorum, pro quacunque coniugum aetate," *ibid.*

57. "Commentatio de utilissima ac commodissima directione potentiarum frictionibus mechanicis adhibendarum," *ibid.,* **13,** 1768 (1769).

58. "Disquisitiones analyticae de novo problemate coniecturali," *ibid.,* **14,** 1769, pt. 1 (1770).

59. "Mensura sortis ad fortuitam successionem rerum naturaliter contingentium applicata," *ibid.;* "Continuatio," *ibid.,* **15,** 1770 (1771).

60. "Commentationes physico-mechanicae de frictionibus," *ibid.,* **14,** 1769, pt. 1 (1770).

61. "Examen physico-mechanicum de motu mixto qui laminis elasticis a percussione simul imprimitur," *ibid.,* **15,** 1770 (1771).

62. "De summationibus serierum quarundam incongrue veris," *ibid.,* **16,** 1771 (1772).

63. "De vibrationibus chordarum," *ibid.*

64. "De indole singulari serierum infinitarum quas sinus vel cosinus angulorum arithmetice progredientium formant, earumque summatione et usu," *ibid.,* **17,** 1772 (1773).

65. "Expositio theoretica singularis machinae hydraulicae," *ibid.*

66. "Theoria elementaria serierum, ex sinibus atque

cosinibus arcuum arithmetice progredientium diversimode compositarum, dilucidata," *ibid.,* **18,** 1773 (1774).

67. "Vera determinatio centri oscillationis in corporibus qualibuscunque filo flexili suspensis eiusque ab regula communi discrepantia," *ibid.*

68. "Commentatio physico-mechanica generalior principii de coexistentia vibrationum simplicium haud perturbatarum in systemate composito," *ibid.,* **19,** 1774 (1775).

69. "Commentatio physico-mechanica specialior de motibus reciprocis compositis," *ibid.*

70. "Adversaria analytica miscellanea de fractionibus continuis," *ibid.,* **20,** 1775 (1776).

71. "Disquisitiones ulteriores de indole fractionum continuarum," *ibid.*

72. "Diiudicatio maxime probabilis plurium observationum discrepantium atque verisimillima inductio inde formanda," in *AP,* 1777, pt. 1 (1778).

73. "Specimen philosophicum de compensationibus horologicis, et veriori mensura temporis," *ibid.,* pt. 2 (1780).

74. "Sur la cause des vents," in Berlin Academy's *Recueil des prix* without Bernoulli's name. Also at University of Basel, LIa753E5.

75. "Oratio physiologica de vita," with German trans. and historical essays, ed. O. Spiess and F. Verzár, in *Verhandlungen der Naturforschenden Gesellschaft Basel,* **52** (1940/1941), 189–266. Also at University of Basel, LIa-753E18.

The library of the University of Basel has most of Bernoulli's original MSS, and photocopies of the rest.

76. "Methodus isoperimetricorum ad novam problematum classem promota," LIa751C3. From all curves of equal length lying between two fixed points, to find the ones for which $\int R^m ds$ is a minimum, where R^m is the mth power of the radius of curvature R and ds is the arc element. This was first satisfactorily solved by Euler.

77. "Solutio problematis inveniendi curvam, quae cum aliis data sit tatuochrona" (1729), LIa751C4. Solution to the problem of finding the curves in which the oscillations of a center of mass moving in a vacuum are isochronous regardless of starting point. Also treated by Euler.

78. "De legibus motus mixti variati, quo corpus sphaericum super plano aspero progredietur," LIaC19.

79. "Discours sur la cause et la nature de la pesantur," LIa752D2, submitted in the 1728 prize competition of the Paris Academy. The prize was awarded to Bilfinger.

80. "Quelle est la meilleure méthode d'observer les hauteurs sur mer par le soleil et par les étoiles," LIa752D3, submitted in the 1729 prize competition of the Paris Academy. The prize was awarded to Bouguer.

81. "Recherches mécaniques et astronomiques sur la théorie de Saturne et de Jupiter," LIa33, submitted in the 1748 prize competition of the Paris Academy. The prize was awarded to Euler.

82. An outline of the *Hydrodynamica* (1733) is in Archives of the Academy of Sciences, Leningrad; a photocopy is in Basel.

Biographical works on Bernoulli include Daniel II Bernoulli, "Vita Danielis Bernoulli," in *Acta Helvetica,* **9** (1787), 1–32, a memorial address with an almost complete bibliography of printed works; "Die Basler Mathematiker Bernoulli und Leonhard Euler. Vorträge von Fr. Burckhardt u.a.," in *Verhandlungen der Naturforschenden Gesellschaft Basel,* **7** (1884), appendix; Marquis de Condorcet, "Éloge de M. Bernoulli," in *Hist. Paris,* 1782 (1785), 82–107, and in *Oeuvres de Condorcet,* II (Paris, 1847), 545–585; *Gedenkbuch der Familie Bernoulli zum 300. Jahrestage ihrer Aufnahme in das Basler Bürgerrecht. 1622, 1922* (Basel, 1922); Peter Merian, *Die Mathematiker Bernoulli. Jubelschrift zur 4. Säcularfeier der Universität Basel, 6 September 1860* (Basel, 1860); Otto Spiess, *Basel anno 1760* (Basel, 1936); "Daniel Bernoulli," in Eduard Fueter, ed., *Grosse Schweizer Forscher* (Zurich, 1939), pp. 110–112; "Johann Bernoulli und seine Soehne," in *Atlantis* (1940), 663–669; "Die Mathematikerfamilie Bernoulli," in Martin Huerlimann, ed., *Grosse Schweizer,* 2nd ed. (Zurich, 1942), pp. 112–119; and "Bernoulli, Basler Gelehrtenfamilie," in *Neue deutsche Biographie* (1955), pp. 128–131; and Rudolf Wolf, "Daniel Bernoulli von Basel, 1700–1782," in *Biographien zur Kulturgeschichte der Schweiz,* 3rd ser. (Zurich, 1860), pp. 151–202.

Hans Straub

BERNOULLI, JAKOB (JACQUES) I (*b.* Basel, Switzerland, 27 December 1654; *d.* Basel, 16 August 1705), *mathematics, mechanics, astronomy.*

Bernoulli came from a line of merchants. His grandfather, Jakob Bernoulli, was a druggist from Amsterdam who became a citizen of Basel in 1622 through marriage. His father, Nikolaus Bernoulli, took over the thriving drug business and became a member of the town council and a magistrate; his mother, Margaretha Schönauer, was the daughter of a banker and town councillor. Jakob was married in 1684 to Judith Stupanus, the daughter of a wealthy pharmacist; their son Nikolaus became a town councillor and master of the artists' guild.

Bernoulli received his master of arts in philosophy in 1671, and a licentiate in theology in 1676; meanwhile, he studied mathematics and astronomy against the will of his father. In 1676 he went as a tutor to Geneva, where in 1677 he began his informative scientific diary, *Meditationes;* he then spent two years in France, familiarizing himself with the methodological and scientific opinions of Descartes and his followers, among whom was Nicolas Malebranche. Bernoulli's second educational journey, in 1681–1682, took him to the Netherlands, where he met mathematicians and scientists, especially Jan Hudde, and to England, where he met Robert Boyle and Robert Hooke. The scientific result of these journeys was his inadequate theory of comets (1682) and a theory of gravity that was highly regarded by his contemporaries (1683).

After returning to Basel, Bernoulli conducted experimental lectures, concerning the mechanics of solid and liquid bodies, from 1683 on. He sent reports on scientific problems of the day to the *Journal des sçavans* and the *Acta eruditorum,* and worked his way through the principal mathematical work of those days, *Geometria,* the Latin edition of Descartes's *Géométrie,* which had been edited and provided with notes and supplements by Frans van Schooten (2nd ed., Amsterdam, 1659–1661). As a result of this work, Bernoulli contributed articles on algebraic subjects to the *Acta eruditorum.* His outstanding achievement was the division of a triangle into four equal parts by means of two straight lines perpendicular to each other (1687). After these contributions had been extended and supplemented, they were published as an appendix to the *Geometria* (4th ed., 1695).

In four disputations published from 1684 to 1686, Bernoulli presented formal logical studies that tended toward the sophistical. His first publication on probability theory dates from 1685. By working with the pertinent writings of John Wallis (those of 1656, 1659, and 1670–1671) and Isaac Barrow (1669–1670), concerning mathematical, optical, and mechanical subjects, Bernoulli was led to problems in infinitesimal geometry.

In the meantime his younger brother Johann began attending the University of Basel after an unsuccessful apprenticeship as a salesman. As respondent to one of Jakob's scholarly logic debates, Johann earned his master of arts degree in 1685 and, by order of his father, studied medicine. Simultaneously, however, he secretly studied mathematics under his brother, becoming well versed in the fundamentals of the field. In 1687 Jakob became professor of mathematics at Basel, and with his brother he studied the publications of Leibniz and of Ehrenfried Walther von Tschirnhaus in *Acta eruditorum* (1682–1686), which had in essence been limited to examples and intimations of infinitesimal mathematics and its application to mechanics and dynamics. After much effort, Bernoulli was able to make himself master of these new methods, which he erroneously believed to be merely a computational formalism for Barrow's geometrical treatment of infinitesimals. His mathematical studies reached a first peak about 1689 with the beginnings of a theory of series, the law of large numbers in probability theory, and the special stress on complete induction.

Bernoulli showed his mastery of the Leibnizian calculus with his analysis (in May 1690) of the solutions given by Huygens in 1687 and by Leibniz in 1689 to the problem of the curve of constant descent in a gravitational field. (It was in that analysis that

the term "integral" was first used in its present mathematical sense.) The determination of the curve of constant descent had been posed as a problem by Leibniz in 1687. As a counterproblem Bernoulli raised the determination of the shape of the catenary, to which he had, perhaps, been directed by Albert Girard's notes to the *Oeuvres* of Simon Stevin (1634); Girard claimed that the catenary is a parabola. Leibniz promptly referred to the significance of this counterproblem, which he had spontaneously solved (1690) and which was later treated by Johann Bernoulli, Huygens, and himself in the *Acta eruditorum* (1691). Jakob, who found himself at that time in difficulties at the university because of his open criticism of university affairs and saw himself being overshadowed by his brother, did not take part directly, but proposed generalizations of the problem, allowing the links of the chain to be elastic or of unequal weight. He also announced a treatise on the *elastica,* the form of a bent elastic beam, which, under certain conditions, satisfies the differential equation $dy/dx = x^2/\sqrt{a^4 - x^4}$. Later he investigated this thoroughly, supposing arbitrary functions of elasticity (1694). In two notable contributions to differential calculus (1691), he examined the parabolic spiral (in polar coordinates: $r = a - b\sqrt{\phi}$, the elliptical integral for the curve length with its characteristic feature of symmetry) and the logarithmic spiral.

In Johann Bernoulli's study concerning the focal line of incident parallel rays of light on a semicircular mirror (1692), there is reference to Jakob's general procedure for determination of evolutes. This procedure is based on the generation of an algebraic curve as the envelope of its circles of curvature, and this procedure is worked out fully in the case of the parabola. Here Bernoulli corrected a mistake made by Leibniz (1686)—the statement that the circle of curvature meets the curve at four coinciding points—but he himself made a mistake in his assertion that the radius of curvature becomes infinite at every point of inflection. This error, corrected in 1693 by G. F. A. de L'Hospital, was the occasion for Bernoulli's removing of the singularity $a^2x^3 = y^5$ in the origin (1697). Almost simultaneously, and independently of each other, the brothers recognized that the form of a sail inflated by the wind is described by $(dx/ds)^3 = a d^2y/dx^2$. Jakob made a preliminary report in 1692 and a thorough one in 1695.

Further investigations concerned evolutes and caustics, first of the logarithmic spiral (*spira mirabilis*) and the parabola (1692), and later of epicycloids (1692) and diacaustic surfaces (1693), this in connection with Johann's similar studies. These last were included in his private instruction to L'Hospital

(1691–1692). Here, for the first time, public reference was made to the *theorema aureum,* which had been developed in the spring of 1692. The theorem, which gives the radius of curvature as $(ds/dx)^3 : (d^2y/dx^2)$, was published in 1694. Bernoulli's solution of the differential equation proposed by Johann Bernoulli (1693), $xdy - ydx/yds = a/b$, was completed by Huygens (1693). In treating the paracentric isochrone, a problem proposed by Leibniz (1689) that leads to the differential equation

$$(xdx + ydy)/\sqrt{y} = (xdy - ydx)/\sqrt{a},$$

Bernoulli separated the variables by substituting

$$x^2 + y^2 = r^2, \qquad ay = rt$$

and was able to relate the solution to the rectification of the *elastica;* later he found the reduction to the rectification of the lemniscate,

$$x^2 + y^2 = a\sqrt{x^2 - y^2}.$$

These and other studies—among which the kinetic-geometrical chord construction for the solution (1696) of $dy/dx = t(x)/a$ and the solution (1696) of the so-called Bernoullian differential equation $y' = p(x)y + q(x)y^n$ (1696) merit special attention—are proof of Bernoulli's careful and critical work on older as well as on contemporary contributions to infinitesimal mathematics and of his perseverance and analytical ability in dealing with special pertinent problems, even those of a mechanical-dynamic nature.

Sensitivity, irritability, a mutual passion for criticism, and an exaggerated need for recognition alienated the brothers, of whom Jakob had the slower but deeper intellect. Johann was more gifted in working with mathematical formulations and was blessed above all with a greater intuitive power and descriptive ability. Johann was appointed a professor at the University of Groningen in 1695, and in 1696 he proposed the problem to determine the curve of quickest descent between two given points, the brachistochrone. In connection with this he replied to the previous gibes of his brother with derisive insinuations. Jakob gave a solution (1697) that was closely related to that given by Leibniz (1697). It is based on the sufficient but not necessary condition that the extreme-value property of the curve in question (a common cycloid) is valid not only for the entire curve but also for all its parts. As a counterproblem Jakob set forth the so-called isoperimetric problem, the determination of that curve of given length between the points $A(-c; 0)$, $B(+c; 0)$ for which $\int_{-c}^{c} y^n dx$ takes a maximum value. Johann, in the *Histoire des ouvrages des savants* (1697), through a misunderstanding of the difficulty of the problem and of its

nature (calculus of variations), gave a solution based on a differential equation of the second degree. A differential equation of the third degree is necessary, however. After showing that a third-degree equation is required (1701), Jakob was able also to furnish the proof, which Johann and Leibniz had been seeking in vain, that the inexpansible and homogeneous catenary is the curve of deepest center of gravity between the points of suspension.

Johann Bernoulli may have comprehended the justification for his brother's argument soon after publication of the dissertation of 1701 (*Analysis magni problematis isoperimetrici*), but he remained silent. Only after Brook Taylor had adopted Jakob's procedure (1715) was he induced to accept Jakob's point of view. In the 1718 series of the *Mémoires* of the Paris Académie des Sciences, of which the brothers had been corresponding members since 1699, Johann gave a presentation, based on Jakob's basic ideas but improved in style and organization. It was not superseded until Leonhard Euler's treatment of the problems of variations (1744).

The antagonism between the brothers soon led to ugly critical remarks. In 1695 Jakob failed to appreciate the significance of Johann's extraordinarily effective series expansion (1694), which is based on iterated integration by parts and leads to a remainder in integral form. On the other hand, Johann, who in 1697 had challenged the criterion for geodetic lines on convex surfaces, complained in the following year that his brother knew how to solve the problem "only" on rotation surfaces. Other items of disagreement were the determination of elementary quadrable segments of the common cycloid and related questions. The brothers argued over this in print in 1699 and 1700. The formulas for the multisection of angles are connected with these problems. In his ingenious use of Wallis' incomplete induction (1656), Jakob presented $2 \cdot \cos n\alpha$ and $2 \cdot \sin n\alpha$ as functions of $2 \cdot \sin \alpha$. This is related to his notes from the winter of 1690–1691, in which, furthermore, the exponential series was derived from the binomial series in a bold but formally unsatisfactory manner.

Jakob Bernoulli's decisive scientific achievement lay not in the formulation of extensive theories, but in the clever and preeminently analytical treatment of individual problems. Behind his particular accomplishments there were, of course, notions of which Bernoulli was deeply convinced, primarily concerning continuity of all processes of nature (*natura non facit saltum*). Although Bernoulli assigned great significance to experimental research, he limited himself—for example, in investigations of mechanics—to a few basic facts to which he tried to cling and on which

48

he sought to base full theories. For this reason his final results were intellectually interesting, and as points of departure they were significant for further investigation by his contemporaries and subsequent generations. Naturally enough, they usually do not conform to more modern conclusions, which rest on far wider foundations. It is to be regretted, however, that Bernoulli's contributions to mechanics are hardly ever mentioned in the standard works.

The theory that seeks to explain natural phenomena by assuming collisions between particles of the ether, developed in the *Dissertatio de gravitate aetheris* (1683), of course does not mean much to a later generation. There are extensive discussions about the center point of oscillation, which had been determined correctly for the first time by Huygens in his *Horologium oscillatorium* (1673), but this was strongly debated by some of the members of the Cartesian school. On this subject Bernoulli expressed his opinions first in 1684, and then in more detail in 1686 and 1691; finally he succeeded in developing a proof from the properties of the lever (1703–1704). Important also is his last work, on the resistance of elastic bodies (1705). Supplementary material from his scientific diary is contained in the appendix to his *Opera*. Additional, but unpublished, material deals with the center of gravity of two uniformly moved bodies, the shape of a cord under the influence of several stretching forces, centrally accelerated motion (in connection with the statements of Newton in his *Principia* [1687]), and the line of action and the collective impulse of infinitely many shocks exerted on a rigid arc in the plane.

In the field of engineering belongs the 1695 treatment of the drawbridge problem (the curve of a sliding weight hanging on a cable that always holds the drawbridge in balance), stemming from Joseph Sauveur and investigated in the same year by L'Hospital and Johann Bernoulli. Leibniz was also interested in the problem. In Bernoulli's published remains, the contour is determined upon which a watch spring is to be developed so that the tension always remains the same for the movements of the watch.

The five dissertations in the *Theory of Series* (1682–1704) contain sixty consecutively numbered propositions. These dissertations show how Bernoulli (at first in close cooperation with his brother) had thoroughly familiarized himself with the appropriate formulations of questions to which he had been led by the conclusions of Leibniz in 1682 (series for $\pi/4$ and log 2) and 1683 (questions dealing with compound interest). Out of this there also came the treatise in which Bernoulli took into account short-term compound interest and was thus led to the exponential

series. He thought that there had been nothing printed concerning the theory of series up until that time, but he was mistaken: most conclusions of the first two dissertations (1689, 1692) were already to be found in Pietro Mengoli (1650), as were the divergence of the harmonic series (Prop. 16) and the sum of the reciprocals of infinitely many figurate numbers (Props. 17–20).

The so-called Bernoullian inequality (Prop. 4), $(1 + x)^n > 1 + nx$, is intended for $x > 0$, n as a whole number > 1. It is taken from Barrow's seventh lecture in the *Lectiones geometricae* (1670). Bernoulli would have been able to find algebraic iteration processes for the solution of equations (Props. 27–35) in James Gregory's *Vera . . . quadratura* (1667). The procedure of proof is still partially incomplete because of inadmissible use of divergent series. At the end of the first dissertation Bernoulli acknowledged that he could not yet sum up $\Sigma_{k=1}^{\infty} k^{-2}$ in closed form (Euler succeeded in doing so first in 1737); but he did know about the majorant $\Sigma_{k=1}^{\infty} 2k^{-1}(k + 1)^{-1}$, which can be summed in elementary terms. In Proposition 24 it is written that $\Sigma_{k=1}^{\infty} (2k - 1)^{-m}/\Sigma_{k=1}^{\infty} (2k)^{-m}$ equals $(2^m - 1)/1$ (m integer > 1), and that $\Sigma_{k=1}^{\infty} k^{-1/2}$ diverges more rapidly than $\Sigma_{k=1}^{\infty} k^{-1}$. Informative theses, based on Bernoulli's earlier studies, were added to the dissertations; and theses 2 and 3 of the second dissertation are based on the still incomplete classification of curves of the third degree according to their shapes into thirty-three different types.

The third dissertation was defended by Jakob Hermann, who wrote Bernoulli's obituary notice in *Acta eruditorum* (1706). In the introduction L'Hospital's *Analyse* is praised. After some introductory propositions, there appear the logarithmic series for the hyperbola quadrature (Prop. 42), the exponential series as the inverse of the logarithmic series (Prop. 43), the geometrical interpretation of

$$\sum_{k=1}^{\infty} k^{-2}x^k$$

(Prop. 44), and the series for the arc of the circle and the sector of conic sections (Props. 45, 46). All of these are carefully and completely presented with reference to the pertinent results of Leibniz (1682; 1691). In 1698 previous work was supplemented by Bernoulli's reflections on the catenary (Prop. 49) and related problems, on the rectification of the parabola (Prop. 41), and on the rectification of the logarithmic curve (Prop. 52).

The last dissertation (1704) was defended by Bernoulli's nephew, Nikolaus I, who helped in the publication of the *Ars conjectandi* (1713) and the reprint of the dissertation on series (1713) and became a prominent authority in the theory of series. In the dissertation Bernoulli first (Prop. 53) praises Wallis' interpolation through incomplete induction. In Proposition 54 the binomial theorem is presented, with examples of fractional exponents, as an already generally known theorem. Probably for this reason there is no reference to Newton's presentation in his letters to Leibniz of 23 June and 3 November 1676, which were made accessible to Bernoulli when they were published in Wallis' *Opera* (Vol. III, 1699). In proposition 55 the method of indeterminate coefficients appears, without reference to Leibniz (1693). Propositions 56–58 and 60 deal with questions related to the *elastica*.

In Proposition 59 it is stated that the series

$$\sum_{k=1}^{\infty} (-1)^{k+1} k^{-1}$$

for log 2 should be replaced by

$$\sum_{k=1}^{\infty} 2^{-k} k^{-1},$$

which converges more rapidly. From the letter to Leibniz of 2 August 1704, we know that in Proposition 59 Bernoulli used an idea of Jean-Christophe Fatio-de-Duillier (1656–1720), an engineer from Geneva, for the improvement of convergence. The procedure was expanded by Euler in the *Institutiones calculi differentialis* (1755) to his so-called series transformation. In the dissertations on series Bernoulli apparently wished to reproduce everything he knew about the subject. In this he was primarily concerned with the careful rendering of the results and not so much with originality.

The *Ars conjectandi* is Bernoulli's most original work, but unfortunately it is incomplete. The first part is basically a first-rate commentary on Huygens' *De ratiociniis in aleae ludo,* which was published as an appendix to van Schooten's *Exercitationes mathematicae* (1657). In the second part Bernoulli deals with the theory of combinations, based on the pertinent contributions of van Schooten (1657), Leibniz (1666), Wallis (1685), and Jean Prestet's *Élémens de mathématiques* (1675; 2nd ed., 1689). The chief result here is the rigid derivation of the exponential series through complete induction by means of the so-called Bernoullian numbers. In the third part Bernoulli gives twenty-four examples, some simple, some very complicated, on the expectation of profit in various games.

The fourth part contains the philosophical thoughts on probability that are especially characteristic of Bernoulli: probability as a measurable degree of certainty; necessity and chance; moral versus mathematical expectation; a priori and a posteriori probability; expectation of winning when the players are divided according to dexterity; regard of all available arguments, their valuation, and their calculable evaluation; law of large numbers, and reference to the *Art de penser* (*Logique de Port Royal,* Antoine Arnauld and Pierre Nicole, eds., 1662). The last section contains a penetrating discussion of *jeu de paume,* a complicated predecessor of tennis that was very popular. This part is Bernoulli's answer to the anonymous gibes occasioned by his debate of 1686 on scholarly logic.

Bernoulli's ideas on the theory of probability have contributed decisively to the further development of the field. They were incorporated in the second edition of Rémond de Montmort's *Essai* (1713) and were considered by Abraham de Moivre in his *Doctrine of Chances* (1718).

Bernoulli greatly advanced algebra, the infinitesimal calculus, the calculus of variations, mechanics, the theory of series, and the theory of probability. He was self-willed, obstinate, aggressive, vindictive, beset by feelings of inferiority, and yet firmly convinced of his own abilities. With these characteristics, he necessarily had to collide with his similarly disposed brother. He nevertheless exerted the most lasting influence on the latter.

Bernoulli was one of the most significant promoters of the formal methods of higher analysis. Astuteness and elegance are seldom found in his method of presentation and expression, but there is a maximum of integrity. The following lines taken from the *Ars conjectandi* (published posthumously in 1713) are not without a certain grace, however, and represent an early statement, made with wit and clarity, of the boundaries of an infinite series.

> Ut non-finitam Seriem finita cöercet,
>   Summula, & in nullo limite limes adest:
> Sic modico immensi vestigia Numinis haerent
>   Corpore, & angusto limite limes abest.
> Cernere in immenso parvum, dic, quanta voluptas!
>   In parvo immensum cernere, quanta, Deum!

> Even as the finite encloses an infinite series
>   And in the unlimited limits appear,
> So the soul of immensity dwells in minutia
>   And in narrowest limits no limits inhere.
> What joy to discern the minute in infinity!
>   The vast to perceive in the small, what divinity!

BIBLIOGRAPHY

I. ORIGINAL WORKS. Bernoulli's most famous single writing is *Ars conjectandi* (Basel, 1713; Brussels, 1968). His *Opera,* G. Cramer, ed. (Geneva, 1744; Brussels, 1968), contains all his scientific writings except the *Neuerfundene*

Anleitung, wie man den Lauff der Comet- oder Schwantz-sternen in gewisse grundmässige Gesätze einrichten und ihre Erscheinung vorhersagen könne (Basel, 1681), as well as a *Prognosticon.* Its contents were incorporated in the *Conamen novi systematis cometarum . . .* (Amsterdam, 1682), which is reproduced in the *Opera* as part 1.

Two collections of letters are printed: those to Leibniz in Leibniz' *Mathematische Schriften,* Vol. III, C. I. Gerhardt, ed. (Halle, 1855; Hildesheim, 1962); those to his brother Johann, in *Der Briefwechsel von Johann Bernoulli,* Vol. I, O. Spiess, ed. (Basel, 1955).

His MSS at the library of the University of Basel are *Reisebüchlein* (1676–1683); *Meditationes, annotationes, animadversiones* (1677–1705); *Stammbuch* (1678–1684); *Tabulae gnomicae. Typus locorum hypersolidorum,* which concerns classification of the curves of the third degree into thirty-three types; *Memorial über die Missbräuche an der Universität* (1691); and *De arte combinatoria* (1692) (all unpublished manuscripts); and "De historia cycloidis" (1701), in *Archiv für Geschichte der Mathematik und Naturwissenschaften,* **10** (1927–1928), 345 ff.

Bernoulli's unpublished manuscripts at the library of the University of Geneva are lectures on the mechanics of solid and liquid bodies, *Acta collegii experimentalis* (1683–1690), parts of which have been transcribed.

The collected works are in preparation. Included are Bernoulli's correspondence with Nicolas Fatio-de-Duillier (1700–1701) and with Otto Mencke (1686, 1689). The most important correspondence with L'Hospital and Pierre Varignon seems to have been lost.

Translations from the *Opera* are in the series Ostwald's Klassiker der Exacten Wissenschaften: *Unendliche Reihen* (1689–1704), translated into German by G. Kowalewski, no. 171 (Leipzig, 1909); and *Abhandlungen über das Gleichgewicht und die Schwingungen der ebenen elastischen Kurven von Jakob Bernoulli (1691, 1694, 1695) und Leonh. Euler (1744),* translated into German by H. Linsenbarth, no. 175 (Leipzig, 1910). Other translations are *Abhandlungen über Variations-Rechnung. Erster Theil: Abhandlungen von Joh. Bernoulli (1696), Jac. Bernoulli (1697) und Leonhard Euler,* translated into German by P. Stäckel, no. 46 (Leipzig, 1894; 2nd ed., 1914); and *Jakob Bernoulli: Wahrscheinlichkeitsrechnung,* translated into German by R. Haussner, nos. 107 and 108 (Leipzig, 1899). The latter was translated into English by Fr. Masères in his *Doctrine of Permutations and Combinations* (London, 1795) and in *Scriptores logarithmici,* Vol. III (London, 1796).

II. SECONDARY LITERATURE. Works concerning Bernoulli and his contributions to mathematics are P. Dietz, "The Origins of the Calculus of Variations in the Works of Jakob Bernoulli," in *Verhandlungen der Naturforschenden Gesellschaft in Basel,* **70** (1959), 81–146, a dissertation presented at the University of Mainz in 1958; J. O. Fleckenstein, *Johann und Jakob Bernoulli* (Basel, 1949), which is supp. 6 to the journal *Elemente der Mathematik;* J. E. Hofmann, *Uber Jakob Bernoullis Beiträge zur Infinitesimalmathematik* (Geneva, 1956), no. 3 in the series *Monographies de l'Enseignement Mathématique;* and O. Spiess, "Bernoulli," in *Neue deutsche Biographie,* II (Berlin, 1955), 128–129,

and "Jakob Bernoulli," *ibid.,* pp. 130–131, which include supplementary bibliographical material.

The poem "On Infinite Series," Helen M. Walker, trans., appears in D. E. Smith, *A Source Book in Mathematics;* repr. by permission Harvard University Press.

J. E. HOFMANN

BERNOULLI, JAKOB (JACQUES) II (*b.* Basel, Switzerland, 17 October 1759; *d.* St. Petersburg, Russia, 15 August 1789), *mathematics.*

Jakob II was Johann II's most gifted son. He graduated in jurisprudence in 1778 but successfully engaged in mathematics and physics. In 1782 he presented a paper (6) to support his candidacy for the chair of his uncle Daniel. The decision (made by drawing lots) was against him, however, and he traveled as secretary of the imperial envoy to Turin and Venice, where he received a call to St. Petersburg. There he married a granddaughter of Euler and published several treatises (2–4) at the Academy. When only thirty years old, he drowned while swimming in the Neva.

BIBLIOGRAPHY

Jakob's writings include (1) "Lettre sur l'élasticité," in *Journal de physique de Rozier,* **21** (1782), 463–467; (2) "Considérations hydrostatiques," in *Nova acta Helvetica,* **1,** 229–237; (3) "Dilucidationes in Comment. L. Euleri de ictu glandium contra tabulam explosarum," in *Nova acta Petropolitana,* **4** (1786), 148–157; (4) "De motu et reactione aquae per tubos mobiles transfluentis," *ibid.,* **6** (1788), 185–196; (5) "Sur l'usage et la théorie d'une machine, qu'on peut nommer instrument ballistique," in *Mém. Acad. Berl.* (1781), pp. 347–376; and (6) *Theses physicae et physico-mathematicae quas vacante cathedra physica die 28 Maii 1782 defendere conabitur Jacobus Bernoulli* (Basel, 1782).

J. O. FLECKENSTEIN

BERNOULLI, JOHANN (JEAN) I (*b.* Basel, Switzerland, 6 August 1667; *d.* Basel, 1 January 1748), *mathematics.*

The tenth child in the family, Johann proved unsuited for a business career, much to his father's sorrow. He therefore received permission in 1683 to enroll at his native city's university, where his brother Jakob (or Jacques), who was twelve years older and who had recently returned from the Netherlands, lectured as *magister artium* on experimental physics. In 1685 Johann, respondent to his brother in a logical disputation, was promoted to *magister artium* and began the study of medicine. He temporarily halted his studies at the licentiate level in 1690, when his first publication appeared, a paper on fermentation processes.[1] (His doctoral dissertation of 1694[2] is a mathematical work despite its medical subject, and reflects the influence of the iatromathematician Borelli.)

Bernoulli privately studied mathematics with the gifted Jakob, who in 1687 had succeeded to the vacant chair of mathematics at the University of Basel. From about this time, both brothers were engrossed in infinitesimal mathematics and were the first to achieve a full understanding of Leibniz' abbreviated presentation of differential calculus.[3] The extraordinary solution[4] of the problem of *catenaria* posed by Jakob Bernoulli (*Acta eruditorum,* June 1691) was Johann's first independently published work, and placed him in the front rank with Huygens, Leibniz, and Newton. Johann spent the greater part of 1691 in Geneva. There he taught differential calculus to J. C. Fatio-de-Duillier (whose brother Nicolas later played a not very praiseworthy role in the Leibniz-Newton priority dispute) and worked on the deepening of his own mathematical knowledge.

In the autumn of 1691 Bernoulli was in Paris, where he won a good place in Malebranche's mathematical circle as a representative of the new Leibnizian calculus, and did so by virtue of a "golden theorem" (stemming actually from Jakob)—the spectacular determination of the radius of curvature of a curve by means of the equation $\rho = dx/ds : d^2y/ds^2$. During this period he also met L'Hospital, then probably France's most gifted mathematician. "Grandseigneur of the science of mathematics"—he corresponded also with Huygens—L'Hospital engaged Bernoulli to initiate him into the secrets of the new infinitesimal calculus. The lessons were given in Paris and sometimes in L'Hospital's country seat at Oucques, and Bernoulli was generously compensated. L'Hospital even induced Bernoulli to continue, for a considerable fee, these lessons by correspondence after the latter's return to Basel. This correspondence[5] subsequently became the basis for the first textbook in differential calculus,[6] which assured L'Hospital's place in the history of mathematics. (Bernoulli's authorship of this work, which was still doubted by Cantor,[7] has been substantiated by the Basel manuscript of the *Differential Calculus* discovered in 1921 by Schafheitlin,[8] as well as by Bernoulli's correspondence with L'Hospital.[9])

In 1692 Bernoulli met Pierre de Varignon, who later became his disciple and close friend. This tie also resulted in a voluminous correspondence.[10] In 1693 Bernoulli began his exchange of letters with Leibniz, which was to grow into the most extensive correspondence ever conducted by the latter.

Bernoulli's most significant results during these years were published in the form of numerous memoirs in *Acta eruditorum* (*AE*) and shorter papers in the *Journal des Sçavans* (*JS*). Bernoulli's two most important achievements were the investigations concerning the function $y = x^x$ and the discovery, in 1694, of a general development in series by means of repeated integration by parts, the series subsequently named after him:

$$\int_0^x y\,dx = xy - \frac{x^2}{2!}y' + \frac{x^3}{3!}y'' - + \cdots$$

(cf. *Addidamentum AE,* 1694, letter to Leibniz of 2 September 1694). This series—whose utility, incidentally, Jakob Bernoulli failed to recognize—is based on the general Leibnizian principle for the differentiation of a product:

$$d^m[f(x)g(x)] = (df + dg)^{(m)} = \sum_{\nu=0}^{m}\binom{m}{\nu}d^{m-\nu}f d^\nu g.$$

This formalism is characteristic of a large part of the Bernoulli-Leibniz correspondence between 1694 and 1696.

Integration being viewed as the inverse operation of differentiation, Bernoulli worked a great deal on the integration of differential equations. This view was generally accepted in the Leibniz circle. In Paris he had already demonstrated the efficacy of Leibniz' calculus by an anonymous solution of "Debeaune's problem" (*JS,* 1692), which had been put to Descartes as the first inverse tangent problem. Five years later he demonstrated that with the aid of the calculus much more complex differential equations could be solved. In connection with Debeaune's problem, Jakob Bernoulli had proposed the general differential equation since called by his name,

$$y' + P(x)y + Q(x)y^n = 0,$$

and had solved it in a rather cumbersome way. Johann, more flexible with regard to formalism, solved this equation by considering the desired final function as the product of two functions, $M(x)$ and $N(x)$. In the resulting equation,

$$\frac{dM}{M} + \frac{dN}{N} + P(x)\,dx + (MN)^{n-1}Q(x)\,dx = 0,$$

the arbitrariness of the functions M and N makes it possible to subject one of them (e.g., M) to the secondary condition

$$\frac{dM}{M} + P(x)\,dx = 0,$$

resulting in $M = \exp[-\int P(x)\,dx]$. This substitution promptly leads to a linear differential equation in N.

Bernoulli's "exponential calculus" is nothing other than the infinitesimal calculus of exponential functions. Nieuwentijt, in a paper criticizing the lack of logical foundations in Leibniz' calculus,[11] pointed out the inapplicability of Leibniz' published differentiation methods to the exponential function x^y. Thereupon Bernoulli developed, in "Principia calculi exponentialium seu percurrentium" (*AE*, 1697), the "exponential calculus," which is based on the equation

$$d(x^y) = x^y \log x\, dy + yx^{y-1}\, dx.$$

Also in 1695 came Bernoulli's summation of the infinite harmonic series

$$\sum_{K=1}^{n} (-1)^{K-1} K^{-1} \binom{n}{K}$$

from the difference scheme, the development of the addition theorems of trigonometric and hyperbolic functions from their differential equations, and the geometric generation of pairs of curves, wherein the sum or difference of the arc lengths can be represented by circular arcs. Neither Johann nor Jacob Bernoulli succeeded in mastering the problem, originated by Mengoli, of the summation of reciprocal squares ($\sum_{k=1}^{\infty} 1:k^2$). This problem was solved only by Johann's greatest pupil, Leonhard Euler.[12]

In 1695 Bernoulli was offered both a professorship at Halle, and, through the intervention of Huygens, the chair of mathematics at Groningen. He eagerly accepted the latter offer, particularly since his hopes of obtaining a chair in Basel were nil as long as his brother Jakob was alive. On 1 September 1695 he departed for Holland with his wife (the former Dorothea Falkner) and seven-month-old Nikolaus, his first son, not without resentment against Jakob, who had begun to retaliate for Johann's earlier boastfulness when he solved the differential equation of the velaria (*JS*, 1692): Jakob termed Johann his pupil, who after all could only repeat what he had learned from his teacher. This cutting injustice was promptly paid back by Johann, now his equal in rank.

In June 1696, Johann posed (in *AE*) the problem of the brachistochrone, i.e., the problem of determining the "curve of quickest descent." Since no solution could be expected before the end of the year, Bernoulli, at Leibniz' request, republished the problem in the form of a leaflet dedicated to *acutissimis qui in toto orbe florent mathematicis* ("the shrewdest mathematicians of all the world") and fixed a six-month limit for its solution. Leibniz solved the problem on the day he received Bernoulli's letter, and correctly predicted a total of only five solutions:

from the two Bernoullis, Newton, Leibniz, and L'Hospital. (It should be noted that it was only through Johann's assistance—by correspondence—that L'Hospital had arrived at his solution.)

This problem publicly demonstrated the difference in the talents of the two brothers. Johann solved the problem by ingenious intuition, which enabled him to reduce the mechanical problem to the optical problem already resolved by means of Fermat's principle of least time. He deduced the differential equation of the cycloid from the law of refraction. Jakob, on the other hand, furnished a detailed but cumbersome analysis, and came upon the roots of a new mathematical discipline, the calculus of variations. Unlike Jakob, Johann failed to perceive that such extreme-value problems differed from the customary ones in that it was no longer the unknown extreme values of a function that were to be determined, but functions that made a certain integral an extreme.[13]

In connection with his solution of the brachistochrone problem, Jakob (*AE*, May 1697) posed a new variational problem, the isoperimetric problem.[14] Johann underestimated the complexity of this problem by failing to perceive its variational character; and he furnished an incomplete solution (wherein the resulting differential equation is one order too low) in *Histoire des ouvrages des savants* (VI, 1697), and thereby brought on himself the merciless criticism of his brother.[15] This was the beginning of alienation and open discord between the brothers—and also the birth of the calculus of variations. A comparison of Jakob's solution (Basel, 1701; *AE*, May 1701) with Johann's analysis of the problem (which he presented through Varignon to the Paris Academy on 1 February 1701) clearly shows Johann's to be inferior. Nevertheless, Jakob was not able to enjoy his triumph, since—for reasons that remain mysterious—the sealed envelope containing Johann's solution was not opened by the Academy until 17 April 1706, the year following Jakob's death.

Soon after publication of Jakob's *Analysis magni problematis isoperimetrici* (1701), Johann must have felt that his brother's judgment was valid, although he never said so. Only after having been stimulated by Taylor's *Methodus incrementorum* (1715) did he produce a precise and formally elegant solution of the isoperimetric problem along the lines of Jakob's ideas (*Mémoires de l'Académie des sciences,* 1718). The concepts set forth in this paper contain the nucleus of modern methods of the calculus of variations. Also in this connection Bernoulli made a discovery pertaining to the variational problem of geodetic lines on convex surfaces: in a letter addressed to Leibniz, dated 26 August 1698, he perceived the characteristic property of geodetic lines, i.e., three consecutive points determine a normal plane of the surface.

Bernoulli's studies on the determination of all rationally quadrable segments of the common cycloid —the "fateful curve of the seventeenth century" (*AE,* July 1699)—in connection with the cyclotomic equation (*AE,* April 1701; more detailed in his correspondence with Moivre[16])—resulted in a systematic treatment of the integrals of rational functions by means of resolution into partial fractions. The general advance in algebraic analysis under Bernoulli's influence is evident in the typical case of the relation

$$2i \arctan x = \log \frac{x - i}{x + i}.$$

Nevertheless, Bernoulli had not yet perceived that such logarithmic expressions may take on infinitely many values.

Immediately after Jakob's death, Johann succeeded him in Basel, although he would undoubtedly have preferred to accept the repeated invitations extended to him by the universities of Utrecht and Leiden (see correspondence of the rector of Utrecht University, Pieter Burman, with Bernoulli's father-in-law, Falkner[17]). Family circumstances, however, caused him to settle in Basel.

Bernoulli's criticism of Taylor's *Methodus incrementorum* was simultaneously an attack upon the method of fluxions, for in 1713 Bernoulli had become involved in the priority dispute between Leibniz and Newton. Following publication of the Royal Society's *Commercium epistolicum* in 1712, Leibniz had no choice but to present his case in public. He released —without naming names—a letter by Bernoulli (dated 7 June 1713) in which Newton was charged with errors stemming from a misinterpretation of the higher differential. Thereupon Newton's followers raised complicated analytical problems, such as the determination of trajectories and the problem of finding the ballistic curve, which Newton had solved only for the law of resistance $R = av$ ($R =$ resistance, $a =$ constant, $v =$ velocity). Bernoulli solved this problem (*AE,* 1719) for the general case ($R = av^n$), thus demonstrating the superiority of Leibniz' differential calculus.

After Newton's death in 1727, Bernoulli was unchallenged as the leading mathematical preceptor to all Europe. Since his return to Basel in 1705, he had devoted himself—in the field of applied mathematics—to theoretical and applied mechanics. In 1714 he published his only book, *Théorie de la manoeuvre des vaisseaux.* Here Bernoulli (as Huygens had done before him) criticizes the navigational theories ad-

vanced in 1679 by the French naval officer Bernard Renau d'Eliçagaray (1652–1719), a friend of Varignon's. In this book Bernoulli exposed the confusion in Cartesian mechanics between force and *vis viva* (now kinetic energy). On 26 February 1715—and not 1717, as stated in the literature because of a printing error in Varignon's *Nouvelle mécanique* (1725)—Bernoulli communicated to Varignon the principle of virtual velocities for the first time in analytical form. In modern notation it is

$$\partial A = \sum_{i=1}^{n} \vec{K}_i \, \partial \vec{s}_i = 0.$$

Since this principle can be derived from the energy principle $A + mv^2/2 = $ const., which Bernoulli applied several times to conservative mechanical systems of central forces, he considered it a second general principle of mechanics—which, however, he had demonstrated only for the statical case. For central forces, Bernoulli applied the *vis viva* equation to the inverse two-body problem, which he for the first time expressed in the form used today for the equation of the orbit (*Mémoires de l'Académie des sciences,* 1710):

$$\varphi(r) = \varphi\left(\frac{1}{u}\right) = \vartheta(u) = \int \frac{c \, du}{\sqrt{2(u + h) - u^2 c^2}} \, .$$

For the corresponding problem of centrally accelerated motion in a resisting medium (*ibid.,* 1711), he solved the differential equation

$$\frac{a}{\rho} \frac{dv}{v} \frac{ds}{d\theta} \pm av^{n-2} \, ds + \frac{dv}{v} = 0$$

($\rho = $ radius of curvature of the orbit) on the premise that $v = M(r)N(r)$, and determined the central force, in accordance with Huygens' formula, from

$$\rho(r) = \frac{v^2}{\rho} \frac{ds}{r d\theta} \, .$$

Newton severely criticized the Cartesian vortex theory in Book II of the *Principia.* Bernoulli's advocacy of the theory delayed the acceptance of Newtonian physics on the Continent. In three prize-winning papers, Bernoulli treated the transmission of momentum (1727), the motions of the planets in aphelion (1730), and the cause of the inclination of the planetary orbits relative to the solar equator (1735). Bernoulli's 1732 work on hydraulics (*Opera,* IV) was generally considered a piece of plagiarism from the hydrodynamics of his son Daniel. Nevertheless, Bernoulli did try to manage without Daniel's formulation of the principle of *vis viva.*

Bernoulli also worked in experimental physics. In several papers (*Mémoires de l'Académie des sciences,* 1701; Basel, 1719), he investigated the phenomenon of the luminous barometer within the framework of contemporary Cartesian physics, although he was unable to furnish a sufficient explanation for the electrical phenomenon of triboluminescence discovered by Picard.

Bernoulli was a member of the royal academies of Paris and Berlin, of the Royal Society, of the St. Petersburg Academy, and the Institute of Bologna. As son-in-law of Alderman Falkner, he not only enjoyed social status in Basel, but also held honorary civic offices there. He became especially well known as a member of the school board through his efforts to reform the humanistic Gymnasium. His temperament might well have led him to a career in politics, but instead it only involved him in scientific polemics with his brother Jakob and in the Leibniz-Newton priority dispute. Even abroad he was unable to curb his "Flemish pugnacity." In 1702, as professor in Groningen, he became involved in quarrels with the theologians, who in turn, because of his views in natural philosophy, accused him of what was then the worst of heresies, Spinozism.

Bernoulli's quarrelsomeness was matched by his passion for communicating. His scientific correspondence comprised about 2,500 letters, exchanged with some 110 scholars.

NOTES

1. *De effervescentia et fermentatione.*
2. *De motu musculorum.*
3. Leibniz, *Nova methodus de maximis et minimis.*
4. J. E. Hofmann, "Vom öffentlichen Bekanntwerden der Leibniz'schen Infinitesimalmathematik."
5. O. Spiess, ed., *Der Briefwechsel von Johann Bernoulli.*
6. L'Hospital, *Analyse des infiniment petits.*
7. Cantor, *Vorlesungen über Geschichte der Mathematik.*
8. *Lectiones de calculo differentialium,* MS Universitätsbibliothek, Basel.
9. O. J. Rebel, *Der Briefwechsel zwischen Johann Bernoulli und dem Marquis de l'Hôpital.*
10. E. J. Fedel, *Johann Bernoullis Briefwechsel mit Varignon aus den Jahren 1692–1702.*
11. *Considerationes secundae circa calculi differentialis principia.*
12. O. Spiess, *Die Summe der reziproken Quadratzahlen.*
13. P. Dietz, *Die Ursprünge der Variationsrechnung bei Jakob Bernoulli;* J. E. Hofmann, *Ueber Jakob Bernoullis Beiträge zur Infinitesimalmathematik.*
14. J. O. Fleckenstein, *Johann und Jakob Bernoulli.*
15. Hofmann, *Ueber Jakob Bernoullis Beiträge zur Infinitesimalmathematik.*
16. K. Wollenschlaeger, *Der mathematische Briefwechsel zwischen Johann I Bernoulli und Abraham de Moivre.*
17. O. Spiess, ed., *Der Briefwechsel von Johann Bernoulli.*

BIBLIOGRAPHY

I. ORIGINAL WORKS. Among Bernoulli's writings are *De effervescentia et fermentatione* (Basel, 1690); *De motu musculorum* (Basel, 1694), his doctoral dissertation; and *Lectiones de calculo differentialium,* MS in library of Univ. of Basel, also trans. and ed. by P. Schafheitlin as no. 211 in Ostwald's Klassiker der Exakten Wissenschaften (Leipzig, 1924). His works were collected as *Opera Johannis Bernoullii,* G. Cramer, ed., 4 vols. (Geneva, 1742). For his correspondence, see Bousquet, ed., *Commercium philosophicum et mathematicum G. Leibnitii et Joh. Bernoullii* (Geneva, 1745); his correspondence with Euler in *Bibliotheca mathematica,* 3rd ser., **4** (1903)–**6** (1905); E. J. Fedel, *Johann Bernoullis Briefwechsel mit Varignon aus den Jahren 1692–1702* (Heidelberg, 1934), dissertation; O. J. Rebel, *Der Briefwechsel zwischen Johann Bernoulli und dem Marquis de l'Hôpital* (Heidelberg, 1934), dissertation; O. Spiess, ed., *Der Briefwechsel von Johann Bernoulli,* I (Basel, 1955); and K. Wollenschlaeger, *Der mathematische Briefwechsel zwischen Johann I Bernoulli und Abraham de Moivre* (Basel, 1932), dissertation, publ. separately in *Verhandlungen der Naturforschenden Gesellschaft in Basel.* Handwritten material is in the library of the University of Basel.

II. SECONDARY LITERATURE. Writings on Bernoulli, on his work, or on background material are Jakob Bernoulli, *Analysis magni problematis isoperimetrici* (Basel, 1701); M. Cantor, *Vorlesungen über Geschichte der Mathematik,* 2nd ed., III (Leipzig, 1901), 207–233; C. Carathéodory, "Basel und der Beginn der Variationsrechnung," in *Festschrift zum 60. Geburtstag von Andreas Speiser* (Zurich, 1945), pp. 1–18; P. Dietz, *Die Ursprünge der Variationsrechnung bei Jakob Bernoulli* (Basel, 1959), dissertation, Univ. of Mainz; J. O. Fleckenstein, "Varignon und die mathematischen Wissenschaften im Zeitalter des Cartesianismus," in *Archives d'histoire des sciences* (1948); and *Johann und Jakob Bernoulli* (Basel, 1949), supp. no. 6 of *Elemente der Mathematik;* J. E. Hofmann, *Ueber Jakob Bernoullis Beiträge zur Infinitesimalmathematik,* no. 3 in the series Monographies de l'Enseignement Mathématique (Geneva, 1956); "Vom öffentlichen Bekanntwerden der Leibniz'schen Infinitesimalmathematik," in *Sitzungsberichte der Oesterreichischen Akademie der Wissenschaften,* no. 8/9 (1966), 237–241; and "Johann Bernoulli, Propagator der Infinitesimalmethoden," in *Praxis der Mathematik,* **9** (1967/1968), 209–212; Guillaume de L'Hospital, *Analyse des infiniment petits* (Paris, 1696); G. Leibniz, "Nova methodus de maximis et minimis," in *Acta eruditorum* (Oct. 1684); B. Nieuwentijt, *Considerationes secundae circa calculi differentialis principia* (Amsterdam, 1696); A. Speiser, "Die Basler Mathematiker," *Neujahrsblatt der G.G.G.,* no. 117 (Basel, 1939); O. Spiess, "Johann B. und seine Söhne," in *Atlantis* (1940), pp. 663 ff.; "Die Summe der reziproken Quadratzahlen," in *Festschrift zum 60. Geburtstag von Andreas Speiser* (Zurich, 1945), pp. 66 ff.; *Die Mathematiker Bernoulli* (Basel, 1948).

E. A. FELLMANN
J. O. FLECKENSTEIN

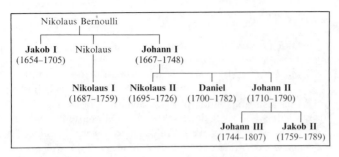

The Bernoullis. Names in boldface are discussed in articles.

BERNOULLI, JOHANN (JEAN) II (*b.* Basel, Switzerland, 28 May 1710; *d.* Basel, 17 July 1790), *mathematics.*

Johann II was perhaps the most successful of Johann I's sons, inasmuch as he succeeded his father in the chair of mathematics after having previously taught rhetoric. In 1727 he obtained the degree of doctor of jurisprudence (1). Subsequently he won the prize of the Paris Academy four times, either by himself or with his father (2–5)—undoubtedly sufficient qualification to make him Johann I's successor. But thereafter his mathematical production dwindled to occasional academic papers and a treatise (6), although he lived to be almost as old as his father. His shyness and frail constitution did not, however, prevent him from engaging in extensive scientific correspondence (about 900 items) and from furthering the publication, in four volumes, of his father's *Opera omnia.* He personified the mathematical genius of his native city in the second half of the eighteenth century. In 1756, after resigning as president of the Berlin Academy, Maupertuis found refuge with him in Basel, where he died in 1759.

BIBLIOGRAPHY

Bernoulli's writings include (1) *De compensationibus* (Basel, 1729), dissertation for the doctor of jurisprudence; (2) "Recherches physiques et géométriques sur la question: Comment se fait la propagation de la lumière," in *Recueil des pièces qui ont remporté les prix de l'Académie royale des sciences,* III (1736); (3) "Discours sur les ancres," *ibid.* (1737); (4) "Discours sur le cabestan," *ibid.,* V (1741); (5) "Nouveaux principes de mécanique et de physique tendans à expliquer la nature et les propriétés de l'Aiman," *ibid.,* V (1743); and (6) "Réponse à une lettre anonyme sur la figure de la terre," in *Journal Helvét.* (1740), pp. 219 *et seq.*

J. O. FLECKENSTEIN

BERNOULLI, JOHANN (JEAN) III (*b.* Basel, Switzerland, 4 November 1744; *d.* Berlin, Germany, 13 July 1807), *mathematics, astronomy.*

The most successful of the sons of Johann II—although his mathematical achievements were insignificant—Johann was a universally knowledgeable child prodigy. At fourteen he obtained the degree of master of jurisprudence, and at twenty he was invited by Frederick II to reorganize the astronomical observatory at the Berlin Academy. His frail health and his encyclopedic inclinations hampered him in his practical scientific activities, however. His treatises are of no particular interest. On the other hand, his travel accounts (1772–1776; 1777–1779; 1781) had a great cultural and historical impact. With Hindenburg he published the *Leipziger Magazin für reine & angewandte Mathematik* from 1776–1789.

Johann was entrusted with the administration of the mathematical estate of the Bernoulli family. The major part of the correspondence was sold to the Stockholm Academy; and its existence there was overlooked until his letters were rediscovered by Gylden at the Stockholm Observatory in 1877. His correspondence, comprising about 2,800 items, exceeded that of Johann I.

BIBLIOGRAPHY

Johann's writings include various essays in *Mém. Acad. Berlin* (1766–1775), as well as astronomical observations and computations, 1767–1807, in *Neue Berliner Ephemeriden* and *Bodes astronomisches Jahrbuch.* Other works are *Recueil pour les astronomes,* 3 vols. (Berlin, 1772–1776); *Liste des astronomes connus actuellement* (Berlin, 1776); *Lettres écrites pendant la cours d'un voyage par l'Allemagne 1774/75,* 3 vols. (Berlin, 1777–1779); "Essai d'une nouvelle méthode de déterminer la diminution séculaire de l'obliquité de l'écliptique," in *Mém. Acad. Berlin* (1779), pp. 211–242; and *Lettres astronomiques* (Berlin, 1781).

J. O. FLECKENSTEIN

BERNOULLI, NIKOLAUS I (*b.* Basel, Switzerland, 21 October 1687; *d.* Basel, 29 November 1759), *mathematics.*

The son of Nikolaus Bernoulli, a Basel alderman and painter, Nikolaus I studied with his two uncles, Jakob I and Johann I, and made rapid progress in mathematics. As early as 1704, studying under Jakob I, he obtained his master's degree by defending Jakob's last thesis on infinite series (1), in which quadratures and rectifications are determined by means of series expansions, arrived at by the method of undetermined coefficients or by interpolation, after Wallis, for binomial expansions. In 1709 he obtained the degree of doctor of jurisprudence (2) with a dissertation on the application of the calculus of probability to questions of law. In 1712 Nikolaus undertook a journey to Holland, England, and

France, where he met Montmort, later his friend and collaborator. He became a member of the Berlin Academy in 1713, of the Royal Society in 1714, and of the Academy of Bologna in 1724. In 1716 he succeeded Hermann as professor of mathematics in Padua, but in 1722 he returned to Basel in order to accept the chair of logic, which he exchanged in 1731 for a professorship in law. He served four times as rector of the University of Basel.

Nikolaus was a gifted but not very productive mathematician. As a result, his most important achievements are hidden throughout his correspondence, which comprises about 560 items. The most important part of his correspondence with Montmort (1710–1712) was published in the latter's *Essai d'analyse sur les jeux de hazard* (2nd ed., Paris, 1713). Here Nikolaus formulated for the first time the problem of probability theory, later known as the St. Petersburg problem.

In his correspondence with Leibniz (1712–1716), Nikolaus discussed questions of convergence and found that the binomial expansion $(1 + x)^n$ diverges for $x > 1$. In his letters to Euler (1742–1743) he criticized Euler's indiscriminate use of divergent series. In this correspondence he also solved the problem of the sum of reciprocal squares $\sum_{\nu=1}^{\infty} 1/\nu^2 = \pi^2/6$, which had confounded Leibniz and Jakob I. His personal copy of the *Opera omnia* of his uncle Jakob, which he had published, contains the proof, which does not require the help of analytical methods.

To his edition of his uncle's *Opera,* he added as an appendix thirty-two articles from Jakob I's diary ("Annotationes et meditationes"). His concern with editing his uncle's works went back to at least 1713, when he published the *Ars conjectandi.*

In the priority quarrel with Newton, Nikolaus sided with his uncle Johann in defending the interests of Leibniz. It was he who pointed out Newton's misunderstanding of the higher-order derivatives (3), which had caused Newton's errors with the inverse problem of central force in a resisting medium (5). He also considered the problem of orthogonal trajectories (6) and Riccati's differential equation (6, 8, 10–12).

BIBLIOGRAPHY

Jakob I's last thesis on infinite series is (1) *De seriebus infinitis earumque usu in quadraturis spatiorum et rectificationibus curvarum* (Basel, 1704). Nikolaus' writings include (2) *De usu artis conjectandi in jure* (Basel, 1709); (3) "Addition au Mém. de Mr. Jean Bernoulli touchant la manière de trouver les forces centrales dans les milieux résistans . . . ," in *Mémoires de l'Académie des Sciences*

(1711), pp. 53–56; (4) "Solutio generalis problematis 15 propositi a D. de Moivre in Transactiones de mensura sortis," in *Philosophical Transactions of the Royal Society,* **29** (1714), 133–144; (5) "Calculus pro invenienda linea curva, quam describit projectile in medio resistente," in *Acta eruditorum* (1719), 224–226; (6) "Modus inveniendi aequationem differentialem completam ex data aequatione differentiali incompleta . . . ," *ibid.,* supp. **7** (1719), pp. 310–859; (7) "Tentamen solutionis generalis problematis de construenda curva, quae alias ordinatim positione datas ad angulos rectos secat," *ibid.* (1719), pp. 295–304; (8) "Novum theorema pro integratione aequationum differentialium secundi gradus, quae nullam constantem differentialem supponunt," *ibid.,* supp. **9** (1720); (9) *Theses logicae de methodo analytica et synthetica* (Basel, 1722); (10) "Annotazioni sopra lo schediasma del Conte Jacopo Riccati etc. coll'annessa soluzione propria del problema inverso delle forze centrali . . . ," in *Giornale de letterati d'Italia,* **20,** 316–351; (11) "Dimostrazione analitica di un teorema, il qual serve per la soluzione del problema proposto nel T.XX. del Giorn. Lett. Ital. . . . ," *ibid.,* **29,** 163–171; and (12) "Òsservazione intorno al teorema proposto dal Conte Jacopo de Fagnano," *ibid.,* pp. 150–163. See also *Athenae Rauricae* (Basel, 1778), pp. 148–151.

J. O. FLECKENSTEIN

BERNOULLI, NIKOLAUS II (*b.* Basel, Switzerland, 6 February 1695; *d.* St. Petersburg, Russia, 31 July 1726), *mathematics.*

Nikolaus was the favorite son of Johann I, whose mediation made it possible for him to enter the University of Basel at the age of thirteen. There he passed the master's examinations at sixteen, and in 1715 he became a licentiate in jurisprudence (1). Nikolaus assisted his father with his correspondence, particularly in the priority quarrel between Leibniz and Newton, during which he drafted the reply to Taylor (6) and supplied valuable contributions to the problem of trajectories (2–4). With his brother Daniel he traveled in France and Italy, where both received and accepted appointments to the St. Petersburg Academy. Within the year, however, he contracted and died of a hectic fever.

BIBLIOGRAPHY

Nikolaus' writings are (1) *Dissertatio de jure detractionis* (Basel, 1715); (2) "Solutio problematis invenire lineam, quae ad angulos rectos secet omnes hyperbolas ejusdem verticis et ejusdem centri," in *Acta eruditorum* (1716), pp. 226–230; (3) "Exercitatio geometrica de trajectoriis orthogonalibus . . . ," *ibid.* (1720), pp. 223–237; (4) "De trajectoriis curvas ordinatim positione datas ad angulos rectos vel alia lege secantibus," *ibid.* (1718), pp. 248–262; (5) "Animadversiones in Jac. Hermanni solutionem propriam duorum problematum geometricorum . . . ," *ibid.,* supp. **8** (1720), pp. 372–389; (6) "Responsio ad Taylori Angli querelas . . . ," *ibid.* (1720), pp. 279–285; (7) "Analysis aequationum quarundam differentialium," in *Comment. Acad. Petrop.*, **1** (1728), 198–207; and (8) "De motu corporum ex percussione," *ibid.*, pp. 121–126.

J. O. FLECKENSTEIN

BERNSTEIN, FELIX (*b.* Halle, Germany, 24 February 1878; *d.* Zurich, Switzerland, 3 December 1956) *mathematics.*

Bernstein's father, Julius, who wrote on electrobiology, studied under DuBois Reymond. Felix studied in Halle with Georg Cantor, a friend of his father, then went to Göttingen to study with Hilbert and Klein. In 1896 he took his Abitur in Halle, then taught mathematics and studied physiology there. He received his Ph.D. at Göttingen in 1901 and his Habilitation in Halle in 1907. He returned to Göttingen in 1911 as associate professor of mathematical statistics. After military service in World War I, he headed the statistical branch of the Office of Rationing and in 1921 became commissioner of finance. Also in 1921 he became full professor and founded the Institute of Mathematical Statistics, where he was director until 1934. He emigrated to the United States, where he became a citizen in 1940. He taught at Columbia, New York, and Syracuse universities. In 1948 he returned to Göttingen.

It was in 1895 or 1896 while a student at the Gymnasium that Bernstein gave the first proof of the equivalence theorem of sets. If each of two sets, A and B, is equivalent to a subset of the other, then A is equivalent to B. This theorem establishes the notion of cardinality and is thus the central theorem in set theory. It bears some similarity to the Eudoxean definition of equal irrationals.

Cantor, who had been working on the equivalence problem, had left for a holiday and Bernstein had volunteered to correct proofs of his book on transcendental numbers. In that interval, the idea for a solution came to Bernstein one morning while shaving. Cantor then worked with the approach for several years before formulating it to his satisfaction. Cantor always gave full credit to Bernstein, who meanwhile had become a student of fine arts at Pisa. He was persuaded to return to mathematics by two professors there who had heard Cantor expound the equation at a mathematical congress. Bernstein retained this interest in the arts, particularly painting and sculpture, throughout his life.

Bernstein's subsequent work in pure and applied mathematics shows great versatility, and includes some of the earliest applications of set theory outside pure mathematics, contributions to isoperimetric problems, convex functions, the Laplace transform, and number theory, as well as set theory itself.

Toward the 1920's Bernstein became increasingly interested in the mathematical treatment of questions in genetics; he was to contribute decisively to the development of population genetics in the analysis of modes of inheritance. The discovery of human blood groups had made possible an entirely new approach to human genetics. In 1924 Bernstein was able to show that the A, B, and O blood groups are inherited on the basis of a set of triple alleles, and not on the basis of two pairs of genes, as had been thought. He compared a population genetic analysis of the frequencies of the four blood groups—numerous records of racially variant blood-group frequencies had been available since the discovery of this phenomenon by L. and H. Hirschfeld—with the expectations for the blood-group frequencies according to the expanded Hardy-Weinberg formula $p^2 : 2pq : q^2$, and found significant and consistent differences. When he applied the same technique to an expectation based on a triple-allelic system of a single locus, the agreement with observation was excellent.

Bernstein also applied the techniques of population genetics to such problems as linkage, to the measures of the degree of inbreeding for individuals and populations, to determination of the presence of recessive inheritance, to a method for deriving genetic ratios based on an a priori expectation, and to use of the development of presbyopia as an indicator of age. He also interpreted the direction of hair whorl and variations in singing voice, as found in different populations, in terms of allelic differences of single pairs of genes, but this interpretation has not withstood the test of time.

BIBLIOGRAPHY

Bernstein's writings are *Untersuchungen aus der Mengenlehre,* dissertation (Halle, 1901); "Ueber die Reihe der transfiniten Ordnungszahlen," in *Mathematische Annalen,* **60,** no. 2 (1905), 187–193; "Die Theorie der reellen Zahlen," in *Jahresbericht der Deutschen Mathematikervereinigung,* **14,** no. 8/9 (1905); "Zur Theorie der trigonometrischen Reihe," in *Bericht. Königliche Sächsische Gesellschaft der Wissenschaften zu Leipzig,* Math.-phys. Klasse, **60** (meeting of 7 Dec. 1908); "Ueber eine Anwendung der Mengenlehre auf ein aus der Theorie der säkularen Störungen herrührendes Problem," in *Mathematische Annalen,* **71** (1909), 417–439 (see also *ibid.,* **72** [1912], 295–296, written with P. Bohl and E. Borel); "Ueber den letzten Fermatschen Lehrsatz," in *Nachrichten von der Gesellschaft der Wissenschaften zu Göttingen,* Math.-phys. Klasse (1910); "Zur Theorie der konvexen Funktionen," in *Mathematische Annalen,* **76,** no. 4 (1915); "Die Mengenlehre Georg Cantors und der Finitismus," in *Jahresbericht der Deutschen Mathematikervereinigung,* **28,** no. 1/6 (1919), 63–78; "Die Theorie der gleichsinnigen Faktoren in der Mendelschen Erblichkeitslehre vom Standpunkt der mathematischen Statistik," in *Zeitschrift für induktive Abstammungs- und Vererbungslehre* (1922);

"Probleme aus der Theorie der Wärmeleitung. I. Mitteilung. Eine neue Methode zur Integration partieller Differentialgleichungen. Der lineare Wärmeleiter mit verschwindender Anfangstemperatur," in *Mathematische Zeitschrift,* **22,** no. 3/4 (1925), written with Gustav Doetsch; "Zusammenfassende Betrachtungen über die erblichen Blutstrukturen des Menschen," in *Zeitschrift für induktive Abstammungs- und Vererbungslehre* (1925); "Ueber die numerische Ermittlung verborgener Periodizitäten," in *Zeitschrift für angewandte Mathematik und Mechanik,* **7** (1927), 441–444; "Fortgesetzte Untersuchungen aus der Theorie der Blutgruppen," in *Zeitschrift für induktive Abstammungs- und Vererbungslehre* (1928); "Ueber Mendelistische Anthropologie," in *Verhandlungen des V. internationalen Kongresses für Vererbungswissenschaft* (Berlin, 1927), repr. in *Zeitschrift für induktive Abstammungs- und Vererbungslehre* (1928); "Ueber die Anwendung der Steinerschen Fläche in der Erblichkeitslehre, insbesondere in der Theorie der Blutgruppen," in *Zeitschrift für angewandte Mathematik und Mechanik,* **9** (1929); "Ueber die Erblichkeit der Blutgruppen," in *Zeitschrift für induktive Abstammungs- und Vererbungslehre* (1930); "Berichtigung zur Arbeit: Zur Grundlegung der Chromosomentheorie der Vererbung beim Menschen mit bes. Berücksichtigung der Blutgruppen," *ibid.* (1931, 1932); "Principles of Probability in Natural Science," in *Journal of Mathematics and Physics,* **14** (Mar. 1935); "The Continuum Problem," in *Proceedings of the National Academy of Sciences,* **24,** no. 2 (Feb. 1938), 101–104; and "Law of Physiologic Aging as Derived from Long Range Data on Refraction of the Human Eye," in *Archives of Ophthalmology,* **34** (Nov.-Dec. 1945), written with Marianne Bernstein. See also Corrado Gini, "Felix Bernstein," in *Revue de l'Institut international de statistique,* **25** (1957), 1–3; English trans. in *Records of the Genetics Society of America,* **60** (1968), 522–523. Bernstein's correspondence with Einstein (1933–1952) is at the Institute for Advanced Study, Princeton, and Houghton Library, Harvard.

HENRY NATHAN

BERNTHSEN, HEINRICH AUGUST (*b.* Krefeld, Prussia, 29 August 1855; *d.* Heidelberg, Germany, 26 November 1931), *chemistry.*

Bernthsen was the son of Heinrich Friedrich Bernthsen, a construction contractor, and Anna Sybilla Terheggen. He first studied mathematics and natural sciences but later turned to chemistry, which he studied in Bonn and then in Heidelberg under Bunsen and Kopp. Uninterested in the work on gases being pursued there, Bernthsen thought of leaving Heidelberg, but remained to study with Kekulé and Victor Meyer. In 1877 he became lecturer-assistant to Kekulé, and two years later he gave his inaugural academic lecture. In 1883 he was appointed extraordinary professor. Bernthsen left the University of Heidelberg in 1887 to join the Badische Anilin- und Sodafabrik as head of the main laboratory; he later headed the patent division. In 1884 he married the

daughter of a judge, Maria Magdelene Haubenschmied. Bernthsen was a member of the Deutsche Chemische Gesellschaft, an editor of the *Jahresbericht über die Fortschritt der Chemie,* and received honorary degrees from the Technische Hochschule in Berlin and the University of Heidelberg.

Bernthsen's chemical work was mainly industrial and dealt with dyes of the acridine and azine groups. He explained the composition of such substances as methylene blue and safranine. He also gave the correct composition of sodium hyposulfite. Although he was generally more interested in analysis than in synthesis, he developed technically feasible processes for producing indigo, rhodamine, and tolyl red. As a patent director, he secured patents for his indigo process, contact sulfuric acid, and lac dye. In the course of obtaining patents for his firm, Bernthsen became active in the movement for a patent law to protect industrial chemical processes and to strengthen the chemical industry.

BIBLIOGRAPHY

I. ORIGINAL WORKS. Among Bernthsen's writings are "Zur Kenntniss der Derivate der Alpha-toluylsäure," in *Berichte der Deutsche chemische Gesellschaft,* **8** (1875), 691–693; "Zur Kenntniss der Amidine und der Thiamide ein basischer organischer hydrosäuren," in *Annalen der Chemie,* **184** (1877), 290–320; **192** (1878), 1–60, 197; **197** (1879), 341–351; "Ueber die Zusammensetzung des unterschwefligsäuren Natrons," in *Berichte der Deutsche chemische Gesellschaft,* **14** (1881), 438–440; "Ueber das Methylenblau," *ibid.,* **16** (1883), 1025–1028, 2896–2904; *Kurzes Lehrbuch der organischen Chemie* (Brunswick, 1887); "Zur Kenntniss der Konstitution der blawen Schwefelfarbstoffe," in *Chemiker-Zeitung,* **32** (1908), 956–957; and *Fünfzig Jahre Tätigkeit in chemischer Wissenschaft und Industrie* (Heidelberg, 1925).

II. SECONDARY LITERATURE. Works on Bernthsen are M. Bodenstein, "H. A. Bernthsen," in *Berichte der Deutsche chemische Gesellschaft,* **65** (1932), 21a; K. Elbs, "August Bernthsen zum 70. Geburtstag," in *Zeitschrift für angewandte Chemie,* **38** (1925), 737–739; K. Holdermann, "August Bernthsen," in *Zeitschrift für Elektrochemie,* **38** (1932), 49, and "August Bernthsen zum Gedächtnis," in *Zeitschrift für angewandte Chemie,* **45** (1932), 141–143; and J. R. Partington, *A History of Chemistry,* IV (1964), 839.

RUTH ANNE GIENAPP

BERT, PAUL (*b.* Auxerre, France, 19 October 1833; *d.* Hanoi, Indochina [now People's Republic of Vietnam], 11 November 1886), *physiology, comparative anatomy, natural history, education.*

Bert was the son of Joseph Bert, lawyer and *conseiller de préfecture* of the Department of Yonne, and of Jeanne Henriette Massy, who was of Scottish extraction on her father's side. He attended the elementary school and the Collège Amyot at Auxerre from 1843 to 1852. In 1853 he went to Paris, where he studied jurisprudence and obtained the licentiate in law. He studied medicine and science from 1857 to 1866, receiving the licentiate in natural sciences in 1860 and the M.D., with a thesis on animal transplantation, in 1863. From 1863 to 1866 Bert was the student and *préparateur* of Claude Bernard at the Collège de France. In 1866 he was awarded the doctorate in natural sciences with the thesis *De la vitalité propre des tissus animaux.* In 1865 he married Josephina Clayton, a Scotswoman.

Bert was professor of zoology and physiology at the Faculté des Sciences of the University of Bordeaux in 1866–1867. In 1868 he replaced Flourens (as *suppléant*) in the chair of comparative physiology at the Muséum d'Histoire Naturelle, and the following year he succeeded Claude Bernard in the chair of physiology at the Sorbonne.

Bert was a member of the Société des Sciences Historiques et Naturelles de l'Yonne, the Société Philomathique de Paris, the Société de Biologie de Paris, and the Académie des Sciences de Paris.

The Franco-Prussian War caused Bert to enter politics. After the capitulation of the French Imperial Army at Sedan in September 1870 and the resignation of Napoleon III, Bert joined the forces led by the Government of National Defense, whose driving force was Léon Gambetta, who was also his personal friend. The following January he became Préfet du Nord and organized a hopeless resistance. After the war, Bert joined the Liberal Republican party and was elected in 1872 to the Chamber of Deputies. His political aims were regeneration and *revanche,* the recovery of the lost eastern provinces, Alsace and Lorraine.

The means of political, economic, industrial, intellectual, and moral regeneration was, for him, a radical reform of education at all levels, and particularly of the elementary schools. As a deputy, as chairman of important parliamentary committees, and as minister of public instruction (from 14 November 1881 to 26 January 1882) he fought for enactment of the laws proposed by Jules Ferry, a well-known liberal politician and influential minister of public instruction. In a masterful report on elementary education presented to the Chamber of Deputies, Bert explained Ferry's principles. He demanded free, compulsory elementary schooling, with a secular program and lay personnel. The schoolteacher must not belong to the clergy. He is the soldier of the secular republic.

The program of the elementary school must include the elements of science, for the sciences sharpen the intellect: the natural sciences develop the power of observation, and the physical sciences sow the seeds of causal thinking. History and paleontology will reveal the gradual development of man from a cave dweller to a culture-bearing, free, republican citizen; thus evolution, and not revolution, will be man's guiding principle. The "Ferry laws" were enacted between 1880 and 1886.

Bert supported secondary education for girls. For many years he lectured at the Sorbonne on biology and zoology to girls aged sixteen to eighteen. Bert also published elementary and secondary textbooks on natural history, zoology, and the physical sciences. These books, which show his remarkable didactic ability, were reprinted many times and translated into English, Italian, and Spanish. It was Bert's firm belief that the principles of science should pervade the whole of society in order to make it better. The *Revues scientifiques,* a periodical that he edited from 1879 to 1885, bears witness to Bert's endeavor to popularize science.

In 1885 the French people were alternately enraged and perplexed by the revolts in Indochina against the new French colonial regime. In the stormy debates in the National Assembly, Bert sternly defended the French colonial expansion in Indochina. Appointed as the first civil governor of Annam and Tonkin in 1886, Bert sought a true partnership between France and Indochina. He eliminated the military interference in administration and "pacified" the areas he governed. In Annam he strengthened the power of the emperor, and in Tonkin he relied on popular forces, which he strengthened by political and social reforms. In addition, he established the Tonkin Academy and founded numerous schools.

In 1855, when still a law student, Bert became a member of the Société des Sciences Historiques et Naturelles de l'Yonne. This marked the beginning of his research. Bert's scientific thought and method were later shaped by three eminent scientists who taught him: Pierre Gratiolet, Henri Milne-Edwards, and Claude Bernard. He studied with Gratiolet in the laboratory of comparative anatomy at the Muséum d'Histoire Naturelle in Paris. There he learned that within the great variety of structures "the problem of life was but one." Bert was also deeply influenced by Milne-Edwards' *Introduction to General Zoology* (1851). This book revealed to him how nature worked through "the law of economy," the "division of labor," and the perfection of functions by means of specialization. It was to Claude Bernard, however, that Bert owed his greatest debt. As a student and collaborator of Bernard, he became acquainted with the methods of experimental physiology, the critical evaluation of experimental findings, and the attempt to describe the basic laws of physiology. Bert was deeply impressed by Bernard's concept of an internal environment.

Bert's scientific activity can be divided into three periods: (1) In the 1860's—as a student of medicine and science, as *préparateur* for Bernard, and as professor of zoology and comparative physiology at Bordeaux and Paris—he dealt with questions of general physiology, plant physiology, and comparative anatomy and physiology. During that period he published important monographs on animal transplantation and the vitality of tissues, and a comprehensive study on the comparative physiology of respiration. (2) After the Franco-Prussian War, he published his magnum opus, *La pression barométrique* (1878). (3) A last period of scientific activity, which was a direct offshoot of his barometric work, dealt with the experimental and clinical study of anesthesia and with the properties of blood at high altitude.

Through all his scientific work we can perceive the constant endeavor to study the phenomena of life in the context of the exterior and interior environment.

Bert's first important work dealt with animal transplantation. This study was not intended to be a contribution to experimental surgery, for transplantation was conceived of as a physiological problem: How can transplanted organs and tissues live in a new environment? Bert succeeded in creating "double monsters," uniting two rats by suturing their skins together. He also implanted the tip of the tail of a young albino rat under the skin of its back; the proximal end of the tail was then cut, so that it formed the tip of the grafted tail. The transplanted tail grew, formed new bone, and reestablished circulation and sensibility—but the direction of the sensory impulse was reversed.

Bert's investigation of the specific vitality of animal tissues was a pure environmental study. He used the transplantation technique as a means of examining the vital resistance of organs and tissues. Isolated tails of rats were exposed to different temperatures and humidities, and to various gases and chemical agents. After these exposures the tails were transplanted under the skins of rats. The transplantation reestablished a physiological internal milieu and made it possible to test the survival of the tail tissues after they had been subjected to various changes of environment. Bert emphasized that the cells and tissues lived their own lives, growing and differentiating independently of any superior vital force as long as they were in a suitable milieu.

During his professorship at Bordeaux, Bert dealt with problems of marine biology and plant physiology. He studied the mechanism of death in marine fishes exposed to fresh water, and he observed the occurrence of *Amphioxus lanceolatus* on the southwestern coast of France. He also published his classic study on the movements of the "sensitive plant" (*Mimosa pudica*). Using ether, he succeeded in differentiating spontaneous movements from induced ones. The spontaneous movements depended on differences of osmotic pressure, which was regulated by light and darkness.

In 1870 Bert published an important work dealing with problems of the comparative physiology of respiration. This monograph was based on his lectures delivered at the Muséum d'Histoire Naturelle, Paris, in the spring semester of 1868. Bert analyzed the anatomical structures and the physiological functions of the respiratory organs of vertebrates and invertebrates. He also dealt with the problem of tissue respiration and showed that the oxygen content of the blood depended on the exterior air pressure. The respiratory movements of aquatic and air-breathing animals were investigated with novel methods and techniques. He studied the nervous regulation of the respiratory rhythm and investigated the respiration of diving animals. He analyzed the death mechanism of asphyxia and succeeded in clearly differentiating true asphyxia caused by lack of oxygen from carbon dioxide poisoning. This work, rich in facts and new views, also provided a critical evaluation of a vast number of problems.

Bert's definitive work, *La pression barométrique*, published in 1878, represented an environmental study on the largest scale. The great questions were How does the changing exterior milieu—the atmospheric pressure—act on an organism? What are the effects of low pressure at high altitudes? How does the high pressure to which the caisson worker is exposed affect the body? How does the blood, the interior milieu, behave under high and low pressures? What is the mechanism of mountain sickness? Bert's interest in these problems arose early in his scientific work. In 1864, while a student of Bernard, he had discussed the question of whether the amount of gases dissolved in the blood depended on atmospheric pressure. The French physician Denis Jourdanet, who had practiced in Mexico and was interested in the biological and medical aspects of high-altitude climates, provided Bert with the necessary means to build the costly apparatus needed for the experimental study of the physiological effects of air pressure. Jourdanet had constructed pressure chambers for therapeutic use and had formulated the hypothesis that the blood contained less oxygen under the low atmospheric pressure of high altitudes, a condition he called barometric anoxemia.

In his first series of experiments Bert studied the following problem: What extremes of air pressure can a living being endure? He first examined the effects of low pressure and found that animals died after the partial pressure of oxygen sank below a critical level that was constant for each species. Bert therefore announced the following principle: Oxygen tension is everything; barometric pressure in itself does nothing or almost nothing.

Bert then studied the physiological effects of pure oxygen or air inhaled under high pressure: the dogs used died with tonoclonic cramps. Thus he discovered the phenomenon of acute oxygen poisoning and was able to state that if the organism receives too little oxygen, it suffocates; and if it is exposed to oxygen at high pressure, it is poisoned. In a second series of experiments he studied the behavior of blood gases under various pressures. These investigations relied on the fundamental studies of blood gases made by Carl Ludwig and his students. Bert found: "The combination of oxygen with hemoglobin is likely to be partially destroyed, to be dissociated at low pressures. . . . Everything seems to indicate that there exists in the neighborhood of normal pressure a point of chemical saturation of the oxyhemoglobin, and that beyond this point there is added to the blood only oxygen dissolved in the serum according to Dalton's law."

Bert also experimented on himself in the pressure chamber in order to study the effects of low atmospheric pressure on the human body. When the pressure was decreased to 400 millimeters of mercury, he noted increased pulse rate, headache, dizziness, darkening of the vision, mental lassitude, and nausea. These complaints disappeared as soon as he inhaled oxygen. Bert also trained balloonists in his pressure chamber and provided them with oxygen bags for their flights to high altitudes. He recognized that when the partial pressure of oxygen is reduced, the altitude sickness of the aviator, the mountain sickness of the alpinist, and the complaints encountered in low-pressure experiments appear. Bert also clarified experimentally the mechanism of decompression sickness (caisson disease). Sudden decompression from high atmospheric pressures (five–ten atmospheres) produced gas bubbles in the blood and tissues. These gas bubbles consisted primarily of nitrogen that had previously been dissolved under high pressure and was liberated by the decompression. Bert emphasized that the duration of compression was of great importance. As prophylaxis for caisson disease he devised a slow, gradual decompression; treatment was recompression and low decompression.

The results of Bert's monumental work were manifold: He fully realized the physiological importance of the partial pressures of the respiratory gases. He described the relationship between the external partial pressure and the behavior of the blood gases. He recognized that mountain sickness and altitude sickness are a consequence of the low partial pressure of oxygen. He introduced oxygen apparatus to avert the dangerous consequences of ascent to high altitudes. He was the first to study in a pressure chamber the conditions of high-altitude ascents. He discovered and described oxygen poisoning, and explained the cause and mechanism of caisson disease.

Bert applied his knowledge of the physiological effects of atmospheric pressure to the field of anesthesia. He reasoned that the pressure at which nitrous oxide induces anesthesia is about one atmosphere. Therefore pure nitrous oxide has to be used, but it induces asphyxia. To avoid this, Bert prepared a mixture of one-sixth oxygen and five-sixths nitrous oxide. He administered this mixture to dogs under the slightly increased pressure of $1\frac{1}{5}$ atmospheres. Thus the blood was supplied with enough oxygen to sustain life and enough nitrous oxide to produce anesthesia. Bert went on to the application of this method to clinical anesthesia, constructing a horse-drawn anesthetic chamber in which the surgeon, his assistants, and the patient could be placed under slightly increased pressure.

In *La pression barométrique* Bert discussed the problem of acclimatization to high altitudes. Is the blood of individuals at high altitudes capable of absorbing more oxygen than the blood of individuals living at sea level? Three possibilities might be conceived: a qualitative change in the hemoglobin, an increase in the hemoglobin content of the red corpuscles, or an increase in the number of erythrocytes. Starting from Bert's hypothesis that people living at high altitudes might possess more red corpuscles, François Viault examined human and animal blood in the Peruvian Andes from 1890 to 1892. He observed on himself and on his companions that the number of erythrocytes increased from 5,000,000 to 7,000,000 within three to five weeks. This important fact, which showed a substantial change of what is normally a biological constant, initiated the systematic high-altitude research in the Alps (conducted by Hugo Kronecker, A. Mosso, and others).

As a true disciple of Bernard, Bert hailed the emergence of physiology as an exact scientific discipline. He refused to adhere to any dogmatic materialism. He emphasized that the cerebral function was indispensable for the production of psychic phenomena. But organized matter could not be the only cause, the sufficient condition, for intellectual manifestations, for this would mean transcending the realm of scientific physiology. Outside of physiology, he believed, remained the immense field of subjective phenomena that cannot be investigated with the methods of physiology.

BIBLIOGRAPHY

I. ORIGINAL WORKS. Bert's scientific works include "De la greffe animale," M.D. thesis (Paris, 1863); *Recherches expérimentales pour servir à l'histoire de la vitalité propre des tissus animaux,* Ph.D. thesis (1866), pub. in *Annales des sciences naturelles* (Zoologie), **5** (Paris, 1866), 123–218; "Recherches sur les mouvements de la sensitive (*Mimosa pudica,* Linn.)," in *Mémoires de la Société des Sciences Physiques et Naturelles de Bordeaux,* **4** (1866), 11–46, also published separately (Paris, 1867); *Leçons sur la physiologie comparée de la respiration . . .* (Paris, 1870); *Recherches expérimentales sur l'influence que les modifications dans la pression barométrique exercent sur les phénomènes de la vie* (Paris, 1873); "Sur la possibilité d'obtenir, à l'aide du protoxyde d'azote, une insensibilité de longue durée, et sur l'innocuité de cet anesthésique," in *Comptes rendus hebdomadaires de l'Académie des Sciences* (Paris), **87** (1878), 728–730; *La pression barométrique, recherches de physiologie expérimentale* (Paris, 1878), his masterpiece, trans. into English by M. A. Hitchcock and F. A. Hitchcock as *Barometric Pressure. Researches in Experimental Physiology* (Columbus, Ohio, 1943); "Anesthésie par le protoxyde d'azote mélangé d'oxygène et employé sous pression," in *Comptes rendus hebdomadaires de l'Académie des Sciences* (Paris), **89** (1879), 132–135; *Leçons de zoologie . . .* (Paris, 1881); *Leçons, discours et conférences* (Paris, 1881), which includes political and educational papers as well as scientific ones; *La première année d'enseignement scientifique, sciences naturelles et physiques: animaux, végétaux, pierres et terrains, physique, chimie, physiologie végétale* (Paris, 1882), trans. by Josephina Clayton (his wife) as *First Year of Scientific Knowledge* (Paris, 1885).

II. SECONDARY LITERATURE. Works on Bert are E. H. Ackerknecht, "Paul Bert's Triumph," in *Bulletin of the History of Medicine,* supp. **3** (1944), 16–31, which treats Bert's place in French intellectual history and includes an important bibliography; E. Bérillon, *L'oeuvre scientifique de Paul Bert* (Paris, 1887); C. A. Culotta, "A History of Respiratory Theory: Lavoisier to Paul Bert, 1777–1880," thesis (Univ. of Wisconsin, 1968), which includes a detailed discussion of Bert's work on blood gases; L. Dubreuil, *Paul Bert* (Paris, 1935), comprehensive biography with special emphasis on Bert's political importance; J. Ducloz, "L'enfance et la jeunesse de Paul Bert," in *Bulletin de la Société des Sciences Historiques et Naturelles de l'Yonne,* **78** (1924), 5–102; and N. Mani, "Paul Bert als Politiker, Pädagog und Begründer der Höhenphysiologie," in *Gesnerus,* **23** (1966), 109–116.

NIKOLAUS MANI

BERTHELOT, PIERRE EUGÈNE MARCELLIN[1] (*b.* Paris, France, 25 October 1827; *d.* Paris, 18 March 1907), *chemistry.*

His father, Jacques Martin Berthelot, had married Ernestine Sophie Claudine Biard in 1824. Marcellin was the second of three children; the first died in infancy. The father, who was from a family of ironsmiths in the region of Orléans, had come to Paris in 1822 to study medicine. After qualifying he spent most of his life tending the sick in the poorer districts of Paris. Only in his heroic work during the cholera epidemic of 1832, about which he wrote a book, did he rise above obscurity. His income was just sufficient to support his family, and his wife, who came from the bourgeoisie, brought only a small dowry.

At the age of eleven Marcellin Berthelot entered the Collège Henri IV in Paris. He showed himself reserved in the extreme but brilliant at his lessons, distinguishing himself particularly in Latin verse. In 1846 he won first prize for philosophy among pupils from *lycées* throughout France. At fourteen Berthelot became a boarder; four years later he met Ernest Renan, whose room was adjacent to his in the *pension.* Renan was twenty-two and was employed as an assistant master. In 1847 Berthelot became *bachelier ès lettres* and then attended courses in the Paris Faculty of Medicine and the Faculty of Science, graduating from the latter in July 1849. He undertook a rigorous program of reading, including languages and the main branches of science. Berthelot asserted his independence by deliberately avoiding the two great educational institutions, the École Polytechnique and the École Normale, the training ground of so many French men of science. He obtained entry to the private laboratory of the chemist Pelouze, where he learned some practical chemistry. In February 1851 there was a vacancy for the post of demonstrator to Balard at the Collège de France. Berthelot accepted the post although it carried only a nominal salary; in his spare time he prepared for his doctorate. On 24 June 1854 he defended his thesis, "Mémoire sur les combinaisons de la glycérine avec les acides et sur la synthèse des principes immédiats des graisses des animaux." Berthelot carried out further studies at the École de Pharmacie in Paris, graduating as a pharmacist on 29 November 1858. He also visited Italy and Germany during this time. Thus, until he was over thirty, Berthelot lived the life of a student, relying on his father for financial support.

Berthelot became a leading advocate in France of science as an ideal with direct moral implications. Although his parents were Roman Catholic and brought him up in the same faith, he reacted to his philosophy course at school by soon questioning the validity of religion and becoming a skeptic. In this he was influenced by Renan but also by his republican feelings, since the Roman Catholic Church in mid-nineteenth-century France was unsympathetic to radical thought. In his later writings Berthelot attacked clerical influence, particularly in education.

On 30 May 1861 Berthelot married Sophie Caroline Niaudet, a girl from a Protestant family and ten years younger than he. His wife was a descendant of the famous clockmaker Breguet. The couple had six children. They were devoted to each other for forty-five years, and within an hour of the death of Sophie Berthelot, Marcellin, who had tended her night and day, also died. A special law was passed to permit Berthelot and his wife to be buried together in the Panthéon.

Berthelot was appointed to a chair of organic chemistry created at the École de Pharmacie on 2 December 1859. The success of his book *La chimie organique fondée sur la synthèse* resulted in his giving a course of lectures at the Collège de France (1863–1864), and on 8 August 1865 this course was attached to a chair of organic chemistry entrusted to Berthelot. From that time until his death, if Berthelot was in Paris he went daily to his laboratory at the Collège de France. Although he gave more attention to research than to teaching, he had a number of distinguished students, including Jungfleisch, Sabatier, and A. Werner. Berthelot was elected to the Académie de Médecine in 1863, and the Académie des Sciences (after three unsuccessful attempts) in 1873. He was made permanent secretary of the latter in 1889. He became a member of the Académie Française in 1901. First nominated as *chevalier* of the Legion of Honor in 1861, he received the highest grade, *Grand-Croix,* in 1900. He was also a member of a large number of foreign academies.

When Paris came under siege in the Franco-Prussian War (1870–1871), Berthelot was made president of the Comité Scientifique pour la Défense de Paris. His activities during the siege of Paris called much attention to himself, and in the election of 1871 he was given a large vote, although he had not put himself forward as a candidate. He first took his seat in the Senate in 1871. In 1874 he was appointed to a War Ministry commission on explosives, and in 1878, when a new commission on explosives was formed, Berthelot was named president. In July 1881 the Senate elected him to a permanent senatorship. Berthelot sat with the parties on the Left and spoke frequently on educational matters. In 1886 he presided over a commission on the laicization of primary education. Berthelot was minister of education from 11 December 1886 to 30 May 1887 in the cabinet of René Goblet. In 1895 he was appointed foreign minister in the cabinet of Léon Bourgeois but resigned after five months because of disagreements over policy on Egypt and the Sudan.

In 1869, on the occasion of the opening of the Suez Canal, Berthelot visited Egypt, the country traditionally associated with the birth of chemistry; but it was not until 1884 that he committed to paper a few ideas on alchemy. Attracted both by the mysticism of the alchemists and by the connection of other parts of their art with the rational science he professed, he began to use his knowledge of Greek to interpret unpublished alchemical manuscripts. Like Hermann Kopp, Berthelot took the view that alchemy had developed as a misunderstanding of the earlier empirical knowledge of Egyptian metalworkers. He and Kopp were the two nineteenth-century figures who were able not only to make outstanding contributions to chemistry but also to undertake an extensive study of its history. Berthelot studied the transmission of ancient alchemy to the Middle Ages. He distinguished a practical tradition, exemplified by the *Liber ignium* of Marcus Graecus, from a theoretical approach transmitted through Syriac and Arabic sources. He argued that the Latin author Geber was distinct from Jābir ibn Ḥayyān. In much of his alchemical studies Berthelot was dependent on his collaborators, who translated the original Syriac and Arabic manuscripts. His interpretation was, therefore, not faultless.

On a more practical plane, Berthelot's analysis of metallic objects from ancient Egypt and Mesopotamia laid the foundations of chemical archaeology. In 1889, to celebrate the centenary of the French Revolution, Berthelot, as secretary of the Académie des Sciences, was called upon to commemorate men of science. He prepared material for a lecture on Lavoisier, and on the basis of this and Grimaux's study of 1888 he produced a book, *La révolution chimique, Lavoisier.* One detects special sympathy by the patriotic nineteenth-century French chemist for the eighteenth-century liberal who had also used his scientific knowledge to help his country. Berthelot's publication of extracts from the laboratory notebooks of Lavoisier, which were in the possession of the Academy, performed a valuable service to the history of science. Whatever criticisms may be leveled at Berthelot's earlier publications on the history of chemistry, this study was an astonishing achievement for a man who was simultaneously carrying out important research on thermochemistry and agricultural chemistry.

In the last years of his life Berthelot published books that suggest his concept of science as an all-embracing philosophy: *Science et philosophie* (1886), *Science et morale* (1897), *Science et education* (1901), *Science et libre pensée* (1905). He regarded it as unreasonable to assign limits to the possible progress of science. He foresaw a Utopia through science that could be realized by the year 2000. In this new world he considered chemistry to have a central place not only because of its almost unlimited powers of synthesis but also through the exploitation of agriculture and natural resources. Berthelot continually fought against clerical influence in education. He wanted a greater place for science in the school curriculum, but not at the expense of classical studies. The moral value of science for Berthelot lay not only in its respect for truth but also in its justification for work. Berthelot, like Claude Bernard, favored a positivistic philosophy. It was in this spirit of accepting only the observable that he regarded atomic and molecular theories with great suspicion.

Berthelot's publications were particularly numerous. Jungfleisch lists 1,600 titles of papers on inorganic, organic, physical, analytical, technical, agricultural, and physiological chemistry, as well as on the history of chemistry. His work in organic chemistry may, however, be singled out as being of special importance; and if his contributions to physical chemistry hold second place, even they may be considered as originating in the context of his interest in the reactions and formation of organic compounds. In order to systematize Berthelot's vast work, covering a period of sixty years, Graebe divided his productive life into four periods:

(1) The first organic period, 1850–1860. This covers his work on alcohols and includes his early work on synthesis. In the fall of 1860 he published his definitive work on organic synthesis.

(2) The second organic period, 1861–1869. This period is characterized by research and synthesis of acetylene, benzene, and aromatic compounds occurring in coal tar. It was in the early 1860's that Berthelot collaborated with Péan de Saint-Gilles on the formation and decomposition of esters, research that constituted a bridge between his interest in organic and physical chemistry. At the end of this period Berthelot was using hydrogen iodide to reduce organic compounds.

(3) The period 1869–1885, which covers Berthelot's most important contributions to thermochemistry.

(4) The period 1885–1907, in which Berthelot's most original work was his contribution to agricultural chemistry and to the history of chemistry.

Berthelot's first publication, read to the Académie des Sciences on 27 May 1850, was concerned with a simple method of liquefying a gas by applying pressure. The choice of this physical topic may have been inspired by his admiration for Regnault, but in the next year his deep interest in organic chemistry

revealed itself. He studied the action of red heat on alcohol and acetic acid. It was already known that at high temperatures alcohol could be transformed into a crystalline solid, naphthalene. Berthelot was able to show that in addition benzene and phenol were formed. Acetic acid at red heat produced naphthalene and benzene. He concluded very significantly that the synthesis (*synthèse*) of naphthalene, benzene, and possibly phenol, could now be considered as an established fact, since they could all be obtained from acetic acid, which in turn could be prepared via the respective stages: carbon disulfide, carbon tetrachloride, trichloracetic acid. This was one of the first examples of the use of the word *synthesis* to denote the production of organic compounds from their elements.[2]

Two compounds to which Berthelot gave considerable attention were oil of turpentine and camphor. It was known that on reaction with hydrochloric acid, turpentine formed a hydrochloride, $C_{10}H_{16} \cdot HCl$. In 1852 Berthelot showed that the reaction could be taken further to produce a product identical with oil of lemon. It was Berthelot who first discovered isomeric changes in oil of turpentine, from which he obtained the solid hydrocarbon camphene. He distinguished what we would now call *d*-pinene and *l*-pinene and *d*-, *l*-, and *dl*-camphene. He found that camphene may be oxidized by chromic acid (or by air in the presence of platinum black) into a camphorlike substance; in 1870 he proved that the product was true camphor.

The classical work on fats had been carried out by Chevreul. In 1853 and 1854, Berthelot established his reputation in this field by his research on the derivatives of glycerin. By heating glycerin with hydrochloric acid and a selection of fatty acids, he obtained compounds of glycerin with acetic, valeric, benzoic, and sebacic acids. He went on to obtain compounds of glycerin with one, two, or three molecules of acid, the other product being contained in natural fats, e.g., tristearin. Thus, with stearic acid and glycerin he obtained successively monostearin, distearin, and tristearin. He also investigated the products of glycerin with other acids, including acetic acid, the reactions for which he formulated as follows (C = 6, O = 8):

$$\text{monoacetin} = \underset{\text{acetic acid}}{C^4H^4O^4} + \underset{\text{glycerin}}{C^6H^8O^6} - \underset{\text{water}}{2HO}$$

$$\text{diacetin} = \underset{\text{acetic acid}}{2C^4H^4O^4} + \underset{\text{glycerin}}{C^6H^8O^6} - \underset{\text{water}}{4HO}$$

$$\text{triacetin} = \underset{\text{acetic acid}}{3C^4H^4O^4} + \underset{\text{glycerin}}{C^6H^8O^6} - \underset{\text{water}}{6HO}$$

Whichever of the above esters was hydrolyzed, the product was glycerin. From these reactions Berthelot concluded that glycerin in organic chemistry corresponded to phosphoric acid in inorganic chemistry as alcohol corresponded to nitric acid. In other words, this was the beginning of the idea that, corresponding to polybasic acids in inorganic chemistry, there were polyatomic alcohols in organic chemistry. Berthelot's younger contemporary and rival, Adolphe Wurtz, is sometimes given credit for this work, although Wurtz's contribution in 1855 was to make the correct analogy between different salts of the same (ortho-phosphoric) acid rather than the three different acids of phosphorus to which Berthelot had referred. Berthelot's most important contribution in his work on glycerin was to introduce the concept (and name) of polyatomic alcohols, but hardly less important was his synthesis of stearin and palmitin, the chief constituents of ordinary hard fats. He also carried out further work on the esterification of glycerin, some of it in collaboration with his pupil S. de Luca.

From glycerin Berthelot turned his attention to sugars and succeeded in isolating several new sugars. He showed that sugars behave partly as polyatomic alcohols (i.e., with the —OH group) and partly as aldehydes (i.e., with the —COH group). With the object of systematizing the confused knowledge of sugars, he divided carbohydrates into three classes: (1) ordinary sugars, which are like either (a) glucose (i.e., monosaccharides) or (b) cane sugar (i.e., polysaccharides); (2) carbohydrates, such as starch, cellulose, etc.; and (3) polysaccharides, which on hydrolysis combine with water to form glucoses:

$$(C^6H^{10}O^5)^n + nH^2O = nC^6H^{12}O^6.$$

Berthelot gave cane sugar the systematic name *saccharose*. From his researches on sugar and alcohol he naturally had an interest in fermentation. He showed that the conversion of cane sugar into invert sugar in fermentation is caused by an enzyme (*ferment glucosique*) present in yeast. He succeeded in obtaining it from an extract of yeast through precipitation by alcohol. Having obtained new sugars and thrown some light on the relation of sugars to other compounds, Berthelot turned to the alcohols. He prepared new alcohols from cholesterol, ethal, Borneo camphor, etc. He gave a definition of alcohols as neutral compounds consisting of carbon, hydrogen, and oxygen, and which with acids had water eliminated to form another neutral compound. The latter was capable of taking up water to form the original alcohol and acid. He was the first to consider the phenols as a group, which he characterized in a similar way.

One of the earliest of Berthelot's triumphs in his program of synthesis was in the preparation of alcohol. This was, of course, traditionally the product of fermentation of sugars with yeast; but in 1854 Berthelot showed that it could be prepared from ethylene. When ethylene was subjected to prolonged and vigorous shaking with sulfuric acid, it dissolved; and when the product was heated with water and distilled, the alcohol passed over. An obvious objection to this preparation was that the ethylene had itself been obtained from alcohol. Berthelot therefore obtained ethylene from coal gas as ethylene iodide by passing the crude gas into a solution containing iodine. Hennel had already suggested in 1828 that ethylene could be converted to alcohol by treatment with sulfuric acid but had been criticized by Liebig, so that most chemists in the mid-nineteenth century regarded the possibility of this conversion as doubtful. The fact that previous work had been done on the subject was emphasized by Chevreul in an attack on Berthelot, who had neglected to mention this. By a similar method Berthelot synthesized isopropyl alcohol from propylene.

Berthelot had now begun his program of general synthesis of organic compounds, and in 1856 he set out to prepare formic acid. He reasoned that formic acid was related to carbon monoxide in the same way as alcohol was to ethylene. As he wrote it:

$$C^2H^2O^4 = C^2O^2 + 2HO$$
$$\text{formic acid}$$

$$C^4H^6O^2 = C^4H^4 + 2HO$$
$$\text{alcohol}$$

As he had produced alcohol by adding water to ethylene in the presence of an acid capable of fixing the alcohol, so he should be able to react water with carbon monoxide in the presence of an alkali to fix the acid product. He accordingly heated moist caustic potash in an atmosphere of carbon monoxide for seventy hours. This produced potassium formate, which, when distilled with sulfuric acid, yielded formic acid in no way different from that occurring naturally in the ant.

Berthelot then synthesized methane (contaminated with a little ethylene) by passing a mixture of carbon disulfide vapor and hydrogen sulfide over red-hot copper. Then in 1857, by reacting methane with chlorine, he obtained methyl chloride, which, on hydrolysis, formed methyl alcohol. There is an obvious parallel between the success of Cannizzaro two years earlier in obtaining benzyl alcohol from toluene via benzyl chloride. Berthelot, nevertheless, achieved the first true synthesis of an aliphatic alcohol, and in 1858 he summarized his achievements in a long table, "Sur la synthèse des carbures d'hydrogène," in which he described his preparation of the hydrocarbons methane, ethylene, propylene, butylene, amylene, ethane, and propane, as well as benzene and naphthalene. The first stage in synthesis, the preparation of these hydrocarbons and their conversion to the corresponding alcohol, was the most difficult stage; but once the alcohol had been prepared, it was possible "to achieve the synthesis of an almost infinite number of organic compounds." Berthelot attacked the idea of a vital force distinguishing organic from inorganic compounds. Chemistry had proceeded up to then by the method of analysis, but by the complementary method of synthesis he claimed to have shown that the forces acting in organic chemistry were no different from those operating in inorganic compounds.

In Berthelot's memoir of 1858 he wrote that "carbon does not combine directly with hydrogen." Before he achieved this he wrote his monumental *Chimie organique fondée sur la synthèse,* in which he presented a review of his work in organic chemistry during the previous ten years. The work begins with an extensive historical introduction, which contains no more than a passing reference to Wöhler's preparation of urea in 1828. One obtains the impression from the book that the author was the first to recognize the importance of synthesis in organic chemistry and that it was he who had undertaken the basic research. The first volume is devoted to a discussion of the synthesis of hydrocarbons and the synthesis of alcohols. In the second volume glycerin and sugars are discussed, and toward the end there is a chapter dealing with the evidence for genuine synthesis and the implications for physiological chemistry. Berthelot emphasized throughout the work that the success of synthesis in organic chemistry meant that the claim of vitalists that vegetable and animal substances were essentially different from those made in the laboratory was no longer tenable. There were not two chemistries but one, and chemical reactions in both the inorganic and organic realms depended ultimately on purely mechanical factors. In his conclusion Berthelot argued that chemistry differed from a descriptive science such as natural history by being creative and that in this it resembled the mathematical sciences.

Acetylene was first prepared in 1836 by Edmund Davy, for whom it was "a new carburet of hydrogen." The gas was then forgotten until it was rediscovered by Berthelot in 1860. Berthelot prepared it by passing either ethylene, or methyl or ethyl alcohol in the vapor state, or ether through a red-hot tube. Alternatively it could be prepared by passing an electric discharge through a mixture of cyanogen and hydrogen. To isolate it he made use of its reaction with ammoniacal cuprous chloride solution to form cuprous acetylide, a substance discovered by Quet in 1858. Berthelot gave the gas the name "acetylene," saying that it was derived from acetyl (C_2H_3—H) in the same way that ethylene was related to ethyl (C_2H_5—H). Having already found that acetylene is the product formed when ethylene or methane is strongly heated or sparked, Berthelot concluded that it is the most stable of hydrocarbons and might therefore be produced by direct combination of carbon and hydrogen. Taking special precautions to ensure the purity of his materials, he therefore passed hydrogen through an electric arc formed between carbon poles. By reduction of acetylene Berthelot obtained ethylene and ethane and, by oxidation, acetic acid and oxalic acid, all reactions of great use to their author in extending his program of synthesis. As the direct reaction of acetylene with chlorine usually produced an explosion, Berthelot and Jungfleisch used antimony chloride (as a negative catalyst) and succeeded in obtaining two addition products.

Berthelot opened a new field when he carried out a systematic investigation of hydrocarbons obtained by heating suitable substances in the temperature range from red to white heat. His most famous experiment was that in which he heated acetylene in a glass tube; polymerization took place, forming benzene with some toluene. This was the first demonstration that it was possible to effect a simple conversion of an aliphatic to an aromatic compound. By passing benzene vapor through a red-hot iron tube filled with broken glass, Berthelot obtained diphenyl. Similarly, he obtained styrolene and naphthalene from benzene and acetylene, and acenaphthene from naphthalene and ethylene. These reactions had the additional value of throwing light on the formation of byproducts in the manufacture of coal gas. For example, Berthelot first discovered acenaphthene ($C_{10}H_6 \cdot C_2H_4$) as described above and then as a constituent of coal tar. He also discovered fluorene ($C_{13}H_{10}$) in crude anthracene and heavy coal-tar oil.

After Berthelot's success in obtaining acetylene directly from carbon and hydrogen, he considered acetylene to be the most important starting point in his whole system of synthesis, since from it could be obtained ethylene, methane, and benzene:

$$C_2H_2 \longrightarrow C_2H_4 \longrightarrow C_2H_6$$
$$C_2H_2 \longrightarrow CH_3 \cdot COOH \longrightarrow CH_4$$
$$3C_2H_6 \longrightarrow C_6H_6$$

Previously Berthelot had deliberately used molecular formulas only, but from 1864, when he gave a lecture on isomerism to the Société Chimique in Paris, he began to develop the molecular implications of his researches. Nevertheless, he remained outside the development of the theory of chemical structure developed by Butlerov and Kekulé and the notational reforms arising from the Karlsruhe Congress of 1860, so that he continued to use the old equivalent notation. He distinguished hydrocarbons according to their degree of saturation, as follows (using modern atomic weights):

carbure complet   e.g., ethane C_2H_6

carbure incomplet   e.g., ethylene $C_2H_4(-)$

carbure incomplet du 2me ordre
     e.g., acetylene $C_2H_2(-)(-)$

According to this system, and wishing to call attention to its method of synthesis, Berthelot considered benzene an incomplete hydrocarbon of the fourth order. This formulation was acceptable for its reduction to hexane but was unsatisfactory in accounting for the majority of benzene's reactions, where it behaves as a saturated hydrocarbon. Berthelot rejected Kekulé's formula for benzene (1865–1866), and it was not until 1897 that he accepted modern structural formulas.

Berthelot's last major research in organic chemistry was the application of hydrogen iodide as a reducing agent—he called it "une méthode universelle d'hydrogenation." His publications on this research covered the period 1867–1870. Although he claimed that they were a continuation of work published in 1855, it must be mentioned that meanwhile Lautemann had already used hydrogen iodide as a reducing agent in organic chemistry. Berthelot, anxious to carry out even the most difficult reductions, was prepared to use concentrated hydriodic acid saturated at 0°C. and heated with the substance to be reduced in an oil bath up to 280°C. He succeeded in reducing a large number of unsaturated aliphatic hydrocarbons. His results with aromatic hydrocarbons were less definite. His study of the mechanism of decomposition of hydrogen iodide provides a further link with his work in physical chemistry.

One of the earliest papers in which Berthelot revealed his interest in physical chemistry was published in 1856. In this he paid special attention to the boiling point, specific gravity, specific heat, heat of combustion, and refractive index of organic compounds, particularly esters. His great contributions to physical chemistry began in the 1860's. Berthelot later explained how he had become interested in physical chemistry:

In a succession of publications for several years I endeavored to compare experimentally the origins of organic compounds with those of inorganic compounds and to formulate general methods of synthesis. To extend my research I considered it appropriate to make a special study of the mechanism of these changes. The experiments which I have published on the laws governing the production of esters were published with this intention. Now I propose to examine what thermal phenomena accompany the formation and the decomposition of organic compounds—in other words the extent of the energy change [*le travail des forces vives*] necessary to bring about their synthesis.[3]

Thus Berthelot's contributions to physical chemistry arose directly from his consuming interest in the synthesis of organic compounds. His studies on chemical equilibrium published in 1862 and 1863 were followed by equally fundamental work on thermochemistry.

Berthelot and L. Péan de Saint-Gilles studied the reaction of alcohols with acids to form the corresponding ester and water. They found that the reaction never went to completion but arrived at an equilibrium state that was independent of the quantities (measured in equivalents) of alcohol, acid, ester, or water present at the beginning. On the other hand, the rate of reaction did depend on the quantities of alcohol and acid present: "The amount of ester produced at each moment is proportional to the product of the active masses [*masses actives*] present."[4] Such a statement appears to be an anticipation of the law of mass action later formulated by Guldberg and Waage. Berthelot attempted a mathematical treatment and drew graphs illustrating the formation and decomposition of esters. Berthelot and Saint-Gilles, however, while appreciating that the equilibrium was affected by the reverse reaction (ester + water), failed to take this into consideration in deriving a general mathematical expression. In their comprehensive experiments they varied temperature, pressure, concentration, and types of alcohols and acids used. They recorded the increase in reaction velocity with rise of temperature, although the final position of equilibrium was found to be almost independent of temperature. As one of their experiments on the effect of mass, one equivalent of ethyl alcohol was reacted with increasing amounts of acetic acid and the various yields of ester formed were recorded. With one equivalent of acid reacting with x equivalents of alcohol the amount of ester formed was also recorded:

X	1	1.5	2	3	4	5.4	12	19
% ester	66.5	77.9	82.8	88.0	90.9	92.0	93.2	95.0

During their investigation of a large number of alcohols Berthelot and Saint-Gilles found that the rate of reaction of borneol with acids was excessively slow, but it was left to Menschutkin to clear up the relation between the constitution of an alcohol and its rate of esterification. When Guldberg and Waage announced their law of mass action, they fully acknowledged their debt to the studies of esterification by Berthelot and Saint-Gilles.

In 1856, in the course of his investigations of esters, Berthelot established that the heat of combustion of an ester is almost exactly equal to the sum of the heats of combustion of the alcohol and acid from which it is formed. In this, as in some later work, he was able to make use of the experimental results of Favre and Silbermann. It was not until 1865, however, that Berthelot seriously turned his attention to problems of thermochemistry. He chose this subject for his lectures at the Collège de France, but he made it clear from the beginning that his interest was really in the heat changes involved in the formation and decomposition of organic compounds so that these could be compared with the comparatively well-known basic thermochemistry of inorganic chemistry as a further basis of comparison between the two branches of the science. It was in these lectures that Berthelot introduced the terms *exothermic* and *endothermic.*

Berthelot enunciated the principle that the heat evolved or absorbed in a chemical change depends only on the initial and final states of the reactants and products, provided no external work is done. This is Berthelot's "second principle," analogous to Hess's law of constant heat summation. He based this principle on the assumption of an equivalence between internal work (*le travail moléculaire*) and heat changes in a chemical reaction (Berthelot's "first principle"). Best known is Berthelot's "third principle," or "law of maximum work," which was first published in its complete form in 1873: "Every chemical change accomplished without the intervention of energy from outside tends toward the production of a body or system of bodies which produce the most heat." In the same publication Berthelot introduced the expression "principle of maximum work." The honor of beginning a new epoch in thermochemistry must be shared between Berthelot and Julius Thomsen, who had arrived in 1853 at essentially the same principle as that established by Berthelot in 1873. The principle was soon recognized as no more than a useful approximation, strictly true only at a temperature of absolute zero; it was superseded by the researches of Helmholtz, Gibbs, and van't Hoff. The accuracy of

Berthelot's thermochemical data has often been criticized, but sometimes it was his arithmetic rather than his experimental work that was at fault. In interpreting nineteenth-century thermochemical data it is also necessary to know that Berthelot's calories refer to water at 0°C., whereas Thomsen, for example, used the range 18°C.–20°C.

Berthelot tried to improve the experimental technique of thermochemistry. Recognizing the unreliability of the mercury calorimeter used by Favre and Silbermann, for example, he used a water calorimeter and either a platinum reaction chamber or one made of glass to facilitate direct observation. By the use of a water jacket, a mechanical stirrer, and a thermometer reading to .005°C., he was able to carry out experiments accurately over small temperature ranges. After more than ten years' research on thermochemistry, Berthelot published a major two-volume work for which he significantly chose the title *Essai de mécanique chimique fondée sur la thermochimie.* In the first volume, entitled *Calorimétrie,* Berthelot gave a general survey of thermochemistry, with particular reference to his own apparatus and results. The second volume, entitled *Mécanique,* was a general account of chemical reactions and decompositions. Berthelot continued his thermochemical research by determining the heats of combustion of gases. He introduced the use of the bomb calorimeter, in which the gas under test was mixed with excess oxygen compressed to 20–25 atmospheres and then sparked. This method enabled him to determine heats of combustion with an accuracy hitherto unattainable. With his bomb calorimeter, Berthelot made a fundamental contribution to both pure and applied chemistry. This was immediately recognized, and a stream of foreign scientists came to Paris to acquire firsthand knowledge from Berthelot of the new thermochemical methods. Through Berthelot, his pupils, and such collaborators as Vieille and Recoura, there was established a substantial body of reliable data of heats of combustion, solution, neutralization, and so on. Berthelot's continuing interest in thermochemistry is suggested by the publication in 1897, when he was seventy years old, of another two-volume work, *Thermochimie. Données et lois numériques.* He had already devoted to practical methods the smaller work *Traité pratique de calorimétrie chimique* (1893).

Almost as important was the fundamental work carried out by Berthelot, largely in collaboration with Jungfleisch, and published in the years 1869 and 1872. Studying the distribution of a solute between two immiscible solvents (e.g., water and benzene) or partially miscible solvents (e.g., water and ether), they were able to enunciate the partition law: "At a given temperature the quantities of solute dissolved by equal volumes of two solvents is a constant." They called this constant the partition coefficient (*coefficient de partage*). Some twenty years later Nernst investigated some apparent exceptions to this law.

While serving as president of the scientific commission appointed to bring any possible aid from science to help in the defense of Paris during the Franco-Prussian War, Berthelot investigated the possibility within the city of extracting saltpeter for gunpowder. Also in November 1870 he presented to the Académie des Sciences three memoirs on explosives, which were published in full in the *Comptes rendus* under the heading "Art militaire." Combining his patriotic duty as a Frenchman and his interest in thermochemistry, Berthelot showed how the power of explosive materials could be quantitatively expressed. He expanded this research in his book *Sur la force des matières explosives d'après la thermochimie,* first published in 1871 but greatly expanded in the two-volume definitive work published as a third edition in 1883. In the intervening period Berthelot had made a particular study of the thermochemistry of nitrogen compounds used as the basis of explosives. But it was his work carried out in collaboration with Vieille that laid the foundations of a new scientific study of the mechanism of explosions. They found that explosions were propagated in a manner in many ways analogous to that of a sound wave. They accordingly introduced the concept of an explosive wave (*onde explosive*), found to have a velocity much greater than that of sound; for example, with an explosive mixture of hydrogen and oxygen the velocity was 2,841 meters per second. The explosive wave was propagated uniformly. Its velocity depended on the nature of the explosive rather than on the material or dimensions of the vessel. Apart from its direct utility, Berthelot found his research on explosives interesting because in it he witnessed natural forces acting under extreme conditions. He insisted that his studies of explosives could have peaceful uses; and in 1896, after accidents with liquid acetylene used for illumination, he and Vieille collaborated in a study to establish how acetylene could be used with safety.

The possibility of finding new reactions by using a silent electric discharge also appealed to Berthelot. He constructed an ozonizer, essentially the same as that devised earlier by Brodie. By passing a silent electric discharge for up to ten hours through a mixture of sulfur dioxide and oxygen, Berthelot discovered persulfuric anhydride, S_2O_7. This is the anhydride of an acid of both theoretical and practical importance for the use of its salts as oxidizing agents.

It was Berthelot's lifelong interest in energy states that prompted his studies of allotropic forms of sulfur, phosphorus, and arsenic, and his discovery in 1891 (simultaneously with Mond and Quincke) of iron carbonyl.

Berthelot's first study of animal heat was in 1865, and in this he followed the thermal tradition of Lavoisier and Laplace. In 1890 he took up the subject again, carefully distinguishing the heat produced in the lungs by the action of oxygen in the blood from the later reaction in which carbon dioxide is produced, and he was able to show that the former reaction produced only one-seventh of the total heat.

In other experiments Berthelot was able to show that the combustion of foodstuffs in the laboratory produced no less heat than in the body. There was therefore no energy to relate to any vital force, a conclusion in which Berthelot took particular satisfaction. He opposed Pasteur's vitalistic interpretation of fermentation, preferring a theory of fermentation fully analogous to that of inorganic catalysts, thus once more championing the view of the strict parallel between the organic and the inorganic.

In 1883 Berthelot founded a research establishment for vegetable chemistry, and during the last twenty-five years of his life he undertook research into aspects of chemistry useful to agriculture. He found that certain carbohydrates, such as cellulose, could be made to absorb nitrogen under the action of a silent electric discharge. By treating them with lime, the absorbed nitrogen was liberated as ammonia. He was able to demonstrate that the fixation of atmospheric nitrogen by sparking is parallel to the natural process in plants. In 1885 he made his first reference to microorganisms capable of fixing nitrogen and in 1893, in collaboration with Guignard, he succeeded in isolating and forming a culture of such bacteria. In accordance with his usual practice, he brought together work done by himself and his collaborators (particularly G. André) and published it in book form. The result was his *Chimie végétale et agricole* (1899). Berthelot's last chemical book, *Traité pratique de l'analyse des gaz* (1906), concerned a subject he had studied practically from the time of his early syntheses of hydrocarbons.

NOTES

1. The spelling "Marcelin" was used on his birth certificate, and Berthelot himself sometimes wrote it that way, particularly toward the end of his life in the signing of official documents.
2. In 1850 Williamson had described the production of ethyl methyl ether from sodium ethoxide and methyl iodide as a "synthesis," *Philosophical Magazine,* 3rd ser., **37** (1850), 350.
3. *Annales de chimie et de physique,* 4th ser., **6** (1865), 290–291.
4. *Ibid,* 3rd ser., **66** (1862), 112.

BIBLIOGRAPHY

I. ORIGINAL WORKS. Berthelot wrote the following books (all published in Paris): *Chimie organique fondée sur la synthèse,* 2 vols. (1860); *De la synthèse en chimie organique. Leçon professée le 16 Mars 1860 à la Société chimique de Paris* (1861); *Sur les principes sucrés. Leçons professées à la Société chimique de Paris* (1863); *Leçons sur les méthodes générales de synthèse en chimie organique, professées au Collège de France* (1864; 1876; 1879; 1880; 1883; 1887; 1891; 1897; 1903; 1910); *Sur l'isomérie. Leçons de chimie professées devant la Société chimique de Paris* (1866); *Sur la force de la poudre et des matières explosives* (1871; 2nd ed., 1872; see also *Sur la force . . .,* 1883); *Traité élémentaire de chimie organique* (1872; later eds. with Jungfleisch, 2 vols., 1881, 1886, 1898 [Vol. I]; 1904 [Vol. II]; new ed. of Vol. I, 1908); *Essai de mécanique chimique fondée sur la thermochimie,* 2 vols. (1879; supp., 1881); *Sur la force des matières explosives d'après la thermochimie,* 3rd ed., 2 vols. (1883); *Les origines de l'alchimie* (1885); *Science et philosophie* (1886; 1905); *Collection des anciens alchimistes Grecs,* 3 vols. (1887-1888); *Introduction à l'étude de la chimie des anciens et du moyen age* (1889); *La révolution chimique. Lavoisier* (1890; 2nd ed., 1902); *Histoire des sciences. La chimie au moyen age,* 3 vols. (1893); *Traité pratique de calorimétrie chimique* (1893; 2nd ed., 1905); *Science et morale* (1897); *Thermochimie. Données et lois numériques,* 2 vols. (1897); *Chaleur animale. Principes chimiques de la production de la chaleur chez les êtres vivants,* 2 vols. (1899); *Chimie végétale et agricole. Station de chimie végétale de Meudon,* 4 vols. (1899); *Les carbures d'hydrogène, 1851-1901. Recherches expérimentales,* 3 vols. (1901); *Science et éducation* (1901); *Science et libre pensée* (1905); *Archéologie et histoire des sciences* (1906); *Traité pratique de l'analyse des gaz* (1906).

A selection from Berthelot's principal research papers is listed below. The order follows that of the text. "Sur un procédé simple et sans danger pour démontrer la liquéfaction des gaz et celle de l'acide carbonique en particulier," in *Comptes rendus,* 30 (1850), 666-667; "Action de la chaleur rouge sur l'alcool et sur l'acide acétique," in *Annales de chimie et de physique,* 33 (1851), 295-302; "Sur le bichlorhydrate d'essence de térébenthine," in *Comptes rendus,* 35 (1852), 736-738; "Sur les diverses sortes d'essence de térébenthine," *ibid.,* 36 (1853), 425-429 (see also *Annales,* 38 [1853], 38; 39 [1853], 5; 40 [1854], 5); "Sur la série camphénique," in *Comptes rendus,* 47 (1858), 266-268; "Sur l'oxidation des carbures d'hydrogène," in *Annales,* 19 (1870), 427-429; "Sur les combinaisons de la glycérine avec les acides," in *Comptes rendus,* 36 (1853), 27-29; "Mémoire sur les combinaisons de la glycérine avec les acides et sur la synthèse des principes immédiats des graisses des animaux," in *Annales,* 41 (1854), 216-319, esp. 296; "Sur quelques matières sucrées," *ibid.,* 46 (1856), 66-89; "Sur divers carbures contenus dans le goudron de houille," *ibid.,* 12 (1867), 195-243; "Sur plusieurs alcools nouveaux. Combinaisons des acides avec la cholesterine, l'éthal, le camphre de Bornéo et la méconine," *ibid.,* 56 (1859), 51-98; "Sur la reproduction de l'alcool par le bicarbure d'hydrogène," in *Comptes rendus,* 40 (1855),

102-106; "Transformation de l'oxyde de carbone en acide formique," *ibid.,* 41 (1855), 955; "Synthèse des carbures d'hydrogène," *ibid.,* 43 (1856), 236-238; "Synthèse de l'esprit-de-bois," *ibid.,* 45 (1857), 916-920; "Sur la synthèse des carbures d'hydrogène," in *Annales,* 53 (1858), 69-208; "Note sur une nouvelle série de composés organiques, le quadricarbure d'hydrogène et ses dérivés," in *Comptes rendus,* 50 (1860), 805-808; "Recherches sur l'acétylène," in *Annales,* 67 (1863), 52-77, esp. 65-68; "Nouvelle méthode pour la synthèse de l'acide oxalique et des acides homologues," in *Comptes rendus,* 64 (1867), 35-38; "Les polymères de l'acétylène. Première partie: Synthèse de la benzine," *ibid.,* 63 (1866), 479-484; "Action de la chaleur sur quelques carbures d'hydrogène," in *Annales,* 9 (1866), 445-469, esp. 454; "Action reciproque des carbures d'hydrogène. Synthèse du styrolene, de la naphtaline, de l'anthracène," *ibid.,* 12 (1867), 5-52; "Sur divers carbures contenus dans le goudron de houille," *ibid.,* 195-243; "Méthode universelle pour réduire et saturer d'hydrogène les composés organiques," in *Comptes rendus,* 64 (1867), 710-715, 760-764, 786-791, 829-832; "Remarques sur quelques propriétés physiques des corps conjugués," in *Annales,* 48 (1856), 332-347; "Recherches sur les affinités:—De la formation et de la décomposition des éthers," *ibid.,* 65 (1862), 385-422; 66, 5-110; 68 (1863), 225-359, esp. 290-291, written with Péan de St. Gilles; "Recherches de thermochimie. Premier mémoire. Sur la chaleur dégagée dans les réactions chimiques," *ibid.,* 6 (1865), 290-328; "Sur la statique des dissolutions salines," in *Comptes rendus,* 76 (1873), 94-98; "Méthode pour mesurer la chaleur de combustion des gaz par detonation," in *Annales,* 23 (1881), 160-187; "Nouvelle méthode pour la mesure de la chaleur de combustion du charbon et des composés organiques," in *Comptes rendus,* 99 (1884), 1097-1103, written with Vieille; "Sur les lois qui président au partage d'un corps entre deux dissolvants (expériences)," *ibid.,* 69 (1869), 338-342, written with Jungfleisch; "L'onde explosive," in *Annales,* 28 (1883), 289-332, written with Vieille; "Sur l'acide persulfurique, nouvel acide oxygéné du soufre," in *Comptes rendus,* 86 (1878), 20-26; "Sur la chaleur animale, chaleur dégagée par l'action de l'oxygène sur le sang," in *Annales,* 20 (1890), 177-202, esp. 199; "Fixation directe de l'azote atmosphérique libre par certains terrains argileux," in *Comptes rendus,* 101 (1885), 775-784; "Nouvelles recherches sur les microorganismes fixateurs de l'azote," in *Annales,* 30 (1893), 419-431.

II. SECONDARY LITERATURE. The following studies describe Berthelot's life and work. The articles by Graebe and Jungfleisch are particularly thorough. H. E. Armstrong, "Marcelin Berthelot and Synthetic Chemistry," in *Journal of the Royal Society of Arts,* 76 (1927-1928), 145-171; A. A. Ashdown, "Marcellin Berthelot," in *Journal of Chemical Education,* 4 (1927), 1217-1232; G. Bredig, "Marcelin Berthelot," in *Zeitschrift für angewandte Chemie,* 20, pt. 1 (1907), 689-694; A. Boutaric, *Marcellin Berthelot* (Paris, 1927); E. Farber, "Berthelot," in *Das Buch der grossen Chemiker,* G. Bugge, ed., II (Berlin, 1930), 190-199, trans. in E. Farber, ed., *Great Chemists* (1961), pp. 677-685; C. Graebe, "Marcelin Berthelot," in *Berichte der Deutschen*

Chemischen Gesellschaft, **41** (1908), IIIB, 4805–4872; E. Jungfleisch, "Notice sur la vie et les travaux de Marcellin Berthelot," in *Bulletin de la Société chimique de France,* **13** (1913), i–cclx; L. Velluz, *Vie de Berthelot* (Paris, 1964); and R. Virtanen, *Marcelin Berthelot. A Study of a Scientist's Public Role,* University of Nebraska Studies, no. 31 (Lincoln, 1965).

M. P. CROSLAND

BERTHIER, PIERRE (*b.* Nemours, France, 3 July 1782; *d.* Paris, France, 24 August 1861), *mineralogy, mining engineering, agricultural chemistry.*

Berthier entered the École Polytechnique in Paris in 1798, where he studied under Monge and Berthollet. On completing his course, he entered the École des Mines in 1801, and was one of the few students who moved with the school to Montier. Schreiber, the head of the school, was an experienced mining engineer who stressed fieldwork, including actual mining experience in the lead and silver mine at Pesey.

Berthier was called in 1806 to the newly completed central laboratory of the Board of Mines. In 1816, after additional field experience, he was appointed professor of assaying and chief of the laboratory at the École des Mines. Even after he retired in 1848, Berthier maintained his laboratory there. He was paralyzed in 1851 as a result of a street accident.

With Arago's warm support, Berthier was elected a member of the Académie des Sciences in 1825, and became a *chevalier* of the Legion of Honor in 1828. In 1859 he was awarded the Grand Gold Medal of the Society of Agriculture of France.

Berthier published more than 150 papers on a wide variety of scientific subjects. Most appeared in *Annales de chimie* and *Annales des mines,* with later papers in Erdmann's *Journal,* Liebig's *Annalen,* and the *Quarterly Journal of Science.* He analyzed kaolin, pioneered in locating deposits of native phosphates for use in agriculture, analyzed dozens of minerals and metalliferous ores, and discovered several new mineral species, including bauxite and Berthierite. Berthier is credited with knowing, before Mitscherlich's work on isomorphism, that substances that are chemically different may have the same crystalline form and may even cocrystallize.

Berthier put forth no new concepts, preferring instead to add to man's stock of chemical and mineralogical facts. His well-known *Traité des essais par la voie sèche* was widely used by mineralogists and mining engineers because his analytical procedures were simple, relatively accurate, and practical. Berthier maintained a lifelong interest in plant chemistry, and his analyses of plant constituents received

some notice, but his importance lies in what he added to French geology, mineralogy, and metallurgy.

BIBLIOGRAPHY

I. ORIGINAL WORKS. Berthier's papers are too numerous to cite individually. A nearly complete list is in the *Catalogue of Scientific Papers,* Royal Society of London, 1st ser., pp. 315–319. The list in Poggendorff, I (1863), 166, is incomplete. Berthier's major work is *Traité des essais par la voie sèche ou des propriétés, de la composition et de l'essai des substances, métalliques et des combustibles,* 2 vols. (Paris, 1834).

II. SECONDARY LITERATURE. The best recent, although somewhat uncritical, account is R. Samuel LaJeunesse, *Grands mineurs français* (Paris, 1948), ch. 7. See also *Dictionnaire de biographie française,* VI (1954), 218; Jerome Nicklès, "Correspondence of Jerome Nicklès...," in *American Journal of Science,* 2nd ser., **32** (1861), 108; École Polytechnique, *Livre du centenaire,* III (Paris, 1897), *passim;* and J. R. Partington, *A History of Chemistry,* IV (London, 1964), 97–98.

LOUIS I. KUSLAN

BERTHOLD, ARNOLD ADOLPHE (*b.* Soest, Germany, 26 February 1803; *d.* Göttingen [?], Germany, 3 February 1861), *physiology.*

Berthold, who came from a simple family of artisans, began his medical studies at Göttingen in 1819 and presented his doctoral thesis in 1823. Following custom, he visited various German and foreign universities—including Berlin in 1824 and Paris in 1825—in order to increase his knowledge of practical medicine and comparative anatomy. Having qualified as *Privatdozent* at Göttingen in 1825, he began his lifetime career there. He was named extraordinary professor of medicine in 1835 and ordinary professor the following year. He was also curator of the zoological collections.

Since he was absorbed in both the practice and the teaching of medicine, Berthold left many and varied published works. As early as 1829 he wrote *Lehrbuch der Physiologie des Menschen und der Thiere,* which was reissued many times. His monographs, articles, and notes were published in medical, scientific, and even literary periodicals. A piece of research done with Bunsen (1834) led to the discovery of hydrated iron oxide as an antidote for arsenic poisoning. Some of his other works dealt with myopia, the length of pregnancy, male hermaphroditism, and the formation of fingernails and hair. His short work commemorating Goethe's centennial in 1849 was one of the first German publications to do justice to Goethe as a naturalist. According to Gurlt, Berthold's life was typical of a nineteenth-century German university professor with a high reputation. Gurlt's article has been reprinted without modification and without special mention of the experiment that made Berthold a forerunner of modern endocrinology.

In 1849 Berthold published "Transplantation der Hoden," a four-page article that in its conciseness was a model for experimental investigation. It is the report of the experiments he performed on six cockerels, using each pair for the removal and transplantation of the testicles. The most remarkable result was the successful grafting of testicles from one cockerel into the abdominal cavity of another, with the cockerel receiving the transplant retaining the secondary sexual characteristics of crowing and combativeness. This article completely escaped notice at the time and remained in oblivion until 1910, when Biedl, in his *Innere Sekretion* (p. 5), demonstrated that Berthold should be considered the first scientist to have shown by experimental means the correlation of a gland with the *milieu intérieur* of an organism. (In reality, as early as 1905 Nussbaum had analyzed and evaluated Berthold's experiments before reproducing them on batrachians.) Since then, all historical accounts have considered Berthold as one of the founders of endocrinology. In reality, he was a forerunner without immediate successor.

Berthold's article was translated into English and commented upon by Rush (1929), Quiring (1944), and Forbes (1949), who speculated upon the origin of this particular experiment. From a study of the text it is apparent that Berthold was preoccupied with the trophic nerves:

> Since, however, transplanted testes are no longer connected with their original innervation, and since, as indicated [above], no specific secretory nerves are present, it follows that the results in question are determined by the productive function of the testes (*productive Verhältniss der Hoden*), i.e., by their action on the bloodstream, and then by corresponding reaction of the blood upon the entire organism, of which, it is true, the nervous system represents a considerable part [Quiring, p. 401].

It was not the first time that Berthold showed interest in the physiology of reproduction. He was the author of an important, comprehensive study of sexual characteristics in Wagner's *Handwörterbuch*, a reference book highly regarded at the time and still of great interest in the history of biology.

BIBLIOGRAPHY

I. ORIGINAL WORKS. Among Berthold's writings are *Erster Abriss der (menschlichen und thierischen) Physiologie* (Göttingen, 1826); *Lehrbuch der Physiologie des Menschen und der Thiere* (Göttingen, 1829); *Das Eisenoxydhydrat, ein Gegengift der arsenigen Säure* (Göttingen, 1834), written with R. G. E. Bunsen; "Geschlechtseigentümlichkeiten," in R. Wagner, *Handwörterbuch der Physiologie*, I (Brunswick, 1842), 597–616; *Lehrbuch der Physiologie für Studierende und Aerzte*, 3rd ed., 2 vols. (Göttingen, 1848); *Am 28 August des J. 100, nach der Geburt Göthes in einem Kreise Göttingischer Verehrer . . .* (Göttingen, 1849); and "Transplantation der Hoden," in *Archiv für Anatomie, Physiologie und wissenschaftliche Medicin. . .* (1849), 42–46, trans. by D. P. Quiring as "Transplantation of Testis," in *Bulletin of the History of Medicine,* **16** (1944), 399–401.

II. SECONDARY LITERATURE. Works on Berthold are A. Biedl, *Innere Sekretion. Ihre physiologische Grundlagen und ihre Bedeutung für die Pathologie* (Berlin–Vienna, 1910); T. R. Forbes, "A. A. Berthold and the First Endocrine Experiment: Some Speculations as to Its Origin," in *Bulletin of the History of Medicine,* **23** (1949), 263–267; E. Gurlt, "A. A. Berthold," in *Biographisches Lexikon der hervorragenden Ärzte aller Zeiten und Völker* (Leipzig–Vienna, 1884; 3rd ed., 1962), I, 501–502; M. Klein, "Goethe et les naturalistes français," in *Publications de la Faculté des lettres, Université de Strasbourg,* **137** (1958), 169–191, see 177; N. Nussbaum, "Innere Sekretion und Nerveneinfluss," in *Ergebnisse der Anatomie und Entwickelungsgeschichte,* **15** (1905), 39–89, esp. 67–68; and H. P. Rush, "A Biographical Sketch of Arnold Adolf Berthold," in *Annals of Medical History,* n.s. **1** (1929), 208–214.

MARC KLEIN

BERTHOLLET, CLAUDE LOUIS (*b.* Talloire, near Annecy, Savoy, 9 December 1748; *d.* Arcueil, France, 6 November 1822), *chemistry.*

Berthollet came from a French family that had emigrated to Savoy during the previous century and had become members of the *noblesse de robe.* The family was, however, in straitened circumstances when Claude Louis was born. He first studied at the *collèges* in Annecy and Chambéry, and later qualified as a physician at the University of Turin in 1768. After this he settled in the Piedmont for four years before moving to Paris in 1772, where he studied chemistry under Macquer and Bucquet while continuing to study medicine. As a Savoyard he could introduce himself to his near-compatriot Tronchin, from Geneva, an associate of the Académie des Sciences, propagator of vaccination in France, and the chief personal physician to the regent, the duke of Orléans. Upon Tronchin's recommendation, the duke had Berthollet appointed private physician to Mme. de Montesson, and allowed him to carry out research in the private laboratory installed by the regent and his son in the Palais Royal. Here Berthollet repeated the experiments on elastic fluids of Lavoisier, Priestley, and Scheele, and met Lavoisier. He qualified as a doctor of medicine at the University of Paris in 1778 and married Marguerite Baur in the same year.

Between 1778 and 1780 Berthollet presented seventeen memoirs to the Academy; these led to his election as a member on 15 April 1780, on the death of Bucquet (Fourcroy opposed him). In 1784, on Macquer's death, he was appointed inspector of dye works and director of Manufacture Nationale des Gobelins. He subsequently collaborated with Lavoisier, Fourcroy, and Guyton de Morveau in the publication of *Méthode de nomenclature chimique* (1787), incorporating the principles of the new chemistry of Lavoisier.

Berthollet flourished under four different political regimes. In 1792 he was appointed a member of the commission for the reform of the monetary system, and in 1793 the Committee of Public Safety made him an important member of the scientific commission concerned with war production, particularly that of munitions. He was appointed to the commission on agriculture and arts on 22 September 1794 and was made a professor at the École Normale. Berthollet was also charged, with his lifelong friends Monge and Guyton de Morveau, with the organization of the École Polytechnique, where for a time he taught animal chemistry. In 1795 he was one of the first members elected to the Institut de France, which replaced the suppressed Academy in 1793.

With the fall of Robespierre and the revolutionaries, Berthollet's star shone even more brightly under Napoleon, who showed a deep admiration and affection for the chemist. In 1796 Napoleon appointed Berthollet and Monge to accompany the commission that was to bring back the great works of Italian art to France. In the execution of this assignment Berthollet developed some of the earliest chemical methods for the restoration of paintings. Two years later, Berthollet and Monge accompanied Napoleon as scientific members of his expedition to Egypt, where they stayed for two years and established an Institute modeled on that of Paris. In 1804 Napoleon made Berthollet a count, senator for Montpellier, administrator of the mint, and *grand officier* of the Légion d'Honneur. After this, Berthollet led a semi-retired existence in the Paris suburb of Arcueil. In 1807 he and Laplace founded the Société d'Arcueil, which met regularly to discuss scientific problems, and published three volumes of its proceedings. In 1811 Berthollet's son, Amédée, committed suicide when his business (manufacturing sodium carbonate according to a new method developed by his father) failed.

In tracing the development of Berthollet's scientific work, it must be emphasized that, for all his original contributions, he was essentially part of the continuous historical tradition of chemistry. Unlike his senior contemporary Lavoisier, Berthollet wanted to improve rather than to revolutionize the basis of the science. Whereas Lavoisier had tried to found a new chemistry deriving from the analysis of its most fundamental principles, Berthollet wanted to reinvigorate the traditional science by synthesizing ideas derived from various sources. He was trained as a physician, as was common for chemists from Paracelsus to Boerhaave and Black; in addition, like them, he not only sought an adequate theoretical explanation of chemical phenomena but also strove to find an immediate application for his ideas. His enthusiastic espousal of the ideals of the French Encyclopedists reinforced this longing to put science at the service of man's practical needs.

From 1778 to 1783 Berthollet sent a large number of memoirs to the Academy, all of them admitting the essential correctness of the phlogiston theory. The main characteristic of his mature work is foreshadowed in these early contributions: the quest for a synthesis that would lead to a more adequate understanding of chemical phenomena, by fusing the divergent principles of Stahl's and Lavoisier's systems while incorporating the eighteenth-century tradition of the chemistry of affinities. His preoccupation with practical applications was revealed in such investigations as the attempt to revive asphyxiated animals with the help of "dephlogisticated air." He also studied the properties of soap, which he explained by assuming different distributions of the respective affinities of its components during their interaction with a solvent, and developed original procedures for making new types of "metallic" soap that would have been useful for medicinal purposes.

Berthollet's earlier investigations were largely concerned with Lavoisier's major preoccupation, the study of gases. In the beginning he opposed Lavoisier's ideas while defending the phlogiston theory. It might help to dispel the popular caricature of the opponents of Lavoisier if we recall some of Berthollet's arguments in this debate.

Tradition. Stahl's doctrine, according to Berthollet, had been so successful in accounting for chemical phenomena that there appeared to be no adequate reason for rejecting it. Berthollet argued that the study of air and other elastic fluids (or gases) had shown the need for modifying this doctrine in some details, with which many of the other chemists agreed, but he did not think that it followed from this that the traditional basis of chemistry was to be swept aside. Lavoisier alone appeared radical, setting out systematically to disprove the existence of phlogiston by a series of experiments, the precision of which Berthollet admired. Berthollet held that Lavoisier's positive contributions could be synthesized with a revised phlogiston theory, while the negative conclusions as to the existence of phlogiston could only be harmful to chemistry, the unity of which would be destroyed by such iconoclasm.

Weight. While later historians have made the whole debate on phlogiston revolve about Lavoisier's so-called crucial experiments showing the increase in weight of substances during combustion, it can be seen by studying Berthollet's earlier work that it was possible to have accepted this part of Lavoisier's work without rejecting Stahl. In a series of experiments on the conversion of sulfur, phosphorus, and arsenic to the corresponding "acids" (oxides), Berthollet confirmed Lavoisier's conclusion that there was an increase in weight in all three cases because "vital air" was added. He also agreed that, notwithstanding earlier investigations in which a diminution in the quantity of the surrounding air had been observed, Lavoisier was the first to have indicated that there was an increase in weight when metals, sulfur, and phosphorus were calcinated. So far from seeing any inconsistency between these ponderable considerations and the phlogiston theory, Berthollet, although a supporter of Stahl, felt justified in criticizing Lavoisier for not having been sufficiently vigorous in insisting on his conclusions (a deficiency attributed, paradoxically enough, to Lavoisier's rejection of the phlogiston theory). While it would be tedious to go into the explanations offered by Berthollet in accordance with Stahl's doctrines, it is historically interesting to list some alleged inconsistencies on Lavoisier's part when dealing with quantitative relations.

(*a*) Lavoisier had maintained that the causticity of metals was due to their combination with "vital air." To Berthollet this was doubly suspect. First, oxygen combined with all the metals in almost the same proportion to form calces, yet the latter varied greatly in their causticity, contrary to what should have happened according to Lavoisier's doctrine. Second, the red precipitate of mercury was a highly caustic calx. This implied that it contained a large amount of oxygen, if causticity were due to the presence of this substance—but in fact the red oxide of mercury contained only a tiny proportion of oxygen, not more than one part to every ten or twelve parts of the metal.

(*b*) When hydrogen and carbon were separately burned, Lavoisier's theory led to inconsistencies with regard to the quantity of oxygen required in each case. For hydrogen, he had to say that the heat generated was due only fractionally to the consumption of the oxygen present—about 1/21 of the total—while the rest was due to hydrogen itself. For the burning of carbon, he attributed all the heat to the oxygen consumed. Berthollet rightly questioned the discrepancy in the quantitative factors involved in the explanation of the same type of chemical phenomenon, combustion.

Accompanying effects. The rejection of phlogiston had led Lavoisier to postulate an alternative set of explanations for the physical phenomena accompanying combustion: production of a flame (heat and light), and the change of physical state from solid or liquid to vapor. This he achieved by postulating a generic principle called caloric, which was also brought in to explain the existence of substances that were permanently in the gaseous state. Berthollet pointed out that not only was it impossible to clarify, even in principle, the exact relationship of heat to light in Lavoisier's system (unlike Stahl's, in which all such phenomena were explained uniformly by the presence of phlogiston), but also that empirical evidence refuted Lavoisier's ideas about such physical effects. Thus carbon, sulfur, and the metals detonated strongly, producing a flame, when distilled with niter. For Lavoisier substances in the solid state, such as all the reactants in these cases, were deprived of caloric and therefore should not have been able to produce a flame giving out heat and light. He had explicitly stated that only substances in the gaseous or liquid state contained sufficient caloric to produce these effects.

Lavoisier had asserted that the physical effects accompanying combustion required a diminution in the volume of the elastic fluids or air present, since caloric was required for the production of the effects. On the other hand, Berthollet pointed out that in cases such as the detonation of a mixture of carbon and gunpowder, an elastic fluid possessing a greater volume than the reactants was actually given out. Consequently this experiment should have been accompanied by the production of cold, since caloric was being absorbed and not given out, contrary to the experimental evidence, which showed that heat and light were produced.

Berthollet had here seized upon the real weakness of Lavoisier's explanation of combustion: that this was not only a chemical phenomenon in which something (oxygen) was absorbed, but also a physical phenomenon in which something (energy or, rather, enthalpy, heat minus entropy) was released. This inquiry was not undertaken, however, until the notion of energy was clarified in the third and fourth decades of the nineteenth century.

The foregoing controversy ushered in Berthollet's preoccupation in his later scientific career, the desire to reformulate the basic principles of chemistry by synthesizing the traditional views with the important new discoveries of Lavoisier, a task that was to take him the better part of twenty years. In the intervening period he rejected the phlogiston theory and embraced Lavoisier's doctrine.

In 1785, after he had set out to test its correctness by performing a large number of experiments on "dephlogisticated marine acid" (chlorine), Berthollet explicitly stated for the first time that "this principle, which Stahl had ingeniously imagined in order to explain a large number of phenomena, and by means of which a genuine relationship could be established between them, namely phlogiston, having sufficed for the needs of chemistry during a long period," had at last become a useless hypothesis. Ironically, the test case chosen was more correctly explained by the followers of Stahl than by Lavoisier. As the name suggested, Stahl correctly thought of chlorine as marine (hydrochloric) acid that had lost phlogiston (often identified with hydrogen), while Lavoisier assumed that he was dealing with oxygenated marine acid (oxygen plus hydrochloric acid).

Berthollet's reasons for agreeing with Lavoisier were summarized in the statement that in the preparation of chlorine, using a mixture of manganese dioxide and hydrochloric acid, the "vital air" of the former could be shown to have combined with the latter, thereby proving that oxygenated marine acid had been formed during the reaction. The actual details of Berthollet's reasoning were rather obscure. He cited the formation of fixed air and common salt (by boiling oxygenated marine acid with soda) as an important detail in this context, without making it clear why and how it was relevant. Three other experiments were also given in support of the Lavoisienne view. All of them, however, appeared to be curiously vitiated by circular reasoning, in that Berthollet assumed that the phenomena observed were best explained by supposing that oxygen was combined with marine acid in chlorine. These experiments were (*a*) the dissolution of metals in a solution of oxygenated marine acid when no gas was given off, because Berthollet thought that the metals combined directly with the oxygen of the acid and therefore did not decompose the water, which otherwise would have given off hydrogen; (*b*) the conversion of hydrogen sulfide gas into vitriolic acid when it was passed through the solution of chlorine, which he explained by assuming that the oxygen of the latter combined with the gas; and (*c*) the transformation of mercury into a corrosive sublimate which contained a large proportion of "vital air" and marine acid, the two supposed ingredients of the solution of chlorine in which the metal was dissolved.

Two points are worth observing here. First, Berthollet was already preoccupied with finding an explanation of phenomena, interpreted on Lavoisienne principles, that would finally be understood through a complete theory of affinity. Thus, he mentioned that it was not a simple elective affinity between the two that caused marine acid to combine with the oxygen of manganese dioxide, but that a more complex distribution of affinities was required to account for the production of oxygenated marine acid. He suggested that marine acid had only a feeble affinity for oxygen and believed that its combination with the "vital air" of the calx was to be explained by a change in the state of manganese, which was easily dissolved by the marine acid, this dissolution being accompanied by an expulsion of oxygen with which the metal was originally combined. In this process a more concentrated form of oxygen, which lost a large part of its caloric when separated from the manganese, was obtained. The more concentrated form of the oxygen helped it to combine with marine acid despite their weak mutual affinities. Berthollet later expanded this explanation into a complete system of chemistry.

Second, Berthollet's change of allegiance from Stahl to Lavoisier was directed more by practical expediency than by any apparent superiority in either the logic or the adequacy of the new system. Lavoisier's ideas enabled Berthollet to give a much simpler explanation of the phenomena, although one that was obviously quite inadequate on many counts. Not only were Berthollet's original objections to Stahl's system not answered, but many new ones also arose in the application of Lavoisier's principles.

A glaring instance was Lavoisier's explanation of the properties of acids as deriving from the presence of oxygen. Berthollet had already shown, in 1778, that hydrogen sulfide, while possessing the characteristics of a feeble acid, did not contain oxygen. He was to reconfirm this discovery nearly twenty years later. But Berthollet's most important contribution in this domain was his analysis of prussic acid (1787), which he correctly showed to be composed of hydrogen, carbon, and nitrogen. Although he did not succeed in determining the relative proportions of its components, he was convinced that it contained no oxygen. Lavoisier himself accepted these results but avoided their theoretical implications by suggesting that prussic acid was perhaps not an acid after all. Several years later Berthollet followed up this investigation with those of some other acids—hydrochloric, uric, boric, and fluoric—all of which contained no oxygen, according to him.

Although Berthollet never rejected Lavoisier's ideas after 1785, he gave indications of how he was eventually going to suggest an alternative explanation that would avoid such difficulties as that of the composition of acids. Lavoisier had explained the properties of substances by referring to the elements of which they were composed; Berthollet proposed to derive them from the relations between their constituents. He did not define acidity in isolation, depending upon the presence of oxygen, but by the interaction between the components of one substance in the presence of another; the former was an acid if it was neutralized by a base, and if together the two gave rise to a series of salts.

A valuable consequence of Berthollet's adherence to Lavoisier's system was the determination of the composition of ammonia, for which he gave the earliest accurate analysis (1785). He had tried in 1778 to explain the origin of this alkali, which had been formed in various experiments concerned with distilling alcohol over ammonium carbonate and other "fixed" alkalies. Lavoisier, who had given an account of this work to the Academy, had enjoined Berthollet not to publish it because it contained serious errors. This advice was accepted, although Berthollet intended to repeat the analysis of ammonia, which he carried out soon after his conversion to Lavoisier's ideas.

Berthollet also acknowledged his debt to Priestley (who had decomposed ammonia in 1775–1777) in the method of the analysis: the passage of an electric current through ammonia. The resulting mixture of gases was exploded with oxygen and its composition was analyzed. From the results obtained it was shown that ammonia was composed of 2.9 volumes of inflammable gas (hydrogen) to 1.1 volumes of moffette (nitrogen).

These investigations into pneumatic chemistry were prompted not only by Berthollet's interest in the theoretical issues raised by the differences between the systems of Lavoisier and Stahl, but also by his interest in the practical implications of these new discoveries. An illustration of this was his study of chlorine, which immediately led to two separate applications. The first of these stemmed from his preparation of potassium chlorate by saturating a concentrated solution of caustic potash with chlorine. A mixture of the chlorate and carbon exploded energetically, leading Berthollet to try to replace niter with potassium chlorate to obtain a more powerful kind of gunpowder. A public experiment was carried out with this new type of gunpowder in 1788, with unexpected results—the director of the plant and four other people were killed on the spot. It appears that this innovation was later effectively used in military operations.

Berthollet was more successful in his other attempt to put the properties of chlorine to practical use. Having observed that chlorine had bleaching properties, he wanted to find a simple technique for introducing it as a bleaching agent for textiles. This was partly prompted by his affiliations with Gobelins, but chiefly (as he often insisted) by the humanitarian ideals inculcated in him by the Encyclopedists. The traditional methods of bleaching, which involved soaking cloth in whey and spreading it in a sunny field, was, he said, wasteful, since it prevented large tracts of land from being tilled. It was a measure of his humanitarian impulse that, unlike such contemporaries as Watt, who amassed large fortunes from their industrial inventions, Berthollet published his technique for the bleaching of textiles by chlorine without bothering to patent it. It is some reflection on the morality of pioneers in the same field that they invited Berthollet to demonstrate the application of his method—which he did gratis—and then tried to patent his discovery of the bleaching liquid, calling it *lye de Javelle.*

Berthollet's method of chlorine bleaching consisted of pouring sulfuric acid on a mixture of six ounces of manganese monoxide and sixteen ounces of salt. This mixture was heated by immersion in boiling water and a bleaching solution was obtained by collecting the chlorine in water (100 quarts for every pound of salt). The cloth to be bleached was first soaked in diluted caustic potash, then washed and eventually immersed in the bleaching solution for three to four hours; the operation was repeated several times. Finally the bleached cloth was washed with soft soap and rinsed in diluted sulfuric acid.

Another of Berthollet's important contributions to the textile industry was the treatise in which he endeavored to place the ancient craft of dyeing on a scientific basis by a systematic discussion of its procedures, coupled with an attempt to find an adequate set of theoretical principles to explain the chemical actions involved. His explanation was that, depending on the variable physical conditions of temperature, quantity of solvent employed, and so forth, when a cloth was dyed the reciprocal affinities of the particles of the dye, the mordants, and the cloth itself were responsible for the kind and quality of dyeing. The colors produced were due to the oxidation of the mordant by the atmosphere.

During his stay in Egypt, Berthollet noticed the apparently inexhaustible source of sodium carbonate constituted by Lake Natron, at the threshold of the desert. He sought an explanation for this natural phenomenon in terms of a chemical theory of affinity that had been maturing in his mind over the years. The ground surface of Egypt, he reflected, was covered with a layer of ordinary salt, while the neighboring mountains of Libya were formed of limestone. If these substances reacted with each other, a double decomposition would have occurred, forming sodium carbonate and calcium chloride. But limestone did not ordinarily react with salt—an explanation in terms of double decomposition for the formation of sodium carbonate in Egypt would therefore have been acceptable only if some special circumstances obtained in this case. Berthollet pointed out that the physical conditions in the area were sufficiently unusual to warrant this assumption. Two factors probably intervened: the high temperatures prevailing in the region and the relatively large quantities of limestone present. When a salt solution filtered slowly through the pores of the limestone, the relatively weak affinities between these two substances were enhanced by the combined effects of the temperature and the enormous mass of limestone. This led to decomposition of the salt, assuring a constant production of sodium carbonate and calcium chloride through double decomposition, as a result of the redistribution of the affinities between the original reactants.

These observations lent a renewed interest to Berthollet's earlier suggestion that such physical conditions as temperature, relative concentration, and quantities of reactants affected the nature and direction of affinities in a chemical reaction. Berthollet read a memoir on the general theory of affinities while he was still in Egypt. This was the starting point of his complete new system of chemistry, first briefly sketched in *Recherches sur les lois de l'affinité* (1801) and later developed into the comprehensive, two-volume *Essai de statique chimique*. Here he attempted to provide a proper basis for chemistry, so that its experimental results could be viewed in the light of theoretical first principles. Berthollet developed a theory and a model adequate for the understanding and the interpretation of the rapidly growing body of chemical knowledge in his time. He was aware that the positive work of constructing a new theory had yet to be performed after the shock of Lavoisier's criticism of the old chemistry.

In his attempt to provide chemistry with an adequate theoretical foundation, Berthollet recognized the importance of the theory of affinity. He pointed out that for lack of a proper critical appraisal of the principles involved, his predecessors' ideas on affinity, as expressed in the construction of "tables of affinity"

through the eighteenth century, were perhaps somewhat crude and immature. The main objection to these tables was that they assumed affinity to be a general force, unaffected by the experimental conditions and always constant. For example, it had been supposed that if two acids, A and B, were considered, the table of affinities could show at a glance which of the two had a stronger affinity for a base, Z. This led to the view that the acid with the stronger affinity could always replace the other acid in a compound with the base, no matter what the experimental conditions might be. Thus if, according to a table of affinities, A had a stronger affinity for Z than B did, then it would always be the case that on adding acid A to a compound of B and Z, all the B would be replaced in the compound by A, giving the substitution product AZ. With the introduction of rudimentary quantitative methods in the latter half of the eighteenth century, this view had been extended to justify the doctrine that all substances combined in constant proportions: two substances always combined in the same proportion because of their fixed affinity for each other.

Before the time of Berthollet, it had already been maintained by various chemists that this view oversimplified the nature of chemical combinations. Lavoisier had shown, for example, that the nature of chemical combination varied with the temperature. Consequently, he pointed out, a separate table of affinity should be constructed for each degree of temperature. Bergman actually constructed tables showing how affinity varied with temperature. In spite of a few efforts along these lines, affinity remained in some sense an "absolute" that could be determined once and for all from a given table, irrespective of any variation of the conditions under which a reaction took place.

Berthollet undertook a thorough examination of the notion of affinity as his predecessors had employed it. His main contribution to its development was the proof that affinity was a relative concept which varied with the physical conditions accompanying an experiment: quantity, temperature, solubility, pressure, and physical state (solid, liquid, or gas) determined the relative force with which one substance attracted another. Berthollet then tried to prove that the proportions in which two substances combined also varied according to the conditions. This led to his famous controversy with J. L. Proust.

According to the *Essai*, there were two main types of forces in nature: gravitation, which accounted for astronomical phenomena, and chemical affinity. It was quite possible that they had a common origin, but they were best treated separately, so as not to lose sight of the very important differences between chemical affinity and astronomical attraction. For, unlike affinity, astronomical attraction operated at such enormous distances that its action could always be considered to be uniform. The shapes, sizes, and specific properties of the molecules composing a particular substance determined the way in which chemical affinity was defined in any given case, so that the exercise of this force was not uniform in all cases. Besides, the results produced were quite different in the two cases. The end product of chemical affinity was always a combination of the substances concerned, whereas no such phenomenon could be associated with astronomical attraction.

To elucidate the nature of chemical combination and affinity, Berthollet employed an explanatory model. This was a mixture of the two main types of methods later utilized by chemists, the atomic and the planetary, although he was not very consistent in the use of either model. Thus, he envisaged substances as composed of minute particles or molecules, which roughly corresponded to atoms. But he also asserted that the proportions in which two substances combined varied continuously, thereby making it impossible to think of these molecules as possessing the most characteristic property of atoms, that of indivisibility. Berthollet's molecules were supposed to be endowed with mutual attraction. The interplay between the total number of forces in a given substance was supposed to lead to the production of a stable system. This was analogous to a planetary system, where the sum of the forces between the heavenly bodies results in a state of equilibrium. The analogy did not extend further than this: there was no central nucleus or sun around which the molecules of the substances revolved as in a planetary system. It is not clear whether the model was a dynamic one.

Berthollet pointed out that a chemical substance represented a state of equilibrium between the forces of its component molecules. Likewise, all chemical combinations were caused by an interplay of the forces of molecules composing the reactants. The manner in which the different forces influenced a chemical reaction required that they be distinguished from each other. The former theories of affinity had failed to take these differences into account because of the belief in a uniform force of affinity. The factors that had to be considered in evaluating these forces, according to Berthollet, were the following:

Chemical affinity. The attraction that different substances had for each other, or that the molecules of the same substance exercised upon one another, ultimately depended on their chemical natures, which determined their chemical affinities for each other. The affinity of one substance for another had an upper, but generally no lower, limit. The proportions in which two substances combined could vary from the smallest part to the maximum. The variation normally was continuous and differed very slightly from one compound to another. When the maximum reciprocal affinities between two substances had been satisfied, they were said to be "saturated" with respect to each other.

Quantity. Given a certain force of attraction between two substances, it was natural to suppose that the larger the quantities used, the greater the force deployed. Since a substance was thought of as an aggregate of minute particles or molecules, it was only natural to suppose that each of these would bear a determinate portion of the total force. In any reaction, therefore, affinity was not a constant force that could be determined once and for all: it would vary according to the quantities of the substances. The nature of the combination resulting from the interaction of two or more substances was not a simple function of their respective affinities, considered independently of their masses: in fact, it was not possible to give any precise meaning to the idea of chemical affinity unless it was associated with that of the masses of the reactants. If a salt was formed by the combination of an acid, A, and a base, Z, then it could not be determined in advance whether another acid, B, would displace A in the given salt. For example, A might have a greater affinity for Z than B had for Z in a given reaction. Nevertheless, if a sufficient quantity of B was added to AZ, a point would be reached at which the joint action of the quantity of B and its (relatively weaker) affinity for Z would start to counteract the affinity of A for the base: part of AZ would be converted into BZ.

The forces responsible for chemical combinations depended on the relative attraction or affinity of one substance for another, according to its chemical nature, as well as on the number of its reacting molecules, measured by its quantity. Instead of considering affinity in isolation, as in the tables of affinity, Berthollet proposed to use a more complex concept combining the idea of affinity with that of the mass of a reacting substance.

The relative affinities of different substances could be measured by comparing the quantities of each that would saturate a given amount of the same substance, provided the physical conditions were constant. The idea of saturation was extended by Berthollet to the maximum quantity of any given substance that would combine with another under given conditions, rather than limiting it to the neutralization of acids by bases and of bases by acids.

Berthollet combined the concept of relative affinity with that of the mass of reactants in a chemical combination. This gave the total force with which a given quantity of a substance reacted with another. Instead of taking the quantity of the affinity by itself, Berthollet suggested the use of a concept such as "effective mass" or "chemical mass" of a substance in given reaction. He added that the use of such terms as "mass" and "affinity," which implies that they have clear meanings, should be abandoned if chemistry did not want to be stunted for lack of properly analyzed theoretical foundations. Instead of saying, for instance, that sulfuric acid had a greater affinity for caustic soda than acetic acid did, it was necessary to take into account how much of either would combine with the alkali under given conditions. This would indicate the total attraction exerted by either acid, under similar conditions, upon the alkali: this idea is properly expressed by the complex concept of chemical mass, representing the product of the power of saturation and its mass. Although this was one of the most important ideas introduced into chemistry by Berthollet, its importance was generally overlooked until the second half of the nineteenth century, when there was a revival of interest in the nature of chemical equilibrium and the physical conditions that affect that equilibrium.

Distance (*cohesion and elasticity*). Besides affinity and quantity, Berthollet pointed out that a chemical reaction was strongly influenced by the physical conditions under which substances were made to react. This was one of his most original contributions: he was probably the first chemist to undertake an exhaustive examination of these conditions. In order to grasp his position, we have to understand the details of the model he constructed to represent the course of a reaction. Although his model was not stated very explicitly, the following appears to be the most coherent interpretation of his ideas on the subject.

It was generally observed that substances increased in volume when they were heated, and contracted when cooled. From this Berthollet inferred a particulate structure of matter: substances were composed of small, discrete particles, invisible to the naked eye, and located at definite distances from each other. The application of heat increased the distances between these particles. If substances had been compact masses or undivided wholes, rather than aggregates of discrete particles, it would have been difficult to account for the expansion of bodies when they were heated and their contraction when cooled. This was particularly true when one attempted to explain how an apparently compact mass, such as a solid, could contract when cooled. The different states of matter were thus explicable by relative increases or decreases in the distances between particles as a substance passed from one state to another.

Berthollet used this model to elucidate the relationship between the physical states of matter and the phenomena of chemical combination. This made him the first chemist to attempt a detailed explanation of chemical reactions in quasi-mechanical terms. He asserted that the affinity between two particles attracted to each other was influenced by their distances. The closer together they were, the more strongly they were attracted. It followed that the following minimum conditions had to be fulfilled before any two substances combined: The chemical nature of each substance had to be such that its molecules attracted, and were attracted by, those of the other. Also, the attraction of the molecules of substance A for substance B had to be greater than the reciprocal attraction of the molecules of A for each other. Conversely, the attraction that the molecules of A exercised upon those of B, combined with the attraction of B for A, had to be powerful enough to overcome the reciprocal affinity between the molecules of B. The closer together the molecules of a given substance, the stronger their reciprocal affinity and, consequently, the more difficult would it be for them to combine with any other substance. In more concrete terms, it was often difficult to make two substances combine in the solid state, although they might combine readily in the liquid state. As solids, their molecules were so closely packed together that the reciprocal affinity was too great to be overcome by the attraction exerted on them by the molecules of another substance. In the liquid state, however, the distances between the molecules of a substance were much greater. There was a corresponding decrease in their reciprocal attractions, which rendered them more susceptible to the affinity exerted by the molecules of another substance. For the same reason, chemical combinations took place more easily between substances in the gaseous state than in either the solid or the liquid state: in a gas the distances between the molecules were, relatively speaking, the greatest, so that their reciprocal attractions were reduced to a minimum. For instance, when steam was passed over iron filings, an oxide of iron was formed more readily than when the metal was immersed in water. The reaction took place with far greater facility when the iron had been powdered into small filings than when a block of the metal was used.

Apart from this effect on the physical states, the relative distances between the molecules of reacting substances explained the role of such factors as temperature and pressure. The most important factor influencing chemical combination was heat. The reason for this was to be sought in the expansive power of caloric, the active principle underlying the effects of heat. Caloric caused substances to expand, thereby increasing the distances between their molecules. The reciprocal attraction between the molecules of any given substance consequently decreased. If a substance in the solid state was heated, it passed, as a general rule, through the liquid to the gaseous state. When the reciprocal affinities between the molecules of the same substance decreased, the molecules were more easily attracted to those of another substance, with which they could then combine. For the same reason, the variation of pressure was important in studying the reactions between gases: distances were inversely proportional to the pressure.

A problem that arose in Berthollet's model, with the important role he assigned to the variation of distance between molecules when they reacted with a given substance, was the explanation of the nature of the force responsible for the increase and the decrease of distances between molecules. So far as the more compact states of matter were concerned, the closeness of the molecules could be attributed to the force of reciprocal attraction. Thus, crystallization was explained as a consequence of the tendency to attain a maximum effect by the symmetrical arrangement of molecules: distances were thus reduced to a minimum, with a corresponding increase in reciprocal affinities. The tendency to cohere as closely as possible was the result of mutual attraction between the molecules, and crystallization was a secondary effect derived from this primary attraction. In fact, the *Essai* distinguished between the different kinds of effects due to affinity. One of them accounted for chemical combination of two or more different substances. The other expressed itself as the reciprocal affinity between the molecules of a given substance: its intensity was measured by the state of cohesion. Obviously the two effects could work in opposite directions during a chemical reaction. Berthollet may therefore be considered to be the originator of crystal chemistry; unlike his contemporary Haüy, he did not have to assume that the internal symmetry of the crystal (i.e., the symmetry around a particular atom or molecule) had to be the same as its external (or macroscopic) symmetry.

Thanks to this model and the accompanying analysis of chemical reactions, especially combination, the analyzed idea of affinity had been replaced by a group of concepts. This had an important influence upon early nineteenth-century chemistry. On the negative side, chemists dispensed with the use of

tables of affinity as guiding hypotheses after the publication of the *Essai*. Nobody could accept such a simplified account of chemical phenomena after Berthollet's criticisms. On the positive side, there were the continued efforts of chemists to provide theories accompanied by adequate models showing the internal workings and structure of chemical substances. This was nowhere more evident than in the development of chemical kinetics and thermochemistry in the late 1850's, with the accompanying clarifications of the notion of mass action and the mechanism of chemical equilibria.

Berthollet's compatriot J. L. Proust had asserted in 1799 that all combinations occurred in definite proportions; in the formation of any chemical compound the same elements were always combined in the same proportions by weight. Berthollet interpreted this to be yet another version of the doctrine of elective affinities and challenged Proust's notions.

From Berthollet's concept of chemical mass it followed that the proportions in which one substance combined with another increased directly with its chemical mass, the "active" quantity in any given reaction. There were a maximum and a minimum proportion in which one substance would combine with another. Between these two limits, which one might call the "threshold" and "saturation" points, the substances would combine in any proportions, depending on their respective quantities, the difference in the proportions being continuous between one extreme and the other. From this it obviously followed that the proportions in which substances combined were not fixed, at least within limits. This was not borne out, however, by the facts in all cases—a point that Berthollet had to concede to Proust. There was at least an appearance that substances combined, in some cases, in definite proportions that could not be made to vary indefinitely. For instance, oxygen and hydrogen did not combine in varying degrees, but in the same proportions, to form water; likewise, ammonia was always formed by the same proportions of nitrogen and hydrogen, as Berthollet himself had been the first to demonstrate. Berthollet did not contest such evidence, although he continued to affirm that in the majority of cases, combinations occurred in conformity with his theory of variability of proportions. For him the problem was to reconcile these two apparently conflicting views of chemical combination. He attempted to do this in two different ways.

On the one hand, Berthollet admitted the existence of fixed proportions. Some substances did combine in only one fixed proportion, but this could be explained quite satisfactorily in accordance with his principles. The reason was to be found in the special conditions under which some combinations took place, so that it was not possible for the substances involved to follow the general law of variability of proportions within limits. On the other hand, he maintained that it was only the poverty or superficiality of experimental observations that had led chemists to attribute a fixity of proportions in combinations where, in fact, there was none to be discovered. A case such as the fixed proportions in which hydrogen and oxygen combined to form water vapor was explicable in terms of the physical state of the resulting combination. The reaction was accompanied by a strong condensation: in fact, with the application of slight pressure the resulting product was converted into the liquid state instead of being gaseous; such a conversion was accompanied by a thousandfold (or more) condensation in the volume of the product. As a result of the condensation, the molecules of water vapor were packed together more closely than the molecules of hydrogen and oxygen surrounding them. The reciprocal affinity between the compound molecules of water vapor was considerably greater, due to their closeness, than that between the individual hydrogen and oxygen molecules surrounding them. Berthollet assumed that it was a special characteristic of such molecules that their combination would be accompanied by condensation only when they came together in a particular proportion, such as 2 : 1 for hydrogen and oxygen. When they were mixed in proportions other than this, they did not form stable compounds for two reasons. First, their distances might be too great for the mutual affinity to be effective. Second, even if they did combine in small quantities, the surrounding molecules, by their physical impact or their opposing affinities (e.g., of cohesion), succeeded in dissociating whatever compound molecules might be produced. This explained why oxygen and hydrogen combined in a fixed proportion to produce water. When they were present in the ratio of 1 : 2, their molecules were so combined that the particles of the product were closer together, and hence relatively isolated, due to condensation.

The arguments advanced by Berthollet and Proust in their controversy (1801–1807) were both empirical and theoretical. The empirical objections revolved about the ability to distinguish between a genuine chemical compound and a mere mixture. Here Proust's intuition was more often correct than Berthollet's. The theoretical difficulties showed Berthollet at an advantage because, unlike Proust, he had worked out a complete set of principles in terms of which he tried to account for all the known chemical phenomena. While Berthollet agreed that there was clearly a difference between such substances as

glass and the metallic oxides, he could discover no criterion by which any definite distinction could be established between them. On Berthollet's model it was quite possible to account for the existence of the substances in which the constituents were always combined in the same proportions, but no difference of principle was involved in distinguishing substances held together by weaker affinities (their combination being unaccompanied by condensation). Proust's reply was a circular one. He was supposed to state a principle by which substances that he considered to be of fixed composition could be distinguished from those that he acknowledged to be of variable composition. His reply took the form that compounds were substances whose constituents always combined in fixed proportions, whereas solutions or mixtures were substances having variable constitutions.

BIBLIOGRAPHY

I. Original Works. Most of Berthollet's papers were published in *Mémoires de l'Académie/Institut* or, after 1789, in *Annales de chimie.* He also wrote three books: *Éléments de l'art de la teinture,* 2 vols. (I, 1791; 2nd ed., 1804); *Recherches sur les lois de l'affinité* (1801); and *Essai de statique chimique,* 2 vols. (1803). Also see *Mémoires de physique et de chimie de la Société d'Arcueil,* 3 vols. (1807–1817).

II. Secondary Literature. No comprehensive study of Berthollet's life and works has been published, but see E. F. Jomard, *Notice sur la vie et les ouvrages de C. L. Berthollet* (Annecy, 1844); and S. C. Kapoor, "Berthollet, Proust, and Proportions," in *Chymia,* **10** (1965), 53–110. Also of value is M. P. Crosland, *The Society of Arcueil. A View of French Science at the Time of Napoleon I* (Cambridge, Mass., 1967).

Satish C. Kapoor

BERTHOLON, PIERRE (*b.* Lyons, France, 28 October 1741; *d.* Lyons, 21 April 1800), *physics.*

Bertholon, a priest of the Lazarist order, spent time in Lyons, Paris, and Béziers before settling in Montpellier, where he was invited by the États de Languedoc to teach all aspects of contemporary science. He held the chair of physics specially created for him in 1784 by the Société Royale des Sciences de Montpellier. His courses were greatly appreciated and were given regularly until the Revolution—and privately even later. Bertholon taught physics at the École Centrale de l'Hérault in 1791 and in Lyons in 1797. His renunciation of the priesthood was probably the cause of his leaving the school in Lyons. He died three years later.

Bertholon's scientific contribution is important both

qualitatively and quantitatively, for it included areas of great diversity—including urban public health, agriculture, aerostatics, and fires. He is particularly well known for his work in physics, especially in electricity. He played the same role in the south of France that the Abbé Nollet played in Paris; that is, he contributed greatly to the development of research in electricity—as much by work and personal experience as by his lectures. Three principal works brought him fame. *De l'électricité des météores* is a study of all atmospheric manifestations, as well as of volcanoes and earthquakes; Bertholon proposed to overcome the latter by sinking metal shafts into the ground. Influenced by his friend Benjamin Franklin, he supplied southern France with lightning rods. *De l'électricité des végétaux* deals with the application of electricity to the growth of plants; for this Bertholon used an electrovegetometer of his own invention. *De l'électricité du corps humain dans l'état de santé et de maladie* classifies all ailments according to their positive or negative electrical reactions. Appropriate therapy is advised for each: positive or negative electricity, electric baths, aigrettes, electric sparks, electric shock, or *impressions de souffle*—"When the face, the back of the hand, or another part of the body the sensitivity of which is not too weakened by touch is brought near an electrified conductor, there is felt the impression of a fresh breeze, of a light breath, or of a cobweb."

BIBLIOGRAPHY

Louis Dulieu, "L'Abbé Bertholon," in *Cahier lyonnais d'histoire de la médecine,* **6,** no. 2 (1961), includes a complete bibliography of Bertholon's works (publications, letters, and manuscripts, as well as writings that have not been found) and a bibliography of secondary literature.

Louis Dulieu

BERTI, GASPARO (*b.* Mantua [?], Italy, *ca.* 1600; *d.* Rome, Italy, 1643), *physics, astronomy.*

Berti seems to have been a native of Mantua who spent most of his life in Rome. He was first mentioned (under the name of Alberti) as a distinguished mathematician who about 1629 collaborated with Francesco Contini in the mapping of the Roman catacombs. Berti's friendship with Luc Holste, Athanasius Kircher, and Rafaello Magiotti suggests that he was born about 1600. In 1636, Holste described him to Nicholas Peiresc as an expert in mathematics and in the construction of mathematical instruments. About the same time, Berti's observations of an eclipse came to the attention of Pierre Gassendi, who spoke of him as young, industrious, and erudite. Berti also refined the earlier observations of Christopher Clavius in order to determine the precise latitude of Rome; this he communicated to the English geographer John Greaves, who called him a celebrated astronomer. In July 1638, Magiotti informed Galileo that Berti had been recommended for a chair of mathematics by Benedetto Castelli, who considered him particularly well versed in the Galilean doctrines. Upon Castelli's death in 1643, Berti was named his successor as professor of mathematics at the Sapienza, but died shortly afterward.

Berti's historical importance, however, is in physics rather than in mathematics or astronomy. It was his experimental apparatus, constructed in Rome sometime between 1640 and 1643, that ultimately led to Evangelista Torricelli's work on atmospheric pressure. Berti's experiment seems to have been inspired by Galileo's *Discorsi* (1638), in which it was asserted that water could not be raised more than eighteen cubits by a lift pump. Berti's apparatus was described and illustrated in Magiotti's letter to Marin Mersenne dated 12 March 1648. It consisted of a lead tube no higher than twenty-two cubits, bent downward at the top and terminating at either end in a valve submerged in a container filled with water. Magiotti stated that Berti, who thought he had refuted Galileo's statement, had improperly measured the distance from the water surface in the upper container to the floor rather than to the water surface in the lower container. Properly measured, Magiotti said, the height was indeed eighteen cubits. He went on to say that in writing of the experiment to Torricelli, he had suggested that if seawater were used, a lower level would result; and it was this suggestion that led to Torricelli's experiments using mercury.

A more elaborate apparatus, which was attached to the façade of Berti's house, is illustrated in Athanasius Kircher's *Musurgia* (1650) and in Gaspar Schott's *Technica curiosa* (1664). It consisted of a lead tube about eleven meters long that terminated at its lower end in a valve and at its upper end in a globe, said to have been originally of copper and later of glass. From the juncture of globe and tube, a second lead tube that terminated in a valve was brought to a window about ten meters above the pavement. At Kircher's suggestion, a bell that could be struck by a hammer activated by an external magnet was enclosed in the globe. The fact that a sound was heard convinced Berti that no vacuum existed; but, as Emanuel Maignan later remarked to him, the attachment of the bell to the tube could communicate the sound to the air outside.

Cornelis De Waard assigned to Berti's experiment

a probable date after the spring of 1639 (when Kircher returned to Rome after a long absence) and before 1642, taking a phrase in Magiotti's letter to imply that Galileo was still living when the experiment was first performed. That implication is questionable, however, for the passage reads: "Il Sig⟨r⟩ Berti credeva con questa esperienza convincere il Sig⟨r⟩ Galileo . . . ," and *convincere* usually meant "refute" rather than "convince." Probably the first experiment was performed not earlier than June 1641, when Torricelli left Rome, for it appears that he was informed of it by letter. The magnetic device escaped mention in both the 1641 and 1643 editions of Kircher's *Magnes,* in which all manner of devices employing magnets were described and illustrated, which suggests that the elaborate apparatus was not earlier than 1643. Thomas Cornelius, recounting various Italian experiments of this kind in his *Progymnasmata physica* (1663), spoke of Berti as professor of mathematics at the Roman Academy when the experiment using the glass globe took place, which also suggests the year 1643. Torricelli's mercury experiments of 1644 probably occurred soon after he received Magiotti's first communication; if so, the various forms of Berti's apparatus may all belong to the period 1642–1643.

BIBLIOGRAPHY

All present knowledge of Berti is from secondary literature, which has been collected by Cornelis De Waard in *L'expérience barométrique. Ses antécédents et ses explications* (Thouars, 1936), pp. 104 ff., 169 ff. Works cited herein and letters mentioning Berti may most readily be found in that work.

STILLMAN DRAKE

BERTIN, LOUIS-ÉMILE (*b.* Nancy, France, 23 March 1840; *d.* La Glacerie, France, 22 October 1924), *naval architecture, hydraulics.*

Bertin was the son of Pierre-Julien Bertin, a hosiery dealer, and Anne-Frédéric Merdier. He entered the École Polytechnique in 1858 and, after graduating in 1860, joined the Naval Engineering Corps. In 1862 Bertin finished his studies at the School of Naval Engineering in Paris. Assigned to the Cherbourg naval district, he proved to be an inventive engineer who had a good sense of the practical but at times lacked critical insight into the mathematical development of his thought.

In 1864 the French navy sought to improve the ventilation of its horse transport ships. In 1865 Bertin presented the winning design, which brought its inventor the Plumey Prize (2,500 francs) of the Academy of Sciences in 1873.

In 1866 Bertin produced artificial waves in calm water and measured the decrease of successive amplitudes. The following year he observed sea swells while aboard a ship, measured the roll by means of a simple recording apparatus he had made, and began to give his observations a mathematical form. In accordance with the course given at the School of Naval Engineering, he at first believed that the swell was a cylindrical surface with a sinusoidal section. At that time French engineers still did not know of Franz von Gerstner's "Theorie der Wellen" (1804), even though it had been republished in 1809 by L. W. Gilbert in the *Annalen der Physik.* English engineers had long known of Gerstner's trochoidal swell, and Bertin obtained his knowledge from references in their publications. Accordingly, he published a series of articles in the *Mémoires de la Société des sciences naturelles de Cherbourg.* In order to complete his observations on shipboard, he ordered an oscillograph with two pendulums of different periods and used it, from 1872, to record swell and roll on the same band.

The Franco-Prussian War led to Bertin's work on compartmentalization, the enduring part of his work as an engineer. In order to limit damage, he suggested protecting the horizontal compartment adjoining the waterline by combining armor plate with a cellular compartment. The first French-built compartmentalized cruiser was based on his plans.

In 1881 Bertin was sent to Brest, where he drew up the plans for a cruiser that attained a speed of eighteen knots, a world record at that time. He was placed at the disposal of the Japanese government from 1886 to 1890, and the ships he helped build enabled the Japanese navy to defeat the Chinese navy in 1894. Upon his return to France, Bertin served in Toulon and then in Rochefort, where he was appointed director of naval construction in 1892. From 1893 to 1895 he directed the School of Naval Engineering, and in November 1895 he became head of the Technical Department of Naval Construction of the Naval Ministry, a post he held until his retirement in 1905.

BIBLIOGRAPHY

I. ORIGINAL WORKS. Among Bertin's books are *Notice sur la marine à vapeur de guerre et de commerce depuis son origine jusqu'en 1874* (Paris, 1875); *État actuel de la marine de guerre* (Paris, 1893), section de l'ingénieur, no. 42A of *Encyclopédie scientifique des aide-mémoire,* M. Léauté, ed.; *La marine des États-Unis* (Paris, 1896); *Chaudières marines, cours de machine à vapeur professé à l'École d'application du génie maritime* (Paris, 1896, 1902), trans. and ed. by Leslie S. Robertson as *Marine Boilers, Their Construction and Working, Dealing More Specially With Tubulous Boilers* (London, 1898, 1906); *Machines marines, cours de machines à vapeur professé à l'École d'application du génie maritime* (Paris, 1899); *Les marines de guerre à l'Exposition Universelle de 1900* (Paris, 1902); *Évolution de la puissance défensive des navires de guerre, avec un complément concernant la stabilité des navires* (Paris, 1907); and *La marine moderne, ancienne histoire et questions neuves* (Paris, 1910, 1914).

His articles appeared in *Mémoires de la Société imperiale* [*nationale* after 1871] *des sciences naturelles* [*et mathématiques* after 1879] *de Cherbourg* between **15** (1869–1870) and **31** (1898–1900); the *Mémoires présentés par divers savants à l'Académie des sciences . . .* between **22**, no. 7 (1876) and **26**, no. 5 (1879); in *Comptes rendus hebdomadaires des séances de l'Académie des sciences* between **69** (26 July 1869) and **158** (2 June 1914).

Also of value are *Notes sur mes travaux scientifiques et maritimes* (Cherbourg, 1879); three vols. with the title *Notice sur les travaux de M. L.-E. Bertin* (Paris, 1884, 1885, 1896); and the autographed album "Projets de navires à flottaison cellulaire (1870–1873)," presented by Bertin to the library of the Institute in 1884.

II. SECONDARY LITERATURE. Works containing information on Bertin are Barrillon, "Émile Bertin," in *Neptunia,* no. 10 (2nd trimester 1948), 22–24; Henri Bouasse, *Houle, rides, seiches et marées* (Paris, 1924); Edgar de Geoffroy, "Bertin (Louis-Émile)," in *Larousse mensuel illustré,* no. 216 (Feb. 1925), 691–692; E. Sauvage, "Louis Émile Bertin," in *Bulletin de la Société d'encouragement pour l'industrie nationale* (June 1925), 438–459; and Togari, *Louis-Émile Bertin. Son rôle dans la création de la Marine japonaise* (n.p., 1935). Also used in preparing this article were the birth records for 1840 and the death records for 1843 of Nancy; the register of students of the École Polytechnique, IX, covering 1855–1862; *Annuaire de la marine et des colonies* for 1863–1870 and 1872–1889; *État du personnel de la marine, décembre 1871* (Versailles, 1871); and *Annuaire de la marine* for 1890–1906.

ARTHUR BIREMBAUT

BERTINI, EUGENIO (*b.* Forlì, Italy, 8 November 1846; *d.* Pisa, Italy, 24 February 1933), *mathematics.*

In 1863 Bertini registered at the University of Bologna, intending to study engineering, but after taking the course taught by Luigi Cremona, he turned to pure mathematics. In 1866 he fought with Garibaldi in the third war for Italian independence. On the advice of Cremona, he resumed his studies and transferred to the University of Pisa, from which he received his degree in mathematics in 1867. During the academic year 1868–1869 he attended the course in Milan taught by L. Cremona, F. Brioschi, and F. Casorati. This course, dealing with Abel's integrals, exerted considerable influence on Bertini's own research.

In 1870 Bertini began his teaching career in the secondary schools of Milan, and in 1872 taught in Rome. There, on the recommendation of Cremona, he was appointed a special lecturer to teach descriptive and projective geometry. In 1875 he accepted the professorship of advanced geometry at the University of Pisa. From 1880 to 1892 he taught at the University of Pavia, and then returned to his former professorship at Pisa, a post he held until his retirement at the age of seventy-five. For the next ten years he taught an elective course in geometrical complements, which he had started as an introductory course to higher geometry.

Bertini's research deals particularly with algebraic geometry and constitutes definite progress in relation to the studies pursued by the school of Cremona. In this connection it is necessary to note that Cremona, having formulated the theory on plane and space transformations that bears his name, availed himself of the same transformations to change higher geometric figures into simpler figures and then apply to the higher figures the properties of the simpler ones. Bertini studied the geometric properties that remain constant during such transformations. He conceived the idea of exploring this field after studying the problem of the classification of plane involutions. In 1877 he succeeded in determining the various types, irreducible from each other, in which the planar involutions may be reduced through Cremona's transformations. His treatises are noteworthy for their order and clarity.

BIBLIOGRAPHY

I. ORIGINAL WORKS. Bertini's works include "La geometria delle serie lineari sopra una curva piana, secondo il metodo algebrico," in *Annali di matematica pura ed applicata,* 2nd ser., **22** (1894), 1–40; *Introduzione alla geometria proiettiva degli iperspazi* (Pisa, 1906; Messina, 1923); and *Complementi di geometria proiettiva* (Bologna, 1927).

II. SECONDARY LITERATURE. More information on Bertini may be found in G. Castelnuovo, "Commemorazione del socio Eugenio Bertini," in *Atti della Reale Accademia nazionale dei Lincei. Rendiconti,* Classe di scienze fisiche, matematiche e naturali, 6th ser., **17** (1933), 745–748; and F. Enriques, *Le matematiche nella storia e nella cultura* (Bologna, 1938), pp. 284, 286, 287, 292.

ETTORE CARRUCCIO

BERTRAND, CHARLES-EUGÈNE, or **Charles-Egmont** (*b.* Paris, France, 2 January 1851; *d.* Lille, France, 10 August 1917), *plant anatomy.*

After studying natural science at the Sorbonne and working for a short time in the laboratory of the

Faculty of Sciences at Paris, Bertrand spent the rest of his scientific career at the University of Lille. He considered himself primarily a botanist, although much of his work furthered geologists' understanding of coals and the evolution of the extinct plants found in them. He concentrated on aberrant or anomalous forms, and by this method discovered the phylogenetic relations not recognizable solely from normal types. Bertrand's monographs on aberrant forms developed primarily the affinities and filiations of thallophytes, higher vascular cryptogams (the ferns and club mosses), and lower phanerogams (the ancient gymnosperms).

Because plant fragments in coal are generally mutilated, Bertrand endeavored to learn the comparative anatomy of the vegetative structures (primarily stems), which are preserved. This anatomical knowledge permitted him to identify numerous forms for which the usual keys—leaf and flower morphology—failed. He was one of the earliest to examine thin sections of the carbonaceous rocks by means of the ordinary light microscope. This new technique contributed further to his success in differentiating the natures and compositions of the various types of coal. He showed bogheads to be primarily accumulations of algae, sometimes with other debris mixed in, all cemented by a primitive, humic paste (*gelée fondamentale*). In cannel coals, spores predominate. The *gelée fondamentale* is also a primary constituent in common coals and in two kinds of bituminous shales (*charbons humiques* and *charbons de purins*).

Part of Bertrand's fame rests on his establishment at the University of Lille of a laboratory famous for both education and research. Many of his studies were produced with collaborators, most notably with his son Paul, a paleobotanist.

BIBLIOGRAPHY

E. Morvillez, *Charles-Eugène Bertrand* (Caen, 1918), is the most complete biography and includes a bibliography of Bertrand's writings. Another bibliography is in Giuseppe de Toni, ed., *Bibliographia algologica universalis* (Forlì, 1932), pp. 247–252. For his geological work in particular, see the following unsigned articles: "L'oeuvre géologique de C. Eg. Bertrand," in *Annales de la Société géologique du nord*, **44** (1919), 47–64 (see also 6–7); and "Célébration du centenaire de Ch. Barrois et Ch.-E. Bertrand et du souvenir de P. Bertrand," *ibid.*, **71** (1951), 135–143.

BERT HANSEN

BERTRAND, GABRIEL (*b*. Paris, France, 17 May 1867; *d*. Paris, 20 June 1962), *biochemistry*.

Gabriel Bertrand introduced into biochemistry both the term "oxidase" and the concept of trace elements. The son of a Paris merchant, Bertrand early showed an interest in the natural sciences, especially in the botanical specimens in the collections of the Muséum d'Histoire Naturelle. After obtaining his baccalaureate degree in 1886, he entered the École de Pharmacie in Paris, at the same time enrolling in Edmond Frémy's courses at the chemical laboratory of the museum.

In 1890 Bertrand was appointed *préparateur* to Albert Arnauld, who had just taken over the course in organic chemistry after the death of his teacher Michel Chevreul; Bertrand held this post for ten years. He had also been noticed by Émile Duclaux, Pasteur's successor at the Institut Pasteur, and in 1900 was appointed to the staff of the recently created institute of biochemistry at the Institute.

Duclaux was professor of biochemistry in the Faculté des Sciences, Paris, although his teaching was done at the Institut Pasteur. After Duclaux's death in 1904, Bertrand was placed in charge of his courses; in 1908, he was named to the vacant chair, a position he held until his retirement in 1937. But retirement for Bertrand did not mean the end of work, and for many years thereafter he remained a familiar sight at the Institute.

Bertrand obtained his doctorate in 1904 with a dissertation that was a study of the conversion of sorbitol (D-sorbitol) into sorbose (L-sorbose), a sugar first identified in the sorb berry. He found ultimately that the conversion depended on the presence of a microbe, *Bacterium xylinum* (i.e., *Aerobacter xylinum*), and that it was an oxidation occurring only in the presence of oxygen.

In the years 1894–1897 Bertrand investigated the process of the darkening and hardening of the latex of lacquer trees. He recognized that the color change was caused by the oxidation of a phenol—laccol—in the presence of another substance, laccase. Other phenolic compounds, he found, underwent similar organic oxidation reactions, also in the presence of substances similar to laccase. In 1896 Bertrand first used the term "oxidase" for these oxidizing enzymes (including tyrosinase, which he had described). During the following year he published several studies of oxidases.

Bertrand made another important advance in the analysis of enzymes when he observed that laccase ash contained a large proportion of manganese. Throughout the last half of the nineteenth century it had been known that plants contained minerals, and in 1860 it was demonstrated that in artificial situations plants could be grown in a water culture containing only metallic salts. Researchers still accepted the presence of minerals in the plant as incidental, however, and thought them the result of the presence of minerals in the soil. Bertrand's work in 1897, and especially his later claim that a lack of manganese caused an interruption of growth, forced a change in thinking on this matter. He concluded that the metal formed an essential part of the enzyme, and, more generally, that a metal might be a necessary functioning part of the oxidative enzyme. From this and similar researches he developed his concept of the trace element, essential for proper metabolism.

During his career Bertrand published hundreds of papers on the organic effects of various metals. In 1911 he showed that the development of the mold *Aspergillis niger* was greatly influenced by the presence of minute amounts of manganese. For such researches Bertrand was forced to develop more precise methods of organic analysis, many of which later came into widespread use.

Bertrand's researches were immediately applied to the elimination of previously undiagnosable pathological conditions, thereafter recognized as the result of deficiencies of trace elements. His work also provided the basis for further elaboration of the enzymatic systems involved in respiration and metabolic processes.

BIBLIOGRAPHY

I. ORIGINAL WORKS. Among Bertrand's articles are "Sur le latex de l'arbre à laque," in *Comptes rendus de l'Académie des sciences* (Paris), **118** (1894), 1215–1218, which also appeared in *Bulletin de la Société chimique de France,* **11** (1894), 717–721; "Sur le latex de l'arbre à laque et sur une nouvelle diastase contenue dans ce latex," in *Comptes rendus de la Société biologique* (Paris), **46** (1894), 478–480; "Sur la présence simultanée de la laccase et de la tyrosinase dans le suc de quelques champignons," in *Comptes rendus de l'Académie des sciences* (Paris), **123** (1896), 463–465; "Sur une nouvelle oxydase, ou ferment soluble oxidant, d'origine végétale," in *Comptes rendus de l'Académie des sciences* (Paris), **122** (1896), 1215–1217, which also appeared in *Bulletin du Muséum d'histoire naturelle* (Paris), **2** (1896), 206–208, and in *Bulletin de la Société chimique* (Paris), **15** (1896), 793–797; "Nouvelles recherches sur les ferments oxidants ou oxidases," in *Annales agronomique,* **23** (1897), 385–399; "Les oxidases ou ferments solubles oxidants," in *Revue scientifique,* 4th ser., **8** (1897), 65–73; "Recherches sur la laccase, nouveau ferment soluble, à propriétés oxydantes," in *Annales de chimie,* **12** (1897), 115–140; "Sur l'emploi favorable du manganèse comme engrais," in *Comptes rendus de l'Académie des sciences* (Paris), **141** (1905), 1255–1257.

With M. Javiller, Bertrand wrote "Influence du manganèse sur le développement de l'*Aspergillis niger,*" "Influence combinée du zinc et du manganèse sur le développement de l'*Aspergillis niger,*" and "Influence du zinc et du manganèse sur la composition minérale de l'*Aspergillis niger,*" all of which appeared in *Comptes rendus de l'Académie des sciences* (Paris), **152** (1911), 225–228, 900–902, and 1337–1340, respectively.

II. SECONDARY LITERATURE. Two biographical memoirs appeared soon after Bertrand's death, one by Y. Raoul in *Bulletin de la Société de chimie biologique,* **44** (1962), 1051–1055, and the other by Marcel Delépine in *Comptes rendus de l'Académie des sciences* (Paris), **255** (1962), 217–222. The former was to be reprinted separately as a pamphlet containing a complete bibliography of Bertrand's works, but has not yet appeared. No other complete bibliographical listings are available, although partial listings may be found in the *Royal Society Catalogue of Scientific Papers,* XIII, and in Poggendorff, V and VI. Bertrand's relationship with the Institut Pasteur is discussed in Albert Delaunay, *L'Institut Pasteur. Des origines à aujourd'hui* (Paris, 1962).

The presence of metallic salts in plants was demonstrated in the nineteenth century, as was their ability to maintain plant life. See W. Knop, "Ueber die Ernährung der Pflanzen durch wässerige Lösungen bei Ausschluss des Bodens," in *Landwirtschaftliche Versuchsstationen,* **2** (1860), 65–99, 270–293; and J. Sachs, "Ueber die Erziehung von Landpflanzen in Wasser," in *Botanisches Zentralblatt,* **18** (1860), 113–117.

ALAN S. KAY

BERTRAND, JOSEPH LOUIS FRANÇOIS (*b.* Paris, France, 11 March 1822; *d.* Paris, 5 April 1900), *mathematics.*

Bertrand's father was Alexandre Bertrand, a writer of popular scientific articles and books. Alexandre had attended the École Polytechnique in Paris with Auguste Comte and Jean Marie Constant Duhamel, and the latter married his sister. When his father died, young Bertrand went to live with the Duhamels. A well-known professor of mathematics at the École Polytechnique, Duhamel was the right man to guide his precocious nephew. At the age of eleven the boy was allowed to attend classes at the École Polytechnique. In 1838, at sixteen, Bertrand took the degrees of bachelor of arts and bachelor of science, and at seventeen he received the doctor of science degree with a thesis in thermomechanics. The same year (1839) he officially entered the École Polytechnique, and in 1841 he entered the École des Mines. Bertrand's first publications date from this period, the first being "Note sur quelques points de la théorie de l'électricité" (1839), which deals with Poisson's equation, $\Delta V = -4\pi\rho$, and the law of Coulomb.

In 1841 Bertrand became a professor of elementary mathematics at the Collège Saint-Louis, a position that he filled until 1848. In May 1842 he and his brother, returning to Paris from a visit to their friends the Aclocques at Versailles, were nearly killed in a railroad accident which left a scar on Bertrand's face. Bertrand married Mlle. Aclocque in 1844, in which year he also became *répétiteur d'analyse* at the École Polytechnique. Three years later he became *examinateur d'admission* at this school and *suppléant* of the physicist Jean-Baptiste Biot at the Collège de France. In 1848, during the revolution, Bertrand served as a captain in the national guard. He published much during these years—in mathematical physics, in mathematical analysis, and in differential geometry. The first of Bertrand's many textbooks, the *Traité d'arithmétique,* appeared in Paris in 1849 and was followed by the *Traité élémentaire d'algèbre* (1850); both were written for secondary schools. They were followed by textbooks for college instruction. Bertrand always knew how to fascinate his readers and his lecture audiences, and his books had a wide appeal because of content and style. In 1853 he edited and annotated the third edition of J. L. Lagrange's *Mécanique analytique.* From the many publications in this period, one, "Mémoire sur le nombre de valeurs . . .," introduces the so-called problem of Bertrand: to find the subgroups of the symmetric groups of lowest possible index. Another publication, "Mémoire sur la théorie des courbes à double courbure" (1850), discusses curves with the property that a linear relation exists between first and second curvature; these are known as curves of Bertrand.

In 1852 Bertrand became professor of special mathematics at the Lycée Henry IV (then Lycée Napoléon). He also taught at the École Normale Supérieure. In 1856 he replaced Jacques Charles François Sturm as professor of analysis at the École Polytechnique, where he became the colleague of Duhamel. He then left secondary education to pursue his academic career. In 1862 he succeeded Biot at the Collège de France. Bertrand held his position at the École Polytechnique until 1895, that at the Collège de France until his death.

In 1856 Bertrand was elected to the Académie des Sciences, where in 1874 he succeeded the geologist Élie de Beaumont as *secrétaire perpétuel.* In 1884 he replaced the chemist Jean-Baptiste Dumas in the Académie Française. These high academic positions, combined with his erudition, his eloquence, and his natural charm, gave him a position of national prominence in the cultural field.

During the Commune of 1871 Bertrand's Paris house was burned, and many of his manuscripts were lost, among them those of the third volume of his textbook on calculus and his book on thermodynamics. He was able to rewrite and publish the latter as *Thermodynamique.* Afterward he lived at Sèvres and then at Viroflay. At his home Bertrand enjoyed being the center of a lively intellectual circle. Many of his pupils became well-known scientists—for instance, Gaston Darboux, who succeeded him as *secrétaire perpétuel.* In his *Leçons sur la théorie générale des surfaces,* Darboux elaborated many results of Bertrand and his mathematical circle.

Bertrand's publications, apart from his textbooks, cover many fields of mathematics. Although his work lacks the fundamental character of that of the great mathematicians of his period, his often elegant studies on the theory of curves and surfaces, of differential equations and their application to analytical mechanics, of probability, and of the theory of errors were widely read. Many of his articles are devoted to subjects in theoretical physics, including capillarity, theory of sound, electricity, hydrodynamics, and even the flight of birds. In his *Calcul des probabilités,* written, like all his books, in an easy and pleasant style, there is a problem in continuous probabilities known as Bertrand's paradox. It deals with the probability that a stick of length $a > 2l$, placed blindly on a circle of radius l, will be cut by the circle in a chord of less than a given length $b < 2l$. It turns out that this probability is undetermined unless specific assumptions are made about what constitute equally likely cases (i.e., what is meant by "placed blindly").

From 1865 until his death Bertrand edited the *Journal des savants.* For this periodical, as for the *Revue des deux mondes,* he wrote articles of a popular nature, many dealing with the history of science. This interest in history of science appears also in the many *éloges* he wrote as *secrétaire perpétuel* of the Academy, among which are biographies of Poncelet, Élie de Beaumont, Lamé, Leverrier, Charles Dupin, Foucault, Poinsot, Chasles, Cauchy, and F. F. Tisserand. He also wrote papers on Viète, Fresnel, Lavoisier, and Comte, and books on d'Alembert and Pascal.

Bertrand spent the later part of his life in the midst of his large family, surrounded by his friends, who were many and distinguished. His son Marcel and his nephews Émile Picard and Paul Appell were his fellow members in the Académie des Sciences. In 1895 his pupils gave him a medal in commemoration of his fifty years of teaching at the École Polytechnique. The influence of Bertrand's work, however, is hardly comparable to that of several of his contemporaries and pupils. Lest it be judged ephemeral, it must be viewed in the context of nineteenth-century Paris and of Bertrand's brilliant academic career, his exalted social position, and the love and respect given him by his many pupils.

BIBLIOGRAPHY

I. Original Works. Bertrand's works include "Note sur quelques points de la théorie de l'électricité," in *Journal de mathématiques pures et appliquées,* **4** (1839), 495–500; "Mémoire sur le nombre de valeurs que peut prendre une fonction quand on y permute les lettres qu'elle renferme," in *Journal de l'École polytechnique,* **30** (1845), 123–140; *Traité d'arithmétique* (Paris, 1849); "Mémoire sur la théorie des courbes à double courbure," in *Journal de mathématiques pures et appliquées,* **15** (1850), 332–350; *Traité élémentaire d'algèbre* (Paris, 1850); *Traité de calcul différentiel et de calcul intégral,* 2 vols. (Paris, 1864–1870); *Les fondateurs de l'astronomie moderne* (Paris, 1867); *Rapport sur les progrès les plus récents de l'analyse mathématique* (Paris, 1867); *L'Académie des sciences et les académiciens de 1666 à 1793* (Paris, 1869); "Considérations relatives à la théorie du vol des oiseaux," in *Comptes rendus de l'Académie des sciences,* **72** (1871), 588–591; *Thermodynamique* (Paris, 1887); *Calcul des probabilités* (Paris, 1889; 2nd ed., 1897); *D'Alembert* (Paris, 1889); *Éloges académiques* (Paris, 1889); *Leçons sur la théorie mathématique de l'électricité* (Paris, 1890); *Pascal* (Paris, 1891); *Éloges académiques, nouvelle série* (Paris, 1902), which has a complete bibliography of Bertrand's works on pp. 387–399.

II. Secondary Literature. Gaston Darboux, "Éloge historique de J. L. F. Bertrand," in Bertrand's *Éloges académiques, nouvelle série,* pp. 8–51, and in Darboux's *Éloges académiques et discours* (Paris, 1912), pp. 1–60. Another source of information is *Comptes rendus de l'Académie,* **130** (1900), 961–978, addresses delivered in the Academy to honor Bertrand and used by G. H. Bryan for his article "Joseph Bertrand," in *Nature,* **61** (1899–1900), 614–616. The library of the Institut de France nos. 2029–2047 comprises correspondence and some papers of Bertrand; 2719 (5) contains "Notes autobiographiques" (information from Henry Nathan)—these are probably the notes used by Darboux in his *Éloge.* Discussion of Bertrand's problem may be found in H. Weber, *Lehrbuch der Algebra,* II (Brunswick, 1899), 154–160. The curves of Bertrand are dealt with in books on differential geometry, e.g., G. Darboux, *Leçons sur la théorie générale des surfaces,* I (Paris, 1887), 13–17, 44–46, and III (Paris, 1894), 313–314.

D. J. Struik

BERTRAND, MARCEL-ALEXANDRE (*b.* Paris, France, 2 July 1847; *d.* Paris, 13 February 1907), *geotectonics, stratigraphy, general geology.*

Bertrand's father was the mathematician Joseph Bertrand. Marcel studied at the École Polytechnique and the École des Mines in Paris. After graduation he worked in the Geological Survey of France, and in 1886 he succeeded his teacher Béguyer de Chancourtois at the École des Mines. In 1896 the Académie des Sciences elected him to the chair Pasteur had held.

Inspired by the writings of Eduard Suess, Bertrand always maintained a concern for what he called the grand problems of general geology. Early in his career he devoted his attention to the general problems of mountain structure while producing a dozen sheets of the geologic map of France. He solved the anomaly of le Beausset (and was awarded the Prix Fontannes by the Geological Society of France for it in 1889) by discovering that the islands of Triassic sediments resting on Cretaceous formations are the eroded remains of an enormous overturned fold. His conception of very large-scale overturned folds and overthrusts related the geological structure of Provence to that of the Alps. Bertrand was the first to conceive of the overthrust structure of the Alps, and by this theory of *grandes nappes* he attempted to connect the structures of the Pyrenees, Provence, and the Alps. His analysis of horizontal crustal compression and the displacements resulting from it won the Prix Vaillant of the Académie des Sciences of the Institut de France in 1890, but the essay was not published until 1908.

Bertrand developed an orogenic wave concept that he used to separate earth history into natural divisions on the basis of successive periods of intense folding and orogeny, each division identified with a chain of mountains. Working from Suess's brilliant synthesis, Bertrand demonstrated in 1887 that the Caledonian, Hercynian, and Alpine deformation produced consecutively those three mountain chains, thus building up the European continent gradually from north to south.

In 1894, at Zurich, Bertrand offered his very original conception of the complete sedimentary cycle with its recurring facies; each cycle represented one of the fundamental deformations. He showed that four kinds of facies are repeated in the different mountain chains, typically gneiss, followed by schistous flysch, then coarse flysch and coarse sandstone. At this time he also added the Huronian orogeny of Precambrian time to the other three deformations. In essaying a mechanism for these orogenies, Bertrand revived, then abandoned, the tetrahedral plan of the earth of Lowthian Green and Michel-Lévy.

BIBLIOGRAPHY

I. Original Works. *Oeuvres géologiques de Marcel Bertrand,* Emmanuel de Margerie, ed., 3 vols. (Paris, 1927–1931), contains all Bertrand's published works except the sheets of the *Carte géologique détaillée de la France* (scale 1:80,000); "Études sur les terrains secondaires et tertiaires dans les provinces de Grenade et de Malaga," in *Mémoires de l'Académie des sciences,* **30** (1899), 377–579;

and the posthumously published "Mémoire sur les refoulements qui ont plissé l'écorce terrestre et sur le rôle des déplacements horizontaux," *ibid.,* **50** (1908), 1–267. It includes Bertrand's own notice of his scientific works to 1894.

II. SECONDARY LITERATURE. The most complete biographical notice is the *éloge* by Pierre Termier, in *Bulletin de la Société géologique de France,* 4th ser., **8** (1908), 163–204, including a bibliography. Other notices are Archibald Geikie, in *The Quarterly Journal of the Geological Society of London,* **61** (1908), 1–liv; W. Kilian and J. Révil, in *Annales de l'Université de Grenoble,* **20** (1908), 15–35; and Otto Wilckens, in *Centralblatt für Mineralogie, Geologie, und Paläontologie* (1909), 499–501. V. V. Beloussov, in his *Basic Problems in Geotectonics* (New York, 1962), pp. 39–43, sets some of Bertrand's contributions in historical perspective.

BERT HANSEN

BERWICK, WILLIAM EDWARD HODGSON (*b.* Dudley Hill, England, 11 March 1888; *d.* Bangor, Wales, 13 May 1944), *mathematics.*

Berwick's total output of original work is relatively small (thirteen papers and a monograph), due in part to ill health, and is concerned primarily with the theory of numbers and related topics, including the theory of equations. A penchant for problems involving numerical computation is reflected throughout his publications.

Much of Berwick's work is concerned with the following problem: Given a simple algebraic extension of the rational field, establish methods for computing its algebraic integers and the ideals they form. In the monograph *Integral Bases* (1927) Berwick made his most significant contribution to the resolution of this problem by developing methods for constructing an integral basis for the algebraic integers in such a field. The theoretical existence of integral bases is easily established but does not afford a practicable computational procedure. Methods for special cases—such as quadratic, cubic, and cyclotomic fields—had already been devised, but Berwick was the first to attack the much more formidable problem of developing methods that would apply to simple algebraic extensions in general. Although his method is not workable in certain exceptional cases, it possesses a wide range of applicability. Its strong numerical orientation, however, kept his work outside the mainstream of developments in algebraic number theory.

Berwick also obtained a necessary and sufficient condition that the general quintic equation be solvable by radicals in the field of its coefficients (1915), and was instrumental in bringing about the publication of tables of reduced ideals in quadratic fields by the British Association for the Advancement of Science (1934).

BIBLIOGRAPHY

Further information on Berwick is in H. Davenport, "W. E. H. Berwick," in *The Journal of the London Mathematical Society,* **21** (1946), 74–80, which contains references to all of Berwick's scientific work. See also the notes by Davenport and E. H. Neville in *Nature,* **154** (1944), 265, 465.

THOMAS HAWKINS

BERZELIUS, JÖNS JACOB (*b.* Väversunda, Östergötland, Sweden, 20 August 1779; *d.* Stockholm, Sweden, 7 August 1848), *chemistry.*

Berzelius came from an old Swedish family. A number of his ancestors had been clergymen. His father, Samuel, a teacher in the Linköping Gymnasium, died when his son was four years old. The mother, Elizabeth Dorothea, two years later married Anders Ekmarck, the pastor at Norrköping and himself the father of five children. Young Berzelius and his sister were raised with the Ekmarck children and educated by Ekmarck and private tutors. In 1788 Berzelius' mother died and within two years Ekmarck remarried. The two Berzelius children were sent to the home of their maternal uncle, Magnus Sjösteen. Young Jöns and his cousins quarreled frequently. Even after he entered the Linköping Gymnasium in 1793 conflicts continued, and to escape them Berzelius took a position in 1794 as tutor on a nearby farm, where he developed a strong interest in collecting and classifying flowers and insects. He had originally intended to become a clergyman, but he instead chose to develop his interest in natural science and decided on a career in medicine. Two years later he began his medical studies at Uppsala, but had to interrupt his work to earn money as a tutor. Fortunately, in 1798 he received a three-year scholarship which permitted him to continue his medical studies.

At this time his oldest stepbrother introduced him to chemistry, of which Jöns knew nothing. Together they studied Girtanner's *Anfangsgründe der antiphlogistischen Chemie.* Thus in his first studies Berzelius learned of the new chemistry which had not yet had much influence on the older Swedish chemists. The professor of chemistry at Uppsala, Johan Afzelius, did not offer much encouragement to Berzelius who therefore began to carry out experiments in his own quarters. During the next summer, lacking any financial support, he intended to stay at the home of an aunt, but her husband did not approve of the young man and sent him to work in a pharmacy at Vadstena. Here he was able to learn glassblowing. Later in the summer his uncle, to get rid of him, introduced him to Sven Hedin, chief physician at the

Medevi mineral springs. He took Berzelius as his assistant for the summer of 1800. There he began his scientific career by analyzing the mineral content of the spring water. At the same time he read of the newly described voltaic pile, the first reliable source of a continuous electric current. He soon built one for himself from sixty pairs of alternating zinc disks and copper coins. In 1802 he used the knowledge thus gained in his doctoral thesis, a study of the effect of the galvanic current on patients with a number of different diseases. He found that the current had no effect on the patients, but his interest in electrical phenomena remained strong.

Sven Hedin was aware of Berzelius' interest in chemistry and soon after the latter's medical degree was granted, Hedin arranged for his appointment as an unpaid assistant to the professor of medicine and pharmacy at the College of Medicine in Stockholm. His duties involved the preparation of artificial mineral waters. He lived in a house owned by Wilhelm Hisinger, a wealthy mine owner with a great interest in mineralogy and chemistry. Berzelius soon began to undertake serious chemical investigations with Hisinger. The electrochemical and mineralogical investigations that the two enthusiasts carried out laid the foundations for Berzelius' future work.

Meanwhile, Berzelius became involved in serious financial difficulties. His only income came from his position as physician to the poor in several Stockholm districts, and the salary was very low. Business ventures that he attempted turned out disastrously, and he was deep in debt. His position improved in 1807 when the professor of medicine and pharmacy died and he was appointed to the post. He now had an increased salary and access to a laboratory. In this period he began to work on the textbooks whose composition strongly influenced the direction of his later career. In 1810 the Medical College became the Karolinska Institutet, an independent medical school, and Berzelius was able to devote most of his time to chemistry. He became a member of the Swedish Academy of Science in 1808 and its president in 1810.

During this time Berzelius began the series of travels abroad through which he became personally acquainted with almost all of the leading chemists of his day. In 1807 he met Hans Christian Oersted, the noted discoverer of electromagnetism. In 1812 he visited England and met all the important British chemists. He was especially anxious to meet Humphry Davy, whose electrochemical researches had been closely related to those of Berzelius and Hisinger. At first Davy and Berzelius got on well, but later some of Berzelius' criticisms of one of Davy's books were reported indirectly to Davy and a coolness developed between the two which was never entirely eliminated. In 1818 Berzelius visited Paris, where he remained for a year, meeting his French colleagues. He spent a part of his time working with Dulong at the home of Berthollet in Arcueil. On his way back to Stockholm he traveled through Germany, where he met the most prominent German chemists. He created such a strong impression there that he was later followed to Sweden by a number of younger chemists who wished to work in his laboratory. These included Eilhard Mitscherlich and Gustav and Heinrich Rose. Friedrich Wöhler, who spent a year (1823–1824) with Berzelius, maintained a close personal friendship with him for the rest of his life, translating many of his important works into German.

While he was in France, Berzelius was elected secretary of the Swedish Academy of Science. This doubled his income and furnished him with an excellent laboratory. He was able to devote himself almost entirely to his laboratory research and to his voluminous writings, including correspondence with scientists all over Europe. He became the recognized authority on chemical questions, although he became involved in a number of polemics, especially with Dumas and Liebig.

Berzelius suffered from poor health through much of his life. He was subject to severe periodic headaches which occurred regularly each month at the time of the new and full moon. In later life these disappeared, but were replaced by attacks of gout and periods of prolonged depression and apathy which interfered with his scientific work. Feeling the need for a more domestic life, he finally decided he should marry, and so in 1835, at the age of fifty-six, he married Elizabeth Poppius, the twenty-four-year-old daughter of one of his old friends. On this occasion the king of Sweden gave him the title of baron. The marriage, which was childless, eased the remaining years of his life.

As he grew older, he became more and more set in his ideas, refusing to accept the newer developments in chemistry which contradicted some of his own theories. He withdrew more and more from the laboratory and spent much time trying to discredit the ideas which the growth of the new field of organic chemistry was forcing upon younger chemists. The last years of his life were not happy ones. At the time of his death he had become a respected figure, but one whose opinions were generally disregarded by his younger colleagues.

Berzelius' most active and productive years were those in which chemistry was beginning to show the full effects of Lavoisier's revolution. The fundamental tools which he created were extremely influential in determining the direction in which the science developed. His achievements were many and varied, and at first glance they seem rather unrelated to each other. Upon closer examination we find an underlying unity of thought and a logical interconnection and development of this thought in most of his work.

He was almost self-taught in chemistry. His first textbook adhered to the antiphlogistic viewpoint. The theories of Lavoisier came late to Sweden, but Berzelius never learned the theory of phlogiston. From the time of one of the earliest experiments that he carried out in his rooms, the preparation of oxygen, he was a firm believer in the essential participation of this element in the constitution of chemical compounds. His first scientific papers were rejected for publication by the Academy because he used the antiphlogistic nomenclature. Berzelius was also heir to the chemical work of his own countrymen. The systematization of chemistry and the interest in the nature of chemical affinity which were characteristic of the work of Torbern Bergman as well as the discovery of new minerals and elements carried out by Scheele and other Scandinavian chemists were certainly influential in determining the direction of Berzelius' theoretical and laboratory studies. These men had developed mineral analysis to a high degree and it is not strange that one of the first pieces of chemical work by Berzelius was the analysis of the composition of Medevi mineral waters.

Aside from these influences of older workers, Berzelius was always keenly aware of the importance of the current literature. He studied it carefully, making critical surveys at first for his own use and later for the benefit of all chemists. His earliest book, published in 1802, was a treatise on galvanism, a review of all the work done up to that time on the action of electricity on salts and minerals. This reflected his early appreciation of the importance of the voltaic pile. It showed his ability to synthesize the literature, and it formed the basis for his pioneering studies with Hisinger.

His association with Hisinger during the first decade of the nineteenth century was particularly fruitful. Through this association Berzelius gained access to the largest voltaic pile in Sweden, owned by the Galvanic Society, of which Hisinger was a leading member. In 1803 Hisinger and Berzelius published the results of their studies on the action of the electric current on a number of sodium, potassium, ammonium, and calcium salts. They found that all the salts were decomposed by electricity. Oxygen, acids, and oxidized bodies accumulated at the positive pole, while combustible bodies, alkalies, and alkaline earths passed to the negative pole. Some acids were converted to a lower oxidation state and passed to the negative pole. Thus the lower oxidation state represented a "combustible body." Similar results were obtained and extended by Humphry Davy in England in the years from 1806 to 1807 and led him to the isolation of the alkali and alkaline earth metals. Berzelius with his friend Pontin also continued this type of work, and in 1808 introduced the use of mercury as the negative electrode. This permitted obtaining amalgams of the metals and even ammonium amalgam.

These studies aroused the interest of Berzelius and Davy in each other and was one of the reasons Berzelius visited England in 1812. More important, they convinced him of the significance of electricity in binding chemical elements together and also strengthened his conviction, gained from reading Lavoisier, that oxygen was an essential constituent not only of all acids, but also of bases as well. From these ideas he was later to develop his dualistic theory of the nature of salts.

Hisinger was not solely interested in electrochemical studies. He had also been interested since boyhood in the minerals found in and around his mines. He had had analyses made of a number of minerals that he had collected, and he himself had analyzed a number of them. Among his minerals was a very heavy stone found near the iron mine of Bastnäs, in which Scheele had vainly tried to find tungsten. Since Berzelius had been carrying out mineral analyses, Hisinger proposed that they should study this mineral together. In 1803 they found that it contained a new element which they named cerium. This was discovered at the same time by Klaproth in Berlin.

While these studies were in progress, Berzelius was acting as assistant in medicine at the School of Medicine in Stockholm. In the course of his work he realized that there were no adequate Swedish textbooks on chemical subjects, and so he decided to prepare such texts himself. The first of these was a book on animal chemistry, published in 1806, which included the results of numerous analyses he had made on animal tissues and fluids. In the course of this work he noted that muscle tissues contain lactic acid, previously found by Scheele in milk. He developed an interest at this time in organic acids, which he later studied in greater detail. After completion of this text, Berzelius turned to the composition of a general textbook of chemistry, and in 1808 he published the first volume of his *Lärbok i kemien*, which was destined to become the most authoritative chemical text of its day. While he was writing it, problems occurred to Berzelius and the search for solutions to these led him to carry out much of the research which occupied his most productive period.

At this time chemists were still debating the questions of whether chemical compounds had a fixed composition. The Berthollet-Proust controversy had nearly been decided in favor of the Proustian view that the composition of salts was invariant, but the actual evidence in support of this view was far from conclusive, largely because of the inadequate number of analyses of salts that existed, and the inaccuracies of many of the analyses which did exist. Furthermore, there was no theoretical reason for assuming a fixed composition. The Berthollet-Proust controversy had Berzelius studied the work of Jeremias Benjamin Richter, who in 1792 had published a work on stoichiometry in which he reported measurements of the amounts of various acids required to neutralize certain bases and of bases to neutralize acids. The work actually demonstrated the law of constant proportions. Berzelius saw that his own analyses, so far as they had been carried, also agreed with this law. He decided to devote himself to the analysis of a large number of salts to confirm or disprove this law. At just this time he learned of the atomic theory of Dalton, which supplied a theoretical basis for the law. Berzelius now realized that he did not yet possess the information needed to complete the second volume of his textbook. For the next four years he carried on his analytical studies and finally summarized them in the second volume, which appeared in 1812. In every subsequent edition he presented further results of his continuing analytical studies. In the meantime, he published many of them in the *Afhandlingar i fysik, kemi och mineralogi,* a journal which Hisinger had founded in 1806 because neither he nor Berzelius were satisfied with the brief form required by the Academy for publication of papers in its *Proceedings.* Berzelius took over publication of later volumes of this journal, and it continued to appear until 1818 when he became secretary of the Academy and could himself have an influence on its policies. Most of Berzelius' important papers were also published in foreign journals, and the various editions of his textbook appeared in German, French, Dutch, Italian, and Spanish. Wöhler's German translations were especially helpful in making his ideas known abroad.

The scientific apparatus and reagents available in Sweden when Berzelius began his work were very inadequate. In consequence, he had to design and build almost everything he needed and to synthesize most of his own reagents. The new forms of apparatus that he built were described in the various editions of his textbook and became standard pieces of equipment in laboratories all over the world. He was especially skillful in the use of the blowpipe, which had been developed in the Scandinavian countries. He utilized it in many of his analytical procedures, and the book that he wrote concerning it popularized its use abroad. It was not until his visit to Paris in 1818 that he was able to secure better materials for his laboratory; he sent home twelve large packing cases of apparatus.

Not only did Berzelius have to design his apparatus, but he also had to work out new analytical methods, and in the planning of such methods he showed his chemical genius. He spent much time in the preliminary work for each analysis, so that when he performed the analysis itself he was sure of his results and seldom felt the need to repeat the work.

He set himself the task "to find the definite and simple proportions in which the constituents of inorganic nature are bound together." In general he based his work on oxygen compounds. His conviction of the importance of oxides had begun with his studies of Lavoisier's work and had been strengthened by his electrochemical experiments. He determined the ratio of metal to oxygen in a number of metallic oxides by reducing the oxide to the metal with hydrogen, or sometimes by converting the metal to its oxide. Similarly he determined the oxygen to sulfur ratio in sulfur dioxide and trioxide. From these results he went on to analyses of sulfates and other salts. He reported his analyses in terms of the positive and negative components; for example, for calcium sulfate as CaO and SO_3. This method of reporting analytical results was long continued by analytical chemists. To his great delight all his analyses fitted into his original assumption of the validity of the law of constant proportions. His results permitted him to determine the atomic weights of the elements he studied, although at first he had no way of determining whether a given value or some multiple of it represented the true atomic weight. When Dulong and Petit in 1819 announced the law that the product of atomic weight and specific heat is a constant, Berzelius recognized that he had a new tool for his purpose, and when Mitscherlich, who later studied with him, published the law of isomorphism in 1820, he saw another. By applying these laws to his own results, he was able to correct his values, and only in the case of the alkali metals did he finally accept values that were double the correct ones. He published revised tables of atomic weights in 1814, 1818, and 1828, and a separate pamphlet was issued in French in 1819 to give wider circulation to his values. In the 1818 table he reported the atomic weights of forty-five of the forty-nine elements then known. Thirty-nine of the determinations were his; the other six were by his students.

The table included the chemical composition of nearly 2,000 compounds.

This work with so many salts of so many elements brought home to him the need for a simple and logical system of symbols to represent the compounds he discussed. His first publication in this field was a pamphlet in French issued in 1811, and he explained his ideas in German and English papers published over the next three years. His basic suggestion was that as the symbol for each element the first letter of the Latin name be chosen, or, if more than one element began with the same letter, the next letter of the name be added to the initial for one of them. The use of letters to represent the names of elements and compounds was not entirely new, but Berzelius introduced a new quantitative concept with his symbols. The letter stood for the atomic weight of the element as well, and so the chemical formulas of the compounds of these elements represented the chemical proportions of the elements in that compound. To indicate these proportions he wrote the appropriate small numbers in the formulas. He placed the numbers as superscripts resembling algebraic exponents in these formulas (e.g., SO^3), a practice that continued to be used in France, although elsewhere the numbers came to be written as subscripts. At first there was some opposition to the use of these formulas, but their advantages eventually came to be recognized and their use became universal. Berzelius later introduced certain modifications which he believed made the formulas simpler. Instead of writing O for oxygen he placed a dot for each oxygen atom above the symbol of the element combined with it, and for a double atom he placed a bar through the letter involved. Thus the symbol for water became Ḣ. Such formulas were not easily set in type, and these innovations did not survive for very long. The basic principles of the Berzelius system have served chemistry well, however.

In accord with the interest that Swedish chemists had long shown in mineralogical studies, Berzelius had from time to time analyzed minerals that came into his hands. As was noted above, the discovery of cerium was the result of such an analysis. However, when he began his systematic studies to establish the law of constant proportions, he worked largely with simple salts. In 1812 he received a gift of a large number of minerals which he later decided to classify. The methods of mineral classification existing at that time were based on appearance and physical properties. These seemed highly unsystematic to Berzelius. He concluded from his analytical experience that a logical classification could be based only on chemical composition. In his original system, first published in 1814, he arranged the minerals in terms of their basic constituents, although he later revised this and placed chief emphasis on the acid component. Like many of Berzelius' innovations, his system of mineral classification was at first received with some hostility, but this was gradually overcome. During his visit to Paris in 1818 he won the approval of Haüy, the leading mineralogist of the day, whose own system was based on physical properties.

Interest in the composition of inorganic substances and even of industrial wastes led to the discovery of a number of new elements in the Berzelius laboratory. He himself discovered selenium and thorium, while, as students working with him, Arfwedsen isolated lithium, Sefström found vanadium, and Mosander discovered a number of rare earth elements.

Berzelius was not only a brilliant laboratory experimenter. He constantly tried to bring together the isolated facts discovered by experiment and to produce a synthesis that could explain the basic problems of his science. The major synthesis of his career was his dualistic theory, by which he believed he had explained the long-discussed problem of the nature of the affinity that held chemical substances together. His analytical work furnished him with numerous examples of salts composed of acid and basic radicals, and his early electrochemical studies suggested to him the mechanism that he sought.

He had found that an electric current splits salts into positive and negative components. Berzelius believed in the two-fluid theory of electricity and he held that electricity was itself a substance. Therefore when a salt was split by a current, the negative electricity combined with the positive component of the salt, while the positive electricity combined with the negative component. This maintained electrical neutrality. When the electricities were not present, the negative and positive components of the salt would combine and neutralize each other. Berzelius built his theory by elaboration of these facts and ideas. He believed that one pole of a magnet could be stronger than the other, and similarly the electricity in a substance might be concentrated at one point in it, leading to a predominance at that point of either negative or positive electricity. This condition of unipolarity determined the electrical behavior of the substance. The intensity of the polarity was another important factor, since the more intense the polarity, the stronger would be the affinity for another substance which would neutralize it. Thus, as he said,

. . . every chemical combination is wholly and solely dependent on two opposing forces, positive and negative electricity, and every chemical compound must be composed of two parts combined by the agency of their electrochemical reaction, since there is no third force. Hence it follows that every compound body, whatever the number of its constituents, can be divided into two parts, one of which is positively and the other negatively electrical [*Essai sur la théorie des proportions chimiques* (1819), p. 98].

Berzelius arranged all the elements in a series of decreasing electronegativity. Since oxygen combined with everything and was liberated at the positive pole, it was obviously the most electronegative element, while potassium was the most electropositive. In compounds, the electrochemical nature of the element combined with oxygen determined the total polarity of the compound. This followed because the amounts of electricity in the two parts of an oxide seldom exactly neutralized each other. Therefore, when oxygen combined with an element, a compound of the "first order" resulted, such as potassium oxide, in which a positive charge remained, due to the strong electropositive character of the potassium. In the case of sulfuric acid (SO_3, the anhydride) a negative charge predominated, for sulfur stood next to oxygen in the table of decreasing electronegativities. If now potassium oxide and sulfuric acid were brought together, potassium sulfate was formed, $KO \cdot SO_3$ as Berzelius would write it, since he doubled the atomic weight of potassium. This would be a compound of the second order. A charge could still remain, since the two parts would not exactly neutralize each other, and another charged salt such as aluminum sulfate could combine with the potassium sulfate to form alum, a third-order compound. Finally, to neutralize completely the various charges, water could be taken up to give the fourth-order compound, hydrated alum.

This theory involves several physical difficulties. Unipolarity, either in magnets or atoms, is not possible, and Berzelius confused quantity and intensity of electricity, a distinction which had been made by Faraday. Berzelius was not well trained in physics, and physicists in general did not pay much attention to his theory. Among chemists, however, it attracted many followers, since it explained so easily the behavior of inorganic substances and since the great authority of Berzelius gave it added weight. It can be seen that it contains many features that were later incorporated into the more modern theories of the structure of polar compounds. Until the discovery of organic compounds, which did not fit readily into the scheme, the dualistic theory dominated the thinking of almost all chemists.

Although the dualistic concept was the most influential of the theoretical syntheses of Berzelius, he drew together other scattered facts and gave generalized definitions of other chemical phenomena upon which much later chemistry developed. These generalizations were made in the course of his compilation of the annual reviews of the progress of chemistry which he published from 1821 until his death.

In the days following the Lavoisier revolution, chemical analyses and syntheses revealed a great variety of new compounds. It was generally assumed that each of these compounds must have an individual composition. Eventually, however, analytical results indicated that quite distinct compounds might have the same chemical composition. The most famous case was the identity in analytical results for the fulminates and cyanates as revealed by Liebig and Wöhler. Berzelius became interested in these strange results and collected a number of other cases in which the same phenomenon was observed. In 1831 he proposed the name isomerism for this phenomenon. The name was chosen by analogy with the term isomorphism used by Mitscherlich for different compounds with the same crystal structures. In 1840 Berzelius suggested the name allotropy for the existence of different forms of the same element.

An even more important generalization was made when Berzelius gathered together a rather large number of cases in which a reaction occurred only when some third substance was present, although this substance seemed to remain unchanged throughout the reaction. In 1835 he suggested that here a new force must exist whose nature was not clear to him. He suggested the name catalytic force and called decomposition of bodies by this force catalysis "as one designates the decomposition of bodies by chemical affinity analysis."

Another term suggested by Berzelius was the word *protein,* which he proposed in a letter to Gerardus Mulder when the latter was investigating these compounds. He derived it from the Greek word *proteios* ("primitive"), since he recognized the prime importance of these compounds.

Berzelius was primarily interested in inorganic chemistry and most of his theoretical ideas were derived from the behavior of inorganic compounds. Nevertheless much of his early work involved analysis of animal products, and he continued to investigate organic compounds throughout most of his life. He developed a form of combustion apparatus which permitted him to analyze a number of carbon compounds, but which required a great amount of time. It took him eighteen months to carry out twenty-one analyses of seven organic acids. Liebig later developed this method to permit much more rapid determinations. However Berzelius was never as happy dealing with organic compounds as he was with inorganic salts. He considered organic chemistry not as the chemistry of carbon compounds, but as the chemistry of the living organism. To the day of his death he remained a vitalist. In the last edition of his textbook he said, "In living nature the elements seem to obey entirely different laws than they do in the dead." His attitude toward the rapidly developing field of organic chemistry became more and more antagonistic in the later years of his life. This fact emphasizes certain characteristics which were always important in his scientific outlook and which significantly determined the course of his work.

Berzelius was essentially a scientific conservative. His great experimental ability and his power to draw together diverse facts to produce important generalizations should not obscure the point that his work was based almost entirely upon the principles that he had learned in the first decade of his scientific activity. At that period chemical investigations were based very largely on the reactions of inorganic compounds, and this explains why Berzelius never really felt at home with organic chemistry. It involved new principles which were not his own. He resisted change when he felt his ideas were being violated. In the first part of his life he could gradually come to accept unpalatable conclusions. Thus, at first he refused to believe in the elementary nature of chlorine and nitrogen, believing them to be oxides of as yet undiscovered radicals. By 1818, however, he admitted that chlorine was an element, and by 1824 he came to the same conclusion for nitrogen.

As he grew older, it became more and more difficult to convince him that any change in his theories was possible. The mass of facts accumulated by the organic chemists alarmed him. Although at first he welcomed the radical theory expounded by Liebig and Wöhler from their investigation of the benzoyl radical, he soon realized that in this radical oxygen was present as a relatively unimportant constituent. This violated the dualistic theory. He tried to write formulas for radicals which could combine with oxygen as did metals or acids, and these formulas became more and more complicated as new facts contradicted them. Eventually all these formulas were rejected by his colleagues. The final blow came with the discovery by Dumas that chlorine could substitute for hydrogen in organic radicals without altering the essential properties of the compounds. Dumas and Laurent expanded the substitution principle into a major feature of organic chemistry. It was impossible for Berzelius to accept this, since for him negative chlorine could not replace positive hydrogen. His whole dualistic theory would collapse if he agreed to such a substitution.

Actually even the organic chemists recognized that they could not account for affinity in the compounds they studied. In developing their theory of types and later structural chemistry they simply represented chemical bonds by brackets or lines and made no attempt to explain what these represented. The relation of the forces holding salts together and those binding carbon to carbon or hydrogen could not be established until the electron theories of chemical bonding began to develop in the twentieth century. Then it was seen that there had been much truth in the Berzelius dualism, at least so far as polar compounds were concerned. This was of no help to Berzelius when he saw his precious theory discarded and his attempts to salvage it patronizingly disregarded by the new organic chemists. His attacks on Liebig, Dumas, and Laurent became more violent and much of the bitterness of his last years resulted from his inability to admit to any modification of the ideas he had developed in his most active years.

The tremendous influence which Berzelius exerted on the chemists of his time came not only from his experimental discoveries and his theoretical interpretations. His voluminous writings were translated into all important European languages and circulated everywhere in the chemical world. He reported his own discoveries in the various editions of his textbook, and he surveyed the whole progress of chemistry in his annual reports, the *Arsberättelser över vetenskapernas framsteg,* which were translated into German, mostly by Wöhler, and were read everywhere. Aside from these formal writings, Berzelius was personally acquainted with almost all the active chemists of Europe and after his visits to them he kept up an extensive correspondence, learning of new developments as they occurred and informing his friends about them even before he described them in his books. Much of his correspondence has been published.

In his own laboratory he worked directly with a succession of young Swedish and foreign students who thus learned his methods and thoughts firsthand and spread them abroad when they left him. Most of them maintained close friendship with him. In his autobiographical notes Berzelius lists twenty-four Swedes and twenty-one foreigners who worked in his laboratory. By the force of his personality, by the skill of his laboratory techniques, and by his power to collect, synthesize, and publicize the chemistry of his day, he exerted an influence on his own time which is still reflected in chemistry more than a century after his death.

BIBLIOGRAPHY

I. ORIGINAL WORKS. The publications of Berzelius were so numerous and appeared in so many editions, translations, and excerpts that a listing of even the major ones would consume a large amount of space. Fortunately a complete bibliography of all works by, and most works about, Berzelius has been compiled by Arne Holmberg, *Bibliografi över Berzelius* (Uppsala–Stockholm), Vol. I (1933), supp. 1 (1936), supp. 2 (1953); Vol. II (1936), supp. 1 (1953). A bibliography of the most important works is also given by J. R. Partington, *A History of Chemistry*, IV (London, 1964), 144–147. Berzelius' own account of some of his work is found in *Jöns Jacob Berzelius Autobiographical Notes*, Olof Larsell, trans. (Baltimore, 1934).

II. SECONDARY LITERATURE. The standard biography is H. G. Söderbaum, *Jac. Berzelius, Levnadsteckning*, 3 vols. (Uppsala, 1929–1931). A detailed account of the most important work is given in H. G. Söderbaum, *Berzelius Werden und Wachsen 1779–1821* (Leipzig, 1899). Useful shorter biographies are Wilhelm Prandtl, *Humphry Davy, Jöns Jacob Berzelius* (Stuttgart, 1948) and J. Erik Jorpes, *Jac. Berzelius, His Life and Work* (Stockholm, 1966).

HENRY M. LEICESTER

BESSEL, FRIEDRICH WILHELM (*b.* Minden, Germany, 22 July 1784; *d.* Königsberg, Germany [now Kaliningrad, U.S.S.R.], 17 March 1846), *astronomy, geodesy, mathematics.*

Bessel's father was a civil servant in Minden; his mother was the daughter of a minister named Schrader from Rheme, Westphalia. Bessel had six sisters and two brothers, both of whom became judges of provincial courts. He attended the Gymnasium in Minden but left after four years, with the intention of becoming a merchant's apprentice. At school he had had difficulty with Latin, and apart from an inclination toward mathematics and physics, he showed no signs of extraordinary talent until he was fifteen. (Later, after studying on his own, Bessel wrote extensively in Latin, apparently without difficulty.)

On 1 January 1799 Bessel became an apprentice to the famous mercantile firm of Kulenkamp in Bremen, where he was to serve for seven years without pay. He rapidly became so proficient in calculation and commercial accounting that after his first year he received a small salary; this was gradually increased, so that he became financially independent of his parents.

Bessel was especially interested in foreign trade, so he devoted his nights to studying geography, Spanish, and English, learning to speak and write the latter language within three months. In order to qualify as cargo officer on a merchant ship, he studied books on ships and practical navigation. The problem of determining the position of a ship at sea with the aid of the sextant stimulated his interest in astronomy, but knowing how to navigate by the stars without deeper insight into the foundations of astronomy did not satisfy him. He therefore began to study as-

tronomy and mathematics, and soon he felt qualified to determine time and longitude by himself.

Bessel made his first time determination with a clock and a sextant that had been built to his specifications. The determination of the longitude of Bremen and the observation of the eclipse of a star by the moon are among his first accurate astronomical exercises. He learned of observations and discoveries through the professional astronomical journals *Monatliche Correspondenz* and *Berliner astronomisches Jahrbuch*, and thus was able to judge the accuracy of his own observations.

In a supplementary volume of the *Berliner astronomisches Jahrbuch* Bessel found Harriot's 1607 observations of Halley's Comet, which he wanted to use to determine its orbit. He had equipped himself for this task by reading Lalande and then Olbers on the easiest and most convenient method of calculating a comet's orbit from several observations. The reduction of Harriot's observations and his own determination of the orbit were presented to Olbers in 1804. With surprise Olbers noted the close agreement of Bessel's results with Halley's calculation of the comet's elliptical elements. He immediately recognized the great achievement of the twenty-year-old apprentice and encouraged him to improve his determination of the comet's orbit by making additional observations. After Bessel had done so, this work was printed, upon Olbers' recommendation, in *Monatliche Correspondenz*. The article, which was on the level of a doctoral dissertation, attracted much attention because of the circumstances under which it had been written. It marks the turning point in Bessel's life; from then on he concentrated on astronomical investigations and celestial mechanics. Later, Olbers claimed that his greatest service to astronomy was having encouraged Bessel to become a professional astronomer.

At the beginning of 1806, before the expiration of his apprenticeship with Kulenkamp, Bessel accepted the position of assistant at Schröter's private observatory in Lilienthal, near Bremen, again on Olbers' recommendation. Schröter, a doctor of law and a wealthy civil servant, was renowned for his observations of the moon and the planets; and as a member of various learned societies, he was in close contact with many scientists. In Lilienthal, Bessel acquired practical experience in observations of comets and planets, with special attention to Saturn and its rings and satellites. At the same time, he studied celestial mechanics more intensively and made further contributions to the determination of cometary orbits. In 1807 Olbers encouraged him to do a reduction of Bradley's observations of the positions of 3,222 stars, which had been made from 1750 to 1762 at the Royal Greenwich Observatory. This task led to one of his greatest achievements.

When Friedrich Wilhelm III of Prussia ordered the construction of an observatory in Königsberg, Bessel was appointed its director and professor of astronomy (1809), on the recommendation of Humboldt. He had previously declined appointments in Leipzig and Greifswald. He took up his new post on 10 May 1810. The title of doctor, a prerequisite for a professorship, had been awarded to him without further formalities by the University of Göttingen after Gauss had proposed it. Gauss had met Bessel in 1807 at Bremen and had recognized his unusual ability.

While the observatory in Königsberg was being built (1810–1813), Bessel made considerable progress in the reduction of Bradley's observations. In 1811 he was awarded the Lalande Prize of the Institut de France for his tables of refraction derived from these observations, and the following year he became a member of the Berlin Academy of Sciences. In 1813 Bessel began observations in Königsberg, primarily of the positions of stars, with the Dollond transit instrument and the Cary circle. The observatory's modest equipment was markedly improved by the acquisition of a Reichenbach-Ertel meridian circle in 1819, a large Fraunhofer-Utzschneider heliometer in 1829, and a Repsold meridian circle in 1841. Bessel remained in Königsberg for the rest of his life, pursuing his research and teaching without interruption, although he often complained about the limited possibilities of observation because of the unfavorable climate. He declined the directorship of the Berlin observatory, fearing greater administrative and social responsibilities, and nominated Encke, who was appointed in his stead. Of Bessel's students, several became important astronomers; Argelander is perhaps the most famous.

Bessel married Johanna Hagen in 1812, and they had two sons and three daughters. The marriage was a happy one, but it was clouded by sickness and by the early death of both sons. Bessel found relaxation from his intensive work in daily walks and in hunting. He corresponded with Olbers, Schumacher (the founder of the *Astronomische Nachrichten*), and Gauss, and left Königsberg only occasionally.

From 1840 on, Bessel's health deteriorated. His last long trip, in 1842, was to England, where he participated in the Congress of the British Association in Manchester. His meeting with important English scientists, including Herschel, impressed him deeply and stimulated him to finish and publish, despite his weakened health, a series of works.

After two years of great suffering, Bessel died of cancer. He was buried near the observatory. Bessel was small and delicate, and in his later years he appeared prematurely aged because of his markedly pale and wrinkled face. This appearance altered, however, as soon as he began to talk; then the force of a strong mind was evidenced in brilliant, rapid speech, and his otherwise rigid expression revealed mildness and friendliness.

Newcomb, in his *Compendium of Spherical Astronomy* (1906), has called Bessel the founder of the German school of practical astronomy. This German school started with astrometry and, after Bessel's death, was expanded to astrophysics by Bunsen and Kirchhoff's discovery of spectral analysis. Foremost among the interests of this school were the construction of precision instruments, the study of all possible instrument errors, and the careful reduction of observations. Bessel's contributions to the theory of astronomical instruments are for the most part restricted to those instruments used for the most accurate measurement of the positions of the stars and planets. The principles he laid down for the determination of errors were later followed so painstakingly by less gifted astronomers that the goal to be achieved—the making of a great number of good observations—was relegated to the background in favor of important investigations relative to the instruments themselves. Such was never Bessel's intention; he was undoubtedly one of the most skillful and diligent observers of his century. His industry is well illustrated by the twenty-one volumes of *Beobachtungen der Königsberger Sternwarte*.

Bessel recognized that Bradley's observations gave a system of very accurate star positions for the epoch 1755 and that this could be utilized in two ways. First, a reference system for the measurement of positions of stars and planets was required. Second, the study of star motions necessitated the determination of accurate positions for the earliest possible epoch. Tobias Mayer had determined fundamental star positions from his own observations around the middle of the eighteenth century, but Bradley was never able to reduce his own numerous observations.

The observations of star positions had to be freed of instrumental errors, insofar as these could be determined from the measurements themselves, and of errors caused by the earth's atmosphere (refraction). The apparent star positions at the time of a particular observation (observation epoch) had to be reduced to a common point in time (mean epoch) so that they would be freed of the effects of the motion of the earth and of the site of observation. For this a knowledge of the precession, the nutation, and the aberration was necessary. Bessel determined the latitude of Greenwich for the mean epoch 1755 and the obliquity of the ecliptic, as well as the constants of precession, nutation, and aberration. To determine precession from proper motions, Bessel used both Bradley's and Piazzi's observations. Bessel's first published work on the constant of precession (1815) was awarded a prize by the Berlin Academy of Sciences.

The positions of Bradley's stars valid for 1755 were published by Bessel as *Fundamenta astronomiae pro anno 1755* (1818). This work also gives the proper motions of the stars, as derived from the observations of Bradley, of Piazzi, and of Bessel himself. It constitutes a milestone in the history of astronomical observations, for until then positions of stars could not be given with comparable accuracy: through Bessel's work, Bradley's observations were made to mark the beginning of modern astrometry. During this investigation Bessel became an admirer of the art of observation as practiced by Bradley; and because Bradley could not evaluate his own observations, Bessel followed and also taught the principle that immediately after an observation, the reduction had to be done by the observer himself. Further, he realized that the accurate determination of the motions of the planets and the stars required continuous observations of their positions until such motions could be used to predict "the positions of the stars . . . for all times with sufficient accuracy."

Later, when many unpublished observations of Bradley's were found and when, about 1860, Airy had made accurate observations of the same stars at the Royal Greenwich Observatory, Auwers improved Bessel's reductions and derived proper motions of better quality. Auwers' star catalog was published in three volumes (1882–1903).

Bessel's first and very important contribution to the improvement of the positions and proper motions of stars consisted of the observations of Maskelyne's thirty-six fundamental stars. As Bradley's successor, Maskelyne had chosen these stars to define the system of right ascensions. Bradley had been able to make differential measurements of positions with such accuracy that the star positions for 1755 and those determined by Airy for 1860 resulted in proper motions with the excellent internal accuracy of about one second of arc per century. Greater difficulties were experienced, however, with the measurement of positions with respect to the vernal equinox as zero point of the right ascensions. The continuously changing position of the vernal equinox had to be determined at the time of the equinoxes from the differences in time of the transits of bright stars and the sun through the meridian. In 1820 Bessel succeeded in determining the position of the vernal equinox with an accuracy

of .01 second by observing both Maskelyne's stars and the sun. This can be verified by measurements made in the twentieth century.

In *Tabulae Regiomontanae* (1830), Bessel published the mean and the apparent positions of thirty-eight stars for the period 1750–1850. He added the two polar stars α and δ Ursae Minoris to Maskelyne's thirty-six fundamental stars. The foundations for the ephemerides were the mean positions for 1755 and the positions derived for 1820 from observations at Königsberg. The position of the vernal equinox for 1822, as determined by Bessel, served as the zero point for counting the right ascensions. Bessel derived the ephemerides of the *Tabulae Regiomontanae* without using a specific value of the constant of precession, for in order to find a third position from two given positions of a star, it is necessary to know only the annual variation of precession, not the value of precession itself. Therefore Bessel's ephemerides are correct (aside from errors in observation) up to and including the first magnitude for the proper motions and up to the second magnitude for precession. Only for the two polar stars did Bessel determine the proper motions and also give the values, since for these stars the terms of higher order in proper motion and precession could not be neglected. In calculating the data of the *Tabulae Regiomontanae* Bessel improved his 1815 determination of the precession by utilizing his Königsberg observations of Bradley's stars.

The star positions given for one century in the *Tabulae Regiomontanae* constitute the first modern reference system for the measurement of the positions of the sun, the moon, the planets, and the stars, and for many decades the Königsberg tables were used as ephemerides. With their aid, all observations of the sun, moon, and planets made since 1750 at the Royal Greenwich Observatory could be reduced; and thus these observations could be used for the theories of planetary orbits.

During observations of the stars α Canis Major (Sirius) and α Canis Minor (Procyon), which are among Maskelyne's fundamental stars, Bessel discovered the variation of their proper motions. He concluded that these stars must have dimmer companions whose masses, however, were large enough to make visible the motions of the brighter double-star components around the center of gravity. Arguing from the variation of the proper motion, more than a hundred years later astronomers discovered stars with extremely low luminosity, called dark companions.

Observing the positions of numerous stars with the Reichenbach meridian circle, Bessel pursued two aims: the determination of the motions of the stars in such a way that their positions could be predicted for all time, and the definition of a reference system for the positions of the stars. Between 1821 and 1833 he determined the positions of approximately 75,000 stars (brighter than ninth magnitude) in zones of declination between $-15°$ and $+45°$. With these observations he also developed the methods for determining instrumental errors, including those of the division of the circle, and eliminated such errors from his observations. He published all measurements in detail, and thus they can be verified. These observations were continued by Argelander, who measured the positions of stars in zones of declination from $+45°$ to $+80°$ and from $-16°$ to $-32°$. The work of Bessel and Argelander encouraged the establishment of two large-scale programs: Argelander's *Bonner Durchmusterung* and the first catalog of the Astronomische Gesellschaft (*AGK 1*) with the positions of the stars of the entire northern sky. The *Bonner Durchmusterung* is a map of the northern sky that contains all stars up to magnitude 9.5, and the catalog is the result of meridian circle observations made at many observatories.

One of Bessel's greatest achievements was the first accurate determination of the distance of a fixed star. At the beginning of the nineteenth century, the approximate radius of the earth's orbit (150,000,000 km.) was known, and there was some idea of the dimensions of the planetary system, although Neptune and Pluto were still unknown. The stars, however, were considered to be so far away that it would be hopeless to try to measure their distances. The triangulation procedure was already known, and for this the diameter of the earth's orbit could serve as the base line, since its length was known. It was also known that the motion of the earth around the sun must be mirrored in a periodic motion of the stars within the period of a year, in such a way that a star at the pole of the ecliptic would describe a circular orbit around the pole, stars at ecliptical latitudes between 0° and 90° would describe ellipses, and stars at the ecliptic would undergo periodic variations of their ecliptical longitudes. This change of position of the stars, as evidenced by the motion of the earth, was considered to be immeasurably small, however. The radius of the circle, of the ellipse, or of the ecliptical segment of arc—the so-called parallax figure—is the parallax of the star; the parallax π is the angle subtended by the radius of the earth's orbit at the position of the star. If this angle π can be measured, then the distance r of a star can be obtained from $\sin \pi = a/r$, where a represents the radius of the earth's orbit. An angle of $\pi = 1''$ corresponds to 206,265 radii of the earth's orbit (or $3.08 \cdot 10^{13}$ km., or 3.26 light-years).

In the first half of the eighteenth century Bradley had attempted to determine the parallaxes of the stars γ Draco and η Ursa Major by measurements of the angular distances of these stars from the zenith (zenith distances). Both stars culminate in the vicinity of the zenith of Greenwich and thus are particularly suitable for the accurate measurement of zenith distances. The "absolute" parallax of the stars—that is, the parallactic change of position with respect to a fixed direction on earth (direction of the plumb line at Greenwich)—should be determinable from the variation of the zenith distances. Bradley found an annual variation with an amplitude of twenty seconds of arc, but the phase was shifted by three months from the expected parallactic change of position. He correctly interpreted the phenomenon as a change in direction—arising from the motion of the earth in its orbit—of the stellar light that reaches the earth, and thus discovered the aberration of light. However, he could not detect parallaxes of the stars, but could only conclude that the parallaxes must be smaller than .50 second of arc for the stars he observed.

As a result of this knowledge of the small size of the parallaxes to be expected, the measuring procedures were changed in later experiments, for the accuracy of the measurements of zenith distances was obviously inadequate for the purpose. The angular distance between two stars very close together on the sphere could be determined much more accurately. If one star of a star pair is very far from the sun and the other is near the sun, then the parallax figure of the nearer star must become visible as a result of frequent measurements of the angular distances between the two stars. It was therefore suggested that astronomers measure "relative" parallaxes, that is, the parallactic changes in the position with reference to other stars that can be assumed to be very far away. Herschel's attempts to measure stellar parallaxes in this way led to the discovery of the physical double stars; he found that the components of most star pairs are near to each other in space, as is shown by their motion around the common center of gravity. Herschel's attempt to determine parallaxes failed, however.

This lack of success led to a search for signs that one of the stars would be especially near, with great brightness of an individual star regarded as an indication of its great nearness. (This assumption would be correct if all stars had the same luminosity and if there were no inhomogeneous interstellar absorption. Since both of these conditions are not fulfilled, the relation between apparent magnitude and distance holds only in the statistical mean.) In determining proper motions, Bessel found that individual stars are marked by especially great motions and that these stars are not among the brightest. He concluded that great proper motions are, in most cases, the result of small star distances. Therefore, in order to determine the parallax, he selected the star with the greatest proper motion known to him (5.2″ per year), a star of magnitude 5.6, which had been designated as 61 Cygni in Flamsteed's star catalog.

To determine the parallax Bessel used the Fraunhofer heliometer, an instrument intended primarily for the measurement of the angular diameter of the sun and the planets. The heliometer is a telescope with an objective that can be rotated around the optical axis. The objective is cut along a diameter; both halves can be shifted along the cutting line and the displacement can be measured very accurately. Each half of the objective acts optically as a complete objective would, so that upon moving the halves, two noncoincident images of one object arise. The distance of two stars, A and B, that are in the field of view is measured by sliding the halves so that the image of A coincides with the image of B produced in the second half; thus the two stars appear as one. In Bessel's day this procedure of coincidence determination permitted more accurate measurements than did the customary micrometer determinations with an ordinary telescope; the latter were used to determine the angular distances of the components of double stars. Further, with the heliometer one could measure greater angular distances than with the micrometer (up to nearly two degrees with Bessel's heliometer). For determining the parallax of 61 Cygni, Bessel selected two comparison stars of magnitude 9–10 at distances of roughly eight and twelve minutes of arc. 61 Cygni is a physical double star whose components differ in brightness by less than one magnitude. The distance of sixteen seconds of arc between the components favored the accuracy of the determination of the parallax because pointing could be carried out with two star images. After observing for eighteen months, by the fall of 1838 Bessel had enough measurements for the determination of a reliable parallax. He found that $\pi = 0.314''$ with a mean error of $\pm 0.020''$. This work was published in the *Astronomische Nachrichten* (1838), the first time the distance of a star became known. Bessel's value for the parallax shows excellent agreement with the results obtained by extensive modern photographical parallax determinations, which have yielded the value $\pi = 0.292''$, with a mean error of $\pm 0.0045''$. The distance of 61 Cygni thus amounts to $6.9 \cdot 10^5$ radii of the earth's orbit, or 10.9 light-years.

Bessel's conjecture that the stars with the greatest proper motions are among the nearest was later proved

correct, and the amount of proper motion has remained a criterion for the choice of stars for parallax programs. Only one year after the completion of Bessel's work, two other successful determinations of parallaxes were made known. F. G. W. Struve in Dorpat determined the parallax of the bright star α Lyra (Vega) by means of micrometric measurements. The value he found, $\pi = 0.262'' \pm 0.037''$ (m.e.) nevertheless deviates considerably from the now reliably known value $\pi = 0.121'' \pm 0.006''$ (m.e.). In addition, Thomas Henderson had observed the bright star α Centaurus at the Cape Observatory and had found a parallax of approximately one second of arc. The reliable value for this today amounts to $\pi = 0.75''$. The pioneering work of Bessel, Henderson, and Struve not only opened up a new area of astronomical research but also laid the foundation for the investigation of the structure of our star system.

Bessel was also an outstanding mathematician whose name became generally known through a special class of functions that have become an indispensable tool in applied mathematics, physics, and engineering. The interest in the functions, which represent a special form of the confluent hypergeometric function, arose in the treatment of the problem of perturbation in the planetary system. The perturbation of the elliptic motion of a planet caused by another planet consists of two components, the direct effect of the perturbing planet and its indirect effect, which arises from the motion of the sun caused by the perturbing planet. Bessel demonstrated that it is appropriate to treat the direct and the indirect perturbations separately, so that in the series development of the indirect perturbation, Bessel functions appear as coefficients. In studying indirect perturbation, Bessel made a systematic investigation of its functions and described its main characteristics. This work appeared in his Berlin treatise of 1824. Special cases of Bessel coefficients had been known for a long time; in a letter to Leibniz in 1703, Jakob Bernoulli mentioned a series that represented a Bessel function of the order 1/3. In addition, in a work on the oscillations of heavy chains (1732) Daniel Bernoulli used Bessel coefficients of the order zero, and in Euler's work on vibrations of a stretched circular membrane (1744) there was a series by means of which $J_n(z)$ was defined. Probably a work by Lagrange on elliptical motion (1769), in which such series appear, had led Bessel to make these investigations. The impulse, however, did not come from pure mathematical interests, but from the necessity of applying such series in the presentation of indirect perturbations. Bessel left few mathematical works that do not have some practical astronomical application.

Like nearly all great astronomers of his era, Bessel was obliged to spend part of his time surveying wherever the government wished. In 1824 he supervised the measurement of a 3,000-meter base line in the Frischen Haff because he liked to spend a day in the fresh air once in a while. In 1830 he was commissioned to carry out triangulation in East Prussia, after Struve had completed the triangulation of the Russian Baltic provinces. Bessel designed a new measuring apparatus for the determination of base lines that was constructed by Repsold; he also developed methods of triangulation by utilizing Gauss's method of least squares. Bessel's measuring apparatus and method of triangulation have been widely used. The triangulation in East Prussia and its junction with the Prussian-Russian chain of triangulation was described in a book written with J. J. Baeyer (1838). From his own triangulations and from those of others, Bessel made an outstanding determination of the shape and dimensions of the earth that won him international acclaim.

Among Bessel's works that contributed to geophysics were his investigations on the length of the simple seconds' pendulum (1826), the length of the seconds' pendulum for Berlin (1835), and the determination of the acceleration of gravity derived from observing the pendulum. Bessel achieved the standardization of the units of length then in use by introducing a standard measure in Prussia, the so-called Toise (1 Toise = 1.949063 meters). The necessity of a standard of length had become apparent to him during his work on triangulation in East Prussia, as did the necessity of an international organization to define the units of measures. This need led to the founding of the International Bureau of Weights and Measures.

BIBLIOGRAPHY

I. ORIGINAL WORKS. A collection of Bessel's numerous papers is *Abhandlungen von Friedrich Wilhelm Bessel,* Rudolf Engelmann, ed., 3 vols. (Leipzig, 1875). Vol. I contains Bessel's account of his youth, "Kurze Erinnerungen an Momente meines Lebens," with a supp. by the ed., 23 papers on the motions of planets, and 28 papers on spherical astronomy. Vol. II contains 25 papers on the theory of astronomical instruments, 29 papers on stellar astronomy, and 19 papers on mathematics. Vol. III contains 11 papers on geodesy, 17 papers on physics (mostly geophysics), and 33 on other subjects. Engelmann's collection is not complete, however. *Beobachtungen der Königsberger Sternwarte,* 21 vols., presents Bessel's observations. Major separate publications are *Fundamenta astronomiae pro anno 1755 deducta ex observationibus viri incomparabilis James Bradley in specola astronomica*

Grenovicensi per anno 1750–1762 institutis (Königsberg, 1818) and *Tabulae Regiomontanae reductionum observationum astronomicarum ab anno 1750 usque ad annum 1850 computatae* (Königsberg, 1830). A complete list of Bessel's publications, presented at the end of *Abhandlungen,* III, has 399 entries, including books and book reviews by Bessel.

II. SECONDARY LITERATURE. A bibliography of sketches of Bessel's life and astronomical works is given in *Abhandlungen,* III, 504. Noteworthy are C. Bruhns, in *Allgemeine deutsche Biographie,* pt. 9 (Leipzig, 1875), 558–567; and Sir William Herschel's addresses delivered to Bessel on presenting honorary medals of the Royal Astronomical Society, in *Monthly Notices of the Royal Astronomical Society,* **1** (1829), 110–113, and **5** (1841), 89. A biography of Bessel in anecdotal style is J. A. Repsold, in *Astronomische Nachrichten,* **210** (1919), 161–214. An excellent review of the first determination of a stellar parallax is H. Strassl, "Die erste Bestimmung einer Fixsternentfernung," in *Naturwissenschaften,* 33rd year (1946), 65–71.

WALTER FRICKE

BESSEY, CHARLES EDWIN (*b.* Milton Township, Ohio, 21 May 1845; *d.* Lincoln, Nebraska, 25 February 1915), *botany, education.*

Protagonist of a leading hypothesis of angiosperm phylogeny that, when revised to admit recent research, will probably stand as the accepted system of classification for flowering plants; advocate of the values of scientific meetings for the communication of ideas; author of the most successful textbook of botany published in the United States between 1880 and 1910; exceptional teacher who carried modern botany and its symbol, the microscope, across the Mississippi and planted them firmly in Iowa and then Nebraska and who had among his 4,000 students an impressive number of prominent biologists of the early twentieth century: Charles Bessey was all of these.

Bessey's father, Adnah Bessey, a schoolteacher of Huguenot ancestry, married Margaret Ellenberger, who had been his pupil, in 1841. Educated in rural schools and at home, Charles Bessey was certified to teach at seventeen. In July 1866 he entered Michigan Agricultural College, where he came under the influence of Albert Nelson Prentiss and William James Beal, two botanists noted for their teaching skills. He graduated from the scientific course in 1869 and remained as an assistant in horticulture, but soon left to inaugurate botany and horticulture at Iowa Agricultural College, Ames. His single room, which held two chairs, a table, bureau, washstand, and bedstead, was to serve as office, library, study, and bedroom for three years. Nevertheless, Bessey opened his first botany class for forty-three sophomores the month after his arrival, using Gray's *Lessons in Botany*. In 1871 he added laboratory work to his undergraduate botany course, using his one Tolles compound microscope.

Bessey took part in the Iowa Farmers' Institute, the first of its kind in the country, and encouraged the launching of the Iowa Academy of Sciences. Asa Gray, attracted by Bessey at the American Association for the Advancement of Science meeting at Dubuque, Iowa, in 1872, persuaded him to go to Harvard that year. In Gray's laboratory he was impressed by the importance of morphology and cell structure in plant systematics, whereas previously he had approached the subject from gross macroscopical characteristics alone. Bessey lectured in botany at the University of California in 1875 at the invitation of President Gilman, and returned to Harvard that same year to study under the mycologist William Farlow. From this inoculation Bessey produced four papers on plant diseases that were among the first published in the United States.

Gray's recommendation that Bessey prepare an American adaptation of Julius von Sachs's *Lehrbuch der Botanik,* under the title *Botany for High Schools and Colleges* (1880), reoriented botanical instruction in this country. Bessey's text introduced cryptogamic botany and physiological plant anatomy into American colleges.

Bessey gave direction to botanical literature through his associate editorship of the *American Naturalist* (1880–1897) and *Science* (1897–1915), two of the most influential journals of the time. He offered the first laboratory course in botany at the University of Minnesota in the summer of 1881, using compound microscopes borrowed from Iowa Agricultural College. After teaching at Ames for fifteen years and after repeated solicitations, he moved to the University of Nebraska as professor of botany in September 1884. Bessey's teaching philosophy was later identified with John Dewey's "science with practice." He wrote of the "relatedness of knowledge" and that "the teacher represents the *life* of the subject." He believed introductory classes should be taught by the best-informed, usually senior, professors. His comradeship with students was demonstrated by "Sem. Bot.," where ten advanced botany students conducted investigations, reported in technical detail—and interspersed their seminars with limericks and refreshment. Bessey's second highly successful text, *Essentials of Botany* (1884), had seven editions by 1896. He wrote more than 150 papers and reviews, but his writings on plant phylogeny, "Evolution and Classification" (1893) and "Phylogenetic Taxonomy of Flowering Plants" (1915), established his permanent place in botany.

Bessey took the two well-known systems of angio-sperm classification, those of Engler and Prantl, and of Bentham and Hooker, and rearranged the families on the basis of twenty-eight "dicta," producing a scheme that by its logic and attractive phyletic patterns has proved effective in teaching systematic botany—a success fostered by the textbook of his student Raymond J. Pool. Thousands of students have come to know "Bessey's cactus," a cartoon-like table suggestive of a many-jointed *Opuntia.* Bessey's dicta were a refinement of Candolle's concept that three principal factors have been effective in the differentiation of the angiosperms: loss, or fusion, or specialization of floral parts from a multimerous, free-membered prototype. The fused and few-membered condition was construed as advanced. Whereas apetalous catkin-bearing genera had been considered primitive by the Engler and Prantl school, Bessey viewed them as derived from petaliferous flowers often borne in conelike clusters. Another concept was that monocotyledonous families have been derived from dicotyledonous forms.

Bessey's deep, modulated voice and his genial, persuasive, generous, perennially enthusiastic manner made him popular as a lecturer and officer in many of the some twenty organizations to which he belonged. His memberships ranged from the State Teachers' Association of Nebraska (president in 1888) and the American Microscopical Society (president in 1903 and 1908) to the American Association for the Advancement of Science, of which he was national president in 1910.

Bessey married Lucy Athearn of West Tisbury, Massachusetts, on Christmas Day 1873. All three sons, Edward, Carl, and Ernst, graduated from the University of Nebraska; Ernst Athearn Bessey became internationally known as a mycologist.

BIBLIOGRAPHY

I. ORIGINAL WORKS. Note references in text. "Evolution and Classification" appeared in *Proceedings of the American Association for the Advancement of Science,* **42** (1894), 237–251; "Phylogenetic Taxonomy of Flowering Plants," in *Annals of the Missouri Botanical Garden,* **2** (1915), 109–164. Bessey's MSS in the University of Nebraska botany department and the Nebraska Historical Society Archives include "Discussion of a Plan of a Scientific Course," read before the Nebraska State Teachers' Association 28 December 1876.

II. SECONDARY LITERATURE. There is no biography of Bessey. Works on him are Ernst Athearn Bessey, "The Teaching of Botany Sixty-five Years Ago," in *Iowa State College Journal of Science,* **9** (1935), 227–233; L. H. Pammel, "Prominent Men I Have Met. Dr. Charles Edwin

Bessey," in *Ames* [Iowa] *Daily Tribune and Evening Times,* 26 Nov. 1927, p. 19, and 17 Dec. 1927, p. 20, a eulogy; Raymond J. Pool, "A Brief Sketch of the Life and Work of Charles Edwin Bessey," in *American Journal of Botany,* **2** (1915), 505–518, with portrait and bibliography; Andrew Denny Rodgers, III, *John Merle Coulter* (Princeton, N.J., 1944), *passim,* with short quotations from Bessey's correspondence; and the anonymous "Some Men Under Dr. Bessey," in [Nebraska] *State Journal* (newspaper).

JOSEPH EWAN

BETANCOURT Y MOLINA, AUGUSTIN DE (*b.* Tenerife, Canary Islands, 1758; *d.* St. Petersburg, Russia, 14 July 1824), *physics, engineering.*

Betancourt was a descendant of the Norman navigator Jean de Béthencourt, who discovered the Canary Islands in 1402. After completing his studies in Paris, he was sent by the Spanish government to France, England, Germany, and Holland to study methods of shipbuilding, navigation, mechanics, and using steam engines. He brought back a number of drawings and models which formed the nucleus of the scientific cabinet of the king of Spain. While in France he submitted two important reports to the Académie des Sciences of Paris. In the first he revealed to the Continent the double-action steam engine, which he had observed in action in England. This memoir led Jacques-Constantin Périer to construct the first double-action steam engine in France.

In the second report (1790), Betancourt gave the results of a series of measurements establishing the relation of temperature and steam pressure. This was the first work of its kind, but Betancourt underestimated the importance of the disturbances caused by the presence of even a minimum quantity of residual air.

After another trip to England to study its mining industry Betancourt returned to France. There he became interested in the optical telegraph invented by Claude Chappe, and constructed a line from Madrid to Cadiz. He was then entrusted with the organization of a school of civil engineering in Spain, and became its inspector general. Disturbances in Spain led him to settle temporarily in Paris, where he became well known, especially for a system of water-saving locks.

In 1808 Betancourt accepted an offer from the Russian government, and on 30 November was made chief of staff of the czar's retinue. In Russia he improved the arms industry and constructed bridges using a new system of arches. In collaboration with Carbonnier, Betancourt built the riding school of Moscow, which was then the largest hall without inner supports; the span of its roof was said to be forty

meters. He also constructed the aqueduct of Taïtzy and set up a state paper industry. In 1810 a school of civil engineering was founded in St. Petersburg, and Betancourt became its inspector as well as a professor; in 1819 he was made director of the Central Administration of Civil Engineering.

On 22 August 1822 Betancourt was summarily retired when an official investigation revealed numerous irregularities in the running of the Central Administration of Civil Engineering. Undoubtedly he was simply caught up in a general wave of reform, for he did not lose the czar's favor. He was in the midst of rebuilding St. Isaac's Cathedral when he died.

BIBLIOGRAPHY

I. ORIGINAL WORKS. Among Betancourt's works are *Mémoire sur la force expansive de la vapeur de l'eau* (Paris, 1790); *Mémoire sur un nouveau système de navigation intérieure* (n.p., n.d. [Paris, *ca.* 1808]); and *Essai sur la composition des machines* (Paris, 1808, 1819), written with Lanz. See also the *Procès-verbaux des séances de l'Académie des Sciences tenues depuis la fondation de l'Institut jusqu'au mois d'août 1835,* 10 vols. (Hendaye, 1910–1921): I, 306 (11 Frimaire an VI); 308, 309, 313 (16 Frimaire an VI); 353 (11 Ventose an VI); 373–375 (21 Germinal an VI); III, 504 (2 March 1807); 563 (17 August 1807); 581–585 (14 September 1807); IV, 77 (13 June 1808); 78 (20 June 1808); 88 (8 August 1808), the presentation of the memoir "Nouveau système de navigation . . ."; 279 (27 November 1809); 286 (5 December 1809), Betancourt's election as corresponding member of the Première Classe of the Institut; 331 (12 March 1810); VI, 233 (3 November 1817); 395 (21 December 1818); 448 (10 May 1819); VII, 9 (31 January 1820); VIII, 131 (30 August 1824, 6 September 1824, the latter a replacement).

II. SECONDARY LITERATURE. Works on Betancourt are Jean-Baptiste-Joseph Delambre, "Rapport sur un nouveau télégraphe de l'invention des citoyens Breguet et Bétancourt" (read 21 Germinal an VI), in *Mémoires de l'Institut,* 1st ser., **3,** pt. 1 (an IX), *Histoire,* 22–32; Sebastián Padrón Acosta, *El ingeniero Augustín de Bethencourt y Molina* (Tenerife, 1958); Jacques Payen, "Bétancourt et l'introduction en France de la machine à vapeur à double effet (1789)," in *Revue d'histoire des sciences,* **20** (1967), 187–198, an edition of "Mémoire sur une machine à vapeur à double effet" (read before the Académie Royale des Sciences on 16 December 1789); and Antonio Ruiz Álvarez, "En torno al ingeniero canario don Augustín de Bethencourt y Molina," in *El museo canario,* nos. 77–84 (1961–1962), 139–147.

JACQUES PAYEN

BETTI, ENRICO (*b.* near Pistoia, Italy, 21 October 1823; *d.* Pisa, Italy, 11 August 1892), *mathematics.*

Since his father died when Betti was very young,

the boy was educated by his mother. At the University of Pisa, from which he received a degree in physical and mathematical sciences, he was a disciple of O. F. Mossotti, under whose leadership he fought in the battle of Curtatone and Montanara during the first war for Italian independence.

After having taught mathematics at a Pistoia high school, in 1865 Betti was offered a professorship at the University of Pisa; he held this post for the rest of his life. He also was rector of the university and director of the teachers college in Pisa. In addition, he was a member of Parliament in 1862 and a senator from 1884. His principal aim, however, was always pure scientific research with a noble philosophical purpose.

In 1874 Betti served for a few months as undersecretary of state for public education. He longed, however, for the academic life, solitary meditation, and discussions with close friends. Among the latter was Riemann, whom Betti had met in Göttingen in 1858, and who subsequently visited him in Pisa.

In algebra, Betti penetrated the ideas of Galois by relating them to the previous research of Ruffini and Abel. He obtained fundamental results on the solubility of algebraic equations by means of radicorational operations. It should be noted that the most important results of Galois's theory are included—without demonstration and in a very concise form—in a letter written in 1832 by Galois to his friend Chevalier on the eve of the duel in which Galois was killed. The letter was published by Liouville in 1846. When Betti was able to demonstrate—on the basis of the theory of substitutions, which he stated anew—the necessary and sufficient conditions for the solution of any algebraic equation through radicorational operations, it was still believed in high mathematical circles that the questions related to Galois's results were obscure and sterile. Among the papers in which Betti sought to demonstrate Galois's statements are "Sulla risoluzione delle equazioni algebriche" (1852) and "Sopra la teorica delle sostituzioni" (1855). They constitute an essential contribution to the development from classical to abstract algebra.

Another area of mathematical thought developed by Betti is that of the theory of functions, particularly of elliptic functions. Betti illustrated—in an original way—the theory of elliptic functions, which is based on the principle of the construction of transcendental entire functions in relation to their zeros by means of infinite products.

Betti published these results in a paper entitled "La teorica delle funzioni ellitiche" (1860–1861). These ideas were further developed by Weierstrass some fifteen years later. However, Betti, who in the mean-time had turned to another theory of elliptic functions—this one inspired by Riemann—did not wish to claim priority. These two methods are linked with the two basic aspects of Betti's mathematical thought: the algebraic mode of thought, which went deep into Galois's research, and the physicomathematical mode of thought, developed under Riemann's influence. Betti, an enthusiastic supporter of theoretical physics, had turned toward the procedures already started in electricity and subsequently applied to analysis.

Among Betti's physicomathematical researches inspired by Riemann are *Teorica della forze newtoniane* (1879) and "Sopra le equazioni di equilibrio dei corpi solidi elastici." A law of reciprocity in elasticity theory, known as Betti's theorem, was demonstrated in 1878. Having mastered the methods by which Green had opened the way to the integration of Laplace's equations, which constitute the basis for the theory of potentials, Betti applied these methods to the study of elasticity and then to the study of heat.

Of particular interest is Betti's research on "analysis situs" in hyperspace, which is discussed in "Sopra gli spazi di un numero qualunque di dimensioni" (1871). This research inspired Poincaré in his studies in this field and originated the term "Betti numbers," which subsequently became common usage for numbers characterizing the connection of a variety.

Betti played an important role in the rebirth of mathematics after the Risorgimento. He loved classical culture, and with Brioschi he championed the return to the teaching of Euclid in secondary schools, for he regarded Euclid's work as a model of discipline and beauty. This led to *Gli elementi d'Euclide* (1889).

His enthusiasm and brilliance made Betti an excellent teacher. At the University of Pisa and at the teachers college, he guided several generations of students toward scientific research, among them the mathematicians U. Dini, L. Bianchi, and V. Volterra.

BIBLIOGRAPHY

I. ORIGINAL WORKS. Betti's collected writings are *Opere matematiche*, R. Accademia dei Lincei, ed., 2 vols. (Milan, 1903–1915). Among his works are "Sulla risoluzione delle equazioni algebriche," in *Annali di scienze matematiche e fisiche*, **3** (1852), 49–115; "Sopra la teorica delle sostituzioni," *ibid.*, **6** (1855), 5–34; "La teorica delle funzioni ellitiche," in *Annali di matematica pura e applicata*, **3** (1860), 65–159, 298–310; **4** (1861), 26–45, 57–70, 297–336; "Sopra gli spazi di un numero qualunque di dimensioni," *ibid.*, 2nd ser., **4** (1871), 140–158; "Sopra le equazioni di equilibrio dei corpi solidi elastici," *ibid.*, **6** (1874), 101–111; *Teorica della forze newtoniane* (Pisa, 1879);

and *Gli elementi d'Euclide con note aggiunte ed esercizi ad uso dei ginnasi e dei licei* (Florence, 1889), written with Brioschi.

II. SECONDARY LITERATURE. Works on Betti are F. Brioschi, "Enrico Betti," in *Annali di matematica pura e applicata,* 2nd ser., **20** (1892), 256, or his *Opere matematiche,* III (Milan, 1904), 41–42; F. Enriques, *Le matematiche nella storia e nella cultura* (Bologna, 1938), pp. 187, 203–204, 222, 224–226; and his article on Betti in *Enciclopedia italiana Treccani,* VI (1930), 834; G. Loria, *Storia delle matematiche,* III (Turin, 1933), 497, 541, 556–557; and V. Volterra, *Saggi scientifici* (Bologna, 1920), pp. 37, 40–41, 46–50, 52–54.

ETTORE CARRUCCIO

BEUDANT, FRANÇOIS-SULPICE (*b.* Paris, France, 5 September 1787; *d.* Paris, 9 December 1850), *mineralogy, geology.*

Educated at the École Polytechnique and the École Normal Supérieure in Paris, Beudant began his career as *répétiteur* at the latter institution, leaving this post to become professor of mathematics at Avignon (1811) and then professor of physics at Marseilles (1813). During these years his primary interests were zoology and paleontology, tastes he acquired while studying with Gilet de Laumont. He studied species of coelenterates and mollusks, trying to determine whether freshwater varieties could adapt to saltwater and whether marine forms could have originated from freshwater fauna. Some of his observations were included in "Mémoire sur la possibilité de faire vivre des mollusques fluviatiles dans les eaux salées et des mollusques marins dans les eaux douces . . . "

Louis XVIII appointed Beudant as assistant director of his cabinet of mineralogy in 1814, charging him with the task of cataloguing the enormous mineralogical collection of the Comte de Bournon, which was to be moved to Paris from England the following year. This work directed Beudant's attention from natural history to mineralogy and geology, with which he was thereafter concerned. In 1818 he was sent by the state on a scientific expedition to Hungary, where he gathered masses of important data that were published in his three-volume *Voyage minéralogique et géologique en Hongrie* (1822).

In 1820 Beudant became professor of mineralogy and physics on the Faculty of Sciences at the Sorbonne but resigned the chair of physics so that Ampère might have it. In 1839 he left the university and became *inspecteur général des études,* which was equivalent to being supervisor for the entire French educational system. He held this position until his death. In 1841 he wrote a grammar of the French language that was favorably received by his contemporaries.

Mineralogical investigations, particularly experiments with carbonates and other salts, revealed to Beudant a principle of the combination of mineral substances that he expressed in Beudant's law. Essentially, he found that some compounds dissolved in the same solution would precipitate together, forming a crystal whose properties they determined in common. The interfacial angles of this new crystal would have a value intermediate between the angles of the original compounds, proportional to the quantity of each. The same idea had been put forth by Robert Boyle in "The Origine of Form and Qualities" (1666). Beudant was rather conservative about the generality of his proposition, although Delafosse enthusiastically maintained that it should apply to all crystals.

The generalization of this idea, the law of isomorphism, was proposed by Mitscherlich in 1819.

BIBLIOGRAPHY

I. ORIGINAL WORKS. Beudant published a great number of papers in geology, mineralogy, zoology, and paleontology. The most important are "Mémoire sur la possibilité de faire vivre des mollusques fluviatiles dans les eaux salées et des mollusques marins dans les eaux douces, considérée sous le rapport de la géologie," in *Journal de physique,* **83** (1816), 268–284, and "Recherches sur les causes qui déterminent les variations des formes cristallines d'une même substance minérale," in *Annales de chimie,* **8** (1818), 5–52. Among his texts are *Voyage minéralogique et géologique en Hongrie pendant l'année 1818,* 3 vols. (Paris, 1822); *Traité élémentaire de minéralogie* (Paris, 1824); *Traité élémentaire de physique* (Paris, 1824); *Nouveaux éléments de grammaire française* (Paris, 1841); and *Cours élémentaire de minéralogie et de géologie* (Paris, 1842).

II. SECONDARY LITERATURE. Articles on Beudant are in *Dictionnaire de biographie française,* V (1951), 358–359, and *Larousse grande dictionnaire du XIX siècle,* III (Paris, 1867), 656.

MARTHA B. KENDALL

BEXON, GABRIEL-LÉOPOLD-CHARLES-AMÉ (*b.* Remiremont, France, 10 March 1747; *d.* Paris, France, 15 February 1784), *biology.*

Bexon, whose short scientific career was closely linked with that of the great French naturalist Buffon, received his early education from his parents, Amé Bexon, a lawyer, and Marthe Pillement. Having shown considerable intelligence, he was sent north to the seminary at Toul to continue his education and prepare for the clergy. He completed the course of study and received his doctorate in theology at the

Faculty of Theology at the University of Besançon in 1766 or 1767, and then returned to Toul to accept a post as a subdeacon. There Bexon spent some time studying canon law; he was ordained in 1772.

Bexon's first published works appeared in 1773. *Catéchisme d'agriculture,* written about 1768, was published anonymously in Paris. It was a very simply written book that dealt as much with morals as with agriculture, and was meant to aid and educate the French peasants. In the same year he also published, under his brother Scipion's name, *Le système de la fertilisation.*

Although Bexon had announced plans for a two-volume history of Lorraine, and had later expanded the project to a four-volume work that would include one volume on the natural history of the province, he ceased work on the project in 1774. Only one volume appeared (*Histoire de Lorraine,* Paris, 1777), and was dedicated to Marie Antoinette. Probably as a result of this dedication, Bexon was appointed canon of Sainte-Chapelle in Paris in 1778 and three years later was elevated to the post of precentor.

From childhood, Bexon had been interested in natural history. He was a keen observer, and knew the mineralogy of the Vosges region especially well. While at Paris for a six-month visit in 1768–1769 he may have tried to meet Buffon, hoping to help with the *Histoire naturelle.* At that time Buffon's collaborator, Guéneau de Montbéliard, overworked and in poor health, was seeking an assistant; and there is an ambiguous reference to "the abbé" in a letter from Buffon to Montbéliard dated 17 May 1769.

Bexon did meet Buffon in 1772. The young man spoke of the influence of Buffon's works on him and of his desire to be of some assistance, which timely offer the naturalist accepted. Soon thereafter Bexon began to supply information and descriptions for various articles in the *Histoire naturelle.*

Although he supplied Buffon with descriptions, Bexon did not actually begin to collaborate in the writing of the articles until 1777. From then on, though, his contributions were numerous, appearing ultimately in six of the nine volumes of the *Histoire des oiseaux* and in the *Histoire des minéraux.*

When he began to assist Buffon, at the age of twenty-five, Bexon's writing style was extremely erudite and flowery, but under Buffon's tutelage it became more concise and exact. He worked hard, inspired by a love of natural history and by the need to support a sick mother and a young sister. Buffon publicly acknowledged his work on the *Histoire des oiseaux* in the preface to the seventh volume (1780), where he noted not only Bexon's scholarly researches but also the "solid reflections and ingenious ideas"

he had supplied. From this collaboration there developed an increasingly close friendship that ended only with Bexon's untimely death.

Despite their friendship, Bexon did not agree with the philosophical aspects of Buffon's system; he regarded it as "an ingenious and learned hypothesis" but not a true system of nature. In an unpublished manuscript on religion in relation to the universe, written about 1773, Bexon said he believed he could use the phenomena that supported the theory to show that nature contradicted Buffon's views.

The complete *Histoire naturelle* consisted of forty-four volumes published over a span of fifty-five years, and it proved to be a very popular and influential work. But although it is known as the product of Buffon's genius and industry, the work could not have appeared had it not been for the aid of Buffon's collaborators—Daubenton, Guéneau de Montbéliard, and the Abbé Bexon.

BIBLIOGRAPHY

I. ORIGINAL WORKS. Bexon's earlier works are quite scarce today. The articles in the *Histoire naturelle* that P. Flourens attributed to Bexon are listed in Flourens's *Des manuscrits de Buffon* (see below), pp. 221–222. Flourens also mentions (p. 58 f.) Bexon's unpublished manuscript on religion and natural history, which he entitles "De la religion par rapport à l'univers." Although no letters from Bexon to Buffon have ever been published, several from master to collaborator have been. One group of letters was originally published in the year VIII (1799–1800) by François de Neufchateau in his journal *Le conservateur,* **1,** 101–146. These were later reprinted by Flourens in *Histoire des travaux et des idées de Buffon* (Paris, 1850), pp. 307–344, and by Henri Nadault de Buffon in *Correspondance inédite de Buffon,* II (Paris, 1860). Several letters were added by Lanessan in his edition of the *Oeuvres complètes de Buffon* (Paris, 1855), XIII and XIV. These letters were reprinted, with Lanessan's notes, in H. M. J. A. P. de Bremond d'Ars, *Un collaborateur de Buffon: l'abbé Bexon* (see below).

II. SECONDARY LITERATURE. Of the several short accounts of Bexon's life that have appeared in the periodical literature, the most useful are E. Buisson, "Un collaborateur de Buffon. L'abbé Bexon—sa vie & ses oeuvres," in *Bulletin de la Société philomatique vosgienne,* **14** (1888–1889), 275–317, and Paillart, "L'abbé Bexon. Étude biographique et littéraire," in *Mémoires de l'Académie de Stanislas* (1867), 195–230. Flourens discusses Bexon in his *Des manuscrits de Buffon* (Paris, 1860), pt. III, ch. 3. Bexon's mother wrote a sympathetic, motherly sketch of her son after his death, and Humbert-Bazile, Buffon's secretary, published it in conjunction with his discussion of Bexon in his *Buffon. Sa famille, ses collaborateurs et ses familiers,* Nadault de Buffon, ed. (Paris, 1863). Relying on all these sources, Bremond d'Ars published the only extensive treatment of Bexon, *Un collaborateur de Buffon: L'abbé Bexon* (Paris, 1936).

ALAN S. KAY

BEYRICH, HEINRICH ERNST (*b.* Berlin, Germany, 31 August 1815; *d.* Berlin, 9 July 1896), *geology, paleontology.*

Beyrich was born into a substantial old Berlin merchant family. After completing his secondary education he entered Berlin University at the age of sixteen in order to study the natural sciences, especially mineralogy under Christian Samuel Weiss. In 1834 he transferred to the University of Bonn, where Goldfuss inspired him to specialize in paleontology. Geology and paleontology were his major interests throughout most of his life.

Before and after completing his studies Beyrich wandered through much of Germany, Switzerland, and Italy in order to add to his knowledge of geology and paleontology. In the course of his travels he met a number of eminent geologists and paleontologists, including Peter Merian, Agassiz, and Studer in Switzerland and Élie de Beaumont, Deshayes, and Brongniart in Paris. He had been recommended to these men by Leopold von Buch and Alexander von Humboldt.

Beyrich received the Ph.D. from Berlin University in 1837 with a Latin dissertation dealing with the goniatites of the Rhenish Schiefergebirge. His choice of topic was greatly influenced by Buch. In 1841 Beyrich was appointed *Privatdozent* at Berlin University, where he spent the rest of his life, holding a wide range of academic and civil service posts. In 1846 he was appointed associate professor, and in 1865 he became full professor. He was made custodian of the geological collections of the Prussian Mining Administration in 1855, and from 1857 he taught mining students.

Beyrich was one of the founders of the German Geological Society in 1848, and for the rest of his life he was one of its most active promoters; from 1872 to 1895 he was its president. In 1853 Beyrich was elected a full member of the Prussian Academy of Sciences, and when the Königliche Geologische Landesanstalt und Bergakademie was founded in Berlin in 1873, he was appointed its scientific director. In the same year he became director of the Museum of Natural History.

In 1842 Beyrich was commissioned by the Prussian Mining Administration to make a geological survey of Silesia. His findings were published in 1844 as "Über die Entwickelung des Flötzgebirges in Schlesien," a masterpiece that combines clear presentation, acute perception, and thoroughness. The work established Beyrich's reputation as a geologist and stratigrapher; its most important chapters deal with Paleozoic, Jurassic, Cretaceous, and Tertiary strata, as well as the tectonic structures of the area. The investigation comparing the malm strata in Poland (Wieluń, Krakow) and in Moravia (Štramberk, Mikulov, Brno), with the Upper Jurassic in southern France and northern Italy, and the clarification of certain stratigraphic problems of the Mesozoic and Tertiary formations in the Carpathians are a continuation of his Silesian studies.

Beyrich's explorations of the north German Tertiary formations were of particular importance. They started around 1847 and occupied him for most of his life, although during his last years he published little on the subject. The most important work on these explorations is "Die Conchylien des norddeutschen Tertiärgebirges" (1852–1855), which remained unfinished. The result of this paleontological study was a classification of the north German Tertiary ("Über den Zusammenhang der norddeutschen Tertiärbildungen . . .," 1855); its best-known result is the conception of an independent Oligocene interpolated between Lyell's Eocene and Miocene. The Oligocene, the north German counterpart of the Belgian Tongrian and the Rupelian, was first mentioned in 1854 and briefly defined in "Über die Stellung der hessischen Tertiärbildungen." Ever since, the Oligocene has occupied its established place in the stratigraphy of the Tertiary.

In his treatise of 1855, Beyrich describes the classification and extension of the north German Tertiary strata. He points out that in Belgium the Tongrian stage immediately follows the Upper Eocene, whereas in the large adjoining sections to the northeast and east there is no Eocene and the north German Tertiary strata cover the older pre-Tertiary strata in transgressive deposits. In 1855 Beyrich wrote:

> Their connected and independent geognostic distribution, independent of the presence of older Eocene Tertiary formations, is the prime reason which prevents me from classifying them—in accordance with Lyell—merely as an upper part of the Eocene Tertiary series. Their diverse formations and their abundance in specific organic remains—the extent of which was first revealed in Germany—have motivated me to consider them, rather, as a separate part of the Tertiary period (i.e., Oligocene) instead of assigning them to the Miocene Tertiary stage, as I had done earlier in accordance with d'Orbigny and other authors ["Über den Zusammenhang . . .," p. 11].

Beyrich's studies resulted in a subdivision of the Oligocene into Lower, Middle, and Upper Oligocene. He defined the marine sands of Egeln, near Magdeburg, as Lower Oligocene (i.e., Lower Tongrian); the septarian clay of the Brandenburg region and the sands of the Stettin area connected to the latter's facies by lateral transition as Middle Oligocene (Rupelian); and the so-called Sternberg rocks and the sedimentary rocks of the same period in central Germany as Upper Oligocene.

Some minor details of Beyrich's concepts have been changed, but his initial theory has been retained in principle. Even today his definition of the Oligocene inspires a flood of discussions and treatises. Beyrich also investigated stratigraphic problems of the Paleozoic in Germany. In 1865 he published a classification of the Permian (Continental Rothliegende [i.e., Lower Permian] and marine Zechstein [i.e., Upper Permian]) on the southern edge of the Harz Mountains. Beyrich then applied his paleontological knowledge to the clarification of the difficult and tectonically complicated conditions of the Devonian in the Harz Mountains.

Beyrich further defended the division of the Carboniferous in Germany into an older marine system (Carboniferous limestone and culm, respectively) and a younger system (with coal-bearing strata) on several occasions. He was especially interested, too, in comparing the development of the Triassic system in Germany with the Alpine Triassic system. He obtained data primarily by the paleontological findings from the Middle Triassic system (i.e., Muschelkalk) of Upper Silesia.

Geological cartography constitutes a large part of Beyrich's scientific work, beginning with his fieldwork in Silesia in 1842. In the 1860's he mapped the Harz Mountains, their northern and southern foothills, and the vicinity of Magdeburg. After that Beyrich emerged as the organizer and coordinator of the geological mapping operations in Prussia and Thuringia, for which he was officially commissioned in 1867 by the Prussian government.

One of Beyrich's most effective organizational accomplishments was the introduction of the 1:25,000 topographic map as the basis for geological mapping in Prussia. The other German geological surveys, and those of some foreign countries, adopted this scale.

The main part of Beyrich's work, reflected in his published papers, concerned studies in paleontology and biostratigraphy. More than half of his total of 205 publications are devoted to this subject. His doctoral thesis dealt with paleontological stratigraphy, in that it treated the Devonian goniatites of the Rhenish Schiefergebirge and their stratigraphic distribution. This part of his scientific work is significant not only in scope but also in diversity. There are many groups of individual or multiple fossil representatives of the animal kingdom that were treated by Beyrich. Among them are Mammalia, Stegocephalia, Pisces, Cystoidea, Crinoidea, Echinoidea, Graptolitoidea, Trilobita, Phyllopoda, Cephalopoda, Gastropoda, Pelecypoda, and Brachiopoda; he also investigated corals, sponges, and trace fossils. He dealt with paleobotanical subjects in a number of short papers on Tertiary and Carboniferous plants.

Nevertheless, Beyrich left few purely paleozoological works. Of these, mention should be made of his papers on the Muschelkalk Crinoidea (1857, 1871), which had been suggested by Johannes Müller's classic studies on the living *Pentacrinus*. In addition to a precise and thorough description of the Crinoidea of the German Muschelkalk, which were then little known, Beyrich furnished important information on the organization of the crinoidean skeleton. Thus he was the first to demonstrate the canals in the plates of the crinoid cup and to discuss the symmetrical principles in the cup structure and their taxonomic value.

In his treatise "Über einige Cephalopoden aus dem Muschelkalk der Alpen" (1866) Beyrich made the first attempt to establish connections between Triassic and Jurassic ammonites. This work also contains the first approaches to the taxonomic classification of the later ammonites and their evolutionary relations.

In "Conchylien des norddeutschen Tertiärgebirges" Beyrich split the rather largely classified genera of the older conchologists into subdivisions, so that their relationships could be better understood. Furthermore, such smaller groups were apt to offer more reliable material in discussions of stratigraphic and paleobiogeographic problems. On the other hand, Beyrich was an enemy of unfounded and wanton classification of species. In the introduction to his doctoral thesis (1837) we find the following passage:

> I was least inclined to imitate the methods of some excellent scholars of great merit who are wont to label everything in the collections available to them and to publicize such names without scruples simply to add to and to decorate synonymics. In classifying new species I shall always proceed with the greatest care. I am absolutely disinclined to consider the authorship of the newest species possible as an accomplishment or as something enviable. It would appear much more deserving to me to do away with useless divisions and to solidify already known facts by more precise observations [pp. 1–2].

If we take into account that Beyrich was only twenty-two when he wrote this paper, we must admire the great maturity of his systematic and taxonomic insights and principles. Yet in questions of zoological nomenclature, he took more liberties. He did not think much of the system of nomenclatural priorities, of the system of generic and specific classification, or of the use of synonymy lists.

Beyrich also investigated vertebrates. His main work in this field is on the catarrhine monkey *Mesopithecus pentelici* of the Lower Pliocene at Pikermi, near Athens. He established its difference from the hominoid *Hylobates* and its close relationship to the cynomorph *Semnopithecus*. He also produced memoranda on the Oligocene *Anthracotherium,* the Pliocene mastodons and rhinoceroses, the Pleistocene elephants and rhinoceroses, the Triassic labyrinthodonts, the Devonian *Pterichthyodes* and Coccosteidae, the Permian *Acanthodes,* the Triassic ganoid *Tholodus,* and the Tertiary selachian *Carcharodon.*

The range of Beyrich's paleontological works was due, first, to the era in which he lived, an era in which the differentiation of the geological sciences did not extend so far as today. At that time important fields, such as paleontology, were still comprehensible for the single scientist. Second, under Beyrich's direction the Berlin Museum started to receive a great volume of paleontological materials from all parts of Germany and many other European countries, the processing of which was entrusted to Beyrich by virtue of his office.

Most of Beyrich's work dealt with the geology and paleontology of Germany and adjoining European countries. During the last third of the nineteenth century, however, Berlin also received paleontological collections from overseas that were of great interest to Beyrich. These included Devonian, Carboniferous, and Cretaceous samples from Tripoli, ammonites of the Upper Malm, and Pelecypoda of the Lower Cretaceous from the Zanzibar coast.

Another collection from the island of Timor contained Permian marine fossils: an early work by Beyrich concerned the Timor fauna made famous by Wanner and others. Beyrich also occupied himself for years with the Cretaceous fauna between Cairo and Suez; he received the materials from the explorer Georg Schweinfurth. He also treated the Himalayan ammonite fauna of the Triassic era, and demonstrated that the Triassic contained not only elements of the European Upper Triassic fauna but also of the Middle Triassic.

BIBLIOGRAPHY

I. ORIGINAL WORKS. Beyrich's major writings are "Beiträge zur Kenntniss der Versteinerungen des rheinischen Übergangsgebirges," in *Abhandlungen der Preussischen*

Akademie der Wissenschaften (1837); *De goniatitis in montibus rhenanis occurrentibus* (Berlin, 1837), his doctoral dissertation; "Über die Entwickelung des Flötzgebirges in Schlesien," in *Karsten's Archiv,* **18** (1844), 3–68; "Untersuchungen über Trilobiten. Als Fortsetzung zu der Abhandlung: Ueber einige böhmische Trilobiten," in *Abhandlungen der Preussischen Akademie der Wissenschaften* (1845), 1–38; "Die Conchylien des norddeutschen Tertiärgebirges," in *Zeitschrift der Deutschen geologischen Gesellschaft,* **5** (1853), 273–358; **6** (1854), 408–500, 726–781; **8** (1856), 21–88; "Über die Stellung der hessischen Tertiärbildungen," in *Verhandlungen der Preussischen Akademie der Wissenschaften* (1854), 640–666; "Über den Zusammenhang der norddeutschen Tertiärbildungen, zur Erläuterung einer geologischen Übersichtskarte," in *Abhandlungen der Preussischen Akademie der Wissenschaften* (*für 1855*), 1–20; "Über die Crinoiden des Muschelkalkes," *ibid.* (*für 1857*), 1–50; "Über eine Kohlenkalk-Fauna von Timor," *ibid.* (*für 1864*), 61–98; "Über einige Cephalopoden aus dem Muschelkalk der Alpen und über verwandte Arten," *ibid.* (*für 1866*), 105–150; and "Über die Basis der Crinoidea brachiata," in *Monatsberichte der Preussischen Akademie der Wissenschaften* (*für 1871*), 33–55.

II. SECONDARY LITERATURE. Works on Beyrich are H. W. Dames, "Gedächtnisrede auf Ernst Beyrich," in *Abhandlungen der Königlichen Akademie der Wissenschaften, Berlin* (1898), 3–11; W. Hauchecorne, "Nekrolog auf E. Beyrich," in *Jahrbuch der Königlichen Preussischen Geologische Landesanstalt und Bergakademie,* **17** (Berlin, 1897), pp. 102–148, with complete bibliography; and E. Koken, *Die Deutsche geologische Gesellschaft in den Jahren 1848–1898 mit einem Lebensabriss von Ernst Beyrich* (Berlin, 1901).

HEINZ TOBIEN

BEZOLD, ALBERT VON (*b.* Ansbach, Bavaria, 7 January 1836; *d.* Würzburg, Bavaria, 2 March 1868), *physiology.*

Bezold's father, Johann Daniel Christoph, was a physician in Rothenburg and Ansbach. After attending secondary school at Ansbach, Albert began to study medicine at the University of Munich in 1853, with the aim of devoting himself to experimental research; he transferred to Würzburg the following year. At the beginning of 1854 he contracted rheumatic endocarditis, which recurred several times. The ailment led to a mitral stenosis, which resulted in his death at the age of thirty-two.

For the completion of his studies Bezold went in the fall of 1857 to Berlin, where he worked under the physiologist Emil Du Bois-Reymond, famous for his electrophysiological investigations; from him Bezold learned to apply physical methods to the study of biological phenomena. In 1858 Du Bois-Reymond made him an assistant in his institute and in 1859, at the age of twenty-three, he received a surprise appointment to the newly created chair of physiology at the University of Jena. Before he assumed his post in Jena, he went for several days to Würzburg to complete the requirements for the M.D. by presenting the dissertation "Über die gekreuzten Wirkungen des Rückenmarks." He had tried by experimental means to elucidate the cross effects of the motor and sensory paths while studying there. In 1865 Bezold was called to the new chair of physiology in Würzburg. Although he was active there for only three years, he made it a renowned center for physiological research.

Bezold's investigations were primarily on the physiology of the nerves and muscles, as well as of the heart. At Würzburg he also made pharmacological investigations, especially on the effects of veratrine, atropine, and curare on the muscles, nerves, and the heart and circulatory system. He confirmed Pflüger's law in *Untersuchungen über die elektrische Erregung der Nerven und Muskeln* (1861) and demonstrated that it is also valid for the direct stimulation of the muscle fiber.

In the physiology of the heart, Bezold occupied himself primarily with the problem of which nerve impulses influence the work of the heart. Galen had assumed that the contractions of the heart were independent of the brain. When Bezold turned to these problems, intracardial sources of heart stimulation had already been described; he localized them in the collection of ganglia in the *septum interatriale* (Bezold's ganglia). Through cleverly planned experiments Bezold settled the controversy over the relationship between the vagus nerve and the heartbeat. He proved that stimulation of the vagus decreases the rate of heartbeat and that it depresses the total energy output of the heart. In the *Untersuchungen über die Innervation des Herzens* (1863), Bezold also presented his experiments on the function of the sympathetic nerve, from which he concluded that stimulation of the cervical sympathetic would increase the frequency and the force of the heartbeat.

Bezold's experiments with veratrine led to the discovery of an effect on the circulation that originated from the heart and was characterized by bradycardia and lowering of the blood pressure. In the heart, veratrine stimulates the sensory ("depressing") vagus fibers and thus stimulates the vagus center. The efferent component of the reflex, via the vagus tract, decreases the heartbeat; through the vasomotor center, it lessens the tone of the vessels and thus decreases the blood pressure. In 1937 the pharmacologist Adolf Jarisch reexamined this regulatory heart-circulation reflex and recognized its general significance; today it has become important, as the "Schonreflex" (protective reflex) of the heart, for the understanding of pathological processes and is known as the Bezold-Jarisch reflex.

BIBLIOGRAPHY

I. ORIGINAL WORKS. Among Bezold's writings are *Untersuchungen über die elektrische Erregung der Nerven und Muskeln* (Leipzig, 1861); *Untersuchungen über die Innervation des Herzens,* 2 vols. (Leipzig, 1863); and "Über die physiologischen Wirkungen des essigsäuren Veratrins," written with L. Hirt, in *Untersuchungen aus dem physiologischen Laboratorium in Würzburg* (Leipzig, 1867), pp. 73–123.

A complete bibliography of his writings is in Robert Herrlinger and Irmgard Krupp, *Albert von Bezold,* pp. 123–124; see also pp. 113–114.

II. SECONDARY LITERATURE. Works on Bezold are Paul Diepgen, *Unvollendete* (Stuttgart, 1960), pp. 34–37; Robert Herrlinger, "Albert von Bezold und die Entdeckung der Innervation des Herzens," in *Von Boerhaave bis Berger,* ed. K. E. Rothschuh (Stuttgart, 1964), pp. 106–120; Robert Herrlinger and Irmgard Krupp, *Albert von Bezold* (Stuttgart, 1964); and Friedrich von Recklinghausen, "Gedächtnisrede auf Albert von Bezold," in *Verhandlungen der physikalisch-medizinischen Gesellschaft in Würzburg,* n.s. **1** (1869), xli–xlviii.

JOHANNES STEUDEL

BEZOUT, ÉTIENNE (*b.* Nemours, France, 31 March 1739; *d.* Basses-Loges, near Fontainebleau, France, 27 September 1783), *mathematics.*

Étienne Bezout, the second son of Pierre Bezout and Hélène-Jeanne Filz, belonged to an old family in the town of Nemours. Both his father and grandfather had held the office of magistrate (*procureur aux baillage et juridiction*) there. Although his father hoped Étienne would succeed him, the young man was strongly drawn to mathematics, particularly through reading the works of Leonhard Euler. His accomplishments were quickly recognized by the Académie des Sciences, which elected him *adjoint* in 1758, and both *associé* and *pensionnaire* in 1768. He married early and happily; although he was reserved and somewhat somber in society, those who knew him well spoke of his great kindness and warm heart.

In 1763, the duc de Choiseul offered Bezout a position as teacher and examiner in mathematical science for young would-be naval officers, the Gardes du Pavillon et de la Marine. By this time, Bezout had become a father and needed the money. In 1768 he added similar duties for the Corps d'Artillerie. Among his published works are the courses of lectures he gave to these students. The orientation of these books is practical, since they were intended to instruct people in the elementary mathematics and mechanics needed for navigation or ballistics. The experience of teaching nonmathematicians shaped the style of the works: Bezout treated geometry before algebra, observing that beginners were not yet familiar enough with mathematical reasoning to understand the force of algebraic demonstrations, although they did appreciate proofs in geometry. He eschewed the frightening terms "axiom," "theorem," "scholium," and tried to avoid arguments that were too close and detailed. Although criticized occasionally for their lack of rigor, his texts were widely used in France. In the early nineteenth century, they were translated into English for use in American schools; one translator, John Farrar, used them to teach the calculus at Harvard University. The obvious practical orientation, as well as the clarity of exposition, made the books especially attractive in America. These translations considerably influenced the form and content of American mathematical education in the nineteenth century.

A conscientious teacher and examiner, Bezout had little time for research and had to limit himself to what was, for his time, a very narrow subject—the theory of equations. His first two papers (1758–1760) were investigations of integration, but by 1762 he was devoting all his research time to algebra. In his mathematical papers, Bezout often followed a "method of simplifying assumptions," concentrating on those specific cases of general problems which could be solved. This approach is central to the conception of Bezout's first paper on algebra, "Sur plusieurs classes d'équations" (1762).

This paper provides a method of solution for certain nth-degree equations. Bezout related the problem of solving nth-degree equations in one unknown to the problem of solving simultaneous equations by elimination: "It is known that a determinate equation can always be viewed as the result of two equations in two unknowns, when one of the unknowns is eliminated."[1] Since an equation can be so formed, Bezout investigated what information could be gained by assuming that it actually was so formed. Such a procedure resembles the eighteenth-century study of the root-coefficient relations in an nth-degree equation by treating it as formed by the multiplication of n linear factors. Now, if one of the two composing equations had some very simple form—for instance, had only the nth-degree term and a constant—Bezout saw that he could determine the form of its solution. Conversely, if the coefficients of a given nth-degree equation in one unknown had the form built up from such a special solution, that nth-degree equation could be solved. Bezout's principal example considers

$$x^n + mx^{n-1} + px^{n-2} + \cdots + M = 0$$

as resulting from the equations

$$y^n + h = 0 \quad \text{and} \quad y = \frac{x + a}{x + b}.$$

The importance of this paper lay in drawing Bezout's attention from the problem of explicitly solving the nth-degree equation—an important concern of eighteenth-century algebraists—to the theory of elimination, the area of his most significant contributions. The central problem of elimination theory for Bezout was this: given n equations in n unknowns, to find and study what Bezout called the resultant equation in one of the unknowns. This equation contains all values of that unknown that occur in solutions of the n given equations. Bezout wanted to find a resultant equation of as small degree as possible, that is, with as few extraneous roots as possible. He wanted also to find its degree, or at least an upper bound on its degree.

In his 1764 paper, "Sur le degré des équations résultantes de l'évanouissement des inconnues," he discussed Euler's method for finding the equation resulting from two equations in two unknowns, and computed an upper bound on its degree.[2] He extended this method to N equations in N unknowns. But, although Euler's method yielded an upper bound on the degree of the resultant equation, Bezout observed that it was too clumsy to use for equations of high degree.

Another procedure, which gives a resultant equation of lower degree (now called the Bezoutiant) is given at the end of the 1764 paper. The equations to be solved are

(1) $\qquad Ax^m + Bx^{m-1} + \cdots + V = 0$
(2) $\qquad A'x^{m'} + B'x^{m'-1} + \cdots + V' = 0$

where $A, A', B, B', \cdots$ are functions of y, and where $m \geqslant m'$. From these, he obtained m polynomials in x, of degree less than or equal to $m-1$, which have among their common solutions the solutions of (1) and (2). For the case $m = m'$, these polynominals are

$$A'(Ax^m + Bx^{m-1} + \cdots + V)$$
$$- A(A'x^{m'} + \cdots + V') = 0,$$
$$(A'x + B')(Ax^m + \cdots + V)$$
$$- (Ax + B)(A'x^{m'} + \cdots + V) = 0, \text{ etc.}$$

He considered these polynomial equations as m linear equations in the unknowns $x, x^2, \cdots, x^{m-1}$. And he observed that (1) and (2) have a common solution if these linear equations do. But when can the linear equations be solved?

At the beginning of this 1764 paper, Bezout had

expressed what we would call a determinant by means of permutations of the coefficients, in what is sometimes called the Table of Bezout. He described the use of this table in solving simultaneous linear equations and, in particular, as a criterion for their solvability. This gave him a criterion for finding the resultant of (1) and (2). J. J. Sylvester, in 1853, explicitly gave the determinant of the coefficients of these m linear equations, and called it the Bezoutiant. The Bezoutiant, considered as a function of y, has as its zeros all the y's that are common solutions of equations (1) and (2).

It was not until 1779 that Bezout published his *Théorie des équations algébriques,* his major work on elimination theory. Its best-known achievement is the statement and proof of Bezout's theorem: "The degree of the final equation resulting from any number of complete equations in the same number of unknowns, and of any degrees, is equal to the product of the degrees of the equations."[3] Bezout, following Euler, defined a complete polynomial as one that contains each possible combination of the unknowns whose degree is no more than the degree of the polynomial. Bezout also computed that the degree of the resultant equation is less than the product of the degrees for various systems of incomplete equations. Here we shall consider only the complete case.

The proof makes one marvel at the ingenuity of Bezout, who, like Euler, not only could manipulate formulas but also had the ability to choose those manipulations that would be fruitful. He was compelled to justify his nth-order results by a naive "induction" from the observed truth of the statements for $1, 2, 3, \cdots$. Also, numbered subscripts had not yet come into use, and the notations available were clumsy.

Here is Bezout's argument. Given n equations in n unknowns, of degrees $t, t', t'', \cdots$. Let us call the equations $P_1(u,x,y,\cdots), P_2(u,x,y,\cdots), \cdots$. (Bezout wrote them $(u \cdots n)^t, (u \cdots n)^{t'}, \cdots$). Suppose now that P_1 is multiplied by an indeterminate polynomial, which we shall designate as Q for definiteness, of degree T. If a Q can be found such that P_1Q involves only the unknown u, P_1Q will be the resultant; Bezout's problem then becomes to compute the smallest possible degree of such a P_1Q.

Bezout stated, and later[4] gave an argument to show, that he could solve the equations

$$P_2(u, x, y, \cdots) = 0, \cdots P_n(u, x, y, \cdots) = 0$$

to determine, respectively, $x^{t'}, y^{t''}, z^{t'''}, \cdots$ in terms of lower powers of the unknowns. Substituting the values for $x^{t'}, y^{t''}, \cdots$ in the product P_1Q would eliminate all the terms divisible by those powers of the unknowns.

The key to Bezout's proof was in counting the number of terms in the final polynomial, P_1Q. Bezout began his book with a derivation, by means of finite differences, of a complicated formula for the number of terms in a complete polynomial in several unknowns which are not divisible by the unknowns to particular powers; that is, for a complete polynomial in $u, x, y, z, \cdots$ of degree T, he gave an expression for the number of terms not divisible by $u^p, x^q, y^r, \cdots$, where $p + q + r + \cdots < T$. Bezout used this formula to compute the number of terms in the polynomials P_1Q and Q which remained after the elimination of $x^{t'}, y^{t''}, \cdots$.

Let us write N (instead of Bezout's complicated expression) for the number of terms remaining in P_1Q, M for those remaining in Q. If the degree of the resultant is to be D, then it will have $D + 1$ terms, since it is an equation in the single unknown u. Then the coefficients of Q must be such that $N - (D + 1)$ terms in the product P_1Q will be annihilated by them. But, since Q or any multiple of Q would have the same effect, one of the coefficients of Q may be taken arbitrarily. Thus, Bezout argued, there were $M - 1$ coefficients at his disposal to annihilate the number of terms beyond $D + 1$ remaining in the product P_1Q. In other words, Bezout had to solve $N - (D + 1)$ linear equations in $M - 1$ unknowns—these unknowns being the coefficients of Q. This can be done if the number of equations equals the number of unknowns, although Bezout did not explicitly state this. Equating $N - (D + 1)$ with $M - 1$, and using his formulas for N and M, Bezout was able to compute that $D = t, t', t'', t''', \cdots$.[5] Bezout briefly noted that his theorem has a geometric interpretation: "The surfaces of three bodies whose nature is expressible by algebraic equations cannot meet each other in more points than there are units in the product of the degrees of the equations."[6] We should note that Bezout did not show that the equations for the coefficients of Q form a consistent, independent set of linear equations, or that extraneous roots can never occur in the resultant equation. Further, the geometric statement must be modified to deal with special cases, since, for instance, three planes can have a straight line in common.

Later on in the work,[7] Bezout discussed another method of finding the resultant equation; this was by finding polynomials, which we may write $Q_1, \cdots, Q_n$, such that

$$P_1Q_1 + P_2Q_2 + \cdots + P_nQ_n = 0$$

is the resultant equation. Each Q_k has indeterminate coefficients, which Bezout explicitly determined for many systems of equations by comparing powers of the unknowns $x, y, z, \cdots$.

Bezout's work on resultants stimulated many investigations in the modern theory of elimination, including Cauchy's refinements of elimination procedure and Sylvester's work on resultants and inertia forms. Bezout's theorem is crucial to the study of the intersection of manifolds in algebraic geometry. In the preface to the *Théorie des équations,* Bezout had complained that algebra was becoming a neglected science. But his accomplishment showed that the fact that his contemporaries could not solve the general equation of nth degree did not mean that there were no fruitful areas of investigation remaining in algebra.

NOTES

1. "Sur plusieurs classes d'équations," 20.
2. For Euler's method, see *Introductio in analysin infinitorum* (Lausanne, 1748), **2**, secs. 483 ff.
3. *Théorie des équations algébriques,* 32.
4. *Ibid.,* 206.
5. *Ibid.,* 32.
6. *Ibid.,* 33.
7. *Ibid.,* 187 ff.

BIBLIOGRAPHY

I. ORIGINAL WORKS. Bezout's major works are the following: "Sur plusieurs classes d'équations de tous les degrés qui admettent une solution algébrique," in *Mémoires de l'Académie royale des sciences* (1762), 17–52; *Cours de mathématiques à l'usage des Gardes du Pavillon et de la Marine,* 6 vols. (Paris, 1764–1769), reprinted many times with slight variations in the title, often translated or revised in parts; "Sur le degré des équations résultantes de l'évanouissement des inconnues," in *Mémoires de l'Académie royale des sciences* (1764), 288–338; "Sur la resolution des équations de tous les degrés," *ibid.* (1765), 533–552; and *Théorie générale des équations algébriques* (Paris, 1779).

II. SECONDARY WORKS. Secondary works are Georges Bouligand, "À une étape décisive de l'algèbre. L'oeuvre scientifique d'Étienne Bezout," in *Revue générale des sciences,* **55** (1948), 121–123; Marquis de Condorcet, "Éloge de M. Bezout," in *Éloges des académiciens de l'Académie Royale des Sciences,* **3** (1799), 322–337; E. Netto and R. Le Vavasseur, "Les fonctions rationnelles," in *Encyclopédie des sciences mathématiques pures et appliquées,* I, pt. 2 (Paris–Leipzig, 1907), 1–232; and Henry S. White, "Bezout's Theory of Resultants and Its Influence on Geometry," in *Bulletin of the American Mathematical Society,* **15** (1909), 325–338.

JUDITH V. GRABINER

BHĀSKARA I (*fl.* 629), *astronomy.*

Bhāskara I, who was one of the leading exponents of Āryabhaṭa I's two systems of astronomy (see Essays V and VI), composed his commentary on the *Āryabhaṭīya* in 629. In this work he mentions Valabhī (Vala, in Saurāṣṭra), Bharukaccha (Broach, in Gujarat), Śivabhāgapura (Śivarājapura, in Saurāṣṭra), and Sthāneśvara (Thanesar, in the Panjab). But in this same work, and in the *Mahābhāskarīya,* Bhāskara constantly speaks of the *Āryabhaṭīya* as the *Āsmakatantra* and its followers as the *Āśmakīyāḥ.* This seems to indicate that he belonged to a school of followers of the *Āryabhaṭīya* which flourished in Aśmaka (probably the Nizamabad District of Andhra Pradesh). It is supposed by Shukla that Bhāskara was born in either Saurāṣṭra or Aśmaka, and later migrated to the other.

Bhāskara is the author of three works: the *Mahābhāskarīya,* the *Laghubhāskarīya,* and the *Āryabhaṭīyabhāṣya.* The first contains eight chapters:

1. On the mean longitudes of the planets.

2. On the correction due to local longitude.

3. On the three problems relating to diurnal motion, and on the conjunctions of the planets with the stars.

4. On the true longitudes of the planets.

5. On solar and lunar eclipses.

6. On heliacal risings and settings, on the lunar crescent, and on the conjunctions of the planets.

7. The parameters according to the *audayaka* (*Āryapakṣa*) and the *ārdharātrika* systems.

8. Examples.

There are two published commentaries on the *Mahābhāskarīya: Bhāsya,* by Govindasvāmin (*fl. ca.* 800–850), on which there is a supercommentary (*Siddhāntadīpikā*) by Parameśvara (*fl.* 1400–1450), and *Karmadīpikā,* by Parameśvara. The anonymous *Prayogaracanā* is unpublished, and no manuscripts are known of the *Govindasvāmya* of Sūryadeva and the *Ṭīkā* of Śrīkaṇṭha. The text was published with the *Karmadīpikā* by Balavanta Rāya Āpṭe, as Ānandāśrama Sanskrit Series, no. 126 (Poona, 1945); with Govindasvāmin's *Bhāsya* and Parameśvara's *Siddhāntadīpikā* by T. S. Kuppanna Sastri, as Madras Government Oriental Series, no. 130 (Madras, 1957); and with an English translation and commentary by Kripa Shankar Shukla (Lucknow, 1960).

The *Laghubhāskarīya* also contains eight chapters:

1. On the mean longitudes of the planets.

2. On the true longitudes of the planets.

3. On the three problems relating to diurnal motion.

4. On lunar eclipses.

5. On solar eclipses.

6. On the visibility of the moon and on its crescent.

7. On the heliacal risings and settings of the planets and on their conjunctions.

8. On the conjunctions of the planets with the stars.

There exist three commentaries on the *Laghubhāskarīya:* Śaṅkaranārāyaṇa's *Vivaraṇa* (869), Udayadivākara's *Sundarī* (1073), and Parameśvara's *Parameśvara.* No manuscripts are known to me of the *Bālaśaṅkara* of Śaṅkara (b. 1494) nor of the *Ṭīkā* of Śrīkaṇṭha. The text was edited with the *Parameśvara* by Balavanta Rāya Āpṭe, as Ānandāśrama Sanskrit Series, no. 128 (Poona, 1946); with the *Vivaraṇa* of Śaṅkaranārāyaṇa, as Trivandrum Sanskrit Series, no. 162 (Trivandrum, 1949); and with an English translation and commentary by Kripa Shankar Shukla (Lucknow, 1963).

BIBLIOGRAPHY

In addition to works listed in the text, readers may consult B. Datta, "The Two Bhāskaras," in *Indian Historical Quarterly,* **6** (1930), 727–736.

DAVID PINGREE

BHĀSKARA II (b. 1115), *astronomy, mathematics.*

Bhāskara II has been one of the most impressive Indian astronomers and mathematicians, not only to modern students of the history of science but also to his contemporaries and immediate successors. An important inscription discovered at Pāṭnā, near Chalisgaon in East Khandesh, Mahārāṣṭra, by Bhāu Dājī, and reedited by F. Kielhorn (*Epigraphia Indica,* **1** [1892], 338–346), records the endowment, by Soïdeva the Nikumbha, on 9 August 1207, of an educational institution (*maṭha*) for the study of Bhāskara's works, beginning with the *Siddhāntaśiromaṇi.* There is further reference in this inscription to Soïdeva's brother and successor, Hemāḍideva, who was a feudatory of the Yādava king of Devagiri, Siṅghaṇa, whose rule began in 1209/1210. The following genealogy is given in the inscription.

Trivikrama belonged to the Śāṇḍilya *gotra*—which indicates that he and his descendants were Brāhmaṇas. His son was Bhāskarabhaṭṭa, who was given the title of Vidyāpati by Bhojarāja (the Paramāra king of Dhārā from *ca.* 995 to *ca.* 1056). The next four generations were respectively Govinda, Prabhākara, Manoratha, and Maheśvara; the last was the father of Bhāskara II. Bhāskara's son, Lakṣmīdhara, was made chief of the Paṇḍitas by Siṅghaṇa's predecessor, Jaitrapāla (1191–1209); and Lakṣmīdhara's son, Caṅgadeva, was the chief

astrologer to Siṅghaṇa himself. It is confirmed in Bhāskara's works—e.g., in the concluding verses of the *Siddhāntaśiromaṇi*—that his father was Maheśvara of the Śāṇḍilya *gotra;* it is further added that he came from the city Vijjaḍaviḍa (Bījāpur in Mysore), which was probably named after the Kalacūri king Vijjala II (1156–1175). If this identification is correct—since the *Siddhāntaśiromaṇi* was written in 1150—Bhāskara II must have been in Vijjala's capital while the latter was still *daṇḍanāyaka* of the Cālukya kings, Jagadekamalla II (1138–1150) and Taila III (1150–1156). We further know from Trivikrama's *Damayantīkathā* that he was the son of Nemāditya (Devāditya?) and the grandson of Śrīdhara; and there exists a popular astrological work by Maheśvara, Bhāskara II's father, entitled *Vṛttaśataka.*

Bhāskara II is the author of at least six works, and possibly of a seventh as well:

1. *Līlāvatī* (see Essay XII).
2. *Bījagaṇita* (see Essay XII).
3. *Siddhāntaśiromaṇi* (see Essay IV).
4. *Vāsanābhāṣya* on the *Siddhāntaśiromaṇi* (see Essay IV).
5. *Karaṇakutūhala* (see Essay IV).
6. *Vivaraṇa* on the *Śiṣyadhīvṛddhidatantra* of Lalla (see Essay V).
7. *Bījopanaya* (see Essay IV).

The *Līlāvatī* and the *Bījagaṇita* are sometimes taken to be parts of the *Siddhāntaśiromaṇi;* the ascription of the *Bījopanaya* to Bhāskara II is questionable.

1. The *Līlāvatī* is a work on mathematics addressed by Bhāskara II to a lady (his daughter or wife?) named Līlāvatī. It contains thirteen chapters:

1. Definitions of terms.
2. Arithmetical operations.
3. Miscellaneous rules.
4. Interest and the like.
5. Arithmetical and geometrical progressions.
6. Plane geometry.
7–10. Solid geometry.
11. On the shadow of a gnomon.
12. Algebra: the pulverizer (*kuṭṭaka*). This is the same as chapter 5 of the *Bījagaṇita.*
13. Combinations of digits.

The *Līlāvatī* has been commented on many times:

1. *Karmapradīpikā* of Nārāyaṇa (*fl.* 1356).
2. *Vyākhyā* of Paraśurāma Miśra (1356).
3. *Vyākhyā* of Parameśvara (*fl.* 1400–1450).
4. *Gaṇitāmṛtasāgarī* of Gaṅgādhara (*ca.* 1420).
5. *Vyākhyā* of Lakṣmīdāsa (*fl.* 1501).
6. *Gaṇitāmṛtakūpikā* of Sūryadāsa (1541). See K. Madhava Krishna Sarma, *Siddha-Bhāratī,* part 2 (Hoshiarpur, 1950), 222–225.

7. *Buddhivilāsinī* of Gaṇeśa (1545). Published. See below, Sanskrit text of the *Līlāvatī* no. 14.

8. *Kriyākramakarī* of Śaṅkara (*fl.* 1556).

9. *Vivaraṇa* of Mahīdhara, alias Mahīdāsa (1587). Published. See below, Sanskrit text of the *Līlāvatī* no. 14.

10. *Mitabhāṣiṇī* of Raṅganātha (1630).

11. *Nisṛṣṭārthadūtī* of Munīśvara, alias Viśvarūpa (1635).

12. *Gaṇitāmṛtalaharī* of Rāmakṛṣṇa (1687). See P. K. Gode, "Date of Gaṇitāmṛtalaharī of Rāmakṛṣṇa," in *Annals of the Bhandarkar Oriental Research Institute,* **11** (1930), 94–95.

13. *Sarvabodhinī* of Śrīdhara (1717).

14. *Udāharaṇa* of Nīlāmbara Jhā (*fl.* 1823).

15. *Ṭīkā* in Kannada of Alasiṅgārya, alias Aliśiṅgarāja.

16. *Vyākhyā* of Bhaveśa.

17. *Udāharaṇa* of Candraśekhara Paṭanāyaka.

18. *Ṭīkā* of Dāmodara(?).

19. *Vilāsa* of Devīsahāya.

20. *Bhūṣaṇa* of Dhaneśvara. Refers to Sūryadāsa (1541).

21. *Ṭīkā* (in vernacular) of Giridhara.

22. *Vyākhyā* of Keśava.

23. *Ṭippaṇa* of Mukunda.

24. *Vṛtti* of Moṣadeva.

25. *Subodhinī* of Rāghava.

26. *Gaṇakabhūṣaṇa* of Rāmacandra, son of Śoṣaṇabhaṭṭa.

27. *Kautukalīlāvatī* of Rāmacandra, son of Vidyādhara.

28. *Ṭippaṇa* of Rāmadatta (?).

29. *Manorañjana* of Rāmakṛṣṇadeva.

30. *Ṭīkā* of Rāmeśvara.

31. *Ṭīkā* of Śrīkaṇṭha.

32. *Gaṇitāmṛtavarṣiṇī* of Sūryamaṇi.

33. *Udāharaṇa* of Vīreśvara. Refers to Lakṣmīdāsa (1501).

34. *Udāharaṇa* of Viśveśvara.

35. *Ṭīkā* of Vṛndāvana (?).

In addition to these and a number of anonymous commentaries, there are others in Marāṭhī and Gujarātī. A modern Sanskrit commentary (aside from those which accompany some of the editions listed below) was published by Candra Śekhara Jhā under the title *Vyaktavilāsa* (Benares, 1924).

There are also numerous editions of the Sanskrit text of the *Līlāvatī*:

1. Calcutta, 1832.

2. Tārānātha Śarman, ed. (Calcutta, 1846).

3. Baptist Mission Press (Calcutta, 1846; 2nd ed., Calcutta, 1876).

4. With the *Vivaraṇa* of Mahīdhara and a Telugu commentary by Taḍakamalla Veṅkaṭa Kṛṣṇarāva, Vāvilla Rāmasvāmin Śāstrin, ed. (Madras, 1863).

5. Jīvānanda Vidyāsāgara, ed. (Calcutta, 1876).

6. Sudhākara Dvivedin, ed. (Benares, 1878).

7. Edited, with his own Sanskrit commentary, by Bāpūdeva Śāstrin (Benares, 1883).

8. Bhuvanacandra Basak, ed. (Calcutta, 1885).

9. Edited as an appendix to Banerji's edition of Colebrooke's translation (Calcutta, 1892; 2nd ed., Calcutta, 1927).

10. Edited, with a Marāṭhī commentary, by Vināyaka Pāṇḍuraṅga Khānāpūrkar (Poona, 1897).

11. Sudhākara Dvivedin, ed., Benares Sanskrit Series, no. 153 (Benares, 1912).

12. Rādhāvallabha, ed. (Calcutta, 1914).

13. Edited, with his own Sanskrit commentary, by Muralīdhara Thākura, as Śrī Harikṛṣṇa Nibandha Maṇimālā Series, no. 3 (Benares, 1928; 2nd ed., Benares, 1938).

14. With *Buddhivilāsinī* of Gaṇeśa and *Vivaraṇa* of Mahīdhara, Dattātreya Āpṭe, ed., Ānandāśrama Sanskrit Series, no. 107, 2 vols. (Poona, 1937).

15. With Sanskrit commentary, edited by Dāmodara Miśra and Payanātha Jhā, as Prācīnācārya Granthāvalī Series, no. 8 (Durbhanga, 1959).

16. With Sanskrit and Hindī commentaries of Laṣaṇa Lāla Jhā, edited by Sureśa Śarman, as Vidyābhavana Saṃskṛta Granthamālā Series, no. 62 (Benares, 1961).

There are also many translations of the *Līlāvatī.* A Kannada version is supposed to have been made by Bhāskara II's contemporary Rājāditya, who flourished, apparently, under the Hoysala king Viṣṇuvardhana (1111–1141). There also exists a Hindī translation, and the various commentaries in Gujarātī, Marāṭhī, and Telugu have already been referred to. Three Persian translations are known. That made by Abū al-Fayḍ Fayḍī at the request of Akbar in 1587 was published at Calcutta in 1827; another was done by Dharma Nārāyan ibn Kalyānmal Kāyath *ca.* 1663 (H. J. J. Winter and A. Mirza, in *Journal of the Asiatic Society of Science,* **18** [1952], 1–10); and the third was made in 1678 by Muḥammad Amīn ibn Shaykh Muḥammad Saʿīd. There are also two English translations. That by J. Taylor was published at Bombay in 1816, and that by H. T. Colebrooke in his *Algebra, With Arithmetic and Mensuration: From the Sanscrit of Brahmegupta and Bháscara* (London, 1817). The latter was republished by Haran Chandra Banerji as *Colebrooke's Translation of the Lilávati* (Calcutta, 1892; 2nd ed., Calcutta, 1927).

2. The *Bījagaṇita,* on algebra, contains twelve chapters:

1. On positive and negative numbers.

2. On zero.

3. On the unknown.

4. On surds.

5. On the pulverizer (*kuṭṭaka*).

6. On indeterminate quadratic equations.

7. On simple equations.

8. On quadratic equations.

9. On equations having more than one unknown.

10. On quadratic equations having more than one unknown.

11. On operations with products of several unknowns.

12. On the author and his work.

The commentaries on the *Bījagaṇita* are all relatively late, and they are far fewer in number than those on the *Līlāvatī*.

1. *Sūryaprakāśa* of Sūryadāsa (1538). See K. Madhava Krishna Sarma, in *Poona Orientalist*, **11** (1946), 54–66, and his article in *Siddha-Bhāratī*, part 2 (Hoshiarpur, 1950), 222–225.

2. *Navāṅkura* (or *Bījapallava*, or *Bījāvataṃsa*, or *Kalpalatāvatāra*) of Kṛṣṇa (1602). See M. M. Patkar, in *Poona Orientalist*, **3** (1938), 169. Published. See below, Sanskrit texts nos. 13 and 16.

3. *Bījaprabodha* of Rāmakṛṣṇa (1687). See P. K. Gode in *Annals of the Bhandarkar Oriental Research Institute*, **10** (1929), 160–161, and **11** (1930), 94–95.

4. *Bālabodhinī* of Kṛpārāma (1792).

5. *Vāsanābhāṣya* of Haridāsa.

6. *Bījālavāla* of Nijānanda.

7. *Kalpalatā* of Paramaśukla (most likely Kṛṣṇa's work?).

8. *Bījavivaraṇa* of Vīreśvara (?).

The Sanskrit text has been frequently published:

1. Calcutta, 1834; rev. ed., Calcutta, 1834.

2. Calcutta, 1846.

3. Partial edition with a German translation by H. Brockhaus, "Über die Algebra des Bhāskara," in *Berichte über die Verhandlungen der Königlich Sächsischen Gesellschaft der Wissenschaften zu Leipzig, Philosophisch-historische Klasse*, **4** (Leipzig, 1852), 1–46.

4. Calcutta, 1853.

5. Gopinātha Pāṭhaka, ed. (Benares, 1864).

6. Bāpūdeva Śāstrin, ed., 2 parts (Calcutta [?], 1875).

7. Jīvānanda Vidyāsāgara, ed. (Calcutta, 1878).

8. Edited, with his own Sanskrit commentary, by Jīvanātha Śarman (Benares, 1885).

9. Edited, with his own Sanskrit commentary, by Sudhākara Dvivedin (Benares, 1888).

10. Edited, with a Marāṭhī translation and com-

mentary, by Vināyaka Pāṇḍuraṅga Khānāpūrkar (Poona, 1913).

11. Edited, with his own Sanskrit commentary, by Rādhāvallabha (Calcutta, 1917).

12. Edited, with Sudhākara Dvivedin's Sanskrit commentary and one of his own, by Muralīdhara Jhā, as Benares Sanskrit Series, no. 154 (Benares, 1927).

13. Edited, with the *Navāṅkura* of Kṛṣṇa, by Dattātreya Āpṭe, as Ānandāśrama Sanskrit Series, no. 99 (Poona, 1930).

14. Edited, with his own Sanskrit and Hindī commentaries, by Durgāprasāda Dvivedin (3rd. ed., Lakṣmaṇapura, 1941; the preface is dated Jayapura, 1916).

15. Edited, with Jīvanātha Śarman's Sanskrit commentary and with his own in Sanskrit and Hindī, by Acyutānanda Jhā, as Kāśī Sanskrit Series, no. 148 (Benares, 1949).

16. Edited, with the *Bījapallava* of Kṛṣṇa, by T. V. Rādhākṛṣṇa Śāstrin, as Tanjore Sarasvati Mahal Series, no. 78 (Tanjore, 1958).

There are two Persian translations of the *Bījagaṇita*, one anonymous and the other by ʿAtā allāh Rashīdī ibn Aḥmad Nādir for Shah Jahan in 1634/1635. An English translation of the latter by E. Strachey, with notes by S. Davis, was published at London in 1813. It was also translated into English directly from the Sanskrit by H. T. Colebrooke in *Algebra, With Arithmetic and Mensuration* . . . (London, 1817).

3. The *Siddhāntaśiromaṇi*, which was written in 1150, consists of two parts—the *Grahagaṇitādhyāya* (or *Gaṇitādhyāya*) and the *Golādhyāya*—which are sometimes preserved singly in the manuscripts. The first part, on mathematical astronomy, contains twelve chapters:

1. On the mean longitudes of the planets.

2. On the true longitudes of the planets.

3. On the three problems involving diurnal motion.

4. On the syzygies.

5. On lunar eclipses.

6. On solar eclipses.

7. On planetary latitudes.

8. On the heliacal risings and settings of the planets.

9. On the lunar crescent.

10. On planetary conjunctions.

11. On conjunctions of the planets with the stars.

12. On the *pātas* of the sun and moon.

The second part, on the sphere, contains thirteen chapters:

1. Praise of (the study of) the sphere.

2. On the nature of the sphere.

3. On cosmography and geography.

4. Principles of planetary mean motion.

5. On the eccentric-epicyclic model of the planets.

6. On the construction of an armillary sphere.

7. Principles of spherical trigonometry.

8. Principles of eclipse calculations.

9. Principles of the calculation of the first and last visibilities of the planets.

10. Principles of the calculation of the lunar crescent.

11. On astronomical instruments.

12. Descriptions of the seasons.

13. On problems of astronomical computations. The chapter on the sine function is placed differently in different editions. The *Golādhyāya*, then, is to a large extent an expansion and explanation of the *Ganitādhyāya*.

The following commentaries on the *Siddhāntaśiromani* are known (besides various anonymous ones):

1. *Mitākṣarā* (or *Vāsanābhāṣya*) of Bhāskara II himself (see **4**, below). Published. See below, under Sanskrit texts.

2. *Ganitatattvacintāmaṇi* of Lakṣmīdāsa (1501).

3. *Śiromaṇiprakāśa* of Gaṇeśa (*b.* 1507). Published in part. See below, Sanskrit text of *Grahaganitādhyāya*, no. 4.

4. *Marīci* of Munīśvara, alias Viśvarūpa (*b.* 1603). Published. See below, under Sanskrit texts.

5. *Ṭīkā* of Rāmakṛṣṇa (*fl.* 1687).

6. *Ṭīkā* of Cakracūḍāmaṇi (?).

7. *Vyākhyā* of Dhaneśvara.

8. *Vyākhyā* of Harihara (?).

9. *Ṭīkā* of Jayalakṣmaṇa (?).

10. *Lakṣmīnāthī* of Lakṣmīnātha Miśra (?).

11. *Bhāṣya* of Maheśvara (?).

12. *Vāsanā* of Mohanadāsa (?).

13. *Vyākhyā* of Raṅganātha.

14. *Ṭīkā* of Vācaspati Miśra (?).

The *Ṭippaṇīvivaraṇa* of Buddhinātha Jhā was published at Benares in 1912.

The list of editions of the text is arranged under three headings: *Siddhāntaśiromaṇi, Grahaganitā-dhyāya,* and *Golādhyāya.*

Siddhāntaśiromaṇi.

1. *Siddhāntaśiromaṇiprakāśa* (of Gaṇeśa?), with a Marāṭhī translation (Bombay, 1837).

2. *Siddhāntaśiromaṇi,* with the *Prakāśa* (of Gaṇeśa?), Rāmacandra, ed. (Madras, 1837).

3. Edited, with the *Vāsanābhāṣya,* by Bāpūdeva Śāstrin (Benares, 1866); revised by Candradeva (Benares, 1891); revised by Gaṇapatideva Śāstrin, as Kāśī Sanskrit Series, no. 72 (Benares, 1929).

4. Edited, with the *Vāsanābhāṣya,* the

Vāsanāvārttika of Nṛsiṃha, and the *Marīci* of Munīśvara, by Muralīdhara Jhā, in *The Pandit,* n.s. **30-38** (1908-1916)—incomplete; the first chapter of the *Grahaganitādhyāya* was reprinted at Benares in 1917.

5. Edited, with a Sanskrit commentary, by Girijāprasāda Dvivedin (Ahmadabad, 1936).

Grahaganitādhyāya.

1. Edited, with the *Mitākṣarā,* by L. Wilkinson (Calcutta, 1842).

2. Edited, with the *Mitākṣarā,* by Jīvānanda Vidyāsāgara (Calcutta, 1881).

3. Edited, with a Marāṭhī translation and commentary, by Vināyaka Pāṇḍuraṅga Khānāpūrkar (Poona, 1913).

4. Edited, with the *Vāsanābhāṣya* and the *Śiromaṇiprakāśa* of Gaṇeśa, by Dattātreya Āpṭe, as Ānandāśrama Sanskrit Series, no. 110, 2 vols. (Poona, 1939-1941).

5. Edited, with the *Vāsanābhāṣya* and his own Sanskrit commentary, by Muralīdhara Ṭhakkura, as Kāśī Sanskrit Series, no. 149 (Benares, 1950)—the first two chapters only.

6. Edited, with the *Vāsanābhāṣya,* the *Marīci* of Munīśvara, and his own Sanskrit and Hindī commentaries, by Kedāradatta Jośī, 3 vols. (Benares, 1961-1964); this edition does not include the *Marīci* on chapter 1.

Golādhyāya.

1. Edited, with the *Mitākṣarā,* by L. Wilkinson (Calcutta, 1842).

2. Calcutta, 1856.

3. Edited, with the *Vāsanābhāṣya,* by Jīvānanda Vidyāsāgara (Calcutta, 1880).

4. Edited, with the *Vāsanābhāṣya* and a Bengali translation, by Rasikamohana Chattopādhyāya (Calcutta, 1887).

5. Edited, with the *Vāsanābhāṣya* and a Bengali translation, in *Aruṇodaya,* **1** (1890), part 6.

6. Edited, with a Marāṭhī translation and commentary, by Vināyaka Pāṇḍuraṅga Khānāpūrkar (Bombay, 1911)—chapters 1-8 only.

7. Edited, with the *Vāsanābhāṣya* and a Hindī commentary, by Girijāprasāda Dvivedin (Lucknow, 1911).

8. Edited, with the *Vāsanābhāṣya* and a Bengali translation, by Rādhāvallabha (Calcutta, 1921).

9. Edited, with the *Vāsanābhāṣya* and the *Marīci* of Munīśvara, by Dattātreya Āpṭe, as Ānandāśrama Sanskrit Series, no. 122, 2 vols. (Poona, 1943-1952).

Aside from the translations into the vernacular mentioned above, I know only of the following two: a Latin translation of the *Grahaganitādhyāya* published by E. Roer in *Journal of the Royal Asiatic*

Society of Bengal, **13** (1844), 53–66, and an English translation of the *Golādhyāya* by L. Wilkinson, revised by Bāpūdeva Śāstrin, as Bibliotheca Indica, no. 32 (Calcutta, 1861), with the Paṇḍit's translation of the *Sūryasiddhānta.* See also L. Wilkinson, "On the Use of the Siddhāntas in the Work of Native Education," in *Journal of the Royal Asiatic Society of Bengal,* **3** (1834), 504–519.

4. The *Vāsanābhāṣya* or *Mitākṣarā* is Bhāskara II's own commentary on the *Siddhāntaśiromaṇi.* A commentary on it, the *Vāsanāvārttika,* was written by Nṛsiṃha of Golagrāma in 1621. Editions of both these works have been listed in the preceding material on the *Siddhāntaśiromaṇi.*

5. The *Karaṇakutūhala,* which is also known as the *Brahmatulya,* the *Grahāgamakutūhala,* and the *Vidagdhabuddhivallabha,* was written in 1183; it gives simpler rules for solving astronomical problems than does the *Siddhāntaśiromaṇi.* There are ten sections:

    1. On the mean longitudes of the planets.
    2. On the true longitudes of the planets.
    3. On the three problems involving diurnal motion.
    4. On lunar eclipses.
    5. On solar eclipses.
    6. On heliacal risings and settings.
    7. On the lunar crescent.
    8. On planetary conjunctions.
    9. On the *pātas* of the sun and moon.
    10. On the syzygies.

There are, aside from the usual quantity of anonymous commentaries on the *Karaṇakutūhala,* eight whose authors' names are known:

    1. *Bhāṣya* of Ekanātha (*ca.* 1370).
    2. *Nārmadī* of Padmanābha (*ca.* 1575).
    3. *Udāharaṇa* of Viśvanātha (1612).
    4. *Gaṇakakumudakaumudī* of Sumatiharṣa Gaṇi (1622). Published. See below.
    5. *Ṭīkā* of Caṇḍīdāsa.
    6. *Brahmatulyasāra* of Keśavārka (?).
    7. *Ṭīkā* of Śaṅkara.
    8. *Ṭīkā* of Soḍhala.

For a set of tables based on the *Karaṇakutūhala,* see David Pingree, "Sanskrit Astronomical Tables in the United States," in *Transactions of the American Philosophical Society,* n.s. **58,** no. 3 (1968), 36–37.

The *Karaṇakutūhala* has twice been edited: by Sudhākara Dvivedin, with his own Sanskrit commentary (Benares, 1881); and, with the *Gaṇakakumadakaumudī* of Sumatiharṣa Gaṇi, by Mādhava Śāstrī Purohita (Bombay, 1902).

6. Bhāskara II's *Vivaraṇa* on the *Śiṣyadhīvṛddhidatantra* of Lalla has not been studied or pub-

lished. There are three manuscripts: in Benares, in Bikaner, and in Ujjain.

7. A short text of fifty-nine verses entitled *Bījopanaya* is attributed to Bhāskara II. The author claims to be that scholar and to have written this work in 1151. A *Tithinirṇayakārikā* published with it is the only other Sanskrit work to mention it; the author of this text claims to be Śrīnivāsa Yajvan, who flourished in Mysore in the second half of the thirteenth century and wrote a *Śuddhidīpikā* and a commentary on the *Karaṇaprakāśa* of Brahmadeva. Both works, despite their acceptance by Mukhopadhyaya and Sengupta, are evidently late forgeries.

Kuppanna Sastri has shown that the *Bījopanaya,* which gives rules for computing a correction to the moon's equation of the center and variation, was most probably forged in south India in the early 1870's to buttress the position of the partisans of the *dṛk* system against those of the *vākya* system. His argument is based on three main points:

(1) The first correction is astronomically invalid and would have appeared so to the author of the *Siddhāntaśiromaṇi.*

(2) The style is completely at variance with Bhāskara's normal method of exposition.

(3) There are oblique references in the *Vāsanābhāṣya,* a commentary accompanying the *Bījopanaya,* which is also alleged to be by Bhāskara II, to Raṅganātha's commentary on the *Sūryasiddhānta,* which was written in 1602 and was published in 1859.

These arguments seem to this writer quite convincing.

The *Bījopanaya* has been published twice: by Cintāmaṇi Raghunāthācārya and Taḍhakamalla Veṅkaṭakṛṣṇa Rāya at Madras in 1876; and by Ekendranāth Ghosh at Lahore in 1926.

BIBLIOGRAPHY

The following bibliography generally excludes articles that deal only in part with Bhāskara II. It is divided into five sections: General, *Līlāvatī, Bījagaṇita, Siddhāntaśiromaṇi,* and *Bījopanaya.* All entries are listed in chronological order.

I. GENERAL. The following deal with Bhāskara II and his works in general: Bhāu Dājī, "Brief Notes on the Age and Authenticity of the Works of Āryabhaṭa, Varāhamihira, Brahmagupta, Bhaṭṭotpala, and Bhāskarāchārya," in *Journal of the Royal Asiatic Society* (1865), 392–418, esp. 410–418; Janārdana Bāḷājī Moḍaka, *Bhāskara Āchārya and His Astronomical System* (n.p., 1887); Sudhākara Dvivedin, *Gaṇakataraṅgiṇī* (Benares, 1933; repr. from *The Pandit,* n.s. **14** [1892]), pp. 34–42; Bāpūdeva Śāstrin, "A

Brief Account of Bhāskara, and of the Works Written, and Discoveries Made, by Him," in *Journal of the Asiatic Society of Bengal,* **62** (1893), 223–229; S. B. Dīkṣita, *Bhāratīya Jyotiḥśāstra* (Poona, 1931; repr. of Poona, 1896), pp. 246–254; G. Thibaut, *Astronomie, Astrologie und Mathematik, Grundriss der indo-arischen Philologie und Altertumskunde,* III, pt. 9 (Strasbourg, 1899), 60; S. K. Ganguly, "Bhāskarāchārya's References to Previous Teachers," in *Bulletin of the Calcutta Mathematical Society,* **18** (1927), 65–76; B. Datta, "The Two Bhāskaras," in *Indian Historical Quarterly,* **6** (1930), 727–736; and Brij Mohan, "The Terminology of Bhāskara," in *Journal of the Oriental Institute, Baroda,* **9** (1959/1960), 17–22.

II. LĪLĀVATĪ. The *Līlāvatī* is discussed in E. Strachey, *Observations on the Mathematical Science of the Hindoos, With Extracts From Persian Translations of the Leelawuttee and Beej Gunnit* (Calcutta, 1805); H. Suter, "Über die Vielecksformel in Bhāskara," in *Verhandlungen des 3. Mathematikerkongresses in Heidelberg* (Leipzig, 1905), pp. 556–561; Sarada Kanta Ganguly, "Bhāskarācārya and Simultaneous Indeterminate Equations of the First Degree," in *Bulletin of the Calcutta Mathematical Society,* **17** (1926), 89–98; M. G. Inamdar, "A Long Forgotten Method," in *Annals of the Bhandarkar Oriental Research Institute,* **9** (1927/1928), 304–308; A. A. Krishnaswami Ayyangar, "Bhaskara and Samclishta Kuttaka," in *Journal of the Indian Mathematical Society,* **18** (1929), 1–7; Saradakanta Ganguli, "Bhāskara and Simultaneous Indeterminate Equations of the First Degree," *ibid.,* **19** (1931/1932), 6–9; A. S. Bhandarkar, " 'Method of False Assumption' of Pacioli, an Italian Mathematician," in *Indian Culture,* **8** (1941/1942), 256–257; K. S. Nagarajan, "Bhaskara's Leelavathi," in *The Aryan Path* (1949), 310–314; D. A. Somayaji, "Bhaskara's Calculations of the Gnomon's Shadow," in *The Mathematics Student,* **18** (1950), 1–8; and Brij Mohan, "The Terminology of Līlāvatī," in *Journal of the Oriental Institute, Baroda,* **8** (1958/1959), 159–168.

III. BĪJAGAṆITA. The *Bījagaṇita* is dealt with in Reuben Burrow, "A Proof That the *Hindoos* Had the *Binomial Theorem,*" in *Asiatick Researches,* **2** (1790), 487–497; A. A. Krishnaswami Ayyangar, "New Light on Bhaskara's Chakravala or Cyclic Method of Solving Indeterminate Equations of the Second Degree in Two Variables," in *Journal of the Indian Mathematical Society,* **18** (1929), 225–248; K. J. Sanjana, "A Brief Analysis of Bhaskara's *Bījagaṇita* With Historical and Critical Notes," *ibid.,* 176–188; and D. H. Potts, "Solution of a Diophantine System Proposed by Bhaskara," in *Bulletin of the Calcutta Mathematical Society,* **38** (1946), 21–24.

IV. SIDDHĀNTAŚIROMAṆI. Works discussing the *Siddhāntaśiromaṇi* are Bapudeva Sastri, "Bhāskara's Knowledge of the Differential Calculus," in *Journal of the Asiatic Society of Bengal,* **27** (1858), 213–216; W. Spottiswoode, "Note on the Supposed Discovery of the Principle of the Differential Calculus by an Indian Astronomer," in *Journal of the Royal Asiatic Society* (1860), 221–222; H. Suter, "Eine indische Methode der Berechnung der Kugeloberfläche," in *Bibliotheca mathematica,*

3rd ser., **9** (1908/1909), 196–199; R. Sewell, "The Siddhanta-siromani," in *Epigraphia Indica,* **15** (1919/1920), 159–245; M. G. Inamdar, "A Formula of Bhaskara for the Chord of a Circle Leading to a Formula for Evaluating Sin α°," in *The Mathematics Student,* **18** (1950), 9–11; and A. A. Krishnaswami Ayyangar, "Remarks on Bhaskara's Approximation to the Sine of an Angle," *ibid.,* 12.

V. BĪJOPANAYA. Further discussion of the *Bījopanaya* can be found in Dhirendranath Mukhopadhyaya, "The Evection and the Variation of the Moon in Hindu Astronomy," in *Bulletin of the Calcutta Mathematical Society,* **22** (1930), 121–132; P. C. Sengupta, "Hindu Luni-solar Astronomy," *ibid.,* **24** (1932), 1–18; and T. S. Kuppanna Sastri, "The Bījopanaya: Is It a Work of Bhāskarācārya?," in *Journal of the Oriental Institute, Baroda,* **8** (1958/1959), 399–409.

DAVID PINGREE

BIAGGIO PELICANI. See **Blasius of Parma.**

BIAŁOBRZESKI, CZESŁAW (*b.* Pošechonje, near Jaroslavl, Russia, 31 August 1878; *d.* Warsaw, Poland, 12 October 1953), *physics, natural philosophy.*

Białobrzeski studied physics at the University of Kiev from 1896 to 1901 and received the *veniam legendi* there in 1907. From 1908 to 1910 he worked in Langevin's laboratory at the Collège de France. After his return to Russia he held the chair of physics and geophysics at the University of Kiev from 1914 to 1919. Białobrzeski assumed the chair of theoretical physics at Warsaw University in 1921 and occupied it for the rest of his life. In 1935 he was appointed to the International Institute of Intellectual Cooperation of the League of Nations, filling the vacancy created by the death of Marie Curie-Skłodowska. Białobrzeski served several terms as president of the Polish Society of Physics, was a vice-president of the International Union of Pure and Applied Physics (1947–1951), and belonged to the Polish Academy of Science, among many others.

Białobrzeski's work may be divided into three periods. From 1900 to 1912 he carried out experimental and theoretical research on the electrical and optical phenomena in fluid and solid dielectrics.

In 1912 Białobrzeski turned his attention to the role of radiation pressure in the equilibrium of the star interior. His paper on this (1913) drew the attention of the Polish physicists Smoluchowski and Natanson but attracted little notice abroad because the journal had only a limited circulation. Other works dealing with radiation pressure were his papers on the mechanism of light absorption (1923–1926). The second period closed with the publication of *La thermodynamique des étoiles* (1931).

In the third period Białobrzeski concentrated on the philosophical problems of physics, mainly on the interpretation of quantum-theory foundations. He initiated and was elected chairman of the international scientific conference in Warsaw (1938) where this problem was discussed by many famous theorists. During World War II, Białobrzeski prepared a three-volume work to be entitled *Podstawy poznawcze fizyki świata atomowego* ("Epistemological Foundations of the Physics of the Atomic World"), in which he developed his philosophical interpretation of the quantum theory. Unfortunately, the manuscripts of the first two volumes were burned during the Warsaw Insurrection (1944). After the war Białobrzeski returned to Warsaw University and started to reconstruct the book. The work, limited to one volume, was finished in 1951 and published in 1956.

BIBLIOGRAPHY

Among Białobrzeski's works are "Sur les théories des diélectriques," in *Le radium,* **9** (1912), 250; "Sur l'équilibre thermodynamique d'une sphère gazeuse libre," in *Bulletin international de l'Académie des sciences de Cracovie,* ser. A (1913), 264–290; "Sur l'absorption vraie de la lumière," in *Annales de physique,* **5** (1926), 215; "Szkic autobiograficzny i uwagi o twórczości naukowej" ("Autobiographical Essay and Remarks on Scientific Work"), in *Nauka Polska* ("Polish Science"), **6** (1927), 49–76, also in *Wybór pism* (see below), pp. 13–48; *La thermodynamique des étoiles* (Paris, 1931); "Sur l'interpretation concrète de la mécanique quantique," in *Revue de métaphysique et de morale,* **41** (1934), 83–103; the introductory discourse in *New Theories in Physics* (Paris, 1939), also published in French (Paris, 1939); *Podstawy poznawcze fizyki świata atomowego* ("Epistemological Foundations of the Physics of the Atomic World"; Warsaw, 1956); and *Wybór pism* ("Selected Papers"; Warsaw, 1964), a selection of philosophical papers, with a bibliography.

W. Scisłowski, "Czesław Białobrzeski (1878–1953)," in *Acta physica Polonica,* **13** (1954), 301–308, an obituary with a bibliography, also appeared in Polish in *Postepy fizyki,* **5**, no. 4 (1954), 413–422.

ANDRZEJ A. TESKE

BIANCHI, LUIGI (*b.* Parma, Italy, 18 January 1856; *d.* Pisa, Italy, 6 June 1928), *mathematics.*

The son of Francesco Saverio Bianchi, a jurist and senator of the kingdom of Italy, Bianchi entered the Scuola Normale Superiore of Pisa after passing a competitive examination in November 1873. He studied under Betti and Dini at the University of Pisa, from which he received his degree in mathematics on 30 November 1877. He remained in Pisa for two additional years, pursuing postgraduate studies. Later he attended the universities of Munich and Göttingen, where he studied chiefly under Klein.

Upon his return to Italy in 1881, Bianchi was appointed professor at the Scuola Normale Superiore of Pisa, and after having taught differential geometry at the University of Pisa, in 1886 he was appointed extraordinary professor of projective geometry on the basis of a competitive examination. During the same year he was also made professor of analytic geometry, a post he held for the rest of his life. By special appointment Bianchi also taught higher mathematics and analysis. After 1918 he was director of the Scuola Normale Superiore. He was a member of many Italian and foreign academies, and a senator of the kingdom of Italy.

Bianchi concentrated on studies and research in metric differential geometry. Among his major results was his discovery of all the geometries of Riemann that allow for a continuous group of movements, that is, those in which a figure may move continuously without undergoing any deformation. These results also found application in Einstein's studies on relativity. In addition, Bianchi devoted himself to the study of non-Euclidean geometries and demonstrated how the study of these geometries may lead to results in Euclidean geometry that, through other means, might have been obtained by more complex methods.

A writer of clear and genial treatises, Bianchi wrote many works on mathematics, among which are some dealing with functions of a variable complex, elliptic functions, and continuous groups of transformations.

BIBLIOGRAPHY

I. ORIGINAL WORKS. *Lezioni di geometria differenziale* (Pisa, 1886; 3rd ed., 1922–1923); *Lezioni sulla teoria dei gruppi di sostituzioni e delle equazioni algebriche secondo Galois* (Pisa, 1900); *Lezioni sulla teoria aritmetica delle forme quadratiche binarie e ternarie* (Pisa, 1912); *Lezioni di geometria analitica* (Pisa, 1915); *Lezioni sulla teoria delle funzioni di variabile complessa e delle funzioni ellittiche* (Pisa, 1916); *Lezioni sulla teoria dei gruppi continui finiti di trasformazioni* (Pisa, 1918); *Lezioni sulla teoria dei numeri algebrici e principii di geometria analitica* (Bologna, 1923). Bianchi's works were collected in *Opere,* Edizioni Cremonese, 11 vols. (Rome, 1952–1959); Vol. I, pt. 1 contains a bibliography and analyses of Bianchi's scientific work by G. Scorza, G. Fubini, A. M. Bedarida, and G. Ricci.

II. SECONDARY LITERATURE. Works on Bianchi are G. Fubini, "Luigi Bianchi e la sua opera scientifica," in *Annali di matematica,* 4th ser., **6** (1928–1929), 45–83, and "Commemorazione di Luigi Bianchi," in *Rendiconti della Accademia nazionale dei Lincei,* Classe di scienze fisiche matematiche e naturali, ser. 6a, **10** (1929), xxxiv–xliv (appendix).

ETTORE CARRUCCIO

BICHAT, MARIE-FRANÇOIS-XAVIER (*b*. Thoirette, Jura, France, 14 November 1771; *d*. Paris, France, 22 July 1802), *surgery, anatomy, physiology.*

The son of Jean-Baptiste Bichat, a physician and graduate of the Faculté de Médecine of Montpellier, who was then practicing in Poncin-en-Bugey, and of Jeanne-Rose Bichat, a cousin of her husband, Bichat studied humanities at the Collège de Nantua, completed the course in rhetoric, and was then sent to the Séminaire Saint-Irénée at Lyons to study philosophy. In 1791 he became the pupil of Marc-Antoine Petit at the Hôtel-Dieu in Lyons, in order to study surgery and anatomy. Three years later he went to Paris, where he was the favorite student and collaborator of Pierre Desault (1738–1795), who had created the surgical clinic at the Hôtel-Dieu, then called Grand Hospice d'Humanité by the Revolutionary powers. Desault was very skillful in the treatment of fractures and was particularly interested in vascular surgery. His death temporarily interrupted the publication of his *Journal de chirurgie,* but Bichat, encouraged by Corvisart, published as its fourth volume observations of Desault that he himself had written up. On 23 June 1796 Bichat founded, with Henri Husson and Guillaume Dupuytren, the Société Médicale d'Émulation, which Cabanis, Corvisart, and Pinel then joined.

At this time Bichat also started a private course in anatomy in a house on the Petite Rue de Grès (today the Rue Cujas). This course was transferred, in 1798, to the Rue des Carmes. It was then that Bichat added demonstrations in physiology using animal vivisections to his teaching of anatomy and of medical operations. Simultaneously he worked on the *Oeuvres chirurgicales de Desault* (1798–1799); wrote up, for the Société Médicale d'Émulation, several reports on surgery; and composed the memoirs *Sur les membranes et leurs rapports généraux d'organisation* (1798), which led to the *Traité des membranes* and *Anatomie générale,* and *Sur les rapports qui existent entre les organes à forme symétrique et ceux à forme irrégulière* (1798), which ushered in *Recherches physiologiques sur la vie et la mort*. In 1801 he was made *médecin expectant* (supernumerary) at the Grand Hospice d'Humanité.

Sensing that his life would not be long, Bichat published, between 1799 and 1801, the three works that made him famous: the *Traité des membranes,* the *Recherches physiologiques sur la vie et la mort,* and the *Anatomie générale*. The *Anatomie descriptive* was left unfinished at his death. Bichat's funeral services were held at Notre-Dame de Paris.

Bichat's most important contribution to modern anatomy consists in the generalization of a theory set forth by Pinel in his *Nosographie philosophique* (1798). Pathology must be based not upon the topographical situations of organs, but upon the structure of the membranes (i.e., of the tissues making up the organs), regardless of the location of the latter in the organism. Bichat recognized his debt to Pinel in the *Traité des membranes* (art. 1, sec. iv), but Magendie, in the preface to its new edition (1827), says that Bichat, by his extension of Pinel's idea, showed "that he was of such a stature as to owe the idea only to himself."

Bichat's best statement on his own method as an anatomist is in the sixth and seventh paragraphs of "Considérations générales," a preface to the *Anatomie générale*. Just as chemistry is the science of elementary bodies, says Bichat, anatomy is the science of elementary tissues, which differ from each other in the composition and the arrangement of their fibers, the combination of which forms organs. General anatomy is the study of the simple organic elements and of the similarly elementary structures. Bichat distinguished twenty-one organized elements, characterized by their textures and their properties. Since it differs from others in its vital properties, each tissue also differs in its diseases because diseases are nothing more than alterations of its vital properties. As a background to the diversity of symptoms and the uneven duration of illnesses, the physician must consider the diversity of the tissues. Therefore, general anatomy should set up a new pathological anatomy, substituting for the descriptive order, generally accepted since Morgagni, a systematic order of the diseases common to each elementary structure, to each tissue.

Bichat distinguished the properties of tissues according to their texture, properties that are retained after death: extensibility, contractility, and the vital properties—organic contractility and sensibility ("insensible" or subliminal) on the one hand, and animal contractility and sensibility ("sensible" or conscious) on the other. Vital properties were, in his eyes, irreducible to physical laws. *Recherches physiologiques sur la vie et la mort* begins with the famous sentence "Life is the ensemble of functions that resist death." Without completely admitting Barthez's vital principle, Bichat was hostile to the traditional medicine of Boerhaave and praised the doctors of the Montpellier school for having "more or less followed the impetus given by Stahl" (animism).

Bichat's ideas had a profound influence not only in medicine but also in philosophy. General anatomy and the pathology of tissues were both transformed and confirmed in the nineteenth century by the development of histology, cytology, and cellular pathology. Claude Bernard, while recognizing that in his time the morphological analysis of organized bodies had decentralized the seat of life "beyond the term fixed by Bichat,"—beyond tissue and down to the cell—wrote, "Modern opinions concerning vital phenomena are based on histology and really have their origin in Bichat's ideas" (*Leçons sur les phénomènes de la vie*, II, 452). No less hostile than Claude Bernard toward all metaphysical vitalism, Auguste Comte expressed his admiration for Bichat insofar as the latter had helped to establish the specificity of a general science of life, at the very time when Lamarck and Treviranus simultaneously invented the term "biology" (1802) in order to denote it. The German philosopher Schopenhauer insisted upon calling himself a disciple of Bichat, as did Cabanis.

BIBLIOGRAPHY

I. ORIGINAL WORKS. Bichat's writings include *Oeuvres chirurgicales de Desault*, 3 vols. (Paris, 1798–1799; new ed., rev. and enl., 1801–1803); *Recherches physiologiques sur la vie et la mort* (Paris, 1800; new ed., with notes by M. Magendie, 1822), translated by F. Gold as *Physiological Researches on Life and Death* (London, 1815); *Traité des membranes en général et de diverses membranes en particulier* (Paris, 1800; new ed., rev. and enl., with notes by M. Magendie, 1827), new ed. with a notice on Bichat's life and works by M. Husson; *Anatomie générale, appliquée à la physiologie et à la médecine*, 4 vols. (Paris, 1801), supplemented by P.-A. Beclard, *Additions à l'Anatomie générale de X. Bichat* (1821), prefaced by a historical note on Bichat by Scipion Pinel; and *Traité d'anatomie descriptive*, 5 vols. (Paris, 1801–1803)—Bichat wrote the first three volumes, Mathieu-François Buisson wrote the fourth, and Philibert-Joseph Roux wrote the fifth.

II. SECONDARY LITERATURE. Works on Bichat are *Bulletin de la Société française d'histoire de la médecine*, **1** (1902; repr. 1967), which contains a series of articles and documents relating to Bichat and was published on the centenary of his death—see especially the articles by R. Blanchard and Émile Gley; J. Coquerelle, *Xavier Bichat* (Paris, 1902); "Bichat," in Dezeimeris, Ollivier, and Raige-Delorme, *Dictionnaire historique de la médecine ancienne et moderne*, I (Paris, 1928), 385–396; Michel Foucault, *Naissance de la clinique* (Paris, 1963), ch. 8; Geneviève Genty, "Bichat, médecin du Grand Hospice d'Humanité," thesis (Paris, 1943); Maurice Genty, "Bichat," in *Biographies médicales*, II (Paris, 1929–1931), 35–36, and "Bichat et son temps," in *Médecine internationale* (1934), nos. 7–12 and (1935), nos. 1–10; M. Laignel-Lavastine, "Sources, principes, sillage et critique de l'oeuvre de Bichat," in *Bulletin de la Société française de philosophie*, **46** (1952), 1; and Entralgo P. Lain, "Sensualism and Vitalism in Bichat's *Anatomie générale*," in *Journal of the History of Medicine and Allied Sciences*, **3** (1948), 47–64.

GEORGES CANGUILHEM

BICKERTON, ALEXANDER WILLIAM (*b*. Alton, England, 1842; *d*. London, England, 22 January 1929), *cosmology, natural philosophy*.

An orphan at an early age, Bickerton was given an engineering education by his uncle in Bridgwater; however, he found the time an engineer had to spend outdoors was too much for his health, and turned his attention to science. In 1864 he established a small factory in the Cotswolds to develop his woodworking inventions; three years later he organized technical classes at Birmingham while studying at the Royal School of Mines in London. In 1870 Bickerton accepted a post on the staff of the Hartley Institution in Southampton, and was later appointed lecturer in science at Winchester College. In the meantime his publications on the relation between electricity and heat attracted some attention, and as a result he was invited to accept the professorship of physics and chemistry at Canterbury College in Christchurch, New Zealand, a post he held until 1903.

Most of the scientific work for which Bickerton is remembered was carried out during his New Zealand years, and is characterized by originality and boldness of approach. His theory of the build-up of celestial bodies by collisions, published in a number of papers from 1880 on, attracted considerable attention and some hostility in astronomical quarters. Among his scientific papers are "On a New Relation of Heat and Electricity"; "On Temporary and Variable Stars," containing an outline of a view that novae originate by collisions of two stars in space; "On the Problem of Stellar Collisions"; "On the Origin of Double Stars, of Nebulae and of the Solar System"; and "On Agencies Tending to Alter the Eccentricities of Planetary Orbits."

BIBLIOGRAPHY

Bickerton's books include *Materials for Lessons in Elementary Science* (1883), *A New Story of the Stars* (1894), *Some Recent Evidence in Favour of Impacts* (1894), *The Romance of the Earth* (1900), *The Romance of the Heavens* (1901), and *The Birth of Worlds and Systems* (1911).

ZDENĚK KOPAL

BIDDER, FRIEDRICH HEINRICH (*b*. Kurland, Russia, 28 October 1810; *d*. Dorpat, Russia [now Tartu, Estonian S.S.R.], 27 August 1894), *anatomy, physiology*.

His father, Ernst Christian, was an agriculturalist. In 1834 Friedrich received a medical degree from the University of Dorpat and was appointed professor extraordinary and prosector in anatomy. Before assuming his duties, Bidder spent a year in Berlin studying anatomy with J. Müller, C. Ehrenberg, J. Henle, and F. Schlemm. His itinerary included a tour of the research facilities at Dresden, Halle, and Leipzig. He became a full professor of anatomy in 1842, and accepted the new chair of physiology and pathology in 1843. He served as dean of the medical faculty from 1843 to 1845 and as rector of the university in 1858. Bidder found an outlet for his humanitarian interests in the social problems of the city. He helped to found a public baby nursery and was president of Dorpat's Hülfsverein, a charitable institution. Bidder received the Karl E. von Baer medal in 1879 for his contributions to biology.

Bidder was not a gifted teacher in the conventional sense, but he was highly successful at providing fruitful research topics for his students. Over seventy-five dissertations were completed under his direction. Bidder's superior grasp of scientific literature and his command of anatomy, physiology, histology, and embryology enabled him to single out important questions for study. With few exceptions, Bidder selected his research topics from published differences of opinion. This habit reveals Bidder's confidence and broad knowledge of biology; it also explains why Bidder's name is not associated with any major innovations. His best-known works were published in cooperation with other scientists. Bidder's partnerships brought to light his most innovative and creative abilities. When Bidder did reach out to creative projects, his efforts frequently were not appreciated. His contributions to intermediary metabolism and nerve physiology were too advanced for the majority of his contemporaries.

In 1852, Bidder and Carl Schmidt, a student of Justus von Liebig, published their classic *Verdauungssäfte und der Stoffwechsel*. The treatise was a brilliant extension of the concepts suggested by Liebig's *Animal Chemistry* (1842). *Verdauungssäfte* was the first major publication on intermediary metabolism (*intermediären Kreislauf*). Bidder examined the effects of digestive juices (salivary, pancreatic, biliary, intestinal, and gastric) on foodstuffs. He elucidated the chemical changes induced by enzymes and the effects of nervous control on the secretion of digestive juices. He was able to show that bile was not an excretion but a secretion serving a physiological function.

Bidder opened the bile duct and inserted a cannula to draw off the biliary liquids. By controlling the diet of cats fitted with this fistula, he found that digestive disturbances occurred when the nutrients contained large amounts of fat. The chemical composition of bile was not similar to fat but to carbohydrate or protein. Yet, further fistula experiments indicated that bile did have some unknown complex function in fat metabolism. When bile flowed into the gut, animals could digest more fat than when bile was removed through the fistula. Oil mixed with bile rose higher in capillary tubes than did untreated samples of oil. *In vitro* mixtures of fat, water, and bile rendered the fat water-soluble and neutral to litmus paper. Fat globules in the lymphatics, however, were acidic. Bidder hypothesized that bile aids in the absorption of fats in the stomach and is then reabsorbed in the gut. By comparing the concentrations of bile salts in the feces and urine with the concentration of bile in the gut, Bidder concluded that such a hypothesis was correct.

Although Bidder worked on a wide spectrum of problems, he maintained a persistent interest in the physiology and anatomy of the nervous system. He provided an improved description of the rods and cones of the retina (1839), repeated many of Flourens's experiments on the regeneration of sectioned nerves, investigated olfaction, and was successful in joining the severed ends of the lingual and hypoglossal nerves (although the crossed fibers were not functional). He cataloged all of the inhibitory nerves known to him and traced their fibers to the autonomic ganglia (1871). He discovered the auriculoventricular and interauricular ganglion cells in the hearts of frogs ("Bidder's ganglia") and demonstrated that the ganglia contained fibers of the vagus nerve.

His best-known work on the nervous system was the outcome of collaboration. Bidder and Alfred W. Volkmann made an extensive histological study of the autonomic nervous system and the spinal cord. They demonstrated that although certain fibers of the intercostal nerve were unmyelinated, they were genuine nerve fibers. They established the general rule that postganglionic fibers are not covered by a myelin sheath. Detailed numerical analysis revealed that the peripheral nerves contained more fibers than could be accounted for by the spinal cord and the sympathetic chain alone. This discovery provided anatomical evidence for the theory of double innervation of the organs from (1) the sympathetic chain and (2) the ganglionic system.

In 1857 Bidder and Carl Kupffer made a histological study of the embryonic and adult spinal cord in order to settle the dispute whether or not the spinal nerve fibers are continuous with the gray matter of the brain. Employing a new chromic-acid stain developed by Schroeder van der Kolk, the authors were able to show the continuity of cord fibers with cell bodies located in the gray matter of the brain. Bidder was an advocate of the neuron theory.

Bidder's name is also associated with an organ found only in certain frogs (*Bufonides*). It is a circular mass of tissue located slightly ventral to, and between, the kidneys and testes. Bidder suspected it was testicular tissue; following an interesting array of speculative theories, it was shown in the twentieth century to be endocrine tissue.

A great deal of Bidder's work was excellent but his virtuosity and diverse interests prevented its recognition. His tendency to resolve minor conflicts rather than carry out sustained research on any one topic reduced his impact. His interest in controversies, however, increases the historical value of his work. Bidder's discussions of opposing views provide insights into the research problems of the nineteenth century.

BIBLIOGRAPHY

I. ORIGINAL WORKS. Bidder's most important monographs are *De graviditatis vi medicatrice* (Dorpat, 1834); *Neurologische Beobachtungen* (Dorpat, 1836); *Neue Beobachtungen über die Bewegungen des weichen Gaumens und über den Geruchssinn* (Dorpat, 1838); "Reichen," in Rudolph Wagner's *Handwörterbuch der Physiologie,* II (Brunswick, 1844), 916–926; *Vergleichend-anatomische Untersuchungen über den Harn und die Geschlechtswerkzeuge der nachten Amphibien* (Dorpat, 1846); and *Zur Lehre von dem Verhältniss der Ganglienkörper zu den Nervenfasern. Neue Beiträge* (Leipzig, 1847). His most famous joint efforts are *Untersuchungen über die Textur des Rückenmarks und die Entwicklung seiner Formelelemente* (Leipzig, 1857), with C. Kupffer; *Verdauungssäfte und der Stoffwechsel. Eine physiologisch-chemische Untersuchung* (Leipzig, 1852), with C. Schmidt; and *Die Selbständigkeit des sympathischen Nervensystems, durch anatomische Untersuchung nachgewiesen* (Leipzig, 1842), with A. W. Volkmann. The bulk of Bidder's papers appeared in the *Archiv für Anatomie, Physiologie und wissenschaftliche Medicin,* Johannes Müller, ed. Those relevant to the subjects discussed are the following: "Zur Anatomie der Retina" (1839), pp. 371–388, (1841), pp. 248–262; "Versuche über die Möglichkeit des Zusammenheilens functionell verschiedener Nervenfasern" (1842), pp. 102–120; "Versuche zur Bestimmung der Chylusmenge die durch den Ductus thoracicus dem Blute zugeführt wird" (1845), pp. 45–60; "Ueber functionell verschiedene und räumlich getrennte Nervencentra im Froschherzen" (1852), pp. 163–177; "Erfolge von Nervendurchschneidung an einem Frosch" (1865), pp. 67–79; "Beobachtung doppelsinniger Leitung im Nervus lingualis nach Vereinigung desselben mit dem N. hypoglossus" (1865), pp. 246–260; "Ueber die Unterschiede in den Beziehungen des Pfeilgifts zu verschiedenen Abtheilungen des Nervensystems" (1865), pp. 337–359; "Zur näheren Kenntniss des Froschherzens und seiner Nerven" (1866), pp. 1–25; "Die Endigungsweise der Herzzweige des Nervus

vagus beim Frosch" (1868), pp. 1–50; "Einige Bemerkungen über Hemmungsnerven und Hemmungscentren" (1871), pp. 447–472; and "Erfahrungen über die functionelle Selbständigkeit des sympathetischen Nervensystems" (1841), pp. 359–380, with A. Volkmann.

II. SECONDARY LITERATURE. Further biographical details can be found in the *Allgemeine deutsche Biographie,* XLVI (Leipzig, 1902), 538–540; *Leopoldina,* **3,** nos. 17–18 (1894), 145, 162; and the *Saint Petersburger medicinische Wochenschrift,* **19** (1894), 314–315. Bidder's nutritional research is discussed in Graham Lusk, "A History of Metabolism," in *Endocrinology and Metabolism,* III (New York, 1922), 3–78; Nikolaus Mani, *Die historischen Grundlagen der Leberforschung II Teil. Die Geschichte der Leberforschung von Galen bis Claude Bernard* (Basel, 1967); Elmer V. McCollum, *A History of Nutrition* (Boston, 1957); and Fritz Lieben, *Geschichte der physiologischen Chemie* (Leipzig, 1935). An enumeration of Bidder's minor contributions can be found in Karl Rothschuh, *Entwicklungsgeschichte physiologischen Probleme in Tabellenform* (Munich, 1952). References to Bidder's research on the nervous system are inadequate. Brief discussions appear in John Langley, *The Autonomic Nervous System* (Cambridge, 1921); R. Herrlinger, "Albert von Bezold und die Entedeckung der Innervation des Herzens," in K. Rothschuh, ed., *Von Boerhaave bis Berger* (Stuttgart, 1964), pp. 106–120; and V. Kruta, "G. Prochaska and J. E. Purkyně's Contributions to Neurophysiology," *ibid.,* pp. 134–156. A history of the interpretations given to Bidder's organ can be traced through H. King, "The Structure and Development of Bidder's Organ in Bufo Levtiginosus," in *Journal of Morphology,* **19** (1908), 439–465; W. Harms, "Untersuchungen über das Biddersche Organ der männlichen und weiblichen Kröten," in *Zeitschrift für Anatomie und Entwicklungsgeschichte,* **62** (1921), 1–38; and N. Takahashi, "Biological and Anatomical Studies of the Nuptial Excrescence and Bidder's Organ of the Toad," in *Endocrinology,* **7** (1923), 302–304.

CHARLES A. CULOTTA

BIELA, WILHELM VON (*b.* Rossla, Stolberg am Harz, Germany, 19 March 1782; *d.* Venice, Italy, 18 February 1856), *astronomy.*

Descended from a Bohemian noble family, Biela was educated at the school for pages of the Elector of Saxony. In 1802 he became a cadet in an Austrian infantry regiment and fought in the Napoleonic Wars. He became a lieutenant in 1809 and later was promoted to captain. During the battle of Leipzig (1813) he was wounded, and while recuperating in Prague, attended the astronomical lectures of Alois David. Subsequently he became a very successful amateur astronomer. In 1825 the army transferred him to Naples, and in 1832 he became local governor of Rovigo, Italy. Biela suffered a stroke in 1844 and retired two years later with the rank of major. His health led him to move to Venice, and since he could no longer devote himself to astronomy, he turned to art and became a well-known connoisseur.

Biela made many valuable astronomical observations, mainly of comets and meteors; the most remarkable was that of a comet in 1826. On 27 February of that year, in Josefstadt, Bohemia, he saw a comet and recognized it as one already seen in 1772 and 1805. On 14 March he found its period to be six years and nine months, a discovery that made him famous throughout Europe. Other astronomers confirmed that his comet was indeed identical with those of 1772 and 1805—a determination difficult to establish because of the disturbances caused by Jupiter. The most remarkable phenomenon appeared when the comet returned in 1846: it separated into two parts of the same shape but of changing intensity. It was last seen in 1852. E. Weiss of Vienna has pointed out that fragments of Biela's comet constitute the periodic shower of certain meteors, the Andromedides.

Biela was respected by all astronomers of his time, and he corresponded with many of them. A crater on the moon was named for him, and the Andromedides are now called Bielides or Belides. His only published work shows the influence of the romantic philosophy of nature prevalent at the time.

BIBLIOGRAPHY

Biela's only published work is *Die zweite grosse Weltenkraft nebst Ideen über einige Geheimnisse der physischen Astronomie oder Andeutungen zu einer Theorie der Tangentialkraft* (Prague, 1836).

Works on Biela or his contributions are Alois David, *Geschichte des Kometen, den Hauptmann von Biela entdeckte* (Prague, 1827); Josef von Hepperger, "Bahnbestimmung des Bielaschen Kometen," in *Sitzungsberichte der mathematisch-naturwissenschaftlichen Classe der Kaiserlichen Akademie der Wissenschaften, Wien,* **107,** pt. 2a (1898), 377–489 and **109,** pt. 2a (1900), 299–382, 623–655; J. Hirtenfeld and H. Meynert, eds., *Österreichisches Militär-Konversations-Lexikon,* I (Vienna, 1851), p. 410; and Constant von Wurzbach, *Biographisches Lexikon des Kaiserthums Oesterreich,* I (Vienna, 1856), 388–390.

JOSEF MAYERHÖFER

THOMAS WIDORN

BIENEWITZ, PETER. See **Apian, Peter.**

BIESTERFELD, J. H. See **Bisterfeld, Johann Heinrich.**

BIGOURDAN, CAMILLE GUILLAUME (*b.* Sistels, Tarn-et-Garonne, France, 6 April 1851; *d.* Paris,

France, 28 February 1932), *astronomy, history of science.*

Bigourdan was born to a peasant family from the Bigorre (whence his surname) and from them inherited his passion for work, his strong character, and his deep religious convictions. His teacher, Félix Tisserand, introduced him to the study of astronomy in Toulouse, and in 1879 summoned him to the Paris Observatory, where Bigourdan was astronomer-in-chief from 1897 to 1925. He was a member of the Bureau des Longitudes (1903), of the Académie des Sciences (1904), and, for his work in meteorology, of the Académie d'Agriculture (1924).

Bigourdan was a remarkable observer, and most of his contributions to astronomy were visual surveys of position: meridian observations and equatorial observations of double stars, asteroids, comets, and especially nebulae. He also perfected instruments and methods. His catalog of the positions of 6,380 nebulae brought him the gold medal of the Royal Astronomical Society in 1919.

His work has been made partially obsolete by technical progress, but his studies on time and the history of astronomy are still of value. After serving as a promoter, with Gustave Ferrié, of the International Congress on Time (1912), Bigourdan became the director of the Bureau International de l'Heure when it was created in 1919. He deliberately oriented this originally technical agency toward science. His emphasis on science, although it led to his dismissal in 1929, greatly benefited astronomy.

Moreover, Bigourdan left numerous authoritative historical studies. He also discovered various lost manuscripts, particularly those of the French astronomer J. A. G. Pingré, which allowed him to publish his *Annales célestes du XVIIe siècle* in 1901.

BIBLIOGRAPHY

I. ORIGINAL WORKS. Bigourdan's astronomical works include "Sur l'équation personnelle dans les mesures d'étoiles doubles" (doctoral thesis, 1886), in *Annales de l'Observatoire de Paris* (*Mémoires*), **19** (1889), C1–C74; *Observations de nébuleuses et d'amas stellaires, 1884–1909,* 6 vols. (Paris, 1899–1917); "Détermination de la différence de longitude entre les méridiens de Greenwich et de Paris, exécutée en 1902," written with F. Lancelin, in *Annales de l'Observatoire de Paris* (*Mémoires*), **26** (1910), B1–B214; and *Gnomonique ou Traité théorique et pratique de la construction des cadrans solaires* (Paris, 1921).

In the history of science Bigourdan published *Le système métrique des poids et mesures* (Paris, 1901); *De l'origine à la formation de l'Observatoire de Paris,* Vol. I of *Histoire de l'astronomie d'observation et des observatoires en France*

(Paris, 1918); "Un institut d'optique à Paris au XVIIIᵉ siècle," in *Comptes rendus du Congrès des Sociétés Savantes en 1921, Sciences* (Paris, 1922), pp. 19–74; and *De la fondation de l'Académie à la fin du XVIIIᵉ siècle,* Vol. II of *Histoire de l'astronomie de l'observation et des observatoires en France* (Paris, 1930).

His articles in the *Bulletin astronomique* include "Honoré Flaugergues, sa vie et ses travaux," in **1** (1884), 569–576, and **2** (1885), 151–156, 491–500; "Histoire des observatoires de l'École Militaire," in **4** (1887), 497–504, and **5** (1888), 30–40; and "La prolongation de la méridienne de Paris, de Barcelone aux Baléares, d'après des correspondances inédites de Méchain, de Biot et d'Arago," in **17** (1900), 348–368, 390–400, 467–480.

Articles in the *Annuaire du Bureau des Longitudes* are "Le jour et ses divisions" (1914), B1–B107; "Le calendrier babylonien" (1917), A1–A20; "Le calendrier égyptien" (1918), A1–A42; "Les comètes, liste chronologique de celles qui ont paru depuis l'origine à 1900" (1927), A1–A76; and "Le Bureau des Longitudes. Son histoire et ses travaux depuis l'origine (1795)" (1928), A1–A72; (1929), C1–C92; (1930), A1–A110; (1931), A1–A151; (1932), A1–A117; and (1933), A1–A91.

II. SECONDARY LITERATURE. Works on Bigourdan are A. Collard, "L'astronome G. Bigourdan," in *Ciel et terre,* **48** (1932), 165–167; F. W. Dyson, "G. Bigourdan," in *Monthly Notices of the Astronomical Society,* **93** (1933), 233–234; and P. A. MacMahon, "Address . . . on the Award of the Gold Medal . . . to G. Bigourdan," *ibid.,* **79** (1919), 306–314.

JACQUES R. LEVY

BILHARZ, THEODOR (*b.* Sigmaringen, Germany, 23 March 1825; *d.* Cairo, Egypt, 9 May 1862), *anatomy, zoology.*

Bilharz' name is perpetuated in the name of the disease bilharziasis, which he described in 1851; the following year he discovered its cause. The son of Anton Bilharz, a counsellor of the exchequer, and Elisa Fehr, Bilharz grew up in the small, *biedermeierlich* south German city of Sigmaringen. He attended the secondary school there, and having developed a particular interest in natural history during his school years, he entered the University of Freiburg im Breisgau in 1843. There he studied under Friedrich Arnold, who gave him insight into anatomical research. In 1845 Bilharz accompanied Arnold to the University of Tübingen, where in 1849 he passed the state examination and received the M.D.

At Tübingen he became acquainted with Wilhelm Griesinger, who in 1850 asked Bilharz to accompany him to Egypt when he was named director of the Egyptian Department of Hygiene. Bilharz began as Griesinger's assistant; and after Griesinger's return to Germany in 1852, he was promoted to chief physician of a medical department. In 1856 he became professor of descriptive anatomy at the Cairo medical school, Kasr-el Aïn. There he had the opportunity to perform numerous dissections and thus discovered peculiar pathological changes—white extuberances of cancerous aspect—in the mucous membranes of the bladder, intestines, ureters, and seminal glands. He found that the cause of the changes was a hitherto unknown trematode, which he described to his zoology teacher Carl Theodor von Siebold in nine letters from 1 May 1851 to 1 January 1853; Siebold published these letters in 1853. In them Bilharz gave not only a detailed anatomical description of the parasite and the anatomical changes produced by it, but he also supplied excellent diagrams of a pair of the copulating flatworms, which he called *Distomum haematobium,* and diagrams of the eggs. In these diagrams he also depicted the *Schistosomum mansoni,* which was not mentioned again until 1907, when Sambon named it for his teacher Patrick Manson. The terms "bilharzia" and "bilharziasis" were coined by Heinrich Meckel von Hemsbach, who introduced them into scientific nomenclature in 1856, two years before David F. Weinand introduced the term "schistosoma."

Bilharz never considered this consequential discovery to be as important as the work he did on the electrical organ of the thunderfish; in 1857 he wrote a monograph on the latter subject. He took his only European vacation in 1858, at which time he had the chance to report on his zoological research. In the following years Bilharz occupied himself with investigating native fauna—he gave the first description of a Nile fish, the *Alestes macrolepidotus*—and with anthropological and etymological activities.

In 1862 Bilharz accompanied the German explorer Ernst von Coburg-Gotha to Ethiopia; there, while treating a patient, he contracted typhoid fever and, just after his return to Cairo, he died. His discovery of the agent of tropical hematuria, which was by no means recognized by Bilharz' contemporaries as epoch-making, was nevertheless to introduce a new era of tropical parasitology; and it initiated a successful fight against an illness that still infects millions of the earth's inhabitants.

BIBLIOGRAPHY

I. ORIGINAL WORKS. Among Bilharz' writings are "Ein Beitrag zur *Helminthographia humana,* aus brieflichen Mittheilungen des Dr. Bilharz in Cairo," in *Zeitschrift für wissenschaftliche Zoologie,* **4** (1853), 53–76, table 5; 454–456, table 17; "*Distomum haematobium* und sein Verhältnis zu gewissen pathologischen Veränderungen der menschlichen Harnorgane," in *Wiener medizinische Wochenschrift,* **6**

(1856), 49–52, 65–68; "Über *Pentastomum constrictum,*" in *Zeitschrift für wissenschaftliche Zoologie,* **7** (1856), 329–330, table 17; *Das electrische Organ des Zitterwelses. Anatomisch beschrieben* (Leipzig, 1857); and "Über die Eingeweidewürmer Ägyptens," in *Zeitschrift der kaiserlich-königlichen Gesellschaft der Ärzte Wien,* **14** (1858), 447–448.

II. SECONDARY LITERATURE. The most comprehensive biography is E. Senn, *Theodor Bilharz: Ein deutsches Forscherleben in Ägypten, 1825–1862,* Schriften des Deutschen Auslandsinstituts Stuttgart, series D, V (Stuttgart, 1931); see also H. Ben-Amram, "L'histoire de la draconculose et de la bilharziose et leur incidence économique et sociale," thesis (Rennes, 1959); A. Bilharz, "Theodor Bilharz," in *Archiv für die Geschichte der Naturwissenschaften und der Technik,* **8** (1918), 232–236; J. Roos, "Theodor Bilharz," M.D. dissertation (Würzburg, 1929); and H. Schadewaldt, "Theodor Bilharz," in *Deutsche medizinische Wochenschrift,* **80** (1955), 1053–1055; "Die Erstbeschreibung und -abbildung von *Bilharzia haematobia* und *mansoni* durch Theodor Bilharz," in *Zeitschrift für Tropenmedizin und Parasitologie,* **4** (1953), 410–414; "Theodor Bilharz, einer der Begründer deutscher tropenmedizinischer Forschung," in *Münchener medizinische Wochenschrift,* **104** (1962), 1730–1734; "Theodor Bilharz und die Bilharziose," in *Berliner medizinische Zeitschrift,* **14** (1963), 244–250; and "Unveröffentlichte Zeichnungen aus dem Nachlass von Theodor Bilharz," in *Münchener medizinische Wochenschrift,* Bildbeilage 7 (1963).

HANS SCHADEWALDT

BILLINGS, ELKANAH (*b.* Billings Bridge, near Bytown [now Ottawa], Ontario, 5 May 1820; *d.* Montreal, Quebec, 14 June 1876), *paleontology.*

Billings' father, Bradish, was a farmer in comfortable circumstances. After attending several private schools, Elkanah entered St. Lawrence Academy at Potsdam, New York, in 1837. Returning to Bytown in 1839, he enrolled in the Law Society of Upper Canada and was admitted to the bar in 1844. For eight years he practiced law in Toronto, Renfrew, and Bytown, but in 1852 he virtually gave up his law practice to become editor of the Bytown *Citizen;* in it he began to publish popular articles on natural history, including local fossils, of which he had a large collection. His proficiency in paleontology earned him membership in the Canadian Institute of Toronto in 1854, and his first scientific papers were read before that institute the same year and were published in its *Journal.*

In 1856 Billings wrote: "I have abandoned my [legal] profession, and intend to devote the rest of my life to the study of Natural History." He thereupon began publication of the *Canadian Naturalist & Geologist,* and for the first year (1856) was its owner, editor, and sole contributor. Although he relinquished

official connection with that journal in 1857, he subsequently contributed at least forty articles to it.

Billings' love for natural history may have been fostered by his eldest brother, who became an accomplished botanist and entomologist; as a paleontologist he was entirely self-taught. While living in Toronto he doubtless had access to scientific libraries and collections. Indeed, he could not have written his first two papers (1854) without detailed knowledge of the morphology of echinoderms, both living and fossil, of the available literature on that subject, and of the rules of taxonomic nomenclature. From 1852 on, he corresponded with William E. Logan, director of the Geological Survey of Canada, who considered his scientific attainments of such high caliber that in 1856 he obtained for Billings the post of paleontologist with the Survey. This necessitated his moving to the Survey headquarters in Montreal, where a large collection of fossils, some identified and some not, had accumulated in its museum. Billings plunged at once into the task of identifying and classifying these fossils so that they could form a meaningful public display and could provide a standard against which new collections could be compared. Within two years this task was accomplished.

Except for two scientific papers (1854) concerned with technical descriptions of cystids from the Trenton limestone, Billings' writings prior to the Survey appointment were essays on natural history. After becoming an officer of the Survey, he turned his attention strictly to the exacting scientific descriptions of fossils, mostly from Paleozoic formations of eastern Canada. His first official "Report . . . as Palaeontologist" (1857) contains detailed descriptions of 106 new species belonging to thirty-five genera (of which thirteen were new), a remarkable achievement. His second report (1858) was similarly constituted, and was immediately followed by Decades 3 (1858) and 4 (1859) of the Canadian Organic Remains series, which were concerned almost wholly with fossil echinoderms. Billings began a second series of paleontological works, *Palaeozoic Fossils,* of which he wrote the first volume (1865) and the first part of the second (1874). Among other publications are his 1866 report on fossils from Anticosti Island and his continuing work on Devonian fossils from western Canada and on Silurian (i.e., Ordovician and Silurian) crinoids and cystids. The last category embraced what might be called his specialty. In all he published descriptions of sixty-one new genera and 1,065 new species of fossils. His bibliography contains nearly 200 titles, of which ninety are concerned directly or indirectly with paleontological subjects.

Although he accomplished much in purely descriptive taxonomic paleontology, Billings is also remembered for the stratigraphic interpretation of his identifications. Because of his determination of the age of the rocks of the "Quebec Group" as Beekmantown and Chazy, Logan was able to demonstrate his "great overlap," which is now referred to as Logan's line.

BIBLIOGRAPHY

I. ORIGINAL WORKS. Billings' writings include "Report for the Year 1856 as Palaeontologist," in Geological Survey of Canada, *Report of Progress for 1853–56* (Montreal, 1857), pp. 247–345; "Report for the Year 1857 as Palaeontologist," in Geological Survey of Canada, *Report of Progress for 1857* (Montreal, 1858), pp. 147–192; *Figures and Descriptions,* Canadian Organic Remains, Decade 3 (Montreal, 1858) and Decade 4 (Montreal, 1859); *Palaeozoic Fossils,* Vol. I (Montreal, 1865) and Vol. II, pt. 1 (Montreal, 1874); and "Catalogue of the Fossils of the Island of Anticosti," in *Special Report of the Geological Survey of Canada* (Montreal, 1866), pp. 1–82.

II. SECONDARY LITERATURE. Articles on Billings are H. M. Ami, "Brief Biographical Sketch of Elkanah Billings," in *American Geologist* (May 1901), 265–281; B. E. Walker, "List of the Published Writings of Elkanah Billings, F. G. S.," in *Canadian Record of Science,* **8** (1902), 366–388; and J. F. Whiteaves, "Obituary Notice of Elkanah Billings, F. G. S.," in *Canadian Naturalist,* **8** (1876), 251–261.

T. H. CLARK

BILLROTH, CHRISTIAN ALBERT THEODOR
(*b.* Bergen, on the island of Rügen, Germany, 26 April 1829; *d.* Abbazia, Istria, Italy [now Opatija, Yugoslavia], 6 February 1894), *pathological anatomy.*

Billroth's father, Theodor, was a clergyman; his mother, Christine Nagel, was the daughter of a Berlin *Kammerrat* (counsellor of the exchequer). His father died when Theodor was five, and his mother then moved to Greifswald, where Billroth attended the Gymnasium. He was musically inclined (a family characteristic) and probably for that reason was not an exceptional pupil, even needing tutoring at home; he seemed unable to master languages and mathematics, was not quick-witted, and spoke slowly.

His mother and two professors of medicine in Greifswald, Baum and Seiffert, induced Billroth to become a doctor for financial reasons. He was nevertheless an artist: intuitive, humane, inventive. His home in Vienna later became a musical center where he played second violin or viola and became friends with Johannes Brahms and with the musical theorist and writer Eduard Hanslick. Two of Brahms's string quartets are dedicated to Billroth, and during his last illness Billroth was working on the physio-psychological book *Wer ist musikalisch?,* published by Hanslick in 1896.

During his first semester as a medical student in Greifswald, Billroth studied natural sciences and began the multifaceted activity and careful use of his time that characterized his later years. He followed Baum to the University of Göttingen, where he established a lasting friendship with Georg Meissner. Like Billroth, Meissner was interested in music and a pupil of the physiologist Rudolf Wagner, who taught Billroth microscopy. With Wagner and Meissner, Billroth went to Trieste to study the origin and insertion of the nerves of the torpedo fish. In 1851 he continued his studies at Berlin with Bernard von Langenbeck, Johann Lukas Schönlein, Moritz Romberg, and Ludwig Traube. Traube taught him experimental pathology and encouraged him to write the thesis "De natura et causa pulmonum affectionis quae nervo utroque vago dissecto exoritur." On 30 September 1852 Billroth received his doctorate, and that winter he passed the state medical examination, after which he worked in the ophthalmological clinic of Albrecht von Graefe.

In order to take courses in dermatology with Ferdinand von Hebra, in pathology with Henschel, and in internal medicine with Johann von Oppolzer, Billroth went to Vienna in the spring of 1853. That fall he tried in vain to establish himself as a general practitioner in Berlin, but after a few months he was appointed assistant to Langenbeck at the surgical clinic of Berlin University. He published on pathological histology and in 1856 became *Privatdozent* in surgery and pathological anatomy. Later he lectured on surgery and gave practical demonstrations.

It was in Berlin that Billroth met his wife Christine, daughter of the court physician Edgar M. Michaelis and of Karoline Eunike. They were married in 1858, and of their four daughters and one son, three daughters survived.

In 1860 Billroth was nominated ordinary professor and director of the well-known surgical hospital and clinic at Zurich. He added greatly to its fame and its growth during the seven years he was its director. Modern surgery was in its infancy, and Billroth was especially interested in the causes of wound fever. He insisted on regular temperature-taking and believed that wound fever was caused by a chemical poison produced by some living organism. His *Die allgemeine chirurgische Pathologie und Chirurgie in fünfzig Vorlesungen* (1863) is a classic surgical textbook. Billroth collaborated with Pitha on *Handbuch der Chirurgie* (1865–1868) and with Lücke on *Deutsche Chirurgie* (1879). After having declined calls to

Rostock and Heidelberg, he was appointed professor of surgery and director of the surgical clinic at the University of Vienna, where he remained until his death, much beloved by his students, assistants, and patients.

Billroth excelled as a surgeon, as a teacher, and as a scientist. He performed many hazardous operations successfully because of his great ability and caution. *Chirurgische Klinik* (1869–1876), his collection of clinical experiences and the surgical results of his Zurich and Vienna years, is notable because it reports failures and successes alike. Billroth was one of the first to introduce antisepsis on the Continent and was the first to resect the esophagus (1872), to perform total laryngectomy (1873), and to resect a cancerous pylorus (1881), which caused a great sensation in medical circles. His methods of resection, although modified, remained in use for many years. Plastic surgery, especially of the face, was another of his specialties.

Billroth founded the House of the Society of Physicians in Vienna and the Rudolfinerhaus, a nursing school for which he wrote the handbook *Ueber die Krankenpflege im Hause und im Hospitale* (1881). He was a member of the Academy of Sciences in Vienna and honorary member of thirty-two scientific societies, and a member of the Austrian Herrenhaus (from 1886); he was also honored with sixteen high decorations. His bibliography contains some 150 items.

Billroth kept his robust health until February 1887, when he contracted pneumonia and suffered from cardiac weakness that increased during his last years. His pupils, who spread his teaching all over the Continent, included Czerny, Gussenbauer, Mikulicz, Eiselsberg, Wölfer, Nikoladoni, Hacker, Winiwarter, Gersuny, Salzer, Fraenckel, and Narath.

BIBLIOGRAPHY

I. ORIGINAL WORKS. A full list of Billroth's publications is K. Gussenbauer, in *Wiener klinische Wochenschrift,* **7** (1894), 118–120; and Mikulicz, in *Berliner klinische Wochenschrift,* **31** (1894), 203–205. His thesis was "De natura et causa pulmonum affectionis quae nervo utroque vago dissecto exoritur" (Berlin, 1852). His major publications include "Beobachtungsstudien über Wundfieber und accidentelle Wundkrankheiten," in *Archiv für klinische Chirurgie,* **2** (1861); **6** (1864); **8** (1866); **13** (1872); *Die allgemeine chirurgische Pathologie und Chirurgie in fünfzig Vorlesungen* (Berlin, 1863; 16th ed., 1906), translated as *General Surgical Pathology and Therapy* (1871; 3rd ed., 1880); and *Chirurgische Klinik,* 4 vols. (Berlin, 1869–1879). Billroth collaborated with Pitha on *Handbuch der Chirurgie* (1865–1868) and with Lücke in editing *Deutsche Chirurgie*

(1879). On the study of medicine he wrote *Ueber das Lehren und Lernen der medizinischen Wissenschaften* (Vienna, 1876) and *Aphorismen* to it (Vienna, 1886). His handbook for nurses is *Ueber die Krankenpflege im Hause und im Hospitale* (Vienna, 1881; 7th ed., 1905). Among his special studies are *Chirurgische Briefe aus den Kriegslazarethen in Weissenburg und Mannheim 1870* (Berlin, 1872); *Untersuchungen über die Vegetationsformen von Coccobacteria septica* (Berlin, 1874); and "Offenes Schreiben und Herrn Dr. D. Wittelshöfer" (on the stomach resection performed 29 January 1881), in *Wiener medizinische Wochenschrift,* **31** (1881), cols. 161–165. Billroth's autobiography is in *Wiener klinische Wochenschrift,* **7** (1894), 120–122, and *Wiener medizinische Blätter,* **17** (1894), 91–95. His last work, *Wer ist musikalisch?,* was published by his friend Hanslick (Berlin, 1896).

II. SECONDARY LITERATURE. Articles on Billroth's operations are K. Gussenbauer (on the first laryngectomy, 27 November 1873), in *Archiv für klinische chirurgie,* **17** (1874), 343–356; Mikulicz, "Ueber die Totalexstirpation des Uterus," in *Wiener medizinische Wochenschrift,* **31** (1881), cols. 241–245; and "Zur Resektion des carcinomatösen Magens," *ibid.,* cols. 634–635. J. Mundy, "Ein neues Buch von Th. Billroth," in *Wiener medizinische Wochenschrift,* **31** (1881), cols. 225–229 and 251–254, consists of notes on Billroth's work and a bibliography. On his Zurich period see A. Huber, "Prof. Dr. Theodor Billroth in Zürich 1860–1867," in *Züricher medizinische Geschichte,* I (1924). On Billroth's relation to Brahms see O. G. Billroth, *Billroth und Brahms im Briefwechsel* (Berlin-Vienna, 1935). His genealogoy is P. von Gebhardt, "Ahnentafel berühmter Deutscher, 1," in *Familiengesch. Blätter,* **27** (1929), 88–89. Additional correspondence is in G. Fischer, ed., *Briefe von Theodor Billroth* (Hannover-Leipzig, 1895; 9th ed., 1922); W. von Brunn, *Jugendbriefe Theodor Billroths an Georg Meiszner* (Leipzig, 1941); and *Wiener klinische Wochenschrift,* **7** (1894), 122–124, 425–427.

Obituaries are K. Gussenbauer, in *Wiener klinische Wochenschrift,* **7** (1894), 115–117, with complete bibliography and list of honors; *Jahrbuch der Universität Wien* (Vienna, 1893–1894), pp. 37–46, with Gussenbauer's bibliography and list of honors; and J. Mikulicz, in *Berliner klinische Wochenschrift,* **31** (1894), 199–205, based on R. Gersuny's biography in *Nord und Süd,* **141** (1888), with complete bibliography; see also Oehlschläger, "Jugenderinnerungen an Theodor Billroth," *ibid.,* 229–230; and E. von Bergmann, memorial speech, *ibid.,* 205–207. Additional obituaries are A. von Bardeleben, in *Deutsche medizinische Wochenschrift,* **20** (1894), 145; E. von Bergmann and E. Gurlt, in *Archiv für klinische Chirurgie* (1894); M. Benedikt, in *International klinische Rundschau,* **8** (1894), 184–189; E. Hanslick, in *Deutsche Rundschau,* **20** (1893–1894), 274–277; E. Kappeler, in *Archiv für klinische chirurgie* (1894), 161; Sozin, in *Korr. bl. für Schweizer Ärtze* (1894), 129; and A. Wölfer, in *Archiv für klinische Chirurgie* (1894), *Wiener medizinische Wochenschrift* (1894), 339, and *Zentralblatt für Chirurgie* (1894).

There is a biography by R. Gersuny (Berlin-Leipzig-Munich, 1922). See also I. Fischer, *Billroth und* seine Zeitgenossen (Berlin-Vienna, 1929); H. Fischer, in *Medical Life*, **37** (1930), 432–440; J. C. Hemmeter, in *Johns Hopkins Hospital Bulletin*, **11** (1900), 297–317; R. E. Weise, in *Annals of Medical History*, **10** (1928), 278–286; *Mitteilungen für die Geschichte der Medizin und Naturwissenschaften*, **28**, 238–314; and Winiwarter, in *Wiener klinische Wochenschrift*, **20** (1909), 309–312.

H. ENGEL

BILLY, JACQUES DE (*b.* Compiègne, France, 18 March 1602; *d.* Dijon, France, 14 January 1679), *mathematics, astronomy.*

A Jesuit, Billy spent his teaching career in the *collèges* of the society's administrative province of Champagne—Pont à Mousson, Rheims, and Dijon. He taught either theology or mathematics, depending on which was needed. In 1629–1630 he taught mathematics at Pont à Mousson while he was still a theology student and not yet ordained a priest. Billy taught mathematics at Rheims from 1631 to 1633. Around this time he became a close friend of Claude Gaspar Bachet de Méziriac, the commentator on Diophantus who introduced him to indeterminate analysis.

Billy became master of studies and professor of theology at the Collège de Dijon, where one of his students was Jacques Ozanam, whom he taught privately because there was no chair of mathematics at the *collège,* and in whom he instilled a profound love for calculus. Finally, a professorship having been created in mathematics, he taught his favorite subject from 1665 to 1668.

An active correspondence between Billy and Fermat began before 1659, of which one letter remains. Some of Billy's writings originated in this exchange, including parts of the *Doctrinae analyticae inventum novum,* through which his name is still known to number theorists. It is an elaborate study of the techniques of indeterminate analysis used by Fermat and, on the whole, it explains them correctly. From it one can guess at Fermat's general line of activity in a field in which there are few pertinent documents.

In astronomy Billy published numerical tables applicable to the three important theories (Ptolemy, Brahe, Copernicus) of the time. There are also a study on comets and several critiques against forensic astrology.

BIBLIOGRAPHY

I. ORIGINAL WORKS. Billy's works are *Abrégé des préceptes d'algèbre* (Rheims, 1637); *Le siège de Landrecy* (Paris, 1637); *Nova geometriae clavis algebra* (Paris, 1643); *Tabulae Lodoicae, seu eclipseon doctrina* (Dijon, 1656);

Tractatus de proportione harmonica (Paris, 1658); *Diophantus geometria, sive opus contextum ex arithmetica et geometria simul. . . .* (Paris, 1660); *Opus astronomicum* (Dijon, 1661); *Discours de la comète qui a paru l'an 1665 au mois d'avril* (Paris, 1665); *Crisis astronomica de motu cometarum* (Dijon, 1666); *Diophanti redivivi pars prior . . . pars posterior* (Lyons, 1670); and "Doctrinae analyticae inventum novum, collectum a R. P. Jacobo de Billy . . . ex variis epistolis quas ad eum diversis temporibus misit D. P. de Fermat. . . .," a study in Samuel Fermat, *Diophanti Alexandrini arithmeticorum libri sex* (Toulouse, 1670), Latin text and German trans. by P. von Schaewen (Berlin, 1910), French trans. by Paul Tannery, in *Oeuvres de Fermat,* III (Paris, 1896).

The Dijon municipal library owns several of Billy's autograph manuscripts. Paul Tannery wished to publish the part of the correspondence concerned with indeterminate analysis, but he died before he could carry out the project.

II. SECONDARY LITERATURE. There is a notice on Billy in R. P. Niceron, *Mémoires pour servir à l'histoire des hommes illustres dans la république des lettres,* XL (Paris, 1739), 232–244.

JEAN ITARD

BINET, ALFRED (*b.* Nice, France, 8 July 1857; *d.* Paris, France, 18 October 1911), *psychology.*

The career of Binet, the founder of French experimental psychology, developed on the periphery of the traditional institutions and established frameworks. His training was unusual: he was a licentiate in jurisprudence and a doctor of natural sciences, but he did no teaching, with the exception of a course in psychology at the University of Bucharest, where he had been invited in 1895. In 1892 he was named assistant director of the Laboratory of Physiological Psychology created at the Sorbonne in 1889 and directed by Henri Beaunis. In 1895 Binet and Beaunis founded the first French journal of psychology, *Année psychologique.* In the same year he succeeded Beaunis at the Laboratory, now connected with the École Pratique des Hautes Études, where he worked until his death.

About 1900 Binet's experiments began to go beyond the somewhat narrow framework of the Laboratory. He interviewed children in nurseries, schools, and school camps. In 1898 he and Ferdinand Buisson founded the Société Libre pour l'Étude Psychologique de l'Enfant, which after his death became the Société Alfred Binet. In 1904 the minister for public education appointed Binet to a commission for the study of problems connected with the education of retarded children. This appointment resulted in a series of studies leading to the creation of the metric intelligence scale.

Young Binet was interested in psychiatry; frequented the Salpêtrière, where Charcot taught; and studied Hippolyte Taine, Théodule Ribot, and John Stuart Mill. His first book, *Psychologie du raisonnement* (1886), based on experiments with hypnosis, related the reasoning process to an organization of images and taught associationism. It also touched off a long series of researches into intelligence and the thought process. To this series belong his studies on mathematical prodigies and chess players, retarded children, and especially the *Étude expérimentale de l'intelligence* (1903), which completed his break with Ribot's associationism. The work described experiments with and observations of Binet's two daughters, to whom he presented simple problems, insisting on justification of the replies as well as on the solution itself. These experiments demonstrated the impossibility of translating reasoning in sensory terms and proved the unity and activity of thought and its independence with respect to images. R. S. Woodworth and K. Bühler were to arrive at analogous results in 1907. Binet's study was also a fine in-depth investigation of the individual differences of the two subjects; the experiment was made complete by introspection and was oriented toward a qualitative typology.

Binet expanded the idea of experimentation in psychology. Although he worked on esthesiometric thresholds, tactile sensibility, and optic illusions, he preferred to proceed by means of questionnaires, investigations, and personal interviews rather than the complicated apparatus and artificial techniques of the laboratory. His research covered much ground: he wrote on personality changes, suggestibility, intellectual fatigue, and graphology.

His major contribution to psychology consists in the introduction of new methods of measuring intelligence. When expansion of the educational system created the need to find criteria for detecting the mentally defective, Binet, at the government's request, pursued his former work on the evaluation of intelligence in children. He proposed the metric intelligence scale, based on the idea of classifying the subjects according to the observed differences between individual performances. In 1905 Binet drew up a whole series of tests: a large number of short, varied problems related to daily situations, bringing into play "superior processes" such as memory and ratiocination. The series was arranged according to mental levels, and the measure of intelligence was established by comparison of the results and their classification. A revision of the scale in 1908 resulted in an important innovation: assuming that intelligence increases with age, Binet ranked the tests in accordance with age levels corresponding to performances by the average child. The mental age (the age

the child attains on the scale) was distinguished from chronological age. This latter work of Binet, in collaboration with Théodore Simon, enjoyed wide popularity. It was translated, adapted, imitated, and administered on a large scale. It was the beginning of a new era in testing.

BIBLIOGRAPHY

I. ORIGINAL WORKS. Binet's writings include *La psychologie du raisonnement, recherches expérimentales par l'hypnotisme* (Paris, 1886); *Le magnétisme animal* (Paris, 1887), written with C. Féré; *On Double Consciousness* (Chicago, 1889); *Les altérations de la personnalité* (Paris, 1892); *Contribution à l'étude du système nerveux sous-intestinal des insectes* (Paris, 1894), his thesis at the Sorbonne; *Introduction à la psychologie expérimentale* (Paris, 1894), written with P. Courtier and V. Henri; *Psychologie des grands calculateurs et joueurs d'échecs* (Paris, 1894); *La fatigue intellectuelle* (Paris, 1898), written with V. Henri; *La suggestibilité* (Paris, 1900); *L'étude expérimentale de l'intelligence* (Paris, 1903); *L'âme et le corps* (Paris, 1906); and *Les idées modernes sur les enfants* (Paris, 1911). Binet and Simon's tests were published in *Année psychologique,* **11** (1905–1906), 163–336, and **14** (1908–1909), 1–94, and in the final form, in their *La mesure du développement de l'intelligence chez les enfants* (Paris, 1911), which has gone through numerous editions.

II. SECONDARY LITERATURE. Works on Binet are F.-L. Bertrand, *Alfred Binet et son oeuvre* (Paris, 1930), which contains a bibliography of Binet's writings; R. Martin, *Alfred Binet* (Paris, 1924); E. Varon, *The Development of Alfred Binet's Psychology,* Psychological Monographs, Vol. XLVI (Lancaster, N. Y., 1935); and F. Zuza, *Alfred Binet et la pédagogie expérimentale* (Paris, 1948), with a bibliography of Binet's works and secondary literature.

JAN SEBESTIK

BION, NICOLAS (*b. ca.* 1652; *d.* 1733), *instrumentation.*

There are almost no biographical data available on Bion. His workshop was located on the Quai de l'Horloge in Paris, at the sign of the Quart de Cercle or of the Soleil d'Or. This last sign may have been simply that of his printer–bookseller, Boudot. Bion had the title of king's engineer for mathematical instruments, and his name was often mentioned in his time; however, very few of his instruments are extant and no important technical innovations can be attributed to him. Undoubtedly he was extremely clever and had excellent manufacturing facilities at his disposal. Less specialized than most of his colleagues, he seems to have made globes, sundials, mathematical instruments, and mechanical machines with equal accuracy.

Also unlike his colleagues, Bion published several

works, and they as well as his instruments were probably responsible for his fame. Two pamphlets concern a sphere and globes executed for the dauphin and a celestial planisphere constructed to reflect the most recent observations made by the members of the Académie des Sciences. He also published three important treatises on globes and cosmography, on astrolabes, and on precision instruments in general. These writings had great success and went into many editions, the most recent of which were printed under the supervision of Bion's son after his father's death.

The *Traité de la construction . . . des instruments mathématiques* gives a fairly complete list of instruments normally constructed during the first quarter of the eighteenth century. It should, however, be noted that some of the instruments described by Bion—such as astrolabes, marine astrolabes, the jacob staff, and the Davis quadrant—were no longer used. There are deficiencies also in the descriptions of eyeglasses, microscopes, and micrometers. Bion apparently did not wish to have his instruments copied by others. All of his treatises were more for the user and the amateur than for the manufacturer.

The extant instruments are sundials of the Butterfield type, a pair of calipers, a proportional compass, an artillery calibrating compass, a theodolite, a graphometer, and a water level with pinnules of a special type.

BIBLIOGRAPHY

Bion's writings are *L'usage des globes célestes et terrestres et des sphères suivant les différents systèmes du monde, précédé d'un traité de cosmographie* (Paris, 1699; 6th ed., 1751); *L'usage des astrolabes tant universels que particuliers accompagné d'un traité qui en explique la construction* (Paris, 1702); *Description de la sphère et des globes dédiés et présentés à Mgr. le Dauphin* (Paris, 1704); *Description et usage du planisphère céleste nouvellement construit suivant les dernières observations de Messieurs de l'Académie des Sciences* (Paris, 1708); and *Traité de la construction et des principaux usages des instruments de mathématiques* (Paris, 1709; 5th ed., 1752).

Another work on Bion is Maurice Daumas, *Les instruments scientifiques aux XVIIe et XVIIIe siècles* (Paris, 1953), pp. 109–110.

JACQUES PAYEN

BIOT, JEAN-BAPTISTE (*b.* Paris, France, 21 April 1774; *d.* Paris, 3 February 1862), *physics.*

Biot's father, Joseph, was of Lorraine peasant stock; he had risen on the social scale and held a post in the treasury. Jean-Baptiste attended the Collège Louis-le-Grand in Paris and distinguished himself in the classical curriculum. About 1791, he left the school and took private lessons in mathematics. His father intended him to have a career in commerce, but Biot rebelled and, taking an opportunity provided by the Revolutionary Wars, volunteered for the army, enlisting as a gunner in September 1792. Biot, who had not abandoned mathematics, took the entrance examination for the École des Ponts et Chaussées and was accepted in January 1794. Shortly afterward, the École Polytechnique was founded; Biot transferred to it, and was appointed section leader of a group of students in November 1794. At the École Polytechnique, Biot's outstanding ability drew the attention of the director, Monge.

Under the influence of the ideas of the Revolution, and later of Laplace, Biot became skeptical of all belief in a personal God. Yet in 1825, in Rome, he sought and obtained a personal audience with Pope Leo XII and he became increasingly attracted to the religion of his childhood. In 1846 he made a formal return to the Roman Catholic Church. In 1797 Biot married the daughter (then aged sixteen) of Antoine François Brisson of Beauvais, *inspecteur général du commerce et des manufactures,* whose son was Biot's friend at the École Polytechnique. Biot instructed his wife in science and mathematics, and since she was a competent linguist, she was able to collaborate with him in a translation into French of E. G. Fischer's physics textbook.

Biot was closely associated with many of the institutions for education and research that were a prominent feature of France after the Revolution. On graduation from the École Polytechnique in 1797, he was appointed professor of mathematics at the École Centrale of the Oise department at Beauvais. From 1799 he was entrance examiner for the École Polytechnique, a post he retained when, in 1800, he was appointed professor of mathematical physics at the Collège de France. Under the patronage of Laplace, Biot was given the post of assistant astronomer at the Bureau des Longitudes in 1806. When the University of France was established by Napoleon in 1808, Biot was appointed professor of astronomy at the Paris Faculté des Sciences. From 1816 to 1826, however, while retaining the official title of professor of astronomy, he agreed to teach physics related to his own research and gave courses on light, sound, and magnetism. From 1840 until his retirement in 1849, Biot was dean of the Faculty.

Biot joined the Société Philomathique in Paris in 1801. His association with Laplace and Berthollet at about this time qualifies him for consideration as one of the original members of the Société d'Arcueil. In 1800 he was elected a nonresident member of the

First Class of the Institute, and when, in 1803, a vacancy for full membership in the mathematics section occurred, he was elected. Biot was unsuccessful in his candidature for the post of permanent secretary of the Académie des Sciences on the death of Delambre in 1822, but was elected vice-president of the Academy in 1835. By virtue of his research on Egyptian, Babylonian, and Chinese astronomy, Biot was elected a member of the Académie des Inscriptions et Belles Lettres in 1841. In 1856 he received the honor, unusual for a man of science, of election to the Académie Française. Biot, who in his youth had detested the Jacobins, also had little sympathy for Napoleon. Upon the restoration of the Bourbons in 1814, Biot was awarded the Legion of Honor (*chevalier*) for his services to science and education, and was successively promoted to officer (1823) and commander (1849) of the order. Unlike many of his contemporaries in France, Biot took no part in politics, living a long and active life devoted almost entirely to scientific research.

Throughout his life Biot made contributions to literature beyond those expected of a man of science. His *Essai sur l'histoire générale des sciences pendant la révolution française* was published in 1803. When in 1812 the Académie Française proposed the subject of Montaigne for a prize, Biot's essay received an honorable mention. He was commissioned to write a hundred-page biography of Newton for the *Biographie universelle*. Biot was well known as an ardent follower of Newton. In 1813, in the *Journal de physique,* he described Newton as a person "whose conceptions seem to have surpassed the limits of thought of mortal man" (p. 131). Biot continued: "Words fail to convey the profound impression of astonishment and respect which one experiences in studying the work of this admirable observer of nature." In the biography, Biot's solution to the problem of the interrelation of Newton's natural philosophy and his theology was to suggest that all the original scientific work had been completed early in Newton's life and that he had become seriously interested in theology only after mental illness. Biot later took issue with Brewster's interpretation of Newton. In the last years of his life, Biot wrote appreciations of Gay-Lussac and Cauchy.

Biot was the author of several important textbooks. His *Traité élémentaire d'astronomie physique* was the source from which Sir George Airy, later British astronomer royal, learned his basic astronomy; and he claimed that he had acquired his interest in the subject through Biot's work. Biot's *Traité de physique* (1816) constitutes a comprehensive account of contemporary physics, including not only recent original research by himself (e.g., on polarization) but also the recent and often unpublished work of his associates, particularly Laplace, Gay-Lussac, and Dulong.

Although Biot's first publications were in mathematics, he soon came strongly under the influence of Laplace, whose advice he followed in the application of analysis to physical problems. In 1802 Biot demonstrated that the attraction of an ellipsoid at an external point might be deduced by simple differentiations from a particular expression, which is theoretically known when the attraction is known for all points situated in the plane of one of the principal sections. Biot's memoir, however, constituted little more than a commentary on the earlier writings of Legendre and Laplace. His introduction to Laplace had come about through his offer to read the proofs of Laplace's *Mécanique céleste*. Laplace encouraged Biot to undertake experimental investigation of a wide range of problems, many of which constituted a deliberate extension of the Newtonian framework of science. This can be seen particularly in Biot's research on refraction, polarization of light, and sound. If we were to select any one branch of physics to which Biot made the most important contribution, the choice would be the polarization of light, but, since none of his contributions in this field occurred before 1812, it will be convenient to deal first with his varied contributions to other branches of physical science.

An unusual piece of research at the beginning of Biot's career was concerned with a meteorite said to have fallen from the sky at l'Aigle in the Orne department on 26 April 1803. Biot was ordered by Chaptal, minister of the interior, to confirm the report. Shortly before, M. A. Pictet had called attention to reports of meteorites—reports that many rationalists had dismissed as superstitious. Biot questioned people in the locality where the meteoric stones had fallen. Various specimens were examined and compared with the composition of the ground from which they had been taken, and some were subjected to chemical analysis. Biot's report, read to the First Class of the Institute on 18 July 1803, marks the beginning of a general recognition in France of the reality of meteorites.

In the years 1804–1809 Biot undertook several scientific projects in collaboration with other men, notably fellow members of the Arcueil group or the Bureau des Longitudes. On 24 August 1804, Biot made a balloon ascent with Gay-Lussac. The ascent was notable in that it was undertaken entirely for purposes of scientific research and had the approval of the French government. The primary purpose was to find whether the magnetic intensity of the earth decreased at great altitudes, as had been suggested by Horace de Saussure's experiments in the Alps. From their experiments, in which they timed the oscillations of a magnetized needle at various altitudes, Biot and Gay-Lussac concluded that up to 4,000 meters there was no change.

Biot undertook further work on magnetism in collaboration with Humboldt, who furnished much of the data used in their joint memoir (1804). Biot attempted to derive general laws governing inclination, using as a basis the hypothesis of an infinitely small magnet situated at the center of the earth and placed perpendicular to the magnetic equator. The theoretical values for the inclination agreed well with Humboldt's readings, particularly in the northern hemisphere. Anomalies were attributed to purely local factors.

Biot collaborated with Arago in 1805–1806 in research which, in a typically Newtonian spirit, they presented with the title "Mémoire sur les affinités des corps pour la lumière, et, particulièrement sur les forces réfringentes des différents gaz" (1806). The accurate determination of the refractive indices of various gases and vapors at different pressures was a legitimate subject of study for two members of the Bureau des Longitudes who were concerned with astronomical observations. In another memoir, Biot reported on the refraction of light rays near the horizon. He made a full study of mirages, taking the subject beyond Wollaston's work of 1803. In later years, in a succession of memoirs, Biot made further contributions to the subject of atmospheric refraction. As regards the joint work with Arago, however, not the least important part of Biot's research was the accurate determination of the densities of gases weighed in glass globes. The values obtained were part of the data used by Gay-Lussac to establish his law of combining volumes of gases. When Gay-Lussac and the chemist Thenard carried out combustion analyses of organic compounds, they calculated their results with atomic weights for carbon and hydrogen deduced from the density measurements of Biot and Arago. Prout also used the data of Biot and Arago to support his hypothesis.

In 1807 Biot collaborated with Thenard in a thorough comparative study of rhombic aragonite and hexagonal calcite. Apart from obvious chemical tests, they compared the refractivity not only of the crystals but also of their solutions in hydrochloric acid and of the carbon dioxide evolved from each. They concluded that aragonite and calcite were composed of the same chemical elements in the same proportions, but with a different arrangement of the molecules, which resulted in physically different substances. This was one of the earliest examples of what was later called dimorphism.

Biot made a number of contributions to the determination of the velocity of sound. His first memoir in 1802 mentions that he had undertaken this research at the instigation of Laplace. Biot was not then able to measure directly the tiny temperature changes produced by sound waves, although Laplace believed that this was the key to solving the discrepancy between Newton's formula for the velocity of sound and the value obtained in practice. In 1807 Biot carried out more experiments at Arcueil on the transmission of sound through vapors. In 1808 the extensive laying of water mains in Paris gave Biot the opportunity of carrying out further experiments on sound. With cast iron pipes forming a continuous length of 951 meters Biot determined, by repeated experiments, the time interval between transmission and reception of sound through the pipe and through the air. Knowing the velocity of sound in air under the temperature and pressure conditions of the experiment, Biot was able to compare the two velocities. One factor limiting the accuracy of the experiments was the presence of lead, used to join the iron pipes. Biot therefore did not give an explicit value for the velocity of sound in cast iron, but concluded that it was 10.5 times that in air—a value that was long considered authoritative because of the difficulty of direct determination.

In 1806 Biot and Arago were sent by the Bureau des Longitudes to determine the arc of the meridian in Spain and the Balearic Islands, a task begun by Méchain but left incomplete at his death in 1804. The post-Revolution metric system was based on the idea of a "natural" unit, the meter, which was supposed to be exactly one ten-millionth of a meridian quadrant of the earth. Méchain and Delambre had made measurements over a meridian arc of 10° stretching from Dunkirk to Barcelona, and from their readings the length of the standard meter was obtained. It was proposed that this should be redetermined with greater accuracy by extending measurements farther south to the Balearic Islands. Special difficulties in triangulation were encountered because of the distance of the islands from the mainland, but these difficulties were eventually overcome. Biot presented a report on this expedition to the Institute in 1810. Meanwhile, in the company of Mathieu, Biot measured the length of the seconds pendulum at Bordeaux and at Dunkirk. In 1817, Biot took part in another expedition, this time to Scotland and the Shetland Islands in order to confirm the geodesic work that had recently been undertaken by the British under Colonel Mudge. In 1818, Biot was again in Dunkirk; in 1824 and 1825 he went to Italy and Sicily with his son, and then revisited Formentera and Barcelona to correct his earlier geodesic measurements. From a comparison of his determinations, Biot concluded that the weight of a given body is not the same on all points with the same latitude, nor is its variation uniform along a particular meridian. This work established the necessity of revising the generally accepted simple ellipsoid theory of the earth.

During Biot's visit to Spain in 1806 and 1807 for geodesic work, he carried out other experimental work which is not generally known. He made a special study of the composition of the air contained in the swim bladders of fish found off the islands of Ibiza and Formentera. He can claim credit for recording the extremely high proportion of oxygen in the swim bladders of certain fish which live at great depths. He found a maximum of 87 percent oxygen, a figure that agrees well with the modern value. Another series of experiments that he carried out in the Mediterranean on the compression of gases (published in 1809) is of some theoretical interest. He lowered mixtures of gases, in appropriate proportions, to great depths to see whether they would combine to form the corresponding compounds. Combination did not take place up to pressures of about thirty atmospheres, and he was able to conclude that even when the pressure was increased thirty times and the distance between the molecules was correspondingly reduced, the molecules were still too far apart to exercise their chemical affinity.

In 1804 Biot carried out an experimental investigation of the conductivity of metal bars by maintaining one end at a known high temperature and taking readings of thermometers placed in holes along the bar. He was able to report the significant result that the steady-state temperature decreased exponentially along the length of the bar. He saw that this could be explained in terms of a balance of loss of heat at the surface and transfer of heat along the bar, which he analyzed in terms of adjacent pairs of cross-sectional areas. Unfortunately he was unable to present the differential equation corresponding to his physical model because of his inability to find plausible physical reasons for dividing a second difference of temperatures by the square of the infinitesimal element of length. Hence he could not convert his second difference into a second derivative. This was later achieved by Fourier. (I owe the above analysis of Biot's work on conductivity to Dr. J. R. Ravetz.)

In 1813 Biot attempted to derive a general formula for the expansion of liquids, and in 1815 he made a critical examination of Newton's law of cooling. Biot's friend Delaroche had already shown that at high temperatures heat losses were greater than the simple proportionality suggested by Newton. Biot proposed the equation

$$t = aT + bT^3,$$

where t represents the heat loss, T is the difference in temperature between the hot body and its surroundings, and a and b are constants. While considering Biot's early work on heat, it will be convenient to mention two later contributions. Biot proposed a general formula for the pressure, p, of a saturated vapor:

$$\log p = a + b\alpha^\theta + c\beta^\theta,$$

where a, b, c, α, and β are determined by means of five experiments and θ is the temperature measured from a convenient zero, such as the lowest temperature in the five experiments used to determine the constants. Biot's formula was a considerable improvement in generality and precision over the earlier formula of Delaroche. Biot also derived a formula (occasionally referred to as Biot's law) relating the intensity of solar radiation to the thickness of the atmosphere. If I is intensity of radiation of incident beam and I' is intensity of radiation transmitted through a thickness, t, of a medium whose coefficient of absorption is k, then

$$I' = Ie^{-kt}.$$

In 1800, the announcement of the voltaic pile aroused general interest; and when Volta came to Paris in 1801, the official report of the committee appointed by the First Class to examine Volta's work was edited and presented by Biot. On 14 August 1801, Biot read a memoir to the First Class describing his study of the "movement of the galvanic fluid," based on the hypothesis of Laplace that it consisted of mutually repellent particles. In collaboration with Cuvier, Biot investigated the chemistry of the voltaic pile. Expanding the work of W. H. Pepys, they confirmed that the voltaic pile in action absorbed oxygen, which they measured. They found that as long as any oxygen remained to be absorbed, the voltaic pile was still active, but with decreasing intensity. Nevertheless, the pile continued to function in the exhausted receiver of an air pump.

In the first edition of his physics textbook, Biot adopted a theory of electrolytic decomposition that was substantially that of Grotthus, but he later suggested that the liquid undergoing decomposition is most positive at the positive pole and most negative at the negative pole. When a particle of a salt is decomposed at the negative pole, the latter communicates a strong negative charge to the acid part. By repulsion from the surrounding negatively charged particles and by attraction toward the positive pole, it moves toward the latter. Only at the poles does decomposition take place. Biot made a brief study of the distribution of electricity on the surface of irregular spheroids, an extension of earlier work by Laplace. By analysis, he also established that when a Leyden jar is discharged by successive contacts, the losses of electricity form a geometrical progression.

News of Oersted's discovery of the connection between magnetism and electricity was brought to Paris by Arago in September 1820. Immediately the Paris scientists, including Ampère, began to explore the subject. Biot was away at the time, but on his return he was said to be working day and night to make up for lost time. He presented the result of his research with Savart to the Academy on 30 October 1820. They had measured the rate of oscillation of a suspended magnet placed at various distances from a conductor carrying a current. They were thus able to show that the magnetic force acts at right angles to the perpendicular joining the point considered to the conductor, and that its intensity is inversely proportional to the distance (Biot and Savart's law).

After this review of Biot's miscellaneous contributions to science, we must turn to the field in which he did his most important work—the study of polarization of light, the research for which Biot was awarded the Rumford Medal in 1840 by the Royal Society of London. The polarization of light by reflection had been discovered by Malus in Paris in the fall of 1808. This was of fundamental importance in the history of optics, since it showed that a phenomenon that had previously been observed in a few crystalline substances, such as Iceland spar, was a general property of light. Malus's discovery opened up an entirely new field of research, and no one was stimulated more than his two associates in the Arcueil group, Arago and Biot. In August 1811, Arago announced that he had found that white light polarized by reflection could, on passing through certain crystals, be split into two differently colored beams.

Biot repeated Arago's experiments and established the relationship between the thicknesses of the crystal plates and the colors produced. He observed that for perpendicular incidence, the colors seen correspond to those seen by reflection and transmission in thin films of air; and he concluded that the thicknesses at which the colors appeared were proportional to the thickness of the air gap that gave the same color on Newton's scale. These thicknesses depended on the nature of the crystal, but were always much greater than the thicknesses of thin films of air that gave the same tint. Biot found that the colors disappeared if the plate was extremely thin, and there was also an upper limit—for example, no colors would be seen

if the thickness of a plate of gypsum was greater than 0.45 mm. In this research the exact measurement of the plate was of the utmost importance, and Biot was fortunate in being able to use the spherometer, newly invented by Cauchoix. He began his research by taking eleven plates of gypsum, varying in thickness from 0.087 mm. to 0.345 mm. He determined the color produced by each and compared it with the color on Newton's scale. For oblique incidence, Biot found that the color depended on the thickness of the crystal traversed by the refracted ray and varied as the square of the sine of the angle that the direction of the ray formed with the optical axis.

Biot's interpretation of his results was in terms of a repulsive force that caused polarization by acting on the particles of light. This conception was first worked out in detail in a memoir presented to the First Class on 30 November 1812. The discovery of polarization had greatly encouraged Laplace, Biot, and others who supported a corpuscular theory of light. Malus had been successful in deriving the fundamental cosine law of polarization on such a model. To explain the complementary polarization in crystalline plates, Biot developed a theory of "mobile polarization." The particles of a polarized ray were supposed to preserve their original polarization until they reached a certain depth in the crystal, when they began to oscillate around their center of gravity so that the axes of polarization were carried alternately to each side of the axes of the crystal. The period was considered to vary with the color (as in Newton's theory of fits). When the ray emerged from the crystal, oscillation stopped, and the ray assumed "fixed polarization," in which the axes of the particles were arranged in two perpendicular directions. The theory was plausible up to a point, but Biot had considerable difficulty in accounting for the difference in the effect of thin and thick plates on polarized light. In 1841 Biot considered that he had found a new phenomenon of polarization, which was dependent on the existence of different layers in the crystal and which he called lamellar polarization.

In 1812 Biot observed that the rotation of polarized light produced by a plate of quartz decreased progressively with change of color from violet to red. In a paper read to the Academy on 22 September 1818, Biot was able to announce what has become known as Biot's law of rotatory dispersion and would now be expressed by the equation

$$\alpha = \frac{k}{\lambda^2},$$

where α is the rotation and λ is the wavelength. For Biot, however, it was "la loi de rotation réciproque aux quarrés des longueurs des accès" (the law of rotation in inverse proportion to the squares of the lengths of the fits)—a reminder that Biot did not accept a wave theory, but followed Newton's theory of "fits." Biot also found that the amount of rotation is proportional to the thickness of the crystalline plate traversed by the ray and that the rotation effected by two plates is the algebraic sum of the rotations produced by each separately.

Biot deduced his law without the use of monochromatic light, and his wavelength values given in the graph (Fig. 1) are Newton's values for the boundaries of different colors.

The horizontal axis in Biot's graph represents the square of the wavelength of light, and the vertical axis denotes the thickness of the plates of quartz required to produce rotations of 180°, 360°, 540°, etc., in light of a given color. Biot found that the same

law apparently applied equally to liquids. Later, when trying to compensate levorotatory turpentine against dextrorotatory oil of lemon, Biot observed that exact compensation of all rays was not possible. The amendment to the expression of Biot's law that this implied was not achieved until much later (Drude, 1898; Lowry and Dickson, 1913).

In 1814, Biot found that in certain crystalline substances the refractive index was less for the ordinary ray than for the extraordinary ray, unlike the standard doubly refracting substance calcite. Huygens had explained the formation of the extraordinary ray in calcite by the construction of an oblate spheroid. Biot modified this construction, drawing a prolate spheroid to describe the new phenomenon.

Tourmaline was known to be a doubly refracting substance. In 1815 Biot found that a plate of a certain thickness of tourmaline crystal cut parallel to the axis

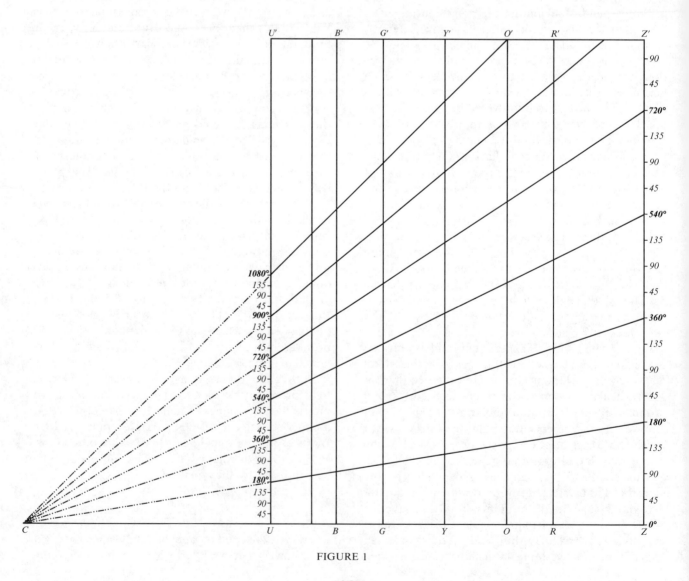

FIGURE 1

had the property of transmitting only light polarized in one plane. If light was allowed to pass through two plates and the second was rotated until it was perpendicular to the first, the light would gradually be extinguished. Biot's difficulty in explaining the action of polarized light on certain crystals such as calcium sulfate was overcome only when Brewster distinguished biaxial crystals from uniaxial ones. Nevertheless, in 1818 Biot did clearly distinguish a uniaxial form of magnesia mica from the more common biaxial types, and his work is commemorated in the name biotite given to a type of mica by J. F. L. Hausmann in 1847.

Until 1815 it had been assumed that only in the solid state did substances have the effect of rotating the plane of polarized light. It was Biot who discovered that certain crystalline solids, which had no effect on polarized light, did have an effect when in solution. On 23 October 1815, Biot announced to the First Class that the property of rotating the plane of polarized light was shared by liquids. Turpentine placed in a long tube with plane glass at each end exhibited a similar property, although to a less marked degree. Within a week he found a similar effect with oil of laurel and oil of lemon. Biot appreciated the immense importance of his discovery—that since this was a property of liquids, it must be a property of the molecules. To confirm this, he demonstrated that the effect on polarized light applied equally to turpentine in the liquid and vapor states.

In 1811 Arago found that in plates of quartz, the colors polarized along the axis were different from those he had studied in other crystals. When they were analyzed through a prism of Iceland spar, he found that the two images had complementary colors and changed through the spectrum as the prism was rotated. Biot repeated these experiments and found that while in some quartz crystals the tints descended in the scale of colors by turning the analyzing prism from left to right, in others the same effect was obtained by turning the prism from right to left. He thus distinguished what he called right-handed and left-handed quartz. Biot found that liquids also had opposite effects on polarized light. If liquids that rotated the plane of polarization in opposite directions were mixed in suitable proportions, the effect was canceled out. For this effect, Biot introduced the term *compensé* ("compensated"). He found that the rotation of the plane of polarization for a given liquid was proportional to its concentration or, with a given concentration, to the length of the tube containing the solution. It was Biot who introduced the practice of denoting the effect of polarized light on a liquid or solution by the value of the rotation produced by a column of standard length. Biot's major contribution to instrumentation in polarimetry was to design one of the first polariscopes.

In 1816 Biot suggested that the equal effect of polarized light on respective solutions of cane sugar and beet sugar constituted an additional proof of their identity. After 1820 he put his optical research aside, and for the next twelve years his work was mainly in astronomy and electricity. From 1832 he resumed his optical research with renewed vigor, going back to his earlier work and carrying out comparative tests on sugars. In 1833, working with Persoz, Biot found that when cane sugar was heated with dilute sulfuric acid, a chemical change took place; this was revealed by the solution's rotating the plane of polarized light to the left instead of to the right. He described the effect as "inversion," a term which is still used. In further collaboration with Persoz, Biot studied the conversion of starch by dilute acids into sugar and a gum which, from its effect on polarized light, they named dextrine. Biot introduced the polarimetric method of quantitative estimation of sugar remaining in molasses.

In 1832 Biot recorded the property of a solution of tartaric acid of rotating polarized light, and he remarked on the anomalous dispersion it gave, the rotation being greater for "less refrangible rays." In 1836 he presented to the Academy a memoir devoted entirely to the study of the rotatory power of tartaric acid under different conditions. He stressed that tartaric acid constituted an outstanding exception to his law of inverse squares (of wavelength). Biot accordingly divided optically active substances into two classes, those that obeyed his law and those that did not. He observed the crystalline forms of some salts of tartaric acid, but it was left to Pasteur to show the relationship between the crystalline form and the effect on polarized light. Biot was, significantly, an ardent champion of Pasteur at the beginning of his career. Pasteur, for his part, felt that Biot's work on the rotation of polarized light by liquids constituted a valuable scientific tool that had been hitherto unjustly neglected by chemists.

BIBLIOGRAPHY

I. Original Works. *The Royal Society Catalogue of Scientific Periodicals* lists some three hundred memoirs by Biot published in scientific periodicals. It also lists fifteen memoirs written with collaborators.

A few of Biot's most significant memoirs are "Mémoire sur la propagation de la chaleur, et sur un moyen simple et exact de mesurer les hautes températures," in *Journal des mines,* **17** (1804), 203–224; "Mémoire sur les affinités des corps pour la lumière, et, particulièrement sur les forces réfringentes des différents gaz" (written with Arago), in *Mémoires de l'Institut,* **7** (1806), 301–385; "Mémoire sur l'analyse comparée de l'arragonite et du carbonate de chaux rhomboïdal" (written with Thenard), in *Mémoires de la Société d' Arcueil,* **2** (1809), 176–206; "Mémoire sur des nouveaux rapports qui existent entre la réflexion et la polarisation de la lumière par les corps cristallisés," in *Mémoires de l'Institut* (1811), 135–280; "Sur une loi remarquable qui s'observe dans les oscillations des particules lumineuses, lorsquelles traversent obliquement des lames minces de chaux sulfatée ou de cristal de roche, taillées parallèlement à l'axe de cristallisation," in *Société philomatique, Bulletin* (1815), 149–156; "Phénomènes de polarisation successive, observés dans des fluides homogènes," in *Société philomatique, Bulletin* (1815), 190–192; "Mémoire sur les rotations que certaines substances impriment aux axes de polarisation des rayons lumineux," in *Mémoires de l'Académie des sciences de l'Institut,* **2** (1817), 41–136; "Note sur le magnétisme de la pile de Volta" (written with Savart), in *Annales de chimie et de physique,* **15** (1820), 222–223; "Mémoire sur les modifications que la fécule et la gomme subissent sous l'influence des acides" (written with Persoz), in *Annales de chimie et de physique,* **52** (1833), 72–90; "Mémoire sur la polarisation circulaire et sur ses applications à la chimie organique," in *Mémoires de l'Académie des sciences de l'Institut,* **13** (1835), 39–175; "Méthodes mathématiques et expérimentales, pour discerner les mélanges et les combinaisons définies ou non définies qui agissent sur la lumière polarisée; suivies d'applications aux combinaisons de l'acide tartrique avec l'eau, l'alcool, et l'esprit de bois," in *Mémoires de l'Académie des sciences de l'Institut,* **15** (1838), 93–279.

The following are the principal books by Biot: *Analyse du Traité de mécanique céleste de P. S. Laplace* (Paris, 1801); *Traité analytique des courbes et des surfaces du second degré* (Paris, 1802; 8th ed., 1834), translated by Frances H. Smith as *An Elementary Treatise on Analytical Geometry; Translated From the French and Adapted to the Present State of Mathematical Instruction in the Colleges of the United States* (New York–London, 1840); *Traité élémentaire d'astronomie physique,* 2 vols. (Paris, 1802; 3rd ed., 10 vols., 1841–1857); *Essai sur l'histoire générale des sciences pendant la Révolution française* (Paris, 1803); *Recherches sur les réfractions extraordinaires qui ont lieu près de l'horizon* (Paris, 1810); *Recherches expérimentales et mathématiques sur les mouvements des molécules de la lumière autour de leur centre de gravité* (Paris, 1814); *Traité de physique,* 4 vols. (Paris, 1816); *Précis élémentaire de physique expérimentale,* 2 vols. (Paris, 1817; 3rd ed., 1824), translated as *J. B. Biot's . . . Anfangsgrunde der Erfahrungs—Naturlehre, aus dem französisch übersetzt von F. Wolff,* 2 vols. (Berlin, 1819); *Recueil d'observations géodésiques, astronomiques et physiques, executées par ordre du Bureau des Longitudes de France en Espagne, en France, en Angleterre et en Écosse pour déterminer la variation de la pesanteur et des degrés terrestres sur le prolongement du méridien de Paris . . . rédigé par MM Biot et Arago* (Paris, 1821); *Recherches sur plusieurs points de l'astronomie égyptienne, appliquées*

aux monuments astronomiques trouvés en Égypte (Paris, 1823); *Notions élémentaires de statique* (Paris, 1828); *Instructions pratiques sur l'observation et la mesure des propriétés optiques appelées rotatoires, avec l'exposé succinct de leur application à la chimie médicale, scientifique et industrielle* (Paris, 1845); *Mélanges scientifiques et littéraires,* 3 vols. (Paris, 1859); *Études sur l'astronomie indienne* (Paris, 1859); *Études sur l'astronomie indienne et sur l'astronomie chinoise* (Paris, 1862).

II. SECONDARY LITERATURE. There is an anonymous biography of Biot in American Academy of Arts and Sciences, *Proceedings,* **6** (1862–1865), 16–23. Other works on Biot are D. Brewster, "Optics," in *Encyclopaedia Britannica,* 8th ed. (Edinburgh, 1858), XVI; M. P. Crosland, *The Society of Arcueil. A View of French Science at the Time of Napoleon I* (Cambridge, Mass. 1967); F. Lefort, "Un savant chrétien. J. B. Biot," in *Le correspondant,* n.s. **36** (1867), 955–995; T. M. Lowry, "Optical Rotatory Dispersion. A Tribute to the Memory of Jean Baptiste Biot (1774–1862)," in *Nature,* **117** (1926), 271–275; E. Mach, *The Principles of Physical Optics* (New York, 1926), ch. X; P. F. de Mottelay, *Bibliographical History of Electricity and Magnetism* (London, 1922); C. E. Picard, "La vie et l'oeuvre de Jean Baptiste Biot," in *Éloges et discours académiques* (Paris, 1931), pp. 221–287; D. Sidersky, "Le centenaire du premier polarimètre," in Association des Chimistes, *Bulletin de l'Association des chimistes* (Jan.–Feb. 1940).

M. P. CROSLAND

BIRD, JOHN (*b.* Bishop Auckland, England, 1709; *d.* London, England, 31 March 1776), *mathematics, optics.*

Bird was an eminent maker of mathematical instruments; the precision of his products made possible considerable advancement of practical astronomy. Almost nothing is known of his early life, which was spent in northeast England. He worked as a weaver and developed an interest and a proficiency in metal engraving. He also divided and engraved clock dial plates, demonstrating considerable skill in this work. About 1740 Bird moved to London, where he was employed by the well-known maker of astronomical and navigational instruments, Jonathan Sisson. Shortly after his arrival in London, Bird devised an instrument for finding the latitude at sea. He soon came to the attention of George Graham, another maker of astronomical instruments and clocks. Graham, impressed with the accuracy and quality of his work, assisted Bird in establishing his own shop for the making of mathematical instruments in 1745.

The nature of the instruments first produced by Bird at his own shop is not precisely known, but among them were portable quadrants with radii ranging from ten to twenty-four inches, as well as transit instruments ranging in length from one foot to five feet.

Bird received his first major order in 1749. The astronomer royal, James Bradley, had applied to and received from George II a grant of £1,000 for the construction of new instruments for the Royal Observatory at Greenwich. Bradley thereupon commissioned Bird to produce a movable forty-inch-radius quadrant cast in brass. He labored three years over the production of this instrument, which required 2,000 screws and weighed nearly 800 kilograms. Bird provided against error due to temperature change, and when the instrument was completed in June 1750 it was found to be a mere .05 greater than 90°.

This achievement brought Bird considerable fame, and by means of the instrument Bradley was able to achieve numerous important astronomical observations. When the Imperial Academy of Science at St. Petersburg remodeled its observatory within the next several years, Bird was employed to provide the instruments, including an eight-foot quadrant. He produced similar instruments for the Royal Observatory at Paris and for the Naval Observatory at Cádiz. In 1754 Bird constructed a six-foot mural quadrant that was presented by the king to Tobias Mayer at the University of Göttingen. Other important instruments made by Bird were two eight-foot quadrants for Radcliffe Observatory at Oxford. In 1758 and 1759 he was commissioned to produce a standard yard measure for the committees of the House of Commons and an apparatus for determining capacity measures.

During this same period Bird was engaged in commercial production of reflecting telescopes, barometers, thermometers, octants, and drafting instruments. He produced the reflecting circles, designed by Tobias Mayer, that were tested by Captain John Campbell for the Admiralty in 1757–1759, and in 1773 he was called upon by the Board of Longitude to examine the sextant scale divided by Jesse Ramsden's new dividing engine.

Bird's methods for the accurate division of scientific instruments were made available to others in two published works in which he described his methods and tools.

Not only were Bird's instruments the most accurately graduated prior to the invention of the dividing engine, but by his example and his publications others were enabled to achieve greater precision in scientific instrumentation.

BIBLIOGRAPHY

I. ORIGINAL WORKS. Bird's published writings are *The Method of Dividing Astronomical Instruments* (London, 1767) and *The Method of Constructing Mural Quadrants* (London, 1768).

II. SECONDARY LITERATURE. Writings on Bird or his work are Jean Bernoulli, *Lettres astronomiques* (Berlin, 1771), pp. 107–108, 126–129; C. Doris Hellman, "John Bird (1709–1776) Mathematical Instrument-Maker in the Strand," in *Isis,* **17** (1932), 127–153; Thomas Hornsby, ed., *Astronomical Observations Made at the Royal Observatory at Greenwich, From the Year MDCCL, to the Year MDCCLXII, by the Rev. James Bradley,* I (Oxford, 1798), Preface, ii–iii, vii, xiv, xv; Henry C. King, *The History of the Telescope* (London, 1955), pp. 115–118; Pierre-Charles Le Monnier, "Description et usage des principaux instruments d'astronomie où l'on traite de leur stabilité, de leur fabrique, et de l'art de les diviser," in *Description des arts et métiers,* **42** (1774), 1–60, plates I–XIV; and W. Ludlam, *An Introduction and Notes, on Mr. Bird's Method of Dividing Astronomical Instruments, to Which Is Added a Vocabulary of English and French Astronomical Terms* (London, 1786).

SILVIO A. BEDINI

BIRGE, EDWARD ASAHEL (*b.* Troy, New York, 7 September 1851; *d.* Madison, Wisconsin, 9 June 1950), *limnology.*

Birge's father, Edward White Birge, was an English-born carpenter of limited means; his mother was Ann Stevens, of Troy, New York. After completing grammar school and high school in Troy, Birge attended Williams College from 1869 to 1873. There Mark Hopkins (philosophy), John Bascom (English literature), and Sanford Tenney (natural history) were his most influential teachers. Following Tenney's advice, Birge next studied zoology under Louis Agassiz at the Museum of Comparative Zoology in Cambridge, Massachusetts, during the fall of 1873.

After Agassiz's death in December 1873, Birge transferred to the Harvard Graduate School. In January 1876 he assumed the position of instructor in natural history and curator of cabinet at the University of Wisconsin, at the behest of John Bascom, then president of that institution. In 1878 Birge returned briefly to Harvard for his doctoral examination and to receive his doctorate. He became professor of biology at Wisconsin in 1879, and soon thereafter he married Anna Grant, a friend since childhood. Accompanied by her, he spent the year 1880–1881 at the University of Leipzig in postdoctoral study under the physiologist Ludwig and under Gaule, from whom he learned the newly devised paraffin sectioning technique, which he introduced to Wisconsin upon his return.

Birge spent the rest of his life at the University of Wisconsin, where he was professor and chairman of zoology (1879–1911), dean of the College of Letters and Science (1891–1918), acting president (1900–1903), president (1918–1925), and president emeritus (1925–1950). In addition he served the state of Wisconsin as commissioner of fisheries (1895–1915), director of the Geological and Natural History Survey (1879–1919), and member of the Conservation Commission (1908–1915). He received honorary degrees from Williams College; the universities of Pittsburgh, Wisconsin, and Missouri; and Rensselaer Polytechnic Institute.

At Wisconsin, Birge continued his undergraduate interest in water fleas by studying plankton crustaceans in Lake Mendota in relation to depth, light, temperature, and currents. This led him to the independent discovery of the thermocline (observed only shortly before in European lakes) and to a study of dissolved oxygen at different depths and times, its depletion by respiration of organisms, and its interchange with carbon dioxide at the water surface. Birge came to regard the lake as a kind of superorganism with a physiology of its own—respiring, metabolizing, and exchanging matter and energy with its environment.

In 1900 Birge brought Chancey Juday to Wisconsin and, greatly assisted by him, in 1911 published a major monograph on the dissolved gases in lakes and their biological significance. From this time on, Birge turned his attention more toward the physical factors in lakes, especially heat distribution and light penetration, while Juday specialized in lake chemistry and biological productivity. Their collaboration continued, and most of their work was published jointly.

From a study of the Finger Lakes of New York in 1910 and comparable investigations in Wisconsin, Birge worked out the principle of the "heat budget" of lakes; he also studied the thermal exchange between lake water and air in considerable detail, including the very significant role of wind in the process. Impressed by the importance of sunlight in the heat budget and in productivity, Birge next began a detailed study of light penetration into inland lake waters. With the help of physicists, he devised an instrument, which he named the pyrlimnometer, that was in principle a delicate thermocouple capable of recording (as heat) light delivered by the sun to different lake depths. Later, equipped with filters, it was used to study separately the penetration of light of different wavelengths into the water.

Upon Birge's retirement from the presidency of the University of Wisconsin, he and Juday established the Trout Lake Limnological Laboratory near Minocqua, in Vilas County, Wisconsin, a region of numerous lakes. Thus Birge was able to extend his studies of light penetration to many lakes with different intensities of water coloration. The results were published in a series of reports written with Juday and H. J. James between 1929 and 1933. Birge showed that the upper meter of water absorbs nearly all the infrared and a large part of the ultraviolet light. The depth of penetration of the remaining rays was shown to depend on the amount of suspended matter (e.g., algae and sediment) and dissolved stains (e.g., humic acids from bogs). The amount and kind of organic matter was presented in a comprehensive report in 1922.

Birge was the first great American limnologist. Ably assisted by Juday, he gave the initial and major impetus to the development of limnology in the United States, just as Forel, the founder of limnology, had done a generation before in Europe.

BIBLIOGRAPHY

I. ORIGINAL WORKS. Birge's writings include "The Inland Lakes of Wisconsin. The Dissolved Gases," in *Bulletin of the Wisconsin Geological and Natural History Survey*, science ser. 7, no. 22 (1911), written with Juday; "The Water Fleas (*Cladocera*)," in Ward and Whipple, *Fresh-Water Biology* (New York, 1918), pp. 676–740; and "The Inland Lakes of Wisconsin. The Plankton," in *Bulletin of the Wisconsin Geological and Natural History Survey*, science ser. 13, no. 64 (1922), written with Juday.

II. SECONDARY LITERATURE. John L. Brooks et al., "Edward Asahel Birge (1851–1950)," in *Archiv für Hydrobiologie*, **45** (1951), 235–243, includes a nearly complete bibliography of Birge's works; G. C. Sellery, *E. A. Birge* (Madison, Wis., 1956), contains an appraisal of Birge as a limnologist by C. H. Mortimer.

LOWELL E. NOLAND

BIRINGUCCIO, VANNOCCIO (*b.* Siena, Italy, 20 October 1480; *d.* Rome [?], Italy, *ca.* 1539), *metallurgy.*

The son of Lucrezia and Paolo Biringuccio, the latter an architect and public servant, Biringuccio traveled as a young man throughout Italy and Germany, inspecting metallurgical operations. After running an iron mine and forge at Boccheggiano for Pandolfo Petrucci, he was appointed to a post with the arsenal at Siena and in 1513 directed the mint. In 1516, after the fall of the Petrucci family, he was exiled by the Republic of Siena on a charge of having debased the coinage. Biringuccio returned with the Petruccis in 1523, and was again exiled in 1526. Thereafter he served the Venetian and Florentine republics, and cast cannon and built fortifications for the Este and Farnese families. In 1531, with political peace, he returned once more to Siena, this time in honor, as senator and, succeeding Baldassare Peruzzi, as architect and director of building construction at the Duomo. He later moved to Rome. In 1538 Biringuccio was appointed head of the papal foundry and director of papal munitions, but he died soon after, probably in Rome and certainly before 30 April 1540.

Biringuccio's reputation derives from a single work, his *Pirotechnia,* published posthumously in 1540. The work is divided into ten books, which deal with (1) metallic ores; (2) the "semiminerals" (including mercury, sulfur, alum, arsenic, vitriol, several pigments, gems, and glass); (3) assaying and preparing ores for smelting; (4) the parting of gold and silver, both with nitric acid and with antimony sulfide or sulfur; (5) alloys of gold, silver, copper, lead, and tin; (6) the art of casting large statues and guns; (7) furnaces and methods of melting metals; (8) the making of small castings; (9) miscellaneous pyrotechnical operations (including alchemy; the distillation of acids, alcohol, and other substances; the working of a mint "both honestly and with profit"; the goldsmith, silversmith, and ironsmith; the pewterer; wire-drawing; mirror-making; pottery; and bricks); and (10) the making of saltpeter, gunpowder, and fireworks for warfare and celebration. Virtually all of Biringuccio's descriptions are original. He is important in art history for his description of the peculiarly Renaissance arts of casting medallions, statues, statuettes, and bells. His account of typecasting, given in considerable detail, is the earliest known. The *Pirotechnia* contains eighty-three woodcuts, the most useful being those depicting furnaces for distillation, bellows mechanisms, and devices for boring cannon and drawing wire.

As the first comprehensive account of the fire-using arts to be printed, the *Pirotechnia* is a prime source on many practical aspects of inorganic chemistry. Biringuccio emphasizes the adaptation of minerals and metals to use—their alloying, working, and especially the art of casting, of which he writes in great detail. In this area he is far better than the two other sixteenth-century authors with whom he is inevitably compared, Georgius Agricola and Lazarus Ercker. Although Agricola excels on mining and smelting, his famed sections on glass, steel, and the purification of salts by crystallization are in fact taken nearly verbatim from the *Pirotechnia.*

Biringuccio's approach is in strong conflict with that of the alchemists, whose work he evaluates in eleven pages of almost modern criticism, distinguishing their practical achievements from their theoretical motivations. His interest in theoretical questions is limited to the repetition of an essentially Aristotelian view of the origins of metallic ores and the nature of metals, with a rather forced extension to account for the observed increase in weight of lead when it is turned to litharge.

Biringuccio has been called one of the principal exponents of the experimental method, for he states that "It is necessary to find the true method by doing it again and again, continually varying the procedure and then stopping at the best" and "I have no knowledge other than what I have seen with my own eyes." He gives quantitative information wherever appropriate. He was certain that the failure of an operation was due to ignorance or carelessness, not to either ill luck or occult influences: Fortune could be made to favor the foundryman by paying careful attention to details. Biringuccio's method, however, is not that of the scientist, for none of his operations is planned to test theory or even reflects the conscious application of it. He represents the strain of practical chemistry that had to develop and to be merged with philosophy before it could become science. Yet the enjoyment of the diverse properties of matter and the careful recording of a large number of substances and types of reactions that had been established by various craftsmen were just as necessary as the works of the philosophers, and in some sense were nearer the truth.

BIBLIOGRAPHY

I. ORIGINAL WORKS. Biringuccio's only work was *De la pirotechnia. Libri .X. dove ampiamente si tratta non solo di ogni sorte & diuersita di miniere, ma anchora quanto si ricerca intorno à la prattica di quelle cose di quel che si appartiene à l'arte de la fusione ouer gitto de metalli come d'ogni altra cosa simile à questa* (Venice, 1540; repr. 1550, 1558, 1559; Bologna, 1678). Books I and II only were reprinted with an important introduction by A. Mieli (Bari, 1914). There is a French translation by Jacques Vincent (Paris, 1556; repr. 1572, 1627); a German translation by Otto Johannsen (Brunswick, 1925); and an English translation by C. S. Smith and M. T. Gnudi (New York, 1942; repr. 1943, 1959; Cambridge, Mass., 1966).

II. SECONDARY LITERATURE. Icilio Guareschi, *Enciclopedia di chimica,* XX (Turin, 1903–1904), supplemento annuale, 419 ff.; Aldo Mieli, "Vannoccio Biringuccio e il metodo sperimentale," in *Isis,* **2** (1914), 90–99; and "Vannoccio Biringuccio," in *Gli scienziati italiani dall'inizio del medio evo ai nostri giorni,* I (Rome, 1921), pt. 1; Otto Johannsen, "Vannoccio Biringuccio," in Günther Bugge, ed., *Das Buch der grossen Chemiker,* I (Berlin, 1929), 70–84; M. T. Gnudi and C. S. Smith, *Of Typecasting in the Sixteenth Century* (New Haven, 1941). See also the introductions to the 1914 Italian edition and to the German and English translations listed above.

C. S. SMITH

BIRKHOFF, GEORGE DAVID (*b.* Overisel, Michigan, 21 March 1884; *d.* Cambridge, Massachusetts, 12 November 1944), *mathematics.*

The son of a physician, Birkhoff studied at the Lewis Institute (now Illinois Institute of Technology) in Chicago from 1896 to 1902. After a year at the University of Chicago he went to Harvard, where he received the A.B. in 1905. Returning to the University of Chicago, he was awarded the Ph.D., *summa cum laude,* in 1907. His graduate study at the University of Chicago was followed by two years as an instructor in the University of Wisconsin. In 1908 he was married to Margaret Elizabeth Grafius.

In 1909 Birkhoff went to Princeton as a preceptor and in 1911 was promoted to a full professorship in response to a call from Harvard. The following year he accepted an assistant professorship at Harvard, where he became professor in 1919, Perkins professor in 1932, and dean of the Faculty of Arts and Sciences from 1935 to 1939. As Perkins professor, the major part of his academic life was devoted to mathematical research and direction of graduate students.

Birkhoff was very generally regarded, both in the United States and abroad, as the leading American mathematician of his day. Honors came early in life and from all over the world. He was president of the American Mathematical Society in 1925 and of the American Association for the Advancement of Science in 1937.

Of Birkhoff's teachers, Maxime Bôcher of Harvard and E. H. Moore of the University of Chicago undoubtedly influenced him most. He was introduced by Bôcher to classical analysis and algebra. From Moore he learned of "general analysis." There are indications that Birkhoff preferred the approach of Bôcher to that of Moore. Through his reading, Birkhoff made Henri Poincaré his teacher and took over Poincaré's problems in differential equations and celestial mechanics. Like Moore, Birkhoff was a pioneer among those who felt that American mathematics had come of age.

He had many close friends among his colleagues in Europe. With Hadamard he shared a deep interest and understanding of Poincaré. Neils Nörlund and Birkhoff had common ground in their study of difference equations. Between Levi-Città and Birkhoff there were deep ties of friendship cemented by their common interest in the problem of three bodies. The correspondence of Sir Edmund Whittaker and Birkhoff on the existence of periodic orbits in dynamics was intense, illuminating, and friendly.

Birkhoff's thesis was concerned with asymptotic expansions, boundary-value problems, and the Sturm-Liouville theory. Nonself-adjoint operators

$$L(z) = \frac{d^n z}{dx^n} + * + p_2(x)\frac{d^{n-2} z}{dx^{n-2}} + \cdots + p_n(x)z$$

$$(a \leq x \leq b)$$

were introduced with continuous coefficients and n boundary conditions, $U_i(u) = 0$, $i = 1, \cdots, n$, linear and homogeneous in u and its first n-1 derivatives at $x = a$ and $x = b$. Birkhoff defined an operator $M(z)$ adjoint to $L(z)$, and boundary conditions $V_i(v) = 0$, $i = 1, \cdots, n$ adjoint to the conditions $U_i(u) = 0$. For $n > 2$ he introduced a parameter λ, as in the classical Sturm-Liouville equations, and, with suitable conditions on the matrix of coefficients in the boundary conditions, obtained an expansion of a prescribed real function $x \to f(x)$, "piecewise" of class C^1. This expansion was shown to converge essentially as does the classical Fourier expansion. This work of depth admitted extension both by Birkhoff and by such pupils as Rudolph Langer and Marshall Stone; he collaborated with Langer on "The Boundary Problems and Developments . . ." (1923).

Birkhoff next devoted his attention to linear differential equations, difference equations, and the generalized Riemann problem. With Gauss, Riemann, and Poincaré showing the way, second-order differential equations of the Fuchsian type with regular singular points have become central in conformal mapping, in the theory of automorphic functions, and in mathematical physics, including quantum mechanics. Linear differential systems with irregular singular points appeared as a challenging new field, and Birkhoff turned to it.

Thomé had used formal solutions; Poincaré and Jakob Horn, asymptotic expansions; Hilbert and Josef Plemelj, unknown to Birkhoff, had solved one of the relevant matrix problems; and Ebenezer Cunningham had generalized Poincaré's use of Laplace's transformation. It remained for Birkhoff to formulate a program of so vast a scope that it is still an object of study today.

Among analytic systems with a finite number of irregular singular points with prescribed ranks, Birkhoff defined a "canonical system" and a notion of the "equivalence" of singular points. Under the title "generalized Riemann problem," Birkhoff sought to construct a system of linear differential equations of the first order with prescribed singular points and a given monodromy group. That he carried his program as far as he did is remarkable. The total resources of modern function space analysis are now involved.

Carmichael's thesis, done under Birkhoff's supervision at Princeton in 1911, was perhaps the first significant contribution on difference equations in America. Birkhoff extended his notion of a "generalized Riemann problem" to systems of difference equations. In "Analytic Theory of Singular Difference Equations," he collaborated with Trjitzinsky in an extension and modification of earlier work.

Birkhoff's major interest in analysis was in dynamical systems. He wished to extend the work of Poincaré, particularly in celestial mechanics. One can divide his dynamics into formal and nonformal dynamics. The nonformal portion includes the metrical and topological aspects.

Birkhoff was concerned with a real, analytic Hamiltonian or Pfaffian system. A periodic orbit gives rise, after a simple transformation, to a "generalized equilibrium point" at which the "equations of variation" are independent of t. First-order formal stability at such a point requires that the characteristic multipliers at the point be purely imaginary. Formal trigonometric stability is then defined. It is a major result of Birkhoff's work that under the limitations on generality presupposed by Poincaré, first-order formal stability at a generalized equilibrium point implies formal trigonometric stability.

Possibly the most dramatic event in Birkhoff's mathematical life came when he proved Poincaré's "last geometric theorem." In "Sur un théorème de géométrie" (1912), Poincaré had enunciated a theorem of great importance for the restricted problem of three bodies, acknowledging his inability to prove this theorem except in special cases. The young Birkhoff formulated this theorem in "Proof of Poincaré's Geometric Theorem" (1913, p. 14):

> Let us suppose that a continuous one-to-one transformation T takes the ring R, formed by concentric circles C_a and C_b of radii a and b ($a > b > 0$), into itself in such a way as to advance the points of C_a in a positive sense, and the points of C_b in a negative sense, and at the same time preserve areas. *Then there are at least two invariant points.*

Birkhoff's proof of this theorem was one of the most exciting mathematical events of the era.

In 1912, in "Quelques théorèmes sur le mouvement des systèmes dynamiques," Birkhoff introduced his novel conceptions of minimal or recurrent sets of motions and established their existence under general conditions. This was the beginning of a new era in the theory of dynamical systems. Birkhoff continued by introducing the concepts of wandering, central, and transitive motions.

Metric transitivity, as defined by Birkhoff and Paul Smith in "Structure Analysis of Surface Transformations" (1928), requires that the only sets that are invariant under the "flow" in phase space be sets of measure zero or measure of the space. Metric transitivity implies topological transitivity (i.e., the existence of a transitive motion). Great problems abound and are today the object of research. On a compact regular analytic manifold it is not known, even today, whether topological transitivity implies metric transitivity.

From these concepts of Birkhoff's the main body of modern dynamics has emerged, together with such branches as symbolic dynamics and topological dynamics. Other concepts of Birkhoff's, his minimax principle and his theorem on the fixed points of surface transformations, have motivated some of the greatest advances in global analysis and topology.

One of Birkhoff's theorems of major current interest in his "ergodic theorem." Following an idea of Bernard Koopman, Von Neumann established his "mean ergodic theorem" in 1931. Stimulated by these ideas, Birkhoff presented his famous "pointwise ergodic theorem." As formulated by Khintchine, Birkhoff's theorem takes the form "The space M is assumed to have a finite measure m invariant under the flow. Let f be integrable over M and let P be a point of M. Then

$$\lim_{T \to \infty} \frac{\int_0^T f(P_t)\, dt}{T}$$

exists for almost all P on M."

Birkhoff thought critically for many years about the foundations of relativity and quantum mechanics. His philosophical and scientific ideas found vivid expression in "Electricity as a Fluid" (1938), where he described a "perfect fluid" that he proposed as a model from which to deduce the observed spectrum of hydrogen without postulating "energy levels." In "El concepto matemático . . ." (1944) he formulated a theory of gravitation in flat space-time, and deduced from it the three "crucial effects." Both of these models were consistent with special relativity; both avoided the general curvilinear coordinates basic to Einstein's general relativity but always considered by Birkhoff to be unnecessary and difficult to interpret experimentally.

Although Birkhoff's physical models may be controversial, his original critiques and interpretations are stimulating and illuminating.

Birkhoff wrote on many subjects besides those of his major works; for example, he devised a significant formula for the ways of coloring a map. At sixteen, he began a correspondence with H. S. Vandiver, who was eighteen, on number theory. A significant paper resulted in 1904.

Another paper, written in collaboration with his colleague Oliver Kellogg (1922), was one of the openers of the age of function spaces. Schauder and Leray acknowledged this paper as an inspiration for their later, more powerful theorem.

In 1929 Birkhoff and Ralph Beatley joined in writing a textbook on elementary geometry, which they called "basic geometry." After a period of revision and development, the pedagogical conceptions of this book have been widely adopted in current teaching of high school geometry.

Birkhoff's lifelong interest in music and the arts culminated in his book *Aesthetic Measure* (1933), in preparation for which he had spent a year traveling around the world, observing objects of art, ornaments, tiles, and vases, and recording impressions of music and poetry.

BIBLIOGRAPHY

I. ORIGINAL WORKS. Among Birkhoff's works are "On the Integral Divisors of $a^n - b^n$," in *Annals of Mathematics,* **5** (1904), 173–180, written with H. S. Vandiver; "On the Asymptotic Character of the Solutions of Certain Differential Equations Containing a Parameter," in *Transactions of the American Mathematical Society,* **9** (1908), 219–231; "Boundary Values and Expansion Problems of Ordinary Linear Differential Equations," *ibid.,* 373–395; "Quelques théorèmes sur le mouvement des systèmes dynamiques," in *Bulletin de la Société mathématique de France,* **40** (1912), 305–323; "Proof of Poincaré's Geometric Theorem," in *Transactions of the American Mathematical Society,* **14** (1913), 14–22; "Invariant Points in Function Space," *ibid.,* **23** (1922), 96–115, written with O. D. Kellogg; "The Boundary Problems and Developments Associated With a System of Ordinary Linear Differential Equations of First Order," in *Proceedings of the American Academy of Arts and Sciences,* **58** (1923), 49–128, written with R. E. Langer; "Structure Analysis of Surface Transformations," in *Journal de mathématiques pures et appliquées,* 9th ser., **7** (1928), 345–379, written with P. A. Smith; "On the Number of Ways of Coloring a Map," in *Proceedings of the Edinburgh Mathematical Society,* 2nd ser., **2** (1930), 83–91; *Aesthetic Measure* (Cambridge, Mass., 1933); "Analytic Theory of Singular Difference Equations," in *Acta mathematica,* **60** (1933), 1–89, written with W. J. Trjitzinsky; "Electricity as a Fluid," in *Journal of the Franklin Institute,* **226** (1938), 315–325; *Basic Geometry* (1940), written with Ralph Beatley; "El concepto matemático de tiempo y la gravitación," in *Boletín de la Sociedad matemática mexicana,* **1** (1944), 1–24; and *Dynamical Systems,* rev. ed. (Providence, R.I., 1966), with introduction, bibliography, and footnotes by Jürgen Moser.

His works have been brought together by the American Mathematical Society as *Collected Mathematical Works of George David Birkhoff,* 3 vols. (Providence, R.I., 1950).

II. SECONDARY LITERATURE. Works on Birkhoff are American Mathematical Society Semicentennial Publications, I, *History* (Providence, R.I., 1938), p. 212, a list of his honors up to 1938; P. Masani, "On a Result of G. D. Birkhoff on Linear Differential Systems," in *Proceedings of the American Mathematical Society,* **10** (1959), 696–698; and H. L. Turrittin, "Reduction of Ordinary Differential Equations to the Birkhoff Canonical Form," in *Transactions of the American Mathematical Society,* **107** (1963), 485–507.

MARSTON MORSE

BIRMINGHAM, JOHN (*b.* probably at Millbrook, near Tuam, Ireland, *ca.* 1816; *d.* Millbrook, 7 September 1884), *astronomy.*

Birmingham was a country gentleman and amateur astronomer who first became prominent when, on 12 May 1866, while walking home from a friend's house, he noticed in Corona Borealis a new star of the second magnitude, later termed T Coronae. This nova was the brightest since that of 1604 and the first to be identified with an existing star: it had been listed in the *Bonn Durchmusterung* as of magnitude 9.5, and by the beginning of June 1866 it had returned to the ninth magnitude. It was also the first nova to be subjected to spectroscopic examination, and William Huggins' visual and spectroscopic observations showed that it consisted of a star surrounded by a shell of hydrogen. T Coronae is remarkable for the fluctuations in the decline of its brightness and for its recurrence in 1946.

In 1872, at the suggestion of T. W. Webb, Birmingham undertook the revision of the catalog of red stars assembled in 1866 by H. C. F. C. Schjellerup. This task occupied him for four years. His catalog of 658 red stars, supplemented by numerous spectroscopic observations, was presented to the Royal Irish Academy on 26 June 1876, and was recognized by the award of the Academy's Cunningham Medal in 1884. On his deathbed Birmingham requested that Webb produce a revision of the catalog, but the task was undertaken by T. E. Espin and completed by him in 1888.

In the 1870's Birmingham published a number of papers on the members of the solar system, especially on features of the moon and of Jupiter. He was also a man of many parts: musician, linguist, antiquarian, poet, and the author of several geological papers. At the time of his death he was an inspector of applications for loans under the Land Law (Ireland) Act. He never married.

BIBLIOGRAPHY

I. ORIGINAL WORKS. Birmingham's most substantial publication is "Catalogue of Red Stars," in *Transactions of the Royal Irish Academy,* **26** (1879), 249–354; rev. by T. E. Espin, in Royal Irish Academy's *Cunningham Memoirs,* no. 5 (Dublin, 1890). Minor astronomical papers published by Birmingham between 1869 and his death are scattered among a number of English-language astronomical and scientific journals. For details see the Royal Society's *Catalogue of Scientific Papers,* I (1867), 388; VII (1877), 178; IX (1891), 246–247; XIII (1914), 567; J. C. Houzeau and A. Lancaster, *Bibliographie générale de l'astronomie jusqu'en 1880,* new ed., II (London, 1964); and Poggendorff, III (1898), 133. Details of his discovery of T Coronae are given in *Monthly Notices of the Royal Astronomical Society,* **26,** no. 8 (1866), 310.

II. SECONDARY LITERATURE. Birmingham has attracted almost no attention. The main sources are two unsatisfactory items in the (now extinct) *Tuam News* (12 Sept. 1884). More accessible are the biographical sketch by Agnes Clerke in *Dictionary of National Biography,* V, 85–86; and the obituary notice in *Astronomische Nachrichten,* **110,** no. 2632 (1885), 255.

M. A. HOSKIN

BIRT, WILLIAM RADCLIFF (*b.* Southwark [London], England, 15 July 1804; *d.* Leytonstone, England, 14 December 1881), *astronomy.*

Birt's capacity for the measurement and analysis of observational data, exhibited in his early studies of the brightness fluctuations of the stars β-Lyra and α-Cassiopeia, which he submitted for publication in the *Memoirs of the Royal Astronomical Society,* so impressed the society's president, Sir John Herschel, that from 1839 to 1843 he employed Birt as his assistant in the arrangement and reduction of numerous series of barometric measurements. In the course of this work, Birt discovered large fluctuations in the readings that lent strong support to Herschel's view that well-defined atmospheric waves were produced by contrary winds blowing across Britain and western Europe.

The results of Birt's subsequent research on this and other meteorological phenomena, including electrical measurements made at the Kew Observatory, are in the annual reports of the British Association (1844–1849) and in a series of articles in the *Philosophical Magazine* (1846–1850). By this time Birt had become convinced that the height of the column of mercury in the barometer was a reliable index for forecasting the occurrence of storms and for obtaining one's position relative to the center of a "revolving storm" (cyclone). The manner in which such data can be used by ships' captains to provide working rules for steering away from the storm center is described in his *Handbook on the Laws of Storms* (1853).

Shortly after being elected a fellow of the Royal Astronomical Society on 14 January 1859, Birt began making systematic observations of sunspots, solar rotation, and lunar markings. It is for his work in selenography—he was the Selenographical Society's first president—which he pursued between 1861 and 1866 at Dr. John Lee's observatory at Hartwell, Bed-

fordshire, and thereafter (with the aid of his own seven-and-one-half-inch equatorial reflector) at his small private observatory in Waltham's Town, Essex, that he is now best remembered. As secretary of the Lunar Committee for Mapping the Surface of the Moon, set up by the British Association in 1865 to revise and supplement Beer and Mädler's lunar map, he wrote the annual reports up to 1869, introducing in them his own notation for the identification of such small features as craterlets, mountains, and rills.

As an aid to classifying the brightness of the lunar markings, Birt used a scale of lunar tints consisting of twenty-four shades of a single pigment—his homochromoscope—which he describes at the end of his monograph *Mare Serenitatis* (1869). His careful comparison and measurement of photograms and numerous telescopic observations by other leading selenographic experts strengthened his previous conviction that there was a "secular variation of tint" on the floor of the crater Plato, such as might have been caused by eruptive action or chemical activity on the moon. He was always very conscious, however, of the provisional nature of his conclusions, and later expressed the belief that any speculations on such physical changes would require the support of terrestrial analogies from chemistry, geology, and mineralogy.

After 1873 his health began to fail, and in 1877 he stopped observing altogether. Two years later he presented to the Royal Astronomical Society twelve manuscript volumes containing the portion of the lunar catalog that had been completed. He died, after a further rapid decline in his already poor health, at the age of seventy-seven.

BIBLIOGRAPHY

Among Birt's works are *Handbook on the Laws of Storms* (Liverpool, 1853) and *Mare Serenitatis* (London, 1869). Most of his publications are in the *Monthly Notices of the Royal Astronomical Society* (1859–1872), the *British Association Reports* (1859–1870), and the *Philosophical Magazine* (1846–1880). A detailed list, with an indication of each work's contents, is in Poggendorff, III (1898), 134. In addition, there are ninety-five letters from Birt to Sir John Herschel among the latter's unpublished correspondence at the Royal Society, London.

An obituary notice is in *Monthly Notices of the Royal Astronomical Society,* **42** (1882), 142–144.

ERIC G. FORBES

AL-BĪRŪNĪ (or **Bērūnī**), **ABŪ RAYHĀN** (or **Abu'l-Rayhān**) **MUHAMMAD IBN AHMAD** (*b.* Khwārazm [now Kara-Kalpakskaya A.S.S.R.], 4 September

973; *d.* Ghazna [?] [now Ghazni, Afghanistan], after 1050), *astronomy, mathematics, geography, history.*

Bīrūnī was born and grew up in the region south of the Aral Sea, known in ancient and medieval times as Khwārazm. The town of his birth now bears his name. The site was in the environs (*bīrūn,* hence his appellation) of Kāth, then one of the two principal cities of the region, located (in the modern Kara-Kalpakskaya A.S.S.R.) on the right bank of the Amu Dar'ya (the ancient Oxus) and northeast of Khīva. The second capital city of Khwārazm was Jurjāniyya (modern Kunya-Urgench, Turkmen S.S.R.), on the opposite side of the river and northwest of Khīva. There also Abū Rayhān spent a good deal of time during the early part of his life. About his ancestry and childhood nothing is known. In verses ridiculing a certain poet (Yāqūt, p. 189; trans., *Beiträge,* LX, p. 62) he claims ignorance of his own father's identity, but the statement may have been rhetorical. He very early commenced scientific studies and was taught by the eminent Khwārazmian astronomer and mathematician Abū Nasr Mansūr. At the age of seventeen he used a ring graduated in halves of a degree to observe the meridian solar altitude at Kāth, thus inferring its terrestrial latitude (*Tahdīd,* 249:7). Four years later he had made plans to carry out a series of such determinations and had prepared a ring fifteen cubits in diameter, together with supplementary equipment. There was, however, time only for an observation of the summer solstice of 995, made at a village south of Kāth and across the Oxus from it. At this time, civil war broke out. Bīrūnī went into hiding and shortly had to flee the country (*Tahdīd,* 87:3, 109:6–110:11). "After I had barely settled down for a few years," he writes, "I was permitted by the Lord of Time to go back home, but I was compelled to participate in worldly affairs, which excited the envy of fools, but which made the wise pity me."

Since these "worldly affairs" essentially affected not only Bīrūnī's personal well-being but also his scientific work, it is necessary to introduce the names of six princely dynasties with which he became directly involved.

(1) The ancient title of Khwārazmshāh had long been held by the lord of Kāth, a member of the Banū 'Irāq. Abū Nasr was a prince of this house (Krause, p. 3). In 995, however, the emir of Jurjāniyya attacked his suzerain, captured and killed him, and seized the title for himself (*Chahār Maqāla,* p. 241). It was this disturbance that caused Bīrūnī's flight.

(2) For well over a century the Khwārazmshāhs had been dominated by the Sāmānids, a royal house of Zoroastrian origin but early converted to Islam. The Sāmānid capital was in Bukhara, about two

hundred miles southeast of Khīva, from whence the dynasty ruled in its heyday an area comprising roughly all of present Afghanistan, Transoxiana, and Iran. In Bīrūnī's youth this empire was rapidly breaking up. Nevertheless, in a poem written much later (Yāqūt, p. 187; trans., *Beiträge,* LX, p. 61) he names as his first patron Mansūr II, almost the last of the Sāmānid line, who reigned from 997 to 999.

(3) Much farther to the west flourished the Buwayhid dynasty, which had originated in the highlands south of the Caspian and extended its domain south to the Persian Gulf and, by 945, west over Mesopotamia.

(4) Set precariously between the Sāmānids and the Buwayhids was the Ziyārid state, based in Gurgān, a city just back of the southeast corner of the Caspian shore.

(5) All these competing dynasties were menaced, and eventually absorbed, by the swift expansion of another kingdom, that of the Ghaznavids, named from Ghazna, their base in east-central Afghanistan. Sultan Mahmūd, son of a Turkish slave and the second and greatest of the line, was two years older than Bīrūnī. By 1020 he had carved out a realm extending a thousand miles north and south, and twice as far east and west.

(6) Over these kaleidoscopic shifts there presided at Baghdad the spectral figure of the Abbasid caliph, retaining only the shadow of power over these fragments of his ancestors' empire. Playing a role somewhat analogous to that of the medieval popes, he was accorded a strange religious respect by the temporal princes of Islam. Upon them the successive caliphs conferred prestige by investing them with honorific titles and robes of honor.

To which or from which of these kingdoms Bīrūnī fled in 995 is now uncertain. It may have been then that he went to Rayy, near modern Teheran. In the *Chronology* (p. 338) he quotes a ribald poem on the tribulations of penury, and to illustrate it states that he was once in Rayy, bereft of a royal patron and in miserable circumstances. A local astrologer chose to ridicule his views on some technical matter because of his poverty. Later, when his circumstances improved, the same man became friendly.

At the command of the Buwayhid prince, Fakhr al-Dawla, the astronomer al-Khujandī built a large mural sextant on a mountain above Rayy. With this Fakhrī sextant, named for the ruler, he observed meridian transits during 994. Bīrūnī wrote a treatise describing this instrument (*Sextant*) and a detailed account of the observations (*Tahdīd,* 101:20–108:19). Part of his information was obtained from al-Khujandī in person, and since the latter died about

1000 (Suter, p. 74), the conversation between the two cannot have been long after the observations.

There is some reason for thinking that Abū Rayḥān also was in the Caspian province of Gīlān about this time. He dedicated a book (*RG* 7) to the Ispahbad (Persian for "ruler," or "commander") of Gīlān, Marzubān ibn Rustam, who was connected with the Ziyārids. In the *Chronology,* completed about 1000 (trans., pp. 47, 191), he mentions having been in the presence of this individual, perhaps the same Ispahbad who sheltered Firdawsī, the epic poet of Iran, from the wrath of Sulṭān Maḥmūd (Browne, pp. 79, 135).

Regardless of where he had been, Bīrūnī was back in Kāth by 997, for on 24 May of that year he observed a lunar eclipse there (Oppolzer 3403), having previously arranged with Abu'l-Wafā' that the latter should simultaneously observe the same event from Baghdad (*Taḥdīd,* 250:11, gives only the year; but Oppolzer 3404, on 17 November 997, was invisible from both cities). The time difference so obtained enabled them to calculate the difference in longitude between the two stations.

This year saw the beginning of the short reign of the Sāmānid Manṣūr II. If Bīrūnī ever resided at his court in Bukhara (as Bīrūnī's poem mentioned above may imply), it probably was at this time. Meantime, the ruler of Gurgān, the Ziyārid Qābūs, had been expelled from his lands, and at Bukhara he sought support for a return to power. He succeeded in reestablishing himself at Gurgān, and Bīrūnī either accompanied him or followed almost immediately thereafter, for about 1000 Bīrūnī dedicated to Qābūs his earliest extant major work, the *Chronology* (text, p. xxiv). This was by no means his first book, for in it he refers incidentally to seven others already completed, none of which are extant. Their titles indicate that he had already broken ground in the fields he later continued to cultivate, for one (*RG* 34) is on decimal computation, one (*RG* 46) on the astrolabe, one (*RG* 146) on astronomical observations, three (*RG* 42, 99, 148) on astrology, and two (*RG* 161, 162) are histories. By this time he also had engaged in an acrimonious correspondence with the brilliant Bukharan philosopher and physician Avicenna on the nature and transmission of heat and light. Bīrūnī refers to him (*Chronology,* text, p. 257) as "the youth." The appellation, coming from an individual still in his twenties, may seem less condescending when it is realized that the precocious Avicenna was still in his teens.

In the *Taḥdīd* (214:15–215:3), after describing the measurement of a degree along a terrestrial meridian made at the direction of the Caliph Ma'mūn, Bīrūnī writes of his own abortive project to repeat the operation. A suitable tract of land was chosen between Gurgān and the land of the Oghuz Turks (in the deserts east of the Caspian?), but the patron, presumably Qābūs, lost interest.

The end of Abū Rayḥān's sojourn at the Ziyārid court can be fixed within precise limits, for in 1003 he observed two lunar eclipses from Gurgān, one on 19 February and the other on 14 August. On 4 June of the following year he observed a third lunar eclipse (*Canon,* pp. 740, 741), but this one from Jurjāniyya. Hence, sometime in the interim he had returned to his homeland, high in favor with the reigning Khwārazmshāh. This was now a certain Abu'l-'Abbās Ma'mūn, a son of the usurper to the title mentioned above. Both Ma'mun and a brother who preceded him on the throne had married sisters of the ever more powerful and truculent Sultan Maḥmūd of Ghazna.

The bounty of the shah enabled Bīrūnī to set up at Jurjāniyya an instrument, apparently a large ring fixed in the meridian plane, which in gratitude he called the Shāhiyya ring (*Canon,* 612:5). He reports in various places in the *Taḥdīd* and the *Canon* some fifteen solar meridian transit observations at Jurjāniyya, the first the summer solstice of 7 June 1016, the last on 7 December of the same year. It was probably during this interlude of prosperity and royal favor that he had a hemisphere constructed, ten cubits in diameter, to be used as a plotting device for the graphical solution of geodetic problems (*Taḥdīd,* 38:6).

Meanwhile, Khwārazmian political affairs, in which Bīrūnī was closely involved, had been building up to a climax. The Caliph Qādir conferred upon Ma'mūn an honorific title and dispatched an envoy bearing the insignia of the award. The shah was frightened lest Maḥmūd take offense at his accepting the honor conferred directly and not through Maḥmūd as implied overlord. Ma'mūn therefore sent Bīrūnī west into the desert to intercept the embassy, take delivery of the objects, and thus forestall a public investiture.

In 1014 Maḥmūd let it be understood to Ma'mūn that he wanted his own name inserted into the *khuṭba,* the Friday prayer for the faithful and for the reigning monarch. Ma'mūn convened an assembly of the notables, proposing that he accede to this demand, but the chiefs refused to allow him to do so, realizing that it meant the end of the region's autonomy. Ma'mūn then sent to them Bīrūnī, who, "with tongue of silver and of gold," convinced them that their liege was only testing them by his request and that the *khuṭba* would not be changed. At this,

Maḥmūd dispatched an insulting ultimatum to the shah, demanding that he keep his nobles in line, or he, Maḥmūd, would do it himself. The hapless Ma'mūn introduced the sultan's name into the *khuṭba* in the provincial mosques, but not those of Jurjāniyya and Kāth. Thereupon the Khwārazmian army revolted and killed Ma'mūn. This was all Maḥmūd needed. He marched into Khwārazm with ample forces, obtained the delivery of his sister, the Khwārazmshāh's widow, took Kāth, on 3 July 1017, cruelly executed the insurgent leaders, and set one of his officers on the throne. The surviving princes of the local dynasty were carried off to imprisonment in various parts of his domain (Barthold, pp. 275–279).

Much of our knowledge of these events is from Bīrūnī's extensive history of his native land, a work that has been lost except for fragments incorporated into other histories. As for Abū Rayḥān himself, he also was led off by the conqueror, partly, no doubt, to grace the sultan's court but also to remove an active partisan of the native rulers from the scene. He is next heard of in a village near Kabul, depressed and in miserable circumstances, but hard at work on the *Taḥdīd* (119:1–12). On 14 October 1018 he wanted to take the solar altitude, but had no instrument. He therefore laid out a graduated arc on the back of a calculating board (*takht*) and, with a plumb line, used it as an improvised quadrant. On the basis of the results obtained, he calculated the latitude of the locality.

The next firm date at our disposal is 8 April 1019, when he observed a solar eclipse from Lamghān (modern Laghman?), north of Kabul. He uses this, and the lunar eclipse mentioned below, to comment sarcastically upon the ignorance of the local astronomers.

Sachau has shown (*India*, trans., I, xi) that Bīrūnī's relations with Maḥmūd were never good, although the stories in the *Chahār Maqāla* (text, pp. 57–59) alleging cruel and arbitrary treatment of the savant by the sultan are doubtless apocryphal. It is evident that Abū Rayḥān received some sort of official support for his work, for in the *Canon* (p. 609) he writes of having determined the latitude of Ghazna by a series of observations carried out between 1018 and 1020 with an instrument he calls the Yamīnī ring. A title bestowed upon Sultan Maḥmūd by the caliph was Yamīn al-Dawla ("Right Hand of the State"). No doubt this ring was a monumental installation named, as was the custom, for the ruler patron.

It is also clear that Bīrūnī's interests in Sanskrit and in Indian civilization are due to his having become an involuntary resident of an empire that had by then expanded well into the Indian subcontinent. Already in 1002 Maḥmūd had conquered the district of Waihand, on the Indus east of Ghazna. By 1010 he had subjugated Multan and Bhatinda, the latter 300 miles east of the Indus. Twice repulsed (in 1015 and 1021) from the borders of Kashmir, by 1022 he had penetrated and subdued the Ganges valley to a point not far west of Benares. In 1026 Maḥmūd led a raid due south from Ghazna all the way to the Indian Ocean. From Somnāth, at the tip of the Kathiawar Peninsula, he carried off immensely valuable booty, as well as fragments of the phallic idol in the temple. One of the pieces was laid at the entrance to the Ghazna mosque, to be used as a footscraper by the worshipers (*India*, trans., II, 103; Nāzim, ch. 8).

Abū Rayḥān profited from these events by travel and residence in various parts of India. The names of many of the places he saw are known, but no dates can be given for his visits. They were confined to the Punjab and the borders of Kashmir. Sachau (*India*, text, p. xii) lists some eleven Indian towns whose latitudes Bīrūnī reports as personally determined by him. Bīrūnī himself writes that while living (in detention?) at Nandana Fort, he used a nearby mountain to estimate the earth's diameter (*Taḥdīd*, 222:10). The installation at Nandana, taken by Maḥmūd in 1014, commanded the route by which he, the Moghuls after him, and Alexander the Great long before, penetrated the Indus valley. Bīrūnī's temporary residence overlooked the site where, in the face of King Poros and his elephants, Alexander effected his famous crossing of the Jhelum River, the classical Hydaspes (Stein).

It is also clear that Bīrūnī spent a great deal of time at Ghazna. The cluster of recorded observations made by him there commences with a series of meridian solar transits covering the summer solstice of 1019, and includes the lunar eclipse on 16 September of the same year (*Taḥdīd*, 291:9). He continued to observe equinoxes and solstices at Ghazna, the last being the winter solstice of 1021. In fact, this is the latest of Bīrūnī's observations that has been preserved. At about this time, according to Barani (*Canon*, III, vii), he completed his treatise on *Shadows*.

In 1024 the ruler of the Volga Turks sent an embassy to Ghazna. These people had trade relations with inhabitants of the polar regions, and Bīrūnī questioned members of the mission to supplement his knowledge of these lands. One of the ambassadors asserted in the sultan's presence that in the far north the sun sometimes did not set for days on end. Maḥmūd at first angrily put this down as heresy, but Abū Rayḥān convinced him that the report was both credible and reasonable (*Commemoration Volume*, p. 235; Yāqūt).

By the late summer of 1027 the treatise on *Chords* was completed (according to the Patna MS). During the same year a Chinese and Uighur Turkish embassy came to Ghazna, and from this mission Bīrūnī obtained geographical information on the Far East which he later incorporated into the *Canon* (*Commemoration Volume*, p. 234).

In 1030 Sultan Maḥmūd died, and the succession was disputed between two of his sons for a short period. Bīrūnī finished the *India* during this interim and, perhaps because of the uncertain political situation, refrained from dedicating it to any particular patron. Within the year Masʿūd, the elder son, won the crown. His accession brought about a drastic improvement in the situation of his most famous scientist, and Bīrūnī named the *Canon* for the new ruler amid "a farrago of high-sounding words" in the preface (*India*, trans., I, xii).

Perhaps it was the change of regime that enabled him to revisit his native land. By whatever means, he made at least one trip back, for in the *Bibliography* he writes that for over forty years he had sought a certain Manichaean work, a copy of which he at length procured while in Khwārazm (*Chronology*, text, p. xxxvi). In the same source Bīrūnī relates that after he was fifty years old he suffered from a series of serious illnesses, and in his distress inquired of several astrologers concerning the length of his life. Their answers diverged wildly, and some were patently absurd. At the end of his sixty-first (lunar?) year he began improving, and had a dream in which he was seeking the new moon. As its crescent disappeared, a voice told him that he would behold 170 more of the same.

Masʿūd was murdered by his officers and succeeded by his son Mawdūd in 1040. During Mawdūd's eight-year reign, Bīrūnī wrote the *Dastūr* (*RG* 167) and the *Gems*. Of his subsequent activities we have no knowledge, save that in the *Pharmacology* (p. 7) he notes having passed his eightieth (lunar?) year; his eyesight and hearing are failing, but he is still hard at work with the assistance of a collaborator. Thus the date of his death given by Ghaḍanfar as 13 December 1048 is incorrect; Bīrūnī outlasted his third Ghaznavid patron and achieved the life-span foretold in his dream.

When he was sixty-three years old, Bīrūnī prepared a bibliography of the works of the physician Muḥammad ibn Zakariyya al-Rāzī, to which he appended a list of his own books. This runs to 113 titles (not counting twenty-five additional treatises written "in his name" by friends), partially arranged by subject matter and occasionally with a brief indication of the contents. Most of the entries also give the

length of the particular manuscript in folios. The list is incomplete, for Abū Rayḥān lived at least fourteen years after this, working until he died. Moreover, seven additional works by him are extant and many more are named, some in his own writings and others in a variety of sources. All told, these come to 146. The reckoning is uncertain, for some titles counted separately may be synonyms, and additional items may well turn up in the future.

There is a wide range in size of the treatises. Several amount to only ten folios each, while, at the other extreme, three lost astronomical works run to 360, 550, and 600 folios respectively. Largest of all is the *India,* at 700 folios. The English translation of the latter, incidentally, takes up 654 pages of small type, so that one of Bīrūnī's folios is roughly equivalent to a modern printed page. The mean length of the seventy-nine books of known size is very nearly ninety folios. Assuming that the same holds for all 146 works, it follows that Bīrūnī's total output is on the order of 13,000 folios (or pages), consisting for the most part of highly technical material, including numerical tables, the results of involved computations, and analyses of materials from multifarious sources—a formidable accomplishment indeed.

The classification attempted in the table below is only approximate; for instance, a book placed in the geographical category could legitimately be classed as primarily geodetic, and so on. Practically nothing Bīrūnī wrote confines itself strictly to a single subject, and in many cases where the title alone survives, an informed guess is our only recourse. Nevertheless the table gives a reasonable breakdown of the man's activity. In the second column a "major work" has been taken arbitrarily as anything of 200 folios or more. The third and fourth columns show, respectively, the compositions known to exist in manuscript form and the numbers of these that have thus far been printed. Roughly four-fifths of Bīrūnī's work has vanished beyond hope of recovery. Of what has survived, about half has been published. Most of the latter (with the notable exception of the *Canon*) has been translated into other languages and has received some attention from modern scholars.

The table also clearly reveals both scope and areas of concentration. Bīrūnī's interests were very wide and deep, and he labored in almost all the branches of science known in his time. He was not ignorant of philosophy and the speculative disciplines, but his bent was strongly toward the study of observable phenomena, in nature and in man. Within the sciences themselves he was attracted by those fields then susceptible of mathematical analysis. He did serious work in mineralogy, pharmacology, and philology, subjects where numbers played little part; but about half his total output is in astronomy, astrology, and related subjects, the exact sciences par excellence of those days. Mathematics in its own right came next, but it was invariably applied mathematics.

CLASSIFICATION OF BĪRŪNĪ'S WORKS

	Works		Major Works	Extant	Published
Astronomy	35		8	4	3
On astrolabes	4	62		2	
Astrology	23		1	3	2
Chronology	5		1	1	1
Time measurement	2				
Geography	9		1	1	1
Geodesy and		19			
Mapping Theory	10			1	
Mathematics					
Arithmetic	8			1	1
Geometry	5	15		1	1
Trigonometry	2			1	
Mechanics	2			1	
Medicine and					
Pharmacology	2		1	1	
Meteorology	1				
Mineralogy and					
Gems	2			1	1
History	4				
India	2		1	1	1
Religion and					
Philosophy	3			1	1
Literary	16				
Magic	2			1	
Unclassified	9		1	1	1
Total	146		14	22	13

Below are brief descriptions of most of Bīrūnī's works that are still available. They are our best sources for estimating the extent and significance of his accomplishments.

The *Chronology.* The day, being the most apparent and fundamental chronological unit, is the subject of the first chapter. Bīrūnī discusses the advantages of various calendric epochs—sunset or sunrise (horizon-based), noon or midnight (meridian-based) —and names the systems that use each. Next the several varieties of year are defined—lunar, solar, lunisolar, Julian, and Persian—and the notion of intercalation is introduced. Chapter 3 defines and discusses the eras of the Creation, the Flood, Nabonassar, Philip Arrhidaeus, Alexander, Augustus, Antoninus, Diocletian, the Hegira, Yazdigird, the Caliph Muʿtaḍid, the pre-Islamic Arabs, and Bīrūnī's native Khwārazm. Chapter 4 discusses the Alexander legend, giving sundry examples of pedigrees, forged

and otherwise. Next are lists of the month names, with variants, used by the Persians, Soghdians, Khwārazmians, Egyptians, Westerners (Spaniards?), Greeks, Jews, Syrians, pre-Islamic Arabs, Muslims, Indians, and Turks. In this chapter, the fifth, Bīrūnī commences his very extensive description of the Jewish calendar. (Except for the work of al-Khwārizmī, another Muslim, his is the earliest extant scientific discussion of this calendar.)

Chapter 6 culminates with a table (trans., p. 133) giving the intervals in days between each pair of the eras named above. This is preceded, however, by chronological and regnal tables in years (sometimes with months and days) for the Jewish patriarchs and kings; the Assyrians, Babylonians and Persians; the Pharaohs, Ptolemies, Caesars, and Byzantine emperors; the mythical Iranian kings; and the Achaemenid, Parthian, and Sasanian dynasties. Where tables from different sources conflict, all are given in full, and there are digressions on the length of human life and the enumeration of chessboard moves.

Chapter 7 continues the exhaustive discussion of the Jewish calendar, but includes a derivation of the solar parameters, a table of planetary names, and the Mujarrad table giving the initial weekdays of the mean (thirty-year cycle) lunar year.

Chapter 8 is on the religions of various pseudo prophets, the most prominent being the Sabians (or Mandaeans, alleged to be followers of Būdhāsaf = Bodhisattva!), Zoroastrians, Manichaeans, and adherents of Mazdak.

The remaining half of the book (save the last chapter) describes the festivals and fasts of the following peoples: Chapter 9, the Persians; 10, the Soghdians; 11 and 12, the Khwārazmians; 13, the Greeks (including material from Sinān ibn Thābit ibn Qurra on the parapegmatists); 14, the Jews; 15, the Melchite Christians; 16, the Jewish Passover and Christian Lent; 17, the Nestorian Christians; 18, the Magians and Sabians; 19, the pre-Islamic Arabs; 20, the Muslims. The concluding chapter, 21, gives tables and descriptive matter on the lunar mansions, followed by explanations of stereographic projection and other plane mappings of the sphere.

The *Astrolabe.* Amid the plethora of medieval treatises on the astrolabe, this is one of the few of real value. It describes in detail not only the construction of the standard astrolabe but also special tools used in the process. Numerical tables are given for laying out the families of circles engraved on the plates fitting into the instrument. Descriptions are also given of the numerous unusual types of astrolabes that had already been developed in Bīrūnī's time. As for the underlying theory, not only are the techniques and properties of the standard stereographic projection presented, but also those of certain nonstereographic and nonorthogonal mappings of the sphere upon the plane.

The *Sextant.* This two-page treatise describes the giant mural instrument for observing meridian transits built by al-Khujandī at Rayy for Fakhr al-Dawla, and perhaps seen by al-Bīrūnī, although he does not say so.

The *Taḥdīd.* The central theme is the determination of geographical coordinates of localities. In particular, Bīrūnī sets out to calculate the longitudinal difference between Baghdad and Ghazna. Several preliminary problems present themselves: latitude determinations, inclination of the ecliptic, the distribution of land masses and their formation, length of a degree along the terrestrial meridian, and differences in terrestrial longitudes from eclipse observations. Techniques and observations used by Bīrūnī and by others are reported. Application is made of a theorem of Ptolemy's that gives the longitudinal difference between two places in terms of the latitude of each and the great circle distance between them. The latter was estimated from caravan routes and lengths of stages. Successive computations then yield the differences in longitude between Baghdad, Rayy, Jurjāniyya, Balkh, and Ghazna, and likewise along a southern traverse including Shiraz and Zaranj. The final result is in error by only eighteen minutes of arc.

The *Densities.* By means of an ingenious form of balance exploiting Archimedes' principle, Bīrūnī worked out a technique for ascertaining the specific gravity of a solid of irregular shape. He reports very precise specific gravity determinations for eight metals, fifteen other solids (mostly precious or semiprecious stones), and six liquids.

The *Shadows.* As its full title indicates, this is a comprehensive presentation of all topics known to Bīrūnī to be connected with shadows. Of the total of thirty chapters, the first three contain philosophical notions about the nature of light, shade, and reflection. There are many citations from the Arabic poets descriptive of kinds of shadows.

Chapter 4 shows that the plane path traced in a day by the end point of a gnomon shadow is a conic. The next two chapters discuss the properties of shadows cast in light emanating from celestial objects. Chapters 7 and 8 define the shadow functions (tangent and cotangent) and explain the origins of the gnomon divisions used in various cultures: the Hellenistic 60, Indian 12, Muslim 7 or 6-1/2. The succeeding three chapters explain rules for converting between functions expressed in different gnomon

lengths and for conversions into the other trigonometric functions (sine, secant, and their cofunctions, together with *their* various parameters), and vice versa. Chapter 12 gives tangent-cotangent tables for the four standard gnomon lengths and discusses interpolation. The next two chapters explain how to engrave the shadow functions on astrolabes. There follows, in Chapter 15, a discussion of gnomon shadows cast on planes other than horizontal, and on curved surfaces. Chapters 16 and 17 consider the effect of solar declination and local latitude on the meridian shadow length. A number of nontrigonometric approximate Indian rules are given. Chapters 18–21 list a variety of meridian-determination methods (including one from the lost *Analemma* of the first-century B.C. Diodorus). Chapter 22 is on daylight length and rising times of the signs as functions of the local latitude and the season. Here and in the next two chapters (on determining the time of day from shadows) rules are reproduced from numerous Indian, Sasanian, and early Islamic documents, many no longer extant. Some early Muslim rules are in Arabic doggerel written in imitation of Sanskrit *slokas.* Chapters 25 and 26 define the time of the Muslim daily prayers, some in terms of shadow lengths. Chapter 27 shows that in many situations on the celestial sphere, Menelaus' theorem gives relations between shadow functions. The concluding three chapters describe Indian and early Islamic techniques for calculating terrestrial and celestial distances by the use of shadows.

The *Chords.* The book begins by stating the following theorem: A, B, and C, three points on a circle, are so situated that $AB > BC$. From D, the midpoint of arc AC, drop a perpendicular, DE, to the chord AB. Then the foot of the perpendicular bisects the broken line ABC. There follow a number of proofs of this theorem, attributed to sundry Greek and Islamic mathematicians, some otherwise unknown to the literature. A second theorem, that in the configuration above, $\overline{AD}^2 = \overline{AB} \times \overline{BC} + \overline{BD}^2$, is also followed by a long series of proofs. The same thing is done for the expression $\triangle ADC - \triangle ABC = \overline{DE} \times \overline{EB}$. Then comes a set of metric relations between chords, based on the foregoing and leading up to propositions useful for calculating a table of chords (or sines).

The *Patañjali.* Cast in the form of a series of questions put by a hermit student and the answers given by a sage, this book deals with such philosophical and mystical topics as liberation of the soul and its detachment from the external world, the attributes of God, the power of spirit over the body, and the composition of the universe.

The *Tafhīm.* A manual of instruction in astrology,

well over half of the book is taken up with preliminaries to the main subject. Persian and Arabic versions are extant, both apparently prepared by Bīrūnī himself. It is arranged in the form of questions and answers. There are five chapters in all, the first (thirty-three pages in the Persian edition) on geometry, ending with Menelaus' theorem on the sphere. The second (twenty-three pages) is on numbers, computation, and algebra. Chapter 3, the longest (229 pages), deals with geography, cosmology, and astronomy. From it a complete technical vocabulary may be obtained, as well as sets of numerical parameters, some of them uncommon. The next chapter (thirty-one pages) describes the astrolabe, its theory and application. Only the last chapter (223 pages) is on astrology as such, but it is complete and detailed.

The *India.* The book commences with a prefatory chapter in which the author states that the subject is difficult because Sanskrit is not easy; there are extreme differences between Indians and non-Indians; and Indian fear and distrust has been exacerbated by Muslim conquests. The book will not be polemical and, when appropriate, Indian customs and beliefs will be compared with cognate ones of the Greeks.

Chapters 2–8 are on religion and philosophy: the nature of God, the soul, matter, mysticism, paradise, and hell. Chapters 9, 10, and 11 describe, respectively, the Hindu castes, laws concerning marriage, and the construction of idols. Chapters 12, 13, and 14 are on categories of literature: sacred, grammatical, and astronomical. The latter gives a table of contents of the *Brāhmasphuṭasiddhānta.* Chapter 15 presents tables of metrological units and gives various approximations to the number π. The next two chapters are on Indian systems of writing, number names, chess rules, and superstitions. Chapter 18 is geographical; in particular, sixteen itineraries are given with the distances in *farsakhs* between successive stages. Chapters 19–30 present astronomical and cosmological nomenclature, legends, and theories. Chapter 31 cites the geodetic parameters used by various astronomers, and the latitudes (observed by Bīrūnī) of a number of Indian cities. Chapters 32–53 are on Indian notions of time, including detailed definitions of the hierarchies of enormous cycles—the *yugas, kalpas,* and so on—interspersed with accounts of sundry religious legends. Calendric procedures are given in great profusion. Chapters 54–59 are astronomical, dealing with the computation of mean planetary positions, the sizes and distances of the planets, heliacal risings, and eclipses. The remainder of the book is largely astrological, but includes chapters on rites, pilgrimages, diet, lawsuits, fasts, and festivals.

The *Ghurra.* This is an example of an Indian *karana,* a handbook enabling the user to solve all the standard astronomical problems of his time, with the emphasis on actual computation rather than on theory. Hence it resembles an Islamic *zīj* (astronomical handbook). Topics include calendric rules; length of daylight; determination of the astrological lords of the year, month, day, and hour; mean and true positions of the sun, moon, and planets; time of day; local latitude; solar and lunar eclipses; and visibility conditions for the moon and the planets. Bīrūnī has added worked-out examples, in particular, conversions from the Šaka calendar into the Hegira, Yazdigird, and Greek (so-called era of Alexander) calendars. Otherwise, he states, in his translation he has made no changes.

In general, the methods are those common to medieval Indian astronomy, but the parameters are not identical with any extant Sanskrit document. For instance, the radius of the defining circle for the sine function is 200 minutes, and the increment of arc, the *kardaja,* is ten degrees.

The *Canon.* This most comprehensive of Bīrūnī's extant astronomical works contains detailed numerical tables for solving all the standard problems of the medieval astronomer-astrologer. But it also has much more in the way of observation reports and derivations than the typical *zīj.* It is organized in eleven treatises (*maqāla*) that are further subdivided into chapters and sections.

Treatises 1 and 2 set forth and discuss general cosmological principles (that the earth and heavens are spherical, that the earth is stationary, etc.), units of time measurement, calendars, and regnal and chronological tables. This covers much of the ground gone over in the *Chronology,* but the chapter on the Indian calendar is additional.

Treatises 3 and 4 are on plane and spherical trigonometry respectively. There are tables of all the standard trigonometric functions, more extensive and precise than preceding or contemporary tables. Methods of solving many problems of spherical astronomy appear, together with tables of ancillary functions: oblique ascensions, declinations, and so on.

Treatise 5, on geodesy and mathematical geography, reworks much of the subject matter of the *Taḥdīd.* A table gives the geographical coordinates of localities.

Treatises 6 and 7 are on the sun and moon, respectively. Here (and with planetary theory farther on) the abstract models are essentially Ptolemaic, but many parameters are independently derived on the basis of all available observations (including Bīrūnī's own).

Treatise 8 treats of eclipse computations and the first visibility of the lunar crescent.

Treatise 9, on the fixed stars, includes a star table with 1,029 entries (cf. Ptolemy's 1,022). Magnitudes according to Ptolemy and to al-Ṣūfī are given.

The next treatise is on the planets, with tables and text for calculating longitudes, latitudes, stations, visibility, distances, and apparent diameters.

The concluding treatise is on astrological operations, describing various doctrines for calculating the astrological mansions, projection of the rays, the *taysīr,* the sectors (*niṭāqāt*), transits, and the curious cycles apparently developed by Abū Maʿshar.

The *Transits.* This book describes the various categories of astrological phenomena to which the term *mamarr* (transit or passage) was attached. One planet was said to transit another if it passed the other planet in celestial longitude, or celestial latitude, or in its relative distance from the earth. The notion seems to have been developed by astrologers using non-Ptolemaic astronomical doctrines described in documents no longer extant. Hence the main interest of the work is the assistance it gives toward the reconstruction of these lost Indian, Sasanian, and early Islamic theories.

The *Gems.* The work is organized in two parts, the first being on precious and semiprecious stones, the second on metals. Bīrūnī brings together material from Hellenistic, Roman, Syriac, Indian, and Islamic sources, supplemented by his own observations. In addition to descriptions of the physical properties of the various substances, there are very extensive etymological discussions of the technical terminology in many languages and dialects, and numerous illustrative quotations from Arabic poetry. The principal mines and sources of supply are cited. Relative weights of the metals with respect to gold are given, and there are tables showing the prices of pearls and emeralds as functions of size.

The *Pharmacology.* The book commences with an introduction in five chapters. The first presents an etymology for the Arabic word for druggist. The second gives technical terminology for categories of drugs. The next chapter is on the general theory of medicaments. In the fourth and fifth chapters Bīrūnī states his preference for Arabic over Persian as a language of science, and he names polyglot dictionaries available to him.

The main body of the work is an alphabetical listing of drugs comprising about 720 articles. For a typical entry the name of the substance is given in Arabic, Greek, Syriac, Persian, and an Indian language, and sometimes also in one or more less common languages or dialects: Hebrew, Khwārazmian, Tokharian,

Zabuli, and so on. There follows a full presentation of the Arabic variants and synonyms, liberally illustrated with quotations from the Arabic poets. The substance is described, its place or places of origin named, and its therapeutic properties given, although Bīrūnī disclaims medical competence on his own part. Sources are fully and critically mentioned.

Abū Rayḥān's dominant trait was a passion for objective knowledge. In pursuit of this he early began studying languages. His mother tongue was Khwārazmian, an Iranian language in which, he wrote, it would be as strange to encounter a scientific concept as to see a camel on a roof gutter (*mīzāb*) or a giraffe among thoroughbred horses (*ʿirāb,* an example of rhymed prose). Therefore he acquired a deep knowledge of both Arabic and Persian. The former, in spite of the ambiguity of its written characters, he esteemed a proper vehicle for the conveyance of science, whereas the latter he deemed fit only for the recital of bedtime stories (*al-asmār al-layliyya*) and legends of the kings (*al-akhbār al-kisrawiyya,* more rhymed prose; *Pharmacology,* p. 40). Of Greek, Syriac, and Hebrew he attained at least sufficient knowledge to use dictionaries in these languages. His command of Sanskrit, on the other hand, reached the point where, with the aid of *pandits,* he was able to translate several Indian scientific works into Arabic, and vice versa. He took obvious delight in Arabic poetry, composed verses himself, and liberally interlarded his writings with quotations from the classics.

Thus equipped, he made full use of all the documents that came to his hand (many of which have since disappeared), exercising a critical faculty that extended from the minutiae of textual emendations to the analysis of scientific theories. A strong sense of history permeates all his writings, making them prime sources for studying the work of his predecessors, as well as his own and that of his contemporaries.

Bīrūnī's pursuit of the truth was not confined to the written or spoken word. He had a strong penchant for firsthand investigation of natural phenomena, exercised at times under very trying circumstances. Along with this went an ingenuity in the devising of instruments and a flair for precision in observations. Because of this feeling for accuracy, and because of a well-founded fear of losing precision in the course of calculations, he tended to prefer observational methods that yielded direct results, as against techniques requiring extensive reduction by computation.

Speculation played a small role in his thinking; he was in full command of the best scientific theories of his time, but he was not profoundly original or a constructor of new theories. His attitude toward astrology has been debated. He spent a great deal of time in serious study of the subject, but Krause (p. 10) has collected passages in which Bīrūnī not only heaps ridicule upon ignorant or unscrupulous astrological practitioners, but indicates disbelief in the basic tenets of this pseudo science. Krause also reminds us that there were many centuries when the casting of horoscopes was the only way by which an astronomer could support himself in the exercise of his profession.

As for religion, Bīrūnī was doubtless a sincere Muslim, but there is no firm evidence of his having been an adherent of any particular sect within the faith. In the *Chronology* (trans., pp. 79, 326), written at the court of Qābūs, are passages that have been interpreted as betraying a Shīʿi (hence anti-Arab and pro-Persian) bent. On the other hand, the *Pharmacology,* compiled under Ghaznavid patronage, represents the author as an orthodox Sunnī. Probably these two situations reflect no more than the fact that the two patrons were Shīʿi and Sunnī, respectively. From time to time Bīrūnī inveighs harshly against various groups, but the criticism is of particular acts or attitudes, not of the group as such. Thus his strictures against the Arab conquerors of Khwārazm were called forth, not because they were Arab, or alien, but because they destroyed ancient books. Concerning the Christian doctrine of forgiveness he writes, "Upon my life, this is a noble philosophy, but the people of this world are not all philosophers. . . . And indeed, ever since Constantine the Victorious became a Christian, both sword and whip have ever been employed" (*India,* trans., II, 161).

In these, and in most matters, Bīrūnī had a remarkably open mind, but his tolerance was not extended to the dilettante, the fool, or the bigot. Upon such he exercised a broad and often crude sarcasm. Upon his showing an instrument for setting the times of prayer to a certain religious legalist, the latter objected that it had engraved upon it the names of the Byzantine months, and this constituted an imitation of the infidels. "The Byzantines also eat food," stated Abū Rayḥān. "Then do not imitate them in this!" and he ejected the fellow forthwith (*Shadows,* 37:9).

Such were the life, labors, and character of a man known to his contemporaries as the Master (*al-Ustādh*). Unknown in the medieval West, except perhaps by the garbled name Maître Aliboron, his name and fame have been secure in his own lands from his time until the present.

BIBLIOGRAPHY

The standard bibliographical work on Bīrūnī is D. J. Boilot, "L'oeuvre d'al-Beruni. Essai bibliographique," in *Mélanges de l'Institut dominicain d'études orientales,* **2** (1955), 161–256; and "Corrigenda et addenda," *ibid.,* **3** (1956), 391–396; no attempt has been made here to duplicate it. A good deal of material has, of course, appeared since it was published in 1955.

For points of view somewhat different from that expressed in the text, see Boilot's article on al-Bīrūnī in the new ed. of the *Encyclopaedia of Islam,* Krause's paper (cited below), and Sachau's prefaces to the text and to the translation of the *Chronology* and the *India. RG* stands for "Répertoire général," the numbered listing of Bīrūnī's works in Boilot.

I. ORIGINAL WORKS. Following are Bīrūnī's extant major works, listed alphabetically.

Astrolabe (*RG* 46). The Arabic title is *Kitāb fī istīʿāb al-wujūh fī ṣanʿat al-asṭurlāb.* Several MSS exist (see Boilot), but the text has not been published. Sections of it have, however, been translated and studied.

Bibliography (*RG* 168). Bīrūnī calls this *Risāla fī fihrist kutub Muḥammad b. Zakariyyāʾ al-Rāzī.* The text was published by Paul Kraus as *Épître de Bērūnī contenant le répertoire des Ouvrages de Muḥammad b. Zakarīyā ar-Rāzī* (Paris, 1936). The text of the part giving Bīrūnī's own bibliography appears in the text edition of the *Chronology,* pp. xxxviii–xxxxviiii. It is translated into German in Wiedemann's "Beiträge," LX.

Canon (*RG* 104). The Arabic text has been published as *al-Qānūn al-Masʿūdī* (Canon Masudicus), 3 vols. (Hyderabad-Dn., 1954–1956). References in the article are to page and line of the printed text, pagination of which is continuous, not commencing anew with each volume. A Russian translation, in preparation by P. G. Bulgakov, M. M. Rozhanskaya, and B. A. Rozenfeld, will be Vol. V of the *Selected Works.*

Chords (*RG* 64). There are three MS versions of this work: (1) Leiden Or. 513(5) = CCO 1012; (2) Bankipore Arabic MS 2468/42 = Patna 2,336,2519/40; (3) Murat Molla (Istanbul) 1396. The Leiden version has been published in translation and with a commentary, both by H. Suter, as "Das Buch der Auffindung der Sehnen im Kreise . . . ," in *Bibliotheca mathematica,* **11** (1910), 11–78. The text of version (2) has been published as the first of the four *Rasāʾil* (Arabic for *treatises*). This contains, however, extraneous material, part of which is probably not by Bīrūnī, and part probably a fragment of *RG* 11. Many topics in (2) and (3) are missing from (1), and those parts that are in common are in drastically different orders. Two recensions by Bīrūnī himself are indicated. See H. Hermelink, in *Zentralblatt für Mathematik und ihre Grenzgebiete,* **54** (1956), 3; and A. S. Saidan, in *Islamic Culture,* **34** (1960), 173–175. Many of the additional sections in (2) and (3) are described by E. S. Kennedy and Ahmad Muruwwa in *Journal of Near Eastern Studies,* **17** (1958), 112–121. A composite Arabic text based on (2) and (3) was published by A. S. Demerdash as *Istikhrāj al-awtār fiʾl-dāʾira* (Cairo, 1965). There is a Russian translation by C. A. Krasnova and L. A. Karpova, with commentary by B. A. Rosenfeld and C. A. Krasnova: *Iz istorii nauki i texniki v stranax Vostoka,* III (Moscow, 1963).

Chronology (*RG* 105). In Arabic this is *al-Āthār al-bāqiya min al-qurūn al-khāliya.* It was edited by E. Sachau as *Chronologie orientalischer Voelker von Albērūnī* (Leipzig, 1878, 1923; repr. Baghdad, 1963). The parts missing from Sachau's text are given by K. Garbers and J. Fück in J. Fück, ed., *Documenta Islamica inedita* (Berlin, 1952), pp. 45–98. It was translated into English by Sachau as *The Chronology of Ancient Nations* (London, 1879). The Russian translation by M. A. Sal'e, *Pamyatniki minuvshikh pokolenii,* is Vol. I of the *Selected Works* (Tashkent, 1957).

Densities (*RG* 63). This work's Arabic title is *Maqāla fi'l-nisab allatī bayn al-filizzāt wa'l-jawāhir fi'l-ḥajm* ("Treatise on the Ratios Between the Volumes of Metals and Jewels"). The text has never been published, but portions of it have been taken over by other authors and have been studied in modern times.

Gems (*RG* 156). Known as the *Kitāb al-jamāhir fī maʿrifat al-jawāhir,* this text was edited by F. Krenkow (Hyderabad-Dn., 1936). Krenkow also translated the text, but only the chapter on pearls has been published (see Boilot). There is, however, a translation by A. M. Belenskii—*Mineralogiya* (Moscow, 1963).

Ghurra. The *Ghurrat al-zījāt* is Bīrūnī's Arabic translation of the Sanskrit astronomical handbook called *Karaṇatilaka* (forehead caste mark of the Karaṇas), by one Vijayanandin or Vijaya Nanda. The original text is not extant, but a MS of the translation is in the Dargah Library of Pir Muhammad Shah, Ahmadabad. Portions of the Arabic text, with English translation, and a commentary were published in installments by Sayyid Samad Husain Rizvi in *Islamic Culture,* **37** (1963), 112–130, 223–245, and **39** (1965), 1–26, 137–180. Another text, translation, and commentary, by M. F. Qureshi, exist in typescript but have not been published.

India (*RG* 93). Also known as *Kitāb fī taḥqīq ma li'l-Hind . . .,* this was edited by E. Sachau (London, 1888). A later edition has been published by the Osmania Oriental Publications Bureau (Hyderabad-Dn., 1958). Translated by E. Sachau as *Al-Beruni's India,* 2 vols. (London, 1910). Translated into Russian by A. B. Khalidov and Y. N. Zavadovskii as Vol. II of *Selected Works* (Tashkent, 1963).

Patañjali (*RG* 98). Bīrūnī's Arabic translation of this Sanskrit work is extant only in an incomplete MS edited by H. Ritter as "Al-Bīrūnī's Übersetzung des Yoga-Sūtra des Patañjali," in *Oriens,* **9** (1956), 165–200. See Boilot.

Pharmacology (*RG* 158). The Arabic title of this is *Kitāb al-ṣaydala fi 'l-ṭibb.* There is no edition of the entire work. M. Meyerhof translated it into German, but of this only the introduction has been published, together with the corresponding part of the Arabic text and an extremely valuable foreword and commentary: "Das Vorwort zur Drogenkunde des Bērūnī," in *Quellen und Studien zur Geschichte der Naturwissenschaften,* **3** (1932), 157–208. A Russian translation, in preparation by U. I. Kazimov, will be Vol. IV of *Selected Works.*

Rasā 'ilu-l-Bīrūnī. This is the Arabic text of *RG* 64, 15, 45, and 38, published by Osmania Oriental Publications Bureau (Hyderabad-Dn., 1948).

Selected Works (*Izbrannye proizvedeniya*). Bīrūnī's extant works are being published in Russian by the Academy of Sciences of the Uzbek S.S.R. Volumes in print or in preparation are listed by individual titles.

Sextant (*RG* 169). The *Ḥikāyat al-ālāt al-musammāt al-suds al-fakhrī* ("Account of the Instrument Known as the Fakhrī Sextant") is MS 223, pp. 10–11, of the Univ. of St. Joseph, Beirut. It was edited by L. Cheikho in *Al-Mashriq,* **11** (1908), 68–69. With minor changes, this small treatise was copied without acknowledgment by Abu'l-Ḥasan al-Marrākushī as part of a larger work. Text and French translation appear in L. A. Sédillot, "Les instruments astronomiques des arabes," in *Mémoires . . . à l'Académie royale des inscriptions . . .,* 1st ser., **1** (1844), 202–206.

Shadows (*RG* 15). The text has been published as the second of the *Rasā 'il* with the title *Kitāb fī ifrād al-maqāl fī amr al-ẓilāl* ("The Exhaustive Treatise on Shadows"). An English translation has been made by E. S. Kennedy, but publication awaits completion of the commentary. References to the *Shadows* made in the article are to page and line of the published text.

Tafhīm (*RG* 73). This is the *Kitāb al-tafhīm li-awāʾil ṣināʿat al-tanjīm.* R. Ramsay Wright published an edition of the Arabic text with English translation as *The Book of Instruction in the Art of Astrology* (London, 1934). Bīrūnī's Persian version was published by Jalāl Humāʾi (Teheran, 1940).

Taḥdīd (*RG* 19). The Arabic title is *Taḥdīd nihāyāt al-amākin li-tashīḥ masāfāt al-masākin,* and the work is extant in the unique Istanbul MS Fatih 3386. The Arabic text was published by P. Bulgakov as a special number of the Arab League journal, *Majallat maʿhad al-makhṭūṭāt al-ʿarabiyya* (Cairo, 1962). Translated into Russian by P. G. Bulgakov as *Geodeziya,* Vol. III of *Selected Works* (Tashkent, 1966). An English translation by Jamil Ali is *The Determination of the Coordinates of Cities, al-Bīrūnī's Taḥdīd al-Amākin* (Beirut, 1967). References in the article to the *Taḥdīd* are to page and line of the published text.

Transits (*RG* 45). In Arabic this is *Tamhīd al-mustaqarr li-taḥqīq maʿnā al-mamarr* ("Smoothing the Basis for an Investigation of the Meaning of Transits"). The text has been published as the third of the *Rasāʾil.* A translation by Mohammad Saffouri and Adnan Ifran, with commentary by E. S. Kennedy, is *Al-Bīrūnī on Transits* (Beirut, 1959).

II. SECONDARY LITERATURE. Works referred to parenthetically in the text, by author and page, are W. Barthold, *Turkestan Down to the Mongol Invasion,* 2nd ed. (London, 1928); E. G. Browne, *A Literary History of Persia,* II (Cambridge, 1928); *Chahār Maqāla of Aḥmad ibn ʿAli an-Niẓāmī al-ʿArūḍī as-Samarqandī,* Mirza Muḥammad ibn ʿAbd'l-Wahhāb, ed. (Leiden–London, 1910); Iran Society, *Al-Bīrūnī Commemoration Volume, A. H. 362–A. H. 1362* (Calcutta, 1951); Max Krause, "Al-Biruni. Ein iranischer Forscher des Mittelalters," in *Der Islam,* **26** (1940), 1–15; Muḥammad Nāẓim, *The Life and Times of Sulṭān Muḥmūd of Ghazna* (Cambridge, 1931); Aurel Stein, "The Site of Alexander's Passing of the Hydaspes and the Battle With Poros," in *Geographical Journal,* **80** (1932), 31–46; Heinrich Suter, "Die Mathematiker und Astronomen der

Araber . . .," in *Abhandlungen zur Geschichte der mathematischen Wissenschaften . . .,* X (Leipzig, 1900); Eilhard Wiedemann et al., "Beiträge zur Geschichte der Naturwissenschaften," in *Sitzungsberichte der Physikalisch-medizinischen Sozietät in Erlangen;* and Yāqūt al-Rūmī, Shihāb al-Dīn, Abū ʿAbdallāh, *Muʿjam al-udabāʾ* (= *Irshād al-arīb ilā maʿrifat al-adīb*), XVII (Cairo, 1936–1938).

E. S. KENNEDY

BISCHOF, CARL GUSTAV CHRISTOPH (*b.* Wörth, near Nuremberg, Germany, 18 January 1792; *d.* Bonn, Germany, 29 November 1870), *chemistry, geology.*

Bischof probably acquired his interest in natural sciences from his father, a teacher of natural history and geography. He attended the University of Erlangen, where he obtained the doctorate and became *Privatdozent* in chemistry and physics. He received much early encouragement and inspiration from Nees von Esenbeck and Goldfuss. With the latter he published a two-volume physical and statistical description of the Fichtelgebirge, a mountain range near Nuremberg (1817). His next work, written with Esenbeck and Rothe, dealt with the evolution of plants. His first independent book was the *Lehrbuch der Stöchiometrie* (1819). In the same year he was called to the newly founded University of Bonn, where he became professor of chemistry and technology.

For the rest of his life he concentrated on the chemical changes accompanying geological processes, first in the Rhineland and later in other German areas and even foreign regions. His main interest at first was the volcanic phenomena of the Eifel and neighboring areas—specifically, the springs in these areas, which he interpreted as being largely of volcanic origin. In 1824 he published *Die vulkanischen Mineralquellen Deutschlands und Frankreichs,* a work that aroused much interest and led to his being considered one of the main defenders of volcanistic theories (as opposed to the neptunistic). He corroborated his ideas on the origin of springs by a case study of the mineral spring of Roisdorf (1826).

The chief work of his volcanistic period was *Wärmelehre des Innern unseres Endkörpers* (1837). In this work he presented a critical compilation of all that was known at the time, together with many of his own observations on the thermal properties of the earth's surface, including observations made in mines. From these he derived his theories of thermal gradients, which were essentially correct, and his ideas on the origin of volcanism and the heat required for his kind of "metamorphic" transformation. He concluded that the observed heat gradients explained satisfactorily all known volcanic activity, as well as springs and earthquakes. The evidence presented in this book was used immediately to support the plutonist theories dominant at the time. It also included experimental evidence; the volume reduction observed during the cooling of melted basalt inspired Élie de Beaumont to propose that folded mountain chains arose from wrinkling of the surface of the contracting earth, assuming that the earth had once been in a state of fusion.

Bischof had an interest in and talent for communicating his ideas to the general public. This is shown in his popular lectures, many of which are collected in *Populäre Vorlesungen über naturwissenschaftliche Gegenstände* (1842–1843) and *Populäre Briefe an eine gebildete Dame über das gesammte Gebiet der Naturwissenschaften* (1848–1849). Bischof was not only a gifted experimentalist; he also had a flair for translating scientific knowledge into practical use. For example he was the first to harness the HCO_3 springs in the volcanic areas of the Niederrhein for industrial purposes. He also promoted the recovery of copper from very low-grade ores by an inexpensive leaching and "cementation" method.

In 1848 Bischof began to publish his *Lehrbuch der chemischen und physikalischen Geologie,* the main source of his fame. The second edition, with a supplement, ran to 3,005 pages, and was published between 1863 and 1871. It was in many ways a continuation rather than a new edition. This enormous work soon became the standard geochemical text. It appeared to support a new school of thought, the "neoneptunistic." At the beginning of the first volume Bischof was still a plutonist and opposed neptunist views, although somewhat hesitantly. The intensive studies he made in connection with the *Lehrbuch,* however, convinced him more and more of the validity of the role of surface waters.

Bischof was and remained in many ways basically a laboratory chemist, despite his great interest in natural phenomena. This was his weakness in many instances, and the reason for his strong adherence to plutonism-volcanism until about 1846 and his fervent advocation of opposite views after this time.

Just as the exaggeration of the magmatic-hydrothermal theory of the formation of rocks and ore deposits was in part caused by one-sided experiments between 1900 and 1960, Bischof during his time exaggerated the role of water in his experiments, extrapolating from laboratory results to natural phenomena without adequate support from observations in nature. In this manner he assumed numerous transformations of sediments to crystalline schists, gneisses, and granites—and even basic igneous rocks—often without even discussing field relations. He rejected plutonic metamorphism and assumed that all metamorphic processes were caused by hydrochemical ("neptunic or katogene") reactions, i.e., changes at surface temperatures caused by a continuous flow of water through the rocks and introduction and subtraction of material. He was, in this respect, a forerunner of the extreme transformationists who, a century later, insisted upon similar transformations of sediments into various igneous rocks, usually without knowing of Bischof's work. He also believed that ore veins had been formed from descending solutions or by lateral secretion.

Bischof's work benefited several branches of geology and promoted a more scientific approach to many geological problems, such as the use of analogies with experiments (even though his geological theories were often proposed on the basis of laboratory evidence rather than field relationships). In 1849 he introduced the so-called oxygen coefficient into chemical comparisons of rocks by using the ratio between oxygen in bases and oxygen in SiO_2. He offered experimental evidence for causes of landslides in 1846 and 1863.

Bischof determined the relationship of gypsum and anhydrite and proposed a connection between crystallization and climate in the Dead Sea. He recognized that gypsum is not the end of salt deposition, but the beginning of a new sequence. He argued that since gypsum and halite are the first salts to precipitate during evaporation of seawater, calcareous sediments must have been formed through the action of organisms. He also recognized that dolomitization must have taken place and assumed magnesium bicarbonate to be a major cause. Bischof did not succeed in reproducing dolomitization in the laboratory, but was able to explain cavernous limestones and the lack of fossil shells in dolomites by leaching of dolomites. He also found, in 1864, that phosphoric acid accumulates in bones, shells, and the soft parts of animals. As early as 1829 Bischof had obtained melnikowite precipitates experimentally, and in 1863–1866 he discussed the different reactions by which gypsum or anhydrite forms in one case, and in another native sulfur is precipitated. Thus he contributed to the fundamental knowledge that decades later led to the suggestion that massive stratiform sulfide deposits may be sedimentary or exhalative-sedimentary.

BIBLIOGRAPHY

I. ORIGINAL WORKS. Bischof's major work is *Lehrbuch der chemischen und physikalischen Geologie,* 3 vols. (Bonn, 1846–1855; 2nd. ed., 3 vols. and supp., 1863–1871), trans. into English by B. H. Paul and J. Drummond as *Elements of Chemical and Physical Geology,* 3 vols. (London, 1854–1859). His other writings include *Physikalisch-statistische Beschreibung des Fichtelgebirges,* written with A. Goldfuss, 2 vols. (Nuremburg, 1817); with Nees von Esenbeck and Rothe, *Die Entwicklung der Pflanzensubstanz* (Erlangen, 1819); *Lehrbuch der Stöchiometrie* (Bonn, 1819); *Die vulkanischen Mineralquellen Deutschlands und Frankreichs* (Bonn, 1824); *Die Mineralquellen von Roisdorf* (Bonn, 1826); *Wärmelehre des Innern unseres Erdkörpers* (Bonn, 1837), also trans. into English (1844); and *Populäre Vorlesungen über naturwissenschaftliche Gegenstände* (Bonn, 1842–1843).

II. SECONDARY LITERATURE. Works on Bischof are F. Behrend and G. Berg, *Chemische Geologie* (Stuttgart, 1927); V. von Cotta, *Geologie der Gegenwart,* I (Leipzig, 1866), 61–62, 347–372; W. Fischer, *Gesteins- und Lagerstättenbildung im Wandel der wissenschaftlichen Anschauung* (Stuttgart, 1961); C. W. von Gümbel, in *Allgemeine Deutsche Biographie,* II, 665–669; Poggendorff, I, 202; and K. von Zittel, trans. by Maria M. Ogilvie-Gordon, *History of Geology and Paleontology* (London, 1901).

G. C. AMSTUTZ

BISCHOFF, GOTTLIEB WILHELM (*b.* Dürkheim an der Hardt, Germany, 21 May 1797; *d.* Heidelberg, Germany, 11 September 1854), *botany.*

Bischoff, who came from a family of pharmacists, was introduced to botany in Kaiserslautern by Wilhelm Koch, the author of *Synopsis florae Germanicae et Helveticae,* while he was studying graphic arts. In 1819 he entered the Academy of the Creative Arts at Munich, in order to perfect his drawing techniques; but in 1821, at the University of Erlangen, he turned to the study of botany and chemistry. Here he published his first work, *Die botanische Kunstsprache* (1822), in which his interest in terminology is apparent.

The botanist and scientific explorer Philipp von Martius encouraged Bischoff to continue his studies in Munich. In the first volume of Martius' *Nova genera et species plantarum* (1824), nearly all the drawings of plants, which Martius had brought back with him from his explorations in Brazil, were done by Bischoff. After completing his studies, Bischoff managed his father's pharmacy for about a year, but in 1824 he went to Heidelberg to study botany. By 1825 he had established himself there as a *Privatdozent;* in 1833 he became extraordinary professor, and in 1839 full professor of botany and director of the Botanical Garden of Heidelberg, which he completely reorganized.

Bischoff grew up in the era of *Naturphilosophie* in

German science, which still bore the imprint of Schelling, but he sought to dissociate himself from speculative discussion of nature. His investigations of cryptogams showed him to be an excellent observer, and his work on Characeae and on the Archegoniatae deserves special mention. He collected great amounts of the material that enabled Wilhelm Hofmeister to trace the development of lichens and ferns and to elucidate the relations between the life cycles and life histories of different plant phyla. In this connection his unfinished work *Die kryptogamischen Gewaechse* (1828), with its experiments on spores, must be mentioned because it prepared the way for Hofmeister's discovery of the alternation of generations. In *De hepaticis,* Bischoff described the reproductive organs of liverworts, calling the male sexual organs "antheridia" and the female ones "fruit-germs"; only later did he call the latter "archegonia." He also studied the Selaginellales and the Equisetales, which are interesting from the point of view of organic evolution.

In *Lehrbuch der Botanik* (1834–1840) Bischoff presented in detail the morphology and physiology of plants. He concurred with Goethe's assumption that all parts and organs of the plant originate from the leaf. The *Lehrbuch* contains a synopsis of plant pathology and one of the first histories of botany. In a clear and orderly fashion, Bischoff treats botanical knowledge up to 1837, when Matthias Schleiden's cell theory was published. His *Handbuch der botanischen Terminologie und Systemkunde* (1833–1844) is still of value as a general survey of the plant kingdom, for orientation on the development of botanical terminology, and for its discussion of numerous synonyms and the history of plant classification. Several of the nearly 4,000 figures that Bischoff himself drew for the book are still used in textbooks.

BIBLIOGRAPHY

I. ORIGINAL WORKS. A complete bibliography of Bischoff's works is in G. A. Pritzel, *Thesaurus literaturae botanicae,* new ed. (Leipzig, 1872), pp. 27–28. Among his works are *Die botanische Kunstsprache* (Nuremberg, 1822); *Die kryptogamischen Gewaechse,* 2 pts. (Nuremberg, 1828); *Handbuch der botanischen Terminologie und Systemkunde,* 3 vols. (Nuremberg, 1833–1844); and *Lehrbuch der Botanik,* 3 vols. (Stuttgart, 1834–1840).

II. SECONDARY LITERATURE. Works on Bischoff are *Botanische Zeitung,* **12** (1854), no. 39; Martin Möbius, *Geschichte der Botanik* (Jena, 1937), pp. 85, 122, 135, 152; *Neue deutsche Biographie,* II (Berlin, 1955), 263; Claus Nissen, *Die botanische Buchillustration,* 2 vols. (Stuttgart, 1951), see especially I, 217, and list of works illustrated by Bischoff in II, 16, nos. 165–167; 117, no. 1288;

M. Seubert, "G. W. Bischoff," in *Badische Biographien,* I (Heidelberg, 1875), 86; and E. Stübler, *Geschichte der medizinischen Fakultät der Universität Heidelberg* (Heidelberg, 1926), pp. 267, 274, 297.

JOHANNES STEUDEL

BISCHOFF, THEODOR LUDWIG WILHELM (*b.* Hannover, Germany, 28 October 1807; *d.* Munich, Germany, 5 December 1882), *comparative anatomy, physiology.*

Theodor's father, Ernst Christian Heinrich Bischoff, was a physician and a follower of Schelling's *Naturphilosophie.* He was a romantically stern and pious man. In 1806, after having served for several years as professor of physiology at the Medizinisch-Chirurgische Kollegium in Berlin, he divorced his first wife and married Juliane Hufeland, née Amelung, the wife of Christoph Wilhelm Hufeland. He then left Berlin and waited in Hannover for a new post. There Theodor, his only son by this second marriage, was born. After serving as a physician in Barmen-Elberfeld (now Wuppertal) and as an army physician during the "War of Liberation" against Napoleon (1813–1814), the elder Bischoff was appointed associate professor of pharmacology and forensic medicine at the reconstituted University of Bonn in 1818.

Theodor Bischoff was reared in a financially secure, cultured, and strict Protestant atmosphere. He attended the Gymnasium in Bonn and in 1825 passed the *Maturitätsprüfung.* After attending a special class in Gotha, he began his medical studies in Bonn (1826–1829). He attended the lectures given by Nees von Esenbeck, Friedrich Nasse, Philipp Franz von Walther, and Johannes Müller, all men with strong leanings toward speculative *Naturphilosophie.* This influence on Bischoff waned in 1830, when he continued his studies in Heidelberg under F. Carl Nägeli, Friedrich Tiedemann, and F. Arnold, and learned their preference for empirical research. His M.D. thesis (1832), completed under Tiedemann and Arnold, concerned the areas innervated by the *nervus accessorius willisii* in mammals, birds, and reptiles.

After passing the state medical examinations in Berlin, Bischoff interned at the University of Berlin's maternity clinic. He also attended lectures on comparative anatomy by Johannes Müller, who had taken over the Berlin chair of anatomy and physiology for the summer semester of 1833. Bischoff became lecturer in physiology at Bonn in September 1833. For the summer semester of 1836 he went to Heidelberg as lecturer in comparative and pathological anatomy. On 18 April 1839 he married the oldest daughter of Friedrich Tiedemann, his teacher in Heidelberg. In February 1843, he was appointed professor of anatomy and physiology at Heidelberg. From the fall of 1843 until 1854 he also taught in Giessen, first physiology and, from 1844, anatomy as well. In 1854, upon Liebig's recommendation, he received an invitation to teach the same subjects in Munich. He turned over his lectures in physiology to Carl Voit in 1863 and retired fifteen years later, at the age of seventy-one. He died 5 December 1882, after an intestinal perforation.

Bischoff received many decorations and honors. He was a member of the scientific academies of Berlin, Vienna, St. Petersburg, and Munich, the Royal Society of London, and the Kaiserlich Leopoldinischen Carolinischen Deutschen Akademie, as well as honorary member of many associations and societies.

Bischoff's scientific work began with zoological and botanical research. Next he wrote on physiological and physiological-chemical subjects. His most important works concerned the embryology of mammals and of man. While he worked in Munich, he stressed anthropological research. In addition, throughout his life he remained interested in general problems of natural philosophy and religion.

Bischoff's student papers on *Helix pomatia* and dragonflies were never printed, but his Ph.D. thesis on the spiral vessels of plants was published. His interest in physiological problems is shown in his treatises on the electric nerve currents (1841) and the reabsorption of narcotic toxins (1846). Also of interest is his attempt to determine the amount of blood in a fresh corpse (1856). In Heidelberg he conducted —jointly with P. G. von Jolly—the first, but relatively unsatisfactory, experiments concerning blood-bound oxygen and carbon dioxide.

Although Bischoff was neither inclined toward nor particularly gifted in vivisection, he developed an interest in embryology while an intern in Berlin. In this field he achieved excellent results. He began with research on the human fetal membrane and was able to demonstrate the existence of the decidual vessels and the amniotic epithelium. In Bonn he was the first to lecture before a large audience on the history of embryology. His interest in this subject had been aroused by the work of Karl Ernst von Baer, a disciple of Ignaz Döllinger, on the existence of the mammalian ovum in the graafian follicle (1827) and on the subsequent formation of embryonic epithelia through a process of segmentation.

In 1835 Bischoff showed the canine ovum moving through the fallopian tube. This achievement formed the starting point for a series of related research. At the Freiburg convention of natural scientists and physicians (1838) he reported on the presence of sperm in the peritoneal sac of the ovary of a bitch some twenty hours after copulation. He deduced that the follicle was made to burst by the entering sperm. Next, in answer to a contest problem posed by the Berlin Academy of Sciences, he investigated the first phases of mammalian development in the rabbit. Bischoff was the first to clarify the successive division of the mammalian ovum and the first subsequent segmentation processes. He also demonstrated that the embryonic vesicle consists of cells. This finding enabled him to establish the connection between embryology and the then new science of cytology.

In this work there arose controversies regarding the choice between epigenesis and evolution. Bischoff considered the embryo to be a liquid-filled vesicle, and attempted to derive the cellular substance of which the embryo is formed from the cellular beads of the morula. His history of the development of the rabbit egg was printed in 1842, that of the canine egg in 1845, of the guinea pig egg in 1852, and of the doe egg in 1854. His *Entwicklungsgeschichte der Säugetiere und des Menschen* was also translated into French, as were several other of his works.

At that time many scientists were doing research on embryology, with the result that all sorts of controversies and priority feuds arose. In 1843 Bischoff was compelled to correct his former views when it was found that ovulation occurs periodically, independent of copulation, and that subsequently the ovum commences its movement through the fallopian tube. His paper on this appeared in 1844 in German and in 1847 in English. His studies on the embryology of the guinea pig showed some remarkable differences from all prior observations. Of importance in this connection was Bischoff's clarification of the fertilization process in the doe: after being fertilized in August, the egg moves within a few days through the fallopian tube into the uterus and remains there for four and a half months without any further important development. In 1854 Bischoff had to correct his former views on one not unimportant point. It had been found, particularly by Barry (1843), that in the fertilization process the sperm penetrates the mammalian ovum. This had been disputed by Bischoff.

In 1859 Darwin's *On the Origin of Species* appeared. Bischoff then decided to change his field of research, and began to occupy himself more and more with anthropological investigations. While he approved the theory of selection in principle, he considered the general theory of the origin of species to be insufficiently substantiated, since it did not furnish any explanation for the general and individual origin of life. This criticism resulted from Bischoff's assumption that specific vital forces are active in development whereby "individual immortal basic causes" unite with matter. Although he closely related the activity of the soul with that of the brain, he nevertheless felt that the immortal basic cause of the individual continued to exist after death halted the soul's activity. Such ideas are also expressed in his posthumously published treatise *Gedanken eines Naturforschers über die Natur und über die Religion* (1878).

In connection with the question of man's place in the living world, Bischoff investigated the cerebral convolutions (1868) and the weight of the human brain (1880). He also studied the weight of the brain in relation to sex, body weight, body size, age, and race, and found no relationship between intelligence and brain weight. He did, however, consider the possibility of certain parallels. Bischoff's studies on the anatomy of the anthropoid apes also belong to this group of works.

Among Bischoff's physiological works particular mention should be made of those resulting from collaboration with or suggestions made by the great chemist Justus Liebig. In 1843, when Bischoff came to Giessen, Liebig's influential books *Die Chemie in ihrer Anwendung auf Agrikultur und Physiologie* (1840) and *Die Tierchemie . . .* (1842) had been published. The wealth of ideas formulated in these works stimulated Bischoff's interest in the metabolic processes. At that time scientists were beginning to investigate more closely the transformation of certain foods and their metabolic products. Liebig considered urea to be the measure of the protein metabolism in the tissue, whereas F. H. Bidder and Carl Schmidt interpreted it merely as the result of the actual amount of protein in the food. In Bischoff's *Der Harn als Maass des Stoffwechsels* (Giessen, 1853), which was dedicated to Liebig, the findings of Bidder and Schmidt are corroborated, but the production of urea is ascribed to the metabolism of nitrogen-rich substances. Liebig's theory of nitrogen-containing structural foodstuffs and respiratory nitrogen-free foods is substantiated. In Munich, Bischoff had Carl Voit as his collaborator. They jointly investigated nutrition in carnivores (1860) under conditions of starvation, an all-meat diet, and an all-fat diet, among others. Fat and similar substances were considered to be heat producers, nitrogen-containing substances to be energy producers. Subsequently, Voit founded the Munich school of metabolic physiology.

Bischoff had a gift for oratory and was a popular banquet speaker and eulogist. His eulogies of Johannes Müller and Friedrich Tiedemann and his address honoring Liebig's achievements in physiology are valuable documents in the history of natural science.

Somewhat peculiar is Bischoff's brusque rejection of women as university students in *Das Studium und die Ausübung der Medizin durch Frauen, beleuchtet durch Dr. Th. L. W. v. Bischoff* (1872). According to him, unqualified, half-trained female "artisans" impede and "most disastrously" disrupt the further development of medicine. Bischoff's life was filled with unflinching scientific labor, involving experiments, preparations, observations, comparisons, readings, and weighings. He investigated many subjects and left indelible marks on many areas. His vigor and his strong will, his firm character and integrity are testified to by all biographers.

BIBLIOGRAPHY

I. ORIGINAL WORKS. Complete bibliographies of Bischoff's works are in the biographies by Kupffer and Sudhoff (see below). Among his major writings are "Berichte über die Fortschritte der Physiologie," in Johannes Müller, ed., *Archiv für Anatomie, Physiologie und wissenschaftliche Medizin für die Jahre 1839–1847; Entwicklungsgeschichte des Kaninchens* (Brunswick, 1842); *Entwicklungsgeschichte der Säugetiere und des Menschen,* Vol. VII of Samuel Thomas von Sömmering's *Vom Baue des menschlichen Körpers,* new ed. (Leipzig, 1843); *Beweis der von der Begattung unabhängigen periodischen Reifung und Loslösung der Eier der Säugetiere und des Menschen* (Giessen, 1844); *Entwicklungsgeschichte des Hundeeies* (Brunswick, 1845); *Entwicklungsgeschichte des Reheies* (Giessen, 1854); *Die Gesetze der Ernährung des Fleischfressers durch neue Untersuchungen festgestellt* (Leipzig–Heidelberg, 1860), written with Carl Voit; "Über die Bildung des Säugetier-Eies und seine Stellung in der Zellenlehre," in *Sitzungsberichte der Königlichen Bayerischen Akademie der Wissenschaften zu München,* **1** (1863), 242; *Das Studium und die Ausübung der Medizin durch Frauen* (Munich, 1872); *Gedanken eines Naturforschers über die Natur des Menschen und über die Religion* (Bonn, 1878); and *Das Hirngewicht des Menschen. Eine Studie* (Bonn, 1880).

II. SECONDARY LITERATURE. Biographies of Bischoff with full bibliographies are Carl Kupffer, *Gedächtnisrede auf Theodor L. W. von Bischoff, 28.III.1884* (Munich, 1884); and Karl Sudhoff, "Bischoff, Theodor Ludwig Wilhelm (von)," in *Hessische Lebensbilder,* **3** (1928), 1–11. See also *Allgemeine Deutsche Biographie,* XLVI (1902), 570; *Almanach der Königlichen Bayerischen Akademie der Wissenschaften* (1875), pp. 182–187 (1878), pp. 133–134; *Biographisches Lexikon der hervorragenden Arzte,* 2nd ed. (Berlin-Vienna, 1929–1935), II, 550–551; and *Index Catalogue of the Library of the Surgeon-General's Office, U.S. Army, Washington,* 1st ser., II (Washington, D.C., 1881), 72–73; 2nd ser., II (Washington, D.C., 1897), 353–354.

K. E. ROTHSCHUH

BISTERFELD, JOHANN HEINRICH (*b*. Siegen, Germany, *ca*. 1605; *d*. Weissenburg, Transylvania [after 1715 Karlsburg; now Alba Iulia, Rumania], 16 February 1655), *philosophy, theology.*

Bisterfeld's father, Johann, was a minister and professor of theology. He published a book on Ramist dialectics, and died while attending the Synod of Dort in 1619. The mother's maiden name was Schickard, and she appears to have been a sister of Martin Schickard, also of Siegen, professor of jurisprudence at Heidelberg and later at Deventer. As a student at the reformed University of Herborn, Bisterfeld studied under Comenius' teacher, Johann Heinrich Alsted, whose *Encyclopaedia* he is said to have known by heart at the age of sixteen. He may have studied in England in the middle 1620's, but in any case he matriculated at the University of Leiden on 3 November 1626, where he made the acquaintance of André Rivet, with whom he later corresponded. In 1628 Bisterfeld traveled in the Netherlands. He married Alsted's daughter Anna, but it is not known exactly when or where.

Early in 1629, Bisterfeld and Alsted were invited by Gabriel Bethlen, prince of Transylvania, to join the newly established (1622) academy in Weissenburg. Under the pressure of the dislocations caused by the Thirty Years' War in Nassau, they accepted the call to Transylvania, where Bisterfeld was professor of philosophy and theology until his death. His successor was the French-English traveler and divine Isaac Basire. During the late 1630's and the early 1640's, Bisterfeld also performed diplomatic duties for György Rákóczy I, in order to secure an alliance with France and Sweden against the Holy Roman Empire. Owing to the hesitation of Sweden, this alliance was not effected until 1643. In late July 1638, he arrived in Paris, where he conferred with Marin Mersenne. He spent the remainder of the year in western Europe, including Hamburg and Amsterdam.

Owing to the efforts of Rivet, Bisterfeld received a call to the University of Leiden, a position that would have satisfied his desire to return to "the more cultivated parts of Europe," but Rákóczy did not wish to lose so useful a man. Bisterfeld corresponded with Samuel Hartlib, John Dury, Theodore Haak, and others of their circle, and like them he looked forward to the union of the divided Protestant churches. A projected visit to Hartlib in London late in 1638 did not occur, but Bisterfeld's name figures in Hartlib's plans for an office of correspondence.

The philosophical basis of Bisterfeld's thought was the Ramism that reigned at the University of Herborn, and he also had much in common with Alsted and Comenius on other points: He shared their respect for Bacon, Ramón Lull, and Campanella, as well as their chiliasm and their belief in the universal harmony of all creation; universal knowledge, or pansophy, was their common aim. The Trinity was the source, norm, and end of all order. Philosophy was the pedagogue to theology, and Scriptures were the foundation of philosophy. Bisterfeld differed from Alsted and Comenius in his greater insight into the philosophical requirements of the system that would reveal the universal harmony and thus put man in control of nature. "Whatever is most true in philosophy," Bisterfeld said, "is also most useful in practice." He criticized Bacon and Campanella for failing to pay sufficient attention to Lull's *Ars magna,* and Comenius for ignoring metaphysics, for the lack of a strict method that would tie his system together.

Bisterfeld was strongly impressed by the need for a consistent terminology and precise definitions. It was not enough to "open the door to languages," as Comenius had done; Bisterfeld's nomenclature would be a "new door" ("nomenclator meus sit porta linguarum reformata"). On this point, Bisterfeld may have influenced projects for a philosophical language through his *Alphabeti philosophici libri tres* (1661). But he also saw the need to go beyond terminology. He realized more fully than his contemporaries the value of an *ars combinatoria,* or a logic of relations, as an Ariadne thread to serve as a guide in the labyrinth of the encyclopedia of knowledge. It was the chief aim of philosophy to reduce all the principles of particular areas of knowledge to the fewest possible common principles; it was the soul of practical theology to demonstrate that all things could be referred back to God.

This aspect of Bisterfeld's work had a strong effect on the young Leibniz, who read and commented upon Bisterfeld's most important philosophical writings during his student years at Leipzig. He noted that the *Philosophiae primae seminarium* was "a most brilliant little work whose equal in this kind I have not seen," and called the *Phosphorus catholicus, seu artis meditandi epitome* "a most ingenious little book." Together with the *Elementorum logicorum libri tres,* both were published at Leiden in 1657, but the *Phosphorus* had already been separately printed at Breda in 1649. In these works Leibniz seems first to have encountered the idea of universal harmony and the suggestion of a mathematical mode of logical calculation. Among the passages he especially noted was the statement that "logic is nothing but a mirror of relations." Leibniz remarked that "Bisterfeld proceeds in metaphysics almost like Bacon in physics," and he was fully aware of the affinity between Lull's and Bisterfeld's art of combinations. In his *De uno Deo, Patre, Filio ac Spiritu Sancto, mysterium pietatis* (Leiden, 1639; Amsterdam, 1645) Bisterfeld provided a detailed critique of Socinianism, directed against a work by Johannes Crellius.

BIBLIOGRAPHY

I. ORIGINAL WORKS. Bisterfeld's works are sparsely represented in even the largest libraries. His posthumous works were published as *Bisterfeldius redivivus*, 2 vols. (Leiden, 1661). Vol. I contains *Alphabeti philosophici*. Excerpts with Leibniz' comments are in Leibniz, *Sämtliche Schriften und Briefe*, 6th ser., *Philosophische Schriften*, I (Darmstadt, 1930), 151–161. An important letter to Hartlib is in J. Kvačala, ed., *Die pädagogische Reform des Comenius in Deutschland*, I, *Texte* (Berlin, 1903), 112–118 (*Monumenta Germaniae paedagogica*, XXVI).

II. SECONDARY LITERATURE. J. Kvačala, "Johann Heinrich Bisterfeld," in *Ungarische Revue*, **13** (1893), 40–59, 171–197, is diffuse and not entirely reliable. See also Kvačala, in *Die pädagogische Reform des Comenius in Deutschland*, II, *Historischer Überblick, Bibliographie, Namen und Sachregister* (Berlin, 1904), Index (*Monumenta Germaniae paedagogica*, XXXII). Interesting information about Bisterfeld in general and his work as a diplomat in particular is in I. Hudita, *Histoire des relations diplomatiques entre la France et la Transylvanie au XVIIe siècle (1635–1683)* (Paris, 1927), esp. pp. 43–47, 57–58. See also Hudita, ed., *Répertoire des documents concernant les négociations diplomatiques entre la France et la Transylvanie au XVIIe siècle (1636–1683)* (Paris, 1926), esp. pp. 61–68. Leroy E. Loemker offers a good account of Bisterfeld's philosophical position in "Leibniz and the Herborn Encyclopedists," in *Journal of the History of Ideas*, **22** (1961), 323–338. See also Paolo Rossi, *Clavis universalis* (Milan, 1960), pp. 197–200, 238–239. Much useful information can be gathered from letters and editorial notes in *Correspondance du P. Marin Mersenne*, Cornélis de Waard, ed. (Paris, 1932–   ), by tracing Bisterfeld through the indexes to VII (1962), VIII (1963), IX (1965), and X (1967), so far the last published.

HANS AARSLEFF

AL-BIṬRŪJĪ AL-ISHBĪLĪ, ABŪ ISḤĀQ, also known as **Alpetragius** (his surname probably derives from Pedroche, Spain, near Cordoba; *fl.* Seville, *ca.* 1190), *astronomy, natural philosophy.*

Al-Biṭrūjī's only extant work is *De motibus celorum*, originally written in Arabic. He was a contemporary of Ibn Rushd (Averroës), and his astronomical system aroused much interest among such Christian natural philosophers as Albertus Magnus, Robert Grosseteste, and Roger Bacon. For a detailed study of his life and work, see Supplement.

BIZZOZERO, GIULIO CESARE (*b.* Varese, near Milan, Italy, 20 March 1846; *d.* Turin, Italy, 8 April 1901), *histology.*

Bizzozero was the son of Felice Bizzozero, a small manufacturer, and Carolina Veratti. After studying classics at Milan, he went to Pavia for medical study and received the M.D. in 1866. Upon graduation, Bizzozero enlisted in the army of Garibaldi as a military physician. The war against Austria ended within the year, and in 1867 he was appointed substitute professor of general pathology at the University of Pavia. He taught there until 1872; in that year Bizzozero became ordinary professor of general pathology at Turin. For almost thirty years he was a well-known teacher and researcher at Turin. Among the more illustrious of his pupils were Camillo Bozzolo, Camillo Golgi, Pio Foà, Gaetano Salvioli, and Cesare Sacerdotti. The school of Bizzozero in Turin and, later, the school of Alessandro Lustig in Florence were the first Italian teaching centers in general pathology.

A man of wide medical learning, Bizzozero was among the first to understand the importance to medicine of the microscope. In 1878 he instituted an annual course in clinical microscopy as an aid to the exhaustive study of sick persons. In 1879 he published his *Manuale di microscopia clinica*, later reprinted many times and translated into German. Three years before, Bizzozero had founded the *Archivio per le scienze mediche*. Stricken, after 1890, by a distressing debility of the right eye, he was unable to continue his own microscopical observations. He therefore devoted himself more and more to writing works that would develop an awareness of the need for public health measures. As early as 1883 Bizzozero set forth his program: the defense of mankind against infectious diseases. In 1890, he was appointed a senator.

Bizzozero was trained in scientific research by Eusebio Oehl, director of the laboratory of experimental physiology of Pavia University, and by Paolo Mantegazza, who in 1861 had founded the laboratory of general pathology at Pavia, the first in Italy. But the direction of Bizzozero's research was morphological. From his first paper in 1862 to his last paper in 1900, he was, essentially, a histologist. Bizzozero was so convinced of the importance of microscopical morphology that in 1880 he began, on his own initiative, to teach normal histology at Turin, a free course which he continued until his death.

Bizzozero was, undoubtedly, one of the outstanding histologists of his time. Of highest importance were his works on epithelial tissues—for instance, his studies of stratified squamous epithelium in 1870 and 1886—but even earlier there was his paper of 1864,

"Delle cellule cigliate del reticolo malpighiano dell'epidermide," in which he demonstrated the connections between the cells of the Malpighian layer. In collaboration with Gaetano Salvioli, Bizzozero defined the mesothelium of the great serous membranes (pleural and peritoneal) as a continuous layer, without stomata, separating the serous cavity from the lymphatic vessels. He also published studies (1888–1893) on the gastric and intestinal glands of mammals. Bizzozero thus contributed decisively to epithelial histology, fundamental knowledge of which had begun with the work of Jacob Henle.

Equally important were Bizzozero's observations on connective tissue: he illustrated (1865, 1872, 1873) the cells of the meninges and showed that tumors of the meninges (meningiomas) arise from the connective tissue cells of the meninges proper, not from the endothelial cells of blood vessels. In the structure of tendons he showed that the tendinous cells are in direct contact with the collagenous fibers. Special attention should be given to his work on the structure of the lymphatic glands. In 1872 Bizzozero clarified the peculiarity of cytoplasmatic relations between the cells and the fibers of the reticulum: "The cells are only applied on the reticulum, and they do not take integrated part in the constitution of reticulum." Bizzozero also anticipated actual knowledge of the "reticular cell," the staminal element of lymph nodes which he first called *cellula del reticolo*. He defined the important morphological problem of the sinuses of lymph nodes, demonstrating the existence of endothelial cells (reticuloendothelial cells) on the internal wall of sinuses. In 1868 Bizzozero demonstrated the erythrocytopoietic function of bone marrow, which he illustrated in several works, and in 1869 he discovered the megakaryocytes.

Bizzozero then undertook the solution of very important biological problems. In his first experimental work (1866) he demonstrated that granulation tissue originated as a consequence of the proliferative activity of mobile cells of loose connective tissue proper—cells which, for their high potentiality, he termed "embryonal cells" of loose connective tissue. Bizzozero also delineated the morphology of these "wandering polyblastic cells," knowledge of which was developed from 1902 to 1906 by the Russian histologist Alexander Maximov. Moreover, in this work Bizzozero demonstrated that in the neoproduction of granulation tissue there is also a neoproduction of capillaries; he described their development from compact cellular cordons in which a vascular cavity appears secondarily. This important observation was confirmed in 1871 by Julius Arnold in his *Experimentelle Untersuchungen über die Blutcapillaren*.

With his work on granulation tissue, Bizzozero was, with Virchow and Cohnheim, among the pioneers in the modern study of inflammation. Bizzozero's observations went further, however, for he first illustrated the power of ingestion (phagocytosis) as characteristic of the great mobile cells of loose connective tissue, of similar cells of bone marrow, and of the reticular cells of lymph nodes—a capacity to ingest damaged cells as well as the products of an inflammatory process: "The *celluliferous* cells [macrophages containing dead leukocytes] are large cells of connectival origin, which introduce into the proper contractile protoplasm the pus corpuscles [leukocytes]. . . . The finding of these celluliferous cells is not interpretable as a process of 'endogenesis' [as Rudolf Virchow then thought in his erroneous doctrine of endogenous cell formation], but as the effect of ingestion of pus corpuscles. . . . These great *celluliferous* cells unquestionably possess the power to devour pus corpuscles, erythrocytes, pigment granules."

In 1873 Bizzozero affirmed, ten years before Metchnikoff, that connective tissue cells (reticular cells of lymph nodes) act also against infection: "These cells, which contain granules of cinnebar or China ink 2–3 days after the injection in the subcutaneous connective tissue of those substances, constitute a very intricate labyrinth, through which the liquid carried from the lymphatic vessels is obliged to flow, and the corpuscles carried by the liquid, coming in contact with the protoplasm of reticular cells, are ingested by them. This fact is, perhaps, the cause of stoppage of some infections in the lymph nodes in which the lymphatic vessels arrive from the part overwhelmed by infection's products."

Although the name of Bizzozero is not linked with phagocytosis today, it is associated with the platelets. Knowing that Max Schultze in 1865 had described the presence, in the blood of healthy man, of irregular accumulations which he termed "granular formations," Bizzozero demonstrated in 1882 that they were normal elements of the blood and that they were linked to the phenomenon of thrombosis—hence their name of "thrombocytes."

BIBLIOGRAPHY

I. ORIGINAL WORKS. Bizzozero's works are collected in *Le opere scientifiche di Giulio Bizzozero*, 2 vols. (Milan, 1905), with a preface by Camillo Golgi, which contains all of Bizzozero's works, complete with original illustrations. For Bizzozero's research on thrombocytes, see especially his "Di un nuovo elemento morfologico del sangue e della sua importanza nella trombosi e nella coagulazione del sangue," in *Archives italiennes de biologie*, **2** (1882),

345–365; and "Ueber einen neuen Formbestandteil des Blutes, und dessen Rolle bei der Thrombose und der Blutgerinnung," in *Virchow's Archiv,* **90** (1882), 261–280. He also published a short book: *Di un nuovo elemento morfologico del sangue . . .* (Milan, 1883).

II. SECONDARY LITERATURE. Works on Bizzozero are R. Fusari, "Giulio Bizzozero," in *Monitore zoologico italiano,* **12** (1901), 103–107, a complete list of his work; and P. Franceschini, "La conoscenza dei tessuti connettivi nelle ricerche di Giulio Bizzozero," in *Physis,* **4** (1962), 227–267, which includes an extensive bibliography.

PIETRO FRANCESCHINI

BJERKNES, CARL ANTON (*b.* Christiania [later Kristiania, now Oslo], Norway, 24 October 1825; *d.* Kristiania, 20 March 1903), *mathematics, physics.*

Bjerknes was the son of Abraham Isaksen Bjerknes, a veterinarian, and Elen Birgitte Holmen. Both of his parents were of peasant stock, and throughout his life Bjerknes retained strong ties to his relatives in the country. The father, who as the youngest son did not inherit any land, died in 1838, leaving his widow and three children in straitened circumstances. In 1844 Bjerknes entered the University of Christiania and completed his undergraduate studies in 1848 with a degree in mining engineering. After several years at the Kongsberg silver mines (1848–1852) and as a mathematics teacher (1852–1854), he was awarded a fellowship that enabled him to study mathematics in Göttingen and Paris (1856–1857). The lectures of Dirichlet made a great impression on him and turned his interest to hydrodynamics, which later became the main subject of his research.

In 1859 Bjerknes married Aletta Koren, daughter of a minister. Two years later he was appointed lecturer in applied mathematics at the University of Christiania and was promoted to professor in 1866; in 1869 the professorship was converted to a chair of pure mathematics.

Bjerknes had a delightful personality and was an excellent teacher who was greatly respected by his students for his personal qualities and outstanding lectures. As the years passed, however, he showed an increasing tendency to professional isolation and a fear of publishing the results of his research, which was concerned mainly with hydrodynamic problems. Apart from the very close cooperation with his son Vilhelm, he lived for the most part in his own world. At one point Vilhelm had to extricate himself from this collaboration in order to avoid the danger of unproductive isolation. Nevertheless, in many fields he contributed to the elucidation and continuation of his father's theories.

Bjerknes had been particularly impressed by Dirichlet's demonstration that, according to the principles of hydrodynamics, a ball can move at a constant speed and without external force through ideal (frictionless) fluids, i.e., without the fluid's offering resistance to the ball's movement. Earlier, he had been greatly influenced by Leonhard Euler's *Lettres à une princesse d'Allemagne,* in which Euler opposed the concept of certain forces, such as Newtonian gravity, which are presumed to work at a distance rather than through an overall encompassing medium or ether. One of the strongest objections to the ether theory had always been the difficulty in understanding that according to the principle of inertia, a body not influenced by force should be able to move through such a medium without resistance, but in his lectures Dirichlet had proved that this was possible for movements in the frictionless fluids of hydrodynamics.

Slowly, Bjerknes developed the notion that it was possible, on the basis of hydrodynamics, to form a general theory of the forces active between the solid elements and the influence of the forces on the movements of those elements. First he studied the movement of a ball of variable volume through frictionless fluid according to the method of mathematical physics, and was thus led to further calculations of the simultaneous movements of two such balls. In this way he arrived at the historical conclusion, in 1875, that two harmoniously pulsating balls moving through frictionless fluid react as though they were electrically charged, i.e., they attract or repel one another with a force similar to that of Coulomb's law: they repulse one another when performing harmoniously pulsating oscillations in opposite phases (i.e., when one has maximum volume and the other's volume is minimal); conversely, they attract each other when oscillating in the same phase, thus attaining maximum or minimum volume at the same time.

This important discovery was followed by a number of tests that further stressed the analogy between the movement of bodies in frictionless fluids and the phenomena of electrodynamics. This research, which Bjerknes carried out in collaboration with his son, was substantiated by experiments that drew considerable attention at the electrical exhibition held in Paris in 1881.

Bjerknes' goal was now to develop this analogy to include Maxwell's general theory for electrodynamic phenomena, but despite his intensive efforts he did not attain this goal. His "hydrodynamic picture of the world" and his efforts to explain the electromagnetic forces through hydrodynamics are today more a fascinating analogy than a basic physical theory, yet through this research Bjerknes attained a great insight into hydrodynamic phenomena and thus anticipated later developments in several fields. It is especially noteworthy that through his efforts to describe the action of a magnetic field on an electric current he came to the conclusion that a cylinder rotating in a moving fluid is influenced by a force of the type that today is known as the hydrodynamic transverse force.

Shortly before his father's death Vilhelm Bjerknes published a work on long-range hydrodynamic forces as formulated in his father's theories. In it he explains and clarifies the important results of his father's research.

BIBLIOGRAPHY

I. ORIGINAL WORKS. Among Bjerknes' writings are *Niels Henrik Abel, en skildring af hans liv og videnskabelige virksomhed* (Stockholm, 1880) and *Hydrodynamische Fernkräfte. Fünf Abhandlungen,* no. 195 in the series Ostwald's Klassiker der exakten Wissenschaften (Leipzig, 1915).

II. SECONDARY LITERATURE. Works on Bjerknes are Vilhelm Bjerknes, *Vorlesungen über hydrodynamische Fernkräfte nach C. A. Bjerknes's Theorie,* 2 vols. (Leipzig, 1900–1902); "Til minde om professor Carl Anton Bjerknes," in *Forhandlinger i Videnskabs-selskabet i Kristiania,* no. 7 (1903), 7–24; *Fields of Force* (New York, 1906); *Die Kraftfelder* (Brunswick, 1909); and *C. A. Bjerknes. Hans liv og arbejde* (Oslo, 1925), translated as *Carl Anton Bjerknes. Sein Leben und seine Arbeit* (Berlin, 1933); Elling Holst, "C. A. Bjerknes som matematiker," in *Det Kongelige Frederika Universitet 1811–1911,* II (Kristiania, 1911); and Holtsmark, in *Norsk biografisk leksikon,* I, 581–583.

MOGENS PIHL

BJERKNES, VILHELM FRIMANN KOREN (*b.* Christiania [later Kristiania, now Oslo], Norway, 14 March 1862; *d.* Oslo, 9 April 1951), *physics, geophysics.*

Bjerknes was the son of Carl Anton Bjerknes and Aletta Koren. His life and scientific activities were strongly influenced by his father; even in boyhood he became interested in the elder Bjerknes' hydrodynamic research, especially in the experimental verification of his father's discovery of the generation of forces between pulsating and rotating bodies in ideal (frictionless) fluids. His collaboration with his father was also necessary because the elder Bjerknes, who had never received any formal training in experimental physics, was rather impractical. It should be noted, however, that at an early age Bjerknes was able to give an independent, even critical, evaluation of his father's research. On the other hand, he defended his father's memory with great devotion and gave a clearer and more general explanation of his theoretical thinking in *Vorlesungen über hydrodynamische Fernkräfte nach C. A. Bjerknes' Theorie* and in *Die Kraftfelder.*

Bjerknes began his scientific studies at the University of Kristiania in 1880 and in 1888 received the M.S. During the last years of his studies he decided to cease collaborating with his father, a decision that must certainly have been very difficult to make but is a tribute to the maturity and independence with which Bjerknes regarded his possibilities for scientific research. In spite of his great devotion to his father, he was fully aware of the drawbacks to the elder Bjerknes' scientific isolation and one-sidedness, and feared that he himself could become a victim of the same circumstances. At this time he decided that after completing his education in mathematics and physics and obtaining a position that afforded him comparative peace and security, he would complete his father's work as far as possible.

After completing his studies, Bjerknes went to Paris on a state fellowship; there he attended Henri Poincaré's lectures on electrodynamics, during which Heinrich Hertz's studies on the diffusion of electrical waves were mentioned. He then went to Bonn, where he worked for nearly two years as Hertz's assistant and first scientific collaborator. For the rest of his life he remained a close friend of the Hertz family and helped Hertz's widow and daughter in 1933, when they had to flee the Nazis and seek refuge in England. This collaboration with Hertz resulted in some very important scientific publications on resonance in oscillatory circuits; and the theoretical and experimental resonance curves discovered by Bjerknes, along with a work by Poincaré, were of considerable importance for the understanding and final proof of Hertz's revolutionary experiments.

After his return to Norway, Bjerknes continued his studies, obtaining the Ph.D. in 1892 on the basis of the dissertation "Elektricitetsbevaegelsen i Hertz's primaere leder." It was his research in this field that especially qualified him for appointment as lecturer in applied mechanics at Stockholm's Högskola (School of Engineering) in 1893 and his appointment as professor of applied mechanics and mathematical physics at the University of Stockholm in 1895. Even though he abandoned experimental research in this field fairly soon, he retained a deep interest in the problems of electrodynamics for the rest of his life.

During the following years Bjerknes worked on his father's theories of hydrodynamic forces, which he succeeded in explaining in a simpler form than that based on his father's calculations of such specific examples as forces between pulsating balls in frictionless fluids. These investigations resulted in the two-volume work *Vorlesungen über hydrodynamische Fernkräfte nach C. A. Bjerknes's Theorie* (1900–1902). Later he often returned to the problem of force fields, which he treated in a simple, clear-cut fashion in two books published in 1906 and 1909.

During the period of his hydrodynamic studies, Bjerknes generalized on the well-known propositions of Lord Kelvin and Hermann Helmholtz concerning the so-called velocities of circulation and the conservation of the circular vortex. He then applied this generalization to the movements in the atmosphere and the ocean.

Bjerknes' generalization depended on the introduction of a broader interpretation of the concept of fluids than that normally used in classical hydrodynamic theory, which assumes that a unique relationship exists between pressure and the specific volume (the volume of a unit mass). He perceived the fluids as thermodynamic systems, which made it necessary to renounce such an unambiguous relationship, and was led to the formulation of the theory of physical hydrodynamics.

In this connection, however, reference should be made to the contribution made in 1896 by L. Silberstein, at that time unknown to Bjerknes, who developed one of Bjerknes' two circulation theorems without comprehending its far-reaching implications. The atmospheric movements that cause weather changes result from the radiation of heat from the sun, and the atmosphere thus works as a sort of thermodynamic heat engine that is constantly converting heat to mechanical energy; it also emits heat because of the friction resulting from atmospheric movements. It is therefore necessary, when atmospheric movements are described, to produce the synthesis of classical hydrodynamics and thermodynamics that results from the formulation of the theory of physical hydrodynamics.

Although he realized that it would not be completed in the near future, Bjerknes planned an ambitious program as the final goal of this research: he hoped to be able, with the help of the hydrodynamic and thermodynamic theories, to use knowledge of the present conditions of the atmosphere and hydrosphere to calculate their future conditions. During a visit to the United States in 1905 he presented these plans, and received from the Carnegie Foundation an annual stipend to support his research in this field. The grant continued until 1941.

During his period as professor at the University of Stockholm, Bjerknes began collaboration with various scientists, for which he was eminently suited because of his stimulating intellect and deep understanding of his associates' need for independent development and research. Of special importance was his collaboration with J. W. Sandström, with whom he wrote the first volume of *Dynamic Meteorology and Hydrography* (1910). The second volume (1911), dealing with kinematics, was written with Th. Hesselberg and O. Devik. The projected third volume, dealing with dynamics, was completed by associates, but he lived to see its publication in 1951. In 1933 he coauthored a book with his son, Jack, and a friend of the son, H. Solberg, *Physikalische Hydrodynamik mit Anwendung auf die dynamische Meteorologie.*

After his return from Stockholm in 1907 Bjerknes became professor of applied mechanics and mathematical physics at the University of Kristiania, where he collaborated with Sandström, Hesselberg, Devik, and H. U. Sverdrup in developing dynamic meteorology. In 1912, when he was offered the professorship of geophysics at the University of Leipzig and the chairmanship of the newly organized geophysical institute, he decided to accept the offer, in the hope of better prospects. The new institute was started under the best possible conditions: Hesselberg and Sverdrup followed him to Leipzig, and a few years later both his son and Solberg joined them.

A visit from Fridtjof Nansen resulted in an offer to Bjerknes to take over a professorship at the University of Bergen and to start a geophysical institute there. He decided to accept the offer after assuring himself that the institute in Leipzig would be carried on. Bjerknes was fifty-five when he started working in Bergen, and he remained there until 1926. His years in Bergen were perhaps the most productive of his life. His main collaborators were again his son and Solberg; later they were joined by S. Rosseland, T. Bergeron, E. Bjørkdal, C. Rossby, and E. Palmén. Bjerknes himself continued to play an active role in both the practical implementation of extensive meteorological services and the work on theoretical meteorological problems. From this period came his now classic work *On the Dynamics of the Circular Vortex With Applications to the Atmosphere and to Atmospheric Vortex and Wave Motion* (1921). One of his finest books, it contains a clear explanation of the most important basic ideas in his research.

After his appointment as professor of applied mechanics and mathematical physics at the University of Oslo in 1926, Bjerknes continued his studies in dynamic meteorology in cooperation with Solberg, J. Holmboe, C. L. Godske, and E. Høiland. He became involved in the teaching of theoretical physics, but remained within the limits of classical physics, and in 1929 he published a small book on vector analysis and kinematics as the first part of a textbook in theoretical physics. The next volume planned, which was to include an explanation of the elder Bjerknes' "hydromagnetic" theory, was never completed. Despite intensive efforts, Bjerknes and Høiland never succeeded in finding a satisfactory formulation of this theory, which had occupied Bjerknes from his earliest years. This was a problem from which he could not, and would not, disengage himself.

BIBLIOGRAPHY

I. ORIGINAL WORKS. A bibliography of Bjerknes' works is in *Geofysiske publikationer,* **24** (1962), 26–37. Among his writings are "Über die Dämpfung schneller electrischer Schwingungen," in *Annalen der Physik,* **44** (1891), 74–79; and "Über electrische Resonanz," *ibid.,* **55** (1895), 121–169; *Über die Bildung von Cirkulationsbewegungen und Wirbeln in reibungslosen Flüssigkeiten,* no. 5 in *Skrifter udgivet af Videnskabsselskabet i Christiania,* I (1898), 1–29; "Über einen hydrodynamischen Fundamentalsatz und seine Anwendung besonders auf die Mechanik der Atmosfäre und des Weltmeeres," in *Kungliga Svenska vetenskapsakademiens handlingar,* **31,** no. 4 (1898–1899), 1–35; *Vorlesungen über hydrodynamische Fernkräfte nach C. A. Bjerknes's Theorie,* 2 vols. (Leipzig, 1900–1902); *Fields of Force* (New York, 1906); *Die Kraftfelder* (Brunswick, 1909); *Dynamic Meteorology and Hydrography,* 2 vols. (Washington, D.C., 1910–1911), Vol. I written with J. W. Sandström and Vol. II with Th. Hesselberg and O. Devik, also translated as *Dynamische Meteorologie und Hydrographie,* 2 vols. (Brunswick, 1912–1913); *On the Dynamics of the Circular Vortex With Applications to the Atmosphere and to Atmospheric Vortex and Wave Motion* (Kristiania, 1921); *C. A. Bjerknes. Hans liv og arbejde* (Oslo, 1925), translated as *C. A. Bjerknes. Sein Leben und seine Arbeit* (Berlin, 1933); *Teoretisk fysik* (Oslo, 1929); and *Physikalische Hydrodynamik mit Anwendung auf die dynamische Meteorologie* (Berlin, 1933), written with J. Bjerknes, H. Solberg, and T. Bergeron, translated as *Hydrodynamique physique avec applications à la météorologie dynamique* (Paris, 1934).

II. SECONDARY LITERATURE. Works on Bjerknes are T. Bergeron, "Vilhelm Bjerknes," in *Småskrifter udgivet af Universitet i Bergen,* no. 11 (1962), 7–30; T. Bergeron, O. Devik, and C. L. Godske, "Vilhelm Bjerknes," in *Geofysiske publikationer,* **24** (1962), 6–25; T. Hesselberg, in *Norsk biografisk leksikon,* I, 584–588; and Harald Wergeland, "Vilhelm Bjerknes," in *Det kongelige Norske videnskabers selskabs forhandlinger,* **24,** no. 16 (1951), 74–78.

MOGENS PIHL

BJERRUM, NIELS JANNIKSEN (*b.* Copenhagen, Denmark, 11 March 1879; *d.* Copenhagen, 30 September 1958), *chemistry, physics, history of science.*

Bjerrum was the son of the well-known ophthalmologist and university professor Jannik Petersen Bjerrum and Anna Johansen, and the nephew of Kirstine Bjerrum Meyer, who edited the works of Oersted (1920) and wrote an important treatise on the history of the concept of temperature, *Temperaturbegrebets udvikling* (1909). He completed his doctorate at the University of Copenhagen in 1908 under S. M. Jørgensen with a dissertation entitled *Studier over basiske kromiforbindelser: Bidrag til hydrolysens teori* (Copenhagen, 1908). He had begun to lecture on elementary inorganic chemistry in 1907, and in 1914 he was made professor of chemistry at the Royal Veterinary and Agricultural College in Copenhagen, a post he held until his retirement in 1949. During the early years of his career he studied with Robert Luther in Leipzig (1905), Alfred Werner in Zurich (1907), Jean Perrin in Paris (1910), and Walther Nernst in Berlin (1911). In 1907 Bjerrum married Ellen Emilie Dreyer. Their son Jannik has been professor of chemistry at the University of Copenhagen since 1948. In many ways his work is an extension of his father's.

In 1928 Bjerrum was awarded the Oersted Medal. He was a member of the Carlsberg Foundation, the Rask–Oersted Foundation, and the Committee of the Solvay Institute for Chemistry, as well as the academies of science of Denmark, Norway, Sweden, Vienna, and New York. His honorary memberships were in the chemical societies of Belgium, the Netherlands, Sweden, and Switzerland.

Noteworthy for the historian of science, primarily because it presents the point of view of a prominent theoretical and experimental physical chemist, is Bjerrum's *Fysik og kemi* (in *Det nittende aarhundrede,* **18**[1925], 71–192). This volume gives his interpretation of various late nineteenth-century developments in atomic theory, thermodynamics, the electromagnetic theory of light, relativity and the physics of the ether, and the structure of the atom. In a lecture of 1922, "Kemiens udvikling i det 19. aarhundrede," Bjerrum emphasized the historical importance of the "mathematization" of chemistry, the exploitation of the atomic theory for stereochemical considerations, and the coordination of science and technology, of theory and practice.

Bjerrum's contributions to chemical physics, an outgrowth of his work with Nernst in Berlin, were made primarily in four papers (1911–1914). They deal with the application of the kinetic and quantum theories, and employ information obtained from absorption measurements in the infrared to elucidate the constitution and the optical and thermal properties of matter. His theoretical studies on specific heat as a function of temperature for gases represent advances over the specific heat studies that had been made for solids by Einstein, Nernst, and Lindemann. In this work Bjerrum succeeded in demonstrating the interdependence of specific heats and the spectrum as required by the quantum theory. The infrared absorption spectra of water vapor were further related, on the quantum hypothesis, to line broadening caused by molecular rotational frequencies that vary discontinuously and to radiating atoms that do not rotate—thus providing agreement with specific heat investigations which suggest that the rotational energy of atoms must be very small.

In a number of more general papers, such as "Nyere undersøgelser over atomernes bevaegelser med saerligt henblik paa kvantehypotesen" (1915) and "Moderne atomlaere og kvanteteori" (1919), Bjerrum revealed an extraordinarily keen appreciation of the significance of the quantum theory for atomic-molecular problems in general.

Physical chemistry, theoretical as well as analytical, was Bjerrum's lifelong interest and is the subject of the major part of his publications. As early as 1909 Bjerrum proposed a new form for the electrolytic dissociation theory. In 1916, at the sixteenth meeting of Scandinavian scientists in Kristiana (Oslo), he presented, in a most convincing form, his now-celebrated view on the dissociation of strong electrolytes—according to which some acids and hydroxyl compounds, and most salts, are almost completely dissociated into ions in the dissolved state. Arrhenius, who was chairman of the meeting, had won the Nobel Prize in chemistry in 1903 for his electrolytic theory of dissociation, and was not willing to accept this extension of his own theory. Bjerrum's view, according to which the "anomalous" behavior of strong electrolytes should be interpreted in terms of interionic forces, was extended in a second paper in 1916 and then explored in depth in a series of studies (1920–1932) on electrolytic dissociation theory. These studies also dealt with the activity and distribution coefficients of ions, osmotic pressure, association of ions, the Debye-Hückel theory, and the solubility of gases.

The currently accepted method of introducing activity coefficients into the expression for the velocity of a chemical reaction was published in 1923 by the Danish chemist J. N. Brønsted in a paper entitled "Zur Theorie der chemischen Reaktionsgeschwindigkeit." Bjerrum's papers on chemical kinetics and the discussions that took place between the two men appeared in the Zeitschrift für Chemie of 1923–1925. Bjerrum's view on how his own new theories regarding acids, bases, and salts developed within the context of the older views of Brønsted is treated admirably in a lecture delivered before the Danish Chemical Society in 1931 ("Syrer, salte og baser").

Bjerrum's joint paper on Brownian motion with Jean Perrin, "L'agitation moléculaire dans les fluides visqueux" (1911), is noteworthy for having demonstrated that the equilibrium distribution of particles is independent of the viscosity of the fluid in the gravitational field.

The study of buffer mixtures and indicators, and the measurement of the hydrogen ion concentration of solutions, was an early and continuing interest. In a paper of 1905 Bjerrum introduced his well-known extrapolation method for the elimination of the diffusion potential. The theory upon which this study rests was developed further and improved by E. A. Guggenheim, who worked with Bjerrum for a number of years. In his study of the theory and the sources of error in acidimetric and alkalimetric titrations and of buffer solutions (Die Theorie der alkalimetrischen und azidimetrischen Titrierungen [Stuttgart, 1914]), Bjerrum showed how to determine the end point of a titration and how to estimate the error that accompanies the choice of pH values associated with the use of a particular indicator.

Bjerrum's contributions to the theory of acids and bases include a novel method of using the experimentally determined strength constants of different acidic and basic groups in a molecule to establish the constitution and dissociation constants of ampholytes, particularly of amino acids. In a paper of 1923 he applied the notion of the different strength constants of polybasic acids to the determination of molecular distances. On the practical side, Bjerrum's contributions to the needs of a country where agriculture is of paramount importance are seen in his papers devoted to the factors that determine the pH, and therefore the reaction, of the soil, the hardness of water, and the general application of physicochemical measurements to agricultural problems.

In his study of the coordination chemistry of inorganic complexes, Bjerrum went beyond the classical methods of analysis and synthesis of his teacher Jørgensen by emphasizing the importance of physicochemical principles. In 1906 Bjerrum published a comprehensive 120-page study on the chromic chlorides (Studier over kromiklorid) that revealed the existence and mode of isolation of the previously unknown chromium monochloropentaquo complex $[Cr\ Cl\ (H_2O)_5]^{++}$. Two years later, in his doctoral dissertation on the theory of hydrolysis of chromium compounds, he investigated the formation and relation between "truly basic" and "latently basic" complexes. Bjerrum's papers on the complex chromium and gold salts span a period of more than forty years.

Bjerrum also made substantial contributions to colloid theory, concentrating on the study of substances with high molecular weights, their colloidal properties (e.g., charge), and the preparation of collodion membranes with reproducible permeabilities (1924–1927).

The function of the thiocyanate group as a ligand in chromium and gold compounds was the subject of two significant papers of 1915 and 1918. In the second paper Bjerrum and Aage Kirschner proposed a sequence of reactions that explained the overall complex kinetics of the aqueous decomposition of dithiocyanogen—a compound the preparation of which from nonaqueous solution was first accomplished by Erik G. Söderbäck in 1919. In 1949 Bjerrum investigated the gold chloride complexes.

During the last decade of his life, Bjerrum extended his interest in problems of molecular structure to the study of ice. This work, summarized in *Structure and Properties of Ice* (1951), treats the position of the hydrogen atoms and their zero-point entropy, changes in configuration, ionization and "molecular turns," and the proton-jump conductivity of ice and water.

On the frontier where physics and chemistry interact, Niels Bjerrum achieved world renown during the early decades of this century. He published about eighty scientific papers (not including translations) and ten books and monographs. Best known among the latter is his *Laerbog i uorganisk kemi* (1916–1917), which went through six editions and was translated into English, German, and Russian.

BIBLIOGRAPHY

Niels Bjerrum, *Selected Papers,* ed. by friends and co-workers on the occasion of his seventieth birthday (Copenhagen, 1949), contains, besides the 27 selected papers trans. into English, J. A. Christiansen, "A Survey of the Scientific Papers of Niels Bjerrum," a foreword by Niels Bohr, and a bibliography of Bjerrum's scientific publications, books, and papers (1903–1948), items 1–92; J. A. Christiansen, "Niels Bjerrum," in *Fysisk tidsskrift,* **57** (1959), 24–36, appends a list of Bjerrum's publications (items 93–103) that appeared after the publication of the papers listed in *Selected Papers;* Aksel Tovborg Jensen, "Niels Bjerrum," in *Oversigt over det Kongelige Danske Videnskabernes Selskabs Virksomhed* (1958/1959), 99–113, is an excellent short account of his life. See also Stig Veibel, "N. J. Bjerrum," in *Dansk biografisk leksikon,* III (1934), 183–185; and *Kemien i Danmark,* II, *Dansk kemisk bibliografi 1880– 1935* (Copenhagen, 1943), 61–67, which contains a list of Bjerrum's 80 publications that appeared 1908–1935.

ERWIN N. HIEBERT

BLACK, DAVIDSON (*b.* Toronto, Canada, 25 July 1884; *d.* Peking, China, 15 March 1934), *anatomy, anthropology.*

By tradition Black's family followed the law (his father was Queen's Counsel), but early in life he showed a marked interest in biology and natural history. In 1903 he enrolled in the medical school of the University of Toronto, graduating in 1909 with M.D. and M.A. degrees. His first post was at Western Reserve University, Cleveland, Ohio, as anatomist; there he and T. Wingate Todd built up the museum of comparative anthropology and anatomy begun in 1893 by C. A. Hamann.

In 1914 Black went to Manchester, England, on sabbatical leave to study advanced anthropology under Grafton Elliot Smith. There he learned also to make casts, a skill that later proved valuable to him. Under Smith's auspices in London he met Arthur Keith, Arthur Smith Woodward, and Frederick Wood Jones. From London he went to Amsterdam to study neuroanatomy under C. V. Ariëns Kappers at the Central Institute of Brain Research. On his return to Cleveland he wrote his paper "Brain in Primitive Man" (1915).

Through his experience abroad and as the result of his intensive study of a treatise by William Diller Matthews, *Climate and Evolution,* Black became convinced that Asia had probably been the realm of early man and the center of dispersal of land mammals. In 1920 he had the opportunity to explore this theory when he accepted an appointment as anatomist and neurologist at the Peking Union Medical College. Always a tireless worker, in China, Black launched his brilliant career, teaching, writing, conducting field expeditions, and gathering about him as friends and co-workers such men as J. Gunnar Andersson, Wong Wen-hao, V. K. Ting, A. W. Grabau, C. C. Young, G. B. Barbour, and Teilhard de Chardin. His search for hominid fossils took him to eastern Mongolia, the Gobi Desert, Siam, Honan, and Kansu, and finally concentrated twenty-five miles from Peking at Chou K'ou-tien in the western hills. Here extensive excavations were undertaken by the archaeologists, and Black studied all fossil-bearing material in the laboratories of the Peking Union Medical College.

In 1927 a well-preserved left molar was recovered, and after study Black pronounced a new hominid genus, which he named *Sinanthropus pekinensis* Black and Zansky. Under the archaeologist W. C. Pei the site yielded numerous teeth and pieces of jawbone with teeth *in situ*. On 1 December 1929, in a cave seventy feet below working level, an almost complete skull cap was found in an environment of extinct animal bones, crudely chipped stones, and man-made fires. In 1930 a second skull was recovered. Black himself freed the skulls of their heavy coating of travertine, made casts, and wrote his series of reports on the discovery, morphology, and environment of *Sinanthropus pekinensis*. Among the scientists who flocked to Peking to consult with Black and to examine the fossils and the cave excavations were Henri Breuil, Walter Granger, Alĕs Hrdečha, and G. Elliot Smith.

In 1934 he died suddenly while working on *Sinanthropus pekinensis*.

Black was a fellow of the Royal Society of London and of the Geological Society of America, and an honorary member of the National Academy of Sciences, Washington, as well as of other scientific organizations.

BIBLIOGRAPHY

Black's works include "Brain in Primitive Man," in *Cleveland Medical Journal*, **14** (Mar. 1915), 177–185; "On a Lower Molar Hominid Tooth From the Chou Kou Tien Deposit," in *Palaeontologia Sinica*, ser. D, **7**, fasc. 1 (Nov. 1927), 1–28; "*Sinanthropus pekinensis*: The Recovery of Further Fossil Remains of This Early Hominid From the Chou Kou Tien Deposit," in *Science*, **69**, no. 1800 (June 1929), 674–676; "Preliminary Notice of the Discovery of an Adult *Sinanthropus* Skull at Choukoutien," in *Bulletin of the Geological Society of China*, **8**, no. 3 (1930), 207–230; "Notice of the Recovery of a Second Adult *Sinanthropus* Skull Specimen," *ibid.*, **9**, no. 2 (1930), 97–100; "On an Adolescent Skull of *Sinanthropus pekinensis* in Comparison With an Adult Skull of the Same Species and With Other Hominid Skulls, Recent and Fossil," in *Palaeontologica Sinica*, ser. D, **7**, fasc. 2 (1931), 1–144; "Present State of Knowledge Concerning the Morphology of *Sinanthropus*," in *Proceedings of the Fifth Pacific Science Congress* (Vancouver, B.C., 1933); "Fossil Man in China: The Choukoutien Cave Deposits With a Synopsis of Our Present Knowledge of the Late Cenozoic in China," in *Memoirs of the Geological Survey of China*, ser. A, no. 11 (1933), 1–168, written with Teilhard de Chardin, C. C. Young, and W. C. Pei; and "On the Discovery, Morphology and Environment of *Sinanthropus pekinensis*," in *Philosophical Transactions of the Royal Society*, **B223** (1934), 57–120.

See also Dora Hood, *Davidson Black, a Biography* (Toronto, 1964).

DORA HOOD

BLACK, JAMES (*b.* Scotland, *ca.* 1787; *d.* Edinburgh, Scotland, 30 April 1867), *medicine, geology.*

During the Napoleonic Wars, Black was a naval surgeon and served in the West Indies. Subsequently he practiced as a physician in Newton Stewart, Bolton, Manchester (1839–1849), and again in Bolton, until his retirement to Edinburgh in 1856.

In the course of caring for his patients, Black made careful observations and collected data that he related, through his wide medical reading, to medical theory and to social conditions. *Capillary Circulation* (1825) reports his only experimental work, mainly on capillaries in the feet of ducks and frogs, in which he repeated the experiments of John Thomson. In this work, Black describes the blood as moving faster in the arteries than in the veins, the anastomoses, and the effects of ammonia in decreasing the diameter of the blood vessels and speeding the flow of blood—which is followed by passive dilation, congestion, and typical symptoms of inflammation. He further reviews the literature and discusses the nature of inflammation, its causes and conditions, its symptoms, and its relation to capillary circulation. Black's work was not highly regarded by his colleagues, but he continued it in 1826 with *The Nature of Fever*, which similarly reviewed alternative hypotheses and contemporary practice.

In the spring of 1829, Black visited the United States, where he studied medical practice and education and wrote the appreciative and critical sketch of the state of medicine in America, which included descriptions of the institutions that he had visited in New York and Philadelphia. On returning to Bolton, he embarked on an extensive survey of that town, relating the physical and social environment to health and habits and noting industrial changes; this was published in 1837 as "Sketch of Bolton."

Black was also a competent amateur geologist and paleontologist and published a number of papers on these subjects; they were mostly local and descriptive, but they also dealt with such topics as submerged forests and coal formation. A founding member of the council of the Manchester Geological Society, he delivered an address to it, "On Some Objects and Uses of Geological Research," in which he discussed the development of organisms toward perfection in terms of successive acts of a Creative Power, in relation to changes in the environment.

Black was active and well loved in many spheres. His family life seems to have been happy and devout, and one of his sons also became a naval surgeon. He was a founder-member of the British Association in 1831, lectured on geology at the Bolton Mechanics

Institute, and served on the committee whose work resulted in the establishment of one of the first municipal public libraries (in Bolton) in 1853. In 1854 a lithographed portrait of Black was published by public subscription.

BIBLIOGRAPHY

I. ORIGINAL WORKS. Black's medical works include *A Short Inquiry Into the Capillary Circulation of the Blood* . . . (London, 1825); *A Comparative View of the More Intimate Nature of Fever* . . . (London, 1826); "A Sketch of the State of Medicine, and of Medical Schools and Institutions, in the United States of America . . .," in *New England Medical and Surgical Journal,* **1** (1830–1831), 209–219, 301–313, 398–409; and "A Medico-topographical, Geological, and Statistical Sketch of Bolton and Its Neighbourhood," in *Transactions of the Provincial Medical and Surgical Association,* **5** (1837), 125–224 and map.

His geological works include "On Some Objects and Uses of Geological Research," in *Transactions of the Manchester Geological Society,* **1** (1841), 1–34; and *An Eclectic View of Coal Formation* (Manchester, 1847).

II. SECONDARY LITERATURE. There is an anonymous "Obituary" in *British Medical Journal,* **1** (25 May 1867), 623. Signed works on Black are A. Sparke, "James Black," in *Bibliographia Boltonensis* (Manchester, 1913), pp. 26–27; C. W. Sutton, "James Black," in *Dictionary of National Biography,* V (London, 1886), pp. 106–107, which gives additional references; and P. A. Whittle, "James Black," in *Bolton-le-Moors* (Bolton, 1855), p. 392.

DIANA SIMPKINS

BLACK, JOSEPH (*b.* Bordeaux, France, 16 April 1728; *d.* Edinburgh, Scotland, 6 December 1799), *chemistry, physics, medicine.*

A founder of modern quantitative chemistry and discoverer of latent and specific heats, Joseph Black, although born in France, was by blood a pure Scot. His father, John Black, was a native of Belfast; his mother, Margaret Gordon, was the daughter of an Aberdeen man who, like John Black, had settled in Bordeaux as a factor, or commission merchant, in the wine trade.

John Black and Margaret Gordon were married at Bordeaux in 1716, apparently in the Catholic faith. Joseph, fourth of their twelve children, was first educated by his mother. At the age of twelve he was sent to Belfast, where he learned the rudiments of Latin and Greek in a private school. About 1744 he crossed the North Channel to attend, as did so many Ulster Scots, the University of Glasgow. Here he followed the standard curriculum until, pressed by his father to choose a profession, he elected medicine. At this point he began the study of anatomy and

attended the lectures in chemistry recently inaugurated by William Cullen. These lectures were the decisive influence on Black's career; chemistry captivated him, and for three years he served as Cullen's assistant. So began a close friendship that lasted until Cullen's death.

In 1752 Black left Glasgow for the more prestigious University of Edinburgh, which boasted on its medical faculty the great anatomist Alexander Monro *primus,* the physiologist Robert Whytt, and Charles Alston, a botanist and chemist who lectured on *materia medica.* Black gained less, he said, from their lectures than from the bedside clinical instruction provided by the university's Royal Infirmary. Alston's lectures pleased him most, although he found him deficient in chemical knowledge, a matter of concern, for, as he wrote to Cullen, "no branch should be more cultivated in a medical college."[1]

In 1754 Black received the M.D. with his now historic dissertation *De humore acido a cibis orto et magnesia alba.* The next year, before the Philosophical Society of Edinburgh, he described the chemical experiments, considerably expanded, that had formed the second half of his dissertation. This classic paper—the chief basis of Black's scientific renown and his only major publication—appeared in 1756 in the Society's *Essays and Observations* under the title "Experiments Upon Magnesia Alba, Quicklime, and Some Other Alcaline Substances." Here Black demonstrated that an aeriform fluid that he called "fixed air" (carbon dioxide gas) was a quantitative constituent of such alkaline substances as *magnesia alba,* lime, potash, and soda.

The same year, 1756, brought Cullen to Edinburgh as professor of chemistry, and saw Black—at the age of twenty-eight Cullen's outstanding student—replace him in Glasgow. Here Black spent the next ten years. Although this period is sparsely documented, we know that he soon emerged as a gifted and effective teacher. His course in chemistry, launched in 1757–1758, proved so popular that many students, some with no particular relish for the subject, pressed to attend. Alongside his teaching, Black carried on an active and demanding medical practice; and since Glasgow, unlike Edinburgh, was administered by its faculty, he was constantly pressed upon by multifarious college duties. Yet it was at Glasgow that he developed his ideas about latent and specific heats—the second of his major scientific achievements—and carried out experiments, alone or with his students, to confirm his theories. These important discoveries he could never be induced to publish.

In 1766 Black received the call to Edinburgh. William Cullen relinquished the chair of chemistry to succeed Robert Whytt as Professor of the Institutes of Medicine, and Black took over Cullen's chair. At Edinburgh he was destined to remain. Although his duties were less onerous than at Glasgow—he limited his medical practice to the care of a few close friends like David Hume—his period of scientific creativity was at an end. Two short papers on insignificant subjects were his only publications. The teaching of chemistry now became his central concern; here, as at Glasgow, he became an idol to the medical students and to many others as well. Each year, from October to May, he delivered a series of more than a hundred lectures, and sometimes offered a course during the summer months.

Black's pedagogic achievement at least equals that of his great French contemporary, G.-F. Rouelle. Although he had no student of the stature of Lavoisier, there were many of great ability. His audience was surprisingly cosmopolitan; although French students were rare, men came from Germany, Switzerland, Scandinavia, and from as far away as Russia and America, attracted by the reputation of the Scotch medical schools and of Black himself. Lorenz Crell, known as editor of early chemical journals, was one of his German students. To Edinburgh from the American colonies came such men as James McClurg, later a successful Richmond physician, and the still more famous Benjamin Rush. Black's British students were no less gifted. At Glasgow there were John Robison, who was to bring out in 1803 his master's *Lectures on Chemistry*, and William Irvine, Black's collaborator in the work on specific heat. His Edinburgh students included Thomas Charles Hope, who succeeded him in 1797; Daniel Rutherford, the discoverer of nitrogen; and John McLean, who emigrated to America in 1795, where he became Princeton's first professor of chemistry. Among the last to hear Black lecture were Thomas Young, the versatile physician, physicist, and linguist; the elegant and prolific Henry Brougham; and Thomas Thomson, chemist and pioneer historian of chemistry.

There are several contemporary descriptions of the appearance, personality, and lecturing skills of this great teacher. On the platform he was an immaculate figure; his voice was low but so clear that he was heard without difficulty by an audience of several hundred. His style was simple, his tone conversational, far different from Rouelle's flamboyance. He spoke extemporaneously from the scantiest of notes; yet his lectures, of which numerous manuscript versions by his students have survived, were models of order and precision: the facts and experiments led a listener by imperceptible degrees to the theories and principles by which he explained them. Vivid accounts of his own discoveries, and demonstration experiments conducted with unvarying success were the highlights of his performance. He kept abreast of the progress of chemistry: through the years the outline of the lectures remained the same, but new material was added as chemistry, that "opening science," as he called it, steadily advanced; and Black told his students of new discoveries and theories, and of the men who had made them.

Black was a typical valetudinarian; never robust, he suffered all his life from chronic ill health, perhaps pulmonary in origin. With only limited reserves of energy, he nevertheless managed by careful diet and moderate exercise—hours of walking were part of his regimen—to husband his strength. In his prime, as the portrait by David Martin depicts him, he was a handsome man; and even in old age his appearance was impressive. Henry Cockburn, who saw Black in his last years, gives the following description:

> He was a striking and beautiful person; tall, very thin, and cadaverously pale; his hair carefully powdered, though there was little of it except what was collected in a long thin queue; his eyes dark, clear, and large, like deep pools of pure water. He wore black speckless clothes, silk stockings, silver buckles, and either a slim green silk umbrella, or a genteel brown cane. The general frame and air were feeble and slender. The wildest boy respected Black. No lad could be irreverent towards a man so pale, so gentle, so elegant, and so illustrious. So he glided, like a spirit, through our rather mischievous sportiveness, unharmed.[2]

And so we see him, on one of his increasingly rare strolls, pictured by the sharp eye of the caricaturist John Kay: slim, slightly stooped, an intent and pensive figure.

Black never married, but he was no recluse. Calm, self-possessed, gentle, and a trifle diffident, he nevertheless enjoyed conviviality. Until at last his health failed him, he frequented, besides the Philosophical Society and the Royal Society of Edinburgh which replaced it in 1783, those informal clubs for which Edinburgh was famous: the Select, the Poker, and the Oyster. The Oyster, a weekly dining club, was his favorite; indeed, with his two closest friends, Adam Smith and the geologist James Hutton, he had founded it. Other members were his cousin Adam Ferguson, William Cullen, Dugald Stewart, John Playfair, and James Hall: in a word, the scientific luminaries of that remarkable Scotch Renaissance. Since the rising industrialists of the region were often at table—John Roebuck, Lord Dundonald, for example, and visitors like Henry Cort—the Oyster might be compared with the famous Lunar Society of

Birmingham, for discussion often turned on the role of science in technological progress.

One after another Black's friends passed from the scene: Cullen in 1790 and William Robertson, principal of the University of Edinburgh, in 1793. Adam Smith was the first of the triumvirate to die, but increasing infirmity afflicted the two surviving members. Hutton, wasted by years and illness, fell gravely ill in the winter of 1796/97. Black, too, found his feeble strength waning. He gave his last full course of lectures in 1795–1796; but, aware of his debility, he chose Thomas Charles Hope as his assistant and eventual successor. The next year his health worsened, and Hope in effect took over. For a time Black's health improved slightly, and he lingered on two more years. The manner of his death was so peaceful, in a way so characteristic of his methodical and undramatic life, that it has been several times recounted. Curiously, the early authorities on Black's life are mistaken about the date of his death, variously given as November 26 (Adam Ferguson, John Robison, and Lord Brougham) and November 10 (Thomson, also quoted by his modern biographer, Sir William Ramsay). But a letter from Robison to James Watt settles the point: Black died on 6 December 1799, in his seventy-second year.[3]

By scrupulous frugality, Black had quietly amassed a substantial competence, something in excess of £20,000. In his will, this sum was divided among his numerous heirs according to an ingenious, and of course mathematical, plan. Black had a certain reputation for parsimony—it is said that he weighed on a balance the guineas his students paid to attend his course—and his biographers have felt obliged to set the record straight by citing instances of his generosity: the loans he made to friends, the poor patients he treated without charge, and even the spaciousness of his house and the plenty of his table, "at which he never improperly declined any company."[4] He was, at the very least, as methodical in his financial affairs as he was in his science, his teaching, and all the other aspects of his life.

Black's investigation of alkaline substances had a medical origin. The presumed efficacy of limewater in dissolving urinary calculi ("the stone") was supported by the researches of two Edinburgh professors, Robert Whytt and Charles Alston. It interested Cullen as well, and Black came to Edinburgh as a medical student with the intention of exploring the subject for his doctoral dissertation.

But at this moment Whytt and Alston were at loggerheads: they disagreed as to the best source, whether cockleshells or limestone, for preparing the quicklime. And they differed as to what occurs when

mild limestone is burned to produce quicklime. Whytt accepted the common view that lime becomes caustic by absorbing a fiery matter during calcination, and thought he had proved it by showing that quicklime newly taken from the fire was the most powerful dissolvent of the stone. Alston, in an important experiment on the solubility of quicklime, showed that this was not the case, and that the causticity must be the property of the lime itself. Both men were aware that on exposure to the air quicklime gradually becomes mild, and that a crust appears on the surface of limewater. For Whytt, this resulted from the escape of fiery matter; but Alston, noting that the crust was heavier than the lime in solution, hinted that foreign matter, perhaps the air or something contained in it, produced the crust. Yet he was more disposed to believe that the insoluble precipitate formed when the quicklime combined with impurities in the water. Black, although he had criticized Alston as a chemist, was soon to profit from his findings.

Preoccupied at first with his medical studies, Black did not come to grips with his chosen problem until late in 1753. When he did so, he found it expedient to avoid any conflict between two of his professors; instead of investigating limewater, he would examine other absorbent earths to discover, if possible, a more powerful lithotriptic agent. He chose a white powder, *magnesia alba,* recently in vogue as a mild purgative. Its preparation and general properties had been described by the German chemist Friedrich Hoffmann; although it resembled the calcareous earths, *magnesia alba* was clearly distinguishable from them.

Black prepared this substance (basic magnesium carbonate) by reacting Epsom salts (magnesium sulfate) with pearl ashes (potassium carbonate). He treated the purified product with various acids, noting that the salts produced differed from the corresponding ones formed with lime. The *magnesia alba,* he observed, effervesced strongly with the acids, much like chalk or limestone.

Could a product similar to quicklime be formed by calcining *magnesia alba?* Would its solutions have the causticity and solvent power of limewater? Black's effort to test this possibility was the turning point of his research. When he strongly heated *magnesia alba,* the product proved to have unexpected properties. To be sure, like quicklime, this *magnesia usta* did not effervesce with acids. But since it was not sensibly caustic or readily soluble in water, it could hardly produce a substitute for limewater.

The properties of this substance now commanded Black's entire attention, notably the marked decrease in weight that resulted when *magnesia alba* changes into *magnesia usta.* What was lost? Using the balance more systematically than any chemist had done before him, he performed a series of quantitative experiments with all the accuracy he could command. Heating three ounces of *magnesia alba* in a retort, he determined that the whitish liquid that distilled over accounted for only a fraction of the weight lost. Tentatively he concluded that the major part must be due to expelled air. Whence came this air? Probably, he thought, from the pearl ashes used in making the *magnesia alba*; for Stephen Hales, he well knew, had shown long before that fixed alkali "certainly abounds in air."[5] If so, upon reconverting *magnesia usta* to the original powder, by combining it with fixed alkali, the original weight should be regained. This he proved to his satisfaction, recovering all but ten grains.

Magnesia usta, he soon found, formed with acids the same salts as *magnesia alba,* although it dissolved without effervescence. Only the presence or absence of air distinguished the two substances: *magnesia alba* loses its air on combining with acids, whereas the *magnesia usta* had evidently lost its air through strong heating before combining with acids.

Could the same process—the loss of combined air—also explain the transformation of lime into quicklime? Tentative experiments suggested something of the sort; but not until the work on magnesia was completed, late in 1753, did he examine this question. When he precipitated quicklime by adding common alkali, the white powder that settled out had all the properties of chalk, and it effervesced with acids. Early in 1754, Black wrote William Cullen that he had observed interesting things about the air produced when chalk was treated with acid: it had a pronounced but not disagreeable odor; it extinguished a candle placed nearby; and "a piece of burning paper, immersed in it, was put out as effectually as if it had been dipped in water."[6] This was an observation clearly worth pursuing. Nevertheless, he could no longer postpone the writing of his Latin dissertation and his preparation for his doctoral examination.

The dissertation is in two parts: the first, dealing with gastric acidity, was clearly added to give medical respectability to the work; Black was never happy about it, and hoped it would pass "without much notice."[7] The second set forth the experiments on *magnesia alba* and the tentative conclusions he drew from them. Nothing of significance was said about other alkaline substances or about what he was to call "fixed air."

The "Experiments," on the other hand, is a longer and more elaborate work. Like his dissertation, it is divided into two parts. In Part I, he recounts the experiments on *magnesia alba*; little or nothing is added, but now his theory is presented without equivocation: this substance he now describes as "a compound of a peculiar earth and fixed air."[8]

In Part II, Black describes experiments that enabled him to generalize the theory and to support his explanation of causticity. When lime is calcined, air is given off in abundance, and the caustic properties of the resulting quicklime do not derive from some fiery matter, but from the lime itself. He showed by experiment, in effect confirming what Alston had already done, that all of a given amount of quicklime, not merely a part of it, is capable of solution, if enough water is used. Thus the mysterious property of causticity is associated with a definite chemical entity having a definite solubility. As in the case of magnesia, he showed that calcareous earth combines with the same quantity of acid whether it is in the form of chalk (combined with air) or of quicklime. Again quicklime, made from a measured weight of chalk, when saturated with a fixed alkali can be converted into a fine powder nearly equal in weight to the original chalk. The quicklime was evidently saturated with air obtained from the alkali.

Black's theory also explained the production of strong or caustic alkalies (e.g., caustic potash) prepared by boiling quicklime with a solution of a mild alkali. What must occur is not, as chemists thought, that the acrimony of the potash is derived from the lime, but that fixed air is transferred from the mild alkali to the quicklime, thereby uncovering the inherent causticity of the alkali. Careful experiments confirmed this new extension of his doctrine.

A conclusive test of his theory of inherent causticity was Black's demonstration that both quicklime and *magnesia usta* could be produced by the "wet way," without the use of fire. He argued that if caustic alkali is caustic when not combined with "fixed air," it should separate magnesia from combination with acid and deposit it as *magnesia usta*. This he easily demonstrated. He performed a similar experiment with chalk.

An important collateral investigation stemmed from Black's experiments on the solubility of quicklime. Although he established to his satisfaction that it could be almost completely dissolved, he was puzzled not to find a larger residue of insoluble matter, for the air dissolved in water ought to combine with the quicklime to form a small amount of insoluble earth (carbonate). Perhaps the air had been driven off when the water was saturated with quicklime. The rough experiment to test this was performed after he had presented his major results to the Philosophical Society. In the receiver of an air pump, Black placed a small vessel containing four ounces of limewater; alongside it he put an identical vessel containing the same quantity of pure water. When the receiver was exhausted, the same amount of air appeared to bubble from both vessels. Clearly, the limewater contained dissolved air, but not of the kind that combined so readily with quicklime. In his "Experiments" he wrote: "Quicklime therefore does not attract air when in its most ordinary form, but is capable of being joined to one particular species only."[9] This he proposed to call "fixed air," preferring to use a name already familiar "in philosophy," rather than invent a new one. The nature and properties of this substance, he wrote, "will probably be the subject of my further inquiry."[10]

Black had shown that a particular kind of air, different from common air, can be a quantitative constituent of ordinary substances and must enter, as Lavoisier put it later, into their "definition." But he was not destined to make the investigation of such elastic fluids his "future inquiry." This was to be mainly the work of his British disciples—MacBride, Cavendish, Priestley, and Rutherford—and he published nothing further on the subject. Nevertheless, from his *Lectures* and other bits of evidence we learn that his discoveries did not end abruptly with the publication of his "Experiments." He knew that "fixed air" did not support combustion, that it had a density greater than common air, and that its behavior with alkaline substances resembled that of a weak acid. By experiments with birds and small animals, he soon demonstrated that this air would not support life. Using the limewater test, he showed that air expired in respiration consisted mainly of "fixed air"; and likewise that the elastic fluid given off in alcoholic fermentation, like that produced in burning charcoal, was identical with the "fixed air" yielded by mild alkalies when they effervesce with acids.

Black's doctrines did not have the prompt success on the Continent that they enjoyed in Britain. His influence on the early stages of the chemical revolution in France was far less than scholars have imagined; indeed, at first it was negligible. Before 1773 French chemists were unfamiliar with his "Experiments," which had appeared in English in an obscure publication. What they knew of his work derived largely from the arguments advanced against him by the German chemist J. F. Meyer, whose rival theory of *acidum pingue* was for a time widely credited. Black's case would surely have been strengthened had he published the simple experiment, performed at Glasgow in 1757 or 1758, of directly impregnating a solution of caustic alkali with the "fixed air" expelled from chalk or limestone, and so obtaining a product both effervescent and mild.

Black's discoveries concerning heat, the major achievement of his Glasgow period, were originally stimulated by William Cullen. In 1754 Cullen noted a striking phenomenon—the intense cold produced when highly volatile substances like ether evaporate—and he promptly wrote Black about his experiment. At about this time, Black set down in a notebook a curious observation made by Fahrenheit: water can be cooled below the freezing point without congealing; yet if shaken, it suddenly freezes and the thermometer rises abruptly to 32° on Fahrenheit's scale. This, Black speculated, might be due to "heat unnecessary to ice."

Fahrenheit's observation was recorded in Boerhaave's *Elementa chemiae,* a famous work that Cullen, and later Black, recommended in the English version of Peter Shaw to their students. This observation was hard to reconcile with the prevailing view that when water is brought near the freezing point, withdrawal of a small increment of heat must bring prompt solidification. But Fahrenheit's experiment showed that solidification (or liquefaction) required the transfer of substantial quantities of heat: of heat lying concealed and not directly detectable by the thermometer; of heat, to use Black's term, that was *latent.* Upon reflection, Black saw this notion to be quite consistent with commonly observed facts of nature. Snow, for example, requires a considerable time to melt after the surrounding temperature has risen well above the freezing point. A gradual absorption of heat must therefore be taking place, although the temperature of the snow remains unaltered.

Black became convinced of the reality of this latent heat through thoughtful reading and meditation on the familiar phenomena of change of state. He presented his doctrine in his Glasgow lectures, perhaps as early as 1757–1758, before he had performed any experiments of his own. Nor did his doctrine arise from any firmly held theory as to what heat might be.

Not until 1760 did Black carry out his earliest experiments on heat. The first fact to be ascertained was the reliability of the thermometer as a measuring tool. Would a thermometric fluid, having received equal increments of heat, show equal increments of expansion? Ingenious and simple experiments on mixing amounts of hot and cold water (*Lectures,* I, 56–58) convinced him that the scale of expansion of mercury, over that limited range, was indeed a reliable scale of "the various heats, or temperatures of heat."

Crucial to Black's experiments was his recognition of the distinction between *quantity of heat* and *temperature,* between what we sometimes describe as the *extensive* and *intensive* measures of heat. Although not the first to note this distinction, he was the earliest to sense its fundamental importance and to make systematic use of it. In his lectures, which always opened (after certain preliminaries) with a careful discussion of heat, he would tell his students:

> Heat may be considered, either in respect of its quantity, or of its intensity. Thus two lbs. of water, equally heated, must contain double the quantity that one of them does, though the thermometer applied to them separately, or together, stands at precisely the same point, because it requires double the time to heat two lbs. as it does to heat one.[11]

Temperature, of course, is read directly from the thermometer. But how to measure the *quantity* of heat? Black's answer is implied in the above quotation: the time required to warm or cool a body to a given temperature is related to the amount of heat transferred. This elusive quantity required a *dynamic* measurement: the heat gained or lost should be proportional to the temperature and the time of heat flow "taken conjointly."[12] Here, as so often in his career, the influence of Newton is quite apparent. In a famous paper of 1701, Newton had used a dynamic method to estimate temperatures beyond the reach of his linseed oil thermometer. Black in his lectures gave a clear account of Newton's experiments and how his law of cooling (as we now call it) was used to estimate relative temperatures above that of melting tin. Black's own experiments made use of the law of cooling, and it is not hard to imagine that Newton's dynamic method was the key to Black's.

Characteristically, Black was not satisfied to demonstrate qualitatively that there is such a thing as the latent heat of fusion: he proposed to measure it. The method occurred to him in the summer of 1761. First cooling a given mass of water to about 33°F., he would determine the time necessary to raise its temperature one degree, and compare this with the time required to melt the same amount of ice. Conversely, he would compare the time necessary to lower the temperature of a mass of water with the time necessary to freeze it completely. Assuming that both systems received heat from, or gave up heat to, the surrounding air at the same rate, as much heat should be given off in freezing a given amount of water as in melting the same amount of ice. Obliged to wait until winter, Black carried out the experiment in

December 1761 in a large hall adjoining his college rooms. The following April he described his results to his Glasgow colleagues and friends.

Black soon saw that latent heat must play a part in the vaporization of water as well as in the melting of ice. The analogy was so persuasive that as early as 1761—before testing his conjecture by experiment—he presented this version of his doctrine to his students. The success of the freezing experiments soon led him to investigate the latent heat of vaporization. But the method he first employed, a precise analogue of Fahrenheit's observation on supercooled water, was unsuited to measurement. A better method could be modeled on his freezing experiments. Tin vessels, containing measured amounts of water, would be heated on a red-hot cast iron plate. The time necessary to heat the water from 50°F. to the boiling point would be compared with the time necessary for the water to boil away. The chief obstacle was to find a source of heat sufficiently unvarying so that the absorption of heat could be safely measured by the time. A "practical distiller" informed Black that when his furnace was in good order, he could tell, to a pint, the quantity of liquor that he would get in an hour. When Black confirmed this by boiling off small quantities of water on his own laboratory furnace, he was ready for the experiments.[13] These he performed late in 1762. From the average of three experiments he calculated that the heat absorbed in vaporization was equal to that which would have raised the same amount of water to 810°, were this actually possible. This gives a figure of 450 calories per gram for the latent heat of vaporization of water, compared to a modern figure of 539.1 calories per gram. More accurate figures were obtained in later experiments, but several years elapsed before Black took up the subject again.

The second, and closely related, discovery made by Black concerning heat was that different substances have different heat capacities. It is commonly assumed that Black discovered *specific heats* before his work on latent heat. This is not the case. To be sure, the clue, once again, was found in Boerhaave's textbook; Fahrenheit had made certain experiments at his request and had obtained the surprising result that when he mixed equal quantities of mercury and water, each at a different temperature, the mercury exerted far less effect in heating or cooling the mixture than did the water. At first Black was puzzled; but he soon realized that mercury, despite its greater density, must have a smaller store of heat than an equal amount of water at the same temperature. If so, the capacities of bodies to store up heat did not vary with their bulk or density, but in a different fashion "for which no general principle or reason can yet be assigned."[14] Now Black could explain a peculiar effect reported twenty years earlier by George Martine, an authority on thermometers. Martine had placed equal volumes of water and mercury in identical vessels before a fire, and observed that the mercury increased in temperature almost twice as fast as the water. Black saw that since less heat was required to bring mercury up to a given temperature, a thermometer placed in it should rise more rapidly. Not until 1760 did Black perceive the significance of this effect, but he did not pursue the subject; he was principally absorbed with the more striking phenomena of changes of state. In 1764—a year that James Watt made memorable in the history of invention—Black returned to the study of heat. His experimental inquiry into specific heats, and his attempts to obtain a more accurate value for the latent heat of vaporization, were stimulated by the activities of Watt.

The year that Black began his Glasgow lectures, James Watt, a young man of nineteen, skilled in making mathematical instruments, was taken under the wing of the university as what we might today call a technician. He was soon called upon by Black to make things he needed for his experiments. Watt, in turn, after repairing the now-famous model of a Newcomen engine and undertaking experiments to improve its performance, turned to Black to explain an effect he could not comprehend. He was astonished at the large amount of cold water required to condense the steam in the engine cylinder, until Black explained his ideas about latent heat.

Watt was many months, and many experiments, away from hitting upon the historic invention of the separate condenser, and Black may be pardoned for believing that this disclosure inspired Watt's radical improvement of the steam engine, a claim advanced even more strongly by John Robison, who spoke of Watt as Dr. Black's most illustrious pupil. This, in a strict sense, Watt never was; and, despite his lifelong attachment to Black, he later insisted that the invention of the separate condenser had not been suggested by his knowledge of the doctrine of latent heat. But he readily credited Black with having clarified the problems he encountered and with teaching him "to reason and experiment in natural philosophy."[15]

On the other hand, Watt's ingenuity and questioning mind, and the practical problems he raised, revived Black's interest in heat. The problems he now investigated with John Robison and William Irvine were closely related to those Watt needed to elucidate. Irvine was set to work determining a more accurate value of the latent heat of steam. Using a common laboratory still as a water calorimeter, Irvine obtained improved values, although these were not high enough to be really accurate. Black, it should be remarked, never made or used the mythical ice calorimeter associated with his name, although it occurred to him in the spring or summer of 1764 that his knowledge of the latent heat of fusion of ice could be used to measure the latent heat of steam. Plans to put this to the test were set aside when Watt, late in 1764, began to obtain values that Black deemed sufficiently precise. Years later, Black gave the French scientists full credit for the independent invention and first use of an ice calorimeter.[16]

The measurements made at this time by Black, Irvine, and Watt on the specific heats of various substances are the earliest of which we have any trace. Watt seems to have been the first to stress the importance of investigating the subject systematically. He carried out experiments of his own, and Black put Irvine to work on the problem. Using the method of mixtures, Black and Irvine determined the heats communicated to water by a number of different solids. These joint experiments continued until Black left for Edinburgh. After Black's departure, Irvine continued these investigations, but his results were not published in his lifetime.

Joseph Black's view of chemistry, his chemical doctrines, can be derived from his single major paper and from the various versions of his lectures. Chemistry, to him, was a subject with wide practical application to medicine and to the progress of industry. But he insisted, as Cullen had done, that it is a science, albeit an imperfect one, not merely an art: "the study of the effects of heat and mixture, natural or artificial, with a view to the improvements of arts and natural knowledge."[17]

On the question of the "elements" or "principles" of bodies, Black showed a typical caution. He no longer credited the venerable doctrine of the four elements; little could be known about them, for there was no knowledge of the underlying constitution and forces of nature. It was more sensible to group into several classes, as Cullen did, those substances sharing certain distinguishable properties: the salts, earths, inflammable substances, metals, and water. These were not necessarily elementary; of earths there were several sorts that could not be decomposed further; water, Black believed, can on distillation be converted into earth; and there was reason to suspect that salts were not compounds of earth and water, but of an earth and some other unknown substance.

Black invariably devoted the early lectures of his course to the subject of heat, telling his students that Boerhaave, Robert Boyle, and Sir Isaac Newton followed Lord Bacon in believing that heat is caused by motion, and that the French thought heat to be the vibration of an imponderable, elastic fluid. The fluid theory was the one to which he quite definitely leaned, for it seemed to agree best with the phenomena; he could not, for example, readily conceive a motion of particles in dense, solid bodies. But such questions are involved in obscurity. And he told his students: "The way to acquire a just idea of heat is to study the facts."[18]

In discussing problems of combustion, fermentation, and the calcination of metals, Black—until the close of his career—presented a gingerly version of the phlogiston theory, although he generally avoided the term. Air, of course, was required for combustion; but like his contemporaries he invoked the property of elasticity to explain its role. Combustion was caused by the presence of an inflammable principle, for which different substances had a different "elective attraction." Inflammable and combustible substances, including the calcinable metals, were pervaded with this mysterious substance, the nature of which "we are still at a loss to explain." Although little could be said on the subject, heat and light appeared to be the principles of inflammability. There was, however, a fact that was hard to explain and was a strong objection to this theory. When it was possible to collect the product of combustion, or weigh a calcined metal, this product was heavier, despite the loss of the inflammable principle. Possibly this was a kind of matter that defied the general law of gravitation. Yet speculations of this sort were not Black's cup of tea. These doubts almost certainly prepared him to accept, in the main, Lavoisier's discoveries.

As late as 1785 Black was reluctant to adopt the new "French chemistry," a term he heartily disliked. The geologist Sir James Hall was the earliest of Black's circle to sense the winds of change. A visit to Paris in 1786 convinced him, and on his return to Scotland, in the course of long discussions, he brought Thomas Charles Hope around to his opinions. Early in 1788 Hall read before the Royal Society of Edinburgh a paper entitled "A View of M. Lavoisier's New Theory of Chemistry." Black may well have been present—we cannot be sure—but his intimate friend James Hutton was, and defended the phlogistic hypothesis with a paper of his own. Nevertheless, an Italian visitor to England, an admirer of Mme. Lavoisier, wrote her from London in 1788 that Black and Watt ("in my opinion the two best heads in Great Britain") were on the verge of being convinced by Lavoisier's antiphlogistic theory. Soon thereafter, Black began to mention the new doctrine

in his lectures; he wrote to Lavoisier, in a famous letter of October 1790, that he had begun to recommend the new system to his students as simpler, and more in accord with the facts, than the old. Robison's edition of Black's lectures, based on what Black was telling his students between 1792 and about 1796, amply confirms his statement. Yet it is clear that Black strongly disapproved of the new nomenclature, while recognizing that advances in chemistry made some such reform necessary. He objected to obliterating the work of the earlier chemists completely; he felt that the new terms were "evidently contrived to suit the genius of the French language," and he perceived in the new scheme a clever stratagem of the French chemists to give their doctrines "universal currency and authority."[19]

The most interesting and pervasive of Black's doctrines is the theory of chemical affinity. Here his debt to Cullen, and beyond Cullen to Newton's *Opticks,* is clearly evident. Chemical reactions result from the differential or "elective" attraction of chemical individuals for one another. Simple elective attractions are those produced by heat; "double elective attractions," reactions of double decomposition, are chiefly those that take place in solution. Black saw no reason for avoiding the term "attraction," as did the French chemists who spoke instead of "affinities" or "rapports." As Newton had insisted, "attraction" should be taken as a descriptive term, not a causal explanation.

Black's earliest use of this concept, and of Geoffroy's well-known table of affinities, appears in his "Experiments." He employs it to show the differential behavior of alkaline substances toward acids and "fixed air." For Black, as for Cullen, this became a centrally important pedagogical device; invariably he devoted several lectures to elective attractions, describing the table, and referring to it elsewhere when speaking of particular reactions. In the lectures of the early years Black set forth these reactions with the diagrams Cullen had invented, adding numbers to indicate the relative force of attraction between substances:

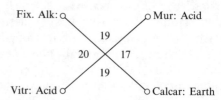

Later, he used a diagram consisting of segmented circles, but without numbers to indicate the relative attractions.[20] The following example illustrates what takes place when a compound of volatile alkali with any acid (e.g., ammonium chloride) reacts with a mild fixed alkali (e.g., sodium carbonate); here the "fixed air" or "mephitic air" represented by **MA** combines with the volatile alkali, ♉; and the fixed alkali, ♌, joins itself to the acid, ≻:

Did Black wish to represent molecules in which the atomic partners are interchanged? Perhaps this occurred to him, but he probably conceived these diagrams primarily as what we call visual aids. Unlike his master, William Cullen, Black did not explicitly link his discussion of elective attractions with the corpuscular or atomic doctrine, about which he had little to say in his lectures.

Robison records that in an early conversation Black "gently and gracefully" checked his disposition to form theories and warned him to reject "even without examination, every hypothetical explanation, as a mere waste of time and ingenuity."[21] Like Newton, whom he so greatly esteemed—at least Newton as he understood him—Black chose not to "deal in conjectures."

In mid-career he told his students, well before the fulfillment of what we call the chemical revolution:

> Upon the whole, Chymistry is as yet but an opening science, closely connected with the usefull and ornamental Arts, and worthy the attention of a liberal mind. And it must always become more and more so: for though it is only of late, that it has been looked upon in that light, the great progress already made in Chymical knowledge, gives us a pleasant prospect of rich additions to it. The Science is now studied on solid and rational grounds. While our knowledge is imperfect, it is apt to run into errour: but Experiment is the thread that will lead us out of the labyrinth.[22]

NOTES

1. Thomson, *Cullen,* I, 573.
2. Henry Cockburn, *Memorials of his Times* (Edinburgh, 1856), pp. 48–49.
3. Muirhead, *Origin and Progress,* II, 261–263.
4. Ferguson, "Minutes," p. 116.
5. *Dissertatio,* p. 272, and "Experiments," p. 17. Compare *Lectures,* II, 63–64.
6. Thomson, *op. cit.,* I, 50.
7. Letter to Cullen, 18 June 1754, *ibid.,* pp. 50–51.
8. "Experiments," p. 25.
9. *Ibid.,* pp. 30–31.
10. *Ibid.,* p. 31.
11. Law, "Notes of Black's Lectures," I, 5.
12. *Ibid.,* p. 18.
13. *Lectures,* I, 157.
14. *Ibid.,* p. 79.

15. Muirhead, *op. cit.,* 264.
16. *Lectures,* I, 175.
17. Cochrane, "Notes From Black's Lectures, 1767/8," p. 3. Black saw no reason to change his definition: see *Lectures,* I, 12–13.
18. Law, *op. cit.,* I, 5. Compare *Lectures,* I, 35.
19. *Lectures,* I, 489–493.
20. Henry Guerlac, "Commentary on the Papers of Cyril Stanley Smith and Marie Boas," in Marshall Clagett, ed., *Critical Problems in the History of Science* (Madison, Wis., 1959), pp. 515–519. For a fuller treatment, see M. P. Crosland, "The Use of Diagrams as Chemical 'Equations' in the Lecture Notes of William Cullen and Joseph Black." Crosland argues that Black wished to illustrate the course of chemical reactions without implying mechanical explanations and that his symbols were generalized expressions for reactions of a similar type.
21. *Lectures,* I, vii.
22. Law, *op. cit.,* III, 88.

BIBLIOGRAPHY

I. ORIGINAL WORKS. Black's published writings are the following:

Dissertatio medica inauguralis . . . (Edinburgh, 1754). Reprinted in *Thesaurus medicus Edinburgensis novus,* II (Edinburgh–London, 1785). English translation by A. Crum Brown in *Journal of Chemical Education,* **12** (1935), 225–228, 268–273, with facsimile of title page and dedication.

"Experiments Upon Magnesia Alba, Quicklime, and Some Other Alcaline Substances," in *Essays and Observations, Physical and Literary. Read Before a Society in Edinburgh,* **2** (1756), 157–225. Republished, together with Cullen's "Essay on the Cold Produced by Evaporating Fluids" (Edinburgh, 1777; repr. 1782). Black's famous paper is most readily available as Alembic Club Reprints no. 1 (Edinburgh, 1898). The first French translation was "Expériences sur la magnésie blanche, la chaux vive, & sur d'autres substances alcalines, par M. Joseph Black, Docteur en Médecine," in *Observations sur la physique,* **1** (1773), 210–220, 261–275. A short summary of Black's work on magnesia had been published in the *Journal de médecine, chirurgie, pharmacie,* **8** (1758), 254–261.

"On the Supposed Effect of Boiling Upon Water in Disposing It to Freeze More Readily," in *Philosophical Transactions of the Royal Society of London,* **65** (1775), 124–129.

"Lettre de M. Joseph Black à M. Lavoisier," in *Annales de chimie,* **8** (1791), 225–229. The English original of Black's letter was printed by Douglas McKie in *Notes and Records of the Royal Society of London,* **7** (1950), 9–11.

"An Analysis of the Water of Some Hot Springs in Iceland," in *Transactions of the Royal Society of Edinburgh,* **3** (1794), 95–126.

Lectures on the Elements of Chemistry, John Robison, ed., 2 vols. (Edinburgh, 1803; American ed., 3 vols., Philadelphia, 1806–1807). Robison omitted Black's introductory lecture and much of the next two or three lectures. In a letter of 16 September 1802 to James Black, Joseph's brother, he tells of his difficulties in putting together, from Dr. Black's sparse notes, a coherent text. He even speaks of having to "manufacture" one lecture; obviously, although Robison's edition probably represents Black's opinions, the language may sometimes be Robison's. The text should be compared with the MS versions. See McKie, *Annals of Science,* **16** (1960), 131–134, 161–170.

"Case of Adam Ferguson, Drawn up by Joseph Black, M.D., in May, 1797," in *Medico-Chirurgical Transactions,* **8** (1816). Cited and summarized by Crowther.

"A Letter From Dr. Black to James Smithson, Esq. Describing a Very Sensible Balance," in *Annals of Philosophy,* n.s. **10** (1825), 52–54. Black's letter is dated 18 September 1790.

The earliest pictures of Black are two ink sketches made by Thomas Cochrane in his notes while attending Black's lectures in 1767–1768 in what appears to be an anatomical theater. Reproduced by McKie in *Annals of Science,* **1** (1936), 110, and in his edition of Cochrane's "Notes 1767/8."

Not much later (*ca.* 1770) is a fine oil by David Martin, the teacher of Sir Henry Raeburn (Collection of the University of Edinburgh). Published by Guerlac in *Isis,* **48** (1957) and by McKie as frontispiece to his edition of the Cochrane "Notes 1767/8."

The most familiar portrait of Black is by Raeburn (Collection of the University of Edinburgh), showing Black at about the age of sixty; often reproduced, sometimes (as in the frontispiece of Robison's edition of Black's *Lectures*) from an inferior engraving.

Roughly contemporaneous with the Raeburn portrait are the sketches of John Kay: one shows Black walking in the country; another places *en face* a birdlike Hutton and a pensive Black. The most interesting is a close view of Black lecturing, spectacles in hand, and before him a spate of scattered notes, a syphon, a burning candle, and a small bird in a cage, all to demonstrate the properties of "fixed air." See *A Series of Portraits and Caricature Etchings by the Late John Kay,* 2 vols. (Edinburgh, 1837), I, pt. 1, 52–57. Reproductions by Ramsay in his biography of Black and, more attractively, by John Read in *Humour and Humanism in Chemistry,* plates 41 and 44–45.

No published census of manuscript versions of Black's lectures exists, but a few may be mentioned, together with their present locations. The earliest is Thomas Cochrane's (Andersonian Library, University of Strathclyde, Glasgow); dating from Black's early teaching at Edinburgh, it is otherwise notable only for the caricatures of Black. Very sketchy, it has recently been published as *Notes from Doctor Black's Lectures on Chemistry 1767/8,* ed. with intro., by Douglas McKie (Cheshire, 1966).

More complete are the closely written set of 120 lectures recorded by James Johnson, 1770 (University of Edinburgh Library), and three volumes bearing the name of Joseph Freyer Rastrick and covering Black's lectures of 1769–1770 (History of Science Collections, Cornell University Library). A fine set is the Beaufoy MS, 1771/72 (University of Saint Andrews). Quite different from others is Alexander Law's "Notes of Doctor Black's Lectures on Chemistry" (University of Edinburgh Library), with fifty-seven lectures from 13 June to 22 December 1775, and elaborate notes

and appendices. For Black's later period there are the notes of George Cayley, 6 vols., 118 lectures, 1785–1786 (York Medical Society) and a similar set, without date or name of original owner, now at University College, London. McKie, in *Annals of Science* (1959), p. 73, believes this set to be contemporary with the Cayley notes.

II. SECONDARY LITERATURE. A brief, uninspired account of a visit to Black in 1784 (largely devoted to a description of Black's portable furnace) is given by the geologist Barthélemy Faujas de Saint-Fond, in his *Voyage en Angleterre et en Ecosse,* II (Paris, 1797), 267–272. The earliest biographical sketch is the short, anonymous account, probably by Alexander Tilloch, in *Philosophical Magazine,* **10** (1801), 157–158. But the most important source is Adam Ferguson, "Minutes of the Life and Character of Joseph Black, M.D.," in *Transactions of the Royal Society of Edinburgh,* **5** (1805), 101–117. Ferguson was Black's cousin and close friend. John Robison, "Editor's Preface," in *Lectures,* I, v–lxvi, draws heavily on Ferguson, yet adds much useful information. Thomas Thomson, *History of Chemistry,* I (London, 1830), ch. 9, relies on Ferguson and Robison, but adds personal impressions. Thomson also contributed short accounts of Black to the *Annals of Philosophy,* **5** (1815), 321–327, and the *Edinburgh Encyclopedia.* Lord Brougham, like Thomson one of Black's last students, devotes an interesting chapter to Black in his *Lives of Philosophers of the Time of George III* (London–Glasgow, 1855), pp. 1–24. George Wilson's brief note, in *Proceedings of the Royal Society of Edinburgh,* **2** (1849), 238, corrects the date of Black's death as given by Ferguson and Robison, citing newspaper accounts and Muirhead.

John Playfair's "Biographical Account of the Late Dr. James Hutton," in *Transactions of the Royal Society of Edinburgh,* **5,** pp. 39–99 of the "History of the Society," has references to Black. James Patrick Muirhead, *Origin and Progress of the Mechanical Inventions of James Watt,* 3 vols. (London, 1854), published numerous letters from Watt to Black, and one from Black to Watt. Also valuable is John Thomson's *Account of the Life, Lectures and Writings of William Cullen, M.D.,* 2 vols. (Edinburgh–London, 1859), with several early letters of Black to Cullen.

For Glasgow University in Black's time, see Henry G. Graham, *Social Life of Scotland in the Eighteenth Century,* II (London, 1899), chs. 12–13; W. Innes Addison, *Roll of Graduates of the University of Glasgow* (Glasgow, 1898) and *Matriculation Album of the University of Glasgow* (Glasgow, 1913). Worth consulting is W. R. Scott, *Adam Smith as Student and Professor* (Glasgow, 1937). Letters of Thomas Reid in *The Works of Thomas Reid, D.D.,* Sir William Hamilton, ed., 7th ed., I (Edinburgh, 1872), 39–50, describe Black's Glasgow lectures. For the chair of chemistry at Glasgow, see Andrew Kent, ed., *An Eighteenth Century Lectureship in Chemistry* (Glasgow, 1950). For Edinburgh, consult Alexander Bower, *History of the University of Edinburgh,* 2 vols. (Edinburgh, 1817); and Sir Alexander Grant, *Story of the University of Edinburgh,* 2 vols. (London, 1884).

Agnes Clarke's article on Black in the *Dictionary of*

National Biography is disappointing and sometimes inaccurate, but gives the correct date for Black's death. The most scientifically eminent of Black's modern biographers is Sir William Ramsay. He first discussed Black's work in his Gases of the Atmosphere (London, 1896), pp. 527–531, and again in his Joseph Black, M.D., A Discourse (Glasgow, 1904). His Life and Letters of Joseph Black, M.D. (London, 1918) is the only full-length biography; published posthumously, it is valuable chiefly for the use made of letters and papers of Black, including an autobiographical sketch, which have been otherwise inaccessible to scholars. Ramsay's book is unsatisfactory; when a scholarly biography is written, and one is badly needed, use will surely be made of Henry Riddell's "The Great Chemist, Joseph Black, His Belfast Friends and Family Connections," in Proceedings of the Belfast Natural History and Philosophical Society, **3** (1919/20), 49–88.

Black is, of course, discussed in the familiar histories or studies of early chemists. Most can be ignored; an exception is Max Speter's "Black," in G. Bugge, Das Buch der grossen Chemiker, 2 vols. (Weinheim, 1929), I, 240–252. J. R. Partington, History of Chemistry, III (London–New York, 1962), 131–143, appraises Black's proficiency as a chemist and gives detailed citations of the literature. John Read has a readable, if not wholly reliable, account of Black in his Humour and Humanism in Chemistry (London, 1947), ch. 8, and a brisk chapter in Kent's Eighteenth Century Lectureship, pp. 78–98. A longer and more informative account of Black, with new insights and some inaccuracies, is J. G. Crowther, Scientists of the Industrial Revolution (London, 1962), pp. 9–92. Archibald and Nan L. Clow, The Chemical Revolution (London, 1952), has a misleading title: it deals with the applications of chemistry to industry in the eighteenth and early nineteenth centuries, and is valuable for many passing references to Black's involvement in such matters.

Douglas McKie's paper on the Cochrane "Notes," in Annals of Science, **1** (1936), 101–110, has been superseded by his edition of that MS. But see his "Some MS Copies of Black's Chemical Lectures," ibid., **15** (1959), 65–73; **16** (1960), 1–9; **18** (1962), 87–97; **21** (1965), 209–255; **23** (1967), 1–33. E. W. J. Neave, "Joseph Black's Lectures on the Elements of Chemistry," in Isis, **25** (1936), 372–390, merely outlines the contents of Robison's edition of Black's Lectures.

Black's influence on the progress of scientific medicine and biology is treated by Heinrich Buess, "Joseph Black und die Anfänge chemischer Experimentalforschung in Biologie und Medizin," in Gesnerus, **13** (1956), 165–189. Henry Guerlac, "Joseph Black and Fixed Air," in Isis, **48** (1957), 124–151, 433–456, attempts to clarify the chronology of Black's early life and to reconstruct the steps in his chemical investigations. Guerlac's Lavoisier, The Crucial Year (Ithaca, N.Y., 1962), pp. 8–35, 68–71, sees Stephen Hales, rather than Joseph Black, as the chief British influence on Lavoisier before 1773. M. P. Crosland has studied Black's teaching symbols in "The Use of Diagrams as Chemical 'Equations' in the Lecture Notes of William Cullen and Joseph Black," in Annals of Science, **15** (1959),

75–90. Twenty-six recently discovered letters by or concerning Black, including twenty-one written to his brother Alexander, have been published by Douglas McKie and David Kennedy, "Some Letters of Joseph Black and Others," in Annals of Science, **16** (1960), 129–170. Included is the important letter by John Robison on the problems encountered in publishing Black's Lectures.

For Black's work on heat, consult Ernst Mach, Die Principien der Wärmelehre (Leipzig, 1896), pp. 153–181; Douglas McKie and Niels H. de V. Heathcote, The Discovery of Specific and Latent Heats (London, 1935), pp. 1–53; and Martin K. Barnett, "The Development of the Concept of Heat From the Fire Principle of Heraclitus Through the Caloric Theory of Joseph Black," in Scientific Monthly, **42** (1946), 165–172, 247–257.

HENRY GUERLAC

BLACKMAN, FREDERICK FROST (b. Lambeth, England, 25 July 1866; d. Cambridge, England, 30 January 1947), plant physiology.

Blackman, whose father was a doctor, was the third child and eldest son in a family of eleven children. His interest in botany may have been started by a set of Sowerby's British Botany that belonged to his book-collecting father; it developed during his school days at Mill Hill, where he started a herbarium. His interest in the plant as a whole remained throughout his life. On leaving school in 1883, Blackman entered St. Bartholomew's Hospital to train as a doctor. Although his studies were highly satisfactory—he graduated B.Sc. in 1885—in 1887 he accepted an opportunity to read science at St. John's College, Cambridge, where he shared rooms with his younger brother, later Professor V. H. Blackman. In 1895 he was elected to fellowship of his college, which he retained until his death. Blackman played an active part in the life of St. John's, where he lived until his marriage in 1917, holding the office of steward from 1908 to 1914. As an undergraduate his interests included music and pictures, and the college later benefited from his advice in aesthetic matters.

Blackman was appointed a university demonstrator in botany in 1891, and continued as a member of the botany school until he retired from his readership in 1936. He assumed a full share of the administrative work of the department, including the extension of the buildings in 1933. Everything he did was done with meticulous care, and he inspired others to do likewise. When, in recognition of the active school of plant physiology that he had developed from the basis laid down by his predecessors S. H. Vines and Francis Darwin, a subdepartment of plant physiology was created in 1931 with the aid of a grant from the Rockefeller Fund, Blackman naturally became its head and, but for technicalities regarding retirement age (resulting from the university statutes of 1926), he would have become the first professor of plant physiology at Cambridge.

Blackman was elected a fellow of the Royal Society in 1908 and was awarded a Royal Medal in 1921. Outside the botany school he served the university in various ways, including many years as a member of the Fitzwilliam Museum Syndicate. From 1901 to 1936 he was a member of the board of the Cambridge Instrument Company, which was started by Horace Darwin, brother of Francis, who had preceded Blackman as reader in botany.

Blackman's published papers are not numerous. Although his mind was quick, he was not hurried over the planning of his experiments and the devising and perfecting of the apparatus; neither was he hurried in the contemplation of the results and the search for a just interpretation of them. He took great pains over the preparation of a paper; such was the standard he set himself that several papers, which in 1935 had reached a stage that would have satisfied many, were not published until after his death. Many papers by pupils in his laboratory bear his stamp if not his name.

Blackman's first two botanical papers appeared in the *Philosophical Transactions of the Royal Society* in 1895. They began a series entitled "Experimental Researches in Vegetable Assimilation and Respiration," which was continued by him and his pupils until 1933. As early as 1832 it had been suggested that most of the gaseous exchange between the leaves of a plant and the surrounding atmosphere takes place through the stomata. For a long time this was much disputed, and it was not accepted until convincing experimental evidence was produced by Blackman. The first paper described an apparatus for measuring the carbon dioxide in gaseous mixtures, of which Blackman said, "The *raison d'être* of the apparatus is to be found in the perfection of details and their adaptation so that all the various processes can be performed with the minimum of error, time, and labour . . ." (p. 487). This approach characterized all his designs.

Blackman was surprised by the high rate at which gases diffused through a septum with many small holes, as compared with that when there were fewer holes with the same total area. His correspondence with his friend R. A. Lehfeldt, a physical chemist, shows that the latter advised him that "The effectiveness of a stoma should be proportional to its diameter but if the stomata were very close more nearly proportional to the area."

In 1904 the third paper in the series, by Blackman's pupil G. L. C. Matthaei, demonstrated that temperature had little effect on the rate of carbon assimilation at low intensities of illumination, while at high intensities the effect was comparable with that of many chemical reactions. This work directed Blackman's thoughts along lines that resulted in the publication of his classic paper "Optima and Limiting Factors" (1905). The terms "limiting factor" and "bottleneck" were not on the lips of biologists at that time. This paper is that of a pioneer in the application of physicochemical ideas to biological problems, a pioneer not unaware of the complexities of such problems. For instance, he says, "Physico-chemical finality is not to be attained in this matter, but special research might at least show how far the recorded optima are real metabolic truths and how far they are illusions of experimentation" (p. 286), and continues ". . . at present our science entirely lacks data that will stand critical analysis from the point of view indicated" (p. 295). The process limiting the rate of carbon assimilation at high intensities of illumination was later named the Blackman reaction by Otto Warburg.

These ideas were developed further in Blackman's presidential address to the Botany Section at the British Association meeting in 1908. It was a bold action to plead for the consideration of vital processes from the point of view of laws governing physicochemical processes at a time when, as he said, in consequence of the teaching of Pfeffer, "The notion that *every* change in which protoplasm takes part is a case of *'reaction'* of an *'irritable'* living substance to a *'stimulus'* overflowed its legitimate bounds and swamped the development of physical-chemical concepts" (p. 2), and continued, "No general treatment of the physiology of plants had yet been attempted in terms of reaction velocity" (p. 17).

Blackman later turned his attention to plant respiration. In 1928 three papers appeared dealing with part of the investigations his pupil P. Parija had made on the effect of the partial pressure of oxygen on the production of carbon dioxide by apples. Other papers were published with the assistance of a former pupil, J. Barker, after Blackman's death. The third paper of 1928, under Blackman's name, gives his interpretation of the complicated set of results analyzed in the previous papers. On the assumption that the products of glycolysis underwent fermentation to carbon dioxide and alcohol or complete oxidation to carbon dioxide and water, he deduced that glycolysis in air is more nearly complete than in nitrogen. Because the production of carbon dioxide in nitrogen was much more than one-third of that in air, he concluded there must be yet another fate for the products of glycolysis, which he called "oxidative anabolism."

His typical conclusion was that this schema provided a "plausible interpretation of all the quantitative variations of CO_2 production observed" (p. 521);

yet other data awaited similar analysis, "after which it may become necessary to take the present schema to pieces and reconstruct it . . ., but at least we shall have consolidated a mass of relations to which any future system must conform" (p. 522). This conclusion was based on an analysis of the behavior of more than twenty individual apples. Under Blackman's guidance his pupils carried out investigations on sugar content, rate of oxygen consumption, and composition of the atmosphere in the intercellular spaces of apples, in an attempt to complete the respiratory picture for this plant organ. Unfortunately, the work did not progress further than doctoral theses, now in the University Library at Cambridge.

Blackman's interests in botany were not confined to plant physiology. In his early years he lectured on the algae and later, with A. G. Tansley, his future brother-in-law, published "Classification of the Green Algae" in the first volume of the *New Phytologist* (1902). His other contribution to biology—certainly not the least—was the effect he had on many who attended his lectures. These were works of art, appreciated more on a second hearing, which they received from many of his research students. A mass of data collected from his wide reading or his own experiments was marshaled to point to his interpretation. The full discipline of Blackman's ways was experienced by those who had the privilege of starting research under his guidance. These students, many of whom proceeded to leading posts in British botany, carried the torch he lit.

BIBLIOGRAPHY

I. ORIGINAL WORKS. Blackman's writings include "Experimental Researches in Vegetable Assimilation and Respiration. I. On a New Method for Investigating the Carbonic Acid Exchange of Plants. II. On the Paths of Gaseous Exchange Between Aërial Leaves and the Atmosphere," in *Philosophical Transactions of the Royal Society,* **B186** (1895), 485–562; "III. On the Effect of Temperature on Carbon-dioxide Assimilation," *ibid.,* **B197** (1904), 47–105 (by G. L. C. Matthaei); "Optima and Limiting Factors," in *Annals of Botany* (London), **19** (1905), 281–295; "The Manifestations of the Principles of Chemical Mechanics in the Living Plant," in *Transactions of Section K of the British Association Meeting* (London, 1908), pp. 1–18; "Analytic Studies in Plant Respiration. I. The Respiration of a Population of Senescent Ripening Apples [written with P. Parija]. II. The Respiration of Apples in Nitrogen and Its Relation to Respiration in Air [by Parija]. III. Formulation of a Catalytic System for the Respiration of Apples and Its Relation to Oxygen [by Blackman alone]," in *Proceedings of the Royal Society,* **B103** (1928), 412–513; and *Analytic Studies in Plant Respiration* (Cambridge, 1954).

II. SECONDARY LITERATURE. Articles on Blackman are G. E. Briggs, in *Obituary Notices of Fellows of the Royal Society,* **5** (May 1948), 651–657; and *Dictionary of National Biography,* supp. for 1941–1950.

G. E. BRIGGS

BLAEU, WILLEM JANSZOON (*b.* Alkmaar [?], Holland, 1571; *d.* Amsterdam, Holland, 21 October 1638), *cartography.*

Before beginning his scientific career, Blaeu was a carpenter and a clerk in the Amsterdam mercantile office of his patrician cousin Cornelius Pieterszoon Hooft. His main interests, however, were astronomy and navigation, so in 1595–1596 he worked with Tycho Brahe at the latter's observatory on the island of Hven, Denmark. He then settled in Amsterdam, where he married Marytje Cornelisdochter. In 1599 Blaeu bought a house on the Y, where he established himself as a merchant of maps and globes, in the making of which he soon became quite proficient.

In constant contact with merchants and navigators, Blaeu was well informed on their latest discoveries. At this time Holland was beginning to send its fleets to Asia, Africa, America, and the Arctic Ocean, and interest in navigation and cartography grew by leaps and bounds. Blaeu's first terrestrial globe dates from 1599; his first celestial globe, from 1602. In 1605 he published his first world map, *Nova universi terrarum orbis mappa;* his sea atlas, *Het Licht der Zeevaert,* appeared in 1608. He moved his shop to the Damrak "in de vergulde Sonnewyzer" ("at the sign of the gilded sundial"), where he also began to publish maps made by others, thus laying the foundation of his once-famous world atlas, *Novus atlas* (1634).

In 1633 Blaeu became the official cartographer of the Dutch East India Company. Four years later he moved his printing plant to the Bloemgracht, where, with its specially designed presses, its foundry of special types, and its rooms for engravers and collectors, it became a showplace.

After Blaeu's death the business was continued by his sons Joan and Cornelis. The Bloemgracht plant continued operations until 1650, and the bookstore at the Damrak remained open. In 1672 a fire destroyed its warehouse, but the firm was in the family until 1695–1696, under the management of Joan's sons Willem, Pieter, and Joan. The establishment was then taken over by J. Van Keulen.

BIBLIOGRAPHY

I. ORIGINAL WORKS. Blaeu's main works are *Nova universi terrarum orbis mappa* (Amsterdam, 1605); *Het Licht der Zeevaert* (Amsterdam, 1608); and *Novus atlas* (Amsterdam, 1634), trans. into Dutch as *Toonneel des*

Aerdrycks, 4 vols. (Amsterdam, 1635–1645), with various later eds. entitled *Atlas major, Le grand atlas,* and *Grooten Atlas,* in 9–12 vols. (Amsterdam, 1662–1665). The various known eds. are listed by Baudet and Stevenson (see below).

II. SECONDARY LITERATURE. Works on Blaeu are P. J. H. Baudet, *Leven en Werken van Willem Jansz. Blaeu* (Utrecht, 1871; supplement, 1872), and *Notice sur la part prise par W. J. Blaeu . . . dans la détermination des longitudes terrestres* (Utrecht, 1875); J. Keuning, "Blaeu's Atlas," in *Imago mundi,* **14** (1959), 74–89; H. Richter, "William Jansz. Blaeu With Tycho Brahe on Hven," *ibid.,* **3** (1939), 53–60; E. L. Stevenson, *William Janszoon Blaeu* (New York, 1914), with facsimile repro. of 1605 world map in 18 sheets, and *Terrestrial and Celestial Globes,* 2 vols. (New Haven, 1921).

D. J. STRUIK

BLAGDEN, CHARLES (*b.* Wooten-under-Edge, Gloucestershire, England, 17[?] April 1748; *d.* Arcueil, France, 26 March 1820), *physical chemistry.*

Virtually nothing is recorded of Blagden's family background or education. He studied medicine at Edinburgh and received the M.D. in 1768. He was elected a fellow of the Royal Society in 1772 and served as a medical officer in the British Army from about 1776 to 1780. From about 1782 to 1789 Blagden was Cavendish's assistant. Neither man ever revealed "the circumstances which brought them together or separated them" (G. Wilson, *Life of Cavendish,* p. 129). Cavendish settled an annuity on Blagden and left him a considerable legacy.

Blagden succeeded Paul Henry Maty as secretary of the Royal Society on 5 May 1784, at a time when the Society was sorely divided over the efficacy of the administration of its president, Sir Joseph Banks. Blagden, Banks's close friend for many years, was elected secretary by a large majority. Both in this capacity and as Cavendish's assistant he became involved in the prolonged "water controversy"—the question of priority in discovering the composition of water, claimed by both Cavendish and James Watt in England and by Lavoisier in France. Blagden admitted responsibility for conveying, quite well meaningly, word of the experiments and conclusions of both Watt and Cavendish to Lavoisier; and he seems to have been careless in overlooking errors of date in the printing of Cavendish's and Watt's papers. There appears, however, to be little ground for the charge, leveled by Muirhead and other supporters of Watt's claims, that Blagden deliberately falsified the evidence in favor of Cavendish.

Blagden's earliest published papers concerned experiments carried out on himself, Banks, and others, to determine the endurance of air temperatures of up to 260° F.—he found the body temperature did not rise by more than one or two degrees. He also wrote a history of the attempts by Cavendish and others to determine the freezing point of mercury. A series of experiments on the supercooling of distilled water and solutions of salts led Blagden to study the effects of dissolved substances, beginning with common salt, on the freezing point of water. His conclusion that the salt lowers the freezing point in the simple inverse ratio of the proportion the water bears to it in the solution has come to be known as Blagden's law, although Richard Watson first discovered the relationship in 1771 (see J. R. Partington, *Text-book of Inorganic Chemistry* [London, 1921], p. 103; and Watson, in *Philosophical Transactions,* **61** [1771], 213–220). Nearly a century elapsed before his results could be integrated into a new theory of solutions initiated by the work of Raoult, Arrhenius, and van't Hoff; in the meantime, Blagden's work was virtually forgotten until Louis de Coppet drew attention to it in 1871.

Blagden spent much of his time in Europe, particularly in France—he was a close friend of Berthollet and other French scientists. Indeed, he had gone to live in Arcueil shortly before he died. He was knighted in 1792.

BIBLIOGRAPHY

I. ORIGINAL WORKS. Blagden published little outside the *Philosophical Transactions of the Royal Society.* Among his papers are "Experiments in a Heated Room," in *Philosophical Transactions,* **65** (1775), 111–128, 484–494; "History of the Congelation of Mercury," *ibid.,* **73** (1783), 329–397; "Experiments on the Cooling of Water Below Its Freezing Point," *ibid.,* **78** (1788), 125–146; and "Experiments on the Effect of Various Substances in Lowering the Point of Congelation of Water," *ibid.,* 277–312.

II. SECONDARY LITERATURE. A biography is F. H. Getman, "Sir Charles Blagden," in *Osiris,* **3** (1937), 69–87. The two sides of the "water controversy," giving diametrically opposed opinions respecting Blagden's integrity, are best studied in J. P. Muirhead, ed., *Correspondence of the Late James Watt on His Theory of the Composition of Water . . .* (London, 1846); and G. Wilson, *The Life of the Honourable Henry Cavendish, Including Abstracts of His Most Important Scientific Papers, and a Critical Inquiry Into the Claims of All the Alleged Discoverers of the Composition of Water* (London, 1851). The latter gives a useful bibliography of the subject and a biographical sketch of Blagden.

E. L. SCOTT

BLAINVILLE, HENRI MARIE DUCROTAY DE (*b.* Arques, France, 12 September 1777; *d.* Paris, France, 1 May 1850), *anatomy, zoology.*

Son of Pierre Ducrotay and Marie Pauger de Blainville, Henri grew up among the lesser but intensely proud Norman nobility. His schooling, interrupted by the French Revolution, recommenced in Rouen and Paris, where he at first studied music, art, and literature. There followed a brief but spectacular dissipation of his patrimony. Reforming himself and pursuing his ferocious desire to learn, Blainville turned to medicine (M.D., Paris, 1808) and then to natural history. Working in Cuvier's laboratory, he soon became an outstanding comparative anatomist and developed further as a remarkably independent thinker. In addition to anatomy, he lectured and published widely on descriptive and taxonomic invertebrate zoology (particularly malacology), comparative osteology, history of science, and the first principles of natural history. About 1810 he began formal instruction (as Cuvier's deputy) in various Parisian institutions (Athénée, Collège de France, Muséum d'Histoire Naturelle). He was named a professor at the Muséum in 1830 and in 1832 was appointed to Cuvier's vacant chair of comparative anatomy. In 1825 he was elected to the Académie des Sciences.

Blainville's lifelong objective in natural history was order. Order in the chaos of existence necessarily could derive only from clearly defined *principes*. These would follow inevitably from what Blainville called *la philosophie chrétienne*. Natural order was simply the unfolding of the Creator's design; that design in turn refocused our regard upon His wisdom and power. There were two roads to God, faith and knowledge, and for Blainville they merged into a single Christian philosophy. Blainville was a believing and, it appears, a practicing Roman Catholic. His religion, however, aimed less at spiritual experience than at an understanding of God's plan of creation, and thus was largely an elaboration on earlier objectives and beliefs of the deists.

God's plan for ordering animals and plants was the long-familiar scale of being, or *série*. Blainville vigorously defended the generalized *série* against Cuvier's attacks. Apparent gaps in the arrangement of existing organisms were nicely filled by fossil forms; Blainville carried out valuable paleontological research to support this proposition. Together, extinct and extant organisms testified to the original fullness, and hence rightness, of God's creation. Such change as occurred, possibly including that of species, was predicated by the divine plan.

Man stood both morally and physically at the summit of the *série,* presenting the standard by which the rank of all other forms was to be decided. Blainville followed Bichat in defining life as a general responsiveness of the organism to, and its persistence amid, ever-varying ambient conditions. This characteristic, *sensibilité,* was to biology what gravitation was to the Newtonian world machine; whether causal or not, it gave meaningful substance to the essential fact of *relation.* Organs of relation (principally the sensory and locomotory parts) thereupon assumed primacy and allowed Blainville to base his intricate and numerous classifications upon external features, those which mediated with the environment.

Blainville's influence was exerted principally through his famous lectures; he was a somewhat unsystematic author and brought few works to true completion. His extreme personal and family pride led him to bitter relations with contemporaries and an unsympathetic view of bourgeois France. Blainville's grand and enduring trinity was God, king, and France. His distaste for egalitarian society and contempt for Republican ideals led him to examine alternative social structures and to discuss favorably various utopian socialistic schemes. In 1813 he became a close acquaintance of Saint-Simon and, about 1824, the friend, disciple, and mentor of Comte, who carefully followed Blainville's most notable lecture series, that on physiology. F. L. P. Gervais, F. A. Pouchet, and H. C. M. Nicard were among Blainville's pupils.

BIBLIOGRAPHY

I. ORIGINAL WORKS. Blainville published over 150 articles and numerous monographs. A topical list of his writings is given in Flourens, pp. xliii–lx. His major publications are *De l'organisation des animaux, ou Principes d'anatomie comparée* (Paris, 1822); *Manuel de malacologie et conchyiologie,* 2 vols. (Paris–Strasbourg, 1825–1827); *Cours de physiologie générale et comparée professé à la Faculté des Sciences de Paris en 1829–1833,* Hollard, ed., 3 vols. (Paris, 1833); *Manuel d'actinologie et de zoophytologie,* 2 vols. (Paris, 1834); *Ostéographie ou description iconographique comparée du squelette et du système dentaire des cinq classes d'animaux vertébrés récents et fossiles pour servir de base à la zoologie et à la géologie* (Paris, 1839–1864)—24 fascicles were issued by Blainville between 1839 and 1850; the 25th was published, with a biographical notice of Blainville and indexes to all parts, by Nicard in 1864; *Histoire des sciences de l'organisation et de leurs progrès, comme base de la philosophie, rédigées d'après ses notes et ses leçons faites à la Sorbonne de 1830 à 1841, avec les développements nécessaires et plusieurs additions,* Maupied, ed., 3 vols. (Paris, 1845)—Maupied severely distorted Blainville's intention and arguments, and made him appear a shrill Catholic apologist (see Nicard, p. 149); *Sur les principes de la zooclassie ou de la classification des animaux* (Paris, 1847); and *Cuvier et [E.] Geoffroy Saint-*

Hilaire: Biographies scientifiques, H. C. M. Nicard, ed. (Paris, 1890), an angry polemic against Cuvier.

Blainville's papers passed to his student and biographer Nicard, and are today in the Bibliothèque Centrale, Muséum National d'Histoire Naturelle, Paris. These MSS include notes for lectures at the Muséum, drafts of published works, and occasional items of poetry and drama. (Information courtesy of Yves Laissus.) No collection of correspondence can be located.

II. SECONDARY LITERATURE. There exists neither a satisfactory study of Blainville's scientific work nor a detailed biography. The only comprehensive view of both is given in H. C. M. [Pol] Nicard, *Étude sur la vie et les travaux de M. Ducrotay de Blainville* (Paris, 1890). Nicard is at once hagiographer and confusing, quite unsystematic expositor. Nevertheless, he knew Blainville well and controlled his literary legacy; hence his volume is a major source and interpretation. In the preface, Nicard lists earlier biographical notices. M. J. P. Flourens wrote the official *éloge* for the Académie des Sciences: "Éloge historique de Marie-Henri [*sic*] Ducrotay de Blainville," in *Mémoires de l'Académie des Sciences, Paris,* **27** (1860), i–lx, also in Flourens's *Recueil des éloges historiques lus dans les séances publiques de l'Académie des Sciences* (Paris, 1856), pp. 285–341. Blainville's early physiological ideas are expounded in C. J. F. B. Dhéré, *De la nutrition considérée anatomiquement et physiologiquement dans la série des animaux, d'après les idées de M. Ducrotay de Blainville* (Paris, 1826). A brief but penetrating estimate of Blainville's religiophilosophical viewpoint is given in H. Gouhier, "La philosophie 'positiviste' et 'chrétienne' de H. de Blainville," in *Revue philosophique,* **131** (1941), 38–69. See also A. Comte, *Cours de philosophie positive,* III (Paris, 1838), leçon 40; P. Ducassé, *Méthode et intuition chez Auguste Comte* (Paris, 1939); and E. Littré, *Auguste Comte et la philosophie positive,* 2nd ed. (Paris, 1864), pp. 632–639, a discussion of Blainville and Maupied.

WILLIAM COLEMAN

BLAIR, PATRICK (*d.* Boston, England, 1728), *botany, biology, medicine.*

Reliable biographical information on Blair is scanty and contradictory; for example, the evidence for his death in 1728 is indirect—he had reached the letter H in his *Pharmaco-Botanologia,* and his death is assumed to have occurred when no further material appeared. Certainly he worked as a surgeon in Dundee, Scotland, for some years; and when, on 27 April 1706, a female Indian elephant died there, Blair dissected it. He presented his findings, with an extensive review of the literature, in a letter to the Royal Society in 1710; this communication was later published separately (1713). In 1712 Blair was elected a fellow of the Royal Society, and the same year was given an honorary M.D. by King's College, Aberdeen. Blair was a Jacobite, and was sentenced to death as

such on 7 July 1716; he was, however, pardoned after he successfully appealed to Sir Hans Sloane and others, such as Richard Mead, to intercede with the authorities on his behalf.

Blair appears to have been a practicing surgeon for most of his life, and in a communication to the Royal Society in 1717 he gave what Caulfield has called "probably the earliest description" of pyloric stenosis on record. But it is for his contributions to botany that Blair is most famous—not so much for his natural interest as a physician in the medicinal properties of plants (although this took up much of his time) but more for his work on plant sexuality. Although Sachs, in his *History of Botany,* felt that in this regard ". . . Patrick Blair . . . did nothing himself, but merely appropriated the general results of Camerarius' observations" (p. 391), Blair was only one of many—including Grew, Ray, Camerarius, and Bradley—who, according to Ritterbush, "escaped the consequences of plant sexuality for the scale of functions by ascribing hermaphroditic generation to plants, which they shared only with the lower animals" (*Overtures to Biology,* p. 117). Blair was an "ovulist" rather than a "pollenist."

Blair was something of a polemicist, and his unpublished manuscript preface bound with Bishop Rawlinson's copy of Bradley's *Philosophical Account of the Works of Nature* was, as Ritterbush has observed, "an exceptionally abusive attack upon Richard Bradley" (*op. cit.,* p. 96).

BIBLIOGRAPHY

I. ORIGINAL WORKS. In addition to several communications in the *Philosophical Transactions of the Royal Society,* Blair wrote the following: *Osteographia Elephantina: or, a Full and exact description of all the bones of an elephant, which died near Dundee, April the 27th. 1706. with their several dimensions: To which are premis'd, 1. An historical account of the natural endowments . . . of elephants . . . 2. A short anatomical account of their parts . . . In a letter to Dr. Hans Sloane* (London, 1713); *Miscellaneous Observations in the Practise of Physick, Anatomy and Surgery. With new and curious remarks in botany* (London, 1718); *Botanick Essays. In two parts. The first containing, the structure of flowers . . . and the second, the generation of plants, etc.* (London, 1720); and *Pharmaco-Botanologia: or, an Alphabetical and classical dissertation on all the British indigenous and garden plants of the new London Dispensatory . . . With many curious and useful remarks from proper observation* (London, 1723–1728).

One of the most important collections of Blair MSS is bound in with Bishop Richard Rawlinson's copy of Richard Bradley's *Philosophical Account of the Works of Nature,* now in the possession of the Bodleian Library, Oxford.

There are several Blair letters dated 1723–1724 as well as a 28-page preface to an apparently unfinished treatise.

II. SECONDARY LITERATURE. For an account of Blair's work on the elephant, see F. J. Cole, *A History of Comparative Anatomy* (London, 1944), pp. 325–328. A good general account of Blair's botanical contributions is P. C. Ritterbush, *Overtures to Biology* (New Haven, 1964), ch. 3. For general biographical information, Ernest Caulfield, "An Early Case of Pyloric Stenosis," in *American Journal of Diseases of Children,* **40** (1930), 1070–1077, is surprisingly informative. See also C. E. Raven, *John Ray* (Cambridge, 1950), pp. 185–186, for an account of Blair's attack on John Ray; and Julius von Sachs, *History of Botany* (Oxford, 1890), p. 391.

L. R. C. AGNEW

BLAISE. See **Blasius of Parma.**

BLANC, ALBERTO-CARLO (*b.* Chambéry, France, 30 July 1906; *d.* Rome, Italy, 3 July 1960), *prehistory, ethnology.*

Blanc was the son of Gian-Alberto Blanc, a distinguished naturalist and professor at the University of Rome and descendant of an old and prominent Savoyard family. When a part of the Duchy of Savoy voted for France in 1860, the Blancs chose to become Italian out of loyalty to the prince of Savoy. These origins endowed the Blanc family with a double culture, French and Italian. Like his father, Alberto-Carlo spoke and wrote the two languages with equal perfection. The family was affluent, owning extensive property, and like almost all Savoyard families they were Catholic. They were on friendly terms with the Abbé Breuil, the prehistorian. Sometimes, too, Cardinal Tisserant would accompany them into the field to take part in excavations in search of traces of prehistoric man.

After distinguishing himself as a student at the universities of Pisa and Rome, Blanc was appointed in 1938 to teach a course in geology at the former institution. He was attracted primarily by the most recent era, the Quaternary, in which man made his appearance. He married in 1939. One month later he wrote to André Cailleux:

> Forgive me for my somewhat tardy reply to your card of 21 February. In order to excuse myself at least partially, may I say that I was married on 20 February and that on the 25th of the same month I had the good luck to discover, in the midst of my honeymoon, a beautiful Neanderthal skull. Naturally, this upset all my plans and at the moment we, i.e., my wife and I, are in the process of digging in the new deposit here.

This letter was dated San Felice Circeo (Littoria), where the Blancs owned property. The Monte Circeo

skull, after thorough examination, quickly became a classic, since it is the most complete Neanderthal skull yet discovered.

Blanc was called to the University of Rome in 1939, where he taught ethnology and human paleontology. In 1957 he was appointed to the chair in paleoethnology. A member of numerous academies and scientific societies, he also was invited to lecture at more than twenty foreign universities. He was elected president of several international commissions and of the Sixth Congress of the International Union of Prehistoric and Protohistoric Sciences. He was the moving spirit behind the magazine *Quaternaria,* dedicated to the natural and cultural history of the Quaternary era.

Toward the end of 1958, while on a field trip in Apulia in southern Italy, Blanc experienced the first attacks of the disease that proved fatal. But to his very last hours, according to his pupil Georges Laplace, he "retained his extraordinary clarity of spirit, his marvelous smile, and the calm strength which he radiated."

An industrious and efficient worker, Blanc published 164 works in the span of twenty-six years. When he started his work in 1934, European prehistory was dominated by the great if somewhat authoritarian figure of the Abbé Breuil. With the latter's aid and advice, yet always preserving his independence, Blanc addressed himself first—and properly so—to the severe but beneficial school of empirical study. He participated in some hundred digs, most of which were performed under his direction. In Italy he discovered some fifty prehistoric deposits and six of the seven known human Neanderthal fossils of that country. Wisely, he studied the geology of these deposits with great care so as to be able to link up their history with that of the rest of Europe. The flora and fauna, which were interpreted by specialists, revealed changes during the course of time, a development to which Blanc turned his attention.

In 1939 Blanc was asked to give a course in the ethnology of presently existing populations, particularly the most primitive peoples. In comparing their type of life with that of prehistoric man of the advanced Paleolithic age, with which he was familiar, Blanc was struck by a clear deterioration in techniques and an ever-narrowing specialization as the transition from ancient to present times was followed. This led him to a new theory concerning the formation of ethnic groupings which he termed *ethnolysis:* starting from an ancient polymorphous ethnic grouping, present ethnic groupings are differentiated by a loss of characteristics:

". . . their centrifugal diffusion, which occurred in very special conditions varying for each people, resulted necessarily in widely differing specializations; certain cultural elements persisted (and developed) in certain peoples while disappearing in others, who in turn preserved certain elements which the former had lost. Thus, by alternative elimination, there occurred a separation, a segregation, a lysis of cultural elements which had coexisted originally [*L'évolution humaine*].

Blanc immediately recognized the potential affinities and generalizations of his idea. What he had assumed concerning ethnic groupings, he very soon extended to animal and plant associations, to the evolution of species, and in particular to the appearance of the human races: from ethnolysis he passed to *cosmolysis,* which embraced the entire physical, biological, and human universe:

Cosmolysis is that universal modality of evolution through which heterogeneous archaic entities and groupings, which in the state of primary blending contain a great number of characteristics and elements, are resolved into more and more homogeneous and distinct entities, through lysis (from the Greek *luo,* "I separate, dissolve") and segregation in each case of characters and elements which coexist in a mixed state in the above mentioned archaic entities and groupings [*Ibid.*].

Objections were raised to Blanc's views on the grounds that human prehistory, like the evolution of the species, shows inverse examples, where we pass from the simple to the complex and more varied elements. Indeed, Blanc himself recognized these phases of growing complexity, for instance, between the beginning and the last part of the Stone Age, between the Abbevillian and the upper Paleolithic periods. From the point of view of ethnology, he placed them within a two-phase cycle, i.e., complexification–simplification—the cycle that was to be repeated several times in the course of the ages. Evidently, however, he was far more interested in the second phase, that of simplification, or lysis, which he had invented himself. Even if, contrary to Blanc, it is supposed that the first phase, i.e., acquisition of new characters and their formation, plays just as important or even more important a role in the evolution of animal and especially plant species, we must recognize that the idea of cosmolysis explains certain facts.

But there was more to it. Like his contemporaries, Blanc saw clearly that his hypothesis and the facts supporting it agreed quite well with Neo-Darwinism, with the mutation-selection theory, and with the corpuscular theory of heredity that has been confirmed by so many other works since his death. In this respect, he was a precursor.

Given to bold generalizations, Blanc went even

further and speculated about the hydrogen atom and its components. Its then apparent simplicity seemed to him deceptive and he saw there "the simplified product of a history proceeding from a fundamental complexity of matter" (*Ibid.*). Even if one finds Blanc's theory of cosmolysis a less than complete explanation for the evolution of the cosmos, life, and mankind, Blanc's idea must still be given credit for its intrinsic strength and greatness.

BIBLIOGRAPHY

I. ORIGINAL WORKS. Blanc's writings include "Le glaciaire considéré au point de vue paléobiologique et géomorphologique," in *L'anthropologie,* **48** (1938); "Il Monte Circeo, le sue grotte paleolitiche ed il suo uomo fossile," in *Bollettino Società geografica italiana* (1939); "Etnolisi—Sui fenomeni di segregazione in biologia ed in etnologia," in *Rivista di antropologia,* **33** (1940); "Cosmolisi—Interpretazione genetico-storica delle entità e degli aggrupamenti biologici ed etnologici," *ibid.,* **34** (1941–1942); "Sviluppo per lisi delle forme distinte," in *Quaderni di sintesi,* **2** (1946); "I paleantropi di Saccopastore e del Circeo," in *Quartär,* **4** (1942); "Etnologia e paleontologia," in *Atti della Società italiana per il progresso delle scienze,* **41** (1943); "L'évolution humaine dans le cadre de la cosmolyse," in *Cahiers de la Faculté de Théologie de l'Université de Lausanne* (1946).

II. SECONDARY LITERATURE. A good account and a discussion of cosmolysis is in Piero Leonardi, *L'evoluzione dei viventi* (Brescia, 1957), also trans. into Spanish as *La evolución biológica* (Madrid, 1957); see pp. 265–276. For the life and works of Blanc, see the note by Georges Laplace in *Bulletin de la Société préhistorique française,* **58** (1961), 515–519.

ANDRÉ CAILLEUX

BLANCHARD, RAOUL (*b.* Orléans, France, 4 September 1877; *d.* Paris, France, 24 March 1965), *geography.*

Blanchard graduated from the École Normale Supérieure in 1900 and passed the *concours d'agrégation* in history and geography. At the time, geography was still mainly in a descriptive stage and an adjunct to history. His thesis on Flanders, presented at the University of Lille in 1906, was one of the first important works of regional geography based on research done *in situ.*

At the time of his appointment as lecturer in 1905, there was not a single student of geography at Grenoble. Through perseverance, teaching ability, and the novelty of his subject, he turned the university into one of the most active centers for geography in France. He later became a professor. In 1940 he was appointed dean of the Faculty of Liberal Arts and taught there until his retirement in 1948.

Blanchard was the creator of French Alpine geography, founding the *Revue de géographie alpine,* which published the first works on the French Alps. For the study of mountains he instituted field observation (he was an indefatigable hiker) together with rigorous arguments for studying structure, to which he gave priority (instead of erosion surfaces).

His twelve volumes on the French Alps are considered his chief work, but his range of interests was much greater. In human geography he produced important studies of cities, works that were novel at the time because they combined a study of the site with that of the development of the city. He prepared the volume on western Asia for the series Géographie Universelle (1929), and later published a general study of North America (1933). He knew North America well. He had been appointed instructor at Harvard in 1917 and was a full professor from 1928 to 1936. He also taught at Chicago, Columbia, and other schools in the United States as well as the universities of Montreal and Laval in Canada.

Blanchard was representative of his epoch in French geography. In time, of course, concepts and methods changed, and he was reproached by some for not having advanced his morphology beyond a somewhat oversimplified determinism.

BIBLIOGRAPHY

I. Original Works. Blanchard's writings include *La Flandre . . .* (Dunkerque, 1906), his thesis; *Geography of France* (Chicago–New York, 1919), written with M. Todd; *Les Alpes françaises* (Paris, 1925); *L'Asie occidentale,* in the series Géographie Universelle, directed by P. Vidal de la Blache and L. Gallois (Paris, 1929); *L'Amérique du Nord: États-Unis, Canada et Alaska* (Paris, 1933); *Grenoble: Étude de géographie urbaine* (Grenoble, 1935); *Les Alpes occidentales,* 12 vols. (Grenoble, 1941–1958); and *A Geography of Europe* (New York, 1944), written with R. E. Crist.

II. Secondary Literature. Biographies of Blanchard are J. Blache, in *Revue de géographie alpine,* **7** (1965), 361–370; P. Dagenais, in *Revue de géographie de Montréal,* **18** (1964), 133–135; D. Faucher, in *Revue de géographie des Pyrénées et du Sud-Ouest,* **13** (1965), 157; and A. Perpillou, in *Acta geographica* (Paris), no. 5 (1965), 1. See also *In Memoriam Raoul Blanchard,* prepared by the Association des Amis de l'Université de Grenoble (Grenoble, 1966).

Juliette Taton

BLASCHKE, WILHELM JOHANN EUGEN (*b.* Graz, Austria, 13 September 1885; *d.* Hamburg, Germany, 17 March 1962), *mathematics.*

Blaschke's father, Josef Blaschke (1852–1917), was professor of descriptive geometry at the Landes-Oberrealschule at Graz. Wilhelm inherited his father's predilection for the geometry of Jakob Steiner and his love of concrete problems. Josef also imparted to the boy a feeling for history and an open-mindedness toward foreign cultures that remained with him throughout his life.

Blaschke began his studies at the Technische Hochschule of Graz and earned his doctorate from the University of Vienna in 1908. For more than a decade afterward he traveled through Europe, seeking contact with many of the leading geometers of his day. He spent some months in Pisa with Luigi Bianchi and a semester in Göttingen, drawn there by Felix Klein, David Hilbert, and Carl Runge. He worked at Bonn with Eduard Study, whose main fields of research were geometry, kinematics, and the theory of invariants. Blaschke became *Privatdozent* at Bonn in 1910, but in the following year he went to the University of Greifswald to join Friedrich Engel, with whom he shared an admiration for the great Norwegian mathematician Sophus Lie.

In 1913 Blaschke accepted an extraordinary professorship at the Deutsche Technische Hochschule in Prague, and in 1915 he moved to Leipzig, where he became a close friend of Gustav Herglotz. Two years later he was made full professor at the University of Königsberg. After a short stay at Tübingen, Blaschke was called in 1919 to the full professorship of mathematics at the University of Hamburg, a position he retained until his retirement in 1953. He also held visiting professorships at Johns Hopkins University, at the University of Chicago, at the University of Istanbul, and at the Humboldt University in Berlin, and lectured at universities all over the world. He was married to Augusta Meta Röttger and had two children.

At Hamburg, Blaschke succeeded within a few years in gaining worldwide recognition for the department of mathematics of the newly founded university, for he was able to attract to Hamburg such well-known mathematicians as Erich Hecke, Emil Artin, and Helmut Hasse. Very soon Hamburg became a center of great mathematical activity and productivity, testimony to which is given by the *Abhandlungen aus dem mathematischen Seminar der Universität Hamburg* and the *Hamburger mathematische Einzelschriften,* both founded by Blaschke.

One of the leading geometers of his time, Blaschke centered most of his research on differential and integral geometry and kinematics. He combined an unusual power of geometrical imagination with a consistent and suggestive use of analytical tools; this gave his publications great conciseness and clarity and, with his charming personality, won him many students and collaborators.

Blaschke made "kinematic mapping" (discovered independently in 1911 by Josef Grünwald), which established a mapping between the group of isometries (motions) in the plane and the three-dimensional point space, a central tool in kinematics; and in an abstract turn given to it by Kurt Reidemeister, it proved very useful in the axiomatic foundation of several geometries. In *Kreis und Kugel* (1916), Blaschke investigated the isoperimetric properties of convex bodies, characterizing circles and spheres as figures of minimal properties. In this he was following methods suggested by Steiner, who had been criticized by Dirichlet for omitting an existence proof. This was first remedied by Weierstrass by means of the calculus of variation, but Blaschke supplied the necessary existence proofs in a fashion closer to the spirit of Steiner.

Blaschke's books on differential geometry soon gained worldwide recognition. The three-volume *Vorlesungen* (1921–1929) put into practice Felix Klein's "Erlangen Program" for differential geometry: Volume I was devoted to classical geometry, Volume II to affine differential geometry (a subject developed by Blaschke and his pupils), and Volume III to the differential geometry of circles and spheres, controlled by the transformation groups of Moebius, Laguerre, and Sophus Lie. (The treatment of projective differential geometry, however, was left to Blaschke's pupil Gerrit Bol.) Furthermore, Blaschke originated topological differential geometry, which studies invariants of differentiable mappings; he collected the results in his books *Geometrie der Gewebe* (1938) and *Einführung in die Geometrie der Waben* (1955). In 1950 Blaschke gave a new, concise exposition of differential geometry based on ideas of E. Cartan.

Inspired by Gustav Herglotz and by some classical problems of geometrical probability (Buffon's needle problem, Crofton's formulas), Blaschke began, about 1935, a series of papers on integral geometry. Because of its relations to convex bodies and kinematics, this field of research was especially to his liking; and many of his students continued his work in this area—Hadwiger, Wu, Chern, and Santaló.

Blaschke received honorary doctorates from the universities of Sofia, Padua, and Greifswald, and the Karlsruhe Technische Hochschule. He was elected corresponding or honorary member of about a dozen European scientific academies.

BIBLIOGRAPHY

I. ORIGINAL WORKS. Blaschke's works include *Kreis und Kugel* (Leipzig, 1916; Berlin, 1956), trans. into Russian (Moscow, 1967); *Vorlesungen über Differentialgeometrie,* 3 vols., I (Berlin, 1921, 1924, 1930, 1945), trans. into Russian (Moscow, 1935); II, rev. by Kurt Reidemeister (Berlin, 1923); III, rev. by G. Thomsen (Berlin, 1929); *Vorlesungen über Integralgeometrie,* 2 vols., I (Leipzig–Berlin, 1935, 1936, 1955), trans. into Russian (Moscow, 1938); II (Leipzig–Berlin, 1937; 3rd ed., 1955—together with Vol. I); *Ebene Kinematik* (Leipzig–Berlin, 1938); *Geometrie der Gewebe,* written with Gerrit Bol (Berlin, 1938); *Einführung in die Differentialgeometrie* (Berlin, 1950; 2nd ed., 1960), written with Hans Reichardt, trans. into Russian (Moscow, 1957); *Einführung in die Geometrie der Waben* (Basel–Stuttgart, 1955), trans. into Russian (Moscow, 1959), trans. into Turkish (Istanbul, 1962); *Ebene Kinematik,* written with H. R. Müller (Munich, 1956); *Reden und Reisen eines Geometers* (Berlin, 1957, 1961); and *Kinematik und Quaternionen* (Berlin, 1960).

II. SECONDARY LITERATURE. The following obituary notices describe Blaschke's life and work in greater detail: Werner Burau, "Wilhelm Blaschkes Leben und Werk," in *Mitteilungen der Mathematischen Gesellschaft in Hamburg,* **9,** no. 2 (1963), 24–40; Otto Haupt, "Nachruf auf Wilhelm Blaschke," in *Jahrbuch 1962 der Akademie der Wissenschaften und der Literatur zu Mainz* (Mainz–Wiesbaden, 1962), pp. 44–51; Erwin Kruppa, "Wilhelm Blaschke," in *Almanach der Österreichischen Akademie der Wissenschaften,* **112** (for 1962) (Vienna, 1963), 419–429; Hans Reichardt, "Wilhelm Blaschke †," in *Jahresbericht der Deutschen Mathematiker-Vereinigung,* **69** (1966), 1–8; and Emanuel Sperner, "Zum Gedenken an Wilhelm Blaschke," in *Abhandlungen aus dem Mathematischen Seminar der Universität Hamburg,* **26** (1963), 111–128 (with a bibliography by W. Burau, to which Reichardt gives an addition).

CHRISTOPH J. SCRIBA

BLASIUS OF PARMA (*b.* Parma, Italy, *ca.* 1345; *d.* Parma, 1416), *natural philosophy.*

Although presumably he was born in Parma, the first known reference to Blasius is found in the records for 1377 of the University of Pavia, where he took his doctorate, perhaps in 1374 (the latter date makes 1345 a plausible birth year). Listed as an examiner in March 1378, Blasius probably left Pavia by October of that year for the University of Bologna, where he remained at least until 1382 (for 1379–1380, he was officially described as a teacher of logic, philosophy, and astrology), and probably through 1383. On 20 May 1384, he agreed to teach at the University of Padua for four years; his name appears in the university records from February 1386 to 11 May 1387 and again on 16 December 1388 as the sponsor of a doctoral candidate who was represented by another scholar, probably because in 1387 Blasius had returned to Bologna as professor of philosophy and astrology for the period 1387–1388. On 29 July 1388, he was appointed a lecturer in natural philosophy at the University of Florence, where he remained until 1389.

During the next decade, when he reached the summit of his career, Blasius was again at Pavia as professor of "mathematical arts and both philosophies" (i.e., moral and natural). His whereabouts between 1400 and 1403 are unknown, but in subsequent years he taught at the University of Pavia (1403–1407) and the University of Padua (1407–1411). In October 1411 he was dismissed from the latter because he lacked students and was deemed no longer fit to teach, conditions that were probably caused by the infirmities of old age. He died five years later.

A sojourn in Paris, where he received his doctorate (so we are told in an explicit to his *Questio de tactu corporum duorum,* which was disputed at Bologna no later than 1388), is mentioned in his *Questiones super tractatum de ponderibus.*[1] It was probably while in Paris that he absorbed the new ideas of the Parisian Scholastics, ideas that he was to disseminate and popularize in Italy.

Blasius was not merely an Aristotelian commentator, but also wrote independent treatises on important scientific topics. Prior to 16 October 1396, when he was compelled to recant unspecified transgressions against the Church[2] (by this time he had probably written the bulk of his extant treatises), Blasius seems to have been a materialist and determinist, accepting as true certain articles condemned at Paris in 1277. In his *Questiones de anima* (Padua, 1385), he denied that the intellective soul was separable from the body, insisting that it was produced from transient matter. It was only by authority of the Church and faith—not by natural reason or evidence—that one ought to believe in its separability. Furthermore, he denied the immortality of the intellective soul while accepting the eternity of the world and a necessary determinism exerted by the celestial bodies and constellations on terrestrial and human events. Such opinions, characteristic of earlier Bolognese Averroists, were probably instrumental in provoking the ecclesiastical authorities. Blasius capitulated and complied swiftly. During 1396–1397, in lectures delivered at Pavia on the *Physics* of Aristotle, he repudiated all these views; and in 1405 he attacked astrological determinism (but not astrology) in his *Iudicium revolutionis anni 1405,* declaring that while the stars influenced men and events, the will of God and human free will could resist if they chose.

Of his numerous scientific treatises, only those on optics, statics, and intension and remission of forms have received more than cursory examinations, resulting in partial or complete modern editions. In addition to relevant discussions on optics in his *Questiones*

de anima and *Meteorologica,* Blasius also wrote *Quaestiones perspectivae* (dated 1390 in one manuscript), a lengthy commentary on some of the propositions of John Peckham's enormously popular thirteenth-century optical treatise, *Communis perspectiva.* Guided by an empiricist outlook derived ultimately from the optics of Ibn al-Haytham (Alhazen), and perhaps influenced by fourteenth-century nominalism, he made visual sensation the basis for human certitude and knowledge; consequently he placed heavy emphasis upon the psychology of perception. Traditional geometric optics was placed in the broader matrix of a theory of knowledge and cognitive perception based on vision.

Blasius composed at least two treatises on statics: one Scholastic, the other longer and non-Scholastic in form. The longer work, *Tractatus de ponderibus,* drew heavily on the thirteenth-century statical treatises associated with the name of Jordanus de Nemore. It was probably from the *Elementa Jordani super demonstrationem ponderum* that Blasius adopted the important concept of "positional gravity," which involved a resolution of forces where the effective "heaviness" or weight of a body in a constrained system is proportional to the directness of its descent as measured by the projection of an arbitrary segment of its path onto the vertical drawn through the fulcrum of a lever or balance. Ignoring straightforward and available definitions of positional gravity, Blasius presents the concept in Pt. I, Supps. 6 and 7, proving in the first of these that "in the case of equal arcs unequally distant from the line of equality (i.e., the line of horizontal balance) that which is a greater distance intercepts less of the vertical [through the axis]," and in the second that "one body is heavier than another by the amount that its movement toward the center [of the world] is straighter."[3] But the more of the vertical cut off by a projected arc, the straighter its descent and, therefore, the greater its positional gravity.

In Pt. II, Prop. 4 (probably based upon *Elementa Jordani,* Th. 2), Blasius misapplied the concept of positional gravity in demonstrating that "when the equal arms of a balance are not parallel with the horizon and equal weights are hung [on their ends], the beam assumes a horizontal position."[4] In Figure 1, let arms *AB* and *BC* be equal but not parallel to the horizon, *DF.* If equal weights are suspended at the ends of the equal arms, the latter would become parallel to the horizon, or *DF.* This will occur because, being heavy bodies, *a* and *c* will seek to descend, *c* to *F* and *a* to *G.* Assuming, quite improperly, that arcs *CF* and *AG* are equal, but unequally distant from *DF,* Blasius applies Pt. I, Supp. 6, to show that arc

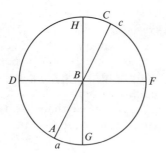

FIGURE 1

CF cuts off more of the vertical along HG than does arc AG, and concludes that c's downward motion to F is more direct than a's to G. Consequently (by Pt. I, Supp. 7, which is not cited but is clearly required), c is positionally heavier than a and will descend, forcing a upward until horizontal equilibrium is attained. The basic error in all this lies in assuming a simultaneous descent for c and a, instead of comparing the descent of c with the ascent of a and the ascent of c with the descent of a. The equality of these two ratios would have yielded the desired proof.

Positional gravity was more felicitously applied in proving the law of the lever (Pt. II, Prop. 10) and in demonstrating that equilibrium is attained on a bent lever when equal weights are suspended on its unequal arms, which terminate equidistantly from the axis of support (Pt. II, Prop. 11). A brief third part concerned the specific gravities of fluids and solids. Dynamic considerations played a role, since Blasius says (Pt. III, Prop. 1) that of two solid bodies descending in water, the heavier will descend more quickly. For comparison of liquids, a hydrometer was used, and its principle was utilized in the comparison of two solid, similarly shaped floating bodies, when Blasius advocated that each be divided into twelve equally spaced parts.

In one of two treatises on intension and remission of forms, *Questiones de latitudinibus formarum,* Blasius included both the English (arithmetic) and French (geometric) fourteenth-century versions of the mean speed theorem[5] that demonstrated the equality of the distances traversed during the same time by one body moving with uniform acceleration and another moving with a uniform speed equal to the velocity acquired by the body in uniform acceleration at the middle instant of its period of acceleration (i.e., $S = V_0 t + at^2/2$, where S is distance; V_0 initial velocity; a acceleration; and t time). Blasius may have introduced both versions into Italy.

Over a long career, Blasius discussed many traditional scientific concepts. Initially an apparent supporter of the impetus theory, he later denied, in his

Questiones super octo libros physicorum (1397), that it could explain acceleration in free fall or the rebound of bodies.[6] He accepted "Bradwardine's function,"[7] which described the relationship between the speeds of two bodies as $F_2/R_2 = (F_1/R_1)^{V_2/V_1}$, where F is motive power; R, the resistive force of the body in motion; and V, velocity. He also reflects common Parisian arguments when he denies the natural existence of vacuum inside or outside the cosmos[8] and then allows (contrary to Aristotle) that if a separate vacuum did exist, bodies could undergo motion and change[9] (in his *Questio de tactu corporum duorum,* Blasius holds that if a vacuum actually existed, many difficulties and contradictions involving the physical contact of two bodies could be resolved).[10]

Although not an original thinker, Blasius sympathetically absorbed the scientific ideas current among the "moderns" at the University of Paris during the fourteenth century. He helped disseminate these in Italy, where they were widely discussed until the time of Galileo.

NOTES

1. Quoted by Marshall Clagett in Marshall Clagett and E. A. Moody, *The Medieval Science of Weights,* pp. 413–414.
2. The brief document has been translated by L. Thorndike in *University Records and Life in the Middle Ages,* pp. 258–259.
3. Clagett and Moody, *op. cit.,* p. 243.
4. *Ibid.,* p. 251. The figure appears on p. 250.
5. Marshall Clagett, *Science of Mechanics,* pp. 402–403.
6. A. Maier, *Zwei Grundprobleme der scholastischen Naturphilosophie,* pp. 271–273.
7. *Questiones super tractatu De proportionibus Thome Berduerdini,* questions 10, 11 (MS Vat. lat. 3012, 151r–153v).
8. *Questiones super octo libros physicorum,* MS Vat. lat. 2159, 120v, c. 1.
9. *Ibid.,* 124r, c. 2–124v, c. 1.
10. G. F. Vescovini, "Problemi di fisica aristotelica in un maestro del XIV secolo: Biagio Pelacani da Parma," in *Rivista di filosofia,* **51** (1960), 196.

BIBLIOGRAPHY

I. ORIGINAL WORKS. All but a few of Blasius' works are unpublished. References to manuscripts of these treatises can be found in L. Thorndike and P. Kibre, "Blasius of Parma," in *A Catalogue of Incipits of Mediaeval Scientific Writings in Latin* (rev. and enl. ed., Cambridge, Mass., 1963), c. 1764; and, in an earlier but more conveniently arranged list, in L. Thorndike, *A History of Magic and Experimental Science,* IV (New York, 1934), app. 40, pp. 652–662.

Questions or commentaries on the following works of Aristotle are extant: *Questiones librorum de caelo et mundo* (Milan, Ambros. P. 120 Sup., 1–69; Bodleian Library, Canonicus Misc. 422, 1–52; Rome, Angelica 592 [F.6.4], 1–34; Rome, Angelica 595 [folio numbers unavailable];

Vienna, National Bibliothek, 2402, 1–63v); *Commentaria in Aristotelis de generatione et corruptione* (Vienna, National Bibliothek, 2402, 99r–123v); *Questiones de anima* (Vat. Chigi O IV 41, 112–217v; Vat. Urbinas lat. 1489, 74–[terminating folio number unavailable]; Bodleian Library, Canonicus Misc. 393, 1–78; Turin 1247 [folio numbers unavailable]); *Questiones in libros metheororum* (Vat. lat. 2160, 63–138v, which is immediately preceded by Blasius' *Conclusiones super libris methaurorum Aristotelis,* a work differing from the *Questiones;* Florence, Ashburnham 112, 1–60; Vat. Chigi O IV 41, 62–108v; University of Chicago 10, 39 ff.). On the *Physics* of Aristotle, Blasius left at least three versions: (1) *Expositio per conclusiones super octo libros physicorum Aristotelis* (Vat. lat. 2159, 1–98v); (2) *Questiones super octo libros physicorum,* preserved, with variant titles, in separate versions, one written in or before 1385 at Padua, of which only the first two books remain (Vat. Chigi O IV 41), and the other, copied in 1397 in Pavia, differing somewhat and embracing all eight books (Vat. lat. 2159, 61r–230r; despite an apparent overlap in pagination with the *Expositio,* it immediately follows the latter with independent pagination beginning with 61r); an incomplete and mutilated version of the Pavia copy is contained in Vat. lat. 3012, 2v–110v; and (3) an arrangement of Buridan's *Questiones super octo phisicorum libros Aristotelis* made around 1396 (Venice, S. Marco X, 103, 83–84).

The two works on statics are *Questiones super tractatum de ponderibus,* containing five questions and known only in a single manuscript (Milan, Ambros. F. 145 Sup., 18r–28r), edited and translated by Father Joseph Brown in a thesis, "The *Scientia de ponderibus* in the Later Middle Ages" (Univ. of Wis., 1967); and *Tractatus de ponderibus,* edited and translated by Marshall Clagett in Marshall Clagett and E. A. Moody, *The Medieval Science of Weights* (Madison, Wis., 1952), pp. 238–279. On intension and remission of forms, Blasius left two treatises, *De intensione et remissione formarum* and *Questiones super tractatu de latitudinibus formarum:* for manuscripts of both, see Marshall Clagett, *The Science of Mechanics in the Middle Ages* (Madison, Wis., 1959), pp. 404, 685–686; the second treatise was published in 1482 and 1486 at Padua and in 1505 at Venice, while part of the third and final question of the second treatise was edited and translated by Clagett in *Science of Mechanics,* pp. 402–408. On problems of motion, he wrote *Questiones super tractatu De proportionibus Thome Berduerdini* (i.e., Bradwardine); for manuscripts, see Marshall Clagett, *The Science of Mechanics in the Middle Ages,* p. 686; and *De motu iuxta mentem Aristotelis* (MS Vat. Barb. 357, 1–16v).

Of the three books and twenty-four questions of the *Quaestiones perspectivae,* Book I, quests. 1–10 were edited by F. Alessio as "Questioni inedite di ottica di Biagio Pelacani da Parma," in *Rivista critica di storia della filosofia,* **16** (1961), 79–110, 188–221; Book I, quests. 14 and 16, and Book III, quest. 3 (*ultima questio*), were edited by G. F. Vescovini as "Le questioni di 'Perspectiva' di Biagio Pelacani da Parma," in *Rinascimento,* **1** (1961), 207–243—the text is preceded by a lengthy discussion of the questions and their historical context on pp. 163–206.

The 1505 edition of *Questiones super tractatu de latitudinibus formarum* includes Blasius' *Questio de tactu corporum duorum,* which is summarized by G. F. Vescovini in "Problemi di fisica aristotelica in un maestro del XIV secolo: Biagio Pelacani da Parma," in *Rivista di filosofia,* **51** (1960), 179–200.

On astronomy, Blasius wrote *Questiones super tractatum sperae Johannis de Sacrobosco* (MS Parma 984) and a *Theorice planetarum* (Vat. lat. 4082, 47r–60v; Venice, S. Marco VIII.69, 175r–216v). The Latin text of the titles of the problems discussed by Blasius in the latter treatise was published by L. Thorndike in *Isis,* **47** (1956), 398–400. Thorndike mentions that the Latin texts of the first three problems and the last were published by G. Boffito and U. Mazzia in *Bibliofilia,* **8** (1907), 372–383, where they are mistakenly ascribed to Peter of Modena. In the same article, Thorndike (*Isis,* **47** [1956], 401–402) cites another astronomical work by Blasius, *Demonstrationes geometrice in theorica planetarum,* printed anonymously by Octavianus Scotus (Venice, 1518), fols. 143r–152v (a possible manuscript version of this treatise is Vat. lat. 3379, 52r–61r, which bears the slightly variant title *Blasii Parmensis demonstrationes geometrie in theoricam planetarum*). An astrological prediction constitutes Blasius' *Iudicium revolutionis anni 1405* (Bibliothèque Nationale MS 7443, 11v–17r).

The diverse treatises cited below conclude the list of Blasius' scientific and philosophic works known thus far: *Questiones super tractatus loyce* [i.e., *logice*] *magistri Petri Hyspani* [i.e., Peter of Spain] (Bodleian Library, Canonicus Misc. 421, 92–222); *Questiones undecim de locis* (Venice, S. Marco X, 208, 82–92); *Queritur utrum spericum tangat planum* (Bodleian Library, Canonicus Misc. 177, 153–154); *Questiones viginti sex predicamentis* (Venice, S. Marco X, 208, 43–82 and perhaps also Vat. Barberini 357); *De motu* (Vat. Barberini 357, 1–16v); *Elenchus questionum Buridani* (i.e., *A Refutation of Questions of Buridan;* Venice, S. Marco X, 103, 83–84); and a *De terminis naturalibus* (Bodleian Library, Canonicus Misc. 393, 78–83), of uncertain attribution. A theological work, *De predestinatione,* has also been preserved (Venice, Bibl. de' Santi Giovanni e Paolo 163).

II. Secondary Literature. There is relatively little literature on Blasius. To what has already been cited, we may add L. Thorndike, *A History of Magic and Experimental Science,* IV, ch. 39; G. F. Vescovini, *Studi sulla prospettiva medievale* (Turin, 1965), ch. 12; A. Maier, *Die Vorläufer Galileis im 14. Jahrhundert* (Rome, 1949; 2nd ed., 1966), pp. 279–299, and *Zwei Grundprobleme der scholastischen Naturphilosophie,* 2nd ed. (Rome, 1951), pp. 270–274; and F. Amodeo, "Appunti su Biagio Pelacani da Parma," in *Atti del IV Congresso Internazionale dei Matematici,* **3** (Rome, 1909), 549–553.

Edward Grant

BLAUW, WILLEM. See **Blaeu, Willem.**

BLAZHKO, SERGEI NIKOLAEVICH (*b.* Khotimsk-Mogilevskaya province, Russia, 17 November 1870; *d.* Moscow, U.S.S.R., 11 February 1956), *astronomy.*

The son of a merchant who had risen from the enserfed peasantry, Blazhko graduated from the Smolensk Gymnasium in 1888 and from the Physics and Mathematics Faculty of Moscow University in 1892. From 1894 to 1915 he was assistant at Moscow Observatory and from 1915 to 1918, an astronomer-observer there. In 1917 he married Maria Ivanovna Ushina, a teacher. In 1918 he became professor of astronomy at Moscow University and from 1920 to 1931 was, simultaneously, director of the observatory. From 1922 to 1931 he was also director of the university's Scientific Research Institute of Astronomy and Geodesy.

Blazhko's pedagogic career began in 1896, when he was entrusted with conducting exercises in practical astronomy at the university. From 1900 to 1918 he taught astronomy at the Women's Pedagogical College, and from 1909 to 1919 at the A. L. Shanyavsky People's University; from 1910 to 1918 he was *Privatdozent* and from 1918 professor at Moscow University; at the latter he held the chair of astronomy from 1931 to 1937 and the chair of astrometry from 1937 to 1953. In 1929 Blazhko was elected an associate member of the Academy of Sciences of the U.S.S.R. He received the title Honored Scientist of the R.S.F.S.R. in 1934 and was twice awarded the Order of Lenin and the Order of the Red Banner of Labor. He also belonged to numerous astronomical societies, both Russian and foreign.

Blazhko's primary sphere of scientific activity was the study of variable stars. In 1895 he began the systematic photographing of the heavens with a special "equatorial camera" built in Germany according to the design of V. K. Cerasky, director of the Moscow Observatory. He hoped, through comparison of plates taken at different times, to discover new variable stars. He also conducted visual observations of variable stars over several decades. Through these observations Blazhko discovered the periodic change of the periods and the shape of light curves of a number of short-period variable stars of the type RR Lyrae, a phenomenon that came to be called the "Blazhko effect." In all, Blazhko investigated more than 200 variable stars, and his valuable series of observations, which covered many years, is still used.

Blazhko's other scientific work involved photographing the sun with a photoheliograph in order to determine the period of its rotation according to the motion of its faculae (1895); obtaining spectra of two meteors in 1904 with an apparatus of his own construction; one of the first detailed investigations of meteor spectra (1907); and one of the first investiga-tions of the spectrum of the eclipsing variable star U Cephei. Blazhko later obtained the spectrum of another meteor, and these three spectra were long among the first five known.

Blazhko devoted special attention to the study of eclipsing binary stars of the Algol type. In 1911, in his dissertation, "O zvezdakh tipa Algolya" ("On Stars of the Algol Type"), he was the first to give a general method for determining the elements of the orbits of eclipsing binaries. He also provided the first analysis of the influence of darkening toward the limb of a star's disk on the shape of the light curve. However, one must note that in 1912–1913 there appeared in the United States a series of articles by H. N. Russel and Harlow Shapley, who, independently of Blazhko, developed methods for studying eclipsing binary stars, not only of the Algol type but of other types as well (β Lyrae and W Ursae Majoris).

In 1919 Blazhko proposed an original photographic method for discovering minor planets—the method of triple exposure on one plate with intervals between exposures and with a shift in the declination of the telescope during the intervals. He devised a number of original instruments: a star spectograph, a blink-microscope, a special magnifying glass for reading the division marks of meridian circles, and a device for eliminating the "stellar magnitude equation" from the times of transit taken with meridian instruments.

BIBLIOGRAPHY

I. ORIGINAL WORKS. Principal works include "On the Spectra of Two Meteors," in *Astrophysical Journal,* **26,** no. 5 (1907), 341–348; "Über der Veränderlichensterne U Cephei," in *Astronomische Nachrichten,* **181** (1909), 295–298; "Étude de l'étoile RW Draconis à période variable et à courbe de la lumière variable," in *Russkii astronomicheskii zhurnal,* **1,** no. 2 (1924); *Kopernik* (Moscow-Leningrad, 1926); "Sur le variable XZ Cygni à période et à courbe de lumière variable," in *Annales de l'Observatoire astronomique de Moscou,* 2nd ser., **8** (1926), 23–41; "Photographische Aufnahmen der kleiner Planeten auf der Universitäts-Sternwarte zu Moskau," in *Astronomische Nachrichten,* **232** (1928), 131–134; "O spektre meteora 1907 g. avgusta 12" ("On the Spectrum of the Meteor of 12 August 1907"), in *Russkii astronomicheskii zhurnal,* **9,** nos. 3–4 (1932), 146–162; *Kurs prakticheskoy astronomii* ("Course of Practical Astronomy"; Moscow-Leningrad, 1938; 3rd ed., 1951); "Istoria Moskovskoy astronomicheskoy observatory v svyazi s istorey prepodavania astronomy v universitete (1824–1920)" ("History of the Moscow Astronomical Observatory in Connection with the Teaching of Astronomy at the University [1824–1920]"), in *Uchenye zapiski MGU,* jubilee ser., no. 58 (1940), 5–106; *Kurs obshchey astronomy* ("Course of General Astronomy";

Moscow–Leningrad, 1947); and *Kurs sfericheskoy astronomy* ("Course of Spherical Astronomy"; Moscow, 1948; 2nd ed., 1954).

II. SECONDARY LITERATURE. Works on Blazhko are B. V. Kukarkin, "Sergei Nikolaevich Blazhko (Necrolog)," in *Peremennye zvezdy*, **11**, no. 2 (1956), 63–64, with portrait; P. G. Kulikovsky, "50-letny yubiley S. N. Blazhko," in *Astronomicheskii kalendar na 1945 god* (Gorki, 1945), pp. 205–207; "Zasluzhenny deyatel nauki S. N. Blazhko. K 80-letiyu so dnya rozhdenia," in *Priroda* (1951), no. 8, 59–61; and "Sergei Nikolaevich Blazhko," in *Astronomicheskii kalendar na 1957 god* (Moscow, 1956), pp. 275–276, with portrait; the obituary "Sergei Nikolaevich Blazhko (1870–1956. Necrolog)," in *Astronomicheskii zhurnal*, **33**, no. 2 (1956), 278–280, with portrait; and V. V. Podobed, "S. N. Blazhko," in *Astronomicheskii tsirkular*, no. 168 (1956), 1–2.

P. G. KULIKOVSKY

BLICHFELDT, HANS FREDERICK (*b.* Illar, Denmark, 9 January 1873; *d.* Palo Alto, California, 16 November 1945), *mathematics.*

The son of Erhard Christoffer Laurentius Blichfeldt, a farmer who came from a long line of ministers, and Nielsine Maria Scholer, Blichfeldt showed unusual mathematical aptitude at an early age. He was assisted in his studies by his father, and in general he did well in all subjects. He passed the university entrance examinations with honors but did not attend because his parents were unable to afford it.

Fortunately for Hans, his family emigrated to the United States when he was fifteen. He spent four years as a laborer on farms and in sawmills in the Midwest and West and two years traveling about the country as a surveyor. His phenomenal ability to do all the surveying computations mentally so impressed his colleagues that they encouraged him to become a mathematician. He entered the recently founded Stanford University in 1894 and received his B.A. in 1896 and his M.A. in 1897. Not having enough money to go to Europe for a doctorate, as was the custom among the better-known mathematicians, he borrowed the money from a Stanford professor, Rufus L. Green, and enrolled in the University of Leipzig, where he studied under the famous mathematician Sophus Lie. In one year he received his doctorate *summa cum laude,* with the dissertation "On a Certain Class of Groups of Transformation in Three-dimensional Space."

During the year 1898 Blichfeldt was employed by Stanford as an instructor. He obtained the rank of full professor in 1913. He accepted the chairmanship of the mathematics department in 1927 and served in that capacity until his retirement in 1938. In addition, Blichfeldt served as a visiting professor at the University of Chicago in the summer of 1911 and at Columbia University during the summers of 1924 and 1925. He was professor emeritus at Stanford until his death.

Blichfeldt was extremely active in the American Mathematical Society and gave numerous talks in many parts of the country on his favorite topics, group theory and number theory. In 1912 he was elected vice-president of the Society.

Blichfeldt's contributions were primarily in the form of articles for the Society publications and European mathematics journals. His lifework was devoted to group theory and number theory. Some of the many topics that he covered were diophantine approximations, orders of linear homogeneous groups, theory of geometry of numbers, approximate solutions of the integers of a set of linear equations, low-velocity angle fire, finite collineation groups, and characteristic roots. In addition, he published the text *Finite Collineation Groups* and coauthored *Theory and Applications of Finite Groups* with G. A. Miller and L. E. Dickson.

During his life Blichfeldt received many honors. In 1920 he was elected to the National Academy of Sciences, which at the time was an achievement for a mathematician. From 1924 to 1927 he was a member of the National Research Council. After he retired from Stanford, the king of Denmark made him a Knight of the Order of the Dannebrog.

Blichfeldt's contributions in group theory and group characteristics are now of considerable importance because of recent applications of Lie groups in the sciences.

BIBLIOGRAPHY

Blichfeldt's works include "On a Certain Class of Groups of Transformation in Three-dimensional Space," in *American Journal of Mathematics*, **22** (1900), 113–120; "On the Determination of the Distance Between Two Points in *m* Dimensional Space," in *Transactions of the American Mathematical Society*, **3** (1902), 467–481; "On the Order of Linear Homogeneous Groups. I," *ibid.*, **4** (1903), 387–397; ". . . II," **5** (1904), 310–325; ". . . III," **7** (1906), 523–529; ". . . IV," **12** (1911), 39–42; "A Theorem Concerning the Invariants of Linear Homogeneous Groups With Some Applications to Substitution Groups," *ibid.*, **5** (1904), 461–466; "Theorems on Simple Groups," *ibid.*, **11** (1910), 1–14; "Finite Groups of Linear Homogeneous Transformations," Part II of *Theory and Applications of Finite Groups* (London-New York, 1916), pp. 17–390, written with G. A. Miller and L. E. Dickson; and *Finite Collineation Groups* (Chicago, 1917).

G. H. MILLER

BLISS, GILBERT AMES (*b.* Chicago, Illinois, 9 May 1876; *d.* Harvey, Illinois, 8 May 1961), *mathematics.*

Gilbert Ames Bliss, the son of George Harrison Bliss and Mary Maria Gilbert, devoted his life to the study of mathematics. Although his scientific interests ranged broadly over the field of analysis, with special emphasis on the basic existence theorems, the focal point of much of his work was the calculus of variations. Prior to World War I he wrote, with Max Mason and A. L. Underhill, on the application of the methods of Weierstrass to a number of problems in the latter subject. He worked in the ballistic laboratory at Aberdeen, Maryland, during the war, and used his knowledge of the calculus of variations to construct new firing tables. In the 1920's his papers encompassed the transformation of Clebsch, proofs of the necessity of the Jacobi condition, multiple integrals, and boundary value problems in his field.

His elementary Carus Monograph on the calculus of variations (1925) was followed, after some twenty years, by his definitive book: *Lectures on the Calculus of Variations* (1946). In this publication Bliss employed the scattered results of mathematicians of past decades, many of whom were his former students, to establish firmly the theoretical foundations of the calculus of variations. He approached his subject from the viewpoint of analysis and covered the use of existence theorems for implicit functions, differential equations, and the analysis of singular points for the transformations of the plane. He improved upon and extended the theories of the problems of Lagrange, Mayer, and Bolza and simplified the proofs of the necessary and sufficient conditions of these problems. He clearly presented the theory of the calculus of variations for cases involving no side conditions. Overall he gave a greater comprehensiveness and generality to the field than had previously existed. As a result of his earlier work as summarized in this book, Bliss may be judged one of the chief architects of the edifice of the calculus of variations.

Bliss's work represents a turning point in American mathematics. With his generation, American mathematics came of age. Previously, most American mathematicians had received their training in, and inspiration from, Europe. From the beginning of his career, Bliss was identified with the University of Chicago. He enrolled there in 1893, one year after the university opened its doors. He received his bachelor's degree in 1897, his master's in 1898, and his doctorate in 1900. Although he began his studies in mathematical astronomy, under the guidance of F. R. Moulton, he soon turned to the study of pure mathematics. E. H. Moore, Oskar Bolza—who aroused his interest in the calculus of variations—and H. Maschke were his instructors.

Bliss spent his apprenticeship as a mathematics instructor at the universities of Minnesota (1900–1902) and Chicago (1903–1904). From 1902 to 1903 he did postgraduate work at the University of Göttingen. Bliss was assistant professor of mathematics at the University of Missouri (1904–1905) and at Princeton (1905–1908). In 1908 he returned to Chicago as an associate professor.

On 15 June 1912, Bliss married Helen Hurd (*d.* 1918). They had two children, Elizabeth and Gilbert, Jr. He married Olive Hunter 12 October 1920.

Bliss taught and worked at the University of Chicago from 1908 to 1941. He was associate professor until 1913, professor from 1913 to 1941, and professor emeritus from 1941. He succeeded Moore as chairman of the mathematics department in 1927 and was Martin A. Ryerson distinguished professor of mathematics from 1933 to 1941. Throughout his career at Chicago he was known for his lively sense of humor and for stressing the importance of a strong union between teaching and fundamental mathematical research.

From 1909 until his death, Bliss exerted a strong influence on the American mathematical scene. He was an associate editor of the *Transactions of the American Mathematical Society* from 1909 to 1916, and from 1921 to 1922 was president of the society. He was elected to the National Academy of Sciences in 1916, and in 1924, with G. D. Birkhoff and Oswald Veblen, he became a member of the awards committee of the newly instituted National Research Fellowships in mathematics. Bliss served on this committee until 1936. In 1925 he received the first Chauvenet Prize awarded by the Mathematical Association of America for his paper "Algebraic Functions and Their Divisors." The following year Bliss was elected a member of the American Philosophical Society. In 1935 he was made a fellow of the American Academy of Arts and Sciences. For many years Bliss served as chairman of the editorial committee established by the Mathematical Association of America for its Carus Monographs, a series of short expository books on mathematics for the layman.

BIBLIOGRAPHY

I. Original Works. Bliss's books are *Fundamental Existence Theorems* (Princeton, 1913); *Calculus of Variations,* Carus Mathematical Monograph No. 1 (Chicago, 1925); *Algebraic Functions,* American Mathematical Society Colloquium Publications, XVI (New York, 1933); *Mathematics for Exterior Ballistics* (New York, 1944); *Lectures on the Calculus of Variations* (Chicago, 1946).

Bliss also wrote many articles: "The Geodesic Lines on the Anchor Ring" (doctoral dissertation), in *Annals of Mathematics,* **4** (1902), 1–21; "The Solutions of Differential Equations of the First Order as Functions of Their Initial Values," *ibid.,* **6** (1905), 49–68; "A Problem of the Calculus of Variations in Which the Integrand Is Discontinuous," in *Transactions of the American Mathematical Society,* **7** (1906), 325–336, written with Max Mason; "A New Proof of Weierstrass' Theorem Concerning the Factorization of a Power Series," in *Bulletin of the American Mathematical Society,* **9** (1910), 356–359; "Generalizations of Geodesic Curvatures and a Theorem of Gauss Concerning Geodesic Triangles," in *American Journal of Mathematics,* **37** (1914), 1–18; "A Note on the Problem of Lagrange in the Calculus of Variations," in *Bulletin of the American Mathematical Society,* **22** (1916), 220–225; "Integral of Lebesgue," *ibid.,* **24** (1917), 1–47; "Solutions of Differential Equations as Functions of the Constants of Integration," *ibid.,* **25** (1918), 15–26; "The Problem of Mayer With Variable End Points," in *Transactions of the American Mathematical Society,* **19** (1918), 305–314; "Functions of Lines in Ballistics," *ibid.,* **21** (1920), 93–106; "Algebraic Functions and Their Divisors," in *Annals of Mathematics,* **26** (1924), 95–124; "The Transformation of Clebsch in the Calculus of Variations," in *Proceedings of the International Congress of 1924 at Toronto,* **1** (1928), 589–603; "The Problem of Lagrange in the Calculus of Variations," in *American Journal of Mathematics,* **52** (1930), 673–744; "The Problem of Bolza in the Calculus of Variations," in *Annals of Mathematics,* **33** (1932), 261–274; "Mathematical Interpretations of Geometrical and Physical Phenomena," in *American Mathematical Monthly,* **40** (1933), 472–480; "The Calculus of Variations for Multiple Integrals," *ibid.,* **49** (1942), 77–89.

II. Secondary Literature. Articles on Bliss are L. M. Graves, "Gilbert Ames Bliss, 1876–1951," in *Bulletin of the American Mathematical Society,* **58** (1952), 251–264 (this article contains a bibliography of Bliss's publications); and Saunders MacLane, "Gilbert Ames Bliss (1876–1951)," in *Yearbook of the American Philosophical Society for 1951,* pp. 288–291.

Ronald S. Calinger

BLISTERFELD, J. H. See **Bisterfeld, Johann Heinrich.**

BLOMSTRAND, CHRISTIAN WILHELM (*b.* Växjö, Sweden, 20 October 1826; *d.* Lund, Sweden, 5 November 1897), *chemistry, mineralogy.*

The son of John Blomstrand, a teacher in a Gymnasium, and his wife Severina Rodhe, Blomstrand originally studied mineralogy at the University of Lund. His interest in chemistry began only after he had been awarded his doctorate (1850) and had been named the first recipient of the Berzelius scholarship. His *Habilitationsschrift* dealt with bromine and iodine compounds of tin. At the University of Lund, Blomstrand became adjunct in chemistry (1856) and later professor of chemistry and mineralogy (1862), a position that he occupied until his retirement (1895). He never married, and except for a brief period in 1861 when he served as mineralogist on an expedition to Spitzbergen, he remained at Lund.

Blomstrand's experimental inorganic research largely concerned the Group VB elements—the so-called earth acids (halides and oxyhalides of niobium, tantalum, molybdenum, and tungsten; heteropoly acids of iodic and periodic acid with chromic, molybdic, and tungstic acids). In addition to his strictly chemical work, he characterized and analyzed many minerals, especially those of the rarer elements or of unknown composition, such as monazite, ilmenite, tantalite, niobite, and euxenite.

Most of Blomstrand's theoretical works (such as those on azoammonium and chain theories) are polemical, but more often conciliatory than inflammatory in tone. Since he lived in Sweden during a period of transition between the older and newer chemistries and since he was a scientific as well as a political conservative, he sought to reconcile Berzelius' dualistic theory with the unitary and type theories. He was opposed to Kekulé's dogma of constant valence and strove to establish a sound and complete theory of variable valence. Blomstrand's chain theory, as modified and developed by Sophus Mads Jørgensen, was the most successful of the numerous attempts to explain the constitution of metalamines. It held sway for roughly a quarter century, until it was displaced by Alfred Werner's coordination theory in 1893.

BIBLIOGRAPHY

I. Original Works. "Zur Frage über die Constitution der Diazoverbindungen," in *Berichte der Deutschen chemischen Gesellschaft,* **8** (1875), 51–55; "Über die Metallammoniake oder die Metallamine," *ibid.,* **4** (1871), 40–52 (an English translation appears in G. B. Kauffman, ed., *Classics in Coordination Chemistry, Part II: Selected Papers* [*1798–1935*], New York [in press]); *Die Chemie der Jetztzeit vom Standpunkte der electrochemischen Auffassung aus Berzelius Lehre entwickelt,* Blomstrand's best-known work (Heidelberg, 1869) made his name known throughout Europe. A complete list of Blomstrand's publications can be found in *Svensk kemisk tidskrift,* **38,** no. 9 (1926), 235–238.

II. Secondary Literature. The entire Sept. 1926 issue of *Svensk kemisk tidskrift,* **38,** no. 9, 234–314, is devoted to articles on various aspects of Blomstrand's career; for obituaries, which include discussions of his works, see P. Klason, *Berichte der Deutschen chemischen Gesellschaft,* **30** (1897), 3227–3241; and E. von Meyer, *Journal für praktische Chemie,* **56** (1897), 397–400.

GEORGE B. KAUFFMAN

BLONDEL, ANDRÉ EUGÈNE (*b.* Chaumont, France, 28 August 1863; *d.* Paris, France, 15 November 1938), *physics, engineering.*

Blondel came from a family of Burgundian magistrates. His mother died when he was nine, and his education was directed entirely by his father. He completed his secondary studies in Dijon, entered the École Polytechnique in 1883, and then attended the École des Ponts et Chaussées. Upon graduation in 1888, he chose assignment to the Service Central des Phares et Balises. Blondel received his degree in mathematical sciences in 1885, and in physical sciences in 1889. He worked in Cornu's laboratory at the École Polytechnique in 1888–1889, and there acquired the knowledge that gave rise to his subsequent discoveries.

Blondel's very first projects drew attention to him, and in 1893 he became a professor of electro-technology at both the École des Mines and the École des Ponts et Chaussées.

During the last few years of the century, Blondel, who until then had enjoyed excellent health, participated in sports, and traveled extensively, was stricken with paralysis of the legs, probably of psychosomatic origin. His father then settled near him in Paris. Bedridden, Blondel devoted himself wholeheartedly to his research and inspired a group of associates who worked in the research laboratory he had set up in Levallois.

His health improved somewhat around 1919, after the death of his father. His physical activity remained extremely limited, however, for the rest of his life.

Blondel indicated that all his work had been suggested by his research for the Service des Phares and his teaching of electrotechnology. This explains the tremendous variety of subjects he dealt with and the large number of his published works (more than 250).

His two main contributions are the oscillograph and the system of photometric units of measurement. Struck by the outdated units of measurement used in photometry, Blondel specified, in 1894, different units for this branch of optics. He introduced the fundamental concept of luminous flux and defined illumination according to the flux received by the unit surface. He thus established a coherent system of units, using as a basis the Violle candle and the meter. The unit of flux, or lumen, was independent of the unit of length. This system was adopted in 1896 by the International Electrical Congress, meeting in Geneva, and it became the system used by the International Illumination Commission and the International Conference on Weights and Measures. It is also included, practically without change, in the international system adopted by the eleventh International Conference on Weights and Measures (1960).

Assigned to study the arc lamps used in lighthouses and their feed, he found deficiencies in the prevailing research methods, which did not allow a worker to see instantly the intensity of alternating currents. After a rather unsatisfactory test using a stroboscopic method (1891) he solved the problem by invention of the oscillograph (1893); he perfected two variations, a "soft iron" version and a bifilar one. For forty years these instruments were the most advanced and the most widely used in the study of variable electric phenomena. From the point of view of the moving band, they have been superseded by the cathode-ray tube oscilloscope, but they are no less valuable because of their simplicity and their small bulk.

BIBLIOGRAPHY

Blondel's oscillograph was patented in April 1897 (no. 266246); the original file is in the Institut National de la Propriété Industrielle, Paris. Among his writings is "La détermination de l'intensité moyenne sphérique des sources de lumière," in *L'éclairage électrique,* **2** (1895), 385–391; **3** (1895), 57–62, 406–414, 538–546, 583–586.

Works on Blondel are Louis de Broglie, *La vie et l'oeuvre d'André Blondel* (Paris, 1944), a lecture given at the annual meeting of the Académie des Sciences, 18 Dec. 1944; *Commemoration de la naissance d'André Blondel, 1863–1938* (Paris, 1963), a brochure published for the ceremony held 15 May 1963 at the Conservatoire National des Arts et Métiers, Paris; and *Commemoration de l'oeuvre d'André Eugène Blondel* (Paris, 1942), a collection of articles by Louis de Broglie, Camille Guitton, Joseph Béthenod, Eugène Darmois, Robert Gibrat, and others.

JACQUES PAYEN

BLONDEL, NICOLAS-FRANÇOIS (*b.* Ribemont, France [baptized there 15 June 1618]; *d.* Paris, France, 21 January 1686), *military engineering, architecture.*

Blondel was the eldest son of François-Guillaume Blondel, master of petitions to the queen mother and king's attorney at the bailiff's court of Vermandois, and of Marie de Louen. At seventeen he became an infantry cadet and fought against the Imperial forces in the Thirty Years' War. Between 1640 and 1652 he held a variety of positions, many of them in naval engineering. In July 1652 he became tutor to the son of Loménie de Brienne, secretary of state for foreign affairs, and spent the next three and a half years traveling through Europe with his pupil. Upon his return to Paris, Blondel succeeded Gassendi as lecturer in mathematics at the Collège Royal (now the Collège de France). From 1657 to 1662 Loménie de Brienne entrusted him with several diplomatic missions; when his period of service was over, he became commissioner general of the navy.

As a member of a commission seeking a harbor for the navy between the Loire and the Gironde, Blondel recommended the site of what is now Rochefort. He directed construction for the region and drew up plans for the town and its fortifications, and for restoration of the Saintes bridge and the Roman arch. Construction was not far advanced when Colbert sent him to the West Indies in July 1666, to look for harbors, make maps, and plan fortifications. The maps he made of Grenada (1:98,000) and Tortuga (1:76,500) in 1667 are in the Bibliothèque Nationale.

From 1640 on, when he was in Paris, Blondel never failed to attend the meetings held regularly by the scholars of the capital prior to the founding of the Royal Academy of Sciences. When he was unable to be present, he relied on correspondence to convey his passionate interest in theoretical discussions of any kind. Thus on 12 August 1657, he wrote at great length to Paul Wurz, one of his Swedish friends, to the effect that Galileo had erred in his *Discorsi e dimostrazioni matematiche* with respect to beams of equal resistance. Actually, the text that he criticized applied correctly to beams fixed at one end only, i.e., to brackets. (Also, he was wrong to claim that in Galileo's thinking beams resting on two supports were involved.)

In 1669 Colbert sponsored Blondel's admission to the Royal Academy of Sciences as a geometer, which in his case meant topographer. In 1671 he was appointed professor at, and director of, the Royal Academy of Architecture, which had just been founded. From then on, he attended the regular meetings of the two academies, gave a course in mathematics at the Collège Royal, and delivered public lectures at the Royal College of Architecture. His work increased in 1672, when Louis XIV put him in charge of public works for the city of Paris, and in 1673, when he was made mathematics tutor to the dauphin.

In the seventeenth century the profession of architect embraced subjects now considered a part of engineering. Blondel devoted part of his lectures at the Royal College of Architecture to geometry, arithmetic, mechanics, gnomonics, the art of fortification, and perspective and stereometry; but he did not feel it necessary to publish corresponding texts. In order to provide an architecture text, in 1673 he brought out a new edition, without the illustrations, of Louis Savot's *L'architecture françoise des bastimens particuliers* (first published in 1624). In his *Cours d'architecture* (1675–1683) he treated both decorative elements and such practical structures as bridges and aqueducts. He presented the concepts of the great architects since Vitruvius and shared the lessons he had learned from wide reading and many visits to most of the major monuments of antiquity, as well as his practical experience and his theories. The *Cours*, illustrated with many remarkable figures, remained the definitive reference work for French architects for more than a century.

While writing the portion of the *Cours* devoted to staircases, Blondel formulated the following rule, approved in 1675 by the Royal Academy of Architecture and still used today: "The length of the normal step of a man walking on level ground is two feet (Paris measure), and the distance between steps on a perpendicular ladder is one foot." From this he deduced that a staircase is well-balanced when $G + 2H =$ two feet (G is the tread of a step and H is the height of a step). This is the origin of the rule of art approved 28 January 1675 by the Royal Academy of Architecture and applied universally ever since; thus it would be proper to call it Blondel's rule.

In 1673, while editing his *Cours*, Blondel published the *Résolution des quatre principaux problèmes d'architecture*. The first problem concerns the sketching of columns; the second and the third, rampant arches. The fourth problem pertains to beams of equal resistance. Here Blondel reproduced his letter of 1657 addressed to Wurz, which he had printed in 1661, as well as his reply to a correspondent who had pointed out to him that his interpretation of Galileo's text was wrong. In his reply Blondel evinced an obstinacy worthy of a better cause and stuck to his criticism.

In 1676 Félibien, secretary of the Royal Academy of Architecture, published a work entitled *Des principes de l'architecture, de la sculpture, de la peinture et des autres arts qui en dependent; avec un dictionnaire des termes propres à chacun de ces arts* which Blondel recommended to his architectural course. Félibien's plates, remarkable in every respect, were to serve as models for the ones in the *Description des arts et métiers,* which was commissioned by order of the king on 16 June 1675. In turn they became the prototypes of the plates in the *Encyclopédie.*

During his infantry service, Blondel had been amazed at the crude methods of the French bombardiers. In 1637, during the siege of Landrecies, he had made the acquaintance of the English engineer Maltus, whom Louis XIII had summoned from Holland to teach his army the use of the mortar. Since Maltus fired by guesswork and then proceeded to make corrections, his first bombs at Landrecies at times fell on Frenchmen. In 1642, at the siege of Collioure, by sheer luck he had destroyed the water supply of the beleaguered Spaniards, who hastened to surrender; thus he appeared to have made progress. In order to improve this method, Blondel scrutinized all the publications on the subject, going even to the extent of reading the two chapters on fireworks inserted in Volume III of the *Cursus seu mundus mathematicus,* by the Jesuit Milliet de Chales. Galileo, who had overlooked air resistance, had demonstrated that the trajectory assumed the form of a parabola. Torricelli had indicated that it was sufficient to make a test shot under a predetermined angle by measuring the range obtained with a given charge and that a range was a function of the sine of the double angles of inclination of the mortar over the horizontal. Blondel too considered air resistance a negligible factor. Bent on arriving at a rule easy to apply, he presented the problem to the Royal Academy of Sciences on 6 March 1677. Buot came up with a solution on 10 March. Roemer offered an even simpler one on 20 March, and on 27 March he proposed to adopt a scaled half-circle as a sighting instrument. Finally, Philippe de La Hire, who did not yet hold membership in the Academy, communicated a third solution through Cassini as intermediary. Blondel used the three solutions for establishing the practical rules which he formulated in his *L'art de jetter les bombes,* the printing of which was delayed until 1683 by order of Louis XIV, who hardly cared to have the enemy profit by it. The French gunners, however, paid no attention to it and continued to use faulty firing tables until 1731 and the publication of the treatise by Belidor, under the title *Le bombardier françois, ou nouvelle methode de jetter les bombes avec precision.*

BIBLIOGRAPHY

I. ORIGINAL WORKS. Blondel's writings include *La solitude royalle ou description de Friderisbourg* (n.p., n.d.; dedicated to the king of Denmark, 5 Jan. 1653); *F. B. Epistola ad P. W., in qua famosa Galilaei propositio discutitur, circa naturam lineae qua trabes secari debent ut sint aequalis ubique resistentiae; et in qua lineam illam non quidem parabolicam, ut ipse Galilaeus arbitratus est, sed ellipticam esse demonstratur* (Paris, 1661); his ed. of Louis Savot's *L'architecture françoise des bastimens particuliers* (Paris, 1673; 2nd ed., enl., 1685), with illustrations and notes by Blondel; *Comparaison de Pindare et d'Horace* (Paris, 1673); *Résolution des quatre principaux problèmes d'architecture* (Paris, 1673); *Cours d'architecture enseigné dans l'Académie Royale d'architecture,* 6 pts. (Paris, 1675–1683); *Histoire du calendrier romain, qui contient son origine et les divers changements qui luy sont arrivez* (Paris, 1682; The Hague, 1684); *L'art de jetter les bombes, et de connoitre l'etendue des coups de volée d'un canon en toutes sortes d'élevations* (Paris, 1683; Amsterdam, 1699); *Cours de mathematique contenant divers traitez composez et enseignez à Monseigneur le Dauphin* (Paris, 1683); and *Nouvelle maniere de fortifier les places* (Paris, 1683; The Hague, 1684, 1711), also trans. into Russian (Moscow, 1711).

II. SECONDARY LITERATURE. Archival sources are the Academy of Sciences, registers 5 (1669), 7 (1675–1679), 9 (1679–1683), and 11 (1683–1686); and the Collège de France, register of professorships. Published sources are Paul Bonnefon, ed., *Mémoires de Louis-Henri de Loménie, comte de Brienne, dit le jeune Brienne, publiés d'après le manuscrit autographe pour la Société d'Histoire de France,* 3 vols. (Paris, 1915–1919), *passim;* Harcourt Brown, *Scientific Organizations in Seventeenth Century France, 1620–1680* (Baltimore, 1934), pp. 12, 92; *Histoire de l'Académie Royale des Sciences,* I (Paris, 1733), 230–236; Henry Lemonnier, *Procès-verbaux de l'Académie Royale d'architecture 1671–1793,* I–II (Paris, 1911–1912), *passim;* Charles Lucas, "François Blondel à Saintes, à Rochefort et aux Antilles (1665–1667)," in *Congrès archéologique de France, LXIe session (La Rochelle-Saintes), 1894* (Caen, 1897), pp. 326–341; and "Relation d'un voyage de Berlin à Constantinople par François Blondel, sieur des Croisettes et de Gallardon (novembre–décembre 1658)," in *Bulletin de géographie historique et descriptive, année 1899* (1900), 111–118; Louis-Placide Mauclaire and C. Vigoureux, *Nicolas-François de Blondel, ingénieur et architecte du roi (1618–1686)* (Laon, n.d. [1938]); and René Mémain, *Le matériel de la marine de guerre sous Louis XIV, Rochefort, arsenal modèle de Colbert (1666–1690),* thesis (Paris, 1936), commercially published without the first three words of the original title (Paris, 1937)—Mémain attributes to Blondel a monograph on the hot springs of Aix-la-Chapelle that was really the work of a contemporary physician of the same name.

ARTHUR BIREMBAUT

BLONDLOT, RENÉ-PROSPER (*b.* Nancy, France, 3 July 1849; *d.* Nancy, 24 November 1930), *physics.*

Son of Nicolas Blondlot, a renowned physiologist and chemist, René Blondlot spent nearly all his life in Nancy, where he taught physics at the local university. Although he was never a member of the Paris Academy of Sciences, he was named *correspondant* for the Section of General Physics in 1894. In addition, the Academy awarded him three of its most important prizes, chiefly for his experimental determinations of the consequences of Maxwell's theories of electromagnetism.

In 1875 John Kerr discovered that birefringence was produced in glass and in other dielectrics subjected to an intense electrical field. With the aid of a movable mirror Blondlot established that in the oscillating discharge of a condenser the time-lag between the electrical phenomenon and the appearance of the Kerr effect is less than 1/40,000 of a second. With Ernest Bichat he observed the same instantaneity for magnetic, rotatory polarization.

Using this same technique of a rapidly rotating mirror, Blondlot measured the velocity of electricity propagated through conducting wires. He sent simultaneous electrical charges through two wires, one of which was 1,800 meters longer than the other. By photographing the light from the resulting sparks successively reflected in a rotating mirror and then measuring the distance between the photographic images, Blondlot established that the speed of electricity in a conductor was nearly the same as that of light.

Blondlot received his greatest notoriety in the controversy over the existence of N rays. In the course of an attempt to polarize the newly discovered X rays, Blondlot claimed in 1903 to have found a new kind of invisible radiation, capable of being emitted not only from cathode ray tubes but also from many luminous sources, most notably from Auer burners. He named this new species of radiation N rays, in honor of the city of Nancy. No fewer than fourteen of his fellow scientists claimed also to have observed these N rays emanating from various animal and vegetable substances and, in the case of Jean Becquerel, from "anesthetized metals." Because of the difficulty of reproducing the experiments, many scientists came to doubt the existence of Blondlot's new ray. In 1904 the editors of the *Revue scientifique* undertook an examination of the entire matter and concluded that the positive experimental results had been the chimerical products of autosuggestion. Blondlot finally recanted and, probably as a result of the scandal, spent the rest of his life in relative obscurity.

BIBLIOGRAPHY

I. ORIGINAL WORKS. Blondlot's writings include *Recherches expérimentales sur la capacité de polarisation voltaïque* (Paris, 1881); *Introduction à l'étude de l'électricité statique* (Paris, 1885), written with Ernest Bichat; *Introduction à l'étude de la thermodynamique* (Paris, 1888); *Sur un électromètre astatique pouvant servir comme wattmètre* (Nancy, 1889), written with P. Curie; and *"N" Rays, A Collection of Papers Communicated to the Academy of Sciences,* J. Garcin, trans. (London, 1905).

II. SECONDARY LITERATURE. A brief *éloge* for Blondlot is in *Comptes rendus hebdomadaires des séances de l'Académie des sciences,* **191** (1 Dec. 1930), 80–81. For the N-ray controversy see *Revue scientifique, 74,* no. 2 (July–Dec. 1904), 73–79, 545–552, 620–625, 705–709, 718–722, 783–785.

J. B. GOUGH

BLUMENBACH, JOHANN FRIEDRICH (*b.* Gotha, Germany, 11 May 1752; *d.* Göttingen, Germany, 22 January 1840), *natural history, anthropology, comparative anatomy.*

Blumenbach was born into a cultured, wealthy Protestant family. His father, Heinrich, was the assistant headmaster at the Gymnasium Ernestinum in Gotha; his mother, Charlotte Eleonore Hedwig Buddeus, was the daughter of a high government official in Gotha and the granddaughter of a Jena theologian. Thus, from a very early age, Blumenbach was exposed to both literature and natural science. After completing his Gymnasium studies in 1769, he studied at the universities of Jena and Göttingen, and received the M.D. in 1775 at the University of Göttingen.

At Jena, Blumenbach attended the lectures of the mineralogist Johann Ernst Immanuel Walch, the author of *Naturgeschichte der Versteinerungen,* which interested him in the study of fossils. In Göttingen, he studied under Christian W. Büttner, who lectured on natural history, beginning with man, and fascinated Blumenbach with vivid accounts of travel and foreign peoples, encouraged him to write his doctoral dissertation, and gave him the impetus to start his widely admired anthropological-ethnographic collection. His dissertation, *De generis humani varietate nativa liber,* which became world-famous and is considered one of the basic works on anthropology, was published for the first time in 1776 (and perhaps as early as 1775). In 1776 he was appointed curator of the natural history collection and extraordinary professor at Göttingen, and in 1778 he was named full professor of medicine.

Through his marriage in 1778 Blumenbach became the son-in-law of Georg Brandes, who held an influential position in the administration of the University of Göttingen, and a brother-in-law of Christian Gottlieb Heyne, the classics scholar. These connections helped strengthen Blumenbach's influence at the university. In 1816 he was appointed *professor primarius* of the Faculty of Medicine. In 1776 he had become a member of the Königliche Societät der Wissenschaften zu Göttingen, and in 1812 he became its permanent secretary. In addition, Blumenbach was either a regular or a corresponding member of more than seventy other academies and scientific organizations, including the Institut de France, the Royal Society and Linnean Society of London, the Königliche Akademie zu Berlin, the Imperial Academy of St. Petersburg, and the American Philosophical Society. He carried on extensive correspondence with scientists, the most noteworthy of whom were Albrecht von Haller, Peter Camper, and Charles Bonnet. He was unusually successful as a teacher, and many of his students who later became famous, such as Karl Ernst von Hoff, claimed that Blumenbach had given them the decisive impetus for the formation of their ideas.

Blumenbach's fame is based mainly on his role in the founding of scientific anthropology. He was one of the first scientists to view man as an object of natural history, and saw in him "the most perfect of all domesticated animals." On the other hand, he gave special emphasis to the gap between man and animal and attacked all political or social abuses of anthropological ideas, in particular the notion that black men were on a lower level of humanity than white men. In his dissertation one can find the first reliable survey of the characteristics and distribution of the human races; its most significant points were included in almost all later anthropological classifications.

Blumenbach's ideas on *Bildungstrieb* (*Nisus formativus*) made a great impression on his contemporaries (and later scientists as well). They are of historical significance because they offered some new arguments in favor of epigenesis to the conflict between it and preformation. However, they were very short-lived and did not exert any lasting influence.

Blumenbach's lectures and his textbook on comparative anatomy were epoch-making. Blumenbach believed that he was the first, at least in Germany, to lecture on comparative anatomy, and that his textbook was the first "to have appeared that dealt with the entire area of *anatome comparata.*" This assumption was most probably correct, but one must not overlook the fact that his *anatome comparata* had less to do with homology than it did with *anatome animata comparata,* i.e., comparative physiological anatomy. Nevertheless, this textbook was without question a milestone in the history of this subject.

The *Handbuch der Naturgeschichte,* which went through many editions and was translated into many languages, exerted an even greater influence on the advancement of science. Although Blumenbach tended to follow the Linnaean system, this work ushered in a new era in natural history. It contains an abundance of new or hitherto insufficiently evaluated morphological and ecological findings, from which Blumenbach drew conclusions that led to a

more modern (biological and evolutionary) concept of the plant and animal kingdoms. He concluded from the spread of certain parasites found only in the domestic pig that such parasites did not exist as long as pigs were not domesticated and that they could therefore not possibly have existed since the creation of the world. Such ideas, revolutionary in their day, were carefully presented in various places in the *Handbuch,* and were demonstrated by concrete examples.

In connection with the morphological analysis and geological dating of fossil plants and animals, Blumenbach developed ideas that were still unknown to most of the scientists of his day and were touched upon by only a few others, such as Soulavie. He came to the conclusion that there had been groups of plants and animals, now extinct, which could not be classified in the system of recent forms of life, and he even attempted to draw up a geological-paleontological time scale.

Blumenbach developed these ideas more deeply and in greater detail in his *Beyträge zur Naturgeschichte,* still a valuable source for historians of science. In addition to some interesting anthropological essays (e.g., on the alleged appearance of a *Homo ferus*), the *Beyträge* contains several essays on the "variability" of nature, a concept that was not understood very well. It also showed that the earth, with all its flora and fauna, had a very long history. Blumenbach was one of the earliest thinkers to recognize the "historicalness" of nature, and therefore occupies an important place in the history of evolution theory.

BIBLIOGRAPHY

I. Original Works. Blumenbach's major writings are *De generis humani varietate nativa liber* (Göttingen, 1776); *Handbuch der Naturgeschichte* (Göttingen, 1779; 12th ed., 1830); "Über den Bildungstrieb (*Nisus formativus*) und seinen Einfluss auf die Generation und Reproduction," in *Göttingisches Magazin der Wissenschaften,* **2** (1780), 240 ff.; "*Über den Bildungstrieb und das Zeugungsgeschäft* (Göttingen, 1781); *Geschichte und Beschreibung der Knochen des menschlichen Körpers* (Göttingen, 1786); *Handbuch der vergleichenden Anatomie* (Göttingen, 1805); and *Beyträge zur Naturgeschichte,* 2 vols. (Göttingen, 1806–1811).

II. Secondary Literature. Works dealing with Blumenbach are Walter Baron, "Evolutionary Ideas in the Writings of J. F. Blumenbach (1752–1840)," in *Ithaca 1962* (Paris, 1962), pp. 945–947; Walter Baron and B. Sticker, "Ansätze zur historischen Denkweise in der Naturforschung an der Wende vom 18. zum 19. Jahrhundert. I: Die Anschauungen Johann Friedrich Blumenbachs über die Geschichtlichkeit der Natur," in *Sudhoffs Archiv,* **47**

(1963), 19–26; K. E. A. von Hoff, *Erinnerung an Blumenbach's Verdienste um die Geologie bey der fünfzigjährigen Jubelfeyer seines Lehramtes am 24. Februar 1826* (Gotha, 1826); K. F. H. Marx, *Zum Andenken an Johann Friedrich Blumenbach, Eine Gedächtniss-Rede gehalten in der Sitzung der Königlichen Societät der Wissenschaften den 8. Februar 1840* (Göttingen, 1840); Hans Plischke, "Johann Friedrich Blumenbachs Einfluss auf die Entdeckungsreisen seiner Zeit," in *Abhandlungen der K. Gesellschaft der Wissenschaften zu Göttingen,* Phil. Hist. Kl. 3rd ser., no. 20 (Göttingen, 1937); and Johann St. Pütter, *Versuch einer academischen Gelehrten-Geschichte an der Georg-Augustus-Universität zu Göttingen,* pt. 2 (1788), pp. 148–149; pt. 3 (1820), pp. 303–307; pt. 4 (1838), pp. 421–424. See also *Allgemeine deutsche Biographie,* II (1875), 748–751.

WALTER BARON

BLYTH, EDWARD (*b.* London, England, 23 December 1810; *d.* London, 27 December 1873), *natural history.*

Like his father, a native of Norfolk, Blyth had a great love of nature and an extraordinary memory. When his father died in 1820, Blyth's mother took charge of the four children and immediately sent Edward, the oldest boy, to Dr. Fennell's school at Wimbledon. Although he displayed great intellectual ability, Blyth was frequently truant to go on scientific expeditions to the nearby woods; and in 1825 Dr. Fennell suggested that Mrs. Blyth, who had planned a university career and ultimately the ministry for her son, should send him to study chemistry in London with a Mr. Keating. Blyth thought the chemist was unsatisfactory, however, and upon coming of age, he invested in a small druggist's business in the town of Tooting, Surrey. This venture was doomed to failure because of his indifference: the real object of his affections was natural history. When not reading extensively in the British Museum, he collected butterflies, stuffed birds, or perhaps studied German—rising as early as three or four in the morning to do so.

While barely managing to exist, Blyth established an excellent reputation as a diligent and accurate naturalist. From 1833 until 1841, he contributed numerous notes and articles to scholarly journals, especially *The Magazine of Natural History.* In 1836 he published an edition of Gilbert White's *The Natural History and Antiquities of Selborne,* and he both translated and annotated the sections on mammals, reptiles, and birds for an edition of Georges Cuvier's *The Animal Kingdom* (1840). As a result of his scientific reputation (and perhaps as a result of an article on sheep in which he discussed Indian specimens), Blyth was offered a job as curator for the Royal Asiatic Society of Bengal. Despite the extremely low pay of 250 rupees per month,[1] he ac-

cepted the offer when his doctor advised him to seek a more salubrious climate for his delicate health. After arriving in Calcutta in September 1841, he worked with great industry for the society and published numerous reports, articles, and monographs, with particular emphasis on mammals and birds. The great English ornithologist John Gould observed that Blyth was "one of the first zoologists of his time, and the founder of the study of that science in India,"[2] and Allan Hume described him as "the greatest of Indian naturalists."[3]

In 1854 Blyth married Mrs. Sutton Hodges, a widow whom he had known in England, but this extremely happy marriage ended in December 1857, when his wife died. This great shock further weakened his chronically frail health, which led him to return to England in the summer of 1862, before learning whether he had received a government pension for his long and faithful service.[4] After returning to his homeland, in addition to his other scholarly works he contributed extensive notes and articles to *Land and Water* and *The Field* under the pseudonym "Zoophilus" or "Z." Unfortunately, Blyth found it extremely difficult, as his early rambling articles indicate, to complete his own longer works, for there were always more facts to be gathered and additional hypotheses to test. Some of these hypotheses were grandiose indeed, and in 1865 he attempted to explain, according to A. R. Wallace, "*everything* by the *Precession of the Equinoxes,*" which led Wallace to comment that Blyth was "certainly very queer."[5]

Nevertheless, his scientific abilities were recognized in 1865 by his election as honorary member of the Asiatic Society of Bengal. He was also a corresponding member of the Zoological Society of London and various other scientific academies and societies in Turin, Norway, Batavia, Moselle, and Philadelphia. He died from heart disease shortly after his sixty-third birthday.

At the time of his death, Blyth was preparing a work entitled "The Origination of Species." Although he was one of the very early converts to Darwin's theory of natural selection,[6] Blyth had long entertained ideas of his own on the species question. In 1835 he published the first of two excellent articles that discuss variation, the struggle for existence, sexual selection, and natural selection in terms that have a Darwinian sound.[7] Although Blyth was equivocal, he conceded that a better adapted organism might survive the struggle for existence and "transmit its superior qualities to a greater number of offspring"; that there is a decided tendency in nature for peculiarities to increase when two similar animals mate; that in human races if nontypical variations were propa gated, they "would become the origin of a new race"; and that by selection man can produce breeds "very unlike the original type." But he also thought that the law allowing differences to be propagated "was intended by Providence to keep up the typical qualities of a species." Without man's intervention, domestic breeds would revert to the original type. In nature simple variations "are generally lost in the course of two or three generations" by the swamping effect of blending inheritance. He erroneously believed that the "original form of a species," not a subsequent modification, was "unquestionably" better adapted to its "natural habits." When he wrote these words, Blyth was not an evolutionist despite limited concessions that species may depart from the original type.[8]

On the other hand, Blyth's perceptive observations on variation in nature bear an interesting resemblance to Darwin's ideas, and, in fact, Darwin's copies of Blyth's articles in *The Magazine of Natural History* are heavily marked. While Blyth was in India, they corresponded frequently, and Blyth's lengthy letters were filled with detailed information, comments, and recommendations. One interesting example occurred in 1855 when he suggested that Darwin read his two papers of 1835 and 1837.[9] Although these articles were obviously germane to Darwin's work, and despite the fact that he had already read, marked, and taken notes on them, he apparently never cited them in print, which is one reason Darwin has been accused of plagiarizing the idea of natural selection from Blyth.[10] While the evidence indicates that Blyth probably did influence Darwin far more than most scholars have previously recognized, no one has established precisely when Darwin first read Blyth's works and what he found of interest there. Blyth no doubt provided Darwin with many insights and possibly reconfirmed certain ideas, but it was Darwin's *On the Origin of Species* which seems to have revolutionized Blyth's ideas on species, and not vice versa.[11] On the other hand, he was keenly aware of the issues in question and warmly recommended A. R. Wallace's first evolutionary paper, "On the Law Which Has Regulated the Introduction of New Species" (Sept. 1855), to Darwin; this was perhaps Darwin's first clear warning of Wallace's work on evolution.[12] Blyth's influence on Darwin, however, does not affect the epochal importance of *On the Origin of Species*.

NOTES

1. Allan Hume was especially bitter about Blyth's low pay. "In Memoriam," in *Stray Feathers,* **2** (1874), n.p.
2. John Gould, *The Birds of Asia* (London, 1850–1853), III, pt. 26, pl. 41.

3. *Op. cit.*
4. His request was at first denied, but through the efforts of Sir P. Cautley and Dr. Falconer, he finally received a pension of £150 per year. See Grote, p. xii.
5. Letter, Alfred Russel Wallace to Alfred Newton, 9 November 1865, at the Balfour Library, Cambridge. The italics are Wallace's. I wish to thank A. J. R. Wallace and R. R. Wallace for permission to quote from their grandfather's MSS.
6. The Asiatic Society of Bengal received a copy of Darwin's *Origin* in February or March 1860. Shortly thereafter Blyth commented favorably on Darwin's views, and at the meetings of November 1860 he staunchly defended his ideas. See *Journal of the Asiatic Society of Bengal,* **29** (1860), 86, 283, 383, 385, 428–438, esp. 436–438.
7. These two articles were published in three parts each in *The Magazine of Natural History.* The first had a different title for each part: pt. 1, "An Attempt to Classify the 'Varieties' of Animals, With Observations on the Marked Seasonal and Other Changes Which Naturally Take Place in Various British Species and Which Do Not Constitute Varieties," in **8** (1835), 40–53; pt. 2 "Observations on the Various Seasonal and Other External Changes Which Regularly Take Place in Birds, More Particularly in Those Which Occur in Britain; With Remarks on Their Great Importance in Indicating the True Affinities of Species; And Upon the Natural System of Arrangement," in **9** (Aug. 1836), 393–409; pt. 3, "Further Remarks on the Affinities of the Feathered Race; and Upon the Nature of Specific Distinctions," *ibid.* (Oct. 1836), 505–514. The second article, "On the Psychological Distinctions Between Man and All Other Animals; And the Consequent Diversity of Human Influence Over the Inferior Ranks of Creation From Any Mutual and Reciprocal Influence Exercised Among the Latter," appeared in three consecutive parts all under the same title, *ibid.,* n.s. **1** (Jan. 1837), 1–9; (Feb. 1837), 77–85; (Mar. 1837), 131–141. Loren Eiseley has reprinted these articles in "Charles Darwin, Edward Blyth, and the Theory of Natural Selection," in *Proceedings of the American Philosophical Society,* **103** (Feb. 1959), 94–158; see 114–150.
8. Blyth (1835), pp. 41, 46–49.
9. The books and MSS referred to are at the University Library, Cambridge. For Blyth's remark see the Darwin Papers, Box 98, MSS headed "Notes on Lyell, Vol. 2, Edit. 1832" (1855), p. 48.
10. See Eiseley, *op. cit.,* 102–103. His arguments are repeated and elaborated in "Darwin, Coleridge, and the Theory of Unconscious Creation," in *Daedalus,* **94,** no. 3 (Summer 1965), 588–602.
11. Darwin to Sir Charles Lyell, (1 June 1860), in *The Life and Letters of Charles Darwin,* II (1887), 314–316; see 316.
12. Letter, Blyth to Darwin, 8 Dec. 1855, Darwin Papers, Box 98, Cambridge University. The letter probably arrived in England about Feb. 1856.

BIBLIOGRAPHY

I. ORIGINAL WORKS. In addition to the articles referred to in the Notes, Blyth wrote an enormous number of notes, articles, and monographs. Many of these are listed by A. Grote with his memoir of Blyth prefaced to "Catalogue of Mammals and Birds of Burma," in *Journal of the Asiatic Society of Bengal,* pt. 2, extra no. (Aug. 1875), iii–xxiv, esp. xvii–xxiv; reprinted without bibliography in Eiseley (1959). Since Grote used Blyth's letters (then in the possession of his sister but now missing) as well as his own recollections, his biography of Blyth is the most valuable one available. Some of Blyth's letters to Darwin are in the Darwin Papers, Cambridge University, but his handwriting was frequently very poor, and these valuable letters have not yet been fully utilized.

II. SECONDARY LITERATURE. Other than Grote's biographical sketch, the obituaries of Blyth—such as those in *Nature,* **9** (1874), 191, and *The Field* (3 Jan. 1874), 3—are of little value. H. D. Geldart's "Notes on the Life and Writings of Edward Blyth" (read 28 Oct. 1879), in *Transactions of the Norfolk and Norwich Naturalists Society,* **3** (1884), 38–46, used Grote exclusively, except for excerpts from and comments on Blyth's 1835 article which Grote omitted. H. M. Vickers also cited this perceptive article by Blyth in "An Apparently Hitherto Unnoticed 'Anticipation' of the Theory of Natural Selection," in *Nature,* **85** (16 Feb. 1911), 510–511. Recently, attention has been directed to Blyth by Eiseley (1959). Perhaps the tenor of the replies to Eiseley has been established by Theodosius Dobzhansky, "Blyth, Darwin, and Natural Selection," in *The American Naturalist,* **93** (1959), 204–206. He stresses that Darwin's thinking processes were not totally free from "subconscious components" and asked, "Might not even Darwin have been mistaken about the sources of some of his ideas?" Eiseley replied in 1965 (see n. 10).

H. LEWIS MCKINNEY

BOAS, FRANZ (*b.* Minden, Germany, 9 July 1858; *d.* New York, N.Y., 21 December 1942), *anthropology.*

Boas exercised considerable influence in the "historical" and "scientific" reorientation of anthropology from about 1890 to 1925, especially in the United States. He made significant contributions to formulation of problems and methods in human growth, linguistics, folklore, art, and the ethnology of the Indians of the Northwest Coast.

Franz was one of six children of Meier Boas, a moderately successful merchant, and Sophie Meyer, who founded a Froebel-type kindergarten in Minden. At an early age he expressed an interest in traveling to far-off lands to study the life and customs of exotic peoples, but as he pursued his studies at the Gymnasium in Minden, mathematics and physics claimed his attention and remained the focus of his graduate work at Heidelberg, Bonn, and Kiel. At twenty-three Boas received a doctorate in physics, but during his university career his desire to travel and to understand "nature as a whole" was renewed, probably by the inspiration of the naturalist and geographer Alexander von Humboldt. A friendship with the geographer Theobald Fischer also influenced the turn to anthropology.

In the fall of 1882 Boas attended meetings of the Berlin Anthropological Society, where he sought out Adolf Bastian, the foremost German ethnologist, and Rudolf Virchow, the famous anthropometrist. From Virchow he received anthropometric training, while Bastian gave him "good advice," encouragement, and friendly assistance. By June 1883 Boas was on his way to Cumberland Sound, Baffin Island, on the polar research schooner *Germania.*

Before setting out for the Arctic, Boas read widely on Eskimo culture and studied the language, apparently intending to gather both geographic and ethnographic materials. He had every reason to be satisfied with his geographic researches, completed in 1884, for he had corrected many misconceptions about Baffin Island and had established the presence of two large lakes (not one) in the interior. He had also charted some 250 miles of coastline, often at great discomfort and hazard to his life. The expedition was crucial in turning Boas to anthropology, for he not only came to admire the Arctic people but also observed firsthand that geographic forces were not the prime determinants of human behavior. Obviously the source of these determinants must be sought elsewhere, and Boas elected to follow Bastian in focusing on mental processes. By the time Boas published *The Central Eskimo* in 1888, ethnological contributions were outrunning the geographic.

On Boas' return to Germany, Bastian found him a position in the Museum für Völkerkunde as an assistant, and in 1886 he became lecturer in geography at the University of Berlin. At the time he was undecided about pursuing a career in Germany, for he felt frustrated by the discrimination and insults to which he had been subject because of his Jewish origins. Still fresh from the Arctic adventure, he was excited by the visit of some Bella Coola Indians "on exhibit" in Berlin. Boas interviewed them and threw himself into the task of ordering the museum's collection of Northwest Coast materials. He soon had managed a shaky financing that would enable him to live among the Indians of the Northwest Coast for a few months. This was the first of thirteen field trips to the area (the last was in 1931). Altogether he spent some two and a half years among the coastal Indians.

Contact with F. W. Putnam during a meeting of the American Association for the Advancement of Science in Buffalo, New York, developed into an enduring association; it provided new opportunities for Boas and influenced his decision to remain in the United States. Putnam arranged for Boas to become a foreign associate member of the Association and in 1887, when Boas appeared at the New York office of *Science* to present an article, he was asked to become assistant editor. This provided a small income, whereupon Boas married Mary Krackowizer of New York, whom he had met in Germany.

A chance meeting with G. Stanley Hall led to an invitation to become a lecturer in anthropology at Clark University, a position Boas held until 1892, when he accepted Putnam's invitation to become his anthropological assistant in charge of exhibits for the Columbian Exposition in Chicago. In that same year, under Boas' direction, the first Ph.D. in anthropology in North America was conferred on A. F. Chamberlain, who became his successor at Clark. Boas stayed with Putnam during the formative years of the Field Museum, which had been founded to house some of the riches of the ethnological exhibits of the Columbian Exposition. When Putnam left in 1894, Boas became curator of anthropology, but difficulties already afoot soon provoked his resignation. Boas was virtually destitute until 1896, when Putnam offered him the post of curator of ethnology and somatology at the American Museum of Natural History. The museum appointment led to a lectureship in physical anthropology at Columbia University, followed by a professorship in 1899 (he retired in 1937). Putnam left the museum in 1901, and Boas assumed the post of curator of anthropology, resigning in 1905.

Boas made a number of basic assumptions about the nature of reality which guided his approach to the phenomenon of man and dictated his contributions to anthropology. The heart of his position was the assumption that reality is structured. It has an inner core that remains relatively stable in the face of altered conditions and an outer form that changes in response to alterations in conditions. The scientific problem lay in uncovering the inner core and in finding to what extent it was affected by variations in external form and vice versa. The only way to uncover this hidden reality, in Boas' view, was to study the variations, and when the variable aspects of range and nature had been determined, it would be possible to describe the inner form with confidence. This empirical-inductive position was not novel, however; it echoed the usual approach of natural scientists at that time. However, the application of an inductive orientation to the "evolutionary" anthropology of the day was bound to produce profound alterations in problem definition, methodology, and results.

Boas' most general and signal contributions to anthropology stem from his earnest desire to make the discipline a rigorous and exact science. No anthropologist at the time seems to have understood so well as Boas the full implications of the inductive position and why the path of deduction then followed in anthropology must lead to erroneous conclusions and perpetuation of a quasi-science. No one pressed more vigorously for abandonment of uncritical inferences and comparisons than Boas, as he systematically destroyed the popular and "scientific" myths of the day. Except for his ethnographic texts, which presented material with little commentary, he used every

review, public lecture, article, introduction, and analytical monograph to rephrase problems and to point out the complex logical and technical operations involved in achieving precise controls and conclusions. Most of the rephrasings picked up the complexities glossed over by uncritical classification: What criteria shall be used when we say that a culture or a custom is simple in comparison with another? How does one classify cultures that are simple in technology, yet complicated in social organization? How can the criteria employed be freed from subjective bias traceable to the relativistic categories of thought found in one's own culture?

Boas has been cited as a destroyer of the "evolutionist" position, but as a scientist, he was the opponent of any kind of speculation. "Evolutionary" anthropology, dominant when Boas entered the discipline, operated deductively with a number of unproved assumptions. It was therefore necessary to replace this pseudo anthropology and its uncritical "comparative method" with a "scientific" anthropology grounded in a critical use of detailed factual materials. Attention should be directed to process—how forms change.

In the course of unseating "evolutionary" anthropology, Boas contributed some basic principles to practice. (1) All classifications are relative, and the scientist must realize this when he seeks to conceptualize his findings. The error of forcing the logic of one's own categories of thought on the "primitive" must be avoided, or the product of research will be useless. (2) Classification is no substitute for process—for what actually happens. Data must be collected in such detail that the operation will record variations in forms and the causes of these variations, thereby offering clues to process. (3) For items to be comparable, they must show similarities not only in outward form but also in their histories. One must beware of using functional analogies produced by convergence as if they were homologies—which alone can pass muster for scientific comparison. (4) Coincidence must not be taken for causal connection. Boas' classic target here was the alleged relation linking race, language, and culture. With well-known historic illustrations, supplemented by examples out of his own fieldwork, Boas was able to show that race, language, and culture varied independently of each other. (5) No system of explanation can account for everything, yet there can be no scientific explanation at all without accounting for the variations insofar as objective controls can be brought to bear. (6) Valid interpretations are possible only when derived from a relevant context. To seek the source of religious behavior in a "contemplation" of nature is suspect because more likely sources for religious ideas are the "feelings" and "imaginative play" accompanying "social experience." (7) The data compared must be quantified before "laws" can be extracted.

The determination to reorient anthropology came gradually, as Boas developed field experience and began to analyze his Northwest Coast materials, especially myths and legends, in the light of their geographic distributions. He paid special tribute to the geographer Friedrich Ratzel in supplying this methodological lead. As early as 1888 he called attention to the great importance of "cultural contact" for the cultural development of "primitive" societies, but he was nevertheless willing to admit that comparative evolutionary studies were sound enough to allow the conclusion that the "human mind develops everywhere according to the same laws." By 1896, however, Boas was convinced that the "vain endeavor to construct a uniform systematic history of the evolution of culture" with the aid of the traditional "comparative method" had to be renounced if anthropology were to become a science. The method he advocated was the "historical."

The historical method would uncover process through a meticulous consideration of individual forms and their variations. By plotting culture "elements" in space, historic relations could be outlined and psychological processes recovered, as one noted alterations in form and meaning made by individual peoples when reinterpreting borrowed traits according to their own social traditions. Properly applied, the historical method would disclose how diverse elements had been accumulated and ordered in a kind of unity not only within particular tribes but also within areas where they seemed to be especially characteristic. Once the variations had been disposed of, the local culture base would be recognizable. In applying the historical method, it was best to begin with a limited area and gradually extend one's controlled comparisons outward. Certain uniformities could be found in the cultures of societies located in geographic areas suggesting a kind of natural culture region with which to begin.

At all times the investigator must stay within the limits of his data and avoid inferences that have not been checked against facts. The occurrence of similar forms in adjacent tribes usually indicated that diffusion had taken place, but the researcher must consider other contingencies. Likewise, a wide and continuous distribution would suggest an ancient history for the culture element in question, but an age-area principle could not be applied to ethnographic facts as a blanket rule, because historic exceptions were known. To hold that distant tribes with similar customs were linked historically would not be reasonable unless a more or less continuous chain of distribution of elements held in common could be traced, or migration were likely. Boas thus admitted that parallel developments in widely separated areas could occur and that such might owe their similarities to identical psychological processes. In this regard he noted how language elements were distributed so irregularly throughout the world that some phonetic, morphological, and classificatory features must owe their similarities to "psychological causes" operating in the context of a limited number of alternatives. Such an admission would not lend wholehearted support to a unilinear type of development, however. Boas' viewpoint stressed a generalized psychological base, an impulse as it were, that could be expressed in any number of ways. Therefore, differentiation was more expectable than restriction to a single response. It followed that no unilinear parallelism based on identical psychological processes could be granted until the strictures of historic contact, migration, and convergence had been satisfied. The inner form thus would stand in relief only as the variations had been accounted for.

By the time Boas wrote the introduction to the *Handbook of American Indian Languages* (1911), he was able to outline the essentials of his position. He intended to establish the "historical method" and then move on to "problems of cultural dynamics, of integration of culture and of the interaction between individual and society." All these efforts were programmed to uncover the ultimate anthropological reality—the psychological laws governing human behavior.

Once convinced that he was on the track of a scientific anthropology, Boas drove himself tirelessly and systematically toward the goal. He preferred to operate in the context of problems and methods rather than to formulate neat definitions for the classification and manipulation of data. In consequence, the conceptual language of anthropology owes little to Boas, who rarely offered a definition of culture, the alleged hallmark of anthropology. He was, however, a prime contributor to the emergence of the culture concept, stressing the weight of "social tradition" as a molding force in human behavior. Boas thus was in the vanguard of those who recognized that what individuals experienced and learned as members of a society represented a basic problem for social science investigation, and that the key to human behavior would be found in sociopsychological processes rather than in common human psychological tendencies stressed by evolutionists when reconstructing the stages of culture growth.

From his field experience Boas was convinced that the sociopsychological processes linking primitive men to social tradition were basically little different from those of civilized societies. By 1930 two of his students, Ruth Benedict and Margaret Mead, were breaking new ground in learning how cultures stimulate the development of unique personality types.

Boas distinguished three important problems to which anthropology must be directed: "reconstruction of human history"—biological, linguistic, and social; "determination of types of historical phenomena and their consequences"; and the "dynamics of change." In starting with problems and methods, it is apparent that Boas wished to reorient his readers to the intricate processes involved in human phenomena. Aware of his own deficiencies in ethnographic details, he had no wish to forestall a scientific advance by presenting an elaborate explanatory system. Classifications and explanations must come later—after full description of the phenomena. Ultimate understandings would emerge as psychological processes came to light, but Boas wanted to make it clear that anthropology, since it dealt with "secondary [historic] features," could not use psychology as a jumping-off point. Rather, "The psychological problem [of inner form] is contained in the results of the historic inquiry."

When casting about for the most sensitive indicators of psychological processes, Boas was drawn to language, for in the "linguistic categories" of specific languages he hoped to uncover the "unconscious" impulses that had given rise to the "fundamental ideas of language." Since ethnology could be considered the "science dealing with mental phenomena of the life of the peoples of the world," the two stood to gain by mutual cooperation, and Boas encouraged the thought that linguistics would make a greater contribution to anthropology as a part of ethnology than were it to go its own way. The study of folklore was selected as the second center of investigation because Boas had observed that "nothing seems to travel as readily as fanciful tales." Some elements were widely distributed throughout North America, and there were obvious connections between the mythology of northeast Asia and that of the Northwest Coast. Boas labored for thirty years to produce a compilation of folklore elements from throughout the Northwest Coast area and from which he could extract the psychological processes by which they had been altered during their spread and reinterpretation. Many of the conclusions advanced in the voluminous quantitative study *Tsimshian Mythology* (1916) had been foreshadowed in shorter papers on how folklore traits drawn from a number of sources accumulated

gradually and were integrated into the folklore traditions and narrative styles of a particular tribe.

Boas' theory of culture history, and his view of the integration of culture, owed much to his pursuit of mythological elements on the Northwest Coast. Since mythic elements could be disengaged from narratives and plots and widely diffused, he became convinced that culture seldom, if ever, diffused as a complex and unified whole. He therefore opposed the *Kulturkreislehre*, who assembled disparate elements into "culture complexes" and treated them as if they had diffused *en bloc.* Again he found confirmation for his notion that any cultural tradition was a complicated product of intricate historical and psychological processes rooted in the "social life of the people." Under such circumstances one could hardly expect mythology to be a "direct reflex of the contemplation of nature" and to present a uniform "organic growth" expressing uniformities in the human mind. It was true that incidents and actors in folktales had a way of becoming associated with natural phenomena (stars, animals, mountains, trees), but these were secondary symbolizations projected through the "play of imagination with the events of human life." In Boas' historical view, individual cultures might develop some tendencies toward becoming a unity, but none could become a thoroughly consistent and integrated whole.

Myths also proved a useful touchstone for language studies, literary analysis, and world view. When they were dictated by the informant and recorded phonetically in text, a corpus of materials was available for linguistic analysis. Myths often contained theological, philosophical, and scientific explanations, and thus provided an important channel to primitive thought. But although Boas was aware of the uses to which myths could be put in understanding the psychology of a people in their choice of metaphor, in the symbolism expressed in the personalities of the actors, and in explanations of events, he never explored these dimensions in any depth.

Art was the third medium Boas used to demonstrate that human phenomena are highly variable and have complex historic, social, and psychological roots. He confounded those who interpreted the history of art as an evolution from a realistic to an abstract representation by pointing out that within any society there may be found examples of both realistic and geometric design. Actually, an utterly realistic treatment did not occur in primitive art. The intent might be representative, but "symbolic forms" or conventions representing heads, legs, and other body details normally were used, and the primitive artist knew very well that he had not succeeded in duplicating what he portrayed. The imaginative stimulus the artist received from materials and tools while approaching the task within a traditional context provided the key to much of the variation in the artistic expression of any tribe, and Boas cautioned against assuming that conventionalization was so rigid that no variations could occur in primitive art.

Stress on the interplay of tradition, material, tool, and technical virtuosity channeled Boas' focus on process in art, and helped him explain why individual variations and stylistic conventionalizations must occur. Only the total context of any art could provide relevant and valid source materials for scientific generalization. The elements of art could be explained much better by the rhythmic and alternating movements involved in technical operations than by contemplation of nature's analogues. Uniformities found in any art were a product of stylistic traditions, which limited human inventiveness, and not the consequence of uniform responses by the human mind.

Physical anthropology constituted Boas' fourth instrument to promote a scientific anthropology. It seemed highly probable to him that the predominance of heredity over environment assured a great permanence of human types. However, the problem of physical variability had to be dealt with first because variability was not a matter of biology but the result of unknown influences on the more permanent characteristics. The plotting of human types according to their geographic distribution therefore must be the first order of business, paying due attention to variations stemming from "mixture," the fixing of types through inbreeding and isolation, and differentiations that followed natural tendencies for succeeding generations to vary from parental stems. Subtle physiological changes induced by nutrition and mode of life could not be ignored, because man, like other domesticated species, was not impervious to such influences. The races that anthropologists so easily assumed to be homogeneous and stable in type actually masked a considerable heterogeneity that would become apparent—and significant—as the complex histories of the "genetic lines" of which they were composed came to the surface. When the historic analysis had revealed the range and sources of variations, the permanent features would stand clear and would allow definition of local "genetic" types instead of "ecotypes." Boas was soon convinced that historic complexities in anatomical variation were such that it would be futile to expect to find any "pure races" or an original race from which a rigid genealogy could be plotted.

Boas set out to document physical variation by investigating alterations that might be due to rates of growth, intermixture, and mode of life. Knowing individuals differed in their maturational rates, he asked how these uneven rates influenced distributions and classifications. He admitted that development of a dynamic set of growth statistics would increase the complexities of research problems, but a vast and interesting array of challenging questions would open up as "physiological changes in the individual and the types and variabilities of these changes become accessible to investigation."

Measurement of nearly 18,000 immigrants in New York provided Boas with a controversial study in physical variation. In the New World urban setting, round-headed east European Jews became more long-headed, while south Italians, long-headed to begin with, became more short-headed. Two ethnic groups with differing head shapes thus tended to converge in a new urban environment. The changes were registered in children, and native-born children registered a greater shift in stature and in head form than did foreign-born. The study excited anthropometrists, who had viewed physical types as quite stable and had relied on head form as a special index for racial classification and history. Although Boas drew on his data to emphasize the "plasticity" of human types, he realized that genetic organization had not been altered. Nonetheless the study called attention to the dynamic relation linking type and environment, and cautioned against the ready acceptance of external form as a mirror image of genetic type.

Stimulated by an experimental study in which "pure lines" of beans had shown variations apparently traceable to both inheritance and environment, Boas attempted to determine the relative importance of these variables for human physical types. By statistical correlation of variance "averages" among "fraternities" comprising a line of descent, he hoped to measure the heterogeneity of the line. Boas was sufficiently knowledgeable in mathematics to attempt innovations in correlational procedures to suit his anthropometric data, but these *ad hoc* efforts had no lasting effect, for Pearson and Fisher developed more sophisticated techniques.

With variations the key to his initial scientific purpose, Boas turned to controlled observation and recording of detailed information as prerequisites to the critical and "exhaustive" analysis that must take place. There could be no substitute for facts, which alone could lead to the understanding of forms and processes. Hence, the somewhat natural history approach of Boas to the collection of anthropological data. The anthropologist must be a fieldworker, but he could not approach his task with preconceived ideas. A properly trained anthropologist would observe, but above all he would depend on the people he was studying to supply their own categories of thought for sifting the phenomena. It followed that the interview would not be directed, except incidentally. The informant would be presented with a problem, but he then would be left free to develop his narrative and to follow the lead of his own interests. A fieldworker should strive for language mastery, but in the absence of the language skill, texts dictated by the informant in his own language could be recorded phonetically.

Boas recommended the collection of comparable materials from several informants, but he made no real effort to increase the number of his informants in the hope of increasing reliability. Instead, he customarily sought control through numbers by expanding the list of elements. The historic emphasis tended to focus attention on issues other than the determination of culture patterns or literary style, but Boas' treatment invariably turned to the presence or absence of literary expression in riddles, moralizing fables, epic poetry, and the like to show that primitive literature had a history and was not a natural and spontaneous product of common human mental processes.

In reviewing Boas' substantive contributions to anthropology, it is apparent that the key to his success lay in a determined massing of data, a quick grasp of the ramifications of problems, and a perceptive utilization of leads furnished by investigators in related fields.

A restrained and businesslike tone governed Boas' relations with his Indian informants, preventing him from probing their offbeat habits, but his methodical record established a platform upon which subsequent investigators could build. His publications, often with ethnographic notes, concerned a remarkable number of Pacific Coast and inland tribes: Northern and Southern Kwakiutl, Bella Coola, Tsimshian, Chinook, Tillamook, Kathlamet, Kutenai, and a Keresan-speaking pueblo of the Southwest.

In his search for the inner world of the primitive, Boas pioneered in using those who had been raised among the Indians or who were their close associates. He went to some pains to train George Hunt, a mixed-blood Tlingit reared among the Kwakiutl, in phonetic transcription, and Hunt rewarded him with many pages of text, including mythology, potlatch narratives, and Kwakiutl recipes. James Teit, a sheepherder, also was encouraged to publish extensively on the Thompson and other Indians of the plateau region.

Boas was largely responsible for conduct of the Jesup North Pacific Expedition (1897–1900), which sought answers to cultural relations linking Siberian and Northwest Coast peoples. His efforts to establish an international school of American archaeology and ethnology dedicated to research in Mexico ultimately failed as a result of the revolution that erupted in 1910, but during his directorship (1911–1912), Boas was able to acquaint the school and the Mexican appointee, Manuel Gamio, with the uses of stratigraphy and typology in formulating a sequence of cultures for the Valley of Mexico. The interest Boas generated in folklore was passed on to students and associates, including William Jones, George A. Dorsey, Edward Sapir, Pliny E. Goddard, Robert H. Lowie, Melville J. Herskovits, Gladys Reichard, and Melville Jacobs. During his editorship (1908–1924) of the *Journal of American Folk-Lore,* Boas assured an outlet for the texts and translations he had inspired. Publications of the American Ethnological Society and of the Columbia University series Contributions to Anthropology also were utilized for folklore works until 1940.

In linguistics, Boas' accomplishments parallel his achievements in folklore. A self-taught linguist, he used his editorship of the *Handbook of American Indian Languages* (1911–1941) to outline a proper study of language and to provide "grammatical sketches" that would serve as models for future research. His intent was to achieve uniform presentations, treating the grammar "as though an intelligent Indian was going to develop the forms of his own thoughts by analysis of his own form of speech." In 1917 Boas took the initiative in launching the *International Journal of American Linguistics,* which he not only edited until his death but also maintained with personal funds.

In physical anthropology, Boas' pioneer efforts led to discovery of individual variability in "tempo of growth." He was the first to initiate human growth studies in North America, and his was the first chart of standardized heights and weights for American children according to chronological age, corrected statistically to give the standard deviation or variance at each half-year interval. Boas also pointed out the close correlation between the rate of physiological development and mental development; physiologically advanced subjects made better scores in school work and on psychological tests than did physiologically retarded subjects of the same age.

Boas has been criticized for impeding the progress of American anthropology for two or more decades by training a generation of students antagonistic to evolutionary problems and by exercising his commanding influence in publication and organizational channels. Such an accusation is difficult to evaluate, but the existence on the Continent of a similar anti-evolutionary drive leaves the charge open to doubt. The basic limitations of Boas' approach stem from his determined efforts to overturn "speculative" evolutionary theories of the origin and development of human thought and culture. He concentrated on investigation of "mental phenomena" in art, language, and mythology and thus was diverted from the study of cultures in their organizational and functional operations. Preoccupation with historic processes also led to an indifference to the processes of change then taking place in Indian cultures. His methodological practice of reducing larger culture units to elemental parts in order to trace historic relations prevented full appreciation of the stylistic integration that cultures might exhibit. Once Boas developed his position, he was little inclined to change course or to go beyond the scientific boundaries he had outlined. This inflexibility is apparent in an inclination to press more rigorous and searching methods upon unfinished research rather than to open up new problems.

Besides holding honorary and regular memberships in many national anthropological organizations in Europe and in the United States, Boas held a number of presidencies, including those of the American Anthropological Association (1907, 1908), New York Academy of Science (1910), American Folklore Society (1931), and American Association for the Advancement of Science (1931). From 1901 to 1919 he served as honorary philologist to the Bureau of American Ethnology.

BIBLIOGRAPHY

I. ORIGINAL WORKS. A select list of papers (some revised or condensed) assembled by Boas in *Race, Language and Culture* (New York, 1940) offers an excellent introduction to the range of his activities and his theoretical position. This can be supplemented with such reprints as *The Central Eskimo* (1888; repr. Lincoln, Neb., 1964); *The Mind of Primitive Man* (1911; repr. New York, 1963); and *Primitive Art* (1927; repr. New York, 1955). In *Kwakiutl Ethnography* (Chicago, 1966), Helen Codere has added a number of published selections to an unfinished manuscript in order to furnish a representative account of Boas' description of this people. The raw data for Boas' study of bodily changes in immigrants was published as *Materials for the Study of Inheritance in Man* (New York, 1928). He was also contributing editor of the *Handbook of American Indian Languages,* 4 vols. (Washington, D.C.–New York, 1911–1941). A full bibliography is in Memoir 61 (see below).

II. SECONDARY LITERATURE. Two sympathetic views of Boas appeared as memoirs of the *American Anthropologist: Franz Boas, 1858–1942,* Memoir 61 (Menasha, 1943), with contributions by A. L. Kroeber, Ruth Benedict, and others; and Walter Goldschmidt, ed., *The Anthropology of Franz Boas: Essays on the Centennial of His Birth,* Memoir 89 (Menasha, 1959). In June Helm, ed., *Pioneers of American Anthropology,* American Ethnological Society, Monograph 43 (Seattle, 1966), Ross Parmenter describes the friendship of Boas and Zelia Nuttall, the celebrated Mexican researcher, and Robert Rohner presents diary and correspondence material relevant to Boas' fieldwork on the Northwest Coast. M. J. Herskovits, *Franz Boas: The Science of Man in the Making* (New York, 1953), offers the appreciative reflections of a student, viewing Boas as teacher, scientist, and citizen of the world. More critical treatments can be found in Leslie White, *The Ethnography and Ethnology of Franz Boas,* Texas Memorial Museum Bulletin 6 (Austin, Tex., 1963), and *The Social Organization of Ethnological Theory,* Rice University Studies, Monograph in Cultural Anthropology, **52,** no. 4 (Houston, Tex., 1966). For Boas' contribution to the concept of culture, see George Stocking, Jr., "Franz Boas and the Culture Concept in Historic Perspective," in *American Anthropologist,* **68,** no. 4 (1966), 867–882.

FRED W. VOGET

BOBILLIER, ÉTIENNE (*b.* Lons-le-Saulnier, France, 17 April 1798; *d.* Châlons-sur-Marne, France, 22 March 1840), *geometry, mechanics.*

Étienne Bobillier was the second son of Ignace Bobillier, a merchant, who died when Étienne was seven years old. He and his brother Marie André were raised by their mother, a wallpaper merchant. Étienne attended the local secondary school and seemed inclined toward literary studies, in which he won awards. Until he was sixteen, he showed no interest in mathematics, but then his brother, a student at the *lycée* of Besançon, was accepted by the École Polytechnique. Étienne resolved to follow this example and, shutting himself up with the books left behind by his brother and aided from time to time by advice from him, he completed the course in special mathematics. He then presented himself for the competitive entrance examination in 1817. The examiner, well known for his severity, put Étienne first on his list. He was admitted to the École Polytechnique as fourth in rank.

Bobillier finished his first year eighth out of sixty-four students. Because of financial needs, he took a leave of absence from the École Polytechnique in October 1818, in order to become an instructor in mathematics at the École des Arts et Métiers at Châlons. The young instructor soon showed a remarkable gift for teaching; exhibiting a rapid judgment, lively mind, lucid language, and strength of character that impressed, captivated, and subdued his students.[1] He taught trigonometry, statics, analytic geometry, descriptive geometry, practical mechanics, physics, and chemistry.

In 1829 Bobillier, who saw no future where he was, applied for a university post and, upon Poisson's recommendation, he became professor of special mathematics at the Collège Royal of Amiens. The minister of commerce, however, who had authority over all Écoles des Arts et Métiers, named him director of studies at the École of Angers. He took up his post there on 1 January 1830. After the July Revolution, civil war broke out again in the west and Bobillier, a volunteer in the National Guard of Angers, fought in a rather hard month-long campaign against the Chouans.

In 1832 his post as director of studies was abolished, and Bobillier returned to Châlons as instructor-in-chief in mathematics. By 1834 he was full professor, a rank he retained until his death. He also held a professorship of special mathematics at the municipal high school. Bobillier was named Chevalier of the Legion of Honor in 1839.

In 1836 he became seriously ill; but the following year he married. In spite of a recurring illness, Bobillier refused to interrupt his work, even though he finally was confined to bed. Imprudently, he resumed his teaching and other activities too soon—a decision that hastened his death in 1840.

Bobillier became known to the scientific world particularly through his contributions to the *Annales de Gergonne.*[2] The first, in August 1826, was a modest solution of certain problems posed to the readers by the editor. In April 1830 he went on to demonstrate the principle of virtual velocities for machines in equilibrium. He also contributed to Quetelet's *Correspondance mathématique et physique,*[3] to the *Mémoires de l'Académie de Caen,*[4] and to a few provincial journals[5]—probably a total of some forty writings. He also edited for his students a book of elementary algebra and a complete course in geometry. His courses in mechanics and physics were written out in autograph.[6]

At the time of his death, Bobillier was working on a dissertation concerning the geometric laws of motion[7] that he meant to present as a report before the Académie des Sciences. Some of the passages in his course in geometry are probably an early outline for this.

Because of his isolation in the provinces, Bobillier had few direct contacts with the scientific world of his time. His premature departure from the École

Polytechnique prevented him from forming the lasting friendships that are one of the principal hallmarks of that famous school.

He knew Poncelet, however, through correspondence and a close relationship that began in 1828–1829.[8] Unfortunately, their correspondence seems to have been lost. Bobillier never met J. D. Gergonne, the editor of the *Annales,* nor Adolphe Quetelet and Michel Chasles. In spite of the coincidence of their statements and interests, and of the fact that questions of priority arose between them, they do not seem to have corresponded directly. In fact, in the notice that Chasles wrote on Bobillier, he made a serious error as to the date of his death.[9]

There is even more reason to think that Bobillier had neither contact nor correspondence with Jakob Steiner and Julius Plücker, his emulators on the staff of the *Annales.* It should be noted that Gergonne often edited the articles of his collaborators to suit himself, which makes it difficult to judge them definitively.

Loyal to Gaspard Monge's ideas, Bobillier treated geometric problems in a way akin to both analytic geometry and projective geometry. He first set up a problem in the form of an equation in a particular case, simple enough so that the analytic geometry of his time could deal with it. Then, through a transformation by reciprocal polars, he obtained the dual. In this respect he was a disciple of Gergonne.

Such a method was hardly suitable for treating metric proportion. In 1824 Poncelet presented to the Académie des Sciences a report in which he solved the difficulty by taking the sphere as the quadric of reference.[10] Since his report had not been published, he then, at Gergonne's insistence, gave a somewhat sybilline sketch of it in the *Annales.*[11] Upon reading this, Chasles and Bobillier rediscovered Poncelet's method; and Bobillier was first to publish it.[12]

In his course in geometry Bobillier was the very first to use transformation by reciprocal polars relative to a circle in order to provide an elementary study of conic sections. For a more sophisticated point of view of the same field, his following proposition may be cited: The polar circles of a fixed point of a conic section, relative to all the triangles inscribed in the curve, meet at the same point.[13]

Bobillier is best known, however, for his studies of successive polars of curves or algebraic surfaces, and for his abridged notation. He stated, following Monge, that the tangents drawn from a point to a plane curve of order m have their points of contact on a curve of order $m - 1$, which he called the polar of the point. He made analogous statements concerning space.

In a series of studies[14] Bobillier showed that the polars of collinear points have $(m - 1)^2$ points in common. The polars of a point relative to a linear pencil (an expression that came into use after his time) of curves of order m form a pencil of order $(m - 1)$. If the point describes a straight line, the $(m - 1)^2$ points at the base of this pencil describe a curve of order $2(m - 1)$.

In considering the successive polars of the same points that are of order $(m - 1)$, $(m - 2) \cdots 1$, one is led to the following theorem: The polar of order n of a point P is the locus of all points whose polars of order $(m - n)$ pass through P. Plücker agreed with Bobillier in stating this theorem, which he had mentioned to Gergonne without any proof.

In May 1828 there appeared in the *Annales* Bobillier's essay on a new mode of research on the properties of space.[15] The method of research that he expounded was, he stated, susceptible to various applications, which he hoped to publish in successive issues.

With A a linear function of two coordinates and a a constant, $A = 0$ is the equation of a straight line. $ABC = 0$ is the equation of the extensions of the sides of a triangle, known as the equation of the triangle; and $aBC + bCA + cAB = 0$ is the general equation of a conic section circumscribed about the triangle. Its tangents at the vertices are $bC + cB = 0$; $cA + aC = 0$; $aB + bA = 0$. In fact, the straight line $bC + cB = 0$ meets the conic only at the point $B = 0$, $C = 0$. The triangle circumscribed about the conic, the points of contact being the vertices of the first triangle, has for its "equation" $(bC + cB)(cA + aC)(aB + bA) = 0$. The straight line passing through the three points of intersection of the tangents with the corresponding sides of the original triangle is $bcA + caB + abC = 0$. The lines of junction of the corresponding vertices of the two triangles—$cB - bC = 0$; $aC - cA = 0$; $bA - aB = 0$—are concurrent. It is then a trivial matter to show that this point and this line are pole and polar of each other in relation to the conic section in question.

Bobillier showed by a similar and clever process that, in space, if a tetrahedron is inscribed in a quadric, then another tetrahedron can be circumscribed about the same quadric so that the points of contact of its faces are the vertices of the original tetrahedron. The oppositely placed faces of the two tetrahedrons cut each other in four lines, and their oppositely placed vertices are joined by four other lines. The straight lines of each group belong to the same quadric.

Aware of the value of his method, Bobillier applied it in the 1 June 1828 number of the *Annales* to some elementary geometric propositions.[16] In particular, he obtained from it the known proposition concerning the chords common to a circle and a conic section, and the theorems of Pascal and C. J. Brianchon. The efficacy of the method may be judged by these simple examples.

In statics, in which Bobillier was particularly concerned with catenaries, his report "De l'équilibre de la chaînette sur une surface courbe"[17] should be mentioned. This problem was taken over by F. Minding in 1835, by C. Gudermann in 1834 and 1846, by P. Appell in 1885, and by A. G. Greenhill in 1897. In spite of a few minor errors in computation, the work remains most elegant.

Bobillier's demonstration of the principle of virtual velocities[18] consisted in substituting "for any ordinary machine, whose character can be changed in an infinite number of ways, the winch, whose conditions of equilibrium are so well known and that, at least for the infinitely small deviation that we can estimate in its equilibrium, remains exactly the same." His method is extremely clever.

In kinematics there seem to be no known traces of the work Bobillier was doing toward the end of his life, although the passages in his book on geometry that treat this subject are still extant.[19] Two theorems and one problem are particularly in evidence: All movement of a triangle on a plane can be produced by rolling a certain line over another fixed line, the triangle being invariably linked to the first line. If a triangle, *abc*, moves in such a fashion that the sides *ab* and *ac* constantly touch two circles, the envelope of the third side is also a circle; and the centers of the three envelopes determine a new circle that includes all the instantaneous centers of rotation. Bobillier then went on to pose the problem of how to determine the corresponding center of curvature in the path of the third vertex, *c*, when given the centers of curvature at points *a* and *b* of the paths described by vertices *a* and *b* of triangle *abc*. The construction he gave of this center is known as the Bobillier construction.

NOTES

1. Obituary, in *Almanach du département de la Marne* (1841), pp. 316–320.
2. In vols. **17–20.**
3. Vol. **4** (1828).
4. "De la courbe nommée chaînette," 1831.
5. E.g., *Recueil des travaux de la Société d'agriculture, commerce, sciences et arts de la Marne:* "Notes sur les puits à bascule" (1826) and "Note sur le principe de Roberval" (1834).
6. From the obituary of 1841.
7. From the obituary. However, Bobillier notes this memoir in the 10th ed. of the *Géométrie*, p. 208.

8. Poncelet, *Applications d'analyse et de géométrie*, II, 486. Poncelet calls Bobillier "an intelligent and singularly active mind."

9. *Rapport sur les progrès de la géométrie*, pp. 65–68. "We owe remarkable researches to Bobillier, a distinguished geometer who gave hopes of great achievements for mathematical sciences, from which he was snatched in 1832 at the age of thirty-five."

10. See, e.g., *Applications*, p. 529.

11. Vol. **17**, pp. 265–272; see also vol. **18**, pp. 125–149.

12. *Annales*, **18**, pp. 185–202, "Démonstration de divers théorèmes de géométrie." "In writing the above article we have used only the contents of M. Poncelet's letter" (the letter in vol. **17**). On p. 269 of vol. **18**, Chasles said: "It was only yesterday evening that your number of January 1828 reached me. I have read M. Bobillier's report with considerable eagerness . . . but I must say that the special case in which a sphere is taken . . . has come to me only through the reading of the analysis in the report" (report by Poncelet, vol. **17**, p. 265).

13. "Mémoire sur l'hyperbole équilatère."

14. *Annales*, vol. **18**, pp. 89, 157, 253; vol. **19**, pp. 106, 138, 302.

15. Vol. **18**, pp. 321–339.

16. "Philosophie mathématique. Démonstration nouvelle de quelques propriétés des lignes du second ordre."

17. *Annales*, vol. **20**, pp. 153–175.

18. *Ibid.*, pp. 285–288.

19. *Cours de géometrie*, 10th ed., pp. 204–208.

BIBLIOGRAPHY

I. Original Works. Among Bobillier's writings are "Notes sur les puits à bascule," in *Recueil des travaux de la société d'agriculture, commerce, sciences et arts de la Marne* (1826); material in *Correspondance mathématique et physique*, **4** (1828); "De la courbe nommée chaînette," in *Mémoires de l'Académie de Caen* (1831); "Note sur le principe de Roberval," in *Recueil des travaux . . .* (1834); *Cours de géométrie* (10th ed., Paris, 1850); and *Principes d'algèbre* (6th ed., Paris, 1865).

Many of his articles were published in the *Annales de mathématiques pures et appliquées:* "Démonstration de divers théorèmes de géométrie," **18**, 185–202; "En rédigeant l'article qu'on vient de lire . . . ," *ibid.*, 269; an article on polars of collinear points, *ibid.*, 89, 159, 253, and **19**, 106, 138, 302; an article on new methods of research on the properties of space, **18**, 321–339; "Philosophie mathématique. Démonstration nouvelle de quelques propriétés des lignes du second ordre," *ibid.*, 359–367; "Mémoire sur l'hyperbole équilatère," **19**, 349–359; "De l'équilibre de la chaînette sur une surface courbe," **20**, 153–175; and an article on the principle of virtual velocity, *ibid.*, 285–288.

II. Secondary Literature. Works providing more information on Bobillier and his accomplishments are Michel Chasles, *Rapport sur les progrès de la géométrie* (Paris, 1870), pp. 65–68, excellent from a mathematical standpoint; J. L. Coolidge, *A History of Geometrical Methods* (Oxford, 1940; New York, 1963), p. 143, which perpetuates Chasles's error on the date of Bobillier's death; and Poncelet's *Applications d'analyse et de géométrie* (Paris, 1864), II, 486, 529; and an article on the sphere as the quadric of reference, in *Annales de mathématiques pures et appliquées*, **17**, 265–272, and **18**, 125–149. There is an obituary of Bobillier in *Almanach du département de la Marne* (1841), 316–320.

J. Itard

BOCHART DE SARON, JEAN-BAPTISTE-GASPARD (*b.* Paris, France, 16 January 1730; *d.* Paris, 20 April 1794), *astronomy.*

Bochart, whom Laplace aptly described as an enlightened amateur scientist, followed family tradition by choosing a legal career and by having a keen interest in astronomy. His prominent name, his scientific competence, and his generosity to academicians (e.g., he financed the publication of Laplace's *Théorie du movement et de la figure élliptique des planètes*) led to his election, in 1779, as an honorary member of the Académie des Sciences and to his appointment, after the death of Camus, as one of the directors of Jacques Dominique Cassini's project, the Carte de France. He approached science with concrete images rather than with abstractions, through mechanics rather than through mathematics, and with practical interests rather than with theoretical ones.

At an early age Bochart successfully tried his hand at making reflecting telescopes, and astronomical instruments then became his passion. He spared neither his free time nor his wealth in creating one of Europe's largest and finest collections. Renowned craftsmen constructed his instruments: for example, John Dolland made an early achromatic telescope for him, and Jesse Ramsden, a very accurate degree-cutting machine. Bochart placed his collection at the disposition of the academicians: Jean Delambre, Pierre Méchain, Guillaume Le Gentil, Achille Dionis du Séjour, and Charles Messier.

Bochart calculated the orbits of comets on the basis of observations furnished by his long-time collaborator, Messier. The observed positions were few and close together; this presented special difficulties for Bochart, who used Boscovich's method, which he simplified and mechanized in his calculations. His predictions vastly aided Messier in finding the comets after they had disappeared behind the sun. The supposition that Bochart was forced to make when trying to calculate the orbit of Herschel's comet—discovered in 1781—played an important role in identifying it as the planet Uranus. He supposed that it followed a circular orbit with a radius equal to twelve times the distance of Saturn from the sun. This proved accurate, and Laplace then calculated the precise elliptical orbit. Bochart, who became the First President of the Parliament of Paris a few months before the outbreak of the French Revolution, lost his head during the reign of terror of the Committee of Public Safety.

BIBLIOGRAPHY

I. Original Works. Bochart's astronomical observations are in *Mémoires de l'Académie Royale des Sciences, année 1769,* 421, 429; *ibid., année 1770,* 232; *ibid., année 1774,* 19; *ibid., année 1775,* 217; and *ibid., année 1776,* 450. An orbital calculation is "Comète observée en 1779," in *Connaissance des Temps pour . . . 1782* (Paris, 1779), 395.

II. Secondary Literature. More information on Bochart and his work may be found in Jacques Dominique Cassini, "Découverte de la planète Herschel," in *Connaissance des temps . . . 1786* (Paris, 1783), 3–4; Cassini published his eulogy of Bochart separately and in *Mémoires pour servir à l'histoire des sciènces* (Paris, 1810), 373–391; J.-J. DeLalande, "Histoire de l'astronomie pour 1794," in *Connaissance des temps . . . pour l'année sextile VIIe* (Paris, 1797), 282–318, esp. 310–311 for a eulogy of Bochart, which was also published in DeLalande's *Bibliographie astronomique* (Paris, 1803), 752–754. Boscovich gives his method in "De orbitis cometarum determinandis, ope trium observationum parum a se invicem remotarum," in *Mémoires de mathématique et de physique* (Académie des Sciences), VI (Paris, 1774), 198–215, 401–434. Detailed information on Bochart's family is given in P. Humbert, "Les astronomes français de 1610 à 1667," in Société d'Études Scientifiques et Archéologiques de Draguignan, *Mémoires,* **63** (1942), 1–72; L. Moréri, *s.v.* Bochart, *Le grand dictionnaire historique* (Paris, 1759).

Robert M. McKeon

BÔCHER, MAXIME (*b.* Boston, Massachusetts, 28 August 1867; *d.* Cambridge, Massachusetts, 12 September 1918), *mathematics.*

Maxime Bôcher was the son of Ferdinand Bôcher, the first professor of modern languages at the Massachusetts Institute of Technology, and Caroline Little, of Boston. He entered Harvard in 1883, specializing in mathematics and natural science under W. E. Byerly, B. O. Peirce, and J. M. Peirce. He was elected to Phi Beta Kappa upon his graduation in 1888. Bôcher then went to Göttingen as a traveling fellow to audit the lectures of Felix Klein, Schönflies, Schur, Schwarz, and Voigt. Encouraged by Klein, he wrote a tract that won the prize in a competition sponsored by the Philosophical Faculty at Göttingen in 1891. It also served as his doctoral dissertation (1891), and was published as a book (1894) with an introduction by Klein.

In 1891 Bôcher returned to Harvard as an instructor in mathematics, and rose through the ranks to a professorship in 1904. In 1913 he was an exchange professor at the Sorbonne for a three-month period beginning in November.

He served the mathematical community unstintingly as a member of the editorial staff of the *Annals of Mathematics* in 1896–1900, 1901–1907, and 1911–1914; as vice-president, in 1902, and as president, in 1909 and 1910, of the American Mathematical Society; and as editor of the Society's *Transactions* in 1908, 1909, and 1911–1913. Under Klein's leadership as president of the International Commission on the Teaching of Mathematics, Bôcher served as chairman of the American Committee on Graduate Work in Universities, which published the report "Graduate Work in Mathematics in Universities and in Other Institutions of Like Grade in the United States" in the *Bulletin of the U.S. Bureau of Education,* no. 6 (1911). He was an invited speaker at the St. Louis Congress of Mathematicians in 1904 and at the Fifth International Congress of Mathematicians, Cambridge, England, in 1912, where his paper dealt with boundary problems in one dimension.

Bôcher was a prolific contributor to mathematical journals on the theory of differential equations and related questions. His research topics included systems of linear differential equations of the first order, singular points of functions satisfying partial differential equations of the elliptic type, exposition of the work of Jacques Sturm on algebraic and differential equations, boundary problems, and George Green's functions for linear differential and difference equations, and the theorems of oscillation of Sturm and Klein.

He was a member of the National Academy of Sciences and the American Philosophical Society, and was a fellow of the American Academy of Arts and Sciences.

BIBLIOGRAPHY

I. Original Works. Bôcher's books are *Über die Reihenentwickelungen der Potentialtheorie* (Leipzig, 1894), his doctoral thesis; *Introduction to Higher Algebra* (New York, 1907); *An Introduction to the Study of Integral Equations,* Cambridge Tracts in Mathematics and Mathematical Physics, no. 10 (Cambridge, 1909); *Plane Analytic Geometry With Introductory Chapters on the Differential Calculus* (New York, 1915); *Trigonometry With the Theory and Use of Logarithms* (New York, 1915), written with H. D. Gaylord; *Leçons sur les méthodes de Sturm dans la théorie des équations différentielles linéaires et leurs développements modernes,* delivered at the Sorbonne in 1913–1914, G. Julia, ed. (Paris, 1916), in the Borel series.

His numerous papers are listed in Poggendorff, V, 129. A more complete list is found in G. D. Birkhoff, "The Scientific Work of Maxime Bôcher," in *Bulletin of the American Mathematical Society,* **25,** no. 5 (1919), 197–216.

II. Secondary Literature. Further biographical detail may be found in W. F. Osgood, "The Life and Services of Maxime Bôcher," in *Bulletin of the American Mathematical Society,* **25,** no. 8 (1919), 337–350; in "Maxime

Bôcher," in *Science*, n.s. **48,** no. 1248 (29 Nov. 1918), 534–535, repr. by the National Academy of Sciences in its *Annual Report* for 1918, pp. 49–50, and also by *The Harvard University Gazette* (22 Oct. 1918), p. 14; in the *Lebenslauf* in his doctoral thesis (gift copy presented by Bôcher to Widener Library, Harvard University, 19 Sept. 1891); in *Who's Who in Science* (1912), p. 53; in *American Men of Science* (1910), p. 47; and in the "Reports of Meetings," in *Bulletin of the American Mathematical Society,* **17** (1910–1911), 77, 277, 507.

CAROLYN EISELE

BOCK, JEROME (*b.* Heidesbach, or Heidelsheim, Germany, 1498; *d.* Hornbach, Germany, 21 February 1554), *botany.*

Bock (also known as Hieronymus Tragus) was one of the three "German fathers of botany." Along with Otto Brunfels and Leonhard Fuchs, he represented the transition from late medieval botany, with its philological scholasticism, to early modern botany, with its demand that descriptions and illustrations be derived from nature.

Melchior Adam, Bock's first biographer, provides the earliest, and in some cases the only, information on his career. His birthplace is debatable, but internal evidence indicates that his adult life was spent in the Saar. His parents, Heinrich and Margarethe (maiden name unrecorded), apparently wished their son to enter a cloister. Where he received his early schooling is unknown. He may have attended the University of Heidelberg; but whether he studied medicine, philosophy, or theology is uncertain, for there is no record that he received a degree. In January 1523 Bock accepted a position in Zweibrücken; and on 25 January 1523 he married Eva, daughter of Heinrich and Margarethe Victor. He remained in Zweibrücken until 1533, when he accepted a canonry at the Benedictine church of St. Fabian in nearby Hornbach. The growing religious unrest forced Bock, who had become a follower of Luther, to leave Hornbach in August 1550. For a short period he acted as personal physician to the Landgraf Philipp II of Nassau, whose garden he is said to have supervised and to whom his *Kreuterbuch* was dedicated. In 1551 he returned to Hornbach, where he died three years later, probably of consumption. Bock's memorial tablet at St. Fabian's, where he was buried, was later discovered by Adam, who preserved a transcript of it.

The first result of Bock's botanizing excursions, dating from his years at Hornbach and conducted, he states, while dressed as a peasant, is the short tract *De herbarum quarundam nomenclaturis.* As the title suggests, it is concerned primarily with nomenclature—more specifically, with relating Greek and Latin names to local plants. Despite the lexicographical orientation, the brief entries indicate a personal acquaintance with plants, and Bock is not afraid to admit that he has never seen some of the plants mentioned by the ancients.

The appearance of Bock's *Neu Kreütterbuch* (1539) marked a new beginning in botany. It was some time, however, before his departure from tradition found general acceptance. Written in the vernacular, lacking illustrations, and sandwiched between the better-known writings of Brunfels and Fuchs, it was soon lost from sight. Only with the publication in 1546 of the first illustrated edition, and bolstered by the Latin translation of 1552, did Bock's position become assured. As the first to describe the local flora, Bock has been credited with discovering many new species. The lack of illustrations turned out to be a blessing in disguise, for it forced Bock to describe plants in such a manner that they could be recognized by a reader whose botanical knowledge was limited to local species and their vernacular names.

Bock's lasting contributions to botany, commemorated by Charles Plumier, who named the genus *Tragia* (*Euphorbiaceae*) in his honor, were the result of a happy union of talent and perseverance. By combining personal observation and precise description with an attempt to establish taxonomic relationships on a new basis, Bock broke sharply with the past. Being neither a physician, at least in the ordinary sense, nor a university scholar, he looked at plant life with the eyes of a true amateur, unencumbered by the necessity of finding a therapeutic rationale or a classical antecedent for every plant.

The third German edition (1551), from which the Latin translation was made, may be considered Bock's final statement. Despite such additions as a preface and an index, the text and illustrations remained essentially unaltered in successive editions. It will be convenient first to note the general format of the *Kreuterbuch*. The descriptions of approximately seven hundred plants and trees are arranged in three parts. The first two deal with herbs, monocotyledons, and cryptogams, while the third part treats of shrubs and trees. Each of the more than four hundred chapter divisions follows a set formula: "On Names," "On the Power and Effect," "Internal Uses," and "External Uses." Prior to the section on nomenclature, which contains Greek, Latin, and Arabic synonyms, there is an untitled section in which the plant is described. It is this material upon which Bock's reputation depends. Innocent of the sexuality of plants and the taxonomic significance of the reproductive organs, Bock necessarily based the descriptions upon the morphological characteristics of the vegetative portions. The descriptions usually contain the following information: the general aspect, including height, sometimes expressed in the form of a comparison with another, better-known plant; remarks on the foliage, including any noteworthy shape, texture, odor, or color; and miscellaneous observations concerning root systems, time of flowering, and economic uses. By establishing marks useful for field identification—the presence of milky sap or stipules, the distinction between various underground parts or between terete and quadrangular stems—Bock was the first modern botanist to teach the importance of fine structure. Although this momentarily diverted attention from the potential significance of floral organs, it stimulated inquiry until optical aids changed the conception of plant anatomy.

Floral structure was not ignored, however, and it is here that Bock's powers of observing and recording details are most apparent. He described the stamen, noting that it was typically composed of two parts, the filament and anther, and that while the number of stamens varied, their number was constant for a given species. This description, one of the earliest in botanical literature, is matched by his account of the pistil, which he correctly noted was composed of stigma and style. Another remarkable observation was his recognition that species of the birch family (*Betulaceae*) have, in addition to the familiar, tassel-like aments, other, quite inconspicuous flower clusters. In neither case, however, was Bock able to identify them with the staminate catkins or the pistillate inflorescence recognized today.

Passing from the blossom to the subsequent seed or fruit, another side of Bock's ability is revealed. As the first to describe the lily of the valley (*Convallaria majalis* L.), his account is the more noteworthy because it calls attention to the fruit, which he likens to red coral (fol. 204v, 1577 ed.). Ever searching for more accurate information or a confirmation of his suspicions, he planted the downy catkins of a willow. He was pleased to see them germinate, which demonstrated that they were seeds (fol. 380v). An even more determined effort was his nightly vigil to collect seeds from a fern (*Osmunda regalis* L.). Naturally he failed, but by collecting some of the ejected sporangia without resorting to incantations or other superstitious practices (fol. 194v), he made a major, if unappreciated, step forward.

The larger, drupaceous fruits, which served many domestic purposes, did not require so exacting a description. On the other hand, in order to illustrate a fruit-bearing tree in the same naturalistic manner as herbs, a different technique was demanded. In conjunction with David Kandel, the Strasbourg artist whom Bock employed, a workable solution was found for some thirty trees. The woodcuts depict the charac-teristic leaf, the shape of the fruit (often disproportionately enlarged and placed in inset), and various genre scenes representing the symbolic value or economic use of the tree in question.

In his efforts to observe native plants, Bock traveled widely in the Rhineland and elsewhere, often supplying the names of towns where he encountered unusual plants. He recorded such ecological and phenological data as would provide a more accurate account, including habitat, occurrence of weeds, and time of budding. Not all of his observations were made in the field, however, for he mentions his friends' gardens, some of which he visited or from some of which he received specimens.

As a consequence of his wide knowledge, it was inevitable that Bock made some effort at classification. Expressly rejecting an alphabetical arrangement, he made the fullest possible use of relating plants in terms of similarity of form, corolla shape, and formation of seed capsules. Because of his ignorance of plant sexuality, his efforts have only historical interest today. Nevertheless, by indicating a method based upon more than one criterion, he provided guidelines for succeeding generations of taxonomists.

By focusing attention on the plants themselves and by daring to question the high authority of Dioscorides and other classical writers, Bock laid down methodological canons whose future importance transcended even his own accomplishments. The rapid development of botany in the latter half of the sixteenth century owed much to the schoolteacher from Zweibrücken who led the exodus from the library into the fields.

BIBLIOGRAPHY

I. ORIGINAL WORKS. Because of serious discrepancies in the accounts of Nissen, Pritzel, and Roth (1899), and in the absence of any bibliography of Bock's published writings, the following list may be useful.

Bock's major work is, of course, the *Kreuterbuch*. It first appeared as *Neu Kreütterbuch von Underscheydt, Würckung und Namen der Kreutter, so in teutschen Landen wachsen* . . . (Strasbourg, 1539). The second German ed., the first to be illustrated, was *Kreuterbuch. Darin Underscheid, Würckung und Namen der Kreuter, so in deutschen Landen wachsen* . . . (Strasbourg, 1546); there was an enlarged edition entitled *Kreuterbuch, darinn Underscheidt, Namen und Würckung der Kreuter, Stauden, Hecken und Beuman, sammt ihren Früchten, so in deutschen Landen wachsen* . . . (Strasbourg, 1551, 1556, 1560, 1565, 1572, 1574). Melchior Sebizius edited *Kreütterbuch, darin Underscheidt, Nammen und Würckung der Kreütter* . . . (Strasbourg, 1577, 1580, 1586, 1587, 1595, 1630; facsimile reprint, Munich, 1964, essentially a reprint of the 1551 ed.). The

Speisskammer has been added as pt. 4, and there is new material in the preliminary leaves.

Others works by Bock are *De herbarum quarundam nomenclaturis,* in Otto Brunfels, *Herbarum vivae eicones ad naturae imitationem . . .,* II (Strasbourg, 1531–1532), the ninth of twelve tracts appearing under the title *De vera herbarum cognitione appendix* (the pagination varies with the ed. of Brunfels); *Hieronymi herbarii Apodixis Germanica, ex qua facile vulgares herbas omnes licebit perdiscere, ibid.,* the last tract of the collection *De vera herbarum . . .*—although sometimes ascribed to Bock, it may well be the work of Hieronymus Brunschwig (1450–1512); *Der vollen brüder ordern. Diss buchlein zeyget an was der wein würcke inn denen so ihn missbrauchen* (Strasbourg, ca. 1540); *Kurtz regiment für das grausam Haupt, wehe und breune, vor die Gemein und Armes heuflin hin und wider im Wasgaw und Westereich . . .* (Strasbourg, 1544); *Regiment für alle zufallende kranckheit das Leibs auch wie man die Leibsgebrechen so jetzund vorhanden sol abschaffen* (n. p., 1544; *vide* Roth [1899], p. 67, no. 3); *Regiment für das Hauptweh,* in *Artzneybuch köstlich für mancherley Kranckheit des gantzen Leibs . . .* (Erfurt, 1546; Frankfurt, ca. 1549; Nurcmberg, 1549 [the title of the collection is different]; Königsberg, 1565); *Bader Ordnung . . . aufs den Hochgelerten Hippocrate und Barptholomeo Montagnana, sampt andern auffs kürtzest, allen frommen Badern zu Trost, ins Teutsch gestelt* (Strasbourg, 1550); *Teutsche Speisskammer* (Strasbourg, 1550, 1555); *Verae atque ad vivum expressae imagines omnium herbarum, fructicum et arborum . . .* (Strasbourg, 1550, 1553); *De stirpium maxime earum quae in Germania nostra nascuntur . . . interprete Davide Kybero . . .* (Strasbourg, 1552), a translation of the 1551 edition of the *Kreuterbuch* that also contains tracts by Conrad Gesner and Benedict Textor; *Artzneibüchlin,* in Johannes Dryander, *New Artznei und Practicierbüchlin* (Frankfurt, 1557, 1572).

II. SECONDARY LITERATURE. Works concerning Bock are Melchior Adam, *Vitae Germanorum medicorum* (Heidelberg, 1620), pp. 67–72; J.-E. Gérock, "Les illustrations de David Kandel dans le *Kreuterbuch* de Tragus," in *Archives alsaciennes d'histoire de l'art* (1931), 137–148, with eleven figures; Edward Lee Greene, *Landmarks of Botanical History,* Smithsonian Miscellaneous Collections, LIV (Washington, D.C., 1909), 220–262; Heinrich Marzell, "Das Buchsbaum-bild im Kräuterbuch (1551) des Hieronymus Bock," in *Sudhoffs Archiv,* **38** (1954), 97–103; Louis Masson, "Le 'Livre de plantes' de Tragus," in *Aesculape,* **24** (1934), 301–310, ten figures; Ernst H. F. Meyer, *Geschichte der Botanik,* IV (Königsberg, 1857), 303–309; Claus Nissen, *Die botanische Buchillustration,* II, nos. 182–184 (Stuttgart, 1951); Georg A. Pritzel, *Thesaurus literaturae botanicae,* 2nd ed. (Milan, 1950), facsimile reproduction of Leipzig 1872 ed., nos. 864–868; F. W. E. Roth, "Hieronymus Bock, genannt Tragus (1498–1554)," in *Botanisches Centralblatt,* **74** (1898), 265–271, 313–318, 344–347; and "Hieronymus Bock, genannt Tragus, Prediger, Arzt und Botaniker 1498 bis 1554," in *Mitteilungen des historischen Vereins der Pfalz,* **23** (1899), 25–74; Kurt Sprengel, *Geschichte der Botanik,* I (Altenberg–Leipzig,

1817), 269–272, which contains a list of 109 plants first described by Bock.

JERRY STANNARD

BODE, JOHANN ELERT (*b.* Hamburg, Germany, 19 January 1747; *d.* Berlin, Germany, 23 November 1826), *astronomy.*

Bode, the son of a commercial accounting teacher and the nephew of the well-known writing master and mathematic master Jürgen Elert Kruse of Hamburg, had a great love for practical calculations throughout his life. This, with his pedagogical abilities, made him an excellent teacher of astronomy. He studied astronomy by himself and was strongly stimulated in his studies by the Hamburg scholars J. A. Reimarus and J. G. Büsch, as well as by the poet Friedrich Klopstock. They encouraged the nineteen-year-old to publish his famous *Anleitung zur Kenntnis des gestirnten Himmels* (1768), which was in print for nearly a hundred years and won innumerable adherents to astronomy.

In 1772 Johann Lambert summoned Bode to the astronomical observatory of the Berlin Academy as an arithmetician, to help in the publication of accurate ephemerides. The sale of astronomical almanacs was one of the chief sources of income for the Academy; because of their low degree of accuracy, however, the almanacs were not selling well. The new *Astronomisches Jahrbuch für 1776* was compiled under Bode's direction and published in 1774. He continued to do the calculations for and publish each successive annual volume until that of 1829 (published in 1826). Bode's almanacs were soon greatly esteemed. Aside from the ephemerides, the *Jahrbücher* also contained scientific news on observations and discoveries around the world.

In 1786 Bode was appointed royal astronomer, director of the astronomical observatory, and member of the Berlin Academy. He was active in these positions for nearly forty years, until his retirement in 1825. In spite of the renovations which he arranged, the observatory, situated on the roof of a five-story tower of the Academy building, could not compete with those of Paris and London; it was equipped only for modest investigations of comets, planets, double stars, and so forth.

Bode's literary activity more than made up for the observatory's deficiencies. Besides his tables, his two sky atlases were for a long time indispensable tools for astronomers: the *Vorstellung der Gestirne,* which, according to the example set by John Flamsteed's atlas, contained more than 5,000 stars; and the *Uranographia,* which surpassed all its predecessors by listing over 17,000 stars and containing, for the first time, the nebulae, star clusters, and double stars discovered by William Herschel.

Bode was almost the only writer to support the then not widely known ideas of Kant, Lambert, and Herschel on the infinity of space, the infinite number of inhabited worlds, and the continuous birth and passing away of stars according to natural laws. He also made public in 1772 the relation first established by the Wittenberg professor Johann Daniel Titius, the Titius-Bode series, according to which the distances between the large planets are in a nearly regular geometric progression. The minor planets, unknown at that time, and the planet discovered in 1781 by Herschel fit well into this series. (If one substitutes for Venus, Earth, Mars, Jupiter, and Saturn the numbers $n = 0, 1, 2, 4, 5$, then the distance is obtained from the relation $A = 0.4 + 2^n \cdot 0.3$ [astronomical units]. For Mercury the last member is zero, for the minor planets $n = 3$, for Uranus $n = 6$.) Bode gave the name Uranus to Herschel's newly discovered planet.

Bode married three times: his first two wives were nieces of the Berlin astronomer Christine Kirch; the third was the niece of the chemist Andreas Marggraf. He was a member of the Royal Society of London, as well as of the academies of Berlin, St. Petersburg, Stockholm, Copenhagen, and Göttingen.

BIBLIOGRAPHY

I. ORIGINAL WORKS. Bode's writings include *Anleitung zur Kenntnis des gestirnten Himmels* (Hamburg, 1768; 11th ed., Berlin, 1858); *Sammlung astronomischer Tafeln* (Berlin, 1776); *Kurzgefasste Erläuterung der Sternkunde* (Berlin, 1778; 3rd ed., Berlin, 1808); *Dialogen über die Mehrheit der Welten* (Berlin, 1780; 3rd ed., Berlin, 1798), a translation of Fontenelle's *Entretiens sur la pluralité des mondes; Vorstellung der Gestirne auf 34 Kupfertafeln nebst Fixsternverzeichnis* (Berlin–Stralsund, 1782; 2nd ed., Berlin, 1805); *Allgemeine Betrachtungen über das Weltgebäude* (Berlin, 1801; 3rd ed., Berlin, 1807); and *Uranographia sive astrorum descriptio* (Berlin, 1801), 20 folios with the catalog *Allgemeine Beschreibung und Nachweisung der Gestirne*.

II. SECONDARY LITERATURE. Biographical material on Bode is in Encke, an obituary speech, in *Abhandlungen* of the Berlin Academy (1827); Poggendorff, I; *Allgemeine deutsche Biographie*, III, 1; and Schroeder, *Lexikon der hamburgischen Schriftsteller*, I (Hamburg, 1851), 282 ff., with an extensive list of his works.

BERNHARD STICKER

BODENHEIMER, FRITZ SIMON (b. Cologne, Germany, 6 June 1897; d. London, England, 4 October 1959), *entomology, zoology, history of science.*

The son of Max Yitshaq Bodenheimer and Rosa Dalberg, Bodenheimer studied zoology in Frankfurt and subsequently specialized in agricultural entomology. In 1921 he received his doctorate from the University of Bonn and in 1922 emigrated to Palestine, to become head of the department of entomology at the agricultural experiment station of the Jewish Agency. He moved to The Hebrew University in Jerusalem in 1928; he was promoted to professor in 1931 and remained there until his retirement in 1953. Bodenheimer also served as adviser on applied entomology in Europe, the Near East, South Africa, and Australia.

Bodenheimer contributed significantly to knowledge of the Palestinian fauna, with emphasis on insects. At the time, his *Animal Life in Palestine* (1935) was the generally accepted survey of the subject, but his main interest was the ecological relations of animals, chiefly moles, insect pests, and parasitic insects. He correlated variations in dormancy and rhythms of development, and in the population of pests, parasites, and hosts, with a wide range of climatic and biotic environmental factors in the widely divergent climates of Palestine. In collaboration with many others, Bodenheimer studied particular adaptations to different habitats, and it was one of his major accomplishments that he stimulated the interest of so many. In animal ecology he was a pioneer of more than regional importance, giving particular attention to theoretical concepts. Bodenheimer's special interest in citrus, the major crop of Israel, is reflected in his exhaustive *Citrus Entomology in the Middle East* (1951).

Beginning with his two-volume study *Materialien zur Geschichte der Entomologie* (1928–1929), Bodenheimer showed great interest in the history of biology. His numerous studies emphasized early Hebrew works and the biology and biologists of the Near East. Major works in this field are *Animal Life in Bible Lands* (1949–1956) and *History of Biology* (1958). In all of these a very wide range of relevant source material is brought together, but the final version lacks lucidity and polish.

From 1947 Bodenheimer was increasingly active in organizations related to the history of science. From 1950 to 1953 he was vice-president, and from 1953 to 1956 president, of the Académie Internationale d'Histoire des Sciences.

BIBLIOGRAPHY

Bodenheimer's writings include *Materialien zur Geschichte der Entomologie bis Linné*, 2 vols. (Berlin, 1928–1929); *Die Schädlingsfauna Palästinas* (Berlin, 1930); *Animal Life in Palestine* (Jerusalem, 1935); *Animal Life in Bible Lands,* 2 vols. (Jerusalem, 1949–1956); *Citrus Entomology in the Middle East* (The Hague, 1951); *Précis d'écologie animale* (Paris, 1954); *Animal Ecology To-day* (The Hague, 1958); *History of Biology* (London, 1958); and his autobiography, *A Biologist in Israel* (Jerusalem, 1959), which includes a bibliography of 420 items.

JACOB LORCH

BODENSTEIN, ADAM OF. See **Adam of Bodenstein.**

BOË, FRANZ DE LA. See **Sylvius, Franciscus.**

BOEHME, JACOB (*b.* Alt Seidenberg, near Görlitz, Germany, 1575; *d.* Görlitz, 17 November 1624), *theology, mysticism.*

Boehme was the fourth child of Jacob and Ursula Boehme. The father belonged to a well-to-do, old family of German-speaking farmers. A prominent man in the village, he held lay offices in the local church. As a boy Boehme herded cattle with neighboring farm boys, attended the village school, and was given a Lutheran upbringing. At fourteen he was apprenticed to the village cobbler, perhaps owing to his delicate health, and three years later he set out on his journeyman travels. Around 1595 he returned to Görlitz, where in 1599 he became a citizen of the town, set up as a master cobbler, and married Catharina Kuntzschmann, with whom he had four sons. Her father was a butcher, and her family was prosperous and influential in city affairs. Boehme was now enabled to buy a house in Görlitz, where he spent the remainder of his life, interrupted only by visits to his spiritual friends among the nobility of the region and by travels on business to the Leipzig fair and to Prague. In 1613 he gave up cobbling and entered the cloth trade, but his later years were clouded by financial distress caused by inflation and the beginning of the Thirty Years' War.

In 1600 occurred the decisive event in Boehme's life. Through the chief pastor of Görlitz, Martin Moller, he had recently been exposed to the great tradition of German mysticism: to Johann Tauler, Heinrich Suso, and Jan Van Ruysbroeck, among others. He went through a period of anxious search for insight. The result was the profound mystical experience that shaped his life and inspired the writings that form the culmination of the German mystical tradition. One day while sitting in his room in a state of melancholy, his eyes by chance caught the sunlight reflected from a pewter dish. His soul was immediately ushered into a mystical vision, and he maintained that the innermost part of the secrets of

nature as well as the true nature of good and evil were revealed to him. In a quarter of an hour, he saw and knew more than he could have learned by years of study in the universities. At first full of doubt, he soon became convinced that he had received the gift of vision. When it was upon him, he could penetrate into the very heart and being of all things in creation. But it was twelve years before Boehme recorded this experience and its fruits in his first work, *Aurora oder Morgenröthe im Aufgang.* Meant only as a private record, and left unfinished, it soon began to circulate in manuscript copies among his friends and thus came to the attention of the local church authorities, who took a hostile view of the work and enjoined him to desist from any further writing.

Devout and humble, Boehme did not write until 1618, when the spirit again urged him so strongly that he could no longer remain silent. The rest of his works, amounting to more than thirty items, were written during the remaining years of his life—some short, some long, many as responses to questions from friends or as polemics against opponents. He was again called to account by the authorities, but no action was taken, and he died quietly in his house in Görlitz as a member of the Lutheran church to which he had belonged all his life. Among his most important works are *Von den drei Principien Göttliches Wesens, Vom dreifachen Leben des Menschen, De signatura rerum oder von der Geburt und Bezeichnung aller Wesen, Mysterium magnum oder Erklärung über das erste Buch Mosis, Der Weg zu Christo* (the only work published [1624] during Boehme's lifetime), and *Von der Gnaden-Wahl oder von dem Willen Gottes über die Menschen,* which Boehme considered his greatest work.

In all his works Boehme spoke as a prophet; he believed that God had chosen him to reveal to mankind what lay hidden. Convinced that he wrote under direct inspiration, he claimed that he changed nothing once it was written. The obscurity of his style is the expression of his mode of insight—full of bold metaphors, alchemical terms, number symbolism, and Neoplatonic conceptions—and reveals his background in Luther, Paracelsus, Kaspar Schwenckfeld, and Valentin Weigel. Boehme deprecated book learning, distrusted reason and the disputes of the theologians, and was fond of saying, as did Weigel, that all knowledge was revealed within him as in a book, even the Bible, so long as he had Christ's spirit in him. All the same, it is evident that Boehme was by no means unfamiliar with the thought of his spiritual forerunners, whose work he no doubt knew both from his own reading and from conversations with his visionary friends. Having had only elementary schooling, Boehme was largely self-taught. Later in life he often voiced regret that he had not learned Latin, and his writing in German earned him the title of *teutonicus philosophus.* His friend and first biographer, Abraham von Frankenberg, observed that Boehme's eyes were sky-blue and shone like the windows of Solomon's Temple.

Boehme's philosophy cannot be reduced to brief, systematic statement; it was not conceived in terms that would permit such reduction, and his own conceptions shifted in the course of time—the *Aurora,* he later said, was a work of his spiritual childhood. An understanding of his philosophy can, however, be gained by noting a few of its fundamental concerns. Boehme's initial problem was the existence of evil and the concealment of God from the world of man. His answer revealed a cosmic drama, with opposition of light and dark, spirit and body, love and wrath, joy and pain, eternity and time. All things visible were emanations of things invisible; the hidden God lay revealed under the visible creation. The "out-breathed" or "outspoken" invisible power, the Word, had called forth creation. Nature is the language of God to man, if only man will read it aright. The undifferentiated *Ungrund,* or nothingness (in the English translations, the "abyss"), is like an eye that seeks an object in order to become aware of itself, a mirror of images whose possibilities suggest the actualities of nature. By a repeated process of reflecting, willing, and creating, God's self-knowledge finds expression in nature, which is thus ordered according to the heavenly wisdom, the eternal Sophia.

Placed in time and body after the Fall, man's path to regeneration is renewed revelation of the secrets of nature. Adam's first and decisive fall occurred when he fell asleep and lost the direct insight into creation that he had hitherto possessed, being purely spiritual and ever awake. Sinfulness is caused by a perverse imagination, whose consequence is inadequate or false knowledge. The promise of again placing good over evil, light over dark, was given in the obedience and suffering of Christ, a reversal of Adam's course into time, history, and body. By placing cosmogony at the center of his theology, Boehme reveals his debt to the Lutheran tradition, and especially to Paracelsus and the Protestant mystics of the sixteenth century. But he is more explicit and detailed than his predecessors. One far-reaching effect of this theology was profound reverence for nature and closeness to it. Nature is given positive reality; its study gains justification; its observation—if rightly used as an avenue to the invisible realm beyond it—is an act of devotion. To the seventeenth century this view was more widely derived from Boehme rather than directly from Paracelsus. To Boehme a meadow in bloom with flowers was a mystical opening.

Adam, the first man, was created in the image of God; as the microcosm, he had the macrocosm in him. Although not original with Boehme, this view is prominent in his philosophy. It was in this sense that he wrote out of the book "which I myself am," not from other books or the instruction of the learned. He also claimed another special gift, direct insight into the language of nature, the language Adam spoke when he named the animals in the Garden of Eden. He even felt that he had direct access to new truths through his God-given insight into the interpretation of the sounds and forms of his own language, which, like all other languages, reveals the divine plan to the truly inspired mind. The language of nature had been adumbrated before Boehme, but he carried his doctrine on this point far beyond any previous speculation. This doctrine proved especially influential during the seventeenth century, often occurring in isolation from his theology. The observation was made more than once that the hoped-for philosophical language would copy the function of Adam's language, thus recapturing a measure of the insight into nature and the unity of knowledge that had been lost. Boehme was a mystic, but his mysticism did not advocate withdrawal from the world; on the contrary, his way was spiritual immersion in it.

After his death, Boehme's manuscripts were carefully collected and taken to Holland, where the first published versions appeared. From Holland his works, printed or in manuscript copies, passed into England, where they were all published in English versions between 1644 and 1663, many for the first time in any language. Thus Boehme was first discovered in England, where he had a wide and varied influence, most clearly among the Quakers. In Germany he did not gain prominence until he was taken up by the Pietists during the eighteenth century. He had a strong impact on German Romantic thought and later gained a position of eminence in post-Kantian idealism, in large measure through the French translations of Louis-Claude de Saint-Martin (1743–1803). Boehme has also had a strong and enduring influence on Russian writers.

BIBLIOGRAPHY

I. ORIGINAL WORKS. The MSS are described in Werner Buddecke, *Verzeichnis von Jakob Böhme-Handschriften* (Göttingen, 1934) (= *Hainbergschriften,* 1). The German eds. are listed in Buddecke, *Die Jakob Böhme-Ausgaben. Ein beschreibendes Verzeichnis,* pt. 1 (Göttingen, 1937) (= *Hainbergschriften,* 5), and the translations in pt. 2

(Göttingen, 1957) (= *Arbeiten aus der Staats- und Universitätsbibliothek,* Göttingen, N.F. 2). The best collected edition is *Theosophia revelata,* J. W. Ueberfeld, ed., 21 pts. (Amsterdam, 1730); repr. in 11 vols., Will-Erich Peuckert, ed. (Stuttgart, 1955–1961). K. W. Schiebler's edition of *Jakob Böhme's Sämmtliche Werke,* 7 vols. (Leipzig, 1831–1847), has a poor text and cannot be used for serious work.

The recently discovered autograph copies of Boehme MSS have been ed. by Werner Buddecke in Jacob Böhme, *Die Urschriften,* 2 vols. (Stuttgart, 1963–1966). The first collected English edition is *The Works of Jacob Behmen,* G. Ward and T. Langcake, eds., 4 vols. (London, 1764–1781), usually called William Law's edition. With the exception of *Der Weg zu Christo* (from the 1775 version of G. Moreton), this ed. reprints the seventeenth-century versions of J. Sparrow, J. Ellistone, and H. Blunden.

II. SECONDARY LITERATURE. There has been a continual flow of Boehme literature since the 1640's; only the most important can be mentioned here: Gottfried Arnold, *Unparteyische Kirchen- und Ketzer-Historie,* 2 vols. in 4 pts. (Frankfurt, 1699–1700), pt. 2, 656–682; Franz von Baader, "Vorlesungen über J. Böhme's Theologumena und Philosopheme," in *Gesammelte Schriften zur Naturphilosophie,* Franz Hoffmann, ed. (Leipzig, 1852), pp. 357–432; and *Vorlesungen und Erläuterungen zu Jacob Böhmes Lehre,* Julius Hamberger, ed. (Leipzig, 1855)—Vols. III and XIII, respectively, in Franz von Baader, *Sämmtliche Werke,* Franz Hoffman and Julius Hamberger, eds., 16 vols. (Leipzig, 1851–1860); R. Jecht, "Die Lebensumstände Jacob Böhmes," in his ed. of *Jacob Böhme, Gedenkgabe der Stadt Görlitz* (Görlitz, 1924), pp. 7–75; and Will-Erich Peuckert, *Das Leben Jacob Böhmes,* 2nd ed., rev., in Vol. X of the reprint edition of *Theosophia revelata* listed above. The best full exposition is A. Koyré, *La philosophie de Jacob Boehme* (Paris, 1929). On particular aspects of Boehme, the following works are useful: Ernst Benz, *Der Vollkommene Mensch nach Jacob Böhme* (Stuttgart, 1937); *Der Prophet Jakob Boehme, ein Studie über den Typus nachreformatorischen Prophetentums* (Akademie der Wissenschaften und der Literatur. Mainz. Abhandlungen der Geistes- und Sozialwissenschaftlichen Klasse, 1959, no. 3); "Zur metaphysischen Begründung der Sprache bei Jacob Böhme," in *Dichtung und Volkstum,* **37** (1936), 340–357; "Zur Sprachalchimie der deutschen Barockmystik," *ibid.,* 482–498; "Die Sprachtheologie der Reformationszeit," in *Studium Generale,* **4** (Apr. 1951), 204–213; "Die Geschichtsmetaphysik Jakob Böhmes," in *Deutsche Vierteljahrsschrift für Literaturwissenschaft und Geistesgeschichte,* **13** (1935), 421–455; M. L. Bailey, *Milton and Jakob Boehme* (New York, 1914); Heinrich Bornkamm, *Luther und Böhme* (Bonn, 1925); Emanuel Hirsch, *Geschichte der neuern evangelischen Theologie im Zusammenhang mit den allgemeinen Bewegungen des europäischen Denkens,* II (Gütersloh, 1951), 208–255; Serge Hutin, *Les disciples anglais de Jacob Boehme aux XVIIe et XVIIIe siècles* (Paris, 1960); Wolfgang Kayser, "Böhmes Natursprachenlehre und ihre Grundlagen," in *Euphorion,* **31** (1930), 521–562, an especially useful and stimulating study; Peter

Schäublin, *Zur Sprache Jakob Boehmes* (Winterthur, 1963); Wilhelm Struck, *Der Einfluss Jakob Boehmes auf die englische Literatur* (Berlin, 1936); and Nils Thune, *The Behemenists and the Philadelphians, a Contribution to the Study of English Mysticism in the 17th and 18th Centuries* (Uppsala, 1948). The following three works offer good introductions: Howard H. Brinton, *The Mystic Will* (New York, 1930); Rufus M. Jones, *Spiritual Reformers in the 16th and 17th Centuries* (New York, 1914; Beacon Paperback, 1959); and John Joseph Stoudt, *Sunrise to Eternity, a Study in Jacob Boehme's Life and Thought* (Philadelphia, 1957).

HANS AARSLEFF

BOERHAAVE, HERMANN (*b.* Voorhout, Netherlands, 31 December 1668; *d.* Leiden, Netherlands, 23 September 1738), *medicine, botany, chemistry.*

Boerhaave was the son of the Reverend Jacobus Boerhaave and of his second wife, Hagar Daelder. The boy's mother died when he was five years old; his father then married Eva Dubois, the daughter of a Leiden clergyman, who proved to be a devoted stepmother. The elder Boerhaave personally saw to his son's upbringing, supervising his physical, as well as moral and intellectual, education. Boerhaave also spent three years in the grammar school in Leiden.

Boerhaave's father died in 1683; in accordance with his wish, Boerhaave applied himself to the study of theology and philosophy upon his matriculation at the University of Leiden in 1684. As a student, Boerhaave distinguished himself by a series of five disputations, three of which dealt with the human mind. He also delivered an oration on Cicero's view of Epicurus' concept of the *summum bonum,* for which the governors of the university awarded him a gold medal. He earned a degree in philosophy in 1690 with a thesis on the distinction of mind from body.

Upon graduation, Boerhaave continued to study theology. At the same time, however, his interest began to turn to medicine. He attended the yearly public dissections conducted by Anton Nuck and independently studied the works of Hippocrates, Vesalius, Fallopio, Bartholin, and Sydenham. In 1693 he took a medical degree at the academy of Harderwijk, having presented a thesis, *De utilitate explorandorum in aegris excrementorum ut signorum.*

Unfounded rumors had raised the suspicion that Boerhaave was a secret adherent of Spinoza; since such an allegation could only damage his ecclesiastical career, he turned definitely to a medical one. He settled in Leiden, augmenting the income from his small practice by giving lessons in mathematics. In 1701 he was appointed a lecturer in medicine by the University of Leiden; in his inaugural public address he advocated the study of the works of Hippocrates. He lectured on the *institutiones medicae* at the university, and gave private lectures on the theory and practice of medicine (for which he was paid by his students). In addition, he began to lecture on chemistry at the request of foreign (probably English) students.

Boerhaave restored the declining prestige of the Faculty of Medicine at Leiden, and in 1703 he was offered a professorship at the University of Groningen. He rejected the offer, and the governors of Leiden, anxious to retain him, promised him the first chair to become vacant there. At the same time, he was authorized to give an academic oration. This address, *De usu ratiocinii mechanici in medicina,* was his iatromechanistic credo.

Boerhaave's lectures became the basis for several textbooks. His *Institutiones medicae* was published in 1708 and his *Aphorismi de cognoscendis et curandis morbis* appeared in 1709. These books were largely responsible for Boerhaave's European reputation; indeed, pirated editions appeared both in the original Latin and in modern languages.

In 1709 the chair of botany and medicine fell vacant at Leiden, and Boerhaave was immediately appointed to it. He thus entered into a new field of science. As professor of botany, he was ex officio supervisor of the university's botanical garden and was given an official residence and an allowance for foreign correspondence and the exchange of seeds and plants. He hastily drew up a new catalog of plants for the garden (*Index plantarum,* 1710)—the previous list dated from 1687. During the next ten years, Boerhaave made extensive additions to the botanical garden, and the second edition of his catalog (*Index alter,* 1720) listed 5,846 species, more than 2,000 more than his earlier index. Although he was totally untrained in botany, Boerhaave recognized the need for a new system of classification; he was aware of his own limitations, however, and made no attempt to provide one. Rather, he helped Linnaeus in every way that he could.

On 14 September 1710 Boerhaave married Maria Drolenvaux, the daughter of a rich merchant. They had four children, of whom one daughter, Maria Joanna, lived to adulthood.

In 1714 Boerhaave was appointed vice-chancellor of the university, a position that he again filled in 1730. The same year he was, perhaps by his own request, charged with clinical teaching, which in Leiden had been practiced since 1637 but had fallen into neglect. Boerhaave revivified bedside teaching (for which two six-bed wards, one for men and one for women, of the Caecilia Hospital were reserved) and raised it to new heights, attracting students from all over Europe. His oration *De comparando certo in physicis,* made 8 February 1715, marked the end of his first term as vice-chancellor.

When Le Mort, the professor of chemistry, died in 1718, Boerhaave was chosen to succeed him, and for the next ten years he held simultaneously three of the five chairs that constituted the whole of Leiden's Faculty of Medicine. Boerhaave assumed his new duties with the oration *De chemia suos errores expurgante.* He lectured with extraordinary zeal and energy, four or five hours a day, until he was halted by a severe, painful illness (which he himself diagnosed as *lumbago rheumatica,* as described by Sydenham) that confined him to bed for five months in 1722. In January 1723 the students and citizens of Leiden celebrated his recovery and return to teaching with illuminations of the university building and a large part of the city.

Boerhaave was undeniably a great teacher. His lecture room was crowded with students from several countries (all lectures were given in Latin, of which Boerhaave had an easy mastery). Often many students had to stand, and some young noblemen were known to hire men to get to the classroom early to reserve their seats. Haller called Boerhaave *communis Europae praeceptor;* in the years of his tenure, 1,919 students were enrolled in the Medical Faculty, of whom 659 came from English-speaking countries. Many of his students copied out their lecture notes at home, to be copied by others in some instances; several exercise books containing unpublished lectures are thus extant, in addition to his lectures as published by some pupils.

Boerhaave's influence spread throughout Europe. His textbooks were published in Great Britain, France, Germany, and Italy, among other countries, and his students transmitted his teachings (even to later generations, since after Boerhaave's death Haller published a seven-volume edition of the *Institutiones,* and Gerard van Swieten published a five-volume commentary on his *Aphorismi*). The medical faculties of the universities of Vienna, Göttingen, and Edinburgh were begun or reformed after the system that Boerhaave instituted at Leiden. Indeed, the modern medical curriculum—with its emphasis on natural science, anatomy, physiology, pathology, and, in particular, clinical training—owes much to Boerhaave. His little book *Atrocis, nec descripti prius, morbi historia* (1724) made a twofold contribution to medicine. In this case history—that of Baronet Wassenaer, Admiral of the Republic, who ate a heavy meal, experienced severe chest pains, and died the next day—Boerhaave, who performed the postmortem examination himself, made the first diagnosis of a spontaneous rupture of the esophagus. Moreover, in his presentation, he established the classic form for a morbid history—anamnesis, physical examination, diagnosis, history of the disease, and autopsy findings.

Boerhaave's interest in, and influence on, systematism extended to the synthesis of older and newer theories of medicine. Medical science was in a state of confusion at the beginning of the eighteenth century; the heritage of old Greek medicine was still honored, but no attempt had been made to reconcile it with the medical discoveries of the previous century. Boerhaave attempted to build a comprehensive medical doctrine. To this end he published a new edition of the works of Aretaeus of Cappadocia, which he furnished with a valuable critical apparatus, and reedited or wrote new prefaces for editions of the works of Prospero Alpini, Eustachius, Bellini, Carolus, and Nicholas Piso. With his younger colleague B. S. Albinus he reedited the *Opera omnia* of the great anatomist Vesalius. Another work on anatomy is his *Opusculum anatomicum de fabrica glandularum* (1722), in which he defended Malpighi's concept of the structure of the glands against that of Fredrik Ruysch.

He further collected the iatromechanical theories of the late seventeenth century—most notably those of Willis, Baglivi, Borelli, and Bellini—and merged them into a creative synthesis. For example, Boerhaave, like Baglivi, stressed the pathological and therapeutic significance of mechanically deranged body fibers, but at the same time he joined Borelli's and Bellini's iatromechanical theories to Willis' detailed discussions of nervous disorders. In attempting to develop a satisfactory theory of a self-regulating bodily machine, Boerhaave manipulated the older ideas—without, however, entire success.

Accepting Leeuwenhoek's faulty interpretation of the process of hemolysis, he held that, in the blood, the red globules could be broken up into six yellow globules, which in turn consisted of six very small pellucid spherules, thought to be made up of particles too small to be visible through the microscope. Boerhaave made appropriate distinction of orders of vessels into sanguiferous arteries, of which the smallest would admit one red globule; serous capillaries, of which the smallest would admit one yellow globule; lymphatic vessels, which would admit nothing larger than the pellucid spherules; and a series of still smaller vessels, in descending order of diameter, which would admit specific particles. In his attempt to explain the vital phenomena, Boerhaave rejected the hypotheses of the iatrochemical school and used hydraulic and mechanical principles, taking into account the velocity of the blood, the angle and diameter of the vessels, the size and shape of the particles, the viscosity of the blood, and so forth. He interpreted inflammation, for example, as the result of stagnation of the blood in the smallest capillaries combined with an increased velocity of the blood, leading to increased pressure on the obstructing matter. Moreover, the smallest elements of the walls of the vessels might be too rigid and stiff or too weak and lax, which could produce pathological conditions. (Since Boerhaave could not know of the microbiological causes of disease, he attributed many of them to insufficient digestion of food, from which an acid or alkaline putrid acrimony might arise, for which he advised proper therapeutic measures.) Thus he devised a doctrine that was generally accepted for some time—although it failed him as an explanation for specific secretion of the glands.

Boerhaave belongs, with Stahl and Hoffmann, to the great systematists of the early eighteenth century. Apparently there was a demand for a comprehensive and consistent medical system for the mass of new facts and observations that he had scrupulously collected and attempted to arrange appropriately.

How much his system was appreciated by his pupils may be gathered from the words of one of them, his biographer W. Burton: "It will now perhaps be universally granted that our professor has indeed supplied us with the best system from an unparallel'd fund of medical learning happily digested."

Boerhaave was influenced by the philosophy of Descartes, but more so by the great English scientists Boyle and Newton. He accepted the corpuscular theory of the structure of matter. In accordance with this, his system was essentially mechanistic, although he acted eclectically and introduced chemical and other viewpoints. Nevertheless, although he avoided the extreme one-sidedness of the iatrochemical school, against which he campaigned, he tried to understand the vital processes and phenomena in the body, using an inanimate model.

In this respect he was in direct opposition to the animistic organicism of his contemporary Stahl, who understood the distinction between an organism and a mechanism better than Boerhaave. (Boerhaave never entered into a discussion with Stahl and he did not even mention Stahl's well-known phlogiston theory in his chemical textbook.)

The inconsistencies of Boerhaave's system generated discussion that eventually led, however, to the statement of the problems of animal heat and irritability. Boerhaave's contribution to physiology thus came as an almost accidental side effect; but there is no doubt that his work served as the stimulus to physiological research of the latter half of the eighteenth century, and that he contributed significantly to the discussion of the important problems that his work raised. Boerhaave's lasting influence on medicine does not lie so much in his system, or in new discoveries, but in his teaching. In his clinical instruction he indoctrinated his pupils with the old Hippocratic method of bedside observation and taught them to act methodically in the examination of their patients.

Boerhaave's most important contributions to science, perhaps, were made in chemistry—paradoxically, since his medical system was mainly based on mechanics and he did not think that chemistry was yet an adult science. He introduced exact, quantitative methods into chemistry by measuring temperature and using the best available balances made by Fahrenheit; indeed, he may be considered the founder of physical chemistry as well as a contributor to pneumatic chemistry and biochemistry. He was an indefatigable experimenter, exhibiting an unbelievable tenacity in his experiments on mercury. He introduced biochemical demonstrations into his chemical courses for medical students.

Boerhaave was the first to obtain urea, by a procedure that took more than a year, and to discover its diuretic properties, as well as its cooling effect when dissolved in water. He demonstrated that water could be obtained by condensation from burning alcohol, and described a rapid method of making vinegar—sometimes called Boerhaave's method.

Boerhaave's attitude toward alchemy was somewhat ambiguous; he did not dogmatically deny a priori the possibility of the transmutation of metals, but examined it in a series of painstaking experiments that lasted over a period of many years. He purified his mercury specimens by forcing them through leather and then washing them in seawater. In one experiment, he used a fulling mill to shake a specimen of mercury, enclosed in a glass bottle, for a period of eight and one-half months; he then distilled it sixty-one times. Other specimens of mercury were variously heated for fifteen and one-half years, boiled 511 times, or mixed with gold and then distilled 877 times. Gold remained gold, and mercury, mercury; he did, however, obtain mercury with the specific weight of 14.1 as the result of one of these year-long experiments. These experiments were published in the *Philosophical Transactions of the Royal Society of London* (1734–1736).

After his illness of 1722, Boerhaave realized he must take care of his health. In 1724 he bought an estate near Leiden, where he spent his leisure time arranging a great private botanical garden. He pub-

lished, at the dying author's request, a splendid edition of Sebastian Vaillant's *Botanicon Parisiense* in 1727. In 1729 he resigned his professorships of botany and chemistry; on this occasion he made a public oration in which he took a retrospective view on his career and thanked many botanists throughout Europe who had helped him in enriching the garden. He continued, however, to lecture on the theory and the practice of medicine and to give clinical demonstrations until the year of his death. When a spurious edition of his chemical lecture notes was published in 1724 as *Institutiones et experimenta chemiae* (translated into English by Shaw and Chambers under the captivating title *The New Chemistry*), he felt impelled to publish a textbook on chemistry, the *Elementa chemiae*, which was later translated into English and French and remained the authoritative chemical manual for decades. In addition he published the papers of Jan Swammerdam, which he had bought in Paris, in both Dutch and Latin, as *Biblia naturae* (2 volumes, 1737–1738).

His popularity was now at its highest; he was created a foreign member of the Académie Royale des Sciences of Paris (1728) and elected a fellow of the Royal Society of London (1730); the czarina of Russia invited him to become her court physician, and royalty and members of the nobility sought his advice. He was now the most famous man of science in Europe and was considered an oracle. The secret of his influence lay in the conjunction of a universal scholarship with a cheerful personality and impeccable character.

In autumn 1737 Boerhaave began to show symptoms of serious heart failure. Dyspnea forced him to interrupt a bedside lecture in April 1738; he made his will, advised the governors of the university about the choice of his successor, and died in his house in Leiden, after an illness of several months. He was buried in St. Peter's Church, and the whole scholarly community of Europe mourned him. On 4 November 1738 his friend Albertus Schultens delivered a eulogy based, in part, upon autobiographical notes left by Boerhaave.

BIBLIOGRAPHY

I. Original Works. Boerhaave's *Atrocis, nec descripti prius, morbi historia*, a facsimile of the first edition (1724) and first French translation, with an introduction by G. A. Lindeboom, appeared as the ninth volume in the series Dutch Classics in the History of Science (Nieuwkoop, 1964).

A complete list of all works written, edited, or provided with a preface by Boerhaave, as well as of the works based on his textbooks, lectures, etc., is in G. A. Lindeboom, *Bibliographia Boerhaaviana* (Leiden, 1959). See also Lindeboom, ed., *Boerhaave's Correspondence,* 2 vols. (Leiden, 1962–1964).

II. SECONDARY LITERATURE. Further works on Boerhaave are [W. Burton] *An Account of the Life and Writings of Herman Boerhaave* (London, 1743); F. W. Gibbs, "Boerhaave and the Botanists," in *Annals of Science,* **13** (1957), 47–61, and "Boerhaave's Chemical Writings," in *Ambix,* **6** (1958), 117–135; F. R. Jevons, "Boerhaave's Biochemistry," in *Medical History,* **6** (1962), 343–362; Lester S. King, *The Medical World of the Eighteenth Century* (Chicago, 1958), chs. 3 and 4, and *The Growth of Medical Thought* (Chicago, 1963), pp. 177–185; G. A. Lindeboom, *Iconographia Boerhaavii* (Leiden, 1963), and *Herman Boerhaave. The Man and His Work* (London, 1968); M. Maty, *Essai sur le caractère du grand médecin ou Éloge de Mr. Herman Boerhaave* (Cologne, 1747); and D. Schoute et al., *Memorialia Hermanni Boerhaave* (Haarlem, 1939), lectures given at the 1938 international Boerhaave commemoration.

G. A. LINDEBOOM

BOETHIUS, ANICIUS MANLIUS SEVERINUS

(*b.* Rome [?], *ca.* 480; *d.* near Pavia, Italy, 524/525), *logic, mathematics, music, theology, philosophy.*

Very little is known of Boethius' life before his downfall, imprisonment, and execution (522–525). He belonged to one of the more eminent families of the Roman aristocracy, the Anicii, to which two emperors and perhaps also Pope Gregory the Great belonged. Manlius Boethius, consul for 487, may have been his father, and a prefect of the *praetorium* for 454 may have been his grandfather. Indirect evidence suggests an approximate date for Boethius' birth: he was younger than the writer Ennodius (*b.* 475), his distant relative and friend; he considered himself not old in 523; and he achieved public eminence in 510. His appointment to the honorific title of consul in 510, while he was writing a commentary on Aristotle's *Categories;* his presence in Rome in 522, when he delivered a speech in the Senate before King Theodoric, who had just made Boethius' two sons consuls; his imprisonment in or near Pavia in 522/523; and his death there two years later are well documented. All other chronological data are hypothetical, including his appointment to one of the highest offices in the Roman Gothic kingdom, the *magisterium officiorum,* which gave him some measure of control over state affairs.

For a long time it was taken for granted that Boethius studied in Athens because of a statement made in Theodoric's name by Cassiodorus that in fact suggests a contrary conclusion: "You [Boethius] have penetrated *from a distance* the schools of Athens"

(italics author's).[1] Many now accept the view that he studied under Ammonius in Alexandria; the hypothesis is based on a vague possibility that a prefect of Alexandria *ca.* 476 named Boetios was Boethius' father and on the close connection of many passages in the two philosophers' works.[2] But common doctrines most often derive from common sources, and books travel more easily than men. There is no reason to believe that Boethius ever left Italy.

When still young Boethius lost his father, but acquired the powerful and inspiring protection of Q. Aurelius Memmius Symmachus, a member of an eminent Roman family that combined public authority with great culture. Symmachus may well have provided Boethius with his first knowledge of fourth-century Greco-Latin learning and with the encouragement to bring it up to date. Symmachus' daughter, Rusticiana, became Boethius' wife and bore him two sons, Boethius and Symmachus. Theodoric flattered him for his learning, and asked his advice when the king of France wanted a harper and when the king of Burgundy wanted a water clock and a sundial. Whether Theodoric appointed him to high office because of his special abilities or in order to strengthen his hold on the Roman nobility we cannot know; but he certainly did not take into account Boethius' solidarity with other members of the Senate and his attachment to the idea of the Roman Empire and Roman "freedom," nor did he realize that collaboration does not necessarily mean submission and renunciation. In 522, when Boethius defended Albinus against the charge of betraying the Gothic king for the Roman emperor, Theodoric took his revenge: he ordered Boethius' imprisonment and death.

Boethius left no perceptible mark on politics and statesmanship. His death inspired many to consider him a martyr,[3] but hagiography does not lead to proper appreciation of a man's work. On the other hand, centuries after his death Boethius was responsible for what he probably achieved in a very small measure during his lifetime: the spread of encyclopedic learning. He became the broadcaster of much Greek knowledge to many generations who used Latin and, through them, to many others. Several factors converged to produce this result: basic among them are the body of works that he translated, elaborated, or adapted from the Greek and his own writings, in which he probably exercised somewhat more independent judgment.

Here again we must be cautious. Much has been made of Boethius' grand plan to leave behind, in Latin, the achievements of the Greek past, but he did not outline any such plan. His interests were varied; he had some acquaintance with the general scheme of the lay encyclopedia of knowledge dominating the Greek schools and cultural life of his time, and with the new developments of Christian doctrine. However, for two areas of knowledge he outlined a vague scheme. The first was the basic doctrines of philosophy: "I shall translate and comment upon as many works by Aristotle and Plato as I can get hold of, and I shall try to show that their philosophies agree."[4] This echoes a plan first suggested by Plotinus' forerunner Ammonius Saccas and partly carried out by Plotinus' faithful pupil Porphyry. It is particularly important because it can be shown more than once that Boethius is repeating his source almost literally, even where the translation is disguised; and Porphyry was often his source. It must also be noted that Boethius speaks of writings of which he can "get hold," thus hinting that he was not working where works of Aristotle and Plato were easily obtained.

Another partial plan is suggested by the introductory section of Boethius' *Arithmetic,* dedicated to his father-in-law.[5] There he says that he intends to produce a handbook for each of the four mathematical disciplines—arithmetic, music, geometry, astronomy—which he calls the *quadrivium,* probably the first time this word was used. This led, by analogy, to the term *trivium* for the disciplines dealing with words instead of with numbers or magnitudes. Here again one ought to be cautious and not interpret the intention as a definite plan: the four disciplines were linked in the Greek tradition from which Boethius drew his material.[6] Nor should one be drawn by the flattering letter of Theodoric/Cassiodorus (*ca.* 507–513) into believing that what is written there described works already composed rather than Boethius' knowledge and an ability to discuss matters contained in Greek works.

We know too little about schools and intellectual life when Boethius was young to be able to infer what he learned from whom, or how and where he learned it. We can only try to find out from his works what may have contributed to their composition. The two elements that seem to emerge from such an inquiry are the Roman intellectual life of the latter fourth century and the Greek scholastic tradition as it appeared in the fifth century.

A few books, possibly very few, written in fourth-century Rome had come into Boethius' hands: books of logic or on the line between logic and rhetoric. He may have learned more from his father-in-law, one of whose ancestors had been a member of the learned circles of *ca.* 360–380. Representing that period in Boethius' works are Marius Victorinus, African and pagan by birth, Roman and Christian by adoption; Vettius Agorius Praetextatus, the leader of the pagan revival; Albinus; and Themistius, the eminent Greek rhetorician, philosopher, and teacher of many Romans, including Agorius, in Constantinople. Cicero should be added, because he was the great Roman of that period, master and inspirer of these revivalists.

Boethius possessed, at least in part, Victorinus' Latin adaptation of Porphyry's *Isagoge* and used it for his shorter commentary, in dialogue form, on this work. Victorinus may even have encouraged him to present as his original work what he was actually adapting from the Greek: Victorinus had done this in the *Isagoge* and Boethius did it in several of his "original" works of logic and, perhaps, of theology. Victorinus may also have been the source for other writings by Boethius, if we accept as authentic one of the two basic versions of Cassiodorus' *Institutiones:* there[7] Victorinus is credited with a translation of Aristotle's *Categories* and *De interpretatione,* commentaries on the *Categories* and Cicero's *Topica,* and a *De syllogismis hypotheticis.* In any case, Victorinus provided an example of how to spread Greek culture among Latin-speaking people.

Boethius may have known one work by Agorius Praetextatus: his Latin version of Themistius' paraphrase of Aristotle's *Analytics,* but he is rather ambiguous: he may simply have known that such a version existed. Of Albinus, Boethius knew that he had written something on logic. It may be suspected that Albinus was in fact responsible for the Latin version of Themistius' exposition of the *Categories,* which, from *ca.* 780, was ascribed to St. Augustine;[8] but Boethius was not familiar with it. The connection with Themistius appears to be indirect. Apart from Agorius' (and Albinus'?) dependence on Themistius, this idea seems to be confirmed by the place that Themistius' doctrines concerning the "topics," or types of logical and rhetorical arguments, have in Boethius' work; Themistius' classification of topics is discussed by Boethius as a parallel to Cicero's classification and analysis of them.

Greece and the Greek world still had active and organized centers of higher studies and well-stocked libraries. Boethius may never have gone near them, but he could try to obtain some of the books used there, most probably in Athens, by students and teachers. There is no mention in his works[9] of contemporary Greek scholars or philosophers, nor of those of the two or three previous generations. The most modern man he mentions is Proclus' teacher Syrianus (first half of the fifth century). More than once mention is made of Iamblichus, a Neoplatonist of the first half of the fourth century, whose intellectual legacy passed, after three generations, to Proclus, a Constantinopolitan who headed the Athenian school in the decades immediately preceding Boethius' birth. Recent studies have strengthened the hypothesis that the few books from which Boethius derived his knowledge of Greek philosophy and science came from Athenian circles.

When it is maintained, with a great wealth of quotations and parallel passages, that Boethius was a pupil of Ammonius,[10] master in Alexandria, nothing more is shown than that what Ammonius had learned from his masters in Athens, especially from Proclus, had also reached Boethius. The detailed analysis of the Porphyrian and Aristotelian commentaries of Boethius made by J. Shiel leaves little doubt that his conclusions are right: Boethius possessed one volume of the Greek *Organon,* in which the logical texts of Porphyry and Aristotle were surrounded by a rich collection of passages extracted from the main commentaries of the third and fourth centuries. All the quotations from and references to Porphyry, Iamblichus, Themistius, and Alexander of Aphrodisias are secondhand. Wherever it is possible to check, they are also found in the corresponding extant Greek commentaries. Even quotations from other works of Aristotle, not commented upon by Boethius, come from these selections of Greek commentaries.

In general, considering the nature of most of Boethius' writings, one would do well to discount even internal references to "past" works: some of these references may come from the original Greek works[11] or—as happens with many writers—may be expressions based on the author's wishful thought that, by the time one work is finished, others will also be completed, so that the reader will be able to take the whole series of works in a definite systematic order linked by cross-references. Consequently, it is reasonable to consider as works surely written by Boethius those which are extant and cannot easily be denied as his. Doubts still remain regarding the actual "Boethian" form of several of these works: double recensions suggest that early editors took more freedom than we should like in reshaping the works of the man they intended to glorify. This might even lead us to suggest that Boethius' name was soon added to works not his own, as was done in later times.

The existing works include a considerable body of logical writings: translations, commentaries, and independent treatises.[12] We still have the translations of (1) Porphyry's *Isagoge* (*ca.* 507), in two slightly different versions; (2) Aristotle's *Categories* (before 510), in one uniform, quite polished recension and in a mixture of parts of this recension with parts of a rougher rendering (perhaps Boethius' own, incompletely preserved); (3) Aristotle's *De interpretatione* (before 513), in three slightly different forms; (4) Aristotle's *Prior Analytics* (before 520), like the *Categories,* in one polished recension and in a mixture of parts of this with parts of a more primitive (perhaps Boethius' original) rendering; (5) Aristotle's *Topics* (before 520), in a uniform, unpolished edition and one small section from a more finished text; (6) Aristotle's *Sophistical Refutations* (before 520), in one recension (another existing recension is probably the result of the mixture of the usual recension by Boethius with some elements of a twelfth-century translation or revision by James of Venice). The suggestion that a Latin collection of passages from Greek commentaries on *Prior Analytics* was also translated by Boethius may have to be discarded, and there is only scanty evidence that he translated the *Posterior Analytics.* The translations, especially if one considers only the less finished recensions as undoubtedly authentic, suggest that Boethius' knowledge of Greek was by no means excellent.

The logical works commented upon by Boethius are (1, 2) Porphyry's *Isagoge:* one commentary (*ca.* 505), in the form of a dialogue, is based on some sections of Victorinus' adaptation, and another (*ca.* 508), in five books, is based on Boethius' own translation; (3) Aristotle's *Categories* (509–511), on the basis of Boethius' translation, with a second commentary perhaps intended but probably never written;[13] (4, 5) Aristotle's *De interpretatione* (513–516), a shorter commentary in two books and a longer one in six, both based on Boethius' translation; (6) Cicero's *Topics* (*ca.* 522), preserved incomplete, in seven books. A commentary on Aristotle's *Topics* is mentioned by Boethius, but it is not known whether it was ever written.

The "independent" logical works are (1) *On Categorical Syllogism* (*ca.* 505–506), in two books; (2) *On Division* (*ca.* 507); (3) *On Hypothetical Syllogisms* (*ca.* 518), in three books; (4) *Prolegomena* (*ca.* 523), known in the Middle Ages as *Antepraedicamenta* and, from 1492 on, as *Introductio in syllogismos categoricos;* and (5) *De differentiis topicis* (*ca.* 523). (*On Definitions,* a treatise ascribed to Boethius from the twelfth to the nineteenth centuries, is the work of Marius Victorinus. Small rhetorical treatises published as independent works are extracts or adaptations from the *De differentiis topicis.*)[14]

Two works by Boethius on disciplines of the *quadrivium* still exist: the *Arithmetic,* in two books, and the *Music,* in five. No agreement has been reached by scholars on the status of the various recensions of a *Geometry* that bear Boethius' name in many manuscripts and editions and were quoted as his for several centuries; it is quite possible that they include at least some sections originally written by him as translations of and adaptations from Euclid. None of the texts on astronomy that have been tentatively connected with Boethius can be ascribed to him unless new evidence comes to light.

Boethius' writings on theology are confined to two short pamphlets, *On the Trinity* and *On the Two Natures and One Person of Christ,* and the briefly argued answers to two questions, *Are "Father," "Son," "Holy Spirit" Predicated Substantially of "God"* and *How Can Substances Be Good in Virtue of Their Existence, Without Being "Goods" qua Substances* (*Quomodo Substantiae . . .,* often known as *De hebdomadibus*).[15]

All these writings are obviously didactic or scholastic. The same character is shared, but veiled in a literary form, by Boethius' one personal, original, and attractive work, the *Consolation of Philosophy* (523–524), written in verse and prose while he was awaiting execution.

Among the books most frequently—and erroneously—ascribed to him are Dominic González' (or Gundissalinus') *De unitate et uno* (twelfth century) and Thomas of Erfurt's *De disciplina scholarium* (thirteenth century). Translations from Aristotle (*Metaphysics, Ethics,* etc.) made in the twelfth century were occasionally attributed to Boethius from the twelfth to the sixteenth centuries; more persistent was the attribution, from 1510 to the early twentieth century, of the translation by James of Venice of the *Posterior Analytics* (ca. 1140).

Originality is rare in Boethius' works. Even where the sources of the doctrines expounded cannot be traced back exactly to a particular author, it can easily be assumed that he was following a definite model. It is also clear, especially in advanced logic, mathematics, and theology, that his preparation, and possibly his linguistic knowledge, was not sufficient for him to pass on all the best that was available to him. But, considering the enormous influence that his works exerted on the revival of learning from the late eighth to the thirteenth centuries, it is important to delineate the doctrines he expounded. We shall not include, however, those contained in those works of Aristotle that he translated.

Two points from the commentaries on Porphyry—which go back mainly to the commentaries of the Porphyrian school itself as it continued, particularly in Athens—deserve special mention. One concerns the Aristotelian divisions of philosophy, and more especially the general plan of logic.[16] Boethius' texts contributed more than anything else to popularization of those divisions. Philosophy, as the encyclopedia of knowledge, is divided into two parts: the theoretical (speculative) sciences and the practical sciences. The first is tripartite: it contains the sciences of nature that consider things material and changeable (physical sciences in a wide sense); those that consider the same things abstracted from movement and matter (mathematical, or "intelligible," sciences); and those that consider things immaterial and unchangeable ("theology" or, later, metaphysics). The second part contains the sciences that deal with action, in relation either to the individual (ethics), or to the family ("economics"), or to social life (politics). Logic is the science of persuasive argument, composed of several propositions; it is the science of syllogism in its general form, or in its applications in common discussion, or in its application to demonstration. This main part of logic must be preceded by a study of individual propositions, and this, in its turn, by the study of individual terms or classes of terms.

The other point concerns what came to be known as the problem of universals.[17] Porphyry had only mentioned its difficulties; Boethius treated some of them and suggested solutions. Especially important are his distinction between "things as they are" and "things as they are conceived" and his mention of the theory of *indifferentia,* a half-way solution that simultaneously allows for and denies the presence of, in things outside the mind, the common element that characterizes universality. This became the doctrine of one of the main schools of thought of the early twelfth century.

In the commentary to the *Categories,* derived largely from the two commentaries by Porphyry, one finds such statements as "A sign of continuity in a body is this: if one part of it is put in motion, the whole body is put in motion, and, if a body which is a whole is moved, at least other parts near those which are set in motion will be moved; as if I push a stick touching one extreme, the other parts of the stick will be moved as that extreme."[18] The commentaries on *De interpretatione* contain interesting analyses of the meanings of necessity[19] and—a source of interminable meditation and discussion—the different aspects of the so-called problem of future contingents:[20] Is a future event, which is not foreseeable on the basis of a known law of nature, such that a proposition describing it is bound to be true or false?

The *De divisione,* covering one of the main sections of logic as detailed by Porphyry at the beginning of the *Isagoge* and possibly based on a similar treatise of the Roman or Athenian school of the fourth or fifth century, contains a classification and partial analysis of the kinds of distinctions that must be considered when inquiring into one's subject matter. It propounds the elements for a methodical approach to scientific inquiry. Four kinds of "division" are listed: (1) division of a genus according to fundamental, substantial, different features and according to species, which are determined by at least some of these differences—this is indispensable for achieving satisfactory definitions; (2) division of a whole into its constituent parts, so that precision in accounting for the nature and structure of the whole may be attained; (3) "division of words," i.e., classification of the different meanings or functions of individual words, in order to avoid confusion and sophistry; and (4) "division of accidents," i.e., classification of some feature that may belong, but not essentially, to many different things or kinds of things (the blue of the sea, the blue of a wall, etc.), which will aid in understanding the relationship between accidental features and the essential nature of things.

The *Prolegomena* (*Introductio ad syllogismos categoricos*), which may go back, directly or indirectly, to a similar introduction by Porphyry and is mentioned by Boethius in his first commentary on the *Isagoge,*[21] restates and expands Aristotelian doctrines on noun and verb, but concentrates mainly on the relationships between propositions that are quantified in the subject and either positive or negative in the subject and/or the predicate. This is a later and more extensive treatment of what had appeared as the first book of *De syllogismis categoricis,* the second book of which is a rather poor synthesis, with the addition of a few mechanically constructed combinations, of the first part of Aristotle's *Prior Analytics*. This work most probably also reflects an elementary textbook of Porphyrian origin.

In *De syllogismis hypotheticis* the basic formulation of the Theophrastian syllogism ("If A then B; if B then C; therefore, if A then C") is played upon through a multiplication of formulas resulting from the insertion of the negative at different places in the premise. The importance of this is limited because A, B, C, must stand for nouns; thus, we fall directly back into the nonhypothetical syllogism. The Stoic hypothetical syllogism had its role in this work, as well as in the commentary on the *Topics* of Cicero, but with no original contribution. The one element that may be useful for an analysis of scientific method is the distinction between accidental connection or coincidence ("Fire being warm, the heavens are spherical") and natural connection ("There being man, there is animal" and, more compelling, "If the Earth comes in between, there follows an eclipse of the moon"), technically termed by Boethius *consequentia secundum accidens* and *consequentia naturae* (the latter being either *non per positionem terminorum* or *per positionem terminorum*).

The commentary on Cicero's *Topics* and the *De differentiis topicis* deal with the kinds of arguments used to persuade, either in a purely theoretical context or in a practical one, i.e., in dialectical or rhetorical arguments. The second work includes most that is important, from a methodological point of view, in the first. It is a systematic exposition of the nature of individual propositions (categorical and hypothetical), questions, theses, and rhetorical "hypotheses," and of connected propositions (such as syllogisms); and then of the headings under which arguments can be classified according to Themistius and Cicero. The importance of such a work lies mainly in its provision of the tools for a critical evaluation of arguments used in discussion and exposition of theories and facts. Thus, distinctions are made between arguments based on definitions, on descriptions, on similarities, on different interpretations of words, on assertions valid for whole classes (and therefore for subclasses), on regular causality, on contradiction, on authority, and on parallelism of situations.

The theological treatises must be considered here because of their role in training several generations, from the ninth century to the thirteenth, to apply the concepts developed by philosophy as a basis for clear thinking to fields where acceptance of dogmatic statement would have appeared more apposite. In *On the Trinity* and, within narrower limits, in the question on the predication of the three Persons to the subject "God," Boethius tries to explain the apparently absurd equation "one = three" by using the distinctions of Porphyry's (and Aristotle's) classes of predicates (genus, species, difference, accident, property) and the ten Aristotelian categories (substance, quantity, quality, relation, etc.). He was, of course, not the inventor of rational theology: *On the Trinity,* which reflects one of the revolutionary trends in Greek theology, is perhaps no more than a disguised translation. But his exposition of the problem and the attempt to locate the absurdity, or possibly the validity, of a statement within the intellectual framework of his time give him an eminent position in the progress toward clarity and exercise of critical power.

The short work on goodness of beings (*Quomodo substantiae . . .*) also claims more than an antiquarian interest. In this writing Boethius set out to solve an eminently nonmathematical problem with something of a mathematical method, and thereby, through many centuries, trained students to organize their thoughts and apply their powers of deduction: "Just as is the custom in mathematics and other disciplines, I begin with a series of definitions and axioms or postulates, from which all the rest will be derived."[22] The *Quomodo* is also important for the neat distinction between essence (*esse*) and existence (*quod est*), which may have a distant echo in the distinction between hypothesis and verification.

The treatise *Two Natures and One Person in Christ* provides us with, among other things, an analysis of the meanings that *natura* has in different contexts. The four meanings are set forth in these formulas: "Nature is to be found in things that can somehow be grasped by our mind"; "Nature (of substances) is what can bring about or be the recipient of an effect"; "Nature (of bodily substances) is the principle of movement per se, not accidentally"; and "Nature is the specific difference giving a definite thing its form." With the definition of *persona*—which became traditional in theology and is at the basis of most of our usages of "person"—Boethius also contributed to the establishment of the technical distinction between *personalis* and *confusa* in the context of the development of the medieval and modern theory of "supposition." For this second purpose, Boethius' definition ("Person is the individual substance of a rational nature") lost the connotation "rational," preserving above all the element of individuality.

The mathematical works by Boethius reproduced Greek works. Although it is not as clear as it has been thought, partly on the basis of what Boethius himself says, exactly which Greek works were reproduced,[23] it is clear that the neo-Pythagorean theory of number as the very divine essence of the world is the view around which the four sciences of the quadrivium are developed. Number, qua multitude considered in itself, is the subject matter of arithmetic; qua multitude applied to something else (relations between numbers?), the subject matter of music; qua magnitude without movement, of geometry; and qua magnitude with movement, of astronomy. The *Arithmetic* develops here and there what was too concise in Nicomachus and abbreviates what was too diffuse. Further, it passes on to the Latin reader many of the basic terms and concepts of arithmetical theory: prime and composite numbers, proportionality, *numeri figurati* (linear, triangular, etc.; pyramidal and other solid numbers), and ten different kinds of *medietates* (arithmetical, geometrical, harmonic, counterharmonic, etc.). His interest in proportions is perhaps connected with the story according to which, while in prison, he thought out a game based on number relations. Here it is noticeable, however, that his understanding of arithmetic, and possibly of Greek, was limited: the more advanced propositions and proofs in Nicomachus, such as the proposition that cubic numbers can be expressed as the successive

sums of odd numbers and the proposition expressing the relation between triangular numbers and the polygonal numbers of polygons with *n* sides, are missing from the *Arithmetic*. He does not, however, miss such elementary things as the multiplication table up to ten.

The *Music* is a continuation of the *Arithmetic*, which contains several elements and terms more appropriate for the treatment of speculative, purely arithmetical music. But, before he comes to this, the very essence of the second science of the quadrivium, Boethius reminds us of the Platonic view that, unlike the other "mathematical sciences," which have only a theoretical value, music has a moral value as well. He also distinguishes the three kinds of music in which number relationships express themselves: the music of the universe (each of the heavens has its special chord), the music of human nature (which harmonizes man's bodily and psychic activities), and the music of some instruments. The third is the only one that, although deteriorated because of its involvement in matter, can be heard. Most of the book is devoted to a lengthy catalog of somewhat classified number relations, most of them with their technical terms and with some description of the nature of the sounds corresponding to them. But, music being considered as science, most of what the musicologist, the artistic composer, and the practicing player would consider essential to the understanding of what music is, is beyond Boethius' grasp.

Boethius' *Geometry*, which is mentioned in Cassiodorus' *Institutiones*, may well have been very different from any of the texts, varied in extent and, in many cases, with different contents, that appeared under his name during the Middle Ages. There is very little more of a geometrical nature in the most ancient manuscripts ascribed to Boethius than Euclid's definitions (from Book I) and some propositions (from Books III and IV) without the proofs. But, as part of the *Geometry*, there is the description of the abacus, the elementary computer based on a decimal system with the individual numbers classified under the headings *numeri incompositi*—the *digiti* (1–9) and the *articuli* (10, 20, · · · , 100, etc.)—and *compositi* (11–19, 21–29, · · · , 101–109, etc.), and there are rules for multiplication and division.

One additional contribution to mathematics that reached the Middle Ages through Boethius is in his commentaries to Porphyry (a sign that his knowledge of such matters is secondhand): the formula $\dfrac{n(n-1)}{2}$ for the number of possible combinations of two elements in a class of *n* elements.[24]

The *De consolatione philosophiae*, considered from

the doctrinal point of view, is on the whole a restatement of the eclectic Neoplatonic cosmology. Three aspects may be usefully emphasized, because this book contributed in large measure to impressing them into the minds of philosophers and scientists, and of the world at large. (1) Independently of any revelation, the mind can achieve certainty about the existence of God, his goodness, and his power of ruling over the universe. (2) The universe is ordered according to unbroken chains of causes and effects, where necessity, under supervision and determination by God, would be apparent to an all-knowing mind and where chance is nothing more than the coincidental intersection of distinct lines of causation. (3) The order of the universe includes a descent from the first cause to the lowest effects and a return from the lowest ends to the highest beginning. Causality, in the more restricted modern sense, and teleology have preserved a stronger hold on the minds of many generations because of the enormous popularity, until the sixteenth century, of the *Consolatio.* But Boethius' insistence on the possibility of combining freedom of the will with God's eternally present knowledge of the order he willed engaged scholars in theological subtleties more than in a scientific approach to research or organization of knowledge.

NOTES

1. Cassiodorus, *Variae* I.45.3.
2. P. Courcelle, *Les lettres grecques,* p. 299, n. 1.
3. E.g., Dante, *Divine Comedy, Paradiso* X.124–129.
4. *Second Commentary on De interpretatione,* Meiser, ed., pp. 79–80.
5. *Arithmetic,* Friedlein, ed., p. 3.
6. See esp. Iamblichus' *Commentary on Nicomachus' Arithmetic,* E. Pistelli, ed. (Leipzig, 1894), pp. 5–8.
7. Cassiodorus, *Institutiones* I.iii.18, R. A. B. Mynors, ed. (Oxford, 1937), p. 128.
8. *Categoriae,* in *Aristoteles Latinus* I.1–5 (Bruges, 1961), p. lxxviii.
9. There is no foundation for the view held by Courcelle in *Les lettres grecques* (p. 278) that *audivimus* in Boethius' *Second Commentary,* Meiser, ed., p. 361, line 9, should be read "Ammonius."
10. Courcelle, pp. 270–277.
11. See J. Shiel, "Boethius' Commentaries on Aristotle," *passim.*
12. For the dates of the logical works I follow De Rijk, "On the Chronology." Some of the views I express here on the question of second recensions are at variance with hypotheses I put forward in the past.
13. But see P. Hadot, in *Archives d'histoire doctrinale et littéraire du moyen âge.*
14. A. Mai "discovered" these texts in MS Vat. lat. 8591; they are part of a collection of Boethian logical texts made in Constantinople ca. 530, of which many copies exist.
15. Views have been expressed by competent scholars both for and against the authenticity of a fifth theological text, the *De fide Catholica,* which seems to have intruded itself, anonymously, at some later stage into the collection of the other four. The arguments in favor seem unsatisfactory.
16. G. Schepss and S. Brandt, eds., pp. 7–10.
17. *Ibid.,* pp. 23–32, 159–167.
18. *Patrologia Latina,* LXIV, cols. 204–205.
19. E.g., in the *Second Commentary on the De interpretatione,* Meiser, ed., pp. 241 ff.
20. *Ibid.,* pp. 190–230.
21. Schepss and Brandt, p. 15.
22. H. F. Stewart and E. K. Rand, eds., p. 40.
23. Very close similarities can be noticed between Boethius and Nicomachus' commentator Iamblichus.
24. Schepss and Brandt, pp. 118–120, 319–321.

BIBLIOGRAPHY

I. ORIGINAL WORKS. The first ed. meant to contain all the works of Boethius was brought out by Iohannes and Gregorius de Gregoriis, with the scholarly collaboration of Nicolaus Iudecus (Venice, 1491–1492; repr. 1498–1499); it did not include the translations of *Prior Analytics, Topics,* and *Sophistical Refutations* but did contain the pseudepigrapha *On Definition, De unitate et uno,* and *De disciplina scholarium.* A complete ed. (Basel, 1546, 1570), with the pseudepigrapha and the non-Boethian translation of *Posterior Analytics* includes the translations missing from the Venice collection reproduced from a text rev. by Jacques Lefèvre d'Étaples (Paris, 1503), which was based on the Greek, under the supervision and with the collaboration of Heinrich Lorit; for the logical works (except the uncommented translations) and for the theological treatises this ed. depends on Giulio Marziano Rota's ed. (Venice, 1537). J. P. Migne, ed., *Patrologia Latina,* LXIII and LXIV, contains all the works of the 1570 ed., some of them from more recently published texts, and some fragments wrongly thought to be new discoveries. Both the Corpus Scriptorum Ecclesiasticorum Latinorum and the Corpus Christianorum include complete editions of Boethius in their plans. In Vol. 48 of the former (Vienna, 1906), G. Schepss and S. Brandt edited the two *Commentaries on Porphyry,* and in Vol. 67 (Vienna, 1934), W. (Guillelmus) Weinberg edited the *Consolatio philosophiae;* in Vol. 94 of the latter (Turnhout, Belgium, 1957), L. Bieler edited the *Consolatio.*

Critical editions of the translations are being done by L. Minio-Paluello, partly with the collaboration of B. G. Dod, as part of the *Aristoteles Latinus,* a section of the Corpus Philosophorum Medii Aevi (Bruges–Brussels–Paris): I, pts. 1–2, *Categoriae* (1961); III, pts. 1–2, *Analytica priora* (1962); II, pt. 1, *De interpretatione* (1965); I, pt. 6, Porphyry's *Isagoge* (1966); V, pts. 1–2, *Topica* (1969); and VI, pt. 1, *Elenchi sophistici* (in preparation).

Among the earliest eds. are *Consolatio philosophiae* (Savigliano, ca. 1471)—at least sixty-two Latin eds. of the work were printed before 1501; *Analytica priora* (Louvain, 1475); *Second Commentary on Porphyry, Commentary on Categories,* text of *De interpretatione* (Naples, ca. 1476); all the translations (Augsburg, 1479); *De differentiis topicis* and *In Ciceronis Topica commentarium* (Rome, 1484); *De institutione arithmetica* (Augsburg, 1488); *De Trinitate, Utrum Pater . . ., Quomodo substantiae* (Venice, 1489); and the doubtful *De fide Catholica* (Leiden, 1656).

Among the recent eds. not mentioned above, the following are important: *In Ciceronis Topica commentarium,*

I. G. Baiter, ed., in Cicero's *Opera,* I. C. Orelli and I. G. Baiter, eds., I (Zurich, 1833)—this ed. also contains the short section discovered and published by C. B. Hase in *Johannis Laurentii Lydi, De ostentis* (Paris, 1823), pp. 341–356; *De institutione arithmetica, De institutione musica, Geometria,* G. Friedlein, ed. (Leipzig, 1867); *Opera theologica,* R. Peiper, ed. (Leipzig, 1871); *Commentaries on the De interpretatione,* C. Meiser, ed. (Leipzig, 1877–1880); *De divisione,* in an appendix to L. Davidson, *The Logic of Definition* (London, 1885); *The Theological Tractates,* with English translation by H. F. Stewart and E. K. Rand, and *The Consolation of Philosophy,* with English translation by I. T. [John Thorpe?], rev. by H. F. Stewart (London–Cambridge, Mass., 1936). A fragment, believed by the ed. to come from Boethius' *Second Commentary to the Categories,* was published by P. Hadot in *Archives d'histoire doctrinale et littéraire du moyen âge,* **34** (1960), 10–27.

II. SECONDARY LITERATURE. Extensive bibliographies on Boethius can be found in L. Bieler's ed. of the *Consolatio* (see above), pp. xvi–xxvi; P. Courcelle, *Les lettres grecques en occident de Macrobe à Cassiodore,* 2nd ed. (Paris, 1948), pp. 401–415, and *La consolation de philosophie dans la tradition littéraire* (Paris, 1967), pp. 383–402 and, for the commentaries on the *Consolatio,* pp. 403–438; M. Cappuyns, "Boèce," in *Dictionnaire d'histoire et de géographie ecclésiastique,* IX (1937), cols. 349–380; B. Geyer, *Die patristische und scholastische Philosophie,* Vol. II of Friedrich Ueberweg's *Grundriss der Geschichte der Philosophie,* 11th ed. (Berlin, 1928), pp. 133, 669–670; C. Leonardi, L. Minio-Paluello, U. Pizzani, and P. Courcelle, "Boezio," in *Dizionario biografico degli italiani,* XII (in press); and A. Momigliano, "Cassiodorus and Italian Culture of His Time," in *Proceedings of the British Academy,* **41** (1955), 227–245.

Besides the above-mentioned works by Cappuyns, Courcelle (*Les lettres . . .*), and Momigliano, see the following on Boethius' life and work in general: H. M. Barrett, *Boethius, Some Aspects of His Times and Works* (Cambridge, 1940); M. Grabmann, *Geschichte der scholastischen Methode,* I (Freiburg, 1909), 148–177; M. Manitius, *Geschichte der lateinischen Literatur des Mittelalters,* I (Munich, 1911), 22–36; A. Momigliano, "Gli Anicii e la storiografia latina del VI secolo," in *Rendiconti dell'Accademia nazionale dei Lincei, classe scienze morali,* 8th ser., **9** (1956), 279–297; B. G. Picotti, "Il Senato Romano e il processo di Boezio," in *Archivio storico italiano,* 7th ser., **15** (1931), 205–228; E. K. Rand, *Founders of the Middle Ages* (Cambridge, Mass., 1928), pp. 135–180; and H. Usener, *Anecdoton Holderi* (Bonn, 1877).

On the influence of Boethius see R. Murari, *Dante e Boezio* (Bologna, 1905); and H. R. Patch, *The Tradition of Boethius: A Study of His Importance in Mediaeval Culture* (New York–Oxford, 1935).

On Boethius' logical works (sources, chronology, translations, theories, influences) see L. Bidez, "Boèce et Porphyre," in *Revue belge de philologie et d'histoire,* **2** (1923), 189 ff.; I. M. Bocheński, *Formale Logik* (Freiburg–Munich, 1956), translated by I. Thomas (Notre Dame, Ind., 1961); L. M. De Rijk, "On the Chronology of Boethius's Works

on Logic," in *Vivarium,* **2** (1964), 1–49, 125–162, which supersedes all previous studies on the subject; K. Dürr, *The Propositional Logic of Boethius* (Amsterdam, 1951); W. Kneale and M. Kneale, *The Development of Logic,* (Oxford, 1962), pp. 189–198; L. Minio-Paluello, "Iacobus Veneticus Grecus, Canonist and Translator of Aristotle," in *Traditio,* **8** (1952), 265–304, and "Les traductions et les commentaires aristotéliciens de Boèce," in *Texte und Untersuchungen zur Geschichte der altchristlichen Literatur,* Vol. 64 of Studia Patristica (1957), pp. 358–365; C. Prantl, *Geschichte der Logik im Abendlande,* I (Leipzig, 1855; repr. Graz, 1955), 679–721; A. N. Prior, "The Logic of Negative Terms in Boethius," in *Franciscan Studies,* **13** (1953), 1–6; J. Shiel, "Boethius' Commentaries on Aristotle," in *Mediaeval and Renaissance Studies,* **4** (1958), 217–244; and A. Van de Vyver, "Les étapes du développement philosophique du haut moyen âge," in *Revue belge de philologie et d'histoire,* **8** (1929), 425–452.

Also see the prefaces to Minio-Paluello's eds. of Boethius' works listed above; however, some of the views expressed in this article are new, and will be discussed in future writings. The previous literature on the authorship of the translations is discussed in full in these prefaces.

For the theological treatises see, besides Usener's *Anecdoton Holderi,* V. Schurr, *Die Trinitätslehre des Boethius im Lichte der skytischen Kontroversen* (Paderborn, 1935). The latest discussion of the authenticity of *De fide Catholica,* with references to the previous works on the subject, is W. Bark, "Boethius's Fourth Tractate: The So-called 'De Fide Catholica,'" in *Harvard Theological Review,* **59** (1946), 55–69. For the influence of the treatises in the Middle Ages, see M. Grabmann, *Die theologische Erkenntnis- und Einleitungslehre des heiligen Thomas auf Grund seiner Schrift In Boethium De Trinitate* (Fribourg, 1948); and N. M. Haring's editions of *A Commentary on Boethius' De hebdomadibus by Clarenbaldus of Arras* and *The Commentaries of Gilbert, Bishop of Poitiers on the Two Boethian Opuscula Sacra on the Holy Trinity,* in *Nine Mediaeval Texts,* Vol. I of Studies and Texts, published by the Pontifical Institute of Mediaeval Studies (Toronto, 1955), pp. 1–96.

On the mathematical works, including the *De musica,* see M. Cantor, *Vorlesungen über Geschichte der Mathematik,* 3rd ed., I (Leipzig, 1907), 573–585, which contains references to previous works, especially Friedlein's; J. L. Heiberg, in *Philologus,* **43,** 507–519; F. T. Koppen, "Notiz über die Zahlwörter im Abacus des Boethius," in *Bulletin de l'Académie des sciences de St. Pétersbourg,* **35** (1892), 31–48; O. Paul, *Boethius, fünf Bücher über die Musik aus dem lateinischen . . . übertragen und . . . sachlich erklärt* (Leipzig, 1872); G. Pietzsch, *Die Klassifikation der Musik von Boetius bis Ugolino von Orvieto* (Halle, 1929); U. Pizzani, "Studi sulle fonti del *De institutione musica* di Boezio," in *Sacris erudiri,* **16** (1965), 5–164; H. Potiron, *Boèce théoricien de la musique grecque* (Paris, 1961); P. Tannery, "Notes sur la pseudo-géométrie de Boèce," in *Bibliotheca mathematica,* **3** (1900), 39–50; and R. Wagner, "Boethius," in *Die Musik in Geschichte und Gegenwart,* II (Kassel–Basel, 1952), cols. 49–57.

All the relevant bibliography for the *De consolatione*, its sources, doctrines, diffusion, and influence, is in the edition by Bieler and in Courcelle's *La consolation*.

A good source for recent bibliography is Menso Folkerts' critical edition of the two-book version of Boethius' *Geometry, Boethius Geometrie II: Ein mathematisches Lehrbuch des Mittelalters* (Göttingen, 1967), doctoral dissertation.

LORENZO MINIO-PALUELLO

BOETIUS DE BOODT, ANSELMUS. See **Boodt, Anselm de.**

BOETTGER, RUDOLPH CHRISTIAN VON. See **Böttger, Rudolf Christian von.**

BOGUSLAVSKY, PALM HEINRICH LUDWIG VON (*b.* Magdeburg, Prussia, 7 September 1789; *d.* Breslau, Prussia, 5 June 1851), *astronomy.*

The son of a Prussian captain, Boguslavsky attended the Dom School in Magdeburg, then entered the Prussian military service; after military training he took part in the campaign against Napoleon in 1813–1815. After his discharge he lived on his estate in Silesia, where, as an amateur, he occupied himself with astronomical observations. In 1831 he became a senior astronomer at the astronomical observatory in Breslau; in 1836 he was appointed extraordinary professor at the University of Breslau, and in 1843 became director of the astronomical observatory there. He was concerned primarily with the observation and orbit computation of comets, meteor groups, planets, and solar eclipses, and he also contributed to the *Berliner academischen Sternkarten*.

His son Heinrich Georg (1827–1884) was an oceanographer and hydrographer in Berlin.

BIBLIOGRAPHY

Biographical articles on Boguslavsky are in *Allgemeine deutsche Biographie*, III, 58; and Poggendorff, I, 225.

BERNHARD STICKER

BÖHEIM, MARTIN. See **Behaim, Martin.**

BOHL, PIERS (*b.* Walka, Livonia [now Latvian S.S.R.], 23 October 1865; *d.* Riga, Latvia, 25 December 1921), *mathematics.*

The son of George Bohl, a merchant, Piers Bohl first studied in his native city and then at a German Gymnasium in Viljandi, Estonia. In 1884 he entered the department of physics and mathematics at the University of Dorpat, Estonia, from which he grad-

uated in 1887 with a candidate's degree in mathematics (equivalent to a master's degree in the United States), having won a gold medal for a competitive essay on the theory of invariants of linear differential equations (1886). Bohl defended dissertations in applied mathematics for his master's degree in 1893 (equivalent to a doctorate in the United States) and for his doctorate in 1900. (The doctorate, a degree that can be gained only after the candidate has done outstanding work in his chosen field, allows the holder to be called professor.) He received both of these advanced degrees from Dorpat. From 1895 Bohl taught at Riga Polytechnic Institute (from 1900 with the rank of professor); and when the institute was evacuated to Moscow at the beginning of World War I, he accompanied it. He returned to Riga in 1919 and was appointed professor at the University of Latvia, which had been founded that year. Two years later he died of a cerebral hemorrhage.

In his master's dissertation, Bohl was the first to introduce and to study that class of functions (more general than ordinary periodic functions) which in 1903 were named quasi-periodic by the French mathematician E. Esclangon, who discovered them later than, but independently of, Bohl. Finite sums of periodic functions with, generally speaking, incommensurable periods (of the type $\sin x + \sin \sqrt{2}x + \sin \sqrt{3}x$) are an example. Harald Bohr's concept of almost-periodic functions is the further generalization of this class.

In his doctoral dissertation, Bohl, following Henri Poincaré and A. Kneser, presented a new development of topological methods of systems of differential equations of the first order. To the investigation of the existence and properties of the integrals of these systems, he applied a series of theorems, which he developed and proved, concerning points that remain fixed for continuous mappings of n-dimensional sets of points. L. Brouwer's famous theorem on the existence of a fixed point under the condition of the mapping of a sphere onto itself is easily obtained as a consequence of one of the propositions completely demonstrated in Bohl's "Über die Bewegung. . . ." Bohl's topological theorems did not, however, attract the attention of contemporary mathematicians.

Studying one problem of the theory of secular perturbations (1909), Bohl encountered the question of the uniform distribution of the fractional parts of functions satisfying certain conditions. The theorem he developed was also developed independently by H. Weyl and W. Sierpinski; it was generalized by Weyl in 1916. Later the theory of the distribution of fractional parts of functions became a large part of number theory.

BIBLIOGRAPHY

I. ORIGINAL WORKS. For Bohl's early work, see *Theorie und Anwendung der Invarianten der linearen Differentialgleichungen* (Dorpat, 1886), which manuscript is in the Historical Archive of the Estonian S. S. R., Tartu; and *Über die Darstellung von Funktionen einer Variablen durch trigonometrische Reihen mit mehreren einer Variablen proportionalen Argumenten* (Dorpat, 1893), his master's dissertation. His doctoral dissertation, "O Nekotorykh Differentsialnykh Uravneniakh Obshchego Kharaktera, Primenimykh v Mekhanike" ("On Some Differential Equations of a General Character, Applicable in Mechanics;" Yurev, 1900), is also available in French as "Sur certaines équations différentielles d'un type général utilisables en mécanique," in *Bulletin de la Société mathématique de France*, **38** (1910), 1–134. See also "Über die Bewegung eines mechanischen Systems in der Nähe einer Gleichgewichtslage," in *Journal für reine und angewandte Mathematik*, **127** (1904), 179–276; and "Über ein in der Theorie der säkularen Störungen vorkommendes Problem," *ibid.*, **135** (1909), 189–283.

II. SECONDARY LITERATURE. For further information on Bohl, see A. Kneser and A. Meder, "Piers Bohl zum Gedächtnis," in *Jahresbericht der Deutschen Mathematiker-vereinigung*, **33** (1925), 25–32. A complete bibliography of Bohl's work and of literature devoted to him appears in A. D. Myshkis and I. M. Rabinovich, eds., *P. G. Bohl. Izbrannye Trudy* ("P. G. Bohl, Selected Works"; Riga, 1961), biography and analysis of scientific activity, pp. 5–29.

A. P. YOUSCHKEVITCH

BOHN, JOHANNES (*b.* Leipzig, Germany, 20 July 1640; *d.* Leipzig, 19 December 1718), *physiology, medicine.*

Bohn was the son of a wealthy merchant family. He studied medicine in Jena and Leipzig and about 1665 received the doctorate from the medical school of Leipzig. From 1663 to 1665 he traveled through Denmark, Holland, England, Switzerland, France, and possibly Italy. In 1668 he was named professor of anatomy and surgery at Leipzig; in 1690 he became municipal physician; and the following year he was appointed professor of practical medicine. Bohn was a critical, truth-loving man who was so careful of his scientific reputation that on his deathbed he arranged for the destruction of all his unpublished writings.

Bohn's accomplishments are in three areas: anatomy and physiology, iatrochemistry, and forensic medicine. His twenty-six *Exercitationes physiologicarum* appeared at irregular intervals from 1668 on; these are doctoral dissertations, written by Bohn and disputed by various candidates for the doctorate. Most of the *Exercitationes* appeared in 1668; the rest appeared from time to time until about 1677. They were later reprinted as a whole in a pirated edition. Only a few copies of the work are available. Bohn later reworked these dissertations into a completely new composition which appeared in 1680 as *Circulus anatomicus-physiologicus seu Oeconomia corporis animalis,* and was dedicated to Malpighi.

The *Exercitationes* and the *Circulus* show Bohn to have been an expert on the then new anatomical and physiological discoveries. He cites contemporary authors almost exclusively and thereby proves himself one of the innovators in physiology who completely forsook the Galenic tradition. He describes and discusses all major functions of the body. He complements the knowledge gained from the literature with numerous firsthand experiments, for example, experiments on bile and the biliary tract, lymph ducts, heart contractions, pancreatic secretion, on the conjectured swelling of ligated nerves, and artificial perfusion of an excised kidney.

Bohn's basic attitude was mechanistic in that he gave predominantly physical interpretations of vital processes. He especially esteemed Malpighi, Borelli, and Boyle. Bohn had an excellent knowledge of iatrochemistry as well, but he maintained a critical position against this doctrine. He condemned the ancient theory of qualities as unsuitable to the explanation of chemical processes. Wherever possible, he referred to Jan van Helmont's theories of the *fermentum* and to those of Sylvius on the *acidum* and the *alcali.* In his opinion, the process of digestion cannot be explained without the theories of iatrochemistry; with the help of *spiritus* and *sal volatile,* a fermentative transformation of food into chyle takes place. But he opposed a general explanation of physiological findings and clinical observations exclusively by these theories, and especially in his *De alcali et acidi insufficientia* (1675) he explicates this attitude.

Bohn contributed several significant works to forensic medicine. He is considered one of the founders of this discipline and one of the initiators of forensic autopsy.

BIBLIOGRAPHY

I. ORIGINAL WORKS. Bohn's works on forensic medicine are not mentioned here. Among his other works are *Disputatio de sudore* (Leipzig, 1661), his dissertation, sponsored by Johannes Michaelis; *Exercitationum physiologicarum XXVI* (Leipzig, 1668–1677); *Circulus anatomicus-physiologicus, seu Oeconomia corporis animalis, hoc est cogitata functionum animalium, notissimarum formalitatem et causas concernantia* (Leipzig, 1680, 1686), dedicated to Malpighi and consisting of thirty progymnasmata and eleven other dissertations; *Observationes quaedem anatomica circa structuram vasorum biliarum et motuum bilis spectantes* (Leipzig, 1682); *Dissertationes chymico-physicae, chymiae finem, instrumenta et operationes frequentiones explicantes*... (Leipzig, 1685, 1696); and *De duumviratu hypochondriacorum* (Leipzig, 1689), a polemic against Sylvius.

II. SECONDARY LITERATURE. There is no biography of Bohn, but further information on him and his work may be found in A. von Haller, *Bibliotheca anatomica,* I (1774), pp. 497–499; M. Neuburger, "Deutsche Experimentalphysiologen des 17. Jahrhunderts," in *Deutsche medizinische Wochenschrift,* **23** (1897), 483–486; and J. C. Rosenmüller, *De viris quibusdam in Academia Lipsiensi Anatomes peritia in clavuerunt,* III (1816), 7–9.

Also see *Allgemeine deutsche Biographie,* III (Leipzig, 1876), 81–99, with an incomplete bibliography; A. von Haller, *Bibliotheca medicinae practicae,* 4 vols., III (Basel, 1778), 87–88, with a list of forty-two dissertations; *Biographie médicale,* I (Paris, 1855), 539–540; *Biographie universelle ancienne et moderne,* IV (Paris, 1843), 553; and *Biographische Lexikon des hervorragenden Ärzte aller Zeiten und Länder,* 2nd ed., I (Berlin–Vienna, 1929), 606–607.

K. ROTHSCHUH

BOHR, HARALD (*b.* Copenhagen, Denmark, 22 April 1887; *d.* Copenhagen, 22 January 1951), *mathematics.*

Bohr's father was the distinguished physiologist Christian Bohr; his mother, a daughter of the prominent financier, politician, and philanthropist D. B. Adler. In the home he and his elder brother Niels imbibed a deep love of science. At the age of seventeen Bohr entered the University of Copenhagen. Of his teachers, he felt the closest kinship to H. G. Zeuthen, but the most decisive factor in his development as a mathematician was his study of Jordan's *Cours d'analyse* and Dirichlet's *Vorlesungen über Zahlentheorie* with Dedekind's supplements. In his later student years, his interests centered on analysis. After his master's examination he went to study with Landau in Göttingen. This center of mathematics became like a second home to Bohr, and he returned there often. During the years before World War I, he also came into close contact with Hardy and Littlewood, and he often went to Cambridge and Oxford to study.

After obtaining his doctor's degree in 1910, Bohr joined the faculty of the University of Copenhagen. In 1915 he was appointed professor at the College of Technology, a position he retained until returning in 1930 to the University of Copenhagen, where he headed the newly founded Institute of Mathematics. Bohr was one of the leading analysts of his time, and he exerted an extraordinary influence both in inter-

national mathematical circles and in the academic life of his own country. As a teacher he was greatly admired and loved. When the rise of Nazism in Germany in 1933 endangered the academic community, among others, Bohr was among the first to offer help. His close personal relations with colleagues in many countries enabled him to help in finding new homes for those scientists who were either forced to leave Germany or who chose to do so, and he turned all his energies to this task. He himself did not escape exile in the latter part of World War II, when he was compelled to take refuge in Sweden.

Bohr's contribution to mathematics is one of great unity. His first comprehensive investigation, which formed the subject of his doctor's thesis, was concerned with the application of Cesàro summability to Dirichlet series. In a number of later papers he studied other aspects of the theory of Dirichlet series, in particular the distribution of the values of functions represented by such series. His method consists in a combination of arithmetic, geometric, and function-theoretic considerations. His collaboration with Landau was concentrated mainly on the theory of the Riemann zeta-function. It culminated in the so-called Bohr-Landau theorem (1914), concerning the distribution of its zeros. In later papers Bohr gave a detailed study of the distribution of its values in the half plane to the right of the critical line.

The problem of which functions may be represented by Dirichlet series led Bohr to his main achievement, the theory of almost periodic functions, on which the greater part of his later work is concentrated. If a Dirichlet series is considered on a vertical line in the complex plane, it reduces to a trigonometric series. It was therefore natural to consider more generally the problem of which functions of a real variable can be represented by such a series, i.e., can be formed by superposition of pure oscillations. In the special case where the frequencies of the oscillations are integers, the answer is given in the classical theory of Fourier series of periodic functions. Whereas hitherto in the theory of Dirichlet series one had always worked with frequencies forming a monotonic sequence, Bohr discovered that in order to obtain an answer to the problem one would have to consider series with quite arbitrary frequencies. The answer was obtained by introducing the notion of almost periodicity. The theory was published in three papers in *Acta mathematica* (1924–1926), and numerous mathematicians joined in the work on its simplification and extension. Thus Weyl and Wiener connected it with the classical theories of integral equations and Fourier integrals, and Bochner developed a summation method for Bohr-Fourier series generalizing Fejér's theorem. Stepanoff, Wiener, and Besicovitch studied generalizations depending on the Lebesgue integral. Other aspects of the theory were studied by Favard, Wintner, and many others. In the 1930's Von Neumann succeeded in extending the theory to functions on arbitrary groups, and it thus found a central place in contemporary mathematics.

BIBLIOGRAPHY

Bohr's *Collected Mathematical Works,* 3 vols. (Copenhagen, 1952), contain all his mathematical writings, with the exception of elementary articles and textbooks in Danish. An English translation of an autobiographical lecture appears as a preface to this edition.

Obituaries of Bohr include those by S. Bochner, in *Bulletin of the American Mathematical Society,* **58** (1952), 72–75; B. Jessen, in *Acta mathematica,* **86** (1951), i–xxiii, repr. as supp. S. 25 in Bohr's *Collected Mathematical Works,* III, supp. 163–176; O. Neugebauer, in *Year Book 1952 of the American Philosophical Society* (1953), pp. 307–311; N. E. Nørlund, in *Oversigt over det Kongelige Danske Videnskabernes Selskabs Virksomhed, 1950–1951* (1951), pp. 61–67, in Danish; O. Perron, in *Jahresbericht der Deutschen Mathematiker-vereinigung,* **55** (1952), pt. 1, 77–88; and E. C. Titchmarsh, in *Journal of the London Mathematical Society,* **28** (1953), 113–115.

BØRGE JESSEN

BOHR, NIELS HENRIK DAVID (*b.* Copenhagen, Denmark, 7 October 1885; *d.* Copenhagen, 18 November 1962), *atomic and nuclear physics, epistemology.*

A tradition common to many pioneers in science has been the combination of achievement in actual discovery of natural laws with philosophical reflection on the nature of scientific thinking and the foundations of scientific truth. This combination is essential to such scientists in the sense that epistemological considerations played a decisive part in the success of their investigations and that, conversely, the results of the latter led them to deeper understanding of the theory of knowledge. Niels Bohr in particular was very conscious of this twofold aspect of his scientific activity, deep-rooted as it was in the environment in which he grew up and received his education.

The family in which Bohr was the second of three children belonged to the well-to-do intellectual circles of Copenhagen; his father, Christian Bohr, was a talented professor of physiology at the University of Copenhagen; his mother, Ellen Adler, came from a wealthy Jewish family that was prominent in such varied activities as banking, politics, classical philology, and progressive pedagogy. The parents allowed the children's native gifts the fullest development, and

formal education was supplemented at every stage by example and encouragement at home. Niels was not as brilliant a pupil as his younger brother Harald, who became an eminent mathematician; they both, however, showed interests in other fields, including sports. At the University of Copenhagen, Niels stood out as an unusually perceptive investigator. His first research project, a precision measurement of the surface tension of water by the observation of a regularly vibrating jet, was completed in 1906, when he was still a student, and it won him a gold medal from the Academy of Sciences. It is a mature piece of work, remarkable for the care and thoroughness with which both the experimental and theoretical parts of the problem were handled.

Bohr's doctoral dissertation, *Studier over metallernes elektrontheori* (1911), was a purely theoretical work that again exhibited a mastery of the vast subject he had chosen, the electron theory of metals. This theory, which pictures the metallic state as a gas of electrons moving more or less freely in the potential created by the positively charged atoms disposed in a regular lattice, accounted qualitatively for the most varied properties of metals; but it ran into many difficulties as soon as a quantitative treatment was attempted on the basis of then accepted principles of classical electrodynamics.

In order to throw light on the nature of these difficulties, Bohr developed general methods allowing him to derive the main features of the phenomena from the fundamental assumptions in a very direct way. He could thus clearly exhibit the fundamental nature of the failures of the theory, which were in fact attributed to an insufficiency of the classical principles themselves. Thus, he showed that the magnetic properties of the metals could in no way be derived from a consistent application of these principles. The rigor of his analysis gave him, at this early stage, the firm conviction of the necessity of a radical departure from classical electrodynamics for the description of atomic phenomena.

The study of physics, even carried to such unusual depth, did not absorb all of Bohr's energy; his intellectual curiosity knew no bounds. With his characteristic earnestness and thoroughness he took up the hints that circumstances offered as starting points for highly original philosophical reflections. His father's scientific work concentrated on the quantitative analysis of physical processes underlying the physiological functions; the school which he founded and which was brilliantly continued by his pupils still flourishes in modernized form. The type of problem that Christian Bohr was investigating required the closest attention to the elaboration of refined techniques of physical measurement, and simultaneously raised profound philosophical questions about the relationship between physical and biological phenomena.

During Niels's adolescence, the philosophical trend in scientific circles was a reaction against the mechanistic materialism of the preceding generation. In the liberal atmosphere surrounding Christian Bohr's friends, a group to which the philosopher Harald Höffding belonged, this reaction took a moderate and thoughtful form, however. Bohr, the master of the investigation of the physical basis of the physiological processes, insisted on the practical necessity of considering these processes also from the teleological point of view in order to arrive at a complete description. Niels and Harald Bohr were admitted as silent listeners to the philosophical conversations of their father and his friends, and this first confrontation with the epistemological problem of biology, in which apparently conflicting views were found equally indispensable for a full understanding of the phenomena, made a lasting impression upon Niels's mind.

He also soon came to share the negative attitude of the progressive bourgeoisie, to which his family belonged, toward the church and religious beliefs in general; but it is characteristic of his independence of judgment that he arrived at this conclusion only after he had convinced himself that the church upheld doctrines that were logically untenable and shunned the pressing task, at the time preoccupying all liberal minds, of alleviating a still widespread pauperism. His approach to social and philosophical questions, even at such an early stage, was marked by the same logical rigor and breadth of vision as his scientific thinking.

It was in the course of his meditations on the human condition that, considering the role of language as a means of communication, he first came across a situation of great generality whose recognition was the source of his later decisive contribution to the epistemology of physics. He was struck by the fact that the same word is currently used to denote a state of consciousness and the concomitant behavior of the body. In trying to describe this fundamental ambiguity of every word referring to mental activity, Bohr had recourse to an analogy drawn from the mathematical theory of multivalued functions: each such word, he said, belongs to several "planes of objectivity," and we must be careful not to allow them to glide from one plane of objectivity to another. However, it is an inherent property of language that there is only one word for the different aspects of a given psychical activity. There is no point in trying to remove such ambiguities; we must recognize their existence and live with them.

After finishing his studies in Copenhagen, Bohr went to Cambridge, hoping to pursue his work on electron theory under the guidance of J. J. Thomson. Unfortunately, Thomson had lost interest in the subject, and failed to appreciate the importance of Bohr's dissertation, which the latter showed him in an English translation he had been at great pains to make; this was turned down by the Cambridge Philosophical Society as too long and too expensive to print, and Bohr's further attempts to get it published were equally abortive.

This grievous disappointment did not prevent Bohr from making the most of his stay in Cambridge, but as soon as he conveniently could, he moved to Manchester, where Ernest Rutherford had established a flourishing laboratory. There, from March to July 1912, working with utmost concentration, he laid the foundations of his greatest achievement in physics, the theory of atomic constitution. It would be difficult to imagine two temperaments more different than those of Bohr and Rutherford; but this first contact initiated, besides a new epoch in science, a lifelong friendship, compounded of filial affection on Bohr's part and of warm cordiality, tinged with respect, on the part of the jovial New Zealander. With his shrewd judgment of people, Rutherford soon sensed the genius in the shy, unassuming young man, and his immense strength, imaginative insight, and directness of approach were an inspiration to Bohr.

Toward the end of 1910, Rutherford had proposed a "nuclear" model of the atom in order to account for the large-angle scattering of α rays observed in his laboratory. Since the discovery of the electron as a carrier of an elementary unit of negative electric charge, the atom was thought of as a system of a certain number of electrons, kept together by an equivalent positive charge, somehow attached to the massive substance of the atom (the electron itself being nearly two thousand times lighter than the lightest atom). If this positive charge were spread over the whole atom, the α rays, or positively charged helium atoms, impinging upon it would generally undergo small deviations from their courses; the frequent occurrence of large-angle deviations suggested direct collisions with a strongly concentrated positive substance. A quantitative check fully confirmed this inference and revealed that the massive, positively charged nucleus of the atom had linear dimensions a hundred thousand times smaller than those of the whole atomic structure.

Bohr eagerly took up the new model and soon recognized its far-reaching implications. In particular, he pointed out that the nuclear model of the atom implied a sharp separation between the chemical properties, ascribed to the peripheral electrons, and the radioactive properties, which affected the nucleus itself. This immediately suggested a close relation between the atomic number, which indicates the position of an element in Mendeleev's periodic table, and the number of its electrons, or its nuclear charge, which should thus be more fundamental than its atomic weight. Indeed, the periodic table showed one or two irregularities in the sequence of atomic weights, and it became increasingly difficult to accommodate in it the newly discovered radioactive products; Bohr showed how all these anomalies could be eliminated if one admitted the occurrence of atomic nuclei of the same charge but different mass, so that there could be more than one species of atom occupying the same place in the periodic table. Somewhat later, the name "isotope" was given to these chemically indistinguishable atomic species of different weights.

According to the nuclear model, radioactive transformations had to be conceived as actual transmutations of the atomic nucleus. Thus, Bohr argued, by the emission of an α ray, the nucleus lost two units of charge and became an isotope of the element two places back in the periodic table. In β decay, on the other hand, the emission of an electron resulted in the gain of one unit of charge, and the product nucleus occupied the next higher place in the periodic table. Simple as it may seem, the inference leading to these "displacement laws" of radioactive elements was far from obvious at that time.

The only person in the laboratory who followed Bohr's thoughts with deep interest and genuine understanding, and who was able to help him in the discussion of the empirical information, was a young Hungarian chemist, Georg von Hevesy, who was himself on the verge of discovering the use of isotopes as tracers, which brought him fame. Indeed, Rutherford himself, insensible to the logical cogency of Bohr's argument, dissuaded him from publishing such hazardous deductions from his own atomic model, to which he was not prepared to ascribe the fundamental significance that Bohr gave it; and when, a few months later, the displacement laws could be discerned by mere inspection of the accumulated experimental evidence, Kasimir Fajans (one of those who then enunciated them) so little understood their meaning that he actually presented them as evidence against the Rutherford atomic model.

Bohr's survey of the implications of Rutherford's atomic model did not stop at the recognition of the existence of a relation between the atomic number (which summarizes the whole physicochemical behavior of the element) and the number of electrons

in the atom. He resolutely attacked the much harder problem of determining the exact nature of this relation, which amounts to a dynamic analysis of the atomic structure represented by the nuclear model. Following J. J. Thomson's example, Bohr assumed that the electrons would be symmetrically distributed around the nucleus in concentric circular rings. He had then to face the problem, not present in Thomson's model, of how to account for the stability of such ring configurations, which could not be maintained by the electrostatic forces alone.

Bohr had become convinced, from his study of the behavior of electrons in metals, that the validity of classical electrodynamics would be subject to a fundamental limitation in the atomic domain, and he had no doubt that this limitation would somehow be governed by Planck's quantum of action; he knew already how to quantize the motion of a harmonic oscillator, i.e., to select from the infinity of possible motions a discrete series characterized by energy values increasing by finite steps of magnitude $h\nu$, where h is Planck's universal constant and ν the frequency of the oscillator. One could try to apply a similar quantization to the motions of an atom's electrons, whose frequencies might be identified with the resonance frequencies observed in the scattering of light by the atom.

Thus, an allowed state of motion characterized by a frequency ω_n would have a binding energy of the form $W_n = Knh\omega_n$, where n is an integer numbering the state and K is some numerical factor that could possibly depend on the type of motion. Such a formula could be combined with the relation given by the classical theory between the binding energy and the amplitude of the motion, in order to obtain a relation between the amplitude of motion, whose order of magnitude is known from various evidence about the atomic dimensions, and the corresponding resonance frequency, which is obtained from optical measurements. It was easy to ascertain that the numerical value of Planck's constant, entering such a relation, did lead to the expected orders of magnitude; but this rough check, however encouraging, was clearly insufficient to establish the precise form of the quantum condition.

At this juncture, Bohr obtained a much deeper insight into the problem by a brilliant piece of work, which he—working, as he said, "day and night"—completed with astonishing speed. The problem was one of immediate interest for Rutherford's laboratory: in their passage through a material medium, α particles continually lose energy by ionizing the atoms they encounter, at a rate depending on their velocity. This energy loss limits the depth to which the particles can penetrate into the medium, and the relation between this depth, or range, and the velocity offers a way of determining this velocity. What Bohr did was to analyze the ionizing process on the basis of the Rutherford model of the atom and thus express the rate of energy loss in terms of the velocity by a much more accurate formula than had so far been achieved—a formula, in fact, to which modern quantum mechanics adds only nonessential refinements.

Bohr's interest in atomic collision problems never faltered. In the early 1930's, when the modern theory of these processes was being elaborated, especially by Hans Bethe, Felix Bloch, and E. J. Williams, he took an active part in the work, a good deal of which took place in Copenhagen; and as late as 1948 he wrote a masterly synthesis of the whole subject, in which one still finds, in modernized form, the arguments of his early analysis.

The success of this analysis showed him, however, that the classical theory, while completely failing to account for the stability of the periodic motions of the atomic electrons, could deal with undiminished power with the aperiodic motions of charged particles traversing a region in which there is an electric field. This means that, however radical the break with classical ideas implied by the existence of the quantum of action, one must expect a gradual merging of the quantum theory into the classical one for motions of lower and lower frequencies. Moreover, one may expect that the effect of a very slow and gradual modification of the forces acting on or within an atomic system will be correctly estimated by the classical theory.

These were shrewd points, which Bohr used skillfully and which he eventually developed into powerful heuristic principles. An immediate application of the second principle helped him to discuss simple models of atomic and molecular structures, which reproduced, at least in order of magnitude, a number of features derived from various experiments and thus further illustrated the fruitfulness of the Rutherford atomic model. Indeed, this model was the first to permit a clear-cut distinction to be made between atom and molecule—a molecule being defined as a system with more than one nucleus—and thereby to open the way to an understanding of the nature of chemical binding. The models studied by Bohr were characterized by the arrangement of the electrons in one or more ring configurations, disposed around the nucleus as the common center in an atom, or symmetrically with respect to the nuclei in molecules. While the absolute dimensions of these configurations depended on quantum conditions that he could only roughly guess, their stability, owing to the argument

mentioned above, could be examined by classical methods; thus, he could explain why hydrogen could form a diatomic molecule, while helium could not.

Although these considerations were crude—and are completely superseded by the modern conceptions—they were remarkably successful; in fact, they do embody an important feature of the chemical bond that is part of the modern theory: the fact that this bond is due to the formation of a configuration of electrons shared by the combining atoms. The hydrogen molecule, for instance, was well represented by a ring of two electrons perpendicular to the line joining the two nuclei.

With regard to the determination of the states of motion allowed by the quantum condition mentioned above, Bohr found that the Rutherford model leads to remarkably simple results, at least for the type of configuration he considered. In general, the classical theory of the motion furnishes an additional relation between the binding energy and the frequency, which allows one to eliminate the frequency from the quantum condition and thus obtain for the binding energy W_n an expression depending only on the integer n, with a coefficient that, besides Planck's constant, contains the parameters characterizing the system and the type of motion. Thus, to take the simplest example of the hydrogen atom, consisting of a singly charged nucleus and an electron of mass m and charge e, the classical theory shows that there is proportionality between W_n^3 and ω_n^2; this leads, for the allowed states of binding, to the very simple law $W_n = A/n^2$, and the precise value of the coefficient A is $\pi^2 e^4 m/2K^2 h^2$; only the numerical factor K remains in doubt.

When he left Manchester in July 1912, Bohr was filled with ideas and projects for further exploration of this world of atoms that was displaying such wide prospects; but he had another reason to be in high spirits. Since 1911, shortly before his departure for England, he had been engaged to Margrethe Nørlund, a young woman of great charm and sensibility. The marriage took place in Copenhagen on 1 August 1912 and was a happy and harmonious union. Margrethe's role was not an easy one. Bohr was of a sensitive nature, and constantly needed the stimulus of sympathy and understanding. When children came —six sons, two of whom died young—Bohr took very seriously his duties as paterfamilias. His wife adapted herself without apparent effort to the part of hostess, and evenings at the Bohr home were distinguished by warm cordiality and exhilarating conversation.

In the autumn of 1912, Bohr took up the duties of assistant at the University of Copenhagen; he fulfilled them conscientiously, and used the privilege extended to holders of the doctorate of giving a free course of lectures. At the same time, he started to write up the account of his Manchester ideas. Then, at the beginning of 1913, the orientation of his thought took a sudden turn toward the problem of atomic radiation, which rapidly led him to the decisive step in the process of incorporating the quantum of action into the theory of atomic constitution. The rest of the academic year was spent reconstructing the whole theory upon the new foundation and expounding it in a large treatise, which was immediately published, in three parts, in the *Philosophical Magazine*.

It had been known since Kirchhoff's pioneering work that the spectral composition of the light emitted by atoms is characteristic for the chemical species; a whole science of spectroscopy had developed on this principle and a great deal of extremely accurate material had been accumulated. Obviously, the tables of wavelengths of the characteristic spectral lines must contain very precise information on the structure of the emitting atoms; but since atomic spectra consist of apparently capricious sequences of thousands of lines, it seemed hopeless to try to decipher such complicated codes. It therefore came as a great surprise to Bohr to learn from a casual conversation with a colleague that spectroscopists had managed to discover regularities behind the chaos.

In particular, J. R. Rydberg, of the nearby University of Lund, had found a very simple and remarkable formula expressing the frequencies of several "series" of spectral lines which recurred, with different values of the parameters, in the spectra of different atoms. The striking feature of Rydberg's formula was that the frequencies were represented by differences of two terms, each of which depended in a simple way on a number which could take a sequence of integral values; a series corresponded to the sequence obtained by keeping one of the terms fixed and varying the other.

Thus, the frequencies ν_{nm} of the lines of the hydrogen spectrum could be represented in the simplest possible form in terms of two integers as

$$\nu_{nm} = R\left(\frac{1}{n^2} - \frac{1}{m^2}\right),$$

with a single parameter, R, of accurately known numerical value. As soon as Bohr saw this formula, he immediately recognized that it gave him the missing clue to the correct way to introduce the quantum of action into the description of atomic systems.

The formal similarity between the terms of the Rydberg formula R/n^2 and the expression for the energies $W_n = A/n^2$ of the possible stationary states of the atom suggested to him, in the spirit of Planck's conception of the quanta of radiation, that the emission by the atom of light of frequency ν_{nm} occurred in the form of single quanta of energy $h\nu_{nm}$; Rydberg's formula then indicated that in this process the atom passed from an initial stationary state W_n to another stationary state, W_m. An immediate control of this interpretation offered itself: according to it, the value of Rydberg's constant should be given by $Rh = A$, that is, by $R = \pi^2 e^4 m/2K^2 h^3$. Inserting in this expression the known values of e, m, and h, and taking the value $1/2$ for K (which would give the correct binding energy W_n for a harmonic oscillator of frequency ω_n), Bohr obtained a value of R as near the experimental one as the errors in the determinations of the other constants allowed.

However convincing such a stringent quantitative test could appear, there was in this new conception of the radiation process a feature that must be considered so unusual as to be almost unthinkable: the frequencies ν_{nm} of the emitted light did not coincide with any of the allowed frequencies of revolution ω_n of the electrons or their harmonics—a feature of the classical theory of radiation so immediate and elementary that it seemed impossible to abandon it.

That Bohr was not deterred by this consideration was due essentially to the dialectical turn of mind he had acquired in his youthful philosophical reflections. The conflict between the classical picture of the atomic phenomena and their quantal features was so acute that no hopes (such as those Planck was still expressing) could be entertained of solving it by reducing the latter to the former; one had, rather, to accept the coexistence of these two aspects of experience, and the real problem was to integrate them into a rational synthesis. Bohr later said that the clue offered by Rydberg's formula was so transparent as to lead uniquely to the quantal description of the radiation process he proposed; this gave him the conviction that it was right, in spite of the radical break with classical ideas that it implied.

In order to clinch the argument, however, Bohr went a very important step further. He knew that the quantal behavior of a system, whatever it was, had to satisfy the requirement of going over to the corresponding classical behavior in the limiting case of motions involving large numbers of quanta of action. Applying this test to his interpretation of Rydberg's formula, Bohr found that the condition could be fulfilled only by ascribing the value $1/2$ to the numerical coefficient K, for which the right value of Rydberg's constant was obtained. Indeed, for large values of the number n, the frequencies $\nu_{n,n+p}$ are then seen to tend to the values of the frequency of revolution, $\omega_n = 2R/n^3$, and its successive harmonics, $p\omega_n$. Thus, as Bohr expressed it, "the most beautiful analogy" was established—in the sense just indicated—between classical electrodynamics and the quantum theory of radiation.

In his great papers of 1913, Bohr presented his theory as being founded upon two postulates, whose formulation he refined in later papers. The first postulate enunciates the existence of stationary states of an atomic system, the behavior of which may be described in terms of classical mechanics; the second postulate states that the transition of the system from one stationary state to another is a nonclassical process, accompanied by the emission or absorption of one quantum of homogeneous radiation, whose frequency is connected with its energy by Planck's equation. As for the principle by which the possible stationary states are selected, Bohr was still very far from a general formulation; indeed, he was keenly aware of the necessity of extending the investigation to configurations other than the simple ones to which he had restricted himself. The search for sufficiently general quantum conditions defining the stationary states of atomic systems was going to be a major problem in the following period of development of the theory.

A statement in Bohr's first paper gave rise to a controversy that soon ended in triumph for the new theory and in no small degree contributed to its swift acceptance. On the strength of his interpretation of Rydberg's formula, Bohr had pointed out that a certain series of lines attributed to hydrogen ought actually to be ascribed to helium: it had been fitted to the formula for hydrogen with half-integral values of the numbers n,m; in Bohr's view, which required integral values for these numbers, this could only mean that the Rydberg constant for this series was four times that for hydrogen, corresponding to a doubly charged nucleus. The experienced spectroscopist Alfred Fowler received the suggestion with understandable skepticism, but control experiments, which were at once performed in Rutherford's laboratory, confirmed Bohr's prediction. Fowler's last-ditch resistance, in the form of the pointed objection that Rydberg's coefficient for the contested series was not exactly $4R$ (R being the hydrogen value), was brilliantly countered by Bohr: he showed that the slight difference was to be expected as an effect of the motion of the nucleus, which he had neglected in his first approximation.

There is no doubt that this dramatic incident was decisive in convincing Rutherford and Fowler that there was something after all in this young foreigner's

theorizing. This was also James Jeans's attitude when, in the report of Bohr's work he gave at the British Association meeting at Birmingham in September 1913, he pointed out that the only justification of Bohr's postulates "is the very weighty one of success." At Göttingen, that center of mathematics and physics, where the sense of propriety was strong, the prevailing impression was one of scandal, or at least bewilderment, in the face of the undeserved success of such high-handed disregard of the canons of formal logic; but the significance of Bohr's ideas did not escape those who had themselves most searchingly pondered the problems of quantum theory, Albert Einstein and Arnold Sommerfeld.

No one realized more keenly than Bohr himself the provisional character of his first conclusions, and above all the need for a deeper analysis of the logical relationship between the classical and quantal aspects of the atomic phenomena that were embodied in the two postulates. At the same time, he was faced with an overwhelming program of generalizing the theory and unfolding all its consequences. He was naturally more and more dissatisfied with his job at the university, which left him little time for research and (since he had mainly to deal with medical students) little hope of turning out pupils able to assist him in his work.

The academic authorities were slow in realizing that an exceptional situation had arisen, and when Rutherford offered him a lectureship in Manchester, Bohr was glad to avail himself of the opportunity to pursue his work under the most favorable conditions. He remained in Manchester for two years. In the meantime, the Danish authorities had moved to offer Bohr a professorship, which he accepted; and three years later, thanks to the active intervention of a group of friends, who donated the ground, they were at last persuaded to build Bohr a laboratory: this was the famous Institute for Theoretical Physics, of which he was director for the rest of his life. The founding of the institute came just in time to keep Bohr in his native country, for Rutherford, who had just been called to the directorship of the Cavendish Laboratory in Cambridge, had already invited Bohr to join him.

The new institute was meant to be primarily a physical laboratory; what was termed "theoretical physics" would now be called "fundamental physics." Bohr did not draw any sharp distinction between theoretical and experimental research; on the contrary, he visualized these two aspects of research as supporting and inspiring each other, and he wanted the laboratory equipped so as to make it possible to test new theoretical developments or conjectures by appropriate experiments. He managed to put this conception into effect; the experimental investigations carried out at the institute have not been numerous, but have always been of high quality—some of them, indeed, of pioneering importance—and all have been directly relevant to the theoretical questions under consideration. In order to keep up with the changing outlook of current theory, it was imperative to expand and even to renew the experimental equipment in order to adapt it to entirely new lines of research; this Bohr did with remarkable foresight as well as persuasive tenacity in securing the necessary funds.

Bohr's atomic theory inaugurated two of the most adventurous decades in the history of science, a period in which the efforts of the elite among the younger generation of physicists were concentrated on the numerous problems raised by the theory and on experimental investigations that further stimulated the theoretical developments or provided the required proof of theoretical predictions. Three experimental advances that furthered the progress of the theory were made as early as 1913 and 1914. The domain of X-ray spectroscopy was opened up by H. G. J. Moseley's brilliant work in Manchester, and its significance for atomic theory, on the basis of Bohr's ideas, was pointed out by Walther Kossel. The experiments of James Franck and Gustav Hertz on the excitation of radiation from atoms by collisions with electrons, and those of Johannes Stark on the modification of the atomic spectra by strong electric fields, offered a new approach to the study of the dynamical behavior of atomic systems; their interpretation was soon outlined by Bohr himself.

Optical spectroscopy, whose importance had been suddenly enhanced, was actively developed, especially by the school established at Tübingen under Friedrich Paschen's leadership; with his collaborators Ernst Back, Alfred Landé, and others, Paschen analyzed in great detail the fine structure of the line spectra and the further splitting of the lines under the action of magnetic fields of increasing strength, and he formulated the regularities obeyed by the frequencies and intensities of the lines in terms of sets of quantum numbers attached to the spectroscopic terms and taking integral or half-integral values.

On the theoretical side, too, the scene was rapidly changing. The isolation in which Bohr had hitherto found himself gave way to a lively collaboration with a growing number of fellow workers all striving toward the common goal, freely exchanging ideas, discussing results and conjectures, sharing the thrill of success and the expectation of further progress. By tacit consent, Bohr was the leader to whom all turned for guidance and inspiration. There were other great schools of theoretical physics, the foremost being

those newly established by Sommerfeld in Munich and by Max Born in Göttingen; they pursued their own lines of research, always keeping in close contact with the Copenhagen group. The first to join Bohr in Copenhagen was a young Dutchman, H. A. Kramers, who arrived in 1916 and for the next ten years was Bohr's tireless assistant and talented collaborator. During this period, many others came to Bohr's institute; among them was Bohr's faithful friend Hevesy, as well as younger men—Oskar Klein, Wolfgang Pauli, and Werner Heisenberg.

The first of the main problems requiring consideration was the generalization of the quantum conditions defining the stationary states. Bohr did not at first attempt to make use of the general methods of classical mechanics; this was not his way of tackling problems. He preferred to handle concrete cases and to develop ingenious arguments which, although lacking generality, had the advantage of clearly bringing out the physical features of essential importance. In the present instance, he again started from the premise that slow deformations of a system would not change its quantal state, and developed it into a principle of mechanical transformability, which proved quite efficient within a limited scope. The idea was to transform one type of motion continuously into another by slow variation of some parameter; if the determination of the stationary states had been accomplished for one of the two motions, one could derive stationary states, by such a transformation, for the other. To this end, one could take advantage of the existence of dynamical quantities, the adiabatic invariants, which have the property of remaining unchanged under slow mechanical transformations.

As early as 1911, Paul Ehrenfest had emphasized the important role played by adiabatic invariants in the quantum theory of radiation in thermodynamic equilibrium; but neither he nor Bohr at first succeeded in extending this conception to modes of motion more complicated than simple periodic ones. Decisive progress in this problem was made by Sommerfeld, who at the end of 1915 succeeded in formulating a full set of quantum conditions for the general Keplerian motion, including even the relativistic precession of the elliptic trajectory. Sommerfeld's work not only supplied an explanation (a partial one as it turned out) of the fine doublet structure of the lines of the hydrogen spectrum, but showed the way to the desired generalization of the rules of quantization to more complex atomic systems, whose motions were not simply periodic.

Bohr eagerly followed this new line of attack; he now made full use of the powerful methods of Hamiltonian dynamics, especially in the form adapted to the wide class of motions known as multiply periodic, to which the motions of the electrons in atoms belonged. It was fortunate that Kramers, skilled in the relevant techniques, was at hand to help him; even so, it took years of strenuous effort to bring the work to completion. In their general form, the quantum conditions stated that a certain set of adiabatic invariants should be integral multiples of Planck's constant; but in the process of establishing this result, a formidable hurdle was the occurrence of "degeneracies" of the motions into simple periodic ones, which led to ambiguities in the formulation of the corresponding quantum conditions. This difficulty was eventually overcome by another ingenious application of the principle of mechanical transformability.

The theory of multiply periodic systems offered the possibility of a more rational treatment of the question which Bohr had tackled in his very first reflections on the nuclear atomic model: the gradual building up of atoms of increasing complexity and the origin of the periodicities in the atomic structures revealed by Mendeleev's table. The starting point was the consideration of the individual stationary orbits of each single electron in the electrostatic field of the nucleus, "screened" by the average field of the other electrons; the residual interaction of the electrons could then be treated by the perturbation methods originally developed for use by astronomers. For those spectra originating from quantum transitions of a single electron, usually the most weakly bound one, the quantum conditions provided a characterization directly comparable with the specification of the spectroscopic terms by quantum numbers.

The confrontation of the theory with the relevant spectroscopic evidence led to partial success: the main features of the empirical term sequences were well reproduced by the theory, and the spectroscopic quantum numbers on which these features depended accordingly acquired a simple mechanical interpretation (except for the occurrence of half-integral values, which appeared as an arbitrary modification of the quantum conditions); but the finer structure of the term sequences presented a complexity for which the atomic model offered no mechanical counterpart.

In spite of this imperfection, the model could be expected to give reliable guidance at least in the investigation of the broader outlines of atomic structures. The primitive ring configurations of Bohr's previous attempt were now replaced by groupings of individual electron orbits in "shells" specified by definite sets of quantum numbers, according to rules that were inferred from the spectroscopic data. This conception of the shell structure of atomic systems did

not merely account for the main classification of the stationary states; its scope could be extended to include the interpretation of the empirical rules established by the spectroscopists for the intensities of the quantal transitions between these states. This was a much more difficult problem than that of the formulation of quantum conditions for the stationary states; the complete breakdown of classical electrodynamics, reflected in Bohr's quantum postulates, seemed at first to remove the very foundation on which a comprehensive theory of atomic radiation could rest. It was in taking up this challenge that Bohr was led to one of his most powerful conceptions: the idea of a general correspondence between the classical and the quantal descriptions of the atomic phenomena.

Bohr seized upon the only link between the emission of light in a quantal transition and the classical process of radiation: the requirement that the classical description should be valid in the limiting case of transitions between states with very large quantum numbers. If the atom were treated as a multiply periodic system, its states of motion could be represented as superpositions of harmonic oscillations of specified frequencies and their integral multiples, each occurring with a definite amplitude; it was indeed possible to verify that the frequencies of quantal transitions between states of very large quantum numbers tended to become equal to those multiples of the classical frequencies given by the differences between these quantum numbers; in the limit of large quantum numbers, then, the classical amplitudes could be used directly to calculate the intensities of the quantal transitions. Bohr boldly postulated that such a correspondence should persist, at least approximately, even for transitions between states of small quantum numbers; in other words, the amplitudes of the harmonics of the classical motion should in all cases give an estimate of the corresponding quantal amplitudes.

The power of this correspondence argument was immediately illustrated by the application Kramers made of it, in a brilliant paper, to the splitting of the hydrogen lines in an electric field. Not only did the correspondence argument, for want of a more precise formulation, play an indispensable part in the interpretation of the spectroscopic data, but it eventually gave the decisive clue to the mathematical structure of a consistent quantum mechanics.

By 1918 Bohr had visualized, at least in outline, the whole theory of atomic phenomena, whose main points have been presented in the preceding sections. He of course realized that he was still very far from a logically consistent framework wide enough to incorporate both the quantum postulates and those aspects of classical mechanics and electrodynamics that seemed to retain some validity. Nevertheless, he at once started writing up a synthetic exposition of his arguments and of all the evidence upon which they could have any bearing; in testing how well he could summarize what was known, he found occasion to check the soundness of his ideas and to improve their formulation. In the present case, however, he could hardly keep pace with the growth of the subject; the paper he had in mind at the beginning developed into a four-part treatise, "On the Theory of Line Spectra," publication of which dragged over four years without being completed; the first three parts appeared between 1918 and 1922, and the fourth, unfortunately, was never published. Thus, the full impact of Bohr's views remained confined to the small but brilliant circle of his disciples, who indeed managed better than their master to make them more widely known by the prompter publication of their own results.

Bohr's theory of the periodic system of the elements, based essentially on the analysis of the evidence of the spectra, renewed the science of chemistry by putting at the chemists' disposal rational spectroscopic methods much more refined than the traditional ones. This was dramatically illustrated in 1922, by the identification, at Bohr's institute, of the element with atomic number 72. This discovery was made by Dirk Coster and Hevesy, under the direct guidance of Bohr's theoretical predictions of the properties of this element; they gave it the name "hafnium," from the latinized name of Copenhagen. The conclusive results were obtained just in time to be announced by Bohr in the address he delivered when he received the Nobel Prize in physics for that year.

There was never any question of Bohr's resting on his well-deserved laurels. He did not allow the apparent triumph of the quantum theory of atomic systems to mislead him into believing that the model used to describe these systems—simple point charges interacting by electrostatic forces according to the laws of classical mechanics—bore any close resemblance to reality. In fact, the fine structure of the spectroscopic classification manifested an essential insufficiency of this model, whose nature was not yet elucidated; but above all, the peculiar character of the correspondence between the quantal radiation processes and their classical counterpart strongly suggested that the classical model was no more than an auxiliary framework in the application of quantum conditions and correspondence considerations.

After Kramers had succeeded in extending the scope of the correspondence argument to the theory of optical dispersion—thus rounding off a treatment of the interaction of atomic systems with radiation that accounted for all emission, absorption, and scattering processes—Bohr ventured to propose a systematic formulation of the whole theory, in which what he called the virtual character of the classical model was emphasized. In this he was aided by Kramers and a young American visitor, J. C. Slater, and the new theory was published in 1924 under the authorship of all three. The most striking feature of this remarkable paper, "The Quantum Theory of Radiation," was the renunciation of the classical form of causality in favor of a purely statistical description. Even the distribution of energy and momentum between the radiation field and the "virtual oscillators" constituting the atomic systems was assumed to be statistical, the conservation laws being fulfilled only on the average. This was going too far: the paper was hardly in print before A. H. Compton and A. W. Simon had established by direct experiment the strict conservation of energy and momentum in an individual process of interaction between atom and radiation. Nevertheless, this short-lived attempt exerted a profound influence on the course of events; what remained after its failure was the conviction that the classical mode of description of the atomic processes had to be entirely relinquished.

This conviction was strengthened by the outcome of the other line of investigation most actively pursued in Copenhagen in these years, the search for the missing dynamic element of the atomic model. Pauli approached this arduous problem by trying to unravel the spectroscopic rules governing the fine structure of the terms and the splitting of the spectral lines in an external magnetic field—the anomalous Zeeman effect. He at length recognized that the entire problem could be simplified by attributing to the individual stationary states of each electron an additional quantum number, susceptible to two values only and combining with the other quantum numbers according to definite rules.

This conclusion at once threw light on the systematics of the shell structure of the elements, which Bohr had left incomplete, but which had lately been improved by E. C. Stoner. In fact, Pauli was able, in 1925, to formulate the simple underlying principle of this systematics: each stationary state—including the specification of the new quantum number—cannot be occupied by more than one electron. This exclusion principle has since received considerable extension, and has in fact turned out to be one of the most fundamental in nature. In the same year, decisive progress was made in the interpretation of the new quantum number by two of Ehrenfest's young pupils, S. A. Goudsmit and G. E. Uhlenbeck: they pointed out that the new quantum number could be ascribed to a proper rotation, or spin, of the electron, and that an intrinsic magnetic moment, related to the spin, could then account for the anomalous Zeeman effect. However, the quantization of the spin was at variance with that expressed by the quantum conditions; this circumstance, as well as the exclusion principle, which obviously was quite unaccountable in classical terms, showed in the most striking fashion that not only the radiation field but also the atomic constituents were out of reach of the conceptions of classical physics.

The crisis to which the attempt to treat the atom as a classical dynamic system had led did not last long. By the summer of 1925 Heisenberg had found the clue to the construction of a consistent mathematical scheme embodying the quantum postulates. This momentous progress was the direct outcome of the investigation of the optical dispersion theory initiated by Kramers. Heisenberg had taken an active part in this work and had been much impressed by the stand taken by Bohr, Kramers, and Slater. If classical conceptions could no longer be relied upon to supply at least a framework for the quantum theory, he concluded, what must be looked for is an abstract formal scheme expressing only relations between directly observable quantities, like the stationary states and the amplitudes whose absolute squares should express the probabilities of quantal transitions between these states. The correspondence between classical and quantal amplitudes established in the theory of dispersion, envisaged from this point of view, took the shape of a set of algebraic rules that these quantal amplitudes had to obey and that defined an algorism adapted to the rational formulation of laws of motion and quantum conditions, as well as the precise calculation of radiation amplitudes.

Heisenberg's program was eagerly taken up in Göttingen, where Born immediately recognized that the noncommutative algebra involved in Heisenberg's relations was the matrix calculus; at the same time, a young Cambridge physicist, P. A. M. Dirac, was developing even more abstract and elegant methods. While in the high places of mathematics the formal scheme of the new quantum mechanics was thus being built up, a more critical attitude prevailed in Copenhagen. Pauli pointed out that by limiting the observable quantities to stationary states and radiation amplitudes, Heisenberg was unduly restricting the scope of the theory, since it was an essential part of the correspondence argument that the new theory should contain as limiting case, for large quantum

numbers, the more detailed description of the motion in classical terms.

The fulfillment of this essential requirement necessitated a considerable extension of the mathematical framework of the theory, allowing it to accommodate both discontinuous and continuous aspects of the atomic phenomena. The decisive contribution was unexpectedly made by the "outsiders," Louis de Broglie and Erwin Schrödinger, who were exploring the conjecture that the constituents of matter might, like radiation, be governed by a law of propagation of continuous wave fields.

Although the idea in this one-sided form was at once seen to be untenable, it nevertheless provided the missing element; as Born especially emphasized, the wave fields associated with the particles give the probability distributions of the variables specifying the state of motion of these particles. Thus, the required formal completion of quantum mechanics could be carried out at the beginning of 1927, when Dirac indicated the most general representation of the operators belonging to the physical quantities, and the way to pass at will from any representation to any other according to definite prescriptions which guaranteed the fulfillment of all correspondence requirements. However, such classical features of the motion of particles as a sequence of positions forming a uniquely determined trajectory appeared only as a limiting case of a more general mode of description that was essentially statistical.

The quantum conditions were found to impose a peculiar restriction on the statistical distributions of the values of physical quantities. If, as a consequence of these conditions, the operators representing two such quantities do not commute, the average spreads in the assignment of the values they may take under given circumstances are reciprocal; their product exceeds a limit that depends on the degree of noncommutation and is proportional to Planck's constant. Thus, if in definite experimental circumstances the position of an electron, relative to some fixed frame of reference, is confined within narrow limits, its momentum will have a correspondingly wide range of possible values, each with its definite probability of occurrence, depending on the experimental conditions.

Heisenberg, who in 1927 discovered these remarkable indeterminacy relations, realized their epistemological significance. In fact, the novelty of quantum mechanics in this respect is that it allows for the possibility of using all classical concepts, even though their precise determinations may be mutually exclusive—as is the case with the concept of a particle localized at a point in space and time, and that of a wave field of precisely given momentum and energy, whose space-time extension is infinite. Indeterminacy relations between such concepts, then, indicate to what extent they may be used concurrently in statistical statements. Heisenberg saw that the origin of these reciprocal limitations must lie in quantal features of the processes in which the quantities in question are observable, and he attempted to analyze such idealized processes of observation from this point of view.

This was the occasion for Bohr to reenter the scene. His role so far had been to inspire and orient the creative efforts of the younger men, especially Heisenberg and Pauli, and he could legitimately consider the new theory as the attainment of the goal toward which he had so long been striving. On the one hand, the radical break with classical physical theories, which he had felt to be inescapable from the very beginning, was now formally accomplished by the substitution of abstract relations between operators for the simple numerical relations of classical physics. On the other hand, the abstract character of the new formalism made it at last possible to fulfill the requirement he had always emphasized: not to sacrifice any aspect of the phenomena, but to retain every element of the classical description within the limits suggested by experience.

The peculiar form of limitation of the validity of classical concepts expressed by the indeterminacy relations demanded a more thorough analysis than that which Heisenberg had initiated, however. For this challenging task Bohr was, of course, not unprepared. The occurrence of conflicting, yet equally indispensable, representations of the phenomena evoked the ambiguities of mental processes over which he had pondered in his student days. Now, however, similar dilemmas confronted him in an incomparably simpler form, for the description of atomic phenomena operated with only a few physical idealizations. Bohr hoped that the study of such a transparent case would lead him to a formulation of the epistemological situation that was sufficiently general to be applicable to the deeper problems of life and mind, and he devoted all his energy to it. Although he very soon was able to elucidate the essential features, he spent most of the following decade patiently refining the formulation of the fundamental ideas and exploring all their implications.

In any investigation of the scope of physical concepts, the method to follow is prescribed by the nature of the problem; one has to go back to the definition of the concepts by means of apparatus—real or idealized—suited to the measurement of the physical quantities they represent. The analysis of such measuring operations should then reveal any limitation in the use of these concepts resulting from the laws of physics. This had been the method followed by Einstein in establishing the relativity of simultaneity; the same method was followed by Heisenberg and Bohr to elucidate the indeterminacy relations. It emerged from Bohr's analysis that the decisive element brought in by the quantum of action is what he called the individual character of quantal processes: any such process—for instance, the emission of radiation by an atom—occurs as a whole; it is well defined only when it is completed, and it cannot be subdivided like the processes dealt with in classical physics, which involve immense numbers of quanta, into a sequence of gradual changes of the system.

In particular, the measurement of a physical quantity pertaining to an atomic system can be regarded as completed only when its result has been recorded as some permanent mark left upon a registering device. Such a recording cannot be performed without some irreversible loss of control of the quantal interaction between the atomic system and the apparatus. Thus, if we record the position of an electron by a spot on a rigidly fixed photographic plate, we lose the possibility of ascertaining the exchange of momentum between the electron and the plate. Conversely, an apparatus suited to the determination of the momentum of the electron must include a mobile part, completely disconnected from the rigid frame of spatial reference, whose position, when it exchanges momentum with the electron, therefore necessarily escapes our control. Here is the root of the mutual exclusion of the application of such concepts as position and momentum in the extreme case of their ideally precise determination. More generally, by relaxing the accuracy requirements, it is possible to limit the reciprocal exclusion to the extent indicated by the indeterminacy relations, thus allowing for the concurrent use of the two concepts in a description that is then necessarily statistical.

It thus appears that in order to reach full clarity in such a novel situation, the very notion of physical phenomenon is first of all in need of a more careful definition that embodies the individuality or wholeness typical of quantal processes. This is achieved by inserting in the definition the explicit specification of all the relevant experimental arrangement, including the recording devices. Between phenomena occurring under such strictly specified conditions of observation, there may then arise the type of mutual exclusion for which an indeterminacy relation is the formal expression.

It is this relationship of mutual exclusion between two phenomena that Bohr designated as complementarity; by this he wanted to stress that two complementary phenomena belong to aspects of our experience which, although mutually exclusive, are nevertheless indispensable for a full account of experience. The introduction of the notion of complementarity finally solved the problem of the consistent incorporation of the quantum of action into the conceptual framework of physics—the problem with which Bohr had struggled so long. Complementarity was not an arbitrary creation of Bohr's mind, but the precise expression, won after patient efforts demanding a tremendous concentration, of a state of affairs entirely grounded in nature's laws, one that, according to Bohr's familiar exhortation, had to be learned only from nature. It consecrated the recognition of a statistical form of causality as the only possible link between phenomena presenting quantal individuality, but made it plain that the statistical mode of description of quantum mechanics was perfectly adapted to these phenomena and gave an exhaustive account of all their observable aspects.

From the epistemological point of view, the discovery of the new type of logical relationship that complementarity represents is a major advance that radically changes our whole view of the role and meaning of science. In contrast with the nineteenth-century ideal of a description of the phenomena from which every reference to their observation would be eliminated, we now have the much wider and truer prospect of an account of the phenomena in which due regard is paid to the conditions under which they can actually be observed—thereby securing the full objectivity of the description, since the description is based on purely physical operations intelligible and verifiable by all observers. The role of the classical concepts in this description is obviously essential, since those concepts are the only ones adapted to our capabilities of observation and unambiguous communication.

In order to establish a link between these concepts and the behavior of atomic systems, we have to use measuring instruments composed—like ourselves—of large numbers of atoms, and this requirement unavoidably leads to complementary relations and a statistical type of causality. These are the main lines of the new structure of scientific thought that gradually unfolded itself as Bohr, with uncompromising consistency, pursued his epistemological analysis to its limits. That some of the greatest representatives of the type of physical thinking with which he was so decisively breaking refused to follow him is understandable; that Einstein should be among them was always a matter of surprise and regret to Bohr. On the other hand, the progress of his work owed much

to Einstein's opposition; indeed, its successive stages are marked by the refutation of Einstein's subtle objections. Bohr himself retraced the dramatic course of this long controversy in an article of 1949, which marks the nearest he ever came to a systematic exposition of his argumentation.

The role of complementarity in quantum mechanics is above all to provide a logical frame sufficiently wide to ensure the consistent application of classical concepts whose unrestricted use would lead to contradictions. Obviously, such a function is of universal scope, and an occasion soon presented itself to put its usefulness to the test. In the early 1930's the extension of the mathematical methods of quantum mechanics to electrodynamics was beset with considerable formal difficulties, which raised doubts regarding the possibility of upholding the concept of the electromagnetic field in quantum theory.

This was clearly a point of crucial importance, since it bore upon the fundamental issue of a possible limit to the validity of the correspondence argument, hitherto unchallenged. According to Bohr's point of view, one had to inquire whether every component of the electromagnetic field could, in principle, be measured with unlimited accuracy, and whether the measurements of more than one component were subject only to the reciprocal limitations resulting from their complementary relationships. Bohr took up this investigation, which occupied him and Leon Rosenfeld during most of the period from 1931 to 1933. He succeeded in devising idealized measuring procedures, satisfying all requirements of relativity, by means of which all consequences of the quantization of the electromagnetic field could be confirmed. In view of the significance of the issue at stake, this work had a wider repercussion than its immediate effect of establishing the consistency of quantum electrodynamics: it showed how essential a part Bohr's epistemological standpoint played in the conception of the quantum phenomena.

By the middle 1930's the main interest had shifted, in Copenhagen as elsewhere, to the rapidly expanding field of nuclear physics. On the theoretical side, the results of the experiments on the reactions induced by the impact of slow neutrons on nuclei, carried out by Enrico Fermi and his school at Rome, created a critical situation. In discussing the processes involving the impact of charged particles, α particles or protons, on a nucleus, it had been found sufficient to represent the effect of the forces acting between the nucleus and the impinging particle schematically by an attractive potential well extending over the volume of the nucleus; to this was added the repulsive electrostatic potential, forming a coulomb barrier around the nucleus. It was therefore natural to analyze the neutron reactions with the help of the same potential, without the coulomb barrier; and it was a surprise that this model did not even qualitatively account for the observed effects. In particular, it was impossible on this basis to understand the very large probabilities with which the capture of the neutron by the nucleus occurred for a sequence of resonance energies.

Faced with this puzzling problem, Bohr proceeded to look for cases of capture processes occurring under a simpler form than in the range of low energies, in which they appeared to be tied to resonance conditions. As it happened, he had only to return to James Chadwick's earliest experiments, performed with neutrons of higher energy; he noticed that the different reactions induced by these neutrons all occurred at any energy with about the same probability, whose order of magnitude indicated that almost every neutron hitting the nucleus was captured by it. This strikingly simple result suggested to him a reaction mechanism radically different from the distortion of neutron waves by a potential well; indeed, in contrast with the quantal character of the latter model, the analogy Bohr proposed was completely classical. He visualized the nucleus as an assembly of nucleons held together by short-range forces, and thus, in effect, behaving like the assembly of the molecules forming a droplet of liquid.

The energy of a particle impinging upon such a system of similar particles moving about and continually colliding with each other will be rapidly distributed among all of them, with the result that none has enough energy to leave the system: the impinging neutron is captured, and a "compound nucleus" is thus formed in a state of high excitation. This state will subsist during a time that is long on the nuclear scale, i.e., which corresponds to many crossings of the nuclear volume by any single nucleon. It will decay as soon as some random fluctuation in the energy distribution has concentrated a sufficient amount of energy on some nucleon, or group of nucleons, to allow it to escape, a process comparable to evaporation from the heated droplet. It was also easy to understand that the density of possible states of the compound nucleus would rapidly increase with the energy of excitation; this explained the absence of resonance effects at high energies as well as their presence in the low-energy range.

Bohr's "droplet model" of nuclear reactions, refined in various ways since it was proposed in 1936, still holds as the adequate mode of description of one of the most important types of nuclear processes. It is of course an idealized model, and its basic assumptions are not always sufficiently fulfilled to ensure its

validity. Thus, another type of reaction has been found to occur, in which the interaction of the impinging particle with a single mode of motion of the target nucleus leads directly to a transfer of energy large enough to complete the process, without formation of a compound nucleus; these "direct interaction" processes are successfully treated with the help of the old method of the potential well, in which provision is made for the possibility of capture by a formal trick imitating the way in which the absorption of light is taken into account in classical optics. Compound nucleus and "optical" potential have now shed all apparent opposition and are blended into a comprehensive theory.

The most important application of Bohr's theory was the interpretation of nuclear fission. This is a type of reaction that may be initiated by impact of a neutron on a very heavy nucleus: the compound nucleus formed by the capture of the neutron has so little stability that it can split into two fragments of about the same mass and charge. It was Otto Hahn's and Fritz Strassmann's chemical identification of such fragments as decay products of uranium under neutron bombardment that led O. R. Frisch and Lise Meitner to recognize that the fission mechanism was the only conceivable interpretation.

The first experiments actually showing the emission of the fragments were performed in Copenhagen by Frisch in January 1939. By then Bohr had left for the United States, where he had been invited to spend a few months at Princeton. It was on his departure that he had heard of Frisch's idea and project; during the voyage and shortly after his arrival, also in January 1939, he outlined the whole theory of the process. In the following months, this theory was refined and elaborated in great detail, with J. A. Wheeler's collaboration.

A point that at first seemed surprising was that such a splitting of the nucleus into two parts, obviously initiated by a relative oscillation of these parts with increasing amplitude, could occur with a probability comparable to that of more familiar processes, such as the emission of a γ ray, which results from a stable motion affecting only a very few nucleons. As Bohr pointed out, however, this is a direct consequence of the statistical law of energy distribution among the various modes of motion of the compound nucleus. It seemed harder to explain the differences in the efficacy of slow and fast neutrons in inducing fission in different nuclei, but Bohr solved this problem as soon as he was confronted with the experimental data. By one of his most brilliant feats of rigorous induction from experiment, he unraveled the complex case of uranium, concluding that only the rare isotope of mass number 235 was fissile by slow neutrons, while the abundant isotope of mass 238 was not; and he showed by a very simple argument that this difference of behavior was due solely to the fact that the numbers of neutrons in the two isotopes were odd and even, respectively.

The discovery that the highly unstable fission fragments emitted neutrons immediately raised the question of the possibility of a chain reaction leading to the liberation of huge amounts of energy of nuclear origin. The answer to this question was soon found, and, coming as it did at a critical moment in the social and political evolution of the world, the unfolding of its consequences was precipitated with unprecedented violence. If this was a fateful development in the history of mankind, it also deeply affected Bohr's individual fate.

The work with fission, continued after his return to Copenhagen during the first three years of World War II, was the last piece of research he carried to completion in the quiet and serene atmosphere he had himself done so much to create. Only much later, during the last two summers of his life, did he for a while manage to concentrate again on a phenomenon very near to those with which he had started his scientific career: the superconductivity of metals, in which the quantum of action manifests itself, so to speak, by macroscopic effects. He tried, without success, to put the somewhat abstract theory of these effects on a more physical basis. In 1943, however, he was dragged into the turmoil of the war, and when he later came back to Copenhagen, he had to cope with profoundly changed conditions of scientific work that banished from his institute the intimacy of bygone years.

Bohr did not fare well among statesmen. In their eyes his candor and directness appeared strange and suspicious, and his clearsightedness was beyond their grasp. The physicists who were desperately striving, under great moral and intellectual stress, toward the dark goal of the nuclear weapon, felt the need of calling Bohr to their support. Bohr was transported in 1943 from Copenhagen to England, through Sweden, not without danger to his life, and was suddenly faced, to his surprise and dismay, with the advanced stage of a project he had deemed beyond the realm of technical accomplishment. Although he did take part, both in England and in the United States, in discussions of the physical problems related to the development of nuclear weapons, his main concern was to make the statesmen, as well as the physicists, aware of the political and human implications of the new source of power.

It is a striking example of his optimism that, besides the obvious dangers, he also stressed the potential advantages of the situation: the existence of a weapon equally threatening to all nations, he argued, offered a unique opportunity for reaching a universal agreement never to use it, which could become the foundation of an era of lasting peace. The condition for setting up such an agreement, he added with his customary logic, was universal knowledge of the issue. More concretely, he urged the Western leaders to initiate contacts with the Russians, with the view of creating a climate conducive to the establishment of peaceful relations and mutual confidence between the West and the East. Although these thoughtful considerations were appreciated by some of the men in key positions, his attempts to put them before Roosevelt and Churchill ended in failure. The fulfillment of his darkest predictions in the following years did not prevent him from persevering, and in 1950 he decided to publish an open letter to the United Nations, in which he repeated his plea for an "open world" as a precondition for peace. The timing of such an appeal was the worst possible; but it is now as relevant as ever, and may still perhaps find a response some day.

Apart from this unhappy excursion into the realm of world politics, Bohr devoted much time and energy to the more immediate tasks he was called upon to fulfill. In Denmark, the expansion of his institute occupied him to the last, and he also took a leading part in the foundation and organization, in 1955, of a Danish establishment for the constructive application of nuclear energy. When the European Center for Nuclear Research was founded in 1952, its theoretical division was installed in Bohr's institute, until it could move nearer the experimental divisions at Geneva in 1957; it was then replaced in Copenhagen by a similar institution of more restricted scope, the Nordic Institute for Theoretical Atomic Physics, created with Bohr's participation by the five Nordic governments to accommodate young theoretical physicists from those countries. In these years of unprecedented expansion of scientific research all over the world, Bohr's advice and support were sought on many occasions, and never in vain. He was more than ever a public figure, and honors were conferred on him from every quarter.

Unaffected by this lionization, Bohr made the best of it. An invitation to give a lecture was the occasion for him to orient his thought toward the particular aspect of science that would be familiar to his audience, and to reflect on the possible bearing on it of the new epistemological conceptions he had developed in quantum theory. Thus, in the 1930's he had given a lecture entitled "Light and Life" before a congress of phototherapists, and had spoken of the complementary features of human cultures in an assembly of anthropologists. In the postwar period, he went on in this vein and expressed thoughts about the human condition which for him were inseparable from a proper understanding of the aim and meaning of science. His writings on such topics were collected in three books, published in 1934, 1958, and 1963. These have been translated into several languages, and one must hope that in spite of the difficulty of style they may exert the same influence on the philosophical attitude of coming generations as on the minds of those who heard Bohr himself.

In fact, the form of publication of Bohr's essays is not felicitous. The books contain some repetition, especially in elementary expositions of the physical background, and the main points are often suggested to the reader rather than plainly stated. Involved sentences try to embrace all the shades of an uncommonly subtle dialectical form of thinking. Such obstacles ought not to deter those who are genuinely concerned with the problems, but the unprepared audiences to whom the message was addressed have too often failed to appreciate its true character. Bohr put an enormous amount of work into the composition of his essays, and they contain the most carefully weighed expression of his philosophy.

Bohr's essays strikingly illustrate the continuity of his thought. He was striving all the time to find more precise formulations and to disclose new aspects of the complementary relationships he was exploring, but the basic conception remained the same in all essentials from his youth to his last days. Critics endeavoring to trace foreign influences on his thinking are quite misguided: he was no doubt interested when analogies were pointed out to him between his own conceptions and those of others, but such comparisons never led to any modification of his argumentation—this argumentation, in contrast with the other, was so solidly founded in the analysis of the clear and precise situation offered by the development of quantum theory that there was no need for any firmer foundation. Indeed, Bohr repeatedly stressed the fortunate circumstance that the simplicity of the physical issue made it possible for him to arrive at an adequate formulation of the relations of complementarity he perceived in all aspects of human knowledge.

The domain in which complementary situations manifest themselves most immediately is the realm of psychical phenomena—which had been the starting point for Bohr's early observations. He was now able to express in terms of complementarity the peculiar relation between the description of our emotions as revealed by our behavior and our consciousness of them; in such considerations he liked to imagine (on slender evidence, it must be said) that sayings of ancient philosophers and prophets were groping expressions for complementary aspects of human existence.

In the development of human societies, Bohr emphasized the dominant role of tradition over the complementary aspect of hereditary transmission in determining the essential elements of what we call culture; this he held in opposition to the racial theories then propagated in Germany. Nearer to physics, he pointed out that the two modes of description of biological phenomena which were usually put in absolute opposition to each other—the physical and chemical analysis on the one hand, the functional analysis on the other—ought to be regarded as complementary. Altogether, he saw in complementarity a rational means of avoiding the exclusion of any line of thought that had in any way proved fruitful, and of always keeping an open mind to new possibilities of development.

In his last years, he followed with the deepest satisfaction the spectacular advance of molecular biology. In the last essay he wrote, "Light and Life Revisited," he made it clear that in upholding the use of functional concepts in biology, he did not have in mind any insuperable limitation of the scope of the physical description; on the contrary, he saw in the recent progress the unlimited prospect of a full account of biological processes in physical terms, without prejudice to an equally full account of their functional aspect.

The origin of Bohr's epistemological ideas in a purely scientific situation confers on them the character of scientific soundness and certainty. Bohr was always careful to stress both the necessity, in epistemological investigations, of divesting oneself of all preconceived opinions and of seeking guidance exclusively in the data of experience and the equally stringent necessity of recognizing in every case the limitations inherent in the concepts used in the account of the phenomena. In order to understand the unique significance of his contribution to epistemology, it is necessary to realize that complementarity is a logical relationship, referring to our way of describing and communicating our experience of a universe in which we occupy the singular position of being at the same time, and inseparably, spectators and actors. Far from excluding any aspect of the universe from our reach, complementarity enables us, so far as we can judge, to account for all aspects of the phenomena—comprehensively, rationally, and objectively. By the rigor of his rational thinking, the universality of his outlook, and his deep humanity, Bohr ranks among the fortunate few to whom it has been given to help the human mind take a decisive step toward a fuller harmony with nature.

BIBLIOGRAPHY

This biography is based mainly on personal experience and conversations with Niels Bohr and his closest collaborators, as well as on the correspondence and documents in the Niels Bohr Archive in Copenhagen. Detailed biographical material is published in *Niels Bohr, His Life and Work as Seen by His Friends and Colleagues,* S. Rozental, ed. (Amsterdam, 1967). See also the report of the Niels Bohr Memorial Session held in Washington, D.C., on 22 April 1963 in *Physics Today,* **16,** no. 10 (Oct. 1963), 21–62; and an earlier essay of a more personal character: L. Rosenfeld, *Niels Bohr: An Essay* (Amsterdam, 1945; rev. ed., 1961). There is much autobiographical material in Bohr's Rutherford memorial lecture, "Reminiscences of the Founder of Nuclear Science and of Some Developments Based on His Work," in *Proceedings of the Physical Society of London,* **78** (1961), 1083–1115.

A full bibliography of Bohr's publications may be found in *Nuclear Physics,* **41** (1963), 7–12. The main items are the following: *Studier over metallernes elektrontheori* (Copenhagen, 1911); *On the Constitution of Atoms and Molecules,* papers of 1913 reprinted from the *Philosophical Magazine* with an introduction by L. Rosenfeld (Copenhagen, 1963); "On the Quantum Theory of Line Spectra," pts. I–III, in *Det Kgl. Danske Videnskabernes Selskabs Skrifter, naturvidenskabelig-matematisk Afdeling,* **4,** no. 1 (1918–1922); "The Structure of the Atoms," his Nobel lecture, in *Nature,* **112** (1923), 29–44; "The Quantum Theory of Radiation," with H. A. Kramers and J. C. Slater, in *Philosophical Magazine,* **47** (1924), 785–802; "Zur Frage der Messbarkeit der elektromagnetischen Feldgrössen," with L. Rosenfeld, in *Det Kgl. Danske Videnskabernes Selskab, matematisk-fysiske Meddelelser,* **12,** no. 8 (1933); "Neutron Capture and Nuclear Constitution," in *Nature,* **136** (1936), 344–348, 351; "The Mechanism of Nuclear Fission," with J. A. Wheeler, in *Physical Review,* **56** (1939), 426–450; and "The Penetration of Atomic Particles Through Matter," in *Det Kgl. Danske Videnskabernes Selskab, matematisk-fysiske Meddelelser,* **18,** no. 8 (1948).

The three volumes of collected essays are *Atomic Theory and the Description of Nature* (Cambridge, 1934; repr. 1961); *Atomic Physics and Human Knowledge* (New York, 1958); and *Essays 1958–1962 on Atomic Physics and Human Knowledge* (New York, 1963).

LEON ROSENFELD

BOISBAUDRAN, PAUL ÉMILE LECOQ DE (called **François**) (*b.* Cognac, France, 18 April 1838; *d.* Paris, France, 28 May 1912), *chemistry.*

Boisbaudran's family belonged to the ancient Protestant nobility of Poitou and Angoumois, and had been wealthy prior to the religious persecutions of the seventeenth and eighteenth centuries. When Boisbaudran was born, however, his father and uncle were coproprietors of a wine business at Cognac. His mother, daughter of an army officer, was a learned woman who taught her son classics, history, and foreign languages. Although he had no formal schooling, Boisbaudran worked through the course books of the École Polytechnique, performing experiments in a home laboratory equipped by his uncle.

At the age of twenty, Boisbaudran began to work for the family company, traveling through the Continent and England on business. He continued to study chemistry and physics in his spare time; as business prospered, he was allowed to spend more time on scientific work. His early research concerned supersaturation of salt solutions, conditions of crystallization, and crystalline shapes.

Boisbaudran is best known for his work on spectroscopic methods of elementary analysis. In a volume entitled *Spectres lumineux* (1874), he reported the results of extensive and refined spectral examinations of thirty-five elements. This work was undertaken to test several generalizations relating spectral wavelength to atomic weight. In this work, Boisbaudran held that the various kinds of spectra for the elements were related to the various motions (rotation, vibration, and translation) of the molecules. He believed, however, that the displacement of the lines in related elements did not correspond to the magnitude of the molecular forces (as some chemists held) but to the mass of the molecules.

In 1875 Boisbaudran spectroscopically discovered a new element, gallium, which he found in zinc blende from a mine in Hautes-Pyrénées. Continuing his work in Wurtz's laboratory in Paris, he was able to obtain the free metal by electrolysis of a solution of the hydroxide in potassium hydroxide. Gallium, Boisbaudran realized, was the "eka-aluminum" predicted by Mendeleev, and was the first of Mendeleev's predicted elements to be isolated. Boisbaudran's finding thus provided valuable evidence for the validity of Mendeleev's periodic classification of the elements.

In 1879 Boisbaudran began spectroscopic experimentation with the rare earth elements, research that he pursued for several decades. Collaborating with John Lawrence Smith, he showed that didymium from cerite differed from that coming from samarskite. This discrepancy led to the discovery of samarium. In 1885 the elements that have since been named dysprosium, terbium, and europium were identified from their phosphorescent spectra. Such phosphorescent bands were produced by making the liquid under consideration the positive pole, instead of the negative pole, when a line spectrum is formed. In 1886 a new element, later called gadolinium, was detected in earths yielding samarium. In 1904 Boisbaudran used "Z" to denote an element contained in earth separated from impure terbium and subsequently identified as pure terbium.

After 1895 Boisbaudran's scientific work decreased considerably because of failing health and family and business concerns. He married late in life, but no further information about his family can be found. Boisbaudran was a winner of the Cross of the Legion of Honor, the 1879 Davy Medal (for his discovery of gallium), and the Prix Lacaze. He was a corresponding member of the chemistry section of the French Academy of Sciences, and a foreign member of the Chemical Society of London.

BIBLIOGRAPHY

I. ORIGINAL WORKS. Among Boisbaudran's writings are *Spectres lumineux: Spectres prismatiques et enlonguers d'ondes destinés aux recherches de chimie minérale,* 2 vols. (Paris, 1874), and an article on gallium, in E. Fremy, ed., *Encyclopédie chimique,* XVI (Paris, 1884), 201–222. The discovery of gallium is reported in *Comptes rendus de l'Académie des sciences,* **81** (1875), 493–495; *Philosophical Magazine,* 5th ser., **2** (1876), 398–400; and *Annales de chimie et de physique,* **10** (1877), 100–141. A crucial work is *Analyse spectrale appliquée aux recherches de chimie minérale,* 2 vols. (Paris, 1923), written with A. Gramont, which contains a biographical sketch of Boisbaudran and a complete bibliography, pp. xi–liv; most of his publications are also listed in Poggendorff.

II. SECONDARY LITERATURE. Works on Boisbaudran are M. A. Gramont, "Lecoq de Boisbaudran: Son oeuvre et ses idées," in *Revue scientifique,* **51,** pt. 1 (25 Jan. 1913), 97–109; W. Ramsay, obituary, in *Journal of the Chemical Society,* **103** (1913), 742–746; Urbain, obituary, in *Chemische Zeitung,* **36** (1912), 923–933; and Mary Elvira Weeks, *Discovery of the Elements,* 6th ed. (Easton, Pa., 1956), esp. chs. 25, 26.

SUSAN G. SCHACHER

BOISLAURENT, FRANÇOIS BUDAN DE. See **Budan de Boislaurent, François.**

BOLK, LODEWIJK, usually called **LOUIS** (*b.* Overschie, the Netherlands, 10 December 1866; *d.* Amsterdam, the Netherlands, 17 June 1930), *anatomy.*

Bolk's parents wanted him to study for the ministry, but he preferred to attend medical school, which they could not afford. In 1888 he managed, with financial aid, to matriculate as a medical student at Amsterdam University, where he became an assistant to Georg Ruge, professor of human anatomy, after he had passed his final medical examination in October 1896. When Ruge retired, Bolk was appointed his successor in February 1898, because he had proved himself an exponent of functional, as against descriptive, anatomy.

In the anatomical periodical *Petrus Camper,* which Bolk and Winkler started in 1900, Bolk published on the comparative anatomy of the cerebellum and its nerves. The localization of muscle coordination in the cerebellum proved to have clinical implications. Leiden University awarded him an honorary doctorate in 1902, but he refused the professorship of anatomy there—Amsterdam University had started to build a new anatomical institute adapted to his needs and ideas.

The removal of an old graveyard near the institute induced Bolk to study human skulls and teeth. His research dealt with the ontogeny of the teeth, left-handedness and right-handedness, the length of the body in different races (Nordic and Alpine in the Netherlands), color of the eyes and hair, endocrinology, and general ontogeny.

Bolk was knighted (1918) while *rector magnificus* of Amsterdam University. Here, for the first time, he put forward his ideas of "fetalization." The theory was fully expounded during the Anatomical Congress in Freiburg (1926) and was published as "Das Problem der Menschwerdung." It considers man to be a neotenic ape and states that the retention of many fetal characters in the adult is caused by the endocrine glands. The theory indicates that, in a way, apes are more specialized from an evolutionary point of view than man is.

On the occasion of his silver jubilee as professor, Bolk was made a Commander in the Order of Orange-Nassau, and in 1927 he received the Swedish Retzius Medal for his work on the cerebellum. At the time of his death all professors of anatomy in the Netherlands and the East Indies had been his pupils.

BIBLIOGRAPHY

I. ORIGINAL WORKS. For a nearly complete list of Bolk's 179 publications, see A. J. P. Van Den Broek, "Louis Bolk," in *Gegenbaurs morphologisches Jahrbuch,* **65** (1931), 497–516. Additional papers are nos. 1, 3, 11, 12, 21, 24, 44, 47, and 67 in the list in A. J. Van Bork-Feltkamp, "Anthropological Research in the Netherlands," in *Verhandelingen van de Koninklijke Nederlandse Akademie van Wetenschappen, afdeling Natuurkunde,* **37** (1938), 1–166, list on 137–139. Bolk's earlier works include papers on the problems of segmental anatomy, in *Morphologisches Jahrbuch* (1894–1900); "Das Cerebellum der Säugetiere," in *Petrus Camper,* **3** and **4,** also published separately (1906); and *Odontologische Studien,* 2 vols. (1913–1914). The "fetalization theory" is in *Hersenen en Cultuur* (1918; 3rd ed., 1932); "The Part Played by the Endocrine Glands in the Evolution of Man," in *The Lancet* (10 September 1921); and "Das Problem der Menschwerdung," in *25sten Versammlung der anatomischen Gesellschaft in Freiburg* (Jena, 1926). A posthumous work is his contribution to *Handbuch der vergleichenden Anatomie der Wirbeltiere* (Berlin–Vienna, 1931).

II. SECONDARY LITERATURE. The best biography is that of A. J. P. Van Den Broek (see above). Additional data are in Van Bork-Feltcamp (see above). Obituaries are C. U. Ariëns Kappers, in *Psychologische en Neurologische Bladen,* **4** (1930), 1–6; J. A. J. Barge, in *Jaarboek van de Maatschappy de Nederlandsche Letterkunde* (1935–1936); Brouwer, *Verslagen der Koninklijke Nederlandsche Akademie van Wetenschappen,* **6** (1930); *Gedenkboek Universiteit van Amsterdam* (Amsterdam, 1932), pp. 188, 554; W. A. Mijsberg, in *Geneeskundig Tijdschrift voor Nederlandsch-Indië,* **2** (1930), 737–738; C. A. J. Quant, in *The Lancet* (12 July 1930), 76; A. J. P. Van Den Broek, in *Nederlandsch Tijdschrift voor Geneeskunde* (1930); and F. A. F. C. Went, "In Memoriam Lodewijk Bolk," in *Verslagen der Koninklijke Nederlandsche Akademie van Wetenschappen,* **39** (1930), 1–7.

H. ENGEL

BOLOS OF MENDES, also known as **Bolos the Democritean** (*b.* Mendes, Egypt; *fl. ca.* 200 B.C.), *biology.*

The dates of Bolos of Mendes cannot be established with certainty, and nothing seems to be known of his life. The extent of his influence cannot be estimated easily because of his deliberate policy of passing off his writings under Democritus' name. Obvious pseudo-Democritean writings, in turn, were later attributed to Bolos. Together these writings, none of which has survived, constitute a complex literary tradition in which the original contributions cannot be disentangled from later additions. Bolos was widely read in antiquity, when his reputation rivaled that of Aristotle as an authority in natural history. Judging from the titles of his lost writings, he wrote on a wide range of subjects. Some evidence exists that he was one of the principal early sources for the later encyclopedic tradition in which natural history and the lore of marvels are indistinguishable. Only a few fragments actually bear his name, thus precluding a detailed reconstruction of any of his writings. It is doubtful that he was systematic in collecting his data or that he made original contributions. The evidence suggests, rather, that he collected a large and diverse body of information, largely supernatural in nature, that could be put to the various purposes suited to the exploitation of the irrational in Hellenistic times.

Bolos' best-known and most influential work was entitled Φυσικὰ δυναμερά ("Natural Properties"). Sometimes known as Περὶ συμπαθειῶν καὶ ἀντιπαθειῶν ("On Sympathies and Antipathies"), it was an attempt to categorize observed biological and ecological relationships and to explain them in terms of the supposed, conscious "loves and hates" existing between the entities in question. Animals, plants, and minerals, each associated with its particular astral god, were believed to be invested with miraculous powers. As a result, magical ritual and religious invocations tended to take the place of causal explanation based on empirical observation and description. Bolos' influence is seen most clearly in Aelian, Pliny, and the anonymous Hermetic writers.

Other works ascribed to Bolos include Χειρόκμητα ("Things Made or Performed by Hand"), which is usually cited under the name of Democritus. It dealt with medical and magical herbs and various agricultural practices. His Περὶ λιθῶν ("On Stones") was a catalog of precious and semiprecious stones in which their supernatural powers were described in a manner foreshadowing the later lapidaries. Another of his writings, Περὶ γεωργίας ("On Farming"), was known to Columella and was frequently consulted by the writers of the *Geoponica.* Two other writings ascribed to him, but whose precise titles and meaning are unclear, are "Concerning Miracles" and an astrological tract, "On the Signs of the Sun and Moon." Portions of late magical papyri containing alchemical texts and directions for technological processes, some of which have been attributed to Pseudo-Democritus, may derive ultimately from Bolos' lost Βαφικά ("Things Dyed or Gilded"). He is also credited with separate writings on medicine, magical herbs, military tactics, and ethics, and Περὶ Ἰουδαίων ("History of the Jews"), but few identifiable fragments remain.

BIBLIOGRAPHY

Further discussion of Bolos or his work can be found in Hermann Diels, *Antike Technik,* 2nd ed. (Leipzig, 1920), pp. 127–138; Wilhelm Kroll, "Bolos und Demokritos," in *Hermes,* **69** (1934), 228–232; E. H. F. Meyer, *Geschichte der Botanik,* I (Königsberg, 1854), 277–284; Eugen Oder, "Beiträge zur Geschichte der Landwirthschaft bei den Griechen," in *Rheinisches Museum für Philologie,* **45** (1890), 58–99; and Max Wellmann, "Bolos aus Mendes," in Pauly-Wissowa, *Real-Encyclopädie,* III, cols. 676–677; "Die Georgika des Demokritos," in *Abhandlungen der Preussischen Akademie der Wissenschaft, Philosophisch-Historische Klasse,* no. 4 (1921), which contains an edition of the eighty-two fragments surviving from Bolos' writings on agriculture; and "Die ΦΥΣΙΚΑ des Bolos Demokritos und der Magier Anaxilaos aus Larissa," *ibid.*, no. 7 (1928), the fundamental study.

JERRY STANNARD

BOLOTOV, ANDREI TIMOFEEVICH (*b.* Dvoryaninovo, Tula oblast, Russia, 18 October 1738; *d.* Dvoryaninovo, 15 October 1833), *agronomy, biology.*

Bolotov's father was an army officer, and in 1755 Bolotov, too, entered military service. From 1757 to 1761 he was a translator attached to the Russian military governor at Königsberg, and in 1761–1762 he was adjutant to the St. Petersburg chief of police. Bolotov retired in 1762, moved to the country, and devoted himself to agronomy and botany, especially horticulture.

Characteristic of Bolotov's work was a striving not only to explore a given problem and to explain a given phenomenon arising from his practical experience and experimentation, but also to discover general biological laws and to find means of influencing the development of plants for practical purposes.

Bolotov's most noteworthy achievement was the defense and further development, during a period when the humus theory of soil fertility reigned supreme, of the theory that plants need mineral nourishment. Following Linnaeus, Camerarius, Koelreuter, and Christian Sprengel, Bolotov advocated, developed, and propagandized the field concept at a time when the majority of botanists approached it with distrust or completely ignored it as unfounded.

Bolotov considered the union of male and female sexual elements as the condition of fertilization and development of a new plant from a seed bud. He noted the quantity of pollen necessary for normal fertilization, the widespread occurrence in nature of cross-pollination, and the role of wind and insects in the latter process. His research on these questions covered the period 1778 to 1823. In 1778, fifteen years before Sprengel, Bolotov gave a sufficiently precise description of dichogamy—the maturation of the pistil and the stamen of bisexual plants at different times, which ensures cross-pollination in these plants.

In an article on the hazel nut (1804) and in several earlier works, such as "O semenakh" ("On Seeds," 1780), he noted the role of multiple pollination in increasing the fitness for survival of a species; thus, he anticipated the general features of Thomas Knight's discovery, which later received an explanation in the works of Darwin and of I. V. Michurin. Bolotov saw in cross-pollination and intraspecific and interspecific

hybridization some of the sources of the multiplicity of forms in nature. Much attention is given in his works to the formative influence of environmental conditions; in many of these works elements of an ecological approach to the analysis of phenomena in the plant world are clearly shown.

Bolotov was the author of more than 300 works, a prodigious output.

BIBLIOGRAPHY

Many of Bolotov's works were brought together in *Izbrannye sochinenia po agronomy, pludovodstvu, lesovodstvu, botanike* ("Selected Works on Agronomy, Fruit Growing, Forestry, and Botany," Moscow, 1952).

Works on Bolotov are A. P. Berdyshev, *A. T. Bolotov—perv russky ucheny agronom* ("A. T. Bolotov—The First Russian Agronomist," Moscow, 1949); *Istoria estestvoznania v Rossy* ("The History of Natural Science in Russia"), I, pt. 1 (Moscow, 1957), 475–478; I. M. Polyakov, "Istoria otkrytia dikhogamy i rol russkikh uchenykh v etom otkryty" ("History of the Discovery of Dichogamy and the Role of Russian Scientists in This Discovery"), in *Uspekhi sovremennoi biologii,* **30,** no. 2 (1950), 291–306; and I. M. Polyakov and A. P. Berdyshev, "A. T. Bolotov i ego trudy v oblasti selskokhozyaystvennoy i biologicheskoy nauki" ("A. T. Bolotov and His Works in the Fields of the Agricultural and Biological Sciences"), in Bolotov's *Izbrannye sochinenia.*

S. R. MIKULINSKY

BOLTWOOD, BERTRAM BORDEN (*b.* Amherst, Massachusetts, 27 July 1870; *d.* Hancock Point, Maine, 14/15 August 1927), *radiochemistry.*

Scientists spent the first several years following Henri Becquerel's discovery of radioactivity in 1896 largely in studying the physical properties of the radiations. By 1904, however, enough radioelements had been found to shift their interest to the bodies emitting these radiations. In the chemical identification of the radioelements and in positioning them in proper sequence in the decay series, Boltwood was an equal among such first-generation radiochemists as Otto Hahn, Frederick Soddy, Friedrich Giesel, and Herbert N. McCoy.

Boltwood's paternal ancestors came to America from Great Britain in the mid-seventeenth century. They settled in New England, where for several generations they were farmers, millers, and blacksmiths. One was able to work his way through Williams College, graduating in 1814 and later becoming a lawyer in Amherst, Massachusetts. This was Lucius Boltwood, Bertram's grandfather, who was active in the founding of Amherst College and served as its secretary from 1828 to 1864. He also was a candidate for the governorship of Massachusetts in 1841. Bertram's father, Thomas Kast Boltwood, graduated from Yale College in 1864, received a degree from the Albany Law School in 1866, and practiced his profession until his untimely death in 1872. Bertram, an only child, thereafter was raised entirely by his mother, Margaret Van Hoesen Boltwood, in her native village of Castleton-on-Hudson, New York. The Van Hoesens, of Dutch stock, were among the early settlers of Rensselaer County during the seventeenth century.

Bertram Boltwood grew up in comfortable surroundings, attended a private school, and from 1879 to 1889 prepared at the Albany Academy for Yale. The intellectual stature of his family, represented by cousin Ralph Waldo Emerson and uncle Charles U. Shepard, a professor of mineralogy at Amherst, presumably had great influence on him, although his childhood was characterized more by fun and practical jokes than by scholarship. Nevertheless, he entered Yale's Sheffield Scientific School in 1889, majoring in chemistry. Upon completion of the three-year course, Boltwood took highest honors in his subject and then departed for two years of advanced study at the Ludwig-Maximilian University in Munich. The training he received there under Alexander Krüss in special analytical methods and in the rare earths was to prove valuable in later years.

Boltwood returned to Yale in 1894 as a laboratory assistant in analytical chemistry and also to pursue graduate research. His work on double salts was accomplished under the direction of Horace L. Wells, who became his thesis adviser. In 1896 Boltwood spent a semester at the University of Leipzig, where he studied physical chemistry in Ostwald's laboratory, and then returned to Yale, where he received the Ph.D. in June 1897. A strong attachment brought Boltwood back to Europe several times in later years; his exuberant personality, his lifelong bachelorhood, and his height—well over six feet—made him both distinctive and welcome there.

Following graduation, Boltwood remained at the Sheffield Scientific School as an instructor in analytical chemistry, a position he had assumed a year earlier; later he was an instructor in physical chemistry. Until 1900, when he established a private laboratory as a consulting chemist, Boltwood devoted himself to perfecting laboratory apparatus and techniques and supplying teaching materials for students. He devised a simple automatic Sprengel pump, a new form of water blast, a lead fume pipe for the Kjeldahl nitrogen determination apparatus, and, somewhat later, Boltwax, a wax with low melting point, useful

for vacuum seals. He also translated German texts on physical chemistry and quantitative analysis by electrolysis. Boltwood's eager acquisition of new techniques made him a storehouse of information upon which his colleagues often drew. In later years he conducted demonstration classes in laboratory arts for research students.

In 1900 Boltwood left Yale and, with Joseph Hyde Pratt, also a Sheffield graduate, established a partnership: Pratt and Boltwood, Consulting Mining Engineers and Chemists. Pratt worked in the field, mostly in the Carolinas, and sent ore samples for analysis to Boltwood's private laboratory in downtown New Haven. Many of these samples contained rare earth elements and uranium and thorium, with which they commonly are associated. In 1896 Becquerel had discovered the radioactivity of uranium, and in 1898 Gerhard C. Schmidt and Marie Curie independently found thorium to be radioactive. It was perhaps inevitable that Boltwood's interest would turn in this direction, considering his early training in the analysis of rare earths, his inclination toward analytical and physical chemistry, his current familiarity with such ores as monazite and uraninite, and the challenge offered to his laboratory skill by work in radioactivity. He was not a total stranger to radioactivity, moreover, for in a senior thesis written under his direction in 1899, a student had repeated the Curies' separation process for radium and had narrowly missed the discovery of actinium in the pitchblende residues. Upon André Debierne's announcement of this new radioelement, Boltwood tested his student's substance and confirmed actinium's presence.

In April 1904, Boltwood began research on radioactivity. Not long before, Ernest Rutherford and Frederick Soddy had advanced a revolutionary new interpretation of this phenomenon: that radioactive atoms decay and transmute into other elements. While the evidence supporting this theory already was impressive, Boltwood reasoned that he could more strongly confirm it by showing a constant ratio between the amounts of radium and uranium in unaltered minerals. Such uniformity in composition would have to be accepted as proof of a genetic relationship, wherein the uranium decayed in several steps to form radium, which in turn decayed to form a series of several daughter products.

Boltwood quickly saw that the minute traces of radium, with chemical properties of its own, would be difficult to separate and test quantitatively. He therefore chose to measure radium's first daughter product, emanation, as an indication of the amount of radium present. Emanation, being chemically inert and a gas, required only mechanical separation; its activity thus was easier to measure. Within a few months, Boltwood's gas-tight gold-leaf electroscope yielded data showing that the activity of radium emanation was directly proportional to the amount of uranium in each of his samples. Rutherford, delighted with this news, encouraged Boltwood to perform the same tests on minerals with much smaller percentages of uranium. Yet even with this further confirmation in hand, Boltwood decided that direct proof that uranium decays into radium was desirable—he would try to "grow" radium.

One of the steps in this effort was to determine the equilibrium amount of radium. Boltwood's voluminous correspondence with Rutherford had ripened into a warm friendship; and the two collaborated, by mail, in the 1906 announcement that "the quantity of radium associated with one gram of uranium in a radio-active mineral is equal to approximately 3.8×10^{-7} gram." (The figure accepted today is 3.42×10^{-7} gram.) But Boltwood's attempts to grow radium were unsuccessful. Only one product between uranium and radium was then known; this was uranium X, whose short half-life should allow detectable quantities of radium to form within reasonable time limits. Yet even after more than a year, he was unable to observe any radium emanation in his uranium solution. Since Boltwood's faith in the disintegration theory did not waver, he concluded that there must be a long-lived decay product between uranium X and radium that was preventing the rapid accumulation of the daughter product.

Boltwood's search for this "parent of radium" was interrupted by his appointment as assistant professor of physics at Yale College. During the summer of 1906 he moved his apparatus into the Sloane Physics Laboratory and prepared to undertake his new academic duties. These responsibilities proved more extensive than anticipated, since, owing to the illness of the laboratory's director, Boltwood's close friend Henry A. Bumstead, he was left in charge of extensive renovations in the old building. Resumption of the search led him to Debierne's actinium, and Boltwood indeed believed for a while that he had properly placed actinium in the decay series. Among others, Soddy in Glasgow was working on the same problem, and the two carried on a heated controversy in the pages of *Nature*. Rutherford also was disinclined to accept actinium as the parent of radium and based his objection on the relative activities of actinium's products, a field in which Boltwood and McCoy had done basic work. The activities of many radioelements had been determined, relative to that of uranium, and

if those of actinium and its products were added to those of uranium, uranium X, radium, radium emanation, radium A, B, and so on, the total would be far greater than that of the mineral which supposedly contained them all in secular equilibrium.

Further investigation showed Boltwood that his difficulty lay in accepting Debierne's work on actinium as correct. In fact, there were other constituents in the Frenchman's radioelement, one of them having chemical properties similar to those of thorium. It was this substance, named "ionium" by Boltwood in 1907, that was the immediate parent of radium. He had now proved that ionium grows radium; that uranium grows ionium had still to be shown to complete the direct proof of this relationship. Tests a few years later were unsuccessful due to the small quantity of ionium accumulated. Finally, in 1919, Soddy conclusively proved this relationship, using uranium purified many years earlier.

An outgrowth of this work was a superior method for the determination of the half-life of radium. Under Boltwood's direction during the 1913–1914 academic year, a Norwegian chemist, Ellen Gleditsch, who had previously worked in Madame Curie's laboratory, obtained a value of slightly under 1,700 years. Another result of this intensive study of the chemistry of the radioelements was the realization that many of these substances, which differed in type and intensity of radiation, nevertheless could not be separated chemically. Thus, beginning about 1907, Boltwood, as well as Hahn, McCoy, and most other radiochemists, recognized the inseparability of, for example, thorium, radiothorium, ionium, and uranium X. But it was not until 1913 that Kasimir Fajans and Soddy declared them to be chemically identical isotopes, and explained the decay sequence by the group displacement laws.

Just as the radioelements were related, Boltwood's research activities bore logical connections. His first foray into the intricacies of the radioactive decay series in 1904 soon led him to examine the question of the inactive end products. Earlier analyses of uranium minerals showed that lead invariably appeared with the uranium. Between 1905 and 1907 Boltwood extended these observations and noted further that the geologically older minerals contained higher proportions of lead, as would be expected if this end product were accumulating over the ages. The thorium series was less well understood, and Boltwood at first doubted that it ended in lead, while actinium was not then recognized as forming part of a distinct series.

A direct result of this work was a striking application of science, the method of radioactive dating of rocks. If the rate of formation of an inactive decay product could be determined, the total amount found in a mineral would immediately yield its age. Both lead and helium (believed by most to be the alpha particle) were seen as suitable elements and, indeed, served in radioactive dating techniques. The helium method, pioneered in England by R. J. Strutt (later the fourth Baron Rayleigh), could not, however, give more than a minimum age because a variable portion of the gas would have escaped from the rock. But the lead method, developed by Boltwood in 1907, proved satisfactory and is still in use today. In effect, Boltwood reversed his procedure of confirming the accuracy of lead:uranium ratios by the accepted geological ages of the source rocks, and used these ratios to date the rocks. Because most geologists, under the influence of Lord Kelvin's nineteenth-century pronouncements, inclined toward an age of the earth measured in tens of millions of years, Boltwood's claim for a billion-year span was met with some skepticism. However, the subsequent work of Arthur Holmes, an understanding of isotopes, and the increasing accuracy of decay constants and analyses finally brought widespread acceptance of this method in the 1930's.

Boltwood's major contributions lay in the understanding of the uranium decay series. Still, he was able to suggest, with Rutherford in 1905, that actinium is genetically related to uranium, though not in the same chain as radium, while in the thorium series he almost beat Hahn to the discovery of mesothorium in 1907. His other significant service to the study of radioactivity was to bring greater precision and advanced techniques into the laboratory, as in his insistence that only by complete dissolution and boiling of the mineral could all the emanation be extracted from radioactive bodies.

Boltwood remained at Yale the rest of his life, except for the academic year 1909–1910, when he accepted an invitation to Rutherford's laboratory at the University of Manchester. Yale, fearing that he would remain in England indefinitely, offered Boltwood a full professorship in radiochemistry. This appointment brought him back to New Haven, but it also marked the end of his research career. Heavy academic duties, including supervision of construction of the new Sloane Physics Laboratory and unsuccessful efforts to obtain large quantities of radioactive minerals for research, seem to have taken all his time and energy. His stature as the foremost authority on radioactivity in the United States brought him membership in the National Academy of Sciences, the American Philosophical Society, and other organizations, but it also brought him numerous requests from prospectors, mine owners, speculators, chemical refiners, and wholesalers to analyze samples, devise separation processes, and find financial backing (from wealthy Yale alumni) for various projects. These efforts probably helped stimulate the production of radium, in which the United States led the world by about 1915, although they did not appreciably aid the progress of science.

In 1918 Boltwood was appointed director of the Yale College chemical laboratory and presided over the consolidation of the Yale and Sheffield chemistry departments. To cement this union, the new Sterling Chemistry Laboratory was proposed, and Boltwood was placed in charge of its design. He completed it successfully, but the strain of this effort caused a breakdown in his health from which he never fully recovered. Periods of severe depression alternated with his more customary cheerful spirits, and resulted in his suicide during the summer of 1927.

Boltwood's influence in radioactivity was widespread—through his published papers, correspondence, and personal contacts, for he trained surprisingly few research students. Part of his success stemmed from his close association with Rutherford, but like Rutherford's other chemical collaborators, Soddy and Hahn, he was eminently capable of major contributions in his own right.

BIBLIOGRAPHY

I. ORIGINAL WORKS. A reasonably complete list of Boltwood's publications is in Alois F. Kovarik's sketch of him in *Biographical Memoirs of the National Academy of Sciences,* **14** (1930), 69–96. His unpublished correspondence, papers, and laboratory notebooks are preserved in the Manuscript Room, Yale University Library. His extensive correspondence with Rutherford is in the Rutherford Collection, Manuscript Room, Cambridge University Library.

II. SECONDARY LITERATURE. In addition to Kovarik's memoir (see above), the following obituary notices offer information about Boltwood: *Yale Alumni Weekly,* **37** (7 Oct. 1927), 65; Kovarik, in *Yale Scientific Magazine,* **2** (Nov. 1927), 25, 44, 46; Rutherford, in *Nature,* **121** (14 Jan. 1928), 64–65; Kovarik, in *American Journal of Science,* **15** (Mar. 1928), 188–198.

LAWRENCE BADASH

BOLTZMANN, LUDWIG (*b*. Vienna, Austria, 20 February 1844; *d*. Duino, near Trieste, 5 September 1906), *physics.*

Boltzmann's father, Ludwig, was a civil servant (*Kaiserlich-Königlich Cameral-Concipist*); his mother was Katherina Pauernfeind. He was educated at Linz and Vienna, receiving his doctorate in 1867 from the

University of Vienna, where he had studied with Josef Stefan. Boltzmann held professorships at the universities of Graz, Vienna, Munich, and Leipzig. In 1876 he married Henrietta von Aigentler, who bore him four children.

Distribution Law. The first stimulus for Boltzmann's researches came from teachers and colleagues at the University of Vienna, especially Stefan and Josef Loschmidt. In a lecture Stefan suggested the problem in electrical theory whose solution constituted Boltzmann's first published paper (1865);[1] he also published a few papers on kinetic theory and did important experimental work on gases and radiation that provided the basis for some of Boltzmann's theories. Loschmidt (also in 1865) accomplished the first reliable estimate of molecular sizes with the help of the Clausius-Maxwell kinetic theory. Although Loschmidt was later to dispute Boltzmann's interpretation of the second law of thermodynamics, the problem of finding quantitative relations between atomic magnitudes and observable physical quantities was a common interest of both men.

Boltzmann began his lifelong study of the atomic theory of matter by seeking to establish a direct connection between the second law of thermodynamics and the mechanical principle of least action (1866). Although Clausius, Szily, and others later worked along similar lines, and Boltzmann himself returned to the subject in his elaboration of Helmholtz' theory of monocyclic systems (1884), the analogy with purely mechanical principles seemed insufficient for a complete interpretation of the second law. The missing element was the statistical approach to atomic motion that had already been introduced by the British physicist James Clerk Maxwell.[2] Boltzmann's first acquaintance with Maxwell's writings on kinetic theory is indicated by his paper on thermal equilibrium (1868). In this paper, he extended Maxwell's theory of the distribution of energy among colliding gas molecules, treating the case when external forces are present. The result was a new exponential formula for molecular distribution, now known as the "Boltzmann factor" and basic to all modern calculations in statistical mechanics. To understand how Boltzmann arrived at this result, we must first review the work of Maxwell on which it is based.

Maxwell, in his first paper on kinetic theory (1859), had pointed out that the collisions of gas molecules would not simply tend to equalize all their speeds but, on the contrary, would produce a range of different speeds. Most of the observable properties of a gas could be calculated if one knew, instead of the positions and velocities of all the molecules at any given time, only the average number of molecules having various positions and velocities. In many cases it seems reasonable to assume that the gas is spatially uniform, that is, the average number of molecules is the same at different places in the gas. The problem is then to determine the velocity distribution function $f(v)$, defined so that $f(v)\,dv$ is the average number of molecules having speeds between v and $v + dv$.

Maxwell argued that $f(v)$ should be a function that depends only on the magnitude of v, and that the velocity components resolved along the three coordinate axes should be statistically independent. Hence, he inferred that

$$f(v) = (N/\alpha^3\pi^{3/2})e^{-(v^2/\alpha^2)}, \qquad (1)$$

where N is the total number of molecules, and α^2 is inversely proportional to the absolute temperature.

In his long memoir of 1866, Maxwell admitted that the assumptions used in his previous derivation of the distribution law "may appear precarious"; he offered another derivation in which the velocities of two colliding molecules, rather than the velocity components of a single molecule, were assumed to be statistically independent. This means that one can express the joint distribution function for the probability that molecule 1 has velocity v_1, while at the same time molecule 2 has velocity v_2, as the product of the probabilities of these two separate events:

$$F(v_1, v_2) = f(v_1)f(v_2). \qquad (2)$$

To derive the distribution function itself, Maxwell argued that the equilibrium state would be reached when the number of collisions in which two molecules with initial velocities (v_1, v_2) rebound with final velocities (v_1', v_2') is equal to the number of collisions in which two molecules with initial velocities (v_1', v_2') rebound with final velocities (v_1, v_2); from this condition it follows that

$$F(v_1, v_2) = F(v_1', v_2'). \qquad (3)$$

By combining this equation with that for the conservation of energy (in the case when no forces act),

$$\frac{1}{2}m_1 v_1^2 + \frac{1}{2}m_2 v_2^2 = \frac{1}{2}m_1 v_1'^2 + \frac{1}{2}m_2 v_2'^2, \quad (4)$$

Maxwell deduced (as before) that

$$f(v_1) = (N/\alpha^3\pi^{3/2})\,e^{-(v^2/\alpha^2)}. \qquad (5)$$

This type of reasoning about velocity distribution functions was repeatedly used and generalized by Boltzmann in his own works on kinetic theory. He began, in his 1868 paper, by considering the case in which one of the particles of a system is acted on by a force with a corresponding potential function,

$V(x)$. The condition of conservation of energy would then be

$$\frac{1}{2} m_1 v_1{}^2 + V(x_1) + \frac{1}{2} m_2 v_2{}^2$$

$$= \frac{1}{2} m_1 v_1{}'^2 + V(x_1') + \frac{1}{2} m_2 v_2{}'^2, \quad (6)$$

and Boltzmann could then apply Maxwell's procedure to deduce the distribution function

$$f(v) = (\text{const.}) \, e^{-h(mv^2/2 + V[x])}. \quad (7)$$

The constant factor h could be related to the absolute temperature of the gas, as Maxwell and Clausius had done, by comparing the theoretical pressure of the gas with the experimental relation between pressure and temperature (Gay-Lussac's law). In modern notation, h is equivalent to $1/kT$, where k is a constant, now called Boltzmann's constant, and T is the absolute temperature on the Kelvin scale.

The physical meaning of the Maxwell-Boltzmann distribution law is that the energy ($E = mv^2/2 + V[x]$) of a molecule is most likely to be roughly equal to kT; much larger or much smaller energies occur with small but finite probability.

In the same paper of 1868, Boltzmann presented another derivation of the Maxwell distribution law that was independent of any assumptions about collisions between molecules. He simply assumed that there is a fixed total amount of energy to be distributed among a finite number of molecules, in such a way that all combinations of energies are equally probable. (More precisely, he assumed uniform distribution in momentum space.) By regarding the total energy as being divided into small but finite quanta, he could treat this as a problem of combinatorial analysis. He obtained a rather complicated formula that reduced to the Maxwell velocity-distribution law in the limit of an infinite number of molecules and infinitesimal energy quanta.

The device of starting with finite energy quanta and then letting them become infinitesimal is not essential to such a derivation, but it reveals an interesting feature of Boltzmann's mathematical approach. Boltzmann asserted on several occasions that a derivation based on infinite or infinitesimal quantities is not really rigorous unless it can also be carried through with finite quantities. While this prejudice kept him from appreciating and using some of the developments in pure mathematics that appeared toward the end of the nineteenth century, it also had the curious effect of making some of his equations for energy distribution and transfer look similar to those of modern quantum theory. (This is perhaps not quite

accidental, since Planck and other early quantum theorists were familiar with Boltzmann's works and used many of his techniques.)

Transport Equation and H-*theorem.* Although Maxwell and Boltzmann had succeeded in finding the correct distribution laws by assuming that the gas is in an equilibrium state, they thought that the kinetic theory should also be able to show that a gas will actually tend toward an equilibrium state if it is not there already. Maxwell had made only fragmentary attempts to solve this problem; Boltzmann devoted several long papers to establishing a general solution.

Approach to equilibrium is a special case of a general phenomenon: dissipation of energy and increase of entropy. It was Boltzmann's achievement to show in detail how thermodynamic entropy is related to the statistical distribution of molecular configurations, and how increasing entropy corresponds to increasing randomness on the molecular level. This was a peculiar and unexpected relationship, for macroscopic irreversibility seemed to contradict the fundamental reversibility of Newtonian mechanics, which was still assumed to apply to molecular collisions. Boltzmann's attempts to resolve this contradiction formed part of the debate on the validity of the atomic theory in the 1890's. Seen in this context, the proof of the distribution law has even more significance than the law itself.

Boltzmann's major work on the approach to equilibrium (and on transport processes in gases in general) was published in 1872. This paper, like that of 1868, took Maxwell's theory as the starting point. Boltzmann first derived an equation for the rate of change in the number of molecules having a given energy, x, resulting from collisions between molecules. He considered a typical collision between two molecules with energies x and x' before the collision, and energies ξ and $x + x' - \xi$ after the collision. Such a collision reduces by one the number of molecules with energy x; the number of such collisions is assumed to be proportional to the number of molecules with energy x, and also to the number of molecules with energy x'. Boltzmann used here, without any comment, Maxwell's assumption of statistical independence of the velocities of two colliding molecules (eq. 2); later it was recognized that there might be valid grounds for objecting to this assumption.[3] With this assumption, the decrease in $f(x)$ will be equal to the product, $f(x)f(x')$, multiplied by an appropriate factor for the collision probability and integrated over all values of x'. Similarly, the increase in $f(x)$ may be attributed to inverse collisions in which the molecules have energies ξ and $x + x' - \xi$ before the collision, and x and x' after the collision. By such arguments Boltzmann arrived at the equation

$$\frac{\partial f}{\partial t} = \int_0^\infty \int_0^{x+x'} \left[\frac{f(\xi,t)f(x+x'-\xi,t)}{\sqrt{\xi}\sqrt{x+x'-\xi}} - \frac{f(x,t)f(x',t)}{\sqrt{x}\sqrt{x'}} \right]$$
$$\sqrt{xx'}\psi(x,x',\xi)\, dx'\, d\xi. \quad (8)$$

(This is a special case of the general Boltzmann transport equation [eq. 9]; terms describing the effect of external forces and nonuniformities on the change of f are here omitted. The square root expressions in the denominators, which do not appear in the form of the equation generally used, result from the fact that energy rather than velocity is the variable.)

One additional assumption involved in this derivation should be mentioned: the collision probability function, $\psi(x, x',\xi)$, is the same for both the direct and inverse collisions; that is, the collision is perfectly reversible.

Following Maxwell's 1866 development of the transport equations, Boltzmann showed how the diffusion, viscosity, and heat conduction coefficients of a gas could be calculated by solving the general transport equation

$$\frac{\partial f}{\partial t} + \xi \frac{\partial f}{\partial x} + \eta \frac{\partial f}{\partial y} + \zeta \frac{\partial f}{\partial z} + X \frac{\partial f}{\partial \xi} + Y \frac{\partial f}{\partial \eta}$$
$$+ Z \frac{\partial f}{\partial \zeta} + \int d\omega_1 \int b\, db \int d\phi V(ff_1 - f'f_1') = 0, \quad (9)$$

where (ξ,η,ζ) are components of the velocity of a particle and (X,Y,Z) are components of the force acting on it, and V, ϕ, b, and ω_1 are variables characterizing the relative motion of the two molecules during a collision. (Values of the function f for velocities of the two molecules before and after the collision are indicated by f, f_1, f', and f_1'.)

It is difficult to obtain exact solutions of Boltzmann's transport equation except when the molecules interact with inverse fifth-power forces, a case for which Maxwell had found an important simplification.[4] Boltzmann made several attempts to develop accurate approximations for other force laws, but this problem was not satisfactorily solved until the work of S. Chapman and D. Enskog in 1916–1917. Boltzmann's equation is now frequently used in modern research on fluids, plasmas, and neutron transport.

If the velocity distribution function is Maxwellian, then the integral on the right-hand side of eq. 8 vanishes identically for all values of the variables, and we find $\partial f/\partial t = 0$. In other words, once the Maxwellian state has been reached, no further change in the velocity distribution function can occur.

So far this is simply an elaboration of the previous arguments of Maxwell and of Boltzmann himself, but

now, with an explicit formula for $\partial f/\partial t$, Boltzmann was able to go further and show that $f(x)$ probably tends toward the Maxwell form. He did this by introducing a function, E, depending on $f(x)$

$$E = \int_0^\infty f(x,t)\left\{ \log\left[\frac{f(x,t)}{\sqrt{x}} \right] - 1 \right\} dx \quad (10)$$

and showing that E always decreases unless f has the Maxwellian form:

$$\frac{dE}{dt} < 0 \text{ if } f \neq (\text{const.})\ \sqrt{x}e^{-hx},$$
$$\frac{dE}{dt} = 0 \text{ if } f = (\text{const.})\sqrt{x}e^{-hx}. \quad (11)$$

(The proof is straightforward and relies simply on the fact that the quantity $(a - b) \log b/a$ is always negative if a and b are real positive numbers.) Boltzmann also noted that in the Maxwellian state E is essentially the same as the thermodynamic entropy (aside from a constant factor). Thus Boltzmann's "H-function" (the notation was changed from E to H in the 1890's) provides an extension of the definition of entropy to nonequilibrium states not covered by the thermodynamic definition.

The theorem that H always decreases for nonequilibrium systems was called "Boltzmann's minimum theorem" in the nineteenth century, and now goes by the name "Boltzmann's H-theorem." (It has not yet been proved rigorously except with certain specializing assumptions.)

Reversibility and Recurrence Paradoxes. The H-theorem raised some difficult questions about the nature of irreversibility in physical systems, in particular the so-called "reversibility paradox" and "recurrence paradox." (The modern terminology goes back only to the Ehrenfests' 1911 article, in which the words *Umkehreinwand* and *Wiederkehreinwand* were introduced.) The reversibility paradox, first discussed by Lord Kelvin (1874) and brought to Boltzmann's attention by Loschmidt, is based on the apparent contradiction between one of the basic premises of Boltzmann's derivation—the reversibility of individual collisions—and the irreversibility predicted by the theorem itself for a system of many molecules. Of course there must be such a contradiction between any molecular theory based on Newtonian mechanics and the general principle of dissipation of energy, but Boltzmann's work was the first to reveal this inconsistency explicitly.

Boltzmann's initial response (1877) to the reversibility paradox was the suggestion that the irreversibility of processes in the real world is not a consequence of the equations of motion and the form of the intermolecular force law but, rather, seems to be a result of the initial conditions. For some unusual initial conditions the system might in fact decrease its entropy (increase the value of H) as time progresses; such initial conditions could be constructed simply by reversing all the velocities of the molecules in an equilibrium state known to have evolved from a nonequilibrium state. But, Boltzmann asserts, there are infinitely many more initial states that evolve with increasing entropy, simply because the great majority of all possible states are equilibrium states. Moreover, the entropy would also be almost certain to increase if one picked an initial state at random and followed it backward in time instead of forward.

The recurrence paradox arises from a theorem in mechanics first published by Poincaré in 1890. According to this theorem, any mechanical system constrained to move in a finite volume with fixed total energy must eventually return to any specified initial configuration. If a certain value of the entropy is associated with every configuration of the system (a disputable assumption), then the entropy cannot continually increase with time, but must eventually decrease in order to return to its initial value. Therefore the H-theorem cannot always be valid.

Poincaré, and later Zermelo (1896), argued that the recurrence theorem makes any mechanical model, such as the kinetic theory, incompatible with the second law of thermodynamics; and since, it was asserted, the second law is a strictly valid induction from experience, one must reject the mechanistic viewpoint.

Boltzmann replied that the recurrence theorem does not contradict the H-theorem, but is completely in harmony with it. The equilibrium state is not a single configuration but, rather, a collection of the overwhelming majority of possible configurations, characterized by the Maxwell-Boltzmann distribution. From the statistical viewpoint, the recurrence of some particular initial state is a fluctuation that is almost certain to occur if one waits long enough; the point is that the probability of such a fluctuation is so small that one would have to wait an immensely long time before observing a recurrence of the initial state. Thus the mechanical viewpoint does not lead to any consequences that are actually in disagreement with experience. For those who are concerned about the cosmological consequences of the second law—the so-called "heat death" corresponding to the final attainment of a state of maximum disorder when all irreversible processes have run their course —Boltzmann suggested the following idea. The universe as a whole is in a state of thermal equilibrium, and there is no distinction between forward and backward directions of time. However, within small regions, such as individual galaxies, there will be noticeable fluctuations that include ordered states corresponding to the existence of life. A living being in such a galaxy will distinguish the direction of time for which entropy increases (processes going from ordered to disordered states) from the opposite direction; in other words, the concept of "direction of time" is statistical or even subjective, and is determined by the direction in which entropy happens to be increasing. Thus, the statement "Entropy increases with time" is a tautology, and yet the subjective time directions in different parts of the universe may be different. In this way local irreversible processes would be compatible with cosmic reversibility and recurrence. (Boltzmann's concept of alternating time directions has recently been revived in connection with theories of oscillating universes.)

Statistical Mechanics and Ergodic Hypothesis. Having followed Boltzmann's work on irreversible processes into some of the controversies of the 1890's, let us now return to his contributions to the theory of systems in thermal equilibrium (for which the term "statistical mechanics" was introduced by J. Willard Gibbs).

It would be possible (as is in fact done in many modern texts) to take the Maxwell-Boltzmann distribution function (eq. 7) as the basic postulate for calculating all the equilibrium properties of a system. Boltzmann, however, preferred another approach that seemed to rest on more general grounds than the dynamics of bimolecular collisions in low-density gases. The new method was in part a by-product of his discussion of the reversibility paradox, and is first hinted at in connection with the relative frequency of equilibrium, as opposed to nonequilibrium, configurations of molecules: "One could even calculate, from the relative numbers of the different distributions, their probabilities, which might lead to an interesting method for the calculation of thermal equilibrium."[5] This remark was quickly followed up in the same year (1877) in a paper in which the famous relation between entropy and probability,

$$S = k \log W,$$

was developed and applied. In this equation, W is the number of possible molecular configurations ("microstates," in modern terminology) corresponding to a given macroscopic state of the system.[6] (To make this expression meaningful, microstates have to be defined with respect to finite cells in phase space; the size of these cells introduces an arbitrary additive constant in S which can be determined from quantum theory.)

The new formula for entropy—from which formulas for all other thermodynamic quantities could be deduced—was based on the assumption of equal a priori probability of all microstates of the system (that is, all microstates that have the same total energy). As noted above, Boltzmann had already proved in 1868 that such an assumption implies the Maxwell velocity distribution for an ideal gas of noninteracting particles; it also implies the Maxwell-Boltzmann distribution for certain special cases in which external forces are present. But the assumption itself demanded some justification beyond its inherent plausibility. For this purpose, Boltzmann and Maxwell introduced what is now called the "ergodic hypothesis," the assumption that a single system will eventually pass through all possible microstates.

There has been considerable confusion about what Maxwell and Boltzmann really meant by ergodic systems. It appears that they did not have in mind completely deterministic mechanical systems following a single trajectory unaffected by external conditions; the ergodic property was to be attributed to some random element, or at least to collisions with a boundary. In fact, when Boltzmann first introduced the words *Ergoden* and *ergodische,* he used them not for single systems but for collections of similar systems with the same energy but different initial conditions. In these papers of 1884 and 1887, Boltzmann was continuing his earlier analysis of mechanical analogies for the second law of thermodynamics and also developing what is now (since Gibbs) known as ensemble theory. Here again, Boltzmann was following a trail blazed by Maxwell, who had introduced the ensemble concept in his 1879 paper. But while Maxwell never got past the restriction that all systems in the ensemble must have the same energy, Boltzmann suggested more general possibilities and Gibbs ultimately showed that it is most useful to consider ensembles in which not only the energy but also the number of particles can have any value, with a specified probability.

The Maxwell-Boltzmann ergodic hypothesis led to considerable controversy on the mathematical question of the possible existence of dynamical systems that pass through all possible configurations. The controversy came to a head with the publication of the Ehrenfests' article in 1911, in which it was suggested that while ergodic systems are probably nonexistent, "quasi-ergodic" systems that pass "as close as one likes" to every possible state might still be found. Shortly after this, two mathematicians, Rosenthal and Plancherel, used some recent results of Cantor and Brouwer on the dimensionality of sets of points to prove that strictly ergodic systems are

indeed impossible. Since then, there have been many attempts to discover whether physical systems can be ergodic; "ergodic theory" has become a lively branch of modern mathematics, although it now seems to be of little interest to physicists.

After expending a large amount of effort in the 1880's on elaborate but mostly fruitless attempts to determine transport properties of gases, Boltzmann returned to the calculation of equilibrium properties in the 1890's. He was encouraged by the progress made by Dutch researchers—J. D. van der Waals, H. A. Lorentz, J. H. van't Hoff, and others—in applying kinetic methods to dense gases and osmotic solutions. He felt obliged to correct and extend their calculations, as in the case of virial coefficients of gases composed of elastic spheres. The success of these applications of kinetic theory also gave him more ammunition for his battle with the energeticists (see below).

Other Scientific Work. Although Boltzmann's contributions to kinetic theory were the fruits of an effort sustained over a period of forty years, and are mainly responsible for his reputation as a theoretical physicist, they account, numerically, for only about half of his publications. The rest are so diverse in nature—ranging over the fields of physics, chemistry, mathematics, and philosophy—that it would be useless to try to describe or even list them here. Only one common characteristic seems evident: most of what Boltzmann wrote in science represents some kind of interaction with other scientists or with his students. All of his books originated as lecture notes; in attempting to explain a subject on the elementary level, Boltzmann frequently developed valuable new insights, although he often succumbed to unnecessary verbosity. He scrutinized the major physics journals and frequently found articles that inspired him to dash off a correction, design a new experiment, or rework a theoretical calculation to account for new data.

Soon after he started to follow Maxwell's work on kinetic theory, Boltzmann began to study the electromagnetic theory of his Scottish colleague. In 1872, he published the first report of a comprehensive experimental study of dielectrics, conducted in the laboratories of Helmholtz in Berlin and of Töpler in Graz. A primary aim of this research was to test Maxwell's prediction that the index of refraction of a substance should be the geometric mean of its dielectric constant and its magnetic permeability ($i = \sqrt{\epsilon\mu}$). Boltzmann confirmed this prediction for solid insulators and (more accurately) for gases. He also confirmed the further prediction that if the speed of light (and hence the index of refraction) varies with direction in an anisotropic crystal, then the dielectric constant must also vary with direction.

During the next few years, Boltzmann began experimental work in diamagnetism while continuing his theoretical research in kinetic theory. He proposed a new theory of elastic aftereffects, in which the stress on a material at a given time depends on its previous deformation history.

In 1883, as a result of preparing an abstract of H. T. Eddy's paper (on radiant heat as a possible exception to the second law of thermodynamics) for Wiedemann's *Beiblätter,* Boltzmann learned of a work by the Italian physicist Adolfo Bartoli on radiation pressure. Bartoli's reasoning stimulated Boltzmann to work out a theoretical derivation, based on the second law of thermodynamics and Maxwell's electromagnetic theory, of the fourth-power law previously found experimentally by Stefan:

(radiation energy) $\propto$ (absolute temperature)^4.

Although at the time the "Stefan-Boltzmann law" for radiation seemed to be an isolated result with no further consequences, it did at least show a possible connection between thermodynamics and electromagnetism that was exploited in the later quantum theory. In the 1920's it was applied by Eddington and others in explaining the equilibrium of stellar atmospheres.

In the 1890's Boltzmann again revived his interest in electromagnetic theory, perhaps as a result of Hertz's experiments, which he repeated before a large audience in Graz. He published his *Vorlesungen über Maxwells Theorie . . .* in 1891 and 1893, along with some papers in which he suggested new mechanical models to illustrate the field equations. In 1895 he published an annotated German edition of Maxwell's paper "On Faraday's Lines of Force" in Ostwald's *Klassiker der exakten Wissenschaften.* Boltzmann was partly responsible for the eventual acceptance of Maxwell's theory on the Continent, although he did not advance the theory itself as much as did Lorentz, nor did he grapple with the difficult problems that ultimately led to Einstein's theory of relativity.

Defense of the Atomic Viewpoint. Throughout his career, even in his works on subjects other than kinetic theory, Boltzmann was concerned with the mathematical problems arising from the atomic nature of matter. Thus, an early paper with the title "Über die Integrale linearer Differentialgleichungen mit periodischen Koeffizienten" (1868) turned out to be an investigation of the validity of Cauchy's theorem on this subject, which is needed to justify the application of the equations for an elastic continuum to a crystalline solid in which the local properties vary

periodically from one atom to the next. Every time someone published new data on the specific heats of gases, Boltzmann felt obliged to worry again about the distribution of energy among the internal motions of polyatomic molecules.

Until the 1890's, it seemed to be generally agreed among physicists that matter *is* composed of atoms, and Boltzmann's concern about the consistency of atomic theories may have seemed excessive. But toward the end of the century, the various paradoxes—specific heats, reversibility, and recurrence—were taken more seriously as defects of atomism, and Boltzmann found himself cast in the role of principal defender of the kinetic theory and of the atomistic-mechanical viewpoint in general. Previously he had not been much involved in controversy—with the exception of a short dispute with O. E. Meyer, who, ironically, had accused Boltzmann of proposing a theory of elasticity that was inconsistent with the atomic nature of matter. But now Boltzmann found himself almost completely deserted by Continental scientists; his principal supporters were in England.

In retrospect it seems that the criticisms of kinetic theory in this period were motivated not primarily by technical problems, such as specific heats of polyatomic molecules but, rather, by a general philosophical reaction against mechanistic or "materialistic" science and a preference for empirical or phenomenological theories, as opposed to atomic models. The leaders of this reaction, in the physical sciences, were Ernst Mach, Wilhelm Ostwald, Pierre Duhem, and Georg Helm. Mach recognized that atomic hypotheses could be useful in science but insisted, even as late as 1912, that atoms must not be considered to have a real existence. Ostwald, Duhem, and Helm, on the other hand, wanted to replace atomic theories by "energetics" (a generalized thermodynamics); they denied that kinetic theories had any value at all, even as hypotheses.

In the first volume of his *Vorlesungen über Gastheorie* (1896), Boltzmann presented a vigorous argument for the kinetic theory:

> Experience teaches that one will be led to new discoveries almost exclusively by means of special mechanical models. . . . Indeed, since the history of science shows how often epistemological generalizations have turned out to be false, may it not turn out that the present "modern" distaste for special representations, as well as the distinction between qualitatively different forms of energy, will have been a retrogression? Who sees the future? Let us have free scope for all directions of research; away with all dogmatism, either atomistic or anti-atomistic! In describing the theory of gases as a mechanical *analogy,* we have already indicated, by

the choice of this word, how far removed we are from that viewpoint which would see in visible matter the true properties of the smallest particles of the body [Brush trans., p. 26].

In the foreword to the second volume of this book (1898), Boltzmann seemed rather more conscious of his failure to convert other scientists to acceptance of the kinetic theory. He noted that attacks on the theory had been increasing, but added:

I am convinced that these attacks are merely based on a misunderstanding, and that the role of gas theory in science has not yet been played out. The abundance of results agreeing with experiment which van der Waals has derived from it purely deductively, I have tried to make clear in this book. More recently, gas theory has also provided suggestions that one could not obtain in any other way. From the theory of the ratio of specific heats, Ramsay inferred the atomic weight of argon and thereby its place in the system of chemical elements— which he subsequently proved, by the discovery of neon, was in fact correct. . . .

In my opinion it would be a great tragedy for science if the theory of gases were temporarily thrown into oblivion because of a momentary hostile attitude toward it, as was for example the wave theory because of Newton's authority.

I am conscious of being only an individual struggling weakly against the stream of time. But it still remains in my power to contribute in such a way that, when the theory of gases is again revived, not too much will have to be rediscovered. . . [Ibid., pp. 215–216].

Boltzmann and Ostwald, although on good personal terms, engaged in bitter scientific debates during this period; at one point even Mach thought the argument was becoming too violent, and proposed a reconciliation of mechanistic and phenomenological physics.[7] While teaching at Leipzig with Ostwald during the period 1900–1902, Boltzmann was undergoing periods of mental depression and made one attempt at suicide. He returned to Vienna in 1902, where he succeeded himself as professor of theoretical physics and also lectured on the philosophy of science, replacing Ernst Mach, who had to retire for reasons of health. In 1904 he went to the United States to attend the World's Fair at St. Louis, where he lectured on applied mathematics, and also visited Berkeley and Stanford. He later described his experiences on this trip in a satirical article, "Reise eines deutschen Professors ins Eldorado." But despite his travels and discussions with scientific colleagues, he somehow failed to realize that the new discoveries in radiation and atomic physics occurring at the turn of the century were going to vindicate his own theories, even if in somewhat altered form. The real cause of Boltzmann's

suicide in 1906 will never be known; but insofar as despair over the rejection of his lifework by the scientific community may have been a contributing factor (as has sometimes been suggested without much evidence), it is certainly one of the most tragic ironies in the history of science that Boltzmann ended his life just before the existence of atoms was finally established (to the satisfaction of most scientists) by experiments on Brownian motion guided by a kinetic-statistical theory of molecular motion.

NOTES

1. All of Boltzmann's publications for which only the year is given may be found in his *Wissenschaftliche Abhandlungen.*
2. See *The Scientific Papers of James Clerk Maxwell* (Cambridge, 1890). The 1859 and 1866 papers of Maxwell, together with other papers by Clausius, Boltzmann, Kelvin, Poincaré, and Zermelo (cited by year in this article) may be found in S. G. Brush, ed., *Kinetic Theory,* 2 vols. (Oxford, 1965–1966).
3. See Boltzmann, *Vorlesungen über Gastheorie,* I, §3; P. and T. Ehrenfest, "Begriffliche Grundlagen der statistischen Aufassung in der Mechanik."
4. It was in reference to this result of Maxwell's that Boltzmann wrote his oft-quoted comparison of styles in theoretical physics and styles in music, dramatizing the almost magical disappearance of V from the integrand of eq. 9 when the words "let $n = 5$" were pronounced (*Populäre Schriften,* p. 51).
5. Brush, *Kinetic Theory,* II, 192.
6. This formula for S is clearly related to Boltzmann's earlier expression for the H-function (eq. 9). If we know that the system has probability W_i of being in macrostate i, with given values of W_i for all i, then the expectation value of the entropy can be calculated from

$$S = k \Sigma W \log W$$

with an appropriate interpretation of the summation or integration.
7. *Die Principien der Wärmelehre, historisch-kritisch entwickelt* (Leipzig, 1896), pp. 362 ff.

BIBLIOGRAPHY

I. ORIGINAL WORKS. The technical papers that originally appeared in various periodicals have been reprinted in Boltzmann's *Wissenschaftliche Abhandlungen,* F. Hasenöhrl, ed., 3 vols. (Leipzig, 1909). Lectures and articles of general interest are collected in *Populäre Schriften* (Leipzig, 1905). A review article written with J. Nabl, "Kinetische Theorie der Materie," was published in *Encyklopädie der mathematischen Wissenschaften,* V, pt. 1 (Leipzig, 1905), art. V8. Other works are *Vorlesungen über Maxwells Theorie der Elektrizität und des Lichtes,* 2 vols. (Leipzig, 1891–1893); his ed. of Maxwell's "On Faraday's Lines of Force," *Ueber Faraday's Kraftlinien* (Leipzig, 1895), with 31 pages of notes by Boltzmann; *Vorlesungen über Gastheorie,* 2 vols. (Leipzig, 1896–1898), trans. by S. G. Brush, with introduction, notes, and bibliography, as *Lectures on Gas Theory* (Berkeley, 1964); *Vorlesungen über die Principe der Mechanik,* 3 vols. (Leipzig, 1897–1920); and

Über die Prinzipien der Mechanik, Zwei akademische Antrittsreden (Leipzig, 1903). Books based on Boltzmann's lectures are Charles Emerson Curry, *Theory of Electricity and Magnetism* (London, 1897), with a preface by Boltzmann; and Hugo Buchholz, *Das mechanische Potential,* published with *Die Theorie der Figur der Erde* (Leipzig, 1908).

II. SECONDARY LITERATURE. Works on Boltzmann are Engelbert Broda, *Ludwig Boltzmann: Mensch, Physiker, Philosoph* (Berlin, 1955); S. G. Brush, "Foundations of Statistical Mechanics 1845–1915," in *Archive for History of Exact Sciences,* **4** (1967), 145–183; René Dugas, *La théorie physique au sens de Boltzmann et ses prolongements modernes* (Neuchâtel, 1959); P. and T. Ehrenfest, "Begriffliche Grundlagen der statistischen Auffassung in der Mechanik," in *Encyklopädie der mathematischen Wissenschaften,* IV, pt. 32 (Leipzig, 1911), trans. by M. J. Moravcsik as *The Conceptual Foundations of the Statistical Approach in Mechanics* (Ithaca, N.Y., 1959); and G. Jaeger, "Ludwig Boltzmann," in *Neue Österreichische Biographie 1815–1918,* pt. 1, *Biographien,* II (Vienna, 1925), 117–137. Other articles on Boltzmann are listed in the bibliography of the Brush trans. of *Lectures on Gas Theory.*

STEPHEN G. BRUSH

BOLYAI, FARKAS (WOLFGANG) (*b.* 9 February 1775, Bolya [German, Bell], near Nagyszeven [German, Hermannstadt], Transylvania, Hungary [now Sibiu, Rumania]; *d.* 20 November 1856, Marosvásárhely, Transylvania, Hungary [now Târgu-Mureş, Rumania]), *mathematics.*

Farkas Bolyai was the son of Gáspár (Kasper) Bolyai and Christina Vajna (von Páva) Bolyai. Bolya was the hereditary estate of the noble family of Bolyai de Bolya, which was mentioned as early as the thirteenth and fourteenth centuries. By the time of Gáspár it had been reduced to a small holding, but Gáspár added another holding (which belonged to his wife's family) in Domáld, near Marosvásárhely. He enjoyed a reputation as an industrious and intelligent landholder of strong character.

Young Farkas received an education at the Evangelical-Reformed College in Nagyszeven, where he stayed from 1781 to 1796. He excelled in many fields, especially in mathematics, and showed interest in theology, painting, and the stage. In 1796, he traveled to Germany, going first to Jena and then, with a fellow student at Nagyszeven, Baron Simon Kemény, entered the University of Göttingen, where he studied until 1799. Among his teachers were the astronomer Felix Seyffer and the mathematician Abraham Gotthelf Kästner. It was at this time that Bolyai began his lifelong friendship with Carl Friedrich Gauss, also a student at Göttingen, who already was intensely engaged in mathematical research. From this period dates Bolyai's interest in the foundations of geometry,

especially in the so-called Euclidean or parallel axiom, to which Kästner and Seyffer, as well as Gauss, were devoting attention. Bolyai maintained a correspondence with Gauss that, with interruptions, lasted all their lives.

After his return to Transylvania, Bolyai became a superintendent in the house of the Keménys in Koloszvár (German, Klausenburg; now Cluj, Rumania). In 1801 he married Susanna von Árkos, the daughter of a surgeon. His wife was talented but sickly and nervous, and the marriage was not a happy one. The couple settled in Domáld, where Bolyai farmed from 1801 to 1804. In 1802 their son, János, was born, at the von Árkos home in Koloszvár.

In 1804, Farkas accepted the position of professor of mathematics, physics, and chemistry at the Evangelical-Reformed College at Marosvásárhely, where he taught until his retirement in 1853. During this half century he was known as a patient and kind teacher, but one who lacked the faculty of easily transmitting to others his own scientific enthusiasm, despite the emphasis he placed on correct mathematical education. Meanwhile, he continued his research, concentrating on the theory of parallels. He sent a manuscript on this subject, *Theoria parallelarum,* with an attempt to prove the Euclidean axiom, to Gauss in 1804. The reasoning, however, satisfied neither Gauss nor himself; and Bolyai continued to work on it and on the foundations of mathematics in general.

The Euclidean axiom, which appears as the fifth postulate in Book I of Euclid's *Elements,* is equivalent to the statement that through a given point outside a given line only one parallel can be drawn to the line. It is also equivalent to the statement that there exists a triangle in which the sum of the three angles is equal to two right angles and, hence, that all triangles have this property. Attempts to prove this axiom—that is, to deduce it from other, more obvious, assumptions—began in antiquity. These attempts were always unsatisfactory, however, and the nature of the axiom had remained a challenge to mathematicians. Bolyai, working in almost total scientific isolation, often despaired while trying to understand it.

During such moments of discouragement, he sought consolation in poetry, music, and writing for the stage. In 1817, his *Öt Szomorujátek, Irta egy Hazafi* ("Five Tragedies, Written by a Patriot") was entered in a contest. The following year, another play, *A Párisi Par* ("The Paris Process"), appeared. Bolyai's wife died in 1821, and in 1824 he married Theresia Nagy, the daughter of an iron merchant in Marosvásárhely. They had one son, Gregor.

Farkas began to interest himself in mathematics

again when his son János evinced unusual mathematical talent. In 1829 Bolyai finished his principal work, but because of technical and financial problems it was not published until 1832–1833. It appeared in two volumes, with the title *Tentamen juventutem studiosam in elementa matheseos purae, elementaris ac sublimioris, methodo intuitiva, evidentiaque huic propria, introducendi, cum appendice triplici* ("An Attempt to Introduce Studious Youth Into the Elements of Pure Mathematics, by an Intuitive Method and Appropriate Evidence, With a Threefold Appendix"). While writing the *Tentamen,* Bolyai had his first difficulties with his son János. In spite of warnings from his father to avoid any preoccupation with Euclid's axiom, János not only insisted on studying the theory of parallels, but also developed an entirely unorthodox system of geometry based on the rejection of the parallel axiom, something with which his father could not agree. However, despite misgivings, Bolyai added his son's paper to the first volume and thus, unwittingly, gave it immortality. In 1834, a Hungarian version of Volume I was published.

The *Tentamen* itself, the fundamental ideas of which may date back to Bolyai's Göttingen days, is an attempt at a rigorous and systematic foundation of geometry (Volume I) and of arithmetic, algebra, and analysis (Volume II). The huge work shows the critical spirit of a man who recognized, as did few of his contemporaries, many weaknesses in the mathematics of his day, but was not able to reach a fully satisfactory solution of them. Nevertheless, when it is remembered that Bolyai worked in almost total isolation, the *Tentamen* is a most remarkable witness to the sharpness of his mind and to his perseverance. In many respects, he can be taken as a precursor of Gottlob Frege, Pasch, and Georg Cantor; but, as with many pioneers, he did not enjoy the credit that accrued to those who followed him.

The *Tentamen* was almost totally unappreciated by Bolyai's contemporaries, although Gauss expressed his pleasure at finding "everywhere thoroughness and independence." Disappointed and again a widower, the sensitive man found little consolation in the equally disappointed János, who after his retirement from military service had come to live in Marosvásárhely. The two men often clashed. In 1837 both entered a contest on complex numbers sponsored by the Jablonow Society in Leipzig. The elder Bolyai's contribution was taken essentially from his *Tentamen.* When no prize was awarded to either of them, their disillusionment grew; but whereas the son sank more and more into melancholy, the father—poetic, musical, and venerable—remained an outstanding, although somewhat eccentric, citizen of the provincial town, who was often consulted on technical questions. Both men also wrote much on a theory of salvation for mankind, and both returned occasionally to mathematics. Besides some elementary books, Bolyai published a summary of his *Tentamen* in German as *Kurzer Grundriss eines Versuches* (1851); after retiring from college teaching, and after having heard of Gauss's death, he wrote *Abschied von der Erde.* He died after suffering several strokes.

BIBLIOGRAPHY

I. ORIGINAL WORKS. Among Bolyai's works are *Az arithmetica Eleje* ("Elements of Arithmetic," Marosvásárhely, 1830); *Ürtan elemei kerdóknek* ("Elements of the Theory of Space for Beginners," Marosvásárhely, 1850–1851); and *Kurzer Grundriss eines Versuches* (Marosvásárhely, 1851). His major work, the *Tentamen,* was published in Latin in 2 vols. (Marosvásárhely, 1832–1833); 2nd ed., Vol. I (Budapest, 1897), Vol. II (Budapest, 1904), with an additional volume of figures.

II. SECONDARY LITERATURE. For information on Bolyai, see P. Stäckel, *W. und J. Bolyai, Geometrische Untersuchungen,* 2 vols. (Leipzig, 1913): the first volume is biographical; the second contains German translations of the theory of parallels of 1804, parts of the *Tentamen,* and the *Kurzer . . . Versuches.* The correspondence between Bolyai and Gauss is found in *Briefwechsel zwischen C. F. Gauss und W. Bolyai* (Leipzig, 1899). Further biographical material may be found in L. David, *A két Bólyai élete és munkássága* ("Life and Work of the Two Bolyais," Budapest, 1923) and "Die beiden Bolyai," supp. to *Elemente der Mathematik,* no. 11 (1951). A memorial work, *Bolyai Farkas 1856–1956,* was published in Târgu-Mureş in 1957. A stage play by Lászlo Németh, "A két Bólyai" ("The Two Bolyais") was first produced in 1962, and is collected in the author's *Változatok egy témara* (Budapest, 1961). See also K. R. Biermann, "Ein Brief von Wolfgang Bolyai," in *Mathematische Nachrichten,* **32** (1966), 341–346.

D. STRUIK

BOLYAI, JÁNOS (JOHANN) (*b.* 15 December 1802, Koloszvár [German, Klausenburg], Transylvania, Hungary [now Cluj, Rumania]; *d.* 27 January 1860, Marosvásárhely, Hungary [now Târgu-Mureş, Rumania]), *mathematics.*

The son of Farkas (Wolfgang) Bolyai and Susanna von Árkos Bolyai, János Bolyai received his early education in Marosvásárhely, where his father was professor of mathematics, physics, and chemistry at Evangelical-Reformed College. The precocious lad was first taught by his father and showed early proficiency not only in mathematics but also in other fields, such as music. He mastered the violin at an early age. From 1815 to 1818, he studied at the college where his father taught. The elder Bolyai had hopes that the son would go on to Göttingen to study with his friend Gauss, but he did not. In 1818 János entered the imperial engineering academy in Vienna, where he received a military education; he remained there until 1822.

From his father, János had inherited an interest in the theory of parallels; but in 1820 his father warned him against trying to prove the Euclidean axiom that there can be only one parallel to a line through a point outside of it:

> You should not tempt the parallels in this way, I know this way until its end—I also have measured this bottomless night, I have lost in it every light, every joy of my life— . . . You should shy away from it as if from lewd intercourse, it can deprive you of all your leisure, your health, your peace of mind and your entire happiness.— This infinite darkness might perhaps absorb a thousand giant Newtonian towers, it will never be light on earth, and the miserable human race will never have something absolutely pure, not even geometry . . . [Stäckel, pp. 76–77].

In the same year, however, János began to think in a direction that led him ultimately to a non-Euclidean geometry. He profited by conversations with Karl Szász, governor in the house of Count Alexis Teleki. In 1823, after vain attempts to prove the Euclidean axiom, he found his way by assuming that a geometry can be constructed without the parallel axiom; and he began to construct such a geometry. "From nothing I have created another entirely new world," he jubilantly wrote his father in a letter of 3 November 1823. By this time János had finished his courses at the academy and had entered upon a military career, beginning as a sublieutenant. His duties took him first to Temesvár (now Timişoara, Rumania), in 1823–1826, then to Arad (Rumania), in 1826–1830, and finally to Lemberg (now Lvov, W. Ukraine), where in 1832 he was promoted to lieutenant second class. During his military service, he was often plagued with intermittent fever, but he built up a reputation as a dashing officer who dueled readily. In 1833 he was pensioned off as a semi-invalid, and he returned to his father's home in Marosvásárhely.

While visiting his father in February 1825, János had shown him a manuscript that contained his theory of absolute space, that is, a space in which, in a plane through a point P and a line l not through P there exists a pencil of lines through P which does not intersect l. When this pencil reduces to one line, the space satisfies the Euclidean axiom. Farkas Bolyai could not accept this geometry, mainly because it depended on an arbitrary constant, but he finally decided to send his son's manuscript to Gauss. The first letter (20 June 1831) went unanswered, but Gauss did answer a second letter (16 January 1832). In this famous reply, dated 6 March 1832 and directed to his "old, unforgettable friend," Gauss said:

> Now something about the work of your son. You will probably be shocked for a moment when I begin by saying *that I cannot praise it,* but I cannot do anything else, since to praise it would be to praise myself. The whole content of the paper, the path that your son has taken, and the results to which he has been led, agree almost everywhere with my own meditations, which have occupied me in part already for 30–35 years. Indeed, I am extremely astonished. . . .

Further on, after mentioning that there had been a time when he had been inclined to write such a paper himself, Gauss continued, "Hence I am quite amazed, that now I have been saved the trouble, and I am very glad indeed that it is exactly the son of my ancient friend who has preceded me in such a remarkable way." Gauss ended with some minor remarks, among them a challenge to János to determine, in his geometry, the volume of a tetrahedron, and a critique of Kant's theory of space.

It is now known from Gauss's diaries and from some of his letters that he was not exaggerating; but for János the letter was a terrible blow, since it robbed him of the priority. Even after he became convinced that Gauss spoke the truth, he felt that Gauss had done wrong in remaining silent about his discovery. Nevertheless, he allowed his father to publish his manuscript, which appeared as an appendix to the elder Bolyai's *Tentamen* (1832), under the title "Appendix scientiam spatii absolute veram exhibens" ("Appendix Explaining the Absolutely True Science of Space"). This classic essay of twenty-four pages, which contains János' system of non-Euclidean geometry, is the only work of his published in his lifetime. Gauss's letters had such a discouraging influence on him that he withdrew into himself more and more, and for long periods he did hardly any mathematics. Disappointment grew when his essay evoked no response from other mathematicians.

After his retirement from the army, János lived with his father, who was then a widower. This arrangement lasted only a short time, however. Tension grew between father and son, who were both disappointed at the poor reception given their work, and János withdrew to the small family estate at Domáld, visiting Marosvásárhely only occasionally. In 1834 he contracted an irregular marriage with Rosalie von Orbán. The couple had three children, the first born in 1837.

In an attempt to reestablish themselves in mathematics, both father and son participated in the Jablonow Society prize contest in 1837. The subject was the rigorous geometric construction of imaginary quantities, at that time a subject to which many mathematicians (for example, Augustin Cauchy, W. R. Hamilton, and Gauss) were paying attention. The Bolyais' solutions were too involved to gain a prize, but János' solution resembled that of Hamilton, which was published about the same time, although in simpler terms, and which considered complex numbers as ordered pairs of real numbers. Again the Bolyais had failed to obtain due recognition. János continued to do mathematical work, however, some of it strong and some, because of his isolation, very weak. His best work was that on his absolute geometry, on the relation between absolute trigonometry and spherical trigonometry, and on the volume of the tetrahedron in absolute space. On the last subject, there are notes written as late as 1856. Nikolai Lobachevski's *Geometrische Untersuchungen zur Theorie der Parallellinien* (1840), which reached him through his father in 1848, worked as a powerful challenge, for it established independently the same type of geometry that he had discovered. In his later days he occasionally worried about the possibility of contradictions in his absolute geometry—a real difficulty that was not overcome until Beltrami did so later in the nineteenth century. János also worked on a salvation theory, which stressed that no individual happiness can exist without a universal happiness and that no virtue is possible without knowledge.

János' father died in 1856 and his relationship with Rosalie ended at about the same time, thus depriving him of two of his few intimate contacts. However, in the four years left to him, he did have his good moments. He could write enthusiastically about the ballet performances of the Vienna Opera and compose some beautiful lines to the memory of his mother. He died after a protracted illness, and was buried in the Evangelical-Reformed Cemetery in Marosvásárhely.

The "Appendix" was practically forgotten until Richard Baltzer discussed the work of Bolyai and Lobachevski in the second edition of his *Elemente der Mathematik* (1867). Jules Houel, a correspondent of Baltzer's, then translated Lobachevski's book into French (1867) and did the same with Bolyai's "Appendix" (1868). Full recognition came with the work of Eugenio Beltrami (1868) and Felix Klein (1871).

BIBLIOGRAPHY

In addition to the works cited in the article on Farkas Bolyai, see the English translation of the "Appendix," with an introduction, by G. B. Halsted (Austin, Texas, 1891; new ed., Chicago–London, 1914), reprinted in R. Donola, *Non-Euclidean Geometry* (reprinted New York, 1955). There are accounts of Bolyai's geometry in the many books on non-Euclidean geometry. See D. M. Y. Sommerville, *Bibliography of Non-Euclidean Geometry* (London, 1911). Further material may be found in I. Tóth, *Bolyai János élete és miive* ("Life and Work of Johann Bolyai," Bucharest, 1953); *János Bolyai Appendix* (Bucharest, 1954), in Rumanian.

E. Sarlóska, "János Bolyai, the Soldier," in *Magyar tudományos akadémia Matematikai es fizikai osztályanak kőzleményei*, **15** (1965), 341–387, contains a documentary study of Bolyai's army life.

D. J. STRUIK

BOLZA, OSKAR (*b.* Bergzabern, Germany, 12 May 1857; *d.* Freiberg im Breisgau, Germany, 5 July 1942), *mathematics.*

Bolza's principal mathematical investigations covered three topics: the reduction of hyperelliptic integrals to elliptic integrals, elliptic and hyperelliptic functions, and the calculus of variations. On the first two topics Bolza proved an able follower of his teachers Karl Weierstrass and Felix Klein. In the realm of reduction problems he worked chiefly on third-degree and fourth-degree transformations. He stressed elliptic theory and often reformulated it as a special case of the hyperelliptic theory in his papers on hyperelliptic θ, σ, and ζ functions. On the third topic his book *Lectures on the Calculus of Variations* (1904) presented the most recent contributions of Weierstrass, Adolf Kneser, and David Hilbert, as well as his own comments. In this book and other writings he added to the theory in the plane and the problem of Lagrange with fixed end points. He extended and applied existence theorems for implicit functions and for solutions to differential equations. Bolza's most significant single contribution was the unification of the problems of Lagrange and Mayer into his more general problem of Bolza. This problem was the fifth, classical necessary condition for a minimum to appear. Leonhard Euler, Adrien-Marie Legendre, Karl Jacobi, and Karl Weierstrass had formulated the previous four. The problem of Bolza in parametric form is to find in a class of arcs $y_i(x)$, where ($i = 1, \cdots, n; x_0 < x < x_1$) which satisfy equations of the form

$$\phi_\beta(y, y') = 0 \qquad (\beta = 1, \cdots, p)$$
$$\psi_\mu(y) = 0 \qquad (\mu = 1, \cdots, q)$$

and end conditions of the form

$$J_\gamma[y(x_0), y(x_1)] - 0 \qquad (\gamma = 1, \cdots, r),$$

one that minimizes a sum of the form

$$I = G[y(x_0), y(x_1)] + \int_{x_0}^{x_1} f(y, y') \, dx.$$

In this formulation the problem of Mayer with variable end points is the problem of Bolza with its integrand function f identically zero, while the problem of Lagrange with variable end points is the case when G is absent from I.

The son of Emil Bolza and Luise König, Bolza displayed a variety of interests during his youth. At the Gymnasium in Freiburg, he eagerly studied languages and comparative philology, but when he entered the University of Berlin in 1875, he decided to study physics under Kirchhoff and Helmholtz. After tiring of experimental work, in 1878 Bolza switched to the study of pure mathematics. The chief mentor for his mathematical studies at Berlin was Weierstrass, who was particularly interested in the calculus of variations and strongly influenced the course of Bolza's research. From 1878 to 1880 Bolza's studies led him from Berlin to Strasbourg, back to Berlin, and then to Göttingen. After deciding that he wanted to teach, either in a Gymnasium or a university, he interrupted his mathematical studies from 1880 to 1883 in order to prepare for and pass the Staatsexamen, a prerequisite for Gymnasium teaching. From 1883 to 1885 Bolza returned to his mathematical studies, working privately on his doctoral dissertation at the University of Freiburg. After Felix Klein accepted his dissertation on hyperelliptic integrals, he received his doctorate from Göttingen in June 1886. He followed this with a year's private seminar with Klein in Göttingen.

After completing his studies, two reasons prompted Bolza to abandon his teaching plans and go to the United States. Friends complained of the lack of time allowed for research in German schools; second, he was not robust and feared that Gymnasium teaching would be too strenuous for him. He had been rejected for military service in 1887. Bolza arrived in the United States in 1888, and in January 1889 he became reader in mathematics at Johns Hopkins University. In October of that same year he advanced to associate professor in mathematics at Clark University. On 1 January 1893, Bolza became associate professor of mathematics at the newly founded University at Chicago. He advanced to full professor in the following year.

After 1898 Bolza felt a growing desire to return to Germany. In 1908 the death of Heinrich Maschke, an old college friend and a colleague at Chicago, severed perhaps the strongest bond that kept him in America. In addition, he felt that America had made great strides in the training of scholars and believed that he should step aside for the increasing number of young American-trained teachers. In 1910, when he left the University of Chicago, he was given the title of nonresident professor of mathematics.

Upon his return to Germany, Bolza studied various subjects. He accepted the position of honorary professor of mathematics at the University of Freiburg, but in a few years World War I turned his prime interest from mathematics to religious psychology and languages, especially Sanskrit. He had grown up in a pre-World War I Europe in which people believed no major war could occur again: all problems would be resolved by reason. The trauma of World War I shook the foundations of thought for many, including Bolza; he turned to religious psychology and Sanskrit in search of answers on how to establish a better society. Bolza studied Sanskrit so that he could read firsthand the literature concerning the religious systems of India. His new interests prompted him to interrupt his mathematical research in 1922 and his class lectures in 1926. Bolza became more and more engrossed in psychological research, and he devoted full time to it from 1926 until 1929. The result of this work was *Glaubenlose Religion,* which he published in 1931 under the pseudonym F. H. Marneck.

In his final years Bolza remained an active academician. He returned to lecturing on mathematics at the University of Freiburg from 1929 to 1933, when he retired. After his retirement he continued to publish papers on mathematics and religious psychology. At the request of friends he wrote a brief autobiography, *Aus meinen Leben.* As late as 1939 Bolza wrote to friends of his interest in studying the foundations of geometry.

BIBLIOGRAPHY

I. ORIGINAL WORKS. Books by Bolza include *Lectures on the Calculus of Variations* (Chicago, 1904); *Vorlesungen über Variationsrechnung, Umgearbeitete und stark Vermehrte deutsche Ausgabe der "Lectures on the Calculus of Variations"* (Leipzig, 1908); "Gauss und die Variationsrechnung," in Gauss's *Werke,* X, pt. 2, 5 (Göttingen, 1922); *Aus meinen Leben* (privately published, 1936); *Glaubenlose Religion* (Munich, 1931), published under the pseudonym F. H. Marneck; *Meister Eckehart als Mystiker, eine religions-psychologische Studie* (Munich, 1938). See also "Elliptic Functions," a handwritten record of a course given at the University of Chicago, probably in the winter quarter, 1901; and "Lectures on Integral Equations," in W. V. Lovitt, *Linear Integral Equations* (New York, 1924).

For articles by Bolza see "Über die Reduction hyperelliptischen Integrale erste Ordnung und erster Gattung auf elliptische durch eine Transformation vierten Grades," in *Mathematical Annals,* **28** (1887), 447–456; "On Binary

Sextics With Linear Transformations Into Themselves," in *American Journal of Mathematics*, **10** (1888), 47–70; "The Elliptic Function Considered as a Special Case of the Hyperelliptic Function," in *Transactions of the American Mathematical Society,* **1** (1900), 53–65; "Weierstrass' Theorem and Kneser's Theorem on Transversals for the Most General Case of an Extremum of a Simple Definite Integral," *ibid.,* **7** (1906), 459–488; "Die Lagrangeschen Multiplikatorenregel in die Variationsrechnung für den Fall von gemischten Bedingung bei variabeln Endpunkten," in *Mathematical Annals,* **64** (1907), 370–387; "Heinrich Maschke, His Life and Work," in *Bulletin of the American Mathematical Society,* **15** (1908), 85–95; "Über den Hilbertischen Unabhangigkeitssatz beim Lagrangeschen Variationsproblem," in *Rendiconti del Circolo matematico di Palermo,* **31** (1911), 257–272; "Über zwei Eulersche Aufgaben aus der Variationsrechnung," in *Annali di matematica pura ed applicata,* **20** (1913), 245–255; "Einführung in E. H. Moore's *General Analysis* und deren Anwendung auf die Verallgemeinerung der Theorie der linearen Integralgleichungen," in *Jahrbuch Deutschen Mathematische Verein,* **23** (1914), 248–303; "Der singuläre Fall der Reduction hyperelliptische Integrale erster Ordnung auf elliptische durch Transformation dritten Grades," in *Mathematical Annals,* **111** (1935), 477–500.

II. Secondary Literature. An article on Bolza is G. A. Bliss, "Oskar Bolza—In Memoriam," in *Bulletin of the American Mathematical Society,* **50** (1944), 478–489, which contains a chronological list of Bolza's writings.

Ronald S. Calinger

BOLZANO, BERNARD (*b.* Prague, Czechoslovakia, 5 October 1781; *d.* Prague, 18 December 1848), *philosophy, mathematics, logic, religion, ethics.*

Bolzano was born in one of the oldest quarters of Prague and was baptized Bernardus Placidus Johann Nepomuk. His mother, Caecilia Maurer, daughter of a hardware tradesman in Prague, was a pious woman with an inclination to the religious life. At the age of twenty-two she married the elder Bernard Bolzano, an Italian immigrant who earned a modest living as an art dealer. The father was a widely read man with an active social conscience, and felt responsible for the well-being of his fellow men. He put his ideas into practice and took an active part in founding an orphanage in Prague.

Bernard was the fourth of twelve children, ten of whom died before reaching adulthood. Of delicate health, he had a quiet disposition, although he was easily irritated and very sensitive to injustice.

From 1791 to 1796 he was a pupil in the Piarist Gymnasium, and in 1796 he entered the philosophical faculty of the University of Prague, where he followed courses in philosophy, physics, and mathematics. His interest in mathematics was stimulated by reading A.

G. Kästner's *Anfangsgründe der Mathematik,* because Kästner took care to prove statements which were commonly understood as evident in order to make clear the assumptions on which they depended.

The benevolence of the environment in which Bernard Bolzano was reared, both at home and in school, influenced his entire life. In fact, he raised to the supreme principle of moral conduct the precept always to choose that action, of all possible actions, which best furthers the commonweal.

After having finished his studies in philosophy in 1800, Bolzano entered the theological faculty. These studies did not strengthen his belief or resolve his doubts concerning the truth and divinity of Christian religion, but he found a solution in his professor's statement that a doctrine may be considered justified if one is able to show that faith in it yields moral profit. This made it possible for Bolzano to reconcile religion with his ethical views and to consider Catholicism the perfect religion.

In 1805 Emperor Franz I of Austria, of which Czechoslovakia was then a part, decided that a chair in the philosophy of religion would be established in each university. The reasons for this were mainly political. The emperor feared the fruits of enlightenment embodied in the French Revolution, and therefore was sympathetic to the Catholic restoration that joined issue with the spirit of freethinking which had spread over Bohemia. Bolzano, who had taken orders in 1804, was called to the new chair at the University of Prague in 1805.

Spiritually, Bolzano belonged to the Enlightenment. Both his religious and social views made him quite unsuitable for the intended task, and difficulties were inevitable. His appointment was received in Vienna with suspicion, and it was not approved until 1807.

Bolzano's lectures, in which he expounded his own views on religion, were enthusiastically received by his students; in particular, his edifying Sunday speeches (*Erbauungsreden,* in *Gesammelte Schriften,* I) to the students were warmly applauded. He was respected by his colleagues, and in 1815 became a member of the Königlichen Böhmischen Gesellschaft der Wissenschaften and, in 1818, dean of the Prague philosophical faculty.

In the struggle between the Catholic restoration and the Enlightenment action against Bolzano was postponed until 1816, when a charge was brought against him at the court in Vienna; his dismissal was issued on 24 December 1819. He was forbidden to publish and was put under police supervision. Bolzano repeatedly refused to recant the heresies of which he was accused, and in 1825 the action came to an end through the intervention of the influential nationalist leader J. Dobrovsky.

From 1823 on, Bolzano spent summers on the estate of his friend J. Hoffmann, near the village of Těchobuz in southern Bohemia. He lived there permanently from 1831 until the death of Mrs. Hoffmann in 1842. Then he returned to Prague, where he continued his mathematical and philosophical studies until his death.

Though Bolzano's career was concerned mainly with social, ethical, and religious questions, he was irresistibly attracted by philosophy, methodology of science, and especially mathematics and logic. His philosophical education—which acquainted him with the Greeks and with Wolff, Leibniz, and Descartes—convinced him of the necessity of forming clear concepts and of sound reasoning, starting from irreducible first principles and using only intrinsic properties of defined concepts. Such methods could not take into account properties alien to their definition, such as geometrical evidence (see *Beyträge*). On occasion he applied these principles with remarkable results; on other occasions, however, his philosophical approach, particularly to mathematics, led him to introduce insufficiently founded and incorrect assumptions. Such was the case in *Die drey Probleme,* which was explicitly intended to lead to a completely new theory of space—which, of course, it failed to do. In the domain of mathematical analysis, however, Bolzano's struggle for clear concepts did lead to profound results that, unfortunately, did not attract the attention of the mathematical world or influence the development of mathematics.

Around the turn of the nineteenth century, mathematicians in Europe were concerned with two major problems. The first was the status of Euclid's parallel postulate, and the second was the problem of providing a solid foundation for mathematical analysis, so as to remove the so-called scandal of the infinitesimals. Bolzano tried his hand at both problems, with varying success.

In 1804 he published his *Betrachtungen über einige Gegenstände der Elementargeometrie,* in which he tried to base the theory of triangles and parallels on a theory of lines, without having recourse to theorems of the plane. The full development of this theory of lines was postponed—and although Bolzano often returned to the theory of parallels (without success), his linear theory was never completed.

In the course of the following years, Bolzano became acquainted with the extensive work done in the **theory** of parallels, such as that of A. M. Legendre and **F. K. Schweikart**. There are no indications that he ever knew of the final breakthrough to non-

Euclidean geometry by Nikolai Lobachevski and János Bolyai, although the latter's work was published in 1832 in Hungary. Bolzano's own manuscript "Anti-Euklid" follows a different line of thought and is devoted mainly to methodological criticism of Euclid's *Elements*. In fact, in his methodological principles he went so far as to require definitions of such geometrical notions as those of (simple closed) curve, surface, and dimension (see *Die drey Probleme; Ueber Haltung;* "Geometrische Begriffe"; and E. Winter, *Die historische Bedeutung*), and to require proofs of such seemingly evident statements as "A simple closed curve divides the plane into two parts," which is now known as the Jordan curve theorem. Indeed, the discussion in "Anti-Euklid" confirms the opinion held by H. Hornich that Bolzano was the first to state this as a theorem (requiring proof). The problems raised in this connection by Bolzano found their final solution at the end of the nineteenth century and the beginning of the twentieth in that branch of mathematics called topology (for a discussion of these matters, see Berg, *Bolzano's Logic*).

It should be emphasized that Bolzano was not the only one, or even the first, to be concerned with the problem of rigorous proofs in mathematics. A curious fact, however, is that although many of the mathematicians actively interested in the problem of the foundation of mathematical analysis surpassed him in mathematical skill, Bolzano overcame them decisively in the foundation of analysis, in which as early as 1817 (see *Rein analytischer Beweis*) he obtained fundamental results, which were completed in 1832–1835 in his theory of real numbers (see Rychlík, *Theorie der reellen Zahlen*).

The introduction of infinitesimals by Newton and Leibniz in the seventeenth century met with violent resistance from philosophers and mathematicians, and vivid discussions on infinitesimal quantities went on throughout the eighteenth century. Bishop Berkeley's attack in *The Analyst* (1734) is well known. Although Leibniz himself did not consider the existence of infinitesimals to be well founded, and held that their use could be avoided, he admitted them as ideal quantities, which could be handled in calculations like ordinary quantities (except that they equal their finite multiples). These arithmetical properties, however, formed the weak point in the theory of infinitesimals because of the lack of an exhaustive description of the real number system, which was accomplished only in the second half of the nineteenth century. How badly the general laws of arithmetic were understood may be illustrated by the problem of division by zero. This problem kept Bolzano busy from 1815 on, and he never fully got to the bottom of it, as can be seen, for instance, in §34 of his *Paradoxien des Unendlichen,* where he admits identities of the form $A/0 = A/0$, despite his knowledge of Ohm's solution.

To overcome the difficulties presented by infinitesimals, Lagrange proposed to base analysis on the existence of Taylor's expansion for functions, and this attitude was widely accepted for a time. Bolzano did not escape its influence, and made extensive studies on Taylor's theorem (see *Der binomische Lehrsatz* and "Miscellanea mathematica").

A different position was held by d'Alembert, who proposed to found differential calculus on the notion of limit and contended that differential calculus does not treat of infinitely small quantities, but of limits of finite quantities.

Certainly d'Alembert's opinion impressed his contemporaries, and many attempts, such as Lagrange's, were made to free differential calculus from infinitesimals. The first successful attempt was made by Bolzano in his *Rein analytischer Beweis* (1817), which is devoted to a proof of the important theorem which states that if for two continuous functions f and ϕ we have $f(\alpha) < \phi(\alpha)$ and $f(\beta) > \phi(\beta)$, then there is an x between α and β such that $f(x) = \phi(x)$.

Bolzano argues that a sound proof of this theorem presupposes a sound definition of continuous function. In his introduction he gives such a definition, which is important because it is the first that does not involve infinitesimals, and is, essentially, the one used up to now. In the more accurate formulation of Volume I of the *Functionenlehre,* it reads: If $F(x + \Delta x) - Fx$ in absolute value becomes less than an arbitrary given fraction $1/N$, if one takes Δx small enough, and remains so, the smaller one takes Δx, the function Fx is said to be continuous (in x). Bolzano also distinguishes between right and left continuity.

It should be noted that in 1821 Cauchy, in his *Cours d'analyse,* adopted a different definition: $f(x + \alpha) - f(x)$ infinitely small for all infinitely small α. Because of its elegance, this definition was generally accepted.

In his proof of the theorem in the *Rein analytischer Beweis,* Bolzano uses a lemma that later proved to be the cornerstone of the theory of real numbers. He was fully aware of the paramount importance of this theorem, and he formulated it with great generality, as follows: If a property M does not hold for all values of a variable x, but does hold for all x less than a certain u, then there is a quantity U, which is the greatest of all those for which it holds that all x less than it have property M.

In modern terminology, U is the greatest lower bound of the (nonempty) set of x for which M does not hold.

Though the two theorems mentioned above already bear witness to the outstanding content of the *Rein analytischer Beweis,* it contains another theorem of equal importance, which is known as Cauchy's condition of convergence. Bolzano devotes a whole section to it and proves that if a sequence $F_1(x)$, $F_2(x)$, $F_3(x)$, $\cdots$, $F_n(x)$, $\cdots$, $F_{n+r}(x)$ is such that the difference between the nth term $F_n(x)$ and every later one $F_{n+r}(x)$ remains less than any given quantity if only n has been taken large enough, then there is a fixed quantity, and only one, to which the terms approach—as near as one likes, if one continues the sequence far enough.

The proofs of these theorems are incomplete, and were bound to be so, because complete proofs would require a precise notion of quantity (real number), which Bolzano did not have at that time. He was aware of at least some of the difficulties involved, because his methodology, as expounded in the *Beyträge,* demanded the systematic development of a theory of real numbers that should logically precede his theory of real functions.

A fairly complete theory of real functions is contained in Bolzano's *Functionenlehre,* including many of the fundamental results that were rediscovered in the second half of the nineteenth century through the work of K. Th. Weierstrass and many others.

In the first part, concerning continuous functions, it is shown that a function Fx which is unbounded on the closed interval $[a, b]$ cannot be continuous on $[a, b]$. The proof uses the so-called Bolzano-Weierstrass theorem that a bounded infinite point set has an accumulation point. For this theorem Bolzano refers to his own work, in which up to now it has not been found.

Functions continuous on a closed interval attain there a maximal and a minimal value. Bolzano sharply distinguishes between continuity and the property of assuming intermediate values, and proves that continuous functions assume all values intermediate between any two function values, while the converse is shown not to be true.

In §13 Bolzano notices a property of continuous functions which is rather close to uniform continuity, a notion which is due to E. Heine (1870, 1872). In connection with the function

$$Fx = \frac{1}{1 - x},$$

which is continuous on (0, 1), he observes that though a function may be continuous on the open interval (a, b), it does not follow that a real number e, independent of x in (a, b), exists, such that one need not choose $\Delta x < e$ in order that $F(x + \Delta x) - Fx < 1/N$. Indeed, if in the example x approaches 1, the Δx has to be taken increasingly smaller in order that $F(x + \Delta x) - Fx < 1/N$.

As K. Rychlík has pointed out in his commentary in Volume I of the *Schriften*, the said property is weaker than uniform continuity. One may be tempted, however, to assume that Bolzano intended uniform continuity and that only the formulation is defective. The more so, when in "Verbesserungen und Zusätze" we find the correct theorem: If the function Fx is continuous on the closed interval $[a, b]$, then there exists a (real) number e such that for all x in $[a, b]$ the Δx need not be chosen $< e$ in order that $F(x + \Delta x) - Fx <$ a given number $1/N$. Further reading reveals, however, that Bolzano had no clear notion of uniform continuity after all.

Careful attention is paid to the connection between monotonicity and continuity. Thereby the following correction to §79 of the *Functionenlehre* in "Verbesserungen und Zusätze" should not remain unnoticed: If the (real) function Fx increases (or decreases) steadily on the closed interval $[a, b]$, then Fx is continuous on $[a, b]$, with the exception of a set of isolated values of x which may be infinite or finite.

The most remarkable result of the *Functionenlehre*, however, is the construction in §75 of the so-called Bolzano function. There Bolzano constructs a function as the limit of a sequence of continuous functions which is continuous on the closed interval $[0, 1]$ such that it is in no subinterval monotone. The importance of this function, however, derives from another property—its nondifferentiability.

The second part of the *Functionenlehre* is devoted to derivatives. Particular emphasis is laid on the distinction between continuity and differentiability. Bolzano shows that the above-mentioned function, though continuous in $[0, 1]$—which is not proved—fails to be differentiable on an everywhere dense subset of $[0, 1]$. In fact, the function is nowhere differentiable on $[0, 1]$. This example preceded by some forty years that of Weierstrass, who in 1875 published a different example of a nowhere differentiable continuous function which roused wide interest and even indignation.

Bolzano erroneously believed that his function was continuous because it was the limit of continuous functions; in explanation it may be remarked that Cauchy made the same error. Apparently Bolzano was not aware of a counterexample given by N. H. Abel in 1826.

Though the second part of the *Functionenlehre*

contains many interesting results, it contains as many errors, such as the statement that the derivative of an infinite series is the sum of the derivatives of its terms, and the conclusion that the limit of a sequence of continuous functions again is a continuous function. Both errors tie up with the notion of uniformity and therefore are explainable; the following error is less easy to understand. In 1829 Cauchy put forward the function

$$C(x) = e^{-1/x^2} \text{ (to be completed by } C[0] = 0)$$

as an example of a function, different from zero for all $x \neq 0$, having all its derivatives zero for $x = 0$ and, hence, not admitting a Taylor expansion in the neighborhood of $x = 0$. Bolzano knew of this example in 1831 (see "Miscellanea mathematica," p. 1999), yet in the *Functionenlehre*, §89, we find the following theorem:

If $F^{n+r}a = 0$ for $r > 0$, then

$$F(a + h) = Fa + h \cdot F'a + \frac{h^2}{2} F''a + \cdots +$$

$$\frac{h^n}{2.3 \cdots n} F^n a,$$

which is clearly refuted by Cauchy's example.

The firm base on which the theory of functions was to rest, according to Bolzano's methodology—the theory of quantities (real numbers)—was completed in 1832–1835. Like most of Bolzano's mathematical work, it remained in manuscript and was published for the first time only in 1962 (see Rychlík, *Theorie der reellen Zahlen*). As a result, this bold enterprise failed to exercise any influence on the development of mathematics, which in the second half of the nineteenth century independently took the same course.

Real numbers occur in Bolzano's writings under the name of measurable infinite number-expressions. They make their appearance in "Miscellanea mathematica," pt. 22, p. 2000–2001 (1832), in connection with the geometric progression, which has inspired many interesting ideas. The representation of the sum S of an infinite geometric progression as given in the footnote to §18 of the *Paradoxien des Unendlichen* is paradigmatic.

Bolzano's idea is that descriptions of (real) numbers make sense only if they permit determination of the numbers described with an arbitrarily high degree of precision by means of rational numbers. In general, these descriptions require an infinite number of arithmetical operations to be carried out—for instance, the sum S of a geometric progression. These are the infinite number expressions with which Bolzano is concerned. If the results obtained by carrying out only a finite number of operations is always positive, the number expression is called positive.

An infinite number expression S is called measurable (or determinable by approximation) if to any natural number q there is an integer p, such that

$$S = \frac{p}{q} + P_1 = \frac{p+1}{q} - P_2,$$

where P_1 and P_2 are positive (infinite) number expressions.

Infinitely small numbers are those for which all $p = 0$, i.e., those S for which

$$S = P_1 = \frac{1}{q} - P_2,$$

as well as their opposites.

An essential requirement is that measurable numbers differing in an infinitesimal number have to be considered as equal. Therefore, equality is defined by equality of p for all q in the above representation of infinite number expressions. On the basis of these definitions, Bolzano completed his systematic exposition of the theory of real numbers and, thereby, of mathematical analysis.

The elaboration is not quite satisfactory because of many errors due to his insufficient mathematical skill (for interpretations and evaluation of Bolzano's theory, see Laugwitz, "Bemerkungen"; van Rootselaar, "Bolzano's Theory of Real Numbers"; Rychlík, *Theorie der reellen Zahlen*).

The essential differences between Bolzano's incomplete theory of real numbers and those of, for instance, Weierstrass and Georg Cantor are marked by the shift from intensional meaning, in Bolzano's work, toward a general tendency to extensionality, and, above all, by the possibility of creating new mathematical objects by means of definition by abstraction, based on equivalence relations, of which Bolzano was unaware. These differences also appear clearly in his *Paradoxien des Unendlichen,* which contains many interesting fragments of general set theory.

The existence of infinite sets is proved in a way similar to that followed by Richard Dedekind in his memoir *Was sind und was sollen die Zahlen* (1887). Most noteworthy, however, is that in §20 of *Paradoxien des Unendlichen* Bolzano is at the border of cardinal arithmetic, a border which he is unable to cross. There he notices a property of infinite sets: that they may be brought into one-to-one correspondence with a proper subset. In fact, he observes that this will always be the case with infinite sets. That two sets may be brought into one-to-one correspondence is no reason for him to consider them to be composed

of the same number of elements (*Paradoxien des Unendlichen,* §21), however, and he sees no reason to consider such sets as equal. On the contrary, in order for two sets to be considered as equal, he argues, they must be defined on the same basis (*gleiche Bestimmungsgründe haben*). Needless to say, this is too vague to be dealt with mathematically. Here again, we see that Bolzano halts at a point where application of the method of definition by abstraction would have opened entirely new fields of knowledge.

Precisely that property of infinite sets noticed by Bolzano was afterward used by Dedekind as a definition of the infinite (1882). The introduction of equivalence classes of sets under one-to-one correspondence was fully exploited by Cantor in his theory of cardinals, a very important chapter of general set theory.

Bolzano planned to elaborate the methodology begun in his *Beyträge* and to develop it into a complete theory of science, of which a treatise on logic was to form the cornerstone. From 1820 on, he worked steadily on it, and his four-volume treatise *Wissenschaftslehre* appeared in 1837. The plan of the *Wissenschaftslehre* appears clearly from the following subdivision (see Kambartel, *Bernard Bolzano's Grundlegung der Logik,* pp. 14–17):

(1) Fundamental theory: proof of the existence of abstract truths and of the human ability to judge.

(2) Elementary theory: theory of abstract ideas, propositions, true propositions, and deductions.

(3) Theory of knowledge: condition of the human faculty of judgment.

(4) Heuristics: rules to be observed in human thought in the search for truths.

(5) Proper theory of science: rules to be observed in the division of the set of truths into separate sciences and in their exposition in truly scientific treatises.

The work did not induce a complete revision of science, as Bolzano hoped, but, on the contrary, remained unnoticed and did not exercise perceptible influence on the development of logic. Some of the innovations in logic contained in the first two volumes did attract attention, as well as excessive praise— notably from Edmund Husserl and Heinrich Scholz (see Berg, *op. cit.*; Kambartel, *op. cit.*; and the literature cited in them).

The rise of logical semantics, initiated by Alfred Tarski in the 1930's, has led to a revival of the study of Bolzano's logic in the light of modern logic (see Berg, *op. cit.*) and of his theory of an ideal language.

The heart of Bolzano's logic is formed by his concepts of (abstract) proposition (*Satz an sich*), abstract idea (*Vorstellung an sich*), truth, and the notions of derivability (*Ableitbarkeit*) and entailment (*Abfolge*).

These notions may be explained with the help of Bolzano's example:

(*a*) Cajus is a human being.

(*b*) All human beings have immortal souls.

(*c*) Cajus has an immortal soul.

First of all, (*a*) expresses an abstract proposition, which in itself has no real existence, but is something to which (*a*) refers and which is either true or false. An abstract proposition may be expressed in many ways linguistically, and it is said to be true if it asserts something as it actually is ("*so wie es ist*," *Wissenschaftslehre*, §25).

Bolzano argues that any proposition may be expressed in the normal form "*A* has *b*." For instance,

(*a′*) Cajus has human existence

is the normal form of the proposition expressed by (*a*).

Parts of propositions which are not themselves propositions are (abstract) ideas; for example, in (*a′*) the expression "human existence" refers to an abstract idea.

Between abstract propositions there exist relations, among which those of consistency and derivability are of paramount importance. Propositions $A, B, C, \cdots$ are called consistent with respect to the common ideas $i, j, \cdots$ if there are ideas $i′, j′, \cdots$ which, after substitution for $i, j, \cdots$ turn the propositions $A, B, C, \cdots$ into simultaneously true propositions $A′, B′, C′, \cdots$. Propositions $A′, B′, C′, \cdots$ are called derivable from $A, B, C, \cdots$ with respect to the ideas $i, j, \cdots$ whenever $A, B, C, \cdots, A′, B′, C′, \cdots$ are consistent with respect to $i, j, \cdots$ and if any substitution $i′, j′, \cdots$ for $i, j, \cdots$ that turns $A, B, C, \cdots$ into true propositions also turns $A′, B′, C′, \cdots$ into true propositions. According to Bolzano, (*c*) is derivable from (*a*) and (*b*).

The relation of entailment (*Abfolge*) may subsist between true propositions, and refers to the situation that A is true because $A_1, A_2, \cdots$ are true. The treatment of this notion, however, is rather unsatisfactory (see Berg, *op. cit.;* Buhl, *Ableitbarkeit und Abfolge;* Kambartel, *op. cit.* for details).

The resemblance that many of the concepts introduced by Bolzano bear to modern logic has led to the opinion that Bolzano may be considered a true precursor of modern logic. (For a detailed account, consult Berg, *op. cit.;* and Kambartel, *op. cit.;* for Bolzano's philosophy, Fujita, *Borutsāno no tetsugaku* ["Bolzano's Philosophy"]).

BIBLIOGRAPHY

I. ORIGINAL WORKS. Bolzano's published works include the following: *Betrachtungen über einige Gegenstände der Elementargeometrie* (Prague, 1804; repr. *Schriften,* V); *Beyträge zu einer begründeteren Darstellung der Mathematik* (Prague, 1810; new ed. by H. Fels, Paderborn, 1926); *Der binomische Lehrsatz und als Folgerung aus ihm der polynomische und die Reihen, die zur Berechnung der Logarithmen und Exponentialgrössen dienen, genauer als bisher erwiesen* (Prague, 1816); *Die drey Probleme der Rectification, der Complanation und der Cubirung, ohne Betrachtung des unendlich Kleinen, ohne die Annahmen des Archimedes, und ohne irgend eine nicht streng erweisliche Voraussetzung gelöst; zugleich als Probe einer gänzlichen Umstaltung der Raumwissenschaft, allen Mathematikern zur Prüfung vorgelegt* (Leipzig, 1817; repr. *Schriften,* V); *Rein analytischer Beweis des Lehrsatzes, dass zwischen je zwey Werthen, die ein entgegengesetztes Resultat gewähren, wenigstens eine reelle Wurzel der Gleichung liege* (Prague, 1817; new ed. in Ostwald's Klassiker der exakten Wissenschaften, no. 153 [Leipzig, 1905]; also in Kolman, *Bernard Bolzano*).

Also *Lebensbeschreibung des Dr. Bernard Bolzano mit einigen seiner ungedruckten Aufsätze und dem Bildnisse des Verfassers*, ed. M. J. Fesl (Sulzbach, 1836; repr. Vienna, 1875), an autobiography; *Wissenschaftslehre, Versuch einer ausführlichen und grösstentheils neuen Darstellung der Logik mit steter Rücksicht auf deren bisherigen Bearbeiter*, 4 vols. (Sulzbach, 1837; new ed. by A. Höfler and W. Schultz, 4 vols. [Leipzig, 1914–1931]; also in *Gesammelte Schriften*); *Paradoxien des Unendlichen*, ed. F. Přihonsky (Leipzig, 1851; English ed. by D. A. Steele, *Paradoxes of the Infinite* [New Haven, 1950]); *Ueber Haltung, Richtung, Krümmung und Schnörkelung bei Linien sowohl als Flächen sammt einigen verwandten Begriffen* (ed. in *Schriften,* V).

There are two collections of Bolzano's works. One is *Gesammelte Schriften*, 12 vols. (Vienna, 1882); the contents are as follows: I: *Erbauungsreden;* II: *Athanasia oder Gründe für die Unsterblichkeit der Seele;* III–VI: *Lehrbuch der Religionswissenschaft;* VII–X: *Wissenschaftslehre;* XI: *Dr. Bolzano und seine Gegner;* XII: *Bolzano's Wissenschaftslehre und Religionswissenschaft in beurteilender Übersicht*. The other is *Bernard Bolzano's Schriften*, 5 vols., ed. Königlichen Böhmischen Gesellschaft der Wissenschaften (Prague, 1930–1948), which contains the following: I: *Functionenlehre*, ed. K. Rychlik (1930); II: *Zahlentheorie*, ed. K. Rychlik (1931); III: *Von dem besten Staate*, ed. A. Kowalewski (1932); IV: *Der Briefwechsel B. Bolzano's mit F. Exner*, ed. E. Winter (1935); V: *Mémoires géometriques*, ed. J. Vojtech (1948).

Additional works are in manuscript: "Anti-Euklid" (fragment), in Österreichische Nationalbibliothek, Vienna, Handschriftensammlung, Series Nova, 3459, section 5 (also edited in Večerka, "Bernard Bolzano's Anti-Euklid"); "Geometrische Begriffe, die jeder kennt und nicht kennt," in Österreichische Nationalbibliothek, Vienna, Handschriftensammlung, Series Nova, 3459, sections 3b and 3c; "Miscellanea mathematica," 1–24, in Österreichische Nationalbibliothek, Vienna, Handschriftensammlung, Series Nova, 3453–3455; and "Verbesserungen und Zusätze zu dem Abschnitt von der Differenzialrechnung," in Österreichische Nationalbibliothek, Vienna, Handschriftensammlung, Series Nova, 3472, section 7.

II. Secondary Literature. Works on Bolzano include J. Berg, *Bolzano's Logic* (Stockholm, 1962), which has an extensive bibliography; and "Bolzano's Theory of an Ideal Language," in R. E. Olson, ed., *Contemporary Philosophy in Scandinavia* (Baltimore, in press); G. Buhl, "Ableitbarkeit und Abfolge in der Wissenschaftstheorie Bolzanos," in *Kantstudien*, **83** (1961); H. Fels, *Bernard Bolzano, sein Leben und sein Werk* (Leipzig, 1929), which includes a Bolzano bibliography; I. Fujita, *Borutsāno no tetsugaku* ("Bolzano's Philosophy") (Tokyo, 1963); H. Hornich, "Ueber eine Handschrift aus dem Nachlass von B. Bolzano," in *Anzeiger. Osterreichische Akademie der Wissenschaften, mathematische-naturwissenschaftliche Klasse,* no. 2 (1961); F. Kambartel, *Bernard Bolzano's Grundlegung der Logik* (Hamburg, 1963), which includes selections from Wissenschaftslehre I and II and an excellent introduction; A. Kolman, *Bernard Bolzano* (Berlin, 1963), which has an extensive bibliography; D. Laugwitz, "Bemerkungen zu Bolzanos Grössenlehre," in *Archive for History of Exact Sciences,* **2** (1964), 398–409; B. van Rootselaar, "Bolzano's Theory of Real Numbers," in *Archive for History of Exact Sciences,* **2** (1964), 168–180; K. Rychlik, *Theorie der reellen Zahlen im Bolzanos handschriftlichen Nachlasse* (Prague, 1962); K. Večerka, "Bernard Bolzano's Anti-Euklid," in *Sbornik pro dějiny přirodnich věd a teckniky (Acta historiae rerum naturalium nec non technicarum,* Prague), **11** (1967), 203–216; E. Winter, *Leben und geistige Entwicklung des Sozialethikers und Mathematikers B. Bolzano 1781–1848,* Hallische Monographien, no. 14 (Halle, 1949), which has a Bolzano bibliography; *Die historische Bedeutung der Frühbegriffe B. Bolzano's* (Berlin, 1964); and *Wissenschaft und Religion im Vormärz. Der Briefwechsel Bernard Bolzanos mit Michael Josef Fesl 1822–1848* (Berlin, 1965); and E. Winter, P. Funk, J. Berg, *Bernard Bolzano, Ein Denker und Erzieher in Österreichischen Vormärz* (Vienna, 1967).

B. van Rootselaar

BOMBELLI, RAFAEL (*b.* Bologna, Italy, January 1526; *d.* 1572), *algebra.*

Rafael Bombelli's family came from Borgo Panigale, a suburb three miles north of Bologna. The original family name was Mazzoli. The Mazzolis, who seem to have been small landowners, adopted the name Bombelli early in the sixteenth century. Some of them were supporters of the Bentivoglio faction. An unsuccessful conspiracy to restore the Bentivoglio *signoria* in 1508 resulted in the execution of seven men, among whom was Giovanni Mazzoli, Rafael Bombelli's great-grandfather. Giovanni Mazzoli's property was confiscated but was later restored to his grandchildren. One of them was Antonio Mazzoli, alias Bombelli, who later became a wool merchant and moved to Bologna. There he married Diamante Scudleri, the daughter of a tailor. Six children were born to this marriage, of whom the eldest son was Rafael Bombelli.

All that is known about Bombelli's education is that his teacher (*precettore*) was Pier Francesco Clementi of Corinaldo, an engineer-architect. It has been suggested that Bombelli might have studied at the University of Bologna, but this seems unlikely when one considers his family background and the nature of his profession. He spent the greater part of his working life as an engineer-architect in the service of his patron, Monsignor Alessandro Rufini, a Roman nobleman. Rufini was *cameriere* and favorite of Pope Paul III, and later was bishop of Melfi. The major engineering project on which Bombelli was employed was the reclamation of the marshes of the Val di Chiana. It was at a time when the reclamation work had been suspended that he wrote his treatise on algebra in the peaceful atmosphere of his patron's villa in Rome. His professional engagements seem to have delayed the completion of the book, but the more important part of it was published in 1572. His death soon afterward prevented the publishing of the remainder of the work. It was not published until 1929.

Bombelli's teacher, Pier Francesco Clementi, was employed by the Apostolic Camera (*ca.* 1548) in draining the marshes of the Topino River at Foligno (100 miles from Rome). It is not known whether Bombelli himself worked in Foligno; but by 1551 he had begun to work for Rufini in the reclamation of the Val di Chiana marshes. Rufini began to take an interest in this project in 1549, when the rights of reclamation of that part of the marshes which belonged to the Papal States were obtained by his nominee. Evidence of Bombelli's activity is found in the record relating to the marking out and settlement of the boundaries of the reclaimed land. The work of reclamation was interrupted sometime between 1555 and 1560. By 1560 Bombelli had returned to the Val di Chiana, and his work there ended in that year. In 1561 he was in Rome, where he took part in the unsuccessful attempt to repair the Ponte Santa Maria, one of the bridges over the Tiber.

Bombelli's work in the Val di Chiana earned him a reputation as an engineer, and led to his being one of the consultants on a proposed project for draining a part of the Pontine Marshes during the reign of Pius IV (1559–1565). The historian Nicolai, in his *De' bonificamenti delle Terre Pontine* (1800), says that the work was to have been directed by Rafael Bombelli, "famous among hydraulic engineers for having successfully drained the marshes of the Val di Chiana." The project was not realized, however.

Rafael Bombelli grew up in an Italy that was active in the production of works on practical arithmetic. Luca Pacioli, author of the *Summa di arithmetica, geometria, . . .* (1494), had lectured at Bologna at the beginning of the century. So had Scipione dal Ferro, a citizen of Bologna and one of the foremost mathematicians of the time. Their successors, Cardano, Tartaglia, and Ferrari, who were attempting the solution of the cubic and biquadratic equations, lived and worked in the neighboring cities of northern Italy. Cardano's *Practica arithmeticae* was published in 1539 and was followed in 1545 by his great treatise on algebra, the *Ars magna,* which gave the methods of dal Ferro and Ferrari for solving the cubic and biquadratic equations, respectively. In 1546 the controversy between Cardano and Tartaglia became public with the appearance of the latter's *Quesiti et inventioni diverse.* Copies of the *Cartelli di matematica disfida* (1547–1548), exchanged between Ferrari and Tartaglia, were circulated in the principal cities of Italy. Such was the climate in which Bombelli conceived the idea of writing a treatise on algebra. He felt that none of his predecessors except Cardano had explored the subject in depth; but Cardano, he thought, had not been clear in his exposition. He therefore decided to write a book that would enable anyone to master the subject without the aid of any other text. The work, written between 1557 and 1560, was a systematic and logical exposition of the subject in five parts, or books. In Book I, Bombelli dealt with the definitions of the elementary concepts (powers, roots, binomials, trinomials) and applications of the fundamental operations. In Book II he introduced algebraic powers and notation, and then went on to deal with the solution of equations of the first, second, third, and fourth degrees. Bombelli considered only equations with positive coefficients, thus adhering to the practice of his contemporaries. He was therefore obliged to deal with a large number of cases: five types of quadratic equations, seven cubic, and forty-two biquadratic. For each type of equation, he gave the rule for solution and illustrated the rule with examples. Bombelli feared that the examples given in Book II would not be sufficient for a beginner who wished to master the subject, so he decided to include in Book III a series of problems by which the student would be taken, in stages, through the various operations of algebra. For this purpose he chose problems that were common to books on practical arithmetic of his day. Many of them were "applied problems"— that is, problems that had denominate numbers— and not mere exercises in manipulating symbols. They were often woven into incidents that could have occurred in the marketplace or tavern. Books IV and V formed the geometrical portion of the work. Book IV contained the application of geometrical methods to algebra, *algebra linearia,* and Book V was devoted to the application of algebraic methods to the solution of geometrical problems. Unfortunately, Bombelli was unable to complete the work as he had originally planned, in particular Books IV and V.

He had the opportunity, however, of studying a codex of Diophantus' *Arithmetic* in the Vatican Library during a visit to Rome. It was shown to him by Antonio Maria Pazzi, *lector ad mathematicam* at the University of Rome. They set out to translate the manuscript, but circumstances prevented them from completing the work. The changes that Bombelli made in the first three books of his *Algebra* show evidence of the influence of Diophantus. At the end of Book III, Bombelli said that the geometrical part, Books IV and V, was not yet ready for the publisher, but that it would follow shortly. His death prevented his keeping the promise. It was only in 1923 that the manuscript of the *Algebra* was rediscovered by Ettore Bortolotti in the Biblioteca Comunale dell'Archiginnasio in Bologna.

In his *Algebra,* Bombelli gave a comprehensive account of the existing knowledge of the subject, enriching it with his own contributions. Cardano had observed that the general rule given by dal Ferro could not be applied in solving the so-called irreducible case of the cubic equation, but Bombelli's skill in operating with "imaginary numbers" enabled him to demonstrate the applicability of the rule even in this case. Because of the special nature and importance of these imaginary quantities, he took great care to make the reader familiar with them by introducing them early in his work—at the end of Book I. He said he had found "un altra sorte di radice cuba legata" ("another kind of cube root of an aggregate") different from the others. This was the cube root of a complex number occurring in the solution of the irreducible case of the cubic equation. He called the square roots of a negative quantity *più di meno* and *meno di meno* (that is, *p. di m.* 10, *m. di m.* 10 for $+\sqrt{-10}, -\sqrt{-10}$). Having pointed out that the complex root is always accompanied by its conjugate, he set out the rules for operating with complex numbers and gave examples showing their application. Here he showed himself to be far ahead of his time, for his treatment was almost that followed today. Bombelli also pointed out that the problem of trisecting an angle could be reduced to that of solving the irreducible case of the cubic equation (this was illustrated in Book V). Although he made no significant contribution to the solution of the biquadratic equation, he showed the application of Ferrari's rule to every possible case.

In Book III of the printed version of the *Algebra* one finds no trace of the influence that practical arithmetics originally had on Bombelli. He said in the preface that he had deviated from the custom of those authors of arithmetics who stated their problems in the guise of human actions; his intention was to teach the "higher arithmetic." The problems of applied arithmetic that were originally included in Book III were left out of the published work; by doing so, Bombelli helped to raise algebra to the status of an independent discipline. In place of these applied problems he introduced a number of abstract problems, of which 143 were taken from the *Arithmetic* of Diophantus. Although Bombelli did not distinguish Diophantus' problems from his own, he acknowledged that he had borrowed freely from the *Arithmetic*. He was in fact the first to popularize the work of Diophantus in the West.

Apart from the solution of the irreducible case of the cubic equation, the most significant contribution Bombelli made to algebra was in the notation he adopted. He represented the powers of the unknown quantity by a semicircle inside which the exponent was placed: $\underset{\smile}{1}$ for the modern x, $\underset{\smile}{2}$ for x^2, and $5\underset{\smile}{1}$ or $5^{\underset{\smile}{1}}$ for $5x$. In the printed work the semicircle was reduced to an arc: $\underset{\smile}{1}$, $\underset{\smile}{2}$, $5^{\underset{\smile}{1}}$. The zero exponent, $\underset{\smile}{0}$, was used in the manuscript, $48\underset{\smile}{0}$ for 48, but was omitted from the published work. The notation $R\lfloor\underline{\quad}$ was used in the manuscript in applying the radical to the aggregate of two or more terms: $R\lfloor 4pR6\rfloor$ for $\sqrt{4+\sqrt{6}}$. He even used the radical sign as a double bracket: $R^3\lfloor 2pR\lfloor 0m121\rfloor\rfloor$ for $\sqrt[3]{2+\sqrt{0-121}}$. In the printed work the horizontal line was broken, and R, R^3 became Rq, Rc: for example, $Rc\lfloor 2pRq\lfloor 0m121\rfloor\rfloor$.

Although incomplete, Books IV and V of the *Algebra* reveal Bombelli's versatility as a geometer. He had reduced some of the arithmetical problems of Book III to an abstract form and had interpreted them geometrically. He did not feel obliged to give geometrical proofs for the correctness of the results that he had obtained by algebraic methods. In doing so, he had broken away from a long-established tradition. The linear representation of powers, the use of the unit segment, and the representation of a point by "orthogonal coordinates" are some of the noteworthy features of this part of the work.

Bombelli was the last of the algebraists of Renaissance Italy. The influence that his *Algebra* had in the Low Countries is attested to by Simon Stevin and Adrien Romain. In the course of a short historical survey of the solution of equations, in his *Arithmetique,* Stevin referred to Bombelli as "great arithmetician of our time." He used a slightly modified form of Bombelli's notation for the powers of the unknown. While giving Bombelli due credit, he stressed the superiority of his notation to that of the Cossists. About a century later Leibniz, while teaching himself mathematics, used Bombelli's *Algebra* as a guide to the study of cubic equations. His correspondence with Huygens shows the keen interest these two men took in the work of the Italian mathematicians of the Renaissance. In the words of Leibniz, Bombelli was an "outstanding master of the analytical art."

BIBLIOGRAPHY

I. ORIGINAL WORKS. Versions of the *Algebra* are *L'algebra di Rafaello Bombello, cittadino bolognese,* in Biblioteca Comunale dell'Archiginnasio in Bologna, Codex B.1569; *L'algebra* (Bologna, 1572); and Ettore Bortolotti, ed., *L'algebra, opera di Rafael Bombelli da Bologna, Libri IV e V* (Bologna, 1929).

II. SECONDARY LITERATURE. For references to earlier literature, see S. A. Jayawardene, "Unpublished Documents Relating to Rafael Bombelli in the Archives of Bologna," in *Isis,* **54** (1963), 391–395, and "Rafael Bombelli, Engineer-Architect: Some Unpublished Documents of the Apostolic Camera," *ibid.,* **56** (1965), 298–306.

S. A. JAYAWARDENE

BONANNI, FILIPPO. See **Buonanni, Filippo.**

BONAPARTE, LUCIEN JULES LAURENT, called **Charles Lucien** (*b.* Paris, France, 24 May 1803; *d.* Paris, 29 July 1857), *zoology.*

Bonaparte was the son of Napoleon's younger brother, Lucien (1775–1840). His mother was Alexandrine de Bleschamp. In 1822 Bonaparte married his cousin Zénaïde Charlotte Julie, the daughter of the king of Naples and Spain. They had twelve children.

Soon after his marriage Bonaparte went to the United States, where he started a brilliant career as a naturalist. At twenty-five he returned to Europe and settled in Italy, beginning his great political activity. He advocated the organization of scientific congresses, which in Italy also served as the opportunity for meetings of independents and reformers. Upon the accession of Pope Pius IX, Bonaparte became a member of the pope's party. Next he was a member of the Radical party and of the supreme junta that seized power in the Roman states. After the pope's flight to Gaeta in November 1848, Bonaparte became a deputy from Viterbo; having been made vice-president of the Assemblée Nationale Romaine, he was also on the commission to draw up the constitution. When the Italian republic fell and French troops marched into Rome, he left Italy and went to France. He was not allowed to remain in Marseilles and therefore continued his trip. In Orléans he was arrested but released, then fled to Le Hâvre, where he took a boat for England. In 1850, once again allowed to live in Paris, Bonaparte left politics and turned exclusively to his scientific work. He had begun with a few essays in botany, but his zoological research became very important. While in the United States he had published numerous ornithological notes in the *Journal of the Philadelphia Academy of Sciences* and applied himself to the continuation of Wilson's work on birds.

As early as 1831 Bonaparte became interested in the great principles of classification and was critical of Cuvier's concepts. He classified the Insectivora before the Rodentia and separated the Chiroptera from the Primates. Besides the morphological characteristics, he considered the physiological data, such as, in the case of birds, "the perfect or imperfect condition" of the chicks at birth. Bonaparte raised the Batrachia to the rank of a subclass. He then united the saurians and the ophidians (Reptilia), placing the iguanodons at the head, as "the most perfect of cold-blooded and air-breathing animals."

In ichthyology Bonaparte made use of the location, the structure, and the relationships of the branchiae in the classification of fishes. He made a distinction between two new "sections," the Physostomi and the Physoclisti, according to whether or not the alimentary canal (the branchiae, according to Bonaparte) communicates with the air bladder. In general classifications, Bonaparte carefully reconstituted the synonymy of a species, minutely described its coloration (often illustrated by beautiful plates), and considered both its behavior and its geographic distribution.

Bonaparte tried to establish our knowledge of various zoological groups once and for all, and published numerous synopses, conspectuses, and catalogs. He urged zoologists to study local fauna and conceived the writing of a general work on the fauna of France, *Histoire naturelle générale et particulière des animaux qui vivent en France.* This was to be carried out in collaboration with Victor Meunier, and its prospectus is dated 1857. Bonaparte's death prevented the realization of the project. He had visited many museums in the United States and Europe and was deeply interested in the Muséum d'Histoire Naturelle. He hoped for the creation of a special gallery in which to exhibit the native fauna, with an accompanying professorship for its study. The teaching of natural sciences, according to him, should affect the study of agriculture and be conceived so as "to reach those who do not have the time to study."

Bonaparte had many friends abroad. In the United States he had known Audubon, and in Leiden he was friendly with J. C. Temminck and H. Shlegel. He was also a member of numerous learned societies and academies: the Philadelphia Academy of Natural Sciences, the Academy of Sciences and Literature of Baltimore, the Academy of Sciences of Berlin, and the Royal Academy of Turin. In 1839 Agassiz defeated Bonaparte for election to corresponding membership of the Académie des Sciences de Paris by one vote, but Bonaparte was elected on 18 March 1844.

Bonaparte left his personal library to the Muséum d'Histoire Naturelle. It contains works on the natural sciences, meteorology, history, and politics. It also includes his extensive correspondence, not yet cataloged.

BIBLIOGRAPHY

I. ORIGINAL WORKS. Bonaparte's writings include *American Ornithology or the Natural History of Birds Inhabiting the United States,* 4 vols. (Philadelphia, 1825–1833); *Observations on the Nomenclature of Wilson's Ornithology* (Philadelphia, 1826); "The Genera of North American Birds and a Synopsis of the Species to Be Found Within the Territory of the United States," in *Annales du Lycée New-York,* **11** (1826–1828); "Tableau comparatif des ornithologies de Rome et de Philadelphie," in *Nouveau journal des savants* (1827); "*Fauna Italica,*" *Iconographia della fauna italica per le quattro classi degli animali vertebrati,* 3 vols. (Rome, 1832–1841); "Amphibia Europaea ad systema nostrum vertebratorum ordinata," in *Memorie della Accademia delle scienze di Torino* (1840), 385–456; "Monographia leuciscorum Europaeorum," in *Actes du Congrès zoologique Pisa* (1840), p. 150; "Systema vertebratorum," in *Transactions of the Linnean Society of London,* **18** (1840), 247–305; *Catalogo metodico degli ucelli europei* (Bologna, 1842); *Catalogo metodico dei mammiferi europei* (Milan, 1845); *Catalogo metodico dei pesci europei* (Naples, 1846); *Conspectus systematis erpetologiae et amphibiologiae* (Leiden, 1850); *Conspectus systematis ornithologiae* (Leiden, 1850; rev., enl. ed., 1854); *Monographie des Loxiens* (Leiden-Düsseldorf, 1850), written with H. Schlegel; *Conspectus generum avium,* 2 vols. (Leiden, 1850–1857); *Discours, allocutions et opinions de Charles Louis Prince Bonaparte dans le Conseil des Députés et l'Assemblée Constituante de Rome en 1848 et 1849* (Leiden, 1857); and *Iconographie des pigeons non figurés par Madame Knip* [Mlle. Pauline de Courcelles], 2 vols. (Paris, 1857).

II. SECONDARY LITERATURE. Works on Bonaparte are Élie de Beaumont, *Notice sur les travaux scientifiques de S. A. le Prince Charles Louis Bonaparte (Réflexions sur ce travail par M. Richard du Cantal)* (1886); *Biographie du Prince Charles Bonaparte, Prince de Canino, fils de Lucien,* J. P. Jules Pautet, trans.; and the article in *Dictionnaire de biographie française,* VI (Paris, 1954). There is also an anonymous notice on Bonaparte's works in *Revue et magazine de zoologie,* **11** (1850).

G. PETIT

BONAVENTURA, FEDERIGO (*b.* Ancona, Italy, 24 August 1555; *d.* Urbino [?], Italy, March 1602), *meteorology.*

Bonaventura was the son of Pietro Bonaventura, an officer in the army of the duke of Urbino and a poet, and of Leonora Landriani of Milan. Upon his father's death in 1565 Federigo, supported by the duke of Urbino, went to Rome, where he was educated with Francesco Maria della Rovere at the house of Cardinal Giulio della Rovere. At the age of eighteen he returned to Urbino. Following the accession of Duke Francesco Maria II in 1574, Federigo met with even greater favor. He continued his studies at Urbino, particularly Greek mathematics and natural philosophy. In addition to his scholarly activities, Bonaventura served as Urbino's ambassador to several European courts. His marriage to Pantasilea, countess of Carpegna, in 1577 produced twelve children, including Pietro, who became bishop of Cesena, and Francesco Maria, who wrote several literary works.

Bonaventura's most important scientific writings deal with meteorology. As yet they have not been carefully studied, nor has their significance in sixteenth-century meteorological thought been determined. They include *De causa ventorum motus* (1592), in which he argues, in opposition to many later interpreters, that there is no basic disagreement between the theory of winds of Aristotle and that of Theophrastus; *Pro Theophrasto atque Alexandro Aphrodisiensi . . . apologia* (1592), in which he again attempts to defend the ancient Peripatetic meteorological theories against such modern interpreters as Francesco Vimercato; *Anemologiae pars prior* (1593), which is essentially a Latin translation of Theophrastus' *De ventis* and *De signis,* with long and detailed commentaries on the two works; and *Quomodo calor a sole corporibus coelestibus producatur secundum Aristotelem* (1627), in which he argues that Aristotle held that the sun's heat is transferred to other bodies through motion, rather than through light.

All of Bonaventura's writings on meteorology are marked by an attempt to determine the precise meaning of the ancient texts through philological techniques, with apparently little effort being made to utilize experience and observation to verify their truth. He also wrote works on medical subjects (especially *De natura partus octomestris* [1596]) and

political philosophy, and translated into Latin works of Themistius and of Ptolemy (*Inerrantium stellarum apparitiones ac significationum collectio* [1592]).

BIBLIOGRAPHY

I. ORIGINAL WORKS. Lists of Bonaventura's published works are available in Mazzuchelli, Narducci, and Vecchietti (see below). See also the following MSS: Biblioteca Ambrosiana, Milan, Q. 118 sup., and S. 87 sup.; Biblioteca Oliveriana, Pesaro, 1494, 1500, 1503, 1509; and Vat. urb. lat. 1333, 1349. Among his published works are *De causa ventorum motus* (Urbino, 1592; Venice, 1594); *Inerrantium stellarum apparitiones ac significationum collectio* (Urbino, 1592), a Latin translation of the work by Ptolemy, with a long commentary; *Pro Theophrasto atque Alexandro Aphrodisiensi . . . apologia* (Urbino, 1592; repr. Venice, 1594); *Anemologiae pars prior* (Urbino, 1593), repr. as *Meteorologicae affectiones* (Venice, 1594); *De natura partus octomestris* (Urbino, 1596, 1600; Frankfurt, 1601, 1612; Venice, 1602); and a collection of *Opuscula* (Urbino, 1627), which contains, among others, *Quomodo calor a sole corporibus coelestibus producatur secundum Aristotelem.*

II. SECONDARY LITERATURE. Writings on Bonaventura or his work are Giammaria Mazzuchelli, *Gli scrittori d'Italia,* II (Brescia, 1760), 1563–1564; M. Michaud, *Biographie universelle,* 2nd ed. (Paris, 1880), IV, 687; Enrico Narducci, *Giunte all'opera . . . del . . . Mazzuchelli* (Rome, 1894), p. 95; and P. Vecchietti and T. Vecchietti, *Biblioteca picena,* III (Osimo, 1796), 1–6.

See also *Degli uomini illustri di Urbino commentario del P. Carlo Grossi con aggiunte scritte dal conte Pompeo Gherardi* (Urbino, 1856), pp. 72–78.

CHARLES B. SCHMITT

BONCOMPAGNI, BALDASSARRE (*b.* Rome, Italy, 10 May 1821; *d.* Rome, 13 April 1894), *history of mathematics, history of physics.*

The son of Luigi Boncompagni, prince of Piombino, and of Maria Maddalena Odelscalchi, Baldassarre was a student of Barnaba Tortolini, the noted mathematician. In 1843 *Crelle's Journal* published the results of mathematical analyses obtained by Boncompagni, who afterward concentrated mainly on the history of mathematics and of physics. His works in this field include one on the development of the study of physics in Italy during the sixteenth and seventeenth centuries, as well as publications concerning Guido Bonatti, Plato of Tivoli, Gerard of Cremona, and Gerard of Sabbionetta.

From these studies, Boncompagni was led to examine the works of Leonardo Fibonacci, about whom little was known at that time. By means of numerous accurate works, he made known Fibonacci's importance in the history of mathematics, illustrating his life and works in the accurate edition of the *Scritti di Leonardo Pisano* (1857–1862).

In order to meet the requirements of his scientific publications, Boncompagni established his own printing plant, called "delle Scienze Matematiche e Fisiche." For forty years he assumed full financial responsibility for the entire cost of its operation, freely granting to other scientists the privilege of using its facilities. The plant published important documents on the history of science, such as the papers on challenging mathematics between Ferrari and Tartaglia and the unpublished letters of Lagrange and Gauss. In order to have a specialized journal for his favorite studies, in 1868 Boncompagni undertook the publication of *Bullettino di bibliografia e di storia delle scienze matematiche e fisiche*. Known as *Bullettino Boncompagni,* it ceased publication in 1887.

Boncompagni was among the first thirty members of the Pontifical Academy of the New Lincei, which was founded in 1847 by Pope Pius IX, who desired to reactivitate the academy founded by Federico Ccsi, of which Galileo had also been a member. He published the transactions of the Academy, from volume XXIV to volume XLVII, at his own expense. He was faithful to it even after the Italian government established the Lincei Academy. Boncompagni came to the assistance of needy scholars and students, assigning them to well-paying tasks in transcription and in translation, thus leaving behind him the memory of an enlightened and generous patronage.

BIBLIOGRAPHY

I. ORIGINAL WORKS. There are 209 works listed in "Catalogo degli scritti del Principe D. Baldassarre Boncompagni," I. Galli, ed., in *Atti dell'Accademia Pontificia dei nuovi Lincei,* **47** (1893–1894), 171–186. Among these works are "Recherches sur les intégrales définies" in *Crelle's Journal,* **25** (1843); "Studi intorno ad alcuni avanzamenti della fisica in Italia nei secoli XVI e XVII," in *Giornale Arcadico di scienze, lettere ed arti* (1846); "Della vita e delle opere di Guido Bonatti astrologo ed astronomo del secolo decimo terzo," *ibid.* (1851); "Delle versioni fatte da Platone Tiburtino traduttore del secolo duodecimo," in *Atti dell'Accademia Pontificia dei nuovi Lincei,* **4** (1850–1851), 247–286; "Della vita e delle opere di Gherardo Cremonese traduttore del secolo duodecimo e di Gherardo da Sabbionetta astronomo del secolo decimoterzo," *ibid.,* 387–493; and *Saggio intorno ad alcune opere di Leonardo Pisano* (Rome, 1854).

II. SECONDARY LITERATURE. Works on Boncompagni are M. S. De Rossi, "Commemorazione del socio ordinario Principe D. Baldassarre Boncompagni," in *Atti dell'Accademia Pontificia dei nuovi Lincei,* **47** (1893–1894), 131–134; A. Favaro, "Don Baldassarre Boncompagni e la storia delle Scienze Matematiche e Fisiche," in *Atti del R. Istituto veneto di scienze, lettere ed arti,* 7th ser., **6** (1894–1895), 509–521; and I. Galli, "Elogio del Principe Don Baldassarre Boncompagni," in *Atti dell'Accademia Pontificia dei nuovi Lincei,* **47** (1893–1894), 161–170.

ETTORE CARRUCCIO

BOND, GEORGE PHILLIPS (*b.* Dorchester, Massachusetts, 20 May 1825; *d.* Cambridge, Massachusetts, 17 February 1865), *astronomy.*

The third son of William Cranch Bond and Selina Cranch, George Bond grew up in an environment focused on Harvard Observatory, where his father was the first director. The scientific collaboration with his father began so early that it is often difficult to separate their contributions. At the age of twenty-three he assisted in the observations of Saturn that led to his discovery of the satellite Hyperion. Two years later, he found Saturn's crepe ring. Hence Bond was the natural choice for director of Harvard Observatory when his father died in 1859.

The selection was not unchallenged, however, for Benjamin Peirce, the top mathematical astronomer in the country, also aspired to the directorship. The resulting antagonism with Peirce and his scientific clique hampered Bond in many ways and embittered his career. A serious and uncompromising man, Bond believed that this rivalry cost him a place when the National Academy of Sciences was incorporated in 1863.

Bond's principal observations were carried out with the observatory's 15-inch refractor, which, until it was surpassed in 1862, ranked with the Pulkovo instrument as the largest refractor in the world. His comprehensive and handsomely illustrated monograph on Donati's Comet of 1858, in *Annals of the Harvard College Observatory* (1862), won widespread acclaim and in 1865 brought him the gold medal of the Royal Astronomical Society, the first ever awarded to an American.

George Bond directed the observatory a scant six years—he died of tuberculosis at the age of thirty-nine. He had undertaken an intense investigation of the Orion Nebula, but his health broke before he could complete it. The memoir was published posthumously. His remarkable drawing of the nebula can be favorably compared with modern photographs. In 1860 Bond reported on the comparative brightness of the sun, moon, and Jupiter, a fundamental research that has placed "Bond albedo" in the contemporary astronomical vocabulary.

Bond's most enduring fame, however, rests on his enthusiastic experimentation with stellar photography and his perceptive anticipation of its potential; in

1857 he wrote, "There is nothing, then, so extravagant in predicting a future application of photography on a most magnificent scale. . . . What more admirable method can be imagined for the study of the orbits of the fixed stars and for resolving the problem of their annual parallax?" His pioneering daguerreotype work, undertaken from 1847 to 1851 in cooperation with his father, resulted in the first photograph of a star, Vega. Bond's 1857 experiments with wet collodion photography achieved still greater success. With considerable justification Edward S. Holden called him "the father of celestial photography."

BIBLIOGRAPHY

I. ORIGINAL WORKS. Among his writings are "Account of the Great Comet of 1858," in *Annals of the Harvard College Observatory,* **3** (1862); "Observations upon the Great Nebula of Orion," *ibid.,* **5** (1867). Selections from Bond's diaries during his trips abroad in 1851 and 1863, and from his correspondence, as well as an extensive bibliography, appear in Edward S. Holden, *Memorials of William Cranch Bond and George Phillips Bond* (San Francisco, 1897); bound copies of the Bond correspondence used by Holden are in the Lick Observatory Library. See also "Diary of the Two Bonds: 1846–1849," Bessie Z. Jones, ed., in *Harvard Library Bulletin,* **15** (1967), 368–386, and **16** (1968), 49–71, 178–207,

II. SECONDARY LITERATURE. An extensive evaluation of Bond's scientific work is given in the Royal Astronomical Society presidential address by Warren De La Rue, in the Society's *Monthly Notices,* **35** (1865), 125–137. See also Dorrit Hoffleit, *Some Firsts in Astronomical Photography* (Cambridge, Mass., 1950).

OWEN GINGERICH

BOND, WILLIAM CRANCH (*b.* Falmouth [now Portland], Maine, 9 September 1789; *d.* Cambridge, Massachusetts, 29 January 1859), *astronomy.*

His father, William Bond, a fiery Cornishman, and his strict, forceful mother, Hannah Cranch, emigrated to Massachusetts in 1786; soon after William Cranch Bond's birth, their lumber export business failed and the family moved to Boston, where they opened a clockmaking shop. Bond's youth was spent in the hardship of poverty, and he was obliged to leave school at an early age. His rare mechanical ability proved invaluable in the shop, where he constructed a chronometer at the age of fifteen.

A total solar eclipse in 1806 fixed Bond's attention on astronomy, and at the age of twenty-one he independently found the Comet of 1811. The parlor of the first house he owned, in Dorchester, was converted into an observatory complete with granite pier and, in the ceiling, a meridian opening. As an expert

clockmaker, he rated the chronometers for numerous expeditions to determine longitudes in the eastern United States. Bond married his cousin Selina Cranch in 1819; she bore him four sons and two daughters. After her death in 1831, Bond married her elder sister, Mary Roope Cranch.

In 1815, when Bond traveled to Europe, he was commissioned by Harvard to examine instruments and observatories in England. Proposals for a meteorological and astronomical observatory at Harvard came to naught, however, until 1839, when Bond was invited to transfer his own equipment to Dana House in Cambridge and to serve (without salary) as astronomical observer to Harvard University.

The great sun-grazing comet of 1843 aroused an immense latent interest in astronomy, and some ninety societies and individuals subscribed $25,730 for the building of a large telescope at Harvard. A 15-inch refractor, equal in size to the largest in the world, was ordered from Munich and mounted in Cambridge in June 1847. Bond's mechanical ingenuity manifested itself in the construction of the dome, in the remarkable observing chair, and in the regulating device that made the chronograph a precision instrument.

With the 15-inch telescope Bond undertook elaborate studies of the Orion Nebula and of the planet Saturn, and during his administration the daguerreotype process was first used to photograph stars. Bond was a modest, retiring, and deeply religious man. An accurate evaluation of his abilities was given by Benjamin Peirce: "In his original investigations he naturally restrained himself to those forms of observation which were fully within reach of his own resources. . . . He consequently availed himself less of the remarkable capacity of his instrument for delicate and refined measurements than of its exquisite optical qualities."

BIBLIOGRAPHY

Bond's writings include "History and Description of the Astronomical Observatory of Harvard College," in *Harvard College Observatory Annals,* **1,** pt. 1 (1856); "Observations on the Planet Saturn," *ibid.,* **2,** pt. 1 (1857); and "Diary of the Two Bonds: 1846–1849," Bessie Z. Jones, ed., in *Harvard Library Bulletin,* **15** (1967), 368–386, and **16** (1968), 49–71, 178–207. An extensive bibliography and much original material are given in Edward S. Holden, *Memorials of William Cranch Bond and George Phillips Bond* (San Francisco, 1897). An early photographic portrait appears as the frontispiece of *Harvard College Observatory Annals,* **7** (1871).

OWEN GINGERICH

BONNET, CHARLES (*b.* Geneva, Switzerland, 13 March 1720; *d.* Geneva, 20 May 1793), *natural history, biology, philosophy.*

Bonnet was the son of Pierre Bonnet, whose family lived at Thônex, near Geneva, and of Anne-Marie Lullin. A mediocre student, he was gifted neither in languages nor in mathematics; moreover, he was hindered by increasing deafness, which exposed him to the taunts of his playmates. His father decided to engage a private tutor, a Dr. Laget, who played a prominent part in stimulating the boy's early interest in the natural sciences. In 1736, Bonnet avidly read Pluche's study *Le spectacle de la nature.* The following year, he read the memoirs of René Réaumur and began a correspondence with the great scientist. These were decisive steps. In 1738 he submitted an essay on entomology to the Academy of Sciences of Paris. His father, however, looked unfavorably upon a career in the natural sciences, and Bonnet agreed to study law instead. In 1744 he received a doctorate in law. He married Jeanne-Marie de La Rive in 1756 and retired to his wife's estate at Genthod, near Geneva. All his life Bonnet had to contend with precarious health. In addition to being deaf, he became, while still young, almost completely blind, then began to suffer severe asthma attacks.

Bonnet is considered one of the fathers of modern biology. He is distinguished for both his experimental research and his philosophy, which exerted a profound influence upon the naturalists of the eighteenth and nineteenth centuries.

Bonnet was twenty-six when he made his first and greatest discovery, the parthenogenesis of the aphid. He very carefully raised a female spindle-tree aphid, then observed that she produced ninety-five offspring without mating. Virginal generation was therefore possible. Bonnet wrote a note for the Academy of Sciences, which, on Bernard de Fontenelle's recommendation, appointed Bonnet a corresponding member. Then, taking up Réaumur's (1712) and Abraham Trembley's (1740) research on regeneration, Bonnet began his observation of rainwater worms of the species *lumbriculus.* He demonstrated that one of these worms, cut into twenty-six pieces, would become twenty-six perfectly constituted new worms. In 1745, he published several monographs on this subject. Bonnet devoted another work to the regeneration of a snail's head (1769), and in 1777 he dealt with regeneration of the limbs of a triton.

Bonnet's research induced him to study the breathing of caterpillars and the locomotion of ants. In 1745 he published the comprehensive *Traité d'insectologie,* a work that entitles him to consideration as an early

exponent of experimental entomology. After 1750 Bonnet published only a few studies on zoology; henceforth, his research was in plant physiology. In the preface to a remarkable work entitled *Recherches sur l'usage des feuilles dans les plantes* (1754), Bonnet wrote:

> Insects held my attention for some years. The strenuousness with which I worked on this study strained my eyes to such an extent that I was forced to interrupt it. Deprived thus of what had so far been my greatest pleasure, I tried to console myself by changing subjects. I then turned toward the physics of plants—a matter less animated, less fertile in discoveries but of a more generally recognized usefulness.

In the *Recherches,* Bonnet grouped five memoirs, all of which were of prime importance for plant biology: He precisely described the characteristics of the nutrition of leaves and of their transpiratory phenomena. Although he did not know the kinds of gases (oxygen and carbon dioxide) produced and absorbed by green leaves exposed to light, Bonnet made very careful observations on their production. For his masterly experimentation, Bonnet should be considered one of the first naturalists to investigate experimentally the question of photosynthesis. He studied the movement of leaves and discovered the epinastic phenomena; he observed the position of leaves on the axis of the stalk and collected a great many anatomic facts; he returned to experiments on etiolation, on the movement of the sap, and on teratology. Bonnet could no longer observe with his own eyes, so he surrounded himself with collaborators, all of whom later became distinguished naturalists—for example, François Huber, the bee specialist, and Jean Sénebier, famous for his research on photosynthesis. The collaborators performed innumerable tests on hybridation of corn, wheat, and darnel. Bonnet was opposed to the theory of transmutation of the species and may rightly be considered a forerunner of Lamarckism through his definition of his original concept of the "chain of beings"—parts of which, it is true, he had borrowed from Leibniz.

Next Bonnet turned to philosophy and methodology. A true theoretician of biology, he exercised an enormous influence in this field and maintained a correspondence with almost all the scientists of his time. He published works that caused a considerable stir—among them *Essai de psychologie* (1754) and *Essai analytique sur les facultés de l'âme* (1759). Bonnet then returned to theoretical biology, publishing *Considérations sur les corps organisés* (1761), *Contemplation de la nature* (1764), *Palin-*

génésie philosophique (1769), and *Recherches philosophiques sur les preuves du christianisme* (1771).

Bonnet was an enthusiastic champion of preformation, the theory postulating that the animal already existed in miniature in the germ cell. His discovery of parthenogenesis was, to him, proof that the female germ cell contains the preformed individual. Thus Bonnet became a fervent partisan of ovism. Many other naturalists, such as Albrecht von Haller, supported this thesis—a surprising one nowadays—of preformation. Yet Bonnet was less doctrinaire than his colleagues; he supported, for example, a very elastic thesis of the germ cell, which, according to him, was not only "an organized body reduced in size . . ." but "every kind of original preformation out of which may result an organic whole, as of his immediate principle." This theory, which Bonnet christened "palingenesis," set forth the functional and structural notion of the cell, which was not stated formally until a hundred years later.

Bonnet was not only a remarkable experimentalist in his younger years and a theoretician with fertile ideas: he was the instigator of a whole series of fundamental experiments. His extraordinary imagination suggested projects that his poor eyesight prevented him from carrying out; he treated these projects in numerous works, and above all he discussed them with his many correspondents. For instance, he suggested to Lazzaro Spallanzani that he carry out experiments on artificial insemination.

Mention must be made of the importance of Bonnet's methodological work. Of course he was—particularly in his more important writings—a theoretician, but a theoretician who experimented widely. Every observed fact, every proposed theory, gave Bonnet the opportunity to suggest the technique best suited for progress toward a solution. In his voluminous correspondence (he purportedly wrote over 700 letters annually), philosophical treatises, notes submitted to the Academy, and his most important monographs, he showed himself constantly preoccupied with methodological problems. Well before Claude Bernard, Bonnet attributed a preponderant role in scientific research to the "art of observing." The quality of his work fluctuates greatly, so it is not surprising that Bonnet is not appreciated without reservations. He sought after truth with courage and perseverance, was distrustful of his own hypotheses, and was ready to accept the conclusions of his strongest adversaries, if convinced that they were right. The personal qualities of Charles Bonnet, as much as his writings, justify the extraordinary reputation that he enjoyed in his lifetime and that survives today.

BIBLIOGRAPHY

I. ORIGINAL WORKS. Bonnet's writings include *Traité d'insectologie* (Paris, 1745); *Essai de psychologie* (Leiden, 1754); *Recherches sur l'usage des feuilles dans les plantes* (Leiden, 1754); *Essai analytique sur les facultés de l'âme* (Copenhagen, 1760); *Considérations sur les corps organisés* (Amsterdam, 1762); *Contemplation de la nature,* 2 vols. (Amsterdam, 1764); *La palingénésie philosophique* (Geneva, 1769); *Recherches philosophiques sur les preuves du christianisme* (Geneva, 1771); and *Oeuvre d'histoire naturelle et de philosophie,* 8 vols. in 4°, 18 vols. in 8° (Neuchâtel, 1779–1783).

II. SECONDARY LITERATURE. Works on Bonnet include G. Bonnet, *Charles Bonnet* (Paris, 1929), a thesis; R. de Caraman, *Ch. Bonnet, philosophe et naturaliste, sa vie et ses oeuvres* (Paris, 1859); E. Claparède, *La psychologie animale de Ch. Bonnet* (Geneva, 1909); J. Rostand, *Esquisse d'une histoire de la biologie: Un préformationniste—Ch. Bonnet* (Paris, 1945), pp. 65–80, and *Hommes d'autrefois et d'aujourd'hui: Ch. Bonnet* (Paris, 1966), pp. 7–45; R. Savioz, *La philosophie de Ch. Bonnet de Genève* (Paris, 1948), and *Mémoires autobiographiques de Ch. Bonnet de Genève* (Paris, 1948); A. Schubert, *Die Psychologie von Bonnet und Tetens* (Zurich, 1909), a thesis; and J. Trembley, *Mémoire pour servir à l'histoire de la vie et des ouvrages de M. Ch. Bonnet* (Bern, 1794).

P. E. PILET

BONNET, PIERRE-OSSIAN (*b.* Montpellier, France, 22 December 1819; *d.* Paris, France, 22 June 1892), *mathematics.*

Bonnet was the son of Pierre Bonnet, *commis banquier,* and Magdelaine Messac. After attending the College of Montpellier, he in 1838 entered the École Polytechnique in Paris, where he studied at the École des Ponts et des Chaussées. After graduation, however, he declined an engineering position, preferring teaching and research. In 1844 he became *répétiteur* at the École Polytechnique, augmenting his income by private tutoring. In a paper of 1843 he published convergence criteria of series with positive terms, among them logarithmic criteria. Another paper on series was honored by the Brussels Academy of Sciences and was published in 1849. By that time Bonnet had, starting in 1844 with the paper "Sur quelques propriétés générales des surfaces," begun to publish that series of papers on differential geometry on which his fame is based. The merit of this work was recognized by the Académie des Sciences when it elected Bonnet to membership in 1862 to replace Biot.

In 1868 Bonnet became Michel Chasles's *suppléant* at the École Polytechnique in the latter's course on higher geometry, and in 1871 he became director of studies there. He also taught at the École Normale Supérieure. In 1878 he obtained a chair at the Sorbonne, succeeding the astronomer Leverrier, and in 1883 he succeeded Liouville as a member of the Bureau des Longitudes. Married and the father of three sons, he always lived the quiet and unpretentious life of a scholar.

Bonnet's favorite field was the differential geometry of curves and surfaces, a field opened by Euler, Monge, and Gauss, but at that time lacking systematic treatment and offering wide fields of research. Between 1840 and 1850 this challenge was taken up by Bonnet and a group of younger French mathematicians—Serret, Frenet, Bertrand, and Puiseux—but it was Bonnet who most consistently continued in this field. In the "Mémoire sur la théorie générale des surfaces," presented in 1844 to the Académie, Bonnet introduced the concepts of geodesic curvature and torsion, and proved a series of theorems concerning them. One of these is the formula for the line integral of the geodesic curvature along a closed curve on a surface, known as the Gauss-Bonnet theorem (Gauss had published only a special case). Bonnet also showed the invariance of the geodesic curvature under bending of the surface.

From 1844 to 1867 Bonnet wrote a series of papers on differential geometry of surfaces. Special attention should be given to the "Mémoire sur la théorie des surfaces applicables sur une surface donnée" (1865–1867), written as a solution for a prize contest announced by the Académie in 1859: i.e., to find all surfaces of a given linear element. The problem is sometimes associated with Edouard Bour, who wrote a competing memoir (1862). The third entrant was Delfino Codazzi. Bonnet, in his contribution, showed the importance of certain formulas introduced in 1859 by Codazzi, formulas now taken as part of the so-called Gauss-Codazzi relations. He also showed the role these formulas play in the existence theorem for surfaces, if first and second fundamental forms are given. Bour, in his paper, came to similar conclusions.

In these and other papers Bonnet stressed the usefulness of special coordinate systems on a surface, such as isothermic and tangential coordinates; studied special curves, such as lines of curvature with constant geodesic curvature (1867); and investigated the conditions under which geodesic lines are the shortest connection between two points on a surface. He also paid much attention to minimal surfaces—for instance, those applicable on each other—and surfaces of constant total and constant mean curvature (1853).

Bonnet also published works on geodesy and cartography, theory of series (convergence criteria), al-gebra, rational mechanics, and mathematical physics. In 1871 he gave a definition of the limit for functions of a real variable.

BIBLIOGRAPHY

I. ORIGINAL WORKS. Among Bonnet's papers are "Note sur la convergence et divergence des séries," in *Journal de mathématiques pures et appliquées,* **8** (1843), 73–109; "Sur quelques propriétés générales des surfaces et des lignes tracées sur les surfaces," in *Comptes rendus de l'Académie des Sciences,* **14** (1844); "Mémoire sur la théorie générale des surfaces," in *Journal de l'École Polytechnique,* **32** (1848), 1–46; "Sur la théorie des séries," in *Mémoires couronnés de l'Académie de Bruxelles,* **22** (1849); "Mémoire sur l'emploi d'un nouveau système de variables dans l'étude des propriétés des surfaces courbes," in *Journal de mathématiques pures et appliquées,* ser. 2, **5** (1860), 153–266; and "Mémoire sur la théorie des surfaces applicables sur une surface donnée," in *Journal de l'École Polytechnique,* **41** (1865), 201–230, and **42** (1867), 1–151.

Bonnet's most important papers are mainly in the *Journal de l'École Polytechnique,* the *Journal de mathématiques pures et appliquées,* and the *Comptes rendus de l'Académie des Sciences.* He also wrote a *Mécanique élémentaire* (Paris, 1858). Bonnet's papers have never been collected, but the essence of his work on the theory of surfaces can be found in Gaston Darboux's *Leçons sur la théorie générale des surfaces,* 4 vols. (Paris, 1887–1896), *passim.*

II. SECONDARY LITERATURE. Works on Bonnet are P. Appell, "Notice sur la vie et les travaux de Pierre Ossian Bonnet," in *Comptes rendus de l'Académie des Sciences,* **117** (1893), 1013–1024; Michel Chasles, *Rapport sur les progrès de la géometrie en France* (Paris, 1870), pp. 199–214; and A. Franceschini, "Bonnet," in *Dictionnaire de biographie française,* Vol. VI (Paris, 1954).

D. J. STRUIK

BONNEY, THOMAS GEORGE (*b.* Rugeley, England, 27 July 1833; *d.* Cambridge, England, 10 December 1923), *geology.*

One of the last links with the heroic age of geology, Bonney was contemporary with Sedgwick, Murchison, Lyell, and Darwin during his early professional life; his pupils included Sollas, Marr, Watts, Teall, and Strahan.

The eldest of ten children born to Rev. Thomas Bonney, headmaster of the Rugeley grammar school, and his wife, the daughter of Edward Smith, Bonney graduated from St. John's College, Cambridge, as Twelfth Wrangler in the Mathematical Tripos in 1856 and obtained a second-class in the Classical Tripos; illness frustrated his intention to sit for the Theological Tripos as well. After five years as mathematics master at Westminster School, during which time he was ordained a priest and elected to a fellowship at St. John's, Bonney was recalled to the college as junior dean in 1861, becoming tutor in 1868 and also college lecturer in geology, a subject in which he was until then a self-taught amateur. At that time there were no university lectureships, but Bonney shouldered the main responsibility for university as well as college teaching in the subject during Sedgwick's declining years and exerted a powerful influence in molding the Cambridge school of geology.

Surprisingly, Bonney was not elected to succeed Sedgwick, and in 1877 he accepted the Yates-Goldschmidt professorship of geology at University College, London. At first conducting his part-time professorial duties from Cambridge, he became secretary of the British Association for the Advancement of Science in 1881, and this brought with it a sufficient increase of income for him to set up housekeeping in London with his sister. As secretary of the Association, Bonney organized its first meeting outside Britain (in Montreal), and during his London residence he became president of its Geological Section. (In 1910 he became president of the Association.) He also served, successively, as secretary and president of the Geological Society of London, and president of the Mineralogical Society. The Royal Society elected him a fellow in 1878; and he also became honorary canon of Manchester, Whitehall Preacher (at the Chapel Royal), and a Hulsean and Rede's lecturer at Cambridge. Bonney resigned his chair in 1901, but remained "doing fairly lucrative work" as one of the regular writers for a newspaper, the *Standard,* until 1905, when he returned to Cambridge.

Lack of formal education in geology may have contributed to that independence of outlook which constantly questioned dogma, and his insistence on proof may be attributed to his mathematical training. Bonney would accept no theory until it was exhaustively proved, and he inevitably became a formidable controversialist in nineteenth-century geology: "Fine phrases unsupported by facts prove to be no better than cheques without a balance at the bank." He was profoundly impressed by the *Origin of Species,* and in one of his presidential addresses lamented that the science of mineralogy "still needs its Darwin"; nevertheless, many contemporary theories and hypotheses came under his stricture.

Bonney's interest in glaciology was lifelong; it was the subject of his second paper in the *Geological Magazine* and of his presidential address to the British Association in 1910. Observations on valley glaciers made during his frequent visits to Switzerland caused him to dispute the efficacy of ice as an erosive agent,

and he remained unconvinced of the formation of cirques by plucking action or of more than superficial modification of river valleys by moving ice. While tarns and lakelets might be formed by glacial excavation, he could not accept the latter as the origin of larger lakes. Neither could he believe that the Scandinavian ice sheet ever reached the shores of Britain; he contended, *inter alia,* that the deep coastal trough bordering Norway must have afforded an easy path to the Arctic Ocean. His final presidential address, on the British "drift," marshaled his difficulties in accepting the land-ice hypothesis without venturing an alternative explanation.

Bonney was a leading figure in the early days of British petrography, following closely in the wake of Zirkel and Rosenbusch. Important studies on basic and ultrabasic igneous rocks led naturally to his demonstration of the true character of British serpentines, and in one of his numerous papers on these rocks he strongly contested Sterry Hunt's views of their sedimentary origin. As an authority on the Archaean rocks of England and Wales, Bonney became involved in the heated discussions of the age of the Eastern Gneiss (Moinian) of the northwest Highlands of Scotland, and of the age and relations of the metamorphic rocks of the Alps. Rocks for identification and analysis were sent to him from all over the world, and several of his later papers concern the parent rock (eclogite) of the diamond in South Africa. His petrological interests were not, however, confined to igneous rocks, and one of his presidential addresses (British Association, 1886), "The Application of Microscopic Analysis to Discovering the Physical Geography of Bygone Ages," is a remarkable pioneering achievement in what is now called sedimentology. Bonney was also chairman of the Coral Reef Committee of the Royal Society, which organized and reported on the Funafuti boring.

Bonney was by training and by temperament an exceedingly versatile man: geologist, mathematician, theologian, and classicist; an alpine traveler and a noted climber; a journalist and writer on architecture and scenery; and a draftsman of great merit. But above all, he was an outstanding teacher, and it was through teaching as much as through his prolific writings that he exercised such influence upon nineteenth-century geology. He was one of the first to introduce the examination of thin slices of rocks under the microscope, to lay emphasis on practical work in the laboratory and in the field, and, by his severely critical attitude, to compel his students to seek the facts and the evidence underlying any theory. As a teacher, he put out all his talents at compound interest.

BIBLIOGRAPHY

I. ORIGINAL WORKS. Bonney was the author of more than 200 scientific papers; a dozen books on geology, travel, architecture, and theology; several volumes of sermons; and a great many newspaper articles. The majority of his scientific papers are in *Quarterly Journal of the Geological Society of London, Geological Magazine,* and *Mineralogical Magazine.* His books include *The Story of Our Planet* (London, 1893); *Charles Lyell and Modern Geology* (London, 1895); *Ice Work Past and Present* (London, 1896); *Volcanoes* (London, 1899); *The Building of the Alps* (London, 1912); and *Memories of a Long Life* (Cambridge, 1921).

II. SECONDARY LITERATURE. Articles on Bonney are "Eminent Living Geologists: The Rev. Professor T. G. Bonney," in *Geological Magazine,* **38** (1901), 385–400, which contains a full catalog of his scientific papers to 1901; and W. W. Watts, an obituary notice in *Proceedings of the Royal Society,* **B99** (1926), xvii-xxvii.

O. M. B. BULMAN

BONNIER, GASTON (*b.* Paris, France, 1853; *d.* Paris, 30 December 1922), *botany.*

During the last half of the nineteenth century, botany changed from a descriptive science to an experimental one; Gaston Bonnier was one of the botanists responsible for this transformation. A fervent advocate of the experimental approach, he made several discoveries that, although not revolutionary, were of considerable importance to the growth of the science.

Bonnier's father and grandfather were both professors of law, but he seems to have been interested in botany from the very beginning of his academic career. He was a student at the École Normale in Paris and taught there for several years before succeeding to the chair of botany at the Sorbonne in 1887.

In 1879, Bonnier received the D.S.N. upon publication of his thesis, an anatomical and physiological study of the nectary organs. This thesis earned him the Prix de Physiologie Expérimentale of the Paris Academy of Sciences (of which he was made a member in 1896). Traditionally the nectary organs had been one of the best weapons in the arsenal of teleology; it was said that they existed solely to produce nectar, which itself existed solely to attract bees, the agents of cross-fertilization. Bonnier demonstrated the absurdity of this argument by proving that in many species of plants a bee can easily collect nectar without going near the pollen-carrying stamens. He proved that the nectaries are important to the plant chiefly because they store the excess sugar that is needed during periods of increased physiological activity, nectar being only an incidental by-product of transpiration.

The decade after the publication of his thesis was the most productive period of Bonnier's career. Between 1880 and 1882 he studied (in collaboration with Philippe van Tieghem) the physiological activity of seeds, grains, and bulbs and discovered that they are not physiologically "dead," as had been thought. From 1883 to 1885, he published several lengthy papers on plant respiration in collaboration with Louis Mangin. At that time botanists had just begun to realize that respiration and photosynthesis are different processes. Bonnier and Mangin were particularly interested in the relationship between respiration and various environmental conditions. Of their many discoveries, the most crucial was the fact that respiration proceeds most rapidly in the absence of light. They also developed an apparatus for determining the ratio of carbon dioxide discharged to oxygen absorbed at any given moment and demonstrated that this ratio remains constant for each species, no matter what the respiration rate may be at any given time. Finally, toward the end of this decade, Bonnier branched out into an entirely different field—lichenology—and settled a long-standing botanical debate by proving that lichens are composed of two symbiotic forms, an alga and a fungus, the latter reproducing by means of spores.

Bonnier became increasingly involved in administrative affairs. He was deeply concerned about the development of botany as a discipline, and he had helped to organize a separate faculty in natural sciences while at the École Normale. Botanical facilities were inadequate at the Sorbonne, and in 1889 Bonnier was able to remedy that situation by founding and directing the Laboratoire de Biologie Végétale at Fontainebleu. In 1890 he took on the additional burden of editing a new botanical journal, *Revue générale de botanique.* He continued in that post, as well as in the others, until his sudden death.

From 1890 to 1922, most of Bonnier's scientific work was concerned with the relationship between structure and environment. In this connection he studied the differences between alpine and arctic plants, and attempted to find correlations with differences in their habitats. He also undertook studies of heat production and pressure transmission in plants. During this time he wrote extensively, producing several botanical handbooks, a few popularizations, a basic text, and, finally, his twelve-volume masterpiece, *Flore complète . . . de France, Suisse et Belgique,* some of which was published posthumously.

BIBLIOGRAPHY

Articles by Bonnier include "Les nectaires," in *Annales des sciences naturelles (botanique)*, 6th series, **8** (1879), 5–212, and "Recherches sur la vie ralentie et sur la vie latente," in *Bulletin de la Société Botanique de France,* **27** (1880), 83–88, 116–122; **29** (1882), 25–29, 149–153. A summary of the work done by Bonnier and Mangin can be found in "La fonction respiratoire chez les végétaux," in *Annales des sciences naturelles (botanique)*, 7th series, **2** (1885), 365–380; the original papers are in *Annales,* 6th series, **17** (1884), 210–306; **18** (1884), 293–381; **19** (1885), 217–255. A representative article of several on alpine and arctic plants is "Les plantes arctiques comparées aux mêmes espèces des Alpes et des Pyrénnées," in *Revue générale de botanique,* **6** (1894), 505–527. Of his books, other than practical guides such as *Nouvelle flore pour la détermination facile des plantes* (Paris, 1887), the most important are his text, *Cours de botanique,* 2 vols. (Paris, 1901), and his *Flore complète illustrée et en couleurs de France, Suisse et Belgique,* 12 vols. (Paris, 1912–1934).

For a discussion of Bonnier's work, see M. H. Jumelle, "L'oeuvre scientifique de Gaston Bonnier," in *Revue générale de botanique,* **36** (1924), 289–307.

RUTH SCHWARTZ COWAN

BONOMO, GIOVAN COSIMO (*b.* Leghorn, Italy, 30 November 1666; *d.* Florence, Italy, 13 January 1696), *medicine.*

Bonomo received the doctorate in philosophy and medicine at the University of Pisa on 22 June 1682. The following year, on 18 December, he passed the qualifying examination to practice his profession in Tuscany. He often served as physician in the galleys of Grand Duke Cosimo III, which had Leghorn as their home port.

Bonomo belonged to the biological school that originated with Galileo. Inspired by the research that had enabled his teacher Francesco Redi to disprove the theory of the spontaneous generation of insects in 1668, and availing himself of Giacinto Cestoni's skill with the microscope, Bonomo, in his *Osservazioni intorno a' pellicelli del corpo umano* (1687), affirmed that scabies is caused by mites. As a matter of fact, the mites of patients suffering from scabies had been known for some time, but they were considered a consequence and not a cause of the disease.

With the aid of the microscope, it was demonstrated that this arachnid reproduces by means of eggs and that it possesses an oral apparatus with which it penetrates the skin. Hence, Bonomo resolved to adopt local therapy aimed at killing the mites, instead of the general therapy that had previously been used. The results thus obtained enabled him to conclude that the mites were the cause of the disease. It followed that scabies is transmitted by the mites from a victim to a healthy person. Therefore, it is a "live" infection, of which Bonomo's work constituted the first clinical and experimental proof.

In April 1691, the grand duke appointed Bonomo physician to his daughter Anna Maria Luisa, who had married the elector of the Rhenish Palatinate, Johann Wilhelm. In this capacity he accompanied her to Düsseldorf and remained there until the end of 1694 or the beginning of 1695, when he was obliged, for reasons of health, to return to Florence.

BIBLIOGRAPHY

Bonomo's major work is *Osservazioni intorno a' pellicelli del corpo umano* (Florence, 1687).

Works on Bonomo are Luigi Belloni, "Le 'contagium vivum' avant Pasteur," in *Les conférences du Palais de la Découverte* (Paris, 1961), pp. 10–11; "La medicazione topica nella scoperta della etiologia acarica della scabbia," in *Simposi clinici,* **1** (1964), xxi–xxvi; "I secoli italiani della dottrina del contagio vivo," *ibid.,* **4** (1967), liii; Ugo Faucci, "Contributo alla storia della scabbia," in *Rivista di storia delle scienze mediche e naturali,* **22** (1931), 153–170, 198–215, 257–371, 441–475; and C. Lombardo, "Giovan Cosimo Bonomo a Pisa," *ibid.,* **29** (1938), 97–121.

LUIGI BELLONI

BONVICINO (also known as **BONVOISIN** or **BUONVICINO**), **COSTANZO BENEDETTO** (*b.* Centallo, near Cuneo, Piedmont, 1739; *d.* Turin, Italy, 25 January 1812), *chemistry.*

The comfortable circumstances of Bonvicino's family enabled him to attend the University of Turin, from which he received his degree in medicine on 14 April 1764. In 1778 he was admitted to the College of Physicians. Bonvicino devoted himself almost entirely to chemistry, however, at first under the guidance of V. A. Gioanetti, and acquired considerable scientific prestige. In 1783 he became a member of the Academy of Sciences of Turin, of which he was president in 1801–1802. In 1800 he was made professor of pharmaceutical chemistry and of the natural history of drugs at the University of Turin.

Bonvicino's first scientific work dealt with a qualitative and a partially quantitative analysis of a Piedmontese mineral called *pierre hydrophane,* known for its characteristic of losing its opacity when immersed in water. A similar mineral had previously been examined by Torbern Bergman.

Bonvicino's activity in the Academy of Sciences of Turin during the reign of the house of Savoy (1783–1798) was in analytical and industrial chemistry. He was placed in charge of the Academy's laboratory and of the analysis of mineral waters and waters used by dyers. He also supervised the production and control of common salts and wrote reports on new proposals dealing largely with the textile and dye industries and with metallurgy.

In opposition to the views held by Lavoisier, in November 1787 Bonvicino requested the Turin Academy to take an official position against the chemical nomenclature proposed by the French scientist. This occurred at a time when the Academy was totally dedicated to a defense of the theories of Stahl. On 12 February 1792 Bonvicino reported to the Academy on some experiments he had performed in support of the phlogiston theory. His scientific production during this period achieved some noteworthy results that are recorded in all the chemistry manuals of the period. His memoir on the isolation of phosphoric acid from calcium phosphate by means of ammonium carbonate dates from 1784–1785; previously, oxalic acid, which gave only a partial purification, had been used.

The writings that give us a truer measure of Bonvicino's scientific achievements are those related to the analysis of water and salts. The predominance of the experimental aspect freed the scientist from the rigid interpretative schemes that were based on preconceived theories.

In his carefully conducted studies, Bonvicino made use of a technique that aimed at the solution of problems of quantitative determination, that is, problems that dealt with the loss of weight, due to mechanical and physical factors, of substances being examined. The results yielded a great mass of data that appeared in scientific publications during the second half of the eighteenth century while being subjected to further experimental verification. From these studies he wrote his *Elementi di chimica,* a rich compendium of his university courses.

During the Napoleonic period (1798–1812), Bonvicino was especially interested in a more rational exploitation of the mineral resources of the Piedmont, a problem he tackled from all points of view: the importance of the mineral wealth (the survey and analysis of minerals were the chief objects of his studies during this period); the development of the industries that processed the minerals (to be achieved by the gradual replacement of the miners, who used crude techniques, by students who had been trained in technical schools, the establishment of which Bonvicino strongly advocated); the updating of mining legislation (still on a feudal basis, mining laws presented a serious obstacle to free enterprise); and combustibles (besides the traditional source, wood, he also searched for coal deposits).

BIBLIOGRAPHY

I. ORIGINAL WORKS. A fairly complete bibliography of Bonvicino's published works may be found in G. G. Bonino, *Biografia medica piemontese* (see below), II, 593–596. Books not included by Bonino are *Sulle cagioni recenti della minor produzione in bozzoli e in sete nel Piemonte e sui mezzi di rimediarvi* (Turin, 1802); *Elementi di chimica farmaceutica e di storia naturale e preparazione de' rimedi,* 2 vols. (Turin, 1804–1810); *Pensieri sulla cura dell'epizozia che regna ora in Piemonte* (Turin, 1805); *Storia di quattro persone che morirono avvelenate dai funghi* (Turin, n.d.); *Mémoire présenté à la Commission du Grand Conseil d'Administration de l'Université de Turin nommée pour examiner tout ce qui a rapport aux examens de médecine* (Turin, n.d.); and *Memorie ed istruzioni sui mezzi di minorare i danni delle carestie nel Piemonte per mezzo della dilatata coltura dei pomi di terra, volgarmente detti tartifle* (Turin, n.d.). Articles omitted by Bonino are "Rouissage du chanvre," in *Memorie dell'Accademia delle scienze di Torino,* **10** (1793), xxxii–xxxix; "Note sur la diopside," in *Journal des mines,* **20,** no. 115 (1812), 65 ff.; and "Sur la formation de l'hydrophane et du cacholong," *ibid.,* no. 118 (1812), 305 ff.

Unpublished memoirs and MSS of Bonvicino, preserved in the library of the Academy of Sciences of Turin, are "Sulla maniera di trarre il sale catartico dal scisto delle montagne di Sallances" (1784), MS collection 485; "Analisi di alcune acque naturali relativamente alle tinture" (1790), MS collection 540; "Analyse de la teinture tonique dite les Gouttes de Bestouscheff" (1791), MS collection 443–445; and "Parere su un saggio di Vitriolo della nuova fabbrica eretta in Carouge dal sig. De Voiseray" (1791), MS collection 2331.

Also preserved in the same library are reports prepared by Bonvicino and the commissions of which he was a member at the request of private industry, and six letters from Bonvicino to colleagues in the Academy. Bonvicino's activity in the Academy of Sciences is recorded in its MS minutes, preserved in its library.

II. SECONDARY LITERATURE. An obituary notice on Bonvicino is A. Garmagnano, *Clarissimi Benedicti Bonvicini chimiae pharmaceuticae professoris medicae facultatis praesidis laudatio* (Turin, 1812). A longer sketch of his life is G. G. Bonino, in his *Biografia medica piemontese,* II (Turin, 1825), 585–596. Its information was republished in a biography of Bonvicino by I. Guareschi, in *Supplemento annuale alla Enciclopedia di chimica* (Turin, 1910), pp. 445–453. For information on the scientific milieu in which Bonvicino worked, the only worthwhile work is still that of I. Guareschi, "La chimica in Italia dal 1750 al 1800," in *Supplemento annuale alla Enciclopedia di chimica* (Turin, 1909), pp. 327–378.

GIORGIO PEDROCCO

BOODT, ANSELMUS BOETIUS DE (*b.* Bruges, Belgium, *ca.* 1550; *d.* Bruges, 21 June 1632), *mineralogy.*

Boodt was the son of Willem de Boodt and Johanna Voet, daughter of a famous lawyer. The Boodts were a noble Roman Catholic family, and Anselmus was destined for a career in the administration of his native town. He probably took his first university degree in civil and canon law at Louvain. After this he studied medicine under Thomas Erastus at Heidelberg and obtained his M.D. in Padua.

From 1583 Boodt lived in Bohemia as physician to Wilhelm Rosenberg, the burgrave of Prague. He was on very friendly terms with the imperial physician, Thadeus Hayek, a well-known Bohemian naturalist and historian. With Hayek and Nicolas Barnaud he made some alchemical experiments that are mentioned in Barnaud's "In aenigmaticum"; however, he was critical of alchemy and a decided opponent of Paracelsus. Besides being a polyglot and poet, Boodt drew and painted flowers, animals, and minerals. He made many mineralogical excursions in Germany, Silesia, and Bohemia.

On 11 February 1584, while he was in Prague, Boodt was appointed canon of St. Donat's Church in Bruges; he held this post until 1595 without leaving Prague. On 1 January 1604 he was nominated physician in ordinary to Rudolf II and retained this position until the death of the emperor in 1612. Under the influence of Rudolf, a devoted collector of all curiosities, Boodt began in 1604 to write his chief work, *Gemmarum et lapidum historia* (1609).

In 1612 Boodt returned to Bruges, where he spent the rest of his life as a town councillor. He never married. On 17 October 1630 he made his will, bequeathing to the Jesuits of Bruges the sums that Rudolf II owed him. His next of kin received his books, pictures, instruments, and collections of minerals.

In his *Gemmarum et lapidum historia* Boodt made the first attempt at a systematic description of minerals, dividing the minerals into great and small, rare and common, hard and soft, combustible and incombustible, transparent and opaque. He uses a scale of hardness expressed in three degrees and notes the crystalline forms of some minerals (triangular, quadratic, and hexangular). Boodt criticizes some of the views of Aristotle, Pliny, Paracelsus, and others, but accepts the existence of the four elements and three principles, although he also mentions atoms. He enumerates about 600 minerals that he knows from personal observation, and describes their properties, values, imitations, and medical applications. There are also tables of values of diamonds according to their size and a short description of the polishing of precious stones. Boodt cites nineteen authors and,

besides the minerals known to him, gives a list of 233 minerals whose names he knows from Pliny and Bartholomeus Anglicus, among others.

BIBLIOGRAPHY

I. ORIGINAL WORKS. Boodt's only published work is *Gemmarum et lapidum historia* (Hanau, 1609); 2nd ed., prepared by Adrianus Toll, M.D. (Leiden, 1636); the 3rd ed., also prepared by Toll (Leiden, 1649), has as supplement John de Laet's *De gemmis et lapidibus librii II et Theophrastus' Liber de lapidibus Graece et Latine cum brevibus notis* (1647). The *Gemmarum* was translated into French by Jean Bachou as *Le parfaict joaillier, ou histoire des pierreries* (Lyons, 1644, 1649).

II. SECONDARY LITERATURE. Writings on Boodt or his work are Nicolas Barnaud, "In aenigmaticum quoddam epitaphium Bononiae," in *Theatrum chemicum*, III (Strasbourg, 1613), 787; O. Delepierre, *Biographie des hommes remarquables de la Flandre occidentale*, I (Bruges, 1843), 31–35; G. Dewalque, *Biographie nationale de Belgique*, IV, 814–816; F. M. Evans, *Magic Jewels* (Oxford, 1922), p. 154; F. V. Goethals, *Lectures relatives à l'histoire des sciences, des arts, des lettres, des moeurs et de la politique en Belgique* (Brussels, 1838), pp. 98–105; J. E. Hiller, in *Annales Guebhard-Severine*, 11 (1935), 74; in *Archeion*, 15 (1933), 348–368; in *Quellen und Studien zur Geschichte der Naturwissenschaften und der Medizin*, 8 (Berlin, 1942), 1–125; and in *Fortschritte der Mineralogie, Krystallographie und Petrographie* (Stuttgart), 17 (1932), 418–419; E. Hoefer, ed., *Nouvelle biographie universelle*, VI (Paris, 1853), 665; F. M. Jaeger, in *Chemisch Weekblad*, 15 (1918), 628, and in *Historische Studiën. Bijdragen tot de Kennis van de Geschiedenis der Natuurwetenschappen in Nederlanden gedurende de 16e en 17e Eeuw* (Groningen, 1919), pp. 99–149; *Nieuw Nederlands biografisch Woordenboek*, VI, 151–152; Oesterreichische Nationalbibliothek, Vienna, MS 14724, p. 133; J. R. Partington, *A History of Chemistry*, II (London, 1961), 101–102; "Testament olographe d'Anselme de Boodt, conseiller-pensionnaire de Bruges, 1630," in *Annales de la Société d'émulation*, 2nd ser., 11 (1861), 370–383; and A. J. J. Van de Velde, "*Rede op de hulde Anselmus Boetius de Boodt de Brugge op 20 november 1932*, names de Academie," in *Koninklijke Vlamsche Academie voor Wetenschappen, Letteren en Schoone Kunsten van België, Klasse der Wetenschappen. Verslagen en Mededeelingen* (Brussels, Nov. 1932), pp. 1505–1507.

WŁODZIMIERZ HUBICKI

BOOLE, GEORGE (*b.* Lincoln, England, 1815; *d.* Cork, Ireland, 1864), *mathematics.*

George Boole was the son of John Boole, a cobbler whose chief interests lay in mathematics and the making of optical instruments, in which his son learned to assist at an early age. The father was not a good businessman, however, and the decline in his business had a serious effect on his son's future. The boy went to an elementary school and for a short time to a commercial school, but beyond this he educated himself, encouraged in mathematics by his father and helped in learning Latin by William Brooke, the proprietor of a large and scholarly circulating library. He acquired a knowledge of Greek, French, and German by his own efforts, and showed some promise as a classical scholar; a translation in verse of Meleager's "Ode to the Spring" was printed in a local paper and drew comments on the precocity of a boy of fourteen. He seems to have thought of taking holy orders, but at the age of fifteen he began teaching, soon setting up a school of his own in Lincoln.

In 1834 the Mechanics Institution was founded in Lincoln, and the president, a local squire, passed Royal Society publications on to the institution's reading room, of which John Boole became curator. George, who now devoted his scanty leisure to the study of mathematics, had access to the reading room, and grappled, almost unaided, with Newton's *Principia* and Lagrange's *Mécanique analytique,* gaining such a local reputation that at the age of nineteen he was asked to give an address on Newton to mark the presentation of a bust of Newton, also a Lincolnshire man, to the Institution. This address, printed in 1835, was Boole's first scientific publication. In 1840 he began to contribute to the recently founded *Cambridge Mathematical Journal* and also to the Royal Society, which awarded him a Royal Medal in 1844 for his papers on operators in analysis; he was elected a fellow of the Royal Society in 1857.

In 1849, Boole, on the advice of friends, applied for the professorship of mathematics in the newly established Queen's College, Cork, and was appointed in spite of his not holding any university degree. At Cork, although his teaching load was heavy, he found more time and facilities for research. In 1855 he married Mary Everest, the niece of a professor of Greek in Queen's College and of Sir George Everest, after whom Mount Everest was named.

Boole was a clear and conscientious teacher, as his textbooks show. In 1864 his health began to fail, and his concern for his students may have hastened his death, since he walked through rain to a class and lectured in wet clothes, which led to a fatal illness.

Boole's scientific writings consist of some fifty papers, two textbooks, and two volumes dealing with mathematical logic. The two textbooks, on differential equations (1859) and finite differences (1860), remained in use in the United Kingdom until the end of the century. They contain much of Boole's original work, reproducing and extending material published in his research papers. In the former book, so much use is made of the differential operator D that the method is often referred to as Boole's, although it is in fact much older than Boole. Both books exhibit a great technical skill in the handling of operators: in the volume on finite differences, an account is given of the operators π and ρ, first introduced in Boole's Royal Society papers. The basic operators of this calculus, Δ and E, are defined by the equations

$$\Delta u_x = u_{x+1} - u_x, \; E \, u_x = u_{x+1};$$

Boole then defines his new operators by the operational equations

$$\pi = x\Delta, \; \rho = xE,$$

and shows how they can be used to solve certain types of linear difference equations with coefficients depending on the independent variable. These operators have since been generalized by L. M. Milne-Thomson.

In papers in the *Cambridge Mathematical Journal* in 1841 and 1843, Boole dealt with linear transformations. He showed that if the linear transformation

$$x = pX + qY, y = rX + sY$$

is applied to the binary quadratic form

$$ax^2 + 2hxy + by^2$$

to yield the binary quadratic form

$$AX^2 + 2HXY + BY^2,$$

then $\quad AB - H^2 = (ps - qr)^2 (ab - h^2).$

The algebraic fact had been partly perceived by Lagrange and by Gauss, but Boole's argument drew attention to the (relative) invariance of the discriminant $ab - h^2$, and also to the absolute invariants of the transformation. This was the starting point of the theory of invariants, so rapidly and extensively developed in the second half of the nineteenth century; Boole himself, however, took no part in this development.

Other papers dealt with differential equations, and the majority of those published after 1850 studied the theory of probability, closely connected with Boole's work on mathematical logic. In all his writings, Boole exhibited considerable technical skill, but his facility in dealing with symbolic operators did not delude him into an undue reliance on analogy, a fault of the contemporary British school of symbolic analysis. E. H. Neville has remarked that mathematicians of that school treated operators with the most reckless disrespect, and in consequence could solve problems beyond the power not merely of their predecessors at the beginning of the century but of their inhibited successors at the end of the century, obtaining many remarkable and frequently correct formulas but ignoring conditions of validity.

Boole greatly increased the power of the operational calculus, but seldom allowed himself to be carried away by technical success: at a time when the need for precise and unambiguous definitions was often ignored, he was striving, although perhaps not always with complete success, to make his foundations secure. There is a clear and explicit, although later, statement of his position in his *Investigation of the Laws of Thought;* there are, he says, two indispensable conditions for the employment of symbolic operators: "First, that from the sense once conventionally established, we never, in the same process of reasoning, depart; secondly, that the laws by which the process is conducted be founded exclusively upon the above fixed sense or meaning of the symbols employed." With the technical skill and the desire for logical precision there is also the beginning of the recognition of the nonnumerical variable as a genuine part of mathematics. The development of this notion in Boole's later and most important work appears to have been stimulated almost accidentally by a logical controversy.

Sir William Hamilton, the Scottish philosopher (not to be confused with the Irish mathematician Sir William Rowan Hamilton), picked a logical quarrel with Boole's friend Augustus De Morgan, the acute and high-minded professor of mathematics at University College, London. De Morgan's serious, significant contributions to logic were derided by Hamilton, on the grounds that the study of mathematics was both dangerous and useless—no mathematician could contribute anything of importance to the superior domain of logic. Boole, in the preface to his *Mathematical Analysis of Logic* (1847), demonstrated that, on Hamilton's own principles, logic would form no part of philosophy. He asserted that in a true classification, logic should not be associated with metaphysics, but with mathematics. He then offered his essay as a construction, in symbolic terms, of logic as a doctrine, like geometry, resting upon a groundwork of acceptable axioms.

The reduction of Aristotelian logic to an algebraic calculus had been more than once attempted; Leibniz had produced a scheme of some promise. If the proposition "All A is B" is written in the form A/B, and "All B is C" in the form B/C, then it is tempting to remove the common factor B from numerator and denominator and arrive at A/C, to be correctly interpreted as the conclusion "All A is C." Any attempt to extend this triviality encountered difficulties: Boole's predecessors had tried to force the algebra of real numbers onto logic, and since they had not envisaged a plurality of algebras, it was believed that only if the elementary properties of the symbols implied formal rules identical with those of the algebra of real numbers could the subject be regarded as a valid part of mathematics. Boole recognized that he had created a new branch of mathematics, but it is not clear whether he appreciated that he had devised a new algebra. He appears not to have known that geometries other than Euclidean could be constructed; but he knew of Rowan Hamilton's quaternions, an algebra of quadruplets in which products are noncommutative, for one of his minor papers (1848) deals with some quaternion matters. Grassmann's similar, if more general, work in the *Ausdehnungslehre* (1844) seems to have been unknown. Boole, then, knew of an algebra similar to, but not identical with, the algebra of real numbers.

If we consider a set U, the universal set or the universe of discourse, often denoted by 1 in Boole's work, subsets can be specified by elective operators $x, y, \cdots$, so that xU is the subset of U whose elements have the property defining the operator x. Thus, if U is the set of inhabitants of New York, we can select those who are, say, male by an elective operator x and denote the set of male inhabitants of New York by xU. Similarly, the left-handed inhabitants of New York may be denoted by yU, and blue-eyed inhabitants by zU, and so on. The elective operators may be applied successively. Thus we may first select all the males and from these all the left-handers by the symbolism $y(xU)$; if we first select all the left-handers and from these all the males, we have the symbolism $x(yU)$. Since in each case the final set is the same, that of all left-handed males, we can write $y(xU) = x(yU)$, or, since the universe of discourse U is understood throughout, simply write $yx = xy$. The analogy with the commutative algebraic product is clear. The associative law for products, $x(yz) = (xy)z$, can be verified at once in this interpretation, since each side denotes the set of those who are at once male, left-handed, and blue-eyed; Boole uses this without bothering to give any explicit justification. He was careful, however, to remark that although an analogy exists, the evidence on which the laws are based in his work is not related to the evidence on which the laws of the algebra of real numbers are based. To select the set of males from the set of males is merely to arrive at the set of males; thus the definition of the operator x leads to the idempotent law $x(xU) = xU$, or $x^2 = x$, the first break with ordinary algebra.

The product or intersection operation can also be regarded as a symbolic expression of the logical concept of conjunction by means of the conjunctive "and," since xy will denote the set of those inhabitants of New York who are at once male *and* left-handed.

If xU is the subset of males in the universal set U, it is natural to write the set of nonmales, that which remains when the set of males is subtracted from U, as $U - xU$, or, briefly, $1 - x$. This set, the complement of x relative to U, which Boole for brevity denoted by $\bar{x}$, can be regarded as arising from the application of the logical negation "not" to the set x. Addition has not yet been defined, but Boole did not hesitate to rewrite the equation $\bar{x} = 1 - x$ in the form $x + \bar{x} = 1$, implying that the universal set is made up of the elements of the subset x or of the subset not-x; this suggests that the sign $+$ is the symbol for the connective "or." But the word "or" in English usage has an inclusive and an exclusive sense: "either . . . or . . . and possibly both" and "either . . . or . . . but not both." Boole chose the exclusive sense, and so did not allow the symbolism $x + y$ unless the sets x, y were mutually exclusive.

Modern usage takes $x + y$ for the union or logical sum, the set of elements belonging to at least one of x,y: this union Boole included in his symbolism as $x + \bar{x}y$. Kneale suggests that Boole's choice of the exclusive sense for the symbol $+$ was caused by a desire to use the minus sign $(-)$ as the inverse of the plus sign $(+)$. If y is contained in x, $x - y$ can consistently denote those elements of x which are not elements of y—the complement of y relative to x—but if $+$ is used in the inclusive sense, then the equations $x = y + z$, $x = y + w$ do not imply $z = w$, so that $x - y$ is essentially indeterminate. Alternately, a use of the idempotent law implies that

$$(x - y)^2 = x - y,$$

and a further application of this law suggests that from

$$x^2 - 2xy + y^2 = x - y$$

it follows that

$$x - 2xy + y = x - y$$

and, hence, that $y = xy$; this is a symbolic statement that y is a subset of x. Boole was thus led to the use of the sign $+$ in the exclusive sense, with the sign $-$ as its inverse.

The idempotent law $x^2 = x$ is expressed in the form $x(1 - x) = 0$, but it is not altogether clear whether Boole regarded this as a deduction or as a formulation of the fundamental Aristotelian principle that a prop-

osition cannot be simultaneously true and false. Some of the obscurity is due to the fact that Boole does not always make clear whether he is dealing with sets, or with propositions, or with an abstract calculus of which sets and propositions are representations.

Much of the 1847 tract on the mathematical analysis of logic is devoted to symbolic expressions for the forms of the classical Aristotelian propositions and the moods of the syllogism. The universal propositions "All X's are Y's," "No X's are Y's" take the forms $x(1 - y) = 0$, $xy = 0$. The particular propositions "Some X's are Y's," "Some X's are not Y's" do not take what might appear to be the natural forms $xy \neq 0$, $x(1 - y) \neq 0$, possibly because Boole wished to avoid inequalities and to work entirely in terms of equations. He therefore introduced an elective symbol, v; any elements common to x and y constitute a subset v; which, he says, is "indefinite in every respect but this"—that it has some members. The two particular propositions he wrote in the forms $xy = v$, $x(1 - y) = v$. This ill-defined symbol needs careful handling when the moods and figures of the syllogism are discussed. Thus the premises "All Y's are X's," "No Z's are Y's" give the equations $y = vx$, $0 = zy$, with the inference $0 = vzx$ to be interpreted as "Some X's are not Z's." Boole explains that it would be incorrect to interpret $0 = vzx$ as "Some Z's are not X's" because the equation $y = vx$ fixes the interpretation of vx as "Some X's" and "v is regarded as the representation of 'Some' only with respect to the class X."

A similar obscurity is encountered when an attempt is made to define division. If $z = xy$, what inferences can be drawn about x, in the hope of defining the quotient z/y? Since z is the intersection of x and y, and thus is contained in y, $yz = z$; thus x, which contains z, contains yz. Any other element of x that is not in z cannot be in y, and hence x is made up of yz and an indeterminate set of which all that can be said is that its elements belong neither to y nor to z, and thus belong to the intersection of $1 - y$ and $1 - z$. Thus

$$z/y = yz + \text{an indefinite portion of } (1 - y)(1 - z).$$

Boole gave this result as a special case of his general expansion formula, and his argument is typical of that used to establish the general theorem. From $y + \bar{y} = 1$, $z + \bar{z} = 1$, it follows that $yz + y\bar{z} + \bar{y}z + \bar{y}\bar{z} = 1$, that is, the universe of discourse is the sum of the subsets yz, $y\bar{z}$, $\bar{y}z$, $\bar{y}\bar{z}$. Hence, any subset whatsoever will be at most a sum of elements from each of these four subsets; thus

$$z/y = Ayz + By\bar{z} + C\bar{y}z + D\bar{y}\bar{z},$$

with coefficients A, B, C, D to be determined. First, set $y = 1$, $z = 1$, so that $\bar{y} = \bar{z} = 0$; then $A = 1$. Next, set $y = 1$, $z = 0$, so that $\bar{y} = 0$, $\bar{z} = 1$; then $B = 0$. Third, set $y = 0$, $z = 1$, so that $\bar{y} = 1$, $\bar{z} = 0$; if the term in $\bar{y}z$ were present, then C would have to be infinite; hence, the term in $\bar{y}z$ cannot appear. Finally, if $y = z = 0$, the coefficient D is of the form $0/0$, which is indeterminate. This asserts the possible presence of an indefinite portion of the set $\bar{y}\bar{z}$. Thus, as before,

$$z/y = yz + \text{ an indefinite portion of } \bar{y}\bar{z},$$

or, as Boole frequently wrote it,

$$\frac{z}{y} = yz + \frac{0}{0}(1 - y)(1 - z).$$

Schröder showed that the introduction of division is unnecessary. But the concept of the "development" of a function of the elective symbols is fundamental to Boole's logical operations and occupies a prominent place in his great work on mathematical logic, the *Investigation of the Laws of Thought*. If $f(x)$ involves x and the algebraic signs, then it must denote a subset of the universe of discourse and must therefore be made up of elements from x and $\bar{x}$. Thus

$$f(x) = Ax + B\bar{x},$$

where the coefficients A and B are determined by giving x the values of 0 and 1. Thus

$$f(x) = f(1)x + f(0)(1 - x),$$

which in the *Mathematical Analysis of Logic* Boole regards as a special case of MacLaurin's theorem, although he dropped this analogy in the *Investigation of the Laws of Thought*. A repeated application of this method to an expression $f(x,y)$ containing two elective symbols yields

$$f(x,y) = f(1,1)\,xy + f(1,0)\,x(1 - y)$$
$$+ f(0,1)(1 - x)y + f(0,0)(1 - x)(1 - y),$$

and more general formulas can be written down by induction. Logical problems which can be expressed in terms of elective symbols may then be reduced to standard forms expediting their solution.

Boole's logical calculus is not a two-valued algebra, although the distinction is not always clearly drawn in his own work. The principles of his calculus, as a calculus of sets, are nowhere set out by him in a formal table, but are assumed, sometimes implicitly, and are, save one, analogous to the algebraic rules governing real numbers:

$$xy = yx$$
$$x + y = y + x$$
$$x(y + z) = xy + xz$$
$$x(y - z) = xy - xz.$$

If $x = y$, then

$$xz = yz$$
$$x + z = y + z$$
$$x - z = y - z.$$

$$x(1 - x) = 0.$$

Of these, only the last has no analogue in the algebra of real numbers. These principles suffice for the calculus of sets. But Boole observes that in algebra the last principle is an equation whose only roots are $x = 0$, $x = 1$. In the calculus of sets this would assert that any set is either the null set or the universal set. Boole added this numerical interpretation in order to establish a two-valued algebra, of which one representation would be a calculus of propositions in which the truth of a proposition X is denoted by $x = 1$ and its falsehood by $x = 0$: the truth-value of a conjunction "X and Y" will be given by xy, and of an exclusive disjunction "X or Y" by $x + y$. The distinction between propositions and propositional functions, not drawn by Boole, was made later by C. S. Peirce and Schröder.

The use of $x + y$ to denote the exclusive sense of "or" led to difficulties, such as the impossibility of interpreting $1 + x$ and $x + x$, which Boole surmounted with considerable ingenuity. But Jevons, in his *Pure Logic* (1864), used the plus sign in its inclusive (and/or) sense, a use followed by Venn and C. S. Peirce and since then generally adopted. Peirce and Schröder emphasized that the inclusive interpretation permits a duality between sum and product, and they also showed that the concepts of subtraction and division are superfluous and can be discarded. With the use of $x + y$ to denote "either x or y or both," the expression $x + x$ presents no difficulty, being just x, while $1 + x$ is the universal set 1. The duality of the two operations of sum and product exemplified by the equations $xx = x$, $x + x = x$ can now be carried further: the formulas

$$xy + xz = x(y + z), \qquad (x + y)(x + z) = x + yz$$

are duals, since one can be derived from the other by an interchange of sum with product. This duality is clearer if these operations are denoted by the special symbols $\cap$, $\cup$ now in general use for product and sum, that is, for intersection and union. In this notation, the preceding equations are written

$$(x \cup y) \cap (x \cup z) = x \cup (y \cap z),$$
$$(x \cap y) \cup (x \cap z) = x \cap (y \cup z).$$

With the inclusive interpretation, the system can now be shown to obey the dual rules of De Morgan:

$$\overline{xy} = \bar{x} + \bar{y}, \qquad \overline{x + y} = \bar{x}\bar{y}.$$

In the *Investigation of the Laws of Thought,* the calculus is applied to the theory of probability. If $P(X) = x$ is the probability of an event X, then if events X, Y are independent, $P(X$ and $Y) = xy$, while if X and Y are mutually exclusive, $P(X$ or $Y) = x + y$. The principles laid down above are satisfied, except for the additional numerical principle in which the allowable values of x are 0 and 1, which is not satisfied. A clear and precise symbolism enabled Boole to detect and correct flaws in earlier work on probability theory.

E. V. Huntington in 1904 gave a set of independent axioms on which Boole's apparatus can be constructed, and various equivalent sets have been exhibited. One formulation postulates two binary operations $\cup$, $\cap$ (union and intersection) which have the commutative and distributive properties:

$$x \cup y = y \cup x, \qquad x \cap y = y \cap x,$$
$$x \cup (y \cap z) = (x \cup y) \cap (x \cup z),$$
$$x \cap (y \cup z) = (x \cap y) \cup (x \cap z);$$

further, there are two distinct elements, 0 and 1, such that for all x

$$x \cup 0 = x, \qquad x \cap 1 = x;$$

also, for any x, there is an element $\bar{x}$ (the complement) for which

$$x \cup \bar{x} = 1, \qquad x \cap \bar{x} = 0.$$

The system so defined is self-dual, since the set of axioms remains unchanged if $\cup$ and $\cap$ are interchanged when 0, 1 are also interchanged. The associative laws for union and intersection are not required as axioms, since they can be deduced from the given set.

If intersection and complement are taken as the basic operations, with the associative law $x \cap (y \cap z) = (x \cap y) \cap z$ now an axiom and the relation between the basic operations given by the statements

if $x \cap \bar{y} = z \cap \bar{z}$ for some z, then $x \cap y = x$,
if $x \cap y = x$, then $x \cap \bar{y} = z \cap \bar{z}$ for any z,

then union can now be defined in terms of intersection and complement by the equation

$$x \cup y = \overline{\bar{x} \cap \bar{y}},$$

0 can be defined as $x \cap \bar{x}$, and 1 as the complement of 0. The two systems are then equivalent.

The theory of lattices may be regarded as a generalization. A lattice is a system with operations $\cup$, $\cap$ having the commutative, distributive, and associative properties. Thus every Boolean algebra is a lattice; the converse is not true. The lattice concept is wider than the Boolean, and embraces interpretations for which Boolean algebra is not appropriate.

Boole's two-valued algebra has recently been applied to the design of electric circuits containing simple switches, relays, and control elements. In particular, it has a wide field of application in the design of high-speed computers using the binary system of digital numeration.

BIBLIOGRAPHY

I. ORIGINAL WORKS. Boole's papers include "Researches on the Theory of Analytical Transformations, With a Special Application to the Reduction of the General Equation of the Second Order," in *Cambridge Mathematical Journal,* **2** (1841), 64–73; "On a General Method in Analysis," in *Philosophical Transactions of the Royal Society of London,* **134** (1844), 225–282. *An Address on the Genius and Discoveries of Sir Isaac Newton* was published in Lincoln in 1835.

His textbooks are *Treatise on Differential Equations* (Cambridge, 1859, and later editions); a posthumous *Supplementary Volume* (Cambridge, 1865), compiled from Boole's notes by Isaac Todhunter, and containing a list of Boole's publications; *Treatise on the Calculus of Finite Differences* (Cambridge, 1860, and later editions).

On mathematical logic: *The Mathematical Analysis of Logic, Being an Essay Towards a Calculus of Deductive Reasoning* (Cambridge, 1847; repr. Oxford, 1948, and in Boole's *Collected Logical Works,* I, Chicago–London, 1916); *An Investigation of the Laws of Thought, on Which Are Founded the Mathematical Theories of Logic and Probability* (London, 1854; repr. New York, 1951, and in Boole's *Collected Logical Works,* II, Chicago–London, 1916).

II. SECONDARY LITERATURE. E. V. Huntington, "Sets of Independent Postulates for the Algebra of Logic," in *Transactions of the American Mathematical Society,* **5** (1904), 208–309; E. V. Huntington, "Postulates for the Algebra of Logic," in *Transactions of the American Mathematical Society,* **35** (1933), 274–304; W. Kneale, "Boole and the Revival of Logic," in *Mind,* **57** (1948), 149–175, which contains a useful bibliography; W. Kneale, "Boole and the Algebra of Logic," in *Notes and Records of the Royal Society of London,* **12** (1956), 53–63; Sir Geoffrey Taylor, "George Boole, 1815–1864," *ibid.,* 44–52, which gives an account of Boole's life by his grandson.

T. A. A. BROADBENT

BORCH, OLUF. See **Borrichius, Olaus.**

BORCHARDT, CARL WILHELM (*b.* Berlin, Germany, 22 February 1817; *d.* Rudersdorf, near Berlin, 27 June 1880), *mathematics.*

The son of Moritz Borchardt, a wealthy and respected Jewish merchant, and Emma Heilborn, Borchardt had among his private tutors the mathematicians J. Plücker and J. Steiner. From 1836 he studied at the University of Berlin with Dirichlet, and from 1839 at the University of Königsberg with Bessel, F. Neumann, and Jacobi. In his doctoral thesis (1843; unpublished and now lost), written under the supervision of Jacobi, he dealt with certain systems of nonlinear differential equations. In 1846–1847 he was in Paris, where he met Chasles, Hermite, and Liouville. Borchardt became a *Privatdozent* at the University of Berlin in 1848, and a member of the Berliner Akademie der Wissenschaften in 1855. He married Rosa Oppenheim. Very poor health interrupted his teaching for years; nevertheless, from 1856 to 1880 he edited, as Crelle's successor, Volumes **57–90** of the celebrated *Crelle's Journal für die reine und angewandte Mathematik,* upholding its high standard of mathematical scholarship.

Borchardt became known as a mathematician through his first publication (1846), in which he generalized a result obtained by Kummer concerning the equation that determines the secular disturbances of the planets (characteristic equation, or secular equation). By means of determinants Borchardt proved that in this case Sturm's functions can be represented as a sum of squares. From this it follows that the roots of the characteristic equation are real. In several further papers Borchardt applied the theory of determinants to algebraic questions, mostly in connection with symmetric functions, the theory of elimination, and interpolation. Another group of his papers dealt with the arithmetic-geometric mean (AGM). Gauss and Lagrange had established its connection with the complete elliptic integral of the first class. Borchardt, starting from the functional equation for the limit value of the AGM, derived a linear differential equation of the second order, the differential equation of the complete, first-class elliptic integral. He also studied a variant process of the AGM connected with the circular functions, and the generalization of the AGM to four elements and its relation to hyperelliptic integrals. Other papers dealt with problems of maxima and the theory of elasticity.

BIBLIOGRAPHY

I. Original Works. Borchardt's *Gesammelte Werke,* G. Hettner, ed. (Berlin, 1888), contains 25 papers and some short communications. His works are listed in Poggendorff, I, 238; III, 162; IV, 158.

II. Secondary Literature. Works on Borchardt are Maurice d'Ocagne, "C. W. Borchardt et son oeuvre," in *Revue des questions scientifiques* (Jan. 1890), also repr. separately (Brussels, 1890); and Max Steck, in *Neue deutsche Biographie,* II (Berlin, 1955), 456.

Christoph J. Scriba

BORDA, JEAN-CHARLES (*b.* Dax, France, 4 May 1733; *d.* Paris, France, 19 February 1799), *physics, mathematics.*

Borda was the tenth child and the sixth son of the sixteen children of Jean-Antoine de Borda and Jeanne-Marie-Thérèse de Lacroix. His parents were both of the nobility, and his parental ancestors had been in the military since the early seventeenth century. Borda began his studies at the Collège des Barnabites at Dax, then continued at the Jesuit Collège de la Flèche. He entered the École du Génie de Mézières in 1758 and finished the two-year course in one year. Borda scorned religion, at least in his youth, and he never married. While commanding a flotilla of six ships in the Antilles in 1782, Borda was taken prisoner by the English. After this misfortune, his health declined steadily. He was elected a member of the Paris Académie des Sciences in 1756 (and of its successor, the Institut de France), the Académie de Bordeaux in 1767, the Académie de Marine in 1769, and the Bureau des Longitudes in 1795. Borda is a major figure in the history of the French navy. He attained the rank of *capitaine de vaisseau,* participated in several scientific voyages and in the American Revolution, and in 1784 was named *inspecteur des constructions et de l'École des Ingénieurs de vaisseau.*

Borda's most important contributions are his work in fluid mechanics and his development and use of instruments for navigation, geodesy, and the determination of weights and measures. In a series of theoretical and experimental memoirs he studied fluid flow reactions and fluid resistance as applied to artillery, ships, scientific instruments, and hydraulic wheels and pumps. Specifically, he demonstrated that Newton's theory of fluid resistance was untenable and that the resistance is proportional to the square of the fluid velocity and to the sine of the angle of incidence. He introduced the Borda mouthpiece and calculated the coefficient of fluid contraction from an orifice. Borda's use of the principle of conservation of *vis viva* was important as a precursor of Lazare Carnot's work in mechanics.

Borda's development of a surveying instrument, the *cercle de réflexion,* contributed to the French success in measuring the length of the meridional arc. He participated in the work on a standard system of weights and measures, and designed the platinum standard meter and the standard seconds pendulum. He contributed memoirs on the calculus of variations and, in connection with his *cercle de réflexion,* developed a series of trigonometric tables. Borda's importance to science lies in his skillful use of calculus and experiment, unifying them in diverse areas of physics. This led Biot to state that one owes to Borda and Coulomb the renaissance of exact physics in eighteenth-century France.

BIBLIOGRAPHY

I. ORIGINAL WORKS. A complete bibliography of Borda's memoirs is contained in Mascart (see below). His various papers on fluid mechanics are contained in the *Mémoires de l'Académie des sciences* for the years 1763 and 1766–1769. For a description of his *cercle de réflexion,* see *Description et usage du cercle de réflexion avec différentes méthodes pour calculer les observations nautiques, par le Chevalier de Borda* (Paris, 1787; 4th ed., 1816).

II. SECONDARY LITERATURE. The most important treatment of Borda's work is the massive 800-page study by Jean Mascart, *La vie et les travaux du Chevalier Jean-Charles de Borda,* published as a volume of the *Annales de l'Université de Lyon,* n.s., **2,** Droit, Lettres, fasc. 33 (Lyons-Paris, 1919). The best contemporary essay is S. F. Lacroix, *Éloge historique de Jean-Charles Borda* (Paris, *ca.* 1800). For a recent summary of Borda's work in fluid mechanics, see R. Dugas, *Histoire de la mécanique* (Paris, 1950), pp. 292–300.

C. STEWART GILLMOR

BORDET, JULES (*b.* Soignies, Belgium, 13 June 1870; *d.* Brussels, Belgium, 6 April 1961), *bacteriology, immunology.*

Bordet established the basis of humoral immunity and founded serology. The second son of a schoolteacher, he was an outstanding student at the Athénée Royal of Brussels and received the M.D. from the University of Brussels in 1892. He had begun his research even before finishing his medical studies.

In 1894, thanks to a scholarship awarded by the Belgian government, Bordet went to Paris to work in Élie Metchnikoff's laboratory at the Institut Pasteur. It was there, between the ages of twenty-five and thirty, that he made his principal discoveries in humoral immunity.

Married in 1899, he had two daughters and one son, Paul, who also worked in experimental medicine.

In Metchnikoff's laboratory Bordet studied the mechanics of bacteriolysis, a phenomenon consisting in the lysis of cholera vibrios injected into the peritoneum of immunized animals and recently discovered by R. Pfeiffer and Issaeff (1894). Bordet reached the conclusion that bacteriolysis was due to the action of two substances: a specific antibody that he called the sensibilizer, which was resistant to heat of 55°C. and present in serum from immunized animals, and a nonspecific, thermolabile substance, which is found in serum from both unvaccinated and vaccinated animals. He identified this substance as Büchner's "alexin," which later became known as "complement." Bordet then demonstrated that the mode of action of the hemolytic serums is absolutely analogous to that of the bacteriolytic ones.

As early as 1895 Bordet underscored the specific character of the agglutination of the *Vibriocomma* (Asiatic cholera bacillus) through anticholeric immunoserum. By using hemolytic serums, he extended the idea of antigenic specificity to the constitution of the cells, and by using precipitating serums, to the proteins of the various animal species.

Famous at thirty, Bordet in 1901 accepted the directorship of the Institut Antirabique et Bactériologique, which had just been founded in Brussels and which in 1903 was renamed the Institut Pasteur du Brabant. There he continued his research on immunity and demonstrated that if an antibody has the ability to unite with an antigen, the alexin can be absorbed only by the complex antigen-antibody, that is, the antigen "sensitized" by the antibody. This complex antigen-antibody can bring about the fixation of the alexin of fresh serum, and because of this, the alexin can no longer cause the lysis of red corpuscles sensitized by the hemolysin. This is the alexin-fixation reaction (the complement-fixation reaction), which Bordet and his brother-in-law Octave Gengou applied in 1901 to the serodiagnosis of typhoid fever, carbuncle, hog cholera, and other diseases and which makes it possible to trace the antibody in the patient's serum. This reaction was taken up again by Wassermann in the diagnosis of syphilis, and more recently has been used in the diagnosis of virus infections.

In his interpretation of the mechanism of the union of antigen and antibody, Bordet compared this union to adsorption phenomenon, while Ehrlich defended the theory of a union by definite proportions. The further work of Heidelberger and his school confirmed Bordet's concept.

In 1906, while carrying out research in different directions, Bordet and Gengou discovered the whooping cough bacillus and extracted an endotoxin, prepared a vaccine, and, with Sleeswijk, studied the antigenic variability of the bacillus. In 1909 Bordet isolated the germ of bovine peripneumonia and that of avian diphtheria. From 1901 to 1920 he studied blood coagulation and, from 1920 on, bacteriophages.

All of his research was conducted while Bordet bore

300

the heavy duties of directing the Institut Pasteur du Brabant (until 1940, when his son Paul succeeded him) and teaching at the Faculty of Medicine of the Free University of Brussels, where he occupied the chair of bacteriology from 1907 to 1935. Besides this, he went to Paris every year to lecture on immunity in the microbiology course at the Institut Pasteur, where he was made president of the Conseil Scientifique in 1935.

Bordet's work on humoral immunity, which made possible the application of serological techniques to diagnosis and control of infectious diseases, brought him many international awards, including the Nobel Prize in medicine for 1919, as well as the highest academic distinctions and honors.

BIBLIOGRAPHY

I. ORIGINAL WORKS. Most of Bordet's papers were published in *Annales de l'Institut Pasteur.* His major work is *Traité de l'immunité dans les maladies infectieuses* (Paris, 1920, 1939). Many documents, such as laboratory notebooks, are preserved in the Musée Jules Bordet, at the Institut Pasteur du Brabant, in Brussels.

II. SECONDARY LITERATURE. Works on Bordet are J. Beumer, "Jules Bordet 1870–1961," in *Journal of General Microbiology,* **29** (1962), 1–13; Paul Bordet, "Jules Bordet," in *Florilège des sciences en Belgique pendant le XIX siècle et le début du XXe* (Brussels, 1968), pp. 1036–1067; A. M. Dalcq, "Notice biographique sur J. Bordet," in *Bulletin de l'Académie royale de médecine de Belgique,* 7th ser., **1** (1961), 352–365; and "Jules Jean Vincent Bordet," in Blakiston's *New Gould Medical Dictionary,* N. L. Hoerr and Arthur Osol, eds., 2nd ed. (New York–Toronto–London, 1956). Also of value is the "Volume jubilaire de Jules Bordet," *Annales de l'Institut Pasteur,* **79,** no. 5 (1950), 479–520.

JEAN VIEUCHANGE

BORDEU, THÉOPHILE DE (*b.* Izeste, France, 22 February 1722; *d.* Paris, France, 23 November 1776), *medicine.*

The son and grandson of Béarnese physicians, Bordeu studied medicine at Montpellier, where he received his medical degree on 10 November 1743. He returned to Montpellier for further study in 1745 and in 1746 went to Paris, where he devoted his time particularly to the clinical examination of patients at La Charité hospital. He returned to Béarn, to Pau, in 1749. There he held, in succession, the posts of intendant and superintendent of the mineral waters of Aquitaine, drawing attention to their therapeutic value through a newspaper that he had founded. His reputation led to his appointment as a corresponding member of the Royal Academy of Sciences of Paris in 1747.

In 1752 Bordeu went again to Paris, with the intention of practicing medicine, but he had to observe the regulation that only graduates of the Paris Faculty of Medicine could practice in the capital. He renewed his studies and was again graduated M.D. on 7 October 1754. At this time he became an attending physician at La Charité. Very soon he attracted a large practice, which aroused jealousy. As the result of a conspiracy organized by Michel-Philippe Bouvart, his name was removed from the list of Paris physicians in 1761, an act that had the effect of preventing him from practicing. Bordeu defended himself vigorously and was reinstated in 1764. His success with patients continued, and he cared for many important persons, among them Madame Du Barry.

The part that Bordeu played in the history of thermalism has caused him to be considered the founder of modern hydrotherapy. It was through him that the baths of the Pyrenees became known throughout the south of France and even in Paris. His *Journal de Barèges* and his *Recherches sur les maladies chroniques* (1775) are among his most important contributions to this field.

Bordeu also studied anatomy. In 1747 he published "Recherches anatomiques sur les articulations des os de la face," which won him membership in the Academy of Sciences. He had already written a treatise on the formation of chyle, *Chylificationis historia* (1742), that foreshadowed his important work on the glands, *Recherches anatomiques sur les différentes positions des glandes et sur leur action* (1752). In the latter he announced the double innervation (trophic and functional) of glands and organs, thus proving the existence of secretory nerves. He offered in evidence a local increase in the circulation when a gland is in action and, while emphasizing the influence of the imagination, showed that excretion is due to the gland itself and not the surrounding muscles. This shows the importance of Bordeu as a precursor in the science that in the twentieth century came to be distinguished as endocrinology. Finally, he demonstrated that secretion is the active elaboration of a new product separate from the constituents of the blood. Bordeu completed this sequence with his famous *Recherches sur le tissu muqueux* (1767), in which he described connective tissue—under the name of mucous tissue—showing its role in exchanges, the phenomena of nutrition, and the mechanical equilibrium of organs and tissues.

Semeiology also interested Bordeu. His "Recherches sur les crises" appeared in Diderot's *Encyclopédie,* as did his "Recherches sur le pouls par rapport aux crises." Both mark him as a clinician of the first order who knew how to obtain a large number of diagnostic and prognostic facts from an examination. Since Bordeu, physicians take the pulse by applying the tips of the four fingers to the hollow of the radius.

Bordeu played a great role in the history of medical theories. His thesis at Montpellier, *De sensu generice considerato* ("On the Senses, Considered Generically," 1742), suggested the direction in which he was heading. His other works demonstrated that he considered each organ as having its own life and believed that life was the sum of all these "little lives" of the organs. Coordination among them was due to the mucous tissues. The organs and glands were set in motion by an irritation termed "sensibility." These ideas opened the way for vitalism, a doctrine of which the school of Montpellier became the champion.

BIBLIOGRAPHY

I. ORIGINAL WORKS. Works by Bordeu are *Chylificationis historia* (Montpellier, 1742); *De sensu generice considerato* (Montpellier, 1742); "Recherches anatomiques sur les articulations des os de la face," in Académie Royale des Sciences, *Mémoires des savants étrangers,* II (Paris, 1747); *Recherches anatomiques sur les différentes positions des glandes et sur leur action* (Paris, 1752); *Recherches sur le tissu muqueux* (Paris, 1767); and *Recherches sur les maladies chroniques* (Paris, 1775). His works are collected in A. Richerand, ed., *Oeuvres complètes de Théophile de Bordeu, précédées d'une notice sur sa vie et ses ouvrages,* 2 vols. (Paris, 1818).

Many of Bordeu's previously unpublished works have been brought to light by Lucien Cornet: *Théophile de Bordeu* (Bordeaux, 1922); "Une consultation médicale au XVIIIe siècle," in *Bulletin de la Société des Sciences, Lettres et Arts de Pau,* 3rd ser., **10** (1949), 44–50; "Le procès de Théophile de Bordeu, documents inédits," *ibid.,* **14** (1953), 139–143; "Lettres inédites de Théophile de Bordeu, présentées et commentées par le Dr. Lucien Cornet," *ibid.,* **15** (1954), 24–29; "Nouvelles lettres inédites de Théophile de Bordeu (1753)," *ibid.,* **18** (1957), 65–70; "Lettres inédites de Théophile de Bordeu (1746)," *ibid.,* **20** (1959), 49–67; "Un ami de Théophile de Bordeu: Le médecin Jean de Brumont-Disse," *ibid.,* **21** (1960), 39–52; "Lettres inédites de Théophile de Bordeu (1746). Deuxième séjour à Montpellier," in *Journal de médecine de Bordeaux,* no. 136 (May 1960), 501–520; "Lettres inédites de Théophile de Bordeu (1747). Premier séjour à Paris," *ibid.,* no. 137 (September-November 1960), 1302–1404; "Lettres inédites de Théophile de Bordeu (1748). L'année de Versailles," *ibid.,* no. 138 (May 1961), 1475–1483; "Lettres de Théophile de Bordeu (1749). Fin du séjour à Versailles et retour à Pau," *ibid.,* no. 140 (February 1963), 328–337; "Lettres inédites de Théophile de Bordeu (1749)," in *Journal de médecine de Bordeaux,* no. 10 (October 1963); "Théophile de Bordeu et Madame de Sorbério," in *Revue régionaliste des Pyrénées,* no. 161–162 (January-June 1964), 6–22;

"Lettres inédites de Théophile de Bordeu (1750). Deuxième séjour à Pau et tournée d'inspection en Bigorre," in *Revue régionaliste des Pyrénées,* no. 59 (1964), 155–164, and no. 60 (October-December 1964), 277–287; and "Théophile de Bordeu, le biologiste," in *Bulletin de la Société des Sciences, Lettres et Arts de Pau,* 4th ser., **1** (1966), 123–125.

II. SECONDARY LITERATURE. Works on Bordeu are E. Forgue, *Théophile de Bordeu, fondateur de l'hydrologie, précurseur de la biologie moderne (1722–1776)* (Paris, 1937); J. J. Gardane, *Éloge historique de Bordeu* (Paris, 1777); F. Granel, *Un médecin du XVIIIe siècle aux conceptions biologiques modernes: Théophile de Bordeu (1722–1776), docteur de Montpellier et de Paris* (Montpellier, 1964); and P. Roussel, *Éloge historique de Théophile de Bordeu* (Paris, 1778).

LOUIS DULIEU

BORDONE DELLA SCALA, GIULIO. See **Scaliger, Julius Caesar.**

BOREL, ÉMILE (FÉLIX-ÉDOUARD-JUSTIN) (*b.* Saint-Affrique, Aveyron, France, 7 January 1871; *d.* Paris, France, 3 February 1956), *mathematics.*

Borel's father, Honoré, son of an artisan, was a Protestant village pastor. His mother, Émilie Teissié-Solier, came of a local merchant family. In 1882, already known as a prodigy, he left his father's school for the *lycée* at nearby Montauban. In Paris as a scholarship student preparing for the university, he entered the family circle of G. Darboux through friendship with his son, saw the "good life" of a leading mathematician, and set his heart on it. In 1889, after winning first place in the École Polytechnique, the École Normale Supérieure, and the general competitions, Borel chose the gateway to teaching and research, in spite of the blandishments of a special representative of the École Polytechnique.

Fifty years later Borel's colleagues celebrated the jubilee of his entrance to the École Normale, rightly considering that as the beginning of his scientific career. Indeed, he published two papers during his first year and appears to have established there his lifetime pattern of intensely serious and well-organized activity. He embraced an agnostic, scientific, and rational outlook that implied a responsible interest in all aspects of human affairs, and the extensive friendships of his undergraduate days helped make possible his broad cultural and political influence in later life. First in the class of 1893, he was promptly invited to teach at the University of Lille, where he wrote his thesis and twenty-two papers in three years before being called back to the École Normale, where publications, honors, and responsibilities piled up rapidly.

In 1901 Borel married Paul Appell's eldest daughter, Marguerite, who had interested him for some time but had only then turned seventeen. She wrote more than thirty novels (as Camille Marbo), was president of the Société des Gens de Lettres, and both assisted and complemented her husband's many-sided activity. They had no children but adopted Fernand Lebeau, son of the older of Borel's two sisters, after the early death of his parents. In 1906 they used money from one of Émile's prizes to launch *La revue du mois,* which appealed successfully to a very broad circle until the war and economic crisis killed it in 1920. During this period Borel's publications and activities showed a progressive broadening of interest from pure mathematics to applications and public affairs. Without seeming to diminish his mathematical creativity, he wrote texts and popularizations, edited several distinguished series of books, contributed to popular magazines and the daily press, played leading roles in professional and university affairs, and maintained acquaintances ranging from poets to industrialists.

Such an implausible level of activity was possible because Borel's uncommon intelligence and vigor were accompanied by efficient organization and self-discipline. He could be kind and generous of his time and energy in meeting his official or self-imposed obligations. He was even ready to risk his status for a good cause. But he had no time for "small talk" or trivial activity, seemed formidable and even rude to outsiders, and with increasing age grew more impatient with would-be wasters of his time. His lectures displayed his mind at work rather than a finished exposition, and his teaching consisted primarily in directing his students' efforts.

In 1909 Borel occupied the chair of theory of functions, newly created for him at the Sorbonne, and began thirty-two years on the University Council, representing the Faculty of Science. In 1910 he entered what he called the happiest time of his life as vice-director of the École Normale in charge of science students, but World War I cut it short. His service in sound location at the front (while Marguerite headed a hospital), and in organizing research and development in the War Office under his old friend Paul Painlevé, turned his interests more than ever toward applications. After the war he could not be happy again at the École Normale. There were "too many ghosts in the hallways," including that of his adopted son. At his request he moved to the chair of probability and mathematical physics at the Sorbonne and maintained only honorary connections with the École Normale. The era was closed by the longest of his many trips abroad, including five months in China

with Painlevé, and by his election to the Academy in 1921.

While continuing his flow of publications and his lectures in mathematics, Borel now moved rapidly into politics as mayor of Saint-Affrique (with Marguerite presiding over the Jury Femina), councillor of the Aveyron district, Radical and Radical-Socialist member of the Chamber of Deputies (1924–1936), and minister of the navy (1925). Important scientific legislation, the founding of the Centre National de la Recherche Scientifique, and several ships named after mathematicians are traceable to his initiative. He helped plan and raise funds for the Institut Henri Poincaré and served as director from its founding in 1928 until his death.

Retired from politics in 1936 and from the Sorbonne in 1940, Borel still had the vigor to produce more than fifty additional books and papers, to participate in the Resistance in his native village, to which he returned after a brief imprisonment by the Germans in 1940, and to travel extensively. The breadth of his services was recognized by such honors as the presidency of the Academy (1934), the Grand Cross of the Legion of Honor (1950), the first gold medal of the Centre Nationale de la Recherche Scientifique (1955), the Croix de Guerre (1918), and the Resistance Medal (1945). A fall on the boat while returning from giving a paper at a meeting of the International Institute of Statistics in Brazil in 1955 hastened his death the following year at eighty-five.

Borel's undergraduate publications showed virtuosity in solving his elders' problems rather than great originality, but a "big idea" was incubating. Already in 1891 he was "extrêmement séduit" by Georg Cantor, whose "romantic spirit" mixed explosively with Borel's rigorous training in classical analysis and geometry. By sensing both the power and danger of set concepts, Borel anticipated the unifying themes of his lifework and of much mathematics in the twentieth century. In his thesis of 1894, which Collingwood rightly calls "an important mathematical event," can be found the ideas with which he initiated the modern theories of functions of a real variable, measure, divergent series, nonanalytic continuation, denumerable probability, Diophantine approximation, and the metrical distribution theory of values of analytical functions. All are related to Cantorian ideas, especially to the notion of a denumerable set. This is obvious for the two most famous results in the thesis, the Heine-Borel covering theorem (misnamed later by Schoenflies) and the proof that a denumerable set is of measure zero. The first asserted that if a denumerable set of open intervals covers a bounded set of points on a line, then a finite subset of the intervals is sufficient to cover. The second involves implicitly the extension of measure from finite sets of intervals to a very large class of point sets, now known as Borel-measurable sets.

Borel exploited his first insights in many directions. His *Leçons* of 1898 and other works laid the basis of measure theory so solidly that in that field the letter *B* means Borel. In 1905 he noticed that probability language was convenient for talking about measure of point sets, and in 1909 he introduced probability on a denumerable set of events, thus filling the gap between traditional finite and "geometrical" (continuous) probability. In the same paper he proved a special case of his strong law of large numbers. But Borel remained skeptical of the actual infinite beyond the denumerable and of nonconstructive definitions. Much of his work was motivated by finitistic ideas, and his last book (1952) discussed his observation that most real numbers must be "inaccessible," since with a finite alphabet we can name at most a denumerable subset. By this caution he avoided some of the pitfalls into which others fell, but he also was barred from the fruits of greater daring. It was Lebesgue, Baire, Fréchet, and others who pushed set and measure theoretic ideas more boldly and so opened the way to the abstract analysis of the mid-twentieth century.

Other motivations are visible in Borel's work: the challenge of unsolved classical problems and visible gaps, an early and increasing admiration for Cauchy, an interest in physical and social problems, all tinged strongly with French patriotism. Often his solutions opened whole fields for exploitation by others. His "elementary" proof of Picard's theorem in 1896 not only created a sensation because the problem had resisted all attacks for eighteen years, but also established methods and posed problems that set the theme of complex function theory for a generation. Borel's work on divergent series in 1899 filled the gap between convergent and divergent series. His work on monogenic functions (summed up in his monograph of 1917) showed the primacy of Cauchy's idea of the existence of the derivative over the Weierstrassian notion of series expansion and filled the gap between analytic and very discontinuous functions.

Before World War I, Borel had worked out most of his original ideas, and thereafter his scientific publications were largely the development and application of earlier ideas and the solution of minor problems. A major exception is the series of papers on game theory (1921–1927) in which he was the first to define games of strategy and to consider best strategies, mixed strategies, symmetric games, infinite games, and applications to war and economics. He proved the minimax theorem for three players, after some doubts for five and seven, and finally (1927) conjectured its truth a year before John von Neumann independently first took up the subject and proved the general theorem. Although Borel's papers were overlooked until after von Neumann's work was well known, he must be considered the inventor, if not the founder, of game theory.

Borel's innovations are essential in twentieth-century analysis and probability, but his research methods belong rather to the nineteenth. He abjured generalization except when it was forced on him. He was motivated by specific problems and applications. He disliked formalism ("pure symbolism turning about itself"), logicism, and intuitionism (both too removed from the physical reality that he thought should guide mathematics). Borel was the most successful mathematician of his generation in using specific problems and results as scientific parables pointing the way to broad theories that still remain fertile.

BIBLIOGRAPHY

I. Original Works. A complete scientific bibliography to 1939 appears in *Selecta. Jubilé scientifique de M. Émile Borel* (Paris, 1940), and is extended to 1956 in the biographies by Collingwood and Fréchet, which also analyze Borel's work in detail. The papers in the *Selecta* are in part more representative of the commentators' interests than of Borel's most significant work, but a complete collected works is in preparation. His writings on philosophical questions, pedagogy, and social problems are well covered in *Émile Borel, philosophe et homme d'action. Pages choisies et presentées par Maurice Fréchet* (Paris, 1967). Borel's own analysis of his work appears in his *Notice sur les travaux scientifiques* (Paris, 1912) and his *Supplément (1921) à la Notice (1912)*, in the *Selecta*. Very revealing also are his "Documents autobiographiques," in *Organon* (Warsaw), **1** (1936), 34–42, repr. in *Selecta*, and "Allocution," in *Notices et discours de l'Académie des Sciences,* **2** (1949), 350–359.

Of more than 300 scientific publications the most notable are his thesis, "Sur quelques points de la théorie des fonctions," in *Annales de l'École Normale*, 3rd ser., **12** (1895), 9–55; "Démonstration élémentaire d'un théorème de M. Picard sur les fonctions entières," in *Comptes rendus de l'Académie des Sciences,* **122** (1896), 1045–1048; "Fondements de la théorie des séries divergentes sommables," in *Journal de mathématique*, 5th ser., **2** (1896), 103–122; "Sur les zéros des fonctions entières," in *Acta mathematica*, **20** (1897), 357–396; *Leçons sur la théorie des fonctions* (Paris, 1898; 4th ed., 1950), his most influential book; "Mémoire sur les séries divergentes," in *Annales de l'École Normale*, 3rd ser., **16** (1899), 9–131, which won a grand prize of the Academy and led to over 200 papers by others during the following two decades; *Leçons sur les fonctions entières*

(Paris, 1900; 2nd ed., 1921), an exposition of the work growing out of his paper on the Picard theorem; *Leçons sur les séries divergentes* (Paris, 1901; 2nd ed., 1928); *Leçons sur les fonctions de variables réeles et les développements en séries de polynomes* (Paris, 1905; 2nd ed., 1928); "Les probabilités dénombrables et leurs applications arithmétiques," in *Rendiconti del Circolo Matematico di Palermo,* **27** (1909), 247–270; *Le hasard* (Paris, 1914), probably his best popularization; "I. Aggregates of Zero Measure. II. Monogenic Uniform Non-analytic Functions," in *Rice Institute Pamphlet,* 4th ser., **1** (1917), 1–52; *Leçons sur les fonctions monogènes uniformes d'un variable complexe* (Paris, 1917), the definitive exposition of his work in this area; "La théorie du jeu et les équational intégrales à noyau symétrique," in *Comptes rendus de l'Académie des Sciences,* **173** (1921), 1302–1308—this and two later notes (1924, 1927) on game theory appear in translation with commentary by Fréchet and von Neumann in *Econometrica,* **21** (1953), 95–125; *Méthodes et problèmes de la théorie des fonctions* (Paris, 1922), a collection winding up his work in that area; *La politique républicaine* (Paris, 1924), his most substantial political work; *Principes et formules classiques du calcul des probabilités* (Paris, 1925), the first fascicle of the Traité; *Valeur pratique et philosophique des probabilités* (Paris, 1939), the last fascicle of the Traité; *Théorie mathématique du bridge à la portée de tous* (Paris, 1940), written with A. Cheron; *Le jeu, la chance et les théories scientifiques modernes* (Paris, 1941); "Sur l'emploie du théorème de Bernoulli pour faciliter le calcul d'une infinité de coefficients—Application au problème de l'attente à un quichet," in *Comptes rendus de l'Académie des Sciences,* **214** (1942), 425–456, his last original contribution to probability theory; *Les probabilités et la vie* (Paris, 1943), another fine popularization with later editions and translations; *Éléments de la théorie des ensembles* (Paris, 1949), a summation containing some new results; and *Les nombres inaccessibles* (Paris, 1952), his last book.

Series that he edited (always contributing substantially also) include Collection de Monographies sur la Théorie de Fonctions (Paris, 1898–1952)—sometimes called the Borel tracts, this totaled over fifty volumes, ten by Borel himself —and Cours de Mathématiques (Paris, 1903–1912), a series of elementary texts designed to cover various curricula, usually written with collaborators. Other series include La Nouvelle Collection Scientifique (1910–1922), thirty-five popularizations; Bibliothèque d'Éducation par la Science (Paris, 1924–1946), high-level popularizations for the educated layman; Traité de Calcul des Probabilités et de Ses Applications (Paris, 1925–1938), eighteen fascicles in four volumes, intended to cover the whole field as it had developed since 1875; Collection de Physique Mathématique (Paris, 1928–1950); and Collection de Monographies des Probabilités et de Leurs Applications (Paris, 1937–1950), seven volumes intended to supplement the Traité by current research.

II. SECONDARY LITERATURE. Along with the material in the *Selecta* and Borel's autobiographical writings cited above, the best sources are "Jubilé scientifique de M. Émile Borel . . . 14 janvier 1940," in *Notices et discours de l'Académie des Sciences,* **2** (1949), 324–359; L. de Broglie, *ibid.,* **4** (1957), 1–24; E. F. Collingwood, in *Journal of the London Mathematical Society,* **34** (1959), 488–512, and **35** (1960), 384; M. Fréchet, "La vie et l'oeuvre d'Émile Borel," in *Enseignement mathématique,* 2nd ser., **11** (1965), 1–95; M. Loève, "Integration and Measure," in *Encyclopaedia Britannica* (1965); and P. Montel, in *Comptes rendus de l'Académie des Sciences,* **242** (1965), 848–850.

KENNETH O. MAY

BOREL, PIERRE (*b.* Castres, Languedoc, France, *ca.* 1620; *d.* Paris, France, 1671), *history, medicine, chemistry.*

Borel studied at Montpellier and returned to Castres as an M.D. in 1641. Besides practicing medicine, he collected rarities, plants, antiquities, and minerals from the town and countryside of Castres. In 1645 he published a catalog of his collection and expanded it in 1649 to include the history and Roman inscriptions of the area.

About the end of 1653 Borel moved to Paris, where he received the title of *médecin ordinaire du roy.* Again, he became very active as a collector. He assembled some 4,000 manuscripts and books of the Hermetic philosophers or chemists and published a catalog in Paris in 1653. A collection of linguistic antiquities listed in alphabetical order (1655) was the basis for Favre's greatly enlarged *Dictionnaire du vieux François,* published in 1882. Borel also studied reports about the telescope; in his book of 1656 he cites Zacharias Janssen (1590) as the first inventor and Hans Lipperhey (1608) as the second. He also describes a "polemoscope," a 1637 invention designed for looking around corners, which is particularly useful in warfare. He appended to this book an account of a hundred medicophysical observations with the microscope.

Among his original contributions to medicine are the statement that cataract is a darkening of the crystalline lens and the recommendation of the use of concave mirrors in the diagnostic examination of the nose and throat. Borel is credited with the first description of brain concussions.

His last work seems to have been a *Hortus* (1667), which listed plants with known uses in medicine.

BIBLIOGRAPHY

I. ORIGINAL WORKS. *Catalogue des raretés du cabinet de P. Borel* (Castres, 1645); *Les antiquités, raretés, plantes, minéraux et autres choses considérables de la ville et comté de Castres, d'Albigeois, et des lieux qui sont à ses environs etc.* (Castres, 1649); *Bibliotheca chimica, seu catalogus librorum philosophicorum hermeticorum, in quo quatuor millia circiter auctorum chimicorum . . . usque ad annum 1653 continentur* (Paris, 1654); *De vero telescopii inventore, cum brevi omnium conspicillorum historia etc., accessit etiam centuria observationum microscopicarum* (The Hague, 1655, 1656); *Hortus, seu armamentarium simplicium plantarum et animalium ad artem medicam spectantium, cum brevi eorum etymologia, descriptione, loco, tempore et viribus* (Castres, 1667); *Petri Borelli historiarum et observationum medico-physicarum centuriae IV* (Frankfurt–Leipzig, 1676), with additions by Arnold de Boot and L. Cattier.

II. SECONDARY LITERATURE. See articles by R. P. Niceron, in *Mémoires pour servir à l'histoire des hommes illustres dans la république des lettres avec un catalogue résumé de leurs ouvrages,* XXXVI (Paris, 1736), 218–224 (Niceron gives certain biographical dates that have been proved wrong); August Hirsch, ed., *Biographisches Lexikon der hervorragenden Ärzte,* 2nd. ed., I (Berlin, 1929), 632; and Mme. Puech-Milhau, in *Revue du Tarn,* 4th ser., no. 7 (1936), 279–280.

EDUARD FARBER

BORELLI, GIOVANNI ALFONSO (*b.* Naples, Italy, January 1608; *d.* Rome, Italy, 31 December 1679), *astronomy, epidemiology, mathematics, physiology (iatromechanics), physics, volcanology.*

Borelli is not as widely known or appreciated as perhaps he should be. What reputation he has is based upon his mechanics, including celestial mechanics, and his physiology or iatromechanics. The former, unfortunately, was quickly and completely overshadowed by the work of Isaac Newton; and his iatromechanics, although important and influential, was too much informed by what proved to be a relatively sterile systematic bias to bear much immediate fruit. Accordingly historians have undervalued his place in the development of the sciences in the seventeenth century, and they have paid little attention to his career or his personality. (There has been no lengthy treatment of his life since the eighteenth century, and important and elementary biographical information is still hard to come by.) But he was highly respected by his contemporaries. He read widely, and he drew his scientific inspiration from a broad spectrum of the heroes and near-heroes of the early seventeenth century: such men as Galileo Galilei, William Harvey, Johannes Kepler, and Santorio Santorio. He worked on many problems, contributed significantly to all the topics he touched, and in fact played an important part in establishing and extending the new experimental-mathematical philosophy. He was brilliant enough scientifically to be very much ahead of his time, even if he was not quite brilliant enough nor free enough from other commitments to produce general synthetic solutions in his fields of interest which would be either successful or entirely convincing.

During the century prior to Borelli's birth, Italians had been in the forefront of the late Renaissance effort to translate and master the Alexandrian astronomers, mathematicians, and physiologists. By the end of that century many had learned all they could from the past and had begun to strike out on their own. Galileo's telescopic discoveries only dramatically underscored the fact that major innovations were underway in all fields of natural philosophy. And they also indicated that the Italians could be expected to continue playing a leading role in these new enterprises. But during Borelli's lifetime the world saw Galileo condemned for his innovations, the Lincei persecuted, the Cimento disbanded, and the Investiganti of Naples suspended. It also saw the death, in the decade of the 1640's, of many of Galileo's most talented disciples: Benedetto Castelli, Bonaventura Cavalieri, Vincenzo Renieri, and Evangelista Torricelli. Borelli's Italy rejoiced over the conversion of Queen Christina of Sweden and perhaps was as much interested in the fact of Nicholas Steno's conversion as in his scientific accomplishments. Moreover, it was a politically fragmented Italy, portions of which were absorbed in struggles to throw off oppressive foreign domination. And later on its best investigators, for example, Marcello Malpighi and Gian Domenico Cassini, had to find recognition and support north of the Alps. In sum, the new philosophy faced distracting competition and even open hostility from several quarters, and in the long run the Italians could find neither the wherewithal nor the enthusiasm to support science in the ways it was beginning to be supported elsewhere. Borelli's career, then, is an illuminating record of an original scientist who was also politically active in Counter-Reformation Italy. Borelli himself ended his life in political exile in Rome—poverty stricken, teaching elementary mathematics.

Borelli's birth was not auspicious. As part of their rule of southern Italy at the turn of the century, the Spanish maintained military garrisons in the three principal fortresses of Naples. On 28 January 1608, a Spanish infantryman, Miguel Alonso, stationed at Castel Nuovo, witnessed the baptism of his first son, Giovanni Francesco Antonio. The mother was a local woman by the name of Laura Porrello (variously spelled in the records as *porrello, porrella, borrella, borriello, borrelli*). The couple went on to have one daughter and four more sons, including a Filippo baptized 9 March 1614. In later years both Giovanni and Filippo used Borelli as a family name; Giovanni dropped two of his baptismal names but retained an

Italianized version of his father's name in their place. Why they did this perhaps can be guessed from the circumstances of their early years.

In November 1614 Tommaso Campanella was returned to Castel St. Elmo, where he had previously been confined. Meanwhile Miguel Alonso had been ordered to Castel St. Elmo. Just after Campanella's return, Miguel became implicated in some serious offense and was arrested along with several other persons. Although it is not known for certain what the alleged crime was, responsible sources suggest that there may have been a conspiracy to free Campanella. In any case the interrogations and trial took place in secret, and during the summer of 1615 Miguel was found guilty and sentenced to the galleys. Upon his certification that he was unable to serve in the galleys the sentence was commuted to exile. Miguel seems to have gone to Rome, and it has usually been supposed that this was the occasion for young Borelli's presence there and eventual contact with Benedetto Castelli. But now we know that Miguel did not remain in exile. He appealed his case and was exonerated. In April 1617 he returned to duty at Castel St. Elmo, where he stayed until he died in 1624. Laura Porrello possibly remained attached to Castel St. Elmo in some capacity, for at her death in 1640 she was buried, as Miguel had been, at the church serving the fortress.

We can guess that sometime before 1626 young Borelli came to the attention of Campanella; there was no lack of opportunity. In 1616 the latter was given a few months of at-large detention in Castel Nuovo (he may have written his *Defense of Galileo* at this time), but he was back in the dungeon of Castel St. Elmo when Miguel returned from exile. In May 1618 he was again sent to Castel Nuovo, where he had a relatively easy imprisonment; he was able to write, see friends, and even have students. It is possible that Borelli was among these, and it is also possible that Borelli received some medical training at the University of Naples in this period, although we have no published records to that effect. In 1626 Campanella was taken to Rome, where he was fully liberated in 1628. Five years later a disciple, under duress, implicated Campanella in a plot to assassinate the Spanish viceroy in Naples. Under great pressures Campanella fled Italy for Paris, in 1634, taking Filippo Borelli with him. There Filippo helped to edit and publish various of Campanella's works, and in at least one he appears as *nipote ed amanuense dello autore*. What happened to Filippo later is not known, but a letter of another of Campanella's disciples in 1657 connects Giovanni Alfonso with information concerning several hundred copies of Campanella's books left at the Dominican convent of Santa Maria Sopra Minerva and also indicates that Giovanni had a brother, a "P. Tomaso filosofo." It has been suggested that on Campanella's death, in 1639, Filippo entered orders and took the name Tommaso.

We do not know when Borelli himself went to Rome. Anytime after 1628 he could have resumed whatever relationship he had established in Naples with Campanella; and it is quite possible that Campanella in turn introduced him to Castelli. In any case he became a student of Castelli along with Torricelli. He must have been in Rome through the period of the publication of Galileo's *Dialogo* and the subsequent trial. Although he did not meet Galileo, he probably had access to all the ins and outs of the affair through both his mentors. And possibly it was during this period that he acquired a copy of calculations or tables made by Galileo concerning the Medici planets (the moons of Jupiter), calculations which were not among the papers inherited by Vincenzo Viviani at Galileo's death and which Viviani requested a copy of in 1643. After Campanella left Rome, Borelli continued for a while with Castelli. In 1635, or shortly thereafter, Castelli's recommendation obtained for Borelli the public lectureship in mathematics in Messina, Sicily. And Castelli continued to look after Borelli's welfare. In 1640, when the mathematics chair at the University of Pisa became vacant, he wrote two letters to Galileo praising Borelli very highly, calling him in one *huomo di grandissimo ingegno e sapere, versatissimo nelle dottrine di V.S. molto Ill.re e tutto tutto* NOSTRI ORDINIS. Galileo's choice, however, was Vincenzo Renieri who then held the position until his death in 1647. Borelli would eventually obtain the post, but not until 1656.

Meanwhile Borelli made his way in Messina. The city had had little to boast of since the death of Francesco Maurolico in 1575. In the 1630's, however, there was an effort toward a political and intellectual revival which included an attempt to improve substantially the city's university. The people backing these moves were among the same who formed the Accademia della Fucina in 1639, a group of the young, enlightened nobility and merchant class, jealous of its political rights and beginning to grow restless under the restrictions of Spanish rule. The Fucina itself became a forum for both political and intellectual discussion, and in 1642 it came under the direct protection of the Messinese senate. It is not clear when Borelli became a member, but his talents as a public lecturer of mathematics were already highly appreciated. In 1642 the senate provided him with ample funds and sent him on a mission to leading universities to hire away good teachers, especially in law and medicine. We can guess that on this trip Borelli stopped in Naples to see Marco Aurelio Severino, perhaps renewing an old association. He must have visited Castelli in Rome. We know that he visited Tuscany, but unfortunately too late to see Galileo. But he did spend some time in Florence, and while there he met both Viviani and Prince Leopold, the youngest brother of the grand duke. After Florence he went on to Bologna where he very favorably impressed Bonaventura Cavalieri. Then he was off to Padua and eventually Venice where he planned to catch a ship back to Messina. Among the topics of discussion in Florence must have been the work of Santorio, for in Venice he bought a copy of *De statica medicina* and mailed it back to Viviani along with other items of scientific interest. By 1643, then, even though he had not yet published, he was beginning to be known in Italy, and what evidence we have indicates that he had already exposed himself to the studies that were to concern him for the rest of his life: mathematics, physiology, and planetary astronomy.

From 1643 to 1656 Borelli remained in Sicily, so far as we know; he published two works and possibly had a hand in a third. The first developed out of a dispute that may have had some polemic roots in the political and intellectual rivalry between Messina and Palermo. In 1644 a Pietro Emmanuele of Palermo published a *Lettera intorno alla soluzione di un problema geometrico*. This was attacked, so he followed it a year later with a *Lettera in difesa di un problema geometrico*. In the second, at least, Borelli's reputation was impugned, and Borelli replied in the *Discorso del Signor Gio: Alfonso Borelli, accademico della Fucina e professore delle scienze matematiche nello Studio della nobile città di Messina, nel quale si manifestano le falsità, e gli errori, contenuti nella difesa del Problema Geometrico, risoluto dal R. D. Pietro Emmanuele* (Messina, 1646). The Fucina also reacted to protect both itself and Borelli by encouraging the publication of several pamphlets. In one of them, Daniele Spinola's *Il Crivello* (Macerata, 1647), the resolution of the original problem was provided by Giovanni Ventimiglia, a student and a friend of Borelli.

As this controversy died down, Sicily was invaded by an epidemic of fevers. Messina was especially hard hit and the senate encouraged its local *dotti* to try to discern its causes. One study that resulted was Borelli's *On the causes of the malignant fevers of Sicily in the years 1647 and 1648 . . .*; to which he added a section entitled *And at the end the digestion of food is treated by a new method* (Cosenza, 1649). During his investigation of the epidemic Borelli had visited other cities, observed autopsies, and noted in detail the circumstances under which the disease was prevalent. He concluded that in no way were the fevers caused by meteorological conditions or astrological influences, but were probably caused by something getting into the body from the outside. Since this thing seemed to be chemical, Borelli prescribed a chemical remedy, sulfur, and for this recommendation he acknowledged the counsel of his friend and colleague Pietro Castelli (*d.* 1661). In the addendum he again disclosed a chemical approach; he characterized digestion as the action of a *succo acido corrosivo* turning food into a liquid form. Borelli would repeat and expand this particular inquiry during his stay in Pisa.

In 1650 Borelli was considered for the chair of mathematics at Bologna. Cavalieri had died in 1647 and the authorities there wished to fill the post with someone equally able. Accordingly they made inquiries concerning Borelli and received strong endorsements for him as the best mathematician in Italy after Cavalieri. They also learned that Borelli was a trifle capricious and had a leaning toward the "moderns," Copernicus and Galileo (*il Gubernico et il Galileo*). Whether or not this latter was a factor, Borelli was passed over and the chair went to Gian Domenico Cassini. So Borelli remained in Messina and was there when Maurolico's *Emendatio et restitutio conicorum Apollonii Pergaei* was finally published in 1654. The original of the *Conics* of Apollonius had contained eight books, but the sixteenth century possessed only the texts of the first four. Maurolico had attempted to reconstruct Books V and VI. The extent of Borelli's connection with this project is not certain. We do know that he had composed a digest of the first four books before he left Messina. On this account alone he would have been prepared for an opportunity that presented itself when he later arrived in Pisa. Sometime previously the Medici had acquired an Arabic manuscript which seemed to contain all the original eight books. As early as 1645 Michelangelo Ricci had corresponded with Torricelli about the possibility of translating and publishing it, but with no results. Somehow Borelli had learned of it, however, for just a month after his inaugural lecture at Pisa, in the spring of 1656, he wrote to Leopold suggesting that with the aid of someone who knew Arabic he could edit these "most eagerly awaited" last four books. This led, in 1658, to a long summer's collaboration in Rome with the Maronite scholar Abraham Ecchellensis during which the two substantially completed an edition of Books V, VI, and VII. (It turned out that Book VIII was missing from the manuscript.) After many frustrating delays the work finally saw print in 1661 along with an appended Archimedean *Liber assumptorum* taken from another manuscript.

We must presume that in the years before Borelli left Messina he was already in touch with what would become a very important group in Naples. Tommaso Cornelio and Leonardo Di Capoa had both studied with Marco Aurelio Severino. On Severino's urging Cornelio had traveled for several years and had studied with such leading innovators of northern Italy as Ricci, Torricelli, and Cavalieri. When he came back to Naples in 1649 he brought with him the works of Galileo, Descartes, Gassendi, Bacon, Harvey, and Boyle, among others; and he and a lawyer named Francesco d'Andrea started an informal gathering which met to hear the results of its members' investigations. As it gained notoriety, the group faced various pressures, among them political, and in 1663 expediency compelled it to organize formally as the Accademia degli Investiganti under the protection of Andrea Concublet, the marchese d'Arena. All the while it pursued its physical, chemical, and physiological inquiries; corresponded with individuals and groups in other cities; and from time to time received distinguished visitors. Marcello Malpighi, for instance, had been at Pisa from 1656 to 1659 and then went to Bologna. In 1662 Borelli recommended him for the chair that had become vacant with the death of Pietro Castelli in Messina, and on his way south in the fall of that year Malpighi was warmly entertained by Cornelio and Di Capoa. From at least the time of his return to Naples, Cornelio had devoted himself to physiological experimentation in the new mathematical-mechanical manner. He became a professor of mathematics at the University of Naples in 1653. By 1656 his old teacher Severino had persuaded Cornelio to publish his investigations and speculations; delays occurred, unfortunately, but when his *Progymnasmata physica* appeared in 1663 one section of it carried a dedication to Borelli. For Borelli's part, almost immediately upon his arrival in Pisa he established a flourishing anatomical laboratory in his own house. Here he collaborated with and taught many talented students of the various disciplines of anatomy from Marcello Malpighi, at the beginning of his stay, to Lorenzo Bellini and Carlo Fracassati, in his last few years. Here also he nurtured his great iatromechanical project, a work on the movements of animals. He probably had had such an endeavor in mind before he came; in 1659 he could already complain of having to put it aside because of the work on Apollonius. By 1659, of course, Borelli had become involved in many things, not the least of them the experimental investigations of the Accademia del Cimento.

One year after Borelli arrived in Tuscany the Accademia del Cimento held its first session; the year Borelli left, the Cimento quietly died. Indeed, Borelli seems to have been the principal animus of the academy, but lest he appear the sole mover, we should recall the documentation, especially for the extensive experimental work performed during this Galilean epoch, in Giovanni Targioni Tozzetti's *Atti e memorie inedite dell'Accademia del Cimento e notizie aneddote dei progresse delle scienze in Toscana*. In fact the Tuscan court had been thoroughly infected by Galileo's ideas and those of his pupils. Grand Duke Ferdinand II, from the time of his accession to power in 1628 until his death in 1670, maintained a personal laboratory as did Prince Leopold. From the time of the death of the Master, Galileo, informal gatherings met at the court and presented and discussed experiments. At first Torricelli was the most prominent figure; after his death in 1647 Viviani presided over the activities.

Then, possibly under the crystallizing influence of Borelli, Leopold asked for and received permission from Ferdinand to organize formally an academy for purely experimental research. Under Leopold's aegis it met for the first time in June of 1657. Among its more distinguished members, besides Borelli and Viviani, were Antonio Oliva (*d.* 1668), Carlo Rinaldini (*d.* 1698), and Francesco Redi (*d.* 1697). Nicholas Steno arrived in Florence in 1666 and soon thereafter joined the group. Lorenzo Magalotti, after attending the University of Pisa as a student, was appointed secretary in 1660. The Cimento had adopted a policy of submerging the identities of its members and presenting itself as a group. Accordingly, when Magalotti brought out the *Saggi di naturali esperienzi fatte nell'Accademia del Cimento* in 1666–1667, it appeared anonymously and refrained from identifying the individual contributions of the members. Actually the *Saggi* presented only part of the work performed; it tended to emphasize the identification and description of physical phenomena and the perfecting of measuring techniques. It failed to present other interesting investigations, including some potentially controversial observations and discussions of comets.

During the life of the Cimento dissension appeared among the membership; Borelli may have originated some of it. He seems to have chafed under the requirement of anonymity, and by all accounts he was a touchy person to get along with under any circumstances. In any case, toward the end of 1666 and just after the publication of his important work on the theory of the motions of the moons of Jupiter, Borelli made his decision to leave Tuscany and return to Messina. In 1667 Leopold was created a cardinal and thus had some of his energies diverted. Rinaldini moved on to the University of Padua, and Antonio Oliva went to Rome where he came under the suspicion of the Inquisition and died by throwing himself from a window of one of its prisons. In December of 1667 Steno converted to Catholicism and shortly thereafter set out on a series of journeys. How or whether any of these events may have been connected is not known with any degree of certainty. But at this point the Cimento effectively ceased to function, even though it apparently was not formally dissolved, and even though Prince, now Cardinal, Leopold continued to direct some experimental work until he died in 1675. As far as Borelli was concerned, he had been, and afterward remained, on excellent terms with Leopold; and Leopold maintained his high regard for Borelli.

Besides his involvement with the Cimento and his own laboratory, Borelli had had other things to keep him busy during these years in Tuscany, among them his teaching duties. He was by no means the usual sort of professor. Nor did he bother to cultivate the finer graces of that calling. His first lectures at Pisa, for instance, were something of a disaster. He lacked any particular eloquence and was long-winded and dull. The students reacted with catcalls and agitation, once forcing him to stop before finishing his lesson. Very quickly, however, he demonstrated his capabilities, and his lack of Tuscan oratorical polish probably became less of a barrier. Then, in connection with his post, he prepared for publication of his *Euclides restitutus.* Not one to be overawed by canonical texts, he frankly stated that although Euclid had done an excellent job in compiling his *Elements,* these nevertheless could be repetitive and prolix, and it was time to put the material together in a clearer and more concise package. While he was about it, Borelli took the opportunity not only to reexamine the parallel postulate and propose his own version but also to try to establish the theory of proportions on firmer grounds. The Latin edition of this work appeared in 1658. Five years later his student Domenico Magni undertook the task of providing a "Euclid for the layman" by editing out most of Borelli's technical commentary and shortening and translating the remainder into Italian. Both works apparently were very well received. In subsequent editions of the Latin version, Borelli's short summary of Apollonius and other brief analyses appeared.

One of the more notable events during Borelli's stay in Pisa had been the appearance of a comet in late 1664. Borelli immediately took up the vigil and kept very close track of it throughout December and until the beginning of February 1665. Out of this came

a small paper, which he published in the form of a letter addressed to Stefano degli Angeli, a mathematician at the University of Padua. Borelli showed that, no matter which interpretation one preferred, Ptolemaic, Tychonic, or Keplerian, one had to admit that the comet changed in its absolute distance from the earth. This fact raised obvious difficulties for the first two systems, and Borelli argued that it presented difficulties for the Keplerian also. He went on to show that his parallax measurements proved the comet to be above the moon, at least toward the end of the observations presented here. This was touchy material, and Borelli published under the pseudonym of Pier Maria Mutoli. His interest in comets continued into the spring. In early May he wrote Leopold that he believed that the true motion of a comet *then* visible could in no ways be accounted for by means of a straight line but rather by a curve very similar to a parabola. And he proposed to demonstrate it, not only by calculation, but also with some kind of mechanical device. Borelli apparently built this instrument; unfortunately, neither it nor any description of it remains.

During the summer of 1665 Borelli established an astronomical observatory in the fortress of San Miniato, a pleasant site on a hill a short distance from Florence. Here he used an excellent Campani telescope and some instruments of his own design to try to determine with extreme accuracy the motions of Jupiter's satellites. From this work came his *Theoricae mediceorum planetarum ex causis physicis deductae* (1666), in which, among other things, he explained how the elliptical orbits of planetary bodies could be understood in terms of three types of action. In the first place, a planetary body has a tendency toward a central body and would move toward that central body if no other factors intervened. Then, a central body, such as the sun, sends out rays and as that body rotates the rays also rotate. The cumulative effect of the impacts of these seemingly corporeal rays is to impart to the planet a motion around the central body. This motion in revolution thus produces a centrifugal tendency which balances the original centripetal one and thereby establishes the planet in a given mean orbit. Small self-correcting fluctuations account qualitatively for the observed ellipses. There are some obvious difficulties in accommodating these proposals to the satellites of the major planets, and it is clear that Borelli had much more in mind than just explaining the motions of the moons of Jupiter. The Copernican implications of his scheme, however, could be masked by seeming to focus attention on Jupiter.

Meanwhile, as time allowed, Borelli continued his

anatomical research. He collaborated with Lorenzo Bellini in an investigation of the structure of the kidney, and in 1664 this resulted in a short piece entitled *De renum usu judicum.* And he also produced two major studies which were not only exercises in pure mechanics but also, in the eyes of Borelli himself, necessary introductions to what he would consider to be his most important work, the *De motu animalium.* Respectively, these were *De vi percussionis* (1667) and *De motionibus naturalibus a gravitate pendentibus* (1670). Both cover considerably more subject matter than their titles indicate. In the first, for instance, Borelli discusses percussion in detail, some general problems of motion, gravity, magnetism, the motion of fluids, the vibrations of bodies, and pendular motion, to cite just a few items. Likewise, in the second, he argues against positive levity, discusses the Torricellian experiment, takes up siphons, pumps, and the nature of fluidity, tries to understand the expansion of water while freezing, and deals with fermentation and other chemical processes. When we consider that all this was the product of years of experimental and theoretical investigation, we should not wonder that he objected to giving it over to be brought out anonymously by the Cimento just because he happened to present a good deal of it before that society. To the apparent displeasure of Leopold, Borelli published *De vi percussionis* in Bologna. And in the early summer of 1667 he set out once more to Messina.

On the way he passed through Rome and stopped for the summer in Naples. While there he was the guest of the Investiganti for whom he repeated many of the experiments he had performed at the Cimento. And he also repeated for his own edification some work that the Investiganti had accomplished independently. As a result of this visit, Concublet provided for the publication of *De motionibus naturalibus,* for which Borelli reciprocated by writing a warm dedication to him. Back in Messina, Borelli resumed his chair in mathematics. Stefano degli Angeli had raised some objections to parts of *De vi percussionis,* so in 1668 Borelli wrote the short *Risposta;* one of the problems concerned the deviation toward the east of a body dropped from a tower. In 1669 there occurred a major eruption of Etna and Borelli took the occasion to observe it closely, making notes on the topography of the mountain, the locations of the flows, and the nature of the various materials ejected, and offering some reasoned speculations of the sources of the heat powering the display. These he published in the *Historia et meteorologia incendii Aetnaei anni 1669.* Meanwhile he tried to return to his long delayed *De motu animalium.*

Borelli did not confine himself only to the sciences.

He had always taken a great interest in the public affairs of Messina. For example, while he was in Tuscany he helped to procure a copy of a manuscript the Messinese wished to publish. The work in question was the *Storia della guerra di Troja* by Guido Giudici delle Collone. A Latin version had been found among the papers of Maurolico, but it was known that the Accademia della Crusca had cited an Italian translation in Florence. At the request of the Messinese senate and with the aid of Borelli a copy was made in 1659. The Fucina published it in 1665 with a dedication to the senate. When Borelli returned from Pisa, then, he was coming home. And even though he was nearing sixty, he seems to have taken up an active political role. Agitation had been growing between the local citizens and their Spanish overlords. This led in 1674 to an open revolt. With some assistance from the French the struggle continued until 1678 when the French decided to leave the city, taking with them many of the city's leaders and (among other things) ensuring the closing of the Fucina. But trouble had brewed even before 1674. Borelli himself was thought to have provided the ideological inspiration for a party of republicans. In 1672 the Spanish Conservatore del Regno managed to stir up riots against the party, during which the home of Carlo Di Gregorio, which served as the meeting place for the Fucina, was burned. Borelli was declared a rebel and a price was placed on his head. He left very quickly and seems to have gone directly to Rome. One of his current projects also became a casualty. He had been into the papers of Maurolico and was publishing the latter's edition of the works of Archimedes when in 1672 the Spanish confiscated the nearly completed printing.

When Borelli arrived in Rome he was by no means unknown to that city. Besides his years of study there and several visits during the intervening period, he also knew and had corresponded frequently with Michelangelo Ricci and from its beginning the *Giornale de' Letterati* had published news of his scientific accomplishments: abstracts of his longer works and complete versions of a few shorter pieces. It is not surprising, then, that he would come to the attention of Queen Christina and come under her somewhat erratic patronage. Christina had been the only legal offspring of Gustavus Adolphus of Sweden. She had received an excellent education and undertook many projects, among them the creation of a learned academy in Stockholm. One of her first acts after her spectacular conversion to Catholicism was to attempt to start an academy in Rome, this in early 1656. Unfortunately, political and financial problems occupied her attention for many years. Finally, in

1674, she launched her Accademia Reale. Borelli appeared twice before it in 1675—in February when he spoke on the construction of the triremes of the ancients and again in April when he discussed Etna, this time including considerations resulting from a climb to the rim of the volcano in 1671. Christina also patronized another, more scientific group, known variously as the Accademia dell'Esperienza or the Accademia Fisica-matematica. It was organized in July of 1677 under the leadership of Giovanni Giustino Ciampini, who was also connected with the *Giornale de' Letterati*. Its membership included Borelli and an old friend and disciple, Lucantonio Porzio. But recognition apparently did not entail too much tangible support, and Borelli began to look farther afield for that. Cassini had been in Paris for several years and had become a member of the Royal Academy of Sciences. In 1676 Borelli wrote him complaining of the extreme circumstances to which he had been reduced by his enemies and the lack of quiet which was interfering with the completion of his works; he hinted that he too would like to serve the Most Christian King. By February 1677, negotiations were under way. A year later he had hopeful news, but he wrote that he was too old to travel to Paris. Instead he would send his work on the motion of animals to be printed there with a dedication to the king. In May of 1678 he still hoped for his election to the Royal Academy, but since he did not wish to trust his only copy of *De motu animalium* to the mails, he wrote that he needed time to have another made. Actually it is unlikely that he ever was elected to the Academy. A short time previously he had been robbed of all his possessions by a servant. Lacking adequate means, he had accepted the hospitality of the fathers of the Casa di S. Pantaleo and had entered their house on 13 September 1677. For the last two years of his life he taught mathematics at its Scuole Pie. Apparently he never sent a copy of his manuscript to Paris. Then in late 1679 Queen Christina agreed to bear the printing costs and Borelli dedicated the *De motu animalium* to her. He died in December, however, and his benefactor at the convent, P. Giovanni di Gesù, accepted the responsibility of seeing this last and most important work through the press. Volume I, treating of external motions, or the motions produced by the muscles, appeared in 1680. Volume II, dealing with internal motions, such as the movements of the muscles themselves, circulation, respiration, the secretion of fluids, and nervous activity, appeared in late 1681. A simple stone in the wall of the Church of S. Pantaleo recalls: *Joh. Alphonso Borellio, neapolitano, philosopho medico et matematico, clarissimo, . . .*

BIBLIOGRAPHY

I. ORIGINAL WORKS. Borelli's major writings are *Discorso . . . nel quale si manifestano le falsità e gli errori contenuti nella difesa del problema geometrico risoluto dal R. D. Pietro Emmanuele* (Messina, 1646); *Delle cagioni delle febbri maligne di Sicilia negli anni 1647 e 1648, . . . Ed in fine si tratta della digestione di cibi con nuovo metodo* (Cosenza, 1649); *Euclides restitutus* (Pisa, 1658); *Apollonius Pergaeus Conicorum lib. v. vi. vii. paraphraste Abalphato Asphahanensi nunc primum editi. Additus in calce Archimedis assumptorum liber, ex codicibus Arabicis m. ss. . . . Abrahamus Eccellensis . . . latinos reddidit* (Florence, 1661); *Euclide rinnovato* (Bologna, 1663); *Del movimento della cometa apparsa il mese di Dicembre 1664* (Pisa, 1665); *Theoricae mediceorum planetarum ex causis physicis deductae* (Florence, 1666); *De vi percussioni liber* (Bologna, 1667); *Risposta . . . alle considerazioni fatte sopra alcuni luoghi del suo libro della forza della percossa del R. P. F. Stefano de gl' Angeli* (Messina, 1668); *De motionibus naturalibus a gravitate pendentibus, liber* (Regio Iulio [Reggio di Calabria], Bologna, 1670); *Istoria et meteorologia incendii Aetnaei anni 1669 . . . accessit. Responsio ad censuras Rev. P. Honorati Fabri contra librum auctoris De vi percussionis* (Regio Iulio [Reggio di Calabria], 1670); *Elementa conica Apollonii Pergaei, et Archimedis opera, nova et breviori methodo demonstrata* (Rome, 1679); *De motu animalium . . . Opum Posthumum. Pars prima* (Rome, 1680), *Pars altera* (Rome, 1681); and "Discorso sopra la laguna di Venezia. Relazione sopra lo stagno di Pisa. Supplemento da aggiungersi alla proposizione seconda del secondo libro del P. Castelli, ecc.," in *Raccolta d'autori che trattano del moto dell'acque,* IV (Florence, 1765), 15–63.

Shorter tracts and less important works appeared in various issues of *Giornale de' Letterati;* Borelli, et al., *Tetras anatomicarum epistolarum de lingua, et cerebro* (Bologna, 1665); Marcello Malpighi, *Opera posthuma* (London, 1697); and Giovanni Targione Tozzetti, *Atti e memorie inedite dell' Accademia del Cimento,* 3 vols. (Florence, 1780).

The collections of the libraries of Florence, especially the Galileiana of the Biblioteca Nazionale, contain a great deal of unpublished correspondence to, from, and relating to Borelli. Other Italian libraries, and perhaps French and English ones, must still have a great deal of unrecognized and unpublished Borelli materials. The following have made many Borelli letters available: Howard B. Adelmann, *Marcello Malpighi and the Evolution of Embryology,* 5 vols. (Ithaca, N. Y., 1966); Giovanni Arenaprimo di Montechiaro, "Gio: Alfonso Borelli a Marcello Malpighi," in *Studi di medicina legale e varii . . . in onore di Giuseppe Ziino ecc.* (Messina, 1907), pp. 467–475; Vincenzo Busacchi and Giordano Muratori, "Giovanni Alfonso Borelli e lo Studio di Bologna," in *Bollettino delle scienze mediche* [Società di Bologna], **136** (1964), 86–90; Modestino Del Gaizo, *Alcune lettere di Giovanni Alfonso Borrelli, dirette una al Malpighi, le altre al Magliabechi* (Naples, 1886); "Contributo allo studio della vita e delle opere di Giovanni Alfonso Borrelli," *Atti dell'Accademia Pontaniana, Napoli,*

20 (1890), 1–48; "Una lettera di G. A. Borelli ed alcune indagini di pneumatica da lui compiute," in *Memorie della Pontificia Accademia Romana dei Nuovi Lincei,* 21 (1903), 61–78; "Note di storia della vulcanologia," *Memoria* no. 5 in *Atti dell'Accademia Pontaniana, Napoli,* 36 (1906); "Evangelista Torricelli e Giovanni Alfonso Borrelli. Appunti raccolti nel compiersi il terzo secolo dalla loro nascita," in *Rivista di fisica, matematica e scienze naturali* (Pavia), 17 (1908), 385–402; "L'opera scientifica di G. A. Borelli e la Scuola di Roma nel secolo XVII," in *Memorie della Pontificia Accademia Romana dei Nuovi Lincei,* 27 (1909), 275–307; and "Di una lettera inedita di G. A. Borelli diretta a M. Malpighi," in *Atti dell'Accademia Pontaniana, Napoli,* 49 (1919), 29–40; Tullio Derenzini, "Alcune lettere di Giovanni Alfonso Borelli ad Alessandro Marchetti," in *Physis,* 1 (1959), 224–243; and "Alcune lettere di Giovanni Alfonso Borelli a Gian Domenico Cassini," *ibid.,* 2 (1960), 235–241; Angelo Fabroni, *Lettere inedite di uomini illustri,* 2 vols. (Florence, 1773–1775); Giovanni Giovannozzi, "La versione borelliana dei *Conici* di Apollonio," in *Memorie della Pontificia Accademia Romana dei Nuovi Lincei,* 2nd ser., 2 (1916), 1–32; *Lettere inedite di Giovanni Alfonso Borelli al P. Angelo [Morelli] di S. Domenico sulla versione di Apollonio* (Florence, 1916); "Carte Borelliane nell' Archivio Generale delle Scuole Pie a Roma," in *Atti della Pontificia Accademia Romana dei Nuovi Lincei,* 72 (1918–1919), 81–86; and "Una lettera di Famiano Michelini a Giovanni Alfonso Borelli," *ibid.,* 80 (1926–1927), 315–319; Ugo Morini and Luigi Ferrari, *Autografi e codici di lettori dell'Ateneo Pisano esposti in occasione dell' XI congresso di medicina interna* (Pisa, 1902), pp. 19–23; Giuseppe Mosca, *Vita di Lucantonio Porzio pubblico primario cattedratico di Notomia* (Naples, 1765); Luigi Tenca, "Le relazioni fra Giovanni Alfonso Borelli e Vincenzio Viviani," in *Rendiconti dell'Istituto Lombardo di scienze e lettere, Milano,* 90 (1956), 107–121; and Giambatista Tondini, *Delle lettere di uomini illustri* (Macerata, 1782).

Among the translations of portions of the *De motu animalium* are Max Mengeringhausen, *Die Bewegung der Tiere,* no. 221 in *Ostwald's Klassiker der exakten Wissenschaften* (Leipzig, 1927); and T. O'B. Hubbard and J. H. Ledoboer, *The Flight of Birds,* Royal Aeronautical Society of London, Aeronautical Classics, no. 6 (London, 1911).

II. Secondary Literature. The most extensive treatment of Borelli's life is in Angelo Fabroni, *Vitae italorum doctrina excellentium,* II (Pisa, 1778), 222–324. More recently, Gustavo Barbensi, *Borelli* (Trieste, 1947), and Tullio Derenzini, "Giovanni Alfonso Borelli, fisico," in *Celebrazione dell'Accademia del Cimento nel tricentenario della fondazione* (Pisa, 1958), pp. 35–52, offer useful shorter treatments. Luigi Amabile, in his *Fra Tommaso Campanella ne' castelli di Napoli, in Roma ed in Parigi,* 2 vols. (Naples, 1887), published the documents pertaining to Borelli's birth and family and possible connections with Campanella, II, 361–369. Both Max H. Fisch, "The Academy of the Investigators," in E. A. Underwood, ed., *Science, Medicine and History: Essays . . . in Honor of Charles Singer,* I (Oxford, 1953), 521–563; and Nicola Badaloni, *Introduzione a*

G. B. Vico (Milan, 1961), provide much information about the Investiganti and Borelli's relation to it. Howard B. Adelmann's work on Malpighi (cited above) is indispensable for Borelli's life and work after he came to Pisa. For the Fucina and Borelli's connection with the Messina revolt one can begin with Giacomo Nigido-Dionisi, *L'Accademia della Fucina di Messina (1639–1678) ne' suoi rapporti con la storia della cultura in Sicilia* (Catania, 1903), and Giuseppe Olivà, "Abolizione e rinacimento della Università di Messina," in *CCCL Anniversario della Università di Messina* (Messina, 1900), Parte Prima, 209–365.

Borelli's celestial mechanics have been studied in Angus Armitage, "'Borell's Hypothesis' and the Rise of Celestial Mechanics," in *Annals of Science,* 6 (1950), 268–282; Alexandre Koyré, "La mécanique céleste de J. A. Borelli," in *Revue d'histoire des sciences et de leurs applications,* 5 (1952), 101–138; "La gravitation universelle de Kepler à Newton," in *Archives internationales d'histoire des sciences,* 4 (1954), 638–653; "A Documentary History of the Problem of Fall from Kepler to Newton," in *Transactions of the American Philosophical Society,* n.s. 45 (1955), 327–395; and *La révolution astronomique: Copernique, Kepler, Borelli* (Paris, 1961); and Charles Serrus, "La mécanique de J.-A. Borelli et la notion d'attraction," in *Revue d'histoire des sciences et de leur applications,* 1 (1947), 9–25. His physics have been examined in particular in Pierre Varignon, *Projet d'une nouvelle mechanique, avec Un Examen de l'opinion de M. Borelli, sur les propriétez des poids suspendus par des cordes* (Paris, 1687); Giovanni Antonio Amedeo Plana, "Mémoire sur la découverte de la loi du choc direct des corps durs publiée en 1667 par Alphonse Borelli . . .," in *Memorie della Reale Accademia delle scienze di Torino,* 2nd ser., 6 (1844), esp. 1–37; and J. MacLean, "De historische ontwikkeling der stootwetten van Aristoteles tot Huygens," a dissertation (Amsterdam, 1959).

Various particular aspects of Borelli's life and work, as well as additional bibliographical sources, are given in the following: Gustavo Barbensi, "Di una diversa soluzione di un problema di meccanica muscolare da parte di due medici matematici," in *Rivista di storia delle scienze mediche e naturali, Siena,* 29 (1938), 168–180; Pietro Capparoni, "Sulla patria di Giovanni Alfonso Borelli," *ibid.,* 22 (1931), 53–63; Modestino Del Gaizo, *Studii di Giovanni Alfonso Borrelli sulla pressione atmosferica, con note illustrative intorno alla vita ed alle opere di lui* (Naples, 1886); "Di un' antica indagine sul calore animale," in *Atti della R. Accademia medico-chirurgica di Napoli,* 49 (1895), 378–394; "Di un' opera di G. A. Borelli sulla eruzione dell' Etna del 1669 e di Adriano Auzout corrispondente, in Roma, del Borelli," in *Atti della Pontificia Accademia Romana dei Nuovi Lincei,* 60 (1906–1907), 111–117; "Qualche ricordo di Giovanni Alfonso Borelli in Firenze," in *Studium: Rivista universitaria mensile,* 2 (1907), 234–238; "Giovanni Alfonso Borrelli e la sua opera *De motu animalium,* discorso," in *Atti della R. Accademia medico-chirurgica di Napoli,* 62 (1908), 147–169; "Il *De motu animalium* di G. A. Borelli studiato in rapporto del *De motu cordis et sanguinis* di G. Harvey," *ibid.,* 67 (1914), 195–227; and "Ipotesi di antiche fisiologi e specialmente di Giovanni Alfonso Borelli sulla esistenza del succo nervoso," *ibid.,* **69** (1916), 85–108; Giovanni Battista De Toni, "Per la conoscenza delle opinioni sulla ascesa dei liquidi nelle piante," in *Rivista di fisica, matematica e scienze naturali,* **3** (1901), 199–203; Pietro Franceschini, "L'apparato motore nello studio di Borelli e di Stenone," in *Rivista di storia delle scienze mediche e naturali,* **42** (1951), 1–15; Giovanni Giovanozzi, "La patria di Gio. Alfonso Borelli," in *Atti della Pontificia Accademia Romana dei Nuovi Lincei,* **79** (1925–1926), 61–66; Raymond Hierons and Alfred Meyer, "Willis's Place in the History of Muscle Physiology," in *Proceedings of the Royal Society of Medicine,* **57** (1964), 687–692; Michelangelo Macrì, "Lettere d'illustre autori de' secoli XVII e XVIII," in *Nuova Biblioteca Analitica di scienze, lettere ed arti,* **14** (1819), letters 1 and 1 *bis,* 349–353; and Giuseppe Ziino, "G. A. Borelli medico e igienista," in *CCCL Anniversario della Università di Messina* (Messina, 1900), Parte Seconda, 3–40.

THOMAS B. SETTLE

BORGOGNONI OF LUCCA, THEODORIC (*b.* Parma or Lucca, Italy, *ca.* 1205; *d.* Bologna, Italy, 1298), *medicine, surgery.*

It is quite certain that Theodoric's father was Hugh of Lucca, a pioneer of Italian surgery, although this is contested by some. When Theodoric was nine, the family moved to Bologna, the medical capital of medieval Europe, where he became a Dominican friar in 1226. Under his father's tutelage he learned medicine and surgery, and taught these subjects at the University of Bologna for thirty-three years. As a Dominican he first was appointed a papal penitentiary of Pope Innocent IV, then was named bishop of Bitonto (1262), and finally bishop of Cervia (1266). He never took possession of the see of Bitonto, however, and resided but little in Cervia, his usual home being Bologna. His practice of surgery there, unusual for a friar and a bishop, was necessary for his teaching; it provides an apt illustration of the varied occupations of churchmen in the Middle Ages.

Theodoric wrote treatises on mineral salts and on the sublimation of arsenic, both of which have been lost. He wrote also on falconry and on the veterinary science of horses (*Practica equorum*). His most famous work, however, is his *Surgery* (*Chirurgia,* sometimes entitled *Filia principis*), begun while he was a papal penitentiary. He seems to have prepared three redactions before arriving at a definitive version, which he released in 1266 at the urging of a fellow Dominican, Andrew of Abalat, bishop of Valencia, whom he had met in Rome and to whom he dedicated the work. The first printed edition was produced at Venice in 1498, and other editions followed in 1499, 1513, 1519, and 1546. During the medieval period the work was translated into several vernaculars, and in 1955 an English translation was published by two doctors, E. Campbell and J. Colton, whose war service led them to new developments in surgery that they found adumbrated in the work of Theodoric.

The four books of the *Surgery* cover the subject in both its general and its specific aspects. Theodoric recorded, in the main, the surgical knowledge and practices of Hugh of Lucca, appealing over fifty times to the latter's authority. He verified Hugh's work by his own experience, however, and supplemented it where it was incomplete or defective. In his quite modern discussion of fractures and dislocations, for example, he states: "In this book I have not been willing to include anything which I have not tested; nor did I wish my book to seem to contain more of another than of me" (Campbell and Colton, I, 218). He also claimed to be well acquainted with "the accounts of the ancients, especially of Galen. . ." (*ibid.,* I, 4).

Theodoric's methods were progressive, particularly his advocacy of antiseptic surgery at a time when everyone held the theory of "laudable pus." He condemned the use of unguents and poultices to generate pus in a wound, holding that this, far from being necessary for healing, actually impedes nature, prolongs sickness, prevents the uniting and consolidating of the wound, deforms the part, and impedes cicatrization. He described his own revolutionary methods in detail, outlining procedures for cleaning the wound, eliminating dead tissue and foreign matter, accurately reapproximating the wound walls, using stitches where necessary, and adequately protecting the area.

Theodoric also improved techniques for preparing and using soporific sponges to induce sleep before surgery. The sponges were impregnated with a mixture of narcotics, dried out, and stored for use. When needed, they were immersed in hot water, wrung out, and held to the nose of the patient, who was instructed to breathe deeply. Other of his practices include the use of mercurial ointments in the treatment of some skin diseases and the sparing application of cautery.

Unfortunately for the history of medicine, aseptic surgery died with Theodoric's pupil, Henri of Mondeville, the father of French surgery. Despite persecution by his contemporaries, Mondeville applied his master's methods in France and found them most successful. His principal adversary, however, was Guy de Chauliac, whose *Chirurgia magna* (1363) became a standard textbook for centuries. Guy unjustly accused Theodoric of plagiarism and otherwise denigrated him; he particularly rejected Theodoric's aseptic treatment of wounds and thereby perpetuated the older methods. The result was that, with few exceptions (e.g., Mondeville, Ambrose Paré, and Richard Wiseman), one studying the history of surgery encounters "a gulf of 'laudable pus' centuries wide" (*ibid.*, I, xxix).

BIBLIOGRAPHY

I. ORIGINAL WORKS. Theodoric's major work, the *Chirurgia* (Venice, 1498, 1499, 1513, 1519, 1546), was translated by E. Campbell and J. Colton as *The Surgery of Theodoric* (New York: Vol. I, 1955; Vol. II, 1960). The translators' use of modern technical terms occasionally gives a distorted and anachronistic view of Theodoric's contributions. The *Practica equorum* appeared as *El libro de los caballos: Tratado de albeitería del siglo XIII,* G. Sachs, ed. (Madrid, 1936).

II. SECONDARY LITERATURE. Theodoric or his work is discussed in H. F. Garrison, *An Introduction to the History of Medicine,* 4th ed. (Philadelphia, 1929), pp. 153–155; Adalberto Pazzini, "Borgognoni, Teodorico," in *Enciclopedia cattolica* (Rome, 1949–1954), II, cols. 1923–1924, with additional bibliography; Jacques Quétif and Jacques Échard, *Scriptores ordinis praedicatorum,* 2 vols. (Paris, 1719–1721; repr. New York, 1959), I, 354–355, in which the authors confuse Theodoric with Theodoricus Catalanus; and George Sarton, *Introduction to the History of Science,* II, pt. 2 (Baltimore, 1931), 654–656.

WILLIAM A. WALLACE, O.P.

BORN, IGNAZ EDLER VON (*b.* Karlsburg, Transylvania [now Alba Iulia, Rumania], 26 December 1742; *d.* Vienna, Austria, 24 July 1791), *mineralogy.*

Born was descended from a noble German family. At the age of thirteen he began his studies in Vienna with the Jesuits, who induced him to join their order; he remained a member for only about sixteen months, however. Leaving Vienna and the Jesuits behind, Born went to Prague to study jurisprudence. After completing his education there, he traveled in Germany, the Low Countries, and France. Upon his return to Prague, he took up the study of natural history and mining, and in 1770 he joined the department of mines and the mint. In the same year Born visited the principal mines of Hungary and Transylvania. His account of this expedition is preserved in a series of lively and interesting letters addressed to the mineralogist J. J. Ferber. In 1774 Ferber published these letters in a work that later appeared in English, French, and Italian editions. During his visit to a mine at Felso-Banya, Born suffered an accident that nearly cost him his life. He descended into the mine too soon after the fires used to detach the ore had been extinguished, and inhaled a dangerously large quantity of arsenical vapors. This unfortunate occurrence seriously affected Born's health and may well have shortened his life. Upon returning to Prague, he was appointed counselor of mines.

In 1772 Born published *Lithophylacium Bornianum,* a description of his own collection. This work attracted the favorable attention of mineralogists, and Born was soon admitted to various learned societies throughout Europe. In 1776 Empress Maria Theresa, having heard of his reputation, called him to Vienna to arrange and describe the imperial collection. He completed a portion of this task, but after the empress' death in 1780 it was discontinued. In 1779 Born was raised to the office of counselor of the court chamber in the department of mines and the mint.

Born's interests and activities extended into fields other than mineralogy and mining. While in Prague, he had helped to found a literary and philosophical society that was the forerunner of the Bohemian Scientific Society. In Vienna, Born was active in the secret fraternity of Freemasons. After Joseph II's accession to the throne, this brotherhood was allowed to pursue its anticlerical activities with greater freedom. In 1783 Born published *Specimen monachologiae,* a vicious satire against monks in which the various orders were classified according to a system modeled after Linnaeus'.

Besides preparing catalogs of fossil and mineral collections, works of classification, and descriptions of mines and mining equipment, Born invented an amalgamation process for removing gold and silver from various ores. Since the process did not require the usual melting of the ore, its use effected a considerable saving of fuel. A trial of the process in the presence of observers was made at Selmeczbánya (German, Schemnitz), Hungary (now Banská Štiavnica, Czechoslovakia). In 1786 Born published his description of it. The process was adopted in copper mines throughout Hungary, and Born was given a share of the savings occasioned by its use.

BIBLIOGRAPHY

I. ORIGINAL WORKS. Born's writings include *Briefe über mineralogische Gegenstände, auf seiner Reise durch das Temeswarer Bannat, Siebenbürgen, Ober- und Nieder-Hungarn, an den Herausgeber derselben geschreiben,* J. J. Ferber, ed. (Frankfurt–Leipzig, 1774), trans. by R. E. Raspe as *Travels Through the Bannat of Temeswar, Transylvania, and Hungary in the Year 1770, etc.* (London, 1777); *Lithophylacium Bornianum,* 2 vols. (Prague, 1772–1775); *Effigies virorum eruditorum atque artificum Bohemiae et Moraviae, etc.,* 2 vols. (Prague, 1773–1775); *Index rerum naturalium Musei Caesarei Vindobonensis* (Vienna, 1778); *Ueber das*

Anquicken der gold- und silberhältigen Erze, Rohsteine, Schwarzkupfer und Hüttenspeise (Vienna, 1786), trans. by R. E. Raspe as *Baron Inigo Born's New Process of Amalgamation of Gold and Silver Ores, and Other Metallic Mixtures* (London, 1791); and *Bergbaukunde,* 2 vols. (Leipzig, 1789–1790), written with F. W. H. von Trebra. Born edited the *Abhandlungen einer Privatgesellschaft in Böhmen,* 6 vols. (Prague, 1775–1784). He published in L. Crell's *Chemische Annalen* some reports of Matteo Tondi's alleged reduction of the alkaline earths (1790), pt. 2, no. 12, 483–485; (1791), pt. 1, no. 1, 3–10; no. 2, 99–100; no. 5, 387–389.

II. SECONDARY LITERATURE. An excellent biography with numerous and useful citations is Baur's article on Born in *Allgemeine Encyclopädie der Wissenschaften und Künste,* J. G. Ersch and J. G. Gruber, eds., XII (Leipzig, 1824), 38–40. The detailed English biography in Alexander Chalmers' *Biographical Dictionary,* VI (1812), 123–127, is based on an account given in Robert Townson's *Travels in Hungary* (London, 1797). For further biographical sources, see the list in Paul Mayer's article on Born in *Neue deutsche Biographie,* II (1953). In addition, see Robert Keil, *Wiener Freunde, 1784–1808, Beitraege zur Jugendgeschichte der deutsch-oesterreichischen Literatur* (Vienna, 1883), pp. 33–36, which contains three of Born's letters.

J. B. GOUGH

BORODIN, ALEKSANDR PORFIREVICH (*b.* St. Petersburg, Russia, 12 November 1833; *d.* St. Petersburg, 27 February 1887), *chemistry.*

Borodin was the illegitimate son of the wife of an army doctor and an Imeretian prince. Between 1850 and 1856 he studied at the St. Petersburg Academy of Medicine and Surgery, where he did his first research in chemistry under the direction of N. N. Zinin. In 1858 he defended his doctoral thesis, "Ob analogy fosfornoy myshyakovoy kislot v khimicheskom i toksikologicheskom otnosheny" ("Analogy of Phosphoric and Arsenic Acids from the Chemical and Toxicological Viewpoints"). From 1859 to 1862, he traveled in Italy, Germany, France, and Switzerland, sometimes in the company of Mendeleev and I. M. Sechonov. As a member of the Russian delegation, Borodin took part in the work of the First International Congress of Chemists in Karlsruhe in 1860. From 1864, Borodin held a professorship at the Academy of Medicine and Surgery, where he helped to found medical courses for women.

Borodin's most important research work was done in organic and physiological chemistry. He was among the first to obtain fluorine benzol, and in 1861 he developed a method for the fluorination of organic compounds. In 1869 he proposed a method for obtaining bromine-producing fatty acids by the action of bromine on the silver salts of acids.

A number of Borodin's studies (1863–1874) were devoted to an investigation of the polymerization and condensation of aldehydes. By the action of metallic sodium on valeric aldehyde ($C_5H_{10}O$), Borodin obtained the condensation products $C_{10}H_{18}O$ and $C_{20}H_{38}O_3$. He also showed that from valeric aldehyde, valeric acid and amyl alcohol are formed:

$$C_4H_9C\overset{H}{\underset{O}{}} \xrightarrow{O_2} C_4H_9-C\overset{OH}{\underset{O}{}}$$

$$C_4H_9C\overset{H}{\underset{O}{}} \xrightarrow[\text{NaOH}]{H_2} C_4H_9CH_2OH$$

In studying the condensation products of acetaldehyde, Borodin found a substance having two kinds of alcohol aldehydes (aldol), which, dehydrating rapidly, turn into crotonic aldehyde:

$$2CH_3-C\overset{H}{\underset{O}{}} \longrightarrow CH_3-CHOH-CH_2-C\overset{H}{\underset{O}{}} \longrightarrow$$

$$CH_3-CH=CH-C\overset{H}{\underset{O}{}}$$

This aldol condensation reaction was subsequently employed by I. I. Ostromyslensky to obtain butadiene from alcohol.

In 1876 Borodin developed an azotometric method and apparatus for the quantitative determination of urea by measuring the amount of elementary nitrogen that is extracted from the urea by the action of excess sodium bromate ($Br_2 + NaOH$). The Borodin method was widely adopted in biochemical and clinical laboratories.

Borodin's name is also well-known in music, chiefly as the composer of *Prince Igor,* the first heroic opera on a Russian theme (completed by Rimsky-Korsakov), and of the B-minor symphony and numerous songs.

BIBLIOGRAPHY

I. ORIGINAL WORKS. Borodin's writings include "Über die Einwirkung des Natriums auf Veraldehyd," in *Bulletin de l'Académie imperiale des sciences de St.-Pétersbourg, phys.-math. classe,* **7** (1869), 463–474; "O poucheny produkta uplotnenia obyknovennogo aldegida" ("On Polymerization and Condensation Products of Common Aldehydes"), in *Zhurnal Russkogo khimicheskogo obshchestva,* **6** (1872), 209; "Über einen neuen Abkömmling des Valeraldehyds," in *Berichte der deutschen* *chemischen Gesellschaft zu Berlin,* **6** (1873), 982–985; Borodin's major work on his methods of azometric measurement of nitrogen in urea is *Uproshchenni azometrichesky sposob opredelenia azota v primeneny k klinicheskomu opredelenniiu metamorfozy azotistykh veshchestv v organizme s sovremennoy tochki zrenia* (St. Petersburg, 1886).

II. SECONDARY LITERATURE. For biographical information, see N. A. Figurovski and Y. I. Soloviev, *Aleksandr Porfirevich Borodin* (Moscow–Leningrad, 1950).

Y. I. SOLOVIEV

BORREL, JEAN. See **Buteo, Johannes.**

BORRICHIUS (or **BORCH**), **OLAUS** (*b.* Nørre Bork, in Ribe, Denmark, 7 April 1626; *d.* Copenhagen, Denmark, 13 October 1690), *chemistry.*

The son of Oluf Clusen, a rector, Borrichius went to school in Ribe and entered the University of Copenhagen in 1644 to study medicine under Thomas Bartholin, Olaus Worm, and Simon Pauli. He remained a close friend of Bartholin until the latter's death in 1680. Borrichius was a teacher at the chief grammar school in Copenhagen for a time, won fame as a physician during the plague epidemic of 1654, and became tutor to the sons of Joachim Gersdorf, the lord high steward (*Rigshofmester*), in 1655. In 1660 Borrichius was appointed *professor ordinarius* of philology and *professor extraordinarius* of botany and chemistry. The posts were supernumerary until vacancies occurred.

Later in 1660 Borrichius was granted permission by the university to absent himself for two years in order to prepare himself for these posts by study and travel in other countries. He was joined at Hamburg by Gersdorf's sons. Borrichius' diary of the tour, and his correspondence with Bartholin, provide an interesting picture of European intellectual life during the period. He visited Germany, the Netherlands, France, England, and Italy. Among those he met were Sylvius, Swammerdam, Boyle, Petit, Redi, and Gui Patin. Borrichius received the M.D. at Angers in 1664. He gathered much information on the Hermetic sciences during the tour, and was greatly impressed by the Italian alchemist Giuseppe Francesco Borri.

His tour having lasted six years by then, Borrichius was reminded that his university posts could not be kept vacant indefinitely, and he began his journey home from Italy, reaching Copenhagen in 1664. He assumed the posts that he was to hold for nearly thirty years, becoming famous for his polymath erudition and establishing a large and profitable medical practice (he was royal physician to Frederick III and Christian V). He was twice *rector magnificus* at his university, and in 1686 was appointed counselor to the Supreme Court of Justice and in 1689 to the Royal Chancellery. He never married, and before his death bequeathed his house as a *collegium mediceum,* to lodge six students.

Borrichius was famous in his own time as a physician, as a polemicist and defender of Hermeticism, and as a prolific writer on chemical, botanical, and philological topics. His histories of the development of chemistry are among his best-known works. His travels and meetings with other European natural philosophers had not weakened his allegiance to the revived Hermeticism of the sixteenth and seventeenth centuries. Borrichius was prepared to concede that there was no healing virtue in words, seals, and images; astral influences, if they really existed, consisted of balsamic exhalations. At the same time, he believed in the existence of the philosophers' stone. Of all profane sciences, chemistry came closest to the contemplation of divinity in nature, and hence to Scripture. Opposing Athanasius Kircher's views, and especially those of Hermann Conring, Borrichius defended the genuineness and antiquity attributed to the Emerald Table and the Hermetic writings. He also accepted as authentic the alchemical works ascribed to such authors as Democritus, Albert the Great, Arnald of Villanova, Ramón Lull, and Nicolas Flamel. He opposed Conring's views that Paracelsian principles had no use in medicine and that chemistry was better employed in perfecting pharmacy than in presuming to correct physiology and pathology.

BIBLIOGRAPHY

I. ORIGINAL WORKS. Borrichius' chief works are *Docimastice metallica* (Copenhagen, 1660); *De ortu et progressu chemiae dissertatio* (Copenhagen, 1668); *Lingua pharmacopoeorum sive de accurata vocabulorum in pharmacopoliis usitatorum pronunciatione* (Copenhagen, 1670); *Hermetis, Aegyptiorum et chemicorum sapientia, ab Hermanni Conringii animadversionibus vindicata* (Copenhagen, 1674); *De somno et somniferis maxime papavereis* (Copenhagen, 1680); *De usu plantarum indigenarum in medicina* (Copenhagen, 1688); and *Conspectus scriptorum chemicorum illustriorum libellus posthumus, cui prefixa historia vitae auctoris ab ipso conscripta* (Copenhagen, 1697). *De ortu* and *Conspectus* repr. in J. L. Manget, *Bibliotheca chemica curiosa* (Geneva, 1702), I, 1–53. Borrichius also published numerous works on general and Latin philology. His botanical observations in the *Acta Hafniensia* were collected by S. Lyntrup in his *Orationes academicae in duos tomos distributae,* 2 vols. (Copenhagen, 1714). "Nitrum non inflammari" appeared in the *Acta Hafniensia,* **5** (1680), 213–216. Borrichius' autobiographical sketch was ed. by F. Rostgaard in his *Vitae selectae quorundam eruditissimorum ac illustrium virorum* (Bratislava, 1711), pp. 276–294.

II. SECONDARY LITERATURE. Various aspects of Borrichius' life and work are discussed in *Dansk biografisk leksikon* (Copenhagen, 1934), III, 454–462; a fuller treatment is E. F. Koch, *Oluf Borch* (Copenhagen, 1866). Supplementary details are in Lenglet du Fresnoy, *Histoire de la philosophie hermétique* (The Hague, 1742), I, 417–422; C. S. Petersen, *Den danske litteratur* (Copenhagen, 1929), pp. 669–682, 1041; Lynn Thorndike, *A History of Magic and Experimental Science* (New York, 1958), VII, 318–320; VIII, 344–346; and E. Warburg, *Subacute and Chronic Pericardial and Myocardial Lesions* (Copenhagen–London, 1938), esp. pp. 13–14, trans. by H. Anderson and G. Seidelin.

P. M. RATTANSI

BORRIES, BODO VON (*b.* Herford, Westphalia, Germany, 22 May 1905; *d.* Cologne, Germany, 17 July 1956), *electron microscopy.*

The scion of a long line of distinguished civil servants, and on his mother's side related to the Kamps of the pioneering metallurgical firm of Kamp & Harkort, Borries wavered between the law and engineering. He opted for a career in technology and studied electrical engineering in Karlsruhe, Danzig, and Munich. In 1930 he became an assistant to Adolf Wilhelm Matthias at the High-Voltage Institute of the Technische Hochschule in Berlin, where a group of young engineers led by Max Knoll sought to develop the electronic oscilloscope (picture tube) into a technological tool. In 1931, almost simultaneously, G. R. Rüdenberg, research director at the Siemens laboratories in Berlin, applied for the first patent on an electron microscope; Knoll and his group reported their work on it at the Technische Hochschule; and Ernst Brüche described *his* efforts on a similar instrument in the laboratories of the AEG firm in Berlin. Borries was thus involved in electron microscopy from its conception and later contributed significantly to it himself.

In 1937, feeling the need for better facilities than were likely to be made available at an engineering college, Borries and his colleague (and later brother-in-law) Ernst Ruska persuaded Siemens to support their project, thereby giving that firm a lead in the field that persists to the present day. Two years later they put the first transmission electron microscope on the market. Borries not only had a leading part in the design of this instrument and became the outstanding expert in the associated photographic techniques, but also extended its use beyond very thin specimens to metallic surfaces by reflection methods. In 1941 he was awarded the silver Leibniz Medal of the Prussian Academy of Sciences.

After World War II, Borries formed (with joint government and private support) an electron-microscopy institute in Düsseldorf and became its first director, helped to found the German Society for Electron Microscopy, and (after 1953) served as professor of electron optics at the Technische Hochschule in Aachen. Throughout his career he promoted the introduction of electron microscopy into many fields of research, notably in the life sciences. He also took a prominent part in securing public support for science. In 1949 Borries was appointed honorary professor at the Düsseldorf Medical Academy, and in the same year he published a text, *Die Übermikroskopie.* He was the principal organizer of the International Federation of Electron Microscope Societies and served as its president from its foundation in 1954 until his death.

BIBLIOGRAPHY

A list of Borries' publications through 1952 appears in Poggendorff's *Biographisch-literarisches Handwörterbuch,* VIIa. An important text is *Die Übermikroskopie* (Aulendorf, 1949). Partial autobiographies are in *Physikalische Zeitschrift,* **45** (1944), 316, and in *Frequenz,* **2** (1948), 267.

An obituary by Ernst Ruska is in *Zeitschrift der wissenschaftlichen Mikroskopie,* **63** (1956), 129, and in *Electron Microscopy* (the proceedings of the 1956 Stockholm conference on the subject), where it is followed by a further appreciation by V. E. Cosslett. For an account of Borries' place in the development of electron microscopy, see L. Marton, *Early History of the Electron Microscope* (San Francisco, 1968).

CHARLES SÜSSKIND

BORTKIEWICZ (or **Bortkewitsch**), **LADISLAUS** (or **Vladislav**) **JOSEPHOWITSCH** (*b.* St. Petersburg, Russia, 7 August 1868; *d.* Berlin, Germany, 15 July 1931), *mathematics.*

Bortkiewicz's mother was Helene von Rokicka; his father was Joseph Ivanowitsch Bortkewitsch, a member of the gentry from the Kovno [now Kaunas] province of Russia who was a colonel, an instructor in artillery and mathematics, a notary, and an author of several textbooks on elementary mathematics and works in economics and finance.

Bortkiewicz graduated from the Faculty of Law of the University of St. Petersburg in 1890 and took a postgraduate course in political economy and statistics. He also studied at Strasbourg (1891–1892) under G. F. Knapp, at Göttingen (1892) under W. Lexis, and at Vienna and Leipzig. In 1893 he defended his doctoral dissertation in philosophy at Göttingen. Bortkiewicz was a *Privatdozent* in Strasbourg and lectured in actuarial science and theoretical statistics in 1895–1897; in 1897–1901 he was a clerk in the general office of the Railway Pension Committee in St. Petersburg. Simultaneously, from 1899 to December 1900, he taught statistics at the prestigious Alexandrowsky Lyceum. In 1901 he became extraordinary professor of statistics at the University of Berlin, where he spent the rest of his life, becoming ordinary professor of statistics and political economy in 1920. He was a member of the Swedish Academy of Sciences, the Royal Statistical Society, the American Statistical Association, and the International Statistical Institute.

Bortkiewicz's publications concern population and statistical theory; mathematical statistics; and application of the latter and of probabilities to statistics, to actuarial science, and to political economy. Following Lexis' reasoning, Bortkiewicz was a proponent (almost the only one) of connecting statistics with the theory of probabilities and mathematical statistics. This idea was featured in an empirical "law of small numbers" (law of rare events, which formerly, beginning with Jakob I Bernoulli, were considered "morally" impossible and were discarded as such): The small numbers of events in large series of trials are stable in time; oscillations of the numbers of such events are accounted for by the Lexis criterion (Q quotient). The most important feature of this law, contrary to Bortkiewicz's opinion, appeared to be its connection with the Poisson law of large numbers and the popularization of the Poisson distribution. The Q quotient was regularly used by Bortkiewicz (who, moreover, deduced its expectation and standard deviation) in the same way that the x^2 criterion is used now. His other works in the theory of probabilities and mathematical statistics pertain to radioactivity, the theory of runs, and order statistics (he was a pioneer in the latter).

Noting the concrete and social nature of statistical deductions, Bortkiewicz recommended that legislation be based on them. His works are distinguished by independent opinions (dissenting with V. J. Buniakowsky, G. F. Knapp, M. E. L. Walras, and others), rigorous deductions, and voluminous references of international scope. At the same time, being comprehensive and not accompanied by a summary, they make hard reading.

Bortkiewicz was one of the main representatives of the "Continental direction" in mathematical statistics and its application to statistics, but he left no monographs, and the German scientists were only marginally interested in his works. He did not create a school but was closely associated with A. A. Tschuprow.

His last days were marred by a heated argument with Gini, an Italian statistician, who accused Bortkiewicz of plagiarism. Original correspondence on this alleged plagiarism is appended to Andersson's obituary (see bibliography).

BIBLIOGRAPHY

I. ORIGINAL WORKS. The only more or less comprehensive enumeration of approximately 100 of Bortkiewicz's works is in the obituary by T. Andersson (see below). These works include a few books, papers (including rather lengthy ones in various journals), and reviews. Seven of his papers (1889–1910) are in Russian; the other works are almost exclusively in German. Among his writings are *Das Gesetz der kleinen Zahlen* (Leipzig, 1898); *Die radioaktive Strahlung als Gegenstand wahrscheinlichkeitstheoretischer Untersuchungen* (Berlin, 1913); *Die Iterationen* (Berlin, 1917); "Die Variabilitätsbreite beim Gauschen Fehlergesetz," in *Nordisk statistisk tidskrift,* **1** (1922), 11–38, 193–220; and "Variationsbreite und mittlerer Fehler," in *Sitzungsberichte der Berliner mathematischen Gesellschaft,* **21** (1922), 3–11.

Three of his papers are available in English trans. by the W.P.A., published in the early 1940's together with trans. of related works, notably those of W. Lexis: "Kritische Betrachtungen zur theoretischen Statistik" (1894–1896), trans. as "Critical Comments on the Theory of Statistics"; "Homogeneität und Stabilität in der Statistik" (1918), trans. as "Homogeneity and Stability in Statistics"; and "Das Helmertsche Verteilungsgesetz für die Quadratsumme zufälliger Beobachtungsfehler" (1918), trans. as "Helmert's Law of Distribution for the Sum of Squares of Random Errors of Observation." The W.P.A. trans. are accompanied by a short bibliography of Bortkiewicz's works. At least two of his works in economics are also available in English.

Information about the St. Petersburg period of Bortkiewicz's life and about his father is in the U.S.S.R. State Historical Archives, Leningrad. Information about his life in Berlin is in the archives of the Humboldt University, Berlin.

II. SECONDARY LITERATURE. Information on the life and works of Bortkiewicz (with reference to his obituaries) is in *Kürschners deutscher Gelehrten-Kalender* (Berlin–Leipzig, 1931), 274; *Reichshandbuch der deutschen Gesellschaft,* I, *Handbuch der Persönlichkeiten in Wort und Bild* (Berlin, 1930), 188, with portrait; *Neue deutsche Biographie,* II (Berlin, 1955), 478; and Poggendorff, VI, pt. 1. The most comprehensive obituary is T. Andersson, in *Nordisk statistisk tidskrift,* **10** (1931), 1–16, published simultaneously in English in *Nordic Statistical Journal,* **3** (1931), 9–26. The latest published biography is E. J. Gumbel, in *International Encyclopedia of the Social Sciences* (New York, 1968).

O. B. SHEYNIN

BORTOLOTTI, ETTORE (*b.* Bologna, Italy, 6 March 1866; *d.* Bologna, 17 February 1947), *mathematics, history of mathematics.*

A disciple of Salvatore Pincherle, Bortolotti received his degree in mathematics *summa cum laude* from the University of Bologna in 1889. He was a university assistant until 1891, when he was appointed professor at the lyceum of Modica, Sicily. After completing postgraduate studies in Paris in 1892–1893, he taught in Rome from 1893 to 1900. In the latter year Bortolotti was appointed professor of infinitesimal calculus at the University of Modena, where he taught analysis and rational mechanics. He was dean of the Faculty of Science from 1913 to 1919, the year in which he assumed the professorship of analytical geometry at the University of Bologna. He retired in 1936.

Bortolotti's early studies were devoted to topology, whereas his later works in pure mathematics dealt largely with analysis: calculus of finite differences, the general theory of distributive operations, the algorithm of continuous fractions and its generalizations, the order of infinity of functions, the convergence of infinite algorithms, summation and asymptotic behavior of series and of improper integrals.

Bortolotti's interest in the history of mathematics was clear in his early works on topology; it increased during his stay in Rome, when he was an associate of the physicist and mathematician Valentino Cerruti; and it was fully developed in Modena, when he made deep studies of Paolo Ruffini's manuscripts. His first published historical work was "Influenza dell'opera matematica di Paolo Ruffini sullo svolgimento delle teorie algebriche" (1902). He later edited Ruffini's *Opere matematiche* (1953–1954). Bortolotti gradually widened the scope of his studies to include more remote times. The period in the seventeenth century during which infinitesimal analysis was developed was the subject of profound studies by Bortolotti, who revealed the importance of Torricelli's infinitesimal results while vindicating Cataldi's claim to the discovery of continuous fractions.

Bortolotti also studied the work of Leonardo Fibonacci and of Scipione Dal Ferro, Nicolò Tartaglia, Girolamo Cardano, Ludovico Ferrari, and Rafael Bombelli. He found and published (1929), with an introduction and notes, the manuscript of books IV and V of Bombelli's *L'algebra*. Among his other contributions is the objective reconstruction of the argumentations of the Sumerian, Assyrian, Babylonian, and Egyptian mathematicians.

BIBLIOGRAPHY

I. ORIGINAL WORKS. Bortolotti's works total more than 220, and lists of them may be found in the appendixes to the articles by Bompiani and Segre (see below). Among his works are "Influenza dell'opera matematica di Paolo Ruffini sullo svolgimento delle teorie algebriche," in *Annuario della R. Università di Modena, 1902–1903*, pp. 21–77; *Lezioni di geometrica analitica*, 2 vols. (Bologna, 1923); *Studi e ricerche sulla storia della matematica in Italia nei secoli XVI e XVII* (Bologna, 1928); *I cartelli di matematica disfida e la personalità psichica e morale di Girolamo Cardano* (Imola, 1933); and *La storia della matematica nella Università di Bologna* (Bologna, 1947). He also edited Books IV and V of *L'algebra, opera di Rafael Bombelli da Bologna* (Bologna, 1929) and Ruffini's *Opere matematiche*, 3 vols. (Rome, 1953–1954).

II. SECONDARY LITERATURE. Works on Bortolotti are E. Bompiani, "In ricordo di Ettore Bortolotti," in *Atti e memorie dell'Accademia di scienze, lettere e arti di Modena*, 5th ser., **7** (1947); E. Carruccio, "Ettore Bortolotti," in *Periodico di matematiche*, 4th ser., **26** (1948), and "Commemorazione di Ettore Bortolotti," in *Atti della Società italiana di scienze fisiche e matematiche "Mathesis"* (1952); and B. Segre, "Ettore Bortolotti—commemorazione," in *Rendiconti dell'Accademia delle scienze dell'Istituto di Bologna, classe di scienze fisiche*, n.s. **52** (1949), 47–86.

ETTORE CARRUCCIO

BORY DE SAINT-VINCENT, JEAN BAPTISTE GEORGES MARIE (*b*. Agen, France, 6 July 1778; *d*. Paris, France, 22 December 1846), *biology*.

As a young boy, Bory de Saint-Vincent wandered about his native Guyenne to escape the persecution by the Jacobins with which his father was threatened. He became interested in plants and insects and, by the age of sixteen, was corresponding with established French naturalists. He was conscripted into the army in 1799 and remained an officer until his retirement in 1840. He was seconded to Baudin's expedition to Australia in 1801 but, officially owing to ill health, spent the year in the Mascarene Islands. After returning to France, Bory de Saint-Vincent saw army service in Germany and Spain. In 1815 he entered politics as deputy for Lot-et-Garonne; this led, at the fall of Napoleon, to his banishment from France between 1816 and 1820. He edited the *Dictionnaire classique de l'histoire naturelle* from 1822 to 1831 and, in 1829, led the scientific section of the French government's expedition to the Peloponnese. He was elected to the Académie des Sciences in 1834 and, in 1840, led an official scientific expedition to Algeria.

Bory de Saint-Vincent is remembered as the leader of successful botanical collecting expeditions and for his contributions to the theory, principles, and knowledge of island faunas; the zoogeography of the seas; and the classification of man.

As a disciple of Buffon and Lamarck, Bory de Saint-Vincent accepted the idea of change in the natural world. He suggested that in earlier times the oceans had covered the globe and that fish were, therefore, the most ancient inhabitants. The continents emerged in their turn and, finally and recently, the volcanic islands. He accepted spontaneous generation but believed that, after initial creations, species changed under the influence of the environment. Developing the ideas of Buffon, he argued that, on continents, species were relatively old and fixed in type but, on recent volcanic islands, such as Réunion, they were still in a state of flux, polymorphic. After many generations the stability of the environment would lead to a stable monomorphic species.

Bory de Saint-Vincent was the first to notice that oceanic islands were without amphibia and speculated on the reasons for flightless birds occurring independently on different islands. While accepting that the Mascarene Islands had emerged from the ocean, he considered the Atlantic islands to be the remains of a continent, the lost Atlantis. He referred to the formation of coral reefs and, many years later, he attempted one of the first biogeographical classifications of the oceans. Further developments along these lines were not made in France, but later devolved upon the English naturalists.

Bory de Saint-Vincent was also prominent in studies on the classification of the races of man by physical characteristics, which had begun with Blumenbach. In 1827 he divided man into fifteen species on the basis of the combined value of all the physical characteristics then known. The fifteen species were grouped into two major types, those with straight hair and those with crinkly hair. Treating man as a creation no different from other animals, Bory de Saint-Vincent unified his biological ideas by suggesting that the species of man were probably created at different times in different parts of the world, some of them on former islands that have become part of Asia.

BIBLIOGRAPHY

Bory de Saint-Vincent's most important writings are *Essais sur les îles fortunées et l'atlantique atlantide* (Paris, 1802); *Voyage dans les quatre principales îles des mers d'Afrique*, 3 vols. (Paris, 1803); *L'homme: Essai zoologique sur le genre humain* (Paris, 1827); *Relation du voyage de la commission scientifique du Morée*, 2 vols. (Paris, 1836–1838); *Exploration scientifique de l'Algérie*, 2 vols. (Paris, 1846–1847); and many contributions to the *Dictionnaire classique de l'histoire naturelle, Encyclopédie moderne,* and *Encyclopédie méthodique.*

There is no biography, but see P. Romieux, *Les carnets de Bory de Saint-Vincent 1813–1815* (Paris, 1934).

WILMA GEORGE

BOSC, LOUIS AUGUSTIN GUILLAUME, known in his youth as **Dantic** (*b.* Paris, France, 29 January 1759; *d.* Paris, 10 July 1828), *natural history, agronomy.*

Bosc belonged to a Protestant family. His father, Paul Bosc d'Antic, the son of a surgeon from the Tarn, was a master glassmaker and a doctor who was acquainted with most of the great naturalists of his time; his mother, Marie d'Hangest, daughter and sister of generals, died when Louis was only two years old. Bosc was interested in nature as early as six or seven, but little is known of what he studied at the Collège de Dijon. At eighteen he became secretary of the Intendance des Postes in Paris. He assiduously followed the courses at the Jardin du Roi, directed by Buffon, where Thouin and Antoine-Laurent de Jussieu taught. A large and varied group, often enamored of and nurtured on Rousseau, met there regularly. He made friends among them, notably with Desfontaines and Mme. Roland, wife of a future member of the National Convention.

The period between 1780 and 1796 was of particular importance for Bosc, who showed equal enthusiasm for science and politics. He took an extremely active part in the Revolution: he became secretary of the Club des Jacobins in 1791 and postmaster general under the Girondist ministry in 1792. After the fall of the Girondins, he took refuge in the ancient priory of Ste. Radegonde, in the forest of Montmorency, from September 1793 to July 1794. Later, an affair of the heart led him into emotional difficulties which he decided to cure by taking a long voyage. He sailed for Charleston, South Carolina, on 8 July 1796.

Bosc's work had already become well known. His first publication was an article in the Abbé Rozier's *Journal de physique* (1784). Bosc was then only twenty-five, but his article was a masterpiece. The insect described was an unusual cochineal, a new genus that was to become the type for the subfamily *Ortheziinae.* In 1792 Bosc named the genus *Ripiphorus,* a coleopteron that was later elevated to the rank of family by Thomson (1864).

Between 1790 and 1792 Bosc published numerous notes on insects, mollusks, birds, and plants. Along with Broussonet, l'Héritier, and a few others, he founded the Société Linnéenne de Paris. Bosc became more and more interested in sciences applicable to agriculture, and in 1796 he agreed to substitute for Thouin and the Abbé Tessier to write on agriculture for the *Encyclopédie de Panckoucke;* thanks to his skill, Volume IV follows the preceding ones without interruption. Again asked to help in 1810, Bosc edited, almost single-handed, the three last volumes of the monumental work (1813–1821). He also collaborated on the *Nouveau dictionnaire d'histoire naturelle* (1803–1804) and on the *Nouveau cours complet d'agriculture théorique et pratique* (1809), and was one of the leading editors of the *Annales de l'agriculture française* from 1811 to 1828.

Little is known of Bosc's stay in America, where he arrived 14 October 1796, except that he made at least two journeys: one to Wilmington, Delaware, where he was named vice-consul, and the other to the border of Tennessee, on which latter trip he gathered material of great scientific value: "500 kinds of seeds; two new quadrupeds, 15 birds, 20 or so reptiles; shells; about 30 fish; 150 zoophytes, worms, or mollusks; 1,200 insects; all these objects described and drawn from life" (Poiret in Lamarck, *Encyclopédie*, VIII [1808], 716–718).

The ocean voyage from Bordeaux to Charleston offered Bosc the opportunity for making several discoveries, notably of two genera: *Tentacularia*, a *Cestoda tetrarhynchus*, which was classed as a family by Poche, and *Oscana*, a mollusk.

In 1797 Bosc was named consul in New York. He never fulfilled his functions as consul, for President John Adams refused the *exaequatur*. His publications indicate that he was scientifically active. In November 1798, Bosc returned to Paris, where he married his cousin, Suzanne Bosc, the following year. He was made inspector of gardens and nurseries in 1803 and a member of the Institute in 1806. He concluded his career by succeeding Thouin as professor at the Muséum d'Histoire Naturelle in 1825.

Bosc's return to France was followed by the publication of a work of considerable importance, *The Natural History of Worms, Shellfish and Crustaceans* (1802), in which new species, and even an annelid genus, *Polydora*, are described.

Bosc never ceased to be interested in worms, and he was responsible for naming the genera *Capsala* (Platyhelminthes), a fish parasite that became the type of a family and even of an order; *Hepatoxylon* (1811), a genus later raised to a family by Dollfus (1940); *Dipodium* (1812), *Thalazia* (1819), and *Nematoda*.

Bosc devoted himself to science without thought of personal gain. The greater part of his collections, often containing descriptions and drawings made from life, were distributed among such specialists as J. C. Fabricius, G. A. Oliver, Latreille, Jean Daudin, Lacépède, and Jean Louis Poiret. It is known, for instance, according to Harper (1940), that he collaborated with Daudin in the latter's research on tree frogs and frogs, and that he discovered, described, drew, and named species from the Carolinas: *Hyla squirella, H. femoralis, H. lateralis, H. ovularis,* and *Rana clamitans.* The same is true of the turtles *Testudo odorata, T. reticularia,* and *T. serrata* and of the lizard *Stellio undulatus.* In arachnology and botany he left some unpublished papers that at the time would have brought to light a great number of new taxa. His arachnology of the Carolinas was published by Walckenaer (1805–1837); but a manuscript on the spiders of the forest of Montmorency (1793), in which eighty-nine species, most of them unknown until then, were described, named, and drawn, remained unpublished.

In botany Bosc discovered, defined, drew, and named numerous species, particularly some *Gramineae,* but most of his names, going unpublished, fell into synonymy. One of the most noteworthy species, published in 1807, is *Hydrocharis spongiosa* Bosc, a type of the American genus *Limnobium* L. C. Richard, which he had discovered in the Carolinas.

An accomplished naturalist who was important in his own time, Bosc was interested in the factual and practical side of science rather than in theories, although perhaps he was the first to point out the idea of "biological competition" in agriculture. He was a typical product of the eighteenth century, and he perhaps became too involved in encyclopedism. A man of good will and truth, Bosc probably will be remembered as a remarkable artisan of natural history and a pioneer in practical natural history.

BIBLIOGRAPHY

I. ORIGINAL WORKS. Bosc's numerous articles and notes appeared in *Transactions of the Linnean Society* (London); *Bulletin de la Société philomatique; Annales du Muséum d'histoire naturelle; Journal des mines; Journal de physique; Mémoires de l'Institut; Annales de chimie; Journal d'histoire naturelle;* and *Annales de l'agriculture française.* His first article was "Description de l'*Orthezia characias,*" in *Journal de physique,* **24** (1784), 171–173. The journal he kept while crossing Spain on foot, in the course of his return voyage from America to France, appeared as "Voyage en Espagne, à travers les royaumes de Galice, Léon, Castille vieille et Biscaye," in *Magazin encyclopédique,* **6** (1800), I, 448–493.

Bosc also was largely responsible for the last three volumes of the *Encyclopédie de Panckoucke* (Paris, 1813–1821) and collaborated on the *Nouveau dictionnaire d'histoire naturelle,* 24 vols. (Paris, 1803–1804; 2nd ed., 36 vols., 1816–1819) and the *Nouveau cours complet d'agriculture théorique et pratique,* 13 vols. (Paris, 1809; new ed., 16 vols., 1821–1823). His books include *Histoire naturelle des vers,* 3 vols. (Paris, 1802); *Histoire naturelle des coquilles,* 5 vols. (Paris, 1802); and *Histoire naturelle des crustacés,* 2 vols. (Paris, 1802). Outside science, Bosc published *Appel à l'impartiale postérité* (Paris, 1795), the memoirs of Mme. Roland, with introduction and notes by Bosc and letters from Mme. Roland to Bosc, written between 1782 and 1791.

Work left unpublished by Bosc—at least that signed by him—is represented by a set of MSS preserved in the library of the Muséum National d'Histoire Naturelle in Paris. It consists of the following: a 16-page notebook entitled "Araignées d'Amérique," in which there are descriptions and drawings of 25 species and five plates (MS 841); *Tableau des Aranéides* (Paris, 1805) and *Histoire naturelle des insectes* (Paris, 1837), both published by Walckenaer; a notebook entitled "Araignées de la forêt de Montmorency décrites et dessinées pendant que j'étais caché à Radegonde lors de la Terreur," dated Sept. 1793 (MS 872); a notebook of 100 pages (1788) containing a "Cenaculum insectorum," in the manner of Linnaeus, and descriptions of new species of insects; a mineralogical line drawing, and four-color wash drawing of the *Sphex scutellata* (MS 873); a "Flora Caroliniana," a catalog of the plants observed, with ecological, phenological, and geographical notes, descriptions of new species, principally of the *Gramineae* (these last pages were to be part of his "Agrostographie"), and ten pages dating from 1788, with descriptions of species from various places (MS 874); an "Agrostographie carolinienne," 54 leaves containing the description of *Hydrocharis spongiosa* and that of a plant of the Indies (1791), *Oryza aristata* (MS 875); a notebook of 120 original wash drawings of *Gramineae* and *Cyperaceae*—the illustrations for the "Agrostographie carolinienne" (MS 876); and a list of the seeds of 200 plants of the South China Sea and of Carolina (MS 569), given to the Museum in 1799.

There are also documents on Bosc in the archives of the Académie des Sciences, in particular a heliogravure portrait by Boilly (1821); in the archives of the Institute (Fonds Cuvier 3157); in the Laboratoire de Phanérogamie of the Muséum National (Paris); in the historical library of the city of Paris (MSS 1007, 1008, 1009), an autobiography and a journal of the voyage to the United States; and in the Archives Nationales in Paris (AJ 15, 569).

II. SECONDARY LITERATURE. No critical analysis of Bosc's work has ever been published. The eulogies or notices of Georges Cuvier (*Mémoires du Muséum national d'histoire naturelle,* **18** [1829], 69–92), A. F. de Silvestre (*Mémoires de la Société royale et centrale d'agriculture,* **1** [1829], lxxxi–cvii), and Degérando have no pretensions to being such; they have, however, the value of biographical documents by men who had been Bosc's friends. The memoir of Auguste Rey, "Le naturaliste Bosc. Un Girondin herborisant," in *Revue de l'histoire de Versailles et de Seine et Oise* (1900), 241–277 (1901), 17–42; and C. Perroud, "Le roman d'un Girondin," in *Revue du dix-huitième siècle,* **1** (1916), 57–77, can be read with pleasure and, since they rely on numerous unpublished documents, make an important contribution to our knowledge of Bosc and of his political and private life (his relations with the Rolands, his love for Mme. Roland and then for her daughter, Eudora). Scientific appraisals are in L. Berland, "Voyageurs d'autrefois et insectes historiques," in *Livre du centenaire de la Société entomologique de France* (1932), pp. 157–166; and

especially in Francis Harper, "Some Works of Bartram, Daudin, Latreille and Sonnini and Their Bearing Upon North American Herpetological Nomenclature," in *American Midland Naturalist,* **23,** no. 3 (1940), 692–723. The indications on the arachnological work of Bosc are given by P. Bonnet, in *Bibliographia araneorum,* I (Toulouse, 1945), 278.

JEAN-FRANÇOIS LEROY

BOSCH, CARL (*b.* Cologne, Germany, 27 August 1874; *d.* Heidelberg, Germany, 26 April 1940), *chemistry.*

The Bosch family was of Swabian peasant stock. Two sons had already left the farm by the second half of the nineteenth century, however. The younger of the two founded an electrotechnical firm that later became world-famous. The older of the two, Carl, opened a gas and plumbing equipment business.

The latter gave his name to his son, the eldest of his children. From his very young days, the boy showed a special talent for the natural sciences and technology, which his father ardently encouraged. The younger Bosch was not allowed to study chemistry until he had first completed a few semesters of training in an ironworks and then a few semesters of mechanical engineering. He then turned to chemistry in Leipzig, where, after four years, he obtained his doctorate. He applied for a position at the Badische Anilin- und Sodafabrik in Ludwigshafen and was accepted. Quite by chance, he touched upon the field that would subsequently become his lifework. Wilhelm Ostwald, a winner of the Nobel Prize, claimed to have found a process for obtaining ammonia from nitrogen and hydrogen by conduction over fine wire. Bosch was entrusted with supplementary testing and quickly found the error that Ostwald had made and the reasons for it. Thereupon he was given the task of studying the nitrogen question further. Only the use of atmospheric nitrogen was feasible. Use of the electric arc was not possible because of the high current requirements in Germany. Numerous experiments were carried out. New problems continually arose, but there were so many positive results that a breakthrough appeared near.

In 1909 Fritz Haber of Karlsruhe began work on the synthesis of ammonia, employing unusually high pressures and temperatures. Bosch undertook to transform Haber's laboratory experiments into large-scale technological ones, which in turn developed into a huge industry within five years. Haber's technically unsuitable catalysts had to be replaced. After thousands of experiments, iron with admixed alkaline material proved to be especially suitable. Equipment had to be built that would be capable of withstanding high pressures and temperatures. The furnaces, which were first heated from outside, lasted only a few days; the iron lost its carbon content, and thereby its steel properties, because of the hydrogen, brittle iron carbide being the result. Bosch devised a twin tube that allowed the hydrogen to escape through tiny openings. After numerous experiments, he found a solution to the heat problem whereby he introduced the uncombined gases into the furnace and, accordingly, produced an oxyhydrogen flame, the temperature of which could be regulated according to the quantity of oxygen added. The combination of nitrogen and metals was attempted. The work was conducted at enormous cost. The gases required for synthesis had to be obtained in an exceptionally pure state. The separation of the ammonia that had formed from the untransformed gases was extremely difficult. Bosch's technological know-how proved itself to be the most outstanding aspect of his work, and he was always personally active in all practical testing of his equipment.

Bosch also undertook another experiment in high-pressure engineering, an attempt to obtain urea from ammonium carbamate, and developed the methanol synthesis used in the manufacture of formaldehyde. Finally he tackled the problem of carbon hydrogenation and the production of synthetic rubber.

In 1925 Bosch became president of I. G. Farben, and in 1931 he shared the Nobel Prize with Bergius "for the discovery and development of chemical high-pressure methods."

He died after a long illness.

BIBLIOGRAPHY

I. ORIGINAL WORKS. Bosch's writings include "Verfahren zur Darstellung von Bariumoxyd und von Cyaniden," in *Chemisches Zentralblatt* (1907), 1999; "Verfahren zur Darstellung von Mono- und Dichlorhydrin aus Glycerin und gasförmiger Salzsäure," *ibid.* (1908), 1655; "Verfahren zur Darstellung von Aluminiumstickstoffverbindungen," *ibid.* (1912), 865; "Verfahren zur Herstellung von Ammoniak aus seinen Elementen mit Hilfe von Katalysatoren," *ibid.* (1913), 195; "Stickstoff in Wirtschaft und Technik," in *Naturwissenschaften,* **8** (1920), 867–868; "Entwicklung der chemischen Hochdrucktechnik bei dem Aufbau der neuen Ammoniakindustrie," in *Les prix Nobel en 1931* (Stockholm, 1933); and "Probleme grosstechnischer Hydrierungs-Verfahren," in *Chemische Fabrik,* **7** (1934), 1–10.

II. SECONDARY LITERATURE. For a discussion of Bosch's life and work, see K. Holdermann, "Carl Bosch, 1874–1940," in *Chemische Berichte,* **90** (1957), xix–xxxix; and Carl Krauch, "Carl Bosch zum 60 Gedächtnis," in *Angewandte Chemie,* **53** (1940), 285–288; see also the unsigned tribute "Carl Bosch zum 60 Geburtstag," *ibid.,* **47** (1934), 593–594.

GÜNTHER KERSTEIN

BOSE, GEORG MATTHIAS (*b.* Leipzig, Germany, 22 September 1710; *d.* Magdeburg, Germany, 17 September 1761), *electricity.*

Bose, a merchant's son, was educated at the University of Leipzig, where he concentrated on philosophy, mathematics, and languages. In 1727 he received the M.A. and joined the philosophy faculty as a junior lecturer (*Assessor*) in mathematics and physics. In 1738 he accepted the chair of natural philosophy (*Naturlehre*) at the University of Wittenberg, where he remained until 1760, when the Prussians carried him off to Magdeburg as a hostage of war.

Although at Leipzig Bose had written only on eclipses, sound, and the errors of physicians, he had begun to study electricity, inspired by the weak papers of J. J. Schilling in the *Miscellanea Berolinensa.* Through ignorance of Du Fay's work, however, he had not progressed far by 1738, as is apparent from his inaugural oration, which is interesting chiefly for its attack on action-at-a-distance. Probably at Wittenberg, and independently of C. A. Hausen of Leipzig, he revived Hauksbee's electrical machine, and added a "prime conductor" that greatly enhanced its power.

From 1742 to 1745, after close reading of Du Fay, Bose vigorously and successfully promoted the study of electricity in Germany, where it had never before been cultivated extensively. To this purpose he produced wonderful displays with his electrical machine, a German poem, a French tract, and several high-flown Latin commentaries. This mixture of literary polish and striking demonstration, this "writing sublimely of wonderful things," as a contemporary electrician put it, was Bose's great contribution, for he thereby indirectly brought about the central event in the early history of electricity, the discovery of the condenser (1745–1746). Musschenbroek began the experiments that culminated in the Leyden jar by repeating an impressive Bosean demonstration; and J. G. von Kleist very likely hit on his form of the condenser while attempting to ignite spirits by means of sparks, a recently successful experiment first urged by Bose. In theory Bose followed the Nollet system, much of which he invented independently.

Bose promoted himself as assiduously as he did electricity, urging his work on the royal societies of France and England, the Grand Mufti of Istanbul, and the Pope of Rome. His fulsome praises of Benedict XIV annoyed the Wittenberg theologians, whose rancor culminated, in 1750, in an attempt to expurgate a little tract of Bose's on eclipses (astronomy was his other scientific subject), raising a storm that ultimately involved Frederick the Great and the Royal Society of London.

BIBLIOGRAPHY

I. ORIGINAL WORKS. Bose's chief works on electricity are *Tentamina electrica in academiis regiis Londensi et Parisina primum habita omni studio repetita quae novis aliquot accessionibus locupletavit Georg Matthias Bose* (Wittenberg, 1744), a reprint of his inaugural oration and two other pamphlets, and *Recherches sur la cause et sur la véritable téorie de l'électricité* (Wittenberg, 1745), which contains his Nolletesque theory. Poggendorff, and Jöcher, *Algemeines gelehrten Lexikon,* I (Leipzig, 1784), cols. 2098–2099, give bibliographies of Bose's printed works; his manuscripts apparently were lost in the Thirty Years' War.

II. SECONDARY LITERATURE. For a biography of Bose, see Bose's *Tentamina,* especially pp. 48–53; Jöcher, *loc. cit;* and A. Mercati, "Il fisico tedesco Giorgio Mattia Bose e Benedetto XIV," in *Acta pontificiae academiae scientiarum,* **15** (1952), 57–70. For Bose's works on electricity, see J. L. Heilbron, "G. M. Bose: The Prime Mover in the Invention of the Leyden Jar?," in *Isis,* **57** (1966), 264–267; and Joseph Priestley, *The History and Present State of Electricity,* 3rd ed. (London, 1775), I, 87–88, 93–94.

JOHN L. HEILBRON

BOSE, JAGADIS CHUNDER (*b.* Mymensingh, Bengal, India [now East Pakistan], 30 November 1858; *d.* Giridih, Bengal, India, 23 November 1937), *physics, comparative physiology.*

The son of a deputy magistrate, Bose studied at St. Xavier's, a Jesuit college in Calcutta, and then went to London to study medicine. He transferred to Cambridge University after receiving a scholarship to Christ's College and graduated in natural science in 1884. He was immediately appointed to the professorship of physics at Presidency College, Calcutta, where he remained until his retirement in 1915.

Bose first attracted worldwide attention in 1895 with his meticulous experiments on the quasi-optical properties of very short radio waves, which led him to design and construct some fine generating apparatus. His improvements in the coherer, a tube of iron filings widely used as an early form of radio detector, were of both scientific and technological importance, and led him to formulate a more general theory of the properties of contact-sensitive substances that figures in the history of solid-state physics.

Bose was struck by the way in which the responses of certain inorganic substances to various stimuli resembled biological response. That observation led him to compare the behaviors of animal and plant tissue, a study that occupied him for the rest of his life. His papers and lectures on these subjects fell short of general acceptance, however. In 1901 and again in 1904, his papers were rejected by the Royal Society, partly because of the philosophical terms in which they were couched. Today, when biophysics is a generally recognized discipline and comparative physiology rests on a more scientific basis, the idea that animal and plant tissues exhibit similar responses seems less controversial and may even be taken as foreshadowing Norbert Wiener's cybernetics. Bose aroused general admiration, however, for the extremely sensitive automatic recorders he devised to measure plant growth with great precision and for the way in which he used them to accumulate records of microscopic changes caused by various stimuli.

Bose was knighted in 1917 and was elected a fellow of the Royal Society in 1920, the first Indian physicist so honored. In 1915 he had retired from government service to organize the Bose Research Institute, which he founded in 1917, largely with his own considerable fortune but also with contributions from private well-wishers and from the government of India. He was a great friend of another famous Bengali, the Nobel Prize-winning author Rabindranath Tagore. Bose was happily married for fifty years to Abala Das, daughter of the Calcutta lawyer and political leader Durga Mohan Das.

BIBLIOGRAPHY

I. ORIGINAL WORKS. Bose's books are *Response in the Living and Non-living* (London, 1902); *Plant Response as a Means of Physiological Investigation* (London, 1906); *Electro-physiology: A Physico-physiological Study* (London, 1907); *Researches on Irritability of Plants* (London, 1913); *Life Movements in Plants* (Calcutta, 1918); *The Physiology of the Ascent of Sap* (London, 1923); *The Physiology of Photosynthesis* (London, 1924); *Mechanism of Plants* (London, 1926); *Plant Autographs and Their Revelations* (London, 1927); and *Tropic Movement of Plants* (London, 1929).

His *Collected Physical Papers* (London, 1927) were compiled by Bose himself.

II. SECONDARY LITERATURE. A contemporary biography is Sir Patrick Geddes, *The Life and Work of Sir Jagadis C. Bose* (London, 1920). An obituary by M. N. Saha appears in *Obituary Notices of Fellows of the Royal Society,* **3** (1940), 2–12. An appreciation by S. K. Mitra is in *Journal of the British Institution of Radio Engineers,* **18** (1958), 661; an announcement of an annual award named in Bose's honor appears in an earlier issue of the same journal: **13** (1953), 130.

CHARLES SÜSSKIND

BOŠKOVIĆ, RUDJER J. (*b.* Dubrovnik, Yugoslavia, 18 May 1711; *d.* Milan, Italy, 13 February 1787), *natural philosophy, mathematics, astronomy, physics, geodesy.*

Bošković was perhaps the last polymath to figure in an important way in the history of science, and his career was in consequence something of an anachronism and presents something of an enigma. He stands between the natural philosophy of Newton and Leibniz at one extreme and Faraday and field theory at the other, but too far from both for the connection either forward or backward to appear a coherent one. A somewhat isolated figure, he belonged to no definite eighteenth-century tradition. Croatian by birth, he became a Jesuit; and like many intellectuals from the Dalmatian cities, he was drawn to Italy and lived the first part of his career in Rome. A man of the Enlightenment, he sometimes gives the effect of a Renaissance scholar moving about Europe from place to place for reasons of circumstance and patronage and departing on great journeys at critical junctures. As will appear from consulting his bibliography, he published in the mode of an earlier time. He wrote treatises on whole sciences, and at certain periods in his life composed several such works in the course of a year. Nevertheless, his reputation has been rather that of a forerunner than a survival. His doctrine of atomism which modified the massy corpuscles of Newtonian natural philosophy into immaterial centers of force appeared to foretell, and there are historical reasons to believe that it actually influenced, the basic position of nineteenth-century field physics in regard to the relations between space and matter.

Life. Bošković was the son of Nikola Bošković, a merchant of Dubrovnik, and Paula Bettera, the daughter of Bartolomeo Bettera, a merchant originally from Bergamo, Italy. The family was of average means and was noted for its literary interests and accomplishments. Bošković began his education in the Jesuit college of Dubrovnik and continued it in Rome, first at the novitiate of Sant'Andrea, which he entered in 1725 at the age of fourteen, and later at the Collegium Romanum. He was extraordinarily sharp of mind, comprehensive in intelligence, and tireless in application—in short, an outstanding student. He learned science in a way characteristic of his later career, through independent study of mathematics, physics, astronomy, and geodesy. In 1735 he began studying Newton's *Opticks* and the *Principia* at the Collegium Romanum, where he made himself an enthusiastic propagator of the new natural philosophy. The exact sciences were what always appealed to him—in the first instance mathematics. In 1740,

although he had not yet completed his theological studies, he was appointed professor of mathematics at the Collegium Romanum. That event largely determined the course of his career. Teaching interested him in its methods as well as for its content. In this respect, as in others, his spirit was progressive. He published a textbook of his teaching in 1754—*Elementa universae matheseos*—of which the third and final volume contains an original theory of conic sections.

During this period of his life Bošković undertook, as was customary among qualified clergymen of his time, several practical and diplomatic commissions for lay or ecclesiastical authorities. The cupola of St. Peter's having developed alarming fissures, a commission was appointed consisting of Bošković and two fellow "mathematicians," F. Jacquier and Th. Le Seur, to investigate the causes and make recommendations. Bošković drafted the report which, analyzing the problem in theoretical terms, achieved—despite certain errors—the reputation of a minor classic in architectural statics. Thereafter the papal government entrusted the planning for draining of the Pontine marshes to Bošković. He composed a series of memoirs on the practice of hydraulic engineering, on regulation of the flow of the Tiber and other streams, and on harborworks. He did a plan for the harbor at Rimini in 1764 and for that at Savona in 1771.

Archaeology also interested Bošković. In 1743 he discovered and excavated an ancient Roman villa above Frascati in Tusculum, and in 1746 published a description of a sundial that had been among the finds. In 1750 he also published a critical study of the Augustan obelisk in the Campo Marzio. In 1757 Bošković undertook the most important of his several diplomatic missions, representing the interests of the Republic of Lucca before the Hapsburg court in Vienna in a dispute with Tuscany over water rights. He won the case, and in the intervals of tending to its ramifications, he also while in Vienna completed his major work in the field of natural philosophy, *Philosophiae naturalis theoria,* which appeared in the autumn of 1758.

As the years went by, Bošković fell out of sympathy with certain policies of his ecclesiastical superiors. He resented their rejection of proposals he had advanced looking to the modernization of education both in method and in subject matter. He disliked the Vatican's reaction to the persecution of his order in Portugal. He was disappointed by the negative attitude that a number of Jesuit philosophers—Peripatetics he thought them to be—adopted toward his own system of natural philosophy. It seemed time for a move. The Academy of Sciences in Paris had long since elected him to corresponding membership—he was correspondent of Dortous de Mairan—on the publication in 1738 of his discourse on the aurora borealis. His superiors gave him permission, and in 1759 Bošković set off on his travels, going first to Paris.

There he remained for six months, well received in aristocratic, scientific, and literary circles. He came to know members of the Academy of Sciences at first hand. A diplomatic intervention on behalf of his native city of Dubrovnik took him to the court at Versailles. He decided not to remain in Paris, however, and in 1760 crossed over to London, where again his reputation had preceded him among literary and scientific people. He had discussions with representatives of the Church of England; met Benjamin Franklin, who showed him electrical experiments; and visited Oxford and Cambridge. On 15 January 1761 the Royal Society elected him a fellow, and in recognition of the honor, he dedicated to it a poem on eclipses of the sun and moon. He then lent his weight to efforts to persuade the Society to organize an expedition for the purpose of observing the transit of Venus in June 1761.

Bošković had planned to make such observations himself in Istanbul but, dependent in his plans on a companion, Correr, the new Venetian ambassador to Istanbul, Bošković arrived too late for observation. He made a trip through Flanders, Holland, the court of Stanislas in Nancy, and various centers in Germany. Once in Istanbul, he fell dangerously ill and had to remain there for seven months of recuperation. Partially recovered he set off again, this time in the company of the British ambassador, and traveled through Bulgaria and Moldavia, and went on alone from there to Poland. In Warsaw he was received in ecclesiastical and diplomatic circles. The Czartoryski and the Poniatowski connections took him up. His *Diary* of the trips he made through Bulgaria and Moldavia amounts to a systematic description of the country. It was published in Italian in 1784, having already been translated into French and German. From Poland, finally, he returned to Rome—by way of Silesia, Austria, and Venice—arriving there in November 1763 after an absence of over four years which marked a stage in his life.

Back in Italy, Bošković found a situation in Pavia, where at the end of the year he won election as professor of mathematics at the university, revived under Austrian administration. He organized both his own lectures and his department realistically, with an emphasis upon applied mathematics. At Pavia he concentrated his own efforts mainly in the field of optics and the improvement of telescopic lenses, and

played a leading role in the organization of the Jesuit observatory at Brera near Milan in 1764. Had his program been carried out and the instruments he advocated installed, the observatory would have been one of the most elaborate in Europe. Remembering his interest in the transits of Venus, the Royal Society invited him to lead an expedition to California for the purpose of observing the second of the famous pair of transits, that of 1769. Unfortunately political conditions prevented that trip. In 1770 he moved his work to the department of optics and astronomy at the Scuole Palatine in Milan. As time went on, he provoked opposition among his colleagues at the observatory. In 1772 the court in Vienna yielded to the demands of the majority and relieved Bošković of "concern" for the observatory. In despair he resigned his professorship also. All his world was dissolving: the next year, 1773, the pope suppressed the Jesuit order.

By now Bošković was in his sixty-third year. Influential friends urged him to repair to Paris. There a post was arranged for him as director of optics for the navy, and he even became a subject of the French crown. In Paris during this, the last productive period of his life, he worked mainly on problems of optics and astronomy. It may be that his nature was a little contentious, for there too disputes attended him, one with the young Laplace over Bošković's early method (1746) of determining the path of a comet, another with the Abbé Alexis de Rochon over priorities in the invention of a type of micrometer and megameter consisting of pairs of rotating prisms. The device became important in the design of geodetic telemeters. In search of health and tranquillity, Bošković spent the greater part of each year in the country residing at the estates of one or another of his friends.

In 1782 Bošković received leave to return again to Italy in order to ready his French and Latin manuscripts for the press. He settled in Bassano, and there in 1785 the printing firm of the brothers Remondini brought out his five-volume *Opera*. The preparation of those writings and the strain of proofreading told on Bošković's health. Once again he set out to travel, although only in Italy, in order to visit old friends. He found a cordial welcome in Milan, where former opponents were inclined to let bygones be bygones, and settled down in the Brera observatory, which he had founded, to work on the notes for the third volume of Benedict Stay's poem *Philosophiae recentioris versibus traditae libri X,* on Newtonian natural philosophy. His mental powers were leaving him, however. Forgetfulness, anxiety, fear for his scientific reputation grew upon him, and it was clear that his mind was failing. He mercifully died of a lung ailment before the decline reached an extreme and was buried in the church of Santa Maria Podone in Milan, where, however, all trace of his tomb has been lost.

Bošković's interests were more manifold than was at all normal, even in the eighteenth century, for one who participated deeply in the actual work of science. For purposes of clarity, they may be grouped under the headings of the instrumental sciences of astronomy, optics, and geodesy, and the abstract subjects of mathematics, mechanics, and natural philosophy. It must be appreciated, however, that such a classification is a mere convenience. Bošković's work in the former trio exhibited a consistent penchant for the invention or improvement of instruments of observation as well as for recognition and compensation of procedural errors. In the second, theoretical set of sciences, his writings develop a highly individual point of view. All his work, finally, may be read as physical essays in the working out of an epistemology and metaphysic that styled his career in a way, again, not at all characteristic of his century.

Instrumental Sciences. Bošković's earliest (1736) publication was a description of methods for the determination of the elements of the rotation of the sun on its axis from three observations of a single sunspot. In 1737 there followed an exposition of a graphical method for the resolution on a plane of problems in spherical trigonometry and the treatment of an actual problem in the transit of Mercury. In 1739, two years after the treatise on the aurora borealis, Bošković published an account of the principle of the circular micrometer based on the idea that the circular aperture of the objective may serve for determination of the times at which a celestial body enters and leaves the field of vision of a telescope; these values, when compared with those of a known star, give the relative positions of the two bodies.

From these specific matters, Bošković turned his attention in astronomy to a comprehensive survey of the theoretical foundations and instrumental practice and resources of practical, observational astronomy, and in the years 1742 through 1744 he published a series of works that deal with these matters in a spirit of *severioris critices leges.*

Thereafter, Bošković took up the study of comets. A widely read work of 1746 offered his opinions on a number of questions concerning the nature of comets. In it he proposed his first method—that much later criticized adversely by Laplace—for the determination of parabolic orbits. The procedure was essentially similar to that afterward introduced by J. H. Lambert (1761). Bošković's method, developed in

Volume III of his *Opera pertinenta ad opticam et astronomiam* (1785), comes close to the classic method of H. W. Olbers (1797). An interesting treatise of 1749 concerns the determination of an elliptical orbit by means of a construction previously employed for resolving the reflection of a light ray from a spherical mirror. Bošković employed this method again in 1756, in a treatise discussing the reciprocal perturbations of Jupiter and Saturn, which he entered in a competition on the subject set by the Academy of Sciences in Paris. The winner was Leonhard Euler; Bošković received an honorable mention.

Bošković's interest in optics seems to have developed in the first instance out of his astronomical concerns. As early as 1747 he was discussing the tenuity, or rarity, of sunlight, apparently with the old question in mind of the hypothetical materiality of light, and at the same time attempted to estimate the density of a solar atmosphere, supposing it to reach as far as the earth. Having reflected on the problems of light, Bošković published in 1748 a treatise (in two parts) of a broadly critical nature. The central Newtonian positions in optics did not at all appear to him to be securely established. It is perhaps the most interesting feature of his critical attitude that he regarded rectilinear propagation as an unproved hypothesis, a question on which he dwelt in detail. Some other aspects of optical phenomena he thought hidden or unclear even after Newton's discoveries. Discussing phenomena of parallax, he drew attention to the distance of fixed stars in dimensions of light-years. He formulated, and was the first to do so, a general photometric law of illumination and enounced the law of emission of light known under Lambert's name. He was critical of Newton's account of colors arising from the passage of light through thin plates involving the ether and periodicity, and he provided an alternative interpretation in the spirit of his own theory of natural forces, of which more below.

In his later years at Pavia and at the observatory in Brera, he concentrated his attention on the improvement of lenses and optical devices. A series of five discourses on dioptrics (1767) treats of achromatic lenses and offers an impressive example of Bošković's experimental dexterity and accuracy, most notably in respect of measurements of the reflection and dispersion of light by means of his vitrometer. Having confirmed that two-lens arrangements will recombine only two spectral colors, he recommended a composition involving three or more lenses. He also stressed the importance of the eyepiece in achromatic telescopes. In the actual fabrication of lenses he worked with Stephen Conti of Lucca, who manufactured them according to his specifications and assisted in performing the optical experiments.

At Brera he worked intensively on methods for verifying and rectifying astronomical instruments and improved or invented a number of them, of which accounts later appeared in Volume IV of his *Opera*. Perhaps the most ingenious were a leveler that determined the plane of the edge of a quadrant and a micrometric wedge. To ensure that the border of an astronomical quadrant would be on the same level as the plane passing through the center, Bošković made use of a sort of surveying device. In a canal filled with water leading around the border of the quadrant and along one of its radii floated a small boat with a wire mast hooked at the top. Its point nearly touched the border of the instrument, permitting the measurement of small distances between the point and the water level, thus revealing the true form of the so-called plane of the quadrant. Bošković's micrometric wedge is a metallic wedge truncated on the thin side, which he used to measure the distance between two planes by inserting it into the opening between them and noting the corresponding number on the scale engraved on its side. He also thought that it ought to be possible to decide between Newton's emission theory of light and the wave theory by observations of aberration of light from the fixed stars, first through an ordinary telescope and then through a telescope filled with water. It was his further prediction that observation would detect an aberration of light from terrestrial sources: in these matters research in the nineteenth century failed to confirm his expectations.

Bošković's work in meteorology and geophysics was closely related to astronomical concerns. In 1753 he advanced the idea that the moon was probably enveloped not by an atmosphere like that of the earth, but rather by a concentric layer of some homogeneous, extremely transparent fluid. As to our own atmosphere and its behavior, or misbehavior, he investigated a tornado that devastated Rome in June 1749 and attempted to interpret its effects in terms of Stephen Hales's theory of "fixed air"—it was ever his way to try connecting phenomena in one domain with famous developments in the science of another; his mind ranged over the whole of physical science with more or less cogency, but never without imagination.

It was his idea that mountains had originated from the undulation of rock strata under the influence of subterranean fires, and developing this notion in 1742 led him to the concept of compensation of strata, which could be taken as basis of the later theory of isostasy. He also conceived the idea of a kind of gravimeter for measuring gravitation even in the ocean. At the same time, he proposed a method for determining the mean density of the earth by measuring the incremental attraction of masses of water at high tide by the deviation of a pendulum situated in the proximity.

Early in his career his interest was drawn to the problem of the size and shape of the earth, an issue intensively discussed in the first half of the eighteenth century, since its resolution was thought to be crucial in an eventual choice between a Cartesian cosmology of vortices, which predicted an earth slightly elongated at the poles, and a Newtonian one of inertial motion under attractive forces, in which case the globe should be slightly flattened. In 1739 Bošković initiated a critical investigation of existing measurements of the length of a degree along the meridian. It appeared to him that in addition to cosmological effects, superficial inequalities and irregularities of structure and density beneath the surface might well affect and distort measurements of distance along a meridian, modify the length of the second pendulum at a given locality, and bias the apparent direction of the vertical.

Bošković always promoted international cooperation in geodesy. On his initiative, meridians were measured in Austria, Piedmont, and Pennsylvania, and he himself readily collaborated with an English colleague, Christopher Maire, rector of the English Jesuit College in Rome, in surveying the length of two degrees of the meridian between Rome and Rimini. The onerous work took three years. Its results confirmed, among other things, the geodetic consequences of unevenness in the earth's strata, the possibility of determining surface irregularities by such measurements, as well as the deviation of meridians and parallels from a properly spherical shape. The report on these measurements came out in Rome at the end of 1755. Bošković employed novel methods for measuring the base line in his surveys, and he developed an exact theory of errors and learned to employ his instruments to the most accurate effect. The earliest device for verifying the points of division on the edge of such an instrument originated with Bošković, who determined from the inequalities of their chords that the circular arcs on the border of the instrument, although theoretically equal, were not in fact so. Having determined errors of division corresponding to 60° by comparing the chord with the radius of the instrument, he proceeded by bisecting to angles of 30°, 15°, and finally 5°. The method of compensating errors being applicable to astronomical as well as geodetic observations, he took an important step toward the newer practical astronomy, which for most astronomers begins with Friedrich Bessel. In the French edition of his report on measurements of the meridian, Bošković included the first theory of the combination of observations based on a minimum principle for determining their most suitable values, making use of absolute values instead of their squares, as Gauss later did in his classical method of least squares.

Abstract Sciences. Science in general took its lead in physics from Newton and in mathematical analysis from Leibniz, and it was at the root of Bošković's idiosyncrasy that, whether deliberately or not, he took the opposite tack in both respects. Mathematics had always attracted him. Instead of the calculus as developed by the great analysts among his own contemporaries—d'Alembert, the Bernoullis, and Euler—he preferred the geometric method of infinitely small magnitudes "which Newton almost always used," as he said, and which embodied the "power of geometry." He particularly applied it to problems of differential geometry, terrestrial and celestial mechanics, and practical astronomy. In 1740 he studied the properties of osculatory circles, and in 1741 devoted an entire treatise to the nature of the infinitely great and small magnitudes employed in that method. He relied upon it also in a few problems of classical mechanics: in 1740 he studied the motion of a material point and in 1743 was the first to solve the problem of the body of greatest attraction.

In mechanics (as in optics), however, his allegiance to Newton was qualified. True, he annotated Stay's elegant Latin verses on Newtonian natural philosophy, the *Philosophiae,* published in Rome in three volumes, the first in 1755. Nevertheless, his heterodoxy in mechanics began to be apparent at least as early as 1745, when he published an important discourse on the subject of living force (*vis viva*). He there put forward the view that the speed of a movement is to be computed from the *actio momentanea* of the force that generates it. Attacking the problem of the generation (*generatio*) of velocity in a new way, by distinguishing between actual and potential velocity and by introducing subtle conceptions in connection with the notion of force, he reduced the famous debate over the true measure of force, whether it be momentum (*mv*) or *vis viva* (mv^2), to the status of a mere argument "over titles." This discourse contained the first statement of Bošković's universal force law.

That law was inspired partly by Leibniz' law of continuity and partly by the famous thirty-first query with which Newton concluded the fourth and final edition of his *Opticks.* There Newton raised speculatively the question whether there might not exist both attractive and repulsive forces alternately operative between the particles of matter. From this idea Bošković proceeded by way of an analysis of collision of bodies to the enunciation of a "universal law of forces" between elements of matter, the force being alternately attractive or repulsive, depending upon the distance by which they are separated. As that distance diminishes toward zero, repulsion predominates and grows infinite so as to render direct contact between particles impossible. A fundamental role is played by the points of equilibrium between the attractive and repulsive forces. Bošković called such points "boundaries" (*limes,* the Latin singular). Some of them are points of stable equilibrium for the particles in them and others are points of unstable equilibrium. The behavior of these boundaries and the areas between them enabled Bošković to interpret cohesion, impenetrability, extension, and many physical and chemical properties of matter, including its emission of light.

It was because of its consequences for the constitution of matter that the law of forces was particularly important. In Bošković's natural philosophy the "first elements" of matter became mere points—real, homogeneous, simple, indivisible, without extension, and distinguished from geometric points only by their possession of inertia and their mutual interaction. Extended matter then becomes the dynamic configuration of a finite number of centers of interaction. Many historians have seen in Bošković's derivation of matter from forces an anticipation of the concept of the field, an anticipation still more clearly formulated very much later by Faraday in 1844. Matter, then, is not a continuum, but a discontinuum. Mass is the number of points in the volume, and drops out of consideration as an independent entity. In the special case of high-speed particles, Bošković even envisaged the penetrability of matter.

The principle of inertia itself did not escape his criticism. It was impossible in his view to prove it or indeed to prove any metaphysical principle to be true of physical reality a priori. But neither could it be proved a posteriori as Newtonians were wont to do from "the phenomenon of movement." Bošković emphasized the necessity of defining the space to which the principle relates. Since he held it to be impossible to distinguish absolute from relative motion by direct observation and without invoking "unproven physical hypotheses," he introduced the notion that inertia as it is observed is relative to a space chosen to include all the bodies in the universe that are within range of our senses, i.e., all the subjects of all our experiments and observations. The translation of that space as a whole can have no effect on the motion of a body within it, on its rotation at a given angle, or on its contraction or dilation if there is a simultaneous and equivalent contraction or dilation of the scale of forces. From these considerations Bošković concluded that experiment and observation could never decide whether inertia is relative or absolute.

It must not be supposed, however, that his natural philosophy represents a reversion to a Leibnizian metaphysic. He was in fact as skeptical and critical of the principle of sufficient reason or final causes as of that of inertia. In general Bošković was convinced that we know nothing so far as the absolute is concerned and just as little of what is relative. He often emphasized the impotence of the human mind, and spoke more than once of the imaginability of beings with a geometry different from ours. He had a clear understanding of the hypothetical-deductive nature of geometry, especially of the Euclidean fifth postulate about parallels. In his view our universe is no more than a grain of sand in a horde of other universes. There might well be other spaces quite unconnected to our own and other times that run some different course.

Sharp in thought, bold in spirit, independent in judgment, zealous to be exact, Bošković was a man of eighteenth-century European science in some respects and far ahead of his time in others. Among his works are writings that still repay study, and not only from a historical point of view.

BIBLIOGRAPHY

I. ORIGINAL WORKS. Bošković's more important treatises and works include *De maculis solaribus* (Rome, 1736); *De Mercurii novissimo infra solem transitu* (Rome, 1737); *Trigonometriae sphaericae constructio* (Rome, 1737); *De aurora boreali* (Rome, 1738); *De novo telescopii usu ad objecta caelestia determinanda* (Rome, 1739); *Dissertatio de telluris figura* (Rome, 1739); *De circulis osculatoribus* (Rome, 1740); *De motu corporum projectorum in spatio non resistente* (Rome, 1740); *De inaequalitate gravitatis in diversis terrae locis* (Rome, 1741); *De natura, & usu infinitorum & infinite parvorum* (Rome, 1741); *De annuis fixarum aberrationibus* (Rome, 1742); *De observationibus astronomicis, et quo pertingat earundem certitudo* (Rome, 1742); *Disquisitio in universam astronomiam* (Rome, 1742); *Parere di tre mattematici sopra i danni, che si sono trovati nella cupola di S. Pietro sul fine dell'anno MDCCXLII, dato per ordine di Nostro Signore Papa Benedetto XIV* (Rome, 1742).

Later works are *De motu corporis attracti in centrum immobile viribus decrescentibus in ratione distantiarum reciproca duplicata in spatiis non resistentibus* (Rome, 1743);

"Problema mecanicum de solido maximae attractionis solutum a P. Rogerio Josepho Boscovich," in *Memorie sopra la fisica e istoria naturale di diversi valentuomini,* **1** (Lucca, 1743), 63–88; *Nova methodus adhibendi phasium observationes in eclipsibus lunaribus ad exercendam geometriam, et promovendam astronomiam* (Rome, 1744); *De viribus vivis* (Rome, 1745); *De cometis* (Rome, 1746); *Dissertatio de maris aestu* (Rome, 1747); *Dissertazione della tenuità della luce solare* (Rome, 1747); *Dissertationis de lumine pars prima* (Rome, 1748); *Dissertationis de lumine pars secunda* (Rome, 1748); *De determinanda orbita planetae ope catoptricae ex datis vi, celeritate et directione motus in dato puncto* (Rome, 1749); *Sopra il turbine, che la notte tra gli 11 e 12 di giugno del 1749 daneggiò una gran parte di Roma* (Rome, 1749).

During the 1750's Bošković wrote *De lunae atmosphaera* (Rome, 1753); *De continuitatis lege et ejus consectariis pertinentibus ad prima materiae elementa eorumque vires* (Rome, 1754); *Elementa universae matheseos,* 3 vols. (Rome, 1754); *De lege virium in natura existentium* (Rome, 1755); *De litteraria expeditione per Pontificiam ditionem ad dimetiendos duos meridiani gradus et corrigendam mappam geographicam jussu, et auspiciis Benedicti XIV. Pont. Max. suscepto a Patribus Societ. Jesu Christophoro Maire et Rogerio Josepho Boscovich* (Rome, 1755), trans. into French as *Voyage astronomique et géographique dans l'état de l'Église, entrepris par l'ordre et sous les auspices du Pape Benoît XIV, pour mesurer deux degrés du méridien et corriger la carte dans l'état ecclésiastique par les PP. Maire et Boscovich, traduit du latin* (Paris, 1770); *De inaequalitatibus, quas Saturnus et Jupiter sibi mutuo videntur inducere, praesertim circa tempus conjunctionis* (Rome, 1756); "De materiae divisibilitate, et principiis corporum" (1748), in *Memorie sopra la fisica . . .,* IV (Lucca, 1757); *Philosophiae naturalis theoria redacta ad unicam legem virium in natura existentium* (Vienna, 1758).

Bošković's works in his last quarter-century include *De solis ac lunae defectibus libri V* (London, 1760); *Dissertationes quinque ad dioptricam pertinentes* (Vienna, 1767); *Les éclipses* (Paris, 1779); *Giornale di un viaggio da Constantinopoli in Polonia, dell'Abate R. G. Boscovich* (Bassano, 1784); and *Rogerii Josephi Boscovich Opera pertinentia ad opticam et astronomiam maxima ex parte nova et omnia hucusque inedita in V tomos distributa* (Bassano, 1785).

II. SECONDARY LITERATURE. The earlier works are F. Ricca, *Elogio storico dell'abate Ruggiero Giuseppe Boscovich* (Milan, 1789); M. Oster, *Roger Joseph Boscovich als Naturphilosoph,* dissertation (Cologne, 1909); V. Varićak, "L'oeuvre mathématique de Bošković," in *Rad* (Zagreb), **181, 185, 190, 193** (1910–1912), condensed by Ž. Marković, in *Bulletin des travaux de la classe des sciences mathématiques et naturelles de l'Académie yougoslave de Zagreb,* **1** (1914), 1–24.

See also D. Nedelkovitch, *La philosophie naturelle et relativiste de R. J. Boscovich* (Paris, 1922); *A Theory of Natural Philosophy, Put Forward and Explained by Roger Joseph Boscovich, S. J.,* Lat.-Eng. ed. with trans. by J. M. Child from 1st Venetian ed. (1763), with a short life of Bošković (Chicago–London, 1922); V. Varićak, "Latin-English Edition of Bošković's Work *Theoria philosophiae naturalis,*" in

Bulletin des travaux de la classe des sciences mathématiques naturelles de l'Académie de Zagreb, **19-20** (1923-1924), 45–102; L. Čermelj, "Roger Joseph Boscovich als Relativist," in *Archiv für Geschichte der Mathematik, der Naturwissenschaft und der Technik, 2,* no. 4 (1929), 424–444; and H. V. Gill, *Roger Boscovich, Forerunner in Modern Physical Theories* (Dublin, 1941).

A bibliography of publications on Bošković in English, French, German, and Italian up to 1961 is in L. L. Whyte, ed., *Roger Joseph Boscovich, S.J., F.R.S., 1711–1787, Studies of His Life and Work on the 250th Anniversary of His Birth* (London, 1961), which contains articles on Bošković by E. Hill, L. L. Whyte, Ž. Marković, L. P. Williams, R. E. Schofield, Z. Kopal, J. F. Scott, C. A. Ronan, and Churchill Eisenhart. G. Arrighi has published newly found correspondence between Bošković and G. A. Arnolfini of Lucca in *Quaderni della rivista "La provincia di Lucca,"* **3** (1963), **5** (1965), *Studi scientifici;* and correspondence between Bošković and G. A. Slop, the Pisan astronomer, in *Studi trentini di scienze storiche,* **43,** no. 3 (1964), 209–242; he also published a study on Bošković's good friend Conti, "Scienziati lucchesi del settecento: Giovan Stefano Conti," in *La provincia di Lucca,* **2,** no. 3 (July–Sept. 1962), 31–44. R. Hahn, "The Boscovich Archives at Berkeley," in *Isis,* **56,** no. 183 (Spring 1965), 70–78, reports on the literary legacy of Bošković that has been at the University of California since 1962. Of the recent articles, we should mention P. Costabel's "Le *De viribus vivis* de R. Boscovic ou de la vertu des querelles de mots," in *Archives internationales d'histoire des sciences,* **14,** nos. 54–55 (Jan.–June 1961), 3–12.

The literature on Bošković in Yugoslavia is abundant. On the occasion of the centenary of his death, the Yugoslav Academy of Arts and Sciences in Zagreb issued a collection of works on Bošković in its publication *Rad,* **87, 88,** and **90** (1887–1888), including his correspondence from the archives of the observatory at Brera as transcribed by G. V. Schiaparelli. The latter correspondence was reprinted in *Publicazioni del R. Osservatorio astronomico di Milano–Merate,* n.s. no. 2 (1938). *Gradja za život i rad Rudjera Boškovića* ("Material Concerning the Life and Work of Rudjer Bošković"), 2 vols. (Zagreb, 1950–1957), is a separate publication of the Yugoslav Academy of Arts and Sciences.

Other publications on Bošković in languages other than Serbo-Croatian are in *Actes du symposium international R. J. Bošković 1958* (Belgrade–Zagreb–Ljubljana, 1959); *Actes du symposium international R. J. Bošković 1961* (Belgrade–Zagreb–Ljubljana, 1962); and *Atti del convegno internazionale celebrativo del 250° anniversario della nascita di R. G. Boscovich e del 200° anniversario della fondazione dell' Osservatorio di Brera* (Milan, 1963).

Studies on Bošković have made considerable advances in Yugoslavia, as shown by the works of V. Varićak, in the Yugoslav Academy's *Rad;* B. Truhelka, in various reviews, based on unpublished material on Bošković, especially correspondence with his brothers; S. Hondl, in *Almanah Bošković* ("The Bošković Almanac") of the Croatian Society of Natural Science; J. Majcen; Ž. Marković; Ž. Dadić; D. M. Grmek; and others. Mention

should also be made of D. Nedeljković's numerous articles in reviews, as well as in the publications of the Serbian Academy of Arts and Sciences in Belgrade; and of the works of S. Ristić, D. Nikolić, and others. A comprehensive general bibliography up to 1956 can be found in "Bošković," in *Enciklopedija Jugoslavije* ("The Encyclopedia of Yugoslavia"), II (Zagreb, 1956).

Željko Marković

BOSS, LEWIS (*b.* Providence, Rhode Island, 26 October 1846; *d.* Albany, New York, 5 October 1912), *positional astronomy.*

Boss, who was honored by the Royal Astronomical Society for his "long-term work on the positions and proper motions of fundamental stars," had little, if any, academic training for this work. As a student at Dartmouth College he followed a classical course, but also frequented the observatory, where he learned to handle astronomical instruments and to reduce observations. After graduation he worked as a clerk in various government offices in Washington, D.C., and frequented the U.S. Naval Observatory, from which he borrowed small astronomical instruments.

In 1872 Boss was appointed assistant astronomer for the survey of the 49th parallel, between the United States and Canada; his job during the next four years was to locate, by celestial observations, latitude stations from which the surveyors could work. His observations with a zenith telescope led him to realize that latitude determinations can be no more accurate than the stellar declinations on which they are based. Therefore, while the survey was still in progress, Boss developed a homogeneous system of declinations, as free as possible from systematic errors resulting from faulty observations and methods of reduction. From a comparison of numerous star catalogs he devised tables for the systematic correction of each, as well as a new catalog of the declinations and proper motions of 500 stars, which was adopted by the American Ephemeris in 1883.

In 1876 Boss became director of the Dudley Observatory, a position he held for the rest of his life. His first major project at Albany, New York, was observation and reduction of a zone for Leipzig's Astronomische Gesellschaft. By determining his magnitude equation and investigating the flexure of, and division corrections needed by, each of the two circles of the Pistor and Martins meridian circle, Boss was able to keep his probable errors to less than $\pm 0''.6$ for each observation—well within the limits expected for the society's catalog. Although he had little assistance, and started ten years after the work on many other zones was begun, Boss was the first to finish his zone. A comparison of the zone results with earlier observations led him to realize the need for an extensive analysis and comparison of the many available star catalogs, in order to make a reliable determination of proper motions.

The outcome of this study—financed in large part by the Carnegie Institution, which in 1906 established a department of meridian astrometry under the direction of Boss, and later of his son Benjamin—was published in numerous papers and four great catalogs. The *Preliminary General Catalogue* (1910) included information on 6,188 bright stars. The *San Luis Catalogue* (1928) was based on observations of 15,333 stars made with the great meridian circle of the Dudley Observatory, moved temporarily to Argentina. The *Albany Catalogue* (1931) included 20,811 stars observed with the meridian circle at Albany, and the *General Catalogue of 33,342 Stars* (1937) contains "the standard positions and proper motions of all stars brighter than the seventh magnitude, extending from the north to the south pole, and some thousands of additional fainter stars promising to yield reasonably accurate proper motions."

While at Dudley, Boss undertook several other projects. He computed the orbits of many comets and in 1881, under the pseudonym of Hipparchus III, won the $200 Warner Prize for the best essay on comets. The following year Boss took charge of the U.S. government party sent to Santiago, Chile, to observe and photograph the transit of Venus. From 1883 to 1906 he served as superintendent of weights and measures of the state of New York. In 1893 Boss moved the Dudley Observatory to a more astronomically advantageous location in Albany. With experience gained from editing and managing the daily Albany *Morning Express,* in 1897 Boss became associate editor, and in 1909 editor, of the *Astronomical Journal.*

Among Boss's many honors besides the gold medal of the Royal Astronomical Society (1905), the Lalande Prize of the Paris Académie des Sciences (1911), and membership in the National Academy of Sciences (1889), the Königlich Preussischen Akademie der Wissenschaften (1910), and the Imperial Academy of Sciences of St. Petersburg.

BIBLIOGRAPHY

Among Boss's writings is his prize-winning essay, "Comets: Their Composition, Purpose, and Effect Upon the Earth," in *History and Work of the Warner Observatory, 1883-6* (Rochester, N.Y., 1887), pp. 25–30. The four star catalogs, all prepared at the Dudley Observatory, Albany, N.Y., and published by the Carnegie Institution of Washington, D.C., are the *Preliminary General Catalogue of 6188 Stars* (1910), by Lewis Boss; *San Luis Catalogue of 15,333 Stars* (1928), by Lewis and Benjamin Boss; *Albany Catalogue of 20,811 Stars* (1931), and *General Catalogue of 33,342 Stars* (1936–1937), both the work of Benjamin Boss.

Additional information and a bibliography of his scientific writings is in Benjamin Boss, "Biographical Memoir of Lewis Boss, 1846–1912," in *Biographical Memoirs of the National Academy of Sciences,* IX (Washington, D.C., 1920), 239–260. Also of value is "Address Delivered by the President, H. H. Turner, on Presenting the Gold Medal of the Society to Professor Lewis Boss," in *Monthly Notices of the Royal Astronomical Society,* **65** (1904–1905), 412–425.

DEBORAH JEAN WARNER

BOSSE, ABRAHAM (*b.* Tours, France, 1602; *d.* Paris, France, 14 February 1676), *geometry, graphic techniques.*

The son of Louis Bosse, a tailor, and Marie Martinet, Bosse settled in Paris around 1625 and worked as a draftsman and engraver. In 1632 he married Catherine Sarrabat; four of their children lived to adulthood.

His drafting technique was obviously derived from the *méthode universelle* of perspective, presented by Girard Desargues as early as 1636. Bosse became Desargues's most ardent propagandist, and it was through his efforts that Desargues's methods achieved some success among artists of the seventeenth century and spread to foreign countries.

The art world of the seventeenth century was split into vigorously warring factions. Bosse sided with Desargues, who was conducting violent polemics, and in 1643 published two treatises, *La pratique du trait à preuves de Mr. Desargues . . . pour la coupe des pierres en l'architecture* and *Manière universelle de Mr. Desargues . . . pour poser l'essieu & placer les heures et autres choses aux cadrans au soleil,* which were complex expositions of two essays that Desargues had written in 1640 on the cutting of stone and on gnomonics.

In 1648 Bosse published a third tract, *Manière universelle de Mr. Desargues pour pratiquer la perspective,* which included several texts by Desargues himself, some of which had not been published before. Among them was the famous theorem on perspective triangles. In 1653 the work was amplified by a demonstration of the application of perspective to curvilinear surfaces. Bosse followed this work with several others dealing with particular applications, and became involved in controversies that eventually cost him his membership in the Académie Royale de Peinture et de Sculpture.

Bosse also illustrated books, particularly works on

science. These included such various works as Glaser's *Traité de la chymie,* M. Cureau de La Chambre's *Traité de la lumière,* Moyse Charas's *Pharmacopée royale,* and a series of botanical plates for Dodart's *Mémoires pour servir à l'histoire des plantes.*

BIBLIOGRAPHY

I. ORIGINAL WORKS. Bosse's writings include *La pratique du trait à preuves de Mr. Desargues, Lyonnois, pour la coupe des pierres en l'architecture* (Paris, 1643), also translated into German (Nuremberg, 1699); *La manière universelle de Mr. Desargues, Lyonnois, pour poser l'essieu & placer les heures et autres choses aux cadrans au soleil* (Paris, 1643), also translated into English (London, 1659); *Traité des manières de graver en taille douce sur l'airin par le moyen des eaux fortes et des vernix durs et mols* (Paris, 1645), often reissued and translated; *Manière universelle de Mr. Desargues pour pratiquer la perspective par petit-pied, comme le géometral. Ensemble les places et proportions des fortes et foibles touches, teintes et couleurs* (Paris, 1648), also translated into Dutch (Amsterdam, 1664, 1686); *Moyen universel pour pratiquer la perspective sur les tableaux ou surfaces irrégulières. Ensemble quelques particularitez concernant cet art & celuy de la gravure en taille douce* (Paris, 1653), also translated into Dutch (Amsterdam, 1664, 1686); *Représentations géométrales de plusieurs parties de bastiments faites par les reigles de l'architecture antique* (Paris, 1659); *Traité des manières de dessiner les ordres de l'architecture antique en toutes leurs parties* (Paris, 1664); *Traité des pratiques géométrales et perspectives enseignées dans l'Académie Royale de la Peinture et Sculpture* (Paris, 1665); *Le peintre converty aux précises et universelles regles de son art* (Paris, 1667); *Regle universelle pour décrire toutes sortes d'arcs rampans sur des points donnez de sujetion* (Paris, 1672); *Catalogue des traitez que le Sieur Bosse a mis au jour* (Paris, 1674); and *Recueil des plantes gravées par ordre du roi Louis XIV,* 3 vols. (Paris, n.d.), with N. Robert and L. Chatillon.

A more complete bibliography is in J. C. Brunet (see below) and especially in A. Blum, *Abraham Bosse et la société française du dix-septième siècle,* pp. 217–227, which also reproduces numerous documents relating to the polemic between Bosse and the Académie Royale de Peinture et de Sculpture.

II. SECONDARY LITERATURE. Bosse or his work is discussed in the following (listed chronologically): J. and L. G. Michaud, *Biographie universelle,* new ed., V (Paris, 1843), 124–125; J. M. B. Renouvier, *Des types et des manières des maîtres-graveurs,* XVI–XVII, pt. 2 (Montpellier, 1856), 117–123; P. J. Mariette, *Abecedario,* P. de Chennevières and A. de Montaiglon, eds., II (Paris, 1851–1853), 159–161; G. Duplessis, *Catalogue de l'oeuvre d'A. Bosse* (Paris, 1859); J. C. Brunet, *Manuel du libraire,* I (Paris, 1860), cols. 1126–1129; F. Hoefer, *Nouvelle biographie générale,* IV (1862), cols. 786–787; G. Poudra, *Oeuvres de Desargues,* 2 vols. (Paris, 1864), I, 352–493, II, 1–113, and *Histoire de la perspective* (Paris, 1864); A. Jal,

Dictionnaire critique de biographie et d'histoire, 2nd ed. (Paris, 1872), pp. 348–352; A. de Montaiglon, *Procès-verbaux de l'Académie Royale de Peinture et de Sculpture, 1648–1792,* I (Paris, 1875), *passim;* E. Haag and E. Haag, *La France protestante,* 2nd ed., II (Paris, 1879), cols. 922–928; A. Valabrègue, *Abraham Bosse* (Paris, 1892); G. C. Williamson, *Bryan's Dictionary of Painters and Engravers,* new ed., I (London, 1903), 174; U. Thieme and F. Becker, *Allgemeines Lexikon der bildenden Künstler,* IV (Leipzig, 1910), 402–403; A. Fontaine, *L'art dans l'ancienne France. Académiciens d'autrefois* (Paris, 1914), pp. 67–114; F. Amodeo, "Lo sviluppo della prospettiva in Francia nel secolo XVII," in *Atti dell'Accademia Pontaniana,* **63** (Naples, 1933); A. Blum, *Abraham Bosse et la société française du dix-septième siècle* (Paris, 1924), and *L'oeuvre gravé d'Abraham Bosse* (Paris, 1925); R. Taton, *L'oeuvre mathématique de Desargues* (Paris, 1951), see Index H, and "La première oeuvre géométrique de Philippe de La Hire," in *Revue d'histoire des sciences,* **6** (1953), 93–111; M. L. Blumer, in *Dictionnaire de biographie française,* VI (Paris, 1954), cols. 1146–1147; F. Bénézit, *Dictionnaire des peintres, sculpteurs, dessinateurs et graveurs,* new ed., II (Paris, 1961), 33–34; and A. Kondo, "Abraham Bosse et Poussin devant les problèmes de l'espace et du temps," in *Annales,* 23rd year (1968), 127–135.

A more complete bibliography relating to the artistic aspect of Bosse's work is given by A. Blum in *Abraham Bosse et la société française du dix-septième siècle,* pp. 213–221; more precise references to the geometrical aspect of the problems are in R. Taton, *L'oeuvre mathématique de Desargues,* pp. 70–71.

RENÉ TATON

BOSSUT, CHARLES (*b.* Tartaras, Rhône-et-Loire, France, 11 August 1730; *d.* Paris, France, 14 January 1814), *mathematics, mechanics.*

Bossut was the son of Barthélemy Bossut and Jeanne Thonnerine. His father died when Charles was six months of age, and the boy was raised by a paternal uncle. He entered the Jesuit Collège de Lyon at fourteen and was a student of Père Béraud, a mathematician whose pupils included Jean Étienne Montucla and Joseph Jérome Lalande. Bossut took minor ecclesiastical orders and was an *abbé* until 1792. He was aided in his professional formation by d'Alembert, Clairaut, and the Abbé Charles Étienne Louis Camus. Bossut never married, was without family, and, according to some, lived his last years as a misanthrope.

Bossut's importance for the history of science lies in his role as a major contributor to European scientific education. His career began in 1752, when he was appointed as professor of mathematics at the École du Génie at Mézières. He remained as professor until 1768, then continued as examiner of students until 1794. His other teaching post was from 1775 to 1780, in the chair of hydrodynamics established by Turgot at the Louvre. For a time he was also examiner of students at the École Polytechnique. Bossut wrote a series of textbooks that appeared in several French and foreign-language editions and won wide acceptance from the 1770's until the early years of the Empire. The texts of Bossut and Étienne Bézout best represent the emergence in the eighteenth century of a standardized, rigorous system of engineering physics textbooks. In France, for example, Bossut's course was used at the Benedictine Collège de Sorèze, the Collège de France, the École du Génie, the École des Ponts et Chaussées, and the École des Mines. He also wrote a history of mathematics that achieved popularity, but never the scholarly recognition of Montucla's history. He edited the works of Pascal, contributed to the *Encyclopédie méthodique,* and aided d'Alembert in editing contributions to Diderot's *Encyclopédie.*

Bossut was one of a very few whom d'Alembert took as students, and as such he was admitted as a *correspondant* to the Académie des Sciences on 12 May 1753; subsequently, he rose to *géomètre, mécanicien,* and *mathématicien.* In 1761, 1762, and 1765 he won or shared prizes given by the Academy for memoirs on mechanics applied to the operation of ships and on the resistance of the ether in planetary motions. He won additional prizes for his mechanics memoirs from the academies of Lyons and Toulouse, and was elected to the scientific academies of Bologna, Turin, and St. Petersburg. In 1775 he participated with d'Alembert and Condorcet in a well-known series of experiments on fluid resistance. Never more than a minor mathematician or physicist, Bossut is nevertheless one of the important figures in the history of physics and engineering education.

BIBLIOGRAPHY

I. ORIGINAL WORKS. Most of Bossut's memoirs appeared in the *Mémoires* and publications of the Académie des Sciences, Paris. Some of these were reissued in the collection *Mémoires de mathématiques, concernant la navigation, l'astronomie physique, l'histoire . . . par Charles Bossut* (Paris, 1812). His first textbook, a volume that does not figure in the catalogs of most major libraries, is *Traité élémentaire de méchanique et de dinamique appliqué principalement aux mouvemens des machines* (Charleville, 1763). The various editions of his textbooks (*Cours de mathématiques, Traité élémentaire d'arithmétique,* and others) are cited in the general catalogs of the Bibliothèque Nationale and the British Museum. The first edition of his history of mathematics is *Essai sur l'histoire générale des mathématiques,* 2 vols. (Paris, 1802). For Bossut's edition of Pascal's works see Blaise Pascal, *Oeuvres complètes,* 5 vols. (The Hague, 1779).

II. SECONDARY LITERATURE. For a short biography, see M. E. Doublet, "L'abbé Bossut," in *Bulletin des sciences mathématiques,* 2nd ser., **38** (1914), 93–96, 121–125, 158–160, 186–190, 220–224. See also the *éloge* by J. B. J. Delambre in *Mémoires de l'Académie Royale des Sciences de l'Institut de France—Année 1816,* **1** (1818), xci-cii. For Bossut's career at Mézières and for his general influence on education, see René Taton, ed., *Enseignement et diffusion des sciences en France au XVIIIe siècle* (Paris, 1964); Vol. XI of the series Histoire de la Pensée. Bossut's appointment to the chair of hydrodynamics is discussed in Roger Hahn, "The Chair of Hydrodynamics in Paris, 1775–1791: A Creation of Turgot," in *Acts of the Xth International Congress of the History of Science (Ithaca)* (Paris, 1964), pp. 751–754. A convenient summary of Bossut's work in fluid resistance is in René Dugas, *A History of Mechanics,* J. R. Maddox, trans. (Neuchâtel, 1955), pp. 313–316. On the question of whether Bossut was a Jesuit, see Thomas F. Mulcrone, S.J., "A Note on the Mathematician Abbé Charles Bossut," in *Bulletin—American Association of Jesuit Scientists,* **42** (1965), 16–19.

C. STEWART GILLMOR

BOSTOCK, JOHN (*b.* Liverpool, England, 1773; *d.* London, England, 6 August 1846), *medical chemistry.*

The only child of Edinburgh graduate and Liverpool physician John Bostock the elder (*d.* 1774), John grew up amid the town's active and prospering Protestant Dissenter community. Bostock enjoyed a good education, and in 1792 he attended Joseph Priestley's chemical lecture course at Hackney College. Further study with an apothecary and attendance at the Liverpool General Dispensary preceded his move to Edinburgh University in the autumn of 1794. Fellow medical students there included Alexandre Marcet, Thomas Thomson, and Thomas Young; and Bostock was elected a member, then a president, of the Medical Society. Graduating M.D. in 1798 with a thesis on the secretion of bile, he returned to Liverpool as a physician to the General Dispensary. Soon active in the town, Bostock played key roles in the formation of the Botanic Garden, the Fever Hospital, and the Philosophical and Literary Society, and he was also quickly launched on a career of prolific scientific publication.

For several years Bostock wrote nearly all the medical and scientific articles in the *Monthly Review,* as well as publishing a host of original papers in the *Edinburgh Medical and Surgical Journal,* Nicholson's *Journal of Natural Philosophy, Chemistry and the Arts,* and the *Transactions* of the London Medico-Chirurgical Society. An early interest in physiology led to his *Essay on Respiration* (Liverpool, 1804). This was well received, a commission to write many medical articles for David Brewster's *Edinburgh Encyclopaedia* being one result. Bostock's most successful work was his three-volume *Elementary System of Physiology* (1824–1827), which enjoyed wide popularity and reached a fourth edition in 1844. He also wrote critical pamphlets on the new Edinburgh (Liverpool, 1807) and London (London, 1811) pharmacopoeias, a *History . . . of Galvanism* (London, 1818), a *History of Medicine* (London, 1835), and an incomplete translation of Pliny's *Natural History* (London, 1828).

Bostock developed an extensive medical and chemical consulting practice; and in 1817, having made a secure fortune, he abandoned Liverpool medicine for London and the full-time pursuit of science in its social and administrative, as much as its laboratory, context. He soon succeeded his friend Marcet as chemical lecturer at Guy's Hospital, and (with A. Aikin) he for two years edited the *Annals of Philosophy,* following Thomas Thomson's move to Glasgow. His widening scientific activity is reflected in his work as secretary of the Geological Society (president, 1826), council member of the Royal Society (vice-president, 1832), and treasurer of the Medico-Chirurgical Society. He also served on the councils of the Linnean, Zoological, and Horticultural societies, and the Royal Society of Literature.

Bostock's primary research interest was medical chemistry and he made valuable contributions to, *inter alia,* the study of body fluids and urinary components. He also gave the first complete description of hay fever. Equally interesting to the historian is the way that Bostock's career typifies the scientific research and administrative opportunities available to the British medical man of the early nineteenth century. In this respect his life forms an instructive contrast to the activities of such less formally educated and more industrially oriented contemporary chemists as, say, Friedrich Accum and William Nicholson.

BIBLIOGRAPHY

I. ORIGINAL WORKS. Bostock's separately published writings are *Essay on Respiration* (Liverpool, 1804); *Remarks on the Reforms of the Pharmaceutical Nomenclature; and Particularly on That Adopted by the Edinburgh College* (Liverpool, 1807); *Remarks on the Nomenclature of the New London Pharmacopoeia* (London, 1810); *History . . . of Galvanism* (London, 1818); *Elementary System of Physiology,* 3 vols. (London, 1824–1827; 3rd ed., 1 vol., London, 1876); and *History of Medicine* (London, 1835). He also was responsible for a partial translation of Pliny's *Natural History* (London, 1828). An incomplete list of Bostock's multitudinous journal and encyclopedia contributions is in T. J. Pettigrew, *Biographical Memoirs of the Most Celebrated Physicians,* III (London, 1839), sec. 4, 1–20.

II. SECONDARY LITERATURE. Further information on Bostock's life may be gleaned from the obituary notice in *Proceedings of the Royal Society,* **5** (1846), 636–638. His medical chemistry is evaluated in J. R. Partington, *A History of Chemistry,* III (London, 1962), 711–712.

ARNOLD THACKRAY

BOTALLO, LEONARDO (*b.* Asti, Italy, *ca.* 1519; *d.* Chenonceaux or Blois, 1587/1588), *medicine.*

Botallo studied and obtained a degree in medicine at the University of Pavia, after which he continued his studies for a time at the University of Padua under Gabriele Falloppio. He then practiced medicine in Asti. He joined the French forces in Italy, at least by 1544, since he refers to his participation in the battle of Ceresole as a military surgeon. He was already located in Paris as one of the physicians of Charles IX in 1560, the same year in which his surgical treatise *De curandis vulneribus sclopettorum* was published in Lyons. Based partly on earlier, similar treatises and partly on Botallo's own experiences as a military surgeon, the work is notable chiefly for its support of the opinion, first advocated in print by Ambroise Paré in 1545, that gunshot wounds were not envenomed and ought to receive mild rather than harsh treatment. It was also concerned with the neurological effects of cranial injuries and the indications for treatment. The work was frequently and widely reprinted.

Botallo's name appears in the eponymous nomenclature of anatomy through the terms Botallo's duct (*ductus arteriosus*) and Botallo's foramen (*foramen ovale cordis*), actually incorrect attributions of discovery based upon Botallo's brief note "Vena arteriarum nutrix a nullo antea notata" in his *De catarrho commentarius.* Assertions that this note appeared in an earlier form entitled *De foramine ovalis dissertatio* appear to be incorrect. Through accidental discovery of the above structures in the calf's heart and thereafter in various other animals, and in particular through his chance observation of persistence of the *foramen ovale,* Botallo was led to believe that the blood's passage from the right to the left side of the heart was by way of this opening rather than through the imaginary pores of the cardiac septum, as (incorrectly) proposed by Galen, or through the lungs, as (correctly) proposed by Realdo Colombo. Actually, Botallo's "discovery" had been mentioned in the second century by Galen and, more recently, the *ductus arteriosus* by Falloppio, in 1561 and the *ductus arteriosus* and *foramen ovale* by Vesalius, also in 1561. Botallo may therefore be credited merely with independent rediscovery; but, since he did not dissect the fetus, he failed to recognize the true significance of the structures, although it appears to have been known to Galen and was later reemphasized by William Harvey.

A second note in *De catarrho commentarius,* entitled "Addita est in fine monstrorum renum figura, nuper in cadavere repertorum," provides a careful description, accompanied by a detailed illustration, of an instance of fused kidneys with horseshoe configuration. This anomaly, too, had previously been observed and described, although more briefly, by Berengario da Carpi in the *Isagogae* (1522, f. 17v). Nonetheless, Botallo's detailed account indicates his interest in anatomy and his not inconsiderable ability as a dissector and observer.

Botallo's other writings are of lesser significance. The *Luis venereae curandae ratione* was characteristic of its time; but *De incidendae venae, cutis scarificandae et hirudinum applicandarum modo,* through presentation of Botallo's independent, anti-Galenic opinions regarding venesection, gained him the enmity of the conservative Parisian physicians. Botallo believed in the therapeutic value of liberal bloodletting. The *Commentariola duo, alter de medici, alter de aegroti munere,* dealing with medical ethics and the physician-patient relationship in general, reveals Botallo as skeptical of the value of astrology to medicine.

Throughout his years in the French royal medical service, Botallo enjoyed the favor and confidence of the queen mother, Catherine de' Medici, in part perhaps because of their common Italian origin. In any event, she was instrumental in having his services transferred to her favorite son, the duke of Anjou, later Henry III. It was during this latter service, and no doubt as a reflection of his fame as a military surgeon, that Botallo was temporarily disengaged in 1575 to undertake the care and treatment of Henry I of Lorraine, duke of Guise, who had received a gunshot wound of the cheek and ear. Botallo was, at least professionally, inactive during the final years of his life as a result of sickness, most likely the effects of malaria. He died probably at Chenonceaux or Blois, but there is no available information as to where he was buried.

BIBLIOGRAPHY

I. ORIGINAL WORKS. Botallo's works include *De curandis vulneribus sclopettorum* (Lyons, 1560, 1564, 1566, 1575, 1583), translated into German as *Von den Schuss-Wunden, und wie dieselben zu heilen* (Nuremberg, 1676); *De foramine ovalis dissertatio* (Lyons, 1561); *Luis venerae curandae ratione* (Paris, 1563); *De catarrho commentarius* (Paris, 1564); *Commentariola duo, alter de medici, alter de aegroti munere* (Lyons, 1565); and *De incidendae venae, cutis scarificandae et hirudinum applicandarum modo* (Lyons, 1565).

II. SECONDARY LITERATURE. In addition to the few autobiographical remarks in Botallo's writings, there is a biographical study by Leonardo Carerj, *Leonardo Botallo Astese, medico regio* (Asti, 1954). Also see John A. Benjamin and Dorothy M. Schullian, "Observations on Fused Kidneys With Horseshoe Configuration: The Contribution of Leonardo Botallo (1564)," in *Journal of the History of Medicine and Allied Sciences,* **5** (1950), 315–326; Leonardo Carerj, "Leonardo Botallo, il foro ovale e il dotto arterioso," in *Minerva medica,* **2** (1955), varia, 789–795; K. J. Franklin, "A Survey of the Growth of Knowledge About Certain Parts of the Foetal Cardio-vascular Apparatus, and About the Foetal Circulation, in Man and in Some Other Mammals. Part I: Galen to Harvey," in *Annals of Science,* **5** (1941), 57–89; E. J. Gurlt, *Geschichte der Chirurgie und ihrer Ausübung,* II (Berlin, 1898), 403–415; Antonio Nitto, "Considerazioni medico-storiche sulla fossa ovale, il legamento arterioso e la priorità della loro scoperta tra Leonardo Botallo e Giulio Cesare Aranzio," in *Policlinico, sez. med.,* **68** (1961), 299–312; and Mario Truffi, "Leonardo Botallo sifilografo," in *Minerva medica,* **46** (1955), varia, 34–42.

C. D. O'MALLEY

BOTHE, WALTHER WILHELM GEORG (*b.* Oranienburg, Germany, 8 June 1891; *d.* Heidelberg, Germany, 8 February 1957), *physics.*

Bothe's father was Fritz Bothe, a merchant. During the period 1908 to 1912 Walther studied at the University of Berlin, where his training included not only physics and mathematics but also chemistry. In 1914 he obtained his doctorate under Max Planck for a study of the molecular theory of refraction, reflection, dispersion, and extinction, a subject he continued to study from 1915 to 1920 while a prisoner of war in Russia. On his return to Germany, where he married Barbara Below, Bothe accepted Hans Geiger's invitation to work at the radioactivity laboratory of the Physikalische-Technische Reichsanstalt. Much of his early experimental work was done with Geiger, and Bothe later said that it was Geiger who had initiated him into his researches in physics. Bothe taught physics at the University of Berlin from 1920 until he accepted the professorship of physics at Giessen in 1931. In 1934 he became director of the Max Planck Institute at Heidelberg, where he remained until his death. Bothe received the Max Planck Prize, and he shared the Nobel Prize with Max Born in 1954.

From 1921 to 1924 Bothe was active in both theoretical and experimental work on the scattering of alpha and beta rays. He devised a statistical theory for processes involving multiple scattering through small angles, a phenomenon far more complex than large-angle scattering. Bothe studied multiple scattering of electrons experimentally by tracking the trajectories on photographic plates. He also deduced mathematical expressions for the relationships between scattering angle and foil thickness, using a nuclear model of the atom.

Among the topics that Bothe studied in 1924 was the ejection of electrons by X rays, and it was in connection with this phenomenon that he and Geiger performed an important experiment. In an effort to reconcile the particulate and wavelike properties of radiation, Bohr, Kramers, and Slater in 1924 formulated a new quantum theory of radiation. According to their hypothesis, momentum and energy are conserved only statistically in interactions between radiation and matter. Bothe and Geiger suggested that this could be tested experimentally by examining individual Compton collisions. Bothe introduced a modification into the Geiger counter that made it appropriate for use in coincidence experiments (a very novel procedure in 1924). Using two counters, they studied the coincidences between the scattered X ray and the recoiling electron. Correlating photons with electrons, Bothe and Geiger found a coincidence rate of one in eleven; since the chance coincidence rate for the situation was 10^{-5}, the experimental results contradicted the theoretical predictions and indicated small-scale conservation of energy and momentum.

With Werner Kolhörster, Bothe used the coincidence method again in 1929 to demonstrate that cosmic rays might be particles. Ever since their discovery in 1912, physicists had assumed that cosmic rays were high-energy photons, and Millikan's hypothesis that they were released during the elementary synthesis of elements by fusion in the atmosphere was especially popular. In the experiments of Bothe and Kolhörster, two Geiger counters were separated by about 4 cm. of gold; and in order for a photon to produce a pulse in a counter, it would have to undergo a Compton collision and produce an ionizing electron. The known probability of Compton collisions and the average energy of the photons indicated that coincidences between the two counters were highly improbable. The high coincidence rate in the experiment, approximately 75 percent of the original single-counter rate, therefore indicated that the cosmic radiation might well be particulate.

Nuclear transmutation was a third new topic of interest for physicists at this time. In 1919 Rutherford had produced oxygen nuclei and protons by bombarding nitrogen with alpha particles, and during the following decade various laboratories worked on this type of transmutation. Bothe took up the subject in 1926, and in the following years he studied the transmutation, via alpha particles, of boron to carbon. He was among the early users of the electronic counter to detect the protons in this type of reaction. With H. Becker he searched systematically for a gamma radiation accompanying the transmutation; its existence seemed reasonable because there were some light elements, such as beryllium, that did not disintegrate. In these experiments there were problems in finding a suitable source of alpha particles (one that did not produce other radiations as well), and Bothe himself finally prepared one by a process of chemical extraction. In 1930 Bothe and Becker detected a highly penetrative radiation from beryllium bombarded by alpha particles, and they assumed that it was gamma radiation. Bothe estimated the photon energy from the degree of absorption of the secondary electrons. When physicists studied this "beryllium radiation," estimating its energy constituted a problem, for it varied greatly according to the substance used as absorber. Chadwick later suggested that the radiation was particulate and consisted of a new particle, the neutron.

After detecting the new radiation in 1930, Bothe continued studying nuclear transmutation, and made coincidence measurements on the products of the reaction between alpha particles and beryllium. When he became director of the Max Planck Institute, he was involved in the construction of a Van de Graaf accelerator there and in the planning of a cyclotron which was finally completed in 1943. After the outbreak of World War II he did much work on uranium and on neutron transport theory, and he was one of the foremost scientists of Germany's "uranium project" for nuclear energy.

In the early 1950's Bothe again dealt with questions of electron scattering and cosmic rays, and of beta and gamma spectra. Thus, over a period of more than thirty years Bothe studied a broad range of highly relevant problems in a variety of ways.

BIBLIOGRAPHY

I. ORIGINAL WORKS. A few of Bothe's more important papers are "Ein Weg zur experimentellen Nachprüfung der Theorie von Bohr, Kramers, und Slater," in *Zeitschrift für Physik*, **26** (1924), 44 (written with H. Geiger); "Ueber das Wesen des Comptoneffekts; ein experimentellen Beitrag zur Theorie der Strahlung," in *Zeitschrift für Physik*, **32** (1925), 639–663 (written with H. Geiger); "Das Wesen der Hoehenstrahlung," in *Zeitschrift für Physik*, **56** (1929), 75–77 (written with W. Kolhörster); "Kunstliche

Erregung von Kern-γ-Strahlen," in *Zeitschrift für Physik,* **66** (1930), 289–306 (written with H. Becker). Other papers are listed in *Science Abstracts.*

II. SECONDARY LITERATURE. Discussions of Bothe's work on scattering are in E. Rutherford, J. Chadwick, and C. D. Ellis, *Radiations from Radioactive Substances* (Cambridge, 1930), pp. 209–212, 219–220, 237–238. In connection with the work on cosmic rays, see Bruno Rossi, *Cosmic Rays* (New York, 1964), pp. 30–42. The context of the early research on nuclear transformations may be found in M. Korsunsky, *The Atomic Nucleus* (New York, 1965), pp. 137–145. M. Jammer, *The Conceptual Development of Quantum Mechanics* (New York, 1966), pp. 181–188, and B. L. Van der Waerden, ed., *Sources of Quantum Mechanics* (Amsterdam, 1967), pp. 12–14, discuss the significance of the 1925 experiment. There is an account of Bothe's role in the German uranium project in S. A. Goudsmit, *Alsos* (New York, 1947). Obituaries are Lise Meitner, in *Nature,* **179** (1957), 654–655; and R. Fleischmann, in *Die Naturwissenschaften,* **44** (1957), 457–460.

SIGALIA DOSTROVSKY

BOTTAZZI, FILIPPO (*b.* Diso, Apulia, Italy, 23 December 1867; *d.* Diso, 19 September 1941), *physiology.*

Bottazzi received the M.D. in Rome in 1893. Early the following year, he became an assistant in physiology at the Institute of Higher, Practical, and Postgraduate Studies of Florence (now the University of Florence), and in 1896, *Privatdozent* in physiology. In 1902 he was appointed director of the Institute of Physiology of the University of Genoa, and in 1905, director of the Institute of Physiology of the University of Naples. While in Naples he was actively engaged in research at the Zoological Station, and from 1915 to 1923 he was director of its department of physiology.

Through his association with his teacher Giulio Fano, Bottazzi belonged to the physiological school of Luigi Luciani, which, with the school of Angelo Mosso, was at that time the most important in Italy. Bottazzi's work in physiology is distinguished by its close relationship to biological and physical chemistry. This relationship became evident in his *Elementi di chimica fisica* (1906) and his *Trattato di chimica fisiologica* (1898–1899), which for decades were the standard works on the subjects in Italy. Their recognition abroad is attested by H. Boruttau's German translation, *Physiologische Chemie für Studierende und Ärzte* (1901).

In 1894 Bottazzi's earliest investigations demonstrated the diminution of the osmotic resistance experienced by the red corpuscles as they travel through the splenic cycle: this is the hemocatatonistic function of the spleen, which is a part of the wider function of the splenic hemocatheresis and, more generally, of the endothelial reticulum.

A similar chemicophysical basis was demonstrated by his research on the osmotic pressure of the organic liquids, which Bottazzi explored in nearly all classes of animals, from the invertebrates to man (and in the case of man, also in a variety of physiological and pathological conditions). He arrived at the conclusion that homeo-osmosis—that is, the constancy of the osmotic pressure of the *milieu intérieur* (blood and interstitial liquids)—must be considered a relatively late philogenetic acquisition. Homeo-osmosis does not occur in aquatic invertebrates or in cartilaginous fishes, animals whose internal liquids are nearly in osmotic equilibrium with the water of the external environment. In the teleost fishes there is the beginning of a limited or conditioned osmotic independence of the external environment. In all other vertebrates, starting with the amphibians, the osmotic pressure of the blood becomes an absolute physical constant, like the body temperature of birds and mammals. Therefore, Bottazzi distinguished between homeo-osmotic and poikilosmotic animals, in much the same way that one distinguishes between homeothermic and poikilothermic animals.

The same broadly comparative criterion that was adopted in these investigations was also used in those that Bottazzi and his collaborators conducted on the smooth and striated muscles. Thanks to this work, Bottazzi was able to attribute an essentially tonic function to the sarcoplasm, that is, to the part of the muscle cell that is not differentiated into myofibril. According to the theory formulated by Bottazzi, the fibrillar formations serve for the rapid clonic contraction, while the sarcoplasm serves for the slow and sustained tonic contraction. Therefore, the more readily a muscle attains tonus, the more sarcoplasm it contains (following this decreasing scale: smooth muscle, red striated muscle, and pale striated muscle). The sarcoplasm, which is less irritable and slower to contract, carries out the simple function of tone. The fibrillar anisotropic substance, which is more irritable, is capable of rapid movements and usually is more fully developed, depending on how quickly the muscle reacts to stimuli.

In the area of practical studies, special mention must be made of Bottazzi's research on the physiology of nutrition. This work was presented in *Fisiologia dell'alimentazione con speciale riguardo all'alimentazione delle classi povere* (1910), written with G. Jappelli, and *L'alimentazione dell'uomo* (1919).

Bottazzi translated Michael Foster's *Treatise on Physiology* in 1899. He and Foster, who in 1894 invited him to teach a practical course on physiology for advanced students at Cambridge, shared a keen interest in the history of science. The main results of this interest were numerous studies that Bottazzi wrote on Leonardo da Vinci from 1902 to the eve of his death, and the new national edition of the works of Lazzaro Spallanzani (1932–1936).

BIBLIOGRAPHY

I. ORIGINAL WORKS. Bottazzi's writings include *Trattato di chimica fisiologica,* 2 vols. (Milan, 1898–1899), trans. by H. Boruttau as *Physiologische Chemie für Studierende und Ärzte* (Leipzig-Vienna, 1901); *Trattato di fisiologia,* 4 vols. (Milan, 1899), a translation of Michael Foster's *Treatise on Physiology; Elementi di chimica fisica* (Milan, 1906), with G. Jappelli; *Fisiologia dell'alimentazione con speciale riguardo all'alimentazione delle classi povere* (Milan, 1910); and *L'alimentazione dell'uomo* (Naples, 1919). He was also an editor of the new national edition of *Le opere di Lazzaro Spallanzani,* 6 vols. (Milan, 1932–1936).

II. SECONDARY LITERATURE. A fundamental source for this article was Pietro Rondoni's "Filippo Bottazzi," a discerning commemoration of the man and his work, in *Annuario della Reale Accademia d'Italia,* **14** (1941–1942), 156–169. Complete lists of Bottazzi's writings, with brief biographical sketches, are in *Annuario della Reale Accademia d'Italia,* **1,** (1929), 87–102; and in *Annuario della Pontificia Accademia delle scienze,* **1** (1936–1937), 159–185.

LUIGI BELLONI

BÖTTGER, RUDOLPH CHRISTIAN (*b.* Aschersleben, Germany, 28 April 1806; *d.* Frankfurt am Main, Germany, 29 April 1881), *chemistry.*

Böttger entered the University of Halle in 1824 to study theology. Despite the time devoted to theology and philosophy, he regularly attended the chemistry lectures of Johann S. C. Schweigger, thus fulfilling a youthful ambition to study chemistry.

On leaving the university in 1828, he continued his interest in chemistry while employed as a tutor, corresponding with Schweigger and pursuing a program of research in his spare time. By 1835 he had published a dozen papers in the leading German chemical periodicals. In that year, he was called to the Physicalischer Verein at Frankfurt am Main, where he taught physics and chemistry for the rest of his long, active life. In 1837 he received the doctorate from the University of Halle, and in 1841 he married Christiane Harpke.

Böttger's interests in chemistry were far-reaching, and he was particularly interested in practical applications of research. For example, he invented a useful kindling apparatus modified from Schönbein's lamp, and in 1841 he developed an electroforming process for reproducing illustrations which was widely used.

Böttger showed in 1843 that nickel could be plated on other metals by electrodeposition, although the technology of the times was insufficient for commercial use, and in 1845 he produced high-purity iron by electrolytic deposition.

Böttger has been credited with the independent discovery of guncotton, first announced by Schönbein in March 1846. He had collaborated with Heinrich Will at Giessen on styphnic (hydroxypicric) acid and was able to advise Schönbein in July 1846 that the preparation of guncotton could be improved by use of a mixture of sulfuric and nitric acid.

Shortly after Anton Schrötter discovered red phosphorus in 1847, Böttger introduced match heads covered with a mixture of potassium chlorate, red lead, and a gum, to be used on match boxes whose striking surfaces were coated with red phosphorus.

Böttger investigated silver and copper acetylides, chrome alum, chromic oxide, and lampic acid. He devised tests for nitrites and chlorates still used today. He also studied the reduction of palladous chloride, and the chemistry of indium, thallium, and cesium.

Böttger was an able chemist, a "skillful experimenter whose tact in manipulation is well known" (Jerome Nicklès, "Correspondence . . . 1858"). His work, mainly qualitative, was ingenious and accurate. He rarely pursued a particular topic in depth, turning instead to the other aspects of chemistry that intrigued him, nor was he prone to hypothesize or to frame chemical theories, preferring instead the daily routine of laboratory experimentation.

BIBLIOGRAPHY

I. ORIGINAL WORKS. Böttger wrote more than one hundred papers. Convenient but incomplete lists may be found in Poggendorff, *Biographisch-literarisches Handwörterbuch,* I (1963), 150–151; and in the Royal Society's *Catalogue of Scientific Papers,* X (1867), 508–511; VII (1877), 223–224; IX, 303. In 1846 he founded the *Polytechnisches Notizblatt* and edited it for 35 years. One of his major works was the three-volume *Beiträge zur Physik und Chemie* (Frankfurt am Main, 1838–1846).

II. SECONDARY LITERATURE. Theodor Petersen gives a brief description of Böttger's life and work in *Berichte der Chemischen Gesellschaft,* **14** (1881), 2913–2919. Robert Knott's biography in *Allgemeine deutsche Biographie,* XLVII, 143–144, is based on Petersen. J. R. Partington, *A History of Chemistry,* IV (London, 1964), 196, gives a concise summary of Böttger's work. See also Jerome Nicklès, "Correspondence of Jerome Nicklès, Dated Paris, Oct. 26th, 1858," in *American Journal of Science,* 27 (1859), no. 79, 121. There is no detailed study of Böttger's life.

LOUIS I. KUSLAN

BOUCHER DE CRÈVECOEUR DE PERTHES, JACQUES (*b.* Rethel, Ardennes, France, 10 September 1788; *d.* Abbeville, France, 5 August 1868), *archaeology.*

Boucher was the director of the customshouse at Abbeville, where he spent his leisure time in archaeological pursuits. Here he found evidences of Stone Age cultures, which he reported and interpreted in a series of monographs. Although rejected by the scientific society of Abbeville, his work came to the attention of a group of eminent British scientists (including Lyell), who in 1859 visited the sites of his excavations and supported his conclusions. For a complete study of his scientific accomplishments and writings see Supplement.

BOUÉ, AMI (diminutive of **Amédée**) (*b.* Hamburg, Germany, 16 March 1794; *d.* Vöslau, Austria, 21 November 1881), *geology.*

Boué's father, Jean-Henri, was a shipbuilder and shipowner; his mother, Suzanne de Chapeaurouge, was the daughter of an Alsatian merchant who had settled in Hamburg. Orphaned at the age of eleven, he was taken in by relatives in Geneva and later went to Paris, where he was educated by an uncle, Antoine Odier, a banker. He showed no particular interest in either business or bookkeeping, and convinced his relatives that he should not enter the family shipping business. He also had no inclination for the intricacies of the law, and he therefore interrupted his legal studies (which his uncle had urged him to begin) when, at the age of twenty, he came into an inheritance.

Boué decided at the beginning of 1814 to study medicine at the University of Edinburgh and learned English. Through the influence of one of his teachers, the mineralogist Robert Jameson, he became interested in geology and botany, which he studied while traveling throughout Scotland. In 1815 he published in the *Edinburgh Philosophical Journal* an unsigned article on a crystallized hyacinth found in gneiss along the Caledonian Canal. On 15 September 1817, he received the M.D. after having defended a thesis on the botanical geography of Scotland.

Although Boué continued his medical studies in Paris (1818–1819), Berlin (1820), and Vienna (1821), he decided to devote himself exclusively to geology and traveled through the Auvergne, the south of France, and almost all of Germany, Austria-Hungary, and Italy. His *Essai géologique sur l'Écosse,* dedicated to Jameson, appeared in 1820. Two years later he published "Mémoire géologique sur l'Allemagne." His observations on the geology of Germany were completed in *Geognotisches Gemälde von Deutschland* (1829), in which he discussed the ages of the various mountain chains.

In 1830 Boué went to Paris and, with his friends Constant Prevost, Paul Deshayes, and Jules Desnoyers, founded the Société Géologique de France, of which he became president in 1835. In 1830–1831, in collaboration with Jobert and Claude Rozet, he published *Journal de géologie,* followed by *Mémoires géologiques et paléontologiques* in 1832. Boué expressed his views on geological controversies with complete candor, and he did not fail to criticize the authorities whenever he felt they were wrong. Thus, he reproved Cuvier for having refused (1823) to acknowledge the age of the human skeleton he had found among fossil mammal remains at the base of the loess at Lahr (near Strasbourg). He also censured Élie de Beaumont for his daring theory on the pentagonal network.

From 1830 to 1834 Boué published bibliographical notices on geological progress in foreign countries in the *Bulletin de la Société géologique de France.* In 1833, when Paul-Émile Botta published his "Observations sur le Liban et l'Anti-Liban," Boué remarked in a note to the article that, according to the fossils sent back by the author, the three stages distinguished in Lebanon correspond to the Lower Cretaceous, greensand, and Jurassic limestone.

In 1835 Boué left Paris for Vienna, just after the first volume of his *Guide du géologue-voyageur, sur le modèle de l'Agenda geognostica de M. Léonhard* had been published. (Leonhard's work was published in 1829.) Boué expanded the subject considerably, discussing preparations and preliminary instructions for geological trips, physical geography, and various aspects of geology.

In 1836, 1837, and 1838 Boué made three trips to European Turkey to study the resources of the country and its people. *La Turquie d'Europe* was published in 1840.

When he returned to Austria, Boué bought property at Vöslau, where during the summer he worked in his gardens and vineyards. In 1845 his synthesis of geological knowledge, *Essai de carte géologique du globe terrestre,* was published. He was elected a member of the Academy of Sciences of Vienna in 1849.

BIBLIOGRAPHY

Boué's principal publications are *Dissertatio inauguralis de methodo floram cujusdam conducendi, exempli é flora Scoticâ, etc., ductis, illustrata* (Edinburgh, 1817); *Essai géologique sur l'Écosse* (Paris, n.d. [1820]); "Mémoire géologique sur l'Allemagne," in *Journal de physique, de chimie et d'histoire naturelle,* **94** (1822), 297–312, 345–378; **95** (1822), 31–48, 88–112, 173–200, 275–304; "Allgemeine geologische Beobachtungen über die Entstehung der Gebirge Schottlands," in *Taschenbuch für die gesammte Mineralogie von Karl Cäsar Leonhard* (Frankfurt, 1823), pp. 239–362; *Geognotisches Gemälde von Deutschland mit Rücksicht auf die Gebirgsbeschaffenheit nachbarlicher Staaten* (Frankfurt, 1829); "Des progrès de la géologie," in *Bulletin de la Société géologique de France,* **1** (1831), 71–75, 94–97, 105–124; *Journal de géologie,* 3 vols. (Paris, 1830–1831); "Résumé des progrès de la géologie en 1831," in *Bulletin de la Société géologique de France,* **2** (1832), 133–218; *Mémoires géologiques et paléontologiques* (Paris, 1832); "Résumé des progrès de la géologie, et de quelques unes de ses principales applications, pendant l'année 1832," in *Bulletin de la Société géologique de France,* **3** (1833), ii–clxxxviii; "Résumé des progrès des sciences géologiques pendant l'année 1833," *ibid.,* **5** (1834); *Guide du géologue-voyageur, sur le modèle de l'Agenda geognostica de M. Léonhard,* 2 vols. (Paris, 1835–1836); *La Turquie d'Europe . . .,* 4 vols. (Paris, 1840); *Essai de carte géologique du globe terrestre* (Paris, 1845); *Der ganze Zweck und der hohe Nutzen der Geologie in allgemeiner und in specieller Rücksicht auf die Oesterreichischen Staaten und ihre Völker* (Vienna, 1851); *Sur l'établissement de bonnes routes et surtout de chemins de fer dans la Turquie d'Europe* (Vienna, 1852); and *Ueber die Nothwendigkeit einer Reform des bergmännischen Unterrichtes in Österreich . . .* (Vienna, 1869).

A secondary source is Franz Ritter von Hauer, "Zur Erinnerung an Dr. Ami Boué," in *Jahrbuch der Kaiserlich-Königlichen geologischen Bundesanstalt* (Vienna), **32** (1882), 1–6.

ARTHUR BIREMBAUT

BOUELLES, CHARLES DE. See **Bouvelles, Charles.**

BOUGAINVILLE, LOUIS ANTOINE DE (*b.* Paris, France, 11 November 1729; *d.* Paris, 31 August 1811), *geography, mathematics.*

Bougainville was the son of a notary, Pierre-Yves de Bougainville. To escape his father's profession he joined the army, saw service with Montcalm in Canada, and, on his own initiative, founded a French colony in the Falkland Islands in 1764. Two years later he was commissioned to sail around the world. On his return he received many honors and was promoted in both the army and the navy. He saw further service in North America. He married into a naval family in 1780 and had four children. Despite his royalist sympathies, he survived the Terror. He escaped the massacre of Paris and lived quietly for the rest of his life. He was an associate of the Académie des Sciences, a member of the Legion of Honor, a count of the empire under Napoleon, and a senator. He was buried with full honors in the Panthéon.

Bougainville's contributions to science were twofold: he began a career in mathematics but achieved his greatest fame as an explorer. At the completion of his schooling he came under the influence of d'Alembert, and as a result he wrote the *Traité du calcul-intégral* during 1752. L'Hospital had written the first textbook on calculus in 1696. Bougainville's contribution was to extend L'Hospital's treatise to cover the integral calculus and to bring the differential calculus up to date. He brought such clarity and order to the subject, as well as incorporating new work, that he achieved immediate recognition. The Académie des Sciences noticed the work in January 1753. It was published the following year, and at the beginning of 1756, it brought Bougainville election to the Royal Society of London. A further volume was published in 1756, and this was the end of his career as a mathematician.

At the end of 1766 Bougainville left Nantes in the frigate *La Boudeuse.* After handing over the Falkland Island colony to Spain in 1767, he called at Rio de Janeiro to meet his supply ship. On board was the botanist Commerson. Among the plants Commerson collected around Rio de Janeiro was a climbing shrub with large purple-red bracts which he named bougainvillea.

The two ships left the Falkland Islands in July 1767 and sailed through the Strait of Magellan. By the end of March 1768, Bougainville was discovering new islands in the Pacific archipelago of Tuamotu. He sailed on to Tahiti, only to find that La Nouvelle Cythère, as he named it, had been discovered eight months earlier by Samuel Wallis. Sailing west, he almost reached the Great Barrier Reef but turned north without exploring further. Bougainville sailed through an archipelago that he named the Louisiade, and discovered two of the Treasury Islands before reaching the Solomons. On 1 July he left the west coast of Choiseul Island and for three days sailed along "a new coast which is of an astonishing height." This is now Bougainville Island, and the strait between it and Choiseul is Bougainville Strait.

Putting into the Moluccas, Bougainville found a "species of wild cat that carries her young in a pocket below her belly," and thus confirmed what Buffon had doubted, that pouched mammals exist in the East Indies. In 1771 Bougainville published the best-selling *Voyage autour du monde.*

It has been said by Frenchmen that, in spite of his mathematical abilities, Bougainville was no great navigator. But he was the first Frenchman to sail around the world. His voyage took three years and, in an age when the death rate for sailors was high, he lost only seven men. He named new islands in the Solomons and the Tuamotu Archipelago; and he was the first to make systematic astronomical observations of longitude, providing valuable charts for future sailors in the Pacific.

Bougainville's attitude toward exploration can be summed up in his own words: "But geography is a science of facts; one cannot speculate from an armchair without the risk of making mistakes which are often corrected only at the expense of the sailors" (*Voyage autour du monde,* p. 210).

BIBLIOGRAPHY

I. ORIGINAL WORKS. Bougainville's works include *Traité du calcul-intégral, pour servir de suite à l'analyse des infiniment-petits du Marquis de l'Hôpital,* 2 vols. (Paris, 1754–1756); *Voyage autour du monde* (Paris, 1771); "Journal de l'expédition d'Amérique . . .," in *Rapport de l'archiviste de la province de Québec* (1923/1924), pp. 204–393, trans. and ed. by E. P. Hamilton as *Adventure in the Wilderness, the American Journals 1756–60* (Norman, Okla., 1964). MSS are in the Archives Nationales, Paris.

II. SECONDARY LITERATURE. Works on Bougainville are J. Dorsenne, *La vie de Bougainville* (Paris, 1930); and J. Lefranc, *Bougainville et ses compagnons* (Paris, 1929).

WILMA GEORGE

BOUGUER, PIERRE (*b.* Croisic, France, 16 February 1698; *d.* Paris, France, 15 August 1758), *geodesy, hydrography, physics.*

The son of Jean Bouguer, royal professor of hydrography, Pierre Bouguer was a prodigy who at the age of fifteen, upon the death of his father, applied for and obtained the professorship. He quickly became the leading French theoretical authority on all things nautical, and by the time he was twenty-nine had won three prizes for essays on subjects set by the Académie Royale des Sciences: on the masts of ships (1727), on the best way of observing the altitudes of stars at sea (1729), and on the observation at sea of the magnetic declination (1731). In 1731 Bouguer was made an associate geometrician of the Académie Royale, and in 1735 he became a full Academician. In the same year he was sent, with Charles Marie de La Condamine, Louis Godin, and Joseph de Jussieu, on the celebrated expedition to Peru that was to measure an arc of the meridian near the equator.

Bouguer's work on this expedition, from which he did not return until 1744, was of high quality. Apart from the main geodetic program, he did an astonishing amount of other scientific work, measuring the dilatation of various solids by making use of the large range of temperatures found in the Cordillera, investigating the phenomena of atmospheric refraction and the measurement of heights with the barometer, devising a new type of ship's log, and undertaking a number of other researches, in spite of the very difficult physical conditions under which the geodetic measurements had to be carried out. The results of these measurements were published formally in 1749 as *La figure de la terre, déterminée par les observations de Messieurs De la Condamine et Bouguer.* . . .

His work on naval architecture and navigation produced *Traité du navire* (1746), *Nouveau traité de navigation* (1753), and *De la manoeuvre des vaisseaux* . . . (1757), as well as several papers in the *Mémoires* of the Academy. These treatises seem to have been very useful to the naval services of the time—for example, his early paper on "lines of pursuit" (1732), one of several that display his considerable mathematical ability. He was also good with instruments, as is shown by his invention of the heliometer in 1748.

Nevertheless, in the twentieth century Pierre Bouguer is probably best known as the father of photometry, in spite of the fact that the subject seems to have been a part-time occupation, a hobby to which he returned in the last years of his life.

His interest in the measurement of light dates from about 1721, when J. J. d'Ortous de Mairan proposed a problem that necessitated a knowledge of the relative amount of light from the sun at two altitudes. Bouguer succeeded in making such a measurement of the light from the full moon on 23 November 1725, by comparing it with that of a candle.

Bouguer's achievement was to see that the eye could be used, not as a meter but as a null indicator, i.e., to establish the equality of brightness of two adjacent surfaces. He then made use of the law of inverse squares, first clearly set forth by Kepler. In his *Essai d'optique sur la gradation de la lumière* (1729), he showed how to compare lights in this way; he then went on to deal with the transmission of light through partly transparent substances. In the latter part of the *Essai,* Bouguer published the second of his great optical discoveries, often called Bouguer's law: In a medium of uniform transparency the light remaining in a collimated beam is an exponential function of the length of its path in the medium. This law was restated by J. H. Lambert in his *Photometria* (1760) and, perhaps because of the great rarity of copies of Bouguer's *Essai,* is sometimes unjustifiably referred to as Lambert's law.

Just before he died, Bouguer completed a much larger book on photometry, the *Traité d'optique sur la gradation de la lumière,* published posthumously

(1760) by his friend the Abbé Nicolas Louis de la Caille. The *Traité* goes far beyond the *Essai*, describing a number of ingenious kinds of photometers, including a method of goniophotometry, and even attempting an elaborate theory of the reflection of light from rough surfaces, although this was not successful. The third and last part of the book, however, gives a valid elementary theory of the horizontal visual range through an obscuring atmosphere, arriving at a law, usually credited to H. Koschmieder, considered to belong to the twentieth century. It is fair to consider Pierre Bouguer not only the inventor of the photometer but also the founder of an important branch of atmospheric optics. The eighteenth century is not an outstanding epoch in the history of optics, but Bouguer's contribution to that science is notable by any standard.

BIBLIOGRAPHY

I. ORIGINAL WORKS. Works by Bouguer are *Essai d'optique sur la gradation de la lumière* (Paris, 1729), reprinted in the series *Les maîtres de la pensée scientifique* (Paris, 1921); "De la méthode d'observer exactement sur mer la hauteur des astres" (1729), in *Recueil des pièces qui ont remporté le prix de l'Académie Royale des Sciences . . . ,* **2,** no. 4 (1732); "De la mâture des vaisseaux" (1727), *ibid.,* **1,** no. 8 (1732); "De la méthode d'observer en mer la déclinaison de la boussole" (1731), *ibid.,* **2,** no. 6 (1732); *Traité du navire, de sa construction, et de ses mouvemens* (Paris, 1746); *La figure de la terre, déterminée par les observations de Messieurs De la Condamine et Bouguer, de l'Académie Royale des Sciences, envoyés par ordre du Roy au Pérou pour observer aux environs de l'équateur . . .* (Paris, 1749); *Nouveau traité de navigation, contenant la théorie et la pratique du pilotage* (Paris, 1753); *De la manoeuvre des vaisseaux . . .* (Paris, 1757); and *Traité d'optique sur la gradation de la lumière: Ouvrage posthume . . . publié par M. l'Abbé de la Caille . . .* (Paris, 1760), Latin translation by Joachim Richtenburg (Vienna, 1762), Russian translation by N. A. Tolstoy and P. P. Feofilov with a commentary by A. A. Gershun (Moscow, 1950), English translation with introduction and notes by W. E. K. Middleton (Toronto, 1961). Bouguer also contributed more than thirty articles to the *Mémoires de l'Académie Royale des Sciences* and the *Journal des Sçavans.*

II. SECONDARY LITERATURE. Works on Bouguer include Jean Paul Grandjean de Fouchy, "Éloge de M. Bouguer," in *Histoire de l'Académie Royale des Sciences, Paris* (1758), 127–136; Roland Lamontagne, *La vie et l'oeuvre de Pierre Bouguer* (Montreal and Paris, 1964), a short memoir dealing mainly with Bouguer's relations with the Americas that contains a brief list of manuscript sources on pp. 95–96; M. Prevost and R. d'Amat, eds., *Dictionnaire de biographie française* (Paris, 1954), VI, 1298; and Vasco Ronchi, *Storia della luce* (Bologna, 1952), Ch. 6.

W. E. KNOWLES MIDDLETON

BOUILLES, CHARLES DE. See **Bouvelles, Charles.**

BOUIN, POL ANDRÉ (*b.* Vendresse, Ardennes, France, 11 June 1870; *d.* Vendresse, 5 February 1962), *biology.*

Pol Bouin, the son and grandson of veterinary surgeons, grew up in the Ardennes, where at that time horse-breeding was a flourishing occupation. He would often tell how his interest in testicular physiology and pathology arose out of his father's method of treating cryptorchism in horses and pigs. While a student at Nancy he was attracted to the study of histology by the example of his teacher, Auguste Prenant. As early as 1895, he fixed his attention on phenomena of degeneration in the testes, to which problem he devoted what became a significant medical thesis in 1897. He began working with Paul Ancel in 1903, a collaboration which, developing through thirty years of close friendship and fruitful cooperation, laid the fundamental groundwork for the rapid development of reproductive endocrinology. Bouin and Ancel performed many types of operations on laboratory animals, and to test their experimental results relied on the morphology of the gonads, the genital tract, and secondary sexual characteristics. They were pioneers in the physiology of reproduction long before the isolation (around 1930) of sex hormones.

It will be useful to summarize here their essential discoveries, which continue to be valid. By employing convergent techniques, Bouin and Ancel elucidated the dual function of the testis: in the first place, gametogenesis (the production of semen) in the interior of the seminiferous tubules; and in the second place, the secretion of hormones in the interstitial gland located between the seminiferous tubules. They demonstrated that the interstitial gland controls the secondary sexual characteristics in the male. What used to be called the interstitial theory attracted few supporters at the outset, a large number of specialists remaining faithful to the old superstition about the importance of the "seed" in male potency. Violent controversies set in, reaching their peak between 1920 and 1925, which is to say on the eve of the discovery of male hormones. Some biologists, for example Champy in Paris and Stieve in Halle, bitterly resisted the demonstration that the interstitial gland is the source of male hormones. On the other side Steinach in Vienna and Lipschütz in Dorpat (currently living in Santiago de Chile) defended the theory ardently. Those extinct disputes are worth recalling as an essential chapter in the history of testicular endocrinology.

The same pattern appears in the work of Bouin

and Ancel on ovarian physiology and especially on the corpus luteum. Among the original techniques they devised was that of using a male rabbit which had been rendered sterile (while remaining potent) by ligaturing of the ductus deferens. During sexual intercourse the rabbit was able to rupture the graafian follicle of the female, inducing ovulation. Between 1909 and 1911 the two colleagues demonstrated irrefutably that in the absence of fertilization the corpus luteum, through an internal secretion, controls the readying of the uterine mucosa for implantation of the ovum as well as the morphogenetic development of the mammary glands. The effect is particularly striking in female rabbits. These results gave rise to intensive discussion until progesterone, the hormone secreted by the corpus luteum, was isolated by the epoch-making work of George Corner and Willard Allen in 1929.

From time to time Bouin worked in the field of cytology, which he enjoyed enormously. His drawings of spermatogenesis in myriopods would in themselves have assured his distinguished reputation among cytologists in the early decades of the century.

Concern for teaching in the university played a great part in Bouin's scientific life. From the outset he collaborated with his own teacher, Prenant, in the preparation of a *Traité d'histologie* (published in two volumes, 1904–1911). A very full work for the time, completely illustrated by the authors themselves, it continues to be valuable as a source of iconography. Bouin's own *Éléments d'histologie* (two volumes, 1929–1932), a sumptuously illustrated work, is now out of print, but the illustrations have been reproduced in more recent standard publications. In his textbooks Bouin laid down the main lines of the lectures in which he attempted to exhibit correlations between structure and function. As a professor he enjoyed enormous prestige. He spoke without notes in a quiet voice and in a manner both familiar and confidential, illustrating his account with drawings of great elegance. His course made an indelible impression on his audience and was the starting point of a significant number of scientific careers.

It was in laboratory work, however, which he always followed closely, that Bouin picked out his future disciples, in the course of conversations back and forth across the microscope. Having no confidence in selecting talent by competitions, he preferred to choose his collaborators directly. He encouraged students with a flair for research to begin work in his laboratory at as youthful an age as possible. There he watched over them, particularly at the outset, with paternal solicitude. Bouin thus trained numerous disciples who have made their mark on

various levels of the university structure. The full list would be excessively long, but a few may be named: Rémy Collin at Nancy, Max Aron at Strasbourg, Robert Courrier (permanent secretary of the Academy of Science in Paris) at the Collège de France, Jacques Benoit of the Collège de France, and the author of the present article at Strasbourg.

Bouin's laboratory was for many years, and particularly during his time at Strasbourg, the starting point of a number of significant discoveries: Courrier's work on folliculin (estrone) in the female and on the physiology of periodic sexual activity in the testicle; the discovery by Stricker and Grüter in 1928 of the lactogenic hormone produced by the anterior pituitary; and the numerous works of Max Aron on comparative endocrinology. From 1925 on, the pace of Bouin's personal work slowed, but he continued to participate with undiminished vigilance in the work of his students, though without adding his signature to theirs except in very rare instances. More and more he effaced himself in the work of his collaborators while continuing to give them the benefit of his illuminating advice almost to the day of his death. His involvement in teaching reached far beyond the institute of which he was director. In consequence of a number of trips abroad he had developed very definite views on the way to recruit talent for scientific research, particularly in biology. The Rockefeller Foundation showed him great confidence through the years, especially in nominating candidates for fellowships. In 1927 the Foundation made him an extremely important grant for the construction of an entirely up-to-date Institute of Histology. Around 1930 Bouin was one of the promoters of the Caisse Nationale des Sciences, forerunner of the present Centre National de la Recherche Scientifique in Paris.

Bouin's gentle nature and surpassing kindness remained alive in the memory of all who knew him. Severe in his judgment of himself and of his collaborators when it came to the publication of scientific results, he liked to converse at length on the progress of an experiment when it was under way and, if need be, to defend its results with his authority. Having done his duty in every respect during the war of 1914–1918 and again throughout the black years from 1939 to 1945, his tolerance in political and religious matters was as complete as it was uncommon. He became a world-famous scientist during his years at Strasbourg, from 1919 to 1939 and then lost all his possessions in the course of World War II. Thereupon, he retired to his native village and devoted himself to forestry, a subject of great interest to him and one in which he had attained considerable reputation.

During the years of his retirement, from 1945 to

1962, Bouin remained continually in touch with the progress of biology, partly by keeping up with the journals and partly by correcting in a manner both kindly and precise the manuscripts that his former students continued to submit to his judgment. The example he gave of a scientist entirely devoted to disinterested research and the compassionate nature that led him to participate in the personal life of all his acquaintanceship remain a vivid memory among the numerous disciples who count themselves among the school of Bouin.

Bouin's academic career was distinguished. He was *préparateur d'histologie* in the Faculty of Medicine in Nancy in 1892 and received the *docteur en médecine* there in 1897 (when he also served as *chef des travaux d'histologie*), becoming *professeur-agrégé* of anatomy (histology) in 1898. He was *professeur titulaire* of histology and pathological anatomy at the medical school of Algiers in 1907, *professeur titulaire* of histology of the Faculty of Medicine in Nancy in 1908, and held the same position at Strasbourg in 1918.

Bouin received many honors. He was a commander of the Legion of Honor; *membre titulaire* of the Academy of Sciences of Paris and of the Académie Nationale de Médecine, Paris; and honorary foreign member of the Royal Belgian Academy of Medicine of Brussels. He won a number of prizes, including the Prince Albert I of Monaco award of the Academy of Medicine (which he shared with Paul Ancel) in 1937, the Prix de la Fondation Singer-Polignac (again with Paul Ancel) in 1951, and the gold medal of the Centre National de la Recherche Scientifique, Paris, in 1961.

BIBLIOGRAPHY

R. Courrier's *Notice sur la vie et les travaux de Pol Bouin* (Paris, 1962) contains a complete list of Bouin's scientific writings, in addition to extensive biographical information, a list of biographical sources, and a portrait.

MARC KLEIN

BOULE, MARCELLIN (*b.* Montsalvy, Cantal, France, 1 January 1861; *d.* Montsalvy, 4 July 1942), *human paleontology, geology.*

Boule was born at the beginning of the decade during which the stratified deposits of the caves of southern France were to support in abundance the long record of man's newly discovered antiquity. It was in Cantal and in its neighboring departments that, one after another, the sites were excavated to help complete the record of ancient man. Time and place set the intellectual scene for the emergence of Boule as the central figure in French prehistoric studies during the half century of his scientific career.

In 1880 Boule enrolled at the Faculty of Sciences of Toulouse, and during the succeeding four years earned the *licence* in natural sciences and then that in physical science. He completed his work toward the coveted teaching certificate in 1887 at the Museum of Natural History in Paris, to which he had received a scholarship the preceding year. The museum was then at its height as a center of research in the natural sciences, particularly geology. The influence of the museum pushed Boule in the direction of research rather than that of teaching, for which he had prepared. Although his initial interests were in the problems of stratigraphy and petrography, it was finally to paleontology that he devoted most of his efforts and to which he made his most significant research contributions. His personal association with Louis Lartet and Albert Gaudry, and particularly the enthusiasm of the latter, were responsible for the final localization of his interests. Returning to Paris in 1890 after a brief teaching assignment at Clermont-Ferrand, Boule received a doctorate in natural sciences in 1892. In the same year he became *préparateur* in the paleontological laboratory of the museum in preference to the offer of a professorship at Montpellier. An assistant to Gaudry, he succeeded him as professor of paleontology at the museum, a position he held until his retirement in 1936.

Although he began as a geologist, with much of his research in descriptive geology and stratigraphy, Boule's importance lies in the role he played in the establishment of prehistory or paleoanthropology in France. Boule himself illustrates the close kinship between geology and prehistory in the period of the latter's emergence, for in its beginnings prehistory was an extension of geology, drawing from it both its methodology and its scientific status and pretensions. No geologist working in the general region of the Auvergne could ignore the importance of the stratified deposits which provided the record of a vastly expanded time period for human existence and which made of man a species to be fitted into a geological context.

As Boule's interests in nonhuman paleontology were shaped by his association with Lartet and Gaudry, so his archaeological and paleoanthropological work was stimulated by his close association and friendship at Toulouse with Émile Cartailhac, already a distinguished prehistoric archaeologist. As early as 1884 Boule described a prehistoric flint mine; and in 1887 he made stratigraphic studies of several newly discovered Mousterian rock shelters and, with

Cartailhac, published a monograph on the Grotte de Reilhac which demonstrated that the transition from the Paleolithic to the Neolithic was such as to cast doubt on the classical thesis of a clear break between the two epochs. These early studies foreshadowed an increasing concern with the earlier phases of human prehistory; and to both the collection and the interpretation of data Boule was able to bring a unique synthetic approach that was the product of his training and competence in stratigraphic geology, paleontology, and prehistoric archaeology. It was this total view of prehistoric man that formed the essential orientation of *Les hommes fossiles: Éléments de paléontologie humaine,* which was first published in 1921 and for a quarter of a century remained, in its several editions, the essential synthesis in paleoanthropology. Here he brought together the archaeological, geological, and zoological data in order to provide the record of human achievement and adaptation through the changing landscapes of the geological past. In this he saw human evolution in the proper sense, not as a series of anatomical stages alone but rather as the continuing process by which the human form adapted to a constantly changing ecology.

Although its extensive approach was adumbrated by his earlier work with Cartailhac, *Les hommes fossiles* was built upon two decades of detailed work in paleoanthropology, of which the central accomplishment was the three-part monograph "L'homme fossile de la Chapelle-aux-Saints" (1911–1913). Its anatomical account of the most complete specimen of Neanderthal man then known set the standard for such detailed descriptions. Now known to be wrong in several particulars, Boule's precise anatomical definition of Neanderthal man and the subsequent reconstruction provided the authority for his view that such specimens represented a distinct species population more primitive or apelike than modern man, one that could not be ancestrally related to him. It was this view that long formed the cornerstone of the theoretical structure of French paleoanthropology.

Apart from his many contributions to the rapidly expanding body of knowledge relating to man's past, Boule was one of the most distinguished statesmen and persuasive representatives of this burgeoning science. One of the founders of *L'anthropologie* in 1890, he was its editor from 1893 to 1920. When Albert I of Monaco, impressed by the remains uncovered in his own principality, founded the Institute for Human Paleontology in Paris in 1914, Boule was appointed its first director, a position he held until his death. The institute, whose opening was delayed until 1920 by World War I, served as the center and instigator of research in human paleontology. Boule

founded its *Archives* and was the active agent of its researches. As editor of the two most distinguished journals in the field and as director of the only research establishment devoted solely to its pursuit, Boule set the tone and the tempo for Old World paleoanthropology in the first third of this century.

BIBLIOGRAPHY

I. ORIGINAL WORKS. Works by Boule include "Essai de paléontologie stratigraphique de l'homme," in *Revue de anthropologie* (1888–1889); *Les grottes de Grimaldi: Géologie et paléontologie* (Monaco, 1906–1910); "L'homme fossile de la Chapelle-aux-Saints," in *Annales de paléontologie,* **6–8**; *Les hommes fossiles: Éléments de paléontologie humaine* (Paris, 1921).

II. SECONDARY LITERATURE. See "Jubilé de M. Marcellin Boule," in *L'anthropologie,* **47** (1937), 583–648, which includes a complete bibliography through 1937; and H. V. Vallois, "Marcellin Boule," *ibid.,* **50** (1946), 203–210.

J. GRUBER

BOULLANGER, NICOLAS-ANTOINE (*b.* Paris, France, 11 November 1722; *d.* Paris, 1 September 1759), *geology, philosophy.*

Boullanger received a classical education at Beauvais College in Paris, then studied mathematics and architecture. He later joined the army as an engineer. In 1745 he joined the Département des Ponts et Chaussées as superintendent of works. He was appointed deputy engineer in 1749; two years later he was posted to the generality of Paris and, sometime afterward, to that of Tours.

As early as 1745 Boullanger became interested in the morphology of the Marne Basin and the peculiarities of the sedimentary terrains under its workings. In the freestone quarries he paid special attention to the soft layers, but he mistook the oolites for the eggs or embryos of shellfish. He continued his observations as far as the confluence of the Seine and the Marne, where he mistook oolites for the Miliola fossils of the rough limestone of the Paris region. He published these observations in the *Mercure de France* for June 1753. Although this work was republished several times, no naturalist before d'Archiac (1862) pointed out its errors.

Boullanger collected his observations and comments on the courses of the Loire and the Marne in "Anecdotes de la nature." Having determined that most of the strata involved had been formed from the remains of marine shellfish, he deduced that these deposits had originally constituted continuous layers in the ocean and that, after the sea had retreated, they had been eroded by the currents. He did not, however, study stratigraphic succession or identify the fossils in the various layers. Instead of adhering to observation, he sought to develop a theory on the formation of the earth that involved a universal deluge. Boullanger retained Descartes's hypothesis of a subterranean stratum of water, and attributed the deluge to a vast and sudden eruption of this water through springs. Since he could not assume the age of the earth to be that assigned it by biblical chronology, he read and cited the ancient authors in their original languages.

The modern distribution of the oceans suggested a complementary hypothesis, that of the elasticity of the earth's strata, half of which had, by bending, caused the elevation of the other half. Boullanger presented this hypothesis in his *Mémoire sur une nouvelle mappemonde* (1753) and *Nouvelle mappemonde dédiée au progrès de nos connoissances* (1753). His map showed two hemispheres: the terrestrial, with its center near Paris, comprised most of the continents; the maritime contained practically all the oceans. Boullanger considered the lifting or sinking of the earth's strata to have been the origin of the universal deluge. He expressed this theory in the article "Déluge" in the *Encyclopédie* (1754). Boullanger retired in 1758 with the rank of engineer.

BIBLIOGRAPHY

I. ORIGINAL WORKS. Boullanger's published works are *Mémoire sur une nouvelle mappemonde* (Paris, 1753); *Nouvelle mappemonde dédiée au progrès de nos connoissances* (Paris–Nuremberg, 1753; Paris, 1760); "Déluge," in *L'encyclopédie,* IV (1754); *Recherches sur l'origine du despotisme oriental* (Geneva, 1761; Paris, 1763; Amsterdam, 1766); and *L'antiquité dévoilée par ses usages . . .* (Amsterdam, 1766). An unpublished MS is "Anecdotes de la nature sur l'origine des vallées, des montagnes et des autres irrégularités extérieures et intérieures du globe de la terre . . .," Library of the Muséum d'Histoire Naturelle, MS 869. It was sent to Buffon about 1750, and he borrowed part of the description of the Langres Plateau for *Les époques de la nature* (the borrowing was reported to the *Journal de littérature, science et arts* by Nicolas Gobet). The MS was abstracted by Nicolas Desmarest as part of the article "Géographie physique," in *L'encyclopédie méthodique,* I (Paris, an III), 8–28.

II. SECONDARY LITERATURE. The basic source of information on Boullanger's life is the anonymous introduction to *L'antiquité dévoilée* (Amsterdam, 1766), attributed to Diderot by Grimm (see *Correspondance littéraire, philosophique et critique par Grimm,* Maurice Tourneux, ed., VI [Paris, 1878], 468). Other sources are John Hampton, *Nicolas-Antoine Boulanger et la science de son temps* (Geneva–Lille, 1955); and Jacques Roger, "Un manuscrit inédit perdu et retrouvé: Les *Anecdotes de la Nature* de Nicolas-Antoine Boulanger," in *Revue des sciences humaines* (Lille) (July–Sept. 1953), 231–254.

ARTHUR BIREMBAUT

BOULLIAU, ISMAEL (*b.* Loudun, France, 28 September 1605; *d.* Paris, France, 25 November 1694), *mathematics, astronomy.*

Boulliau was born of Calvinist parents, but he became a Roman Catholic at the age of twenty-one. About four years later, he was ordained a priest. His early studies had been in law and the humanities, but upon settling in Paris in 1633, he resumed an early interest in astronomical observation, a taste he had shared with his father. Thereafter, he pursued a predominantly scientific career, becoming known as Clarissimus Bullialdus. In addition to the usual French and Latin spellings of his name, there were such variants as Bouillaud, Boulliaud, and Bulliald.

The Galilean storm broke during the very year that Boulliau joined the Parisian scientific circle. A recent convert both to Catholicism and to science, he nevertheless joined his friend Gassendi in support of Galileo. Boulliau's publication of the *Philolaus* in 1639 placed him squarely in the Copernican camp, although not yet as a Keplerian. In assuming that the sun stood still, so that he could retain uniform circular motions, Copernicus had been right for the wrong reason. So it was with Boulliau. In the *Philolaus,* Boulliau went further than Copernicus in suggesting the resolution of rectilinear accelerated motion in free fall into two uniform circular components. His law of fall (equivalent to $s = k$ vers t) is in close agreement with the definitive Galilean formulation for small intervals of time only.

In 1645 Boulliau published his most significant scientific work, a more accomplished heliocentric treatise entitled *Astronomia philolaica.* He had now become one of the very few astronomers to accept the ellipticity of orbits, but he categorically rejected all those suggestions of variation in celestial forces which had made Kepler's *Astronomia nova* of 1609 more revolutionary, in a sense, than the work of Copernicus. As against Kepler's astrophysics, Boulliau preferred a geometrical astronomy which saved uniformity of circular motion. He asserted, however, that *if* a planetary moving force did in fact exist, then it should vary inversely as the square of the distance—and not, as Kepler had held, inversely as the first power. The inverse-square hypothesis, which Boulliau published in his *Astronomia philolaica,* evidently had been carried over from his *De natura lucis* of 1638, in which the inverse-square law for intensity of illumination, used earlier by Kepler, had appeared.

Rejecting all dynamic hypotheses, including the inverse-square hypothesis in astronomy, Boulliau proposed instead a kinematic representation of planetary motion in which a planet moved along a linear element of an oblique cone while the element in turn revolved uniformly about the axis of the cone. In this way, he reconciled ellipticity of orbits with uniformity of circular motion. Seth Ward modified the scheme shortly afterward in a hypothesis by which the motion of the planet is uniform as seen from the "blind" focus of the ellipse.

The *Astronomia philolaica* was one of the most important treatises written in the period between Kepler and Newton. In his *Principia,* Newton referred to Boulliau's inverse-square hypothesis and praised the accuracy of his tables (Bk. 3, Phen. 4). Boulliau was also highly regarded as a mathematician. Before he was thirty, he had prepared the first printed edition (1644) of the *Arithmetica* of Theon of Smyrna; in his fifties, he published (besides several minor works) the *De lineis spiralibus* (1657), a work inspired by Archimedes; and when he was more than seventy-five years of age, he published a ponderous *Opus novum ad arithmeticam infinitorum* (1682), purporting to clarify the *Arithmetica infinitorum* of Wallis. The mathematical works of Boulliau had little influence on the development of the subject, however, because they were old-fashioned. He evidently failed to see the significance of the Cartesian contributions, whether to mathematics or to science, and seems pointedly to have avoided mentioning Descartes's name. Boulliau's astronomical observations at Paris covered over half a century, but it has been ungenerously said that Boulliau's only permanent contribution to science is the word "evection" in astronomy. Nevertheless, it was Boulliau who, in his *Ad astronomos monita duo* of 1667, first established the periodicity of a variable star, Mira Ceti. His explanation of the phenomenon as a rotating semiluminous body or "half sun" was incorrect, but his estimate of the period as 333 days was accurate, exceeding by less than two days that determined since then.

Boulliau was one of the last reputable scholars to maintain confidence in astrology. Among the works he edited were the *Astronomicon* of Marcus Manilius (1655) and the *De judicandi facultate* of Ptolemy (1667). Despite all his publications, Boulliau's contribution to science should perhaps be measured less by his treatises and ideas than by his scientific activity. He rivaled Mersenne as a correspondent. He served as librarian, first to the brothers du Puy, then to de Thou, French ambassador in Holland, and ultimately to the Bibliothèque Royale in Paris. There he joined the groups which gave rise to the Académie des Sciences. Although never elected to the Academy, in 1663 he was among the first foreign associates elected to the Royal Society of London. It was to Boulliau that Huygens first entrusted his secret of the rings of Saturn and to him that he sent his earliest pendulum clocks. The distribution in Paris of Huygens' *Systema saturnium* (1658) was entrusted to Boulliau; and it was through Boulliau that Pascal's *Lettres d'Amos Dettonville* (1658–1659) went to English and Dutch mathematicians. Prince Leopold in Italy and Hevelius in Danzig depended upon Boulliau to keep them informed of scientific news from Paris, although at times Boulliau was himself traveling to England or Poland or the Levant, seeking out manuscripts, books, and information.

BIBLIOGRAPHY

I. ORIGINAL WORKS. Thirty-nine volumes containing Boulliau's unedited papers and correspondence are to be found in Paris (Bibliothèque Nationale, fonds franc. 13019–13058). His published works include *De natura lucis* (Paris, 1638); *Philolaus* (Amsterdam, 1639); *Astronomia philolaica* (Paris, 1645); *De lineis spiralibus* (Paris, 1657); *Ad astronomos monita duo* (Paris, 1667); *Opus novum ad arithmeticam infinitorum* (Paris, 1682). He also edited works of Theon of Smyrna (Paris, 1644), Ptolemy, and Marcus Manilius, as noted in the text.

II. SECONDARY LITERATURE. There is no biography of Boulliau. Some information on his life and work may be found in G. Bigourdan, *Histoire de l'astronomie d'observation et des observatoires en France,* pt. 1 (Paris, 1918), and in J. P. Niceron, *Mémoires pour servir à l'histoire des hommes illustres dans la république des lettres* (Paris, 1727–1745), Vols. I, X.

CARL B. BOYER

BOUQUET, JEAN-CLAUDE (*b.* Morteau, Doubs, France, 7 September 1819; *d.* Paris, France, 9 September 1885), *mathematics*.

After entering the École Normale Supérieure in 1839, Bouquet became a professor at the lycée of Marseilles. He received the *doctorat ès sciences* in 1842, presenting a thesis on the variation of double integrals, and was appointed professor at the Faculté des Sciences of Lyons. There he found his school friend Charles Briot, with whom he collaborated throughout his career.

Bouquet taught special mathematics at the Lycée Bonaparte (now the Lycée Condorcet) from 1852 to 1858, then at the Lycée Louis-le-Grand until 1867. After serving as *maître de conférence* at the École Normale Supérieure and *répétiteur* at the École Polytechnique, Bouquet succeeded J. A. Serret in the chair of differential and integral calculus at the Sorbonne (1874–1884). He was elected to the Académie des Sciences in 1875.

After his thesis Bouquet took up differential geometry, writing a memoir on the systems of straight lines of space and one on orthogonal surfaces that was basic to the important research carried on successively by Ossian Bonnet, Gaston Darboux, Maurice Levy, and Arthur Cayley.

From 1853 on, Bouquet's name is generally associated with that of his friend Briot. Their joint scientific work was a profound study and clarification of the analytic work of Augustin Cauchy. In a memoir that has remained famous since 1853, they proposed to establish precisely the conditions that a function must fulfill in order to be developable into an entire series. They also perfected the analysis by which Cauchy had, for the first time, established the existence of the integral of a differential equation. They opened the way to research on singular points and showed their importance for knowledge of the integral. Their works of 1859 and 1875 on elliptic functions finally brought out the great force of Cauchy's analytic methods.

The mathematical activity of Bouquet and Briot was equaled by remarkable teaching activity. Bouquet, who was as fond of teaching as of science, taught Jules Tannery. Collaborating with Briot, he produced several textbooks that went into numerous printings.

BIBLIOGRAPHY

I. ORIGINAL WORKS. Bouquet's works include "Sur la variation des intégrales doubles," doctoral thesis (Faculté des Sciences, Paris, 1842); "Remarques sur les systèmes de droites dans l'espace," in *Journal des mathématiques pures et appliquées*, 1st ser., **11** (1846), 125 ff.; "Note sur les surfaces orthogonales," *ibid.*, 446 ff.; *Mémoire sur les propriétés d'un système de droites* (Lyons, 1848); "Sur la courbure des surfaces," a note in Cournot's *Traité de la théorie des fonctions* (Paris, 1857), along with other, lesser notes by Bouquet and Briot; "Mémoire sur la théorie des intégrales ultra-elliptiques," in a shorter version in *Comptes rendus des séances de l'Académie des sciences* (1868), which led to a report by J. A. Serret on 4 July 1870, in *Recueil des savants étrangers*, pp. 417–470; *Notice sur les travaux mathématiques de M. Bouquet* (Paris, 1870); "Sur l'intégration d'un système d'équations différentielles totales simultanées du Ier ordre," in *Bulletin des sciences mathématiques et astronomiques*, 3 (1872), 265 ff.; "Note sur le calcul des accélérations des divers ordres dans le mouvement d'un point sur une courbe gauche," in *Annales scientifiques de l'École normale supérieure*, 2nd ser., 3 (1874).

Works written in collaboration with Charles Briot are "Note sur le développement des fonctions en séries con-

vergentes, ordonnées suivant les puissances croissantes de la variable," in *Comptes rendus des séances de l'Académie des sciences,* **36** (1853), 334; "Recherches sur les séries ordonnées suivant les puissances croissantes d'une variable imaginaire," *ibid.*, 264 ff.; "Recherches sur les propriétés des fonctions définies par des équations différentielles," *ibid.*, **39** (1854), *séance* of 21 August; "Additions au mémoire précédent," *ibid.*—Cauchy's report on this memoir, *ibid.*, **40** (1855), 567 ff.; "Recherches sur les fonctions doublement périodiques," *ibid.*, **40** (1855), 342 ff.; "Mémoire sur l'intégration des équations différentielles au moyen des fonctions elliptiques," *ibid.*, **41** (1855), 1229, with Cauchy's report in **43** (1856), 27, *séance* of 7 July 1856. All these memoirs, divided into three distinct parts, form, "with certain modifications," the *Journal de l'École polytechnique,* **36** (1856).

Other works are *Théorie des fonctions doublement périodiques et en particulier des fonctions elliptiques* (Paris, 1859), also translated into German (Halle, 1862); *Leçons de géométrie analytique* (Paris, 1875); *Leçons nouvelles de trigonométrie* (Paris, 1875); and *Théorie des fonctions elliptiques* (Paris, 1875).

II. SECONDARY LITERATURE. Works on Bouquet are Michel Chasles, *Rapport sur les progrès de la géométrie* (Paris, 1870), pp. 214–215; G. H. Halphen, "Notice nécrologique sur Bouquet," in *Comptes rendus des séances de l'Académie des sciences,* **102**, no. 23 (7 June 1886); and Jules Tannery, "Notice nécrologique sur Bouquet," in *Mémorial de l'Association des anciens élèves de l'École normale* (Paris, 1885).

JEAN ITARD

BOUR, EDMOND (*b.* Gray, Haute-Saône, France, 19 May 1832; *d.* Paris, France, 9 March 1866), *mathematics, analytical mechanics, celestial mechanics.*

Bour, the son of Joseph Bour and Gabrielle Jeunet, came from a rather modest provincial family. After receiving his secondary education in Gray and Dijon, he was admitted in 1850 to the École Polytechnique, from which he graduated first in his class in 1852; he then continued his studies at the École des Mines in Paris. At this time he worked on the paper "Sur l'intégration des équations différentielles de la mécanique analytique," which he read before the Académie des Sciences of Paris on 5 March 1855. He also wrote two theses in celestial mechanics, one on the three-body problem and the other on the theory of attraction, which he set forth brilliantly before the Faculté des Sciences in Paris on 3 December 1855.

In July 1855 Bour became both a mining engineer and professor of mechanics and mining at the École des Mines of Saint-Étienne, but he returned to Paris at the end of 1859 as lecturer in descriptive geometry at the École Polytechnique. The following year he was appointed professor at the École des Mines, and professor of mechanics at the École Polytechnique in 1861. Also in 1861 he received the grand prize in mathematics awarded by the Académie des Sciences for his paper "Théorie de la déformation des surfaces." In April 1862 Bour was a candidate for membership in the Académie des Sciences but was defeated by Ossian Bonnet. Disappointed by this failure, he concentrated entirely on his course in mechanics at the École Polytechnique.

Although Bour died of an incurable disease at the age of thirty-four, he left valuable works in mathematical analysis, algebra, infinitesimal geometry, theoretical and applied mechanics, and celestial mechanics. In mechanics his essential contributions dealt with differential equations in dynamics, the theme of his first memoir and of another published in 1862; the analytical study of the composition of movements (1865); and the reduction of the three-body problem to the plane case. In infinitesimal geometry his memoir on the deformation of surfaces, in line with the analogous studies of Bonnet and Codazzi, contained several theorems on ruled surfaces and minimal surfaces; but in its printed version this work does not include the test for the integration of the problem's equations in the case of surfaces of revolution, which had enabled Bour to surpass the other competitors for the Academy's grand prize.

BIBLIOGRAPHY

I. ORIGINAL WORKS. A nearly complete list of Bour's published works is given in Poggendorff and the *Catalogue of Scientific Papers* (see below). Among his works are "Sur l'intégration des équations différentielles de la mécanique analytique," in *Journal de mathématiques pures et appliquées,* **20** (1855), 185–202; his theses for the *docteur-ès-sciences* were published separately (Paris, 1855) as *Thèses présentées à la Faculté des Sciences à Paris pour obtenir le grade de docteur-ès-sciences.* . . . and reproduced in *Journal de l'École polytechnique,* **21,** cahier 36 (1856), 35–84; "Théorie de la déformation des surfaces," *ibid.,* **22,** cahier 39 (1862), 1–148; *Cours de mathématiques et machines,* 3 vols. (Paris, 1865–1874); and *Lettres choisies,* Joseph Bertin and Charles Godard, eds. (Gray, 1905).

II. SECONDARY LITERATURE. Works on Bour are M. Chasles, *Rapport sur les progrès de la géométrie* (Paris, 1870), pp. 211, 295, 325–327; M. D'Ocagne, *Histoire abrégée des sciences mathématiques* (Paris, 1955), p. 300; A. Franceschini, in *Dictionnaire de biographie française,* VI (1954), col. 1383; A. de Lapparent, in *École Polytechnique. Livre du centenaire,* I (Paris, 1895), 143–145; "Notice biographique sur Edmond Bour," in *Nouvelles annales de mathématiques,* 2nd ser., **6** (1867), 145–157; Poggendorff, III, 172–173; and Royal Society of London, *Catalogue of Scientific Papers, 1800–1863,* I (1867), 532.

RENÉ TATON

BOURBAKI, NICOLAS. Bourbaki is the collective pseudonym of an influential group of mathematicians, almost all French, who since the late 1930's have been engaged in writing what is intended to be a definitive survey of all of mathematics, or at least of all those parts of the subject which Bourbaki considers worthy of the name. The work appears in installments that are usually from 100 to 300 pages long. The first appeared in 1939 and the thirty-third in 1967; many intervening installments have been extensively revised and reissued. The selection of topics is very different from that in a traditional introduction to mathematics. In Bourbaki's arrangement, mathematics begins with set theory, which is followed, in order, by (abstract) algebra, general topology, functions of a real variable (including ordinary calculus), topological vector spaces, and the general theory of integration. To some extent the order is forced by the logical dependence of each topic on its predecessors. Bourbaki has not yet reached the other parts of mathematics. Although the work as a whole is called *Elements of Mathematics,* no one could read it without at least two years of college mathematics as preparation, and further mathematical study would be an advantage.

The exact composition of the Bourbaki group varies from year to year and has been deliberately kept mysterious. The project was begun by a number of brilliant young mathematicians who had made important contributions to mathematics in their own right. At the beginning they made no particular attempt at secrecy. With the passage of time, however, they seem to have become more and more enamored of their joke, and have often tried to persuade people that there is indeed an individual named N. Bourbaki, who writes the books. Indeed, Bourbaki once applied for membership in the American Mathematical Society, but was rejected on the ground that he was not an individual. The original group included H. Cartan, C. Chevalley, J. Dieudonné, and A. Weil (all of whom are among the most eminent mathematicians of their generation). Many younger French mathematicians have joined the group, which is understood to have ten to twenty members at any one time and has included two or three Americans. The founding members are said to have agreed to retire at the age of fifty, and are believed to have done so, although with some reluctance.

The origin of the name Nicolas Bourbaki is obscure. The use of a collective pseudonym was presumably intended to obviate title pages with long and changing lists of names and to provide a simple way of referring to the project. The family name appears to be that of General Charles-Denis-Sauter Bourbaki (1816–

1897), a statue of whom stands in Nancy, where several members of the group once taught. Possibly the Christian name was supposed to suggest St. Nicholas bringing presents to the mathematical world.

In the early days Bourbaki published articles in mathematical journals, as any mathematician would. He soon gave that up, however, and his reputation rests on his books. People who are unsympathetic to the "new mathematics" introduced into the schools since 1960 accuse Bourbaki of having inspired that movement. The accusation is probably unjustified, although aspects of his work bear a superficial resemblance to less attractive aspects of new mathematics. Bourbaki himself does not intend his approach to be used even in college teaching. Rather, it is meant to improve a mathematician's understanding of his subject after he has learned the fundamentals and to serve as a guide to research.

The most obvious aspects of Bourbaki's work are his insistence on a strict adherence to the axiomatic approach to mathematics and his use of an individual and (originally) unconventional terminology (much of which has since become widely accepted). The former is the more important. Any mathematical theory starts, in principle, from a set of axioms and deduces consequences from them (although many subjects, such as elementary algebra, are rarely presented to students in this way). In classical axiomatic theories, such as Euclidean geometry or Peano's theory of the integers, one attempts to find a set of axioms that precisely characterize the theory. Such an axiomatization is valuable in showing the logical arrangement of the subject, but the clarification so achieved is confined to the one subject, and often seems like quibbling.

A good deal of the new mathematics consists of introducing such axiomatizations of elementary parts of mathematics at an early stage of the curriculum, in the hope of facilitating understanding. Bourbaki's axiomatization is in a different spirit. His axioms are for parts of mathematics with the widest possible scope, which he calls structures. A mathematical structure consists, in principle, of a set of objects of unspecified nature, and of certain relationships among them. For example, the structure called a group consists of a set of elements such that any two can be combined to give a third. The way in which this is done must be subject to suitable axioms. The structure called an order consists of a set of elements with a relationship between any two of them, corresponding abstractly to the statement (for numbers) that one is greater than the other.

Having studied a structure, one may add axioms to make it more special (finite group or commutative group, for example). One can combine two structures, assuming that the objects considered satisfy the axioms of both (obtaining, for example, the theory of ordered groups). By proceeding in this way, one obtains more and more complicated structures, and often more and more interesting mathematics. Bourbaki, then, organizes mathematics as an arrangement of structures, the more complex growing out of the simpler.

There are great advantages in dealing with mathematics in this way. A theorem, once proved for an abstract structure, is immediately applicable to any realization of the structure, that is, to any mathematical system that satisfies the axioms. Thus, for example, a theorem about abstract groups will yield results (which superficially may look quite different) about groups of numbers, groups of matrices, or groups of permutations. Again, once it is recognized that the theory of measure and the theory of probability are realizations of a common set of axioms, all results in either theory can be reinterpreted in the other. Historically, in fact, these two theories were developed independently of each other for many years before their equivalence was recognized. Bourbaki tries to make each part of mathematics as general as possible in order to obtain the widest possible domain of applicability. His detractors object that he loses contact with the actual content of the subject, so that students who have studied only his approach are likely to know only general theorems without specific instances. Of course, the choice of an axiom system is never arbitrary. Bourbaki's collaborators are well aware of the concrete theories they are generalizing, and select their axioms accordingly.

Bourbaki has been influential for a number of reasons. For one thing, he gave the first systematic account of some topics that previously had been available only in scattered articles. His orderly and very general approach, his insistence on precision of terminology and of argument, his advocacy of the axiomatic method, all had a strong appeal to pure mathematicians, who in any case were proceeding in the same direction. Since mathematicians had to learn Bourbaki's terminology in order to read his work, that terminology has become widely known and has changed much of the vocabulary of research. The effect of the work in the development of mathematics has been fully commensurate with the great effort that has gone into it.

BIBLIOGRAPHY

I. ORIGINAL WORKS. Works by Bourbaki include "The Architecture of Mathematics," in *American Mathematical*

Monthly, **57** (1950), 221–232; and *Éléments de mathé-matique,* many numbers in the series Actualités Scientifiques et Industrielles (Paris, 1939–   ).

II. SECONDARY LITERATURE. André Delachet, "L'école Bourbaki," in *L'analyse mathématique* (Paris, 1949), pp. 113–116; Paul R. Halmos, "Nicolas Bourbaki," in *Scientific American,* **196,** no. 5 (May 1957), 88–99.

R. P. BOAS, JR.

BOURDELIN, CLAUDE (*b.* at or near Villefranche-sur-Saône, France, *ca.* 1621; *d.* Paris, France, 15 October 1699), *chemistry.*

Bourdelin was a beneficiary of the tradition of iatrochemistry. After having completed his apprenticeship under master apothecaries in Paris, he purchased an apothecary's license from the house of "Monsieur" (the king's younger brother, Gaston of Orléans). For some twenty years he held two offices in the house of Monsieur: that of assistant apothecary to Monsieur and the officers of his household, and that of apothecary to the stable.

Bourdelin's scientific knowledge and his title assured him prominent patients and powerful connections, which helped him become a member of the Royal Academy of Sciences when it was created in 1666. Next to the meeting room of the Academicians he installed the Academy's laboratory, equipped with furnaces and a distilling apparatus. There he began research in chemical analysis in March 1667 and continued until the end of 1686.

Initially, Bourdelin worked with Samuel Cottereau-Duclos, the king's physician, on the analysis of mineral waters. Subsequently, he devoted himself to the chemical analysis of plant, animal, and mineral substances; of normal and pathological organic liquids; and of water, both fresh and salt. His usual technique for plant analysis was distillation, which enabled him to find phlegm, an acid or corrosive spirit, volatile salt, oil, solid salt (sometimes in the form of tartar and sometimes of marine salt), and a residue, the *caput mortuum* of the alchemists.

Bourdelin kept careful records of his expenses, for which the Academy reimbursed him at the beginning of each year, and of the authorized transfers of chemical substances to other chemists or naturalists of the Academy. As of 1 January 1687, he was authorized to work at home, and until his death he continued his experimentation and his record-keeping.

Although Bourdelin's scientific work is of no value today, his importance lies in his having made clear to some of his contemporaries and to his successors that progress in chemical knowledge required use of less antiquated experimental methods and the elaboration of hypotheses as guidelines for research.

BIBLIOGRAPHY

I. ORIGINAL WORKS. The archives of the Academy of Sciences contain eleven handwritten records of plant analyses carried out by Bourdelin between 14 June 1672 and 2 September 1699. The Bibliothèque Nationale (MS dept.) has other handwritten records kept by Bourdelin, which had been stolen by Libri: the record of expenses incurred for the laboratory between 1667 and 1699, n.a. fr. 5147; and fourteen records of analyses covering 1667–1668, n.a. fr. 5133, and 1672–1699, n.a. fr. 5134–5146. Some of Bourdelin's findings are published in *Histoire de l'Académie royale des sciences* (Paris, 1733), I, 27–35, 345–346; II, 9–10, 26–27, 68.

II. SECONDARY LITERATURE. Bourdelin's laboratory at the Academy is illustrated by Sébastien Leclerc in *Mémoires pour servir à l'histoire des plantes dressés par M. Dodart* (Paris, 1676). Valuable articles are A. Birembaut, "Le laboratoire de l'Académie royale des sciences," in *Revue d'histoire des sciences* (1969); and Paul Dorveaux, "Les grands pharmaciens. Apothicaires membres de l'Académie royale des sciences," in *Bulletin de la Société d'histoire de la pharmacie* (Aug. 1929), 290–298.

ARTHUR BIREMBAUT

BOURDELOT, PIERRE MICHON (*b.* Sens, France, 2 February 1610; *d.* Paris, France, 9 February 1685), *medicine, dissemination of science.*

Bourdelot was the son of Maximilien Michon, a barber-surgeon, and of Anne Bourdelot. About 1629 he began medical studies in Paris, where he had two uncles on his mother's side: Edmé Bourdelot, physician to Louis XIII, and Jean Bourdelot, a jurist and distinguished Hellenist. They adopted young Michon in 1634 and obtained for him the right to bear the name of Bourdelot. They also introduced him into the intellectual life of Paris. François de Noailles made Bourdelot his physician and took him to Rome in 1634.

When he returned to Paris in 1638, Bourdelot entered the service of Prince Henri II de Condé. He accompanied the latter on his campaigns in Spain, taking advantage of stopovers in Paris to pass his medical examinations. Having earned the title of king's physician, he settled in Paris early in 1642 and became the Condé family's physician. Eager for fame, he founded the Académie Bourdelot, the biweekly meetings of which were attended by nobles, men of letters, philosophers, and devotees of new fashions who were more critical than knowledgeable. Men interested in science also came, as did such truly learned men as Roberval, Gassendi, and Étienne and Blaise Pascal. Although Bourdelot often allowed the reading of extravagant theses at the meetings, his academy, along with that of Mersenne, played an important role in spreading scientific ideas in Paris. During the winter of 1647/1648 several new experiments on the vacuum were presented and discussed.

Upon the death of Henri II de Condé in 1646, Bourdelot served his son, Louis II. Political disorders brought on by the Fronde interrupted the academy's activity, and when Condé was arrested in January 1650, Bourdelot followed the dowager duchess into hiding. In October 1651, Bourdelot left the Condé family to go to Sweden as physician to Queen Christina, over whom he gained marked influence. This success aroused widespread animosity, and in June 1653 the queen was persuaded to send Bourdelot back to France.

In Paris, he obtained the living of the abbey of Massay, in Berry, which gave him the right to the title *abbé*. When the Great Condé returned from exile in 1659, Bourdelot again became his physician. His success with a treatment for gout, which he had previously developed, brought him new fame with numerous noted patients, among them Mme. de Sévigné. Early in 1664 he resumed the meetings of his academy. These meetings were attended by future members of the Académie Royale, by foreign scholars passing through, by violent partisans of Descartes or Gassendi, and by all sorts of alchemists and visionaries who espoused ideas of the past. The academy continued to meet more or less regularly until 1684. Although its *Conférences,* published in 1672 by Le Gallois, do not give a good idea of its standards and scientific level, references to some of its meetings in contemporary journals, memoirs, and correspondence indicate that interesting experimental work was often done there and that the academy helped to arouse sympathy for and interest in science. This was possible largely because the secrecy surrounding the work of the Académie Royale prevented it from participating in the dissemination of scientific knowledge.

Bourdelot's writings have little more than anecdotal interest. A possible exception is his history of music, published posthumously by his two nephews, which has recently been studied by historians of music. Nevertheless, Bourdelot played an important role in the scientific life of Paris between 1640 and 1680, providing material assistance and a means of diffusing experimental results. He also created a climate favorable to science in influential circles that participated in foreign intellectual exchanges, particularly with Italy.

BIBLIOGRAPHY

I. ORIGINAL WORKS. Bourdelot's writings are *Recherches et observations sur les vipères* . . . (Paris, 1671); *Ré-*
ponse . . . à la lettre de Boccone . . . sur l'embrasement du Mont Etna, n.p., n.d. (Paris, 1671); *Histoire de la musique et de ses effets* . . . (Paris, 1715), written with P. Bonnet-Bourdelot and J. Bonnet; and *Histoire générale de la danse sacrée et profane* . . . (Paris, 1732), written with J. Bonnet.

II. SECONDARY LITERATURE. Works on Bourdelot are Roman d'Amat, in *Dictionnaire de biographie française,* VI (1954), cols. 1439–1440; A. Cabanès, *Dans les coulisses de l'histoire* (Paris, 1923), pp. 93–123; A. Chérest, *Un médecin du grand monde au XVIIe siècle* (*Pierre Michon, devenu par adoption Pierre Bourdelot*), n.p., n.d. (Auxerre, 1861); Henri Chérot, *Trois éducations princières au dix-septième siècle* (Lille, 1896), pp. 120–132, 248; R. J. Denichou, *Un médecin du grand siècle: L'abbé Bourdelot* (Paris, 1928); F. Halévy, *Souvenirs et portraits* (Paris, 1861), pp. 87–115; H. Brown, *Scientific Organizations in Seventeenth Century France* (Bâltimore, 1933), pp. 111–112, 161, 165, 231–253, 296; P. Le Gallois, ed., *Conversations de l'Académie de Monsieur l'Abbé Bourdelot . . .* (Paris, 1672, 1673), and *Conversations académiques, tirées de l'Académie de M. l'Abbé Bourdelot* (Paris, 1674); Jean Lemoine and André Lichtenberger, *Trois familiers du grand Condé* (Paris, 1908), pp. 1–138; René Pintard, "Autour de Pascal. L'Académie Bourdelot et le problème du vide," in *Mélanges . . . offerts à Daniel Mornet* (Paris, 1951), pp. 73–81; D. Riesman, "Bourdelot, a Physician of Queen Christina of Sweden," in *Annals of Medical History,* **9** (1937), 191; and D. C. Vischer, *Der musikgeschichtliche Traktat des Pierre Bourdelot (1610–1684* [*sic*]) (Bern, 1947), dissertation (available at Bibliothèque Nationale, Paris, 8° θ *Bern.* ph. 2062).

Numerous references are in *Correspondance du P. Marin Mersenne,* C. de Waard, ed., 10 vols. (Paris, 1932–　　　); René Pintard, *Le libertinage érudit en France dans la première partie du XVIIe siècle,* 2 vols. (Paris, 1943); and René Taton, *Les origines de l'Académie Royale des Sciences* (Paris, 1966).

RENÉ TATON

BOURDON, EUGÈNE (*b.* Paris, France, 8 April 1808; *d.* Paris, 29 September 1884), *instrumentation.*

Bourdon was the son of a merchant who, expecting him to enter business, sent him to Nuremberg for two years in order to learn German. After his father's death in 1830, however, Bourdon spent two years in an optician's shop and, in 1832, set up his own instrument and machine shop. He moved in 1835 to 71 Faubourg du Temple, and continued to work there until 1872, when his sons took over its management. After his retirement, Bourdon continued to perform experiments that interested him. He died from a fall that occurred while he was testing an anemometer he had designed, one that employed a venturi tube.

In 1832 Bourdon made and presented to the Société d'Encouragement pour l'Industrie Nationale a model of a steam engine with glass cylinders; subsequently

changes. He went further in *Experiments and Considerations Touching Colours,* for here he not only described various ways of producing color changes, such as the conversion of a blue vegetable solution to red or green, but he also emphatically indicated a *use* for these color changes: chemical classification and identification. It had long been known that some acids turned the blue syrup of violets red; Boyle claimed to be the first to realize that all acids did so and that those substances that did not do so were not acids—a bold but useful distinction. Similarly, he claimed to be the first to note that alkalies—all alkalies—turned syrup of violets green. This left him with three classes of salts: acid, alkali, and those that were neither. He reinforced this empirically derived classification by observing that the blue opalescence of the yellow solution of *lignum nephriticum* (a South American wood with supposed medical virtues and of considerable optical interest) was destroyed when the solution was acidified and could be restored by the addition of alkali; he also used this reaction to determine the relative strength of acidic and alkaline solutions.

These tests also allowed Boyle to determine the purity of chemicals bought from apothecary shops. He discovered a further useful color change when he demonstrated (what he claimed to have deduced) that different alkalies give differently colored precipitates with mercury sublimate (mercuric chloride): "vegetable alkalis" (potassium carbonate and possibly sodium carbonate) give an orange precipitate (a form of mercuric oxide), while "animal alkalis" (ammonia compounds) give a white precipitate (ammonium-mercury chloride). Since limewater was already known to turn cloudy when a solution of niter was added, Boyle now had the ability to distinguish among all the common alkalies.

The importance of these tests, upon which Boyle placed absolute reliance, was very great, for in the late seventeenth century there was still great confusion over the identity, not to mention the composition, of various simple substances. He found it necessary on the one hand to insist that all salts were not common salt, but on the other that salt of tartar, hartshorn, and vegetable alkali were all one salt—a point not always appreciated by his contemporaries. Many chemists of his day insisted upon sweeping generalizations like that of the acid-alkali hypothesis (which claimed that all substances were either acids or alkalies, and all reactions therefore were neutralizations), or carelessly confused end products with starting materials.

Boyle was unique in realizing the continual need for meticulous care in examining purity, testing composition, and searching for chemical differences and similarities. How methodical he could be is amply demonstrated in his two investigations into the nature of phosphorus, *The Aerial Noctiluca* (1680) and *New Experiments and Observations Made Upon the Icy Noctiluca* (1682), in the course of which he discovered the chief chemical and physical properties of phosphorus and phosphoric acid, and in *Short Memoirs for the Natural Experimental History of Mineral Waters* (1685), an admirable set of analytical directions. Few other chemists of his day seem to have been sufficiently patient to elaborate an analytical procedure, although it became commonplace in the next century.

To his work on acids and alkalies Boyle added a host of other specific tests: for copper by the blue color of its solutions; for silver by its ability to form silver chloride, with its characteristic blackening over time; for sulfur and various mineral acids by their characteristic reactions. Some of these were new, others had been known for years or even centuries. These tests enabled him to discuss the composition of substances in what can only be called positivistic terms—that is, in terms of empirically determined components rather than in terms of metaphysical, a priori "elements." This was perhaps Boyle's greatest contribution to chemistry, for while few chemists followed him in altogether dispensing with elements, most chemists saw the utility of distinguishing in analysis between empirically verifiable components and a priori elements. When this occurred generally, the way lay open for the immense strides to be made in analytical chemistry in the eighteenth century. And although it was quite possible to practice this sort of chemistry without paying more than lip service to the corpuscular philosophy (a physicist's theory, after all), there is no doubt that Boyle was helped by his adherence to the corpuscular philosophy, and perhaps would never have formulated his new chemical method without his habit of thinking in corpuscular terms. And it is not without significance that this more physical way of approaching chemistry was introduced into France by William Homberg, who worked for some time in Boyle's laboratory.

In his lifetime Boyle was honored not only as an original chemist and physicist, a great exponent of English experimental philosophy, and a pillar of the Royal Society, but also as a prolific writer on natural theology, the point where religion and science met. He was a truly devout man and scrupulous to the point of having a tender conscience where oaths were concerned, as he explained in declining to serve as president of the Royal Society in 1680—presumably because he thought he might be subject to the provisions of the Test Act, as a public officer. Yet he was not dogmatic, and he was a devoted scientific investigator. Fortunately, he experienced no conflicts of conscience; for him a God who could create a mechanical universe—who could create matter in motion, obeying certain laws out of which the universe as we know it could come into being in an orderly fashion—was far more to be admired and worshiped than a God who created a universe without scientific law.

Boyle's God stands in the same relation to the watchmaker as the watchmaker might to an untutored savage who thinks a watch is a living creature because its hands move. Boyle never tired of writing on this subject, finding his thoughts becoming more devout the more he studied the wonders of nature. Not all his numerous books on religious subjects were an offshoot of his scientific endeavors, but many were, and it was these that were influential. At his death Boyle left a sum of money to found the Boyle lectures (really sermons), intended for the confutation of atheism; and his contemporaries immediately concluded that he meant the arguments against atheism to be drawn from the scientific advances of his day. Hence the first and most famous series of Boyle lectures, by Robert Bentley, was filled with arguments and illustrations drawn from Bentley's discussions with Newton. Subsequent Boyle lectures (by Clarke, Whiston, Woodward, Derham, and others) followed this pattern to produce what is thought of as a characteristically eighteenth-century form of "natural" religion, much less formal and theological than anything usual in Boyle's day.

In this, as in so much else, Boyle set the tone and inspired the methods of thought widely accepted by the next two generations. In large part this was so because eighteenth-century Newtonians found that Boyle's opinions, discoveries, and scientific method usefully supplemented those of Newton. Modern historians see this as a sign of Boyle's very real influence upon Newton; Newtonians saw it as a proof that the "new science" was a product of the English methods proclaimed in the charter of the Royal Society.

BIBLIOGRAPHY

The standard edition of Boyle's works is Thomas Birch, ed., *The Works of the Honourable Robert Boyle,* 5 vols. (London, 1744), 2nd ed., 6 vols. (London, 1772); available in a facs. repr. Separate works are *The Sceptical Chymist* (London–New York, 1911, and later eds.) and *Experiments and Considerations Touching Colours* (repr. New York, 1964). John Fulton, *A Bibliography of the Honourable Robert Boyle,* 2nd ed. (Oxford, 1961), is a complete bibliography of works by Boyle and of secondary sources to 1960. M. B. Hall, ed., *Robert Boyle on Natural Philosophy* (Bloomington, Ind., 1965), contains an introductory discussion of his work and long illustrative excerpts from his writings. Boyle's manuscripts are preserved chiefly in the archives of the Royal Society. Many letters to Boyle are published in Birch's edition cited above; his letters to Oldenburg are published in A. R. and M. B. Hall, eds., *The Correspondence of Henry Oldenburg* (Madison, Wis., 1965–   ).

MARIE BOAS HALL

BRACE, DE WITT BRISTOL (*b.* Wilson, New York, 5 January 1859; *d.* Lincoln, Nebraska, 2 October 1905), *optics.*

De Witt Bristol Brace, professor of physics and specialist in optics at the University of Nebraska, is best remembered for his experimental test of the Lorentz-Fitzgerald contraction hypothesis in 1904. He received his bachelor's degree from Boston University in 1881 and went on to graduate study at the Massachusetts Institute of Technology and Johns Hopkins University. In 1883 his admiration for Kirchhoff and Helmholtz took him to Berlin, where he wrote a doctoral dissertation on magneto-optical effects. In 1888, shortly after his return to the United States, Brace became professor of physics at the University of Nebraska, where he remained until his death. He was survived by his wife of four years, Elizabeth Russell Wing of Massachusetts. Brace was a member of the American and British Associations for the Advancement of Science.

For eight years Brace was the entire physics department at Nebraska and had little opportunity for research until the faculty was expanded in 1896; then, however, he began a series of experiments on the effect of various external factors, such as magnetism (the "Faraday effect"), pressure, and strain on light passing through a transparent medium. In the course of this research he invented several new instruments, including a spectropolariscope and a spectrophotometer which today bear his name.

In 1904 the opportunity arose for Brace to apply the extremely sensitive optical techniques he had developed to one of the crucial problems of his day. Two years earlier, Lord Rayleigh had proposed that the Lorentz-Fitzgerald contraction, if it existed, might produce an observable double refraction in a moving transparent medium. Rayleigh made experiments in which he failed to find the predicted effect, but his work was not quite accurate enough to be conclusive. Brace pointed this out and reconducted the investigation in his own laboratory, establishing beyond a doubt the absence of double refraction caused by movement of the refracting medium through the ether. This did not disprove the contraction hypothesis, but Brace at first believed that it did. Joseph Larmor showed that double refraction need not result from Lorentz contraction if matter is composed of electrically charged particles that contract in the same proportion as large bodies; he thus saved the Lorentz hypothesis and gave the electron its status as a fundamental particle of matter.

Brace remained interested in the problems of contraction, and in 1905 performed experiments to detect higher-order effects that might clear up the issue. Death kept him from seeing its final resolution in Einstein's theory of special relativity.

BIBLIOGRAPHY

I. ORIGINAL WORKS. Articles by Brace include "Observations on Light Propagated in a Dielectric Normal to the 1 Lines of Force," in *Philosophical Magazine,* **44** (1897), 342–349; "Description of a New Spectrophotometer . . .," *ibid.,* **48** (1899), 420–430; "Observation on the Circular Components in the 'Faraday Effect,'" in *Nature,* **62** (1900), 368–369; "A Sensitive-strip Spectropolariscope," in *Philosophical Magazine,* n.s. **5** (1903), 9; "Double Refraction in Matter Moving Through the Ether," *ibid.,* n.s. **7** (1904), 317–328; and three articles on tests for the "ether drift," *ibid.,* n.s. **10** (1905).

II. SECONDARY LITERATURE. Articles relating to Brace's work are Joseph Larmor, "On the Ascertained Absence of Effects of Motion Through the Aether . . .," in *Philosophical Magazine,* n.s. **7** (1904), 621; and Lord Rayleigh, "Does Motion Through the Aether Cause Double Refraction?," *ibid.* (1902).

The only substantial biographical material about Brace is found in the *Dictionary of American Biography,* II, 540, and *Who's Who in America, 1903–1905.* His role in the Lorentz contraction controversy is discussed in Florian Cajori, *A History of Physics* (repr. New York, 1962), p. 378. For bibliography, see the *Royal Society Catalogue of Scientific Papers,* XIII, 757, and Poggendorff, V, 157.

EUGENE FRANKEL

BRACHET, ALBERT (*b.* Liège, Belgium, 1 January 1869; *d.* Brussels, Belgium, 27 December 1930), *embryology.*

Brachet was the only son of Auguste Brachet, an industrialist, and Louise Despac. A shy, sensitive, and self-contained child, he was a mediocre student in elementary school, and later said that his last years in secondary school reduced him to intellectual torpor. Having decided to study medicine, he entered the University of Liège and at once came under the influence of its celebrated professor of zoology,

Edouard Van Beneden, whose lectures awoke young Brachet's enthusiastic interest in the form and embryonic development of animals and man. At Van Beneden's invitation, Brachet worked in the zoology laboratory in 1887–1888, learning the techniques of biological research. Thereafter, while still a medical student, he became histological preparator in the laboratory of Auguste Swaen, professor of anatomy. There Brachet began research on the development of the long bones of birds, which resulted in his first scientific publication, a monograph that appeared in 1893. In 1894 he was granted the M.D. by the University of Liège. About this time, at Swaen's suggestion Brachet, in spite of his youth, boldly attacked some of the most difficult problems in vertebrate embryology—the earliest development of the liver and pancreas, and of the diaphragm and the pleuroperitoneal cavity. He published these studies in various French and German scientific periodicals from 1895 to 1897.

When Brachet received his medical degree at the age of twenty-five, his parents wished him to practice medicine and even furnished an office for him, but by this time he had become thoroughly committed to morphological research. Winning a government traveling fellowship, he went to the University of Edinburgh, a famous center of human anatomy and the art of dissection; then to the laboratory of Ernst Gaupp at Freiburg im Breisgau to study the embryology of the head; and finally to Gustav Born at Breslau, where he learned the important new method of constructing enlarged models of embryos and their internal structures by the wax-plate method devised by Born. In Born's laboratory he also observed newly introduced procedures of experimental embryology, by which Born and other embryologists were beginning to analyze embryonic development by excising parts of early embryos (chiefly of frogs and salamanders), transplanting limb buds, and producing local damage in the ovum and in early embryos.

After Brachet's return to Liège in 1895, as assistant in anatomy, he collaborated with Swaen in studies involving the whole vertebrate group, seeking to compare the early development of the heart, blood vessels, and urinary organs in mammals with that in the lower vertebrates (fishes and amphibians), where the pattern often is simpler and more comprehensible.

This broadened experience in comparative morphology turned Brachet's attention to the earliest stages of development—the first cleavage of the fertilized ovum, the early differentiation of the embryonic area, and the beginnings of the head and trunk. These phenomena, complex and obscure in the higher animals, are susceptible of analysis by comparison with simpler processes of the same nature in lower vertebrates and in the invertebrates. For example, there is a well-known stage when the embryo has the form of a hollow ball so deeply indented at one point that the indented region coats the inside of the outer wall and the embryo is thus a two-layer sphere with an opening to the outside, the blastopore. The region of the blastopore becomes the active center of development of the embryonic body. In the vertebrates, however, the gastrula stage is so modified that it is recognizable only with difficulty. Brachet made notable contributions to the solution of this problem.

Another and even more recondite set of problems includes such questions as the extent to which embryonic structures are laid down in the undivided ovum, and the relation of the first planes of its cleavage to the axis of symmetry of the embryo. Such problems as these are to be solved partly by descriptive studies, partly by experiment. Brachet devoted himself to them for the rest of his life, by his own researches and those of the numerous advanced students who came to work with him. To the elucidation of these problems he brought the great variety of insights and methods he had gained at several centers of biological research. Among his most important studies at this time was an analysis of gastrulation in amphibians, in which he pointed out the similarity of this process to that of more primitive animals.

In 1900 Brachet was promoted to the post of *chef des travaux pratiques,* i.e., director of the students' laboratory work, from which the next step would be to a professorial chair. No such vacancy was likely to occur soon in Liège, however, and when the chair of anatomy at Brussels became vacant in 1904 Brachet, with Van Beneden's warm support, was appointed professor of anatomy and embryology, and director of the Institute of Anatomy of the University of Brussels.

Soon after he settled at Brussels, Brachet married Marguerite Guchez, a gifted student whose devotion and talents freed him from domestic concerns and allowed him to attack with undivided attention the tasks of building up the anatomy department, conducting his own intricate investigations, and guiding the numerous research students who came to his laboratory. The Brachets had two children, one of them the distinguished biochemist Jean Brachet.

At this time Brachet's experimental work was directed chiefly to analyzing the inherent capacities of the egg cell and the way in which these potentialities of future development become localized in the embryo. For example, by pricking one of the two cells into which the fertilized ovum first divides, or similarly damaging one or another region of an embryo a few days old, he could discover the rate at which the embryonic cells differentiate into organ-forming areas. Brachet also devoted much time to the study of ova, chiefly of the frog, into which, by experimental treatment, he caused more than one sperm cell to enter, with resulting abnormalities of development that threw light on the normal process. He was one of the first to confirm the possibility of causing unfertilized ova to develop parthenogenetically by mechanical stimulation, and he pioneered in experimental attempts to cultivate the mammalian embryo (rabbit) *in vitro.*

The results of his varied and persistent investigations have become integral to today's biological thought. It is to Brachet, perhaps more than to any other biologist of his time, that we owe our understanding that the fertilized egg cell, of whatever animal species, is neither a diminutive model that has only to unfold and enlarge in order to become an adult (preformation) nor an undifferentiated mass to be transformed by some mysterious force into adult tissues and organs (epigenesis) but, rather, a packet of living materials endowed with potentialities of growth and differentiation under inherent physical and chemical influences.

When World War I broke out, Brachet was in France, at a seaside laboratory in Roscoff, Brittany, and was not able to return to Brussels. He was invited to join the Medical Faculty of the University of Paris as adjunct professor of anatomy and embryology. On frequent trips to the Belgian front he gave anatomical instruction to the hospital surgeons. A highly successful series of special lectures at the Collège de France led to publication of an important book embodying Brachet's findings of the previous decade, *L'oeuf et les facteurs de l'ontogénèse* (1917).

After the war Brachet returned to Brussels, bearing with him the high regard of French colleagues, a regard soon expressed by corresponding membership in the Institut de France and an honorary doctorate from the University of Paris. Resuming his chair at Brussels, he published a valuable textbook, *Traité d'embryologie des vertèbres* (1921). After the death of Van Beneden he assumed the editorship of the influential journal *Archives de biologie.* Always working on the fundamental problems of the organization of the ovum, the nature of fertilization, and the pattern of segmentation and early embryonic differentiation, he had at his side a succession of young biologists who, inspired by his enthusiasm and intellectual force, went on to become professors of anatomy, embryology, or zoology in universities of Belgium, France, and Portugal. From 1923 to 1926, Brachet was rector of the University of Brussels. In the winter of 1928/1929 he made a successful lecture tour of the United States.

BIBLIOGRAPHY

I. ORIGINAL WORKS. Among Brachet's writings are *L'oeuf et les facteurs de l'ontogénèse* (Paris, 1917) and *Traité d'embryologie des vertèbres* (Paris, 1921).

II. SECONDARY LITERATURE. Works on Brachet are A. Celestino da Costa, "L'oeuvre embryologique d'Albert Brachet," in *Bulletin de la Société Portugaise des Sciences Naturelles,* **11** (1932), 179–220, with a list of Brachet's publications; A. M. Dalcq, "Albert Brachet, 1869–1930," in *Florilège des sciences en Belgique* (Brussels, 1968), pp. 991–1013, with a portrait; Giuseppe Levi, "Commemorazione del socio straniero, Albert Brachet," in *Rendiconti della Reale Accademia dei Lincei,* **14** (1931), 382–393, with a list of Brachet's publications; and H. de Winiwarter, *Notice d'Albert Brachet* (Brussels, 1932), with a portrait and list of publications.

GEORGE W. CORNER

BRACONNOT, HENRI (*b.* Commercy, France, 29 May 1780; *d.* Nancy, France, 13 January 1855), *chemistry.*

Braconnot, a pioneer in the study of plant and animal chemistry, was the son of Gabriel Braconnot, a lawyer. His father died in 1787, leaving to his widow's care not only Henri but also a younger brother, André. His mother enrolled Henri in the Collège de Commercy, a Benedictine school, but he rebelled against its strict discipline and she was forced to remove him and find private means of education. Meanwhile, Mme. Braconnot remarried. This compounded Henri's unhappiness, for his stepfather, a physician named Huvet, disliked him from the start and did all he could to come between the boy and his mother.

Braconnot was apprenticed to an apothecary in Nancy at the age of thirteen; two years later he went to Strasbourg as a pharmacist in a military hospital. He remained in Strasbourg until 1801, studying scientific and medical subjects at a number of institutions, the most important being the École Centrale du Département du Bas-Rhin. In 1801 he went to Paris to complete his scientific education, attending courses in chemistry, biology, and geology given by such luminaries as Fourcroy and Étienne Geoffroy Saint-Hilaire. He made a good impression on his teachers, and reinforced it by his first major paper, "Recherches sur la force assimilatrice dans les végétaux" (1807). Through the influence of Fourcroy, Braconnot secured the directorship of the Jardin Botanique de

Nancy, a position he held for the rest of his career. In 1823 he was elected a corresponding member of the Institut de France. After the death of his stepfather, Braconnot lived with his mother, to whom he was very much attached, until her death in 1843. He was highly skeptical of physicians and suffered greatly from an untreated cancer of the stomach, which caused his death.

Braconnot lived a life of great simplicity; his only amusements were literature and the theater. In adult life, in contrast to his childhood, he was retiring and painfully shy. He never married.

Braconnot's research was conditioned by his double interest in botany and chemistry. What made it original and fruitful was not any novelty of theory or experimental method but, rather, the scope and detail of his investigations, carried out over thirty years. Added to this were his experimental facility and his sensitivity to the presence of previously unknown substances. A guiding theme, exhibited in his first major paper in bizarre, romantic chemical terms and in a number of subsequent important papers, was the attempt to elucidate the steps by which the complex organic constituents of plants were synthesized from simple inorganic substances.

Braconnot's methods of analysis remained conservative—similar, for example, to those outlined by Fourcroy in 1801 for organic substances. Although he made careful gravimetric determinations of the initial and final products collectively, he almost never attempted quantitative analysis of the individual products that he had discovered. He was generally content to describe such substances as new "vegetable acids" or "animal substances" and to give some idea of their chemical properties.

From 1807 to 1819, Braconnot analyzed a variety of plant and animal substances. One of his first discoveries was a cellulose substance in mushrooms, which he named fungine. He began a study of fatty substances, which he analyzed physically into combinations of a solid tallow and a liquid oil, but his work was superseded by that of Chevreul. In 1817 he published a paper in which he disproved the idea, enunciated by Fourcroy and held by Berzelius, among others, that there existed a unitary "extractive principle" in vegetable substances. The following year, he published his discovery of ellagic acid in nutgalls.

Beginning in 1819, Braconnot embarked upon a series of researches on the effects of sulfuric acid on wood and ligneous fibers. He discovered that sawdust, when treated with concentrated sulfuric acid, was converted into a gum that, in turn, was convertible into sugar and what he called a "vegeto-sulfuric" ulmin—a substance discovered by Vauquelin in 1797.

He went on to study the effects of sulfuric acid on animal substances: gelatin, muscle fibers, and wool. In the case of gelatin, he discovered a sugar-like substance which he called *sucre de gélatine,* later named glycocoll (glycine). Sulfuric acid converted wool and muscle fiber to a white substance he called leucine.

In the 1820's and 1830's Braconnot's discoveries included pectic acid; legumin, a substance discovered in beans that he thought was analogous to albumin; populin (benzoylsalicin), discovered in the bark of the aspen; pyrogallic acid, produced from the heating of gallic acid; and xyloïdine (nitrocellulose), produced by the action of concentrated nitric acid on potato starch or wood. Various substances he thought he had discovered turned out subsequently to be identical with known ones; these included *aposépédine,* a product obtained from the distillation of the liquid of petrified cheese, later shown to be leucine, and equisetic acid, shown to be maleic acid.

BIBLIOGRAPHY

I. Original Works. For a listing of Braconnot's many articles, see the Royal Society of London's *Catalogue of Scientific Papers,* I (1867), 557–561. The biography by M. J. Nicklès (see below) contains a bibliography taken from Braconnot's own catalog of his works. He published no large monographs or textbooks. His papers include: "Recherche sur la force assimilatrice dans les végétaux," in *Annales de chimie,* 1st ser., **61** (1807), 187–246; "Mémoire sur le principe extractif et sur les extraits en général," in *Journal de physique, de chimie, d'histoire naturelle et des arts,* **84** (1817), 267–296, 325–349; "Observations sur la préparation et la purification de l'acide gallique, et sur l'existence d'un acide nouveau dans la noix de galle," in *Annales de chimie et de physique,* 2nd ser., **9** (1818), 181–189; "Mémoire sur la conversion du corps ligneux en gomme, en sucre, et en un acide d'une nature particulière, par le moyen de l'acide sulfurique; conversion de la même substance ligneuse en ulmine par la potasse," *ibid.,* **12** (1819), 172–195; "Mémoire sur la conversion des matières animales en nouvelles substances par le moyen de l'acide sulfurique," *ibid.,* **13** (1820), 113–125; "Recherches sur un nouvel acide universellement répandu dans tous les végétaux," *ibid.,* **28** (1825), 173–178; "Nouvelles observations sur l'acide pectique," *ibid.,* **30** (1825), 96–102.

II. Secondary Literature. M. J. Nicklès, "Braconnot sa vie et ses travaux," in *Mémoires de l'Académie Stanislaus* (1855), xxiii–cxlix, is the only extensive biography of Braconnot. Although a bit melodramatic about his personal life, it is very thorough in its account of his scientific career. See also E. Frémy, *Encyclopédie chimique,* I (Paris, 1882), fasc. 1, 101–103; M. J. Nicklès, "Correspondence of J. Nicklès," in *American Journal of Science,* 2nd ser., **21**

(1856), 118–119; and J. R. Partington, *A History of Chemistry,* IV (London, 1964), 251.

Seymour H. Mauskopf

BRADFORD, JOSHUA TAYLOR (*b.* Bracken County, Kentucky, 9 December 1818; *d.* Augusta, Kentucky, 31 October 1871), *medicine, surgery.*

The son of early Kentucky settlers, William and Elizabeth Johnson Bradford, Joshua attended Augusta College and began the study of medicine in the Augusta surgery of his brother Jonathan. He was graduated M.D. in 1839 from the medical school of the University of Transylvania in Lexington, Kentucky. After establishing a practice in Augusta, Bradford spent most of his professional life there and in the nearby Kentucky and Ohio counties. On 4 February 1845 he married Sarah Emily Armstrong of Augusta; their children were William and Emily.

A believer that boiled water was the best surgical dressing, Bradford became the most successful early ovariotomist of record, on either side of the Atlantic. In the early 1840's, when the newly developed operation for the removal of ovarian tumor was being abandoned because of its danger, Bradford operated on thirty patients, with only three deaths. No other surgeon of his time approached Bradford's accomplishment of keeping deaths at no more than 10 percent of all patients operated on.

During the Civil War, Bradford served as a brigade surgeon with the Union Army. In this period, especially in 1861–1862, he developed improved techniques of camp sanitation and battlefield surgery.

Ironically, in the later years of his practice Bradford devised and prescribed an ointment effective against skin cancer—but died at fifty-two of cancer of the liver.

BIBLIOGRAPHY

I. Original Works. Among Bradford's writings are "Selections From a Report on Ovariotomy," presented to the annual meeting of the Kentucky State Medical Society, Louisville, Ky. (Apr. 1857), printed in *Kentucky Medical Journal,* **15** (1917), 142–165; "Report of Cases of Ovariotomy Occurring in Kentucky in 1857," in *Medical News* (Aug. 1859); and "Complete Rupture of the Perineum of Ten Years' Standing, Successfully Operated on," in *Lancet and Obstetrics* (Feb. 1869).

An unpublished diary covering 1862 is on deposit in the Division of Manuscripts, Library of Congress. Unpublished letters and notes are in the possession of Helen Yoder Johnson (Mrs. Howard Johnson), Mrs. Charles Bradford, and the writer.

II. SECONDARY LITERATURE. Works on Bradford include W. W. Anderson, "Dr. Joshua Taylor Bradford," in *Kentucky Medical Journal*, **15** (1917), 140–142; Charles Clay, *Results of the Operation for Extirpation of Diseased Ovaria* (Manchester, 1848); E. R. Peaslee, *Ovarian Tumors, Their Pathology, Diagnosis and Treatment* (New York, 1872); and "Early History of Ovariotomy in Kentucky," in J. N. McCormack, ed., *Some of the Medical Pioneers of Kentucky* (Bowling Green, Ky., 1917), pp. 105–107.

See also *Cyclopedia of American Medical Biography; Dictionary of National Biography;* and *Biographical Encyclopedia of Kentucky* (Cincinnati, 1876).

CHARLES O'NEILL

BRADLEY, JAMES (*b*. Sherbourne, Gloucestershire, England, March 1693; *d*. Chalford, Gloucestershire, England, 13 July 1762), *astronomy.*

The Bradley family has been traced as far back as the fourteenth century at Bradley Castle, near Wolsingham, Durham, but a branch to which James Bradley's father belonged had moved south to Gloucestershire. James, the third son of William Bradley and his wife Jane Pound, was intended for the Church. His father's income was limited, however, and his education was helped financially by his uncle, the Reverend James Pound, rector of Wanstead, Essex, who was then one of the ablest amateur astronomers in England and who fostered his nephew's fondness for astronomy. Bradley was educated at Northleach Grammar School and at Balliol College, Oxford, which he entered in 1711 and from which he received his B.A. in 1714 and his M.A. in 1717; upon his appointment as astronomer royal in 1742, Oxford awarded him an honorary D.D.

Bradley was ordained in 1719 and installed as vicar of Bridstow, near Ross, Monmouthshire, by the bishop of Hereford, who also presented him with an additional sinecure living and soon after made him his chaplain. A distinguished career in the Church seemed in prospect for the clever young scholar; but Bradley, whose parochial duties were very light, was able to continue his visits to Wanstead and to take part in his uncle's astronomical observations.

Pound had introduced his nephew to a friend of his, the eminent astronomer Edmund Halley, and in 1716 Bradley had made accurate and prompt observations of Mars and certain nebulae at Halley's request. A year later Halley drew the special attention of the Royal Society to Bradley's erudition, ability, and industry, predicting that he would advance astronomical studies. In 1718 Bradley was elected a fellow of the Royal Society. Three years later he was appointed to the Savilian professorship of astronomy at Oxford and resigned his livings and gave up his

prospects in the Church, since he did not believe he could do full justice to two different employments; his Oxford appointment made astronomy no longer a spare-time hobby. Many years later, for the same reason, he refused the living of Greenwich as a means of supplementing his meager salary of £100 per year as astronomer royal.

When Halley died in 1742, Bradley was appointed—as Halley had wished—to succeed him as astronomer royal; and he held that office with great distinction for twenty years until his death. In 1744 Bradley married Susannah Peach of Chalford, Gloucestershire. There was one daughter of the marriage, born in 1745. Bradley's wife died in 1757. Bradley was humane, benevolent, and kind; a good son and an affectionate husband and father. He was very abstemious. Apart from an attack of smallpox in 1717, he seems to have enjoyed excellent health for most of his life. A hard worker, he was able to endure long hours of observing and intensive calculating with no apparent ill effects. In the last few years of his life, partly through overwork, Bradley's health gradually deteriorated, and he began to suffer from severe headaches. By 1761 he became unfit for regular work, and was obsessed by the unfounded fear that his brain was giving way. He was cared for by his deceased wife's family until he died of an abdominal inflammation.

Bradley was a fellow of the Royal Society for over forty years, and by 1748 his brilliant discoveries and work at the Royal Observatory brought him preeminence among both English and foreign astronomers. He was elected a member of the Académie Royale des Sciences and of the academies of Berlin, Bologna, and St. Petersburg.

Bradley's celebrated discovery of the aberration of light is a good example of the way in which his accuracy, industry, and clarity of perception could extract an unforeseen success from an apparent failure. Since the stars should appear to be very slightly displaced in direction because of the earth's annual motion round the sun, these parallactic displacements would, if measurable, reveal the distances of the stars. Robert Hooke had unsuccessfully attempted this in 1669, and in 1725 Samuel Molyneux, a wealthy amateur astronomer, tried to better Hooke's effort to measure the parallax of the star Gamma Draconis by means of an improved twenty-four-foot zenith sector, made by George Graham and erected at Molyneux's house at Kew. He invited his friend Bradley to join in the observations. Gamma Draconis, passing almost through the zenith, was chosen to avoid refraction and to have the telescope fixed vertically, so that it could easily be checked. Within a few days Molyneux and Bradley detected a small but increasing deviation of the star, a displacement too large and in the wrong direction to be due to its parallax. Having verified the accuracy of the instrument, they carefully measured the deviations of Gamma Draconis, finding that they went through a cycle in the course of a year and that a similar effect occurred with other stars.

Molyneux gave up the observations but Bradley continued, using a smaller and more convenient sector made by Graham that would take in a greater number of stars; this was erected at Wanstead in 1727. Bradley tested numerous hypotheses to explain the effect, but none of them would fit. One story tells that he obtained the clue when on a pleasure trip on the Thames by noticing that every time the boat put about, the vane at the masthead shifted slightly; the sailors assured him that the wind direction had not changed—the shift of the vane was due to the boat's change of direction. Bradley concluded that the phenomenon he had observed in the stars was due to the combined effect of the velocity of light and the orbital motion of the earth. He verified this by calculation, and presented an account of the work and his discovery of the aberration of light to the Royal Society in 1729, in the form of a long letter to Halley, then astronomer royal. In this paper Bradley stated that if the parallax of any of the stars he observed had been as great as one second of arc, he would have detected it, and concluded that their parallaxes were much smaller than had been hitherto supposed. He was quite correct: there are only twenty-one stars with parallaxes exceeding $0''.25$, and that of Gamma Draconis is approximately $0''.017$. The discovery not only provided an essential correction for star positions but was also the first direct observational proof of the Copernican theory that the earth moves round the sun.

In 1727 Bradley had noticed a small "annual change of declination in some of the fixed stars" for which neither precession nor aberration completely accounted, so he continued to observe the stars involved with his zenith sector. He found that stars of the right ascension near 0 hours and 12 hours were affected differently than were those near right ascension 6 hours and 18 hours. By 1732 he had guessed the real cause, suspecting "that the Moon's action upon the equatorial parts of the earth might produce these effects. . . ." He felt confident that a complete cycle of these displacements of the stars due to the moon's action would correspond to the period (nineteen years) of the revolution of the nodes of the moon's orbit, so he continued the observations for

twenty years, finding at the end of nineteen "that the stars returned into the same positions again, as if there had been no alteration at all in the inclination of the earth's axis. . . ."

Since this effect on star positions arose from a slow nodding of the earth's axis due to the moon's attraction, Bradley called it "nutation." In 1748 he announced the results to the Royal Society in a very long letter to his patron and friend the Earl of Macclesfield, himself a keen amateur astronomer. The paper contained much geometrical discussion and tables of precession, aberration, and nutation for several stars for the years 1727–1747. (At current values, aberration ranges from zero to $20''.4958$, nutation from zero to $9''.210$.) Bradley further improved the exact determination of star positions by deriving practical rules for refraction from elaborate calculations, introducing corrections for air temperature and barometric pressure.

On becoming astronomer royal, Bradley tested, adjusted, and had repairs made on the astronomical equipment at Greenwich Royal Observatory. Then, with one assistant, he embarked on an intensive program of star observations. He found, however, that Halley's instruments had developed defects that caused observational errors. He managed to obtain a grant of £1,000 from the Admiralty, and by 1750 had thoroughly reequipped the observatory; the chief additions were two mural quadrants and a transit instrument, all made by John Bird, a pupil of Graham's. As a result, the massive program of observations (at least 60,000) made at Greenwich from 1750 to 1762 attained a very high standard of accuracy, sufficient to make them useful to modern astronomers.

Throughout his adult life Bradley made many observations of bodies in the solar system as well as of stars. With his uncle, in 1719 he had derived an improved value for the solar parallax from observations of Mars. He observed and calculated the elements of several comets, and published short papers on three. In one paper (1726) Bradley derived the longitudes of Lisbon and New York from differences in the observed times of eclipses of one of Jupiter's bright satellites. He was the only astronomer to record the reappearance of Saturn's ring in 1730 from the edgewise phase. He made laudable attempts at the very difficult feat of measuring the diameters of Venus, Mars, Jupiter, and of Saturn and its ring system, a task that taxed the resources of astronomers with much larger and better telescopes a century and a half later.

As befitted an astronomer royal, Bradley was

keenly interested in the accurate measurement of time. In the early 1730's Graham experimented in London with a clock whose pendulum beat sidereal seconds, and gave Bradley the results. The clock was then sent to Jamaica and tested on the transits of certain stars, with the times and temperatures recorded. From these data Bradley worked out a correction for the higher temperatures in Jamaica and deduced a slowing of the clock by 1 minute, 58 seconds per day due to lower gravity near the equator. From Newton's theory of the relation between latitude and gravity, Bradley derived the same slowing. He then worked out a table, for each five degrees of latitude, of the lengths required for pendulums that would keep the same time as one 39.126 inches long in London, and reported the results of the investigation to the Royal Society in 1734. One use Bradley made of his new quadrants at the observatory after 1750 was to determine accurately the latitude of Greenwich. His value, $+51°\ 28'\ 38\ 1/2''$, exceeds the current one by only $1''.3$, and is closer than those derived by two of his successors.

The Royal Observatory had been founded to assist navigation—to increase the safety of ships on ocean voyages by prescribing better methods of finding longitude at sea. Bradley recognized the importance for navigation of magnetic observations, so he included magnetic instruments among his new equipment. In 1755 the Admiralty asked Bradley to examine and report on the usefulness of Tobias Mayer's new lunar tables for finding longitude at sea. After comparing them with more than 230 Greenwich observations, and doing many calculations, Bradley reported in 1756 that, subject to trials on shipboard, the tables should give the longitude to within $1/2°$. Observations made at sea proved less encouraging, however, so in 1759 and 1760 Bradley compared Mayer's tables with many more observations and worked out detailed corrections for them by laborious and intricate calculations. In 1760 he reported that the difficulty of finding longitude by this method was not insuperable, and that the corrected tables should give it with an error of less than $1°$.

Bradley was a brilliant original thinker, a very skillful observer, and a thoroughly practical astronomer who exercised unremitting care in examining the errors of his instruments and in insuring their accurate adjustment. The value of his star observations increases with time, for they provide a firm starting point for long-term investigations of stellar motions. Without his two great discoveries and his work on refraction, it is difficult to see how later progress by others in the determination of star positions, distances, and motions would have been possible.

BIBLIOGRAPHY

I. ORIGINAL WORKS. Bradley's writings are collected in *Miscellaneous Works and Correspondence of the Rev. James Bradley, D.D., F.R.S., Astronomer Royal . . .,* S. P. Rigaud, ed. (Oxford, 1832), and *Supplement to Dr. Bradley's Miscellaneous Works . . .,* S. P. Rigaud, ed. (Oxford, 1833). The 1832 volume includes all Bradley's papers and memoranda; his zenith star observations of 1725–1747; a selection of his observations of sun, moon, planets, and comets of 1715–1742 and some observations of them made at Greenwich between 1743 and 1748 and in 1759; the best of his astronomical correspondence; and a detailed biography by Rigaud. The 1833 supplement includes a discussion of Bradley's refraction calculations. (A reprint of this valuable collection is in preparation.)

Bradley's papers are printed in the *Philosophical Transactions of the Royal Society* as follows: "Comet of 1723," **33**, No. 382 (1723), 41; "Longitudes by Eclipses of a Jupiter Satellite," **34**, No. 394 (1726), 85; "Aberration," **35**, No. 406 (1729), 637; "Pendulum Experiments in London and Jamaica," **38**, No. 432 (1734), 302; "Comet of 1737," **40**, No. 446 (1737) 111; "Nutation," **45**, No. 485 (1748), 1; and "Comet of 1757," **50** (1757), 408.

Much of Bradley's work was recorded in *Astronomical Observations Made at the Royal Observatory at Greenwich From the Year 1750 to the Year 1762 by the Rev. James Bradley D.D., Astronomer Royal . . .,* 2 vols. (Oxford, 1798–1805), the second volume of which includes observations by Bradley's successor, Nathaniel Bliss.

II. SECONDARY LITERATURE. Writings on Bradley or his work are G. F. A. Auwers, *Neue Reduktion der Bradley'schen Beobachtungen aus den Jahren 1750 bis 1762,* 3 vols. (St. Petersburg, 1882–1903), a reduction of all the Greenwich observations of 1750–1762 that shows which of the observations were made by the four men who were, in succession, Bradley's assistant; an earlier reduction is F. W. Bessel, *Fundamenta astronomiae pro anno MDCCLV deducta ex observationibus viri incomparabilis, James Bradley, in specula astronomica Grenovicensi per annos 1750–1762 institutis* (Königsberg, 1818), which contains a catalogue of more than 3,000 stars, based on Bradley's observations.

See also the article on Bradley in *Dictionary of National Biography;* G. Abetti, *The History of Astronomy* (London, 1954); A. F. O'D. Alexander, *The Planet Saturn* (London, 1962), and *The Planet Uranus* (London, 1965); and H. Spencer Jones, *The Royal Observatory, Greenwich* (London, 1948), which contains an excellent short account on pp. 10–13, by a recent astronomer royal, of Bradley's outstanding achievements.

A. F. O'D. ALEXANDER

BRADLEY, RICHARD (*d.* Cambridge, England, 5 November 1732), *botany.*

Bradley's main scientific contributions were his studies on the movement of sap and on the sexual reproduction of plants. His experiments, particularly on trees, led him to consider sap as circulating in some way; from his work on tulips and hazel he drew analogies with animal reproduction, and emphasized the significance of pollination and the importance of insects in fertilization. He then went on to discuss the novel idea of cross-fertilization and the production of different strains. This work was published in his *New Improvements of Planting and Gardening* (1717) and *A General Treatise on Husbandry and Gardening* (1724).

Bradley was a prolific science writer, producing more than twenty botanical works, as well as writing on the plague at Marseilles in 1720, advocating cleanliness and a "wholesome diet" as prophylactics. His style was clear and readable, and his reputation immense; indeed, his publications did much to encourage a scientific approach to gardening and husbandry. Bradley claimed to have invented the kaleidoscope, which he used for preparing symmetrical designs for formal gardens, thus anticipating the claims of Sir David Brewster by some ninety years. He also strongly advocated the use of steam to power the irrigation of gardens and farmland.

On 1 December 1712 Bradley was elected a fellow of the Royal Society of London, and on 10 November 1724 was appointed to the chair of botany at Cambridge University. It is said he obtained the latter by claiming a verbal recommendation from the botanist William Sherard (1659–1728) and promising to provide a botanic garden at his own expense. He provided no garden and was unfamiliar with Latin and Greek, and, because of some supposed scandal, there was a petition to remove him. It proved of no avail, and he died in office. Sir Joseph Banks and other botanists named genera to commemorate him.

BIBLIOGRAPHY

Bradley's books include *The Gentleman and Farmer's Kalendar, Directing What Is Necessary to Be Done Every Month* (London, 1718); *New Improvements of Planting and Gardening, Both Philosophical and Practical; Explaining the Motion of the Sap and Generation of Plants* (London, 1718); *The Plague at Marseilles Consider'd: With Remarks Upon the Plague in General* (London, 1721); *Precautions Against Infection: Containing Many Observations Necessary to Be Considered, at This Time, on Account of the Dreadful Plague in France* (London, 1722); *A General Treatise on Husbandry and Gardening,* 2 vols. (London, 1724); *Philo-*

sophical Account of the Works of Nature* (London, 1725); *A Survey of Ancient Husbandry and Gardening* (London, 1725); *The Country Gentleman and Farmer's Monthly Director* (London, 1726); *A Complete Body of Husbandry* (London-Dublin, 1727); *Dictionarium botanicum,* 2 vols. (London, 1728); *The Riches of a Hop-Garden Explain'd* (London, 1729); *A Course of Lectures Upon the Materia Medica* (London, 1730); and *Collected Writings on Succulent Plants,* with an introduction by G. D. Rowley, a facsimile ed. (London, 1946).

His scientific papers include "Motion of Sap in Vegetables," in *Philosophical Transactions of the Royal Society,* **29** (1716), 486–490; and "Some Microscopical Observations and Curious Remarks on the Vegetation and Exceeding Quick Propagation of Moldiness of the Substance of a Melon," *ibid.,* 490–492.

A work containing information on Bradley is Richard Pulteney, *Historical and Biographical Sketches of the Progress of Botany in England,* II (London, 1790), 129–133.

COLIN A. RONAN

BRADWARDINE, THOMAS (*b.* England, *ca.* 1290–1300; *d.* Lambeth, England, 26 August 1349), *mathematics, natural philosophy, theology.*

Both the date and place of Bradwardine's birth are uncertain, although record of his early connection with Hartfield, Sussex, has often been taken as suggestive. His own reference (*De causa Dei,* p. 559) to his father's present residence in Chichester is too late to be relevant.

Our knowledge of Bradwardine's academic career begins with the notice of his inscription as fellow of Balliol College in August 1321. Two years later we find him as fellow of Merton College, a position he presumably held until 1335. We also have evidence of a number of other university positions during this period. The succession of his Oxford degrees would seem to be the following: B.A. by August 1321; M.A. by about 1323; B.Th. by 1333; D.Th. by 1348.

Bradwardine's ecclesiastical involvement appears to have begun with his papal appointment as canon of Lincoln in September 1333, although his less official entry about 1335 into the coterie of Richard de Bury, then bishop of Durham, was probably of greater importance in determining the remainder of his career. For not only did this latter move place Bradwardine in more intimate contact with some of the more engaging theologians in England, but it also may well have proved to be of some effect in introducing him to the court of Edward III. Indeed, shortly after his appointment as chancellor of St. Paul's, London (19 September 1337), we find him as chaplain, and perhaps confessor, to the king (about 1338/1339). We know that he accompanied Edward's retinue, perhaps to Flanders, but certainly to France during the campaign of 1346. In point of fact, it was late in that year, in France, that Bradwardine delivered his (still extant) *Sermo epinicius* in the presence of the king, the occasion being the commemoration of the battles of Crécy and Neville's Cross. The closeness of his ties with Edward might also be inferred from the fact that the king annulled Bradwardine's first election to the archbishopric of Canterbury (31 August 1348). He was, however, elected a second time (4 June 1349), apparently without Edward's opposition, and consecrated at Avignon approximately a month later (10 July 1349). Bradwardine immediately returned to England, where, after scarcely more than a month as archbishop, he fell before the then raging plague and died at the residence of the bishop of Rochester in Lambeth, 26 August 1349.

Although our evidence is not absolutely conclusive, it seems highly probable that Bradwardine composed all of his philosophical and mathematical works between the onset of his regency in arts at Oxford and approximately 1335.

Early Logical Works. In spite of the lack of any direct testimony, it is nevertheless a reasonable assumption that the early logical works were the result of a youthful Bradwardine first trying his hand at a kind of activity common, even expected, among recent arts graduates in the earlier fourteenth century. A number of these logical treatises ascribed to Bradwardine are undoubtedly spurious, but at least two seem, to judge in terms of present evidence, to be most probably genuine: *De insolubilibus* and *De incipit et desinit.* Neither of these works has been edited or studied, yet the likelihood is not great that they will eventually reveal themselves to be much more than expositions of the *opinio communis* concerning their subjects. Both treatises are of course relevant to the history of medieval logic, but the *De incipit et desinit*, like the many other fourteenth-century tracts dealing with the same topic, had a direct bearing upon current problems in natural philosophy as well. For the medieval works grouped under this title (or under the alternative *De primo et ultimo instanti*) address themselves to the problem of ascribing what we would call intrinsic or extrinsic boundaries to physical changes or processes occurring within the continuum of time. Thus, to cite the fundamental assumption of Bradwardine's *De incipit* (an assumption shared by almost all his contemporaries), the duration of the existence of a permanent entity (*res permanens*) that lasts through some temporal interval is marked by the fact that it possesses a first instant of being (*primum instans in esse*) but no last

instant of being (*ultimum instans in esse*); its termination is signified, rather, by an extrinsic boundary, a first instant of nonbeing (*primum instans in non esse*).

Tractatus de proportionibus velocitatum in motibus. It is this work, composed in 1328, that has firmly established Bradwardine's position within the history of science. As its title indicates, it treats of the "ratios of speeds in motions," a description of the contents of the *Tractatus* that becomes more properly revealing once one identifies the basic problem Bradwardine set out to resolve: How can one correctly relate a variation in the speeds of a mobile (expressed, as in the work's title, as a "ratio of speeds") to a variation in the causes, which is to say the forces and resistances, determining these speeds? The proper answer to this question is, without doubt, the fundamental concern of the *Tractatus de proportionibus.* In Bradwardine's own words, to find the correct solution is to come upon the *vera sententia de proportione velocitatum in motibus, in comparatione ad moventium et motorum potentias* (Crosby ed., p. 64).

Answers to this question were, Bradwardine points out, already at hand. Yet they failed, he argues, to resolve the problem satisfactorily. Basically, their failure lay in that they would generate results which were inconsistent with the "postulate" of Scholastic-Aristotelian natural philosophy which stipulated that motion could ensue only when the motive power exceeded the power of resistance: when, to use modern symbols, $F > R$. Thus, for example, one unsatisfactory answer was that implied by Aristotle. For, although Aristotle was certainly not conscious of Bradwardine's problem as such and, it can be argued, never had as a goal the firm establishment of any mathematical relation as obtaining for the variables involved, the medieval natural philosopher took much of what he had to say in the *De caelo* and the *Physics* (especially in bk. VII, ch. 5) to entail what is now most frequently represented by $V \propto F/R$. But this will not do. For, if one begins with a given $F_1 > R_1$, and if one continually doubles the resistance (i.e., $R_2 = 2R_1$, $R_3 = 2R_2$, etc.), then F_1 as a randomly given mover will be of infinite capacity (*quelibet potentia motiva localiter esset infinita* [*ibid.*, p. 98]). In our terms, what Bradwardine intends by his argument is that, under the continual doubling of the resistance, if we hold F_1 constant, then at some point one will reach $R_n > F_1$, which, on grounds of the suggested resolution represented by $V \propto F/R$, implies that some "value" would still obtain for V; this in turn violates the "motion only when $F > R$" postulate. Therefore, $V \propto F/R$ is an unacceptable answer to his problem. Bradwardine also sets forth related arguments against other possible answers, which are usually symbolized by $V \propto F - R$ and $V \propto (F - R)/R$.

The correct solution, in Bradwardine's estimation, is that "the ratio of the speeds of motions follows the ratio of the motive powers to the resistive powers and vice versa, or, to put the same thing in other words: the ratios of the moving powers to the resistive powers are respectively proportional to the speeds of the motions, and vice versa. And," he concludes, "geometric proportionality is that meant here" (*Proportio velocitatum in motibus sequitur proportionem potentiarum moventium ad potentias resistivas, et etiam econtrario. Vel sic sub aliis verbis, eadem sententia remanente: Proportiones potentiarum moventium ad potentias resistivas, et velocitates in motibus, eodem ordine proportionales existunt, et similiter econtrario. Et hoc de geometrica proportionalitate intelligas* [*ibid.*, p. 112]).

Given just this much, it is not at all immediately clear what Bradwardine had in mind. His intentions reveal themselves only when one begins to examine his succeeding conclusions and, especially, the examples he uses to support them. If we generalize what we then discover, we can, in modern terms, say that his solution to his problem of the corresponding "ratios" of speeds, forces, and resistances is that speeds vary arithmetically while the ratios of forces to resistances determining these speeds vary geometrically. That is, to use symbols, for the series $V/n, \cdots V/3, V/2, V, 2V, 3V, \cdots nV$, we have the corresponding series $(F/R)^{1/n}, \cdots (F/R)^{1/3}, (F/R)^{1/2}, F/R, (F/R)^2, (F/R)^3, \cdots (F/R)^n$. Or, straying an even greater distance from Bradwardine himself, we can arrive at the now fairly traditional formulations of his so-called "dynamical law":

$$(F_1/R_1)^{V_2/V_1} = F_2/R_2 \text{ or } V = \log_a F/R,$$
$$\text{where } a = F_1/R_1.$$

Furthermore, if we continue our modern way of putting Bradwardine's solution to his problem, we can more easily express the advantage it had over the medieval alternatives cited above. In essence, this advantage lay in the fact that Bradwardine's "function" allowed one to continue deriving "values" for V, since such values—the repeated halving of V, for example—never correspond to a case of $R > F$ (as was the case with $V \propto F/R$); they correspond, rather, to the repeated taking of roots of F/R, and if the initial $F_1 > R_1$ (as is always assumed), then for any such root $F_n/R_n = (F_1/R_1)^{1/n}$, F_n is always greater than R_n. With this in view, it would seem that Bradwardine's most notable accomplishment lay in discovering a mathematical relation governing speeds, forces, and resistances that fit more adequately than others the Aristotelian-Scholastic postulates of motion involved in the problem he set out to resolve.

It is of the utmost importance to note, however, that almost all we have said in expounding Bradwardine's function goes well beyond what one finds in the text of the *Tractatus de proportionibus* itself. Notions of arithmetic versus geometric increase or of the exponential character of his "function" may well translate his intentions into our way of thinking, but they also simultaneously tend to mislead. Thus, to speak of exponents at once implies or suggests a mathematical sophistication that is not in Bradwardine, and also obscures the relative simplicity of his manner of expressing (by example, to be sure) his "function." This simplicity derives from the symmetrical use of the relevant terminology: If one *doubles* a speed, he would say, then one *doubles* the ratio of force to resistance, and if one *halves* the speed, then one *halves* the ratio. Although we feel constrained to note that doubling or halving a *ratio* amounts, in our terms, to squaring or taking a square root, such an addendum was unnecessary for Bradwardine, since for him the effect of applying such operations as doubling or halving to ratios was unambiguous. To double A/B always gave—again in our terms—$(A/B)^2$, and never $2(A/B)$. What is more, the examples Bradwardine utilizes to express his function deal *merely* with doubling and halving, a factor which makes it evident that he was still at a considerable remove from the general exponential function so often invoked in explaining the crux of the *Tractatus de proportionibus*.

This limitation not only derived from the fact that the relevant material in Aristotle so often spoke of doubles and halves, but may also be related to the possible origin of Bradwardine's function itself. The locale of this origin is medieval pharmacology, where we find discussion of a problem similar to Bradwardine's; in place of investigating the corresponding variations between the variables of motion, we have instead to do with an inquiry into the connection of variables within a compound medicine and its effects. Given any such medicine, how is a variation in the strength (*gradus*) of its effect related to the variation of the relative strengths (*virtutes*) of the opposing qualities (such as hot-cold or bitter-sweet) within the medicine which determine that effect? As early as the ninth century, the Arab philosopher al-Kindī replied to this question by stipulating that while the *gradus* of the effect increases arithmetically, the ratio of the opposing *virtutes* increases geometrically (where this geometric increase follows the progression of successive "doubling," that is, squaring). Now, not only was the pertinent text of al-Kindī translated into Latin, but the essence of his answer to this pharmacological puzzle was analyzed, developed, and, in a way, even popularized by the late thirteenth-century physician and alchemist Arnald of Villanova. From Arnald's work al-Kindī's "function" found its way into early fourteenth-century pharmacological works, even into the *Trifolium* of Simon Bredon, fellow Mertonian of Bradwardine in the 1330's.

Now it is certainly possible, indeed even probable, that Bradwardine may have appropriated his function from the al-Kindian tradition (a borrowing that may also have occurred in the case of the use of "exponential" relations within certain fourteenth-century alchemical tracts as well). But even admitting this, Bradwardine did a good deal more than simply transfer the function from the realm of compound medicines to the context of his problem of motion. For, quite unlike his pharmacological forerunners, he developed the mathematics behind his function by axiomatically connecting it with the whole medieval mathematics of ratios as he knew it. Thus, the entire first chapter (there are but four) of his *Tractatus* is devoted to setting forth the mathematical framework required for his function. A beginning exposition of the standard Boethian division of particular numerical ratios (e.g., *sesquialtera, superpartiens,* etc.) furnishes him with the terminology with which he was to operate. Second, and of far greater insight and importance, he axiomatically tabulated the substance of the medieval notion of composed ratios. That is, to use our terms, A/C is composed of (*componitur ex*) $A/B \cdot B/C$. Furthermore (and here lies the specific connection with Bradwardine's function), when $A/B = B/C$, then $A/C = (A/B)^2$. Or, as Bradwardine stated in general, if $a_1/a_2 = a_2/a_3 = \cdots a_{n-1}/a_n$, then $a_1/a_n = a_1/a_2 \cdot a_2/a_3 \cdots a_{n-1}/a_n$ and $a_1/a_n = (a_1/a_2)^{n-1}$. In point of fact, the insertion of geometric means and the addition of continuous proportionals that are here manipulated were Bradwardine's, and the standard medieval, way of dealing with what are (for us), respectively, the roots and powers involved in his function.

The fact that Bradwardine was thus able to state in its general form the medieval mathematics behind his function suggests that, although his expression of the function itself in mathematical terms was never general, this was due to his inability to formulate such a general mathematical statement. The best he could do was, perhaps, to give his function in the rather opaque, and certainly mathematically ambiguous, form we have quoted *in extenso* above, and then merely to express the mathematics of it all by way of example.

Proper generalization of Bradwardine's function had to await, it seems, his successors. Hence, John

Dumbleton, like Bradwardine a Mertonian, gave a more general interpretation of the function through a more systematic investigation of its connections with the composition of ratios (see his *Summa de logicis et naturalibus,* pt. III, chs. 6–7). He also extended Bradwardine by translating him, as it were, into the then current language of the latitude of forms (that is, equal latitudes of motion [V] always correspond to equal latitudes of ratio [F/R], where the corresponding "scales" of such latitudes are, respectively, what we would term arithmetic and geometric).

A further development of Bradwardine, in many ways the most brilliant, occurred in the *Liber calculationum* of Richard Swineshead, yet another Mertonian successor. In *Tractatus* XIV (entitled *De motu locali*) of this work, Swineshead elaborates his predecessor's function by setting forth some fifty-odd rules that, assuming Bradwardine to be correct, specify which different *kinds* of change (uniform, difform, uniformly difform, and so on) in F/R obtain relative to corresponding variations in V. Swineshead also extended Bradwardine in *Tractatus* XI (*De loco elementi*) of his *Liber calculationum,* where, in what is something of a fourteenth-century mathematical tour de force, he applies his function to the problem of the motion of a long, thin, heavy body (*corpus columare*) near the center of the universe.

Another significant medieval development of Bradwardine's function was effected by Nicole Oresme in his *De proportionibus proportionum.* Here one observes an extension of the mathematics implicit in the function into a whole new "calculus of ratios" in which rules are prescribed for dealing with what are, for us, rational and irrational exponents. Moreover, Oresme then applies this calculus to the problem of the possible incommensurability of heavenly motions and the consequences of such a possibility for astrological prediction.

Many other Scholastic legatees of the Bradwardinian tradition could be cited as well, but, unlike the three we have mentioned above, most appear to have concerned themselves chiefly with rather belabored expositions of what Bradwardine meant, although a few, such as Blasius of Parma and Giovanni Marliani, produced somewhat unimpressive dissents from his opinion.

One should not close an account of the *Tractatus de proportionibus* without some mention of its final chapter. Here, in effect, Bradwardine attacks the question of the appropriate measure of a body in uniform motion, a matter that becomes problematic when rotational movement is considered. Again investigating, and rejecting, proposed alternative solutions to his question, he argues that the proper measure must be determined by the fastest-moving point of the mobile at issue. Once more his decision bore fruit, especially in his English successors. The resulting "fastest moving point rule" gave birth to an extensive literature treating of the sophisms that arise when one attempts to apply the rule to bodies undergoing condensation and rarefaction or generation and corruption. (The work of the Mertonian William Heytesbury furnishes the best example of this literature.)

Tractatus de continuo. In book VI of the *Physics,* Aristotle had formulated a battery of arguments designed to refute, once and for all, the possible composition of any continuum out of indivisibles. Like all Aristotelian positions, this received ample confirmation and elaboration in the works of his Scholastic commentators. Yet two features of the medieval involvement with this particular segment of *Physics* VI are especially important as background to Bradwardine's entry, with his *Tractatus de continuo,* into what was soon to become a heated controversy among natural philosophers. To begin with, from the end of the thirteenth century on, Scholastic support and refortification of Aristotle's anti-indivisibilist position almost always included a series of mathematical arguments that did not appear in the *Physics* itself or in the standard commentary on it afforded by Averroës. Considerable impetus and authority were given to the inclusion of such arguments by the fact that Duns Scotus had seen fit to feature them, as it were, in his own pro-Aristotelian treatment of the "continuum composition" problem in book II of his *Commentary on the Sentences.* The second important medieval move in the history of this problem occurred in the early years of the fourteenth century, when we witness the eruption of anti-Aristotelian, proindivisibilist sentiments. These two factors alone do much to explain the nature and purpose of Bradwardine's treatise, for he wrote it (sometime after 1328, since it refers to his *Tractatus de proportionibus*) to combat the rising tide of atomism, or indivisibilism, as personified by its two earliest adherents: Henry of Harclay (chancellor of Oxford in 1312) and Walter Chatton (an English Franciscan, *fl. ca.* 1323). Furthermore, in attacking the atomistic views of his two adversaries, Bradwardine used as his most lethal ammunition the appeal to mathematical arguments that, as we have noted, were by now standard Scholastic fare. But he developed this application of mathematics to the problem at issue far beyond that of his predecessors.

The *Tractatus de continuo* was, first of all, mathematical in form as well as content, for it was modeled on the axiomatic pattern of Euclid's *Elements,* beginning with twenty-four "Definitions" and ten "Suppositions," and continuing with 151 "Conclusions" or "Propositions," each of them directly critical of the atomist position. These "Conclusions" purport to reveal the absurdity of atomism in all branches of knowledge (to wit: arithmetic, geometry, music, astronomy, optics, medicine, natural philosophy, metaphysics, logic, grammar, rhetoric, and ethics), but the nucleus of it all lies in the geometrical arguments Bradwardine brought to bear upon his opponents.

To understand, even in outline, the substance and success of what Bradwardine here accomplished, one should note at the outset that the atomism he was combating was, at bottom, mathematical. The position of the fourteenth-century atomistic thinker consisted in maintaining that extended continua were composed of nonextended indivisibles, of points. Given this, Bradwardine astutely saw fit to expand the mathematical arguments that were already popular weapons in opposing those of atomist persuasion. Such arguments can be characterized as attempts to reveal contradictions between geometry and atomism, in which the revelation takes place when assorted techniques of radial and parallel projection are applied to the most rudimentary of geometrical figures. For example, parallels drawn between all the "point-atoms" in opposite sides of a square will destroy the incommensurability of the diagonal, while the construction of all the radii of two concentric circles will, if both are composed of extensionless indivisibles, entail the absurdity that they have equal circumferences. In applying these and related arguments, Bradwardine effectively demolished the atomist contentions of his opponents, at least when they maintained that the atoms composing geometrical lines, surfaces, and solids were finite in number or, if not that, were in immediate contact with one another. His success (and that of others who employed similar arguments against atomism) was not as notable, however, in the case of an opponent who held that continua were composed of an infinity of indivisibles between any two of which there is always another. Here the argument by geometrical projection faltered due to a failure—which Bradwardine shared—to comprehend the one-to-one correspondence among infinite sets and their proper subsets (although this property was properly appreciated, it seems, by Gregory of Rimini in the 1340's).

The major accomplishment of Bradwardine's *Tractatus de continuo* lay, however, in yet another mathematical refutation of his atomist antagonist. To realize the substance of what he here intended, we should initially note that Aristotle's own arguments in *Physics* VI against indivisibilism made it abundantly clear that the major problem for any prospective mathematical atomist was to account for the connection or contact of the indivisibles he maintained could compose continua. As if to grant his opponent all benefit of the doubt, Bradwardine suggests that this problematic contact of point-atoms might appropriately be interpreted in terms of the eminently respectable geometrical notion of superposition (*superpositio*), a respectability guaranteed for the medieval geometer in the application of this notion within Euclid's proof of his fundamental theorems of congruence (*Elements* I, 4 and 8; III, 24). However, immediately after this concession to the opposing view, Bradwardine strikes back and, in a sequence of propositions, conclusively reveals that the superposition of any two geometrical entities systematically excludes their forming a single continuum. Consequently, the urgently needed contact of atoms is geometrically inadmissible.

Finally, as if to reveal his awareness of the mathematical basis of his whole *Tractatus,* toward its conclusion Bradwardine puts himself the question of whether, in using geometry as the base of his refutation of atomism, he had perhaps not begged the very question at issue; does not geometry *assume* the denial of atomism from the outset? He replies by carefully pointing out that while some kinds of atomism are, by assumption, denied in geometry, others are not. And he explains why and how. In our terms, he has attempted to point out just which continuity assumptions are independent of the axioms and postulates, both expressed and tacit, of Euclid's *Elements* and which are not. That he realized the pertinence of such an issue to the substance of the medieval continuum controversy is certainly much to his credit.

Geometria speculativa and *Arithmetica speculativa.* These two mathematical works, about which we lack information concerning the date of composition, are both elementary compendia of their subjects and were intended, it seems plausible to claim, for arts students who may have wished to learn something of the quadrivium, but with a minimal exposure to mathematical niceties. The *Arithmetica* is the briefer of the two and appears to be little more than the extraction of the barest essentials of Boethian arithmetic. More interesting, both to us and to the medievals themselves, to judge from the far greater number of extant manuscripts, is the *Geometria speculativa.* From the mathematical point of view, it contains little of startling interest, although it does include elementary materials not developed in Euclid's *Elements* (e.g., stellar polygons, isoperimetry, the filling of space by touching polyhedra [*impletio loci*], and so on). Of

greater significance would seem to be Bradwardine's concern with relating the mathematics being expounded to philosophy, even to selecting his mathematical material on the basis of its potential philosophical relevance. Such a guiding principle was surely in Bradwardine's mind when he saw fit to have his compendium treat of such philosophically pregnant matters as the horn angle, the incommensurability of the diagonal of a square, and the puzzle of the possible inequality of infinites. Indeed, it is precisely to passages of the *Geometria* dealing with such questions that we find reference in numerous later authors, such as Luis Coronel and John Major. Such authors were fundamentally philosophers—philosophers, moreover, with little mathematical expertise—and it would seem fair to conclude that Bradwardine had just this type of audience in view when he composed his *Geometria.*

Theological Works. Bradwardine's earliest venture into theology is perhaps represented by his treatment of the problem of predestination, extant in a *questio* entitled *De futuris contingentibus.* His major theological work, indeed the *magnum opus* of his whole career, is the massive *De causa Dei contra Pelagium et de virtute causarum ad suos Mertonenses,* completed about 1344. Its primary burden was to overturn the contemporary emphasis upon free will, found in the writings of those with marked nominalist tendencies (the "Pelagians" of the title), and to reestablish the primacy of the Divine Will. Although this reaffirmation of a determinist solution to the problem of free will is not of much direct concern to the history of science, brief excursions into sections of the *De causa Dei* have revealed that it is not without interest for the development of late medieval natural philosophy. Thus, to cite but two instances, Bradwardine discusses the problem of an extramundane void space and, within the context of rejecting the possible eternity of the world, again struggles with the issue of unequal infinites. One is tempted to suggest that closer study of the *De causa Dei* will reveal that Bradwardine's theological efforts contain yet other matters of importance for the history of science.

BIBLIOGRAPHY

I. Life. The fundamental point of departure is A. B. Emden, *A Bibliographical Register of the University of Oxford to A.D. 1500,* I (Oxford, 1957), 244–246. One may also profitably consult H. A. Obermann, *Archbishop Thomas Bradwardine, a Fourteenth Century Augustinian* (Utrecht, 1958), pp. 10–22. The *Sermo epinicius* has been edited with a brief introduction by H. A. Obermann and J. A. Weisheipl, in *Archives d'histoire doctrinale et littéraire du moyen age,* **25** (1958), 295–329.

II. WRITINGS AND DOCTRINE. The most complete bibliography of the editions and MSS of Bradwardine's works is to be found in the unpublished thesis of J. A. Weisheipl, "Early Fourteenth-Century Physics and the Merton 'School'" (Oxford, 1957), Bodl. Libr. MS D. Phil. d.1776.

Logical Works. The unedited *De insolubilibus* is extant in at least twelve MSS, including Erfurt, Amplon. 8° 76, 6r–21v and Vat. lat. 2154, 13r–24r. For the equally unedited *De incipit et desinit:* Vat. lat. 3066, 49v–52r and Vat. lat. 2154, 24r–29v. Although Bradwardine's treatises are not considered, the kinds of problems they bear upon are dealt with (for the *De insolubilibus*) in I. M. Bochenski, *A History of Formal Logic* (Notre Dame, Ind., 1961), pp. 237–251; and (for the *De incipit*) in Curtis Wilson, *William Heytesbury. Medieval Logic and the Rise of Mathematical Physics* (Madison, Wis., 1956), pp. 29–56. A variety of other logical writings, although often ascribed to Bradwardine in MSS, are most likely spurious; they are too numerous to mention here.

Tractatus de proportionibus velocitatum in motibus. This has been edited and translated, together with an introduction, by H. Lamar Crosby as *Thomas of Bradwardine. His Tractatus de Proportionibus. Its Significance for the Development of Mathematical Physics* (Madison, Wis., 1955). Corrections to some of Crosby's views can be found in Edward Grant, ed., *Nicole Oresme. De proportionibus proportionum and Ad pauca respicientes* (Madison, Wis., 1966), pp. 14–24, a volume that also contains a text, translation, and analysis of Oresme's extension of the mathematics of Bradwardine's "function." For the problem and doctrine of Bradwardine's *Tractatus* one should also note Marshall Clagett, *The Science of Mechanics in the Middle Ages* (Madison, Wis., 1959), pp. 215–216, 220–222, 421–503; and Anneliese Maier, *Die Vorläufer Galileis im 14. Jahrhundert,* 2nd ed. (Rome, 1966), pp. 81–110. Discussion of some of the factors of the above-cited application of Bradwardine's function in the *Liber calculationum* of Richard Swineshead can be found in John E. Murdoch, *"Mathesis in philosophiam scholasticam introducta:* The Rise and Development of the Application of Mathematics in Fourteenth Century Philosophy and Theology," in *Acts of the IVth International Congress of Medieval Philosophy, Montreal, 1967* (in press); and M. A. Hoskin and A. G. Molland, "Swineshead on Falling Bodies: An Example of Fourteenth Century Physics," in *British Journal for the History of Science,* **3** (1966), 150–182, which contains an edition of the text of Swineshead's *De loco elementi* (*Tractatus* XI of his *Liber calculationum*). For a new interpretation of how Bradwardine's function should be understood, see A. G. Molland, "The Geometrical Background to the 'Merton School': An Exploration Into the Application of Mathematics to Natural Philosophy in the Fourteenth Century," in *British Journal for the History of Science,* **4** (1968), 108–125. A brief discussion and citation of the relevant texts in Dumbleton that treat of Bradwardine will appear in an article by John Murdoch in a forthcoming volume of the *Boston University Studies in the Philosophy of Science.* Finally, the issue of the probable pharmacological origin of Bradwardine's function is treated in Michael McVaugh,

"Arnald of Villanova and Bradwardine's Law," in *Isis,* **58** (1967), 56–64.

Tractatus de continuo. The as yet unpublished text of this treatise was first indicated, and extracts given, in Maximilian Curtze, "Über die Handschrift R. 4° 2: Problematum Euclidis Explicatio, des Königl. Gymnasial Bibliothek zu Thorn," in *Zeitschrift für Mathematik und Physik,* **13** (1868), Hist.-lit. Abt., 85–91. A second article giving a partial analysis of the contents of the *Tractatus* is Edward Stamm, "Tractatus de Continuo von Thomas Bradwardina," in *Isis,* **26** (1936), 13–32, while V. P. Zoubov gives a transcription of the enunciations of the definitions, suppositions, and propositions of the *Tractatus,* with an accompanying analysis of the whole, in "Traktat Bradvardina 'O Kontinuume,'" in *Istoriko-matematicheskie Issledovaniia,* **13** (1960), 385–440. A critical edition of the text has been made from the two extant MSS (Torun, Gymn. Bibl. R. 4° 2, pp. 153–192; Erfurt, Amplon. 4° 385, 17r–48r) by John Murdoch and will appear in a forthcoming volume on mathematics and the continuum problem in the later Middle Ages. Some indication of the issues dealt with in the *De continuo* can be found in Anneliese Maier, *Die Vorläufer Galileis im 14. Jahrhundert,* 2nd ed. (Rome, 1966), pp. 155–179; and John Murdoch, *"Rationes mathematice." Un aspect du rapport des mathématiques et de la philosophie au moyen age,* Conférence, Palais de la Découverte (Paris, 1961), pp. 22–35; "Superposition, Congruence and Continuity in the Middle Ages," in *Mélanges Koyré,* I (Paris, 1964), 416–441; and "Two Questions on the Continuum: Walter Chatton (?), O.F.M. and Adam Wodeham, O.F.M.," in *Franciscan Studies,* **26** (1966), 212–288, written with E. A. Synan.

Mathematical Compendia. The *Arithmetica speculativa* was first printed in Paris, 1495, and reprinted many times during the fifteenth and sixteenth centuries. The *Geometria speculativa* (Paris, 1495) was also republished, and has recently been edited by A. G. Molland in his unpublished doctoral thesis, "*Geometria speculativa* of Thomas Bradwardine: Text with Critical Discussion" (Cambridge, 1967); cf. Molland's "The Geometrical Background to the 'Merton School,'" cited above. Brief consideration of the *Geometria* can also be found in Moritz Cantor, *Vorlesungen über Geschichte der Mathematik,* 2nd ed., II (Leipzig, 1913), 114–118. One might note that a good deal of Bradwardine's *Geometria* was repeated in a fifteenth-century *Geometria* by one Wigandus Durnheimer (MS Vienna, Nat. Bib. 5257, 1r–89v). The *Rithmomachia* ascribed to Bradwardine (MSS Erfurt, Amplon. 4° 2, 38r–63r; Vat. Pal. lat. 1380, 189r–230v) is most probably spurious.

The *Questio de futuris contingentibus* was edited by B. Xiberta as "Fragments d'una qüestió inèdita de Tomas Bradwardine," in *Beiträge zur Geschichte der Philosophie des Mittelalters,* Supp. **3,** 1169–1180. The *editio princeps* of the *De causa Dei* at the hand of Henry Savile (London, 1618) has recently been reprinted (Frankfurt, 1964). The basic works dealing with Bradwardine's theology are Gordon Leff, *Bradwardine and the Pelagians* (Cambridge, 1957), and H. A. Obermann, *Archbishop Thomas Bradwardine. A Fourteenth Century Augustinian* (Utrecht,

1958), whose bibliographies give almost all other relevant literature. For the discussion of void space and infinity in the *De causa Dei,* see Alexandre Koyré, "Le vide et l'espace infini au XIVe siècle," in *Archives d'histoire doctrinale et littéraire du moyen age,* **17** (1949), 45–91; and John Murdoch, *"Rationes mathematice"* (see above), pp. 15–22. Also of value are A. Combes and F. Ruello, "Jean de Ripa I Sent. Dist. XXXVII: De modo inexistendi divine essentie in omnibus creaturis," in *Traditio,* **23** (1967), 191–267; and Edward Grant, "Medieval and Seventeenth-Century Conceptions of an Infinite Void Space Beyond the Cosmos," in *Isis* (in press).

If one disregards the various epitomes of the *De causa Dei,* it would appear that the only remaining work which may well be genuine is a *Tractatus de meditatione* ascribed to Bradwardine (MSS Vienna, Nat. Bibl. 4487, 305r–315r; Vienna, Schottenkloster 321, 122r–131v). Of the numerous other works that are in all probability spurious, it will suffice to mention the *Sentence Commentary* in MS Troyes 505 and the *Questiones physice* in MS Vat. Pal. lat. 1049, which is not by Bradwardine but, apparently, by one Thomas of Prague.

JOHN E. MURDOCH

BRAGG, WILLIAM HENRY (*b.* Westward, Cumberland, England, 2 July 1862; *d.* London, England, 12 March 1942), *physics.*

Born on his father's farm near Wigton, Bragg was the eldest child of Robert John Bragg, former officer in the merchant marine, and Mary Wood, daughter of the vicar of the parish of Westward. His mother died when he was seven. A bachelor uncle, William Bragg, a pharmacist and the dominant member of the family, then took his namesake to live with him. After six years Bragg's father removed his son from the uncle's house in Market Harborough (50 miles northeast of Cambridge) and sent him to King William's College, a public school on the Isle of Man. Bragg continued, however, to return to Market Harborough during vacations even after he had gone up to Cambridge, and to look forward to his uncle's pride in his accomplishments.

Bragg was always at the top of his school class, quiet and rather unsocial but tall, strong, and good at competitive sports. Having outstripped his schoolmates, he made little progress in his final year, 1880–1881. "But a much more effective cause for my stagnation was the wave of religious experience that swept over the upper classes of the school during that year. . . . we were terribly frightened and absorbed; we could think of little else."[1] The mature Bragg preserved his composure by refusing to take literally the biblical threat of eternal damnation, although he retained his faith and his abhorrence of atheism.

Bragg entered Trinity College on a minor scholarship, obtaining a major scholarship the following year. Beginning his work at Cambridge in the long vacation, July and August 1881, he went up every "long" afterward. Under Routh's coaching he read mathematics, and only mathematics, "all the morning, from about five to seven in the afternoon, and an hour or so every evening" for three years, coming out third wrangler in Part I of the mathematical tripos in 1884. "I never expected anything so high. . . . I was fairly lifted into a new world. I had new confidence: I was extraordinarily happy."[2] Bragg obtained first-class honors in Part III of the mathematical tripos in 1885, and left Cambridge at the end of that year upon being appointed to succeed Horace Lamb as professor of mathematics and physics at the University of Adelaide. Although in his last year at Cambridge Bragg attended lectures by J. J. Thomson at the Cavendish Laboratory, at the time of his appointment his physical studies had not included any electricity; he subsequently attempted Maxwell's *Treatise* only after reading more elementary texts.

At Cambridge, Bragg published nothing; in his first eighteen years at Adelaide (1886–1904) he published three minor papers on electrostatics and the energy of the electromagnetic field. Rather, his efforts were invested in the development of a marvelously, indeed beguilingly, simple and comprehensible style of public and classroom exposition, in the affairs of his university, and in those of the Australasian Association for the Advancement of Science. One of the Australian notables by virtue of his office, in 1889 he married the daughter of the postmaster and government astronomer, Charles Todd, and fell in with the extensive but relaxed social life and out-of-doors recreations. His elder son, William Lawrence, caddied for his father, a fine golfer; his daughter was a devoted companion.

This is not the sort of life that brings election to the Royal Society of London (1907), the Bakerian lectureship (1915), the Nobel Prize in physics (1915), the Rumford Medal of the Royal Society (1916), sixteen honorary doctorates (1914–1939), presidency of the Royal Society (1935–1940), and membership in numerous foreign academies, including those of Paris, Washington, Copenhagen, and Amsterdam. The new life began, at age forty-one, in 1903–1904.

In 1903 Bragg was once again president of Section A (astronomy, mathematics, and physics) of the Australasian Association for the Advancement of Science. His presidential address, delivered at Dunedin, New Zealand, on 7 January 1904, was entitled "On Some Recent Advances in the Theory of the Ionization of Gases."[3] Conscious that he was addressing Rutherford's "friends and kindred," and possibly stimulated by the unavoidable comparison between his own accomplishments and those of the younger man, Bragg gave a highly critical review of the field, finding fault with much of the work that had been done and with many of the assumptions upon which it rested. The most damaging criticism was directed toward the work on the scattering and absorption of the ionizing radiations (α, β, and γ rays) by matter—the atoms of which Bragg, following the most modern views, supposed to consist of "thousands of electrons." The absorption of the particulate β and α rays had, unjustifiably, been assumed analogous to the exponential decrease in intensity of a wave traversing an absorbing medium; moreover, the analogy confounded energy flux and particle flux. "The exponential law is not applicable to this kind of radiation. . . . 'Amount of radiation' is not a term with definite meaning."[4] If an exponential law seemed to hold, that was only because of the superposition of a variety of factors—principally the broad spectrum of initial velocities of the particles and the scattering of the particles by the absorber.

In the spring of 1904, "through the generosity of a constant friend of the University of Adelaide"[5] and with the assistance of R. D. Kleeman, Bragg began experiments on the absorption of α particles emitted by a radium bromide source. Early in September, Bragg and Kleeman reported the results of a rather thorough investigation that combined simple experiments and highly ingenious analysis.[6] The α particles fell into a few groups, each of which had a definite range, and thus a definite initial velocity. Each group corresponded to a different radioactive species in the source, so that the measurement of α particle ranges soon became an invaluable tool in identifying radioactive substances.

For the next two and a half years, until the spring of 1907, Bragg followed up this line of investigation very vigorously, publishing a paper every few months. Then, as in 1903–1904, a highly critical review paper ("On the Properties and Natures of Various Electric Radiations")[7] heralded a reorientation of his interest. Again the title was a misnomer, for the main point of the paper was to present arguments supporting "the possibility that the γ and X rays may be of a material nature," specifically neutral pairs consisting of an electron and an α particle. (This was a full year before Rutherford and Geiger found the α particle to be doubly charged.)

As early as January 1904, Bragg had expressed doubts about the identity of γ rays and X rays—the latter just then being rather convincingly shown to have many properties of transverse ether pulses.[8] Now, considering γ and X rays to be of the same nature, he declared the evidence in favor of the ether pulse theory to be "overrated," and emphasized that the theory was unable to account for the large quantity of energy and momentum that remained in the ray regardless of the distance from its source, and that could all be delivered to a single electron. During the following five years Bragg backed off somewhat from this concrete model of the γ ray, emphasizing its "corpuscular" rather than its "material" nature,[9] but did not abandon the general concept of an electron-with-its-charge-neutralized until after the discovery of X-ray diffraction in 1912. Thus, initially without being aware of the views of Einstein and Stark, Bragg became the first, and remained the foremost, English-language advocate of a view of X rays that stressed their "quantal" properties.[10]

Barkla answered Bragg's challenge,[11] and their exchange of body blows over the distribution of scattered X rays initiated a continuing feud. Thereafter, Bragg's experiments, and the controversy, focused upon a remarkable inference that Bragg drew from his neutral-pair theory: the ionization accompanying the passage of X rays and γ rays through matter is not produced by the direct action of these rays, but is entirely a secondary effect occurring only after the ray has been converted into a high-speed electron (through removal of the neutralizing positive charge).

Until the spring of 1911 the available data were ambiguous, and the opponents numerous.[12] However, the first result to come out of C. T. R. Wilson's cloud chamber was a clear demonstration that the exposure of a gas to a beam of X rays did not produce a diffuse homogeneous fogging, but a large number of short wiggly lines; ionization occurred only along the path of the photoelectron.[13] Bragg's inference became—and has remained—the accepted view of the interaction of high-frequency radiation with matter. And yet, just as Bragg's contention was receiving striking experimental support, the theory from which he derived it seemed to be decisively refuted by the discovery of an interference phenomenon accompanying the passage of an X-ray beam through a crystal.

But to pick up the biographical thread: Bragg's star rose rapidly after his first publications. In 1907, nominated by Horace Lamb and supported by Rutherford, he was elected to the Royal Society of London; in 1908 he was appointed Cavendish professor of physics at the University of Leeds, returning to England in March 1909. The first year or two at Leeds were not happy ones; the removal from Australia, the lack of solid scientific results, and the sniping criticism of his work by Barkla undermined Bragg's self-esteem. Things brightened in 1911–1912 with the vindication of his views on ionization by X rays, and of his views on the scattering of β and α particles (in this latter question he was closely allied with Rutherford against J. J. Thomson).[14] After completing a detailed account of his researches and views, *Studies in Radioactivity* (1912), Bragg was on the lookout for a new problem.[15]

During the summer and fall of 1912 the Laue-Friedrich-Knipping phenomenon was, naturally, *the* subject of discussion. After some initial success in construing the photographs on the corpuscular hypothesis,[16] Bragg and his son convinced themselves that a wave interpretation was unavoidable. This transition was smoothed by the instrumentalist epistemology that Bragg had adopted in the course of his corpuscular hypothesis campaign: "Theories were no more . . . than familiar and useful tools."[17] Early in November, William Lawrence, working at the Cavendish, showed how the Laue phenomenon might be regarded as a *reflection* of electromagnetic radiation in the incident beam from those planes in the crystal that were especially densely studded with atoms, and he derived the famous Bragg relation, $n\lambda = 2d \sin \theta$, connecting the wavelength of the X ray with the glancing angle at which such a reflection could occur.[18]

The younger Bragg's paper was entitled "The Diffraction of Short Electromagnetic Waves"—not "X rays"—for he wished to hold open the possibility that X rays (as known especially by their ionizing properties) were nevertheless his father's corpuscles, the diffracted-reflected entity affecting the photographic plates being merely the *Bremsstrahlung* necessarily accompanying the stopping of cathode rays in the X-ray tube. Despite the uncertainty whether the reflected rays could ionize—and even despite some counter evidence—Bragg's epistemology did not allow him to see the issue any longer as either/or: "The problem then becomes, it seems to me, not to decide between two theories of X-rays, but to find, as I have said elsewhere, one theory which possesses the capacities of both."[19]

In January 1913 Bragg succeeded in detecting the reflected rays with an ionization chamber,[20] and by March he had constructed the first X-ray spectrometer. Initially Bragg used it to investigate the spectral distribution of the X rays, relations between wavelength and Planck's constant, the atomic weight of emitter and absorber, and so on.[21] But very quickly he adopted his son's interest in the inversion of the Bragg relation: using a known wavelength in order to determine *d*, the distances between the atomic planes, and thus the structure, of the crystal mounted in the spectrometer. Apart from specifying general

symmetry conditions, before June 1912 it had not been possible to give the actual arrangement of the constituent atoms of any crystal. Laue's assignment of a simple cubic lattice to zinc sulfide had been corrected by William Lawrence to face-centered cubic, and he went on to analyze the crystal structure of the alkali halides on the basis of "Laue diagrams" that he had made at Cambridge. The spectrometer first served to confirm these structures and to determine the absolute values of the lattice spacings, and then was applied to more difficult cases.[22] By the end of 1913 the Braggs had reduced the problem of crystal structure analysis to a standard procedure.

In 1915 Bragg moved to London as Quain professor of physics at University College, and throughout the war he continued to direct some crystal analyses. He had, however, already become involved in war work, and it soon took almost all of his time. Bragg was a member of the panel of scientific experts attached to the Central Committee of the Board of Invention and Research, an institution created by Lord Fisher in July 1915 to aid the navy by screening inventions and sponsoring research. In April 1916, with the title of resident director of research and a staff of two physicists and a mechanic, Bragg was installed at the Naval Experiment Station at Hawkcraig to work on submarine detection. No satisfactory cooperation could be obtained here because of intraservice rivalry between Fisherites and anti-Fisherites, and at the end of 1916 the work was transferred to Harwich, with much loss of time and momentum.[23] "It was," Andrade opined, "probably in acknowledgment of his war work, as well as of his scientific eminence, that Bragg was made a C.B.E. in 1917 and was knighted as a K.B.E. in 1920."[24]

"The outbreak of war," Bragg asserted in 1920, "practically put a stop to the work with the spectroscope [i.e., X-ray spectrometer], which had been commenced in England, and we have fallen behind other countries which have been able to push on with it."[25] Of the two sorts of work—measuring λ (or, more generally, the properties of X rays and the X-ray spectra emitted by atoms) and measuring *d* (or, more generally, the structure of various crystals)—the first and more fundamental task, although pioneered by Barkla, Bragg, and Moseley, was largely abandoned in Britain after the war. Bragg assumed his duties at the University of London and began gathering a research school about himself. In 1923, when Bragg became head of the Royal Institution, this young and energetic group was installed in the previously moribund Davy-Faraday Research Laboratory. Their work, following a tacit agreement between Bragg and his son, was confined to the analysis of organic crystals. And in this field, which has now become so fundamental to molecular biology, Bragg put Britain way out in front.[26]

Bragg was president of the Royal Society at a very difficult time (1935–1940). He was one of the three Britons who had been members of the Deutsche Physikalische Gesellschaft since before World War I, and he now welcomed "certain ambiguous advances from learned bodies in Nazi Germany, and he did his best to further ostensible plans for an understanding between the two countries which, in his goodness of heart, he took at their face value."[27] Then the Royal Society was caught in the cross currents of agitation over the study of the social relations of science and the assertion of the social responsibility of science. Finally, there was the war, with its innumerable committees and councils—and air raids. Bragg's mind remained keen, even for scientific questions, but his energy began to fail. On 10 March 1942 he "had to take to his bed: two days later he was dead."[28]

NOTES

1. Autobiographical note quoted by Bragg and Caroe, pp. 171–172.
2. *Ibid.,* p. 173.
3. *Report of the Australasian A.A.S.* (Dunedin), **10** (1904), 47–77.
4. *Ibid.,* p. 69. Bragg's critique, contrary to the usual account, was not limited to (and thus not derived from the peculiar constitution of) the α particle.
5. Bragg, *Studies in Radioactivity* (London, 1912), p. 5.
6. Bragg, "On the Absorption of X-rays, and on the Classification of the X-rays of Radium," in *Philosophical Magazine,* 6th ser., **8** (Dec. 1904), 719–725; Bragg and Kleeman, "On the Ionization Curves of Radium," *ibid.,* 726–738. Dated 8 September 1904.
7. *Philosophical Magazine,* 6th ser., **14** (Oct. 1907), 429–449. Read before the Royal Society of South Australia, 7 May and 4 June 1907.
8. See the article on Barkla in *D.S.B.*
9. Bragg, "The Consequences of the Corpuscular Hypothesis of γ and X-rays, and the Range of β Rays," in *Philosophical Magazine,* 6th ser., **20** (Sept. 1910), 385–416; *Studies in Radioactivity.*
10. Not everyone shut his mind to this new gospel: "Personally, I have long been a convert to Professor Bragg's views on the nature of X-rays. . . ." H. L. Callendar, "Presidential Address, Section A," in *Report of the British A.A.S.* (Dundee, 1912), p. 396. Cf. Russell McCormmach, "J. J. Thomson and the Structure of Light," in *British Journal for the History of Science,* **3** (1967), 362–387.
11. See the article on Barkla in *D.S.B.*
12. Bragg, in *Philosophical Magazine,* 6th ser., **20** (Sept. 1910), 385–416; **22** (July 1911), 222–223; and **23** (Apr. 1912), 647–650.
13. C. T. R. Wilson, "On a Method of Making Visible the Paths of Ionising Particles Through a Gas," in *Proceedings of the Royal Society of London,* **85A** (9 June 1911), 285–288. Received 19 April 1911.
14. J. L. Heilbron, "The Scattering of α and β Particles and Rutherford's Atom," in *Archive for History of Exact Sciences,* **4** (1968), 247–307.

15. R. A. Millikan, *Autobiography* (London, 1951), pp. 95, 99.
16. Bragg, "X-rays and Crystals," in *Nature,* **90** (24 Oct. 1912), 219; dated 18 October. P. P. Ewald, "William Henry Bragg and the New Crystallography," in *Nature,* **195** (28 July 1962), 320–325. P. Forman, "On the Discovery of the Diffraction of X-rays by Crystals: Why Munich, Which X-rays?," in *Acts du XIIe Congrès International d'Histoire des Sciences, Paris, 1968.*
17. Bragg, "Radiations Old and New," in *Report of the British A.A.S.* (Dundee, 1912), pp. 750–753.
18. W. L. Bragg, "The Diffraction of Short Electromagnetic Waves by a Crystal," in *Proceedings of the Cambridge Philosophical Society,* **17** (14 Feb. 1913), 43–57. Read 11 November 1912.
19. Bragg, "X-rays and Crystals," in *Nature,* **90** (28 Nov. 1912), 360–361. The "elsewhere" may refer to *Studies in Radioactivity,* p. 192.
20. Bragg, "X-rays and Crystals," in *Nature,* **90** (23 Jan. 1913), 572. Dated 17 January.
21. W. H. Bragg and W. L. Bragg, "The Reflection of X-rays by Crystals," in *Proceedings of the Royal Society of London,* **88A** (1 July 1913), 428–438, received 7 April 1913; W. H. Bragg, "The Reflection of X-rays by Crystals (II)," *ibid.,* **89A** (22 Sept. 1913), 246–248, received 21 June 1913.
22. W. H. Bragg and W. L. Bragg, "The Structure of Diamond," *ibid.* (22 Sept. 1913), 277–291, received 30 July.
23. J. J. Thomson, *Recollections and Reflections* (New York, 1937), pp. 207–210.
24. Andrade, "William Henry Bragg 1869–1942," p. 284.
25. Bragg, as president of the Physical Society of London, opening a discussion on X-ray spectra, in *Proceedings of the Physical Society of London,* **33** (1920), 1.
26. Articles by J. M. Robertson, J. D. Bernal, and K. Lonsdale, in P. P. Ewald, ed., *Fifty Years of X-Ray Diffraction* (Utrecht, 1962).
27. Andrade, *op. cit.,* p. 290.
28. *Ibid.*

BIBLIOGRAPHY

I. ORIGINAL WORKS. An excellent chronological bibliography prepared by K. Lonsdale is appended to the article by Andrade (see below). Bragg's papers at the Royal Institution, London, include a collection of offprints, research notebooks covering 1903–1913, summaries of some literature read, a few MS drafts (notably that of *Studies in Radioactivity*), miscellaneous scientific correspondence after 1920, a very little scientific correspondence before 1920, autobiographical notes on his youth, and family correspondence. Bragg's correspondence with Rutherford, some fifty letters written between 1904 and 1915, is in the Rutherford papers at the Cambridge University Library; copies are available at the Royal Institution and at McGill University, Montreal. The locations of thirty-one letters from Bragg to other correspondents are given in T. S. Kuhn *et al., Sources for History of Quantum Physics* (Philadelphia, 1967), p. 26.

II. SECONDARY LITERATURE. Works on Bragg are E. N. da C. Andrade, "William Henry Bragg 1869–1942," in *Obituary Notices of Fellows of the Royal Society of London,* **4** (1943), 277–300; Sir Lawrence Bragg and Mrs. G. M. Caroe (Gwendolen Bragg), "Sir William Bragg, F.R.S.," in *Notes and Records of the Royal Society of London,* **16** (1961), 169–182; and *Who Was Who 1941–1950* (London, 1952), p. 134.

PAUL FORMAN

BRAHE, TYCHO (*b.* Skåne, Denmark [now in Sweden], 14 December 1546; *d.* Prague, Czechoslovakia, 24 October 1601), *astronomy.*

The second child and eldest son of Otto Brahe and his wife, Beate Bille, Tycho (Danish, Tyge) was born at the family seat, Knudstrup. He had five sisters and five brothers, including his still-born twin. Otto Brahe was a privy councillor and later became governor of Helsingborg Castle. Probably Tycho and Christine, whose last name is unknown and who was not of noble family, were never formally married, but they lived together from about 1573 to the end of his life. They had five daughters and three sons; their daughter Elizabeth married Tycho's assistant, Franz Gansneb Tengnagel von Camp. Tycho's best observing was done on the island of Hven from 1576 to 1597. His observations of the nova of 1572 and several comets forced abandonment of the traditional celestial spheres, and his observations of Mars enabled Kepler to discover the laws of planetary motion. Information about his observatory and observational techniques was widely disseminated, and his geoheliocentric system gained numerous supporters.

Tycho was brought up by his paternal uncle, Jörgen Brahe, and from the age of seven was taught Latin and the preparatory subjects by a tutor. From April 1559 to February 1562 he attended the Lutheran University of Copenhagen, where the theologians and faculty were under the influence of Melanchthon as well as Aristotle and the Scholastics. Tycho must have begun his studies in the Faculty of Philosophy by applying himself first to the *trivium;* his study of the arts probably began under the lecturers in pedagogy, who emphasized the writing and speaking of Latin. No doubt he received instruction in the articles of faith from the Lutheran catechism on Sunday mornings. He must have studied Greek grammar and Greek and Latin literature, and probably also dialectic, attending lectures in Greek on Aristotle's *Dialectics* and lectures on the Latin rhetorical works and on Roman epistolary authors.

Since his family was a noble one, Tycho did not need a university degree to establish himself in a profession. Therefore, he must have entered on the study of the *quadrivium* as soon as he was able, without waiting to earn a degree. Ethics and singing were included in the university *quadrivium;* and at the chapter house of the cathedral, students practiced and heard lectures on singing. Also available were lectures on harmonic theory, a mathematical discipline since the time of Pythagoras. From the lectures on the natural sciences and philosophy that Tycho may also have heard, he would have emerged as a convinced Aristotelian. By 1560 he was, no doubt,

studying arithmetic, then Sacrobosco's *Sphaera* and Peter Apian's *Cosmographia.* His copies of the *Sphaera,* a medical handbook, an herbal, Gemma Frisius' edition of Apian's *Cosmographia,* and Regiomontanus' *Tabulae directionum* are preserved.

In 1561 and 1562 Tycho was probably attending lectures on Aristotle's *Physics,* Euclid's *Elements,* Ptolemy's theory of the planets, and on astrology, which united astronomy with medicine. Tycho made friends with, and later wrote an epitaph for, Hans Fransden, from Ribe in Jutland (Johannes Franciscus Ripensis), who lectured on Hippocrates and Galen as well as on mathematics, acted as physician to the king, and prepared an annual astrological almanac. Tycho also made friends with Johannes Pratensis, who later became professor of medicine and whose copy of Ptolemy's *Almagest* Tycho probably inherited in 1576. On 21 August 1560 the occurrence at the predicted time of a solar eclipse, although only partial in Copenhagen, turned Tycho toward observational astronomy, which was not part of the university curriculum. He immediately obtained a copy of Stadius' *Ephemerides,* which is based on the *Prutenic Tables* and, consequently, on the Copernican system.

So that he would be parted from friends interested in science and would study law, a necessary part of the education of a member of the nobility, Tycho's uncle sent him to the University of Leipzig, where he arrived in March 1562. With him, as tutor, went Anders Sörensen Vedel, only four years his senior. Vedel had spent less than a year attending lectures on divinity and studying history at the university, but he was later to distinguish himself as a historian. Except for two short visits, Tycho remained away from his homeland until 1570.

At Leipzig, although Vedel tried to keep his charge busy with the study of law, Tycho's interest in astronomy was not to be thwarted; and as late as May 1564 he was pursuing it secretly, while Vedel slept. During the daytime he attended to the studies prescribed by his uncle. He used what money he could save for the purchase of astronomical books, tables, and instruments. Not content with Stadius' *Ephemerides,* he also obtained the *Alphonsine Tables* and *Prutenic Tables,* and used the *Ephemerides* of Giovanni Battista Carellus (1557). To learn the constellations, he secretly used a globe no bigger than a fist.

A conjunction of Saturn and Jupiter in August 1563 was later regarded by Tycho as the turning point in his career. Although equipped with only a pair of compasses, he recorded his observations relative to it. The discrepancy between the time of the observed closest approach of the two planets and that computed from the tables, about a month using the *Al-*

phonsine Tables and a few days by the *Prutenic Tables,* greatly impressed Tycho. On 1 May 1564 he began observing with a radius, or cross staff, consisting of a three-foot arm along which could slide the center of an arm of half that length. Both arms were graduated. Bartholomaeus Scultetus subdivided the instrument for him by means of transversals. There was a fixed sight at the end of the longer arm that he held near his eye. To measure the angular distance between two objects, Tycho set the shorter arm at any graduation of the longer arm and moved a sight along the shorter arm until he saw the two objects through it and a sight at the center of the transversal arm. The required angle was then obtained from the graduations and a table of tangents. This instrument was not very accurate, but, since he could not get money from Vedel for a new one, he made a table of corrections to apply to it.

Tycho left Leipzig 17 May 1565 and traveled to Copenhagen via Wittenberg and Rostock. Of his family, only his mother's brother, Steen Bille, showed any sympathy for his scientific interests. When his Uncle Jörgen died, 21 June 1565, there was no longer any reason for him to remain at home. He reached Wittenberg 15 April 1566 and began studies under Caspar Peucer. He left after five months, however, arriving in Rostock 24 September and matriculating at the university soon thereafter.

On 29 December 1566, in an unfortunate duel with another Danish nobleman, part of Tycho's nose was cut off. This he replaced by what was long thought to be a composition of gold and silver, but probably had considerable copper content. When his tomb was opened 24 June 1901, a bright green stain was found on the skull at the upper end of the nasal opening.

At Rostock, Tycho met several men devoted to astrology and alchemy as well as to medicine and mathematics. He observed a lunar eclipse 28 October 1566 and a partial solar eclipse 9 April 1567. That summer he visited home, but was back in Rostock by 1 January 1568. He immediately began observations, although without an instrument until he used the cross staff 19 January. His last recorded observation in Rostock was 9 February. On 14 May 1568, King Frederick II of Denmark formally promised Tycho the first vacant canonry in the cathedral chapter at Roskilde, Zealand. He matriculated at the University of Basel in 1568 and, probably early in 1569, went to Augsburg via Lauingen in Swabia, where he met the astronomer Cyprian Leowitz. He entered into the intellectual life of Augsburg, where he made his first observation 14 April.

Among Tycho's friends in Augsburg were Johann Baptist Hainzel, an alderman, and his brother, Paul, the burgomaster,[1] who helped Tycho arrange for the manufacture of a wooden quadrant, suspended at the center, with a radius of about nineteen feet. The divisions marked on the arc and a plumb line gave altitude measures. Tycho does not seem to have used it himself, and it was destroyed in a storm in December 1574. Tycho also designed a portable sextant, which he used and gave to Paul Hainzel, and ordered a five-foot globe. His last recorded observation in Augsburg was made in Hainzel's presence 16 May 1570. At Augsburg he argued with Ramus, who advocated constructing a new astronomy based entirely on logic and mathematics, without recourse to any hypothesis. They agreed on the need for new and accurate observations before attempting to explain the celestial motions, and it is obvious that Tycho was aware of the need for good instruments to obtain those observations. He returned home in 1570, probably because of his father's poor health. On the way, in Ingolstadt, he met Philip Apian, son of Peter.

Although at his father's death, 9 May 1571, he and his brother Steen inherited Knudstrup, Tycho soon moved to Heridsvad Abbey, the home of his uncle, Steen Bille, where he devoted himself to chemical experiments until 11 November 1572. After sunset on that day, almost directly overhead, in the constellation Cassiopeia he noticed a star shining more brightly than all the others and immediately realized it had not been there before.

To measure the star's angular distances from the neighboring stars in Cassiopeia, Tycho used a sextant similar to the one he had left with Hainzel. The two arms of seasoned walnut, less influenced by climate than other woods and lighter than metal, were joined by a bronze hinge. A 30° arc, graduated by individual minutes, without transversals, was fixed to one arm; the other arm could slide along the arc. Square metal sights with holes through the centers were attached at the ends of the arms. Tycho later described this instrument in the *Mechanica* and in the *Progymnasmata,* by itself and, as used for the nova observations, in the plane of the meridian, pointing out a window with the end of the arm where the arc was fixed (this time a 60° arc) resting on the sill while the end of that arm, near the joining of the two arms, rested on a post some five feet inside the window. To make sure that this arm was horizontal, it was moved until a plumb line, hanging from the end of the arc, touched a mark in the middle of the arm. The plumb line would show any change in position of the instrument, thus indicating the correction to be made to the observation.

To make sure that observations made the same night were made under the same conditions, Tycho left the instrument clamped in position between such observations. He measured the angular distance of the new star, at both upper and lower culmination, from the star Schedar (α Cassiopeia), which crossed the meridian at nearly the same time, and found no parallax. He measured the distance of the nova from nine stars in Cassiopeia and found no variation between observations. Had the new star been as close to the earth as the moon, a parallax of 58'30" would have been found. Tycho observed the star until the end of March 1574, when it ceased to be visible. His records of its variations in color and magnitude identify it as a supernova. At first clear white, with the magnitude of Venus at its brightest, it grew yellowish and diminished in brightness to that of Jupiter. By February and March it was of the first magnitude and reddish, in April and May of the second magnitude and lead-colored. Thereafter its color did not change. By August it was a third-magnitude star, fourth-magnitude by October, hardly more than fifth-magnitude at the turn of the year, and sixth-magnitude or less in February 1574.

Tycho concluded that the phenomenon was not an atmospheric exhalation and was not attached to the sphere of a planet, since it did not move contrary to the direction of the diurnal rotation, but that it was situated in the region of the fixed stars. He called it a star, not a comet, because, as the ancients asserted, comets are generated in the upper regions of the air, not in the heavens. He noted that it twinkled like a star and did not have a tail like a comet. It could not be a comet with its tail turned away from the earth because Peter Apian and Gemma Frisius had shown that the tail of a comet is turned away from the sun. Tycho thought it not impossible that the star would again cease to be visible, as he wrote in a brief tract published in 1573, while the star was still visible. This tract, dedicated to Johannes Pratensis, at whose urging it had been printed, included an exchange of letters between the latter and Tycho, a section on the astrological significations of the star, the introduction to an astrological calendar, and that part of the calendar dealing with the lunar eclipse of December 1573.

All over Europe scholars observed the star. Some, using crude observational procedures, such as holding a thread before their eyes, assured themselves that the newcomer did not move relative to certain known fixed stars. Such observations, showing the star to be supralunar, were widely appreciated as necessitating an alteration in cosmological theories.

Tycho's scholarly treatise concerning the star, the *Progymnasmata* (1602), was the first volume of a proposed trilogy. The second chapter on planets having been printed and paged first, there was space in chapter 1 to describe the lunar theory, the complexity of which delayed publication of the volume. The work reprinted most of Tycho's 1573 tract and gave his carefully compiled observations of the nova, discussing its position in space and its expected annual parallax if the Copernican system were true. Tycho attempted to calculate the real diameters of the sun, moon, planets, and the nova from his measurements of their apparent diameters. He estimated the maximum distance of Saturn as 12,300 semidiameters of the earth, the distance of the fixed stars as 14,000—not all at the same distance—and that of the new star as 13,000. His estimate of the real diameter of the new star at its first appearance was 7-1/8 times that of the earth. He assigned the diminution in light to actual decrease in size. Galileo pointed out[2] the impossibly enormous sizes of the stars if Tycho's estimates of their diameters were correct. The *Progymnasmata* also reprinted, summarized, or criticized the works on the nova by others. Tycho deplored Hagecius' use of clocks because of their inaccuracy. Because he was unable to observe the star with his sextant at upper culmination, Tycho used Hainzel's observations made at Augsburg with the big quadrant.

Tycho's observations of the nova were separately recorded. His journal of observations skips from one made at Helsingborg 30 December 1570 to his entries of three distance measurements between the nova and known fixed stars made with a parallax instrument 10 May 1573. There are entries for 14 August and for a lunar eclipse observed at Knudstrup 8 December, for observations at Heridsvad in March and April 1574, and at Copenhagen at the end of April and May. None appear for 1575.

In September 1574, in the first lecture of his course for young noblemen at the University of Copenhagen, Tycho spoke of the skill of Copernicus, whose system, although not in accord with physical principles, was mathematically admirable and did not make the absurd assumptions of the ancients, who let certain bodies move irregularly in respect to the centers of the epicycles and eccentrics. Doubtless, Tycho had Copernicus' rejection of Ptolemy's equant in mind. The influence of the stars on nature—seasons, tides, weather—seemed obvious. If forewarned, thanks to astrology, men could conquer the influence of the stars on themselves, but Tycho had reservations concerning public calamities.

Soon after completing these lectures, early in 1575, and wondering where to settle permanently, Tycho went first to Kassel, where he visited Landgrave William IV. The two men, convinced of the need for systematic observations, observed together for more than a week, Tycho with some of his own portable instruments and the landgrave with his quadrants and torqueta. They made an accurate determination of the position of Spica. Their discussion of the retardation of the sun near sunset spurred Tycho later to study refraction at low altitudes. The landgrave was so impressed by Tycho's ability that he suggested to the Danish monarch that something be done to enable Tycho to pursue his astronomical studies in his native land.

Tycho's next stop was Frankfurt am Main. There, at the book fair, he purchased many pamphlets on the recent nova. He next journeyed to Venice via Basel, where he contemplated settling, and then returned to Augsburg, inquiring about instruments he had ordered during his previous visit. Wherever he went, he met the leading astronomers and, whenever possible, inspected their astronomical instruments.

At Regensburg, where the future emperor, Rudolph II, was crowned King of the Romans, 1 November 1575, Tycho met Rudolph's physician, Hagecius, who had written an excellent book on the nova of 1572. From him Tycho obtained a copy of Copernicus' *Commentariolus* and a copy of a letter from Hieronymo Mugnoz to Hagecius about the new star. It is probable that at the same time Tycho presented Hagecius with a copy of his tract on that star. At Saalfeld, on the return journey, Tycho saw the manuscripts of Erasmus Rheinhold, who had prepared the *Prutenic Tables*. In Wittenberg, Tycho examined the wooden parallactic instrument, or triquetrum, with which Wolfgang Schuler had observed the nova after his earlier observations with Johannes Praetorius, made with an old wooden quadrant, had resulted in the finding of a large parallax that was inconsonant with the results obtained by the landgrave.

Tycho reached home near the end of 1575. In February 1576, possibly because of the landgrave's recommendation, King Frederick II offered him the island of Hven in the Danish Sound and asked him to erect suitable buildings and construct instruments there. Tycho accepted, feeling that he could thus obtain in his native land the desired quiet and convenience. He immediately visited the island, and on 23 May a document was signed by the king conferring and granting in fee the island and its tenants and servants, with the rent therefrom; there was also the obligation to govern it in accordance with the law and to attend to the welfare of the inhabitants. Tycho was also given sufficient funds to augment his own, in order to erect a suitable residence and other buildings necessary to his work, and certain landholdings, the income from which, together with his own fortune,

made it possible for him to lead an almost regal existence. From time to time additional sources of income were made available.

The island is roughly 2,000 acres in area. The inhabitants lived in a village near the northern coast and tilled about forty farms in common. Near the center of the island, at the highest point, about 160 feet above sea level, Tycho began construction of Uraniborg (heavenly castle), the edifice that was to be his home and observatory for more than twenty years. He made one observation of Mars in October 1576 and began observations of the sun 14 December. Although he probably moved into the building that winter, it was not completed until 1580, and even thereafter additions and alterations were made. On the island were the workshops of the artisans who constructed his instruments, a windmill, a paper mill begun in 1590 and completed in 1592, which could also be used to grind corn and prepare hides, and nearly sixty fishponds, one of which, for the use of the mill, was secured by a large dam.

The main building was erected exactly in the center of a square enclosure the walls of which were about 255 feet long, eighteen and one-half feet high, and seventeen feet wide at the base. At the center of each wall was a semicircular bend about seventy-six feet in diameter that enclosed a pavilion. There were gates at the eastern and western corners, and above the gates were kennels in which two English watchdogs were kept to warn of arrivals. At the northern corner was a small house for servants in the same Gothic-Renaissance style as the main house. A similar house at the southern corner housed a printing office. The press was installed in 1584. Four roads directed exactly to the cardinal points led from the main house to the gates and houses. Within the enclosure were herbaries and flower gardens and about 300 trees of various species.

The main house, too, was exactly square, its four walls, about fifty-one feet long and thirty-eight feet high, facing the four points of the sky. The rounded towers added on the south and north were eighteen and one-half feet in diameter, with eight and one-half-foot galleries encircling them. From the ground to the Pegasus weathervane, the house measured about sixty-four feet. Beneath the entire house was a cellar more than ten feet deep, divided into many rooms, and beneath the towers were the well and arrangements for storing food. The original four corridors on the ground floor, which met at right angles, were later reduced to three so as to make possible the establishment, behind the furnace, of a small chemical laboratory, thereby lessening the need to go down to the large subterranean one. There were a fountain that could turn, sending water in all directions, and pipes and pumping apparatus to distribute water to rooms on both floors. On the ground floor there were also a library, a kitchen, a table for collaborators in each corner of the building, and spare bedrooms. The observatories were on the upper level, the larger southern and northern ones containing several of the important, large instruments—such as the azimuthal semicircle, Ptolemaic rulers, brass sextant and azimuthal quadrant, and parallactic rulers that also showed azimuths. An octagonal gallery contained one of the globes on which an instrument could be placed and turned in all directions. At the very top of the house were eight bedrooms for assistants.

About a hundred feet south and slightly east of Uraniborg a separate observatory, Stjerneborg (castle of the stars), constructed about 1584, housed additional instruments in five subterranean rooms. Stone columns outside could be used to support Ptolemaic rulers or the portable armillae. There were also places for globes on which instruments could be placed and turned. In this building was a study with only the vaulted roof and the top of the walls above ground. On the ceiling was depicted the Tychonic system, and on the walls were the portraits of Timocharis, Hipparchus, Ptolemy, al-Battānī, King Alfonso X of Castile, Copernicus, Tycho, and the still unborn but hoped-for descendant, Tychonides, each with a legend beneath it—that for Tychonides expressing the wish that he would be worthy of his great ancestor.

The accuracy of the observations depended on the instruments and the care with which they were used. Although Tycho's were without magnification, error was minimized by their huge size and by the graduations carefully marked on them to facilitate angular measurements on the celestial sphere, altitudes, and azimuths. Tycho checked instruments against each other and corrected for instrumental errors. Unfortunately, he considered refraction negligible at altitudes above 70°. He observed regularly and achieved an accuracy within a fraction of a minute of arc, an

accuracy unsurpassed from the time of Hipparchus to the invention of the telescope.

In the library was the globe, almost five feet in diameter, ordered from Augsburg. Tycho filled the cracks, restored the spherical shape with pieces of parchment, tested it for two years to see whether it would retain its shape and whether it would withstand the seasonal temperature changes, then covered it with brass sheets and again had it smoothed. On it he engraved the zodiac and the equator with their poles and, using transversal points, divided each degree of these circles into sixty minutes. The globe could be turned on an axis through its poles inside the meridian and horizon circles that were mounted on it and that were divided into degrees and minutes. A vertical brass quadrant marked in degrees and minutes indicated altitudes as well as azimuths along the horizon. On this globe, over the years, Tycho marked the exact positions, referred to the year 1600, of the fixed stars that he observed. He also investigated the planet motions with reference to this globe.

In the southwest room on the ground floor at Uraniborg, affixed to a wall in the plane of the meridian, was Tycho's most famous instrument, the mural quadrant with a radius of about six feet. The degrees marked off on its arc were so far apart that each minute was divided by transversal points into six subdivisions of ten seconds each, making it possible to read off measurements of five seconds. In a wall pointing exactly east and west, and over the center of the quadrant, was a square hole that could be opened and closed and that contained a brass cylinder along both sides of which the observer could sight, using one of two pinnules on the quadrant. Each pinnule had a square plane the width of which was exactly equal to the diameter of the cylinder. Each side of the plane had a slit for use in determining a star's altitude and meridian transit at the same time. To determine the altitude alone, which was done to the sixth of a minute, an observer looked through the upper and lower slits and the corresponding sides

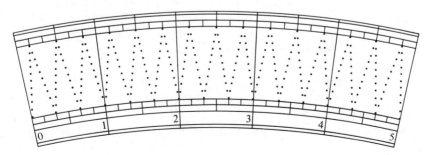

FIGURE 1.   Method of dividing an arc by transversal points, proof of the applicability of the method to curved lines and pinnules for use in sighting.

of the cylinder, and an assistant entered the reading on the record. A third person watched two clocks when the observer at the pinnule signaled, and the time was noted in the ledger. Two clocks that gave seconds as accurately as possible and could be checked against each other were necessary. Tycho had four. Elsewhere he expressed his distrust of clocks, preferring to check the time by observation. Despite his faith in this quadrant, he also consulted other large instruments.

Inside the quadrant's arc, for ornamental purposes, was painted a life-size portrait of Tycho seated at a table, with arm outstretched as though pointing to the cylinder. In a niche in the wall, above and near the head, was a brass globe fitted with interior wheels. It could turn to imitate diurnal rotation and to show the paths of the sun and moon and the lunar phases.

The smaller southern observatory housed a brass armillary instrument with four armillae, or rings; the smaller northern observatory, another with three armillae. In the northern tower were the sextant with which one observer could measure distances, the bipartite arc for measuring small angular distances, and the sextant with which Tycho had observed the nova. Among his other instruments were several smaller quadrants and sextants of various designs for various purposes, an astronomical radius, an astronomical ring, a small astrolabe, an azimuth semicircle, and some parallactic or ruler instruments, one of which had belonged to Copernicus.

In these fantastically ornate but exceedingly useful observatories, Tycho watched the skies and trained his assistants. Some of the larger instruments could not have been used without their aid. Among these assistants were Peter Jacobsøn Flemløs, Longomontanus, Elias Olsen, Gellius Sascerides (who stayed six years), Otto Islandus (Oddur Einarsson, who was bishop in Iceland), and Willem Blaeu (who later made excellent maps and globes). Paul Wittich, who was an assistant at Uraniborg in 1580 and who, at Kassel in 1584, described Tycho's instruments, including the transversal divisions, so impressed the landgrave that he had his instrument maker, Joost Bürgi, alter his instruments to conform to the description. Wittich was probably largely responsible for the development of the prosthaphaeretic method (from πρόσθεσις [addition] and ἀφαίρεσις [subtraction]) for simplifying trigonometrical computations by replacing multiplications and divisions with additions and subtractions. This is the basis of the set of rules for solving plane and spherical triangles, *Triangulorum planorum et sphaericorum praxis arithmetica,* drawn up, without proof, by Tycho and made available in numerous

manuscript copies for the use of his assistants. Wittich also revealed this method at Kassel, to the annoyance of Tycho; he was even more annoyed, however, by the inclusion of the first two rules in a book by Nicolai Reymers Bär (Ursus), printed at Strasbourg in 1588. Afterward the method was further developed by other mathematicians. Ursus had visited Hven in 1584. Tycho was also visited by members of the nobility, possibly including Frederick II and certainly Frederick's son, the future Christian IV (1592), as well as James VI of Scotland, the future James I of England (1590). Christoph Rothmann, Landgrave William IV's mathematician, was there from 1 August to 1 September 1590.

Although Tycho saw no objection to the adoption of the Gregorian calendar by the Protestant world, since questions of theology were not involved, he does not seem to have used it until early 1599, when, on the Continent, he began to date his letters in the new style. His first observation so dated was made 22 July of that year.

From Hven, Tycho carried on a vast correspondence that kept alive the personal contacts made in his student days, apprised the scholarly world of his work, and provided him with the observations of others for comparison with his own. Although Tycho and William IV never met again after Tycho's 1575 visit to Kassel, in later years they exchanged letters, sending each other records of their observations. The correspondence, including letters between Tycho and Rothmann, was printed at Uraniborg in 1596. It begins with data concerning the comet of 1585 and largely concerns the techniques of observation, the instruments used, and their divisions. Appended is a description of Hven, with its observatories and instruments. The majority of Tycho's other letters, written between 14 January 1568 and 30 April 1601, first appeared in print in the *Opera omnia.* They provide a survey of observational astronomy in the last three decades of the sixteenth century, having achieved that dissemination of ideas which is now the province of learned journals.

Shortly after sunset on 13 November 1577, Tycho noticed, for the first time, a large comet with a very long tail. Although he later heard that the comet had been seen in the Northwendic Sea on 9 November, in his opinion it had begun with the new moon that had occurred shortly before, on 10 November at one hour after midnight.[3] He observed from 13 November to 26 January, by which time it was barely distinguishable. He used a radius and a sextant, and occasionally a quadrant with an azimuth circle—the larger instruments were not yet all installed. He fixed the quadrant in the meridian. Shortly after the comet ceased to be visible, he described it in a short German tract, first published in 1922.[4]

Five hours after noon on 13 November, Tycho found the comet 26°50′ from the bright star in Aquila and 21°40′ from the lowest star in the horn of Capricorn, toward which the tail was stretched. Using trigonometry, he computed the comet's position as 7°15′ in Capricorn, with a declination of 8°20′ north of the ecliptic. In the next twenty-four hours it moved 3°30′ in its circle. Having found it moved more rapidly in the beginning, Tycho decided it had moved 4° in its circle each of the days before he saw it, at new moon having been near the ecliptic beneath the twenty-fifth degree of Sagittarius in the line of the Milky Way, which he considered the place whence comets usually come. He traced the comet's path from west to east. It had described a quarter of a great circle from the twenty-fifth degree of Sagittarius in the ecliptic, intersecting the equator at an angle of 34° at a point 300°40′ from the vernal equinox. Its rate of motion gradually decreased, so that in the end it moved only 20′ in a day, or 4°20′ from 13 January to 26 January. Its tail, 22° long in the beginning, gradually became smaller and shorter, and could scarcely be seen in January. Tycho used the direction of comets' tails as evidence that the tails are merely solar rays transmitted through the head of the comet, an argument against Aristotle's theory of the formation of the tail out of "dry fatness."

In the first chapter of the untitled German tract, Tycho described Aristotle's theory of comets and objected to it on the grounds that the star in Cassiopeia four years before had been supralunar, having had no parallax and having remained stationary like the fixed stars, for which reasons many had abandoned the Aristotelian theory in favor of the belief that something new can be born in the heavens. Tycho suggested that other comets could be born there, and are not composed of dryness and fatness pulled up from the earth. He said that Aristotle's proof had been based on meditation, not mathematical observation or demonstration, whereas comets *are* generated in the heavens.

Tycho referred frequently to his still incomplete Latin work on the same phenomenon, considering the two works as serving different purposes. The German one was intended for a wider audience than could be reached by a work in Danish, but it was meant for a less skilled audience than the one for whom the Latin work was written. Because it could reach only the literate, the German work would have an intelligent audience, but not one expected to be trained in mathematics. Repeatedly Tycho referred to the mathematical explanations in the Latin work, which the "masters" could read and understand. Indeed, the numerical values, such an important part of the Latin work, are almost entirely absent from the German. Tycho's main objective was to determine the comet's distance from the earth as a means of refuting Aristotle. He was also concerned with the comet's physical appearance—color, magnitude, and the direction of the tail.

As clearly as anything he wrote, this tract shows Tycho as a product of his times. Breaking with established tradition, he knew exactly where he stood in the historical development of astronomy. Moreover, the tract demonstrates how early and how fully he understood the implications of his break and stresses his insistence on putting observation above deduction by reasoning. Emphasis is placed on the comet's lack of parallax and the resultant untenability of the so-called Aristotelian doctrine of solid spheres in an unchanging heaven. It hints at Tycho's own system of the universe, on which he was already working. It deals at length with the astrological implications of this fiery sign, but secondarily to the observational revelations.

De mundi aetherei recentioribus phaenomenis (1588), on the comet of 1577, the second volume of Tycho's proposed trilogy, was printed on his own press and is profusely illustrated with useful diagrams. Chapter 1 records in detail, day by day, each of Tycho's observations of the comet. The next chapter gives his positional data, computed from his observations, for the comparison stars used in observing the comet. In chapter 3 the comet's latitude and longitude for each day are derived by means of spherical trigonometry, using observed angular distances of the comet from certain fixed stars. The diagrams, but not the mathematical steps, are reproduced. Chapter 4 treats the comet's right ascension and declination with respect to the equator, and chapter 5 deals with the portion of a circle described by the comet, ending with a table of its daily motion, latitude, longitude, right ascension, and declination (first southern, then northern) for 9 November 1577 to 26 January 1578. Chapter 6 treats the comet's parallax as a measure of its distance from the earth and states that the comet was in the etherial rather than the elementary region and moved in a great circle. Tycho's observations of the comet's angular distance from certain fixed stars are compared with those of other observers in other localities. Chapter 7 deals with past writings about the direction of comet tails and with the 1577 comet's tail, which was directed away from Venus. In chapter

9, however, Tycho states his opinion that this was an illusion, since it would seem more likely that the tail be directed away from the sun. Chapter 8 discusses the comet's position in regard to the planetary spheres.

Since his observations of the nova of 1572 and the comet of 1577 had made him discard the reality of the spheres, Tycho included a description of his own geoheliocentric system of the universe. The comet, whose greatest elongation from the sun was 60°, moved about that body in a circle outside that of Venus, that part of the circle where Tycho observed the comet being closer to the earth than Venus was. Moreover, the comet's orbit, inclined to the ecliptic

at an angle of 29° 15', was not a true circle, but an oval. Chapter 9 is concerned with the actual size of the comet and its tail, the diameter of the head being 3/14 of the diameter of the earth, and the length of the tail in November being ninety-six semidiameters if turned from Venus. The tenth and last chapter summarizes in detail the observations of others, both of those who found the comet supralunar and those who thought they found it sublunar.

At least six later comets were visible to the naked eye before Tycho left his island. The comets of 1580, 1582, 1585, and 1590 were supposed to be treated in the third volume of the trilogy, but that volume was never written. Tycho observed the comet of 1580

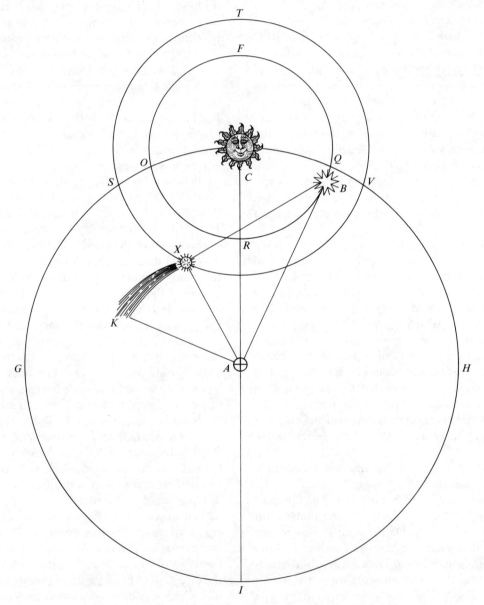

FIGURE 2:   The position of the comet of 1577 relative to the sun and Venus.

from 10 October to 25 November and again on 13 December, after it had passed perihelion. On 12, 17, and 18 May 1582 he observed another comet. By 1585 his major astronomical instruments, including a large armillary instrument at Stjerneborg, had been installed. His excellent observations of the tailless comet visible in October and November of that year appeared in 1586 in the first book printed on the island, the *Diarium astrologicum et metheorologicum* of his assistant Elias Olsen. They were more fully preserved in manuscript and studied in detail in the nineteenth century. The comet of 1590 was observed at Hven the end of February and the beginning of March, whereas that of 1593 was not observed at Hven but at Zerbst in Anhalt (Seruesta Anhaldinorum) by one of Tycho's former students, Christiernus Johannis Ripensis. Tycho saw the comet of 1596 in Copenhagen

on 14, 15, and 16 July. More complete observations were made at Uraniborg on 18, 21, 24, and 27 July.

Hinted at in the German tract on the comet of 1577, probably first worked out by 1583, and first described in print in the 1588 Latin work on the comet of 1577, the Tychonic system was never presented in detail. In it the earth is at rest in the center of the universe, and there is still need for a sphere of fixed stars revolving in twenty-four hours. The planets circle the sun while the sun circles the earth. The orbits of Mercury and Venus intersect the orbit of the sun in two places but do not encompass the earth. The orbit of Mars also twice intersects that of the sun, but encloses the earth and its orbiting moon. The orbits of Jupiter and Saturn enclose the entire path of the sun.

Tycho prized parts of the Copernican doctrine or

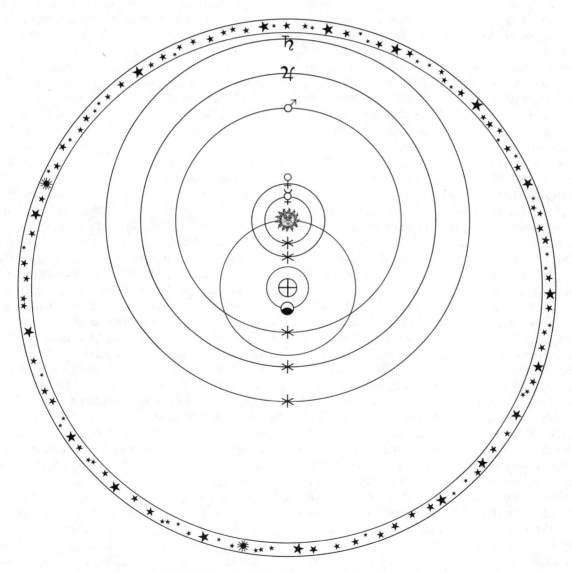

FIGURE 3.   Tycho Brahe's system of the universe.

at least acknowledged the abilities of its originator, but could not bring himself to accept a sun-centered universe. His reluctance to do so can be ascribed partly to his respect for Scripture and partly to his feeling of common sense, but largely to his inability to conceive of a universe so immense that an observer as accurate as he knew himself to be could not detect any stellar parallax, the necessary consequence of the earth's motion around the sun. The Tychonic system was timely and gained acceptance in many quarters. It did not bring its author into conflict with the theologians, yet it cared for observed phenomena, including the motion of comets through space, which had necessitated Tycho's rejection of the Aristotelian spheres. It could account for the phases of Venus, first observed by Galileo and not explicable by the Aristotelian and Ptolemaic schemes.

Just as Tycho was only one of a number of observers who stressed the supralunar position of comets and novae, so his compromise theory of the universe was only one of a number that accentuated the abandonment of Aristotelian tradition and helped prepare men to accept the Copernican doctrine. It is only natural, especially in the light of Tycho's arrogant ways, that there was some feeling of rivalry, especially toward Ursus, who described a similar system.

In Aristotelian theory the planets were attached to spheres and rotated with them. The destruction of these solid orbs made it necessary to find a cause for the motion of the planets, and this cause was provided by the next generations of astronomers and physicists, the sun assuming an importance not accorded to it by either Copernicus or Tycho. Undoubtedly the traditional crystalline spheres would eventually have been discarded without the aid of Tycho's work, but he speeded up the change.

Tycho presented his cosmologic views in his introduction to a pamphlet[5] on weather forecasting by his assistant Flemløs. To explain how the heavens influenced matters on earth and so could be used for prognostication, Tycho described his cosmology, but focused not so much on the system as on the way the heavens affect the earth. He maintained the concept of "free will" while conceding celestial influence. Accepting three elements—earth, water, and air—he theorized that air is the instrument by which the celestial region influences the terrestrial, with the animals and plants therein and, to a lesser extent, men (some more than others). Thus he voiced disagreement with traditional concepts while maintaining the validity of astrology and distinguishing it from astronomy.

Elsewhere Tycho criticized astrologers who drew improper conclusions based on superstition and error rather than astrology itself, which he considered a science for which both accurate knowledge of the course of the stars and experience gained from signs seen in the elementary world were needed. From the lunar eclipse observed during his stay in Leipzig, he predicted the wet weather that followed. Also while in Leipzig, he calculated Caspar Peucer's horoscope, predicting the misfortunes that befell him, as well as his reestablishment. In Rostock, from the lunar eclipse of 28 October 1566, Tycho predicted the death of the aged Sultan Suleiman the Magnificent, but later learned that Suleiman had died before the eclipse. Tycho calculated horoscopes for the three sons of Frederick II; but, although he continued to prepare annual prognostications for his ruler, by 1588, if not earlier, he held them of little importance, preferring to devote himself to the restoration of astronomy.

The German tract on the comet of 1577 stressed the comet's astrological significance, whereas the Latin work did not. In the *Progymnasmata,* the main part of the 1573 nova tract was reprinted, but not the section on the star's signification. These differences can, no doubt, be explained by the differences in the intended audience and a change in Tycho's point of view. Yet as late as 1598, in the autobiography included in the *Mechanica,* he said that both natural and judicial astrology are more reliable than one would think, provided the times are correctly determined and the paths of the celestial bodies and their entrances into the separate divisions of the sky are used in accordance with the observed sky, and their directions of motion and revolution are properly computed. He indicated that he had developed a method for this that he did not care to divulge.

In the *Astrologia,* Flemløs gave 399 short rules for weather prediction from the appearance of the sky, sun, moon, and stars, or animal behavior. However, the daily weather record kept at Hven from 1 October 1582 to 22 April 1597 was not published until the nineteenth century. It recorded the arrivals and departures of Tycho, his visitors, and students or assistants; and, although no instruments were used and precise times were not entered, it provided useful meteorological information for the area—frequency of wind, rain, snow, fog, hail, thunder, halos, and aurorae, and whether the sky was clear, semiclear, or covered. Some estimates were made of the force of winds.

Tycho's main occupation on Hven was the redetermination of the positions of the fixed stars and the observation of the planets, the sun, and the moon for the purpose of improving the theory of their motions. For six years, beginning in 1582, the distance between Venus and the sun was measured with the triangular sextant, which required two observers. Simultaneously the altitudes, and occasionally the azimuths, of Venus and the sun were measured. The distance of Venus from selected bright stars near the zodiac was measured with the same sextant after sunset, altitudes and declinations also being noted. The motions of Venus and the sun between daytime and nighttime observations were considered in calculating the positions of the observed stars. A star's declination was measured directly, but the difference in right ascension between the sun and a star was obtained by trigonometry. Using the right ascension of the sun as given in the tables, the right ascension of the star could be found. The stars were connected with α Arietis by distance measures. By suitable selection of observations, minimizing the effects of parallax and refraction, he determined the right ascension of α Arietis, and with this as reference, he determined the right ascensions of eight standard stars. Later he added three stars near the zodiac.

In determining the position of another star, a meridian quadrant or armillary was used to measure the declination, and a sextant was used to measure the distance from a known star. For the complete determination, two or three standard stars were used as reference. Included in the *Progymnasmata* (1602), before the section on the nova, are revisions of the solar and lunar theories and a catalog giving the positions of 777 fixed stars. Having indicated familiarity with the work of his predecessors, Tycho, using diagrams, described his observational methods and depicted the instruments used. In his later years he brought the list of stars to 1,000 by the less careful determination of the positions of 223 additional ones. The *Tabulae Rudolphinae,* prepared by Kepler in accordance with his modification of the Copernican system but on the basis of Tycho's observations, did not appear until 1627. Included were logarithm and other tables, the most significant of which were those of the positions of the sun and moon and five planets, and of 1,000 fixed stars calculated for the year 1600.

When he had more instruments, Tycho used several quadrants simultaneously for repeated observations of the sun's meridian altitude, begun 14 December 1576. From March 1582 he mostly used the great mural quadrant. He determined the equinoxes for the years 1584–1588, using the time when the sun was 45° from the equinoxes to determine the position of the apogee and the eccentricity of the orbit instead of using the solstice, the exact moment of which was difficult to find. He believed that the sun moved uniformly in an eccentric circle, but by 1591 he might have noted, from the motion of Mars, another inequality due to that eccentricity. He considered his

tables of the sun's motion to be accurate within 10″ or at most 20″. His values were 95°30′ for the longitude of the apogee with an annual motion of 45″ and .03584 for the eccentricity of the orbit, the greatest equation of center being 2°3′15″.

The difference in the colatitude as determined from his solar observations and his observations of the polestar led Tycho to investigate the effects of refraction, using the armillae at Stjerneborg, and to compose a refraction table. Unfortunately, he assumed the value of 3′ for the sun's horizontal parallax. He also composed a refraction table for the stars. He erred in believing refraction negligible at 45° and over, but made a step forward in determining the refraction for an observation and in correcting the instruments.

Tycho's handling of the lunar theory illustrates not only the accuracy of his observations and his awareness of the need to observe over long periods of time and over the whole course of the moon's orbit, but also his computational prowess and talent for theory construction. His discoveries of new inequalities in both longitude and latitude stem from his efforts at accurate determination of eclipses and his interest in parallax. Making approximately 300 observations of the moon in different parts of its orbit from 1582 to 1595, he noted its position relative to known fixed stars, observing in such a way as to minimize the effects of parallax. In the *Progymnasmata* he recorded twenty-one lunar and nine solar eclipses.[6] At his death all the important lunar perturbations, with the exception of the secular variation of the mean motion, were known.

Tycho made his first discovery regarding the moon's motion in 1587, when preparing his observations of the comet of 1577 for publication. The comet's position obtained from observations of its stellar distances differed by 21′ from the position computed from the lunar distance, suggesting some error in his theoretical position of the moon. Four lunar observations in August 1587 confirmed his suspicion that the inclination of the lunar orbit was 5°15′ instead of the previously accepted 5° of Ptolemy. When Tycho announced this finding in his book on the comet of 1577,[7] he expressly interpreted it as due to a long-term change rather than as a correction of Ptolemy. But in 1595 he discovered that the inclination varied in the short term—that he had, by chance, observed the moon's latitude in quadrature, whereas previous interest in the moon's latitude had been focused on syzygy, where eclipses occur. To account for the semimonthly fluctuations of the inclination, Tycho let the pole of the lunar orbit describe a circle twice a month to bring the pole 5° from the ecliptic when the moon was in syzygy and 5° 15′ from it in quadrature, and also to provide a smooth variation in between. Since this device implied an oscillation of the nodes along the ecliptic, Tycho sought and found empirical evidence that such an oscillation did occur, thus making what has been described as "a true deductive discovery." From his determinations of the extreme values for the inclination of the orbit (4° 58′ 30″ to 5° 17′ 30″), Tycho deduced a value of 1° 46′ for this nodal oscillation.

Tycho's discovery of the third inequality in longitude began with his observation of the lunar eclipse of 1590, in which the moon reached opposition about an hour before the time he had computed. By 1595 he had isolated the cause of his difficulty and had determined the approximate value of the "variation," the discovery of which he announced in his *Mechanica*.[8] During the winter of 1598–1599 another refractory eclipse led him to a fourth inequality—the so-called annual equation with a period of a solar year.

Tycho's theory was put into its finished form by Longomontanus in 1601 and was published in the *Progymnasmata*. In it the first inequality (4° 58′ 27″) was represented by a double epicycle, while the second appeared in the form of a hypocycle by means of which the center of the deferent was made to pass through the earth twice a month at syzygies and to reach its greatest distance from the earth at quadratures. The third inequality (40′ 30″) was accounted for by letting the center of the large epicycle librate on the deferent in the period of half a synodic revolution. Since these mechanisms left no room for the fourth inequality, Longomontanus introduced it—only partially, and to Tycho's expressed displeasure—by dispensing with the anomalistic component of the equation of time.

After the death of Frederick II, 4 April 1588, Tycho gradually lost the favor he had enjoyed at court. His own personality had much to do with this. He was arrogant, haughty with members of the royal family, neglectful of the welfare of the tenants on Hven, and careless in the maintenance of the public buildings on his fiefs. Although Hven had been conferred on him for life and he had some inherited wealth, the maintenance of his buildings and instruments required additional funds. Young King Christian IV, after gaining majority, did not seem to find that the astronomical work warranted the large expenditures. A quarrel with his former pupil Gellius Sascerides, who was engaged to his daughter Magdalene, put Tycho in an unpleasant light and may have contributed to his desire to leave Denmark. Besides, he may have wanted more opportunity for intellectual inter-

course than he had on his island; and he may have hoped for patronage from Emperor Rudolph II, of whose interest in alchemy and astronomy he must have been aware through his correspondence with Hagecius and with Vice-Chancellor Curtius, who had written describing Clavius' method of dividing instruments, which was similar to Vernier's later, more practical one.

After 15 March 1597, the date of the last observation at Hven, Tycho's instruments, printing press, chemical apparatus, and portable possessions were transported to his house in Copenhagen; the mural quadrant and three other large instruments were left behind. Little is known of Tycho's activities in Copenhagen. Early in June he sailed for Rostock with his instruments, press, and other belongings, as well as his family and entourage, including Tengnagel, who had come to Hven in 1595.

On 10 June 1597 the Roskilde prebend was conferred on another. Tycho made an unsuccessful attempt at reconciliation with Christian IV and in October, at the invitation of Heinrich Rantzov, took up residence in the castle at Wandsbeck, near Hamburg. There he continued his efforts to have the king permanently endow Uraniborg. Tycho began observing again 21 October, using only a radius until February 1598, when he got some of his better instruments together. He observed the solar eclipse of 25 February 1598, and later he received records of observations by others and information that it had been observed from beginning to end at Hven. He observed two lunar eclipses and some meridian altitudes but concentrated on the planets. He was assisted by the mathematician Johannes Müller, from Brandenburg, who had visited Hven in 1596 and whom the electress of Brandenburg had asked Tycho to train in chemistry and the preparation of medicines. Among visitors at Wandsbeck was the astronomer David Fabricius.

Tycho now completed the *Mechanica* and dedicated it to Emperor Rudolph II. Excellent woodcuts accompany Tycho's descriptions of his globe and of each of his instruments and its use. Also included are descriptions of Hven and its buildings, the instrument sights and the method of dividing by transversals, and a brief autobiography. From Wandsbeck he distributed a large number of manuscript copies of his star catalog, also dedicated to Rudolph.

Tycho's eldest son brought the emperor the catalog, the *Mechanica*, and a letter expressing the hope that the astronomer could complete his work under the emperor. At the same time Tycho sent bound copies of the catalog to scholars and influential people, including Christian IV, to whom he also addressed a respectful letter. On 24 March 1598 Tycho wrote to Longomontanus, inquiring about Wittich's books and manuscripts, and asked if Longomontanus had seen a recent publication by Ursus that Tycho did not consider deserving of refutation. He requested Longomontanus to join him at Wandsbeck, perhaps to continue work on the lunar theory.

Rantzov asked the elector of Cologne to use influence with the emperor and to try to interest Barwitz, the Austrian privy councillor, in Tycho's cause. Tycho himself wrote to Hagecius, hoping he would influence the emperor and the vice-chancellor. He also investigated the possibilities for settling in the Netherlands. Shortly after the middle of September 1598, having been assured that he would be welcome in Prague, Tycho left Wandsbeck with his sons, his students, and a few instruments. Longomontanus reached Wandsbeck after Tycho's departure but accompanied Tycho's ladies as far as Magdeburg. He returned to Denmark, however, and did not rejoin Tycho until January 1600. An epidemic of pestilence and dysentery in Prague caused Tycho to remain in Dresden. The first week in December, he moved to Wittenberg.

On Tycho's arrival in Prague in June 1599, he was escorted to the home of the late Vice-Chancellor Curtius and was soon granted an audience by the emperor, who arranged for him to receive financial support. Tycho had only a few instruments with him, but tried to display them in the same splendid setting they had had at Uraniborg. He never again got his instruments properly set up, nor did he make any important observations. He observed the end of a partial solar eclipse 22 July. He did not want to remain in the city of Prague and soon took up residence in the castle of Benatky, one of those offered him by the emperor. On a hill above the Iser about twenty-two miles northeast of Prague, Benatky had unobstructed views of the skies. It was small, but he altered it to fit his needs, building a laboratory and an observatory and planning to set up the instruments in separate rooms. He had difficulty, however, in obtaining the necessary funds. His family arrived, and he sent his eldest son for the four large instruments left at Hven. These, as well as the instruments and books that Tycho had brought with him as far as Magdeburg, were delayed in transit, the latter not arriving in Prague until November 1600.

Tycho's assistants in Bohemia included Longomontanus (January 1600–4 August 1601), David Fabricius (June 1601), Johannes Müller (March 1600–Spring 1601), and Melchior Joestelius, a mathematician from Wittenberg, who returned there before June 1600 but who was probably responsible for completing the solution of triangles by the prosthaphaeretic method that Tycho said he and Joestelius had done together. The assistant most important for the future of astronomy, Johannes Kepler, a firm believer in the heliocentric system, arrived at Benatky 3 February 1600. Longomontanus was working on Mars, but that planet was eventually turned over to Kepler. The relations between Tycho and Kepler were frequently strained.

In the summer of 1600 Tycho moved to Prague and set up his instruments in the Belvedere, a villa belonging to the emperor and close to the castle. Kepler, who had returned to Graz to settle his affairs and call for his family, arrived in Prague in October. Until April 1601 he was mostly engaged, at Tycho's behest, in a refutation of Ursus, although the latter had died in August 1600. Because of Tycho's death, the refutation remained unpublished until the nineteenth century. The emperor bought Curtius' house, and Tycho took possession of it in February 1601. The Kepler family moved there, too, although Kepler returned to Graz on business from April to August 1601. Kepler worked on the theories of Mercury, Venus, and Mars, and tried to persuade Tycho of the impossibility of describing the motion of the sun (or of the earth) as uniform in an eccentric circle.

Tycho died after an illness of eleven days, probably caused by prostate difficulties. He was buried with pomp on 4 November in the Tyn Church in the city's main square, his tomb marked by an upright slab bearing a life-size raised image of him with an inscription. On his deathbed, Tycho begged Kepler to complete the *Rudolphine Tables* as quickly as possible and expressed the wish that their theory be demonstrated in accordance with the Tychonic system.

Kepler did not obtain Tycho's instruments. They were stored beneath the Curtius house, and their subsequent fate is uncertain. The great globe was placed in the Round Tower in Copenhagen in the middle of the seventeenth century, having first been in Silesia, at Rosenborg Castle in Denmark, and at the University of Denmark. Some instruments must have been carried off during the Thirty Years' War, for they were discovered in a castle in Sweden in the twentieth century. Kepler had difficulties with Tengnagel and Tycho's other heirs over the records and publications. Giving him due credit, Kepler used the records of Tycho's observations, especially of Mars, to derive the laws of planetary motion, announcing the first two in his *Astronomia nova* (1609) and the third in the *Harmonices mundi libri V* (1619). He published Tycho's *Progymnasmata* (1602), already partly printed at Uraniborg, and the *Tabulae Rudolphinae* (1627), in conformity with a heliocentric system.

Thus Tycho's accurate observations of the positions of the sun, moon, stars, and planets provided the basis for refinements of the Copernican doctrine. Had the observations been as accurate as Tycho considered them, or less accurate than they actually were, the history of astronomy would have been different. But they provided the suitable degree of accuracy at the critical time. A discrepancy of 8′ of arc between theory and observation led Kepler to his reformation of astronomy.

NOTES

1. *Opera omnia*, II, 342–343; V, 81. Dreyer (1890), p. 30, erred in describing the Hainzels.
2. *Dialogue Concerning the Two Chief World Systems*, Stillman Drake, trans. (Los Angeles, 1962), pp. 358 ff.
3. Pingré, *Cométographie*, I (1783), 511, says the comet was seen in Peru as early as 1 November.
4. *Opera omnia*, IV².
5. *En Elementisch oc Jordisch ASTROLOGIA*, 1591.
6. *Opera omnia*, II, 98.
7. *Ibid.*, IV, 42.
8. *Ibid.*, V, 111.
9. A broadside, item 3026 in Zinner, *Geschichte und Bibliographie* (Leipzig, 1941), is not included.

BIBLIOGRAPHY

There is no complete bibliography of the vast literature dealing with Tycho Brahe's life or work. Neither is there a printed list of his writings. Also lacking is a bibliography of works with references to him. Nevertheless, the following bibliography is selective.

The impact of Tycho's work must be studied in the writings, both printed and MS, of his contemporaries and immediate followers. Works by and about Kepler, including the recent ones, are of particular importance, but the writings of less important seventeenth-century men also give evidence of the influence of Tycho in Europe and the East. In honor of the 300th anniversary of Tycho's death (1901) and the 400th anniversary of his birth (1946) much was written, both scholarly and popular.

I. ORIGINAL WORKS. Tycho's writings were collected as *Tychonis Brahe Dani Opera omnia*, J. L. E. Dreyer, ed., 15 vols. (Copenhagen, 1913–1929), which includes charts, diagrams, facsimiles, maps, portraits, and tables, and is copiously annotated.[9]

His books on the nova of 1572 are *De nova et nullius aevi memoria prius visa stella* . . . (Copenhagen, 1573; facs. ed., 1901), trans. into Danish by Otto Gelsted as *Tyge Brahe: Den ny stjerne* (*1572*) . . . (Lemvig, 1923), and partially trans. into English by John H. Walden in Harlow Shapley and Helen E. Howarth, eds., *A Source Book in Astronomy* (New York–London, 1929), pp. 13–19; and *Astronomiae instauratae progymnasmata* . . . (Prague, 1602; Frankfurt, 1610). A number of seventeenth-century tracts summarized or translated excerpts from the *Progymnasmata,* the most important work on the 1572 nova.

Tycho's book on the comet of 1577 is *De mundi aetherei recentioribus phaenomenis* . . . (Uraniborg, 1588; Prague, 1603; Frankfurt, 1610). Part of ch. 8 is trans. in Marie Boas and A. Rupert Hall, "Tycho Brahe's System of the World," in *Occasional Notes of the Royal Astronomical Society,* **3,** no. 21 (1959), 252–263 (trans. on pp. 257–263).

His letters have been brought together by Dreyer in the *Opera omnia* and in *Tychonis Brahe Dani Epistolarum astronomicarum* . . . (Uraniborg, 1596; Nuremberg, 1601; Frankfurt, 1610); *Tychonis Brahei et ad eum doctorum virorum epistolae . . .,* F. R. Friis, ed., 2 vols. (Copenhagen, 1876–1909); and Wilhelm Norlind, *Ur Tycho Brahes brevväxling, från Latinet* (Lund, 1926), 23 letters trans. into Swedish, with notes, and "Några Anteckningar till Tycho Brahes brevväxling," in *Nordisk astronomisk tidsskrift* (1956), no. 2, 51–55.

Tycho's book on his instruments is *Astronomiae instauratae mechanica* (Wandsbeck, 1598; Nuremberg, 1602), also in facsimile of 1598 ed., B. Hasselberg, ed. (Stockholm, 1901); the 1598 ed. was printed in Rantzov's castle near Hamburg on Tycho's press by Philip de Ohr. For trans., see Hans Raeder, Elis Strömgren, and Bengt Strömgren, eds. and trans., *Tycho Brahe's Description of His Instruments and Scientific Work* . . . (Copenhagen, 1946).

His tables are in *Historia coelestis . . .,* Lucius Barrettus (pseud. of Albert Curtz), ed. (Augsburg, 1666). Kepler's *Tabulae Rudolphinae . . .* (Ulm, 1627) are based on Tycho's observations and the Copernican–Keplerian system of the universe.

The MS material has been thoroughly used by Dreyer in his biography and in the *Opera omnia,* and MSS for which he gives bibliographical details will not be described here. For his biography Norlind has examined MS sources; see also his "On Some Manuscripts Concerning Tycho Brahe," in *The Observatory,* **78,** no. 903 (1958), 73–75. It is impossible to list here MSS in which Tycho is discussed, e.g., Gregoriana (Rome) 530, ff. 208–211, a letter dated 26 January 1601 from Magini in Mantua to Clavius in Rome, which speaks of Tycho's book on the star of 1572; or Ambrosiana (Milan) D 246 inf. 83r, the fragment of a letter from Padua in 1592 that speaks of Tycho and Galileo, who had just begun his lectures there. Nor need the location and description of the presentation copies of the *Mechanica* and the MS copies of the catalog of stars be listed.

Letter no. 102 (*Opera omnia,* XIV, 68) is at The Historical Society of Pennsylvania, Philadelphia, as are an undated autograph and an autograph dated 10 August 1594. Presumably the letter from Tycho to T. Saville (1 December 1590) cited in the British Museum's Sloane Collection catalog of 1782 is the same as the museum's Harleian 6995, 40 used by Dreyer (*Opera omnia,* VII, 283–285). The same catalog of the Sloane Collection lists an MS of the *De mundi aetherei. . . .* Two letters in the hand of a sixteenth-century scribe are in the possession of the author: Tycho to Caspar Peucer, 13 September 1588, 23 leaves (*Opera omnia,* VII, 127–141), and Caspar Peucer to Tycho, 10 May 1589, 9 leaves (*Opera omnia,* VII, 184–191).

An interesting summary (from Padua) of Tycho's work, *Epitome de restitutione motuum solis ac lunae, et de nova stella anni 1572,* is preserved in Venice (Marciana lat. Cl. VIII, Cod. XXXVII, 3493). In Milan (Ambrosiana D 246 inf. 84r–87v) there are part of the *Mechanica* and epigrams to Scaliger.

II. SECONDARY LITERATURE. The best single treatment of Tycho's life and work is J. L. E. Dreyer, *Tycho Brahe, a Picture of Scientific Life and Work in the Sixteenth Century* (Edinburgh, 1890; repr. New York, 1963). This is based on and cites the sources available in 1890, and forms the basis for this article. Except for Tycho's major publications, works cited by Dreyer are not in this bibliography, although they include much material to which the reader may want to refer, such as Gassendi's biography of Tycho (1654) and Tycho's *Opera omnia* (1648), which presents the *Progymnasmata* and *De mundi aetherei* rather than the complete works. More recent biographies are John Allyne Gade, *The Life and Times of Tycho Brahe* (Princeton, 1947), with a bibliography that, while not selective, includes some useful items that appeared after Dreyer's work; and Wilhelm Norlind, *Tycho Brahe. Mannen och verket. Efter Gassendi overs. med kommentar* (Lund, 1951); *Tycho Brahe* (Stockholm, 1963); and *Tycho Brahe. En biografi. Med nya bidrag belysande hans liv och verk* (Lund, in press), with a summary in German.

The island of Hven is discussed in the anonymous "Stjerneborg," in *Nordisk astronomisk tidsskrift,* n.s. **20,** no. 3 (1939), 79–99; Francis Beckett and Charles Christensen, *Uraniborg og Stjaerneborg* (Copenhagen–London, 1921), text in Danish, summary and explanation of plates in English, title also given in English: *Tycho Brahe's Uraniborg and Stjerneborg on the Island of Hveen;* C. L. V. Charlier, *Utgräfningarna af Tycho Brahes observatorier på ön Hven sommaren 1901,* which is *Acta universitatis Lundensis. Lunds universitets årsskrift,* **37,** afdeln. 2, no. 8 (Lund, 1901); John Christianson, "The Celestial Palace of Tycho Brahe," in *Scientific American,* **204,** no. 2 (1961), 118–128; Charles D. Humberd, "Tycho Brahe's Island," in *Popular Astronomy,* **45** (1937), 118–125, which reproduces the Cologne map of 1586 and translates the Latin explanations inserted on the map and its back; William Lengert, *Tycho Brahe-tryck* (Malmö, 1940); N. A. Møller Nicolaisen, "Et Tycho Brahe-minde paa Hven," in *Nordisk astronomisk tidsskrift,* n.s. **11,** no. 3 (1930), 122–128; "Tycho Brahes mølledaemning paa Hven," *ibid.,* no. 4, 173–175; "Tycho Brahes papirmølle," *ibid.,* n.s. **14,** no. 3 (1933), 85–95; and *Tycho Brahes papirmølle paa Hven. Udgravningen 1933–34 og forsøg til rekonstruktion . . .* (Copenhagen, 1946); Harald Mortensen, "Johannes Mejers kort over øen Hven," in *Skäne årsbok* (1925), pp. 9–16; and "Et Tycho Brahe-minde paa Hven," in *Nordisk astronomisk tidsskrift,* n.s. **11,** no. 4 (1930), 172–173; and Lauritz Nielsen, *Tycho Brahes bogtrykkeri. En bibliografisk-boghistorisk undersøgelse* (Copenhagen, 1946).

Works treating other subjects are Joseph Ashbrook, "Tycho Brahe's Nose," in the column "Astronomical Scrapbook," in *Sky and Telescope,* **29,** no. 6 (1965), 353, 358; F. Burckhardt, *Zur Erinnerung an Tycho Brahe 1546– 1601 . . .* (Basel, 1901); John Christianson, "Tycho Brahe at the University of Copenhagen, 1559–1562," in *Isis,* **58**

(1967), 198–203; and "Tycho Brahe's Cosmology From the 'Astrologia' of 1591," *ibid.,* **59** (1968), 312–318; J. L. E. Dreyer, "Note on Tycho Brahe's Opinion About the Solar Parallax," in *Monthly Notices of the Royal Astronomical Society,* **71,** no. 1 (1910), 74–76; "The Place of Tycho Brahe in the History of Astronomy," in *Scientia,* **25,** no. 83–3 (Mar. 1919), 177–185; and "On Tycho Brahe's Manual of Trigonometry," in *The Observatory,* no. 498 (Mar. 1916), 127–131; Antonio Favaro, "Ticone Brahe e la corte di Toscana," in *Archivio storico italiano,* 5th series, **3** (1889); Edvard Gotfredsen, "Tycho Brahes sidste sygdom og død" ("Tycho Brahe's Last Disease and Death"), in *Københavns Universitets medicinsk-historiske museum: Årsberetning 1955–1956;* Poul Hauberg, "Tycho Brahes opskrifter paa Laegemidler," in *Dansk tidskrift for farmaci,* **1,** no. 7, 205–212; C. Doris Hellman, "Was Tycho Brahe as Influential as He Thought?," in *British Journal for the History of Science,* **1,** pt. 4, no. 4 (Dec. 1963), 295–324; Flora Kleinschnitzová, "Ex Bibliotheca Tychoniana Collegii Soc. Jesu Pragae ad S. Clementem," in *Nordisk tidskrift för bok- och biblioteksväsen,* **20** (1933), 73–97; Wilhelm Krebs, "Facsimile einer eigenhändigen Zeichnung Tycho's de Brahe von dem grossen Kometen 1577," in *Das Weltall. Illustrierte Zeitschrift für Astronomie und verwandte Gebiete,* **12** (1911), 52–53; Martha List, *Des handschriftliche Nachlass der Astronomen Johannes Kepler und Tycho Brahe* (Munich, 1961); Knud Lundmark, "Tycho Brahe och astrofysiken," in *Nordisk astronomisk tidsskrift,* n.s. **11,** no. 3 (1930), 89–112; and "Om Tycho Brahes liv och gärning," in *Cassiopeia* (1945), 14–47; N. A. Møller Nicolaisen, "Nicholai Raimarus Ursus contra Tycho Brahe," *ibid.* (1942), 81–91; Harald Mortensen, "Portraeter af Tycho Brahe," *ibid.* (1946), 52–77; "Tycho Brahe i Wandsbek," in *Astronomiska sällskapet Tycho Brahe ärsbok* (1945), pp. 94–98; and "Tychoniana," in *Cassiopeia* (1948), 21–27; Wilhelm Norlind, "Tycho Brahe och Thaddaeus Hagecius: Ur en brevväxling," *ibid.* (1939), 122–130; "A Hitherto Unpublished Letter From Tycho Brahe to Christopher Clavius," in *Observatory,* **74** (1954), 20–23; and "Tycho-Brahé et ses rapports avec l'Italie," C. Cardot, trans., in *Scientia,* **49** (1955), 47–61; Eiler Nystrøm, "Tyge Brahes Brud med Faedrelandet," in *Festskrift til Kristian Erslev, dem 28. Decbr. 1927 . . .,* pp. 291–320; and *Epistolae et acta ad vitam Tychonis Brahe pertinentia* (Copenhagen, 1928); Wilhelm Prandtl, "Die Bibliothek des Tycho Brahe," in *Philobiblon; eine Zeitschrift für Bücherliebhaber,* **5,** no. 8 (1932), 291–300, no. 9, 321–330; Hans Raeder, "Tycho Brahe og hans korrespondenter," in *Edda,* **27** (1927), 250–264; Edward Rosen, "Kepler's Defense of Tycho Brahe Against Ursus," in *Popular Astronomy,* **54,** no. 8 (1946), 1–8; Henrik Sandblad, "En Tycho Brahe-notis," in *Lychnos* (1937), 366–368, concerning the duel in which Tycho lost part of his nose; Aydin Sayili, "Islam and the Rise of Seventeenth Century Science," in *Belleten* (Ankara), **22,** no. 87 (July 1958), 353–368; and "Tycho Brahe Sistemi Hakkinda XVII. Asir Başlarina ait Farça Bir Yazma. An Early Seventeenth Century Persian Manuscript on the Tychonic System," in *Anatolia revue annuelle de l'Institut d'Archéologie de l'Université d'Ankara,* **3** (1958),

79–87; Ottomar Schiller, "Tycho Brahe à Prague," in *Archeion,* **22,** no. 4 (12 Feb. 1941), 372–375; H. C. F. G. Schjellerup, "Tycho Brahes Original-Observationer, benyttede til Banebestemmelse af Cometen 1580," in *Det Kongelige Danske Videnskabernes selskab. Skrifter. 5 Raekke, Naturvidenskabelig og Matematisk Afdeling.* IV (1856–1859), pp. 1–39; Christine Schofield, "The Geoheliocentric Mathematical Hypothesis in Sixteenth Century Planetary Theory," in *British Journal for the History of Science,* **2** (1965), 291–296; Harold Spencer Jones, "Tycho Brahe (1546–1601)," in *Nature,* **158** (1946), 856–861; E. S., "Tycho Brahes Immatrikulation ved Universitetet i Leipzig vinterhalvaaret 1561–62," in *Nordisk astronomisk tidsskrift,* n.s. **9,** no. 2 (1928), 41–42; Elis Strömgren, "Tycho Brahes sekstanter," *ibid.,* n.s. **14,** no. 2 (1933), 69–75; F. J. Studnička, ed., *Prager Tychoniana* (Prague, 1901); and, as author, *Bericht über die astrologischen Studien des Reformators der beobachtenden Astronomie Tycho Brahe. . . .* (Prague, 1901); Sevim Tekeli, "Nasirüddin, Takiyüddin ve sesi" ("The Comparison Instruments of Taqi al Din and Tycho Brahe"), in *Ankara Üniversitesi, Coğrafya Fakültesi Dergisi,* **16** (1958), 301–393; and "Solar Parameters and Certain Observational Methods of Taqî al Dîn and Tycho Brahe," in *Proceedings of the 10th International Congress of the History of Science,* I (Paris, 1964), 623–626; and Victor E. Thoren, "Tycho Brahe on the Lunar Theory," doctoral dissertation (Indiana Univ., 1965); "Tycho and Kepler on the Lunar Theory," in *Publications of the Astronomical Society of the Pacific,* **79,** no. 470 (Oct. 1967), 482–489; "An Early Instance of Deductive Discovery: Tycho Brahe's Lunar Theory," in *Isis,* **58** (1967), 19–36; and "Tycho Brahe's Discovery of the Variation," in *Centaurus,* **12,** no. 3 (1967), 151–166.

Dr. Thoren kindly advised on the description of Tycho's lunar theory and helped in the final draft of that section.

C. DORIS HELLMAN

BRAHMADEVA (*fl. ca.* 1092), *astronomy.*

Brahmadeva was the son of Candrabudha (or Śrīcandra, or Candrabhaṭṭa), a Brāhmaṇa of Mathurā (or Madhurā). The epoch date of his only work, the *Karaṇaprakāśa,* is Thursday, 11 March 1092. The work contains nine chapters:

1. On the mean longitudes of the planets.
2. On *tithis* and so on.
3. On the true longitudes of the star-planets.
4. On the three problems relating to diurnal motion.
5. On lunar eclipses.
6. On solar eclipses.
7. On heliacal risings and settings.
8. On the lunar crescent.
9. On planetary conjunctions and latitudes.

The work is based on the *Āryabhaṭīya* of Āryabhaṭa I, with modifications proposed by Lalla (see Essay V).

It was particularly popular in Madras, Mysore, and Mahārāṣṭra.

There are commentaries on the *Karaṇaprakāśa* by Amareśa in the Kannada language (*Karṇāṭabhāṣāvyākhyāna*); by Brahmaśarman (*Vyākhyā*); by Dāmodara, the pupil of Padmanābha (*fl. ca.* 1575) (*Vṛtti*); by Govinda, the son of Viśvanātha Tāmbe (*Vivṛtti*); by Sampatkumāra (*Vyākhyā*); and by Śrīnivāsa Yajvan (*Prabhā, ca.* 1275). There also exist an *Udāharaṇa,* once (probably erroneously) ascribed to Viśvanātha of Benares (*fl.* 1612–1630), and a *Sadvāsanā* by Sudhākara Dvivedin, who published the *Karaṇaprakāśa* along with this commentary as the twenty-third work in the Chowkhambā Sanskrit Series (Benares, 1899).

BIBLIOGRAPHY

Additional works concerning Brahmadeva are Ś. B. Dīkṣita, *Bhāratīya Jyotiḥśāstra* (Poona, 1896; repr. Poona, 1931), pp. 240–243; and Sudhākara Dvivedin, *Gaṇakataraṅgiṇī* (Benares, 1933; repr. from *The Pandit,* n.s. **14** [1892]), pp. 31–33.

DAVID PINGREE

BRAHMAGUPTA (*b.* 598; *d.* after 665), *astronomy.*

Brahmagupta was the son of Jiṣṇugupta; the names compounded with -*gupta* may indicate membership in the Vaiśya caste. At the age of thirty, he composed the *Brāhmasphuṭasiddhānta;* the reigning king was Vyāghramukha of the Cāpavaṃśa of the Gurjaras, and we know from the account of the Chinese pilgrim Hiuan-tsang (641) that the capital of the Gurjaras was Bhillamāla (modern Bhinmal, near Mt. Abu in Rajasthan). In the colophon of chapter 24 of the *Brāhmasphuṭasiddhānta* and in Pṛthūdakasvāmin's commentary on the *Khaṇḍakhādyaka,* Brahmagupta's second work, he is called Bhillamālācārya—the teacher from Bhillamāla. The *Khaṇḍakhādyaka* uses as its epoch Sunday, 15 March 665. Both of these treatises were studied primarily in Rajasthan, Gujarat, Madhya Pradesh, Uttar Pradesh, Bihar, Nepal, Panjab, and Kashmir.

The *Brāhmasphuṭasiddhānta,* whose planetary parameters are mainly derived from the *Paitāmahasiddhānta* of the *Viṣṇudharmottarapurāṇa* (see Essays IV and XII), contains twenty-four chapters, with the twenty-fifth appended in some manuscripts.

1. On the mean longitudes of the planets.
2. On the true longitudes of the planets.
3. On the three problems relating to diurnal motion.
4. On lunar eclipses.
5. On solar eclipses.
6. On heliacal risings and settings.

7. On the lunar crescent.

8. On the lunar "shadow."

9. On planetary conjunctions.

10. On conjunctions of the planets with stars. (These ten chapters form the *Daśādhyāyī*, which sometimes is found independently in manuscripts.)

11. Examination of previous treatises on astronomy.

12. On mathematics.

13. Additions to chapter 1.

14. Additions to chapter 2.

15. Additions to chapter 3.

16. Additions to chapters 4 and 5.

17. Additions to chapter 7.

18. On algebra.

19. On the gnomon.

20. On meters.

21. On the sphere.

22. On instruments.

23. On measurements.

24. Summary of contents.

25. Versified tables.

A commentary (*Vāsanābhāṣya*) on the *Brāhmasphuṭasiddhānta* was written by Pṛthūdakasvāmin of Kurukṣetra (*fl.* 864); unfortunately, it is but imperfectly preserved in the few surviving manuscripts. There are also an anonymous commentary (*Ṭīkā*) on the *Daśādhyāyī* in Vishveshvarananda Vedic Research Institute 2363, and another (*Ṭīkā*) on chapter 18 in two manuscripts (one in Calcutta, one in London). A translation of chapters 12 and 18 was presented by H. T. Colebrooke in *Algebra, With Arithmetic and Mensuration: From the Sanscrit of Brahmegupta and Bháscara* (London, 1817); the preface is reprinted in Colebrooke's *Miscellaneous Essays,* II (Madras, 1872), 417–531. The whole text, with his own Sanskrit commentary, was edited by Sudhākara Dvivedin (Benares, 1902; reprinted from *The Pandit,* n.s. **23** [1901] and **24** [1902]). It was published again by Ram Swarup Sharma in four volumes (New Delhi, 1966): Volume I contains a sometimes useful introduction by Satya Prakash (pp. 1–344) and the text based on five of some fifteen known manuscripts; Volume II, chapters 1–9 with fragments of Pṛthūdakasvāmin's commentary on chapters 1–3, excerpts from Sudhākara Dvivedin's commentary, a new Sanskrit commentary (*Vijñānabhāṣya*), and a Hindi explanation; Volume III, chapters 10–16 with the same commentaries, save that, in this volume, the available portions of Pṛthūdakasvāmin's *Vāsanābhāṣya* are strangely missing; and Volume IV, chapters 17–25 as in Volume III, but with Pṛthūdakasvāmin's commentary on chapter 21 added as an appendix.

The *Khaṇḍakhādyaka* is the best-known treatise on the *ārddharātrika* system (see Essay VI). It contains eight chapters:

1. On computing the *tithis* and *nakṣatras.*

2. On the longitudes of the planets.

3. On the three problems relating to diurnal motion.

4. On lunar eclipses.

5. On solar eclipses.

6. On heliacal risings and settings.

7. On the lunar crescent.

8. On planetary conjunctions.

There also exists an appendix (*Uttarakhaṇḍakhādyaka*). If one follows Utpala (*fl.* 966), it contained three chapters:

1. On the conjunctions of the planets with the stars.

2. On the *pātas* of the sun and moon.

3. On the projection of eclipses.

But the (incomplete) manuscript tradition of Pṛthūdakasvāmin's version provides only one chapter in the *Uttarakhaṇḍakhādyaka*—one in which new methods of approximation are presented; and Āmarāja (*fl.* 1180) cites various verses from the *Uttara* which are found in neither Pṛthūdakasvāmin nor Utpala. The whole problem of the extent and authorship of the *Uttarakhaṇḍakhādyaka* needs to be extensively investigated.

Commentaries have been written on the *Khaṇḍakhādyaka* by the following scholars: Lalla; Pṛthūdakasvāmin (*Vivaraṇa*, 864); Utpala (*Vivṛti*, 966); Someśvara (1040); Āmarāja (*Vāsanābhāṣya*, 1180); Śrīdatta, the son of Nageśvara Miśra Mahopādhyāya (*Ṭīkā*); and Yāmaṭa (*Ṭīkā*); as well as a number of anonymous *ṭīkās* and *udāharaṇas*. No manuscripts of the commentaries of Lalla and of Someśvara have yet been identified. Dīkṣita also refers to a *ṭīkā* by one Varuṇa. The version and commentary of Āmarāja was edited by Babua Miśra (Calcutta, 1925). The recension and commentary of Pṛthūdakasvāmin were edited by P. C. Sengupta (Calcutta, 1941); an appendix contains additional verses from the version of Utpala. Sengupta has also translated Pṛthūdakasvāmin's version into English (Calcutta, 1934).

BIBLIOGRAPHY

Virtually every paper on every aspect of Indian astronomy discusses Brahmagupta. Listed here are only those which deal especially with his works. For the others, the reader is referred to David Pingree, *A Census of the Exact Sciences in Sanskrit* (Philadelphia, 1969). It should also be noted that the early 'Abbāsid *Zīj al-Arkand* and *Zīj al-Sindhind* have been said to be versions, respectively, of the *Khaṇḍakhādyaka* and of the *Brāhmasphuṭasiddhānta;*

but in neither case is a direct influence discernible. See David Pingree, "The Fragments of the Works of Ya'qūb ibn Ṭāriq," in *Journal of Near Eastern Studies,* **27** (1968), 97–125; and E. S. Kennedy and David Pingree, *The Kitab 'ilal al-zījāt of al-Hāshimī.* Both of Brahmagupta's works were known in Sanskrit to al-Bīrūnī; he cites them extensively, especially in *India* and in *Al-Qānūn al-Mas'ūdī.* In his catalog of his own works, D. J. Boilot, ed., in *Mélanges de l'Institut Dominicain d'études Orientales du Caire,* **2** (1955), 161–256, al-Bīrūnī includes *Translation of What Is in the Brahmasiddhānta of the Method of Calculation* (*RG* 40).

Additional literature on Brahmagupta includes B. Datta, "On the Supposed Indebtedness of Brahmagupta to *Chiuchang Suan-shu,*" in *Bulletin of the Calcutta Mathematical Society,* **22** (1930), 39–51; Ś. B. Dīkṣita, *Bhāratīya Jyotiḥśāstra* (Poona, 1931; repr. of Poona, 1896), pp. 216–227; Sudhākara Dvivedin, *Gaṇakataraṅgiṇī* (Benares, 1933; repr. from *The Pandit,* n.s. **14** [1892]), pp. 18–19; P. C. Sengupta, "Brahmagupta on Interpolation," in *Bulletin of the Calcutta Mathematical Society,* **23** (1931), 125–128; M. Simon, "Zu Brahmegupta diophantischen Gleichungen zweiten Grades," in *Archiv der Mathematik und Physik,* **20** (1913), 280–281; G. Thibaut, *Astronomie, Astrologie und Mathematik, Grundriss der indo-arischen Philologie und Altertumskunde,* III, pt. 9 (Strasbourg, 1899), 58–59; and H. Weissenborn, "Das Trapez bei Euclid, Heron und Brahmagupta," in *Abhandlungen zur Geschichte der Mathematik,* **12** (1879), 167–184.

Reference should also be made to the solar and lunar tables based on the *Brāhmasphuṭasiddhānta* that were published by R. Sewell in *Epigraphia Indica,* **17** (1923–1924), 123–187, 205–290.

DAVID PINGREE

BRAIKENRIDGE (BRAKENRIDGE), WILLIAM (*b. ca.* 1700; *d.* 30 July 1762), *mathematics.*

The precise date and the place of Braikenridge's birth are not known. He lived in a period of intense mathematical activity, one that abounded in illustrious mathematicians: the Bernoullis, Maclaurin, and Brook Taylor, to name but a few. Newton was still in his prime, but his interest in mathematics had begun to wane (no doubt as a result of his important duties as master of the Mint).

The main lines of development in mathematics at this time were the extension and systematization of the calculus, the further study of the theory of equations, and a revival of interest in geometry. It was in the last of these that Braikenridge excelled, and it is upon his *Exercitatio geometrica de descriptione linearum curvarum* (1733) that his reputation mainly rests.

This work is divided into three parts, and its scope is indicated by their titles: "De descriptione curvarum primi generis seu linearum ordinis secundi," "De descriptione linearum cujuscunque ordinis ope linearum ordinis inferioris," and "Ubi describuntur sectiones conicae ope plurium rectarum circa polos moventium."

The study of the properties of curves has always been an inexhaustible subject of speculation and research among geometers. Colin Maclaurin had already published his *Geometria organica* (1720), which contained an elegant investigation of curves of the second order by regarding them as generated by the intersection of lines and angles turning about fixed points, or poles. Many of Maclaurin's theorems were discovered independently by Braikenridge, notably the Braikenridge-Maclaurin theorem: If the sides of a polygon are restricted so that they pass through fixed points while all the vertices except one lie on fixed straight lines, the free vertex will describe a conic or a straight line. A general statement of this appeared in 1735 in the *Philosophical Transactions* (no. 436), and a dispute at once arose regarding priority. Braikenridge, in the Preface to the *Exercitatio,* maintained that as early as 1726, when he was living in Edinburgh, he had discovered many of the propositions contained in that work and had actually discussed some of them with his contemporaries, including Maclaurin. There followed a lively correspondence between the two men which, however, it would be profitless to discuss here.

About the middle of the century, interest in the geometry of curves began to languish. It was revived, however, when a group of French mathematicians—Monge, Carnot, Poncelet—by employing projective methods, gave the study a fresh impetus.

Braikenridge was a noted theologian, and for many years he was rector of St. Michael's, Bassishaw, London. On 6 February 1752 he was elected fellow of the Royal Society of Antiquaries, and on 9 November of the same year he became a fellow of the Royal Society.

Braikenridge contributed a number of papers to the *Philosophical Transactions.* Their titles reflect the wide range of his interests: "A General Method of Describing Curves, by the Intersection of Right Lines, Moving About Points in a Given Plane"; "A Letter . . . Concerning the Number of Inhabitants Within the London Bills of Mortality"; "A Letter . . . Concerning the Method of Constructing a Table for the Probabilities of Life in London"; "A Letter . . . Concerning the Number of People in England"; "A Letter . . . Concerning the Present Increase of the People in Britain and Ireland"; "A Letter . . . Containing an Answer to the Account of the Numbers and Increase of the People of England by the Rev. Mr. Forster"; "A Letter Containing the Sections of a Solid, Hitherto not Considered by Geometers."

BIBLIOGRAPHY

Works concerning Braikenridge are Moritz Cantor, *Vorlesungen über Geschichte der Mathematik,* III (Leipzig, 1894–1898), 761–766, 773; and J. F. Montucla, *Histoire des mathématiques,* III (Paris, 1799–1802), 87.

J. F. SCOTT

BRAMER, BENJAMIN (*b.* Felsberg, Germany, *ca.* February 1588; *d.* Ziegenhain, Germany, 17 March 1652), *mathematics.*

After the death of his father in 1591, Bramer was taken as a foster son into the home of his sister and her husband, the court clockmaker Joost Bürgi, in Kassel. His brother-in-law tutored Bramer and awakened his passion for mathematics, which was later combined with his architectural abilities. When Bürgi left Kassel in 1604, Bramer accompanied him to the imperial court at Prague; he returned to Kassel in 1609. In 1612 Landgrave Moritz of Hesse-Kassel appointed Bramer the master builder of the court in Marburg, and he was naturalized there on 16 February 1625. (Since 1620 he had been directing the construction of fortifications at the castle and in the town.) In the same year he was consultant to the count of Solms at the fortress of Rheinfels. From 1630 to 1634, Bramer was in charge of the fortifications in Kassel, and in November 1635 he was appointed princely master builder and treasurer of the important Hessian fortress of Ziegenhain.

In his first publication on the calculation of sines (1614), Bramer's talents are evident. In a work on the vacuum (1617), we can see his wide-ranging interests, but no particular field of concentration. The problem of empty space, which had been under active investigation since classical times, was of special topical interest in the seventeenth century. On this matter Bramer held the views of Tommaso Campanella, the contemporary and follower of Galileo.

The problem of central perspective obtained by means of instruments, which had been taken up by Leone Battista Alberti in 1435 and for which instruments had been designed by Albrecht Dürer in 1525 and by Bürgi in 1604, was further developed by Bramer in 1630 by means of a device that enabled one to draw accurate geometrical perspectives true to nature. He described his method in his *Trigonometria planorum* (1617). In 1651 Bramer contributed to the completion of the instruments for triangulation with the semicirculus: he used an inclined ruler, in order to determine simultaneously the sighted point and its inclination; the instrument, however, differed little from a similar one described by Leonhard Zubler in 1607. Another form of this instrument was mounted on a calibrated plate to determine angulation; Bramer used this for the solution of planar triangles.

We know very little of Bramer's architectural achievements. From advice he gave in 1618 to Count Christian von Waldeck, we know of a plan for construction of a new church for the city of Wildungen. Although the project was not undertaken because of the Thirty Years' War, it is of special importance because it is one of the earliest plans to introduce central church construction into Protestant German church architecture.

BIBLIOGRAPHY

I. ORIGINAL WORKS. *Problema, wie aus bekannt gegebenem sinu eines Grades, Minuten oder Sekunden alle folgenden sinus aufs leichteste zu finden und der canon sinuum zu absolvieren sei* (Marburg, 1614); *Beschreibung und Unterricht, wie allerlei Teilungen zu den mathematischen Instrumenten zu verfertigen, neben dem Gebrauch eines neuen Proportional-Instrumentes* (Marburg, 1615); *Bericht und Gebrauch eines Proportional-Lineals, neben kurzem Unterricht eines Parallel-Instrumentes* (Marburg, 1617); *Kurze Meldung vom Vacuo oder leerem Orte, neben anderen wunderbaren und subtilen Quaestionen, desgleichen Nic. Cusani Dialogus von Waag und Gewicht* (Marburg, 1617); *Trigonometria planorum mechanica oder Unterricht und Beschreibung eines neuen und sehr bequemen geometrischen Instrumentes zu allerhand Abmessung* (Marburg, 1617); *Etliche geometrische Quaestiones, so mehrerteils bisher nicht üblich gewesen. Solviert und beschrieben* (Marburg, 1618); *Beschreibung eines sehr leichten Perspektiv- und Grundreissenden Instrumentes auf einem Stande: auf Joh. Faulhabers, Ingenieurs zu Ulm, weitere Continuation seines mathematischen Kunstspiegels geordnet* (Kassel, 1630); *Appollonius Cattus oder Kern der ganzen Geometrie,* 3 vols. (Kassel, 1634–1684); *Benjamin Brameri Bericht zu Meister Jobsten seligen geometrischen Triangularinstrument* (Kassel, 1648); *Kurzer Bericht zu einem Semicirculo, damit in allen Triangeln in einer Observation nicht allein die drei latera, sondern auch die drei Winkel eines Triangels zu finden* (Augsburg, 1651); *Von Wasserwerken* (unpub. MS Math. 4°27), National Library, Kassel.

II. SECONDARY LITERATURE. Johann Heinrich Zedler, *Universal-Lexikon,* IV (Halle-Leipzig, 1733), 997; Christian Gottlieb Jöcher, *Allgemeines Gelehrten-Lexikon,* I (Leipzig, 1750), 1328; Friedrich Wilhelm Strieder, *Grundlage zu einer hessischen Gelehrten- und Schriftstellergeschichte,* I (Göttingen, 1781), 521 ff.; *Nouvelles annales de mathématiques* (*Bulletin de bibliographie*) (Paris, 1858), 75 ff.; Wolfgang Medding, "Das Projekt einer Zentralkirche des hessischen Hofbaumeisters Benjamin Bramer", in *Hessenland, Heimatzeitschrift für Kurhessen,* **49** (Marburg, 1938), 82 ff.; Karl Justi, "Das Marburger Schloss," *Veröffentlichungen der Historischen Kommission für Hessen und Waldeck,* XXI (Marburg, 1942), 94, 98, 105.

PAUL A. KIRCHVOGEL

BRANDE, WILLIAM THOMAS (*b.* London, England, 11 January 1788; *d.* London, 11 February 1866), *chemistry.*

Brande's father was proprietor of the Brande Pharmacy in Arlington Street, London (Friedrich Accum, one of the pioneers of coal-gas lighting, became his assistant in 1793). The Brandes were apothecaries to George III, and operated shops in both London and Hannover. His family moved to Chiswick when he was about fourteen, and Brande became acquainted with Charles Hatchett, who was keenly interested in chemistry and mineralogy. Hatchett allowed him to help in his laboratory and encouraged him to study the classification of rocks and ores. The mineralogical series with which Brande later illustrated his lectures at the Royal Institution originated in specimens given to him by Hatchett.

Brande was a pupil at the Anatomical School in Windmill Street and began to study chemistry at St. George's Hospital (*ca.* 1804). At about this time he seems to have been befriended by Sir Everard Home, who, as one of the trustees of the Hunterian collection at the Royal College of Surgeons, later entrusted Brande with the analysis of calculi selected from the collection. Brande submitted the report, with observations by Home, to the Royal Society (*Philosophical Transactions,* **98** [1808], 223–243) and was elected a fellow in 1809.

It appears that quite early in life Brande became acquainted with Davy and attended his lectures at the Royal Institution. In 1808 he himself began lecturing on chemistry and pharmacy at London medical schools. In 1812 he became superintendent of chemical operations at Apothecaries' Hall, and the following year succeeded Davy as professor of chemistry at the Royal Institution, in which post he remained until 1852. When Faraday returned from Europe in 1815, he began to assist Brande in the laboratory and, from 1824, as a lecturer. Thus the two men were associated for many years, both in teaching and chemical investigations and in editing the *Quarterly Journal of Science, Literature and the Arts,* published at the Royal Institution, to which they both made many contributions. In 1823 Brande was consulted by the government with a view to obtaining a more coherent metal for dies; his report, which led to improvements and economy at the Mint, led also to his appointment as superintendent of the dies department and, later, as chief officer of the coinage department.

Brande was an indefatigable lecturer and prolific writer, and published many papers on his investigations, but it is difficult to point to any that led to significant progress in chemistry. He was awarded the Copley Medal in 1813 for papers in which he showed, contrary to the prevailing belief, that alcohol was present in fermented liquors as such and not produced as a result of distillation; he also ascertained the alcohol content of many wines. In 1819 Brande examined a substance thought to be benzoic acid, but almost certainly naphthalene, then unknown, and carried out experiments which indicated that it contained no oxygen. He suggested that it was a binary compound of carbon and hydrogen, but carried out no analysis to confirm this or to determine the proportions. In the same year, in a paper read to the Royal Society on the inflammable gases from coal and oil, he inferred that there existed "no definite compound of carbon and hydrogen except that called olefiant gas" (*Philosophical Transactions,* **110** [1820], 11–28), although Dalton and William Henry had clearly distinguished methane and olefiant gas some fifteen years before.

BIBLIOGRAPHY

Brande wrote textbooks based on his courses of lectures on geology, chemistry, and pharmacy: *Outlines of Geology* (London, 1817, 1829); *A Manual of Chemistry: Containing the Principal Facts of the Science Arranged in the Order in Which They Are Discussed and Illustrated in the Lectures at the Royal Institution* (London, 1819; 6th ed., 1848; American ed., New York[?], 1829); *A Manual of Pharmacy* (London, 1825, 1833). He also wrote *Chemistry* (London, 1863), with A. S. Taylor, terming it "especially adapted for students." Among his other reference works should be noted *A Dictionary of Materia Medica and Practical Pharmacy, Including a Translation of the Formulae of the London Pharmacopoeia* (London, 1839), said to have been invaluable to medical students of his day. His catholic interests are evidenced by his editorship of *A Dictionary of Science, Literature and Art* (London, 1842, 1852, 1853; New York, 1847; rev. ed., 3 vols., 1865–1867, 1875). Most of his papers are listed in *The Royal Society Catalogue of Scientific Papers,* I (1867), 564–566. Those omitted are of little importance, and most are in the *Quarterly Journal,* in the period 1816–1830.

No adequate biography exists; the best biographical sketch is the obituary notice in *Proceedings of the Royal Society,* **16** (1868), ii–viii.

E. L. SCOTT

BRANDES, HEINRICH WILHELM (*b.* Groden, near Cuxhafen, Germany, 22 July 1777; *d.* Leipzig, Germany, 17 May 1834), *astronomy, physics.*

The son of a Protestant preacher, Albert Georg Brandes, Heinrich attended the grammar school in Ottendorf and studied natural sciences in Göttingen under A. G. Kästner and G. C. Lichtenberg. From

1801 to 1811 he was a dike official in Eckwarden near Jadebusen, duchy of Oldenburg, and in 1811 he was appointed professor of mathematics at Breslau. He became professor of physics at Leipzig in 1826.

As a student Brandes, with J. F. Benzenberg, developed the method of the corresponding observations of shooting stars. With its aid they determined the planetary velocity and the height, at the border of the atmosphere, of shooting stars. He also discovered the periodicity of the August meteorites. Later, Brandes occupied himself with practical and theoretical problems of the refractions of rays and other problems of atmospheric optics, as well as with the theory of comet tails. His books were widely read and contributed much to the popularization of astronomy.

BIBLIOGRAPHY

I. ORIGINAL WORKS. Among Brandes' more technical writings are *Versuche die Entfernung, die Geschwindigkeit und die Bahnen der Sternschnuppen zu bestimmen* (Hamburg, 1800), written with J. F. Benzenberg; *Beobachtungen und theoretische Untersuchungen über die Strahlenbrechung* (Oldenburg, 1807); *Beitrag zur Theorie der Cometenschweife* (1812); *Lehrbuch der Gesetze des Gleichgewichts und der Bewegung fester und flüssiger Körper* (Leipzig, 1817–1818); and *Beiträge zur Witterungskunde* (Leipzig, 1820).

His popular works include *Die vornehmsten Lehren der Astronomie deutlich dargestellt in Briefen an eine Freundin,* 4 vols. (Leipzig, 1811–1816); and *Unterhaltungen für Freunde der Physik und Astronomie,* 3 vols. (1825–1829).

II. SECONDARY LITERATURE. Additional works on Brandes are Poggendorff, I, 277–278; and H. Schröder, *Lexikon der hamburgischen Schriftsteller,* I (1851), 368, with a list of works.

BERNHARD STICKER

BRANDT, GEORG (*b.* Riddarhyttan, Sweden, 21 July 1694; *d.* Stockholm, Sweden, 29 April 1768), *chemistry, mineralogy.*

A son of Jurgen Brandt, a mineowner and former pharmacist, and Katarina Ysing, Brandt inherited his father's interest in chemistry and metallurgy, and as a child was allowed to participate in his father's experiments in these fields. He continued his studies at Uppsala University and worked for the Council of Mines. Convinced that he needed a more extensive background in the natural sciences, he decided to go abroad. Brandt arrived in Leiden in 1721 and became a pupil of Boerhaave. During three years of intensive study he acquired an extensive knowledge of chemistry, and his medical studies led to the M.D. from

Rheims in 1726. On his way back to Sweden he stopped in the Harz Mountains, where he studied mining and smelting, and when he arrived home, he was made director of the chemical laboratory of the Council of Mines. He was named warden of the Royal Mint in 1730. In 1747 he became associate member of the Council of Mines and in 1757 was named a member.

The Laboratorium Chymium Holmiense, where Urban Hiärne had produced his great lifework, had gradually declined because of Hiärne's advanced age and a lack of funds. When its work was resumed in new offices at the Royal Mint, Brandt's original contributions and his leadership, as well as the work of his collaborators, Henrik Scheffer and Axel Cronstedt, laid the groundwork for the eminence that Swedish chemistry achieved under such scientists as Bergman, Scheele, and Berzelius. Brandt's reputation as a chemical experimenter and as a teacher led to an offer of the chair of chemistry (which he refused) when it was established at the University of Uppsala in 1750.

Besides being an able administrator and chief of the laboratory, and making valuable contributions as a chemist, Brandt did outstanding research on arsenic. His findings, published in 1733, constitute the first detailed treatise on various arsenic compounds, their composition, and their solubility in various media. Henckel had begun to analyze the little that was known about arsenic, but Brandt's research clearly established its metallic nature and proved that white arsenic (arsenious oxide) was an oxide of this metal.

Brandt continued his metallurgical investigations, the results of which he published in a dissertation on semimetals (1735); in addition to mercury, antimony, bismuth, arsenic, and zinc, he dealt with cobalt, which he here showed to be a distinct metal. It is mainly for the discovery of this element that Brandt is known in the history of chemistry. In 1748 he published more findings on cobalt, describing his production of it as a regulus by reducing cobalt pyrite with charcoal, and declaring that it was magnetic.

Brandt also published findings on the difference between soda and potash (1746) and the methods of producing sulfuric acid, nitric acid, and hydrochloric acid (1741, 1743), and the ability of aqua fortis to dissolve gold, provided the gold was alloyed with a certain quantity of silver. In an article on the metallurgy of iron (1751), he proved that thermal brittleness was due to the sulfur content of the iron. However, he stated erroneously that arsenic was the cause of cold brittleness. Before Bergman, Brandt observed that the carbon content of steel was greater than that of cast iron.

BIBLIOGRAPHY

I. ORIGINAL WORKS. Brandt's writings are "De arsenico observationes," in *Acta literaria et scientiarum Sveciae*, **3** (1733), 39–43; "Dissertatio de semimetallis," *ibid.*, **4** (1735), 1–10; "De vitriolo albo," *ibid.*, 10–12; "Acta laboratorii chymici," in *Kongliga Svenska vetenskapsakademiens handlingar*, **2** (1741), 49–63; "Continuation," *ibid.*, **4** (1743), 89–105: "Rön och anmärkningar angående en synnerlig färg-cobolt," *ibid.*, **7** (1746), 119–130; "Rön och anmärkningar angående åtskilnaden emellan soda och pottaska," *ibid.*, 289–290; "Rön och anmärkningar angående det flyktige alcaliske saltet," *ibid.*, **8** (1747), 301–308; "Nytt rön angående gulds uplösning uti skedvatten," *ibid.*, **9** (1748), 45–54; "Cobalti nova species examinata et descripta," in *Acta Regiae Societatis Scientiarum Upsaliensis*, 1st ser., **3** (1748), 33–41; "Rön och försök angåendejärn, des förhållande mot andra kroppar, samt rödbräckt och kallbräckt järns egenskaper och förbättring," in *Kongliga Svenska vetenskapsakademiens handlingar*, **12** (1751), 205–214; "Några rön och anmärkningar angående köks-salt och dess syra," *ibid.*, **14** (1753), 295–312 and **15** (1754), 53–68; and *Tal om färg-cobolter, hållit för Kongliga vetenskaps academien vid praesidii nedläggande den 30 juli 1760* (Stockholm, 1760).

Some of Brandt's writings are in *Abhandlungen der Königlichen Schwedischen Akademie der Wissenschaften* (Hamburg, 1749–1753), and in *Recueil des mémoires les plus intéressants de chymie et d'histoire naturelle, contenus dans les Actes de l'Académie d'Upsal, et dans les Mémoires de l'Académie royale des sciences de Stockholm; publiés depuis 1720 jusqu'en 1760*, trans. by Augustin Roux and Paul-Henri, baron d'Holbach, 2 vols. (Paris, 1764).

There are bibliographies of Brandt's work in *Svenskt Biografiskt Lexikon*, V (Stockholm, 1925), 788–789; in J. R. Partington, *A History of Chemistry*, III (London–New York, 1962), 168–169; and Poggendorff, I, 280.

II. SECONDARY LITERATURE. Works on Brandt are Johan Axel Almquist, *Bergskollegium* (Stockholm, 1909); and Torbern Bergman, *Åminnelsetal öfver Georg Brandt* (Stockholm, 1769).

UNO BOKLUND

BRANDT, JOHANN FRIEDRICH (*b.* Jüterbog, Germany, 25 May 1802; *d.* Baths of Merreküll, Finland, 15 July 1879), *zoology, paleontology, botany.*

Brandt was carefully educated by his parents at Jüterbog, in the Prussian province of Brandenburg, where his father was a successful surgeon. From his great-uncle Hensius he early acquired a liking for botany, which captivated his attention for more than twenty years.

After graduating from the Gymnasium at Jüterbog, Brandt attended the Lyceum of Wittenberg, studying classics, and in 1821 entered the University of Berlin to study medicine. Although he had such famous teachers as Karl A. Rudolphi, Karl A. F. Kluge, and Albrecht von Gräfe, he still preferred botany, zoology, and even mineralogy. During his first vacation he visited the Harz Mountains with his fellow student Julius T. C. Ratzeburg; having won the prize for an essay on respiration, he was able to continue his travels the following year through the Riesengebirge.

The lectures of Martin H. K. Lichtenstein stimulated Brandt's interest in zoology and prompted him to visit many museums of anatomy. At that time he became a protégé of Rudolphi, to whom he was secretary for a short time. In 1825 he published *Flora Berolinensis*, based on his previous field trips. On 24 June 1826, having defended his thesis, "Observationes anatomicae de mammalium quorundam vocis instrumento," he obtained the M.D. During the same summer he passed his state examinations and became a licensed surgeon.

After having been assistant to the famous surgeon Ernst L. Heim for a short time, Brandt became assistant at the Anatomical Institute of the University of Berlin in 1827. He at once began, in collaboration with his friend Ratzeburg, to work on the first volume of their *Medizinische Zoologie* (1829–1833). This work, considered one of the major achievements of his career, was an enumeration and description of the animals used in the preparation of medical drugs.

Brandt was accepted as *Privatdozent* at the University of Berlin in 1828 and lectured on several subjects, among which medical botany and pharmacology were his favorites. Although he wrote many articles for the *Encyklopädische Lexikon*, his research continued to be more in botany than in zoology, as shown by *Deutschlands phanerogamische Giftgewächse* (1828) and *Tabellar Uebersicht d. offizin. Gewächse nach d. Linn. Sexualsystem u. d. natürl. System* (1829), each the first part of a major work. In 1830 he began writing monographs on the myriopods and *Oniscidae*, as well as a memoir on mammals based on Friedrich Bürde's *Abbildungen merkwürdiger Säugethiere*.

However, in spite of his achievements Brandt was unable to find any permanent position in Germany, and like many other German scientists of that time he emigrated to Russia. He left Berlin in 1831, and through the influence of Humboldt and Rudolphi he was appointed an associate member of the Academy of Sciences of St. Petersburg, as well as an assistant in its zoological museum, of which he later became director. In that position he succeeded another German scientist, K. E. von Baer, who a few years earlier had left his position in Königsberg to join the Academy of St. Petersburg, but found life in that town not to his liking and returned home.

Brandt was elected an ordinary member of the

Academy in 1833, a position he held until his death. Besides his activity in the Academy, he taught at the Central Pedagogical Institute, a teacher's college, and for eighteen years (1851–1869) was professor of zoology at the Military-Medical Academy, where army surgeons were trained.

Many honors were bestowed upon Brandt: he became an Imperial Russian Councillor, received the title of "Excellency," was invested with several distinguished orders, and was elected honorary member of many academies and scientific societies of Europe.

The fiftieth anniversary of his doctoral degree was celebrated in January 1876 with great pomp and with the participation of many of his students and friends. A special medal was struck for the occasion, and the Brandt Prize was established to reward outstanding zoological works. At this time Brandt's published scientific writings numbered 318, and their distribution among different scientific disciplines illustrates the extent of his knowledge: 176 publications are zoological, twenty-four relate to comparative anatomy, thirty-five are paleontological, eleven deal with geographical zoology, and the remainder pertain to archaeological zoology, botany, and various other subjects.

Russia presented Brandt with unusual opportunities for original studies. His activity followed two major lines: research and collecting specimens. He had found the zoological collection of the Academy very incomplete, and he undertook its development by means of many scientific expeditions financed by the Academy. He went to the Crimea, to Bessarabia, and to Nicolayev in search of the mammoth, and to the Caucasus to study its fish. For comparative purposes, and also to keep in touch with the progress of science in the West, he visited and studied in the museums of many European countries. Brandt collected not only animals during his travels but also books, making the zoological division of the Academy's library outstanding.

Although Brandt's zoological publications are remarkable, his fame is based essentially on his paleontological writings, which relate to the fossil Mammalia. The most important of these, *Untersuchungen über die fossilen und subfossilen Cetaceen Europa's*, gives a complete account of all the European Cetacea known until 1873. This work includes descriptions of species of *Cetotherium, Pachyacanthus, Cetotheriopsis, Cetotheriomorphus, Delphinapterus, Heterodelphis, Schizodelphis, Champsodelphis, Squalodon, Zeuglodon,* and other remarkable types. Several important memoirs relate to *Elasmotherium, Dinotherium, Rhytina,* the elk, and the mammoth. Another monograph is devoted to the characters of the Sirenia,

and their relations to different orders. Brandt also made important contributions to the knowledge of the osteology and structure of many other groups of mammals.

As a paleontologist, Brandt ranks among the best. His exhaustive and lucid monographs were written with a full understanding of their philosophical implications, an attitude very close to the final aim of scientific research.

BIBLIOGRAPHY

Among Brandt's writings are *Flora Berolinensis, sive descriptio plantarum phanerogamarum circa Berolinum sponte crescentium vel in agris cultarum additis filicibus et charis* (Berlin, 1824); *Deutschlands phanerogamische Giftgewächse* (Berlin, 1828); *Tabellar Uebersicht d. offizin. Gewächse nach d. Linn. Sexualsystem u. d. natürl. System* (Berlin, 1829); *Medizinische Zoologie oder getreue Darstellung und Beschreibung der Thiere die in der Arzneimittellehre in Betracht kommen,* 2 vols. (Berlin, 1829–1833), written with J. T. C. Ratzeburg; *Uebersicht d. Charactere d. Familien d. offizin. Gewächse nach R. Brown, De Candolle, Jussieu, . . .* (Berlin, 1830); *Deutschlands kryptogamische Giftgewächse* (Berlin, 1838), published together with *Deutschlands phanerogamische Giftgewächse* as *Abbildung und Beschreibung der in Deutschland wild wachsenden und in Gärten in freien ausdauernden Giftgewächse, nach natürlichen Familien erläutert, mit Beiträgen von P. Phoebus und J. T. C. Ratzeburg* (Berlin, 1838); *Symbolae Sirenologicae quibus praecipue Rhytinae historia naturalis illustratur* (St. Petersburg, 1846); *Symbolae Sirenologicae . . .,* fasc. 2 and 3 (St. Petersburg, 1861–1868), also *Mémoires de l'Académie impériale des sciences, St. Pétersbourg,* 7th ser., **12,** no. 1; *Untersuchungen über die fossilen und subfossilen Cetaceen Europa's mit Beiträgen von Van Beneden, Cornalia, Gastaldi, Quenstedt, und Paulson, nebst einem geologischen Anhange von Barbot de Marny, G. von Helmersen, A. Goebel und Th. Fuchs, ibid.,* **20,** no. 1 (1873); *Ergänzungen, ibid.,* **21,** no. 6 (1874); and *Bericht über die Fortschritte, welche die zoologischen Wissenschaften den von der kaiserlichen Akademie der Wissenschaften zu St. Petersburg von 1831 bis 1879 herausgegeben Schriften verdanken* (St. Petersburg, 1879).

An index to Brandt's works is *J. F. Brandtii index operum omnium* (St. Petersburg, 1876), issued as a *Festschrift*.

ALBERT V. CAROZZI

BRASHEAR, JOHN ALFRED (*b.* Brownsville, Pennsylvania, 24 November 1840; *d.* Pittsburgh, Pennsylvania, 8 April 1920), *astrophysical instruments.*

Until 1881 John Brashear was a mechanic in a Pittsburgh steel mill by day and an amateur telescope maker by night; then, through the encouragement of Samuel Pierpont Langley and with the financial support of William Thaw, he was able to establish a workshop for making astronomical and physical instruments. During the next half century the John A. Brashear Company produced many of the major instruments used throughout the world for astrophysical research. In 1926 the Brashear concern was bought by J. W. Fecker of Cleveland, Ohio.

Besides a dozen or so workmen and "Uncle John," as he was widely and affectionately known, the Brashear Company employed two notable "associates": Charles Sheldon Hastings, an optical physicist at Yale University, and James McDowell, Brashear's son-in-law. Hastings computed the curves of most of the objective lenses figured in the Brashear shops, and McDowell did most of the actual optical work.

The Brashear Company began at an opportune time: establishing large and well-equipped astronomical observatories had recently become an acceptable philanthropy; and astronomers, beginning to study the quality and quantity, as well as the position, of starlight, needed new types of instruments. Since most other astronomical instrument makers concentrated on equatorial refracting and reflecting telescopes or apparatus for terrestrial and celestial surveying, Brashear, who was adept at producing special-purpose instruments, had few competitors.

Among Brashear's astronomical contributions were an improved and soon widely used process for silvering glass mirrors; the concave metal mirrors on which Henry A. Rowland ruled diffraction gratings; spectroscopes for use with the large refractors at the Allegheny, Lick, Princeton, and Yerkes observatories; George Ellery Hale's first spectroheliograph for photographing solar prominences; the optical parts of the interferometer with which Albert A. Michelson measured the standard meter; a 16-inch-aperture, double-photographic doublet for Max Wolf at Heidelberg; and numerous telescope objectives—both lenses and mirrors—culminating in the 72-inch-aperture primary mirror for the Dominion Observatory in Canada.

BIBLIOGRAPHY

The best original sources for an instrument maker are the instruments themselves; and the best secondary sources are the articles about the instruments and the work done with them published in scientific journals and in observatory annals.

I. Original Works. Brashear wrote many short articles, frequently describing his techniques and work in progress, which were published in most of the contemporary astronomical journals; the first, "Hints on Silvering Specula, Periscopic Eyepieces, & c.," appeared in *English Mechanic,* **31** (1880), 327. Also of value is his *The Autobiography of a Man Who Loved the Stars* (Boston, 1925).

II. Secondary Literature. Among the numerous published accounts of Brashear's life that stress his many educational and philanthropic activities, mention may be made of J. S. Plaskett, "James B. McDowell, an Appreciation," in *Journal, Royal Astronomical Society, Canada,* **18** (1924), 185–193; and Frank Schlesinger, "John Alfred Brashear, 1840–1920," in *Popular Astronomy,* **28** (1926), 373–379.

Deborah Jean Warner

BRASHMAN, NIKOLAI DMITRIEVICH (*b.* Rassnova, near Brno, Czechoslovakia, 14 June 1796; *d.* Moscow, Russia, 13 May 1866), *mathematics, mechanics.*

Although Brashman's family was of limited means, he was able to study at the University of Vienna and the Vienna Polytechnical Institute by working as a tutor. In 1824 he went to Russia, and after a short stay in St. Petersburg he obtained the post of assistant professor of physicomathematical sciences at the University of Kazan, where he taught mathematics and mechanics.

In 1834 Brashman accepted a professorship of applied mathematics (mechanics) at the University of Moscow. Here he became known as a gifted scientist and teacher, and laid the foundations of instruction in both theoretical and practical mechanics.

In his lectures on mechanics and in his articles Brashman not only tried to show the achievements of this science, but also worked out its most difficult sections. He also prepared textbooks for Russian institutions of higher education. His texts on mathematics and mechanics reflect the state of science at that time, and his proofs of important theorems show originality, clarity, and comprehensiveness. Brashman wrote one of the best analytical geometry texts of his time, for which the Academy of Sciences awarded him the entire Demidov Prize in 1836.

In 1837 Brashman published the textbook on mechanics, *Teoria ravnovesia tel tverdykh i zhidkikh,* which contains an original presentation of problems of statics and hydrostatics. Upon the recommendation of Ostrogradski, this work also brought Brashman the full Demidov Prize.

In 1859 Brashman published a textbook, *Teoreticheskaya mekhanika* ("Theoretical Mechanics"), dealing with the theories of equilibrium and the motion of a point and of a system of points.

In addition to texts, Brashman wrote articles on various problems in mathematics and mechanics. Brashman's memoirs on mathematics were intended for those interested in the progress of the mathematical sciences, and dealt with the latest results of Russian and foreign scientists.

More important are Brashman's memoranda on mechanics. "O prilozhenii printsipa naimenshego deystvia k opredeleniu obema vody na Vodoslive" ("On the Application of the Principle of Minimum Action to the Determination of Water Volume in a Spillway," 1861), which was published in both Russian and foreign periodicals, drew much favorable attention.

Also in 1861 Brashman published "Note concernant la pression des wagons sur les rails droits et des courants d'eau sur la rive droite du mouvement en vertu de la rotation de la terre" (*Comptes rendus de l'Académie des sciences*, **53**, [1861], 370–376). With the aid of general equations, he tried to prove in this article that the rotation of the earth invariably imposes a pressure on the right rail of a railroad track as a train travels over it and on the right bank of a river as the current moves along it, no matter in what direction the train is moving or the river is flowing, provided this force is a single one (i.e., the motion must be rectilinear and uniform).

Another article of considerable interest is his "Printsip naimenshego deystvia" ("Principle of Minimum Action") that appeared in *Mélanges mathématiques et astronomiques* (**1** [1859], 26–31).

Brashman was not only an important scientist but also an excellent teacher. His students included such prominent scientists as P. L. Chebyshev, I. I. Somov, and other talented specialists in mathematics and mechanics. He founded the Moscow Mathematical Society and its journal, *Matematicheskiy sbornik* ("Mathematical Symposium"), the first issue of which appeared in the year of his death. This journal was equal to the best European publications in its scientific value and wide range of contents.

For his distinguished services to science, Brashman was elected a corresponding member of the Petersburg Academy of Sciences in 1855.

BIBLIOGRAPHY

I. Original Works. Brashman's writings include *Kurs analiticheskoy geometrii* ("Course in Analytical Geometry"; Moscow, 1836); *Teoria ravnovesia tel tverdykh i zhidkikh. Statika i gidrostatika* ("Theory of Equilibrium of Solid and Liquid Bodies. Statics and Hydrostatics"; Moscow, 1837); and *Teoreticheskaya mekhanika* ("Theoretical Mechanics"; Moscow, 1859).

II. Secondary Literature. Works on Brashman are A. Davidov, *Biograficheskiy slovar professorov i prepodovateley Moskvskogo Universiteta* ("Biographical Dictionary of Professors and Teachers at Moscow University"), I (Moscow, 1855), 206; *Matematicheskiy sbornik*, **1** (1866); A. T. Grigorian, *Ocherki istorii mekaniki v Rossii* ("Essays on the History of Mechanics in Russia"; Moscow, 1961), pp. 96–107; I. I. Liholetov and S. H. Yanovskaya, "Iz istorii prepodavaniya mehaniki v Moskovskom Universitete" ("From History of Teaching Mechanics at Moscow University"), in *Istoriko-matematicheskie issledovaniya*, **8**, 294–368; M. Viyodski, "Matematika i eyo deyateli v Moskovskom Universitete vo vtoroy polovine XIX V." ("Mathematics and Its Representatives at Moscow University in the Second Half of the Nineteenth Century"), *ibid.*, **1**, 141–149.

A. T. Grigorian

BRAUN, ALEXANDER CARL HEINRICH (*b.* Regensburg, Germany, 10 May 1805; *d.* Berlin, Germany, 29 March 1877), *botany, philosophy.*

For more than twenty-five years Braun was professor of botany and director of the botanical gardens at the University of Berlin, and during his lifetime was the most highly regarded botanist of the *Naturphilosoph* school. His father, also named Alexander, was a civil servant with scientific interests ranging over mineralogy, physics, and astronomy; his mother, Henriette Mayer, was the daughter of a mathematics professor and former priest. Braun himself married twice. In 1835 he married Mathilde Zimmer, who died in 1843 after the birth of their sixth child. Five children survived to adulthood; two of the daughters from this marriage married the German botanists Robert Caspary and Georg Mettenius. In 1844 Braun married Adele Messmer, who bore him five more children, four surviving to adulthood. She died just a few months after her husband.

Braun grew up in Karlsruhe in Baden, surrounded by the natural beauty of the Schwarzwald area, and his interest in natural history and botany developed at a very early age. After private tutoring, in 1816 he entered the Karlsruhe Lyceum, where he was still a student when he published his first paper at the age of sixteen. While still a youth Braun also discovered several new species of cryptogams that now bear his name: *Chara braunii, Orthotrichum braunii,* and *Aspidium braunii.* In 1824 he enrolled at the University of Heidelberg, and in compliance with his father's wishes, neglected the liberal arts in order to take up the study of medicine. Medicine was soon pushed into the background, however, as his abiding interest in botany took pride of place. At Heidelberg he became intimate friends with Carl Friedrich Schimper and Louis Agassiz. Braun and Agassiz became inseparable friends, and Agassiz eventually married Braun's sister, Silly. In 1827 Braun and Agassiz went to the University of Munich, attracted especially by the fame of the *Naturphilosophen* Oken and Schelling, and in 1829 both went on to

Tübingen, where they received their doctorates in that same year.

After four more years of study and travel, including an eventful stay in Paris, Braun accepted a position as teacher of botany and zoology at the newly instituted polytechnic school in Karlsruhe. He remained there until 1846, when he accepted the chair of botany at Freiburg im Breisgau. In 1849 he was elected prorector of the Freiburg *Hochschule,* and his diplomatic leadership during the Baden revolution did much to keep the school free from political strife. In 1850 Justus von Liebig persuaded Braun to move to Giessen, but he had hardly established residence there before he was offered a position at the prestigious University of Berlin. After some hesitation, Braun accepted the call to the big city in 1851, and he remained there the rest of his life.

Braun made his early reputation in botanical circles by his work on the arrangement of leaves. With his old Heidelberg friend Carl Schimper, Braun established the doctrine of spiral phyllotaxis, according to which growth in a stem has an upward direction in a spiral line such that the leaves are arranged on the stem according to fixed geometrical rules. Most anomalies were accounted for by a formula expressed as a simple continuous fraction. Between 1830 and 1835, Braun and Schimper each published articles introducing and explicating the doctrine, and their friendship was interrupted by Schimper's rather bellicose claim that he deserved full credit for the theory. Years later, a reconciliation was achieved. Braun, in particular, extended the theory between 1840 and 1860.

The Schimper-Braun theory focused attention on the important question of the relative positions of plant organs and inspired much of the detailed work of the opposing school of "genetic morphologists," such as Wilhelm Hofmeister, who claimed that the developmental history of plants proved the theory erroneous. Numerous exceptions to the rule certainly did exist, but it served as a roughly valid description of the arrangement of leaves on adult plants, and under its influence the morphological examination and comparison of plants, inflorescences, and vegetative shoots reached unprecedented completeness.

Braun's most important single work was *Betrachtungen über die Erscheinung der Verjüngung in der Natur,* written while he was at Freiburg and originally delivered as the prorectorial address for 1849; its publication was delayed until 1851 by Braun's call to Giessen and by the Baden revolution. Braun's chief object in this volume is to show that the phenomenon of "rejuvenescence" distinguishes the organic realm from the inorganic. His diffuse concept of rejuvenescence was an extension of

Goethe's doctrine of metamorphosis and included a consideration of developmental history, reproduction, and the dissolution of formed structures. More significant than his particular object, however, were Braun's contributions to the morphology of plants, to the biology of freshwater algae, and especially to a reconstruction of the cell theory. In this last area, his investigation of algae swarm-spores led him to oppose Schleiden and Schwann's emphasis on the cell wall, and to insist instead that the cell contents were the site of all the physiological activities of the cell. In passages remarkable for the beauty of their language, Braun suggested dramatically that the cell wall was in fact the structure that *entombed* the true life of the cell, and that could eventually destroy that life by interfering with "rejuvenescence."

It is clear from Braun's other important general work, "Das Individuum der Pflanze in seinem Verhältnisse zur Species" (1853), that he believed in the transmutation of species, but he insisted that the process was teleological. He objected strongly to Darwin's principle of natural selection and to any "mechanistic" explanations that substituted blind external forces for inner direction and purpose. Braun's general interpretations were always colored by his *Naturphilosophie,* and his work may have exerted less influence than it deserved because he was so obviously and genuinely a *Naturphilosoph* at a time when that mode of looking at nature was becoming unfashionable.

While at Berlin, Braun had among his pupils Anton De Bary and A. W. Eichler, his successor, who extended the natural system of classification developed by Braun for the ordering of the university's botanical gardens. In 1879 a bust of Braun was unveiled and placed in the botanical gardens.

BIBLIOGRAPHY

I. ORIGINAL WORKS. The Schimper-Braun theory of phyllotaxis was developed primarily in the following series of papers: Carl Schimper, "Beschreibung des Symphytum Zeyheri und seiner zwei deutschen Verwandten . . .," in *Magazin für Pharmacie von Geiger,* **3** (1830); Braun, "Vergleichende Untersuchung über die Ordnung der Schuppen an den Tannenzapfen, als Einleitung zur Untersuchung der Blattstellung überhaupt," in *Nova acta Academiae Caeserae Leopoldino Carolinae Germanicae naturae curiosorum,* **15** (1831), 195–402; and Braun, "Dr. Carl Schimper's Vorträge über die *Moglichkeit eines wissenschaftlichen Verständniss der Blattstellung,"* in *Flora, oder allgemeine botanische Zeitung* (Regensburg), **18** (1835), 145–192. See also report of the meeting of 20 Sept. 1834, *ibid.,* 7–8. For Braun's reasonableness on the question of priority, see his "Nachträgliche Erläuterungen zu meinem

Aufsatz in Nr. 10, 11, und 12 der 'Flora' laufenden Jahres über Dr. Schimper's Vorträge," *ibid.*, 737–746; Schimper's intemperate reply is *ibid.*, 748–758.

Braun's most important general works were *Betrachtungen über die Erscheinung der Verjüngung in der Natur, insbesondere in der Lebens- und Bildungsgeschichte der Pflanze* . . . (Leipzig, 1851), translated by Arthur Henfrey, in *Ray Society Botanical and Physiological Memoirs* (1853), vii–xxvi, 1–341; and "Das Individuum der Pflanze in seiner Verhältnisse zur Species, Generationsfolge, Generationswechsel und Generationstheilung der Pflanze," in *Abhandlungen der Königlichen Akademie der Wissenschaften Berlin* (Physikalische Klasse) (1853), 19–122, translated by Charles Francis Stone, in *The American Journal of Science and Arts* (*Silliman's Journal*), 2nd ser., **19** (1855), 297–318; **20** (1855), 181–200; and **21** (1856), 58–79.

For Braun's contributions in the area of systematics, see Paul F. A. Ascherson, *Flora der Provinz Brandenburg* . . . (*Nebst einer Übersicht des natürlichen Pflanzensystems nach A. Braun*) (Berlin, 1864); and A. W. Eichler, *Blüthendiagramme*, 2 vols. (Leipzig, 1875–1878), *passim* (esp. I, 3, 9, 11, 14–30). Most of Braun's scientific papers are listed in the *Royal Society Catalogue of Scientific Papers* (I, 582–585; VII, 248; IX, 333–334). Braun left behind an extensive manuscript collection that the Berlin Royal Academy of Sciences acquired in 1879 and turned over to the Royal Herbarium for safekeeping.

II. SECONDARY LITERATURE. Of the existing sketches of Braun's life and work, those written by his daughter Cecilie, though uncritical, are the most interesting because they are illuminated by Braun's personal letters. See Cecilie Mettenius, "Alexander Braun," in *Leopoldina,* **13** (1877), 50–60, 66–72; and *Alexander Braun's Leben nach seinem handschriftlichen Nachlass dargestellt* . . . (Berlin, 1882). The best source for estimating Braun's place in the history of botany is Ferdinand G. Julius von Sachs, *Geschichte der Botanik vom 16. Jahrhundert bis 1860* (Munich, 1875), pp. 185–195. It should be pointed out, however, that Sachs is unsympathetic toward Braun's general mode of viewing nature. For other sketches of Braun's life and work, see Robert Caspary, in *Flora*, n.s. **25** (1877), 433–442, 449–457, 465–471, 497–507, 513–519; A. W. Eichler, in *Leopoldina,* **15** (1879), 163–165; *Allgemeine deutsche Biographie*, XLVII (1903), 186–193; and *Neue deutsche Biographie*, II (1953), 548.

GERALD L. GEISON

BRAUN, FERDINAND (*b.* Fulda, Germany, 6 June 1850; *d.* Brooklyn, New York, 20 June 1918), *physics.*

Braun studied at the University of Marburg and received his doctorate from the University of Berlin in 1872 with a dissertation on the vibrations of elastic rods and strings. He later did work in thermodynamics, but his major accomplishments were in electricity. Braun began his career as assistant to Quincke at Würzburg and later held positions at Leipzig, Marburg, Karlsruhe, and Tübingen; during the short period that he spent at Tübingen he founded the Physical Institute. From 1880 to 1883 Braun was at Strasbourg, and he returned there permanently in 1895 to become professor of physics and director of the Physical Institute. He was called to the United States to testify in litigation involving radio broadcasting and then was detained when the United States entered World War I. He died in a Brooklyn hospital on 20 June 1918.

Although his contributions were all in the realm of pure science, in 1909 Braun shared the Nobel Prize in physics with Marconi for his practical contributions to wireless telegraphy. The work recognized by the Nobel committee was his fundamental modification of Marconi's transmitting system. Braun was first drawn to the study of wireless transmission by the puzzle of why it was so difficult to increase the range of the transmitter over 15 kilometers. It seemed to Braun that the range should easily be increased by increasing the power of the transmitter. His study of Hertz oscillators indicated that the attempt to increase the power output by increasing the length of the spark gap eventually reached a limit at which the spark caused a decrease in output. The solution, Braun thought, was to produce a sparkless antenna circuit. The power from the transmitter was coupled magnetically to the antenna circuit by a transformer effect instead of having the antenna directly in the power circuit. The principle has been applied to all such transmissions, including radio, radar, and television. A patent was granted on this circuit in 1899. Braun also developed an antenna that directed the transmission of electric waves in one direction.

In 1874 Braun published the results of his research on mineral metal sulfides. He found that these crystals conducted electric currents in only one direction. This information was important in electrical research and in measuring another property of substances, the electrical conductivity, but Braun's discovery did not have immediate practical application. In the early twentieth century the principle that Braun had discovered was employed in crystal radio receivers.

The first oscilloscope, or Braun tube, was introduced in 1897. In order to study high-frequency alternating currents Braun used the alternating voltage to move the electron beam within the cathode tube. The trace on the face of the cathode tube represented the amplitude and frequency of the alternating-current voltage. He then produced a graph of this trace by use of a rotating mirror. The Braun tube was a valuable laboratory instrument, and modifications of it are a basic device in electronic testing and research. The principle of the Braun tube, moving an electron beam by means of alternating voltage, is the principle on which all television tubes operate.

BIBLIOGRAPHY

I. ORIGINAL WORKS. Among Braun's writings are "On the Conduction of Current Through Sulpho-Metals," in *Poggendorff's Annalen*, **102** (1874), 550, and *Drahtlose Telegraphie durch Wasser und Luft* (Leipzig, 1901). His papers are listed in the Royal Society's *Catalogue of Scientific Papers, 1884–1900* (1914), pp. 773–774, and in the *International Catalogue of Scientific Literature* (1902), p. 65; (1904) p. 74; (1907) pp. 83–84; (1908) p. 85; (1912) pp. 74–75; (1917) pp. 73–74.

II. SECONDARY LITERATURE. There are no full-length biographies of Braun. He is best remembered by biographers for his Nobel Prize, although his most important work was in pure science. Some information is given in N. de V. Heathcote, *Nobel Prize Winners in Physics 1901–1950* (New York, 1953), pp. 81–86; and *The Nobel Prize-Winners and the Nobel Foundation 1901–1937* (Zurich, 1938), pp. 52–53.

HAROLD I. SHARLIN

BRAUNER, BOHUSLAV (*b.* Prague, Czechoslovakia, 8 May 1855; *d.* Prague, 15 February 1935), *chemistry*.

Brauner's family was one of comfortable means, extensive education, and wide-ranging scientific and political interests. His father, Dr. Francis Brauner, was a lawyer and Czech political leader. His mother, Augusta, was the daughter of Karl August Neumann, professor of chemistry at the Prague Technical College. As a boy, Brauner came into contact with the many scientific, intellectual, and political personalities who frequented his parents' home. He showed an early interest in science and studied at the Czech Polytechnical High School. In 1873 he entered the University of Prague, where he attended courses in chemistry and performed research in metal analysis and organic preparations. Preferring inorganic chemistry, Brauner went to Heidelberg in 1878 to train under Robert Bunsen. He returned to Prague in 1880 to take his doctorate and, still widening his experience, he went to England in that year to work under the direction of Sir Henry Roscoe at Owen's College, Manchester.

In 1882 Brauner was appointed lecturer in chemistry in the Czech branch of the Charles University, Prague, where gradually his influence made English the language of the laboratory. He rose to docent in 1885, assistant professor in 1890, and full professor in 1897. Brauner played a large part in the construction of a chemical institute, completed in 1904, at the university. He retired from academic life in 1925.

Brauner led an active and athletic life, playing soccer, bicycling, skiing, and hiking. He suffered a thrombosis in 1922, but his health was good and he was not incapacitated. In 1886 Brauner married Ludmilla, the adopted daughter of Professor Safarik, his predecessor in Prague. The couple had two sons and one daughter.

Brauner's scientific work consisted largely in the exemplification and perfection in the laboratory of Mendeleev's periodic law and system of classifying the chemical elements, published in 1869. Since the system had many problems—gaps, too many elements for one position, atomic weight anomalies—it was not immediately popular. It was supported in 1875, however, by the discovery of "eka-aluminum" (now called gallium), which Mendeleev had predicted. This finding, by Boisbaudran, convinced Brauner of the validity of Mendeleev's system. Brauner, who was to become a correspondent and friend of Mendeleev's, chose as his life's work the "experimental examination of the problems connected with Mendeleev's system. . . ."

Brauner began this work by trying to ascertain the atomic weight of beryllium. Its accepted value, 13, was not in agreement with its position in Mendeleev's periodic table. Working with John I. Watts, Brauner showed that previous measurements were faulty and that the correct value was 9, thus confirming beryllium's place at the head of Group II, as Mendeleev had insisted. Thereafter, Brauner devoted most of his attention to determining the proper positions of the rare earth elements. While working on cerium, he produced $CeF_4 \cdot H_2O$ and was thus the first to prepare a salt of quadrivalent cerium. From this compound, he computed cerium's atomic weight to be 141.1, placing it in Group IV. Later he revised this value to 140.22. From crude ceria, Brauner isolated compounds of other elements and determined their atomic weights: lanthanum, 138.92; praseodymium, 140.9; neodymium, 144.3; samarium, 150.4. This time-consuming, exacting analysis enabled Brauner to conclude that the rare earths are a group of closely related elements which occupy a single place in the periodic table—that between number 57 (lanthanum) and number 72 (hafnium). Brauner also studied the atomic weight of tellurium, an anomaly because it was thought to be 128, higher than that of iodine (127), although it was supposed to precede iodine in Mendeleev's system. Brauner's value of 127.61 for tellurium, although still an inversion, remains the accepted value today. In the decades following 1900, Brauner revised his earlier atomic weight results and also investigated the weights of tin and thorium.

In other areas, Brauner discussed the position of hydrogen in the periodic table, supporting Mendeleev's assignment of this element to the place above Group I. During his experiments with cerium he became the first to prepare elemental fluorine by a chemical method. When a fluorine-oxygen compound was produced in 1927, he speculated on its composition. When inert gases were discovered in the atmosphere, Brauner was reluctant to recognize them as new elements: he suggested that helium might be an allotrope of hydrogen and that argon might be triatomic nitrogen. He was also interested in astronomy and physiology, and made small contributions in both fields.

In 1888 Brauner advocated the adoption of oxygen ($=16$) instead of hydrogen ($=1$) as the standard for calculating atomic weights. Reviving an argument previously put forth by Marignac, Brauner claimed that the change would be advantageous, since most of the elements would then have atomic weights closer to a whole number. An international commission for atomic weights formed, at Brauner's suggestion, to consider the problem decided to use the oxygen standard, starting in 1904. Brauner was a member of the International Committee on Chemical Elements from 1921 to 1930. In 1921 he made a number of proposals regarding the naming of terms—atomic weight, for example—which had become clouded as a result of the discovery of isotopes. Brauner called for the formation of a subcommittee for atomic weights and served as its first president.

Brauner was invited to contribute sections on atomic weights for Abegg's *Handbuch der anorganischen Chemie,* and he spent the years between 1904 and 1913 preparing and translating the necessary materials. In 1906 he wrote a section on rare earth elements for a revised edition of Mendeleev's *Principles of Chemistry*. He collaborated with his assistant, H. Krepelka, on a textbook of qualitative analysis (in Czech) published in 1919. Additionally, he supported a journal, *Collection of Czechoslovak Chemical Communications,* which was founded in 1929 and was edited by E. Votocek and J. Heyrovsky. As a cultural "neo-Slavist," Brauner popularized Mendeleev's ideas and wrote for English journals on the contributions of other Russian scientists.

Brauner received numerous honors in his lifetime. He was a prominent member of the Czech Academy of Science and of Czechslovakia's National Research Council. He was recognized by scientific societies in England, the United States, France, Poland, Russia, and Austria; and he was honored and decorated by the governments of Austria, Yugoslavia, and France.

BIBLIOGRAPHY

I. ORIGINAL WORKS. An autobiographical source is "D. I. Mendeleef," in *Pokrokova revue* (1907). It was translated into English in *Collection of Czechoslovak Chemical Communications,* **2** (1930), 219–243; a bibliography of Brauner's scientific papers precedes it on pp. 212–218.

A selected list of Brauner's scientific papers includes the following: "On the Atomic Weight of Beryllium," in *Philosophical Magazine,* **11** (1881), 65–72; "On the Specific Volumes of Oxides," *ibid.,* 60–65, written with John I. Watts; "Contributions to the Chemistry of Rare Earth-Metals," in *Journal of the Chemical Society,* **41** (1882), 68–79; **43** (1883), 278–289; **47** (1885), 879–897; "The Atomic Weight of Tellurium," in *Journal of the Russian Physical-Chemical Society* (1883); "The Standard of Atomic Weights, I," in *Chemical News,* **58** (1888), 307–308; "Experimental Researches on the Periodic Law. I. Tellurium," in *Journal of the Chemical Society,* **55** (1889), 382–411; "Experimental Studies on the Periodic Law," in *Chemicke listy,* **14** (1889), 1–30; "The Standard of Atomic Weights, II," in *Chemische Berichte,* **22** (1889), 1186–1192; "Fluoplumbates and Free Fluorine," in *Journal of the Chemical Society,* **65** (1894), 393–402, and in Czech Academy, *Bulletin,* **18** (1894), 1–9; "Atomic Weight of Tellurium," in *Journal of the Chemical Society,* **67** (1895), 549–551; "Note on the Gases of the Helium and Argon Type," in *Chemical News,* **71** (1895), 271; "On the Compound Nature of Cerium," in Czech Academy, *Bulletin,* **2** (1895), 1–6; "Contributions to the Chemistry of Thorium; The Atomic Weight of Thorium; On the Compound Nature of Cerium; On Praseodymium and Neodymium," in *Proceedings of the Chemical Society,* **14** (1898), 67–72; "Contributions to the Chemistry of Thorium. Comparative Researches on the Oxalates of the Rare Earths," in *Journal of the Chemical Society,* **73** (1898), 951–985.

Also see "On the Atomic Weight of Lanthanum and the Error of the Sulphate Method for the Determination of the 'Equivalent' of the Rare Earths [written with F. Pavlicek]; On the Atomic Weight of Praseodymium; On Praseodymium Tetroxide and Peroxide; Note on Neodymium; Contribution to the Chemistry of Thorium," in *Proceedings of the Chemical Society,* **17** (1901), 63–68; "On the Position of Hydrogen in the Periodic System," in *Chemical News,* **84** (1901), 233–234, and in Czech Academy, *Bulletin,* **34** (1901), 1–4; "Revision of the Atomic Weight of Tin," in *Journal of the American Chemical Society,* **42** (1920), 917–925, written with H. Krepelka; "The New International Commission on Chemical Elements," in *Chemical News,* **123** (1921), 230–232; "Atomic Weight of Silver," in *Nature,* **119** (1927), 348, 526; "Oxide of Fluorine or Fluoride of Oxygen?," *ibid.,* **120** (1927), 842;

"Some Physiologico-optical Experiments," in Czech Academy, *Bulletin,* **38** (1929).

II. SECONDARY LITERATURE. Two biographical sources are Gerald Druce, *Two Czech Chemists, Bohuslav Brauner and Frantisek Wald* (London, 1944), pp. 4–44, bibliography pp. 62–65; and "Obituary of Bohuslav Brauner," in *Nature,* **135** (30 Mar. 1935), 497–498.

SUSAN G. SCHACHER

BRAUNMÜHL, ANTON VON (*b.* Tiflis, Russia, 12 December 1853; *d.* Munich, Germany, 7 March 1908), *history of mathematics.*

Braunmühl, descended from the old Bavarian nobility, was the son of the famous architect Anton von Braunmühl (1820–1858), who had studied with Fr. Gärtner, and Anna Maria Schlenz (1823–1892). In 1879 he married Franziska Stölzl (1853–1917), who bore him two daughters.

After the sudden death of his father, Braunmühl grew up in Munich and enrolled in its university in 1873. There he attended lectures on astronomy by Johann Lamont, on physics by Philipp Jolly, on the history of literature by Michael Bernays, and on cultural history by Wilhelm Riehl; he also studied mathematics under Ludwig Seidel, Gustav Bauer, and Friedrich Narr. At the Munich Technical University, Braunmühl studied further under Alexander Brill, Felix Klein, and Johann Bischoff. In 1888 he was appointed extraordinary professor of mathematics at the Technical University and was promoted to ordinary professor of mathematics in 1892. He was recognized as a scientist and was held in extraordinary esteem as a teacher. Braunmühl's lectures on the history of mathematics, given regularly after 1893, were unique in that they were offered without credit, as were the seminars on the history of mathematics that were given right after the lectures. These lectures and seminars stimulated Wilhelm Kutta, Axel Bjoernbo, and Carl Wallner, among others, to undertake independent work in the history of mathematics.

At the turn of the century Braunmühl, Moritz Cantor, Maximilian Curtze, and Sigmund Günther were leading authorities on the history of mathematics in Germany. Braunmühl's contributions, pertaining especially to the history of trigonometry, surpass those of many of his contemporaries in thorough study of sources, complete reflection of previous literature, and precise presentation of specific details, as well as in their critical evaluation.

BIBLIOGRAPHY

I. ORIGINAL WORKS. Braunmühl's writings include *Chr. Scheiner als Mathematiker, Physiker und Astronom*

(Bamberg, 1891); "Beiträge zur Geschichte der Trigonometrie," in *Nova acta Leopoldina,* **71** (1897), 1–30; and *Vorlesungen über Geschichte der Trigonometrie,* 2 vols. (Leipzig, 1900–1903).

Preliminary studies by Braunmühl were utilized after his death by H. Wieleitner in *Geschichte der Mathematik,* II, pt. 1 (Leipzig, 1911).

II. SECONDARY LITERATURE. Works on Braunmühl are S. Günther, "Anton von Braunmühl," in *Mitteilungen zur Geschichte der Medizin und der Naturwissenschaften,* **7** (1908), 362–367; J. E. Hofmann, "Anton von Braunmühl," in *Neue deutsche Biographie,* II (1955), 560; and H. Wieleitner, "Zum Gedächtnis Anton von Braunmühls," in *Bibliotheca mathematica,* 3rd ser., **11** (1910), 316–330, with a portrait and a bibliography.

JOSEPH E. HOFMANN

BRAVAIS, AUGUSTE (*b.* Annonay, France, 23 August 1811; *d.* Le Chesnay, France, 30 March 1863), *botany, physics, astronomy, crystallography.*

The ninth of ten children born to François-Victor Bravais, a physician, and Aurélie-Adelaïde Thomé, Auguste completed his classical education at the Collège Stanislas, Paris, in 1827, winning honorable mention in mathematics in the general competitive examination. He returned to Annonay and failed the polytechnical examination in 1828; after a year of special mathematics at the Collège St.-Louis, he won first prize in mathematics in the general competitive examination and was accepted at the École Polytechnique. He led his class at the end of his first year, and his academic record the next year made him eligible for any technical corps except mining. Since he wanted to participate in exploration, he chose to enter the navy. Bravais became a first-class cadet in 1831 and was sent to the Toulon naval district. He shipped out in January 1832, sailed the Mediterranean, and that April was assigned to map the coast of Algeria.

During his leaves in Annonay (November 1833–June 1835) Bravais studied plant organography with his brother Louis and his friend Charles Martins. The brothers' publications on this subject won them membership in the Société Philomathique de Paris. Auguste's work on shipboard led him to consider various methods of nautical surveying and the stability of ships; these were the subjects of his doctoral thesis, which he presented at Lyons in 1837. Shortly afterward he and Martins were assigned to the Commission Scientifique du Nord and sailed in June 1838 with the expedition that landed near North Cape.

While in Norwegian Lapland (August 1838 to September 1839), Bravais and the other physicists from the commission made numerous observations in astronomy, meteorology, and terrestrial magnetism. He noticed that a certain alga, *Fucus vesiculosus,* formed a yellowish area whose upper limit always occurred at the same height above sea level; this served as a point of reference for surveying ancient shorelines. Accompanied by Martins, Bravais returned overland to Paris, arriving in January 1840. Immediately afterward the navy assigned him to publish an account of the expedition; the volumes dealing with meteorology, terrestrial magnetism, the northern lights, botanical geography, astronomy, and hydrography are largely his own work.

The navy authorized Bravais to teach astronomy at the Faculté des Sciences at Lyons in 1841. Shortly after his arrival, he and J. Fournet founded the Commission Hydrométrique, des Orages et Météorologique de Lyon. Bravais climbed the Faulhorn in 1841 and 1842 with his brother Louis and Martins, in order to make observations. He was elected a member of the Académie Royale des Sciences, Belles-Lettres et Arts de Lyon in 1844, and later that year the minister of public education entrusted him with a scientific mission in the Alps: accompanied by Martins, he climbed Mont Blanc and, once again, the Faulhorn. During his stay in Lyons he wrote the segment of *Patria* that contains a survey of verified falls of meteorites in France from 1198 until 1842.

Bravais was appointed professor of physics at the École Polytechnique in 1845. Among the many communications he sent to the Académie des Sciences, the ones that drew the most attention were those of 1848 and 1849. They concerned reticular groupings and were based on speculations arising from a paper by Delafosse on the meaning of hemihedrism in crystals. The Academy admitted Bravais to its geography and navigation section in 1854.

The relationship between the external forms of crystals and an internal periodic corpuscular structure had been discussed by Kepler, Descartes, Hooke, and Huygens but was neglected in the eighteenth century for studies of the external forms alone, in the manner of Steno's *Prodromus*. After Romé de l'Isle and Haüy, attention was again directed to internal structure seen as the repetition of fundamental polyhedral nuclei. In an exhaustive study of the properties of lattices (1848), Bravais derived the fourteen possible arrangements of points in space. The Bravais lattices effectively combined earlier concepts of periodicity with Haüy's law of rational intercepts. (A. J. Shaler's English translation of this fundamental paper was published in 1949 as Memoir no. 1 of the Crystallographic Society of America.)

In the *Études cristallographiques* (1866), Bravais concentrated on the relationships between the ideal lattice and the material crystal. He proposed to locate his lattice points at the centers of gravity of congruent molecular polyhedra (p. 196). His analysis of the symmetry of the molecular polyhedra (the thirty-two point groups had been derived by Hessel in 1830) led to the later derivation of the space groups by Barlow, Schönflies, and Fedorow. The Bravais molecular polyhedra, repeated in parallel orientation in the same way as the points of the Bravais lattice, constituted the modern representation of the atomic structure of crystals. Each peak of the polyhedron was "a center or pole of force" (*ibid.*), which Cauchy, in a remarkable report introducing the paper, understood as an atom of a particular species. The Bravais rule that referred the most prominent planes of the crystal and the planes of cleavage to the nets of the Bravais lattice with the greatest reticular density (concentration of points) afforded a method by which mineralogists could determine lattice types and, in the simplest cases, structures. Bravais himself applied the rule to most mineral species with excellent but not perfect results.

Only constant work allowed Bravais to carry on extensive studies of the many different subjects that aroused his curiosity. He was, however, not able to bear such a strain indefinitely. He became seriously ill in 1856, resigned from his post at the École Polytechnique, and retired from the navy in 1857. His wife tried to nurse him back to health; on his death she withdrew to the Convent des Clarisses, in Versailles, of which she was founder.

Bravais's work, continued by Friedel, Tammann, Barlow, and others, provided the mathematical and conceptual basis for the determination of crystal structures after Laue's discovery of X-ray diffraction in 1911. The intensive development of solid-state physics in recent times has publicized Bravais's role in the origin of the application of the mathematics of symmetry and groups to the theory of solids.

BIBLIOGRAPHY

I. ORIGINAL WORKS. A complete bibliography is in Poggendorff, I, 283–284; III, 184–185; IV, 177. Bravais's books include *Thèse d'astronomie sur les méthodes employées dans les levés sous voiles . . .* (Paris, 1837); *Über die geometrische Anordnung der Blätter und der Blüthenstände . . .* (Breslau, 1839); *Sur l'équilibre des corps flottants . . .* (Paris, 1840); *Le Mont Blanc* (Paris, n.d. [1854]); and *Études cristallographiques* (Paris, 1866). He also contributed to many works or was a coauthor: "Étoiles," in P. Leroux and J. Reynaud, eds., *Encyclopédie nouvelle . . .*, V (Paris, 1839), 100–109, also published separately as *Les étoiles ou résumé d'astronomie stellaire* (Paris, 1844); the notes for L. F. Kaemtz's *Cours complet de météorologie . . .*, trans. and annotated by Charles Martins (Paris, 1843); the following volumes (as coauthor) of *Voyages de la*

Commission scientifique du Nord . . .: Météorologie, 3 vols. (1844–1855); *Magnétisme terrestre,* 3 vols. (1843–1850); *Aurores boréales* (1845); *Géographie physique, géographie botanique, botanique et physiologie,* 2 vols. (1844–1846); and *Astronomie et hydrographie* (n.d.); three letters in A. Quételet, *Lettres à S. A. R. le duc régnant de Saxe-Cobourg et Gotha, sur la théorie des probabilités . . .* (Brussels, 1846), pp. 412–424; and two chapters in *Patria. La France ancienne et moderne . . .,* 2 vols. (Paris, 1847), I, 1–142, 143–176.

Bravais's papers appeared in many journals: the *Annales des sciences naturelles,* from 2nd ser., Botanique, 7 (1837) to 3rd ser., Botanique, 3 (1845); *Comptes rendus hebdomadaires . . . de l'Académie des sciences,* from 4 (1837) to 42 (1856); *Mémoires présentés par divers savants . . .,* 9 (1846); *Nouveaux mémoires de l'Académie royale des sciences et belles-lettres de Bruxelles,* 14 (1841); *Mémoires couronnés . . . par l'Académie royale . . . de Bruxelles,* 15, pt. 2 (1843); *Annales de chimie et de physique,* from 3rd ser., 5 (1842) to 3rd ser., 46 (1856); *Journal de mathématiques pures ou appliquées . . .,* from 8 (1843) to 2nd ser., 1 (1856); *Journal de l'École polytechnique,* from 30th cahier, 18 (1845) to 34th cahier, 20 (1851); the reports of the 6th and 9th sessions of the Congrès scientifique de France (1839, 1842); the *Bibliothèque universelle de Genève* (1845); *Le moniteur universel* (18 Sept. 1844); *Revue scientifique et industrielle,* 2nd ser., 4 (Jan. 1845); and *Annuaire météorologique de la France* for the years 1849–1853.

His communications to the Société Philomathique are condensed in *L'Institut; 1re section, Sciences mathématiques, physiques et naturelles* between 15 Feb. 1837 and 29 Nov. 1854; they are reproduced, after correction of misprints, in *Extraits des procès-verbaux . . .* of the Society for the years 1837–1854.

Lithographed copies of Bravais's courses are École Polytechnique, 2nd div., "Cours de physique, 1re année d'étude 1847–1848"; École Polytechnique, 1st div., "Cours de physique, 2e année 1848–1849"; École Impériale Polytechnique, 2nd div., "Sommaire du cours de physique, 1re année" (1853–1854); École Impériale Polytechnique, 1st div., "Sommaire du cours de physique, 2e année" (1853–1854); and École Impériale Polytechnique, 1st div., "Sommaire du cours de physique, 2e année" (1854–1855).

II. SECONDARY LITERATURE. Writings on Bravais or his work are J.-M.-J. Bouillat, "Auguste Bravais, voyageur et savant (1811–1863)," in *Les contemporains,* no. 588 (Paris, 1904); Élie de Beaumont, *Éloge historique d'Auguste Bravais* (Paris, 1865), trans. in the *Smithsonian Report* for 1869, pp. 145–168; J. Fournet, "Rapport sur trois mémoires de Bravais," in *Annales des sciences physiques et naturelles . . .,* 9 (1846), lxxi–lxxv; and Jean Messié, "Auguste Bravais, savant annonéen," in *Revue du Vivarais,* 67 (Jan.–Mar. 1963), 7–11.

Charles Martins wrote several works concerning his work with Bravais: "Lettre sur le voyage aux terres arctiques," in *Revue médicale française et étrangère* (1838), 4, 433–438; "Observations sur les migrations et les moeurs des Lemmings," in *Revue zoologique . . .* (July 1840), 193–206; "Une ascension au Faulhorn," in *Revue médicale française*

et étrangère (1841), 4, 209–214; "Un hivernage scientifique en Laponie," in *Revue indépendante* (25 Dec. 1843), 483–511; "Ascension au Mont Blanc par MM. Martins, Bravais et Lepileur," in *L'illustration, journal universel* (5 Oct. 1844), 68–74; "Deux ascensions scientifiques au Mont-Blanc. . .," in *Revue des deux mondes* (15 Mar. 1865), 377–411; and *Du Spitzberg au Sahara* (Paris, 1866).

Archival materials are Bravais's dossiers in the Service Historique de la Marine and in the Archives de l'Académie des Sciences; the *Registre de matricule des élèves,* V (1820–1830), 353, at the École Polytechnique, Secrétariat de la bibliothèque; and the archives of the Société Philomathique, cartons 123–134, at the Bibliothèque de la Sorbonne.

Also used in preparing this article were *Distribution générale des prix aux élèves des collèges royaux de Paris et de Versailles* for 1827 (p. 23) and for 1829 (p. 10); Auguste Bérard, *Description nautique des côtes de l'Algérie . . .* (Paris, 1837); "Instructions pour l'expédition scientifique qui se rend dans le Nord de l'Europe," in *Comptes rendus de l'Académie des sciences,* 6 (23 Apr. 1838), 526–571; (30 Apr. 1838), 585; (13 May 1838), 673; (21 May 1838), 704; Gabriel Delafosse, "Recherches sur la cristallisation, considérée sous les rapports physiques et mathématiques," in *Mémoires présentés par divers savants . . .,* 8 (1843), 641–690, first presented in the *Comptes rendus,* 11 (31 Aug. 1840), 394–400, and reported by Beudant, *ibid.,* 12 (25 Jan. 1841), 205–210; Abbé Filhol, *Histoire religieuse et civile d'Annonay,* IV (Annonay, 1882); *Mémoires de l'Académie royale des sciences, belles-lettres et arts de Lyon. Section des lettres et arts,* 1 (1845), 111; and *Notice des travaux scientifiques de M. A. Bravais* (Paris, 1851, 1854).

ARTHUR BIREMBAUT

BREDIKHIN, FEDOR ALEKSANDROVICH (*b.* Nikolaev, Russia, 8 December 1831; *d.* St. Petersburg, Russia, 14 May 1904), *astronomy.*

Bredikhin was born into an old aristocratic family. His father, Aleksandr Fedorovich, served with the Black Sea fleet; his mother, Antonida Ivanovna Rogulya, was the sister of an admiral. Since all of his uncles were naval officers, it was expected that Bredikhin would join the navy, but science attracted him instead. Until 1845 he studied at home; at Solonikha, his father's estate near Kherson; and at the Lycée Richelieu in Odessa. In 1851 he enrolled in the Physics and Mathematics Faculty of Moscow University, where he became interested in physics and, in his final year, astronomy.

After graduation in 1855 Bredikhin stayed at the university in order to prepare for a teaching career. He also worked at the Moscow observatory, which was directed by the famous "discoverer of comets," K. G. Schweizer. In 1858–1859 Bredikhin began teaching a lecture course at the university.

He married Anna Dmitrievna Bogolovskaya in

1860; their son, Aleksandr, died tragically in 1888. Bredikhin was a man of great general learning. He knew several European languages, possessed undoubted literary ability, loved music, and played the violin.

In 1861 he published his first scientific paper, "Quelques mots sur les queues des comètes" (on the bright comet Donati of 1858, which, in addition to its primary tail, had several weaker, more distorted ones). At this time he began his series of remarkable investigations of the nature of comets, which was the major work of his life. In 1862 Bredikhin defended his master's thesis, *O khvostakh komet* ("On the Tails of Comets"), and in 1863 was designated deputy extraordinary professor in the astronomy department. He defended his doctoral dissertation, "Vozmushchenia komet, nezavisyashchie ot planetnykh prityazheny" ("Comet Perturbations Which Are Not Caused by Attractions by the Planets") in 1865 and became an ordinary professor. In 1867–1868 Bredikhin visited Italy, where the first classification of star spectra had been developed and where, generally speaking, astrophysics was rapidly being developed. There he studied the technique of spectroscopic observation with Secchi, Tacchini, and other astrophysicists.

After Schweizer's death in 1873, Bredikhin became director of the Moscow observatory, where his fundamental works on the study of comets were written. In 1890, following Struve's retirement, he was elected an academician and director of the Pulkovo observatory. He left this post in 1895 for reasons of health and settled in St. Petersburg, where he worked on his investigations of comets for the rest of his life.

Notwithstanding the great significance that Bredikhin ascribed to theoretical investigations, he was a tireless observer. His observations encompassed all major aspects of contemporary astronomy, including even gravimetry—which was then inextricably associated with astronomy. He observed comets with astronomical instruments and studied them, systematically drawing their heads and tails and the contours of solar protuberances and chromosphere; he also observed comets and gaseous nebulae with the aid of a spectroscope, as well as meteor showers and zodiacal light. In addition, he observed the surfaces of Jupiter (especially the famous red spot) and Mars and conducted gravimetric measurements at various locations in Russia.

Although the fundamental peculiarities of changes in the apparent shape of comets as they approach the sun had been noted before Bredikhin, he was the one who developed the so-called mechanical theory of a comet's form. This theory's basic premise, which

dates as far back as Kepler and, ultimately, to Olbers, consists in the fact that particles of matter, which fly off from the core of the comet at a certain initial velocity, are repelled by the sun and, under the influence of that repulsion, move along hyperbolic trajectories. Bredikhin developed a method for determining the value of the repulsive accelerations (which are designated by $1 + \mu$) in the tails of comets and classified the tails according to these values: tails of type I are formed by forces of repulsion eleven to eighteen times greater than the sun's gravity and are inclined directly away from the sun, although they bend slightly in the direction opposite to the motion of the comet; the tails of type II are wider and strongly bent, their repulsive accelerations encompassing all values from 0.7 to 2.2 times the force of the sun's gravity; all tails of type III are even more inclined in the direction opposite that of the comet's motion, with an acceleration from 0.1 to 0.3. The fourth (anomalous) type of tail, which is directed straight toward the sun, is seen very rarely and always with a tail of one of the first three types. Some comets have tails of two or three different types.

Bredikhin's theory, which was based on an analysis of observations of about forty comets taken by himself and by others, explains several details of the structure of comets: the transverse bands in type II tails (synchrons), the motion of individual clumps in tails, and the form of a comet's head. The physical ideas on which the theory of comet forms was based included the supposition of an interaction between the sun's electric charge and the like charge of the gaseous molecules emanating from the comet's core. The values of the repulsive accelerations of these molecules are inversely proportional to their molecular weights. From this, Bredikhin proposed that hydrogen molecules predominate in type I tails; that hydrocarbons and light metals, such as sodium, predominate in type II tails; and that type III tails might contain molecules of iron. Observations by Bredikhin and his contemporaries seemingly confirmed such an explanation. It has since been proved that type I tails contain ionized molecules of carbon monoxide and nitrogen, while tails of type II and type III consist of neutral gases and dust particles that reflect the rays of the sun.

Through the detailed study of spectra of the heads and tails of comets, the mechanical theory of comet forms has to a large degree been retained. The classification of comet tails proposed by Bredikhin and developed by S. V. Orlov has also retained its significance. Certain peculiarities of comets that follow Bredikhin's theory, such as the parabolic shape of a comet's head, rotation of the core, and the complex movements of clumps of matter in comets' tails, have received exhaustive explanations in contemporary investigations. Bredikhin foresaw in the physical theory of comets the possibility of the influence of electrical forces, which also takes place in the interaction of solar corpuscular streams (so-called solar wind) and particles of matter in a comet.

He also devoted much attention to the study of meteor showers, which he believed occurred as the result of the earth's intersection of a swarm of particles from the tails of comets, these particles having been gradually scattered along the comets' orbits.

Bredikhin directed special attention to astrophysical investigations. In connection with his systematic observations of the sun, he developed a theory of the movement of matter in sunspots and in the rays of the solar corona. He proposed descending and ascending gaseous streams as the explanation of the formation of faculae and spots. On the basis of measurements made from photographs of eight total solar eclipses, taken between 1870 and 1896 by various astronomers, Bredikhin correctly noted the intimate connection between coronal eruptions and chromospheric protuberances and the absence of a direct connection between coronal eruptions and sunspots. Having determined that coronal eruptions are not radial, but somewhat curved, Bredikhin sought a similarity between them and the form of comet tails, proposing that a certain repulsive force also acts on coronal eruptions. He also applied his mechanical theory of comet forms to the analysis of coronal rays. Only in recent times has the physics of the corona received solid substantiation in the theory of the interaction between magnetic fields and a plasma, which is what the solar corona actually is.

Bredikhin's activity as director of the Pulkovo observatory greatly influenced the development of astronomy in Russia. Having replaced Struve, who strictly limited his contacts with other Russian university observatories and who preferred to have Germans and Swedes on the observatory staff, Bredikhin undertook a tour of all Russian observatories in order to familiarize himself with their activities, to help them obtain necessary instruments, to determine the most urgent scientific problems, and to attract the most talented young Russian astronomers to Pulkovo. In his first director's report (for 1889–1891) he wrote: "To the alumni of all Russian universities . . . must be afforded, within the limits of possibility, free access [to the observatory] . . . the recruitment abroad of scientists for its staff must and can be stopped forever." This greatly encouraged Russian astronomers, who now became frequent guests at Pulkovo, learned how to observe, and some-

times remained permanently. Before Bredikhin, there were only two Russians (A. A. Belopolsky and M. N. Morin) on the fifteen-man staff at Pulkovo; soon after he assumed the directorship, the number rose to nine.

Bredikhin was a charter member of the Mathematical Society in Moscow and an active member of the Moscow Society of Natural Scientists. In 1890 he was chairman of the Russian Astronomical Society, which had just been founded in St. Petersburg. He also belonged to many foreign organizations: the Astronomische Gesellschaft, Deutsche Akademie der Naturforscher Leopoldina, the Royal Astronomical Society, the Italian Society of Spectroscopists, and the Bureau des Longitudes in Paris. In 1892 he was awarded an honorary doctorate by the University of Padua.

BIBLIOGRAPHY

I. ORIGINAL WORKS. Bredikhin's complete bibliography of published works contains over 200 titles. Among them are *O khvostakh komet* ("On the Tails of Comets"; Moscow, 1862; 2nd ed., K. D. Pokrovsky, ed., Moscow, 1934), 2nd ed. includes a biographical essay and a bibliography; "Protsess Galileya po novym dokumentam" ("The Trials of Galileo According to New Documents"), in *Russkii vestnik,* **92**, no. 4 (1871), 405–414; "O solnechnoy korone" ("On the Solar Corona"), in *Izvestiya Imperatorskoi akademii nauk,* 5th ser., **9**, no. 3 (1898), 179–207; *Prof. Th. Bredikhin's mechanische Untersuchungen über Cometenformen in systematischer Darstellung,* R. Jaegermann, compil. (Leipzig, 1903), a systematic survey of all of Bredikhin's papers on the mechanical theory of comet forms, compiled under his supervision; *Études sur l'origine des météores cosmiques et la formation de leurs courants* (St. Petersburg, 1903), Bredikhin's survey monograph on meteor showers; and *Etyudy o meteorakh* ("Studies on Meteors"), S. V. Orlov, ed. (Moscow, 1954), with an article and commentary by A. D. Dubyago. His correspondence is in the archives, Academy of Sciences, Leningrad.

II. SECONDARY LITERATURE. Works on Bredikhin include A. A. Belopolsky, "Fedor Aleksandrovich Bredikhin," in *Izvestiya Imperatorskoi akademii nauk,* 5th ser., **21**, no. 2 (1904), i–iv, and "Fizicheskoe stroenie kometnykh khvostov" ("The Physical Structure of Comet Tails"), in *Russkii astronomicheskii kalendar na 1927 god* ("Russian Astronomical Calendar for 1927"; Nizhni Novgorod, 1926), pp. 137–161; S. K. Kostinsky, "Fedor Aleksandrovich Bredikhin," in *Russkii astronomicheskii kalendar na 1905 god* ("Russian Astronomical Calendar for 1905"; Nizhni Novgorod, 1904), pp. 2–29; B. Y. Levin, "F. A. Bredikhin," in *Lyudi russkoy nauki* ("People of Russian Science"; Moscow, 1961), pp. 141–151; O. A. Melnikov, "Fedor Aleksandrovich Bredikhin (k 125-letiyu so dnya rozhdenia)" ("Fedor Aleksandrovich Bredikhin: On the 125th Anniversary of the Day of His Birth"), in

Izvestiya GAO v Pulkove, **20**, pt. 6, no. 159 (1958), 1–27; N. I. Nevskaya, *Fedor Aleksandrovich Bredikhin (1831–1904)* (Moscow–Leningrad, 1964), the most complete biography, with a bibliography of all his works and of 252 secondary sources; S. V. Orlov, *Fedor Aleksandrovich Bredikhin (1831–1904)* (Moscow, 1948), a biographical essay with a bibliography of both original and secondary works; and K. D. Pokrovsky, "Teoria kometnykh form" ("The Theory of Comet Forms"), in *Russkii astronomicheskii kalendar na 1905 god* ("Russian Astronomical Calendar for 1905"; Nizhni Novgorod, 1904), pp. 35–51.

P. G. KULIKOVSKY

BREDON, SIMON (*b.* Winchcomb, England, *ca.* 1300; *d. ca.* 1372), *mathematics, astronomy, medicine.*

Originally a fellow of Balliol College, Oxford, Bredon moved to Merton College and was a fellow there in 1330, becoming junior proctor of the university in 1337 and keeper of the Langton chest about 1339. In 1348 he left Merton to become vicar of Rustington, Sussex, and thereafter held a succession of church appointments. His will, probated in 1372, listed the contents of his library, which covered theology, law, medicine, mathematics, and astronomy, as well as grammar and dialectic.

Bredon's earliest writings were concerned with philosophy, but he soon turned to mathematics and produced an explanation of Boethius' *Arithmetic.* This he split up into two parts, the first dealing with numbers, including multiplication, the second concerned with geometrical figures—triangles, squares, pentagons, hexagons, etc. In his copy of William Rede's astronomical tables for 1341–1344 he jotted down five conclusions on square numbers, which he considered useful for the squaring of the circle. These were followed by two criticisms of statements made by Vitello in his book on perspective, which Bredon dubbed "marvellous but false." His possession of Richard of Wallingford's book on sines and John Maudith's table of chords shows him to have taken an interest in trigonometry, but his own writings on these subjects have not survived except for a few brief notes; therefore it is not possible to assess his contribution in this field.

Bredon's works on astronomy are better attested. He wrote a treatise on the use of the astrolabe, giving detailed instructions how to find the altitude, degree, and declination of the sun; the latitude of any region; the degree of eclipse; and so on. The opening paragraph, entitled "Nomina instrumentorum," is not his work, but a borrowing from Messehallach. His *Theorica planetarum,* sometimes attributed to Walter Brytte, a contemporary at Merton, sometimes to Gerardo da Sabbionetta, is largely a paraphrase of the latter's treatise, although it lacks the two final sections on the latitude of the planets and the invection of the aspects of the planets. The text *De equationibus planetarum* formerly ascribed to Bredon has been shown to belong to Chaucer.

Bredon wrote a commentary on the first three books of Ptolemy's *Almagest.* No complete copy survives, but the work can be reconstructed from two incomplete manuscripts, both of which were annotated by Thomas Allen and John Dee. According to a marginal note in MS Digby 179, Bredon also made a new translation of Ptolemy's *Quadrepartitum,* probably to be identified with the *Astronomia judiciaria* mentioned in John Bale's *Index Britanniae scriptorum.* This translation is inserted into the lower margins of the version done by Egidius de Thebaldis of Parma, a copy of which was in Bredon's library. He drew up tables for the declination of the sun and the ascension of the signs and gave the longitude of Oxford as 14°56′. Bale ascribes three other works to him—*Super introductorio Alcabitii, Astronomia calculatoria,* and *Astronomia judiciaria*—without giving incipits.

Bredon's most ambitious work was the *Trifolium,* a medical compilation modeled on Avicenna's *Canon.* Only one-twelfth of it survives, dealing with the prognostication of disease from feces and urine, and with the composition of medicines. He was physician to Richard, Earl of Arundel, in 1355 and treated Joanna, Queen of Scots, in 1358.

BIBLIOGRAPHY

I. ORIGINAL WORKS. Bredon's writings are *Questiones in X libros Ethicorum Aristotelis:* Vienna, Bibl. Monast. B.V.M. ad Scotos MS 278.

De arithmetica: Oxford, Digby MS 98, fols. 109–117; Digby MS 147, fols. 92–103; Corpus Christi Coll. MS 118, fols. 101–118; Cambridge, Univ. Lib. MS Ee.iii, 61, fols. 92–101; Univ. of Alabama, MS 1, fols. 1–16; Boston Public Lib. MS 1531. On the last, see Margaret Munsterberg, "An Unpublished Mathematical Treatise by Simon Bredon," in *More Books, The Bulletin of the Boston Public Library,* **19** (1944), 411.

Conclusiones quinque de numero quadrato: Digby MS 178, fols. 11v–14.

Massa compoti (of Alexandre de Ville Dieu, not of Grosseteste, as ascribed by Bale): Digby MS 98, fols. 11–21, "bene correctus secundum sententiam Bredone."

Theorica planetarum: London, British Museum Egerton MS 847, fols. 104–122; Egerton MS 889; Oxford, Digby MS 48; Digby MS 93; Digby MS 98. The following MSS listed by Lynn Thorndike do not contain Bredon's work, but the treatise by Gerardo da Sabbionetta: London, B.M. Royal 12 C.ix; Royal 12 C.xvii; Royal 12 E.xxv; Oxford, Digby MS 47; Digby MS 168; Digby MS 207.

Commentum . . . Almagesti: Oxford, Digby MS 168, fols. 21–39; Digby MS 178, fols. 42–87; Cambridge, Univ. Lib. Ee.iii, 61, art. 8.

Astrolabii usus et declaracio: London, B.M. Harl. 321, fols. 24v–28.

Liber Quadrepartiti Ptolemei: Digby MS 179. See Axel Anthon Björnbo, "Die Mittelalterlichen lateinischen Übersetzungen aus dem Griechischen auf dem Gebiete der mathematischen Wissenschaften," in *Archiv für Geschichte der Naturwissenschaften und der Technik,* **1** (1909), 391 ff.

Trifolium: Oxford, Digby MS 160, fols. 102–223.

Bredon is quoted in Thomas Werkworth, *Tractatus de motu octavae spherae* (1396): Digby MS 97, fol. 143.

Two letters addressed to him are in London, B.M. Royal 12 D.xi, fols. 25r, 35r. His longitude for Oxford is in Royal 12 D.v, fol. 50r.

II. Secondary Literature. Full biographical details are in A. B. Emden, *A Biographical Register of the University of Oxford to A.D. 1500,* I (Oxford, 1957), 257–258; R. T. Gunther, *Early Science in Oxford,* II (Oxford, 1923), 52–55; and Lynn Thorndike, *A History of Magic and Experimental Science,* III (New York, 1934), 521–522. See also C. H. Talbot, "Simon Bredon (*c.* 1300–1372), Physician, Mathematician and Astronomer," in *British Journal of the History of Science,* **1** (1962–1963), 19–30; and J. A. Weisheipl, "Early 14th Century Physics and the Merton School," Bodl. Lib. MS D.Phil. d.1776. A list of the contents of Bredon's library is in F. M. Powicke, *The Mediaeval Books of Merton College* (Oxford, 1931), pp. 82–86, 138–142.

C. H. Talbot

BREFELD, JULIUS OSCAR (*b.* Telgte, Germany, 19 August 1839; *d.* Schlachtensee, near Berlin, Germany, 7 January 1925), *mycology.*

Oscar Brefeld, a founder of modern mycology, developed pure culture techniques and a comparative morphological approach in the study and classification of fungi, pioneered in researches on the cereal smuts, and published, over a period of forty years, a monumental fifteen-part treatise on his observations.

He was born in a small town near Münster in Westphalia, the third of four children (two sons and two daughters) of Wilhelm Brefeld, a prosperous pharmacist, and his wife Franziska Povel. Since the elder son, Ludwig, had studied law (in due course he became Prussian minister of trade and commerce), the other son was expected to follow his father's profession. Accordingly, having attended school in Telgte and completed his year of military service, Oscar studied pharmacy at Breslau for a year and a half. After a similar period at Berlin, he passed his state examination in pharmacy in 1863; but preferring chemistry and botany, he went to Heidelberg to work

under Bunsen and Hofmeister. There he obtained his Ph.D. in June 1864, for a thesis entitled "Chlor- und Bromgehalt des Meerwassers."

Soon afterward, a severe attack of pneumonia forced Brefeld to give up chemistry. During convalescence in Italy he studied art. On his return, he managed the family pharmacy for a short period, meanwhile beginning private researches on fungi. In 1868, financially assisted by his father, he went to the Botanical Institute at Halle to work with Anton de Bary. Two years later the Franco-Prussian war intervened, and Brefeld was drafted as an army pharmacist. During the siege of Paris he contracted typhoid fever, which proved nearly fatal. He was invalided home early in 1871 after devoted nursing by a French pharmacist. When fully recovered, he returned briefly to Halle. Thence, after short periods in Munich and at the Botanical Institute at Würzburg, he went in 1873 to Berlin, where he became *Privatdozent* in botany in 1875.

In 1878, shortly after becoming professor of botany at the Forest Academy of Eberswalde, near Berlin, Brefeld suffered a grave mishap. While examining a forestry class in chilly rain, he caught a severe cold, accompanied by ocular inflammation. This culminated in retinal detachment, glaucoma, and surgical removal of the left eye. He spent two years in Italy recuperating, again studying art. In the autumn of 1881 he resumed his researches, having appealed in vain to the Agriculture Ministry for an assistant because of his damaged eyesight.

In 1884 Brefeld became professor of botany at the Royal Botanical Institute and director of the botanical garden at Münster, where he spent fourteen very productive years. His work attracted international attention, and various foreign academies elected him to membership. He traveled in England, France, and Spain, meeting many scientists, including Pasteur, with whom he later corresponded. In 1896 he was appointed *Geheimer Regierungsrat* and in 1897 was made a corresponding member of the Berlin Academy of Sciences.

Brefeld married Elizabeth Godendahl, daughter of a Münster merchant, in 1896. Two years later he succeeded Ferdinand Cohn at Breslau as professor of botany and director of the Institute of Plant Physiology. In 1902, when Brefeld was sixty-three, his wife died shortly after the birth of their only child, Walter. His remaining eye now developed glaucoma, and in 1905 Brefeld became unable to teach; but with his assistant, R. Falck, he continued to work until increasingly defective vision forced his resignation in 1907.

Brefeld moved with his young son and housekeeper to Berlin, where he became completely blind in 1910, following an unsuccessful eye operation. Thereupon he withdrew to a property he owned in Berlin-Lichterfelde. In 1918, wartime malnutrition compelled him to enter a Rhineland sanatorium until 1924, when he moved to a nursing home at Schlachtensee, near Berlin. He remained completely alert mentally until his death from a colonic disorder.

Brefeld's blunt individuality and caustic wit were effective when used in defense of freedom of speech or for prescient warnings against the political philosophy that led Germany into World War I. His scientific reports, however, were sometimes needlessly polemic—a fault that his personal misfortunes, his single-minded attachment to mycology, and the lasting loyalty of his assistants helped to mitigate.

Brefeld's first publication concerned a new species of Myxomycetes, *Dictyostelium mucoroides* (1869). After the war, he began those detailed investigations into the developmental history and systematic relationships of fungi that became his lifework. Part I of his *Botanische Untersuchungen über Schimmelpilze* (1872), respectfully dedicated to his teacher de Bary, described three fungal species of Zygomycetes, a subclass characterized by Brefeld himself. The preface states a fundamental tenet that was to govern all his work: "The developmental history of a mold is deduced completely from the culture of the individual spore."

A report on *Penicillium,* completed while Brefeld was at Würzburg, appeared in 1874. In Berlin he published his first monograph on the Basidiomycetes (1877). These works, which constituted Parts II and III of his *Untersuchungen,* reflect his changing views on the sexuality of fungal fruit bodies. Part IV (1881) included observations on various species of Zygomycetes and Ascomycetes, and on *Bacillus subtilis,* preceded by a section expounding methods he had developed since 1869 for microscopic observations of pure fungal cultures. The same volume reveals an open feud with de Bary, whose claim—eventually verified—that the higher fungi exhibit sexuality Brefeld henceforth denied. Thirty years later, he intransigently wrote: "It is a basic error to correlate fungal pleomorphy with sexuality in the higher fungi, which has been construed after the pattern of the algae, but in reality does not exist." Meanwhile, he reacted unduly whenever criticisms of his work appeared in de Bary's journal, the *Botanische Zeitung.*

Brefeld's cultural methods (first outlined in 1874) stressed the heat sterilization of culture media, glassware, utensils, and instruments; precautions to exclude dust-borne contaminants; and use of sufficiently diluted inocula to permit single-spore transfers. He invented or adapted such useful devices as "capillary

culture chambers" and the "hanging drop" (generally known as the Van Tieghem cell). As basal nutrient medium, he favored decoctions of fresh manure from a herbivore; but solid sterile dung, or bread soaked in dung decoction, prevented dispersal and facilitated study of growing cultures. Addition of gelatin to the fluid medium permitted freer manipulation of preparations under the microscope and reduced evaporation of the medium without altering its transparency or nutritive qualities.

On returning to work after losing his eye, Brefeld made thousands of cultures, employing the foregoing techniques, to demonstrate that parasitic fungi, such as the cereal smuts, might be grown saprophytically. He cultivated over twenty species of *Ustilago,* as well as the potato blight fungus and other usually parasitic species. His observations led him to conclude that the yeasts, hitherto classified as "sprouting fungi," were conidial forms of higher fungal genera. These findings were reported in Part V of the *Untersuchungen* (1883), Part VI, on certain Myxomycetes and Entomophthorales, followed (1884), under the final version of the main title, *Untersuchungen aus dem Gesammtgebiete der Mykologie.*

At Münster, Brefeld had two young assistants, G. Istvánffy and O. J. Olsen (later known as O. Sopp), for Parts VII (1888) and VIII (1889), which reported further studies on over 200 species of Basidiomycetes. De Bary was again attacked in Part VII. He had suggested that Brefeld's views on the "sprouting fungi" were too sweeping and that yeasts were rudimentary forms of Ascomycetes.

F. von Tavel, and to a lesser extent G. Lindau, assisted Brefeld with Parts IX and X, on the Ascomycetes (1891), of which more than 400 species were studied. Work was resumed on the cereal smuts, over sixty species being cultivated on laboratory media. Growing "inexhaustible quantities" of specific fungi saprophytically, Brefeld investigated the mechanisms of infection, and the temporal and nidal variations in host susceptibility, in several smut diseases. These climactic efforts were reported in Parts XI and XII (1895).

Despite the many difficulties, domestic and administrative, that beset Brefeld in Breslau, Part XIII appeared in 1905, with Falck as co-author, describing the mechanisms of blossom infection by smut fungi and their natural modes of dissemination. In retirement, the totally blind mycologist dictated a recapitulation of his cultural methods and also reported miscellaneous earlier observations, published in Part XIV (1908). In Part XV (1912), he reverted mainly to smuts and smut diseases. A sixteenth part in manuscript was never published. Falck, in his obituary tribute, summarizes the contents of each part of the *Untersuchungen.*

Brefeld's manifold contributions to microbiology included a cultural methodology that was far more precise than Pasteur's and stimulated Koch to improve upon the gelatinized nutrient medium he is wrongly credited with having initiated. The indirect practical benefits to North America of Brefeld's work on smuts, especially of wheat, are inestimable. His *Untersuchungen,* often termed the "Bible of mycology," contains some errors of observation and interpretation, as well as unduly repetitive and vituperative passages; but the work is replete with classic observations, novel findings, and prophetic conjectures, recorded in colorful style. The accurate artistry of innumerable drawings on over 100 folio plates ensures the continuing reproduction of samples in mycology texts. Unfortunately, only fragments of his writings are available in English. Brefeld's extraordinary experimental skill, combined with rare patience and complete dedication, inspired A. H. R. Buller to call him "one of the ablest botanists of the nineteenth century."

BIBLIOGRAPHY

I. ORIGINAL WORKS. The first seven pts. of Brefeld's chief work, *Botanische Untersuchungen . . .,* were published between 1872 and 1888 in Leipzig, and the last eight pts. between 1889 and 1912 in Münster. The dates and titles are, under the general title *Botanische Untersuchungen über Schimmelpilze,* I, *Zygomyceten* (1872); II, *Penicillium* (1874); III, *Basidiomyceten I* (1877); IV, *Culturmethoden . . .* (1881); as *Botanische Untersuchungen über Hefenpilze,* V, *Die Brandpilze I* (1883); and, under the comprehensive title *Untersuchungen aus dem Gesammtgebiete der Mykologie,* VI, *Myxomyceten I. Entomophthoreen II* (1884); VII, *Basidiomyceten II* (1888), written with G. Istvánffy and O. J. Olsen; VIII, *Basidiomyceten III* (1889); IX, *Die Hemiasci und die Ascomyceten* (1891), written with F. von Tavel and G. Lindau; X, *Ascomyceten* (1891), written with F. von Tavel; XI, *Die Brandpilze II* (1895), written with G. Istvanffi; XII, *Hemibasidii. Brandpilze III* (1895); XIII, *Brandpilze (Hemibasidii) IV* (1905), written with Richard Falck; XIV, *Die Kultur der Pilze* (1908); and XV, *Die Brandpilze und die Brandkrankheiten V* (1912). Falck, in his obituary on Brefeld, lists twenty-six additional publications in various German scientific journals, the earliest being "Dictyostelium mucoroides, ein neuer Organismus aus der Verwandtschaft der Myxomyceten," in *Abhandlungen herausgegeben von der Senckenbergischen naturforschenden Gesellschaft,* 7 (1869), 85–107.

Translations of works by Brefeld are "Recent Investigations of Smut Fungi and Smut Diseases," trans. from *Nachrichten aus dem Klub der Landwirthe zu Berlin,* nos. 220–222 (1888) by Erwin F. Smith in *Journal of Mycology,*

6 (1890), 1–9, 59–71, 153–164; and *Investigations in the General Field of Mycology, Blossom Infection by Smuts and Natural Distribution of Smut Diseases,* pt. 13, written with R. Falck, trans. by Frances Dorrance (n.p., 1912).

II. SECONDARY LITERATURE. Obituaries are W. Brefeld, "Oscar Brefeld, 'Ein Leben für die Mykologie,'" an unpub. memoir by his son (personal communication, Sept. 1967); R. Falck, "Oskar Brefeld," in *Botanisches Archiv,* 11 (1925), 1–25; M. Kienitz, "Zum Gedächtnis. Dr. Oskar Brefeld, Professor der Botanik an der Forstakademie Eberswalde in den Jahren 1.10.1878–1884," in *Zeitschrift für Forst- und Jagdwesen,* 57 (1925), 709–711; F. Rosen, "Das pflanzenphysiologische Institut," in *Festschrift zur Feier des hundertjährigen Bestehens der Universität Breslau* (Breslau, 1911), see pp. 496–498; and O. Sopp, "Minnetale over prof. dr. Oscar Brefeld," in *Norske Videnskaps-Akademi i Oslo Årbok* (1925), pp. 83–86.

Important references to Brefeld's work are A. H. R. Buller, *Researches on Fungi,* 7 vols. (I–VI, London, 1909–1934; VII, Toronto, 1950; all repr. New York, 1958); A. De Bary, *Comparative Morphology and Biology of the Fungi. Mycetozoa and Bacteria,* Henry E. F. Garnsey, trans. (Oxford, 1887; repr. New York, 1966), pp. 256–257, 272, 295–297; M. Möbius, *Geschichte der Botanik. Von der ersten Anfängen bis zur Gegenwart* (Jena, 1937), pp. 74, 102–106; and J. Ramsbottom, "The Expanding Knowledge of Mycology Since Linnaeus," in *Proceedings of the Linnean Society of London,* 151 (1939), 280–367.

CLAUDE E. DOLMAN

BREGUET, LOUIS FRANÇOIS CLÉMENT (*b.* Paris, France, 22 December 1804; *d.* Paris, 27 October 1883), *instrumentation.*

Breguet's career can be understood only with a knowledge of the milieu in which he lived. His grandfather, Abraham, from Neuchâtel, was one of the best-known clockmakers of Paris; his shop was established as early as 1775. Louis's father, Antoine, became Abraham's partner in 1807. After spending some time in Neuchâtel with his godfather when he was about eight, Louis was apprenticed to Perrelet, in Versailles, for two years, and then joined his father and grandfather. From 1824 to 1827 he worked with Barral in Geneva in order to improve his craft, and upon his return to Paris worked on naval chronometers. His father, having little interest in business, withdrew more and more to the country. Finally, in 1833, the enterprise was organized into a company and turned over to Louis and two other partners, one of whom was a cousin.

After 1830 Breguet turned to making electrical instruments, particularly precision apparatus. His first electric clocks date from 1839. In 1840 he devised a thermometer that registered temperature electrically, and recorded a temperature of −42° at Kazan, Russia. Work on induced currents with Antoine

Masson in 1842 led to the creation of a genuine induction coil, a feat later ascribed to Heinrich Ruhmkorff in 1851. In 1843 Breguet created, for François Arago, an apparatus with a revolving mirror, which could attain a speed of 9,000 rps, for measuring the speed of light. It was used in Fizeau's experiments.

Breguet was then named designer-manufacturer to the Bureau des Longitudes. He gave a definitive form to the Wheatstone dial plate adopted by French railroads and constructed the Foy-Breguet instrument used in the French telegraphic system. In 1856 Breguet's firm made the first clocks to transmit time electrically.

In 1873 his son, Antoine, became his partner, and the Breguets turned to electrotechnics, then in its infancy: the company produced Daniell and Leclanché batteries, arc lamps, and Gramme dynamos. Metal thermometers, barometers, and manometers were made, as were very small, experimental aluminum helicopters for Antoine Pénaud, a pioneer in aeronautics. In 1876 Cornelius Roosevelt, representing Bell in Paris, put the Breguet firm in charge of setting up the French telephone system. The first Exposition Internationale d'Électricité was opened in Paris in 1881, and Antoine Breguet was the director of the installation services. Before his death Antoine collaborated with Charles Richet in founding the *Revue scientifique* (1881), an important journal during the next forty years. (It was the famous "Revue rose.") On 1 January 1882, Louis Breguet retired, leaving his son in charge, but within two years both had died.

BIBLIOGRAPHY

I. ORIGINAL WORKS. Breguet's most important works are *Mémoire sur l'induction . . . (. . . présenté à l'Académie des Sciences le 23 août 1841)* (Paris, 1842), written with Antoine Masson; *Télégraphie électrique, son avenir. Poste aux lettres électriques. Journaux électriques; suivi d'un aperçu théorique de télégraphie* (Paris, 1849), written with Victor de Seré; *Manuel de la télégraphie électrique à l'usage des employés des chemins de fer* (Paris, 1851; 2nd ed., 1853; 3rd ed., 1856; 4th ed., 1862); *Notice sur les appareils magnéto-électriques brevetés de Breguet et sur leur application à l'explosion des torpilles et des mines en général* (Paris, 1869); and *Catalogue illustré. Appareils et matériaux pour la télégraphie électrique, instruments divers, électricité, physique, mécanique, météorologie, physiologie, etc.* (Paris, 1873). For lists of his publications, see *Tables générales des comptes-rendus des séances de l'Académie des sciences . . . 1835–1850* (Paris, 1853), pp. 99–100; *1851–1865* (Paris, 1870), p. 74; *1866–1880* (Paris, 1888), p. 88; *1881–1895* (Paris, 1900), p. 96; and *Catalogue général des livres imprimés . . . de la Bibliothèque nationale,* XIX (Paris, 1904), cols. 165–167.

II. SECONDARY LITERATURE. Works on Breguet are

Claude Breguet, "La maison Breguet," in *Annuaire de la Société historique du quatorzième arrondissement de Paris* (1962), pp. 65–92; E. Ferret, *Les Breguet* (Paris, n.d. [*ca.* 1890]); and E. de Jonquières, "Notice sur la vie et les travaux de Louis Breguet," in *Comptes-rendus de l'Académie des sciences,* **103** (5 July 1886), 5–14, repr. in Ferret, pp. 60–79.

JACQUES PAYEN

BREISLAK, SCIPIONE (*b.* Rome, Italy, 17 August 1750; *d.* Milan, Italy, 15 February 1826), *geology, natural history.*

Breislak, of German extraction, was a priest who devoted much of his life to the teaching of the natural sciences. He had become interested in them during his early youth, which he spent in Sicily; and when he returned to Rome, he improved his scientific knowledge under Giovanni Fortis and Pietro Petrini. He later published reports on the natural resources of Latium: the Tolfa, Allumiere, Oriolo Romano, Latera, and Civitavecchia mines; underground water; and agriculture.

Subsequently, Breislak moved to Naples, where he taught at the military academy. In 1794 he witnessed the eruption of Vesuvius and made direct observations at the Campi Flegrei and Pozzuoli sulfur mines. *Topografia fisica della Campania* (1798), which discusses his studies on volcanoes, includes his opinion that the volcanic systems of Latium and Campania were connected at the ancient Roccamonfina volcano, whose eruptive apparatus and lava he had discovered. An adherent of Plutonist theory, Breislak believed that metamorphic rocks originated during consolidation of the earth's crust.

Considered one of the founders of volcanology in Italy, Breislak was the first to determine that basaltic rocks were of extrusive origin; he also emphasized that the tufaceous deposits of Campania originated under water, and he reconstructed the evolution of Vesuvius. He concluded that for a long time Monte Somma, the second summit of Vesuvius, had been the sole volcanic apparatus; that it had been active until A.D. 79; that it had been eroded; and that the cone of Vesuvius had been set upon it within historical time.

Breislak's final move was to Milan, where he was inspector of the niter factory and cooperated with the Biblioteca Italiana. He also investigated the mineral resources of most of Lombardy, particularly the building-stone and clay quarries of Brianza and the gold placers of the Tessin, Adda, and Serio rivers. In addition, he studied the igneous rocks on the western shore of Lake Maggiore, between Intra (Verbania) and Arona.

BIBLIOGRAPHY

Breislak's writings are *Topografia fisica della Campania* (1798); *Voyages physiques et lithologiques dans la Campanie, suivis d'une mémoire sur la constitution physique de Rome* (1801), also trans. into German (Leipzig, 1802); *Descrizione geologica della provincia di Milano* (1822); and *Traité sur la structure interne du globe* (1822).

VINCENZO FRANCANI

BREITHAUPT, JOHANN FRIEDRICH AUGUST (*b.* Probstzella, Germany, 18 May 1791; *d.* Freiberg, Germany, 22 September 1873), *mineralogy.*

The son of a local official in Probstzella, Breithaupt received his early education in the schools and Gymnasium of Saalfeld, to which his father was transferred. He was influenced by the mining activity that still thrived near Saalfeld, and decided to prepare for a career in this field. After a year and a half of study at the University of Jena, he went to the Bergakademie at Freiberg, Saxony, in the spring of 1811. There he came under the influence of Abraham Gottlob Werner, who had been a dominant figure at the Bergakademie for nearly forty years.

In 1813 Breithaupt was appointed to the post of assistant teacher (*Hülfslehrer*) in the Mining Academy and also to the post of gem inspector (*Edelstein-Inspektor*); he was the last to hold the latter office. Werner died in 1817 and was succeeded as professor of mineralogy by Friedrich Mohs. When Mohs left for Vienna in 1826, Breithaupt became his successor; he held the post until his retirement in 1866. He became blind soon after he retired, and this put an end to his mineralogical activity, which had been intensive until then.

Breithaupt was married in 1816 to Agnes Ulrike Winkler, the daughter of an official of a *Blaufarbenwerk* (a plant for producing cobalt blue) in the Erzgebirge near Freiberg. They had three daughters and one son, Hermann, who studied at the Bergakademie, was imprisoned for his political activities in 1848, and later became a mining official in Spain.

Breithaupt's first major publication was the completion of C. A. S. Hoffmann's four-volume *Handbuch der Mineralogie* (1816–1818), which was the only authorized publication of Werner's mineral system. Systematic mineralogy—the recognition and classification of minerals and the discovery of new species—was Breithaupt's chief interest. The so-called natural classifications, derived ultimately from Linnaeus, were still much in vogue, and for Breithaupt the elaboration of such a classification was a principal objective for many decades. The result appeared in his *Vollständiges Handbuch der Mineralogie* of which only three of a planned four volumes appeared (1836, 1841, and 1847). Although the work remained incomplete, the taxonomy of minerals into classes, orders, genera, and species and the associated nomenclature show that Breithaupt resisted the introduction of the chemical and crystal-chemical classifications which have since been generally adopted. In contrast, J. D. Dana, who in the first two editions of his *System of Mineralogy* (1837, 1844) had also attempted a "natural classification" with binomial nomenclature, wrote in the preface to the third edition (1850): "To change is always seeming fickleness. But not to change with the advance of science, is worse, it is persistence in error. . . ." Succeeding editions of his work developed into the standard in the field.

Breithaupt was a great observer and named many minerals. His successor and author of a necrology, Albin Weissbach, listed forty-seven, but the true number is even greater. Of the more than eighty mineral names devised by Breithaupt, about half—including such important ones as monazite, phlogopite, and orthoclase—are still regarded as referring to valid species. The naming of new minerals may, however, be regarded as merely a phase of Breithaupt's interest in systematics. Of more fundamental importance was his early work *Über die Echtheit der Krystalle* (1815), on pseudomorphs, which, although they had been recognized as such by Werner, had not received the attention they deserved. Breithaupt also was the first to distinguish amorphous minerals, which he referred to as "porodine," a term long since obsolete.

Breithaupt's greatest contribution by far to mineralogy and to the study of ore deposits was his little book *Die Paragenesis der Mineralien* (1849). Although others had noticed that there is some regularity in the association of different minerals, he was the first to make a comprehensive study of such regularities and to emphasize the importance of age relations among associated minerals. In the year following the publication of the *Paragenesis,* but only occasionally thereafter, Breithaupt gave a formal course of lectures on paragenesis. The importance of his work in this field was immediately recognized. For instance, in the first volume of the English edition of his *Textbook of Chemical and Physical Geology* (1859), Gustav Bischoff based his discussion of mineral associations largely on Breithaupt's work and referred extensively to the *Paragenesis* and related publications. The term "paragenesis" has been generally adopted, and the importance of the relations to which it applies is now recognized by all mineralogists and geologists. The Breithaupt Colloquium held at Freiberg in 1966 to commemorate the one hundred seventy-fifth anniversary of his birth (and, incidentally, just a century after his retirement) was entitled "Problems of Paragenesis in Mineralogy, Geochemistry, Petrology, and Ore Geology."

Breithaupt was a great teacher, and during his last years his students included many Americans: Eugene Hilgard, George Brush, Arnold Hague, and Raphael Pumpelly, among others. In 1865 twenty-four of the fifty-one newly admitted students at the Bergakademie were from the United States, and it is probable that most of them studied under Breithaupt. Pumpelly gave a warm and amusing account of his years at Freiberg (1856–1859), with several references to Breithaupt, of whom he wrote: "Breithaupt was already old; he was, however, one of the fathers of Mineralogy, and an inspiring lecturer. He taught crystallography without mathematics. . . . He created in his students an interest in crystal forms and systems, that I did not find later in the mathematical treatment under Alvin Weissbach" (*My Reminiscences,* 1918).

Breithaupt summarized his views at the end of his career in a contribution to the *Festschrift* published for the centennial of the Bergakademie (1866). From this it is clear that mineral classification was still his chief interest. He had failed to follow the progress of crystallography and had persisted in recognizing only four crystal systems and in attempting to develop the forms of hexagonal and tetragonal crystals from those of the isometric system by means of his *Progressions-Theorie.* He appears to have been unaware of Johann F. C. Hessel's work (1830) on the derivation of the crystallographic symmetry classes and of Bravais's (1850) on translation lattices. Although he knew of the work of William Whewell and William H. Miller (1839), he did not favor their approach. Breithaupt apparently failed to recognize his own most valuable contribution: in summarizing his career, he did not so much as mention the paragenesis of minerals.

BIBLIOGRAPHY

I. Original Works. A complete bibliography of Breithaupt's publications is in Poggendorff's *Biographisch-litterarisches Handwörterbuch,* I (1863), 290, and III (1898), 187.

His elaboration of "natural classifications" was *Vollständiges Handbuch der Mineralogie,* 3 vols. (Dresden-Leipzig, 1836–1847). The first volume is a textbook of mineralogy and also contains an extended statement of Breithaupt's views on mineral classification; Vols. II and III comprise systematic description of species, arranged by classes, orders, and genera. A planned fourth volume was never published. *Die Paragenesis der Mine-*

ralien (Freiberg, 1849) carries the subtitle *Mineralogisch, geognostisch und chemisch beleuchtet, mit besonderer Rücksicht auf Bergbau.* This little book of 274 pages is by far Breithaupt's most important and lasting contribution.

II. Secondary Literature. Franz von Kobell, *Geschichte der Mineralogie, von 1650–1860* (Munich, 1864), is the most comprehensive history of mineralogy by a slightly younger (1803–1882) contemporary of Breithaupt's. It contains many critical comments but only incidental reference to paragenesis.

See also H. J. Rösler, "August Breithaupt—sein Leben und Werk," in *Freiberger Forschungshefte,* C-230 (Leipzig, 1968), 9–19.

B. Voland, "Über die Entwicklung der Mineralsystematik in der ersten Hälfte des 19. Jahrhunderts durch die Schüler Werners," in *Abraham Gottlob Werner Gedenkschrift, Freiberger Forschungshefte,* C-223 (Leipzig, 1967), 179–190, is a discussion of the development of mineral systematics that considers Breithaupt's contributions in the light of the advances made by his contemporaries.

Also of value are Raphael Pumpelly, *My Reminiscences,* I (New York, 1918); and *Festschrift zum hundertjärigen Jubiläum der Königl. Sächs. Bergakademie zu Freiberg am 30. Juli 1866* (Dresden, 1866).

Adolf Pabst

BREMIKER, CARL (*b.* Hagen, Germany, 23 February 1804; *d.* Berlin, Germany, 26 March 1877), *astronomy, geodesy.*

After a long period as a geometer with the Rhine-Westphalian Land Survey, Bremiker went to the royal observatory in Berlin, where he served as a mathematician, an observer, and an editor of widely used astronomical and mathematical tables. At the suggestion of Friedrich Bessel, several observatories jointly published the *Berliner academischen Sternkarten*; and from 1841 to 1859 Bremiker observed and calculated the hours 6, 9, 13, 17, and 21 for these atlases. He also took part in the calculations for the *Berliner astronomische Jahrbuch*; and from 1850 to 1877 he edited the *Nautische Jahrbuch.*

In 1868 Bremiker was appointed *Sektionschef* in the Prussian Geodetic Institute. He became well known for his *Logarithmisch-trigonometrische Handbuch* (1856), which, by the advent of the calculating machine, had gone through forty editions. Easy to use and offering an accuracy never before attained, it served as an indispensable tool for generations of calculators. Bremiker's tables, however, with centesimal arguments, were less popular.

BIBLIOGRAPHY

I. Original Works. Bremiker's publications include *Logarithmisch-trigonometrische Handbuch mit 7 Dezimal-*

stellen (Berlin, 1856); *Logarithmisch-trigonometrische Tafeln mit 6 Dezimalstellen* (Berlin, 1862); *Logarithmisch-trigonometrische Tafeln mit 5 Dezimalstellen* (Berlin, 1872); and *Tafel vierstelliger Logarithmen* (Berlin, 1874). Bremiker was also editor of the second edition of Crelle's *Rechentafeln* (Berlin, 1864) and of the tenth and eleventh editions of Bode's *Anleitung zur Kenntnis des gestirnten Himmels* (Berlin, 1844, 1858).

II. SECONDARY LITERATURE. Further information can be found under Bremiker's name in *Neue deutsche Biographie,* I, 582; Poggendorff, I and III; an obituary notice is L. A. Winnecke, in *Monthly Notices of the Royal Astronomical Society,* London, **38** (1878), 151.

BERNHARD STICKER

BRENDEL, OTTO RUDOLF MARTIN (*b.* Niederschönhausen, near Berlin, Germany, 12 August 1862; *d.* Freiburg, Germany, 6 September 1939), *astronomy.*

After extensive study of astronomy and mathematics from 1883 to 1889 in Berlin, Munich, London, Paris, and primarily in Stockholm, Brendel received his doctorate for "Anwendung der Gyldenschen absoluten Störungstheorie auf die Breitenstörungen" (1890); all his life Brendel considered himself a disciple of Gyldén, the Swedish expert on celestial mechanics. This aspect of celestial mechanics—the development of mathematical methods to consider the influence of perturbations upon the computation of orbits—remained his chief concern.

Whereas the problem of the computation of the orbits of the major planets had been nearly solved by the great works of Laplace, Lagrange, Olbers, and Gauss in the eighteenth and nineteenth centuries, the great number of hard-to-observe minor planets presented the theoreticians with entirely new tasks. A complete consideration of the perturbations to which the minor planets are subjected by Jupiter and Saturn would require considerable time. The methods developed by Brendel are based on consideration of only the largest perturbations and their tabulation, so that the minor planets will not again be lost; this way, at least over a certain period (say about one hundred years) they can again be looked for and identified with certainty during their opposition.

After his habilitation in Greifswald in 1892, Brendel devoted himself to this work. He became extraordinary professor of theoretical astronomy at Göttingen in 1898, also undertaking the teaching of insurance mathematics and geodesy in 1902. In 1907 he went to the Academy of Commerce, Frankfurt-am-Main, where he was lecturer in mathematics and insurance mathematics, and the next year he became director of the newly founded observatory of

the Physikalische Verein there. With the support of various foreign academies, he established the Internationales Planeteninstitut, which was to develop his ideas. In 1914 he was appointed ordinary professor at the University of Frankfurt and director of the university's observatory. In 1927 he retired.

Brendel's book on the minor planets was awarded the Prix Damoiseau by the Paris Academy. He participated in the publication of Gauss's works (Volumes VII–XII, 1898–1929).

BIBLIOGRAPHY

I. ORIGINAL WORKS. Brendel's writings include *Theorie der kleinen Planeten,* 4 vols. (Göttingen, 1897–1911); *Theorie des Mondes* (Göttingen, 1905); and *Theorie der grossen Planeten,* 2 vols. (Kiel, 1930–1933).

II. SECONDARY LITERATURE. Obituaries of Brendel are *Astronomische Nachrichten,* **270** (1940), 248; *Neue deutsche Biographie,* II, 584; and Poggendorff, IV (1904), 180; V (1926), 164–165; VI (1936), 322.

BERNHARD STICKER

BRESCHET, GILBERT (*b.* Clermont-Ferrand, France, 7 July 1783; *d.* Paris, France, 10 May 1845), *anatomy.*

Breschet became a nonresident medical student in the Paris hospitals in 1808, a resident student in 1809, and a doctor in 1812. Seven years later he was appointed surgeon and head of anatomical studies at the Hôtel Dieu, and he was *agrégé* in 1825. In 1832 he was elected to the Académie de Médicine, and in 1835 to the Académie des Sciences; the next year he became professor of anatomy at the Faculty of Medicine. Breschet attained this chair after a three-month competitive examination (14 April–9 July 1836) that was decided by a jury composed of professors from the Faculty and members of the Institute and was passionately followed by hundreds of students. He was elected on the third ballot, defeating Pierre-Paul Broc, a mere professor at the École Pratique who was far less learned than he but was an idol of the students. Infuriated by the outcome, the students attempted to sack the Faculty of Medicine. The damage was estimated at 5,000 francs, and the disturbance ended in court, where the journalist Fabre, a supporter of Broc, was fined 500 francs. (Fabre was fined for having published a newspaper, between 1835 and 1836, without having put up the security demanded by law and for having moved the paper's printing plant without having informed the authorities.) He took his revenge by writing *Orfilaïade,* a verse pamphlet illustrated by Daumier

442

that was directed against Mathéo Orfila, the dean of the Faculty of Medicine.

Although Breschet studied under the renowned surgeon Guillaume Dupuytren, whose chair he filled at the Faculty of Medicine, he was never a first-rate surgeon and left no special mark on the history of that discipline. His great and methodical capacity for work was particularly applied to the study of human anatomy, comparative anatomy, and the natural sciences. This work was carried on in collaboration with an excellent team: Milne-Edwards, Vavasseur, Villermé, Roussel de Vauzème, Rayer, Bogros, and Raspail. Breschet was the only doctor-naturalist at the Faculty of Medicine, which was essentially oriented to clinical and practical medicine and to human anatomy. Because of this, at the time of Breschet's death, the Faculty rejected an offer made by the minister of education to create a chair of comparative anatomy.

Breschet became known through his work on the veins of the spine and the human skull, his contributions to knowledge of the auditory system in vertebrates, and his knowledge of the arterial plexuses of the Cetacea, showing their adaptation for diving. With Roussel de Vauzème he discovered the sweat glands, and in 1818 he coined the word *phlebitis* to designate an inflammation that soon dominated medico-surgical pathology. Breschet also studied the human ovum and that of other vertebrates.

Having a genuine gift for languages and being keenly aware of what was being done in foreign countries, Breschet translated into French the classic works of Meckel, Heusinger, Hodgson, Kaltenbrunner, Gimbernait, von Baer, Gottfried Treviranus, Rathke, Pander, Jacobson, and Joseph Arnold. His fame was much greater elsewhere in Europe than in France. At the Stuttgart Congress, for instance, the whole assembly gave him a standing ovation. Breschet was a member of the Académie des Curieux de la Nature, as well as academies and learned societies throughout Europe.

The death of his parents (1842, 1845) deeply grieved Breschet and brought on a slight stroke whose effects progressively became so severe that he died.

BIBLIOGRAPHY

I. ORIGINAL WORKS. A complete bibliography of Breschet's works is in Huard (see below). His principal works are "Recherches sur les hydropisies actives en général et sur l'hydropisie du tissu cellulaire en particulier," *Thèse de Paris*, no. 173 (1812); *Essai sur les veines du rachis* (Paris, 1819); *Recherches historiques et expérimentales sur la formation du cal* (Paris, 1819); *Traité des maladies des*

artères et des veines par J. Hogdon, which Breschet translated, adding notes, 2 vols. (Paris, 1819), the work in which he uses *phlébite* for the first time; *Mémoire sur une nouvelle espèce de grossesse extra-utérine* (Paris, 1826); *Recherches anatomiques, physiologiques et pathologiques sur le système veineux*, 30 parts (Paris, 1828); four papers on the structure of the hearing organ in fish, read to the Académie Royale des Sciences, 13 August 1832; "De la structure de l'organe de l'ouïe et particulièrement de celle du labyrinthe chez l'homme et les mammifères," presented to the Académie Royale des Sciences in 1832; *Traité des maladies des enfants*, 2 vols. (Paris, 1832); and *Études anatomiques, physiologiques et pathologiques de l'oeuf dans l'espèce humaine et dans quelques unes des principales familles des animaux vertébrés* (Paris, 1833).

II. SECONDARY LITERATURE. Breschet was the subject of a eulogy by Royer-Collard: "Éloge du 3 novembre 1845 à la Faculté de médecine de Paris," in *Moniteur universel* (14–15 May 1845). Other memorial notices are in *Archives générales de médecine*, **2** (1845), 257–342; *Bulletin de l'Académie de médecine*, **10** (1844–1845), 680–685; *Gazette médicale de Paris*, **2** (1845), 301–316; and *Journal des connaissances médicales pratiques et de pharmacologie*, **12** (1844–1845), 305–306.

See also P. Balme and G. Dastugue, "Gilbert Breschet," in *Clermont médical*, no. 44 (1961), 89–115; A. Corlieu, *Centenaire de la Faculté de médecine de Paris, 1794–1894* (Paris, 1896), pp. 250–256; L. Delhoume, *Dupuytren* (Paris, 1935), pp. 243–260; P. Huard, "Gilbert Breschet," in *Comptes rendus du Congrès des Sociétés savantes Clermont-Ferrand*, **3** (1963), 117–128, which contains Breschet's complete bibliography; and A. Thierry, "Gilbert Breschet," in *Le temps* (6 Dec. 1928).

PIERRE HUARD

BRET, JEAN JACQUES (*b.* Mercuriol, Drôme, France, 25 September 1781; *d.* Grenoble, France, 29 January 1819), *mathematics.*

He was the son of Jacques Bret, a notary. After passing the entrance examinations given at Lyons, Bret entered the École Polytechnique on 22 November 1800 and was admitted to the course of preparation for civil engineering (Service des Ponts et Chaussées). Unfortunately, because of poor health, he did not complete his studies, but was forced to take a leave of absence from October 1802 to November 1803. The school administration offered to let him stay a fourth year on condition that he take the examinations. He was definitely removed from the rolls in December 1803.

In 1804 Bret became professor of transcendental mathematics at the lycée in Grenoble, and from 8 October 1811 until his death, he was professor at the Faculté des Sciences in the same city, having became *docteur ès sciences* on 10 March 1812.

There are some twenty publications by Bret in the

Annales de mathématiques de Gergonne, a note in the *Correspondance* of the École Polytechnique, and a memoir in the latter's journal. Most of his articles deal with analytical geometry on plane surfaces and in space, notably with the theory of conics and quadrics. He sets forth, for example, the third-degree equation that determines the length of the axes of a central quadric.

In this research the cumbersome techniques of the time are unpleasantly obvious. By way of exception, a study on the squares of the distance between a point in space and fixed points is remarkable for its simplicity, elegance, and generality.

Other works have a bearing on the theory of algebraic equations, particularly upon the limitation of real roots, a subject in style at the time. Bret also worked on the theory of elimination, where he used the greatest common divisor of polynomials in order to establish Bézout's theorem on the degree of the polynomial resultant.

Bret became involved in a long polemic with J. B. E. Dubourguet in the *Annales de Gergonne.* This had to do with the demonstration of the fundamental theorem that an algebraic equation admits a number of roots equal to its degree.

BIBLIOGRAPHY

Among Bret's works are "Sur la méthode du plus grand commun diviseur appliquée à l'élimination," in *Journal de l'École polytechnique,* **15** (1809), 162–197; and "Sur les équations du quatrième degré," in *Correspondance de l'École polytechnique,* **2** (1811), 217–219.

Of particular note, all in *Annales de mathématiques,* are "Recherche des longueurs des axes principaux dans les surfaces du second ordre qui ont un centre," **2** (1812), 33–38; "Recherche de la position des axes principaux dans les surfaces du second ordre," *ibid.,* 144–152; "Discussion de l'équation du second degré entre deux variables," *ibid.,* 218–223; "Démonstration de quelques théorèmes relatifs au quadrilatère," *ibid.,* 310–318; "Théorie de l'élimination entre deux équations de degrés quelconques, fondée sur la théorie du plus grand commun diviseur," **3** (1812), 13–18; "Démonstration du principe qui sert de fondement au calcul des fonctions symétriques et de la formule binomiale de Newton," **4** (1813), 25–28; "Théorèmes nouveaux sur les limites des racines des équations numériques," **6** (1815), 112–122; and "Théorie générale des fractions continues," **11** (1818), 37–51.

An article on Bret is Niels Nielsen, "Bret," in *Géomètres français sous la Révolution* (Copenhagen, 1929), pp. 31–37.

JEAN ITARD

BRETHREN OF PURITY. See **Ikhwān al-Ṣafā'.**

BRETONNEAU, PIERRE (*b.* St.-Georges-sur-Cher, France, 3 April 1778; *d.* Passy, France, 18 February 1862), *medicine.*

Bretonneau's father, Pierre, a master surgeon, and his mother, Elisabeth Lecomte, came from the old bourgeois class. Almost completely uneducated during his childhood (at nine he was not yet able to read), he was sent to the École de Santé in Paris in 1795. There he attended the clinical lectures of Corvisart. He abandoned his studies, however, after being unjustifiably failed on an examination (1801) and became a public health officer in Chenonceaux. His skill gained Bretonneau reputation, and he was made chief physician at the Tours hospital; to qualify for this position he took his final examinations and completed his doctoral thesis in 1815. In 1838 he left this post and the directorship of the École de Santé in Tours, to dedicate himself to practicing medicine among the poor. After a first marriage to a woman twenty-five years his senior he married at the age of seventy-eight a girl of eighteen.

A man of many interests, Bretonneau constructed hydraulic hammers, barometers, and thermometers; sculpted and drew; and studied the habits of bees and ants. He was a first-rate botanist and horticulturist, and wrote a treatise on plant grafting; his garden in Palluau was famous throughout Europe. Independent, proud yet modest, and disdainful of honors, Bretonneau was a dedicated physician and an able therapist. His lectures, which were in the Hippocratic tradition, made a profound impression upon his students.

Bretonneau's outlook made him a member of the school of Paris, which considered the lesion to be the trace that makes possible the definition, classification, and comprehension of an illness. Physical signs, as direct mediators of the lesion, were preferred to epiphenomenal symptoms. G. L. Bayle introduced the concept of a definite development of a lesion that defines its specificity instead of altering it. Laennec illustrated this idea by showing that the various lesions of "phthisis" were in fact the gross signs of a specific disease, tuberculosis. In 1801 Bayle defended the unicity of smallpox (either discrete or confluent) in his M.D. thesis, arguing for the existence of cross-contagiousness. This thesis, which described two kinds of smallpox, had many repercussions.

The concepts of specificity and contagiousness, as well as the previous work on smallpox, later gave Bretonneau a model for introducing those concepts into his work on two diseases of the mucous membranes of the digestive and respiratory tracts. He demonstrated that, just as the skin could show a great many reactions, the mucous membranes did not have only one response to all pathogens. Against an excessively narrow concept of tissue pathology, inherited from Bichat, he proposed the concept of specific inflammation.

As early as 1819 Bretonneau individualized typhoid fever (called *dothinentérie*). He demonstrated the localization of the lesion on Peyer's patches in the ileum, the cyclic development of the lesion (each phase had formerly been described as a separate disease), and the uniqueness of the various fevers, and defended the concept of a specific transmissible agent. In his memoir of 1829, Bretonneau described the course of a typhoid epidemic at Chenonceaux and showed the role of contact in its propagation. He observed that the disease was endemic to Paris because the chains of transmission were frequently broken by the immunity of those who had already had the disease. He suspected that besides typhoid there was a petechial pyrexia with a more rapid course (exanthematic typhus) that was often mistaken for typhoid.

From 1818 to 1820 a diphtheria epidemic raged in Tours. In 1821 Bretonneau published his observations. He individualized the disease through analysis of the characteristics of the false membrane, showing its primary tonsillar localization and its possible nasopharyngeal, auricular, and laryngeal ("croup") extensions, which lead to asphyxia. He showed that certain (toxic) symptoms exist regardless of localization, and deduced that the specificity of the inflammation—much more than the type of tissue in which it occurs—is responsible for the disturbance of functions that every inflammatory lesion produces: duration, severity, and danger of most fevers depend on the specificity of the inflammation.

Concerned with preventing fatal asphyxia, Bretonneau, after two experiments that failed in man but succeeded in a dog, performed a successful tracheotomy on a four-year-old girl in July 1825. The operation, the first to be performed on a croup patient, was made possible by his manual dexterity and his ingenuity (he invented the double cannula). He was convinced that diphtheria was contagious and that it was transmitted by drinking glasses. He tried, in vain, to infect animals with the disease. He defended the idea of specific therapy, but his experimental work in that area proved fruitless.

BIBLIOGRAPHY

I. ORIGINAL WORKS. Bretonneau published his works sparingly, some time after they had been written, and at the insistence of his students: *De l'utilité de la compression et, en particulier, de l'efficacité du bandage de Theden dans les inflammations idiopathiques de la peau* (Paris, 1815), his M.D. thesis; *Des inflammations spéciales du tissu muqueux et en particulier de la diphthérite ou inflammation pelliculaire* (Paris, 1826), which consists of two papers read to the Académie Royale de Médecine on 26 June and 6 August 1821; "Notice sur la contagion de la dothinentérie, lue à l'Académie royale de médecine le 7 juillet 1829," in *Archives générales de médecine*, **21** (1829), 57–78; and *Traités de la dothinentérie et de la spécificité,* L. Dubreuil-Chambardel, ed. (Paris, 1922), which contains a biography and an analysis of the work.

II. SECONDARY LITERATURE. Works on Bretonneau are E. Apert, "Bretonneau," in *Biographies médicales,* no. 6 (1938); J. D. Rolleston, "Bretonneau: His Life and Work," in *Proceedings of the Royal Society of Medicine, Section of the History of Medicine,* **18** (1924); and P. Triaire, *Bretonneau et ses correspondants* (Paris, 1892). Useful explanations of Bretonneau's ideas may be found in Trousseau's complete works.

ALAIN ROUSSEAU

BREUER, JOSEF (*b.* Vienna, Austria, 15 January 1842; *d.* Vienna, 20 June 1925), *medicine, physiology, psychoanalysis.*

Breuer's father, Leopold (1791–1872), was a teacher of religion employed by the Jewish community of Vienna, and Breuer described him as belonging to "that generation of [Eastern European] Jews which was the first to step out of the intellectual ghetto into the air of the Western world." Breuer's mother died when he was about four, and he was raised by her mother. His father tutored him until he was eight, and he then entered the Akademisches Gymnasium of Vienna, from which he graduated in 1858. After a year of general university studies, Breuer entered the medical school of the University of Vienna in 1859 and completed his medical studies in 1867. In the same year, immediately after passing his doctoral examination, he became assistant to the internist Johann Oppolzer. When Oppolzer died in 1871, Breuer relinquished his assistantship and entered private practice.

In 1875 Breuer qualified as *Privatdozent* in internal medicine. He resigned the position on 7 July 1885, however, apparently because he felt he had been improperly denied access to patients for teaching purposes; he also refused to let the surgeon C. A. T. Billroth nominate him for the title of professor extraordinarius. His formal relationship to the medical faculty was thus tenuous and strained; yet he was considered one of the best physicians and scientists in Vienna. His practice was his chief interest, and although he once referred to himself as a "general practitioner," he was what today would be called an internist. Some idea of his reputation can be gathered from the fact that among his patients were many of the professors on the medical faculty, as well as Sigmund Freud and the prime minister of Hungary. He was elected to the Viennese Academy of Science in 1894 upon the nomination of three of its most distinguished members: the physicist Ernst Mach and the physiologists Ewald Hering and Sigmund Exner.

Breuer married Matilda Altmann on 20 May 1868, and she bore him five children: Robert, Bertha Hammerschlag, Margaret Schiff, Hans, and Dora. When faced with deportation by the Nazis, Dora committed suicide; she did not kill herself in the United States as stated by Ernest Jones. Breuer's granddaughter Hanna Schiff was killed by the Nazis. The remainder of his descendants live in England, Canada, and the United States.

Breuer was a skeptic in matters of religion. In espousing the views of Fechner and Goethe, he referred to himself as one of "the many intellectuals who have religious needs and find themselves utterly unable to satisfy them within the faith of popular religion." In his will he expressed the wish (which was followed) that he be cremated, a wish inconsistent with conventional Jewish religious practices.

Breuer was one of the great physiologists of the nineteenth century. He had no pupils and no permanent affiliation with a university or institute, which may explain why his fame today is not in proportion to his achievements. His first major scientific study led to the discovery of the reflex regulation of respiration. The work was conducted at the military medical school of Vienna (the Josephinum) in collaboration with its professor of physiology, Ewald Hering, and the results were published in 1868. By the very simple device of occluding the trachea at the end of an inspiration or expiration, Breuer and Hering were able to show that the lung contains receptors that detect the degree to which it is stretched. When the lung is distended by inspiration, nerve impulses arise in the lung and are transmitted to the brain via the vagus nerve; these impulses reflexly initiate expiration. When the lung is deflated, other receptors are stimulated and their impulses, also arriving in the brain via the vagus nerve, reflexly initiate the next inspiration. The whole mechanism, called by Hering and Breuer the "self-regulation" of respiration, was one of the first "feedback" mechanisms to be demonstrated in the mammal. Writing thirty years later, E. H. Starling said that these experiments caused "a complete revolution in our idea of the relationship of the vagus to the respiratory movements." At the time Starling wrote, doubt had been cast on the existence of the receptors that respond to deflation of the lung, but the existence of such receptors was demonstrated

by A. S. Paintal in the 1950's. The picture of the reflex regulation of respiration drawn by Hering and Breuer remains in all essentials the view held today, and the underlying reflex is still known as the Hering-Breuer reflex.

After completing his work with Hering, Breuer began his long series of investigations of the function of the labyrinth, remarkable for their importance and even more remarkable because he conducted them privately, working in his own home and supported only by fees from his medical practice. We now know that the inner ear is a double organ, part of which is the organ of hearing and part of which detects movement of the head and its position in space. The structures of the inner ear are small and delicate, and are placed deep within the skull, which makes them inaccessible to easy experimental investigation. Breuer's first studies of the labyrinth were concerned with the semicircular canals. These canals, three on each side of the head, are filled with fluid (endolymph). The angular position of the canals suggests that they might have something to do with the detection of movement; crude experiments carried out in the 1820's by M. J. P. Flourens showed that injury to them produced disturbances of motor function in animals. In 1870 Friedrich Goltz suggested that the semicircular canals were the sense organs that detect the position of the head relative to the gravitational field; he believed that the lowest part of each canal would, as the result of the weight of the endolymph, be stimulated by pressure.

Our understanding of the function of the semicircular canals dates from the insight that it is not pressure but a tendency of the endolymph to flow within the canals during motion of the head that stimulates the receptors in the ampullae at the end of the canals. This insight was based on Goltz's suggestion of 1870 and was reached essentially simultaneously by Breuer, Mach, and by the Edinburgh chemist A. Crum Brown. (Their initial communications were made on 14 November 1873, 6 November 1873, and 19 January 1874, respectively.) During movement of the head, the endolymph of the canals moves, but its angular rotation lags slightly behind that of the head and results in stimulation of the receptors at the ends of the canals. Thus, the semicircular canals respond to angular acceleration.

Breuer's first article (printed in 1874) concerned mainly the explanation and interpretation of previous observation on animals and humans (thereby resembling Mach's article); his second article, which appeared a year later, reported the results of many experiments, carried out chiefly on pigeons. In those experiments Breuer developed evidence for his theory, which has stood virtually unchallenged since that time. In addition he called attention to the importance of another receptor system connected with posture (also located in the inner ear), the otolith system. The otolith is a minute solid body whose movement stimulates receptors in the utricle, another part of the inner ear. Breuer suggested, and accumulated evidence to show, that the otoliths and the hair cells of the utricle are static position receptors that provide information about the orientation of the head in the gravitational field as well as information about linear acceleration. We are indebted to Breuer for the clear-cut analysis of the differing functions of the semicircular canals and the otoliths, as well as a clear-cut depiction of the relationship of the labyrinthine reflexes to optical nystagmus. His results were by no means immediately accepted and, in particular, his work on the otoliths was not generally known or accepted as late as 1900. Yet it was correct, and today is recognized as the foundation of our knowledge of the sensory receptors for sensations of posture and movement.

Thus, Breuer deserves the credit for two fundamental and far-reaching advances in mammalian physiology: the Hering-Breuer reflex and the elucidation of the function of the labyrinth. His scientific techniques included mastery of physiological experiment and of delicate surgery, as well as the use of histological techniques. Above all, he was a remarkably patient and accurate observer. It must not be supposed that this means he was a passive recorder of events, for observation always implies full awareness of the relation of the facts observed to their meaning for theory and interpretation, as well as the ability to suspend judgment and retain a multitude of observations pending an intellectual survey and rearrangement of them. It has been suggested that Breuer was in some ways less scientific in his psychoanalytic reporting than in his physiological research, but the same sort of active observation, active accumulation of facts, and active suspension of final judgment until the facts arranged themselves into meaningful patterns is entirely characteristic of his investigations of psychopathology.

The general impression that Breuer published little and infrequently is true in the sense that he published relatively few major scientific articles at relatively long intervals; but quite apart from the importance of these articles, it should be noted that some of them were very long and detailed. His purely physiological articles, published over a forty-year period, numbered about twenty and comprised more than five hundred pages.

In the summer of 1880, while attending a man who was seriously ill with a peripleuritic abscess, Breuer observed the onset of a serious psychological disturbance in the man's daughter, "Anna O.," who was also his patient. Her symptoms were later summarized by Freud as follows:

> Her illness lasted for over two years, and in the course of it she developed a series of physical and psychological disturbances which decidedly deserved to be taken seriously. She suffered from a rigid paralysis, accompanied by loss of sensation, of both extremities on the right side of her body; and the same trouble from time to time affected her left side. Her eye movements were disturbed and her power of vision was subject to numerous restrictions. She had difficulties over the posture of her head; she had a severe nervous cough. She had an aversion to taking nourishment, and on one occasion she was several weeks unable to drink in spite of a tormenting thirst. Her powers of speech were reduced, even to the point of her being unable to speak or understand her native language. Finally, she was subject to conditions of *absence,* of confusion, of delirium and of alterations of her whole personality . . . [*The Complete Psychological Works of Sigmund Freud,* XI, 10].

Breuer noted that Anna O. showed two markedly different states of consciousness each day: during one she seemed relatively normal, during the other she was "clouded." He also found that if, during her normal state, she could be induced to tell him the fantasies that occupied her during her clouded state her restlessness was greatly reduced. To facilitate this "catharsis," he began to hypnotize her. Far more importantly, he eventually noted that under special circumstances of recall she would trace a series of memories back over time until she reached the memory of a "traumatic" episode that had been transformed into a symptom. After seeing several of her symptoms vanish as the result of this sort of recall, Breuer began to visit her twice a day in order to have time for more intensive and frequent hypnosis. He gradually succeeded in relieving all of her symptoms by this process of catharsis.

From his treatment of Anna O., Breuer arrived at two conclusions of fundamental importance: (1) that the symptoms of his patient were the result of "affective ideas, deprived of the normal reaction" which remained embedded in the unconscious, and (2) that the symptoms vanished when the unconscious causes of them became conscious through being verbalized. These two observations form the cornerstone upon which psychoanalysis was later built.

Breuer did not publish or publicize the results of his treatment of Anna O. He did, however, discuss them with Freud, and the cathartic treatment resumed when Freud began to use it under Breuer's guidance. For several years Breuer and Freud jointly explored this form of psychotherapy. Only Freud treated patients, but he and Breuer continually discussed the results and implications of the treatment. Freud first used the cathartic method in either 1888 or 1889. The practical and theoretical conclusions they reached through their collaboration were published in an article in 1893 and as a book (*Studien über Hysterie*) in 1895. The publication of the book very nearly coincided with the end of their collaboration—and of their friendship. The contribution of Breuer and Freud to the development of psychoanalysis may be stated as follows: (*a*) Breuer discovered that neurotic symptoms arise from unconscious processes. (*b*) Breuer discovered that neurotic symptoms disappear when the unconscious processes become conscious. (*c*) These major discoveries were communicated by Breuer to Freud. (*d*) The first serious attempt to explore the implications of these discoveries was made by Freud and Breuer working in close collaboration. (*e*) Breuer was not anxious to pursue these studies, and the major development must be attributed to Freud after he separated from Breuer, during a period when Breuer apparently did not seek another collaborator but dropped the subject. (*f*) If by psychoanalysis we mean a discipline relying on the technique of free association, psychoanalysis was solely Freud's discovery.

Breuer, in writing the theoretical chapter of the *Studien über Hysterie,* advanced a number of very important concepts, among them one rejected by Freud but now regarded as very important: that the hypnoid state and varying levels of consciousness are of great importance in normal and abnormal mental functioning. Other theoretical concepts usually attributed to Breuer include the distinction between the primary and secondary processes, the concept of hallucination as a regression from imagery to perception, and the suggestion that perception and memory cannot be performed by the same psychic apparatus. The "principle of constancy" was first mentioned by Breuer, but he attributed it to Freud. This concept is fundamental to the development of psychoanalytic theory, and one cannot but wonder whether Breuer did not in fact play an important role in formulating it. Breuer after all, formulated the notion of "feedback" in the respiratory cycle and studied the sense organ that plays a key role in postural balance in animals; and the homeostatic devices involved in those systems are very reminiscent of the "principle of constancy."

In his obituary of Breuer, Freud spoke with regret of the fact that Breuer's brilliance had been directed toward the problems of psychopathology for only a brief time. Although Breuer actually dealt with that subject during a period of nearly fifteen years, he did so in a way that deserves special examination. In the first place, Breuer's mastery of hypnosis and his readiness to use it in treating Anna O. may, as Professor E. H. Ackerknecht has suggested to me, indicate that Breuer had more interest in psychopathology than the average internist even before the case of Anna O.; this supposition receives confirmation from the time Breuer spent on that case and from the care with which he recorded it. In the opinion of Freud it was the case of Anna O. that also caused Breuer to draw back from psychotherapy for a time, since the case had, near its end, an unexpected and disturbing result: Anna O. formed a strong attachment to Breuer, an attachment that had a definitely sexual quality. Freud believed that this upset Breuer and prevented him from again practicing "deep" psychotherapy. Breuer himself stated that after the case of Anna O. he gave up treating patients in this manner since such treatment could not be carried out by a physician subject to the demands of a busy general practice. The resumption of that sort of psychotherapy, which was to evolve into psychoanalysis, was undertaken jointly by Breuer and Freud about five years after the case of Anna O., but the treatment of patients was solely in the hands of Freud, with Breuer taking part only in discussion of the techniques and the results of treatment.

There were for a time a few "Breuerians," i.e., physicians who used Breuer's original cathartic therapy without Freud's amplifications. Breuer does not seem to have been a "Breuerian" in this sense, although letters by him in the Medizinhistorisches Institut der Universität Zürich indicate that his handling of psychiatric patients remained very sophisticated. On the whole it is probably correct to say that while Breuer was persuaded intellectually of the validity and importance of the new concepts and techniques that developed from his own work and from the work he did jointly with Freud, he was equally dismayed by the recurrent intrusion of sexuality into the subject. In a sense, therefore, Breuer's anxiety over Anna O.'s reaction to him may be taken as a symbol of the reasons for his ambivalence toward the subject, but only as a symbol, since he returned, via collaboration with Freud, to the very same subject, and did not finally split with Freud until thirteen years after the Anna O. episode. Whatever the nature of Breuer's interest in the subject and whatever the reasons for his ambivalence, nothing can minimize the fact that his treatment of Anna O. can convincingly be regarded as the first modern example of "deep psychotherapy" carried out over a prolonged period of time.

Although they had been very close for many years, Freud and Breuer separated in 1896 and never spoke again. Whatever the roots of this break were in the character of the Breuer and Freud relationship, the quarrels that led up to it grew out of their work on psychotherapy. It was a difficult period for Freud, who felt, among other things, that Breuer was ambivalent about the value of their work, ambivalent about publishing it, and ambivalent about publicly supporting him. Interestingly enough, their final quarrel seems to have concerned a matter in which Breuer was right and Freud only later found himself to be wrong, the question of the reality of the memories of having been seduced in early childhood, which had occurred in many patients. Breuer did refuse to back Freud in his belief that nearly all their patients had experienced such seductions; when Freud finally realized that such memories were memories not of real events but of childhood fantasies, he made one of his most important discoveries.

There is no possibility of meaningfully exploring the dynamics of their relationship at this date. It may be simplest to say that for a long time Freud needed Breuer and depended upon him; he then came to need him less and to depend on him less; and eventually he had a positive need to break with him, which he did thoroughly and in a way that left bad feelings. We cannot say even that much about Breuer; all we can say is that he was very fond of Freud for a long time and deeply wounded by the break. That a relationship between an older and a younger man, first full of warm and close friendship and then of turmoil, should have accompanied the birth of psychoanalysis may have been inevitable. That relationship should not be allowed to obscure the brilliant intellectual and observational contributions each made to the founding of modern psychoanalysis, psychiatry, and psychotherapy, nor should it obscure the fact that their long and close collaboration was an integral part of the creation of psychoanalysis.

Breuer was friendly with many of the most brilliant intellects of his time. He sustained a long correspondence with Franz Brentano, was a close friend of the poet Maria von Ebner-Eschenbach, and was on friendly terms with Mach, whom he had met at the time of their simultaneous work on the labyrinth. His opinion on literary and philosophical questions seems to have been widely respected and often sought. His correspondence with Maria von Ebner-Eschenbach has been preserved as has part of the Brentano-Breuer correspondence. Breuer had a considerable command of languages, and it is interesting to note that his treatment of Anna O. was for a long period conducted in English. The eulogies published after his death all emphasize that the range and depth of his cultural interests were as unusual and important as his medical and scientific accomplishments.

BIBLIOGRAPHY

I. ORIGINAL WORKS. No detailed bibliography of Breuer's publications has ever been assembled. His scientific articles can easily be traced through the usual guides to the medical literature and, in particular, in the *Cumulative Author Index to Psychological Index . . .* (Boston, 1960). The Hering-Breuer reflex is described in "Die Selbststeuerung der Athmung durch den *Nervus vagus*," in *Sitzungsb. d. k. Akad. d. Wissensch., Math.-naturwissensch. Klasse,* Abtheilung II, **58** (1868), 909–937. Breuer's first major articles on the labyrinth were "Ueber die Funktion der Bogengänge des Ohrlabyrinths," in *Medizinische Jahrbücher,* 2nd series, **4** (1874), 72–124, and "Beiträge zur Lehre vom statischen Sinne (Gleichgewichtsorgan, Vestibularapparat des Ohrlabyrinths)," *ibid.,* **5** (1875), 87–156. Breuer's other articles on the labyrinth may be found with the aid of the bibliographic sources mentioned above or in the bibliographies to the article by Roth and the book by Camis cited below.

Breuer's publications in the sphere of psychopathology, both written with Freud, were "Ueber den psychischen Mechanismus hysterischer Phänomene (Vorläufige Mittheilung)," in *Neurologisches Centralblatt,* **12** (1893), 4–10, 43–47; and *Studien über Hysterie* (Leipzig-Vienna, 1895). The preliminary communication and the book are readily available in English translation as Vol. II of *The Standard Edition of the Complete Psychological Works of Sigmund Freud,* James Strachey, ed. (London, 1955).

Breuer's brief autobiography, entitled simply *Curriculum vitae,* presumably was published in Vienna in 1925 (the actual publication bears no place, date, or publisher's name). There is a photographic copy of this rare pamphlet in the library of the New York Academy of Medicine.

II. SECONDARY LITERATURE. Apart from the brief autobiography mentioned above, the most detailed sketch of Breuer's life is found in Hans Horst Meyer, "Josef Breuer," in Anton Bettelheim, ed., *Neue Österreichische Biographie* (Zurich-Leipzig-Vienna, 1928), V, 30–47. Additional useful information is found in obituaries: Sigmund Freud, in *Internationale Zeitschrift für Psychoanalysis,* **11** (1925), 255–256, translated in *The Complete Psychological Works of Sigmund Freud,* XIX, 279–280; A. de Kleyn, in *Acta Otolaryngologica,* **10** (1927), 167–171; and A. Kreidl, in *Wiener medizinische Wochenschrift* (1925), 1616–1618. The function of the labyrinth is a difficult and obscure topic. A good survey of it, as well as a clear idea of how pervasive and enduring Breuer's contributions to the subject were, can be gained from Mario Camis, *The Physiology of the Vestibular Apparatus* (Oxford, 1930).

Freud's many comments on Breuer's role in the history of psychoanalysis may easily be located in *The Complete Psychological Works of Sigmund Freud.* The evaluation of the contributions of Breuer and Freud to the founding of psychoanalysis follows that in P. F. Cranefield, "Josef Breuer's Evaluation of His Contributions to Psychoanalysis," in *International Journal of Psychoanalysis*, **39** (1958), 319–322; the same article contains an important letter from Breuer discussing his contributions to psychoanalysis. A good deal of information about Breuer and Freud is found in Ernest Jones, *The Life and Work of Sigmund Freud* (Vol. I, New York, 1953). Jones's book must be used with care, however, since its impressive quantity of information is not always matched by accuracy either of fact or of interpretation.

The most recent detailed biographical article on Breuer is E. H. Ackerknecht, "Josef Breuer," in *Neue Österreichische Biographie ab 1815* (Vienna-Munich-Zurich, 1963), XV, 126–130. Other recent articles include J. E. Gedo et al., "Studies on Hysteria; a Methodological Evaluation," in *Journal of the American Psychoanalytic Association*, **12** (1964), 734–751; N. Roth, "The Place of Josef Breuer in Medical History," in *Comprehensive Psychiatry*, **5** (1964), 322–326; N. Schlessinger et al., "The Scientific Style of Breuer and Freud in the Origins of Psychoanalysis," in *Journal of the American Psychoanalytic Association*, **15** (1967), 404–422; and J. Sullivan, "From Breuer to Freud," in *Psychoanalysis and the Psychoanalytic Review*, **46** (1959), 69–90. The Gedo and Schlessinger articles attempt to evaluate Breuer's scientific and "cognitive" style via a consideration of what little is known of his life and via an analysis of a few of his publications. In these articles one finds a rather mechanical analysis of psychological propositions in terms of their "remoteness from concrete clinical data." The authors' remark that Breuer's work was "limited by deficient scientific reality testing" suffices to show the dangers of the use of unsophisticated ideas about the nature of scientific reasoning and creativity.

For help in obtaining previously unpublished information contained in the above biography of Breuer, I am deeply indebted to the late Dr. Walter Federn, to Professor Erwin H. Ackerknecht, to Professor Erna Lesky, to Dr. Kurt Eissler and to Breuer's granddaughter, Mrs. Felix Ungar.

PAUL F. CRANEFIELD

BREUIL, HENRI ÉDOUARD PROSPER (*b.* Mortain, Manche, France, 28 February 1877; *d.* L'Isle-Adam, Seine-et-Oise, France, 14 August 1961), *prehistory*.

Breuil, the son of farmers, entered the Séminaire St.-Sulpice, Paris, in 1897 and was ordained a priest in 1900. From an early age he displayed a great interest in natural history, particularly geology and human paleontology. He was lecturer in prehistory and ethnography at the University of Fribourg from 1905 to 1910; professor of prehistoric ethnography at the Institut de Paléontologie Humaine, Paris, from 1910; and professor of prehistory at the Collège de France from 1929 to 1947. He was elected a member of the Institut de France in 1938, was a gold medalist of the American Academy of Science and of the Society of Antiquaries of London, and was awarded the Huxley Memorial Medal and the Prestwich Medal for Geology. He was a member of nineteen foreign societies and academies and received honorary degrees from Oxford, Cambridge, Edinburgh, Cape Town, Lisbon, and Fribourg.

Breuil did original research on the Paleolithic period in Europe, China, and South Africa, and was for years the doyen of Paleolithic studies. His first contact with this field was through Émile Cartailhac, professor of prehistoric archaeology at Toulouse, and he was present at the field meetings in the Dordogne in 1901 when Les Combarelles and Font de Gaume were discovered. He was also at La Mouthe when the authenticity of Paleolithic cave art was accepted. He journeyed to Altamira with Cartailhac the following year. From then on, one of his main contributions to the development of archaeology was his painstaking recording and analysis of Paleolithic cave art. He was closely associated with the discovery of Tuc d'Audoubert in 1912 and Les Trois-Frères in 1916, and was the first archaeologist to visit and describe Lascaux in 1940.

Breuil's other main contribution to prehistoric archaeology was his reclassification of Paleolithic industries, which began with his classic paper "Les subdivisions du paléolithique supérieur et leur signification," given at the Geneva Congress of Prehistoric and Protohistoric Sciences in 1912. He was not so successful when he strayed away from his Paleolithic studies to write about the art of the megalith builders of France and Iberia, and toward the end of his life he became involved in controversies about the interpretation of paintings in South Africa and the authenticity of those found at Rouffignac, France, in 1956.

BIBLIOGRAPHY

I. ORIGINAL WORKS. Breuil's writings include *La caverne d'Altamira, à Santillane près Santander,* written with Émile Cartailhac (Monaco, 1906); *La caverne de Font-de-Gaume aux Eyzies* (*Dordogne*), written with L. Capitan and D. Peyrony (Monaco, 1910); *Les Combarelles aux Eyzies* (*Dordogne*), written with L. Capitan and D. Peyrony (Paris, 1924); *Rock Paintings of Southern Andalusia*, written with M. C. Burkitt and Montagu Pollock (Oxford, 1929); *Les peintures rupestres schématiques de la péninsule ibérique,* 4 vols. (Paris, 1933–1935); *Beyond the Bounds of History* (London, 1949); *Les hommes de la pierre ancienne,* written with R. Lantier (Paris, 1951); *Quatre cents siècles d'art pariétal; les cavernes ornées de l'âge du renne* (Montignac,

1952); and *The White Lady of Brandberg*, written with Mary Boyle and E. R. Scherz (London, 1955).

II. SECONDARY LITERATURE. *Hommage à l'abbé Henri Breuil pour son quatre-vingtième anniversaire* (Paris, 1957) includes a complete bibliography of Breuil's writings. See also Mary Boyle et al., "Recollections of the Abbé Breuil," in *Antiquity*, **12** (1963); A. H. Brodrick, *The Abbé Breuil, Prehistorian* (London, 1963); and N. Skrotzky, *L'Abbé Breuil* (Paris, 1964).

GLYN DANIEL

BREWSTER, DAVID (*b.* Jedburgh, Scotland, 11 December 1781; *d.* Allerly, Melrose, Scotland, 10 February 1868), *optics.*

The son of Margaret Key and James Brewster, rector of the Jedburgh grammar school, David entered the University of Edinburgh in 1794. Although he completed the prescribed courses, like other students of the time, he did not take a bachelor's degree. Continuing at the university as a divinity student, he was awarded an honorary M.A. in 1800, and in 1804 he was licensed to preach in the Church of Scotland, although he was never ordained a minister. As an evangelical, Brewster became an adviser to the leaders of the Disruption; was long a friend of its leader, Thomas Chalmers; and became a member of the Free Church of Scotland in 1843.

His formal training, however, was but one part of his education. As amanuensis to Dr. Thomas Somerville, scholar, author, and minister of Jedburgh, Brewster acquired in his youth the writing and editing skills that later were his principal source of income. More important, as a child Brewster began to learn about physical science from his father's manuscript notes from the University of Aberdeen. Encouraged and assisted by the "peasant astronomer" James Veitch, Brewster built sundials, microscopes, and telescopes. While studying at the university, he continued to build instruments and to exchange astronomical observations with Veitch. Stimulated by his classmate Henry Brougham, Brewster began his experimental researches on light about 1798. While he believed that he had disproved part of Newton's explanation of the "inflexion" of light, Brewster did not then, or ever, abandon the emission theory of light.

Brewster's income depended on his literary, rather than his scientific, efforts. He was a private tutor from 1799 to 1807; he edited the *Edinburgh Magazine* and *Scots Magazine* from 1802 to 1806, the *Edinburgh Encyclopaedia* from 1807 to 1830, and various scientific journals from 1819 to the end of his life; and he was author of numerous popular books and articles. Throughout his life he lost no opportunity to deplore the lack of paid careers for British scientists, and his repeated failure to obtain a professorship only increased his calls for reform. Moreover, he aggressively promoted scientific education for all groups in society, in the hope that wider knowledge of science would lead to its increased prestige. A leader in the establishment of the Edinburgh School of Arts (1821), the Royal Scottish Society of Arts (1821), and the British Association for the Advancement of Science (1831), Brewster attempted to create organizations, as well as opinions, that would diffuse and promote science and education. Yet with all his effort, he failed in his great plan to establish a national institute patterned on the ideas of Bacon and Newton and on the practice of France.

Brewster was a reform Whig. His political friends, including Brougham, Charles Grant (Lord Glenelg), and the earl of Buchan (who recommended Brewster for the LL.D. at Aberdeen), were sources of influence and honor. Not only did Brougham, as lord chancellor, assist in founding the British Association, but he also obtained pensions and knighthoods for Brewster (1832) and other scientists. Through political influence, two of Brewster's sons became East Indian officials, and in 1838 Brewster was relieved of financial worries by the government's gift of the principalship of the United Colleges of St. Salvator and St. Leonard, St. Andrews. In 1859 Brewster was elected principal of the University of Edinburgh by the town council. Shortly thereafter he also became vice-chancellor.

Brewster received numerous honors, including the LL.D. from Aberdeen (1807) and the D.C.L. from Oxford (1832), fellowships of the Royal Societies of Edinburgh and of London, and a foreign associateship in the French Institute (1848). His awards included the Copley, Rumford, and Royal medals of the Royal Society of London and the Keith Prize of the Royal Society of Edinburgh.

Brewster's interest in astronomy and instruments is evident in his first major scientific publication, *A Treatise on New Philosophical Instruments* (1813). In the latter sections of the book, he reported his determinations of the refractive and dispersive powers of nearly two hundred substances that he had made in a quest for improvement of achromatic telescopes. Not until much of this work was completed did he learn, in 1811 or 1812, that in 1808 Malus had discovered that reflected light acquired the same polarization as one of the doubly refracted beams in Iceland spar. This information, combined with his discoveries that doubly refracting bodies have two dispersive powers, that the single beam transmitted by agate is polarized, and that noncrystallized bodies such as mica "depolarise" light, shifted Brewster's concern from instruments back to optical theory. In exploring the consequences of these experiments, Brewster followed four separate but related lines of research.

Since Brewster had found that light was partially polarized by oblique refraction in mica, he attempted to determine the law of this polarization in the simpler case of successive refractions by a pile of thin glass plates. By the end of 1813 he had concluded that "the number of plates in any parcel, multiplied by the tangent of the angle, at which it [completely] polarises light, is a constant quantity." More important, since "the pencil of light polarised by transmission [comports] itself, in every respect like one of the pencils formed by double refraction," study of the physical optics of transparent bodies ought to enable philosophers "to unfold the secrets of double refraction, to explain the forms and structure of crystallised bodies, and to develope the nature and properties of that etherial matter, which . . . performs . . . a capital part in the operations of the material world." Nothing less was at stake than understanding of the structure of organized matter and the nature of light. In conceiving of his results in this way, Brewster defined much of his optical career for the next twenty years.

Brewster's second line of study was a search for the law of polarization by reflection. While Malus had concluded that the "polarising angle neither follows the order of the refractive powers nor that of the dispersive forces," Brewster was not convinced. His own "measures for *water* and the *precious stones* afforded a surprising coincidence between the indices of refraction and the polarising angles; but the results for glass formed an exception, and resisted every method of classification." Persisting, however, he concluded that chemical changes on the surface of the glass had obscured the general law that "the index of refraction is the tangent of the angle of polarisation."

Moreover, analogously to polarization by successive refractions, successive reflections at any angle continuously increased the quantity of polarized light in the beam. Now making explicit his adoption of an emission theory of light, Brewster explained his results in terms of "polarising forces" that acted on light particles "in every state of POSITIVE and NEGATIVE polarisation from particles completely polarised to particles not polarised at all." Each successive reflection brought the particles nearer to complete polarization. By late 1829 he was convinced not only that his particulate theory was the simplest possible, but also that it was fully adequate to account quantitatively for the intensities and resultant angles of polarization of reflected and refracted light. He regarded "all the various phenomena of polarisation of light by reflexion and refraction as brought under the dominion of laws as well determined as those which regulate the motions of the planets." In 1816 he received the Copley Medal, in 1819 two Rumford Medals, and in 1831 a Royal Medal for the papers in which he announced these discoveries. On the popular level, Brewster's reputation was established in 1816 by the fad for his kaleidoscope. Its invention was a direct result of his studies of the theory of polarization by multiple reflections.

Study of metallic reflection was a third line of Brewster's research. Using successive reflections to increase the degree of polarization, he concluded that light reflected by metals was neither plane nor circularly—but elliptically—polarized. Moreover, from his results he deduced laws that not only accurately predicted the quantities and angles of polarization of light, but also were the foundation for theoretical researches on metallic reflection by MacCullagh and Cauchy.

Pursuing his fourth line of research, Brewster created the new fields of optical mineralogy and photoelasticity. In 1813, while studying the "depolarising" action of topaz, he observed two sets of elliptical rings (interference patterns) centered on axes in the topaz that were apparently inclined at 65°. He interpreted this to mean that topaz must have two axes, not one, of double refraction, an entirely unexpected result. After many laborious experiments he was able in 1819 to group all but a few of hundreds of minerals and crystals into mutually consistent optical and mineralogical categories: the primitive form determined the number of axes of double refraction.

During these investigations, Brewster quite unexpectedly observed that heat and pressure could produce or change a doubly refracting structure in uncrystallized, crystallized, or organic bodies. Moreover, from the geometry of the interference patterns he deduced equations that permitted him to predict the shapes, numbers, and colors of patterns that would be produced by changes in configuration, temperature, pressure, and method of observation.

In an attempt to improve colored eyeglasses and microscopy, in 1821 Brewster began an intensive study of absorption spectroscopy. Ironically, these researches led Newton's biographer to a profound dissent from Newton's doctrine of colors and to a strong reaffirmation of a "Newtonian" emission theory of light.

In examining a blue eyeglass spectroscopically, he concluded that it caused extreme eye fatigue by transmitting only red and blue rays. Their differing refrangibilities prevented the eye from accommodating to one focal distance. Less than a year later, in utilizing his new monochromatic microscope illuminator in studies of other absorption spectra, Brewster entered the debate over the number of colors in the spectrum: was it seven, as Newton had held, or four, as Wollaston believed, or some other number? Since Wollaston had asserted that yellow was merely a combination of green and red, Brewster first examined the solar spectrum with red- and green-absorbing glasses. Rather than the reds and greens vanishing, "the space from which the colours were absorbed was in both cases occupied by *yellow* light. . . ." While this established, for Brewster, the separate existence of yellow, it invalidated Newton's identification of color and refrangibility, for "*Yellow* light . . . has its *most* refrangible rays mixed with *green* light of equal refrangibility, and its *least* refrangible rays mixed with *red* light of equal refrangibility." Further experiments in which he examined a salted candle flame with these filters led Brewster to the startling conclusion that, while yellow light had an independent existence, "the prism is incapable of decomposing that part of the spectrum which [yellow] occupies." By 1831 he had extended this interpretation to the entire spectrum. However formed, it "consists of *three* spectra of *equal length, beginning and terminating at the same points,* viz. a *red* spectrum, a *yellow* spectrum, and a *blue* spectrum." Moreover, a certain amount of undecomposable white light exists at every point in the spectrum.

This line of research led to Brewster's most effective defense of an emission theory of light. In an attempt to establish techniques for optical chemical analysis, he turned to a detailed examination of the action on the spectrum of plant juices, gases, and the earth's atmosphere. Not only did he succeed in identifying bodies by their characteristic dark lines, but he also added some 1,600 dark lines to Fraunhofer's 354. Impressed by the extremely selective absorption of light by "nitrous acid gas," he concluded that he could "form no conception of a simple elastic medium so modified by the particles of the body which contains it, as to make such an extraordinary selection of the undulations which it stops or transmits."

After the 1830's Brewster directed his attention to such subjects as photography, stereoscopy, and the physiology of vision. At the same time he began to emphasize his writing rather than his editing. His biographies of Newton, Galileo, Tycho Brahe, and Kepler; his numerous articles for the *Encyclopaedia Britannica*; and his hundred or more major essay reviews were, for the most part, written after 1830. His time for research was limited, and he had largely achieved his original research goals. Also, optics was increasingly dominated by an unwelcome theory, the undulatory theory of light.

Brewster never wholeheartedly accepted the undulatory theory. His most honored papers had either been based on the emission theory or had been directed to its defense. Using that theory, he had derived mathematical laws that successfully explained and predicted phenomena. At the same time he frankly admitted his admiration for the "singular power of [the undulatory theory] to explain some of the most perplexing phaenomena of optics. . . ." However, "the power of a theory . . . to explain and predict facts is by no means a test of its truth. . . ." But more important for Brewster, the undulatory theory, based on a hypothetical ether that he could not conceive of in physical terms, and that in principle never could be observed, was fatally "defective as a *physical* representation of the phaenomena of light. . . ."

While Brewster repeatedly asserted the value of theory and was a competent, if not brilliant, mathematician, he was above all an experimenter. And the experimenter, not the theoretician, could achieve "true" knowledge. Most important, however, Brewster believed that the undulatory theory, a mere speculation, had been raised to the level of an assumed Truth. As a devout evangelical Presbyterian who believed in the unity of truth, he felt that such unbridled speculation in physics had profoundly serious implications for religion. To him, "Speculation engenders doubt, and doubt is frequently the parent either of apathy or impiety." To accept the undulatory theory of light would have required Brewster to abandon his deepest convictions about man's ability to know the world and man's duty to serve God.

BIBLIOGRAPHY

I. ORIGINAL WORKS. There is no collected edition of Brewster's works, nor is there a bibliography. An appendix in the *Home Life* (see below) reproduces the bibliography of his articles given in the Royal Society's *Catalogue of Scientific Papers.* It is very incomplete and has some inaccuracies. Many other articles by him of varying importance are printed in the various periodicals that he edited, especially the *Edinburgh Philosophical Journal* and the *Edinburgh Journal of Science.* Most of his anonymous, and very revealing, essay reviews are listed in Walter E. Houghton, ed., *The Wellesley Index to Victorian Periodicals, 1824–1900,* Vol. I (Toronto, 1966). Brewster's own collection of papers was accidentally burned early in the twentieth century, but important manuscript collections of his letters exist at the British Museum (Charles Babbage and MacVey Napier papers), University College, London (Henry, Lord Brougham papers), the National Library of Scotland, and the Royal Society, London.

Information in this article is based upon the following of Brewster's own books and articles: *A Treatise on New Philosophical Instruments, for Various Purposes in the Arts and Sciences. With Experiments on Light and Colours* (Edinburgh-London, 1813); "On some properties of Light," in *Philosophical Transactions of the Royal Society,* **103** (1813), 101–109; "On the affections of Light transmitted through crystallized bodies," *ibid.,* **104** (1814), 187–218; "On the Polarisation of Light by oblique transmission through all Bodies, whether crystallized or uncrystallized," *ibid.,* 219–230; "Results of some recent experiments on the properties impressed upon Light by the action of Glass raised to different temperatures, and cooled under different circumstances," *ibid.,* 436–439; "Experiments on the depolarisation of light as exhibited by various mineral, animal, and vegetable bodies, with a reference of the phenomena to the general principles of polarisation," *ibid.,* **105** (1815), 29–53; "On the effects of simple pressure in producing that species of crystallization which forms two oppositely polarised images, and exhibits the complementary colours by polarised light," *ibid.,* 60–64; "On the laws which regulate the polarisation of Light by reflexion from transparent bodies," *ibid.,* 125–159; "On new properties of heat as exhibited in its propagation along glass plates," *ibid.,* **106** (1816), 46–114; "On the communication of the structure of doubly-refracting crystals to glass, muriate of soda, fluor spar, and other substances by mechanical compression and dilation," *ibid.,* 156–178; "On the Effects of Compression and Dilation in altering the Polarising Structure of Doubly Refracting Crystals," in *Transactions of the Royal Society of Edinburgh,* **8** (1818), 281–286; "On the Laws which regulate the Distribution of the Polarising Force in Plates, Tubes, and Cylinders of Glass that have received the Polarising Structure," *ibid.,* 353–372; "On the laws of Polarisation and double refraction in regularly crystallised bodies," in *Philosophical Transactions of the Royal Society,* **108** (1818), 199–273; "On the connection between the primitive forms of Crystals, and the number of their Axes of double refraction," in *Memoirs* of the Edinburgh Wernerian Society, **3** (1817–1820), 50–74; *A Treatise on the Kaleidoscope* (Edinburgh, 1819); "Observations on Vision through Coloured Glasses, and on their application to Telescopes and Microscopes of great magnitude," in *Edinburgh Philosophical Journal,* **6** (1822), 102–107; "Description of a Monochromatic Lamp for Microscopical purposes, &c. with Remarks of the Absorption of the Prismatic Rays by coloured Media," in *Transactions of the Royal Society of Edinburgh,* **9** (1823), 433–444; "On the production of regular double refraction in the molecules of bodies by simple pressure; with observations on the origin of the doubly refracting structure," in *Philosophical Transactions of the Royal Society,* **120** (1830), 87–96; "On the law of the partial polarisation of light by reflexion," *ibid.,* 69–84; "On the laws of the polarisation of light by refraction," *ibid.,* 133–144; "On the Phenomena and Laws of Elliptic Polarisation, as exhibited in the Action of Metals upon Light," *ibid.,* 287–326; "Decline of Science in England and Patent laws," in *Quarterly Review,* **43** (Oct. 1830), 305–342; "Observations on the Decline of Science in England," in *Edinburgh Journal of Science,* **5** (July 1831), 1–16; "On a New Analysis of Solar Light, indicating three Primary Colours, forming Coincident Spectra of equal length," in *Transactions of the Royal Society of Edinburgh,* **12** (1834), 123–136; "Report on the recent Progress of

Optics," in *Report of the British Association for the Advancement of Science* (1831–1832), pp. 308–322; "Observations of the Absorption of Specific Rays, in reference to the Undulatory Theory of Light," in *London, Edinburgh, and Dublin Philosophical Magazine and Journal of Science,* **2** (1833), 360–363; "Life and Correspondence of Sir James Edward Smith," in *Edinburgh Review,* **57** (April 1833), 39–69 (the quotation on "Speculation" is on p. 41); "Observations on the Lines of the Solar Spectrum, and on those produced by the Earth's Atmosphere, and by the action of Nitrous Acid Gas," in *Transactions of the Royal Society of Edinburgh,* **12** (1834), 519–530; "On the Colours of Natural Bodies," *ibid.,* 538–545; "Arago *Éloge historique de Baron Fourier,*" in *North British Review,* **4** (Feb. 1846), 380–412 (contains his plan for a national institute, particularly pp. 410–412); and *Memoirs of the Life, Writings, and Discoveries of Sir Isaac Newton,* 2 vols. (Edinburgh, 1855). The *Memoirs . . . of Sir Isaac Newton* has been reprinted in 2 volumes (New York, 1965) with a useful introduction by Richard S. Westfall that discusses both Brewster's life and his interpretation of Newton.

II. SECONDARY LITERATURE. Very little contemporary secondary material on Brewster exists, nor is he adequately treated in the standard histories of optics. The only biography that exists is Margaret Maria Gordon (Brewster), *The Home Life of Sir David Brewster,* 2nd ed., rev. (Edinburgh, 1870). While it is generally reliable, it is frankly directed to the "unscientific" reader. It is the source for most of his obituaries. Brewster himself particularly recommended, as very full and complete, the article "Brewster, David," in the *Biographie universelle et portative des contemporaines,* V (Paris, 1836), 77–81. It has biographical information not included in the *Home Life.*

The standard, although very one-sided, histories of the emission-undulation controversy are Ernst Mach, *The Principles of Physical Optics, an Historical and Philosophical Treatment* (New York, 1953); Vasco Ronchi, *Histoire de la lumière,* translated by Juliette Taton (Paris, 1956); and Sir Edmund Whittaker, *A History of the Theories of Aether and Electricity,* Vol. I, *The Classical Theories* (New York, 1960). Brewster's work in optical mineralogy is briefly treated in John G. Burke, *Origins of the Science of Crystals* (Berkeley, 1965). A very useful discussion of Brewster's optical theories is contained in Henry Steffens, "The Development of Newtonian Optics in England, 1738–1831" (unpublished Master's dissertation, Cornell University, 1965).

EDGAR W. MORSE

BRIANCHON, CHARLES-JULIEN (*b.* Sèvres, France, 19 December 1783; *d.* Versailles, France, 29 April 1864), *mathematics.*

There appears to be no record of Brianchon's early years. He entered the École Polytechnique in 1804 and was a pupil of the noted geometer Gaspard Monge. While a student there, he published his first paper, "Sur les surfaces courbes du second degré"

(1806), which contained the famous theorem named after him.

Brianchon graduated first in his class in 1808 and became a lieutenant in artillery in the armies of Napoleon. He took part in the Peninsular campaigns, serving in Spain and Portugal, and is said to have distinguished himself both in bravery and ability. The rigors of his army service affected his health, and after the cessation of hostilities in 1813, Brianchon applied for a teaching position. He was finally appointed professor at the Artillery School of the Royal Guard in 1818.

By this time he had published several works in geometry, including "Sur les surfaces courbes du second degré" (1816), *Mémoire sur les lignes du second ordre* (1817), *Application de la théorie des transversales* (1818), and "Solution de plusieurs problèmes de géométrie" (1818).

Brianchon's teaching duties apparently affected both his output and his interests. In 1820 there appeared "Recherches sur la détermination d'une hyperbole équilatère, au moyen de quatres conditions données," written with Poncelet. It is notable for containing the nine-point circle theorem and is an instance of the many times this theorem has been rediscovered by independent investigators. At any rate, this paper contains the first complete proof of the theorem and the first use of the term "nine-point circle."

Brianchon's next publication, "Description du laboratoire de chimie de l'École d'Artillerie de la Garde Royale" (1822), indicates his change of interests. Two works appeared in 1823: "Des courbes de raccordement" and *Mémoire sur la poudre à tirer*. His last known work, *Essai chimique sur les réactions foudroyantes,* appeared in 1825. Brianchon ceased writing after 1825 and devoted all his time to teaching. Details of his personal life are singularly scarce.

Brianchon's fame rests ultimately on one theorem. In 1639 Pascal had proved that "If all the vertices of a hexagon lie on a circle, and if the opposite sides intersect, then the points of intersection lie on a line." He then boldly extended this result to a hexagon inscribed in any conic, since he recognized that his theorem was projective in nature. Oddly enough, it took 167 years before someone else—Brianchon—realized that since the theorem is projective in nature, its dual should also be true. Simply stated, Brianchon's theorem is "If all the sides of a hexagon are tangent to a conic, then the diagonals joining opposite vertices are concurrent." The theorem is useful in the study of the properties of conics and—if the hexagon is specialized in various ways—for the study of properties of pentagons, quadrilaterals, and triangles.

BIBLIOGRAPHY

Brianchon's writings are "Sur les surfaces courbes du second degré," in *Journal de l'École Polytechnique* (1806); "Sur les surfaces courbes du second degré," *ibid.* (1816); *Mémoire sur les lignes du second ordre* (Paris, 1817); *Application de la théorie des transversales* (Paris, 1818); "Solutions de plusieurs problèmes de géométrie," in *Journal de l'École Polytechnique,* **4** (1818); "Recherches sur la détermination d'une hyperbole équilatère, au moyen de quatres conditions données," *ibid.* (1820); "Description du laboratoire de chimie de l'École d'Artillerie de la Garde Royale," in *Annales de l'industrie nationale* (1822); "Des courbes de raccordement," in *Journal de l'École Polytechnique,* **12** (1823); *Mémoire sur la poudre à tirer* (Paris, 1823); and *Essai chimique sur les réactions foudroyantes* (Paris, 1825).

S. L. GREITZER

BRIDGES, CALVIN BLACKMAN (*b.* Schuyler Falls, New York, 11 January 1889; *d.* Los Angeles, California, 27 December 1938), *genetics.*

Calvin Blackman Bridges was the only child of Leonard Victor Bridges and Charlotte Amelia Blackman. His mother died when Calvin was two years old and his father a year later, so the boy was brought up by his paternal grandmother. When he was fourteen, he was sent to Plattsburg to attend high school. Because of his deficient primary school training and because he worked to help support himself, he did not graduate from high school until he was twenty. His record was good enough, however, for him to be offered scholarships at both Cornell and Columbia. He chose the latter and entered as a freshman in 1909. His record at Columbia was outstanding, and he graduated in three years in spite of largely supporting himself by outside work and, in the last half of the period, spending much of his time and energy on research with *Drosophila.* In 1912 he married Gertrude Ives. The couple had four children.

In his freshman year Bridges and the writer (then a sophomore) took the beginning course in zoology, which was given (for the only time during his twenty-four years at Columbia) by T. H. Morgan. This was the beginning of a very close association, among the three of us, which lasted until Bridges' death.

Morgan's work on the genetics of *Drosophila* began in earnest in the summer of 1910; and in the academic year 1910–1911 Bridges and I were given desks in his laboratory, a room 16 by 23 feet, which came to be known as "the fly room." Here the three of us reared *Drosophila* for the next seventeen years. A steady succession of American and foreign doctoral and postdoctoral students also had desks there. From 1915 Bridges was a research associate of the Carnegie Institution of Washington, and in 1928 he moved from Columbia to the California Institute of Technology, where he spent the rest of his life.

At the beginning of this period the techniques were unsatisfactory, and Bridges was largely responsible for their improvement. He introduced the use of binocular microscopes instead of hand lenses, he developed dependable temperature controls, and he played the largest part in the improvement and standardization of the culture bottles and media.

The working material for the group was the series of mutant types, and their detection and isolation was one of the major concerns of all of us—but Bridges was so good at this that he contributed many more mutants than did the rest of us. He also had the skill and patience required to organize this material into a coherent body of detailed information and to produce a series of carefully planned and useful stocks of combinations of mutant types. This information and many of these stocks are still basic and are in constant use by students of *Drosophila*.

In the early work there were often found a few exceptions to the usual rules for the inheritance of sex-linked genes, and Bridges undertook a study of them. He published an account in 1913, giving the phenomenon the name "nondisjunction." No satisfactory scheme emerged, until he made a microscopical study of the chromosomes (1914). Here, and especially in his doctoral thesis (1916), he produced a brilliant and characteristically detailed and convincing account that constituted a proof of the correctness of the chromosome theory of heredity.

Bridges followed this with a study of nondisjunction of the small fourth chromosome, and this led to his development of the idea of genic balance, which has played a large part in all later interpretations of the way in which genes influence development. The classic work in this field is Bridges' study of the determination of sex in *Drosophila*, which he based on his combined genetical and cytological study of the offspring of triploid females—again a brilliantly conceived and convincingly thorough piece of work.

Bridges had made himself the outstanding authority on the cytology of *Drosophila*, but the chromosomes were very small and so lacking in structural detail that they could be used effectively in the analysis of only a few of the many chromosome rearrangements that were discovered by genetic means—most of them by Bridges.

In 1933 the work of Heitz and Bauer, and of Painter, showed that the chromosomes of the salivary glands of *Drosophila* and of some other flies had a wealth of structural detail far greater than that known in the chromosomes of any other organism. Bridges threw himself into the study of these chromosomes, and produced a series of drawings of them that are still the standards of reference. This, like much of his work, required great patience, accurate observation, technical skill and ingenuity, and an understanding of what was important. In 1936 he was elected to the National Academy of Sciences.

Bridges was a friendly and generous person. Politically he was rather far to the left—a circumstance related to his visit to Russia in 1931–1932. In his personal and social relations he was a nonconformist, largely as a matter of principle.

BIBLIOGRAPHY

I. ORIGINAL WORKS. Bridges wrote about 125 scientific papers, not counting numerous notes in *Drosophila Information Service,* which he and M. Demerec compiled and edited from 1934 to 1939. The list that follows is a selection of the more important contributions. A fuller listing may be found in Morgan's biography (1941; see below).

Bridges' works include "Dilution Effects and Bicolorism in Certain Eye Colors of *Drosophila,*" in *Journal of Experimental Zoology,* **15** (1913), 429–466, written with T. H. Morgan; "Non-disjunction of the Sex Chromosomes of *Drosophila,*" in *Science,* **37** (1913), 112–113; "Direct Proof Through Non-disjunction That the Sex-linked Genes of *Drosophila* Are Borne by the X-Chromosomes," in *Science,* **40** (1914), 107–109; "A Linkage Variation in *Drosophila,*" in *Journal of Experimental Zoology,* **19** (1915), 1–21; *The Mechanism of Mendelian Heredity* (New York, 1915), written with T. H. Morgan, A. H. Sturtevant, and H. J. Muller; "Non-disjunction as Proof of the Chromosome Theory of Heredity," in *Genetics,* **1** (1916), 1–52, 107–163; *Sex-linked Inheritance in Drosophila,* Carnegie Institution of Washington publication 237 (Washington, 1916); "Deficiency," in *Genetics,* **2** (1917), 445–465; "The Constitution of the Germinal Material in Relation to Heredity," in *Carnegie Institution of Washington Year Book,* XV-XXXVIII (Washington, 1917–1939)—written with T. H. Morgan and A. H. Sturtevant through 1929 and with T. H. Morgan and J. Schultz from 1930, these annual reports give a picture of the work in progress and are often the only published accounts of ideas and experiments.

Also see "The Second-chromosome Group of Mutant Characters" and "The Origin of Gynandromorphs," Carnegie Institution of Washington publication 278 (Washington, 1919), pp. 1–304; "Specific Modifiers of Eosin Eye Color in *Drosophila melanogaster,*" in *Journal of Experimental Zoology,* **28** (1919), 337–384; "Proof of Non-disjunction of the Fourth Chromosome of *Drosophila melanogaster,*" in *Science,* **53** (1921), 308; "Triploid Intersexes in *Drosophila melanogaster,*" in *Science,* **54** (1921), 252–254; *The Third-*

chromosome Group of Mutant Characters of Drosophila melanogaster, Carnegie Institution of Washington publication 327 (Washington, 1923), written with T. H. Morgan; "The Translocation of a Section of Chromosome II on Chromosome III," in *Anatomical Record,* **24** (1923), 426–427; "Crossing Over in the X-Chromosomes of Triploid Females of *Drosophila melanogaster,*" in *Genetics,* **10** (1925), 418–441, written with E. G. Anderson; "The Genetics of *Drosophila,*" in *Bibliographia genetica,* **2** (1925), 1–262, written with T. H. Morgan and A. H. Sturtevant; "Sex in Relation to Chromosomes and Genes," in *American Naturalist,* **59** (1925), 127–137; "Some Physicochemical Aspects of Life, Mutation, and Evolution," in *Colloid Chemistry,* **2** (1928), 9–58, written with J. Alexander; "The Genetic Conception of Life," an address given before the Academy of Science, Leningrad (1931); "Specific Suppressors in *Drosophila,*" in *Proceedings of the 6th International Congress on Genetics,* II (1932), 12–14; "The Mutants and Linkage Data of Chromosome Four of *Drosophila melanogaster,*" in *Biologicheskii zhurnal* (Moscow), **4** (1935), 401–420; "Salivary Chromosome Maps—With a Key to the Banding of the Chromosomes of *Drosophila melanogaster,*" in *Journal of Heredity,* **26** (1935), 60–64; "The Bar 'Gene' a Duplication," in *Science,* **83** (1936), 210–211; "A Revised Map of the Salivary Gland X-Chromosome of *Drosophila melanogaster,*" in *Journal of Heredity,* **29** (1938), 11–13; "A New Map of the Second Chromosome. A Revised Map of the Right Limb of the Second Chromosome of *Drosophila melanogaster,*" ibid., **30** (1939), 475–476, written with P. N. Bridges; *The Mutants of Drosophila melanogaster,* Carnegie Institution of Washington publication 552 (Washington, 1944), completed and edited by K. S. Brehme.

II. SECONDARY LITERATURE. T. H. Morgan wrote four biographical accounts of Bridges: in *Science,* **89** (1939), 118–119; in *Journal of Heredity,* **30** (1939), 355–358; in *Genetics,* **25** (1940), i–v; and in *Biographical Memoirs. National Academy of Sciences,* **22** (1941), 31–48. The last three have two different photographs of Bridges, and the last has a full bibliography.

Additional biographical works are H. J. Muller, in *Nature,* **143** (1939), 191–192; J. Schultz, an unsigned article in *National Cyclopedia of American Biography,* **30** (1943), 374; A. H. Sturtevant, in *Biological Bulletin,* **79** (1940), 24. There are also numerous references in A. H. Sturtevant, *A History of Genetics* (New York, 1965).

A. H. STURTEVANT

BRIDGMAN, PERCY WILLIAMS (*b.* Cambridge, Massachusetts, 21 April 1882; *d.* Randolph, New Hampshire, 20 August 1961), *physics, philosophy of science.*

Bridgman was the only son of Raymond Landon Bridgman, a newspaper correspondent and the author of a number of books on public affairs, and Ann Maria Williams Bridgman. The family moved to Newton, Massachusetts, where Percy attended the public schools until he entered Harvard College in 1900. He graduated with a B.A., *summa cum laude,* in 1904, with rigorous training in physics and mathematics. He remained at Harvard for his M.A. (1905) and Ph.D. (1908) in physics, whereupon he was immediately appointed research fellow in the department of physics, then instructor in 1910. In 1912 Bridgman married Olive Ware. The couple had two children. Bridgman was appointed assistant professor in 1913, professor in 1919, Hollis professor of mathematics and natural philosophy in 1926, Higgins university professor in 1950, and professor emeritus in 1954.

Percy Bridgman's penetrating analytical thought and physical intuition, fertile imagination for mechanical detail, and exceptional dexterity in manipulating equipment defined a clear channel of activity into which he threw himself with untiring energy and singleness of purpose. He was an individualist of the most determined stamp, and refused to be diverted by faculty business, by the demands of society, or by any personal weakness from his main interest: his scholarly activity as experimenter, teacher, and critic of the basic concepts of physical science.

While avoiding almost all university committees, Bridgman was an active member of the American Academy of Arts and Sciences, and served on the board of editors of its journal, *Daedalus.* He was fond of music, and also pursued a number of other avocations—all with concentration and perfectionism, whether it was chess, handball, gardening, mountain climbing, or photography. But play was never allowed to interfere with the main business of his life; his unremitting activity was reflected in the high and steady output of papers on physics and philosophy of science. Bridgman wrote about six papers a year, many with such titles as "The Resistance of 72 Elements, Alloys, and Compounds to 100,000 kg/cm²." His lifetime total was over 260 papers, in addition to thirteen books that were largely the products of his summers at Randolph, New Hampshire. All of his writing is remarkably personal and often in the first person singular; whether the subject is the polymorphism of bismuth or the duties of intelligent individuals in an unintelligent society, the characteristic Bridgman tone and quality are immediately evident.

Bridgman, a man of generosity and integrity, was regarded with affection and admiration by his associates. His honors included the Rumford Medal of the American Academy of Arts and Sciences, the Cresson Medal of the Franklin Institute, the Roozeboom Medal of the Royal Academy of Sciences of Amsterdam, the Bingham Medal of the Society of Rheology, the Comstock Prize of the National Academy of Sciences, the New York Award of the Research Corporation of America, and, "for the invention of an apparatus to produce extremely high pressures, and for the discoveries he made therewith in the field of high-pressure physics," the Nobel Prize in physics for 1946. He was president of the American Physical Society in 1942, a member of the National Academy of Sciences, a fellow of the American Academy of Arts and Sciences, a foreign member of academies of science in England, Mexico, and India, and the holder of honorary degrees from six universities.

Bridgman's early papers give no explanation of the origins of his interest in high pressures. He may have been influenced by Theodore Richards, who had measured the compressibility of elements, or by Wallace Sabine, with whom he took a research course in heat and light for four years.

His first three papers, published in the *Proceedings of the American Academy of Arts and Sciences* (Vol. **44**, 1908-1909), laid the foundation for most of his later work. The maximum pressure attained—6,500 atmospheres—was not much higher than was currently used by other investigators, and was inefficiently produced with a screw compressor turned with a six-foot wrench. Bridgman's first concern appears to have been the establishment of an adequate pressure scale rather than the production of drastically higher pressures. He developed the free-piston gauge, or pressure balance, used earlier by Amagat, and introduced a more convenient secondary gauge based on the effect of pressure upon the electrical resistance of mercury (the subject of his Ph.D. thesis).

The new design of a leakproof pressure seal or "packing"—later called the "unsupported area seal," and the key to so much subsequent achievement—appears in the discussion of the free-piston gauge, with scarcely a suggestion of its importance. Indeed, Bridgman later explained (*American Scientist*, Vol. **31**, 1943) that the self-sealing feature of his first high-pressure packing was incidental to the design of a closure for the pressure vessel that could be rapidly assembled or taken apart; the basic advantages of the scheme were realized only afterward. In his brief autobiographical remarks in a questionnaire filed with the National Academy of Sciences, under the heading "Discoveries Which You Regard as Most Important," Bridgman wrote: "Doubtless the most influential single discovery was that of a method of producing high hydrostatic pressure without leak. The discovery of the method had a strong element of accident."

In principle, the construction insures that the sealing gasket, of rubber or soft metal, is restrained on the upper, or low-pressure, side of the vessel by a

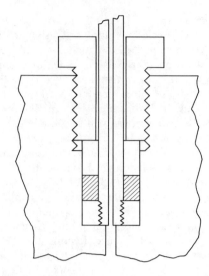

FIGURE 1. The general scheme of the packing, by which pressure in the soft packing materials is automatically maintained at a fixed percentage higher than in the liquid.

fixed surface the area of which is somewhat smaller than that acting on the other side of the packing. Hence, the latter is always compressed to a pressure higher than that to be confined inside the vessel (Fig. 1); the high pressure itself is used to tighten the packing; and the ultimate limitation becomes the strength of the metal parts. It was mainly this advance that allowed Bridgman to open up a virgin field for experimental exploration.

The third paper of the early series gives new measurements of the compressibility of steel, mercury, and glass. We recognize already the characteristic Bridgman style: the evident pleasure in the manipulations of shop and laboratory; the meticulous pursuit of the numerous corrections; the experiments with homely mixtures of mercury, molasses, glycerine, and marine glue. None of these early measurements proved to be definitive; the absolute gauge was soon improved, the mercury gauge was discarded in favor of a manganin wire gauge, and the compressibilities were revised. But his rapid succession of publications quickly transformed the field of high-pressure research.

By 1910 the equipment had been completely redesigned. The screw compressor was replaced by a hydraulic ram, and the new packing was systematically exploited. For the first time, pressures of the order of 20,000 atmospheres and more are reported. Bridgman remarks: "The magnitude of the fluid pressure mentioned here requires brief comment, because without a word of explanation it may seem so large as to cast discredit on the accuracy of all the data." The techniques to be used for the next twenty years had been substantially perfected and were described more fully in the paper "The Technique of High Pressure Experimenting" (1914).

Bridgman had the good fortune to begin his experiments at a time when metallurgical advances were providing steels of unprecedentedly high strength; his achievement of still higher pressures in the 1930's was made possible by the development of the cobalt-bonded tungsten carbides. The leakproof packing would have been of little value with Amagat's steel, but the new alloys permitted spectacular increases of the useful pressure range. Bridgman settled on an electric-furnace chrome-vanadium steel (equivalent to the present AISI 6150) for most of his pressure vessels and connecting tubes. It is not a deep-hardening steel, and in pressure vessels the size of Bridgman's, four or five inches in diameter, the interior remains relatively soft. This condition is advantageous in pressure vessels because the elastic limit is reached first in the ductile material near the bore, which can be stretched appreciably without rupture; at the same time, the expansion of the inner part transmits the load to the strong outer parts, which are inefficiently stressed so long as the whole cylinder remains in the elastic range. Thus Bridgman found that pressures far in excess of predictions based on simple elastic criteria could be contained.

The maximum fluid pressure for routine measurements of the mechanical, electrical, and thermal properties of matter was gradually raised to 30,000 atmospheres. Still higher pressures, to an estimated 400,000 atmospheres, were finally obtained in quasi-fluid systems.

At pressures above 3,000 atmospheres, Bridgman was in a realm of physical conditions new to the physicist; the instruments for measuring pressure had to be devised and calibrated, and novel methods developed for making other kinds of physical measurements. Bridgman's fifty years of concentrated effort, characterized both by the magnitude of the pressures employed and by the range of phenomena investigated, have provided a large part of all the measurements now used in this field and form the basis for most of the recent advances in high-pressure technology. His *The Physics of High Pressure* (1931) has remained the basic work in this field.

Much of Bridgman's work was done while the theory of the solid state was still in its infancy, and the interpretation of many of his measurements has become possible only in recent years. The massive treasure of data that he left has proved invaluable for the development of solid state physics. Also among Bridgman's achievements were an early method of refining by zone melting, the discovery of polymorphism of many materials at high pressures (including ice at high temperature), and a new electrical effect (internal Peltier heat) in metal crystals. His investigations had great geophysical significance, for they proved that drastic alterations in the physical properties and crystal structure of rock material must take place under the high pressures that prevail in the earth's interior. He lived to see the artificial production of many natural high-pressure mineral forms, such as diamond, coesite, and jadeite, by techniques based on his discoveries.

From the beginning, Bridgman, working alone or with his long-time machinist Charles Chase and research assistant Leonard Abbott, made much of his own apparatus. In almost any weather, he would arrive on his bicycle as soon as the workshop opened at 8:00 A.M. His papers contain many useful bits of shop lore and throw light on the amount of labor and persistence underlying his studies: "It is easy, if all precautions are observed, to drill a hole . . . seventeen inches long in from seven to eight hours." They also helped greatly in the adaptation of his techniques all over the world when, after 1945, there was a great rise of interest in experimental high-pressure work.

One may surmise that Bridgman thoroughly enjoyed the complete personal control he exercised over his equipment. The manipulations took him from pumps to measuring apparatus—usually a set of direct current electrical bridges or potentiometers requiring telescopic observation of galvanometer deflections—to the notebook in which notations were made in his private shorthand, and back to the pumps for a new cycle. All of this was performed as fast as the various thermal and pressure lags would permit, sometimes on a fixed schedule that covered several hours. Much philosophical debate has taken place over the meaning of his term *operational*, but his original meaning must have been closely related to the manifold physical activities of his laboratory, with every adjustment and every measurement dictated by his own mind and controlled by his own muscles.

This desire for full personal involvement in the experiment probably also accounted for Bridgman's reluctance to do joint research or to take on thesis students. He rarely had more than two at one time; the record shows fourteen doctoral theses on high-pressure topics, in addition to several on other subjects that he supervised. He was usually most pleased when least consulted, but was always willing to listen to interesting findings or to put his mind to real difficulties.

Bridgman's mechanical genius was reflected in the essential simplicity of his apparatus and his mastery of manipulative techniques. Whether machining a miniature mechanical part, blowing glass, preparing samples of intractable materials, drawing wires, sealing off volatile liquids, purifying chemicals, or growing simple crystals of unprecedented size, he accomplished his purpose with rapidity, a minimum of equipment, and a remarkably low budget.

During World War I, Bridgman moved with his family to New London, Connecticut, where he engaged in the development of sound-detection systems for antisubmarine warfare. He also developed an application of his high-pressure laboratory technique for the prestraining of one-piece gun barrels. Just before World War II, Bridgman's libertarian outlook and his fear of "the misuse of scientific information" caused him to close his laboratory to "citizens of any totalitarian state." After the war broke out, he undertook a series of studies for the Watertown Arsenal on the plastic flow of steel under high pressure, a consideration related to the problem of the strength of armor plate. For the Manhattan Project he measured the compressibility of uranium and plutonium.

Bridgman's lectures were at first baffling to many students. He spoke quickly, and in spurts, with little regard for clear enunciation. Nevertheless, his basic lucidity of thought and his way of coming to grips with the subject forced students to think deeply for themselves. This was reinforced by problem sets of legendary brevity and difficulty.

One of Bridgman's early teaching assignments—giving two advanced courses in electrodynamics, suddenly thrust upon him in 1914 by the death of B. O. Peirce—turned out to be the genesis of his active interest in philosophy of science. Years later he commented on the obscurity of the underlying conceptual situation that he found in electrodynamics, and the intellectual distress that it caused him. His efforts to meet the logical problems in this area led to a critical examination of the logical structure of physics, until he could say, "I was able to think the situation through to my own satisfaction."

Bridgman's first publication in this area (1916) dealt with dimensional analysis, and he returned to this subject in his first book, *Dimensional Analysis* (1922). It was the first systematic and critical exposition of the principles involved and was characterized by a rigorous analysis of the mental operations involved in dimensional reasoning; the analysis was based on his demand that the equations of physics be given unambiguous meanings by interpreting the letter symbols for the different physical quantities as placeholders for the numbers that form the measure-values of the physical quantities, rather than as place-holders for "physical quantities" formed by multiplying each measure-value by the corresponding physical unit.

Bridgman's success in thinking his way through the confusions of dimensional analysis encouraged him to turn his attention to the larger task of eliminating similar confusions in the broader field of physics proper. He was deeply impressed by Einstein's demonstration of the meaninglessness of the conception of absolute simultaneity between events at different places, noting that the proof involved an analysis of the operations of synchronization of clocks at different locations. That the basic concepts of time and space had been seriously misconceived, that sloppy thinking and uncritical usage of language were revealed at the core of physics, seemed to Bridgman to call for a critical reexamination of the conceptual structure of physics as a whole. In order to circumvent the word traps of ordinary speech, he proposed to use physical and mental operations as the measure of meaning.

The resulting "operational" point of view was brilliantly argued, in simple but stark statements, in *The Logic of Modern Physics* (1927): "In general, we mean by any concept nothing more than a set of operations; the concept is synonymous with the corresponding set of operations. . . . If a specific question has meaning, it must be possible to find operations by which answers may be given to it. It will be found in many cases that the operations cannot exist, and the question therefore has no meaning."

This volume was of immense value to a generation of scientists then facing the apparent paradoxes of a new world of atoms and quanta that flatly refused to follow the rules of common sense. In due course *The Logic of Modern Physics* was followed by other books and papers extending and deepening Bridgman's critical examination of the concepts and theories of physics. His Princeton University lectures were published as *The Nature of Physical Theory* (1936). *The Nature of Thermodynamics* followed (1941), and *A Sophisticate's Primer of Relativity* appeared posthumously (1962). In these studies Bridgman drew attention away from the apparent precision of the mathematical equations of physics and the seemingly rigorous logic of axiomatically constructed theories, and turned it to the matrix of crude observations and approximate verbal explanations from which the symbols and equations derive their significance. His relentless probing exposed a surprising penumbra of uncertainty regarding the interpretation of the symbols in thermodynamics (for example, in different physical situations) and the limits of applicability of its concepts. Through these books, and through papers on the application of his ideas to other areas of science and even to the social sciences, Bridgman's influence spread far beyond the field of physics.

His philosophic point of view is usually classified with the positivism of Stallo, Mach, Charles Peirce, William James, and the Vienna Circle; but it derived from his own experience and maintained its individual line.

Bridgman's influence was strongest among scientists, who found his point of view congenial; and much of what he had to say is commonly accepted among them today. His philosophic writing was, to be sure, always iconoclastic and stimulated a good deal of controversy, especially among some philosophers of science. Operational analysis, he had to explain, was proposed as an aid to clear thinking, and not as a solution for all the problems of philosophy. Again and again he expressed his dislike of the word "operationalism," with its implication of an associated dogma. Similarly, he gave repeated evidence in operational terms of the seriousness with which he advocated almost ruthless intellectual integrity. One case in point was the publication in 1959 of his own appraisal of *The Logic of Modern Physics.*

A final affirmation of this ideal is to be found in the circumstances of his death. In the essay "The Struggle for Intellectual Integrity" (1933), he had dealt with the choice of death if one's probable future is irreversibly and predominantly painful. This choice came to him with suddenness in his eightieth year, and he was not a man to think one thing and do another. After careful diagnosis by undoubted authorities, he found that, in his own words,

> . . . the disease [Paget's disease] has run its normal course, and has now turned into a well-developed cancer for which apparently nothing can be done. . . . In the meantime there is considerable pain, and the doctors here do not offer much prospect that it can be made better. . . . I would like to take advantage of the situation in which I find myself to establish a general principle, namely, that when the ultimate end is as inevitable as it now appears to be, the individual has a right to ask his doctor to end it for him.

Unable to make such arrangements, and finding that his limbs were rapidly losing mobility, Bridgman felt obliged to take action himself. He left behind a two-sentence note:

> It isn't decent for Society to make a man do this thing himself. Probably this is the last day I will be able to do it myself. P.W.B.

A day after his death, Harvard University Press received one of the last things Bridgman must have written, the index for the collection of his complete scientific papers. His ashes were buried in the garden of his beloved summer home.

BIBLIOGRAPHY

I. Original Works. Among Bridgman's writings are *Dimensional Analysis* (New Haven, 1922; rev. ed., 1931); *A Condensed Collection of Thermodynamic Formulas* (Cambridge, Mass., 1925); *The Logic of Modern Physics* (New York, 1927); *The Physics of High Pressure* (New York, 1931; new impression with suppl., London, 1949); *The Thermodynamics of Electrical Phenomena in Metals* (New York, 1934); *The Nature of Physical Theory* (Princeton, 1936); *The Intelligent Individual and Society* (New York, 1938); *The Nature of Thermodynamics* (Cambridge, Mass., 1941); *Reflections of a Physicist* (New York, 1950, 1955); *The Nature of Some of Our Physical Concepts* (New York, 1952); *Studies in Large Plastic Flow and Fracture, With Special Emphasis on the Effects of Hydrostatic Pressure* (New York, 1952); *The Way Things Are* (Cambridge, Mass., 1959); *The Thermodynamics of Electrical Phenomena in Metals and a Condensed Collection of Thermodynamic Formulas,* rev. ed. (New York, 1961); *A Sophisticate's Primer of Relativity* (Middletown, Conn., 1962); and *Collected Experimental Papers of P. W. Bridgman*, 7 vols. (Cambridge, Mass., 1964).

Bridgman's books on philosophy of science and his *Collected Experimental Papers* contain reprints of almost all his published papers. A fairly complete listing of individual titles is also contained in the obituary note of the Royal Society of London (1962). Further biographical details are in a booklet of essays presented at the memorial meeting at Harvard University, 24 October 1961, and in an obituary volume of the National Academy of Sciences (in press). Much of the material in this article was drawn from these sources and Birch et al. Bridgman's documentary *Nachlass* is largely at the Harvard University Archives; some materials (e.g., scientific data and laboratory books) are kept in Lyman Laboratory, Harvard University, and at the Center for History and Philosophy of Science, American Institute of Physics, New York City. Much of his equipment is still in use in research laboratories at Harvard, but some items are at the Smithsonian Institution, Washington, D.C.

II. Secondary Literature. Among the works on Bridgman are Francis Birch, Roger Hickman, Gerald Holton, and Edwin C. Kemble, "Percy Williams Bridgman," in Faculty of Arts and Sciences, *Harvard University Gazette* (31 March 1962); and Philipp Frank, *The Validation of Scientific Theories* (Boston, 1956), ch. 2.

Edwin C. Kemble
Francis Birch
Gerald Holton

BRIGGS, HENRY (*b.* Warleywood, Yorkshire, England, February 1561; *d.* Oxford, England, 26 January 1630), *mathematics.*

Although J. Mede of Christ's College, Cambridge, wrote on 6 February 1630, "Mr. Henry Briggs of Oxford, the great mathematician, is lately dead, at

74 years of age," implying thereby that Briggs was born about 1556, it seems that he was in error. The Halifax parish register gives the 1561 date.

After a local grammar schooling in Greek and Latin, Briggs went to St. John's College, Cambridge, about 1577, and was admitted as a scholar on 5 November 1579. He received the B.A. in 1581 and the M.A. in 1585, became examiner and lecturer in mathematics in 1592, and soon afterward was appointed Dr. Linacre's reader of the physic (medicine) lecture. He had been elected fellow of his college in 1589.

Early in 1596 Briggs became the first professor of geometry at the newly founded Gresham College in London. He first worked on navigation and composed a table for the finding of the height of the pole, the magnetic declination being given. By 1609 he was in correspondence with James Ussher, later the famous archbishop of Armagh; from one of Briggs's letters we learn that he was studying eclipses in 1610. By 10 March 1615, however, he was entirely engaged in the study of logarithms, the subject for which he is renowned: "Neper, lord of Markinston, hath set my head and hands a work with his new and admirable logarithms. I hope to see him this summer, if it please God, for I never saw book, which pleased me better, and made me more wonder."

Briggs at once applied his energies to the advancement of logarithms and to lecturing on them at Gresham College. He soon proposed a modification of the scale of logarithms from Napier's hyperbolic form, a change that Napier discussed with Briggs, who went to Edinburgh for a month's visit after completing his lectures in the summer of 1616. One result of these exchanges was that Briggs saw E. Wright's translation of Napier's *Canon mirificus* through the press, Wright having died. To the work Briggs added a preface and some material of his own—"A description of an instrument table to find the part proportional, devised by Mr. Edward Wright" (1616).

Briggs's *Logarithmorum chilias prima* is dated 1617; in the preface, which mentions the recent death of Napier, the change from the hyperbolic form of logarithms is justified and the publication of Napier's *Rhabdologia* foretold. That work duly appeared in 1619, with comments by Briggs himself on the new form of logarithms and on the solution of spherical triangles.

The parts taken by Napier and Briggs in developing logarithms were described by the latter in his *Arithmetica logarithmica* (1624). The proposals there recorded do not yield common logarithms: for if R is the radius, Briggs suggested that $\log R = 0$ and $\log R/10 = 10^{10}$. Napier, having abandoned the hyper-

bolic form in which

$$\text{Nap. log } y = 10^7 \log_e \frac{10^7}{y},$$

proposed an improvement whereby $\log 1 = 0$ and $\log R = 10^{10}$. Later, Briggs replaced $\log R = 10^{10}$ with $\log 10 = 1$. Brigg's key words are:

> I myself, when expounding this doctrine publicly in London to my auditors in Gresham College, remarked that it would be much more convenient that 0 should be kept for the logarithm of the whole sine (as in the *Canon Mirificus*), but that the logarithm of the tenth part of the same whole sine, that is to say 5 degrees 44 minutes and 21 seconds should be 10,000,000,000. And concerning that matter I wrote immediately to the author himself.

Later, however, in Edinburgh, Napier suggested to Briggs "that 0 should be the logarithm of unity and 10,000,000,000 that of the whole sine; which I could not but admit," says Briggs, "was by far the most convenient."

Briggs's edition of Euclid's *Elements* (Books I–VI), printed without the editor's name, was published in London in 1620. In the previous year Sir Henry Saville had invited Briggs to become professor of geometry at Oxford, where he took up his duties at Merton College in January 1620. In his last lecture, Saville introduced Briggs with the words, "Trado lampadem successori meo, doctissimo viro, qui vos ad intima geometriae mysteria perducet." Tactfully Briggs began his lecture course where Saville had left off, at the ninth proposition of Euclid.

His next achievement was the *Arithmetica logarithmica*, which included thirty thousand logarithms, those from 1 to 20,000 and those from 90,000 to 100,000. The work contains a dissertation on the nature and use of logarithms and proposes a scheme for dividing among several hands the calculation of the intermediate numbers from 20,000 to 90,000. Briggs even offered to supply paper specially divided into columns for the purpose. Chapters 12 and 13 of the introduction explain the principles of the method of constructing logarithms by interpolation from differences, an interesting forerunner of the *Canonotechnia* of Roger Cotes. A second edition of the *Arithmetica,* completed by Adrian Vlacq (or Flack), contained the intermediate seventy chiliads and appeared in 1628.

Vlacq also printed Briggs's tables of logarithmic sines and tangents. The responsibility for seeing this work through the press was entrusted by Briggs, when dying, to his friend Henry Gellibrand, then professor of astronomy at Gresham College, who added a preface explaining the application of logarithms to plane and spherical trigonometry. The work was published in 1633 as *Trigonometria Britannica sive de doctrina triangulorum.*

Briggs was an amiable man, much liked by his contemporaries. Unlike Napier, he scorned astrology, thinking it to be "a system of groundless conceits." His last years were spent at Merton College, Oxford, where he died. Some Greek elegiacs were written for him by his Merton colleague Henry Jacob; they end with the statement that not even death has put a stop to his skill, for his soul still astronomizes while his body measures the earth. Oughtred called him "the mirrour of the age for excellent skill in geometry," and Isaac Barrow expressed in his inaugural lecture at Gresham College the sincere gratitude of mathematical contemporaries to Briggs for his outstanding work on logarithms. The interest of this brilliant man extended to the problem of a northwest passage to the South Seas, on which he wrote a treatise (1622), and to the relative merits of the ancients and moderns.

BIBLIOGRAPHY

I. ORIGINAL WORKS. Briggs's contributions, and the rest of Napier's *Canon mirificus,* were published at London in 1616 and reprinted in 1618; his own *Logarithmorum chilias prima* soon followed the original edition of the *Canon* (London, 1617). Briggs also added comments to Napier's *Rhabdologia* (Edinburgh, 1619) and edited a version of Euclid's *Elements,* Books I–VI (London, 1620), although his name did not appear as editor. His interest in a northwest passage to the South Seas was expressed in a treatise on the subject (London, 1622). A major work by Briggs was *Arithmetica logarithmica* (London, 1624); a second edition, completed by Adrian Vlacq (Gouda, 1628), contained the intermediate seventy chiliads. The relative merits of the ancients and moderns were discussed in *Mathematica ab antiquis minus cognita,* published in the second edition of G. Hakewill's *Apologie* (1630). Briggs's last work was *Trigonometria Britannica sive de doctrina triangulorum* (Gouda, 1633).

II. SECONDARY LITERATURE. Works concerning Briggs are D. M. Hallowes, "Henry Briggs, Mathematician," in *Transactions of the Halifax Antiquarian Society* (1962), 79–92; Christopher Hill, *Intellectual Origins of the English Revolution* (Oxford, 1965), p. 38, where it is claimed that "significant though Briggs was as a mathematician in his own right, his greatest importance was as a contact and public relations man"; C. Hutton, *Mathematical Tables,* 5th ed. (London, 1811), pp. 33–37, and *A Philosophical and Mathematical Dictionary,* I (London, 1815), 254–255; F. Maseres, ed., *Scriptores logarithmici,* I (London, 1791), lxxvi ff. (on Briggs's abacus ΠΑΓΧΡΗΣΤΟΣ and binomials, see especially p. lxviii); Thomas Smith, biography of Briggs, in his *Vitae quorundam eruditissimorum et illustrium*

virorum (1707), translated into English by J. T. Foxell in A. J. Thompson, *Logarithmetica Britannica,* I (Cambridge, 1952), lxvii–lxxvii; H. W. Turnbull, a study of Briggs's work on finite differences, in *Proceedings of the Edinburgh Mathematical Society,* 2nd ser., **3** (1933), 164–170; J. Ward, biography of Briggs, in *The Lives of the Professors of Gresham College* (London, 1740), pp. 120–129, which includes a list of Briggs's writings, both published and unpublished; D. T. Whiteside, "Patterns of Mathematical Thought in the Later Seventeenth Century," in *Archive for the History of the Exact Sciences,* **1** (1961), 232–236.

G. HUXLEY

BRIGHT, RICHARD (*b.* Bristol, England, 28 September 1789; *d.* London, England, 16 December 1858), *medicine.*

The first third of the nineteenth century brought forth an extraordinary flowering of medical talent, among whom Bright was a truly outstanding figure. Born to a well-to-do banking family and always in comfortable circumstances, he could devote himself wholly to medical and scientific pursuits without the financial cares of the less fortunate. He began the study of medicine at the University of Edinburgh in 1808. In his second year he interrupted his studies for a few months to serve as a naturalist accompanying Sir George Mackenzie on an expedition to Iceland. Bright contributed notes on botany and zoology. In 1810 he transferred his medical studies to Guy's Hospital in London, where he spent two years. During this first association with Guy's, he became deeply interested in pathology and postmortem examinations. In 1812 Bright returned to Edinburgh for a year, receiving his degree on 13 September 1813. He then spent two terms at Cambridge, but he left quite dissatisfied and embarked on a period of foreign study and travel, particularly in Berlin, Vienna, and Hungary. He described his travels in *Travels From Vienna Through Lower Hungary,* one of the outstanding travel books of the nineteenth century. Of particular interest are his descriptions of Hungary and the Congress of Vienna in 1814.

Returning to England, he became a licentiate of the London College of Physicians in 1816, but he was not elected a fellow until 1832. In 1820 he was appointed assistant physician at Guy's Hospital, and in 1824 was promoted to physician. For approximately twenty years, besides teaching medicine, he engaged in extensive clinical research and clinico-pathological correlations, always of the highest quality. After 1843 he no longer published papers, and from about 1840 until his death, he was absorbed in practice rather than in research.

We associate Bright's name particularly with kidney

disease, but we must not ignore his broad interest in the entire field of internal medicine. He had an intense concern with diseases of the nervous system, and he also contributed to the knowledge of visceral disease, especially of the pancreas, duodenum, and liver. He wrote extensively on abdominal tumors; and in his textbook, written with Thomas Addison, he provided the first accurate account of appendicitis. In all his investigations he combined meticulous clinical observation and careful postmortem studies.

Bright was, in essence, a "naturalist" who had unrivaled powers of observation. He noted the clinical phenomena of disease. Then, in the tradition of Morgagni, Gaspard Bayle, and Laënnec, he tried to correlate these clinical findings with the postmortem observations, but always on an empirical level. Remote causes or occult properties interested him not at all. Instead, he dealt with facts of observation. Gathering his observations not only at the bedside but also at the autopsy table and in the rudimentary clinical laboratory, he tried to correlate the data so that meaningful patterns might emerge. Constantly he tried to bring together, as he expressed it, "such facts as seem to throw light upon each other." As a scientist he could select, from the total profusion of facts, those which somehow belonged together.

In 1827 Bright published his first great contribution to kidney disease. He had clearly perceived that some relationships existed between three separate features. Each of them was familiar, but he was the first to connect them; and then he provided overwhelming evidence that the association was valid. He pointed out that of all patients with edema, some showed albumin in their urine, and this albumin was coagulable by heat; futhermore, these patients showed structural changes in their kidneys. Bright, with convincing clinico-pathological studies, thus identified a new disease pattern that tied together pathologic changes, chemical alterations, and clinical signs. Later research showed that this pattern was not a simple disease entity but, rather, a congeries of enormous complexity. But this does not detract from the brilliance of Bright's initial correlation.

Bright, one of the great men of Guy's, helped the hospital attain its enviable position among British medical educational institutions. He helped found the journal *Guy's Hospital Reports,* which, as a ready medium of publication, proved important in encouraging research, and he did much to popularize the then-new concept of bedside teaching. A neglected part of medical education, he realized, was meaningful contact between the students and the patients. He helped reform the educational methods so that, in 1837, *Guy's Hospital Reports* (**2,** v-vi) could proudly declare that "each student who has passed three months in the clinical wards is ready to admit that that period has proved the most profitable portion of his medical education. . . . Under the guidance of the experienced physicians, the student is instructed how to make observations upon the sick, and to interpret the signs of disease. . . ." Then, because of his efforts, the hospital set aside special wards for clinical research.

Bright, standing at a crossroads in medical history, exemplified the transition between old and new—the tradition, on the one hand, that emphasized observation with virtually unaided senses and, on the other hand, the tradition that emphasized laboratory and other technical aids to knowledge. Bright used the most rudimentary technique to identify albumin in the urine, heating a sample in a teaspoon held over a candle flame. But even at that very time analytic tools of considerable precision were being forged for chemistry, physiology, anatomy, and pathology. These new tools, and the theories to which they gave rise, became increasingly significant in medicine. Bright ceased active research just at the threshold of a new era.

His link with the past is clearly visible in the textbook written with Addison, which looks backward rather than forward. But his spirit of inquiry, respect for evidence, careful observation, critical acumen, and restrained conclusions, all were forward-looking.

BIBLIOGRAPHY

I. ORIGINAL WORKS. A complete bibliography of Bright's work, compiled by William Hill, is in *Guy's Hospital Gazette,* **64** (1950), 456 ff.; repr. in *Guy's Hospital Reports,* **107** (1958), 531 ff. This includes books, journal articles, and critical and biographical studies.

Among the books by Bright are *Travels From Vienna Through Lower Hungary; With Some Remarks on the State of Vienna During the Congress in the Year 1814* (Edinburgh, 1818); and *Reports of Medical Cases Selected With a View of Illustrating the Symptoms and Cure of Diseases With a Reference to Morbid Anatomy,* 2 vols. (London, 1827–1831). With Thomas Addison he wrote *Elements of the Practice of Medicine* (London, 1839). A collection of his papers on renal disease is the reprint *Original Papers of Richard Bright on Renal Disease,* A. A. Osman, ed. (London, 1937). His numerous periodical publications are listed in Hill's bibliography.

II. SECONDARY LITERATURE. A vast amount of secondary material deals with Bright. Worthy of special mention are B. Chance, "Richard Bright, Traveler and Artist," in *Bulletin of the History of Medicine,* **8** (1940), 909 ff.; F. H. Garrison, "Richard Bright's Travels in Lower Hungary, a Physician's Holiday," in *Johns Hopkins Hospital Bulletin,* **23** (1914), 173 ff.; Sir William Hale-White, "Bright's Observations Other Than Those on Renal Disease," in *Guy's Hospital Reports,* **71** (1921), 143 ff., repr. *ibid.,* **107** (1958), 308 ff.; and "Richard Bright and His Discovery of the Disease Bearing His Name," *ibid.,* **71** (1921), 1 ff., repr. *ibid.,* **107** (1958), 294 ff.; R. M. Kark, "A Prospect of Richard Bright on the Centenary of His Death, December 16, 1958," in *American Journal of Medicine,* **25** (1958), 819 ff.; Samuel Wilks and G. T. Bettany, *Biographical History of Guy's Hospital* (London, 1892).

LESTER S. KING

BRILL, ALEXANDER WILHELM VON (*b.* Darmstadt, Germany, 20 September 1842; *d.* Tübingen, Germany, 8 June 1935), *mathematics.*

Brill, the nephew of the geometer Christian Wiener, was a student of Alfred Clebsch at both the Politechnikum in Karlsruhe and at the University of Giessen. He graduated in 1864 and passed his *Habilitation* in 1867. From then until 1869 he was a *Dozent* at Giessen; from 1869 to 1875, a professor at the Politechnikum in Darmstadt; and from 1875 to 1884, a professor at the Politechnikum in Munich, where he worked with Felix Klein and was influenced by him. From 1884 to 1918, when he retired, Brill was a professor at the University of Tübingen. He worked primarily on the theory of algebraic functions and algebraic geometry, characteristically using algebraic methods, striving to avoid transcendental methods and aiming at "Weierstrassian strictness" of exposition. The systematic study of those properties of algebraic functions which are invariant under birational transformations is contained in his fundamental work, written with Max Noether (1874). In it many of the results obtained by Riemann and by Clebsch and Gordan, using transcendental means, are substantiated by algebraic-geometrical methods. Also noteworthy are his papers on three-dimensional algebraic curves (1907) and on pseudospherical three-dimensional space (1885), where the impossibility of putting such a space into a Euclidean four-dimensional space and the possibility of its being placed in a Euclidean five-dimensional space are proved.

At the end of the last century, Brill published a series of articles on methodology of mathematics, participated—following Klein—in the movement to reform its teaching, and was an initiator of the use of models of geometrical figures in teaching; many such models were prepared under his guidance.

Brill also wrote on the theory of determinants, on the theory of elimination, on the theory of elliptic functions, on some special curves and surfaces, and on the singularities of planar and spatial algebraic curves. He was also concerned with theoretical mechanics. In *Vorlesungen über allgemeine Mechanik*

(1928) and *Vorlesungen über algebraische Kurven und algebraische Functionen* (1925) Brill, who was then retired, summed up his scientific and pedagogical career.

Brill's survey of the development of the theory of algebraic functions ("Die Entwicklung der Theorie der algebraischen Functionen in älterer und neurer Zeit," 1894), which was written with Noether, has significance for the history of mathematics. His last work, published when he was eighty-seven, dealt with Kepler's *New Astronomy.*

BIBLIOGRAPHY

I. ORIGINAL WORKS. Among Brill's writings are "Ueber die algebraische Functionen und ihre Anwendung in der Geometrie," in *Mathematische Annalen,* **7** (1874), 269–370, written with Max Noether; "Bemerkungen ueber pseudophärischen Mannigfaltigkeiten," *ibid.,* **26** (1885), 300–303; "Die Entwicklung der Theorie der algebraischen Functionen in älterer und neurer Zeit," in *Jahresbericht der Deutschen Mathematiker-Vereinigung,* **3** (1894), 107–566, written with Max Noether; "Ueber algebraische Raumkurven," in *Mathematische Annalen,* **64** (1907), 289–324; *Vorlesungen über algebraische Kurven und algebraische Functionen* (Brunswick, 1925); and *Vorlesungen über allgemeine Mechanik* (Munich–Berlin, 1928). For a more complete list see Poggendorff.

II. SECONDARY LITERATURE. See S. Finsterwalder, "Alexander von Brill. Ein Lebensbild," in *Mathematische Annalen,* **112** (1936), 653–663; and F. Severi, "Alexander von Brill," in *Jahresbericht der Deutschen Mathematiker-Vereinigung,* **31** (1922), 89–96

J. B. POGREBYSSKY

BRILLOUIN, MARCEL LOUIS (*b.* Melle, Deux-Sèvres, France, 19 December 1854; *d.* Paris, France, 16 June 1948), *mathematics, physics.*

Brillouin came from a middle-class family. His father was a painter, and the family lived in Paris, where Marcel studied at the Lycée Condorcet. They moved back to Melle during the Franco-Prussian War, and he spent the years 1870 and 1871 reading all the books on philosophy he could find in his grandfather's big library. Back in Paris in 1872, he brilliantly passed his baccalaureate the following year and became a student at the École Normale Supérieure (1874–1878). He then was an assistant, at the Collège de France, to the well-known physicist Mascart, whose daughter he later married. In 1881 Brillouin obtained doctorates in both mathematics and physics. He spent the next several years, as assistant professor of physics, at the universities of Nancy, Dijon, and Toulouse. Brillouin returned to the

École Normale Supérieure in 1888, when he married Charlotte Mascart. From 1900 on, he was professor of mathematical physics at the Collège de France until his retirement in 1931. He became a member of the Académie des Sciences de Paris in 1921.

Brillouin was a prominent theoretical physicist, but he was also a very skillful experimenter. He always had a laboratory and a large library nearby. In his teaching he always outlined the history of the subject and organized a seminar on the history and philosophy of physics for all his students. He had a great influence on the formation and careers of such students as Perrin, Langevin, Villat, Pérès, A. Foch, his son Léon, and J. Coulomb. He also maintained friendly personal relations with many foreign scientists, including Kelvin, Lorentz, Planck, and Sommerfeld.

In his long career Brillouin published more than 200 papers and books. He was a great admirer of Kelvin's lectures and wrote a preface and notes for their translation (1893); he also provided notes for a book of translations of original papers on meteorology (1900), a subject in which he was always highly interested. His interest in the kinetic theory of gases, liquids, and solids is reflected in his contribution of a preface and many notes to the French translation of Boltzmann's book (1902). This was followed by a book on viscosity (1906–1907) and a number of papers on kinetic theory and thermodynamics of liquids (isotropic or anisotropic) and solids, plasticity, and melting conditions. A book on the propagation of electricity (1904) included a complete calculation of proper vibrations for a metallic ellipsoid, a problem that became later of great importance for ultrashort wavelengths.

About 1900 Brillouin spent considerable time building a new model of the Eötvös balance and testing it in the Simplon Tunnel, which was opened in 1906. This is described in a long paper published by the Académie des Sciences in 1908. The Brillouin balance was later used for oil prospecting.

There followed a series of important papers on Helmholtz' flow and surfaces of discontinuity, with applications to hydrodynamics and hydraulic problems, and a long paper on the stability of airplanes.

From 1918 to 1922, and later, Brillouin tried to find an explanation of Bohr's condition of stable atom trajectories and their n, l, m quantum numbers. He attempted to use retarded actions of unknown nature (rather similar to de Broglie waves) and obtained stability conditions containing some sort of quantum numbers. Similar conditions were used later by de Broglie and modified by Schrödinger.

A few papers on the problem of an electromagnetic source in uniaxial or biaxial crystals are of interest for crystal optics. From 1925 on, most of Brillouin's research centered on physics of the earth, especially tides, and was published in the Academy's *Comptes rendus.* He also lectured on these subjects at the Collège de France and the Institut Poincaré (1930). His lectures on tides were edited by J. Coulomb, but most of them remained unpublished. Brillouin discussed a variety of mathematical problems in connection with tides, especially problems of varying boundary conditions, and transformations of spherical harmonics from one polar axis to another, the idea being to use, for tides, an axis of coordinates running through continental regions.

The interests of this wide-ranging, open-minded scientist extended from the history of science to the physics of the earth and the atom.

BIBLIOGRAPHY

I. ORIGINAL WORKS. Books that Brillouin wrote or contributed to are *Conférences de Lord Kelvin,* Lugol, trans. (Paris, 1893), ed., preface, and notes by Brillouin; *Mémoires originaux sur la théorie de la circulation de l'atmosphere* (Paris, 1900), notes by Brillouin; *Théorie cinétique des gaz de Boltzmann* (Paris, 1902), notes and preface by Brillouin; pt. 2, *Sur la condition de l'état permanent. Sur la tendance apparente à l'irréversibilité d'après Gibbs* (Paris, 1905); *Propagation de l'électricité, histoire et théorie* (Paris, 1904); and *Leçons sur la viscosité des liquides et des gaz,* 2 vols. (Paris, 1906–1907).

Papers of special importance are "Vents contigus et nuages," in *Mémoires du Bureau central météorologique* (1897), also in *Annales de chimie et physique,* **12** (1897), 145 ff.; "L'ellipticité du geoïde dans le tunnel du Simplon," in *Mémoires présentés par divers savants à l'Académie des sciences de l'Institut de France,* **23** (1908); "Stabilité des aéroplanes, surface métacentrique, planeurs, etc.," in *Revue de mécanique* (1909–1910); "Surfaces de glissement d'Helmholtz et resistance des fluides," in *Annales de chimie et physique,* **23** (1911); "Structure des cristaux et anisotropie des molécules," Solvay Congress, 1913; "Milieux biaxes," in *Comptes rendus de l'Académie des sciences,* **165** (1917) and **166** (1918); "Sources électromagnétiques dans les milieux uniaxes," in *Bulletin des sciences mathématiques,* **42** (1918); "Actions mécaniques à hérédité discontinue, essai de l'atome à quanta," in *Comptes rendus de l'Académie des sciences,* **168** (1919), **171** (1920), **173** (1921), and **174** (1922); and "Atome de Bohr, fonction de Lagrange circumnucléaire," in *Journal de physique,* **3** (1922).

II. SECONDARY LITERATURE. Works on Brillouin are H. Villat, ed., *Jubilé de M. Brillouin pour son 80ème anniversaire,* 2 vols. (Paris, 1935); and H. Villat, *Titres et travaux scientifiques* (Paris, 1930), pp. 8, 10, 19–21, 25–26, and "Notice nécrologique sur Marcel Brillouin," in *Comptes rendus de l'Académie des sciences,* **226,** no. 25 (1948), 2029.

L. BRILLOUIN

BRINELL, JOHAN AUGUST (*b.* Bringetofta, Sweden, 21 June 1849; *d.* Stockholm, Sweden, 17 November 1925), *metallurgy, materials testing.*

While he is best known as the originator of a standard procedure for determining the hardness of a metal, Brinell also did significant work on the metallurgy of steel.

The son of Johannes Månsson, a farmer, and Katarina Jonasdotter, Brinell graduated in 1871 from the technical school in Borås and was employed in the Swedish iron industry for some fifty years. From 1882 to 1903 he was chief engineer of the Fagersta Ironworks, where his most original scientific work was done. He was chief engineer of Jernkontoret, an iron industry association, from 1903 to 1914, and chairman of the board of Fagersta from 1915 to 1923. He was a member of the Swedish Academy of Science and the (British) Iron and Steel Institute; and he received the Polhem Medal in 1900, the Bessemer Medal in 1907, and an honorary Ph.D. from Uppsala in 1907, as well as many other awards. He was married in 1880 to Selma Nilsson.

Brinell's first studies at Fagersta, which were concerned with changes of the internal structure of steel as it was heated or cooled, compared the appearance of steel fracture surfaces in a very large number of experiments. His first major paper (1885) was, in the opinion of Cyril Stanley Smith, ". . . a monument of imaginative and careful work and shows how much can be learned about steel without knowledge of its microstructure." However, the work of Floris Osmond, in which microstructure was identified through microscopic examination of etched surfaces, published also in 1885, overshadowed Brinell's work. Nevertheless, Osmond's conclusions probably were accepted more readily by the iron industry because they were reinforced by Brinell's, which were based upon an observational procedure well known in the shop.

Brinell's apparatus for testing the hardness of a material was first displayed in 1900 at the Paris Exposition. A hardened steel ball, 10 millimeters in diameter, is pressed into the test surface under a heavy load (up to 3,000 kilograms). The Brinell hardness number (Bhn) in kg/mm^2 is calculated by dividing the load by the area of indentation. This procedure, not essentially modified, is still one of the most widely used tests of hardness.

BIBLIOGRAPHY

I. ORIGINAL WORKS. Brinell's works include "Om ståls texturförändringar under uppvärmning och afkylning," in *Jern-kontorets annaler,* n.s., **40** (1885), 9–38, published in German in *Stahl und Eisen,* **5** (1885), 611–620, and abstracted in English as "Changes in the Texture of Steel on Heating and Cooling," in *Journal of the Iron and Steel Institute* (1885), no. 1, 365–367; and "Sätt att bestämma kroppars hårdhet jämte några tillämpningar af detsamma," in *Teknisk tidskrift,* **30** (1900) [section on mechanics], 69–87; English version prepared by Axel Wahlberg as "Brinell's Method of Determining Hardness and Other Properties of Iron and Steel," in *Journal of the Iron and Steel Institute* (1901), no. 1, 243–298, and no. 2, 234–271.

II. SECONDARY LITERATURE. The best biographical sketch is in *Svenskt biografiskt lexicon,* VI (Stockholm, 1926), 236–241, which includes a list of Brinell's works. An obituary notice is in *Journal of the Iron and Steel Institute* (1926), no. 1, 482–483. A critical analysis of Brinell's 1885 paper is in Henry M. Howe, *The Metallurgy of Steel* (New York, 1890), pp. 170–175. See also Cyril Stanley Smith, *History of Metallography* (Chicago, 1960), esp. pp. 119–121.

EUGENE S. FERGUSON

BRING, ERLAND SAMUEL (*b.* Ausås, Kristianstad, Sweden, 19 August 1736; *d.* Lund, Sweden, 20 May 1798), *mathematics.*

The son of Iöns Bring, a clergyman, and Christina Elisabeth Lagerlöf, Erland Bring studied jurisprudence at Lund University from 1750 to 1757. Beginning in 1762 he was a reader at Lund and from 1779 a professor. He taught history at the university, although his favorite field was mathematics. In the university library are preserved eight volumes of his manuscript compositions on various questions of algebra, geometry, mathematical analysis, and astronomy, and commentaries on the work of L'Hospital, Christian von Wolf, Leonhard Euler, and other scholars.

In 1786 Bring's *Meletemata* was published. Like many eighteenth-century mathematicians, he attempted to solve equations of higher than fourth degree in radicals by means of reduction into binomial form, employing the transformation of the unknown quantity first proposed by Tschirnhaus (1683). Bring succeeded in reducing a general fifth-degree equation to the trinomial form $x^5 + px + q = 0$, using a transformation whose coefficients are defined by equations of not higher than the third degree. This remarkable result received practically no attention at the time and was obtained independently by George Birch Jerrard in his *Mathematical Researches* (1832–1835). Shortly thereafter, Sir William R. Hamilton demonstrated (1836) that with the aid of this operation a general fifth-degree equation reduces to any of four trinomial forms. It is not known whether Bring hoped to solve the fifth-degree equation in radicals with the aid of his transformation; Jerrard retained this hope, even though Niels Abel proved (1824–1826) that such a solution is impossible for a general fifth-degree equation.

In 1837 Bring's nephew, the historian Ebbe Samuel Bring, tried unsuccessfully to attract the attention of mathematicians to the algebraic investigations of his uncle. The deep significance of the Bring-Jerrard transformation was ascertained only after Charles Hermite (1858) used the above-mentioned trinomial form for the solution of fifth-degree equations with the aid of elliptic modular functions, thereby laying the foundations for new methods of studying and solving equations of higher degrees with the aid of transcendental functions.

Hermite cited only Jerrard, calling his result the most important event in the theory of fifth-degree equations since Abel. Shortly thereafter, in 1861, the scholarly world also recognized Bring's merits, mainly through the efforts of Carl J. D. Hill, professor of mathematics at Lund University.

BIBLIOGRAPHY

Bring's major work is *Meletemata quaedam mathematica circa transformationem aequationum algebraicarum* (Lund, 1786).

Writings on Bring include Moritz Cantor, *Vorlesungen über Geschichte der Matematik,* IV (Leipzig, 1908), 130–132; C. J. D. Hill, "Nagra ord om Erland Sam. Brings reduktion af 5te gradens equation," in *Öfversigt af Kongelige vetenskapsakademiens förhandlingar* (1861), pp. 317–355; Felix Klein, *Vorlesungen über die Ikosaeder* (Leipzig, 1884), pp. 143–144, 207–209, 244; and *Svenska män och kvinnor biografisk uppslagsbok,* I (Stockholm, 1942), 466.

A. P. YOUSCHKEVITCH

BRINKLEY, JOHN (*b.* Woodbridge, England, 1763; *d.* Dublin, Ireland, 14 September 1835), *astronomy, mathematics.*

Brinkley, whose greatest contribution was his researches into stellar parallaxes, received his early education at Woodbridge Grammar School and with a Mr. Tilney of Harleston. He went on to Caius College, Cambridge, and received his B.A. as senior wrangler and first Smith's Prizeman in 1788. During his senior year he was assistant to N. Maskelyne at Greenwich and was fellow of his college from 1788 to 1792. Upon Maskelyne's personal recommendation he was appointed Andrews professor of astronomy at Dublin University, 11 December 1790. The following year Brinkley was ordained a priest at Lincoln and received his M.A. at Cambridge. In 1792 he was incorporated at Dublin and elected first astronomer royal for Ireland. He proceeded D.D. (Dublin) 1806.

Between 1790 and 1808 he prepared the excellent textbook *Elements of Plane Astronomy,* published in 1808, and ten mathematical papers, some with direct application to celestial astronomy.

Upon acquiring a splendid eight-foot meridian circle in 1808, Brinkley attempted to determine the long-sought parallax of the fixed stars, with a view to determining their distances. Two years later he announced the detection of an annual (double) parallax for α Lyrae of 2″.52, and in 1814 similarly large values of 2″.0, 5″.5, 2″.2, and 2″.1 for the stars α Lyrae, α Aquilae, Arcturus, and α Cygni, respectively. The validity of these measurements was disputed in the literature for fourteen years by Pond, who was unable to deduce analogous results with Greenwich instruments. This controversy, by necessitating repeated tests of the observations, was of great value in stimulating the study of previously unappreciated factors affecting the measurements. Brinkley's results, although now themselves discredited, thus led to the later successful detection of stellar parallaxes.

Among Brinkley's other major work was the publication of a new theory of astronomical refractions (1815), estimation of the obliquity of the ecliptic (1819), determination of north polar distances of the principal fixed stars (1815, 1824), and determination of the precession of the equinoxes (1828). He also used the south polar distances of certain fixed stars observed by Sir Thomas Brisbane at Paramatta, New South Wales, to investigate the accuracy of separate determinations by himself and by Bessel of their north polar distances (1826). His astronomical career ended with his elevation, after numerous ecclesiastical preferments, as bishop of Cloyne, 28 September 1826.

Brinkley's honors were many. Fellowship of the Royal Society (1803) was followed by the Conyngham Medal of the Royal Irish Academy, for his essay on investigations relating to the mean motion of the lunar perigee (1817). He was also awarded the Copley Medal of the Royal Society (1824) for his scientific achievements and his approximations to the solution of the parallax problem. He was president of the Royal Irish Academy from 1822 to 1835, vice-president of the Astronomical Society from 1825 to 1827, and its president from 1831 to 1833.

BIBLIOGRAPHY

I. ORIGINAL WORKS. Brinkley's various observations at Greenwich (1787–1788) are distributed through Maskelyne's *Astronomical Observations Made at the Royal Observatory, Greenwich,* **3** (1799), starting with an entry for 23 Sept. 1787. Maskelyne appends the letters *JB* to

Brinkley's observations; those by Brinkley's contemporary John Bumpstead appear under Bumpstead's full name.

His elementary astronomical textbook was compiled from lectures given between 1799 and 1808 to undergraduates at Dublin University. The earliest record of the course is *Synopsis of Astronomical Lectures to Commence October 29, 1799 at Philosophical School, Trinity College, Dublin* (Dublin, 1799). The finished book, *Elements of Plane Astronomy* (Dublin, 1808), was prepared at the request of the board of the college when the acquisition of a meridian circle diverted Brinkley's efforts to practical astronomy. The book went through five editions subject to his revision during his lifetime; a sixth edition was edited and revised by Thomas Luby (Dublin, 1845), and two further editions were revised and partly rewritten by J. W. Stubbs and F. Brünnow (London, 1874, 1886).

Ten mathematical papers of considerable elegance were published between 1800 and 1818, nine in the *Transactions of the Royal Irish Academy* (see its Index) and one in *Philosophical Transactions of the Royal Society*, 97 (1807), 114–132.

Eighteen significant astronomical papers on various subjects appeared between 1810 and 1828, eight in *Transactions of the Royal Irish Academy*, eight in *Philosophical Transactions of the Royal Society*, and two in *Memoirs of the Astronomical Society* (see Royal Society's *Catalogue of Scientific Papers*, I [1867], 627–629). Those relating to the parallax question include the following: Brinkley's original announcement of his detection of the annual (double) parallax of α Lyrae, communicated to the Royal Society by Maskelyne, in *Philosophical Transactions*, 100 (1810), 204. The 1814 report of similar and even larger results for other stars, in *Transactions of the Royal Irish Academy*, 12 (1815), 33–75. Discordance with Pond's results suggested to be due to uncertainty of elements used in reduction of Greenwich observations, in *Philosophical Transactions*, 108 (1818), 275–302. Results of further observations introducing a determination of the constant of aberration and of that of lunar nutation, *ibid.*, 109 (1819), 241–248, and 111 (1821), 327–360. An instrumental investigation of the effect of solar nutation cited to exhibit the competence of his equipment to detect the larger quantity of parallax, first reported to the Royal Irish Academy in 1822, in *Transactions of the Royal Irish Academy*, 14 (1825), 3–37. Disengagement from Greenwich results of a parallax for α Lyrae not differing sensibly from that measured at Dublin, in *Memoirs of the Royal Astronomical Society*, 1, pt. 2 (1822), 329–340. Reassertion of parallax of α Lyrae and attempt to form a correct estimate of the absolute and relative degrees of accuracy of the Dublin and Greenwich instruments, in *Philosophical Transactions*, 114 (1824), 471–498.

Among the ten remaining catalogued papers, see *Transactions of the Royal Irish Academy*, 13 (1818), 25–51, containing an essay on investigations relative to the mean motion of the lunar perigee, which was awarded the Conyngham Medal of the Academy; and *Memoirs of the Astronomical Society*, 2, pt. 1 (1826), 105–123, containing Brisbane's Paramatta observations. There are several minor

references in the *Quarterly Journal of Science, Literature and the Arts*, 9 (1820), 164–167; 11 (1821), 364–370, 370–372; and 12 (1822), 151–154; and in *Astronomische Nachrichten*, 3 (1825), cols. 105–106; 4 (1826), cols. 101–104; and 5 (1827), cols. 131–138.

II. SECONDARY LITERATURE. An excellent discussion of Brinkley's life and work is contained in *Dictionary of National Biography*, VI (1886); a more general account is in Sir Robert Ball, *Great Astronomers* (London, 1895), pp. 233–246. A synopsis of the contents of his Royal Society papers is in *Proceedings of the Royal Society*, 3 (1835), 354–355. An account of the oration by Sir Humphry Davy upon the award to Brinkley of the Copley Medal may be found in *Philosophical Magazine*, 64 (1824), 459–462.

Obituaries are Henry Cotton, *Fasti ecclesiae Hibernicae*, I (Dublin, 1851), 307–309; *Gentlemen's Magazine*, 11 (1835), 547; Rev. J. B. Leslie, *Clogher Clergy and Parishes* (Enniskillen, 1929), p. 47; and *Memoirs of the Royal Astronomical Society*, 9 (1836), 281–282.

SUSAN M. P. MCKENNA

BRIOSCHI, FRANCESCO (*b.* Milan, Italy, 22 December 1824; *d.* Milan, 14 December 1897), *mathematics, hydraulics.*

Brioschi graduated in 1845 from the University of Pavia, where he was a student of Antonio Bordoni. From 1852 to 1861 he was a professor of applied mathematics there, teaching theoretical mechanics, civil architecture, and hydraulics. He was the general secretary of the Ministry of Education in 1861–1862, a senator from 1865, and, from 1870 until 1882, a member of the Executive Council of the Ministry of Education. In 1863 Brioschi organized the Istituto Tecnico Superiore in Milan, serving as director and professor of mathematics and hydraulics until his death. From 1884 he was president of the Accademia Nazionale dei Lincei.

From the beginning of his career, Brioschi strove to overcome the backwardness of Italian mathematics, to popularize new scientific trends, and to raise the quality of the teaching of mathematics in secondary schools and universities. He published many essays and reviews, and participated in the organization of the journal *Annali di matematica pura ed applicata*, heading its editorial staff from 1867 until his death (until 1877 in conjunction with Cremona). He also helped to organize the journal *Politecnico*.

In his original papers Brioschi appears as a virtuoso in computation, as an analyst, and as an algebraist. In the works of his most fruitful decade (1851–1860) he widely applied and developed the still new theory of determinants. His *Teoria dei determinanti* (1854) was the first nonelementary statement of the theory and its basic applications. Brioschi devoted several important papers, following Caley, Sylvester, and

Hermite, to the then developing theory of forms of two or more variables, which Hermite termed ". . . one of the major mathematical achievements of our time." He applied exclusively algebraic means of solution to such questions as the deduction of equations in partial derivatives for the discriminant of a binary form and for the resultant of two such forms. A significant part of his results in this area was included in a monograph published in the first four volumes of *Annali di matematica.*

In these same years Brioschi added new results to the theory of the transformation of elliptic and Abelian functions. In his greatest achievement, following Hermite and simultaneously with Kronecker, he applied elliptical modular functions to the solution of fifth-degree equations. At the same time, Brioschi popularized Gauss's theory of surfaces in Italy and brought forth, in connection with this, geometric papers.

During the 1860's and 1870's Brioschi continued his work in algebra and analysis in traditional directions, using the Weierstrass theory of elliptic functions. From these viewpoints, he addressed himself to the theory of differential equations and, in the 1880's, to the theory of hyperelliptic functions. His second great achievement relates to this latter period: the solution of sixth-degree equations with the aid of hyperelliptic functions.

Brioschi did not propound any strikingly new ideas in mathematics, nor did he discover any new fields. "I am only a calculator," he humbly characterized himself. However, he was a brilliant analyst with algebraic propensities and possessed a rare mobility of thought that responded to new ideas from their very inception. This enabled him to enrich science with new results for half a century.

Along with Betti, Brioschi began a new epoch in the history of Italian mathematics, leading it out of its provincial backwardness. He was the teacher of its most outstanding representatives in the next generation, among them Casorati, Cremona, and Beltrami.

In mechanics Brioschi dealt with problems of statics, proving Moebius' results by analytic means; with the integration of equations in dynamics, according to Jacobi's method; with hydrostatics; and with hydrodynamics. His work as a hydraulic engineer was significant, although it is reflected comparatively little in his publications. Brioschi used the findings of a series of major projects or participated in the projects' development—for example, in the regulation of the Po and Tiber (which goals remained unaccomplished). Two more of Brioschi's works should be mentioned: with Betti he brought out a treatment of the first six

books of Euclid's *Elements* for secondary schools, and he edited Leonardo da Vinci's *Codice Atlantico,* an important source for the history of science and technology.

An adherent of pure mathematics, Brioschi highly valued its significance in application and allotted to it a significant place in technical education, emphasizing the great role of the latter in the development of national industry. At the same time he insisted on the value of the humanities and, simultaneously with his founding of the Politechnicum, he organized the Accademia Scientifica-Litteraria in Milan.

In addition to the publication of the *Codice Atlantico,* Brioschi produced several important articles on contemporary mathematicians.

BIBLIOGRAPHY

I. ORIGINAL WORKS. Many of Brioschi's writings have been brought together in Ascoli *et al.,* eds., *Opere,* 5 vols. (Milan, 1901–1908). Among individual works of note are *Teoria dei determinanti* (Pavia, 1854) and "La teoria dei covarianti e degli invarianti delle forme binarie, e le sue principali applicazioni," in *Annali di matematica,* **1** (1858), 269–309, 549–561; **2** (1859), 82–85, 265–277; **3** (1860), 160–168; **4** (1861), 186–194.

II. SECONDARY LITERATURE. The fullest characterization of Brioschi's scientific work is in M. Noether, "Francesco Brioschi," in *Mathematische Annalen,* **50** (1898), 477–491. On Brioschi as an engineer, see E. Paladini, "Commemorazione di F. Brioschi," in *Atti del Collegio degli ingegneri ed architetti* (Milan), **30** (1898). See also E. Beltrami's obituary notice of Brioschi in *Annali di matematica,* 2nd ser., **26** (1897), 340–342; Charles Hermite, "Notice sur M. F. Brioschi," in *Comptes rendus de l'Académie des sciences,* **125** (1897), 1139–1141; and the speeches given at Brioschi's funeral, in *Reale Istituto tecnico superiore, programma 1891–1898* (Milan, 1898).

JOSEPH POGREBYSSKY

BRIOT, CHARLES AUGUSTE (*b.* St.-Hippolyte, France, 19 July 1817; *d.* Bourg-d'Ault, France, 20 September 1882), *mathematics, physics.*

Briot's father, Auguste, a merchant at St.-Hippolyte, had a considerable reputation in the tanning trade. Charles, the eldest of a large family, became a teacher after an accident that left him with a stiff arm. He was sent to Paris and in only five years attained a remarkable level of scholarship. When he entered the École Normale Supérieure in 1838, he was ranked second. Three years later he completed the course and received his *agrégation* in mathematics with the highest rank. In March 1842 he received his doctorate of science, having presented his thesis on the movement of a solid body round a fixed point. This brilliant success lit the way for a group of young men from his native Franche-Comté: Claude Bouquet, L. E. Bertin, and Louis Pasteur.

Briot devoted himself to teaching, first as a professor at the Orléans Lycée and afterward at the University of Lyons, where he reencountered his friend Claude Bouquet. In 1851 he moved to Paris, where he taught the course in *mathématiques speciales* (preparation for the École Normale Supérieure and the École Polytechnique) at the Lycée Bonaparte and later at the Lycée Saint-Louis, as well as acting as substitute at both the École Polytechnique and the Faculté des Sciences for the courses in mechanical engineering and surveying (1850), calculus (1853), and mechanics and astronomy (1855). From 1864 on, he was a professor at the Sorbonne and at the École Normale Supérieure. In his courses he particularly stressed the relation between thermodynamics and rational mechanics.

Briot's studies on heat, light, and electricity were based on the hypothesis of the existence in the ether of imponderable molecules acting upon each other, as well as upon the ponderable molecules of matter. Particularly in his study of the crystalline medium, he linked his findings to Pasteur's experimental work on the dissymmetry of crystals. These studies, which were conducted from a mathematical point of view, led to the simplification of methods for integral calculus and the advance of the theories of elliptic and Abelian functions. To honor him for this work, the Göttingen Academy named him a corresponding member.

A large part of Briot's activity was devoted to the writing of textbooks for students, so that he and Bouquet could provide them with a library of basic books on arithmetic, algebra, calculus, geometry, analytical geometry, and mechanics. These books were published in numerous editions and for many years contributed to establishing the level of mathematics teaching in France. Briot also published, with Bouquet, an important work on elliptic functions (1875) and, alone, a treatise on Abelian functions (1879). The Académie des Sciences awarded Briot the Poncelet Prize in 1882 for his work in mathematics.

BIBLIOGRAPHY

Briot's works include "Recherches sur la théorie des fonctions," in *Journal de l'École Polytechnique* (1859), also published as an independent work (Paris, 1859); *Théorie des fonctions doublement périodiques,* written with Bouquet, 2 vols. (Paris, 1859; 2nd ed., 1875); *Essai sur la théorie mathématique de la lumière* (Paris, 1864); *Théorie mécanique de la chaleur* (Paris, 1869); *Théorie des fonctions elliptiques,* written with Bouquet (Paris, 1875); and *Théorie des fonctions abéliennes* (Paris, 1879).

LUCIENNE FÉLIX

BRISBANE, THOMAS (*b.* Brisbane House, Ayrshire, Scotland, 23 July 1773; *d.* Brisbane House, 27 January 1860), *astronomy.*

Although himself an able practical astronomer, Brisbane is better remembered as a munificent patron of science through his founding and equipping of Paramatta (astronomical) and Makerstoun (magnetic) observatories, the personal remuneration of their observers, and the provision of support for the publication of their findings.

Brisbane was descended from the distinguished Brisbane family of Bishopton. His early education was under tutors at home; he then studied at Edinburgh University and at Kensington Academy, where he attended lectures on astronomy and mathematics. Brisbane was gazetted an ensign in 1789 and progressively advanced to the rank of general (1841). He saw active service in Europe, the West Indies, and Canada. He was elected a fellow of the Royal Society in 1810, corresponding member of the Paris Institute in 1816 (for protecting its premises from military attack earlier that year), vice-president of the Astronomical Society in 1827, president of the Royal Society of Edinburgh in 1833, and honorary member of the Royal Irish Academy in 1836. He received honorary degrees from Edinburgh (1824), Oxford (1832), and Cambridge (1833); was created baronet in 1836; and was made G.C.B. in 1837.

Brisbane's decision to master practical astronomy came on his first voyage to the West Indies (1795), when an error of the ship's commander in taking the longitude resulted in their being almost wrecked. Retired on half pay for health reasons from 1805 to 1810, he built an observatory at Brisbane House in 1808 and became skilled in the use of astronomical instruments. This was the second of two observatories then in Scotland and the foremost in equipment, having a four-and-a-half-foot transit and an altitude and azimuth instrument (both by Troughton), a mural circle, and an equatorial. During the Peninsular campaigns (1812–1813) Brisbane took regular observations with a pocket sextant and, while serving in France (1815–1818), computed a set of tables for determining apparent time with a sextant from the altitudes of the sun and stars. These tables, commissioned by the Duke of Wellington and published privately by the army in 1818, also formed the subject of his first scientific contribution to the Royal Society of Edinburgh.

Appointed governor of New South Wales in 1821, Brisbane decided to establish, at his own expense, an observatory at Paramatta, in order to promote knowledge of the then little-known stars of the Southern Hemisphere. The observatory, equipped with a five-and-a-half-foot transit and a two-foot mural circle by Troughton and other instruments, opened on 2 May 1822 under his personal direction, with Charles Rümker and James Dunlop as observers. The importance of this station was underlined a month later by Dunlop's rediscovery, in its predicted place (invisible from Europe), of Encke's comet, thus establishing the existence of comets of short period and providing information on their spatial motions. Besides standard astronomical observations, the greatest effort at Paramatta was the cataloging of 7,385 stars between 1822 and 1826 ("Brisbane Catalogue," 1835). Unfortunately, the inherent unsteadiness of the transit instrument used in this program has since caused the catalog to prove largely useless. Brisbane's provision of an observatory in the Southern Hemisphere was honored by the award of the gold medal of the Astronomical Society in 1828.

When he returned to Scotland, Brisbane built and equipped another observatory at Makerstoun in 1826, making astronomical observations there until about 1847. It is noteworthy, since he later supported a worldwide effort—instigated by Humboldt in 1837 and undertaken by the British and other national governments, the East India Company, and private enterprise in 1839—to elucidate the problems of terrestrial magnetism, that a personal letter from him to the Royal Society of Edinburgh dated as early as 15 March 1830 regrets that the taking of magnetic measurements should be neglected in Britain. His support of the international cooperation took the form of personally founding and equipping a magnetic observatory at Makerstoun in 1841, thus filling the need, in view of its extreme northwesterly position in Europe, of taking magnetic measurements in Scotland. The results obtained at this station under the director John Allan Broun now constitute the most valuable fruits of Brisbane's patronage of science. His philanthropy in its establishment and maintenance, and in the dissemination of its results, was honored by the award of the Keith Medal of the Royal Society of Edinburgh in 1848.

Other benefactions included the founding of two medals for reward of scientific merit—one to be awarded by the Royal Society of Edinburgh, the other by the Scottish Society of Arts—and the endowment of Brisbane Academy, Ayrshire.

BIBLIOGRAPHY

I. ORIGINAL WORKS. Brisbane's writings include *Tables for Determining the Apparent Time From the Altitudes of the Sun and Stars* (France, 1818); "A Method for Determining the Time . . .," in *Transactions of the Royal Society of Edinburgh*, **8**, pt. 2 (1818), 497–506; and papers on the repeating reflecting circle and on a method of determining the latitude by a sextant or circle, *ibid.*, **9** (1823), 97–102 and 227–234, respectively. *Memoirs of General Sir T. M. Brisbane* (Edinburgh, 1860) contains personally compiled accounts of his military campaigns.

A great variety of observations made at Paramatta between 1822 and 1826 by Brisbane and/or his assistants were forwarded by Brisbane for publication in the journals of the Royal Society of London and the Royal Society of Edinburgh, and in Schumacher's *Astronomical Notices* (see *Royal Society Catalogue of Scientific Papers*, I, 632–633). A large collection of assorted Paramatta observations, compiled by Charles Rümker, appeared in *Philosophical Transactions of the Royal Society*, **119**, pt. 3 (1829), 1–152. The bulk of Brisbane's personal observations are contained in *A Catalogue of 7385 Stars Chiefly in the Southern Hemisphere . . .* ("The Brisbane Catalogue"), compiled by William Richardson (London, 1835).

A variety of observations, mainly planetary, made at Makerstoun astronomical observatory by Brisbane and/or his assistants, appeared in *Monthly Notices of the Royal Astronomical Society*, vols. **1, 2, 4, 7, 8** and in *Memoirs of the Royal Astronomical Society*, vols. **4, 5, 9, 10.**

A personal letter from Brisbane to the Royal Society of Edinburgh concerning the taking of magnetic measurements in Britain is published in *Transactions of the Royal Society of Edinburgh*, **12** (1834), 1–2. Significant results obtained at Makerstoun magnetic observatory by John A. Broun and his staff, published at the joint expense of Brisbane and the Royal Society of Edinburgh, appear in *Transactions of the Royal Society of Edinburgh*, **17** (1845)–**19** (1850) and in a supplement to **22** (1861) published after Brisbane's death.

II. SECONDARY LITERATURE. Writings on Brisbane include A. Bryson, "Memoir of General Sir Thomas Makdougall Brisbane, GCB . . .," in *Transactions of the Royal Society of Edinburgh*, **22** (1861), 589–605, which contains many anecdotes of his military campaigns and a complete quotation of Herschel's presentation address (see below); Fraser, *Genealogical Table of Sir T. M. Brisbane* (Edinburgh, 1840); and the original address by Sir John Herschel upon the presentation of the gold medal of the Astronomical Society to Brisbane, in *Memoirs of the Astronomical Society*, **3** (1829), 399–407.

A general account of his life and work appear in the *Dictionary of National Biography*, VI (1886). Obituaries are in *Gentlemen's Magazine*, pt. 1 (1860), 298–302; *Monthly Notices of the Royal Astronomical Society*, **21** (1861), 98–100; and *Proceedings of the Royal Society*, **11** (1862), iii–vii.

SUSAN M. P. McKENNA

BRISSON, BARNABÉ (*b.* Lyons, France, 11 October 1777; *d.* Nevers, France, 25 September 1828), *hydraulic engineering, mathematics.*

The son of Antoine-François Brisson, inspector of commerce and manufacture for the financial district of Lyons, Brisson studied at the Collège Oratorien de Juilly and was admitted to the École des Ponts et Chaussées in December 1793. A year later, at the newly founded École Centrale des Travaux Publics (the future École Polytechnique), he became one of the brilliant team of aspiring instructors and was highly thought of by Gaspard Monge. In December 1796, upon graduation from this school, he was admitted to the Corps des Ponts et Chaussées, where he remained for the rest of his career.

After completing his professional training at the École des Ponts et Chaussées in May 1798, Brisson specialized in the design and construction of ship canals. In 1802 he and his colleague Pierre-Louis Dupuis-Torcy presented a brilliant memoir based on applying methods of descriptive geometry to the determination of crest lines and of thalwegs, as well as establishing the course of the canals. After having been the civil engineer for the department of Doubs, he collaborated from 1802 to 1809 in the construction of the Canal de St.-Quentin, and then in the extension of the dikes and canals of the department of l'Escaut (until 1814). Appointed professor of stereometry and construction at the École des Ponts et Chaussées in 1820, he later assumed the additional duties of inspector for the school (from 1821) and secretary of the Conseil Royal des Ponts et Chaussées (from 1824).

Brisson remained one of Monge's favorite disciples, and his marriage in 1808 to Anne-Constance Huart, the latter's niece, strengthened his admiration and affection for the famous geometer. In 1820 he edited the fourth edition of Monge's *Géométrie descriptive* and finished off the work with two previously unpublished chapters on the theory of shadows and on perspective, which he revised with great care. But his favorite field of study was the theory of partial differential equations. Brisson drew up two important reports on this subject. One was read before the Académie des Sciences by Biot, his fellow student at the École Polytechnique and his brother-in-law. This paper was published in 1808. The other was read in 1823 and was not published. The main idea in these reports was the application of functional calculus, through symbols, to the solution of certain kinds of linear differential equations and of linear equations with finite differences.

The 1823 report was the object of lively discussion in 1825 before the Academy and was approved of by Cauchy, who, although he had some reservations about the validity of some of the symbols used and the equations obtained, emphasized the elegance of the method and the importance of the objects to which they were applied. Cauchy followed the way opened by Brisson, who thus became one of those who developed the methods of functional calculus.

BIBLIOGRAPHY

I. ORIGINAL WORKS. Brisson's writings include "Essai sur l'art de projeter les canaux de navigation," in *Journal de l'École polytechnique,* **7,** no. 14 (Apr. 1808), 262–288; "Mémoire sur l'intégration des équations différentielles partielles," *ibid.,* 191–261; *Notice historique sur Gaspard Monge* (Paris, 1818); *Nouvelle collection de 530 dessins ou feuilles de textes relatifs à l'art de l'ingénieur et lithographiés . . . sous la direction de M. Brisson,* 2 vols. (Paris, 1821–1825); and *Essai sur le système général de navigation intérieure de la France* (Paris, 1829).

II. SECONDARY LITERATURE. Biographical sketches of Brisson are A. Debauve, *Les travaux publics et les ingénieurs des ponts et chaussées depuis le XVIIe siècle* (Paris, 1893), pp. 381–382; *École polytechnique—Livre du centenaire,* III (Paris, 1895), 62–64, *passim.;* F. Hoefer, in *Nouvelle biographie générale,* VII (1863), cols. 436–437; H. Massiani, in *Dictionnaire de biographie française,* VII (1956), col. 364; J. and L. G. Michaud, *Biographie universelle,* new ed., V (1843), 565–567; *Le moniteur* (19 Oct. 1828); N. Nielsen, *Géomètres français sous la Révolution* (Paris, 1937), pp. 37–38, 83–84; J. Petot, *Histoire de l'Administration des ponts et chaussées (1599–1815)* (Paris, 1955); S. Pincherle, "Opérations fonctionnelles," in *Encyclopédie des sciences mathématiques,* II, fasc. 26, 10; Poggendorff, III (1898), col. 196; and *Procès verbaux de l'Académie des sciences,* VIII (Hendaye, 1918), 223–226.

RENÉ TATON

BRISSON, MATHURIN-JACQUES (*b.* Fontenay-le-Comte, Vendée, France, 30 April 1723; *d.* Brouessy, Commune of Magny-les-Hameaux, near Versailles, France, 23 June 1806), *physics, natural history.*

Brisson was the eldest son of Mathurin Brisson, who was named *président des traites* at Fontenay in 1726, and of Louise-Gabrielle Jourdain. He belonged to one of the most famous families of the legal nobility of the Poitou, particularly known for the jurist Barnabé Brisson, *président à mortier* of the Parliament of Paris who was executed in 1591 by the Holy League. Brisson was also related to the illustrious naturalist Réaumur; Catherine Brisson, his father's sister, had married Réaumur's younger brother.

After completing his early studies at the Collège de Fontenay in 1737–1738, Brisson finished a year of philosophy at the Collège de Poitiers and then turned to theology. He passed his baccalaureate in theology in 1744, and after taking minor orders he was allowed to continue his studies at the St.-Sulpice Seminary in Paris in 1745. But in 1747, just when he was to be elevated to the subdiaconate, he renounced that vocation and soon returned to his family.

Brisson then resumed the study of natural history, which he had begun with Réaumur when the latter spent his vacation on his country estate in the Poitou. In October 1749 Réaumur engaged him as caretaker and demonstrator of his own collection of natural history, as a successor to the Abbé Menou, who had recently died. This position, paying 600 livres a year, was underwritten by the Académie des Sciences, to which Réaumur had donated his collection. This post was a responsible one, for the attendant not only had to classify and care for Réaumur's collections, but also to help him during his observations and experiments and to be his main collaborator and confidant. Brisson's first research was thus set in the line of fire of the great rivalry between Réaumur and Buffon.

Buffon had attempted in his *Histoire naturelle* to give a general description of the animal world based upon the collections in the Cabinet du Roi, and Réaumur desired to launch a similar enterprise in extending the six volumes of his *Mémoires pour servir à l'histoire des insectes* by using the innumerable observations he had made or gathered from his correspondents, as well as the numerous specimens in his collection. Brisson was to play the principal role in this project. After having translated J. T. Klein's *Système du règne animal* (1754), he published the *Règne animal* (1756), a bilingual work with Latin and French texts printed side by side. In it he announced the project, presented the classification, and dealt with the study of the first two classes: quadrupeds and cetaceans. He then went on to the study of ornithology, in which specialty Réaumur's collection was extremely rich. But after Réaumur's death in October 1757, Brisson had to give up the care of his collection, which the Academy had transferred to the Cabinet du Roi, under the supervision of Buffon and Daubenton. Allowed for a time to continue his research, Brisson soon found himself denied access to the collections, but in spite of this he managed to publish the six volumes of his *Ornithologie* (1760).

This work, also in Latin and French, contained the descriptions of 1,500 species of birds, grouped into 115 genera, twenty-six orders, and two classes (distinguished by the presence or absence of webbed feet). There were 220 plates by F.-N. Martinet of 500 birds, many of which had never been illustrated before. In spite of its insufficient classification, this work, essentially didactic and written without style or pictorial research, was one of the most complete treatises in ornithology before the *Histoire des oiseaux* of Buffon, Guéneau de Montbeillard, and Gabriel Bexon.

Brisson, who had continued to receive the payment previously allocated to him as caretaker of Réaumur's collection, was elected an adjoint fellow in botany of the Academy in 1759 and royal censor in 1760. But forever deprived of any access to direct documentation and being the target of Buffon and his colleagues' hostility, he understood that it was useless to continue his work as a naturalist. On the advice of the Abbé Nollet he then turned to experimental physics, to which he devoted all his time.

In 1768 Nollet, who had been appointed to the Collège de La Fère, had Brisson named his deputy professor and successor to the chair of experimental physics that had been created for him at the Collège de Navarre in 1753. In 1770, a few months before his death, Nollet also arranged that Brisson be named his successor as "master of physics and natural history to the children of France," which position put him in touch with the royal family and assured him a comfortable living. In 1771 Brisson translated Priestley's *History of Electricity* and took this occasion to defend passionately Nollet's point of view against that of Franklin. The few memoirs he presented before the Academy concerned physics: the measurement of density, refraction, burning mirrors, barometers, magnetism, and atmospheric electricity.

But it was as a botanist that Brisson was made an associate member of the Academy in 1779 and a supernumerary pensioner in 1782, and it was only upon the reorganization of 1785 that he became a pensioner of the new section of general physics. In the meantime he had published the *Dictionnaire raisonné de physique* (1781), which was a fair presentation of various aspects of physics at that time but was soon out of date in spite of some additions in 1784, at the time of the first aerostatic experiments. His *Pesanteur spécifique des corps* (1787) was of more lasting interest, for many rather precise experimental data were included. In 1789 Brisson published in his *Traité élémentaire ou Principes de physique* the essentials of his courses given both privately and at the Collège de Navarre. The success of these courses is borne out by the testimony of a Russian traveler, P. I. Strakhov, who attended them from 1785 to 1787 and was enthusiastic over the presentation of Brisson's new discoveries on gases. On his return to Russia, Strakhov organized a course in experimental physics at the University of Moscow and published a Russian translation of the *Traité élémentaire*.

Brisson married Marie-Denise Foliot de Foucherolles in 1775. Their son, Louis-Antoine, who had the king and queen as godparents, died when he was only ten. There also were two daughters. Brisson was financially comfortable until the Revolution. In 1792 he became a member of the Academy commission entrusted with preparations for setting up the metric system, but was removed in 1793. He was reinstated after Thermidor and also was on the first list of professors for the new *écoles centrales*. In December 1795 he was appointed resident member of the experimental physics section, first class, of the Institut National. From 1796 on, Brisson taught experimental physics and chemistry at the Collège des Quatre Nations and published well-conceived, up-to-date manuals. He also published, but prematurely, several lessons on the comparison between the new and old units of measure, a subject he had studied when on the Commission of Weights and Measures. In 1801 Chaptal saw to it that his salary as professor was continued, but his appointment to the professorship at the Lycée Bonaparte in 1805 was purely honorific. He died only a short time later, following a stroke that made his last months very painful.

Having undergone an involuntary adjustment in his activities, Brisson carried on two successive careers, a short one as a naturalist and Réaumur's collaborator, and a longer one as Nollet's disciple and the disseminator of the ideas of experimental physics. His rather considerable influence was due to his teaching and his works, which, although not containing any original or important discoveries, were nevertheless an excellent means of spreading the scientific knowledge of the time. Probably his creative contribution would have been more important had not Buffon opposed the pursuit of his work as a naturalist.

BIBLIOGRAPHY

I. ORIGINAL WORKS. Brisson's writings include *Regnum animale in classes IX distributum . . . Le règne animal divisé en IX classes . . .* (Paris, 1756; Leiden, 1780); *Ornithologia, sive Synopsis methodica sistens avium divisionem in ordines . . . Ornithologie ou Méthode contenant la division des oiseaux en ordres . . .*, 6 vols. (Paris, 1760; Latin part reissued, 2 vols., Leiden, 1763, and 1 vol., Paris, 1788); *Dictionnaire raisonné de physique,* 3 vols. (Paris, 1781; 2nd ed., Paris, 1800); *Observations sur les nouvelles découvertes aérostatiques . . .* (Paris, 1784); *Pesanteur spécifique des corps* (Paris, 1787); *Traité élémentaire ou Principes de physique . . .,* 3 vols. (Paris, 1789; 3rd ed., 1800), translated into Russian by P. I. Strakhov, 3 vols. (Moscow, 1801–1802; 2nd ed., incomplete, 2 vols., 1812), also translated into Georgian (Tiflis, 1812); *Rapport sur la vérification du mètre*

qui doit servir d'étalon (Paris, 1795), with Borda; *Instruction sur les nouveaux poids et mesures* (Paris, 1795); *Réduction des mesures et poids nouveaux* (Paris, 1799), new ed. entitled *Instruction sur les mesures et poids nouveaux* (1800); *Principes élémentaires de l'histoire naturelle et chymique des substances minérales* (Paris, 1797); and *Élémens ou principes physico-chymiques, destinés à servir de suite aux principes de physique à l'usage des écoles centrales* (Paris, 1800, 1803).

Manuscripts are "Dossier Brisson, Mathurin Jacques," in Archives de l'Académie des Sciences, Paris; and Library of the Institut National, Paris, MS 2041, no. 90: "Principales étapes de la vie de M. Brisson écrites par lui-même," among the papers of J. M. Delambre, transmitted by J. Bertrand.

II. SECONDARY LITERATURE. Works on Brisson are Roman d'Amat, in *Dictionnaire de biographie française,* VII (1956), cols. 366–367; H. Beauchet-Filleau and Paul Beauchet-Filleau, in *Dictionnaire historique et généalogique des familles du Poitou,* 2nd ed., II (Poitiers, 1895), 6, and III (1905), 389; A. Birembaut, "Les liens de famille entre Réaumur et Brisson, son dernier élève," in *Revue d'histoire des sciences,* **9** (1958), 167–169, and in the collection *La vie et l'oeuvre de Réaumur* (Paris, 1962), pp. 168–170; J. B. Delambre, in *Mémoires de la classe des sciences mathématiques et physiques de l'Institut National de France* (2nd semester 1806), pp. 184–205; Constant Merland, "Mathurin-Jacques Brisson," in *Biographies vendéennes,* II (Nantes, 1883), 1–47, partially repr. in *Revue de la Société Littéraire Historique et Généalogique de la Vendée* (1st trimestre 1883), pp. 145–161, and (2nd trimestre 1883), pp. 37–46; Poggendorff, I (1863), col. 301; J. Quérard, *La France littéraire,* I (Paris, 1827), 518–519, in which two notices on B. Brisson are wrongly attributed to M. J. Brisson; R. Taton, ed., *L'enseignement et la diffusion des sciences en France au XVIIIe siècle* (Paris, 1964), esp. pp. 158, 630–632, 640, 648; J. Torlais, *L'abbé Nollet* (Paris, 1954), esp. pp. 234–236, and *Réaumur,* rev. ed. (Paris, 1961), esp. pp. 343–345; and M. G. Th. Villenave, in F. Hoefer, *Nouvelle biographie générale,* VII (1863), cols. 437–438.

RENÉ TATON

BRITTEN, JAMES (*b.* London, England, 3 May 1846; *d.* Brentford, Middlesex, England, 8 October 1924), *botany.*

Britten was very clearly interested in botany as a child, and for many years he continued to be a keen field botanist, contributing notes on various flora of English counties. He has acknowledged the pleasures he derived from Anne Pratt's *Flowering Plants, Grasses, Sedges and Ferns of Great Britain* (1855). He was educated privately and for five years, until he reached the age of twenty-three, he resided with a doctor at High Wycombe, Buckinghamshire, as a preliminary to entering the medical profession. During that period he was secretary to the local natural history society and editor of its magazine. But he abandoned his medical studies when, in 1869, he was appointed assistant in the herbarium of the Royal Botanic Gardens, Kew. The following year he attended a course of lectures at University College, London, given by Daniel Oliver, who at the time was also keeper of the herbarium at Kew. Oliver befriended Britten and greatly encouraged him in botanical field pursuits, which they often shared. On Oliver's advice, Britten applied for a post in the department of botany in the British Museum, then still part of the main establishment at Bloomsbury. Much to the resentment of Sir Joseph Hooker, he left Kew, after only two years, to begin what was to be an industrious and somewhat tempestuous career at the museum, where his irascible, controversial temperament was often an embarrassment to his colleagues.

Britten's most important contributions to botany concerned historical, literary, and biographical aspects of the subject; and his long obituary notices, notes, and anecdotal comments, occasionally caustic and vituperative, are most valuable reference sources. With G. S. Boulger he produced in 1893 the invaluable *Biographical Index of British and Irish Botanists,* to which three supplements were published before his death and of which a second edition, revised by Dr. A. R. Rendle, appeared in 1931. His interest in old English dialects and folklore was manifest in many of his writings, especially in two works published by the English Dialect Society: the important *Dictionary of English Plant Names* (1878–1886), compiled in cooperation with Robert Holland, and his reprint, with notes, of William Turner's *The Names of Herbes* (1881). He also published a number of articles of ephemeral importance in the popular scientific press of the day.

The herbarium of Sir Hans Sloane, probably the most extensive collection of plants in existence in the seventeenth or eighteenth century, was purchased when the British Museum was founded in 1753. Britten's historical bent led him, throughout his service in the museum, to accumulate a mass of data, written on slips of paper, that until recently were the only commentary to the contents of Sloane's 265 bound volumes of *exsiccatae*. They formed the basis of *The Sloane Herbarium,* an annotated list of the *horti sicci* composing the herbarium, with biographical accounts of the principal contributors revised and edited by J. E. Dandy; it was published by the trustees of the British Museum in 1958.

Britten's first published work was a short note on locations of rare plants, mostly in the Thames Valley in Irvine's *The Phytologist* (November 1862). In the following year he contributed the paper "Rare and Exotic Plants at Kew Bridge, Surrey" to the first volume of the *Journal of Botany,* which he later edited for almost forty-five years. He used his editorial prerogative in a highly individualistic manner, and never missed an opportunity to express his candid criticism, deserved or not. This attitude caused resentment, and he was quick to pounce on any apparent shortcomings of the authorities and publications of Kew, which became a constant target for his pungent comments. Relations between his department and Kew were severely strained after an official committee recommended in 1901 that the herbarium collections at the museum should be transferred to Kew. Britten did not conceal his malice toward Kew, particularly criticizing the administration and making personal attacks on members of the staff. His indiscretions led to legal action, and he was obliged in the same year to make a public apology to the Kew authorities and to pay a donation to an agreed charity. His wit was always evident and perhaps one of his best efforts concerned the supposed demise of the regular numbers of the Kew *Bulletin* in a period when four appendices appeared and he surmised that the main publication had developed appendicitis!

Britten was elected a fellow of the Linnean Society in 1870. He was admitted to the Roman Catholic Church when he was twenty-one, and with his characteristic energy threw himself with zest into the propaganda activities of his adopted church. He was particularly prominent in the work of the Catholic Truth Society, and for this and other religious causes he was made a Knight of the Order of St. Gregory in 1897 by Pope Leo XIII and promoted to Knight Commander in 1917.

BIBLIOGRAPHY

Among Britten's writings are "Rare and Exotic Plants at Kew Bridge, Surrey," in *Journal of Botany,* **1** (1863), 375–376; *Dictionary of English Plant Names* (London, 1878–1886), compiled with Robert Holland; editing and notes for a reprint of William Turner's *The Names of Herbes* (London, 1881); *Biographical Index of British and Irish Botanists* (London, 1893), written with G. S. Boulger, 2nd ed., rev. by A. R. Rendle (London, 1931); and "The History of Aiton's 'Hortus Kewensis,'" in *Journal of Botany,* **50** (1912), supp. 3. A bibliography accompanies the obituary notice by A. B. Rendle, in *Journal of Botany,* **62,** 337.

GEORGE TAYLOR

BRITTON, NATHANIEL LORD (*b.* Staten Island, New York, 15 January 1859; *d.* New York, N. Y., 25 June 1934), *botany.*

Britton was educated at the School of Mines of Columbia College, from which he graduated as Engineer of Mines in 1879. He then became assistant in geology at Columbia and later served as botanist and assistant geologist for the Geological Survey of New Jersey for five years. Although his early training was in geology and mining, botanical interests dominated his career. Thus, in 1887 he returned to Columbia as instructor in botany and geology, in 1890 became adjunct professor of botany, and in 1891 was made professor of botany.

He was married in 1885 to Elizabeth Gertrude Knight (1858–1934), herself a botanist of distinction. Best known in her own specialty of bryology, she also was a constant helper in her husband's work.

Britton is best known for his role in the establishment and development of the New York Botanical Garden, a process set in motion in 1888, on a visit to the Royal Botanic Gardens at Kew, England, when his wife asked, "Why couldn't we have something like this in New York?" This question led to the formation of a committee by the Torrey Botanical Club to consider the establishment of a botanical garden in New York City. It appealed for funds in 1889, and in 1891 the New York Legislature chartered the New York Botanical Garden Corporation, which by July 1895 had persuaded the city to set aside 250 acres in Bronx Park for the development of the garden. In 1896 Britton was formally appointed director in chief, and during the next thirty-three years, thanks to his enthusiasm, initiative, drive, and organizing ability, this undeveloped area without buildings or roads became a garden with greenhouses, laboratories, library, and herbarium that make it one of the world's great botanical institutions.

Britton's own interests were primarily taxonomic and concerned mainly with the plants of eastern North America and the West Indies. His adherence to the American Code of Nomenclature and his extremely narrow generic concept have made much of the nomenclature of his works obsolete, but their value as comprehensive descriptive surveys remains. Britton founded the *Bulletin of the New York Botanical Garden* in 1896, the garden's *Journal* and *Memoirs* in 1900, and *Addisonia* in 1916, as well as the *North American Flora* in 1905. The periodical *Brittonia* (founded in 1931) is named for him, as are the plant genera *Brittonamra*, *Brittonastrom*, and *Brittonella*. *Bryobrittonia* commemorates Mrs. Britton.

BIBLIOGRAPHY

I. ORIGINAL WORKS. The most important of Britton's publications are *Illustrated Flora of the Northern United States, Canada and the British Possessions,* 3 vols. (New York, 1896–1898; 2nd ed., 1913), written with Addison Brown; *Manual of the Flora of the Northern States and Canada* (New York, 1901; 2nd ed., 1905); accounts of various families in *North American Flora* (1905–1930); *North American Trees* (New York, 1908), written with J. A. Shafer; *Flora of Bermuda* (New York, 1918); *The Cactaceae,* 4 vols. (Washington, D.C., 1919–1923), written with J. N. Rose; *The Bahama Flora* (New York, 1920), written with C. F. Millspaugh; and *Botany of Porto Rico and the Virgin Islands,* 2 vols. (New York, 1923–1930), written with P. Wilson.

II. SECONDARY LITERATURE. Writings on Britton are H. A. Gleason, "The Scientific Work of Nathaniel Lord Britton," in *Proceedings of the American Philosophical Society,* **104** (1960), 205–226; M. A. Howe, "Nathaniel Lord Britton," in *Journal of the New York Botanical Garden,* **35** (1934), 169–180; E. D. Merrill, "Biographical Memoir of Nathaniel Lord Britton," in *Biographical Memoirs of the National Academy of Sciences of the United States of America,* **295** (1934), 147–202; and T. A. Sprague, "Nathaniel Lord Britton," in *Kew Bulletin* (1934), 275–279.

WILLIAM T. STEARN

BROCA, PIERRE PAUL (*b.* Sainte Foy-la-Grande, near Bordeaux, France, 28 June 1824; *d.* Paris, France, 8 July 1880), *medicine, anthropology.*

Broca was the son of a Huguenot doctor, Benjamin Broca; his mother was the daughter of a Protestant preacher. At the local college he received a *bachelier ès lettres* and diplomas in mathematics and physical sciences. He entered the University of Paris medical school in 1841, and in an unusually short time became *externe* (1843), *interne* (1844), and prosector of anatomy (1848); he received the M.D. degree in 1849.

Broca's graduate studies were in pathology, anatomy, and surgery, and in 1853 he became assistant professor at the Faculty of Medicine and surgeon of the Central Bureau. He was an active figure in the Anatomical Society of Paris and in the Society of Surgery. His interests later turned to anthropology, and he was one of the most outstanding pioneers of the new discipline. During this period he held important posts in the hospitals of Paris, finally serving as surgeon to the Necker Hospital. In 1867 Broca was elected to the chair of *pathologie externe* at the Faculty of Medicine, and the following year he became professor of clinical surgery. He was elected a life member of the French Senate, representing science, six months before he died. He also received many honors in the medical and scientific world, and at his death was vice-president of the French Academy of Medicine.

Broca's versatility was noteworthy and his knowledge wide; his bibliography reveals the breadth of his scientific and clinical work. He made important contributions to anatomy, pathology, surgery, cerebral function, and other areas of medicine, and to anthropology. He showed deep interest in all his work and approached each problem with enthusiasm and thoroughness. He married the daughter of a Paris physician named Lugol.

Broca published several minor papers on anatomy, as well as *La splanchnologie,* a volume of the *Atlas d'anatomie descriptive.* During the twenty years or so after his graduation, he wrote extensively on pathology, including a two-volume work on tumors, *Traité des tumeurs.* These studies were closely associated with his contributions to surgery, which also appeared in his publications of this period; a book on strangulated hernia (1853) and one on aneurysm (1856) demonstrated his theoretical and practical knowledge of surgery.

Broca is, however, better known for his role in the discovery of cortical localization in the brain. This concept had begun with the phrenologists earlier in the century; but the majority of physicians, following Pierre Flourens, denied it. In a famous discussion in Paris in 1861, Broca was able to provide the essential link in the argument favoring the localization of speech function in the left inferior frontal gyrus (since known as Broca's convolution); an aphasic patient was found to have a lesion there. Much later it was shown that the lesion was not so precisely located as Broca had claimed, but his evidence was nevertheless a significant step toward proving that the cerebral hemisphere has localized areas of function, although precise parcelation is no longer accepted. He published extensively on cerebral localization and on normal, comparative, and pathological anatomy of the brain.

Broca's equally important labors were in anthropology, which field he helped to create. In 1847 he served on a commission to report on excavations in the cemetery of the Celestins, and this led him to study craniology and ethnology. These subjects suited him best, for they allowed him to use his anatomical and mathematical skills as well as his diversified knowledge; and his synthetic abilities were necessary to coordinate the wide range of data presented. He was mainly responsible for the formation of the Société d'Anthropologie de Paris in 1859, of the *Revue d'anthropologie* in 1872, and of the École d'Anthropologie in 1876. At this time anthropology was considered by both church and government to be sinister and subversive, but Broca surmounted all opposition and eventually established it securely. He invented at least twenty-seven instruments for the more accurate study of craniology, and he helped to standardize methods. Between 1850 and his death he published 223 papers and monographs on general anthropology, ethnology, physical anthropology, and other aspects of the field.

BIBLIOGRAPHY

I. ORIGINAL WORKS. S. Pozzi compiled a bibliography of Broca's writings according to subject (see below); this has been reprinted by Huard (see below). In addition, *Mémoires sur le cerveau de l'homme et des primates,* S. Pozzi, ed. (Paris, 1888), contains a wide selection of his papers. Broca's books are *La splanchnologie* (Paris, 1850–1866), Vol. III of Broca, C. Bonamy, E. Beau, *Atlas d'anatomie descriptive du corps humain* (Paris, 1844–1866); *De l'étranglement dans les hernies abdominales et des affections qui peuvent le simuler* (Paris, 1853; 2nd ed., 1856); *Des anévrismes et de leur traitement* (Paris, 1856); and *Traité des tumeurs,* 2 vols. (Paris, 1866–1869).

II. SECONDARY LITERATURE. There is an obituary of Broca in *The Lancet* (1880), **2**, 153–154. Articles on Broca (listed chronologically) include J. R. C., "Paul Broca of Paris," in *Edinburgh Medical Journal,* **26** (1880), 186–192; "Paul Broca, Honorary Member," in *Journal of the Anthropological Institute* (London), **10** (1880–1881), 242–261; S. Pozzi, "Biographie-bibliographie," in *Revue scientifique,* 3rd ser., **2** (1881), 2–12; *Popular Science Monthly,* **20** (1881–1882), 261–266; R. Fletcher, "Paul Broca and the French School of Anthropology," 15 April 1882, in *Saturday Lectures No. 6* (Washington, D.C., 1882), and in Fletcher's *Miscellaneous Papers 1882–1913* (Washington, D.C.); S. Zaborowski, "La psychologie et les travaux de M. Broca," in *Revue internationale des sciences biologiques,* **10** (1882), 141–159; M. Genty, "Broca (Paul) (1824–1880)," in *Les biographies médicales,* **9** (1935), 209–224, with portraits and references to iconography; K. Goldstein, "Pierre Paul Broca (1824–1880)," in W. Haymaker, ed., *Founders of Neurology* (Springfield, Ill., 1953), pp. 259–263; and P. Huard, "Paul Broca (1824–1880)," in *Revue d'histoire des sciences,* **14** (1961), 47–86, which contains references to most of the secondary material on Broca and the Pozzi bibliography of 1881.

EDWIN CLARKE

BROCARD, PIERRE RENÉ JEAN-BAPTISTE HENRI (*b.* Vignot, France, 12 May 1845; *d.* Bar-le-Duc, France, 16 January 1922), *mathematics, meteorology.*

Henri Brocard, born in a small, unpretentious town in northeastern France, was the son of Jean Sebastien and Elizabeth Auguste Liouville Brocard. No record has been found of brothers, sisters, or other close relatives, and Brocard never married. He is now known chiefly for his work in the geometry of the triangle, but he is also remembered as a French army officer and a meteorologist.

For some time, knowledge of Brocard's life fell far short of knowledge about the Brocard configuration, on which his renown rests. This was remedied by an autobiographical account published in 1894 at Bar-le-Duc. It covers the first fifty years of Brocard's life and tells of his mathematical and scientific publications and activities. He sent a copy of this pamphlet to the Smithsonian Institution shortly after its publication.

Brocard received his early education at the *lycée* of Marseilles, and the *lycée* and academy of Strasbourg. He attended the École Polytechnique from 1865 to 1867, and then joined the Corps of Engineers of the French army. It is known that he was a prisoner of war at Sedan in 1870, but for the most part his army career was devoted to teaching and research rather than to active combat. He became a life member of the newly organized Société Mathématique de France in 1873, and in 1875 he was made a life member of the Association Française pour l'Avancement des Sciences and of the Société Météorologique de France. For several years after 1874 he was assigned to service in north Africa, chiefly in Algiers and Oran. He was a co-founder of the Meteorological Institute at Algiers.

As a member of the local committee for the tenth session of the Association Française pour l'Avancement des Sciences, which met in Algiers in 1881, he presented a paper entitled "Étude d'un nouveau cercle du plan du triangle." It was in this paper that he announced the discovery of the circle that is now known by his name. In 1884 he returned to Montpellier, where he had taught for a short time after his graduation from the École Polytechnique.

There followed appointments to many government commissions and many scientific honors. Brocard served with the Meteorological Commission at Montpellier, Grenoble, and Bar-le-Duc. In 1894 he became a member of the Society of Letters, Sciences, and Arts of Bar-le-Duc; and it is through the publications of this society that one can follow the activities of the last twenty-six years of his life. His scientific and mathematical publications began when he was about twenty-three, and over the years showed him to be an indefatigable correspondent with the editors of mathematical and scientific journals. Brocard contributed to *Nouvelles annales de mathématiques, Bulletin de la Société mathématique de France, Mathesis, Zeitschrift für mathematischen und naturwissenschaftlichen Unterricht, Educational Times, El progreso matemático, L'intermédiaire des mathématiciens,* and many others. In his autobiography, a brief descriptive paragraph of about three or four lines is devoted to each journal, giving the names of the editors, the dates of publication, etc. These paragraphs provide a succinct and handy source of information, particularly for journals that later ceased publication.

Brocard's most extensive publication was a large, two-part work entitled *Notes de bibliographie des courbes géométriques,* followed by *Courbes géométriques remarquables,* which appeared under the joint authorship of Brocard and T. Lemoyne. The first part of the earlier work appeared in 1897, and the second in 1899. Probably no more than about fifty copies of this work were prepared, lithographed in the printscript of the author, and privately distributed. The *Notes* may be regarded as a source book of geometric curves, with a painstakingly prepared index containing more than a thousand named curves. The text consists of brief descriptive paragraphs, with diagrams and equations of these curves. About twenty years later, Volume I of the projected three-volume work *Courbes géométriques remarquables* was published in Paris. In 1967 both Volume II and a new edition of Volume I were published in Paris. *Courbes géométriques remarquables* is described as an outgrowth of *Notes de bibliographie des courbes géométriques.*

During the latter part of his life, Brocard made his home in Bar-le-Duc. He lived completely alone and rarely had visitors. He obviously enjoyed his membership in, and his work as librarian of, the Society of Letters, Sciences, and Arts of Bar-le-Duc, although he had declined the honor of becoming president. Largely through his efforts, one of the streets of Bar-le-Duc was named for Louis Joblot, a native Barisian who was an acknowledged but almost forgotten pioneer in the field of microscopy. When he retired from the army in 1910, Brocard was a lieutenant colonel and an officer in the Legion of Honor. In his retirement he spent much of his time making astronomical observations with a small telescope in the garden behind his house. Every fourth year he took a long trip to the meetings of the International Congress of Mathematicians.

The unit of mathematical theory identified as the Brocard configuration is founded upon two points, O and O', in a triangle ABC such that the angles OAB, OBC, and OCA, and the angles $O'BA$, $O'CB$, and $O'AC$ are equal. Brocard readily admitted that he had no claim to priority in the discovery of the existence of these points. Yet his influence upon his contemporaries was so great that the points O and O' are now universally recognized as the Brocard points of a triangle.

Of the several solutions available for the construction of the Brocard points of a triangle, the most striking and familiar is one in which circles are drawn as follows: A circle tangent to side AB of triangle ABC at A and passing through C; a second circle tangent to BC at B and passing through A; and a third circle tangent to CA at C and passing through B. It is easily proved that these three circles are concurrent at a point O which satisfies the above conditions (see Figure 1). Point O' is obtained in a similar manner, after a slight modification in procedure. The angle OAB (angle W) is called the Brocard angle of triangle ABC, and it is a simple matter to prove that

$$\cot W = \cot A + \cot B + \cot C.$$

Obviously a similar relation holds for angle $O'BA$ (angle W'). Brocard's truly original contribution to the theory of the geometry of the triangle was his discovery of the circle drawn on the line segment PK as diameter, where P is the circumcenter of the triangle and K is its symmedian point. This circle, called the Brocard circle of a triangle, passes through the points O and O' and has many additional interesting geometric properties.

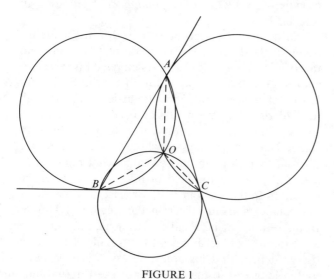

FIGURE 1

During the last decade of the nineteenth century several books were published about the Brocard configuration. The subject, which stirred the imagination and aroused the enthusiasm of many mathematicians in different parts of the world, has remained a pleasant and fruitful topic for discussion.

On 16 January 1922, Brocard was found dead at his desk. In accordance with his specific request, he was buried in the small cemetery at Vignot, next to his father and mother.

BIBLIOGRAPHY

I. ORIGINAL WORKS. Brocard's works are *Notice sur les titres et travaux scientifiques* (Bar-le-Duc, 1895); *Notes de bibliographie des courbes géométriques,* 2 vols. (Bar-le-Duc, 1897–1899); and *Courbes géométriques remarquables,* 2 vols. (I, Paris, 1920; new ed., 1967; II, Paris, 1967), written with T. Lemoyne.

II. SECONDARY LITERATURE. Works dealing with Brocard are Lucien Braye, "De New York à Bar-le-Duc sur les traces d'un théorème," in *L'est républicain,* Bar-le-Duc sec. (16 Sept. 1951); A. Emmerich, *Die Brocard'schen Gebilde* (Berlin, 1891); Laura Guggenbuhl, *Concerning Notes de bibliographie des courbes géométriques* (New York, 1951), and "Henri Brocard and the Geometry of the Triangle," in *Mathematical Gazette,* **32,** no. 322 (Dec. 1953), 241–243, also in *Proceedings of the International Congress of Mathematicians,* II (Amsterdam, 1954), 420–421; and Roger A. Johnson, *Modern Geometry* (Boston, 1929), chs. 12, 16, 17, 18; repr. in paperback as *Advanced Euclidean Geometry* (New York, 1960). Illustration courtesy of *Mathematical Gazette.*

LAURA GUGGENBUHL

BROCCHI, GIOVANNI BATTISTA (*b.* Bassano, Italy, 18 February 1772; *d.* Khartoum, Sudan, 23 September 1826), *geology.*[1]

The son of Cornelio Brocchi and Lucrezia Verci, Brocchi was brought up in a cultured and well-to-do family, educated in the classics and literature, but also brought into contact with the outdoors by his father's enthusiasm for hunting. He early developed strong interests in antiquities and natural history despite parental discouragement. Sent to Padua to study law, he preferred the botanical garden and the lectures in botany by Bonato. Upon the death of his father, Brocchi went to Rome, where he studied the art and monuments of antiquity intensively for six months. Returning to Bassano, he devoted himself to literature, especially the works of Dante, for two years.

In 1802 Brocchi became an instructor in natural history at the Gymnasium in Brescia. Appointment in 1808 as inspector of mines at Milan gave him the opportunity to travel widely through Italy, making extensive notes and collecting numerous specimens. The return of Lombardy to Austria in 1814 deprived Brocchi of this position but did not lessen his activity, which henceforth was centered in Rome. In 1821 he was restored to his position, but meanwhile he had accepted the invitation of the viceroy of Egypt to conduct a survey of the mineral resources of that country and to organize its mining industry. In the fall of 1822 he sailed from Trieste to Egypt, whence he made excursions up the Nile, and into Syria and Palestine. In 1826, as he was preparing to return to

Italy, he contracted bubonic plague,[2] of which he died. His last journals and collections are preserved in the Museo Civico, Bassano.

Brocchi published five major books and contributed about seventy articles to various journals. He wrote upon an amazing range of subjects from antiquities to zoology in carefully documented papers replete with classical references. While at Brescia he published zoological and mineralogical articles; these include observations on the anatomy of insect eyes and on infusoria. In 1818 he conducted experiments on "night air" at Rome, in the hope of finding the cause of malaria; the results were negative but duly reported.[3] He also published several articles on recent shells.

Brocchi's most significant contributions, however, were in the field of geology. Early papers such as the memoir on the Val di Fassa consist largely of mineralogical descriptions. The geognostical introduction to this report shows no appreciation of the stratigraphic significance of fossils beyond the uncertain importance of their presence or absence as criteria for "secondary" or "transition" rocks. The origin of basalts interbedded with limestones is discussed in terms of the Neptunist-Vulcanist controversy.

Brocchi's masterpiece is the *Conchiologia fossile subappennina* (1814). It opens with an eighty-page survey of paleontological studies in Italy—a mine of historical data that has been freely used by Lyell, Zittel, and other writers. The nature of the fossiliferous deposits at various localities is described in detail.[4] He noted the contrast between the Subapennine fossils, most of which could be identified with the living Mediterranean fauna, and those described by Lamarck from the Paris Basin, most of which were extinct. The possibility that deposits of greater antiquity have fewer living species is mentioned, but not examined in detail.[5] A chapter is devoted to the problem of extinction.

This work was completed before William Smith had published on the stratigraphic significance of fossils, and probably before Brocchi had read Cuvier and Brongniart's *Essai sur la géographie minéralogique des environs de Paris* (1811).[6] It is doubtful whether knowledge of the chronological value of fossil shells would have enabled Brocchi to reach more fundamental conclusions concerning these deposits; nearly all his collections were from Pliocene deposits—indeed, they constitute much of the "type" Pliocene of Lyell.[7]

Brocchi made his first extensive exploration of central Italy in 1811–1812. In the course of this he witnessed an eruption of Vesuvius and was thus able to compare the state of its crater before and after an eruption. From 1814 until 1820 his work was largely

on volcanic rocks, especially tuffs. Impressed by the wide distribution of certain ash beds around Rome and Naples, he argued for the submarine deposition of these even in the face of included freshwater fossils, which he insisted were transported. His descriptions of these deposits are thorough and accurate, and his *De stato fisico del suolo de Roma* (1820) still contains valuable data on the geology of the Imperial City.[8] The *Catologo ragionato* (1817) likewise is a useful compendium of original observations on the geology of central and southern Italy.

Brocchi's scientific contribution lay in the accurate observation, recording, and incorporation into the current body of scientific knowledge of a vast amount of new data from very diverse fields, especially the earth sciences.

NOTES

1. Or natural history. Brocchi's journals cover many phases of natural history and archaeology.
2. Malaria, according to Sacchi.
3. It is interesting to note that at this date Brocchi believed that the agent responsible for malarial infection entered the body through the skin rather than by respiration, and he concluded (without, however, suspecting the mosquito) that the draining of swamps would reduce the incidence of the disease.
4. Most of the collection is preserved in the Museo Civico di Storia Naturale di Milano. Its recent history has been summarized by Carla Rossi Ronchetti.
5. Lyell refers repeatedly to Brocchi's work in the *Principles of Geology*. He criticizes him (III, 18-19) for correlating the Subapennine formations with those of the Paris Basin despite the faunal dissimilarity. To what extent Brocchi's suggestive but incomplete analysis of this matter influenced Lyell in formulating his subdivision of the Tertiary into epochs is unknown; it certainly was noted and accorded a place in the exposition of this principle.
6. Brocchi in 1817 alludes to Cuvier's work and to Tertiary fossils "buried at the time of the great revolutions of the earth."
7. Lyell, *Principles of Geology*, 2nd ed. (1833), III, 157, refers various Subapennine beds of Brocchi to the Miocene, Older Pliocene, and Newer Pliocene. Ronchetti, p. 7, considers all but one of the localities of the Brocchi collection to be Pliocene.
8. *Cf.* Clerici, p. xciii.

BIBLIOGRAPHY

I. ORIGINAL WORKS. Lists of Brocchi's writings are in *Biographie universelle ancienne et moderne, deuxième edition, par MM Michaud*, V (Paris, 1880), 580–584; and Giuseppi Roberti, in G. I. Ferrazzi, ed., *Atti della festa commemorativa il primo centenario della nascita di Giambattista Brocchi celebratosi in Bassano il xv ottobre MDCCCLXXII* ... (Bassano, 1873; repr. Milan, 1874, 1881), pp. 37–42 (the pagination is different in each edition). Meli (see below) cites several other, less extensive lists and adds a number of previously unnoticed titles.

Brocchi's major works are *Memoria mineralogica sulla*

valle di Fassa in Tirolo (Milan, 1811); *Conchiologia fossile subappennina, con osservazioni geologiche sugli Appennini e sul suolo adiacente di G. Brocchi . . .,* 2 vols. (Milan, 1814); *Catologo ragionato di una raccolta di rocca deposto con ordine geografico per servire alla geognosia dell'Italia* (Milan, 1817); *Dello stato fisico del suolo di Roma . . .* (Rome, 1820); and *Giornale delle osservazioni fatte ne viaggi in Egitto, nella Siria e nella Nubia di G. B. Brocchi,* 5 vols. (Bassano, 1841–1843).

II. Secondary Literature. Works on Brocchi are Giambattista Baseggio, "Della vita e degli studii di Giambattista Brocchi," in G. I. Ferrazzi, ed., *Di Bassano e dei Bassanesi illustri . . .* (Bassano, 1847), pp. 325–358, with portrait; Enrico Clerici, "In occasione del centenario dell'opera di Giovanni Battista Brocchi *Dello stato fisico del suolo di Roma,*" in *Bollettino della Società geologica italiana,* **38** (1919), lxxxiii–xciii; Giovanni Larber, *Elogio storico di Gio. Batt. Brocchi bassanese compilato dal suo concittadino Giovanni Larber* (Padua, 1828), with portrait, the most detailed source; Romolo Meli, "Una lettera inedita dell'insigne naturalista Giambattista Brocchi," in *Bollettino della Società zoologica italiana,* 2nd ser., **7** (1906), 303–323, which contains references to numerous other biographies; Carla Rossi Ronchetti, "I tipi della *Conchiologia fossile subappennina di G. Brocchi,*" in *Rivista italiana di paleontologia e stratigrafia,* supp. **57–62** (1951–1956); Defendente Sacchi, "Elogio biografico," in *Annali universali di statistica,* no. 144 (Feb. 1828), 132 ff.; and Antonio Stoppani, "Elogio di Giambattista Brocchi letto in occasione del primo centenario celebratosi in Bassano al di 15 ottobre 1872," in G. I. Ferrazzi, ed., *Atti di festa commemorativa il primo centenario della nascità di Giambattista Brocchi celebratosi in Bassano il xv ottobre MDCCCLXXII . . .* (Bassano, 1873; repr. Milan, 1874, 1881), pp. 42 ff. (pagination is different in each edition).

Joseph T. Gregory

BROCHANT DE VILLIERS, ANDRÉ-JEAN-FRANÇOIS-MARIE (*b.* Villiers, near Mantes, France, 6 August 1772; *d.* Paris, France, 16 May 1840), *geology, mineralogy.*

Brochant warmly supported Haüy's theories of crystal structure, although his writings also aided in the diffusion in France of A. G. Werner's mineral classification and nomenclature, as well as the early crystallographic ideas of C. S. Weiss. Wernerian principles influenced his geological memoirs, which were concerned exclusively with the geology of the Alps. He devoted the last two decades of his life primarily to the preparation of a geological map of France.

Brochant studied mineralogy under Werner at Freiberg from 1791 to 1793 and entered the newly organized École des Mines in 1794. He became an engineer in the Agence des Mines in 1800 and an editor of the *Journal des mines* in 1801. In 1804 Brochant was named professor of geology and min-

eralogy at the École des Mines, at that time located at Pesey in Tarentaise (Savoy), and he retained this post when the school was moved to Paris in 1815. He then became a member of the Académie des Sciences, a director of the glass factory at St. Gobain, and later inspector general of mines.

Stimulated by the publication of Greenough's geological map of England in 1822, Brochant pleaded successfully for the preparation of a similar map for France. After a preliminary journey to England with his collaborators, Armand Dufrénoy and Élie de Beaumont, to establish procedures, Brochant began the survey in 1825. During the next ten years, Élie and Dufrénoy made summer field trips, the former in eastern France and the latter in western France, while Brochant supervised the work and compiled the results. He published a report on the project, "Notice sur la carte géologique générale de la France," in the *Comptes rendus* (**1** [1835], 423–429). The completed map was published in 1841, after his death.

BIBLIOGRAPHY

I. Original Works. Brochant published two works in mineralogy: *Traité élémentaire de minéralogie, suivant les principes du professeur Werner,* 2 vols. (Paris, 1801–1802; 1808); and *De la cristallisation considérée géométriquement et physiquement, ou traité abrégée de cristallographie* (Strasbourg, 1819). In geology he compiled a four-volume *Mémoires pour servir à une description géologique de la France* (Paris, 1830–1838); published posthumously was his *Explication de la carte géologique de la France, rédigée sous la direction de M. Brochant de Villiers . . . par M. M. Dufrénoy et Élie de Beaumont* (Paris, 1841). Brochant's most important geological memoirs were "Observations géologiques sur les terrains de transition qui se trouvent dans la Tarentaise et autres parties des Alpes," in *Journal des mines,* **23** (1808), 321–380; "Observations sur les terrains de gypse ancien qui se trouvent dans les Alpes," in *Annales des mines,* **2** (1817), 256–300; and "Considérations sur la place qui doivent occuper les roches granitoides du Mont Blanc et d'autres cimes centrales des Alpes dans l'ordre d'antériorité des terrains primitif," *ibid.,* **4** (1819), 283–300.

II. Secondary Works. Brochant is discussed in Alfred Lacroix, "Notice historique sur le troisième fauteuil de la section de minéralogie," in *Académie des sciences—séance publique annuelle du lundi 17 décembre 1928* (Paris, 1928), pp. 18–26.

John G. Burke

BRODIE, BENJAMIN COLLINS (*b.* Winterslow, Wiltshire, England, 8 June 1783; *d.* Broome Park, Betchworth, Surrey, England, 21 October 1862), *physiology, surgery.*

Benjamin Collins Brodie was the third son of Rev. Peter Brodie, the rector of the parish of Winterslow, and Sarah Collins, the daughter of a banker from Salisbury. He received his early education at home, being taught—along with his elder sister and brothers—by his father. At eighteen he went to London to study medicine and began attending the anatomy lectures of John Abernethy, pupil and "disciple" of John Hunter, at St. Bartholomew's Hospital. He entered the Windmill Street School of Anatomy in 1802 and in 1803 became surgical pupil of Everard Home, enrolling in St. George's Hospital in June 1804. His father had died in March of that year, leaving the family in difficult circumstances. Fortunately, through his uncle, Thomas Denman, a distinguished obstetrician, Brodie became known to, and was helped by, many of the prominent medical men in London at that time. In May 1805, he was appointed house surgeon at St. George's, and on 18 October of that year was admitted as a member of the Royal College of Surgeons. He became assistant surgeon at St. George's in March 1808, and began lecturing on surgery at the Windmill Street School of Anatomy. He continued this course until 1830. In 1816 he married Anne Sellon; they had four children, one of whom, Benjamin, Jr., became a famous chemist.

Brodie's scientific career has three aspects: he was a researcher, a surgeon and general practitioner, and a member of the medical establishment. He contributed six papers to the *Philosophical Transactions* of the Royal Society between 1809 and 1814. These papers, whose significance is further discussed below, were not so much concerned with physiological theory as with factual experimental reporting. They were widely recognized, however, and their impact on the Royal Society gave Brodie considerable professional prominence. He was elected a member of the Royal Society, at the age of twenty-six, in February 1810, and in 1811 was awarded the Copley Medal, the youngest member ever to receive it. He was the Croonian Lecturer in 1810 and 1813, and in 1858 was elected president. He was the first surgeon to hold this post.

Around 1811, however, Brodie realized that he would, in effect, have to choose between research and teaching on the one hand, and surgery and private practice on the other. Thus, from that time on, his written publications consisted mainly of reports to the clinical journals. He had kept scrupulous case notes from the very beginning of his career, and sixteen volumes of these were given to St. George's Hospital by his grandson. His book *Diseases of the Joints* was extremely influential and ran to five editions, the first appearing in 1818 and the last in 1850. It is based on an analysis of case histories, and Brodie clearly preferred to retain limbs rather than amputate them. In an age when women were cosseted indoors, he was a pioneer advocate of fresh air and exercise, and was able to demonstrate how many joint afflictions probably had hysterical origins.

Brodie was an excellent diagnostician and built up a flourishing practice as the leading surgical consultant. He became personal surgeon to King George IV in 1828, and president of the Royal College of Surgeons in 1844, having been an active and reforming member of its council for a number of years. He was knighted in 1834, and three years later, as he felt befitted his new station in life, bought a landed estate in Surrey.

In retrospect, Brodie's six papers in the *Philosophical Transactions* seem to have had an impact out of all proportion to their theoretical importance, for the positive contributions were slight. In trying to assess why this was so, we find important clues to the prevailing state of physiological thought at the time. With the exception of the first paper, which was communicated to the Society by Everard Home, all the papers were communicated by, or first read to, the Society for Promoting Knowledge of Animal Chemistry, which had been founded *ca.* 1802 and of which Humphry Davy was a member. Physiology in England had received a great stimulus from the work of Joseph Black, Joseph Priestley, and Adair Crawford, and was also influenced—to a lesser extent—by the researches of Lavoisier and Laplace on respiration and animal heat. In particular Adair Crawford's theory of animal heat production, based both on the chemical theory of respiration and on Black's theory of specific heats, apparently tied up in a neat theoretical form one of the most perplexing problems of animal physiology. The importance of the new chemistry for physiology was recognized by the founding of the society mentioned above and by active research by many medical men.

Brodie's papers made their impact almost solely because the empirical results they presented challenged the whole chemical theory of animal heat, with respiration (and, by implication, combustion) as the actual source of heat production. The results were indisputable. He destroyed the animal's brain by pithing, decapitation, or poisoning, yet maintained respiration and heartbeat artificially—managing to do this for periods up to two and a half hours, getting the appropriate changes of color in the blood. If respiratory changes were the immediate cause of the heat in animals, then the temperature of the animals should be maintained. This did not occur. Moreover, if Brodie inactivated the higher cerebral centers by poisoning, then gradually allowed the animal to recover, as the "sensibility" was recovered, the animal also recovered the power of generating heat, until it could counteract the loss of heat due to the cold of the surrounding atmosphere. Brodie concluded, rightly, (1) that the presence of the brain was not *directly* necessary to the action of the heart, but only indirectly, by maintaining the "life" of the organs; (2) that if the brain is destroyed, the secretory functions and heat production are totally impaired; (3) that respiratory changes could not, *of themselves,* be responsible for animal heat production, because in his experimental situation, with the respiration artificially maintained, if the air inspired was cooler than the natural temperature of the animal, the effect of the artificial respiration was to diminish the heat of the animal, not increase it; (4) that the phenomena of animal heat production were very complicated indeed, and it was relevant to question whether, in view of the multifarious processes going on in the animal body, one was justified in attributing only one of them to animal heat production.

In later years Brodie was often represented as producing a "nervous" theory of animal heat (e.g., by Claude Bernard, in his *La chaleur animale,* 1876, p. 290). This is wrong: there is little trace of any theory in these papers. Brodie said, at the end of the 1812 paper, that he did not wish to advance some opinions, but simply to state facts. In certain aspects his empirical findings needed to be explained, and his doubts were genuine ones. Animal heat production and maintenance are controlled by nervous centers. It was recognition of this fact that led Bernard so many years later to quote Brodie. But it was the direct conflict between Brodie's results and the contemporary chemical ideas that made Brodie's papers the sensation they were.

BIBLIOGRAPHY

The Works of Sir B. C. Brodie, collected by C. Hawkins, includes an autobiographical sketch in the first volume, pp. 1–116.

Biographies of Brodie are T. Holmes, *Sir Benjamin Brodie* (1898); and William R. LeFanu, "Sir Benjamin Brodie, F.R.S. (1783–1862)," in *Notes and Records of the Royal Society of London,* **19,** no. 1 (1964), 42–52. See also G. J. Goodfield, *The Growth of Scientific Physiology* (1960), especially ch. 4, pp. 76–99.

G. J. GOODFIELD

BRODIE, BENJAMIN COLLINS, JR. (*b.* London, England, 5 February 1817; *d.* Torquay, England, 24 November 1880), *chemistry.*

Brodie was the son of a prominent surgeon who was also a president of the Royal Society. He attended Harrow and then entered Balliol College, Oxford. After graduating, he left England for the famous chemical laboratory at Giessen, where, in the summer of 1845, he worked under Liebig on the analysis of beeswax. He returned home about 1847 and continued his chemical studies in a private laboratory in London. He was elected a fellow of the Royal Society in 1849. In 1855 he became Waynflete professor of chemistry at Oxford, a position he held until 1873.

Brodie pursued the investigation of waxes that he had begun in Germany, and soon discovered and named cerotic acid, cerotin, and melissic acid. A few years later he examined graphite and discovered graphitic acid. Brodie discussed his findings in this early work in terms of the accepted language of the atomic theory, representing the new substances by conventional formulas. He soon abandoned this treatment, however.

Brodie is interesting as a historical figure on account of his drastic proposals for an alternative approach to chemistry, which he first put before the Royal Society in 1866. By this time a skeptical attitude toward Dalton's atomic theory had arisen. Chemists welcomed the convenience of the theory as a summary of known facts and employed its language, but they were not prepared to discuss the question of the ultimate divisibility of matter. It was sufficient to know that certain masses were undivided by the powers of chemistry. Gerhardt's statement that chemical formulas did not represent actual atomic arrangements, but merely the relations between substances in chemical change, was quoted with approval by Brodie, who placed it at the head of his 1866 paper.

This was not enough for Brodie, however. The advertisement in a chemical journal of a set of balls and wires as an aid to the study of chemical combination was evidence enough for him that the science of chemistry had gone "off the rails of philosophy." Only an accumulation of errors could have produced such a "bathos." He therefore proposed to substitute an exact language, free from any association with Dalton's theory. He said that this would be independent of any hypothesis on the nature of matter. Its symbols would simply express the facts.

The new symbols represented chemical operations performed on space. Brodie regarded the method as an applied algebra and stated that he had been guided by algebraic procedures in geometry and logic. He made several references to George Boole. Just as in geometry a symbol could indicate the operation on a unit of length by which a line was generated, so in chemistry a symbol could represent an operation on space which produced a weight. All substances were considered to be perfect gases. The chemical unit of ponderable matter was that which occupied 1,000 cc., the "unit of space," at standard temperature and pressure. The symbol α represented the operation on the unit of space from which a unit of hydrogen resulted. The compound that resulted from the successive operations α and χ had the symbol $\alpha\chi$. The logarithmic relation $\alpha + \chi = \alpha\chi$ was one of the fundamental equations of the calculus.

In Brodie's scheme only weight would be considered; the form, color, and other properties of matter would be neglected. He would investigate the distribution of weights in chemical change by the appropriate symbols, which were derived with the aid of experimental data on gases. For example, the decomposition of water vapor, in which two volumes of water produce two volumes of hydrogen and one of oxygen, was represented as $2\phi = 2\phi_1 + \phi_2$, where ϕ, ϕ_1, ϕ_2 are the symbols of units of water, hydrogen, and oxygen, respectively.

Let $\phi = \alpha^m \xi^n$, $\phi_1 = \alpha$, $\phi_2 = \alpha^p \xi^q$, where α and ξ symbolize uncompounded component weights (prime factors) and m, n, p, and q are positive integers. By the fundamental logarithmic relationship $2\phi = 2\phi_1\phi_2$, or $(\alpha^m \xi^n)^2 = \alpha^2 \alpha^p \xi^q$, so that $2m = 2 + p$ and $2n = q$. Density data were brought into the calculation and the equations were finally solved, subject to the condition of a minimum number of prime factors. The unique solution is then $m = 1$, $n = 1$, $p = 0$, $q = 2$. The symbol for the unit of oxygen is therefore ξ^2 and that for water is $\alpha\xi$.

Symbols of other substances, obtained by similar reasoning, formed three distinct classes. One group of substances, typified by hydrogen, was represented by a single letter, that is, only one operation on space was needed for their production. The group of double-lettered symbols indicated that two successive operations were needed to form the substances concerned. This was the case with oxygen, ξ^2 or $\xi\xi$. The third group included chlorine, $\alpha\chi^2$; nitrogen, $\alpha\upsilon^2$; and hydrogen peroxide, $\alpha\xi^2$. These substances could not be formed by fewer than three operations on the unit of space.

Brodie discussed the exciting implications of his symbols. Substances regarded as elements were of the same symbolic form as substances known to be compounds. The ξ^2 component of hydrogen peroxide indicated the presence of oxygen. But how were χ^2, υ^2, and ξ to be interpreted? They could not be dismissed as imaginary units, since, Brodie argued, they had been discovered in the course of analysis. Nor would he positively claim that υ, ξ, and χ represented real pieces of matter. He preferred to call such symbols "ideal."

He could not exclude the possibility that chlorine, for example, might really consist of α and χ. This view that the elements might be compounded had earlier been favored by Davy and Prout. Brodie speculated that at some remote age, when the earth was very hot, the simple forms of matter—α, χ, ξ, and so forth—might have existed. As the earth's temperature fell, these formed combinations, like chlorine, which were so stable that they never decomposed again. The primitive materials might still exist, however, on the sun and in distant nebulae. The independent existence of χ and υ might be detected by a spectroscopic examination of heavenly bodies. Indeed, a recent investigation of a nebula had shown its spectrum to be like that of nitrogen, but less complex. Later, he also seized on some erroneous experiments on chlorine as evidence of its compound nature.

Brodie failed to convince chemists to abandon their association with atoms. The weakness of the calculus of operations was indicated by Naquet, who translated Brodie's work into French. He pointed out that there was no way to distinguish isomers, compounds such as acetaldehyde and ethylene oxide, which have identical compositions by weight but exhibit completely different properties. This anomaly was readily explained on the atomic theory by assuming different arrangements of particles in space. Brodie had restricted his treatment to weights, and without additional suppositions his calculus could not explain the different qualitative changes resulting from identical gravimetric redistributions. The spatial properties of matter assumed particular importance in the stereochemistry of Van't Hoff and Le Bel, which appeared in 1874. The excess explanatory capacity of the atomic theory won the confidence of chemists, and Brodie's calculus was never adopted. It was left as a curious relic of the positivistic tendencies of nineteenth-century chemistry.

BIBLIOGRAPHY

I. ORIGINAL WORKS. Some of Brodie's scientific papers include "An Investigation on the Chemical Nature of Wax," in *Philosophical Transactions,* **138** (1848), 147–158, 159–170, and **139** (1849), 91–108; "On the Atomic Weight of Graphite," *ibid.,* **149** (1859), 249–259; "On the Decomposition of the Simple Weight X Effected by Victor Meyer," in *Journal of the Chemical Society,* **35** (1879), 673–682. A list of Brodie's other scientific papers is given in the *Royal Society Catalogue of Scientific Papers.*

Brodie presented his calculus in "The Calculus of Chemical Operations," in *Philosophical Transactions,* **156** (1866), 781–859, and **167** (1877), 35–116. A less formal exposition, together with some reactions to the calculus, can be found in Brodie's "On the Mode of Representation

Afforded by the Chemical Calculus, as Contrasted With the Atomic Theory," in *Chemical News,* **15** (1867), 295–305. A corrected version of this was later published as *Ideal Chemistry* (London, 1880).

II. SECONDARY LITERATURE. Brodie's work has been discussed in W. H. Brock and D. M. Knight's "The Atomic Debates: Memorable and Interesting Evenings in the Life of the Chemical Society," in *Isis,* **56** (1965), 5–25; and in W. V. Farrar's "Sir B. C. Brodie and his Calculus of Chemical Operations," in *Chymia,* **9** (1964), 169–179. A. Naquet's critique of Brodie's calculus may be found in "Considerations on the Two Memoirs of Sir B. C. Brodie on the Calculus of Chemical Operations," in *Philosophical Magazine,* **7** (1879), 418–432. Some additional information on Brodie can be found in J. R. Partington, *A History of Chemistry,* IV (London, 1961–   ), 425–427.

D. C. GOODMAN

BRØGGER, WALDEMAR CHRISTOPHER (*b.* Christiania [now Oslo], Norway, 10 November 1851; *d.* Oslo, 17 February 1940), *geology.*

Brøgger was the son of Anton Wilhelm Brøgger, a well-known book publisher, and Oline Marie Bjerring. In 1878 he married Antonie Siewers; their home fostered musical and other cultural interests. After receiving his baccalaureate in 1869, Brøgger took the *examen philosophicum* (a year's study of various subjects, before specialization) at the University of Oslo in 1870 and then began to study zoology. In 1872 he published a paper on the mollusks of the Oslo Fjord. Attracted by Theodor Kjerulf, professor of mineralogy and geology, he transferred to that department and in 1876 received an appointment as assistant at the mineralogical institute of the university. At the same time he served as assistant at the Geological Survey of Norway. During this period he was president of the Norwegian Students' Union several times.

During the next few years, Brøgger engaged in rich and varied scientific activity, and in 1881 he was appointed professor at the newly established Stockholms Høgskola (now the University of Stockholm). There he founded a first-class mineralogical institute, as well as teaching and conducting research that dealt mainly with the geology and mineralogy of Norway.

In 1890 Brøgger succeeded Kjerulf as professor at the University of Oslo, remaining in this post until 1917. His main contribution was in the study of the Permian eruptive rocks of the Oslo district; and he was one of the pioneers in the theory of magmatic differentiation. Besides his writings, Brøgger published, alone or in collaboration with his colleague and successor J. Schetelig, numerous geological maps. Among other important petrographical works is that

on carbonatite-containing rocks at Fen, in the Precambrian just outside the Oslo district, published when he was nearly seventy; this type of rock was one of his main concerns. Outstanding among Brøgger's many mineralogical contributions is his long treatise on pegmatite minerals from Langesund Fjord, in the southernmost part of the Oslo district (1890).

Other scientific achievements are Brøgger's important studies on tectonic geology, Cambro-Silurian stratigraphy and paleontology, and Quaternary geology. At the presentation of the Wollaston Medal to Brøgger in 1911, the president of the Geological Society of London said: "In an age of specialization he is a specialist, but a specialist in almost every branch of science. That it should have fallen to one man to do so much and so well almost passes belief."

With his forceful personality and his administrative ability, Brøgger became a leader in Norwegian scientific circles at an early stage in his career. For a time he was rector of the University of Oslo, and for many years served as president of the Norwegian Academy of Science in Oslo. By establishing a number of scientific endowments, largely through private and institutional subscriptions, and through his untiring efforts to achieve recognition for scientific research, he had a strong and lasting effect on the history of his country. Between 1900 and 1906 he was a member of the Storting (parliament), where he championed the cause of science.

BIBLIOGRAPHY

I. ORIGINAL WORKS. Brøgger's writings include "Bidrag til Kristianiafjordens molluskfauna," in *Nyt magazin for naturvidenskaberne,* **19** (1872), 103–145; "Om paradoxidesskifrene ved krekling," *ibid.,* **24** (1878), 18–88; "Die silurischen Etagen 2 und 3 im Kristianiagebiet und auf Eker, ihre Gliederung, Fossilien, Schichtenstörungen und Contactmetamorphosen," in *Universitätsprogramm,* 2nd sem. 1882, no. 8; "Spaltenverwerfungen in der Gegend Langesund-Skien," in *Nyt magazin for naturvidenskaberne,* **28** (1884), 253–419; "Ueber die Bildungsgeschichte des Kristianiafjords. Ein Beitrag zum Verständnis der Fjord- und Seebildung in Skandinavien," *ibid.,* **30** (1886), 99–231; "Die Mineralien der Syenitpegmatitgänge der südnorwegischen Augit- und Nephelinsyenite. Mit zahlreichen chemisch-analytischen Beiträgen von P. T. Cleve," in *Zeitschrift für Krystallographie und Mineralogie,* **16** (1890); "Die Eruptivgesteine des Kristianiagebietes. I. Die Gesteine der Grorudit-Tinguait-Serie," in *Videnskaps-selskapets skrifter, mathematisk-naturvidenskapelig klasse* (1894), no. 4; "Die Eruptivgesteine des Kristianiagebietes. II. Die Eruptionsfolge der triadischen Eruptivgesteine bei Predazzo in Südtyrol," *ibid.* (1895), no. 7; "Die Eruptivgesteine des Kristianiagebietes. III. Das Ganggefolge des Laurdalits," *ibid.* (1897), no. 6; "Om de senglaciale og postglaciale nivåforandringer i Kristaniafeltet (molluskfaunan). . .," in *Norges geologiske undersøgelse,* no. 31 (1900); "Strandliniens beliggenhed under stenalderen i det sydøstlige Norge," *ibid.,* no. 41 (1905); "Die Eruptivgesteine des Kristianiagebietes. IV. Das Fengebiet in Telemark, Norwegen. . .," in *Videnskaps-selskapets skrifter, mathematisk-naturvidenskapelig klasse* (1920), no. 9; *Geologisk oversikts-kart over Kristianiafeltet. 1:250,000 . . .* (Christiania, 1923); "Die Explosionsbreccie bei Sevaldrud zwischen Randsfjord und Sperillen," in *Norsk geologisk tidsskrift,* **11** (1930), 281–346; "Die Eruptivgesteine des Oslogebietes. V. Der grosse Hurumvulkan," in *Norske Videnskaps-Akademi i Oslo, matematisk-naturvidenskapelig klasse* (1930), no. 6; "Die Eruptivgesteine des Oslogebietes. VI. Über verschiedene Ganggesteine des Oslogebietes," *ibid.* (1931), no. 7; "Essexitrekkens erupsjoner. Den elste vulkanske virksomhet i Oslofeltet," in *Norges geologiske undersøkelse,* no. 138 (1933); "Die Eruptivgesteine des Oslogebietes. VII. Die chemische Zusammensetzung der Eruptivgesteine des Oslogebietes. . .," in *Norske Videnskaps-Akademi i Oslo, matematisk-naturvidenskapelig klasse* (1933), no. 1; "On Several Archäan Rocks From the South Coast of Norway. I. Nodular Granites From the Environs of Kragerø," *ibid.,* no. 8; and "On Several Archäan Rocks From the South Coast of Norway. II. The South Norwegian Hyperites and Their Metamorphism," *ibid.* (1934), no. 1.

II. Secondary Literature. A special publication, *Ved W. C. Brøggers bortgang* (Oslo, 1940), was printed and distributed by Norske Videnskaps-Akademi soon after Brøgger's death. It contains, besides biographical data, speeches delivered at the memorial held in the university auditorium on the day of his funeral and at a special meeting of the Academy of Science and Letters in Oslo on 8 March 1940. The scientific achievements of Brøgger are discussed by V. M. Goldschmidt and Olaf Holtedahl in that publication, which also contains a complete bibliography by W. P. Sommerfeldt on pp. 57–105.

Olaf Holtedahl

BROGLIE,   LOUIS-CÉSAR-VICTOR-MAURICE DE (*b.* Paris, France, 27 April 1875; *d.* Neuilly-sur-Seine, France, 14 July 1960), *physics.*

The ancient family of de Broglie has long supplied France with soldiers, diplomats, and politicians of the highest rank. Maurice, the second of the family's five children and the eldest son, was virtually required by tradition to follow a diplomatic or a military career. After some discussion, his grandfather, the head of the family, allowed him to enter the navy, the most technically demanding of the services, which Maurice preferred partly from a taste for the sea and partly from a youthful but sincere interest in the physical sciences. After a brilliant career at the École Navale (1893–1895), he was assigned to the Mediterranean Squadron, which he improved by installing the first French shipboard wireless. Simultaneously he continued his education in the schools of Toulon and Marseilles University, from which he emerged with the *licence ès science* in 1900.

As Maurice de Broglie's experience of science and technology grew, he thought to resign his commission and to follow a career in physics. In 1898 he broached the possibility to his family. His grandfather was scandalized. "Science," said the old duke, "an old lady content with the attractions of old men," was no career for a de Broglie. A compromise was effected: Maurice fitted out a room of the family mansion as a laboratory and returned to the navy. There he so distinguished himself in wireless work that it appeared that he might add lustre to the family name by following his bent; and so he was able, after the death of his grandfather, to convert the furlough he obtained for his marriage in 1904 into an indefinite leave, which lasted until he formally resigned his commission in 1908. During those years he prepared for his new career, first at the observatory at Meudon, where he studied spectroscopy with Deslandres, and then at the Collège de France, where, in 1908, he successfully defended a thesis on ionic mobilities which he had prepared under the direction of Paul Langevin.

The "ions" of de Broglie's thesis were charged particles of smoke and dust floating about in a gas. His research involved two central problems in the physics of the time, the mechanism of ionization and the measurement of Brownian motion. He accordingly had occasion to work with and to improve upon some of the most advanced techniques then employed in studying and producing ionization; in particular, he tried to improve the capricious apparatus used to generate X rays. The research was done in his own home, in his unusually well-equipped private laboratory, in which he was to work for the rest of his long career.

For a time de Broglie pursued themes suggested by his thesis, either alone or in collaboration with his old teacher from Toulon, L. Brizard. These investigations, which were gradually moving him out of the mainstream of physics, ended in 1912, when Laue and the Braggs showed that X rays could yield diffraction patterns. De Broglie immediately took up the study of X-ray spectra, which became the chief field of his researches and the subject of his most notable discoveries. His earliest important contribution was the "method of the rotating crystal," an application, and perhaps an independent discovery, of the "focusing effect" first described by the Braggs. With this technique, which eliminates spurious spectral lines arising from local imperfections in the face of the diffracting crystal, de Broglie explored the X-ray emission spectra at the same time that Moseley was preparing his classic papers (1913–1914); but whereas the latter breathlessly mapped a few high-frequency lines from many elements, and so arrived at his well-known formulas, the former, proceeding more cautiously, investigated a wider spectrum in only a few metals and found no regularities. World War I interrupted the work of both men. Moseley enlisted in the army and was killed at Gallipoli; de Broglie rejoined the navy and labored on submarine communications, for which he received a medal. Meanwhile the neutral Swedes, particularly Manne Siegbahn, advanced the study of X-ray spectra by extending Moseley's mapping with the aid of de Broglie's technique.

After the war de Broglie returned to his laboratory and the X rays, attending primarily to their absorption spectra, which he had briefly examined during a period of leave in 1916. Then he had made the capital discovery of the third L absorption edge, a matter of great theoretical interest; now he began (partly with the aid of A. Dauvillier) a careful study of the fine structure of the various edges. This investigation led naturally to the exploration of "corpuscular spectra," i.e., of the velocities of photoelectrons released by X rays of ·a given frequency v. These "spectra" can reveal the various absorption edges of an atom, as the difference between hv and the kinetic energy of a liberated electron gives the energy of the absorption edge with which that electron was associated before its release. De Broglie was joined in these researches, which date from 1921–1922, by his brother Louis, still undecided about whether to settle on physics for his own career; and their cooperation proved very helpful in refining Bohr's specification of the substructure of the various atomic shells. Subsequently (1924) the brothers briefly studied an analogous phenomenon together, the Compton effect.

In the mid-1920's de Broglie began to direct more of his energy toward the leadership of his laboratory. Although he continued to work on diverse problems involving X rays, he began increasingly to look for other aspects of the interaction between radiation and matter. His private laboratory, kept up to date, began to attract students, some of whom were to become leaders of French science. The first group included (besides Dauvillier and L. de Broglie) J. Thibaud, J. Trillat, F. Dupré la Tour, and L. Leprince-Ringuet. After their initiation they tended to follow their own interests, and so de Broglie's private laboratory became the scene of pioneering studies in nuclear physics and cosmic radiation. He followed this work closely, initially (1930–1932) as a collaborator, and subsequently as an influential, inspirational master. In this last role, for which his informed, independent, natural —we might even say aristocratic—authority admirably prepared him, he made what are perhaps his most lasting contributions to French physics.

Many honors came to de Broglie, among which we may single out election to the Académie des Sciences (1924), the Académie Française (1934), the French Atomic Energy Commission, the Académie de Marine, and the Institut Océanographique. He also played a part in international science—for example, through his participation in the Solvay Congresses and his successful early texts on X rays, one of which was translated into English. He was also, perhaps, the last representative of a type that has contributed mightily to the advancement of science—the wealthy independent experimentalist who could follow what he pleased as far as his energies and ability might carry him.

BIBLIOGRAPHY

I. ORIGINAL WORKS. Adequate although not exhaustive bibliographies of de Broglie's writings will be found in W. Wilson, "Maurice, Le Duc de Broglie," in *Biographical Memoirs of Fellows of the Royal Society* (1961), pp. 31–36, and in P. Lépine, *Notice sur la vie et les travaux de Maurice de Broglie* (Paris, 1962). Among the most important of his papers are "Sur un nouveau procédé permettant d'obtenir la photographie des spectres de raies des rayons de Röntgen," in *Comptes rendus de l'Académie des sciences,* **157** (1913), 924–926 (the rotating crystal); "Sur un système de bandes d'absorption correspondant aux rayons L . . .," *ibid.,* **162** (1916), 352–354 (the third L edge); "Les spectres corpusculaires . . ." (several articles alone and in collaboration with L. de Broglie), *ibid.,* **172** (1921), 274–275, 527–529, 746–748, 806–808; **173** (1921), 939–941, 1157–1160; **174** (1922), 939–941; **175** (1922), 1139–1141; and "La relation $hv = E$ dans les phénomènes photo-électriques . . .," in *Atomes et électrons* (Paris, 1923), pp. 80–130 (de Broglie's report to the Solvay Congress of 1921). Among the most important of his books are *La théorie du rayonnement et les quanta* (Paris, 1912), M. de Broglie and P. Langevin, eds. (the proceedings of the first Solvay Congress); *Les rayons X* (Paris, 1922); and, with L. de Broglie, *Introduction à la physique des rayons X et gamma* (Paris, 1928).

II. SECONDARY LITERATURE. For biographical material, see Lépine's *Notice;* the obituaries by Wilson (*op. cit.*); L. Leprince-Ringuet, *Comptes rendus de l'Académie des sciences,* **251,** pt. 3 (1961), 297–303; and R. Sudre, *Revue de deux mondes* (July–Aug. 1960), 577–582; family reminiscences by the Comtesse de Pange (Pauline de Broglie), "Comment j'ai vu 1900," *ibid.* (Apr. 1962), pp. 548–557; and L. de Broglie, *Savants et découverts* (Paris, 1951), pp. 298–305; and the romantic dynastic history by La Varende, *Les Broglie* (Paris, 1950), pp. 265–320. For some assessment of de Broglie's work on X rays and its background, see

J. L. Heilbron, "The Kossel-Sommerfeld Theory and the Ring Atom," in *Isis*, **58** (1967), 451–485, and works there cited.

ADRIENNE R. WEILL-BRUNSCHVICG
JOHN L. HEILBRON

BROILI, FERDINAND (*b.* Mühlbach, Germany, 11 April 1874; *d.* Mühlbach, 30 April 1946), *paleontology, geology.*

Ferdinand was the son of J. B. Broili, squire of the castle of Mühlbach, near Karlstadt. The family was of Italian origin; an ancestor had emigrated in 1741 from Treviso to Würzburg. At first Broili attended the village school in Mühlbach, then the Gymnasium in Würzburg. While still a child he collected his first fossils from a shell limestone quarry on the estate and thus became interested in geology. In 1894 Broili began his study of the natural sciences at the University of Würzburg. The following year he transferred to Munich, where he was the favorite student of the internationally renowned paleontologist Karl von Zittel, under whose guidance he received the Ph.D. in 1898 with a thesis on paleontology.

In 1899, after one year as assistant in the geology department of the Technische Hochschule of Munich, Broili became Zittel's assistant at the State Paleontological Collection in Munich, which at that time was probably the most significant of its kind. In 1901 Zittel sent him to Texas, where he and the American Charles Sternberg successfully collected and investigated amphibian and reptile fossils of the Permian era. For nearly a decade he occupied himself extensively with the Permian fauna in Texas, and between 1904 and 1913 he published several works on his investigations, especially on saurians.

In 1903 Broili qualified as academic lecturer under Zittel, again with a paleontological work. When Zittel died in 1904, Broili was appointed the custodian of the State Paleontological Collection; in 1909 he became curator, and after 1908 he had the title of professor. In 1919 he was appointed director of the Institute for Paleontology and Historical Geology of the University of Munich, as well as director of the State Paleontological Collection. In 1904 he had married Emma Morneburg of Passau; they had one son and one daughter.

Broili's main concern was paleontology, but he also participated in the geological investigations in the Bavarian Alps.

In paleontology, Broili was very active in many areas. In 1919 he began extensive investigations of the many fossils of the laminated lime formations of the upper Malm at Sonthofen and in Eichstätt, in Upper Franconia, most of them unique surviving fossils. He was especially successful in his investigations of winged reptiles, demonstrating that they had hairy coverings, and thus were warm-blooded, as well as that they had webbed skin and a pecten on the crown of the head. Broili did not allow his duties as director of the State Paleontological Collection to restrict his scientific work, but conducted many-faceted investigations of the Paleozoic and Mesozoic eras. At the beginning of his career he had been active outside central Europe, and he continued foreign investigations in the 1920's and the 1930's, stimulating and organizing several expeditions to the Karroo formation in South Africa and also taking a leading part in the evaluation of the findings. In his investigations of fossils Broili did not stop at description and systematic explanation, but also attempted to depict the main life habits of the animals involved and succeeded in working out several excellent descriptions.

Broili also worked on the methodology of paleontology and developed fundamental methods for the evaluation of fossil deposits. As director of the State Paleontological Collection he continued Zittel's pioneer work and greatly enlarged the collection, both in general scope and in number of local Bavarian specimens.

In 1930 he became editor-in-chief of the journal *Paleontographica,* and was coeditor of several other journals. He resigned from his various positions in 1939 and gave his full time to his private investigations, especially to devising a unified description of the amphibia, which remained unfinished. In 1943 Broili left Munich, where he had been active for so long, and spent the rest of his life in Mühlbach.

BIBLIOGRAPHY

I. ORIGINAL WORKS. Although Broili's writings have been only partially published, more than 100 scientific works, predominantly in paleontology, are in print. These are dispersed in many different journals and cover the years 1899–1942. Among his published works should be mentioned *Die permischen Brachiopoden von Timor* (Stuttgart, 1916); *Paläozoologie* (*Systematik*) (Berlin–Leipzig 1921); "Ein neuer Fund von Plesiosaurus aus dem Malm Frankens," in *Abhandlungen der Bayerischen Akademie der Wissenschaften,* Math.-nat. Abt., **30** (1926), Abh. 8; "Über Gemündina Stürtzi Traquair," *ibid.,* n.s. (1930), Abh. 6. The majority of Broili's publications are found in *Sitzungsberichte der Bayerischen Akademie der Wissenschaften.* Math.-nat. Abt. For bibliography, see Dehm and Schroeder (below).

II. SECONDARY LITERATURE. Broili's life and works were

discussed during his lifetime by W. D. Matthew, "Notes on the Scientific Museums of Europe," in *Natural History,* **21** (1921), 184–190. Obituary notices include R. Dehm and J. Schroeder, "Ferdinand Broili 1874–1946," in *Neues Jahrbuch für Mineralogie, Geologie und Paläontologie,* Sec. B (1945–1948), pp. 257–271, with complete bibliography; B. Peyer, "Ferdinand Broili 1874–1946," in *Verhandlungen der Schweizerischen naturforschenden Gesellschaft* (1946), 358–360; and D. M. S. Watson, "Ferdinand Broili," in *Nature,* **158** (1946).

HANS BAUMGÄRTEL

BROMELL, MAGNUS VON, known as **Bromelius** before being raised to the nobility in 1726 (*b.* Stockholm, Sweden, 26 March 1679; *d.* Stockholm, 26 March 1731), *geology.*

Bromell was the son of Olof Bromelius, a prominent physician and botanist in Göteborg. During his studies in Holland, England, and France from 1697 to 1704, he acquired a thorough knowledge of medicine, anatomy, chemistry, and botany; in 1703 he became a doctor of medicine in Rheims. Returning to Sweden, Bromell practiced as a physician in Stockholm, where he periodically gave lectures as a professor of anatomy and in 1724 was elected head of the Collegium Medicum; about 1715 he lectured, as a medical professor, in natural history at the University of Uppsala. An accomplished chemist and mineralogist, he became associated with the Board of Mines, where, in 1720, he was named assessor and, in 1724, head of the chemical laboratory.

Bromell earned his scientific renown as a geologist and mineralogist. He was a passionate collector whose great natural history cabinet—some of it inherited from his father—contained a beautiful collection of ore, mineral, and fossil specimens that he described and partly illustrated. In his small *Mineralogia* (1730), which was also translated into German and which had considerable influence, Bromell classified, to a degree, minerals according to their chemical characteristics and thus became a forerunner of A. F. Cronstedt and later eighteenth-century mineralogists. His "Litographiae Svecanae specimen secundum" (1727–1730) is a pioneer paleontological work that describes a multitude of Swedish animal and plant fossils, including trilobites, ammonites, and corals from Gotland limestone.

BIBLIOGRAPHY

I. ORIGINAL WORKS. Bromell's writings are "Litographiae Svecanae specimen primum et secundum," in *Acta literaria Sveciae,* II (1725-1729) and III (1730-1734); and *Mineralogia eller inledning til nödig kunskap at igenkienna . . . allahanda bergarter* (Stockholm, 1730), trans. into German as *Mineralogia et litographia Svecana,* (Stockholm–Leipzig, 1740).

II. SECONDARY LITERATURE. Works on Bromell are O. Hult, "Några anmärkningar om Olof och Magnus Bromelius," in *Svenska Linnésällskapets Årsskrift* (1926); and G. Regnéll, "On the Position of Palaeontology and Historical Geology in Sweden Before 1800," in *Arkiv för mineralogi och geologi,* **1,** no. 1 (1950), 1–64.

STEN LINDROTH

BROMWICH, THOMAS JOHN I'ANSON (*b.* Wolverhampton, England, 8 February 1875; *d.* Northhampton, England, 26 August 1929), *mathematics.*

Bromwich, whose father was a woolen draper, received his early education in Wolverhampton and in Durban, South Africa, where the family immigrated. He entered Cambridge in October 1892 as a pensioner of Saint John's College and graduated three years later as senior wrangler in a class that included E. T. Whittaker and J. H. Grace. He obtained a fellowship in 1897 but left Cambridge in 1902 to become professor of mathematics at Queen's College, Galway. Bromwich returned in 1907 as permanent lecturer in mathematics at Saint John's College and received the Sc.D. in 1909. He was elected to the Royal Society in 1906 and was active in the London Mathematical Society, serving as its secretary (1911–1919) and vice-president (1919, 1920). The first two decades of Bromwich's career were distinguished by numerous publications and vigorous teaching, but mental affliction led to diminished productivity in his later years and eventually to suicide.

Described by G. H. Hardy as the "best pure mathematician among the applied mathematicians at Cambridge, and the best applied mathematician among the pure mathematicians," Bromwich was well known for his precision, mastery of technique, and skill in algebraic manipulation. But Hardy also described Bromwich as lacking the power of "thinking vaguely" and Bromwich's work as "a little wanting in imagination."

The author of two books, two pamphlets, and some eighty papers, Bromwich is best known for his encyclopedic *Introduction to the Theory of Infinite Series* (1908). Although this book has been praised for its richness of detail and its abundance of examples, it has also been criticized for defects in its general structure—for example, its frequent failure to set off and to emphasize fundamental ideas. The book, based on Bromwich's lectures at Galway, incorporates many of his own researches separately published between 1903 and 1908.

Another series of researches culminated in Bromwich's Cambridge Tract, *Quadratic Forms and Their Classification by Means of Invariant Factors* (1906). In these publications Bromwich's creative powers are most fully evident, for in them he both introduced English readers to Kronecker's ideas and methods in the theory of quadratic and bilinear forms and advanced the knowledge of these forms.

Bromwich's first publication, as well as many later papers, was in applied mathematics. Especially under the influence of George Stokes, Bromwich did significant work in the mathematics of electromagnetism and of other subjects as well (including lawn tennis). Most memorable is a series of papers that began in 1916 with "Normal Coordinates in Dynamical Systems." In this paper Bromwich indicated how Oliver Heaviside's much criticized calculus of symbolic operators could be developed in a manner acceptable to pure mathematicians by treating his operators as contour integrals.

BIBLIOGRAPHY

I. ORIGINAL WORKS. Bromwich's two books are *Quadratic Forms and Their Classification by Means of Invariant Factors* (Cambridge, 1906); and *An Introduction to the Theory of Infinite Series* (London, 1908, 1926). For a bibliography of his papers, see below.

II. SECONDARY LITERATURE. Articles on Bromwich are G. H. Hardy, "Thomas John I'Anson Bromwich," in *Journal of the London Mathematical Society,* **5** (1930), 209–220; and Harold Jeffreys, "Bromwich's Work on Operational Methods," *ibid.,* 220–223. See also G. H. Hardy, *Proceedings of the Royal Society of London* (Section A), **129** (1930), i–x. All three of the above articles include bibliographies; the last includes a portrait. See also "Bromwich," in *Alumni Cantabrigienses,* Part II, Vol. I (Cambridge, 1940), 392.

MICHAEL J. CROWE

BRONGNIART, ADOLPHE-THÉODORE (*b.* Paris, France, 14 January 1801; *d.* Paris, 18 February 1876), *paleobotany, plant anatomy, plant taxonomy.*

The son of Alexandre Brongniart, the eminent geologist, Brongniart was trained by his father and at an early age collaborated with him. He rapidly showed signs of being a superior student, and his gifts were so carefully developed that he became a precocious scientist capable of an immediately high level of work.

Nothing in Brongniart's life indicates the slightest hesitation in his pursuit of science. Between 1817 and 1828 he was able to attend to his studies and his initiation in science while carrying on original research. In 1818 he was registered for courses in medicine, but they constituted only a fraction of his occupations; two years later he published his first report, on a new genus of crustacean. After this youthful attempt Brongniart hoped to reach the level of the great biological movements of his time: research on the primary divisions of the vegetable kingdom, anatomy and taxonomic anatomy (following the work of Mirbel and Candolle), and the theory of generalized plant sexuality. The progress already made in these fields, as well as that in geology and botanical geography (he had early acquired a knowledge of tropical flora), heralded a new science of which Brongniart was to be the architect: the comparative morphology of living and fossil plants.

In 1822 Brongniart published his first important memoir, on the classification and distribution of fossil plants. In it he conceived of paleobotany as a part of botany and gave it a theoretical value of prime importance for biology as well as for geology. Coming as he did after such scholars as Ernst Schlotheim and Kaspar von Sternberg, Brongniart was not entirely an innovator, but his study did show an assurance previously unknown.

The masterworks of 1828, the *Prodrome* and the *Histoire des végétaux fossiles,* mainly confirmed and extended his early ideas, giving them foundation and breadth of perspective. The *Histoire,* which he had hoped to continue in a second volume (only the first parts appeared in 1837), was a long, methodical, detailed, and precise study that clearly showed Brongniart's two concerns: nomenclature and illustration. Its general principles and theoretical views were expressed in condensed form in the *Prodrome,* to striking effect. In it Brongniart recognized the existence of four successive periods of vegetation, each characterized geologically. Three were particularly well characterized: the first, extending to the end of the Carboniferous, by the vascular cryptogams; the third, covering the Jurassic and the Cretaceous, by ferns and the gymnosperms; the fourth, which was the Tertiary, by the dicotyledons.

Brongniart then divided the vegetable kingdom into six classes: Agamae (thallophytes), cellular cryptogams (liverworts and mosses, i.e., Hepaticae and Muscae), vascular cryptogams, and three classes of phanerogams: gymnosperms, monocotyledonous angiosperms, and dicotyledonous angiosperms. This excellent classification clearly indicated modern views, but unfortunately, for unknown reasons, Brongniart did not follow it in his later publications. For the first time gymnosperms were taken as a class and correctly placed among the phanerogams. After more than a century, the cotyledons were no longer the major criterion for classification.

Although Brongniart agreed with Cuvier's theories of fixity of species and cycles in the history of the earth, he, like Candolle before him, accepted the law of organic improvement of plants, adding to it a fundamental geological dimension. The sequence went from the structural simplicity of the Carboniferous plant life, to the intermediary structure of the Jurassic gymnosperms, to the dicotyledons of the Tertiary and the modern flora. This work led to the biological chain formulated by Hofmeister in 1851. Brongniart noted both the phenomena of extinction, which affected the genera and even the classes of the Carboniferous flora, and the correspondence between changes in fauna and flora and changes in climate.

At twenty-seven, Brongniart seemed to have reached the zenith of his creative power. The year before, he had passed his *agrégation* in medicine and had published a valuable memoir on the fertilization of phanerogams that followed up Amici's early research. The improvements in the microscope (notably by Amici) finally made possible the direct study of fertilization, so Brongniart decided to repeat Vaucher's investigation (which had been attempted by Brongniart's great-uncle, Romain Coquebert, as early as 1794): to follow the process of fertilization all the way to the fusion of the male and female germ cells. Brongniart's text confirmed and generalized the existence of the pollen tube; he also named the embryo sac and adopted the theory of epigenesis. But, most importantly, he confusedly provided a description of two fundamental discoveries: the existence of the tetrads, which appear during male sporogenesis, and the distinction between the fertilized egg and the seed. This work led to a new understanding of classification and of the alternation of generations.

In 1824 Brongniart and a few colleagues founded the *Annales des sciences naturelles.* He succeeded Desfontaines as professor of botany at the Muséum d'Histoire Naturelle in 1833, and the following year he was elected a member of the Académie des Sciences.

Until 1849 Brongniart extended his researches to the whole plant world, past and present, including taxonomy, anatomy, or biology. He was one of the first, after H. T. M. Witham's work of 1833, to use thin sections in paleobotany (inaugurated by W. Nicol). His most notable use of thin sections was in his famous anatomical observations on the *Sigillaria* (1839), a genus in the class of plants proper to the Primary era and related to the lycopods.

In 1849 Brongniart's article "Végétaux fossiles" appeared in d'Orbigny's *Dictionnaire universel d'histoire naturelle* and was also printed separately under the title *Tableau des genres de végétaux fossiles . . . .* This was the first attempt at a synthesis of paleobotany: the inventory of fossil genera as a whole and the place of these genera in natural classification.

After 1849 Brongniart's activity turned more and more to the systematic study of living plants, particularly the Neo-Caledonian flora: Proteaceae, Eleocarpaceae, Saxifragaceae, Cunoniaceae, Myrtaceae, Pittosporaceae, Dilleniaceae, Umbelliferae, Epacridaceae, palms, conifers, and so on. His articles appeared in such journals as *Annales des sciences naturelles, Annales du Muséum, Archives du Muséum, Comptes rendus de l'Académie des sciences,* and *Bulletin de la Société botanique de France.* Yet he had not abandoned his interest in paleobotany. A quarter of a century after the *Tableau des genres . . . fossiles,* his pupil Grand' Eury sent him some silicified seeds from Grand' Croix, near St. Étienne. Brongniart went to work with enthusiasm and made a last great discovery: the pollen chamber in fossil cycads, a structure that he and Bernard Renault also found in a living cycad species from Mexico—*Ceratozamia brongniart*—in 1846.

Brongniart was one of the greatest French botanists of the nineteenth century, and his work exerted a major influence on the progress of botany. It is possible, however, that he was too much influenced by Cuvier and too little by Lamarck, for the theoretical aspect of his work may not quite equal its descriptive excellence.

BIBLIOGRAPHY

I. ORIGINAL WORKS. Brongniart's writings include "Mémoire sur le *Limnadia,* nouveau genre de crustacés," in *Mémoires du Muséum d'histoire naturelle,* **6** (1820), 83–92; "Description d'un nouveau genre de fougère nommée *Ceratopteris,*" in *Bulletin de la Société philomatique de Paris,* **7** (1821), 184–187; "Sur la classification et la distribution des végétaux fossiles en général, et sur ceux des terrains de sédiment supérieur en particulier," in *Mémoires du Muséum d'histoire naturelle,* **8** (1822), 203–240, 297–348, also published separately; *Essai d'une classification naturelle des champignons, ou Tableau méthodique des genres rapportés jusqu'à cette famille* (Paris, 1825); "Recherches sur la génération et le développement de l'embryon dans les végétaux phanérogames," in *Bulletin de la Société philomatique de Paris,* **12** (1826), 170–175, and in *Annales des sciences naturelles,* **12** (1827), 14–53, 145–172, 225–296; "Mémoire sur la famille des Rhamnés," *ibid.,* **10** (1827), 320–386, his thesis for the M.D.; *Prodrome d'une histoire des végétaux fossiles* (Paris, 1828); *Histoire des végétaux fossiles, ou Recherches botaniques et géologiques sur les végétaux renfermés dans les diverses couches du globe,*

2 vols. (Paris, 1837); *Botanique du voyage de la Coquille pendant 4 années 1822–1825* . . . (Paris, 1829), with atlas; "Recherches sur la structure et les fonctions des feuilles," in *Annales des sciences naturelles,* **21** (1830), 420–458; "Observations sur la structure intérieure du *Sigillaria elegans,* comparée à celle des *Lepidodendron* et des *Stigmaria,* et à celle des végétaux vivants," in *Archives du Muséum d'histoire naturelle,* **1** (1839), 405–460; "Note sur un nouveau genre de Cycadées du Mexique," in *Annales des sciences naturelles,* 3rd ser., **5** (1846), 1–9; "Végétaux fossiles," in d'Orbigny's *Dictionnaire universel d'histoire naturelle,* XIII (Paris, 1849), 52–173, published separately as *Tableau des genres de végétaux fossiles considérés sous le point de vue de leur classification botanique et de leur distribution géologique* (Paris, 1849); *Rapport sur les progrès de la botanique phytographique* (Paris, 1868); "Études sur les graines fossiles trouvées à l'état silicifié dans le terrain houiller de Saint-Étienne," in *Annales des sciences naturelles,* section botanique, **20** (1874), 234, and *Comptes rendus de l'Académie des sciences,* **79** (1874), 343, 427, 497; "Sur la structure de l'ovule et da la graine des Cycadées, comparée à celle de diverses graines fossiles du terrain houiller," *ibid.,* **81** (1875), 305–307; and *Recherches sur les graines fossiles silicifiées* (Paris, 1881), with a preface on Brongniart's work by J.-B. Dumas.

II. SECONDARY LITERATURE. For biographical information on Brongniart, see Maxime Cornu, "Éloge de M. Ad. Brongniart," in *Revue scientifique,* **10** (1876), 564–574; Pierre Duchartre et al., *Discours prononcé le 21 fevrier 1876 sur la tombe de M. Adolphe Brongniart* (Paris, 1876), with a good list of works by Brongniart; J.-B. Dumas, "Les Brongniart, généalogies et compléments biographiques" (1923–1924), MS in Bibliothèque du Muséum d'histoire naturelle, and "La famille d'un échevin de Paris: Les Hazon. Généalogies et compléments historiques" (1939), MS in Bibliothèque du Muséum d'histoire naturelle; L. de Launay, *Une grande famille de savants: Les Brongniart* (Paris, 1940); and G. de Saporta, "Étude sur la vie et les travaux paléontologiques d'Adolphe Brongniart," in *Bulletin de la Société botanique de France,* 7th ser., **4** (1876), 373, meeting of 20 April.

Brongniart's work is discussed in Cornu and in Saporta (above). See also P. Bertrand, "La chaire d'anatomie comparée des végétaux vivants et fossiles du Muséum," in *Bulletin du Muséum d'histoire naturelle,* 2nd ser., **13**, no. 5 (1941), 369–391; K. Chester, "Fossil Plant Taxonomy," in W. B. Turrill, ed., *Vistas in Botany,* IV (Oxford, 1964), 238–297; F. H. Knowlton, *Fossil Wood and Lignite of the Potomac Formation,* U.S. Geological Survey Bulletin no. 56 (Washington, D. C., 1889), pp. 1–72; and L. F. Ward, "Historical Review of Paleobotanical Discovery," in U.S. Geological Survey, *Fifth Annual Report (1883–1885)* (Washington, D.C., 1885), pp. 368–425.

JEAN-FRANÇOIS LEROY

BRONGNIART, ALEXANDRE (*b.* Paris, France, 5 February 1770; *d.* Paris, 7 October 1847), *geology.*

Brongniart was the son of the distinguished Parisian architect Alexandre-Théodore Brongniart (1739–1813) and Anne-Louise Degremont. He studied at the École des Mines and later at the École de Médecine, and for a time acted as assistant to his uncle Antoine-Louis Brongniart (1742–1804), who was then professor of chemistry at the Jardin des Plantes. After serving as *aide-pharmacien* in the French forces in the Pyrenees, Brongniart returned to Paris; in 1794 he was appointed *ingénieur des mines,* and in 1797 became professor of natural history at the École Centrale des Quatre-Nations. In 1818 he was appointed *ingénieur en chef des mines,* and in 1822 he succeeded R. J. Haüy as professor of mineralogy at the Muséum d'Histoire Naturelle. Brongniart was elected a member of the Académie des Sciences in 1815. Most of his life was spent in Paris, in teaching, research, and administration. He is said to have been exceptionally helpful and generous to his students, for whom he opened his collections every Sunday; and he attracted distinguished gatherings of scientists at his evening *salons.*

Brongniart traveled widely in western Europe and published geological papers on areas ranging from Sweden to Italy. As a young man, immediately after the Revolution, he visited England to learn the techniques of the ceramics industry, and in 1800 he was appointed director of the Sèvres porcelain factory, a post he held until his death. The problems of ceramic technology occupied his attention increasingly toward the end of his life, and his last major work was the two-volume *Traité des arts céramiques* (1844). He married Cécile, daughter of the statesman-scientist Charles-Étienne Coquebert de Montbret; their only son was the botanist and paleobotanist Adolphe-Théodore Brongniart.

Brongniart's earliest scientific papers (the first was published in 1791) were on various zoological and mineralogical subjects. In the former he was strongly influenced by Georges Cuvier, who was almost exactly his contemporary. For example, his "Essai d'une classification naturelle des reptiles" (1800) emphasized the prime importance of careful comparative anatomy, and on that basis he divided the class Reptilia into four groups. He recognized, however, that one group, the batrachians, was significantly different from all the others, especially in the reproductive organs, and that this distinction was far more important than the more striking difference between the limbless snakes and the rest. In 1804 Pierre Latreille elevated the batrachians into a separate class, the amphibians; Brongniart's grouping of the true reptiles—into chelonians, saurians, and ophidians—has been retained, in essence, in modern systematics.

In 1807 Brongniart published *Traité élémentaire de minéralogie,* a work commissioned as a textbook for his and Haüy's courses at the Faculté des Sciences and the Muséum d'Histoire Naturelle. He adopted a simple scheme of classification based mainly on physical properties, but he also made extensive use of Haüy's crystallographic work. Like other mineralogists at this time, he could not easily distinguish some fine-grained rocks from true simple minerals; but he classed clay and basalt, for example, as *fausses espèces,* recognizing that they were *mélangées* although they were too fine-grained to be analyzed. He emphasized the importance of studying the modes of occurrence of minerals as well as their properties, but firmly avoided any discussion of their origins as being too speculative. The sole exception to this was a review of divergent opinions on the aqueous or igneous origin of basalt, but even in this he avoided expressing his own opinion.

Brongniart's early studies in zoology and mineralogy coalesced in the geological work that made him famous throughout the scientific world. Cuvier had already begun his series of spectacular reconstructions of extinct mammals from the Paris region; these fossils clearly belonged to several distinct periods, but he needed a reliable clue to their relative ages. He and Brongniart therefore collaborated in surveying the region and determining the order of the strata in which the fossils had been found. About 1804 they began a series of traverses of the region, and on 11 April 1808 they read their "Essai sur la géographie minéralogique des environs de Paris" before the Institute. The paper was first published in June 1808; a greatly enlarged version, accompanied by a large colored geological map and several horizontal sections, appeared in 1811. With characteristic modesty Brongniart allowed his name to appear after Cuvier's on this memoir, although the geological work seems to have been largely Brongniart's.

Nine "formations" (distinctive rock units) were recognized in the initial version of the work (the details were amended and amplified later, but the main conclusions were unchanged). The oldest was the Chalk, which underlay a series of later strata that had been deposited in succession, in a kind of gulf or embayment. There was no transition between the lowest, an unfossiliferous clay (*Argile plastique*), and the underlying Chalk; and at one locality a conglomerate showed that the Chalk must have been lithified before the clay was deposited, thus implying a long interval of time. The next formation, the *Calcaire grossier,* a series of limestones, was remarkable in that several subordinate beds could be distinguished over a very wide area, invariably in the same order, by their distinctive assemblages of fossils: "This is a mark of recognition which up to the present time has not deceived us." Marine mollusks were the most abundant fossils (they were described and classified by Lamarck), but they were totally distinct from those of the Chalk. Slow deposition in a calm sea was inferred from the regular stratification and the excellent preservation of the fossils. These limestones passed laterally into an unfossiliferous *Calcaire silicieux.* Both these formations were succeeded by the *Formation gypseuse,* a series of marls and beds of gypsum; among the latter were those quarried for plaster at Montmartre, from which Cuvier was obtaining many of his most remarkable fossil vertebrates— mammals, birds, and reptiles that were unknown not merely in species but even in genus. Rare shells of freshwater genera, and a total absence of marine shells, confirmed the probable freshwater origin of the series; but the following *Sables et Grès marins,* with abundant marine fossils, indicated a return to marine conditions. Yet after an unfossiliferous sandstone (*Grès sans coquilliers*), a *Terrain d'eau douce* with freshwater shells demonstrated that the conditions had changed yet again. This was the highest, and therefore the youngest, of the regularly stratified deposits. But after all of these, and after the excavation of valleys, a superficial *Limon d'aterrissement* had been deposited; and in the valleys these deposits had yielded a fauna of extinct species of elephants, antelopes, and so forth. Although very modern in comparison with all the other deposits, this last formation clearly originated before historic times.

The significance of Brongniart's stratigraphy of the Paris "basin" was quickly recognized. The general nature of stratified sedimentary rocks and the importance of observing their order of superposition were commonplaces in geology before his time. The highest, and therefore most recent, stratified deposit that could be recognized over a wide area was the Chalk, however; only "superficial" deposits, assumed to be relatively recent in origin, were thought to overlie it. Brongniart's work proved that above the Chalk was a complex series of stratified rocks, many of them evidently formed by very slow deposition. By implication, therefore, the time that must have elapsed since the end of the Chalk period was greatly extended. This extension of geological time was the first important effect of the stratigraphy.

Second, the strata showed an alternation between marine and freshwater conditions, countering the earlier assumption that all stratified rocks had been deposited in a gradually shrinking ocean. So important was this conclusion that Brongniart devoted a separate memoir, "Sur les terrains qui paraissent avoir été formés sous l'eau douce" (1810), to arguing it in detail, demonstrating the close analogies between living freshwater mollusks and the corresponding fossils. In this memoir he also described similar freshwater deposits from far outside the Paris region, even into central France (Cantal), arguing that such extensive deposits were not improbable when compared with the Great Lakes of the present day.

This alternation of marine and freshwater conditions implied a broadly cyclic, or at least a repetitive, character for this part of geological time. With his usual caution, Brongniart refrained from speculating on the causes of these changes; but his stratigraphical conclusions were certainly influential in molding Cuvier's geological theory, as first set out in the *Discours préliminaire* to the *Recherches sur les ossemens fossiles* (1812). Here Cuvier used the rather sharp breaks between the successive marine and freshwater formations in support of his hypothesis of *sudden* changes of sea level as a cause of the extinction of terrestrial faunas; but the recurrence of broadly similar marine and freshwater shells in the successive formations proved that these sudden "revolutions" must have been local in their effects.

The third important feature of Brongniart's stratigraphy was his use of fossils for the detailed correlation of strata. Previously it had been normal practice to use the lithology, physical position, and fossil content of a formation, with varying relative emphases, as criteria for recognizing it in widely separated areas. But Brongniart's work demonstrated the value of precisely collected and identified fossils as criteria for tracing a detailed series of strata, which might differ little in either lithology or physical position, across an extensive area. It was the precision with which the method was applied that was original.

A similar detailed use of fossils had been made some years previously by William Smith, who worked on the Jurassic strata around Bath and subsequently over a wide area of England. Smith's work was known to English geologists, but its validity could not be assessed by the scientific community as a whole until, some years later than Brongniart, he eventually published his geological map (1815) and the illustrations of the characteristic fossils on which it was based (1816–1819). Although Smith's work had strict priority, Brongniart's independent discovery of the value of fossils as a tool for stratigraphy was the first to be published, and therefore had the greater influence on the direction of geological research. Brongniart's method was rapidly and successfully applied in other areas, not only to the recognition of similar strata younger than the Chalk but also to the analogous problems of the older strata.

Brongniart himself played an important part in this development. By 1822 he had traveled widely enough to be able to describe strata, equivalent to those of the Paris region, from many different parts of Europe: these descriptions were inserted in the *Description géologique des environs de Paris* (a new edition of the *Géographie minéralogique*). But this extension of Tertiary stratigraphy over a wider geographical area posed the fundamental problem of geological facies. Rocks of the same age could not be expected to have the same lithological characters if they were deposited under different conditions in different areas. Even within the Paris region Brongniart had recognized lateral changes in lithology; on a wider scale the changes became more general. It was this that led him to stress the primacy of fossil evidence over that of lithology as a criterion for age, wherever the two sources of evidence were found to conflict. Thus he recognized that the London Clay of southern England must be the approximate equivalent in age of the *Calcaire grossier* of Paris, although they were totally different in appearance: they occurred in equivalent positions in the succession and, more important, they had similar fossils.

Brongniart defended this methodology in a special memoir, "Sur les caractères zoologiques des formations" (1821), in which its validity was argued with reference to the strata equivalent to the Chalk. His most striking example was his discovery of fossils identical to those of the Chalk (more precisely, of the Greensand) in a hard black limestone outcropping more than 2,000 meters above sea level on a mountain in the Savoy Alps. This countered the earlier belief that strata formed at a single period should be relatively constant in both lithology and physical position; but Brongniart described a series of other occurrences of rocks with similar fossils, which bridged the vast difference between this black Alpine rock and the more usual appearance and position of the Chalk.

The *Mémoire sur les terrains de sédiment supérieurs* (1823) described a similar variety of occurrence, in rocks containing fossils identical to those of the Tertiary strata around Paris. Like the "Chalk" rocks, these varied greatly in lithology, some being associated with volcanic rocks; they might be flat-lying or highly folded; and they might outcrop at any altitude from sea level to the summits of the Alps. In spite of this variety, Brongniart maintained, it was far more satisfactory to accept them all as being of the same age than to assign the Alpine occurrences to a much earlier epoch simply on the grounds of their position. Although such differences of altitude seemed vast by human standards, he argued, they amounted to no more than a millimeter on a two-meter model globe, and should be kept in proportion. Nevertheless, although Brongniart himself drew no such conclusion, his demonstration of such spectacular elevation of relatively recent strata later acted as powerful evidence for a greatly expanded time scale for the earth's history in the hands of those who, like Charles Lyell, believed that the elevation had occurred slowly and gradually.

Brongniart's empirical demonstration of the primacy of fossils as criteria for correlation was justified by a theoretical argument that owed much to Cuvier's catastrophism. Brongniart stressed the relative uniformity of faunas in different areas at the present time, and argued that whereas a whole fauna might be suddenly and drastically extinguished, leaving few survivors, it would take a long time for a new fauna to replace it; each major period should therefore be characterized by a distinctive set of fossils, although a few might be common to more than one period. Brongniart recognized the possible effects of climatic differences on the faunas of a single period and the need to discount "derived" fossils, but he concluded that fossil evidence must be given greater weight than lithology or physical position in determining the relative age of a deposit: only *superposition évidente* could have priority over fossils.

In the oldest "Transition" (broadly, Lower Paleozoic) strata then known, great interest was aroused by the fossil trilobites, a class totally unknown in more recent strata, which nevertheless seemed remarkably "highly organized" for such an ancient period of earth history. In the *Histoire naturelle des crustacés fossiles* (1822) Brongniart published the first full-length study of the trilobites (in the same volume, A.-G. Desmarest made a similar study of the fossil remains of true crustaceans). He classified a wide variety of species from many parts of Europe and even from America, and attempted to group them according to their relative age. For the latter objective he had insufficient evidence from superposition and was misled, for example, by the undisturbed and "young-looking" strata in Scandinavia; but his systematic work on trilobites was an important contribution to the later unraveling of Paleozoic stratigraphy.

Among the distinctive rocks that were commonly thought to antedate even these earliest fossils were the coarse-grained crystalline rocks granite and gabbro (the evidence for an intrusive origin for granite was well-known but generally regarded as atypical; most lists of strata still showed granite as the earliest rock of all). It was therefore an important conclusion when Brongniart showed, in the "Gisement des ophiolites" (1821), that in the Apennines some gabbros and similar rocks actually overlay normal limestones and detrital rocks, and were not

the oldest rocks in the area. Brongniart had no fossil evidence of the age of the sediments but thought the limestones resembled those of the Jura (i.e., of Mesozoic age). This implied that at least one crystalline rock analogous to granite could have been formed much later than the earliest fossils. This conclusion served to emphasize still further the fallibility of distinctive rock types as criteria of particular periods of earth history.

Throughout his life Brongniart remained reluctant to speculate on the causes of the phenomena he described. This is best illustrated by his contribution to one of the most difficult geological puzzles of the period, the phenomena of erratic blocks and striated pavements. These features, which in the 1840s were recognized as the effects of slowly moving ice sheets, were in the 1820s generally attributed to some sudden "diluvial" action in the geologically recent past. In his paper "Blocs des roches des terrains de transport" (1828), Brongniart gave careful descriptions of erratic blocks and striated pavements, and also the first detailed account (with a map) of the sinuous ridges or eskers that cross the low-lying glaciated areas of Sweden. He pointed out that all these features indicated some extremely powerful transporting process acting from the north and able to move large blocks of rock even across the Baltic; but he offered this as a simple conclusion from his observations and refused to speculate further on the nature of the remarkable force involved.

Brongniart's last major geological work, the *Tableau des terrains qui composent l'écorce du globe* (1829), was the culmination of his life's work on the ordered classification and interpretation of rocks. It had a disappointing reception, however, and exerted little influence on the further development of geology. Yet in some ways it was an attempt to tackle problems that are still important in modern geology. Having recognized that the rocks formed at one period might be widely different in appearance, he was concerned to establish a system of nomenclature for the periods of earth history, which would not involve misleading references to particular rock types: in other words, he was trying to distinguish time units from rock units. But this led him to propose a cumbersome and largely novel nomenclature that was difficult to remember and therefore failed to win general acceptance. Moreover, although he expressed the intention of avoiding theorizing about causes, he implicitly accepted the theories of others. Thus, his first and main division of time was between a *période jovienne* (Recent) and a *période saturnienne* (all earlier time); the sedimentary rocks of the latter were divided into *Terrains clysmiens* ("diluvial" or glacial deposits) and *Terrains izemiens* (all other sediments). This implied a distinction between past and present and a unique role for the most recent "revolution" (the glacial period), which were far more questionable than he seemed to realize. On the other hand, he made a clear distinction between the *Terrains stratifiés ou Neptuniens,* which invariably occurred in the same order, and the *Terrains massifs ou Typhoniens* (broadly, igneous rocks), which might be intercalated at any point in the series. He also emphasized that the stratigraphical succession would have to be pieced together from many sections in different areas, correlated as well as possible with each other; and he urged the desirability of fixing type sections for the definition of stratal terms ("une suite de terrains admis comme type ou module"). In such ways he recognized the problems of stratigraphical geology more perceptively than most of his contemporaries. Although his classification had little effect on their thinking, he provided valuable lists of characteristic fossils for each *terrain* from the most recent back to the *Terrains hemilysiens* (pre-Carboniferous); and these were widely used in the rapidly developing stratigraphical research of the following decade.

In retrospect, Brongniart's influence on nineteenth-century geology might have been greater if he had been less cautious about theorizing. In spite of that limitation, however, the influence of his early work on the Paris region, and its later extension to the whole of western Europe, is difficult to overestimate. For this careful stratigraphical work provided the principal model on which much of the exceptionally productive geological research of the period 1810–1840 was based; and thus it lay at the root of the greatest achievement of early nineteenth-century geology, the elucidation of the main outlines of the history of the earth and of life on earth.

BIBLIOGRAPHY

Brongniart's principal writings are "Essai d'une classification naturelle des reptiles," in *Bulletin de la Société Philomathique,* **2** (1800), 81–82, 89–91, and *Mémoires de l'Institut de France,* **1** (1806), 587–637; *Traité élémentaire de minéralogie,* 2 vols. (Paris, 1807); "Essai sur la géographie minéralogique des environs de Paris," written with Georges Cuvier, in *Journal des mines,* **23,** no. 138 (June 1808), 421–458; *Annales du Muséum d'Histoire Naturelle,* **11** (1808), 293–326; *Mémoires de l'Institut Impérial de France,* année 1810 (1811), 1–274, also issued separately (Paris, 1811); and in Cuvier's *Recherches sur les ossemens fossiles,* I (Paris, 1812); "Sur les terrains qui paraissent avoir été formés sous l'eau douce," in *Annales du Muséum d'Histoire Naturelle,* **15** (1810), 357–405; "Eau (minéralogie et géognosie)," in *Dictionnaire des sciences naturelles,* **14** (1819), 1–62; "Sur le gisement des ophiolites (roches à base de serpentine), des euphotides, etc., dans quelques parties des Apennins," in *Annales des mines,* **6** (1821), 177–238; "Sur les caractères zoologiques des formations, avec application de ces caractères à la détermination de quelques terrains de craie," *ibid.,* 537–572; *Description géologique des environs de Paris,* written with Georges Cuvier, new ed. with additions by Brongniart (Paris, 1822); *Histoire naturelle des crustacés fossiles, sous les rapports zoologiques et géologiques,* written with A.-G. Desmarest (Paris, 1822); *Mémoire sur les terrains de sédiment supérieurs calcaro-trappéens du Vicentin, et sur quelques terrains d'Italie, de France, d'Allemagne, etc., qui peuvent se rapporter à la même époque* (Paris, 1823); *Classification et caractères minéralogiques des roches homogènes et hétérogènes* (Paris, 1827); "Notice sur les blocs des roches des terrains de transport en Suède," in *Annales des sciences naturelles,* **14** (1828), 5–22; *Tableau des terrains qui composent l'écorce du globe, ou Essai sur la structure de la partie connue de la terre* (Paris, 1829), also published as "Théorie de la structure de l'écorce du globe," in *Dictionnaire des sciences naturelles,* **54** (1829), 1–256; and *Des volcans et des terrains volcaniques* (Paris, 1829), also published as "Volcans," in *Dictionnaire des sciences naturelles,* **54** (1829), 334–446.

The most important biography is Louis de Launay, *Une grande famille de savants. Les Brongniart* (Paris, 1940). This is chiefly devoted to Alexandre Brongniart, and makes use of MS sources. It includes an extensive, though incomplete, list of Brongniart's published works.

M. J. S. RUDWICK

BRONN, HEINRICH GEORG (*b.* Ziegelhausen bei Heidelberg, Germany, 3 March 1800; *d.* Heidelberg, Germany, 5 July 1862), *paleontology, zoology.*

Fundamental systematic works in paleontology are Bronn's most enduring contribution to science. He was educated in public administration and natural science in Heidelberg, after which he traveled through northern Italy and southern France, making paleontological investigations. In 1833 he became professor of natural science at Heidelberg. For a short time Louis Agassiz studied with Bronn, and decades later returned to purchase Bronn's collection of fossils for the newly established Museum of Comparative Zoology at Harvard.

In 1831 Bronn's *Italiens Tertiär-Gebilde* distinguished different series of Tertiary strata on the principle that in successively more recent strata the number of extinct species diminishes while the number of modern species increases. His divisions correspond closely to those established as Eocene, Miocene, Older Pliocene, and Newer Pliocene in Lyell's *Principles of Geology* (1833). Bronn's extensive *Lethaea geognostica,* which sought to establish a chronological sequence of fossil organisms, summarized all that was then known in stratigraphy and paleontology. His later *Index palaeontologicus* and the *Lethaea geognostica* were for decades the chief reference works in paleontology. At his death Bronn left unfinished the enormous undertaking of systematizing the whole animal kingdom, recent as well as fossil forms, in *Die Klassen und Ordnungen des Thier-reichs.* From 1830 he edited the *Jahrbuch für Mineralogie, Geognosie, Geologie, und Petrefackten-kunde* with Karl von Leonhard, and later with Georg von Leonhard. Soon after Darwin's *Origin of Species* appeared in 1859, Bronn, at Darwin's suggestion, was responsible for having it translated into German.

Near the end of his life Bronn synthesized developmental laws of nature from his extensive and detailed paleontological studies. He rejected theories of development like those of Lamarck and Geoffroy St. Hilaire and the modification of one species into another. Instead, he explained progressive development, "the successive appearance of forms with more and more complicated organization," by a law of creation operating according to a definite plan. Species became extinct and were replaced by improved ones within the limits imposed by the external "conditions of existence." Bronn used this phrase, borrowed from Cuvier, to indicate adaptation to environment rather than functional harmony, as Cuvier had. The conditions of existence, which exerted a negative, restricting effect on organic development, varied as the earth's crust evolved. During the formation of the crust the primitive ocean was first modified by the appearance of islands and mountain chains, which then grew into continents, thereby dividing the earth's surface into separate oceans, inland seas, lakes, coastal plains, and mountain ranges. The organic kingdoms adhered to a law of *terripetal evolution* analogous to the development of the crust. At first entirely pelagic, they evolved littoral, then coastal, and finally continental forms.

Bronn emphasized that these changes, both organic and inorganic, had been gradual and continuous. Like many of his contemporaries, he criticized the new glacial theory of Louis Agassiz because it appeared to introduce a catastrophe into the history of the world. Bronn argued that the gaps then acknowledged in the paleontological and stratigraphic succession were illusory and would be bridged by the future discovery of transitional species and intermediate formations. (Compare Darwin's similar essay, "On the Imperfection of the Geological Record," *Origin of Species,* ch. 9.)

Bronn's contributions, both theoretical and systematic, were recognized by his contemporaries. In 1857 his "Laws of Evolution of the Organic World" won the prize of the Academy of Sciences of Paris. For the *Handbuch einer Geschichte der Natur* he was awarded a prize medal by the Scientific Society of Haarlem, and in 1861 he received the Wollaston Prize of the Geological Society of London.

BIBLIOGRAPHY

Bronn's main theoretical contributions are found in his *Untersuchungen über die Entwicklungsgeschichte der organischen Welt während der Bildungszeit unserer Erdoberfläche* (Stuttgart, 1858), of which the last chapter, summarizing the whole work, is available in English in *The Annals and Magazine of Natural History,* 3rd ser., **4** (1859), 81–90, 175–184.

His more important classificatory and descriptive works are *Gaea Heidelbergensis, oder Mineralogische Beschreibung der Gegend von Heidelberg* (Heidelberg, 1830); *Italiens Tertiär-Gebilde und deren organischen Einschlüsse* (Heidelberg, 1831); *Lethaea geognostica, oder Abbildungen und Beschreibungen der für die Gebirgs-Formationen bezeichnendsten Versteinerungen,* 2 vols. (Stuttgart, 1835–1838 and several later editions); *Handbuch einer Geschichte der Natur,* 2 vols. (Stuttgart, 1841–1843), which includes as its third part *Index palaeontologicus,* a systematic reduction of all published lists of fossils; and *Die Klassen und Ordnungen des Thier-reichs, wissenschaftlich dargestellt in Wort und Bild* (begun in 1859 and continued after Bronn's death), paleontology and zoology combined in one system.

For a list of his memoirs, see the *Royal Society Catalogue of Scientific Papers,* I, VII.

For biographical information, see the obituary notice in the *Quarterly Journal of the Geological Society of London,* **19** (1863), xxxii–xxxiii.

BERT HANSEN

BRØNSTED, JOHANNES NICOLAUS (*b.* Varde, Denmark, 22 February 1879; *d.* Copenhagen, Denmark, 17 December 1947), *chemistry.*

Brønsted entered the Faculty of Chemical Engineering at the Technical University of Denmark in 1897. Two years later he received his degree, then left the Technical University and entered the Faculty of Natural Sciences at the University of Copenhagen, from which he obtained the M.S. in chemistry in 1902. After a period of nonchemical research he was appointed assistant at the university's chemical laboratory in 1905, and from then on he was attached to the university, serving as professor of physical chemistry from 1908.

Since the conclusion of Julius Thomsen's studies on thermochemistry in 1886, physical chemistry had been somewhat neglected in Denmark, although the work of Ostwald, Arrhenius, and Nernst was followed up in most other countries. Brønsted took over Thomsen's idea of determining chemical affinity by measuring the maximum work of a chemical process, but instead of using calorimetric determinations, he used electromotive force measurements for galvanic cells, which give correct values at room temperature, whereas the calorimetric method gives values that are in error by an amount proportional to the entropy changes taking place for the process in the chemical reaction. He published the results in a series of thirteen monographs on chemical affinity (1906–1921). He defended the third paper of this series, on the affinity of mixing in binary systems, for the Ph.D. at the University of Copenhagen in 1908.

Other aspects of physical chemistry aroused Brønsted's interest after 1913: not only the determination of specific heats but also the determination of affinity constants, published in a series of studies on solubility (1921–1923) and on the specific interaction of ions (1921–1927). These studies evoked considerable interest among physical chemists, especially in the United States and in England, and from 1921 to about 1935 Brønsted's laboratory was crowded with foreign guests desiring to study under his guidance. The poor laboratory conditions were considerably improved when the International Education Board offered to defray the expenses connected with the building of a new Institute of Physical Chemistry, provided the Danish government would take over the operation of the institute, which began operation in 1930. Famous among these studies is a paper, written with V. K. la Mer, on the relation between activity coefficients and the ionic strength of the solution, a relation derived theoretically at the same time by P. Debye and E. Hückel.

Other achievements, too, deserve to be mentioned: Brønsted's definition of acids and bases (1923), simultaneously suggested in almost identical form by T. M. Lowry and in an extended version by G. N. Lewis; his studies on catalysis (1924–1933); and his work on the separation of isotopes of mercury and chlorine (1920–1922, 1929), done with G. von Hevesy.

In 1912 Brønsted published a short manual of physical chemistry, based on the thermodynamic cycle of Carnot. Before 1936, when a new edition had to be written, Brønsted had become convinced of the superiority of J. W. Gibbs's approach to thermodynamics, and the new, substantially enlarged edition was based on Gibbs's ideas.

Brønsted was unhappy with the classical formulation of the laws of thermodynamics, according to which heat is not directly comparable to other forms of energy. To him, heat, like other forms of energy, can be considered as composed of a quantity factor (the entropy) and an intensity factor (the temperature). In this way it was possible to formulate the first law of thermodynamics as a work principle, whereas the second law was broadened to a heat-and-equivalence principle, including also irreversible reactions. A characteristic of this approach is that it relates thermodynamics to physical concepts rather than to mathematical complexities.

Brønsted's formulations, especially his use of the principles "work" and "heat," were not approved by the physicists, and angry discussions took place. He tried to concrete his principles in later works (1940, 1941, 1946), but no agreement had been reached by the time of his death.

BIBLIOGRAPHY

A list of 103 of Brønsted's 120 papers (i.e., those written up to 1935) is in Stig Veibel, *Kemien i Danmark,* II (Copenhagen, 1943), 80–88. Among his works are *Blandingsaffiniteten: binaere Systemer* (Copenhagen, 1908); *Grundrids af den fysiske kemi* (Copenhagen, 1912), new ed. entitled *Laerebog i fysisk kemi* (Copenhagen, 1936, 1943), also trans. into English by R. P. Bell (London, 1937); "Einige Bemerkungen über den Begriff der Säuren und Basen," in *Recueil des travaux chimiques des Pays-Bas et de la Belgique,* **42** (1923), 718–728; "The Activity Coefficients of Ions in Very Dilute Solutions," in *Journal of the American Chemical Society,* **46** (1924), 555–573, written with V. K. la Mer; "Die katalytische Zersetzung des Nitramids und ihre physikalisch-chemische Bedeutung," in *Zeitschrift für physikalische Chemie,* **108** (1924), 185–235, written with K. Pedersen; *Om syre- og basekatalyse* (Copenhagen, 1926); "The Fundamental Principles of Energetics," in *Philosophical Magazine,* 7th ser., **29** (1940), 449–470; "On the Concept of Heat," in *Kongelige Danske Videnskabernes Selskabs Skrifter,* **19**, no. 8 (1941), 79 ff.; and *Principer og problemer i energetiken* (Copenhagen, 1946), trans. by R. P. Bell as *Principles and Problems in Energetics* (New York, 1955), with an enthusiastic foreword by V. K. la Mer. There is an obituary with a complete bibliography by J. A. Christiansen in *Oversigt Danske Videnskabernes Selskabs 1948–1949,* pp. 57–79.

STIG VEIBEL

BROOKS, ALFRED HULSE (*b.* Ann Arbor, Michigan, 18 July 1871; *d.* Washington, D.C., 22 November 1924), *geology.*

Brooks was the dominant figure in the early geological exploration of Alaska and in the formulation of general concepts of the geological framework and natural resources of that then remote region; later he brought about the first applications of geology to military problems and thus became one of the founders of engineering geology.

Brooks was the only son and second of the four children of a self-made and distinguished mining engineer and geologist, Thomas Benton Brooks, who was well known for his studies of the iron and copper deposits of the Upper Peninsula of Michigan. Brooks received his elementary and secondary education in the Newburgh, New York, schools and from private tutors. He studied engineering in the Polytechnik Institut of Stuttgart (1889) and of Munich (1890), then entered Harvard in 1891. At Harvard he studied under Nathaniel Southgate Shaler and William Morris Davis, graduating with the B.S. in 1894.

His father's wide acquaintance among geologists had provided opportunities for Brooks's employment as a junior member of topographical mapping parties of the U. S. Geological Survey in Vermont (1888) and northern Michigan (1889), and in a geological party investigating potential iron ore lands in northern Michigan (1891). While recuperating from an illness in 1893, he undertook some independent geological studies of the Georgia coastal plain and in the region around Newburgh. He joined the Geological Survey as a geologist and petrographer in 1894, and spent the next several years studying various parts of the Appalachian Mountains. In August 1897, Brooks attended the VIIth International Geological Congress in St. Petersburg and then was in Paris until the following spring to study at the École des Mines of the Sorbonne, under Alfred Lacroix, F. A. Fouqué, Charles Bertrand, and Louis de Launay. His studies were terminated by a cable calling him to Washington, to participate in the geological exploration of Alaska.

This exploration had barely begun when Brooks first joined the Geological Survey. Although Alaska had been purchased in 1867, the federal government virtually ignored the new territory for several decades. Interest quickened, however, with the discovery of gold in Alaska and northwestern Canada in the 1880's; and beginning in 1890, several Survey geologists were attached to exploration parties under military, Coast and Geodetic Survey, or National Geographic Society sponsorship. Charles W. Hayes, one of these early geological explorers, was Brooks's first chief after his permanent appointment to the Survey. Working under him, Brooks became strongly attracted to the idea that he might assist in the sound scientific development of a new region.

The years 1898–1902 saw a series of long and difficult, but scientifically rewarding, treks through remote and largely unexplored parts of Alaska. During this period, Brooks undertook the exploration of hundreds of thousands of square miles. The geological exploration was conducted under the severest of difficulties; the geologist was required to divide his time between chopping trail for his pack train, assisting in the construction of rafts or boats when unfordable streams were encountered, and assisting in the preparation of a sketchy topographical map of the country through which he was passing—all the while attempting to gain some understanding of the significance of the rocks and topography.

Nevertheless, Brooks, during these years, discovered and named Rainy Pass, now an avenue of air transportation from Anchorage through the Alaska Range to northwestern Alaska; predicted and guided the discovery of the gold bonanzas of the ancient beaches buried beneath the coastal plain at Nome, discovered the tin placers of western Alaska, defined and described a scheme for the physiographical division of Alaska that has only recently been elaborated upon, and laid a general conceptual framework for future studies of the geological history and topographical development of Alaska that was beautifully expressed in his monograph "The Geography and Geology of Alaska" (1913).

In 1903 Brooks was placed in charge of the newly formed Alaska Branch of the United States Geological Survey. Annual trips to Alaska between 1903 and 1917 acquainted him with nearly every corner of the territory, and in the meantime his corps of dedicated but highly individualistic geologists completed much of the reconnaissance phase of the mapping of Alaska. His blending of scientific and practical interests resulted, during this period, in a series of major works on the coal and metal resources of Alaska and their possible development and use, and on possible railroad routes in Alaska. He was appointed vice-chairman of the Alaska Railroad Commission when it was formed by Congress in 1912, and he played a major role in bringing into existence Alaska's major artery of ground transportation, the Alaska Railroad, completed in 1923.

Shortly after World War I broke out, Brooks became impressed with the contributions that geologists might make to the planning of fortifications, to the evaluation of terrain on which fighting might occur, and to the location of adequate water and fuel for military bases and transport systems. His interest led to the creation of the Geologic Section of the American Expeditionary Force, of which he assumed charge in France in 1918. After the Armistice, he was assigned to the American Commission to Negotiate Peace, for which he prepared an analysis of the mineral industries of Lorraine, the Saar, Luxembourg, and Belgium.

Brooks resumed his post as chief Alaskan geologist of the U.S. Geological Survey in 1919, and retained that position until he died at his desk five years later. During those last years he was at work on a history of Alaska and a description of its resources and geological development, finally published as *Blazing Alaska's Trails* (1953).

Brooks was a small man, but vigorous and active throughout most of his life. A dark, closely trimmed full beard lent distinction to his rather serious features. In love with nature and the outdoors, he was a bit of a romantic, yet a man of unflinching intellectual honesty and almost excessive humility. He wrote with style and clarity, and devoted a good deal of his time to attempts to improve the prose produced by his Alaskan colleagues and subordinates. He led a quietly happy family life with his wife, the former Mabel Baker, whom he married in Washington, D.C., in 1903, and his two children, Mary and Benton.

Brooks served as president of the Geological Society of Washington in 1910–1911 and of the Washington Academy of Science in 1921. In 1913 he was awarded the Charles P. Daly Medal of the American Geographical Society and the Conrad Malte-Brun Medal of the Société de Géographie (Paris) for the excellence and importance of his work in the exploration and mapping of Alaska. In 1920 Colgate University conferred an honorary D. Sc. upon him. His name will always be associated with Alaska; after his death, the chain of rugged mountains that marks the continuation of the Rocky Mountains across northern Alaska was named the Brooks Range. In addition, the largest peak in the tin district of the western Seward Peninsula bears his name, as does a large river in southwestern Alaska.

BIBLIOGRAPHY

A complete list of Brooks's ninety-odd books, scientific papers, administrative reports, and popular articles is given in the lengthy memorial by Phillip S. Smith (see below). His most enduring work consists of the long series of monographic reports of his explorations in Alaska that appeared as parts of the Annual Reports, as Professional Papers, or as Bulletins of the U.S. Geological Survey; his monograph, *The Geography and Geology of Alaska: A Summary of Existing Knowledge,* U.S. Geological Survey Professional Paper 45 (Washington, D.C., 1906); and his book, *Blazing Alaska's Trails,* B. L. Fryxell, ed. (1953).

Other writings include short papers, written early in his career, on local geological problems in the Appalachians; notes calling attention to the occurrence of stream tin near the Bering Strait; notes on glacial phenomena; climber's notes on Mount McKinley and other peaks in the Alaska

Range; summaries of gold, silver, copper, lead, zinc, tin, coal, and petroleum resources of Alaska; reports on transportation problems in Alaska; a lengthy series of reports on water supply, mining, and excavation problems in the areas of interest to the American Expeditionary Force during World War I; and several papers, written near the end of his life, on the future of Alaska and on the role there and elsewhere of applied geology.

Brooks's Alaskan field notes are in the Alaskan Geology Branch of the U.S. Geological Survey, Menlo Park, Calif.

A detailed account of Brooks's life is Phillip S. Smith, "Memorial to Alfred Hulse Brooks," in *Bulletin of the Geological Society of America,* 37 (1926), 15–48. Smith was Brooks's colleague and successor.

DAVID HOPKINS

BROOKS, WILLIAM KEITH (*b.* Cleveland, Ohio, 25 March 1848; *d.* Baltimore, Maryland, 12 November 1908), *zoology, embryology.*

William was the son of Oliver Allen Brooks, a Cleveland merchant, and Ellenora Bradbury Kingsley. Shy, retiring, and gentle, and afflicted by a congenital heart defect that became progressively more limiting in later life, Brooks inherited from his mother artistic skill, a studious, idealistic nature, and a dislike for the obvious and the trivial.

A graduate of Williams College in 1870, Brooks enrolled in the first Anderson School of Natural History on Penikese Island, Massachusetts, in the summer of 1873. There he met Louis Agassiz and at once developed the interest in marine organisms that was to determine the course of his career. He completed a thesis under Alexander Agassiz and, in June 1875, received the third Ph.D. conferred at Harvard. He was then appointed associate in biology at the new Johns Hopkins University to serve under H. Newell Martin, who was himself fresh from graduate study in physiology at Cambridge with T. H. Huxley and Michael Foster. Martin and Brooks developed a major new venture in graduate education (modeled on the German university) that led the way in the vigorous growth of American biology during the last quarter of the century. Much of Brooks's contribution to this venture and the bulk of his research were accomplished at the Chesapeake Zoological Laboratory, a movable marine station established each summer between 1878 and 1906 at various points along the Atlantic coast and in the West Indies. He spent his entire academic career at Johns Hopkins, rising in 1894 to the biology department chairmanship, a position he retained until his death.

Brooks was a descriptive evolutionary morphologist with a strong bias toward studies of whole organisms in their natural environment. A keen observer and an indefatigable amateur philosopher, he was among the late nineteenth-century morphologists who, accepting the transforming power of function in the tradition of Cuvier, added the new insights permitted by the Darwinian concept of the organism as a historical being. Organic form viewed both as a living record of its own ancestry and as dynamically adaptable to new circumstances of life lay at the center of Brooks's thought. For him, life was adjustment, or it was nothing; fitness, the adaptive response, was paramount: "The thing to be explained is not the structure of organisms, but the fitness of this structure for the needs of living things in the world in which they pass their lives." For him, nature was a language that a rational being may read. His reading led him to believe with Aristotle that the "essence of a living thing is not what it is made of or what it does, but why it does it"—in a word, its purpose. In thus withholding judgment on analytical, reductionist approaches, Brooks became unable to look beyond the confines of his own generation, and he did not participate in the transformation of morphology from a comparative to a causal science. Yet his more noteworthy students were somehow stimulated to share fully in that transformation; four of them—E. B. Wilson, T. H. Morgan, E. G. Conklin, and R. G. Harrison—laid the groundwork of much of modern cytology, embryology, and genetics.

In both their substance and their manner of presentation, Brooks's descriptions of the embryology, morphology, and life habits of marine invertebrates were memorable for their scope, their meticulousness, their wealth of illustration, their evolutionary insight, and their charming literary style. His studies of the pelagic tunicate *Salpa* are classic. He described the complex development of *Salpa* buds with unexcelled clarity and demonstrated their relation to the buds of the sessile tunicates. He discovered the remarkable fact that the *Salpa* embryo is first fashioned from follicle cells that are later replaced by regular blastomeres. Brooks's investigations of crustacea were equally fundamental. In several species he identified and collated the larval stages, a major morphological feat. He was the first to follow an entire crustacean life history from a single egg; this he did for *Lucifer,* remarkable among the arthropods in having an egg that cleaves totally. Brooks's descriptions of the life histories of several hydroid coelenterates still stand out for the perceptiveness with which their phylogenetic relationships were assessed. Yet he could also be wrong: his theory of heredity based on pangenesis had only the merit of stimulating others to explore more deeply. And he could be ineffective: an extensive analysis of the practical aspects of conserving the Chesapeake Bay oyster won him only temporary notice despite the accuracy of his warnings.

In his own mind at least, Brooks's forte was his capacity to identify and reflect on the great issues of biology. His many papers and especially his major book, *The Foundations of Zoology,* are laden with rhetorical speculation which, as Brooks cautioned his readers, might at first seem obscure but "may, on review, be found consistent and intelligible."

BIBLIOGRAPHY

I. ORIGINAL WORKS. A complete list of Brooks's works is given in the biographical memoir by Conklin (see below). Collections of his scientific and popular articles are in the Eisenhower Library of the Johns Hopkins University and in the library of the Marine Biological Laboratory, Woods Hole, Massachusetts. His major publications were "The Genus *Salpa*," in *Memoirs of the Biological Laboratory of the Johns Hopkins University,* **2** (1893), 1–303; and *The Foundations of Zoology* (New York, 1899).

Only fragments of Brooks's correspondence are known to exist; these are in the Brooks Papers and the D. C. Gilman Papers at Johns Hopkins, and in the Alexander Agassiz Papers at the Museum of Comparative Zoology, Harvard University. No systematic search through other possible locations has been made.

II. SECONDARY LITERATURE. There is no full biography of Brooks. Brief contemporary accounts of his life were prepared by two former students, E. A. Andrews and E. G. Conklin, the latter's essay being the more valuable. These and other helpful sources are E. A. Andrews, "William Keith Brooks, Zoologist," in D. S. Jordan, ed., *Leading American Men of Science* (New York, 1910), pp. 427–455; E. G. Conklin, "Biographical Memoir of William Keith Brooks, 1848–1908," in *Biographical Memoirs of the National Academy of Sciences,* **7** (1913), 23–88; D. M. McCullough, "William Keith Brooks and American Biology in Transition: A Re-evaluation of His Influence" (thesis, Harvard Univ., 1967); C. P. Swanson, "A History of Biology at the Johns Hopkins University," in *Bios,* **22** (1951), 223–262; and "William Keith Brooks. A Sketch of His Life by Some of His Former Pupils and Associates," in *Journal of Experimental Zoology,* **9** (1910), 1–52.

M. V. EDDS, JR.

BROOKS, WILLIAM ROBERT (*b.* Maidstone, England, 11 June 1844; *d.* Geneva, New York, 3 May 1921), *astronomy.*

Brooks was an enthusiastic and successful observer, whose efforts led to the discovery of twenty-seven comets. His family emigrated to Darien, New York, in 1857, and he spent the remainder of his life in western New York state. While still a child, Brooks developed an interest in observational astronomy and a talent for mechanical construction. In 1858 he built his first telescope, and with it observed Donati's comet. In his father's church, three years later, Brooks delivered the first of his many popular astronomical lectures.

In 1870 Brooks, now married to Mary E. Smith, settled in Phelps, New York, where he worked briefly as a commercial photographer. He increasingly devoted attention to construction and use of telescopes, making a two-inch refractor and reflectors of five and nine inches aperture. His garden, with this portable apparatus, became the Red House Observatory. Here, on 4 October 1881, Brooks discovered his first comet (comet 1881 F, Brooks-Denning), and during the subsequent seven years he found ten more. In 1888 he moved to Geneva, New York, to take charge of the newly established Smith Observatory. Even though this observatory, according to the wishes of its founder, was open for the entertainment and instruction of visitors every clear weekday evening, Brooks was able to discover sixteen more comets—most of them in the morning sky. In addition to his work at the observatory, Brooks was professor of astronomy at Hobart College from 1900, and at William Smith College from 1908 as well.

Brooks belonged to the Royal Astronomical Society, the Liverpool Astronomical Society, the American Association for the Advancement of Science, and the British Astronomical Association. For his cometary discoveries he was widely acclaimed by his contemporaries. He many times won the Warner Prize, and later the Donohoe Medal of the Astronomical Society of the Pacific, for discoveries of unexpected comets. He also won the Prix Lalande of the Académie des Sciences (1900); a gold medal at the Louisiana Purchase Exposition in St. Louis (1904); and a gold medal from the Astronomical Society of Mexico.

BIBLIOGRAPHY

I. ORIGINAL WORKS. Brooks's announcements of his discoveries and observations were published in astronomical journals (e.g., *Monthly Notices of the Royal Astronomical Society*), general scientific journals (e.g., *Scientific American*), and in local New York newspapers.

II. SECONDARY LITERATURE. See Brooks's obituary in *Monthly Notices of the Royal Astronomical Society,* **82** (1922), 246–247. Articles on him are in *The National Cyclopaedia of American Biography,* V (1894), 197–198; and in *Dictionary of American Biography,* III (1929), 91–92.

DEBORAH JEAN WARNER

BROOM, ROBERT (*b.* Paisley, Scotland, 30 November 1866; *d.* Pretoria, South Africa, 6 April 1951), *paleontology.*

Broom was the second son of John Broom, a designer of calicoes and Paisley shawls who, during Robert's youth, was engaged in business in Glasgow, and of Agnes Shearer Broom. As a child he was afflicted with asthma and other respiratory troubles that obliged him to spend a year at the seaside, where he was introduced to marine biology by a retired army officer, John Leavach. His father, an enthusiastic amateur botanist, encouraged the boy's interest in natural history, as did contact with well-known Glasgow naturalists who frequently visited at the Broom summer home near Linlithgow. Also from his father Robert acquired his facility at drawing and a liberal religious viewpoint. His mother belonged to the strict sect of the Plymouth Brethren; he retained strong religious beliefs throughout his life and sought to explain organic evolution as the result of some cosmic intelligence or plan.

Broom attended Hutcheson's Grammar School in Glasgow and in 1883 entered the University of Glasgow, where he assisted in the chemistry laboratory. He attended lectures of Sir William Thomson (later Lord Kelvin), but was most strongly influenced by the botanist F. O. Bower and the anatomist John Cleland, who introduced him to the work of Sir Richard Owen and the embryological researches of W. K. Parker. In 1889 he received his medical degree, and in 1892 he went to Australia. The following year he married Mary Baillie in Sydney, whence she had followed him from Scotland. After a brief visit with his father in Scotland in 1896, he went to South Africa early in 1897, and thereafter made it his home.

Broom practiced medicine most of his life, frequently in remote rural communities. From 1903 until 1910 he held the professorship of zoology and geology at Victoria College (now Stellenbosch University), and from 1934 until his death he served as curator of paleontology at the Transvaal Museum in Pretoria. Throughout his life, his research was interspersed with the duties of his profession. Fossil collecting was his hobby; he also collected paintings and stamps, and played chess.

His major contributions were to the study of the origin of mammals and the structure of their skulls, to the evolutionary history and classification of Permian and Triassic reptiles, and to the discovery and interpretation of the earliest human fossils. Broom also wrote extensively on the cause and mechanism of evolution and on a variety of other topics in zoology, anthropology, medicine, chemistry, and philately.

Broom was invited to deliver the Croonian lecture of the Royal Society of London in 1913; in 1920 he was elected to fellowship in the society. Numerous honorary doctorates, honorary fellowships in distinguished academies, and medals were awarded to him in his later years; these have been listed by Watson and by Cooke.

Between his medical calls in the back country of Australia, Broom began to investigate the anatomy of the native marsupials and primitive egg-laying monotremes. Scientific opinion at the time was divided over the question of the origin of mammals. Cope, Owen, and H. G. Seeley had pointed out many similarities between the structure of mammals and the fossil anomodont and theridont reptiles of South Africa.[1] Most zoologists preferred to derive mammals directly from the amphibians, as advocated by T. H. Huxley.

One of Broom's earliest investigations was of the development of Jacobson's organ, a sensory structure in the nose. In monotremes this organ is supported by a pair of small bones that Broom identified with the paired prevomers near the front of the palate of reptiles. In nine papers published between 1895 and 1935 he sought to demonstrate by both embryonic relationships and features of the palate of mammal-like reptiles that the reptilian and mammalian vomers were not homologous. He relied largely upon the similarity between the anomodont reptile *Dicynodon* and the platypus, despite his recognition that both of these were side branches from the main line of mammalian evolution.[2] Aside from the vomer question, Broom's interest in Jacobson's organ was renewed whenever young specimens of uncommon animals reached his hands, and he reported his findings in fifteen papers published between 1895 and 1939.

In the course of studying the embryonic development of the skull in the Australian phalanger *Trichosurus,*[3] Broom discovered that the mammalian alisphenoid bone does not form in the wall of the braincase but arises from the palate, like the slender epipterygoid bone of lizards and other reptiles (1907). Subsequently he was able to show this transition from reptilian to mammalian condition in the skulls of various mammal-like reptiles. In some ways this is the most important of Broom's many contributions to vertebrate morphology.

One of the major problems of mammalian development was the origin of the three auditory ossicles;[4] some students regarded them as the result of fragmentation of the single sound-transmitting bone in the ear of reptiles or amphibians; others followed Reichert (1838) and Huxley in the view that the outer two bones of the mammalian chain represent the articular and quadrate bones that form the jaw articulation of reptiles. Broom accepted the former view in 1890 and 1904, but in 1911 he reversed his opinion[5] and in 1912 showed how the bones of anomodont reptiles conformed to Reichert's theory. He presented a series of reconstructions suggesting the mode of transition from the reptilian to the mammalian ear condition.[6] In 1936 Broom turned again to this question, and noted that evidence of an extracolumellar portion of the stapes in therocephalians confirmed his earlier opinion that the tympanic membrane of mammal-like reptiles lay close behind the quadrate bone, rather than embedded in the notch of the angular.

Other morphological questions that engaged Broom's attention included the homologies of the variable number of bones at the base of the brain in mammals; the homologies of the coracoid bones of the shoulder region; the peculiarities of the epiphyses at the ends of the metacarpal and metatarsal bones; the arrangement of the reptilian tarsal bones; and the homologies of skull arches in lizards—a problem in which his conclusions based upon embryological and morphological studies were spectacularly confirmed in 1934 when Parrington described a Triassic ancestral lizard, *Prolacerta,* that shows an extremely early stage in the disappearance of the lower temporal arch.

But it is for the study of the fossil reptiles of the Karroo that Broom is most famous. In 1896, when he visited his father in Britain, he had an opportunity to examine Seeley's collection of African fossils at the British Museum. Grasping their significance for the problem of the origin of mammals, he sailed for South Africa.

The nineteenth-century work on the mammal-like reptiles by Owen and Seeley was largely descriptive. It had revealed considerable variety in the fauna and had shown that some of these fossils approached the structure of mammals more closely than did other known reptiles. Broom revealed the details of the skull structure of these animals by splitting specimens with a chisel or by sawing cross sections through them. He thus placed their classification on a firm morphological basis; showed that they had been derived from the pelycosaurs of North America, which his contemporaries had classified with the lizard-like *Sphenodon* of New Zealand; and demonstrated how the distinctive mammalian structures had arisen within the therapsid suborders. As early as 1902 he recognized that certain carnivorous therapsids had a far more primitive palate than the extremely mammal-like cynodonts that Seeley had discovered, and he proposed the suborder *Therocephalia* for the earlier forms. By 1905 he had arrived at the basic groupings of the African forms, which, modified in the light of later discoveries, are still used.

Beginning in 1905, Broom attempted to record the stratigraphic occurrence of various fossils in the thick sequence of Karroo deposits. Building on the earlier work of Seeley, he established the standard sequence of faunal zones. This attention to stratigraphy as well as to morphological detail prepared him for the brilliant synthesis of the African and North American Permian faunas at which he arrived after seeing the American fossils on a brief visit to the American Museum of Natural History in 1910.

Broom described an unbelievably large number of fossils. His brief descriptions are generally accompanied by rather unfinished sketches, which are praised by some of his colleagues for their remarkable fidelity and intuition, and condemned by others for showing structures still concealed by matrix and for carrying restoration beyond available evidence.[7]

These studies of Karroo fossils, as well as the embryological investigations of mammalian development, formed the basis for numerous essays on the origin of mammals, the earliest in 1908; the Croonian lecture to the Royal Society of London in 1913; and *The Origin of the Human Skeleton* and *The Mammal-like Reptiles of South Africa.*

Following World War I, Broom practiced medicine at Douglas, remote from fossil beds and contact with other scientists. It was during this period that his attention first turned to the problems of prehistoric man in Africa and the physical characteristics of African races.[8] His 1923 survey of the Hottentots, Bushmen, and extinct Korana race defined many of the problems of South African anthropology and laid the basis for subsequent work, which has not confirmed his views of racial relationships.[9]

When Raymond Dart announced the discovery of the *Australopithecus* skull at Taungs in 1925, Broom, on the basis of a personal examination of the specimen, supported Dart's conclusions as to its human relationships and phylogenetic importance. After he had been appointed curator of paleontology in the Transvaal Museum, he turned his attention to the Pleistocene cave deposits, hoping to find additional evidence of early man. This quest was rewarded by the discovery of *Plesianthropus* within a week of Broom's first visit to Sterkfontein. Other important specimens were found at Kromdraai in 1938, at Sterkfontein again in 1946–1947, and at Swartkrans in 1949. Broom has given a vivid account of the excitement of these discoveries in *Finding the Missing Link* (1950). He not only collected, but also prepared, illustrated, and described, this material when he was well over eighty years of age.[10]

Between the time Broom retired from medical practice in 1928 and his appointment at the Transvaal Museum in 1934, he wrote three books. *The Origin of the Human Skeleton* (1930) provides a rapid survey of vertebrate evolution, followed by discussion of the transition from reptiles to mammals, as shown by the mammal-like reptiles, and of the various problems that had held his attention for so many years. A final chapter deals with the inadequacies of the views of Lamarck and Darwin as explanations of evolution. *The Mammal-like Reptiles of South Africa and the Origin of Mammals* (1932) is illustrated with several hundred of Broom's rough restorations, which he defends as more useful than exact portrayal of fragmentary specimens. The problem of evolution and its causation is explored more boldly.

In *The Coming of Man. Was It Accident or Design?* (1933) Broom criticizes various theories of evolution in detail. He was fully conversant with the current ideas of geneticists on the roles of mutation and natural selection in producing evolution, but like most paleontologists of the early 1930's he found this view inadequate. Opposing Darwinism, Lamarckism, and evolution by mutation, Broom suggested there must have been intelligent spiritual agencies behind evolution. Arguing that all major evolutionary advances have come from small, generalized animals, and that all living animals are now too specialized to give rise to a completely new type of life, he concluded that further evolutionary advance is impossible. Evolution was over with the appearance of man, its latest and highest product, which must have been the goal of a directing Intelligence.

Educated at the height of the wave of evolutionary biology that followed Darwin's work, Broom devoted his career to unraveling the phylogenetic problems that the nineteenth-century morphologists and paleontologists had raised. His unrelenting drive to investigate these problems, his complete willingness to travel to regions where critical material might be found, and especially his unusual talent for combining paleontological and embryological research enabled him to contribute more to the story of mammalian origins than all his contemporaries together. When one considers his good fortune in obtaining fossil hominids at Sterkfontein within a week of his first inquiry there, as well as at other localities, is it any wonder that he strongly believed in his own Special Providence?

NOTES

1. E. D. Cope (1870v); R. Owen (1876, 1880); H. G. Seeley (1889–1895). These and other paleontological works cited below are listed in the bibliographies of vertebrate paleontology by Camp and by Romer, or in O. P. Hay, *Bibliography*

and Catalogue of the Fossil Vertebrates of North America, U.S. Geological Survey Bulletin 179 (Washington, D.C., 1902).

2. Broom's illustrations of the palate of *Thrinaxodon* ("The Vomer-Parasphenoid Question," *Annals of the Transvaal Museum,* **18** [1935], 23–31) clearly show the closer similarity of cynodont prevomers to the mammalian vomer, which F. R. Parrington emphasized in support of the opposite interpretation in the same year. In 1940 Parrington and Westoll virtually laid Broom's contention to rest.

3. This study was commenced in 1898 and published in 1909 ("Observations on the Development of the Marsupial Skull," *Proceedings of the Linnean Society of New South Wales,* **34**, 192–219). It remains a principal source of knowledge of the development of diprotodont marsupials, according to G. R. DeBeer: *The Development of the Vertebrate Skull* (Oxford, 1939), p. 305.

4. The bones in question in mammals and their homologues are the *malleus = articular* bone of reptilian lower jaw; the *incus = quadrate* bone, which forms the articulation for the jaw at the back of the skull; and the *stapes =* part of the single auditory ossicle of reptiles, etc.

5. "On the Structure of the Skull in Cynodont Reptiles," *Proceedings of the Zoological Society of London* (1911), p. 917; full article on pp. 893–925.

6. Broom's study was published a year before the exhaustive analysis of embryological and morphological relationships of the auditory ossicles by Ernst Gaupp, "Die Reichertsche Theorie (Hammer-, Amboss- und Kieferfrage)," *Archiv für Anatomie und Entwickelungsgeschichte; anatomische Abteilung des Archives für Anatomie und Physiologie . . .,* Supplement-Band 1912.

7. Besides these studies of primitive reptiles, Broom wrote numerous articles on fossil fishes, amphibians, and dinosaurs of South Africa. While in Australia he had collected and described fossil marsupials from Pleistocene caves. In 1909 he first discussed African Pleistocene mammals, and after 1935 devoted much attention to these fossils (at least twenty-three papers). He demonstrated that the faunas associated with early fossil hominids were largely of extinct species and differed considerably from site to site.

8. This interest stemmed from the discovery of the Boskop skull by Fitzsimons in 1913.

9. L. H. Wells (1952) terms this "one of the most important landmarks in Anthropological research in South Africa."

10. Broom and Schepers (1946); Broom and Robinson (1952); and numerous short papers in various journals.

BIBLIOGRAPHY

I. ORIGINAL WORKS. Lists of Broom's publications are given in R. Broom, "Bibliography of R. Broom, M.D., D.Sc., F.R.S.," in A. L. DuToit, ed., *Robert Broom Commemorative Volume* (Capetown, 1948), pp. 243–256; C. L. Camp *et al., Bibliography of Fossil Vertebrates,* Geological Society of America, Special Papers, no. 27 (1940), covering 1928–1933; no. 42 (1942), covering 1934–1938; Memoirs, no. 37 (1949), covering 1939–1943; no. 57 (1953), covering 1944–1948; and no. 84 (1961), covering 1949–1953; A. S. Romer *et al., Bibliography of Fossil Vertebrates Exclusive of North America 1507–1927,* Geological Society of America, Memoirs, no. 87 (1962), I; and D. M. S. Watson (see below).

Broom's major papers and books are "On the Origin of Mammals," in *Philosophical Transactions of the Royal Society of London,* **206B** (1914), 1–48, the Croonian lecture of 1913; *The Origin of the Human Skeleton; an Introduction to Human Osteology* (London, 1930); *The Mammal-like Reptiles of South Africa and the Origin of Mammals* (London, 1932); *The Coming of Man. Was It Accident or Design?* (London, 1933); *The South African Fossil Ape-Men. The Australopithecinae,* Transvaal Museum Memoir no. 2 (Pretoria, 1946), written with G. W. H. Schepers; *Finding the Missing Link* (London, 1950); and *Swartkrans Ape-Man, Paranthropus crassidens,* Transvaal Museum Memoir no. 6 (Pretoria, 1952), written with J. T. Robinson.

II. SECONDARY LITERATURE. Biographical sketches are Raymond A. Dart, "Robert Broom—His Life and Work," in *South African Journal of Science,* **48** (1951), 3–19; Austin Roberts, "Historical Account of Dr. Robert Broom and His Labours in the Interest of Science," in A. L. DuToit, ed., *Robert Broom Commemorative Volume* (Capetown, 1948), pp. 5–15, an account basic to most published obituaries—see also notes on Broom by Field Marshall J. C. Smuts, W. K. Gregory, and L. H. Wells in the same volume; and D. M. S. Watson, "Robert Broom," in *Obituary Notices of Fellows of the Royal Society,* VIII (1952), 37–70.

Twelve other obituaries are listed in Camp *et al.* (1961). Of these, Sonia Cole, in *South African Archeological Bulletin,* **6** (1951), 51, portrays his personality; H. B. S. Cooke, in *South African Journal of Science,* **47** (1951), 277–278, and R. F. Ewer, in *Natal University Journal of Science,* **8** (1952), 27–29, give additional insight into his methods, attitudes toward scientific research, and religious beliefs; W. E. Le Gros Clark, in *Nature,* **167** (1951), 752, provides an excellent, concise evaluation and tribute; and L. H. Wells, in *Royal Society of Edinburgh Yearbook, 1950–1951* (1952), pp. 9–12, evaluates his contributions to anthropology.

JOSEPH T. GREGORY

BROSCIUS, JOANNES. See **Brożek, Jan.**

BROUNCKER, WILLIAM (*b.* 1620; *d.* Westminster, London, England, 5 April 1684), *mathematics.*

Brouncker's father was Sir William Brouncker, who was created viscount of Castle Lyons, Ireland, in September 1645; the father died the same November, and was succeeded by the son. The title passed to William's brother Henry in 1684, and since both were unmarried, became extinct when Henry died in 1687. William's mother was Winefrid, daughter of William Leigh of Newenham, Warwickshire.

Brouncker entered Oxford University at the age of sixteen and showed proficiency in mathematics, languages, and medicine. He received the degree of Doctor of Physick in 1647, and for the next few years devoted himself mainly to mathematics.

He held several offices of prominence: Member of Parliament for Westbury in 1660, president of Gresham College from 1664 to 1667, commissioner for the navy from 1664 to 1668, comptroller of the treasurer's accounts from 1668 to 1679, and master of St. Catherine's Hospital near the Tower from 1681 to 1684.

Brouncker was the king's nominee for president of the Royal Society, and he was appointed without opposition—at a time when there were many talented scientists. He was reappointed annually, and he guarded his position zealously, possibly holding on to it for too long. He resigned in 1677, in effect at the suggestion of an election, and was succeeded by Sir Joseph Williamson. He was an enthusiastic supporter of the society's bias toward experimentation and was very energetic in suggesting and assessing experimental work until Hooke took over that job. Sprat's history records two experiments performed by Brouncker, one on the increase of weight in metals due to burning and the other on the recoil of guns.

His major scientific work was undoubtedly in mathematics. Much of his work was done in correspondence with John Wallis and was published in the latter's books.

One of Wallis' major achievements was an expression for π in the form of an infinite product, recorded in his *Arithmetica infinitorum.* This book states that Brouncker was asked to give an alternative expression, which he did in terms of continued fractions, first used by Cataldi in 1613, as

$$\frac{4}{\pi} = 1 + \frac{1^2}{2+} \frac{3^2}{2+} \frac{5^2}{2+} \cdots,$$

from which he calculated π correct to ten decimal places.

In an exchange of letters between Fermat and Wallis, the French mathematician had proposed for general solution the Diophantine equation $ax^2 + 1 = y^2$. Brouncker was able to supply an answer equivalent to $x = 2r/r^2 - a, y = r^2 + a/r^2 - a$, where r is any integer, as well as another answer in terms of continued fractions.

A paper in the *Philosophical Transactions* (**3** [1668], 753–764) gives a solution by Brouncker of the quadrature of a rectangular hyperbola. He arrived at a result equivalent to

$$\int_0^1 \frac{dx}{1+x} = \frac{1}{1 \cdot 2} + \frac{1}{3 \cdot 4} + \frac{1}{5 \cdot 6} + \cdots$$

or

$$1 - \frac{1}{2} + \frac{1}{3} - \frac{1}{4} + \cdots,$$

and found similar infinite series related to this problem. In order to calculate the sum, he discussed the convergence of the series and was able to compute it as 0.69314709, recognizing this number as proportional to log 2. By varying the problem slightly, he was able to show that 2.302585 was proportional to log 10.

Brouncker also improved Neile's method for rectifying the semicubical parabola $ay^2 = x^3$ and made at least three attempts to prove Huygen's assertion that the cycloidal pendulum was isochronous. A letter from Collins to James Gregory indicates that Brouncker knew how to "turn the square root into an infinite series," possibly an allusion to the binomial series.

Brouncker was a close associate of Samuel Pepys, socially and professionally, and is mentioned many times in the *Diary.* Pepys valued his friendship highly, but sometimes doubted his professional ability. Brouncker shared with Pepys an interest in music, and his only published book is a translation (1653) of Descartes's *Musicae compendium* with notes as long as the work itself, including a mathematical attempt to divide the diapason into seventeen equal semitones.

His fame as a mathematician rests largely on an ability to solve problems set by others. If he had devoted himself more fully to his own studies, he would undoubtedly have been one of the best mathematicians during a period in which talent abounded.

A portrait by Sir Peter Lely is in the possession of the Royal Society.

BIBLIOGRAPHY

Works concerning Brouncker or his work are E. S. de Beer, ed., *The Diary of John Evelyn,* III (Oxford, 1955), 285–286, 332, 353; T. Birch, *History of the Royal Society,* Vol. I (London, 1756–1757); Lord Braybrooke, ed., *Diary and Correspondence of Samuel Pepys* (London, 1865); Sir B. Burke, *Extinct Peerages* (London, 1883), p. 78; M. H. Nicolson, *Pepys' Diary and the New Science* (Charlottesville, Va., 1965), pp. 11, 28–29, 109, 135; H. W. Robinson and W. Adams, *The Diary of Robert Hooke* (London, 1935); J. F. Scott and Sir Harold Hartley, "William Viscount Brouncker," in *Notes and Records of the Royal Society,* **15** (1960–1961), 147–156; T. Sprat, *History of the Royal Society* (London, 1667), pp. 57, 136, 228–229; J. Wallis, *Arithmetica infinitorum* (Oxford, 1656), p. 181; *Tractatus duo* (Oxford, 1659), p. 92; *A Treatise of Algebra* (Oxford, 1685), p. 363; D. T. Whiteside, "Brouncker's Mathematical Papers," in *Notes and Records of the Royal Society,* **15** (1960–1961), 157; A. à Wood, *Athenae oxonienses,* P. Bliss, ed. (London, 1820), p. 98.

JOHN DUBBEY

BROUSSAIS, FRANÇOIS JOSEPH VICTOR (*b.* near St. Malo, France, 17 December 1772; *d.* Paris, France, 18 November 1838), *medicine.*

Broussais began his medical studies at the Hôtel Dieu of St.-Malo and later attended the École de

Chirurgie Navale at Brest. After twice shipping out as a naval surgeon, he had earned the amount of money needed to finish his medical studies in Paris, where he received his doctor's degree in 1802. He then entered the Service de Santé Militaire, where he showed himself to be a conscientious doctor, continuously observing, writing, and teaching. His *Traité des phlegmasies* was written in Spain, from which he returned in 1814 to become a professor at the Val de Grâce. Broussais gave a course in practical medicine that attracted large, enthusiastic classes. His success lasted until his physiological doctrine—based in large part on therapeutic bleeding—was rejected by his students. It was also proved wrong by the 1832 outbreak of cholera, which Broussais treated, with catastrophic results, as acute gastroenteritis.

In 1834 the last number of the *Annales de la médecine physiologique* (which he had founded in 1822) was published, and in a last effort to regain his popularity, Broussais exploited the phrenology that Gall had made fashionable some twenty years earlier. In 1836, however, he began a type of teaching that questioned the immortality of the soul and the existence of God; it led to such violent scenes that the police were called in to restore order, and Broussais decided to leave the Val de Grâce. He became ill soon thereafter and died two years later of cancer.

The 1830 revolution had been a great boon to Broussais. Through government influence he had been exempted from the usual requirements and had won a chair at the Faculty of Medicine, a seat in the Académie des Sciences Morales et Politiques, the rank of inspector general in the Service de Santé Militaire, and had been made a commander of the Legion of Honor.

Broussais lived at a time when a monistic system of pathology was still possible. His was a kind of "Brownism" in reverse, in which the phenomena of illness are different from those of health only in intensity. Laennec and Bretonneau opposed a doctrine inspired by old theories of deep pathological states that did not admit individual illnesses, and desperately fought the idea of specificity, localization, or contagion. Everything came under the heading of gastroenteritis, and consequently was treated by repeated bleedings and debilitating diets. These had disastrous effects on patients who were hemorrhaging or who suffered from cancer, malaria, or syphilis. Some have wished to see Broussais's ideas on the nonspecific states of inflammation as making him a precursor in this field.

Nevertheless, by rejecting Pinel's concept of essential fevers and by affirming that fevers are only reactions to certain given inflammations, Broussais eliminated the ontological notion of illness, which considered pathological and physiological phenomena to be entirely different and saw diseases as real entities, independent of the organism. But even if there had been an idea in his system, one would look in vain for a method.

BIBLIOGRAPHY

I. ORIGINAL WORKS. Broussais's writings include "Recherches sur la fièvre hectique considérée comme dépendante d'une lésion d'action des différents systèmes, sans vice organique" (Paris, 1804), his thesis; *Histoire des phlegmasies ou inflammations chroniques* (Paris, 1808, 1816, 1822, 1826, 1828, 1838), also translated into Spanish by Suárez Pantigo (Madrid, 1828) and English by Isaac Hays and R. Eglesfeld Griffith (Philadelphia, 1831); *Examen de la doctrine médicale généralement adoptée et des systèmes modernes de nosologie,* 4 vols. (Paris, 1816–1834), also translated into Spanish by G. Lanuza (Madrid, 1822) and German by Reulin (Bern, 1820); articles in *Annales de la médecine physiologique* (1822–1834); *Traité de physiologie appliqué à la pathologie* (Paris, 1822); *Catéchisme de la médecine physiologique* (Paris, 1824), also translated into Spanish and into English by I. Hays and R. E. Griffith (Philadelphia, 1832); *Commentaire des propositions de pathologie consignées dans l'examen des doctrines médicales* (Paris, 1824); *De l'irritation et de la folie* (Paris, 1828); *Du cholera-morbus epidémique observé et traité* (Paris, 1832); *Mémoire sur la physiologie de la médecine* (Paris, 1832); *Mémoire sur l'influence que les travaux des médecins physiologistes ont exercé sur l'état de la médecine en France* (Paris, 1832); *Cours de pathologie et de thérapeutiques générales,* 4 vols. (Paris, 1833–1835); *Mémoire sur l'association du physique et du moral* (Paris, 1834); and *Cours de phrénologie fait à la Faculté de médecine de Paris* (Paris, 1836). Forty-five pages of Broussais's autograph letters are at the Faculty of Medicine, Paris, MS 402.

II. SECONDARY LITERATURE. Writings on Broussais are E. Ackerknecht, "La médecine à Paris entre 1800 et 1850," in *Conférences du Palais de la Découverte* (Paris, 1958), pp. 9–13, 17–18, 20, and *Medicine at the Paris Hospitals, 1794–1848* (Baltimore, 1967), pp. 10, 26, 28, 58, 61–79, 96, 212; J.-Z. Amussat, "Relation de la dernière maladie de Broussais," in *Gazette médicale de Paris,* **4,** no. 49 (7 Dec. 1838), 769–774; L. Babonneix, "Le centenaire de la mort de Broussais," in *Gazette des hôpitaux,* **111** (24 Sept. 1938), 1221–1228; E. Beaugrand, "Broussais," in *Dictionnaire Dechambre,* XI (1878), 160–163; L. J. Begin, *Application de la doctrine physiologique à la chirurgie* (Paris, 1823), pp. viii, 17–18, 61, 87, 93, 96; Michel Bert, "Essai sur Broussais" (Paris, 1957), thesis; P. Busquet, "Broussais," in *Les biographies médicales,* I (Paris, 1927–1928), 53–65, 69–80, with an important bibliography; Georges Canguilhem, "Essai sur quelques problèmes concernant le normal et le pathologique" (Strasbourg, 1943), thesis; J. des Cilleuls, "À propos du centenaire de la mort de Broussais," in *Revue du Service de santé militaire,* **110,** no. 4 (1939), 645–655;

Cornilleau, "À propos de Broussais," in *Progrès médical* (26 Nov. 1938), 1604–1609, 1610; F. Dubois, "Éloge de Broussais," in *Mémoires de l'Académie de médecine de Paris,* **14** (1849), 1–28; L. Duplais, *Histoire complète de Broussais* (Paris, 1891); H. Folet, "Broussais et le broussaisisme," in *France médicale,* **54** (1907), 137–157, 217–237, also in *Bulletin de la Société française d'histoire de la médecine de Paris,* **52** (1806), 239–314; M. Foucault, *Naissance de la clinique* (Paris, 1966), 175, 179, 186–187, 189–191, 194–196; Pierre Huard, "Broussais et nous," in *Ouest médical,* no. 15 (1959), 463–466; Jeanne Huet, "Broussais et son oeuvre" (thesis, Paris, 1938); René Laennec, *Traité de l'auscultation médiate* . . . (Paris, 1826), pp. xx–xxxii; P.-C. Louis, *Recherches sur les effets de la saignée dans quelques maladies inflammatoires* . . . (Paris, 1835); J. L. H. P. Peisse, *Les médecins contemporains* (Paris, 1827), pp. 1–56; J. Rochard, *Histoire de la chirurgie française au XIXème siècle* (Paris, 1875), pp. 114–123, 251–252; and R. Villey, *Réflexions sur la médecine d'hier et de demain* (Paris, 1966), pp. 53, 55, 56.

PIERRE HUARD

BROUSSONET (or **Broussonnet** or **Broussounet**), **PIERRE-AUGUSTE-MARIE** (*b.* Montpellier, France, 19 January 1761; *d.* Montpellier, 27 July 1807), *zoology, botany.*

Broussonet's family belonged to the bourgeoisie. His father, François Broussounet (des Terrasses), was a physician who taught at Montpellier's school of medicine. Broussonet's brother Victor studied there and later became its dean. Henri Fouquet, a professor at the medical school, was a relative, as was Jean Chaptal, who subsequently became minister of the interior.

As a child, Broussonet was a passionate collector of natural history specimens, cluttering his father's house with his finds. He also excelled in classical studies in Montpellier, Montélimar, and Toulouse. He thus was headed toward medical studies, which were both a family tradition and, at that time, the only avenue leading to the study of the natural sciences. Antoine Gouan, a convinced Linnaean, taught at the medical school, and apparently it was from him that Broussonet learned of Linnaeus' work. From then on, he worked to have it accepted. His thesis on respiration, which he defended in 1778, marked the end of Broussonet's formal studies. He received the doctorate on 27 May 1779, at the age of eighteen.

Broussonet's thesis was unanimously praised and seemed to the professors of the University of Montpellier to justify exceptional treatment. They asked that he, despite his youth, be made his father's successor when the latter retired. The request was not granted, although Broussonet himself went to Paris to plead his cause. His failure was compensated for, however, by the friendships he established with the Paris scholars who made it possible for him to continue and extend the studies on fish that he had begun at Montpellier as early as 1779. Although the Paris ichthyological collections surpassed those that Broussonet had worked with until then, they were not complete enough for the work on which he planned to spend the greater part of his time. Consequently, he went to England to seek the specimens needed for the morphological and systematic work he had in mind.

London, which he reached in 1780, offered Broussonet all he could wish for: an active scientific community; naturalists already won over to Linnaeus' ideas; collections rich in new species; and a friend, Sir Joseph Banks, whose devotion never lagged. In addition, he was elected to the Royal Society.

Banks had brought back from Cook's first expedition a considerable number of exotic fish, which he turned over to Broussonet for study, thereby making it possible for Broussonet to start his *Ichthyologia,* which was to contain descriptions of 1,200 species. The first ten sections, in which he noted the important discovery of the *pseudobranchia,* were published in 1782. They were the only ones.

When he returned to France, Broussonet finished his "Notes ichthyologiques," and in 1785 he presented six of them before the Académie des Sciences. Their merit and the support of Daubenton, who, although anti-Linnaean, was kindly disposed toward Broussonet, resulted in his election to the Academy. The following year he presented a memoir on the *voilier,* his last work in ichthyology.

The growing unrest among the common people led to Broussonet's decision to abandon ichthyology. Many thought that the improvement of agricultural production, both in quality and in quantity, might appease those seeking reforms. Berthier de Sauvigny, the administrator of Paris, for whose food supply he was responsible, was one who held that view. He had met Broussonet while in England to study methods of cultivation and animal husbandry and had renewed the acquaintance in France. Berthier, who had revived the Société d'Agriculture, persuaded Broussonet to become its secretary. In addition, Daubenton, who in 1783 had accepted the chair of rural economy at the Alfort Veterinary School, passed on this heavy responsibility to his young friend.

Broussonet, having become an agronomist, tried to fulfill the duties of his new offices. Between 1785 and 1788 he regularly published short notices, both signed and anonymous, for the use of farmers. Many of them have not been identified to this day. Unfortunately, this work came too late, and the Revolution put an end to Broussonet's agricultural efforts.

In 1789 Broussonet, then twenty-eight, enthusiastically welcomed revolutionary ideas, as was characteristic of his generation. He turned to politics, but soon realized its dangers. His friend Berthier, held responsible for the current famine, was killed before his eyes. Broussonet, knowing he was in danger, fled Paris. In 1792 he took refuge in Montpellier but was accused of federalism and thrown in jail. He remained there only a few days, but his liberty was still precarious after his release. He therefore left Montpellier for Bagnères-de-Bigorre to join his brother, then a doctor in the army of Pyrénées-Orientales. On 19 July 1794 he crossed the Spanish border.

Although Broussonet was penniless, he was warmly received by the botanists Ortega and Cavanilles in Madrid, Gordon in Jerez, and José Correa de Serra in Lisbon. Sir Joseph Banks continued to be interested in him and helped him financially. But he was not accepted by the French who had emigrated earlier and looked upon him as a revolutionary. Having become friends with Simpson, American consul in Gibraltar, Broussonet accompanied him as physician on a diplomatic mission to Morocco, where he studied the flora.

In 1795, when he voluntarily returned to France, Broussonet's name was removed from the list of political refugees, and he regained possession of his property. Elected to the Institut in 1796, he requested appointment as a *voyageur de l'Institut,* stating that he wished to return to Morocco to continue his research. In 1797, therefore, he was named vice-consul at Mogador, a post created for him. There he carried on his work of collecting and describing plants and animals, as well as attending to his consular duties.

In 1799 Mogador was threatened by the plague. On 8 July, Broussonet sailed with his family to the Canary Islands, where he was made commissioner of commercial relations. He continued his collecting and observations, writing of them to Cavanilles, Charles l'Heritier, and Humboldt. The local authorities forbade him to travel, however, and Broussonet decided to leave his post. He asked to be sent to the Cape of Good Hope, where he hoped to create a botanical garden.

Chaptal, then minister of the interior, supported his young relative's request, and Broussonet was therefore named commissioner of commercial relations to the Cape on 15 October 1802. He returned to France in 1803 to prepare for this new assignment, only to learn that Chaptal had changed his mind and had had him made professor at the medical school of Montpellier, to succeed Gouan. He had to accept.

Broussonet took up his new position at once. His title, besides its teaching duties, gave him charge of the botanical garden of Montpellier. He restored its former layout and, helped financially by Chaptal, built a greenhouse, dug ponds, and enlarged the collections, of which he published a list in 1805. This was the *Elenchus plantarum horti botanici Monspeliensis,* the first in a projected series of works that promised to be considerable. But it was also the last. Broussonet was preparing to describe the 1,500 species collected at Tenerife when he suffered a stroke that caused a gradually worsening aphasia. On 17 August 1806 he notified the director of the medical school that he must resign his post, and a year later, he suffered a final stroke that caused his death.

BIBLIOGRAPHY

I. ORIGINAL WORKS. Among Broussonet's works are *Variae positiones circa respirationem* (Montpellier, 1778), his doctoral thesis; "Mémoire sur les différentes espèces de chiens de mer," in *Mémoires de l'Académie royale des sciences* (1780), 641–680; *Ichthyologia sistens piscium descriptiones et icones* (London–Paris–Vienna–Leipzig–Leiden, 1782); "Essai de comparaison entre les mouvements des animaux et ceux des plantes et description d'une espèce de sainfoin dont les feuilles sont dans un mouvement perpétuel," in *Mémoires de l'Académie royale des sciences* (1784), 609–621; "Mémoire sur le voilier, espèce de poisson peu connue qui se trouve dans les mers des Indes," *ibid.* (1786), 450–455; "Observations sur la régénération de quelques parties des corps des poissons," *ibid.* (1786), 684–688; and *Elenchus plantarum horti botanici Monspeliensis* (Montpellier, 1805).

II. SECONDARY LITERATURE. Works on Broussonet are F. Aubouy, "Auguste Broussonet et la flore de Montpellier," in *Annales de la Société d'horticulture et d'histoire naturelle de l'Hérault,* 2nd ser., **29** (1897–1898), 139–161; J. Caillé, "Un vice-consul de France au Maroc: Auguste Broussonet," in *Revue de l'Institut Napoléon,* no. 89 (1963), 157–166, which cites his writings during his period of travel; A. P. de Candolle, *Éloge de Mr. Auguste Broussonet prononcé dans la séance publique de l'École de médecine le 4 janvier 1809* (Montpellier, 1809), which cites his agricultural and botanical works; J. Castelnau, *Mémoire historique et biographique sur l'ancienne Société royale des sciences et lettres de Montpellier* (Montpellier, 1838), which cites his botanical works; G. Cuvier, "Éloge historique de Pierre-Auguste-Marie Broussonet," in *Éloges historiques des membres de l'Académie royale des sciences,* I (Strasbourg–Paris, 1819), 311–342; H. Dehérain, *Dans l'Atlantique* (Paris, 1912), which cites works from his period of travel; Durand, "Vie de M. Broussonet," Bibliothèque du Muséum National d'Histoire Naturelle de Paris, MS 1991-pièce 242–8 ff., a sketch, several times reworked, of a biography that seems never to have been completed (extracts in Roumeguère, below); E. Fournier, "Broussonet (Pierre Marie Auguste)," in M. H. Baillon, *Dictionnaire de botanique,* I (Paris, 1876), 499–500; F. Granel, "Un grand naturaliste montpelliérain P. M. Auguste Broussonet (1761–1807)," in *Pages médico-historiques montpelliéraines* (Montpellier, 1964), pp. 119–130; and "Les étapes scientifiques d'Auguste Broussonet," in *Monspeliensis Hippocrates*, no. 37 (1967), 25–34; H. Harant and G. Vidal, "Á propos du nom de Broussonet," *ibid.*, no. 8 (1960), 23–26; C. Martins, *Le jardin des plantes de Montpellier. Essai historique et descriptif* (Montpellier, 1854), pp. 1–91; L. Passy, *Histoire de la Société nationale d'agriculture de France*, I (Paris, 1912); G. Roumeguère, "Correspondance de Broussonet avec Alex. de Humboldt au sujet de l'histoire naturelle des îles Canaries," in *Mémoires de la Société nationale des sciences naturelles de Cherbourg*, **18** (1874), 304–317, which cites works written during his period of travel; and A. Thiébaut de Berneaud, *Éloge de Broussonet, premier fondateur de la Société Linnéenne de Paris* (Paris, 1824).

JEAN MOTTE

BROUWER, DIRK (*b*. Rotterdam, Netherlands, 1 September 1902; *d*. New Haven, Connecticut, 31 January 1966), *astronomy*.

Brouwer worked mainly in celestial mechanics, making elegant theoretical contributions but also pioneering in the use of high-speed digital computers to solve its problems with previously unattainable accuracy.

The fourth of six children born to Martinus Brouwer, a government employee, and his wife Louisa van Wamelen, Brouwer studied under Willem de Sitter at the University of Leiden, where he received the Ph.D. in 1927. He then came to the United States on a year's fellowship and joined the faculty of Yale University in 1928 as an instructor. In 1941 he became both professor of astronomy and successor to Frank Schlesinger as director of the Yale Observatory, posts he held, together with the editorship of the *Astronomical Journal*, for the rest of his life. He was married in 1928 to Johanna de Graaf and became an American citizen in 1937.

At Yale, Brouwer first assisted Ernest William Brown in his search for differences between predicted and observed positions of the moon that would reveal changes in the earth's rotation. In 1930 he found that some of the differences were due to incorrectly located reference stars. To get better positions for these stars, Brouwer turned to asteroids (1935), later investigating the origins of these small bodies in a paper extending the membership in Hirayama's families (1951) and one on Kirkwood's gaps (1963).

After Brown's death in 1938, Brouwer took up more general orbital problems. Papers he published in 1938 and 1946 formed the basis for a direct determination of planetary positions by stepwise numerical integration, which was realized in *Coordinates of the Five Outer Planets, 1653–2060* (1951), written jointly with Wallace John Eckert and Gerald Maurice Clemence. This was the first astronomical problem to be solved through use of a high-speed computer. In an address delivered in 1955, when he was awarded the Gold Medal of the Royal Astronomical Society for his outstanding contributions to celestial mechanics, Brouwer outlined the way in which computer techniques were changing his field.

The advent of artificial earth satellites in 1957 provided an application for two theoretical papers Brouwer had written in 1946 and 1947, and led to two more (1959, 1961) of significant merit.

Brouwer was an active member of the International Astronomical Union and influential in its adoption of a new set of fundamental astronomical constants in 1964. It was he who suggested the name for Ephemeris Time (1950) and provided data on the way it diverged from Universal Time between 1820 and 1950 (1952). He was elected to the National Academy of Sciences in 1951, given an honorary D.Sc. by the University of La Plata in 1961, and awarded the Bruce Medal of the Astronomical Society of the Pacific in 1966.

BIBLIOGRAPHY

I. ORIGINAL WORKS. Brouwer's dissertation, "Diskussie van de Waarnemingen van Satellieten I, II en III van Jupiter," was presented to the University of Leiden in April 1927 and incorporated into the paper "Discussion of Observations of Jupiter's Satellites Made at Johannesburg in the Years 1908–1926," in *Annalen van de Sterrewacht te Leiden*, **16** (1928), 7–99, with erratum on 99. Other works, referred to in text, are "Discussion of the Annual Term in the Residuals in the Moon's Longitude," in *Astronomical Journal*, **40** (1930), 161–168; "On the Determination of Systematic Corrections to Star Positions from Observations of Minor Planets," *ibid.*, **44** (1935), 57–63; "The Use of Rectangular Coordinates in the Differential Correction of Orbits," *ibid.*, **46** (1938), 125–132, written with W. J. Eckert; "On the Accumulation of Errors in Numerical Integration," *ibid.*, 149–153; "Integration of the Equations of General Planetary Theory in Rectangular Coordinates," *ibid.*, **51** (1946), 37–43; "The Motion of a Particle of Negligible Mass Under the Gravitational Attraction of a Spheroid," *ibid.*, 223–231; "A Survey of the Dynamics of Close Binary Systems," *ibid.*, **52** (1947), 57–63; *Coordinates of the Five Outer Planets, 1653–2060*, Vol. XII of *Astronomical Papers Prepared for Use of the American Ephemeris and Nautical Almanac* (Washington, D.C., 1951), written with W. J. Eckert and G. M. Clemence; "Families of Minor Planets and Related Distributional Problems," in *Astronomical Journal*, **55** (1951), 162–163; "A Study of the Changes in the Rate of Rotation of the Earth," *ibid.*, **57** (1952), 125–146; "The Motions of the Outer Planets," in *Monthly*

Notices of the Royal Astronomical Society (London), **115** (1956), 221–235, the George Darwin lecture delivered by Brouwer on 6 April 1955, when he received the Gold Medal of the Royal Astronomical Society; "Solution of the Problem of the Artificial Satellite Without Drag," in *Astronomical Journal,* **64** (1959), 378–397, errata in **65** (1960), 108; "Theoretical Evaluation of Atmospheric Drag Effects in the Motion of an Artificial Satellite," *ibid.,* **66** (1961), 193–225, appendix on 264–265, both written with Gen-ichiro Hori; and "The Problem of the Kirkwood Gaps in the Asteroid Belt," *ibid.,* **68** (1963), 152–159.

Brouwer also wrote two textbooks: *Spherographical Navigation* (New York, 1944), written with Frederic W. Keator and Drury A. McMillen; and *Methods of Celestial Mechanics* (New York, 1961), written with Gerald M. Clemence.

A complete list of Brouwer's works, with 156 items, follows Clemence's biographical memoir (see below).

II. SECONDARY LITERATURE. The adoption of Brouwer's suggestion that time based on the year rather than the day be called Ephemeris Time is recorded in "Colloque Internationale sur les Constantes Fondamentales de l'Astronomie, Procès-verbaux des séances," in *Bulletin astronomique,* 2nd ser., **15** (1949 for 1950), 283; the citation made by John Jackson in announcing the award of the Gold Medal of the Royal Astronomical Society to Brouwer was printed in *Monthly Notices of the Royal Astronomical Society,* **115** (1955), 199–202; Louis George Henyey's "Posthumous Award of the Bruce Gold Medal to Professor Dirk Brouwer" appeared in *Publications of the Astronomical Society of the Pacific,* **78** (1966), 194–197; John Michael Anthony Danby wrote an obituary notice in *Quarterly Journal of the Royal Astronomical Society* (London), **8** (1967), 84–88; and Gerald Maurice Clemence's memoir can be found in *Biographical Memoirs. National Academy of Sciences,* Washington, **40** (1969), with portrait and list of publications.

SALLY H. DIEKE

BROUWER, LUITZEN EGBERTUS JAN (*b.* Overschie, Netherlands, 27 February 1881; *d.* Blaricum, Netherlands, 2 December 1966), *mathematics.*

Brouwer first showed his unusual intellectual abilities by finishing high school in the North Holland town of Hoorn at the age of fourteen. In the next two years he mastered the Greek and Latin required for admission to the university, and passed the entrance examination at the municipal Gymnasium in Haarlem, where the family had moved in the meantime. In the same year, 1897, he entered the University of Amsterdam, where he studied mathematics until 1904. He quickly mastered the current mathematics, and, to the admiration of his professor, D. J. Korteweg, he obtained some results on continuous motions in four-dimensional space that were published in the reports of the Royal Academy of Science

in Amsterdam in 1904. Through his own reading, as well as through the stimulating lectures of Gerrit Mannoury, he became acquainted with topology and the foundations of mathematics. His great interest in philosophy, especially in mysticism, led him to develop a personal view of human activity and society that he expounded in *Leven, Kunst, en Mystiek* ("Life, Art, and Mysticism"; 1905), where he considers as one of the important moving principles in human activity the transition from goal to means, which after some repetitions may result in activities opposed to the original goal.

Brouwer reacted vigorously to the debate between Russell and Poincaré on the logical foundations of mathematics. These reactions were expressed in his doctoral thesis, *Over de Grondslagen der Wiskunde* ("On the Foundations of Mathematics"; 1907). In general he sided with Poincaré in his opposition to Russell's and Hilbert's ideas about the foundations of mathematics. He strongly disagreed with Poincaré, however, in his opinion on mathematical existence. To Brouwer, mathematical existence did not mean freedom from contradiction, as Poincaré maintained, but intuitive constructibility.

Brouwer conceived of mathematics as a free activity of the mind constructing mathematical objects, starting from self-evident primitive notions (primordial intuition). Formal logic had its *raison d'être* as a means of describing regularities in the systems thus constructed. It had no value whatsoever for the foundation of mathematics, and the postulation of absolute validity of logical principles was questionable. This held in particular for the principle of the excluded third, briefly expressed by $A \vee \neg A$—that is, A or not A—which he identified with Hilbert's statement of the solvability of every mathematical problem. The axiomatic foundation of mathematics, whether or not supplemented by a consistency proof as envisaged by Hilbert, was mercilessly rejected; and he argued that Hilbert would not be able to prove the consistency of arithmetic while keeping to his finitary program. But even if Hilbert succeeded, Brouwer continued, this would not ensure the existence (in Brouwer's sense) of a mathematical system described by the axioms.

In 1908 Brouwer returned to the question in *Over de Onbetrouwbaarheid der logische Principes* ("On the Untrustworthiness of the Logical Principles") and—probably under the influence of Mannoury's review of his thesis—rejected the principle of the excluded third, even for his constructive conception of mathematics (afterward called intuitionistic mathematics).

Brouwer's mathematical activity was influenced by Hilbert's address on mathematical problems at the

Second International Congress of Mathematicians in Paris (1900) and by Schoenflies' report on the development of set theory. From 1907 to 1912 Brouwer engaged in a great deal of research, much of it yielding fundamental results. In 1907 he attacked Hilbert's formidable fifth problem, to treat the theory of continuous groups independently of assumptions on differentiability, but with fragmentary results. Definitive results for compact groups were obtained much later by John von Neumann in 1934 and for locally compact groups in 1952 by A. M. Gleason and D. Montgomery and L. Zippin.

In connection with this problem—a natural consequence of Klein's Erlanger program—Brouwer discovered the plane translation theorem, which gives a homotopic characterization of the topological mappings of the Cartesian plane, and his first fixed point theorem, which states that any orientation preserving one-to-one continuous (topological) mapping of the two-dimensional sphere into itself leaves invariant at least one point (fixed point). He generalized this theorem to spheres of higher dimension. In particular, the theorem that any continuous mapping of the n-dimensional ball into itself has a fixed point, generalized by J. Schauder in 1930 to continuous operators on Banach spaces, has proved to be of great importance in numerical mathematics.

The existence of one-to-one correspondences between numerical spaces R_n for different n, shown by Cantor, together with Peano's subsequent example (1890) of a continuous mapping of the unit segment onto the square, had induced mathematicians to conjecture that topological mappings of numerical spaces R_n would preserve the number n (dimension). In 1910 Brouwer proved this conjecture for arbitrary n.

His method of simplicial approximation of continuous mappings (that is, approximation by piecewise linear mappings) and the notion of degree of a mapping, a number depending on the equivalence class of continuous deformations of a topological mapping (homotopy class), proved to be powerful enough to solve the most important invariance problems, such as that of the notion of n-dimensional domain (solved by Brouwer) and that of the invariance of Betti numbers (solved by J. W. Alexander).

Finally, mention may be made of his discovery of indecomposable continua in the plane (1910) as common boundary of denumerably many, simply connected domains; of his proof of the generalization to n-dimensional space of the Jordan curve theorem (1912); and of his definition of dimension of topological spaces (1913).

In 1912 Brouwer was appointed a professor of mathematics at the University of Amsterdam, and in the same year he was elected a member of the Royal Netherlands Academy of Science. His inaugural address was not on topology, as one might have expected, but on intuitionism and formalism.

He again took up the question of the foundations of mathematics. There was no progress, however, in the reconstruction of mathematics according to intuitionistic principles, the stumbling block apparently being a satisfactory notion of the constructive continuum. The first appearance of such a notion was in his review (1914) of the Schoenflies-Hahn report on the development of set theory. In the following years he scrutinized the problem of a constructive foundation of set theory and came fully to realize the role of the principle of the excluded third. In 1918 he published a set theory independent of this logical principle; it was followed in 1919 by a constructive theory of measure and in 1923 by a theory of functions. The difficulty involved in a constructive theory of sets is that in contrast with axiomatic set theory, the notion of set cannot be taken as primitive, but must be explained. In Brouwer's theory this is accomplished by the introduction of the notion of free-choice sequence, that is, an infinitely proceeding sequence of choices from a set of objects (e.g., natural numbers) for which the set of all possible choices is specified by a law. Moreover, after every choice, restrictions may be added for future possible choices. The specifying law is called a spread, and the ever-unfinished free-choice sequences it allows are called its elements. The spread is called finitary if it allows only choices from a finite number of possibilities. In particular, the intuitionistic continuum can be looked upon as given by a finitary spread. By interpreting the statement "All elements of a spread have property p" to mean "I have a construction that enables me to decide, after a finite number of choices of the choice sequence α, that it has property p," and by reflection on the nature of such a construction, Brouwer derived his so-called fundamental theorem on finitary spreads (the fan theorem). This theorem asserts that if an integer-valued function, f, has been defined on a finitary spread, S, then a natural number, n, can be computed such that, for any two free-choice sequences, α and β, of S that coincide in their first n choices, we have $f(\alpha) = f(\beta)$.

This theorem, whose proof is still not quite accepted, enabled Brouwer to derive results that diverge strongly from what is known from ordinary mathematics, e.g., the indecomposability of the intuitionistic continuum and the uniform continuity of real functions defined on it.

From 1923 on, Brouwer repeatedly elucidated the role of the principle of the excluded third in mathematics and tried to convince mathematicians that it must be rejected as a valid means of proof. In this connection, that the principle is noncontradictory, that is, that $\neg\neg(A \vee \neg A)$ holds, is a serious disadvantage. Using the fan theorem, however, he succeeded in showing that what he called the general principle of the excluded third is contradictory, that is, there are properties for which it is contradictory that for all elements of a finitary spread, the property either holds or does not hold—briefly, $\neg(\forall \alpha)$ $(P(\alpha) \vee \neg P(\alpha))$ holds.

In the late 1920's the attention of logicians was drawn to Brouwer's logic, and its relation to classical logic was investigated. The breakdown of Hilbert's foundational program through the decisive work of Kurt Gödel and the rise of the theory of recursive functions has ultimately led to a revival of the study of intuitionistic foundations of mathematics, mainly through the pioneering work of S. C. Kleene after World War II. It centers on a formal description of intuitionistic analysis, a major problem in today's foundational research.

Although Brouwer did not succeed in converting mathematicians, his work received international recognition. He held honorary degrees from various universities, including Oslo (1929) and Cambridge (1955). He was elected to membership in many scientific societies, such as the German Academy of Science, Berlin (1919); the American Philosophical Society, Philadelphia (1943); and the Royal Society of London (1948).

BIBLIOGRAPHY

I. ORIGINAL WORKS. "Over een splitsing van de continue beweging om een vast punt 0 van R_4 in twee continue bewegingen om 0 van R_3's," in *Verslagen. Koninklijke akademie van wetenschappen te Amsterdam,* **12** (1904), 819–839; *Leven, Kunst en Mystiek* (Delft, 1905); *Over de Grondslagen der Wiskunde* (Amsterdam, 1907); "Over de onbetrouwbaarheid der logische principes," in *Tijdschrift voor wijsbegeerte,* **2** (1908), 152–158; "Die Theorie der endlichen Kontinuierlichen Gruppen, unabhängig von den Axiomen von Lie (erste Mitteilung)," in *Mathematische Annalen,* **67** (1909), 246–267, and ". . . (zweite Mitteilung)," *ibid.,* **69** (1910), 181–203; "Zur Analysis Situs," *ibid.,* **68** (1910), 422–434; "Beweis des Jordanschen Kurvensatzes," *ibid.,* **69** (1910), 169–175; "Beweis des Jordanschen Satzes für den n-dimensionalen Raum," *ibid.,* **71** (1912), 314–319; "Über eineindeutige, stetige Transformationen von Flächen in sich," *ibid.,* **69** (1910), 176–180; "Beweis der Invarianz der Dimensionenzahl," *ibid.,* **70** (1911), 161–165; "Über Abbildung von Mannigfaltigkeiten," *ibid.,* **71** (1912), 97–115, 598; "Beweis der Invarianz des n-dimensionalen Gebietes," *ibid.,* **71** (1912), 305–313; "Zur Invarianz des n-dimensionalen Gebiets," *ibid.,* **72** (1912), 55–56; "Beweis des ebenen Translationssatzes," *ibid.,* **72** (1912), 37–54;

"Beweis der Invarianz der geschlossene Kurve," *ibid.,* **72** (1912), 422–425; "Über den natürlichen Dimensionsbegriff," in *Journal für die reine und angewandte Mathematik,* **142** (1913), 146–152; "Intuitionism and Formalism," in *Bulletin of the American Mathematical Society,* **20** (1913), 81–96; review of A. Schoenflies and H. Hahn, *Die Entwicklung der Mengenlehre und ihrer Anwendungen. Erste Hälfte. Allgemeine Theorie der unendlichen Mengen und Theorie der Punktmengen* (Leipzig–Berlin, 1913), in *Jahresbericht der Deutschen Mathematikervereinigung,* **23** (1914), 78–83; "Begründung der Mengenlehre unabhängig vom logischen Satz vom ausgeschlossenen Dritten. Erster Teil: Allgemeine Mengenlehre," in *Verhandelingen Koninklijke akademie van wetenschappen te Amsterdam,* **12,** no. 5 (1918), 1–43, and ". . . II. Theorie der Punktmengen," *ibid.,* **12,** no. 7 (1919), 1–33; "Begründung der Funktionenlehre unabhängig vom logischen Satz vom ausgeschlossenen Dritten. Erster Teil: Stetigkeit, Messbarkeit, Derivierbarkeit," *ibid.,* **13,** no. 2 (1923), 1–24; "Intuitionistische Einführung des Dimensionsbegriffes," in *Proceedings. Koninklijke akademie van wetenschappen te Amsterdam,* **29** (1926), 855–863; "Über Definitionsbereiche von Funktionen," in *Mathematische Annalen,* **97** (1927), 60–75; "Essentially Negative Properties," in *Proceedings. Koninklijke akademie van wetenschappen te Amsterdam,* **51** (1948), 963–964; "Consciousness, Philosophy and Mathematics," in *Proceedings of the Tenth International Congress of Philosophy,* I (Amsterdam, 1949), 1235–1249. For Brouwer's topological work, consult the book by Alexandroff and Hopf listed below. Extensive bibliographies of his foundational work may be found in the books by Heyting and Van Heijenoort (see below). A complete edition of Brouwer's work is planned by the Dutch Mathematical Society.

II. SECONDARY LITERATURE. Brouwer or his work is discussed in P. Alexandroff and H. Hopf, *Topologie* (Berlin, 1935), *passim;* P. Benacerraf and H. Putnam, *Philosophy of Mathematics* (Englewood Cliffs, N.J., 1964), pp. 66–84; J. van Heijenoort, *From Frege to Gödel, a Source Book in Mathematical Logic, 1879–1931* (Cambridge, Mass., 1967), pp. 334–345, 446–463, 490–492; A. Heyting, *Intuitionism, An Introduction* (Amsterdam, 1965), *passim;* S. Lefschetz, *Introduction to Topology* (Princeton, N.J., 1949), pp. 1–26, 117–131; S. C. Kleene and R. E. Vesley, *The Foundations of Intuitionistic Mathematics* (Amsterdam, 1965); G. Kreisel, "Functions, Ordinals, Species," in *Logic, Methodology and Philosophy of Science,* III, ed. B. van Rootselaar and J. F. Staal (Amsterdam, 1968), pp. 145–159; J. Myhill, "Formal Systems of Intuitionistic Analysis I," *ibid.,* pp. 161–178; A. S. Troelstra, "The Theory of Choice Sequences," *ibid.,* pp. 201–223.

B. VAN ROOTSELAAR

BROWN, ALEXANDER CRUM (*b.* Edinburgh, Scotland, 26 March 1838; *d.* Edinburgh, 28 October 1922), *chemistry, physiology.*

Crum Brown was the son of Rev. John Brown, a Secessionist minister, and Margaret Crum. His uncle, Walter Crum, was a chemist and fellow of the Royal

Society. He went to Mill Hill School, London, and then Edinburgh University, where he studied medicine. He graduated in 1858, and in 1861 he received the M.D. for a thesis on the theory of chemical combination. In 1862 Crum Brown was awarded a D.Sc. by London University. In the same year, he went to Germany, where he worked under Kolbe and carried out what was probably the first synthesis of adipic acid. Returning to Edinburgh in 1863, he was appointed lecturer in chemistry, and professor in 1869, a position he held until 1908. He became a fellow of the Royal Society in 1879, and from 1891 to 1893 was president of the Chemical Society.

In his thesis Crum Brown discussed chemical structure and the application of mathematics to chemistry, the subjects that interested him most. The first attempts had just been made to represent the structure of compounds by various types of graphical formulas. Crum Brown was dissatisfied with the unwieldy diagrams invented by Kekulé and proposed a more convenient scheme. He represented constituent atoms by circles drawn around the usual letter symbols, with a number of lines proceeding from them according to their valences. He indicated atomic linkage by lines, first dotted and later solid. In 1865 he invented the symbol, which is still in use, of two parallel lines for a double bond. Unknown to him, Archibald Couper, also in Edinburgh, had already employed a similar system of letters and lines. However, the graphical formulas used today most resemble those suggested by Crum Brown. The latter were popularized by Frankland, who adopted them throughout his lectures.

Also in structural chemistry, in 1892 Crum Brown put forward the rule determining the positions of groups entering the benzene ring in the substitution of monoderivatives.

He was convinced that chemistry would one day achieve the perfection of a mathematical science. This was a common feeling, earlier expressed by Davy and Herschel. Crum Brown was inspired by a belief in the unified plan of the Creation. This would be revealed in the ultimate reduction of all physical sciences to dynamics. Just as optics and heat, in their maturity, had become branches of applied mathematics, so he expected chemistry, undeveloped and fragmentary, eventually to be mathematically deducible from mechanics. In 1867 he showed how some steps might be taken in this direction. Regarding chemical substances as operands and the processes performed on them as operators, he derived mathematical expressions corresponding to successive chemical substitutions. Although it was similar in appearance to Brodie's more elaborate operational calculus, his work was distinct from this. He thought Brodie

had been "too severe" on chemists who had used atomic language, and insisted that it could be adopted without believing in atoms. He stated that while physics showed matter to be molecular, chemists could use atomic symbols without concern for their physical significance. Even if matter were proved to be continuous, his graphical formulas would still be useful. This common pragmatic attitude to the atomic theory was soon attacked by Williamson. Crum Brown's address to the British Association in 1874 was a more realistic approach. Neither he nor Brodie made any lasting contribution to mathematical chemistry.

In physiology Crum Brown was fascinated by the effects produced through rotation. He investigated the various sensations of vertigo in a subject rotating blindfolded on a table. He correctly related these to the motion of liquid in the semicircular canals of the inner ear. His work appeared soon after Mach and Breuer had given the same explanation, but his theory was an advance in that it detailed the coordinated action of the canals on both sides of the head, which Mach had rejected. Crum Brown stated for the first time that the two horizontal canals differed from one another in that they received their stimuli from motions in opposite directions, thus allowing a blindfolded subject to distinguish between rotations to the left and to the right. To illustrate his theory, he constructed a mechanical model. This consisted of two heavy wheels fixed horizontally and side by side so that they could rotate about a vertical axis in opposite directions. Each wheel had a stop that prevented its rotation beyond a certain point in one direction; its rotation in the opposite direction was restricted by the stretching of a spring. The axle of each wheel contained an adjustable stopcock, which widened as the springs stretched and through which gas was passed from pipes and ignited.

Crum Brown said that the wheels represented the horizontal canals of the ear and that the inertia of the former corresponded with the inertia of the fluid contained in the latter. The stretching of the springs represented the stretching of the ampullae. The variation in the flames that occurred as the springs stretched and relaxed during the rotation of the wheels illustrated the different brain messages transmitted by the nerves as their endings were stimulated by the stretching of the ampullae during the rotation of the body.

BIBLIOGRAPHY

Crum Brown's papers are listed in the *Royal Society Catalogue of Scientific Papers*, I (1867); V (1877); IX (1891); XIII (1914). His other publications are in the *British Mu-*

seum General Catalogue of Printed Books, XXVII (1965). Ernst Mach discussed Crum Brown's work on rotation in his *Grundlinien der Lehre von den Bewegungsempfindungen* (Leipzig, 1875), pp. 104–110. Some useful information can be found in the obituary notice "Alexander Crum Brown," in *Journal of the Chemical Society,* **123** (1923), 3422–3431. See also David F. Larder, "Alexander Crum Brown and his Doctoral Thesis of 1861," in *Ambix,* **14** (1967), 112–132.

D. C. GOODMAN

BROWN, ERNEST WILLIAM (*b*. Hull, England, 29 November 1866; *d*. New Haven, Connecticut, 22 July 1938), *celestial mechanics.*

Brown, who early excelled in mathematics, won a scholarship to Christ's College, Cambridge, in 1884, and maintained close ties with that school throughout his life: he received a B.A. as sixth wrangler (1887), M.A. (1891), and D.Sc. (1897); and was a fellow (1889–1895) and honorary fellow (1911–1938). In 1891 he moved to the United States to become instructor in, and later (1893) professor of, mathematics at Haverford College. Brown went to Yale University in 1907—largely because Yale agreed to support the computing and publishing of his lunar tables—and he remained there as professor, Sterling professor of mathematics (1921–1931), the first Josiah Willard Gibbs professor of mathematics (1931–1932), and professor emeritus. Among his many honors were those from the Royal Society of London (fellowship, 1897; Royal Medal, 1914) and from the National Academy of Sciences (membership, 1923; Watson Medal, 1937).

While still a student, Brown was encouraged by his professor, George Howard Darwin, to study George Hill's papers on lunar theory, and from that time on, he devoted himself to reconciling lunar theory and observations by finding "in the most accurate way and by the shortest path the complete effect of the law of gravitation applied to the moon" ("Cosmical Physics," p. 185). By 1908 Brown had worked out, and published in five papers, his theory of the motion of the moon. Following Hill's example, he attacked the moon's motion as an idealized problem of three bodies, assuming the sun, earth, and moon to be spherical and the center of the earth-moon system to move in an elliptical orbit about the sun; he then considered inequalities resulting from the actual figures of the earth and moon, and the direct and indirect gravitational attractions of the other planets.

Brown's main objective was a new, accurate calculation of each coefficient in longitude, latitude, and parallax as great as one-hundredth of a second of arc, and the result was not to be in error by more than that amount; in fact, he included many terms with coefficients one order of magnitude smaller. Among the few lunar motions he could not account for by gravitation was the relatively large fluctuation in mean longitude; after rejecting numerous other possibilities, he explained this apparent deviation by irregular variations in the earth's rate of rotation.

After developing his lunar theory Brown proceeded, with the assistance of Henry B. Hedrick, to use it to construct new tables of the moon's motion. The numerical values of the constants were obtained by comparing the theory with 150 years of Greenwich observations, as analyzed by Philip H. Cowell. The tables were designed for actual computation of the moon's position, and in 1923 they were adopted by most national ephemerides. Although Brown included nearly 1,500 terms—nearly five times as many as had Peter Andreas Hansen, the author of the previously used table—the format of his tables made them as convenient to use as were Hansen's. The remainder of his work concerned the interaction of other members of the solar system, such as the Trojan group of asteroids, and the negligible gravitational attraction of Pluto for Uranus and Neptune.

BIBLIOGRAPHY

A bibliography of Brown's writings is in Schlesinger and Brouwer (see below). Among his works are "Theory of the Motion of the Moon," in *Memoirs of the Royal Astronomical Society,* **53** (1896–1899), 39–116, 163–202; **54** (1899–1901), 1–63; **57** (1908), 51–145; and **59** (1910), 1–103; *Tables of the Motion of the Moon,* 3 vols. (New Haven, 1919); and "Cosmical Physics," in *Nature,* **94** (1914), 184–190.

A biographical source is Frank Schlesinger and Dirk Brouwer, "Biographical Memoir of Ernest William Brown," in *Biographical Memoirs of the National Academy of Sciences,* XXI (Washington, D.C., 1939), 243–273.

DEBORAH JEAN WARNER

BROWN, ROBERT (*b*. Montrose, Scotland, 21 December 1773; *d*. London, England, 10 June 1858), *botany.*

In the eighteenth century the honorific designation of *princeps botanicorum* was bestowed by contemporary scholars first, it would seem, upon William Sherard (1659–1728) and later upon Linnaeus. Early in the nineteenth century Alexander von Humboldt referred to Robert Brown as *botanicorum facile princeps,* a designation both apt and just, for he rose intellectually above both these predecessors; although his works, like theirs, deal primarily with taxonomy and hence nomenclature, they embody many profound observations on morphology, embryology, and plant geography. Brown was the first to investigate the continuous erratic motion (now known as the "Brownian movement") of minute particles suspended in a fluid. He noted the streaming of protoplasm and recognized the nucleus as an essential part of the living cell. He demonstrated the lack of an ovary around the ovule in Coniferae and similar plants, thus detecting the fundamental distinction between gymnosperms and angiosperms. His extraordinarily minute and critical study of floral and seed structure in a great diversity of plants, during their development as well as in their mature state, greatly improved the classification of plants into families and genera. Through his emphasis on deep-seated characters, as Martius stated in 1859, he "detected similarity when concealed and he separated that which had merely the appearance of likeness; he sympathetically demonstrated the hidden relations between the most diversified forms." Brown's influence upon his contemporaries was far-reaching. Moreover, largely through his farsightedness, the department of botany at the British Museum, London, came into existence to provide a national botanical collection available to the public and to grow into an internationally important center of phytotaxonomic research. Thus the reputation that Brown acquired during his lifetime as one of the greatest of botanists has proved well founded.

Brown's father, the Rev. James Brown, was a Scottish Episcopalian clergyman of strong independent views; the son inherited and retained his intellectual honesty and sturdiness of character, but lost his uncompromising religious faith. After education at Marischal College, Aberdeen, and the University of Edinburgh, where he completed his medical studies, young Brown joined the Fifeshire Regiment of Fencibles in 1795 as an ensign, with the duties of surgeon's mate, and accompanied the regiment to Ireland. He was then twenty-one. How he spent his time there during the next five years is evident from such entries as the following in his diary for 13 January 1800:

> At breakfast read part of the rules concerning the genders of German nouns in Wendelborn's grammar. After breakfast transcribed into my botanical common place book part of my notes on Sloane's Herbarium of Jamaican Ferns. Attended the Hospital from one till three o'clock, saw about 30 outpatients. Dined with Mr. Thor of the Aberdeen Fencibles and remained with him till half past eleven o'clock. Drank about a pint of port in negus. Conversation various. . . . About twelve o'clock finished the transcription of my notes on Sir Hans Sloane's Ferns. This transcription has not afforded me one new idea on the subject of Filices.

And so these day-to-day records go on: "At breakfast read the rules on the nouns of the first declension in Wendelborn's German grammar"; "At breakfast endeavoured to commit to memory rules concerning the German numerals"; "After breakfast, the auxiliary German verb *Können* To be able."

German nouns, German adjectives, German verbs, the structure of mosses and ferns, the examination of blood under the microscope, medical textbooks—such were the matters to which young Brown industriously gave his time, for his official duties seem to have been light. In these diary entries Brown's later scientific development is implicit; here is expressed that wandering curiosity and that determination to master a subject detail by detail which led to his eminence. Science later gained a rich reward through his knowledge of German acquired during this period of rigorous self-education; in 1841, for example, he brought C. K. Sprengel's then little-appreciated *Das entdeckte Geheimniss der Natur im Bau und in der Befruchtung der Blumen* (1793) to the attention of Charles Darwin.

In October 1798, apparently while in London on a recruiting mission, Brown was introduced by José Correa da Serra, then in exile from Portugal, to Sir Joseph Banks, whose house at Soho Square, with its rich library and herbarium, was the botanical center not simply of London but of Britain. Significantly, Correa referred to Brown as "a Scotchman, fit to pursue an object with constance and cold mind." Thus Brown came to the notice of Banks, who had ever an eye for talent, and of Banks's erudite botanist-librarian Jonas Dryander; he obviously impressed them both by his zeal and ability.

Accordingly, when in December 1800 plans had matured at the Admiralty for a voyage, commanded by Matthew Flinders, to survey the southern and northern coasts of Australia—or New Holland, as the continent was then called—Banks, whose opinion carried much weight at the Admiralty, offered Brown a recommendation for the post of naturalist aboard Flinders' ship, the *Investigator,* at a salary of £420, then a very substantial sum. Brown immediately accepted this attractive offer. He came to London and, until the sailing of the *Investigator* on 18 July 1801, spent his time studying the specimens, illustrations, and literature about New Holland plants available at Banks's house. He and Flinders were both twenty-seven years old; the botanical draughtsman, Ferdinand Bauer, was forty-one.

The *Investigator* stopped at the Cape of Good Hope, a region rich in Proteaceae, to which Brown later gave much attention, then sailed for the southwestern corner of Western Australia. Landing there on 8 December 1801, at King George Sound, both Brown and Bauer were challenged by the astonishing floral richness of this region, its plants in their diversity and strangeness far exceeding anything previously seen. Three weeks there yielded some 500 species, almost all of them new to science. No adequate guides for their classification then existed. Brown's task was to study their structure intimately, to group them into genera and species, and to make detailed descriptions. From Lucky Bay, which yielded 100 more species, the *Investigator* sailed eastward along the southern coast of Australia, passed through the Bass Strait, and turned northward to Port Jackson, which it reached on 8 May 1802. Here they stayed for twelve weeks. The *Investigator* now sailed northward along the east coast to Cape York and into the Gulf of Carpentaria. Unfortunately the *Investigator* had been damp, leaky, and unsound from the start, and Flinders dared not continue his survey. He sailed to Timor for provisions, then returned to Port Jackson, arriving there on 8 June 1803. Fortunately, Brown and Bauer stayed behind when Flinders set out in another ship on his unlucky return voyage to England, which he did not reach until 1810. Brown spent ten months in Tasmania; Bauer went to Norfolk Island. They reached England aboard the repaired *Investigator* in October 1805, bringing with them specimens of nearly 4,000 species of plants as well as numerous drawings and zoological specimens. Thereupon Banks recommended to the Admiralty that Brown prepare for publication "a succinct account" of his plants and that he receive a government salary while so doing, at the same time selecting representative specimens for the public collection (i.e., the British Museum). This task kept Brown busy during the next five years. By 6 January 1810 he had described nearly 2,200 species, over 1,700 of which were new (including 140 new genera), and had selected about 2,800 specimens.

Concurrently with this botanical activity Brown served the Linnean Society of London as "Clerk, Librarian and Housekeeper" from 1806 to 1822. In 1810 Jonas Dryander died, and Banks appointed Brown to succeed him as librarian and curator at Soho Square. He held these posts until Banks's death on 19 June 1820, when he became his own master, for Banks had bequeathed to "my infatigable and intelligent librarian Robert Brown" an annuity of £200 and the life tenancy of his Soho Square house, with the use and enjoyment of its library and collections; on Brown's death these were to pass to the trustees of the British Museum. Brown did not wait that long for their transfer into national keeping. In 1827 he bargained with the trustees for their immediate transfer, stating that if the trustees agreed to form an independent botanical department in the British Museum, he would be willing to take charge of it, his own status to be that of an underlibrarian (the title then of the head or keeper of each separate museum department). The trustees accepted this reasonable stipulation, and Brown spent the winter of 1827/1828 moving the Banksian collections from Soho Square to Montague House, the old British Museum, Bloomsbury. Thereby Brown secured, for the first time in Britain, a nationally owned botanical collection available to the public; he remained in charge from 1827 to 1858. As his assistant he had John Joseph Bennett (1801–1876), who had trained as an apothecary and surgeon. Upon this pair fell the whole business of the department; as W. Carruthers said in 1876, "It is hard to realize that this time [1827/1828] and for eight more years all the work of the department, even the merest manual drudgery, had to be performed by Mr. Bennett or Mr. Brown."

The period from 1806 to 1820 (i.e., from the return of the *Investigator* to the death of Banks) was that of Brown's greatest creative endeavor; it was the period during which he worked under Banks's fatherly eye. After 1828 he published comparatively little. His earlier work relates to the flora of Australia but leads to other matters, linked more by Brown's methods of investigation—which took him from one problem to another, each receiving detailed methodical treatment—than by any general plan.

On Flinders' voyage Brown had many opportunities to study, in the living state, members of the family Proteaceae, which is well represented in Australia and South Africa; back in England, Banks, James Edward Smith, and others made their herbaria available to him. George Hibbert's collection of living plants, skillfully grown by his gardener Joseph Knight, and his herbarium, formed at the Cape of Good Hope by his collector David Niven, were particularly rich in Proteaceae; the same collections also attracted the attention of Richard Anthony Salisbury. On this material Brown based his classic paper "On the Proteaceae of Jussieu," which was read to the Linnean Society of London on 17 January 1809, Salisbury being present, but not published in the Society's *Transactions* (**10**, 15–226) until February 1810. It is notable not only for its clear exposition of the morphology of these plants, for its proposal of a new classification into genera based largely on floral details hitherto uninvestigated, for the definition of these genera and of the species, but also for its observations on geographical distribution and, surprisingly introduced, the androecium of Asclepiadaceae. Sharp practice by Salisbury deprived Brown of priority of publication.

In August 1809 there appeared Joseph Knight's book *On the Cultivation of the Plants Belonging to the Natural Order of Proteēae* [*sic*], described by Bishop Goodenough in December 1809 as "Salisbury's surreptitious anticipation of Brown's paper on the New Holland plants, under the name and disguise of Mr. Hibbert's gardener!" In it, described under other names, were genera and species known to have been recorded in Brown's manuscript; the preface acknowledged Salisbury's participation.

April 1810 saw the publication of Brown's paper "On the Asclepiadeae," subsequently issued in *Memoirs of the Wernerian Natural History Society* (**1** [1811], 12–78), in which he separated the family Asclepiadaceae from the Apocynaceae by the character of its pollen, and the first and only volume of his *Prodromus florae Novae Hollandiae*. His intent was "to include the generic and specific characters of all the plants known to be natives of New Holland." He himself paid the cost of printing. He gave twenty-four copies to leading botanists and learned societies and hoped, no doubt, to sell the rest of the 250 copies printed; in fact he sold twenty-four. The volume started with a survey of the ferns (Filices) and their allies, then dealt with Gramineae and other monocotyledons, followed by families of dicotyledons—among them Proteaceae, Scrophulariaceae, Apocynaceae, and Asclepiadaceae—covering 464 genera and some 2,000 species. It remains a work of fundamental importance for Australian botany. The two dominant systems of plant classification then were Linnaeus', based on the number of floral parts and frankly artificial but often convenient, and A. L. de Jussieu's, based on a wider range of characters and thereby bringing together plants that agreed more closely in the sum of their characters. Brown found neither system satisfactory when dealing with the bewildering variety of Australian plants. In the *Prodromus* he adopted, contrary to prevailing usage, a modified form of Jussieu's more natural system, amending the definition of families and genera, adding many new to science, and inserting a multitude of firsthand observations based not simply on Australian plants but also on plants from elsewhere. It immediately won the esteem of eminent contemporary botanists—as one said, "Everything here is new . . . and every part abounds with observations equally original and useful"—but it did not appeal to the book-buying public. Sadly disappointed by the sale of the first volume, Brown seems to have discontinued work on the second after 1817 and this volume, which would have covered Leguminosae, Myrtaceae, Compositae, and other families, was never published. The loss to science would have been greater had not

Brown incorporated some of this material in memoirs, appended to books of travel, that often were based upon fragmentary specimens which, as Martius remarked, could have been made so important and fruitful only by a genius like Brown.

The most important of these memoirs is probably the "General Remarks, Geographical and Systematical, on the Botany of Terra Australis," published in Flinders' *A Voyage to Terra Australis* (1814). Here Brown estimated the Australian species known to him at about 4,200 species, with the number of dicotyledons more than three times the number of monocotyledons, and established the families Pittosporaceae, Cunoniaceae, Rhizophoraceae, Celastraceae, Haloragaceae, and Stackhousiaceae. Brown's observations also found expression in further papers published by the Linnean Society, notably "Observations on the natural family of plants called Compositae" (*Transactions of the Linnean Society of London,* **12** [1818], 76–142), and in contributions to the second edition of Aiton's *Hortus Kewensis* (1812), notably on Cruciferae, Leguminosae, Myrtaceae, and Orchidaceae. Of great importance for the development of classification was not so much the precision of Brown's descriptions as his perception of relationships and his statements of the evidence for them, which led to the concept that certain characters had not so much absolute as relative worth, being constant, and hence valuable, in some groups but varying in others. Prophetically, he noted in 1810, from a consideration of the shape of pollen in Proteaceae and other families, that "it may be consulted with advantage in fixing our notions of the limits of genera." He introduced other characters, such as aestivation of the flower, into generic descriptions, deriving largely, it would seem, from his interest in ascertaining the early state and the development of organs. The cumulative effect of so many minute observations perspicuously correlated gave Brown's publications their high authority. Necessarily he left it to others to exploit by further investigation the lines of inquiry he indicated. After the disappointment of the *Prodromus* he turned from major works of synthesis and made important information available almost in a casual manner, as digressions or appendages to memoirs only remotely connected with it.

Thus, in 1831 Brown published as a pamphlet for private distribution his "Observations on the Organs and Mode of Fecundation in Orchideae and Asclepiadeae" (reprinted in *Transactions of the Linnean Society of London,* **16** [1833], 685–742), which contains his observations that the pollen of orchids, when placed upon the stigma, emits pollen tubes traceable into the ovary. Embedded in this paper is a discovery highly relevant to the cell theory, and thus to the development of cytology. Regarding the leaves of orchids, Brown stated:

> In each cell of the epidermis of a great part of this family, especially of those with membranous leaves, a single circular areola, generally somewhat more opake than the membrane of the cell, is observable. . . . only one areola belongs to each cell. . . . This areola, or nucleus of the cell as perhaps it might be termed, is not confined to the epidermis, being also found not only in the pubescence of the surface particularly when jointed, as in Cypripedium, but in many cases in the parenchyma or internal cells of the tissue. . . . The nucleus of the cell is not confined to the Orchideae but is equally manifest in many other Monocotyledonous families; and I have even found it, hitherto however in very few cases, in the epidermis of Dicotyledonous plants ["Observations" (1831), 19–21; *Transactions* (1833), 710–712].

A few earlier botanists evidently had observed the presence of this nucleus in some cells, as Brown himself points out, but he was the first specially to demonstrate its general occurrence in living cells and to give it the name "nucleus."

Brown's curiously incidental method of making known an important discovery resulting from long research is exemplified in a paper entitled "Character and Description of Kingia," appended to P. P. King's *Narrative of a Survey of the Intertropical and Western Coasts of Australia* (II [London, 1827], 536–563). Here occurs the remark

> It would entirely remove the doubts that may exist respecting the point of impregnation, if cases could be produced where the ovarium was either altogether wanting, or so imperfectly formed, that the ovulum itself became directly exposed to the action of the pollen. . . . such, I believe, is the real explanation of the structure of Cycadeae, of Coniferae, of Ephedra, and even of Gnetum [p. 555].

Brown then got rid of objections to this view that the ovule in these plants was not contained within an ovary, but he left it for others to emphasize that it established a fundamental difference between the gymnosperms (as they were later named) and the other flowering plants (i.e., the angiosperms). It remained for Hofmeister's later investigations to indicate the relevance of gymnospermy and angiospermy to the theory of the alternation of generations.

In the course of these microscopical explorations Brown passed from the study of the ovule to that of pollen grains, and thus came to investigate the phenomenon known as the Brownian movement, i.e., the continuous motion of minute particles suspended in a fluid, which results from their being bombarded by molecules in like continuous motion.

In June 1827, when examining pollen grains of *Clarkia pulchella,* Brown observed particles suspended in a fluid within the grain which were evidently moving, and he concluded that their motions "arose neither from currents in the fluid nor from its gradual evaporation but belonged to the particle itself." He thereupon extended his observations, as was his wont, to numerous species belonging to many families of plants, and found such motion in the particles of all fresh pollen. This led him to inquire whether the property continued after the death of the pollen; he then found it even in herbarium specimens preserved for not less than a century. Ultimately, after examining powdered pit coal and glass, numerous rocks, and metals in a finely divided state, Brown stated that such active particles occurred in every mineral he could reduce to a powder sufficiently fine to be suspended in water. He published these results in 1828, in a privately printed pamphlet entitled *A Brief Account of Microscopical Observations Made in the Months of June, July and August, 1827, on the Particles Contained in the Pollen of Plants; and on the General Existence of Active Molecules in Organic and Inorganic Bodies.* He took care to point out that the motion of the particles within had earlier been "obscurely seen by Needham, and distinctly by Gleichen," but to Brown belongs the credit for establishing such motion as a property not simply of living pollen but of all minute particles, inorganic as well as organic, suspended in a fluid. Here again it remained for others to carry Brown's work much further and to demonstrate its relevance to the kinetic theory of gases.

The last of Brown's work was contained in Brown and Bennett's *Plante rariores Javanicae,* published in four parts (London, 1838, 1840, 1844, 1852). This scholarly book had an unfortunate history. Among the collections that came into Brown's hands was one made in Java by Thomas Horsfield between 1802 and 1818, comprising 2,196 species, according to Brown's statement, and thus representing a large part of the flora of a little-known area of great phytogeographical interest. Horsfield evidently hoped for a complete enumeration of his material classified and named by Brown, who, however, merely selected for publication "those subjects which appeared to possess the greatest interest, either on account of their novelty, or of their peculiarity of structure." Brown's other activities and his annual eleven-week holidays severely handicapped his participation in this work, and in any event the routine publication of new species had no more interest for him. He liked to let an inquiry lead him from fact to fact until he had obtained a general view and was in a position to write a monograph, for which he then substituted a synopsis or which he condensed into a footnote. The *Plantae rariores Javanicae* ultimately had only 258 pages and fifty plates, and Brown himself wrote on only thirty of Horsfield's 2,196 species. It must have been a severe disappointment to Horsfield. This led John Lindley to write that

> . . . it was the misfortune of Dr. Horsfield to place the very important collections of plants which he formed in Java in the early part of this century, in the hands of a gentleman who to an extensive acquaintance with the details of systematical botany adds habits of procrastination, concerning which we shall only say that they are fortunately unparalleled in the annals of natural history. . . . The fatigue of describing these 50 in the course of 30 years was moreover found to be so excessive that a second editor had to be added to the first, in order that the Herculean labour might be accomplished [*Gardener's Chronicle*, 1852 (26 June), 406–407].

Despite this, Brown's contributions are important for the supplementary observations that he added. Thus, to the description of *Loxonia acuminata* he appended a long essay on the classification of Gesneriaceae and a synopsis of the genera and species of the Cyrtandreae; under the description of *Pterocymbium javanicum* he inserted a synopsis of Sterculiaceae.

Brown's mind became so richly endowed over the years with the details of so much painstaking and critical investigation into the characters of plants, most of it done with the aid of the microscope, that to make them available in a major work of synthesis would have been a task beyond his industry, even had it been congenial to his temperament; instead, he adopted his peculiar, almost haphazard, presentation of isolated parts of his special information, which makes almost all his publications pregnant with unexpected indications of fruitful inquiry. Charles Darwin, who knew Brown well and for many years spent Sunday mornings in discussion with him, stated:

> He seemed to me to be chiefly remarkable for the minuteness of his observations and their perfect accuracy. He never propounded to me any large scientific views in biology. His knowledge was extraordinarily great, and much died with him, owing to his excessive fear of never making a mistake. He poured out his knowledge to me in the most unreserved manner, yet was strangely jealous on some points [*Autobiography of Charles Darwin*, N. Barlow, ed., (London, 1958), p. 103].

Brown never married and had no near relatives. He lived in the house left him by Banks from 1821 until his death; he died in the room that had been Banks's library. His many excursions abroad made him personally well known, particularly in Germany, where he was much esteemed. According to Martius, "he sat whole nights in his arm-chair, reading and thinking." Despite his quiet manner and unobtrusive way of life, Brown was by no means a recluse and was, according to Asa Gray, "very fond of gossip at his own fireside," and, according to W. J. Hooker, "really fond of society and calculated to shine in it; and to my certain knowledge, never so happy as when he is in it." His somewhat feminine but far from effeminate disposition had none of Lindley's aggressiveness; contemporaries who knew him well testify to his tenderness and kindness, but the company he liked was essentially that of his peers. He kept remote from controversy and public affairs, had no contact with university students, and went his tranquil way in continuous search of truth, untroubled by economic difficulties and a multiplicity of duties such as beset Lindley. He refused three professorships. Brown never lost interest in the plants of his Scottish homeland and often returned to Montrose; at the age of eighty, in 1853, he ascended Lochnagar, a mountain on which he had botanized just sixty years earlier.

Brown became a fellow of the Linnean Society in 1822, and was its president from 1849 to 1853; he was elected a fellow of the Royal Society in 1810. Numerous academies honored him with election to foreign membership. It was to the Munich Academy of Sciences that his friend of many years, Carl von Martius, delivered in 1859 a eulogy that both in the original German and in Henfrey's translation still provides an excellent general appreciation of Brown's character and his scientific achievements.

BIBLIOGRAPHY

I. ORIGINAL WORKS. Robert Brown's contributions to learned periodicals, travel books (in which they form botanical appendices), and so on are brought together in *The Miscellaneous Botanical Works of Robert Brown*, J. J. Bennett, ed., 2 vols. (London, 1866–1867); a less complete collection is the earlier *Robert Brown's vermischte botanische Schriften*, C. G. Nees von Esenbeck, ed., 5 vols. (Nuremberg, 1825–1834), which includes a reprint, with different pagination, of his *Prodromus florae Novae Hollandiae* (London, 1810), originally issued in an edition of 250 copies and republished in facsimile, with an introduction by W. T. Stearn, as no. 5 in the series Historiae Naturalis Classica (1960).

His unpublished MSS, notes, diary, and correspondence are at the British Museum (Natural History), South Kensington, London.

II. SECONDARY LITERATURE. Works on Brown are N. T. Burbidge, "Robert Brown's Australian Collecting Localities," in *Proceedings of the Linnean Society of New South Wales,* **80** (1956), 229–233; W. Carruthers, "John Joseph Bennett," in *Journal of Botany, British and Foreign,* **14** (1876), 97–105; J. B. Farmer, "Robert Brown, 1773–1858," in F. W. Oliver, *Makers of British Botany* (London, 1912), pp. 108–125; J. D. Hooker, "Eulogium on Robert Brown," in *Proceedings of the Linnean Society of London,* sess. 1887–1888 (1890), 5–67; C. F. P. von Martius, "Robert Brown, eine akademische Denkrede," in *Flora* (Regensburg), **42** (1859), 10–15, 25–31, repr. in Martius, *Akademische Denkreden* (Leipzig, 1866), pp. 365–381, and in *Annals and Magazine of Natural History,* 3rd ser., **3** (1859), 321–331, A. Henfrey, trans.; J. Ramsbottom, "Robert Brown, botanicorum facile princeps," in *Proceedings of the Linnean Society of London,* **144** (1932), 17–36; and W. T. Stearn, *Three Prefaces on Linnaeus and Robert Brown* (Weinheim, 1962).

WILLIAM T. STEARN

BROWNE, THOMAS (*b.* London, England, 19 October 1605; *d.* Norwich, England, 19 October 1682), *general science, natural history.*

Browne's father, Thomas, had come in early manhood from Chester to London, where he was a mercer, or silk merchant. There he married Anne Garroway of Acton, Middlesex. They lived in comfortable circumstances with their son and four daughters. The senior Browne died when his son was eight years old, but left ample means for his education. Accordingly, the boy was sent to Winchester College in 1615. William of Wykeham's famous school was Anglican and Royalist, and provided a sound classical education that was a good foundation for the erudition acquired by Browne in later years. He remained at Winchester for eight years, until, in 1623, he proceeded to Oxford. He matriculated at Broadgates Hall, which soon afterward was upgraded to become Pembroke College. Browne, although only a freshman, was called upon to deliver a Latin oration at the inauguration ceremony.

Winchester would have afforded the boy little or no opportunity for study of the natural sciences, so it was probably during the school holidays that he began to acquire his knowledge of natural history. At Oxford a chair of anatomy had just been established in addition to several other chairs of physical sciences. Browne's teachers included Dr. Clayton, an anatomist and Regius professor of medicine, and Dr. Thomas Lushington, a mathematician and clergyman. Clayton's influence directed Browne's attention to the study of medicine and human anatomy, but this could not begin seriously until he had taken his M.A. in philosophy in 1629. He then left Oxford and spent some weeks in Ireland with his stepfather, Sir Thomas Dutton, before proceeding to Montpellier for full training in medicine.

It is not known how long Browne remained in France, but his travels in several European countries cannot have occupied less than four years. He probably spent some time in Padua, but his final goal was Leiden, where he defended his thesis and received his M.D. in December 1633. During these travels he studied many subjects besides medicine, absorbing information of all kinds and acquiring knowledge of several modern languages.

English regulations required a medical man with a foreign degree to practice for four years with an established doctor before being allowed to have his M.D. by incorporation at Oxford or Cambridge. It is probable that Browne spent these years of apprenticeship somewhere in Oxfordshire, but no details are known. He took his M.D. at Oxford on 10 July 1637 and was then, at the age of thirty-two, free to practice anywhere that he chose. It was during these four years that Browne wrote his most famous book, *Religio medici,* which was not published until 1642. Influenced, it is believed, by his Oxford friend Dr. Lushington, Browne moved in 1637 to the East Anglia city of Norwich and established himself there as a physician. In 1641 he married Dorothy Mileham, from a neighboring village. They had twelve children, only four of whom survived their parents. Edward, the eldest son, became a well-known physician in London and was president of the College of Physicians in 1704. Browne was knighted in 1671 by King Charles II, who was visiting Norwich and wished to honor its most distinguished citizen.

Browne's *Religio medici* describes the religion and philosophy of a tolerant, humorous, and latitudinarian mind. He did not, however, expose in it much of his attitude toward the rapidly expanding world of science. Yet throughout his apprenticeship and first years in Norwich he must have been reading widely in travel, philosophy, medicine, and science, and compiling the notebooks from which he quarried his next, very long book, *Pseudodoxia epidemica: or, Enquiries Into Very Many Received Tenents, And Commonly Presumed Truths* (1646). In this he sought to dispel popular ignorance about many matters in history, folklore, philology, science, medicine, natural history, and embryology. He was, thus, to be designated an "enquirer after truth" rather than a "scientist" (a term not yet invented), his field of inquiry being as wide as all human knowledge. He accepted the authority of William Harvey, one of the first great experimental scientists, and told a young correspondent: "Be sure you make yourself master of Dr. Harvey's piece, *De circulatione sanguinis,* which discovery I prefer to that of Columbus." Browne conducted many experiments in physics, electricity (a word of his own coining), biology, and comparative anatomy, dissecting animals, birds, fishes, reptiles, worms, and insects. He became an acknowledged authority on the plants, animals, birds, and fishes of East Anglia. Many of his experiments are mentioned in his *Pseudodoxia epidemica* and his letters. Others, such as investigations of bubbles, and of coagulation, freezing, and other properties of matter remained in the privacy of his notebooks.

Throughout his active life Browne lived on the fringe of the scientific world. His profession was medicine; his hobbies were science and natural history. He was an earnest amateur and never, as far as is known, left Norfolk for London. He was elected a fellow of the College of Physicians, but was never a fellow of the Royal Society of London, nor did he betray any desire for this kind of recognition. His elaborate and highly latinized prose style was very different from the much more austere style deliberately adopted by the fellows of the Royal Society. He was content to correspond with various fellows, such as Henry Oldenburg (secretary of the Society), John Ray, Christopher Merrett, and the diarist John Evelyn, and occasionally to send communications through his son Edward.

He was deeply interested in archaeology; one of his most famous books was *Hydriotaphia, or, Urne-buriall* (1658), occasioned by the discovery of some supposed Roman (really Saxon) burial urns near Norwich. He corresponded with other eminent antiquaries, such as Sir William Dugdale, Elias Ashmole, and John Aubrey. With these manifold interests and occupations, it is not surprising that Browne is remembered as a learned man and a literary artist rather than for any important contributions to contemporary science. His qualities served to foster a general interest in science and, above all, to illuminate thought by truth concerning the material world.

BIBLIOGRAPHY

I. ORIGINAL WORKS. Browne's works are *Religio medici* (pirated ed., London, 1642; 1st authorized ed., London, 1643), which also appeared in *Religio medici and Other (Shorter) Works,* L. C. Martin, ed. (Oxford, 1964); *Pseudodoxia epidemica* (London, 1646); *Hydriotaphia, or, Urne-buriall* (London, 1658); *The Garden of Cyrus* (London, 1658); *Miscellany Tracts* (London, 1684); *A Letter to a Friend* (London, 1690); *Posthumous Works* (London, 1712), with a biography by John Whitefoot; and *Christian Morals* (London, 1716), later edited, with a biography, by Samuel Johnson (London, 1756).

The following are collections of Browne's works: *The Works of Sir Thomas Browne,* Thomas Tenison, ed. (London, 1686); *Sir Thomas Browne's Works,* Simon Wilkin, ed., 4 vols. (London, 1836), which contains much biographical and scholarly material; *The Works of Sir Thomas Browne,* Sir Geoffrey Keynes, ed., 6 vols. (London, 1928–1932; rev. and enl. ed., 4 vols., London, 1964), which is the standard and only complete edition and includes miscellaneous material from manuscripts and letters.

II. SECONDARY LITERATURE. Works on Browne are Joan Bennett, *Sir Thomas Browne* (Cambridge, 1962); Jeremiah Finch, *Sir Thomas Browne* (New York, 1950); Sir Edmund Gosse, *Sir Thomas Browne* (London, 1905); F. L. Huntley, *Sir Thomas Browne* (Ann Arbor, Mich., 1962); Sir Geoffrey Keynes, *A Bibliography of Sir Thomas Browne* (Cambridge, 1924; 2nd rev. ed., Oxford, 1968); and Oliver Leroy, *Le chevalier Thomas Browne* (Paris, 1931).

GEOFFREY KEYNES

BROWNRIGG, WILLIAM (*b.* High Close Hall, Cumberland, England, 24 March 1711; *d.* Ormathwaite, Cumberland, England, 6 January 1800), *chemistry.*

Except that he came of a family long settled in western Cumberland and apparently was apprenticed to a physician in Whitehaven some time before 1733, little is known of Brownrigg's background and early life. He studied in London and later at Leiden, where he obtained his doctorate in medicine in 1737. He took up practice in Whitehaven and in 1741 married Mary Spedding, daughter of the steward of the Lowther estates.

Sir James Lowther owned the Whitehaven collieries and encouraged Brownrigg in his investigations into the damps arising in the mines, arranging for firedamp to be piped from the mines to Brownrigg's laboratory as a source of heat. Brownrigg believed that a greater knowledge of these exhalations would help to decrease the mortality among miners; he also thought that there might be a connection between them and epidemic diseases, and that some would prove to be the same as the "elastic spirits" with which mineral waters were impregnated and to which the latter owed their properties. Four papers under the general heading "Of Damps" were communicated through Lowther to the Royal Society in 1741 and 1742, and led to Brownrigg's being elected a fellow in May 1742.

An extract from these was published in the *Philosophical Transactions* for 1765, as an addendum to a paper (for which Brownrigg received the Copley Medal) in which he virtually showed that the gas which could be expelled from water from Pouhon and other mineral springs on heating, and the expulsion of which led to the precipitation of dissolved solids, was identical with the chokedamp (i.e., carbon dioxide) of the mines.

Further experiments and observations made, he says, at about the same time, but not presented until 1774, dealt in detail with the dependence of the solubility of calcareous earths and iron salts on the dissolved gas. In the meantime (as he acknowledged), papers on the solubility of these earths and salts had been given respectively by Cavendish (1767) and Timothy Lane (1769). He expressed the view "that the mephitic air and martial earth, contained in the Pouhon water, strongly attract each other, and uniting together, form a concrete soluble in water" (*Philosophical Transactions,* 1774, p. 363), that is, calcium and ferrous carbonates combine with water and carbon dioxide to form bicarbonates, which are soluble.

Brownrigg's views on gases and his technique in handling them were an advance on those of Hales, and it seems not unjustifiable to claim a place for him in the direct line of British pneumatic chemists that includes Black, Cavendish, and Priestley. Particularly noteworthy is his opinion, contrary to that generally held in his day, that "two elastic fluids, altho' they both possess a repulsive quality, may yet in their other qualities, differ as much as inelastic fluids are found to differ; as water, for example, differs from oil of vitriol" (*Philosophical Transactions,* 1765, p. 238).

Much of the other work of Brownrigg, a man of wide interests, merits attention; here we mention only his investigation, the first by a trained European scientist, of platinum, specimens of which had been brought to England in 1741 by his brother-in-law, Charles Wood. A more thorough and accurate examination of the metal was made during the next decade by Scheffer in Sweden and Lewis in England.

BIBLIOGRAPHY

I. ORIGINAL WORKS. Brownrigg wrote two treatises of contemporary importance, both showing evidence of a painstaking accumulation of facts and a profound grasp of the problems involved. *The Art of Making Common Salt* (London, 1748) advocated the extensive manufacture of salt by the evaporation of seawater at selected sites on the east coast of England, with a view to drastically decreasing its price (almost all of it was imported and heavily taxed). An abstract by William Watson appeared in *Philosophical Transactions,* **45** (1748), 351–372. *Considerations on the Means of Preventing the Communication of Pestilential Contagion, and of Eradicating It in Infected Places* (London, 1771) dealt with measures that Brownrigg thought should be adopted if the plague, which had appeared in Europe that year, should spread to Britain.

The papers to which reference has been made in the text are "Of Damps," a series of four papers read 16 April 1741, 11 March 1742, 8 April 1742, 13 May 1742; the first was entitled "Some Observations Upon the Several Damps in the Coal Mines Near Whitehaven" and the others dealt, respectively, with the possible relations of these damps to epidemics, mineral waters, and the nature of common air. They were never published as a whole, but the MSS are preserved in the archives of the Royal Society. The other papers are "Several Papers Concerning a New Semi-metal Called Platina," in *Philosophical Transactions,* **46** (1749/1750), 584–596; "An Experimental Inquiry Into the Mineral Elastic Spirit, or Air, Contained in Spa Water; as Well as Into the Mephitic Qualities of That Spirit," *ibid.,* **55** (1765), 218–235, the extract from "Of Damps" following on 236–243; and "Continuation of an Experimental Inquiry, Concerning the Nature of the Mineral Elastic Spirit or Air Contained in the Pouhon Water, and Other Acidulae," *ibid.,* **64** (1774), 357–371.

II. SECONDARY LITERATURE. The contemporary biography by Joshua Dixon, his pupil, is eulogistic but invaluable: *The Literary Life of William Brownrigg, M.D., F.R.S., to Which is Added an Account of the Coal Mines Near Whitehaven* (London, 1801). J. Russel-Wood, in *Annals of Science,* **6** (1950), 186–196, 436–447; **7** (1951), 77–94, 199–206, gives a biographical sketch and discusses Brownrigg's published and unpublished work. J. McDonald, *A History of Platinum From the Earliest Times to the Eighteen-eighties* (London, 1960), appraises the work of Brownrigg, Scheffer, and Lewis, *inter alia,* on platinum.

E. L. SCOTT

BROWN-SÉQUARD, CHARLES-ÉDOUARD (*b.* Port Louis, Mauritius, 8 April 1817; *d.* Paris, France, 1 April 1894), *physiology.*

Brown-Séquard's father, Charles Edward Brown, was an American naval officer; his mother, Charlotte Séquard, was French. He was a British subject by birth, but just before becoming professor at the Collège de France in 1878 he became a French citizen. Brown-Séquard's early life was difficult. Born after his father's death, he was raised by his mother in modest circumstances. After receiving the M.D. in Paris on 3 January 1846, he became the protégé of Pierre Rayer and began his work at the hospital of La Charité. Although he always remained interested in the life of Paris and its scientific movements, Brown-Séquard had a restless nature and was unable to have a real home. Before receiving his degree he had lived in Mauritius, and after completing his studies he went to the United States. He practiced medicine and gave private courses in Philadelphia, New York, and Boston in 1852. After his marriage to Ellen Fletcher of Boston, Brown-Séquard returned to Paris and later settled again in Port Louis, where he distinguished himself in fighting a cholera epidemic. In 1854 he accepted a professorship at Virginia Medical College in Richmond, but returned to Paris in 1855. In 1856 he again gave courses in Boston but suddenly decided to live in England, where he taught at the Royal College of Surgeons in 1858. As a physician at the National Hospital for the Paralysed and Epileptics, Brown-Séquard lived in London from 1860 to 1863. He was elected to the Royal Society in 1861. Becoming in England more and more the slave of professional practice, he fled to Paris and then to Boston. The Harvard Medical School offered him the chair of physiology and pathology, which he held from 1864 to 1867. After the death of his wife in 1867, Brown-Séquard again left the United States for Paris, where he gave the course in comparative and experimental pathology at the Faculté de Médecine in 1869. The following year he settled once more in the United States, and married Maria Carlisle, of Cincinnati, Ohio, in 1872. His activity then knew no respite: he founded a new journal and organized a physiology laboratory in New York; lectured in Boston, London, Dublin, and Paris; and wrote several dozen scientific articles. His travel increased after his second wife's early death, and in 1877, in Geneva, he married an Englishwoman named Emma Dakin.

The high point of Brown-Séquard's scientific career came following the death of Claude Bernard, when he became professor of medicine at the Collège de France (3 August 1878). He retained this post until his death but divided his time between Paris and Nice, leaving his assistant, Arsène d'Arsonval, to give the winter courses. In 1886 Brown-Séquard was elected to the Académie des Sciences.

While a student, Brown-Séquard had become particularly interested in problems of physiology and conducted original experiments on gastric digestion and the function of the spinal cord. In his thesis he suggested that in the spinal cord sensations are transmitted through the gray matter rather than through the dorsal columns. He also described the phenomenon that was later known as traumatic spinal shock: marked diminution in reflex activity immediately after sectioning of the cord and its subsequent recovery and even hyperesthesia. His thesis contained the germ of his main discovery: crossing of the nervous pathways for conduction of sensation in the spinal cord. In 1849 he showed that transverse hemisection of the cord produces motor paralysis and hyperesthesia on the corresponding side and anesthesia on the opposite side of the body below the lesion (the Brown-Séquard syndrome). This discovery was confirmed by clinical observations and proved to be very useful in the diagnosis of neurological lesions.

In August 1852, while in the United States, Brown-Séquard demonstrated the existence of vasoconstrictive nerves. When the severed cervical sympathetic trunk was stimulated by a galvanic current, the phenomena described by Claude Bernard as following the section (increasing of the skin temperature, dilation of the blood vessels, sensitization) disappeared and were replaced by inverse phenomena. It must be said that Bernard carried out the same experiment independently of Brown-Séquard and obtained the same results. It should be noted that the publication of Brown-Séquard's experiment preceded that of Bernard's; but, on the other hand, Brown-Séquard's research in this field was directly inspired by Bernard's work and was the logical sequel to the experiments made public just before Brown-Séquard left Paris.

Other of Brown-Séquard's investigations concerned the artificial production of epileptic states through lesions of the nervous system (1856), and the irrigation of dead muscles with warm blood. He was successful in reviving the head of a dog eight minutes after decapitation by injecting oxygenated blood into the arterial trunks. In 1862 he demonstrated that section of the vagus nerve brought on dilation of the coronary arteries. Brown-Séquard was mistaken in opposing the theory of cerebral localizations, but it should be emphasized that we are greatly indebted to him for elaboration of the concept of nervous inhibition.

Brown-Séquard was a pioneer of endocrinology. In 1856 he proved that removal of the adrenal glands always caused death in animals. He also prolonged the survival of animals after adrenalectomy by injecting normal blood. These researches are open to criticism, and this is even more true of his celebrated experiments on "rejuvenation." On 1 June 1889 Brown-Séquard presented a sensational report to the Société de Biologie: he believed that he had "rejuvenated" himself with subcutaneous injections of a liquid extracted from the testicles of freshly killed guinea pigs and dogs. His extravagant claim stimulated the development of modern organotherapy and exerted a strong influence on later research on sex hormones.

Brown-Séquard's investigations show more intuition than critical sense. His great achievement was understanding that through "internal secretion" the cells become dependent on one another by means of a mechanism other than the action of the nervous system (1891).

BIBLIOGRAPHY

I. ORIGINAL WORKS. A good bibliography is in his autobiographic *Notice sur les travaux scientifiques de M. C. É. Brown-Séquard* (Paris, 1883). His principal books are *Recherches et expériences sur la physiologie de la moelle épinière* (Paris, 1846), his thesis; *Recherches expérimentales sur la transmission croisée des impressions dans la moelle épinière* (Paris, 1855); *Researches in Epilepsy* (Boston, 1857); and *Course of Lectures on the Physiology and Pathology of the Central Nervous System* (Philadelphia, 1860). The most important results of Brown-Séquard's research were first printed in the reports of the Société de Biologie (Paris) and of the Académie des Sciences. For instance, the *Comptes rendus de la Société de biologie* contains the description of the experimental hemisection of the spinal cord, **1** (1850), 70–73; and the first mention of the effects produced in man by injections of testicular extract, **41** (1889), 415–422. His fundamental experiments on the adrenal glands were published in the *Comptes rendus de l'Académie des sciences,* **43** (1856), 422–425, 542–546. He also published a great many articles in the three journals of which he was founder and editor: *Journal de la physiologie de l'homme et des animaux, Archives de physiologie normale et pathologique,* and *Archives of Scientific and Practical Medicine.*

II. SECONDARY LITERATURE. Works on Brown-Séquard are M. Berthelot, *Notice sur la vie et les travaux de M. Brown-Séquard* (Paris, 1904); E. Dupuy, "Notice sur M. le Pr. Brown Séquard, ancien président de la Société de biologie," in *Mémoires de la Société de biologie,* **46** (1894), 759–770; E. Gley, "C.-E. Brown-Séquard," in *Archives de physiologie normale et pathologique,* 5th ser., **6** (1894), 501–516; J. M. D. Olmsted, *Charles-Edouard Brown-Séquard. A Nineteenth Century Neurologist and Endocrinologist* (Baltimore, 1946); and F. A. Rouget, *Brown-Séquard et son oeuvre* (Port Louis, Mauritius, 1930). An interesting set of letters between Brown-Séquard and his assistant d'Arsonval is in L. Delhoume, *De Claude Bernard à d'Arsonval* (Paris, 1939).

M. D. GRMEK

BROŻEK (or **Broscius**), **JAN** (*b.* Kurzelow, near Sieradz, Poland, November 1585; *d.* Krakow, Poland, 21 November 1652), *mathematics.*

Brożek's father, Jakub, was an educated landowner who taught his son the art of writing and the principles of geometry. Jan went to primary school in Kurzelow and then to the University of Krakow, where he passed his baccalaureate in March 1605. Among his professors were Stanislaw Jacobeius and Walenty Fontanus. In March 1610 Brożek won the rank of *magister,* and in 1611 he was ordained a priest. His contacts with Adriaan Van Roomen (Romanus), an eminent Belgian mathematician then in Krakow, greatly influenced his studies.

Early in 1614 Brożek became a professor at the Collegium Minus of the University of Krakow, where he was assigned the chair of astrology, and in 1619 at the Collegium Maius. In 1618 he traveled to Torun, Danzig, and Frombork to gather material on Copernicus. In 1620, at Innsbruck, he met the astronomer Christoph Scheiner. From June 1620 to

June 1624 Brożek studied medicine in Padua, receiving his doctorate in medicine in 1624, and was physician to the bishop of Krakow until the autumn of 1625. In 1625 the University of Krakow elected him professor of rhetoric, and in 1629 he gave up his chair in astrology because he had received higher ecclesiastical orders and had become canon of St. Anne's church. He then passed his baccalaureate in theology and became professor of that discipline.

In 1630 Brożek gave up his chair of rhetoric, and from April 1631 to December 1638 he was director of the library of the Collegium Maius. He became active in organizing the teaching of "practical geometry," which was entrusted to his favorite pupil, Pawel Herka, with some supervision on his part during 1635 and 1636. In 1639 Brozek presented his library to the University of Krakow, along with a substantial sum for the purchase of additional books and instruments. He gave up his professorship and the apartment at the Collegium Maius, as well as the canonry of the church of St. Florent, and moved to Międzyrzecze. In 1648, however, he returned to Krakow, where he received the master of theology. In February 1650 he became doctor of theology, and rector of his university in 1652.

Brożek's loyalty to the University of Krakow, one of his strongest characteristics, even surpassed his attachment to the Catholic Church. On the side of the university he took part in the fight against the Jesuit domination of schools, sending reports to Rome and making ten trips to Warsaw (1627–1635) in order to defend the university's rights. In the course of his struggle he answered a letter from a priest, Nicolas Lęczycki, by publishing (1626) a satirical dialogue, *Gratis,* which was soon burned in the public square of Krakow. It provoked a long answer from a priest, Frédéric Szembek, entitled *Gratis plebański gratis wyćwiczony* ("The Priests' *Gratis* Gratuitously Beaten," Poznan, 1627).

It was Brożek's hope to write the history of the University of Krakow, showing its role in the general development of science and education in Poland, but fragments of manuscripts are all that remain. The most important are "De antiquitate litterarum in Polonia" and an excellent biography of Stanislaw Grzepski, Polish geometer and philologist of the sixteenth century. In spite of his being enlightened and erudite, a partisan of progress who was active in reforming the teaching of mathematics, Brożek was not free from astrological prejudices or belief in the magical properties of numbers and their relation to medicine.

Brożek was the author of more than thirty publications. The ones concerning Copernicus, and particularly those dealing with mathematics, which won him the reputation of being the greatest Polish mathematician of his time, are of considerable interest. Among the first are the poem *Septem sidera* (of doubtful authenticity) and many Copernican documents but, unfortunately, not the letters by and about Copernicus that Brożek collected but did not publish that are now lost. In the second group are his purely mathematical works and opuscules, the most important being *Arithmetica integrorum* (1620), a new didactic manual, in which logarithms, then recently discovered, were introduced in schools; *Aristoteles et Euclides defensus contra Petrum Ramum* (1638), reissued in 1652 and 1699 under the title *Apologia pro Aristotele . . .*; a dissertation containing original research on the star-shaped polygons; and two treatises entitled *De numeris perfectis* (1637, 1638), which brought new results, at the time, on perfect numbers and amicable numbers. There one finds the basic theorem of the elementary theory of numbers, better known as Fermat's theorem, which was published in 1670 (also without its proof).

Jan Brożek should not be confused with Nicolas Brożek, nephew or grandson of his sister (*b*. Kurzelow, *ca*. 1635; *d*. Krakow, 1676). His scientific and ecclesiastic career was very similar to Jan Brożek's, but with only slight results, even for his epoch.

BIBLIOGRAPHY

I. ORIGINAL WORKS. Among Brożek's more than thirty writings are *Arithmetica integrorum* (Krakow, 1620); *Gratis* (Krakow, 1626); *De numeris perfectis* (Krakow, 1637, 1638); and *Aristoteles et Euclides defensus contra Petrum Ramum* (Krakow, 1638), reissued as *Apologia pro Aristotele . . .* (Krakow, 1652, 1699). Available in manuscript are parts of "De antiquitate litterarum in Polonia" and the "Biography of Stanislaw Grzepski," and Brożek's diary for 1636–1643, with gaps, in the form of notes written in the margins of the "Ephemerides" edited by L. Eichstadt, MS Bibl. Jagellone, Krakow, sign. Mathesis 513 R.X.VI.12.

II. SECONDARY LITERATURE. Works dealing with Brożek are M. A. Baraniecki, *Arytmetyka*, 2nd ed. (Warsaw, 1894), pp. 43–49; H. Barycz, *Wstęp i przypisy do nowego wydania Gratisa* ("Introduction and Commentaries on the New Edition of *Gratis*"), Vol. LXXXII of Biblioteka Pisarzy Polskich (Krakow, 1929); and "Pierwszy historyk nauki i kultury w Polsce" ("First Historian of Science and Culture in Poland"), in *Księga pamiątkowa ku czci W. Sobieskiego* ("Commemorative Volume in Honor of W. Sobieski"; Krakow, 1932); A. Birkenmajer, "Brożek (Broscius), Jan," in *Polski słownik biograficzny* ("Dictionary of Polish Biography"), II (Krakow, 1937), 1–3; L. A. Birkenmajer, *Nikołaj Kopernik*, I (Krakow, 1900); A. Favaro, *Tito-Livio Burattini* (Venice, 1896), p. 74; W. Konczyńska, *Zarys historii Biblioteki Jagiellońskiej* ("Sketch of the History of the Jagel-

lone Library"; Krakow, 1923); J. Krókowski, *De septem sideribus, quae Nicolao Copernico vulgo tribuuntur* (Krakow, 1926); a monograph on Brożek, with portrait; Z. Mysłakowski, *Walerian Magni*, Vol. LI A of *Rozprawy Wydziału matematyczno-przyrodniczego PAU* (Krakow, 1911); K. Piekarski, *Ex libris Jana Brożka*, Vol. III of *Silva rerum* (Krakow, 1927); E. Stamm, "Z historii matematyki XVII wieku w Polsce" ("History of Mathematics in the Seventeenth Century in Poland"), in *Wiadomości matematyczne*, **40** (1935); and S. Temberski, *Roczniki* (Krakow, 1897). A long monograph on Brożek, containing a reproduction of his portrait at the University of Krakow, was published by J. N. Franke (Krakow, 1884).

B. KNASTER

BRUCE, DAVID (*b*. Melbourne, Australia, 29 May 1855; *d*. London, England, 27 November 1931), *microbiology*.

Sir David Bruce won renown for discovering the bacterial cause of Malta fever and for extensive and fruitful researches on trypanosomiasis. His professional life was spent in the Army Medical Service and the Royal Army Medical Corps, in which he attained the special rank of surgeon general.

Bruce was the only son of Scottish parents who emigrated to Australia during the gold rush of the early 1850's. His father, David Bruce, accompanied by his wife, Jane Hamilton, left Edinburgh to install a crushing plant in a Victoria goldfield. When he was five years old, David's parents returned with him to Scotland and settled in Stirling. He attended high school until the age of fourteen, and then worked for a warehouse firm in Manchester. His ambition to become a professional athlete, for which his great physique fitted him, was frustrated by an attack of pneumonia at age seventeen. He resumed his studies, and in 1876 gained admission to the University of Edinburgh.

As a schoolboy, Bruce became an enthusiastic ornithologist and planned a university course in zoology. After he had completed his first year at Edinburgh as medalist in natural history, a physician friend persuaded him to study medicine. After graduating M.B., C.M. in 1881, he assisted a doctor in Reigate, where he met Mary Elizabeth Steele, six years his senior, daughter of the previous owner of the practice. They were married in 1883. Although childless, their marriage was singularly fortunate, for Mary Bruce proved an indispensable helpmate—domestically, socially, and scientifically.

Bruce found general practice uncongenial, and in August 1883 was commissioned surgeon captain in the Army Medical Service. The following year, he was posted to Malta, where he and his wife were quartered in the Valetta Hospital, which had no facilities for research. Impressed by Koch's recent discovery of the tubercle bacillus, Bruce decided to investigate Malta fever, which annually hospitalized around a hundred soldiers of the British garrison, for an average of three months. He purchased a microscope, and late in 1886 he found "enormous numbers of single micrococci" in the spleen of a fatally ill patient. Splenic pulp from four later patients, inoculated into Koch's nutrient agar, yielded cultures of a slowly growing "micrococcus." Bruce reported these findings in September 1887. Subsequent publications described further properties of the organism, for which he proposed the name *Micrococcus melitensis.* Its bacillary morphology was unrecognized until Bang isolated *Bacillus abortus* in 1897. In 1920, on the suggestion of Feusier and Meyer, the generic term *Brucella* was adopted for these closely related microorganisms. The epidemiology of the disease remained a mystery until 1905, when T. Zammit, a Maltese member of the Commission for the Investigation of Mediterranean Fever, the twelve-man team of experts headed by Bruce, implicated goat's milk as the disseminating vehicle. The disease was conquered when goat's milk was eliminated from the diet of the Malta garrison. The eponymous term "brucellosis" has now replaced such names as Malta, Mediterranean, and undulant fever.

Between these two phases of his work in Malta, Bruce had become famous for trypanosomiasis researches in Zululand and Uganda. After departing from Malta in 1889, he spent his leave discoursing in Koch's laboratory while his wife acquired the latest techniques in microscopy, staining, and media making. He then taught pathology at the Army Medical School at Netley, introducing the experimental attitudes and bacteriological methods of Pasteur, Lister, and Koch. In 1894 Bruce was posted to Natal, whose governor (a former lieutenant governor of Malta) asked him to investigate an epizootic, nagana, that was affecting cattle in northern Zululand.

After trekking for five weeks by ox wagon, the Bruces arrived at Ubombo, where they lived for two months in a wattle-and-daub hut, using the veranda as a laboratory. Bacteriological examinations of affected oxen proved negative; but intensive microscopic study of blood specimens revealed a motile, vibrating hematozoon, which Bruce later concluded was a trypanosome. The relationship of this parasite to nagana was demonstrated by inoculating blood from infected cattle into healthy horses and dogs: they became acutely ill, and their blood swarmed with hematozoa. The natural mode of transmission of the disease was revealed, as Bruce explained in his

Croonian lectures (1915), when two oxen and several dogs, sent into a low-lying "fly belt" for a fortnight, acquired this same parasite in their blood. He was now convinced that nagana was identical with the "tsetse fly disease" described by Livingstone in 1858, and that this fly transmitted the causal trypanosome.

Recalled temporarily to Natal, the Bruces returned in September 1895 to their isolated hut in the Zululand bush and stayed almost two years. Bruce's *Preliminary Report* was published in December 1895 and was followed early in 1897 by his *Further Report.* These classic documents described the hematozoa of nagana, established the tsetse fly *Glossina morsitans* as the vector, and implicated regional wild game, such as antelope and buffalo (themselves immune and unaffected), as the trypanosomal reservoir. Living trypanosome samples were forwarded to the Royal Society, which in 1899 elected Bruce a fellow and published a paper by Plimmer and Bradford characterizing the parasite and naming it *Trypanosoma brucei.*

During the Boer War, Bruce distinguished himself at the siege of Ladysmith, where he directed a hospital and performed successful surgery. Mary Bruce received the Royal Red Cross for devoted work with the wounded, particularly as nursing sister in her husband's operating theater. They returned home in October 1901.

In 1903 Bruce was chosen to head the Royal Society's Sleeping Sickness Commission to Uganda. On behalf of the Foreign Office and at Patrick Manson's urging, the Society had organized a similar commission to investigate an epidemic in Uganda in 1902, but its activities were uncoordinated and two members had returned home. The Bruces reached Entebbe in March 1903, with Dr. David Nabarro and a sergeant technician, and met the remaining representative of the first commission, a young bacteriologist, Dr. Aldo Castellani.

Manson's tentative suggestion of *Filaria perstans* as the causal agent had proved untenable; but Castellani had recently noted trypanosomes in cerebrospinal fluid taken from five victims. Previously, he had grown streptococci from the cerebrospinal fluid and heart blood of more than thirty. Well aware of the potential significance of the trypanosomes, although perturbed at the conflicting evidence, Castellani did not wish to be ridiculed by Bruce, still less by Nabarro, who was little older than he and whose appointment to supersede him he resented. He therefore imparted his observations to Bruce, on condition that he (Castellani) should temporarily continue searching for trypanosomes, that he should then publish his findings as sole author, and that

Nabarro should not be informed. When Castellani left Entebbe three weeks later, he had demonstrated trypanosomes in twenty additional cases. He had also taught Bruce the techniques of lumbar puncture and of examining the cerebrospinal fluid for trypanosomes.

In 1902 Dutton had reported *Trypanosoma gambiense* in the blood of a febrile Englishman in Gambia and in 1903 Baker had diagnosed similar cases of trypanosome fever in Uganda, but no connection between these conditions and Uganda sleeping sickness was then suspected. At first Bruce appeared skeptical about trypanosomes as causal agents of human sleeping sickness, but soon after Castellani's departure Bruce and Nabarro amassed convincing evidence that this disease was caused by *T. gambiense* inoculated by the tsetse fly *Glossina palpalis.* Bruce returned to England in August 1903. The "Progress Report" by himself and Nabarro, sent from Entebbe, acknowledged Castellani's discovery; but the "Further Report," written by Bruce, Nabarro, and E. D. W. Greig, betrays a changed attitude that encouraged his supporters, particularly the zoologist Ray Lankester, to minimize Castellani's contribution. The resulting dispute broke into print intermittently for several years. Despite Nabarro's magnanimous support of Castellani, most subsequent accounts displayed bias, which J. N. P. Davies' carefully documented articles (1962, 1968) should help to correct.

From 1908 to 1910, Bruce rejoined the Royal Society's continuing commission in Uganda, where he directed researches into conditions governing the transmissibility of *T. gambiense* by *Glossina palpalis,* and studied cattle and game as potential reservoirs of the parasite. In 1911, he was appointed director of another Sleeping Sickness Commission, to investigate trypanosomiasis in Nyasaland. During the next two years, *T. rhodesiense* was identified as the main regional pathogen and *Glossina morsitans* as its vector; and certain other trypanosomal species pathogenic to domestic animals were characterized. The commission concluded that *T. rhodesiense* and *T. brucei* were identical, but this view was disputed by contemporary German authorities and now has no adherents.

From 1914 until he retired in 1919, Bruce was commandant of the Royal Army Medical College. During the war, his administrative abilities were fully utilized, especially as director of scientific research and as chairman of committees for the study of tetanus and trench fever. In his last years, Bruce suffered recurrent lung infections and wintered in Madeira. He died of cancer in his seventy-seventh year. His wife, who accompanied her husband on all his arduous trips, working self-effacingly beside him as technician, microscopist, and draftswoman, predeceased him by four days. On his deathbed, Bruce requested that her outstanding assistance should always be emphasized in any biographical account of him.

Bruce's many distinctions included fellowship in the Royal Society and the Royal College of Physicians of London; honorary doctorates from the universities of Glasgow, Liverpool, Dublin, and Toronto; honorary memberships in several foreign academies and societies; and numerous medals of honor. He was appointed C.B. in 1905, knighted in 1908, and made K.C.B. in 1918. His abrupt manner, blunt speech, and egotistical personality endeared him to few; but his great energies and talents were dedicated to mankind's health and welfare, and he died poor. Moreover, he had vision and compassion, as shown in this closing passage from his presidential address, "Prevention of Disease," to the British Association, meeting at Toronto in 1924:

> We are all children of one Father. The advance of knowledge in the causation and prevention of disease is not for the benefit of any one country, but for all—for the lonely African native, deserted by his tribe, dying in the jungle of sleeping sickness, or the Indian or Chinese coolie dying miserably of beri-beri, just as much as for the citizens of our own towns.

BIBLIOGRAPHY

I. ORIGINAL WORKS. Bruce's major works are "Note on the Discovery of a Microorganism in Malta Fever," in *Practitioner,* **39** (1887), 161–170; "The Micrococcus of Malta Fever," *ibid.,* **40** (1888), 241–249; *Preliminary Report on the Tsetse Fly Disease or Nagana in Zululand* (Durban, 1895); *Further Report on the Tsetse Fly Disease or Nagana in Zululand* (London, 1897); "Progress Report on Sleeping Sickness in Uganda," in *Reports of the Sleeping Sickness Commission of the Royal Society,* **1** (1903), 11–88, written with David Nabarro; "Further Report on Sleeping Sickness in Uganda," *ibid.,* **4** (1903), 1–87, written with D. Nabarro and E. D. W. Greig; "The Development of *Trypanosoma gambiense* in *Glossina palpalis,*" in *Proceedings of the Royal Society,* **81B** (1909), 405–414, written with A. E. Hamerton, H. R. Bateman, and F. P. Mackie; "The Morphology of the Trypanosome Causing Disease in Man in Nyasaland," *ibid.,* **85** (1912), 423–433, written with D. Harvey, A. E. Hamerton, J. B. Davey, and Lady Bruce; "Trypanosomes Causing Disease in Man and Domestic Animals in Central Africa" (the Croonian lectures), in *Lancet* (1915), **1,** 1323–1330; (1915), **2,** 1–6, 55–63, 109–115; and "Prevention of Disease," in *Science,* **60** (1924), 109–124.

II. SECONDARY LITERATURE. Unsigned obituaries of Bruce are "Sir David Bruce, K.C.B., M. D. Edin., F.R.S.," in *Lancet* (1931), **2,** 1270–1271; and "Major-General Sir

David Bruce, K.C.B., LL.D., D.Sc., F.R.C.P., F.R.S.," in *Journal of the Royal Army Medical Corps,* **58** (1932), 1–4. Signed ones are J. R. Bradford, "Sir David Bruce—1855–1931," in *Obituary Notices of Fellows of the Royal Society,* **1** (1932), 79–85; S. R. Christophers, "Bruce, Sir David (1855–1931)," in *Dictionary of National Biography 1931–1940,* pp. 108–110; A. E. Hamerton, "Major-General Sir David Bruce, K.C.B., D.Sc., LL.D., F.R.C.P., F.R.S., Late A.M.S.," in *Transactions of the Royal Society of Tropical Medicine and Hygiene,* **25** (1931–1932), 305–312; D.W.T. and W.J.T., "Major-General Sir David Bruce, K.C.B., F.R.S.," in *Nature,* **129** (1932), 84–86; and C. M. Wenyon, "Major-General Sir David Bruce, K.C.B., F.R.S.," *ibid.,* 86–88.

Further information on Bruce's work can be found in M. T. Ashcroft, "A Critical Review of the Epidemiology of Human Trypanosomiasis in Africa," in *Tropical Disease Bulletin,* **56** (1959), 1073–1093; Aldo Castellani, "On the Discovery of a Species of Trypanosoma in the Cerebrospinal Fluid of Cases of Sleeping Sickness," in *Proceedings of the Royal Society,* **71** (1902–1903), 501–508; J. N. P. Davies, "The Cause of Sleeping Sickness? Part II," in *East African Medical Journal,* **39** (1962), 145–160, and "Informed Speculation on the Cause of Sleeping Sickness 1898–1903," in *Medical History,* **12** (1968), 200–204; E. R. Lankester, "Nature's Revenges: The Sleeping Sickness," in *The Kingdom of Man* (London, 1907), pp. 159–191; *Reports of the Commission Appointed by the Admiralty, the War Office, and the Civil Government of Malta for the Investigation of Mediterranean Fever, Under the Supervision of an Advisory Committee of the Royal Society,* 7 pts. in 1 vol. (London, 1905–1907); *Reports of the Sleeping Sickness Commission, Royal Society, Nos. 1–11* (London, 1903–1911); and C. Wilcocks, "Trypanosomiasis," in *Aspects of Medical Investigation in Africa* (London, 1962), pp. 59–90.

There is also enlightening correspondence. Two collections of letters are those of Bruce, in the library of the Royal Society of Tropical Medicine and Hygiene, London, and those of R. U. Moffat, former principal medical officer for Uganda, in Makerere College Library, Kampala. Individual letters are Aldo Castellani, ". . . The Exact History of how Colonel Bruce was Acquainted With My Observations . . .," in *The Times,* 8 July 1908, p. 18; E. Ray Lankester, in *The Times,* 19 Nov. 1903, p. 2, and 14 Aug. 1913, p. 4; and in *British Medical Journal,* 11 Aug. 1917, p. 198, and 17 Sept. 1917, pp. 402–403; and David Nabarro, in *Journal of Tropical Medicine and Hygiene,* 15 July 1908, pp. 224–225; *The Times,* 22 Sept. 1913, p. 21; and *British Medical Journal,* 15 Sept. 1917, pp. 374–375, and 6 Oct. 1917, pp. 467–468.

CLAUDE E. DOLMAN

BRUCE, JAMES (*b.* Stirlingshire, Scotland, 14 December 1730; *d.* Stirlingshire, 27 April 1794), *exploration.*

Bruce was the second son of David Bruce of Kinnaird and Marion Graham of Airth. Born into a wealthy landowning family, he entertained the idea of becoming a Church of England clergyman. He then decided instead to read for the Scottish bar, but later gave up legal studies also. Bruce served as consul-general at Algiers from March 1763 until the summer of 1765. He remained in Africa until March 1773, and then spent over a year in France, where he met the naturalist Buffon, and Italy before returning to London in June 1774.

Bruce's most notable journey began in Alexandria in June 1768. He traveled extensively in Ethiopia and claimed to have been the first European to find the source of the Nile (on 4 November 1770); he did know that the Portuguese Jesuit missionary Pedro Páez had been there more than a century earlier. Bruce actually had found the source of the Blue Nile, the main tributary of the White Nile, which it joins below Khartoum. He knew of the White Nile, although he denied that it was the major branch.

James Bruce is justly considered an explorer rather than an adventurer because of his scientific approach. Before setting out, he learned as much as was known about the geography, customs, and languages of the area. Bruce brought back drawings of buildings, fauna, and flora; collected seeds; and kept precise meteorological and astronomical records. His writings are generally accurate, and their embellishment in personal details is easily recognized. His *Travels,* which were not composed for more than twelve years after he left Africa, were written for a general rather than an academic audience; yet the grotesque and exotic material combined with his difficult and vain personality to arouse an adverse public reaction verging on disbelief. He retired to his estate during his last years and died at Kinnaird House.

BIBLIOGRAPHY

Bruce's *Travels to Discover the Source of the Nile in the Years 1768, 1769, 1770, 1771, 1772, and 1773,* 5 vols. (Edinburgh–London, 1790), was promptly translated into French and German. The second and third editions (8 vols., Edinburgh, 1804–1805; 1813) contain material from Bruce's notes that was added by Alexander Murray. There is a modern abridgment, *Travels to Discover the Source of the Nile by James Bruce,* C. F. Beckingham, ed. (Edinburgh, 1964), which offers a scholarly biographical introduction.

BERT HANSEN

BRÜCKE, ERNST WILHELM VON (*b.* Berlin, Germany, 6 September 1819; *d.* Vienna, Austria, 7 January 1892), *physiology.*

Brücke was the son of a painter and thought of

following his father's profession. Even though he became a doctor instead, he dealt throughout his life so intensively with questions concerning the theory of art that they form an integral part of his work.

In 1838 Brücke began studying at Berlin. His final teacher, whose assistant Brücke became in 1843, was the physiologist Johannes Müller, whose circle of friends and colleagues at that time included Hermann von Helmholtz, Emil Du Bois-Reymond, and, indirectly, Carl Ludwig. Brücke formed lifelong friendships with all these men. In 1842, on the basis of his dissertation *De diffusione humorum per septa mortua et viva*, he was graduated as a doctor of medicine and surgery. In this dissertation he tried to prove that the phenomena of osmosis are not to be related to any sort of uncertain vital force, but to weighable and measurable, repelling and attracting physiochemical forces; what he sought to prove was a part of the program of the new physical physiology. Du Bois-Reymond formulated it in the following way: "Brücke and I, we have sworn to each other to validate the basic truth that in an organism no other forces have any effect than the common physiochemical ones. . . . "

For this program the eye was an especially suitable subject for investigation. Brücke examined optical media, afterimages, stereoscopic vision, and the reflection of light from the backgrounds of the eyes of vertebrates; he also discovered the ciliary muscle named after him. His *Anatomical Description of the Human Eye* (1847) has become the standard anatomical-histological work for contemporary oculists.

His research on luminescence in animal eyes and his method of causing luminescence at will in the human eye created the foundation on which Helmholtz continued his work and which led to the invention of the ophthalmoscope in 1851 by the latter. How close Brücke himself had come to this discovery was later attested by Helmholtz: "He had merely neglected to ask himself to which optical image the rays reflected from the luminescent eye belonged."

In 1848 Brücke became professor of physiology at the University of Königsberg. The following year, he went in the same capacity to Vienna, where he founded a school for physiologists that eventually extended far beyond the borders of Austria and there worked until his death. Brücke's laboratory trained the Austrian physiologists S. Exner, A. Rollett, E. von Fleischl-Marxow, M. von Vintschgau, and A. Kreidl; the German W. Kühne; the Swede F. Holmgren; the Englishman T. Lauder-Brunton; and the Russians Elie Cyon, N. von Kowalewsky, and I. M. Setchenoff. Freud, who worked there from 1876

to 1882, considered Brücke the most highly respected teacher and the greatest authority in the field he had ever met.

Here, in the major city of a polyglot country, Brücke had an unusual opportunity to study linguistic and vocal physiology. To determine each sound of an arbitrary language in his own (alphabetical) characters and thereby to give a phonetic transcription was the aim of his *Characteristics of the Physiology and Taxonomy of Linguistic Sounds* (1856). With the aid of a labiograph he made the first attempt to measure exactly the length of strongly and weakly accented syllables in verse. He recorded the results of these measurements in a monograph, *The Physiological Bases of New High German Poetic Art* (1871). There is unmistakably evident here a typical endeavor of the times (primarily to analyze the effect of a work of art rationally, i.e., by means of scientific methods), as is also the case in Brücke's writings concerning the theory of art, which deal with the determination of the classical ideal of beauty. Such analysis was alien to Billroth's intuitively synthetic comprehension of art: "It is as if one wanted to describe how a good apple tastes; one has to eat it himself; if he does not then recognize it, he should stay with potatoes."

It is said that Brücke was one of the most versatile physiologists of his day. His *Lectures On Physiology* (1873–1874) confirms this; in it he added something of his own to almost every chapter. The diversity of his interests made limited specialization alien to Brücke. His investigations included the physiology of digestion; from 1850, Brücke studied the digestive tract microscopically and recognized the structures designated as Peyer's "glands" as the places where the lymphocytes develop. He explained the mechanism of the transfer of chyle by means of the contraction of intestinal villi. In his work on chyle, which was published for the most part in the *Proceedings of the Imperial Academy of Sciences in Vienna*, he encountered an abundance of questions concerning the reabsorption of fats, carbohydrates, and proteins. As a result, he developed many biochemical concepts. Brücke introduced the terms "achroödextrin" and "erythrodextrin" into physiological chemistry; he discovered that blood did not coagulate in uninjured vessels, and he became a pioneer in enzyme research through his experiments on peptic digestion. With these experiments he endeavored to produce the purest possible pepsin solutions. He tried to combine pepsin mechanically with small solid bodies such as calcium phosphate, sulfur, and cholesterol, and subsequently to extract it again from its adsorbates. He succeeded herein along two possible avenues of approach (through precipitation of calcium phosphate with water, or by treating the cholesterol precipitation with ether). But in order to reach his goal he needed control reactions for further purification, which he did not have.

Brücke is generally honored as a microscopist without it being pointed out that his microscopic investigations invariably grew out of his physiological inquiries and were determined by them. The investigation of function was his chief aim when he observed the flow of protoplasm in the stinging hairs of nettles or molecular motion in the salivary particles or—a classic example of the synthesis of histological, physical, and experimental methods—when he explained the changing of a chameleon's colors by the momentary shifting movements of the skin's pigment cells. Such diverse studies on the function of the most varied cells led Brücke to criticize the mechanistic ideas of structure in the cell theory of Matthias Schleiden and Theodor Schwann—of the cell as a shell formed by the cell membrane. While he and Max Schultze, a histologist from Bonn, were beginning to distinguish protoplasm as an essential cellular component, he was at the same time paving the way for a biological theory of cells in his investigation entitled *The Elementary Organisms*. In 1867–1868, with his experiments on the possibility of electric stimulation of muscles, Brücke moved into the specialty of his friend Du Bois-Reymond, i.e., general nerve and muscle physiology. Du Bois-Reymond held that the stimulating effect of an electric current depended solely on how fast such a current was increased in the stimulated organ and not on how long the stimulus lasted. Brücke, on the other hand, observed that in curare-treated frog muscles a current that was increased too slowly remained ineffective. Accordingly, in electric stimulation the time factor had to be considered no less than the amount of current. In regarding the stimulus as a function of the "distance from the normal state," Brücke arrived at a new concept of the law on stimulation, which approximates modern concepts more closely than the first formulation by Du Bois-Reymond.

When Brücke resigned his teaching position in 1890, he had 143 publications to his credit. The range of this output is made evident by the number of different areas of work: physics, plant physiology, microscopic anatomy, physiological chemistry, physiological optics, and purely experimental physiology. He received many honors from numerous academies, including the highest Prussian order, the Order of Merit, and Austrian ennoblement. Such acclaim left untouched the genuine inner modesty of this great researcher, who was interested only in examining the events of nature with a view to their objective regularities.

BIBLIOGRAPHY

I. ORIGINAL WORKS. Note references in text. A complete list of works is in E. Th. Brücke, *Ernst Brücke* (Vienna, 1928). The most important works and the most important secondary literature are in E. Lesky, "Die Wiener medizinische Schule im 19. Jahrhundert," in *Studien zur Geschichte der Universität Wien*, VI (Graz–Cologne, 1965), 258 ff.

II. SECONDARY LITERATURE. See S. Exner, "Ernst von Brücke und die moderne Physiologie," in *Wiener klinische Wochenschrift*, 3 (1890), 807–812; obituary by A. Kreidl, *ibid.*, 5 (1892), 21 f. See also E. Suess, in *Almanach der Akademie der Wissenschaften in Wien*, 42 (1892), 184–189; and E. Brücke, "Ernst Wilhelm Brücke," in *Neue Österreichische Biographie*, V (1928), 66–73.

ERNA LESKY

BRUHNS, KARL CHRISTIAN (*b.* Plön, Germany, 22 November 1830; *d.* Leipzig, Germany, 25 July 1881), *astronomy*.

Trained as a locksmith, Bruhns came to Berlin in 1851 and worked as a mechanic. Even then his aim was to become an astronomer, and Encke, the director of the Berlin Observatory, to whom Bruhns had been recommended for his mathematical skill by a professor in Altona, recognized his great mathematical talents. After a year of carrying out complicated calculations for Encke in addition to his regular work, Bruhns was made an assistant at the observatory. In 1856 he graduated from the university with the thesis *De planetis minoribus inter Jovem et Martem circa solem versantibus* (Berlin, 1856), and in 1859 became lecturer in astronomy at the University of Berlin. Two years later he was appointed assistant professor of astronomy in Leipzig, becoming professor in 1868; he remained director of the observatory from 1860 until his death. In 1877–1878 he was rector of the university.

Under the influence of Encke, Bruhns's activities in his younger years centered on theoretical astronomy. Having made observations with the equatorial and meridian circle in Berlin, he took a greater interest in observational astronomy in Leipzig. His first act in Leipzig was to replace the antiquated observatory in the tower of the old castle in the middle of the town with a new one at the outskirts of the town. It was well equipped—for its time even excellently equipped. From 1900 on, however, it shared the lot of many German university observatories: unfavorable location and obsolete instruments; a place for teaching and training, but not a center of practical research.

Bruhns paid little attention to astrophysics, which flourished in the last years of his life. He was more interested in the fields related to astronomy—geodesy and meteorology—which he advanced considerably. He served on the Kommission für Mitteleuropäische Gradmessung and held the chair in the astronomical section of the Preussisches Geodätisches Institut in addition to his regular duties. Much of his work concerned the determinations of longitude between his observatory and Berlin, Vienna, Paris, and Munich, and other, less important, places.

More important was Bruhns's contribution to meteorology. In cooperation with Buys-Ballot and Jellinek, among others, he organized uniform world-wide weather observations and undertook to supervise those in his vicinity. He further tried to arouse the interest of agricultural circles, in particular, in a regular weather forecast in *Über das meteorologische Bureau für Witterungsprognosen im Königreich Sachsen* (1879).

Bruhns published such popular works as *Atlas der Astronomie* (1872) and also discussed natural science in general. It is therefore no surprise that he was responsible for the first comprehensive biography of Alexander von Humboldt (1871), which was supported by Loewenberg and Carus, among others. In this publication he wrote on Humboldt's work in astronomy, geodesy, and mathematics. Bruhns knew Humboldt personally and thought it urgent to memorialize this genius who was important to natural science in general. He also wrote a biography of Encke that gives an excellent insight into the work of the observatories in the first half of the nineteenth century. His historical concerns are displayed in other works, especially *Die astronomische Strahlenbrechung in ihrer historischen Entwicklung* (1861), written six years earlier.

In view of his wide-ranging interests, it is no wonder that Bruhns was influential. His superior lectures attracted many students; he was also active in the Astronomische Gesellschaft and initiated and equipped the first German astronomical expeditions. Unfortunately, these manifold tasks prevented him from promoting theoretical astronomy to the extent his outstanding ability would have allowed. His energy and enthusiasm, however, had a strong effect on his colleagues, which should not be ignored.

BIBLIOGRAPHY

Bruhns's books include *Geschichte und Beschreibung der Leipziger Sternwarte* (Leipzig, 1861); *Die astronomische Strahlenbrechung in ihrer historischen Entwicklung* (Leipzig, 1861); *Längendifferenz-Bestimmung Berlin-Leipzig* (Leipzig, 1865), written with W. Förster; *Längendifferenz-Bestimmung Leipzig-Gotha* (Leipzig, 1866),

written with A. Auwers; *John Franz Encke, sein Leben und Wirken* (Leipzig, 1869); *Neues log.-trig. Handbuch auf 7 Dezimalen* (Leipzig, 1870); *Alexander von Humboldt, eine wissenschaftliche Biographie . . .,* 3 vols. (Leipzig, 1871); *Astronomisch-geodätische Arbeiten in den Jahren 1867–1875,* 4 vols. (Leipzig, 1871–1876; 1882); *Atlas der Astronomie* (Leipzig, 1872); and *Über das meteorologische Bureau für Witterungsprognosen im Königreich Sachsen* (Leipzig, 1879).

Numerous short papers appeared in *Astronomische Nachrichten,* **35–67** (1852–1866); in *Leipzig Gesellschaft der Wissenschaften. Berichte* (1872, 1878); and in *Leipzig Gesellschaft der Wissenschaften. Abhandlungen* (1873). Bruhns was also editor of 3 vols. in *Publicationen des Preussischen Geodätischen Institutes* (1871–1874); of 12 vols. of *Resultate aus den meteorologischen Beobachtungen, angestellt an meheren Orten im Königreich Sachsen* (Leipzig, 1866–1880); and of *Kalendar und statistisches Jahrbuch für Sachsen* (Leipzig, 1872–1882).

H. C. FREIESLEBEN

BRUMPT, ÉMILE (*b.* Paris, France, 7 March 1877; *d.* Paris, 7 July 1951), *parasitology.*

Brumpt began his career as *préparateur* at the Paris Faculté des Sciences in 1895 and became an assistant professor at the Faculté de Médecine in 1906. In that year he received his doctorate and in 1907 passed the *agrégation.* In 1919 he followed Raphael Blanchard as full professor in the chair of parasitology and was elected a member of the Académie de Médecine.

His rapid rise and brilliant career are explained by his works and his scientific authority. Brumpt, both a zoologist and a physician, was the originator of medical parasitology in France. He was led to this by experimental research and his numerous trips and sojourns in tropical lands. In 1901, as a second-year student, he made a two-year journey across Africa, from Abyssinia to the Congo, with the Bourg de Bozas mission. He later returned to Africa and also visited South America (especially Brazil) and the Far East.

Brumpt was a remarkable teacher and a most talented experimenter. He was the first to demonstrate, in 1904, the existence of a developmental phase in leeches of the trypanosomes of batrachians and fish. This cycle was noted again in 1911 in the tsetse carriers of *T. gambiense,* the agent causing sleeping sickness. In 1912, in Brazil, Brumpt described the life history of *Trypanosoma cruzi*—the agent of Chagas' disease—in *Triatoma,* a blood-sucking hemipteran. Chagas thought that trypanosoma developed in the general cavity and in the salivary glands of the bug and that infection was conveyed through biting. Brumpt in 1912 described the entire cycle and showed that the disease was transmitted through the feces, which infected the bite wound.

Brumpt studied all groups of parasites with his habitual thoroughness: the trypanosomes in Africa and South America, the *Piroplasma canis* and *bigemina,* the filariae, the *Bilharzia,* as well as the biology of the active and passive vectors of those parasites. His publications on mycology are important. He also studied recurrent fevers and exanthematic typhus. It was at this time, in 1933, that he contracted Rocky Mountain spotted fever and nearly died. In 1935 he brought back from Ceylon a strain of *Plasmodium gallinaceum,* the use of which has been of invaluable help in the chemotherapy of human malaria.

In 1923 Brumpt proved that the cysts of amoebas found in numerous individuals really belong to a nonpathogenic species that differs genetically from the dysenteric amoeba. In the same year, he founded the *Annales de parasitologie humaine et comparée.* He died just after completing the sixth edition of his *Précis de parasitologie.* He trained many students who continue the parasitological tradition that he began.

BIBLIOGRAPHY

Among Brumpt's 376 publications, the principal ones are his doctoral thesis in science, "Réproduction des hirudinées," in *Mémoires de la Société zoologique de France,* **33** (1901); *Mission de Bourg de Bozas. De la Mer Rouge à l'Atlantique à travers l'Afrique tropicale* (Paris, 1903); "Maladie du sommeil expérimentale chez les singes d'Asie et d'Afrique," in *Comptes rendus de la Société de biologie,* **56** (1904), 569; "Expériences relatives au mode de transmission des trypanosomes par les hirudinées," *ibid.,* **61** (1906), 77; *Les mycétomes,* his M.D. thesis (Paris, 1906); "Évolution de *Trypanosoma cruzi* chez *Conorhinus megistus, Cimex lectularius* et *Ornithodorus moubata,*" in *Bulletin de la Société de pathologie exotique,* **6** (1913), 752–758; "Le xénodiagnostic," *ibid.,* **7** (1914), 706–710; "Les piroplasmes des bovidés et leurs hôtes vecteurs," *ibid.,* **13** (1920), 416–460; "Recherches sur la bilharziose au Maroc," *ibid.,* **15** (1922), 632–641; "Les anophèles de Corse," in *Bulletin de l'Académie de médecine,* **93** (1925); "Réalisation expérimentale du cycle complet de *Schistosoma haematobium,*" in *Annales de parasitologie humaine et comparée,* **6** (1928), 440–446; "La ponte des schistosomes," *ibid.,* **8** (1930), 263–292; "Transmission de la fièvre exanthématique de Marseille par *Rhipicephalus sanguineus,*" in *Comptes rendus de l'Académie des sciences* (1930), 1028; "Épreuve de l'immunité croisée dans les fièvres exanthématiques," in *Comptes rendus de la Société de biologie,* **90** (1932), 1197; "Sensibilité du spermophile au Kala Azar chinois," *ibid.* (1935), 21–23; "La tularémie et ses hôtes vecteurs," in *Meditsinskaya parazitologia i parazitarnye bolezni* (1935), 23–28; "Une nouvelle fièvre récurrente humaine découverte à Babylone (Iraq)," in *Comptes rendus de l'Académie des sciences* (1939), 2029; "Étude épidémiologique de la fièvre récurrente des hauts plateaux mexicains," in *Annales de parasitologie humaine et comparée,* **17** (1939), 275–286; "Filarioses et éléphantiasis," in *Annales de la Société belge de médecine tropicale* (1947), 103; and *Précis de parasitologie,* 6th ed. (Paris, 1949).

HENRI GALLIARD

BRUNELLESCHI, FILIPPO (*b.* Florence, Italy, 1377; *d.* Florence, 16 April 1446), *architecture, engineering, geometry.*

While Brunelleschi was undoubtedly the first great Renaissance architect, it remains difficult to assess his importance to the history of science, and in particular to the development of a systematic mathematical perspective. Most of what is known of his life and work is derived from Vasari's *Lives of the Artists,* a book perhaps more notable for its charm than for its accuracy.

Brunelleschi was born into comfortable circumstances; his father, Ser Brunellesco di Lippo Lapi, was a notary and his mother, Giuliana, was a member of the noble Spini family. He had to abandon his formal education at an early age, but showed so much artistic talent that his father apprenticed him to a goldsmith.

Here, according to Vasari, "having become skilled in setting stones, and in niello work, and in the science of the motion of weights and wheels, not content with this, there awoke within him a great desire for the study of sculpture." It may well be that the mechanical knowledge gained in his apprenticeship aided Brunelleschi in the design and construction of engineering devices; certainly, he made some remarkable clocks.

The relationship between the craft of the goldsmith and the art of the sculptor in the fifteenth century is defined by the competition, open to both sculptors and goldsmiths, held in Florence in 1401 for the design of a pair of doors for the baptistery of the church of S. Giovanni. The sculptor Lorenzo Ghiberti won the commission and Brunelleschi, who had also submitted a design, went to Rome with the sculptor Donatello to study architecture. From 1402 to 1418, Brunelleschi lived alternately in Rome and Florence. It was perhaps during this period that, during one of his residences in Florence, he met Paolo dal Pozzo Toscanelli and learned geometry from him. He may also have learned some of the principles of perspective from Toscanelli; at any rate, Vasari states that he not only studied perspective, but also taught it to his friend Masaccio.

Vasari also tells of a meeting of architects and engineers in Florence in 1407 for the purpose of determining how to complete the cathedral of Sta. Maria del Fiore. The medieval architects of the building had intended a dome to be built over the crossing of the cathedral, but the problem of how to erect such a dome had never been solved. Brunelleschi entered the open competition for the design of the dome in 1418—Vasari says that he had already built a model for it—and won. He undertook the work in partnership with his rival Ghiberti, but the latter withdrew from the project. Brunelleschi worked on the cathedral dome from 1420 until his death, just after the lantern had been begun. He did not, as some sources suggest, rediscover the dome, but rather he invented a technique for building it without scaffolding.

Besides his work on the cathedral, Brunelleschi designed notable secular buildings—of which the Ospedale degli Innocenti is perhaps the outstanding example—and carried out military commissions. He may have drawn the plans for the fortress of Milan, constructed by the Sforzas; in 1415 he fortified the Ponte a Mare, and in 1435 he worked on the fortress of Vicopisano. Brunelleschi also worked on the fortification of the old citadel of Pisa and furnished the plans and built the model for the fortifications of the port of Pesaro. It seems likely that he always started such work with the construction of small-scale models; certainly he used such a model for the double dome of Sta. Maria del Fiore.

While many authors have considered Brunelleschi's chief scientific contribution to be his pioneering work in perspective (Vasari even credits him with the invention of monocular perspective), recent research has assigned him a more modest part. As an architect, Brunelleschi was certainly concerned with mathematical proportion, and from this an interest in the theory of perspective may well have been born. And in the Florence of the time, marked as it was by a self-consciously Academic exchange of ideas among artists and scientists, perspective would almost undoubtedly have been a subject for discussion; we know, for example, that Paolo Ucello was simultaneously at work on the problem, and eventually published a treatise on perspective projection that almost certainly incorporated many of Brunelleschi's ideas.

Brunelleschi's initial experiment in perspective may have been his ingenious painting of the baptistery as viewed from the porch of the cathedral. This painting, carefully rendered in perspective, was mounted on a thick wooden panel. A hole was then drilled through the panel at precisely the point that represented the eye of the artist. The aperture was, at the back of the panel, approximately the size of a lentil and widened to an opening about the size of a ducat at the front. The painting was placed to face a perpendicular arrangement of mirrors; when the viewer placed his eye to the hole at the back of the painting, he saw, through an optical illusion, the scene in three dimensions. (Brunelleschi made a second such picture showing the palace of the Seigniory, while Alberti made one of St. Mark's Square in Venice.)

Brunelleschi thus demonstrated his knowledge of conical projection and vanishing points, although it is possible that the concept of the optic box was Toscanelli's, and that Brunelleschi simply made it a reality. In any event, the idea of such a device, known to the ancients, may well have been drawn from the common scientific fund of the fifteenth century.

BIBLIOGRAPHY

The bibliography on Brunelleschi is not very extensive. His life as reported by Vasari has been followed by nearly all of his biographers. The most complete work so far is Venturi, *Brunelleschi* (Rome, 1923). Studies dealing with perspective are G. C. Argan, "The Architecture of Brunelleschi and the Origins of the Perspective Theory in the Fifteenth Century," in *Journal of Warburg and Courtauld Institutes* (1946); and J. B. Lemoine, "Brunelleschi et Ptolémée. Les origines géographiques de la 'boîte d'optique,'" in *Gazette des beaux arts* (1958). One might also consult Francastel, "Naissance d'un espace. Mythes et géométrie du quattrocento," in *Revue d'esthétique.*

BERTRAND GILLE

BRUNFELS, OTTO (*b.* Mainz, Germany, *ca.* 1489; *d.* Bern, Switzerland, 23 [?] November 1534), *botany.*

The earliest of the three "German fathers of botany" (the others being Jerome Bock and Leonhard Fuchs), Brunfels pioneered the dramatically sudden emancipation of botany from medieval herbalism.

Otto was the son of Johann Brunfels, a cooper; his mother's name is unknown. He received his early education locally and the master of arts degree at the University of Mainz in 1508/1509. Subsequently, he entered the Carthusian monastery in Strasbourg. He remained there until 1521, when, aided by Ulrich von Hutten, one of Luther's principal defenders, he fled the monastery and the Catholic faith as well. For the next three years he served as a pastor in Steinau and engaged in theological controversy. He returned to Strasbourg in 1524 and opened his own school. That same year he married Dorothea Heilgenhensin, who later helped to prepare his manuscripts for posthumous publication. There is no record of children from the marriage. He soon demonstrated his interest in medicine by editing and translating various older medical texts and by writing one of the earliest medical bibliographies, the *Catalogus* (1530). In that same year, Volume I of the *Herbarum vivae eicones* appeared, a book destined to change the direction of botany. Between 1530 and 1532, Brunfels supervised the publication of Volume II of the *Herbarum* and Volume I of the *Contrafayt Kreüterbuoch* while writing several other books. About this time, he moved to Basel, where he received the doctor of medicine degree from the university in late 1532. On 3 October 1533 he was appointed town physician in Bern for a period of six years. Approximately a year later, he fell seriously ill and died, possibly from diphtheria. He is commemorated by the genus *Brunfelsia* (*Solanaceae*), named in his honor by Charles Plumier in 1703.

Through his early theological and pedagogical writings and his wide correspondence (still unedited), Brunfels became associated with the local Strasbourg humanists, one of whom, Johann Schott, printed many of his books. Presumably it was through Schott that he became acquainted with the artist Hans Weiditz, whose name is inseparably linked with the *Herbarum.*

Judged by modern standards, the *Herbarum* is a curious combination of the old and the new. The text is a typical late-medieval collection of extracts uncritically compiled from earlier writings and possessing little independent value. The illustrations, on the other hand, are detailed, accurate renderings of plants executed with a realism that revolutionized botanical iconography. Most subsequent sixteenth-century herbals are the direct descendants of a method first enunciated under Brunfels' guidance. The impact of his contribution and the scientific value of the *Herbarum* would have been incalculably greater if the descriptions, like the illustrations, had been taken from nature.

The *Herbarum* is divided into *rhapsodiae* (chapters), each of which is devoted to one plant. The text, essentially a series of verbatim quotations from older authorities, is thematically connected by a concern to identify therapeutically useful plants. For this purpose, classical Greek and Latin names are correlated with the German vernacular names. The plants are not arranged in any systematic order, for it was not Brunfels' intention to propose a classification. Nevertheless, the arrangement is not alphabetical, and related species often appear on successive folios. Most of the plants described (approximately 230 species) were indigenous to Strasbourg and its environs. Over forty species were first described by Brunfels. Exotica, frequently encountered in the incunabula herbals, are ignored.

The bulk of each *rhapsodia* is devoted to the medicinal properties of the plant. Pertinent information includes preparation, administration, and dosage of some specified portion of the plant, time of collection, and the ailments for which the prepared drug was reputedly beneficial. Pharmacological uses are expressed in terms of the Galenic doctrine of "grades" and "temperaments." This information is derived almost exclusively from Brunfels' sources, forty-seven of whom are listed (sig. A iii^v). His main authorities were classical, principally Dioscorides, Pliny, and Galen, although medieval, Arabic, and especially contemporary Italian writers are also cited. Following the extracts from his authorities, there is often a section entitled "Iudicium nostrum" ("My Opinion"), in which Brunfels presents his own evidence.

At the end of Volume II of the *Herbarum* are twelve tracts, collectively entitled "De vera herbarum cognitione appendix," edited by Brunfels. The tracts are devoted primarily to the nomenclature of plants known to the ancients. Both Bock and Fuchs first appear as authors in this collection.

The *Contrafayt Kreüterbuoch,* a German adaptation, not a translation, of the *Herbarum,* was undertaken by Brunfels before the *Herbarum* was completed and prior to his departure for Basel. All but sixteen of the illustrations of the *Herbarum* are repeated, with about fifty additional figures. Not all of the plants discussed are illustrated, however. Altogether about 260 species were depicted in the *Herbarum* and the *Kreüterbuoch.* The text of the latter is better organized, the text being arranged under sectional headings dealing with nomenclature, appearance and form, habitat, time of collecting, and medical uses and properties. The long verbatim extracts were abandoned, although their content was closely paraphrased. Like the *Herbarum,* the *Kreüterbuoch* remained incomplete at Brunfels' death.

One other botanical work was published under Brunfels' name, the posthumous *In Dioscoridis historiam.* It is a series of illustrations taken from the same wood blocks as those used for the *Herbarum,* presented without preface or text save the plants' names, which appear alongside the illustrations.

The three volumes of the *Herbarum* contain 238 woodcut illustrations, ranging in size from full folio to small text figures and normally illustrating the text of the facing or adjacent folio. The illustrations, a happy combination of scientific accuracy and aesthetic charm, were designed by Weiditz, who also cut the majority of the blocks. Despite the width of even the finest lines (less than 200μ) and the fact that Weiditz had had no previous botanical training, details of floral structure and vegetative organography are readily apparent. Moreover, careful attention was given to the general appearance of the plant and its typical habitat. Usually the entire plant is depicted, all portions (root, stem, leaves, blossom, and fruit) receiving equal attention, even though, for example, the function of the stamens or the taxonomic importance of foliaceous bracts was then unappreciated. Leaves damaged by insects, broken petioles and bent stems, and blossoms in different stages of development leave no doubt that the illustrations were based on living plants. The drooping appearance of some stems and leaves suggests that the plant was dug up entire and had begun to wilt when illustrated. Owing largely to the fidelity of the woodcuts, the great majority of the plants discussed by Brunfels have been identified with reasonable certainty.

While Brunfels must be given credit for planning an illustrated herbal and overseeing its preparation, credit is also due Weiditz for executing the realistic illustrations. The *Herbarum* is the first printed botanical book in which scientific value can be assigned to the illustrations. Weiditz' contributions were noted by Brunfels, and through his appreciative comments in the preface of the *Kreüterbuoch* the artist assumes, for the first time, a recognized place in botanical literature. Some of Weiditz' watercolor drawings that served as the originals for the wood blocks of the *Herbarum* were discovered by Rytz in Bern. Their publication demonstrated that the success of the *Herbarum* was, in large measure, the result of Weiditz' participation.

Although Brunfels' other writings were of less scientific importance, they deserve a brief note because they were typical of the times and, contributing to his reputation, they facilitated the acceptance of his botanical work. Leaving aside his theological and pedagogical writings (about twenty-eight separate publications), his nonbotanical work was principally in medicine and pharamacology. In the former, Brunfels was active as a translator (Lanfranchi, Paul of Aegina, Galen) and as an editor (Dioscorides, Fries, Tanstetter, Serapion, and others). He was no less industrious in compiling practical texts designed for the use of physicians and apothecaries, which contained prescriptions and related pharmacological matter and were usually well indexed for ready reference. His most important pharmacological work was the *Reformation der Apotecken*. Originally written in Strasbourg, it was enlarged to serve as a city ordinance for apothecaries in Bern. It contains one of the earliest Swiss dispensatories. Brunfels' passion for compiling and organizing reference material, already evident in the "Appendix de usu et administratione simplicium" (*Herbarum*, I, fols. 273–329) was fully exhibited in his 'Ονομαστικόν, a comprehensive dictionary containing a wealth of material related to medicine, botany, alchemy, and metrology. One other writing, the *De diffinitionibus,* is of interest because of its criticism of astrology.

BIBLIOGRAPHY

I. ORIGINAL WORKS. Brunfels was active as editor, translator, and author.

His own works are *Von allerhandt apoteckischen Confectionen, Lattwergen, Oel, Pillulen, Trencken, Trociscen, Zuckerscheiblin, Salben und Pflastern . . .* (Strasbourg, ca. 1530; repr. Frankfurt, 1552); *Catalogus illustrium medicorum, sive de primis medicinae scriptoribus. . .* (Strasbourg, 1530); *Herbarum vivae eicones ad naturae imitationem, summa cum diligentia et artificio effigiatae, una cum effectibus earundem . . .,* 3 vols.: I (Strasbourg, 1530; repr. 1532, 1536, 1537; with II and III, 1539); II (Strasbourg, 1531 [colophon, 1532], 1536, 1537, 1539); III, Michael Heer, ed. (Strasbourg, 1536, 1537, 1539, 1540); *Contrafayt Kreüterbuoch,* 2 vols.: I (Strasbourg, 1532; repr., with different title, Strasbourg, 1534, 1539; with II, Frankfurt, 1546, 1551); II (Strasbourg, 1537, 1540); facsimile repr. of I (1532) and II (1537) (Munich, 1964); *Theses seu communes loci totius rei medicae. Item. De usu pharmacorum, deque artificio suppressam alvum ciendi, liber* (Strasbourg, 1532); *De diffinitionibus et terminis astrologiae libellus isagogicus,* in Julius Firmicus Maternus, *Ad Mavortium Lollianum astronomicon libri VIII* (Basel, 1533; repr. Basel, 1551); *Jatrion medicamentorum simplicium continens remedia omnium morborum quae tam hominibus quam pecudibus accidere possunt . . .,* 4 vols. (Strasbourg, 1533); 'Ονομαστικόν *medicinae . . .* (Strasbourg, 1534; repr., with different title, Strasbourg, 1543 [colophon, 1544]); *Weiber und Kinder Apoteck,* 2 vols. (Strasbourg, ca. 1534; *Annotationes in quatuor evangelia et acta apostolorum* (Strasbourg, 1535), which, besides autobiographical material, has the only known authentic portrait of Brunfels on the reverse of the title page; *Reformation der Apotecken . . .* (Strasbourg, 1536); *Epitome medices summam totius medicinae complectens . . .* (Paris, 1540); and *In Dioscoridis historiam herbarum certissima adaptatio . . .* (Strasbourg, 1543).

Among the works he edited are Alessandro Benedetti, *Anatomice; sive, De hystoria corporis humani libri quinque* (Strasbourg, 1528); Dioscorides, *Pharmacorum simplicium, reique medicae libri VIII, Jo. Ruellio interprete . . .* (Strasbourg, 1529); Lorenz Fries, *Spiegel der Artzney . . .* (Strasbourg, 1529; repr., with slightly changed title, Strasbourg, 1532, 1546); Georg Tanstetter von Thannau, *Artificium de applicatione astrologie ad medicinam, deque convenientia earundem . . .* (Strasbourg, 1531); and *Neotericorum aliquot medicorum introductiones* (Strasbourg, 1533).

He translated Guido Lanfranchi, *Kleyne Wundartznei . . . auss fürbit des Gregorii Flüguss . . .* (Strasbourg, 1528; repr. with slightly changed title, Strasbourg, 1529; Erfurt, 1529; Zwickau, 1529; Cöllen, n.d. [after 1529]; Frankfurt, 1552, 1569; Magdeburg, n.d. [not before 1578]); Paul of

Aegina, *Pharmaca simplicia, Othone Brunfelsio interprete. Idem, De ratione victus Gulielmo Copo Basiliensi interprete* . . . (Strasbourg, 1531); *In hoc volumine continentur* . . . *Joan. Serapionis Arabis de simplicibus medicinis* . . . *Averrois Arabis de eisdem liber eximius. Rasis filii Zachariae de eisdem opusculum.* . . . *Incerti item autoris de centaureo libellus hactenus Galeno inscriptus. Dictionum Arabicarum juxta atque Latinarum index valde necessarius* . . . (Strasbourg, 1531); and Galen, *De ossibus ad tyrones* (Padua, 1551).

II. SECONDARY LITERATURE. Biographical data on Brunfels' life are meager and ultimately derive from the preface of the *Annotationes in quatuor evangelia.* Supplementing them are many references to him and his religious activities in contemporary theological and humanistic writings.

The following concern his scientific work: H. Christ, "Otto Brunfels und seine *Herbarum vivae eicones.* Ein botanischer Reformator des XVI. Jahrhunderts," in *Verhandlungen der Naturforschenden Gesellschaft in Basel,* **38** (1927), 1–11; A. H. Church, "Brunfels and Fuchs," in *Journal of Botany, British and Foreign,* **57** (1919), 233–244; F. A. Flückiger, "Otto Brunfels, Fragment zur Geschichte der Botanik und Pharmacie," in *Archiv der Pharmacie,* **212** (1878), 493–514; Friedrich Kirschleger, *Flore d'Alsace et des contrées limitrophes,* II (Strasbourg–Paris, 1857), xiii–xvii, which contains the identifications of 106 species figured in the *Herbarum;* E. H. F. Meyer, *Geschichte der Botanik,* IV (Königsberg, 1857), 295–303; Claus Nissen, *Die botanische Buchillustration,* 2 vols. (Stuttgart, 1951), I, 40–44; II, nos. 257–261; F. W. E. Roth, "Otto Brunfels. Nach seinem Leben und litterarischen Wirken geschildert." in *Zeitschrift für die Geschichte des Oberrheins,* n.s. **9** (1894), 284–320; "Die Schriften des Otto Brunfels. 1519–1536," in *Jahrbuch für Geschichte, Sprache und Literatur Elsass-Lothringens,* **16** (1900), 257–288, the best bibliography of Brunfels' writings, 49 publications plus 3 dubious ones, but still incomplete; and "Otto Brunfels 1489–1534. Ein deutscher Botaniker," in *Botanische Zeitung,* **58** (1900), 191–232, a well-documented biographical study; Alfred Schmid, "Zwei seltene Kräuterbücher aus dem vierten Dezennium des sechzehnten Jahrhunderts," in *Schweizerischen Gutenbergmuseum,* no. 3 (1936), 160–180, the only study of the quarto eds. of the *Kreüterbuoch* (1534, 1539, 1540, 1551) and the best bibliographical analysis of the complex dating of the various eds. and vols. of the *Herbarum;* Thomas Archibald Sprague, "The Herbal of Otto Brunfels," in *Journal of the Linnean Society* (London), **48** (1928), 79–124, a fundamental study with modern identifications of the plants figured by Brunfels; and Kurt Sprengel, *Geschichte der Botanik,* I (Altenburg–Leipzig, 1817), 258–262, containing the identifications, some dubious, of 131 species figured by Brunfels.

Also of interest are Karl Hartfelder, "Otto Brunfels als Verteidiger Huttens," in *Zeitschrift für die Geschichte des Oberrheins,* n.s. **8** (1893), 565–578; Heinrich Röttinger, "Hans Weiditz, der strassburger Holzschnittzeichner," in *Elsass–Lothringisches Jahrbuch,* **16** (1937), 75–125; Walther Rytz, *Pflanzenaquarelle des Hans Weiditz aus dem Jahre*

1529. Die Originale zu den Holzschnitten im Brunfels'schen Kräuterbuch (Bern, 1936); and Erich Sanwald, *Otto Brunfels 1488–1534. Ein Beitrag zur Geschichte des Humanismus und der Reformation. I. Hälfte 1488–1524* (Bottrop, Germany, 1932).

JERRY STANNARD

BRUNHES, JEAN (*b.* Toulouse, France, 25 October 1869; *d.* Boulogne-sur-Seine, France, 25 April 1930), *geography.*

Brunhes came from a family of university professors: both his father, Julien, and his older brother, Bernard, were professors of physics. Jean entered the École Normale Supérieure in 1889, and in 1892 he graduated and passed the *agrégation* in history and geography. His faculty adviser was Vidal de la Blache. On a scholarship from the Thiers Foundation from 1892 to 1896, he completed his education by taking courses in law, mining, and agriculture. He found his true vocation in geography when he wrote the thesis "L'irrigation, ses conditions géographiques . . . dans la péninsule ibérique et l'Afrique du Nord," which he defended in 1902.

Brunhes was named professor of general geography at the University of Fribourg in 1896, and in 1908 he was appointed to give a course in human geography at the University of Lausanne. He continued to work in human geography, a science that did not then exist in France.

In his *Anthropogéographie* the German geographer Friedrich Ratzel attempted to explain man in terms of nature and to make history and culture dependent on geography. In contrast, Brunhes saw in nature "not a tyrannical fatalism, but an infinite wealth of possibilities among which man has the power to choose" (S. Charléty, *Notes sur la vie et les travaux de M. J. Brunhes* [Paris, 1932], p. 13). He also believed that there is no social determinism whose laws can be ascertained. In his great work, *Géographie humaine* (1910), Brunhes presented the first attempt to coordinate the geographical phenomena resulting from the activities of man. It was illustrated with numerous photographs. In 1912 the Collège de France created a chair of human geography for him.

A member of the Académie des Sciences Morales et Politiques since 1927, Brunhes died suddenly of a stroke just after he and his daughter, Mme. Raymond Delamarre, had published *Les races,* a small, richly illustrated book.

Certain geographers have reproached Brunhes for having extended geography to cover all forms of human activity; others have criticized him for having limited the study of geography to what is "photographable." Nevertheless, he gave a decisive impetus to human geography.

BIBLIOGRAPHY

I. ORIGINAL WORKS. Brunhes's writings include *La géographie humaine. Essai de classification positive. Principes et exemples* (Paris, 1910, 1912, 1925), trans. into English (Chicago–New York, 1920); 2 vols. in G. Hanotaux's *Histoire de la nation française:* I. *Géographie humaine de la France* (Paris, 1926), and II, *Géographie politique et géographie du travail* (Paris, 1926), written with P. Deffontaines; and *Les races* (Paris, 1930), written with his daughter, Mme. Raymond Delamarre. He also translated Isaiah Bowman's *The New World* as *Le monde nouveau. Tableau général de géographie politique universelle* (Paris, 1928).

II. SECONDARY LITERATURE. Biographies of Brunhes are A. Allix, in *Les études rhodaniennes,* **6** (1930), 340–342; M. Boule, in *L'anthropologie,* **40** (1930), 514–515; V. Châtelain, in *Dictionnaire de biographie française,* fasc. 39 (1955), cols. 554–555; D. Faucher, in *Revue de géographie des Pyrénées et du Sud-ouest,* **1** (1930), 514–515; E. de Martonne, in *Annales de géographie,* **39** (1930), 549–553; and G. Vallaux, in *La géographie,* **34** (1930), 237–239.

JULIETTE TATON

BRUNO, GIORDANO (*b.* Nola, Italy, 1548; *d.* Rome, Italy, 17 February 1600), *philosophy.*

Bruno's baptismal name was Filippo; he took the name Giordano, by which he is always known, on entering the Dominican order. His father, Giovanni, was a soldier, and probably a man of fairly good position; his mother, Fraulissa Savolino, has been conjectured to have been of German descent, although there is no real evidence. Hardly anything is known of Bruno's early years in Nola, a small town near Naples.

At the age of fifteen, Bruno entered the Dominican order and became an inmate of the great Dominican convent in Naples. Here he acquired a grounding in Scholastic philosophy and the reverence for Thomas Aquinas (who had lived and taught in the Naples convent) that he professed throughout his life. Here, too, he became proficient in the art of memory, for which the Dominicans were noted, and was taken to Rome to display his mnemonic skill to Pope Pius V. Another influence which he may have come under in these early years was that of the famous natural magician and scientist Giambattista della Porta, who in 1560 had established in Naples his academy for investigating the secrets of nature. Bruno was formed during these years in Naples: his mind and character never lost the imprint of his training as a friar; and it was as a passionate ex-friar that he wandered over Europe, combining philosophical speculation with a religious mission evolved through deep immersion in Renaissance magic and its Hermetic sources.

Bruno's religion was the moving force behind both his wandering career and his philosophical and cosmic speculations. He believed that he was reviving the magical religion of the ancient Egyptians, a religion older than Judaism or Christianity, which these inferior religions had suppressed but of which he prophesied the imminent return. It included a belief in the magical animation of all nature, which the magus could learn how to tap and to use, and a belief in metempsychosis. The historical origins of Bruno's "Egyptianism" and the printed sources whence he derived it are now clear, owing to the work done by scholars in fairly recent years on the Hermetic core of Renaissance Neoplatonism.

As propagated by Marsilio Ficino, Renaissance Neoplatonism included a firm belief that both Plato and his followers had been inspired by a tradition of *prisca theologia,* or pristine and pure theology, which had come down to them from the teachings of Hermes Trismegistus, a mythical Egyptian sage, and other figures supposedly of extreme antiquity. This belief rested on the misdating of certain late antique texts, of which the most important were the *Asclepius* and the *Corpus Hermeticum,* which were supposed to have been written by Hermes Trismegistus himself.

Ficino believed that these texts contained authentic revelations about ancient Egyptian religion and that in them their supposed author prophesied the coming of Christianity—and, hence, could take on sanctity as a Gentile prophet. The scraps of Platonic notions incorporated by the late antique Gnostic writers of the Hermetic texts were, for Ficino, evidence that these ancient "Egyptian" teachings were the pristine source at which Plato and the Neoplatonists had drunk. These beliefs could be supported from works of some Church Fathers, notably Lactantius. Nor were they peculiar to Ficino; on the contrary, the whole Renaissance Neoplatonic movement contained this Hermetic core, and the religious magic, or theurgy, taught by Hermes Trismegistus, particularly in the *Asclepius,* seemed corroborated by the intensive Renaissance study of the later Neoplatonists, such as Porphyry and Iamblichus. As a pious Christian, Ficino was encouraged by the sanctity of Hermes Trismegistus as a Gentile prophet to embark on the astral magic described in the *Asclepius,* which lies behind his own work on astral magic, the *De vita coelitus comparanda,* although he did this hesitantly and timidly, in fear of the Church's embargo on magic.

The extreme boldness and fearlessness that characterized Giordano Bruno are nowhere more apparent than in his choice of a religion. Discarding the belief in Hermes as a Gentile prophet, which sanctified the Hermetic writings for pious Christian Neoplatonists, Bruno accepted the pseudo-Egyptian religion described in the Hermetic texts as the true religion; he interpreted the lament in the *Asclepius* over the decay of Egypt and her magical worship as a lament for the true Egyptian religion, which had been suppressed by Christianity, although various signs and portents were announcing its return.

Among these signs was the heliocentricity announced by Copernicus—and it must be confessed that Copernicus himself did something to encourage such an interpretation of his discovery when, at a crucial point in his work, just after the diagram showing the new sun-centered system, he referred to Hermes Trismegistus on the sun as a visible god (a quotation from the *Asclepius*). In his defense of Copernicanism against the Aristotelians of Oxford, Bruno presented Copernicus as "only a mathematician" who had not understood the true inwardness of his discovery as he, Bruno, understood it—as portending a return to magical insight into living nature. In support of the movement of the earth, Bruno quoted a passage from one of the treatises of the *Corpus Hermeticum,* which states that the earth moves because it is alive.

The magical animism that permeates Bruno's philosophy of nature, his vision of the living earth moving round the sun, of an infinite universe of innumerable worlds moving like great animals in space, is inseparably connected with his pseudo-Egyptian religion. It is universal animism which makes possible the activities of the magus and justifies the techniques by which he attempts to operate on nature. Bruno aspired to become such a magus, using the techniques described in the *De occulta philosophia* of Henry Cornelius Agrippa von Nettesheim, a work that was itself the product of the Hermetic core within Renaissance Neoplatonism.

It is one of the most extraordinary features of Bruno's outlook that he seems to have believed that his religion could somehow be incorporated within a Catholic framework in the coming new dispensation. He never lost his respect for Thomas Aquinas, and his preaching of his new religion retained traces of Dominican preacher's training. Although Christ was for him a benevolent magus, as were Thomas Aquinas, Paracelsus, Ramón Lull, and Giordano Bruno himself, he proclaimed in the *Spaccio della bestia trionfante* that Christ was to remain in heaven as an example of a good life.

While still in the convent in Naples, he fell under suspicion of heresy and proceedings were instituted against him. The suspicion against him seems to have been of Arian tendencies; possibly his full "Egyptian" program was not yet developed. To avoid the process against him he left Naples in 1576. He went first to Rome, where he fell into new difficulties, from which he escaped by abandoning the Dominican habit and fleeing from Italy. Now began his long odyssey through France, England, Germany. He went first to Geneva, where he soon got into trouble and acquired a strong dislike of Calvinism.

From about 1579 to 1581 he was in Toulouse, where he lectured in the university on, among other things, the *Sphere* of Sacrobosco. From Toulouse he went to Paris; here his public lectures attracted the attention of King Henry III. His first published work, the *De umbris idearum* (Paris, 1582), is dedicated to Henry. It is an example of his transformation of the art of memory into a deeply magical art, and its title is taken from that of a magical book mentioned in the necromantic commentary on the *Sphere* of Sacrobosco by Cecco d'Ascoli, an author whom Bruno greatly admired. Bruno thus came before the world in his first Parisian period as a magician teaching some extremely abstruse art of memory that apparently gained the interest and approval of the king of France, who gave him letters of recommendation to the French ambassador in England. This is the first indication of some mysterious political, or politico-religious, undercurrent in Bruno's activities and movements.

Bruno crossed the Channel to England early in 1583; the royal letters of recommendation had the desired effect, for the French ambassador, Michel de Mauvissière, received him into the French embassy, where during the two years of his stay in England he lived as a "gentleman" attached to the embassy. He states that he often accompanied the ambassador to court and saw Queen Elizabeth, whom he addresses as "divine" in his works, an epithet that he had to try to explain away to the Inquisitors. The ambassadorial protection enabled Bruno to publish his extremely provocative works, in which he criticized Reformation Oxford as inferior in philosophical learning to the Oxford of the Middle Ages and attacked the whole social order of Elizabethan England for having destroyed, without adequately replacing, the institutions of Catholic times. His books were published clandestinely, with false imprints, by John Charlewood. As was to be expected, they aroused tumults against the bold ex-friar that were sometimes so violent that he dared not go outside the embassy.

Bruno opened his campaign in England with one of his obscure works on the magic art of memory, the *Triginta sigilli*; hidden away at the end of it there is a passionate advocacy of a new religion based on love, art, magic, and mathesis; it begins with an abusive dedication to the vice-chancellor and doctors of Oxford. This would seem to have been a strange preparation for his visit to Oxford in the train of the Polish prince, Albert Laski. A newly discovered source, first published in 1960, has thrown much light on Bruno's famous advocacy of the Copernican theory to the recalcitrant Aristotelians of Oxford. It appears that after Laski's party had left, Bruno returned to Oxford and delivered lectures that consisted mainly of quotation from Ficino's book on astral magic, *De vita coelitus comparanda,* with which he associated the opinion of Copernicus "that the earth did go round and the heavens did stand still." Bruno's unacknowledged quotations from Ficino were detected by some of his auditors, as is recounted in the newly discovered report of his speech. This new information about Bruno's Oxford lectures is external confirmation of what can also be clearly deduced from his works; that for Bruno, Copernican heliocentricity was associated with his magical and animist view of nature.

The brilliant dialogues in Italian that Bruno published while in England have been the most widely read of his works and were the main foundation for his reputation as a bold philosopher breaking out of the closed medieval universe into a new vision of the cosmos. This reputation is by no means undeserved, although it now has to be formulated in more accurate historical terms than those used by his nineteenth-century admirers, who were unaware that their hero was a magician and knew nothing of the complex political and religious situation in Elizabethan England, the scene of these exploits. In the *Cena de le ceneri* (1584) he defends Copernican heliocentricity against two Oxford "pedants." The angry protests that this attack aroused are described in *De la causa, principio e uno* (1584); Bruno here offers a slight apology for his attack on Oxford—but in the form of professing admiration for the friars of pre-Reformation Oxford, with whom he unfavorably compares their Protestant successors. This can have done little to improve the situation, and the censor can have been prevented from taking action against the book only because it was dedicated to the French ambassador. In the *De l'infinito, universo e mondi* (1584), Bruno sets forth his remarkable vision of an infinite universe and innumerable worlds infused with divine life.

In the *Spaccio della bestia trionfante* (1584), he turns to the moral, as apart from the physical or philosophical, side of his message, and outlines a universal moral and religious reform. The curious form of this work, which is based on the constellations

from which vices are said to be expelled, to be replaced by virtues, is related to Bruno's adaptations of the art of memory. The *Cabala del cavallo Pegaseo* (1585) is an obscure discussion of the Jewish cabala. In the *De gli eroici furori* (1585), Bruno expresses himself in a sequence of beautiful poems followed by commentaries explaining their philosophic and mystical meanings. This book is dedicated to Philip Sidney, as is the *Spaccio della bestia trionfante*. All the other Italian dialogues, with the exception of the *Cabala del cavallo Pegaseo,* are dedicated to the French ambassador. One is left wondering how far the extraordinary philosophical, magical, and religious views that Bruno propagated from the safety of the French embassy were acceptable to the distinguished persons to whom he dedicated these books. They are all full of Hermetic influences and are bound up with the complex religious, or politico-religious, mission for which he seems to have believed that he had the support of the king of France and to which the French ambassador seems to have lent his protection.

Meanwhile, in France the Catholic League was rising in power; Henry III's position grew precarious; Mauvissière, the liberal ambassador, was recalled, and late in 1585 Bruno returned to Paris in his train. Immediately he began to talk and to publish, expounding his philosophy in an address delivered by a disciple in the Collège de Cambrai, which was tumultuously received. The king's support was indirectly withdrawn; and Bruno made himself notorious in a quarrel about a compass with Fabrizio Mordente, which may have had a political background. Paris became too dangerous for him, and in 1586 he fled, this time toward Germany.

At Wittenberg he felt happy for a time: the university allowed him to lecture, and he found that he greatly preferred German Lutherans to English Calvinists. Here he wrote a number of works, particularly on Lullism, which he believed that he understood better than Lull himself. But eventually here also trouble started, and after delivering a moving farewell oration to the doctors of Wittenberg, he went on to Prague, where he dedicated to Emperor Rudolph II his *Articuli adversus mathematicos* (1588), in which he professed to be strongly against mathematics. This book is illustrated with magical diagrams. In the Preface he urges the emperor to lead a movement of religious toleration and philanthropy. Yet even Rudolph, who collected strange people at his court, did not extend a warm welcome to Bruno; he gave him a little money, but no position, and Bruno wandered on to Helmstedt. Here he found support from Henry Julius of Brunswick-Wolfenbüttel, who may have been in sympathy with his ideas; at any rate, he allowed Bruno to deliver an oration on his recently deceased father which echoed the moral and religious program of the *Spaccio della bestia trionfante.* While at Helmstedt, Bruno was busily writing; the *De magia* and other works on magic preserved in the Noroff manuscript may have been written during this period. Henry Julius possibly gave him money toward the publication of the Latin poems that he had been writing during his travels; and Bruno went on to Frankfurt to supervise their printing.

The *De immenso et innumerabilibus,* the *De triplici minimo et mensura,* and the *De monade numero et figura* were published in 1591. In these poems, written in a style imitating that of Lucretius, Bruno expounded for the last time his philosophical and cosmological meditations, mingled, as in the works published in England, with powerful Hermetic influences. His last published work, also published in 1591 by Wechel at Frankfurt, was a book on the magic art of memory dedicated to the alchemist and magician Johannes Hainzell.

While at Frankfurt, Bruno received, through an Italian bookseller who came to the Frankfurt fair, an invitation from Zuan Mocenigo, a Venetian nobleman, to come to Venice and teach him the secrets of his art of memory. He accepted, and in August 1591, he returned to Italy, going first to Padua and then to Venice. There can be little doubt that Bruno believed, like many others at the time, that the conversion of Henry IV of France was a sign of vast impending religious changes in Rome, and that he and his mysterious mission would be well received in the approaching new dispensation. That he had no idea that he was running into danger is shown by the curious fact that he took with him the manuscript of a book that he intended to dedicate to Pope Clement VIII.

Bruno's reception in Italy was tragically other than he had expected. Mocenigo informed against him, and he was arrested and incarcerated in the prisons of the Inquisition in Venice. There followed a long trial, at the end of which Bruno recanted his heresies and threw himself on the mercy of the inquisitors. He had to be sent on to Rome for another trial, however, and there his case dragged on for eight years of imprisonment and interrogation. After some wavering, he finally refused to recant any of his views, with the result that he was burned alive as a dangerous heretic on the Campo de' Fiori in Rome.

The grounds on which Bruno was sentenced are unknown, for the *processo,* or official document containing the sentence, is irretrievably lost. It formed part of a mass of archives that were transported, by order of Napoleon, from Rome to Paris, where they

were pulped. From the reports of the interrogations, it is, however, possible to form an idea of the drift of the case against him. To his major theological heresy, the denial of the divinity of the Second Person of the Trinity, was added suspicion of diabolical magical practices. It was probably mainly as a magician that Bruno was burned, and as the propagator throughout Europe of some mysterious magico-religious movement. This movement may have been in the nature of a secret Hermetic sect, and may be connected with the origins of Rosicrucianism or of Freemasonry. If any philosophical or cosmological points were included in his condemnation, these would have been inextricably bound up with his "Egyptianism."

The legend that the nineteenth century built around Bruno as the hero who, unlike Galileo, refused to retract his belief that the earth moves is entirely without foundation. Bruno's case may, however, have affected the attitude of the Church toward the Copernican hypothesis and may have encouraged the Inquisition's suspicion of Galileo. Although Galileo accepted the Copernican world view on entirely different grounds from Bruno, there are curious formal resemblances between his *Dialogo dei due massime sistemi del mondo,* in which the pedantic Simplicius takes the Aristotelian side, and Bruno's *Cena de le ceneri,* in which the Oxford pedants oppose the "new philosophy."

The history of Bruno's reputation is instructive. Abhorred by Marin Mersenne as an impious deist, he was more favorably mentioned by Kepler. Rumors of his diabolism seem to have been circulated, and were mentioned even by Pierre Bayle in one of the footnotes to his contemptuous article on Bruno. The eighteenth-century deist John Toland revived interest in some of his works. It was not until about the mid-nineteenth century that a revival on a large scale began to gather strength and the legend of the martyr for modern science was invented—of the man who died, not for any religious belief, but solely for his acceptance of the Copernican theory and his bold vision of an infinite universe and innumerable worlds. Statues in his honor proliferated in Italy; the literature on him became immense.

In the late nineteenth and early twentieth centuries, Giordano Bruno was one of the most widely known, and most frequently written about, philosophers of the Italian Renaissance. His ideas, isolated from their historical context, were interpreted in terms of the then dominant type of history of philosophy, for example, by Giovanni Gentile, and the large areas in his writings that are not intelligible in terms of straight philosophical thinking were neglected or ignored. Leo Olschki was probably one of the first to notice that no coherent philosophical system could be drawn from Bruno's works through this approach; and Antonio Corsano emphasized the magical ingredients in Bruno's thought and the politico-religious aspects of his activities. It is, however, the work that has been done in recent years on the Renaissance Hermetic tradition that has at last made it possible to place Bruno within a context in which his philosophy, his magic, and his religion can all be seen as belonging to an outlook that, however strange, makes historical sense.

Now that Giordano Bruno has been, as it were, found out as a Hermetic magician of a most extreme type, is he therefore to be rejected as of no serious importance in the history of thought? This is not the right way to pose the question. Rather, it should be recognized that Renaissance magic, and that turning toward the world as a revelation of the divine that is the motive force in the "religion of the world" that inspired Bruno, was itself a preparation or a stage in the great movement that, running out of the Renaissance into the seventeenth century, gradually shed its irrational characteristics for the genuinely scientific approach to the world. Bruno's leap upward through the spheres into an infinite universe, although it is to be interpreted as the experience of a Gnostic magician, was at the same time an exercise of speculative imagination presaging the advent of new world views. Although Bruno infused the innumerable worlds of which he had learned from Lucretius with magical animism, this was in itself a remarkable vision of a vastly extended universe through which ran one law. We can accept Bruno's Renaissance vision as prophetic of coming world views, although formulated within a very strange frame of reference.

Again, Bruno's atomism, derived from his study of Lucretius through magical interpretation of Lucretius in such a writer as Palingenius, whose *Zodiacus vitae* was one of Bruno's inspirations, may have stimulated the attention of other thinkers. The Renaissance interpretation of Lucretius, which was begun by Ficino, is a stage in the history of atomism which has not yet been adequately examined. When that history comes to be written, Bruno's magically animated atoms may be found to hold some transitional place in it.

Another example of Bruno's thought as a presage of scientific discovery is his remarkable intuition about the circular movement of the blood, which he based on parallelism between man and the universe; he believed that "spirit" is the driving force that moves the blood, the same spirit that is diffused through the universe and that Plato defined as "number which moves in a circle." Hence, the movement of the blood within the body, said Bruno, is circular, diffused from the heart in a circular movement.

One of the closest connections between Bruno and a seventeenth-century scientific philosopher is that which can be discerned in the influence of Bruno's *Cena de le ceneri* on William Gilbert's *De magnete.* The magnet is always mentioned in textbooks on magic as an example of the occult sympathies in action; and Bruno, when defending his animistic version of heliocentricity, brought in the magnet. Gilbert's language when defending heliocentricity in the *De magnete* is extremely close to that of Bruno; like Bruno, he cites Hermes and others who stated that there is a universal life in nature when he is arguing in favor of earth movement. The magnetic philosophy that Gilbert extended to the whole universe seems most closely allied to that of Bruno, and it is not surprising that Francis Bacon should have listed Gilbert with Bruno as proud and fantastic magi of whom he strongly disapproved.

Even the strangest and most formidably obscure of Bruno's works, those on his magic arts of memory, can be seen to presage, on the Hermetic plane, seventeenth-century strivings after method. Bruno aimed at arranging magically activated images of the stars in memory in such a way as to draw magical powers into the psyche. These systems were of an incredible complexity, involving combinations of memory images with the revolving wheels of Lull to form ways of grasping everything in the universe at once and in all possible combinations. Bruno's Hermetic computers, if one may be permitted to call them such, were almost certainly known to Leibniz, who was also familiar with the art of memory and with Lullism. When introducing his universal calculus, Leibniz uses language that is remarkably similar to that in which Bruno introduced his art of memory to the doctors of Oxford. The many curious connections between Bruno and Leibniz may, when fully explored, form one of the best means of watching the transitions from Renaissance occultism to seventeenth-century science.

Within that view of the history of thought in which the Renaissance magus is seen as the immediate precursor of the seventeenth-century scientist, Giordano Bruno holds a significant place, and his tragic death early in the first year of the new century must still arrest our attention as symbolic of a great turning point in human history.

BIBLIOGRAPHY

I. ORIGINAL WORKS. Bruno's Latin works are in *Opera latine,* Francisco Fiorentino, Vittorio Imbriani, C. M.

Tallarigo, Felice Tocco, and Girolamo Vitelli, eds., 3 vols. (Naples–Florence, 1879–1891), also in a facsimile reprint (Stuttgart–Bad Cannstatt, 1962). Latin works discovered and published since this edition are *Due dialoghi sconosciuti e due dialoghi noti,* Giovanni Aquilecchia, ed. (Rome, 1957); and *Praelectiones geometricae e ars deformationum,* Giovanni Aquilecchia, ed. (Rome, 1964). The Italian works are collected in *Dialoghi italiani,* Giovanni Gentile, ed., revised by Giovanni Aquilecchia (Florence, 1957), which contains all the Italian dialogues in one volume; one of the works, *La cena de le ceneri,* has been published separately with intro. and notes by Giovanni Aquilecchia (Turin, 1955).

Translations of Bruno's works include "Concerning the Cause, Principle, and One," Dorothea W. Singer, trans., in Sidney Greenberg, *The Infinite in Giordano Bruno* (New York, 1950), pp. 77 ff.; "On the Infinite Universe and Worlds," Dorothea W. Singer, trans., in her *Giordano Bruno, His Life and Thought* (New York, 1950), pp. 227 ff.; *Des fureurs héroïques,* Paul-Henri Michel, trans. (Paris, 1954); *The Expulsion of the Triumphant Beast,* Arthur D. Imerti, trans. (New Brunswick, New Jersey, 1964); *Giordano Bruno's "The Heroic Frenzies,"* Paul Eugene Memo, trans. (Chapel Hill, N.C., 1964).

II. SECONDARY LITERATURE. A bibliography of Bruno's works and of books and articles on him up to and including 1950 is Virgilio Salvestrini and Luigi Firpo, *Bibliografia di Giordano Bruno (1582–1950)* (Florence, 1958). Documentary sources on his life are Vincenzo Spampanato, ed., *Documenti della vita di Giordano Bruno* (Florence, 1933); and Angelo Mercati, ed., *Il sommario del processo di Giordano Bruno* (Vatican City, 1942). The standard biography is Vincenzo Spampanato, *Vita di Giordano Bruno* (Messina, 1921); on the trial, see Luigi Firpo, *Il processo di Giordano Bruno* (Naples, 1949).

The following brief selection from a vast literature includes books illustrative of the history of Bruno's reputation: Domenico Berti, *La vita di Giordano Bruno da Nola* (Florence, 1867); Felice Tocco, *Le opere latine di G. Bruno* (Florence, 1889), and *Le fonti più recenti del Bruno* (Rome, 1892); J. Lewis McIntyre, *Giordano Bruno* (London, 1903); Giovanni Gentile, *Giordano Bruno e il pensiero del Rinascimento* (Florence, 1920); Leo Olschki, *Giordano Bruno* (Halle, 1924), also translated into Italian (Bari, 1927); Ernst Cassirer, *Individuum und Kosmos in der Philosophie der Renaissance* (Berlin–Leipzig, 1927), also translated into English by Mario Domandi (New York, 1963); Antonio Corsano, *Il pensiero di Giordano Bruno* (Florence, 1940); Eugenio Garin, *La filosofia* (Milan, 1947); Walter Pagel, "Giordano Bruno: The Philosophy of Circles and the Circular Movement of the Blood," in *Journal of the History of Medicine and Allied Sciences,* **6** (1951), 116–125; Alexandre Koyré, *From the Closed World to the Infinite Universe* (Baltimore, 1957); Paolo Rossi, *Clavis universalis* (Milan, 1960), pp. 109–134; Paul-Henri Michel, *La cosmologie de Giordano Bruno* (Paris, 1962); Paul Oskar Kristeller, *Eight Philosophers of the Italian Renaissance* (Palo Alto, Calif., 1964), pp. 127–144.

This article is based on my books, *Giordano Bruno and the Hermetic Tradition* (Chicago, 1964), and *The Art of Memory* (Chicago, 1966). On Bruno, Gilbert, and Bacon, see my essay "The Hermetic Tradition in Renaissance Science," in *Art, Science, and History in the Renaissance,* Charles S. Singleton, ed. (Baltimore, 1968), pp. 255–274.

FRANCES A. YATES

BRUNSCHVICG, LÉON (*b.* Paris, France, 10 November 1869; *d.* Aix-les-Bains, France, 18 January 1944), *philosophy.*

Brunschvicg, of Alsatian origin, achieved a brilliant record at the Lycée Condorcet. There his fellow students included Marcel Proust; Célestin Bouglé, a future sociologist; Xavier Léon; and Élie Halévy, in whose home, around 1885, he met Victor Hugo, Leconte de Lisle, and Bizet. His professor of philosophy, A. Darlu, taught him technical precision and severe self-criticism, and gave him a living example of a sage. In 1888 Brunschvicg entered the École Normale Supérieure with Halévy and Bouglé. At the Sorbonne he took the courses taught by Victor Brochard and Émile Boutroux; it was with the latter's son Pierre that he became more deeply interested in mathematics. He soon obtained the *licence ès lettres,* and received the *licence ès sciences* in 1891.

With F. Gazier, Brunschvicg and Boutroux prepared a fourteen-volume edition of Pascal (1904–1914). Brunschvicg also founded, with Xavier Léon and Élie Halévy, the *Revue de métaphysique et de morale* (1893). On 30 August 1891 he passed his *agrégation* in philosophy and became a professor at the *lycée* of Lorient until 1893; he then taught at the *lycée* of Tours until 1895 and at that of Rouen until 1900. On 29 March 1897 he presented his doctoral theses: *Qua ratione Aristoteles metaphysicam vim syllogismo inesse demonstraverit* and *La modalité du jugement.* In the same year he published the *Pensées* and *Opuscules* of Pascal.

In 1899 Brunschvicg married Cécile Kahn, who was active in social work and served as undersecretary for education in 1936–1937. They had four children, the first of whom died young. Brunschvicg was made professor at the Lycée Condorcet in 1900, and in 1903 succeeded Bergson at the Première Supérieure of the Lycée Henri IV. Promoted to the Sorbonne in 1909, he taught there and at the École Normale Supérieure for thirty years. After World War I he was elected to the Académie des Sciences Morales et Politiques (1919), of which he became president in 1932. Also in 1919 he founded the Societas Spinozana and received an honorary doctorate from the University of Durham. During World War II, Nazi persecution forced Brunschvicg to leave Paris. He took refuge first at Aix-en-Provence, then in the departments of Gers and Gard, and finally at Aix-les-Bains, where he died.

When Brunschvicg's own doctrine was taking shape, between 1886 and 1896, a reaction was becoming apparent against the dominant influence of Taine and Renan, the eclecticism of Victor Cousin, and psychology like that of Théodule Ribot. Drawing interest were Félix Ravaisson, Charles Renouvier, Jules Lachelier, Émile Boutroux, and, among the classical philosophers, Plato—whom Brunschvicg always opposed to Aristotle and, consequently, to Scholasticism—and Descartes, whose *Géométrie* he preferred to his *Cogito*. Spinoza was his favorite; he also commented on Pascal, with whom, by his own admission, he shared not a single idea except in science; Kant, who was then little known; and even Fichte. On the other hand, Brunschvicg always refused to follow Schelling and Hegel in their *Naturphilosophie*. Thus his critical idealism was already heralded, with the support of the reflective method.

This idealism had already taken form in *La modalité du jugement*. Judgment is an action that defines the mind. Far from finding the concept already present, as an image or a quasi thing—in any event, as a datum—as is supposed by conceptualists, especially Aristotle, the mind creates the concept through syntheses that form the basis of analysis. At times judgment asserts an intrinsic relationship between ideas, and it must then be classified at the level of *necessity*, in the modality of *interiority*. At times it repeatedly asserts a being as an externality, and then it must be classified at the level of the *real*, in the modality of *exteriority*. Finally, at times, knowing that it does not produce sensation, judgment discovers that it alone can produce at least all intelligible reality, and it must then be classified at the level of the *possible* in a *mixed* modality. This means that in the immanence of the mind, the only valuable knowledge is that which unites the interiority of thought with the exteriority of experience. Going from theory to practice, judgment again turns to *necessity* when the activity of the mind includes the conditions of its satisfaction, to *reality* when it does not include them, and to *possibility* when it feels in harmony with the external world.

If judgment is an action, it can be known only by its work, and this implies a method of *historical verification*. In this and this only, Brunschvicg is linked to the Port-Royal *Logic*, to Bernard Fontenelle, and to positivism. Verification is necessary in order not to lose oneself in the verbalism of the a priori. History tests doctrines. It chooses, disengaging from confusion the primitive—the infantile, the adolescent,

in short, the irrational—which it eliminates, and the innovator, the progress of conscience, of which it draws the curve. As from the complexity of facts scientists determine a law, both real and ideal, so Brunschvicg's historico-reflective method endeavors to unsnarl from the tangle of factual history a normative history in which intelligence never ceases to prove the legitimacy of its victory over all empiricism.

Of all the mind's works, is not science, especially mathematics, the best expression of its rationality? Consequently, Brunschvicg began by following the developments of mathematical philosophy in *Les étapes de la philosophie mathématique*. This was not the history of mathematics, but the underlining of its essential innovations: (1) Pythagorean dogmatism, which was shattered by the discovery of irrationals; (2) with Plato, the consciousness of operative dynamism; (3) with Aristotle, the appearance of formal logic; (4) the sinking of mathematics in the Middle Ages into syllogistic deduction; (5) the renaissance of Platonism with Descartes and the invention of analytical geometry; (6) the crisis brought about by infinitesimal calculus; (7) the revolution brought on by non-Euclidean geometries, symbolic logic, renewed intuitionism, and relativity. With what result? According to a contemporary mathematician, André Lichnérowicz, in *Les étapes*, which is probably the last book to treat of "mathematical philosophy," Brunschvicg foresaw the resolutely nonontological orientation and the unification—through the study of algebraic-topological structures—of today's mathematics.

L'expérience humaine et la causalité physique rediscovers, throughout the history of the accepted concepts of nature, the stages corresponding to those of mathematical philosophy, from the most primitive to the most recent, proceeding through Platonism, Aristotelianism, Scholasticism, Cartesian mechanism, and Hegelianism. "There is only one Universe" should be the only correct statement concerning causality for anyone who would defend the value of rational experience against the scorn of empiricism; the constantly unforeseen progress, always free and yet always linked, of physics. Once more the thesis of critical idealism is confirmed.

The lesson drawn from the history of science in *Les étapes* and *L'expérience humaine*, and from the studies on Pascal, Spinoza, and Descartes, was completed by the history of philosophy in Brunschvicg's third masterwork, *Le progrès de la conscience dans la philosophie occidentale*. Raymond Aron has shown how Brunschvicg, by depending on science without falling into positivism, by turning to historical development without losing timelessness, reached an ethics of creative man, free and rational—man at his highest, "equal to his own idea of himself."

Brunschvicg's influence is easily recognized in all types of thinkers: such moralists as René Le Senne and Georges Bastide, such aestheticians as Joseph Segond and Valentin Feldmann, and particularly such epistemologists as Gaston Bachelard, Robert Blanché, Jean Cavaillès, Alexandre Koyré, and Albert Lautmann.

BIBLIOGRAPHY

I. ORIGINAL WORKS. Brunschvicg's books are *La modalité du jugement* (Paris, 1897), 3rd ed., enl., entitled *La vertu métaphysique du syllogisme selon Aristote* (Paris, 1964); *Introduction à la vie de l'esprit* (Paris, 1900); *L'idéalisme contemporain* (Paris, 1905); *Les étapes de la philosophie mathématique* (Paris, 1912); *Nature et liberté* (Paris, 1921); *L'expérience humaine et la causalité physique* (Paris, 1922); *La génie de Pascal* (Paris, 1924); *Spinoza et ses contemporains* (Paris, 1924); *Le progrès de la conscience dans la philosophie occidentale* (Paris, 1927); *De la connaissance de soi* (Paris, 1931); *Pascal* (Paris, 1932); *Les âges de l'intelligence* (Paris, 1934); *La raison et la religion* (Paris, 1939); *Descartes et Pascal, lecteurs de Montaigne* (Neuchâtel, 1942); *Héritage des mots, héritage des idées* (Paris, 1945); *L'esprit européen* (Neuchâtel, 1947); *Agenda retrouvé, 1892–1942* (Paris, 1948); and *La philosophie de l'esprit* (Paris, 1950). His articles were collected as *Écrits philosophiques*, 3 vols. (Paris, 1949–1958).

II. SECONDARY LITERATURE. Works on Brunschvicg include Marcel Deschoux, *La philosophie de Léon Brunschvicg* (Paris, 1949); Bernard Elevitch, in *Encyclopedia of Philosophy* (New York, 1967), I, 408–409; Martial Gueroult, "Brunschvicg et l'histoire de la philosophie," in *Bulletin de la Société française de philosophie,* **48,** no. 1 (1954); and D. Parodi, *La philosophie contemporaine en France* (Paris, 1925), pp. 420–424, 427–431. Also of value are a special number of *Revue de métaphysique et de morale,* **55,** no. 1–2 (1945); and a commemoration of the fiftieth anniversary of the publication of *Les étapes* by A. Lichnérowicz, A. Koyré, R. P. Dubarle, and J. Wahl, *Bulletin de la Société française de philosophie,* **57,** no. 2 (1963).

YVON BELAVAL

BRUNSCHWIG (also **Brunswyck** or **Braunschweig**), **HIERONYMUS** (*b.* Strasbourg, France, *ca.* 1450; *d.* Strasbourg, *ca.* 1512), *surgery.*

After receiving an education in surgery, Brunschwig traveled extensively through Alsace, Swabia, Bavaria, Franconia, and the Rhineland as far as Cologne, practicing surgery and acquiring experience in the preparation of medicines, specifically in the technique of distillation. It is frequently stated that he studied

medicine in Bologna, Padua, and Paris, but this assertion cannot be verified. Brunschwig himself never mentions this training, which at that time was unusual for a surgeon, and he clearly differentiates his sphere of activity from that of the physicians.

Brunschwig probably did not become a military surgeon like his contemporary Hans von Gersdorff; and his often-mentioned participation in the campaign against Charles the Bold and in the battle of Morat in 1476 is doubtful. His reports on these engagements, as well as on battle wounds, are largely secondhand. His employment as a surgeon in Strasbourg apparently left him enough time to become a writer and to continue his travels.

Brunschwig's works concern anatomy, treatment of wounds, and, in pharmacy, the preparation of medicines and simples. Written in German, they are directed primarily to barber-surgeons, barbers, and surgeons; his surgical texts, however, are also directed to laymen. They reveal an intense preoccupation with the medical tradition of Lanfranchi, Guglielmo Saliceti, Guy de Chauliac, Henri de Mondeville, and, through them, the tradition of Galen, Avicenna, Rhazes, Mesue, and Abul Kasim, all of whom Brunschwig cites as sources of his own knowledge. On the basis of his own experience, however, Brunschwig was also able to criticize their work. With its traditional knowledge, critical spirit, and careful citations, his *Cirurgia* (1497) is different from the books of his German-speaking forerunners, such as the *Buch der Bündth-Ertznei* (*ca.* 1460) by the empiricist Heinrich von Pfalzpaint (or Pfolspeunt) and the *Chirurgie* (1481) of Johann Schenk of Würzburg. In the treatment of wounds, fractures, and luxations, and in trepanation and amputation, he made extensive use of traditional methods.

Brunschwig's *Cirurgia,* which has become an important cultural-historical source for medicine and pharmacy because of its excellent illustrations, represents a substantial step forward in the German surgery of that time, which lagged behind that of Italy and France. On the other hand, his *Anathomia* (1497), which shows some familiarity with dissection, had little lasting effect; and his discussion of pestilence (1500), which contains an early description of syphilis, had no lasting effect at all. The *Liber de arte distillandi, de simplicibus* (1500) reveals greater originality; primarily because of the description, complemented by abundant illustrations, of chemical and distillation apparatus, this book became a pharmaceutical–technical handbook that was the authority far into the sixteenth century. Appended to it was a compilation of illnesses "a capite ad calcem," along with a list of vegetable distillates indicated for each case of illness. The *Liber de arte distillandi, de compositis* (1507) contains, among other things, a "Thesaurus pauperum" that—especially as it appeared in the 1512 edition—was often reprinted and became a model for later pharmacopeias for poor people.

Because of their completeness Brunschwig's compilations of the technical terms adaptable to pharmacy in the early sixteenth century and his records of his experience in the treatment of gunshot wounds and in surgery are noteworthy accomplishments. Even if they are not the first of their kind, they still represent an important link between the Middle Ages and modern times.

BIBLIOGRAPHY

I. ORIGINAL WORKS. Brunschwig's works are *Anathomia ossium corporis humani* (Strasbourg, 1497); *Buch der Cirurgia, Hantwirckung der Wundartzney* (Strasbourg, 1497), also trans. into English (London, 1525); *Liber de arte distillandi, de simplicibus. Das Buch der rechten Kunst zu distillieren die eintzigen Ding* (Strasbourg, 1500), also issued as *Medicinarius,* with a section entitled "De compositis" (Strasbourg, 1505); and *Liber pestilentialis de venenis epidemie. Das Buch der Vergift der Pestilentz* (Strasbourg, 1500). The *Medicinarius* was reissued as *Liber de arte distillandi, de compositis. Das Buch der waren Kunst zu distillieren die Composita* (Strasbourg, 1507) had a section entitled "Thesaurus pauperum," also published separately as *Apoteck für den gemainen Man* (Strasbourg, 1507). The works on distillation were reissued under the title *Grosses Buch der Destillation* (Strasbourg, 1512).

II. SECONDARY LITERATURE. Works on Brunschwig are A. Brunschwig, "Hieronymus Brunschwig," in *Annals of Medical History,* n.s. **1** (1929), 640–644; Gerhard Eis, "H. Brunschwig," in *Neue deutsche Biographie,* II (1965), 688; H. W. Grabert, "Nomina anatomica bei den deutschen Wundärzten Hieronymus Brunschwig und Hans von Gersdorff," dissertation (Leipzig, 1943); F. Hommel, "H. Brunschwig," in *Archiv für Geschichte der Mathematik, der Naturwissenschaften und der Technik,* **10** (1927), 155–157; G. Klein, Facsimile of the *Cirurgia* with introduction (Munich, 1911); H. E. Sigerist, Facsimile of the *Cirurgia,* with study on Hieronymus Brunschwig (Milan, 1923); J. Steudel, "Brunschwigs Anatomie," in *Grenzgebiete der Medizin,* **1** (1948), 249 f.; K. Sudhoff, "Brunschwigs Anatomie," in *Archiv für Geschichte der Medizin,* **1** (1907), 41–66, with a facsimile of the *Anathomia, ibid.,* 141–156; and *Deutsche medizinische Inkunabeln,* Vols. II and III in the series Studien zur Geschichte der Medizin (Leipzig, 1908); and F. Wieger, *Geschichte der Medizin und ihrer Lehranstalten in Strassburg vom Jahre 1497 bis zum Jahre 1872* (Strasbourg, 1885), pp. 4–15.

R. SCHMITZ

BRUNTON, THOMAS LAUDER (*b.* Roxburgh, Scotland, 14 March 1844; *d.* London, England, 16 September 1916), *physiology, pharmacology.*

Brunton received his formal scientific education at the University of Edinburgh (B.Sc., 1867; M.D., 1868; D.Sc., 1870) and spent two years in Continental laboratories, including Ludwig's in Leipzig and Kühne's in Amsterdam. In 1870 he settled in London, and the next year he was appointed lecturer in materia medica and casualty physician to St. Bartholomew's Hospital. He became assistant physician there in 1875, physician in 1897, and consulting physician in 1904. Brunton was made a fellow of the Royal Society in 1874 and of the Royal College of Physicians in 1876, was knighted in 1900, and was awarded a baronetcy in 1909.

Brunton's lifelong research interests centered on the physiology and therapy of the cardiovascular and digestive systems. In his prize-winning thesis of 1866 he investigated the pharmacological properties of digitalis, a subject to which he frequently returned. He also studied the cardiotonic actions of casca bark (*Erythrophloeum guineense,* 1880). In 1905 Brunton delivered a series of lectures at the University of London, later published as *Therapeutics of the Circulation* (1908), in which he took the many facets of his own work in the field and placed them in the context of contemporary knowledge.

While working as a resident physician in 1867, Brunton discovered that amyl nitrite, a drug already studied experimentally by Frederick Guthrie, Benjamin Richardson, and Arthur Gamgee, was useful in the relief of angina pectoris. Brunton's physiological approach to therapeutics is nowhere better illustrated than in this episode. His measurements of blood pressure led him to believe that angina is caused by transient bouts of hypertension and that amyl nitrite, the first known vasodilator, should therefore be efficacious. That angina was subsequently found to be caused by ischemia rather than hypertension did not invalidate Brunton's reasoning; the drug is still used in the treatment of this condition.

Brunton's early research on digestive physiology may be found in Burdon-Sanderson's *Handbook for the Physiological Laboratory* (1873). The section on digestion, written by Brunton, contains literally hundreds of experiments, each performed or verified by Brunton himself. His scattered papers on the digestive tract were collected in two later volumes (1886, 1901).

His work in digestive physiology led Brunton to consider diabetes, a condition that he tried to treat by the oral administration of raw muscle, in the hope that the glycolytic properties of the muscle would correct the faulty carbohydrate metabolism (1874). His concept of "organotherapy" was vindicated within fifteen years by the successful treatment of myxedema with thyroid extract.

In addition to his original contributions, Brunton played an important role in the development of pharmacology into an independent and rigorous science. In his Goulstonian lectures of 1877 (*Pharmacology and Therapeutics*) he surveyed the history and contemporary state of pharmacological research. His Croonian lectures of 1889 (*An Introduction to Modern Therapeutics*) dealt with molecular pharmacology in discussing the relation of chemical structure to physiological action. *A Textbook of Pharmacology, Therapeutics and Materia Medica* (1885), the first comprehensive treatise on pharmacology, remains his most important work. In it Brunton abandoned the traditional discussions of classic materia medica and emphasized the physiological actions of pure drugs. Immediately accepted as authoritative, the book was translated into French, German, Italian, and Spanish. Brunton's *Lectures on the Actions of Medicines* (1897), using the same basic structure, went through three editions in as many years.

Brunton was known to his contemporaries as a therapeutic activist. He envisioned an almost unlimited ability of scientific pharmacology to eradicate and prevent disease, stressing at the same time other modes of prophylaxis and therapy, such as massage, baths, exercise, and improved public health services.

BIBLIOGRAPHY

I. ORIGINAL WORKS. Brunton's writings include "On the Use of Nitrite of Amyl in Angina Pectoris," in *Lancet* (1867), **2**, 97–98; *On Digitalis* (London, 1868); the section on digestion in Burdon-Sanderson, *Handbook for the Physiological Laboratory,* I (London, 1873); "The Pathology and Treatment of Diabetes Mellitus," in *British Medical Journal* (1874), **1**, 1–3; *Pharmacology and Therapeutics* (London, 1880); *The Bible and Science* (London, 1881); *A Textbook of Pharmacology, Therapeutics and Materia Medica* (London, 1885); *On Disorders of Digestion* (London, 1886); *An Introduction to Modern Therapeutics* (London, 1892); *Modern Developments of Harvey's Work* (London, 1894), the Harveian oration for 1894; *Lectures on the Actions of Medicines* (1897); and *On Disorders of Assimilation, Digestion, etc.* (London, 1901). Brunton also edited the 3rd ed. of Murchison's *Clinical Lectures on Diseases of the Liver* (London, 1885).

II. SECONDARY LITERATURE. Additional biographical material may be found in the following obituary notices: *British Medical Journal* (1916), **2**, 440–442; *Lancet* (1916), **2**, 572–575; and *Proceedings of the Royal Society,* **89B** (1917), 44–48. For consideration of two relatively minor aspects of Brunton's work, see Fielding H. Garrison, "Sir Thomas Lauder Brunton, M.D. (1844–1916). An Apostle of Preparedness," in *The Military Surgeon,* **40** (1917), 369–377; H. Meade, *A History of Thoracic Surgery* (Springfield, Ill., 1961), pp. 430–458.

WILLIAM F. BYNUM

BRYAN, KIRK (*b.* Albuquerque, New Mexico, 22 July 1888; *d.* Cody, Wyoming, 21 August 1950), *geology, geomorphology.*

Bryan's parents taught in the Presbyterian Indian School in Albuquerque, and his father later practiced law in the same town. He was educated in the Albuquerque public schools and received the B.A. from the University of New Mexico in 1909. Then, with the encouragement of H. E. Gregory, professor of geology at Yale, Bryan entered that college as an undergraduate, receiving his second B.A. in 1910 and the Ph.D. in 1920.

From 1912 to 1926 Bryan was associated with the U.S. Geological Survey. During part of this time, 1914–1917, he also served as instructor in geology at Yale, and in 1918–1919 he was a private in the Army Corps of Engineers. Later he received a commission and served as a second lieutenant in the geological section of the army general headquarters. From 1926 until his death, Bryan taught at Harvard. He belonged to the American Association for the Advancement of Science, the Geological Society of America, the Geological Society of Washington, the Geological Society of Boston, the American Academy of Arts and Sciences, and the Society of Military Geologists, holding office in most of these societies. He was married in 1923 to Mary MacArthur.

The Kirk Bryan Fund, established by the Geological Society of America in 1951, supports the Kirk Bryan Award, which is presented, usually annually, for a significant published contribution to geomorphology or a related field.

Bryan's professional career began in the Geological Survey, where he was assigned to work on the general problem of water resources, particularly of groundwater in arid and semiarid regions. This work was directly predicated on the assumption that it had an immediate human use, and this "human" orientation played a significant role in his professional studies for the rest of his life. Bryan's work for the Geological Survey produced a series of publications reflecting his concern with irrigation, dam sites, and groundwater resources. His subsequent writings deal more with man's history, an interest reflected in his collaboration with archaeologists. Throughout, however, he was equally concerned with the processes of earth change and with geological history.

Bryan's first major contribution was the description of arid-climate landforms and discussion of the processes that led to them (1922); the fieldwork and basic observation had been carried out in the fall and early winter of 1917. The study "Erosion and Sedimentation in the Papago Country, Arizona" was a by-product of a survey of watering places in that desert area. In this work the use of the term "pediment" to describe a plain formed at the foot of mountains through the processes of erosion normal to the desert was firmly established in the literature of landscape evolution, although Bryan adapted the term from a 1912 report on desert surfaces by Sidney Paige. Bryan envisaged the pediment as a slope of transportation veneered with a few inches or a few feet of debris in transit. He indicated that the slope of the pediment surface changed with the size of particle in transport. He planned and, with the help of several graduate students, partially completed a field program to study the geology and geomorphology of the Rio Grande depression in New Mexico and southern Colorado. Among other things, this program focused on the history and origin of pediments.

Bryan's recognition of the pediment as an important topographic element in an arid landscape reflected his own geographic heritage. He was raised in the semiarid New Mexico Territory and spent the first fifteen years of his professional career in arid and semiarid lands, so it is not surprising that he was sensitive to the effects and changes of climate. In his initial study of pediments (1922) Bryan postulates the possibility that an increase in rainfall accounted for the dissection of many of the pediments in the Papago country. In that same study he outlined what came to be known as the "alluvial chronology" of late Pleistocene and Recent time in the southwest. The alternate periods of alluviation and erosion that he described were attributed to fluctuations of climate marked particularly by changes in effectiveness of precipitation. The assumption that climatic changes have taken place within the immediate geological past demanded the evidence for those changes. This Bryan sought in the soils, recent sediments, and landscape. In 1943 he and C. C. Albritton, Jr., published "Soil Phenomena as Evidence of Climatic Change," a paper that summarizes his interest in soils as climatic indicators and marks the beginning of a general interest in paleosoils on the part of geologists.

In 1932 Bryan wrote on the use of pollen in the reconstruction of North American Pleistocene climate. The technique had been used for many years in Europe but was little known in America; his continued interest was instrumental in establishing palynology in the United States. Likewise, Bryan

early recognized that the rigorous climate existing beyond the margins of the Pleistocene glaciers must have expressed itself by extensive frost action in the upper few feet of soil and rock. The effect of modern periglacial climate, as well as that of the past, was being reported by European workers in the 1920's. It was Bryan who drew the attention of American workers to the possibility of using frost forms as stratigraphic and environmental indicators. In 1946 he published a summary statement of the nature of the frost action on soils and coined the term "cryopedology" for its study.

In 1924 and 1925, Bryan participated in an archaeological expedition of the Smithsonian Institution to the Chaco Canyon in New Mexico. His goal was to determine, if possible, the conditions that led to the rise and fall of prehistoric Pueblo Bonito. The field study was the beginning of a long association between Bryan and archaeologists, although the report was not published until 1954. In 1940 the "Geologic Antiquity of the Lindenmeier Site" in Colorado was published in collaboration with L. L. Ray. It and the report on Sandia Cave, New Mexico (1941), stand as models of the application of geological studies to archaeological sites and problems.

BIBLIOGRAPHY

I. ORIGINAL WORKS. See "Erosion and Sedimentation in the Papago Country, Arizona with a Sketch of the Geology," in *Bulletin of the United States Geological Survey,* no. 730 (1922), pp. 19–90. A list of publications is in Esper S. Larsen, Jr., "Memorial to Kirk Bryan," in *Proceedings of the Geological Society of America, Annual Report for 1950* (1951), pp. 91–96. Posthumous publications and some minor earlier papers are not listed.

II. SECONDARY LITERATURE. Brief accounts of Bryan's life are given by Larsen (see above) and Frederick Johnson, "Kirk Bryan—1888–1950," in *American Antiquity,* **16** (1951), 253.

SHELDON JUDSON

BRYSON OF HERACLEA, *mathematics.*

The name Bryson occurs several times in Aristotle's writings. A "sophist" Bryson is mentioned as the son of Herodorus of Heraclea (*Historia animalium* VI, 5, 563a7; IX, 11, 615a10); is blamed for his "sophistic" (in fact, probably proto-Stoic) assertion that there is no such thing as "indecent" language (*Rhetoric* III, 2, 1405b9); and is blamed for his "eristic" and "sophistic" method of squaring the circle (*Posterior Analytics* I, 9, 75b40; *De sophisticis elenchis* 11, 171b16–172a4). A Bryson is named in Plato's *Epistles* XIII, where Polyxenus, teacher of Helicon, is designated as his ἑταῖρος. Theopompus, in Athen XI, 508d (*Fragmenta graecorum historicorum,* F. Jacoby, ed., 115F259), accuses Plato of having plagiarized Bryson's *Diatribes*. It seems preferable (in agreement with Natorp rather than Zeller) to assume that all the above passages refer to one and the same person, although it must be admitted that other biographical information preserved in Diogenes Laertius and Suda, particularly concerning his relations with Socrates, Euclid of Megara, Cleinomachus, Plato, Pyrrho, and Theodorus the Atheist, contain some chronological contradictions.

With the help of ancient commentators Bryson's method of squaring the circle, criticized by Aristotle, can tentatively be reconstructed as follows. We start with two squares, one inscribed in a circle, the other circumscribed. Then we construct successively regular circumscribed and inscribed polygons the perimeters of which approach as closely as we like to the circumference of the circle. Thus, exhausting the area by which one square is larger and the other is smaller than the circle, we eventually make the areas of the larger and the smaller polygon coincide, which means that their areas at that time equal the area of the circle, since the polygon and the circle are both smaller than the circumscribed polygons and greater than all the inscribed figures. Since we can always construct a square equal in area to a polygon, the problem of how to square the circle is solved.

Aristotle criticizes this method because it is based on a principle that is too general and not peculiar to the matter at hand. By this he probably means a principle like this: If two quantities, one larger and one smaller than a third, become equal, they also become equal to that third quantity. It seems that Bryson's method can, however, be defended against Aristotle and can be considered another anticipation of the principles underlying the method of exhaustion.

No other opinions of Bryson have survived. But since his name is linked with that of Polyxenus, who has been credited with having been the first to put forward the so-called third-man argument, and since Plato was accused of plagiarizing Bryson, it could be that there were similarities between Bryson's *Diatribes* and Plato's *Parmenides,* in which the third-man argument is indeed discussed.

BIBLIOGRAPHY

In addition to the sources listed in the text, see John Philoponus, *In Analytica posteriora,* XIII, pt. 3 of *Commentaria in Aristotelem Graeca* (1909), 111–115, 149; Pseudo-Alexander of Aphrodisias, *In Sophisticos elenchos,* III, pt. 3 of *Commentaria . . .* (1898), 76, 90, 92; and Themistius, *Analytica posteriora paraphrasis,* V, pt. 1 of *Commentaria . . .* (1900), 19–20.

Literature that deals with Bryson includes T. L. Heath, *A History of Greek Mathematics,* I (Oxford, 1921), 223–225; Paul Natorp, "Bryson 2," in Pauly-Wissowa, *Real-Encyclopädie,* III, pt. 1 (1897); W. D. Ross, *Aristotle's Prior and Posterior Analytics* (Oxford, 1949), 536 f.; and E. Zeller, *Die Philosophie der Griechen,* 5th ed., II, pt. 1 (1922), esp. 250, n. 4.

A different interpretation of Bryson's attempt to square the circle, according to which he proposed establishing some kind of proportion between the circumscribed and the inscribed circles as extremes, the middle term being the square in question, is in A. Wasserstein, "Some Early Greek Attempts to Square the Circle," in *Phronesis,* 4 (1959), 92–100, esp. 95–100.

PHILIP MERLAN

BRYTTE (also **Britte, Brit,** or **Brute**), **WALTER** (*fl.* Oxford, England, second half of fourteenth century), *astronomy.*

According to Brodrick,[1] Brytte was elected to a fellowship at Merton College in 1377, and in MS Digby 15 (fol. 96v) he is mentioned as *quondam socius collegii de Merton*. This is almost all we know about his life, except for an old tradition related by Wood[2] that he was a follower of John Wycliffe (Bachelor of Merton 1356) and author of the book *De auferendis clero possessionibus*. For this reason R. L. Poole[3] and later A. B. Emden[4] tentatively identified him with a layman who in 1391 was tried for heresy before the bishop of Hereford. Nothing has come to light to substantiate this, however, so we are left with the only work that can safely be ascribed to him: the treatise *Theorica planetarum secundum dominum Walterum Brytte,* with the incipit *Circulus eccentricus, et egresse cuspidis, et ingredientis centri idem sunt,* which is extant in at least eight manuscripts, four of which were identified by Bjørnbo.[5]

Many elementary manuals of planetary theory are known under the title *Theorica planetarum,* and Brytte's treatise has often been confused not only with a similar text by an earlier Mertonian scholar, Simon Bredon (fellow 1330, *d.* 1372), but also with the extremely popular and much used "Old *Theorica planetarum,*" which has been ascribed to Gerard of Cremona and to many other authors but is actually an anonymous textbook written in the second half of the thirteenth century. Recent research has cleared up much of the confusion, with the result[6] that although the "Old *Theorica*" certainly served as a prototype for Brytte, his work is neither a simple copy among several hundred others nor a commentary of the usual kind. It must be regarded as a revised version with strongly individual features. Of course, Brytte retained many of the notions current at his time, such as the "physical" doctrine of ethereal spheres as guidance mechanisms for the planets. On the other hand, he rearranged the traditional matter in a more logical way, dealing more fully with the theory of Venus and discarding the very confused chapter on latitudes in the "Old *Theorica*."

The most interesting characteristic of the *Theorica* is Brytte's obvious efforts to remedy some of the worst errors of the older text by means of his insight into kinematics, which clearly stemmed from the great Merton school of mechanics that flourished before and at the middle of the century. Where the "Old *Theorica*" spoke simply of "motion," Brytte carefully distinguishes between "physical" (i.e., linear) velocity and "astronomical" (i.e., angular) velocity, just as he is familiar with both uniform and nonuniform motion and with the composition of angular velocities. These kinematic concepts are applied to planetary theory, with the consequence that Brytte, unlike the author of the old text, is able to state the correct condition for a planet's being stationary: that the apparent angular velocity of the planet on the epicycle is equal but opposite to the angular velocity of the epicycle center on the deferent. Yet he is unable to deduce a correct geometrical construction from this principle, so that like most medieval astronomers, he determines the stationary points by means of tangents from the center of the earth to the epicycle. Thus his efforts to make the new kinematics useful to astronomy did not result in a correction of this persistent error in the elementary teaching of astronomy in the Middle Ages.[7]

NOTES

1. P. 219.
2. I, 475.
3. II, 1266.
4. Pp. 270–271.
5. Pp. 112 ff.
6. Pedersen, "The *Theorica planetarum* Literature," pp. 225 ff.
7. An ed. of Brytte's treatise by Pedersen is in press.

BIBLIOGRAPHY

The MSS of the *Theorica planetarum* are British Museum Egerton 847, 104v–122v, and Egerton 889, 7r–17r; and Bodleian Library Bodl. 300, 45r–53v; Digby 15, 58v–96v; Digby 48, 96r–112v; Digby 93, 37r–51v; Digby 98, 132r–145r; and Wood D.8, 93–112r. An edition of the *Theorica* is Pedersen, *Theorica Planetarum. Texts and Studies in Mediaeval Astronomy* (in press).

Works dealing with Brytte or the *Theorica* are A. A.

Bjørnbo, "Walter Brytte's *Theorica planetarum,*" in *Biblioteca mathematica,* **6** (1905), 112 ff.; G. H. Brodrick, *Memorials of Merton College* (Oxford, 1885), p. 219; A. B. Emden, *A Biographical Register of the University of Oxford to A. D. 1500* (Oxford, 1957), I, 270 f.; Olaf Pedersen, "The *Theorica planetarum* Literature of the Middle Ages," in *Classica et mediaevalia,* **23** (1962), 225 ff.; R. L. Poole, in *Dictionary of National Biography,* II, 1266; and Anthony à Wood, *Antiquities of Oxford* (Oxford, 1786), I, 475.

OLAF PEDERSEN

BUACHE, PHILIPPE (*b.* Paris, France, 7 February 1700; *d.* Paris, 24 January 1773), *cartography*.

Buache studied with and continued the work of Guillaume Delisle. Initially charged with classifying the maps, plans, and journals in the Naval Archives, he was appointed chief royal geographer in 1729. The next year the Royal Academy of Sciences elected him assistant geographer, a post created for him and held by him until his death. In 1755 Buache was appointed geography tutor to the children of the duke of Burgundy. For their use he had a globe built and atlases compiled; these items are preserved in the Bibliothèque Nationale.

Buache's *Essai de géographie physique . . .,* presented to the Academy in 1752 and published soon afterward, contains most of his ideas as well as new methods, some of which later exerted great influence. He abandoned descriptive work and began theoretical study of the structure of the globe, of which, he believed, the mountain chains are the "bones." At the same time Buache brought out the *Carte physique et profil du canal de la Manche,* in which he showed the underwater configurations by means of contour lines designating ten-fathom intervals.

Buache's theory of submarine basins and "backbones," which was based on the *Carte,* became well known. However, Dainville has pointed out the prior use of such hydrographic contour lines by the Dutch engineer Nicolas Cruquius, who in 1729 made quite an accurate map of the underwater contours of the mouth of the Meuse. Buache must have known of this map through his work in the Naval Archives. Whatever the case may be, after Buache hydrographers perfected the presentation of underwater contours; the technique was later applied to land contours.

In *Considérations . . . sur les nouvelles découvertes au nord de la Grande Mer* (1753) Buache went so far as to posit the existence of Alaska and a connection between America and Asia—"because of the direction of the capes, the mountains, the rivers, and the glaciers."

Although his theories on river basins, basins on the ocean floor, and such were at times overgeneralized, and thus hindered the progress of geography, Buache's work as a whole constituted a definite contribution to cartography.

BIBLIOGRAPHY

I. ORIGINAL WORKS. Buache's works are *Parallèle des fleuves des quatre parties du monde* (Paris, 1751); *Essai de géographie physique* . . . (Paris, 1752); *Carte physique et profil du canal de la Manche* (Paris, 1752); *Considérations géographiques et physiques sur les nouvelles découvertes au nord de la Grande Mer*, 3 vols. (Paris, 1753); *Mémoire sur les différentes idées qu'on a eues de la traversée de la Mer glaciale arctique et sur les communications ou jonctions qu'on a supposées entre diverses rivières* (Paris, 1754); *Cartes et tables de la géographie physique ou naturelle* (Paris, 1754); and *Mémoire sur le comète qui a été observé en 1531, 1607, 1682* . . . (Paris, 1757). Available in MS is "Mémoires et notes sur les tremblements de terre," Bibliothèque Nationale, Paris, Dept. des Manuscrits, f. fr. n. acq. 20236–20237.

II. SECONDARY LITERATURE. Works on Buache are H. Balmer, *Beiträge zur Geschichte der Kenntnis des Erdmagnetismus* (Aarau, Switzerland, 1956); F. de Dainville, "De la profondeur à l'altitude. Des origines marines de l'expression cartographique du relief terrestre par cotes et courbes de niveau," in Michel Mollat, ed., *Le navire et l'économie maritime du moyen-âge au XVIIIe siècle principalement en Méditerranée* (Paris, 1958), pp. 195–213; L. Drapeyron, *Les origines de la reforme de l'enseignement géographique en France. Les deux Buache* (Paris, 1888); Grandjean de Fouchy, *Histoire de l'Académie des sciences*, II (Paris, 1772), 135; F. Hoeffer, in *Nouvelle biographie générale*, VII (Paris, 1855), cols. 676–678; F. Marouis, in *Dictionnaire de biographie française*, VII (1956), cols. 591–592; and Poggendorff, I, 323–324.

JULIETTE TATON

BUCH, [CHRISTIAN] LEOPOLD VON (*b.* Stolpe, Germany, 25 April 1774; *d.* Berlin, Germany, 4 March 1853), *geology.*

Buch, one of thirteen children, was born on his family's estate, about ninety kilometers from Berlin. His father, a *Geheimer Legationsrath* in the Prussian civil service, spent his leisure time in composing a history of Brandenburg, where the Buchs had been known since the twelfth century. His mother was a member of the famous Arnim family. The family fortune must have been considerable, since Buch's scientific zeal and passion for travel were never seriously hampered by a need to work for a living. Indeed, throughout his life he wandered extensively

and almost at will, much in the manner of the medieval traveling scholar.

When he was fifteen years old, Buch was sent to Berlin to study mineralogy and chemistry for a semester; these studies were a requirement for admission to the Bergakademie, the school of mining at Freiberg. Buch spent three years at the Bergakademie under the guidance of A. G. Werner, who taught lithology and geognosy, and was the chief exponent of the Neptunist theory of the origin of rocks. Alexander von Humboldt was a fellow student, and he and Buch became lifelong friends. After another year's training in law and government (and incidentally mineralogy) at the universities of Halle and Göttingen, in 1796 Buch entered the Prussian civil service as an inspector of mines, and was commissioned to make a geognostic survey of Silesia.

Although Buch performed some early chemical experiments, like most of Werner's followers he disliked chemistry, whose "despotic aid was in no way requisite for the independence of mineralogy." Chemistry could at most be of practical value, and the furnace only provided information on the commercial value of ores. The real scientific interest of minerals lay in their intrinsic substance or essence, with which true mineralogy ought to concern itself. Buch stated that, in accordance with what he had learned from Werner, the specific gravity, hardness, cleavage, color, and luster of a mineral belong to the mineral itself, whereas chemical reaction was no more than an external activity (although he did use muriatic acid to distinguish between calcite and dolomite).

A good exposition of Werner's natural system was given by D. L. G. Karsten, another of Werner's students, who arranged a well-known private mineral collection and provided a two-volume *catalogue raisonné* for it. Both collection and catalog were bought by the Irish Academy of Science, whose president, Richard Kirwan, exclaimed, according to Buch, that only by studying both had he come to understand true, and hence Wernerian, mineralogy, whereupon he abandoned his chemical pursuits.

Chronology was of little importance for the Wernerians. Their willful neglect of this fundamental aspect of geology is partly explained by the tenets of *Naturphilosophie*—although few geologists actually studied Schelling and Hegel, whose philosophy may be considered an attempt to state explicitly what was implied in the thought of many during the Enlightenment. What Hegel has to say about geology is particularly revealing. He derides the simple conclusion that of two superimposed layers, the higher must be of later formation. It is, he says, as if someone looking at a house concludes with profound wisdom that the ground floor was built first, then the next floor above it, and so on. Hegel goes on to say:

> But this sequence contains something more profound. The meaning and the spirit of the process is the intrinsic connection, the necessary relation of these formations whereunto the "after-each-other" adds nothing at all. The general law of this sequence of formations has to be understood, without need for any form of history; this is the essential. . . . It is Werner's great merit, to have drawn attention to this sequence and to have looked at it throughout with correct eyes. The internal connection exists in the present as a "side-by-side" and this connection must depend upon the quality, upon the content of these things themselves; the history of the earth is thus on the one side empirical, on the other side concluded from empirical data. To determine how things were millions of years ago (and there we may be free with years) is not interesting, the real interest being confined to that which is now in existence—to this system of different formations [*Encyklopädie,* pt. 2, sec. 339, *Zusatz*].

Naturphilosophie here warns against finding out how the things that we see came into being, but encourages us to study the relationships of such things to see whether and how they are mutually consistent (whatever consistency may mean in the case of geological formations). These mysterious relationships had some part in Buch's unflagging interest in superposition; they led him to his conclusion, on seeing a formation of micaceous schist that was covered in one exposure by slate and in another by sandstone, that the schist could not in these two occurrences be the same, although he could find no perceptible difference.

Naturphilosophie was not wholly harmful in its influence on geology, however. The idea of plumbing essential relationships stimulated geologists in their task of classifying minerals. It must likewise have acted as a stimulus to field geology, holding out as it did a promise of revealing the portentous thoughts of the World-Spirit. Werner's students were so zealous, d'Aubuisson tells us, that they ran out into the mountains as soon as classes were over. Moreover, for practical purposes the skill to recognize and name rocks and ascertain their distribution in the terrain, coupled with the empirical knowledge of their association with ores, coal, and salt, led to excellent results.

The Neptunist system was in principle very rigid. The primitive rocks—granite, gneiss, and micaceous schist—were deposited in that order, followed by the Floetz formations, graywacke shale, and limestone. These were followed by alluvial deposits and the products of volcanoes. In practice, this theory was a bit more flexible; Werner modified it by adding a

transition formation that fell between the primitive rocks and the Floetz formations and that could consist of slate, marble, and so forth. He also allowed for intercalation of all kinds of rocks as accidental formations. Only the independent (*selbständige*) formations had to conform to the rule that, according to Werner, must hold everywhere on earth. He believed that this rule about the right order of succession was necessarily true and did not have to be verified outside Saxony. This is in line with the claim of *Naturphilosophie* that all significant features in nature could be deduced by logic alone.

The height at which a given formation occurred was of great importance, and on many of his trips Buch measured relevant heights. "With due precautions," he said, "I have hardly any trouble in carrying a mercury barometer with me."

In principle, only granite could form the highest mountains, being the first rock deposited before the universal ocean withdrew from its highest level. As the water level descended, other formations would be deposited in their proper order. What is most baffling today is the notion that mountains and valleys were at once formed in their present shapes and positions. Steep precipices of a granite mountain were ascribed to chaotic conditions in the primitive ocean. Gneiss could be deposited—like crystals in a beaker—on such a vertical wall, its streaky pattern thus standing on end.

This was the original and most rigid version of Werner's theory: many incongruous elements had to be brought in to account for what is actually observed. Buch showed great versatility and some originality in applying a startling number of odd and drastic devices. He was grievously unlucky in these guesses and rightly complained of frequent disappointments.

An awkwardness existed in the trees that were found in some coal mines, standing upright in the position in which they had grown. Since this could not have happened under water, the Neptunists were forced to postulate a previous withdrawal of the ocean from all altitudes above those of the mines, and then to hypothesize a return of the water to explain younger formations covering high mountains. They supposed that large caverns in the earth could in turn drain and fill the oceans. Buch found this idea attractive, since the currents in the ocean could then be used to explain the positions of some rock formations.

One of Buch's earliest geognostic publications (1797) dealt with Landeck, a corner of Silesia on the Bohemian frontier. The paper aroused a great deal of interest, and was translated into French and English. In it, Buch provides a Neptunist interpretation of the region and states that the area, which is enclosed on all sides by mountains, had once been a lake. The lake had overflowed at a point on the east side of the area, and a valley, through which the Landeck area is now drained, had been carved out there by erosion. (In later years Buch stubbornly denied that erosion could modify a landscape to any extent.) The mountains that surround Landeck are composed of primitive gneiss and micaceous schist, except for a gap in the northern corner of the area. In this corner, the mountains are composed of the same Floetz sandstone that covers the Landeck plain. In the northern corner the sandstone reaches 500 meters above its general level. Buch's conjecture was that the sandstone came down from the north and in rushing through the gap in the primitive mountains dammed itself up to such a height.

Leaving Silesia in September 1798, Buch set out on a walking trip to see Italy, especially Vesuvius, with his own eyes. While the Napoleonic Wars halted him for six months on the northern side of the Alps, he explored the Tyrol, often accompanied by his friend Humboldt. They found a good agreement with Werner's teaching: peaks of granite in the highest central part, flanked by gneiss, schist, and limestone in due order. When, in the spring, Buch was able to cross the Brenner Pass, he expected to find a symmetrical arrangement on the south side; to his distress, he found quite different formations with enormous masses of porphyry and dolomite. The whole of Werner's beautiful system collapsed in confusion. The device that Buch had used to impose order on Landeck came to his aid once again. Clearly, what had happened was that on the northern side of the pass the formation flood that had brought the gneiss, schist, and limestone had come, naturally enough, from the north and had been stopped by the granite mountains, which, according to Werner's theory, were already there at their full height. The floods from the south had arrived independently, at other stages in the workings of nature, and hence had brought other rocks. (This idea of damming up the formation floods recurs several times in Buch's work—for example, the Trauf, an escarpment that forms the northwest edge of the Jurassic system in southern Germany and consists of coral limestone, was considered to have been an effective barrier reef against floods coming from the south.)

As Buch continued his journey through Italy, the French campaign kept him near Rome for eight months. The imposing remains of extinct volcanoes proved so refractory to Werner's theory that Buch complained, "Only two days at Vesuvius and all this confusion could be set right." But when he had wandered around Naples for eight weeks, he wrote to the editor of *Gilberts Annalen der Physik:* "My friends who believed that after seeing a real volcano, I now could say something definite on the diverging opinions about our basalts, will be badly disappointed." This refers to the battle over basalt, in which almost everyone with a claim to general knowledge took sides—although, regarding the whole of Werner's and Hutton's conflicting theories, basalts generally were not the most important issue.

From 1801 to 1803 Buch worked near Neuchâtel, and in the spring of 1802 he went to the Auvergne for six weeks. This excursion has often been referred to as momentous in bringing about Buch's alleged conversion to Volcanism and even to Plutonism.

In Auvergne, Buch found hemispherical elevations that lacked craters and correctly interpreted them as tholoids. He was struck by the frequent eruptions of lava near the bases of volcanoes, and theorized that the flanks of the volcanoes were formed by thin strata of the country rock that had been heaved up to form the tentlike roof of a large conical cavern filled with lava. Buch's theory of elevation craters was given considerable attention in the nineteenth century, although today it is not easy to see exactly what he meant by the term. He sometimes asserted that the elevation crater is not a volcano, and that it lacks a crater proper; a volcano may be spoken of properly only when a crater, and thus a direct channel to the interior of the earth, is present.

In Italy, Buch had looked in vain for the combustible matter—coal or pyrite—that, according to Werner, was necessary for volcanic action. In Auvergne he found volcanoes standing immediately upon granite, which precluded the presence of other, deeper rocks. He was then led to suppose that the granite was transformed into Domite, a type of trachyte found in the Puy de Dome, and then in a continuous action changed into basaltic lava.

Buch's observational work was much better than his speculations. Both in Auvergne and elsewhere he recorded hundreds of observations with touching fidelity, irrespective of the hypothesis that each might support or gainsay. His writings, however, flatly contradict the facile assertion that he arrived in Auvergne a Neptunist and returned a Volcanist. While it is true that he had to concede that the basalt in Auvergne had been a lava, he could not believe that the basalt in Saxony was of the same origin. (Today these basalts are interpreted as sills, according to Hutton's ideas on subterranean or Plutonic lava intrusions. Buch could not have held this view because of the important amount of erosion necessary for these sills to be visible.) In Saxony, such mighty masses of lava would have had to have been the product of correspondingly large volcanoes, the absence of which compelled him to seek other origins because Buch, like other Neptunists, excluded the action of erosion. (Werner himself was more tolerant in this respect and did not object to the concept of an original continuous deposit of basalt that later came to be separated by erosional valleys.)

Buch further objected to the idea of a lava flow stopping at the edge of a flat hill top. Why, he asked, did it not flow on down the slope? He judged the steep precipices, formed by columns of basalt tens of meters long, to be as alien to the idea of the end of a lava flow as to the (in his opinion) exclusively gentle slopes made by erosion. The Neptunist concept of precipitation—which could often be sharply delimited by a slight change in the substratum—explained this phenomenon to Buch's satisfaction. This provides a striking example of Buch's persistent denial of the effectiveness of erosion, a denial that would seem to have become more stubborn with the years if we compare this with his 1797 explication of the valley that drains Landeck.

Buch published the results of his explorations in *Geognostische Beobachtungen auf Reisen durch Deutschland und Italien,* which appeared in two volumes (1802, 1809). This work was primarily of geological interest, although much of Buch's later work may be considered a part of the *belles lettres* of the century. (An earlier short work on kreuzstein, or harmotone, is the only one that has crystallography as its main subject.)

In 1806, when he was thirty-two, Buch was elected a member of the Royal Academy of Berlin. (In due time he was elected a member of academies in Paris, London, and Vienna, while scores of minor scientific societies competed to offer him honorary memberships.) In his speech to the Berlin Academy, Buch spoke of fundamental principles. He presented the interplay of natural forces as one great onward movement that unified all things, from the crystallization of granite to the highest strivings of human intelligence. No deity of any description played any part in this scheme of things; religion was apparently the least of Buch's concerns.

Buch continued his travels, making a lengthy visit to Scandinavia in 1806 and 1807. In 1815 and 1816 he visited the Canary Islands. In two books of travels, *Reise durch Norwegen und Lappland* (1810) and *Physikalische Beschreibung der Canarischen Inseln* (1825), geological data occupy a modest space among Buch's observations of the landscape, climate, flora and fauna, customs of the inhabitants, and vicissitudes of the road.

His explorations in the southern Alps had suggested to Buch that the towering height of the Dolomites might be the result of upheaval, for which he sought the active agent in porphyry, including monozite. He concluded that the magnesia in which this rock is rich would also have been active in transforming the original limestone into dolomite. Buch thus came to visualize great subterranean activities; in 1815, in the Canary Islands, he ingeniously demonstrated the interdependence of the volcanoes of the archipelago. Since two volcanoes never happened to erupt at the same time, he argued that eruption of one volcano relieved the pressure upon the others. Hence, the masses of material underneath all the volcanoes may be seen as filling one cavern, by which the volcanoes are interconnected.

Buch's view of effective subterranean masses soon grew to encompass the whole world. A treatise of 1842, "Ueber Granit und Gneiss," illustrates the later development of Buch's Wernerism and demonstrates the remarkable consequences to which his refusal to consider erosion led him. Granite and gneiss cover southern Finland and Sweden, while, stretching west from Leningrad, Silurian strata cover the Baltic provinces and the isle of Gotland, and further recur in six small patches in Sweden. These latter, we would now say, represent erosional remnants, hills a few hundred meters high in which the Silurian strata are horizontal, as they are in Gotland and Russia. Buch was compelled to admit that the Silurian strata in the six hills are a remnant of a continuous sedimentary layer that had once extended through a large part of Sweden, but was hard pressed to account for the disappearance of the greater part of it. Since he could not explain it by erosion, he had recourse to the novel, non-Wernerian principle of metamorphism. In Buch's conjecture, the Silurian rocks had been metamorphosed by the action, probably vaporous, of the underlying granite. This theory further served to explain the scarcity of granite in Scandinavia, about which Buch had worried a great deal; the granite is largely hidden from view by the large Silurian deposits that now appear as gneiss.

The basalt-topped hills were pushed up by a subterranean mass of basalt; the sheets on top thus represent possibly fiery extrusions. Buch thought that the lower mass of basalt, situated between the granite and Silurian layers, served as a shield against the metamorphic action of the granite and thus preserved the Silurian layers in their original state. He concluded that the large parts of Russia covered with nonmetamorphosed Silurian strata must be underlain by basalt.

Buch dismissed as nonsense the notions that the

stairlike outline of these hills could be due to weathering and erosion and that the relative height of the hills was due to protection by the more resistant basalt. On one of the six hills, Kinnekulle, no more than a mere dot of basalt is visible at the center of the uppermost Silurian layer; how, Buch asked, could this have afforded protection? Buch conceded that the Silurian layers in all the hills had once been continuous, but it is hard to understand what, in his view, had become of the missing segments. He believed that these rocks must have been reworked by metamorphism, but how they were brought down from the steps of the staircase is not explained. His account of the area is marked by his keen eye for geological observation, trained by Werner, but as soon as past events are brought into consideration, all is dim and misty. It might have been better if Buch had followed Hegel's advice and refrained from peering into the past.

Buch's somewhat more actualistic view of sedimentation displayed in this instance may have been the result of his extensive work in stratigraphy and paleontology. This had been Buch's chief concern since his fiftieth year. His first major achievement in this field consisted in distinguishing between ammonites and nautilidae (1829, 1839) and observing the intricate suture lines in the former. The Wernerians, trained as they were in thoughtful observation and accurate description, soon felt at home with paleontology. We may imagine Buch's delight as he contemplated the orderly succession of faunas, the characteristics of each being in part determined by its predecessors, and all bearing witness to that great onward movement he had enlarged upon in the speech to the Berlin Academy. He was aware of the value of guide fossils, and sought to enhance their usefulness with careful drawings and descriptions.

Buch's fundamental error, perhaps, sprang from the prejudice, older than *Naturphilosophie,* that the earth was specially created as a place for man to dwell in. Hegel's comparison with a house is suggestive of how time could be considered unimportant and of motives that might lead to the rejection of all except the barest minimum of erosion. It is not very interesting to know how long it took to build the house; we are interested in its present state and whether it is appropriate to our purpose. If we have to think about the building operations, they surely were carried out according to a plan. It would serve no purpose to build some walls with the intention of demolishing them in order to use the stones for other walls. The idea of a purposeful construction bars access to the principle of uniformity.

For more than forty years Buch brooded over one specific problem, that of the erratic blocks distributed over Germany. In southern Silesia, Buch had determined that these blocks came from the Sudets; he had identified the kinds of rocks and observed that they became smaller as they occurred in more northern locations. Farther north, there appeared blocks of a kind alien to the mountains of the south; Buch could but conclude that these, strewn copiously over half of Holland and large parts of Germany and Poland, had been brought from Scandinavia. He was, however, unable to decide what agent had brought them, and felt inclined to attribute their presence to the then rather commonly invoked mighty floods. In Switzerland, erratic blocks from the Alps were found in the Juras, at an appreciable height above the intervening Swiss plain; J. A. de Luc held that these had been hurled through the air by great explosions. Buch objected to this unlikely device, and judged it impossible that blocks from Scandinavia should have been so hurled across the Baltic Sea far into Germany. He further opposed the notion (propounded by Hutton and Agassiz and Charpentier) that the blocks had been transported by glaciers, and explained the polished surfaces of the rocks by ascribing them to differential movements along the curved, onion-like partings generally seen in fresh granite. He even contended that he had seen the same polish on a surface just laid bare by quarrying. Swedish scientists had measured the striae on the polished rocks to determine the direction of transport; Buch halfheartedly agreed with them, since they cited currents as a moving agent. But when Berzelius, among others, concluded that the masses carried southward must represent an appreciable erosion of the land in Sweden, Buch would have no part of it. However intimately he might have come to know the enormous volumes of Mesozoic and Tertiary formations, he apparently refused to give a moment's thought to where these masses might have come from.

Buch, who never married, referred to himself as a wandering hermit. In later years, however, he got over his shyness in joining larger parties, and when, around 1830, scientists began to meet on field trips, Buch regularly attended. For example, in his last summer he met with naturalists in Koblenz, with Swiss geologists in Sion, with French scientists in Metz, and with German naturalists in Wiesbaden. He then traveled through Basel to Le Puy, where he had an appointment with Daubrée to study the basalt of the Vivarais. In the late autumn he was in Paris with Mitscherlich and Rose; Rose accompanied him back to Berlin by train.

Buch was free of all anxiety that others should unfairly profit from his accomplishments; in the accounts of his journeys, each new phenomenon is recorded either as a matter of fact or as something first observed by a companion. An extreme example was his anonymous publication, in 1826, of a geological map of Germany, which embodies the solid core of a long life's work. This map, which covered forty-two sheets, was reissued in 1842.

Buch's many published works are written in a highly attractive style that easily conveys his thoughts. His proficiency as an author enables us to learn of the theory and practice of Neptunism, about which the scarce and constrained writings of Werner tell us little.

A week before his death Buch was with friends in the Humanist Club until late at night. He complained of nothing more severe than chilblained toes, but the next day his first serious illness set in. He was buried in the Buch family vault at Stolpe.

BIBLIOGRAPHY

I. ORIGINAL WORKS. Buch's works were brought together as *L. von Buch's gesammelte Schriften*, J. Ewald, J. Roth, H. Eck, and W. Dames, eds., 4 vols. (Berlin, 1867–1885); Vol. I contains J. Ewald, "Leopold von Buch's Leben und Wirken bis zum Jahre 1806," pp. v–xlviii. For the individual works listed below, the Roman and Arabic numerals in parentheses refer to the *Gesammelte Schriften*. Among Buch's works are "Mineralogische Beschreibung der Karlsbader Gegend," in *Kohler und Hoffmann Bergmannisches Journal*, **5**, pt. 2 (1792), 383–424 (I, 3–23); *Beobachtungen über den Kreuzstein* (Leipzig, 1794) (I, 24–35); *Versuch einer mineralogischen Beschreibung von Landeck* (Breslau, 1797) (I, 38–72); "Considérations sur le granite," in *Journal de physique*, **49** (1799), 206–213 (I, 101–108); "Mémoire sur la formation de la leucite," *ibid.*, 262–270 (I, 109–117); "Sur les volcans," in *Bibliographia Britannica*, **16** (1801), 227–249 (I, 132–142); *Geognostische Beobachtungen auf Reisen durch Deutschland und Italien*, 2 vols. (Berlin, 1802–1809) (I, 143–523); "Ueber das Fortschreiten der Bildungen in der Natur," his inaugural address to the Königliche Akademie der Wissenschaften (17 April 1806) (II, 4–12); *Reise nach Norwegen und Lappland* (Berlin, 1810) (II, 109–563); *Physikalische Beschreibung der Canarischen Inseln* (Berlin, 1825) (III, 225–646); and "Ueber Granit und Gneiss," in *Abhandlungen der Königlichen Akademie der Wissenschaften, Berlin* (1844) (IV, pt. 2, 717–738).

His unpublished diary is in the possession of the geology department of the University of Berlin. Buch was responsible for the first geological map of Germany (1826), although it does not bear his name. Consisting of forty-two sheets, it was the first map of this sort to cover a fairly large area of Europe. It was reissued in 1842.

II. SECONDARY LITERATURE. Works on Buch are H. von Dechen, *Leopold von Buch; sein Einfluss auf die Entwick-*
lung der Geognosie (Bonn, 1853); S. Günther, *A. v. Humboldt, L. v. Buch* (*P*), Vol. XXXIX in the series Geisteshelden (Führende Geister) (Berlin, 1900), 185–271; W. Haidinger, "Zur Erinnerung an Leopold von Buch," in *Jahrbuch der Kaiserlich-Königlichen Geologischen Reichsanstalt*, IV (1853), 207–220; G. F. Hegel, *Encyklopädie der philosophischen Wissenschaften*, 3rd ed. (Heidelberg, 1830), pt. 2, "Die Philosophie der Natur," sec. 339, *Zusatz*; H. Hölder, *Geologie und Paläontologie in Texten und Geschichte*, II, pt. 11 of Orbis Academicus; Problemgeschichten der Wissenschaft (Freiburg im Breisgau, 1960), see Index; and R. Hooykaas, *The Principle of Uniformity in Geology, Biology and Theology*, 2nd ed. (Leiden, 1963).

W. NIEUWENKAMP

BUCHANAN, JOHN YOUNG (*b.* Glasgow, Scotland, 20 February 1844; *d.* London, England, 16 October 1925), *oceanography, chemistry.*

After graduating from the University of Glasgow in 1863, Buchanan studied chemistry at Marburg, Bonn, Leipzig, and Paris. He was appointed chemist, physicist, and geologist of the pioneering oceanographic expedition of the *Challenger* (1872–1876), under the leadership of Charles Wyville Thomson. He continued his oceanographic studies in his private laboratory in Edinburgh and on ocean cruises. Among his few long-time scientific associates was Prince Albert of Monaco.

Buchanan's participation in the *Challenger* expedition shaped his scientific career. His work dealt mainly with the design and improvement of oceanographic instruments and observational methods, and with data collection—essential aspects of a young science. His research, always original and based on observations, was carried out with the utmost thoroughness and precision.

Certain of Buchanan's publications contain important generalizations. He prepared the first reliable surface salinity map of the oceans. His analysis of spatial and seasonal distributions of salinity and temperature contradicted the widely adopted thermal circulation theory originated by Humboldt. Buchanan's observations and speculations (1877, 1886) on thermohaline circulations in vertical planes were utilized and confirmed—at least for the subtropics—by J. W. Sandström in 1908 and A. Merz in 1925. Buchanan demonstrated that vertical currents from submarine sources supplied the cold surface water that is generally observed along the western shores of continents. He produced many valuable and meticulous studies on the physical and chemical properties of seawater and sea ice, and on the constitution, formation, and distribution of concretionary deposits of iron and manganese oxides discovered by the *Challenger*. He received much attention for his demonstration of the inorganic nature of the gelatinous deep-sea deposit that leading naturalists had thought to be a protoplasmic slime. Buchanan's studies in limnology helped to establish the generality of the temperature stratification of temperate lakes and the concept of the thermocline (1886), and he pioneered in quantitative studies of seasonal variations in heat content of lakes. Buchanan seems to have worked purely for his own satisfaction, accomplishing more than his publications would indicate.

BIBLIOGRAPHY

I. ORIGINAL WORKS. Buchanan collected most of his publications in four volumes: *Experimental Researches on the Specific Gravity and the Displacement of Some Saline Solutions* (Edinburgh, 1912); *Scientific Papers, Vol. 1* (Cambridge, 1913); *Comptes rendus of Observation and Reasoning* (Cambridge, 1917); and *Accounts Rendered of Work Done and Things Seen* (Cambridge, 1919). Most of his papers published in scientific journals (more than 100) are listed in the Royal Society of London's *Catalogue of Scientific Papers* (London, 1867–1925), VII, 291; IX, 386–387; XIII, 885. Some of Buchanan's more important papers are "On the Distribution of Salt in the Ocean, as Indicated by the Specific Gravity of Its Waters," in *Journal of the Geographical Society* (London), **47** (1877), 72–86; "On the Distribution of Temperature in Loch Lomond During the Autumn of 1885," in *Proceedings of the Royal Society of Edinburgh,* **13** (1886), 403–428; and "On Similarities in the Physical Geography of the Great Oceans," in *Proceedings of the Geographical Society* (London), **8** (1886), 753–768.

II. SECONDARY LITERATURE. Information on Buchanan's life may be found in two obituaries: in *Proceedings of the Royal Society of London,* **110A** (1926), xii–xiii; and *Proceedings of the Royal Society of Edinburgh,* **45** (1925), 364–367.

GISELA KUTZBACH

BUCHER, WALTER HERMAN (*b.* Akron, Ohio, 12 March 1889; *d.* Houston, Texas, 17 February 1965), *geology.*

During his youth Bucher and his parents, Maria Gebhardt and August Bucher, moved to Germany. He received his higher education at the University of Heidelberg, graduating in 1911 with a Ph.D. in geology. He returned to America shortly thereafter and joined the faculty of the University of Cincinnati, remaining there for twenty-seven years. In 1940 Bucher moved to Columbia University as professor of structural geology and became chairman of the geology department in 1950. Six years later, having gained emeritus status, he became a consultant to Humble Oil and Refining Company in Houston, Texas, where he was employed until his death.

While at the University of Cincinnati, Bucher produced his major work, *The Deformation of the Earth's Crust* (1933), an explanation of the origin of orogenic belts. It was an attempt to compile "all essential geological facts of a general nature that bear on the problem of crustal deformation and to derive from them inductively a hypothetical picture of the mechanics of diastrophism. . . ." Tectonic observations were presented as carefully worded generalizations that Bucher called laws; interpretations of the laws were designated opinions. From these building blocks he constructed a theory. Bucher's thesis combines the contraction of the earth by cooling with gravitational forces. Such contraction leads to global fractures of the crust. These lesions, which rise from several hundred kilometers under the earth's mantle and penetrate the crust, allow the escape of heat, volatile substances, and water vapor to the surface. This weakens the crust along the fractures, causing it to form welts and furrows.

Bucher believed that the earth's crust was divided by associated linear swells and basins, or welts and furrows (crustal folds). Since the thickened sediments of furrows, uplifted in the formation of welts, are subjected to the force of gravity, their own weight and the resulting stress cause them to buckle into folds or to collapse and spread, producing the sort of rock deformation associated with mountainous regions. Support for his theory was provided by numerous experiments with Plasticine, glass (Christmas tree ornaments), and wax models of the earth, which he described in his book.

Extensions of this theory are presented in "The Role of Gravity in Orogenesis" (1956) and in the symposium *The Crust of the Earth* (1955). In addition to these broad theoretical endeavors, Bucher published many articles on geophysics and structural geology. His essential reliance on radical forces for the orogenic mechanism is now less popular than theories of primarily tangential movement, such as a continental drift.

Bucher received an honorary D.Sc. from Princeton University in 1947, the William Bowie Medal of the American Geophysical Union in 1955, and the Penrose Medal of the Geological Society of America in 1960. He served as president of the Geological Society of America in 1954 and was a member of the National Academy of Sciences, the American Association for the Advancement of Science, and the Ohio Academy of Sciences.

Bucher married Hannah E. Schmid in 1914. They had two sons and two daughters.

BIBLIOGRAPHY

I. ORIGINAL WORKS. Among Bucher's works are *The Deformation of the Earth's Crust* (Princeton, 1933; repr. New York, 1957); "The Role of Gravity in Orogenesis," in *Bulletin of the Geological Society of America,* **67** (1956), 1295–1318; "Deformation in Orogenic Belts," in *The Crust of the Earth,* Arie Poldervaart, ed. (New York, 1955), pp. 343–368; "Continental Drift Versus Land Bridges," in *Bulletin of the American Museum of Natural History,* **99**, no. 8 (1952), 72–258; "Volcanic Explosions and Over-thrusts," in *Transactions of the American Geophysical Union, 14th Annual Meeting* (1933), pp. 238–242; and "Geologic Structure and Orogenic History of Venezuela," in *Memoirs. Geological Society of America,* no. 49.

II. SECONDARY LITERATURE. Details of Bucher's life are in *Current Biography,* XVIII (1957), 84–86; and in *Modern Men of Science,* J. Green, ed. (New York, 1966), pp. 74–75.

MARTHA B. KENDALL

BUCHERER, ALFRED HEINRICH (*b.* Cologne, Germany, 9 July 1863; *d.* Bonn, Germany, 16 April 1927), *physics.*

Alfred Bucherer's father, Heinrich Bucherer, was the owner of a chemical factory in Cologne. His mother was English. Besides his interests in chemistry and technology, the father was an art lover; his wife was a devotee of music and languages, and it was in such surroundings that Alfred Bucherer was raised. He exhibited unusual talents, not only in mathematics and the physical sciences but in philology as well. He was a man who was equally at home in the worlds of physics, chemistry, technology, and philology.

After spending the year 1884 studying at the Technische Hochschule at Hannover, Bucherer went to the United States, to Johns Hopkins, where he continued his technical studies and at the same time studied philology. At Hopkins he came under the influence of the chemist Ira Remsen. While studying under Remsen he became engrossed in thermo-dynamics, and as a direct consequence of this work, he obtained a patent for the separation of aluminum from its sulfide. He returned to the United States in 1893 to spend a year studying at Cornell. His intimate knowledge of vector analysis stemmed from his work during this period.

Bucherer completed his formal education in 1895 under Braun at Strasbourg. His topic, the effects of magnetic fields on the electromotive force, was another direct outcome of his work with Remsen. Bucherer continued traveling and studying for the next three years, spending part of that time with Ostwald at Leipzig.

He became *Privatdozent* at Bonn in 1899 and was to remain connected with Bonn in one way or another until his death. In 1912 he became professor of physics and in 1923 honorary professor of physics. He maintained an active laboratory in the university until his death.

With his arrival at Bonn, Bucherer's academic interests became modified. He discontinued his work in physical chemistry and became more and more involved in problems in physics. Besides his patent for the separation of aluminum from its sulfide, a second patent, which Bucherer had obtained in 1892, was for the transmission of pictures by wireless. It seems that it was in the interest of making this patent more fruitful that Bucherer turned to the questions of the nature of the electron and the effects of the motion of bodies on electromagnetic phenomena.

In 1904 in a monograph on the theory of electrons (*Mathematische Einführung in die Elektronentheorie*) Bucherer produced his own theory of the moving electron, which rivaled the theories of Max Abraham and H. A. Lorentz. Whereas Abraham's theory was predicated on a rigid electron and Lorentz's theory was predicated on an electron that contracted in the direction of motion, Bucherer's theory assumed an electron that contracted, but in such a way as to maintain a constant volume. According to Bucherer, the contraction was such that the moving electron became an ellipse with axes given by $as^{1/3}$, $as^{-1/6}$, $as^{-1/6}$ where a is the radius of the spherical resting electron and s is given by $(1 - v^2/c^2)$, v being the velocity of the electron and c being the velocity of light. This led to a prediction for the transverse mass of the moving electron which was midway between the predictions of Abraham and Lorentz and Einstein, Einstein's special theory of relativity making predictions for the transverse mass identical with those of Lorentz.

This result was within the range of experimental values for the mass of the moving electron obtained by Wilhelm Kaufmann in 1906. Kaufmann's data, however, which he held to be in favor of the Abraham theory, were suspect. In order to settle the question, Bucherer decided to undertake his own measurements of the specific mass of the electron. This he did in 1908. In a remarkable and abrupt turnabout, he concluded that the data he obtained supported not his own theory but the theory of Einstein.

Bucherer's lively and polemical style was responsible for his often getting into arguments in the literature. This was especially true of his work on the mass of the moving electron. While he ended up supporting Einstein in this case, he was never completely happy with the relativistic formulation of physics.

BIBLIOGRAPHY

I. Original Works. *Die Wirkung des Magnetismus auf die elektromotorische Kraft,* diss. (Leipzig, 1896); *Mathematische Einführung in die Elektronentheorie* (Leipzig, 1904); "Das deformiert Elecktron und die Theorie des Elektromagnetismus," in *Physikalische Zeitschrift,* **6** (1905), 833–834; "On a New Principle of Relativity in Electromagnetism," in *Philosophical Magazine,* **13** (1907), 413–429; "Messungen an Becquerelstrahlen. Die experimentelle Bestätigung der Lorentz-Einsteinschen Theorie," in *Deutsche physikalische Gesellschaft, Verhandlungen,* **10** (1908), 688–699; "On the Principle of Relativity and the Electromagnetic Mass of the Electron, A Reply to Mr. Cunningham," in *Philosophical Magazine,* **15** (1908), 316–318; "On the Principle of Relativity. A Reply to Mr. Cunningham," *ibid.,* **16** (1908), 939–940; "Antwort auf die Kritik von Besterlmeyer bezüglich meiner experimentelle Bestätigung des Relativitätsprinzips," in *Annalen der Physik,* **30** (1909), 974–986; "Gravitation und Quantentheorie I," *ibid.,* **68** (1922), 1–10; "Gravitation und Quantentheorie II," *ibid.,* 546–551.

II. Secondary Literature. Max Abraham, *Theorie der Elektrizität,* 2 vols. (Leipzig, 1904–1905); Stanley Goldberg, "Early Response to Einstein's Special Theory of Relativity," unpublished doctoral thesis (Harvard University, 1968), chs. 1, 2; E.T. Whittaker, *A History of the Theories of Aether and Electricity,* 2 vols. (New York, 1960).

Stanley Goldberg

BUCHNER, EDUARD (*b.* Munich, Germany, 20 May 1860; *d.* Focsani, Rumania, 13 August 1917), *chemistry.*

Buchner came from an old Bavarian family of scholars. His father, Ernst, was professor of forensic medicine and obstetrics as well as editor of the *Ärztliches Intelligenzblatt* (later *Münchener medizinische Wochenschrift*).

Upon graduating from the Realgymnasium in Munich, he served in the field artillery and then studied chemistry at the Technische Hochschule in Munich. After a short while, however, Buchner had to abandon his studies because of financial problems; for four years he worked in canneries in Munich and in Mombach. In 1884, with the assistance of his brother Hans, he was able to resume his chemical studies, this time at the organic section of the chemical laboratory of the Bavarian Academy of Sciences in Munich, under Adolf von Baeyer. Buchner's first work in organic preparative chemistry resulted from the suggestions of Theodor Curtius, an assistant in the organic section, and were done under his direction; his work with Curtius on the chemistry of diazoacetic ester led to a warm friendship.

While studying chemistry, Buchner also worked at the Institute for Plant Physiology, under Karl von Nägeli. Here he became interested in the problems of alcoholic fermentation, the subject of his first publication (1886). In this paper he arrived at the significant conclusion that, contrary to Pasteur's contention, the absence of oxygen is not a necessary prerequisite for fermentation.

Buchner obtained his doctorate in 1888 under Baeyer and in 1890 was appointed his teaching assistant. He became *Privatdozent* the following year. His *Habilitationsschrift* dealt with research on pyrazole, the five-membered heterocyclic derivative of antipyrine. Baeyer procured the funds for him to set up his own laboratory for fermentation chemistry, but up to 1893 Buchner published only one other paper on the physiology of fermentation, a comparative study of the behavior of fumaric and maleic acids.

In 1893 Buchner succeeded Curtius as head of the Section for Analytical Chemistry at the University of Kiel, and in 1895 he was appointed associate professor there. The following year he became professor of analytical pharmaceutical chemistry in Tübingen and, while there, published his pioneering work, *Alkoholische Gärung ohne Hefezellen* (1897).

In 1898 Buchner accepted an appointment as full professor of general chemistry at the College of Agriculture in Berlin and simultaneously he became director of the Institute for the Fermentation Industry.

In 1900 he married Lotte Stahl, daughter of a Tübingen mathematician. Two sons and one daughter resulted from this marriage.

Scientifically, Buchner's Berlin years were his most productive period, especially in the field of the biochemistry of the fermentation process. Nevertheless, he felt that professionally he was not able to develop his knowledge: he missed teaching in a chemical institute of a university. He had to wait a long time before he was invited to teach at a university, however, perhaps because of having insulted the Ministry of Education official in charge of the Prussian academic institutions. After being kept waiting by the official, Buchner rebuked him by pointing to his pocket watch.

After receiving the Nobel Prize in chemistry in 1907 for his work on cell-free fermentation, Buchner was appointed to the chair of physiological chemistry at the University of Breslau in 1909. Two years later he was invited to Würzburg, an invitation that he accepted with alacrity. In Würzburg he was also able to pursue his hobbies of hunting and mountain climbing.

Politically, Buchner was an admirer of Bismarck and a follower of the National Liberal party. He volunteered for active duty at the outbreak of World War I and in August 1914 was sent to the front as a captain of an ammunition supply unit. He was promoted to major in 1916, but in that same year he was called back to Würzburg to resume teaching. In 1917 he again volunteered for front-line duty and was sent to Rumania, where on 11 August he was wounded by shrapnel and died two days later.

Central to Buchner's experimental work are three papers published in 1897, which dealt with his sensational discovery of cell-free fermentation, the turning point for the study of enzymes. In the history of enzymology, it is quite proper to differentiate between the pre-Buchner and post-Buchner periods.

The basis of fermentation chemistry or enzyme chemistry is Berzelius' thesis[1] that all reactions in living organisms are initiated and regulated by catalysts. This of course applies to the processes of fermentation and putrefaction. Originally, the terms "ferment" and "enzyme" designated primarily fermentation, but also putrefaction or a gas-producing agent. Hence, it is understandable that "the history of the conversion of fermentable sugar through yeast overlaps with the history of fermentation processes in general. It was this outstandingly important process in the production of alcoholic beverages that was the prototype of fermentation processes."[2]

Between 1830 and 1860, prior to Buchner's discovery, two theories of fermentation had divided the scientific world into vitalists and mechanists. Liebig was the exponent of the mechanists. According to his theory,[3] formulated in 1839, yeast causes fermentation because, as a body in a state of continuous decomposition, it stimulates the sugar molecules to decompose into alcohol and carbon dioxide. Although this appeared to be analogous to the then known fermentative reactions (the decomposition of amygdalin into hydrocyanic acid and sugar by emulsion, the proteolysis of egg albumin by pepsin, the decomposition of starch by diastase), Liebig refused to ascribe the character of a "catalytic force," as formulated by Berzelius, to the action of yeast.

The mechanical theory of fermentation, regardless of whether it interpreted yeast as an expression of a "catalytic force" or of a body in the process of decomposing, clashed with Pasteur's vitalistic interpretation of the fermentation process. In 1836 Cagniard de la Tour had reported to the Paris Academy that yeast consisted of living organisms, and the next year Theodor Schwann and Friedrich Kützing were to reach the same conclusion. On the basis of these discoveries and as a result of his own observations and experiments, Pasteur (between 1857 and 1860) formulated the thesis that alcoholic fermentation was an expression of the vital action of the yeast fungi and inseparably bound to this physiological action.[4] Fermentation was therefore not a catalytic but a vital process. Then Pasteur differentiated between soluble enzymes, which can be separated from

the vital processes and therefore are also effective outside the organism, and ferments, which are inseparably bound to a living organism and its vitality. Although such researchers as Berthelot, Traube, and Hoppe-Seyler subsequently defended the view that there were active enzymes in living cells, comparable to those acting outside the cell, Pasteur's view of the unalterable connection between vitality and fermentation nevertheless found acceptance among the majority of scientists.

Efforts to isolate the fermentation-producing agent from yeast cells remained unsuccessful. This was the problem Buchner attacked anew in 1893 at the suggestion of his brother Hans, who was engaged in the extraction of powerful pathogenic substances from bacteria. The problem was how to obtain cell fluid, in as pure a state as possible, for therapeutic research. Buchner's approach was an attempt to destroy yeast cells in order to extract their fluid. At the suggestion of Martin Hahn, an assistant of his brother's, Buchner pulverized yeast in a mortar with one part quartz sand and one-fifth part diatomaceous earth; this became a thick paste within a few minutes. This paste was then wrapped in canvas and subjected to a pressure of ninety kilograms per square centimeter, yielding 500 milliliters of fluid from 1,000 grams yeast.

At first, hardly any thought was given to the idea of producing fermentation with the fluid expressed from yeast. Rather, Buchner and Hahn concentrated on preserving the easily decomposed fluid by adding concentrated sucrose solution. In 1896 Buchner discovered that this mixture soon exhibited lively gas formation. In 1897 he published three papers on the results of these first experiments with cell-free alcoholic fermentation, and by the end of 1902 he had published fifteen more papers on the same subject. In 1903 the first comprehensive presentation of his achievements was published by the two Buchners and Martin Hahn as *Die Zymase-Gärung*.

Buchner called the active, fermentation-producing agent of the expressed fluid "zymase." This eliminated the previously valid distinction between the soluble enzymes, effective outside the cell, and the ferments, whose effectiveness is linked to cell structure and cell activity. Accordingly, life and fermentation are not unalterably bound to each other. Instead, fermentation is a chemical, enzymatically catalyzed process.

Through painstaking experimentation, Buchner defended this basic fact against various objections, particularly those of the physiologist Max Rubner, the biochemist Hans von Euler-Chelpin, and the botanist Wilhelm Ruhland. He was able to prove that neither the few intact yeast cells present in the expressed fluid nor any "living plasma particles" cause the sugar conversion that occurs in the fermentation process. Addition of a mixture of alcohol and ether to the fluid yielded a precipitate that could be preserved as a powder without the zymase losing its effectiveness. He also obtained a fully fermentative dry substance by killing the yeast cells through addition of alcohol or acetone to a yeast suspension. He gave the name "zymin" to the substance after it had been washed and dried with ether. It cannot grow but is fully effective in fermentation experiments.

Further test series concerned the chemical properties of the expressed fluid, a yellow-brown opalescent liquid which becomes ineffective after prolonged storage because a proteolytic enzyme, endotryptase, destroys the active principle. The fluid also loses its effectiveness through heating, which causes precipitation of its proteins. In addition to zymase, the fluid contains a series of other enzymes: catalase (diluted fluid decomposes hydrogen peroxide into water and oxygen) as well as enzymes resulting from the splitting of disaccharides and polysaccharides. From the latter, Buchner deduced that not only the simple sugars, glucose and fructose, but also complex sugars, maltose and saccharose (disaccharides), as well as the polysaccharide glycogen, are fermentable. Attempts to separate these enzymes from the zymase were almost always unsuccessful.

In the chemistry of alcoholic fermentation, Buchner and his assistants were confronted by three questions between 1904 and 1917: (1) Is zymase a homogeneous enzyme? (2) In order to be active, does it require the presence of additional substances, especially the presence of a coenzyme? (3) What is the chemical nature of the intermediate products in the decomposition of sugar into alcohol and carbon dioxide? In the experimental treatment of these questions, however, Buchner and his assistants were no longer alone. The scientific world had realized that the fluid expressed from yeast could be profitably studied in order to explain the chemistry of fermentation. The English scientists A. Harden and W. J. Young had been working on these problems since 1904, the St. Petersburg botanist L. Ivanov since 1906, and A. von Lebedev since 1910; and if the reaction chain of alcoholic fermentation can be considered fully explained, it is because of their successful initial experiments. In this phase Buchner and his circle of co-workers take second place. True, they kept up with these developments, but they were not the pacesetters.

Buchner suggested that lactic acid, methylglyoxal, glyceraldehyde, and dihydroxyacetone were the intermediate products of the fermentation of alcohol. It has only recently been established that glyceralde-

hyde and dihydroxyacetone, in the form of the respective phosphates, are indeed intermediate products of the decomposition of sugar by yeast as well as by animal cells.

Buchner was also interested in the fermentation phenomena of other microorganisms. In 1902 he published a paper on the enzymes of *Monilia candida* and other milk-sugar yeasts. In subsequent years he worked on acetous fermentation (1906), butyrous fermentation (1908), and citrous fermentation (1909). These investigations also served to confirm that characteristic life phenomena can be attributed to the regularities of enzyme-catalyzed chemical reactions.

Buchner's revolutionary discoveries in biochemistry overshadow his work in preparative organic chemistry. The starting point of these studies was the synthesis of diazoacetic ester, by means of which Curtius discovered a new group of compounds in 1883. This aliphatic diazo compound is highly reactive. Buchner experimented systematically with it, first under the direction of Curtius and later independently. Between 1885 and 1905 he published forty-eight papers treating preparation of nitrogenous compounds, especially pyrazole, as products of the action of diazoacetic ester on unsaturated acid esters, and the synthesis of trimethylene carboxylic acids by adding diazoacetic ester to fumaric acid ester and heating the mixture. With the aid of brucine salts, Buchner separated the resulting racemic mixture into its two enantiomorphic substances. The papers also concerned the products of reaction of diazoacetic ester with aromatic hydrocarbons (benzene, toluene, *m*-xylene, and *p*-xylene). The result of this work was the synthesis of cycloheptatriene and cycloheptane-carboxylic acid. Thus a new direction was furnished for a synthesis of compounds in the cycloheptane series.

After 1905 Buchner published only six additional studies in preparative organic chemistry, an indication that biochemical problems fully occupied his time.

NOTES

1. *Jahresberichte von Berzelius,* **15** (1835), 245.
2. Carl Oppenheimer, *Die Fermente und ihre Wirkungen,* 2nd ed. (Leipzig, 1903), p. 302.
3. *Annalen der Chemie,* **30** (1839), 362.
4. *Annales de chimie et de physique,* **58** (1860), 323.

BIBLIOGRAPHY

I. ORIGINAL WORKS. All of Buchner's writings are listed in Poggendorff, IV, 200, and V, 182; references to his obituaries are in VI, 362. A complete bibliography can also be found in the appendix to C. Harries, in *Berichte der Deutschen chemischen Gesellschaft,* **50** (1917), 1843–1876. For a comprehensive presentation of his work on fermentation, see *Die Zymase-Gärung* (Munich, 1903), written with Hans Buchner and Martin Hahn; "Alkoholische Gärung des Zuckers," in *Bulletin de la Société chimique de France,* **7** (1910), 1–22; "Neuere Ansichten über die Zymase," in *Sitzungsberichte der Physikalisch-medizinischen Gesellschaft zu Würzburg* (1917), written with S. Skraup; and "Cell-free Fermentation," in *Nobel Lectures Chemistry 1901–1921* (Amsterdam–London–New York, 1966).

II. SECONDARY LITERATURE. The history of the fermentation problem is discussed in M. Delbrück and A. Schrohe, *Hefe, Gärung und Fäulnis* (Berlin, 1904); C. Graebe, *Geschichte der organischen Chemie* (Berlin, 1920); F. F. Nord, in F. F. Nord and R. Weidenhagen, *Handbuch der Enzymologie* (Leipzig, 1940); and C. Oppenheimer, *Die Fermente und ihre Wirkungen,* 4th ed. (Leipzig, 1913). For further literature see the works cited in the notes.

On Buchner's work in organic synthesis, see P. Walden, *Geschichte der organischen Chemie seit 1880* (Berlin, 1941). The history of the discovery of cell-free fermentation is discussed in M. von Gruber, in *Münchener medizinische Wochenschrift* (1907).

HERBERT SCHRIEFERS

BUCHNER, FRIEDRICH KARL CHRISTIAN LUDWIG (*b.* Darmstadt, Germany, 29 March 1824; *d.* Darmstadt, 1 May 1899), *medicine, philosophy, history of science.*

Büchner was the most influential nineteenth-century German representative of a consistent materialism. His major accomplishments were the dissemination of the methods and results of the natural sciences and his work against a dogmatic, metaphysically determined formation of the consciousness. That this could not always take place without biased presentations and controversies, sometimes at the expense of scientific exactness and thoroughness, does not greatly lessen the value of Büchner's works.

The third son of Ernst Büchner, a physician who later became archducal medical adviser, and of Karoline Reuss, Ludwig Büchner entered the University of Giessen in 1842 and studied physics, chemistry, botany, mineralogy, and philosophy. Later, following his father's wishes, he studied medicine and passed the examinations of this faculty in 1848. In the same year, he graduated after submitting "Beiträge zur Hall'schen Lehre von einem excito-motorischen Nervensystem" as his dissertation. One of its theses was that the personal soul is unthinkable without its material substrate.

As a student Büchner took an active part in the republican attempts at reform by working for democratic journals. He also published the *Nachgelassene*

Schriften (1850) and a biography of his brother Georg, the famous playwright.

After further studies in Würzburg, with Rudolf Virchow, and in Vienna, Büchner became academic lecturer in Tübingen in 1854. In the following years he lectured on physical diagnosis, medical encyclopedia, and forensic medicine. He also served as assistant physician at the university clinic and as a medicolegal expert.

In 1854 the convention of German natural scientists and physicians was held in Tübingen, and Büchner wrote a series of reports on the talks given at the convention. These reports and Jacob Moleschott's *Kreislauf des Lebens* (1852) stimulated him to write his first and most famous work, *Kraft und Stoff* (1855). Büchner ascribed the favorable reception of this book (the first edition was sold out within a few weeks) to the public's weariness of discussions of politics and literature. In the wake of the 1848 revolution his ideas were considered dangerous by the clergy and the conservative elements, who therefore forced Büchner to resign his university lectureship. The kind and extent of the polemic conducted becomes evident on reading the Preface and Notes in subsequent editions of *Kraft und Stoff,* as well as the collection of essays *Aus Natur und Wissenschaft* (1862).

In the following years Büchner practiced medicine in Darmstadt and published numerous works meant to disseminate knowledge of the natural sciences. These, like *Kraft und Stoff,* were translated into many languages. In 1860 he married Sophie Thomas. He made lecture tours of Germany and the United States (1874), and in 1881 he founded the Deutschen Freidenkerbund.

As guidelines for his works Büchner took Lamettrie's statement that experience and observation must be our sole guides and Bernhard Cotta's opinion that the empirical investigation of nature has no purpose but to find the truth, regardless of whether it seems reassuring or hopeless, beautiful or ugly, logical or inconsistent, wise or foolish.

Starting with the old materialism and the writings of Ludwig Feuerbach, Büchner tried to base the theses of the materialistic concept of the world on empirical foundations and thus to prepare fruitful philosophical discussions. He defined force as "expression for the cause of a possible or an actual movement." Physics, as the science of forces (mechanical force, gravity, heat), revealed that forces are inseparable from matter. Force and matter could not be destroyed; they were one and the same thing, seen from different aspects. There could be no force without matter, and no matter without force.

To consider matter as of minimal value when

compared with the spiritual is meaningless, according to Büchner, for without an exact knowledge of matter and its laws, no insight is conceivable. Elements unchangeable in quantity and quality combine to form the various inorganic and organic substances, and their transformations take place according to laws; the laws of nature are unchangeable and generally valid. The attempt to "elucidate" phenomena by arbitrarily naming them is to be avoided (for instance, the "vital force" assumed by Justus Liebig). Büchner justifiably conceived the works of Darwin (for the dissemination of which he is especially to be lauded), Lyell, Kirchhoff, and Haeckel, and also the development of chemistry (which forced elimination of the abrupt barrier between inorganic and organic combinations), to be confirmations of his point of view.

BIBLIOGRAPHY

I. ORIGINAL WORKS. Büchner's writings include *Kraft und Stoff. Empirisch-naturphilosophische Studien* (Frankfurt am Main, 1855), later entitled *Kraft und Stoff oder Grundzüge der natürlichen Weltordnung* (20th ed., Leipzig, 1902), also trans. as *Force and Matter, or Principles of the Natural Order of the Universe* (London, 1884; repr. New York, 1950); *Natur und Geist* (Frankfurt am Main, 1857; 3rd ed., Leipzig, 1876); *Physiologische Bilder*, I (Leipzig, 1861; 3rd ed., 1886), II (Leipzig, 1875; new ed., 1886); *Aus Natur und Wissenschaft, Studien, Kritiken und Abhandlungen*, I (Leipzig, 1862; 3rd ed., 1874), II (Leipzig, 1884); *Sechs Vorlesungen über die Darwin'sche Theorie von der Verwandlung der Arten und die erste Entstehung der Organismenwelt* (Leipzig, 1868), later entitled *Die Darwin'sche Theorie von der Entstehung und Umwandlung der Lebewelt* (5th ed., Leipzig, 1890); *Der Mensch und seine Stellung in der Natur in Vergangenheit, Gegenwart und Zukunft* (Leipzig, 1869; 3rd ed., 1889), trans. as *Man in the Past, Present and Future* (London, 1872); *Der Gottesbegriff und dessen Bedeutung in der Gegenwart* (Leipzig, 1874), 3rd ed. entitled *Gott und die Wissenschaft* (Leipzig, 1897); *Aus dem Geistesleben der Thiere, oder Staaten und Thaten der Kleinen* (Berlin, 1876; 4th ed., Leipzig, 1897), trans. as *Mind in Animals* (London, 1903); *Liebe und Liebes-Leben in der Thierwelt* (Leipzig, 1879, 1885); *Licht und Leben* (Leipzig, 1882, 1897); *Die Macht der Vererbung und ihr Einfluss auf den moralischen und geistigen Fortschritt der Menschheit* (Leipzig, 1882, 1909); *Der Fortschritt in Natur und Geschichte im Lichte der Darwinschen Theorie* (Stuttgart, 1884); *Der neue Hamlet* (Zurich, 1885; new ed., Giessen, 1901), written under the pseudonym Karl Ludwig; *Über religiöse und wissenschaftliche Weltanschauung* (Leipzig, 1887); *Thatsachen und Theorien aus dem naturwissenschaftlichen Leben der Gegenwart* (Berlin, 1887); *Das künftige Leben und die moderne Wissenschaft* (Leipzig, 1889); *Fremdes und Eigenes aus dem geistigen Leben der Gegenwart* (Leipzig, 1890); *Das goldene Zeitalter, oder das Leben vor*

der Geschichte (Berlin, 1891); *Das Buch vom langen Leben, oder, die Lehre von der Dauer und Erhaltung des Lebens (Makrobiotik)* (Leipzig, 1892), *Darwinismus und Sozialismus, oder der Kampf um das Dasein und die moderne Gesellschaft* (Leipzig, 1894; 2nd ed., Stuttgart, 1906); *Am Sterbelager des Jahrhunderts* (Giessen, 1898, 1900); *Im Dienste der Wahrheit* (Giessen, 1900), with a biography of the author by Alex Büchner; and *Last Words on Materialism and Kindred Subjects* (London, 1901).

II. SECONDARY LITERATURE. Writings dealing with Büchner or his work are Arthur Drews, *Die deutsche Spekulation seit Kant*, II (Leipzig, 1895), 267–281; Julius Frauenstädt, *Der Materialismus. Seine Wahrheit und sein Irrthum. Eine Erwiderung auf Louis Büchner's "Kraft und Stoff"* (Leipzig, 1856); and Friedrich Albert Lange, *Geschichte des Materialismus und Kritik seiner Bedeutung in der Gegenwart*, II (5th ed., Leipzig, 1896), 89–97, also available in English as *The History of Materialism* (London, 1877–1879).

JOACHIM THIELE

BUCHOLZ, CHRISTIAN FRIEDRICH (*b.* Eisleben, Germany, 19 September 1770; *d.* Erfurt, Germany, 9 June 1818), *chemistry, pharmacy.*

Bucholz's father, an obscure apothecary, died in 1775, leaving his five-year-old son the heir to a pharmacy. Two years later the boy's mother married an eminent Erfurt pharmacist named Voigt. At a very early age, Bucholz showed a talent and liking for chemical research, and under the tutelage of his stepfather and his uncle, the pharmaceutical chemist W. H. S. Bucholz, he rapidly acquired the background and training necessary to his chosen profession. In 1784 he was sent to Kassel as apprentice to the pharmacist Karl Wilhelm Fiedler. There he not only learned his professional duties but also taught himself languages and natural science. Bucholz left Kassel in 1789 and went to Ochsenfurt, Franconia, where for two years he worked as an apothecary's assistant; he then moved to Mulhouse, where for three years he was an associate in an apothecary. Here he completed his first publication, a paper on the crystallization of barium acetate, in 1794. Toward the end of 1794 he returned to Erfurt and took over the pharmacy he had inherited from his father. The following year, he married and began raising a family.

Bucholz's researches in chemistry were primarily of an analytical nature. Altogether, he published over a hundred articles in German chemical and pharmaceutical journals, including Scherer's *Neues allgemeines Journal der Chemie*, Trommsdorff's *Journal der Pharmacie*, Gehlen's *Journal für die Chemie und Physik* (of which he was, after 1804, one of the editors), and Schweigger's *Journal für Chemie*. Some of his more important papers were translated into French (primarily for the *Annales de chimie* and the *Journal des mines*) and English (especially for *Nicholson's Journal*). In addition, he published several books on chemistry and pharmacy.

Bucholz made a few important, but no primary, contributions to chemistry. He investigated in detail some of the more obscure compounds of sulfur. He made extensive analyses of the salts of molybdenum, tungsten, and tin, and he extracted uranium compounds from pitchblende. Bucholz distinguished strontium and barium oxides from the hydroxides of those metals by showing that the former were infusable whereas the latter were not. He investigated methods for the separation of copper and silver, iron and manganese, nickel and cobalt, and magnesium and calcium. He also carried out numerous mineral analyses and investigated several organic compounds, including camphoric acid, which he identified.

In 1808 Bucholz received the doctorate in pharmacy from the University of Rinteln, and in 1809 the University of Erfurt awarded him a Ph.D. and a position as *Assessor* (assistant) at its College of Medicine. In the following year, he was made a professor at the University of Erfurt and was given a place on the Faculty of Philosophy. He became privy councillor in the tiny principality of Schwarzburg-Sonderhausen in 1815. Bucholz was also active in the attempt to improve the lot of his fellow pharmacists. He cooperated with F. A. C. Gren in the establishment of an institution for retired pharmacists, and with J. B. Trommsdorff he founded an apothecaries' syndicate.

Bucholz's health began to fail about 1813. During the occupation of Erfurt by the French in 1813, he was cruelly imprisoned with about thirty of his fellow citizens and held for ransom. As a result, he suffered further illnesses and, finally, total blindness. In his final years he was able to publish only with the aid of his student Rudolph Brandes. Since Bucholz made no signal contribution to his field, his reputation, which had been founded on a large number of exacting analytical researches, did not survive long after him.

BIBLIOGRAPHY

I. Original Works. Two of Bucholz's books are *Beiträge zur Erweiterung und Berichtigung der Chemie,* 3 vols. (Erfurt, 1799–1802); and *Grundriss der Pharmacie mit vorzüglicher Hinsicht auf die pharmaceutische Chemie* (Erfurt, 1802). See list in Schreger article (below).

II. Secondary Literature. A detailed and heavily footnoted life of Bucholz, containing references to works not cited in the usual sources, is T. Schreger, in J. G. Ersch and J. G. Gruber, eds., *Allgemeine Encyclopädie der Wissenschaften und Künste,* XIII (Leipzig, 1824), 303–305. See also Poggendorff, I, 330; and J. R. Partington, *A History of Chemistry,* III (London, 1962), 581–582.

J. B. Gough

BUCKINGHAM, EDGAR (*b*. Philadelphia, Pennsylvania, 8 July 1867; *d*. Washington, D.C., 29 April 1940), *physics.*

Edgar Buckingham is best known for his early work on thermodynamics and for his later study of dimensional theory. Especially attracted to problems that could not be solved by pure calculation but required experimentation as well, he showed more clearly than anyone before him how the planning and interpretation of experiments can be facilitated by the method of dimensions, later called dimensional analysis.

Buckingham graduated from Harvard in 1887 and received his Ph.D. at Leipzig in 1893. His book on thermodynamics (1900) carried the reader from the simplest facts of temperature measurement through the equilibrium of heterogeneous systems. It was a powerful contribution toward clarifying the subject.

Between 1891 and 1901, Buckingham taught physics at Harvard, Bryn Mawr College, and the University of Wisconsin. He entered government service as an assistant physicist in the Bureau of Soils in 1902, and was employed by the Bureau of Standards from 1905 until his retirement in 1937. His natural ability as a teacher and lecturer was evident in his lectures on technical thermodynamics in the Naval Postgraduate School at Annapolis (1911–1912) and in the graduate program of the National Bureau of Standards (1912–1913). Buckingham served as a technical expert on the Council of National Defense during World War I and as an associate scientific attaché with the United States Embassy in Rome from 1918 to 1919. He was frequently called upon during his years with the Bureau of Standards to advise the Navy Department on steam turbine and propeller research.

Buckingham's treatise on thermodynamics was followed by fifty or more scientific papers. The subjects included soil physics, properties of gases, blackbody radiation, acoustics, fluid mechanics, and dimensions. Several of his noteworthy papers on dimensional theory, listed in the bibliography, awakened a lively interest in dimensional methods and their practical application.

He pointed out the advantages of dimensionless variables and how to generalize empirical equations. His frequently cited "pi-theorem" serves to reduce the number of independent variables and shows how to experiment on geometrically similar models so as to satisfy the most general requirements of physical as well as dynamic similarity.

BIBLIOGRAPHY

Buckingham's works include *An Outline of the Theory of Thermodynamics* (New York, 1900); "Physically Similar Systems . . .," in *Physical Review,* 2nd ser., **4** (1914), 345–376; "Windage Resistance of Steam Turbine Wheels," in *Bulletin of the Bureau of Standards,* **10** (1914), 191–234; "Model Experiments and the Forms of Empirical Equations," in *Transactions of the American Society of Mechanical Engineers,* **37** (1915), 263–296; "Notes on the Methods of Dimensions," in *Philosophical Magazine,* 7th ser., **42** (1921), 696–719; see also **48** (1924), 141–145; and "Dimensional Analysis of Model Propeller Tests," in *Journal of the American Society of Naval Engineers,* **48** (1936), 147–198.

MAYO DYER HERSEY

BUCKLAND, WILLIAM (*b.* Axminster, England, 12 March 1784; *d.* Islip, England, 14 August 1856), *geology, paleontology.*

Buckland's father, Charles, was rector of Templeton and Trusham; his mother, Elizabeth, was the daughter of a landed proprietor established in Devon since the seventeenth century. Buckland became interested in rocks and fossil shells by playing among them in the valley of the River Axe, in local quarries, and at the seashore around Lyme Regis. He went on collecting rambles with his father, who had a taste for ammonites and related shells. He also collected birds' eggs and observed the habits of fishes. At Winchester School, Buckland was a good Latin student and became familiar with chalk formations through the common practice of digging for field mice in nearby chalk pits. Environment, family, church, and school experiences had fixed his interest on natural history by the time he entered Oxford.

Buckland won a competitive examination for a scholarship at Corpus Christi College in 1801. He graduated in 1804 with a very good examination and continued in residence, supporting himself on his scholarship and by taking pupils. He was elected fellow of his college and admitted to holy orders in 1809.

Around this period some people at Oxford, partly stimulated by John Kidd's lectures, were showing an interest in geology. Among the members of this group, besides Buckland, were J. J. and W. D. Conybeare, Charles Daubeny, John and Philip Duncan, and W. J. Broderip. The latter introduced Buckland to fieldwork, and Buckland considered the younger man to be his tutor in geology. Broderip was knowledgeable in conchology; he had been instructed by Joseph Townsend, a friend of William Smith. Buckland was thus initiated into the new fossil geology at the beginning of his career. Broderip remained a close friend and scientific adviser; and Buckland sought out Townsend and Benjamin Richardson, another friend of Smith's, on his trips from Oxford to Axminster. Other early geological friends were Henry De la Beche, who grew up at Lyme Regis, and George Greenough.

In 1813 Buckland was elected reader in mineralogy to succeed Kidd, and became a fellow of the Geological Society of London; his first publication, of sorts, was in 1814. His lectures included geology, and were well received. He was appointed to a new readership (not professorship) in geology in 1818; the motives and politics of this endowment have not been satisfactorily explained. Thereafter Buckland usually styled himself "professor" on matter published in London and "reader" on matter published in Oxford. He gave two sets of lectures yearly until 1849, and usually prepared new lecture notes each year. Apparently he emphasized causal explanations of the visible phenomena. Buckland was an active participant in town affairs; among other matters, he was instrumental in introducing gas lighting in 1818, and became chairman of the Oxford gas company.

From 1808 to 1815 Buckland made geological tours of England and other parts of the British Isles. In 1816, with Greenough and W. D. Conybeare, he began his European tours, which eventually took him to Germany, Poland, Austria, Italy, Switzerland, and France. Cuvier visited him at Oxford in 1818, and he visited Cuvier at Paris several times.

Buckland was awarded the Copley Medal of the Royal Society for his cavern researches published in 1822. He was president of the Geological Society of London in 1824–1825 and again in 1840–1841, and was a member of the Council of the Royal Society from 1827 to 1849. In 1825 he accepted a country parsonage in the gift of his college, presumably so that he could marry, but was then appointed a canon of Christ Church, Oxford. These canonries were among the richest governmental rewards for academic distinction without serious administrative responsibilities. Buckland married Mary Morland, of Sheepstead House, near Abingdon, Berkshire, the same year; they had five children who survived. Mary Buckland assisted her husband with his writing, and by drawing illustrations and reconstructing fossils according to his instructions.

Buckland's Continental trips, especially those of 1826 and 1827, made him aware of German attempts to found an annual meeting of scientific men. His public prominence and international contacts made him an obvious choice for president and host of the second meeting of the British Association for the Advancement of Science in 1832 (the first full scientific meeting, since the one in 1831 was largely devoted to organizing the association). He was thereafter active at its meetings. He played a major role in the establishment of the Museum of Practical Geology and affiliated activities of his friend De la Beche.

Buckland was made dean of Westminster in 1845 by the Tory prime minister, Robert Peel, an admirer of his work. He left Oxford for London rather willingly, feeling that he had tried for forty-four years to spread a taste for science at the university, and had failed. We can see that he had raised geology to rank alongside the more prominent sciences, such as anatomy, and had helped foster the growing interest in science that led to its inclusion in the examination curriculum and to the building of the Oxford Museum, in the next decade. His own scientific work, while perhaps ultimately not so significant as that being done in physics at Cambridge, and not leading to an "Oxford school of geology," did give Oxford an international name in science that it did not have in humanistic or biblical scholarship.

As dean of Westminster, Buckland was a vigorous administrator, repairing the physical deterioration of the abbey and the school, and restoring the school scholastically to the status of an effective modern educational institution. He took an interest in local sanitary reform. Basically a liberal in Anglican Church politics, he took particular pleasure in acting as host for the consecration of four missionary bishops, a move opposed by some factions.

Buckland held the rectory of Islip, seven miles from Oxford, as a country home. He was a useful rector, and retired there when, at the end of 1849, he contracted a mysterious illness characterized by apathy and depression. He died seven years later. The autopsy showed that damage to the base of the skull caused by a carriage accident in Germany thirty years before had developed into an advanced state of decay. His wife died a year later, apparently as a result of the same accident. Both were buried in Islip churchyard.

Personally, Buckland was characterized by great energy. His whole life and his household were organized around geology. His unfailing sense of humor puzzled and annoyed some of his more Victorian-minded colleagues, such as Charles Lyell and Adam Sedgwick, and led John Henry Newman to distrust geology altogether. On the other hand, John Ruskin found him stimulating, as did most of his other auditors. Buckland took religion and geology seriously, but took himself, other geologists, and most geological theories much less so.

Buckland was not given to synthesis or system building, and there is a danger of attributing too much importance to his general theoretical positions, which were often derivative. His importance lay, rather, in helping to redefine the nature and method of a geological explanation. British stratigraphers before about 1815 had often been satisfied with a tracing of the strata (largely the secondary formations) or had gone all the way to a total system of geological dynamics. Following the suggestion of Cuvier in the first edition of his *Discours préliminaire,* Buckland and other geologists wished to produce detailed explanations that would in effect constitute a geological history, period by period, of the events in a given locality. To help in doing so, Buckland transferred Cuvier's method of reconstructing fossil animals to geology proper: that is, he tried to reason from the analogies of the existing world (Cuvier's *création actuelle*) to the events of a past world, even though Cuvier himself had cast some doubt on the validity of this process in geology.

Nothing is more characteristic of Buckland's papers than the use of some immediately observable contemporary analogy—the habits of modern hyenas, the cavities formed by air bubbles in clay, the geographical locus of modern animals. His method also differed from that of Cuvier's *Discours* in not depending primarily on paleontological evidence. Although Buckland was one of Cuvier's great admirers and seemingly enjoyed correcting him on all kinds of specific points, his own method was to bring together stratigraphical, petrological, dynamic, and paleontological reasoning and observations on modern forms and habits of life to explain the phenomena of a given locality. This is well exemplified by his paper of 1830, written with De la Beche, on the geology of the neighborhood of Weymouth, which utilizes the techniques he had developed over the previous fifteen years.[1]

In paleontology Buckland's most interesting work was on still-existing forms, such as hyenas and bears, and on marine shells. Cuvier was not much interested in conchology, the study of variations among shells per se; Buckland's emphasis, while including the organisms that produced the shells, was perhaps a bit more in the conchological tradition of William Smith and James Sowerby.

Buckland was thus one of the men, perhaps the ablest and probably the most acute, who built a typically "British" geology, based on careful local stratigraphy and local dynamic explanations but revivified by the addition of fossil evidence. This British geology was recognizable as such for the next half-century, regardless of the ideological conceptions (uniformitarian or catastrophist, for example) of the geologist involved. Cuvier was delighted by the results and, in later editions of the *Discours,* cited Buckland as one of the men who had brought into being the new geology whose possibility was only indicated in the *Discours* itself.

On more particular points of theory, the actual sources of Buckland's positions cannot be documented without a study of private papers, although his relation to some positions of his predecessors can be indicated. The processes postulated by James Hutton for cyclic continent building were dismissed by many geologists as being unobservable and therefore of little scientific importance. Buckland ignored such hypothetical views in classifying theorists. He used "Huttonian" to designate those who explained the earth's surface features by the slow action of atmospheric agents. He opposed such geologists; he believed many surface features had originated in local elevations and dislocations. Like many British geologists, he agreed with Saussure, who had used the idea of a massive "debacle" of water to explain much valley excavation, distribution of alluvial gravel, and erratic transport. Unlike Kidd, but like Townsend, Buckland felt that geology could present positive evidence of this debacle. Like Townsend, he was inclined toward volcanism, although it played no essential role in his own work, and he was effective in spreading a taste for the study of volcanic action among British scientists in the 1820's. Basically anti-Wernerian, he nevertheless looked for a worldwide identification of equivalent strata. In opposition to Cuvier's suggestion, Buckland believed that the current continents are more or less permanent, and were dry land before the most recent catastrophe or debacle; their period of submersion under the ocean had been long before that. Like other geologists who had studied the false alarms of the eighteenth century, he was skeptical of evidence of fossil human bones in strata earlier than the debacle; he examined new cases carefully but concluded each time that the point was not proved.

If no other indication is given, the dates in the text of this essay indicate when the paper in question was read, not when it was published. Buckland's first important paper was read in 1816, when he was thirty-two years old. He modified his own theorizing concerning the cause of the deluge to take into account Louis Agassiz's glacial hypothesis in 1840, when he was fifty-six. A slow starter, he was alert throughout his career to adopt or react to new proposals.

The paper of 1816, on specimens from the plastic clay, shows a number of Buckland's continuing themes.[2] He was studying the British equivalent of the *argile plastique* of Cuvier and Alexandre Brongniart, located above the chalk and below the London clay. They had suggested, on the basis of the differences in organic remains, that each deposition was of immense scale and that a long time had intervened between the formations. Buckland confirmed the identity of the French and English formations on the basis of petrographical considerations, and the long periods involved on the basis of the processes that could have produced the rocks and fossils of the strata. The chalk must first have been consolidated; its breaking up produced chalk pebbles whose smoothness and roundness argued the long-continued action of water. These pebbles were mixed in with the clay before or during its solidification, and there were periods of repose long enough for myriad oysters whose shells were attached to the pebbles to live and die undisturbed. Thus Buckland added comparative geology, local dynamics, and stratigraphical and petrological considerations to the French fossil approach, in order to show the long periods involved and the great time gaps in this part of the geological record.

In 1819 Buckland delivered his inaugural lecture as reader in geology; it was published as *Vindiciae geologicae* in 1820. Ostensibly it was an orthodox presentation of geology as useful to religion, in response to an evangelical Presbyterian, the famous Scots preacher Thomas Chalmers, who objected not to geology but to any kind of natural theology. Geology extends the reign of final causes, said Buckland; it shows the existence of a recent universal deluge and the recent origin of man; geological epochs came (as Chalmers suggested) before the creation story, so there is no need to reconcile them with the biblical days of creation. Actually, the lecture was a careful definition of an independent position. Buckland corrected Cuvier as to the permanence of modern continents. He surrounded quotations from Deluc with a text that contradicted Deluc on the essential point: Buckland insisted that the world was made for all its inhabitants, not for man alone. He contradicted Kidd, now his faculty colleague, who believed that the Mosaic flood had taken place miraculously but had left no geological traces; Buckland maintained the reverse. His real antagonist, however, was an evangelical within the Anglican Church, John Bird Sumner, who was later archbishop of Canterbury. By careful selection Buckland made it appear that Sumner's position supported his own; but Buckland carefully avoided asserting a physical miracle, his "creative interference" being always a final, not an efficient, cause. Buckland's insistence on the actual evidence of a deluge was partly an answer to Sumner's insistence that the Mosaic records were much more reliable than geological evidence. The notion that with Buckland "Cuvier and orthodoxy were triumphant"[3] is an old one, but incorrect. For years one of Buckland's roles was to keep room clear for an independent evaluation of scientific evidence within the Anglican community, in spite of increasing pressures from Evangelicalism and, later, from Tractarianism.

Buckland's major work on the geological evidence for a recent deluge was his paper on the quartz pebbles of Lickey Hill in Worcestershire, read later in 1819.[4] By tracing the distribution of these pebbles as far east as London, he thought to trace the path of the deluge and to show that some valleys were scooped out by its waters. This paper, with its conclusion that such superficial gravel appears in similar circumstances all over the world, represents the high point of Buckland's belief in universal formations and universal events; and it is presented in unusually dogmatic form as compared to Buckland's more usual qualified assertions. In the next year came his first published dealings with non-European rocks, in a brief paper on resemblances between specimens from Madagascar and New South Wales, and English rocks; here he was more moderate.[5] In his major paper on the structure of the Alps in 1821, Buckland showed that formations on the flanks of the mountains were the equivalent of certain secondary formations in England, and that there is a regular order of succession in Alpine districts identical with that of England.[6] The table of equivalents annexed to the paper is quite useful. This paper was perhaps the most important work on the Alps between J. G. Ebel's treatise of 1808 and the work of Sedgwick and Murchison around 1830.

Buckland continued his dynamic and stratigraphic researches, and his important summary was "On the Formation of the Valley of Kingsclere," read in 1825.[7] As opposed to simple Huttonian erosion, he believed in the multicausal origin of valleys, in which elevation, fracture, diluvial currents, and erosion had all played their parts. And he noted that the Savoy Alps had been elevated from the ocean floor since the deposition of the Tertiary strata.

In 1822 Buckland published his study of the fossil bones found in Kirkdale Cave in Yorkshire, and in 1823 expanded it into a full-scale treatise, *Reliquiae diluvianae*.[8] The dedication to Bishop Shute Barrington, Sumner's patron in the church, pointedly hoped it would no longer be asserted that there is no geo-

logical evidence for a universal deluge but reminded the bishop that the deluge's physical cause was still unknown.

The deluge, however, was not the important novelty of the *Reliquiae*. Buckland considered the work his "hyena story,"[9] for he proposed that the cave had been the den of hyenas; he not only found fossil feces ("coprolites") and tooth-marked bones, but also made observations on the habits of modern hyenas. His important conclusion was that species of animals that now exist together only in the tropics had coexisted in northern Europe with species still in existence, and that this demonstrated a tropical climate in antediluvial times, before the deluge buried the bones in a layer of mud. Further, the bones and caves showed that Europe had then been dry land much as it is now. In another cave Buckland found a human skeleton which, since the cave showed signs of human disturbance in historic time, he took to be postdiluvial.

The *Reliquiae* was well received for its scientific content, although some critics felt that Buckland had pushed the use of analogies from "modern causes" too far. Cuvier was very pleased with it, although he did not fully concede the general validity of Buckland's reasoning concerning climate. Buckland's assertion of the reality of the Mosaic flood as shown by paleontological evidence, although praised by his friend Edward Copleston in the *Quarterly Review,* was widely attacked by other critics and even by James Smithson, who, somewhat confused, was under the impression that he was defending Buckland from an attack by Granville Penn.[10] Buckland's Oxford students found the idea more amusing than convincing. Apparently it needed only to be stated clearly and fully in order to seem unconvincing; in the popular *Conversations on Geology* of 1828 the instructress says that Penn's theory was "no less fanciful than Mr. Buckland's."[11] Buckland's geological evidence for a large-scale force or agent acting in geologically recent times remained intact, but he quietly abandoned its identification with the Mosaic flood. For several years he intended a second volume of the *Reliquiae* but never published it because he could propose no convincing physical cause of the debacle.

Buckland took part in the giant saurian hunt of the 1820's, perhaps more as a follower than as a leader, although he deserves much of the credit for the Megalosaurus. He sometimes acted as geological intermediary between the discoverer in the field (who was often a layman) and such expert anatomists as Clift and Broderip. His own striking contribution was his paper (1829) on the coprolites of Ichthyosauri.[12] These coprolites permitted the reconstruction of a soft internal organ of an extinct species and indicated the species' eating habits; they proved that carnivorous "warfare" had always been a law of nature "to maintain the balance of creation." Further, their preservation in vast amounts furnished a "geological chronometer" of a period of undisturbed accumulation at the bottom of a sea. Coprolites should therefore be looked for in all periods when vertebrates had existed.

Toward the end of 1828 Charles Lyell and Roderick Murchison mounted an attack on Buckland's valley-formation ideas. W. D. Conybeare labeled the debate one between "fluvialists" (Lyell and Murchison) and "diluvialists" (Conybeare and Buckland). Since the fluvialists could only show that the particular valleys they discussed were caused by erosion, and since Buckland himself had shown that some valleys were so caused, the debate was inconclusive; and Lyell agreed to Buckland's general position in his *Principles of Geology*.[13] The debate seems, however, to have stimulated Adam Sedgwick and William Hopkins to extend Buckland's notions of elevation and dislocation as agents of surface formation in the 1830's and 1840's.

In 1830 Buckland was nominated to write the geological work in a series of books on natural theology that stemmed from the will of the eccentric eighth earl of Bridgewater; the final contracts were signed in 1832. We may assume that most of his energies until 1836 were directed to this project. Thus his celebrated explanations of the habits of the fossil *Megatherium* and the present-day sloth were devoted to showing how perfect their organization is for their mode of life,[14] and the same examples reappear in the Bridgewater treatise.

Buckland was in the middle of a general conservative revolt at Oxford, led by the Tractarians, and his treatise was completed during the spring of 1836, when their fierce opposition to the appointment of the mildly liberal R. D. Hampden as professor of divinity drove a wedge of bitterness into the Anglican Church. Buckland's natural theology conceded nothing to the new religious challenge. His position was essentially the same as in 1819, but its theological liberalism was by now more obvious, and was presented at length and without subterfuge. Buckland particularly emphasized William Paley's position that the world was not made for man alone but for the pleasure of all species of life; in relation to the object to be attained, all organic mechanisms are equally good, are evidence of beneficent adaptation. He reasserted that it is futile to try to reconcile geological epochs with the days of creation in Genesis, and now openly renounced the identification of his geological

deluge with the Mosaic flood. He went no further toward admitting miracles as physical causes than he had done in 1819. The final cause of successive organic systems, he said, is the purpose of maintaining the greatest possible amount of life on earth at all times. He was insistent that the past was regulated by the same laws and processes as the present, and showed the same kind of ecological balance. This demonstrated the unity of the Deity (whereas the Tractarians were fond of saying that natural theology tended to polytheism).

As a geological system Buckland chose, possibly borrowing from De la Beche and Conybeare, progressive development from an initially hot earth, with discontinuous assemblages of organic life being created and dying out. To express a secular development while simultaneously rejecting continuous progress and transmutation, he deliberately kept the rhetoric of the Great Chain of Being, but with missing links or gaps in the present creation being filled up by fossil organisms from past time periods. This was a noteworthy change from Cuvier, and a major step in the conversion of a balanced Malthusian ecology into a system maintaining its balance while it changed over time. Although not agreeing with Charles Lyell's uniformitarianism, Buckland cited the *Principles of Geology* with respect. The treatise was the major general view of paleontology produced in Britain in the period; Buckland's own new contributions were on mollusks, especially the mechanical contrivances (for example, syphons) used in chambered shells.

His Bridgewater treatise was Buckland's last sustained independent scientific work. He became increasingly interested in Roman archaeology and in the practical applications of geology, particularly the drainage of farms and the use of manures. He spread knowledge of Liebig's work but also advocated the widespread use of the natural phosphates contained in the large beds of coprolites he had identified.

In the period from 1838 to 1840 Buckland at last found a physical cause for a geologically recent catastrophe. Louis Agassiz convinced him that much of his evidence constituted signs of widespread glaciation. He and Agassiz delivered papers to the Geological Society of London in November 1840 on glaciation in Britain, and Buckland gave two more in 1840 and 1841.[15] He did not agree completely with Agassiz, however. He thus had the opportunity, of which he took full advantage in his presidential address to the Geological Society in 1841,[16] to accuse both the "glacialists" (Agassiz) and the "diluvialists" (by whom he meant especially Roderick Murchison) of "extreme opinions." He himself compromised by asserting the influence both of glaciers and of the torrents of water released as the glaciers melted, and of the icebergs drifted along in waters. An explanation for Saussure's debacle had been found at last.

NOTES

1. "On the geology of the neighbourhood of Weymouth and the adjacent parts of the coast of Devon," in *Transactions of the Geological Society of London,* 2nd ser., **4** (1836), 1–46.
2. "Description of a series of specimens from the plastic clay near Reading, Berks," *ibid.,* **4** (1817), 277–304.
3. Robert Knox, *The Races of Men* (London, 1850), p. 170.
4. "Description of the quartz rock of the Lickey Hill in Worcestershire, and of the strata immediately surrounding it; with considerations on the evidence of a recent deluge, afforded by the gravel beds of Warwickshire and Oxfordshire, and the valley of the Thames from Oxford downwards to London; and an Appendix, containing analogous proofs of diluvian action. Collected from various authorities," in *Transactions of the Geological Society of London,* **5** (1821), 506–544.
5. "Notice on the geological structure of a part of the island of Madagascar, founded on a collection transmitted to the Right Honourable the Earl Bathurst, by Governor Farquhar, in the year 1819; with observations on some specimens from the interior of New South Wales," *ibid.,* 476–481.
6. "Notice of a paper laid before the Geological Society on the structure of the Alps and adjoining parts of the continent, and their relation to the secondary and transition rocks of England," in *Annals of Philosophy,* n.s. **1** (Jan.–June 1821), 450–468.
7. "On the formation of the valley of Kingsclere and other valleys by the elevation of the strata that enclose them; and on the evidence of the original continuity of the basins of London and Hampshire," in *Transactions of the Geological Society of London,* 2nd ser., **2** (1829), 119–130.
8. "Account of an assemblage of fossil teeth and bones of elephant, rhinoceros, hippopotamus, bear, tiger, and hyaena, and sixteen other animals discovered in a cave at Kirkdale, Yorkshire, in the year 1821: with a comparative view of five similar caverns in various parts of England, and others on the continent," in *Philosophical Transactions,* **112** (1822), 171–235; *Reliquiae diluvianae* (London, 1823). The article constitutes the first section of the book. The first edition of the book sold out and there was a second edition in 1824.
9. North, "Paviland Cave," p. 103.
10. [Edward Copleston], "Buckland-*Reliquiae Diluvianae*," in *Quarterly Review,* **23** (1823), 138–165; James Smithson, "Some observations on Mr. Penn's theory concerning the formation of the Kirkdale cave," in *Annals of Philosophy,* n.s. **8** (July–Dec. 1824), 50–60.
11. (London, 1828), p. 341.
12. "On the discovery of coprolites, or fossil faeces, in the lias at Lyme Regis, and in other formations," in *Transactions of the Geological Society of London,* 2nd ser., **3** (1835), 223–236.
13. (London, 1830), I, 171–172.
14. "On the adaptation of the structure of the sloths to their peculiar mode of life," in *Transactions of the Linnean Society,* **17** (1837), 17–28.
15. "Memoir on the evidences of glaciers in Scotland and the north of England," in *Proceedings of the Geological Society of London,* **3** (1838–1842), 332–337; "Second part of memoir on the evidence of glaciers in Scotland and the north of England," *ibid.,* 345–348; "On the glacio-diluvial phenomena in Snowdonia and the adjacent part of north Wales," *ibid.,* 579–584.
16. "Presidential Address for 1841," *ibid.,* 509–516.

BIBLIOGRAPHY

I. ORIGINAL WORKS. The list of articles in the Royal Society's *Catalogue of Scientific Papers* (I, 702–705) is handy but incomplete, and the serious student will want to use the list of publications given by Francis Buckland in his "Memoir of the Author," printed in the 1858 ed. of Buckland's Bridgewater treatise, *Geology and Mineralogy Considered With Reference to Natural Theology.* This is in infuriating disarray but is tolerably complete; it lists most but not all abstracts in the British Association's *Reports,* three separately printed sermons, and the forty-odd brief abstracts in the *Proceedings* of the Ashmolean Society of Oxford (most of which are of little importance). The following corrections may be noted (numbers are those of Francis Buckland's list): No. 3: 1823; 2nd ed., 1824. No. 14: n.s. **10** (1825); same as no. 55. No. 19: unlocated, but probably same as no. 68. No. 24: published at the end of William Phillips, *A Selection of Facts From the Best Authors, Arranged so as to Form an Outline of the Geology of England and Wales* (London, 1818). No. 56: same as no. 70. No. 58: same as no. 1. No. 69: II (1814)—this is not listed by any author's name but is "compiled by the Secretaries." No. 29 in the Ashmolean Society list is not in its *Proceedings* but is no. 45 in the main list. To this list may be added "Notice of a Series of Specimens From Mr. Johnson's Granite Quarries," in *Reports of the British Association,* **11** (1841), trans. sect., 64; "Notice of Perforations in Limestone," *ibid.,* **12** (1842), trans. sect., 57; and "On the Cause of the General Presence of Phosphorus in Strata," *ibid.,* **19** (1849), trans. sect., 67.

Francis Buckland refers to a paper entitled "On the Coasts of the North of Ireland." This appears to be W. D. Conybeare, "Descriptive Notes Referring to the Outline of Sections Presented by a Part of the Coasts of Antrim and Derry. . . . From the Joint Observations of the Rev. W. Buckland," in *Transactions* of the *Geological Society of London,* **3** (1816), 196–216. It is not Buckland's "first important paper."

There seem to be no major collections of Buckland's papers. The most interesting group may be that at the National Museum of Wales, Cardiff. There are some papers, especially letters to Buckland, in the Devon County Record Office, Exeter. Christ Church has 21 letters and some material on Buckland's career as canon. The University Museum, Oxford, has MS material relating to his lecture notes and publication drafts. The Bodleian Library has 46 scattered letters, and there are about 35 in the Whewell papers, Trinity College, Cambridge. There are probably others elsewhere.

II. SECONDARY LITERATURE. The biographical material presented by his children—Francis Buckland's "Memoir," cited above, and Anna B. Gordon, *The Life and Correspondence of William Buckland* (New York, 1894)—is indispensable but incomplete and sometimes vague. Two articles by F. J. North are based on unpublished materials: "Paviland Cave, the 'Red Lady', the Deluge, and William Buckland," in *Annals of Science,* **5** (1942), 91–128; and "Centenary of the Glacial Theory," in *Proceedings of the*

Geologists' Association, **54** (1943), 1–28. Otherwise there has been no serious treatment of Buckland's work, nor does any general work place it adequately in the history of geology. Attempts, both inadequate, to place Buckland's theological position in its contemporary intellectual setting are Reijier Hooykaas, *Natural Law and Divine Miracle* (London, 1959), pp. 147, 190–201; and W. F. Cannon, "Problem of Miracles in the 1830's," in *Victorian Studies,* **4** (1960), 5–32. The latter incorrectly attributes to Buckland a statement of belief in physical miracles.

WALTER F. CANNON

BUCQUET, JEAN-BAPTISTE MICHEL (*b.* Paris, France, 18 February 1746; *d.* Paris, 24 January 1780), *chemistry.*

Bucquet, the son of a lawyer, was destined for the bar by his father, but he soon abandoned jurisprudence to take up the study of medicine. He became *docteur-régent* at the Faculty of Medicine in Paris in 1770, by which time chemistry had become his main interest. His enthusiasm for it was prompted by the conviction that chemistry held the key to natural history and medicine, and, toward the end of 1770, he began to give courses in which he linked natural history and chemistry.

From 1775 to 1777, Bucquet was professor of pharmacy at the Faculty of Medicine and, at the death of Augustin Roux in 1776, he was elected to succeed him in the chair of chemistry. Bucquet gave his first public course in chemistry in 1777. He continued the chemistry courses he had been giving in private laboratories, in addition to those at the Faculty, until ill health forced him to abandon these private courses in the autumn of 1779. At different times during his career Bucquet also gave courses in botany, physiology, anatomy, hygiene, and medicine.

In February 1777 Bucquet was elected associate of the Société Royale de Médecine, and he took the place of Sage as *adjoint-chimiste* at the Academy of Sciences in Paris on 14 January 1778.

Bucquet's aim was to integrate chemistry with all the subjects related to it, but he found the published work in chemistry too unreliable to act as a basis for his project. Accordingly, he decided to repeat much of the experimental work that had already been done. Because of this decision and his early death, he made little original contribution to chemistry.

Among Bucquet's published works are memoirs on the analysis of minerals and the chemistry of gases. He modified David Macbride's apparatus to produce and investigate "fixed air" (carbon dioxide) and showed that quicklime would react with it only in aqueous solution. Later, after collecting and measuring the carbon dioxide produced by heating marble,

and confirming that carbon dioxide was acid, he suggested the name *acide crayeux* for it. Bucquet's work on gases was probably of great use to Lavoisier.

The memoirs on ethers and the analysis of blood, which Bucquet read, were never published. Of these, the one on ethers, read in March 1777, is the more interesting. In addition to the memoirs, Bucquet published books on mineral and plant chemistry. His work on plant chemistry seems to be the first detailed account of that branch of chemistry to be published.

From 1777 on Bucquet worked with Lavoisier on a number of topics, and the project they conceived of repeating most of the fundamental experiments done in chemistry up to that time, in preparation for an early draft of Lavoisier's *Traité de chimie,* was so perfectly in keeping with Bucquet's avowed aims that there is little doubt that he was the originator. Bucquet seems to have been the first to teach Lavoisier's theory, which he began to include in his courses as early as 1778.

BIBLIOGRAPHY

I. ORIGINAL WORKS. Bucquet's doctoral theses, which are to be found in the Arsénal Library, Paris, are "An digestio alimentorum, vera digestio chymica?" (Jan. 1769); "An recèns nato, lac recèns enixae matris?" (Mar. 1769); "An in febre malignâ balneum?" (Jan. 1770); and "An in partu difficili, sola manus instrumentum?" (Mar. 1770).

His books are *Introduction à l'étude des corps naturels tirés du règne minéral,* 2 vols. (Paris, 1771); and *Introduction à l'étude des corps naturels tirés du règne végétal,* 2 vols. (Paris, 1773).

Published memoirs are "Premier mémoire sur plusieurs combinaisons salines de l'arsenic," in *Mémoires mathématiques et physiques . . . par divers savans,* **9** (1780), 643–658; "Seconde mémoire sur les combinaisons salines de l'arsenic," *ibid.,* 659–672; "Expériences physicochimiques sur l'air qui se dégage des corps dans le temps de leur décomposition, et qu'on connoît sous le nom vulgaire d'air fixé," *ibid.,* **7** (1776), 1–17; "Sur quelques circonstances qui accompagnent la dissolution du sel ammoniac par la chaux vive, par les matières métalliques et par leurs chaux, relativement aux propriétés attribuées à l'air fixé," *ibid.,* **9** (1780), 563–575; "Analyse de la zéolite," *ibid.,* 576–592; "Sur l'analyse de l'opium," in *Mémoires de la Société royale de médecine* (1779), 399–404; *Sur la manière dont les animaux sont affectés par différens fluides aériformes méphitiques, et sur les moyens de remédier aux effets de ces fluides; précédé d'une histoire abrégée de ces différens fluides aériformes ou gaz* (Paris, 1778); and *Rapport sur l'analyse du rob antisyphillitique du Sr. Laffecteur* (Paris, 1779).

Manuscript works are a letter to an unknown person, concerning the preparation of different kinds of ether, dated Paris, 22 Aug. 1773, Bibliothèque de Besançon, MS

1441, fols. 135–136; "Sur les moyens d'obtenir facilement les éthers nitreux et marin," Académie des Sciences, MS in dossier for 19 Mar. 1777; "Analyse chimique du sang," Académie des Sciences, MS in Bucquet dossier; "Notice abrégée de différents mémoires de chymie dont quelquesuns sont les fruits de mes recherches particulières, les autres ont été faits et rédigés conjointement avec Mr. Lavoisier," Académie des Sciences, MS in Bucquet dossier. Manuscript records of courses given by Bucquet are "Précis des leçons de chimie de feu M. Bucquet," Bibliothèque de St. Brieuc, MS 106; "Analyse du cours de phisiologie de M. Bucquet, commencé le 17 août 1773, fini le 21 octobre 1773," Bibliothèque de Rheims, MS 1021 (N fonds); and Lavoisier's laboratory notebooks, Vol. V, Académie des Sciences, MS.

II. SECONDARY LITERATURE. Works on Bucquet are E. McDonald, *Jean-Baptiste Michel Bucquet (1746–1780)— His Life and Work,* M.Sc. dissertation (Univ. of London, 1965); and "The Collaboration of Bucquet and Lavoisier," in *Ambix,* **13,** no. 2 (1966), 74–83.

E. MCDONALD

BUDAN DE BOISLAURENT, FERDINAND FRANÇOIS DÉSIRÉ (worked Paris, France, *ca.* 1800–at least 1853), *mathematics.*

Almost nothing is known of Budan's life except for the information he provided on the title pages of his published works. He was a doctor of medicine and an amateur mathematician. He was educated in the classics and occasionally quoted Virgil and Horace in his works. A royalist, he published a Latin ode on the birth of the posthumous son of the duke of Burgundy. Budan was named *chevalier* of the Legion of Honor in 1814. He held the post of inspector general of studies at the University of Paris for over twenty years; this post may have been responsible for his interest in finding mathematical methods that would be easy for beginning students to use.

Budan is known in the theory of equations as one of the independent discoverers of the rule of Budan and Fourier, which gives necessary conditions for a polynomial equation to have n real roots between two given real numbers. He announced his discovery of the rule and described its use in a paper read to the Institut de France in 1803 and published the paper, with explanatory notes, as *Nouvelle méthode pour la résolution des équations numériques d'un degré quelconque,* in 1807.

Budan's definitive formulation of his rule was the following: "If an equation in x has n roots between zero and some positive number $p,$ the transformed equation in $(x - p)$ must have at least n fewer variations [in sign] than the original" ("Appendice," p. 89). The "transformed equation in $(x - p)$" is the original polynomial equation developed in powers of $(x - p).$

In modern notation: let $P(x) = 0$ be the given polynomial equation, and let $G(x - p) = P(x).$ Then $G(x - p) = 0$ is what Budan called the "transformed equation in $(x - p)$." The term "variation in sign" is borrowed from Descartes's rule of signs: No polynomial can have more positive roots than there are variations in sign in the successive terms of that polynomial. Indeed, Budan appears to have been led to his rule by Descartes's rule of signs.

Budan's first formulation of his rule assumed that all the roots of the original equation were real. In this case, Budan's rule tells exactly how many roots there are between zero and $p,$ just as Descartes's rule gives the exact number of positive roots in the same case. Budan stated that, for the case of all real roots, his rule could be derived from Descartes's rule. It is not difficult to reconstruct such a derivation, even though Budan did not give it, once one has observed that when x is between zero and $p,$ $(x - p)$ is negative.

The need for a rule such as his was suggested to Budan by Lagrange's *Traité de la résolution des équations numériques* (1767). This seems to have been almost the only nonelementary work Budan had read, and it influenced him greatly. He quoted Lagrange to show that it would be useful to give the rules for solving numerical equations entirely by means of arithmetic, referring to algebra only if absolutely necessary. Budan's goal was to solve Lagrange's problem—between which real numbers do real roots lie? —purely by methods of elementary arithmetic. Accordingly, the chief concern of Budan's *Nouvelle méthode* was to give the reader a mechanical process for calculating the coefficients of the transformed equation in $(x - p).$ He did not appeal to the theory of finite differences or to the calculus for these coefficients, preferring to give them "by means of simple additions and subtractions."[1] His *Nouvelle méthode* includes many specific numerical examples in which the coefficients are calculated and the number of sign changes in the polynomials P and G are compared; he intended this to be a simple and practical procedure.

In 1811 Budan presented a proof for his rule to the Institute; he published the proof, along with a reprint of his original article, as *Appendice à la nouvelle méthode,* in 1822. A. M. Legendre, reporting to the Institute in 1811 on Budan's rule and its proof, recognized the utility of being able to know that there could be no real roots between two given real numbers. Apparently unaware of the prior work of Joseph Fourier, he stated that the result was new.[2] Legendre added that the proof given by Budan was valid only after certain gaps were filled, notably the assumption without proof of Segner's theorem (1756): If $P(x)$ is multiplied by $(x - a)$, the number of variations in sign in the product polynomial is at least one greater than that in $P(x)$. Budan himself did not appreciate the force of this objection; he protested that there was nothing wrong with using a known result, although in fact he assumed it without stating it and, until Legendre's remark, did not seem to realize that the proof needed it.

Budan's success in discovering a correct rule and giving a reasonably satisfactory proof of it shows that, at the beginning of the nineteenth century, it was still possible for one without systematic training in mathematics to contribute to its progress; but mathematics was giving increasing attention to rigor and precision of statement, qualities slighted in Budan's work. The professionals were about to take over. Fourier's simultaneous and independent discovery, using derivatives, exemplifies the powerful methods available to one thoroughly schooled in mathematics. J. C. F. Sturm (1836) gave a necessary *and sufficient* condition for a root to lie between two bounds, thus completely solving the theoretical problem of how many roots lie between given limits. Yet Budan's rule remains the most convenient for computation, although finding bounds on roots is no longer the major business of the algebraist.

NOTES

1. This method is fairly efficient. It is equivalent to the use of successive synthetic divisions, a method often discussed in works on the theory of equations. See, e.g., W. S. Burnside and A. W. Panton, *The Theory of Equations* (Dublin–London, 1928), I, 10 ff., 64 ff.
2. Fourier taught his version of the rule before 1797, although it was not published until after his death, in 1831. Fourier's version is: If $f(x)$ is a polynomial of degree n, the number of real roots of $f(x) = 0$ lying between a and b cannot exceed the difference in the number of changes in sign in the sequence $f(b), f'(b), f''(b), \cdots, f^n(b)$ and that of the sequence $f(a), f'(a), f''(a), \cdots, f^n(a)$. See J. Fourier, *Analyse des équations déterminées*, pp. 98–100; on his priority, see the "Avertissement" to that work by C. L. M. H. Navier, p. xxi. Although the formulations of Budan and Fourier are equivalent, the great difference in conception argues for independence of discovery.

BIBLIOGRAPHY

I. ORIGINAL WORKS. Budan's writings are *Nouvelle méthode pour la résolution des équations numériques d'un degré quelconque* (Paris, 1807), and "Appendice à la nouvelle méthode," in *Nouvelle méthode pour la résolution des équations numériques d'un degré quelconque, revue, augmentée d'un appendice, et suivie d'un apperçu concernant les suites syntagmatiques* (Paris, 1822).

II. SECONDARY LITERATURE. Additional information may be found in J. Fourier, *Analyse des équations déterminées* (Paris, 1830 [*sic*]), which includes C. L. M. H. Navier, "Avertissement de l'éditeur," pp. i–xxiv, dated 1 July 1831; and F. N. W. Moigno, "Note sur la détermination du nombre des racines réelles ou imaginaires d'une équation numérique, comprises entre des limites données. Theorèmes de Rolle, de Budan ou de Fourier, de Descartes, de Sturm et de Cauchy," in *Journal de mathématiques pures et appliquées,* **5** (1840), 75–94.

JUDITH V. GRABINER

BUDD, WILLIAM (*b.* North Tawton, Devon, England, 14 September 1811; *d.* Clevedon, Somerset, England, 9 January 1880), *medicine, epidemiology.*

Budd, a pioneer epidemiologist and a precursor of the Pasteurian germ theory of disease, was born in a small town near the northern edge of Dartmoor. His father, Samuel Budd, practiced surgery there; his mother was the former Catherine Wreford, who came of an old Devon family. Of their ten children, William was the fifth of nine sons. The children received their primary education at home, but their father had inherited landed property from his grandfather, an Anglican clergyman, and could afford to send all his sons to good universities. Six of them graduated as doctors of medicine, three from Edinburgh and three from Cambridge. Two of Budd's older brothers, George and Richard, became fellows of the Royal College of Physicians of London, and George and William were elected fellows of the Royal Society.

Budd's professional training was unusually prolonged, partly because of two severe illnesses. An attack of typhoid fever interrupted his medical studies at the École de Médecine in Paris, where he intermittently spent three and a half years between October 1828 and September 1837, under such well-known teachers as Broussais, Cruveilhier, Lisfranc, Louis, Orfila, and Ricord. During the intervals Budd assisted his father, except for the winter of 1835–1836, when he attended the Middlesex Hospital, London. In the autumn of 1837, he went to Edinburgh University to complete the courses required for the M.D. degree. He graduated in August 1838, returned to general practice at North Tawton for about eighteen months, and was then appointed assistant physician to the Dreadnought, the Seaman's Hospital at Greenwich. An illness, which Budd apparently considered a second attack of typhoid, forced his early resignation. In 1841 he moved to Bristol and spent the rest of his working life in that city.

Budd was appointed physician to St. Peter's Hospital in 1842, lecturer in medicine at Bristol Medical College in 1845, and physician to the Bristol Royal Infirmary in 1847. He founded the Bristol Microscopical and Pathological societies, and served ten years as councillor of the Bath and Bristol branch of the Provincial Medical and Surgical Association, becoming its president in 1855–1856. Thereafter he played an active part on the Council of the British Medical Association until 1866. Budd gave valuable evidence before the Health of Towns Commission in 1841, and before the Royal Sanitary Commission in 1869. For the originality and importance of his views on infection and epidemiology, he was elected a fellow of the Royal Society in 1871.

In 1847, Budd married Caroline Mary Hilton, daughter of a landowner in Kent; they had three sons and six daughters. Although he had a healthy appearance and a fine physique, Budd was subject to attacks of intense headache, and in later life to bouts of nervous exhaustion, attributed to overwork. In 1873 he suffered a stroke, which left him an invalid. He died at the little seaside town of Clevedon, and was buried in Arnos Vale Cemetery, Bristol.

Budd's distinction as a physician was fostered by his family environment, thorough training, and warm humanity. His medicopolitical involvements, like his epidemiological and sanitary investigations, were governed by a sense of obligation to extend medical knowledge and to improve the public health. His forthrightly expressed convictions on the communicability and prevention of certain diseases, involving a scientific approach and inductive logic, set him apart from his medical contemporaries. John Tyndall praised him as "a man of the highest genius," whose "doctrines are now everywhere victorious, each succeeding discovery furnishing an illustration of his marvellous prescience."

Budd's essential doctrine, that specific infective agents determine the epidemic phenomena of communicable diseases, stemmed from his investigations of over eighty cases of typhoid at North Tawton in 1839–1840. Long before he described this epidemic in 1859, he had concluded that the "poisons" of typhoid and cholera multiply in the victim's intestines and are "cast off." His pamphlet *Malignant Cholera* (1849), which declared this disease to be waterborne, appeared about a month after the comparable *Mode of Communication of Cholera* by John Snow, whose priority he fully acknowledged. Budd's report erred in claiming that a fungus was the causal agent. Among preventive measures he stressed disinfection of the patient's excreta as well as purification of the water supply. His regimen successfully curbed the spread of cholera in Bristol during the 1866 outbreak.

In his classic monograph *Typhoid Fever* (1873), Budd integrated and expanded several previously published papers. This scholarly, fearless, and occa-sionally scornful document impressively marshaled the evidence that indicates the disease is contagious. Many of his medical contemporaries were "non-contagionists" or "miasmatists," who buttressed their beliefs by the outmoded but die-hard dogma of spontaneous generation, or by such fashionable fallacies as Murchison's pythogenic theory (that the intestinal fevers arise *de novo* from filth and neglect), and Pettenkofer's theory that the specific agents of typhoid and cholera are not infective until they have undergone metamorphosis in suitable soil. Budd's task was especially difficult because these heresies were popular within the ranks of the "sanitary reformers," whose main objectives he ardently supported.

In the 1860's Budd applied his principles *mutatis mutandis* to other communicable diseases of man, such as diphtheria, scarlet fever, and tuberculosis. His masterly report of a sheep-pox epizootic and reviews of rinderpest and hog cholera reinforced his basic contentions: that each specific agent of contagion multiplies at certain sites within the sick host, is eliminated and transported by definite routes, and can be destroyed or interrupted in its passage to other susceptible hosts.

Budd invested his contagious agents with fairly precise properties—e.g., reproducibility and relative lack of resistance to heat and to disinfectants—but he was almost as noncommittal about their exact nature as Fracastoro (unknown to him) had been about *contagium vivum* three centuries before. There were formidable impediments to accurate visualization of these microbic agents, including the high heat resistance of some species of sporulating contaminants, the inadequate resolving power of available microscopes, and the lack of solid nutrient media on which pathogenic bacteria could be grown and differentiated. Within a few years of Budd's death, these deficiencies had been remedied by Pasteur, Abbe, and Koch.

BIBLIOGRAPHY

I. ORIGINAL WORKS. Budd's works include "Remarks on the Pathology and Causes of Cancer," in *Lancet,* **2** (1841–1842), 266–270, 295–298; *Malignant Cholera: Its Mode of Propagation and Its Prevention* (London, 1849); "On Intestinal Fever: Its Mode of Propagation," in *Lancet,* **2** (1856), 694–695; "Intestinal Fever Essentially Contagious," *ibid.,* **2** (1859), 4–5, 28–30, 55–56, 80–82; "On Intestinal Fever," *ibid.,* 131–133, 207–210, 432–433, 458–459, and **1** (1860), 187–190, 239–240; "Diphtheria," in *British Medical Journal,* **1** (1861), 575–579; "Observations on Typhoid or Intestinal Fever: The Pythogenic Theory," *ibid.,* **2** (1861), 457–459, 485–487, 523–525, 549–551, 575–577, 604–605, 625–627; "On the Occurrence (Hitherto Unnoticed) of Malignant Pustule in England," in *Lancet,*

2 (1862), 164–165; "Variola Ovina, Sheep's Small-Pox; or the Laws of Contagious Epidemics Illustrated by an Experimental Type," in *British Medical Journal*, 2 (1863), 141–150; "Investigation of Epidemic and Epizootic Diseases," *ibid.*, 2 (1864), 354–357; "The Siberian Cattle Plague; or, the Typhoid Fever of the Ox," *ibid.*, 2 (1865), 169–179; "Typhoid (Intestinal) Fever in the Pig," *ibid.*, 81–87; "Asiatic Cholera in Bristol in 1866," *ibid.*, 1 (1867), 413–420; "Memorandum on the Nature and the Mode of Propagation of Phthisis," in *Lancet*, 2 (1867), 451–452; "Scarlet Fever, and Its Prevention," in *British Medical Journal*, 1 (1869), 23–24; and *Typhoid Fever; Its Nature, Mode of Spreading, and Prevention* (London, 1873; New York, 1931).

II. SECONDARY LITERATURE. Works on Budd include G. T. Bettany, "Budd, William," in *Dictionary of National Biography*, VII (1886), 220–221; W. Michell Clarke, "William Budd, M.D., F.R.S., 'In Memoriam,' " in *British Medical Journal*, 1 (1880), 163–166; E. W. Goodall, *William Budd, M.D. Edin., F.R.S.* (Bristol, 1936); and W. C. Rucker, "William Budd, F.R.S., Pioneer Epidemiologist," in *Bulletin of the Johns Hopkins Hospital*, 28 (1916), 208–215.

CLAUDE E. DOLMAN

BUERG, JOHANN TOBIAS. See **Bürg, Johann Tobias.**

BUFFON, GEORGES-LOUIS LECLERC, COMTE DE (*b.* Montbard, France, 7 September 1707; *d.* Paris, France, 16 April 1788); *natural history.*

Buffon was the son of Benjamin-François Leclerc and Anne-Cristine Marlin, both of whom came from the bourgeoisie. Anne Marlin was related to a rich financier whose money enabled Benjamin to become, in 1717, lord of Buffon and of Montbard, and *conseiller* to the Burgundian parliament. Georges-Louis, the naturalist, was the eldest of five children, of whom three others entered the church, where two of them rose to high position. In 1717 the Leclerc family moved to a fine house in Dijon, where they occupied an important place in society. The intellectual life of that provincial capital was active but not oriented toward science at that particular time.

Georges-Louis was a pupil at the Collège des Jésuites in Dijon from 1717 to 1723. He was only an average student, although he distinguished himself by his bent for mathematics. His father undoubtedly wanted him to have a legal career, and he did study law in Dijon between 1723 and 1726. As early as 1727, however, he became friendly with the young Swiss mathematician Gabriel Cramer, a professor at the University of Geneva. In 1728 he went to Angers, where he may have studied medicine and botany, as well as mathematics, with Père de Landreville, pro-

fessor at the Collège de l'Oratoire. A duel forced him to leave Angers in October 1730, and he embarked on a long journey through Southern France and Italy with a young English nobleman, the duke of Kingston, and his tutor, Nathaniel Hickman, an obscure member of the Royal Society.

Buffon returned to France in 1732 and, despite his father's opposition, obtained his mother's fortune (she had died during his absence). At the same time, he began to make himself known in Parisian political and scientific circles. His first works on the tensile strength of timber were written at the request of the minister of the navy, Maurepas, who was seeking to improve the construction of war vessels. Buffon's *Mémoire sur le jeu du franc-carreau,* a study of probability theory, contributed to his admission to the Académie Royale des Sciences as *adjoint-méchanicien* on 9 January 1734. For six years he divided his time among finance (his fortune soon became considerable); research in botany and forestry (he wrote several dissertations and translated Stephen Hales's *Vegetable Statiks* into French in 1734); and mathematics (he wrote dissertations and in 1740 translated Newton's *The Method of Fluxions and Infinite Series* into French from the English translation of the original Latin manuscript, published in 1736 by John Colson). At the end of this time he also became interested in chemistry and biology and conducted some microscopic research on animal reproduction. In June 1739 he became an *académicien-associé* and transferred from the mechanical to the botanical section. That July, through the influence of Maurepas, he succeeded Dufay as *intendant* of the Jardin du Roi.

Each spring, from 1740 on, Buffon left Paris for Montbard, to administer his estates, continue his research, and edit his writings. His robust constitution allowed him to adhere to a well-organized schedule: he arose at dawn and spent the morning at his work, and the afternoon at his business affairs. For fifty years, Buffon spent the summer on his estate, returning to Paris in the fall. At the end of this time, he had doubled the area of the Jardin du Roi, enriched its collections, and enlarged its buildings considerably; moreover, he himself had become rich, having been showered with pensions and having increased his landholdings. He had published the thirty-six volumes of *Histoire naturelle* and was famous throughout Europe and even in America; he was a member of the Académie Royale des Sciences, the Académie Française, the Royal Society of London, and the academies of Berlin and St. Petersburg, among others. Catherine II bestowed gifts upon him, and Louis XV made him Comte de Buffon and commissioned the sculptor Augustin Pajou to do a bust of him.

In 1752 Buffon, scarcely inclined to be governed by his feelings, nevertheless married for love. His wife, Françoise de Saint-Belin-Malain, a pretty girl of twenty, was of gentle birth although poor. Mme. de Buffon led a retiring life and died young, in 1769, leaving a five-year-old son. Toward the end of his life, Buffon developed a Platonic affection for the wife of the famous Swiss financier Jacques Necker. His most serious personal worries were caused by his son, an unstable spendthrift, who was to die on the guillotine during the Terror.

In addition to his scientific works, Buffon published several speeches delivered before the Académie Française, of which only one—*Discours sur le style*, delivered on 25 August 1753, the day of his acceptance—is significant. This speech is of interest not only for the literary ideas that it contains, but also for its embodiment of Buffon's conception of the value of the original work of the scholar, which, according to him, lies less in the discovery of facts than in their organization and presentation.

Buffon's works may be grouped into two main categories, the *Mémoires* presented to the Académie des Sciences and the *Histoire naturelle*. The *Mémoires*, which appeared between 1737 and 1752, deal with mathematics (theory of probability), astronomy (the law of attraction), physics (optics), plant physics (tensile strength of wood), forestry, physiology, and pyrotechnics (aerial rockets). Buffon considered most of these subjects again in the *Supplément à l'Histoire Naturelle* (I, II, IV, 1774–1777; for a complete description, see Hanks, pp. 275–281).

Buffon's works appeared in many editions throughout the eighteenth and nineteenth centuries; the list, with an analysis of each original edition, may be found in the bibliography by E. Genet-Varcin and J. Roger (1954). One must emphasize the importance of the chronology of the various texts, since Buffon's ideas evolved considerably as he assembled his great work.

It was probably his interest in mathematics that first drew Buffon toward science. He was reputed to have already discovered Newton's binomial theorem by himself when he was twenty years old; at this time he became associated with Gabriel Cramer. Their correspondence deals with all types of problems—mechanics, geometry, probability, theory of numbers, differential and integral calculus.

Buffon's first original work was the memoir *Sur le jeu de franc-carreau*, which introduced differential and integral calculus into the theory of probability by extending the latter to the field of surfaces. The study

of the Petersburg paradox led Buffon to certain moral considerations that he clarified in the *Essai d'arithmétique morale* (published in the *Supplément*, IV, 1777). In that work, as well as in his memoirs on the tensile strength of wood and his research in the cooling of planets, Buffon obviously considered mathematics more as a means of clarifying the idea of reality than as an autonomous and abstract discipline. His reasoning is that of an engineer, a moralist, or a philosopher, rather than that of a pure mathematician. This is why he refused to accept the notion of infinity, which he considered to be no more than *une idée de privation,* and why, in his discussion with Clairaut on the law of attraction (1745), he insisted that a simple force ought to be represented by a simple algebraic formula. It was this "realism" that prevented him from becoming a pure mathematician. In fairness, however, it must be pointed out that, with Clairaut and Maupertuis, he was one of the first French disciples of Newton.

A philosopher as well as a naturalist, Buffon throughout his works made observations on the nature and value of science. His most important writing on this subject is the *Discours sur la manière d'étudier et de traiter l'histoire naturelle* (1749), but the *Théorie de la terre* and the *Histoire des animaux* of the same date are also significant.

Breaking with the spirit of his time, Buffon attempted to separate science from metaphysical and religious ideas. As a disciple of Locke he denied idealistic metaphysics, stating that mental abstractions can never become principles of either existence or real knowledge; these can come only as the results of sensation. He thereby also brushed aside Plato, Leibniz, and Malebranche. He also rejected teleological reasoning and the idea of God's direct intervention in nature (herewith abandoning Newton): "In physics one must, to the best of one's ability, refrain from turning to causes outside of Nature" (*Théorie de la terre, preuves,* art. V).

Buffon was particularly sensitive to the disorder that appeared to rule nature: "It would appear that everything that can be, is" (*Sur la manière . . .*). He found fault with classifiers, especially Linnaeus, for trying to imprison nature within an artificial system, since man cannot even hope to understand nature completely. Only in mathematics is there evident truth because that particular science is man-made. Physics deals only with the probable. Buffon did not fall into the pit of skepticism, however. He thought that man should construct a science not based on certitudes but derived from nature.

As time went on, Buffon's ideas changed. In the two *Vues de la nature* (*Histoire naturelle,* XII, XIII,

1764 and 1765), he seems to admit that man is actually capable of ascertaining fundamental laws of nature, and in the *Époques de la nature* (1779) he shows how the history of the earth obeys these laws.

Buffon viewed the study of the earth as a necessary prerequisite to zoology and botany and in 1749 wrote the *Histoire et théorie de la terre,* followed by nineteen chapters of *preuves*. He returned to this subject in the *Supplément* (II, V) and devoted his last work to mineralogy.

In the *Théorie de la terre,* Buffon, like most of his contemporaries, states neptunian views. He has no hesitations about animal or plant fossils or the stratigraphic principles set forth by Sténon. The presence of sea fossils and sedimentation of rock beds indicate former submersion of present continents, of which the topography, shaped under the water by ocean currents, is diminished by erosion and the action of the waters that carry earth to the sea. No explanation of the reemergence of formerly submerged continents is offered. Buffon resolutely refused to accept the notion of catastrophes, including the biblical flood, which many of his contemporaries upheld. He offered several hypotheses (such as subsidence of the ground or earthquakes) to account for the displacement of the sea, but he insisted that such changes "came about naturally." Buffon was an advocate of "real causes": "In order to judge what has happened, or even what will happen, one need only examine what is happening. . . . Events which occur every day, movements which succeed each other and repeat themselves without interruption, constant and constantly reiterated operations, those are our causes and our reasons" (*Oeuvres philosophiques,* p. 56A).

On the other hand, in his cosmogony Buffon also rejected slow causes. According to Newton, planets and their movement had been created directly by God: this was the only possible explanation of the circumstance that the six planets then known revolved in the same direction, in concentric orbits, and almost on the same plane. Buffon's cosmogony was designed to replace the intervention of God by means of a natural phenomenon, a "cause whose effect is in accord with the laws of mechanics." He then hypothesized that a comet, hitting the sun tangentially, had projected into a space a mass of liquids and gases equal to 1/650 of the sun's mass. These materials were then diffused according to their densities and reassembled as spheres which necessarily revolved in the same direction and on almost the same plane. These spheres turn on their own axis by virtue of the obliquity of the impact of the comet on the sun; as they coalesced, they assumed the form of spheroids flattened on both poles. Centrifugal force, due to their rapid rotation, tore from these spheres the material that then became the satellites of the new planets.

This cosmogony, one of the first based on Newtonian celestial mechanics, is remarkable for its coherence. It is founded on the then generally accepted idea that comets are very dense stars, at least at their nucleus. But it also raises some serious difficulties, which were brought to light by Euler: according to the laws of mechanics, the material torn from the sun should have fallen back into it after the first revolution; the densest planets should be farthest away from the sun; and the planetary orbits should always coincide at the point of initial impact. Finally, as early as 1770, it became apparent that comets had a very low density, which destroyed the impact hypothesis.

Not only did Buffon retain this hypothesis, but he also made it the basis for a new theory of the earth, published in 1779 as *Époques de la nature.* In 1749, in the *Théorie de la terre,* Buffon juxtaposed a plutonian cosmogony and a neptunian theory of the earth. In 1767, however, Buffon became convinced (probably by Jean-Jacques d'Ortous de Mairan's *Dissertation sur la glace* of 1749 and *Nouvelles recherches sur la cause générale du chaud en été et du froid en hiver* of 1767) of the existence of a heat peculiar to the terrestrial globe. He saw it as the residue of primitive solar heat and immediately undertook large-scale experiments on the cooling period of globes of varying materials and diameters. He extrapolated the results of his experiments, published in Volumes I and II of the *Supplément,* in order to calculate the time required for the cooling of the earth and other planets.

The *Époques de la nature* presents a plutonian history of the earth—a piece was torn from the sun, the mass took form, the moon was torn from it by centrifugal force, and then the globe solidified during the first epoch. In the course of this solidification, primitive mountains, composed of "vitreous" matter, and mineral deposits were formed (marking the second epoch). The earth cooled, and water vapors and volatile materials condensed and covered the surface of the globe to a great depth. The waters were soon populated with marine life and displaced the "primitive vitreous material," which was pulverized and subjected to intense chemical activity. Sedimentary soil was thus formed, derived from rocks composed of primitive vitreous matter, from calcareous shells, or from organic debris, especially vegetable debris such as coal. In the meantime, the water burst through the vaults of vast subterranean caverns formed during the cooling period; as it rushed in, its level gradually dropped (third epoch). The burning of the accumulated combustible materials then produced volcanos and earthquakes, the land that emerged was shaped in relief by the eroding force of the waters (fourth epoch). The appearance of animal life (fifth epoch) preceded the final separation of the continents from one another and gave its present configuration to the surface of the earth (sixth epoch) over which man now rules (seventh epoch).

This work is of considerable interest because it offers a history of nature, combining geology with biology, and particularly because of Buffon's attempt to establish a universal chronology. From his experiments on cooling, he estimated the age of the earth to be 75,000 years. This figure is considerable in comparison to contemporary views which set the creation of the world at 4000–6000 B.C. In studying sedimentation phenomena, however, Buffon discovered the need for much more time and estimated a period of as long as 3,000,000 years. That he abandoned that figure (which appears only in the manuscript) to return to the originally published figure of 75,000 years, was due to his fear of being misunderstood by his readers. He himself thought that "the more we extend time, the closer we shall be to the truth" (*Époques de la nature*, p. 40).

The *Époques de la nature* contains a great deal of mineralogical material that was restated and elaborated in the *Histoire naturelle des minéraux*. Buffon's work on mineralogy was handicapped by its date of appearance, immediately before the work of Lavoisier, Haüy, and J. B. L. Romé de l'Isle. Although it was soon out of date, Buffon's book does contain some interesting notions, particularly that of the "genesis of minerals," that is, the concept that present rocks are the result of profound transformations brought about by physical and chemical agents. Buffon did not have a clear concept of metamorphic rocks, however. It is also noteworthy that Buffon was one of the first to consider coal, "the pyritous and bituminous matter," and all of the mineral oils as products of the decomposition of organic matter.

In the second volume of the *Histoire naturelle* (1749), Buffon offers a short treatise on general biology entitled *Histoire des animaux*. He takes up this subject again in the *Discours sur la nature des animaux* (*Histoire naturelle*, IV [1753]) and in a great many later texts. Although he deals with nutrition and development in these, he is most interested in reproduction. This, of course, was a question much discussed at that time, but for Buffon reproduction represented the essential property of living matter.

Buffon rejected the then widely accepted theory of the preexistence and preformation of embryos. He spurned its dependence on the direct intervention of God and held it to be incapable of explaining hered-

ity. He further refuted the connected theories of ovism and animalculism because no one had actually seen the egg of a viviparous animal and because spermatozoa were not "animalcules," but rather aggregates of living matter that were also to be found in female sexual organs. On the latter point Buffon was the victim of erroneous observations made during the course of a series of experiments conducted, with Needham's help, in the spring of 1748.

The essentials of Buffon's theory of reproduction may be found long before this date, however. He set forth the principle of epigenesis because it exists in nature and allows heredity to be understood. Buffon revived the ideas of certain physicians of the late seventeenth century who were faithful to an old tradition, and assumed that nutritive matter was first used to nourish the living being and then was utilized in the reproduction process when growth was completed. After being ingested, the nutritive matter received a particular imprint from each organ, which acted as a matrix in the reconstitution of that organ in the embryo. But Buffon departs from his predecessors on two points: (1) he sees the action of these molds as capable of modifying the nutritive substance internally, due to "penetrating forces" (conceived of on the basis of Newtonian attraction), and (2) he considers nutritive material to be already living. Buffon also conceived of living universal matter composed of "organic molecules," which are a sort of living atom. His thinking was therefore formed by a mechanistic tradition, complicated by Newton's influence, and balanced by a tendency toward vitalist concepts.

This tendency diminished as time passed. In 1779, in the *Époques de la nature*, Buffon dealt with the appearance of life on the earth—that is, the appearance of living matter, or organic molecules. He explained that organic molecules were born through the action of heat on "aqueous, oily, and ductile" substances suitable to the formation of living matter. The physicochemical conditions that made such formation possible were peculiar to that period of the earth's history; consequently spontaneous generation of living matter and organized living creatures can no longer occur. Buffon thus resolved the contradiction in his text of 1749, in which he maintained that while living matter was totally different from the original matter, nevertheless "life and animation, instead of being a metaphysical point in being, is a physical property of matter" (*Oeuvres philosophiques*, 238A–B).

In 1749 Buffon saw nothing short of disorder in nature. The only notion that corresponded to reality was the idea of species, to which he gave a purely biological definition: "One should consider as being of the same species that which by means of copulation perpetuates itself and preserves the similarity of that species" (*ibid., Histoire des animaux*, p. 236A). If the product of such mating is sterile, as is the mule, the parents are of different species. Any other criterion, particularly resemblance, is insufficient "because the mule resembles the horse more than the water spaniel resembles the greyhound" (*ibid., Histoire naturelle de l'âne*, p. 356A).

If the species exists in nature, the family does not: ". . . one must not forget that these *families* are our creation, we have devised them only to comfort our own minds" (*ibid.*, p. 355B). All classification is therefore arbitrary and has no merit other than convenience. Buffon violently attacked Linneaus and praised Tournefort. He himself followed an order that he believed to be "easier, pleasanter, and more useful" than any other, without being any more arbitrary— "taking the objects that are the most interesting to us because of their relation to us, and gradually moving toward those that are more distant" (*ibid., Sur la manière . . .*, p. 17B). In the order Buffon followed, the dog follows the horse because, in reality, the dog "is accustomed, in fact, to [so] follow" (*ibid.*, p. 18A). Buffon's order is formed by a philosophical bias rather than by science.

For Buffon to admit the concept of family, it would have to correspond to a reality. Thus:

> If these families really existed, they could have been formed only through the crossing, successive variation, and degeneration of original species; and if one once concedes that there are families of both plants and animals, that the donkey is of the horse family and only differs because it has degenerated, one could also say that the monkey is a member of the family of man and is merely a degenerated man, that man and monkey have a common origin just like the horse and mule, that each family . . . has only one founder and even that all animals came from one single animal which, with the passage of time, by simultaneously perfecting itself and degenerating, produced all of the races of the other animals [*ibid., Histoire naturelle de l'âne*, p. 355B–356A].

Because he rejected the concept of family and denied the value of making classifications, Buffon also rejected, at the beginning of his work, the hypothesis of generalized transformism offered by Maupertuis in 1751 in the *Système de la nature*. Buffon's theory of reproduction and the role he attributes to the "internal mold," as the guardian of the form of the species, prevented him from being a transformist.

This same theory of reproduction did not prevent Buffon from believing in the appearance of varieties within a species, however. Buffon believed in the heredity of acquired characteristics; climate, food, and domestication modify the animal type. From his exhaustive research for the *Histoire naturelle des quadrupèdes,* Buffon came to the conclusion that it was necessary to reintroduce the notion of family. But he attributes to this word—or to the word *genus,* which he also uses—a special meaning: a family consists of animals which although separated by "nature," instinct, life style, or geographical habitat are nevertheless able to produce viable young (that is, animals which belong biologically to the same species, e.g., the wolf and the dog). What the naturalist terms species and family, then, will thus become, for the biologist, variety and species. Buffon was thus able to write, in 1766, the essay *De la dégénération des animaux*—in which he showed himself to be a forerunner of Lamarck—while he continued to affirm the permanence of species in the two *Vues de la nature* (1764–1765) and *Époques de la nature* (1779).

Buffon's final point of view concerning the history of living beings can be summarized as follows: No sooner were organic molecules formed than they spontaneously grouped themselves to form living organisms. Many of these organisms have since disappeared, either because they were unable to subsist or because they were unable to reproduce. The others, which responded successfully to the essential demands of life, retained a basically similar constitution— Buffon affirms unity in the plan of animals' composition and, in variations on that plan, the principle of the subordination of organs. Since the earth was very hot and "nature was in its first stage of activity," the first creatures able to survive were extremely large. The earth's cooling drove them from the North Pole toward the equator and then finally caused their extinction. Buffon offered this in explanation of the giant fossils discovered in Europe and North America, which he studied at length (to the point of becoming one of the founders of paleontology). The organic molecules which were left free in the northern regions formed smaller creatures which in turn moved toward the equator, and then a third and fourth generation, which also moved south. Originating in Siberia, these animal species spread out to southern Europe and Africa, and toward southern Asia and North America. Only South America had an original fauna, different from that of other continents.

In the process of migration, the species varied in response to environment. There are few varieties of the large mammals because they reproduce slowly. The smaller mammals (rodents, for example) offer a large number of varieties because they are very prolific. The same is true of birds. Going back to the basic types, quadrupeds may be divided into thirteen separate species and twenty-five genera. But Buffon was not a transformist, because he believed that these thirty-eight primitive types arose spontaneously and simultaneously from an assembly of organic molecules.

As a naturalist and as a paleontologist Buffon was forced to uphold the variability of animal form; as a biologist he had to admit the permanence of hereditary types. He was never able to resolve this difficulty, although he stated the problem quite clearly.

"Love of the study of nature," Buffon wrote, "implies, in the human mind, two attributes which appear to be opposed, the broad outlook of an ardent spirit that grasps everything in one glance and the minute attention of a hard-working instinct that concentrates on only one point" (*ibid.*, *Sur la manière . . .*, p. 7A). Buffon liked to deal with great biological and zoological problems, but his work is above all a detailed description of quadrupeds, birds, and minerals. To him, the "true method" is "the complete description and exact history of each thing in particular" (*ibid.*, p. 14B). This "history" goes beyond simple morphological description:

> The history of one animal should be . . . that of the entire species of that particular animal; it ought to include their procreation, gestation period, the time of birth, number of offspring, the care given by the mother and father, their education, their instincts, their habitats, their diet, the manner in which they procure food, their habits, their wiles, their hunting methods [*ibid.*, 16A–B].

Physiological characteristics allow species separated by habitat or mores to be grouped together biologically; conversely, the habitats or habits of each animal permit distinctions between species or varieties. The description should also include a study of animal psychology, in particular that of social species (as monkeys and beavers). Buffon's method became more and more comparative, and in some works, he drew up genealogical tables of the varieties of each species. Buffon tried always to observe personally the animals he discussed. Nevertheless, pure description became boring to him, and he entrusted it to his associates.

In the *Histoire naturelle de l'homme,* published in 1749 (*Histoire naturelle*, II, III), and in many of his other works as well, Buffon studied the human species by the same methods that he applied to animal species, including the psychological, moral, and intellectual life of man. At the same time that he proclaimed the absolute superiority that the ability to reason gives man over animals, he demonstrated how the physiological organization and development of the sensory organs make reasoning possible. Through-

out his work Buffon specifies that reason developed only through language, that language grew out of life in society, and that social life was necessitated by man's slow physiological growth (since man is dependent on his mother long after birth). For the same reason, the elephant is the most intelligent of animals, while social life makes beavers capable of astonishing work.

It was, therefore, as a physiologist and as a naturalist that Buffon studied man and his reason; and it was as a biologist that he affirmed the unity of the human species. Aside from a few safe formulas, theology never comes into the picture. According to the *Époques de la nature*—and, in particular according to its manuscript—it is clear that the human species has had the same history as the animals. Buffon even explains that the first men, born on an earth that was still hot, were black, capable of withstanding tropical temperatures. Through the use of the resources of his intelligence and because of the invention of fire, clothes, and tools, man was able to adapt himself to all climates, as animals could not. Man is therefore the master of nature; and he can become so to an even greater degree if he begins to understand "that science is his true glory, and peace his true happiness" (*Époques de la nature,* p. 220).

Buffon's work is of exceptional importance because of its diversity, richness, originality, and influence. Buffon was among the first to create an autonomous science, free of any theological influence. He emphasized the importance of natural history and the great length of geological time. He envisioned the nature of science and understood the roles of paleontology, zoological geography, and animal psychology. He realized both the necessity of transformism and its difficulties. Although his cosmogony was inadequate and his theory of animal reproduction was weak, and although he did not understand the problem of classification, he did establish the intellectual framework within which most naturalists up to Darwin worked.

BIBLIOGRAPHY

I. ORIGINAL WORKS. See *Oeuvres complètes de Buffon,* J. L. Lanessan, ed., followed by Buffon's correspondence, 14 vols. (Paris, 1884–1885), still considered to be the best edition; *Oeuvres philosophiques de Buffon,* J. Piveteau, ed. (Paris, 1954), which contains a bibliography by Mme. E. Genet-Varcin and J. Roger that lists most works on Buffon published before 1954; and *Les Époques de la nature,* critical ed. by J. Roger (Paris, 1962), with an introduction, reproduction of the MS, notes, scientific vocabulary, and bibliography.

II. SECONDARY LITERATURE. Works on Buffon, both with more recent bibliographies, are L. Hanks, *Buffon avant l'histoire naturelle* (Paris, 1966); and J. Roger, *Les sciences de la vie dans la pensée française du 18e siècle* (Paris, 1964), pp. 527–584.

JACQUES ROGER

BULLER, ARTHUR HENRY REGINALD (*b.* Birmingham, England, 19 August 1874; *d.* Winnipeg, Canada, 3 July 1944), *mycology.*

Buller was the son of A. G. Buller, magistrate and county councillor for Birmingham. He received the B.Sc. from the University of London in 1896 and the Ph.D. from the University of Leipzig in 1899. He began his professional career at the International Marine Biological Station, Naples, in 1900–1901, then served as lecturer and demonstrator in botany at the University of Birmingham from 1901 to 1904. From 1904 to 1936 he was professor of botany at the University of Manitoba. Buller belonged to many societies, among them the British Association, British Mycological Society (president, 1913), Botanical Society of America (president, 1928), Canadian Phytopathological Society (president, 1920), Royal Society of Canada (president, 1927–1928), and Royal Society of London. He was the recipient of the Flavelle Medal (Royal Society of Canada, 1929), the Natural History Society of Manitoba Medal (1936), and the Royal Medal (Royal Society of London, 1937), as well as honorary degrees from the University of Manitoba (1924), University of Saskatchewan (1928), University of Pennsylvania (1933), and University of Calcutta (1938).

When Buller arrived at the University of Manitoba in 1904, he threw himself energetically into the founding and development of its department of botany, which soon became one of the most outstanding in Canada. He was also one of a small group of its original staff who were largely responsible for guiding the young institution during the early years of its rapid development. Later, when his worldwide reputation would have underwritten a large graduate section, he steadfastly refused to accept more than the few graduates whose work he could personally supervise; but that select group justified him and themselves by their subsequent contributions. One aspect of his position at Winnipeg which appealed greatly to Buller was that his commitments there permitted him to spend three or four months each year at Birmingham, where he worked in the laboratories or the library, or studied nature in the woods and fields. In later years his summers were spent at the herbarium of the Royal Botanical Gardens, Kew, to whose library he bequeathed his miscellaneous manuscripts. Buller retired to England in 1936, but in 1939,

while he was attending the International Congress of Microbiology in New York, World War II broke out and left him stranded there. He returned to Winnipeg, where once again he became deeply immersed in his researches. Indeed, when in 1944 he developed a brain tumor, his chief worry during the weeks of hopeless struggle was that all of his planned researches had not been completed.

Early in his career Buller had published several articles in scientific journals, and by 1909 he had accumulated enough material for a book to be entitled *Researches on Fungi.* He submitted his manuscript to a society, but was informed it would have to be reduced by about half before it could be published. To him this meant sheer mutilation of his manuscript, and accordingly he arranged for its publication at his own expense. Between 1909 and 1934, five additional volumes of *Researches on Fungi* were published in like manner, although the later ones were subsidized in part by the National Research Council of Canada. The Royal Society of Canada sponsored the posthumous publication in 1950 of Volume VII. These volumes were characterized, as were all his writings and all his lectures, by a unique style. The ease, clarity, precision, and grace with which Buller expressed his thoughts reveal a fine feeling for the choice and meaning of words, and reflect his familiarity with and love of English classics—the works of Milton and Shakespeare being especially beloved and extensively memorized.

The *Researches on Fungi* is both Buller's *magnum opus* and a magnificent memorial to his genius and his zeal. It will long remain a primary reference source for mycologists and biologists concerned with the problems of spore production and liberation in the fungi, of social organization within that group, and of epidemiology in general. While the independent manner of its publication ensured an eminently readable, profusely and beautifully illustrated series of volumes, it greatly limited the number that could be printed. Consequently, even before the last volume was published, the early ones were already collector's items, and complete sets were available in relatively few libraries. This unfortunate condition was remedied in part by the reissue in 1959 of Volumes I to VI. Unfortunately, this reprint is also out of print, and only Volume VII (1950) is presently available. Buller bequeathed his magnificent personal scientific library of over 2,000 volumes to the Canada Department of Agriculture Laboratory of Plant Pathology at Winnipeg.

The casual reader of the *Researches* might underestimate the substance of Buller's contributions. His meticulous, detailed, and thoroughgoing investigation of fungus after fungus might suggest that he was lost in details, whereas he was making very sure of his foundation before he compared individuals and combined them in groups on the basis of both form and function. At times, too, his reasoning sounds teleological until one remembers he was an ardent Darwinian and realizes that he was really looking for adaptations in form and function that made certain species successful and certain groups dominant. This approach and his insistence on dealing only with living materials paid handsome dividends in new discoveries in mycology and interpretations of older ones, and exerted a profound influence in many other areas of biological research. While each of his volumes makes a significant and original contribution in and beyond the area of its chief concern, Volume VII, which deals with sexuality in rust fungi, is monumental. In it is the distillation of all he had learned about that group of fungi, which constitutes the greatest threat to mankind's food supply: it also contains basic contributions in many fields of research, including epidemiology, genetics, and plant breeding, as well as mycology and plant pathology.

In 1941 Buller published in the *Botanical Review* a classic article entitled "The Diploid Cell and the Diploidisation Process in Plants and Animals With Special Reference to Higher Fungi." In it he reviewed the phenomenal advances during the previous decade in the knowledge of social organization and sexuality in the higher fungi, and integrated the new discoveries into the framework of previous knowledge so as to highlight areas still to be explored. This, too, is Buller at his best; and fortunately the article is generally available.

To students of the fungi, regardless of which of their many aspects may be their concern, Buller's works will always be a basic point of reference. Fungal taxonomy has been all but revolutionized by his insistence that only living material be utilized. His studies on the bionomics of production, liberation, and dispersal of fungus spores has made a science of epidemiology. His studies on sexuality of fungi and others which these have spawned have at long last resolved many problems in fungal phylogeny and given to it a rational foundation on which unity and perhaps even continuity may some day rest.

Of Buller's many and varied contributions, few had the impact and significance of his discovery that, in the Hymenomycetes, a haploid mycelium could be diploidized by a dikaryotic diploid mycelium. In homage to him, Quintanilha (1939) designated this as the Buller phenomenon. Within four years this had stimulated investigation in so many areas that Buller, in his diploid cell article, redefines the Buller phenomenon as "in Basidiomycetes and Ascomycetes, the diploidisation of a unisexual mycelium or the unisexual rudiment of a fructification by a bisexual mycelium." Convincing supporting evidence in the Uredinales was supplied by Craigie's demonstration (1927) of the function of the pycnia of the rust fungi and by a number of subsequent contributions by Newton, Johnson, and Brown (1930–1940) in their studies with *Puccinia graminis*. These established the scientific foundation for the origin and variation of rust races and hence also of the problem of rust control.

BIBLIOGRAPHY

I. ORIGINAL WORKS. A complete list of Buller's publications accompanies the biographical sketch by W. F. Hanna, C. W. Lowe, and E. C. Stakman in *Phytopathology*, **35** (1945), 577–584. Fairly complete lists are in *Who's Who in Canada* (1938–1939) and *Who's Who* (1944). *Researches on Fungi*, I–VI, was published in London (1909–1934); VII, G. R. Bisby, ed., was published in Toronto (1950); it was reissued (I–VI only) in New York (1959). His major article is "The Diploid Cell and the Diploidisation Process in Plants and Animals With Special Reference to Higher Fungi," in *Botanical Review*, **7** (1941), 334–345.

II. SECONDARY LITERATURE. A. M. Brown, "The Sexual Behaviour of Several Plant Rusts," in *Canadian Journal of Research*, **18** (1940), 18–26; J. H. Craigie, "Discovery of the Function of the Pycnia of the Rust Fungi," in *Nature*, **120** (1927), 765–767; T. Johnson and Margaret Newton, "Crossing and Selfing Studies with Physiological Races of Oat Stem Rust," in *Canadian Journal of Research*, C, **18** (1940), 54–67; Margaret Newton, T. Johnson, and A. M. Brown, "A Study of the Inheritance of Spore Colour and Pathogenecity in Crosses Between Physiological Forms of *Puccinia graminis tritici*," in *Scientific Agriculture*, **10** (1930), 775–798; and A. Quintanilha, "Contribution à l'étude génétique du phénomène de Buller," in *Comptes rendus de l'Académie des sciences*, **205** (1937), 745.

D. L. BAILEY

BULLIALDUS, ISMAËL. See **Boulliau, Ismaël.**

BULLOCH, WILLIAM (*b.* Aberdeen, Scotland, 19 August 1868; *d.* London, England, 11 February 1941), *bacteriology.*

Bulloch participated in the early development of medical bacteriology in Britain and won lasting recognition as a historian of that science. He came from a plain-living, scholarly Aberdonian family. His father, John Bulloch, an accountant, and his mother, Mary Malcolm, had two sons and two daughters. William was the younger son, and with his brother,

John Malcolm, a distinguished London journalist, shared the literary talent, with predilections for history and genealogy, that their father and grandfather had displayed. He attended Aberdeen Grammar School until 1884, and King's College, Aberdeen, for two years before enrolling in medicine at Marischal College. After graduating in 1890 with highest honors, he studied pathology at Aberdeen, Leipzig, and various other European medical centers before returning briefly to Aberdeen in 1894 to present a prize-winning M.D. thesis. Courses in bacteriology at the Pasteur Institute from Émile Roux, Elie Metchnikoff, and Émile Duclaux were followed by a short assistantship to Victor Horsley, professor of pathology at University College, London.

In July 1895 Bulloch took charge of the serum laboratories at the British (later, Lister) Institute of Preventive Medicine. In 1897 he was appointed bacteriologist to the London Hospital and lecturer on bacteriology and pathological chemistry to its medical school, and in 1919 became Goldsmith's professor of bacteriology at the University of London. After officially retiring in 1934, Bulloch served these institutions as consulting bacteriologist and emeritus professor. He was elected a fellow of the Royal Society in 1913, received an honorary LL.D. from Aberdeen in 1920, and held the following lectureships: Horace Dobell (Royal College of Physicians, 1910), Tyndall (Royal Institution, 1922), and Heath Clark (University of London, 1937). Bulloch's brief marriage in 1901 to Anna Molbo, a Danish pianist, was dissolved. In 1923 he married Irene Adelaide Baker, widow of an Australian cricketer, who survived him. In his last years, Bulloch suffered from paralysis agitans. He died in the London Hospital following a minor operation.

Although a lackadaisical administrator, Bulloch was an unforgettable lecturer. A clever mimic and raconteur, he enjoyed dramatizing the foibles and accomplishments of famous bacteriologists at home and abroad, many of whom were personal friends. His knowledgeableness and scrupulosity brought him membership on various technical advisory committees and the chairmanship, in 1932, of the Lister Institute's board of governors. His bibliography totals more than a hundred titles, and he was generally sole author. The earliest reports, dating from 1892, were histoneurological; but the scope soon broadened to include, for instance, descriptions of a new anaerobic jar or bacterial filter, and investigations of such contemporary problems as Ehrlich's diphtheria toxin "spectra" and antitoxin assay, Almroth Wright's opsonins and vaccine therapy (especially as related to tuberculosis), and the Wassermann test for syphilis. This work, always carefully performed and meticu-

lously recorded, seldom revealed new knowledge of signal importance, but facilitated critical appraisal of others' claims. After 1910 Bulloch deserted the laboratory for the library, and his publications were mainly painstaking reviews of hereditary diseases, notably hemophilia, whose genetics fascinated him, and tributes to distinguished bacteriologists. His innate compilatory and historical talents were best expressed in his contributions to monographs on diphtheria (1923) and surgical catgut (1929), and to *A System of Bacteriology* (1929–1931), culminating in his scholarly masterpiece, *The History of Bacteriology* (1938).

BIBLIOGRAPHY

I. Original Works. The fullest bibliography of Bulloch's works (106 items) is that provided by his stepson-in-law, Clifford Dobell, as an appendix (pp. 842–853) to the detailed obituary by a former pupil, J. C. G. Ledingham (see below). Among the more original and characteristic publications are "Hyaline Degeneration of the Spinal Cord," in *Brain*, 15 (1892), 411–413; "A Contribution to the Study of Diphtheria Toxin," in *Transactions of the Jenner Institute of Preventive Medicine*, 2nd ser. (1899), 46–55; "A Simple Apparatus for Obtaining Plate Cultures or Surface Growths of Obligate Anaerobes," in *Zentralblatt für Bakteriologie, Parasitenkunde, Infektionskrankheiten und Hygiene*, Abt. I, 27 (1900), 140–142; "The Chemical Constitution of the Tubercle Bacillus," in *Journal of Hygiene*, 4 (1904), 1–10, written with J. J. R. MacLeod; "On the Relation of the Suprarenal Capsules to the Sexual Organs," in *Transactions of the Pathological Society of London*, 56 (1905), 189–208, written with J. H. Sequeira; "The Principles Underlying the Treatment of Bacterial Diseases by the Inoculation of Corresponding Vaccines," in *Practitioner*, 75 (1905), 589–610; "On the Transmission of Air and Micro-organisms Through Berkefeld Filters," in *Journal of Hygiene*, 9 (1909), 35–45, written with A. J. Craw; *The Problem of Pulmonary Tuberculosis Considered From the Standpoint of Infection* (London, 1910); "L'Abbate Spallanzani. 1729–1799," in *Parasitology*, 14 (1922), 409–412; *Diphtheria: Its Bacteriology, Pathology and Immunology* (London, 1923), written with Frederick W. Andrewes, S. R. Douglas, Georges Dreyer, *et. al.*; *The Preparation of Catgut for Surgical Use*, Medical Research Council Special Report Series, no. 138 (London, 1929), written with L. H. Lampitt and J. H. Burhill; "History of Bacteriology," in *A System of Bacteriology in Relation to Medicine*, I (London, 1930), 15–103 (Bulloch was chairman of the committee that prepared this nine-volume work, and contributed numerous articles besides the opening chapter); and *The History of Bacteriology* (London, 1938; repr. 1960).

Bulloch's authoritative contributions on rheumatic fever, plague, tuberculosis, and relapsing fever, in Clifford Allbutt and Humphry Davy Rolleston, *A System of Medicine*

(London, 1899), survived several editions. He also wrote articles for Karl Pearson's *Treasury of Human Inheritance* (London, 1912), including those on diabetes insipidus, angioneurotic edema, and (with P. Fildes) hemophilia.

Bulloch's purely historical writings include tributes to Spallanzani, Pasteur, Koch, and Lister. Among his more notable obituaries are those on Emanuel Klein, Charles Creighton, Sir Alexander Ogston, Waldemar Haffkine, Shibasaburo Kitasato, Sir William Watson Cheyne, Émile Roux, and Theobald Smith.

II. Secondary Literature. Obituaries include P. Fildes, "William Bulloch. 1868–1941," in *Journal of Pathology and Bacteriology,* **53** (1941), 297–308; J. C. G. Ledingham, "William Bulloch. 1868–1941," in *Obituary Notices of Fellows of the Royal Society,* **3** (1941), 819–843; J. McIntosh, "Prof. William Bulloch, F. R. S.," in *Nature,* **147** (1941), 504–505; and H. M. Turnbull, "Professor William Bulloch M.D., LL.D., F.R.S.," in *British Medical Journal* (1941), **1,** 341–342. Other references to Bulloch's life and work are C. E. Dolman, "Tidbits of Bacteriological History," in *Canadian Journal of Public Health,* 53 (1962), 269–278; and "Paul Ehrlich and William Bulloch: A Correspondence and Friendship (1896–1914)," in *Clio medica,* 3 (1968), 65–84.

Claude E. Dolman

BUNGE, GUSTAV VON (*b.* Dorpat [now Tartu], Estonia, 19 January 1844; *d.* Basel, Switzerland, 5 November 1920), *physiology.*

Gustav von Bunge was the son of the botanist Alexander von Bunge. The father was famous for his botanical studies in Siberia and China. The elder Bunge's teaching and research work in Dorpat (1836–1867) made the university a center for the study of botany.

At Dorpat the biochemist Carl Schmidt became both teacher and promoter of Gustav von Bunge as a student of chemistry, and employed him as an assistant in his laboratory from 1872 onward. Schmidt inspired him, above all else, to tackle the problem of mineral (inorganic) metabolism, and as early as 1874, the year in which he received his doctorate and became an associate professor of physiology, the first significant results of his experiments in that area appeared.

In 1882 Bunge received the degree of doctor of medicine at the University of Leipzig; in 1884 the University of Kiev made him an honorary doctor. In 1885 he answered an invitation from the University of Basel, became extraordinary professor of medicine there, and then in 1886, an ordinary professor of physiological chemistry. His most famous student was the physiologist Emil Abderhalden.

Gustav von Bunge showed himself to be a first-rate nutritional scientist, especially in recognizing the sig-

nificance of mineral salts in the diet of men and other mammals, together with all of its consequences. One of his very first studies on this subject created quite a stir. It dealt with the role of salt (sodium chloride) in nourishment and concluded that herbivores absorb large excesses of potassium in their diet. This led to an increased salt elimination, and thereby to considerable salt losses to the organism and to a correspondingly increased salt requirement. It was typical of Bunge not to remain content after discovering certain physiological facts, and he immediately set about to figure out their ethnological, social, and sociological significance. He found that hunting and nomadic peoples who lived exclusively on meat did not require salt, and actually detested it, whereas vegetarians could not exist without adding salt to their diet. Therefore, the discovery and utilization of salt must have played a very important role in man's transition from a nomad into a settled agricultural worker.

An important point that Bunge brought up in his studies on the role of salt in man's diet was of a phylogenetic nature: The high salt content of vertebrates attests to their descent from inhabitants of the sea. Since ontogenesis repeats phylogenesis, then the salt content of a vertebrate should be greater the younger the vertebrate is; this was precisely what Bunge was able to demonstrate when he determined the salt content in the cartilage of cattle embryos in the various stages of their development and in that of calves of various ages.

Another area in which Bunge did research was the quantitative analysis of mineral substances in nutrients and the related question of the necessity of a constant supply of salt for the full-grown organism. In connection with the question of the quantitative and qualitative mineral requirement of growing and fully grown organisms there are the analyses of ash constituents in the milk of humans and various other animal species, undertaken first by Bunge in Dorpat and then in Basel by his student Abderhalden. Great differences were found above all in lime content. This phenomenon was explained by the fact that the lime content of the milk and the growth rate of the animal in question were in each case positively correlated.

Proceeding from the importance of lime supply in the development of the organism, two works appeared between 1901 and 1904 concerning "increased sugar consumption and its dangers." Limitation of the supply of calcareous nutrients (calcareous fruits) in favor of consumption of chemically pure sugar was viewed as the reason for arrested skeletal development and for the frequent appearance of tooth decay.

One of the main problems posed in Bunge's studies had to do with iron metabolism. The crux of his study on this subject was his search for the ferriferous precursors of red blood pigment in food. According to the theory of the inability of animal cells to synthesize complicated organic compounds, prevalent at that time, it was unthinkable that inorganic iron found in food and resorbed from it might be utilized to synthesize hemoglobin in the organism. Accordingly, Bunge attempted to isolate an organic iron complex out of egg yolk. In Basel, Bunge's students Häusermann and Abderhalden conducted feeding experiments on anemic animals by adding a hemoglobin, hematin, or iron salt supplement to an iron-poor diet. The results of these experiments seemed to confirm Bunge's thesis: only organically bound iron can be resorbed and utilized for the building up of hemoglobin. Iron-rich vegetables similarly have a favorable influence on laboratory-produced anemia, which meant to Bunge that even in these nutrients, iron is found organically bound.

A prerequisite for conducting such studies on iron metabolism was the control of a perfectly functioning method of quantitative iron determination. With the help of this method, Jaquet and Bunge, in 1889, succeeded in determining the iron content of blood pigment exactly and thereby establishing a firm basis for the elucidation of the constitution of the hemoglobin molecule.

Although nutritional physiology formed the center of Bunge's research activities, there is another area of his work that should not be overlooked. This is the study he pursued with Schmiedeberg on the synthesis of hippuric acid from benzoic acid and glycine in the animal organism (1876). This is important not only because it recognized the kidneys as the site of synthesis, but also because the methods he applied to investigation of metabolic problems are still used universally.

To most of the public, composed of physicians and laymen, Gustav von Bunge is known primarily for such monographs as *Vegetarianismus* and *Die Alcoholfrage.* He saw in alcohol consumption the roots of many types of illnesses and social evils. It is therefore not astonishing that he became a champion of the abstinence movement and a great enthusiast of the prohibition laws in the United States.

BIBLIOGRAPHY

I. ORIGINAL WORKS. Bunge's works are *Der Vegetarianismus* (Berlin, 1885; 2nd ed., 1900); *Die Alcoholfrage* (Leipzig, 1887), trans. into 12 languages; *Die zunehmende Unfähigkeit der Frauen, ihre Kinder zu stillen, die Ursache dieser Unfähigkeit, die Mittel zur Verhütung* (Munich, 1900;

4th ed., 1905), trans. into French; *Lehrbuch der physiologischen und pathologischen Chemie* (Leipzig, 1887; 4th ed., 1898), trans. into five languages; *Lehrbuch der Physiologie des Menschen,* 2 vols. (Leipzig, 1901; 2nd ed., 1905); *Lehrbuch der organischen Chemie für Mediziner* (1906); all other works are registered in Poggendorff, III, 214; IV, 204–205; and VI, 369, which also contains the eulogies on Bunge.

II. SECONDARY LITERATURE. A biography of Bunge, together with an assessment of his work, is that by C. M. McCay, in *Journal of Nutrition,* **49** (1953), 3–19. See also F. Lieben, *Geschichte der physiologischen Chemie* (Leipzig–Vienna, 1935); K. E. Rothschuh, *Geschichte der Physiologie* (Berlin–Göttingen–Heidelberg, 1953).

HERBERT SCHRIEFERS

BUNSEN, ROBERT WILHELM EBERHARD (*b.* Göttingen, Germany, 31 March 1811; *d.* Heidelberg, Germany, 16 August 1899), *chemistry.*

Bunsen was the youngest of four sons born to Christian Bunsen, chief librarian and professor of modern languages at the University of Göttingen. His ancestors on his father's side had lived in Arolsen, where many of them held public office, frequently as master of the mint; his mother was the daughter of a British-Hanoverian officer named Quensel.

Bunsen began school in Göttingen but transferred to the Gymnasium at Holzminden, from which he graduated in 1828. Returning to Göttingen, Bunsen entered the university, where he studied chemistry, physics, mineralogy, and mathematics. His chemistry teacher was Friedrich Stromeyer, who had discovered cadmium in 1817. Bunsen received his doctorate in 1830, presenting a thesis in physics: "Enumeratio ac descriptio hygrometrorum."

Aided by a grant from the Hanoverian government, Bunsen toured Europe from 1830 to 1833, visiting factories, laboratories, and places of geologic interest. In May 1832, he saw a new steam engine in K. A. Henschel's machinery factory in Kassel. Later that year, in Berlin, he studied Christian Weiss's geognostic and mineralogic collections; met Friedlieb Runge, the discoverer of aniline, and Gustav Rose; and worked in Heinrich Rose's laboratory. He visited Justus Liebig in Giessen and met Eilhard Mitscherlich in Bonn for a geological trip through the Eifel Plateau. In September 1832, Bunsen arrived in Paris. There he worked in Gay-Lussac's laboratory and met such prominent scientists as Jules Reiset, Henri-Victor Regnault, Théophile Pelouze, and César Despretz. While in France, Bunsen visited the porcelain works at Sèvres. From May to July 1833, he traveled to Vienna, where he toured several industrial plants.

In the fall of 1833 Bunsen became *Privatdozent* at

the University of Göttingen. He succeeded Friedrich Wöhler at the Polytechnic School in Kassel in January 1836. In October 1838 he was appointed professor extraordinarius of chemistry at the University of Marburg and became professor ordinarius four years later. Bunsen spent part of 1851 at Breslau, where he became acquainted with Gustav Kirchhoff, with whom he later did important research in spectroscopy. In 1852 he succeeded Leopold Gmelin at the University of Heidelberg. Although offered a position as Mitscherlich's successor at the University of Berlin in 1863, Bunsen remained at Heidelberg until he retired in 1889, at the age of seventy-eight. A laboratory, constructed for him by the government of Baden, was completed in the summer of 1855; there Bunsen did his research and guided the work of numerous young men who became well-known scientists during the second half of the nineteenth century. Bunsen never married; his teaching and research consumed most of his time, and he traveled widely, either alone or with friends.

Bunsen was a most devoted teacher. He presented 100 hours of lectures during each of seventy-four semesters in a course entitled "Allgemeine Experimentalchemie." The lectures, which changed little through the years, were concerned with inorganic chemistry; organic chemistry was excluded. Theoretical aspects were at a minimum: neither Avogadro's hypothesis nor the periodic law of the elements— developed by his own students, Dmitri Mendeleev and Lothar Meyer—was mentioned. In his research, as in his teaching, Bunsen emphasized the experimental side of science. He enjoyed designing apparatus and, being a skilled glassblower, he frequently made his own glassware. He was also an expert crystallographer. Bunsen developed and improved several pieces of laboratory equipment, including the Bunsen burner, the Bunsen battery, an ice calorimeter, a vapor calorimeter, a filter pump, and a thermopile.

A man of wide scientific interests, Bunsen did some early research in organic chemistry but later abandoned this field and concentrated on inorganic chemistry. His most important work was the development of a variety of analytical techniques for the identification, separation, and measurement of inorganic substances. Throughout his life Bunsen gave much attention to geology. He was also interested in the application of experimental science to industrial problems.

His first research was on the insolubility of metal salts of arsenious acid, carried out in 1834. While involved in this work, Bunsen discovered that hydrated ferric oxide could be used as an antidote for arsenic poisoning. The ferric oxide is effective, he explained, because it combines with arsenic to form ferrous arsenite, a compound insoluble in both water and body fluids. This finding, still used today, was Bunsen's only venture into physiological chemistry. In other early research, he analyzed a sample of allophane, an aluminum silicate, taken from a lignite bed near Bonn. In 1835 and 1836 Bunsen set forth the compositions and crystal measurements of a new series of double cyanides, showing, for example, that ammonium ferrocyanide and potassium ferrocyanide are isomorphous. He also discovered the double salt of ammonium ferrocyanide and ammonium chloride.

Bunsen's only work in organic chemistry was an investigation of compounds of cacodyl, an arsenic-containing organic compound, the results of which appeared in five papers published between 1837 and 1842. In 1843, Bunsen lost the use of his right eye in an explosion of cacodyl cyanide. The first known cacodyl compound, alkarsine, had been prepared in 1760 by L. C. Cadet de Gassicourt, by distilling a mixture of dry arsenious oxide and potassium acetate. Alkarsine is a highly reactive, poisonous, spontaneously inflammable substance having heavy brown fumes and a nauseating odor. Its chemical composition was shown by Bunsen to be $C_4H_{12}As_2O$, as Berzelius had suggested. Berzelius called the compound *kakodyl oxide* (from the Greek κακόδης, "stinking"). Bunsen conducted a detailed study of cacodyl derivatives, obtaining the chloride, iodide, cyanide, and fluoride by reacting concentrated acids with the oxide. Using vapor density techniques, he determined the molecular formulas of the derivatives and realized that the cacodyl radical, $C_4H_{12}As_2$, was preserved as an "unchangeable member" through the numerous reactions. This conclusion supported the radical theory of organic compounds advocated by Liebig and Berzelius. Bunsen put forth further evidence for the radical theory when he isolated the free radical by heating the chloride with zinc in an atmosphere of carbon dioxide. After presenting his papers, Bunsen withdrew from the controversy over the merits of the radical theory and turned to inorganic chemistry. It remained for his students, Adolph Kolbe and Edward Frankland, to show in 1853 that cacodyl compounds contain dimethylarsenic, As $(CH_3)_2$, and for Auguste Cahours and Jean Riche to demonstrate that free cacodyl is $As_2(CH_3)_4$. Finally, in 1858 Adolph von Baeyer, another of Bunsen's students, clarified the relationships among the members of the cacodyl series.

Between 1838 and 1846, Bunsen developed methods for the study of gases while he was investigating the industrial production of cast iron in Germany and, in collaboration with Lyon Playfair, in England. He demonstrated the inefficiency of the process: in the charcoal-burning German furnaces, over 50 percent of the heat of the fuel used was lost in the escaping gases; worse, in the coal-burning English furnaces, over 80 percent was lost. Valuable by-products, such as ammonia, went unrealized and were among the gases lost to the atmosphere. Further, it was accidentally discovered that potassium cyanide was formed from potassium carbonate and atmospheric nitrogen at high temperatures. In an 1845 paper, "On the Gases Evolved From Iron Furnaces With Reference to the Smelting of Iron," Bunsen and Playfair suggested techniques that could recycle gases through the furnace, thereby utilizing heat otherwise lost. They also discussed ways by which valuable escaping materials could be retrieved.

Bunsen compiled his research on the phenomena of gases into his only book, *Gasometrische Methoden* (1857). This work brought gas analysis to a level of accuracy and simplicity reached earlier by gravimetric and titrimetric techniques. Dividing the book into six parts, Bunsen presented methods of collecting, preserving, and measuring gases; techniques of eudiometric analysis; new processes for determining the specific gravities of gases; results of investigations on the absorption of gases in water and alcohol using an absorptiometer he himself had devised; and results of experiments on gaseous diffusion and combustion. On the problem of gaseous absorption, Bunsen, assisted by several students, showed the experimental limits within which Henry's law of pressures and Dalton's law of partial pressures are valid.

Greatly interested in geology, Bunsen accompanied a scientific expedition to Iceland in 1846, the year after the eruption of the volcano Hekla. The expedition, sponsored by the Danish government, lasted three and one-half months and included Sartorius von Waltershausen and Bergman, both from Marburg, and Alfred DesCloizeaux, a French mineralogist. Bunsen collected gases emitted from the volcanic openings and studied the action of these gases on volcanic rocks. He performed extensive chemical analyses of eruptive rocks, insisting that instead of determining what minerals were in a rock, the chemical composition of the rock as a whole should be ascertained. Bunsen concluded that volcanic rocks are mixtures, in varying proportions, of two extreme kinds of rock: one kind acidic and rich in silica (trachytic), the other kind basic and less rich in silica (pyroxenic). He thought that the formation of different kinds of rock could be traced to their differences in melting-point behavior under pressure. Although this explanation is no longer accepted, his observations contrib-

uted a great deal to the development of modern petrology. Bunsen also explored geysers, and at the Great Geyser made daring temperature measurements at several depths shortly before it erupted. He found that the temperature of water in the geyser tube, although high, did not reach the boiling point for a particular depth and corresponding pressure. He concluded that the driving force for eruption is supplied by steam that enters the tube under great pressure from volcanic vents at the bottom. As the steam lifts the column of water, the effective pressure above the water is reduced. This change in the water's depth results in a lowering of the boiling point and enables the already hot water to boil.

Through the 1840's and 1850's Bunsen made a number of improvements in the galvanic battery. In 1841 he made a battery, known since as the Bunsen battery, with carbon, instead of the more expensive platinum or copper, as the negative pole. To prevent disintegration of the carbon pole by the nitric acid electrolyte, Bunsen treated the carbon, a mixture of coal and coke, with high heat. Forming a battery from forty-four subunits, Bunsen was able to generate a light of great intensity. Later, he made a battery with chromic acid instead of nitric acid, as well as one with zinc and carbon plates in chromic acid. In 1852 Bunsen began to use electrochemical techniques to isolate pure metals in quantities sufficient for determining their physical and chemical properties. He prepared chromium from a solution of the chloride and magnesium from the fused chloride. He pressed magnesium into wire and used it as a light source in his subsequent photochemical experiments. Commercial manufacture of magnesium was also undertaken and the element came into general use as a brilliant illuminating agent.

In the mid-1850's Bunsen prepared sodium and aluminum from their molten chlorides. With the assistance of Augustus Matthiessen, Bunsen isolated lithium and several alkaline earth metals—barium, calcium, and strontium—from their fused chlorides. Bunsen, with William Hillebrand and T. H. Norton, prepared the rare earth metals of the cerium group—cerium, lanthanum, and didymium. To obtain the specific heats of these rare elements, Bunsen devised a sensitive ice calorimeter that measured the volume rather than the mass of the ice melted and required only a small sample of the metal. From the specific heats, the atomic weights of these elements and the formulas of their compounds were calculated. Finally, the Bunsen battery made possible the electrolysis of a variety of organic compounds and the isolation of organic radicals by Kolbe and Frankland, who began their work under Bunsen's direction in Marburg.

Between 1852 and 1862 Bunsen collaborated with Sir Henry Roscoe on photochemical research involving the chemical combination of equal volumes of hydrogen and chlorine when they were illuminated. For this experiment they altered a reaction vessel devised by John Draper in 1843. Bunsen and Roscoe found that for some time after the experiment started—a time they called the induction period—no reaction took place; then the reaction rate slowly increased until a constant rate, proportional to the intensity of the light source used, was reached. The effect of the incident light was related to the wavelength and followed a law of inverse squares. Further, the illumination of chlorine alone before it entered the reaction chamber did not alter the length of the induction period. While variations of temperature within the range 18°–26° had little effect on the reaction, the presence of oxygen appeared to have a catalytic effect. Bunsen and Roscoe determined that the energy of the light radiated by the sun in one minute is equivalent to the energy needed for the conversion of 25×10^{12} cubic miles of a hydrogen-chlorine mixture into hydrogen chloride.

Bunsen developed his well-known burner during the 1850's, building upon the inventions of Aimé Argand and Michael Faraday. The Bunsen burner, with its nonluminous flame, quickly supplanted the blowpipe flame in the dry tests of analytical chemistry. Bunsen used his burner to identify metals and their salts by their characteristic colored flames. Other experiments with the burner yielded data for melting points and rate of volatility of salts.

In the 1860's Bunsen and Kirchhoff worked together to develop the field of spectroscopy. Kirchhoff realized in 1859 that when colored flames of heated materials, which usually give bright, sharp emission spectra, are placed in the path of an intense light source, they absorb light of the same wavelength that they otherwise emit, and produce characteristic absorption spectra. Bunsen saw that analyses of absorption spectra could be made in order to determine the composition of celestial and terrestrial matter. He further predicted that spectral analysis could aid in the discovery of new elements that might exist in too small quantities or be too similar to known elements to be identifiable by traditional chemical techniques. Spectral analysis led to Bunsen and Kirchhoff's announcement in 1860 of a new alkali metal, cesium, detected in a few drops of the alkaline residue from an analysis of mineral water obtained from Durkheim. The element was named cesium (from the Latin caesius, "sky blue") because of its brilliant blue spectral lines. Cesium salts had previously been mistaken for compounds of potassium. The following year the element rubidium (from the Latin rubidus, "dark red") was detected from the spectrum of a few grains of the mineral lepidolite. By comparison, forty tons of mineral water were needed to yield 16.5 grams of cesium chloride and rubidium chloride that could be used in the chemical investigation of the compounds of these new elements. In 1862 Bunsen succeeded in isolating metallic rubidium by heating a mixture of the carbonate and charcoal. During the years that followed, several other elements were identified by spectroscopic methods: thallium (Crookes, 1861), indium (Reich and Richter, 1863), gallium (Lecoq de Boisbaudran, 1875), scandium (Nilson, 1879), and germanium (Winkler, 1886).

Bunsen was concerned with a variety of additional analytic work. In 1853 he developed a technique for the volumetric determination of free iodine using sulfurous acid. In 1868 he worked out methods for separating the several metals—palladium, ruthenium, iridium, and rhodium—that remain in ores after the extraction of platinum; as part of this project Bunsen constructed a filter pump for washing precipitates. With the assistance of Victor Meyer, he conducted a government-sponsored study of the mineral water of Baden; results were published in 1871. He described the spark spectra of the rare earths in 1875. Late in his life Bunsen used a steam calorimeter that he had built to measure the specific heats of platinum, glass, and water.

Bunsen was honored by several European scientific societies. In 1842 he was elected a foreign member of the Chemical Society of London. He became a corresponding member of the Académie des Sciences in 1853, and a foreign member in 1882. He was named a foreign fellow of the Royal Society of London in 1858 and received its Copley Medal in 1860; Bunsen and Kirchhoff received the first Davy Medal in 1877. Finally, Bunsen's scientific contributions to industry were recognized by the English Society of Arts, which awarded him the Albert Medal in 1898.

BIBLIOGRAPHY

I. ORIGINAL WORKS. Bunsen's writings include *Gasometrische Methoden* (Brunswick, 1857; enl. ed., 1877), trans. by Henry E. Roscoe as *Gasometry; Comprising the Leading Physical and Chemical Properties of Gases* (London, 1857); *Photochemical Researches,* 5 pts. (London, 1858–1863), written with Henry E. Roscoe and pub. in German as *Photochemische Untersuchungen* (Leipzig, 1892); *Chemische Analyse durch Spectralbeobachtungen* (Vienna, 1860), written with Kirchhoff; and *Gesammelte Abhandlungen,* Wilhelm Ostwald and Ernst Bodenstein, eds., 3 vols. (Leipzig, 1904). Also of interest, all in Klassiker der exacten Wissenschaften, are *Untersuchungen über die Kakodylreihe,* Adolf von Baeyer, ed., no. 27; and *Photochemische Untersuchungen,* W. Ostwald, ed., nos. 34, 38.

II. Secondary Literature. Works on Bunsen are Theodore Curtin's article in Eduard Farber, *Great Chemists* (New York, 1961), pp. 575–581, a trans. from *Journal für praktische Chemie* (1900); O. Fuchs's article in F. D. G. Bugge, *Das Buch der grossen Chemiker,* II (Berlin, 1930), 78–91; Georg Lockemann, *Robert Wilhelm Bunsen. Lebensbild eines deutschen Forschers* (Stuttgart, 1949); Ralph E. Oesper, "Robert Wilhelm Bunsen," in *Journal of Chemical Education,* **4** (1927), 431–439; W. Ostwald's article in *Zeitschrift für Elektrochemie,* **7** (1900), 608–618; J. R. Partington, *A History of Chemistry,* IV (London, 1964), 281–293; H. Rheinboldt, "Bunsens Vorlesung über allgemeine Experimentalchemie," in *Chymia,* **3** (1950), 223–241; Henry E. Roscoe, "Bunsen Memorial Lecture" (delivered 29 Mar. 1900), in *Journal of the Chemical Society,* **77,** pt. 1 (1900), 513–554; and *Bunseniana. Eine Sammlung von humoristischen Geschichten aus den Leben von Robert Bunsen* (Heidelberg, 1904).

Susan G. Schacher

BUONAMICI, FRANCESCO (*b.* Florence, Italy, first half of the sixteenth century; *d.* 1603), *medicine, natural philosophy.*

Buonamici's importance for the history of science derives less from his career as a Florentine physician than from his having taught physics at Pisa while Galileo was a student there, giving credence to the theory that Galileo's *Juvenilia* (commonly held to be class notes written in 1584) were based on Buonamici's lectures. A copy of Buonamici's *De motu* was in Galileo's personal library; there are resemblances between portions of this and Galileo's early writings (including Galileo's *De motu,* composed *ca.* 1590), although Galileo later attacked Buonamici's teachings in a discourse printed in 1612.

The full title of Buonamici's principal work is *De motu libri X, quibus generalia naturalis philosophiae principia summo studio collecta continentur, necnon universae quaestiones ad libros De Physico audito, de Caelo, de Ortu et Interitu pertinentes explicantur; multa item Aristotelis loca explanantur, et Graecorum, Averrois, aliorumque doctorum sententiae ad theses peripateticas diriguntur* ("Ten books on motion, in which are contained general principles of natural philosophy, culled with great care, and in which are worked out all questions relating to [Aristotle's] books of the *Physics, On the Heavens,* and *On Generation and Corruption;* again, many texts of Aristotle are explained, and the opinions of the Greeks, of Averroës, and of other doctors are brought to bear on Peripatetic theses"). The work, as the title indicates, is more than a treatise on motion; it is a com-

plete course in natural philosophy, consisting of 1,031 closely packed pages in the folio edition of 1591. The principles around which Buonamici organizes his course are the four causes of motion, or change; the various types of motion (straight-line, alteration, growth, and so forth); and the relation of motion to the heavenly bodies.

In content the work is a masterpiece of Renaissance eclecticism: Buonamici takes cognizance of the humanist tradition (for example, interspersing Greek poetry throughout the text) as well as of such Greek commentators on Aristotle as Alexander of Aphrodisias and John Philoponus; he cites approvingly such mathematicians as Archimedes (at some length), Nicomachus, and Campanus; he gives attention to the Paduan Averroists and the various Platonic and Neoplatonic schools, ranging from the Arabs to the Florentine Academy; and he continually casts a respectful eye toward Thomas Aquinas, Duns Scotus, and other Scholastics. His citation of authorities is pretentious but generally inaccurate, leading one to believe that he relied heavily on secondary sources; teachings he ascribes to Aquinas, for example, may have been held by contemporary Thomists but certainly are not to be found in Aquinas. He occasionally mentions such fourteenth-century thinkers as Walter Burley, the Calculator (Richard Swineshead), and Albert of Saxony, but he reports their contributions to the development of mechanics only superficially. Among his immediate predecessors and contemporaries, those who attract his attention include Alessandro Achillini and Ludovico Buccaferrea, both of Bologna; Francesco Vicomercati of Milan; Scaliger; and Cardano. Throughout his work Buonamici reveals himself generally as an orthodox and traditional Aristotelian who cites the views of "moderns" (*iuniores*) mainly to refute them.

In mechanics, Buonamici held a theory of "self-expending" (as opposed to inertial) impetus, and accepted uncritically the pretended initial acceleration of projectiles. He rejected the theory of "accidental" gravity invoked by Parisian terminists to explain the acceleration of bodies in free fall, maintaining that Aristotle's explanation is sufficient. Similarly, he preferred Aristotle's rules for calculating the ratios of motions, velocities, and distances of travel to those of Albert of Saxony and other "Latins," which he seems to have drawn from Achillini's *In quaestione de motuum proportionibus,* a work of which he is generally critical.

BIBLIOGRAPHY

I. Original Works. *De motu libri X* (Florence, 1591), in folio, is quite rare; a copy is in the library of Princeton

University; the work also exists in a 1592 edition, in quarto. Alexandre Koyré provides most of the Latin text, with French translation, of Bk. 4, chs. 37–38, dealing with the increase of speed in natural motion; of Bk. 5, chs. 35–36, dealing with projectile motion; and small portions of the Latin text of Bk. 1, chs. 10–11, dealing with the relation of mathematics to physical science, in his *Études galiléennes* (Paris, 1939), pp. 18–41, 267–268, 279. There is a copy of *Discorsi poetica nella Accademia fiorentina in difesa d'Aristotile . . . xix di settembre 1587* (Florence, 1597) in the Bibliothèque Nationale, Paris. Copies of *De alimento libri V* (Florence, 1603) are in the British Museum, London, and the Bibliothèque Nationale, Paris; *ibid.,* Venice, 1604.

II. SECONDARY LITERATURE. Some details concerning Buonamici may be found in J. H. Zedler, ed., *Grosses vollständiges Universal-Lexikon,* IV (Halle–Leipzig, 1733; repr. Graz, 1961), col. 569; Antonio Favaro, ed., *Le opere di Galileo Galilei,* Edizione Nazionale (Florence, 1890), esp. I, 9–13; and I. E. Drabkin and Stillman Drake, eds., *Galileo Galilei: On Motion and On Mechanics,* Univ. of Wisconsin Publications in Medieval Science, no. 5 (Madison, Wis., 1960), provides an English translation of Galileo's *De motu,* with introduction and notes by Drabkin, who calls attention to possible influences of Buonamici on Galileo, pp. 10, 49n, 55n, 78n, 79n; see also the bibliography on pp. 10–11.

WILLIAM A. WALLACE, O.P.

BUONANNI, FILIPPO (*b.* Rome, Italy, 7 January 1638; *d.* Rome, 30 March 1725), *natural sciences.*

Buonanni, one of the most learned Jesuits of his time, was a pupil of Athanasius Kircher, and in 1680 succeeded his master as teacher of mathematics at the Collegium Romanum; in 1698, he was appointed curator of the Kircherian Museum, which he described in his *Museum Collegii Romani Kircherianum* (1709).

Erudite in a number of fields, including numismatics and ecclesiastical history (writing on both subjects), Buonanni made extensive studies in the natural sciences; he constructed his own microscope with three lenses (according to Tortona's system), which proved to be an ingenious mechanism for continual observation. In his *Ricreazione dell'occhio e della mente nell'osservazione della chiocciole* (1681), a work valuable for its many illustrations of shells, he explicitly affirmed his belief in the spontaneous generation of mollusks and rekindled the controversy over generation that had flared in 1671 between Kircher and Francesco Redi. Buonanni's position was anachronistic, since the Aristotelian theory of spontaneous generation had been disproved by Redi in his *Esperienze intorno alla generazione degli insetti* (1668) and by Marcello Malpighi, who had demonstrated the pathogenesis of oak galls from the development of fertilized insect eggs in his *Anatome plantarum* (1679).

Buonanni made no personal observations on the phenomenon of generation in the lower animals; neither had he understood the validity of Nicolaus Steno's declaration that "the oysters and other shells originate from the eggs, not from putrescence," or the statement of the English naturalist Martin Lister that "snails are generated by coition, which we observed often in many of their kinds." He based his belief in the spontaneous generation of mollusks partly on the authority of Aristotle and Kircher and partly on a report by Camillo Picchi of Ancona that "the conches called 'Ballani' (mollusks of the kind *Balanus*) live only in some rocks and not in others," but principally upon an anatomical error of his own; he was convinced, as he stated in his *Ricreazione,* that the mollusks had no hearts. If this were so, they had no blood; Aristotle had written that no bloodless animal is oviparous, and that "all conches are generated spontaneously by the mud—oysters by dirty mud, the others by sandy mud." Convinced that the conches were heartless and bloodless, Buonanni believed that both observation and authority supported the idea of spontaneous generation.

Two years after the publication of the *Ricreazione,* Antonio Felice Marsili, archdeacon of Bologna, brought out his own *Relazione sul ritrovamento dell'uova di chiocciole,* in which he described and, indeed, provided drawings of the eggs of snails, some of which visibly contained minuscule snails. Redi, because of Buonanni's opposition to his conclusions on the oviparous generation of insects, harshly criticized Buonanni in his *Osservazioni* (1684), pointing out his rival's error regarding the absence of the heart in snails (the existence of which Redi demonstrated) and asserting, further, that all snails had hearts. He was ruthless in his exposure of Buonanni's mistakes in methodology and ridiculed Buonanni's attempts to demonstrate spontaneous generation of insects from putrefied hyacinth flowers and to establish that certain putrefied flowers or leaves generated only certain kinds of insects.

Buonanni replied (1691) to Redi's criticism, but his reply was judged by contemporaries as inadequate, and indeed inane. On the other hand, it should be recorded that he did deny the existence of the mythical *remora,* the reality of which had been accepted from Aristotle and Pliny right down to Girolamo Cardano in the mid-sixteenth century. His rational classification of shells was novel and useful. The quality of his illustrations of various insects was excellent—particularly those of the fly, louse, mite, flea, and mosquito. Indeed, his drawings of the *Culex pipiens* (common house mosquito) are the best of the seventeenth century.

BUONO

BIBLIOGRAPHY

I. ORIGINAL WORKS. Buonanni's works include *Ricreazione dell'occhio e della mente nell'osservazione delle chiocciole* (Rome, 1681); *Recreatio mentis et oculi in observatione animalium testaceorum,* the 2nd ed. of the *Ricreazione* (Rome, 1684); *Observationes circa viventia quae in rebus non viventibus reperiuntur* (Rome, 1691), which has as an appendix the *Micrographia curiosa, sive rerum minutissimarum observationes, quae, ope microscopii recognitae, ad vivum exprimuntur a Patre Philippo Bonanni Societatis Jesu sacerdote,* containing some interesting observations on early microscopes and a precise description of his own microscope; and *Museum Collegii Romani Kircherianum descriptum* (Rome, 1709).

II. SECONDARY LITERATURE. Works on Buonanni are J. A. Battarra, *Rerum naturalium historia existentium in Museo Kircheriano edita iam a P. Phil. Bonanni, nunc vero novo methodo cum notis illustrata ac observationibus locupletata a Johanne Antonio Battarra,* 2 vols. (Rome, 1773–1782); Ugo Faucci, "Contributo alla storia della dottrina parassitaria delle infezioni," in *Rivista di storia delle scienze mediche e naturali,* **26** (1935), 136–193, which includes Kircher's biological views (see note 29, p. 147, and note 30, p. 183); A. Neviani, "Un episodio della lotta fra spontaneisti ed ovulisti. Il Padre Filippo Buonanni e l'Abate Anton Felice Marsili," in *Rivista di storia delle scienze mediche e naturali,* **26** (1935), 211–232; and F. Redi, *Osservazioni intorno agli animali viventi che si trovano negli animali viventi* (Florence, 1684), pp. 58–88. The Buonanni microscope is illustrated in Clay and Court, *History of the Microscope* (London, 1932), pp. 41–44, 84–86.

PIETRO FRANCESCHINI

BUONO, PAOLO DEL (*b.* Florence, Italy, 26 October 1625; *d.* Poland, 1659), *mechanics, physics.*

Del Buono was the son of Leonido and Bartolomea Andreini. He studied in Pisa under Famiano Michelini. A member of the Poor Regular Clerics of the Mother of God of the Pious Schools (Scolopi), Del Buono was learned but eccentric. He deserves much credit, however, for arousing a keen interest in the sciences in his pupil Leopoldo de' Medici, who in 1657–1667 was a patron of the Cimento Academy. Del Buono received his degree in Pisa in 1649. Six years later he went to Germany in the service of Emperor Ferdinand III, who appointed him president of the mint and offered him honors and rich prizes if he could devise a mechanism to draw water from mines. In order to make practical studies, in 1657 and 1658 he visited the imperial mines in the Carpathians, accompanied by Geminiano Montanari, a doctor of jurisprudence whom he had instructed in the sciences. The death of the emperor and the disturbances that broke out in Germany made it necessary for him to go to Poland, where he died about a year later.

Del Buono is included among the correspondents of the Cimento Academy, along with his older brother, Father Candido, and his younger brother, Anton Maria.

Several of Del Buono's letters to Prince Leopold are extant. In them he refers to the observations, made in several observatories, of a comet that was then visible. His contributions include an instrument to demonstrate the incompressibility of water and a communication from Vienna that states that water enclosed in glass vials with very thin necks generates air in amounts dependent on the temperature of the environment. The Cimento Academy confirmed this phenomenon, concerning which Giovanni Borelli and Viviani gave conflicting explanations.

BIBLIOGRAPHY

Lettere inedite di uomini illustri, Angelo Fabronio, ed., I (Florence, 1773), 94, 151, 200.

Angelo Fabronio, *Vitae italorum doctrina excellentium qui saeculis XVII et XVIII fioruerunt,* 12 vols. (Pisa, 1778–1785).

Notizie degli aggrandimenti delle scienze fisiche accaduti in Toscana nel corso degli anni LX del sec. XVII°, raccolte dal dottor Giovanni Targioni Tazzetti, G. Bouchard, ed., I (Florence, 1780), 182, 519; II, pt. 1 (Florence, 1780), 309.

Raffaele Caverni, *Storia del metodo sperimentale in Italia* 6 vols. (Florence, 1891–1892), II, 263.

Galileo Galilei: Le opere, Antonio Favaro *et al.,* eds., national ed., 20 vols. (Florence, 1890–1909; reprinted 1929–1939). There is a short biography of Del Buono in XX, pt. 6; see also XXV, pt. 3, 352, letter from Ward to Galileo of 7 September 1641.

A. NATUCCI

BUONVICINO, COSTANZO BENEDETTO. See **Bonvicino, Costanzo Benedetto.**

BUOT, JACQUES (*d. ca.* 1675), *astronomy, physics, geometry.*

Buot was a member of the Académie des Sciences of Paris from the time it was founded in 1666; as such he received an annual stipend of 1,200 livres.

Buot probably was present on 16 July 1667 when Huygens observed the exact hour at which the diameter of the ring of Saturn seemed to be parallel to the horizon. From this observation, Huygens calculated the inclination of the ring to the equator as 8°58' and to the ecliptic as 31°22'; Buot found a value of 31°38'35" for the latter. (As early as 1659, in his *Systema Saturnium,* Huygens had attempted to determine these values as precisely as possible.) On 15 August 1667 Huygens, Jean Picard, Jean Richer, and

Buot repeated the experiment and obtained values of 9°32'50'' and 32°0'.

Buot made a further contribution to astronomy by inventing the *équerre azimutale,* an instrument for finding the intersection of the meridian with a horizontal plane. He was also active as a physicist, once again drawing upon Huygens' work. In 1667 Huygens and members of the Accademia del Cimento in Florence had made experiments to determine the forces that cause water to expand on congelation. In 1670 Buot repeated these experiments for water and oils and observed that the congealing of water differs from that of oils. In 1669 Buot joined Huygens and others in the discussions on the causes of gravitation that were held by the Academy; in his *mémoire* of 21 August 1669, he showed himself opposed to the action-at-a-distance theory.

Condorcet wrote that Buot died in 1675. A letter from Olaus Römer to Huygens, dated 30 December 1677, states, however, *Dominus Buot post aliquot mensium morbum fato appropinquare creditur—*"After an illness of some months Mr. Buot seems to feel that his end is drawing near." The *Comptes des bâtiments* gives the date of his last stipend as 10 June 1676.

BIBLIOGRAPHY

I. ORIGINAL WORKS. Buot's writings include *Usage de la roüe de proportion, avec un traité d'arithmétique* (Paris, 1647); and "Équerre azimutale," in Gallon, ed., *Machines et inventions approuvées par l'Académie royale des sciences depuis 1666 jusqu'en 1701,* I (Paris, 1735), 67–70.

II. SECONDARY LITERATURE. Works dealing with Buot are Condorcet, *Éloges des Académiens de l'Académie royale des sciences, morts depuis 1666 jusqu'en 1699* (Paris, 1773), p. 157; J. Guiffrey, *Comptes des bâtiments du roi, sous le règne de Louis XIV,* I (Paris, 1881), 163, 227, 299, 378, 449, 565, 650, 782, 856; Panckoucke, ed., *Histoire de l'Académie royale des sciences 1666 à 1698,* I (Paris, 1777), 121; and *Oeuvres complètes de Christiaan Huygens,* VI (The Hague, 1895), 58–66, 139–142, 143–147; VIII (The Hague, 1899), 54; XV (The Hague, 1925), 43, 93, 94, 478; XIX (The Hague, 1937), 182, 183, 344, 630; E. Maindron, *L'Académie des sciences* (Paris, 1888), p. 98; and D. Shapeley, "Pre-Huygenian Observations of Saturn's Rings," in *Isis,* **40** (1949), 12–17.

H. L. L. BUSARD

BURALI-FORTI, CESARE (*b.* Arezzo, Italy, 13 August 1861; *d.* Turin, Italy, 21 January 1931), *mathematics.*

After obtaining his degree from the University of Pisa in December 1884, Burali-Forti taught at the Scuola Tecnica in Augusta, Sicily. In 1887 he moved to Turin after winning a competition for extraordinary professor at the Accademia Militare di Artiglieria e Genio. In Turin he also taught at the Scuola Tecnica Sommeiller until 1914. He remained at the Accademia Militare, teaching analytical projective geometry, until his death. He was named ordinary professor in 1906 and held a prominent position on the faculty; in 1927 he was the only ordinary among twenty-five civilian professors.

After an early attempt to obtain the *libera docenza* failed because of the antagonism to the new methods of vector analysis on the part of some members of the examining committee, he never again attempted to obtain it and thus never held a permanent university position. (The *libera docenza* gave official permission to teach at a university and was required before entering a competition for a university chair.) He was assistant to Giuseppe Peano at the University of Turin during the years 1894–1896, but he had come under Peano's influence earlier, however, and had given a series of informal lectures at the university on mathematical logic (1893–1894). These were published in 1894. Many of Burali-Forti's publications were highly polemical, but to his family and his friends he was kind and gentle. He loved music; Bach and Beethoven were his favorite composers. He was not a member of any academy. Always an independent thinker, he asked that he not be given a religious funeral.

The name Burali-Forti has remained famous for the antinomy he discovered in 1897 in his critique of Georg Cantor's theory of transfinite ordinal numbers. The critique begins: "The principal purpose of this note is to show that there exist *transfinite numbers* (or *ordinal types*) a, b, such that a is neither equal to, nor less than, nor greater than b." Essentially, the antinomy may be formulated as follows: To every class of ordinal numbers there corresponds an ordinal number which is greater than any element of the class. Consider the class of all ordinal numbers. It follows that it possesses an ordinal number that is greater than every ordinal number. This result went almost unnoticed until Bertrand Russell published a similar antinomy in 1903. It should be noted, however, that Cantor was already aware of the Burali-Forti antinomy in 1895 and had written of it to David Hilbert in 1896.

Burali-Forti was one of the earliest popularizers of Peano's discoveries in mathematical logic. In 1919 he published a greatly enlarged edition of the *Logica mathematica,* which contained many original contributions. He also contributed much to Peano's famous *Formulaire de mathématiques* project, especially with his study of the foundations of mathematics (1893).

Burali-Forti's most valuable mathematical contributions were his studies devoted to the foundations of vector analysis and to linear transformations and their various applications, especially in differential geometry. A long collaboration with Roberto Marcolongo was very productive. They published a series of articles in the *Rendiconti del Circolo matematico di Palermo* on the unification of vectorial notation that included a full analysis, along critical and historical lines, of all the notations that had been proposed for a minimal system. There followed a book treating the fundamentals of vector analysis (1909), which was almost immediately translated into French. Their proposals for a unified system of vectorial notation, published in *L'enseignement mathématique* in 1909, gave rise to a polemic with various followers of Josiah Gibbs and Sir William Hamilton that lasted into the following year and consisted of letters, responses, and opinions contributed by Burali-Forti and Marcolongo, Peano, G. Comberiac, H. C. F. Timerding, Felix Klein, E. B. Wilson, C. G. Knott, Alexander Macfarlane, E. Carvallo, and E. Jahnke. The differences in notation continued, however, and the Italian school, while quite productive, tended to remain somewhat isolated from developments elsewhere. Also in 1909 Burali-Forti and Marcolongo began their collaboration in the study of linear transformations of vectors.

Burali-Forti's introduction of the notion of the derivative of a vector with respect to a point allowed him to unify and greatly simplify the foundations of vector analysis. The use of one simple linear operator led to new applications of the theory of vector analysis, as well as to improved treatment of operators previously introduced, such as Lorentz transformations, gradients, and rotors, and resulted in the publication (1912–1913) of two volumes treating linear transformations and their applications. Burali-Forti was able to apply the theory to the mechanics of continuous bodies, optics, hydrodynamics, statics, and various problems of mechanics, always refining methods, simplifying proofs, and discovering new and useful properties. He did not live to see the completion of his dream, a small encyclopedia of vector analysis and its applications. The part dealing with differential projective geometry (1930) was Burali-Forti's last work.

The long collaboration with Marcolongo—their friends called them "the vectorial binomial"—was partly broken by their divergent views on the theory of relativity, the importance of which Burali-Forti never understood. With Tommaso Boggio he published a critique (1924) in which he meant "to consider Relativity under its mathematical aspect, wishing to point out how arbitrary and irrational are its founda-

tions." "We wish," he wrote in the preface, "to shake Relativity in all its apparent foundations, and we have reason for hoping that we have succeeded in doing it." At the end he stated: "Here then is our conclusion. Philosophy may be able to justify the space-time of Relativity, but mathematics, experimental science, and common sense can justify it NOT AT ALL."

Burali-Forti had a strong dislike for coordinates. In 1929, in the second edition of the *Analisi vettoriale generale,* written with Marcolongo, we find: "The criteria of this work . . . are not different from those with which we began our study in 1909, namely, an absolute treatment of all physical, mechanical, and geometrical problems, independent of any system of coordinates whatsoever."

BIBLIOGRAPHY

Besides his scientific publications, Burali-Forti wrote many school texts. In all, his publications total more than two hundred.

No complete list of the works of Burali-Forti has been published, but the following may be considered representative: *Teoria delle grandezze* (Turin, 1893); *Logica matematica* (Milan, 1894; 2nd ed., rev., Milan, 1919); "Una questione sui numeri transfiniti," in *Rendiconti del Circolo matematico di Palermo,* **11** (1897), 154–164; *Lezioni di geometria metrico-proiettiva* (Turin, 1904); *Elementi di calcolo vettoriale, con numerose applicazioni alla geometria, alla meccanica e alla fisica-matematica,* written with R. Marcolongo (Turin, 1909), translated into French by S. Lattès as *Éléments de calcul vectoriel, avec de nombreuses applications à la géometrie, à la mécanique et à la physique mathématique* (Paris, 1910); "Notations rationelles pour le système vectoriel minimum," in *L'enseignement mathématique,* **11** (1909), 41–45, written with Marcolongo; *Omografie vettoriali con applicazioni alle derivate rispetto ad un punto ed alla fisica-matematica* (Turin, 1909), written with Marcolongo; *Analyse vectorielle générale,* 2 vols. (Pavia, 1912–1913), written with Marcolongo; *Espaces courbes. Critique de la relativité* (Turin, 1924), written with Tommaso Boggio; and *Analisi vettoriale generale e applicazioni,* Vol. II, *Geometria differenziale* (Bologna, 1930), written with P. Burgatti and Tommaso Boggio.

A work dealing with Burali-Forti is Roberto Marcolongo, "Cesare Burali-Forti," in *Bollettino dell'Unione matematica italiana,* **10** (1931), 182–185.

HUBERT C. KENNEDY

BURDACH, KARL FRIEDRICH (*b.* Leipzig, Germany, 12 June 1776; *d.* Königsberg, Germany, 16 July 1847), *physiology.*

Karl Friedrich Burdach was a natural scientist typical of his peers and representative of a distinctive period in the intellectual life of Germany—the Romantic age of the early nineteenth century. Guided by the tenets of *Naturphilosophie,* he made significant contributions, particularly to neuroanatomy. His penchant for extreme systematization, however, led him to publish in his many treatises much that later workers ridiculed as "unscientific."

The son of a physician, Burdach was encouraged to undertake medical training. He began his studies at the University of Leipzig in 1793, and received the doctor of philosophy degree five years later, in August 1798. He was still unqualified to practice, however, since at the time Leipzig did not offer any clinical training. He therefore proceeded to Vienna and the great clinician Johann Peter Frank, not only because of Frank's fame and the attractions of Vienna, but also because of Frank's reported adherence to the Brunonian system of medicine, which had fascinated the young man during his training. At the end of his year at Vienna, Burdach began to search—unsuccessfully—for an academic position. He formally became doctor of medicine at Leipzig in June 1799, and settled down to private practice and private lecturing while awaiting a university appointment. During this period, which lasted until 1811, the young man turned to medical writing as a means of supplementing his income and making his name known.

In these early works, Burdach's sympathies for the Romantic philosophy of nature were obvious. Although he took issue with the *Naturphilosophen* followers of Lorenz Oken and Friedrich Schelling on points of method and doctrine, his works represent an attempt to understand particular aspects of the natural world as integral parts of a coherent whole. Burdach was guided by the credo "Those who have thoughts have always seen more than those who merely wanted to see with their own eyes"; he approached the study of natural phenomena with the conviction that nature is totality and unity, an "Idea" of which all individual things are partial representations and in which they participate. To know nature, then, one needs both to study and to reflect on individual phenomena.

In the summer of 1811 Burdach was finally successful in obtaining a position, as professor of anatomy, physiology, and forensic medicine at the University of Dorpat. The new-found financial freedom and increased spare time allowed him finally to begin firsthand research, and his choice of areas reflected his general interests. He began intensive studies of the anatomy of the human brain, and of embryonic and animal brains; he also studied in detail the classic works on embryology, in particular those from Harvey to the treatise by Johann von Autenrieth (1797). It is clear that both interests, which were to

remain with Burdach, stemmed from his conviction that a knowledge of the world could be obtained only through a rational intellectual analysis of all aspects of man's life.

Animated by the desire to return to Germany, in 1813–1814 Burdach applied for and received the recently vacated professorship of anatomy at Königsberg. In addition to an increase in salary and the title *Hofrath,* he obtained permission to create and head an anatomical institute. For the prosectorship of the institute he finally obtained the services of Karl Ernst von Baer, a former student of his at Dorpat, and the Konigliche Anatomische Anstalt in Königsberg was formally opened on 13 November 1817. Within it, Burdach taught courses in anatomy, physiology, propaedeutics, and the life of the fetus, and Baer—treated more as colleague than assistant—dealt with zootomy, human anatomy, and fetal physiology. The research carried out at the institute was reported in yearly *Berichte.*

The first results of Burdach's examination of the nervous system were reported in the first *Bericht* (1818), and a complete treatment of the subject appeared in *Vom Baue und Leben des Gehirns,* the first volume of which appeared in 1819. The work is one of the best examples of the approach taken by the Romantic physiologists. It seeks to be comprehensive and systematic; Burdach maintained that one must know formation and life in order to arrive at the real goal, knowledge of the nature of the total entity. He argued that every coherent, systematic work must proceed in accordance with a particular point of view, and he stated his explicitly: Nature is a unity, of which all phenomena, including those of the mind, partake, and an ideal *Sein* lies at the base of all appearances. It follows that since both reason and appearances are aspects of nature, examination of that unity may proceed by two paths: observation and reflection, or, as Burdach put it, "contemplating ideas in their necessity and demonstrating them in their reality."

Burdach sought to show that the nervous system was itself a unity and not just a conglomerate of various anatomical structures. He examined the parts of the brain with the intent of delineating the systems of which they were elements, and was particularly interested in such integrating structures as the intracerebral connecting paths and the fiber tracts. This entailed a thorough anatomical examination, and Burdach provided precise and detailed descriptions of many parts and their relationships. He is credited with the explanation of the relationship of the olive to its surrounding region and with the clarification of the nature of many of the fasciculi. In the anatomical nomenclature prior to the Basel reforms, the column of fibers in the spinal cord now called the fasciculus cuneatus and first described by Burdach was known as the "column of Burdach." He described several of the nuclei of the thalamus and pointed out the division of the lentiform nucleus of the corpus striatum into putamen and globus pallidus. Among the other structures he named were the claustrum, the brachium conjunctivum (superior cerebellar peduncle), and the cuneus. But such observational advances were only way stations on the path to knowledge of the "Idea" of the brain. Thus, a third volume, *Leben des Gehirns,* followed the first two anatomical volumes. This was an attempt at a physiology of the central nervous system, and contained discussions of meningitis, the function of the cerebral gyri, and the connection of the corpora quadrigemina with the hearing function.

In 1820 the question arose of the desirability of producing a second edition of Burdach's 1810 physiology text. Instead, he chose to embark on a more ambitious course—a large-scale treatment of all physiology as a whole:

> I gave myself the task of deriving the life of man primarily from the nature of life as known through consideration of the entire organic realm. This knowledge, however, was to come from a view of the whole of nature derived from the combining of the appearances and the laws of organic and inorganic life [*Rückblick auf mein Leben*, p. 336].

Burdach thus intended to produce a statement of the nature of life through a study of all the individual appearances that are a part of life—a science of experience, based on nondemonstrable premises. For the work he recruited collaborators and assistants in many ancillary fields: Baer, Heinrich Rathke, the botanist C. W. Eysenhardt, and, for later volumes, Johannes Müller, Theodor von Siebold, and Gabriel Valentin. The first volume of the text, *Die Physiologie als Erfahrungswissenschaft,* appeared in 1826.

The form of the work was determined for Burdach by his conception of physiology. The ultimate goal of physiology was knowledge of the human spirit (*Geist*); but the essence (*Wesen*) of anything, he pointed out, takes root only in the whole of reality, and only therein will it be completely known. Thus, in order to know man, physiology must view the whole of nature and consider all the phenomena of the world. For Burdach, as for many of his contemporaries, physiology was no longer a study of the functions or uses of organic parts, but of life and its appearances. "Physiology is therefore the apex of all natural science, the point of unity of the knowledge of reality."

The ten volumes of the work were to be divided between a consideration of life in general as an ordered progression of processes and the particular functions and processes of living things. However, the death of Burdach's wife in 1838 forced the termination of the work after the sixth volume, with the complete plan unrealized. The first three volumes published dealt with the history of *Leben an sich*: generation (Vol. 1); embryonic life, beginning with the development of the egg (Vols. II and III); and independent life, leading to death (Vol. III). The incomplete second part treated some of the systems involved in the vegetative aspects of life: the blood, the blood circulation, and the associated functions of nutrition and secretion. Incomplete though it was, the text was for a time well known and widely read.

The discussion of development in Volume II, prepared largely by Baer and Rathke, indirectly had a profound influence on the development of modern embryology. Baer submitted an entire treatise on the formation of the embryo; Burdach, however—a man convinced of the correctness of his ideas—rearranged the discussion to conform to his larger system, breaking up Baer's treatise and distributing it throughout the work. Baer, angry at what he considered such cavalier treatment, proceeded to publish his treatise separately, as the first part of his *Ueber Entwicklungsgeschichte der Thiere* (1828), the first great work of modern embryology.

The last of Burdach's major works was the four-volume *Blicke ins Leben* (1842–1848). It was written as the culmination of one aspect of Burdach's scientific life, and it exhibited characteristic elements of the thought not only of Burdach but also of most Romantics. Burdach was always optimistic about the unlimited knowledge of the human spirit available to man; he strove always to be rational, since reason is an aspect of the Absolute; and he strove always for universality, seeking true knowledge, which is knowledge of the totality of nature. With the preparation that Burdach believed he had obtained from a lifetime's properly directed scientific work, he sought to show the reader how, from a scientific point of view, all existence can and must be seen as a harmonious whole. Thus, religion and morality must properly be based on science. The first two volumes (1842) continued Burdach's developmental studies by seeking to set forth a phylogeny of the human soul. The third (1844) contained general reflections on human existence, and the last volume, published posthumously in 1848, was Burdach's autobiography.

BIBLIOGRAPHY

I. ORIGINAL WORKS. A fairly complete bibliography of Burdach's works, compiled by T. H. Bast, is available in *Annals of Medical History*, **10** (1928), 45–46. Much of Burdach's observational work was published in the seven *Berichte von der Königlichen Anatomischen Anstalt zu Königsberg* (1818–1824). His treatise on the brain and spinal cord, *Vom Baue und Leben des Gehirns*, was published in three vols. (Leipzig, 1819–1826). *Die Physiologie als Erfahrungswissenschaft* appeared in six vols. (I, Leipzig, 1826; II–VI, Königsberg, 1828–1840). There was also a 2nd ed. of vols. I–III (Königsberg, 1835–1838). Some of the views that had not appeared in this work were included in *Anthropologie für das gebildete Publicum* (Stuttgart, 1837). Burdach's autobiography, *Rückblick auf mein Leben*, was completed and published posthumously as the fourth and last vol. of *Blicke ins Leben* (Leipzig, 1842–1848). A. W. Meyer has published translations of two of Burdach's embryological treatises, prefaced by a short biographical sketch, in his *Human Generation. Conclusions of Burdach, Döllinger and von Baer* (Stanford, 1956), pp. 3–25. These are *De primis momentis formationis foetus* (1814) and *De foetu humano adnotationes anatomicae* (1828). This choice is somewhat puzzling: the first treatise, as Meyer himself notes, was written to satisfy the requirements for joining the faculty at Königsberg; Burdach called it "a premature birth" and refused to send out copies of it. The second essay was a vehicle for congratulations to Sömmering from the Königsberg faculty on the occasion of his jubilee, and the few observations are drawn from the second vol. of Burdach's *Die Physiologie als Erfahrungswissenschaft*. The embryological content of that work, however, has not yet been examined.

II. SECONDARY LITERATURE. Theodor Bast published a biographical sketch of Burdach, drawn entirely from his autobiography, in *Annals of Medical History*, **10** (1928), 34–46. Sections of the autobiography were reprinted (without the original pagination) in Erich Ebstein, ed., *Ärzte-Memoiren aus vier Jahrhunderten* (Berlin, 1923), pp. 158–165. A portrait of Burdach faces p. 158. The interaction of Burdach's researches and reflections is briefly but interestingly discussed in H. B. Picard, "Philosophie und Forschung bei K. F. Burdach," in *Medizinische Monatsschrift* (Stuttgart), **5** (1951), 125–128. Meyer (see above) briefly discusses the controversy between Burdach and Baer that led to publication of the latter's *Ueber Entwicklungsgeschichte der Thiere*. Baer's point of view is given in his *Nachrichten über Leben und Schriften des Herrn Geheimrathes Doctor Karl Ernst von Baer mitgetheilt von ihm selbst* (St. Petersburg, 1865), pp. 417–419, 455–471.

ALAN S. KAY

BURDENKO, NICOLAI NILOVICH (*b.* Kamenka [near Penza], Russia, 3 June 1876; *d.* Moscow, U.S.S.R., 11 November 1964), *neurology.*

Burdenko was born into a village family of some education. From 1886 to 1890 he studied at the Penza parochial school and from 1891 to 1897 was a student in the seminary there. In 1897 he entered the Tomsk University Medical School, transferring in 1901 to the Yuriev University (Tartu) Medical School, from which he graduated in 1906. As a student he was greatly influenced by the ideas of Nikolai Ivanovich Pirogov and the works of Pavlov. In 1903 he joined V.G. Tsego Manteifel's surgical clinic. In 1909, after presenting his doctoral thesis "Materialy k voprosu o posledstviakh perevyazki venaeportae" ("Data on the Effects of Dressing the Venae Portae"), he was employed in laboratories, clinics, hospitals, and libraries in Germany and Switzerland. He learned the surgical methods of August Bier, O. Hildebrandt, F. Krause, and Hermann Oppenheim. Under Constantin von Monakow, in Zurich, he studied anatomy and the histology of the central nervous system and neurological surgery.

After 1910 he held the chair of assistant professor of surgery at Yuriev University and became adjunct professor of surgery and anatomy. After the death of Tsego Manteifel, in 1917, he was professor ordinarius at the school's surgical clinic. From 1918 to 1923 he headed the Voronezh Medical Institute's surgical clinic and in 1923 he was appointed to the chair of anatomy and surgery at the Moscow State University Institute. From 1924 to the end of his life he devoted himself to organizing the clinic's neurological department. After 1929 he was director of the neurological clinic of the Health Ministry's roentgenology institute; this was the precursor of the Central Neurosurgical Institute (founded in 1934), which is today the Burdenko Neurosurgical Institute of the U.S.S.R. Academy of Medical Sciences. Burdenko further made use of his experiences in three wars (the Russo-Japanese War, World War I, and World War II) to lay the basis for Soviet military field surgery.

Burdenko wrote more than 300 articles on clinical and theoretical medicine. His earliest clinicoexperimental research was concerned with the physiology of the liver, the duodenum, the stomach, and the pancreas; his later work deals with a wide variety of problems in anatomy, physiology, biochemistry, histology, and pathology. He was a pioneer of Soviet neurosurgery (and especially of an important school of surgery that is marked by its readiness to experiment) and the teacher of the first generation of Soviet neurosurgeons. He made contributions to the oncology of the central nervous system and the vegetative nervous system; to the pathology and circulation of the blood and the fluids, edema and swelling of the brain; and to the operative treatment of various serious conditions of the nervous system.

Burdenko held many honorary posts. After 1939 he was Chairman of the Board of Directors of the U.S.S.R. College of Surgeons and after 1937, Chairman of the Medical Sciences Council of the U.S.S.R. Ministry of Health. He was the first president of the U.S.S.R. Academy of Medical Sciences, editor of *Sovremennaya khirurgiya* from 1944 to 1946, editor of *Neirokhirurgia,* and a member of the editorial board of *Khirurgiya* and *Voenno-meditsinskii zhurnal.* He was an honorary member of the International Society of Surgeons, the British Royal Society of Surgeons, and the Paris Academy of Surgeons. He was a deputy to the Supreme Soviet of the U.S.S.R. and a Hero of Socialist Labor.

BIBLIOGRAPHY

Burdenko's collected works have been published in seven volumes (Moscow, 1950–1952). See also C. M. Bagdasarian, *Nikolai Nilovich Burdenko* (Moscow, 1954).

N. A. GRIGORIAN

BURDON-SANDERSON, JOHN SCOTT (*b.* Jesmond, near Newcastle-on-Tyne, England, 21 December 1828; *d.* Oxford, England, 23 November 1905), *pathology, physiology.*

Burdon-Sanderson was the son of Richard Burdon, a onetime Oxford don who later severed his connection with the Church of England and became active in evangelical work, and Elizabeth Sanderson, daughter of Sir James Sanderson, M.P., a London merchant who was twice lord mayor of London. Intended by his family for the law, he early developed an interest in natural science. After private instruction at home, he entered the University of Edinburgh in 1847 to study medicine. Among the teachers who influenced him there were John Balfour, John Goodsir, and John Hughes Bennett. Upon graduating M.D. in 1851, he was awarded a gold medal for his thesis on the metamorphosis of the colored blood corpuscles.

In the autumn of 1851, Burdon-Sanderson traveled to Paris, where he studied chemistry in the laboratories of Charles Gerhardt and Charles Wurtz, and attended Claude Bernard's lectures on physiology. Late in 1852, he settled in London to practice medicine, and in August 1853 married Ghetal Herschell, daughter of the Rev. Ridley Herschell. In the same year, he was appointed medical registrar at St. Mary's Hospital, where he later served as lecturer on botany (1854–1855) and on medical jurisprudence

(1855–1862). He was medical officer of health for the parish of Paddington from 1856 to 1867 and inspector for the medical department of the Privy Council from 1860 to 1865. The duties attached to these positions led him into his first work in pathology. He also served on the staffs of the Brompton Hospital for Consumption (1859–1863; 1865–1871) and Middlesex Hospital (1863–1870).

About 1870, Burdon-Sanderson resigned his hospital appointments in order to devote himself exclusively to scientific research. In the same year, he was appointed professor of practical physiology and histology at University College, London, succeeding Michael Foster. In 1874 he succeeded William Sharpey as Jodrell professor of human physiology at University College and remained there until 1882, when he became the first occupant of the Waynflete chair of physiology at Oxford University. His appointment met with violent opposition from antivivisectionists at Oxford (he was notorious for having coauthored a guide to vivisection), and funds for a laboratory of physiology were secured only with great effort. He resigned the Waynflete chair in 1895 to become Regius professor of medicine at Oxford. During his tenure in this chair, several essential reforms were achieved, including the creation of a complete course in pathology and bacteriology.

Burdon-Sanderson's reputation in pathology resulted primarily from his pioneer experimental investigations of contagious diseases and the infective processes. These began with his demonstration in 1865 of the particulate nature of the infective agent in cattle plague and with his confirmation in 1867 of Jean Villemin's experiments on the inoculability of tuberculosis in animals. Although he was generally considered one of the leading exponents in England of the germ theory of disease, there is an ambiguity in his views that makes it difficult to summarize his position simply. In 1869, in his widely discussed work, "On the Intimate Pathology of Contagion," he confirmed Auguste Chauveau's conclusion that the contagium in vaccine lymph was particulate, since the aqueous portion of the lymph was inactive while the solid portion was active. At the same time, he suggested that the infective particles were probably "organised beings" which owed their pathogenicity to their organic development. But when he later demonstrated that bacteria were invariably present in septicemia and pyemia, he avoided the conclusion that the bacteria were directly causative; and as late as 1877 he held that "there is but one case [splenic fever] in which the existence of a disease germ has been established" (*Nature,* **17** [1877], 86). His cautious attitude toward the germ theory resulted from the conflicting nature of the evidence then available, and from his own tendency toward theoretical skepticism. Although not unreasonable, this caution obscures his position as a prophet of the germ theory.

In physiology Burdon-Sanderson's earliest work dealt with the effects of respiratory movements on the circulation (Croonian lecture, 1867), but he later devoted himself almost exclusively to electrophysiological investigations, most notably those on the leaf of *Dionaea muscipula* (the Venus's-flytrap). While experimenting on insectivorous plants for Charles Darwin in 1873, he found that a pronounced electrical current accompanied the familiar closing of the flytrap leaf after stimulation of its excitable hairs. He suggested that this current was indicative of rapidly propagated molecular changes in the leaf cells, and compared this process with the corresponding process in active animal muscle.

Like Michael Foster at Cambridge, Burdon-Sanderson was an important force in establishing physiology as an independent discipline in England. He urged the adoption of the experimental approach to pathology as well as to physiology, and from 1871 to 1878 he was professor superintendent of the newly created Brown Institution, the first laboratory for pathology in England. Among his students at University College, London, were William Bayliss, Francis Gotch, Victor Horsley, William Osler, and G. J. Romanes. The group that later worked under him at Oxford—although it included Gotch—was, in general, less eminent.

Burdon-Sanderson's versatile achievements brought him numerous honors. He was elected a fellow of the Royal Society in 1867, delivered its Croonian lecture on three occasions (1867, 1877, and 1899), and was awarded its Royal Medal in 1883. The Royal College of Physicians elected him a fellow in 1871, appointed him Harveian orator in 1878, and awarded him the Baly Medal in 1880. He was president of the British Association for the Advancement of Science in 1893, and was created a baronet in 1899.

BIBLIOGRAPHY

I. ORIGINAL WORKS. Burdon-Sanderson's early position on the germ theory of disease is best revealed in the following papers: "Introductory Report on the Intimate Pathology of Contagion," in *Twelfth Report of the Medical Officer of the Privy Council* [1869], Parliamentary Papers (London, 1870), pp. 229–256; "Preparations Showing the Results of Certain Experimental Inquiries Relating to the Nature of the Infective Agent in Pyaemia," in *Transactions of the Pathological Society,* **23** (1872), 303–308; remarks

in the discussion on the germ theory of disease, *ibid.,* **26** (1875), 284–289; and "The Occurrence of Organic Forms in Connection With Contagious and Infective Diseases," in *British Medical Journal* (1875), **1,** 69–71, 199–201, 403–405, 435–437. For his later position, see his "Croonian Lectures on the Progress of Discovery Relating to the Origin and Nature of the Infective Diseases," in *Lancet* (1891), **2,** 1027–1032, 1083–1088, 1149–1154, 1207–1211. The first announcement of his discovery of the electric current in *Dionaea* appeared in the *Report of the British Association,* **43** (1873), 133. The most elaborate accounts of his electrophysiological investigations of *Dionaea* and of muscle are "On the Electromotive Properties of the Leaf of Dionaea in the Excited and Unexcited States [1881]," in *Philosophical Transactions of the Royal Society,* **173** (1882), 1–55; and "Croonian Lecture on the Relation of Motion in Animals and Plants to the Electrical Phenomena Which Are Associated With It [1899]," in *Proceedings of the Royal Society,* **65** (1900), 37–64. Burdon-Sanderson edited *Handbook for the Physiological Laboratory* (London, 1873), which he wrote with E. Klein, Michael Foster, and T. Lauder Brunton; this was the first work of its kind in English. His letters and private papers are deposited in the library of University College, London.

II. SECONDARY LITERATURE. The basic source for Burdon-Sanderson's life and work is Lady [Ghetal] Burdon-Sanderson's *Sir John Burdon Sanderson: A Memoir, With Selections From His Papers and Addresses,* completed and edited by his nephew, J. S. Haldane, and his niece, E. S. Haldane (Oxford, 1911). This valuable book is an unusually good example of its genre. For other accounts, see Arthur MacNalty, in *Proceedings of the Royal Society of Medicine* (London), **47** (1954), 754–758; *British Medical Journal* (1905), **2,** 1481–1492; *Lancet* (1905), **2,** 1652–1655; and Francis Gotch, in *Dictionary of National Biography,* supp. 2, I (1912), 267–269; *Proceedings of the Royal Society* (London), **79B** (1907), iii–xviii; and *Nature,* **73** (1905–1906), 127–129. When seeking references to Burdon-Sanderson in indexes and catalogs, check under both surnames.

GERALD L. GEISON

BÜRG, JOHANN TOBIAS (*b.* Vienna, Austria, 24 December 1766; *d.* Wiesenau, Austria, 25 November 1834), *astronomy.*

His parents were poor, and Bürg was destined to become a craftsman, until he received a fellowship from the Imperial Commission for Education. At Vienna University he studied mathematics and astronomy under Triesnecker and Hell. He became a physics teacher at the lyceum of Klagenfurt, Carinthia, in 1791, and the following year he returned to Vienna as assistant at the university observatory. After 1802 Bürg traveled. For two years he was calculator at the observatory of Seebergen, near Gotha, under Zach. After his return to Vienna he worked again at the university observatory and became professor of mathematics and astronomy at the university in 1806. In the same year, in a competition of the Paris Academy, he received an award for his new lunar ephemerides (see below). He became a knight of the Order of Leopold in 1808. From that year on, Bürg suffered from progressive deafness. In 1813 he took a leave of absence from his teaching duties. He hoped to succeed Triesnecker at the observatory, but Littrow was chosen. Bürg retired in 1819. He lived in Vienna until 1825, when he moved to Wiesenau, Carinthia. He never married.

Bürg began his practical astronomical observations while still a university student. After his appointment as professor he cooperated in, and was adviser to, the survey of Austria, but his main interest was calculation. He worked on the Viennese ephemerides for many years and was coeditor with Triesnecker.

Bürg was one of the leading calculating astronomers of his time, and his most important work was the recalculation of lunar ephemerides. He improved Laplace's perturbation theory of the complicated motion of the moon by adding more terms of the perturbation function: for the influence of the sun—considered not as the central body but as a disturbing one—and for the oblateness of the earth. Thus he found in the secular motion of the moon a term with a period of about 180 years and an amount of 13.8″. Taking this into consideration and making use of more recent observations, Bürg's lunar ephemerides proved to be much more accurate than those of his predecessors. From 1813 to 1820 they formed the basis of the lunar ephemerides in the *Nautical Almanach* of the British Admiralty.

BIBLIOGRAPHY

I. ORIGINAL WORKS. Bürg's writings include *Ephemerides astronomicae anni 1794 [–1806] a Francesco de Paula Triesnecker . . . et Joanne Burg . . . supputatae* (Vienna, 1793–1805); and "Tables de la lune," in *Tables astronomiques publiées par le Bureau des Longitudes* (Paris, 1806). Shorter articles are in *Berliner astronomisches Jahrbuch; Monatliche Korrespondenz zur Beförderung der Erd- und Himmelskunde;* and *Zeitschrift für Astronomie und verwandte Wissenschaften.*

II. SECONDARY LITERATURE. Further information on Bürg may be found in Johann Volkamer von Ehrenberg, "Johann Tobias von Bürg," in *Carinthia,* **25** (1836), 66 ff.; Johann Steinmayr, S. J., "Die Geschichte der Universitäts-Sternwarte," pt. 3 (Vienna, *ca.* 1935), MS at observatory of University of Vienna; Constant von Wurzbach, in *Biographisches Lexikon des Kaiserthums Oesterreich,* pt. 2 (1857), 196–198; and Martin Wutte, "Zum Gedächtnis des Astronomen J. T. Bürg," in *Carinthia,* **124** (1934), 143 ff.

JOSEF MAYERHÖFER
THOMAS WIDORN

BURGER, HERMAN CAREL (*b.* Utrecht, Netherlands, 1 June 1893; *d.* Utrecht, 27 December 1965), *physics, medical physics.*

Burger was named for his father, a first engineer with the Dutch navy; his mother was Jeanne Marie Cecile Docen. He received his primary education at the school of the Moravian Brothers, Zeist, the Netherlands, and attended the State High School in Utrecht, from which he graduated in 1911. In 1912 Burger matriculated at the University of Utrecht, where he studied physics and mathematics. He received his Ph.D., *cum laude,* 31 May 1918. From 1918 until 1920 he was an assistant in theoretical physics in Utrecht, and for the following two years he worked at the physical laboratory of Philips Industries. In 1922 he returned to the University of Utrecht as chief assistant in the physics department, and from 1927 to 1950 he had the title of lecturer. From 1950 until 1963 Burger was professor of medical physics in Utrecht. (In 1926 Burger had been offered a professorship at the University of Delft; he preferred, however, to remain in Utrecht.) Burger was awarded an honorary doctorate by the University of Nijmegen in 1963, and in 1964 he received the Einthoven Medal from the University of Leiden. He was the first Dutch citizen to deliver the Einthoven lecture.

Burger had great faith in men, and was extremely loyal and helpful to friends and students; he was mild in his judgments when he could see good intentions but very hard when there were none. He had vacillated between physics and mathematics or medicine as a course of study; he chose the former but was delighted when he was able to combine all of them. His precise mind was receptive to all types of information, but he was strongly opposed to every form of superstition and unscientific reporting (see his reports on the divining rod, 1930 and 1960).

Burger's work, aside from his teaching, can be divided into two parts: that before and that after World War II. Before the war the emphasis in Burger's work was on intensity measurements and spectral analysis. In connection with the atomic theory of Bohr and Rutherford, the determination of the intensity of spectral lines was an important factor in further theoretical development. In Utrecht in the 1920's this kind of experimentation flourished as the result of the development of measuring apparatus: Moll and Burger refined apparatus for detecting radiation— vacuum thermoelements, the bolometer, and the galvanometer; Burger and Van Cittert experimented with high-resolution apparatus and studied the influence of the apparatus on the shape of the spectral line; and Ornstein and Burger worked with intensity measurements themselves. During this period Burger also worked on liquid crystals.

Burger's younger brother, Eduard, had been a physician for a number of years, and the scientific contact between them was very close. In 1938 Burger's brother made a study on vectorcardiography that led Burger to the second period of his lifework, medical physics. His interest was primarily in electrocardiography, especially vectorcardiography. He also studied heart sounds, ballistocardiography, pumping rate of the heart per minute, stenosis, and pulse-wave reproduction. Besides conducting these studies, for many years Burger taught elementary physics to medical, veterinary, and dental students, covering the application of physics to medicine. In later years his teaching was extended to medical students who were more advanced in their studies.

Under Burger's leadership the Foundation for Biophysics flourished in the postwar years because he established various working groups. He was also successful in bringing together professors and staff members interested in medical physics by organizing colloquia.

BIBLIOGRAPHY

Burger's works include *Oplossen en Groeien van Kristallen,* his Ph.D. dissertation (Utrecht, 1918); *Het Onderwijs in de Natuurkunde aan Studenten in de Geneeskunde* (Utrecht, 1923); *Leerboek der Natuurkunde,* 3 vols. (Groningen, 1920–1936), written with W. J. H. Moll; *Voorlopige Beschrijvingen van een vijftigtal natuurkundige Leerlingenproeven* (Groningen, 1929), written with W. Reindersma *et al.; Natuurkundige proeven voor Leerlingen* (Groningen, 1934–1937), written with W. Reindersma *et al.; Objektive Spektralphotometrie* (Brunswick, 1932), written with W. H. J. Moll and L. S. Ornstein; and *Heart and Vector* (1969), ed. by H. W. Julius, Jr. A survey of Burger's work from 1908 to 1933 appears in a work dedicated to Ornstein by his colleagues and pupils (Utrecht, 1933). This work contains approximately 50 titles of articles written by Burger, alone or with others, such as "Beziehung zwischen inneren Quantenzahlen und Intensitäten von Mehrfachlinien," in *Zeitschrift für Physik,* **23** (1924), 258–266, written with H. B. Dorgelo, also in the Ornstein survey, p. 98; *Leerboek der Natuurkunde,* R. Kronig, ed. (1947; 6th ed., 1962), ch. 12, trans. as *Textbook of Physics* (1954; 2nd ed., 1959), ch. 12; *Medische Physica* (Paris, 1949), written with G. C. E. Burger; *Grensgebied* (Amsterdam, 1952); "Het begrip Arbeid in Natuurkunde, Fysiologie, en Geneeskunde," Einthoven lecture, 1964, in *Leidse Voordrachten,* **42**. Burger also published a great many articles in scientific journals.

J. G. VAN CITTERT-EYMERS

BURGERSDIJK (or Burgersdicius), FRANK (*b.* Lier, near Delft, Netherlands, 3 May 1590; *d.* Leiden, Netherlands, 19 February 1635), *natural philosophy.*

Burgersdijk was a farmer's son. His elementary and secondary studies were at the Latin school of Amersfoort and the Delft Gymnasium (1606–1610). He matriculated at Leiden University as a philosophy student on 6 May 1610. Burgersdijk was a distinguished student, obtaining his doctorate in 1614. His first appointment was at the Protestant academy of Saumur, where he was professor of philosophy from 1616 to 1619; here he composed his *Idea philosophiae naturalis.* He returned to Leiden as professor of logic and ethics, and delivered his inaugural lecture, "De fructu et utilitate logices," in 1620. Burgersdijk was promoted to the chair of philosophy in 1628 and became a leading figure at the university, serving as rector (1629, 1630, 1634) and writing influential textbooks on natural philosophy, metaphysics, logic, ethics, and politics.

The reason for Burgersdijk's popularity is apparent from his first book, *Idea philosophiae naturalis,* which became the model for his later writings. His treatment of his subjects is clear, logical, concise, and well organized. The method of ordered studies that he adopted was designed to impart "solid erudition." Proceeding by his method of definition and division, he explored the whole of natural philosophy in twenty-six disputations, each of which was a series of *theses* that could be further studied by consulting the *pro* and *contra* authorities listed. A further collection of disputations was given in the *Collegium physicum.* On the problems of natural philosophy Burgersdijk was strikingly insensitive to the new science, being content to draw upon the neo-Scholastic commentators of the late sixteenth century. Although a Protestant himself, Burgersdijk drew predominantly from Catholic sources, showing particular liking for the Iberian authors Suárez, Periera, and Toletus, and the Coimbra commentaries. His conservatism in astronomy is illustrated by his simplified edition of Sacrobosco's *Sphaera.* He showed greater originality in the *Institutionum logicarum,* his most popular work, in which he sought a compromise between Aristotelian and Ramist logic, regarding as particularly important the roles of division and definition, which he considered equal to syllogism and method.

Burgersdijk's highly successful textbooks made him the dominant figure in the final stage of Dutch Scholasticism, and his authority extended to England and Germany. His textbooks held their place in the universities long after his ideas had been eclipsed by Cartesianism.

BIBLIOGRAPHY

I. ORIGINAL WORKS. Burgersdijk's works relating to natural philosophy are *Idea philosophiae naturalis sive methodus definitionum et controversarum physicarum* (Leiden, 1622; at least eight later eds.); *Institutionum logicarum libri duo* (Leiden, 1626; at least twenty-seven eds., nine pub. in England); *Institutionum logicarum synopsis* (Leiden, 1626), almost as popular as the preceding work; *Sphaera J. de Sacro-Bosco decreto in usum scholarum ejusdem provinciae recensita ut et latinitus et methodus emendata sit* (Leiden, 1626; two later eds.); *Collegium physicum in quo tota philosophia naturalis aliquot disputationibus perspicue et compendiose explicatur* (Leiden, 1632; four later eds.); and *Institutionum metaphysicarum libri II* (Leiden, 1640).

II. SECONDARY LITERATURE. Of the works given below, Dibon, Risse, and Wundt give assistance toward the compilation of a bibliography of Burgersdijk's words. See Paul Dibon, *La philosophie néerlandaise au siècle d'or*, I (Amsterdam, 1954), 96–120, 150–153; *Nieuw Nederlandisch biographisch Woordenboek, 1911–1937*, VII, 229; Wilhelm Risse, *Die Logik der Neuzeit,* I (Stuttgart–Bad Cannstatt, 1964), 515–520; and Max Wundt, *Die deutsch Schulmetaphysik des 17 Jahrhunderts* (Tübingen, 1939), pp. 87–89.

CHARLES WEBSTER

BÜRGI, JOOST (*b.* Liechtenstein, 28 February 1552; *d.* Kassel, Germany, 31 January 1632), *mathematics, astronomy.*

There is no precise account of Bürgi's youth. Most likely he received no systematic education, for he did not even know Latin, the scientific language of his time. From 1579 he was the court watchmaker to Duke Wilhelm IV, and he probably completed his education while working in the duke's observatory at Kassel. There he worked on the construction of several instruments, especially astronomical ones, and made astronomical observations, developing his skill, inventiveness, and accuracy. Bürgi also improved instruments for use in practical geometry. His proportional compasses competed with those of Galileo for priority, although both were probably no more than an improvement of devices already in use.

The fame of Bürgi's instruments, which made possible more accurate astronomical observations in the observatory at Kassel, drew the attention of scientists assembled at the court of Emperor Rudolf II, who tried to establish a science center in Prague and to enlist prominent European scientists. After the death of Wilhelm IV, Bürgi entered the service of Rudolf and became his court watchmaker, also holding this position under Rudolf's successors Matthias and Ferdinand II. He lived in Prague from about 1603 and became assistant to and computer for Kepler, who was working on the results of astronomical observations made by Tycho Brahe. Even after the imperial court moved to Vienna and the leading foreign scientists left Prague, and the Bohemian anti-Hapsburg

revolt was defeated (1620), Bürgi remained in Prague. Here he became scientifically isolated, which lessened the favorable response to his results. Shortly before his death (probably as late as 1631) Bürgi returned to Kassel.

In mathematics Bürgi was by no means a theoretician, but an indefatigable and inventive computer whose help Kepler appreciated. Bürgi's manuscript "Arithmetics" was taken to Pulkovo with Kepler's unpublished papers. In this manuscript Bürgi uses (probably independently of Stevin) the decimal point and sometimes substitutes a small arc for it. Starting from the method known as *regula falsi,* Bürgi also elaborated the method of approximate calculation of the roots of algebraic equations of higher degree. The need to make the tables of sines more precise led him to undertake this problem. His tables of sines, which have the difference 2″, were never published, and not even the manuscript exists.

The computation of the tables of sines and the elaboration of astronomical data led Bürgi to an easier method of multiplying large numbers. From about 1584 he was engaged, like several other astronomers and computers in the sixteenth century, in the improvement of "prosthaphairesis," the method of converting multiplication into addition by means of trigonometrical formulas—for example, $\sin \alpha \cdot \sin \beta = 1/2[\cos(\alpha - \beta) - \cos(\alpha + \beta)]$. Later, possibly at the end of the 1580's, the idea of logarithms occurred to him. Although he did not know Stifel's *Arithmetica integra,* in which the idea of comparing arithmetic and geometric progression is outlined, Bürgi learned of it from other sources. He had computed the tables of logarithms before his arrival in Prague, but he did not publish them until 1620, under the title *Arithmetische und geometrische Progress-Tabulen, sambt gründlichem Unterricht, wie solche nützlich in allerley Rechnungen zu gebrauchen, und verstanden werden sol;* however, the instruction promised in the title remained in manuscript.

The geometrical progression begins with the value 100,000,000 and has the quotient 1.0001. A term of the arithmetical progression 0, 10, 20, 30, 40, ··· corresponds to each term of the geometrical series. The tables extend to the value 1,000,000,000 in the geometrical progression, with the corresponding value 230,270,022 in the arithmetic progression. Consequently, Bürgi's logarithms correspond to our so-called natural logarithms with the base e. By their arrangement they are in fact antilogarithmic, for the basic progressions are logarithms. This circumstance could have made the use and spread of the tables more difficult, but the fate of Bürgi's work was influenced much more by the disintegration of the scientific and cultural center in Prague after 1620. The Prague edition of the tables remained almost unnoticed, and only a few copies were saved; probably the only complete copy is kept, together with the handwritten "instruction," in the library at Danzig. Thus, Bürgi's greatest discovery had no apparent influence on the development of science.

BIBLIOGRAPHY

Bürgi's only published work is *Arithmetische und geometrische Progress-Tabulen, sambt gründlichem Unterricht, wie solche nützlich in allerley Rechnungen zu gebrauchen, und verstanden werden sol* (Prague, 1620), repr. in H. R. Gieswald, *Justus Byrg als Mathematiker und dessen Einleitung zu seinen Logarithmen* (Danzig, 1856).

There is neither a detailed biography nor an analysis of Bürgi's scientific work. Basic bibliographic data in the following works can be of use: G. Vetter, "Dějiny matematických věd v českých zemích od založení university v r. 1348 až do r. 1620" ("History of Mathematics in the Bohemian Lands From the Foundation of the University in 1348 until 1620"), in *Sborník pro dějiny přírodních věd a techniky* (Prague), **4** (1958), 87–88; and "Kratkii obzor razvitija matematiki v cheshtskikh zemliakh do Belogorskoi bitvy," in *Istoriko-matematicheskie issledovaniya,* **11** (Moscow, 1958), 49, 512; E. Voellmy, "Jost Bürgi und die Logarithmen," in *Beihefte zur Zeitschrift für Elemente der Mathematik,* no. 5 (1948); and E. Zinner, *Deutsche und niederländische astronomische Instrumente des 11.–18. Jahrhunderts* (Munich, 1956), pp. 268–276.

LUBOŠ NOVÝ

BURIDAN, JEAN (*b.* Béthune, France, *ca.* 1295; *d.* Paris, France, *ca.* 1358), *philosophy, logic, physics.*

Although Jean Buridan was the most distinguished and influential teacher of natural philosophy at the University of Paris in the fourteenth century, little is known of his personal life. He was born in the diocese of Arras, and went as a young cleric to study at the University of Paris, where he was first enrolled as a student in the College of Cardinal Lemoine and later became a member of the College of Navarre. It is probable that he obtained his master of arts degree soon after 1320, since a document dated 2 February 1328 mentions him as rector of the university in that year. Other documents, relating to benefices whose revenues provided his financial support and bearing the dates 1329, 1330, 1342, and 1348, describe him as a "very distinguished man," as a "celebrated philosopher," and as "lecturing at Paris on the books of natural, metaphysical, and moral philosophy." Two passages in his own writings indicate that at some date prior to 1334 he made a visit to the papal court at Avignon and, on the way, climbed Mt. Ventoux in order to make some meteorological observations.

In 1340 Buridan was rector of the university for a second time, and in that year he signed a statute of the faculty of arts which censured certain masters for the practice of construing texts in a literal sense rather than in accordance with the intentions of the authors, warning that this practice gave rise to "intolerable errors not only in philosophy but with respect to Sacred Scripture." One of the articles of censure bears on a statement known to have been made by Nicolaus of Autrecourt, whose skeptical views on causal inferences were attacked by Buridan in his own writings. The last documentary mention of Buridan occurs in a statute dated 12 July 1358, where his name appears as witness to an agreement between the Picard and English nations of the university. It is not unlikely that he fell victim to the Black Plague, which in 1358 took the lives of many of those who had managed to survive its first outbreak in 1349.

Buridan was a secular cleric rather than a member of a religious order, and he remained on the faculty of arts to the end of his life without, apparently, seeking to obtain a degree in theology. In his lifetime he was held in high esteem by his colleagues, students, and ecclesiastical superiors; and for nearly two centuries after his death his teachings in natural philosophy and logic were of paramount influence in the universities of northern and eastern Europe. A document in the archives of the University of Cologne, dated 24 December 1425, speaks of the preceding century as "the age of Buridan," and when George Lockert, in 1516, edited one of Buridan's works, he stated that Buridan still ruled the study of physics at Paris. In later centuries, the story of the ass who starved to death because he could not choose between two equally desirable bundles of hay was attributed to Buridan, and another story, presumably legendary but perpetuated by the poet François Villon, related that Buridan had been involved in scandalous relations with the wife of Philip V of France and had, by the king's order, been tied in a sack and thrown into the Seine.

The extant writings of Buridan consist of the lectures he gave on subjects in the curriculum of the faculty of arts at Paris. In the fourteenth century this curriculum was based largely on study of the treatises of Aristotle, along with the *Summulae logicales* of Peter of Spain and other medieval textbooks of grammar, mathematics, and astronomy. Buridan composed his own textbook of logic, *Summula de dialectica,* as a "modern" revision and amplification of the text of Peter of Spain; he also wrote two treatises on advanced topics of logic, entitled *Consequentiae* and *Sophismata,* which are among the most interesting contributions to late medieval logic. All of his other works are in the form of commentaries and of critical books of *Questions* on the principal treatises of the Aristotelian corpus. The literal commentaries are extant only in the unpublished manuscript versions, but the books of *Questions* on Aristotle's *Physics, Metaphysics, De anima, Parva naturalia, Nicomachean ethics,* and *Politics* were published, along with Buridan's writings in logic, after the invention of the printing press.

The only modern edition of a work by Buridan is that of his *Questions* on Aristotle's *De caelo et mundo,* previously unedited, which appeared in 1942. Most of the printed editions represent the lectures Buridan gave during the last part of his teaching career, although earlier versions are to be found among the unpublished manuscript materials. Until a critical study of the manuscripts is made, however, there is no sure way of determining any order of composition for Buridan's works, nor of tracing the development of his thought over the thirty-odd years of his academic career.

Buridan made significant and original contributions to logic and physics, but as a philosopher of science he was historically important in two respects. First, he vindicated natural philosophy as a respectable study in its own right. Second, he defined the objectives and methodology of scientific enterprise in a manner that warranted its autonomy with respect to dogmatic theology and metaphysics; this achievement was intimately connected with the fourteenth-century movement known as nominalism and with the controversies precipitated at the universities of Oxford and Paris by the doctrines associated with William of Ockham. Buridan's own philosophical position was thoroughly nominalistic, and indeed very similar to that of Jean de Mirecourt, a theologian of Paris whose teachings were condemned in 1347 by the chancellor of the university and the faculty of theology. Buridan himself was able to escape the charges of theological skepticism that were directed against his fellow nominalists of the theological faculty. He owed his good fortune in part, no doubt, to his prudence and diplomacy. Primarily, however, he could ward off criticism for the fundamental reason that he employed the logical and epistemological doctrines of nominalism in a methodological, rather than a metaphysical, way in formulating the character and the evidential foundations of natural philosophy.

The formal logic presented in Buridan's *Summula de dialectica* is closely related, in topical structure and terminology, to the so-called terminist logic of the thirteenth century, represented by the textbooks of

William of Sherwood and Peter of Spain. Although it presupposes the nominalist thesis that general terms are signs of individuals, and not of common natures existing in individuals, it does not exhibit any strong evidence of direct influence by the logical writings of Ockham; it may well have been developed independently of such influence on the basis of the modern logic (*logica moderna*) already well established in the arts faculties of Oxford and Paris. The doctrine of the supposition of terms, basic to this logic, is used in defining the functions of logical operators or syncategorematic signs in determining the truth conditions of categorical propositions of various forms and in formulating the laws of syllogistic inference, both assertoric and modal. Treatises on topical arguments, fallacies, and the demonstrative syllogism conclude the work.

Buridan's *Sophismata,* designed to constitute a ninth part of the *Summula,* apparently was written much later in his life, for it contains criticisms of the theory of propositional meanings, or *complexe significabilia,* which Gregory of Rimini introduced in 1344. This work presents a fully developed analysis of meaning and truth which corresponds closely to that of Ockham's *Summa logicae.* It goes beyond the work of Ockham, however, in presenting original and highly advanced treatments of the problem of the nonsubstitutivity of terms occurring in intensional contexts and the problem of self-referential propositions represented by the Liar paradox. Buridan's treatment of these problems exhibits a level of logical insight and skill not equaled until very recent times. His treatise *Consequentiae,* which develops the whole theory of inference on the basis of propositional logic, marks another high point of medieval logic, the significance of which has been appreciated only in the twentieth century.

Buridan's philosophy of science is formulated in his *Questions* on the *Metaphysics,* and is applied to the concepts and problems of natural science in his *Questions* on the *Physics.* The Aristotelian definition of science as knowledge of universal and necessary conclusions by demonstration from necessary, evident, indemonstrable premises is accepted. A sharp distinction is made, however, between premises in which the necessity is determined by logical criteria or by stipulated meaning of the terms, and those in which evidence rests on empirical confirmation and which are called necessary in a conditional sense, or "on the supposition of the common course of nature." Only in the latter sense do the principles of the natural sciences have evidence and necessity.

These principles are not immediately evident; indeed we may be in doubt concerning them for a long time.

But they are called principles because they are indemonstrable, and cannot be deduced from other premises nor be proved by any formal procedure; but they are accepted because they have been observed to be true in many instances and to be false in none.[1]

The significance of this theory of scientific evidence lies in its rejection of the thesis, held by most of the scholastic commentators on Aristotle, that the principles of physics are established by metaphysics and that they are necessary in the sense that their contradictories are logically or metaphysically impossible. This metaphysical interpretation of Aristotelian physics led Bishop Étienne Tempier and the faculty of theology at Paris to condemn, in 1277, doctrines taught by members of the arts faculty as truths necessary to philosophy, although contradictory to dogmas of the Christian faith. By construing the principles of the sciences of nature as inductive generalizations whose evidence is conditional on the hypothesis of the common course of nature, Buridan was able to concede the absolute possibility of supernatural interference with the natural causal order, and yet to exclude such supernatural cases as irrelevant to the purposes and methodological procedures of the scientific enterprise. Nicolaus of Autrecourt, demanding that scientific principles have absolute necessity and certainty, had argued that a science of nature based on causal laws established by inductive generalization had no evidence whatsoever, since it could not be known in any given instance whether or not God was producing an effect without a natural cause. Buridan refers to Nicolaus' position in these words:

> It has hereby been shown that very evil things are being said by certain ones who seek to undermine the natural and moral sciences because absolute evidence is not possessed by most of their principles and conclusions, it being supernaturally possible for them to be rendered false. For in these sciences absolutely unconditional evidence is not required, and it is enough if we have conditional or hypothetical evidence of the kind described above.[2]

The conception of scientific enterprise formulated by Buridan as a means of justifying its pursuit within the framework of the Christian doctrine of divine omnipotence is the conception within which science has operated since the late seventeenth century. To make science compatible with Christian dogma, Buridan had to break its traditional ties with metaphysics and define its principles methodologically, in terms of their value in "saving the phenomena." He still encountered some theological difficulties in applying this method within the domain of physics, as did Galileo three centuries later; but after the time of Buridan, natural philosophy had its own legitimacy

and ceased to be either only a handmaiden of theology or a mere exposition of the doctrines of Aristotle.

The *Questions* composed by Buridan on problems raised in Aristotle's *Physics* and *De caelo et mundo* exhibit his application of these criteria of scientific method and evidence to the critical evaluation of Aristotle's theories and arguments and to the diverse interpretations of them offered by Greek, Moslem, and Christian scholastic commentators. The general scheme and conceptual framework of analysis, within which Aristotle's physics and cosmology are formulated, is accepted by Buridan as the working hypothesis, so to speak, of natural philosophy. But the scheme is not sacrosanct, and Buridan not infrequently entertains alternative assumptions as being not only logically possible but also possibly preferable in accounting for the observed phenomena. While the authority of Aristotle had often been challenged on the ground that his positions contradicted Christian doctrine, it had come, in Buridan's time, to be challenged on grounds of inadequacy as a scientific account of observed facts. Buridan's major significance in the historical development of physics arises from just such a challenge with respect to Aristotle's dynamic theory of local motion and from his proposal of an alternative dynamics which came to be known as the impetus theory.

An obvious weakness of Aristotle's dynamics is its inability to account for projectile motions, such as the upward motion of a stone thrown into the air after it has left the hand of the thrower. According to the assumptions of Aristotelian physics, such a motion, being violent and contrary to the natural movement of the stone toward the earth, required an external moving cause continuously in contact with it. Since the only body in contact with it is the air, Aristotle supposed that in some way the air pushes or pulls such a body upward. This feeble explanation drew criticism in antiquity and from medieval Moslem commentators and gave rise to a theory that the violent action of the thrower impresses on the stone a temporary disposition, of a qualitative sort, which causes it to move for a short time in the direction contrary to its nature. This disposition was called an impressed virtue (*virtus impressa*), and it was held to be self-expending and quickly used up because of its separation from its source. Franciscus de Marchia, a Franciscan theologian who taught at Paris around 1322, gave a full presentation of this theory, and it is likely that Buridan was influenced by it.

In treating of the problem of projectile motion in his *Questions* on Aristotle's *Physics* (VIII, question 12), Buridan expounded Aristotle's theory of propulsion by the air and rejected it with arguments similar to those that Marchia had used. His own solution was in some respects like that of Marchia, but in one crucial point it was strikingly different. The tendency of the projectile to continue moving in the direction in which it is propelled, which Buridan calls *impetus* rather than *virtus impressa,* is described as a permanent power of motion, which would continue unchanged if it were not opposed by the gravity of the projectile and the resistance of the air. "This impetus," he says in another discussion given in his *Questions* on the *Metaphysics,* "would endure forever [ad infinitum] if it were not diminished and corrupted by an opposed resistance or by something tending to an opposed motion."[3]

The suggestion given here of the inertial principle fundamental to modern mechanics is striking, as are some further uses that Buridan makes of the impetus concept in explaining the accelerated velocity of free fall, the vibration of plucked strings, the bouncing of balls, and the everlasting rotational movements ascribed to the celestial spheres by Greek astronomy. Buridan defines impetus in a quantitative manner, as a function of the "quantity of matter" of the body and of the velocity of its motion; thus, he seems to conceive of impetus as equivalent to what in classical mechanics is called momentum, defined as the product of mass and velocity. In treating the action of gravity in the case of freely falling bodies, Buridan construes this action as one imparting successive increments of impetus to the body during its fall.

> It must be imagined that a heavy body acquires from its primary mover, namely from its gravity, not merely motion, but also, with that motion, a certain impetus such as is able to move that body along with the natural constant gravity. And because the impetus is acquired commensurately with motion, it follows that the faster the motion, the greater and stronger is the impetus. Thus the heavy body is moved initially only by its natural gravity, and hence slowly; but it is then moved by that same gravity as well as by the impetus already acquired, and thus it is . . . continuously accelerated to the end.[4]

The effect of a force, such as gravity, is thus conceived of as a production of successive increments of impetus, or of velocity in the mass acted upon, throughout the fall. It is a short step from this to the modern definition of force as that which changes the velocity of the body acted upon, implying the correlative principle that a body in uniform motion is under the action of no force. Buridan does not quite take this step, since he retains the Aristotelian assumption that a constant cause must produce a constant effect, and ascribes the increase in velocity to the addition of impetus as an added cause acting along with the gravity.

Yet his theory obviously requires a distinction between impetus as a "conserving cause" of motion and gravity as a "producing cause" of the motion conserved by the impetus; his failure to draw the consequence of this distinction was perhaps because he did not attempt a mathematical analysis involving the concept of instantaneous velocities added continuously with time. Whether Buridan construed the acceleration as uniform with respect to time elapsed, or with respect to distance traversed, is not clear. He probably regarded the two functions as equivalent, a view that, however impossible from a mathematical point of view, was retained into the seventeenth century, when Descartes and Galileo (in his letter to Sarpi of 1604) sought to prove that velocity increases in proportion to time elapsed from the premise that velocity increases in direct proportion to distance of fall.

Buridan's concept of impetus is further distinguished from the modern inertial concept by the fact that he construes rotational motion at uniform angular velocity as due to a rotational impetus analogous to the rectilinear impetus involved in projectile motion. Galileo did likewise, and was in this respect nearer to Buridan than to Newton. But Buridan makes a striking use of his impetus concept, in its rotational sense, by arguing that since the celestial spheres posited by the astronomers encounter no external resistance to the rotational movements and have no internal tendency toward a place of rest (such as heavy and light bodies have), their uniform rotational motions are purely inertial and require no causes acting on them to maintain their motions. There is, therefore, no need to posit immaterial intelligences as unmoved movers of the heavenly spheres, in the manner that Aristotle and his commentators supposed. "For it could be said that God, in creating the world, set each celestial orb in motion . . . and, in setting them in motion, he gave them an impetus capable of keeping them in motion without there being any need of his moving them any more."[5] It was in this way, Buridan adds, that God rested on the seventh day and committed the motions of the bodies he had created to those bodies themselves.

It is clear that Buridan's impetus theory marked a significant step toward the dynamics of Galileo and Newton, and an important stage in the gradual dissolution of Aristotelian physics and cosmology. Buridan did not, however, exploit the potentially revolutionary implications of his analysis of projectile motion and gravitational acceleration, or generalize his impetus theory into a theory of universal inertial mechanics. Thus, in discussing the argument of Aristotle against the possibility of motion in a void, Buridan accepted the principle that the velocity of a natural motion in a corporeal medium is determined by the ratio of the motive force to the resisting force of the medium, so that if there were no resisting medium, the motion would be instantaneous. This is scarcely consistent with the analysis of gravitational acceleration as finite increments of impetus given to the falling body by its gravity, and Buridan made no effort to harmonize these two different approaches within a common theory.

In a question bearing on the *De caelo et mundo,* Buridan asks whether it can be proved that the earth is at rest, with the celestial spheres rotating around it, as Aristotle supposed. He states that many people of his time held it to be probable that the earth rotates on its own axis once a day and that the stellar sphere is at rest. And he adds that it is "indisputably true that if the facts were as this theory supposes, everything in the heavens would appear to us just as it now appears."[6] In support of the hypothesis, he invokes the principle that it is better to account for the observed phenomena by fewer assumptions or by the simplest theory, and argues that since the earth is a small body and the outer sphere is a very large one, it is more reasonable to attribute the rotation to the earth than to suppose the enormously faster rotation of the much larger sphere. After giving this and other arguments in favor of the theory of diurnal rotation of the earth, Buridan makes it quite clear that they cannot be refuted by any of the traditional arguments purporting to prove that the earth is at rest. He says that for his part he chooses to hold that the earth is at rest and the heavens in motion; and he offers, as a "persuasion" for this view, the argument that a projectile thrown straight upward from the earth's surface will fall back to the same spot from which it was thrown.

This argument does not seem consistent with Buridan's own impetus theory, unless he had in mind a point made later by his pupil Albert of Saxony, who held that the lateral impetus shared by the projectile with that of the surface of the rotating earth would be insufficient to carry it over the greater arc which it would have to traverse, when projected outward from the earth's surface, in order to fall back at the same spot. Not only Albert of Saxony, but also another pupil of Buridan's, Nicole Oresme, took over this discussion of the earth's rotation; Oresme concluded that it is impossible to prove either side of the question, since the motion is purely relative. Oresme said that he accepted the view that the earth is at rest, but only because this seemed to be assumed by the Bible. It is of interest to note that when Copernicus was a student at Cracow, Buridan's works in

physics were required reading in the curriculum of that university.

While rejecting the theory of the diurnal rotation of the earth, Buridan says that the earth is not immobile at the center of the world, and proves it as follows: Because the dry land protruding from the ocean is mostly on one side of the earth, the center of volume of the earth does not coincide with its center of gravity. The earth, however, is the center of the world in the sense that its center of gravity is equidistant from the inner surface of the celestial spheres. But this center of gravity is continuously altered by the erosion of the dry land, which slowly gets washed into the sea; and consequently the whole mass of the earth slowly shifts from the wet side to the dry side in order to keep its center of gravity at the center of the universe.

Buridan's significance in the history of science lies more in the questions he raised than in the answers he gave to them, although in some cases his answers opened up new theoretical possibilities that were undoubtedly influential in the rise of modern mechanics in the seventeenth century. The impetus theory was taken over by Buridan's pupils and was made known throughout central Europe, although in a degenerate form that fused it with the older theory of a self-expending *virtus impressa* and introduced a number of confusions and errors that Buridan himself had avoided. It was in this degenerate form that it was conveyed to Galileo by his teacher Buonamici, so that Galileo had to take the step that Buridan had taken three centuries earlier when he discarded Marchia's theory of the self-expending impressed force in favor of impetus as an enduring condition only changed or diminished by opposed forces. Buridan's application of the impetus concept to the analysis of free fall, although retained and made known by Albert of Saxony, was forgotten by most of the later teachers of physics, even when they retained the concept in dealing with projectile motion.

Even when Buridan's specific contributions to physical problems were forgotten, however, the influence of his conception of scientific evidence and method remained operative; and it may be said that the idea of mechanics, in the modern sense, became established in early modern times through the work of Buridan and of his contemporaries. In particular, Buridan may be credited with eliminating explanations in terms of final causes from the domain of physics, which he does very explicitly in his *Questions* on the *Physics* (II, questions 7 and 13) and in his *Questions* on the *De caelo et mundo* (II, question 8). The mechanistic conception of nature, construed as a methodological assumption more than as a metaphysical thesis, emerged in the fourteenth century as a natural development from Buridan's philosophy of science. He was not an experimental scientist or a mathematical physicist; but as a philosopher of science he did much to clear the way for, and to point the way to, the development of modern science in these directions.

NOTES

1. *Qu. in Metaph.* II, Qu. 2 (1518), fol. 9*v.*
2. *Ibid.,* Qu. 1 (1518), fol. 9*r.*
3. *Ibid.* (1518), fol. 73*r.*
4. *Qu. De caelo et mundo* (1942), 180.
5. *Qu. in Phys.* VIII, Qu. 12, fol. 121*r.*
6. *Qu. De caelo et mundo* II, Qu. 22 (1942), 227.

BIBLIOGRAPHY

I. ORIGINAL WORKS. Early editions of Buridan's works include *Quaestiones super decem libros Ethicorum Aristotelis ad Nicomacheum* (Paris, 1489, 1513, 1518; Oxford, 1637); *Sophismata Buridani* (Paris, 1489, 1491, 1493; best ed., by Antonius Denidel and Nicolaus de Barra, Paris, *ca.* 1495), trans. by Theodore K. Scott as *John Buridan: Sophisms on Meaning and Truth* (New York, 1966); *Summula de dialectica,* with commentary by John Dorp of Leiden (Lyons, 1490, 1493, 1495, 1510; as *Perutile compendium totius logicae Joannis Buridani,* Venice, 1499; there are many later editions); *Consequentiae Buridani* (Paris, 1493, 1495, 1499); *Quaestiones super octo physicorum libros Aristotelis,* ed. Johannes Dullaert Gandavensis (Paris, 1509, 1516); *Quaestiones et decisiones physicales insignium virorum . . . ,* ed. Georgius Lockert (Paris, 1516), which contains *Quaestiones in libros De anima, De sensu et sensato, De memoria et reminiscentia, De somno et vigilia, De longitudine et brevitate vitae,* and *De iuventate et senectute; In Metaphysicen Aristotelis quaestiones argutissimae magistri Joannis Buridani* (Paris, 1518); and *Quaestiones in octo libros Politicorum* (Paris, 1530; Oxford, 1640).

More recent editions are *Quaestiones super libris quattuor De caelo et mundo,* ed. E. A. Moody (Cambridge, Mass., 1942); and "Giovanni Buridano, Tractatus de suppositionibus," ed. Maria Eleina Reina, in *Rivista critica di storia della filosofia,* 12 (1957), 175–208, 323–352.

A list of manuscripts of Buridan's works, both published and unpublished, can be found in Edmond Faral, "Jean Buridan, Notes sur les manuscrits, les éditions et le contenu de ses oeuvres," in *Archives d'histoire doctrinale et littéraire du moyen-âge,* 15 (1946), 1–53.

II. SECONDARY LITERATURE. Works on Buridan include J. Bulliot, "Jean Buridan et le mouvement de la terre," in *Revue de philosophie,* 25 (1914), 5–24; Marshall Clagett, *The Science of Mechanics in the Middle Ages* (Madison, Wis., 1959), 505–599; E. J. Dijksterhuis, *De Mechanisering van het Werelbild* (Amsterdam, 1915); Pierre Duhem, *Études sur Léonard de Vinci,* II (Paris, 1909), 379–384, 420–423, 431–438, and III (Paris, 1913), 1–259, 279–286, 350–360;

Pierre Duhem, *Le système du monde,* Vols. VI–VII (Paris, 1954–1958); and Edmond Faral, "Jean Buridan, maître-ès-arts de l'Université de Paris," in *Histoire littéraire de la France,* 38 (1949), 462–605.

The following works of Anneliese Maier should be consulted: *Die Vorläufer Galileis im 14. Jahrhundert* (Rome, 1949); *Zwei Grundprobleme der scholastischen Naturphilosophie,* 2nd ed. (Rome, 1951), 201–235; *Metaphysische Hintergründe der spätscholastischen Naturphilosophie* (Rome, 1955), 300–335, 348 ff., 384 ff.; "Die naturphilosophische Bedeutung der scholastischen Impetustheorie," in *Scholastik,* 30 (1955), 321–343; and *Zwischen Philosophie und Mechanik* (Rome, 1958), 332–339.

Also of value are works by E. A. Moody: "John Buridan on the Habitability of the Earth," in *Speculum,* 16 (1941), 415–425; "Ockham, Buridan, and Nicholas of Autrecourt," in *Franciscan Studies,* 7 (June, 1947), 113–146; "Galileo and Avempace; The Dynamics of the Leaning Tower Experiment," in *Journal of the History of Ideas,* 12 (1951), 163–193, 375–422; *Truth and Consequence in Medieval Logic* (Amsterdam, 1953); and "Buridan and a Dilemma of Nominalism," in *Harry Austryn Wolfson Jubilee Volume* (Jerusalem, 1965), 577–596.

Additional works concerning Buridan are Karl Prantl, *Geschichte der Logik im Abendlande,* IV (Leipzig, 1870), 14–38; A. N. Prior, "Some Problems of Self-reference in John Buridan," in *Proceedings of the British Academy,* 48 (1962), 281–296; Maria Eleina Reina, "Note sulla psicologia di Buridano," in *Arti grafiche grisetti* (Milan, 1959), p. 9 ff.; Maria Eleina Reina, "Il problema del linguaggio in Buridano," in *Rivista critica di storia della filosofia,* 15 (1959), 367–417, and 15 (1960), 141–165, 238–264; Theodore K. Scott, "John Buridan on the Objects of Demonstrative Science," in *Speculum,* 40 (1965), 654–673; G. Federici Vescovini, "La concezione della natura di Giovanni Buridano," in *La filosofia della natura nel medievo: Atti del III Congresso Internazionale di Filosofia Medievale* (Milan, 1964); G. Federici Vescovini, *Studi sulla prospettiva medievale* (Turin, 1965), 137–163; James J. Walsh, "Buridan and Seneca," in *Journal of the History of Ideas,* 27 (1966), 23–40; and James J. Walsh, "Nominalism and the Ethics: Some Remarks About Buridan's Commentary," in *Journal of the History of Philosophy,* 4 (1966), 1–13.

ERNEST A. MOODY

BURLEY, WALTER (*b.* England, *ca.* 1275; *d. ca.* 1345?), *logic, natural philosophy.*

A colophon to the final version of Burley's commentary on the *Logica vetus* that is dated 1337 stipulates that the work was composed in its author's sixty-second year, a factor which places Burley's birth about 1275, possibly in one of the two towns named Burley in Yorkshire. Although almost nothing is known of his youth, it seems most reasonable to presume that Burley began his studies at Oxford sometime during the last decade of the thirteenth century, for two works dated 1301 and 1302 already designate him as a master in arts. He may have at this time also been a fellow of Merton College, although the first definite connection we have with Merton derives from the bursorial roll of 1305. It seems very probable that, during his regency in arts at Oxford, Burley composed his earliest versions (later to be revised and expanded) of expositions on almost all of Aristotle's works in logic, natural philosophy, and moral philosophy.

In 1310 (at the latest) we find him in Paris, where he began his theological studies, a stage in his career first definitely documented by his citation as *doctor sacre theologie* in a colophon dated 1324. We lack further information concerning Burley's work as a theologian and have thus far not been able to discover his *Commentary on the Sentences,* but his first treatise dealing with the intension and remission of forms (the so-called *Tractatus primus* written between 1320-1327) indicates that Thomas Wilton, also a former fellow of Merton College, was at some point his *socius* and master of theology in Paris.

Burley most likely remained in Paris until 28 February 1327, at which time he was appointed as an envoy of Edward III to the papal court. Indeed, the official connections Burley then began with English public figures seems to have continued, intermittently, for the remainder of his years. Yet this is not to say that this new phase to his career marked the end, or even a substantial lessening, of his academic pursuits. For, just as during his years in Paris, he continued to set down numerous logical and philosophical works.

In Bologna in 1341 for a disputation and again at Avignon in November 1343, our last document mentioning Burley is a register revealing his acquisition of a rectory in Kent on 19 June 1344. It seems unlikely that he lived much beyond that date.

In terms of the literary activity, Burley remained, throughout his career, fundamentally an arts graduate. For, if one sets aside his lost work on the *Sentences,* all but one or two items among the formidable mass of his writings deal with logic and philosophy. He was, to begin with, an Aristotelian commentator with a vengeance, composing two—sometimes three—different versions of a commentary on a single work.

Thus, he wrote commentaries on all of the Aristotelian logical books, including Porphyry's *Isagoge* and the *Liber de sex principiis* ascribed to Gilbert de la Porrée, apparently formulating an initial version of his comments during his earlier Oxford period. Many of the commentaries were later revised, the final version of his complete *Expositio super artem veterem* being written only in 1337. To this already

very substantial body of logical literature one must add Burley's numerous opuscula and treatises on the so-called *parva logicalia* (which constituted, in large part, medieval additions to the logic of Aristotle) and, in particular, the two redactions of his magnum opus in logic, the *De puritate artis logicae.* The earlier, shorter version of this work appears to have been composed (in incomplete form) before the appearance (*ca.* 1324) of the *Summa logicae* of William of Ockham. Indeed, the second version (1325-1328) of Burley's treatise can in many respects be viewed as a reply to some of Ockham's contentions. Yet it is not merely as an anti-Ockhamist tract that Burley's revised *De puritate artis logicae* is of importance; for, at least in the view of its modern editor, Philotheus Boehner, its implicit subsumption of syllogistic under the more general theory of consequences involving unanalyzed propositions strongly suggests Burley to have been a logician of appreciable competence.

As concerns natural philosophy once again one must begin by noting the extensive roster of Burley's Aristotelian commentaries: *Expositiones* or *Questiones* (and again often in multiple versions) on the *Physica, De caelo, De generatione et corruptione, Meteorologica, De anima, Problemata, Parva naturalia, De motu animalium, Metaphysica* (possibly), plus Averroës' *De substantia orbis.* Of these commentaries, those on the *Physics* are undoubtedly the most important. The earliest version appears to have been written at some point before 1316. Apparently also of an early date are the separate *Questiones* on the *Physics,* extant in a single (incomplete) manuscript (MS Basel, Universitätsbibliothek F.V.12). We are more fully informed, however, concerning Burley's definitive version of his comments on the *Physics:* Book I was finished in Paris in 1324, Books II-VI, again in Paris, by 1327, while the final redaction of Books VII-VIII was written between 1334-1337. Although this last effort which Burley devoted to the exposition and analysis of Aristotle's *Physics* contains, it appears, the most significant of his contributions to natural philosophy, at least some of what he has to offer here also appears, in more elaborate form, within various independent treatises and opuscula, most notably the *Tractatus de formis* (which contains his criticisms of Ockham's identification of substance and quantity), the quodlibetal question *De primo et ultimo instanti,* and his two treatises on the intension and remission of forms (resp. the *Tractatus primus* and *Tractatus secundus*).

Burley turned to moral philosophy and varia rather late in his life, completing his exposition of Aristotle's *Ethics* in 1333-1334 and of the *Politics* in 1340-1343. His immensely popular history of philosophers, the

De vita et moribus philosophorum, also appears to derive from the early 1340's.

If one excludes Burley's importance within the history of formal logic, it is clear that the very small segment of his voluminous work within natural philosophy which has hitherto received careful study makes any estimate of his significance for the history of later medieval science radically incomplete and tentative. Yet in spite of this, there are two of Burley's contentions which, to judge merely from the attention they received within the works of other medieval natural philosophers, are surely to prove of more than ephemeral importance: his view of the proper limits to be assigned temporal processes through the ascription of first or last instants, and his view of the nature of motion.

De primo et ultimo instanti. This brief treatise was a quodlibetal question disputed by Burley at Toulouse sometime before 1327. Its subject derived directly from Aristotle's discussion in the *Physics* (Book VI, ch. 5, and Book VIII, ch. 8) of the problem of first and last moments within a given change which occurred over a given time interval. Without unraveling Aristotle's treatment and proposed resolution of the puzzle, suffice it to say that Burley (and a multitude of other Scholastics as well) correctly focused upon the relevant variables within the problem as stated by Aristotle when they formulated what was to become a standard distinction concerning the first and last instants of a given temporal process or change. Briefly, the change may be limited intrinsically at both ends (*incipit et desinit*) by an instant which belongs to the temporal interval covering the change in question (*primum vel ultimum instans esse*), or limited extrinsically by a first or last instant which does not belong to the appropriate interval (*primum vel ultimum instans non esse*), in which case, of course, an *ultimum instans non esse* immediately precedes the time denominated by the change, while a *primum instans non esse* immediately follows it. Given this, in four *regulae* Burley explicitly stipulates the mutual exclusiveness of intrinsic and extrinsic instants as limits for both the beginning and the completion of a given change. The problem is then, however, to decide just which kinds of things can be said to begin or end by possessing just which kind (intrinsic or extrinsic) of limit. Although the many cases that Burley considers in making this decision involve complexities that cannot be expounded here, some notion of his procedure should be indicated. Thus, a *res successiva,* a given temporal motion or change or a time interval itself, can, following Aristotle, only have extrinsic limits at both ends. The same is true of a *res permanens* which depends in its being on a

res successiva (the truth of the proposition "Socrates is running" is Burley's example of such a *res permanens,* since it depends on the existence of a given run of Socrates, which is a *res successiva*). On the other hand, intrinsic limits are naturally appropriate when a given permanent thing itself has only instantaneous existence. And Burley goes on to ascribe what he feels to be proper limits to other cases of *res permanentes* undergoing change, specifically to changes explicable in terms of the intension and remission of forms as viewed under his own special theory concerning the nature of such intension and remission.

On the Nature of Motion. Indeed, to state Burley's view with respect to this much discussed problem of motion is, in effect, to set forth the basis of his theory of intension and remission of forms, a theory most elaborately developed and expressed in his *Tractatus primus.* Some theorists, of whom Ockham is a prominent but not the earliest example, had argued that motion consists of nothing else besides the mobile and the place, quality, or quantity that it successively acquires. The form successively acquired was called (1) the *forma fluens.* Burley admitted that motion could be viewed in this way, but said that it could also be considered as (2) a flux (*fluxus formae*), or (3) a successive quantity. As a flux, motion was the acquisition of the terminus of motion or the transmutation by means of which the terminus was acquired. As a successive quantity, motion was the measure of the motion taken in the second (*fluxus*) sense. Motion as a *forma fluens* belonged to the same category as its terminus (place, quality, or quantity); as a *fluxus formae* it either belonged to or constituted the whole of the Aristotelian category of passions or affectations; as a successive quantity, it belonged to the category of quantity.

Why was it necessary, according to Burley, that motion be not only a *forma fluens* but also a flux? For Burley, places, qualities, and quantities were individual and simple. A body at a given time had only one place, one quality of a given type, and one quantity of a given type. In the case of quality, for example, the not uncommon assumption that a compound simultaneously contained both hot and cold combining to produce a single sensible result was to Burley not only false, it was self-contradictory. Qualities like hot and cold were for him sensible by definition. They were not hypothetical underlying realities without separate effect. It followed further, on this view, that one quality could not be part of another. Similar conclusions could be applied to place and quantity.

Since this was Burley's view of the forms to be acquired, his conception of the *forma fluens* theory could not be the same as Ockham's. Ockham spoke of *the* form acquired or *the* terminus of motion, and assumed that this final form somehow contained the forms acquired along the way. A single form was acquired part by part, and this was the *forma fluens.* For Burley, every instant of motion corresponded to a different form, and these forms were neither part of, nor contained in, the terminus of motion. The *forma fluens* was any one of these instantaneous forms, but no one of these forms could represent the whole motion since the terminus (with successive acquisition assumed) represented the whole motion for Ockham. To represent more than an instant within a motion, Burley needed another existential referent. He might have chosen the entire collection of instantaneous *formae fluentes.* Durand of St. Pourcain did just this, saying that the continuity of these forms unified them (cf. Maier, *Studien,* II, 70–73). For Burley, the forms, analogous to points, could not be continuous, and hence could not be treated as a unity. As a unified referent he chose instead the means by which the forms were acquired, i.e., the transmutation or flux. The *forma fluens* conception, he admitted, was truer to the physical reality of motion (*entitas rei*), but the *fluxus formae* conception was truer to the significance of the term "motion" (*Physics,* Bk. III, Text 4).

If Burley's views of motion are typical of other parts of his physics as yet less well-known, the motivation for his conclusions was not simply a willingness to multiply entities beyond necessity, but also his view of forms as empirical, simple, and separate, and his refusal to assume hypothetical connections between them.

BIBLIOGRAPHY

I. Life and Writings. The three fundamental articles are A. B. Emden, *Biographical Register of the University of Oxford to A.D. 1500* (Oxford, 1957–1959), I, 312–314; Conor Martin, "Walter Burley," in *Oxford Studies Presented to Daniel Callus* (Oxford, 1964), pp. 194–230; J. A. Weisheipl, "Ockham and Some Mertonians," in *Mediaeval Studies,* **30** (1968), 174–188. Father Weisheipl will also shortly publish his "Repertorium Mertoniense" in a future volume of *Mediaeval Studies,* which will include a complete list, together with editions and extant manuscripts, of Burley's works. Cf. Zofia Wlodek, "Les traités de Walter Burleigh dans les manuscrits des bibliothèques en Pologne," in *Mediaevalia philosophica polonorum,* **11** (1963), 152–156.

II. Texts of Burley's Works. *Early Printed Editions and Manuscripts.* (Note: Only selected early editions and a few MSS of works important within the history of science are listed; see Weisheipl's forthcoming "Repertorium

Mertoniense" for a complete listing.) *Expositio super artem veterem* (final version completed 1337; Venice eds., 1481, 1497, 1519). *Expositio super libros duos posteriorum analyticorum* (Venice, 1497). *De sophismatibus,* in St. Bonaventure, *Opera omnia* (Bassani, 1767), cols. 467 ff. *Tractatus de universalibus realibus* (Venice, 1492–1493). *Expositio librorum physicorum,* the definitive version (Venice, 1482, 1491, 1501; Bologna, 1589). *Expositio librorum physicorum* (early, pre-1316 version), MS Cambridge, Gonville & Caius 448/409, pp. 172–543. *Questiones super libros physicorum* (also early; incomplete, Books I–IV only), MS Basel, Universitätbibliothek, F.V.12, ff. 108r–171v. *Expositiones super libros:* (1) *De anima;* (2) *Parva naturalis;* (3) *De generatione et corruptione;* (4) *De caelo;* (5) *De substantia orbis;* (6) *De motu animalium,* MS Vat. lat. 2151, 1r–108v, 149r–256r. *Tractatus de formis,* MSS Vat. lat. 2151, ff. 131r–148r; Vat. lat. 2146, ff. 235r–244v. *Tractatus de potentiis anime,* MS Vat. lat. 2146, ff. 252v–256v. *Tractatus primus* (*sive Tractatus de activitate, unitate et augmento formarum activarum habentium contraria suscipientium magis et minus*), MS Vat. lat. 817, ff. 203r–223r. *Tractatus secundus* (*sive Tractatus de intensione et remissione formarum*) (Venice, 1496). *Opuscula varia,* MSS Vat. lat. 2146, ff. 235r–256v; Lambeth Palace 70, ff. 8r–306r. *Expositio librorum ethicorum* (Venice, 1481, 1500).

Modern Editions. See *De puritate artis logicae, Tractatus longior, With a Revised Edition of the Tractatus Brevior,* Philotheus Boehner, ed. (New York, 1955). Part 1 of Tractatus II of the *Tractatus longior* was republished, together with a brief introduction and an English translation, by Ivan Boh, "Burleigh: On Conditional Hypothetical Propositions," in *Franciscan Studies,* **23** (1963), 4–67. "*De primo et ultimo instanti,*" Herman and Charlotte Shapiro, eds., in *Archiv für Geschichte der Philosophie,* **47** (1965), 157–173. Shapiro has also published—sometimes with the assistance of others—editions of the following brief opuscula of Burley: "*De relativis,*" in *Franciscan Studies,* **22** (1962), 155–171; "*De qualitatibus,*" in *Franziskanische Studien,* **45** (1963), 256–260; "*De ente,*" in *Manuscripta,* **7** (1963), 103–108; "*De deo, natura et arte,*" in *Medievalia et humanistica,* **15** (1963), 86–90; "*De diffinitione,*" in *Mediaeval Studies,* **27** (1965), 337–340; "*De potentia activa et passiva,*" in *Modern Schoolman,* **43** (1966), 179–182; "*De toto et parte,*" in *Archives d'histoire doctrinale et littéraire du moyen age,* **24** (1966), 299–303; and "*De sensibus,*" in *Mitteilungen des Grabmann-Institutes der Universität München,* Heft 13 (Munich, 1966). *De vita et moribus philosophorum* has been edited, together with an old Spanish translation, by H. Knüst (Tübingen, 1886); cf. J. O. Stigall, "The Manuscript Tradition of the *De vita et moribus philosophorum* of Walter Burley," in *Medievalia et humanistica,* **11** (1957), 44–57, and J. N. Hough, "Platus, Student of Cicero, and Walter Burley," *ibid.,* pp. 58–68.

III. Secondary Literature. A. *Logic.* Philotheus Boehner's high opinion of Burley as a logician is succinctly stated in his *Medieval Logic. An Outline of Its Development from 1250 to c. 1400* (Chicago, 1952), pp. 44–51, 84–89. This opinion, seemingly shared by Boh (*vide supra et infra*), has been criticized by L. Minio-Paluello in *Oxford Maga-*

zine, **71** (1953), 200–201. A. N. Prior, "On Some *Consequentiae* in Walter Burleigh," in *New Scholasticism,* **27** (1953), 433–446. Ivan Boh, "A Study in Burleigh: *Tractatus de regulis generalibus consequentiarum,*" in *Notre Dame Journal of Formal Logic,* **3** (1962), 83–101; "Walter Burleigh's Hypothetical Syllogistic," *ibid.,* **4** (1963), 241–269; "An Examination of Some Proofs in Burleigh's Propositional Logic," in *New Scholasticism,* **38** (1964), 44–60. Ernest A. Moody, *Truth and Consequence in Mediaeval Logic* (Amsterdam, 1953), *passim.*

B. *Natural Philosophy and General.* Absolutely fundamental are the five volumes of Anneliese Maier's *Studien zur Naturphilosophie der Spätscholastik,* which contain numerable expositions and analyses of aspects of Burley's work within the context of similar material as treated by other late medieval Scholastics: I. *Die Vorläufer Galileis im 14. Jahrhundert,* 2nd ed. (Rome, 1966); II. *Zwei Grundprobleme der scholastischen Naturphilosophie,* 3rd ed. (Rome, 1968); III. *An der Grenze von Scholastik und Naturwissenschaft,* 2nd ed. (Rome, 1952); IV. *Metaphysische Hintergrunde der spätscholastischen Naturphilosophie* (Rome, 1955); V. *Zwischen Philosphie und Mechanik* (Rome, 1958).

Other articles by Maier also directly concerned with Burley are "Zu Walter Burleys Politik-Kommentar," in *Recherches de théologie ancienne et médiévale,* **14** (1947), 332–336; "Zu einigen Problemen der Ockhamforschung," in *Archivum Franciscanum historicum,* **46** (1953), 181–194; "Handschriftliches zu Wilhelm Ockham und Walter Burley," *ibid.,* **48** (1955), 234–251; "Ein unbeachteter 'Averroist' des XIV Jahrhunderts: Walter Burley," in *Medioevo e rinascimento: Studi in onore di Bruno Nardi* (Florence, 1955), 477–499; "Zu Walter Burleys Traktat *De intensione et remissione formarum,*" in *Franciscan Studies,* **25** (1965), 293–321. Maier has included these articles (save the last), together with addenda, in her *Ausgehendes Mittelalter: Gesammelte Aufsätze zur Geistesgeschichte des 14. Jahrhunderts,* 2 vols. (Rome, 1964–1967).

See also L. Baudry, "Les rapports des Guillaume d'Occam et de Walter Burleigh," in *Archives d'histoire doctrinale et littéraire du moyen âge,* **9** (1934), 155–173; A. Koyré, "Le vide et l'espace infini au XIVᵉ siècle," *ibid.,* **17** (1949), 75–80; three articles of S. H. Thomson, "Walter Burley's Commentary on the *Politics* of Aristotle," in *Mélanges Auguste Pelzer* (Louvain, 1947), pp. 557–579; "An Unnoticed *Questio theologica* of Walter Burley," in *Medievalia et humanistica,* **6** (1950), 84–88; "Unnoticed *questiones* of Walter Burley on the *Physics,*" in *Mitteilungen des Instituts für österreichische Geschichtsforschung,* **62** (1954), 390–405; four articles by H. Shapiro, "Walter Burley and Text F1," (i.e. of Book IV of the *Physics*), in *Traditio,* **16** (1960), 395–404; "Walter Burley and the Intension and Remission of Forms," in *Speculum,* **34** (1959), 413–427 (use in conjunction with Anneliese Maier on the same topic); "A Note on Walter Burley's Exaggerated Realism," in *Franciscan Studies,* **20** (1960), 205–214; "More on the 'Exaggeration' of Burley's Realism," in *Manuscripta,* **6** (1962), 94–98.

A brief treatment of Burley's view on first and last instants, together with other late Scholastic discussion of the same problem, may be found in Curtis Wilson, *William Heytesbury: Medieval Logic and the Rise of Mathematical Physics* (Madison, Wis., 1956), pp. 29–56.

JOHN MURDOCH
EDITH SYLLA

BURNET, THOMAS (*b.* Croft, Yorkshire, England, *ca.* 1635; *d.* London, England, 27 September 1715), *cosmogony, geology.*

Little is known of Burnet's family or early childhood other than that his father was John Burnet and that he attended the Freeschool of Northallerton, where he attracted the attention of his teacher, Thomas Smelt. In 1651 he was admitted as pensioner at Clare Hall, Cambridge. Although he was officially a student of William Owtram's and closely associated with John Tillotson, he was most influenced by Ralph Cudworth. Sometime after receiving his bachelor's degree in 1655, Burnet followed Cudworth to Christ's College, became a fellow of the college in 1657, and received his M.A. in 1658. He became proctor of the college in 1667 and remained listed as a fellow until 1678, even though he was not in residence all of that time. While at Cambridge, Burnet worked closely with the Cambridge Platonists, especially Cudworth and Henry More. In 1671 he went abroad as governor to the young earl of Wiltshire and later made a second tour of Europe with the grandson of the duke of Ormonde, the earl of Orrery. During his travels he started writing his theory of the earth, the first two parts of which he completed soon after his return to England.

Burnet published these two parts, the books concerning the Deluge and Paradise, under the title *Telluris theoria sacra* in Latin in 1681 and produced an English version in 1684. The immediate reaction to the book was favorable. Many praised the style and thought; a few questioned the theory. In 1685 Burnet, who had been ordained in the Anglican Church, was made Master of the Charterhouse at the recommendation of the duke of Ormonde. In 1686, when the king tried to have Andrew Popham, a Roman Catholic, admitted as a pensioner at the Charterhouse, Burnet effectively opposed the appointment. Shortly after William III ascended the throne, Burnet became chaplain in ordinary to the king and clerk of the closet.

During this time *The Sacred Theory of the Earth* had become a subject of controversy. Christianus Wagner, Herbert Crofts, bishop of Hereford, and Erasmus Warren attacked it. For a time Burnet ignored the criticisms, then answered by expanding his theory. In 1689 he published a new Latin edition containing two additional books, and in 1691 he completed a similar English edition that included his "Review of the Theory of the Earth" and a reply to Warren's objections.

The Sacred Theory of the Earth was Burnet's attempt to combine the idealism of the Cambridge Platonists, Scripture, and an explanation of the features of the earth's surface in order to account for the past and present states of the earth and to offer a prophecy about its future. He believed that there were four major events in the earth's history: its origin from chaos, the universal deluge, the universal conflagration, and the consummation of all things. The first two of these had already happened; the last two were yet to come. These four events divided the history of the earth into three periods. The first, from the Creation to the Deluge, Burnet described as the state of paradise and the antedeluvian world. This earth differed in form and constitution from the present earth. Its surface, which covered the waters and a great abyss, was smooth, regular, and uniform, without mountains or seas. The material of this surface was moist, oily earth suitable to sustain living things. When the surface caved into the abyss, an event due to the continued drying action of the sun, and was no longer smooth, the fluctuations of the waters over this irregular earth caused the universal deluge. This marked the end of the first period.

The second, or present, era was for Burnet the age between the Deluge and the Conflagration. During this time the surface and interior of the earth undergo slow but continual change and thus, when the time comes in the plan of Divine Providence, the earth will be ready and able to burn. The final period, that of the millennium, is the era following the universal conflagration, when there will be a new heaven and a new earth in which the blessed will enjoy a life of peace and tranquillity. At the end of the millennium the earth will be changed into a bright star, and the consummation of all things predicted in Scripture will be fulfilled.

The expanded theory occasioned more controversy and more replies from Burnet. In 1692 he attempted to reconcile the account of creation in Genesis with his theory in *Archaelogiae philosophicae*. This book aroused such opposition for its allegorical treatment of Scripture that Burnet, although he had dedicated the book to William III, was forced to resign his position at court. He retired to the Charterhouse and remained there until his death. He was buried in the vault of the Charterhouse chapel a week after he died.

Burnet spent the last years of his life writing in defense of his theory. Most of the attacks upon it were upon religious grounds. Warren, Crofts, John Beaumont, and others accused him of a too liberal or allegorical interpretation of Scripture or of eliminating the necessity of God's working in the universe. John Keill, however, attacked the Cartesian mechanical basis of the theory and refuted it in terms of Newtonian mechanics. In his replies to these, Burnet either reiterated his own interpretation of Scripture or, when unable to refute a logical or mathematical argument against him, pointed out a minor inconsistency in his opponent's work. In his later, minor writings, he applied his method of scriptural interpretation to theological questions.

Burnet's importance in the history of scientific thought is due less to his theory itself than to certain aspects of it that became standards in the then growing science of geology. For more than a hundred years after Burnet, writers discussing the origin of and changes in the surface of the earth felt impelled to reconcile their theories with the account of creation in Genesis. His emphasis on the importance of the Deluge and on the explanation of the formation of mountains continued in geologic writings. Finally, Burnet's style was such that *The Sacred Theory of the Earth* was considered readable long after his death and the ideas expressed in it were widely disseminated. Whether accepted or ridiculed, the theory helped popularize the idea that the features of the earth's surface were constantly changing.

BIBLIOGRAPHY

I. ORIGINAL WORKS. Burnet's writings are *Telluris theoria sacra* (London, 1681, 1689, 1702; Amsterdam, 1694, 1699), trans. into English as *The Sacred Theory of the Earth* (London, 1684, 1690–1691, 1697); and *Archaelogiae philosophicae* (London, 1692, 1728), trans. into English as *The Ancient Doctrine Concerning the Originals of Things* (London, 1692 [this contains only chs. 7–10 of Bk. 2], 1729, 1736).

Burnet's replies to works about *The Sacred Theory of the Earth* are *An Answer to the Exceptions Made by Mr. Erasmus Warren, Against the Sacred Theory of the Earth* (London, 1690); *A Review of the Theory of the Earth and of Its Proofs: Especially in Reference to Scripture* (London, 1690); *A Short Consideration of Mr. Erasmus Warren's Defense of His Exceptions Against the Theory of the Earth. In a Letter to a Friend* (London, 1691); *Some Reflections Upon the Short Considerations of the Defense of the Exceptions Against the Theory of the Earth* (London, 1692); and *Reflections Upon the Theory of the Earth, Occasioned by a Late Examination of It. In a Letter to a Friend* (London, 1699).

Other works by Burnet are *De statu mortuorum et resurgentium liber. Accessit epistola (ad virum clarissimum A.B.) circa libellum de archaelogiis philosophicis* (London,

1720), trans. into English as *Of the State of the Dead and of Those That Are to Rise* (partial ed., London, 1727; complete ed., 1728); *De fide et officiis Christianorum* (London, 1722), trans. into English as *The Faith and Duties of Christians* (London, 1728); *De future Judaeorum restauratione* (London, 1727), an appendix to the 1727 ed. of *De statu mortuorum*; and *A Re-survey of the Mosaic System of the Creation With Rules for the Right Judging and Interpreting of Scripture. In Two Letters to a Friend* (London, 1728).

Works published anonymously but credited to Burnet are *Remarks Upon an Essay Concerning Humane Understanding* [by J. Locke] *in a Letter Addressed to the Author* (London, 1697); *Second Remarks Upon an Essay Concerning Human Understanding. In a Letter Address'd to the Author: Being a Vindication of the First Remarks Against the Answer of Mr. Locke at the End of His Reply to the Bishop of Worcester* (London, 1697); *Third Remarks Upon an Essay Concerning Human Understanding. In a Letter Addressed to the Author* (London, 1699); and *An Appeal to Common Sense: or, a Sober Vindication of Dr. Woodward's State of Physick. By a Divine of the Church of England* (London, 1719).

II. SECONDARY LITERATURE. Marjorie Hope Nicolson, *Mountain Gloom and Mountain Glory: The Development of the Aesthetics of the Infinite* (Ithaca, N.Y., 1959), pp. 184–270, concerns Burnet and his work.

There is no bibliography on Burnet, but the following works on *The Theory of the Earth* may be helpful: John Beaumont, "Considerations on a Book Entitled The Theory of the Earth Published Some Years Since by the Learned Dr. Burnet," in *Philosophical Transactions*, **17** (Sept. 1693), 888–892, and *A Postscript to a Book Entituled Considerations on Dr. Burnet's Theory of the Earth* (London, 1694); Herbert Crofts, *Some Animadversions Upon a Book Intituled the Theory of the Earth* (London, 1685); Robert Hooke, "Animadversions on Burnet's Theory, 1689," MS at the Royal Society, London; John Keill, *An Examination of Dr. Burnet's Theory of the Earth, Together With Some Remarks on Mr. Whiston's New Theory of the Earth* (Oxford, 1698), and *An Examination of the Reflections on the Theory of the Earth Together With a Defense of the Remarks on Mr. Whiston's New Theory* (Oxford, 1699); Melchoir Leydekker, *M. Leydeckeri de republica Hebraeorum . . . subjicitur archaelogia sacra, qua historia creationis et diluvii Mosica contra Burneti profanam telluris theoriam asseritur* (Amsterdam, 1704); Archibald Lovell, *A Summary of Material Heads Which May Be Enlarged and Improved into a Compleat Answer to Dr. Burnet's Theory of the Earth* (London, 1696); Matthew Mackaile, *Terrae prodromus theoricus. Containing a Short Account of the New System of Order and Gradation, in the World's Creation. By Way of Animadversions Upon Mr. T. Burnet's Theory of His Imaginary Earth, etc.* (Aberdeen, 1691); Robert St. Clair, *The Abyssinian Philosopher Confuted; or, Telluris theoria Neither Sacred nor Agreeable to Reason* (London, 1697); Christianus Wagner, *Animadversions in . . . T. Burnetii Telluris theoriam sacram, etc.* (Leipzig, n.d.); and Erasmus Warren, *Geologia; or, a Discourse Concerning*

the Earth Before the Deluge, Wherein the Form and Properties Ascribed to It, in a Book Intituled the Theory of the Earth, Are Excepted Against and It Is Made to Appear That the Dissolution of That Earth Was Not the Cause of the Universal Flood* (London, 1690); *A Defense of the Discourse Concerning the Earth Before the Flood: Being a Full Reply to a Late Answer to Exceptions Made Against the Theory of the Earth, etc.* (London, 1691); and *Some Reflections Upon the Short Consideration of the Defense of the Exceptions Against the Theory of the Earth* (London, 1692).

SUZANNE KELLY

BURNHAM, SHERBURNE WESLEY (*b.* Thetford, Vermont, 12 December 1838; *d.* Chicago, Illinois, 11 March 1921), *astronomy.*

Burnham, an indefatigable observer, was a self-trained amateur astronomer. His formal education ended with his graduation from Thetford Academy. He then became an accomplished shorthand reporter and, except for six months spent at the Washburn Observatory and four years at the Lick Observatory, worked full time in law courts until his retirement at age sixty-four in 1902. From 1897 to 1914 Burnham was professor of practical astronomy at the Yerkes Observatory, living in Chicago and commuting to Williams Bay, Wisconsin, for two nights each week to use the forty-inch telescope. For his astronomical work Burnham was honored by the Royal Astronomical Society (member, 1874; associate, 1898; Gold Medal, 1894) and the Académie des Sciences (Prix Lalande, 1904), and by Yale (M.A., 1878) and Northwestern (Sc.D., 1915) universities.

Burnham's significant contributions to the study of double stars were the discovery of numerous visual binary systems, the measurement of their separation and position angles, and a critical compilation of information concerning all known northern pairs. When Burnham began observing in 1870, it was commonly assumed that the Struves and Herschels had found most of the binaries visible in the Northern Hemisphere. With the aid of excellent telescope lenses figured by Alvan Clark & Sons, as well as extraordinarily keen vision, Burnham proved otherwise. Indeed, many of his discoveries were new companions of stars or star systems that had already been carefully scrutinized. The culmination of these efforts came in 1900, with the publication of *A General Catalogue of 1290 Double Stars Discovered from 1871 to 1899 by S. W. Burnham.*

Burnham devoted most of his time at large telescopes to measuring difficult pairs—those with separations less than 1″ and those of unequal magnitudes. He also measured the positions of components of suspected binaries relative to background stars for evidence of common proper motion, and hence of physical relation. For these measurements Burnham designed and used a filar micrometer with greatly improved bright-wire illumination. To help identify his discoveries Burnham compiled, and finally published, *A General Catalogue of Double Stars Within 121° of the North Pole*, the first comprehensive and critical survey of all (13,665) known binaries in this region.

BIBLIOGRAPHY

I. ORIGINAL WORKS. Burnham's writings include "The Position Micrometer of the Washburn Observatory," in *English Mechanic,* **34** (1881), 39–40; *A General Catalogue of 1290 Double Stars Discovered from 1871 to 1899 by S. W. Burnham, Arranged in Order of Right Ascension with all the Micrometrical Measures of Each Pair,* Publications of the Yerkes Observatory, no. 1 (Chicago, 1900), the introduction to which includes an often-quoted autobiography and references to his nineteen original lists of double stars; *A General Catalogue of Double Stars Within 121° of the North Pole,* Carnegie Institution Publication no. 5 (Washington, D.C., 1906); and *Measures of Proper Motion Stars Made with the 40-Inch Refractor of the Yerkes Observatory in the Years 1907 to 1912,* Carnegie Institution Publication no. 168 (Washington, D.C., 1913).

II. SECONDARY LITERATURE. Works on Burnham consist primarily of obituaries written by his colleagues and published in various astronomical journals.

DEBORAH JEAN WARNER

BURNSIDE, WILLIAM (*b.* London, England, 2 July 1852; *d.* West Wickham, England, 21 August 1927), *mathematics.*

Burnside's research was in such diverse fields as mathematical physics, complex function theory, geometry, group theory, and the theory of probability. On the basis of his work in the first two fields, he was elected a fellow of the Royal Society in 1893. It was to the theory of groups, however, that he made his most significant contributions. The beginnings of an interest in groups can be detected in papers of 1891 and 1892, in which groups of linear fractional transformations of a complex variable are involved. By 1894 the theory of groups of finite order had become the central concern of much of his research, and for the next twenty years Burnside remained one of the most active contributors to its development. A number of his results have become an integral part of the modern theory of groups and their representations.

With the hope of stirring up interest in group theory in England, Burnside published his *Theory of Groups* in 1897. It was the first treatise on groups in English and also the first to develop the theory from the modern standpoint of abstract groups vis à vis permutation groups, although this approach had already been pioneered by H. Weber in his *Lehrbuch der Algebra* (1896). One topic Burnside excluded from his book was that of linear groups, because it did not seem that any result could be obtained most directly by considering linear transformations. This opinion soon became outdated, however, with G. Frobenius' development of the theory of group representations and characters (1896–1899), and Burnside was one of the first to recognize the importance of Frobenius' ideas and to contribute to their development, simplification, and application.

Using group characters, Burnside was able to prove, for example, that every transitive group of prime degree is either solvable or doubly transitive (1901) and that every group of order $p^a q^b$ (p and q prime) is solvable (1904). The latter result greatly extended results of Sylow ($b = 0$, 1872), Frobenius ($b = 1$, 1895), and Jordan ($b = 2$, 1898). It was also Burnside who discovered that groups of odd order admit no nontrivial real irreducible representations, and he was led by its consequences to suspect that every group of odd order is solvable. W. Feit and J. G. Thompson finally established this in 1962 with a proof that involves, among other things, frequent applications of Burnside's discovery.

Because he was convinced of the important role that representation theory was destined to play in the future advancement of group theory, Burnside devoted considerable space to its systematic presentation in the second edition of *Theory of Groups* (1911). This edition was widely read and is now considered a classic.

BIBLIOGRAPHY

I. ORIGINAL WORKS. Besides *Theory of Groups of Finite Order* (Cambridge, 1897, 1911), Burnside also composed a treatise on probability, *Theory of Probability* (Cambridge, 1928), which was published posthumously.

II. SECONDARY LITERATURE. The only article dealing with Burnside's life and work in any detail is A. R. Forsyth's obituary notice in *Proceedings of the Royal Society,* **117A** (1928), xi–xxv; the emphasis is upon Burnside's early work, and insufficient attention is paid to his role in the development of the theory of groups. Some idea of the latter can be obtained from the historical and survey articles scattered throughout *The Collected Works of George Abram Miller,* 5 vols. (Urbana, Illinois, 1935–1959), esp. II, 1–18 and III, 1–15. See also H. Burkhardt and H. Vogt, "Sur les groupes discontinus," in *Encyclopédie des sciences mathématiques,* I, 1, fasc. 4 (Paris–Leipzig, 1909), 532–616.

THOMAS HAWKINS

BURRAU, CARL JENSEN (*b.* Elsinore, Denmark, 29 July 1867; *d.* Gentofte, Denmark, 8 October 1944), *astronomy, actuarial mathematics.*

Burrau studied mathematics at Copenhagen University and was an assistant astronomer at the university observatory from 1893 to 1898. He subsequently worked as an actuary. From 1906 to 1912 he lectured at Copenhagen University on practical mathematics. In his researches as an astronomer and as an actuary he was a disciple of T. N. Thiele.

In 1892 the Royal Danish Academy, at Thiele's suggestion, presented an astronomical prize problem concerning librations in the *problème restreint* with two equal masses in circular movement round their common center of gravity. In his solution, Burrau was the first to point out that a series of periodic orbits, into which the third (massless) body moves, develops into a limiting orbit of ejection from (or collision with) one of the masses. It was a pioneer achievement, the first step taken in the systematic search for periodic orbits in the three-body problem that was later carried out by E. Strömgren and his pupils (in its early years, in collaboration with Burrau).

Burrau's dissertation (1895) deals with the derivation of the constants of a measuring machine for photographic determination of star positions. He suggested a development of Bessel's classic method and discussed previously proposed simplifications.

The "distance" from these studies to actuarial work is short. In an obituary, Kristensen, who audited Burrau's lectures on practical mathematics, mentions "his ability to combine scientific points of view with instructions for using them in practice." Burrau's little book on actuarial mathematics, *Forsikringsstatistikens Grundlag,* which was originally written as a series of lectures and appeared in Danish, German, and Italian, is a similar attempt to use his mathematical knowledge in the domain in which he worked for most of his life. Later, he and B. Strömgren published a paper on dividing a frequency curve into its components.

BIBLIOGRAPHY

I. ORIGINAL WORKS. Burrau's writings include "Recherches numériques concernant des solutions périodiques d'un cas spécial du problème des trois corps," in *Astronomische Nachrichten,* **135** (1894), 233–240; **136** (1894), 161–174; "Undersøgelser over Instrumentkonstanter

ved Kjøbenhavns Universitets Astronomiske Observatoriums Maaleapparat for fotografiske Plader," his dissertation, Univ. of Copenhagen (1895); a review of Darwin's *Periodic Orbits* (1897), in *Vierteljahrsschrift der Astronomischen Gesellschaft,* **33** (1898), 21–33, containing information on the early development of the three-body problem; "Über einige in Aussicht genommene Berechnungen betreffend einen Spezialfall des Dreikörper-Problems," *ibid.,* **41** (1906), 261–266; and *Forsikringsstatistikens Grundlag* (Copenhagen, 1925), originally published in German in *Wirtschaft und Recht der Versicherung,* **56** (1924). Papers are in *Publikationer og mindre Meddelelser fra Københavns Observatorium* (1913–1934).

II. SECONDARY LITERATURE. Articles on Burrau are *Dansk Biografisk Leksikon,* IV (1934), 365–366; and S. Kristensen, in *Skandinavisk Aktuarietidskrift,* **28** (1945), 128–130.

AXEL V. NIELSEN

BUSCH, AUGUST LUDWIG (*b.* Danzig, Prussia, 7 September 1804; *d.* Königsberg, Prussia, 30 September 1855), *astronomy.*

Busch was introduced to astronomy by Friedrich Bessel, whose assistant he became in 1831. When Bessel died in 1846, Busch succeeded him as director of the astronomical observatory in Königsberg. There he chiefly reduced the observations made by James Bradley in Kew and Wanstead, and from these he deduced improved values for the constants of aberration and nutation. Busch was also a pioneer in astronomical photography; in 1851 he succeeded in taking a daguerreotype of the eclipsed sun. Ill health prevented him from fully developing his work.

BIBLIOGRAPHY

Busch's published works include his *Reduction of the Observations Made by J. Bradley at Kew and Wansted* (Oxford, 1838); and *Verzeichnis sämtlicher Werke Bessels* (Königsberg, 1849). Poggendorff, I, gives a more complete list of his work.

BERNHARD STICKER

BUSK, GEORGE (*b.* St. Petersburg, Russia, 12 August 1807; *d.* London, England, 10 August 1886), *medicine, natural history, anthropology.*

George Busk was the second son of Robert Busk, an English merchant in St. Petersburg, and Jane Westley, daughter of John Westley, customshouse clerk at St. Petersburg. His grandfather, Sir Wadsworth Busk, was attorney general of the Isle of Man, and an uncle was Hans Busk, scholar and minor poet.

Busk received his medical education at St. Thomas'

616

and St. Bartholomew's hospitals, and became a member of the Royal College of Surgeons in 1832. He then became surgeon to the Seaman's Hospital Society, recently founded for the relief of merchant seamen, and served at Greenwich on the hospital ship *Dreadnought,* which had been given to the society by the Admiralty. Although not actually at sea, he made good use of his time and the available clinical material. Busk is credited with having worked out the pathology of cholera and having made important observations on scurvy. A few of his notes on scurvy are still extant at the Royal College of Surgeons, but no direct evidence of work on cholera has been found. It is probable that there has been confusion with Busk's work on fasciolopsiasis, which culminated in his description of the fluke now eponymically styled *Fasciolopsis buski,* the adult stage of which occurs in the small intestine among natives of India and eastern China. The disease causes toxic symptoms and acute diarrhea, and thus may have been termed cholera.

When Busk resigned from the *Dreadnought* in 1854, he apparently retired from active surgical practice and turned to biology and teaching. In 1843 he had been among the first elected to fellowship of the Royal College of Surgeons. From 1856 to 1859 he was Hunterian professor of comparative anatomy at the college, and his lecture notes survive in its archives. They make somewhat dull reading now, but his philosophical approach is shown by remarks on reproduction and sexual physiology in invertebrates. "Time was when the difficulty of the physiologist lay in understanding reproduction without the sexual process. At the present day it seems to me the process is reversed and that the question before us is why is sexual union necessary?" (R.C.S. 275.6.3).

In 1850 Busk was made a fellow of the Royal Society. He was four times its vice-president and received its Royal Medal in 1871. His industry and zeal were enormous. He was president of the Royal College of Surgeons in 1871 and belonged to the Linnean Society (vice-president and zoological secretary), the Geological Society (Lyell Medal, 1878; Wollaston Medal, 1885), the Microscopical Society (foundation member, 1839; president, 1848–1849), the Anthropological Society (president, 1873–1874), and the Zoological Society. He was editor of the *Microscopic Journal* (1842), the *Quarterly Journal of Microscopic Science* (1853–1868), the *Natural History Review* (1861–1865), and the *Journal of the Ethnological Society* (1869–1870). Busk was also a member of the Senate of London University, treasurer of the Royal Institution, and the first Home Office inspector under the Cruelty to Animals Act, a difficult position that he fulfilled with tact and humanity. From 1841, he

contributed some seventy papers to scientific journals.

Busk's two main interests in science were the Bryozoa (Polyzoa) and paleontology. In 1856 he formulated the first scientific arrangement of the Bryozoa, the notes and drawings for which are extant (R.C.S. 275.e.3). In the same year, the name *Buskia* was given to a genus of Bryozoa. His collection is at the Natural History Museum, which also has anthropological material collected by him, notably the Gibraltar cranium, a Neanderthal type he found (but did not recognize as such) in 1868. He was an authority on craniometry, and his opinions were much sought on fossil identification.

Busk is thus to be seen as a classifier and investigator whose work, although it may now appear insignificant, was ancillary to and provided corroborative evidence for, the ideas of Darwin, Lyell, and Richard Owen. He was therefore closely connected with the development of zoology and anthropology. A dull writer and lecturer, he was described as an excellent surgical operator and a man of "unaffected simplicity and gentleness of character."

Busk married his cousin Ellen, daughter of Jacob Hans Busk, on 12 August 1843. His portrait in oils was painted by his daughter in 1884 for the Linnean Society; a copy is in the Royal College of Surgeons. Busk died at his home in Harley Street, London.

BIBLIOGRAPHY

I. ORIGINAL WORKS. Early in his career Busk translated three works of other authors: J. J. S. Steenstrap, *On the Alternation of Generations* (London, 1845); A. Kölliker, *Human Histology,* 2 vols. (London, 1853–1854); and *Wedl's Rudiments of Pathological Histology* (London, 1855), which he also edited. Among his own works are *A Catalogue of the Marine Polyzoa in the British Museum,* 3 vols. (London, 1852–1875); sections on Polyzoa in J. MacGillivray, *Narrative of the Voyage of H. M. S. Rattlesnake* (London, 1852), and W. B. Carpenter, *Catalogue of Mazatlan Shells* (London, 1859); *A Monograph of the Fossil Polyzoa of the Crag* (London, 1859); "Parasites, Venomous Insects and Reptiles," in Holme's *Surgery,* 4 vols. (London, 1860–1864), app.; "On the Caves of Gibraltar in Which Human Remains and Works of Art Have Been Found," in *Transactions of the International Congress of Prehistoric Archaeology* (Norwich meeting, 1868); "On a Method of Graphically Representing the Dimensions and Proportions of the Teeth of Mammals," in *Proceedings of the Royal Society,* 17 (1869/70), 544; "Descriptions of the Animals Found in Brixham Cave," in J. Prestwich, *Report on the Exploration of Brixham Cave* (London, 1873); "Report on the Exploration of the Caves of Borneo, Note on the Bones Collected," in *Proceedings of the Royal Society,* 30

(1879/80), 319; and *Report on the Polyzoa Collected by H. M. S. Challenger,* 2 vols. (London, 1884–1886).

The titles of 73 papers are listed in the Royal Society's *Catalogue of Scientific Papers,* I (1867), VII (1877), IX (1891), and XIII (1914). Busk's papers in MSS are in the library of the Royal College of Surgeons, 275.c.1–12; they are listed in *Annual Report of the Royal College of Surgeons* (1930), p. 18.

II. SECONDARY LITERATURE. Obituary notices are *British Medical Journal* (1886), **2**, 346; *Lancet* (1886), **2**, 313; *Nature,* **34** (1886), 387; *Proceedings of the Linnean Society,* **7** (1886), 36; *Quarterly Journal of the Geological Society* (1886); and *The Times* (11 Aug. 1886).

K. BRYN THOMAS

BUTEO, JOHANNES (*b.* Charpey, Dauphine, France, *ca.* 1492; *d.* Romans-sur-Isère, Dauphine, *ca.* 1564–1572), *mathematics.*

Buteo's father, François, seigneur d'Espenel, is said to have had twenty children. Because he did not wish to be a burden to his parents, Buteo entered the Abbaye de St.-Antoine about 1508. He had so much feeling for languages and mathematics, we are told, that he soon could comprehend Euclid in the original Greek. In 1522 he was sent to Paris, where he studied under Oronce Fine. By 1528 he longed for his monastic life and returned to St.-Antoine; he was abbot during two of his years there. In 1562, during the first of the Wars of Religion, he had to leave the monastery and take refuge with one of his brothers in Romans-sur-Isère. He died there of grief and boredom. His original French name was Jean Borrel (*bourreau* means "executioner," but is also a popular name for the buzzard, and in this last sense is translated as *Buteo*). There were such variants as Boteo, Butéon, and Bateon.

Buteo published his works only after he was sixty years old. The *Opera geometrica* contains fifteen articles on different subjects, the last six showing his interest in law through treatment of such mathematical aspects of jurisprudence as division of land and inheritances. The first nine articles treat mechanical, arithmetical, and geometrical problems. The most original is *Ad problema cubi duplicandi,* in which he refutes Michael Stifel's claim of an exact solution to this problem and gives an approximate one.

This is also the main theme of *De quadratura circuli,* in which Buteo refutes the pretensions of those who claimed to have found the solution of the quadrature, most notably those of his master, Oronce Fine. By contrast, he discusses appreciatively the approximations found by Bryson, Archimedes, and Ptolemy. He also mentions two approximate values for π: 3-17/120 (from Ptolemy) and $\sqrt{10}$ (Indian, although he believed it to be Arab).

In the second part of this work, Buteo criticizes errors of many of his contemporaries, particularly in terminological questions. An interesting point is his proof that the author of the proofs of Euclid's *Elements* was not Theon, as was the current opinion, but Euclid himself. Here, too, are the beginnings of the famous dispute involving Peletier, Clavius, and many others on the angle of contact. In the *Apologia* (1562) Buteo pursued his refutation of Peletier's theories.

Buteo's most important work, the *Logistica,* was divided into five books, of which the first two deal with arithmetic, the third deals with algebra, and the last two present many problems in both fields. Terms such as "million" and "zero," and symbols such as p and m for $+$ and $-$ show Italian influence. There is a good treatment of simultaneous linear equations, with notations borrowed from Stifel; and there are approximations to $\sqrt{a}$ and $\sqrt[3]{a}$ influenced by Chucquet through Estienne de la Roche. The work was not practical enough to be reprinted, however.

Buteo's fame rests only on his books. He has been a solitary figure in his love of mathematics and mechanics, and he wanted to be so. As far as we know, he had no pupils; and his criticism, often excessively sharp, must have estranged other mathematicians.

BIBLIOGRAPHY

I. ORIGINAL WORKS. Buteo's works are *Opera geometrica* (Lyons, 1554; reissued 1559. See *British Museum, General Catalogue of Printed Books* for information on reprinted articles.); *Logistica, quae et arithmetica vulgo dicitur in libros quinque digesta . . . eiusdem ad locum Vitruvij corruptum restitutio* (Lyons, 1559, 1560); *Ad locum Vitruvij corruptum restitutio* was reprinted in J. Polenus, *Exercitationes Vitruvianae primae* (Padua, 1739), and M. Vitruvius, *Architectura,* IV, part 2 (Utini, 1825–1830), 37–43; *De quadratura circuli libri duo . . . Eiusdem annotationum opuscula in errores Campani, Zamberti, Orontij, Peletarij, Io. Penae interpretum Euclidis* (Lyons, 1559); and *Apologia adversus epistolam Jacobi Peletarii depravatoris Elementorum Euclidis* (Lyons, 1562).

II. SECONDARY LITERATURE. There is no biography of Buteo. The best sources for information on his life are J. A. de Thou, *Histoire universelle . . . depuis 1543 jusqu'en 1610* (The Hague, 1740), III, 493; and L. Moréri, *Le grand dictionnaire historique* (Paris, 1759; this edition only). G. Wertheim wrote on *Logistica* in *Bibliotheca mathematica,* **2** (1901), 213–219. On *Opera geometrica* and *De quadratura,* see Moritz Cantor, *Vorlesungen über Geschichte der Mathematik,* II (Leipzig, 1913), 561–563, but with the emendations by G. Eneström, in *Bibliotheca mathematica,* **12** (1912), 253.

J. J. VERDONK

IBN BUṬLĀN, ABU'L-ḤASAN AL-MUKHTĀR IBN ʿABDŪN IBN SAʿDŪN (*b*. Baghdad, *ca*. beginning eleventh century; *d*. Antioch, 460/1068), *medicine*.

He was a Christian physician who first practiced in Baghdad. His master, Abu'l-Faraj ibn al-Ṭayyib, was also a Christian. He taught at a hospital founded in Baghdad by ʿAḍud al-Dawla, who held him in high esteem and who made him study a great many medical works. Ibn Buṭlān also knew well Abu'l-Ḥasan Thābit ibn Ibrāhīm al-Ḥarrāni and felt that the latter had taught him most of the practical medicine he knew.

In 440/1049 he left his native city, and came to Fusṭāṭ, Egypt, by way of al-Raḥba, al-Ruṣāfa, Aleppo, Antioch, and Lattaquié. There he met the physician ʿAlī ibn Riḍwān with whom he engaged in sharp controversy. Then he continued on to Constantinople, where the plague was rampant. From there he returned to Antioch. Finally, tired of his wanderings and disappointed by his associations with ignorant people, he retired to a monastery in that city where he remained as a monk until his death.

Ibn al-Qifṭī has preserved for us an account he made of his trip (which was later used by Yāqūt). In it he displays his curious, observing, open-minded character; in particular, his description of Antioch is both interesting and precise (sites, monuments, fortifications). He furnishes us with a specific recollection of the coexistence between Christians and Moslems in Lattaquié, and of the customs practiced in that city. He shows himself to be hungry for contacts with men of learning in all the lands he visited. But it appears that his somewhat difficult, overbearing personality did not make for prolonged relationships. Ibn al-Qifṭī recalls that in Aleppo he was an utter failure with the Christians, whose community he wanted to dominate and whose religious life he wanted to reform.

But it is his controversy with Ibn Riḍwān (excerpts of which have been preserved by Ibn al-Qifṭī) that proves how cunning and tough he was beneath a facade of gentleness. He reminded his adversary that on Judgment Day his patients would demand justice against their poor physicians and that he would have to face his accusers, who would be much more unmerciful than Ibn Buṭlān himself was. He prayed to God that Ibn Riḍwān should be enlightened.

Among the many questions he dealt with, mention can be made of (1) the difficulty of eradicating prejudices and doubts brought about by a purely book-oriented concept of science; (2) the obligation not to condemn the ancients merely by superficial reflection on seemingly contradictory statements: interesting observations on the logic of interpreting texts and the essence of languages and problems of Galen and Aristotle and obvious inconsistencies in works by Aristotle himself; (3) the discussion with a student of Ibn Riḍwān who, in treating everyday fever, practiced purges to treat blood thickness and bleedings to counteract bile; (4) anomalies in the relationship of food and disease to warm and cold climates (i.e., winter and summer), the internal temperature of the body, and why the need to urinate wakes one up when one dreams that this urge has been satisfied, whereas in an erotic dream there is a discharge of sperm during sleep itself.

In developing this theme, Ibn Buṭlān first grapples with questions of physics (the nature of the attraction of iron to magnets), geometry (Euclid's negative definition of the point), an examination of Aristotle's definition of place (if there is no place outside of this world, then the enveloping sphere moves in local motion but not in one single place). He also defended Ḥunayn ibn Isḥāq against the obtuseness of Ibn Riḍwān.

Thus it is clear that Ibn Buṭlān had scientific and philosophical knowledge that extended beyond his knowledge of medicine. Besides Aristotle and Galen he refers to Themistius, Porphyry, and Anebo. He was part of an era that came out of the era of translations, but which, by means of clinical experimentation and observation, sought to verify, extend, and correct the heritage of the Ancients by applying it according to the tradition introduced a century earlier by Rāzī.

BIBLIOGRAPHY

I. Original Works. *Taqwīm al-ṣiḥḥa* ("Health Tables"), trans. into Latin as *Tacuini sanitatis Elluchasem Elimithar medici* (Strasbourg, 1531–1532), and into German by M. Herum as *Schachtafeln der Gesundheit* (Strasbourg, 1532); *Daʿwat al-aṭibbāʾ* ("Vocation of Physicians"), Bassara Zalzal, ed. (Alexandria, 1901); *Tadbīr al-amrāḍ al-ʿāriḍa ʿala'l-akthar bi'l-aghdhiya al-maʾlūfa wa'l-adwiya al-mawjūda yantafiʿu bihā ruhbān al-adyira wa-man baʿuda min al-madīna* ("Diet for Diseases Caused Mainly by Customary Food, and Current Remedies Practiced by Monks in Monasteries and Other Persons Living Far Away From Cities"), in manuscript form; *Risāla fī shirāʾ al-raqīq wa-tqlīb al-ʿabīd* ("Treatise on the Purchase and Examination of Slaves"), instructions for detecting bodily defects in slaves, in manuscript form; *Maqāla fī anna'l-farrūkh aḥarru min al-farkh* ("Dissertation: the Chick Is Warmer Than the Fledgling").

II. Secondary Literature. Ibn Abī Uṣaybiʿa, Müller, ed., I, 241; Ibn al-Qifṭī, Lippert, ed., p. 294; Leclerc, *Histoire de la médecine arabe*, I, 489; H. Derenbourg, *Vie d'Usama b. Munqid* (anecdotes on Ibn Buṭlān's sense of observation, used by E. G. Browne, *Arabian Medicine,* French trans., *La médecine arabe* [Paris, 1933], pp. 81–82); Brockelmann, *Geschichte der arabischen Literatur* (Leiden, 1943) I, 636, Supp. I, p. 885; *Encylopédie de l'Islam,* article on Ibn Buṭlān; M. Meyerhof and J. Schacht, *The Medico-philosophical Controversy Between Ibn Butlan and Ibn Ridwan, a Contribution to the History of Greek Learning Among the Arabs* (Cairo, 1937); I. Krachkowski, *Izbrannie sochinenia,* IV (Moscow, 1957), 266–267.

R. ARNALDEZ

BUTLEROV, ALEKSANDR MIKHAILOVICH (*b.* Chistopol, Kazanskaya [now Tatarskaya, A.S.S.R.], Russia, 6 September 1828; *d.* Butlerovka, Kazanskaya, Russia, 17 August 1886), *chemistry.*

Butlerov's father, Mikhail Vasilievich Butlerov, a retired lieutenant colonel, and mother, Sofia Mikhailovna, owned part of Butlerovka village. Butlerov received his primary education in a private boarding school, later attended a Gymnasium in Kazan, and studied at Kazan University from 1844 to 1849.

Immediately after graduating from the university, Butlerov began teaching chemistry there, at first (1849–1850) part-time, then as Carl Claus's official assistant; from 1852, after Claus's transfer to Dorpat University, he taught all the chemistry courses in the university. Between 1860 and 1863 he was twice rector of the university. From 1868 to 1885 Butlerov was a professor of chemistry at St. Petersburg University. In 1885, after thirty-five years of service, he retired but continued to teach special lecture courses at the university.

In 1852 he married Nadezhda Mikhailovna Glumilina, niece of the writer S. T. Aksakov.

In 1870 Butlerov was selected a junior scientific assistant of the St. Petersburg Academy of Sciences; the following year he became an associate member, and in 1874 a full member. From 1857 he was a member of the Chemical Society of Paris, and from 1869 of the Russian Chemical Society. He was chairman from 1878 to 1882 of the chemistry section of the Russian Physics and Chemistry Society, formed in 1878 by the merging of the chemistry and physics societies. Butlerov was also an honorary or foreign member of the Chemical Society of London (from 1876), the American Chemical Society (1876), the Czech Chemical Society (1880), the German Chemical Society (1881), the Russian Physics and Chemistry Society (1882), the Russian Technical Society (1885), and of many others.

Butlerov became interested in chemistry while still at boarding school. He experimented independently and once, following an explosion in the boarding school kitchen, was placed in a punishment cell, from which he was led to dinner with a board inscribed "Great Chemist" tied to his chest. This ironic inscription proved to be prophetic. At Kazan University, Butlerov studied chemistry under N. N. Zinin, as well as Claus, the discoveror of ruthenium. Under Zinin's influence Butlerov decided to devote himself to chemistry; at home he built a laboratory, where he prepared isatin, alloxazine, and other organic compounds. After Zinin left for St. Petersburg, however, Butlerov devoted his energies to another of his interests, entomology. His thesis, for which he received the degree of candidate of natural sciences, was published as *Dnevnie babochki Volgo-Uralskoy fauny* ("Diurnal Butterflies of the Volga-Ural Fauna"). He had collected the material for the thesis during his excursions around Kazan and during a trip to the steppes on the east bank of the Volga River and near the Caspian Sea in the spring and summer of 1846.

As early as 1851 Butlerov defended his master's dissertation, "Ob okisleny organicheskikh soedineny" ("On the Oxidation of Organic Compounds"). On the whole, this work was a historical survey. It includes few original thoughts, although his remarks that isomerism is based on molecular structure and that changes in chemical characteristics are associated with structural changes are worthy of mention. Butlerov included with his master's dissertation his first experimental work on the oxidizing action of osmic acid on organic compounds. His doctoral dissertation, "Ob efirnykh maslakh" ("On Essential Oils"), which he defended at Moscow University in 1854, was also mainly a historical survey. Both of these remained in manuscript form and were not published until 1953 in Volume I of his *Sochinenia* ("Works").

Butlerov's teaching ability immediately attracted the attention of both his students and his colleagues, but initially he taught by lecture only—work in the laboratory was not required of his students, and he himself worked there only sporadically. Until 1857, Butlerov devoted much more time to experiments set up in his greenhouses and in fields. He reported on this research in a great many articles and notices, most of which were published in the *Zapiski* ("Notes") of the Kazan Economic Society in which he was active. In this period he made unsuccessful attempts to build a soap factory and to improve the production of phosphorus matches.

In the early 1850's Butlerov adhered to obsolete theoretical views (he, as well as Claus, taught chemistry from a textbook by C. Löwig, the author of one version of the theory of radicals); but in 1854, on Zinin's advice, he familiarized himself with the work of Laurent and Gerhardt and became one of their passionate supporters. However, the greatest changes in his work and thought resulted from his trip abroad in 1857–1858. During his travels Butlerov met such eminent young chemists as Kekulé and Erlenmeyer and spent about half a year in Paris, participating in the meetings of the Paris Chemical Society, which had just been formed, and working for two months in Wurtz's laboratory.

Markovnikov, in his reminiscences of Butlerov, gives this evaluation of the significance of Butlerov's trip:

> He did not have to finish his education, as did most of those [Russians] sent abroad. He had to see, rather, how scientific experts worked, to observe the origin of ideas and to enter into intimate relations with these ideas, which the scientists readily exchanged in personal conversations . . . that were often held privately and not committed to print. . . . With a basic reserve of scientific knowledge, and possessing absolute fluency in French and German, he had no difficulty standing on an equal footing with the young European scientists, and owing to his outstanding abilities, choosing the correct direction [*Zhurnal Russkogo fiziko-khimicheskogo obshchestva*, **19** (1887), supp., 76].

In Wurtz's laboratory Butlerov began his first series of experimental investigations. Discovering a new way to obtain methylene iodide, he studied many derivatives of methylene and their reactions. As a result, he was the first to obtain hexamethylenetetramine (urotropine) and a polymer of formaldehyde which in the presence of limewater is transformed into a sacchariferous substance (containing, as was established by E. Fischer, α-acrose). This was the first complete synthesis of a sacchariferous substance.

On the other hand, Butlerov did not succeed in obtaining either dihydroxymethylene, $CH_2(OH)_2$, or the free methylene radical itself, instead of which he obtained its dimer—ethylene. However, both of these negative results served as material for future generalizations. These investigations showed the trait characteristic of Butlerov's work, the effort to study a reaction in full, not neglecting its by-products. They were usually completed with very small quantities of the substances involved and enabled him to perfect his skill in experimentation.

Work on the methylene series ended in 1861, when Butlerov stated the basic ideas of the theory of chemical structure and directed his experimental investigations toward the verification and support of his new theory. He arrived at the theory of chemical structure through continuous research and a recognition of the unsatisfactory state of theoretical chemistry. Although he developed a theory of types similar to Gerhardt's,

defended it in print, and on returning from abroad employed it as the basis of a lecture course in organic chemistry, he clearly recognized that he must go beyond Gerhardt.

Butlerov attempted to develop Dumas's theory of carbonaceous types, but all conventional viewpoints proved unsatisfactory for the explanation of addition reactions, which he had come across in describing the results of his work on the methylene series. Summing up his research, he arrived at the theory of chemical structure, which, according to Markovnikov, he began to expound in his lectures as early as 1860.

At the end of the 1850's and the beginning of the 1860's, the theoretical side of chemistry did not correspond to the sum of its empirical data and knowledge. Kekulé, Wurtz, and the majority of other chemists adhered to the theory of polyatomic radicals, which was a further development of Gerhardt's theory of types; Kolbe and his school developed a unique theory of carbonaceous types; and Berthelot used "formation equations." Several chemists—for example, Kekulé (in 1861)—began in despair to reject rational formulas, based on one or another theoretical representation, and turned to empirical formulas. At precisely this juncture Butlerov read his paper "O khimicheskom stroeny veshchestv" ("On the Chemical Structure of Substances") at the chemical section of the Congress of German Naturalists and Physicians in Speyer (September 1861).

In this paper (*Sochinenia*, I, 561), Butlerov defined the concept of chemical structure: "Assuming that each *chemical* atom is characterized by a specific and limited quantity of chemical force [affinity], with which it participates in the formation of a substance, I would call this chemical bond or [this] capacity for the mutual union of atoms into a complex substance *chemical structure*."

From this definition it follows that the concept of chemical structure (the term is found in the work of Russian chemists before Butlerov, but it is used in another sense) could be brought forward only after there had been a sufficiently clear definition of the concepts "atom" (the attribute "chemical" left open the question of the possibility of its further separation into "physical" atoms), "valency" (the quantity of an atom's affinity), and "interatomic bond." Thus, the following can be considered as the preconditions for the existence of a theory, within chemistry itself, of chemical structure: (1) sufficiently clear concepts of atomic theory and molecular theory—which was achieved at the Congress of Chemists in Karlsruhe (1860); (2) development of the study of valency in the form ascribed to valency by Kekulé (1857–1858); (3) the creation of the concept of interatomic bond, as it was formulated in the works of Kekulé and Couper (1858).

Butlerov advanced the basic proposition of the classical theory of chemical structure:

> I consider it possible, for the time being, to change the well-known rule—to wit, that the nature of a compound molecule is determined *by the nature, quantity, and arrangement* of elementary component parts—in the following manner: *the chemical nature of a compound molecule is determined by the nature of its component parts, by their quantity, and by their chemical structure* [*Sochinenia,* I, 70].

This proposition, as is evident from its wording, broke with the traditional view that the properties of molecules are determined principally by the nature of the space grouping of atoms in the molecules, by the relative position of the atoms, and by the distances separating the atoms; these problems could not be studied by methods then available. All the remaining propositions of the classical theory of chemical structure are directly or indirectly associated with this proposition.

Butlerov noted means for determining the chemical structure of molecules and formulated the rules that should be followed in this determination. He gave primary importance to those synthetic reactions in which the participating radicals retain their chemical structure. He foresaw the possibility of regrouping but believed that after a detailed study of matter from the point of view of chemical structure, the general laws for regrouping would be deduced.

Leaving open the question of the preferred structural formulas, Butlerov explicitly expressed his opinion about their sense: When the general laws of the relationship between the chemical properties of substances and their chemical structure became known, the corresponding formula would be an expression of all these properties.

The only incorrect proposition in Butlerov's paper was the supposition concerning the possibility of a primordial (i.e., inherent in free atoms) difference in units of affinity (valency). In connection with this hypothesis Butlerov was the first to produce a model of a carbon tetrahedron (it was irregular). Having subjected the hypothesis to experimental verification and having rejected it, he further developed the theory of chemical structure in a long article, "Über die verschiedenen Erklärungsweisen einiger Fälle von Isomerie." However, the propositions stated in the article were implied in his paper delivered at Speyer.

Guided by the propositions he had formulated, Butlerov explained the existence of isomerism, stating that isomers were compounds possessing the same elementary composition but different chemical structure. Discovery of the facts of isomerism, which did not correspond to this definition, led to the establishment by van't Hoff and Le Bel of stereochemistry, which Butlerov did not accept immediately and, when he did, only in part; specifically, he accepted only the explanation of the optical activity of organic compounds as the result of the presence of asymmetric carbon atoms.

Butlerov explained the relationship of the properties of isomers—and of organic compounds in general—to their chemical structure by the existence of "the mutual influence of atoms," which is transmitted along the bonds; as a result of this influence, atoms possess different "chemical values" depending on their structural environment. This general proposition was given concrete expression in the form of many "rules" by Butlerov himself and, especially, by his students Markovnikov and Popov. In this century these rules, as well as the whole concept of atoms' mutual influence, have received an electron interpretation.

Of great importance for the consolidation of the theory of chemical structure was its experimental corroboration in the work of Butlerov's school. Butlerov himself deserves credit for the prediction and proof of positional and skeletal isomerism. Having unexpectedly obtained tertiary butyl alcohol, he was able to decipher its structure and predicted (later proving, with the aid of his students) the existence of its homologues; he also predicted (1864) the existence of two butanes and three pentanes and, later, that of isobutylene. The formulas of his two butanes were represented as follows:

$$\begin{cases} CH_3 \\ CH_2 \\ CH_2 \\ CH_3 \end{cases} \quad \text{and} \quad CH \begin{cases} CH_3 \\ CH_3 \\ CH_3 \end{cases}$$

As early as 1866 Butlerov reported the synthesis of isobutane. Regarding this he wrote (*Sochinenia,* I, 199): "The principle of chemical structure . . . can now serve as the best guide in researching questions related to isomerism. Using this principle, one can predict phenomena that could be neither predicted nor explained by prior [theoretical] viewpoints."

In the second half of the 1860's the nature of unsaturated compounds was still unexplained. A series of investigations conducted by Butlerov, completed at the beginning of the 1870's, led to a conclusion supporting the hypothesis that they contain multiple bonds.

Butlerov's indication that sulfur had a valence of six and his experimental proof of the tetravalence of lead must be considered contributions to the theory of valency. Throughout the 1860's he gave much attention to organometallic compounds and developed methods, widely used by his school, for synthesizing organic zinc compounds.

In order to promulgate the theory of chemical structure throughout organic chemistry, Butlerov published *Vvedenie k polnomu izucheniyu organicheskoy khimy* ("An Introduction to the Complete Study of Organic Chemistry"), the second edition of which was published in German under the title *Lehrbuch der organischen Chemie*. E. von Meyer, a student of Kolbe's school (which rejected the theory of chemical structure), considered the work to be a magnificent textbook on organic chemistry that greatly influenced the development and popularization of the structural theory.

In 1867–1868 Butlerov went abroad to aid in the publication of the German edition of his book; he traveled to Algiers for a rest but nearly perished on the way; the ship encountered a violent storm and was off course for several days, out of control and half swamped. There was another goal of his trip, however. The official decree concerning his mission stated that a purpose of the voyage was to enable him to explain to foreign chemists his right to major participation in the development of contemporary chemistry.

Thus, Butlerov's trip was connected with the defense of his priority. Until then there had been a tendency to credit the creation of structural theory to Kekulé (Couper's name was not yet known), for in 1857–1858 he had stated the theoretical propositions that served as the preconditions for the emergence of the theory of chemical structure, and in 1865 he had quite successfully extended the theory to aromatic compounds. Formulas representing conclusions drawn on the basis of several of Kekulé's (and Couper's) valency rules coincided with structural theory (in 1868 L. Meyer stated this viewpoint very clearly in opposition to Butlerov's attempts to defend his priority), while within the framework of this theory, the deduction of formulas was based on the study of the properties of the relevant molecules, as well as on the valency of atoms. It was forgotten, however, that in his textbook and in many magazine articles, Kekulé used the "theory of polyatomic radicals" even after 1861 and that, having changed his views in 1864, after Butlerov's criticism, stated that in his textbook he "had always given preference to one form of rational formulas, specifically, to the one that reflects the views concerning atom bonds as the method of forming molecules" [*Lehrbuch der organischen Chemie* . . ., II, pt. 2 (1864), 244–245].

Markovnikov immediately spoke out against this historically incorrect contention, but the legend was nonetheless created. Kekulé gave the hint, L. Meyer developed it, and Schorlemmer reproduced it in *The Rise and Development of Organic Chemistry* (London, 1879). In his *Benzolfest* (1890) Kekulé told how the structural theory came to him as he rode on the top of a London bus in 1857 or 1858. Since then, this elegant tale has fluttered along the pages of histories of chemistry and of anniversary articles, although long ago Markovnikov indicated Butlerov's basic role in the creation and initial development of the classic theory of chemical structure.

Butlerov's second great service—this time to chemistry in Russia—was the creation of the first Russian school of chemists. After his return from abroad in 1858, he equipped his laboratory with gas and expanded it; his students had to complete required practical work; and his first "disciples" appeared. Of these, V. V. Markovnikov, A. M. Zaytsev, and A. P. Popov occupied professorial chairs in universities during Butlerov's lifetime. Nonetheless, in the 1860's Butlerov sought to leave Kazan. One reason for this was his unsuccessful term as rector. In March 1860 he had become the last "crown" (i.e., appointed by the imperial government) rector of Kazan University; however, striving not only to institute liberal changes but also to halt student abuse of individual teachers, Butlerov came into severe conflict with the student body. This forced him to request retirement, which was granted in August 1861. Nonetheless, in November 1862 Butlerov became—against his wishes—the first elected rector of the university. The outbreak of a struggle between groups of professors and Butlerov's clash with a trustee led to his retirement in July 1863. He was bitter about the experience and tried to find a position outside Kazan. Only the insistence of his friends (as well as the birth of a son in April 1864) stopped him from departing immediately.

In May 1868 Butlerov was made professor of chemistry at St. Petersburg University. He continued teaching there until 1885, when he retired on pension but continued to give special lecture courses. His followers at St. Petersburg form a prominent group of Russian organic chemists—the most famous being A. E. Favorski and I. L. Kondakov. At various times G. G. Wagner, D. P. Konovalov, and F. M. Flavitsky worked in his laboratory. Butlerov's outstanding characteristic as an instructor was that he taught by example; the students could always observe what he was doing and how he was doing it.

Butlerov was an advocate of higher education for women; he participated in the organization of university courses for women (1878) and lectured to them on inorganic chemistry. He also created laboratory courses in chemistry. In addition, Butlerov delivered in St. Petersburg, as he had earlier done in Kazan, a large number of public lectures, most of which had a chemical-technical basis.

Surprisingly, election to the Academy of Sciences hardly aided Butlerov's scientific activity, since the condition of its laboratory was so deplorable that not until 1882, after thorough repairs, could he transfer his experimental work there. A struggle for the right of Russian scholars to recognition of their service by the academy weakened Butlerov's position within the academy. For example, in 1874 Butlerov and Zinin failed in their advocacy of Mendeleev's candidacy for membership. In November 1880 Mendeleev was again nominated by Butlerov to the seat that became vacant after Zinin's death. The second blackballing of Mendeleev elicited a storm of indignation throughout the nation. At the same time, the faction within the academy that had prevented Mendeleev's election proposed their own candidates for membership and for academic prizes. Thus, in January 1882, after Mendeleev's failure to be elected Beilstein was nominated. Seeing Beilstein as a protégé of that same anti-Russian faction, Butlerov energetically opposed his candidacy and succeeded in depriving Beilstein of the requisite number of votes. In order to attain this end, however, Butlerov was forced to turn directly to public opinion, publishing in the Moscow newspaper *Rus* a long article with the provocative title "Russkaya ili tolko Imperatorskaya Akademia nauk v St.-Petersburge?" ("[Is There] a Russian or Merely an Imperial Academy of Sciences in St. Petersburg?").

The work of Butlerov and his students can be classified as follows:

(1) Research designed to confirm the theory of chemical structure and to synthesize theoretically possible isomers. Trimethyl acetic acid was obtained, its genetic connection with pinacolone alcohol was established, and the correct structure of pinacolone alcohol and pinacol were given. In addition, Butlerov gave a general schema of pinacolin rearrangement. Concluding this set of experiments, pentamethyleth-anol was synthesized.

(2) Investigation of polymerization reactions. The possibility of the polymerization of ethylene (unsuccessfully), propylene, butene-2, and isobutylene was studied. The mechanism of the polymerization of isobutylene was especially carefully studied, since the polymerization stops at the dimer and trimer stages; Butlerov sought to study "the simplest instance of pure and, perhaps, less complex polymerization of the ethylene series of hydrocarbons." His students simultaneously studied the polymerization of amylene.

Thus, Butlerov was the first to begin the systematic study of the mechanism of polymerization reactions based on the theory of chemical structure; this was continued in Russia by his successors and was crowned by S. V. Lebedev's discovery of the industrial means of producing synthetic rubber.

(3) Secondary results of the study of polymerization. During the attempts to polymerize ethylene, the conditions were found under which it could be hydrated to obtain ethanol. Studying the polymerization of isobutylene and amylene, and having found their isomers in the reaction mixture, in 1876 Butlerov generalized that a dynamic equilibrium can exist between two isomeric forms. These ideas concerning reversible equilibrium isomerizations are found in Butlerov's work as early as 1862–1866. Subsequently, C. Laar proposed the term "tautomerism" for these phenomena, but his representation of the mechanism of reversible isomerism proved to be less correct than Butlerov's.

(4) Random research in chemistry (organic, inorganic, and physical) and even in physics. Thus, not considering the a priori rejection of Prout's hypothesis to be possible, Butlerov arrived at the assumption of the possibility of changes in atomic weights—for example, under the influence of luminous rays. The experiments undertaken did not, of course, yield positive results.

To Butlerov's St. Petersburg period belong his statements defending and substantiating the theory of chemical structure, from an epistemological point of view, against attacks by his colleagues Menschutkin and, to a lesser degree, Mendeleev. His speech "Sovremennoe znachenie teory khimicheskogo stroenia" ("The Contemporary Significance of the Theory of Chemical Structure," 1879) and his article "Khimicheskoe stroenie i teoria zameshchenia" ("Chemical Structure and the Theory of Substitution," 1885) include such statements. Against Menschutkin's positivist orientation, which in this instance followed Berthelot, Butlerov advanced the thesis that chemists had not only the right, but also the responsibility, to speak of molecules and atoms as if speaking of things that in fact exist and, doing this, to preserve the conviction that this belief would not become a baseless abstraction.

A course of lectures ("A Historical Essay on the Development of Chemistry in the Last Forty Years"), given by Butlerov in 1879–1880 at St. Petersburg University, served as partial substantiation of the theory of chemical structure. In the end, Menschutkin, who was Butlerov's successor in the chair of organic chemistry at St. Petersburg, changed his position, supporting the theory of chemical structure.

Butlerov was the organizer and propagandist for scientific apiculture in Russia. He published many articles and notices in the Russian and foreign press, and in 1886 founded the *Russkii pchelovodnyi listok* ("Russian Apiculture Leaflet").

While in St. Petersburg, Butlerov had yet another unusual interest—spiritualism. He was convinced that "medium" phenomena could be studied by scientific methods and even spoke on this theme at the seventh conference of Russian naturalists (Odessa, August 1883). However, experiments with mediums, conducted in the presence of a scientific commission, ended in complete failure. Mendeleev, who participated in the commission, later wrote, "Our spiritists obviously do not see deception." P. D. Boborykin, a student of Butlerov's at Kazan University, defined his passion as an "atavism of religiosity."

BIBLIOGRAPHY

I. ORIGINAL WORKS. Butlerov's writings include *Vvedenie k polnomu izucheniyu organicheskoy khimy* ("Introduction to the Complete Study of Organic Chemistry"), 3 pts. (Kazan, 1864–1866), trans. into German as *Lehrbuch der organischen Chemie zur Einführung in das specielle Studium derselben*, 4 pts. (Leipzig, 1867–1868); *Stati po mediumizmu* ("Articles on Mediumism"; St. Petersburg, 1889); *Stati po pchelovodstvu* ("Articles on Apiculture"; St. Petersburg, 1891); *Izbrannye raboty po organicheskoy khimy* ("Selected Works on Organic Chemistry"), in the series Klassiki Nauki (Moscow, 1951), with a bibliography of his chemical works; *Nauchnaya i pedagogicheskaya deyatelnost. Sbornik dokumentov* ("Scientific and Pedagogic Activity. A Collection of Documents"; Moscow, 1961), with a list of the archives where Butlerov's MSS are preserved; and *Centenary of the Theory of Chemical Structure. Collection of Papers of A. M. Butlerov, A. S. Couper, A. Kekulé, and V. V. Markovnikov* (Moscow, 1961). His writings are collected in *Sochinenia* ("Works"), 3 vols. (Moscow, 1953–1958); Volume III includes a complete bibliography of his works.

II. SECONDARY LITERATURE. Works on Butlerov are *A. M. Butlerov, 1828–1928* (Leningrad, 1929), a collection of articles on Butlerov; G. V. Bykov, *Istoria klassicheskoy teory khimicheskogo stroenia* ("History of the Classical Theory of Chemical Structure"; Moscow, 1960); "La correspondance des chimistes étrangers avec A. M. Butlerov," in *Archives internationales d'histoire des sciences,* 14 (1961), 85–97; *Aleksandr Mikhailovich Butlerov* ("A Sketch of his Life and Activity"; Moscow, 1961); and "The Origin of the Theory of Chemical Structure," in *Journal of Chemical Education,* 39 (1962), 220–224; G. V. Bykov and J. Jacques, "Deux pionniers de la chimie moderne, Adolphe Wurtz et Alexandre M. Boutlerov, d'après une correspondance inédite," in *Revue d'histoire des sciences,* 13 (1960), 115–134; G. W. Bykow und L. M. Bekassowa, "Beiträge zur Geschichte der Chemie der 60-er Jahre des XIX. Jahrhunderts. I. Briefwechsel zwischen E. Erlenmeyer und A. M. Butlerow (von 1862 bis 1876)," in *Physis,* 8 (1966), 179–198; "II. F. Beilsteins Briefe an A. M. Butlerow," *ibid.,* 267–285; "III. Die im Briefform verfasste Chronik der Herausgabe eines Lehrbuchs für Chemie," *ibid.,* 10 (1968), 5–24; G. V. Bykov and L. V. Kaminer, *Literatura ob A. M. Butlerove i po istory klassicheskoy teory khimicheskogo stroenia* ("Literature on A. M. Butlerov and the History of the Classical Theory of Chemical Structure"; Moscow, 1962); W. N. Dawydoff, *Über die Entstehung der chemischen Structurlehre unter besonderer Berücksichtigung der Arbeiten von A. M. Butlerov* (Berlin, 1959); J. Jacques, "Boutlerov, Couper et la Société Chimique de Paris (notes pour servir à l'histoire des théories de la structure chimique," in *Bulletin de la Société chimique de France* (1953), 528–530; H. M. Leicester, "Alexander Mikhailovich Butlerov," in *Journal of Chemical Education,* 17 (1940), 203–209; and "Contributions of Butlerov to the Development of Structural Theory," *ibid.,* 36 (1959), 328–329; "Pisma russkikh khimikov k A. M. Butlerovu" ("Letters of Russian Chemists to A. M. Butlerov"), in *Nauchnoe nasledstvo* ("Scientific Heritage"), Vol. IV (Moscow, 1961); and *Zhurnal Russkogo fizikokhimicheskogo obshchestva,* 19, supp. (1887), devoted to speeches and articles by G. G. Gustavson, A. M. Zaytsev, V. V. Markovnikov, and N. A. Menschutkin in memory of Butlerov.

G. V. BYKOV

BÜTSCHLI, OTTO (*b.* Frankfurt am Main, Germany, 3 May 1848; *d.* Heidelberg, Germany, 2 February 1920), *zoology, mineralogy.*

Bütschli was the son of a confectioner whose family had come from Switzerland several generations earlier. As a youth his scientific appetite was whetted by the lectures and the museum of the Senkenbergische Naturforschende Gesellschaft in Frankfurt, which also inspired his fellow townsmen August Weismann and Richard Goldschmidt. Bütschli studied mineralogy, chemistry, and paleontology at the Karlsruhe Polytechnic Institute and in 1865–1866, at the age of seventeen, was assistant to the paleontologist Karl von Zittel. Next he traveled to Heidelberg, where he finished his doctorate in mineralogy, with minors in zoology and chemistry, in 1868.

During his year of required military service, Bütschli had decided that his true interest was zoology, so he went to Leuckart's laboratory in Leipzig for a semester of advanced study after he left the army in 1869. In the fall of the same year Bütschli returned to Frankfurt and was recalled to military service when the Franco-Prussian War broke out in 1870. During this period his first cytological paper, on the structure of the germ strings of insects, appeared. After the war Bütschli went to Kiel, where he was assistant to the taxonomist Möbius for two years; among his works completed there is a monograph on free-living nematodes. These organisms permitted detailed study of the living cell, and Bütschli used them successfully for many years.

Dissatisfied with the situation in Kiel and desiring greater freedom for his promising cytological researches, Bütschli returned to Frankfurt in 1873 and spent the next three years working on the problems of cell division, fertilization, and the conjugation of the ciliates. In 1876 he presented the results of these studies in a thesis for admission as *Privatdozent* at the Karlsruhe Polytechnic Institute. At the age of thirty Bütschli was named professor of zoology and paleontology at Heidelberg. Here he remained for the rest of his life, declining attractive positions elsewhere; he retired in 1918, at the age of seventy, and spent the last two years of his life working on a textbook that he did not finish before his death from influenza.

Bütschli's monograph of 1876, "Studien über die ersten Entwickelungsvorgänge der Eizelle, die Zelltheilung und die Konjugation der Infusorien," was important as a pioneer work in the development of several areas of cytology and cell theory. In it he was the first to identify and order sequentially the stages of nuclear division in several types of animal cells, simultaneously with Strasburger's work on the division of plant cells and several years prior to Flemming's studies on mitosis. Bütschli demonstrated that the polar bodies of eggs arise through atypical cell division, and in studying fertilization he was the first to describe the fertilization cone and to prove that normally only one sperm enters the egg.

The monograph was in press when Oscar Hertwig's paper on fertilization in the sea urchin appeared, and Bütschli added an appendix in which he discussed the significance of Hertwig's discoveries in relation to his own. Hertwig had succeeded where Bütschli had failed, in following the sperm from penetration to karyogamy. However, Bütschli's description of many of the stages of fertilization and cleavage were more complete, at least partially because of differences in material. He clearly illustrated the fusion of male and female pronuclei in the eggs of snails, but did not recognize the true identity of the male pronucleus because of confusion created by the presence of the several micronuclei that make up the female pronucleus and gradually fuse before karyogamy.

The illustrations in this monograph are remarkable, especially when one considers that all of the observations were made on unstained material treated only with one percent acetic acid. (The use of carmine as a biological stain did not become common until a few years later.) Bütschli clearly illustrated dividing cells and noted what he called rodlets (chromosomes) that made up the "nuclear plate," which, in his view, divided into halves at the climax of nuclear division. The failure to illustrate nuclear-division stages earlier than metaphase was possibly due to the lack of staining techniques. Bütschli's illustrations of the zygotene "bouquet" stage and of diakinesis during the first meiotic division of spermatogonia in the roach were excellent.

The second half of this monograph was devoted to the conjugation of the ciliates. In these organisms, unlike most other animals, reproduction and the sexual processes are not closely associated. Bütschli was the first to recognize this and to demonstrate that conjugation was not a reproductive process per se, but a sexual reorganization of the cell similar to fertilization. In his view, both conjugation and fertilization brought about a rejuvenation of the cell, either through some effect of the conjugation process or through the acquisition of new nuclear material. This idea formed the basis of subsequent work on the ciliate life cycle by Maupas, R. Hertwig, Calkins, and others. Bütschli recognized that the micronuclei of ciliates were comparable with the nuclei of other cells and that they divided by similar processes. He was thus able to prove that these organisms were unicellular animals and to disprove the earlier views of their reproduction, which were based upon a false analogy between the ciliates and the Metazoa. Instead of comparing the ciliate with the whole metazoan organism, Bütschli compared it with the cell of the metazoan.

Bütschli drew many conclusions from these studies, some of which were well ahead of his time. Above all, he argued that cells were physicochemical systems and that cellular phenomena could best be understood in physical-chemical terms. As he said in the preface to his 1876 monograph:

Morphology composes only a part of the nature of organic forms. Each form must be capable of being explained in itself from given bases and influences. Only when it is shown that one organic form proceeds from another, and when the conditions of this appearance are known—as is hardly true in a single case today—is a material present in which it might be possible to seek a causal-mechanical explanation [of the phenomenon]. . . . If we conceive of the elementary organisms [cells] as the building blocks of morphology, our understanding of the elementary organisms is altered, for the type of morphological observation of [multi]cellular organisms [previously] used loses its justification and physiological methods come to the fore. The phenomena on and in the elementary organisms can be more precisely stated only through a knowledge of the physical-chemical conditions of their appearance and disappearance [pp. 1–2].

In keeping with this conviction, he proposed a model for cell division based upon changes in surface tension.

It is perhaps strange that Bütschli did no further work in nuclear cytology, especially since the field was making rapid advances at the time; however, his deepest interest was in the cytoplasm as a physical-chemical system. He thus devoted his first ten years at Heidelberg to a detailed study of the protozoa, for he felt that life processes and the nature of protoplasm could be studied in unicellular organisms better than in Metazoa. These studies culminated in a three-volume monograph (1880–1889), a critical review of the whole field that included much original work by Bütschli himself—he checked many dubious points—and a complete bibliography of all previously published work. These volumes, which remain a basic reference work, contain many important theoretical discussions, such as Bütschli's hypothesis of the sexuality of the ciliates, and are still of considerable taxonomic importance.

During the years spent on the protozoan studies, Bütschli approached the structural problems of protoplasm more closely through extensive observation of the protoplasm of rhizopods and ciliates. These studies eventually led to the "alveolar theory of protoplasm," which he first discussed in print in 1889 and summarized in *Untersuchungen über microskopische Schäume und die Struktur des Protoplasmas* (1892). In Bütschli's view, protoplasm had an alveolar structure, being composed of a two-phase system similar to an emulsion. In keeping with his fundamental mechanistic conviction, Bütschli felt that many of the properties of living protoplasm could be explained on a physical-chemical basis, and thus he spent much time experimenting with various emulsions of oil and soapy solutions of mineral salts that appear to duplicate the physical properties of living protoplasm to a remarkable degree. This work represented a great advance, for it was the first theory of protoplasmic structure to have a comprehensive observational and experimental basis, and at the same time it provided a physical-chemical model of the underlying phenomena of colloidal materials. This work was, of course, limited by the techniques and concepts of the time; however, more recent knowledge of protoplasm gained by electron microscopy and by colloid and protein chemistry does not detract from Bütschli's pioneer study, which in respect to observation, analysis, and synthesis remains a model of

scientific work in biology. With this work Bütschli felt that he had traced to the root the problems he had posed for himself as a young man.

Following his extensive work on protoplasm, Bütschli proceeded to examine a great number of organic and inorganic materials of a colloidal nature. These studies were published mainly in *Untersuchungen über Strukturen* (1898), which included 300 photomicrographs. This pioneer use of photomicrography was a reflection of Bütschli's effort to present his observations as realistically and precisely as possible. These studies, together with the work on protoplasm, provided one of the bases of the subsequent development of colloid chemistry. After finishing *Untersuchungen über Strukturen,* Bütschli turned to the analysis of many mineral materials of biological origin: shells of mollusks, crustaceans, and echinoderms; tests of Foraminifera; and the spicules of sponges. In 1900 he published a short book on the structure of coagulated sulfur and a monograph on the structure of natural and synthetic silica gels.

At the age of sixty Bütschli returned to purely zoological work and began to prepare a comprehensive textbook of comparative anatomy, which had long been one of his secret loves. This work was to include both vertebrates and invertebrates; although more than a thousand pages of the work were finally published, Bütschli did not live to finish this ambitious undertaking.

BIBLIOGRAPHY

I. ORIGINAL WORKS. Complete bibliographies of Bütschli's works are in *Die Naturwissenschaften,* **7** (1920), 567–570; and *Sitzungsbericht der Heidelberger Akademie der Wissenschaften,* Kl. B (1920), 13–19. His works include "Studien über die ersten Entwickelungsvorgänge der Eizelle, die Zelltheilung und die Konjugation der Infusorien," in *Abhandlungen der Senkenbergische naturforschende Gesellschaft,* **10** (1876), 1–250; *Protozoa,* I, pts. 1–3, in H. G. Bronn's *Klassen und Ordnen des Thier-Reichs* (Leipzig, 1880–1889); *Untersuchungen über microskopische Schäume und die Struktur des Protoplasmas* (Leipzig, 1892), trans. by E. A. Minchin, A. Black, and C. Black as *Investigations on Microscopic Foams and on Protoplasm* (London, 1894); *Untersuchungen über Strukturen, insbesondere über Strukturen nichtzelliger Erzeugnisse des Organismus und über ihre Beziehungen zu Strukturen welche ausserhalb des Organismus entstehen* (Leipzig, 1898); *Vorlesungen über vergleichende Anatomie,* 2 vols. (Leipzig, 1910–1925); and an untitled autobiography with the running head "Das Lebenswerk Otto Bütschli's," in *Sitzungsbericht der Heidelberger Akademie der Wissenschaften,* Kl. B (1920), 1–12.

II. SECONDARY LITERATURE. Works on Bütschli are H. Freundlich, "Otto Bütschli als Kolloidchemiker," in *Die Naturwissenschaften,* **7** (1920), 562–564; Richard B. Goldschmidt, "Otto Bütschli 1848–1920," *ibid.,* 543–549; "Otto Bütschli, Pioneer of Cytology (1848–1920)," in *Science, Medicine, and History* (Oxford, 1953), pp. 223–232; V. Goldschmidt, "Otto Bütschli's Verhältnis zur Kristallographie und Mineralogie," in *Die Naturwissenschaften,* **7** (1920), 564–567; Clara Hamburger, "Otto Bütschli als Protozoenforscher," *ibid.,* 559–561; Max Hartmann, "Otto Bütschli und das Befruchtungs- und Todproblem," *ibid.,* 555–558; L. Rhumbler, "Otto Bütschli's Wabentheorie," *ibid.,* 549–555; and Josef Speck, "Über Bütschli's Erklärung der karyokinetischen Figur," *ibid.,* 561–562.

JAMES D. BERGER

BUYS BALLOT, CHRISTOPH HENDRIK DIEDERIK (*b.* Kloetinge, Netherlands, 10 October 1817; *d.* Utrecht, Netherlands, 3 February 1890), *meteorology, physical chemistry.*

The son of Anthony Jacobus Buys Ballot, Dutch Reformed minister, and Geertruida Françoise Lix Raaven, Buys Ballot attended the Gymnasium at Zaltbommel and the *Hogeschool* (now University) of Utrecht, where he was active in student affairs before receiving his doctorate in 1844. He became lecturer in mineralogy and geology at Utrecht in 1845, and in 1846 he added theoretical chemistry. In 1847 he was appointed professor of mathematics, and from 1867 until his retirement in 1888 he was professor of physics. In 1854 he founded the Royal Netherlands Meteorological Institute (K.N.M.I.), a world center for atmospheric research, whose chief director he remained until his death. A deeply religious man, noted for his proverbial modesty, Buys Ballot became a prominent lay leader of the Walloon church. He was twice married; five of his eight children survived him. He was elected to the Royal Academy of Sciences of Amsterdam in 1855 and to the Royal Belgian Academy. He was decorated by the Dutch, Austrian, and Prussian governments. As a teacher he wrote textbooks in chemistry, mathematics, and physics.

Although he is best known for the law to which he gave his name, Buys Ballot's principal accomplishments were the shape he gave (with others) to the field of meteorology in its formative years. He started as a chemist and shifted to meteorology when his speculations on the relation between molecular structure and the properties of matter (put forward in 1843 but not published until 1849) were badly received by his teachers for lack of experimental foundations. In meteorology, which was growing in importance as the spread of the telegraph made synoptic observations possible, Buys Ballot labored unceasingly for the widest possible network of simultaneous observations, which he published in a series of yearbooks beginning in 1851. He was a leader of the international meteorological cooperation that began with the Brussels Conference in 1853, and he served as chairman of the International Meteorological Committee from its founding at the Vienna Congress in 1873 until 1879. He also was responsible for Dutch participation in the International Polar Year.

His contributions to meteorology were twofold. First, he suggested that only the deviations from the mean state were important for understanding. Second, in spite of his stated devotion to the motto *Sine hypothesi scientia nulla,* his research consisted chiefly of examining long time series for regularities that he was more concerned to establish than to interpret. This overwhelming preoccupation with data gave his papers a strongly Baconian flavor; he left to others the development of the theoretical side of meteorology, to which his training might have led.

In 1857, noting that on his synoptic charts of the Netherlands the wind blew at right angles to the pressure gradient, Buys Ballot published the fact, later stating it in the form now known as Buys Ballot's law: "When you place yourself in the direction of the wind, . . . you will have at your left the least atmospheric pressure" (British Association for the Advancement of Science, *Transactions,* **32** [1863], 20–21). In this he had been anticipated by James Henry Coffin and William Ferrel, and he failed to explain, as Ferrel had, that the law results from the deflecting force of the earth's rotation. Although his theoretical understanding did not go much beyond Dove's, Buys Ballot left his mark on the science of meteorology as one of its chief organizers.

BIBLIOGRAPHY

A complete list of Buys Ballot's works is an appendix to the biography cited below. On his meteorological work, see his *Beredeneerd register van het Koninklijk Nederlandsch Meteorologisch Instituut* (Utrecht, 1882). His law is first stated in "Note sur les rapports de l'intensité et de la direction du vent avec les écarts simultanées du baromètre," in *Comptes rendus de l'Académie des sciences,* **45** (1857), 765–768. His ideas in physical chemistry were expanded in "Über die Art und Bewegung welche wir Wärme und Electricität nennen," in *Annalen der Physik,* **103** (1858), 240–259.

A biography is Ewoud van Everdingen, *C. H. D. Buys Ballot 1817–1890* (The Hague, 1953), by one of Buys Ballot's successors at the K.N.M.I.

HAROLD L. BURSTYN